Peñín Guide

Gifts you

PREMIUM ACCESS*
TO THE ONLINE GUIDE

FREE CODE
BY ACCESSING HERE

OR IN GUIAPENIN.WINE

CODE: **DISFRUTONES2023**

* Register first as a "Disfrutón" user and then active your premium account with the free code. Access valid for a year. Promotions starts on the 15/11/2022.

© PI&ERRE

All rights reserved. No part of this publication may be reproduced, stored in a retrieval system, or transmitted, in any form or by any means, without the prior permission in writing of PI&ERRE.

Team:
    Director: Carlos González
    Editor in Chief: Javier Luengo
    Tasting team: Carlos González, Javier Luengo, Boris Olivas and Ziyang Zhang
    Texts: Javier Luengo and Carlos González.
    Database manager: Erika Laymuns
    Advertising: Usue Aburto, María del Carmen Hernández and Sara Felix.
    Design, layout and desktop publishing: Luis Salgado.
    Cover image: La Kreateca

PUBLISHED BY: PI&ERRE
Gran Vía, 16 – 3º centro
28013 Madrid
SPAIN
Tel.: 914 119 464 - Fax: 915 159 499
comunicacion@guiapenin.com
www.guiapenin.com

ISBN: 978-84-122402-7-6
Copyright library: M-23927-2022
Printed by: Ulzama

DISTRIBUTED BY: GRUPO COMERCIAL ANAYA
Juan Ignacio Luca de Tena, 15
Tel: 0034 913 938 800
28027 MADRID
SPAIN

DISTRIBUTED BY: ACC Art Books Ltd.
Sandy Lane, Old Martlesham,
Woodbridge,
Suffolk
IP12 4SD, United Kingdom

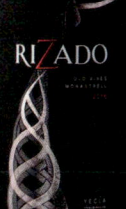

AN ELITE
TO PROMOTE SPANISH WINE
AROUND THE WORLD

There are moments in life when one has the strange sensation that something important is happening. It is clear that we are not living in the most carefree and inconsequential of times, with pandemic, war and global warming in the mix, but it is not the only important thing that is happening, because the passing of time also brings with it changes in many other areas and scenarios, and one of them is wine, which is far more upbeat and motivating.

All the tastings we have carried out this year have shown us a new scenario for Spanish wine. We are not talking about the upward trend in the quality of our wines in recent years, which is indisputable, but rather about the important leap that Spanish wine has made this year, which we are reporting in the pages of this Peñín Guide.

It is this annual journey through the different realities of the wines that we have been making for more than 30 years that has allowed us to observe, from the perspective of the Guide's history, the important moment we are living through. To go through Spanish wine and capture this journey in a massive work such as the one you have in your hands is an intense but very interesting process because of the information that you manage to extract from the large volume of tastings accumulated. Let's look at the events of the year in the Peñín Guide.

> "For the first time in our history two wines have reached the highest possible score, the idyllic 100"

100 POINTS IN THE PEÑÍN GUIDE

For the first time in our history two wines have reached the highest possible score, the idyllic 100. They are **Conde de Aldama Amontillado "Bota NO"** from Bodegas Yuste and **Pedro Ximénez Solera de 1830** from Bodegas Alvear. It has taken more than 30 years for this score to be awarded and it has been awarded to two wines that are undisputed and that represent a very important part of our winemaking history. Don't miss the details of these wines on page 10, where we reveal the reasons that led us to opt for 100 points and where we also take a closer look at the 99-point wines. The number of Podio wines (Podium wins of 95-100 points) has increased in each of their quality levels. This is how we have found 325 new wines, when the trend in past years was 301 (Guide 2022), 269 (Guide 2021), 252 (Guide 2020), 238 (Guide 2019), etc.

This qualitative increase at the top of the Spanish wine pyramid is not surprising, as it is happening in the same way in the rest of the wine producing countries, where there are more and more people obsessed with transcending in the world of wine. The rise of ecology, respect for the environment, for the raw material and the search for a lower impact in the communication between wine and the environment, has meant that much progress has been made in recent years, and we are happy that this is the case! In this elite of Spanish wine, winemaking is becoming more and more local, more direct and less self-conscious. More and more producers are seekinq to adapt to their environment in order to bottle what they interpret as local, which has led us to find different ways of expressing similar environments. For example, today it is possible to access the old Priorat style of wine with strength and forcefulness represented by an interesting Porrera Vi de Vila from Vall Llach, and to find a reference in freshness and subtlety from Priorat such as Les Manyes from Terroir al Limit.

2019
MARBORÉ
Cuvée
CINCO VARIEDADES DE UVA
SELECCIÓN DE VIÑEDOS SINGULARES
UNA DE
SEIS MIL BOTELLAS
0587
SOMONTANO
Denominación de Origen
BODEGA PIRINEOS
WINEinMODERATION
ELEGIR | COMPARTIR | CUIDAR

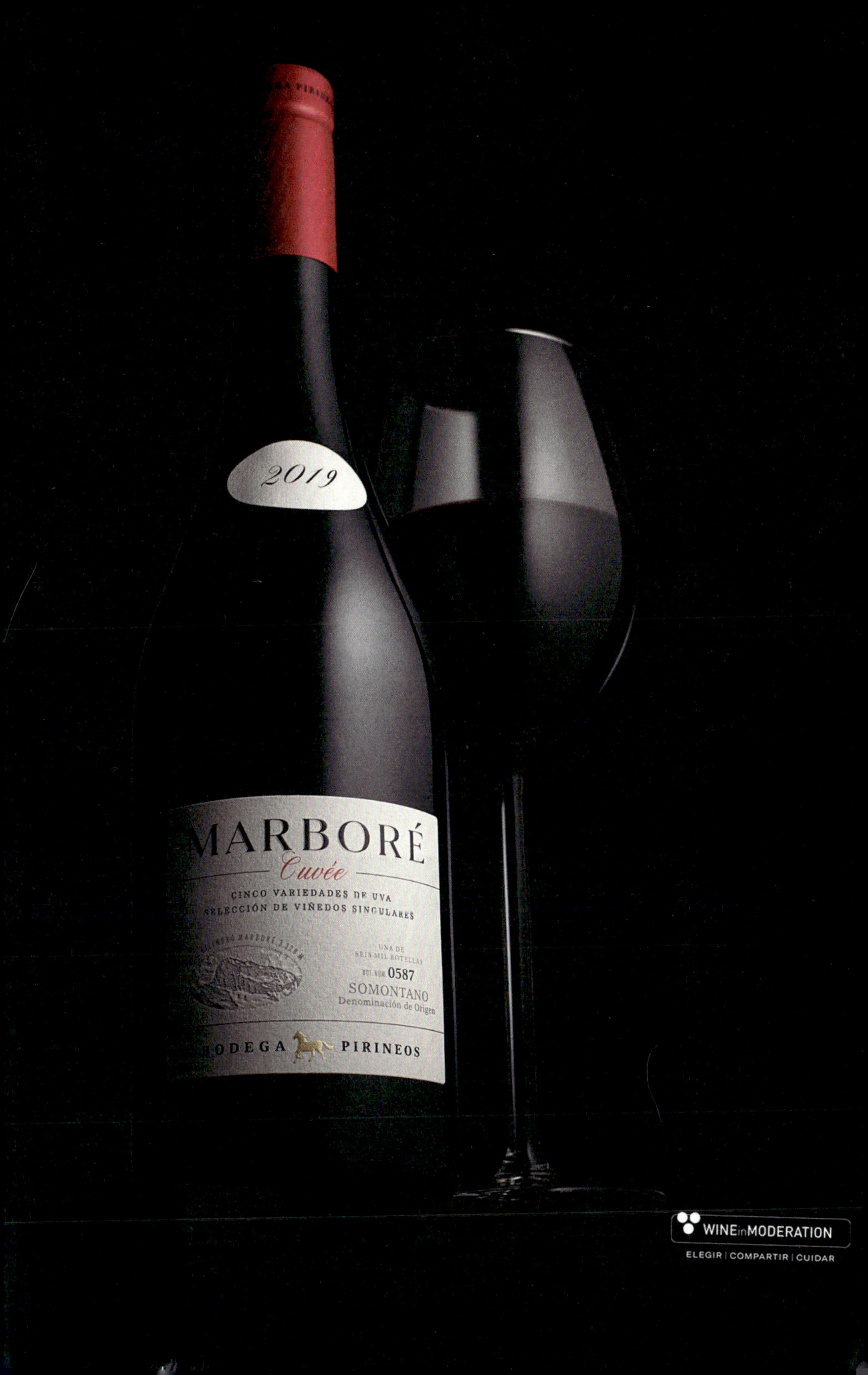

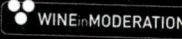

As it is becoming increasingly easier to access the world's wines wherever you are, many winemakers are embracing the new experiences that these wines give them, incorporating part of this winemaking style to the regions where they work. In this creative process they end up adapting the style they like to the conditions of the land where they grow, which allows a new style to be born, inspired by these great wines of the world. There are many examples, such as the Envínate team in their projects in Tenerife and Ribeira Sacra, where an underlying Burgundian style can be appreciated, intermingled with the idiosyncrasies of the island of Tenerife or the slopes of Ribeira Sacra.

The traditional elaborations of the South are serving as inspiration for many winemakers who, broadening their horizons, have wanted to merge the traditional concept of some of their wines with the most modern elaborations. This is how, for example, still wines with a Jerez or Sanlúcar vocation were born, wines touched by the magic of biological ageing but which give great importance in the narrative to the fruit and place of origin. The first example of this typology was born with Equipo Navazos and its Florpower in 2015. Over the years, more interesting elaborations have arrived, such as La Escribana by Luís Pérez or the vibrant Lumière by Muchada-Léclapart.

The fashion for a lighter, fruitier style continues to mark many of the new wines we have tasted. There is a growing number of consumer-oriented works in this style. This has not done anything to eradicate those who play in the opposite vein, i.e. wines with exuberance and structure. These wines continue to be a value and an important part of the Spanish winemaking discourse, especially in reds and in regions where a continental and Mediterranean climate prevails.

Something that characterises this house is that all styles of wine fit in the Peñín Guide, and each of them can represent excellence from different approaches. That is why it is possible to find exuberant wines as well as wines of great subtlety at the top of the Guide's table.

All this play of interpretations and adaptation to the environment is what makes the catalogue of wines available to the consumer even richer, making this magical beverage even more colourful and entertaining.

Every year new producers appear, bringing something new to the wine spectrum. The most daring and groundbreaking, a small selection of them, are the nominees for Vino Revelación, a category that we opened years ago with the aim of rewarding the boldness of innovation and the opening of new paths by winemakers who are committed to their own distinctive idea, even if that idea means working against the tide. This year we have five nominated wines and only one winner. You can find all the information about these wines on page 17.

No one can to say that there is no wine in Spain to suit every consumer's taste. That is what we are here for, to be able to shed some light on the myriad of brands, vintages, varieties and styles of wines that can be found today. Through responsible wine consumption we will find the maximum pleasure in exploring the contrasts of the country's vineyards. It is only necessary to pay attention to what we consume and put it in a geographical context. In this way, we will understand the subtleties, affinities and differences between the wine-growing regions of Spain. Spanish wine already has an elite capable of talking on equal terms with the great wines of the world. Its overwhelming strength will serve to boost the rest of the wines with the Spain brand; all that remains is for the commercial impulse to grow along the same lines as the producer.

A FAMILY BEYOND SUSTAINABILITY

PRESERVING HISTORY, PROTECTING THE EARTH

OVER THIRTY YEARS AGO, WE STARTED A PROFOUNDLY INSPIRING PROJECT: THE RECOVERY OF ANCESTRAL VARIETIES ON THE VERGE OF EXTINCTION. A WAY TO PAY HOMAGE TO CATALONIA'S WINE HERITAGE WHILST ALSO ADDRESSING CLIMATE CHANGE AT THE SAME TIME. TODAY, THE RESULT OF ALL THIS WORK IS CLOS ANCESTRAL, AN UNPRECEDENTED, ECOLOGICAL WINE THAT INCORPORATES THE "MONEU" GRAPE IN ITS BLEND; A VINE RESCUED AND THEN PLANTED IN CASTELL DE LA BLEDA, THE PENEDÈS ESTATE WITH A MORE THAN 2,500 YEARS OF HISTORY. A COMMITMENT TO OUR PAST, BUT ALSO TO OUR FUTURE.

MIGUEL AND MIREIA TORRES MACZASSEK, 5TH GENERATION IN CASTELL DE LA BLEDA (PENEDÈS)

SUMMARY

- The best wines of the Peñín Guide 2023 10
- Revelation wines of 2022 17
- **EL PODIUM (EXCEPTIONAL WINES)** 20
- Wineries and the tasting of the
  wines by *Denominación de Origen* 27
  - Abona 28
  - Alella 31
  - Alicante 34
  - Almansa 46
  - Arabako Txakolina 51
  - Arlanza 54
  - Arribes 57
  - Bierzo 62
  - Binissalem Mallorca 80
  - Bizkaiko Txakolina 83
  - Bullas 89
  - Calatayud 93
  - Campo de Borja 100
  - Cariñena 107
  - Catalunya 114
  - Cava 120
  - Cigales 155
  - Conca de Barberà 162
  - Condado de Huelva 167
  - Costers del Segre 170
  - El Hierro 178
  - Empordà 180
  - Getariako-Txakolina 194
  - Gran Canaria 198
  - Granada 201
  - Jerez-Xérès-Sherry - Manzanilla de
    Sanlúcar de Barrameda 204
  - Jumilla 221
  - La Gomera 234
  - La Mancha 236
  - La Palma 249
  - Lanzarote 251
  - León 254
  - Málaga - Sierras de Málaga 258
  - Manchuela 265
  - Méntrida 272
  - Mondéjar 277
  - Monterrei 278
  - Montilla-Moriles 284
  - Montsant 290
  - Navarra 303
  - Penedès 324
  - Pla de Bages 346
  - Pla i Llevant 350
  - Priorat 353
  - Rías Baixas 377
  - Ribeira Sacra 405
  - Ribeiro 417
  - Ribera del Duero 430
  - Ribera del Guadiana 497
  - Ribera del Júcar 501
  - Rioja 504
  - Rueda 594
  - Somontano 623
  - Tacoronte-Acentejo 630
  - Tarragona 635
  - Terra Alta 640
  - Tierra del Vino de Zamora 652
  - Toro 654
  - Uclés 671
  - Utiel Requena 673
  - Valdeorras 684
  - Valdepeñas 695
  - Valencia 699
  - Valle de Güimar 713
  - Valle de La Orotava 716
  - Vinos de Madrid 720
  - Ycoden-Daute-Isora 729
  - Yecla 731

- VINOS DE PAGO: 735
  - Pago Abadía Retuerta 738
  - Pago Aylés 738
  - Pago Bolandín 739
  - Pago Calzadilla 740
  - Pago Campo de la Guardia 740
  - Pago Casa del Blanco 740
  - Pago Chozas Carrascal 740
  - Pago de Otazu 741
  - Pago de Tharsys 741
  - Pago Dehesa del Carrizal 741
  - Pago Dehesa Peñalba 742
  - Pago El Terrerazo 742
  - Pago El Vicario 742
  - Pago Finca Élez 743
  - Pago Florentino 743
  - Pago Guijoso 743
  - Pago Heredad de Urueña 744
  - Pago La Jaraba 745
  - Pago Los Balagueses 745
  - Pago Los Cerrillos 745
  - Pago Prado de Irache 746
  - Pago Señorío de Arínzano 746
  - Pago Vallegarcía 747
  - Pago Vera de Estenas 747

- VINOS DE CALIDAD / D.O.P. 748
  - VC de Cangas 751
  - VC de Cebreros 751
  - VC de las Islas Canarias 754
  - VC de los Valles de Benavente 756
  - VC de Sierra de Salamanca 757
  - VC de Valtiendas 759

- VINOS DE LA TIERRA / I.G.P. 760
  - VT 3 Riberas 765
  - VT Altiplano de Sierra Nevada 765
  - VT Bajo Aragón 766
  - VT Barbanza e Iria 768
  - VT Cádiz 769
  - VT Castellón 772
  - VT Castilla 773
  - VT Castilla-Campo de Calatrava 799
  - VT Castilla y León 800
  - VT Costa de Cantabria 815
  - VT Eivissa 816
  - VT Extremadura 816
  - VT Formentera 820
  - VT Illa de Menorca 820
  - VT Illes Balears 821
  - VT Laderas del Genil 822
  - VT Liébana 822
  - VT Mallorca 822
  - VT Ribeiras do Morrazo 833
  - VT Ribera del Gállego - Cinco Villas 833
  - VT Ribera del Queiles 833
  - VT Sierra Norte de Sevilla 833
  - VT Val do Miño-Ourense 834
  - VT Valdejalón 834
  - VT Valles de Sadacia 835

- TABLE WINES 836

- SPARKLING WINES 877

- INDEXES 887
  - Organic Wines 888
  - Wineries 908
  - Wines 930

- MAP OF THE DO'S IN SPAIN AND VINOS DE PAGO 1036

- MAP OF VINOS DE LA TIERRA AND VINOS DE CALIDAD 1038

SPANISH WINE

Rías Baixas
HOME OF ALBARIÑO

DENOMINACIÓN DE ORIGEN
Rías Baixas
CONSEJO REGULADOR

THE BEST WINES OF THE PEÑÍN GUIDE 2023

CONDE DE ALDAMA "BOTA NO AMONTILLADO"

Type: fortified white wine / amontillado
Grape variety: palomino
Producer: Bodegas Yuste
Growing Area: DO Jerez

How can a small sip of wine contain so much information? That was the question that assailed us when the amontillado from the Sanlúcar de Sanlúcar winery Yuste reached our glass. The greatness of this wine had already stood out in past editions where it reached 99 points, the highest score until the last edition in the Peñín Guide. However, it arrived with all its strength and splendour and opened the debate on the expressive capacity and the almost infinite projection over time that this type of wine has in its interior. Is it possible to ask for anything more from an amontillado? The answer from our tasting team was no, we were faced with a unique and unbeatable wine. Such is its power and impact that just one drop of this wine can make the taster who is lucky enough to taste it delirious. It comes from an 18th century solera, the Conde de Aldama solera, which began with the purchase of the Aguilar y Cia winery in 1740. A single cask to which access is available only by order, up to the 20 bottles that are extracted from it each year. It is a wine whose evolutionary limit is still unknown and impossible to predict. The amontillado style wines, unique in the world and exclusive heritage of Andalusia, has managed to show us the greatness of some wines that swim at ease in the space-time, and take advantage of this time to leave their mark on the wine. It is precisely this relationship between wine and oxygen that allows all the magic of wine to emerge. However, not all wines are prepared to withstand this incursion of time in the wine's evolution. Very few manage to get the most out of it and for that to happen they have to be prepared. We do not know how this first cask was started, but what we do know is that the raw material used had to be carefully selected in view of its unbeatable evolution. There are true wizards of time in the region, many of them born in past centuries, who have been able to leave us everlasting wines such as this Bota NO by Yuste. Bota NO was simply put in the cask to keep the workers away from this cask, with a clear message NOT to touch.

Aromatically it is so complex that each taster can extract dozens of nuances on the nose. All of them emerge in harmony and with total clarity. The palate is explosive, concentrated, exuberant, and also extremely complex, with sharp sensations that take us to the limit. A single glass of this wine can become a gustatory delirium lasting more than an hour, as it has so much to tell us that we will need time to take it in.

ALVEAR PEDRO XIMÉNEZ SOLERA DE 1830 PX

Type: white / PX
Grape varieties: pedro ximénez
Producer: Alvear
Growing area: DO Montilla - Moriles

Montilla-Moriles is the quintessential appellation for the Pedro Ximénez grape. Although the origin of the sweet wines of this variety in the area is not clear, the fact is that this is where we find some of the most important wines of this type.

The solera of Pedro Ximénez 1830 de Alvear has been with us for almost two centuries. It is a very old PX from the 19th century, that shines for its balance. Imagine a wine with the high sugar concentration of the Pedro Ximénez, but with a vertebrating acidity, capable of counteracting the effect of the sweetness and making it a journey of infinite flavours and a delicate equilibrium. This is this jewel that comes from the oldest solera of the Alvear winery and has been made from sun-raisined grapes, with an eternal oxidative ageing process using the criaderas and soleras system, in which only the very few and very few occasional sacas (extractions for bottling) have been replenished. Fernando Jiménez Alvear, director of the bodega, tells us that it was the fourth generation of the family, in the hands of Enrique de Alvear y Ward, who was responsible for creating this 1830 solera. It is the first sweet wine created in the house, a wine that has endured over time and is expected to continue for many, many years to come.

This is a wine with a long drink for reflection which will stain the glass with an oily, amber-coloured veil and allow us to discover an explosive and rich set of aromas of mocha, cocoa, smoke, incense and a long etcetera. Closing our eyes, we can almost see the Montilla farmhouses with their "pasera" baskets full of bunches of grapes browning in the sun, in a very ancient practice, which brings us back to the rural and artisan essence of our wineries in past centuries. Montilla-Moriles is another of the world's unique winemaking regions. A place capable of creating these jewels in an exceptional way and with a very personal and unique style, which cannot be replicated anywhere else on the planet. Spain has true sacred temples of wine, although we often find it difficult to value and recognise it as it deserves. Alvear is the Montilla bodega that has given Montilla wine the greatest international projection in the world. It has a long family history and has managed to keep the traditional legacy of a whole place intact and to pass it on in its purest form over eight generations.

As with other wines, not all Pedro Ximénez wines have the capacity to age gracefully as this 1830 Solera has done. This proves that heaven is not a question of time, but rather of care, technique and sensitivity in the vineyard and winery, despite the fact that these wines have an apparently very simple, natural and direct winemaking process. It is precisely that sensitivity that ends up reaching us through a wine that has everything to offer, but with a unique and very special elegance. Undoubtedly, this is one of the great wines of the world.

THE 99-POINT WINES

LA RIOJA ALTA GRAN RESERVA 890 2010

La Rioja Alta Gran Reserva 890 2010
Type: red / Gran Reserva
Grape varieties: 95% tempranillo, 3% graciano and 2% mazuelo
Producer: La Rioja Alta
Growing area: DO. Ca. Rioja

For the second consecutive year, La Rioja Alta has managed to position a wine among the most important in the Peñín Guide tastings. Its Gran Reserva 890 represents the excellence of the classic Rioja wines told through the mythical 2010 vintage, for some the great vintage of the 21st century. This is a model of unique style, where the freshness of the wine heralds a slow ageing, with the elegance to which we are accustomed by the house itself. The best Rioja red wine of the year allows us to travel back to the classic Rioja ageing that made the place famous, a type of wine that has transcended our borders and has served to position Spanish wine in the world. Rich and expressive tertiary nuances become the emblem of this wine, where freshness and fruit coexist in perfect harmony with its classic and local essence. The finesse of the tannin and the subtlety with which all the nuances appear in the wine are an example of elegance and balance, and a symbol of the red wines that have most represented the Spanish wine industry beyond our borders.

LOUSAS CAMIÑO NOVO 2020

Type: red
Grape varieties: 85 mencía, 12% alicante
Producer: Envínate
Growing area: Ribeira Sacra

The strength with which new winemakers began to enter Spain just over a decade ago is already leaving its mark on Podio wines. Producers who found in the great wines of the world the inspiration to focus their projects, while each vintage made them realise that this approach to foreign wines would only serve to get them started, being essential to adapt to the terroir and the essence of each place when making wine. The Envínate project is a good example of what we are saying here. Their Lousas Camiño Novo 2020 captures the essence of a place, Ribeira Sacra, but through the eyes of Burgundy wine lovers. An elegant and very wild wine, with all kinds of hints of local flowers and herbs, with minerality as the axis and with a long-lasting and perfectly balanced mouthfeel. This wine comes from an early harvest characterised by an uneven and short distribution of water throughout the year, and by very high temperatures in July, which led to a decrease in production. The appearance of storms in August allowed them to cool the grapes, but did not prevent them from starting to harvest on 1 September, the second earliest harvest of Envínate in the area since they started in 2017, according to what they tell us. In spite of everything, we have found a splendid Lousas Camiño Novo, noticeably more powerful than in its previous vintage, but with freshness, elegance and structure. An example of adaptation in less benevolent climatic years.

SORTE O SORO 2020

Type: white
Grape varieties: godello
Producer: Rafael Palacios
Production area: DO Valdeorras

Rafael Palacios is the great reference for white wines in Spain. He has been for many years now, so it is no coincidence that he has reached this level with a wine that is as unique as it is special. His Sorte o Soro, a wine that comes from a single plot planted in 1978 and located in the Bibei Valley - Santa Cruz do Bolo, reached our glass with an overwhelming force. Despite being a wine still too young, it unfolded its full potential before our eyes, with an uncommon tension and an elegance worthy of admiration. It is a very complex wine in which time will play completely in its favour. Expressiveness, florality and salinity are just some of its assets. But what struck us most about this wine is its complexity. It is a wine that you can dismember in layers. Each layer leads you to a new one, from an oily beginning to a mineral backdrop, leaving a tight and tense finish that will have to be followed as the years go by. Together with Sorte o Soro 2016, it is the best wine created by this restless winemaker and a demonstration of the sensitivity he possesses to make great white wines with a Spanish stamp.

1903 COMA DE CASES GARNATXA VELLES VINYES 2019

Type: red
Grape varieties: 100% Grenache
Producer: Mas Doix
Growing area: DoQ. Priorat

Mas Doix is a winery that has not stopped giving us surprises since we discovered its 1902 Centenary Carignan (96 points) in the 2015 Peñín Guide. Four years later it was crowned with 99 points with its 2014 vintage. Since then, the winery has not stopped showing that nothing is the result of chance and that its way of seeing and interpreting the spectacular Priorat vineyard is also an example of the stylistic diversity that the terroir is capable of producing. 1903 Coma de Cases Garnatxa Velle Vinyes, in addition to being a wine with a very long name, has a little piece of the essence of the classic Priorat, through the production of a wine with strength and structure, but told from the perspective of a new style of winemaking, that is to say, with the presence of well-defined fruit and with very lively floral and mountain herb nuances in its core. If he already broke the mould with his cariñena, now it is his turn to do it with a garnacha of more than 115 years of age, an age that has allowed the grapes to build with singularity and definition all the mineral essence that underlies the slate soils of the place, without the wine being extremely intense and warm despite its 15º of alcohol.

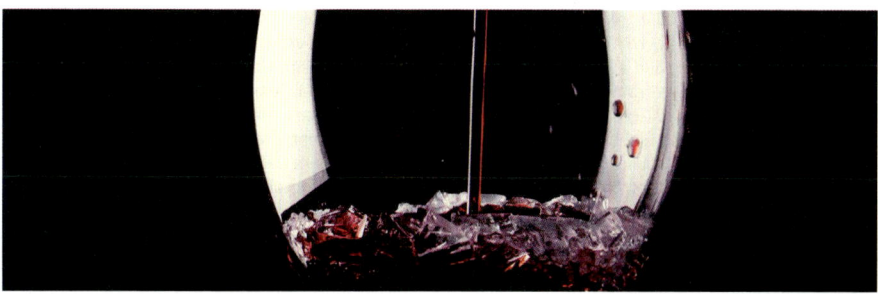

ORDÓÑEZ & CO. Nº4 ESENCIA 2016

Type: sweet white
Grape Varieties: 100% Muscat of Alexandria
Producer: Jorge Ordóñez Málaga
Growing area: Table wine (Malaga)

Jorge Ordoñez is one of the figures who has given Malaga wine the greatest international projection. His vocation as an importer of Spanish wine in the USA has allowed him to show a picture of the quality of Spanish wine to American palates for many years and, of course, Malaga and its wines were in that picture. His Esencia Nº4 is a tribute to the traditional wines of the Axarquía, those wines that as Jorge reminds us were forbidden to fortify until the arrival of wine exports to England and Germany. It is a wine that seeks to approach the traditional wines of Andalusia, but from a modern perspective. This wine comes from the El Panderón vineyard, planted in 1902, on slate and quartz soils, with steep slopes of up to 80%, and which has a northern orientation, where it enjoys afternoon shade, which helps to maintain a refreshing acidity. This is a late harvest wine, which has just been dehydrated in cold chambers instead of drying baskets, which in the eyes of its creator keeps the fruit and acidity purer.

The Malaga Muscatel has always enjoyed a great reputation among wine drinkers for its aromatic strength and unique mouthfeel. In this Esencia Nº4 we can see it in all its bold and expressive arrogance. A wine with a rich acidity that supports its sweetness with great balance. Ordoñez likes to emphasise the fact that this wine has not been fortified, as was done in the past, which has led him to label it as a table wine, making it paradoxical that the best wine from Malaga to date is a table wine. Esencia Nº4 transported us to the Mediterranean identity of Muscat of Alexandria with all its richness of aromas and flavours.

VEGA SICILIA ÚNICO RESERVA ESPECIAL (2009, 2011 Y 2012)

Type: red / Gran Reserva
Grape varieties: 95% tempranillo, 5% cabernet sauvignon
Producer: Bodegas Vega Sicilia
Growing area: DO Ribera del Duero

This is the best Vega wine tasted by the Guide to date. This Único Reserva Especial represents the whole identity of a winery that has become an icon of Spanish wine over the years. It is a wine which is "unique", emphasising its very name, in that all its records lead you to link it indisputably to Vega Sicilia, thanks to its singular and elegant classic essence. It has grapes from three vintages, 2009, 2011 and 2012, the first two rated as excellent by the Ribera del Duero regulatory board, which build the wine's identity. The rest is part of the house's own style where we see a very interesting background of tertiary nuances, still subtle, accompanied by fruit and a silky tannin worthy of the great wines of the world. Time will undoubtedly enrich this wine, which is ready to fight and take advantage of the passing of the years like no other.

This Vega Sicilia is currently the best representation of the classic wines of Ribera del Duero, a category that is not particularly developed, but which has great potential ahead of it. The evidence that Ribera del Duero has a lot to say in the ageing of great wines is undoubtedly to be found in this house.

ENOTECA GRAMONA 2006 BRUT NATURE

Type: sparkling white
Grape Varieties: 60% xarel.lo, 40% macabeo
Producer: Gramona
Growing area: Sparkling wine (Penedés)

Gramona is to sparkling wine as Neil Young is to music, an important and necessary value without which it would be impossible to understand part of the development of local culture, in this case sparkling wine. Since this house culminated all its work by crowning itself as the first sparkling wine to reach 99 points in the 2017 Peñín Guide, with several 98 points in previous editions, its wines have done nothing but increase its prestige. Enoteca Gramona is the crowning achievement of a whole philosophy based on obtaining wines with ageing at the level of the world's great sparkling wines, and when we say this, we are making a direct reference to the great champagnes. Although they could do it from the Brut concept, closer to our French neighbours, it has been from the Brut Nature, much more sincere and direct, how they have managed to reach these 99 points.

We know from experience that these wines gain weight as the years go by, years that come as a given from the very release of the wine (minimum ageing of 160 months) and that from the very first moment they show all their expressive richness. However, subsequent optimal conservation of the bottle will allow us to enjoy some small added subtleties that will make it even greater if possible. As we always say, but it is worth repeating, not all wines are able to withstand this ageing process for years. Only a few are capable of finding in the passage of time the perfect ally to become a great wine, and there is no doubt that Gramona knows exactly how to do it. Enoteca Gramona 2006 is an elegant wine, with an extraordinary liveliness despite its age. It is fine but very complex, with subtle hints of hazelnuts, light touches of cocoa and notes of pastries. The bubbles are as fine as they are persistent, a vehicle to happiness.

FONDILLÓN LUIS XIV 50 AÑOS (TONEL LUNA)

Type: red / fondillón
Grape Varieties: 100% Monastrell
Producer: Colección de Toneles Centenarios
Growing area: DO Alicante

Some of Spain's current wines, the traditional ones, hold the key that allows us to delve into the liquid history of a given area. Such is the case of Alicante's fondillones, wines that have survived to the present day with the same essence as when they were created.

The bodega Colección de Toneles Centenarios left us amazed this year with its wine Fondillón Luis XIV (Tonel Luna), a historic wine of great significance. It is a great wine for its contribution to wine culture. However, not only that, of course, but also on a hedonistic level, capable of taking us on a journey through a myriad of memories and sensations.

Colección de Toneles Centenarios is a small project founded by David Carbonell, from Vins del Comtat, José Ferrero and Regino Ballester. Their Fondillón Luis XIV line comes from unique barrels from wineries that closed their doors in the 1960s after the creation of the Cooperatives in the villages of Canyada and Beneixama. As if they were treasure hunters, their creators have been able to relaunch these old vintages that over the years have been concentrated and refined to such an extent that what has reached our glass has been a real gem.

LA BOTA DE AMONTILLADO (BOTA Nº 109) "BOTA PUNTA"

Type: fortified white wine / amontillado
Grape varieties: palomino
Producer: Equipo Navazos
Growing area: DO Jerez

Equipo Navazos' contribution to the prestige of Andalusian wines in the 21st century is evident. Eduardo Ojeda and Jesús Barquín, experts in the most hidden nooks and crannies of the wines of southern Spain, have rescued great treasures that were hidden in casks in the cellars of Jerez, Sanlúcar, Montilla and El Puerto since 2005, giving them all the voice and prominence that these wines should have. Each new edition of their wines is undoubtedly a piece of history and a vindication of the greatness of wine in this part of Spain. The Bota de Amontillado (Bota nº109) "Bota Punta", is a wine that was kept in a cask of the Albarizas de Trebujena cooperative and has an uncertain origin. As Equipo Navazos himself acknowledges, "the cooperative was founded in 1977, but the wine and the casks that make up this altarillo (small solera) date back to earlier times and it is not entirely clear how they got there". What is certain is that this is a very great sherry wine. Great in that it unites the forcefulness of the very old amontillados, but told not as an overwhelming and crushing force, but from the kind of finesse and edge that comes from its trebujenero origin.

Its arrival in the glass showed us the different stylistic paths that great amontillados can take, bringing you closer to the greatness of these wines from different perspectives. The Navazos team has a gift for this, for with each release they invite us to travel through the different perspectives of the wine of the Marco, with excellence as a backdrop.

1955 SOLERA CINCUENTA ANIVERSARIO

Type: white / PX
Grape Varieties: pedro ximénez
Producer: Pérez Barquero
Growing area: DO Montilla - Moriles

Pérez Barquero left us this year an exceptional PX with the unmistakable seal of Montilla-Moriles and with a very different identity. This is a PX with an average age of 25 to 30 years, of which 250 to 300 bottles are bottled each year. If in other aged PXs we had found explosion and exuberance as a hallmark of identity, this 1955 Solera Cincuenta Aniversario initially caught us by its sensitive austerity. There was an artist's soul in it, wrapped in a certain shyness. A wine that expresses itself slowly but surely. If in Alvear PX Solera 1830, we were amazed by an extrovert and cultured wine, in this Pérez Barquero we were amazed by the subtlety of its character, where a beautiful expression of the place reigned with all its cultural essence. Extreme ageing is not always an argument for achieving excellence and this wine is a good example of what we are saying here.

THE REVELATION WINES

Carlos González (@CarlosGuiaPenin) y Javier Luengo (@JavierGuiaPenin)

CHOLO AGARDA 2015 B

Manuel Formigo is an artisan from Ribeiro who is characterised by seeking the expression of a great wine through simplicity and restraint in the use of elements that might interfere with the final message. Cholo Agarda 2015 is the best example of his contribution to the world of wine, a revealing wine that shows us how a young wine, made in stainless steel tanks and without any ageing on lees or in oak barrels, is capable of becoming a highly complex wine with the passage of time.

Made mainly (85%) with loureiro, and accompanied by treixadura and albariño, this is a wine that seeks the evolutionary limits enhancing only the raw material. Its explosion of aromas and its great freshness opens up a very interesting path for Galician wines, which has already explored bottle ageing but mainly through wines aged on the lees and only in some very specific cases in oak.

LAS ÁNIMAS 2020 T

Young people have come to the world of wine with strength and determination. The best example of this can be found in this small and romantic family project created in Villanueva de Ávila by oenologist Bárbara Requejo and chef Guzmán Sánchez. In 2020 they decided to start Las Pedreras, a project that seeks to recover and create unique vineyards and wines in Gredos. Las Ánimas 2020 is born in the first vintage of this house and it does so in style, with no less than 94 points. The origin of this wine is due to the Garnacha grapes that come from three vineyards with an average age of 80 years, located in the area of Hergüijuelas, east of the village of Villanueva de Ávila. They stand at an altitude of 1,070 metres above sea level and with a northeast orientation, which translates into an extremely cool place, something that Las Animas 2020 perfectly captures. Las Animas is a wine that shows the way for all those who would like to enter the world of wine but cannot find the necessary courage to do so. These young entrepreneurs are an asset for all wine growing areas as they bring with their freshness and determination, a point of view that helps to empower the region.

PUNTO DE FUGA 2019 T

The vibrancy with which this Garnacha arrived in the glass made us wonder why there are still so few wineries in the area that are committed to a crisp, lively Garnacha like Punto de Fuga 2019. This wine was born from the union of two ways of looking at vineyards and wine, that of the Argentinian Matías Michelini and that of the guys from Zorzal, in Navarra. Combining experience and knowledge, they managed to bottle a small plot of just over half a hectare in the Corral de los Altos area, a place with an interesting limestone base that gives the wine its tautness. Viña Zorzal is a very active winery. Always looking for new challenges and projects. The non-conformist character of its creators is clearly marked in this wine, which shows us a Garnacha of exceptional finesse, but with a weighty palate and a long and very complex finish. For the launch of this wine, six different Garnacha wines were tasted, and the one with the best expression was selected.

In recent years we have seen a flourishing of projects that bring together renowned winemakers from different parts of the world. This wine tells us that there are no borders, and that great things can come out of the union of different ways of working and expressing oneself, especially if the people involved have a common sensitivity about what a wine can and should express.

VANDAMA HOYA OSCURA 2021 T

The arrival of this wine was totally unexpected. We had no information about this project, Bodega Vandama, and the first approach to its wines dazzled us. It is a small family winery owned by Diego Cambreleng, with vineyards in the area of the Vandama caldera, in the northeast of the island, barely 10 kilometres away from the city of Las Palmas de Gran Canaria. This wine joins other great examples that are being born today on the island, such as Tamerán or the Bien de Altura project. This wine is the result of the union of the owner of the winery Diego Cambreleng with the winegrower and oenologist Carmelo Peña, a restless young man who is currently expanding his winemaking activity with small projects in which he is putting all his curiosity about wine into practice. The combination of vineyards, people and work is what has allowed the magic to emerge and with it the arrival of new wines that we hope will continue on the rising path. This is an artisanal wine in every sense of the word, of which only 600 bottles have been made in its first and successful 2021 vintage, a particularly complicated year on the island with a drop of up to 60% in its average production due to mildew.

TABUCA 2020 T

There are some wines whose mere presence in a Denominación de Origen (DO) can mark the future of the wines produced there. We believe that this is the case with Tabuca 2020 T. Javier Gil Pejenaute is responsible for this wine, a Riojan from Alfaro who ended up in the small village of Tabuenca, a municipality located in the province of Zaragoza, Aragón, with the aim of creating something unique. And so it was, in 2020, when we had the opportunity to taste his first creation, Las Paradas 2018, a wine that left us crazy for its finesse, balance and complexity. We had an image of Campo de Borja as a more structured wine, where the excesses of oak can be often seen, but where there is exceptional raw material. We were missing a person who could bring that freshness to the winemaking process, and that person is Javier Gil. Tabuca, his second wine, showed us this year that it is not just a lucky coincidence. In this wine, all the essential freshness with which this Riojan winemaker works. A wine that captures part of the essence of a vineyard thanks to the rich forest nuances and minerality that we find in its interior.

KUUSU 2020 T

PEÑÍN GUIDE REVELATION WINE 2023

How would you describe a wine from Toro? It is more than likely that when asked this question, many readers would answer that to talk about Toro is to talk of a wine with a deep colour, a lot of body, the presence of ripe black fruit and perhaps even a sweet finish. We would be talking about pronounced tannins and generous oak. This could well respond to the prevailing style in the DO, with some notable exceptions, where the strength is bound and controlled by the winemakers in a masterful way. However, we hardly find examples such as the wine that has become this year's VINO REVELACIÓN: Kuusu 2020, an elegant, fresh wine, rich in red fruit, flowers and wild herbs. A wine that tames the Toro palate while retaining part of its sapid essence.

Last year we nominated San Román Garnacha, for bringing to Toro the option of creating a less sturdy and fruitier line of wines, this year Kuusu arrived to show us that this option is perfectly possible in Toro with its most important variety, the tinta de Toro. Oxer has been making unique and daring wines for years. He did so initially in Bizkaiko Txakolina, with some vibrant and spectacular wines, he did it again in Rioja Alavesa, with wines that soon entered the top of the Rioja ranks, and now he arrives with a superb table wine, made with grapes from the Toro area.

Kuusu is a wine whose name comes from Basque mythology, a kind of scarecrow-like elf used to ward off evil spirits. Almost all the labels and names of its wines have an artistic-cultural background that is in line with the special sensitivity of the creator of the wines. The grapes that give origin to the wine come from a vineyard of 0.9 hectares planted in 1920 in the town of Villabuena del Puente, on sandy soils with a large amount of quartz. Although fermented in French oak vats and subsequently aged for 12 months in foudres, the presence of wood is very fair and balanced, allowing everything else to come to the surface unhindered. Kuusu, becomes part of a generation of wines that shows the working philosophy of young people in areas with important historical and cultural roots.

PODIUM

PODIUM TRADITIONALS, SWEET AND SPECIAL WINES

WINE	DO	PAGE
100 POINTS		
Alvear Pedro Ximénez Solera 1830 BF PX D	Montilla-Moriles	285
Conde de Aldama "Bota No" BF AM	Jerez	213
99 POINTS		
1955 Pedro Ximenez Solera Cincuentenario BF PX D	Montilla-Moriles	288
Fondillón Luis XIV 50 años (Tonel Luna) T FO	Alicante	42
La Bota de Amontillado (Bota nº 109) "Bota punta" BF AM	Jerez	215
Ordóñez & Co. Nº4 Esencia (sin fortificar) 2016 B D	Table wine	863
98 POINTS		
Jorge Ordóñez & Co. Nº3 Viñas Viejas (sin fortificar) 2018 B D	Málaga y Sierras de Málaga	264
La Bota de Oloroso (Bota nº 108) "Bota NO" BF	Jerez	215
La Saca de Altanza BF PC S	Jerez	205
Oloroso Tradición VORS BF OL S	Jerez	212
Recóndita Armonía 2010 2010 T Solera D	Table wine	848
97 POINTS		
Alvear Pedro Ximénez de Sacristía 1998 BF PX D	Montilla-Moriles	285
Amontillado Tradición VORS BF AM S	Jerez	211
Añada 1975 BF AM	Jerez	216
Coliseo VORS BF AM	Jerez	219
Fernando de Castilla Oloroso Singular BF OL S	Jerez	215
Manuel Aragón BF OL S	Jerez	218
Palo Cortado Tradición VORS BF PC S	Jerez	212
S. Roberto Bota Única 1/1 BF PC	Jerez	207
S. Roberto Bota Única 1/2 BF AM	Jerez	207
Tío Pepe Cuatro Palmas BF AM S	Jerez	216
96 POINTS		
1955 Amontillado Solera Cincuentenario B AM S	Montilla-Moriles	287
1955 Oloroso Solera Cincuentenario B OL S	Montilla-Moriles	288
1955 Palo Cortado Cincuentenario B PC S	Montilla-Moriles	288
1730 VORS BF OL S	Jerez	205
Altanza Colección Roberto Amillo Palo Cortado BF PC S	Jerez	205
Amontillado Fino Imperial BF AM	Jerez	213
Arcos de Moclinejo BF Solera S	Málaga y Sierras de Málaga	261
Brotons Gran Fondillón Reserva 1978 T FO	Alicante	40
Cap de Barbaria Pansit de Formentera B Solera D	VT Formentera	820
Cardenal VORS BF PC S	Jerez	219
Casta Diva Esencia Dulce 2018 B D	Table wine	848
Casta Diva Reserva Real Dulce 2002 B R D	Table wine	848
Castell del Remei Vi Dolç 2019 B D	Costers del Segre	171
Conde de Aldama 1/3 BF OL	Jerez	213
Conde de Aldama 1/8 BF PC S	Jerez	214
Fondillón Ed. Limitada 1959 T FO	Alicante	37
Fondillón Luis XIV 25 años T FO	Alicante	41
Guardianes del Fondillón 1955 T FO	Alicante	44
Harveys Pedro Ximénez VORS BF PX D	Jerez	217
Jorge Ordóñez & Co Nº 2 Victoria Dulce (sin fortificar) 2021 B D	Málaga y Sierras de Málaga	264
La Bota de Manzanilla Pasada Nº103 Magnum BF MZ	Jerez	215
La Bota de Palo Cortado (Bota nº 102) BF PC	Jerez	215
La Cañada BF PX D	Montilla-Moriles	288
La Riva Manzanilla pasada Balbaína Alta BF MZ	Jerez	214
Lustau Amontillado VORS BF AM S	Jerez	218
Manuel Aragón BF PC S	Jerez	218
Montearruit Viejísimo 75 Años Bota 1/1 B AM	Montilla-Moriles	287
Osborne Solera India BF OL S	Jerez	210
Pedro Ximénez Tradición VOS BF PX D	Jerez	212
Quo Vadis VORS BF AM S	Jerez	215
S. Roberto Bota Única 2/2 BF AM	Jerez	207
Solear en Rama saca de verano 2022 BF MZ	Jerez	207
Tío Pepe Tres Palmas BF FI D	Jerez	216
Venerable VORS BF PX D	Jerez	210
Wellington 30 años VORS 50 cl BF PC S	Jerez	210
95 POINTS		
1730 VORS BF AM S	Jerez	205
1730 VORS BF PC S	Jerez	205
1822 Magnum BF MZ	Jerez	206
1822 Solera Fundacional Bota NO BF AM S	Jerez	206
61 Dorado en Rama 2021 BF Solera S	Rueda	597
Altanza Colección Roberto Amillo Amontillado BF AM S	Jerez	205

95 POINTS — PODIUM

WINE	DO	PAGE
Altanza Colección Roberto Amillo Pedro Ximénez BF PX D	Jerez	205
Amontillado 51-1ª VORS BF AM S	Jerez	210
Arcos de Moclinejo BF Solera D	Málaga y Sierras de Málaga	261
Cream Tradición VOS BF CRM	Jerez	211
Bac de les Ginesteres B RC D	Empordà	193
Brotons Gran Fondillon Reserva 1964 T FO	Alicante	40
De Alberto Dorado Verdejo 100% B Solera	Rueda	603
De Muller Garnacha Solera 1926 BF Solera D	Table wine	859
Delgado 1874 B AM S	Montilla-Moriles	286
Don Guido Solera Especial 20 años VOS BF PX D	Jerez	212
Dos Cortados 20 Años BF PC S	Jerez	212
Fernando de Castilla "Amontillado Antique" BF AM S	Jerez	215
Fino Tradición BF FI S	Jerez	212
Harveys Amontillado VORS BF AM S	Jerez	217
Harveys Palo Cortado VORS BF PC MED	Jerez	217
Humboldt 1997 B D	Tacoronte-Acentejo	631
La Gitana Aniversario BF MZ S	Jerez	209
Lustau Almacenista Cayetano del Pino BF PC S	Jerez	217
Lustau Pedro Ximénez VORS BF PX D	Jerez	218
Matusalem VORS BF OL D	Jerez	216
Memòries del Priorat Ranci Cal Marcelino TF RC	Priorat	376
Molino Real 2018 B	Málaga y Sierras de Málaga	262
Noé VORS BF PX D	Jerez	216
Peña del Aguila BF PC S	Jerez	207
Primitivo Quiles Gran Imperial 1892 BF Solera D	Alicante	43
R&G Soleras Olvidadas BF AM	Jerez	211
R&G Soleras Olvidadas BF FI	Jerez	211
Solera Familiar Gutiérrez Colosía BF OL S	Jerez	209
Solera Familiar Gutiérrez Colosía BF PC S	Jerez	209
Somnis de Gerisena - Sol i Resena RD Añejo D	Empordà	184
Tio Pepe Dos Palmas BF FI S	Jerez	216
Triana 30 años VORS 50 cl BF PX D	Jerez	209
V Dulce de Invierno 2019 B D	Table wine	863
Viña Corrales Pago Balbaina BF FI	Jerez	206
Williams & Humbert 2001 BF OL S	Jerez	213

Peñín Guide — SPANISH WINE

PODIUM WHITE

99 POINTS

WINE	DO	PAGE
Sorte O Soro 2020 B	Valdeorras	692

98 POINTS

WINE	DO	PAGE
Chivite Colección 125 1996 B FB	Navarra	318
Maior de Mendoza Finca Las Tablas 2017 B BA	Rias Baixas	396
Mixtura Etiqueta Dorada 2018 B	Table wine	866
Pazo Señorans Selección de Añada 2013 B	Rias Baixas	399

97 POINTS

WINE	DO	PAGE
Suertes del Marqués Vidonia V.P. 2020 B C	Valle de la Orotava	719

96 POINTS

WINE	DO	PAGE
Emilio Rojo 2019 B	Ribeiro	429
Fai un Sol de Carallo 2018 B	Ribeiro	422
Itsasmendi Artizar (botella Borgoña) 2017 B	Bizkaiko Txakolina	85
La Comtesse 2018 B FB	Rias Baixas	398
Lapena 2019 B	Ribeira Sacra	413
Lumière 2019 B	VT Cádiz	771
Nelin 2019 B	Priorat	364
Pedrouzos Magnum 2012 B FB	Valdeorras	693

95 POINTS

WINE	DO	PAGE
As Sortes Val do Bibei 2020 B	Valdeorras	691
Belondrade y Lurton 2020 B FB	Rueda	596
Capellania 2017 B R	Rioja	576
Chivite Colección 125 Vendimia Tardia 2020 B FB D	VT 3 Riberas	765
De los Abuelos Teiró Godello 2020 B	Bierzo	75
El Erial de Valdecañada 2018 B	Bierzo	76
Finca La Terrenal 2018 B	Terra Alta	647
Gran Cruz del Calvario 2020 B	El Hierro	179
Granbazán Don Álvaro de Bazán 2018 B	Rias Baixas	383
José Pariente Finca Las Comas 2018 B	Rueda	606
La Sombrilla 2019 B	Ribeiro	422
La Val Crianza sobre Lías 2016 B C	Rias Baixas	388
Nivarius Finca La Nevera 2017 B	Rioja	541
Ossian 2019 B	VT CastyLe	812
Ossian Capitel 2019 B FB	VT CastyLe	811
Palo Blanco 2020 B	Table wine	861
Parajes del Infierno "El Judas" 2019 B FB	Table wine	840
Qué Bonito Cacareaba 2020 B	Rioja	511
Roda I 2019 B	Rioja	546
Selma de Nin 2018 B	Priorat	365
Sorte Antiga 2020 B	Valdeorras	692
Zárate Tras da Viña 2020 B	Rias Baixas	404

PODIUM RED

WINE	DO	PAGE
99 POINTS		
1903 Coma de Cases Garnatxa Velles Vinyes 2019 T C	Priorat	370
La Rioja Alta Gran Reserva 890 2010 T GR	Rioja	573
Lousas Camiño Novo 2020 T	Ribeira Sacra	414
Vega Sicilia Único Reserva Especial T GR	Ribera del Duero	464
98 POINTS		
Artadi El Carretil 2020 T	Table wine	852
Dominio do Bibei 2019 T	Ribeira Sacra	413
L'Ermita 2020 T C	Priorat	354
La Nieta 2019 T	Rioja	590
Peñas Aladas 2013 T GR	Ribera del Duero	479
Pingus 2020 T	Ribera del Duero	478
Quiñón de Valmira 2020 T FB	Rioja	543
Sierra Cantabria Mágico 2018 T	Rioja	592
97 POINTS		
1902 Tossal d'en Bou Gran Vinya Classificada 2019 T C	Priorat	370
Alabaster 2019 T	Toro	669
Alto de la Estrella 2019 T	Vino de Calidad de Cebreros	753
Aro 2019 T	Rioja	539
Artuke La Condenada 2020 T	Rioja	507
Canta la Perdiz 2017 T R	Ribera del Duero	478
Contador 2020 T	Rioja	511
Dominio de Atauta Llanos del Almendro 2017 T	Ribera del Duero	446
Falcoeira 2019 T	Valdeorras	689
Gran Reserva 904 Selección Especial 2015 T GR	Rioja	573
La Faraona 2020 T BA	Bierzo	73
Las Beatas 2019 T	Rioja	561
Les Aubaguetes 2020 T C	Priorat	354
Les Manyes 2019 T	Priorat	375
Marqués de Riscal 2016 T GR	Rioja	523
Pago de Carraovejas "Cuesta de las Liebres" 2018 T R	Ribera del Duero	487
Pepe Mendoza Fierroca 2020 T	Alicante	43
Pico Ferreira 2020 T	Bierzo	72
Viña El Pisón 2020 T	Table wine	875
96 POINTS		
A Ponte 2020 T	Ribeira Sacra	407
Amancio 2018 T	Rioja	592
Artadi La Poza de Ballesteros 2020 T	Table wine	852
Artadi Quintanilla 2020 T	Table wine	852
Artadi San Lázaro 2020 T	Table wine	853
Barón de Chirel 2017 T	Rioja	522
Casa Castillo Las Gravas 2020 T	Jumilla	224
Casa Castillo Pie Franco 2020 T C	Jumilla	224
Castillo Ygay 2011 T GR	Rioja	576
Cirsion 2019 T R	Rioja	540
Cuentaviñas El Tiznado 2020 T	Rioja	562
Dalmau 2019 T R	Rioja	576
Dominio de Atauta La Roza 2017 T	Ribera del Duero	445
Dominio del Aguila 2018 T R	Ribera del Duero	478
El Jardín de las Iguales 2020 T	VT Valdejalón	834
El Nido 2019 T	Jumilla	225
El Puntido 2019 T	Rioja	590
Finca Dofí 2020 T C	Priorat	354
Finca El Bosque 2019 T	Rioja	592
Finca Meixeman 2020 T	Ribeira Sacra	408
Flor de Pingus 2020 T	Ribera del Duero	478
Galia Le Dean 2019 T	VT CastyLe	802
Gallinas & Focas 2018 T	VT Mallorca	822
La Baixada 2020 T	Priorat	354
La Creu Alta 2018 T BA	Priorat	370
Las Gundiñas La Vizcaína 2019 T	Bierzo	76
Les Tosses 2019 T C	Priorat	375
Macán 2018 T	Rioja	517
Moncerbal 2020 T	Bierzo	73
Nit de Nin La Rodeda 2019 T	Priorat	365
Pradorey Élite 2019 T	Ribera del Duero	471
Propiedad 2020 T	Rioja	543
Quincha Corral 2019 T	Pago El Terrerazo	742
Reserva Real 2017 T	Penedès	335

PODIUM

WINE	DO	PAGE
96 POINTS		
Valbuena 5º 2018 T	Ribera del Duero	464
Vega Sicilia Único 2013 T	Ribera del Duero	464
Victorino 2019 T	Toro	669
Viña Sastre Pago de Santa Cruz 2016 T GR	Ribera del Duero	450
95 POINTS		
4 Kilos 2020 T	VT Mallorca	822
Abadía da Cova Penafión 2018 T BA	Ribeira Sacra	406
Abadía Retuerta Pago Negralada 2017 T BA	Pago Abadía Retuerta	738
Alión 2019 T	Ribera del Duero	467
Arrebatacapas 2019 T	Vino de Calidad de Cebreros	752
Artadi Majadales 2020 T	Table wine	852
Artadi Valdeginés 2020 T	Table wine	853
Artuke El Escolladero 2020 T	Rioja	507
As Caborcas 2019 T	Valdeorras	689
Atteca Armas 2020 T	Calatayud	95
Avrvs 2016 T	Rioja	568
Bernabeleva Carril del Rey 2020 T	Vinos de Madrid	721
Celia Vizcarra 2018 T	Ribera del Duero	466
César Príncipe 2019 T C	Cigales	156
Clio 2019 T	Jumilla	225
Cobrana 2020 T	Bierzo	65
Corteo 2019 T	Jumilla	225
Cuentaviñas 2020 T	Ribera del Duero	476
Cuentaviñas Los Yelsones 2020 T	Rioja	562
Cuevas de Arom Os Cantals 2020 T	Calatayud	97
Dits del Terra 2019 T C	Priorat	374
Doix Costers de Vinyes Velles 2019 T C	Priorat	371
Dominio de Atauta La Mala 2017 T C	Ribera del Duero	445
Dominio de Atauta San Juan 2017 T	Ribera del Duero	446
Dominio de Calogía by José Manuel Pérez Ovejas Cuveé S 2019 T	Ribera del Duero	478
Dominio de Es Carravilla 2019 T	Ribera del Duero	478
Dominio del Pidio 2019 T	Ribera del Duero	479
El Cf de Chozas Carrascal 2017 T	Pago Chozas Carrascal	740
El Rapolao 2020 T	Bierzo	76
El Reventón 2019 T	VT CastyLe	809
El Titán del Bendito 2019 T	Toro	664
El Velado 2019 T	Rioja	561
Espectacle 2019 T C	Montsant	300
Figuero Tinus 2018 T	Ribera del Duero	495
Finca Capeliños 2020 T	Ribeira Sacra	407
Finca Torrea 2018 T R	Rioja	522
Finca Villacreces Specimen Nº2 T	Ribera del Duero	480
Grans Muralles 2018 T R	Conca de Barberà	165
Hacienda Solano Finca Cascorrales 2018 T	Ribera del Duero	482
Hacienda Solano Finca Peñalobera 2019 T	Ribera del Duero	482
Inés Vizcarra 2018 T R	Ribera del Duero	466
Izadi El Regalo 2020 T	Rioja	528
Jiménez-Landi Piélago 2019 T	Méntrida	275
Juan Gil Etiqueta Azul/Blue Label 2020 T	Jumilla	226
Kalamity 2020 T	Rioja	579
Kuusu 2020 T	Table wine	867
La Bovila 2019 T	Table wine	871
La Cueva del Contador 2020 T	Rioja	511
La Viña Escondida 2018 T	Méntrida	274
Lacima 2019 T	Ribeira Sacra	413
Lalomba Finca Iles 2019 T	Rioja	574
Lalomba Finca Valhonta 2018 T	Rioja	574
Las Lamas 2020 T BA	Bierzo	73
Las Suertes 2020 T	Valle de la Orotava	718
Las Tierras de Javier Rodríguez El Teso Alto 2016 T	Toro	660
Las Umbrías 2019 T	Vinos de Madrid	725
Lo Mas D'Edetària 2019 T	Terra Alta	647
Lousas Seoane 2020 T	Table wine	861
Malpuesto 2020 T	Rioja	543
Morca 2019 T	Campo de Borja	104
Pago de los Capellanes Parcela El Picón 2016 T	Ribera del Duero	487
Pago La Jara 2018 T	Toro	663
Peña Caballera 2019 T	Vinos de Madrid	727
Pérez Pascuas Gran Selección 2015 T GR	Ribera del Duero	451
Pico de Luyas 2019 T	Ribera del Duero	462
Pintia 2018 T	Toro	663
Por los Cien 2020 T	Rioja	526
Porrera Vi de Vila de Vall Llach 2020 T C	Priorat	361

PODIUM

WINE	DO	PAGE
95 POINTS		
Prado Enea 2015 T GR	Rioja	539
Proelio La Canal del Rojo 2018 T	Rioja	543
Protos Selección Finca el Grajo Viejo 2018 T	Ribera del Duero	490
Puerto Rubio 2018 T	Rioja	544
Quinta Sardonia QS 2019 T	VT CastyLe	812
Real de Asúa 2019 T	Rioja	563
Remírez de Ganuza 2015 T R	Rioja	545
Renvivas 2019 T	VT CastyLe	814
San Román 2019 T	Toro	668
San Vicente 2018 T BA	Rioja	583
Scala Dei Sant Antoni 2018 T	Priorat	362
Sibila 2018 T	VT Mallorca	832
Sufreiral 2020 T	Bierzo	72
Tadeo Tinaja Cortijo Los Aguilares 2019 T	Málaga y Sierras de Málaga	263
Termanthia 2015 T	Toro	656
Torre Muga 2019 T	Rioja	539
Touran 2019 T	Campo de Borja	104
Vatan 2019 T	Toro	662
Velázquez Colección Artistas Españoles 2011 T R	Rioja	505
Viejas de Izan 2019 T C	Ribera del Duero	492
Viña Arana 2015 T GR	Rioja	573
Viña Ardanza 2016 T R	Rioja	573
Viña Sastre Pesus 2016 T	Ribera del Duero	450
Vivaltus 2017 T	Ribera del Duero	496
Vizcarra Torralvo 2019 T	Ribera del Duero	466
Yo Solo 2020 T FB	Málaga y Sierras de Málaga	261
Ysios Grano a Grano 2017 T	Rioja	559

PODIUM SPARKLING WINES

WINE	DO	PAGE
99 POINTS		
Enoteca Gramona 2006 BE BN	Sparkling Wines	880
98 POINTS		
Enoteca Gramona 2005 BE BN	Sparkling Wines	880
Gramona Celler Batlle 2012 BE BR	Sparkling Wines	880
97 POINTS		
Manuel Raventós 2011 BE BN	Sparkling Wines	883
Mas del Serral 2010 BE BN	Sparkling Wines	885
Mas del Serral 2011 BE BN	Sparkling Wines	885
Rocorva Particular do Rocarodo 2012 BE BN	Sparkling Winoc	884
RNG 20 Magnum 1997 BE BN	Sparkling Wines	882
96 POINTS		
Alta Alella 10 2010 BE BN	Cava	122
Alta Alella 10 2012 BE BN	Cava	122
Gramona TLN 2009 BE BN	Sparkling Wines	881
Llopart Original 1887 Viñas Singulares Les Flandes 2011 BE BN	Sparkling Wines	882
Turo d'en Mota 2008 BE BN	Sparkling Wines	884
95 POINTS		
Agustí Torello Mata Kripta Gran Anyada 2008 BE GR BN	Cava	121
Alta Alella Mirgin Exeo Evolució + 2005 BE GR BN	Cava	122
Alta Alella Mirgin Exeo Paraje Calificado Vallcirera 2016 BE GR BN	Cava	122
Ars Collecta 459 2010 BE GR BR	Cava	135
Ars Collecta La Fideuera Xarel.lo 2011 BE GR BR	Cava	135
Enoteca Personal Manuel Raventós Negra Magnum 2013 BE BN	Sparkling Wines	883
Gramona III Lustros 2013 BE BN	Sparkling Wines	880
Juvé & Camps La Capella 2009 BE GR BN	Cava	141
Manuel Raventós 2015 BE BN	Sparkling Wines	883
María Bernet 2013 BE BN	Sparkling Wines	881
Pere Ventura Gran Vintage Paraje Calificado Can Bas 2015 BE GR BR	Cava	147
Pere Ventura Tresor Anniversary 2017 BE GR BR	Cava	147
Recaredo Serral del Vell 2011 BE BN	Sparkling Wines	883
Torelló Collection 2011 BE BN	Sparkling Wines	884

WINERIES AND TASTING OF WINES BY
DENOMINACIÓN DE ORIGEN

SCORING SYSTEM

95-100 EXCEPTIONAL
The wine excels among those of the same type, vintage and origin. It is in every sense extraordinary. It is full of complexity, with abundant sensory elements both on the nose and on the palate that arise from the combination of soil, grape variety, winemaking and ageing methods; elegant and utterly outstanding, it exceeds average commercial standards and in some cases it may still be unknown to the general public.

90-94 EXCELLENT
Wine with the same attributes as those indicated above but with less exceptional or significant characteristics, or less clarity of nuances.

85-89 VERY GOOD
It stands out thanks to the nuances acquired through a successful winemaking and/or ageing process, or for a pleasant varietal character. A wine with very good features, although it may lack soil or terroir expression.

80 - 84 ACCEPTABLE WINE
Wine which has no outstanding defects nor any virtues. The wine responds fully to the characteristics required of its type and wine growing zone, although it is rather diluted.

70 - 79 WINE NOT TO BE RECOMMENDED
A wine which has an outstanding defect regarding its organoleptic nuances and gives scarce satisfaction to the consumer.

60 - 69 NOT RECOMMENDABLE WINE
A wine which is not acceptable, with noticeable defects which excessively damage it as a whole.

50 - 59 DEFECTIVE WINE
A wine which is not acceptable as enjoyment. It may have serious production defects.

ORGANIC WINES

D.O.P.
DENOMINACIÓN DE ORIGEN PROTEGIDA

I.G.P.
INDICACIÓN GEOGRÁFICA PROTEGIDA

N/A
NOT AVAILABLE

DO. ABONA

CONSEJO REGULADOR

Martín Rodríguez, 9
38588 Porís de Abona - Arico (Santa Cruz de Tenerife)
☎: +34 922 164 241 - Fax: +34 922 164 135
@: vinosdeabona@vinosdeabona.com
www.vinosdeabona.com

LOCATION:

In the southern area of the island of Tenerife, with vineyards which occupy the slopes of the Teide down to the coast. It covers the municipal districts of Adeje, Arona, Vilaflor, San Miguel de Abona, Granadilla de Abona, Arico and Fasnia.

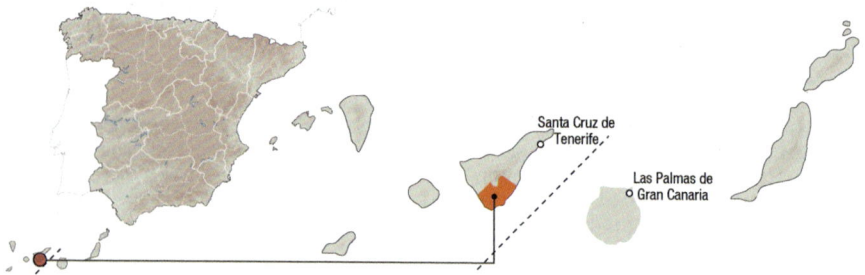

GRAPE VARIETIES:

WHITE: albillo, marmajuelo, forastera blanca, güal, malvasía, moscatel alejandría, sabro, verdello, vijariego, baboso blanco, listán blanco, pedro ximénez and torrontés.

RED: castellana negra, listán negro, malvasía rosada, negramoll, tintilla, baboso negro, cabernet sauvignon, listán prieto, merlot, moscatel negro, pinot noir, ruby cabernet, syrah, tempranillo and vijariego negro.

FIGURES:

Vineyard surface: 818 – **Wine-Growers:** 1,361 – **Wineries:** 20 – **2021 Harvest rating:** Very Good – Production 2021: 848,787 L – **Market percentages:** 95% National - 5% International..

SOIL:

Distinction can be made between the sandy and calcareous soil inland and the more clayey, well drained soil of the higher regions, seeing as they are volcanic. The so-called 'Jable' soil is very typical, and is simply a very fine whitish volcanic sand, used by the local winegrower to cover the vineyards in order to retain humidity in the ground and to prevent weeds from growing. The vineyards are located at altitudes which range between 300 and 1,750 m (the better quality grapes are grown in the higher regions), which determines different grape harvesting dates in a period spanning the beginning of August up to October.

CLIMATE:

Mediterranean on the coastal belt, and gradually cools down inland as a result of the trade winds. Rainfall varies between 350 mm per year on the coast and 550 mm inland. In the highest region, Vilaflor, the vineyards do not benefit from these winds as they face slightly west. Nevertheless, the more than 200 Ha of this small plateau produce wines with an acidity of 8 g/l due to the altitude, but with an alcohol content of 13%, as this area of the island has the longest hours of sunshine.

CLASSIFICATION OF YOUNG WINE HARVESTS PEÑÍNGUIDE

2017	2018	2019	2020	2021
GOOD	GOOD	N/A	GOOD	N/A

ALTOS DE TREVEJOS
Calle la Constitución s/n
38620 San Miguel de Abona
(Santa Cruz de Tenerife)
☎: +34 922 929 294
bodega@altosdetrevejos.com
www.altosdetrevejos.com

Trevejos Organic Wines Listán Blanco 2020 B FB
listán blanco

92 🌱
Balanced, dried flowers. Colour: bright straw. Nose: dried herbs, dried flowers, fine lees, spicy. Palate: flavourful, balanced, fine bitter notes, long.

Trevejos Blanc de Noirs 2017 BE BN
listán prieto

88
Standard, dried herbs, yeasty notes, mineral.

Trevejos Mountain Wines Baboso Negro 2020 T
baboso

91
Colour: cherry, purple rim. Nose: spicy, black fruit, scrubland, mineral. Palate: flavourful, fruity, good acidity.

Trevejos Mountain Wines Listán Blanco & Malvasía 2020 B
listán blanco, malvasía

91
Colour: bright straw. Nose: ripe fruit, fragrant herbs, fine lees, dry stone. Palate: full-bodied, rich, long, good acidity, flavourful.

Trevejos Mountain Wines Listán Prieto 2018 T FB
100% listán prieto

93
Colour: cherry, purple rim. Nose: red berry notes, floral, spicy, black fruit, scrubland, fine reductive notes. Palate: flavourful, good acidity, long.

Trevejos Mountain Wines Vijariego Negro 2020 T
vijariego negro

91
Colour: bright cherry. Nose: fresh fruit, red berry notes, black fruit, earthy notes. Palate: good acidity, spicy, fine tannins.

Trevejos Volcanic Wines Baboso Negro & Syrah 2018 T
baboso, syrah

91
Colour: deep cherry. Nose: dried herbs, creamy oak, fine reductive notes, black fruit. Palate: ripe fruit, spicy, round tannins.

Trevejos Volcanic Wines Blanco Albillo & Verdello 2020 B
albillo, verdello

92
Colour: bright yellow. Nose: powerful, creamy oak, ripe fruit, spicy, dried flowers. Palate: good structure, long, toasty, fine bitter notes.

BODEGA LACASMI
38620 San Miguel de Abona
(Santa Cruz de Tenerife)
☎: +34 922 700 300
Fax: +34 922 700 301
bodega@casanmiguel.com
www.lacasmi.com

Apaga y Vámonos 2020 T
listán negro, castellana, tempranillo, cabernet sauvignon

88
Balanced, herbal, crisp, fruity, tasty.

Apaga y Vámonos 2021 RD SD
listán negro

85

Apaga y Vámonos Afrutado 2021 B SD
listán blanco

85

Chasnero Listán Negro 2021 T
listán negro

86 🌱

BODEGA MENCEY CHASNA
Marta, 3 Chimiche
38594 Granadilla de Abona
(Santa Cruz de Tenerife)
☎: +34 922 777 285
ventas@menceychas.com
www.menceychasna.com

Los Tableros Afrutado 2021 B SS
listán blanco, moscatel, malvasía

85

Mencey Chasna Seco 2021 B
listán blanco

88
Austere, smoky, spicy, mineral, yeasty notes.

Mencey Chasna Semiseco 2021 B SS
albillo, moscatel, listán blanco

85

Mencey de Chasna Afrutado 2021 RD
85

DO ABONA / D.O.P.

DO ABONA / D.O.P.

Mencey de Chasna Afrutado Semidulce 2021 B SD
listán blanco
86

Mencey de Chasna Vijariego Negro 2020 T
vijariego negro
85

BODEGAS REVERÓN

Ctra. Gral. Vilaflor - La Escalona, Los Quemados, 8
38618 Vilaflor (Santa Cruz de Tenerife)
☎: +34 922 725 044
bodegasreveron@hotmail.com
www.bodegareveron.com

Pagos de Reverón 2020 T BA
listán negro, tempranillo, castellana
87 🌱

Pagos de Reverón 2021 B S
listán blanco
86 🌱

Pagos de Reverón 2021 T S
listán negro, tempranillo, castellana
87 🌱

Pagos de Reverón Afrutado 2021 B SD
listán blanco
86

Pagos Reverón Afrutado 2021 RD SD
listán negro
85

Pagos Reverón Malvasia 2019 B S
malvasía
88
Citrus fruit, floral, herbal, tasty.

DO. ALELLA
CONSEJO REGULADOR

Avda. San Mateu, 2 - Masía Can Magarola
08328 Alella (Barcelona)
☎: +34 935 559 153 - Fax: +34 935 405 249
@: doalella@doalella.org
www.doalella.org

LOCATION:

It extends over the regions of El Maresme and el Vallès in Barcelona. It covers the municipal districts of Alella, Argentona, Cabrils, El Masnou, La Roca del Vallès, Martorelles, Montornès del Vallès, Montgat, Orrius, Premià de Dalt, Premià de Mar, Santa Mª de Martorelles, Sant Fost de Campsentelles, Teià, Tiana, Vallromanes, Vilanova del Vallès and Vilasar de Salt. The main feature of this region is the urban environment which surrounds this small stretch of vineyards; in fact, one of the smallest DO's in Spain.

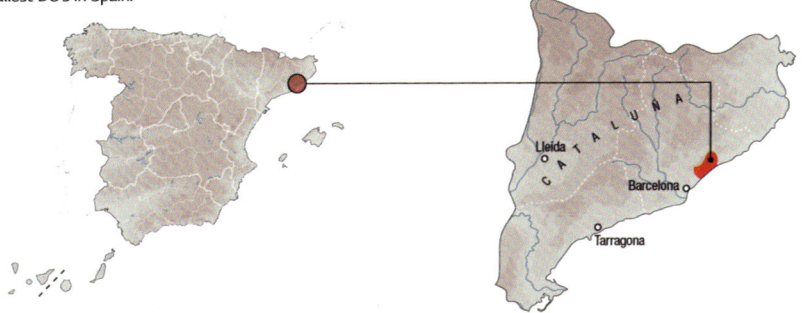

GRAPE VARIETIES:

WHITE: Pansa Blanca (similar to the Xarel·lo from other regions in Catalonia), Garnatxa Blanca, Pansa Rosada, Picapoll, Malvasía, Macabeo, Parellada, Chardonnay, Sauvignon Blanc and Chenin Blanc.

RED (MINORITY): Garnatxa Negra, Ull de Llebre (Tempranillo), Merlot, Pinot Noir, Syrah, Monastrell, Cabernet Sauvignon, Sumoll and Mataró.

FIGURES:

Vineyard surface: 227 – **Wine-Growers:** 56 – **Wineries:** 9 – **2017 Harvest rating:** N/A – **Production 17:** 700,000 L. **Market percentages:** 86% National - 14% International.

SOIL:

Distinction can be made between the clayey soils of the interior slope of the coastal mountain range and the soil situated along the coastline. The latter, known as Sauló, is the most typical. Almost white in colour, it is renowned for it high permeability and great capacity to retain sunlight, which makes for a better ripening of the grapes.

CLIMATE:

A typically Mediterranean microclimate with mild winters and hot dry summers. The coastal hills play an important role, as they protect the vines from cold winds and condense the humidity from the sea.

CLASSIFICATION OF YOUNG WINE HARVESTS PEÑÍNGUIDE

2016	2017	2018	2019	2021
VERY GOOD	VERY GOOD	VERY GOOD	VERY GOOD	N/A

DO ALELLA / D.O.P.

ALTA ALELLA
Camí Baix de Tiana s/n
08328 Alella (Barcelona)
☎: +34 934 693 720
info@altaalella.wine
www.altaalella.wine

AA Cau D'en Genis 2020 B
pansa blanca

90 🌿
Colour: bright straw. Nose: stone fruit, dried herbs, faded flowers, mineral. Palate: flavourful, fruity, balanced.

AA Cau D'en Genis 2021 B
pansa blanca

90 🌿
Colour: bright straw. Nose: citrus fruit, scrubland, white flowers. Palate: fresh, fruity, flavourful, balanced.

AA Lanius 2020 B S
chardonnay

91 🌿
Colour: bright yellow. Nose: ripe fruit, creamy oak, dry nuts, spicy. Palate: flavourful, fresh, powerful, balanced.

AA Orbus 2019 T
syrah

91 🌿
Colour: bright cherry. Nose: ripe fruit, red berry notes, black fruit, black pepper, spicy, dried herbs. Palate: fruity, powerful, fresh, balanced, slightly dry, soft tannins.

AA Parvus Chardonnay 2021 B
chardonnay

89 🌿
Fruity, herbal, tasty, kind finish.

AA Parvus Syrah 2019 T
syrah

89 🌿
Jammy, dried herbs, tasty, powerful, balanced. Nose: black fruit, ripe fruit.

BODEGAS ROURA, J.A. PEREZ ROURA
Valls de Rials, s/n
08328 Alella (Barcelona)
☎: +34 933 527 456
roura@roura.es
www.roura.es

Roura Coupage 2017 T BA
72% garnacha, 12% merlot, 10% syrah, 6% tempranillo

87

Roura Garnatxa Syrah 2017 T C
80% garnacha, 20% syrah

86

Roura Sauvignon Blanc 2021 B
100% sauvignon blanc

86

Roura Tres Ceps 2017 T C
merlot, syrah, tempranillo

86

Roura Xarel.lo 2021 B
100% xarel.lo

86

ELVIWINES
Ctra T-300 Falset-Marça, km 0.97
43775 Marça (Tarragona)
☎: +34 606 186 565
info@elviwines.com
www.elviwines.com

Herenza 2021 B
60% pansa blanca, 40% sauvignon blanc

88
Citrus fruit, floral, herbal, tasty.

MASIA CAN RODA
BV-5006
08106 Santa Maria de Martorelles (Barcelona)
☎: +34 679 196 272
info@cellercanroda.cat
www.cellercanroda.cat

Gran Minguet 2017 BE R BN
100% pansa blanca

91
Colour: bright straw. Nose: fine lees, fragrant herbs, characterful, ripe fruit, dry nuts. Palate: flavourful, good acidity, fine bitter notes.

So de Masia Can Roda Pansa Blanca 2021 B
100% pansa blanca

89 🌿
Sulphur notes, citrus fruit, balanced, yeasty notes, crisp.

Sauló Pansa Blanca 2019 B C
pansa blanca

92

Colour: straw. Nose: powerful, creamy oak, ripe fruit, spicy, wild herbs, white flowers. Palate: rich, good structure, long, toasty, fine bitter notes.

VINS DE LA MEMÒRIA
Montaner 320
08021 Barcelona (Barcelona)
vinsdelamemoria@gmail.com
www.vinsdelamemoria.com

El Badiu 2020 B
pansa blanca

92 🌱

With personality, little interventionist, wild. Colour: yellow. Nose: complex, fine reductive notes, faded flowers, dry nuts. Palate: flavourful, balanced, good acidity.

DO ALELLA / D.O.P.

SEÑORÍO DE LÍBANO
Del Río, s/n
26212 Sajazarra (La Rioja)
☎: +34 941 320 066
bodega@castillodesajazarra.com
www.castillodesajazarra.com

Invita 2021 B
88
Aromatic, pleasant, fruity, tropical, smooth.

Invita Rosé 2021 RD
73% tempranillo, 27% garnacha

88
Standard, crisp, fruity, herbal.

DO. ALICANTE
CONSEJO REGULADOR

Monjas, 6
03002 Alicante
☎: +34 629 513 934 - Fax: +34 965 229 295
@: info@vinosdealicantedop.org
www.vinosalicantedop.org

LOCATION:

In the province of Alicante (covering 51 municipal districts), and a small part of the province of Murcia. The vineyards extend over areas close to the coast (in the surroundings of the capital city of Alicante, and especially in the area of La Marina, a traditional producer of Moscatel), as well as in the interior of the province.

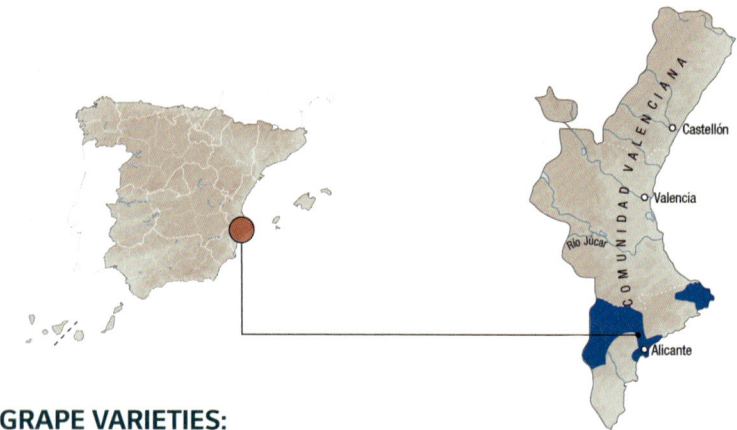

GRAPE VARIETIES:

WHITE: Merseguera (Verdosilla), Moscatel de Alejandría, Macabeo, Planta Fina, Verdil, Airén, Chardonnay, Sauvignon Blanc, Subirat parent, Alarije, Tortosí, Pedro Ximenez, Forcallat Blanca. Plantanova (Tardana), Valencí Blanco, Moscatel de grano menudo, Viognier, Verdejo and Garnacha Blanca.

RED: Monastrell, Garnacha Tinta, Garnacha Tintorera (Alicante bouschet), Bobal, Tempranillo, Cabernet Sauvignon, Merlot, Pinot Noir, Syrah, Petit Verdot, Forcallat Tinta, Bonicaire, Miguel del Arco, Garro, Mandó, Trepat, Valencí Tinto, Cabernet Franc.

FIGURES:

Vineyard surface: 10,320– **Wine-Growers:** 2,294– **Wineries:** 42– **2021 Harvest rating:** Very Good– **Production 21:** 16,505,900 litres – **Market percentages:** 81% National - 19% International.

SOIL:

In general, the majority of the soils in the region are of a dun limestone type, with little clay and hardly any organic matter.

CLIMATE:

Distinction must be made between the vineyards situated closer to the coastline, where the climate is clearly Mediterranean and somewhat more humid, and those inland, which receive continental influences and have a lower level of rainfall.

CLASSIFICATION OF YOUNG WINE HARVESTS PEÑÍNGUIDE

2017	2018	2019	2020	2021
GOOD	VERY GOOD	VERY GOOD	VERY GOOD	VERY GOOD

BODEGA LAS VIRTUDES

Ctra. de Yecla, 27
03400 Villena (Alacant/Alicante)
☎: +34 965 802 187
oficina@virtudes.net
www.bodegavirtudes.com

Casa Ritas 2021 T
100% monastrell

87

Las Virtudes Fortaleza 2021 B FB
moscatel

87

Patojo 2020 T
100% monastrell

88 ♣

Toasty, smoky, standard, fruity, pruney.

Tesoro de Villena Fondillón 1972 T FO
100% monastrell

91

Colour: light mahogany. Nose: toasty, smoky, roasted almonds, dry nuts. Palate: old wood, flavourful, balanced.

Vinalopó 2020 T RB
monastrell, syrah

87

Vinalopó Petit Verdot 2019 T RB
100% petit verdot

89

Pruney, spicy, herbal, dried herbs, powerful, great length.

BODEGAS ARRÁEZ

Pol. 6 Parcela 386 Paraje Ciscar
46630 La Font de la Figuera (València/Valencia)
☎: +34 962 290 031
info@bodegasarraez.com
www.bodegasarraez.com

Bala Perdida 2020 T
100% alicante bouchet (Tinto)

88

Fruity, jammy, herbal, tasty.

BODEGAS BOCOPA

Paraje Les Pedreres, Autovía A-31, km. 200 - 201
03610 Petrer (Alacant/Alicante)
☎: +34 966 950 489
info@bocopa.com
www.bocopa.com

Fuego Lento 2017 T S
70% monastrell, 15% syrah, 15% alicante bouchet (Tinto)

93

Colour: Cherry. Nose: expressive, creamy oak, characterful, sweet spices. Palate: full-bodied, long, great length.

Fuego Lento Monastrell Extremo 2019 T BA
100% monastrell

90

Colour: deep cherry. Nose: ripe fruit, dried herbs, waxy notes, varietal. Palate: ripe fruit, spicy, round tannins, long.

Laudum 2020 T RB
70% monastrell, 30% syrah

88 ♣

Fruity, herbal, spicy.

Laudum Monastrell 2020 T RB
100% monastrell

87

Laudum XII Plus 2017 T C
100% monastrell

89

Floral, fruity, jammy, spicy.

Marina Alta 2021 B
100% moscatel

87

BODEGAS E. MENDOZA

Camino del Romeral, 42
03580 Alfaz del Pi (Alacant/Alicante)
☎: +34 965 888 639
bodegas-mendoza@bodegasmendoza.com
www.bodegasmendoza.com

Enrique Mendoza Chardonnay 2021 B
100% chardonnay

88

Citrus fruit, dried herbs, simple, jammy.

Enrique Mendoza Chardonnay 2021 B FB
100% chardonnay

90

Colour: bright straw. Nose: fruit expression, ripe fruit, floral. Palate: flavourful, fresh, good acidity, fruity aftertaste.

DO ALICANTE / D.O.P.

DO ALICANTE / D.O.P.

Enrique Mendoza Las Quebradas 2020 T C
100% monastrell

93

Colour: bright cherry. Nose: complex, expressive, spicy, mineral, toasty. Palate: elegant, full-bodied, long, great length.

Enrique Mendoza Moscatel de la Marina Dulce 2021 B D
100% moscatel de alejandría

91

Colour: bright yellow. Nose: balsamic herbs, honeyed notes, floral, sweet spices, expressive. Palate: rich, fruity, powerful, flavourful, elegant.

Enrique Mendoza Santa Rosa 2019 T FB
35% monastrell, 35% cabernet sauvignon, 15% merlot, 15% syrah

93

Colour: dark-red cherry. Nose: toasty, spicy, cocoa bean, black fruit, overripe fruit. Palate: flavourful, toasty, fine bitter notes.

BODEGAS FAELO

Camino de los Coves. Partida de Matola, Poligono 3, Nº18
03296 Elche (Alacant/Alicante)
☎: +34 655 856 898
info@vinosladama.com
www.vinosladama.com

L'Alba 2019 T
syrah

87

L'Alba de Faelo 2021 RD
syrah

87

L'Alba del Mar 2021 B
chardonnay

86

La Dama 2017 T C
monastrell, cabernet sauvignon

88

Fruity, herbal, smoky, tasty.

Palma Blanca Dulce 2019 B D
moscatel

88

Pleasant, sweet, dried flowers, tasty.

BODEGAS FRANCISCO GÓMEZ

03400 Villena (Alacant/Alicante)
☎: +34 965 979 195
info@bodegasfranciscogomez.es
www.bodegasfranciscogomez.es

Boca Negra Dulce 2016 T D
monastrell

90

Colour: deep cherry. Nose: black fruit, black liquorice, smoky, spicy. Palate: flavourful, fruity, sweet.

Fruto Noble Joven 2021 T

88

Fruity, herbal, spicy, tasty.

Fruto Noble Roble 2021 T RB

88

Spicy, fruity, jammy.

Fruto Noble Rosado 2021 RD
monastrell, syrah, merlot

87

Fruto Noble Sauvignon Blanc 2021 B
sauvignon blanc

87

Pago Francisco Gómez Fondillón 1988 T FO
monastrell

92

Pruney. Nose: candied fruit, pattiserie, sweet spices, caramel, varnish. Palate: powerful, flavourful.

Quo Vadis Fondillón 1972 T FO
monastrell

94

Representative. Colour: mahogany. Nose: fruit liqueur notes, candied fruit, pattiserie, varnish, expressive, dry nuts. Palate: concentrated, flavourful, long, fine solera notes.

BODEGAS MONÓVAR

Ctra. Monovar-Salinas CV-830 Km 3,2
03640 Monovar (Alacant/Alicante)
☎: +34 965 076 435
info@mgwinesgroup.com
www.mgwinesgroup.com

Borrasca 2018 T
monastrell

89

Fruity, jammy, spicy.

El Caire Monastrell 2019 T
monastrell

89

Herbal, fruity, tasty.

🏆 PODIUM

Fondillón Ed. Limitada 1959 T FO
96
Oxidativ, powerful. Colour: light mahogany. Nose: powerful, complex, dry nuts, toasty, acetaldehyde. Palate: rich, bitter, fine solera notes, long, spicy.

Riesling 2021 B
riesling
88
Pleasant, aromatic, floral, jammy.

BODEGAS MURVIEDRO
Ampliación Pol. El Romeral, s/n
46340 Requena (València/Valencia)
☎: +34 962 329 003
murviedro@murviedro.es
www.murviedro.es

Galeam 2018 T C
100% monastrell
89
Jammy, dried herbs, fruity, balanced.

Galeam Dry Muscat 2021 B
100% moscatel
86

Galeam Monastrell 2021 T
100% monastrell
90 🌿
Colour: cherry, purple rim. Nose: black fruit, dried herbs, smoky, spicy. Palate: fruity, flavourful, balanced, slightly dry, soft tannins.

Sericis Cepas Viejas Monastrell 2018 T R
100% monastrell
90
Colour: bright cherry. Nose: ripe fruit, grassy, spicy, balsamic herbs. Palate: fruity, flavourful, powerful tannins.

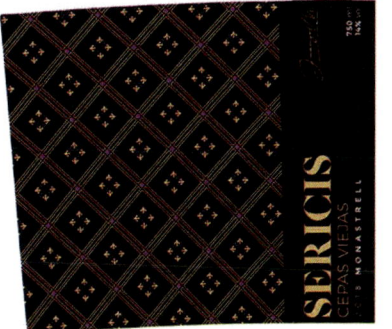

Murviedro Cepas Viejas Monastrell 2018 T R SS
100% monastrell
89
Fruity, herbal, spicy, tasty.

Murviedro Eko 2021 T
100% monastrell
88 🌿
Fruity, spicy, smoky, jammy.

BODEGAS ORTIGOSA
Paraje Ravalet MU 8
03640 Monovar (Alacant/Alicante)
☎: +34 606 457 232
Fax: +34 965 470 083
central@bodegasortigosa.com
www.bodegasortigosa.com

La Pitxotxa Cabernet Sauvignon 2017 T C
cabernet sauvignon
89
Dried herbs, slight reduction, herbal, jammy, tasty.

La Pitxotxa Moscatel de Alejandría 2020 B SD
moscatel de alejandría, macabeo
86

La Pitxotxa Rosé 2020 RD
monastrell
87

Modernitxen Vino Noble de Alicante 2017 TF CRM
88
Jammy, sweety finish, dried herbs, tasty, wild.

BODEGAS PARCENT C.B.
Avda. Denia, 15
03792 Parcent (Alacant/Alicante)
☎: +34 636 536 693
Fax: +34 966 405 173
armando@bodegasparcent.com
www.bodegasparcent.com

CODI/02 2021
garnacha tintorera
88
Fruity, herbal, floral, tasty.

Dolç D'Art Selección de Licor 2017 BF D
moscatel de alejandría
91
Colour: bright golden. Nose: dried fruit, tropical fruit, dried herbs. Palate: sweet, fruity, full-bodied.

DO ALICANTE / D.O.P.

DO ALICANTE / D.O.P.

Auro 2021 B
87

Fruit D'Autor 2017 RF
88
Fruity, sweet, dried herbs, smoky.

Grà D'Or Blanco Seco 2021 B
100% moscatel de alejandría
87

Parcent Giró 2021
giró
87

BODEGAS PINOSO
03650 Pinoso (Alacant/Alicante)
☎: +34 965 477 240
export@bodegaspinoso.com
www.bodegaspinoso.com

Camarillas 2019 T C
100% monastrell
92
Colour: bright cherry. Nose: ripe fruit, fruit preserve, sweet spices. Palate: spicy, round tannins, flavourful, fruity.

Diapiro 2020 B BA
100% merseguera
87

Diapiro 2020 T RB
100% monastrell
88
Aromatic, dried herbs, jammy, spicy, standard.

Pinoso Clásico 2019 T C
100% monastrell
88 🌿
Pruney, dried herbs, tasty, powerful, toasty.

BODEGAS RIKO
Avda Juan Carlos I, 33
03727 Xaló (Alacant/Alicante)
☎: +34 966 480 294
bodegasrikoxalo@gmail.com
www.bodegasrikoxalo.com

Oscar Mestre Giró 2019 T
100% giró
93
Colour: cherry, garnet rim. Nose: red berry notes, ripe fruit, dried flowers, chamomile. Palate: fruity, varietal, spicy, long, great length, good acidity.

Oscar Mestre Insurrecte 2020 B C
moscatel de alejandría
92
With personality. Colour: bright yellow. Nose: ripe fruit, floral, fine lees, balsamic herbs. Palate: long, dry, flavourful.

Renaix de Giró 2020 T
70% giró, 30% syrah
91
Colour: light cherry, garnet rim. Nose: wild herbs, dried flowers, ripe fruit. Palate: flavourful, fruity, balanced, fine bitter notes.

Renaix La Passió 2021 B
100% moscatel de alejandría
90
Defined aromas. Colour: bright straw. Nose: white flowers, white fruit, varietal. Palate: fruity, balanced, easy to drink.

BODEGAS SIERRA DE CABRERAS
La Molineta, 27
03638 Salinas (Alacant/Alicante)
☎: +34 647 515 590
fernando.sirvent@carabibas.com
www.carabibas.com

Carabibas 640 Noches 21 meses 2015 T R S
cabernet sauvignon, merlot, monastrell
89
Spicy, overripe, tasty, fruity.

Carabibas La Viña del Carpintero 2019 T C
monastrell
89
Fruity, jammy, spicy, tasty.

Carabibas VS Vendimia Seleccionada 2019 T C S
merlot, monastrell, cabernet sauvignon
88
Weary, pruney, spicy, toasty.

BODEGAS SIERRA SALINAS

CV 813, Ctra. Villena-Pinoso, km. 18
03400 Villena (Alacant/Alicante)
☎: +34 965 979 786
info@mgwinesgroup.com
www.mgwinesgroup.com

Mira Salinas 2013 T
75% monastrell, 25% cabernet sauvignon

91

Colour: bright cherry. Nose: black fruit, black liquorice, spicy, meaty notes. Palate: flavourful, fruity, balanced, slightly dry, soft tannins.

Mo Salinas 2019 T FB

90

Colour: cherry, purple rim. Nose: ripe fruit, black fruit, spicy, wild herbs. Palate: flavourful, fine bitter notes, round tannins.

Mo Salinas Moscatel Macabeo 2021 B
moscatel, macabeo

88

Aromatic, floral, fruity, jammy, thin, standard.

Puerto Salinas 2013 T R SS
55% monastrell, 35% garnacha tintorera, 10% petit verdot

91

Colour: cherry, garnet rim. Nose: old leather, meaty notes, overripe fruit, dried herbs, spicy. Palate: flavourful, fruity, fresh, good finish, slightly dry, soft tannins.

BODEGAS VIVANZA

Ctra. Jumilla Pinoso, Km. 13 La Alberquilla
30520 Jumilla (Murcia)
☎: +34 966 078 686
agomez@vivanza.es
www.vivanza.es

Lascala 2021 B
sauvignon blanc, verdil

83

Vivanza Elite 2017 T C
monastrell, cabernet sauvignon, syrah

89

Full-bodied, pruney, tasty. Nose: creamy oak, dark chocolate.

Vivanza Gold 2016 T C
monastrell, syrah, pinot noir

88

Full-bodied, tasty, jammy, toasty, dried herbs.

BODEGAS VOLVER

Ctra de Pinoso a Fortuna, s/n
03658 Rodriguillo (Alacant/Alicante)
☎: +34 966 185 624
Fax: +34 965 075 376
export@bodegasvolver.com
www.bodegasvolver.com

Alicante Bouschet by Tarima 2019 T BA
monastrell, alicante bouchet (Tinto)

91

Colour: cherry, garnet rim. Nose: fruit preserve, powerful, scrubland, black liquorice, dark chocolate. Palate: flavourful, long.

Eje Monastrell 2020 T

89

Jammy, spicy, tasty, toasty, great length.

Tarima Hill 2020 T
100% monastrell

90

Colour: very deep cherry. Nose: roasted coffee, aromatic coffee, powerful, ripe fruit, black fruit. Palate: smoky aftertaste, great length, round tannins.

Triga Magnum 2019 T
85% monastrell, 15% cabernet sauvignon

93

Colour: deep cherry. Nose: fruit preserve, black fruit, creamy oak, spicy, cocoa bean, dried herbs. Palate: fruity, good structure, flavourful, balanced, slightly dry, soft tannins.

DO ALICANTE / D.O.P.

DO ALICANTE / D.O.P.

Tarima Hill 2021 B FB
90% chardonnay, 10% merseguera

91
Colour: bright straw. Nose: fruit expression, ripe fruit, floral. Palate: flavourful, fresh, good acidity, fruity aftertaste.

Tarima Selección 2021 T
90% monastrell, 10% syrah

89
Powerful, jammy, hot, toasty.

BODEGAS XALO
Ctra. Xaló Alcalali, s/n
03727 Xaló (Alacant/Alicante)
☎: +34 966 480 034
comercial@bodegasxalo.com
www.bodegasxalo.com

1962 Origen 2018 T
giró

90
Colour: light cherry. Nose: smoky, spicy, dried herbs, red berry notes, ripe fruit. Palate: fruity, flavourful, balanced, slightly dry, soft tannins.

Bahía de Denia 2021 B
moscatel

87
Simple, standard, smooth, kind finish.

Castell D'Aixa 2021 B
moscatel

87

Riu Rau Dulce 2019 B Mistela D
moscatel

91
Colour: golden. Nose: fruit liqueur notes, candied fruit, honeyed notes, faded flowers. Palate: sweet, fruity, rich.

Vall de Xaló 2021 T MC
giró

86

Vall de Xaló Giró Vino de Licor TF D
giró

89
Herbal, full-bodied, dried flowers, great length, tasty. Nose: honeyed notes, candied fruit.

BODEGAS Y VIÑEDOS EL SEQUÉ
El Sequé, 59
03650 Pinoso (Alacant/Alicante)
☎: +34 945 600 119
info@elseque.es
www.elseque.es

El Sequé 2020 T
100% monastrell

93
Colour: deep cherry. Nose: ripe fruit, dried herbs, creamy oak, fruit expression. Palate: powerful, ripe fruit, spicy, round tannins.

El Sequé Dulce 2017 TF C D
100% monastrell

93
Colour: deep cherry, purple rim. Nose: fruit preserve, violet drops, dried herbs, spicy. Palate: flavourful, sweet, balanced, good finish.

BROTONS V & A
Caserío Culebrón, 59
03650 Pinoso (Alacant/Alicante)
☎: +34 965 477 267
info@vinosculebron.com
www.vinosculebron.com

🏆 PODIUM

Brotons Gran Fondillon Reserva 1964 T FO
100% monastrell

95
Colour: light mahogany. Nose: candied fruit, cereal notes, bakery, spicy, dry nuts. Palate: flavourful, long, sweetness, balanced, elegant, complex, fine solera notes.

🏆 PODIUM

Brotons Gran Fondillon Reserva 1978 T FO
100% monastrell

96
Colour: light mahogany. Nose: candied fruit, honeyed notes, dry nuts, acetaldehyde, old leather, sweet spices. Palate: flavourful, full-bodied, fresh, sweetness, good finish, fine solera notes.

Brotons
Gran Fondillon Reserva 1970 T FO
100% monastrell

94

Nose: dried fruit, honeyed notes, faded flowers, spicy. Palate: flavourful, sweetness, balanced, round, old wood.

CASA CORREDOR
Autovía Alicante/Albacete, Salida 168 La Encina
02660 Caudete (Albacete)
☎: +34 966 842 064
info@mgwinesgroup.com
www.mgwinesgroup.com

Semsum 2 2021 B SD
86

CASA SICILIA 1707
Paraje Alcaydias, 4
03660 Novelda (Alacant/Alicante)
☎: +34 965 605 385
administracion@casasicilia1707.es
www.casasicilia1707.es

Cesilia 2020 T RB
petit verdot, cabernet sauvignon

87 🌱

Cesilia Blanc 2021 B
50% moscatel, 30% macabeo, 20% malvasía

88 🌱

Fruity, herbal, jammy, kind finish.

Cesilia La Garnacha 2020 T
100% garnacha

88 🌱

Spicy, jammy, fruity.

Cesilia Rosé 2021 RD
70% monastrell, 15% merlot, 15% syrah

88 🌱

Dried flowers, fruity, herbal.

Cesilia Rosé La Réserve 2021 RD
100% garnacha

90 🌱

Colour: salmon. Nose: ripe fruit, red berry notes, dried herbs. Palate: fruity, fresh, flavourful.

CELLER LES FRESES
Alqueria de Ferrando s/n
03749 Jesús Pobre - Denia (Alacant/Alicante)
☎: +34 682 539 463
celler@lesfreses.com
www.lesfreses.com

Blanc Moscatel Sec Les Freses de Jesús Pobre 2021 B MO
moscatel de alejandría

89

Varietally correct, jammy, defined aromas, pleasant, balanced, floral.

Blanc Sec Ámfora Les Freses de Jesús Pobre 2021 B MO

90

Colour: straw. Nose: white fruit, white flowers, balanced. Palate: flavourful, fruity, good acidity.

Blanc Sec L'Horabona Les Freses de Jesús Pobre 2021 B MO
moscatel de alejandría

88

Jammy, yeasty notes, tasty, great length, varietally correct.

Dolç Les Freses de Jesús Pobre 2021 B MO D

89

Colour: bright yellow. Nose: ripe fruit, candied fruit, honeyed notes, citrus fruit, wild herbs. Palate: flavourful, unctuous, fruity, sweet.

COLECCIÓN DE TONELES CENTENARIOS
Pintor Sorolla, 8
03409 Cañada (Alacant/Alicante)
☎: +34 667 669 287
fondillonluisxiv@gmail.com
www.fondillonluisxiv.com

🏆 **PODIUM**

Fondillón Luis XIV 25 años T FO
100% monastrell

96

Colour: mahogany. Nose: acetaldehyde, varnish, aged wood nuances, creamy oak, candied fruit, pattiserie. Palate: powerful, flavourful, spicy, long, balanced.

Luis XIV Ánforas 2021 T
60% monastrell, 30% arco, 10% bonicaire

90

Wild, little interventionist. Colour: cherry, garnet rim. Nose: fruit preserve, powerful. Palate: long, fruity, easy to drink.

DO ALICANTE / D.O.P.

DO ALICANTE / D.O.P.

🏆 PODIUM

Fondillón
Luis XIV 50 años (Tonel Luna) T FO
100% monastrell

99

With personality, age nuances. Colour: dark mahogany. Nose: candied fruit, fruit liqueur notes, spicy, varnish, acetaldehyde, bakery, complex. Palate: fine solera notes, bitter.

FINCA COLLADO

Ctra. de Salinas a Villena, s/n
03638 Salinas (Alacant/Alicante)
☎: +34 607 510 710
hola@fincacollado.com
www.fincacollado.com

Delit 2019 T
monastrell

92

Colour: cherry, purple rim. Nose: spicy, wild herbs, ripe fruit, balanced. Palate: fruity, good acidity, long, easy to drink.

Fet a Mà 2019 T
monastrell

90

Colour: dark-red cherry. Nose: ripe fruit, dried herbs, creamy oak, fine reductive notes. Palate: ripe fruit, spicy, round tannins.

Finca Collado Messeguera 2019 B
merseguera

92

Defined aromas. Colour: yellow. Nose: expressive, fine lees, ripe fruit, dried flowers. Palate: rich, flavourful, bitter.

Flor Malvés 2020 B
malvasía

91

Colour: yellow. Nose: ripe fruit, fine lees, faded flowers. Palate: good acidity, flavourful, fine bitter notes.

Va de Bo 2019 T C
bobal

91

Colour: deep cherry. Nose: ripe fruit, neat, varietal, balanced, expressive. Palate: round tannins, great length, fruity aftertaste.

HAMMEKEN CELLARS

Llavador, 20
03700 Denia (Alacant/Alicante)
☎: +34 965 791 967
cellars@hammekencellars.com
www.hammekencellars.com

Gran Allegranza 2021 T
100% monastrell

88

Toasty, jammy, spicy, sweet.

JOAN DE LA CASA. VITICULTOR

03720 Benissa (Alacant/Alicante)
☎: +34 670 209 371
info@joandelacasa.com
www.joandelacasa.com

Dissident 2019 T
giró, syrah, cabernet sauvignon

91 🌱

Colour: deep cherry. Nose: dried herbs, meaty notes, spicy, smoky, old leather. Palate: ripe fruit, spicy, round tannins.

GG 2020 T
giró

91 🌱

Colour: light cherry. Nose: red berry notes, fine reductive notes, dried herbs, spicy. Palate: fresh, balanced, slightly dry, soft tannins.

Nimi Gerra 2019 B
moscatel de alejandría

91 🌱

Little interventionist. Colour: bright golden. Nose: dried flowers, characterful, expressive, honeyed notes. Palate: flavourful, fine bitter notes.

Nimi Naturalment Dolç 2017 B FB D
moscatel de alejandría

93

Varietally correct, representative. Colour: old gold, amber rim. Nose: honeyed notes, ripe fruit, floral. Palate: complex, flavourful, good acidity.

Nimi Tossal 2017 B R
moscatel de alejandría

91 🌱

Little interventionist. Colour: old gold. Nose: faded flowers, citrus fruit, ripe fruit, characterful. Palate: flavourful, long.

Terra Fiter 2015 T GR
giró

89 🌱

Fruity, lactic, spicy, tasty.

PEPE MENDOZA CASA AGRÍCOLA

Madrid, 6 2º
03580 Alfaz del Pi (Alacant/Alicante)
☎: +34 688 344 767
info@casaagricola.es
www.casaagricola.es

Pepe Mendoza Casa Agrícola 2020 T
70% monastrell, 27% giró, 3% alicante bouchet (Tinto)

92

Needs time. Colour: cherry, garnet rim. Nose: fruit preserve, powerful, grassy, complex. Palate: flavourful, sweetness, long.

Pepe Mendoza Casa Agrícola 2021 B
40% moscatel, 40% macabeo, 20% airén

91

Colour: bright yellow. Nose: expressive, white flowers, jasmine, dried herbs. Palate: flavourful, fruity, balanced.

Pepe Mendoza Casa Agrícola Pureza 2021 B
100% moscatel

92

Defined aromas. Colour: bright straw. Nose: expressive, white flowers, jasmine, dried herbs, fresh fruit. Palate: flavourful, fruity, balanced.

Pepe Mendoza El Veneno 2020 T BA
100% monastrell

94

Colour: deep cherry. Nose: complex, expressive, spicy, mineral, black fruit. Palate: full-bodied, long, great length, complex.

🏆 PODIUM

Pepe Mendoza Fierroca 2020 T
100% giró

97

Colour: bright cherry. Nose: complex, expressive, spicy, mineral, stone fruit. Palate: full-bodied, long, great length.

Pepe Mendoza Giró de Abargues 2020 T C
100% giró

94

Colour: bright cherry. Nose: complex, expressive, spicy, mineral, ripe fruit. Palate: full-bodied, long, great length, flavourful.

Pepe Mendoza Moscatel Pasa Origen 2016 B D
moscatel

91

Colour: bright yellow. Nose: candied fruit, honeyed notes, wild herbs, tomato, dried flowers, citrus fruit. Palate: flavourful, unctuous, fruity, sweet.

PRIMITIVO QUILES

Mayor, 4
03640 Monovar (Alacant/Alicante)
☎: +34 965 470 099
Fax: +34 966 960 235
info@primitivoquiles.com
www.primitivoquiles.com

Primitivo Quiles Fondillón 1948 T FO
monastrell

93

Colour: mahogany. Nose: candied fruit, fruit liqueur notes, characterful, caramel. Palate: fruity, flavourful, sweet, fine bitter notes.

🏆 PODIUM

Primitivo Quiles Gran Imperial 1892 BF Solera D
100% moscatel

95

Colour: mahogany. Nose: candied fruit, dry nuts, honeyed notes, overripe fruit. Palate: flavourful, old wood, long, sweet.

Primitivo Quiles Monastrell 2017 T C
monastrell

88

Hot, jammy, toasty, powerful.

SHUKHRAT KHAKIMOV & VITICULTORES

Plz Constitución, 7
03550 San Juan de Alicante (Alacant/Alicante)
☎: +34 965 943 090
Fax: +34 965 943 090
a.nogay@winexfood.com
www.winexfood.com

031 Barrica Cabernet Sauvignon 2018 T C
100% cabernet sauvignon

90

Colour: Cherry. Nose: balsamic herbs, sweet spices, scrubland, varietal. Palate: spicy, balsamic, good acidity.

DO ALICANTE / D.O.P.

DO ALICANTE / D.O.P.

031 Barrica
Alicante Bouschet 2018 T BA
87

VICENTE GANDÍA
Ctra. Cheste a Godelleta, s/n
46370 Chiva (València/Valencia)
☎: +34 962 524 242
Fax: +34 962 524 243
info@vicentegandia.com
www.vicentegandia.es

Castillo de Liria Organic 2021 T
monastrell, syrah
86 🌱

Ceremonia
Monastrell Blanc de Noirs 2021 B
monastrell
87

El Miracle Art 2020 T
tempranillo, syrah, monastrell
89
Spicy, fruity, jammy, tasty, toasty.

Puerto
Alicante Aromático 2021 B S
85

Puerto Alicante Selección 2020 T
87

Verema Monastrell
Criado en Ánfora 2020 T
monastrell
90
Colour: deep cherry. Nose: ripe fruit, dried herbs, creamy oak, mineral, earthy notes. Palate: powerful, ripe fruit, spicy, round tannins.

VINOS DE ALGUEÑA
Ctra. Rodriguillo, km. 29,5
03668 Algueña (Alacant/Alicante)
☎: +34 965 476 113
bodega@vinosdealguenya.es
www.vinosdealguenya.es

Casa Jiménez 2019 T C
100% monastrell
86

Flor de Enya 2020 T RB
100% monastrell
88 🌱
Spicy, herbal, rustic.

Fondonet Selección 5 años 2010 T BA D
monastrell
88
Age nuances, sweet, toasty, overripe.

🏆 **PODIUM**
Guardianes del Fondillón 1955 T FO
100% monastrell
96
Colour: light mahogany. Nose: powerful, complex, dry nuts, toasty, acetaldehyde. Palate: rich, long, fine solera notes, spicy.

VINOS SIERRA NORTE
Paraje La Raja, s/n
30520 Jumilla (Murcia)
☎: +34 618 323 119
jumilla@bodegasierranorte.com
www.bodegasierranorte.com

Pasión de Monastrell 2020 T
monastrell
90 🌱
Colour: cherry, purple rim. Nose: red berry notes, ripe fruit, spicy, dried herbs. Palate: flavourful, fruity, balanced.

VINS DEL COMTAT
Turballos, 11
03820 Cocentaina (Alacant/Alicante)
☎: +34 965 593 194
Fax: +34 965 593 590
vinsdelcomtat@gmail.com
www.vinsdelcomtat.com

1921 Monastrell 2019 T
monastrell
93
Full-bodied. Colour: dark-red cherry. Nose: powerful, waxy notes, fruit preserve, black fruit, scrubland. Palate: flavourful, round tannins, spicy, long.

Cristalí Dulce Natural BF D
92
Colour: bright yellow. Nose: honeyed notes, floral, expressive. Palate: rich, fruity, flavourful, balanced.

El Salze 2019 T
89
Overripe, wild, tasty, great length, herbal, spicy.

Montcabrer 2017 T C
90% cabernet sauvignon, 10% monastrell
92
Colour: deep cherry. Nose: ripe fruit, creamy oak, scrubland. Palate: ripe fruit, spicy, round tannins, good structure.

Peña Cadiella 2019 T RB
merlot, monastrell
88
Pleasant, powerful, herbal, jammy, spicy.

Santa Bárbara de Alicante 2019 T RB
monastrell, cabernet sauvignon
88
Jammy, fruity, dried herbs, spicy, tasty.

DO. ALMANSA
CONSEJO REGULADOR

Avda. Carlos III (Apdo. 158)
02640 Almansa (Albacete)
☎: +34 967 340 258 /+34 635 027 519
@: info@denominacion-origen-almansa.com
www.denominacion-origen-almansa.com

LOCATION:

In the South East region of the province of Albacete. It covers the municipal areas of Almansa, Alpera, Bonete, Corral Rubio, Higueruela, Hoya Gonzalo, Pétrola and the municipal district of El Villar de Chinchilla.

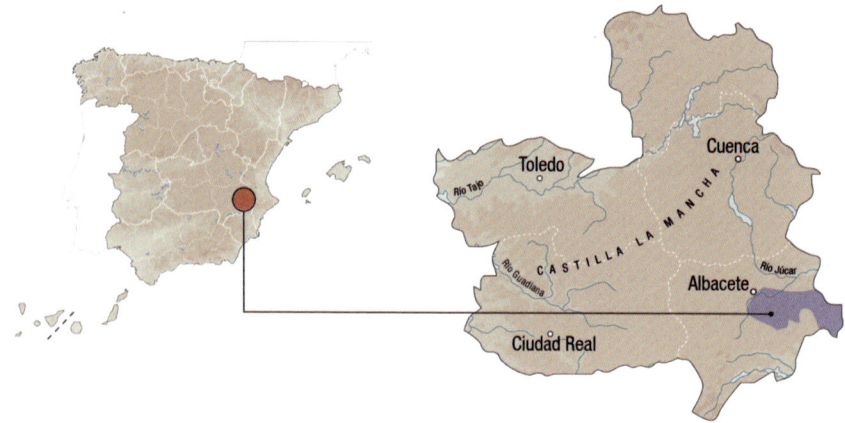

GRAPE VARIETIES:

WHITE: Chardonnay, Moscatel de grano menudo, Verdejo, Macabeo and Sauvignon Blanc.

RED: Garnacha Tintorera (most popular), Cencibel (Tempranillo), Monastrell (second most popular), Syrah, Cabernet Sauvignon, Merlot, Garnacha, Petit Verdot, Cabernet Franc and Pinot Noir.

FIGURES:

Vineyard surface: 9,800 – **Wine-Growers:** 760 – **Wineries:** 12 – **2021 Harvest rating:** Very Good – **Production 21:** 7,174,300 litres – **Market percentages:** 25% National - 75% International.

SOIL:

The soil is limy, poor in organic matter and with some clayey areas. The vineyards are situated at an altitude of about 700 m.

CLIMATE:

Of a continental type, somewhat less extreme than the climate of La Mancha, although the summers are very hot, with temperatures which easily reach 40 °C. Rainfall, on the other hand, is scant, an average of about 350 mm a year. The majority of the vineyards are situated on the plains, although there are a few situated on the slopes.

CLASSIFICATION OF YOUNG WINE HARVESTS PEÑÍNGUIDE

2017	2018	2019	2020	2021
VERY GOOD	VERY GOOD	N/A	VERY GOOD	VERY GOOD

BODEGA DEHESA EL CARRASCAL

Ctra. de Alpera, km. 94.8 - CM-3201
02691 Bonete (Albacete)
☎: +34 967 240 458
Fax: +34 967 210 989
info@dehesaelcarrascal.com
www.vinoselcarrascal.com

Mar García 2019 T
garnacha tintorera

89

Pruney, full-bodied, overripe, dried herbs, powerful. Palate: concentrated.

BODEGA RODRÍGUEZ DE VERA

Pérez Galdós, 3
02003 Albacete (Albacete)
☎: +34 696 168 873
jose@rodriguezdevera.com
www.rodriguezdevera.com

Jumenta Merlot Syrah Garnacha Tintorera 2019 T
merlot, syrah, garnacha tintorera

88

Herbal, jammy, wild, tasty, bitter.

Rodríguez de Vera Serie Limitada Chardonnay 2019 B
chardonnay

90

Colour: straw. Nose: dried herbs, faded flowers, white fruit, cereal notes. Palate: powerful, ripe fruit, balanced.

Rodríguez de Vera 2020 RD
pinot noir

91

Colour: raspberry rose. Nose: elegant, red berry notes, fragrant herbs, fine lees, dried flowers. Palate: spicy, good acidity, fine bitter notes.

Rodríguez de Vera Chardonnay Pequeñas Parcelas 2017 B FB

93

Colour: bright yellow. Nose: powerful, creamy oak, ripe fruit, spicy, toasted bread. Palate: rich, good structure, long, toasty, fine bitter notes, ripe fruit.

Rodríguez de Vera Serie Limitada 2017 T C
merlot, syrah, garnacha tintorera

90

Colour: Cherry. Nose: balsamic herbs, scrubland, black fruit, ripe fruit. Palate: spicy, balsamic, round tannins.

BODEGAS ATALAYA

Ctra. Almansa - Ayora, Km. 1
02640 Almansa (Albacete)
☎: +34 968 435 022
Fax: +34 968 716 051
info@gilfamily.es
www.gilfamily.es

Alaya Tierra 2020 T
garnacha tintorera

92

Colour: deep cherry. Nose: creamy oak, black fruit, ripe fruit. Palate: powerful, spicy, flavourful, sweet tannins.

La Atalaya del Camino 2020 T
85% garnacha tintorera, 15% monastrell

92

Kind finish, balanced. Colour: deep cherry. Nose: wild herbs, sweet spices, toasted bread. Palate: flavourful, ripe fruit, long.

Laya 2021 T
70% garnacha tintorera, 30% monastrell

91

Colour: deep cherry. Nose: dried herbs, creamy oak, ripe fruit, black fruit, black liquorice, balsamic herbs. Palate: powerful, ripe fruit, spicy, round tannins.

BODEGAS CANO

Ctra. CM-3209
02694 Higueruela (Albacete)
☎: +34 690 273 457
adolfo.cano@bodegascano.com
www.bodegascano.com

1860 Selección 2018 T R
100% garnacha tintorera

91

Colour: cherry, garnet rim. Nose: black fruit, ripe fruit, dried herbs, cocoa bean, spicy. Palate: flavourful, long, round tannins.

Cañada del Soto 2018 T C
70% monastrell, 30% garnacha tintorera

87

BODEGAS EL TANINO

Sol, s/n
02696 Hoya Gonzalo (Albacete)
☎: +34 696 946 071
dptotecnico@bodegaseltanino.com
www.bodegaseltanino.com

1752 Garnacha Tintorera Alta Expresión 2017 T C
100% garnacha tintorera

90

Toasty, spicy. Nose: smoky, black fruit, ripe fruit. Palate: flavourful, harsh oak tannins.

DO ALMANSA / D.O.P.

DO ALMANSA / D.O.P.

BODEGAS PIQUERAS
Zapateros, 11
02640 Almansa (Albacete)
☎: +34 967 341 482
info@bodegaspiqueras.es
www.bodegaspiqueras.es

El Abuelo Selección 2017 T
monastrell, syrah, garnacha tintorera

89
Fruity, herbal, spicy, astringent.

Los Losares Garnacha Tintorera 2019 T
garnacha tintorera

90
Colour: cherry, purple rim. Nose: ripe fruit, dried herbs. Palate: fruity, flavourful, slightly dry, soft tannins.

Los Losares Monastrell 2019 T
monastrell

90
Colour: deep cherry. Nose: red berry notes, ripe fruit, dried herbs, spicy. Palate: fruity, flavourful, balanced, powerful tannins.

Piqueras Verdejo Wild Fermented 2021 B FB
verdejo

88
Pleasant, jammy, tasty, tropical.

Piqueras VS 2017 T
monastrell, garnacha tintorera

90
Colour: deep cherry. Nose: black fruit, ripe fruit, grassy, smoky. Palate: flavourful, full-bodied, balanced, slightly dry, soft tannins.

BODEGAS VIRGEN DE BELÉN
Paraje de Belén s/n Apdo. 540
02640 Almansa (Albacete)
☎: +34 655 624 850
bodega@bodegastiopio.com
www.bodegastiopio.com

Bastón 2019 T
petit verdot

88
Full-bodied, spicy, jammy, toasty, wild.

Monóculo 2019 T RB
syrah

89
Full-bodied, jammy, spicy, tasty, varietally correct.

BODEGAS VOLVER
Ctra de Pinoso a Fortuna, s/n
03658 Rodriguillo (Alacant/Alicante)
☎: +34 966 185 624
Fax: +34 965 075 376
export@bodegasvolver.com
www.bodegasvolver.com

La Quinta de Rafa 2021 T
30% syrah, 0% garnacha tintorera

89
Smoky, spicy, herbal, hot, toasty, sweety finish, pruney.

Quinta del 67 2020 T
garnacha tintorera

91
Colour: cherry, garnet rim. Nose: fruit preserve, powerful, sweet spices. Palate: flavourful, long, slightly dry, soft tannins.

DO ALMANSA / D.O.P.

BODEGAS WINOL

Poeta García Carbonell, 8 - Bajo
02005 Albacete (Albacete)
☎: +34 686 136 446
aochoa@bodegaswinol.com
www.bodegaswinol.com

8Lgends Leyenda del Jinete 2017 T C
100% garnacha tintorera

88

Jammy, great length, tasty, spicy, coarse tannins.

BODEGAS Y VIÑEDOS VENTA LA VEGA

Ctra. de Alpera, CM 3201 Km. 98,6
02640 Almansa (Albacete)
☎: +34 965 928 857
info@mgwinesgroup.com
www.mgwinesgroup.com

Adaras Calizo Garnacha Tintorera 2021 T BA
garnacha tintorera

90

Colour: cherry, purple rim. Nose: black fruit, ripe fruit, balsamic herbs, spicy, smoky. Palate: fruity, fresh, flavourful.

Adaras Huella Garnacha Tintorera - Monastrell 2020 T SS
garnacha tintorera, monastrell

91 🌱

Colour: cherry, purple rim. Nose: ripe fruit, spicy, scrubland, dried herbs. Palate: fruity, flavourful, balanced, good finish.

Adaras Kalizo sin Sulfitos 2021 T S
garnacha tintorera

87 🌱

Adaras Lluvia 2021 B SS
verdejo, sauvignon blanc

88 🌱

Pleasant, fruity, yeasty notes, crisp, tasty.

Aldea de Adaras 2021 T BA
garnacha tintorera, syrah

88 🌱

Overripe, spicy, fruity.

L'Entrada P.7 2020 B
sauvignon blanc

91 🌱

Colour: bright straw. Nose: fine lees, spicy, bakery, white fruit, ripe fruit. Palate: rich, long, good acidity, fine bitter notes.

COOP. SANTA QUITERIA - TINTORALBA

Baltasar González Sáez, 34
02694 Higueruela (Albacete)
☎: +34 967 287 012
Fax: +34 967 287 031
direccion@tintoralba.com

Tintoralba 2018 T C
garnacha tintorera, syrah

88

Jammy, toasty, dried herbs, spicy, tasty.

Tintoralba 2021 T RB
100% garnacha tintorera

87

Tintoralba Garnacha Tintorera 2021 T
100% garnacha tintorera

88

Pleasant, fruity, jammy, tasty, smooth.

Tintoralba Sauvignon Blanc - Verdejo 2021 B
sauvignon blanc, verdejo

87

Tintoralba Selección 2017 T C
garnacha tintorera, syrah

90

Colour: bright cherry. Nose: sweet spices, ripe fruit, dark chocolate, black fruit, creamy oak. Palate: fruity, spicy, round tannins.

Tintoralba Syrah 2021 RD
100% syrah

87

PACO MULERO

Partida de la Hoya Torres s/n
30520 Jumilla (Murcia)
☎: +34 676 433 541
info@pacomulero.com
www.pacomulero.com

Prisma Garnacha Tintorera Monastrell 2021 T
75% garnacha tintorera, 25% monastrell

90

Colour: cherry, garnet rim. Nose: fruit preserve, fruit liqueur notes, powerful. Palate: flavourful, sweetness, long.

DO ALMANSA / D.O.P.

Paco Mulero Quince Meses 2020 T
garnacha tintorera

91

Colour: very deep cherry. Nose: roasted coffee, powerful, black fruit, dried herbs, woody. Palate: smoky aftertaste, great length, round tannins.

Esencia Rupestre
Garnacha Tintorera 2016 T C
garnacha tintorera

91

Colour: cherry, garnet rim. Nose: fruit preserve, fruit liqueur notes, powerful, roasted coffee. Palate: flavourful, sweetness, long.

Gold Rupestre
Garnacha Tintorera 2019 T C
garnacha tintorera

91

Colour: cherry, garnet rim. Nose: overripe fruit, creamy oak, hot, black fruit. Palate: pruney, powerful, sweet tannins.

Rupestre de Alpera
Garnacha Tintorera 2018 T R
garnacha tintorera

91

Colour: cherry, garnet rim. Nose: fruit preserve, powerful, dark chocolate, smoky. Palate: flavourful, long.

SPANISH PALATE
Avda. Carlos Latorre, 30
49800 Toro (Zamora)
☎: +34 980 690 643
export@spanishpalate.es
www.spanishpalate.es

Botas de Barro Almansa 2019 T RB
100% garnacha tintorera

87

SANTA CRUZ DE ALPERA SOC. COOP. DE C-L-M
Cooperativa, s/n
02690 Alpera (Albacete)
☎: +34 967 330 108
Fax: +34 967 330 903
laboratorio@bodegasantacruz.com
www.bodegasantacruz.com

Cueva de Chamán
Roble Monastrell 2020 T RB
monastrell

88 🌿

Jammy, toasty, tasty, spicy, great length.

Cueva del Chamán
Garnacha Tintorera 2021 T
garnacha tintorera

91 🌿

Colour: cherry, purple rim. Nose: fruit expression, red berry notes, floral, black fruit. Palate: fruity, flavourful, balanced.

Cueva del Chamán Verdejo 2021 B
verdejo

88 🌿

Aromatic, tropical, tasty, jammy.

DO. ARABAKO TXAKOLINA

CONSEJO REGULADOR

Dionisio Aldama, 7- 1ºD Apdo. 36
01470 Amurrio (Álava)
☎: +34 656 789 372 - +34 945 393 786
@: merino@txakolidealava.eus
www.txakolidealava.com

LOCATION:

It covers the region of Aiara (Ayala), situated in the north west of the province of Alava on the banks of the Nervion river basin. Specifically, it is made up of the municipalities of Amurrio, Artziniega, Aiara (Ayala), Laudio (Llodio) and Okondo.

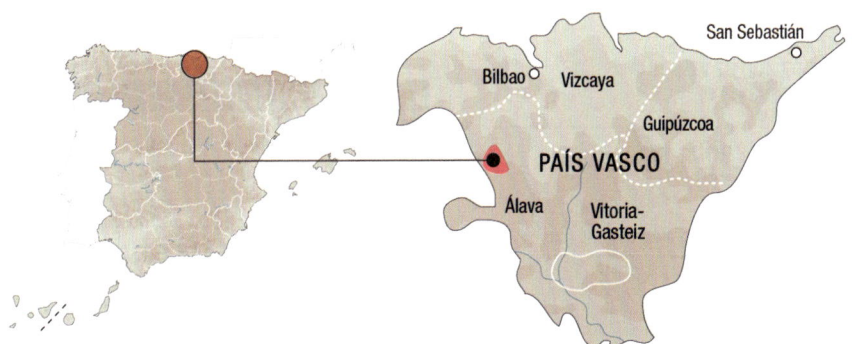

GRAPE VARIETIES:

WHITE:

MAIN: Hondarrabi Zuri.

AUTHORIZED: Petit Manseng, Petit Corbu and Gross Manseng.

FIGURES:

Vineyard surface: 95 – **Wine-Growers:** 43 – **Wineries:** 7 – **2021 Harvest rating:** Excellent   **Production 21:** 315,000 litres – **Market percentages:** 62% National - 38% International.

SOIL:

A great variety of formations are found, ranging from clayey to fundamentally stony, precisely those which to date are producing the best results and where fairly stable grape ripening is achieved.

CLIMATE:

Similar to that of the DO Bizkaiko Txakolina, determined by the influence of the Bay of Biscay, although somewhat less humid and slightly drier and fresher. In fact, the greatest risk in the region stems from frost in the spring. However, it should not be forgotten that part of its vineyards borders on the innermost plantations of the DO Bizkaiko Txakolina.

CLASSIFICATION OF YOUNG WINE HARVESTS PEÑÍNGUIDE

2017	2018	2019	2020	2021
N/A	N/A	N/A	N/A	N/A

DO ARABAKO TXAKOLINA / D.O.P.

ARTOMAÑA TXAKOLINA
Masalarreina, s/n
01468 Artomaña (Araba/Álava)
☎: +34 620 007 452
info@artomanatxakolina.com
www.artomanatxakolina.eus

Eukeni Txakoli 2021 B
80% hondarrabi zuri, 10% gros manseng, 10% petit corbú
88
Crisp, fruity, herbal, simple.

Xarmant Txakoli 2021 B
80% hondarrabi zuri, 20% petit corbú
87

BODEGA ASTOBIZA
Bº Jandiola 16
01409 Okondo (Araba/Álava)
☎: +34 607 400 321
comercial@astobiza.es
www.astobiza.es

Astobiza Rosé 2021 RD
hondarrabi beltza, hondarrabi zuri
88
Crisp, fruity, wild, simple. Palate: slight fizz.

Malkoa Private Collection 2017 B BA S
hondarrabi zuri
94
Colour: yellow. Nose: complex, expressive, floral, fresh, spicy. Palate: rich, long, balanced, fine bitter notes, spicy.

Malkoa Private Collection 2018 B BA S
hondarrabi zuri
94
Colour: yellow. Nose: complex, bakery, fine lees, fresh. Palate: rich, spicy, saline, flavourful, balanced.

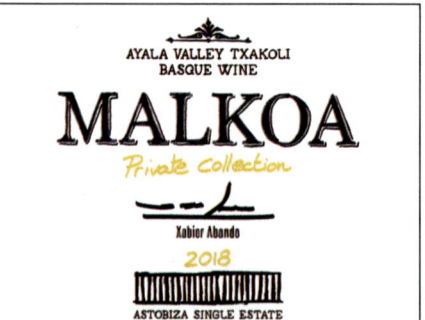

Malkoa Txakoli Edición Limitada 2018 B
hondarrabi zuri
93
Crisp. Colour: yellow. Nose: ripe fruit, fine lees, complex, expressive, floral. Palate: long, good acidity.

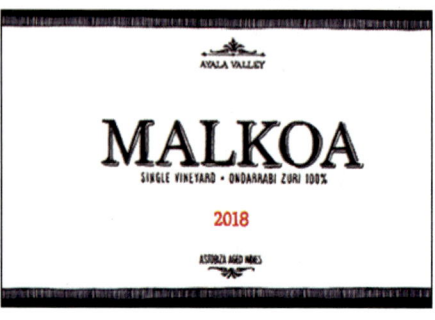

BODEGA BATGARA
Barrio Urtaran, 21
01450 Lezama Amurrio (Araba/Álava)
☎: +34 609 884 826
info@batgara.com
www.batgara.com

Batgara 18 meses 2018 B BA
hondarrabi zuri
93
Colour: bright yellow. Nose: powerful, creamy oak, ripe fruit, spicy, tropical fruit. Palate: rich, good structure, long, toasty, good acidity, mineral.

Batgara Aromas del Sur 2019 B BA
100% hondarrabi zuri
92
Oxidativ, original. Colour: bright straw. Nose: macerated fruit, dry nuts, cereal notes, sweet spices, pattiserie. Palate: flavourful, good structure, good acidity, fine solera notes.

Txakoli Uno 2020 B
95% hondarrabi zuri, 5% riesling
92
Colour: bright straw. Nose: ripe fruit, fine lees, spicy, smoky, mineral. Palate: full-bodied, good acidity, flavourful.

Urtaran 2020 B FB
95% hondarrabi zuri, 5% riesling
93
Colour: bright straw. Nose: fresh fruit, white fruit, citrus fruit, dry stone, wild herbs, toasted bread, fine lees. Palate: good structure, elegant, flavourful.

TANTAKA WINES

Pol. Maskuribai Pabellón Z 12
01470 Amurrio (Araba/Álava)
☎: +34 656 714 709
info@tantaka.eus
www.tantaka.eus

Tantaka 2019 B
hondarrabi zuri

91

Colour: bright straw. Nose: ripe fruit, fragrant herbs, fine lees, smoky. Palate: good acidity, flavourful.

Tantaka 2019 T
hondarrabi beltza

91

Little interventionist, slight oxidation. Colour: light cherry. Nose: red berry notes, red clay notes, wild herbs, animal reductive notes. Palate: light-bodied, mineral, grainy tannins.

Tantaka 2020 T
hondarrabi beltza

91

Little interventionist, slight oxidation. Colour: light cherry. Nose: red berry notes, floral, wild herbs, smoky. Palate: fresh, flavourful, grainy tannins.

Tantaka Diapiro (Lacre Calabaza) 2019 B
80% hondarrabi zuri, 20% petit corbú

93

Colour: bright straw. Nose: fragrant herbs, fine lees, fresh fruit, smoky, tropical fruit. Palate: good acidity, powerful, flavourful, good structure, mineral.

Tantaka Diapiro (Lacre Verde) 2019 B
85% hondarrabi zuri, 15% riesling

94

Colour: bright straw. Nose: ripe fruit, fragrant herbs, fine lees, petrol notes. Palate: full-bodied, long, good acidity, spicy.

VINTAE / ATLANTIS

Caserío Beldio – Barrio Galdea
01400 Llodio (Araba/Álava)
☎: +34 941 483 557
marketing@vintae.com
www.vintae.com

Atlantis Hondarrabi Zuri 2020 B
hondarrabi zuri

89

Pleasant, aromatic, crisp, fruity, tasty. Nose: hint of anise.

DO ARABAKO TXAKOLINA / D.O.P.

DO. ARLANZA
CONSEJO REGULADOR

Ronda de la Cárcel, 4 - Edif. Arco de la Cárcel
09340 Lerma (Burgos)
☎: +34 947 171 046 - Fax: +34 947 171 046
@: info@arlanza.org
www.arlanza.org

LOCATION:

With the medieval city of Lerma at the core of the region, Arlanza occupies the central and southern part of the province of Burgos, on the river valleys of the Arlanza and its subsidiaries, all the way westwards through 13 municipal districts of the province of Palencia until the Pisuerga River is reached.

GRAPE VARIETIES:

WHITE: Albillo and Viura.

RED: Tempranillo, Garnacha, Mencía, Cabernet Sauvignon, Merlot and Petit Verdot.

FIGURES:

Vineyard surface: 350 – **Wine-Growers:** 267 – **Wineries:** 18 – **2021 Harvest rating:** Very Good – **Production 21:** 1,089,558 litres – **Market percentages:** 90% National - 10% International.

SOIL:

Soil in the region is not particularly deep, with soft rocks underneath and good humidity levels. The landscape is one of rolling hills where vines are planted on varied soils, from limestone to calcareous, with abundant granite on certain areas.

CLIMATE:

The climate of this wine region is said to be one of the harshest within Castilla y León, with lower temperatures towards the western areas and rainfall higher on the eastern parts, in the highlands of the province of Soria.

CLASSIFICATION OF YOUNG WINE HARVESTS PEÑÍNGUIDE

2017	2018	2019	2020	2021
N/A	N/A	N/A	N/A	N/A

ARLESE NEGOCIOS
Pol. Ind. de Villalmanzo Parcela 106
09390 Villalmanzo (Burgos)
☎: +34 947 172 866
info@bodegasarlese.com

Almanaque 2017 T C
100% tempranillo
89
Jammy, pruney, tasty, spicy, balanced, pleasant.

Almanaque 2017 T RB
100% tempranillo
87

Almanaque 2020 T
100% tempranillo
88
Fruity, jammy, great length, tasty, coarse tannins.

Almanaque 2021 RD
100% tempranillo
88
Pleasant, fruity, jammy, sweeties, balanced, tasty.

BODEGAS DECORUS
N-1, km 203
09340 Lerma (Burgos)
☎: +34 670 273 809
luroam@hotmail.es

Decorus 2020 T BA S
tempranillo, garnacha, mencía, viura, albillo mayor
90 🌿
Colour: deep cherry. Nose: ripe fruit, dried herbs, creamy oak. Palate: powerful, ripe fruit, spicy, round tannins.

BODEGAS LERMA
Travesía Madrid-Irún, A-1, Km. 203
09340 Lerma (Burgos)
☎: +34 947 177 030
info@tintolerma.com
www.bodegaslerma.com

Gran Lerma Vino de Autor 2016 T R
tempranillo
92
Colour: dark-red cherry, garnet rim. Nose: ripe fruit, aged wood nuances, tobacco, sweet spices. Palate: spicy, round tannins, long.

Lerma 2018 T C
tempranillo
89
Kind finish, classic, toasty, jammy, spicy.

BODEGAS MONTE AMÁN
Ctra. de Lerma a Silos, 5
09348 Castrillo de Solarana (Burgos)
☎: +34 947 173 304
bodegas@monteaman.com
www.monteaman.com

Alto Carmona 2017 T C
100% tempranillo
90
Colour: deep cherry. Nose: dried herbs, creamy oak, black fruit. Palate: ripe fruit, spicy, round tannins.

Alto Carmona 2018 T R
100% tempranillo
91
Colour: deep cherry. Nose: dried herbs, creamy oak, black fruit, meaty notes. Palate: powerful, ripe fruit, spicy, round tannins.

Monte Amán 2017 T C
100% tempranillo
89
Balanced, spicy, creamy, jammy, toasty.

Monte Amán 2018 T RB
tempranillo
88
Creamy, spicy, toasty, coarse tannins.

Monte Amán 2020 T
100% tempranillo
88
Balanced, spicy, fruity, jammy, mineral.

Monte Amán 2021 RD
100% tempranillo
84

BODEGAS VALDESNEROS
Avda. La Paz, 4
34230 Torquemada (Palencia)
☎: +34 979 800 545
Fax: +34 979 800 545
sv@bodegasvaldesneros.com
www.bodegasvaldesneros.com

Cornitero 2021 T MC
tempranillo
87

Eruelo 2016 T C
tempranillo
87

Neros 2021 B
83

DO ARLANZA / D.O.P.

DO ARLANZA / D.O.P.

Neros 2021 RD
tempranillo, garnacha, mencía

87

Valdesneros 2021 RD
tempranillo

88

Jammy, tasty, aromatic, balanced.

BODEGAS Y VIÑEDOS VALTRAVIESO

Finca La Revilla, s/n
47316 Piñel de Arriba (Valladolid)
☎: +34 983 484 030
marketing@valtravieso.com
www.valtravieso.com

Cerro Cerezo 2019 T
tempranillo, mencía, monastrell, bobal, garnacha

93

Colour: cherry, purple rim. Nose: red berry notes, creamy oak, spicy, dried herbs, woody. Palate: good structure, good acidity, balanced, smoky aftertaste, slightly dry, soft tannins.

Las Mamblas 2019 T
tempranillo, mencía, monastrell, bobal, garnacha

92

Colour: cherry, purple rim. Nose: floral, spicy, red berry notes, black fruit, ripe fruit, wild herbs. Palate: flavourful, fruity, good acidity, elegant.

Muniadona 2019 B FB
albillo mayor, viura, cayetana blanca, castellana, alarije

92

Colour: bright yellow. Nose: powerful, creamy oak, ripe fruit, toasty. Palate: rich, good structure, long.

DO. ARRIBES
CONSEJO REGULADOR

Plaza Mayor, 1
49230 Cibanal (Zamora)
☎: +34 669 216 576 - +34 687 846 655
@: info@doarribes.es
www.doarribes.es

LOCATION:

In Las Arribes National Park, it comprises a narrow stretch of land along the southwest of Zamora and northeast of Salamanca. The vineyards occupy the valleys and steep terraces along the river Duero. Just a single municipal district, Fermoselle, has up to 90% of the total vineyard surface.

GRAPE VARIETIES:

WHITE: Malvasía, Verdejo, Albillo and Puesta en cruz.

RED: Bruñal, Juan García, Rufete, Tempranillo, Syrah, Mencía and Garnacha. (preferential); Bastardillo Chico, Gajo Arroba, Mandón and Tinta Jeromo (authorized).

FIGURES:

Vineyard surface: 274 – **Wine-Growers:** 179 – **Wineries:** 21 – **2021 Harvest rating:** N/A – **Production 21:** 783,863 litres – **Market percentages:** 60% National - 40% International.

SOIL:

The region has shallow sandy soils with abundant quartz and stones, even some granite found in the area of Fermoselle. In the territory which is part of the province of Salamanca it is quite noticeable the presence of slate, the kind of rock also featured on the Portuguese part along the Duero, called Douro the other side of the border. The slate subsoil works a splendid thermal regulator capable of accumulating the heat from the sunshine during the day and to slowly release it during the night time.

CLIMATE:

It is marked by a Mediterranean influence. The uneven terrain with abundant slopes forms a climate with important differences depending on the altitude and the location of the vineyard. For instance, in the canyons where there is no frost and the average temperature in winter is 9º and 26º in the summer. In the peneplain, winters are cold and long, and summers are short and hot, while in the river valleys, the "arribe", the temperatures are considerably higher, by about 5ºC higher than the peneplain. In the "arribe" winters are shorter while summers are longer. This territory has such a marked microclimate that it even allows the cultivation of orange and olive trees.

CLASSIFICATION OF YOUNG WINE HARVESTS PEÑÍNGUIDE

2017	2018	2019	2020	2021
VERY GOOD	N/A	N/A	N/A	N/A

DO ARRIBES / D.O.P.

BODEGA ARRIBES DEL DUERO

Ctra. Masueco, s/n
37251 Corporario - Aldeadavila (Salamanca)
☎: +34 923 169 195
Fax: +34 923 169 195
secretaria@bodegasarribesdelduero.com
www.bodegasarribesdelduero.com

Arribes de Vettonia 2018 T R
90
Woody. Colour: dark-red cherry. Nose: toasty, spicy, waxy notes, black fruit. Palate: flavourful, good structure, slightly dry, soft tannins.

Arribes de Vettonia 2019 T C
89
Fruity, spicy, toasty, jammy, tasty.

Arribes de Vettonia 2021 B
86

Arribes de Vettonia 2021 RD
85

Arribes de Vettonia Vendimia Seleccionada 2016 T R
91
Colour: dark-red cherry, garnet rim. Nose: ripe fruit, fruit preserve, aged wood nuances, sweet spices. Palate: spicy, round tannins, long, fruity, flavourful.

Hechanza Real 2019 T C
89
Jammy, rustic, balanced, spicy, pleasant. Nose: black pepper, ripe fruit.

BODEGA FRONTIO

Calle Portilla, 42
49220 Fermoselle (Zamora)
☎: +34 652 460 462
bodegafrontio@gmail.com

Bébeme 2020 T RB
juan garcía
92
Colour: cherry, purple rim. Nose: red berry notes, ripe fruit, wild herbs, scrubland, spicy. Palate: fruity, fresh, flavourful, balanced, good finish.

Corneo 2020 RD RB
juan garcía
90
Little interventionist. Nose: red berry notes, wild herbs, wild herbs, spicy. Palate: fresh, fruity, flavourful, good acidity, balanced.

Follaco 2020 T R
juan garcía, tempranillo
90
Colour: cherry, purple rim. Nose: red berry notes, wild herbs, ripe fruit, wild herbs, spicy. Palate: fresh, fruity, flavourful, good acidity, slightly dry, soft tannins.

Rocoso 2021 T RB
juan garcía
89
Little interventionist, with personality. Colour: cherry, purple rim. Nose: black fruit, spicy, ripe fruit, meaty notes. Palate: flavourful, fresh, fruity, slightly dry, soft tannins.

BODEGA QUINTA LAS VELAS

Humilladero, 44
37248 Ahigal de Los Aceiteros (Salamanca)
☎: +34 619 955 735
enrique@esla.com
www.quintalasvelas.com

Bruñal Quinta las Velas 2019 T C
bruñal
90
Colour: cherry, purple rim. Nose: fruit liqueur notes, fruit preserve, ripe fruit, black fruit, dried herbs, spicy. Palate: fruity, flavourful, fresh, slightly dry, soft tannins.

Origen Bruñal Quinta las Velas 2017 T R
bruñal
91
Rustic. Colour: cherry, garnet rim. Nose: wild herbs, black fruit, fresh fruit, balsamic herbs, expressive, neat. Palate: fresh, fruity, easy to drink.

Quinta las Velas Tempranillo 2019 T C
tempranillo
87

BODEGAS FRANCISCO RODRÍGUEZ GARROTE

Parque, 30
49166 Villalcampo (Zamora)
☎: +34 696 491 471
info@bodegasfrg.com
www.bodegasfrg.es

Begoa 2019 B
malvasía, verdejo
90
Colour: bright straw. Nose: white fruit, ripe fruit, floral, wild herbs. Palate: fruity, fresh, flavourful, good acidity, fine bitter notes.

BODEGAS PASCUAL FERNÁNDEZ - FRONTERA NATURAL
Subida Las Fontanicas 38
49220 Fermoselle (Zamora)
☎: +34 630 027 097
bodega@bodegaspascualfernandez.com
www.sietepeldaños.com

Siete Peldaños 2020 B BA
100% malvasía castellana

90
Colour: bright golden. Nose: ripe fruit, wild herbs, dried herbs, caramel. Palate: fruity, flavourful, full-bodied, balanced, toasty, roasted-coffee aftertaste.

Siete Peldaños Bruñal 2020 T
100% bruñal

92
Colour: cherry, purple rim. Nose: black fruit, ripe fruit, dried herbs, spicy, dark chocolate. Palate: flavourful, fruity, good structure, balanced, spicy, slightly dry, soft tannins.

Siete Peldaños Doña Blanca 2020 B
100% puesta en cruz

90
Colour: bright straw. Nose: fruit expression, white fruit, wild herbs, fine lees. Palate: fruity, full-bodied, flavourful, good acidity, good finish, mineral.

Siete Peldaños Garnacha 2020 T C
100% garnacha

91
Colour: cherry, purple rim. Nose: black fruit, ripe fruit, spicy, creamy oak, floral. Palate: flavourful, fruity, racy, powerful tannins.

Siete Peldaños Malvasía 2020 B
100% malvasía castellana

91
Colour: bright straw. Nose: white fruit, wild herbs, faded flowers, mineral. Palate: fruity, flavourful, balanced, fine bitter notes.

Siete Peldaños Tempranillo 2020 T
100% tempranillo

91
Colour: cherry, purple rim. Nose: ripe fruit, black fruit, spicy, black pepper, dried herbs. Palate: fruity, flavourful, fresh, good structure, balanced, good finish, smoky aftertaste.

BODEGAS PASTRANA
Rumía, 62
49220 Fermoselle (Zamora)
☎: +34 664 546 131
info@bodegaspastrana.es
www.bodegaspastrana.es

Paraje de los Bancales 2018 T R
juan garcía, tinto fino, otras

90
Colour: bright cherry. Nose: black fruit, ripe fruit, dried flowers, spicy, smoky. Palate: fruity, flavourful, good structure, balanced, slightly dry, soft tannins.

BODEGAS VIÑA ROMANA
España, 50
37160 Villarino de Los Aires (Salamanca)
☎: +34 629 756 328
joseluis@vinaromana.com
www.vinaromana.com

Heredad del Viejo Imperio Homenaje 2016 T
100% bruñal

89
Weary, age nuances, pruney, dried herbs, full-bodied.

Winner Premium 2016 T C

90
Colour: deep cherry. Nose: ripe fruit, dried herbs, creamy oak, fruit preserve, spicy. Palate: powerful, ripe fruit, spicy, round tannins, flavourful.

EL HATO Y EL GARABATO
Palazuelo, 4
49230 Formariz de Sayago (Zamora)
☎: +34 658 363 568
jose@elhatoyelgarabato.com
www.elhatoyelgarabato.com

De Buena Jera 2018 T C

92 🌱
Colour: Cherry. Nose: ripe fruit, black fruit, black pepper, fine reductive notes, characterful, expressive. Palate: spicy, fine bitter notes, fine tannins, good acidity.

Ecléctico - Puesta en Cruz Lías B

88
Citrus fruit, fruity, herbal, pronounced acidity.

Otro Cuento 2019 B
90% dona blanca, 10% puesta en cruz, verdejo, godello, albillo

92 🌱
Colour: bright yellow. Nose: ripe fruit, fruit expression, wild herbs, dried flowers, spicy. Palate: fresh, flavourful, balanced, smoky aftertaste.

DO ARRIBES / D.O.P.

DO ARRIBES / D.O.P.

Sin Blanca 2018 T C
92 🌿
With personality, wild, representative. Colour: cherry, garnet rim. Nose: ripe fruit, scrubland, fine reductive notes, wild herbs. Palate: fruity, fine bitter notes, easy to drink.

LA SETERA
Calzada, 7
49232 Fornillos de Fermoselle (Zamora)
☎: +34 676 052 315
info@lasetera.com
www.lasetera.com

La Setera 2019 T C
juan garcía
88
Fruity, pruney, spicy, slight oxidation, toasty.

La Setera 2021 B
malvasía, castellana
85

La Setera 2021 T
juan garcía
89
Fruity, herbal, jammy, tasty, wild.

La Setera Selección Especial 2014 T C
touriga nacional
90
Colour: deep cherry. Nose: black fruit, fruit preserve, violets, scrubland, spicy, toasty. Palate: flavourful, fruity, powerful, balanced, powerful tannins.

La Setera Tinaja Varietales 2015 T RB
juan garcía, mencía, rufete, bastardillo, bruñal, tinta Madrid
90
Colour: deep cherry. Nose: creamy oak, black fruit, wild herbs, spicy. Palate: ripe fruit, spicy, round tannins, sweetness.

OCELLUM DURII
San Juan 56 - 58
49220 Fermoselle (Zamora)
☎: +34 983 390 606
ocellumdurii@hotmail.com
www.bodegasocellumdurii.com

Condado de Fermosel 2015 T C
juan garcía, tempranillo, mencía, rufete, bruñal
91
Colour: dark-red cherry, garnet rim. Nose: ripe fruit, fruit preserve, aged wood nuances, tobacco, sweet spices, wild herbs, earthy notes. Palate: spicy, round tannins, long, great length.

Condado de Fermosel 2016 T C
juan garcía, tempranillo, bruñal, mencía
90
Weary. Colour: dark-red cherry, garnet rim. Nose: aged wood nuances, tobacco, sweet spices, fruit preserve, fruit liqueur notes, wild herbs, toasty, earthy notes. Palate: spicy, round tannins, long, good structure.

Entrelimites La Balanza Bruñal 2015 T
bruñal, tempranillo, garnacha
90
Rustic. Colour: deep cherry. Nose: ripe fruit, dried herbs, creamy oak, toasty. Palate: powerful, ripe fruit, spicy, round tannins.

Entrelimites Limite Natural 2016 T
juan garcía, tempranillo, bruñal, rufete, mencía
91 🌿
Colour: deep cherry. Nose: ripe fruit, dried herbs, creamy oak, earthy notes, tobacco. Palate: powerful, ripe fruit, spicy, round tannins.

Entrelimites Limite Natural 2019 T
tempranillo, juan garcía, malvasía
90
Colour: cherry, purple rim. Nose: fruit expression, floral, spicy, red berry notes. Palate: flavourful, fruity, good acidity, long.

Entrelimites Transitium Durii 2009 T GR
malvasía negra, juan garcía, bruñal, tempranillo, rufete, albillo
91
Colour: deep cherry. Nose: dried herbs, creamy oak, cigar, fruit liqueur notes. Palate: powerful, ripe fruit, spicy, round tannins.

PARDAL Y PUNTO
Fontanicas, 50
49220 Fermoselle (Zamora)
☎: +34 656 252 611
info@bodegapardalypunto.com

Sabaria 2019 T C
100% bruñal
92
Colour: cherry, purple rim. Nose: black fruit, fruit preserve, dried flowers, black pepper, spicy. Palate: fruity, flavourful, good structure, fresh, balanced, good finish, smoky aftertaste, slightly dry, soft tannins.

Salsipuedes 2019 T C
tempranillo, juan garcía, bruñal
90
Colour: deep cherry. Nose: ripe fruit, dried herbs, creamy oak, fruit preserve, black fruit. Palate: powerful, ripe fruit, spicy, round tannins, fresh.

PEÑOS MARTIN MARCOS

San Juan, 21
49220 Fermoselle (Zamora)
☎: +34 639 124 635
vinumromanorum@gmail.com
www.vinumromanorum.com

Romanorum 2016 T R
50% tempranillo, 50% juan garcía

88

Fruity, sweeties, spicy, toasty, wild, astringent.

Romanorum 2018 T C
juan garcía, tempranillo

87

DO. BIERZO

CONSEJO REGULADOR

Mencía, 1
24540 Cacabelos (León)
☎: +34 987 549 408 - Fax: +34 987 547 077
@: info@crdobierzo.es
www.crdobierzo.es

LOCATION:

In the north west of the province of León. It covers 23 municipal areas and occupies several valleys in mountainous terrain and a flat plain at a lower altitude than the plateau of León, with higher temperatures accompanied by more rainfall. It may be considered an area of transition between Galicia, León and Asturias.

GRAPE VARIETIES:

WHITE: Godello, Palomino, Dona Blanca and Malvasia.

RED: Mencía or Negra, Garnacha Tintorera, Estaladiña and Merenzao.

FIGURES:

Vineyard surface: 2,420 – **Wine-Growers:** 1,109 – **Wineries:** 72 – **2021 Harvest rating:** Excellent – **Production 21:** 7,889,893 litres – **Market percentages:** 67% National - 33% International.

SOIL:

In the mountain regions, it is made up of a mixture of fine elements, quartzite and slate. In general, the soil of the DO is humid, dun and slightly acidic. The greater quality indices are associated with the slightly sloped terraces close to the rivers, the half - terraced or steep slopes situated at an altitude of between 450 and 1,000 m.

CLIMATE:

Quite mild and benign, with a certain degree of humidity due to Galician influence, although somewhat dry like Castilla. Thanks to the low altitude, late frost is avoided quite successfully and the grapes are usually harvested one month before the rest of Castilla. The average rainfall per year is 721 mm.

CLASSIFICATION OF YOUNG WINE HARVESTS PEÑÍNGUIDE

2017	2018	2019	2020	2021
GOOD	VERY GOOD	VERY GOOD	VERY GOOD	EXCELLENT

ALMÁZCARA MAJARA
Las Eras, 5
24398 Almázcara (León)
☎: +34 609 322 194
info@almazcaramajara.com
www.almazcaramajara.com

Almázcara Majara 2018 T C
100% mencía
92
Wild, representative. Colour: bright cherry. Nose: waxy notes, ripe fruit, wild herbs, balanced. Palate: fruity, balanced, spicy.

Cobija del Pobre 2021 B
100% godello
90
Colour: straw. Nose: white flowers, wild herbs, white fruit, ripe fruit. Palate: fruity, balanced.

Demasiado Corazón 2019 B FB
100% godello
90
Slight oxidation. Colour: bright yellow. Nose: powerful, ripe fruit, spicy, faded flowers, dried herbs. Palate: rich, good structure, long, fine bitter notes.

Jarabe de Almázcara Majara 2019 T
100% mencía
89
Pleasant, smoky, tasty, rustic.

L'Aphrodisiaque Godello 2021 B
85% godello, 15% jerez
90
Colour: straw. Nose: ripe fruit, dried herbs. Palate: powerful, ripe fruit, good acidity.

L'Aphrodisiaque Mencía 2021 T
100% mencía
88
Kind finish, floral, fruity, simple.

ARTURO GARCÍA VIÑEDOS Y BODEGAS
La Escuela, 3
24516 Toral de los Vados (León)
☎: +34 987 553 000
Fax: +34 987 553 001
info@bodegasarturo.es
www.bodegasarturo.es

Hacienda Elsa Godello 2021 B
88
Pleasant, aromatic, floral, fruity.

Hacienda Elsa Mencía 2020 T
mencía
89
Toasty, tasty, jammy, spicy.

Hacienda Sael Godello 2021 B
godello
88
Kind finish, aromatic, tasty, fruity.

Hacienda Sael Mencía 2020 T
mencía
89
Pleasant, toasty, jammy, spicy.

Solar de Sael Mencía 2018 T C
mencía
88
Pruney, toasty, jammy, spicy.

Valderica Mencía 2021 T
mencía
88
Standard, balsamic herbs, herbal, wild, smooth.

ATTIS BODEGAS Y VIÑEDOS
Lg. Morouzos, 16D - Dena
36967 Meaño (Pontevedra)
☎: +34 986 744 790
administracion@attisbyv.com
www.attisbyv.com

Sangarida Godello 2021 B
100% godello
88
Dried flowers, fruity, kind finish, tasty.

Sangarida La Guiana 2020 B
100% godello
92
Colour: bright straw. Nose: ripe fruit, fragrant herbs, fine lees, toasty, sweet spices. Palate: full-bodied, rich, long, good acidity.

Sangarida La Yegua 2020 B
godello, dona blanca
93
Colour: bright straw. Nose: ripe fruit, dried herbs, faded flowers, scrubland. Palate: powerful, ripe fruit, balanced.

Sangarida Mencía 2021 T
100% mencía
90
Colour: bright cherry. Nose: fruit expression, red berry notes, spicy. Palate: flavourful, fruity, good acidity.

DO BIERZO / D.O.P.

DO BIERZO / D.O.P.

Sangarida Pico Tuerto 2020 T
100% mencía

90

Colour: deep cherry. Nose: ripe fruit, dried herbs, creamy oak, meaty notes, spicy. Palate: powerful, ripe fruit, spicy, flavourful, powerful tannins.

AURELIO FEO VITICULTOR
24491 San Andrés de Montejos (León)
☎: +34 987 401 865
Fax: +34 987 401 867
bodega@bodegafeo.es
www.bodegafeo.es

Buencomiezo Mencia Selección 2017 T
mencía

88

Toasty, smoky, hot, jammy.

Collage 2021 B
godello, doña blanca, palomino

88

Pleasant, aromatic, jammy.

CR Montelios 2015 T
mencía

91

Colour: deep cherry. Nose: ripe fruit, dried herbs, waxy notes, spicy. Palate: ripe fruit, spicy, round tannins.

Cruz de San Andrés 2020 T RB
mencía

90

Colour: Cherry. Nose: balsamic herbs, sweet spices, scrubland. Palate: spicy, balsamic, good acidity.

Metáfora 2019 T
mencía

90

Colour: deep cherry. Nose: ripe fruit, dried herbs, creamy oak. Palate: powerful, ripe fruit, spicy, round tannins.

BODEGA DEL ABAD
Ctra. N-VI, Km 396
24549 Carracedelo (León)
☎: +34 987 562 417
Fax: +34 987 562 428
info@bodegadelabad.com
www.bodegadelabad.com

Abad Dom Bueno 2021 RD
mencía

91

Defined aromas. Colour: raspberry rose. Nose: red berry notes, balanced, expressive. Palate: spicy, fine bitter notes, flavourful.

Abad Dom Bueno Esencia 2021 B
godello

90

Colour: bright straw. Nose: ripe fruit, floral, white flowers, jasmine. Palate: flavourful, fresh, good acidity, fruity aftertaste.

Abad Dom Bueno Godello 2021 B
godello

91

Colour: straw. Nose: ripe fruit, dried herbs, faded flowers. Palate: powerful, ripe fruit, balanced.

Abad Dom Bueno Laderas del Norte 2020 T RB
mencía

89

Pleasant, crisp, fruity, thin, balanced, wild.

Abad Dom Bueno Mencía 2021 T
mencía

88

Fruity, herbal, wild, smooth, balanced.

San Salvador Godello 2019 B FB
godello

91

Colour: bright yellow. Nose: powerful, creamy oak, ripe fruit, spicy, dried herbs. Palate: rich, good structure, toasty, fine bitter notes.

BODEGA PÉREZ CARAMÉS
Peña Picón, s/n
24500 Villafranca del Bierzo (León)
☎: +34 619 782 968
info@perezcarames.com
www.perezcarames.com

Casar de Valdaiga Paraje El Toleiro 2020 T
mencía

91

Colour: deep cherry. Nose: dried herbs, creamy oak, ripe fruit, black fruit. Palate: powerful, ripe fruit, spicy, round tannins.

BODEGA VERÓNICA ORTEGA

24530 Valtuille de Abajo (León)
☎: +34 696 506 485
veronica@veronicaortega.es
www.veronicaortega.es

🏆 PODIUM

Cobrana 2020 T
mencía
95
Taut. Colour: Cherry. Nose: complex, expressive, spicy, mineral, scrubland. Palate: elegant, full-bodied, long, great length.

Kinki 2021 T
91
Subtle, smooth. Colour: light cherry. Nose: medium intensity, wild herbs, red berry notes, ripe fruit. Palate: balanced, easy to drink, good finish.

La Llorona 2021 B
godello
94
Taut. Colour: bright yellow. Nose: powerful, creamy oak, ripe fruit, spicy. Palate: rich, good structure, long, toasty, fine bitter notes.

Quite 2021 T
mencía
93
Colour: cherry, purple rim. Nose: balsamic herbs, red berry notes, scrubland. Palate: balsamic, fruity, balanced.

Roc 2020 T
93
Defined aromas. Colour: bright cherry. Nose: ripe fruit, fruit preserve, neat. Palate: powerful, ripe fruit, spicy, round tannins.

BODEGA Y VIÑEDOS 13 VIÑAS

Campo del Obispo, 13
24492 Cubillos del Sil (León)
Fax: +34 649 312 360
trecevinas@gmail.com
www.13viñas.com

Babu 2021 B
jerez, valenciana, godello
90
Oxidativ. Colour: straw. Nose: dried herbs, faded flowers, white fruit, cereal notes. Palate: powerful, ripe fruit, balanced.

Mingus 2021 T
100% mencía
89
Balanced, spicy, dried flowers, jammy, mineral, tasty.

BODEGA Y VIÑEDOS HEREDAD MORÁN & LÓPEZ

Camino Escaril, 35 Puente Boeza
24401 Ponferrada (León)
☎: +34 676 509 621
heredad.moran.lopez@gmail.com
www.heredadmoranlopez.com

Buleza 2021 RD
100% mencía
86

Heredad 26 2020 T RB
100% mencía
88
Pleasant, toasty, tasty, jammy, spicy.

Heredad 26 Godello 2020 B
100% godello
90
Colour: straw. Nose: ripe fruit, dried herbs, faded flowers. Palate: powerful, ripe fruit, balanced.

Heredad 26 Mencía 2020 T
100% mencía
88
Pleasant, kind finish, crisp, fruity.

Heredad 26 Mencía 2021 T
mencía
91
Colour: cherry, purple rim. Nose: fruit expression, red berry notes, balsamic herbs. Palate: flavourful, fruity, good acidity, long.

Heredad Altos de Talan 2020 B FB
godello
92
Colour: bright yellow. Nose: powerful, creamy oak, ripe fruit, spicy, complex, stone fruit. Palate: rich, good structure, long, toasty, fine bitter notes.

Valdesalas 2021 B
godello
89
Citrus fruit, balanced, herbal, yeasty notes.

DO BIERZO / D.O.P.

DO BIERZO / D.O.P.

BODEGA Y VIÑEDOS HIJA DE ANÍBAL
La Estación, 6
24500 Villafranca del Bierzo (León)
☎: +34 626 512 680
info@anibaldeotero.es
www.anibaldeotero.es

Aníbal de Otero Los Fornos 2016 T
92
Colour: dark-red cherry, garnet rim. Nose: ripe fruit, aged wood nuances, tobacco, sweet spices. Palate: spicy, round tannins, long.

Aníbal de Otero Villa Otero 2017 T
100% mencía
91
Colour: dark-red cherry, garnet rim. Nose: fruit preserve, aged wood nuances, tobacco, sweet spices. Palate: spicy, round tannins, long.

BODEGAS ADRIÁ
Ctra. Antigua Madrid - Coruña, Km. 407
24500 Villafranca del Bierzo (León)
☎: +34 987 540 907
aperez@bodegasadria.com
www.bodegasadria.com

Bodegas Adria Godello 2021 B
godello
90
Colour: bright straw. Nose: fruit expression, ripe fruit, floral. Palate: flavourful, fresh, good acidity, fruity aftertaste.

Bodegas Adria Mencía 2021 T
mencía
89
Balsamic herbs, wild, fruity, standard, herbal, smooth.

Bodegas Adria Silk 2020 T RB
mencía
88
Tasty, jammy, aromatic.

Bodegas Adria Velvet 2019 T
mencía
90
Colour: cherry, garnet rim. Nose: dried herbs, creamy oak, black fruit, ripe fruit. Palate: powerful, spicy, round tannins.

Bodegas Adria Villa El Toleiro 2019 B C
godello
90
With personality, age nuances, oxidativ. Colour: old gold. Nose: acetaldyhde, lees reduction notes, flor yeasts, dry nuts. Palate: flavourful, bitter, spicy.

Etapa 24 2020 B SS
91
Colour: bright yellow. Nose: dried flowers, candied fruit, fine lees, pattiserie. Palate: round, spicy, long, saline.

BODEGAS BERNARDO ÁLVAREZ
San Pedro, 75
24530 Villadecanes (León)
☎: +34 987 562 129
Fax: +34 987 562 129
vinos@bodegasbernardoalvarez.com
www.bodegasbernardoalvarez.com

5 Pulgares 2018 T C
garnacha tintorera
88
Herbal, fruity, jammy, tasty, wild, standard.

Campo Redondo 2018 T RB
mencía
88
Pleasant, fruity, jammy, tasty.

Campo Redondo Godello 2021 B
godello
91
Colour: bright yellow. Nose: powerful, ripe fruit, spicy. Palate: good structure, long, fine bitter notes.

Viña Migarrón 2018 T C
mencía
87

Viña Migarrón 2020 T
mencía
88
Spicy, jammy, tasty.

Viña Migarrón 2021 RD
mencía
88
Pleasant, defined aromas, crisp, fruity, sweeties, lactic, simple.

BODEGAS CANTALOBOS
Avda. Galicia, 187
24411 Ponferrada (León)
☎: +34 619 055 411
info@cantalobos.es
www.cantalobos.es

Cantalobos 2019 T C
mencía
88
Spicy, pruney, fruity, tasty.

Cantalobos 2021 B
100% godello
88
Citrus fruit, balanced, herbal, tasty.

Cantalobos 2021 T
mencía
87

BODEGAS CUATRO PASOS
Santa María, 43
24540 Cacabelos (León)
☎: +34 987 548 089
bierzo@martincodax.com
www.cuatropasos.com

Cuatro Pasos 2020 T
100% mencía
88
Roasted coffee, spicy, jammy, tasty.

Cuatro Pasos Black 2019 T
100% mencía
91
Colour: very deep cherry. Nose: roasted coffee, aromatic coffee, powerful, ripe fruit. Palate: smoky aftertaste, great length, round tannins.

Cuatro Pasos Rosé 2021 RD
100% mencía
88
Pleasant, floral, fruity, tasty.

Martín Sarmiento 2017 T
100% mencía
92
Colour: dark-red cherry, garnet rim. Nose: ripe fruit, aged wood nuances, tobacco, sweet spices. Palate: spicy, round tannins, long.

Pizarras de Otero 2021 T
mencía
90
Colour: cherry, purple rim. Nose: balsamic herbs, red berry notes, scrubland. Palate: balsamic, fruity, balanced.

BODEGAS EMILIO MORO
Ctra. Peñafiel - Valoria, s/n
47315 Pesquera de Duero (Valladolid)
☎: +34 983 878 400
bodega@emiliomoro.com
www.emiliomoro.com

El Polvorete 2021 B
godello
91
Colour: straw. Nose: ripe fruit, dried herbs, faded flowers. Palate: powerful, ripe fruit, balanced.

BODEGAS ESTEFANIA
La Lechería, 3 Dehesas
24390 Ponferrada (León)
☎: +34 987 420 015
info@tilenus.com
www.mqwinesgroup.com

Tilenus Entrecuestas Godello 2020 B FB
godello
93
Colour: bright yellow. Nose: dried flowers, candied fruit, fine lees, pattiserie, toasty. Palate: round, spicy, long, great length.

Tilenus Envejecido en Roble 2019 T RB
91
Colour: deep cherry. Nose: ripe fruit, dried herbs, creamy oak, balsamic herbs. Palate: powerful, ripe fruit, spicy, round tannins.

Tilenus Godello Monteseiros 2021 B
godello
90
Pleasant, aromatic, dried flowers, fruity. Nose: earthy notes, dried herbs. Palate: balanced, fine bitter notes.

Tilenus La Florida 2019 T C
100% mencía
91
Colour: deep cherry. Nose: dried herbs, ripe fruit, neat, wild herbs, waxy notes. Palate: powerful, ripe fruit, spicy, round tannins, varietal.

Tilenus Las Laderas Mencía 2019 T BA
100% mencía
90
Slight reduction, rustic. Colour: cherry, garnet rim. Nose: red berry notes, spicy, wild herbs. Palate: flavourful, fruity, good acidity, long.

Tilenus Pagos de Posada 2017 T BA
90
Colour: Cherry. Nose: balsamic herbs, sweet spices, scrubland, ripe fruit, fruit preserve. Palate: spicy, long, flavourful.

DO BIERZO / D.O.P.

DO BIERZO / D.O.P.

Tilenus Vendimia 2021 T
100% mencía

88

Pleasant, balsamic herbs, fruity, tasty.

BODEGAS GODELIA

Antigua Ctra. N-VI, km. 403,5
24547 Pieros-Cacabelos (León)
☎: +34 987 546 229
Fax: +34 987 548 026
info@godelia.es
www.godelia.es

Godelia Godello 2021 B
90% godello, 10% dona blanca

90

Colour: bright straw. Nose: fruit expression, ripe fruit, floral. Palate: flavourful, fresh, good acidity, fruity aftertaste.

Godelia Mencía 2017 T RB
100% mencía

89

Toasty, jammy, spicy, kind finish.

Godelia Mencía Rosé 2021 RD
100% mencía

88

Pleasant, floral, fruity.

Godelia Selección Godello 2018 B
100% godello

92

Colour: straw. Nose: ripe fruit, dried herbs, faded flowers, toasty. Palate: powerful, ripe fruit, balanced.

Godelia Selección Mencía 2016 T
100% mencía

91

Colour: dark-red cherry, garnet rim. Nose: fruit preserve, aged wood nuances, tobacco, sweet spices, toasty. Palate: spicy, round tannins.

Viernes 2021 T
100% mencía

90

Colour: cherry, purple rim. Nose: fruit expression, red berry notes, floral, spicy. Palate: flavourful, fruity, good acidity, long.

BODEGAS PEIQUE

El Bierzo, s/n
24530 Valtuille de Abajo (León)
☎: +34 987 562 044
bodega@bodegaspeique.com
www.bodegaspeique.com

Luis Peique 2017 T RB
mencía

92

Colour: deep cherry. Nose: ripe fruit, dried herbs, creamy oak, burnt matches, powerful. Palate: powerful, ripe fruit, spicy, round tannins.

Peique Godello 2021 B
godello

89

Standard, crisp, very fruit-driven, herbal, smooth, wild.

Peique Selección Familiar 2018 T
mencía

92

Jammy. Colour: deep cherry. Nose: ripe fruit, dried herbs, spicy. Palate: powerful, ripe fruit, spicy, round tannins.

Peique Mencía 2021 T
mencía

90

Colour: cherry. Nose: balsamic herbs, scrubland. Palate: spicy, balsamic, good acidity.

Peique Ramón Valle 2020 T
mencía

91

Defined aromas, balsamic herbs. Colour: bright cherry. Nose: red berry notes, scrubland, wild herbs, balanced. Palate: easy to drink, ripe fruit.

Peique Viñedos Viejos 2019 T RB
mencía

91

Colour: cherry, garnet rim. Nose: fruit preserve, powerful, dark chocolate, spicy, toasty. Palate: flavourful, sweetness, long.

BODEGAS RODRÍGUEZ Y SANZO
Manuel Azaña, 9
47014 Valladolid (Valladolid)
☎: +34 983 150 150
comunicacion@valsanzo.com
www.rodriguezsanzo.com

Vitis Extrema 2019 T
mencía

90

Colour: cherry, garnet rim. Nose: fruit preserve, powerful, ripe fruit, toasty, aromatic coffee. Palate: flavourful, sweetness, long.

BODEGAS VIÑAS DE VIÑALES
Jaen, 4
24319 Bembibre (León)
☎: +34 609 652 058
bodegasvinasdevinales@gmail.com
www.bodegasvinasdevinales.com

Miliario 2019 T RB

87 ♣

BODEGAS VIORE
Miguel Hernández, 31
47490 Rueda (Valladolid)
☎: +34 941 454 050
Fax: +34 941 454 529
bodega@bodegasriojanas.com
www.bodegasriojanas.com

Viore Mencía 2021 T
100% mencía

88

Pleasant, toasty, powerful, jammy.

BODEGAS Y VIÑEDOS A.A.
Ctra. Puebla de Sanabria, Km. 2
24415 San Lorenzo - Ponferrada (León)
☎: +34 626 452 839
alvarezgarcia75@hotmail.com

Seulalia Godello 2021 B
godello

88

Herbal, dried herbs, tasty, bitter, standard.

Seulalia Mencía 2020 T
mencía

90

Colour: deep cherry. Nose: ripe fruit, dried herbs, creamy oak, toasty. Palate: powerful, ripe fruit, spicy, round tannins.

Seulalia Palomino 2021 B
palomino, dona blanca

89

Pleasant, herbal, wild, balanced, crisp, smooth.

BODEGAS Y VIÑEDOS GANCEDO
Vistalegre, s/n
24548 Quilós (León)
☎: +34 987 134 980
gancedo@bodegasgancedo.com
www.bodegasgancedo.com

Capricho Val de Paxariñas 2021 B
100% godello

91

Colour: bright straw. Nose: fruit expression, ripe fruit, floral. Palate: flavourful, fresh, good acidity, fruity aftertaste.

Capricho Val de Paxariñas 2021 RD
100% mencía

90

Colour: raspberry rose. Nose: red berry notes, floral, fragrant herbs. Palate: light-bodied, spicy, good acidity.

Gancedo 2020 T RB
100% mencía

90

Colour: cherry, garnet rim. Nose: fruit preserve, powerful, toasty, smoky. Palate: flavourful, long.

Herencia del Capricho 2020 B FB
100% godello

92

Colour: bright yellow. Nose: powerful, creamy oak, ripe fruit, spicy, toasty. Palate: good structure, long, toasty, fine bitter notes.

DO BIERZO / D.O.P.

DO BIERZO / D.O.P.

Ucedo 2020 T RB
100% mencía

91
Colour: cherry, garnet rim. Nose: powerful, ripe fruit, fruit preserve, sweet spices, dark chocolate. Palate: flavourful, long, sweet tannins.

Xestal 2015 T RB
89
Nose: roasted coffee, aromatic coffee, powerful, fine reductive notes. Palate: smoky aftertaste, great length, round tannins.

BODEGAS Y VIÑEDOS JOSÉ ANTONIO GARCÍA
El Puente s/n
24530 Valtuille de Abajo (León)
☎: +34 648 070 581

Un Culín
Mencía de Valtuille 2020 T
mencía, dona blanca, godello, garnacha tintorera

88
Toasty, jammy, slight reduction, balsamic herbs.

BODEGAS Y VIÑEDOS LUNA BEBERIDE
Ant. Ctra. Madrid - Coruña, Km. 402
24540 Cacabelos (León)
☎: +34 987 549 002
info@lunabeberide.es
www.lunabeberide.es

Art Luna Beberide 2020 T C
mencía

91
Colour: deep cherry. Nose: ripe fruit, dried herbs, creamy oak. Palate: powerful, ripe fruit, spicy, round tannins.

Finca Luna Beberide 2020 T RB
mencía

92
Colour: cherry, purple rim. Nose: fruit expression, red berry notes, floral, spicy. Palate: flavourful, fruity, good acidity, long.

Luna Beberide Godello 2021 B
godello

90
Colour: bright straw, greenish rim. Nose: fresh fruit, citrus fruit, wild herbs. Palate: fresh, fruity, good acidity, fine bitter notes.

Luna Beberide Mencía 2021 T
mencía

91
Pleasant. Colour: Cherry. Nose: balsamic herbs, sweet spices, scrubland, fruit expression. Palate: spicy, balsamic, good acidity.

Paixar Mencía 2020 T
93
Colour: cherry, purple rim. Nose: floral, spicy, toasty, macerated fruit, ripe fruit. Palate: flavourful, fruity, good acidity, long.

BODEGAS Y VIÑEDOS MERAYO
Ctra. de la Espina, km. 3.5
24491 San Andrés de Montejos (León)
☎: +34 669 372 307
jmerayo@byvmerayo.com
www.bodegasmerayo.com

Aquiana 2019 T C
100% mencía

91
Colour: Cherry. Nose: balsamic herbs, scrubland, dried flowers. Palate: spicy, good acidity, balanced, easy to drink, long.

La Galbana 2019 T R
100% mencía

90
Colour: very deep cherry. Nose: roasted coffee, aromatic coffee, powerful, ripe fruit, black fruit. Palate: smoky aftertaste, great length, round tannins.

Las Tres Filas 2020 T RB
100% mencía

92
Colour: dark-red cherry. Nose: toasty, spicy, cocoa bean, ripe fruit, red berry notes. Palate: flavourful, toasty, fine bitter notes.

Merayo Garnacha Tintorera 2018 T
garnacha tintorera

92
Colour: deep cherry. Nose: ripe fruit, dried herbs, spicy. Palate: ripe fruit, spicy, round tannins, flavourful.

Merayo Godello 2021 B
100% godello

90
Colour: straw. Nose: ripe fruit, dried herbs, faded flowers. Palate: powerful, ripe fruit, balanced.

Merayo Mencía 2021 T
100% mencía

91

Colour: cherry, purple rim. Nose: fruit expression, red berry notes, floral, spicy. Palate: flavourful, fruity, good acidity, long.

BODEGAS Y VIÑEDOS VENTUA
Barbacana, 1
24380 Puente de Domingo (León)
☎: +34 685 521 685
ventua@vinosventua.com
www.vinosventua.com

Ventua 2021 RD
mencía

86

Ventua Mourelo 2018 T C

92

Colour: Cherry. Nose: balsamic herbs, scrubland, spicy, smoky. Palate: spicy, good acidity, flavourful, fine bitter notes, long.

CANTARIÑA
24500 Villafranca del Bierzo (León)
☎: +34 606 075 194
info@vinoscantarina.es
www.vinoscantarina.es

Cantariña 2 Viña de los Pinos 2018 T
mencía, palomino

91

Colour: cherry, garnet rim. Nose: wild herbs, dried herbs, ripe fruit, expressive. Palate: flavourful, balsamic, spicy, easy to drink.

Cantariña 3 El Triángulo 2018 T
mencía, palomino

92

Colour: dark-red cherry, garnet rim. Nose: fruit preserve, aged wood nuances, tobacco, sweet spices, scrubland. Palate: spicy, round tannins, long.

Cantariña 5 Valdeobispo 2018 T
mencía

93

Colour: Cherry. Nose: balsamic herbs, sweet spices, scrubland, ripe fruit, black fruit. Palate: spicy, balsamic, good acidity.

Cantariña 6 Merenzao 2020 T
merenzao

91

Herbal, wild, varietally correct, representative. Colour: Cherry. Nose: balsamic herbs, sweet spices, scrubland. Palate: spicy, balsamic, good acidity, easy to drink.

La Cabeza de Perro 2020 B
palomino, dona blanca

92

Colour: bright yellow. Nose: dried flowers, candied fruit, pattisserie. Palate: round, spicy, great length.

La Cabeza de Perro 2021 RD
mencía

90

Colour: brilliant rose. Nose: red berry notes, floral, fragrant herbs. Palate: light-bodied, spicy, good acidity, fine bitter notes.

CASAR DE BURBIA
Travesía la Constitución, s/n
24549 Carracedelo (León)
☎: +34 987 562 910
info@casardeburbia.com
www.casardeburbia.com

Casar de Burbia 2020 T RB
100% mencía

90 🌱

Colour: dark-red cherry. Nose: toasty, spicy, cocoa bean, ripe fruit. Palate: flavourful, toasty, fine bitter notes.

Casar de Burbia Godello 2021 B
100% godello

91 🌱

Colour: bright straw. Nose: floral, white fruit, ripe fruit, varietal. Palate: flavourful, good acidity, easy to drink, good finish.

Casar Godello 2020 B FB
100% godello

92 🌱

Colour: bright yellow. Nose: powerful, creamy oak, ripe fruit, spicy, fine lees. Palate: good structure, toasty, fine bitter notes, good acidity.

Hombros 2020 T BA
100% mencía

92 🌱

Colour: dark-red cherry, garnet rim. Nose: fruit preserve, aged wood nuances, tobacco, sweet spices. Palate: spicy, round tannins, long.

Tebaida 2020 T RB
100% mencía

92 🌱

Colour: dark-red cherry. Nose: toasty, spicy, cocoa bean, stone fruit, ripe fruit, black fruit. Palate: flavourful, toasty, fine bitter notes.

DO BIERZO / D.O.P.

DO BIERZO / D.O.P.

Tebaida Nemesio 2020 T RB
100% mencía

92

Colour: cherry, garnet rim. Nose: fruit preserve, fruit liqueur notes, powerful, aromatic coffee, roasted coffee. Palate: flavourful, sweetness, long.

Tebaida nº5 2020 T RB
100% mencía

92

Colour: cherry, purple rim. Nose: fruit preserve, black fruit, dried herbs, smoky, dark chocolate, spicy. Palate: fruity, full-bodied, flavourful, good structure, balanced, good finish, slightly dry, soft tannins.

CASTRO VENTOSA
Finca El Barredo, s/n
24530 Valtuille de Abajo (León)
☎: +34 987 562 148
info@castroventosa.com
www.castroventosa.com

El Castro de Valtuille 2021 T
91

Colour: cherry, purple rim. Nose: balsamic herbs, red berry notes, scrubland, fruit expression. Palate: balsamic, fruity, balanced.

El Castro de Valtuille Godello 2020 B BA
92

Colour: bright straw. Nose: ripe fruit, fragrant herbs, fine lees, spicy. Palate: full-bodied, rich, long, good acidity.

Valtuille Cepas Centenarias 2020 T BA
94

Colour: Cherry. Nose: expressive, spicy, wild herbs, scrubland, ripe fruit. Palate: full-bodied, long, great length, ripe fruit.

Valtuille Rapolao 2020 T C
94

Colour: Cherry. Nose: complex, expressive, spicy, mineral, fruit expression, red berry notes. Palate: elegant, full-bodied, long, great length.

Valtuille Vino de Villa 2020 T BA
93

Rustic, with personality. Colour: cherry, garnet rim. Nose: medium intensity, scrubland, wild herbs, dried flowers. Palate: balanced, taut, fine bitter notes.

CÉSAR MÁRQUEZ BODEGAS Y VIÑEDOS
Calle Antigua Ctra. N-VI, 34A
24530 Valtuille de Abajo (León)

El Rapolao Vino de Paraje 2020 T
mencía

94

Colour: Cherry. Nose: balsamic herbs, sweet spices, scrubland, red berry notes, ripe fruit. Palate: spicy, balsamic, good acidity.

La Salvación 2020 B
godello

94

Colour: bright yellow. Nose: ripe fruit, fragrant herbs, fine lees, complex. Palate: full-bodied, rich, long, good acidity.

Parajes Vino de Región 2020 T
85% mencía, 7% alicante bouschet (Tinto), 8% otras

93

Colour: deep cherry. Nose: dried herbs, scrubland, red berry notes, ripe fruit, sweet spices. Palate: powerful, ripe fruit, spicy, round tannins.

🏆 PODIUM

Pico Ferreira 2020 T
85% mencía, 15% alicante bouschet (Tinto)

97

Colour: bright cherry. Nose: complex, expressive, spicy, mineral, fresh fruit, red berry notes, dried herbs, scrubland. Palate: elegant, full-bodied, long, great length.

🏆 PODIUM

Sufreiral 2020 T
85% mencía, 12% alicante bouschet (Tinto), 3% otras

95

Colour: bright cherry. Nose: complex, expressive, spicy, mineral, scrubland, fragrant herbs, red berry notes. Palate: elegant, full-bodied, long, great length.

Valtuille Vino de Villa 2020 T
85% mencía, 10% alicante bouschet (Tinto), 5% otras

93

Colour: deep cherry. Nose: balsamic herbs, scrubland, ripe fruit, red berry notes, spicy. Palate: spicy, balsamic, good acidity.

COMPAÑÍA DE VINOS PEÑA SERRANO (PE ESE)
Ctra. Madrid-Coruña-VI, km 408
24500 Villafranca del Bierzo (León)
☎: 609 128 043
hola@vinospeese.es

Pe Ese 2019 T
mencía
90
Full-bodied, hot. Colour: cherry, garnet rim. Nose: black fruit, fruit preserve, dried herbs, spicy. Palate: flavourful, long.

DESCENDIENTES DE J. PALACIOS
Chao do Pando, 1
24514 Corullón (León)
☎: +34 987 540 821
info@djpalacios.com
www.alvaropalacios.com

🏆 PODIUM
La Faraona 2020 T BA
97% mencía, 3% otras
97
Colour: bright cherry. Nose: neat, red berry notes, elegant, floral, wild herbs. Palate: complex, fruity, elegant, good acidity, taut, racy, round tannins, easy to drink.

🏆 PODIUM
Las Lamas 2020 T BA
95
Elegant, tasty. Colour: bright cherry. Nose: ripe fruit, red berry notes, scrubland, balsamic herbs, sweet spices. Palate: flavourful, fruity, spicy.

🏆 PODIUM
Moncerbal 2020 T
96% mencía, 4% otras
96
Colour: bright cherry. Nose: red berry notes, fresh fruit, scrubland, balsamic herbs, wild herbs. Palate: complex, full-bodied, flavourful, taut.

Pétalos del Bierzo 2020 T
92% mencía, 5% alicante bouchet (Tinto), grao negro, pan y carne, negreda, 3% otras
92
Colour: cherry, garnet rim. Nose: grassy, wild herbs, ripe fruit. Palate: balanced, easy to drink, ripe fruit, aged character.

Pétalos del Bierzo 2021 T
95% mencía, 3% alicante bouchet (Tinto), gran negro, pan y carne, negreda, 2% otras
92
Kind finish, fruity, balanced. Colour: bright cherry. Nose: red berry notes, ripe fruit. Palate: flavourful, fruity, taut, ripe fruit, easy to drink.

Villa de Corullón 2020 T
92% mencía, 1% alicante bouchet (Tinto), 7% otras
94
Aromatic. Colour: Cherry. Nose: balsamic herbs, sweet spices, scrubland, fragrant herbs, red berry notes. Palate: spicy, balsamic, good acidity, long.

DOMINIO DE TARES
Los Barredos, 4
24380 San Román de Bembibre (León)
☎: +34 987 514 550
Fax: +34 987 514 570
info@dominiodetares.com
www.dominiodetares.com

Bembibre 2018 T R
mencía
93
Colour: very deep cherry. Nose: expressive, spicy, mineral, aromatic coffee, roasted coffee. Palate: elegant, full-bodied, long, great length.

Dominio de Tares Cepas Viejas 2018 T C
mencía
91
Colour: deep cherry. Nose: ripe fruit, dried herbs, creamy oak, black fruit. Palate: powerful, ripe fruit, spicy, round tannins.

Dominio de Tares Godello 2021 B FB
godello
92
Colour: bright yellow. Nose: creamy oak, ripe fruit, spicy. Palate: good structure, long, flavourful, balanced.

Tares P. 3 2017 T R
mencía
93
Colour: dark-red cherry, garnet rim. Nose: ripe fruit, fruit preserve, aged wood nuances, tobacco, sweet spices. Palate: spicy, round tannins, long.

DO BIERZO / D.O.P.

DO BIERZO / D.O.P.

DON PEDRONES
Flora, 5
24411 Fuentesnuevas (León)
☎: +34 647 698 485
donpedrones@yahoo.es
www.donpedrones.es

Pethiox 2019 T C
100% mencía
87

Pethiox 2021 T
100% mencía
90
Pleasant, fruity, tasty, jammy.

Pethiox Cielo 2021 B
100% godello
88
Pleasant, fruity, standard, balanced, tasty, jammy.

Pethiox Clarete 2021 RD
50% mencía, 50% dona blanca
87

Pethiox Distinción 2018 T
100% mencía
90
Colour: cherry, garnet rim. Nose: fruit preserve, powerful, dark chocolate, toasty. Palate: flavourful, sweetness, long.

ENCIMA WINES
Era, 35
24413 Molinaseca (León)
☎: +34 987 697 515
info@encimawines.com
www.encimawines.com

El Mago Chalupa 2021 B
60% godello, 40% jerez
88
Pleasant, kind finish, fruity.

El Mago Chalupa 2021 T
100% mencía
87

La Cigüeña Clarete 2021 RD
dona blanca, mencía, palomino
90
Colour: rose. Nose: ripe fruit, hot, faded flowers. Palate: fleshy, flavourful, powerful, ripe fruit.

La Cigüeña Doña Blanca 2021 B
100% dona blanca
90
Defined aromas, with personality. Colour: straw. Nose: floral, neat, medium intensity. Palate: balanced, fine bitter notes, flavourful.

La Cigüeña Godello 2021 B
100% godello
90
Colour: bright straw. Nose: fruit expression, ripe fruit, floral. Palate: flavourful, fresh, good acidity, fruity aftertaste.

La Cigüeña Malvasía 2021 B
100% malvasía
89
Defined aromas, balanced, dried flowers, dried herbs, tasty, standard.

ESTÉVEZ BODEGAS Y VIÑEDOS
Calle Las Flores, s/n
24530 Valtuille de Abajo (León)
info@estevezbodegas.com
www.estevezbodegas.com

Versos de Valtuille 15 meses 2020 T
mencía
92
Colour: cherry, garnet rim. Nose: fruit preserve, powerful, sweet spices, toasty. Palate: flavourful, sweetness, long.

Versos de Valtuille Cepas Centenarias 2020 T RB
mencía
90
Colour: cherry, garnet rim. Nose: ripe fruit, wild herbs, waxy notes. Palate: flavourful, fruity, easy to drink, spicy.

Versos de Valtuille Godello 2020 B FB
godello
91
Colour: bright yellow. Nose: powerful, creamy oak, ripe fruit. Palate: rich, good structure, long, toasty.

LOSADA VINOS DE FINCA
Ctra.a Villafranca LE-713, Km. 12
24540 Cacabelos (León)
☎: +34 987 548 053
bodega@losadavinosdefinca.com
www.losadavinosdefinca.com

Altos de Losada El Cepón 2020 T
100% mencía
92
Smoky. Colour. deep cherry, bright cherry. Nose: black fruit, ripe fruit, lactic notes, spicy. Palate: flavourful, long, ripe fruit, spicy.

Altos de Losada La Bienquerida 2020 T
95% mencía, 5% otras
93
Colour. Cherry. Nose: complex, expressive, spicy, mineral, ripe fruit. Palate: elegant, full-bodied, long, great length.

Altos de Losada La Chana 2020 T
100% merenzao
92
Colour. cherry, purple rim. Nose: fruit expression, red berry notes, floral, spicy. Palate: flavourful, fruity, good acidity, long.

Altos de Losada Mencía 2020 T
100% mencía
92
Colour. cherry, purple rim. Nose: fruit expression, red berry notes, floral, spicy, scrubland. Palate: flavourful, fruity, good acidity, long.

Losada 2020 T
99% mencía, 1% otras
92
Colour. deep cherry. Nose: ripe fruit, dried herbs, creamy oak, raspberry. Palate: powerful, ripe fruit, spicy, round tannins.

Losada Godello 2021 B
100% godello
91
Colour. straw. Nose: ripe fruit, dried herbs, faded flowers, fragrant herbs. Palate: powerful, ripe fruit, balanced.

Villa de San Lorenzo Godello 2019 B
100% godello
92
Colour. bright yellow. Nose: dried flowers, candied fruit, fine lees, pattiserie. Palate: round, spicy, long, great length.

MARIA ZAMARREÑO
Las Salgueras
24413 Molinaseca (León)
☎: +34 639 202 403
mariazv.bierzo@gmail.com
www.mariazamarreño.com

Van Gus Vana 2014 T
mencía
88
Toasty, jammy, spicy, classic.

PAGO DE LOS ABUELOS
24580 Puente de Domingo Florez (León)
☎: +34 692 890 494
bodega@pagodelosabuelos.com
www.pagodelosabuelos.com

🏆 **PODIUM**

De los Abuelos Teiró Godello 2020 B
godello
95
Colour. bright yellow. Nose: dried flowers, candied fruit, fine lees, pattiserie, toasty, ripe fruit. Palate: round, spicy, long, great length.

De los Abuelos Viñas Centenarias 2021 RD
50% mencía, otras
93
Colour. salmon. Nose: sweet spices, red berry notes, fragrant herbs, dried flowers. Palate: full-bodied, flavourful, spicy, sweetness, long.

De los Abuelos Viñedo Barreiros Godello 2020 B FB
godello
92
Colour. bright yellow. Nose: powerful, creamy oak, ripe fruit, spicy. Palate: rich, good structure, long, toasty, fine bitter notes.

De los Abuelos Viñedo Barreiros Mencía 2020 T
mencía
94
Colour. dark-red cherry, garnet rim. Nose: aged wood nuances, tobacco, sweet spices, ripe fruit, black fruit. Palate: spicy, round tannins, long.

De los Abuelos Viñedo Saturno 2021 T
mencía
92
Colour. cherry, purple rim. Nose: balsamic herbs, red berry notes, scrubland, fruit expression. Palate: balsamic, fruity, balanced.

DO BIERZO / D.O.P.

DO BIERZO / D.O.P.

RAÍCES IBÉRICAS
Avda. Mudejar, 61
50340 Maluenda (Zaragoza)
☎: +34 976 893 017
s.richard@raices.wine
www.raicesibericas.com

Raíces Godello 2021 B
100% godello
89
Aromatic, very fruit-driven, simple, fruity, tasty, standard.

Raíces Mencía 2021 T
100% mencía
89
Herbal, jammy, wild, slight reduction, balanced, fruity, smooth.

RAUL PÉREZ BODEGAS Y VIÑEDOS
Bulevar Rey Juan Carlos 1º Rey de España, 11 B
24400 Ponferrada (León)
contacto@raulperez.com
www.raulperez.com

🏆 PODIUM

El Erial de Valdecañada 2018 B
godello
95
Little interventionist. Colour: bright straw. Nose: expressive, ripe fruit, floral, fine lees, mineral, burnt matches. Palate: full-bodied, complex, spicy, long, elegant.

🏆 PODIUM

El Rapolao 2020 T
85% mencía, 10% alicante bouschet (Tinto), 5% otras
95
Colour: deep cherry. Nose: balsamic herbs, sweet spices, scrubland, red berry notes, fruit expression, meaty notes. Palate: spicy, balsamic, good acidity, flavourful.

La del Vivo 2020 B
godello
93
Colour: bright straw. Nose: expressive, white flowers, jasmine, dried herbs. Palate: flavourful, fruity, balanced.

La Poulosa La Vizcaína 2019 T
94
Taut. Colour: bright cherry. Nose: complex, expressive, spicy, mineral, musky notes, faded flowers. Palate: full-bodied, long, great length, balsamic.

La Vitoriana 2020 T
94
Colour: very deep cherry. Nose: complex, expressive, spicy, mineral, red berry notes, fragrant herbs, wild herbs, fine reductive notes. Palate: elegant, full-bodied, long, great length.

🏆 PODIUM

Las Gundiñas La Vizcaína 2019 T
96
Colour: bright cherry. Nose: balsamic herbs, sweet spices, scrubland, red berry notes, expressive. Palate: spicy, balsamic, good acidity.

Ultreia "La Claudina" 2020 B
93
Colour: bright yellow. Nose: dried flowers, candied fruit, fine lees, pattiserie. Palate: round, spicy, long.

Ultreia "Saint Jacques" 2020 T
93
Colour: very deep cherry. Nose: balsamic herbs, sweet spices, scrubland, red berry notes, fragrant herbs, meaty notes. Palate: spicy, balsamic, good acidity, flavourful.

Ultreia Godello 2020 B
godello
94
Kind finish, taut. Colour: straw. Nose: ripe fruit, dried herbs, faded flowers, saffron. Palate: powerful, ripe fruit, balanced, full-bodied.

Ultreia Valtuille 2020 T
93
Colour: bright cherry. Nose: complex, expressive, spicy, mineral, black fruit, earthy notes. Palate: elegant, full-bodied, long, great length, flavourful.

SEÑORÍO DE LOS ARCOS
Ctra.. Caboalles, 332
24191 Villlbalter (León)
☎: +34 987 226 594
info@senoriodelosarcos.es
www.senoriodelosarcos.es

Dobiñon 2020 T
mencía
87

SILVIA MARRAO BARREIRO
☎: +34 616 056 075
banzao@banzao.es
www.banzao.es

Banzao 2018 T
mencía
91
Colour: dark-red cherry. Nose: toasty, spicy, cocoa bean, dark chocolate, ripe fruit, black fruit. Palate: flavourful, toasty, fine bitter notes.

SOTO DEL VICARIO
Ctra. Cacabelos- San Clemente, Pol. Ind. 908
Parcela 155
24547 San Clemente (León)
☎: +34 670 983 534
Fax: +34 926 666 029
sandra.luque@pagodelvicario.com
www.pagodelvicario.com

Soto del Vicario El Origen 2017 T
100% mencía
92
Colour: deep cherry, garnet rim. Nose: aged wood nuances, cocoa bean, cigar, toasty, ripe fruit, black fruit. Palate: flavourful, spicy, toasty, powerful tannins.

Soto del Vicario Go de Godello 2021 B FB
100% godello
91
Colour: bright yellow. Nose: powerful, creamy oak, ripe fruit, spicy. Palate: rich, good structure, long, toasty, fine bitter notes.

Soto del Vicario Godello Especial Cuvée 2020 B
100% godello
92
Colour: bright yellow. Nose: dried flowers, candied fruit, fine lees, pattisserie, toasty. Palate: round, spicy, long, great length.

Soto del Vicario Men de Mencía 2017 T C
100% mencía
91
Colour: dark-red cherry, garnet rim. Nose: fruit preserve, aged wood nuances, tobacco, sweet spices. Palate: spicy, round tannins, long.

Soto del Vicario Men de Mencía Selección de Viñedos 2014 T C
100% mencía
91
Colour: deep cherry, garnet rim. Nose: ripe fruit, cocoa bean, cigar, toasty. Palate: flavourful, spicy, toasty, powerful tannins.

VINOS GUERRA
Avda. Constitución, 106
24540 Cacabelos (León)
☎: +34 987 546 150
info@vinosdelbierzo.com
www.vinosguerra.com

Armas de Guerra 2021 B
dona blanca, palomino
87

Armas de Guerra Godello 2021 B
godello
89
Pleasant, smooth, tasty, jammy.

Armas de Guerra Mencía 2016 T C
mencía
88
Pleasant, toasty, jammy, spicy.

El Valín Mencía 2020 T
mencía
88
Standard, fruity, wild, simple, tasty, balsamic herbs.

DO BIERZO / D.O.P.

DO BIERZO / D.O.P.

Armas de Guerra Mencía 2021 T
mencía

86

Toasty, jammy, spicy.

VINOS VALTUILLE
Promadelo, s/n
24530 Valtuille de Abajo (León)
☎: +34 987 562 165
info@vinosvaltuille.com
www.vinosvaltuille.com

Cabanelas 2018 T
mencía

91

Colour: cherry, garnet rim. Nose: fruit preserve, black fruit, dried herbs, wild herbs. Palate: flavourful, long, round tannins.

Pago de Valdoneje 2021 T
mencía

89

Kind finish, aromatic, fruity, lactic, jammy.

Pago de Valdoneje El Valao 2019 T BA
mencía

91

Colour: Cherry, garnet rim. Nose: balsamic herbs, sweet spices, scrubland. Palate: spicy, balsamic, good acidity.

Pago de Valdoneje Godello 2021 B
godello

89

Crisp, fruity, tasty, jammy.

Pago de Valdoneje La Tellería 2019 T
mencía

90

Colour: Cherry. Nose: balsamic herbs, scrubland, toasty. Palate: spicy, balsamic, good acidity.

Rapolao 2019 T C
mencía

92

Colour: cherry, garnet rim. Nose: balsamic herbs, ripe fruit, scrubland, fine reductive notes. Palate: balsamic, spicy, flavourful.

VIÑAS DEL BIERZO
24410 Camponaraya (León)
☎: +34 987 463 009
vdelbierzo@granbierzo.com
www.granbierzo.com

Gran Bierzo El Culebral 2021 T
80% mencía, 20% palomino

90

Wild, smooth. Colour: cherry, purple rim. Nose: balsamic herbs, scrubland, wild herbs. Palate: fruity, easy to drink, good finish.

Gran Bierzo Mencía 2020 T
mencía

91

Colour: deep cherry. Nose: ripe fruit, dried herbs, creamy oak. Palate: powerful, ripe fruit, spicy, round tannins.

Gran Bierzo Origen 2020 T
95% mencía, 5% estaladiña

91

Colour: dark-red cherry. Nose: toasty, spicy, cocoa bean, ripe fruit. Palate: flavourful, toasty, fine bitter notes.

Valmagaz Mencía 2021 T
mencía

90

Colour: deep cherry. Nose: ripe fruit, dried herbs, creamy oak, red berry notes. Palate: powerful, ripe fruit, spicy, round tannins.

VIÑEDOS SINGULARES

Avda. de La Riera, 11 Nave 1
08960 Sant Just Desvern (Barcelona)
☎: +34 934 807 041
Fax: +34 934 807 076
info@vinedossingulares.com
www.vinedossingulares.com

Corral del Obispo 2020 T RB
mencía

91

Representative. Colour: cherry, garnet rim. Nose: ripe fruit, scrubland, fine reductive notes. Palate: flavourful, spicy, soft tannins.

VIÑEDOS Y BODEGAS PITTACUM

De la Iglesia, 11
24546 Arganza (León)
☎: +34 987 548 054
pittacum@pittacum.com
www.pittacum.com

La Prohibición 2019 T RB
100% garnacha tintorera

91

Colour: very deep cherry. Nose: roasted coffee, aromatic coffee, powerful, red berry notes, ripe fruit. Palate: smoky aftertaste, great length, round tannins.

Petit Pittacum 2021 T
mencía

91

Colour: cherry, purple rim. Nose: fruit expression, red berry notes, floral, spicy. Palate: flavourful, fruity, good acidity, long.

Pittacum 2019 T RB
100% mencía

90

Colour: cherry, garnet rim. Nose: fruit preserve, fruit liqueur notes, powerful, wet leather. Palate: flavourful, sweetness, long.

Pittacum Aurea 2018 T RB
100% mencía

92

Defined aromas, woody. Colour: deep cherry. Nose: ripe fruit, dried herbs, creamy oak. Palate: powerful, ripe fruit, spicy, round tannins.

Pittacum Val de la Osa 2017 T RB
100% mencía

92

Colour: deep cherry. Nose: ripe fruit, dried herbs, creamy oak, black fruit. Palate: powerful, ripe fruit, spicy, round tannins.

DO BIERZO / D.O.P.

DO. BINISSALEM MALLORCA

CONSEJO REGULADOR

Celler de Rei, 9-1º
07350 Binissalem (Mallorca)
☎: +34 971 512 191 - Fax: +34 971 512 191
@: info@binissalemdo.com
www.binissalemdo.com

LOCATION:

In the central region on the island of Majorca. It covers the municipal areas of Santa María del Camí, Binissalem, Sencelles, Consell and Santa Eugenia.

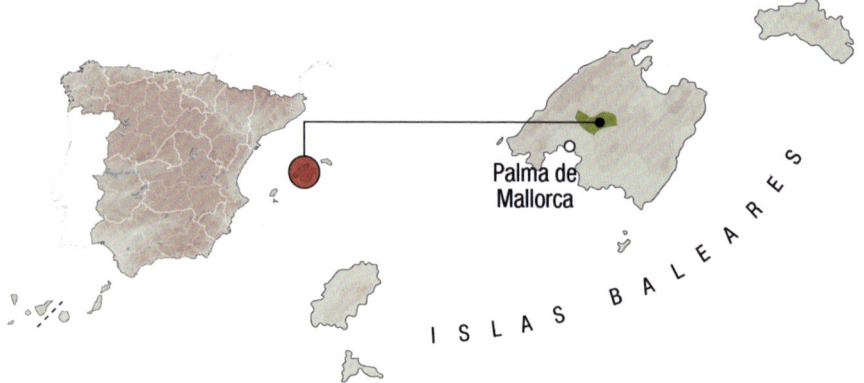

GRAPE VARIETIES:

WHITE: Moll or Prensal Blanc,, Macabeo, Parellada, Moscatel, Giró Ros and Chardonnay.

RED: Manto Negro, Callet, Tempranillo, Syrah, Monastrell, Cabernet Sauvignon, Gorgollassa and Merlot.

FIGURES:

Vineyard surface: 563 – Wine-Growers: 110 – Wineries: 15 – 2021 Harvest rating: N/A – Production 21: 767,129 litres – Market percentages: 86% National - 14% International.

SOIL:

The soil is of a brownish - grey or dun limey type, with limestone crusts on occasions. The slopes are quite gentle, and the vineyards are situated at an altitude ranging from 75 to 200 m.

CLIMATE:

Mild Mediterranean, with dry, hot summers and short winters. The average rainfall per year is around 450 mm. The production region is protected from the northerly winds by the Sierra de Tramuntana and the Sierra de Alfabia mountain ranges.

CLASSIFICATION OF YOUNG WINE HARVESTS PEÑÍNGUIDE

2017	2018	2019	2020	2021
GOOD	GOOD	VERY GOOD	N/A	VERY GOOD

BODEGAS JOSÉ L. FERRER

Conquistador, 103
07350 Binissalem (Illes Balears/Islas Baleares)
☎: +34 971 511 050
Fax: +34 971 870 084
secretaria@vinosferrer.com
www.vinosferrer.com

Ferreret Mantonegro Callet 2021 RD
mantonegro, callet

89

Citrus fruit, herbal, crisp, tasty.

José L. Ferrer 2016 T R
mantonegro, callet, cabernet sauvignon

89

Age nuances, pruney, spicy, dried herbs, balanced, classic.

José L. Ferrer 2018 T C
mantonegro, callet, tempranillo, cabernet sauvignon, syrah

88

Age nuances, balanced, spicy, dried herbs, tasty.

Veritas 2019 BE BN
moll

85

Veritas Blanc 2020 B
moll, chardonnay

90

Colour: bright yellow. Nose: powerful, creamy oak, ripe fruit, spicy, woody. Palate: rich, good structure, long, toasty, fine bitter notes.

Veritas Roig 2021 RD
mantonegro

88

Citrus fruit, dried herbs, crisp, tasty.

FINCA BINIAGUAL

Son Caliú, 18
07181 Calviá (Illes Balears/Islas Baleares)
☎: +34 971 870 111
info@finca-biniagual.com
www.finca-biniagual.com

Finca Biniagual Mantonegro 2020 T FB
100% mantonegro

92

Colour: light cherry. Nose: red berry notes, fruit liqueur notes, meaty notes, fine reductive notes, scrubland. Palate: fleshy, flavourful, elegant.

Finca Biniagual Verán 2019 T BA
40% mantonegro, 30% cabernet sauvignon, 25% syrah, 5% merlot

91

Colour: light cherry. Nose: red berry notes, black fruit, fine reductive notes, cocoa bean, meaty notes. Palate: flavourful, elegant, spicy.

Gran Verán 2019 T BA
45% mantonegro, 55% syrah

94

Colour: cherry, garnet rim. Nose: balanced, complex, ripe fruit, spicy, fine reductive notes. Palate: good structure, flavourful, round tannins, balanced, elegant.

Memòries de Biniagual Negre 2019 T C
52% mantonegro, 32% syrah, 16% cabernet sauvignon

89

Spicy, balanced, herbaceous, full-bodied.

Verán Blanc 2021 B
50% prensal, 50% chardonnay

90

Colour: bright yellow. Nose: ripe fruit, spicy, wild herbs, toasted bread. Palate: good structure, toasty, fine bitter notes.

JAUME DE PUNTIRÓ

Pza. Nova, 23
07320 Santa María del Camí (Illes Balears/Islas Baleares)
☎: +34 606 429 023
pere@vinsjaumedepuntiro.com
www.vinsjaumedepuntiro.com

Buc 2017 T C
mantonegro, cabernet sauvignon

91 🌱

Kind finish. Colour: light cherry, brick rim edge. Nose: ripe fruit, fruit preserve, tobacco, sweet spices, fine reductive notes, toasted bread. Palate: spicy, round tannins, long, flavourful, elegant.

Jaume de Puntiró Blanc 2021 B
prensal

89 🌱

Citrus fruit, balanced, herbal, yeasty notes, tasty, crisp.

Jaume de Puntiró Carmesí 2018 T
mantonegro, syrah, cabernet sauvignon, callet

88 🌱

Balanced, tasty, herbal, spicy.

Jaume de Puntiró Rosat 2021 RD
mantonegro

86 🌱

DO BINISSALEM MALLORCA / D.O.P.

DO BINISSALEM MALLORCA / D.O.P.

Porprat 2016 T
merlot, mantonegro

91 🌱

Colour: straw. Nose: dried herbs, faded flowers, white fruit, citrus fruit. Palate: ripe fruit, balanced, flavourful.

SANTA CATARINA
Ctra. Inca – Sencelles, Km. 3
07140 Sencelles (Illes Balears/Islas Baleares)
☎: +34 971 137 115
info@bodegasantacatarina.com
www.bodegasantacatarina.com

Sta Mantonegro 2020 T
100% mantonegro

91

Colour: light cherry. Nose: red berry notes, scrubland, earthy notes, fine reductive notes. Palate: ripe fruit, spicy, round tannins.

Sta Prensal 2021 B
100% prensal

88

Citrus fruit, balanced, crisp, dried herbs.

VINS NADAL
Ramón Llull, 2
07350 Binissalem (Illes Balears/Islas Baleares)
☎: +34 971 511 058
vinsnadal@vinsnadal.es
www.vinsnadal.es

Albaflor Blanc 2021 B
prensal, mantonegro, moscatel, chardonnay, macabeo

87

Rosat 110 Vins Nadal 2021 RD
mantonegro

87

VINYA TAUJANA
Balanguera, 40
07142 Santa Eugènia (Illes Balears/Islas Baleares)
☎: +34 971 144 494
vinyataujana@gmail.com
www.vinyataujana.es

Lluor Negre 2019 T
48% mantonegro, 33% syrah, 19% merlot

88

Full-bodied, creamy, spicy, full-bodied, roasted coffee.

Torrent Fals 2019 T C
47% mantonegro, 34% syrah, 19% merlot

87

Vinya Taujana Blanc de Blanc 2021 B
59% prensal, 41% chardonnay

85

Vinya Taujana Rosat 2021 RD
100% mantonegro

83

VINYES I VINS CA SA PADRINA
Camí dels Horts, s/n
07140 Sencelles (Illes Balears/Islas Baleares)
☎: +34 646 318 600
Fax: +34 971 874 370
cellermantonegro@gmail.com
www.vinscasapadrina.com

Gran Padrina 2020 T

87

Montnegre 2019 T

88

Age nuances, balanced, spicy, dried herbs, tasty.

DO. BIZKAIKO TXAKOLINA
CONSEJO REGULADOR

Bº Mendibile, 42
48940 Leioa (Bizkaia)
☎: +34 946 076 071 - Fax: +34 946 076 072
@: info@bizkaikotxacolina.eus
www.bizkaikotxakolina.eus

LOCATION:

In the province of Vizcaya. The production region covers both coastal areas and other areas inland.

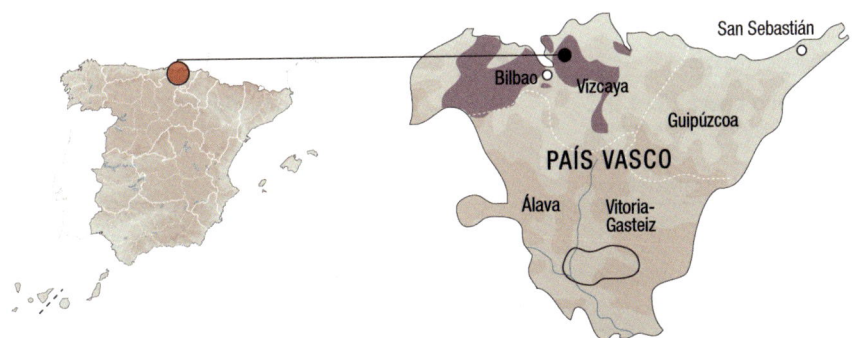

GRAPE VARIETIES:

WHITE: Hondarrabi Zuri, Folle Blanche.

RED: Hondarrabi Beltza, Pinot Noir and Cabernet Franc (Berdesarie).

FIGURES:

Vineyard surface: 430 – **Wine-Growers:** 188 – **Wineries:** 38 – **2021 Harvest rating:** N/A – **Production 21:** 1,557,876 litres – **Market percentages:** 95% National - 5% International.

SOIL:

Mainly clayey, although slightly acidic on occasions, with a fairly high organic matter content.

CLIMATE:

Quite humid and mild due to the influence of the Bay of Biscay which tempers the temperatures. Fairly abundant rainfall, with an average of 1,000 to 1,300 mm per year.

CLASSIFICATION OF YOUNG WINE HARVESTS PEÑÍNGUIDE

2017	2018	2019	2020	2021
VERY GOOD	VERY GOOD	VERY GOOD	VERY GOOD	VERY GOOD

DO BIZKAIKO TXAKOLINA / D.O.P.

AXPE SAGARDOTEGIA TXAKOLINDEGIA

Bº Atxondoa, Caserío Axpe, 13
48270 Markina - Xemein (Bizkaia/Vizcaya)
☎: +34 655 734 625
Fax: +34 946 168 285
axpesagardoa@yahoo.es
www.axpesagardotegia.com

Axpe Txakoli 2020 B
hondarrabi zuri, hondarrabi zerratia, sauvignon blanc, riesling

88

Herbal, citrus fruit, simple, crisp.

BODEGA BERROJA TXAKOLI

Al. de Urkijo, 22 Ppal Dcha.
48008 Bilbao (Bizkaia/Vizcaya)
☎: +34 944 106 254
txakoli@bodegaberroja.com
www.bodegaberroja.com

Aguirrebeko Lasal 2019 B C
hondarrabi zuri, hondarrabi zerratia, riesling

89

Citrus fruit, herbal, jammy, tasty, crisp.

Aguirrebeko Lasal 2020 B C
hondarrabi zuri, riesling, hondarrabi zerratia

87

Txakoli Aguirrebeko 2021 B
hondarrabi zuri, hondarrabi zerratia, riesling

89

Citrus fruit, herbal, tasty, crisp, yeasty notes.

Txakoli Berroja 2019 B
hondarrabi zuri, riesling

91

Colour: bright straw. Nose: fragrant herbs, fine lees, faded flowers. Palate: long, good acidity, balanced, fine bitter notes.

BODEGA MAGALARTE ZAMUDIO

Arteaga Auzoa, 107
48170 Zamudio (Bizkaia/Vizcaya)
☎: +34 695 722 885
kaixo@bodegamagalartezamudio.com
www.bodegamagalartezamudio.com

Aretxabaleta 2019 B
hondarrabi zuri, riesling, izkiota txikia

92

Colour: bright yellow. Nose: toasted bread, toasty, expressive, ripe fruit. Palate: fruity, flavourful, powerful, unctuous.

Aretxabaleta 2020 B
hondarrabi zuri, izkiriota ttppia, riesling

91

Colour: bright yellow. Nose: ripe fruit, creamy oak, faded flowers, expressive. Palate: fresh, flavourful, fruity.

Magalarte Zamudio 2019 B FB
hondarrabi zuri, izkiota txikia

92

Colour: bright yellow. Nose: ripe fruit, fine lees, expressive, spicy, toasted bread. Palate: flavourful, long, balanced.

Magalarte Zamudio 2020 B FB
hondarrabi zuri, izkiriota ttppia

91

Aromatic, floral. Colour: bright yellow. Nose: ripe fruit, spicy, citrus fruit. Palate: flavourful, good acidity, fine bitter notes.

Magalarte Zamudio 2021 B
hondarrabi zuri, hondarrabi zerratia, izkiriota ttppia, riesling, mune mahatsa

89

Floral, crisp, citrus fruit, herbal, tasty.

Zabalondo 2021 B
hondarrabi zuri, hondarrabi zerratia, izkiriota ttppia, riesling

88

Fruity, herbal, simple, crisp.

BODEGA ULIBARRI

Caserío Isuskiza Handi, 1 Barrio Zaldu
48192 Gordexola (Bizkaia/Vizcaya)
☎: +34 665 725 735
ulibarriartzaiak@gmail.com

Artzai 2019 B

91 🌱

Colour: bright straw. Nose: fragrant herbs, fine lees, dry nuts, stone fruit, spicy. Palate: full-bodied, good acidity, flavourful.

Artzai 2020 B

89 🌱

Citrus fruit, crisp, herbal, balanced.

Ulibarri 2019 B

91 🌱

Colour: bright straw, greenish rim. Nose: fresh fruit, citrus fruit, wild herbs. Palate: fresh, fruity, good acidity.

Ulibarri 2020 B

90 🌱

Colour: bright straw, greenish rim. Nose: fresh fruit, citrus fruit, wild herbs, balsamic herbs, fine lees. Palate: fresh, fruity, good acidity.

BODEGAS DE GALDAMES
El Ventorro, 4
48191 Galdames (Bizkaia/Vizcaya)
☎: +34 627 992 063
Fax: +34 946 100 107
info@vinasulibarria.com
www.vinasulibarria.com

Torre de Loizaga Bigarren 2021 B
88
Citrus fruit, herbal, yeasty notes, crisp.

Torre de Loizaga Selección 2020 B
89
Aromatic, slight oxidation, citrus fruit, floral, fruity, smooth.

BODEGAS ITSASMENDI
Barrio Arane, 66
48300 Gernika (Bizkaia/Vizcaya)
☎: +34 946 270 316
info@bodegasitsasmendi.com
www.bodegasitsasmendi.com

Itsasmendi Ados 2020 B
hondarrabi zuri

91
Colour: bright straw, greenish rim. Nose: fresh fruit, citrus fruit, dried flowers, earthy notes, fine lees. Palate: fresh, good acidity, fine bitter notes, rich.

🏆 **PODIUM**

Itsasmendi Artizar (botella Borgoña) 2017 B
hondarrabi zuri

96
Colour: yellow, golden. Nose: expressive, characterful, faded flowers, stone fruit. Palate: round, rich, complex, fine bitter notes.

Itsasmendi nº 7 2020 B
hondarrabi zuri, hondarrabi zerratia, riesling

91
Colour: bright straw. Nose: fresh fruit, citrus fruit, wild herbs. Palate: fresh, good acidity, fine bitter notes.

Itsasmendi Urezti 2018 B
hondarrabi zerratia, izkiriot handi

92
Colour: yellow, pale. Nose: dried flowers, ripe fruit. Palate: fruity, good acidity, easy to drink.

Paradisuak Janeo 2020 RD
pinot noir

92
Colour: raspberry rose. Nose: elegant, dried flowers. Palate: flavourful, saline, balanced, fine bitter notes, fresh.

Paradisuak Jauregi - Abadiño 2018 B
hondarrabi zuri, izkiriot handi

93
Colour: yellow. Nose: ripe fruit, fragrant herbs, fine lees, saline, spicy. Palate: full-bodied, rich, long, good acidity.

Paradisuak Leioa 2019 B
hondarrabi zuri

92
Colour: bright straw. Nose: fresh fruit, fine lees, white fruit. Palate: rich, varietal, fresh, good acidity.

Paradisuak Otxolarre - Morga 2019 B
hondarrabi zuri

93
Colour: bright straw. Nose: ripe fruit, fragrant herbs, fine lees, creamy oak, hint of anise. Palate: full-bodied, rich, long, good acidity.

BODEGAS VIRGEN DE LOREA
Barrio La Flor
48860 Zalla (Bizkaia/Vizcaya)
☎: +34 609 451 875
aitor@muebleselparaiso.es
www.bodegasvirgendelorea.com

Señorío de Otxaran 2017 B
80% hondarrabi zerratia, 10% hondarrabi zuri, 10% sauvignon blanc

91
Colour: bright yellow. Nose: brioche, ripe fruit, dried herbs, sweet spices. Palate: flavourful, fruity, fresh, good finish.

Señorío de Otxaran 2019 B
91
Colour: bright straw. Nose: fragrant herbs, fine lees, stone fruit, mineral. Palate: full-bodied, rich, long, good acidity.

Txakoli Aretxaga 2021 B
20% hondarrabi zuri, 80% hondarrabi zerratia

89
Crisp, herbal, representative, smooth, crisp.

Txakoli Laínoa Magnum 2020 B
20% hondarrabi zuri, 10% folle blanch, 70% hondarrabi zerratia

92
Colour: bright straw, greenish rim. Nose: fresh fruit, citrus fruit, wild herbs, fine lees, dry stone. Palate: fresh, good acidity, great length, mineral.

Txakoli Laínoa Magnum 2021 B
hondarrabi zuri, hondarrabi zerratia, sauvignon blanc

91
Colour: bright straw. Nose: fresh fruit, fine lees, floral, balanced, neat. Palate: fresh, easy to drink, good acidity.

DO BIZKAIKO TXAKOLINA / D.O.P.

DO BIZKAIKO TXAKOLINA / D.O.P.

Txakoli Leyes de Abellaneda Berezia Magnum 2021 B
hondarrabi zerratia, hondarrabi zuri

91
Colour: bright straw. Nose: citrus fruit, wild herbs, fresh, white fruit. Palate: fresh, fruity, good acidity, fine bitter notes.

DE BRINGAS
Molinar, 26
48891 Karrantza Harana (Bizkaia/Vizcaya)
☎: +34 609 776 119
luribc@hotmail.com

De Bringas 2020 B
hondarrabi zuri
88
Kind finish, floral, crisp, fruity.

De Bringas 2021 B
hondarrabi zuri
87

DONIENE GORRONDONA TXAKOLINA
Gibelorratzagako San Pelaio, 1
48130 Bakio (Bizkaia/Vizcaya)
☎: +34 946 194 795
gorrondona@donienegorrondona.com
www.donienegorrondona.com

Doniene 2021 B
90
Colour: bright straw, greenish rim. Nose: fresh fruit, citrus fruit, wild herbs, fine lees. Palate: fresh, fruity, good acidity.

Doniene Apardune 2020 BE BN
90
Colour: bright yellow. Nose: fine lees, balanced, dried herbs, citrus fruit. Palate: good acidity, flavourful, ripe fruit, long.

Doniene XX 2019 B BA
93
Colour: bright yellow. Nose: fine lees, saline, floral, saffron, expressive, neat. Palate: full-bodied, full of life, flavourful, good acidity, spicy.

Gorrondona 2021 B
91
Colour: bright straw. Nose: fragrant herbs, fine lees, white fruit, citrus fruit. Palate: full-bodied, good acidity, flavourful, mineral.

GARENA TXAKOLINA
Barrio Iturriotz,11 (Lamindao)
48141 Dima (Bizkaia/Vizcaya)
☎: +34 946 317 215
info@garenatxakolina.com
www.garenatxakolina.com

Garena 2021 B
80% hondarrabi zerratia, 20% hondarrabi zuri
89
Citrus fruit, crisp, wild, simple, balanced.

Geroa lías 2020 B
100% hondarrabi zerratia
90
Colour: bright yellow. Nose: white fruit, fresh fruit, wild herbs, citrus fruit, fine lees. Palate: varietal, easy to drink.

GARKALDE TXAKOLINA
Goitioltza, 8
48196 Lezama (Bizkaia/Vizcaya)
☎: +34 677 578 664
garkaldetxakolina@hotmail.com

Garkalde Txakolina 2021 B
hondarrabi zuri, petit corbú
87

GORKA IZAGIRRE
Barrio Legina, s/n
48195 Larrabetzu (Bizkaia/Vizcaya)
☎: +34 946 742 706
Fax: +34 946 741 221
txakoli@gorkaizagirre.com
www.gorkaizagirre.com

Ama de Gorka Izagirre 2019 B
100% hondarrabi zerratia
92
Colour: bright straw. Nose: fragrant herbs, fine lees, white fruit, stone fruit, citrus fruit, spicy. Palate: full-bodied, good acidity, balanced.

Ama de Gorka Izagirre Magnum 2015 B
100% hondarrabi zerratia
94
Colour: bright golden. Nose: white fruit, dry nuts, brioche, pattiserie, wild herbs. Palate: full-bodied, flavourful, mineral.

E-Gala 2021 B
89
Fruity, jammy, floral, crisp, tasty.

G22 de Gorka Izagirre 2020 B
100% hondarrabi zerratia

91

Colour: bright yellow. Nose: citrus fruit, white fruit, white flowers, balanced. Palate: fruity, correct, good acidity.

G22 de Gorka Izagirre Magnum 2019 B
100% hondarrabi zerratia

92

Colour: bright yellow. Nose: faded flowers, citrus fruit, fine lees, yeasty notes. Palate: flavourful, easy to drink, good finish, good acidity.

Gorka Izagirre 2021 B
50% hondarrabi zuri, 50% hondarrabi zerratia

89

Citrus fruit, crisp, floral, balanced, wild, smooth.

Zura 2019 B
hondarrabi zerratia

92

Colour: bright yellow. Nose: faded flowers, white flowers, ripe fruit, fine lees, toasted bread. Palate: full-bodied, flavourful, balanced, good acidity, good finish, spicy.

GURE AHALEGINAK
Barrio Ibazurra, 1
48460 Orduña (Bizkaia/Vizcaya)
☎: +34 658 744 181
a_larrazabal@hotmail.com
www.gureahaleginak.com

Ana Vendimia Tardía 2020 B
hondarrabi zuri

93

Colour: bright yellow. Nose: ripe fruit, citrus fruit, floral. Palate: flavourful, good acidity, easy to drink.

Filoxera 2020 T BA
hondarrabi beltza

91

Colour: cherry, purple rim. Nose: red berry notes, spicy, herbaceous, smoky, violets. Palate: flavourful, fruity, good acidity, long.

Gure Ahaleginak 2021 B
hondarrabi zuri, hondarrabi zerratia, gros manseng

87

Mirene 2020 B
hondarrabi zuri

92

Oxidativ. Colour: bright yellow. Nose: ripe fruit, sweet spices, cereal notes, dried herbs, fine lees, smoky. Palate: fresh, fruity, powerful, flavourful, good finish, soft tannins.

HASIBERRIAK WINES
Gametxo, 12
48311 Ibarrangelua (Bizkaia/Vizcaya)
☎: +34 689 604 106
contacto@hasiberriakwine.com
www.hasiberriakwines.com

Arotz 2020 B BA
hondarrabi zuri, hondarrabi zerratia

92

Colour: straw. Nose: white fruit, wild herbs, wild herbs, mineral. Palate: fruity, flavourful, good acidity, balanced.

Nekazari 2020 B C
hondarrabi zuri, hondarrabi zerratia

91

Colour: bright straw. Nose: white fruit, wild herbs. Palate: fruity, fresh, flavourful.

JON ANDER REKALDE
San Roke Bekoa, 11
48150 Sondika (Bizkaia/Vizcaya)
☎: +34 944 458 631
txakoliartxanda@movistar.es

Artxanda 2020 B
hondarrabi zuri

90

Colour: bright straw. Nose: fragrant herbs, fine lees, citrus fruit. Palate: good acidity, balanced, fine bitter notes.

MAGALARTE LEZAMA
B. Garaioltza, 92 B
48196 Lezama (Bizkaia/Vizcaya)
☎: +34 672 249 868
magalarte@icloud.com
www.magalartelezamatxakolina.com

Aitu! 2020 T
hondarrabi beltza

89

Balanced, spicy, with personality, animal funk, dried herbs.

IEUP 2020 B
hondarrabi zuri

90

Colour: bright straw, greenish rim. Nose: fresh fruit, citrus fruit, wild herbs, fine lees, dry stone. Palate: fresh, fruity, good acidity.

IEUP Barrikan 2020 B FB
hondarrabi zuri, izkiriota ttppia

91

Colour: bright yellow. Nose: white fruit, grassy, sweet spices, white flowers. Palate: fresh, fruity, flavourful, full-bodied, spicy.

DO BIZKAIKO TXAKOLINA / D.O.P.

DO BIZKAIKO TXAKOLINA / D.O.P.

IEUP Sobre lías Magnum 2019 B
hondarrabi zuri
93
Colour: bright golden. Nose: ripe fruit, wild herbs, toasty, spicy, fine lees, dried flowers. Palate: fruity, flavourful, fresh, full-bodied.

Magalarte Lezama 2020 B
hondarrabi zuri, mune mahatsa, riesling
88
Kind finish, crisp, smooth, standard.

Sagastibeltza 2020 B
hondarrabi zuri
88
Kind finish, citrus fruit, herbal, tasty.

MERRUTXU
Bº Arboliz, 15
48311 Ibarrangelu (Bizkaia/Vizcaya)
☎: +34 626 140 830
info@merrutxu.com
www.txakolibizkaia.com

Merrutxu 2021 B
88
Floral, crisp, fruity, crisp.

PENÍNSULA VINICULTORES
Hermosilla 29, 1ºA
28008 Madrid (Madrid)
☎: +34 916 039 363
o.moreno@peninsula.wine
www.peninsula.wine

Península Vino Atlántico 2021 B
hondarrabi zerratia
92
Smooth. Colour: bright yellow. Nose: citrus fruit, fresh fruit, wild herbs, grassy. Palate: fresh, good acidity, fleshy.

TALLERI BERRIA UPATEGIA ETA MAHASTIAK
Bº Erroteta, s/n
48115 Morga (Bizkaia/Vizcaya)
☎: +34 946 138 318
admin@bodegatalleri.com

Bitxia Berria 2021 B
87

Gure Aberria 2021 B
hondarrabi zuri, hondarrabi zerratia
89
Crisp, citrus fruit, herbal, crisp, representative, smooth.

Gure Natura 2020 B
hondarrabi zerratia
91
Colour: bright straw. Nose: white fruit, tropical fruit, wild herbs. Palate: fruity, fresh, balanced, good finish.

TX Berria 2020 B
90
Colour: bright yellow. Nose: white flowers, white fruit, balanced, medium intensity. Palate: flavourful, easy to drink, good acidity.

TXAKOLI TXABARRI
Barrio Muñeran, 17
48850 Zalla (Bizkaia/Vizcaya)
☎: +34 625 708 114
txabarri@txakolitxabarri.com
www.txakolitxabarri.com

Txabarri 2021 RD
85

Txabarri 2021 T
85

Txabarri Abeitxa 2020 B
86

Txakoli Txabarri Extra 2021 B
88
Citrus fruit, floral, fruity, herbal, simple.

DO. BULLAS
CONSEJO REGULADOR

Balsa, 26
30180 Bullas (Murcia)
☎: +34 968 652 601 – Fax: +968 652 601
@: consejoregulador@vinosdebullas.es
www.vinosdebullas.es

LOCATION:
In the province of Murcia. It covers the municipal areas of Bullas, Cehegín, Mula Pliego and Ricote, and several vineyards in the vicinity of Calasparra, Caravaca, Moratalla and Lorca.

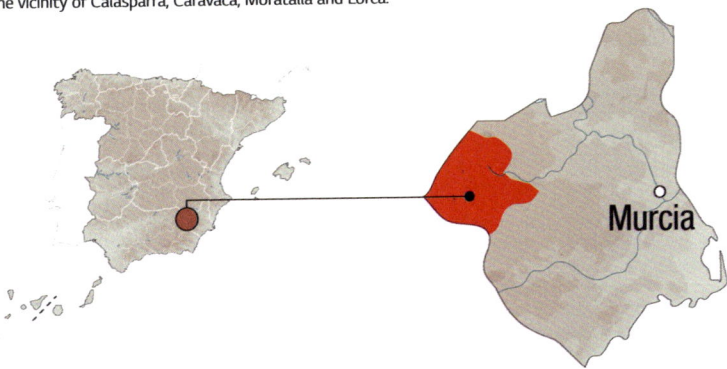

GRAPE VARIETIES:
WHITE: Macabeo (main), Airén, Chardonnay, Malvasía, Moscatel de Grano Menudo and Sauvignon Blanc.

RED: Monastrell (main), Petit Verdot, Tempranillo, Cabernet Sauvignon, Syrah, Merlot, Garnacha and Garnacha Tintorera.

FIGURES:
Vineyard surface: 1,172 – **Wine-Growers:** 252 – **Wineries:** 12 – **2021 Harvest rating:** N/A – **Production 21:** 2,349,000 litres – **Market percentages:** 52% National - 48% International.

SOIL:
Brownish - grey limey soil, with limestone crusts, and alluvial. The terrain is rugged and determined by the layout of the little valleys, each with their own microclimate. Distinction can be made between 3 areas: one to the north north - east with an altitude of 400 – 500 m; another in the central region, situated at an altitude of 500 – 600 m; and the third in the western and north - western region, with the highest altitude (500 – 810 m), the highest concentration of vineyards and the best potential for quality.

CLIMATE:
Mediterranean, with an average annual temperature of 15.6 °C and low rainfall (average of 300 mm per year). The heavy showers and storms which occur in the region are another defining element.

CLASSIFICATION OF YOUNG WINE HARVESTS PEÑÍNGUIDE

2017	2018	2019	2020	2021
GOOD	GOOD	GOOD	VERY GOOD	VERY GOOD

DO BULLAS / D.O.P.

BODEGA BALCONA
Ctra. Carretera de Bullas – Avilés, Km. 9.7 Paraje Aceniche
30180 Bullas (Murcia)
☎: +34 609 104 111
info@bodegabalcona.com
www.bodegabalcona.com

37 Barricas 2018 T C
monastrell
88 🌱
Jammy, fruity, crisp, spicy, pruney.

Mabal 2020 T
monastrell
87 🌱

MaBal Macabeo de Balcona 2019 B
88 🌱
Citrus fruit, dried herbs, tasty, crisp.

Partal Cepas Viejas 2017 T
monastrell
91 🌱
Slight oxidation. Colour: cherry, purple rim. Nose: red berry notes, cocoa bean, sweet spices, raspberry. Palate: fruity, flavourful, balanced.

Partal de Autor 2006 T
60% monastrell, 20% syrah, 10% tempranillo, 10% cabernet sauvignon, merlot
91
Colour: Cherry, garnet rim. Nose: old leather, meaty notes, spicy, smoky, fruit preserve. Palate: fruity, flavourful, slightly dry, soft tannins, balanced.

BODEGA MONASTRELL
Ctra. Bullas-Avilés, km. 9,3 "Valle Aceniche"
30180 Bullas (Murcia)
☎: +34 648 702 412
Fax: +34 968 654 925
info@bodegamonastrell.com
www.bodegamonastrell.com

Almudí 2018 T
88 🌱
Fruity, toasty, simple, crisp, jammy.

Almudí Uno 2017 T C
petit verdot
92
Kind finish, full-bodied. Colour: deep cherry. Nose: dried herbs, creamy oak, black fruit, fine reductive notes. Palate: powerful, ripe fruit, spicy, round tannins.

Chaveo 2019 T C
90 🌱
Colour: deep cherry. Nose: ripe fruit, dried herbs, creamy oak. Palate: spicy, round tannins, balanced.

Salto del Usero 2021 B
macabeo
87

Salto del Usero 2021 T
monastrell
89 🌱
Fruity, vegetal, floral, tasty.

Valché 2019 T C
90 🌱
Colour: deep cherry. Nose: ripe fruit, dried herbs, creamy oak. Palate: spicy, round tannins, balanced.

BODEGA SAN ISIDRO. BULLAS
Pol. Ind. Marimingo, Altiplano, s/n
30180 Bullas (Murcia)
☎: +34 968 654 991
Fax: +34 968 652 160
administracion@bodegasanisidrobullas.com
www.bodegasanisidrobullas.com

Cepas del Zorro 2019 B FB
100% macabeo
86

Cepas del Zorro 2019 T C
85% monastrell, 15% syrah
88
Fruity, spicy, simple.

Cepas del Zorro 2021 RD
80% monastrell, 20% garnacha
89
Nose: fruit expression, red berry notes, lactic notes, floral. Palate: balanced, easy to drink.

Cepas del Zorro Macabeo 2021 B
100% macabeo
87

Cepas del Zorro Monastrell 2020 T
100% monastrell
87

BODEGA TERCIA DE ULEA
Tercia de Ulea, s/n
30440 Moratalla (Murcia)
☎: +34 968 433 213
bodega@terciadeulea.com
www.terciadeulea.com

Mirthya 2020 T
100% monastrell
88
Smoky, jammy, powerful, spicy.

Rambla de Ulea 2021 T
100% monastrell

88

Spicy, fruity, jammy, tasty.

Tercia de Ulea 2020 T C
100% monastrell

86

Travesura Cabernet Sauvignon 2021 T
70% cabernet sauvignon, 30% monastrell

85

Travesura Shiraz 2021 T
100% syrah

88

Colour: cherry, purple rim. Nose: ripe fruit, red berry notes, spicy. Palate: fruity, flavoured, balanced.

Viña Botial 2020 T RB
monastrell, syrah

88

Sweety finish, pruney, fruity, tasty.

BODEGAS CARREÑO
Ginés de Paco, 22
30430 Cehegin (Murcia)
☎: +34 968 740 004
info@bodegascarreno.com
www.bodegascarreno.com

Begastri 2018 T C
60% monastrell, 40% petit verdot

87

Begastri 2021 B
macabeo

86

Begastri 2021 RD
monastrell

85

Marmallejo 2018 T C
petit verdot

88

Pronounced acidity, spicy, dried flowers, dried herbs.

Viña Azeniche Petit Verdot 2018 T
petit verdot

89

Balanced, spicy, smoky, dried herbs.

Viña Azeniche Syrah 2019 T C
syrah

88

Spicy, jammy, dried herbs, toasty.

BODEGAS CONTRERAS
Los Ríos, 2
30812 Avilés de Lorca (Murcia)
☎: +34 685 874 594
info@bodegas-contreras.com
www.bodegas-contreras.com

Sortius Monastrell 2020 T
100% monastrell

87

Sortius Monastrell 2021 T
100% monastrell

88

Fruity, pruney, powerful, dried flowers, standard, tasty, spicy.

Sortius Syrah 2020 T RB
100% syrah

90

Colour: bright cherry. Nose: floral, ripe fruit, balanced. Palate: flavourful, fruity, good acidity.

Uvio C 2019 T C
85% monastrell, 15% syrah

89

Jammy, spicy, roasted coffee, kind finish.

BODEGAS DEL ROSARIO
Avda. de la Libertad, s/n
30180 Bullas (Murcia)
☎: +34 968 652 075
export@bodegasdelrosario.es
www.bodegasdelrosario.es

Inmortalis 2019 T
100% monastrell

88

Spicy, fruity, tasty, wild.

Las Reñas Selección Monastrell Syrah 2018 T C
70% monastrell, 30% syrah

91

Colour: bright cherry. Nose: ripe fruit, black fruit, spicy, dry nuts, wild herbs, tomato. Palate: fresh, flavourful, slightly dry, soft tannins.

Niño de las Uvas Monastrell 2020 T RB
100% monastrell

90

Colour: cherry, purple rim. Nose: powerful, ripe fruit, spicy. Palate: ripe fruit, flavourful, good structure.

Niño de las Uvas Monastrell 2021 RD
monastrell

89

Pleasant, smooth, tasty, fruity.

DO BULLAS / D.O.P.

DO BULLAS / D.O.P.

BODEGAS LAVIA
Paraje Venta del Pino, Parcela 38
30430 Cehegín (Murcia)
info@mgwinesgroup.com
www.bodegaslavia.com

Lavia Finca Paso Malo 2019 T C
monastrell

93 🌱

Colour. Cherry. Nose: complex, expressive, spicy, mineral, balsamic herbs, scrubland. Palate: elegant, full-bodied, long, great length.

Lavia Valle del Aceniche 2019 T C
monastrell

92 🌱

Colour: deep cherry. Nose: ripe fruit, dried herbs, creamy oak, scrubland, balsamic herbs. Palate: powerful, ripe fruit, spicy, round tannins.

Lavia Valle Venta del Pino 2019 T C
monastrell

91 🌱

Colour: bright cherry. Nose: balsamic herbs, sweet spices, scrubland. Palate: spicy, balsamic, good acidity.

Pueblo de Lavia 2019 T
monastrell

89 🌱

Aromatic, pleasant, smooth, tasty, balsamic herbs.

CARRASCALEJO
Finca Carrascalejo, s/n
30180 Bullas (Murcia)
☎: +34 968 652 003
carrascalejo@carrascalejo.com
www.carrascalejo.com

Carrascalejo 2017 T C
60% monastrell, 20% syrah, 20% cabernet sauvignon

87

Carrascalejo 2020 T
100% monastrell

87

Carrascalejo 2021 RD
100% monastrell

86

Rosmarinus 2018 T RB
80% monastrell, 20% syrah

87 🌱

Rosmarinus 2020 T
80% monastrell, 10% tempranillo, 10% syrah

87 🌱

LLANO Y MONTE
Paraje de la Alquibla
30178 Mula (Murcia)
☎: +34 868 087 355
Fax: +34 868 087 388
info@bodegasllanoymonte.com
www.bodegasllanoymonte.com

El Secreto del Abuelo 2018 T C
40% merlot, 25% cabernet sauvignon, 35% syrah

89

Age nuances, pruney, dried herbs, tasty, toasty.

El Secreto del Abuelo 2020 T RB
65% garnacha tintorera, 15% syrah, 20% merlot

86

DO. CALATAYUD
CONSEJO REGULADOR

Ctra. de Valencia, 8
50300 Calatayud (Zaragoza)
☎: +34 976 884 260 - Fax: +34 976 885 912
@: administracion@docalatayud.com
www.docalatayud.com

LOCATION:

It is situated in the western region of the province of Zaragoza, along the foothills of the Sistema Ibérico, outlined by the network of rivers woven by the different tributaries of the Ebro: Jalón, Jiloca, Manubles, Mesa, Piedra and Ribota, and covers 46 municipal areas of the Ebro Valley.

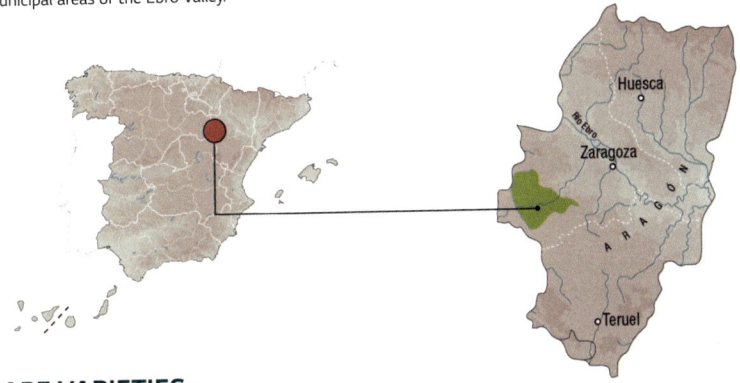

GRAPE VARIETIES:

WHITE: Preferred: Macabeo and Malvasía.

Authorized: Moscatel de Alejandría, Garnacha Blanca, Sauvignon Blanc, Gewürztraminer and Chardonnay.

RED: Preferred: Garnacha Tinta, Tempranillo and Mazuela.

Authorized: Monastrell, Cabernet Sauvignon, Merlot, Bobal and Syrah.

FIGURES:

Vineyard surface: 3,300 – **Wine-Growers:** 860 – **Wineries:** 15 – **2021 Harvest rating:** N/A – **Production 21:** 6,518,700 litres – **Market percentages:** 15% National - 85% International.

SOIL:

In general, the soil has a high limestone content. It is formed by rugged stony materials from the nearby mountain ranges and is on many occasions accompanied by reddish clay. The region is the most rugged in Aragón, and the vineyards are situated at an altitude of between 550 and 880 m.

CLIMATE:

Semi - arid and dry, although somewhat cooler than Cariñena and Borja, with cold winters, an average annual temperature which ranges between 12 and 14 °C, and a period of frost of between 5 and 7 months which greatly affects the production. The average rainfall ranges between 300 – 550 mm per year, with great day/night temperature contrasts during the ripening season.

CLASSIFICATION OF YOUNG WINE HARVESTS PEÑÍNGUIDE

2017	2018	2019	2020	2021
VERY GOOD	VERY GOOD	VERY GOOD	VERY GOOD	VERY GOOD

BODEGA LA DOLORES
Juan José Lorente, 19 bajo
50005 Zaragoza (Zaragoza)
☎: +34 609 251 412
pregunta@bodegaladolores.com
www.bodegaladolores.com

El Hijo de La Dolores 2021 T
garnacha, tempranillo
88
Standard, jammy, simple, wild, tasty.

La Dolores Tinaja 2020 T
garnacha, tempranillo
88
Smoky, toasty, fruity, jammy, spicy.

La Dolores Viñas Viejas 2021 T
garnacha
88
Fruity, jammy, spicy, tasty.

BODEGA SAN GREGORIO
Ctra. Villalengua, s/n
50312 Cervera de la Cañada (Zaragoza)
☎: +34 669 791 152
enologia@bodegasangregorio.com

Armantes Vendimia Seleccionada 2019 T
60% tempranillo, 40% garnacha
90
Colour: bright cherry. Nose: ripe fruit, violets, faded flowers, dried herbs. Palate: fruity, flavourful, good structure, slightly dry, soft tannins.

Tres Ojos Garnacha 2020 T
100% garnacha
88
Spicy, dried herbs, fruity, tasty.

BODEGA VIRGEN DE LA SIERRA S. COOP.
Avda. de la Cooperativa, 21-23
50310 Villarroya de La Sierra (Zaragoza)
☎: +34 976 899 015
manuel@bodegavirgendelasierra.com
www.bodegavirgendelasierra.com

Albada Finca Gemelo 2020 T
96% garnacha, 4% garnacha blanca, bobal
91
Colour: cherry, purple rim. Nose: fruit expression, floral, spicy, red berry notes, ripe fruit. Palate: flavourful, fruity, good acidity, long.

Albada Finca Santos 2020 T
96% garnacha, 4% garnacha blanca, macabeo
90
Taut. Colour: cherry, purple rim. Nose: floral, raspberry, dried herbs. Palate: fruity, flavourful, balanced.

Albada Garnacha Viñas Viejas sobre Lías 2021 T
garnacha
89
Herbal, fruity, spicy, tasty.

Albada Macabeo Viñas Viejas sobre Lías 2021 B
macabeo
87

Albada Paraje La Cañadilla 2019 T
96% garnacha, 4% macabeo, garnacha blanca, bobal, monastrell
91
Colour: deep cherry. Nose: fruit preserve, dried herbs, spicy. Palate: fruity, flavourful, balanced, good finish, slightly dry, soft tannins.

Albada Paraje La Cañadilla 2020 T
garnacha, macabeo, garnacha blanca, bobal, monastrell
91
Colour: cherry, garnet rim. Nose: fruit preserve, overripe fruit, mineral. Palate: flavourful, long, good structure.

Albada Paraje Llano Herrera 2020 T
95% garnacha, 5% macabeo, garnacha blanca, bobal
90
Colour: bright cherry. Nose: ripe fruit, black fruit, dark chocolate, sweet spices, toasty. Palate: flavourful, fruity, balanced, slightly dry, soft tannins.

Cruz de Piedra Selección Especial Garnacha 2020 T
100% garnacha
89
Dried herbs, spicy, toasty, fruity, tasty.

BODEGAS ATECA

Ctra. N-II, s/n
50200 Ateca (Zaragoza)
☎: +34 968 435 022
Fax: +34 968 716 051
info@gilfamily.es
www.gilfamily.es

Atteca 2020 T
100% garnacha

93

Colour: deep cherry. Nose: dried herbs, creamy oak, black fruit. Palate: powerful, ripe fruit, spicy, round tannins.

Atteca Armas 2019 T
100% garnacha

94

Colour: deep cherry. Nose: ripe fruit, dried herbs, creamy oak, black fruit, aromatic coffee. Palate: powerful, ripe fruit, spicy, round tannins.

🏆 PODIUM

Atteca Armas 2020 T
garnacha

95

Colour: very deep cherry. Nose: complex, expressive, spicy, mineral, ripe fruit, red berry notes, toasty, new oak. Palate: full-bodied, long, great length.

Atteca Selección de la Familia 2019 T

90

Colour: cherry, purple rim. Nose: ripe fruit, grassy, dried herbs, spicy. Palate: fruity, flavourful, balanced, good finish.

Atteca Selección de la Familia 2020 T
garnacha

92

Colour: cherry, purple rim. Nose: fruit expression, red berry notes, floral, spicy. Palate: flavourful, fruity, good acidity, long.

Honoro Vera Garnacha 2021 T
100% garnacha

90

Colour: bright cherry. Nose: floral, ripe fruit, balanced. Palate: flavourful, fruity, good acidity, good structure.

BODEGAS AUGUSTA BILBILIS

Carramiedes, s/n
50331 Mara (Zaragoza)
☎: +34 677 547 127
bodegasaugustabilbilis@hotmail.com
www.bodegasaugustabilbilis.com

J de Samitier 2020 T
syrah

90

Colour: cherry, purple rim. Nose: red berry notes, floral, spicy, cocoa bean. Palate: flavourful, fruity, good acidity, long.

Samitier 2020 T RB
garnacha

88

Fruity, floral, crisp, spicy, balanced.

Samitier Garnacha 2019 T
garnacha

91

Colour: bright cherry. Nose: creamy oak, red berry notes, scrubland, floral. Palate: good acidity, spicy, fine tannins.

Samitier Garnacha Blanca 2021 B
garnacha blanca

87

Samitier Syrah 2020 T
syrah

90

Balanced, varietally correct. Colour: bright cherry. Nose: creamy oak, black fruit, wild herbs. Palate: good acidity, spicy, fine tannins.

Segeda de Samitier 2018 T BA
garnacha, syrah

87

DO CALATAYUD / D.O.P.

DO CALATAYUD / D.O.P.

BODEGAS BRECA
Ctra. Monasterio de Piedra, s/n
50219 Munébrega (Zaragoza)
☎: +34 952 504 706
pedidos@jorgeordonez.es
www.jorgeordonez.es

Brega 2018 T
100% garnacha
94
Colour: cherry, garnet rim. Nose: powerful, overripe fruit, toasty, sweet spices, black fruit. Palate: flavourful, sweetness, long.

BODEGAS SAN ALEJANDRO
Ctra. Calatayud - Cariñena, Km. 16,4
50330 Miedes de Aragón (Zaragoza)
☎: +34 976 892 205
Fax: +34 976 890 540
contacto@san-alejandro.com
www.san-alejandro.com

**Baltasar Gracián
Arte de Ingenio 2018 T R**
garnacha, syrah
92
Colour: cherry, purple rim. Nose: fruit preserve, meaty notes, dried herbs, creamy oak. Palate: flavourful, full-bodied, fruity, balanced.

**Baltasar Gracián
Arte de Prudencia 2019 T C**
garnacha, syrah
92
Colour: cherry, purple rim. Nose: ripe fruit, rose petals, faded herbs, dried herbs, spicy. Palate: fruity, good structure, powerful, flavourful, balanced, slightly dry, soft tannins.

**Baltasar Gracián
El Político 2021 T**
garnacha
90
Colour: cherry, purple rim. Nose: ripe fruit, red berry notes, scrubland, spicy, tar, tea leave. Palate: fruity, fresh, flavourful, slightly dry, soft tannins.

**Baltasar Gracián
El Criticón 2021 RD**
garnacha
88
Pleasant, aromatic, fruity, lactic, standard, smooth.

**Baltasar Gracián
El Discreto 2021 B**
viura, garnacha blanca
88
Aromatic, crisp, fruity, jammy, tasty, simple.

**Baltasar Gracián
Viñas Viejas El Héroe 2020 T**
garnacha
92
Colour: deep cherry. Nose: ripe fruit, dried herbs, spicy, creamy oak. Palate: fruity, flavourful, powerful, round tannins, good finish.

Clos Baltasar 2020 T
garnacha
92
Colour: cherry, purple rim. Nose: red berry notes, ripe fruit, faded flowers, rose petals, spicy. Palate: flavourful, fruity, good structure, fresh, slightly dry, soft tannins.

Las Rocas Garnacha 2020 T
garnacha
90
Colour: cherry, purple rim. Nose: ripe fruit, spicy, smoky, dried herbs. Palate: fruity, fresh, good structure, flavourful, slightly dry, soft tannins.

**Las Rocas Garnacha
Viñas Viejas 2020 T**
garnacha
91
Colour: cherry, purple rim. Nose: ripe fruit, red berry notes, violets, faded flowers, dried herbs, spicy. Palate: fruity, flavourful, good structure, balanced, slightly dry, soft tannins.

CUEVAS DE AROM
50330 Miedes de Aragón (Zaragoza)
☎: +34 638 961 395
aiyana@cuevasdearom.com
www.cuevasdearom.com

Cuevas de Arom Altas Parcelas 2020 T
garnacha
92
Colour: deep cherry. Nose: ripe fruit, dried herbs, creamy oak, fruit expression. Palate: powerful, ripe fruit, spicy, round tannins.

Cuevas de Arom As Ladieras 2020 T C
garnacha

94
Colour: deep cherry. Nose: balsamic herbs, sweet spices, scrubland, citrus fruit, ripe fruit, red berry notes. Palate: spicy, balsamic, good acidity.

🏆 PODIUM

Cuevas de Arom Os Cantals 2020 T
garnacha

95
Colour: bright cherry. Nose: complex, expressive, spicy, mineral, red berry notes, citrus fruit, scrubland. Palate: full-bodied, long, great length.

Cuevas de Arom Tuca Negra 2020 T
garnacha

93
Colour: bright cherry. Nose: balsamic herbs, sweet spices, scrubland, red berry notes, chalk. Palate: spicy, balsamic, good acidity.

ESTEBAN CASTEJÓN
Portada, 13
50236 Ibdes (Zaragoza)
☎: +34 610 070 784
bodegasesteban@bodegasesteban.es
www.bodegasesteban.es

180 Noches 2021 B
malvasía

88
Standard, floral, fruity, kind finish, aromatic, simple.

Sargas de Idues Garnacha 2020 T
garnacha

87

Sargas de Idues Garnacha Blanca 2020 B
garnacha blanca

88
Spicy, jammy, dried flowers, fruity, tasty. Nose: caramel.

GALLINA DE PIEL WINES
Escoles, 43
17181 Aiguaviva (Girona/Gerona)
☎: +34 663 594 668
info@gallinadepielwines.com
www.gallinadepielwines.com

Mimetic 2020 T
98% garnacha, 2% otras

91
Colour: cherry, garnet rim. Nose: earthy notes, ripe fruit, dried herbs, wild herbs, waxy notes. Palate: flavourful, fruity, ripe fruit, balanced.

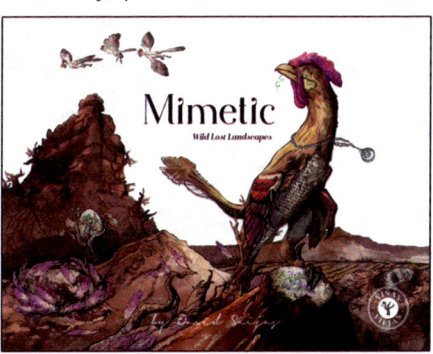

LA GAVACHA
Bulevar Picos de Europa, 20
28701 San Sebastián de los Reyes (Madrid)
☎: +34 646 168 510
info@lagavacha.com
www.lagavacha.com

La Gavacha 2020 T BA
100% garnacha

90
Colour: cherry, purple rim. Nose: ripe fruit, dried herbs, spicy. Palate: fruity, flavourful, good structure, balanced, slightly dry, soft tannins.

LANGA
Ctra. N-II, Km. 241,700
50300 Calatayud (Zaragoza)
☎: +34 976 881 818
Fax: +34 976 884 463
cesar@bodegas-langa.com
www.bodegas-langa.com

Castillo de Ayud 2020 T R
garnacha

90
Colour: cherry, purple rim. Nose: ripe fruit, black fruit, dried herbs, spicy. Palate: fruity, flavourful, slightly dry, soft tannins, toasty.

DO CALATAYUD / D.O.P.

DO CALATAYUD / D.O.P.

Langa Classic 2019 T
garnacha
90
Colour: deep cherry. Nose: ripe fruit, black fruit, creamy oak, spicy, dried herbs. Palate: flavourful, good finish, slightly dry, soft tannins.

Marco Valero Marcial 2019 T
garnacha
90 🌱
Colour: cherry, purple rim. Nose: fruit liqueur notes, spicy, dried herbs. Palate: fruity, powerful, good structure, flavourful, smoky aftertaste, good finish, balanced.

Reyes de Aragón Premium 2021 T BA
garnacha
90
Colour: cherry, purple rim. Nose: ripe fruit, grassy, spicy. Palate: fruity, flavourful, balanced, powerful tannins.

PACO MULERO
Partida de la Hoya Torres s/n
30520 Jumilla (Murcia)
☎: +34 676 433 541
info@pacomulero.com
www.pacomulero.com

Prisma Garnacha 2020 T
100% garnacha
89
Hot, spicy, toasty, jammy, smoky, full-bodied.

PAGOS ALTOS DE ACERED
Fontana de Trevi, 30
50410 Cuarte de Huerva (Zaragoza)
☎: +34 636 474 723
manuel@lajas.es
www.lajas.es

Lajas "Finca el Peñiscal" 2016 T
garnacha, tinta del país, monastrell, macabeo, garnacha blanca, bobal
93
Colour: dark-red cherry. Nose: ripe fruit, sweet spices, toasty, dried herbs. Palate: fruity, flavourful, good finish, slightly dry, soft tannins.

Lajas "Finca el Peñiscal" 2019 T
93% garnacha, 7% monastrell, bobal, garnacha blanca, macabeo
93
Defined aromas. Colour: cherry, purple rim. Nose: red berry notes, ripe fruit, wild herbs. Palate: flavourful, good structure, fruity.

PROYECTO GARNACHAS/VINTAE
Ctra Villalengua, s/n
50312 Cervera de la Cañada (Zaragoza)
☎: +34 608 302 372
marketing@vintae.com
www.vintae.com

La Garnacha Olvidada de Aragón 2020 T
garnacha
89
Fruity, herbal, spicy, tasty.

RAÍCES IBÉRICAS
Avda. Mudejar, 61
50340 Maluenda (Zaragoza)
☎: +34 976 893 017
s.richard@raices.wine
www.raicesibericas.com

Alto Las Pizarras 2021 T BA
85% garnacha, 15% syrah
90
Nose: fruit preserve, dried herbs, spicy. Palate: fruity, good structure, flavourful.

Andrés Alonso Garnacha Syrah 2021 T
80% garnacha, 20% syrah
88
Pruney, spicy, fruity, tasty.

Carlos Rubén Parcela Única 2021 T
100% garnacha
88
Fruity, sweety finish, spicy, tasty, pruney.

Carlos Rubén Sin Duda 2021 T
100% garnacha
90
Colour: cherry, purple rim. Nose: fruit preserve, black fruit, dried herbs, dark chocolate, spicy. Palate: fruity, fresh, flavourful, good structure.

Las Pizarras Fabla 2021 T BA
80% garnacha, 20% syrah
89
Pruney, spicy, fruity, tasty.

Las Pizarras Viña Alarba 2020 T BA
100% garnacha
89
Spicy, fruity, herbal, tasty, jammy.

S. COOP. NIÑO JESÚS DE ANIÑÓN

Las Tablas, s/n
50313 Aniñón (Zaragoza)
☎: +34 976 899 150
produccion@ninojesusaninon.es
www.satninojesus.com

1428 Garnacha Syrah 2020 T
80% garnacha, 20% syrah

90

Colour: cherry, purple rim. Nose: ripe fruit, black fruit, dried herbs, spicy. Palate: fruity, flavourful, balanced, slightly dry, soft tannins.

Aninius 2020 T
88

Smoky, woody, spicy, jammy, tasty, powerful.

Estecillo Macabeo 2021 B
macabeo

86

Legado Estecillo Garnacha & Syrah 2020 T BA
80% garnacha, 20% syrah

87

Legado Garnacha 2020 T BA
garnacha

86

Legado Garnacha Blanca Macabeo 2021 B BA
93% garnacha blanca, 7% macabeo

88

Dried flowers, fruity, standard, balanced, spicy.

Legado Macabeo 2020 B FB
macabeo

89

Standard, spicy, jammy, tasty, simple, yeasty notes. Nose: fine lees. Palate: flavourful, toasty.

WEIN & VINOS GMBH

Hardenbergstr. 9A
10715 Berlin (Berlin)
☎: +49 303 150 6080
info@vinos.de
www.vinos.de

Caliber 2019 T BA
garnacha

93

Colour: dark-red cherry. Nose: toasty, spicy, cocoa bean, ripe fruit, black fruit. Palate: flavourful, toasty, fine bitter notes.

DO CALATAYUD / D.O.P.

DO. CAMPO DE BORJA

CONSEJO REGULADOR

Subida de San Andrés, 6
50570 Ainzón (Zaragoza)
☎: +34 976 852 122 - Fax: +34 976 868 806
@: vinos@docampodeborja.com
www.docampodeborja.com

LOCATION:

The DO Campo de Borja is made up of 16 municipal areas, situated in the north west of the province of Zaragoza and 60 km from the capital city, in an area of transition between the mountains of the Sistema Ibérico (at the foot of the Moncayo) and the Ebro Valley: Agón, Ainzón, Alberite, Albeta, Ambel, Bisimbre, Borja, Bulbuente, Burueta, El Buste, Fuendejalón, Magallón, Malejan, Pozuelo de Aragón, Tabuenca and Vera del Moncayo.

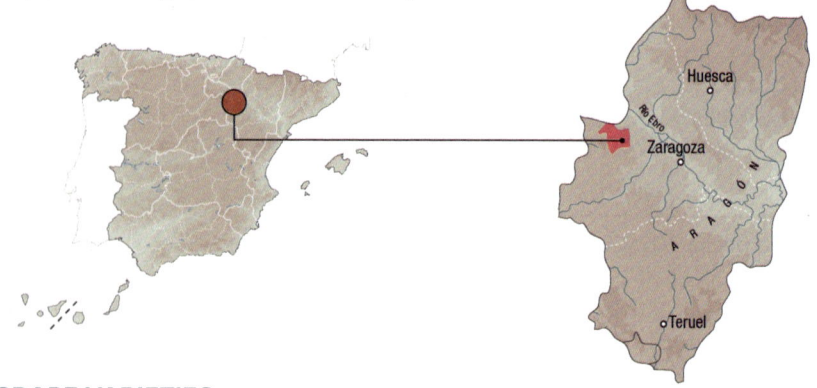

GRAPE VARIETIES:

WHITE: Macabeo, Garnacha Blanca, Moscatel, Chardonnay, Viognier, Sauvignon Blanc and Verdejo.

RED: Garnacha (majority with 75%), Tempranillo, Mazuela, Cabernet Sauvignon, Caladoc, Marselan, Merlot and Syrah.

FIGURES:

Vineyard surface: 6,178 – **Wine-Growers:** 901 – **Wineries:** 19 – **2021 Harvest rating:** Good – **Production 21:** 20,375,600 litres – **Market percentages:** 48% National -52% International.

SOIL:

The most abundant are brownish - grey limey soils, terrace soils and clayey ferrous soils. The vineyards are situated at an altitude of between 350 and 700 m on small slightly rolling hillsides, on terraces of the Huecha river and the Llanos de Plasencia, making up the Somontano del Moncayo.

CLIMATE:

A rather extreme continental climate, with cold winters and dry, hot summers. One of its main characteristics is the influence of the 'Cierzo', a cold and dry north - westerly wind. Rainfall is rather scarce, with an average of between 350 and 450 mm per year.

CLASSIFICATION OF YOUNG WINE HARVESTS PEÑÍNGUIDE

2017	2018	2019	2020	2021
GOOD	GOOD	VERY GOOD	VERY GOOD	VERY GOOD

ARTIGA FUSTEL

Progres, 19 Bajo
08720 Vilafranca del Penedès (Barcelona)
☎: +34 938 182 317
info@artiga-fustel.com
www.artiga-fustel.com

Artiga Garnacha Selección 2021 T
100% garnacha
87

La Bella Garnacha 2020 T
100% garnacha
90
Defined aromas, pruney. Colour: cherry, garnet rim. Nose: fruit preserve, powerful, violets, faded flowers. Palate: flavourful, long, sweet tannins.

La Bestia Garnacha 2019 T RB
100% garnacha
90
Colour: bright cherry. Nose: sweet spices, black fruit, ripe fruit, wild herbs, characterful. Palate: spicy, round tannins.

BODEGAS AINZÓN

Ctra. de Tabuenca ,s/n
50570 Ainzón (Zaragoza)
☎: +34 976 869 696
bodegas@bodegasainzon.es
www.bodegasainzon.es

Flor de Cayus 2019 T BA
garnacha
91
Colour: bright cherry. Nose: red berry notes, spicy, balsamic herbs, dried flowers. Palate: fruity, flavourful, powerful, balanced, round tannins.

Peñazuela Vendimia Seleccionada Garnacha 2020 T RB
100% garnacha
89
Kind finish, fruity, spicy, herbal.

Peñazuela Vendimia Seleccionada Garnacha Blanca 2021 B
100% garnacha blanca
88
Citrus fruit, herbal, tasty, crisp.

Terrazas del Moncayo Garnacha 2018 T BA
garnacha
91
Colour: deep cherry. Nose: ripe fruit, fruit expression, spicy, toasty, wild herbs. Palate: fruity, flavourful, slightly dry, soft tannins.

DO CAMPO DE BORJA / D.O.P.

Viña Ainzón 2018 T R
garnacha
88
Balanced, herbal, jammy, smooth.

Viña Ainzón 2019 T C
garnacha
88
Fruity, dried herbs, jammy, simple.

BODEGAS ALTO MONCAYO

Ctra. Borja - El Buste, CV-606 Km. 1,700
50540 Borja (Zaragoza)
☎: +34 976 868 098
m.arilla@bodegasaltomoncayo.com

Alto Moncayo 2018 T
100% garnacha
94
Colour: deep cherry. Nose: ripe fruit, dried herbs, creamy oak, cocoa bean, sweet spices. Palate: ripe fruit, spicy, round tannins, complex.

Aquilón 2016 T
100% garnacha
94
Colour: cherry, garnet rim. Nose: fruit preserve, powerful, creamy oak, caramel, dark chocolate. Palate: flavourful, long, sweetness, concentrated.

DO CAMPO DE BORJA / D.O.P.

Alto Moncayo Veratón 2019 T
100% garnacha
93
Colour: bright cherry. Nose: ripe fruit, balanced, neat, wild herbs. Palate: fruity, spicy, round tannins, ripe fruit, long.

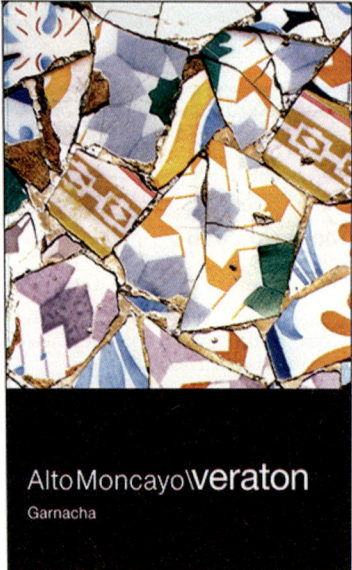

BODEGAS ARAGONESAS
Ctra. Magallón, s/n
50529 Fuendejalón (Zaragoza)
☎: +34 976 862 153
info@bodegasaragonesas.com
www.bodegasaragonesas.com

Aragonia Selección Especial 2019 T
garnacha
90
Colour: Cherry. Nose: woody, new oak, incense. Palate: powerful, good structure, flavourful, toasty.

Coto de Hayas 2019 T C
90
Colour: cherry, garnet rim. Nose: ripe fruit, dried herbs, wild herbs, sweet spices. Palate: ripe fruit, spicy, round tannins.

Fagus de Coto de Hayas 2019 T DA
garnacha
90
Colour: Cherry. Nose: sweet spices, scrubland, woody, black fruit, ripe fruit. Palate: spicy, flavourful, powerful, toasty.

Fagus de Coto de Hayas 2020 T BA
garnacha
90
Colour: cherry, garnet rim. Nose: aged wood nuances, ripe fruit, cocoa bean, toasty, waxy notes. Palate: flavourful, spicy, toasty.

Garnacha Centenaria de Coto de Hayas 2020 T
garnacha
89
Full-bodied, jammy, toasty, tasty, dried herbs, spicy, smoky.

Nabulé 2019 T
garnacha
91
Colour: Cherry. Nose: scrubland, wild herbs, fruit preserve, varietal. Palate: spicy, good acidity, flavourful, easy to drink.

Valdemilanos 2020 T
garnacha
88
Pleasant, fruity, floral, smooth, balanced, varietally correct.

BODEGAS BORSAO
50540 Borja (Zaragoza)
☎: +34 976 867 216
contacto@bodegasborsao.com
www.bodegasborsao.com

Borsao Cabriola 2018 T
55% garnacha, 39% syrah, 5% mazuelo
90
Colour: bright cherry. Nose: fruit preserve, black fruit, spicy, aromatic coffee. Palate: fruity, flavourful, powerful, fresh, good finish, round tannins.

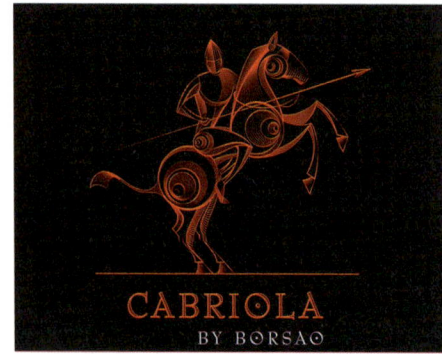

Borsao Berola 2018 T
80% garnacha, 20% syrah

92

Colour: deep cherry. Nose: fruit preserve, spicy, toasty, cocoa bean. Palate: fruity, flavourful, powerful, good finish.

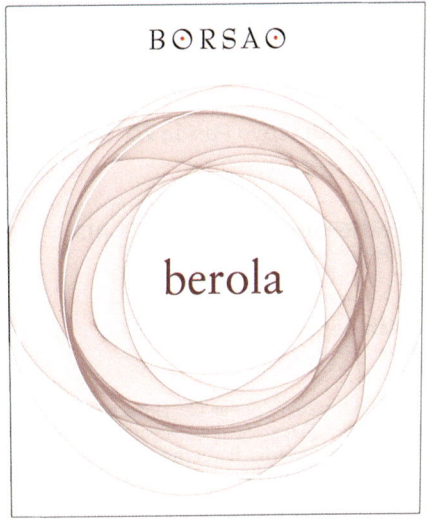

Borsao Bolé 2018 T RB
70% garnacha, 30% syrah

88

Fruity, jammy, tasty, wild.

Borsao Tres Picos 2020 T
100% garnacha

91

Colour: bright cherry. Nose: ripe fruit, wild herbs, toasty, spicy. Palate: fruity, powerful, good structure, slightly dry, soft tannins.

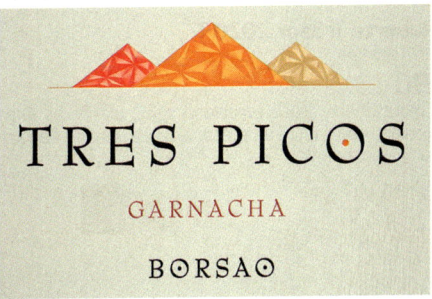

Borsao Selección 2021 B
chardonnay, macabeo

86

Borsao Selección 2021 RD
garnacha

87

Borsao Selección 2021 T
85% garnacha, 10% syrah, 5% tempranillo

90

Colour: cherry, purple rim. Nose: fruit expression, red berry notes, floral, spicy. Palate: flavourful, fruity, slightly dry, soft tannins.

Borsao Zarihs 2018 T
100% garnacha

91

Colour: very deep cherry. Nose: roasted coffee, aromatic coffee, powerful, meaty notes, black fruit. Palate: smoky aftertaste, great length, round tannins.

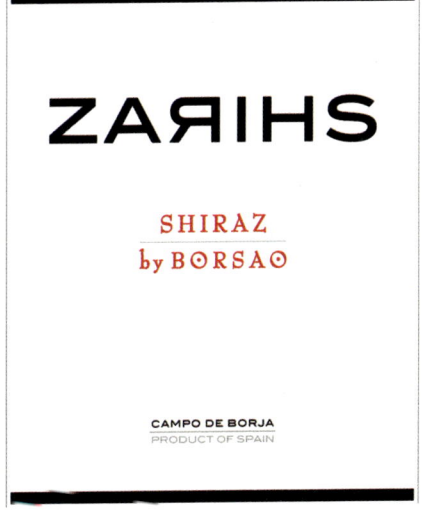

Monte Oton 2020 T
85% garnacha, 15% tempranillo

88

Colour: cherry, purple rim. Nose: fruit expression, red berry notes, floral, spicy, black fruit. Palate: flavourful, fruity, good acidity.

Borsao Clásico 2021 T
garnacha

88

Kind finish, jammy, spicy, floral.

DO CAMPO DE BORJA / D.O.P.

DO CAMPO DE BORJA / D.O.P.

BODEGAS CARLOS VALERO
Castillo de Capúa, 10 Nave 1
50197 Pol. Pla-Za (Zaragoza)
☎: +34 976 180 634
Fax: +34 976 186 326
info@bodegasvalero.com
www.bodegasvalero.com

Garnacha Blanca y Radiante Heredad Carlos Valero 2020 B
100% garnacha blanca

88
Floral, full-bodied, jammy, balanced, smooth.

BODEGAS MORCA
Pol. Molinillo del Fraile s/n
50540 Borja (Zaragoza)
☎: +54 968 435 022
Fax: +34 968 716 051
info@gilfamily.es
www.gilfamily.es

Flor de Morca 2021 T
100% garnacha

91
Colour: cherry, purple rim. Nose: red berry notes, floral, spicy, creamy oak. Palate: flavourful, fruity, good acidity, long.

Godina 2020 T
100% garnacha

93
Colour: deep cherry. Nose: dried herbs, creamy oak, ripe fruit, red berry notes, toasty. Palate: powerful, ripe fruit, spicy, round tannins.

🏆 PODIUM

Morca 2019 T
100% garnacha

95
Colour: deep cherry, garnet rim. Nose: black fruit, ripe fruit, sweet spices, cocoa bean. Palate: flavourful, balanced, round.

🏆 PODIUM

Touran 2019 T
100% garnacha

95
Complex, balanced. Colour: dark-red cherry. Nose: expressive, black fruit, ripe fruit, aromatic coffee, spicy. Palate: good structure, powerful, round tannins, spicy.

BODEGAS ROMÁN
Ctra. Gallur-Agreda, 1
50546 Bulbuente (Zaragoza)
☎: +34 976 852 936
info@bodegasroman.es
www.bodegasroman.es

Portal de Moncayo Ilusión Garnacha 2020 T
garnacha

87

Portal de Moncayo Pasión 2019 T
garnacha

87

Portal de Moncayo Rosé 2021 RD
garnacha

87

Román Cepas Viejas 2017 T
garnacha

92
Colour: bright cherry, purple rim. Nose: ripe fruit, black fruit, creamy oak, spicy, dried herbs, black liquorice. Palate: fruity, powerful, flavourful, balanced, good finish, powerful tannins.

Senda de Hoyas 2021 T
garnacha

87

Senda de Hoyas Orígenes 2020 T RB

88
Fruity, herbal, crisp, simple.

BODEGAS RUBERTE
Avda. de la Paz, 28
50520 Magallón (Zaragoza)
☎: +34 976 858 106
info@bodegasruberte.com
www.gruporuberte.com

Ruberte Tr3sor 2020 T
100% garnacha

90
Colour: Cherry. Nose: scrubland, spicy, wild herbs, dried flowers, ripe fruit, fruit preserve. Palate: spicy, flavourful, easy to drink.

GIL PEJENAUTE
Carretera, 24
50547 Tabuenca (Zaragoza)
☎: +34 677 454 000
info@gilpejnaute.com
www.gilpejnaute.com

Las Paradas 2020 T
garnacha

94

Colour. Cherry. Nose: complex, expressive, spicy, fruit preserve, faded flowers. Palate: elegant, full-bodied, long, great length.

Tabuca 2020 T
garnacha

93

Colour: bright cherry, garnet rim. Nose: red berry notes, fruit preserve, wild herbs, faded flowers, spicy. Palate: flavourful, fruity, long, balanced.

LA GENERAL DE VINOS
Dr Alcay 16-18 estudio D
50006 Zaragoza (Zaragoza)
☎: +34 609 413 578
jp@lageneraldevinos.com
www.lageneraldevinos.com

Furo Garnacha 2019 T
100% garnacha

91

Aromatic. Colour: cherry, garnet rim. Nose: wild herbs, ripe fruit, red berry notes, waxy notes, balanced. Palate: fruity, flavourful, balanced.

LOCOS POR EL VINO
Residencial Paraíso, 1 1ºE
50008 Zaragoza (Zaragoza)
☎: +34 606 233 063
miguel@locosporelvino.com

Azacán Garnacha Shiraz 2018 T
garnacha, syrah

90

Colour: bright cherry. Nose: toasty, spicy, aromatic coffee, ripe fruit. Palate: fruity, flavourful, powerful, round tannins.

Gruñón 2017 T
garnacha, syrah

93

Colour: cherry, purple rim. Nose: fruit expression, red berry notes, floral, spicy, wild herbs. Palate: flavourful, fruity, long, full-bodied.

Zismero 2021 T
garnacha, tempranillo

89

Pruney, fruity, jammy, tasty, floral.

PAGOS DEL MONCAYO
Ctra. Z-372, Km. 1,6
50580 Vera de Moncayo (Zaragoza)
☎: +34 976 900 256
info@pagosdelmoncayo.com
www.pagosdelmoncayo.com

Prados Colección Garnacha 2020 T BA
100% garnacha

91

Colour: bright cherry. Nose: sweet spices, ripe fruit, dark chocolate, dried herbs. Palate: fruity, spicy, round tannins.

Prados Colección Syrah 2020 T
100% syrah

92

Colour: deep cherry. Nose: black fruit, violets, scrubland, cocoa bean. Palate: powerful, ripe fruit, spicy, round tannins.

Prados Fusion Garnacha Syrah 2021 T
65% garnacha, 35% syrah

89

Pleasant, jammy, fruity, dried flowers, balanced, tasty, simple.

Prados Privé 2020 T C
100% syrah

91

Colour: cherry, garnet rim. Nose: fruit preserve, powerful, sweet spices, toasty. Palate: flavourful, long, round tannins, spicy.

DO CAMPO DE BORJA / D.O.P.

DO CAMPO DE BORJA / D.O.P.

PALMERI SICILIA
Cabernet Sauvignon, 3
50547 Tabuenca (Zaragoza)
☎: +34 687 163 015
info@palmerisicilia.com
www.palmerisicilia.es

Palmeri Adán 2015 T GR
100% garnacha

91
Colour: bright cherry. Nose: sweet spices, ripe fruit, dark chocolate, characterful, creamy oak, waxy notes. Palate: fruity, spicy, round tannins.

Palmeri Navalta 2017 T C
100% garnacha

92
Colour: deep cherry. Nose: ripe fruit, dried herbs, spicy, creamy oak. Palate: flavourful, powerful, balanced, slightly dry, soft tannins.

DO. CARIÑENA

CONSEJO REGULADOR

Camino de la Platera, 7
50400 Cariñena (Zaragoza)
☎: +34 976 793 143 / +34 976 793 031 - Fax: +34 976 621 107
@: secretaria@elvinodelaspiedras.es
www.elvinodelaspiedras.es

LOCATION:

In the province of Zaragoza, and occupies the Ebro valley covering 14 municipal areas: Aguarón, Aladrén, Alfamén, Almonacid de la Sierra, Alpartir, Cariñena, Cosuenda, Encinacorba, Longares, Mezalocha, Muel, Paniza, Tosos and Villanueva de Huerva.

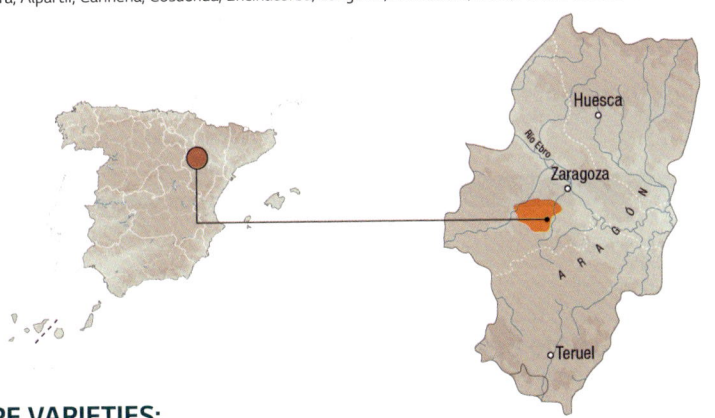

GRAPE VARIETIES:

WHITE: Preferred: Macabeo (majority 20%).

Authorized: Garnacha Blanca, Moscatel de Alejandría, Parellada, Verdejo, Sauvignon Blanc and Chardonnay.

RED: Preferred: Garnacha Tinta (majority 55%), Tempranillo, Cariñena (Mazuela).

Authorized: Juan Ibáñez, Cabernet Sauvignon, Syrah, Monastrell, Vidadillo and Merlot.

FIGURES:

Vineyard surface: 14,110 – **Wine-Growers:** 1,398 – **Wineries:** 32 – **2021 Harvest rating:** Very Good – **Production 21:** 58,137,400 litres – **Market percentages:** 36% National - 64% International.

SOIL:

Mainly poor, either brownish - grey limey soil, or reddish dun soil settled on rocky deposits, or brownish - grey soil settled on alluvial deposits. The vineyards are situated at an altitude of between 400 and 800 m.

CLIMATE:

A continental climate, with cold winters, hot summers and low rainfall. The viticulture is also influenced by the effect of the 'Cierzo'.

CLASSIFICATION OF YOUNG WINE HARVESTS PEÑÍNGUIDE

2017	2018	2019	2020	2021
GOOD	GOOD	VERY GOOD	VERY GOOD	VERY GOOD

DO CARIÑENA / D.O.P.

BIOENOS
Mayor, 88 Bajo
50400 Cariñena (Zaragoza)
☎: +34 639 359 618
bioenos@bioenos.com
www.varietalesantiquos.es

Pulchrum Crespiello Edición Limitada 2012 T R
vidadillo
90
Colour: cherry, garnet rim. Nose: fruit preserve, fruit liqueur notes, powerful, fine reductive notes. Palate: flavourful, sweetness, long.

Pulchrum Crespiello Edición Limitada 2016 T R
90% vidadillo, 10% garnacha
93
Colour: deep cherry. Nose: ripe fruit, creamy oak, fragrant herbs, scrubland. Palate: powerful, ripe fruit, spicy, round tannins.

BODEGAS CARE
Ctra. Aguarón, km 47,100
50400 Cariñena (Zaragoza)
☎: +34 645 515 305
marketing@bodegascare.com
www.bodegasare.com

Care Cariñena Nativa 2020 T C
cariñena
90
Kind finish, jammy, tasty, sweety finish, spicy. Colour: cherry, purple rim. Nose: black fruit, dried herbs, red berry notes, faded flowers. Palate: fruity, good structure, flavourful, fresh.

Care Finca Bancales 2018 T R
100% garnacha
91
Colour: very deep cherry. Nose: black fruit, ripe fruit, spicy, aromatic coffee, cocoa bean. Palate: powerful, flavourful, good finish, slightly dry, soft tannins.

Care Finca Marimú 2020 T BA
100% cariñena
91
Colour: very deep cherry. Nose: black fruit, spicy, creamy oak, dried herbs. Palate: good structure, slightly dry, soft tannins, ripe fruit, flavourful.

Care Garnacha Blanca Nativa 2021 B
garnacha blanca
88
Fruity, dried herbs, dried flowers, tasty, balanced.

Care Garnacha Nativa 2020 T
100% garnacha
89
Fruity, jammy, spicy, dried herbs, tasty.

Care sobre Lías 2021 T
75% garnacha, 25% syrah
88
Fruity, spicy, jammy, balanced.

Care Solidarity Rosé 2021 RD
40% tempranillo, 40% cabernet sauvignon, 20% merlot
88
Crisp, fruity, herbal, sweeties, balanced.

BODEGAS CARLOS VALERO
Castillo de Capúa, 10 Nave 1
50197 Pol. Pla-Za (Zaragoza)
☎: +34 976 180 634
Fax: +34 976 186 326
info@bodegasvalero.com
www.bodegasvalero.com

Heredad X Carlos Valero 2020 T
100% garnacha
86

BODEGAS DEL SEÑORÍO
Afueras, s/n
50108 Almonacid de La Sierra (Zaragoza)
☎: +34 692 471 887
bodegasdelsenorio@gmail.com
www.bodegasdelsenorio.com

Senda Lasarda 2019 T C
tempranillo, garnacha
88
Fruity, spicy, wild, tropical.

Yañoria 2016 T R
garnacha, cabernet sauvignon, tempranillo
89
Pruney, balanced, spicy, dried herbs, full-bodied, tasty.

Yañoria 2017 T C
tempranillo, garnacha, cabernet sauvignon
90
Colour: bright cherry. Nose: black fruit, black pepper, spicy, smoky, dried herbs. Palate: flavourful, fruity, balanced, smoky aftertaste.

BODEGAS ESTEBAN MARTÍN
Camino Virgen de Lagunas, s/n
50461 Alfamen (Zaragoza)
☎: +34 616 903 763
export@estebanmartin.es
www.estebanmartin.com

Esteban Martín 2021 T
garnacha, syrah
88
Spicy, fruity, dried herbs, tasty, crisp.

Niño Mimado Chardonnay 2021 B FB
chardonnay
89
Spicy, fruity, jammy, toasty, tasty.

Niño Mimado Garnacha 2016 T C
garnacha
89
Balanced, toasty, powerful, jammy, full-bodied.

Ulula 2020 T
60% garnacha, 40% syrah
88
Spicy, fruity, jammy.

BODEGAS FERNANDO CASTRO
Paseo Castelar, 70
13730 Santa Cruz de Mudela (Ciudad Real)
☎: +34 926 342 168
info@bodegasfernandocastro.com
www.bodegasfernandocastro.com

Montecruz 2014 T GR
tempranillo
85

Montecruz 2015 T GR
tempranillo
84

BODEGAS HACIENDA MOLLEDA
Ctra. Cariñena-Belchite, 29,3 (A-220, km 29,3)
50154 Tosos (Zaragoza)
☎: +34 976 620 702
haciendamolleda@gmail.com
www.haciendamolleda.com

Finca La Matea Garnacha 2017 T C
garnacha
89
Spicy, fruity, dried herbs, tasty, balanced.

GHM Garnacha + Garnacha 2019 T C
garnacha
87

Gran Hacienda Molleda GHM Cariñena + Cariñena 2016 T C
mazuelo
89
Fruity, spicy, pruney, dried herbs.

Gran Hacienda Molleda GHM Cariñena + Garnacha 2019 T C
garnacha, cariñena
88
Hot, smoky, herbal, fruity, tasty.

Hacienda Molleda 2021 RD
garnacha
84

Hacienda Molleda Cariñena 2021 T
cariñena
89
Floral, fruity, jammy, crisp.

Hacienda Molleda Garnacha 2018 T C
garnacha
88
Pruney, fruity, spicy, tasty.

Hacienda Molleda Garnacha 2020 T RB
garnacha
87

Hacienda Molleda Garnacha 2021 T
garnacha
88
Floral, crisp, fruity, tasty.

Hacienda Molleda Garnacha Blanca 2019 B
garnacha blanca, macabeo
87

Lleda Coupage 2020 T
garnacha, tempranillo
85

BODEGAS HEREDAD ANSÓN
Camino Eras Altas, s/n
50450 Muel (Zaragoza)
☎: +34 606 858 296
info@bodegasheredadanson.com
www.bodegasheredadanson.com

Heredad de Ansón 2017 T C
garnacha, syrah
85

DO CARIÑENA / D.O.P.

DO CARIÑENA / D.O.P.

Heredad de Ansón 2021 B
macabeo
84

Heredad de Ansón 2021 RD
garnacha
85

Heredad de Ansón Merlot Syrah 2021 T
merlot, syrah
86

Legum 2018 T BA
garnacha
83

Liason Garnacha 2021 T
garnacha
85

BODEGAS PANIZA
Ctra. Valencia, Km. 53
50480 Paniza (Zaragoza)
☎: +34 976 622 515
Fax: +34 976 622 958
info@bodegaspaniza.com
www.bodegaspaniza.com

Fabula de Paniza Syrah 2021 T
100% syrah
88
Kind finish, jammy, tasty.

Ibero II 2018 T
90% tempranillo, 5% garnacha, 5% cabernet sauvignon
88
Balanced, spicy, full-bodied, jammy, tasty, toasty.

Jabalí Garnacha & Syrah 2021 T
88
Pleasant, jammy, tasty.

Jabalí Tempranillo & Cabernet 2021 T
50% tempranillo, 50% cabernet sauvignon
88
Kind finish, jammy, tasty, fruity.

Viñas de Paniza Syrah 2020 T
100% syrah
88
Fruity, jammy, spicy, tasty.

Viñas Viejas de Paniza 2020 T
100% garnacha
89
Tasty, jammy, fruity.

BODEGAS SAN VALERO
Ctra. N-330, Km. 450
50400 Cariñena (Zaragoza)
☎: +34 976 620 400
Fax: +34 976 620 398
bsv@sanvalero.com
www.sanvalero.com

8.0.1 T R
cabernet sauvignon, merlot, syrah
89
Spicy, fruity, toasty, balanced.

Particular Chardonnay 2019 B FB
chardonnay
88
Spicy, jammy, tasty, dried flowers, toasty.

Particular Garnacha Blanca 2021 B
garnacha blanca
88
Kind finish, standard, jammy, balanced, dried flowers, wild.

Particular Garnacha Old Vine 2018 T C
garnacha
88
Balanced, spicy, sweety finish, jammy.

Particular Garnacha Viñas Centenarias 2015 T R
garnacha
92
Colour: dark-red cherry. Nose: toasty, spicy, cocoa bean, ripe fruit. Palate: flavourful, toasty, fine bitter notes.

BODEM BODEGAS
Crta. Z-VE-1201 Km 0,3
50108 Almonacid de La Sierra (Zaragoza)
☎: +34 976 780 136
Fax: +34 976 303 035
laboratorio@bodembodegas.com
www.bodembodegas.com

Las Margas Sierra de Algairen Garnacha Blanca 2021 B
garnacha blanca
90
Colour: bright straw. Nose: white fruit, dried herbs, wild herbs. Palate: fruity, flavourful, balanced.

Las Margas Sierra de Algairen Garnacha Tinta 2020 T
garnacha
87

CAMPOS DE LUZ
Avda. Diagonal, 590, 5º - 1
08021 Barcelona (Barcelona)
☎: +34 660 445 464
isabel@vinergia.com
www.vinergia.com

Campos de Luz 2018 T R
garnacha
87

Campos de Luz 2021 B
chardonnay, viura
85

Campos de Luz 2021 RD
garnacha
88
Fruity, mineral, dried herbs, tasty.

Campos de Luz Garnacha 2021 T
garnacha
87

Campos de Luz Revelación 2019 T C
garnacha
87

GRANDES VINOS
Ctra. Valencia Km 45,700
50400 Cariñena (Zaragoza)
☎: +34 976 621 261
Fax: +34 976 621 253
info@grandesvinos.com
www.grandesvinos.com

Anayón Garnacha 2019 T
garnacha
90
Colour: deep cherry. Nose: creamy oak, black fruit, scrubland. Palate: powerful, ripe fruit, spicy, round tannins.

Anayón Selección 2019 T BA
syrah, tempranillo, cabernet sauvignon, cariñena
91
Colour: deep cherry. Nose: dried herbs, creamy oak, black fruit, ripe fruit, roasted coffee. Palate: powerful, ripe fruit, spicy, round tannins.

Beso de Vino Garnacha Viñas Viejas 2020 T
garnacha
85

Corona de Aragón 2019 T C
cabernet sauvignon, tempranillo, garnacha, cariñena
86

Corona de Aragón Sauvignon Blanc 2021 B
sauvignon blanc
88
Varietally correct, tasty, dried herbs, fruity, jammy.

Monasterio de las Viñas 2018 T R
garnacha, tempranillo, cariñena, cabernet sauvignon
87

Monteplogar 2018 T C
86

LIBRE Y SALVAJE
Camino del Bosque, s/n
50108 Almonacid de La Sierra (Zaragoza)
☎: +34 627 445 357
info@libreysalvaje.com
www.libreysalvaje.com

Acecho, Selección del Collado T
91
Colour: bright cherry, garnet rim. Nose: red berry notes, ripe fruit, wild herbs, dried herbs. Palate: good acidity, easy to drink, long.

Camino del Bosque T
garnacha, cariñena
91
Colour: bright cherry, garnet rim. Nose: wild herbs, dried herbs, dried flowers, red berry notes, ripe fruit. Palate: flavourful, balanced, easy to drink.

Libre y Salvaje Cariñena 2020 T
92
Colour: deep cherry, cherry, purple rim. Nose: ripe fruit, dried herbs, red berry notes, spicy. Palate: ripe fruit, spicy, round tannins, good structure.

Libre y Salvaje Clarete 2020 RD
90
Colour: light cherry. Nose: wild herbs, characterful, dried herbs. Palate: correct, fine bitter notes, easy to drink, good finish.

Libre y Salvaje Garnacha 2020 T
90
Colour: deep cherry. Nose: ripe fruit, dried herbs, red berry notes, spicy. Palate: ripe fruit, spicy, round tannins, flavourful.

Libre y Salvaje Garnacha Blanca 2020 B
garnacha blanca
91
Colour: bright straw. Nose: powerful, characterful, toasty, creamy oak, ripe fruit. Palate: flavourful, long, spicy, toasty.

DO CARIÑENA / D.O.P.

DO CARIÑENA / D.O.P.

Libre y Salvaje Narancha 2019 B
91
With personality, oxidativ. Colour: golden, pale. Nose: candied fruit, honeyed notes, faded flowers, wild herbs. Palate: flavourful, long.

Libre y Salvaje Tinajas Antiguas 2019
90
Colour: bright cherry, purple rim. Nose: ripe fruit, wild herbs, grassy, spicy, smoky. Palate: fruity, flavourful, slightly dry, soft tannins.

SOC. COOP. VITIVINÍCOLA DE LONGARES - COVINCA
Ctra. Valencia, s/n
50460 Longares (Zaragoza)
☎: +34 976 142 653
Fax: +34 976 142 402
info@covinca.es
www.covinca.es

Terrai OVC Old Vine Cariñena 2020 T RB
cariñena
90
Colour: cherry, purple rim. Nose: ripe fruit, black fruit, faded flowers, fragrant herbs, spicy. Palate: flavourful, good structure, powerful, balanced.

Terrai OVG 2021 T BA
garnacha
89
Fruity, pruney, sweety finish, tasty, powerful.

Terrai OVG Vendimia Seleccionada 2020 T BA
100% garnacha
89
Fruity, pruney, floral, tasty, crisp.

Torrelongares 2018 T C
50% garnacha, 40% tempranillo, 8% syrah, 2% cariñena
87

Torrelongares 2021 RD
60% tempranillo, 40% garnacha
86

Torrelongares Garnacha 2021 T
100% garnacha
87

SOLAR DE URBEZO
San Valero, 14
50400 Cariñena (Zaragoza)
☎: +34 976 621 968
urbezoecologico@solardeurbezo.es
www.solardeurbezo.es

J. Belmonte 2021 T
merlot
88
Jammy, smooth, tasty, spicy.

Los Cabos de Urbezo 2021 T
garnacha
88
Herbaceous, crisp, fruity, balanced, simple.

Mariola T
91
Colour: cherry, purple rim. Nose: fruit expression, floral, sweet spices. Palate: flavourful, fruity, good acidity, long.

Urbezo 2019 T C
garnacha, cariñena, merlot
87

Urbezo Chardonnay 2021 B
chardonnay
89
Pleasant, aromatic, jammy, tasty, simple, standard.

Viña Urbezo 2021 T
garnacha, syrah
88
Standard, fruity, spicy, toasty.

VINS I LICORS GRAU
Torroella, 163
17200 Palafrugell (Girona/Gerona)
☎: +34 972 301 835
info@vinsilicorsgrau.com
www.grauonline.com

Pillet Blanc 2021 B
macabeo
86

Pillet Negre 2020 T
garnacha
86

Pillet Rosat 2021 RD
50% garnacha, 50% tempranillo
86

VIÑEDOS Y BODEGAS PABLO
Avda. Zaragoza, 16
50108 Almonacid de La Sierra (Zaragoza)
☎: +34 976 627 037
granviu@granviu.com
www.granviu.com

DO CARIÑENA / D.O.P.

Algairen 2020 T
100% garnacha
89
Pruney, spicy, dried herbs, full-bodied, tasty.

Menguante Cariñena Selección 2019 T BA
100% cariñena
88
Hot, jammy, toasty.

Menguante Garnacha 2021 T
100% garnacha
86

Menguante Garnacha Blanca 2020 B
100% garnacha
86

Menguante Garnacha Selección 2019 T
100% garnacha
88
Overripe, hot, pruney, powerful.

Menguante Vidadillo 2019 T
vidadillo
88
Fruity, dried herbs, spicy, overripe, crisp.

DO. CATALUNYA

CONSEJO REGULADOR

Edifici de l`Estació Enológica Passeig Sunyer, 4-6 1º
43202 Reus (Tarragona)
☎: +34 977 328 103 - Fax: +34 977 321 357
@: info@do-catalunya.com
www.do-catalunya.com

LOCATION:

The production area covers the traditional vine - growing Catalonian regions, and practically coincides with the current DOs present in Catalonia plus a few municipal areas with vine - growing vocation.

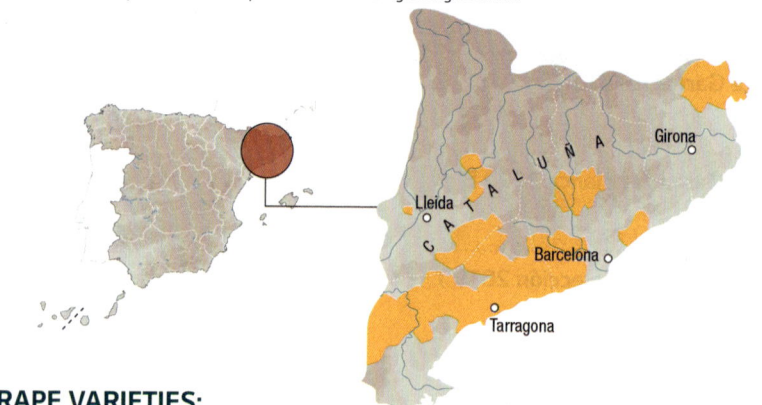

GRAPE VARIETIES:

WHITE: Chardonnay, Garnacha Blanca, Macabeo, Moscatel de Alejandría, Moscatel de Grano Menudo, Parellada, Riesling, Sauvignon Blanc, Xarel·lo. Gewürztraminer, Subirat Parent (Malvasía), Malvasía de Sitges, Picapoll, Pedro Ximénez, Chenin, Riesling, Sauvignon Blanc, Albariño, Sumoll Blanco, Viognier and Vinyater.

RED: Cabernet Franc, Cabernet Sauvignon, Garnacha, Garnacha Peluda, Merlot, Monastrell, Xare.lo Rosat, Pinot Noir, Samsó (Cariñena), Trepat, Sumoll, Ull de Llebre (Tempranillo), Garnacha Tintorera, Marselán, Garnacha Roja (Gris), Petit Verdot, Picapoll Negro and Syrah.

FIGURES:

Vineyard surface: 39,761 – **Wine-Growers:** 4,647 – **Wineries:** 192 – **2021 Harvest rating:** Very Good – **Production 21:** 37,786,334 litres – **Market percentages:** 46% National - 54% International.

CLIMATE AND SOIL:

Depending on the location of the vineyard, the same as those of the Catalonian DO's, whose characteristics are defined in this guide. See Alella, Empordà, Conca de Barberà, Costers del Segre, Montsant, Penedès, Pla de Bages, Priorat, Tarragona and Terra Alta.

CLASSIFICATION OF YOUNG WINE HARVESTS PEÑÍNGUIDE

2017	2018	2019	2020	2021
GOOD	GOOD	VERY GOOD	GOOD	VERY GOOD

BODEGAS CLOS D'AGON
Afores, s/n
17251 Calonge (Girona/Gerona)
☎: +34 972 661 486
info@closdagon.com
www.closdagon.com

Clos D'Agon 2019 T
49% cabernet franc, 38% syrah, 12% petit verdot, 1% cabernet sauvignon

93

Colour: cherry, purple rim. Nose: ripe fruit, wild herbs, spicy, creamy oak, wild herbs, balsamic herbs. Palate: good structure, powerful, fresh, balanced, slightly dry, soft tannins.

Clos D'Agon 2020 B
viognier

92

Colour: bright straw. Nose: floral, white fruit, ripe fruit, wild herbs. Palate: fruity, flavourful, balanced, good acidity, long.

Clos D'Agon Selección Especial 2019 T
57% cabernet franc, 36% petit verdot, 7% cabernet sauvignon

94

Colour: cherry, purple rim. Nose: ripe fruit, black fruit, wild herbs, spicy, smoky, scrubland. Palate: fresh, powerful, flavourful, good structure, balanced, good finish, slightly dry, soft tannins.

Clos D'Agon Viognier 2020 B FB
viognier

94

Colour: bright straw. Nose: stone fruit, ripe fruit, expressive, spicy, mineral. Palate: fruity, flavourful, full-bodied, balanced, good finish.

BODEGAS MASET
Ctra. Vilafranca-Igualada C-15 Km.19
08792 La Granada (Barcelona)
☎: +34 900 200 250
info@maset.com
www.maset.com

Maset del Lleó Syrah 2018 T R
syrah

91

Colour: cherry, purple rim. Nose: black fruit, ripe fruit, toasty, dark chocolate. Palate: good structure, flavourful, balanced, slightly dry, soft tannins.

BODEGAS PUIGGRÒS
Ctra. de Manresa, Km. 13
08711 Òdena (Barcelona)
☎: +34 629 853 587
info@bodegaspuiggros.com
www.bodegaspuiggros.com

Exedra 2021 B
garnacha blanca

91 🌱

Colour: bright straw, greenish rim. Nose: fresh fruit, citrus fruit, wild herbs, floral. Palate: fresh, fruity, good acidity, fine bitter notes.

Exedra 2021 T

92 🌱

Colour: cherry, purple rim. Nose: fruit expression, red berry notes, floral, spicy, earthy notes. Palate: flavourful, fruity, good acidity, long.

Mestre Vila Vell 2019 T
sumoll

89 🌱

Fruity, jammy, tasty, little interventionist.

Sense sentits 2019 T

91

Colour: cherry, garnet rim. Nose: overripe fruit, hot, earthy notes, red clay notes. Palate: pruney, powerful, sweet tannins.

Sentits Blancs 2020 B FB
garnacha blanca

92 🌱

Colour: bright straw. Nose: expressive, white flowers, jasmine, dried herbs, stone fruit. Palate: flavourful, fruity, balanced.

Sentits Negres Garnatxa Negra 2017 T
garnacha

90 🌱

Colour: dark-red cherry, garnet rim. Nose: fruit preserve, aged wood nuances, tobacco, sweet spices, wet leather. Palate: spicy, round tannins, long.

DO CATALUNYA / D.O.P.

DO CATALUNYA / D.O.P.

Signes 2016 T
sumoll, garnacha
92
Colour: deep cherry. Nose: ripe fruit, creamy oak, scrubland, balsamic herbs, grassy. Palate: powerful, ripe fruit, spicy, round tannins.

BODEGUES VISENDRA
Colón, 22
43815 Les Pobles (Tarragona)
☎: +34 639 338 892
info@bodeguesvisendra.com
www.bodeguesvisendra.com

Visendra 2020 T R
86

CA N'ESTRUC
Finca Ca N'Estruc Ctra. C-1414, Km. 1,05
08292 Esparreguera (Barcelona)
☎: +34 937 777 017
Fax: +34 937 772 268
info@canestruc.com
www.canestruc.com

Ca N'Estruc Blanc 2021 B
90
Colour: bright straw, greenish rim. Nose: fresh fruit, citrus fruit, wild herbs, white flowers. Palate: fresh, fruity, good acidity, fine bitter notes.

Ca N'Estruc Negre 2021 T
88
Balanced, spicy, fruity, crisp, toasty.

Ca N'Estruc Rosat 2021 RD
87

Ca N'Estruc Xarel.lo 2021 B
90
Austere, citrus fruit, balanced, tasty. Colour: straw. Nose: white fruit, citrus fruit, dried herbs, dry stone, dry nuts. Palate: flavourful, good structure, fresh, full of life.

Idoia 2018 T
91
Colour: deep cherry. Nose: red berry notes, black fruit, scrubland, earthy notes, toasty. Palate: ripe fruit, spicy, round tannins.

Idoia Blanc 2020 B FB
92
Colour: bright yellow. Nose: creamy oak, ripe fruit, spicy, citrus fruit. Palate: good structure, toasty, fine bitter notes, mineral.

L'Equilibrista Garnatxa 2018 T
92
Colour: deep cherry. Nose: creamy oak, red berry notes, scrubland, fine reductive notes. Palate: powerful, ripe fruit, spicy, round tannins.

L'Equilibrista Negre 2018 T
92
Colour: deep cherry. Nose: red berry notes, black fruit, scrubland, toasty. Palate: ripe fruit, spicy, round tannins.

CAN GRAU VELL
Can Grau Vell, s/n
08781 Hostalets de Pierola (Barcelona)
☎: +34 676 586 933
Fax: +34 932 684 965
info@grauvell.cat
www.grauvell.cat

Alcor 2015 T
92
Colour: deep cherry. Nose: dried herbs, creamy oak, black fruit, tobacco. Palate: ripe fruit, spicy, round tannins.

Quike 2019 RD
90
With personality, yeasty notes, jammy, wild, tasty, powerful, fruity. Nose: macerated fruit.

Super tramp 2017 T
88
Age nuances, pruney, spicy, herbaceous, thin.

Tramp 2017 T
89
Balanced, spicy, age nuances, herbal, jammy, tasty.

CAVES BOHIGAS
Finca Can Maciá
08700 Ódena (Barcelona)
☎: +34 938 048 100
info@bohigas.es
www.fermibohigas.com

Bohigas Garnatxa Negra 2019 T BA
100% garnacha
87

Bohigas Xarel.lo 2021 B
100% xarel.lo
88
Pleasant, aromatic, smooth, fruity.

CAVES ROMAGOSA TORNÉ
Serra de Baix, 24. Alt Penedés
08731 Sant Marti Sarroca (Barcelona)
☎: +34 938 991 353
export@romagosatorne.com
www.romagosatorne.com

Monasterio Real 2019 T BA
88
Smoky, toasty, powerful, classic, spicy, jammy.

CELLER DE CAPÇANES
Llebaria, 9
43776 Capçanes (Tarragona)
☎: +34 977 178 319
cellercapcanes@cellercapcanes.com
www.cellercapcanes.com

Cap Sentit Orange Wine 2021 B
100% garnacha blanca
89
Kind finish, fruity, jammy, tasty, balanced.

Mas Picosa Blanc 2021 B
100% garnacha blanca
86 🌱

Mas Picosa Negre 2021 T
60% garnacha, 20% cabernet sauvignon, 20% syrah
89 🌱
Kind finish, fruity, jammy, standard, balanced.

CELLER GRAU I GRAU
Ctra. C-37, Km. 75,5
08255 Castellfollit del Boix (Barcelona)
☎: +34 938 356 002
info@cellergrauigrau.com
www.cellergrauigrau.com

Seré 2019 T C
garnacha
87

DOMENIO WINES BY CELLERS DOMENYS
Prat de la Riba, 18
43713 St. Jaume dels Domenys (Tarragona)
☎: +34 977 677 135
administracio@cellersdomenys.com
www.domeniowines.com

Capvespre Origen 2021 B
xarel.lo, macabeo, parellada
85

Capvespre Origen 2021 RD
merlot, tempranillo
86

Capvespre Origen 2021 T
merlot, tempranillo
86

Capvespre Sunset 2021 B
xarel.lo, macabeo, chardonnay, moscatel de alejandría
87

Capvespre Sunset 2021 RD
tempranillo, trepat
87

Capvespre Sunset 2021 T
tempranillo, cabernet sauvignon, syrah
87

Domenio Sumoi Negre 2017 T
91
Colour: bright cherry. Nose: red berry notes, ripe fruit, wild herbs, fresh fruit. Palate: fresh, flavourful, slightly dry, soft tannins.

FREIXENET
Joan Sala, 2
08770 Sant Sadurní d'Anoia (Barcelona)
☎: +34 938 917 000
comunicacion@freixenet.es
www.freixenet.es

Freixenet Selección Especial 2021 B
macabeo, sauvignon blanc, chardonnay, moscatel, xarel.lo
87

Freixenet Selección Especial 2021 RD
garnacha, merlot, tempranillo
87

HERETAT MASCORRUBÍ
Camí de Mascorrubí s/n
43714 El Pla de Manlleu - Aigumúrcia (Tarragona)
mascorrubi@gmail.com
www.mascorrubi.cat

Aixarta Blanc 2020 B
garnacha blanca
89 🌱
Spicy, fruity, jammy, tasty.

Captirot 2018 T C
60% garnacha, 40% sumoll
89
Fruity, jammy, tasty, spicy.

Escumoll 2017 BE GR BN
100% sumoll blanc
93
Colour: bright yellow. Nose: fragrant herbs, characterful, ripe fruit, dry nuts, brioche. Palate: powerful, flavourful, good acidity, fine bitter notes.

DO CATALUNYA / D.O.P.

DO CATALUNYA / D.O.P.

InAnnat 2021 RD
100% sumoll
89
Smooth, aromatic, wild, balanced, citrus fruit.

Sumoll Amfora 2017 T R
100% sumoll
91
Colour: Cherry. Nose: balsamic herbs, sweet spices, scrubland. Palate: spicy, balsamic, good acidity.

Teulera 2020 B
100% montonega
88
Citrus fruit, herbal, simple.

MAS DE LA PANSA
Comerç, 2
43422 Barberà de la Conca (Tarragona)
☎: +34 667 894 636
info@masdelapansa.com
www.masdelapansa.com

Mas de la Pansa Parellada 2018 B
100% parellada
92
Colour: bright yellow. Nose: ripe fruit, dried herbs, faded flowers, stone fruit. Palate: powerful, ripe fruit, balanced.

MERCÈ SANGUESA I MILLAN
08786 Capellades (Barcelona)
☎: +34 938 013 532

Pla de Morei 2019 T
90
Colour: deep cherry. Nose: ripe fruit, dried herbs, creamy oak. Palate: powerful, ripe fruit, spicy, round tannins.

PLA DE MOREI
Cami de la Garça s/n
08789 La Torre de Claramunt (Barcelona)
☎: +34 931 313 454
enoturisme@plademorei.com
www.plademorei.com

Filigrana 2020 T
41% tempranillo, 59% merlot
89
Tasty, jammy, spicy, toasty.

Filigrana 2021 B
57,4% garnacha blanca, 42,6% chardonnay
88
Pleasant, aromatic, tasty, smooth.

Saial 2019 B BA
45,3% garnacha blanca, 47,3% chardonnay
89
Fruity, jammy, spicy, tasty.

RAMÓN ROQUETA
Finca Jaumandreu
08259 Fonollosa (Barcelona)
☎: +34 938 743 511
Fax: +34 938 737 204
info@ramonroqueta.com
www.ramonroqueta.com

Ramón Roqueta Garnacha 2021 T
100% garnacha
88
Spicy, fruity, jammy, tasty.

Ramón Roqueta Garnacha Blanca 2021 B
100% garnacha blanca
87

Ramón Roqueta Tempranillo 2021 T
100% tempranillo
88
Fruity, herbal, tasty, wild.

SANT JOSEP VINS
Estació, 2
43785 Bot (Tarragona)
☎: +34 977 428 352
info@santjosepwines.com
www.santjosepvins.com

L'Estació Blanc 2019 B
garnacha blanca, chardonnay, viura
90
Colour: bright yellow. Nose: stone fruit, ripe fruit, lactic notes. Palate: fruity, flavourful, balanced, spicy.

L'Estació Negre 2016 T C
100% cabernet sauvignon
92
Colour: dark-red cherry. Nose: toasty, spicy, cocoa bean, black fruit, dry nuts. Palate: flavourful, toasty, fine bitter notes.

Plana d'en Fonoll Cabernet Sauvignon 2020 T
100% cabernet sauvignon
88
Varietally correct, herbal, fruity, spicy, tasty.

Plana d'en Fonoll Chardonnay 2020 B BA
100% chardonnay
88
Fruity, lactic, jammy, simple.

Plana d'en Fonoll Sauvignon Blanc 2021 B
100% sauvignon blanc

90

Colour: bright straw. Nose: expressive, white flowers, jasmine, dried herbs. Palate: flavourful, fruity, balanced.

Selecció 259 2013 T C
38% mazuelo, 34% cabernet sauvignon, 6% garnacha

92

Colour: cherry, garnet rim. Nose: ripe fruit, scrubland, waxy notes. Palate: flavourful, balsamic, spicy, round tannins.

SEGURA VIUDAS
Ctra. Sant Sadurní a St. Pere de Riudebitlles, Km. 5
08775 Torrelavit (Barcelona)
☎: +34 938 917 070
comunicacion@freixenet.es
www.seguraviudas.com

Segura Viudas Mas d'Aranyó 2016 T C
cabernet sauvignon, merlot, syrah, cariñena, garnacha

90

Colour: Cherry. Nose: balsamic herbs, scrubland, black fruit, ripe fruit. Palate: spicy, round tannins.

Seguras Viudas 2021 RD
55% tempranillo, 45% merlot

87

UNIVERSITAT ROVIRA I VIRGILI
Ctra TV 7211 km 7
43120 Constanti (Tarragona)
☎: +34 977 520 197
pedro.cabanillas@urv.cat
https://www.fe.urv.cat/es/facultad/bodega-mas-dels-frares

Universitat Rovira i Virgili 2021 B
100% moscatel de alejandría

87

VALLFORMOSA
La Sala, 45
08735 Vilobí del Penedès (Barcelona)
☎: +34 938 978 286
berta.grane@vallformosa.com
www.vallformosa.com

Laviña 2021 B
75% macabeo, 25% garnacha blanca

85

Laviña 2021 RD
100% tempranillo

86

Laviña Tempranillo Merlot 2019 T BA
50% tempranillo, 50% merlot

87

VINS DE TALLER
Camí de St. Miquel amb
Ctra. St Tomàs de Fluvià, s/n
17469 Siurana d'Empordà (Girona/Gerona)
☎: +34 972 525 578
Info@vinsdetaller.com
www.vinsdetaller.com

Vins de Taller Baseia 2019 B
100% viognier

91 ♣

Colour: bright straw. Nose: stone fruit, ripe fruit, tropical fruit, dried flowers. Palate: fruity, balanced, mineral, powerful.

Vins de Taller Geum 2020 T C
100% merlot

90 ♣

Colour: cherry, purple rim. Nose: red berry notes, ripe fruit, dried herbs. Palate: flavourful, fruity, spicy, slightly dry, soft tannins.

Vins de Taller Gris 2021 RD
merlot, garnacha gris

90 ♣

Colour: raspberry rose. Nose: red berry notes, floral, fragrant herbs. Palate: good acidity, fine bitter notes, flavourful.

Vins de Taller Larix 2021 B
viognier

91 ♣

Colour: light mahogany. Nose: stone fruit, ripe fruit, wild herbs, dried herbs. Palate: flavourful, balanced, fresh, spicy.

Vins de Taller Phlox 2021 B
viognier, macabeo

88 ♣

Citrus fruit, fruity, herbal, tasty, crisp.

DO CATALUNYA / D.O.P.

DO. CAVA
CONSEJO REGULADOR

Avinguda Tarragona, 24
08720 Vilafranca del Penedès (Barcelona)
☎: +34 938 903 104 - Fax: +34 938 901 567
@: consejo@crcava.es
www.crcava.es

LOCATION:

The defined Cava region covers the sparkling wines produced according to the traditional method of a second fermentation in the bottle of 63 municipalities in the province of Barcelona, 52 in Tarragona, 12 in Lleida and 5 in Girona, as well as those of the municipal areas of Laguardia, Moreda de Álava and Oyón in Álava, Almendralejo in Badajoz, Mendavia and Viana in Navarra, Requena in Valencia, Ainzón and Cariñena in Zaragoza, and a further 18 municipalities of La Rioja.

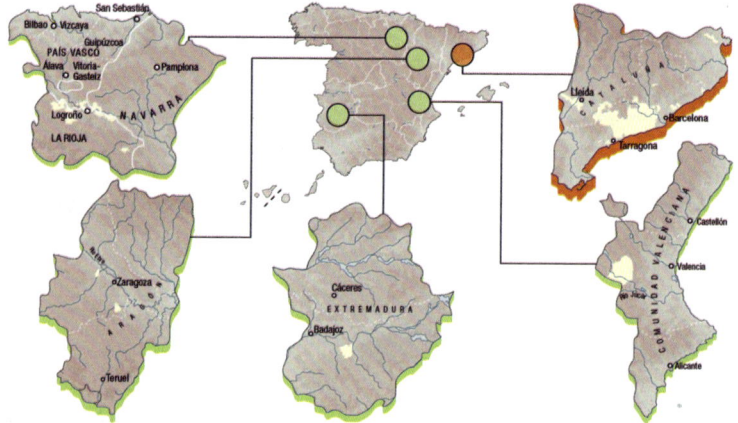

GRAPE VARIETIES:

WHITE: Macabeo (Viura), Xarel.lo, Parellada, Subirat (Malvasía Riojana) and Chardonnay.

RED: Garnacha Tinta, Monastrell, Trepat and Pinot Noir.

FIGURES:

Vineyard surface: 38,133 – Wine-Growers: 6,284 – Wineries: 205 – 2021 Harvest rating: N/A – Production 21: 189,698,811 litres – Market percentages: 29% National - 71% International.

SOIL:

This also depends on each producing region.

CLIMATE:

That of each producing region stated in the previous epigraph. Nevertheless, the region in which the largest part of the production is concentrated (Penedès) has a Mediterranean climate, with some production areas being cooler and situated at a higher altitude.

CLASSIFICATION OF YOUNG WINE HARVESTS PEÑÍNGUIDE

This denomination of origin, due to the wine-making process, does not make available single-year wines indicated by vintage, so the following evaluation refers to the overall quality of the wines that were tasted this year.

AGUSTÍ TORELLÓ MATA

Partida La Serra s/n
08770 Sant Sadurní d'Anoia (Barcelona)
☎: +34 938 911 173
info@agustitorellomata.com
www.agustitorellomata.com

Agustí Torelló Mata 2016 BE GR BN
macabeo, xarel.lo, parellada

91

Colour: bright straw. Nose: fine lees, floral, fragrant herbs, expressive, dry stone. Palate: flavourful, good acidity, fine bead, balanced.

Agustí Torelló Mata 2019 BE R BR
macabeo, xarel.lo, parellada

90

Colour: bright straw. Nose: medium intensity, fresh fruit, dried herbs, fine lees, floral. Palate: fresh, fruity, good acidity.

Agustí Torelló Mata Barrica Gran Reserva 2017 BE GR BN
100% macabeo

91

Colour: bright golden. Nose: fine lees, dry nuts, fragrant herbs, complex, white fruit, floral. Palate: flavourful, good acidity, fine bead, spicy.

Agustí Torello Mata Kripta 2014 BE GR BN
macabeo, xarel.lo, parellada

94

Colour: bright golden. Nose: fine lees, fragrant herbs, ripe fruit, dry nuts, lactic notes, faded flowers. Palate: flavourful, good acidity, fine bead, fine bitter notes, fresh.

🏆 PODIUM

Agustí Torello Mata Kripta Gran Anyada 2008 BE GR BN
macabeo, xarel.lo, parellada

95

Colour: bright golden. Nose: fine lees, dry nuts, complex, dried herbs. Palate: flavourful, good acidity, fine bead, fine bitter notes.

Agustí Torelló Mata Rosat Trepat 2020 RE R BR
trepat

90

Colour: raspberry rose. Nose: red berry notes, fragrant herbs, ripe fruit, dried herbs. Palate: light-bodied, spicy, good acidity, fine bitter notes.

ALDONZA

Ctra. N-430 km 462,3
02612 Munera (Albacete)
☎: +34 967 217 711
info@aldonzagourmet.com
www.aldonzagourmet.com

Aldonza BE BN
xarel.lo, macabeo, parellada

87

Aldonza BE BR
xarel.lo, macabeo, parellada

86

Aldonza BE R BR
chardonnay, xarel.lo, macabeo

90

Colour: bright straw. Nose: fine lees, fragrant herbs, dried flowers. Palate: flavourful, good acidity, balanced, fresh.

Aldonza Rosé RE BR
trepat

87

ALSINA & SARDÁ

Barrio Les Tarumbas, s/n
08733 El Pla del Penedès (Barcelona)
☎: +34 938 988 132
Fax: +34 938 988 132
alsina@alsinasarda.com
www.alsinasarda.com

Alsina & Sardá 2020 BE R BR
xarel.lo, macabeo, parellada

88

Pleasant, aromatic, classic, fruity.

Alsina & Sardá Gran Reserva Especial 2016 BE GR BN
xarel.lo, chardonnay

92

Colour: bright yellow. Nose: characterful, white fruit, ripe fruit, balanced, floral. Palate: flavourful, ripe fruit, long.

Alsina & Sarda Pinoir 2020 RE R BN

88

Balanced, dried flowers, fruity, dried herbs.

Alsina & Sardá Sello 2018 BE GR BN
xarel.lo, macabeo, parellada

89

Standard, balanced, kind finish, jammy, smooth.

DO CAVA / D.O.P.

DO CAVA / D.O.P.

Alsina & Sardá Vestigis Gran Cuvée 2016 BE GR BN
20% xarel.lo, 20% macabeo, 20% parellada, 20% chardonnay, 20% pinot noir

91

Colour: bright yellow. Nose: ripe fruit, fine lees, balanced, dried herbs. Palate: good acidity, flavourful, ripe fruit.

Mas D'Alsina & Sardá 2018 BE R BN
91

Pleasant. Colour: yellow. Nose: ripe fruit, pattiserie, spicy, powerful. Palate: flavourful, rich, easy to drink.

ALTA ALELLA
Camí Baix de Tiana, s/n
08328 Alella (Barcelona)
☎: +34 934 693 720
info@altaalella.wine
www.altaalella.wine

🏆 PODIUM

Alta Alella 10 2010 BE BN
100% chardonnay

96

Colour: bright golden. Nose: fine lees, dry nuts, fragrant herbs, complex, bakery, elegant, expressive. Palate: powerful, flavourful, good acidity, fine bead, fine bitter notes.

🏆 PODIUM

Alta Alella 10 2012 BE BN
chardonnay

96

Colour: bright golden. Nose: caramel, pattiserie, faded flowers, fine lees, candied fruit. Palate: flavourful, rich, fine bitter notes, elegant, fine bead.

Alta Alella Laietà 2017 BE GR BN
xarel.lo, macabeo, chardonnay, pinot noir

93

Colour: yellow. Nose: fine lees, complex, candied fruit, dried flowers, faded flowers, pattiserie. Palate: powerful, flavourful, good acidity, fine bead, fine bitter notes.

Alta Alella Laietà Rosé 2018 RE GR BN
91

Colour: raspberry rose. Nose: balanced, medium intensity, red berry notes, neat, fresh, dried flowers. Palate: fruity, fresh, fine bitter notes, easy to drink.

Alta Alella Mirgin 2018 BE GR BN
92

Colour: bright golden. Nose: fine lees, fragrant herbs, complex, fresh fruit. Palate: powerful, flavourful, good acidity, fine bead.

🏆 PODIUM

Alta Alella Mirgin Exeo Evolució + 2005 BE GR BN
chardonnay, xarel.lo

95

Defined aromas, age nuances. Colour: bright yellow. Nose: caramel, pattiserie, fine lees, balanced, expressive, complex. Palate: flavourful, rich, long, spicy.

🏆 PODIUM

Alta Alella Mirgin Exeo Paraje Calificado Vallcirera 2016 BE GR BN
xarel.lo, chardonnay

95

Colour: bright golden. Nose: fine lees, fragrant herbs, characterful, ripe fruit, dry nuts. Palate: flavourful, good acidity, fine bead, fine bitter notes.

Alta Alella Mirgin Opus Paraje Calificado Vallcirera 2017 BE GR BN
chardonnay, pansa blanca

94

Taut. Colour: bright golden. Nose: fine lees, fragrant herbs, characterful, ripe fruit, dry nuts. Palate: powerful, flavourful, good acidity, fine bead, fine bitter notes.

Alta Alella Mirgin Rosé 2018 RE GR BN
mataró (Blanco)

89

Balanced, dried flowers, crisp, needs time, not representative.

ALTA ALELLA - CELLER DE LES AUS
Cami Baix de Tiana, s/n
08328 Alella (Barcelona)
☎: +34 934 693 720
info@altaalella.wine
www.altaalella.wine

Aus Bruant 2020 BE R BN
xarel.lo

89

Balanced, floral, fruity, jammy, tasty, great length, defined aromas. Palate: easy to drink.

Aus Capsigrany sin sulfitos 2020 BE R BN
xarel.lo vermell

90

Colour: bright yellow, coppery red. Nose: ripe fruit, fine lees, balanced, dried herbs. Palate: good acidity, flavourful, ripe fruit, long.

AVINYÓ

Masia Can Fontanals
08793 Avinyonet del Penedès (Barcelona)
☎: +34 938 970 055
avinyo@avinyo.com
www.avinyo.com

Avinyó 2017 BE R BN
25% macabeo, 70% xarel.lo, 5% parellada

90

Colour: straw. Nose: white flowers, fine lees, fragrant herbs. Palate: flavourful, fruity, fresh.

Avinyó 2019 BE R BR
60% macabeo, 25% xarel.lo, 15% parellada

88

Balanced, spicy, crisp, herbal, fruity.

Avinyó Blanc de Noirs BE R BN
100% pinot noir

90

Colour: bright straw. Nose: fine lees, floral, fragrant herbs, expressive, ripe fruit. Palate: flavourful, good acidity, fine bead, balanced.

Avinyó La Ticota 2015 BE GR BN
xarel.lo

91

Colour: bright golden. Nose: fine lees, fragrant herbs, ripe fruit, dry nuts. Palate: flavourful, good acidity, fine bead, fine bitter notes.

Avinyó Rosé Sublim 2019 RE R BR
100% pinot noir

87

BLANCHER-CAPDEVILA PUJOL

Plaça Pont Romà, Edificio Blancher
08770 Sant Sadurní d'Anoia (Barcelona)
☎: +34 938 183 286
blancher@blancher.es
www.blancher.es

Blancher 2016 BE GR BN
xarel.lo, macabeo, parellada

90

Colour: bright golden. Nose: fine lees, fragrant herbs, ripe fruit, dry nuts. Palate: flavourful, good acidity, fine bead, fine bitter notes.

Blancher BE R BN
xarel.lo, macabeo, parellada

90

Colour: bright straw. Nose: fine lees, floral, fragrant herbs, expressive. Palate: powerful, flavourful, good acidity, fine bead.

Blancher BE R BR
xarel.lo, macabeo, parellada

88

Nose: ripe fruit, balanced, dried herbs. Palate: flavourful, ripe fruit.

Blancher Rosat 2018 RE R BR
garnacha, pinot noir, trepat

87

Capdevila Pujol BE R BN
xarel.lo, macabeo, parellada

89

Citrus fruit, balanced, crisp, floral, herbal, tasty.

Teresa Blancher de la Tieta 2012 BE GR BN
macabeo, parellada, pansa blanca

91

Colour: bright golden. Nose: fine lees, fragrant herbs, characterful, ripe fruit, dry nuts. Palate: powerful, flavourful, good acidity, fine bead, fine bitter notes.

BODEGA JAUME SERRA

Ctra.Vilanova i la Geltrú
a Vilafranca del Penedés, km 2,5
08800 Vilanova i la Geltrú (Barcelona)
☎: +34 938 936 404
jaumeserra@jgc.es
www.garciacarrion.es

Jaume Serra BE R BN
macabeo, parellada, xarel.lo, chardonnay

85

Jaume Serra Chardonnay BE GR BN
chardonnay

88

Citrus fruit, jammy, herbal, full-bodied.

Jaume Serra Orgánico BE BR
macabeo

83

Jaume Serra Pinot Noir Rosé RE BR

86

Jaume Serra Vintage 2019 BE BN
macabeo, parellada, xarel.lo, chardonnay

88

Citrus fruit, crisp, herbal, tasty.

Pata Negra Organic BE BR

85

DO CAVA / D.O.P.

DO CAVA / D.O.P.

BODEGA VERA DE ESTENAS
Ctra. N-III, km. 266 - Paraje La Cabezuela
46300 Utiel (València/Valencia)
☎: +34 962 171 141
estenas@veradeestenas.es
www.veradeestenas.es

Cava Estenas 2020 BE BN
macabeo, chardonnay
88
Fruity, jammy, dried herbs, crisp.

BODEGAS AESSIR
Santa Maria, 76
08340 Barcelona (Barcelona)
☎: +34 937 591 832
Fax: +34 935 207 405
info@aessir.com
www.aessir.com

La Crusset BE R BN
xarel.lo, parellada, macabeo
91
Colour: bright yellow. Nose: ripe fruit, fine lees, balanced, dried herbs, dry nuts. Palate: good acidity, flavourful, ripe fruit.

BODEGAS ARRÁEZ
Pol. 6 Parcela 386 Paraje Ciscar
46630 La Font de la Figuera (València/Valencia)
☎: +34 962 290 031
info@bodegasarraez.com
www.bodegasarraez.com

Sutra by Toni Arraez BE BR
90% macabeo, 10% chardonnay
86

Sutra by Toni Arraez BE R BR
80% macabeo, 20% chardonnay
88
Pleasant, tasty, jammy, spicy.

BODEGAS CA N'ESTELLA
Masia Ca n'Estella, s/n
08635 Sant Esteve Sesrovires (Barcelona)
☎: +34 934 161 387
Fax: +34 934 161 620
a.vidal@fincacanestella.com
www.fincacanestella.com

Rabetllat i Vidal 2018 BE R BN
89
Citrus fruit, dried herbs, tasty, yeasty notes.

Rabetllat i Vidal Rosat 2019 RE R BR
89
Fruity, dried flowers, dried herbs, jammy.

BODEGAS CAPITÀ VIDAL
Ctra. Vilafranca- Igualada, C-15, km 21
08733 El Pla del Penedès (Barcelona)
☎: +34 938 988 630
Fax: +34 938 988 625
administracion@capitavidal.com
www.capitavidal.com

Fuchs de Vidal 2017 BE GR BN
50% xarel.lo, 30% macabeo, 20% parellada
89 🌿
Citrus fruit, balanced, crisp, herbal, yeasty notes.

Fuchs de Vidal BE R BN
40% xarel.lo, 35% macabeo, 25% parellada
90
Colour: bright straw. Nose: fresh fruit, dried herbs, fine lees, floral. Palate: fresh, fruity, flavourful, good acidity.

Fuchs de Vidal Rosé Pinot Noir RE R EBR
100% pinot noir
88 🌿
Pleasant, floral, jammy, tasty.

Fuchs de Vidal Unic BE R BN
50% chardonnay, 35% pinot noir, 15% macabeo, xarel.lo, parellada
89 🌿
Floral, crisp, fruity, jammy.

Gran Fuchs de Vidal BE R BN
70% macabeo, 15% xarel.lo, 15% parellada
90 🌿
Colour: bright straw. Nose: fresh fruit, citrus fruit, fine lees, fragrant herbs. Palate: fresh, fruity, good acidity.

Palau Solá BE BN
40% macabeo, 35% xarel.lo, 25% parellada
86

BODEGAS COVIÑAS
Avda. Rafael Duyos, s/n
46340 Requena (València/Valencia)
☎: +34 962 300 680
covinas@covinas.com
www.covinas.com

Aula BE BN
macabeo
85

Aula BE BR
macabeo
85

Aula BE SS
macabeo
85

Aula Chardonnay BE R BN
88
Fruity, dried flowers, dried herbs, lactic.

Aula Macabeo BE R BN
macabeo
87

Aula Rosé RE BR
garnacha
88
Citrus fruit, full-bodied, jammy, dried flowers.

BODEGAS EMILIO CLEMENTE
Camino de San Blas, s/n
46340 Requena (València/Valencia)
☎: +34 601 410 728
administracion@eclemente.es
www.eclemente.es

Regulus BE BR
macabeo
87

BODEGAS FAUSTINO
Ctra. de Logroño, s/n
01320 Oyón (Araba/Álava)
☎: +34 945 622 500
info@bodegasfaustino.es
www.bodegasfaustino.com

Cava Faustino Art Collection BE R BR
87

Cava Faustino BE R BR
macabeo, chardonnay
86

Cava Faustino Rose RE R BR
garnacha
87

BODEGAS HISPANO SUIZAS
Ctra. N-322, Km. 451,7
46357 El Pontón (València/Valencia)
☎: +34 962 349 370
maria.salinas@bodegashispanosuizas.com
www.bodegashispanosuizas.com

Tantum Ergo Chardonnay Pinot Noir 2019 BE BN
chardonnay, pinot noir
91
Colour: bright yellow. Nose: fine lees, balanced, dried herbs, white fruit. Palate: good acidity, flavourful, ripe fruit.

Tantum Ergo Exclusive Magnum 2011 BE GR BN
chardonnay, pinot noir
91
Colour: bright yellow. Nose: ripe fruit, fine lees, balanced, dried herbs. Palate: good acidity, flavourful, ripe fruit.

Tantum Ergo Pinot Noir Rosé 2019 RE BN
pinot noir
91
Colour: coppery red. Nose: red berry notes, dried flowers, dried herbs. Palate: fresh, fruity, flavourful, balanced.

Tantum Ergo Vintage 2018 BE BN
chardonnay, pinot noir
92
Colour: bright yellow. Nose: ripe fruit, fine lees, balanced, dried herbs, aromatic coffee. Palate: good acidity, flavourful, ripe fruit, long.

BODEGAS LENEUS
Ctra. Alange, km. 2
06200 Almendralejo (Badajoz)
☎: +34 691 340 843
contacto@vinosleneus.com
www.vinosleneus.com

Cava Leneus Selección 2020 BE SS
50% macabeo, 50% parellada
87

BODEGAS MARCELINO DÍAZ
Mecánica, s/n
06200 Almendralejo (Badajoz)
☎: +34 924 677 548
bodega@madiaz.com
www.madiaz.com

Puerta Palma BE EBR
macabeo, parellada
88
Aromatic, fruity, tasty, simple.

Puerta Palma BE R BN
macabeo, parellada
87

Puerta Palma Rosé 2021 RE BR
garnacha
86

DO CAVA / D.O.P.

Peñín Guide | **SPANISH WINE**

DO CAVA / D.O.P.

BODEGAS MARTÍNEZ PAIVA
Ctra. Gijón - Sevilla N-630, Km. 646
Apdo. Correos 87
06200 Almendralejo (Badajoz)
☎: +34 924 671 130
info@bodegasmartinezpaiva.com
www.bodegasmartinezpaiva.com

Paiva 2019 BE R BN
30% chardonnay, 70% macabeo
89
Balanced, slight oxidation, citrus fruit, crisp, tasty, with personality.

Paiva 2020 BE BN
35% chardonnay, 65% macabeo
87

BODEGAS MASET
Ctra. Vilafranca-Igualada C-15 Km.19
08792 La Granada (Barcelona)
☎: +34 900 200 250
info@maset.com
www.maset.com

Maset 1917 2018 BE GR BN
xarel.lo, macabeo, pinot noir
92
Colour: bright golden. Nose: fine lees, fragrant herbs, ripe fruit, dry nuts. Palate: powerful, flavourful, fine bitter notes.

Maset L'avi Pau 2018 BE GR BN
91 🍷
Colour: bright straw. Nose: fine lees, fragrant herbs, expressive, ripe fruit. Palate: powerful, flavourful, good acidity, fine bead.

Maset Nu Brut Rosé 2020 RE BR
trepat, garnacha
88 🍷
Pleasant, standard, dried flowers, crisp, fruity, kind finish.

Maset Vintage 2018 BE R BN
xarel.lo, macabeo, parellada
90
Colour: bright yellow. Nose: ripe fruit, fine lees, balanced, dried herbs. Palate: good acidity, flavourful, ripe fruit, long.

BODEGAS MUGA
Avda. Vizcaya, s/n
26200 Haro (La Rioja)
☎: +34 941 311 825
info@bodegasmuga.com
www.bodegasmuga.com

Conde de Haro 2019 BE R BR
90
Colour: straw. Nose: fragrant herbs, bakery, yeasty notes. Palate: flavourful, long, great length.

Conde de Haro Rosé 2016 RE BR
garnacha
91
Colour: straw. Nose: expressive, dry nuts, fragrant herbs, spicy, brioche, dried flowers. Palate: flavourful, fine bead, balanced.

BODEGAS MURVIEDRO
Ampliación Pol. El Romeral, s/n
46340 Requena (València/Valencia)
☎: +34 962 329 003
murviedro@murviedro.es
www.murviedro.es

Murviedro Arts de Luna Organic BE BN
75% macabeo, 25% chardonnay
86 🍷

Murviedro Arts de Luna Organic BE BR
75% macabeo, 25% chardonnay
87 🍷

Murviedro Arts de Luna Organic Rosé RE BR
100% garnacha
86 🍷

BODEGAS ROMALE
Pol. Ind. Parc. 6, Manzana D - Mecánica s/n
06200 Almendralejo (Badajoz)
☎: +34 924 667 255
romale@romale.com
www.romale.com

Privilegio de Romale BE R BN
macabeo, parellada, xarel.lo
87

Viña Romale Xarel.lo 2021 BE BN
xarel.lo
86

BODEGAS ROURA, J.A. PEREZ ROURA
Valls de Rials, s/n
08328 Alella (Barcelona)
☎: +34 933 527 456
roura@roura.es
www.roura.es

Roura 5* BE EBR
70% xarel.lo, 30% chardonnay
88
Jammy, tasty, smooth. Palate: sweetness.

Roura BE BN
70% xarel.lo, 30% chardonnay
88
Pleasant, aromatic, tasty, simple.

Roura BE BR
70% xarel.lo, 30% chardonnay
89
Pleasant, dried flowers, jammy, tasty.

BODEGAS SANI PRIMAVERA
La Zarza, s/n
06200 Almendralejo (Badajoz)
☎: +34 924 677 917
tienda@bodegassani.com
www.bodegasani.com

Cava Árabe 2020 BE BN
macabeo, parellada
87

Salud de Sani Primavera Cava 2020 BE BR
chardonnay
86

BODEGAS TROBAT
Castello, 10
17780 Garriquella (Girona/Gerona)
☎: +34 972 530 092
xavier.picazo@bmark.es
www.bodegastrobat.com

Celler Trobat 2018 BE R BN
40% xarel.lo, 40% macabeo, 10% parellada, 10% chardonnay
91
Colour: bright yellow. Nose: ripe fruit, fine lees, balanced, dried herbs. Palate: good acidity, flavourful, ripe fruit, long.

Gran Amat 2019 BE BN
40% xarel.lo, 40% macabeo, 20% parellada
88
Citrus fruit, balanced, jammy, tasty, full-bodied, herbal.

Trobat Rosat RE BR
merlot
87

BODEGAS VALDEORITE
Melilla, 13
06360 Fuente del Maestre (Badajoz)
☎: +34 673 038 743
valdeorite@valdeorite.com
www.valdeorite.com

Extrem de Bonaval BE R BR
macabeo, parellada
87

Extrem de Bonaval Rosado RE R BR
garnacha
86

BODEGAS VEGALFARO
Ctra. Pontón - Utiel, Km. 3
46340 Requena (València/Valencia)
☎: +34 962 320 680
info@vegalfaro.com
www.vegalfaro.com

Caprasia 2018 BE R BN
macabeo, chardonnay
88 ♣
Citrus fruit, dried flowers, herbal, jammy, tasty.

Vegalfaro 2017 BE GR BN
macabeo, chardonnay
87 ♣

BODEGAS VEGAMAR
Garcesa, s/n
46175 Calles (València/Valencia)
☎: +34 962 781 443
info@vegamar.es
www.vegamar.es

Cava Esencia Vegamar BE BN
garnacha, chardonnay
90
Colour: bright golden. Nose: dry nuts, fragrant herbs, complex, toasty. Palate: powerful, flavourful, good acidity, fine bead.

Vegamar Privée 18 BE R BN
macabeo, chardonnay
90
Colour: bright golden. Nose: fine lees, characterful, ripe fruit, dry nuts. Palate: powerful, flavourful, good acidity, fine bead, fine bitter notes.

DO CAVA / D.O.P.

DO CAVA / D.O.P.

BODEGAS VILLA CONCHI
Ramon y Cajal 7 1º A
01007 Vitoria-Gasteiz (Araba/Álava)
☎: +34 945 150 589
araex@araex.com
www.araex.com

Villa Conchi Imperial 2018 BE EBR
40% xarel.lo, 30% macabeo, 20% parellada, 10% chardonnay

93

Colour: bright golden. Nose: fragrant herbs, characterful, ripe fruit, dry nuts, candied fruit. Palate: powerful, flavourful, fine bead, sweetness.

BOLET AGRICULTURA ECOLÓGICA
Finca Mas Lluet s/n
08732 Castellví de la Marca (Barcelona)
☎: +34 938 918 153
cavasbolet@cavasbolet.com
www.cavasbolet.com

Bolet Cartoixà 2014 BE GR BN
91 🌿

Colour: bright straw. Nose: fine lees, floral, fragrant herbs, white fruit. Palate: flavourful, balanced, easy to drink, fruity.

Bolet Classic Eco 2019 BE BR
xarel.lo, macabeo, parellada

87 🌿

Bolet Eco 2015 BE GR BN
92 🌿

Colour: bright golden. Nose: fine lees, dry nuts, fragrant herbs, complex, dried herbs. Palate: powerful, flavourful, fine bead, fine bitter notes, fresh.

Bolet Eco 2019 BE R BN
macabeo, xarel.lo, parellada

88 🌿

Standard, jammy, wild, dried herbs.

Bolet Eco 2019 BE R BR
macabeo, xarel.lo, parellada

88 🌿

Citrus fruit, balanced, fruity, jammy, tasty.

Bolet Pinot Noir Rosat 2017 RE R BR
pinot noir

85 🌿

CAL SERRADOR
Montserrat, 87
08770 Sant Sadurní d'Anoia (Barcelona)
☎: +34 938 912 073
info@rosellgallart.com
www.rosellgallart.com

Serra D'Or Chardonnay BE R BN
87

Serra D'Or Coupage Classic BE R BN
87

Serra D'Or Pinot Noir BE R BN
87

Teresa Mata Garriga BE R BN
88

Standard, fruity, jammy, dried flowers.

CANALS I MUNNÉ
Plaza Pau Casals, 6
08770 Sant Sadurní d'Anoia (Barcelona)
☎: +34 938 910 318
Fax: +34 938 911 945
marketing@canalsimunne.com
www.canalsimunne.com

ADN Canals 2017 BE GR BN
90

Colour: bright golden. Nose: fine lees, fragrant herbs, ripe fruit, dry nuts, floral. Palate: flavourful, good acidity, fine bead, fine bitter notes.

Canals & Munné
Ice Sweet White 2020 BE R D
40% macabeo, 30% xarel.lo, 30% parellada

88

Kind finish, aromatic, sweet, floral.

Canals & Munné
Insuperable 2019 BE R BR
40% macabeo, 30% xarel.lo, 30% parellada

91

Colour: bright straw. Nose: fresh fruit, dried herbs, fine lees, floral. Palate: fresh, fruity, flavourful, good acidity.

Canals & Munné X10 2011 BE GR BN
100% xarel.lo

94

Colour: bright golden. Nose: fine lees, fragrant herbs, complex, dry nuts, brioche. Palate: powerful, flavourful, good acidity, fine bead, fine bitter notes.

Lola Rosé Pinot Noir 2019 RE R BR
100% pinot noir

89

Dried flowers, fruity, dried herbs, jammy, tasty.

Pride 2017 BE GR BR
50% xarel.lo, 30% macabeo, 20% parellada

88

Citrus fruit, fruity, simple, crisp, dried flowers, lactic.

CANALS NADAL
Ponent, 2
08733 El Pla del Penedès (Barcelona)
☎: +34 938 988 081
cava@canalsnadal.com
www.canalsnadal.com

**Antoni Canals Nadal
Cupada Selecció 2018 BE GR BR**
40% xarel.lo, 50% macabeo, 10% parellada

91

Colour: straw. Nose: expressive, dry nuts, spicy, floral. Palate: flavourful, fine bead, great length.

Canals Nadal 2017 BE GR BN
50% macabeo, 40% xarel.lo, 10% parellada

90

Colour: bright yellow. Nose: ripe fruit, fine lees, balanced, dried herbs. Palate: good acidity, flavourful, ripe fruit.

Canals Nadal 2019 BE R BN
45% macabeo, 40% xarel.lo, 15% parellada

89

Pleasant, herbal, crisp, fruity, wild, smooth. Palate: fine bitter notes.

Canals Nadal CN 1986 Blanc de Noirs 2017 BE R BR
100% pinot noir

91 ♣

Colour: straw. Nose: ripe fruit, faded flowers, dried flowers, characterful. Palate: flavourful, good acidity, balanced.

Canals Nadal Rosé 2019 RE BR
100% trepat

88 ♣

Kind finish, fruity, standard, simple, floral, citrus fruit.

**Petit Cupada
de Canals Nadal 2019 BE R BN**
60% parellada, 20% macabeo, 20% xarel.lo

89 ♣

Balanced, spicy, tasty, toasty.

CASTELL D'OR
Ctra. de Santes Creus,s/n
43814 Vila-Rodona (Tarragona)
☎: +34 977 459 860
castelldor@castelldor.com
www.castelldor.com

Castell D'Or BE BN
xarel.lo, macabeo, parellada, chardonnay

89

Standard, yeasty notes, dried flowers, dried herbs, citrus fruit, smooth.

Castell D'Or BE C BR
xarel.lo, macabeo, parellada, chardonnay

88

Standard, crisp, floral, kind finish, simple.

Castell D'Or BE GR BN
xarel.lo, macabeo, parellada, chardonnay

89

Pleasant, dried herbs, yeasty notes, crisp. Palate: fine bitter notes, easy to drink.

Castell D'Or Brut Rosat RE BR
trepat

86

Castell D'Or Orgànic BE BR
xarel.lo, macabeo, parellada, chardonnay

87 ♣

Castell D'Or Reserva Imperial BE R BR
parellada, chardonnay

90

Colour: bright yellow. Nose: ripe fruit, fine lees, balanced, dried herbs, bakery. Palate: flavourful, ripe fruit.

CASTELL SANT ANTONI
Passeig del Parc, 13
08770 Sant Sadurní d'Anoia (Barcelona)
☎: +34 938 183 099
cava@castellsantantoni.com
www.castellsantantoni.com

**Castell Sant Antoni
Camí del Sot 2014 BE GR BR**
macabeo, xarel.lo, parellada

92

Colour: bright golden. Nose: fine lees, fragrant herbs, characterful, ripe fruit. Palate: powerful, flavourful, good acidity, fine bead.

**Castell Sant Antoni
Gran Barrica 2015 BE GR BN**
macabeo, xarel.lo, parellada, chardonnay

92

Colour: bright golden. Nose: fine lees, fragrant herbs, ripe fruit, dry nuts, caramel. Palate: flavourful, good acidity, fine bead, fine bitter notes.

DO CAVA / D.O.P.

DO CAVA / D.O.P.

Castell Sant Antoni
Gran Reserva 2012 BE GR BN
macabeo, xarel.lo, parellada, chardonnay
93
Colour: bright golden. Nose: fine lees, fragrant herbs, characterful, ripe fruit, dry nuts. Palate: flavourful, good acidity, fine bead, fine bitter notes.

Castell Sant Antoni Jazz Nature BE R BN
90
Colour: bright straw. Nose: dried herbs, fine lees, floral, ripe fruit. Palate: fresh, fruity, flavourful, good acidity.

Castell Sant Antoni
Jazz Nature Rosé RE R BR
trepat, garnacha
90
Colour: bright straw. Nose: fine lees, floral, fragrant herbs, expressive. Palate: flavourful, good acidity, fine bead, balanced.

Castell Sant Antoni
Torre de L'Homenatge 2005 BE GR BN
xarel.lo, macabeo, parellada
94
Colour: bright golden. Nose: fine lees, fragrant herbs, ripe fruit, dry nuts, brioche. Palate: powerful, flavourful, good acidity, fine bead, fine bitter notes.

CASTELO DE PEDREGOSA
Lluis Maria Vidal, 8
08770 Sant Sadurní d'Anoia (Barcelona)
☎: +34 941 454 050
Fax: +34 941 454 529
bodega@bodegasriojanas.com
www.bodegasriojanas.com

Cum Laude BE R BN
40% xarel.lo, 30% macabeo, 30% parellada
87

CAVA & HOTEL MASTINELL
Ctra. Vilafranca a Sant Martí Sarroca, km. 0,5
08720 Vilafranca del Penedès (Barcelona)
☎: +34 938 170 586
info@mastinell.com
www.mastinell.com

MasTinell Carpe Diem 2016 BE GR BN
30% xarel.lo, 30% parellada, 40% chardonnay
91
Colour: bright golden. Nose: fine lees, dry nuts, fragrant herbs, complex, spicy. Palate: powerful, flavourful, fine bead, fine bitter notes.

MasTinell Cristina 2016 BE GR EBR
10% macabeo, 35% xarel.lo, 35% parellada, 20% chardonnay
91
Colour: bright straw. Nose: dried flowers, dried herbs, citrus fruit, ripe fruit, neat. Palate: fruity, flavourful.

MasTinell Nature 2014 BE GR BN
35% macabeo, 35% xarel.lo, 30% parellada
91
Colour: bright straw. Nose: fine lees, fragrant herbs, characterful, ripe fruit, dry nuts. Palate: powerful, flavourful, good acidity, fine bead, fine bitter notes.

CAVA ANGEL
Vinyals, 161
08223 Terrassa (Barcelona)
☎: +34 937 330 695
botiga@cavaangel.com
www.cavaangel.com

Angel BE R BN
macabeo, xarel.lo, parellada
88
Citrus fruit, crisp, fruity, yeasty notes, creamy.

Angel Cupatge BE R BN
macabeo, xarel.lo, parellada, chardonnay
91
Colour: bright yellow. Nose: ripe fruit, fine lees, balanced, dried herbs, brioche. Palate: good acidity, flavourful, ripe fruit.

Angel Noir Rosat RE BN
pinot noir, garnacha, trepat
85

CAVA ORIOL ROSSELL

Propietat Cal Cassanyes
08732 Sant Marçal (Barcelona)
☎: +34 977 671 061
oriolrossell@oriolrossell.com
www.oriolrossell.com

Oriol Rossell 2019 BE R BN
88 ♣
Kind finish, aromatic, citrus fruit, floral.

Oriol Rossell Ariadna 2016 BE GR BN
xarel.lo, pinot noir
92 ♣
Colour: bright straw. Nose: fine lees, fragrant herbs, expressive, dried herbs, wild herbs. Palate: powerful, flavourful, good acidity, fine bead, balanced.

Oriol Rossell Gran Propietat 2009 BE GR BN
xarel.lo, macabeo, parellada
91
Colour: bright straw. Nose: fine lees, floral, fragrant herbs, expressive. Palate: powerful, flavourful, good acidity, fine bead.

Oriol Rossell Mitic 2017 BE GR BN
macabeo, xarel.lo
91 ♣
Colour: bright yellow. Nose: ripe fruit, fine lees, balanced, dried herbs. Palate: good acidity, flavourful, ripe fruit, long.

Oriol Rossell Reserva de la Propietat 2015 BE GR BN
macabeo, xarel.lo, parellada
92 ♣
Colour: bright straw. Nose: fine lees, floral, fragrant herbs, expressive. Palate: powerful, flavourful, good acidity, fine bead.

Oriol Rossell Reserva de la Propietat Rosé 2016 RE GR BR
pinot noir
90 ♣
Colour: raspberry rose. Nose: elegant, red berry notes, floral, fragrant herbs. Palate: light-bodied, spicy, good acidity, fine bitter notes.

CAVA REVERTÉ

Pss. Estació, 4
43885 Salomo (Tarragona)
☎: +34 630 929 380
Fax: +34 977 629 246
reverte@cavareverte.com
www.cavareverte.com

Cava Reverté "Electe" 2017 BE R BN
88
Floral, fruity, standard, tasty.

Cava Reverté 2016 BE R BN
87

CAVAS BERTHA

Crtra. Sant Sadurní a Vilafranca km. 2,4
08739 Subirats (Barcelona)
☎: +34 938 911 091
cavabertha@cavabertha.com
www.cavabertha.com

Bertha 2019 BE R BN
xarel.lo, macabeo, parellada
90 ♣
Colour: bright yellow. Nose: fine lees, balanced, dried herbs. Palate: good acidity, flavourful, ripe fruit, long.

Bertha Cardús 2018 BE GR BN
xarel.lo, macabeo, parellada
90 ♣
Colour: bright straw. Nose: fine lees, fragrant herbs, pattiserie. Palate: powerful, flavourful, good acidity, fine bead, balanced.

Bertha Lounge 2019 BE BR
xarel.lo, parellada
85 ♣

Bertha Segle XXI 2009 BE GR BN
xarel.lo, macabeo, parellada, chardonnay
91
Age nuances. Colour: bright yellow. Nose: faded flowers, dried flowers, candied fruit, pattiserie, dry nuts, spicy. Palate: slightly evolved, flavourful, long, ripe fruit.

Bertha Segle XXI Rosé 2017 RE GR BR
pinot noir
89
Citrus fruit, standard, herbaceous, yeasty notes.

Ivette 2019 BE R BR
xarel.lo, macabeo, parellada
88 ♣
Kind finish, crisp, floral, simple, smooth.

DO CAVA / D.O.P.

CAVAS HILL

Bonavista, 2
08734 Moja-Olérdola (Barcelona)
☎: +34 938 900 588
Fax: +34 938 170 246
cavashill@cavashill.com
www.cavashill.es

Cavas Hill Cole-lecció Privada BE BN
xarel.lo, parellada

92

Colour: bright golden. Nose: fine lees, fragrant herbs, ripe fruit, bakery. Palate: flavourful, good acidity, fine bead, fine bitter notes.

Cavas Hill Cuvée 1887 BE BN
macabeo, xarel.lo, parellada

88

Pleasant, smooth, simple, very fruit-driven.

Cavas Hill Panot Gaudí BE BN
xarel.lo, macabeo, parellada, chardonnay

90

Colour: bright straw. Nose: fine lees, floral, fragrant herbs, expressive. Palate: powerful, flavourful, good acidity, fine bead, balanced.

Cavas Hill Panot Gaudí BE BR
xarel.lo, macabeo, parellada, chardonnay

88

Defined aromas, yeasty notes, balanced, floral, fruity, smooth.

Cavas Hill Panot Gaudí Coral RE BR
xarel.lo, chardonnay, garnacha

88

Pleasant, aromatic, fruity, jammy.

Conde de Caralt BE BN
xarel.lo, macabeo, parellada

89

Racy, citrus fruit, herbal, tasty.

CAVAS JANÉ SANTACANA

Finca Baldús
08792 Santa Fe del Penedès (Barcelona)
☎: +34 938 988 205
Fax: +34 938 988 205
janesantacana@janesantacana.com
www.janesantacana.com

Baldús 2017 BE GR BN
40% xarel.lo, 50% macabeo, 10% parellada

87

Jane Santacana Etiqueta Blanca 2018 BE R BN
20% xarel.lo, 70% macabeo, 10% parellada

87

Baldús 2018 BE R BN
50% xarel.lo, 25% macabeo, 25% parellada

89

Colour: bright straw. Nose: fine lees, floral, fragrant herbs. Palate: good acidity, balanced, easy to drink.

Jane Santacana 2017 BE GR BN
60% xarel.lo, 30% macabeo, 10% parellada

89

Aromatic, floral, balanced, smooth. Nose: fine lees, pattiserie.

Jane Santacana Etiqueta Cobre 2018 BE R BR
20% xarel.lo, 70% macabeo, 10% parellada

88

Standard, yeasty notes, simple, pleasant.

Jane Santacana Etiqueta Dorada 2018 BE R BN
60% xarel.lo, 30% macabeo, 10% parellada

88

Floral, fruity, crisp, smooth. Palate: fine bitter notes.

CAVAS MIQUEL PONS

08792 La Granada (Barcelona)
☎: +34 938 974 541
Fax: +34 938 974 710
miquelpons@cavamiquelpons.com
www.cavamiquelpons.com

Eulàlia de Pons Cuvée 2018 BE R BR

91

Colour: straw. Nose: expressive, dry nuts, fragrant herbs, spicy. Palate: flavourful, fine bead, long, great length.

Miquel Pons 2019 BE R BN
xarel.lo, macabeo, parellada

89 ♣

Dried herbs, crisp, dried flowers, citrus fruit, yeasty notes, simple.

Miquel Pons Gran Reserva Vintage 2015 BE GR BN
xarel.lo, macabeo, parellada

92

Colour: bright yellow. Nose: ripe fruit, fine lees, balanced, dried herbs, dried flowers. Palate: good acidity, flavourful, ripe fruit.

Miquel Pons Montargull 2015 BE GR BN
xarel.lo, macabeo, parellada

92

Colour: bright straw. Nose: fine lees, floral, fragrant herbs, expressive. Palate: powerful, flavourful, good acidity, fine bead, balanced.

Miquel Pons
Montargull Xarel.lo 2015 BE GR BR
xarel.lo, macabeo
89
Kind finish, aromatic, standard, balanced, fruity, sweety finish.

Nuria Rosé 2019 RE R BR
89
Pleasant, aromatic, fruity, jammy.

CAVES BOHIGAS
Finca Can Macià
08700 Òdena (Barcelona)
☎: +34 938 048 100
info@bohigas.es
www.fermibohigas.com

Bohigas 2020 BE GR EBR
50% xarel.lo, 30% macabeo, 20% parellada
91
Colour: bright yellow. Nose: ripe fruit, fine lees, balanced, dried herbs. Palate: good acidity, flavourful, ripe fruit, long.

Bohigas BE R BN
50% xarel.lo, 30% macabeo, 20% parellada
88
Citrus fruit, crisp, herbal, balanced.

Bohigas BE R BR
50% xarel.lo, 30% macabeo, 20% parellada
88
Citrus fruit, crisp, herbal, full-bodied.

Bohigas Rosat RE BR
100% trepat
88
Citrus fruit, fruity, crisp, balanced.

CAVES NAVERAN
Can Parellada
08735 Torrelavit (Barcelona)
☎: +34 938 988 274
naveran@naveran.com

Naverán Odisea 2019 BE BR
91 ♣
Colour: bright straw. Nose: fine lees, floral, expressive, spicy, ripe fruit. Palate: flavourful, good acidity, fine bead, balanced.

Naveran Perles Blanques 2017 BE GR BR
92 ♣
Colour: bright golden. Nose: fine lees, dry nuts, complex, ripe fruit. Palate: powerful, flavourful, good acidity, fine bead, fine bitter notes.

Naverán Perles D'or 2017 BE GR BR
91 ♣
Colour: bright golden. Nose: dry nuts, dried herbs, complex, spicy, brioche, lees reduction notes, stone fruit. Palate: flavourful, good acidity, fine bead, fine bitter notes.

Naverán Perles Roses Pinot Noir 2019 RE BR
pinot noir
89 ♣
Fruity, lactic, tasty, dried herbs.

CAVES VIDAL I FERRÉ
Nou, 2
43815 Les Pobles (Tarragona)
☎: +34 977 638 554
vidaliferre@vidaliferre.com
www.vidaliferre.com

Vidal i Ferré 2017 BE GR BN
parellada, macabeo, xarel.lo
88
Standard, dried herbs, simple, smooth, crisp.

Vidal i Ferré 2018 BE R BN
macabeo, xarel.lo, parellada
88 ♣
Spicy, dried herbs, jammy, tasty, smooth.

Vidal i Ferré 2020 BE BR
macabeo, xarel.lo, parellada
87 ♣

Vidal i Ferré Blanc de Noirs 2020 BE R BN
pinot noir, xarel.lo
88 ♣
Nose: fine lees, floral, fragrant herbs, expressive. Palate: powerful, flavourful, fine bead.

Vidal i Ferré Rosat 2020 RE BR
pinot noir
87 ♣

CELLER CARLES ANDREU
Sant Sebastià, 19
43423 Pira (Tarragona)
☎: +34 977 887 404
info@cavandreu.com
www.cavandreu.com

Carles Andreu 2018 BE GR BN
78% parellada, 14% macabeo, 8% chardonnay
89
Colour: bright yellow. Nose: ripe fruit, fine lees, balanced, wild herbs. Palate: good acidity, flavourful, ripe fruit, long.

DO CAVA / D.O.P.

DO CAVA / D.O.P.

Carles Andreu 2019 BE BN
86

Carles Andreu Barrica 2017 BE R BN
54% parellada, 23% macabeo, 15% chardonnay, 8% xarel.lo
90
Colour: bright yellow. Nose: fresh fruit, citrus fruit, fine lees, fragrant herbs, spicy. Palate: fresh, fruity, good acidity, easy to drink.

Carles Andreu Rosat 2019 RE BR
100% trepat
90
Dried flowers, dried herbs. Colour: bright yellow. Nose: ripe fruit, fine lees, balanced, dried herbs, dried flowers. Palate: good acidity, flavourful, ripe fruit, good structure.

Carles Andreu Rosat Barrica 2019 RE R BR
91
Colour: coppery red. Nose: expressive, balanced, wild herbs, spicy, dried flowers. Palate: fine bitter notes, balanced, good acidity.

L'Era del Celdoni 2012 BE GR BN
80% parellada, 10% macabeo, 10% chardonnay
91
Colour: bright straw. Nose: fine lees, fragrant herbs, characterful, ripe fruit, dry nuts. Palate: good acidity, fine bead, fine bitter notes.

CELLER JORDI LLUCH
Masia Casa Nova, Barri Les Casetes d'en Raspall
08777 Sant Quintí de Mediona (Barcelona)
☎: +34 938 988 123
vinyaescude@vinyaescude.com
www.vinyaescude.com

Vinya Escudé 523 2018 BE R EBR
macabeo, xarel.lo, parellada
87

Vinya Escudé Daurat 2019 BE R BN
macabeo, xarel.lo, parellada
88
Tropical, jammy, simple.

Vinya Escudé Single Vineyard 2019 BE R BN
macabeo
88
Fruity, dried herbs, simple, lactic.

CELLER PRIVAT
Avda. Barcelona, 78
08720 Vilafranca del Penedès (Barcelona)
☎: +34 938 180 676
Fax: +34 938 180 926
rrpp@pereladachivite.com
www.cellerprivat.com

Privat 2020 BE R BN
40% xarel.lo, 25% macabeo, 25% parellada, 10% chardonnay
92 🌱
Colour: bright golden. Nose: fragrant herbs, characterful, ripe fruit, dry nuts. Palate: powerful, good acidity, fine bead, fine bitter notes.

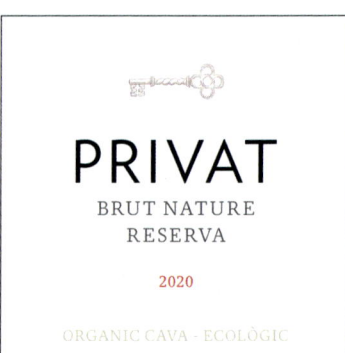

Privat 2020 BE R BR
40% xarel.lo, 25% macabeo, 25% parellada, 10% chardonnay
88 🌱
Aromatic, floral, fruity, simple, standard.

Privat Rosat 2020 RE BN
75% pinot noir, 25% chardonnay
88 🌱
Pleasant, fruity, floral, tasty.

CELLERS CAROL VALLÈS
08739 Subirats (Barcelona)
☎: +34 938 989 078
info@cellerscarol.com
www.cellerscarol.com

Gran Reserva Familiar Millenium 2013 BE GR BR
32% xarel.lo, 22% parellada, 26% macabeo, 20% chardonnay
93
Colour: bright golden. Nose: fine lees, fragrant herbs, characterful, ripe fruit, dry nuts. Palate: powerful, flavourful, good acidity, fine bead, fine bitter notes.

Guillem Carol 2017 BE GR BN
56% xarel.lo, 24% macabeo, 15% chardonnay, 5% pinot noir
90 🌱
Colour: bright straw. Nose: fine lees, floral, fragrant herbs, expressive. Palate: powerful, flavourful, good acidity, fine bead, balanced.

Mallerenga 2017 BE GR BN
40% xarel.lo, 40% chardonnay, 20% xarel.lo
93 🌿
Colour: bright straw. Nose: fine lees, floral, fragrant herbs, expressive. Palate: powerful, flavourful, good acidity, fine bead, balanced.

Rossinyol 2017 BE GR BN
60% chardonnay, 40% pinot noir
90 🌿
Colour: bright straw. Nose: fresh fruit, citrus fruit, fine lees, fragrant herbs. Palate: fresh, fruity, good acidity.

CHOZAS CARRASCAL
Vereda Real
46390 San Antonio de Requena (València/Valencia)
☎: +34 963 410 395
chozas@chozascarrascal.es
www.chozascarrascal.com

Cava Roxanne 2020 BE BR
89 🌿
Aromatic, floral, jammy, balanced, exuberant, tropical.

Chozas Carrascal Ch 2016 BE GR BN
93
Colour: bright golden. Nose: fine lees, fragrant herbs, characterful, ripe fruit. Palate: flavourful, good acidity, fine bead, fine bitter notes.

El Cava de Chozas Carrascal 2018 BE R BN
70% chardonnay, 30% macabeo
91 🌿
Colour: bright straw. Nose: white flowers, white fruit, ripe fruit, yeasty notes. Palate: flavourful, balanced, fine bitter notes.

CODORNÍU
Avda. Jaume Codorníu, s/n
08770 Sant Sadurní d'Anoia (Barcelona)
☎: +34 938 183 232
info@codorniu.es
www.codorniu.es

Anna de Codorníu BE BN
87

Anna de Codorníu BE BR
88
Citrus fruit, balanced, jammy, herbal, tasty.

Anna de Codorníu Blanc de Noirs BE BR
88
Balanced, crisp, herbal, simple.

Anna de Codorníu Rosé RE BR
86

Anna de Codorníu Ecológico Reserva BE R BR
88 🌿
Citrus fruit, crisp, herbal, balanced.

Anna de Codorníu Ice Edition BE SS
86

Anna de Codorníu Ice Edition BE SS
86

🏆 **PODIUM**

Ars Collecta 459 2010 BE GR BR
95
Colour: bright golden. Nose: fine lees, dry nuts, fragrant herbs, complex, toasty. Palate: powerful, flavourful, good acidity, fine bead, fine bitter notes.

Ars Collecta Blanc de Blancs 2018 BE GR BR
92
Colour: bright yellow. Nose: ripe fruit, fine lees, balanced, dried herbs. Palate: good acidity, flavourful, ripe fruit, long.

Ars Collecta El Tros Nou Pinot Noir 2010 BE GR BR
94
Colour: bright yellow. Nose: ripe fruit, fine lees, balanced, dried herbs. Palate: good acidity, flavourful, ripe fruit, long, good structure.

Ars Collecta Jaume Codorníu 2018 BE GR BR
90
Colour: bright yellow. Nose: ripe fruit, fine lees, balanced, dried herbs. Palate: good acidity, ripe fruit, correct.

🏆 **PODIUM**

Ars Collecta La Fideuera Xarel.lo 2011 BE GR BR
95
Colour: bright golden. Nose: fragrant herbs, tea leave, brioche, dry nuts. Palate: powerful, flavourful, good acidity, fine bead, fine bitter notes.

Ars Collecta La Pleta Chardonnay 2011 BE GR BR
94
Colour: bright yellow. Nose: fine lees, balanced, dried herbs, stone fruit, tropical fruit. Palate: good acidity, flavourful, ripe fruit, long.

Codorniu Non Plus Ultra 2018 BE R BR
90 🌿
Colour: bright straw. Nose: fresh fruit, citrus fruit, fine lees, grassy. Palate: fresh, fruity, good acidity.

DO CAVA / D.O.P.

COVIDES VINYES - CELLERS

Finca Prunamala,
Ctra. St. Sadurní a Vilafranca, Km. 1
08770 Sant Sadurní D'Anoia (Barcelona)
☎: +34 938 172 552
marketing@covides.com
www.covides.com

Duc de Foix BE BN
xarel.lo, macabeo, parellada

87

Duc de Foix Reserva Especial BE R BR
chardonnay, xarel.lo, macabeo

91

Kind finish. Colour: bright yellow. Nose: ripe fruit, fine lees, balanced, dried herbs. Palate: good acidity, flavourful, ripe fruit, long.

Terra Terrae BE BR
xarel.lo, macabeo, parellada

87 ♣

Terra Terrae BE R BR
xarel.lo, macabeo, parellada

89 ♣

Balanced, spicy, toasty, tasty, creamy.

Duc de Foix BE R BN
xarel.lo, parellada, chardonnay

88

Pleasant, standard, balanced, crisp, dried flowers.

Duc de Foix BE R BR
xarel.lo, parellada, chardonnay

89

Citrus fruit, lactic, dried herbs, tasty.

CUSCÓ BERGA

Esplugues, 7
08793 Avinyonet del Penedès (Barcelona)
☎: +34 660 829 402
cuscoberga@cuscoberga.com
www.cuscoberga.com

Cuscó Berga 2013 BE GR BN
30% macabeo, 40% xarel.lo, 30% parellada

90 ♣

Colour: bright yellow. Nose: ripe fruit, fine lees, balanced, dried herbs, dry nuts. Palate: flavourful, ripe fruit, good finish.

Cuscó Berga 2019 BE R BR
20% macabeo, 60% xarel.lo, 20% parellada

88

Standard, floral, dried herbs, jammy, wild, smooth.

Cuscó Berga BE GR BR
30% macabeo, 40% xarel.lo, 30% parellada

90 ♣

Colour: bright straw. Nose: fine lees, fragrant herbs, characterful, ripe fruit. Palate: flavourful, good acidity, bitter.

Cuscó Berga BE R BN
89 ♣

Pleasant, crisp, yeasty notes, smooth, balanced, dried herbs.

Cuscó Berga Rosé RE R BR
100% trepat

85

DOMENIO WINES BY CELLERS DOMENYS

Prat de la Riba, 18
43713 St. Jaume dels Domenys (Tarragona)
☎: +34 977 677 135
administracio@cellersdomenys.com
www.domeniowines.com

Tres Naus BE BR
88

Citrus fruit, dried herbs, tasty, yeasty notes.

Tres Naus BE C BN
88

Citrus fruit, balanced, spicy, crisp, herbal, yeasty notes, tasty.

Tres Naus Rosé Intenso RE C BR
87

Tres Naus BE R BR
90
Colour: bright yellow. Nose: ripe fruit, fine lees, balanced, dried herbs. Palate: good acidity, flavourful, ripe fruit.

Tres Naus BE R BN
87
Fruity, citrus fruit, dried herbs, simple.

Tres Naus Rosé Pálido RE BR
88
Floral, fruity, sweety finish, tasty.

DOMINIO DE LA VEGA
Ctra. Madrid - Valencia, Km. 270,6
46390 San Antonio de Requena (València/Valencia)
☎: +34 962 320 570
dv@dominiodelavega.com
www.dominiodelavega.com

Dominio de la Vega Authentique 2019 BE R BN
macabeo, xarel.lo
88
Austere, standard, fruity, simple, dried herbs.

Dominio de la Vega Cuvée Prestige 2017 BE R BN
chardonnay
92
Colour: straw. Nose: expressive, spicy, dried flowers, fine lees. Palate: flavourful, fine bead, long, great length.

Dominio de la Vega Expression 2019 BE R BR
macabeo
88
Standard, dried flowers, fruity, jammy, pleasant. Palate: easy to drink.

Dominio de la Vega No.1 2020 BE BR
macabeo
88 🌱
Fruity, herbal, tasty, dried flowers.

Dominio de la Vega Reserva Especial 2018 BE R BR
macabeo, chardonnay
92
Colour: bright straw. Nose: expressive, dry nuts, fragrant herbs, spicy. Palate: flavourful, long, great length.

Dominio de la Vega Reserva Especial Rosé 2018 RE R BR
pinot noir
90
Colour: brilliant rose. Nose: balsamic herbs, red berry notes, wild herbs. Palate: fresh, fruity, good acidity, flavourful.

ESTEL D'ARGENT
Font-Rubí, 2 4º 1ª
08720 Vilafranca del Penedès (Barcelona)
☎: +34 677 182 347
cava@esteldargent.com
www.esteldargent.com

Estel D'Argent 2018 BE R BN
macabeo, xarel.lo, parellada
90
Colour: bright straw. Nose: medium intensity, fresh fruit, dried herbs, fine lees. Palate: fruity, good acidity, fine bitter notes.

Estel D'Argent Especial 2017 BE R EBR
macabeo, xarel.lo, chardonnay
91
Colour: bright straw. Nose: fine lees, spicy, toasted bread. Palate: powerful, flavourful, good acidity, long.

Estel D'Argent Especial 2018 BE GR EBR
macabeo, xarel.lo, chardonnay
90
Colour: bright straw. Nose: fine lees, floral, fragrant herbs, expressive. Palate: powerful, flavourful, fine bead.

Estel D'Argent Rosé 2019 RE R BN
pinot noir, trepat
88
Jammy, tasty, smooth, dried herbs.

DO CAVA / D.O.P.

FAMILIA AMETLLER

Barri La Rovira Roja, 10
08731 Sant Martí Sarroca (Barcelona)
☎: +34 938 904 735
info@familiaametller.com
www.familiaametller.com

Ametller 2016 BE GR BN
xarel.lo, macabeo, parellada

89

Dried flowers, fruity, dried herbs, jammy.

Ametller 2020 BE BR
xarel.lo, macabeo, parellada

89

Standard, balanced, fruity, dried herbs, dried flowers, tasty.

Ametller 2020 BE R BN
xarel.lo, macabeo, parellada

88

Dried flowers, fruity, tasty, citrus fruit.

Ametller Blanc de Noirs 2020 BE BN
pinot noir

90

Colour: bright straw. Nose: ripe fruit, fine lees, dried herbs, faded flowers. Palate: flavourful, fruity, fresh.

FERRÉ I CATASÚS

Masía Gustems, Crta. Sant Sadurní a Vilafranca, Km.8
08792 La Granada (Barcelona)
☎: +34 938 974 558
administracio@ferreicatasus.com
www.ferreicatasus.com

Cava Ballbé BE BN
macabeo, xarel.lo, parellada

87

Maria Catasús BE R BN
macabeo, chardonnay, xarel.lo

88

Standard, balanced, simple, crisp.

FINCA VALLDOSERA

Masia Les Garrigues, s/n
08734 Olérdola (Barcelona)
☎: +34 938 143 047
general@fincavalldosera.com
www.fincavalldosera.com

Finca Valldosera 2013 BE GR BN
45% xarel.lo, 30% macabeo, 10% parellada, 10% chardonnay, 5% subirat parent

89

Citrus fruit, balanced, toasty, tasty, jammy.

Finca Valldosera 2014 BE R BN
35% xarel.lo, 30% macabeo, 20% parellada, 10% chardonnay, 5% subirat parent

88

Citrus fruit, toasty, age nuances, balanced, yeasty notes.

Finca Valldosera Monastrell 2018 RE R BN
100% monastrell

88

Fruity, jammy, tasty, dried flowers, yeasty notes, balanced.

FREIXA RIGAU

17750 Capmany (Girona/Gerona)
☎: +34 972 549 012
comercial@grupoliveda.com
www.grupoliveda.com

Batec 2014 BE GR BR
pinot noir, xarel.lo

92

Colour: straw. Nose: expressive, dry nuts, fragrant herbs, spicy. Palate: flavourful, fine bead, long, great length.

Gran Rigau Chardonnay BE R BN
chardonnay

91

Colour: bright yellow. Nose: ripe fruit, dried herbs, brioche, citrus fruit. Palate: good acidity, flavourful, ripe fruit.

FREIXENET

Joan Sala, 2
08770 Sant Sadurní d'Anoia (Barcelona)
☎: +34 938 917 000
comunicacion@freixenet.es
www.freixenet.es

Elyssia Gran Cuvée 2020 BE R BR
chardonnay, macabeo, parellada, pinot noir

90

Colour: bright yellow. Nose: ripe fruit, fine lees, balanced, dried herbs, brioche. Palate: good acidity, flavourful, ripe fruit.

Elyssia Pinot Noir Rosé 2020 RE BR
pinot noir

88

Pleasant, fruity, floral, very fruit-driven, crisp.

Freixenet Carta Rosé 2020 RE ES
trepat, garnacha

85

Freixenet
Cordón Negro 2020 BE BR
parellada, macabeo, xarel.lo
88
Pleasant, tasty, smooth, crisp.

Meritum 2016 BE BN
88
Citrus fruit, age nuances, balanced, jammy, tasty.

GATELL
Ctra. T-202, Km-10,5
43884 Bonastre (Tarragona)
☎: +34 609 342 642
director.comercial@cavagatell.com
www.cavagatell.com

Gatell Ambrosía 2017 BE GR BN
xarel.lo, macabeo, parellada
90
Aromatic, thin. Colour: bright yellow. Nose: dry nuts, dried flowers, faded flowers, fine lees, spicy. Palate: fresh, good finish.

Gatell Heritage 2017 BE GR BN
macabeo, xarel.lo, parellada, chardonnay
89
Defined aromas, dried flowers, balanced, spicy, toasty, pleasant, kind finish.

Gatell Initial 2017 BE GR BN
xarel.lo, macabeo, parellada, malvasía
91
Colour: bright yellow. Nose: fine lees, fragrant herbs, characterful, ripe fruit, spicy, toasty, dry nuts. Palate: flavourful, good acidity, fine bitter notes.

Gatell Rosé 2017 RE GR BN
garnacha, pinot noir
88
Citrus fruit, dried flowers, tasty, crisp.

GIRÓ DEL GORNER
Finca Giró del Gorner, s/n
08797 Puigdalber (Barcelona)
☎: +34 938 988 032
gorner@girodelgorner.com
www.girodelgorner.com

Giró del Gorner 2013 BE GR BN
macabeo, xarel.lo, parellada
91
Colour: bright yellow. Nose: ripe fruit, fine lees, balanced, dried herbs, dry nuts. Palate: good acidity, flavourful, ripe fruit.

Giró del Gorner 2013 BE GR BR
macabeo, xarel.lo, parellada
91
Colour: bright yellow. Nose: ripe fruit, balanced, dried herbs, brioche. Palate: good acidity, flavourful, ripe fruit.

Giró del Gorner 2017 BE R BN
macabeo, xarel.lo, parellada
90
Colour: bright yellow. Nose: ripe fruit, fine lees, balanced, dried herbs. Palate: good acidity, flavourful, ripe fruit.

Giró del Gorner 2018 BE R BR
macabeo, xarel.lo, parellada
89
Citrus fruit, crisp, dried herbs, tasty, yeasty notes.

Giró del Gorner Rosat 2020 RE BR
100% pinot noir
90 ♣
Colour: coppery red. Nose: floral, fruit expression, balanced, fresh, medium intensity. Palate: varietal, easy to drink.

GIRÓ RIBOT
Finca El Pont, s/n
08792 Santa Fe del Penedès (Barcelona)
☎: +34 938 974 050
giroribot@giroribot.es
www.giroribot.es

Giró Ribot Avant 2016 BE R BR
xarel.lo, chardonnay, macabeo
92
Defined aromas, with personality. Colour: bright yellow. Nose: fragrant herbs, wild herbs, fine lees, bakery. Palate: flavourful, good acidity, fine bead.

Giró Ribot Excelsus 100 Magnum 2011 BE GR BR
xarel.lo, macabeo, parellada
93
Colour: bright straw. Nose: fine lees, floral, fragrant herbs, expressive. Palate: powerful, flavourful, good acidity, fine bead, balanced.

Giró Ribot Mare 2017 BE GR BN
65% xarel.lo, 30% macabeo, 5% parellada
92
Colour: bright yellow. Nose: fine lees, balanced, dried herbs, citrus fruit, chamomile. Palate: good acidity, flavourful, ripe fruit, long.

DO CAVA / D.O.P.

DO CAVA / D.O.P.

Giró Ribot
Unplugged Rosado 2018 RE R BR
91
Colour: raspberry rose. Nose: elegant, red berry notes, floral, fragrant herbs, fine lees. Palate: light-bodied, spicy, fine bitter notes, sweetness.

Giró Ribot SPur 2017 BE GR BN
50% xarel.lo, 30% chardonnay, 20% parellada
88
Citrus fruit, dried herbs, crisp, simple.

Paul Cheneau 2017 BE R BR
macabeo, xarel.lo, chardonnay
90

Paul Cheneau
Blanc de Blancs BE R BR
90
Colour: bright yellow. Nose: ripe fruit, fine lees, balanced, dried herbs, dry nuts. Palate: good acidity, flavourful, ripe fruit.

HAMMEKEN CELLARS
Llavador, 20
03700 Denia (Alacant/Alicante)
☎: +34 965 791 967
cellars@hammekencellars.com
www.hammekencellars.com

Avent BE BR
80% macabeo, 20% chardonnay
87

Avent Rosé RE BR
100% garnacha
87

JANÉ VENTURA
Ctra. de Calafell, 2
43700 El Vendrell (Tarragona)
☎: +34 977 660 118
janeventura@janeventura.com
www.janeventura.com

Jané Ventura 1914 2012 BE GR BN
xarel.lo, macabeo, parellada
94
Colour: bright golden. Nose: fine lees, fragrant herbs, characterful, ripe fruit, dry nuts. Palate: flavourful, good acidity, fine bead, fine bitter notes.

Jané Ventura Do Major Vinyes Velles 2016 BE GR BN
90
Defined aromas, crisp, needs time. Colour: bright straw. Nose: white flowers, medium intensity. Palate: fresh, easy to drink, fine bitter notes.

Jané Ventura Do Major Vinyes Velles 2017 BE GR BN
xarel.lo, macabeo
92
Colour: bright straw. Nose: white fruit, fine lees, bakery, dried flowers, balanced, neat. Palate: fruity, long, flavourful.

Jané Ventura
Reserva de la Música 2019 BE R BN
90
Colour: bright straw. Nose: fresh fruit, citrus fruit, fine lees, fragrant herbs. Palate: fresh, fruity, good acidity.

Jané Ventura Reserva de la Música Magnum 2017 BE R BN
91
Colour: bright yellow. Nose: fine lees, balanced, dried herbs, citrus fruit, fresh fruit. Palate: good acidity, flavourful, ripe fruit, long.

Jané Ventura Reserva de la Música Rosé 2019 RE R BR
garnacha
88
Pleasant, crisp, fruity, smooth, simple, balanced, floral.

JAUME LLOPART ALEMANY
Cl. Font Rubí, 9
08736 Font-Rubí (Barcelona)
☎: +34 938 979 133
info@jaumellopartalemany.com
www.jaumellopartalemany.com

Aina Jaume Llopart
Alemany Rosado RE R BR
88
Fruity, dried herbs, jammy, tasty, crisp.

Jaume Llopart Alemany 2015 BE GR BN
91
Colour: bright yellow. Nose: fragrant herbs, dried herbs, white fruit, ripe fruit, dried flowers, fine lees. Palate: fruity, flavourful, balanced, good finish.

Jaume Llopart Alemany BE R BN
89
Citrus fruit, dried flowers, dried herbs, tasty.

Jaume Llopart Alemany BE R BR
89
Citrus fruit, dried flowers, herbal, tasty, full of life.

Vinya d'en Ferran
Jaume Llopart Alemany 2012 BE GR BN
91
Colour: bright golden. Nose: dried herbs, fragrant herbs, ripe fruit, dry nuts. Palate: fine bead, balanced, flavourful, fruity, fresh.

JUVÉ & CAMPS
Sant Venat, 1
08770 Sant Sadurní d'Anoia (Barcelona)
☎: +34 938 911 000
juveycamps@juveycamps.com
www.juveycamps.com

Gran Juvé Camps 2016 BE GR BN
xarel.lo, macabeo, chardonnay, parellada
92 ♣
Colour: bright yellow. Nose: ripe fruit, fine lees, balanced, dried herbs, floral. Palate: good acidity, flavourful, ripe fruit.

Juvé & Camps Blanc de Noirs 2010 BE GR BR
100% pinot noir
92
Colour: bright straw. Nose: fine lees, floral, fragrant herbs, expressive. Palate: powerful, flavourful, good acidity, fine bead.

🏆 PODIUM

Juvé & Camps La Capella 2009 BE GR BN
100% xarel.lo
95
With personality, age nuances. Colour: bright yellow. Nose: fine lees, dry nuts, fragrant herbs, complex, lees reduction notes. Palate: powerful, flavourful, good acidity, fine bead, fine bitter notes.

Juvé & Camps La Siberia 2013 RE GR BN
94 ♣
Colour: coppery red, bright. Nose: lees reduction notes, bakery, dry nuts, dried flowers, stone fruit. Palate: flavourful, balanced, long.

Juvé & Camps Milesimé 2008 BE GR BR
100% chardonnay
93
Colour: bright golden. Nose: dry nuts, dried herbs, complex, spicy. Palate: powerful, flavourful, good acidity, fine bead, fine bitter notes.

Juvé & Camps Milesimé 2016 BE R BR
100% chardonnay
93 ♣
Colour: bright golden. Nose: fragrant herbs, characterful, ripe fruit, dry nuts. Palate: flavourful, fine bead, fine bitter notes.

Juvé & Camps
Reserva de la Familia 2018 BE GR BN
xarel.lo, macabeo, chardonnay
91 ♣
Colour: bright straw. Nose: fragrant herbs, expressive, wild herbs. Palate: flavourful, good acidity, fine bead, balanced.

L'ORIGAN
Avernó, 28
08770 Sant Sadurní d'Anoia (Barcelona)
☎: +34 938 183 602
Fax: +34 938 913 461
lorigan@lorigancava.com
www.lorigancava.com

Aire de L'O de L'Origan BE BN
90
Colour: bright yellow. Nose: ripe fruit, fine lees, dried herbs, bakery. Palate: good acidity, flavourful, ripe fruit.

Aire de L'Origan BE BN
92
Full of life. Colour: bright straw. Nose: fine lees, citrus fruit, bakery. Palate: flavourful, good acidity, balanced.

Aire de L'Origan Magnum 2019 BE BN
91
Colour: bright yellow. Nose: ripe fruit, fine lees, balanced, dried herbs. Palate: good acidity, flavourful, ripe fruit, long.

Aire de L'Origan Rosé Magnum 2019 RE BN
88
Racy, standard, dried flowers, herbaceous, yeasty notes.

L'O de L'Origan Rosat RE BN
91
Oxidativ. Colour: bright yellow. Nose: fine lees, balanced, dried herbs, candied fruit, dried flowers. Palate: good acidity, flavourful, ripe fruit.

L'O de L'Origan BE BN
91
Colour: bright golden. Nose: fine lees, characterful, ripe fruit, dry nuts. Palate: powerful, good acidity, fine bead, fine bitter notes.

DO CAVA / D.O.P.

LACRIMA BACCUS
Finca La Porxada s/n
08732 Castellet i la Gornal (Barcelona)
☎: +34 938 918 281
lacrimabaccus@bardinet.es
www.lacrimabaccus.com

Heretat D'Lácrima Baccus 2019 BE R BN
xarel.lo, macabeo, parellada

87 🌱

Heretat D'Lácrima Baccus 2019 BE R BR
xarel.lo, macabeo, parellada

87 🌱

Lácrima Baccus Rosé 2019 RE R BR
garnacha, pinot noir

88 🌱

Jammy, simple, fruity.

Summum Lácrima Baccus 2018 BE R BN
pinot noir, xarel.lo

90 🌱

Colour: bright straw. Nose: ripe fruit, fine lees, dried herbs. Palate: flavourful, good acidity, fine bead.

Summum Lácrima Baccus 2018 BE R BR
pinot noir, xarel.lo

88 🌱

Citrus fruit, fruity, herbal, tasty.

Summum Lácrima Baccus 2019 BE R BN

89 🌱

Citrus fruit, dried herbs, tasty, crisp.

LADRÓN DE LUNAS
Colón, 12
46357 La Portera (València/Valencia)
☎: +34 601 288 998
jorge.ferrer@ladrondelunas.es
www.ladrondelunas.es

Bisila Rosé Brut Nature Prestige BE BN

88

Citrus fruit, crisp, herbal, tasty.

Ladrón de Lunas BE BN
macabeo, chardonnay

85

MARÍA CASANOVAS
Crta. Sant Sadurni a Piera BV-2242 Km. 7,5
08784 Sant Jaume Sesoliveres (Barcelona)
☎: +34 938 910 812
info@mariacasanovas.com
www.mariacasanovas.com

María Casanovas 2019 BE GR BN
chardonnay, pinot noir

92 🌱

Colour: bright straw. Nose: fine lees, fragrant herbs, ripe fruit, dry nuts. Palate: good acidity, fine bitter notes, easy to drink.

María Casanovas XP 2019 BE GR BN
xarel.lo, pinot noir

91 🌱

Colour: bright straw. Nose: fine lees, floral, fragrant herbs, fresh fruit. Palate: good acidity, balanced, easy to drink.

MARIA RIGOL ORDI
Fullerachs, 9
08770 Sant Sadurní d'Anoia (Barcelona)
☎: +34 684 472 424
cava@mariarigolordi.com
www.mariarigolordi.com

Maria Rigol Ordi 2015 BE GR BN
xarel.lo, macabeo, parellada

92 🌱

Colour: bright golden. Nose: dry nuts, fragrant herbs, complex, toasty. Palate: powerful, flavourful, good acidity, fine bead, fine bitter notes.

María Rigol Ordi 2017 BE R BN
xarel.lo, macabeo, parellada

88 🌱

Fruity, jammy, dried herbs, tasty, spicy.

Maria Rigol Ordi Màgnum Cupatge Dos Mil Disset 2017 BE R BN
xarel.lo, macabeo, parellada

91

Colour: bright straw. Nose: medium intensity, fresh fruit, dried herbs, fine lees. Palate: fresh, fruity, good acidity, easy to drink.

María Rigol Ordi Microtiratge #6: Parellada 2016 BE GR BN
parellada, xarel.lo

92

Colour: bright golden. Nose: fine lees, fragrant herbs, ripe fruit, dry nuts, brioche. Palate: flavourful, good acidity, fine bead, fine bitter notes.

María Rigol Ordi Microtiratge #7: Trepat 2020 BE GR BN
parellada, xarel.lo

92

Aromatic. Colour: bright straw. Nose: fine lees, floral, fragrant herbs, expressive. Palate: powerful, flavourful, good acidity, fine bead, balanced.

Maria Rigol Ordi Mil·lenni 2017 BE R BN
xarel.lo, macabeo, parellada, chardonnay

90

Colour: bright straw. Nose: medium intensity, dried herbs, fine lees, floral. Palate: fresh, fruity, flavourful, balanced.

MARQUÉS DE GRIÑÓN FAMILY ESTATES
Finca Casa de Vacas, CM-4015, Km. 23
45692 Malpica del Tajo (Toledo)
☎: +34 925 597 222
adiaz@marquesdegrinon.com

Marqués de Griñón Rosé "La Vie en Rose" RE BR
50% monastrell, 50% pinot noir

88

Pleasant, aromatic, floral, fruity.

MARQUÉS DE LA CONCORDIA FAMILY OF WINES CAVA
Ctra. El Ciego s/n
26350 Cenicero (La Rioja)
aparadiso@marquesdelaconcordia.com
www.marquesdelaconcordia.com

Berberana Gran Tradición BE BR
40% macabeo, 30% xarel.lo, 30% parellada

87

Berberana Gran Tradición BE SS
40% macabeo, 30% xarel.lo, 30% parellada

87

Berberana Gran Tradición RE BR
70% monastrell, 30% pinot noir

87

Marqués de la Concordia MM Selección Especial BE BN
40% macabeo, 30% xarel.lo, 30% parellada

86

Marqués de la Concordia Reserva de la Familia Blanc de Blancs 2018 BE BN
30% chardonnay, 25% macabeo, 25% xarel.lo, 20% parellada

88

Balanced, dried herbs, wild, smooth, bitter.

Marqués de la Concordia Reserva de la Familia Rosé RE BR
70% pinot noir, 30% monastrell

88

Citrus fruit, fruity, dried flowers, tasty.

Marqués de la Concordia Selección Especial Rosé 2019 RE BR
70% monastrell, 30% pinot noir

85

MASCARÓ
Casal, 9
08720 Vilafranca del Penedès (Barcelona)
☎: +34 938 901 628
mascaro@mascaro.es
www.mascaro.es

Cuvée Antonio Mascaró 2014 BE GR BN
parellada, chardonnay, xarel.lo

91

Colour: bright golden. Nose: fine lees, fragrant herbs, ripe fruit, dry nuts, caramel. Palate: powerful, flavourful, good acidity, fine bitter notes.

Mascaró "Ambrosia" 2019 BE R SS
macabeo, xarel.lo, parellada

90

Colour: bright yellow. Nose: ripe fruit, fine lees, balanced, dried herbs. Palate: flavourful, ripe fruit, sweet.

Mascaró Magnum 2019 BE BN
parellada, xarel.lo

91

Colour: bright straw. Nose: dried herbs, spicy, fine lees, neat, fresh. Palate: flavourful, easy to drink, balanced, fine bitter notes.

Mascaró Nigrum 2019 BE R BR
macabeo, xarel.lo, parellada

92

Colour: bright straw. Nose: fine lees, floral, fragrant herbs, expressive. Palate: powerful, flavourful, good acidity, fine bead, balanced.

Mascaró Pure 2019 BE R BN
parellada, xarel.lo

90

Colour: bright straw. Nose: fine lees, floral, fragrant herbs, toasted bread, ripe fruit. Palate: flavourful, fine bead, balanced, fresh.

Mascaró Rubor Aurorae 2019 RE BR
garnacha

90

Colour: bright straw. Nose: fine lees, floral, fragrant herbs. Palate: flavourful, good acidity, fine bead, balanced.

MONT MARÇAL
Finca Manlleu, s/n
08732 Castellví de la Marca (Barcelona)
☎: +34 938 918 281
export@mont-marcal.com
www.mont-marcal.com

Extremarium de Mont Marçal 2019 BE R BN
xarel.lo, macabeo, parellada, chardonnay

87 🌱

Extremarium de Mont Marçal 2019 BE R BR
xarel.lo, macabeo, parellada, chardonnay

90 🌱

Colour: bright straw. Nose: fine lees, floral, fragrant herbs, white fruit. Palate: flavourful, good acidity, fine bitter notes, easy to drink.

Mont Marçal 2019 BE R BN
xarel.lo, macabeo, parellada, chardonnay

88 🌱

Citrus fruit, balanced, standard, fruity, crisp, herbal.

Mont Marçal 2019 BE R BR
xarel.lo, macabeo, parellada, chardonnay

88 🌱

Fruity, dried herbs, spicy, crisp.

Mont Marçal Rosé 2020 RE R BR
pinot noir, garnacha, trepat

87 🌱

MONTESANCO
Casa de la Viña, Ctra. Utiel-Los Isidros km. 7
46340 Requena (València/Valencia)
☎: +34 962 121 626
vinos@montesanco.com
www.montesanco.com

Món Macabeo 2019 BE R BN
100% macabeo

88 🌱

Citrus fruit, tropical, yeasty notes, crisp.

MONTESQUIUS
Rambla de la Generalitat, 1
08770 Sant Sadurní d'Anoia (Barcelona)
☎: +34 938 911 662
info@montesquius.com
www.montesquius.com

Montesquius 1918 Magnum 2004 BE GR BN
68% macabeo, 32% xarel.lo

92

Colour: yellow. Nose: fragrant herbs, characterful, ripe fruit, dry nuts, lees reduction notes. Palate: flavourful, fine bead, fine bitter notes.

Montesquius 2010 BE GR BN
61% xarel.lo, 29% macabeo, 5% parellada, 5% chardonnay

91

Colour: bright yellow. Nose: wild herbs, dried herbs, dried flowers, balanced. Palate: fresh, fruity, easy to drink, fine bitter notes.

Montesquius Colección Privada 2013 BE GR BN
55% xarel.lo, 45% macabeo

92

Colour: yellow. Nose: fine lees, dry nuts, fragrant herbs, complex. Palate: powerful, flavourful, good acidity, fine bead, fine bitter notes.

Montesquius Edición Especial Blanc de Blancs 2015 BE GR BN
75% xarel.lo, 25% macabeo

91

Colour: yellow. Nose: lees reduction notes, characterful, dry nuts, candied fruit, faded flowers. Palate: flavourful, bitter.

Montesquius
La Esencia Rosé 2011 RE GR BN
monastrell, pinot noir, trepat

90

Colour: coppery red. Nose: wild herbs, floral, ripe fruit. Palate: fine bitter notes, correct, easy to drink.

Montesquius
Naturelovers 2018 BE R BN
60% xarel.lo, 25% macabeo, 15% parellada

90

Colour: bright yellow. Nose: ripe fruit, fine lees, dry nuts. Palate: flavourful, long, bitter.

Montesquius Vintage 2018 BE R EBR
38% xarel.lo, 43% macabeo, 19% parellada

90

Colour: yellow. Nose: fine lees, floral, fragrant herbs, balanced. Palate: good acidity, fine bitter notes, easy to drink.

MUSCÀNDIA
Can Rosell de la Llena
08790 Gelida (Barcelona)
☎: +34 625 632 620
info@cavamuscandia.com
www.muscandia.com

Bàlsam BE BR
xarel.lo, macabeo, parellada

88

Crisp, smooth, floral, simple, yeasty notes.

Muscàndia 2016 BE GR BN
xarel.lo, macabeo, parellada

90

Colour: bright golden. Nose: fine lees, fragrant herbs, ripe fruit, bakery. Palate: flavourful, good acidity, fine bead, fine bitter notes.

Muscàndia 2018 BE R EBR
xarel.lo, macabeo, parellada

89

Balanced, crisp, tasty, yeasty notes.

Muscàndia Anhel
Blanc de Noirs 2017 BE GR BN
100% pinot noir

91

Colour: straw. Nose: ripe fruit, fine lees, balanced, dried herbs. Palate: good acidity, balanced, fine bitter notes, easy to drink.

Muscàndia Magnum 2014 BE GR BN
macabeo, xarel.lo, parellada

92

Colour: bright straw. Nose: fine lees, floral, fragrant herbs, expressive. Palate: powerful, flavourful, good acidity, fine bead, balanced.

Muscàndia Rosé
Pinot Noir 2020 RE R EBR
100% pinot noir

87

NADAL
Finca Nadal de la Boadella, s/n
08775 Torrelavit (Barcelona)
☎: +34 938 988 011
comunicacio@nadal.com
www.nadal.com

Nadal Especial 2015 BE GR BN

92

Colour: bright straw. Nose: fresh fruit, citrus fruit, fine lees, fragrant herbs. Palate: fresh, fruity, good acidity, flavourful, good finish.

OLIVER VITICULTORS
Cal Xic de L'Agustí - Can Batista
08770 Subirats (Barcelona)
☎: +34 609 375 242
sadurni@oliverviticultors.com
www.oliverviticultors.com

Gemma 2017 BE GR BN
macabeo, xarel.lo, chardonnay

92

Colour: bright straw. Nose: fine lees, floral, fragrant herbs, expressive. Palate: powerful, flavourful, good acidity, fine bead, balanced.

Oliver Viticultors 2020 BE BN
macabeo, xarel.lo, parellada

89

Pleasant, kind finish, smooth, tasty.

Oliver Viticultors Rosé 2019 RE BN
garnacha, pinot noir

87

Sadurní Oliver 2018 BE R BN
macabeo, xarel.lo, parellada, chardonnay

89

Citrus fruit, balanced, dried herbs, yeasty notes, crisp.

Sadurní Oliver Barrica 2017 BE R BN
macabeo, xarel.lo, parellada, chardonnay

91

Colour: bright yellow. Nose: ripe fruit, balanced, dried herbs, lees reduction notes, toasty. Palate: good acidity, flavourful, ripe fruit, long.

DO CAVA / D.O.P.

Sadurní Oliver Rosat Pinot Noir 2020 RE BN
pinot noir

88
Kind finish, fruity, crisp, tasty.

PAGO DE THARSYS
Ctra. Nacional III, km. 274
46340 Requena (València/Valencia)
☎: +34 962 303 354
bodega@pagodetharsys.com
www.pagodetharsys.com

Carlota Suria Organic BE R BN
100% macabeo

88
Citrus fruit, standard, dried herbs, floral.

Carlota Suria Organic BE R BR
100% macabeo

88
Balanced, crisp, herbal, floral.

Pago de Tharsys 2017 BE GR BN
100% chardonnay

90
Creamy, jammy. Colour: bright straw. Nose: ripe fruit, fine lees, dried herbs, faded flowers. Palate: flavourful, good acidity, fine bead.

Pago de Tharsys Millesime 2018 BE R BR
100% chardonnay

88
Citrus fruit, standard, spicy, floral.

Pago de Tharsys Millesime Barrica 2019 BE R BN
100% chardonnay

91
Colour: straw. Nose: expressive, white flowers, fine lees, fragrant herbs, dried flowers. Palate: flavourful, fruity, toasty.

Pago de Tharsys Millésime Rosé Reserva 2018 RE R BR
100% garnacha

89
Balanced, dried flowers, jammy, tasty.

PARATÓ VINÍCOLA
Can Respall de Renardes s/n
08733 El Pla del Penedès (Barcelona)
☎: +34 938 988 182
info@parato.es
www.parato.es

Ática 2017 BE GR EBR
85% xarel.lo, 15% chardonnay

91
Colour: bright yellow. Nose: ripe fruit, fine lees, balanced, dried herbs, spicy, fine reductive notes. Palate: good acidity, flavourful, ripe fruit.

Ática Rosé 2018 RE R EBR
75% pinot noir, 25% chardonnay

89
Pleasant, kind finish, jammy, tasty.

Elias i Terns 2009 BE R BN
61% xarel.lo, 24% macabeo, chardonnay

93
Colour: bright golden. Nose: fine lees, fragrant herbs, characterful, ripe fruit, dry nuts. Palate: powerful, flavourful, good acidity, fine bead, fine bitter notes.

Finca Renardes 2019 BE R BN
32% macabeo, 33% xarel.lo, 18% parellada, 17% chardonnay

87

Parató 2018 BE R BN
35% macabeo, 30% xarel.lo, 20% parellada, 15% chardonnay

87

Parató 2018 BE R BR
32% macabeo, 36% xarel.lo, 17% parellada, 15% chardonnay

89
Oxidativ, yeasty notes, jammy, dried flowers.

PARÉS BALTÀ
Masía Can Baltá, s/n
08796 Pacs del Penedès (Barcelona)
☎: +34 938 901 399
comunicacio@paresbalta.com
www.paresbalta.com

Parés Baltà Cuvée de Carol 2013 BE GR BN
65% macabeo, 35% chardonnay

93
Age nuances. Colour: old gold. Nose: dry nuts, toasty, toasted bread, faded flowers, lees reduction notes. Palate: rich, flavourful, aged character, long.

Parés Baltà Historic 2018 BE GR BN
macabeo, xarel.lo, parellada

93 🌿

Colour: bright golden. Nose: fine lees, characterful, ripe fruit, dry nuts, sweet spices, pattiserie. Palate: powerful, flavourful, fine bead, fine bitter notes.

Parés Baltà Rosa Cusiné 2017 RE GR BN
100% garnacha

93 🌿

Colour: raspberry rose. Nose: elegant, red berry notes, floral, fragrant herbs. Palate: spicy, good acidity, fine bitter notes.

PARXET
BV- 2429, Km. 23
08793 Sant Sebastià dels Gorgs (Barcelona)
☎: +34 610 486 352
info@glevaestates.com
www.raventoscodorniu.com

Parxet 2019 BE BN
91

Colour: bright yellow. Nose: ripe fruit, fine lees, balanced, dried herbs. Palate: good acidity, flavourful, ripe fruit.

PERE VENTURA
Ctra. de Vilafranca km. 0,4
08770 Sant Sadurní d'Anoia (Barcelona)
☎: +34 938 183 371
Fax: +34 938 912 679
info@pereventura.com
www.pereventuragroup.com

🏆 PODIUM

Pere Ventura Gran Vintage Paraje Calificado Can Bas 2015 BE GR BR
50% macabeo, 50% xarel.lo

95 🌿

Colour: bright golden. Nose: fragrant herbs, characterful, ripe fruit, dry nuts, lees reduction notes. Palate: powerful, flavourful, good acidity, fine bead, fine bitter notes.

Pere Ventura Tresor 2018 BE GR BR
40% xarel.lo, 40% macabeo, 20% parellada

92

Colour: yellow. Nose: fine lees, fragrant herbs, white fruit, ripe fruit, faded flowers. Palate: flavourful, good acidity, fine bitter notes.

🏆 PODIUM

Pere Ventura Tresor Anniversary 2017 BE GR BR
40% macabeo, 40% xarel.lo, 20% parellada

95

Colour: bright golden. Nose: fine lees, dry nuts, fragrant herbs, toasty. Palate: powerful, flavourful, good acidity, fine bead, fine bitter notes.

Pere Ventura Tresor Cuvée Barrique 2018 BE GR BR
60% xarel.lo, 40% chardonnay

92

Needs time. Colour: bright straw. Nose: fine lees, floral, fragrant herbs, expressive. Palate: powerful, flavourful, good acidity, fine bead.

Pere Ventura Vintage 2015 BE GR BR
60% xarel.lo, 40% chardonnay

94

Colour: bright golden. Nose: fine lees, fragrant herbs, ripe fruit, dry nuts. Palate: flavourful, good acidity, fine bead, fine bitter notes.

PERELADA
Avda. Barcelona, 78
08720 Vilafranca del Penedès (Barcelona)
☎: +34 938 180 676
Fax: +34 938 180 926
rrpp@pereladachivite.com
www.perelada.com

Perelada Cuvée Especial 2020 BE BN
45% xarel.lo, 30% macabeo, 20% parellada, chardonnay

90 🌿

Colour: bright straw. Nose: dry nuts, fragrant herbs, spicy. Palate: flavourful, fine bead, long, great length.

DO CAVA / D.O.P.

Perelada Gran Claustro 2017 BE GR BN
60% chardonnay, 40% pinot noir

91

Colour: bright straw. Nose: medium intensity, fresh fruit, dried herbs, fine lees, floral, white fruit. Palate: fresh, fruity, flavourful, good acidity.

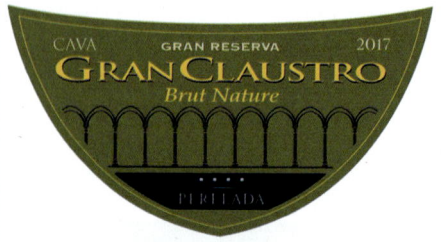

Perelada Gran Claustro Cuvée Especial 2017 BE GR BN
50% pinot noir, 25% chardonnay, 25% chardonnay

92

Colour: bright golden. Nose: fragrant herbs, characterful, ripe fruit, dry nuts. Palate: powerful, flavourful, good acidity, fine bead, fine bitter notes.

Perelada Stars 2020 BE R BN
45% xarel.lo, 20% macabeo, 15% parellada, 10% chardonnay, 10% pinot noir

89

Simple, crisp, fruity, herbal, standard, citrus fruit.

Perelada Stars Touch of Rosé 2020 RE BR
85% garnacha, 15% pinot noir

90 🍷

Colour: raspberry rose. Nose: red berry notes, floral, fragrant herbs. Palate: spicy, good acidity.

PLANAS ALBAREDA
Ctra. a Guardiola, km.3, s/n
08735 Vilobí del Penedès (Barcelona)
☎: +34 607 340 098
info@planasalbareda.com
www.planasalbareda.com

Planas Albareda 2019 BE R BN
macabeo, xarel.lo, chardonnay, parellada

90 🍷

Colour: straw. Nose: dry nuts, fragrant herbs, spicy, brioche. Palate: flavourful, fine bead, great length, fruity.

Planas Albareda 2020 BE BN
xarel.lo, macabeo, parellada

85 🍷

Planas Albareda 2020 BE BR
macabeo, xarel.lo, parellada

87 🍷

Planas Albareda Gran Reserva de L'Avi 2018 BE GR BN
macabeo, xarel.lo, chardonnay, parellada

90 🍷

Colour: bright straw. Nose: fine lees, floral, fragrant herbs. Palate: powerful, flavourful, good acidity, balanced.

Planas Albareda Rosat 2020 RE BR
pinot noir, garnacha

87 🍷

PONY FOODS
36700 Tui (Pontevedra)
☎: +34 698 145 790
info@ponyfoods.es

Gaudir 2021 BE BN
xarel.lo, macabeo, parellada

86

RIMARTS
Avda. Cal Mir, 44
08770 Sant Sadurní d'Anoia (Barcelona)
☎: +34 938 912 775
rimarts@rimarts.net
www.rimarts.net

Martínez Rosé Rimarts 2020 RE BN
garnacha, pinot noir

88 🍷

Aromatic, standard, floral, jammy, simple, fruity.

Rimarts 2017 BE GR BN
xarel.lo, macabeo, parellada, chardonnay

92 🍷

Colour: bright straw. Nose: fine lees, fragrant herbs, characterful, ripe fruit. Palate: good acidity, fine bead, fine bitter notes.

Rimarts 2018 BE R BN
xarel.lo, macabeo, parellada

90 🍷

Colour: straw. Nose: expressive, dry nuts, fragrant herbs, spicy, floral. Palate: flavourful, great length, fruity, spicy.

Rimarts 2019 BE R BR
xarel.lo, macabeo, parellada

90 🍷

Colour: bright straw. Nose: fine lees, floral, fragrant herbs, expressive. Palate: fine bead, balanced, fresh, fruity.

Rimarts Gran Reserva Especial Chardonnay 2017 BE GR BN
chardonnay

92 🍷

Colour: bright straw. Nose: fine lees, complex, spicy, dried flowers, faded flowers. Palate: powerful, flavourful, good acidity, fine bitter notes.

Rimarts Uvae 2011 BE BN
xarel.lo, chardonnay

94

Colour: bright golden. Nose: fine lees, dry nuts, fragrant herbs, complex, toasty. Palate: powerful, flavourful, good acidity, fine bead, fine bitter notes.

ROBERT J. MUR
Rambla de la Generalitat, 9
08770 Sant Sadurní d'Anoia (Barcelona)
☎: +34 938 911 551
info@robertjmur.com
www.robertjmur.com

Robert J. Mur
Ed. Especial 2012 BE GR BN
macabeo, xarel.lo, parellada, chardonnay

90

Slight oxidation. Colour: bright straw. Nose: dried herbs, ripe fruit, dry nuts. Palate: long, rich.

Robert J. Mur Especial
Tradició 2018 BE R BN
macabeo, xarel.lo, parellada

87

Robert J. Mur Especial
Tradició 2018 BE R BR
macabeo, xarel.lo, parellada

87

Robert J. Mur Especial
Tradició Rosé 2018 RE R BN
trepat, monastrell, pinot noir, garnacha

88

Standard, fruity, jammy, floral.

Robert J. Mur Especial
Tradició Rosé 2018 RE R BR
trepat, monastrell

87

Robert J. Mur Millésimé 2016 BE R BN
xarel.lo, macabeo, chardonnay

90

Colour: bright straw. Nose: fine lees, floral, fragrant herbs. Palate: good acidity, balanced, easy to drink.

ROGER GOULART (CVNE)
Major, 6
08635 Sant Esteve Sesrovires (Barcelona)
☎: +34 934 191 000
sac@rogergoulart.com
www.rogergoulart.com

Roger Goulart 2019 BE R BN
40% xarel.lo, 30% macabeo, 30% parellada

90

Colour: bright straw. Nose: fine lees, floral, fragrant herbs, expressive, dried herbs. Palate: powerful, flavourful, good acidity, fine bead, balanced.

Roger Goulart Coral Rosé 2019 RE BR
70% garnacha, 30% pinot noir

88

Fruity, jammy, tasty, floral, kind finish.

Roger Goulart Josep Valls 2017 BE EBR
35% xarel.lo, 15% chardonnay, 15% parellada

89

Aromatic, standard, fruity, dried herbs, great length, tasty.

Roger Goulart Millesimé 2020 BE BR
40% xarel.lo, 30% macabeo, 30% parellada

88

Fruity, dried flowers, dried herbs, jammy.

Roger Goulart
Rosé Millésime 2020 RE BR
85% garnacha, 10% monastrell, 5% pinot noir

88

Kind finish, fruity, tasty, dried flowers, balanced.

Roger Goulart
Semiseco Reserva 2020 BE R SS
40% xarel.lo, 30% macabeo, 30% parellada

88

Citrus fruit, crisp, herbal, jammy, sweetly finish.

ROSELL GALLART
Montserrat, 56
08770 Sant Sadurní d'Anoia (Barcelona)
☎: +34 938 912 073
info@rosellgallart.com
www.rosellgallart.com

Rosell Gallart BE R BN

88

Fruity, lactic, jammy, simple.

DO CAVA / D.O.P.

DO CAVA / D.O.P.

ROVELLATS
Bº La Bleda
08731 Sant Marti Sarroca (Barcelona)
☎: +34 934 880 575
Fax: +34 934 880 819
rovellats@cavasrovellats.com
www.rovellats.com

Rovellats Col.lecció 2016 BE GR EBR
xarel.lo, chardonnay, parellada

92

Colour: straw. Nose: expressive, fragrant herbs, spicy, floral. Palate: flavourful, fine bead, long, great length.

Rovellats Cuvée Especial 2019 BE R BN
macabeo, xarel.lo, parellada

89

Balanced, citrus fruit, dried flowers, dried herbs, tasty.

Rovellats Gran Reserva 2016 BE GR BN
macabeo, xarel.lo, parellada

90

Colour: bright straw. Nose: fine lees, fragrant herbs, ripe fruit, dry nuts, spicy. Palate: powerful, flavourful, good acidity, fine bitter notes.

Rovellats Masia S. XV 2012 BE GR BN
macabeo, xarel.lo, parellada, chardonnay

92

Colour: bright straw. Nose: fine lees, fragrant herbs, dry nuts, pattiserie. Palate: flavourful, good acidity, fine bead, balanced.

Rovellats Reserva Imperial 2019 BE BR
macabeo, xarel.lo, parellada

89

Crisp, dried herbs, wild, smooth, yeasty notes, standard.

Rovellats Reserva Imperial Rosé 2019 RE R
garnacha

88

Dried flowers, fruity, tasty, standard, pleasant, simple.

SEGURA VIUDAS
Ctra. Sant Sadurní a St. Pere de Riudebitlles, Km. 5
08775 Torrelavit (Barcelona)
☎: +34 938 917 070
comunicacion@freixenet.es
www.seguraviudas.com

Segura Viudas BE R BR

89

Herbal, crisp, standard, yeasty notes, simple, smooth.

Segura Viudas Lavit Brut Nature 2020 BE R BN
macabeo, parellada

88

Pleasant, dried flowers, jammy, tasty.

Segura Viudas Vintage 2015 BE GR BR
macabeo, parellada

89

Pleasant, standard, yeasty notes, simple, great length, dried herbs.

SOGAS MASCARÓ
Barri Las Tarumbas, 4
08733 Pla del Penedès (Barcelona)
☎: +34 931 022 212
info@sogasmascaro.com
www.sogasmascaro.com

Sogas Mascaró BE BN
40% macabeo, 30% xarel.lo, 30% parellada

88

Austere, crisp, herbal, smooth.

Sogas Mascaró BE BR
40% macabeo, 30% xarel.lo, 30% parellada

88

Kind finish, standard, dried flowers, dried herbs, simple. Palate: fine bitter notes.

Sogas Mascaró BE R BN
30% macabeo, 40% xarel.lo, 30% parellada

89 🌱

Wild, crisp, herbal, fruity, pleasant, balanced.

SUMARROCA
Masia Molí Coloma. Barri el Rebato s/n
08739 Subirats (Barcelona)
☎: +34 938 911 092
sumarroca@sumarroca.com
www.sumarroca.com

Núria Claverol Allier 2015 BE GR BR
chardonnay

91

Colour: bright yellow. Nose: ripe fruit, fine lees, dried herbs, toasty. Palate: good acidity, flavourful, ripe fruit.

Núria Claverol Blanc de Noirs 2015 BE GR BR
pinot noir

93

Colour: bright golden. Nose: dry nuts, fragrant herbs, complex, toasty. Palate: flavourful, good acidity, fine bead, fine bitter notes.

Núria Claverol
Homenatge 2015 BE GR BR
xarel.lo

93

Colour: bright golden. Nose: fine lees, dry nuts, fragrant herbs, complex, white flowers. Palate: powerful, flavourful, good acidity, fine bead.

Sumarroca 2018 BE GR BN
xarel.lo, macabeo, parellada

90

Colour: bright yellow. Nose: ripe fruit, fine lees, balanced, dried herbs, brioche. Palate: good acidity, flavourful, ripe fruit.

Sumarroca Ecológic 2019 BE R BR
xarel.lo, macabeo, parellada, chardonnay

90

Colour: bright straw. Nose: fine lees, floral, fragrant herbs, expressive, lactic notes, white fruit. Palate: flavourful, good acidity, fine bead, balanced.

TERRA DE FALANIS
Ctra. Campos Felanitx 10.3 Km.
07200 Felanitx (Illes Balears/Islas Baleares)
☎: +34 679 314 406
contactoterradefalanis@gmail.com
www.terradefalanis.com

Tutum Ba 2019 BE BN
xarel.lo, macabeo, parellada

89

Pleasant, crisp, citrus fruit, yeasty notes, smooth, balanced.

TRIAS BATLLE
Pere El Gran, 24
08720 Vilafranca del Penedès (Barcelona)
☎: +34 677 497 892
peptrias@jtrias.com
www.jtrias.com

Trias Batlle 2020 BE R BN
macabeo, xarel.lo, parellada

90

Colour: bright yellow. Nose: ripe fruit, fine lees, balanced, dried herbs. Palate: good acidity, flavourful, ripe fruit, long.

Trias Batlle 2020 BE R BR
macabeo, xarel.lo, parellada

88

Citrus fruit, fruity, dried flowers, tasty.

Trias Batlle Blaué 2018 BE GR BN
xarel.lo, macabeo, parellada, chardonnay

90

Colour: bright straw. Nose: fine lees, floral, fragrant herbs, expressive. Palate: powerful, flavourful, good acidity, fine bead.

Trias Batlle Rosé 2020 RE BR
garnacha, trepat, pinot noir

88

Pleasant, jammy, tasty.

UNIÓN VINÍCOLA DEL ESTE
Construcción 74, Pl. Ind. El Romeral
46340 Requena (València/Valencia)
☎: +34 962 323 343
cava@uveste.es
www.uveste.es

Marevia 2019 BE BR
50% chardonnay, 50% pinot noir

88

Citrus fruit, floral, standard, simple, smooth, kind finish.

Vega Medien Ecológico BE BR
75% macabeo, 25% chardonnay

87

Vega Medien Rosé RE BR
100% garnacha

86

VALLDOLINA VITICULTORS I ELABORADORS
Pl. de la Creu 1 Masia Can Tutusaus
08795 Olesa de Bonesvalls (Barcelona)
☎: +34 938 984 181
info@valldolina.com
www.valldolina.com

Tutusaus 2017 BE GR BN
xarel.lo, macabeo, parellada, chardonnay

90

Colour: bright straw. Nose: fine lees, floral, fragrant herbs, expressive. Palate: powerful, flavourful, good acidity, fine bead.

VallDolina 2018 BE GR BR
xarel.lo, macabeo, parellada, chardonnay

91

Colour: bright golden. Nose: fine lees, dry nuts, fragrant herbs, complex, lactic notes. Palate: flavourful, good acidity, fine bead.

DO CAVA / D.O.P.

DO CAVA / D.O.P.

VallDolina 2019 BE R BN
xarel.lo, macabeo, parellada, chardonnay

90 ♣
Colour: bright yellow. Nose: ripe fruit, fine lees, balanced, dried herbs. Palate: good acidity, flavourful, ripe fruit, long.

VARIAS
Plaça Manuel Raventós, 6
08770 Sant Sadurní d'Anoia (Barcelona)
☎: +34 938 912 763
info@cavavarias.es
www.cavavarias.com

Varias Al·legoria 2015 BE R BN
macabeo, xarel.lo, parellada, chardonnay

91
Colour: bright golden. Nose: characterful, ripe fruit, dry nuts, elegant. Palate: flavourful, fine bitter notes, balanced.

Varias Al·legoria 2015 BE R BR
macabeo, xarel.lo, parellada, chardonnay

90
Colour: straw. Nose: expressive, dry nuts, spicy, ripe fruit. Palate: flavourful, fine bead, fine bitter notes.

Varias Cuvée Clássic 2008 BE GR BN
macabeo, xarel.lo, parellada

90
Colour: bright straw. Nose: dried herbs, fine lees, floral, candied fruit, stone fruit. Palate: flavourful, good acidity.

Varias Cuvée Imperial 2009 BE GR BN
macabeo, xarel.lo, chardonnay

92
Colour: bright straw. Nose: fine lees, floral, fragrant herbs, expressive, white flowers. Palate: powerful, flavourful, good acidity, fine bead.

Varias Edició Limitada Xarel.lo 2008 BE GR BN
100% xarel.lo

93
Colour: bright golden. Nose: fine lees, fragrant herbs, ripe fruit, dry nuts, caramel. Palate: powerful, flavourful, good acidity, fine bead, fine bitter notes.

Varias Genuí 2019 BE BN
macabeo, xarel.lo, parellada

87

VICENTE GANDÍA
Ctra. Cheste a Godelleta, s/n
46370 Chiva (València/Valencia)
☎: +34 962 524 242
Fax: +34 962 524 243
info@vicentegandia.com
www.vicentegandia.es

El Miracle BE BR
macabeo, chardonnay

86

El Miracle Organic BE BR
macabeo, chardonnay

87 ♣

El Miracle Rosé RE BR
garnacha

86

Hoya de Cadenas BE BN
macabeo, chardonnay

85

Hoya de Cadenas BE BR
macabeo, chardonnay

86

Hoya de Cadenas Rosado RE BR
garnacha

86

VID VICA

08770 Sant Sadurní d' Anoia (Barcelona)
☎: +34 666 592 641
Fax: +34 932 510 872
nicole@vidvica.com

Arestel BE BR
40% macabeo, 30% parellada, 30% xarel.lo

85 🌱

Arestel BE SS
40% macabeo, 30% parellada, 30% xarel.lo

85

Arestel Vintage 2019 BE EBR
40% macabeo, 30% parellada, 30% xarel.lo

87

Amorany 2016 BE GR BN
40% macabeo, 30% parellada, 30% xarel.lo

91

Colour: bright yellow. Nose: ripe fruit, fine lees, balanced, dried herbs. Palate: good acidity, flavourful, ripe fruit, long.

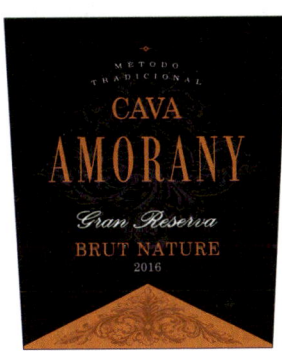

VILARNAU

Ctra. d'Espiells, Km. 1,4 Finca "Can Petit"
08770 Sant Sadurní d'Anoia (Barcelona)
☎: +34 938 912 361
prensa@gonzalezbyass.es
www.vilarnau.es

Albert de Vilarnau Chardonnay Pinot Noir 2015 BE GR BN
50% chardonnay, 50% pinot noir

92

Colour: bright golden. Nose: fine lees, fragrant herbs, ripe fruit, grassy. Palate: flavourful, good acidity, fine bead, fine bitter notes.

Albert de Vilarnau Xarel.lo Castanyer 2016 BE GR BN
xarel.lo

90

Colour: bright golden. Nose: fine lees, ripe fruit, dry nuts, candied fruit, faded flowers. Palate: flavourful, fine bead, fine bitter notes, spicy.

Vilarnau 2019 BE R BN
50% macabeo, 35% parellada, 15% chardonnay

88 🌱

Citrus fruit, fruity, dried herbs, simple.

Vilarnau Brut Rosé Delicat 2020 RE R BR
85% garnacha, 15% pinot noir

88 🌱

Fruity, dried flowers, dried herbs, tasty.

Vilarnau Vintage 2014 BE GR BN
40% macabeo, 30% parellada, 25% chardonnay, 5% pinot noir

92

Colour: bright straw. Nose: fine lees, fragrant herbs, spicy, toasted bread. Palate: flavourful, good acidity, balanced, fine bitter notes.

VINÍCOLA DE NULLES -ADERNATS

Raval de Sant Joan, 7
43887 Nulles (Tarragona)
☎: +34 977 602 622
botiga@vinicoladenulles.com
www.adernats.cat

Adernats 2018 BE R BN
macabeo, xarel.lo, parellada

88

Citrus fruit, crisp, dried herbs, standard.

Adernats 2018 BE R BR
macabeo, xarel.lo, parellada

90

Colour: bright straw. Nose: fine lees, floral, fragrant herbs, expressive. Palate: powerful, flavourful, good acidity, fine bead, balanced.

DO CAVA / D.O.P.

DO CAVA / D.O.P.

Adernats Purn Gran Reserva 2012 BE GR BN
50% macabeo, 50% xarel.lo

93

Colour: bright golden. Nose: fragrant herbs, characterful, ripe fruit, dry nuts. Palate: flavourful, fine bead, fine bitter notes, balanced.

Adernats XC 2015 BE GR BN
100% xarel.lo

91

Age nuances. Colour: bright yellow. Nose: ripe fruit, fine lees, balanced, dried herbs, herbaceous, fine reductive notes. Palate: flavourful, ripe fruit, balanced.

VINS EL CEP

Can Llopart de Les Alzines, Ctra. Espiells
08770 Sant Sadurní d'Anoia (Barcelona)
☎: +34 938 912 353
info@vinselcep.com
www.vinselcep.com

Claror Paraje Calificado Can Prats 2014 BE GR BN
xarel.lo, macabeo, parellada

90

Colour: yellow. Nose: fine lees, dry nuts, ripe fruit, candied fruit. Palate: flavourful, good acidity, fine bead, fine bitter notes.

Clos Gelida 4 Heretats 2017 BE GR BN
90

Standard, crisp, simple, tasty, dried herbs.

Mim Natura Blanc de Noirs 2016 BE GR BN
89

Pleasant, defined aromas, balanced, herbal, dried herbs, great length, tasty.

Mim Natura Pinot Noir Rosado 2019 RE R BR
pinot noir

89 🌿

Crisp, floral, fruity, tasty, balanced, varietally correct.

VIVES AMBRÒS

Mayor, 39
43812 Montferri (Tarragona)
☎: +34 639 521 652
mail@vivesambros.com
www.vivesambros.com

Vives Ambròs 2018 BE GR BN
91

Colour: bright golden. Nose: fine lees, ripe fruit, dry nuts, dried flowers. Palate: powerful, flavourful, good acidity, fine bead, fine bitter notes.

Vives Ambròs 2018 BE R BR
40% xarel.lo, 35% macabeo, 25% parellada

90

Colour: bright straw. Nose: fine lees, fragrant herbs, expressive, spicy, candied fruit. Palate: powerful, flavourful, good acidity, fine bead, balanced.

Vives Ambròs Jujol 2017 BE GR BN
100% xarel.lo

92

Colour: bright golden. Nose: fine lees, fragrant herbs, ripe fruit, dry nuts, toasty, sweet spices. Palate: flavourful, good acidity, fine bead, fine bitter notes.

Vives Ambròs Rosat 2019 RE R BR
40% garnacha, 40% monastrell, 20% pinot noir

88

Dried flowers, lactic, tasty, herbal.

DO. CIGALES
CONSEJO REGULADOR

Corro Vaca, 5
47270 Cigales (Valladolid)
☎: +34 983 580 074 - Fax: +34 983 586 590
@: consejo@do-cigales.es
www.do-cigales.es

LOCATION:

The region stretches to the north of the Duero depression and on both sides of the Pisuerga, bordered by the Cérvalos and the Torozos hills. The vineyards are situated at an altitude of 750 m; the DO extends from part of the municipal area of Valladolid (the wine estate known as 'El Berrocal') to the municipality of Dueñas in Palencia, also including Cabezón de Pisuerga, Cigales, Corcos del Valle, Cubillas de Santa Marte, Fuensaldaña, Mucientes, Quintanilla de Trigueros, San Martín de Valvení, Santovenia de Pisuerga, Trigueros del Valle and Valoria la Buena.

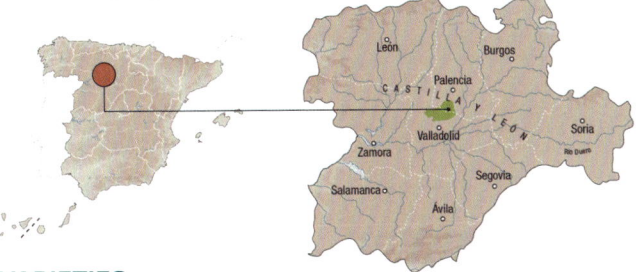

GRAPE VARIETIES:

WHITE: Verdejo, Albillo, Sauvignon Blanc and Viura.

RED: Tinta del País (Tempranillo), Garnacha Tinta, Garnacha Gris, Merlot, Syrah and Cabernet Sauvignon.

FIGURES:

Vineyard surface: 1,890 – **Wine-Growers:** 297 – **Wineries:** 30 – **2021 Harvest rating:** Excellent – **Production 21:** 5,630,381 litres – **Market percentages:** 75% National - 25% International.

SOIL:

The soil is sandy and limy with clay loam which is settled on clay and marl. It has an extremely variable limestone content which, depending on the different regions, ranges between 1% and 35%.

CLIMATE:

The climate is continental with Atlantic influences, and is marked by great contrasts in temperature, both yearly and day/night. The summers are extremely dry; the winters are harsh and prolonged, with frequent frost and fog; rainfall is irregular.

CLASSIFICATION OF YOUNG WINE HARVESTS PEÑÍNGUIDE

2017	2018	2019	2020	2021
GOOD	GOOD	VERY GOOD	VERY GOOD	VERY GOOD

DO CIGALES / D.O.P.

AVELINO VEGAS
Grupo Calvo Sotelo, 8
40460 Santiuste de San Juan Bautista (Segovia)
☎: +34 921 596 002
ana@avelinovegas.com
www.avelinovegas.com

Zarzales 2021 RD
tempranillo, garnacha, verdejo, albillo
89
Pleasant, aromatic, fruity, tasty.

BODEGA CÉSAR PRÍNCIPE
Ctra. Fuensaldaña-Mucientes, s/n
47194 Fuensaldaña (Valladolid)
☎: +34 983 663 123
cesarprincipe@cesarprincipe.es
www.cesarprincipe.es

🏆 PODIUM

César Príncipe 2019 T C
100% tempranillo
95
Colour: very deep cherry. Nose: complex, expressive, spicy, mineral. Palate: elegant, full-bodied, long, great length.

BODEGA COOPERATIVA CIGALES
Las Bodegas, s/n
47270 Cigales (Valladolid)
☎: +34 983 580 135
administracion@bodegacigales.com
www.bodegacigales.com

Gran Torondos 2020 T
tempranillo, garnacha
90
Colour: deep cherry. Nose: ripe fruit, dried herbs, elegant. Palate: powerful, ripe fruit, round tannins.

Gran Torondos 2021 RD
89
Pleasant, aromatic, floral, fruity, jammy.

Parradez 2021 RD
tempranillo, garnacha, albillo, verdejo
88
Kind finish, pleasant, aromatic, smooth.

Torondos Clarete 2021 RD
tempranillo, garnacha, albillo, verdejo
89
Pleasant, aromatic, fruity, jammy.

Torondos Rosé 2021 RD
tempranillo, garnacha, albillo, verdejo
87

BODEGA HIRIART
Tejar, s/n
47270 Cigales (Valladolid)
☎: +34 609 714 402
carlos@bodegahiriart.es
www.bodegahiriart.es

Candiles de Hiriart 2020 T C
89
Toasty, jammy, spicy, fruity.

Hiriart 2016 T C
90
Colour: cherry, garnet rim. Nose: fruit preserve, fruit liqueur notes, powerful, roasted coffee. Palate: flavourful, sweetness, long.

Hiriart 2021 RD
90
Colour: brilliant rose. Nose: fruit expression, red berry notes, floral. Palate: fruity, good acidity, easy to drink.

Hiriart Élite 2021 RD
89
Aromatic, kind finish, smooth, thin.

BODEGA MUSEUM
Ctra. Cigales a Corcos, km. 3
47270 Cigales (Valladolid)
☎: +34 983 581 029
Fax: +34 983 581 030
info@bodegasmuseum.com
www.bodegasmuseum.com

La Renacida 2020 T
garnacha, tempranillo, bobal, mencía, verdejo, albillo mayor
92
Colour: deep cherry. Nose: ripe fruit, dried herbs, dark chocolate, sweet spices, new oak. Palate: powerful, ripe fruit, spicy, round tannins.

Las Musas 2021 RD
verdejo, garnacha, albillo mayor, tempranillo

92

Colour: salmon. Nose: fragrant herbs, elegant, white flowers, dried flowers. Palate: light-bodied, spicy, good acidity, fine bitter notes.

Museum 2018 T R
tempranillo

91

Colour: very deep cherry. Nose: roasted coffee, aromatic coffee, powerful, ripe fruit. Palate: smoky aftertaste, great length, round tannins.

Numerus Clausus 2019 T
tempranillo

93

Colour: bright cherry. Nose: complex, expressive, spicy, mineral, ripe fruit. Palate: elegant, full-bodied, long, great length.

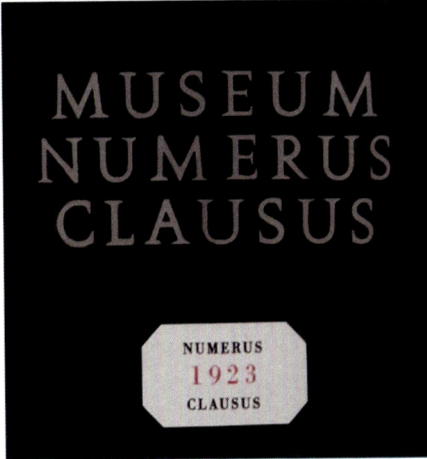

Vinea 2020 T C
tempranillo

91

Colour: deep cherry. Nose: ripe fruit, dried herbs, creamy oak, fruit expression. Palate: powerful, ripe fruit, spicy, round tannins.

Vinea 2021 RD
tempranillo, verdejo

91

Colour: raspberry rose. Nose: elegant, fragrant herbs, faded flowers. Palate: light-bodied, good acidity, fine bitter notes.

BODEGA VALDELOSFRAILES
Camino de Cubillas s/n
47290 Cubillas de Santa Marta (Valladolid)
☎: +34 983 485 028
Fax: +34 983 485 024
valdelosfrailes@matarromera.es
www.valdelosfrailes.es

Valdelosfrailes 2016 T R
tempranillo

93

Colour: dark-red cherry, garnet rim. Nose: ripe fruit, aged wood nuances, tobacco, sweet spices, black fruit. Palate: spicy, round tannins, long.

Valdelosfrailes 2017 T C
100% tempranillo

88

Pruney, powerful, roasted coffee, hot.

Valdelosfrailes Clarete 2021 RD
tempranillo, verdejo, garnacha, merlot, cabernet sauvignon

90

Colour: rose. Nose: ripe fruit, faded flowers. Palate: fleshy, flavourful, ripe fruit.

BODEGAS EMINA
Cº de Cubillas, s/n
47290 Cubillas de Santa Marta (Valladolid)
☎: +34 983 485 028
Fax: +34 983 485 024
valdelosfrailes@matarromera.es
www.valdelosfrailes.es

Emina Rosé 2021 RD
tempranillo, verdejo, garnacha, merlot, cabernet sauvignon

88

Jammy, fruity, tasty, floral.

Emina Rosé Prestigio 2021 RD
tempranillo, verdejo, garnacha tintorera, garnacha gris, albillo, viura

90

Colour: raspberry rose. Nose: elegant, red berry notes, floral, fragrant herbs. Palate: light-bodied, spicy, good acidity, fine bitter notes.

BODEGAS HIJOS DE FÉLIX SALAS
Corrales, s/n
47280 Corcos del Valle (Valladolid)
☎: +34 616 099 148
Fax: +34 983 580 262
bodega@bodegasfelixsalas.com
www.bodegasfelixsalas.com

Félix Salas 2018 T C
tempranillo

88

Pleasant, toasty, tasty, jammy.

DO CIGALES / D.O.P.

Viña Picota 2021 RD
tempranillo, albillo, verdejo, syrah

87

BODEGAS LEZCANO-LACALLE
Ctra. de Valoria, s/n
47282 Trigueros del Valle (Valladolid)
☎: +34 629 280 515
info@lezcano-lacalle.com
www.lezcano-lacalle.com

Lezcano-Lacalle 2017 T R
tempranillo, merlot

92

Colour: dark-red cherry, garnet rim. Nose: aged wood nuances, tobacco, sweet spices. Palate: spicy, round tannins, long.

Lezcano-Lacalle Dú 2015 T
60% tempranillo, 40% merlot

92

Colour: deep cherry, garnet rim. Nose: aged wood nuances, cocoa bean, cigar, toasty, hot. Palate: flavourful, spicy, toasty, powerful tannins.

Maudes 2018 T C
tempranillo, merlot

90

Colour: dark-red cherry. Nose: toasty, spicy, cocoa bean. Palate: flavourful, toasty, fine bitter notes.

BODEGAS MUCY
Ctra. Mucientes-Villalba, km. 1
47194 Mucientes (Valladolid)
☎: +34 692 944 470
info@mucy.es
www.mucy.es

Alpairo 2021 RD
tempranillo, garnacha gris, albillo

88

Elegant, floral, crisp, fruity, thin.

Mucy 2018 T C
tempranillo

89

Toasty, jammy, spicy, tasty.

Mucy 2021 RD
tempranillo, verdejo

88

Pleasant, aromatic, jammy, tasty.

Paño de Lágrimas 2020 T RB
tempranillo

89

Pleasant, spicy, fruity, tasty.

BODEGAS PROTOS
Ctra. VA-VP-4405 km. 3
47290 Cubillas de Santa Marta (Valladolid)
☎: +34 629 024 170
fvillalba@bodegasprotos.com
www.bodegasprotos.com

Aire de Protos 2021 RD
60% tempranillo, 10% garnacha, 10% albillo, 5% verdejo, 5% viura, 5% sauvignon blanc

91

Colour: raspberry rose. Nose: elegant, red berry notes, floral, fragrant herbs. Palate: light-bodied, spicy, good acidity, fine bitter notes.

Protos Clarete 2021 RD
tempranillo, syrah, merlot

92

Colour: brilliant rose. Nose: fruit expression, red berry notes, floral, characterful, expressive. Palate: fruity, good acidity, easy to drink.

Protos Rosé 2021 RD
70% tempranillo, 25% syrah

89 🌿

Pleasant, aromatic, floral, tasty.

BODEGAS SALVUEROS
Ctra. Mucientes Cigales, km. 12,8
47194 Mucientes (Valladolid)
☎: +34 625 115 619
bodegas@salvueros.com
www.salvueros.com

La Guerrera Finca Centenaria 2017 T C
tempranillo

91

Colour: dark-red cherry. Nose: toasty, spicy, cocoa bean. Palate: flavourful, toasty, fine bitter notes.

La Guerrera Finca Centenaria 2019 T MC
tempranillo

93

Colour: Cherry. Nose: balsamic herbs, sweet spices, scrubland, red berry notes, fruit expression, toasty. Palate: spicy, balsamic, good acidity.

Salvueros 2020 T RB
tempranillo

90

Colour: deep cherry. Nose: dried herbs, creamy oak, toasty, black fruit. Palate: powerful, ripe fruit, spicy, round tannins.

Salvueros 2021 RD
tempranillo, albillo, verdejo

91

Colour: brilliant rose, purple rim. Nose: fruit expression, red berry notes, floral, characterful. Palate: balanced, fruity aftertaste.

DO CIGALES / D.O.P.

Salvueros Garnacha Gris 2021 RD
100% garnacha gris

91

Colour: raspberry rose. Nose: elegant, floral, fragrant herbs. Palate: light-bodied, good acidity, fine bitter notes.

BODEGAS SINFORIANO
47194 Mucientes (Valladolid)
☎: +34 983 663 008
sinfo@sinforianobodegas.com
www.sinforianobodegas.com

50 Vendimias de Sinforiano 2015 T
100% tempranillo

93

Colour: deep cherry, garnet rim. Nose: aged wood nuances, ripe fruit, cocoa bean, cigar, toasty. Palate: flavourful, spicy, toasty, powerful tannins.

Liala 2020 RD FB
albillo, garnacha, garnacha gris, garnacha tintorera, verdejo, tempranillo

92

Colour: salmon. Nose: sweet spices, red berry notes, fragrant herbs, dried flowers. Palate: full-bodied, flavourful, spicy, sweetness, long.

Quelías Rosé 2021 RD
50% albillo, 30% garnacha, 10% verdejo, 10% tempranillo

91

Colour: raspberry rose. Nose: elegant, floral, fragrant herbs. Palate: light-bodied, spicy, fine bitter notes.

Sinfo 2020 RD FB
80% tempranillo, 10% verdejo, 10% albillo

90

Toasty, tasty, standard, kind finish. Colour: light cherry.

Sinforiano 2016 T R
100% tempranillo

92

Colour: dark-red cherry. Nose: toasty, cocoa bean, black fruit, sweet spices. Palate: flavourful, toasty, fine bitter notes.

Sinforiano 2018 T C
100% tempranillo

89

Classic, toasty, hot, jammy.

BODEGAS THESAURUS CIGALES
Ctra. Valladolid, 14
47270 Olivares de Duero (Valladolid)
☎: +34 983 250 319
comercial@ciadevinos.com
www.bodegasthesaurus.com

Domine RD S
100% tempranillo

88

Floral, fruity, tasty.

Viña Goy 2021 RD
100% tempranillo

89

Pleasant, fruity, jammy, tasty.

BODEGAS Y VIÑEDOS ALFREDO SANTAMARÍA
Poniente, 18
47290 Cubillas de Santa Marta (Valladolid)
☎: +34 615 052 287
info@bodega-santamaria.com
www.bodega-santamaria.com

Pago el Cordonero Garnacha Albillo 2020 RD FB
garnacha, albillo

89

Spicy, dried flowers, jammy, tasty.

Pago el Cordonero Tempranillo 12 meses 2018 T BA
tempranillo

86

Pago el Cordonero Tempranillo 9 meses 2018 T
tempranillo

88

Slight reduction, pruney, toasty, powerful.

Pago el Cordonero Verdejo 2020 B FB
verdejo

85

Valvinoso Tempranillo 2021 RD
tempranillo

89

Dried flowers, fruity, jammy, tasty.

DO CIGALES / D.O.P.

BODEGAS Y VIÑEDOS ROSAN
47270 Cigales (Valladolid)
☎: +34 983 580 006
rodriguezsanz@telefonica.net

Albéitar 2021 T
tinta del país
87

Rosan 2021 RD
88
Aromatic, floral, fruity, jammy, tasty.

CONCEJO BODEGAS
Ctra. Valoria, Km. 3.6
47200 Valoria La Buena (Valladolid)
☎: +34 983 502 263
info@concejobodegas.com
www.concejobodegas.com

Carredueñas 2019 T C
100% tempranillo
89 ♣
Tasty, powerful, hot, jammy, toasty.

Carredueñas 2020 RD FB
100% tempranillo
90
Colour: light cherry. Nose: powerful, ripe fruit, sweet spices, toasty. Palate: fleshy, flavourful, toasty.

Carredueñas 2021 RD
90
Colour: rose, purple rim. Nose: fruit expression, red berry notes, floral. Palate: fruity, good acidity, easy to drink.

Concejo 2017 T R
100% tempranillo
92 ♣
Colour: cherry, garnet rim. Nose: fruit preserve, fruit liqueur notes, powerful, creamy oak, dark chocolate. Palate: flavourful, sweetness, long.

FRUTOS VILLAR
Camino Los Barreros, s/n
47270 Cigales (Valladolid)
☎: +34 983 586 868
Fax: +34 983 580 180
bodegasfrutosvillar@bodegasfrutosvillar.com
www.bodegasfrutosvillar.com

Conde Ansúrez 2021 RD
100% tempranillo
88
Pleasant, floral, fruity, jammy.

Viña Calderona 2021 RD
merlot, tempranillo
89
Pleasant, aromatic, floral, fruity, crisp.

Viña Cansina 2021 RD
100% tempranillo
88
Aromatic, kind finish, jammy, tasty.

HIJOS DE CRESCENCIA MERINO
47280 Corcos del Valle (Valladolid)
☎: +34 627 269 771
info@bodegashcmerino.com
www.bodegascatajarros.es/com

Viña Catajarros 2021 RD
tempranillo, garnacha, verdejo
90
Colour: rose, purple rim. Nose: fruit expression, red berry notes, floral. Palate: fruity, good acidity, easy to drink.

HIJOS DE RUFINO IGLESIAS
La Canoniga, 25
47194 Mucientes (Valladolid)
☎: +34 983 587 778
Fax: +34 983 587 778
bodega@hijosderufinoiglesias.com
www.hijosderufinoiglesias.com

Carratraviesa 2021 RD
80% tempranillo, 10% garnacha, 10% albillo, verdejo
90
Colour: brilliant rose, purple rim. Nose: red berry notes, lactic notes, floral, characterful, expressive. Palate: balanced, fruity aftertaste.

Millatos 2020 T
100% tempranillo
84

LAGAR DEL DUQUE
Cº de la Ermita
47194 Fuensaldaña (Valladolid)
☎: +34 685 181 363
info@bodegalagardelduque.com
www.bodegalagardelduque.com

Lagar del Duque 2021 RD
tempranillo, verdejo, garnacha
87

OVIDIO GARCÍA
Zona de Bodegas Malpique s/n
47270 Cigales (Valladolid)
☎: +34 628 509 475
patricia@ovidiogarcia.com
www.ovidiogarcia.com

Ovidio García de Autor 2016 T R
100% tempranillo
91
Colour: cherry, purple rim. Nose: fruit expression, floral, spicy, ripe fruit. Palate: flavourful, fruity, good acidity, long.

Ovidio García Esencia 2018 T C
100% tempranillo
89
Nose: roasted coffee, aromatic coffee, powerful, ripe fruit. Palate: smoky aftertaste, great length.

Ovidio García Selección 2020 T RB
100% tempranillo
88
Pleasant, tasty, spicy, fruity.

TRASLANZAS BODEGAS Y VIÑEDOS
Peñaflor, 11
47194 Mucientes (Valladolid)
☎: +34 658 862 416
info@traslanzas.com
www.traslanzas.com

Remolón Rosé 2021 RD PL S
tempranillo, albillo, verdejo, garnacha, viura
91
Colour: raspberry rose. Nose: elegant, red berry notes, floral, fragrant herbs. Palate: light-bodied, spicy, good acidity, fine bitter notes.

Traslanzas 2018 T
tempranillo
92
Colour: deep cherry. Nose: ripe fruit, dried herbs, creamy oak. Palate: powerful, ripe fruit, spicy, round tannins.

Traslanzas 2021 RD
tempranillo, albillo
92
Colour: rose. Nose: ripe fruit, faded flowers, sweet spices. Palate: fleshy, flavourful, powerful, ripe fruit.

Traslanzas V+A 2020 B
verdejo, albillo
91
Colour: bright yellow. Nose: powerful, ripe fruit, spicy. Palate: good structure, long, toasty, fine bitter notes.

Tres Cuestas 2018 T BA S
tempranillo
89
Spicy, toasty, powerful, jammy.

DO CIGALES / D.O.P.

DO. CONCA DE BARBERÀ

CONSEJO REGULADOR

Torre del Portal de Sant Antoni De la Volta, 2
43400 Montblanc
☎: +34 977 926 905 - Fax: +34 977 926 906
@: cr@doconcadebarbera.com
www.doconcadebarbera.com

LOCATION:

In the north of the province of Tarragona with a production area covering 14 municipalities, to which two new ones have recently been added: Savallà del Comtat and Vilanova de Prades.

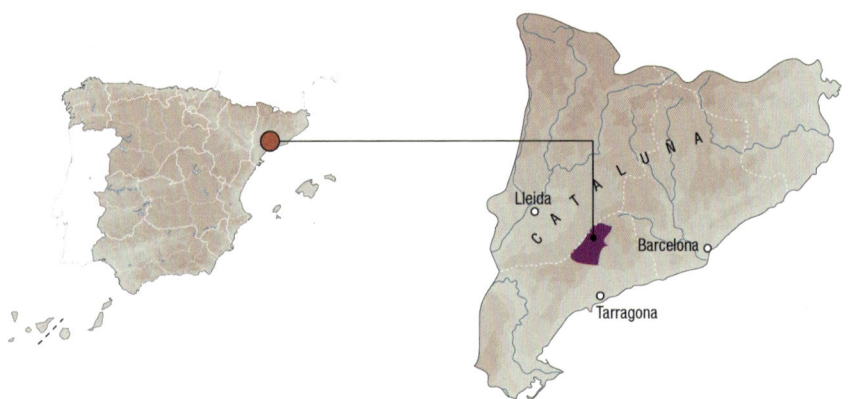

GRAPE VARIETIES:

WHITE: Macabeo, Parellada (majority 3,300 Ha) Chardonnay, Sauvignon Blanc, Viognier, Sumoll Blanc.

RED: Trepat, Ull de Llebre (Tempranillo), Garnatxa, Cabernet Sauvignon, Merlot, Syrah, Pinot Noir, Sumoll Negre.

FIGURES:

Vineyard surface: 2,770 – **Wine-Growers:** 674 – **Wineries:** 24 – **2021 Harvest rating:** Excellent – **Production 21:** 750,000 litres – **Market percentages:** 81% National - 19% International.

SOIL:

The soil is mainly brownish-grey and limy. The vines are cultivated on slopes protected by woodland. An important aspect is the altitude which gives the wines a fresh, light character.

CLIMATE:

Mediterranean and continental influences, as the vineyards occupy a river valley surrounded by mountain ranges without direct contact with the sea.

CLASSIFICATION OF YOUNG WINE HARVESTS PEÑÍNGUIDE

2017	2018	2019	2020	2021
GOOD	GOOD	GOOD	GOOD	VERY GOOD

ABADÍA DE POBLET
Passeig Abat Conill, 6
43448 Poblet (Tarragona)
☎: +34 977 870 358
j.pujol@codorniu.com
www.codorniu.com

Abadía de Poblet Blanc 2019 B
macabeo, parellada

92

Colour: bright straw. Nose: fragrant herbs, fine lees, white fruit, ripe fruit. Palate: full-bodied, long, good acidity, spicy.

Abadía de Poblet Negre 2018 T
trepat, garrut, garnacha

91

Colour: cherry, purple rim. Nose: red berry notes, floral, spicy, black fruit, ripe fruit. Palate: flavourful, fruity, good acidity, long.

El Tossal de la Salut 2019 T
garnacha

94

Colour: cherry, purple rim. Nose: fruit expression, red berry notes, floral, spicy, hint of anise, balsamic herbs, smoky. Palate: flavourful, fruity, good acidity, long.

La Font Voltada 2019 T C
trepat

92

Colour: cherry, purple rim. Nose: red berry notes, floral, spicy, black liquorice, scrubland. Palate: flavourful, fruity, good acidity, long.

BODEGA SANSTRAVÉ
De la Conca, 10
43412 Solivella (Tarragona)
☎: +34 977 892 165
bodega@sanstrave.com
www.sanstrave.com

Camí de Tardor 2021 RD
trepat

86

Llum de Vi Blanc 2021 B
chardonnay, moscatel

88

Kind finish, jammy, full-bodied, floral.

BODEGA VEGA AIXALÁ
De la Font, 11
43439 Vilanova de Prades (Tarragona)
☎: +34 636 519 821
info@vegaaixala.com
www.vegaaixala.com

Vega Aixalà Caliu 2015 T C S
cariñena, syrah, cabernet sauvignon

92

Full-bodied. Colour: black cherry. Nose: sweet spices, dried herbs, tobacco, waxy notes. Palate: flavourful, long, ripe fruit.

Vega Aixalà Carinyena 2015 T C
cariñena

93

Colour: dark-red cherry, garnet rim. Nose: tobacco, black fruit, ripe fruit, dried herbs, wild herbs, earthy notes. Palate: spicy, round tannins, long.

Vega Aixalà Garnatxa Vilanova 2015 T C S
garnacha

91

Wild, spicy. Colour: cherry, garnet rim. Nose: black fruit, waxy notes, balanced. Palate: flavourful, balanced, slightly dry, soft tannins.

Vega Aixalà La Bauma 2019 B
garnacha blanca, chardonnay

89

Jammy, oxidativ. Nose: dry nuts, caramel, faded flowers.

Vega Aixalà La Font dels Aubacs 2018 T BA
pinot noir

91

Slight reduction, with personality. Colour: cherry, garnet rim. Nose: faded flowers, spicy, ripe fruit. Palate: flavourful, easy to drink, long, spicy.

Vega Aixalà Viern 2012 T R S
garnacha, cariñena, cabernet sauvignon, syrah

89

Age nuances, spicy, balanced, toasty, tasty.

CARA NORD CELLER
Plaça Sant Sebastià, 13
25457 El Vilosell (Lleida/Lérida)
☎: +34 973 176 029
hola@caranordceller.com
www.caranordceller.com

Cara Nord Blanc 2021 B

88

Racy, citrus fruit, herbal, standard.

DO CONCA DE BARBERÀ / D.O.P.

DO CONCA DE BARBERÀ / D.O.P.

Cara Nord Negre 2020 T C
garnacha, syrah, garrut
91
Herbaceous, jammy. Colour: cherry, garnet rim. Nose: wild herbs, grassy, ripe fruit, fruit liqueur notes. Palate: flavourful, spicy, ripe fruit.

Cara Nord Trepat Negre 2021 T
90
Colour: cherry, purple rim. Nose: red berry notes, floral, earthy notes, scrubland. Palate: flavourful, fruity, good acidity.

Cara Nord Trepat Rosat 2021 RD
100% trepat
87 🌱

CELLER CARLES ANDREU
Sant Sebastià, 19
43423 Pira (Tarragona)
☎: +34 977 887 404
info@cavandreu.com
www.cavandreu.com

Carles Andreu 12@ 2021 T
100% trepat
91 🌱
Colour: cherry, purple rim. Nose: fruit expression, red berry notes, floral, spicy, cocoa bean. Palate: flavourful, fruity, good acidity.

Carles Andreu Parellada 2020 B
100% parellada
90
Colour: bright straw. Nose: floral, chamomile, balanced. Palate: fresh, fruity, good acidity.

Carles Andreu Trepat 2019 T BA
100% trepat
90
Colour: cherry, purple rim. Nose: spicy, scrubland, black fruit. Palate: ripe fruit, flavourful.

CELLER RENDÉ MASDÉU
Ctra. N-240 km 39,5
43440 L'Esplugá de Francolí (Tarragona)
☎: +34 977 871 361
celler@rendemasdeu.cat
www.rendemasdeu.cat

Arnau de Rendé Masdeu 2020 T C S
syrah
89 🌱
Balanced, spicy, jammy, powerful.

El Follet Rosat 2021 RD
syrah, trepat
88 🌱
Aromatic, fruity, jammy, smooth, wild.

inQuiet de Rendé Masdeu 2021 T
trepat, syrah
87 🌱

La Nimfa Blanc 2021 B
macabeo, garnacha blanca, trepat, parellada
86 🌱

Trepat del Jordiet de Rendé Masdeu 2020 T
trepat
90 🌱
Colour: cherry, purple rim. Nose: fruit expression, red berry notes, floral, spicy, wild herbs. Palate: flavourful, fruity, good acidity.

Vi de Fang 2020 T C
monastrell
90 🌱
Slight reduction. Colour: cherry, purple rim. Nose: fruit expression, floral, spicy, smoky, black fruit. Palate: flavourful, fruity, good acidity, long.

DOMENIO WINES BY CELLERS DOMENYS
Prat de la Riba, 18
43713 St. Jaume dels Domenys (Tarragona)
☎: +34 977 677 135
administracio@cellersdomenys.com
www.domeniowines.com

Anima Nua Cor Viu 2020 B
macabeo, parellada
91 🌱
Colour: bright yellow. Nose: ripe fruit, white fruit, neat, dried flowers. Palate: rich, good structure, long, fine bitter notes.

Anima Nua Cor Viu 2020 T C
tempranillo
89 🌱
Fruity, herbal, spicy, crisp.

Domenio Trepat 2019 T
trepat
89
Slight oxidation, spicy, dried herbs, crisp.

Domenio Ull de Llebre 2019 T
ull de llebre
87

FAMILIA TORRES
Castell de Milmanda - Ctra. N-240, Km 45
43430 Vimbodí (Tarragona)
☎: +34 938 177 400
info@torres.es
www.torres.es

🏆 PODIUM

Grans Muralles 2018 T R
garnacha, cariñena, querol, monastrell, garró
95
Colour: very deep cherry. Nose: complex, expressive, spicy, mineral, dark chocolate, elegant. Palate: elegant, full-bodied, long, great length.

Milmanda 2019 B C
chardonnay
93
Colour: straw. Nose: expressive, white flowers, dried herbs. Palate: flavourful, balanced, long, fine bitter notes.

Sons de Prades 2019 B
chardonnay
92
Colour: bright yellow. Nose: creamy oak, ripe fruit, chamomile, scrubland, sweet spices. Palate: good structure, toasty, fine bitter notes, saline.

JOSEP FORASTER
Camí de L'Ermita de Sant Josep, s/n
43400 Montblanc (Tarragona)
☎: +34 977 860 229
info@josepforaster.com
www.josepforaster.com

Brisat del Coster 2021 B
macabeo
89 🌿
With personality, dried flowers, dried herbs, smooth, little interventionist.

Josep Foraster Blanc Selecció 2021 B
50% garnacha blanca, 40% macabeo, 10% chardonnay
91 🌿
Colour: bright straw. Nose: fresh fruit, wild herbs, dried flowers. Palate: fresh, fruity, good acidity, fine bitter notes.

Josep Foraster Pep 2020 T
91
Colour: cherry, purple rim. Nose: fruit expression, red berry notes, floral, spicy, wild herbs. Palate: flavourful, fruity, good acidity, long.

Josep Foraster Trepat 2021 T
trepat
91 🌿
Colour: cherry, purple rim. Nose: fruit expression, red berry notes, floral, spicy. Palate: flavourful, fruity, good acidity.

Julieta 2020 T
trepat
92 🌿
Colour: cherry, purple rim. Nose: fruit expression, red berry notes, floral, spicy, balsamic herbs, raspberry, ripe fruit. Palate: flavourful, fruity, good acidity, long.

Les Gallinetes 2021 T
89 🌿
Fruity, tasty, crisp, spicy.

MAS DE LA PANSA
Comerç, 2
43422 Barberà de la Conca (Tarragona)
☎: +34 667 894 636
info@masdelapansa.com
www.masdelapansa.com

Mas de la Pansa Dolç de Trepat 2019 T D
92
Yeasty notes. Colour: Rubí, orangey edge. Nose: fruit liqueur notes, dried flowers, caramel, dry nuts. Palate: fruity, flavourful, sweet, fresh.

Mas de la Pansa Escumós Rosat 2018 RE S
100% trepat
89
Balanced, dried flowers, kind finish, jammy, tasty.

Mas de la Pansa Trepat 2018 T
100% trepat
91
Colour: cherry, purple rim. Nose: fruit expression, red berry notes, floral, spicy, balsamic herbs. Palate: flavourful, fruity, good acidity, long.

VINÍCOLA SARRAL I SECCIÓ DE CRÈDIT
Avda. de la Conca, 33
43424 Sarral (Tarragona)
☎: +34 977 890 031
Fax: +34 977 890 136
cavaportell@gmail.com
www.cava-portell.com

Portell 2016 T C
cabernet sauvignon, merlot, ull de llebre
87

DO CONCA DE BARBERÀ / D.O.P.

DO CONCA DE BARBERÀ / D.O.P.

Portell 2016 T R
cabernet sauvignon, merlot, ull de llebre
87

Portell Blanc de Blancs 2021 B S
macabeo, parellada
86

Portell Blanc Semi dolç 2021 BE SD
macabeo, parellada
85

Portell Merlot 2019 T BA
100% merlot
87

Portell Rosat Trepat 2021 RD
100% trepat
86

VINS DE LA MEMÒRIA
Montaner 320
08021 Barcelona (Barcelona)
vinsdelamemoria@gmail.com
www.vinsdelamemoria.com

Pólvora 2020 T
trepat
91
Reduced. Colour: bright cherry, light cherry. Nose: burnt matches, spicy, red berry notes, ripe fruit. Palate: easy to drink, ripe fruit, fine bitter notes, good finish.

DO. CONDADO DE HUELVA / VINO NARANJA DEL CONDADO DE HUELVA

CONSEJO REGULADOR

Plaza Ildefonso Pinto, s/n.
21710 Bollullos Par del Condado (Huelva)
☎: +34 959 410 322 - Fax: +34 959 413 859
@: cr@condadodehuelva.es
www.condadodehuelva.es

LOCATION:

In the south east of Huelva. It occupies the plain of Bajo Guadalquivir. The production area covers the municipal areas of Almonte, Beas, Bollullos Par del Condado, Bonares, Chucena, Gibraleón, Hinojos, La Palma del Condado, Lucena del Puerto, Manzanilla, Moguer, Niebla, Palos de la Frontera, Rociana del Condado, San Juan del Puerto, Villalba del Alcor, Villarrasa and Trigueros.

GRAPE VARIETIES:

WHITE: Zalema, Palomino, Listán de Huelva, Garrido Fino, Moscatel de Alejandría, Pedro Ximénez, Verdejo, Moscatel de Grano Menudo, Colombard, Sauvignon Blanc and Chardonnay.

RED: Merlot, Syrah, Tempranillo, Cabernet Sauvignon and Cabernet Franc.

FIGURES:

Vineyard surface: 2,162 – **Wine-Growers:** 1,022 – **Wineries:** 21 – **2021 Harvest rating:** Very Good – **Production 21:** 13,605,262 litres – **Market percentages:** 95% National - 5% International.

SOIL:

In general, flat and slightly rolling terrain, with fairly neutral soils of medium fertility. The soil is mainly reddish, brownish-grey with alluvium areas in the proximity of the Guadalquivir.

CLIMATE:

Mediterranean in nature, with certain Atlantic influences. The winters and springs are fairly mild, with long hot summers. The average annual temperature is 18 °C, and the average rainfall per year is around 550 mm, with a relative humidity of between 60% and 80%.

CLASSIFICATION OF YOUNG WINE HARVESTS PEÑÍNGUIDE

2017	2018	2019	2020	2021
N/A	N/A	GOOD	N/A	N/A

BODEGA PRIVILEGIO DEL CONDADO

San José, 2
21710 Bollullos Par del Condado (Huelva)
☎: +34 959 410 261
comercial@vinicoladelcondado.com
www.vinicoladelcondado.com

Misterio Condado Viejo BF S
89
Aromatic, age nuances, oxidativ, tasty, representative. Nose: dry nuts.

Misterio Orange Naranja BF Solera D
88
Jammy, powerful. Nose: candied fruit, citrus fruit, pattiserie.

Privilegio del Condado 2021 B S
88
Fruity, herbal, jammy.

VDM Orange Dulce BF AROM D
88
Exuberant, sweety finish, fruity, pruney, full-bodied, citrus fruit.

BODEGAS IGLESIAS

Teniente Merchante, 2
21710 Bollullos Par del Condado (Huelva)
☎: +34 959 410 439
Fax: +34 959 410 463
bodegasiglesias@bodegasiglesias.com
www.bodegasiglesias.com

% UZ Cien x Cien Uva Zalema 2021 B
86

Letrado Solera 1992 BF Solera
100% zalema
89
Aromatic, kind finish, tasty. Nose: caramel, sweet spices, cocoa bean. Palate: fine bitter notes.

Par Vino Naranja BF D
87

Ricahembra Solera 1980 BF Solera D
zalema, pedro ximénez
88
Pruney, creamy, tasty, sweety finish.

UZT Tardía B
zalema
85

BODEGAS INFANTE

Hinojos, 41
21700 La Palma del Condado (Huelva)
☎: +34 959 402 567
dinfante@dinfante.com
www.dinfante.com

Albaleia Colombard 2021 B
100% colombard
88
Tasty, jammy, aromatic, kind finish.

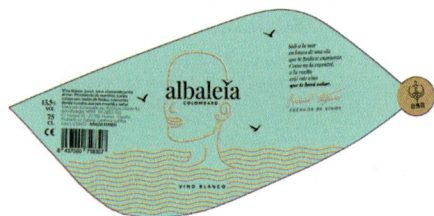

Diosa de las 3 Cabezas 2021 B
40% colombard, 30% zalema, 30% listán blanco
90
Colour: straw. Nose: expressive, white flowers, jasmine, dried herbs. Palate: flavourful, fruity, balanced.

Innocence 2021 B FB
100% colombard
89
Roasted coffee, jammy, tasty, great length.

Tragantia Zalema 2021 B
100% zalema

87

Simple, jammy, thin.

BODEGAS OLIVEROS
Rábida, 12
21710 Bollullos Par del Condado (Huelva)
☎: +34 959 410 057
Fax: +34 959 410 057
oliveros@bodegasoliveros.com
www.bodegasoliveros.com

Oliveros Pedro Ximénez BF PX D
90

Colour: dark mahogany. Nose: fruit liqueur notes, dried fruit, pattiserie, toasty. Palate: sweet, unctuous, powerful.

Oliveros Vino Naranja BF D
pedro ximénez, zalema

89

Citrus fruit, representative, tasty, aromatic, kind finish. Nose: fruit liqueur notes.

BODEGAS SAUCI
Doctor Fleming, 1
21710 Bollullos Par del Condado (Huelva)
☎: +34 959 410 524
sauci@bodegassauci.es
www.bodegassauci.es

Espinapura BF FI ES
palomino, listán blanco

90

Colour: bright yellow. Nose: balanced, yeasty notes, dry nuts, faded flowers. Palate: flavourful, fine bitter notes, long.

Sauci 2019 T C
50% tempranillo, 50% syrah

86

Sauci 2021 B
100% zalema

86

DO CONDADO DE HUELVA / D.O.P.

DO. COSTERS DEL SEGRE

CONSEJO REGULADOR

Complex de la Caparrella, 97
25192 Lleida
☎: +34 973 264 583 - Fax: +34 973 264 583
@: info@costersdelsegre.es
www.costersdelsegre.es

LOCATION:

In the southern regions of Lleida, and a few municipal areas of Tarragona. It covers the sub-regions of: Artesa de Segre, Garrigues, Pallars Jussà, Raimat, Segrià and Valls del Riu Corb.

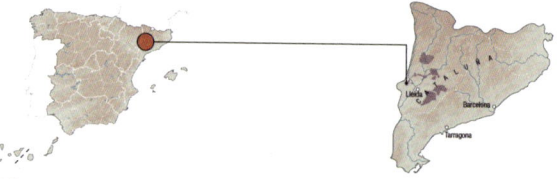

SUB-REGIONS:

Artesa de Segre: Located on the foothills of the Sierra de Montsec, just north of the Noguera region, it has mainly limestone soils. **Urgell:** Located in the central part of the province of Lleida, at an average altitude of 350 meters, its climate is a mix of mediterranean and continental features. **Garrigues:** To the southeast of the province of Lleida, it is a region with a complex topography and marl soils. Its higher altitude is near 700 meters. **Pallars:** Located in the Pyrinees, it is the northernmost sub-zone. Soils are predominantly limestone and its type of climate mediterranean with strong continental influence. **Raimat:** Located in the province of Lleida and with predominantly limestone soils, it has a mediterranean climate with continental features, with predominantly cold winters and very hot summers. **Segrià:** Is the central sub-zone of the DO, with limestone soils. **Valls del Riu Corb:** Located in the southeast of the DO, its climate is primarily mediterranean-continental softened by both the beneficial effect of the sea breezes (called marinada in the region) and "el Seré", a dry sea-bound inland wind.

GRAPE VARIETIES:

WHITE: Preferred: Macabeo, Xarel·lo, Parellada, Chardonnay, Garnacha Blanca, Moscatel de Alenjandría, Malvasía, Gewürztraminer, Riesling and Sauvignon Blanc. Authorised: Albariño, Cheñin, Moscatel de grano menudo and Viognier.

RED: Preferred: Garnacha Negra, Ull de Llebre (Tempranillo), Cabernet Sauvignon, Merlot, Monastrell, Trepat, Samsó, Pinot Noir and Syrah. Authorised: Garnacha tintorera, Gonfaus, Garnacha Peluda, Sumoll and Petit verdot.

FIGURES:

Vineyard surface: 4,182 – **Wine-Growers:** 396 – **Wineries:** 35 – **2021 Harvest rating:** Excellent– **Production 21:** 11,830,145 litres – **Market percentages:** 74% National - 26% International.

SOIL:

The soil is mainly calcareous and granitic in nature. Most of the vineyards are situated on soils with a poor organic matter content, brownish-grey limestone, with a high percentage of limestone and very little clay.

CLIMATE:

Rather dry continental climate in all the sub-regions, with minimum temperatures often dropping below zero in winter, summers with maximum temperatures in excess of 35° on occasions, and fairly low rainfall figures: 385 mm/year in Lleida and 450 mm/year in the remaining regions.

CLASSIFICATION OF YOUNG WINE HARVESTS PEÑÍNGUIDE

2017	2018	2019	2020	2021
VERY GOOD	VERY GOOD	VERY GOOD	VERY GOOD	VERY GOOD

3VVV SINGULAR & WINES
Avda. Tarragona, s/n
25300 Tárrega (Lleida/Lérida)
☎: +34 630 179 962
isidre.ribalta@carviresa.com

Valerna 2020 B FB
chardonnay, macabeo
90
Colour: bright yellow. Nose: powerful, creamy oak, ripe fruit, spicy. Palate: rich, long, toasty, fine bitter notes.

Valerna 2020 T BA
tempranillo, garnacha, cabernet sauvignon
88
Toasty, powerful, hot, jammy.

CAR VINÍCOLAS REUNIDAS
Tarragona, 1
25300 Tárrega (Lleida/Lérida)
☎: +34 973 310 732
carviresa@carviresa.com
www.carviresa.com

3 Setmanes 2021 B
chardonnay, macabeo
88
Citrus fruit, fruity, jammy, spicy, balanced.

3 Setmanes 2021 RD
garnacha
88
Aromatic, fruity, jammy, tasty.

Don Quien 2020 B FB
chardonnay, macabeo
89
Jammy, fruity, smoky, lactic, flat.

Don Quien 2020 T BA
tempranillo, merlot, cabernet sauvignon
90
Colour: deep cherry. Nose: ripe fruit, creamy oak, scrubland. Palate: ripe fruit, spicy, round tannins.

Sot Neral Macabeu 2021 B
macabeo
87

Sot Neral Syrah 2020 T
syrah
88
Balsamic herbs, fruity, jammy, hot, tasty, powerful.

CASTELL DEL REMEI
Finca Castell del Remei s/n
25333 Castell del Remei (Lleida/Lérida)
☎: +34 973 580 200
info@castelldelremei.com
www.castelldelremei.com

Castell del Remei Comtat d'Urgell 2017 B
93
Classic. Colour: bright yellow. Nose: powerful, creamy oak, sweet spices, stone fruit, dried fruit. Palate: rich, good structure, long, toasty, fine bitter notes.

Castell del Remei Gotim Blanc 2021 B
87

Castell del Remei Gotim Bru 2019 T
89
Fruity, jammy, spicy, smoky, herbal.

Castell del Remei Oda Negre 2020 T C
90
Colour: bright cherry. Nose: ripe fruit, black fruit, smoky, toasty, spicy. Palate: fruity, flavourful, smoky aftertaste, slightly dry, soft tannins.

🏆 PODIUM
Castell del Remei Vi Dolç 2019 B D
96
Colour: bright yellow. Nose: acetaldehyde, varnish, candied fruit, citrus fruit, roasted almonds, acetone. Palate: fruity, flavourful, sweet, concentrated.

Garnatxa Castell del Remei 2021 T
90
Colour: cherry, purple rim. Nose: violet drops, fruit expression, ripe fruit, violets, fragrant herbs. Palate: fruity, flavourful, balanced, spicy.

CELLER BATLLIU DE SORT
Bord Batlliu
25568 Olp (Lleida/Lérida)
☎: +34 628 125 473
celler@batlliudesort.cat
www.batlliudesort.cat

Biu Finca de la Borda 2018 B
100% riesling
89
Citrus fruit, balanced, dried flowers, crisp, tasty.

Biu Pinot Noir 2020 T
pinot noir
91
Colour: Cherry. Nose: balsamic herbs, sweet spices, scrubland, red berry notes. Palate: spicy, balsamic, good acidity.

DO COSTERS DEL SEGRE / D.O.P.

DO COSTERS DEL SEGRE / D.O.P.

Finca Les Lleres 2016 B BA
viognier

92

Colour: bright yellow. Nose: ripe fruit, dried herbs, faded flowers, sweet spices. Palate: powerful, ripe fruit, balanced.

Nero de Sort 2020 T
pinot noir

91

Colour: deep cherry. Nose: ripe fruit, dried herbs, creamy oak. Palate: powerful, ripe fruit, spicy, round tannins.

Pantigana 2020 B
garnacha blanca, macabeo

89

Citrus fruit, balanced, spicy, jammy.

Salvavides 2020 T
80% merlot, 15% ull de llebre, 5% garnacha

89

Pruney, powerful, tasty, toasty.

CELLER CERCAVINS
Pol. 8 Parcela 17
25340 Verdú (Lleida/Lérida)
☎: +34 973 348 114
Fax: +34 973 347 197
miquel.bp@@cellercercavins.com
www.cellercercavins.com

Bru de Verdú 14 2017 T
syrah, tempranillo

89

Balanced, spicy, roasted coffee, dried herbs, tasty.

Bru de Verdú 2019 B
garnacha blanca, chardonnay, sauvignon blanc

91

Colour: bright straw. Nose: fragrant herbs, fine lees, white fruit, ripe fruit, waxy notes. Palate: full-bodied, rich, long, good acidity.

Bru de Verdú 2019 T
tempranillo, syrah

87

Guillamina 2020 B
sauvignon blanc, garnacha blanca, gewürztraminer, chardonnay, albariño

87

CELLER DE SANUI
Partida Torres de Sanui –
Antiguo camino de la Cerdera s/n
25193 Lleida/Lérida (Lleida/Lérida)
☎: +34 973 428 163
Fax: +34 973 428 029
pf@melendres.net
www.cellerdesanui.com

Clos De Sanui 2019 T
100% garnacha

89

Herbal, wild, jammy, spicy, hot, tasty.

El Petit De Sanui Blanc 2021 B
55% macabeo, 32% riesling, 13% moscatel grano menudo

85

El Petit De Sanui Negre 2019 T
35% garnacha, 30% tempranillo, 20% merlot, 15% syrah

88

Jammy, tasty, spicy, dried herbs. Nose: black fruit, earthy notes.

Prat d'Hores
Blanc de Sanui 2019 B FB
65% macabeo, 25% riesling, 10% moscatel grano menudo

91

Colour: bright yellow. Nose: powerful, creamy oak, ripe fruit, spicy, faded flowers. Palate: rich, good structure, long, toasty, fine bitter notes.

Prat d'Hores
Negre de Sanui 2019 T
57% garnacha, 23% merlot, 20% ull de llebre

90

Colour: deep cherry. Nose: ripe fruit, dried herbs, creamy oak, smoky, roasted coffee. Palate: ripe fruit, spicy, round tannins.

CELLER MATALLONGA
Raval, 8
25411 Fulleda (Lleida/Lérida)
☎: +34 660 840 791
matallonga60@gmail.com
www.cellermatallonga.cat

Celler Matallonga M 2021 RD
merlot

89

Kind finish, balanced, floral, crisp.

Escorça 2020 B
macabeo

89

Fruity, balanced, jammy, great length, tasty, bitter.

K-Tharsis 2018 T C
cabernet franc

91

Colour: Cherry. Nose: scrubland, dried herbs, fruit preserve, spicy. Palate: spicy, flavourful, good structure.

Matallonga Selecció 2017 T C
80% tempranillo, 20% syrah

87

Vi del Banya 2020 T
80% tempranillo, 20% cabernet franc

87

CELLER MONTSEC
Pol. El Plà s/n
25730 Artesa de Segre (Lleida/Lérida)
☎: +34 629 521 836
jaume.gaspa@coopartesa.com
www.cellermontsec.com

Coop 1958 2021 B
macabeo, garnacha blanca, sauvignon blanc

86

Coop 1958 T
merlot, cabernet sauvignon, ull de llebre

86

Garbuix Origen 2021 RD
ull de llebre, cabernet franc

87

Garbuix Verema Vermella 2021 RD
merlot, ull de llebre, cabernet franc

88

Standard, fruity, jammy, pleasant, tasty.

Montsec Mirapallars 2018 T
cabernet sauvignon, cabernet franc, merlot

88

Powerful, jammy, spicy.

Montsec Sauvignon Blanc 2019 B
sauvignon blanc

87

CELLER PURGATORI
Mas de L'Aranyó,
Ctra. de Bellpuig A Flix C233 – km 59,5
25430 Juneda (Lleida/Lérida)
☎: +34 938 177 400
Fax: +34 938 177 444
info@torres.es
www.torres.es

Purgatori 2019 T BA
cariñena, garnacha, syrah

93

Colour: bright cherry. Nose: fresh fruit, red berry notes, fine reductive notes, meaty notes, scrubland. Palate: good acidity, spicy, fine tannins, fruity.

CELLER VILA CORONA
Camí els Nerets, s/n
25654 Vilamitjana (Lleida/Lérida)
☎: +34 650 529 088
Fax: +34 973 652 638
info@vilacorona.cat
www.vilacorona.cat

Llabustes Chardonnay 2021 B
100% chardonnay

87

Llabustes Merlot 2018 T C
100% merlot

88

Age nuances, balanced, spicy, classic, tasty.

Llabustes Riesling 2020 B
100% riesling

88

Citrus fruit, crisp, fruity, tasty, herbal.

Llabustes Ull de Llebre 2019 T C
100% ull de llebre

87

Tu Rai 2019 T BA
40% monastrell, 30% garnacha, 30% ull de llebre

90

Colour: cherry, purple rim. Nose: red berry notes, floral, spicy, balsamic herbs. Palate: flavourful, fruity, good acidity, long.

DO COSTERS DEL SEGRE / D.O.P.

DO COSTERS DEL SEGRE / D.O.P.

CÉRVOLES CELLER
Avda. Les Garrigues, 26
25471 La Pobla de Cèrvoles (Lleida/Lérida)
☎: +34 973 176 029
info@cervoles.com
www.cervoles.com

Cérvoles Blanc Vinyes Altes de Les Garrigues 2020 B FB
macabeo, chardonnay

90

Colour: bright yellow. Nose: creamy oak, ripe fruit, spicy, woody. Palate: good structure, toasty, fine bitter notes.

Cérvoles Colors Blanc 2021 B

88

Racy, citrus fruit, herbal, standard.

Cérvoles Colors Negre 2020 T C
garnacha, cabernet sauvignon, merlot

90

Defined aromas, balsamic herbs. Nose: black fruit, ripe fruit, faded flowers, characterful, dried herbs. Palate: flavourful, fruity.

Cérvoles Estrats 2019 T

93

Colour: deep cherry. Nose: dried herbs, creamy oak, black fruit, roasted coffee. Palate: powerful, ripe fruit, spicy, round tannins.

Cérvoles Negre 2019 T

91

Colour: bright cherry. Nose: black fruit, ripe fruit, smoky, grassy, spicy. Palate: fruity, flavourful, balanced, round tannins, toasty.

Cérvoles Tossalets 2019 T
syrah

92

Colour: cherry, garnet rim. Nose: powerful, black fruit, scrubland, balsamic herbs, wild herbs, spicy. Palate: flavourful, long, correct, round.

CLOS PONS
Crta. LV-7011, Km. 2.5
25155 L'Albagés (Lleida/Lérida)
☎: +34 973 070 237
Fax: +34 973 730 515
pons.shop@grup-pons.com
www.ponshome.es

Clos Pons Alges 2016 T C
60% garnacha, 25% syrah, 15% tempranillo

91

Colour: deep cherry. Nose: ripe fruit, dried herbs, creamy oak, fruit expression. Palate: powerful, ripe fruit, spicy, round tannins.

Clos Pons Pla del Tet 2018 T C
100% syrah

91

Colour: cherry, garnet rim. Nose: fruit preserve, creamy oak, sweet spices, smoky. Palate: flavourful, long, round tannins.

Clos Pons Roc de Foc 2014 B
100% macabeo

91

Age nuances. Colour: bright yellow. Nose: dry nuts, faded flowers, spicy. Palate: flavourful, long, bitter.

Clos Pons Roc Nu 2012 T R
35% garnacha, 45% cabernet sauvignon, 20% tempranillo

92

Classic. Colour: dark-red cherry, bright ochre rim. Nose: black fruit, tobacco, earthy notes, spicy. Palate: good structure, flavourful, long.

Clos Pons Sisquella 2018 B C
72% garnacha blanca, 28% albariño

90

Colour: bright straw. Nose: medium intensity, dried flowers, balanced. Palate: balanced, fine bitter notes, good finish.

COSTERS DEL SIÓ
Ctra. de Agramunt, Km. 4,2
25600 Balaguer (Lleida/Lérida)
☎: +34 973 424 062
Fax: +34 973 424 112
administracio@costersio.com
www.costersio.com

Alto Siós 2018 T R
syrah, tempranillo, garnacha

92

Colour: cherry, purple rim. Nose: floral, spicy, red berry notes, black fruit, toasty. Palate: flavourful, fruity, good acidity, long, fleshy, warm.

Petit Siós 2021 T RB
tempranillo, garnacha, cabernet sauvignon

90

Colour: deep cherry. Nose: ripe fruit, dried herbs, creamy oak, aromatic coffee, smoky. Palate: ripe fruit, spicy, round tannins.

Siós Cau del Gat 2020 T C
garnacha, tempranillo, syrah

90

Colour: deep cherry. Nose: ripe fruit, dried herbs, creamy oak, woody. Palate: ripe fruit, spicy, round tannins.

Siós Pla del Lladoner 2021 B
garnacha blanca, chardonnay

90
Colour: bright straw. Nose: white fruit, fine lees, floral, neat. Palate: flavourful, long, spicy.

L'OLIVERA
La Plana, s/n
25268 Vallbona de Les Monges (Lleida/Lérida)
☎: +34 973 330 276
comunicacio@olivera.org
www.olivera.org

Agaliu 2020 B FB
100% macabeo

91
Colour: bright yellow. Nose: creamy oak, ripe fruit, spicy. Palate: rich, good structure, toasty, fine bitter notes.

L'Olivera Reserva Superior 2018 BE BN
100% macabeo

91
Colour: bright straw. Nose: fine lees, floral, fragrant herbs, dried flowers, ripe fruit. Palate: flavourful, good acidity, balanced.

Missenyora 2020 B FB S
100% macabeo

90
Colour: bright yellow. Nose: creamy oak, ripe fruit, spicy. Palate: rich, good structure, toasty, fine bitter notes.

Rasim Vi Pansit Naturalmente Dulce 2018 B D
55% garnacha blanca, 27% malvasía, 18% xarel.lo

92
Colour: golden. Nose: candied fruit, dried flowers, dried herbs, honeyed notes. Palate: fruity, fresh, flavourful, powerful.

Rasim Vimadur 2018 T D
100% garnacha

91
Colour: cherry, garnet rim. Nose: spicy, dried herbs, overripe fruit. Palate: flavourful, fruity, balanced, good acidity.

V89 2018 B FB
100% macabeo

92
Colour: bright yellow. Nose: creamy oak, ripe fruit, spicy. Palate: rich, good structure, long, toasty, fine bitter notes, elegant.

MAS BLANCH I JOVÉ
Pol. Ind. 9 Parc. 129 Paratge Llinars
25471 La Pobla de Cèrvoles (Lleida/Lérida)
☎: +34 973 050 018
Fax: +34 973 391 151
sara@masblanchijove.com
www.masblanchijove.com

Petit Blanc Saó 2021 B
70% garnacha blanca, 20% macabeo, 7% viognier, 3% riesling

88
Citrus fruit, crisp, herbal, standard.

Petit Saó 2019 T
50% tempranillo, 30% garnacha, 20% cabernet sauvignon

88
Overripe, powerful, spicy, toasty.

Saó Abrivat 2019 T C
65% garnacha, 15% tempranillo, 11% cabernet sauvignon, 9% syrah

90
Colour: deep cherry. Nose: dried herbs, creamy oak, black fruit. Palate: powerful, ripe fruit, spicy, round tannins.

Saó Blanc 2020 B FB
90% macabeo, 10% garnacha blanca

90
Colour: straw. Nose: dried herbs, faded flowers, white fruit, fine lees. Palate: ripe fruit, balanced, good structure, flavourful.

Saó Expressiu 2019 T R S
70% garnacha, 20% cabernet sauvignon, 10% syrah

92
Colour: Cherry. Nose: balsamic herbs, sweet spices, scrubland, fruit preserve, cocoa bean. Palate: spicy, balsamic, good acidity.

Troballa 2019 T
100% garnacha

90
Colour: cherry, purple rim. Nose: red berry notes, floral, spicy. Palate: flavourful, fruity, good acidity.

MAS RAMONEDA
Crta. L-512, Km. 8
25738 Montmagastre (Lleida/Lérida)
☎: +34 670 237 068
info@masramoneda.com
www.masramoneda.com

Mas Ramoneda Boira Blanc de Noir 2021 B
100% garnacha

89
Citrus fruit, balanced, dried flowers, herbaceous.

DO COSTERS DEL SEGRE / D.O.P.

DO COSTERS DEL SEGRE / D.O.P.

Mas Ramoneda Eriçó 2018 T C
50% merlot, 50% syrah

87

Mas Ramoneda Perdiu 2018 T RB
cabernet sauvignon, merlot, tempranillo

88

Balanced, jammy, tasty, spicy.

Mas Ramoneda Vola Vola Papallona 2021 B
50% garnacha blanca, 50% viognier

88

Pleasant, aromatic, fruity, floral, tasty.

RAIMAT
Passeig Manuel Raventós i Domènech, s/n
25111 Raimat (Lleida/Lérida)
☎: +34 638 486 758
c.escolar@raimat.com
www.15bodegas.com

Raimat El Molí 2018 T C
70% cabernet sauvignon, 30% syrah

90 🌱

Colour: cherry, garnet rim. Nose: scrubland, fine reductive notes, grassy, dried flowers, black fruit. Palate: flavourful, balsamic, spicy.

Raimat El Niu de la Cigonya 2020 B
45% chardonnay, 45% xarel.lo, 10% albariño

90

Colour: bright straw. Nose: ripe fruit, fragrant herbs, fine lees, dried flowers. Palate: good acidity, balanced, fine bitter notes.

Turons de la Pleta 2020 B
100% chardonnay

90

Woody. Colour: bright yellow. Nose: powerful, ripe fruit, spicy, dried flowers. Palate: rich, good structure, toasty.

Turons Vallcorba 2019 T C
100% cabernet sauvignon

90 🌱

Classic, age nuances. Colour: dark-red cherry, garnet rim. Nose: ripe fruit, fruit preserve, aged wood nuances, tobacco, dried flowers. Palate: spicy, round tannins, balanced.

Vol D'Anima de Raimat Blanc 2021 B
45% chardonnay, 40% xarel.lo, 15% albariño

88 🌱

Citrus fruit, crisp, herbal, tasty.

Vol D'Anima de Raimat Rosat 2021 RD
95% garnacha, 5% pinot noir

90 🌱

Smooth. Colour: raspberry rose. Nose: red berry notes, fragrant herbs. Palate: light-bodied, good acidity, fine bitter notes.

TERRER DE PALLARS
Del Vent, 28
25655 Figuerola d'Orcau (Lleida/Lérida)
☎: +34 616 701 080
nuria@terrerdepallars.com
www.terrerdepallars.com

Conca de Tremp 2019 T C
50% merlot, 50% cabernet sauvignon

91

Colour: cherry, garnet rim. Nose: fruit preserve, powerful, dried herbs, scrubland, earthy notes. Palate: flavourful, long, sweet tannins.

Conca de Tremp Blanc 2020 B
macabeo, garnacha blanca

91

Colour: bright straw. Nose: dried flowers, faded flowers, fine lees, yeasty notes. Palate: flavourful, fine bitter notes, spicy.

El Presumit del Pallars 2020 T
syrah, cabernet sauvignon

90

Colour: bright cherry. Nose: ripe fruit, dried herbs, earthy notes. Palate: powerful, ripe fruit, spicy, round tannins.

TOMÁS CUSINÉ
Plaça Sant Sebastià, 13
25457 El Vilosell (Lleida/Lérida)
☎: +34 973 176 029
info@tomascusine.com
www.tomascusine.com

Auzells 2021 B

91 🌱

Austere. Colour: bright straw, greenish rim. Nose: fresh fruit, citrus fruit, wild herbs. Palate: fresh, fruity, good acidity, fine bitter notes.

Finca Comabarra 2019 T C
garnacha, syrah, cabernet sauvignon

92 🌱

Colour: cherry, garnet rim. Nose: ripe fruit, dried herbs, tobacco, creamy oak, powerful. Palate: ripe fruit, spicy, round tannins.

Finca La Serra del Vent 2021 B

91

Colour: bright straw, greenish rim. Nose: fresh fruit, citrus fruit, wild herbs, fine lees, smoky. Palate: fresh, fruity, good acidity, fine bitter notes.

Finca Racons 2017 B
93
Colour: bright straw. Nose: ripe fruit, fruit expression, white fruit, faded flowers, fine lees, mineral. Palate: fruity, full-bodied, flavourful, balanced, unctuous, good finish.

Vilosell 2019 T
91 🌿
Colour: cherry, purple rim. Nose: black fruit, ripe fruit, toasty, black pepper, spicy, smoky. Palate: fruity, flavourful, balanced, fine bitter notes.

VALL DE BALDOMAR
Ctra. de Alós de Balaguer, s/n
25737 Baldomar (Lleida/Lérida)
☎: +34 973 402 205
info@valldebaldomar.com
www.valldebaldomar.com

Baldomà Selecció 2019 T
merlot, cabernet sauvignon, garnacha, monastrell
87

Cristiari 2021 B
60% müller thurgau, 40% incrocio manzoni
87

Cristiari 2021 RD
70% merlot, 30% cabernet sauvignon
88
Fruity, pruney, sweeties, tasty, great length.

Cristiari d'Alòs Merlot 2019 T BA
merlot
88
Pruney, toasty, powerful, overripe.

Petit Baldoma 2020 T
merlot, cabernet sauvignon, garnacha, monastrell
85

Petit Baldomà 2021 B
riesling, garnacha blanca, gewürztraminer
86

VINYA ELS VILARS
Camí de Puiggrós, s/n
25140 Arbeca (Lleida/Lérida)
☎: +34 973 149 144
vinyaelsvilars@vinyaelsvilars.com
www.vinyaelsvilars.cat

Gerard 2018 T R
100% merlot
90
With personality. Colour: deep cherry. Nose: dried herbs, creamy oak, roasted coffee, black fruit, meaty notes. Palate: powerful, ripe fruit, spicy, round tannins.

Leix 2019 T
100% syrah
88
Overripe, hot, powerful, tasty.

Nena 2021 RD
100% syrah
88
Kind finish, standard, fruity, floral, balanced.

Quim 2021 B
macabeo
89
Citrus fruit, balanced, spicy, dried herbs, tasty.

Tallat de Lluna 2019 T
100% syrah
88
Balanced, jammy, spicy, tasty.

Vilars 2019 T RB
syrah, merlot
87

WINES RAMSURT BY MAGI RAVENTOS
Can Majó 18
08173 Sant Cugat (Barcelona)
☎: +34 652 439 922
comercialramsurt@gmail.com
www.ramsurt.com

Vinya Mag Raven 2018 T BA
cabernet sauvignon, merlot
88
Weary, spicy, toasty.

DO. EL HIERRO

CONSEJO REGULADOR

El Hoyo, 1
38911 Municipio de La Frontera (El Hierro)
☎: +34 922 559 622 - Fax: +34 922 559 622
@: doelhierro@doelhierro.es
www.doelhierro.es

LOCATION:

On the island of El Hierro, part of the Canary Islands. The production area covers the whole island, although the main growing regions are Valle del Golfo, Sabinosa, El Pinar and Echedo.

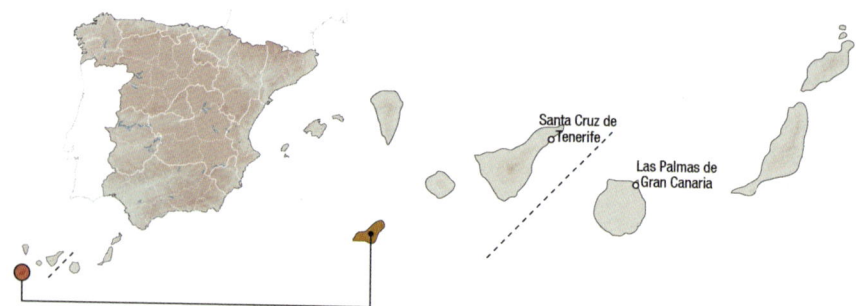

GRAPE VARIETIES:

WHITE: Verijadiego (majority with 50% of all white varieties), Listán Blanca, Bremajuelo, Uval (Gual), Pedro Ximénez, Baboso and Moscatel.

RED: Listán Negro, Negramoll, Baboso Negro and Verijadiego Negro.

FIGURES:

Vineyard surface: 63 – **Wine-Growers:** 238– **Wineries:** 13– 21 Harvest rating: Very Good– **Production 21:** 65,051 litres. **Market percentages:** 95% National - 5% International.

SOIL:

Volcanic in origin, with a good water retention and storage capacity. Although the vineyards were traditionally cultivated in the higher regions, at present most of them are found at low altitudes, resulting in an early ripening of the grapes.

CLIMATE:

Fairly mild in general, although higher levels of humidity are recorded in high mountainous regions. Rainfall is relatively low.

CLASSIFICATION OF YOUNG WINE HARVESTS PEÑÍNGUIDE

2017	2018	2019	2020	2021
N/A	N/A	N/A	N/A	N/A

BIMBACHE VINÍCOLA
Valderde
38900 El Hierro (Santa Cruz de Tenerife)
info@bimbache.com
www.bimbache.es

Bimbache 2020 B
verijadiego blanco, listán blanco, baboso blanco, forastera, gual

93

With personality. Colour: straw. Nose: complex, cereal notes, fine lees, smoky, dried flowers, iodine notes. Palate: powerful, balanced, flavourful, mineral, saline.

Chivo 2020 B
verijadiego blanco, bremajuelo, gual, malvasía, moscatel

92

Little interventionist. Colour: straw. Nose: ripe fruit, floral, cereal notes, fine lees. Palate: balanced, rich, good acidity.

🏆 **PODIUM**

Gran Cruz del Calvario 2020 B
verijadiego blanco, listán blanco, forastera, gual

95

Colour: bright straw. Nose: complex, fine lees, cereal notes, tropical fruit, spicy. Palate: good acidity, flavourful, mineral.

DO EL HIERRO / D.O.P.

DO. EMPORDÀ

CONSEJO REGULADOR

Plaça del Sol, sn
17600 Figueres (Girona)
☎: +34 972 507 513
@: info@doemporda.cat
www.doemporda.cat

LOCATION:

In the far north west of Catalonia, in the province of Girona. The production area covers 40 municipal areas and is situated the slopes of the Rodes and Alberes mountain ranges forming an arch which leads from Cape Creus to what is known as the Garrotxa d'Empordà.

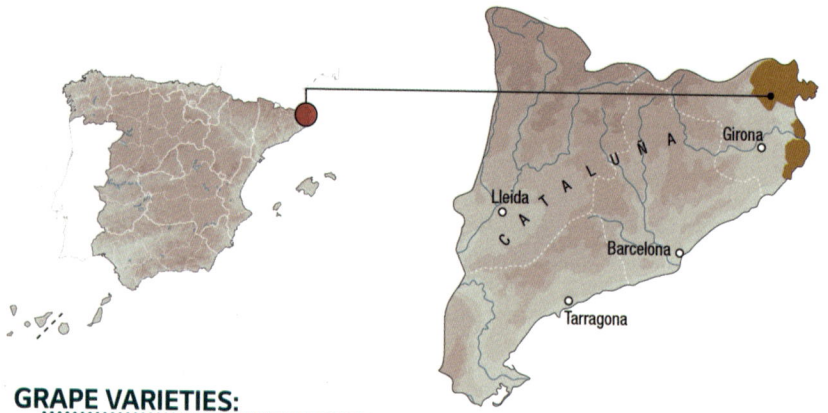

GRAPE VARIETIES:

WHITE: Preferred: Garnacha Blanca, Macabeo (Viura), Cariñena Blanca and Moscatel de Alejandría.

Authorized: Xarel.lo, Chardonnay, Gewürztraminer, Malvasía, Moscatel de Grano Menudo, Picapoll Blanc and Sauvignon Blanc.

RED: Preferred: Cariñena, Garnacha Roja (Lledoner roig), and Garnacha Tinta.

Authorized: Cabernet Sauvignon, Cabernet Franc, Merlot, Monastrell, Tempranillo, Syrah and Garnacha Peluda.

FIGURES:

Vineyard surface: 1,845 – **Wine-Growers:** 254 - **Wineries:** 51 – 21 Harvest rating: N/A – **Production 21:** 5,435,100 litres – **Market percentages:** 92% National - 8% International.

SOIL:

The soil is in general poor, of a granitic nature in the mountainous areas, alluvial in the plains and slaty on the coastal belt.

CLIMATE:

The climatology is conditioned by the 'Tramontana', a strong north wind which affects the vineyards. Furthermore, the winters are mild, with hardly any frost, and the summers hot, although somewhat tempered by the sea breezes. The average rainfall is around 600 mm.

CLASSIFICATION OF YOUNG WINE HARVESTS PEÑÍNGUIDE

2017	2018	2019	2020	2021
GOOD	VERY GOOD	VERY GOOD	VERY GOOD	VERY GOOD

7 MAGNÍFICS
Miquel Torres i Carbó, 6
08720 Vilafranca del Penedès (Barcelona)
☎: +34 938 177 400
7magnifics@7magnifics.com
www.7magnifics.com

Somiadors 2019 T
cariñena, garnacha

90

Colour: cherry, garnet rim. Nose: dried herbs, creamy oak, stone fruit. Palate: ripe fruit, spicy, round tannins, fruity, fleshy.

ALEGRE WINES & SPIRITS
Balmes, 345
08006 Barcelona (Barcelona)
☎: +34 935 641 262
info@alegrews.com
www.alegrews.com

Cala Marquesa 2021 B
garnacha blanca

88

Citrus fruit, fruity, jammy, yeasty notes, balanced.

ALREGI
Pol. Ind. Empordà Internacional – C/Garrotxa, 8
17469 Vilamalla (Girona/Gerona)
☎: +34 972 526 061
alregi@alregi.es
www.alregi.es

Boca Grossa 2019 T
57% garnacha, 33% merlot, 10% syrah

87

Boca Grossa 2021 RD
garnacha

89

Dried flowers, fruity, tasty, dried herbs.

Boca Petita 2021 B
60% macabeo, 40% chardonnay

88

Fruity, herbal, citrus fruit, kind finish.

Boca Petita 2021 RD
40% garnacha, 30% samsó, 30% cabernet sauvignon

87

Boca Petita 2021 T
40% garnacha, 30% samsó, 30% cabernet sauvignon

85

Bufarut 2019 T C
90% garnacha, 10% cabernet sauvignon

86

AV BODEGUERS
Sant Baldiri, 23
17781 Vilamaniscle (Girona/Gerona)
☎: +34 620 006 476
info@avbodeguers.com
www.avbodeguers.com

Elitia Cariñenas Viejas 2018 T
100% cariñena

91 ♣

Colour: deep cherry. Nose: creamy oak, black fruit, scrubland, balsamic herbs. Palate: powerful, ripe fruit, spicy, round tannins.

Elitia Garnatxa d'Empordà B Solera D
garnacha blanca

94

Colour: amber. Nose: candied fruit, dry nuts, sweet spices, aged wood nuances. Palate: fruity, flavourful, sweet, long.

Nereus Garnacha Negra 2018 T
100% garnacha

88 ♣

Full-bodied, balanced, spicy, jammy.

Nereus Selecció 2019 T C
40% syrah, 40% merlot, 20% cariñena

88 ♣

Spicy, crisp, fruity, dried herbs.

Suneus Blanc 2021 B
macabeo

88 ♣

Citrus fruit, crisp, herbal, yeasty notes, balanced.

Suneus Negre 2020 T RB
40% cariñena, 30% cabernet sauvignon, 20% garnacha, 10% syrah

89

Full-bodied, spicy, fruity, powerful, tasty. Nose: woody.

BODEGAS CLOS D'AGON
Afores, s/n
17251 Calonge (Girona/Gerona)
☎: +34 972 661 486
info@closdagon.com
www.closdagon.com

Amic de Clos D'Agon 2019 T
36% cariñena, 35% syrah, 27% garnacha, 2% merlot

90

Colour: deep cherry. Nose: ripe fruit, dried herbs, creamy oak. Palate: ripe fruit, spicy, round tannins, balanced.

DO EMPORDÀ / D.O.P.

DO EMPORDÀ / D.O.P.

Amic de Clos D'Agon 2021 B
53% garnacha blanca, 47% garnacha gris
89
Citrus fruit, balanced, crisp, herbal.

Amic de Clos D'Agon 2021 RD
49% garnacha, 48% cariñena, 3% cabernet franc
89
Balanced, dried flowers, fruity, jammy, yeasty notes, tasty.

Clos D'Agon Syrah Mas Palet 2019 T
100% syrah
94
Colour: cherry, purple rim. Nose: black fruit, wild herbs, spicy, smoky, scrubland, toasty. Palate: fruity, fresh, flavourful, balanced, slightly dry, soft tannins, good finish.

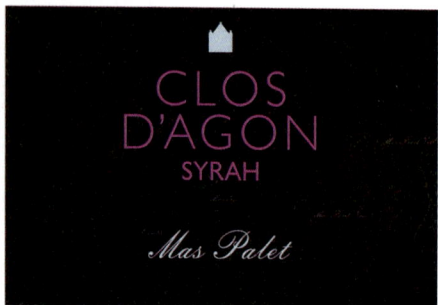

Clos d'Agon Valmaña 2019 T C
40% merlot, 28% syrah, 20% marselan, 8% cabernet franc, 4% petit verdot
91
Creamy. Colour: deep cherry. Nose: dried herbs, creamy oak, black fruit, dark chocolate. Palate: powerful, ripe fruit, spicy, round tannins.

BRUGAROL
Camino de Bell-Lloc, 63
17230 Palamòs (Girona/Gerona)
☎: +34 666 763 540
rrius@brugarol.com
www.brugarol.com

Brisée 2021 B
malvasía
89
Austere, citrus fruit, dried flowers, balanced, tasty.

Malvasía 2020 B
malvasía
88
Citrus fruit, jammy, tropical, smooth.

Mestís 2021 B
garnacha blanca, garnacha gris, malvasía
88
Dried herbs, balanced, full-bodied, citrus fruit.

CELLER ARCHÉ PAGÈS
Sant Climent, 31
17750 Capmany (Girona/Gerona)
☎: +34 626 647 251
bonfill@capmany.com
www.cellerarchepages.com

Bonfill 2019 T C
garnacha, cariñena
89
Colour: cherry, purple rim. Nose: fruit expression, red berry notes, floral, spicy, toasty. Palate: flavourful, fruity, good acidity.

Cartesius Negre 2019 T
garnacha, merlot, cabernet sauvignon
87

Cartesius Rosat 2021 RD
garnacha roja
88
Citrus fruit, balanced, dried flowers, herbal, crisp.

Sàtirs Negre 2019 T C
garnacha, cariñena, syrah, cabernet sauvignon
89
Hot, spicy, jammy, tasty, wild, dried herbs.

Ull de Serp La Closa Carinyena 2019 T C
cariñena
91
Colour: deep cherry. Nose: ripe fruit, dried herbs, creamy oak, black fruit. Palate: powerful, ripe fruit, spicy, round tannins.

Ull de Serp La Closa Macabeu 2019 B FB
macabeo
92
Colour: straw. Nose: dried herbs, faded flowers, stone fruit, toasted bread. Palate: powerful, ripe fruit, balanced, full-bodied, flavourful.

CELLER CASTELLÓ MURPHY
Del Pont, 2
17754 Rabós (Girona/Gerona)
☎: +34 660 665 964
castellojoan10@gmail.com
www.cellercastellomurphy.com

Ixent 2021 RD
100% garnacha
87 ❦

Liber 2021 B
moscatel de alejandría

89

With personality, balanced, spicy, dried flowers, jammy, oxidativ.

Samas 2019 T
50% merlot, 50% samsó

90

Colour: deep cherry. Nose: dried herbs, creamy oak, black fruit, woody. Palate: powerful, ripe fruit, spicy, harsh oak tannins.

CELLER COOPERATIU D'ESPOLLA
Crta. De Roses, s/n
17753 Espolla (Girona/Gerona)
☎: +34 972 563 178
info@celleresspolla.com
www.celleresspolla.com

Babalà – Vi Negre Eixerit 2021 T
15% lledoner roig, 85% lledoner

87

Babalà Vi Blanc Simpàtic 2021 B
75% cariñena blanca, 25% moscatel de alejandría

88

With personality, balanced, dried flowers, herbal, yeasty notes.

Babalà Vi Rosat Alegre 2021 RD
70% lledoner roig, 30% lledoner

87

Clos de les Dòmines 2018 T R
65% cariñena, 35% cabernet sauvignon

88

Roasted coffee, spicy, jammy, powerful.

Clos de les Dòmines Blanc 2021 B FB
30% lledoner roig, 35% lledoner blanco, 20% cariñena blanca, 15% moscatel de alejandría

89

Citrus fruit, balanced, spicy, dried herbs, yeasty notes.

Solera 1931 Espolla Garnatxa d'Empordà Vino Dulce Natural BF Solera D
lledoner blanco, lledoner roig

92

Colour: amber. Nose: dry nuts, dried herbs, candied fruit, sweet spices, aged wood nuances. Palate: good structure, powerful, flavourful.

CELLER COOPERATIU D'ESPOLLA – VINS DE POSTAL
Ctra. Roses, s/n
17753 Espolla (Girona/Gerona)
☎: +34 972 563 178
vinsdepostal@celleresspolla.com
www.vinsdepostal.com

Soliserena Espolla Garnacha D'Empordà BF Solera D
lledoner blanco, lledoner roig

92

Colour: light mahogany. Nose: dried fruit, dry nuts, honeyed notes, spicy, pattiserie. Palate: powerful, flavourful, great length.

Vins de Postal - Coll de Ribera 2014 T BA
100% cariñena

91

Colour: dark-red cherry, garnet rim. Nose: fruit preserve, tobacco, sweet spices, earthy notes. Palate: spicy, round tannins, flavourful.

Vins de Postal – La Creu 2018 B
100% cariñena blanca

89

Citrus fruit, full-bodied, full-bodied, jammy, spicy, woody.

Vins de Postal – La Sureda 2014 T
100% lledoner

92

Colour: dark-red cherry, garnet rim. Nose: fruit preserve, tobacco, black liquorice, scrubland. Palate: spicy, round tannins, balanced.

Vins de Postal – Les Auledes 2017 B
100% lledoner blanco

92

Colour: bright yellow. Nose: powerful, creamy oak, ripe fruit, spicy, woody. Palate: rich, good structure, toasty, fine bitter notes.

Vins de Postal – Les Muntades 2013 T
cariñena

90

Colour: dark-red cherry, garnet rim. Nose: fruit preserve, tobacco, dried herbs, earthy notes. Palate: spicy, round tannins, long.

DO EMPORDÀ / D.O.P.

DO EMPORDÀ / D.O.P.

CELLER GERISENA
17780 Garriguella (Girona/Gerona)
☎: +34 972 530 002
info@cellergerisena.com
www.cellergerisena.com

Blanc de Gerisena 2021 B RB
garnacha blanca

90

Colour: bright yellow. Nose: powerful, creamy oak, ripe fruit, spicy, grassy. Palate: rich, good structure, toasty, fine bitter notes.

Finca Les Roques 2020 B RB
cariñena blanca

90

Colour: bright yellow. Nose: powerful, creamy oak, ripe fruit, spicy, caramel, woody. Palate: rich, good structure, toasty, fine bitter notes.

Negre de Gerisena 2020 T
garnacha

89

Full-bodied, spicy, powerful, tasty, jammy, dried herbs, coarse tannins. Nose: woody.

Rosat de Gerisena 2021 RD
garnacha roja

88

Citrus fruit, crisp, herbal, balanced.

Selecció de Gerisena 2018 T C
garnacha, cabernet sauvignon

91

Colour: deep cherry. Nose: creamy oak, black fruit, scrubland, woody. Palate: powerful, ripe fruit, spicy, round tannins.

🏆 PODIUM

Somnis de Gerisena - Sol i Resena RD Añejo D
garnacha roja

95

Colour: bright cherry, garnet rim. Nose: acetaldehyde, varnish, candied fruit, caramel, sweet spices, dry nuts. Palate: fruity, flavourful, sweet, great length.

CELLER HUGAS DE BATLLE
Francesc Rivera, 28
17466 Colera (Girona/Gerona)
☎: +34 972 389 149
info@cellerhugasdebatlle.com
www.cellerhugasdebatlle.com

30.70 2021 B
70% garnacha blanca, 30% moscatel de alejandría

90

Colour: bright straw. Nose: fragrant herbs, fine lees, white fruit, citrus fruit, wild herbs. Palate: full-bodied, rich, good acidity.

Camí d'en Poca Sang 2021 RD
garnacha

89

Balanced, fruity, jammy, tasty, full-bodied.

Coma Fredosa 2014 T BA
50% garnacha, 40% cabernet sauvignon, 10% cariñena

90

Colour: dark-red cherry, garnet rim. Nose: fruit preserve, tobacco, fine reductive notes, black fruit. Palate: spicy, round tannins, flavourful.

Falguera 2012 T C
60% cariñena, 40% garnacha

91

Colour: dark-red cherry, garnet rim. Nose: fruit preserve, aged wood nuances, tobacco, meaty notes, spicy. Palate: spicy, round tannins, flavourful.

Vall de Molinàs 2013 T RB
70% garnacha, 30% cariñena

88

Age nuances, weary, toasty, tasty, spicy, full-bodied.

Vall de Molinàs 2021 B
100% garnacha blanca

92

Colour: bright yellow. Nose: powerful, creamy oak, ripe fruit, spicy, caramel, woody. Palate: rich, good structure, long, toasty, fine bitter notes.

CELLER JOC
Ctra. Vilajuiga, s/n
17750 Capmany (Girona/Gerona)
☎: +34 607 222 002
info@vinojoc.com
www.vinojoc.com

9 Set 2 2020 T
45% cabernet sauvignon, 29% merlot, 20% garnacha

88

Spicy, tasty, toasty, pruney, full-bodied. Nose: woody.

De Cap a Peus 2021 B
50% garnacha, 50% macabeo

91
Colour: straw. Nose: ripe fruit, dried herbs, faded flowers. Palate: ripe fruit, balanced, full-bodied.

Petardo 2020 T
70% merlot, 25% garnacha, 5% syrah

90
Colour: deep cherry. Nose: dried herbs, red berry notes, black fruit, spicy. Palate: ripe fruit, spicy, round tannins.

CELLER MARIÀ PAGÈS
Pujada, 6
17750 Capmany (Girona/Gerona)
☎: +34 972 549 160
info@cellermpages.com
www.cellermpages.com

Celler Marià Pagès Garnatxa d'Empordà Dulce 2020 B D
garnacha, garnacha blanca

90
Colour: amber. Nose: dry nuts, scrubland, sweet spices, aged wood nuances. Palate: flavourful, full-bodied, good structure.

Celler Marià Pagès Moscat d'Empordà Dulce 2020 B D
moscatel de alejandría

90
Colour: bright yellow. Nose: honeyed notes, floral, sweet spices, dry nuts. Palate: rich, fruity, powerful, flavourful.

Celler Marià Pagès Rosa-T 2021 RD
100% garnacha

87

Celler Marià Pagès Vinya de L'Hort 2021 B
garnacha blanca, garnacha roja

84

Mar de Lluna Moscat 2021 B
100% moscatel de alejandría

87

Taca Negra 2017 T C
garnacha, cariñena, cabernet franc

88
Full-bodied, jammy, herbaceous, powerful, toasty.

CELLER MARTÍ FABRA
Barrio Vic, 26
17751 Sant Climent Sescebes (Girona/Gerona)
☎: +34 972 563 011
info@cellermartifabra.com
www.cellermartifabra.com

Flor D'Albera 2019 B FB
100% moscatel

89
Citrus fruit, floral, jammy, tasty, woody.

L'Oratori 2020 T
50% garnacha, 35% cariñena, 8% tempranillo, 3% cabernet sauvignon, 2% merlot, 2% syrah

90
Colour: deep cherry. Nose: dried herbs, creamy oak, fine reductive notes, black fruit, scrubland. Palate: ripe fruit, spicy, round tannins.

Martí Fabra Selecció Vinyes Velles 2019 T RB
55% cariñena, 45% garnacha

90
Colour: deep cherry. Nose: dried herbs, creamy oak, black fruit, woody. Palate: ripe fruit, spicy, harsh oak tannins.

Masía Carreras Blanc 2019 B FB
40% cariñena blanca, 30% cariñena rosada, 10% garnacha blanca, 10% garnacha rosada, 10% picapoll blanc

91
Colour: bright yellow. Nose: creamy oak, ripe fruit, spicy, woody. Palate: rich, good structure, toasty, fine bitter notes.

Masía Carreras Negre 2019 T
100% cariñena

91
Creamy, pleasant. Colour: deep cherry. Nose: ripe fruit, dried herbs, creamy oak, aromatic coffee, spicy. Palate: powerful, ripe fruit, round tannins.

Masía Pairal Can Carreras Garnatxa de l'Empordà B D
garnacha blanca, garnacha rosada

93
Colour: iodine, amber rim. Nose: complex, dry nuts, creamy oak, varnish, sweet spices. Palate: rich, long, spicy, powerful.

DO EMPORDÀ / D.O.P.

DO EMPORDÀ / D.O.P.

CELLER MARTÍN FAIXÓ
Ctra. de Cadaqués s/n
17488 Cadaqués (Girona/Gerona)
☎: +34 682 107 148
info@martinfaixo.com
w.w.w.martinfaixo.com

MF 2013 T
garnacha, cabernet sauvignon, merlot
88
Age nuances, jammy, powerful, roasted coffee, dried herbs.

Perafita Garnatxa Negra 2021 T
garnacha
89
Subtle, with personality, floral, pruney.

Perafita Picapoll 2021 B
picapoll blanc
87

Perafita Rosat 2021 RD
garnacha, cabernet sauvignon
87

CELLER MASSIS DE L'ALBERA
Plaça Major
17781 Vilamaniscle (Girona/Gerona)
☎: +34 972 549 012
info@massisdelalbera.com

El Visionari 2021 RD
garnacha
90
Colour: salmon. Nose: sweet spices, red berry notes, fragrant herbs, dried flowers. Palate: full-bodied, flavourful, spicy, sweetness, long.

Encanteri 2020 T C
cariñena, garnacha, syrah
90
Colour: deep cherry. Nose: dried herbs, creamy oak, dark chocolate, woody, black fruit. Palate: powerful, ripe fruit, spicy, round tannins.

La Garnatxa D'En Pitu RD GR D
garnacha roja
93
Colour: light mahogany. Nose: aged wood nuances, sweet spices, dry nuts, honeyed notes, woody, varnish. Palate: powerful, flavourful, concentrated, aged character.

CELLER VINÍRIC
Ebre, s/n
17251 Calonge (Girona/Gerona)
☎: +34 635 501 314
david@viniric.cat
www.viniric.cat

Finques Incansables Garnatxa Negre 2019 T BA
garnacha
88
Crisp, spicy, dried herbs, coarse tannins, thin.

Finques Incansables Monastrell 2020 T BA
monastrell
91
Colour: deep cherry. Nose: dried herbs, creamy oak, balsamic herbs, red berry notes, black fruit. Palate: powerful, ripe fruit, spicy, round tannins, taut.

Propòsit 2019 T
garnacha, monastrell, cariñena
88
Fruity, crisp, floral, coarse tannins.

Propòsit Blanc 2021 B
garnacha blanca, malvasía
86

Vella Lola 2021 B
garnacha blanca, macabeo, malvasía
89
Citrus fruit, herbal, yeasty notes, tasty, crisp.

Vella Lola 2021 T
garnacha, cariñena, syrah
87

CELLERS D'EN GUILLA
Camí de Perelada 1
17755 Delfià Rabós d'Empordà (Girona/Gerona)
☎: +34 872 204 054
info@cellersdenguilla.com
www.cellersdenguilla.com

Garnatxa de Cellers d'en Guilla BF D
91
Colour: amber. Nose: dry nuts, creamy oak, macerated fruit, sweet spices. Palate: rich, spicy, old wood, flavourful.

Magenc 2021 B
90
Oxidativ. Colour: straw. Nose: dried herbs, faded flowers, white fruit, fine lees. Palate: powerful, ripe fruit, balanced.

Sa Illa 2017 T R
92
Defined aromas. Colour: cherry, purple rim. Nose: floral, spicy, fine reductive notes, red berry notes, black fruit. Palate: flavourful, fruity, good acidity, long.

Vinya del Metge 2021 RD
89
Oxidativ, dried flowers, crisp, balanced, tasty.

COOPERATIVA DE GARRIGUELLA
Ctra. de Roses, s/n
17780 Garriguella (Girona/Gerona)
☎: +34 972 530 002
info@cooperativagarriguella.com
www.cooperativagarriguella.com

Garriguella Garnatxa D'Empordá Ambré Dulce RD BA D
garnacha roja
89
Balanced, spicy, toasty, dried flowers, sweetish.

Garriguella Garnatxa D'Empordá Robí Dulce Natural T D
garnacha
90
Colour: bright ochre rim. Nose: candied fruit, dry nuts, sweet spices, aged wood nuances. Palate: good structure, fleshy, flavourful.

Garriguella Moscatel D'Empordá Dulce 2021 B D
moscatel de alejandría
88
Slight reduction, herbal, tasty, floral.

Puntils 2018 T C
garnacha, cabernet sauvignon
88
Pruney, balanced, jammy, age nuances.

Puntils Blanc 2021 B
garnacha blanca, moscatel de alejandría
87

Puntils Negre 2021 T
garnacha, cariñena
86

EL CELLER D'EN MARC
Mas Pages, Cami de Bell LLoc, s/n
17230 Palamòs (Girona/Gerona)
☎: +34 639 426 753
elcellerdenmarc@gmail.com
www.elcellerdenmarc.com

Lalut Superior 2018 T C
cabernet sauvignon
87

Lalut Blanc de Noir 2021 B
cabernet sauvignon, tempranillo
87

Lalut Creación Marcus 2017 T C
merlot, cabernet franc
91
Slight reduction. Colour: deep cherry. Nose: dried herbs, creamy oak, black fruit, burnt matches, smoky. Palate: powerful, ripe fruit, spicy, round tannins.

Lalut d'en Marc Magnifico 2017 T
cabernet franc
91
Colour: deep cherry. Nose: creamy oak, red berry notes, black fruit, scrubland. Palate: ripe fruit, spicy, round tannins, balanced.

Lalut Selección Especial 2017 T
syrah, cabernet sauvignon
90
Colour: deep cherry. Nose: dried herbs, creamy oak, black fruit, meaty notes. Palate: powerful, ripe fruit, spicy, round tannins.

EMPORDÀLIA
Ctra. de Roses, s/n
17494 Pau (Girona/Gerona)
☎: +34 972 530 140
Fax: +34 972 530 528
comunicacio@empordalia.com
www.empordalia.com

Antima 2018 T C
garnacha, cariñena
91
Colour: deep cherry. Nose: creamy oak, stone fruit, black fruit, scrubland. Palate: ripe fruit, spicy, round tannins.

Balmeta 2019 T
cariñena
87

DO EMPORDÀ / D.O.P.

DO EMPORDÀ / D.O.P.

Daina 2021 RD
garnacha gris, garnacha roja

88

Citrus fruit, fruity, sweety finish, pleasant, yeasty notes.

Icnos 2018 T C
cariñena, garnacha

92

Creamy. Colour: deep cherry. Nose: dried herbs, creamy oak, black fruit, stone fruit. Palate: powerful, ripe fruit, spicy, round tannins.

Mabre 2021 B
garnacha blanca

88

Citrus fruit, spicy, full-bodied, toasty. Nose: woody.

Sinols Garnatxa RD Solera
garnacha

89

Age nuances, complex, toasty, powerful, tasty.

ESPELT VITICULTORS

Mas Espelt s/n
17493 Vilajuiga (Girona/Gerona)
☎: +34 972 531 727
info@espeltviticultors.com
www.espeltviticultors.com

Espelt ComaBruna 2017 T
100% cariñena

92

Colour: deep cherry. Nose: creamy oak, black fruit, scrubland, cocoa bean, roasted coffee. Palate: powerful, ripe fruit, spicy, grainy tannins.

Espelt Quinze Roures 2021 B FB
60% garnacha blanca, 40% garnacha roja

91 ❦

Colour: straw. Nose: dried herbs, faded flowers, fine lees, white fruit. Palate: ripe fruit, balanced, flavourful.

Espelt Sauló 2021 T
70% garnacha, 30% cariñena

88 ❦

Balanced, fruity, jammy, floral.

Espelt Terres Negres 2019 T
86% cariñena, 14% garnacha

92 ❦

Colour: cherry, purple rim. Nose: fruit expression, red berry notes, floral, spicy, scrubland. Palate: flavourful, fruity, good acidity, long, taut.

Les Elies 2019 T BA
100% garnacha

94 ❦

Defined aromas. Colour: cherry, garnet rim. Nose: fruit expression, floral, spicy, red berry notes, black fruit, wild herbs. Palate: flavourful, fruity, good acidity, elegant, fleshy.

Pardells 2019 B
38% macabeo, 38% lledoner roig, 24% lledoner blanco

91 ❦

Colour: bright yellow. Nose: powerful, creamy oak, ripe fruit, spicy, dried flowers, woody. Palate: rich, good structure, toasty, fine bitter notes.

GALLINA DE PIEL WINES

Escoles, 43
17181 Aiguaviva (Girona/Gerona)
☎: +34 663 594 668
info@gallinadepielwines.com
www.gallinadepielwines.com

Roca del Crit 2020 T
82% cariñena, 18% garnacha

91

Colour: Cherry. Nose: balsamic herbs, sweet spices, scrubland, earthy notes. Palate: spicy, round tannins, flavourful.

MAS LLUNES

Ctra. de Vilajuiga, s/n
17780 Garriguella (Girona/Gerona)
☎: +34 972 552 684
info@masllunes.es
www.masllunes.es

Cercium 2019 T
52% garnacha, 16% cariñena, 16% syrah, 16% cabernet sauvignon

90

Colour: deep cherry. Nose: dried herbs, dark chocolate, fine reductive notes, black fruit. Palate: powerful, ripe fruit, spicy, round tannins.

Empórion 2018 T
58% garnacha, 42% cabernet sauvignon

91

Colour: deep cherry. Nose: dried herbs, creamy oak, black fruit, fine reductive notes. Palate: powerful, ripe fruit, spicy, round tannins.

Finca Butarós 2016 T
70% cariñena, 30% garnacha

91

Colour: deep cherry. Nose: dried herbs, creamy oak, black fruit, roasted coffee. Palate: powerful, ripe fruit, spicy, round tannins.

DO EMPORDÀ / D.O.P.

Nivia 2020 B FB
83% garnacha blanca, 17% macabeo

91

Colour: bright yellow. Nose: powerful, creamy oak, ripe fruit, spicy. Palate: rich, good structure, toasty.

Rhodes 2019 T C
54% cariñena, 30% garnacha, 16% syrah

91

Colour: deep cherry. Nose: dried herbs, creamy oak, black fruit, stone fruit. Palate: ripe fruit, spicy, round tannins.

Singulars Garnatxa Roja 2020 B FB
100% garnacha roja

92

Colour: bright yellow. Nose: powerful, creamy oak, spicy, stone fruit. Palate: rich, good structure, long, toasty, fine bitter notes.

MAS OLLER
Ctra. GI-652, Km. 0,23
17123 Torrent (Girona/Gerona)
☎: +34 972 300 001
info@masoller.es
www.masoller.cat

Mas Oller BlauNit 2021 T
syrah, garnacha

89 🌱

Taut, defined aromas, fruity, crisp, floral.

Mas Oller Malvasía 2019 B Mistela D
malvasía

91 🌱

Colour: bright yellow. Nose: honeyed notes, sweet spices, dried herbs. Palate: rich, fruity, powerful, flavourful.

Mas Oller Mar 2021 B
picapoll blanc, malvasía

91 🌱

Colour: straw. Nose: ripe fruit, dried herbs, faded flowers, citrus fruit. Palate: powerful, ripe fruit, balanced, fleshy.

Mas Oller Plus 2018 T
garnacha, syrah

91 🌱

Colour: deep cherry. Nose: dried herbs, creamy oak, black fruit, cereal notes, fine reductive notes. Palate: powerful, ripe fruit, spicy, round tannins.

Mas Oller Pur 2020 T C
syrah, garnacha, cabernet sauvignon

91 🌱

Colour: deep cherry. Nose: dried herbs, creamy oak, black fruit, cocoa bean. Palate: powerful, ripe fruit, spicy, round tannins.

Mas Oller Syrah de la Muntanya 2019 T C
syrah

91 🌱

Colour: deep cherry. Nose: dried herbs, creamy oak, black fruit, woody. Palate: powerful, ripe fruit, spicy, round tannins.

MASIA SERRA
Dels Solés, 20
17708 Cantallops (Girona/Gerona)
☎: +34 689 703 687
visit@masiaserra.com
www.masiaserra.com

Aroa 2018 T
100% garnacha

88 🌱

Pruney, slight oxidation, smooth, herbal.

Ctònia 2019 B FB
100% garnacha blanca

90 🌱

Colour: bright yellow. Nose: creamy oak, ripe fruit, spicy, balanced. Palate: rich, good structure, toasty, fine bitter notes.

Gneis 2015 T
55% merlot, 35% cabernet sauvignon, 10% garnacha

89

Balanced, spicy, jammy, tasty, toasty, slight reduction.

INO Garnatxa de L'Empordà 2020 RF Solera D

93

Colour: amber, orangey edge. Nose: candied fruit, dry nuts, pattiserie, sweet spices, dried herbs. Palate: fleshy, full-bodied, flavourful.

IO Masia Serra 2018 T R
50% merlot, 40% cabernet franc, 10% garnacha

89 🌱

Full-bodied, balanced, dried herbs, mineral, tasty.

Mosst 2021 B
85% garnacha blanca, 10% garnacha roja, 5% moscatel

87 🌱

DO EMPORDÀ / D.O.P.

OLIVEDA
La Roca, 3
17750 Capmany (Girona/Gerona)
☎: +34 972 549 012
comercial@grupoliveda.com
www.grupoliveda.com

Furot del Monestir 2017 T C
cariñena
90
Colour: deep cherry. Nose: dried herbs, creamy oak, black fruit. Palate: ripe fruit, spicy, round tannins.

Joana Rigau Ros 2021 B
chardonnay
89
Citrus fruit, spicy, dried flowers, full-bodied, jammy, woody.

La Bestia Blanca 2021 B
garnacha blanca
89
Citrus fruit, herbal, crisp, yeasty notes.

La Bestia Negra 2020 T
80% garnacha, 20% cabernet sauvignon
90
Colour: dark-red cherry. Nose: toasty, spicy, cocoa bean, woody. Palate: flavourful, toasty, fine bitter notes, powerful.

Rigau Ros 2015 T R
garnacha, cabernet sauvignon, merlot
90
Colour: dark-red cherry, garnet rim. Nose: ripe fruit, fruit preserve, tobacco, sweet spices. Palate: spicy, round tannins.

PERE GUARDIOLA
Ctra. GI-602, Km. 2,9
17750 Capmany (Girona/Gerona)
☎: +34 972 549 024
vins@pereguardiola.com
www.pereguardiola.com

Anhel d'Empordà 2017 T BA
100% cariñena
89
Full-bodied, dried herbs, jammy, powerful.

Anhel d'Empordà 2021 B
100% garnacha blanca
88
Citrus fruit, herbal, crisp, tasty.

Anhel d'Empordà 2021 RD
100% garnacha
87

Joncària Garnacha 2015 T
100% garnacha
89
Taut, full-bodied, jammy, tasty, powerful, toasty.

Torre de Capmany Moscatel 2021 B
100% moscatel de alejandría
88
Aromatic, varietally correct, tropical, jammy.

Torre de Capmany Viejas Soleras B D
garnacha
92
Colour: amber. Nose: candied fruit, honeyed notes, dry nuts, sweet spices. Palate: flavourful, unctuous, old wood.

PERELADA
Pl. del Carmen, 1
17491 Perelada (Girona/Gerona)
☎: +34 972 538 011
Fax: +34 972 538 277
rrpp@pereladachivite.com

Perelada 5 Finques 2018 T R
32% cabernet sauvignon, 27% garnacha, 22% merlot, 14% samsó
88
Full-bodied, pruney, spicy, roasted coffee, powerful.

Perelada Aires de Garbet 2019 T R
100% garnacha
91
Colour: deep cherry. Nose: dried herbs, creamy oak, aromatic coffee, roasted coffee, woody, red berry notes, black fruit. Palate: powerful, ripe fruit, spicy, round tannins, taut.

Perelada Ex Ex 14 2019 T C
100% garnacha
93
Colour: deep cherry. Nose: dried herbs, creamy oak, wild herbs, black fruit, fruit preserve, fine reductive notes, dark chocolate. Palate: powerful, ripe fruit, spicy, round tannins, elegant.

Perelada Gran Claustro 2015 T C
100% cabernet sauvignon
91
Colour: deep cherry. Nose: dried herbs, creamy oak, black fruit, roasted coffee. Palate: powerful, ripe fruit, spicy, round tannins.

RCR 2019 T C
100% garnacha
92
Colour: deep cherry. Nose: ripe fruit, dried herbs, creamy oak, stone fruit, woody. Palate: powerful, spicy, harsh oak tannins, fleshy, elegant.

Perelada Finca Garbet 2016 T R
100% syrah

94

Colour: cherry, garnet rim. Nose: creamy oak, roasted coffee, black fruit, scrubland, wild herbs. Palate: pruney, powerful, sweet tannins, flavourful.

Perelada Finca Garbet 2017 T R
100% syrah

93

Colour: deep cherry. Nose: creamy oak, black fruit, roasted coffee, meaty notes. Palate: powerful, spicy, round tannins, pruney, flavourful.

Perelada Finca Malaveïna 2019 T
63% merlot, 17% cabernet sauvignon, 14% cabernet franc, 6% garnacha

92

Colour: deep cherry. Nose: dried herbs, creamy oak, wild herbs, black fruit, red berry notes, toasty. Palate: powerful, ripe fruit, spicy, round tannins.

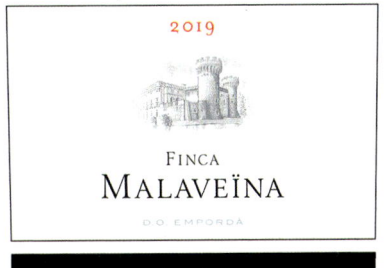

Perelada Garnatxa de l'Empordà 12 Anys Dulce Natural BF Solera D
80% garnacha roja, 20% garnacha blanca

94

Colour: iodine, amber rim. Nose: complex, dry nuts, creamy oak, varnish. Palate: rich, long, spicy, flavourful, old wood.

ROIG PARALS
Les Costes s/n
17752 Mollet de Peralada (Girona/Gerona)
☎: +34 972 634 320
info@roigparals.cat
www.roigparals.cat

Camí de Cormes 2019 T C
cariñena

92 🌱

Colour: deep cherry. Nose: creamy oak, red berry notes, black fruit, scrubland. Palate: powerful, ripe fruit, spicy, round tannins.

Camí de Cormes 2021 B
garnacha blanca, macabeo

92 🌱

Colour: straw. Nose: ripe fruit, dried herbs, faded flowers, fine lees. Palate: powerful, ripe fruit, balanced, flavourful, good finish.

L'Intrús 2019 T
cabernet sauvignon, merlot

89 🌱

Full-bodied, spicy, dried herbs, jammy, roasted coffee.

La Botera 2018 T C
cariñena, garnacha

89 🌱

Balanced, spicy, creamy, full-bodied.

Mallolet 2021 B
macabeo, garnacha blanca

90 🌱

Colour: bright straw. Nose: ripe fruit, fragrant herbs, fine lees, dried flowers. Palate: full-bodied, rich, good acidity, flavourful.

TOCAT DE L'ALA
Carrer de les Costes s/n
17752 Mollet de Peralada (Girona/Gerona)
☎: +34 619 776 948
info@tocatdelala.cat
www.tocatdelala.cat

Tocat de l'Ala 2015 T
92

Colour: deep cherry. Nose: dried herbs, creamy oak, black fruit, scrubland, fine reductive notes. Palate: powerful, ripe fruit, spicy, round tannins.

DO EMPORDÀ / D.O.P.

DO EMPORDÀ / D.O.P.

Tocat de l'Ala Blanc 2021 B
60% garnacha blanca, 40% macabeo

91

Colour: bright straw. Nose: ripe fruit, fragrant herbs, fine lees. Palate: full-bodied, rich, good acidity.

VINS DE RELAT
Ctra. Vilafranca-Sant Jaume
dels Domenys, Km. 8,1
08732 Castellví de la Marca (Barcelona)
☎: +34 938 743 511
info@roquetaorigen.com
www.capcreus.es

Cap de Creus Nacre 2021 B
50% macabeo, 40% lledoner blanco, 10% lledoner roig

87

VINS I LICORS GRAU
Torroella, 163
17200 Palafrugell (Girona/Gerona)
☎: +34 972 301 835
info@vinsilicorsgrau.com
www.grauonline.com

99 Punts 2019 T
80% garnacha, 20% syrah

88

Spicy, herbaceous, jammy, full-bodied.

Malcuat 2021 T
80% garnacha, 10% merlot, 10% syrah

86

Flor d'Empordà 2018 T
70% cariñena, 20% garnacha, 10% syrah

91

Colour: deep cherry. Nose: dried herbs, creamy oak, red berry notes, black fruit. Palate: powerful, ripe fruit, spicy, round tannins.

Murri 2021 B
80% garnacha blanca, 20% macabeo

87

Trapella 2021 RD
100% syrah

87

VINYES D'OLIVARDOTS
Paratge Olivardots, s/n
17750 Capmany (Girona/Gerona)
☎: +34 650 395 627
vdo@olivardots.com
www.olivardots.com

Blanc de Gresa 2020 B FB
45% garnacha blanca, 30% cariñena blanca, 25% garnacha rosada

89 🌱

Citrus fruit, full-bodied, spicy, full-bodied, jammy, woody.

Finca Olivardots Groc D'Anfora 2020 B
50% garnacha blanca, 30% garnacha rosada, 20% macabeo

91 🌱

With personality. Colour: straw. Nose: ripe fruit, dried herbs, faded flowers, bakery, red clay notes. Palate: powerful, ripe fruit, balanced, bitter.

Gresa 2015 T R
50% garnacha, 35% cariñena, 15% syrah

92 🌱

Colour: deep cherry. Nose: creamy oak, black fruit, scrubland, fine reductive notes, meaty notes. Palate: ripe fruit, spicy, round tannins.

Vd'O 1.15 2015 T
100% cariñena

92 🌱

Colour: deep cherry. Nose: black fruit, scrubland, fine reductive notes, mineral, cocoa bean. Palate: powerful, ripe fruit, spicy, round tannins.

Vd'O 2.15 2015 T R
100% cariñena

91 🌱

Colour: garnet rim, cherry, garnet rim. Nose: fruit preserve, tobacco, black liquorice, scrubland. Palate: spicy, round tannins, balanced.

Vd'O 6.19 2019 B
100% cariñena blanca

90 🌱

Colour: bright yellow. Nose: powerful, creamy oak, ripe fruit, spicy, woody. Palate: rich, good structure, toasty.

VINYES DELS ASPRES

Requesens, 7
17708 Cantallops (Girona/Gerona)
☎: +34 619 741 442
dmolas@vinyesdelsaspres.cat
www.vinyesdelsaspres.cat

🏆 PODIUM

Bac de les Ginesteres B RC D
100% garnacha roja

95

With personality. Colour: bright ochre rim. Nose: powerful, smoky, roasted almonds, animal reductive notes, dried fruit, petrol notes. Palate: powerful, flavourful, full-bodied, sweet.

Blanc dels Aspres 2020 B FB

91

Colour: bright yellow. Nose: powerful, creamy oak, ripe fruit, spicy. Palate: rich, good structure, toasty, fine bitter notes.

Negre dels Aspres 2018 T C
48% garnacha, 32% syrah, 20% merlot

91

Colour: cherry, garnet rim. Nose: creamy oak, red berry notes, black fruit, scrubland. Palate: ripe fruit, spicy, round tannins.

S'Alou 2018 T C
66% garnacha, 34% syrah

93

Colour: cherry, garnet rim. Nose: creamy oak, red berry notes, black fruit, fine reductive notes, meaty notes. Palate: powerful, ripe fruit, spicy, round tannins, balanced.

Soques 2018 T R
100% garnacha

92

Colour: deep cherry. Nose: red berry notes, black fruit, scrubland, cocoa bean. Palate: ripe fruit, spicy, round tannins, fleshy.

Xot Blanc 2020 B
48% garnacha blanca, 35% sauvignon blanc, 17% picapoll blanc

92

Colour: bright yellow. Nose: powerful, ripe fruit, spicy, balanced, dried flowers. Palate: rich, good structure, long, toasty, fine bitter notes.

WINES RAMSURT BY MAGI RAVENTOS

Can Majó 18
08173 Sant Cugat (Barcelona)
☎: +34 652 439 922
comercialramsurt@gmail.com
www.ramsurt.com

Claret de Tardor 2020 T
garnacha, cabernet sauvignon

87

DO EMPORDÀ / D.O.P.

DO. GETARIAKO TXAKOLINA

CONSEJO REGULADOR

Parque Aldamar, 4 bajo
20808 Getaria (Gipuzkoa)
☎: +34 943 140 383 - Fax: +34 943 896 030
@: info@getariakotxakolina.eus
www.getariakotxakolina.eus

LOCATION:

Mainly on the coastal belt of the province of Guipuzcoa, covering the vineyards situated in the municipal areas of Aia, Getaria and Zarauz, at a distance of about 25 km from San Sebastián.

GRAPE VARIETIES:

WHITE: Hondarrabi Zuri, Gros Manseng, Riesling, Chardonnay and Petit Courbu.

RED: Hondarrabi Beltza.

FIGURES:

Vineyard surface: 443 – **Wine-Growers:** 105 – **Wineries:** 36 – **2021 Harvest rating:** Good – **Production 21:** 2,699,344 litres – **Market percentages:** 80% National - 20% International.

SOIL:

The vineyards are situated in small valleys and gradual hillsides at altitudes of up to 200 m. They are found on humid brownish-grey limy soil, which are rich in organic matter.

CLIMATE:

Fairly mild, thanks to the influence of the Bay of Biscay. The average annual temperature is 13°C, and the rainfall is plentiful with an average of 1,600 mm per year.

CLASSIFICATION OF YOUNG WINE HARVESTS PEÑÍNGUIDE

2017	2018	2019	2020	2021
GOOD	GOOD	VERY GOOD	VERY GOOD	VERY GOOD

ADUR TXAKOLINA

Eitzaga Auzoa, s/n
20808 Getaria (Gipuzkoa/Guipúzcoa)
☎: +34 629 581 175
info@adurtxakolina.com

Adur sobre Lías 2019 B
100% hondarrabi zuri

89

Rustic, herbal. Colour: yellow. Nose: ebb tide, fine lees, grassy. Palate: fresh, light-bodied.

Adur sobre Lías 2020 B
100% hondarrabi zuri

87

Adur XO 2017 B
100% hondarrabi zuri

91

Crisp, toasty. Colour: yellow. Nose: stone fruit, toasty, creamy oak. Palate: fruity, racy.

AITAREN

Mardubidea, 28 Arroa Behea
20749 Zestoa (Gipuzkoa/Guipúzcoa)
☎: +34 690 053 491
bodega@bodegasaitaren.com
www.bodegasaitaren.com

Aitaren 2019 B FB
100% hondarrabi zuri

91

Colour: straw. Nose: ripe fruit, dried herbs, faded flowers. Palate: powerful, ripe fruit, balanced.

Lurretik 2021 B
100% hondarrabi zuri

90

Colour: bright straw. Nose: ripe fruit, floral. Palate: flavourful, good acidity, fruity aftertaste.

AIZPURUA

Ctra. de Meagas
20208 Getaria (Gipuzkoa/Guipúzcoa)
☎: +34 943 140 696
:txakoliaizpurua@gmail.com
www.txakoliaizpurua.com

Txakoli Aialle 2020 B
hondarrabi zuri

90

Colour: bright straw. Nose: white fruit, ripe fruit, lactic notes, dried herbs. Palate: fresh, fruity, balanced.

Txakoli Aizpurua 2021 B
hondarrabi zuri

87

BODEGA AMEZTOI

Eitzaga, 10
20808 Getaria (Gipuzkoa/Guipúzcoa)
☎: +34 943 140 918
ameztoi@txakoliameztoi.com
www.txakoliameztoi.com

Kirkilla Ameztoi 2017 B
hondarrabi zuri

91

Colour: bright golden. Nose: white fruit, ripe fruit, fine lees. Palate: fruity, balanced, round.

Primus Ameztoi 2020 B
hondarrabi zuri

90

Colour: bright straw. Nose: citrus fruit, fresh fruit, wild herbs, white flowers. Palate: fruity, balanced, fresh.

BODEGA K5

Caserío Estenaga, 16 Bº Andatza
20809 Aia (Gipuzkoa/Guipúzcoa)
☎: +34 943 240 005
bodega@bodegak5.com
www.bodegak5.com

K Pilota 2020 B

90

Colour: bright straw. Nose: fresh fruit, mineral, white flowers. Palate: fruity, fresh, balanced.

K5 2016 B
100% hondarrabi zuri

93

Colour: bright golden. Nose: sweet spices, roasted almonds, toasty, ripe fruit, dried herbs. Palate: fresh, flavourful, balanced, good acidity.

K5 2019 B
hondarrabi zuri

91

Colour: bright straw. Nose: fruit expression, toasted bread, spicy, dried herbs. Palate: fruity, fresh, flavourful.

BODEGAS DE LOS HEREDEROS DEL MARQUÉS DE RISCAL

Torrea, 1
01340 Elciego (Araba/Álava)
☎: +34 945 606 000
Fax: +34 945 606 023
marquesderiscal@marquesderiscal.com
www.marquesderiscal.com

Marqués de Riscal Txakoli 2020 B

91

Colour: bright straw. Nose: expressive, white flowers, jasmine, dried herbs. Palate: flavourful, fruity, balanced.

DO GETARIAKO TXAKOLINA / D.O.P.

DO GETARIAKO TXAKOLINA / D.O.P.

GAINTZA
Bº San Prudencio, 26
20808 Getaria (Gipuzkoa/Guipúzcoa)
☎: +34 943 140 032
info@gaintza.com
www.gaintza.com

Aitako 2019 B AG
90% hondarrabi zuri, 10% chardonnay

90

Colour: bright straw, greenish rim. Nose: fresh fruit, citrus fruit, wild herbs. Palate: fresh, fruity, good acidity, fine bitter notes.

Gaintza 2021 B
90% hondarrabi zuri, 10% gros manseng

88

Citrus fruit, crisp, herbal, yeasty notes.

GARAIKOETXEA TXAKOLINDEGIA
Arana Bailara, 18
20494 Alkiza (Gipuzkoa/Guipúzcoa)
☎: +34 605 758 572
inaziourruzola@gmail.com
www.inaziourruzola.eus

Ernio sobre Lías 2020 B
89

Citrus fruit, fruity, herbal, tasty.

HIRUZTA BODEGA
Barrio Jaizubia, 266
20280 Hondarribia (Gipuzkoa/Guipúzcoa)
☎: +34 943 646 689
info@hiruzta.com
www.hiruzta.com

Hiruzta Berezia 2020 B
100% hondarrabi zuri

91

Colour: bright straw. Nose: white fruit, lactic notes, floral. Palate: fruity, flavourful, balanced.

Hiruzta Rosé 2021 RD
55% hondarrabi beltza, 45% hondarrabi zuri

88

Kind finish, fruity, crisp, simple.

Hiruzta Txakolina 2021 B
100% hondarrabi zuri

90

Colour: bright straw. Nose: dry stone, citrus fruit, dried herbs. Palate: fresh, flavourful, balanced.

REZABAL
Asti Auzoa, 628
20800 Zarautz (Gipuzkoa/Guipúzcoa)
☎: +34 615 748 711
info@txakolirezabal.com
www.txakolirezabal.com

Txakoli Rezabal 2021 B
100% hondarrabi zuri

89

Citrus fruit, herbal, slight fizz.

Txakoli Rezabal Rosé 2021 RD
hondarrabi zuri, hondarrabi beltza

87

TALAI BERRI TXAKOLINA
Talaimendi Auzoa, 728 Junto al Camping Zarautz
20800 Zarautz (Gipuzkoa/Guipúzcoa)
☎: +34 943 132 750
Fax: +34 943 132 750
info@talaiberri.com
www.talaiberri.com

Finca Jakue Txakolina 2021 B
hondarrabi zuri

88

Citrus fruit, crisp, fruity, herbal, balanced.

Talai Berri Rosé 2021 RD
hondarrabi zuri, hondarrabi beltza

87

Talai Berri Txakolina 2021 B
hondarrabi zuri, hondarrabi beltza

87

TXOMIN ETXANIZ
Bº Eitzaga, 21
20808 Getaria (Gipuzkoa/Guipúzcoa)
☎: +34 943 140 702
txakoli@txominetxaniz.com
www.txominetxaniz.com

TX Txomin Etxaniz 2020 B BA
100% hondarrabi zuri

91

Citrus fruit, crisp, herbal. Colour: bright straw. Nose: fragrant herbs, fine lees, fresh fruit. Palate: good acidity, fine bitter notes, balanced, good finish.

Txomin Etxaniz 2021 B
90% hondarrabi zuri, 10% hondarrabi beltza

90

Colour: bright straw. Nose: wild herbs, white fruit, grassy. Palate: fresh, good acidity, fine bitter notes.

Txomin Etxaniz 2021 RD
hondarrabi beltza

87

Txomin Etxaniz Berezia 2020 B
90% hondarrabi zuri, 10% hondarrabi beltza

90

Racy. Colour: bright straw. Nose: fresh fruit, citrus fruit, wild herbs. Palate: fresh, fruity, good acidity, fine bitter notes.

DO. GRAN CANARIA
CONSEJO REGULADOR

Calvo Sotelo, 26
35300 Santa Brígida (Las Palmas)
☎: +34 928 640 462 - Fax: +34 928 640 982
@: crdogc@yahoo.es
www.vinosdegrancanaria.es

LOCATION:

The production region covers 99% of the island of Gran Canaria, as the climate and the conditions of the terrain allow for the cultivation of grapes at altitudes close to sea level up to the highest mountain tops. The DO incorporates all the municipal areas of the island, except for the Tafira Protected Landscape which falls under an independent DO, Monte de Lentiscal, also fully covered in this Guide.

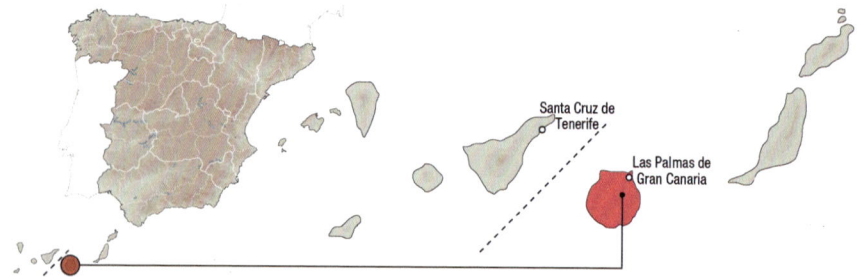

GRAPE VARIETIES:

WHITE: Preferred: Malvasía, Güal, Marmajuelo (Bermejuela), Vijariego, Albillo and Moscatel.

Authorized: Listán Blanco, Burrablanca, Torrontés, Pedro Ximénez, Brebal and Bastardo Blanco.

RED: Preferred: Listán Negro, Negramoll, Tintilla, Malvasía Rosada.

Authorized: Moscatel Negra, Bastardo Negro, Listán Prieto and Vijariego Negro.

FIGURES:

Vineyard surface: 184 – **Wine-Growers:** 289 – **Wineries:** 47 – **2021 Harvest rating:** N/A – **Production 21:** 196,414 litres – **Market percentages:** 95% National - 5% International.

SOIL:

The vineyards are found both in coastal areas and on higher grounds at altitudes of up to 1500 m, resulting in a varied range of soils.

CLIMATE:

As with the other islands of the archipelago, the differences in altitude give rise to several microclimates which create specific characteristics for the cultivation of the vine. Nevertheless, the climate is conditioned by the influence of the trade winds which blow from the east and whose effect is more evident in the higher-lying areas.

CLASSIFICATION OF YOUNG WINE HARVESTS PEÑÍNGUIDE

2017	2018	2019	2020	2021
GOOD	N/A	N/A	N/A	N/A

BIEN DE ALTURA

Ctra. de Bandama, 68
35310 Santa Brígida (Las Palmas)
☎: +34 651 463 133
biendealturavinos@gmail.com

Agan 2020 T
listán negro, otras

94

Colour: cherry, purple rim. Nose: fruit expression, red berry notes, floral, spicy, mineral. Palate: flavourful, fruity, good acidity, taut.

Ikewen 2 años Velo de Flor 2019 B
91

Slight oxidation. Colour: yellow. Nose: flor yeasts, faded flowers, white fruit, ripe fruit, wild herbs, lees reduction notes. Palate: correct, long, fine bitter notes.

Ikewen 2020 T
listán negro, otras

93

Colour: cherry, purple rim. Nose: red berry notes, floral, spicy, earthy notes. Palate: flavourful, fruity, good acidity.

Sansofí 2020 T
91

Slight oxidation. Colour: cherry, purple rim. Nose: red berry notes, floral, citrus fruit. Palate: flavourful, fruity, good acidity.

BODEGA MUNICIPAL DE AGÜIMES

Juan Ramón Jiménez, s/n.
Agüimes (Las Palmas)
☎: +34 928 789 980
ysantana@aguimes.es
www.aguimes.es

Señorío de Agüimes 2021 B D
88

Fruity, tasty, floral, full-bodied, jammy, sweety finish.

Señorío de Agüimes 2021 RD
tintilla, listán negro

85

BODEGA SAN JUAN

Ctra. de Bandama, 68
35310 Santa Brígida (Las Palmas)
☎: +34 626 911 643
info@bodega-sanjuan.com
www.bodega-sanjuan.com

Mocanal 2020 T
85% listán negro, 15% negramoll

93 ✿

Colour: deep cherry. Nose: dried herbs, creamy oak, black fruit, smoky. Palate: powerful, ripe fruit, spicy, round tannins, flavourful.

BODEGA VANDAMA

Ctra. Bandama, 116
35310 Santa Brígida (Las Palmas)
☎: +34 928 352 754
bodegonvandama@bodegonvandama.com

Vandama Hoya Oscura 2021 T
listán negro, negramoll

Nominado Vino Revelación

94

Colour: Cherry. Nose: balsamic herbs, sweet spices, scrubland, red berry notes, black fruit, earthy notes. Palate: spicy, balsamic, good acidity, flavourful.

Vandama Vino de Finca 2021 T
listán negro, negramoll

94

Colour: cherry, purple rim. Nose: red berry notes, floral, spicy, earthy notes, balsamic herbs. Palate: flavourful, fruity, good acidity, good structure.

BODEGA VEGA DE GÁLDAR

Cuesta de la Encarnación
35460 Gáldar (Las Palmas)
☎: +34 605 043 047
lamenora1960@yahoo.es
www.vegadegaldar.com

El Convento de la Vega 2018 T
84 ✿

Vega de Gáldar 2020 T
vijariego negro

82

Viña Amable 12 meses 2017 T
listán negro, castellana

87

Viña Amable 4 meses 2020 T RB
listán negro, castellana

85

DO GRAN CANARIA / D.O.P.

DO GRAN CANARIA / D.O.P.

BODEGA VENTURA
35300 Santa Brígida (Las Palmas)
bodegasventura@gmail.com

Cruz 2020 T C
listán negro
91
Colour: cherry, purple rim. Nose: red berry notes, floral, spicy, smoky. Palate: flavourful, fruity, good acidity, long.

Eidan 2021 B SD
moscatel
87

Eidan 2021 T
listán negro, tintilla
87

BODEGAS LAS TIRAJANAS
Las Lagunas s/n
35290 San Bartolomé de Tirajana (Las Palmas)
☎: +34 928 155 978
info@bodegaslastirajanas.com
www.bodegaslastirajanas.com

Las Tirajanas 2019 T RB
listán negro, castellana, vijariego negro, tintilla
88
Balanced, spicy, floral, fruity, herbal.

Las Tirajanas Malvasía Volcánica 2019 B
malvasía
87

Las Tirajanas Malvasía Volcánica 2021 B FB
malvasía
86

Las Tirajanas Paraje Caldera de Tirajana 2021 T RB
tintilla, vijariego negro
90
Slight oxidation. Colour: cherry, purple rim. Nose: fruit expression, red berry notes, floral, spicy. Palate: fruity, flavourful, balanced.

Las Tirajanas Tinamar 2021 T RB
listán negro
90
Colour: cherry, purple rim. Nose: red berry notes, floral, spicy, smoky. Palate: flavourful, fruity, good acidity.

Las Tirajanas Verijadiego 2019 B
verijadiego
86

RINCÓN DEL GUINIGUADA
35309 Santa Brígida (Las Palmas)
☎: +34 655 472 045
olalladm@hotmail.com

Rincón del Guiniguada 2020 B
listán blanco, verijadiego blanco, marmajuelo, malvasía
86

Rincón del Guiniguada 2020 T
listán negro, castellana, verijadiego, negramoll
87

SEÑORÍO DE CABRERA
Barranco García Ruiz, 5
35200 Telde (Las Palmas)
☎: +34 620 217 695
agustincabrera@sunandbeachhotels.com

Señorío de Cabrera 2019 T
tintilla, listán negro
85

Señorío de Cabrera 2020 T
tintilla, listán negro
87

Señorío de Cabrera 2021 B
malvasía
86

DO. GRANADA

CONSEJO REGULADOR

Cortijo Peinado, Carretera de Fuente Vaqueros s/n, km2
18340 Fuente Vaqueros (Granada)
☎: +34 691 032 409
@: info@dovinosdegranada.es
www.dopvinosdegranada.es

LOCATION:

Located in the southeast of Spain, in Andalusia, in the heart of the Penibetic mountain range, the highest altitude mountain range on the peninsula. It is made up of 168 municipalities in the province of Granada and it has a differentiated sub-zone.

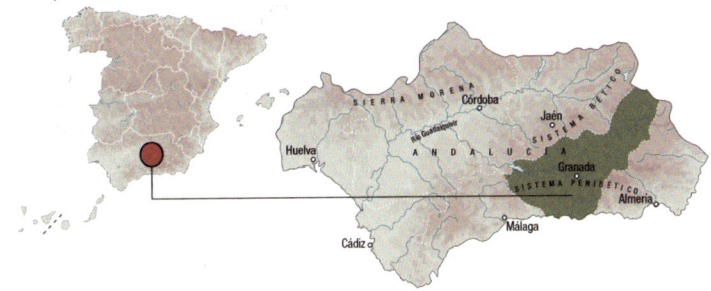

GRAPE VARIETIES:

RED: tempranillo, garnacha, cabernet sauvignon, cabernet franc, merlot, syrah, pinot noir, petit verdot, monastrell and romé.

WHITE: vijiriego, sauvignon blanc, chardonnay, moscatel de grano menudo, moscatel de Alejandría, baladí verdejo, pedro ximénez, palomino and torrontés.

SUBZONES:

It has a complex topography due to the fact that the vineyards are planted mainly in the highest areas of Granada, with an average altitude of around 1,200 m.a.s.l, which gives the area an important thermal amplitude. The soils have variable compositions of slate and clay. The relief of terraced hillsides and slopes means that the work in the vineyards is more labour intensive than in other cases, as it is not possible to mechanise the harvesting process.

FIGURES:

Vineyard surface: 260 – **Wine-Growers:** 50 – **Wineries:** 22 – **2021 Harvest rating:** Very Good – **Production 21:** 430,700 litres.
Market percentages: 79% National - 21% International.

SOILS:

It has a complex topography due to the fact that the vineyards are planted mainly in the highest areas of Granada, with an average altitude of around 1,200 m.a.s.l, which gives the area an important thermal amplitude. The soils have variable compositions of slate and clay. The relief of terraced hillsides and slopes means that the work in the vineyards is more labour intensive than in other cases, as it is not possible to mechanise the harvesting process.

CLIMATE:

Influenced by the Mediterranean and continental climates. The temperature and the cool air currents of the Sierra Nevada affect the development of the grapes. Due to its altitude and its location in the middle of the mountains, the heat summation plays an important role in the growth of the vines and the characteristics of the wines. The average annual rainfall in the area is around 450 mm.

CLASSIFICATION OF YOUNG WINE HARVESTS PEÑÍNGUIDE

2017	2018	2019	2020	2021
GOOD	N/A	N/A	N/A	N/A

DO GRANADA / D.O.P.

BODEGAS AL ZAGAL
Paraje Las Cañaillas, s/n
18518 Cogollos de Guadix (Granada)
☎: +34 958 105 605
info@bodegasalzagal.es
www.bodegasalzagal.es

Rey Zagal 2011 T C
72% tempranillo, 17% cabernet sauvignon, 11% syrah
89
Pruney, fruity, spicy, kind finish, great length, tasty.

Rey Zagal 2013 T R
65% tempranillo, 23% merlot, 12% syrah
89
Slight reduction, pruney, spicy, dried herbs, great length, rustic, tasty.

Rey Zagal Cascamorras 2016 T
69% tempranillo, 27% garnacha, 4% cabernet sauvignon
88
Standard, jammy, simple, wild.

Rey Zagal Sauvignon Blanc 2020 B
100% sauvignon blanc
87

BODEGAS FERNÁNDEZ HERRERO
Ctra. A-4301, Pk. 13.5
18830 Huéscar (Granada)
☎: +34 647 785 367
rfernandez1941@gmail.com

Irving Syrah 2020 T
syrah
89 🌱
Spicy, fruity, jammy, tasty.

BODEGAS FONTEDEI
Doctor Horcajadas, 10
18570 Deifontes (Granada)
☎: +34 958 407 957
info@bodegasfontedei.es
www.bodegasfontedei.es

Aixa 2020 RD
100% merlot
86

Albayda 2020 B FB
chardonnay, sauvignon blanc
90
Fruity, jammy, yeasty notes. Nose: dry nuts, pungent, flor yeasts.

Garnata 2016 T R
garnacha, syrah
89
Classic, pruney, reduced, hot, spicy.

Lindaraja 2019 T RB
70% tempranillo, 30% syrah
89
Spicy, jammy, tasty, great length. Nose: black fruit.

Prado Negro 2016 T C
tempranillo, merlot, garnacha, cabernet sauvignon
88
Full-bodied, pruney, tasty, spicy, smoky, balsamic herbs.

Zacatin sobre Lías 2019 B
100% moscatel de alejandría
89
Aromatic, varietally correct, dried flowers, standard, balanced.

BODEGAS MUÑANA
Ctra. Graena a La Peza, s/n Finca Peñas Prietas
18517 Cortes Y Graena (Granada)
☎: +34 958 670 715
administracion@bodegasmunana.com
www.bodegasmunana.com

**Muñana 1188
Cabernet Sauvignon 2018 T**
cabernet sauvignon
88
Spicy, fruity, jammy, astringent.

BODEGAS SEÑORÍO DE NEVADA
Ctra. de Cónchar, s/n
18659 Villamena (Granada)
☎: +34 958 777 092
Fax: +34 958 107 367
info@senoriodenevada.es
www.senoriodenevada.es

Señorío de Nevada Bronce 2019 T
cabernet sauvignon, merlot, garnacha
87

**Señorío de Nevada
Club de la Barrica 2018 T**
syrah
88
Pruney, full-bodied, dried herbs, spicy, hot. Nose: dark chocolate.

Señorío de Nevada Oro 2019 T
syrah, cabernet sauvignon
86

Señorío de Nevada Plata 2019 T
syrah, cabernet sauvignon, petit verdot
88
Pruney, dried herbs, wild, spicy, full-bodied.

LOS BARRANCOS

Ctra. Cádiar - Albuñol, km. 9,4
18449 Lobras (Granada)
☎: +34 686 387 550
info@lavineria.de
www.los-barrancos.com

Cerro de la Retama 2016 T C
cabernet sauvignon, merlot

89 🌱

Spicy, fruity, tasty, astringent.

Corral de Castro 2014 T R
tempranillo, cabernet sauvignon

88 🌱

Spicy, fruity, jammy, astringent.

Corral de Castro 2017 T
tempranillo, cabernet sauvignon

89 🌱

Herbal, spicy, smoky, fruity, astringent.

Loma de los Felipes 2018 T C
tempranillo, cabernet sauvignon

89 🌱

Spicy, herbal, jammy, tasty, full-bodied. Palate: slightly dry, soft tannins.

XOLAYR

Avda. Andalucía, 1
18659 Cozvijar (Granada)
☎: +34 620 126 514
lopezdelacasa@gmail.com

Xolayr 2019 T BA
70% tempranillo, 30% syrah

86

DO GRANADA / D.O.P.

DO. JEREZ-XÈRÉS-SHERRY- MANZANILLA DE SANLÚCAR DE BARRAMEDA
CONSEJO REGULADOR

Avda. Álvaro Domecq, 2
11405 Jerez de la Frontera (Cádiz)
☎: +34 956 332 050 - Fax: +34 956 338 908
@: vinjerez@sherry.org
www.sherry.org

LOCATION:

In the province of Cádiz. The production area covers the municipal districts of Jerez de la Frontera, El Puerto de Santa María, Chipiona, Trebujena, Rota, Puerto Real, Chiclana de la Frontera and some estates in Lebrija.

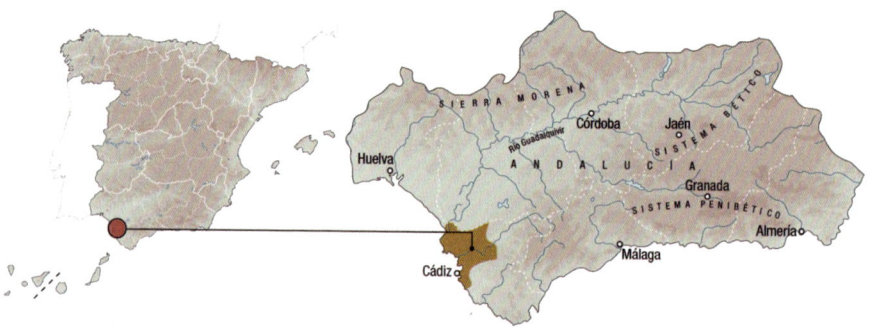

GRAPE VARIETIES:

WHITE: Palomino (90%), Pedro Ximénez, Moscatel, Palomino Fino and Palomino de Jerez.

FIGURES:

Vineyard surface: 7,082 – **Wine-Growers:** 2,203 – **Wineries:** 81 – **2021 Harvest rating:** Very Good – **Production 21:** 39,828,500 litres – **Market percentages:** 67% National - 33% International.

SOIL:

The so-called 'Albariza' soil is a key factor regarding quality. This type of soil is practically white and is rich in calcium carbonate, clay and silica. It is excellent for retaining humidity and storing winter rainfall for the dry summer months. Moreover, this soil determines the so-called 'Jerez superior'. It is found in Jerez de la Frontera, Puerto de Santa María, Sanlúcar de Barrameda and certain areas of Trebujena. The remaining soil, known as 'Zona', is muddy and sandy.arenas.

CLIMATE:

Warm with Atlantic influences. The west winds play an important role, as they provide humidity and help to temper the conditions. The average annual temperature is 17.5°C, with an average rainfall of 600 mm per year.

CLASSIFICATION OF YOUNG WINE HARVESTS PEÑÍNGUIDE

This denomination of origin, due to the wine-making process, does not make available single-year wines indicated by vintage, so the following evaluation refers to the overall quality of the wines that were tasted this year.

DO JEREZ / D.O.P.

ALTANZA - COLECCIÓN R. AMILLO
Asta, 2
11404 Jerez de la Frontera (Cádiz)
☎: +34 618 629 086
altanza@altanza.com
www.altanza.com

🏆 PODIUM

Altanza Colección Roberto Amillo Amontillado BF AM S
100% palomino

95
Colour: iodine, amber rim. Nose: powerful, complex, elegant, dry nuts, toasty, varnish. Palate: rich, fine solera notes, long, elegant.

Altanza Colección Roberto Amillo Oloroso BF OL S
100% palomino

94
Colour: light mahogany. Nose: dry nuts, sweet spices, aged wood nuances, varnish. Palate: complex, flavourful, dry, old wood.

🏆 PODIUM

Altanza Colección Roberto Amillo Palo Cortado BF PC S
100% palomino

96
Colour: light mahogany. Nose: acetaldehyde, pungent, varnish, aged wood nuances, creamy oak, dry nuts. Palate: powerful, flavourful, spicy, long, balanced.

🏆 PODIUM

Altanza Colección Roberto Amillo Pedro Ximénez BF PX D
100% pedro ximénez

95
Colour: mahogany. Nose: powerful, expressive, spicy, dry nuts, incense. Palate: elegant, fine solera notes, toasty, long.

🏆 PODIUM

La Saca de Altanza BF PC S
100% palomino

98
Colour: light mahogany. Nose: acetaldehyde, pungent, varnish, aged wood nuances, powerful, expressive. Palate: flavourful, spicy, long, balanced, fine solera notes.

ÁLVARO DOMECQ
Alamos, 23
11401 Jerez de la Frontera (Cádiz)
☎: +34 956 339 634
alvarodomecqsl@alvarodomecq.com
www.alvarodomecq.com

1730 BF PX D
100% pedro ximénez

94
Representative, complex. Colour: dark mahogany. Nose: powerful, expressive, aromatic coffee, spicy, acetaldehyde. Palate: balanced, fine solera notes, toasty, long.

1730 Fino en Rama BF FI S
100% palomino

94
Balanced, elegant. Nose: pungent, flor yeasts, roasted almonds. Palate: dry, flavourful, long.

🏆 PODIUM

1730 VORS BF AM S
100% palomino

95
Colour: light mahogany. Nose: candied fruit, spicy, varnish, complex, characterful. Palate: fine solera notes, fine bitter notes, round.

🏆 PODIUM

1730 VORS BF OL S
100% palomino

96
Colour: light mahogany. Nose: complex, dry nuts, toasty, acetaldehyde, elegant. Palate: long, fine solera notes, spicy, round, flavourful.

🏆 PODIUM

1730 VORS BF PC S
100% palomino

95
Colour: light mahogany. Nose: acetaldehyde, pungent, varnish, aged wood nuances, creamy oak, spicy, incense. Palate: powerful, flavourful, spicy, long, balanced.

Alburejo BF OL S
100% palomino

91
Colour: iodine, amber rim. Nose: dry nuts, varnish. Palate: spicy, old wood.

Aranda Cream BF CRM
palomino, pedro ximénez

87

DO JEREZ / D.O.P.

La Jaca BF MZ S
100% palomino

91

Colour: bright yellow. Nose: flor yeasts, lees reduction notes, faded flowers, ebb tide. Palate: dry, easy to drink, saline.

La Janda BF FI S
palomino

93

Dried flowers, jammy. Nose: expressive, pungent, yeasty notes. Palate: flavourful, long, full-bodied.

Viña 98 BF PX D
100% pedro ximénez

90

Colour: light mahogany. Nose: fruit liqueur notes, dried fruit, caramel. Palate: flavourful, sweet.

BODEGA SAN FRANCISCO JAVIER
11404 Jerez de la Frontera (Cádiz)
info@pingus.es

🏆 PODIUM

Viña Corrales Pago Balbaina BF FI
palomino

95 🌱

Colour: bright yellow. Nose: pungent, chamomile, dry nuts, faded flowers. Palate: good acidity, bitter, spicy, long.

BODEGAS ARFE
Molino de Viento, 12
11401 Jerez de la Frontera (Cádiz)
☎: +34 665 570 316
clubarfe1767@telefonica.net
www.bodegasarfe.com

De la Cruz de 1767 BF PC S
palomino

94

Colour: light mahogany. Nose: acetaldehyde, pungent, varnish, aged wood nuances. Palate: flavourful, spicy, long, balanced.

BODEGAS ARGÜESO
Mar, 8
11540 Sanlúcar de Barrameda (Cádiz)
☎: +34 956 385 200
comunicacion@argueso.es
www.argueso.es

1822 BF PC S
listán blanco

93

Colour: iodine, amber rim. Nose: powerful, complex, dry nuts, creamy oak, varnish. Palate: rich, long, spicy.

1822 Fino en Rama BF FI S
listán blanco

94

Colour: bright golden. Nose: flor yeasts, pungent, neat, balanced. Palate: good acidity, bitter, spicy, long.

🏆 PODIUM

1822 Magnum BF MZ
listán blanco

95

Colour: bright golden. Nose: dry nuts, powerful, expressive, ripe fruit, faded flowers, flor yeasts. Palate: bitter, long, flavourful, dry.

🏆 PODIUM

1822 Solera Fundacional Bota NO BF AM S
listán blanco

95

Colour: light mahogany. Nose: fruit liqueur notes, spicy, varnish, brioche, acetaldehyde. Palate: fine solera notes, bitter.

La E Argüeso en Rama Magnum BF MZ
listán blanco

94

Colour: old gold. Nose: neat, expressive, dry nuts, ebb tide. Palate: flavourful, long, saline, dry, fine bitter notes, round.

San León Reserva de Familia BF MZ S
listán blanco

93

Colour: golden. Nose: neat, fine lees, pungent. Palate: dry, saline, balanced, fine bitter notes.

BODEGAS BARBADILLO
Luis de Eguilaz, 11
11540 Sanlúcar de Barrameda (Cádiz)
☎: +34 956 385 500
tamara@barbadillo.com
www.barbadillo.com

Barbadillo Cuco BF OL S
palomino

93

Colour: light mahogany. Nose: powerful, complex, dry nuts, toasty, acetaldehyde. Palate: rich, long, fine solera notes, spicy.

Barbadillo La Cilla BF PX D
pedro ximénez

92

Colour: dark mahogany. Nose: complex, fruit liqueur notes, dried fruit, pattiserie, toasty. Palate: sweet, rich, unctuous, powerful.

DO JEREZ / D.O.P.

Barbadillo San Rafael BF MED
palomino, pedro ximénez
91
Colour: iodine, amber rim. Nose: powerful, complex, dry nuts, toasty. Palate: bitter, fine solera notes, long, sweet.

Pastora Pasada BF MZ S
palomino
93
Age nuances. Colour: bright golden. Nose: dried flowers, flor yeasts, lees reduction notes, candied fruit, dry nuts. Palate: old wood, long, flavourful.

Príncipe de Barbadillo BF AM S
palomino
93
Colour: iodine, amber rim. Nose: sweet spices, acetaldehyde, dry nuts. Palate: full-bodied, dry, spicy, long, fine bitter notes.

🏆 **PODIUM**

S. Roberto Bota Única 1/1 BF PC
palomino
97
Colour: light mahogany. Nose: acetaldehyde, pungent, varnish, aged wood nuances, creamy oak. Palate: powerful, flavourful, spicy, long, balanced.

🏆 **PODIUM**

S. Roberto Bota Única 1/2 BF AM
palomino
97
Colour: iodine, amber rim. Nose: elegant, sweet spices, acetaldehyde, dry nuts, pungent, characterful, complex. Palate: full-bodied, dry, spicy, long, fine bitter notes, complex.

🏆 **PODIUM**

S. Roberto Bota Única 2/2 BF AM
palomino
96
Colour: dark mahogany. Nose: candied fruit, fruit liqueur notes, spicy, varnish, lees reduction notes. Palate: fine solera notes, bitter, powerful, flavourful.

Solear Edición Bicentenario BF MZ
palomino
92
Pleasant, aromatic. Colour: bright yellow. Nose: yeasty notes, flor yeasts, ripe fruit. Palate: flavourful, dry.

Solear en Rama Saca de Primavera BF MZ S
94
Colour: bright yellow. Nose: flor yeasts, pungent, dry nuts, dried flowers. Palate: flavourful, dry, classic aged character, long, saline.

🏆 **PODIUM**

Solear en Rama saca de verano 2022 BF MZ
96
Colour: bright yellow. Nose: complex, expressive, pungent, chamomile, lactic notes. Palate: powerful, fresh, fine bitter notes, saline.

BODEGAS CÉSAR FLORIDO
Padre Lerchundi, 35-37
11550 Chipiona (Cádiz)
☎: +34 956 371 285
florido@bodegasflorido.com
www.bodegasflorido.com

César Florido Moscatel Dorado BF MO D
100% moscatel de alejandría
90
Colour: light mahogany. Nose: honeyed notes, floral, sweet spices. Palate: rich, fruity, flavourful.

César Florido Moscatel Pasas BF MO D
100% moscatel de alejandría
91
Colour: mahogany. Nose: sun-drenched nuances, dried fruit, honeyed notes, caramel. Palate: unctuous, flavourful, long.

Cruz del Mar BF AM S
100% palomino
90
Colour: light mahogany. Nose: floral, candied fruit, dry nuts. Palate: correct, balanced, easy to drink.

🏆 **PODIUM**

Peña del Aguila BF PC S
100% palomino
95
Complex, with personality. Colour: light mahogany. Nose: acetaldehyde, pungent, varnish, aged wood nuances. Palate: powerful, flavourful, spicy, long, balanced, fine solera notes.

Peña del Aguila Fino en Rama BF FI S
100% palomino
93
Colour: bright yellow. Nose: flor yeasts, lees reduction notes, pungent, earthy notes. Palate: good acidity, bitter, spicy, long.

DO JEREZ / D.O.P.

Pleamar en Rama BF MZ S
100% palomino
91
Colour: golden. Nose: ripe fruit, lees reduction notes, saline, dried flowers. Palate: flavourful, fine bitter notes, correct.

BODEGAS DIOS BACO

Tecnología, A-14
11407 Jerez de la Frontera (Cádiz)
☎: +34 956 333 337
Fax: +34 956 333 825
info@bodegasdiosbaco.com
www.bodegasdiosbaco.com

Dios Baco BF CRM
100% palomino
91
Colour: iodine, amber rim. Nose: dry nuts, creamy oak, varnish. Palate: rich, long, spicy.

Dios Baco BF AM S
palomino
91
Colour: old gold. Nose: balanced, candied fruit, caramel. Palate: flavourful, long, easy to drink.

Dios Baco BF FI
palomino
91
Colour: yellow. Nose: white fruit, ripe fruit, faded flowers, flor yeasts. Palate: balanced, fine bitter notes, easy to drink, long.

Dios Baco BF PX
pedro ximénez
92
Colour: dark mahogany. Nose: fruit liqueur notes, dried fruit, pattiserie, toasty. Palate: sweet, rich, unctuous, powerful.

Dios Baco Medium BF OL
palomino
89
Balanced, spicy, jammy, oxidativ, tasty, toasty.

BODEGAS GUTIÉRREZ COLOSÍA

Avda. Bajamar, 40
11500 El Puerto de Santa María (Cádiz)
☎: +34 956 852 852
export@gutierrezcolosia.com
www.gutierrezcolosia.com

Gutiérrez Colosía BF AM S
100% palomino
90
Colour: old gold. Nose: balanced, medium intensity, candied fruit, dry nuts. Palate: flavourful, dry.

Gutiérrez Colosía BF FI S
palomino
91
Colour: bright straw. Nose: pungent, flor yeasts, ripe fruit. Palate: flavourful, long, easy to drink.

Gutiérrez Colosía BF OL
100% palomino
91
Colour: iodine, amber rim. Nose: powerful, varnish, caramel, pattiserie. Palate: rich, long, spicy.

Gutiérrez Colosía BF PX D
pedro ximénez
92
Colour: light mahogany. Nose: dried fruit, pattiserie, sweet spices. Palate: unctuous, powerful, flavourful.

Gutiérrez Colosía
Fino en Rama 15 grados BF FI S
100% palomino
92
Colour: bright yellow. Nose: balanced, pungent, flor yeasts. Palate: flavourful, fine bitter notes, long.

Gutiérrez Colosía
Fino en Rama 16 grados BF FI S
100% palomino
93
Colour: yellow, pale. Nose: dry nuts, powerful, flor yeasts. Palate: complex, full-bodied, flavourful.

Sangre y Trabajadero BF OL S
90
Colour: iodine, amber rim. Nose: dry nuts, varnish, pattiserie. Palate: rich, easy to drink.

Solera Familiar
Gutiérrez Colosía BF AM S
94
Colour: old gold, amber rim. Nose: sweet spices, acetaldehyde, dry nuts, incense. Palate: full-bodied, spicy, long, fine bitter notes.

🏆 PODIUM

Solera Familiar Gutiérrez Colosía BF OL S
95
Colour: light mahogany. Nose: complex, dry nuts, toasty, acetaldehyde. Palate: long, fine solera notes, spicy, dry, flavourful.

🏆 PODIUM

Solera Familiar Gutiérrez Colosía BF PC S
95
Complex, with personality. Colour: old gold, amber rim. Nose: powerful, expressive, acetaldehyde, varnish, dry nuts. Palate: flavourful, full-bodied, complex, bitter.

BODEGAS HIDALGO-LA GITANA
Banda de la Playa, 42
11540 Sanlúcar de Barrameda (Cádiz)
☎: +34 956 385 304
bodegashidalgo@lagitana.es
www.lagitana.es

Alameda BF CRM
100% palomino
88
Kind finish, standard, pruney, sweet, smooth, spicy.

Faraón 30 años
VORS 50 cl. BF OL S
100% palomino
93
Colour: light mahogany. Nose: powerful, dry nuts, toasty, roasted almonds. Palate: rich, long, fine solera notes, spicy.

Faraón BF OL S
100% palomino
89
Spicy, jammy, tasty, pruney.

🏆 PODIUM

La Gitana Aniversario BF MZ S
100% palomino
95
Colour: yellow. Nose: complex, characterful, flor yeasts, saline. Palate: dry, fine bitter notes, long.

La Gitana BF MZ S
100% palomino
91
Colour: bright straw. Nose: pungent, flor yeasts, saline. Palate: fresh, fine bitter notes, balanced, easy to drink.

La Gitana en Rama BF MZ S
100% palomino
93
Colour: bright yellow. Nose: flor yeasts, pungent, expressive, balanced. Palate: good acidity, bitter, spicy, long.

Napoleón 30 años
VORS 50 cl. BF AM S
100% palomino
94
Colour: old gold, amber rim. Nose: candied fruit, spicy, varnish. Palate: fine solera notes, bitter.

Napoleón BF AM S
100% palomino
90
Colour: iodine, amber rim. Nose: sweet spices, dry nuts, candied fruit. Palate: dry, spicy, long, fine bitter notes.

Pastrana 50 cl. BF AM
100% palomino
94
Colour: light mahogany. Nose: dry nuts, varnish, characterful, balanced, elegant. Palate: flavourful, dry, long.

Pastrana
Manzanilla Pasada BF MZ S
100% palomino
92
Colour: yellow. Nose: dry nuts, ripe fruit, dried flowers, faded flowers, yeasty notes. Palate: dry, easy to drink, fine bitter notes.

🏆 PODIUM

Triana 30 años VORS 50 cl BF PX D
pedro ximénez
95
Colour: dark mahogany. Nose: complex, aromatic coffee, pattiserie, sweet spices, dark chocolate. Palate: unctuous, concentrated, spicy, toasty, fine solera notes.

Triana Hidalgo
La Gitana BF PX D
pedro ximénez
90
Colour: mahogany. Nose: fruit liqueur notes, dried fruit, toasty. Palate: sweet, rich, unctuous.

Wellington 20 años
VOS 50 cl BF PC S
100% palomino
94
Colour: light mahogany. Nose: dry nuts, balanced, varnish, expressive. Palate: flavourful, dry, long, balanced.

DO JEREZ / D.O.P.

DO JEREZ / D.O.P.

🏆 PODIUM

Wellington 30 años VORS 50 cl BF PC S
100% palomino

96

Colour: dark mahogany. Nose: candied fruit, fruit liqueur notes, spicy, varnish, acetaldehyde, brioche. Palate: fine solera notes, bitter, spicy.

BODEGAS LA CIGARRERA
Pza. Madre de Dios, 4
11540 Sanlúcar de Barrameda (Cádiz)
☎: +34 956 381 285
lacigarrera@bodegaslacigarrera.com
www.bodegaslacigarrera.com

La Cigarrera BF AM ES
100% palomino

90

Colour: iodine, amber rim. Nose: sweet spices, macerated fruit, flor yeasts. Palate: spicy, fruity, light-bodied.

La Cigarrera BF MO D
100% moscatel

91

Full-bodied. Colour: mahogany. Nose: fruit preserve, candied fruit, powerful, ebb tide. Palate: flavourful, good structure, varietal.

La Cigarrera BF MZ ES
100% palomino

89

Kind finish, standard, balanced, dried flowers, jammy.

La Cigarrera BF OL S
100% palomino

91

Colour: old gold, amber rim. Nose: fruit preserve, caramel, sweet spices, varnish. Palate: flavourful, balanced.

La Cigarrera BF PX D
100% pedro ximénez

90

Colour: mahogany. Nose: characterful, dried fruit, powerful, varietal. Palate: flavourful, unctuous, sweet.

La Cigarrera Manzanilla Pasada BF MZ ES
100% palomino

92

Colour: old gold. Nose: violet drops, dry nuts, lees reduction notes, yeasty notes. Palate: flavourful, fruity.

BODEGAS OSBORNE
Fernán Caballero, 7
11500 El Puerto de Santa María (Cádiz)
☎: +34 956 869 000
valeria.morado@osborne.es
www.osborne.es

🏆 PODIUM

Amontillado 51-1ª VORS BF AM S
100% palomino

95

With personality, complex. Colour: iodine, amber rim. Nose: spicy, roasted almonds, dry nuts, varnish. Palate: flavourful, complex, concentrated.

Coquinero en Rama BF FI S
100% palomino

92

Colour: golden, pale. Nose: flor yeasts, lees reduction notes, pungent. Palate: good acidity, bitter, spicy, long.

La Honda Amontillado en Rama BF AM
100% palomino

93

Colour: old gold, amber rim. Nose: spicy, varnish, acetaldehyde, roasted almonds. Palate: fine solera notes, bitter, flavourful.

La Honda BF FI
100% palomino

93

Colour: old gold. Nose: iodine notes, yeasty notes, pungent, dry nuts. Palate: balanced, full-bodied, fine solera notes.

🏆 PODIUM

Osborne Solera India BF OL S
80% palomino, 20% pedro ximénez

96

With personality. Colour: light mahogany. Nose: floral, varnish, sweet spices, candied fruit, dry nuts. Palate: powerful, flavourful, bitter.

🏆 PODIUM

Venerable VORS BF PX D

96

Colour: mahogany. Nose: characterful, expressive, elegant, candied fruit, dried fruit, spicy, aromatic coffee. Palate: full-bodied, complex, long, fine solera notes.

BODEGAS PRIMITIVO COLLANTES

Calle Ancha, 51
11130 Chiclana de la Frontera (Cádiz)
☎: +34 956 400 150
administracion@bodegasprimitivocollantes.com
www.bodegaprimitivocollantes.es

Arroyuelo en Rama BF FI
palomino
92 🏆

Colour: bright yellow. Nose: pungent, chamomile, scrubland. Palate: bitter, spicy, long.

Fossi BF AM
palomino
90 🏆

Colour: iodine, amber rim. Nose: sweet spices, acetaldehyde, dry nuts, candied fruit. Palate: full-bodied, spicy, fruity.

BODEGAS R & G

Ramón y Cajal 7, 1°A
01007 Vitoria-Gasteiz (Araba/Álava)
☎: +34 945 150 589
araex@araex.com
www.araex.com

🏆 **PODIUM**

R&G Soleras Olvidadas BF AM
100% listán blanco
95

Colour: iodine, amber rim. Nose: powerful, complex, elegant, dry nuts, toasty, candied fruit. Palate: rich, bitter, fine solera notes, long, spicy.

🏆 **PODIUM**

R&G Soleras Olvidadas BF FI
100% listán blanco
95

Colour: bright yellow. Nose: flor yeasts, lees reduction notes, pungent, expressive, powerful. Palate: good acidity, bitter, spicy, long.

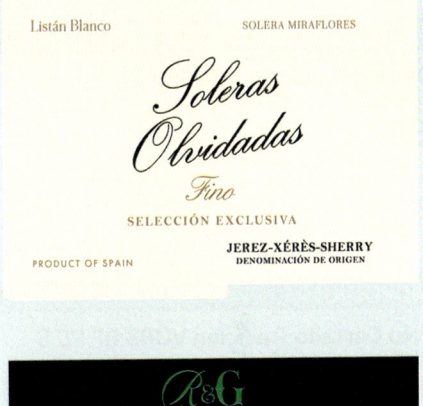

R&G Soleras Olvidadas BF MZ
100% listán blanco
94

Colour: bright yellow. Nose: iodine notes, dry nuts, varnish, candied fruit, fruit liqueur notes. Palate: bitter, spirituous, long.

BODEGAS TRADICIÓN

Cordobeses, 3
11408 Jerez de la Frontera (Cádiz)
☎: +34 956 168 628
Igiglesias@bodegastradicion.com
www.bodegastradicion.com

🏆 **PODIUM**

Amontillado Tradición VORS BF AM S
palomino
97

Colour: iodine, amber rim. Nose: powerful, complex, elegant, dry nuts, characterful, expressive. Palate: rich, long, fine solera notes, spicy, round.

🏆 **PODIUM**

Cream Tradición VOS BF CRM
70% palomino, 30% pedro ximénez
95

Colour: iodine, amber rim. Nose: complex, dry nuts, creamy oak, varnish. Palate: rich, long, spicy, sweet.

DO JEREZ / D.O.P.

DO JEREZ / D.O.P.

🏆 **PODIUM**

Fino Tradición BF FI S
palomino

95

Colour: bright golden. Nose: spicy, charcterful, flor yeasts, faded flowers, dry nuts. Palate: fine solera notes, flavourful, complex.

🏆 **PODIUM**

Oloroso Tradición VORS BF OL S
palomino

98

Colour: dark mahogany. Nose: candied fruit, fruit liqueur notes, spicy, varnish, charcterful, complex, powerful. Palate: fine solera notes, bitter, rich, powerful, fine bitter notes, round.

🏆 **PODIUM**

Palo Cortado Tradición VORS BF PC S
palomino

97

Colour: light mahogany. Nose: acetaldehyde, pungent, varnish, aged wood nuances, creamy oak. Palate: powerful, flavourful, spicy, long, elegant.

🏆 **PODIUM**

Pedro Ximénez Tradición VOS BF PX D
100% pedro ximénez

96

Colour: dark mahogany. Nose: aromatic coffee, spicy, acetaldehyde, dry nuts, varietal, dried fruit. Palate: fine solera notes, toasty, long, sweet, creamy.

BODEGAS WILLIAMS & HUMBERT

Ctra. Madrid-Cádiz, PK. 641,75
11408 Jerez de la Frontera (Cádiz)
☎: +34 956 353 400
Fax: +34 956 353 408
williams@williams-humbert.com
www.williams-humbert.com

Canasta Cream 20 años BF CRM
palomino, pedro ximénez

94

Colour: old gold, amber rim. Nose: candied fruit, expressive, balanced, sweet spices, neat, cocoa bean. Palate: flavourful, sweet, balanced, fine solera notes.

Canasta Cream BF CRM
palomino, pedro ximénez

88

Kind finish, jammy, sweety finish, fruity, toasty.

🏆 **PODIUM**

Don Guido Solera Especial 20 años VOS BF PX D
pedro ximénez

95

Complex, tasty. Colour: dark mahogany. Nose: dried fruit, pattiserie, cocoa bean, aromatic coffee, complex, dry nuts. Palate: flavourful, long, creamy.

Don Zoilo 12 años BF OL S
palomino

90

Kind finish. Colour: bright golden. Nose: candied fruit, dry nuts. Palate: easy to drink, fine bitter notes, correct.

Don Zoilo BF AM S
palomino

92

Colour: old gold, amber rim. Nose: expressive, neat, dry nuts, sweet spices. Palate: flavourful, dry.

Don Zoilo BF FI S
palomino

91

Jammy. Colour: bright golden. Nose: flor yeasts, pungent, charcterful. Palate: flavourful, fine bitter notes.

Don Zoilo BF PC

92

Kind finish. Colour: iodine, amber rim. Nose: varnish, toasty, dry nuts, candied fruit. Palate: flavourful, spicy, long.

Don Zoilo PX BF PX D
pedro ximénez

93

Colour: dark mahogany. Nose: dried fruit, pattiserie, toasty, caramel. Palate: rich, powerful, varietal, unctuous.

🏆 **PODIUM**

Dos Cortados 20 Años BF PC S
palomino

95

Colour: light mahogany. Nose: powerful, complex, dry nuts, acetaldehyde, varnish. Palate: long, fine solera notes, spicy, round, flavourful.

Dry Sack Medium 15 años BF MED
palomino, pedro ximénez

93

Colour: iodine, amber rim. Nose: dry nuts, creamy oak, varnish. Palate: rich, long, spicy.

Jalifa VORS "30 years" BF AM S
palomino

94

Colour: iodine, amber rim. Nose: powerful, complex, elegant, dry nuts, toasty. Palate: rich, long, fine solera notes, spicy.

Williams & Humbert 2001 BF AM S
palomino

94

Colour: dark mahogany. Nose: candied fruit, spicy, varnish, dry nuts. Palate: fine solera notes, bitter, flavourful.

🏆 **PODIUM**

Williams & Humbert 2001 BF OL S
palomino

95

With personality. Colour: light mahogany. Nose: complex, dry nuts, toasty, acetaldehyde. Palate: rich, long, fine solera notes, spicy, round.

Williams & Humbert 2009 BF OL S
palomino

93

Colour: light mahogany. Nose: powerful, dry nuts, toasty, sweet spices, candied fruit. Palate: rich, long, fine solera notes, round.

Williams & Humbert Añada 2002 BF PC S
palomino

92

Colour: amber. Nose: sweet spices, acetaldehyde, dry nuts, roasted almonds. Palate: dry, spicy, long, fine bitter notes, fresh.

BODEGAS Y VIÑEDOS DIEZ MERITO
Ctra. Nac. IV Km. 626,5 Viña El Diablo
11407 Jerez de la Frontera (Cádiz)
☎: +34 956 186 112
joseluis.perez@diezmerito.com
www.diezmerito.com

🏆 **PODIUM**

Amontillado Fino Imperial BF AM
100% palomino

96

Full-bodied, oxidativ, powerful. Colour: amber. Nose: dry nuts, sweet spices, complex, elegant, balanced. Palate: good acidity, fine bitter notes, fine solera notes, great length.

Bertola 12 años BF AM S
100% palomino

92

Colour: iodine, amber rim. Nose: sweet spices, dry nuts, acetaldehyde, yeasty notes. Palate: full-bodied, dry, spicy, long.

Bertola 12 años BF OL S
100% palomino

91

Colour: iodine, amber rim. Nose: dry nuts, varnish, caramel. Palate: rich, spicy, bitter, flavourful.

Bertola 12 años BF PC S
palomino

92

Colour: old gold. Nose: pungent, yeasty notes, dry nuts, varnish, sweet spices. Palate: balanced, flavourful.

Bertola BF CRM
palomino, pedro ximénez

89

Balanced, spicy, full-bodied, toasty, age nuances.

Pemartin BF FI S
100% palomino

86

BODEGAS YUSTE
Ctra. Sanlúcar - Chipiona, km. 93 Bodega Miraflores
11540 Sanlúcar de Barrameda (Cádiz)
☎: +34 956 385 200
comunicacion@bodegasyuste.com
www.bodegasyuste.com

Aurora en Rama Magnum BF MZ
listán blanco

92

Colour: yellow, pale. Nose: pungent, flor yeasts, dry nuts. Palate: flavourful, dry, great length, saline.

🏆 **PODIUM**

Conde de Aldama "Bota No" BF AM
listán blanco

100

Complex, powerful. Colour: light mahogany. Nose: powerful, complex, dry nuts, toasty, acetaldehyde, pungent, saline. Palate: rich, long, fine solera notes, spicy, round.

🏆 **PODIUM**

Conde de Aldama 1/3 BF OL
listán blanco

96

Colour: light mahogany. Nose: powerful, complex, dry nuts, toasty, acetaldehyde. Palate: rich, long, fine solera notes, spicy, round.

DO JEREZ / D.O.P.

DO JEREZ / D.O.P.

🏆 PODIUM

Conde de Aldama 1/8 BF PC S
listán blanco
96
Colour: light mahogany. Nose: acetaldehyde, pungent, varnish, aged wood nuances, creamy oak. Palate: powerful, flavourful, spicy, long, balanced.

Conde de Aldama Magnum Manzanilla Pasada BF MZ
listán blanco
94
Colour: old gold. Nose: characterful, varnish, expressive. Palate: flavourful, fine bitter notes, balanced, spicy, long, fine solera notes.

La Kika Magnum Saca Primavera BF MZ
listán blanco
93
Colour: yellow, pale. Nose: faded flowers, lees reduction notes, yeasty notes, pungent. Palate: flavourful, fine bitter notes.

COOP. ALBARIZAS DE TREBUJENA
Avda. de Jerez s/n
11560 Trebujena (Cádiz)
☎: +34 615 311 024
coopalbarizas@hotmail.com

Castillo de Guzmán BF AM
palomino
88
Aromatic, spicy, tasty, great length, full-bodied, oxidativ.

Castillo de Guzmán BF FI
palomino
88
Kind finish, aromatic, standard, yeasty notes, great length.

Castillo de Guzmán BF OL
palomino
88
Standard, spicy, jammy, tasty, toasty, age nuances.

Castillo de Guzmán BF PC
palomino
87

DE LA RIVA
C. de la Habana, 5
11407 Jerez de la Frontera (Cádiz)

La Riva Manzanilla Fina Miraflores Baja BF MZ
palomino
93
Colour: bright yellow. Nose: complex, expressive, pungent, dry nuts. Palate: powerful, fresh, fine bitter notes, great length.

🏆 PODIUM

La Riva Manzanilla pasada Balbaína Alta BF MZ
96
Colour: bright yellow. Nose: iodine notes, dry nuts, acetaldehyde, chalk. Palate: bitter, long, powerful, fine solera notes.

DELGADO ZULETA
Avda. Rocío Jurado, s/n
11540 Sanlúcar de Barrameda (Cádiz)
☎: +34 956 360 543
tienda@delgadozuleta.com
www.delgadozuleta.com

Goya XL BF MZ S
100% palomino
94
Colour: bright yellow. Nose: flor yeasts, lees reduction notes, pungent, dry nuts, characterful. Palate: bitter, spicy, long.

Las Señoras Medium BF OL MED
palomino, pedro ximénez
91
Colour: iodine, amber rim. Nose: varnish, dry nuts, balanced, neat. Palate: flavourful, easy to drink, long, fine solera notes.

Magnum de Barbiana BF MZ S
92
With personality, representative. Colour: yellow. Nose: pungent, ebb tide, dry nuts, flor yeasts. Palate: flavourful, saline, great length.

Magnum de la Goya BF MZ S
100% palomino
93
Colour: golden, pale. Nose: flor yeasts, chamomile, scrubland, dry nuts. Palate: flavourful, full-bodied.

Monteagudo BF OL S
100% palomino
91
Colour: light mahogany. Nose: balanced, expressive, candied fruit, varnish. Palate: long, fine bitter notes, round.

🏆 PODIUM

Quo Vadis VORS BF AM S
96
With personality, age nuances. Colour: dark mahogany. Nose: candied fruit, fruit liqueur notes, spicy, varnish. Palate: fine solera notes, bitter, complex.

Viejo Zuleta VOS BF AM S
100% palomino
94
Age nuances, aromatic, representative. Colour: old gold, amber rim. Nose: roasted almonds, aged wood nuances, incense, dry nuts. Palate: flavourful, bitter, fine solera notes.

EQUIPO NAVAZOS
11403 Jerez de la Frontera (Cádiz)
equipo@navazos.com
www.equiponavazos.com

🏆 PODIUM

La Bota de Amontillado (Bota nº 109) "Bota punta" BF AM
palomino
99
Colour: dark mahogany. Nose: spicy, varnish, incense, pungent, cedar wood. Palate: fine solera notes, powerful, flavourful, elegant.

🏆 PODIUM

La Bota de Manzanilla Pasada Nº103 Magnum BF MZ
96
Colour: bright golden. Nose: iodine notes, dry nuts, varnish, elegant, expressive. Palate: bitter, long, powerful, saline.

🏆 PODIUM

La Bota de Oloroso (Bota nº 108) "Bota NO" BF
palomino
98
Colour: light mahogany. Nose: powerful, complex, dry nuts, toasty, acetaldehyde. Palate: rich, long, fine solera notes, spicy, round, powerful, great length.

🏆 PODIUM

La Bota de Palo Cortado (Bota nº 102) BF PC
96
Colour: old gold, amber rim. Nose: dry nuts, expressive, candied fruit, characterful, sweet spices, pungent. Palate: balanced, fine bitter notes, dry, long.

FERNANDO DE CASTILLA
Jardinillo, 7-11
11404 Jerez de la Frontera (Cádiz)
☎: +34 956 182 454
Fax: +34 956 182 222
bodegas@fernandodecastilla.com
www.fernandodecastilla.com

🏆 PODIUM

Fernando de Castilla "Amontillado Antique" BF AM S
palomino
95
Colour: iodine, amber rim. Nose: elegant, sweet spices, acetaldehyde, dry nuts. Palate: full-bodied, dry, spicy, long, fine bitter notes.

Fernando de Castilla "Fino Antique" BF FI S
palomino
94
Colour: old gold. Nose: iodine notes, dry nuts, varnish, acetaldehyde, powerful, ripe fruit. Palate: bitter, long, powerful, fine solera notes.

Fernando de Castilla "Palo Cortado Antique" BF PC S
palomino
94
Colour: amber. Nose: acetaldehyde, varnish, aged wood nuances. Palate: flavourful, spicy, long, balanced.

Fernando de Castilla Fino Classic BF FI S
palomino
93
With personality, jammy. Colour: bright yellow. Nose: ripe fruit, yeasty notes, flor yeasts, pungent. Palate: full-bodied, flavourful, fine bitter notes.

🏆 PODIUM

Fernando de Castilla Oloroso Singular BF OL S
palomino
97
Complex, with personality, age nuances. Colour: light mahogany. Nose: powerful, complex, dry nuts, toasty, acetaldehyde. Palate: long, fine solera notes, spicy, round.

DO JEREZ / D.O.P.

Fernando de Castilla
PX Classic BF PX D
pedro ximénez

92

Colour: mahogany. Nose: dried fruit, caramel, aromatic coffee, fruit liqueur notes. Palate: flavourful, full-bodied.

GONZÁLEZ BYASS JEREZ
Manuel María González, 12
11403 Jerez de la Frontera (Cádiz)
☎: +34 956 357 016
prensa@gonzalezbyass.es
www.tiopepe.com

Amontillado
del Duque VORS BF AM S
palomino

94

Representative, tasty. Colour: iodine, amber rim. Nose: neat, dry nuts, candied fruit, varnish. Palate: balanced, long.

🏆 **PODIUM**

Añada 1975 BF AM
palomino

97

Colour: old gold, amber rim. Nose: candied fruit, spicy, varnish, expressive, powerful, acetaldehyde, pattiserie. Palate: complex, concentrated, classic aged character, spicy, great length, round.

Apóstoles VORS BF PC MED
87% palomino, 13% pedro ximénez

92

Colour: light mahogany. Nose: candied fruit, caramel, dried flowers, sweet spices. Palate: flavourful, bitter, sweet.

Leonor BF PC S
palomino

91

Colour: light mahogany. Nose: candied fruit, pattiserie, sweet spices. Palate: flavourful, easy to drink, old wood, great length.

🏆 **PODIUM**

Matusalem VORS BF OL D
75% palomino, 25% pedro ximénez

95

Colour: iodine, amber rim. Nose: dry nuts, varnish, expressive, balanced. Palate: long, spicy, balanced, round.

🏆 **PODIUM**

Noé VORS BF PX D
pedro ximénez

95

Colour: dark mahogany. Nose: powerful, expressive, aromatic coffee, spicy, acetaldehyde, dry nuts. Palate: balanced, elegant, fine solera notes, toasty, long.

Tío Pepe BF FI S
palomino

93

Colour: bright straw. Nose: balanced, fresh, pungent, flor yeasts. Palate: fine bitter notes, long, easy to drink.

🏆 **PODIUM**

Tío Pepe Cuatro Palmas BF AM S
palomino

97

Colour: light mahogany. Nose: powerful, complex, dry nuts, toasty, acetaldehyde, pungent, aged wood nuances, sweet spices. Palate: rich, long, fine solera notes, spicy.

🏆 **PODIUM**

Tío Pepe Dos Palmas BF FI S
palomino

95

Colour: bright golden. Nose: iodine notes, dry nuts, varnish, acetaldehyde, powerful. Palate: bitter, spirituous, long, powerful.

Tío Pepe en Rama BF FI S
palomino

94

Colour: bright yellow. Nose: flor yeasts, lees reduction notes, pungent, white fruit, ripe fruit. Palate: good acidity, bitter, long, classic aged character.

🏆 **PODIUM**

Tío Pepe Tres Palmas BF FI S
palomino

96

Colour: iodine, amber rim. Nose: sweet spices, acetaldehyde, dry nuts, flor yeasts. Palate: full-bodied, dry, spicy, long, fine bitter notes, complex.

Tío Pepe Una Palma BF FI S
palomino

94

Colour: bright golden. Nose: flor yeasts, lees reduction notes, pungent, spicy. Palate: good acidity, bitter, spicy, long.

Viña AB BF AM S
palomino

91

Colour: bright golden. Nose: white fruit, ripe fruit, yeasty notes, pungent. Palate: balanced, fine bitter notes, flavourful, dry.

HARVEYS
Puerta de Rota, s/n
11480 Jerez de la Frontera (Cádiz)
☎: +34 685 082 853
mmartinez@grupoemperador.es
www.harveys.es

Harveys Amontillado Premium BF AM S
palomino

92

Colour: golden. Nose: elegant, balanced, sweet spices, acetaldehyde, pungent. Palate: flavourful, dry, fine bitter notes.

🏆 PODIUM

Harveys Amontillado VORS BF AM S
palomino

95

Colour: iodine, amber rim. Nose: powerful, complex, elegant, dry nuts, toasty. Palate: rich, bitter, fine solera notes, long, spicy.

Harveys Bristol Cream BF CRM
palomino, pedro ximénez

88

Hot, spicy, full-bodied, toasty.

Harveys Fino Premium BF FI S
palomino

92

Jammy. Colour: bright yellow. Nose: balanced, pungent, dry nuts. Palate: flavourful, fine bitter notes, long.

🏆 PODIUM

Harveys Palo Cortado VORS BF PC MED
palomino

95

Colour: light mahogany. Nose: acetaldehyde, pungent, varnish, aged wood nuances. Palate: flavourful, spicy, long, balanced, bitter, full-bodied.

🏆 PODIUM

Harveys Pedro Ximénez VORS BF PX D
pedro ximénez

96 🍃

Colour: dark mahogany. Nose: powerful, expressive, aromatic coffee, spicy, acetaldehyde, dry nuts. Palate: balanced, elegant, fine solera notes, toasty, long.

HIJOS DE RAINERA PÉREZ MARÍN. "LA GUITA"
Misericordia, 1
11540 Sanlúcar de Barrameda (Cádiz)
☎: +34 956 321 004
rvp@grupoestevez.com
www.laguita.com

La Guita BF MZ S
palomino

91

Colour: straw. Nose: dry nuts, white fruit, ripe fruit, lees reduction notes, yeasty notes. Palate: dry, flavourful.

La Guita en Rama BF MZ S
palomino

93

Colour: bright yellow. Nose: complex, expressive, citrus fruit, flor yeasts. Palate: rich, fine bitter notes, flavourful.

LUSTAU
Arcos, 53
11402 Jerez de la Frontera (Cádiz)
☎: +34 956 341 597
lustau@lustau.es
www.lustau.es

La Ina BF FI S
palomino

94

Colour: bright yellow. Nose: flor yeasts, pungent, neat, expressive, chalk. Palate: good acidity, bitter, spicy, long.

Lustau Amontillado del Castillo Antonio Caballero y Sobrinos BF AM S
palomino

94

Representative, oxidativ, complex. Colour: old gold, amber rim. Nose: sweet spices, acetaldehyde, dry nuts, varnish. Palate: full-bodied, dry, spicy, long, fine bitter notes, complex.

🏆 PODIUM

Lustau Almacenista Cayetano del Pino BF PC S
palomino

95

Colour: iodine, amber rim. Nose: powerful, complex, elegant, dry nuts, toasty, varnish. Palate: rich, bitter, fine solera notes, long, spicy.

DO JEREZ / D.O.P.

Lustau Almacenista Manzanilla Pasada Manuel Cuevas Jurado BF MZ
palomino
94
With personality, yeasty notes. Nose: bakery, brioche, lees reduction notes, faded flowers. Palate: complex, flavourful, fine bitter notes.

Lustau Almacenistas Pata de Gallina García Jarana BF OL S
palomino
94
Colour: iodine, amber rim. Nose: complex, dry nuts, varnish, characterful. Palate: long, spicy, flavourful, dry.

🏆 **PODIUM**

Lustau Amontillado VORS BF AM S
palomino
96
Colour: light mahogany. Nose: expressive, balanced, elegant, complex, sweet spices, varnish. Palate: flavourful, full-bodied, fine solera notes, great length.

Lustau Añada Vintage Sherry 2003 BF OL D
palomino
94
Colour: iodine, amber rim. Nose: powerful, complex, dry nuts, creamy oak, varnish. Palate: rich, long, spicy, unctuous, powerful.

Lustau Emperatriz Eugenia BF OL S
palomino
94
Colour: iodine, amber rim. Nose: powerful, complex, dry nuts, creamy oak, varnish. Palate: rich, long, spicy.

Lustau Jarana BF FI
palomino
93
With personality. Nose: burnt matches, dry nuts, flor yeasts, yeasty notes. Palate: flavourful, long, easy to drink.

🏆 **PODIUM**

Lustau Pedro Ximénez VORS BF PX D
pedro ximénez
95
Colour: dark mahogany. Nose: powerful, aromatic coffee, spicy, acetaldehyde, dry nuts. Palate: balanced, fine solera notes, toasty, long.

Lustau San Emilio BF PX D
pedro ximénez
93
Colour: mahogany. Nose: powerful, expressive, spicy, pattiserie, dried fruit. Palate: long, flavourful, great length, unctuous.

MANUEL ARAGÓN
Olivo, 1
11130 Chiclana de la Frontera (Cádiz)
☎: +34 956 400 756
administracion@bodegamanuelaragon.com
www.bodegamanuelaragon.com

Fino Granero en Rama BF FI
palomino
94
Colour: bright yellow. Nose: flor yeasts, lees reduction notes, pungent. Palate: good acidity, bitter, spicy, long.

Hoyo Membrillo Medium BF MED
palomino, pedro ximénez
90
Colour: iodine, amber rim. Nose: dry nuts, creamy oak, aged wood nuances. Palate: rich, spicy, light-bodied, fresh.

Manuel Aragón BF AM S
palomino
94
Colour: iodine, amber rim. Nose: sweet spices, acetaldehyde, dry nuts, citrus fruit. Palate: full-bodied, dry, spicy, long, complex.

🏆 **PODIUM**

Manuel Aragón BF OL S
palomino
97
Complex, with personality, age nuances. Nose: varnish, acetaldehyde, characterful, expressive, powerful. Palate: long, fine solera notes, great length, toasty, complex.

🏆 **PODIUM**

Manuel Aragón BF PC S
palomino
96
Colour: light mahogany. Nose: acetaldehyde, pungent, varnish, aged wood nuances, elegant. Palate: flavourful, spicy, long, balanced, fine solera notes.

Manuel Aragón BF PX D
pedro ximénez
91
Colour: dark mahogany. Nose: fruit liqueur notes, dried fruit, pattiserie, toasty. Palate: unctuous, powerful, sweet.

MARQUÉS DEL REAL TESORO

Ctra. Nacional IV, Km. 640
11408 Jerez de la Frontera (Cádiz)
☎: +34 956 321 004
Fax: +34 956 340 216
info@grupoestevez.com
www.grupoestevez.com

Almirante BF OL S
palomino

91

Colour: light mahogany. Nose: dry nuts, creamy oak, candied fruit. Palate: rich, spicy, balanced.

Del Príncipe BF AM S
palomino

92

Pleasant. Colour: old gold. Nose: spicy, caramel, candied fruit, varnish. Palate: flavourful, dry.

Tío Mateo BF FI S
palomino

90

Colour: bright yellow. Nose: balanced, fresh, expressive, flor yeasts. Palate: flavourful, fine bitter notes.

SÁNCHEZ ROMATE HNOS.

Lealas, 26
11404 Jerez de la Frontera (Cádiz)
☎: +34 956 182 212
Fax: +34 956 185 276
comercial@romate.com
www.romate.com

Encontrado BF OL S
palomino

94

Colour: light mahogany. Nose: dry nuts, acetaldehyde, sweet spices. Palate: rich, long, fine solera notes, spicy, round.

Old & Plus Amontillado VORS BF AM S
palomino

93

Colour: amber. Nose: candied fruit, faded flowers, yeasty notes, varnish. Palate: flavourful, concentrated.

Old & Plus Oloroso BF OL S
palomino

92

Colour: iodine. Nose: candied fruit, caramel, spicy, dry nuts. Palate: flavourful, bitter.

Old & Plus P.X. BF PX D
pedro ximénez

94

Aromatic, complex, representative. Colour: mahogany. Nose: dried fruit, expressive, pattisserie. Palate: unctuous, long, toasty.

Olvidado BF AM S
palomino

94

Colour: iodine, amber rim. Nose: elegant, sweet spices, acetaldehyde, dry nuts. Palate: full-bodied, dry, spicy, long, fine bitter notes.

Regente BF PC S
palomino

92

Colour: light mahogany. Nose: acetaldehyde, varnish, aged wood nuances, creamy oak. Palate: powerful, flavourful, spicy, long, balanced.

VALDESPINO

Ctra. Nacional IV, Km. 640
11404 Jerez de la Frontera (Cádiz)
☎: +34 956 321 004
Fax: +34 956 340 216
info@grupoestevez.com
www.grupoestevez.com

🏆 PODIUM

Cardenal VORS BF PC S

96

Colour: light mahogany. Nose: spicy, varnish, dry nuts. Palate: fine solera notes, bitter, great length, balanced.

🏆 PODIUM

Coliseo VORS BF AM

97

Complex, age nuances. Colour: old gold, amber rim. Nose: spicy, varnish, sweet spices, powerful. Palate: fine solera notes, flavourful, long, bitter.

Deliciosa En Rama Saca Primavera (375cc) BF MZ S
palomino

92

Colour: bright yellow. Nose: flor yeasts, lees reduction notes, pungent. Palate: good acidity, bitter, spicy, long.

El Candado BF PX D
pedro ximénez

92

Colour: mahogany. Nose: complex, fruit liqueur notes, dried fruit, pattisserie, toasty. Palate: sweet, rich, unctuous.

Promesa BF MO D
moscatel de alejandría

92

Colour: iodine, amber rim. Nose: powerful, honeyed notes, candied fruit, floral, expressive, varietal. Palate: flavourful, sweet, fruity, good acidity.

DO JEREZ / D.O.P.

Solera 1842
Medium Sweet VOS BF OL MED
palomino, pedro ximénez

94

Colour: light mahogany. Nose: powerful, complex, dry nuts, toasty, acetaldehyde. Palate: rich, long, fine solera notes, spicy, round, old wood.

Tío Diego BF AM S

91

Colour: old gold. Nose: sweet spices, acetaldehyde, dry nuts. Palate: full-bodied, dry, spicy, long, fine bitter notes.

Viejo CP BF PC S

94

Colour: old gold, amber rim. Nose: elegant, expressive, dry nuts, balanced. Palate: dry, balanced, round.

Ynocente BF FI S
palomino

93

Balanced, representative. Nose: balanced, expressive, pungent. Palate: flavourful, fine bitter notes, long, saline.

DO. JUMILLA
CONSEJO REGULADOR

San Roque, 15
30520 Jumilla (Murcia)
☎: +34 968 781 761 - Fax: +34 968 781 900
@: info@vinosdejumilla.org
www.vinosdejumilla.org

LOCATION:

Midway between the provinces of Murcia and Albacete, this DO spreads over a large region in the southeast of Spain and covers the municipal areas of Jumilla (Murcia) and Fuente Álamo, Albatana, Ontur, Hellín, Tobarra and Montealegre del Castillo (Albacete).

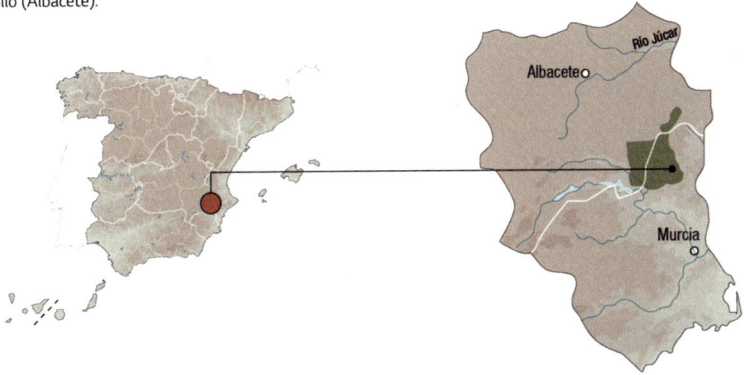

GRAPE VARIETIES:

RED: Monastrell (main), Garnacha Tinta, Garnacha Tintorera, Cencibel (Tempranillo), Cabernet Sauvignon, Merlot, Petit Verdot and Syrah.

WHITE: Airén, Macabeo, Malvasía, Pedro Ximénez, Chardonnay, Sauvignon Blanc, Moscatel de Grano Menudo and Verdejo.

FIGURES:

Vineyard surface: 22,333 – **Wine-Growers:** 1,568 – **Wineries:** 42 – 2021 **Harvest rating:** Excellent – **Production 21:** 39,309,000 litres – **Market percentages:** 30% National - 70% International.

SOIL:

The soil is mainly brownish-grey, brownish-grey limestone and limy. In general, it is poor in organic matter, with great water retention capacity and medium permeability.

CLIMATE:

Continental in nature with Mediterranean influences. It is characterized by its aridity and low rainfall (270 mm) which is mainly concentrated in spring and autumn. The winters are cold and the summers dry and quite hot.

CLASSIFICATION OF YOUNG WINE HARVESTS PEÑÍNGUIDE

2017	2018	2019	2020	2021
VERY GOOD	GOOD	VERY GOOD	VERY GOOD	VERY GOOD

DO JUMILLA / D.O.P.

ALTAMENTE VINOS
Fuente Canónigos, 12
31500 Tudela (Navarra)
☎: +34 650 323 120
admin@altamentevinos.com
www.altamentevinos.com

Altamente Monastrell 2020 T
monastrell

87

Volalto 2020 T BA
monastrell

88
Balanced, spicy, jammy, tasty.

AROMAS EN MI COPA
Luis Martínez Pérez, 2
30520 Jumilla (Murcia)
☎: +34 676 492 477
elisa@aromasenmicopa.com
www.aromasenmicopa.com

Evol 2021 T BA
100% monastrell

90
Colour: deep cherry. Nose: creamy oak, black fruit, scrubland, dark chocolate. Palate: powerful, ripe fruit, spicy, round tannins.

ARTIGA FUSTEL
Progres, 19 Bajo
08720 Vilafranca del Penedès (Barcelona)
☎: +34 938 182 317
info@artiga-fustel.com
www.artiga-fustel.com

Camino de Seda 2021 T
65% syrah, 30% monastrell, 5% petit verdot

87

La Bestia Monastrell 2019 T BA
100% monastrell

90
Colour: deep cherry. Nose: dried herbs, creamy oak, black fruit, tea leave, toasty. Palate: powerful, ripe fruit, spicy, round tannins, warm.

Saleta Monastrell 2020 T RB
100% monastrell

88
Kind finish, dried herbs, jammy, varietally correct, great length.

BODEGA MADRID ROMERO
Ctra. de El Carche, Km 8,3.
30520 Jumilla (Murcia)
☎: +34 611 043 352
info@bodegamadridromero.com
www.bodegamadridromero.com

Acacia Madrid Romero 2020 B FB
macabeo

90
With personality. Nose: cereal notes, dried flowers, faded flowers, yeasty notes, spicy. Palate: flavourful, balanced, fine bitter notes.

Madrid Romero Roble 12 meses 2017 T C
monastrell

91
Colour: deep cherry. Nose: fruit preserve, toasted bread, spicy, expressive. Palate: fleshy, fruity, powerful, balanced, good finish.

Madrid Romero Roble 6 meses 2019 T RB
monastrell, syrah

90
Colour: bright cherry. Nose: ripe fruit, scrubland, smoky, spicy. Palate: fruity, flavourful, balanced.

BODEGAS 1890
Ctra. Murcia s/n
30520 Jumilla (Murcia)
☎: +34 900 189 090
atcliente@jgc.es
www.garciacarrion.com

Mayoral 2016 T C
monastrell

85

Mayoral 2020 T
monastrell

87

Mayoral Reservado T
monastrell, cabernet sauvignon, petit verdot

86

Pata Negra Apasionado Organic T
monastrell

87

Pata Negra Apasionado sin Sulfitos 2020 T
monastrell

86

Pata Negra Apasionado T
monastrell, cabernet sauvignon, petit verdot

88
Fruity, spicy, herbal, tasty.

BODEGAS ALCEÑO
Barrio Iglesias, 34
30520 Jumilla (Murcia)
☎: +34 968 780 142
info@alceno.com
www.alceno.com

Alceño 2021 T
monastrell, syrah, tempranillo
88
Fruity, pleasant, standard, floral, jammy.

Alceño Dulce 2017 T D
monastrell
90
Colour: cherry, garnet rim. Nose: fruit preserve, ripe fruit, spicy, toasty, wild herbs, grassy. Palate: powerful, sweetness.

Alceño Monastrell 12 meses 2019 T
85% monastrell, 7,5% syrah, 7,5% garnacha tintorera
91
Colour: bright cherry. Nose: sweet spices, ripe fruit, dark chocolate, creamy oak. Palate: spicy, round tannins.

Alceño Sauvignon Blanc 2020 B FB
sauvignon blanc
90
Colour: straw. Nose: ripe fruit, dried herbs, faded flowers. Palate: powerful, ripe fruit, balanced.

Alceño Selección 2017 T C
monastrell, syrah, garnacha tintorera, tempranillo
92
Varietally correct. Colour: bright cherry. Nose: sweet spices, dark chocolate, smoky, toasty. Palate: fruity, spicy, round tannins.

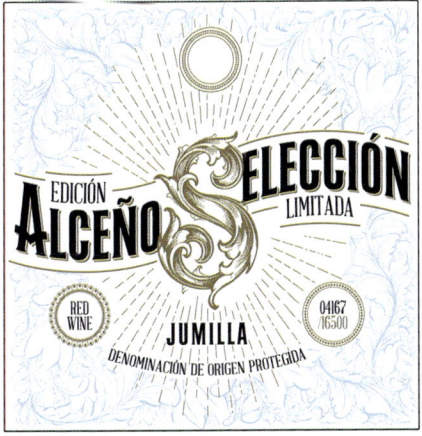

Alceño Syrah 50 Barricas 2020 T
85% syrah, 15% monastrell
89
Roasted coffee, jammy, tasty. Nose: dark chocolate, black fruit, ripe fruit.

Genio Español 12 meses Monastrell 2019 T RB
monastrell
89
Jammy, toasty, spicy, fruity, great length, tasty.

Genio Español 2020 T RB
monastrell
87

BODEGAS ARRÁEZ
Pol. 6 Parcela 386 Paraje Ciscar
46630 La Font de la Figuera (València/Valencia)
☎: +34 962 290 031
info@bodegasarraez.com
www.bodegasarraez.com

Vivir sin Dormir 2020 T RB
100% monastrell
87 ♣

BODEGAS BLEDA
Ctra. Jumilla - Ontur, Km. 2.
30520 Jumilla (Murcia)
☎: +34 968 780 012
vinos@bodegasbleda.com
www.bodegasbleda.com

Castillo de Jumilla 2018 T C
90% monastrell, 10% tempranillo
86

Castillo de Jumilla Monastrell 2020 T
100% monastrell
86

Divus 2018 T RB
100% monastrell
90
Colour: bright cherry. Nose: black fruit, spicy, wild herbs. Palate: balanced, flavourful, fresh.

Pino Doncel 12 Meses 2019 T C
70% monastrell, 30% syrah
89
Toasty, jammy, smoky, tasty. Nose: incense.

Pino Doncel 24 Meses 2017 T C
80% monastrell, 20% cabernet sauvignon
91
Colour: bright cherry. Nose: ripe fruit, black fruit, toasty, dried herbs. Palate: fruity, flavourful, fresh, slightly dry, soft tannins.

DO JUMILLA / D.O.P.

DO JUMILLA / D.O.P.

Pino Doncel Black 2020 T RB
50% monastrell, 30% syrah, 20% petit verdot

87

BODEGAS CARCHELO
Casas de La Hoya, s/n El Carche
30520 Jumilla (Murcia)
☎: +34 968 435 137
info@carchelo.com
www.carchelo.com

Carchelo 2020 T RB
50% monastrell, 35% cabernet sauvignon, 10% syrah, 5% tempranillo

89
Kind finish, balanced, spicy, jammy, toasty.

Carchelo Rosé 2021 RD
monastrell

87 ♣

Carchelo Selecto 2015 T C
monastrell, cabernet sauvignon, syrah, tempranillo

91
Colour: dark-red cherry, garnet rim. Nose: fruit preserve, aged wood nuances, tobacco, sweet spices, black fruit. Palate: spicy, round tannins, long.

Eya 2021 B
verdejo

86 ♣

Eya 2021 RD
monastrell

85 ♣

Eya 2021 T
monastrell

88 ♣
Balanced, spicy, fruity, jammy.

Muri Veteres 2017 T C
monastrell

91
Colour: deep cherry. Nose: dried herbs, creamy oak, black fruit. Palate: ripe fruit, spicy, round tannins.

BODEGAS CASA CASTILLO
Ctra. Jumilla - Hellin, Km. 8
30520 Jumilla (Murcia)
☎: +34 968 781 691
info@casacastillo.es
www.casacastillo.es

Casa Castillo El Molar 2021 T
100% garnacha

93 ♣
Colour: Cherry. Nose: balsamic herbs, sweet spices, scrubland, fruit expression. Palate: spicy, balsamic, good acidity.

Casa Castillo La Tendida 2020 T
85% monastrell, 15% garnacha

93 ♣
Colour: deep cherry. Nose: ripe fruit, dried herbs, creamy oak, candied fruit. Palate: powerful, ripe fruit, spicy, round tannins.

Casa Castillo La Tendida 2021 T
85% monastrell, 15% garnacha blanca

92 ♣
Colour: cherry, purple rim. Nose: red berry notes, floral, spicy, fresh fruit. Palate: flavourful, fruity, good acidity, long.

🏆 PODIUM

Casa Castillo Las Gravas 2020 T
100% monastrell

96 ♣
Colour: deep cherry. Nose: ripe fruit, dried herbs, creamy oak, earthy notes, dry stone. Palate: powerful, ripe fruit, spicy, round tannins.

Casa Castillo Monastrell 2021 T
100% monastrell

93 ♣
Colour: bright cherry. Nose: fruit expression, red berry notes, ripe fruit, sweet spices. Palate: good acidity, spicy, fine tannins.

🏆 PODIUM

Casa Castillo Pie Franco 2020 T C
100% monastrell

96 ♣
Colour: very deep cherry. Nose: complex, expressive, spicy, mineral, red berry notes. Palate: elegant, full-bodied, long, great length.

Valtosca 2021 T
100% syrah

92 ♣
Colour: cherry, purple rim. Nose: fruit expression, red berry notes, floral, powerful. Palate: fruity, flavourful, balanced, full-bodied.

BODEGAS CRAPULA & LANENA

Paraje La Graja
30520 Jumilla (Murcia)
☎: +34 662 380 985
gmartinez@vinocrapula.com
www.crapulawines.com

Cármine 2019 T BA
89
Hot, jammy, tasty, powerful.

Crápula Basado en Hechos Reales 2019 T C
70% monastrell, 15% syrah, garnacha
90
Colour: dark-red cherry. Nose: toasty, spicy, cocoa bean. Palate: flavourful, toasty, fine bitter notes.

Crápula Gold 2019 T
monastrell, syrah
88
Pruney, toasty, powerful, spicy.

Crápula Gold 2020 T
88
Full-bodied, standard, spicy, tasty. Palate: slightly dry, soft tannins.

Crápula Soul 2019 T C
89
Roasted coffee, spicy, tasty, powerful, jammy.

Crápula Soul Edición Limitada 2019 T
monastrell, cabernet sauvignon
89
Woody, toasty, fruity, jammy, tasty.

Gabriel Martínez. Pequeños pasos, grandes Ilusiones T
89
Toasty, smoky, pruney, full-bodied, powerful, tasty.

NdQ Selección 2019 T
monastrell
89
Jammy, toasty, spicy, full-bodied, dried herbs, great length.

BODEGAS EL NIDO

Ctra. de Fuentealamo - Paraje de la Aragona
30550 Jumilla (Murcia)
☎: +34 968 435 023
info@juangil.es
www.gilfamily.es

🏆 PODIUM

Clío 2019 T
70% monastrell, 30% cabernet sauvignon
95
Colour: very deep cherry. Nose: aromatic coffee, powerful, black fruit, ripe fruit, scrubland. Palate: smoky aftertaste, great length, round tannins, concentrated, balanced.

🏆 PODIUM

Corteo 2019 T
100% syrah
95
Colour: very deep cherry. Nose: complex, expressive, spicy, mineral, black fruit, ripe fruit. Palate: full-bodied, long, great length, flavourful.

🏆 PODIUM

El Nido 2019 T
70% cabernet sauvignon, 30% monastrell
96
Full-bodied, balanced. Colour: deep cherry. Nose: black fruit, cocoa bean, dried herbs, complex. Palate: fleshy, round tannins, elegant, round.

BODEGAS JUAN GIL

Ctra. Fuentealamo - Paraje de la Aragona
30520 Jumilla (Murcia)
☎: +34 968 435 022
info@juangil.es
www.gilfamily.es

Albacea Merlot 2021 T
merlot
91
Colour: Cherry. Nose: dried herbs, black fruit, dark chocolate. Palate: powerful, ripe fruit, fine bitter notes.

Comoloco 2021 T
100% monastrell
90
Colour: cherry, purple rim. Nose: red berry notes, floral, spicy, varietal. Palate: flavourful, fruity, good acidity, long.

DO JUMILLA / D.O.P.

DO JUMILLA / D.O.P.

Honoro Vera Monastrell 2021 T
100% monastrell

90

Colour: cherry, purple rim. Nose: fruit expression, red berry notes, floral. Palate: fruity, flavourful, balanced.

Honoro Vera Orgánico 2021 T
100% monastrell

90

Colour: bright cherry. Nose: dried herbs, expressive, black fruit, ripe fruit. Palate: powerful, ripe fruit, fine bitter notes.

Juan Gil 2021 RD
50% tempranillo, 50% syrah

90

Colour: raspberry rose. Nose: floral, balanced, medium intensity, red berry notes, ripe fruit. Palate: good acidity, fine bitter notes, easy to drink.

Juan Gil Etiqueta Amarilla/Yellow Label 2021 T

91

Colour: bright cherry. Nose: floral, balanced, red berry notes. Palate: flavourful, fruity, good acidity.

🏆 PODIUM

Juan Gil Etiqueta Azul/Blue Label 2020 T
60% monastrell, 30% cabernet sauvignon, 10% syrah

95

Colour: dark-red cherry, garnet rim. Nose: aged wood nuances, tobacco, sweet spices, ripe fruit, black fruit, toasty, creamy oak. Palate: spicy, round tannins, long, flavourful.

Juan Gil Moscatel Seco 2021 B
moscatel grano menudo

90

Colour: bright straw. Nose: expressive, white flowers, jasmine, dried herbs, varietal. Palate: flavourful, fruity, balanced.

Juan Gil Petit Verdot 2021 T
petit verdot

92

Colour: cherry, purple rim. Nose: fruit expression, floral, spicy, grassy, black pepper, creamy oak. Palate: flavourful, fruity, good acidity, long.

Juan Gil Selección Bartolomé Abellán 2021 T BA

91

Colour: dark-red cherry. Nose: toasty, spicy, cocoa bean, ripe fruit, black fruit. Palate: flavourful, toasty, fine bitter notes.

Juan Gil Etiqueta Plata/Silver Label 2020 T
100% monastrell

93

Colour: deep cherry. Nose: creamy oak, black fruit, dried fruit, scrubland. Palate: powerful, ripe fruit, spicy, round tannins.

BODEGAS LUZÓN

Ctra. Jumilla-Calasparra, Km. 3,1
30520 Jumilla (Murcia)
☎: +34 968 784 135
info@bodegasluzon.com
www.bodegasluzon.com

Alma de Luzón 2019 T
monastrell, syrah

93

Colour: Cherry, garnet rim. Nose: complex, expressive, spicy, mineral, black fruit. Palate: full-bodied, long, great length.

Altos de Luzón 2020 T
monastrell

92

Colour: very deep cherry. Nose: aromatic coffee, powerful, black fruit, ripe fruit, toasty. Palate: smoky aftertaste, great length, round tannins.

Finca Luzón 2019 T C
monastrell, cabernet sauvignon

91

Colour: cherry, garnet rim. Nose: overripe fruit, creamy oak, hot. Palate: pruney, powerful, sweet tannins.

Finca Luzón 2021 T RB
monastrell

90

Colour: cherry, purple rim. Nose: red berry notes, floral, toasty, dark chocolate. Palate: flavourful, fruity, good acidity, long.

Finca Luzón Monastrell Syrah 2021 T
monastrell, syrah

90

Colour: cherry, purple rim. Nose: powerful, ripe fruit, spicy. Palate: ripe fruit, flavourful, good structure.

Gamellón 2021 T
monastrell, syrah

90

Colour: bright cherry. Nose: floral, ripe fruit, balanced. Palate: flavourful, fruity, good acidity.

Gamellón Viñas Viejas 2019 T BA
monastrell

90

Colour: cherry, garnet rim. Nose: fruit preserve, fruit liqueur notes, powerful, waxy notes. Palate: flavourful, long, sweet tannins.

Luzón Colección Monastrell 2021 T
monastrell

90

Colour: deep cherry. Nose: ripe fruit, dried herbs, spicy. Palate: powerful, ripe fruit, round tannins.

Por Tí 2019 T
monastrell, cabernet sauvignon

93

Colour: deep cherry. Nose: dried herbs, creamy oak, ripe fruit, black fruit, toasty. Palate: powerful, ripe fruit, spicy, round tannins.

BODEGAS OLIVARES

Vereda Real, s/n
30520 Jumilla (Murcia)
☎: +34 968 780 180
correo@bodegasolivares.com
www.bodegasolivares.com

Altos de la Hoya 2020 T BA
90% monastrell, 10% garnacha

88

Balanced, spicy, dried herbs, jammy.

Olivares 2020 T
86

Olivares 2021 RD
100% garnacha

86

BODEGAS SALZILLO

Ctra. Nacional 344, km 57,200
30520 Jumilla (Murcia)
☎: +34 968 782 735
salzillo@bodegassalzillo.com
www.bodegassalzillo.com

Camelot Dulce Monastrell 2019 T D
88

Overripe, representative, tasty, sweety finish, smoky, toasty.

Güertana Sauvignon Blanc 2021 B
sauvignon blanc

88

Aromatic, fruity, dried herbs, citrus fruit, smooth. Nose: stone fruit.

Gúertano Monastrell 2019 T
monastrell

87

Matius Monastrell 2018 T C
monastrell

88

Kind finish, jammy, pruney, tasty, spicy.

Zenizate Cabernet Sauvignon 2020 T
cabernet sauvignon

85

Zenizate Monastrell 2020 T
monastrell

86

BODEGAS SAN DIONISIO, S. COOP.

Ctra. de la Higuera, s/n
02651 Fuentealamo (Albacete)
☎: +34 967 543 032
Fax: +34 967 543 136
sandionisio@bodegassandionisio.es
www.bodegassandionisio.com

Mainetes 12 meses 2018 T C
monastrell

89

Spicy, herbal, tasty, wild, roasted coffee.

Mainetes Verdejo 2021 B FB
verdejo

88

Spicy, jammy, roasted coffee, herbal.

Señorío de Fuenteálamo Monastrell 2020 T RB
monastrell

88

Toasty, standard, smoky, jammy, tasty.

DO JUMILLA / D.O.P.

DO JUMILLA / D.O.P.

Señorío de Fuenteálamo Monastrell 2021 RD
monastrell
87

Señorío de Fuenteálamo Monastrell 2021 T
monastrell
86

Señorío de Fuenteálamo Verdejo 2021 B
verdejo
85

BODEGAS SANTIAGO APÓSTOL
Calle Capataz Santiago, 91
02650 Montealegre del Castillo (Albacete)
☎: +34 967 336 058
Fax: +34 967 336 624
info@bodegassantiagoapostol.com
www.bodegassantiagoapostol.com

Oferente 2018 T C
100% monastrell
87

Oferente 2019 B BA
airén, verdejo
86

Oferente 2020 B
airén, verdejo
85

Oferente Selección 2020 T
100% monastrell
85

BODEGAS SILVANO GARCÍA
Avda. de Murcia, 29
30520 Jumilla (Murcia)
☎: +34 968 780 767
bodegas@silvanogarcia.es
www.silvanogarcia.es

Silvano García 4 meses 2019 T BA
monastrell
86

Silvano García Dulce Monastrell 2020 T D
monastrell
90
Colour: light cherry. Nose: fruit liqueur notes, scrubland, dried flowers, tea leave, tomato. Palate: sweet, rich, flavourful.

Silvano García Etiqueta Negra 2019 T
monastrell
89
Pruney, spicy, roasted coffee, jammy, tasty.

Silvano García Verdejo 2021 B
verdejo
87

Viñahonda 2017 T C
monastrell
90
Balsamic herbs, herbal, jammy. Nose: waxy notes, fine reductive notes, black fruit. Palate: flavourful, long.

Viñahonda Organic + 2019 T BA
monastrell
84

BODEGAS TINTO CORAZÓN
Calle del Duque
30520 Jumilla (Murcia)
☎: +34 610 970 431
info@tintocorazon.com
www.tintocorazon.com

Disfrutar 2019 T R
100% monastrell
89
Creamy, balanced, spicy, herbal, toasty.

Imaginar 2020 T C
50% monastrell, 50% garnacha
88
Spicy, balanced, herbal, toasty.

Olvidar 2021 B MC
100% sauvignon blanc
87

BODEGAS VIÑA ELENA
Paraje Estrecho de Marín, N-344, km 52,5
30520 Jumilla (Murcia)
☎: +34 968 781 340
info@vinaelena.com
www.vinaelena.com

Bruma del Estrecho de Marín "Paraje Marín" 2020 T
monastrell
90
Sweety finish. Colour: light cherry. Nose: ripe fruit, dried herbs, creamy oak. Palate: ripe fruit, spicy, round tannins.

Bruma del Estrecho de Marín "Parcela Mandiles" 2020 T
monastrell

89
Balanced, spicy, tasty, herbal.

Bruma del Estrecho de Marín Finca CQ 2020 T
monastrell

91
Sweety finish. Colour: cherry, garnet rim. Nose: floral, red berry notes, black fruit, ripe fruit, wild herbs. Palate: fleshy, spicy, sweet tannins.

Cucos de la Alberquilla 2020 T
cabernet sauvignon

87

Familia Pacheco Monastrell Orgánico 2020 T
monastrell

87 🌱

Familia Pacheco Selección para Paco 2019 T C

89
Balanced, spicy, jammy, tasty, powerful.

BODEGAS Y VIÑEDOS VALTRAVIESO

Finca La Revilla, s/n
47316 Piñel de Arriba (Valladolid)
☎: +34 983 484 030
marketing@valtravieso.com
www.valtravieso.com

Pie Firme 2019 T R
monastrell

93
Colour: cherry, garnet rim. Nose: ripe fruit, scrubland, waxy notes, earthy notes, dried herbs. Palate: flavourful, spicy, round tannins.

BRUTO

Camino de Tiñosa, 40
30012 Murcia (Murcia)
☎: +34 968 253 533
Fax: +34 968 340 141
fragama.naviser@gmail.com

Bruto 2019 T
100% monastrell

93
Colour: deep cherry. Nose: sweet spices, black fruit, ripe fruit, cocoa bean. Palate: spicy, round tannins, long, powerful.

BSI - BODEGAS SAN ISIDRO

Ctra. Murcia, s/n
30520 Jumilla (Murcia)
☎: +34 968 780 700
info@bsi.es
www.bsi.es

Gémina Cuvée Selección 2019 T C
100% monastrell

90
Colour: very deep cherry. Nose: roasted coffee, aromatic coffee, powerful, black fruit. Palate: smoky aftertaste, great length, round tannins, powerful.

Gémina Finca El Volcán 2018 T
100% monastrell

92
Colour: very deep cherry. Nose: aromatic coffee, powerful, ripe fruit, black fruit, toasted bread. Palate: smoky aftertaste, great length, round tannins.

Gémina Finca La Cabra 2018 T C
100% monastrell

93
Colour: deep cherry. Nose: complex, expressive, spicy, mineral, black fruit. Palate: elegant, full-bodied, long, great length.

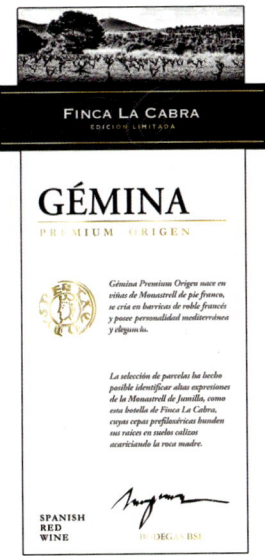

Gémina Finca Los Tomillares 2018 T
100% monastrell

92
Colour: deep cherry. Nose: ripe fruit, dried herbs, creamy oak, earthy notes, red clay notes. Palate: powerful, ripe fruit, spicy, round tannins.

DO JUMILLA / D.O.P.

DO JUMILLA / D.O.P.

Gémina Selección Monastrell 2020 T
monastrell

90

Colour: deep cherry. Nose: ripe fruit, dried herbs, creamy oak. Palate: powerful, ripe fruit, spicy, round tannins.

Sabatacha 2019 T C
monastrell

88

Toasty, powerful, jammy, spicy.

CAMPOS DE RISCA
Avda. Diagonal, 590, 5º 1ª
08021 Barcelona (Barcelona)
☎: +34 935 385 677
vinergia@vinergia.com
www.vinergia.com

Campos de Risca 2020 T
monastrell

86 🌿

DELAMPA
Ctra. Jumilla - Yecla, Km. 79,3
30520 Jumilla (Murcia)
☎: +34 968 759 956
bodegas@delampa.es
www.mrdelampa.com

Delampa 50 años 2018 T C
100% monastrell

88

Full-bodied, spicy, smoky, herbal, jammy, tasty.

Delampa Monastrell 2021 T
100% monastrell

84

Delampa Selección 2021 T
100% monastrell

86

Éxodo 2019 T RB
100% monastrell

87

Éxodo Autor 2018 T C
100% monastrell

88

Spicy, dried herbs, jammy, wild, varietally correct, toasty.

Julián Santos Martínez Edición Centenario 2019 T C
100% monastrell

90

Sweety finish. Colour: deep cherry. Nose: dried herbs, creamy oak, black fruit. Palate: ripe fruit, spicy, round tannins.

DOBLEDEPEREZ MICROBODEGA
Plaza del Ayuntamiento, 7 Bajo A
02653 Albatana (Albacete)
☎: +34 627 588 119
info@dobledeperez.es
www.dobledeperez.es

En Contacto 2021 B
100% verdejo

87

Inaudita 2021 T
100% petit verdot

90

Colour: cherry, purple rim. Nose: fruit expression, red berry notes, spicy, ripe fruit. Palate: flavourful, fruity, fresh, balanced.

EGO BODEGAS
Paraje Hoya de Torres s/n
30520 Jumilla (Murcia)
☎: +34 868 680 939
info@egobodegas.com
www.egobodegas.com

Acuma 2021 T
monastrell, syrah, petit verdot

90

Colour: bright cherry. Nose: sweet spices, ripe fruit, smoky. Palate: fruity, spicy, round tannins.

Don Baffo 2021 T
monastrell, syrah, petit verdot

88

Fruity, herbal, spicy.

Fuerza By Ego 2020 T
monastrell, syrah

89

Hot, spicy, jammy, simple.

Goru 18 M 2018 T
monastrell, cabernet sauvignon

92

Colour: deep cherry. Nose: ripe fruit, dried herbs, creamy oak, waxy notes, sweet spices. Palate: powerful, ripe fruit, spicy, round tannins.

Goru 38 Barrels 2020 T BA
monastrell, cabernet sauvignon

90

Colour: cherry, garnet rim. Nose: fruit preserve, overripe fruit, characterful, smoky, toasty. Palate: flavourful, long.

Infinito 2018 T C
monastrell

90

Colour: bright cherry. Nose: ripe fruit, black fruit, wild herbs. Palate: fruity, flavourful, spicy.

DO JUMILLA / D.O.P.

ESENCIA WINES CELLARS
Ctra. El Carche, Km. 11,5
30520 Jumilla (Murcia)
☎: +34 968 783 035
Fax: +34 968 716 030
info@casadelaermita.com
www.casadelaermita.com

Casa de la Ermita 2020 T C
monastrell, cabernet sauvignon, syrah

90

Powerful. Colour: bright cherry. Nose: sweet spices, dark chocolate, ripe fruit, fruit preserve. Palate: fruity, spicy, round tannins.

Casa de la Ermita Petit Verdot 2019 T C
petit verdot

91

Colour: deep cherry. Nose: dried herbs, creamy oak, black fruit, ripe fruit. Palate: powerful, ripe fruit, spicy, grainy tannins.

Hacienda del Carche Cepas Viejas 2019 T C
monastrell, cabernet sauvignon

88

Balanced, spicy, herbal, jammy, roasted coffee.

Infiltrado 2021 T
syrah, monastrell, garnacha tintorera

88

Kind finish, balanced, spicy, full-bodied, jammy, toasty.

Lunático Monastrell 2020 T
monastrell

88

Pruney, full-bodied, spicy, roasted coffee, creamy.

Taus Fusion 2020 T
syrah

88

Full-bodied, toasty, smoky, pruney.

FINCA BACARA
Calle Rio Taibilla, 13 - 7ºA
30110 Churra (Murcia)
☎: +34 868 680 939
Fax: +34 868 580 285
info@fincabacara.com
www.fincabacara.com

Crazy Grapes 2021 T
monastrell

88 ♣

Smoky, hot, toasty, powerful.

HI 2019 T C
monastrell

92 ♣

Colour: bright cherry. Nose: black fruit, balsamic herbs, dried herbs, spicy. Palate: fruity, flavourful, powerful tannins, balanced.

Tabá 2020 T C
monastrell

92 ♣

Colour: deep cherry. Nose: black fruit, balsamic herbs, dried herbs, spicy. Palate: fruity, flavourful, balanced, slightly dry, soft tannins.

Time Waits for No One 6M 2021 T RB
monastrell

89 ♣

Full-bodied, pruney, spicy, tasty, powerful.

Time Waits for No one Black 2020 T
monastrell

90 ♣

Colour: cherry, purple rim. Nose: black fruit, wild herbs, balsamic herbs, spicy. Palate: fruity, flavourful, smoky aftertaste, slightly dry, soft tannins.

Time Waits for No one Double 2021 T
monastrell

89 ♣

Smoky, jammy, toasty, full-bodied, tasty.

HAMMEKEN CELLARS
Llavador, 20
03700 Denia (Alacant/Alicante)
☎: +34 965 791 967
cellars@hammekencellars.com
www.hammekencellars.com

El Paso Old Vines 2020 T
100% monastrell

89

Powerful, jammy, tasty, spicy.

JOTA JIMÉNEZ WINES
Avda. de Levante, 15
30520 Jumilla (Murcia)
☎: +34 628 801 543
info@jotajimenezwines.com
www.laschovas.com

Las Chovas 2019 T
100% monastrell

91

Colour: deep cherry. Nose: ripe fruit, dried herbs, creamy oak, fine reductive notes. Palate: powerful, ripe fruit, spicy, round tannins.

DO JUMILLA / D.O.P.

ÓSCAR OLMOS VINOS
Paraje Valle Hoya de Torres, s/n
30520 Jumilla (Murcia)
☎: +34 609 663 504
info@oscarolmos.com
www.oscarolmosvinos.com

La Princesa 2020 T BA
monastrell
89
Nose: toasty, cocoa bean, dried fruit. Palate: flavourful, toasty, fine bitter notes.

La Sirvienta 2020 B
100% airén
92
Colour: bright straw. Nose: ripe fruit, fragrant herbs, fine lees, cereal notes, burnt matches. Palate: full-bodied, rich, long, good acidity.

PACO MULERO
Partida de la Hoya Torres s/n
30520 Jumilla (Murcia)
☎: +34 676 433 541
info@pacomulero.com
www.pacomulero.com

Paco Mulero Veinte Meses 2019 T
93
Colour: deep cherry. Nose: smoky, spicy, black fruit, dried herbs, ripe fruit. Palate: flavourful, good structure, balanced, good finish, slightly dry, soft tannins.

Prisma Orgánico 2021 T
monastrell
90
Kind finish, varietally correct. Nose: dried herbs, ripe fruit. Palate: fruity, flavourful, balanced, round tannins.

PARAJES DEL VALLE BODEGAS Y VIÑEDOS
Avda. de Murcia, s/n
30520 Jumilla (Murcia)
☎: +34 968 782 977
info@parajesdelvalle.es
www.parajesdelvalle.es

Parajes del Valle Monastrell 2020 T
monastrell
89
Herbaceous, fruity, jammy, wild, smooth, balanced.

Terraje 2018 T BA
100% monastrell
91
Colour: bright cherry. Nose: fruit preserve, sweet spices, floral. Palate: fresh, fruity, flavourful, varietal.

PITUCO VITICULTOR
Pérez Galdós, 3
02003 Albacete (Albacete)
☎: +34 696 168 873
jose@rodriguezdevera.com
www.rodriguezdevera.com

Pituco MST 2020 T
monastrell, syrah, garnacha tintorera
87

Pituco Paraje las Zorreras 2017 T R
100% monastrell
90
Colour: very deep cherry. Nose: roasted coffee, aromatic coffee, powerful. Palate: smoky aftertaste, great length, round tannins.

SAM'S WINE
Pº Joaquin Garrigues Walker 3, Bajo 1
30007 Murcia (Murcia)
☎: +34 968 979 964
tienda@samwine.es
www.samwine.es

Sam Edición Limitada Jumilla 2018 T
monastrell, cabernet sauvignon, syrah
91
Colour: cherry, purple rim. Nose: black fruit, ripe fruit, smoky. Palate: fruity, powerful, balanced, spicy, smoky aftertaste.

SPANISH PALATE
Avda. Carlos Latorre, 30
49800 Toro (Zamora)
☎: +34 980 690 643
export@spanishpalate.es
www.spanishpalate.es

Botas de Barro Jumilla 2020 T RB
100% monastrell
88
Spicy, toasty, tasty, jammy.

TERRAMAGNA
Ctra. Rincón del Moro, Km 10
02410 Lietor (Albacete)
☎: +34 965 211 955
earellano@grupoterramagna.com
www.grupoterramagna.com

CT en Clave de DO 2019 T RB
monastrell, syrah
90
Colour: deep cherry. Nose: ripe fruit, dried herbs, creamy oak, earthy notes. Palate: powerful, ripe fruit, spicy, round tannins.

VINOS SIERRA NORTE
Paraje La Raja, s/n
30520 Jumilla (Murcia)
☎: +34 618 323 119
jumilla@bodegasierranorte.com
www.bodegasierranorte.com

Equilibrio 2021 T
monastrell, syrah

88

Spicy, balanced, herbal, jammy.

Equilibrio Sauvignon Blanc 2021 B
sauvignon blanc

88

Pleasant, citrus fruit, crisp, herbal, smooth.

Equilibrio-4 2020 T
monastrell

89

Varietally correct, jammy, spicy, balanced, fruity, tasty.

Equilibrio-9 2020 T BA
monastrell

90

Colour: deep cherry. Nose: dried herbs, creamy oak, dark chocolate, black fruit. Palate: ripe fruit, spicy, round tannins.

VIÑAS DEL RÓDAN
Isaac Peral, 6 Bajo C
30520 Jumilla (Murcia)
☎: +34 686 835 565
info@vinasdelrodan.es

Don Carlos 2018 T RB
monastrell

88

Fruity, herbal, spicy, tasty.

Raltollo 24 M 2019 T C
monastrell

91

Colour: bright cherry. Nose: ripe fruit, black fruit, creamy oak, lactic notes, dry stone. Palate: flavourful, balanced, toasty, slightly dry, soft tannins.

VIÑEDOS Y BODEGAS ASENSIO CARCELÉN
RM-714, Km. 8
30520 Jumilla (Murcia)
☎: +34 968 435 543
bodegascarcelen@gmail.com

100 x 100 Monastrell 2020 T
monastrell

84

100 x 100 Syrah 2020 T
syrah

83

Acorde 2020 T
monastrell

86

Pura Sangre 2014 T R
100% monastrell

80

WEIN & VINOS GMBH
Hardenbergstr. 9A
10715 Berlin (Berlin)
☎: +49 303 150 6080
info@vinos.de
www.vinos.de

El Plantonal 2020 T
100% monastrell

91

Colour: deep cherry. Nose: ripe fruit, dried herbs, creamy oak, dark chocolate. Palate: powerful, ripe fruit, spicy, round tannins.

El Plantonal 2021 T
100% monastrell

92

Colour: cherry, purple rim. Nose: red berry notes, floral, spicy, expressive. Palate: flavourful, fruity, good acidity, long.

Mondeo Selección Especial 2020 T BA
80% monastrell, 10% cabernet sauvignon, 10% syrah

91

Colour: Cherry. Nose: ripe fruit, scrubland, black fruit, spicy. Palate: flavourful, fruity, good structure.

Obsesión 2020 T BA
monastrell

93

Colour: dark-red cherry, garnet rim. Nose: fruit preserve, aged wood nuances, tobacco, sweet spices, black fruit, toasty, dark chocolate. Palate: spicy, round tannins, long.

Petit Obsesión 2020 T BA
70% monastrell, 15% cabernet sauvignon, 15% merlot

90

Colour: bright cherry. Nose: red berry notes, wild herbs. Palate: fruity, flavourful, slightly dry, soft tannins.

DO JUMILLA / D.O.P.

DO. LA GOMERA
CONSEJO REGULADOR

Avda. Guillermo Ascanio, 16
38840 Vallehermoso (La Gomera)
☎: +34 922 800 801 - Fax: +34 922 801 146
@: vinoslagomera@gmail.com
www.vinoslagomera.com

LOCATION:

The majority of the vineyards are found in the north of the island, in the vicinity of the towns of Vallehermoso (some 385 Ha) and Hermigua. The remaining vineyards are spread out over Agulo, Valle Gran Rey –near the capital city of La Gomera, San Sebastián– and Alajeró, on the slopes of the Garajonay peak.

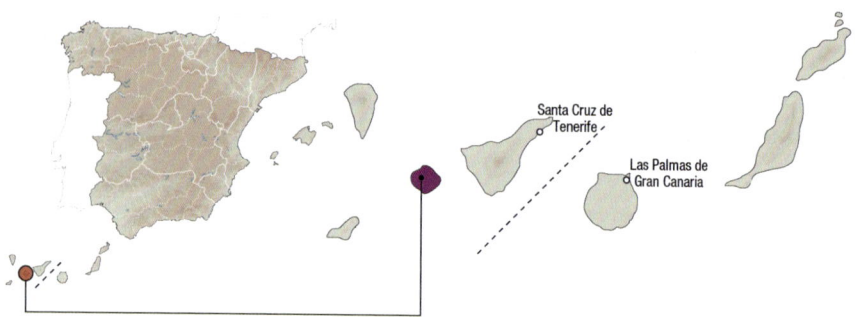

GRAPE VARIETIES:

WHITE: Forastera Gomera (90%), Listán Blanca, Marmajuelo, Malvasía and Pedro Ximenez.

RED: Listán Negra (5%), Negramoll (2%); Experimental: Tintilla Castellana, Cabernet Sauvignon and Rubí Cabernet.

FIGURES:

Vineyard surface: 125 – **Wine-Growers:** 230 – **Wineries:** 20 – **2021 Harvest rating:** Good – **Production 21:** 50,000 litres – **Market percentages:** 100% Nacional.

SOIL:

The most common soil in the higher mountain regions is deep and clayey, while, as one approaches lower altitudes towards the scrubland, the soil is more Mediterranean with a good many stones and terraces similar to those of the Priorat.

CLIMATE:

The island benefits from a subtropical climate together with, as one approaches the higher altitudes of the Garajonay peak, a phenomenon of permanent humidity known as 'mar de nubes' (sea of clouds) caused by the trade winds. This humid air from the north collides with the mountain range, thereby creating a kind of horizontal rain resulting in a specific ecosystem made up of luxuriant valleys. The average temperature is 20°C all year round.

CLASSIFICATION OF YOUNG WINE HARVESTS PEÑÍNGUIDE

2017	2018	2019	2020	2021
N/A	N/A	N/A	N/A	N/A

BODEGA MONTORO, MARIO RODRÍGUEZ MENDOZA

Montoro s/n Hermigua
38820 La Gomera (Santa Cruz de Tenerife)
☎: +34 695 943 245
mariorguezmend@gmail.com

Laderas de Montoro 2020 T
85% listán negro, 10% negramoll, 5% forastera

92

Colour: cherry, garnet rim. Nose: black pepper, meaty notes, fine reductive notes, black fruit, red berry notes. Palate: full of life, fruity, balanced.

Montoro 2021 B
65% tempranillo, 23% merlot, 12% syrah

89

Crisp, citrus fruit, balanced, floral, wild, smooth.

Montoro de Forastera 2021 B FB
100% forastera

91

Colour: bright yellow. Nose: spicy, fine lees, white fruit, balanced, expressive. Palate: long, easy to drink, fine bitter notes.

DO LA GOMERA / D.O.P.

DO. LA MANCHA

CONSEJO REGULADOR

Avda. de Criptana, 73
13600 Alcázar de San Juan (Ciudad Real)
☎: +34 926 541 523 - Fax: +34 926 588 040
@: consejo@lamanchawines.com
www.lamanchawines.com

LOCATION:

On the southern plateau in the provinces of Albacete, Ciudad Real, Cuenca and Toledo. It is the largest wine-growing region in Spain and in the world.

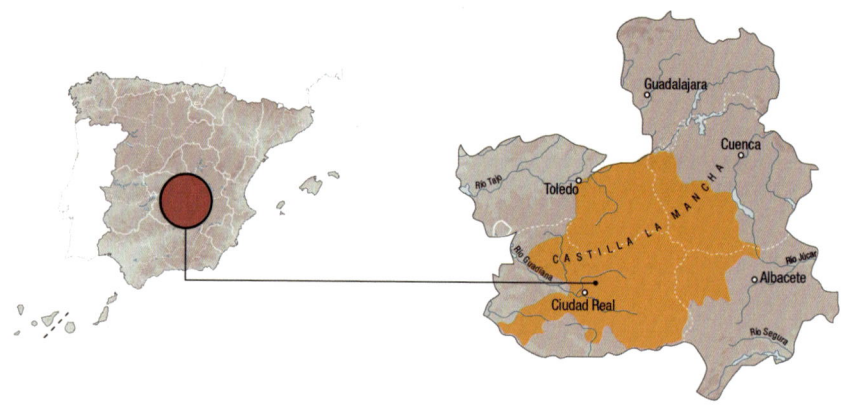

GRAPE VARIETIES:

WHITE: Airén (majority), Macabeo, Pardilla, Chardonnay, Sauvignon Blanc, Verdejo, Moscatel de Grano Menudo, Gewürztraminer, Parellada, Pero Ximénez, Riesling, Torrontés and Moscatel de Alejandría.

RED: Cencibel (majority amongst red varieties), Garnacha, Moravia, Cabernet Sauvignon, Merlot, Syrah, Cabernet Franc, Graciano, Malbec, Mencía, Monastrell, Petit Verdot, Bobal, Garnacha Tintorera and Pinot Noir.

FIGURES:

Vineyard surface: 154,894 – **Wine-Growers:** 14,125 – **Wineries:** 241 – **2021 Harvest rating:** Excellent– **Production 21:** 117,500,000 litres – **Market percentages:** 64% National - 36% International.

SOIL:

The terrain is flat and the vineyards are situated at an altitude of about 700 m above sea level. The soil is generally sandy, limy and clayey.

CLIMATE:

Extreme continental, with temperatures ranging between 40/45°C in summer and –10/12°C in winter. Rather low rainfall, with an average of about 375 mm per year.

CLASSIFICATION OF YOUNG WINE HARVESTS PEÑÍNGUIDE

2017	2018	2019	2020	2021
GOOD	GOOD	GOOD	GOOD	GOOD

BODEGA AMANCIO MENCHERO
Legión, 27
13260 Bolaños de Calatrava (Ciudad Real)
☎: +34 680 402 931
amanciomenchero@hotmail.com
www.bodegamenchero.com

Finca Moriana 2018 T C
cencibel
87

BODEGA LA TERCIA - ORGANIC WINES
Plaza Santa Quiteria, 12
13600 Alcázar de San Juan (Ciudad Real)
☎: +34 610 602 174
bodegalatercia@gmail.com
www.bodegalatercia.com

Yemanueva Airén Pie Franco 2021 B
airén
86

Yemanueva Tempranillo 2019 T
tempranillo
87

Yemaserena 2014 T C
85

Yemaserena 2016 T RB
89
Toasty, hot, classic, jammy, tasty.

Yemaserena 2018 B FB
airén
89
Toasty, smoky, spicy, jammy, tasty. Nose: faded flowers, fine lees, lees reduction notes.

BODEGA Y VIÑAS ALDOBA
Ctra. Alcázar, km. 1
13700 Tomelloso (Ciudad Real)
☎: +34 926 506 534
allozo@allozo.com
www.allozo.com

Aldoba 2021 T MC
100% tempranillo
84

Aldoba Verdejo 2021 B
verdejo
83

BODEGAS AYUSO
Polígono Eras de Santa Lucía, Parcela 35.1
02600 Villarrobledo (Albacete)
☎: +34 967 140 458
comercial@bodegasayuso.es
www.bodegasayuso.es

Benizar Semidulce 2021 B SD
100% macabeo
85

Castillo de Benizar Cabernet Sauvignon 2021 RD
100% cabernet sauvignon
85

Castillo de Benizar Macabeo 2021 B
100% macabeo
86

Castillo de Benizar Tempranillo 2020 T
tempranillo
81

Estola 2014 T GR
65% tempranillo, 35% cabernet sauvignon
88
Classic, hot, spicy, toasty, tasty.

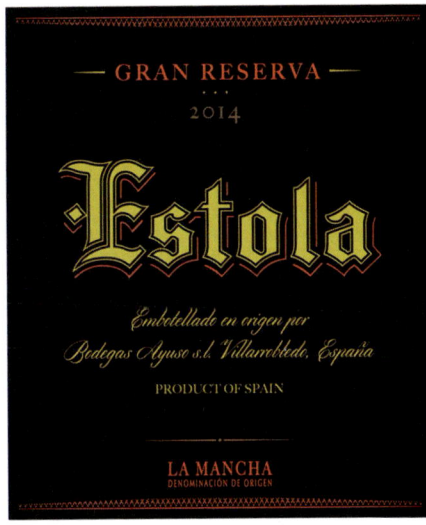

Estola 2018 T C
100% tempranillo
87

Estola 2020 RD
84

DO LA MANCHA / D.O.P.

DO LA MANCHA / D.O.P.

Estola 2020 T RB
85

Estola Selección 2018 T
100% tempranillo
87

Estola Verdejo 2021 B
100% verdejo
86

Finca Los Azares Tempranillo Cabernet Merlot 2017 T
33% tempranillo, 33% cabernet sauvignon, 33% merlot
85

BODEGAS CAMPOS REALES
Castilla La Mancha, 4
16670 El Provencio (Cuenca)
☎: +34 967 166 066
tienda@bodegascamposreales.com
www.bodegascamposreales.com

Campos Reales Cabernet Sauvignon 2019 T RB
100% cabernet sauvignon
88
Full-bodied, balanced, spicy, full-bodied, jammy.

Campos Reales Syrah 2019 T
100% syrah
88
Balanced, spicy, crisp, herbal.

Cánfora Pie Franco 2016 T R
tempranillo
93
Colour: dark-red cherry. Nose: toasty, spicy, cocoa bean, black fruit, ripe fruit. Palate: flavourful, toasty, fine bitter notes.

Canforrales Chardonnay 2021 B
100% chardonnay
87

Canforrales Clásico Tempranillo 2021 T
100% tempranillo
88
Pleasant, fruity, smooth, tasty.

Canforrales Lucía Airén 2021 B
airén
86

Canforrales Petit Verdot 2021 RD
petit verdot
86

Canforrales Selección 2019 T RB
100% tempranillo
88
Nose: overripe fruit, creamy oak, hot. Palate: pruney, powerful, sweet tannins.

Gladium Viñas Viejas 2019 T C
tempranillo
88
Balanced, spicy, fruity, full-bodied, jammy.

BODEGAS CASA ANTONETE
Barrio San José, s/n
02100 Tarazona de La Mancha (Albacete)
☎: +34 967 480 074
Fax: +34 967 480 294
ventas@casaantonete.com
www.casaantonete.com

Casa Antonete Macabeo 2021 B
100% macabeo
85

Casa Antonete Tempranillo 2017 T C
100% tempranillo
87

Casa Antonete Tempranillo 2021 T MC
100% tempranillo
83

Négora Chardonnay 2021 B
100% verdejo
85

Négora Sauvignon Blanc 2021 B
100% sauvignon blanc
86

Négora Verdejo 2021 B
verdejo
85

BODEGAS CENTRO ESPAÑOLAS
Ctra. Alcázar, km. 1
13700 Tomelloso (Ciudad Real)
☎: +34 926 505 653
allozo@allozo.com
www.allozo.com

Allozo 2014 T GR
100% tempranillo
89
Age nuances, elegant, classic, smooth.

Allozo 2017 T R
tempranillo
88
Fruity, floral, spicy, balanced, jammy.

Allozo 2019 T C
85
Spicy, jammy, simple, standard.

Allozo Tempranillo 2021 T
100% tempranillo
84

Allozo Verdejo 2021 B
100% verdejo
84

Flor de Allozo Tempranillo Garnacha 2019 T
50% tempranillo, 50% garnacha
86

BODEGAS EL VINCULO
Avda. Juan Carlos I, s/n
13610 Campo de Criptana (Ciudad Real)
☎: +34 926 563 709
elvinculo@elvinculo.com
www.familiafernandezrivera.com

El Vínculo 2018 T C
100% tempranillo
88
Creamy, balanced, spicy, roasted coffee.

El Vínculo Alejairen 2019 B C
100% airén
88
Citrus fruit, herbal, spicy, toasty, tasty.

El Vínculo Paraje la Golosa 2016 T GR
100% tempranillo
91
Colour: very deep cherry. Nose: roasted coffee, aromatic coffee, powerful, black fruit. Palate: smoky aftertaste, round tannins, flavourful.

BODEGAS ENTREMONTES
45800 Quintanar de la Orden (Toledo)
☎: +34 925 180 237
www.bodegasentremontes.com

Entremontes 2017 T C
tempranillo
86

Entremontes Airén 2021 B
airén
85

Entremontes Selección 2021 B
verdejo, airén
85

Entremontes Tempranillo 2021 T
tempranillo
84

BODEGAS FERNANDO CASTRO
Paseo Castelar, 70
13730 Santa Cruz de Mudela (Ciudad Real)
☎: +34 926 342 168
info@bodegasfernandocastro.com
www.bodegasfernandocastro.com

Castillo Santa Bárbara 2017 T C
tempranillo
88
Toasty, jammy, standard, classic.

Castillo Santa Bárbara 2018 T C
tempranillo
89
Aromatic, spicy, jammy, fruity, tasty, balanced.

Ranguelos 2021 T
tempranillo
86

BODEGAS GARDEL
Toledo, 2
16650 Las Mesas (Cuenca)
☎: +34 627 730 902
omar.gardel@hotmail.com
www.bodegasgardel.com

Dominio de Gardel 2017 T C
tempranillo, syrah
86

DO LA MANCHA / D.O.P.

Dominio de Gardel 2020 T RB
tempranillo, syrah
87

BODEGAS ISIDRO MILAGRO
Avda. del Ebro s/n
26540 Alfaro (La Rioja)
gerardo.export@bodegasisidromilagro.com
www.bodegasisidromilagro.com

Conde de Monterroso 2013 T R
tempranillo
87

Conde de Monterroso 2018 T C
tempranillo
87

Conde de Monterroso 2020 T
tempranillo
84

Libertario 2021 T
86

BODEGAS ISLA
Nuestra Señora de la Paz, 9
13210 Villarta San Juan (Ciudad Real)
☎: +34 926 640 004
manuel@bodegasisla.com
www.bodegasisla.com

Isla Oro Airén 2021 B
100% airén
83

Isla Oro Cabernet Sauvignon 2021 T
100% cabernet sauvignon
84

Isla Oro Garnacha 2021 RD
100% garnacha
83

Isla Oro Macabeo 2021 B
100% macabeo
84

Isla Oro Tempranillo 2018 T C
100% tempranillo
83

Isla Oro Tempranillo 2021 T
100% tempranillo
86

BODEGAS LA REMEDIADORA
Alfredo Atienza, 149
02630 La Roda (Albacete)
☎: +34 967 440 600
export@laremediadora.com
www.laremediadora.com

La Villa Real Cabernet Sauvignon 2018 T C S
100% cabernet sauvignon
85

La Villa Real Macabeo 2021 B
100% macabeo
85

La Villa Real Moscatel 2021 B D
moscatel grano menudo
86

La Villa Real Sauvignon Blanc 2021 B
100% sauvignon blanc
85

La Villa Real Tempranillo 2020 T BA S
100% tempranillo
85

La Villa Real Tempranillo Syrah 2021 T
50% tempranillo, 50% syrah
86

BODEGAS LATÚE
Camino de la Esperilla, s/n
45810 Villanueva de Alcardete (Toledo)
☎: +34 925 166 350
web@latue.com
www.latue.com

Latúe 2021 RD
tempranillo
87

Latúe Airén 2021 B
airén
86

Latúe Cabernet Sauvignon & Syrah 2019 T
cabernet sauvignon, syrah
88
Balanced, herbaceous, thin, spicy.

Pingorote Sauvignon Blanc 2021 B
sauvignon blanc
86

Pingorote Tempranillo 2016 T C
tempranillo
87

Pingorote Tempranillo 2016 T R
tempranillo
89
Hot, classic, toasty, jammy, tasty.

BODEGAS LOZANO
Avda. Reyes Católicos, 156
02600 Villarrobledo (Albacete)
☎: +34 967 141 907
info@bodegas-lozano.com
www.bodegas-lozano.com

Añoranza 2021 RD
tempranillo
85

Añoranza Cabernet Shiraz 2021 T
cabernet sauvignon, syrah
84

Añoranza Sauvignon Blanc 2021 B
sauvignon blanc
86

Oristán 2018 T R
tempranillo, cabernet sauvignon
90
Colour: dark-red cherry, garnet rim. Nose: fruit preserve, aged wood nuances, tobacco, sweet spices. Palate: spicy, round tannins, long.

Oristán 2019 T C
tempranillo, cabernet sauvignon, syrah
89
Pleasant, hot, toasty, jammy.

Oristán Verdejo 2021 B
verdejo
86

BODEGAS NARANJO
Felipe II, 5
13150 Carrión de Calatrava (Ciudad Real)
☎: +34 687 045 574
comercial@bodegasnaranjo.com
www.bodegasnaranjo.com

Casa de la Dehesa 2014 T C
100% tempranillo
88
Classic, elegant, spicy, thin, toasty.

Viña Cuerva 2016 T R
tempranillo
84

Viña Cuerva 2018 T C
tempranillo
86

Viña Cuerva 2021 RD
tempranillo
85

Viña Cuerva 2021 T
tempranillo, syrah
87

Viña Cuerva Airén 2021 B
84

BODEGAS PEDROHERAS
La Mancha, 1
16660 Las Pedroñeras (Cuenca)
☎: +34 967 160 151
compras@pedroheras.es
www.bodegaspedroheras.com

Pedroheras 2015 T R
tempranillo
90
Colour: very deep cherry. Nose: roasted coffee, aromatic coffee, powerful, ripe fruit. Palate: smoky aftertaste, great length, round tannins.

Pedroheras Airén 2021 B
airén
85

Pedroheras Syrah Tempranillo 2021 T S
tempranillo, syrah
88
Pleasant, fruity, sweeties, jammy.

Pedroheras Tempranillo 2021 T
tempranillo
87

Pedroheras Verdejo 2021 B
verdejo
85

DO LA MANCHA / D.O.P.

DO LA MANCHA / D.O.P.

BODEGAS PUENTE DE RUS
Ctra. Almarcha, 50
16600 San Clemente (Cuenca)
☎: +34 969 300 155
exportassistant@puentederus.com
www.puentederus.com

Puente de Rus 2020 T RB
tempranillo, syrah, cabernet sauvignon

86

Puente de Rus 3 Variedades 2020 T
tempranillo, syrah, cabernet sauvignon

86

Puente de Rus La Guacha Blanco 2021 B SD
50% airén, 50% moscatel

85

Puente de Rus Sauvignon Blanc 2021 B
100% sauvignon blanc

84

Puente de Rus Tempranillo 2015 T C
100% tempranillo

87

Puente de Rus Verdejo 2021 B
100% verdejo

85

BODEGAS ROMERO DE ÁVILA SALCEDO
Avda. Constitución, 4
13240 La Solana (Ciudad Real)
☎: +34 926 631 426
media@bodegasromerodeavila.com
www.bodegasromerodeavila.com

Abuelo Paco 2012 T R
tempranillo, syrah

90

Colour: dark-red cherry, garnet rim. Nose: fruit preserve, aged wood nuances, tobacco, sweet spices. Palate: spicy, round tannins, long.

Portento 2012 T R
50% tempranillo

87

Portento 2015 T C
85% tempranillo, 15% cabernet sauvignon

85

Portento Sauvignon Blanc 2021 B
sauvignon blanc

86

Portento Tempranillo 2021 T
tempranillo

84

BODEGAS SIMBOLO
Concepción, 135
13610 Campo de Criptana (Ciudad Real)
☎: +34 926 589 036
comunicacion@bodegassimbolo.com
www.bodegassimbolo.com

Símbolo Airén 2021 B
100% airén

86

Símbolo Chardonnay Selección 2021 B
100% chardonnay

85

Símbolo Petit Verdot 2021 T
100% petit verdot

83

Símbolo Tempranillo 2017 T RB
100% tempranillo

86

Símbolo Tempranillo 2021 T
100% tempranillo

86

Símbolo Verdejo 2021 B
100% verdejo

85

BODEGAS VERDÚGUEZ
Los Hinojosos, 1
45810 Villanueva de Alcardete (Toledo)
☎: +34 925 167 493
export3@bodegasverduguez.com
www.bodegasverduguez.com

Hidalgo Castilla 2016 T GR
90

Colour: cherry, garnet rim. Nose: fruit preserve, fruit liqueur notes, powerful, spicy. Palate: flavourful, balanced, smoky aftertaste, slightly dry, soft tannins

Hidalgo Castilla 2017 T R
tempranillo

90

Colour: deep cherry. Nose: dried herbs, creamy oak, red berry notes, black fruit. Palate: powerful, ripe fruit, spicy, round tannins.

Tierra Imperial 2016 T GR
tempranillo

89

Fruity, pruney, spicy, smoky, tasty, coarse tannins.

Tierra Imperial 2017 T R
tempranillo
88
Standard, spicy, jammy, herbal.

Tierra Imperial 2019 T C
tempranillo
88
Full-bodied, spicy, jammy, powerful, toasty.

Tierra Imperial Oaked Selection 2020 T
60% tempranillo, 20% merlot, 20% syrah
87

BODEGAS Y VIÑEDOS LADERO
Ctra. Alcázar, km. 1
13700 Tomelloso (Ciudad Real)
☎: +34 926 505 653
allozo@allozo.com
www.allozo.com

Ladero 2016 T C
100% tempranillo
85

Ladero 2021 RD
100% tempranillo
84

Ladero Airén Verdejo 2021 B
airén, verdejo
85

Ladero Tempranillo 2021 T
100% tempranillo
87

CASTILLO DE ARESAN
Ctra. N-310, Km.120
02600 Villarrobledo (Albacete)
☎: +34 679 981 697
irene.martinez@castillodearesan.es

Castillo de Aresan Terruño 2019 T
tempranillo, syrah, cabernet franc
91
Colour: dark-red cherry. Nose: toasty, spicy, cocoa bean, ripe fruit, black fruit. Palate: flavourful, toasty, fine bitter notes.

Tradición de Aresan Tempranillo - Cabernet Sauvignon - Merlot 2020 T
tempranillo, cabernet sauvignon, merlot
87

COOP. JESÚS DEL PERDÓN - BODEGAS YUNTERO
P.I. de Manzanares- Ctra. Alcazar s/n
13200 Manzanares (Ciudad Real)
☎: +34 926 610 309
Fax: +34 926 610 516
yuntero@yuntero.com
www.yuntero.com

Epílogo Chardonnay 2019 B BA
100% chardonnay
87

Epílogo Sauvignon Blanc 2021 B
100% sauvignon blanc
88
Kind finish, balsamic herbs, fruity, balanced, varietally correct, wild, smooth.

Mundo de Yuntero Verdejo-Sauvignon Blanc 2021 B
75% verdejo, 25% sauvignon blanc
87

Yuntero 2015 T R
100% tempranillo
90
Colour: deep cherry, garnet rim. Nose: ripe fruit, cocoa bean, cigar, toasty. Palate: flavourful, spicy, toasty, powerful tannins.

Yuntero 68 Vendimia 2015 T R
100% tempranillo
89
Spicy, balanced, toasty, smooth.

DO LA MANCHA / D.O.P.

DO LA MANCHA / D.O.P.

Yuntero Macabeo – Sauvignon Blanc 2021 B
85% macabeo, 15% sauvignon blanc
85

DOMINIO DE BACO
Polígono Industrial Alces. Calle Mencia, s/n
13600 Alcázar de San Juan (Ciudad Real)
☎: +34 926 547 204
sara.rodriguez@dcoop.es
www.dcoop.es

Dominio de Baco 2020 T RB
tempranillo
87

Dominio de Baco Airén 2021 B
airén
87

Dominio de Baco Cabernet 2021 T
cabernet sauvignon
86

Dominio de Baco Syrah 2021 T
syrah
86

Dominio de Baco Tempranillo 2021 T
tempranillo
87

Dominio de Baco Verdejo 2021 B
verdejo
86

EL PROGRESO SDAD. COOP. CLM
Avda. de la Virgen, 89
13670 Villarubia de los Ojos (Ciudad Real)
☎: +34 926 896 135
info@bodegaselprogreso.com
www.bodegaselprogreso.com

Ojos del Guadiana 2018 T R
tempranillo
87

Ojos del Guadiana Airén 2021 B
airén
84

Ojos del Guadiana Chardonnay 2021 B
100% chardonnay
87

Ojos del Guadiana Selección 2020 T BA
33% cabernet sauvignon, 33% merlot, 33% syrah
86

Ojos del Guadiana Tempranillo 2021 T
tempranillo
87

Ojos del Guadiana Verdejo 2021 B
verdejo
85

FAMILIA BASTIDA
Paraje de Titos
13630 Socuéllamos (Ciudad Real)
☎: +34 968 780 142
info@familiabastida.com
www.familiabastida.com

Alceo 2020 T RB
tempranillo
89
Kind finish, toasty, powerful, jammy.

Paraje de Titos 2020 T
garnacha
90
Colour: deep cherry. Nose: ripe fruit, dried herbs, creamy oak. Palate: powerful, ripe fruit, spicy, round tannins.

FÉLIX SOLÍS
Otumba, 2
45840 La Puebla de Almoradiel (Toledo)
☎: +34 925 178 626
fsa@felixsolisavantis.com
www.felixsolisavantis.com

Caliza Merlot Syrah Tempranillo 2021 T
merlot, syrah, tempranillo
87

Caliza Sauvignon Blanc Airén Viura 2021 B
sauvignon blanc, airén, viura
86

Caliza Tempranillo 2021 RD
tempranillo
86

Muchas Manos 2021 RD
tempranillo
84

Muchas Manos Airén 2021 B
airén
85

Muchas Manos Tempranillo 2021 T
tempranillo
86

244 Peñín Guide | SPANISH WINE

FINCA ANTIGUA

Ctra. Quintanar - Los Hinojosos, Km. 11,5
16417 Los Hinojosos (Cuenca)
☎: +34 969 129 700
info@fincaantigua.com
www.fincaantigua.com

Clavis 2014 T R
92
Colour: dark-red cherry, garnet rim. Nose: fruit preserve, aged wood nuances, tobacco, sweet spices. Palate: spicy, round tannins, long.

Finca Antigua 2015 T R
60% merlot, 20% cabernet sauvignon, 20% syrah
88
Jammy, fruity, spicy, balanced.

Finca Antigua Petit Verdot 2019 T
petit verdot
88
Floral, fruity, herbal, standard.

Finca Antigua Syrah 2019 T C
syrah
88
Standard, floral, herbaceous, spicy, bitter.

Finca Antigua Único 2016 T C
50% tempranillo, 20% merlot, 20% cabernet sauvignon, 10% syrah
88
Jammy, toasty, fruity, full-bodied, floral.

Finca Antigua Viura 2021 B
viura
87

HAMMEKEN CELLARS

Llavador, 20
03700 Denia (Alacant/Alicante)
☎: +34 965 791 967
cellars@hammekencellars.com
www.hammekencellars.com

Allegranza Tempranillo-Shiraz 2021 T
tempranillo, syrah
88
Jammy, powerful, full-bodied, roasted coffee.

El Paso Tempranillo Shiraz 2021 T
tempranillo, syrah
89
Jammy, fruity, tasty, smooth, pleasant.

J. GARCIA CARRION LA MANCHA

Guarnicionero, s/n
13250 Daimiel (Ciudad Real)
☎: +34 900 189 090
atcliente@jgc.es
www.garciacarrion.com

Don Luciano 2018 T C
tempranillo
83

Don Luciano 2021 RD
tempranillo
84

Opera Prima Cabernet Sauvignon 2021 T
cabernet sauvignon
85

Opera Prima Chardonnay 2021 B
chardonnay
84

Opera Prima Syrah 2021 T
syrah
86

Opera Prima Verdejo 2021 B
verdejo
84

PUNCTUM BIODYNAMIC FAMILY VINEYARDS

Finca El Fabián, s/n Apdo. Correos 71
16660 Las Pedroñeras (Cuenca)
☎: +34 912 918 326
exportsales@dominiodepunctum.com
www.dominiodepunctum.com

Viento Aliseo 2021 RD
bobal
88 ♣
Pleasant, fruity, floral, sweeties, balanced.

Viento Aliseo Graciano Cabernet Sauvignon 2020 T BA S
50% graciano, 50% cabernet sauvignon
87 ♣

Viento Aliseo Tempranillo Petit Verdot 2021 T S
70% tempranillo, 30% petit verdot
89 ♣
Jammy, spicy, tasty, full-bodied.

DO LA MANCHA / D.O.P.

Viento Aliseo Viognier 2021 B
viognier
88 🌿
Aromatic, dried flowers, jammy, smooth, kind finish, balanced.

SDAD. COOP. CRISTO DE LA VEGA
13630 Socuéllamos (Ciudad Real)
☎: +34 926 530 388
Fax: +34 926 530 024
dptotecnico@bodegascrisve.com
www.bodegascrisve.com

Marqués de Castilla Tempranillo Syrah Merlot 2021 T
tempranillo, syrah, merlot
86

Yugo 2018 T C
tempranillo
85

Yugo 2021 T
tempranillo, syrah, merlot
87

Yugo Airén 2021 B
airén
86

Yugo Garnacha Tempranillo 2021 RD
garnacha, tempranillo
86

Yugo Verdejo 2021 B
verdejo
85

VINÍCOLA DE CASTILLA
Pol. Ind. Calle I, s/n
13200 Manzanares (Ciudad Real)
☎: +34 926 647 800
nacional@vinicoladecastilla.com
www.vinicoladecastilla.com

Guadianeja Paraje Alto Hungrao 2020 T RB
tempranillo
89
Nose: ripe fruit, dried herbs, creamy oak. Palate: ripe fruit, spicy, round tannins.

Guadianeja Paraje Alto Hungrao 2021 B
airén
90
Colour: straw. Nose: dried herbs, faded flowers, sweet spices, ripe fruit, stone fruit. Palate: powerful, ripe fruit, balanced.

Señorío de Guadianeja Macabeo 2021 B
macabeo
86

Señorío de Guadianeja Syrah 2021 T
syrah
88
Balanced, spicy, floral, fruity, smooth.

Señorío de Guadianeja Tempranillo 2021 T S
tempranillo
87

VINÍCOLA DE TOMELLOSO

Ctra. Toledo - Albacete, Km. 130,8
13700 Tomelloso (Ciudad Real)
☎: +34 926 513 004
Fax: +34 926 538 001
vinicola@vinicoladetomelloso.com
www.vinicolatomelloso.com

Añil Fresh 2021 B
50% macabeo, 50% chardonnay

87
Aromatic, dried herbs, wild, smooth. Nose: citrus fruit.

Mantolán BE BN
macabeo, chardonnay

87

Torre de Gazate 2017 T C
50% tempranillo, 50% cabernet sauvignon

85
Standard, herbaceous, jammy, floral.

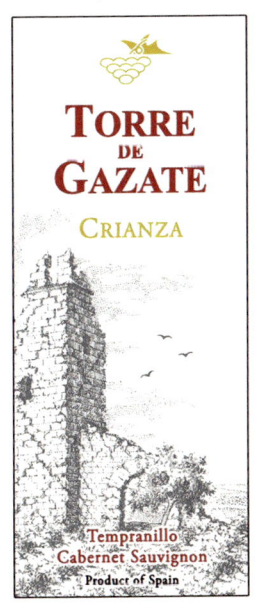

Torre de Gazate Airén 2021 B
100% airén

86

Torre de Gazate Cabernet Sauvignon 2021 RD
100% cabernet sauvignon

86

Torre de Gazate Verdejo 2021 B
100% verdejo

86

VINOS COLOMAN

Goya, 17
13620 Pedro Muñoz (Ciudad Real)
☎: +34 926 586 410
coloman@satcoloman.com
www.satcoloman.com

Besana Real Macabeo 2021 B
macabeo

84

Besana Real Macabeo 2021 B FB
macabeo

85

Besana Real Tempranillo 2018 T C
tempranillo

85

Besana Real Tempranillo 2021 T
tempranillo

87

Besana Real Verdejo 2021 B
verdejo

85

VIÑEDOS Y BODEGAS MUÑOZ

Ctra. Villarrubia, 11
45350 Noblejas (Toledo)
☎: +34 925 140 070
info@bodegasmunoz.com
www.bodegasmunoz.com

Artero 2016 T R
tempranillo, merlot, syrah

88
Full-bodied, creamy, spicy, jammy, roasted coffee.

Artero 2017 T C
tempranillo, merlot, syrah

88
Spicy, full-bodied, jammy, fruity.

DO LA MANCHA / D.O.P.

Artero 2021 RD
tempranillo
86

Artero Macabeo Verdejo 2021 B
macabeo, verdejo
85

Artero Tempranillo 2021 T
tempranillo
86

VIRGEN DE LAS VIÑAS BODEGA Y ALMAZARA
Ctra. Argamasilla de Alba, 1
13700 Tomelloso (Ciudad Real)
☎: +34 926 510 865
comercio.exterior@vinostomillar.com
www.vinostomillar.com

Caballero Hidalgo 2017 T R
tempranillo
87
Balanced, spicy, fruity, jammy, tasty.

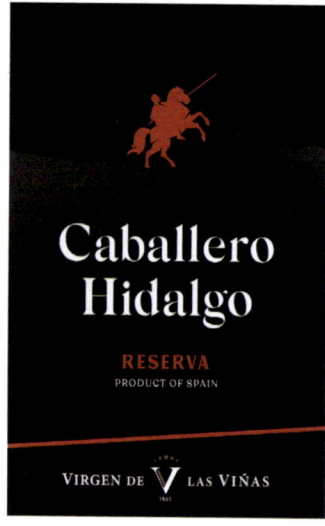

Caballero Hidalgo 2018 T C
tempranillo
86

Tomillar 2017 T R
tempranillo, cabernet sauvignon
85

Tomillar 2018 T C
tempranillo
85

Lienzo Blend 2017 T R
cabernet sauvignon, tempranillo, merlot
89
Herbal, spicy, balanced, hot, jammy, tasty.

Tomillar Airén 2021 B
airén
84

Tomillar Chardonnay 2021 B
chardonnay
88
Aromatic, floral, fruity, balanced, jammy.

Tomillar Merlot 2021 RD
merlot
86

Tomillar Tempranillo 2021 T
tempranillo
85

DO. LA PALMA
CONSEJO REGULADOR

Esteban Acosta Gómez, 7
38740 Fuencaliente (La Palma)
☎: +34 922 444 404
Fax: +34 922 444 432
@: vinoslapalma@vinoslapalma.com
www.vinoslapalma.com

LOCATION:
The production area covers the whole island of San Miguel de La Palma, and is divided into three distinct sub-regions: Hoyo de Mazo, Fuencaliente and Northern La Palma.

SUB-REGIONS:
Hoyo de Mazo: It comprises the municipal districts of Villa de Mazo, Breña Baja, Breña Alta and Santa Cruz de La Palma, at altitudes of between 200 m and 700 m. The vines grow over the terrain on hillsides covered with volcanic stone ('Empedrados') or with volcanic gravel ('Picón Granado'). White and mainly red varieties are grown. **Fuencaliente:** It comprises the municipal districts of Fuencaliente, El Paso, Los Llanos de Aridane and Tazacorte. The vines grow over terrains of volcanic ash at altitudes of between 200 m and 1900 m. The white varieties and the sweet Malvasia stand out. **Northern La Palma:** Situated at an altitude of between 100 m and 200 m, It comprises the municipal areas of Puntallana, San Andrés and Sauces, Barlovento, Garafía, Puntagorda and Tijarafe. The region is richer in vegetation and the vines grow on trellises and using the goblet system. The traditional 'Tea' wines are produced here.

GRAPE VARIETIES:
WHITE: Malvasía, Güal and Verdello (main); Albillo, Bastardo Blanco, Bermejuela, Bujariego, Burra Blanca, Forastera Blanca, Listán Blanco, Moscatel, Pedro Ximénez, Sabro and Torrontés.

RED: Negramol (main), Listán Negro (Almuñeco), Bastardo Negro, Malvasía Rosada, Moscatel Negro, Tintilla, Castellana, Listán Prieto and Vijariego Negro.

FIGURES:
Vineyard surface: 456 – **Wine-Growers:** 823 – **Wineries** 21 – **2021 Harvest rating:** N/A – **Production 21:** 383,579 L.– **Market percentages:** 98% National - 2% International.

SOIL:
The vineyards are situated at altitudes of between 200 m and 1,400 m above sea level in a coastal belt ranging in width which surrounds the whole island. Due to the ragged topography, the vineyards occupy the steep hillsides in the form of small terraces. The soil is mainly of volcanic origin.

CLIMATE:
This is the most north-westerly island of the archipelago of the Canary Islands. Its complex orography with altitudes which reach 2,400 metres above sea level make it a micro-continent with a wide variety of climates. The influence of the anti-cyclone of the Azores and the trade winds condition the thermal variables and the rainfall registered throughout the year. The greatest amount of rainfall is registered in the more easterly and northern parts of the island due to the entry of the trade winds. Throughout the north-east, from Mazo to Barlovento, the climate is milder and fresher, while the western part of the island has drier and warmer weather. The average rainfall increases from the coast as the terrain ascends. The greatest amount of rainfall is in the north and east of the island.

CLASSIFICATION OF YOUNG WINE HARVESTS PEÑÍNGUIDE

2017	2018	2019	2020	2021
GOOD	N/A	N/A	N/A	N/A

DO LA PALMA / D.O.P.

AZUL PERDIDO
La Montaña 2 F
38760 Breña Baja (Santa Cruz de Tenerife)
☎: +34 608 075 809
bodega@azulperdido.com
www.azulperdido.com

Azul Perdido Malvasía 2019 B
malvasía
91
Colour: straw. Nose: ripe fruit, dried herbs, white flowers, creamy oak. Palate: ripe fruit, balanced, mineral, fleshy.

Lapilli 2019 B
malvasía
91
Colour: bright yellow. Nose: creamy oak, ripe fruit, tropical fruit, sweet spices. Palate: rich, toasty, full-bodied.

Maresia del Atlántico 2019 B
listán blanco, malvasía, albillo criollo
90
With personality. Colour: straw. Nose: ripe fruit, dried herbs, faded flowers, red clay notes. Palate: ripe fruit, balanced, full-bodied, rich, mineral.

Maresia del Atlántico 2020 B
listán blanco, negramoll
88
Balanced, spicy, full-bodied, smooth, dried flowers, mineral.

BODEGAS TENEGUÍA
Antonio Frco. Hdez. Santos, 10 Los Canarios
38740 Fuencaliente de la Palma
(Santa Cruz de Tenerife)
☎: +34 922 444 078
Fax: +34 922 444 394
administracion@vinosteneguia.com
www.bodegasteneguia.com

Teneguía 2021 T
negramoll, almuñeco, castellana, vijariego negro, bastardo
86

Teneguía La Gota 2021 B
negramoll, listán blanco, albillo criollo, vijariego blanco
86

Teneguía Malvasía Aromática Naturalmente Dulce 2020 B D
malvasía
91
Colour: bright yellow. Nose: candied fruit, stone fruit, tropical fruit, sweet spices. Palate: flavourful, unctuous, fruity, sweet.

Teneguía Malvasía Dulce 2000 B R
malvasía
94
Colour: amber. Nose: candied fruit, sweet spices, petrol notes, dry nuts. Palate: fruity, flavourful, sweet, powerful.

DAVID LANA GARCÍA VERDUGO
Los Volcanes, 23
38740 Fuencaliente de la Palma
(Santa Cruz de Tenerife)
☎: +34 922 444 427
info@lacasadelvolcan.es
www.lacasadelvolcan.es

La Casa del Volcán 2019 B
vijariego negro, vijariego blanco, bujariego
88
Balanced, dried flowers, smoky, full-bodied, jammy, smooth.

LLANOS NEGROS
Antonio Fco. Hdez. Santos, 10 Los Canarios
38740 Fuencaliente de la Palma
(Santa Cruz de Tenerife)
☎: +34 922 444 078
Fax: +34 922 444 394
info@llanosnegros.com
www.bodegasteneguia.com

Llanos Negros La Batista 2020 B FB
malvasía
92
Colour: straw, golden. Nose: ripe fruit, dried herbs, faded flowers, spicy, smoky. Palate: powerful, ripe fruit, balanced, mineral.

Llanos Negros La Time 2000 B
listán blanco
91
Colour: golden. Nose: ripe fruit, fine lees, waxy notes, smoky, red clay notes. Palate: balanced, flavourful, mineral.

Llanos Negros Los Grillos 2020 T
negramoll
89
Balanced, spicy, herbal, crisp, rustic, wild.

DO. LANZAROTE
CONSEJO REGULADOR

Arrecife, 9
35550 San Bartolomé (Lanzarote)
☎: +34 928 521 313 - Fax: +34 928 521 049
@: info@dolanzarote.com
www.dolanzarote.com

LOCATION:

On the island of Lanzarote. The production area covers the municipal areas of Tinajo, Yaiza, San Bartolomé, Haría and Teguise.

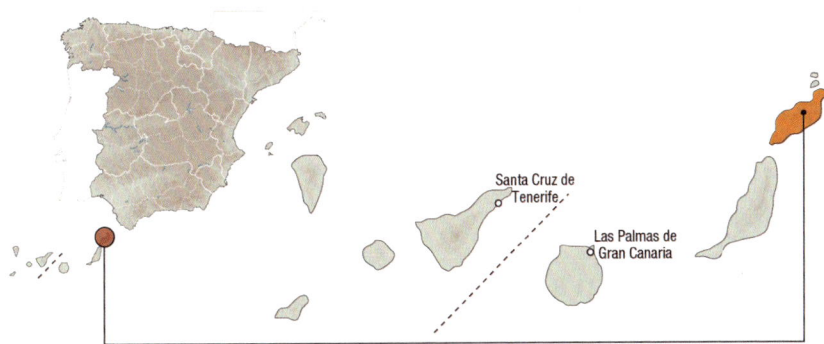

GRAPE VARIETIES:

WHITE: Malvasía (majority 75%), Pedro Ximénez, Diego, Listán Blanco, Moscatel, Burrablanca, Breval.

RED: Listán Negra (15%) and Negramoll.

FIGURES:

Vineyard surface: 1,865 – **Wine-Growers:** 1,822 – **Wineries:** 21 – **2021 Harvest rating:** Very Good – **Production 21:** 1,480,000 litres – **Market percentages:** 97% National - 3% International.

SOIL:

Volcanic in nature (locally known as 'Picón'). In fact, the cultivation of vines is made possible thanks to the ability of the volcanic sand to perfectly retain the water from dew and the scant rainfall. The island is relatively flat (the maximum altitude is 670 m) and the most characteristic form of cultivation is in 'hollows' surrounded by semicircular walls which protect the plants from the wind. This singular trainig system brings about an extremaly low density.

CLIMATE:

Dry subtropical in nature, with low rainfall (about 200 mm per year) which is spread out irregularly throughout the year. On occasions, the Levante wind (easterly), characterised by its low humidity and which carries sand particles from the African continent, causes a considerable increase in the temperatures.

CLASSIFICATION OF YOUNG WINE HARVESTS PEÑÍNGUIDE

2017	2018	2019	2020	2021
VERY GOOD	GOOD	N/A	N/A	N/A

DO LANZAROTE / D.O.P.

BERMEJO
Camino a Los Bermejos, 7
35550 San Bartolomé (Las Palmas)
☎: +34 928 522 463
bodegas@losbermejos.com
www.losbermejos.com

Bermejo Diego Seco 2021 B
89
Citrus fruit, balanced, spicy, herbal, full-bodied, jammy.

Bermejo Listán Negro Maceración Tradicional 2018 T BA
87

Bermejo Malvasia BE BN
91
Colour: bright yellow. Nose: ripe fruit, fine lees, balanced, dried herbs, bakery. Palate: good acidity, flavourful, ripe fruit, long.

Bermejo Malvasia Naturalmente Dulce B D
92
Colour: bright yellow. Nose: candied fruit, honeyed notes, pattiserie, sweet spices. Palate: flavourful, unctuous, fruity, sweet.

Bermejo Malvasía Seco 2021 B FB
89
Balanced, spicy, dried flowers, herbal, full-bodied, jammy.

BODEGA LA FLORIDA
Calle de la Florida, 89
35550 San Bartolomé (Las Palmas)
☎: +34 928 593 001
info@bodegaslaflorida.com

La Florida 2021 RD
listán negro
85

La Florida Malvasía Volcánica 2021 B
malvasía
85

BODEGAS RUBICÓN
Ctra. Teguise - Yaiza, 2
35570 La Geria (Las Palmas)
☎: +34 928 173 708
administracion@bodegasrubicon.com
www.bodegasrubicon.com

Amalia Malvasía Seco 2021 B
88
Austere, jammy, mineral, smooth.

Rubicón 2018 T
listán negro
87

Rubicón Malvasía Seco 2021 B
88
Citrus fruit, herbal, full-bodied, jammy, tasty, mineral.

Rubicón Moscatel Dulce 2020 B D
87

Rubicón Rosado 2020 RD
88
Balanced, dried flowers, fruity, full-bodied, jammy.

Rubicón Semidulce 2021 B SD
88
Citrus fruit, spicy, fruity, slight oxidation, tasty.

EL GRIFO
Ctra. Marzache LZ30, Km. 11
35550 San Bartolomé (Las Palmas)
☎: +34 689 416 663
enologo2@elgrifo.com
www.elgrifo.com

El Grifo Ariana 2019 T
syrah, listán negro
88
Balanced, spicy, smoky, herbaceous.

El Grifo Malvasía Lías 2019 B
malvasía
90
Colour: bright straw. Nose: ripe fruit, fine lees, dried herbs. Palate: full-bodied, rich, long, mineral.

El Grifo Malvasía Seco Colección 2021 B
malvasía
87

El Grifo Moscatel de Ana B D
moscatel de alejandría
92
Colour: bright yellow. Nose: candied fruit, honeyed notes, caramel, sweet spices, dry nuts. Palate: flavourful, unctuous, fruity, sweet.

PURO ROFE

Ctra. Conil, 20
35572 Tías (Las Palmas)
☎: +34 649 628 075
info@purorofe.es
www.purorofe.es

Chibusque 2020 B
93
Slight reduction. Colour: bright straw. Nose: fine lees, faded flowers, characterful, expressive. Palate: full-bodied, long, good acidity.

Chupadero 2020 B
91
Pleasant, balanced, smooth. Nose: medium intensity, neat, spicy, white fruit. Palate: balanced, correct.

Masdache 2020 B
93
Colour: bright straw. Nose: fine lees, wild herbs, expressive. Palate: full-bodied, rich, long, good acidity, spicy.

Mentidero 2020 B
94
Colour: bright straw. Nose: fresh, varietal, fresh fruit, complex, expressive. Palate: long, easy to drink, good acidity, fine bitter notes, saline.

Rofe 2020 B
malvasía
92
Colour: bright straw. Nose: neat, fresh, floral, spicy, dried flowers, lees reduction notes. Palate: flavourful, balanced, easy to drink.

Rofe 2020 T
92
Crisp. Nose: red berry notes, expressive, fresh, medium intensity. Palate: flavourful, easy to drink, round tannins, fresh.

Tilama 2020 B
93
Subtle, elegant, briny. Colour: bright straw. Nose: medium intensity, fresh fruit. Palate: elegant, fine bitter notes, easy to drink.

DO LANZAROTE / D.O.P.

DO. LEÓN

CONSEJO REGULADOR

Edif, Mirador de la Condesa-Complejo de la Isla, s/n.
24200 Valencia de Don Juan (León)
☎: +34 987 751 089 - Fax: +34 987 750 012
@: directortecnico@doleon.es
www.dotierradeleon.es

LOCATION:

In the southeast of Catalonia, in the province of Tarragona. It covers the municipal districts of Arnes, Batea, Bot, Caseres, Corbera d´Ebre, La Fatarella, Gandesa, Horta de Sant Joan, Pinell de Brai, La Pobla de Massaluca, Prat de Comte and Vilalba dels Arcs.

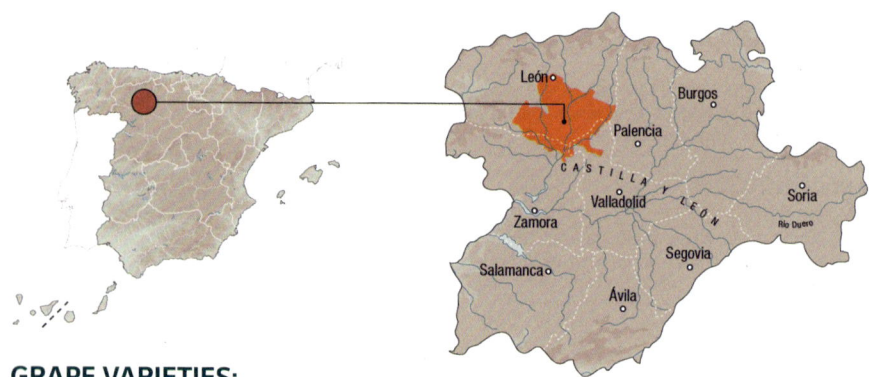

GRAPE VARIETIES:

WHITE: Albarín, Verdejo, Godello, Palomino and Malvasía.

RED: Prieto Picudo, Mencía, Garnacha and Tempranillo.

FIGURES:

Vineyard surface: 1,230 – **Wine-Growers:** 228 – **Wineries:** 42 – 21 Harvest rating: Very Good – **Production 21:** 1,816,179 litres – **Market percentages:** 85% National - 15% International.

SOIL:

High quality for vine cultivation, with good drainage. Most of them are located on alluvial terraces, the brown as well as the limestone soils.

CLIMATE:

The temperatures in the river valleys might suggest a typically Atlantic continental climate, but the elevated position of the León plateau where the vineyards are located favours a very harsh and cold climate. There is a strong contrast between day and night temperatures, with harsh winters, spring frosts and mild summers. Rainfall is concentrated in autumn, reaching 500 mm.

CLASSIFICATION OF YOUNG WINE HARVESTS PEÑÍNGUIDE

2017	2018	2019	2020	2021
VERY GOOD	VERY GOOD	VERY GOOD	VERY GOOD	VERY GOOD

ALBANTO WINES
Jimena Menéndez Pidal, 3
28023 Madrid (Madrid)
☎: +34 670 430 232
antonio@albanto.es

Albanto La Viña de las Flores 2021 RD
prieto picudo,

91

Colour: rose. Nose: ripe fruit, hot, faded flowers, dried herbs. Palate: flavourful, powerful, fruity.

Albanto La Viña de las Flores Lías Finas 2021 B
100% albarín,

91

Colour: bright straw. Nose: ripe fruit, fine lees, faded flowers, dried flowers, balanced. Palate: rich, long, flavourful.

BODEGAS GORDONZELLO
Alto de Santa Marina, s/n
24294 Gordoncillo (León)
☎: +34 987 758 030
info@gordonzello.com
www.gordonzello.com

Gurdos 2021 RD
100% prieto picudo,

89

Exuberant, jammy, powerful, tasty, fruity, floral.

Kyra 2020 B FB
100% albarín,

90

Pleasant. Colour: bright straw. Nose: stone fruit, spicy, toasted bread, citrus fruit. Palate: flavourful, ripe fruit, toasty.

La Costana 2018 T C
100% prieto picudo,

89

Jammy, spicy, dried herbs, tasty, powerful.

Peregrino 14 2016 T C
100% prieto picudo,

90

Colour: deep cherry. Nose: ripe fruit, dried herbs, tobacco, fine reductive notes. Palate: powerful, ripe fruit, spicy, round tannins.

Peregrino Albarín 2021 B
100% albarín,

87

Peregrino Rosé 2021 RD
100% prieto picudo,

88

Fruity, sweeties, floral, balanced, pleasant, smooth.

Peregrino Mil 100 2011 T C
100% prieto picudo,

89

Age nuances, herbal. Nose: old leather, tobacco. Palate: flavourful, slightly dry, soft tannins.

BODEGAS PINCERNA
Ctra. Villada, s/n
24340 Grajal de Campos (León)
☎: +34 679 997 369
alfonso@pincernawines.com
www.bodegaspincerna.com

Pincerna Albarín 2021 B
100% albarín,

88

Aromatic, pleasant, tasty, tropical.

Pincerna Prieto Picudo 2021 RD
prieto picudo,

89

Tasty, jammy, dried flowers.

Pincerna Prieto Picudo 2021 T
100% prieto picudo,

89

Jammy, rustic, tasty, fruity.

Pincerna Sumiller 2020 T
100% prieto picudo,

91

Colour: deep cherry. Nose: ripe fruit, dried herbs, creamy oak, sweet spices. Palate: powerful, ripe fruit, spicy, round tannins.

BODEGAS TAMPESTA
La Socollada, s/n
24230 Valdevimbre (León)
☎: +34 987 304 307
bodegas@tampesta.com
www.tampesta.com

Tampesta Maneki Ed. Especial 2019 B FB
albarín,

91

Dried herbs, rustic. Colour: bright yellow. Nose: balanced, expressive, spicy, wild herbs. Palate: good acidity, fine bitter notes, easy to drink.

Tampesta 2021 RD
100% prieto picudo,

87

Tampesta Albarín 2021 B
100% albarín,

89

Pleasant, floral, smooth, wild, bitter.

DO LEÓN / D.O.P.

DO LEÓN / D.O.P.

Tampesta Golán 2020 RD
100% prieto picudo,
88
Dried flowers, fruity, jammy, tasty, toasty.

Tampesta Golán Edición Limitada 2015 T GR
100% prieto picudo,
91
Colour: dark-red cherry, garnet rim. Nose: fruit preserve, aged wood nuances, tobacco, sweet spices. Palate: spicy, round tannins, long.

Tampesta Neko Godello 2020 B
100% godello,
89
Kind finish, citrus fruit, floral, wild, balanced, defined aromas.

BODEGAS VITALIS
24234 Villamañán (León)
☎: +34 987 131 019
vitalis@bodegasvitalis.com
www.bodegasvitalis.com

Lágrima de Vitalis 2021 RD
100% prieto picudo,
87

Lágrima de Vitalis Albarín 2021 B
100% albarín,
88
Citrus fruit, crisp, herbal, tasty.

Vitalis 2016 T C
100% prieto picudo,
86

Vitalis Selección 2014 T C
100% prieto picudo,
85

LAOSA
Las Cuevas, 8
24232 Ardón (León)
☎: +34 666 217 032
noelia@laosavinos.com
www.laosavinos.com

Grizzly 2019 T
100% prieto picudo,
91
Colour: bright cherry. Nose: balsamic herbs, sweet spices, scrubland, red berry notes. Palate: spicy, balsamic, good acidity, grainy tannins.

Trasto 2020 B BA
100% albarín,
92
Little interventionist. Colour: straw. Nose: expressive, white flowers, jasmine, dried herbs. Palate: flavourful, fruity, balanced.

Trasto 2020 T
100% prieto picudo,
90
Colour: cherry, purple rim. Nose: red berry notes, floral, spicy. Palate: flavourful, fruity, good acidity.

Trasto Finca el Barranco 2020 T
prieto picudo,
92
Colour: cherry, purple rim. Nose: fruit expression, red berry notes, floral, spicy, mineral. Palate: flavourful, fruity, good acidity, long.

PARDEVALLES
Ctra. León s/n
24230 Valdevimbre (León)
☎: +34 987 304 222
info@pardevalles.es
www.pardevalles.es

Pardevalles 2021 RD
100% prieto picudo,
89
Kind finish, aromatic, floral, fruity.

Pardevalles Albarín 2021 B
100% albarín,
88
Tropical, smooth, jammy, floral.

Pardevalles Carroleón 2018 T
100% prieto picudo,
90
Colour: bright cherry. Nose: spicy, cocoa bean, new oak, roasted coffee, ripe fruit. Palate: flavourful, toasty, fine bitter notes.

Pardevalles Carroleón 2020 B FB
albarín,
90
Colour: bright yellow. Nose: dried flowers, candied fruit, fine lees, pattiserie. Palate: round, spicy, long, great length.

Pardevalles Gamonal 2018 T C
100% prieto picudo,
91
Colour: Cherry. Nose: balsamic herbs, sweet spices, scrubland, ripe fruit. Palate: spicy, balsamic, good acidity.

Pardevalles Prieto Picudo 2021 T
100% prieto picudo,

88

Pleasant, balsamic herbs, fruity, tasty.

RAUL PÉREZ BODEGAS Y VIÑEDOS
Bulevar Rey Juan Carlos 1º Rey de España, 11 B
24400 Ponferrada (León)
contacto@raulperez.com
www.raulperez.com

Los Arrotos del pendón 2019 B
albarín,

91

Colour: bright yellow. Nose: dried flowers, candied fruit, fine lees, pattiserie. Palate: round, spicy, long.

SEÑORÍO DE LOS ARCOS
Ctra.. Caboalles, 332
24191 Villlabalter (León)
☎: +34 987 226 594
info@senoriodelosarcos.es
www.senoriodelosarcos.es

El Carriego 2021 RD
prieto picudo,

84

Vega Carriegos 2014 T C
prieto picudo,

84

Vega Carriegos 2016 T MC
prieto picudo,

85

Vega Carriegos 2021 RD
prieto picudo,

84

VILE LA FINCA, BODEGAS Y VIÑEDOS
Crta. Cembranos-Valdevimbre km 6,2
24230 León (León)
☎: +34 987 209 712
lafinca@lafinca.es
www.vilelafinca.es

Don Suero 2018 T C
100% prieto picudo,

86

Don Suero 2020 T RB
100% prieto picudo,

87

Valjunco 2021 RD
100% prieto picudo,

88

Standard, fruity, sweeties, jammy, herbal, lactic, tasty.

Valjunco Albarín 2020 B
100% albarín,

88

Jammy, fruity, tasty, floral, balanced, pleasant.

Vile la Finca 2017 T
100% prieto picudo,

91

Colour: Cherry. Nose: ripe fruit, wild herbs, spicy. Palate: fruity, flavourful, fresh.

DO LEÓN / D.O.P.

DO. MÁLAGA Y SIERRAS DE MÁLAGA

CONSEJO REGULADOR

Plaza de los Viñeros, 1
29008 Málaga
☎: +34 952 227 990 - Fax: +34 952 227 990
@: info@vinomalaga.com
www.vinomalaga.com

LOCATION:

In the province of Málaga. It covers 54 municipal areas along the coast (in the vicinity of Málaga and Estepona) and inland (along the banks of the river Genil), together with the new sub-region of Serranía de Ronda, a region to which the two new municipal districts of Cuevas del Becerro and Cortes de la Frontera have been added.

GRAPE VARIETIES:

WHITE: DO Málaga: Pedro Ximénez, Moscatel, Doradilla and Lairen. **DO Sierras de Málaga:** Chardonnay, Moscatel, Pedro Ximénez, Macabeo, Sauvignon Blanc, Colombard, Garnacha Blanca, Macabeo, Malvasía Aromática, Montúa, Pardina (Jaén Blanco), Perruno, Vermentino, Vijariego Blanco (Bigiriego), Gewürztraminer, Riesling, Verdejo and Viognier.

RED: only **DO Sierras de Málaga:** Romé, Cabernet Sauvignon, Merlot, Syrah, Tempranillo, Petit Verdot, Lemberger (Blaufränkisch), Jaén Tinto, Moscatel Negro, Tinta Velasco (Blasco), Garnacha, Pinot Noir, Tintilla, Monastrell, Graciano and Malbec.

TYPOLOGY OF CLASSIC WINES:

a) LIQUEUR WINES: from 15 to 22% vol.

b) NATURAL SWEET WINES: from 15 to 22 % vol. obtained from the Moscatel or Pedro Ximénez varieties, from musts with a minimum sugar content of 244 grams/litre.

c) NATURALLY SWEET WINES (with the same varieties, over 13% vol. and from musts with 300 grams of sugar/litre) and still wines (from 10 to 15% vol.).

Depending on their ageing:

- MÁLAGA JOVEN: Unaged still wines. - MÁLAGA PÁLIDO: Unaged non-still wines.
- MÁLAGA: Wines aged for between 6 and 24 months. - MÁLAGA NOBLE: Wines aged for between 2 and 3 years.
- MÁLAGA AÑEJO: Wines aged for between 3 and 5 years. - MÁLAGA TRASAÑEJO: Wines aged for over 5 years.

FIGURES:

Vineyard surface: 870 – **Wine-Growers:** 437 – **Wineries:** 45 – **2021 Harvest rating:** Good – **Production 21:** 1,729,121 litres – **Market percentages:** 78% National - 22% International.

SOIL:

It varies from red Mediterranean soil with limestone components in the northern region to decomposing slate on steep slopes of the Axarquía.

CLIMATE:

Varies depending on the production area. In the northern region, the summers are short with high temperatures, and the average rainfall is in the range of 500 mm; in the region of Axarquía, protected from the northerly winds by the mountain ranges and facing south, the climate is somewhat milder due to the influence of the Mediterranean; whilst in the west, the climate can be defined as dry subhumid.

CLASSIFICATION OF YOUNG WINE HARVESTS PEÑÍNGUIDE

2017	2018	2019	2020	2021
GOOD	VERY GOOD	VERY GOOD	N/A	EXCELLENT

BODEGA F. SCHATZ

Finca Sanguijuela s/n
29400 Ronda (Málaga)
☎: +34 952 871 313
bodega@f-schatz.com
www.f-schatz.com

Acinipo 2017 T C
lemberger

91 🌱

Balanced, elegant. Colour: cherry, garnet rim. Nose: red berry notes, ripe fruit, expressive, dried flowers. Palate: good acidity, easy to drink, fruity.

Finca Sanguijuela 2014 T C
34% tempranillo, 33% merlot, 33% cabernet sauvignon

91 🌱

Colour: cherry, garnet rim. Nose: black fruit, ripe fruit, hint of anise, wild herbs. Palate: balanced, long, easy to drink, good acidity.

Schatz Chardonnay 2021 B
chardonnay

90 🌱

With personality. Colour: yellow, pale. Nose: white fruit, ripe fruit, balanced, neat. Palate: flavourful, fruity.

Schatz Petit Verdot 2016 T C
petit verdot

90 🌱

With personality. Colour: cherry, garnet rim. Nose: animal reductive notes, black fruit, spicy. Palate: flavourful, good acidity, balanced.

Schatz Pinot Noir 2016 T C
pinot noir

89 🌱

Aromatic, balsamic herbs, slight reduction, fruity, wild, smooth.

Schatz Rosado 2021 RD
moscatel negro

89 🌱

Wild, jammy, fruity, bitter, tasty.

BODEGA FABIO COULLET

Plaza de la Axarquia, 19
29718 Almáchar (Málaga)
☎: +34 650 202 675
comercial@bodegafabiocoullet.com
www.bodegafabiocoullet.com

Ingénito 2021 T RB
100% garnacha

90

Rustic. Colour: cherry, purple rim. Nose: fruit expression, red berry notes, floral, spicy. Palate: flavourful, fruity, good acidity.

Orange 2021 B
moscatel de alejandría

90

Colour: yellow, straw. Nose: expressive, white flowers, jasmine, dried herbs, citrus fruit, wild herbs, stone fruit. Palate: flavourful, fruity, balanced, good acidity.

Romé 2021 T RB
100% romé

91

Rustic, original. Colour: Cherry, pale. Nose: red berry notes, fruit liqueur notes, black liquorice, scrubland. Palate: flavourful, balanced, grainy tannins.

Secuencial 2021 B
moscatel de alejandría, pedro ximénez, doradilla

90

Colour: bright straw, greenish rim. Nose: fresh fruit, citrus fruit, wild herbs, white flowers. Palate: fresh, fruity, good acidity, fine bitter notes.

Villazo 2021 B
100% moscatel de alejandría

89

Pronounced acidity, herbal, spicy, floral, toasty.

BODEGA GROSS HERMANOS

Duquesa de Parcent, 2 3° Izq.
29001 Málaga (Málaga)
☎: +34 658 079 871
info@bodegasgross.es
www.bodegasgross.es

Bocatinta 2018 T
54,48% merlot, 45,52% cabernet sauvignon

90

Colour: bright cherry. Nose: ripe fruit, black fruit, spicy, green pepper, smoky. Palate: fruity, flavourful, ripe fruit, great length.

Bocatinta 2018 T C
merlot, cabernet sauvignon

89

Fruity, jammy, spicy, tasty.

Ecos 2018 T C
syrah

90 🌱

Colour: cherry, purple rim. Nose: black fruit, fruit preserve, wild herbs, sweet spices. Palate: flavourful, fruity, slightly dry, soft tannins.

Gross 2021 RD
syrah, cabernet sauvignon, garnacha, pinot noir

89 🌱

Kind finish, floral, fruity, tasty.

DO MÁLAGA Y SIERRAS DE MÁLAGA / D.O.P.

DO MÁLAGA Y SIERRAS DE MÁLAGA / D.O.P.

Gross Cabernet Sauvignon 2018 T R
cabernet sauvignon

91 🌱

Colour: very deep cherry. Nose: black fruit, ripe fruit, balsamic herbs, dark chocolate, creamy oak. Palate: flavourful, good structure, grainy tannins.

Gross Petit Verdot 2018 T C
petit verdot, cabernet sauvignon

89 🌱

Fruity, herbal, jammy, spicy, lacks balance.

BODEGA JOAQUÍN FERNÁNDEZ
Finca Los Frutales Paraje de los Frontones s/n
29400 Ronda (Málaga)
☎: +34 951 166 043
info@bodegajf.es
www.bodegajf.es

Finca Los Frutales 2019 RD RB
merlot, syrah, garnacha

87 🌱

Finca Los Frutales Cabernet Sauvignon 2019 T C
cabernet sauvignon

89 🌱

Spicy, fruity, herbal, jammy, varietally correct.

Finca Los Frutales Garnacha 2017 T C
garnacha

89 🌱

Spicy, fruity, dried herbs, tasty, jammy.

Finca Los Frutales Igualado 2017 T
cabernet sauvignon, merlot, garnacha

89 🌱

Fruity, spicy, jammy, wild, tasty.

Finca Los Frutales Merlot Syrah 2019 T C
merlot, syrah

86 🌱

Hacienda de la Vizcondesa 2020 T RB
merlot, cabernet sauvignon, garnacha

89 🌱

Fruity, sweety finish, spicy, balanced.

BODEGA KIENINGER
Paraje Rural los Frontones, 67
29400 Ronda (Málaga)
☎: +34 630 161 156
martin@bodegakieninger.com
www.bodegakieninger.com

7 Vin 2016 T BA
50% blaufraenkisch, 50% zweigelt

90 🌱

Colour: deep cherry. Nose: ripe fruit, dried herbs, creamy oak, wet leather, black fruit. Palate: powerful, ripe fruit, spicy, round tannins.

7 Vin 2017 T BA
50% blaufraenkisch, 50% zweigelt

86 🌱

Maxx 2016 T C
80% garnacha, 20% tintilla de rota

92 🌱

Colour: cherry, garnet rim. Nose: overripe fruit, creamy oak, hot, black fruit. Palate: pruney, powerful, sweet tannins.

Maxx 2017 T C
80% garnacha, 20% tintilla de rota

91 🌱

Colour: dark-red cherry. Nose: toasty, spicy, cocoa bean, dark chocolate. Palate: flavourful, toasty, fine bitter notes.

Vinana 2017 T
40% cabernet sauvignon, 40% cabernet franc, 20% merlot

92 🌱

Colour: deep cherry. Nose: ripe fruit, dried herbs, creamy oak, fruit expression, balsamic herbs, scrubland. Palate: powerful, ripe fruit, spicy, round tannins.

BODEGA LA MELONERA
Paraje Los Frontones, s/n
29400 Ronda (Málaga)
☎: +34 951 194 018
info@lamelonera.com
www.lamelonera.com

Encina del Inglés 2021 T
tintilla de rota, garnacha, cabernet sauvignon

91

With personality. Colour: cherry, purple rim. Nose: black fruit, dried fruit, truffle notes, scrubland. Palate: good structure, flavourful, balanced, ripe fruit.

Payoya Negra 2019 T R
tintilla, romé, syrah, garnacha

92

Colour: deep cherry. Nose: dried herbs, black fruit, fine reductive notes, toasty. Palate: powerful, ripe fruit, spicy, grainy tannins.

DO MÁLAGA Y SIERRAS DE MÁLAGA / D.O.P.

🏆 PODIUM

Yo Solo 2020 T FB
melonera, tintilla

95 🍃

Colour: cherry, garnet rim. Nose: red berry notes, black fruit, dried flowers, scrubland, fine reductive notes. Palate: spicy, round tannins, balanced.

BODEGAS DIMOBE
Ctra. Almachar, s/n
29738 Moclinejo (Málaga)
☎: +34 952 400 594
ignacio@dimobe.es
www.dimobe.es

🏆 PODIUM

Arcos de Moclinejo BF Solera D
pedro ximénez

95

Colour: bright yellow. Nose: bakery, acetaldehyde, sweet spices, dried fruit, expressive, characterful, dry nuts. Palate: sweet, flavourful, powerful, great length, good acidity.

🏆 PODIUM

Arcos de Moclinejo BF Solera S
pedro ximénez

96

Colour: light mahogany. Nose: acetaldehyde, old leather, cereal notes, animal reductive notes, fruit preserve, dried herbs, expressive. Palate: complex, flavourful, balanced, full-bodied.

Dimobe Pajarete BF Trasañejo D
pedro ximénez, moscatel de alejandría

91

Colour: light mahogany. Nose: candied fruit, honeyed notes, floral, expressive, cereal notes, dry nuts. Palate: fresh, flavourful, balanced, good finish.

Dimobe Seco BF Trasañejo S
pedro ximénez

91

Colour: iodine, amber rim. Nose: acetaldehyde, bakery, candied fruit, stone fruit, expressive, sweet spices. Palate: complex, good finish, fresh, bitter.

Piamater
Naturalmente Dulce 2018 B MO D
moscatel de alejandría

92

Colour: golden. Nose: fruit expression, stone fruit, honeyed notes, fragrant herbs, faded flowers. Palate: sweet, fruity, fresh, flavourful, varietal, balanced.

Rujaq Andalusi BF Trasañejo D
moscatel de alejandría

93

Age nuances, original. Colour: bright yellow. Nose: cereal notes, toasty, dried fruit, caramel, honeyed notes, spicy, citrus fruit. Palate: flavourful, sweet, rich, round, good acidity.

BODEGAS LUNARES DE RONDA
Crta. Ronda-El Burgo, Km 1 (A366)
29400 Ronda (Málaga)
☎: +34 649 690 847
vinos@bodegaslunares.com
www.bodegaslunares.com

Altocielo 2016 T R
syrah, graciano, cabernet sauvignon

88

Balsamic herbs, jammy, powerful, tasty, spicy.

Lunares 2018 T
garnacha, syrah

89

Reduced, spicy, sulphur notes, tasty, fruity, crisp.

Lunares 2020 RD MC
merlot, garnacha

87

BODEGAS MÁLAGA VIRGEN
Autovía A-92, Km. 132
29520 Fuente de Piedra (Málaga)
☎: +34 952 319 454
info@bodegasmalagavirgen.com
www.bodegasmalagavirgen.com

Don Juan Pedro Ximénez
Trasañejo 30 años BF PX D
pedro ximénez

94

Colour: dark mahogany. Nose: expressive, aromatic coffee, spicy, acetaldehyde, varnish. Palate: balanced, fine solera notes, long.

Don Salvador Moscatel
Trasañejo 30 años BF MO D
moscatel

94

Colour: iodine, amber rim. Nose: overripe fruit, cocoa bean, aromatic coffee, varnish. Palate: complex, flavourful, varietal, sweet.

Málaga Virgen Dunkel BF PX D
pedro ximénez

88

Balanced, spicy, overripe, toasty, tasty.

DO MÁLAGA Y SIERRAS DE MÁLAGA / D.O.P.

Málaga Virgen Sweet BF PX D
pedro ximénez
86

Marbella Blush Rosé 2020 RD
89
Aromatic, balsamic herbs, dried herbs, jammy, wild.

**Pedro Ximénez
Reserva de Familia BF PX D**
pedro ximénez
91
Colour: dark mahogany. Nose: dried fruit, pattisserie, caramel. Palate: sweet, flavourful, varietal, unctuous.

Pernales Merlot Syrah 2018 T C
merlot, syrah
90 🌱
Colour: deep cherry. Nose: dried herbs, black fruit, ripe fruit, spicy. Palate: powerful, ripe fruit, spicy, round tannins.

Tres Leones Naturalmente Dulce B D
moscatel de alejandría
89
Pleasant, balsamic herbs, floral, varietally correct, smooth, sweet.

BODEGAS QUITAPENAS
Ctra. de Guadalmar, 12
29004 Málaga (Málaga)
☎: +34 952 247 595
bodegas@quitapenas.es
www.quitapenas.es

Málaga Oro Viejo BF Trasañejo D
85% pedro ximénez, 15% moscatel
88
Overripe, oxidativ, toasty, tasty, spicy.

Málaga PX Noble Quitapenas BF PX D
100% pedro ximénez
88
Slight oxidation, overripe, balanced, spicy, full-bodied, standard.

**Pajarete Noble
Quitapenas 2018 BF SD**
80% pedro ximénez, 20% moscatel
86

**Quitapenas
Málaga Dulce 2020 BF PX D**
100% pedro ximénez
88
Spicy, jammy, oxidativ, tasty, toasty, overripe.

Quitapenas Moscatel Dorado 2021 BF D
100% moscatel
89
Citrus fruit, balanced, spicy, floral, full-bodied, hot.

Quitapenas Moscatel Plata 2021 BF D
100% moscatel
84

BODEGAS SEDELLA
29715 Sedella (Málaga)
☎: +34 687 463 082
Fax: +34 967 140 723
info@sedellavinos.com
www.sedellavinos.com

Laderas de Sedella 2018 T
90 🌱
Colour: cherry, garnet rim. Nose: overripe fruit, creamy oak, hot, fragrant herbs, dried herbs. Palate: pruney, powerful, sweet tannins.

Sedella "Las Jacintas" 2019 T
romé
92
Colour: very deep cherry. Nose: balsamic herbs, sweet spices, scrubland, earthy notes. Palate: spicy, balsamic, good acidity.

Sedella 2018 T
92 🌱
Colour: cherry, garnet rim. Nose: fruit preserve, fruit liqueur notes, powerful. Palate: flavourful, sweetness, long.

Vidueños de Sedella 2020 B
moscatel
90 🌱
Colour: bright yellow. Nose: dried flowers, candied fruit, fine lees, pattisserie. Palate: round, spicy, long, great length.

COMPAÑÍA DE VINOS TELMO RODRÍGUEZ
El Monte
01308 Lanciego (Araba/Álava)
☎: +34 945 628 315
contact@telmorodriguez.com
www.telmorodriguez.com

🏆 **PODIUM**

Molino Real 2018 B
moscatel
95
Colour: bright yellow. Nose: honeyed notes, sweet spices, expressive, white fruit, citrus fruit. Palate: rich, fruity, powerful, flavourful, elegant.

Mountain 2019 B
92
Varietally correct, smooth. Colour: straw. Nose: expressive, white flowers, jasmine, dried herbs. Palate: flavourful, fruity, balanced.

MR Dulce 2020 B D
93
Colour: bright yellow. Nose: balsamic herbs, honeyed notes, floral, sweet spices, expressive, citrus fruit. Palate: rich, fruity, powerful, flavourful, elegant.

CORTIJO LOS AGUILARES
Ctra. Ronda a Campillo, km. 35
29400 Ronda (Málaga)
☎: +34 952 874 457
Fax: +34 951 166 000
bodega@cortijolosaguilares.com
www.cortijolosaguilares.com

Cortijo Los Aguilares 2021 RD
71% tempranillo, 18% garnacha, 11% syrah
90
Colour: rose, purple rim. Nose: fruit expression, red berry notes, floral. Palate: fruity, good acidity, easy to drink.

Cortijo Los Aguilares Pago El Espino 2019 T BA
72% petit verdot, 19% syrah, 9% tempranillo
92
Colour: deep cherry. Nose: ripe fruit, dried herbs, creamy oak, earthy notes, truffle notes. Palate: powerful, ripe fruit, spicy, round tannins.

Cortijo Los Aguilares Pinot Noir 2020 T C
pinot noir
94
Colour: Cherry. Nose: balsamic herbs, sweet spices, scrubland, red berry notes. Palate: spicy, balsamic, good acidity.

Cortijo Los Aguilares Pinot Noir 2021 T C
pinot noir
93
Colour: Cherry. Nose: complex, expressive, spicy, mineral, scrubland, fragrant herbs. Palate: elegant, full-bodied, long, great length.

Tadeo Cortijo Los Aguilares 2019 T C
petit verdot
94
Colour: dark-red cherry. Nose: toasty, spicy, cocoa bean, black fruit, powerful, expressive. Palate: flavourful, toasty, fine bitter notes.

🏆 PODIUM

Tadeo Tinaja Cortijo Los Aguilares 2019 T
petit verdot
95
Colour: cherry, garnet rim. Nose: complex, expressive, spicy, mineral, dark chocolate, black fruit. Palate: full-bodied, long, great length, flavourful.

Tadeo Tinaja Cortijo Los Aguilares 2020 T
petit verdot
94
Colour: very deep cherry. Nose: balsamic herbs, sweet spices, scrubland, raspberry, red berry notes. Palate: spicy, balsamic, good acidity.

HUERTO DE LA CONDESA
Calle Genal, 1
29400 Ronda (Málaga)
☎: +34 665 829 423
agumillan95@gmail.com
www.huertodelacondesa.com

Huerto de la Condesa La Hiedra 2020 T RB
75% syrah, 25% garnacha
91
Colour: Cherry. Nose: balsamic herbs, scrubland, red berry notes, ripe fruit. Palate: spicy, good acidity, easy to drink.

Huerto de la Condesa La Palmera 2019 T C
100% syrah
91
Colour: bright cherry. Nose: sweet spices, ripe fruit, dark chocolate, meaty notes. Palate: fruity, spicy, round tannins.

Huerto de la Condesa Los Cipreses 2019 T C
50% syrah, 50% garnacha
91
Colour: cherry, garnet rim. Nose: ripe fruit, red berry notes, black fruit, scrubland, hint of anise, wild herbs. Palate: flavourful, fruity.

Huerto de la Condesa Los Cipreses 2021 RD
100% garnacha
89
Kind finish, aromatic, fruity, jammy, dried herbs.

Huerto de la Condesa Pampaneando 2021 T
100% garnacha
89
Kind finish, aromatic, fruity, dried flowers, wild, smooth.

DO MÁLAGA Y SIERRAS DE MÁLAGA / D.O.P.

JORGE ORDÓÑEZ MÁLAGA

Bartolome Esteban Murillo, 11
29700 Vélez (Málaga)
☎: +34 952 504 706
pedidos@jorgeordonez.es
www.jorgeordonez.es

Botani 2021 B
100% moscatel de alejandría
92
Colour: bright straw. Nose: expressive, white flowers, jasmine, dried herbs, fresh fruit. Palate: flavourful, fruity, balanced.

Botani Garnacha 2021 T
100% garnacha
91
Colour: bright cherry. Nose: balsamic herbs, sweet spices, scrubland, burnt matches, fruit expression. Palate: spicy, balsamic, good acidity.

🏆 PODIUM

Jorge Ordóñez & Co Nº 2 Victoria Dulce (sin fortificar) 2021 B D
100% moscatel de alejandría
96
Colour: bright yellow. Nose: sweet spices, expressive, citrus fruit, white fruit, white flowers. Palate: rich, fruity, powerful, flavourful, elegant.

Botani Nobleza 2021 B
100% moscatel de alejandría
94
Colour: bright yellow. Nose: balsamic herbs, honeyed notes, floral, sweet spices, expressive. Palate: fruity, powerful, flavourful, elegant, good acidity.

🏆 PODIUM

Jorge Ordóñez & Co. Nº3 Viñas Viejas (sin fortificar) 2018 B D
100% moscatel de alejandría
98
Colour: golden. Nose: powerful, honeyed notes, candied fruit, fragrant herbs, acetaldehyde. Palate: flavourful, sweet, fresh, fruity, good acidity.

Jorge Ordóñez & Co Nº 1 Selección Especial Dulce (sin fortificar) 2020 B D
100% moscatel de alejandría
94
Colour: bright yellow. Nose: balsamic herbs, honeyed notes, floral, sweet spices, expressive. Palate: rich, fruity, powerful, flavourful, elegant.

TRES GENERACIONES

Gerald Brenan, 4
29004 Málaga (Málaga)
☎: +34 952 231 519
comercial@laraseleccion.com
www.vinotresgeneraciones.com

El Arquitecto 2018 T BA
cabernet sauvignon, tempranillo, syrah
92
Colour: deep cherry. Nose: ripe fruit, spicy, black liquorice, dried herbs. Palate: fruity, flavourful, fresh, good finish, slightly dry, soft tannins.

El Lero 2018 T BA
tempranillo, cabernet sauvignon, syrah, garnacha
89
Fruity, jammy, toasty.

La Depa 2020 T
garnacha
90
Colour: cherry, purple rim. Nose: ripe fruit, sweet spices, wild herbs. Palate: fruity, flavourful, good finish.

La Depa Rosé 2021 RD
89
Aromatic, floral, fruity, tasty.

DO. MANCHUELA
CONSEJO REGULADOR

Avda. San Agustín, 9
02270 Villamalea (Albacete)
☎: +34 967 090 694 - Fax: +34 967 090 696
@: do@manchuela.wine/ana@manchuela.wine
www.manchuela.wine

LOCATION:

The production area covers the territory situated in the southeast of the province of Cuenca and the northeast of Albacete, between the rivers Júcar and Cabriel. It comprises 70 municipal districts, 26 of which are in Albacete and the rest in Cuenca.

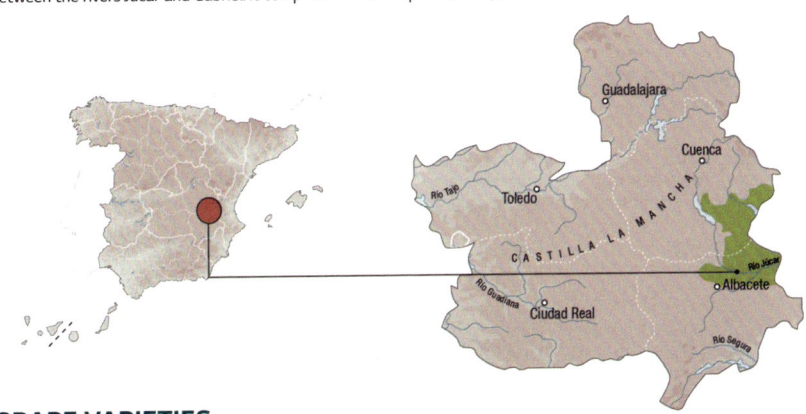

GRAPE VARIETIES:

WHITE: Albillo, Chardonnay, Macabeo, Sauvignon Blanc, Verdejo, Pardillo, Viognier, Garnacha Blanca Moscatél de Alejandría, Tardana and Moscatel de Grano Menudo.

RED: Bobal, Cabernet Sauvignon, Cencibel (Tempranillo), Garnacha, Merlot, Monastrell, Moravia Dulce, Syrah, Garnacha Tintorera, Malbec, Moravia agria, Mazuelo, Graciano, Rojal, Frasco (Tinto Velasco), Petit Verdot, Cabernet Franc, Touriga and Pinot Noir.

FIGURES:

Vineyard surface: 12,478 – **Wine-Growers:** 1,167 – **Wineries:** 38 – **2021 Harvest rating:** Good - **Production 21:** 1,304,2005 litres – **Market percentages:** 13% National - 87% International.

SOIL:

The vineyards are situated at an altitude ranging between 600 and 700 m above sea level. The terrain is mainly flat, except for the ravines outlined by the rivers. Regarding the composition of the terrain, below a clayey surface of gravel or sand, the soil is limy, which is an important quality factor for the region.

CLIMATE:

The climate is continental in nature, with cold winters and hot summers, although the cool and humid winds from the Mediterranean during the summer help to lower the temperatures at night, so creating favourable day-night temperature contrasts for a slow ripening of the grapes.

CLASSIFICATION OF YOUNG WINE HARVESTS PEÑÍNGUIDE

2017	2018	2019	2020	2021
GOOD	GOOD	VERY GOOD	VERY GOOD	VERY GOOD

DO MANCHUELA / D.O.P.

ALTOLANDÓN
Ctra. N-330, km. 242
16330 Landete (Cuenca)
☎: +34 962 302 329
altolandon@altolandon.com
www.altolandon.com

Altolandon by Rosalía 2019 T C
100% garnacha

90 🌿

Colour: deep cherry. Nose: dried herbs, dried flowers, black fruit, red berry notes, fruit preserve. Palate: ripe fruit, spicy, round tannins.

Altolandón Cuencame 2020 T C
bobal, syrah

88 🌿

Balanced, fruity, dried herbs, jammy.

Altolandón Red 2017 T R
syrah, garnacha, cabernet franc

92 🌿

Slight oxidation. Colour: deep cherry. Nose: dried herbs, red berry notes, black fruit, damp earth, spicy. Palate: ripe fruit, spicy, round tannins, good structure.

Biodiverso 2020 B
moscatel, viognier, petit manseng

93 🌿

Oxidativ, with personality, complex. Colour: straw, golden. Nose: white fruit, citrus fruit, mineral, fine lees, cereal notes. Palate: powerful, flavourful, rich, fleshy.

Doña Leo Altolandón 2021 B
100% moscatel grano menudo

90 🌿

Exuberant. Colour: straw. Nose: expressive, jasmine, wild herbs, scrubland. Palate: flavourful, fruity, balanced, powerful.

Mil Historias Bobal 2021 T RB
bobal

90 🌿

Colour: cherry, purple rim. Nose: red berry notes, floral, spicy, wild herbs. Palate: flavourful, fruity, good acidity, balanced.

Mil Historias Malbec 2021 T
malbec

90 🌿

Colour: deep cherry. Nose: dried herbs, black fruit, dried flowers, dark chocolate. Palate: ripe fruit, spicy, round tannins, fleshy.

Rayuelo 2018 T
bobal

92 🌿

Colour: deep cherry. Nose: dried herbs, black fruit, spicy, earthy notes. Palate: powerful, ripe fruit, spicy, round tannins.

BODEGA ANDRÉS INIESTA
Ctra. Fuentealbilla Villamalea, km. 1,5
02260 Fuentealbilla (Albacete)
☎: +34 967 090 650
comunicacion@bodegainiesta.com
www.bodegainiesta.es

Corazón Loco Bobal 2018 T
bobal

88

Spicy, fruity, jammy, dried herbs.

Corazón Loco Premium 2016 T R
syrah, petit verdot, cabernet sauvignon

89

Fruity, spicy, dried herbs, jammy, tasty.

Corazón Loco Selección 2016 T C
syrah, petit verdot, tempranillo, cabernet sauvignon

88

Spicy, fruity, wild, tasty.

Finca El Carril Hechicero 2016 T C
syrah, petit verdot, tempranillo, cabernet sauvignon

90

Colour: cherry, purple rim. Nose: ripe fruit, black fruit, dried herbs, spicy, smoky. Palate: flavourful, powerful, fresh, balanced.

Finca El Carril Paolo Andrea 2016 T
100% bobal

91

Colour: cherry, purple rim. Nose: fruit expression, black fruit, ripe fruit, wild herbs, wild herbs, spicy. Palate: flavourful, powerful, varietal, fresh, balanced.

Finca El Carril Valeria 2019 B
chardonnay

90

Colour: bright yellow. Nose: powerful, creamy oak, ripe fruit, spicy. Palate: rich, good structure, long, toasty, fine bitter notes.

BODEGA PARDO TOLOSA
Villatoya, 26
02215 Alborea (Albacete)
☎: +34 963 517 067
Fax: +34 961 668 213
ventas@bodegapardotolosa.com
www.bodegapardotolosa.com

La Sima 2020 T
tempranillo

87 🌿

Mizaran 2020 B
80% macabeo, 20% moscatel

86 🌿

Mizaran Bobal 2021 RD
100% bobal

87 🌱

Mizaran Tempranillo 2020 T RB
tempranillo

86 🌱

Senda de las Rochas Bobal 2014 T RB
100% bobal

89 🌱

Classic, age nuances. Nose: fruit liqueur notes, old leather.

Senda de las Rochas Tempranillo 2018 T C
tempranillo

87

BODEGA SAN ANTONIO ABAD COOPERATIVA DE CLM
Valencia, 41
02270 Villamalea (Albacete)
☎: +34 967 483 023
Fax: +34 967 483 536
jegj@bodegas-saac.com
www.bodegas-saac.com

Altos del Cabriel 2020 T
bobal, tempranillo

86 🌱

Altos del Cabriel 2021 B
macabeo

87 🌱

Altos del Cabriel 2021 RD
bobal

86 🌱

Gredas Viejas 2017 T RB
syrah

87

Viñamalea 2018 T C
tempranillo, syrah

87

BODEGAS VILLAVID
Niño Jesús, 25
16280 Villarta (Cuenca)
☎: +34 649 361 942
export@villavid.com
www.villavid.com

Villavid Bobal 2021 RD
bobal

87

Villavid Tempranillo 2021 T
tempranillo

86

Villavid Verdejo 2021 B
verdejo

85

BODEGAS VITIVINOS
Camino de Cabezuelas, s/n
02270 Villamalea (Albacete)
☎: +34 967 483 114
Fax: +34 967 483 964
jgarcia@vitivinos.com
www.vitivinos.com

Azua Bobal 2018 T C
bobal

87

Azua Bobal 2018 T RB
bobal

86

Azua Bobal 2020 T
bobal

86

Azua Verdejo 2021 B
verdejo

85

BODEGAS Y VIÑEDOS EL SOLEADO
Pozo Alfaro, 50
16237 Ledaña (Cuenca)
☎: +34 911 994 108
info@elsoleado.com
www.elsoleado.com

La Malquerida Tempranillo- Bobal- Syrah 2016 T BA
tempranillo, bobal, syrah

88

Colour: very deep cherry. Nose: roasted coffee, aromatic coffee, powerful. Palate: smoky aftertaste, great length, round tannins.

BODEGAS Y VIÑEDOS PONCE
Ctra. CM-220, km. 54,500
16230 Villanueva de La Jara (Cuenca)
☎: +34 677 434 523
juanantonio@bodegasponce.es

Clos Lojen 2021 T
bobal

92 🌱

Colour: cherry, purple rim. Nose: fruit expression, red berry notes, floral, dried herbs. Palate: fruity, flavourful, balanced.

DO MANCHUELA / D.O.P.

DO MANCHUELA / D.O.P.

Depaula Ponce 2021 T RB
monastrell

91

Colour: bright cherry. Nose: floral, ripe fruit, balanced, red berry notes. Palate: flavourful, fruity, good acidity.

La Casilla 2020 T RB
bobal

93

Colour: cherry, purple rim. Nose: balsamic herbs, red berry notes, scrubland. Palate: balsamic, fruity, balanced.

La Estrecha 2020 T

93

Colour: cherry, purple rim. Nose: fruit expression, red berry notes, floral, spicy. Palate: flavourful, fruity, good acidity.

P.F. 2020 T
bobal

94

Colour: Cherry. Nose: complex, expressive, spicy, mineral. Palate: elegant, full-bodied, long, great length.

Pino 2020 T RB
bobal

94

Colour: bright cherry. Nose: balsamic herbs, scrubland, wild herbs, elegant, balanced. Palate: spicy, balsamic, good acidity, easy to drink.

CARRIL CRUZADO

Ctra. Iniesta-Villagarcía del Llano km, 13
16236 Villagarcía del Llano (Cuenca)
☎: +34 616 960 992
bodega@carrilcruzado.com
www.carrilcruzado.es

Carril Cruzado Petit Vedot Edición Limitada 2019 T C
100% petit verdot

91

Colour: deep cherry. Nose: dried herbs, creamy oak, black fruit, wild herbs. Palate: powerful, ripe fruit, spicy, grainy tannins.

Carril Cruzado Sauvignon Blanc 2021 B
100% sauvignon blanc

87

CIEN Y PICO WINE

San Francisco, 19
02240 Mahora (Albacete)
☎: +34 610 239 186
luisjimenaz@gmail.com
www.cienypico.com

Cien y Pico Doble Pasta 2019 T
garnacha tintorera

88

Spicy, fruity, jammy, smoky.

Cien y Pico En Vaso 2019 T
bobal

89

Spicy, fruity, herbal, jammy, tasty.

Viña La Ceja 2019 T
50% bobal, 50% garnacha tintorera

88

Spicy, fruity, jammy, smoky, crisp, tasty.

FINCA EL MOLAR

Finca El Molar
02260 Fuentealbilla (Albacete)
☎: +34 647 075 371
info@elmolarderus.com
www.fincaelmolar.com

Finca El Molar 2020 T RB
graciano, syrah

88

Spicy, fruity, tasty, balanced.

Finca El Molar Bobal 2020 T
bobal

88

Pleasant, fruity, balanced, simple, varietally correct.

Finca El Molar Moravia Agria 2020 RD
moravia agria

89

Fruity, spicy, tasty, powerful, varietally correct.

FINCA SANDOVAL

Ctra. CM-3222, Km. 26800
16237 Ledaña (Cuenca)
☎: +34 914 363 636
fincasandoval@fincasandoval.com
www.fincasandoval.com

El Fundamentalista 2020 T
bobal, otras

92

Crisp, fruity, simple. Colour: bright cherry. Nose: red berry notes, fruit expression, balanced. Palate: fruity, easy to drink, long.

DO MANCHUELA / D.O.P.

El Fundamentalista 2021 T
bobal, otras

91 🌱

Colour: light cherry. Nose: fresh fruit, fruit expression, neat. Palate: fruity, good acidity, balanced, correct, easy to drink.

Finca Sandoval 2020 T
syrah, bobal

93 🌱

Colour: bright cherry. Nose: black fruit, fruit preserve, meaty notes, wild herbs. Palate: flavourful, round tannins, spicy.

Salia 2020 T R
syrah, bobal

92 🌱

Pleasant, balanced, spicy. Nose: neat, expressive, red berry notes, ripe fruit, faded flowers. Palate: fruity, flavourful, long.

Signo Bobal 2020 T BA
bobal

93 🌱

Colour: Cherry. Nose: expressive, red berry notes, ripe fruit, dried flowers. Palate: elegant, long, great length, spicy, fruity aftertaste.

LA CEPA DE PELAYO
Batán, 9
02210 Alcala del Júcar (Albacete)
☎: +34 967 099 105
info@bodegalacepadepelayo.com
www.bodegalacepadepelayo.com

Cupido Bobal 2018 T
bobal

90

Slight reduction. Colour: cherry, garnet rim. Nose: ripe fruit, meaty notes, spicy. Palate: flavourful, varietal, balanced, fine bitter notes.

Cupido Macabeo 2018 B BA
macabeo

90

Colour: bright yellow. Nose: dried flowers, candied fruit, fine lees, pattiserie. Palate: round, spicy, long, great length.

La Cepa de Pelayo 2020 B
macabeo

88 🌱

Toasty, powerful, jammy, tasty.

La Cepa de Pelayo Bobal 2019 T
bobal

91 🌱

Colour: Cherry. Nose: scrubland, wild herbs, ripe fruit. Palate: varietal, fruity, easy to drink, spicy.

Ole de Aromas 2019 T
bobal

88 🌱

Spicy, fruity, jammy, astringent.

Ole de Passion 2020 B
macabeo

88 🌱

Aromatic, jammy, tasty, tropical.

LA NIÑA DE CUENCA, BODEGA Y VIÑEDOS PROPIOS
Muela, 1
16237 Ledaña (Cuenca)
☎: +34 629 751 027
info@laninadecuenca.com
www.laninadecuenca.com

Ildania 2017 T
90% bobal, 10% otras

91

Colour: dark-red cherry. Nose: fine reductive notes, black fruit, red clay notes, characterful, fruit preserve. Palate: balanced, flavourful, long.

Orovelo 2019 B
albillo

91 🌱

Colour: bright yellow. Nose: dried flowers, candied fruit, fine lees, pattiserie. Palate: round, spicy, long, great length.

NTRA. SRA. DE LA CABEZA DE CASAS IBÁÑEZ SOC. COOP. DE CLM
Avda. del Vino, 10
02200 Casas Ibáñez (Albacete)
☎: +34 967 460 266
info@coop-cabeza.com
www.coop-cabeza.com

Viaril Cabernet Sauvignon 2020 T
100% cabernet sauvignon

85

Viaril Macabeo 2021 B
100% macabeo

86

Viaril Verdejo Sauvignon Blanc 2021 B
60% macabeo, 40% sauvignon blanc

87

DO MANCHUELA / D.O.P.

PARAJES DEL VALLE BODEGAS Y VIÑEDOS
Avda. de Murcia, s/n
30520 Jumilla (Murcia)
☎: +34 968 782 977
info@parajesdelvalle.es
www.parajesdelvalle.es

Parajes del Valle 2021 RD
100% bobal
88 🌱
Citrus fruit, balanced, fruity, full-bodied.

Parajes del Valle Macabeo 2021 B
89 🌱
Pleasant, aromatic, tasty, fruity.

SAN ANTONIO DE PADUA
San Antonio, 21
16270 Villalpardo (Cuenca)
☎: +34 962 311 002
santopadua@yahoo.es
www.coopsanantoniodepadua.com

Hoya Montés 2021 RD
bobal
85

SAN ISIDRO DE ALBOREA
Extramuros, s/n
02215 Alborea (Albacete)
☎: +34 628 672 290
luisjimenaz@gmail.com
www.cooperativavinoalborea.com

Alterón 2018 T C
bobal
86

Alterón 2021 RD
86

Alterón Macabeo 2019 B
macabeo
87

SEÑORÍO DEL JÚCAR
Pol. Ind. Parc 64-70
02200 Casas Ibáñez (Albacete)
☎: +34 967 460 564
administracion@bsjucar.es
www.senoriodeljucar.com

Cueva Llana Bobal 2021 T
bobal
88 🌱
Fruity, floral, rustic, powerful, tasty.

Cueva Llana Macabeo 2021 B
macabeo
88 🌱
Hot, jammy, tasty.

Cueva Llana Rosé 2021 RD
bobal
88 🌱
Fruity, jammy, powerful.

Cueva Llana Syrah 2021 T
syrah
90 🌱
Colour: cherry, purple rim. Nose: ripe fruit, fruit expression, dried flowers, spicy. Palate: fruity, flavourful, balanced.

Tranco del Lobo 2019 T C
bobal
87 🌱

SHUKHRAT KHAKIMOV & VITICULTORES
Plz Constitución, 7
03550 San Juan de Alicante (Alacant/Alicante)
☎: +34 965 943 090
Fax: +34 965 943 090
a.nogay@winexfood.com
www.winexfood.com

La Bella Ancestral Pet Nat 2020 RE C BN
100% moravia agria
91 🌱
Colour: ruby red. Nose: fresh fruit, fragrant herbs, raspberry, bakery, fine lees. Palate: fresh, fruity, good acidity, flavourful.

La Bella Ancestral Pet Nat 2021 RE BN
100% moravia agria
90 🌱
Colour: ruby red. Nose: fresh fruit, citrus fruit, fine lees, red berry notes. Palate: fresh, fruity, flavourful, good acidity.

SOC. COOP. AGRARIA DE CLM SAN ISIDRO

Príncipe, 153
16220 Quintanar del Rey (Cuenca)
☎: +34 967 495 052
export@bodegasanisidro.es
www.bodegasanisidro.es

Monte de las Mozas Bobal 2021 RD
bobal

87

Monte de las Mozas Macabeo 2021 B
macabeo

87

Quinta Regia Bobal 2020 T
bobal

86

Zaíno Syrah Tempranillo 2020 T
syrah, tempranillo

86

Zaíno Tempranillo 2020 T C
tempranillo

88

Standard, spicy, aromatic, jammy, slight reduction, dried herbs.

VEGA TOLOSA

Pol. Ind. Calle B, 11
02200 Casas Ibáñez (Albacete)
☎: +34 686 908 543
info@vegatolosa.com
www.vegatolosa.com

11 Pinos Bobal 2020 T RB
100% bobal

88

Pruney, spicy, tasty, sweety finish, full-bodied.

Bobal Icon 2020 T RB
100% bobal

89

Aromatic, balsamic herbs, wild, floral, fruity, jammy, pleasant.

Capricho DiVino 2019 BE BN
100% chardonnay

87

Capricho DiVino Sauvignon Blanc 2021 B
100% sauvignon blanc

85

Finca Los Halcones Bobal 2019 T
100% bobal

90

Colour: Cherry. Nose: scrubland, dried herbs, waxy notes, ripe fruit. Palate: spicy, flavourful, slightly dry, soft tannins.

Finca Los Halcones Viognier 2021 B FB

90

Colour: bright yellow. Nose: ripe fruit, dried herbs, faded flowers. Palate: powerful, ripe fruit, balanced.

Vega Tolosa Chardonnay 2018 BE BN
100% chardonnay

88

Citrus fruit, balanced, crisp, herbal, tasty.

DO MANCHUELA / D.O.P.

Peñín Guide | SPANISH WINE

DO. MÉNTRIDA

CONSEJO REGULADOR

Calle Eras de San Francisco, 7
45500 Torrijos (Toledo)
☎: +34 925 785 185 - Fax: +34 925 784 154
@: administracion@domentrida.es
www.domentrida.es

LOCATION:

In the north of the province of Toledo. It borders with the provinces of Ávila and Madrid to the north, with the Tajo to the south, and with the Sierra de San Vicente to the west. It is made up of 51 municipal areas of the province of Toledo.

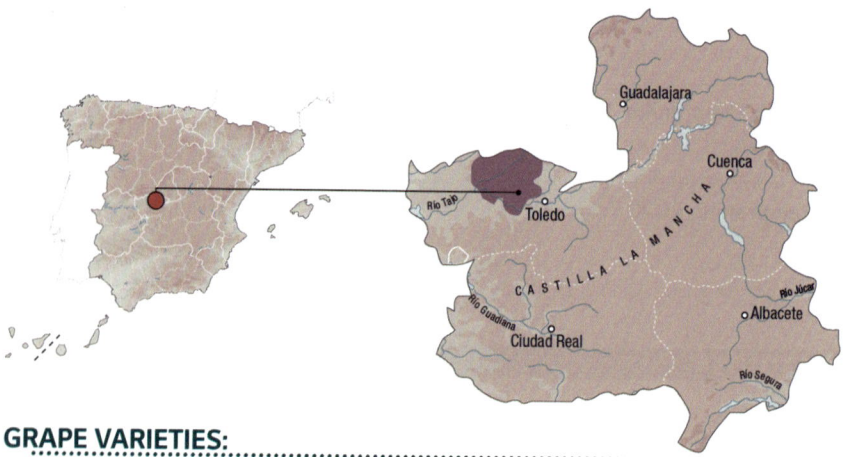

GRAPE VARIETIES:

WHITE: Albillo, Macabeo, Sauvignon Blanc, Chardonnay, Garnacha Blanca and Moscatel de Grano Menudo.

RED: Garnacha (main), Cencibel (Tempranillo), Cabernet Sauvignon, Merlot, Syrah, Petit Verdot, Cabernet Franc, Garnacha Peluda Garnacha Tintorera, Moravia Agria and Graciano.

FIGURES:

Vineyard surface: 5,854 – **Wine-Growers:** 1,260 – **Wineries:** 31 – **2021 Harvest rating:** Very Good – **Production 21:** 14,505,668 litres – **Market percentages:** 75% National - 25% International.

SOIL:

The vineyards are at an altitude of between 400 m and 600 m, although some municipal districts of the Sierra de San Vicente reach an altitude of 800 m. The soil is mainly sandy-clayey, with a medium to loose texture.

CLIMATE:

Continental, dry and extreme, with long, cold winters and hot summers. Late frosts in spring are quite common. The average rainfall is between 300 mm and 500 mm, and is irregularly distributed throughout the year.

CLASSIFICATION OF YOUNG WINE HARVESTS PEÑÍNGUIDE

2017	2018	2019	2020	2021
GOOD	VERY GOOD	VERY GOOD	VERY GOOD	VERY GOOD

BODEGA LÓPEZ CAMPOS
Avda, de Madrid, 24
45940 Valmojado (Toledo)
☎: +34 677 315 470
info@almavid.com
www.almavid.com

Almavid 2018 T R
garnacha, syrah
89
Age nuances, slight reduction, weary, pruney, full-bodied, spicy, dried herbs, jammy.

Almavid Garnacha 2020 T RB
garnacha
88
Spicy, jammy, tasty, standard, kind finish.

Almavid Rosé 2021 RD
garnacha
88
Fruity, jammy, floral, balanced, defined aromas, pleasant.

BODEGA SANTO DOMINGO DE GUZMÁN
Alameda del Fresno, 14
45940 Valmojado (Toledo)
☎: +34 918 170 904
info@santodomingodeguzman.es
www.santodomingodeguzman.es

Valdejuana Syrah 2020 T
syrah
87

BODEGAS ALONSO CUESTA
Pza. de la Constitución, 4
45920 La Torre de Esteban Hambrán (Toledo)
☎: +34 925 795 742
Fax: +34 925 795 742
administracion@alonsocuesta.com
www.alonsocuesta.com

Alonso Cuesta Cállate 2019 T RB
garnacha, syrah
89
Pruney, fruity, dried flowers, standard, kind finish, balanced, jammy.

Alonso Cuesta Cuveé 2018 T BA
garnacha, cabernet sauvignon, petit verdot
91
Colour: Cherry. Nose: balsamic herbs, sweet spices, scrubland, ripe fruit. Palate: spicy, balsamic, good acidity, flavourful.

Alonso Cuesta La Garnacha de Lola 2020 T BA S
garnacha
91
Colour: deep cherry. Nose: dried herbs, dried flowers, ripe fruit, fruit preserve, varietal. Palate: ripe fruit, spicy, flavourful, easy to drink.

Tres Amigos 2018 T BA S
garnacha, cabernet sauvignon, syrah
92
Colour: deep cherry. Nose: ripe fruit, dried herbs, creamy oak. Palate: powerful, ripe fruit, spicy, round tannins.

BODEGAS ARRAYÁN
Finca La Verdosa, s/n
28002 Santa Cruz del Retamar (Toledo)
☎: +34 916 633 131
comercial@arrayan.es
www.arrayan.es

Arrayán Albillo Real 2020 B
albillo real
92
Colour: bright yellow. Nose: faded flowers, dried flowers, spicy, expressive, ripe fruit. Palate: flavourful, rich, long, spicy.

Arrayan Rosado de Garnacha 2021 RD
garnacha peluda
89
Dried flowers, pruney, jammy, dried herbs.

Arrayán Selección 2019 T C
merlot, cabernet sauvignon, syrah, petit verdot
89
Balsamic herbs, dried herbs, jammy, fruity, balanced, spicy, great length, tasty.

Arroyo de Arrayán 2020 B
garnacha gris, garnacha blanca
92
Colour: bright yellow. Nose: powerful, creamy oak, ripe fruit, spicy, dried herbs. Palate: rich, long, fine bitter notes, flavourful.

Estela de Arrayán 2018 T R
merlot, cabernet sauvignon, syrah, petit verdot
92
Colour: Cherry. Nose: balsamic herbs, sweet spices, scrubland, dried herbs. Palate: spicy, good acidity, fruity, flavourful, round tannins.

DO MÉNTRIDA / D.O.P.

DO MÉNTRIDA / D.O.P.

La Suerte de Arrayán 2019 T
garnacha

92

Colour: Cherry. Nose: scrubland, dried flowers, faded flowers, red berry notes, fruit preserve. Palate: spicy, varietal, fruity aftertaste, easy to drink.

BODEGAS ATALAQUE (RODRÍGUEZ DE VERA)
Santa Cruz, 28
45510 Fuensalida (Toledo)
☎: +34 696 168 873
carlos@rodriguezdevera.com
www.rodriguezdevera.com

Atalaque Garnacha del Horcajo 2018 T C
100% garnacha

92

Wild. Colour: cherry, garnet rim. Nose: wild herbs, dried herbs, red berry notes, ripe fruit, balanced. Palate: flavourful, long, easy to drink.

Atalaque Garnacha La Peraleda 2019 T
100% garnacha

90

Reduced. Colour: cherry, garnet rim. Nose: fruit preserve, powerful, black liquorice. Palate: flavourful, long.

Atalaque Moscatel del Horcajo 2018 B D
moscatel grano menudo

90

Colour: bright yellow. Nose: ripe fruit, candied fruit, honeyed notes, faded flowers. Palate: flavourful, unctuous, fruity, sweet.

BODEGAS CANOPY
Herreros, 5
45180 Camarena (Toledo)
☎: +34 619 244 878
achacon@bodegascanopy.com
www.bodegascanopy.com

Castillo de Belarfonso 2020 T RB
100% garnacha

90

Kind finish, smooth. Colour: light cherry. Nose: medium intensity, dried flowers, wild herbs, waxy notes. Palate: easy to drink, good finish.

Ganadero 2021 B
100% verdejo

87

Ganadero 2021 T
100% garnacha

88

Kind finish, simple, wild, jammy, fruity, standard.

🏆 PODIUM

La Viña Escondida 2018 T
100% garnacha

95

Colour: light cherry. Nose: expressive, spicy, mineral, wild herbs, red berry notes, ripe fruit. Palate: full-bodied, long, great length, varietal.

Malpaso 2020 T
100% syrah

92

Slight reduction. Colour: cherry, garnet rim. Nose: ripe fruit, characterful, toasted bread, spicy. Palate: flavourful, fruity.

Tres Patas 2018 T
100% garnacha

93

Colour: Cherry. Nose: scrubland, dried flowers, dried herbs, wild herbs, varietal. Palate: spicy, good acidity, fruity, flavourful, easy to drink.

BODEGAS HIBÉU
Camino Las Ventas, s/n
45920 Torre de Esteban Hambrán (Toledo)
☎: +34 609 027 255
info@bodegashibeu.es
www.bodegashibeu.es

Hibeu 2019 T
tempranillo, syrah

87 🌱

Hibeu 2020 T
tempranillo, syrah

87 🌱

Hibeu Finca La Mineral 2019 T
syrah

89 🌱

Spicy, jammy, tasty, toasty, full-bodied, balanced, fruity.

Peña Hibeu 2018 T
tempranillo, syrah

85 🌱

BODEGAS JIMÉNEZ LANDI

Avda. Solana, 39
45930 Méntrida (Toledo)
☎: +34 918 178 213
info@jimenezlandi.com
www.jimenezlandi.com

Jiménez-Landi Blanco Malvar 2020 B
malvar
92
Oxidativ. Colour: straw. Nose: ripe fruit, dried herbs, faded flowers, white fruit. Palate: powerful, ripe fruit, balanced, flavourful, saline.

Jiménez-Landi El Corralón 2021 T
92
Rustic. Colour: light cherry. Nose: black fruit, ripe fruit, spicy, balsamic herbs. Palate: fleshy, flavourful.

Jiménez-Landi Natural 2021 T
syrah, cabernet sauvignon
92
Colour: deep cherry. Nose: dried herbs, black fruit, floral. Palate: powerful, ripe fruit, spicy, round tannins.

🏆 **PODIUM**

Jiménez-Landi Piélago 2019 T
garnacha
95
Defined aromas, full of life. Colour: cherry, purple rim. Nose: fruit expression, red berry notes, floral, spicy, balsamic herbs. Palate: flavourful, fruity, good acidity, long.

Jiménez-Landi Sotorrondero 2020 T
93
Rustic, tasty. Colour: deep cherry. Nose: dried herbs, creamy oak, black fruit, spicy. Palate: powerful, ripe fruit, spicy, round tannins.

Jiménez-Landi Valdiniebla 2021 RD
91
Colour: rose. Nose: ripe fruit, hot, faded flowers, scrubland. Palate: fleshy, flavourful, ripe fruit.

BODEGAS Y VIÑEDOS TAVERA

Ctra. Valmojado - Toledo, Km. 2
45182 Arcicollar (Toledo)
☎: +34 637 847 777
info@bodegastavera.com
www.bodegastavera.com

Tavera Antiguos Viñedos 2021 RD
100% garnacha
86

Tavera Antiguos Viñedos Garnacha 2019 T RB
100% garnacha
88
Varietally correct, jammy, pruney, dried flowers, balanced, smooth, wild.

Tavera Edición Numerada Syrah 2019 T C
100% syrah
89
Pleasant, dried herbs, jammy, fruity, wild, standard, balanced.

Tavera Syrah Tempranillo 2019 T RB
50% syrah, 50% tempranillo
86

Tavera Vendimia Seleccionada 2017 T C
45% syrah, 45% tempranillo, 10% garnacha
88
Pruney, jammy, tasty, dried herbs, spicy, wild.

COOPERATIVA CONDES DE FUENSALIDA

Avda. San Crispín, 129
45510 Fuensalida (Toledo)
☎: +34 925 784 823
Fax: +34 925 784 823
condesdefuensalida@hotmail.com
www.condesdefuensalida.iespana.es

Condes de Fuensalida 100 años 2020 T
87

Condes de Fuensalida 2019 T C
garnacha, cabernet sauvignon, syrah, tempranillo
84

Condes de Fuensalida 2020 T
garnacha, syrah
85

Condes de Fuensalida 2021 RD
84

Condes de Fuensalida 2021 RE D
84

DO MÉNTRIDA / D.O.P.

DO MÉNTRIDA / D.O.P.

DANIEL LANDI VITICULTOR
Avda de la Constitución, 23
28640 Cadalso de Los Vidrios (Madrid)
☎: +34 918 640 602
info@danilandi.com
www.danilandi.com

Cantos del Diablo 2019 T
garnacha
94
Colour: light cherry. Nose: red berry notes, scrubland, balsamic herbs. Palate: elegant, fine tannins, warm.

Las Uvas de la Ira 2020 T
91
Slight reduction. Colour: light cherry. Nose: red berry notes, rose petals, scrubland, spicy. Palate: spirituous, fleshy.

VIÑEDOS DE CAMARENA, SDAD. COOP. DE CLM
Ctra. Toledo - Valmojado, Km.24.6
45180 Camarena (Toledo)
☎: +34 918 174 347
vdecamarena@hotmail.com
www.vdecamarena.com

Bastión 2021 RD
garnacha
85

Bastión de Camarena 2021 B
verdejo
85

Bastión Garnacha + Syrah 2021 T
garnacha, syrah
87

Bastión Garnacha 2019 T
100% garnacha
90
Pleasant, defined aromas, simple. Colour: cherry, garnet rim. Nose: fragrant herbs, dried flowers, red berry notes, ripe fruit. Palate: varietal, fine bitter notes.

Bastión Selección 2020 T
garnacha, merlot
87

VIÑEDOS Y BODEGAS GONZÁLEZ
Real, 86
45180 Camarena (Toledo)
☎: +34 918 174 063
bodegasgonzalez@yahoo.es
www.vinobispo.com

Señorío del Bispo 2019 T RB
syrah, cabernet sauvignon, merlot
86

Señorío del Bispo 2021 T RB
68% syrah, 32% cabernet sauvignon
86

Viña Bispo 2021 B
50% verdejo, 50% sauvignon blanc
85

Viña Bispo 2021 RD
100% garnacha
86

Viña Bispo 2021 T
68% syrah, 32% cabernet sauvignon
85

DO. MONDÉJAR

CONSEJO REGULADOR

Pza. Mayor, 10
19110 Mondéjar (Guadalajara)
☎: +34 949 385 284 - Fax: +34 949 385 284
@: crdom@crdomondejar.com
www.domondejar.es

LOCATION:

In the southwest of the province of Guadalajara. It is made up of the municipal districts of Albalate de Zorita, Albares, Almoguera, Almonacid de Zorita, Driebes, Escariche, Escopete, Fuenteovilla, Illana, Loranca de Tajuña, Mazuecos, Mondéjar, Pastrana, Pioz, Pozo de Almoguera, Sacedón, Sayatón, Valdeconcha, Yebra and Zorita de los Canes.

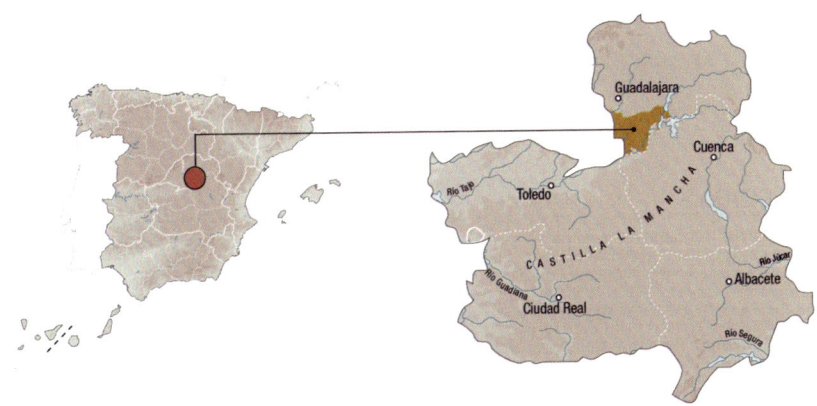

GRAPE VARIETIES:

WHITE (40%): Malvar (majority 80% of white varieties), Macabeo and Torrontés.

RED (60%): Cencibel (Tempranillo – represents 95% of red varieties), Cabernet Sauvignon (5%) and Syrah.

FIGURES:

Vineyard surface: 3.000 – Wine-Growers 11: 300 – Wineries 11: 2 – 20**16 Harvest rating:** N/A – Production 2011: – 421.130 L – Market percentages 2011: 100% National.

SOIL:

The south of the Denomination is characterized by red soil on lime-clayey sediments, and the north (the municipal districts of Anguix, Mondéjar, Sacedón, etc.) has brown limestone soil on lean sandstone and conglomerates.

CLIMATE:

Temperate Mediterranean. The average annual temperature is around 18°C and the average rainfall is 500 mm per year.

CLASSIFICATION OF YOUNG WINE HARVESTS PEÑÍNGUIDE

2017	2018	2019	2020	2021
N/A	N/A	N/A	N/A	N/A

DO. MONTERREI
CONSEJO REGULADOR

Rúa Castelo. 10 Bajo
32600 Verín (Ourense)
☎: +34 988 590 007 - Fax: +34 988 410 634
@: info@domonterrei.com
www.domonterrei.wine

LOCATION:

In the eastern part of the province of Orense, on the border with Portugal. The vineyard covers the valley of Monterrei and includes the municipalities of Verín, Monterrei, Oimbra, Castrelo do Vall, Vilardevós and Riós.

SUB-REGIONS:

Val de Monterrei. This includes the vineyards located in the valley area (hence on flatter terrain). It includes the parishes of Castrelo do Val, Pepín and Nocedo do Val in the municipality of Castrelo do Val; the parishes of Albarellos, Infesta, Monterrei and Vilaza in the municipality of Monterrei; the parishes of Oímbra, Rabal, and San Cibrao in the municipality of Oímbra and the parishes of Abedes, Cabreiroá, Feces de Abaixo, Feces de Cima, Mandín, Mourazos, Pazos, Queizás, A Rasela, Tamagos, Tamaguelos, Tintores, Verín and Vila Maior do Val in the municipality of Verín.

Ladeira de Monterrei. The vineyard on this occasion is located up in the hills. It includes the municipality of Vilardevós, the parishes of Gondulfes and Servoi in the municipality of Castrelo do Val; the parishes of As Chas, Bousés, Vidiferre and A Granxa in the municipality of Oímbra; the parishes of Flariz, Medeiros, Estevesiños and Vences in the municipality of Monterrei; the parish of Queirugás in the municipality of Verín; the parish of Castrelo de Abaixo, the places of Castrelo de Cima, Covelas, O Mourisco, San Paio and A Veiga do Seixo in the parish of Castrelo Cima, the place of Fumaces in the parish of Fumaces, the places of Progo and Pousada in the parish of Progo, the place of Florderrei in the parish of Riós, all in the municipality of Riós.

GRAPE VARIETIES:

WHITE: Dona Blanca, Verdello (Godello), Treixadura (Verdello Louro), Albariño, Caiño Blanco, Loureira and Blanca de Monterrei.

RED: Merenzao (Bastardo), Arauxa (Tempranillo), Mencía, Caiño Tinto and Sousón.

FIGURES:

Vineyard surface: 657 – **Wine-Growers:** 365 – **Wineries:** 28 – **2021 Harvest rating:** N/A – **Production 21:** 4,417,575 litres – **Market percentages:** 95% National - 5% International.

SOIL:

The vineyard extends over the sides of the hills and valleys irrigated by the River Támega and its tributaries. There are three types of soils in the area: slate and shale, granite and sandy, from the degradation of the granite rocks, and sediment type soils.

CLIMATE:

It has a moderate Mediterranean climate with a continental tendency, and influenced by the Atlantic Ocean. Its summers are hot and dry while its winters are cold. The area has considerable temperature fluctuations of up to 30º during the ripening period. It is drier than the rest of Galicia, with maximum temperatures of 35°C in summer and minimum temperatures of -5°C in winter.

CLASSIFICATION OF YOUNG WINE HARVESTS PEÑÍNGUIDE

2017	2018	2019	2020	2021
VERY GOOD	VERY GOOD	VERY GOOD	VERY GOOD	VERY GOOD

ADEGA ABELEDOS

Avda. Portugal, 110 2°A
32600 Verín (Ourense/Orense)
☎: +34 988 414 075
Fax: +34 988 414 075
adegaabeledos@gmail.com

Abeledos 2021 B
85

Abeledos 2021 T
mencía, sousón, caíño
87

Tirabeque 2020 T
mencía
89
Wild, fruity, balanced, standard, tasty, rustic.

Tirabeque 2021 B
godello
90
Colour: bright straw, greenish rim. Nose: citrus fruit, wild herbs, white fruit, fine lees. Palate: fresh, fruity, good acidity, fine bitter notes.

ADEGAS PAZO DAS TAPIAS

Finca As Tapias- Pazos
32600 Verín (Ourense/Orense)
☎: +34 988 261 256
Fax: +34 988 261 264
info@pazodomar.com
www.pazodomar.com

Alma de Autor Godello 2020 B
godello
92
Colour: straw. Nose: ripe fruit, dried herbs, mineral, smoky, white flowers. Palate: powerful, ripe fruit, balanced, fleshy.

Alma de Autor Tinto 2020 T
mencía
89
Balanced, dried flowers, herbal, spicy, fruity.

Alma de Blanco 2021 B
godello
88
Citrus fruit, balanced, crisp, herbal.

Alma de Tinto 2020 T
mencía
87

ADEGAS VALMIÑOR

Estrada A Guarda – Tui n°245. Barrio Portela
36760 O'Rosal (Pontevedra)
☎: +34 986 609 060
valminor@valminorebano.com
www.valminorebano.com

Minius Godello 2021 B
godello
86

AYSV. MENASA

Avda. de Sousas, 56 Bajo
32600 Verín (Ourense/Orense)
☎: +34 988 414 802
magmuszaira@gmail.com
www.menasa.es

Vega de Lucía Godello 2021 B
100% godello
88
Jammy, tasty, fruity, citrus fruit.

Vega de Lucía Mencía 2020 T
100% mencía
89
Pleasant, fruity, tasty, spicy.

BODEGAS CHAVES

Condesa, 3, Barrantes
36636 Ribadumia (Pontevedra)
☎: +34 986 710 015
bodegaschaves@bodegaschaves.com
www.bodegaschaves.com

Silva Daponte Godello 2021 B
100% godello
88
Citrus fruit, crisp, herbal, tasty.

Silva Daponte Mencía 2021 T
100% mencía
88
Fruity, herbal, tasty, balanced.

BODEGAS LADAIRO

Ctra. Ladairo, 42
32613 O Rosal (Ourense/Orense)
☎: +34 988 422 757
info@bodegasladairo.com
www.bodegasladairo.com

J.L. Vilela Ladairo 2019 T C
mencía, arauxa
91
Colour: deep cherry. Nose: dried herbs, black fruit, cocoa bean. Palate: ripe fruit, spicy, round tannins.

DO MONTERREI / D.O.P.

DO MONTERREI / D.O.P.

Ladairo 2019 T BA
mencía, arauxa

90

Colour: deep cherry. Nose: ripe fruit, dried herbs, creamy oak. Palate: ripe fruit, spicy, round tannins.

Ladairo 2020 B FB
godello, treixadura

91

Colour: bright yellow. Nose: sweet spices, fine lees, pattiserie, balanced. Palate: rich, flavourful, fine bitter notes, toasty, good acidity.

Ladairo Colección Familia Godello Treixadura 2021 B
godello, treixadura

90

Colour: bright straw, greenish rim. Nose: fresh fruit, citrus fruit, wild herbs, floral. Palate: fresh, good acidity, fine bitter notes.

Ladairo Colección Familia Mencía y Araúxa 2021 T
mencía, arauxa

89

Crisp, fruity, herbal, balanced.

BODEGAS TRIAY

Rua Ladairo, 36 (O Rosal)
32613 Oímbra (Ourense/Orense)
☎: +34 988 422 776
info@bodegastriay.es
www.bodegastriay.es

Tres Mulleres Godello 2020 B
godello

90

Colour: bright straw. Nose: white fruit, scrubland, white flowers. Palate: fruity, flavourful, varietal.

Tres Mulleres Treixadura 2020 B
treixadura

90

Colour: bright straw. Nose: white fruit, faded flowers, mineral. Palate: fruity, fresh, flavourful, balanced.

Triay 2021 B
godello

90

Colour: bright straw. Nose: white fruit, citrus fruit, white flowers, dried herbs. Palate: fresh, fruity, flavourful, balanced.

Triay 38 2019 T
sousón, mencía

92

With personality, herbal, jammy. Colour: very deep cherry. Nose: ripe fruit, black fruit. Palate: flavourful, fruity, fruity aftertaste, soft tannins.

Triay Mencía 2021 T

90

Colour: cherry, purple rim. Nose: fruit expression, red berry notes, floral, spicy. Palate: flavourful, fruity, good acidity, easy to drink.

BODEGAS Y VIÑEDOS QUINTA DA MURADELLA

Avda. Luis Espada, 99- Entresuelo, dcha.
32600 Verín (Ourense/Orense)
☎: +34 988 411 724
Fax: +34 988 590 427
bodega@muradella.com

Verdello Antiguo 2019 B
verdello

93

Colour: bright straw. Nose: fine lees, citrus fruit, white fruit, dry stone, dried herbs. Palate: full-bodied, rich, long, good acidity, saline, great length.

CASTRO DE LOBARZÁN

Ctra. de Requeixo, 51- Villaza
32618 Monterrei (Ourense/Orense)
☎: +34 988 418 163
lobarzan@gmail.com

Castro de Lobarzán 2021 B
godello, treixadura

88

Standard, floral, fruity, aromatic.

Castro de Lobarzán 2021 T
mencía, arauxa, bastardo

88

Aromatic, standard, fruity, jammy, tasty, wild.

Lobarzán Isaura 2020 B FB
godello

90

Colour: bright yellow. Nose: powerful, ripe fruit, toasty. Palate: rich, good structure, long, fine bitter notes, good acidity.

Lobarzán Isaura 2020 T
mencía, arauxa, bastardo

91

Defined aromas. Colour: deep cherry, purple rim. Nose: characterful, dried herbs, scrubland, ripe fruit. Palate: flavourful, fruity, balsamic.

CREGO E MONAGUILLO
Rua Nova
32618 Salgueira (Ourense/Orense)
☎: +34 699 082 280
tito@cregoemonaguillo.com
www.cregoemonaguillo.com

Crego e Monaguillo Godello 2021 B
godello, treixadura, dona blanca, loureiro, albariño
90
Colour: bright straw. Nose: fruit expression, ripe fruit, floral. Palate: flavourful, fresh, good acidity, fruity aftertaste.

Crego e Monaguillo Mencía 2021 T
88
Pleasant, kind finish, fruity, tasty.

Crego e Monaguillo Treixadura Albariño 2020 B
treixadura, albariño
91
Colour: straw. Nose: expressive, white flowers, jasmine, dried herbs. Palate: flavourful, fruity, balanced.

Father 1943 2019 T BA
91
Colour: cherry, garnet rim. Nose: fruit preserve, fruit liqueur notes, powerful, toasty. Palate: flavourful, sweetness, long.

DANIEL FERNÁNDEZ (ALBA AL-BAR)
Quintá de Arriba, s/n
32698 Queizas Verín (Ourense/Orense)
☎: +34 988 590 864
alberto@bodegasdanielfernandez.com
www.bodegasdanielfernandez.com

Galván 2019 B
godello, treixadura, dona blanca
87

Galván 2020 B
godello, treixadura, albariño, dona blanca
88
Kind finish, aromatic, standard, floral, tasty.

FAUSTO RIVERO PARDO
Avda. do Carregal, 106
32613 Oímbra (Ourense/Orense)
☎: +34 651 488 915
quintasoutullo@gmail.com
www.quintasoutullo.com

Quinta Soutullo 2019 T
mencía
87

Quinta Soutullo 2020 B
godello
87

FRAGAS DO LECER
Touza 21, Vilaza
32618 Monterrei (Ourense/Orense)
☎: +34 616 670 249
comercial@grandespagosgallegos.com
www.fragadocorvo.com

Fraga do Corvo Godello 2021 B
89
Citrus fruit, crisp, fruity, herbal.

Fraga do Corvo Mencía 2021 T
90
Colour: cherry, purple rim. Nose: red berry notes, floral, spicy. Palate: flavourful, fruity, good acidity, long.

Fragas do Lecer 2021 B
godello
90
Colour: bright straw. Nose: fragrant herbs, fine lees, stone fruit. Palate: full-bodied, rich, balanced.

GARGALO
Rua Do Castelo, 59
32619 Verín (Ourense/Orense)
☎: +34 941 454 050
Fax: +34 941 454 529
rrpp@bodegasriojanas.com
www.bodegasriojanas.com

Gargalo Godello 2021 B
100% godello
91
Colour: bright straw. Nose: fruit expression, ripe fruit, floral. Palate: flavourful, fresh, good acidity, fruity aftertaste.

Terra Rubia 3 Uvas 2021 B
treixadura, albariño, godello
88
Pleasant, aromatic, floral, fruity, tasty.

LAGAR DE DEUSES
Ctra. de Lugo – Ourense Km. 50
27500 Chantada (Lugo)
☎: +34 982 454 005
marketing@mendezrojo.com
www.mendezrojo.com

Lagar de Deuses Godello 2021 B
100% godello
88
Citrus fruit, crisp, herbal, balanced.

DO MONTERREI / D.O.P.

DO MONTERREI / D.O.P.

Lagar de Deuses Mencía 2020 T
100% mencía

87

PABLO VIDAL - VINOS CON PERSONALIDAD
Rúa do Miradoiro 8
32004 Ourense/Orense (Ourense/Orense)
☎: +34 609 152 251
pablovidal@vinosconpersonalidad.com
www.vinosconpersonalidad.com

Luxuria 2020 B
85% godello, 10% dona blanca, 5% loureiro

92

Colour: bright straw. Nose: expressive, fine lees, citrus fruit, stone fruit. Palate: full-bodied, complex, long, easy to drink, good acidity.

PAZOS DEL REY
Carrero Blanco, 33
32618 Albarellos de Monterrei (Ourense/Orense)
☎: +34 988 425 959
info@pazosdelrey.com
www.pazosdelrey.com

Pazo de Monterrey Godello 2021 B
100% godello

88

Full-bodied, jammy, tasty, smooth, herbal.

Pazo de Monterrey Mencía 2020 T
100% mencía

90

Colour: cherry, purple rim. Nose: balsamic herbs, red berry notes, scrubland. Palate: balsamic, fruity, balanced.

Pazo de Monterrey Raúl Boo Mencía 2020 T
100% mencía

92

Colour: Cherry. Nose: balsamic herbs, sweet spices, scrubland, red berry notes. Palate: spicy, balsamic, good acidity.

PRIMA VINIA
Soutelo, 3 Goián
36750 Tomiño (Pontevedra)
☎: +34 986 620 137
info@primavinia.com
www.primaviniawines.com

Pájaro Loco Godello 2021 B
godello, treixadura

89

Crisp, herbal, citrus fruit, tasty.

Pájaro Loco Mencía 2021 T
mencía, arauxa

89

Fruity, crisp, herbal, tasty.

SIETE PASOS WINES
Ctra. Nac. 122, km. 311
47300 Peñafiel (Valladolid)
☎: +34 983 881 622
administracion@bodegaspenafiel.com
www.sietepasos.com

La Casa de las Locas Godello 2021 B
godello

90

Colour: bright straw. Nose: citrus fruit, floral, stone fruit. Palate: fresh, good acidity, fine bitter notes, easy to drink.

TERRAS DO CIGARRÓN
Ctra. N-525 Albarellos de Monterrei
32618 Monterrei (Ourense/Orense)
☎: +34 988 418 703
comunicacion@martincodax.com

Mara Martin Godello 2021 B
godello

88

Pleasant, aromatic, herbal, tasty.

Mara Moura 2020 B
95% godello, 5% treixadura

89

Nose: ripe fruit, floral. Palate: flavourful, fresh, good acidity.

O con da Moura 2019 B C
90% godello, 10% treixadura

92

Colour: bright straw. Nose: ripe fruit, fragrant herbs, fine lees. Palate: full-bodied, rich, long, good acidity.

Terras do Cigarrón 2021 B
100% godello

89

Pleasant, aromatic, kind finish, floral.

VIÑA ALMIRANTE

Peroxa, 5
36658 Portas (Pontevedra)
☎: +34 620 294 293
info@vinaalmirante.com
www.vinaalmirante.com

Mar Adentro 2021 B
100% albariño

90

Colour: straw. Nose: expressive, white flowers, jasmine, dried herbs, ripe fruit. Palate: flavourful, fruity, balanced.

Paixón 2021 T

89

Fruity, herbal, balsamic herbs, crisp, tasty.

DO MONTERREI / D.O.P.

DO. MONTILLA - MORILES
CONSEJO REGULADOR

José Padillo Delgado, sn
14550 Montilla (Córdoba)
☎: +34 957 652 110 - Fax: +34 957652 407
@: consejo@montillamoriles.es
www.montillamoriles.es

LOCATION:

To the south of Córdoba. It covers all the vineyards of the municipal districts of Montilla, Moriles, Montalbán, Puente Genil, Montruque, Nueva Carteya and Doña Mencía, and part of the municipal districts of Montemayor, Fernán-Núñez, La Rambla, Santaella, Aguilar de la Frontera, Lucena, Cabra, Baena, Castro del Río and Espejo.

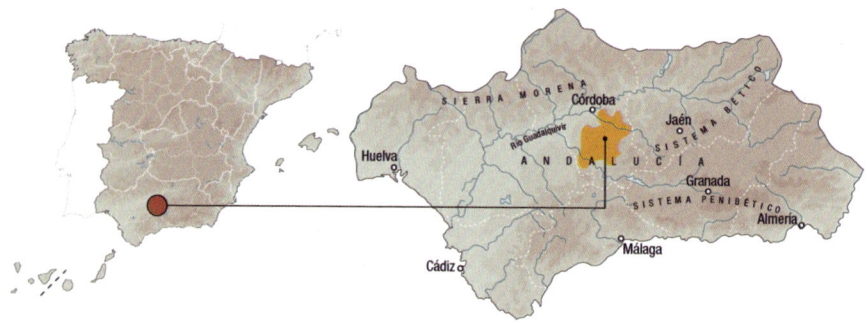

SUB-REGIONS:

We have to differentiate between the vineyards in the flatlands and those in higher areas –such as Sierra de Montilla and Moriles Alto–, prominently limestone soils of higher quality and hardly 2000 hectares planted.

GRAPE VARIETIES:

WHITE: Pedro Ximénez (main variety), Airén, Baladí, Moscatel, Torrontés, Chardonnay, Sauvignon Blanc, Macabeo and Verdejo.

RED: Tempranillo, Syrah and Cabernet Sauvignon.

FIGURES:

Vineyard surface: 4,500 – **Wine-Growers:** 1,600 – **Wineries:** 50 – **2021 Harvest rating:** N/A – **Production 21:** 20,516,600 litres – **Market percentages:** 90% National - 10% International.

SOIL:

The vineyards are situated at an altitude of between 125 m and 640 m. The soils are franc, franc-sandy and, in the higher regions, calcareous ('Albarizas'), which are precisely those of best quality, and which predominate in what is known as the Upper Sub-Region, which includes the municipal districts of Montilla, Moriles, Castro del Río, Cabra and Aguilar de la Frontera.

CLIMATE:

Semi-continental Mediterranean, with long, hot, dry summers and short winters. The average annual temperature is 16.8°C and the average rainfall is between 500 mm and 1,000 mm per year.

CLASSIFICATION OF YOUNG WINE HARVESTS PEÑÍNGUIDE

2017	2018	2019	2020	2021
N/A	GOOD	N/A	N/A	N/A

ALVEAR

María Auxiliadora, 1
14550 Montilla (Córdoba)
☎: +34 957 650 100
alvearsa@alvear.es
www.alvear.es

3 Miradas "Cerro Macho" 2019 B
93
Colour: bright yellow. Nose: characterful, expressive, neat, flor yeasts, balanced, faded flowers. Palate: balanced, fine bitter notes, complex.

3 Miradas Paraje de Río Frío Alto 2018 B
100% pedro ximénez
93
Mineral. Colour: straw, pale. Nose: citrus fruit, white fruit, ripe fruit, bakery, chalk. Palate: flavourful, saline, great length, warm.

3 Miradas Vino de Pueblo 2020 B
100% pedro ximénez
90
Colour: straw, pale. Nose: white fruit, ripe fruit, flor yeasts, dry nuts. Palate: balanced, flavourful, saline.

Alvear Palo Cortado Nº 7 BF PC
100% pedro ximénez
94
Colour: light mahogany. Nose: pungent, varnish, aged wood nuances, creamy oak, sweet spices. Palate: powerful, flavourful, spicy, long, balanced.

Alvear Pedro Ximénez 1927 BF PX D
100% pedro ximénez
93
Colour: dark mahogany. Nose: complex, dried fruit, pattiserie, toasty, tomato. Palate: sweet, rich, unctuous, powerful.

🏆 PODIUM

Alvear Pedro Ximénez de Sacristía 1998 BF PX D
97
Colour: dark mahogany. Nose: powerful, expressive, aromatic coffee, spicy, acetaldehyde, dry nuts, complex, balsamic herbs, characterful. Palate: balanced, fine solera notes, toasty, long.

Carlos VII B AM S
93
Colour: iodine, amber rim. Nose: sweet spices, dry nuts, caramel, pattiserie. Palate: dry, long, fine bitter notes, bitter.

🏆 PODIUM

Alvear Pedro Ximénez Solera 1830 BF PX D
100% pedro ximénez
100
Colour: dark mahogany. Nose: aromatic coffee, caramel, powerful, balanced. Palate: flavourful, creamy, complex, great length, long.

Catón BF OL
100% pedro ximénez
93
Colour: iodine, amber rim. Nose: dry nuts, varnish, sweet spices. Palate: rich, spicy, balanced, concentrated, flavourful.

Fino C.B. BF FI S
100% pedro ximénez
91
Colour: bright straw. Nose: white fruit, flor yeasts, chamomile, scrubland. Palate: flavourful, fine bitter notes, fleshy.

BODEGA SOPLA PONIENTE (RODRÍGUEZ DE VERA)

Pérez Galdós, 3
02003 Albacete (Albacete)
☎: +34 696 168 873
jose@rodriguezdevera.com
www.rodriguezdevera.com

Sopla Poniente Amontillado Viejísimo San Roque 30 B AM
pedro ximénez
92
Colour: light mahogany. Nose: caramel, sweet spices, bakery, dry nuts, candied fruit. Palate: complex, flavourful, good acidity, toasty.

Sopla Poniente Fino en Rama Cerro del Majuelo 20 B FI
pedro ximénez
93
Colour: bright yellow. Nose: flor yeasts, pungent, fine reductive notes, ripe fruit. Palate: flavourful, fresh, good finish.

Sopla Poniente PX Viejísimo El Coscojal 70 BF PX
pedro ximénez
93
Colour: dark mahogany. Nose: honeyed notes, dried fruit, dried herbs, creamy oak. Palate: flavourful, great length, unctuous, sweet.

DO MONTILLA - MORILES / D.O.P.

DO MONTILLA - MORILES / D.O.P.

BODEGAS CRUZ CONDE
Ronda Canillo, 4
14550 Montilla (Córdoba)
☎: +34 957 651 250
Fax: +34 957 653 619
info@bodegascruzconde.es
www.bodegascruzconde.es

Cruz Conde 1902 BF PX D
pedro ximénez
93 ♣
Colour: dark mahogany. Nose: complex, fruit liqueur notes, dried fruit, pattiserie, toasty. Palate: sweet, rich, unctuous, powerful.

Cruz Conde B FI S
100% pedro ximénez
90
Colour: bright yellow. Nose: balanced, fresh, expressive, pungent. Palate: flavourful, fine bitter notes, long.

Cruz Conde BF PX D
100% pedro ximénez
91
Colour: mahogany. Nose: complex, fruit liqueur notes, dried fruit, pattiserie, toasty. Palate: sweet, rich, unctuous.

Donceles Cruz Conde B FI
100% pedro ximénez
89
Kind finish, thin, tasty, saline.

La Tercia B PX D
100% pedro ximénez
88
Spicy, tasty, powerful, sweet.

Soñador Cruz Conde B FI S
100% pedro ximénez
88
Pleasant, spicy, floral, tasty, saline.

BODEGAS DELGADO
Cosano, 2
14500 Puente Genil (Córdoba)
☎: +34 957 600 085
fino@bodegasdelgado.com
www.bodegasdelgado.com

Amon BF AM S
pedro ximénez
94
Colour: iodine, amber rim. Nose: elegant, sweet spices, acetaldehyde, dry nuts. Palate: dry, spicy, fine bitter notes, complex.

🏆 **PODIUM**

Delgado 1874 B AM S
pedro ximénez
95
Colour: iodine, amber rim. Nose: candied fruit, fruit liqueur notes, spicy, varnish, complex, characterful, powerful. Palate: fine solera notes, bitter.

Delgado 1874 B OL S
pedro ximénez
93
Colour: iodine, amber rim. Nose: complex, dry nuts, toasty, aged wood nuances, candied fruit. Palate: rich, long, fine solera notes, spicy.

🏆 **PODIUM**

Delgado 1874 BF PX D
pedro ximénez
95
Colour: dark mahogany. Nose: powerful, expressive, aromatic coffee, spicy, acetaldehyde, dry nuts. Palate: balanced, fine solera notes, toasty, long.

Segunda Bota B FI
pedro ximénez
91
Colour: bright yellow. Nose: fresh, expressive, flor yeasts, pungent. Palate: flavourful, fine bitter notes, long.

BODEGAS MÁLAGA VIRGEN
Autovía A-92, Km. 132
29520 Fuente de Piedra (Málaga)
☎: +34 952 319 454
info@bodegasmalagavirgen.com
www.bodegasmalagavirgen.com

Lagar de Benavides B FI S
pedro ximénez
91
Colour: bright yellow. Nose: balanced, fresh, expressive, flor yeasts. Palate: flavourful, fine bitter notes, long.

CÍA. VINÍCOLA DEL SUR - TOMÁS GARCÍA
Avda. Luis de Góngora y Argote, s/n
14550 Montilla (Córdoba)
☎: +34 957 650 162
Fax: +34 957 650 290
info@vinicoladelsur.com
www.vinicoladelsur.com

Verbenera B FI S
pedro ximénez
87

EQUIPO NAVAZOS
11403 Jerez de la Frontera (Cádiz)
equipo@navazos.com
www.equiponavazos.com

Ovni Pedro Ximenez 2021 B
pedro ximénez
91
Colour: straw. Nose: expressive, white flowers, dried herbs, flor yeasts. Palate: flavourful, balanced, complex.

GRACIA
Avda. Marqués de la Vega de Armijo, 103
14550 Montilla (Córdoba)
☎: +34 957 650 162
Fax: +34 957 652 335
info@bodegasgracia.com
www.bodegasgracia.com

Fino Corredera B FI S
pedro ximénez
87

Gracia Pedro Ximénez Dulce Viejo BF PX D
pedro ximénez
90
Colour: dark mahogany. Nose: spicy, acetaldehyde, dry nuts. Palate: balanced, fine solera notes, toasty.

🏆 PODIUM

Montearruit Viejísimo 75 Años Bota 1/1 B AM
pedro ximénez
96
Colour: dark mahogany. Nose: candied fruit, spicy, varnish, sweet spices, dry nuts. Palate: fine solera notes, bitter, flavourful.

Solera Fina María del Valle en Rama B FI S
91
Colour: bright yellow. Nose: balanced, expressive, pungent, scrubland. Palate: flavourful, fine bitter notes, long, full-bodied.

Tauromaquia Amontillado Viejísimo B AM S
pedro ximénez
92
Colour: iodine, amber rim. Nose: sweet spices, acetaldehyde, dry nuts, dried herbs. Palate: dry, spicy, fine bitter notes.

Solera Fina María del Valle B FI S
pedro ximénez
89
Powerful, herbal, jammy, full-bodied, flat.

Tauromaquia Oloroso Viejísimo B OL S
pedro ximénez
91
Colour: iodine, amber rim. Nose: dry nuts, varnish, sweet spices. Palate: spicy, light-bodied, balanced, fine bitter notes.

Tauromaquia Pedro Ximénez Superior BF PX D
pedro ximénez
92
Colour: dark mahogany. Nose: spicy, acetaldehyde, dry nuts, honeyed notes. Palate: balanced, fine solera notes, toasty, long.

Viñaverde 2021 B SS
pedro ximénez
87

LAGAR DE LA SALUD
Ctra. Córdoba_Málaga, km. 41
14550 Montilla (Córdoba)
☎: +34 659 467 525
info@lagardelasalud.com
www.lagardelasalud.com

Dulas del Lagar de la Salud 2020 B FB
pedro ximénez
91
Colour: yellow, straw. Nose: white fruit, ripe fruit, characterful, powerful, spicy, citrus fruit. Palate: rich, flavourful, long.

PÉREZ BARQUERO
Avda. Andalucía, 27
14550 Montilla (Córdoba)
☎: +34 957 650 500
Fax: +34 957 650 208
info@perezbarquero.com
www.perezbarquero.com

🏆 PODIUM

1955 Amontillado Solera Cincuentenario B AM S
pedro ximénez
96
Colour: iodine, amber rim. Nose: powerful, complex, elegant, dry nuts, toasty. Palate: bitter, fine solera notes, long, spicy.

DO MONTILLA - MORILES / D.O.P.

DO MONTILLA - MORILES / D.O.P.

🏆 PODIUM

1955 Oloroso Solera Cincuentenario B OL S
pedro ximénez

96

Colour: light mahogany. Nose: powerful, complex, dry nuts, toasty, acetaldehyde. Palate: long, fine solera notes, spicy, round.

🏆 PODIUM

1955 Palo Cortado Cincuentenario B PC S
pedro ximénez

96

Colour: light mahogany. Nose: acetaldehyde, pungent, varnish, aged wood nuances, creamy oak. Palate: spicy, long, balanced, flavourful.

🏆 PODIUM

1955 Pedro Ximenez Solera Cincuentenario BF PX D
pedro ximénez

99

Colour: dark mahogany. Nose: powerful, expressive, spicy, acetaldehyde, dry nuts, tomato. Palate: balanced, elegant, fine solera notes, toasty, long.

Fino Los Amigos B FI S
pedro ximénez

87

Fresquito Vino de Tinaja 2020 B
pedro ximénez

90

Colour: straw. Nose: white fruit, flor yeasts, dry nuts, dried flowers, scrubland. Palate: flavourful, balanced, fresh.

Gran Barquero B AM S
pedro ximénez

93

Colour: iodine, amber rim. Nose: sweet spices, acetaldehyde, dry nuts, varnish. Palate: full-bodied, dry, spicy, fine bitter notes.

Gran Barquero B FI S
pedro ximénez

91

Colour: bright yellow. Nose: balanced, fresh, scrubland, flor yeasts. Palate: flavourful, fine bitter notes, balanced.

Gran Barquero B OL S
pedro ximénez

93

Colour: iodine, amber rim. Nose: powerful, dry nuts, varnish, sweet spices. Palate: spicy, flavourful, balanced.

Gran Barquero B PC
pedro ximénez

93

Colour: light mahogany. Nose: acetaldehyde, aged wood nuances, creamy oak. Palate: flavourful, spicy, long, balanced.

Gran Barquero BF PX D
pedro ximénez

93

Colour: dark mahogany. Nose: powerful, expressive, spicy, acetaldehyde, dry nuts, ebb tide. Palate: balanced, toasty, long, unctuous.

Gran Barquero en Rama B FI S

93

Colour: bright yellow. Nose: balanced, fresh, expressive, pungent, scrubland, balsamic herbs. Palate: flavourful, fine bitter notes, long, mineral.

🏆 PODIUM

La Cañada BF PX D
pedro ximénez

96

Colour: dark mahogany. Nose: powerful, expressive, spicy, acetaldehyde, dry nuts. Palate: balanced, elegant, fine solera notes, toasty, long.

Viña Amalia 2021 B SS
pedro ximénez

86

TORO ALBALÁ

Avda. Antonio Sánchez, 1
14920 Aguilar de La Frontera (Córdoba)
☎: +34 957 660 046
Fax: +34 957 661 494
oficina@toroalbala.com
www.toroalbala.com

Don P.X. 1999 BF PX D
pedro ximénez

93

Colour: dark mahogany. Nose: powerful, spicy, dried fruit, dark chocolate. Palate: balanced, toasty, long, sweet.

Don P.X. 2019 BF PX
pedro ximénez

91

Creamy, representative. Colour: light mahogany. Nose: fruit liqueur notes, dried fruit, pattiserie, toasty. Palate: sweet, unctuous, powerful.

Miut El Jabonero 2020 B FB
pedro ximénez

90

Oxidativ, tasty. Colour: yellow. Nose: ripe fruit, spicy, faded flowers. Palate: good structure, long, fine bitter notes.

Poley Amontillado en Rama BF AM
pedro ximénez

93

Colour: bright golden. Nose: acetaldehyde, caramel, sweet spices, neat. Palate: flavourful, dry, long, fine solera notes.

Poley BF CRM
pedro ximénez

92

Colour: dark mahogany. Nose: powerful, dry nuts, creamy oak, varnish, sweet spices. Palate: long, spicy, fleshy.

Poley Fino del Lagar en Rama B FI S
pedro ximénez

92

Colour: bright yellow. Nose: balanced, expressive, pungent, white fruit, dry nuts. Palate: flavourful, fine bitter notes, long.

Poley Fino Pasado en Rama B FI
pedro ximénez

94

Oxidativ, representative. Colour: golden. Nose: flor yeasts, lees reduction notes, pungent, faded flowers, dry nuts. Palate: good acidity, bitter, spicy, long.

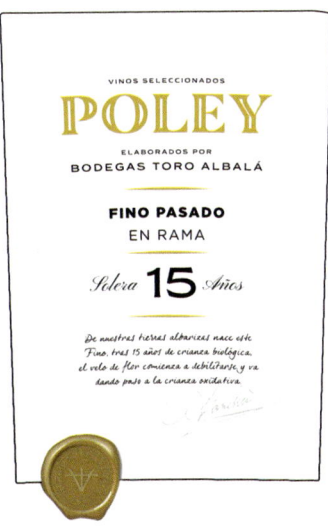

Poley Oloroso en Rama BF OL
pedro ximénez

92

Colour: bright golden. Nose: dry nuts, caramel, sweet spices. Palate: balanced, spicy, dry wood, long.

Poley Palo Cortado en Rama BF PC
pedro ximénez

93

Representative, tasty. Colour: light mahogany. Nose: balanced, expressive, dry nuts, neat, complex. Palate: fine solera notes, old wood.

DO. MONTSANT

CONSEJO REGULADOR

Plaça de la Quartera, 6
43730 Falset (Tarragona)
☎: +34 977 831 742 - Fax: +34 977 830 676
@: info@domontsant.com
www.domontsant.com

LOCATION:

In the region of Priorat (Tarragona). It is made up of Baix Priorat, part of Alt Priorat and various municipal districts of Ribera d'Ebre that were already integrated into the Falset sub-region. In total, 16 municipal districts: La Bisbal de Falset, Cabaces, Capçanes, Cornudella de Montsant, La Figuera, Els Guiamets, Marçá, Margalef, El Masroig, Pradell, La Torre de Fontaubella, Ulldemolins, Falset, El Molar, Darmós and La Serra d'Almos. The vineyards are located at widely variable altitudes, ranging between 200 m to 700 m above sea level.

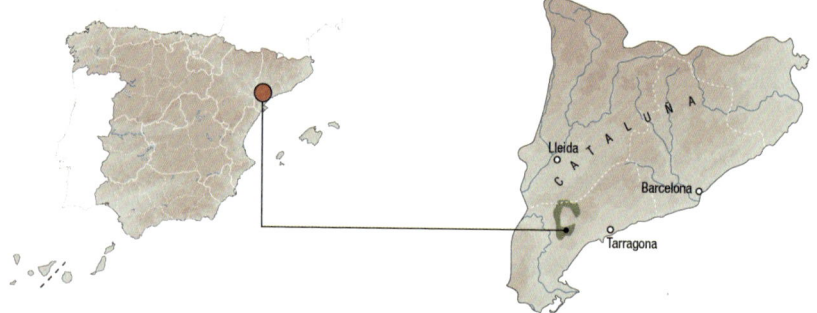

GRAPE VARIETIES:

WHITE: Chardonnay, Garnacha Blanca, Macabeo, Moscatel, Pansal, Parellada.

RED: Cabernet Sauvignon, Cariñena, Garnacha Tinta, Garnacha Peluda, Merlot, Monastrell, Picapoll, Syrah, Tempranillo and Mazuela.

FIGURES:

Vineyard surface: 1,739 – **Wine-Growers:** 551 – **Wineries:** 57 – **2021 Harvest rating:** N/A – **Production 21:** 6,583,485 litres – **Market percentages:** 74% National - 26% International.

SOIL:

There are mainly three types of soil: compact calcareous soils with pebbles on the borders of the DO; granite sands in Falset; and siliceous slate (the same stony slaty soil as Priorat) in certain areas of Falset and Cornudella.

CLIMATE:

Although the vineyards are located in a Mediterranean region, the mountains that surround the region isolate it from the sea to a certain extent, resulting in a somewhat more Continental climate. Due to this, it benefits from the contrasts in day/night temperatures, which is an important factor in the ripening of the grapes. However, it also receives the sea winds, laden with humidity, which help to compensate for the lack of rainfall in the summer. The average rainfall is between 500 and 600 mm per year.

CLASSIFICATION OF YOUNG WINE HARVESTS PEÑÍNGUIDE

2017	2018	2019	2020	2021
GOOD	VERY GOOD	VERY GOOD	VERY GOOD	VERY GOOD

7 MAGNÍFICS
Miquel Torres i Carbó, 6
08720 Vilafranca del Penedès (Barcelona)
☎: +34 938 177 400
7magnifics@7magnifics.com
www.7magnifics.com

El Senat de Montsant 2020 T
garnacha, cariñena, syrah

91

Colour: cherry, garnet rim. Nose: fruit preserve, fruit liqueur notes, powerful, tobacco. Palate: flavourful, sweetness, long.

ACÚSTIC CELLER
Ctra. TV-3002 de Capçanes a Marçà km 3,336
43775 Marçà (Tarragona)
☎: +34 672 432 691
acustic@acusticceller.com
www.acusticceller.com

Acústic Blanc 2019 B
85% garnacha blanca, 10% macabeo, 3% pansal, 2% garnacha roja

91 🌿

Colour: bright yellow. Nose: creamy oak, ripe fruit, spicy, wild herbs, woody. Palate: good structure, toasty, fine bitter notes.

Acústic Blanc 2021 B
93% garnacha blanca, 5% macabeo, 1% pansal, 1% garnacha roja

92 🌿

Colour: straw. Nose: scrubland, white fruit, citrus fruit, dried flowers. Palate: ripe fruit, balanced, flavourful, good finish.

Acústic Negre 2019 T RB
70% cariñena, 30% garnacha

92

Colour: bright cherry. Nose: fresh fruit, creamy oak, red berry notes, scrubland, earthy notes. Palate: good acidity, spicy, round tannins.

Acústic Negre 2020 T RB
70% cariñena, 30% garnacha

92

Colour: cherry, purple rim. Nose: red berry notes, floral, spicy. Palate: flavourful, fruity, good acidity.

Acústic Rosat 2021 RD
100% garnacha

89 🌿

Smooth, fruity, crisp, dried flowers, wild, varietally correct, pleasant.

Braó 2019 T C
90% cariñena, 10% garnacha

93 🌿

Colour: Cherry. Nose: scrubland, wild herbs, spicy, creamy oak. Palate: spicy, flavourful, round tannins, long.

Auditori 2019 T C
100% garnacha

94 🌿

Colour: Cherry. Nose: complex, expressive, spicy, earthy notes, dried herbs, balanced. Palate: full-bodied, long, great length.

Auditori Blanc 2021 B
76% garnacha blanca, 19% xarel.lo, 5% macabeo

92 🌿

Colour: bright straw. Nose: fruit expression, white fruit, balanced, neat. Palate: flavourful, good acidity, fruity aftertaste.

Braó 2018 T C
90% cariñena, 10% garnacha

93 🌿

Colour: Cherry. Nose: expressive, spicy, mineral, ripe fruit, dried herbs, wild herbs. Palate: full-bodied, long, great length.

AGRÍCOLA DE ULLDEMOLINS
43363 Ulldemolins (Tarragona)
☎: +34 977 561 640
agrobotiga@coopulldemolins.com
www.coopulldemolins.com

L'Aniceto 2019 T C
100% garnacha

89

Pruney, full-bodied, jammy, spicy, dried herbs, tasty, great length, woody.

Les Pedrenyeres 2020 B
garnacha blanca, macabeo

88

Standard, jammy, fruity, toasty, spicy, balanced.

Lo Carnisser 2020 B
garnacha blanca

89

Aromatic, dried flowers, dried herbs, tasty, balanced, jammy. Nose: fine lees.

Ssssshhhhhh Ulldemolins 2021 T
garnacha

88

Standard, dried flowers, jammy, herbal, fruity, smooth.

DO MONTSANT / D.O.P.

DO MONTSANT / D.O.P.

ALFREDO ARRIBAS
Sort dels Capellans, 23-25
43730 Falset (Tarragona)
☎: +34 932 531 760
info@portaldelpriorat.com
www.portaldelpriorat.com

Gotes del Montsant 2019 T
garnacha, cariñena
91
Colour: cherry, purple rim. Nose: spicy, ripe fruit, black fruit. Palate: flavourful, good acidity, long.

Tros Blanc Notaria 2016 B FB
garnacha blanca
94 🌿
Oxidativ, reductive, age nuances. Colour: golden. Nose: ripe fruit, dried herbs, dried flowers, fine lees, spicy. Palate: flavourful, long, great length.

Tros Blanc Saleres 2016 B
garnacha blanca
93 🌿
Colour: bright yellow. Nose: expressive, ripe fruit, floral, fine lees, mineral, burnt matches. Palate: full-bodied, complex, spicy, long.

Tros Negre Notaria 2017 T
garnacha
92 🌿
Reductive. Colour: dark-red cherry. Nose: fine reductive notes, wild herbs, dried herbs, balanced. Palate: flavourful, spicy, varietal, good acidity.

Trossos Sants 2019 B
garnacha blanca
92
Colour: bright yellow. Nose: powerful, creamy oak, ripe fruit, spicy, dried flowers, burnt matches. Palate: rich, long, bitter.

Trossos Vells 2019 T BA
samsó
91
Slight reduction. Colour: deep cherry. Nose: ripe fruit, black fruit, toasty, black pepper. Palate: spicy, powerful, pruney.

ANGUERA DOMENECH
43746 Darmos (Tarragona)
☎: +34 654 382 633
angueradomenech@gmail.com
www.vianguera.com

L'Únic 2020 T
monastrell
88
Hot, slight reduction, jammy, toasty.

Reclot 2021 T
cariñena, garnacha, merlot, syrah, tempranillo
89
Jammy, fruity, lactic, herbal, tasty.

Reclot Rosat 2021 RD
garnacha
87

Vinya Gasó 2017 T C
garnacha, samsó
90
Colour: cherry, garnet rim. Nose: fruit preserve, powerful, fine reductive notes, tobacco, dried herbs. Palate: flavourful, long.

ATAVUS PRIORAT
Ctra. 710, Km. 8,3
43737 Gratallops (Tarragona)
☎: +34 977 020 350
info@atavusvines.com
www.atavusvines.com

BK! 2020 B
60% garnacha blanca, 40% macabeo
91
Colour: bright straw, greenish rim. Nose: fresh fruit, citrus fruit, wild herbs, hint of anise, fine lees. Palate: fresh, fruity, good acidity, fine bitter notes.

BK! 2020 T BA
40% cariñena, 30% garnacha peluda, 30% garnacha
90
Colour: cherry, purple rim. Nose: red berry notes, floral, spicy. Palate: flavourful, fruity, good acidity.

CARA NORD CELLER
Plaça Sant Sebastià, 13
25457 El Vilosell (Lleida/Lérida)
☎: +34 973 176 029
hola@caranordceller.com
www.caranordceller.com

Mineral 2020 T C
90
Colour: cherry, purple rim. Nose: black fruit, ripe fruit, black pepper, spicy, dried herbs. Palate: fruity, fresh, good structure, flavourful, balanced, slightly dry, soft tannins.

CASTELL D'OR
Ctra. de Santes Creus,s/n
43814 Vila-Rodona (Tarragona)
☎: +34 977 459 860
castelldor@castelldor.com
www.castelldor.com

Templer 2019 T C
garnacha, cariñena

88

Balsamic herbs, jammy, dried herbs, smooth, wild.

Templer 2021 T
garnacha, cariñena

87

CELLER CAL BESSÓ
Sant Lluis, 16
43777 Els Guiamets (Tarragona)
☎: +34 630 941 959
roquers@roquers.com
www.calbesso.com

Coret 2019 B
garnacha blanca, macabeo, picapoll blanc

90 🍃

Colour: bright straw. Nose: ripe fruit, fragrant herbs, fine lees, spicy. Palate: full-bodied, rich, long, good acidity.

Elvira 2019 B
garnacha blanca

92 🍃

Colour: bright yellow. Nose: powerful, ripe fruit, spicy, toasted bread, fine lees. Palate: rich, good structure, long, fine bitter notes.

La Prunera 2016 T
60% garnacha peluda, 40% cariñena

90 🍃

Colour: deep cherry. Nose: ripe fruit, dried herbs, creamy oak. Palate: powerful, ripe fruit, spicy, round tannins.

CELLER CEDÓ ANGUERA
Ctra. La Serra d'Almos-Darmós, Km. 0,2
43746 La Serra D'Almos (Tarragona)
☎: +34 699 694 728
celler@cedoanguera.com
www.cedoanguera.com

Anexe 2021 T
45% cariñena, 45% garnacha, 10% syrah

87

Anexe Syrah 2020 T
100% syrah

88

Jammy, pruney, slight reduction, tasty.

Anexe Vinyes Velles de Carinyena 2020 T
100% cariñena

88

Roasted coffee, powerful, jammy, spicy.

Clònic 2019 T
60% cariñena, 30% garnacha, 10% syrah

89

Defined aromas, balanced, balsamic herbs, herbal, jammy, tasty.

Clònic Carinyena Vinyas Viejas 2018 T C
100% cariñena

91

Colour: Cherry. Nose: scrubland, earthy notes, spicy, dried herbs, wild herbs. Palate: balsamic, flavourful, round tannins, balanced.

CELLER DE CAPÇANES
Llebaria, 9
43776 Capçanes (Tarragona)
☎: +34 977 178 319
cellercapcanes@cellercapcanes.com
www.cellercapcanes.com

Cabrida 2020 T C
garnacha

93

With personality, reductive. Colour: cherry, garnet rim. Nose: expressive, chamomile, dried flowers, ripe fruit, red berry notes. Palate: flavourful, balanced, long, great length.

Costers del Gravet 2020 T C
garnacha, cabernet sauvignon, cariñena

91

Colour: deep cherry. Nose: ripe fruit, dried herbs, smoky, spicy. Palate: powerful, ripe fruit, good structure, slightly dry, soft tannins.

Els Pájaros 2020 T
100% cariñena

91

Colour: deep cherry. Nose: ripe fruit, dried herbs, wild herbs. Palate: powerful, ripe fruit, spicy, round tannins.

La Nit de Les Garnatxes Limestone 2021 T
100% garnacha

90

Colour: deep cherry. Nose: ripe fruit, dried herbs, dried flowers. Palate: ripe fruit, spicy, fruity, easy to drink.

Mas Tortó 2020 T C
garnacha, cariñena, cabernet sauvignon

90 🍃

Colour: deep cherry. Nose: ripe fruit, dried herbs, earthy notes. Palate: ripe fruit, spicy, round tannins.

DO MONTSANT / D.O.P.

DO MONTSANT / D.O.P.

Mas Tortó Blanc 2021 B
garnacha blanca

91

Colour: golden, pale. Nose: macerated fruit, dried flowers, faded flowers, powerful. Palate: flavourful, balanced, fine bitter notes, good structure.

CELLER DE L'ERA
Mas de las Moreras s/n
43360 Cornudella de Montsant (Tarragona)
☎: +34 977 262 031
jtorres@cellerdelera.com
www.cellerdelera.com

Bri Negre 2015 T R
garnacha, syrah, cariñena

90

Colour: deep cherry. Nose: ripe fruit, dried herbs, toasty, black pepper. Palate: ripe fruit, spicy, round tannins, smoky aftertaste.

Era 2019 T C
cariñena

89

Varietally correct, wild, jammy, tasty, balsamic herbs, slight reduction, spicy, toasty.

MIM 2013 T R
garnacha, cariñena

91

Colour: deep cherry. Nose: ripe fruit, fruit preserve, tobacco, sweet spices, old leather. Palate: spicy, round tannins, long, fruity, full-bodied.

CELLER GRITELLES
Carrer Sant Andreu, 5
43360 Cornudella de Montsant (Tarragona)
☎: +34 637 407 184
celler@gritelles.com
www.gritelles.com

Gritelles Manou 2020 T
garnacha, cariñena

90

Colour: cherry, garnet rim. Nose: dried herbs, black fruit, scrubland, spicy. Palate: powerful, ripe fruit, round tannins.

Gritelles Siurana Brisat 2020 B BA
macabeo

90

Colour: yellow. Nose: animal reductive notes, balanced, expressive, characterful, faded flowers. Palate: flavourful, good acidity, fine bitter notes.

Gritelles Siurana Negre 2018 T BA
60% garnacha, 40% cariñena

92

Colour: deep cherry. Nose: ripe fruit, dried herbs, creamy oak, mineral. Palate: powerful, ripe fruit, spicy, round tannins.

Gritelles Vinyes Velles Carinyena 2019 T C
cariñena

93

Colour: light cherry. Nose: ripe fruit, dried herbs, varietal, balanced, wild herbs. Palate: ripe fruit, spicy, round tannins, good acidity.

Gritelles Vinyes Velles Garnatxa 2019 T C
garnacha

92

Colour: deep cherry. Nose: ripe fruit, dried herbs, creamy oak, fruit preserve. Palate: ripe fruit, spicy, good acidity, balanced.

CELLER MASROIG
Passeig de L'Arbre, 3
43736 El Masroig (Tarragona)
☎: +34 977 825 026
Fax: +34 977 825 315
celler@cellermasroig.com
www.cellermasroig.com

Les Sorts Blanc 2021 B FB
garnacha blanca

88

Citrus fruit, balanced, crisp, herbal, yeasty notes.

Les Sorts Jove 2021 T
garnacha, mazuelo, syrah

91

Colour: cherry, purple rim. Nose: fruit expression, red berry notes, floral, lactic notes. Palate: fruity, good acidity, easy to drink.

Les Sorts Sycar 2019 T
syrah, mazuelo

92

Colour: deep cherry. Nose: ripe fruit, dried herbs, creamy oak. Palate: powerful, ripe fruit, spicy, round tannins.

Les Sorts Vinyes Velles 2018 T C
garnacha, cariñena

92

Colour: very deep cherry. Nose: ripe fruit, dried herbs, creamy oak, black fruit. Palate: powerful, ripe fruit, spicy, round tannins.

Pinyeres Blanc 2021 B
garnacha blanca

88

Citrus fruit, crisp, herbal, tasty.

Pinyeres Negre 2019 T
garnacha, mazuelo

89

Hot, toasty, jammy, spicy.

CELLER PASCONA

Camí dels Fontals, s/n
43730 Falset (Tarragona)
☎: +34 609 291 770
info@pascona.com
www.pascona.com

La Germana de Pascona 2021 B
moscatel grano menudo

89

Citrus fruit, tasty, herbal, crisp.

La Mare de Pascona 2019 T
garnacha

91

Slight reduction. Colour: very deep cherry. Nose: expressive, spicy, mineral, ripe fruit, black fruit, toasty. Palate: full-bodied, long, great length.

Lo Noi del Sac de Pascona 2021 B
merlot

90

Colour: bright straw, greenish rim. Nose: fresh fruit, citrus fruit, wild herbs. Palate: fresh, fruity, good acidity, fine bitter notes.

Maria Ganxa de Pascona 2021 T
cariñena

89

Pleasant, tasty, jammy, fruity.

Trencaclosques de Pascona 2021 RD
87

CELLER RONADELLES

Finca la Plana, s/n
43360 Cornudella del Montsant (Tarragona)
☎: +34 977 821 104
capderuc@ronadelles.com
www.ronadelles.com

Flor Mediterránea 2020 T
garnacha, cariñena

87

CELLERS CAN BLAU

Pl. pol.13 parc. 21 – Ctra. T 734 Kmt. 9
El Molar-Masroig
43736 El Molar (Tarragona)
☎: +34 968 435 022
info@gilfamily.es
www.gilfamily.es

Blau 2020 T
50% cariñena, 25% syrah, 25% garnacha

90

Colour: bright cherry. Nose: ripe fruit, balanced, toasty. Palate: flavourful, fruity, good acidity.

Blauverd 2021 T
garnacha, cariñena

90 🌿

Colour: deep cherry. Nose: sweet spices, scrubland, red berry notes, grassy. Palate: spicy, balsamic, good acidity.

Can Blau 2020 T
40% cariñena, 40% syrah, 20% garnacha

91

Colour: dark-red cherry. Nose: toasty, spicy, cocoa bean, ripe fruit, black fruit. Palate: flavourful, toasty, fine bitter notes.

CELLERS SANT RAFEL

Ctra. La Torre, Km. 1,7
43774 Pradell de la Teixeta (Tarragona)
☎: +34 689 792 305
xavi@cellerssantrafel.com
www.cellerssantrafel.com

Blanca 2021 B
garnacha blanca

86

DO MONTSANT / D.O.P.

Joana 2020 T
garnacha, merlot
88
Kind finish, smooth, tasty, jammy.

Solpost Carinyena 2017 T
cariñena
92
Varietally correct, wild. Colour: cherry, garnet rim. Nose: black fruit, ripe fruit, scrubland, characterful, spicy. Palate: flavourful, sweet tannins, spicy.

Solpost Garnatxa 2016 T C
garnacha
92
Colour: cherry, garnet rim. Nose: ripe fruit, scrubland, dried flowers, balanced, wild herbs. Palate: flavourful, spicy, varietal, easy to drink.

Solpost Origen 2018 T
garnacha, cariñena, cabernet sauvignon
91
Colour: cherry, garnet rim. Nose: powerful, black fruit, fruit preserve, scrubland, dried herbs. Palate: flavourful, long, round tannins.

Xavi 2018 T
garnacha, cariñena
89
Pruney, spicy, toasty, tasty, full-bodied, standard.

CELLERS TERRA I VINS
Av. Falset, 17 Baixos
43206 Reus (Tarragona)
☎: +34 658 567 409
celler@cellersterraivins.com
www.cellersterraivins.com

Bona Nit 2019 T
garnacha, samsó, syrah
88
Spicy, fruity, herbal, jammy, tasty.

Clos del Gos 2020 T
88
Herbal, fruity, spicy, smoky, tasty.

CELLERS UNIÓ
Joan Oliver 16-24
43206 Reus (Tarragona)
☎: +34 977 330 055
Fax: +34 977 330 070
info@cellersunio.com
www.cellersunio.com

Mas dels Mets T
87

Perlat 2020 T RB
garnacha, mazuelo, syrah
89
Jammy, tasty, wild, toasty, pruney, woody.

Perlat Blanc 2021 B
garnacha blanca, macabeo
88
Smoky, toasty, simple, jammy, woody.

Perlat Garnatxa 2020 T BA
garnacha
89
Pruney, powerful, tasty, wild, great length.

Perlat Syrah 2020 T RB
syrah
89
Full-bodied, tasty, wild, jammy, fruity.

CÍA. VITÍCOLA SILEO
Sant Andreu, 17
43360 Cornudella de Montsant (Tarragona)
☎: +34 935 165 043
info@viticolasileo.eu
www.viticolasileo.eu

Costers de Cornudella 2019 T R
garnacha
92
Colour: bright cherry. Nose: red berry notes, ripe fruit, neat, expressive, faded flowers, varietal, mineral. Palate: flavourful, spicy, easy to drink.

Esporreres 2019 T
garnacha
93
Colour: light cherry. Nose: balsamic herbs, scrubland, dried flowers, neat, varietal. Palate: spicy, balsamic, good acidity.

Sileo 2020 T
95% garnacha, 5% cariñena
90
Colour: Cherry. Nose: balsamic herbs, wild herbs, dried flowers. Palate: spicy, balsamic, fruity, varietal.

CINGLES BLAUS
Mas de les Moreres – Afores Cornudella
43360 Cornudella de Montsant (Tarragona)
☎: +34 977 310 382
info@cinglesblaus.com
www.cinglesblaus.com

Cingles Blaus Mas de les Moreres 2019 T
garnacha, cariñena, cabernet sauvignon
88
Pruney, fruity, spicy, jammy.

DO MONTSANT / D.O.P.

Cingles Blaus Mas de les Moreres 2020 B
garnacha blanca, garnacha roja, macabeo

90

Colour: straw. Nose: ripe fruit, dried herbs, faded flowers, fine lees. Palate: powerful, ripe fruit, balanced, mineral.

Cingles Blaus Octubre 2020 T
garnacha, cariñena

89

Fruity, herbal, jammy, tasty, balanced.

Cingles Blaus Octubre 2021 B
garnacha blanca, macabeo, chardonnay

88

Citrus fruit, balanced, spicy, herbal, tasty.

Cingles Blaus Selecció 2019 T C

92

Colour: cherry, purple rim. Nose: red berry notes, floral, spicy, dried herbs, fragrant herbs. Palate: flavourful, fruity, good acidity, long.

CLOS FIGUERAS

Carrer La Font, 38
43737 Gratallops (Tarragona)
☎: +34 977 830 217
info@closfigueras.com
www.closfigueras.info

Poblets del Montsant 2020 T
garnacha, cariñena

88

Austere, jammy, herbal, spicy.

Poblets del Montsant 2021 B
garnacha, macabeo

85

CLOS MOGADOR

Camí Manyetes, s/n
43737 Gratallops (Tarragona)
☎: +34 977 839 171
Fax: +34 977 839 426
closmogador@closmogador.com
www.closmogador.com

Com Tu 2019 T C
100% garnacha

92

Fruity, jammy. Colour: cherry, garnet rim. Nose: ripe fruit, neat, dried flowers. Palate: flavourful, fruity, easy to drink, balanced.

COCA I FITÓ

Avda. Onze de Setembre s/n
43736 El Masroig (Tarragona)
☎: +34 619 776 948
info@cocaifito.cat
www.cocaifito.cat

Coca i Fitó Carinyena 2015 T
100% cariñena

89

Fruity, toasty, pruney, tasty, age nuances.

Coca i Fitó Maragda 2016 T
55% garnacha, 25% cariñena, 20% syrah

89

Spicy, fruity, toasty, pruney, tasty.

Coca i Fitó Natura 2021 T
60% garnacha, 35% syrah, 5% cabernet sauvignon

90 ⚘

Colour: cherry, garnet rim. Nose: fruit preserve, fruit liqueur notes, powerful, toasty, dark chocolate. Palate: flavourful, sweetness, long.

Coca i Fitó Nu 2020 T
100% garnacha

88

Reduced, pruney, full-bodied, tasty, spicy.

Coca i Fitó Rosa 2020 RD
100% syrah

90

Colour: bright cherry. Nose: red berry notes, floral, expressive, ripe fruit. Palate: fruity, flavourful, easy to drink.

Jaspi Negre 2018 T
45% garnacha, 25% cariñena, 15% cabernet sauvignon, 15% syrah

90

Colour: Cherry. Nose: scrubland, balanced, neat. Palate: balsamic, good acidity, ripe fruit, long.

COOPERATIVA FALSET MARÇA

Miquel Barceló, 31
43730 Falset (Tarragona)
☎: +34 977 830 105
info@etim.cat
www.etim.cat

Castell de Falset 2014 T R
garnacha, cariñena, cabernet sauvignon

88

Toasty, smoky, jammy, herbal.

DO MONTSANT / D.O.P.

Ètim Dolça Carinyena 2020 TF D
90
Sweety finish, tasty. Colour: very deep cherry. Nose: black fruit, overripe fruit, dried herbs, hot. Palate: flavourful, sweet.

Ètim El Viatge 2021 T
garnacha, cariñena, syrah
88
Fruity, herbaceous, crisp, bitter.

Ètim L'Antull 2021 B
garnacha blanca, macabeo
87

Ètim La Pausa 2021 RD
garnacha, syrah
87

Ètim Sinestèsics Tardana Negra 2018 TF D
garnacha
91
Colour: cherry, garnet rim. Nose: overripe fruit, dried fruit, characterful, hot. Palate: powerful, flavourful, unctuous.

EDICIONES ILIMITADAS
Carrer Modolell, 56 Local A
08021 Barcelona (Barcelona)
☎: +34 932 531 760
info@edicionesi-limitadas.com

Luno 2020 T
garnacha, cariñena, syrah, cabernet sauvignon, merlot
89
Pruney, powerful, toasty, spicy. Palate: slightly dry, soft tannins.

Luno Blanc 2020 B
garnacha blanca
90
Colour: bright yellow. Nose: white fruit, ripe fruit, dried flowers, faded flowers. Palate: flavourful, balanced, fine bitter notes.

Pell de Gerres 2020 B
93
Dried flowers, little interventionist. Colour: bright straw. Nose: ripe fruit, fragrant herbs, fine lees. Palate: full-bodied, rich, long, good acidity.

Terrícola 2019 T
garnacha, cariñena, syrah, cabernet sauvignon
91
Wild. Colour: bright cherry. Nose: red berry notes, floral, spicy. Palate: flavourful, fruity, good acidity, long.

Terrícola Blanc 2020 B
garnacha blanca
91
Colour: bright straw. Nose: ripe fruit, fragrant herbs, fine lees, smoky. Palate: full-bodied, rich, good acidity.

Violetes de Fang 2018 T
cariñena
92
Colour: deep cherry. Nose: dried herbs, creamy oak, black fruit, earthy notes, burnt matches. Palate: powerful, ripe fruit, spicy, round tannins.

ELVIWINES
Ctra T-300 Falset-Marça, km 0.97
43775 Marça (Tarragona)
☎: +34 606 186 565
info@elviwines.com
www.elviwines.com

Clos Mesorah 2019 T R
42% cariñena, 37% garnacha, 21% syrah
91
Colour: cherry, garnet rim. Nose: fruit liqueur notes, powerful, scrubland, toasty. Palate: flavourful, sweetness, long.

ESTONES VINS
Pol. Ind. Sort dels Capellans, Nave 15
43730 Falset (Tarragona)
☎: +34 666 415 735
vins@estones.cat
www.estones.cat

Coster D'En Fornós 2019 T
90
Colour: cherry, purple rim. Nose: fruit expression, red berry notes, floral, spicy. Palate: flavourful, fruity, good acidity.

Estones GS 2017 T
91
Colour: dark-red cherry. Nose: toasty, spicy, cocoa bean, ripe fruit, black fruit. Palate: flavourful, toasty, fine bitter notes.

Petites Estones Negre 2021 T
89
Fruity, smooth, wild, standard, crisp, herbal.

Set Tota la Vida T
88
Hot, toasty, powerful, overripe.

JOSEP GRAU VITICULTOR
Paratge Les Taules
43776 Capçanes (Tarragona)
☎: +34 657 322 291
celler@josepgrauviticultor.com
www.josepgrauviticultor.com

Granit 2020 B
garnacha blanca
92
Colour: straw. Nose: ripe fruit, faded flowers, waxy notes, spicy, dry stone. Palate: balanced, fleshy, saline.

Granit 2021 B
91
Needs time. Colour: bright straw. Nose: ripe fruit, fine lees, white fruit, grassy. Palate: full-bodied, flavourful, balanced.

La Florens 2019 T
garnacha
94
Colour: cherry, purple rim. Nose: fruit expression, red berry notes, spicy, violets, wild herbs. Palate: flavourful, fruity, good acidity, good structure.

La Florens 2020 T
garnacha
93
Colour: bright cherry. Nose: dried herbs, creamy oak, ripe fruit, red berry notes, earthy notes. Palate: powerful, ripe fruit, spicy, round tannins.

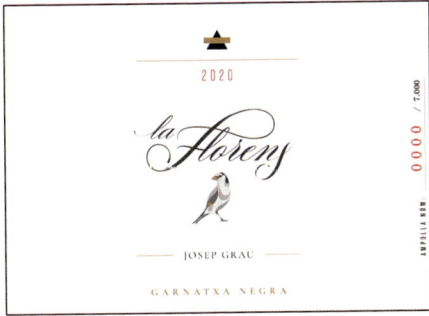

Les Casetes 2020 T
garnacha
94
Colour: cherry, purple rim. Nose: red berry notes, scrubland, lactic notes, spicy, balsamic herbs. Palate: flavourful, fruity, elegant.

Vespres Blanc 2021 B
garnacha blanca
90
Colour: bright yellow. Nose: spicy, dried herbs, white fruit. Palate: flavourful, fleshy, balanced.

Territori 2019 T R
garnacha, cariñena
93
Colour: cherry, purple rim. Nose: fruit expression, red berry notes, spicy, scrubland. Palate: flavourful, good acidity, good structure, balanced.

Una Nit en Globus 2020 T
garnacha, cariñena, syrah
90
Colour: Cherry. Nose: red berry notes, fruit preserve, dried herbs, spicy. Palate: fleshy, balanced, round tannins.

Vespres 2020 T
garnacha, cariñena, syrah
90
Colour: cherry, purple rim. Nose: fruit expression, red berry notes, floral, spicy. Palate: flavourful, fruity, good acidity, long.

LA COVA DELS VINS
Del Bosquet, 5
43730 Falset (Tarragona)
☎: +34 636 395 386
f.perello@lacovadelsvins.com
www.lacovadelsvins.cat

Deler 2021 B
88
Pleasant, fruity, jammy, tasty, standard, great length.

Deler 2021 RD
88
Jammy, fruity, dried herbs, tasty, balanced, kind finish.

Deler Vinyes Seleccionades 2021 T
89
Fruity, jammy, floral, tasty.

Ombra 2019 T BA
garnacha
90
Colour: bright cherry. Nose: red berry notes, spicy, wild herbs. Palate: fruity, good acidity, long, easy to drink.

Ombra 2021 B
garnacha blanca, macabeo
89
Fruity, defined aromas, very fruit-driven, floral, tasty, balanced.

Terròs 2019 T
91
Colour: dark-red cherry, garnet rim. Nose: ripe fruit, aged wood nuances, tobacco, sweet spices. Palate: spicy, round tannins, long.

DO MONTSANT / D.O.P.

DO MONTSANT / D.O.P.

MAS DE LES VINYES
Mas de les Vinyes, s/n
43373 Cabacés (Tarragona)
☎: +34 652 568 848
josep@masdelesvinyes.com
www.masdelesvinyes.com

Traca i Mocador Blanc 2020 B
91
Colour: straw. Nose: ripe fruit, dried herbs, faded flowers, citrus fruit. Palate: powerful, balanced, flavourful.

ORTO VINS
Passeig sde l'Arbre entre 3 y 5
43736 El Masroig (Tarragona)
☎: +34 629 171 246
info@ortovins.com
www.ortovins.com

Les Argiles D'Orto Vins Blanc 2021 B
90% macabeo, 10% garnacha blanca
91 🌿
Colour: straw. Nose: ripe fruit, dried herbs, faded flowers. Palate: powerful, ripe fruit, balanced.

Les Argiles D'Orto Vins Negre 2021 T
90% garnacha, 10% cariñena
90 🌿
Colour: Cherry. Nose: sweet spices, scrubland, ripe fruit, dried flowers. Palate: spicy, balsamic, good acidity.

Les Argiles D'Orto Vins Rosat 2021 RD
55% picapoll negre, garnacha
88 🌿
Citrus fruit, balanced, herbaceous, tasty.

Les Comes D'Orto 2019 T
50% cariñena, 50% garnacha
91
Colour: cherry, garnet rim. Nose: varietal, dried herbs, wild herbs, hot, neat. Palate: flavourful, round tannins, ripe fruit.

Les Tallades de Cal Nicolau 2017 T C
100% picapoll negre
91
Colour: cherry, garnet rim. Nose: fruit preserve, fruit liqueur notes, powerful. Palate: flavourful, long, slightly dry, soft tannins.

Ranci Orto Vins B RC
90% garnacha blanca, 10% macabeo
91
Defined aromas, with personality. Colour: old gold. Nose: dry nuts, candied fruit, sweet spices. Palate: flavourful, fine bitter notes, spicy, long.

PACO MULERO
Partida de la Hoya Torres s/n
30520 Jumilla (Murcia)
☎: +34 676 433 541
info@pacomulero.com
www.pacomulero.com

Puntes de Calnegre 2020 T
cariñena, garnacha, syrah
89
Dried herbs, rustic, toasty, pruney, smoky.

ROUND TABLE
Ramón Freixas, 40 Bx 2
08720 Vilafranca del Penedès (Barcelona)
☎: +34 637 716 322
hola@theroundtabletopics.com
www.roundtable.wine

Round Table 2021 T MC
garnacha, mazuelo
89
Jammy, tasty, fruity, exuberant, herbal.

SPECTACLE VINS
Carretera Bellmunt – Sort del Capellans
43730 Falset (Tarragona)
☎: +34 977 839 171
closmogador@closmogador.com
www.espectaclevins.com

🏆 PODIUM

Espectacle 2019 T C
95 🌿
Colour: bright cherry. Nose: wild herbs, dried flowers, ripe fruit, hint of anise, expressive, red berry notes. Palate: elegant, long, full of life, flavourful.

TERRA DE FALANIS
Ctra. Campos Felanitx 10.3 Km.
07200 Felanitx (Illes Balears/Islas Baleares)
☎: +34 679 314 406
contactoterradefalanis@gmail.com
www.terradefalanis.com

Llenca Plana 2019 T RB
89
Smoky, toasty, jammy, fruity, tasty.

TERROIR SENSE FRONTERES

43775 Marça (Tarragona)
☎: +34 977 839 391
vi@terroir-sense-fronteres.com
www.terroir-al-limit.com

Coreografía Montsant 2021 T
50% garnacha, 25% garnacha peluda, 25% cariñena

92

Aromatic, subtle, wild. Colour: light cherry. Nose: wild herbs, scrubland, red berry notes, ripe fruit, dried flowers, hint of anise. Palate: flavourful, taut, long, easy to drink.

Marcenca 2021 T
100% garnacha

94

Defined aromas, elegant. Colour: light cherry. Nose: chamomile, floral, red berry notes, expressive. Palate: full of life, fruity, fresh, good acidity.

Terroir Sense Fronteres Brisat 2021 B
100% garnacha blanca

91

Dried flowers, fruity, wild. Colour: straw. Nose: dried herbs, spicy, wild herbs. Palate: flavourful, correct, balanced.

Terroir Sense Fronteres Negre 2021 T
100% garnacha

91

Defined aromas, fruity, crisp. Colour: light cherry. Nose: fresh fruit, red berry notes, very fruit-driven, neat. Palate: fresh, fruity, easy to drink.

Vértebra de la Figuera 2021 T
100% garnacha

94

Colour: light cherry. Nose: scrubland, wild herbs, hint of anise, fresh fruit, red berry notes. Palate: full of life, long, easy to drink.

VENUS LA UNIVERSAL

Ctra. Porrera, s/n
43730 Falset (Tarragona)
☎: +34 977 279 189
info@venuslauniversal.com
www.venuslauniversal.com

Dido 2020 T
garnacha, syrah, cariñena, merlot, cabernet sauvignon

92 ♣

Wild, jammy. Colour: cherry, garnet rim. Nose: ripe fruit, fruit expression, fine reductive notes, balsamic herbs. Palate: flavourful, long, easy to drink.

Dido Blanc 2020 B
macabeo, garnacha blanca, cartoixà

93 ♣

Complex, with personality. Colour: straw. Nose: ripe fruit, dried herbs, faded flowers, white fruit, spicy, balsamic herbs, cereal notes. Palate: powerful, ripe fruit, balanced, flavourful.

Dido La Solució Rosa 2020 RD
garnacha blanca, garnacha, garnacha gris, macabeo, cariñena, tempranillo

92

Oxidativ, little interventionist. Colour: rose. Nose: ripe fruit, faded flowers, cereal notes, fine lees. Palate: fleshy, flavourful, powerful, balanced.

Venus 2018 T C
cariñena, garnacha

93 ♣

Defined aromas. Colour: cherry, garnet rim. Nose: wild herbs, dried flowers, scrubland, ripe fruit, expressive. Palate: fruity, full of life, easy to drink, spicy.

Venus de Cartoixà 2018 B
cartoixà

93 ♣

Dried flowers, little interventionist. Colour: bright straw. Nose: ripe fruit, fragrant herbs, mineral, lees reduction notes, white fruit. Palate: full-bodied, rich, long, good acidity.

Venus de la Figuera 2018 T
garnacha

94 ♣

Colour: dark-red cherry. Nose: wild herbs, dried herbs, expressive, ripe fruit, elegant, dried flowers. Palate: fruity, spicy, easy to drink.

VINYES D'EN GABRIEL

Ctra. Darmós - La Serra, s/n
43746 Darmos (Tarragona)
☎: +34 609 989 345
celler@vinyesdengabriel.com
www.vinyesdengabriel.com

Cuvila 2021 T
100% garnacha

90 ♣

Rustic, varietally correct. Nose: wild herbs, dried flowers, neat. Palate: ripe fruit, slightly dry, soft tannins.

L'Heravi 2021 T
cariñena, garnacha, syrah

89 ♣

Standard, fruity, wild, simple, tasty, floral.

DO MONTSANT / D.O.P.

DO MONTSANT / D.O.P.

L'Heravi Blanc de Noir 2021 B
garnacha

89 🌱

Dried herbs, austere. Nose: fine lees, wild herbs, citrus fruit. Palate: fresh, easy to drink.

L´Heravi Selecció 2021 T
syrah, cariñena

89 🌱

Wild, smooth, jammy, fruity, pleasant, dried herbs, balanced.

Mans de Samsó 2020 T
cariñena

92

Colour: Cherry. Nose: balsamic herbs, scrubland, ripe fruit, spicy, wild herbs. Palate: flavourful, fruity, good acidity, fine bitter notes.

VINYES DOMÈNECH

43776 Capçanes (Tarragona)
☎: +34 670 375 828
jidomenech@vinyesdomenech.com
www.vinyesdomenech.com

Bancal del Bosc Blanc 2021 B
garnacha blanca

89

Pleasant, standard, floral, fruity, jammy, balanced.

Empelts 2019 T
100% garnacha peluda

93

Rustic, tasty. Colour: Cherry. Nose: expressive, spicy, mineral, scrubland. Palate: full-bodied, long, great length.

Furvus 2020 T BA
garnacha

91 🌱

Colour: cherry, garnet rim. Nose: fruit preserve, powerful, scrubland, dried herbs. Palate: flavourful, long, round tannins.

Rita 2021 B
garnacha blanca

91 🌱

Colour: bright yellow. Nose: powerful, creamy oak, ripe fruit, sweet spices. Palate: rich, good structure, long, toasty, fine bitter notes.

Teixar 2017 T
100% garnacha peluda

92 🌱

Colour: cherry, garnet rim. Nose: fruit preserve, fruit liqueur notes, powerful, dried herbs, hot. Palate: flavourful, sweetness, long.

Vi D'Amfora 2021 T
100% garnacha

89 🌱

Rustic, tasty, pruney. Nose: earthy notes, characterful.

VIÑEDOS SINGULARES

Avda. de La Riera, 11 Nave 1
08960 Sant Just Desvern (Barcelona)
☎: +34 934 807 041
Fax: +34 934 807 076
info@vinedossingulares.com
www.vinedossingulares.com

El Veïnat 2021 T
garnacha

89

Herbal, jammy, spicy, balanced, tasty, wild, fruity.

DO. NAVARRA
CONSEJO REGULADOR

Rúa Romana, s/n
31390 Olite (Navarra)
☎: +34 948 741 812 - Fax: +34 948 741 776
@: info@navarrawine.com
www.navarrawine.com

LOCATION:

In the province of Navarra. It draws together areas of different climates and soils, which produce wines with diverse characteristics.

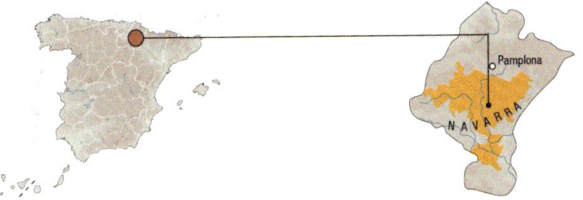

SUB-REGIONS:

Baja Montaña: Situated northeast of Navarra, it comprises 22 municipal districts with around 2,500 Ha under cultivation. **Tierra Estella:** In western central Navarra, it stretches along the Camino de Santiago. It has 1,800 Ha of vineyards in 38 municipal districts. **Valdizarbe:** In central Navarra. It is the key centre of the Camino de Santiago. It comprises 25 municipal districts and has 1,100 Ha of vineyards. **Ribera Alta:** In the area around Olite, it takes in part of central Navarra and the start of the southern region. There are 26 municipal districts and 3,300 Ha of vineyards. **Ribera Baja:** In the south of the province, it is the most important in terms of size (4,600 Ha). It comprises 14 municipal districts.

GRAPE VARIETIES:

WHITE: Chardonnay, Garnacha Blanca, Malvasía, Moscatel de Grano Menudo, Viura and Sauvignon Blanc.

RED: Cabernet Sauvignon, Garnacha Tinta, Graciano, Mazuelo, Merlot, Tempranillo, Syrah and Pinot Noir.

FIGURES:

Vineyard surface: 10,028 – **Wine-Growers:** 2,000 – **Wineries:** 85 – **2021 Harvest rating:** N/A **Production 21:** 37,344,000 litres – **Market percentages:** 72% National - 28% International.

SOIL:

The diversity of the different regions is also reflected in the soil. Reddish or yellowish and stony in the Baja Montaña, brownish-grey limestone and limestone in Valdizarbe and Tierra Estella, limestone and alluvium marl in the Ribera Alta, and brown and grey semi-desert soil, brownish-grey limestone and alluvium in the Ribera Baja.

CLIMATE:

Typical of dry, sub-humid regions in the northern fringe, with average rainfall of between 593 mm and 683 mm per year. The climate in the central region is transitional and changes to drier conditions in southern regions, where the average annual rainfall is a mere 448 mm.

CLASSIFICATION OF YOUNG WINE HARVESTS PEÑÍNGUIDE

2017	2018	2019	2020	2021
GOOD	GOOD	VERY GOOD	VERY GOOD	VERY GOOD

DO NAVARRA / D.O.P.

ALEX
Ctra. Tudela, s/n
31591 Corella (Navarra)
☎: +34 948 782 014
Fax: +34 948 782 164
info@vinosalex.com
www.vinosalex.com

Alex 2018 T C
tempranillo, merlot, graciano
88
Standard, fruity, jammy, wild, herbal.

Alex Garnacha 2021 RD
garnacha
87

Alex Viura 2021 B
viura
86

Ontinar Merlot 2021 RD
merlot
85 ♣

Ontinar Merlot 2021 T
merlot
87 ♣

Ontinar Tempranillo 2021 T
tempranillo
87 ♣

ALIAGA
Avda. de Navarra, 17
31591 Corella (Navarra)
☎: 948 401 321
sales@vinaaliaga.com
www.vinaaliaga.com

Aliaga Doscarlos Sauvignon Blanc 2021 B
100% sauvignon blanc
87

Aliaga Helena Syrah Syrah 2019 T
100% syrah
87

Aliaga Lágrima de Garnacha 2021 RD
100% garnacha
89
Kind finish, aromatic, dried flowers, fruity.

Aliaga Lágrima de Luna 2021 RD
100% garnacha
87

Aliaga Moscatel Vendimia Tardía 2019 B D
100% moscatel grano menudo
88
Pleasant, aromatic, fruity, floral, sweety finish, sweet.

Aliaga Reserva de la Familia 2014 T R
80% tempranillo, 20% cabernet sauvignon
89
Slight reduction, overripe, spicy, fruity.

ANECOOP BODEGAS
Monforte, 1 – entlo.
46010 València/Valencia (València/Valencia)
☎: +34 963 938 500
Fax: +34 963 938 543
anecoopbodegas@anecoop.com
www.anecoopbodegas.com

Dominio de Unx Chardonnay 2021 B
chardonnay
89
Dried flowers, fruity, jammy, yeasty notes, tasty, balanced.

Dominio de Unx Garnacha Blanca Sur Lie 2021 B
garnacha
87

Dominio de Unx Garnacha Old Vines 2020 T
garnacha
91
Colour. Cherry. Nose: balsamic herbs, sweet spices, scrubland, red berry notes. Palate: spicy, balsamic, good acidity.

Dominio de Unx Garnacha Rosado de Lágrima 2021 RD
garnacha
91
Colour. raspberry rose. Nose: elegant, red berry notes, floral. Palate: light-bodied, spicy, good acidity.

La Calma Mágica 2018 T
garnacha
90
Colour. bright cherry. Nose: fruit expression, floral, spicy, violets. Palate: flavourful, fruity, good acidity, easy to drink.

La Calma Mágica 2019 B
garnacha
90
Aromatic, pleasant. Colour. bright straw. Nose: white fruit, floral, medium intensity. Palate: good acidity, easy to drink, fresh.

AROA BODEGAS

Apalaz, 13
26500 Zurukoain (Navarra)
☎: +34 941 483 557
marketing@vintae.com
www.vintae.com

Aroa Jauna 2018 T C
cabernet sauvignon, merlot, tempranillo, garnacha

89 ♣

Fruity, jammy, spicy, tasty.

Aroa Laia 2021 B
garnacha blanca

89 ♣

Fruity, jammy, sweety finish, lactic, tasty.

Aroa Larrosa 2021 RD
garnacha

90 ♣

Colour: brilliant rose, purple rim. Nose: fruit expression, lactic notes, floral. Palate: balanced, fruity aftertaste.

Aroa Mutiko 2020 T
garnacha

88 ♣

Fruity, spicy, jammy, sweety finish, coarse tannins.

Le Naturel 2021 B
garnacha blanca, otras

88 ♣

Fruity, dried flowers, slight oxidation, jammy, tasty.

Le Naturel 2021 T
garnacha, otras

88 ♣

Spicy, herbal, fruity, smoky, jammy.

AZUL Y GARANZA

San Juan, 19
31310 Carcastillo (Navarra)
☎: +34 659 857 979
fernando@azulgaranza.com
www.azulygaranza.com

Garciano de Azul y Garanza 2020 T BA
garnacha, graciano

90 ♣

Wild. Colour: cherry, garnet rim. Nose: ripe fruit, scrubland, dried herbs. Palate: flavourful, balsamic, easy to drink.

Naturaleza Salvaje Garnacha 2020 T
garnacha

90 ♣

Reductive, rustic. Colour: deep cherry, purple rim. Nose: dried herbs, ripe fruit. Palate: flavourful, balsamic, ripe fruit, balanced.

Seis de Azul y Garanza 2019 T
merlot

90 ♣

Colour: cherry, purple rim. Nose: black fruit, ripe fruit, spicy, dried herbs, balsamic herbs. Palate: flavourful, fresh, fruity, good structure, balanced, powerful tannins.

BODEGA CASTILLO DE ENÉRIZ

N- 6010, Km. 7,4
31153 Enériz (Navarra)
☎: +34 948 692 500
info@manzanos.com
www.bodegasmanzanos.com

1864 Castillo de Olite 2016 T R
garnacha, tempranillo, cabernet sauvignon

91

Colour: bright cherry. Nose: black fruit, ripe fruit, dark chocolate, spicy, smoky, dried herbs. Palate: fruity, flavourful, balanced, smoky aftertaste, round tannins.

1864 Castillo de Olite 2018 T C
garnacha, tempranillo

90

Pleasant. Colour: deep cherry. Nose: dried herbs, creamy oak, red berry notes, black fruit. Palate: powerful, ripe fruit, spicy, round tannins.

1864 Castillo de Olite Chardonnay 2021 B FB
100% chardonnay

90

Colour: bright yellow. Nose: powerful, creamy oak, ripe fruit, spicy, white fruit. Palate: good structure, toasty, fine bitter notes.

Castillo de Eneriz Colección 2020 T C
garnacha, cabernet sauvignon, graciano, tempranillo

89

Spicy, fruity, jammy, herbal, toasty.

Reino de Altuzarra 2016 T R
cabernet sauvignon, merlot, tempranillo

89

Balsamic herbs, jammy, spicy, herbal, tasty, great length. Nose: black pepper, spicy, scrubland.

Reino de Altuzarra 2018 T C
garnacha, cabernet sauvignon, merlot, tempranillo

88

Full-bodied, balanced, spicy, jammy, toasty.

DO NAVARRA / D.O.P.

BODEGA CASTILLO DE MONJARDÍN

Viña Rellanada, s/n
31242 Villamayor de Monjardin (Navarra)
☎: +34 948 537 412
contacto@monjardin.es
www.monjardin.es

Castillo de Monjardín Deyo Merlot de Autor 2018 T C
merlot

90

Colour: bright cherry. Nose: ripe fruit, dried herbs, creamy oak, smoky. Palate: spicy, round tannins, fruity, powerful.

Monjardín Chardonnay Especial Superior 2009 B GR
chardonnay

92

Colour: bright golden. Nose: caramel, sweet spices, creamy oak, honeyed notes, candied fruit. Palate: flavourful, rich, aged character, good finish, fresh, fine bitter notes.

Monjardín Tempranillo Especial Superior 2017 T R
tempranillo

90

Colour: cherry, garnet rim. Nose: fruit preserve, black fruit, black pepper, smoky, toasty, dried herbs. Palate: fruity, flavourful, balanced, good finish, slightly dry, soft tannins.

Castillo de Monjardín Chardonnay 2020 B FB
chardonnay

90

Colour: bright yellow. Nose: powerful, spicy, toasted bread, ripe fruit. Palate: rich, long, toasty, fine bitter notes.

BODEGA COSECHEROS REUNIDOS S.COOP

Pza. San Antón, 1
31390 Olite (Navarra)
☎: +34 948 740 067
info@bodegacosecheros.com
www.bodegacosecheros.com

1913 2019 B
87

1913 2021 RD
88
Pleasant, fruity, jammy.

1913 2021 T
88
Fruity, jammy, spicy, coarse tannins.

Gratianvs 2018 T
86

Viña Juguera Selección 2018 T C
86

BODEGA DE LIÉDENA

Ctra. de Jaca s/n
31487 Liédena (Navarra)
☎: +34 948 870 280
info@bodegadeliedena.com
www.bodegadeliedena.com

Ledea 2019 T C
88
Pruney, spicy, toasty, hot, smoky.

Ledea Garnacha 2021 RD
garnacha

87

BODEGA INURRIETA

Ctra. Falces-Miranda de Arga, km. 30
31370 Falces (Navarra)
☎: +34 948 737 309
Fax: +34 948 737 310
info@bodegainurrieta.com
www.bodegainurrieta.com

Altos de Inurrieta 2017 T R
45% graciano, 35% syrah, 10% cabernet sauvignon, 5% merlot, 5% tempranillo

91

Colour: cherry, garnet rim. Nose: fruit preserve, fruit liqueur notes, powerful, spicy, toasty. Palate: flavourful, long, powerful, concentrated.

Inurrieta Mimao 2020 T
100% garnacha

91

Colour: cherry, purple rim. Nose: floral, spicy, ripe fruit, red berry notes. Palate: flavourful, fruity, good acidity, long.

Inurrieta Mimao 2021 B BA
100% garnacha blanca

90

Colour: bright yellow. Nose: ripe fruit, spicy, white flowers. Palate: rich, good structure, long, toasty, fine bitter notes.

Inurrieta Puro Vicio 2019 T
100% syrah

91

Colour: very deep cherry. Nose: roasted coffee, aromatic coffee, powerful, black fruit, ripe fruit. Palate: smoky aftertaste, great length, round tannins.

Inurrieta Cuatrocientos 2019 T C
50% cabernet sauvignon, 20% merlot, 12% syrah, 9% graciano, 5% garnacha, 4% tempranillo

88

Fruity, toasty, jammy, dried herbs, spicy.

Laderas de Inurrieta 2019 T
100% graciano

91

Colour: cherry, purple rim. Nose: fruit expression, ripe fruit, dried herbs, varietal, spicy, incense. Palate: flavourful, fruity, balanced, slightly dry, soft tannins, good finish.

DO NAVARRA / D.O.P.

DO NAVARRA / D.O.P.

BODEGA MARQUÉS MONTECIERZO
San José, 62
31590 Castejón (Navarra)
☎: +34 948 814 414
info@marquesdemontecierzo.com
www.marquesdemontecierzo.com

Emergente 2019 T C
tempranillo, garnacha
88
Balanced, spicy, jammy, tasty.

Emergente Chardonnay 2021 B
100% chardonnay
85

Emergente Moscatel de Grano Menudo 2021 B SD
moscatel
86

Emergente Rosado de Lágrima 2021 RD
garnacha, cabernet sauvignon
88
Citrus fruit, hot, balanced, floral, sweety finish.

Emergente Vino de Autor Garnacha 2016 T R
89
Age nuances, slight reduction, classic, spicy, jammy, powerful, great length.

Emergente Vino de Autor Garnacha 2019 T R
88
Standard, spicy, jammy, herbal, tasty, toasty.

Marqués de Montecierzo Selección Merlot 2016 T RB
88
Balanced, spicy, herbal, jammy, rustic.

Montecierzo Rosé 2021 RD
88
Aromatic, floral, fruity, jammy, tasty.

BODEGA OTAZU
Señorío de Otazu, s/n
31174 Otazu (Navarra)
☎: +34 948 329 200
Fax: +34 948 329 353
lboursier@bodegaotazu.es
www.otazu.com

Otazu Chardonnay 2021 B
100% chardonnay
91
Colour: bright yellow. Nose: ripe fruit, dried herbs, faded flowers, stone fruit. Palate: powerful, ripe fruit, balanced.

Otazu Merlot 2021 RD
100% merlot
89
Jammy, powerful, pleasant, hot.

Otazu Premium Cuvée 2019 T C
35% cabernet sauvignon, 33% merlot, 32% tempranillo
92
Colour: deep cherry. Nose: ripe fruit, dried herbs, creamy oak, grassy. Palate: powerful, ripe fruit, spicy, round tannins.

Otazu Reserva Clásico 2016 T R
40% tempranillo, 40% cabernet sauvignon, 20% merlot
90
Colour: dark-red cherry, garnet rim. Nose: fruit preserve, aged wood nuances, tobacco, sweet spices. Palate: spicy, round tannins, long.

Otazu Tempranillo Rosé 2021 RD
100% tempranillo
90
Colour: raspberry rose. Nose: fruit expression, red berry notes, floral. Palate: fruity, good acidity, easy to drink.

BODEGA PAGO DE CIRSUS
Ctra. de Ablitas a Ribafora, Km. 5
31523 Ablitas (Navarra)
☎: +34 948 386 427
info@pagodecirsus.com
www.pagodecirsus.com

Pago de Cirsus Chardonnay 2021 B
chardonnay
90
Colour: straw. Nose: ripe fruit, faded flowers, balanced. Palate: powerful, ripe fruit, balanced.

Pago de Cirsus Rosé Gran Cuvée Especial 2021 RD FB
91
Colour: raspberry rose. Nose: red berry notes, floral, fragrant herbs. Palate: spicy, good acidity, fine bitter notes.

BODEGA PAGOS DE ARAIZ
Camino de Araiz, s/n
31390 Olite (Navarra)
☎: +34 948 926 963
info@bodegaspagosdearaiz.masaveu.com
www.bodegaspagosdearaiz.com

Blaneo Chardonnay 2021 B FB
100% chardonnay
89
Woody, standard, spicy, fruity, jammy.

Blaneo Syrah 2019 T FB
100% syrah
89
Full-bodied, creamy, full-bodied, jammy, hot, roasted coffee.

Pagos de Araiz Crianza 2019 T C
30% tempranillo, 30% merlot, 20% syrah, 20% cabernet sauvignon
89
Balanced, jammy, tasty, toasty.

Pagos de Araiz Roble 2020 T RB
50% tempranillo, 30% merlot, 20% garnacha
86

Pagos de Araiz Rosado 2021 RD
100% garnacha
88
Fruity, dried herbs, jammy, tasty, kind finish.

Pagos de Araiz Rosé 2021 RD
100% garnacha
88
Fruity, dried herbs, kind finish, jammy, tasty.

BODEGA SAN MARTÍN
Ctra. de Sanguesa, s/n
31495 San Martín de Unx (Navarra)
☎: +34 948 738 294
admon@bodegasanmartin.com
www.bodegasanmartin.com

Alma de Unx 2016 T
garnacha
89
Floral, fruity, sweeties, jammy, simple, tasty, very fruit-driven.

Ilagares 2021 B
chardonnay, viura
86

Ilagares 2021 RD
garnacha
88
Aromatic, floral, fruity, jammy.

Ilagares Garnacha 2021 T
garnacha
87

La Matacalva 2019 T
garnacha
89
Spicy, fruity, jammy, tasty.

Señorío de Unx 2018 T C
tempranillo, garnacha
87

BODEGAS ALCONDE
Ctra. de Calahorra, s/n
31260 Lerin (Navarra)
☎: +34 948 530 058
Fax: +34 948 530 589
manuel@bodegasalconde.com
www.bodegasalconde.com

Alconde 2016 T R
90
Colour: deep cherry, garnet rim. Nose: aged wood nuances, cigar, toasty, fruit liqueur notes. Palate: flavourful, spicy, toasty.

Sardasol 2021 RD
garnacha
88
Pleasant, fruity, floral, jammy.

Sardasol Sauvignon Blanc 2021 B
sauvignon blanc
86

Sardasol Selección Garnacha 2021 T
garnacha
88
Fruity, spicy, herbal, jammy.

Sardasol Selección Garnacha Blanca 2021 B
garnacha blanca
89
Colour: straw. Nose: white fruit, ripe fruit, wild herbs, dried flowers. Palate: fruity, fresh, fruity aftertaste.

DO NAVARRA / D.O.P.

DO NAVARRA / D.O.P.

Sardasol Tempranillo 2019 T C
tempranillo
88
Standard, balanced, jammy, fruity, spicy, simple.

BODEGAS ARMENDÁRIZ
Avda. El Salvador, 9-Bajo
31370 Falces (Navarra)
☎: +34 948 734 235
Fax: +34 948 714 902
info@bodegasarmendariz.com
www.comprarvinodenavarra.com

Armendáriz 2018 T C
46% garnacha, 42% tempranillo, 12% merlot
85

Armendáriz 2020 T
67% tempranillo, 33% garnacha
87

Armendáriz 2021 B
100% chardonnay
85

Armendáriz 2021 RD
52% garnacha, 48% tempranillo
86

Armendáriz Dulce Natural 2021 B D
100% moscatel grano menudo
88
Tasty, hot, sweet, spicy.

BODEGAS BERAMENDI
Ctra. Tafalla, s/n
31495 San Martín de Unx (Navarra)
☎: +34 948 738 262
info@bodegasberamendi.com
www.bodegasberamendi.com

Beramendi 3 Flores Garnacha 2021 RD
100% garnacha
86

Beramendi Chardonnay 2021 B
100% chardonnay
87

Beramendi Edic. Especial Graciano 2019 T
100% graciano
88
Smoky, pruney, hot, herbal, rustic, wild.

Beramendi Edic. Especial Tempranillo 2019 T RB
100% tempranillo
89
Balsamic herbs, standard, fruity, herbal, jammy, tasty, wild, smooth.

Beramendi Garnacha 2021 RD
100% garnacha
89
Kind finish, balanced, floral, fruity, sweety finish.

Beramendi Merlot 2017 T C
100% merlot
87

BODEGAS CAMPOS DE ENANZO S. COOP.
Mayor, 189
31500 Murchante (Navarra)
☎: +34 948 838 030
Fax: +34 948 838 677
enologia@enanazo.com
www.camposenanzo.com

Remonte 2017 T C
garnacha, cabernet sauvignon, graciano
88
Balanced, spicy, dried herbs, toasty, classic.

Remonte 2021 RD
garnacha
86

Remonte Chardonnay 2021 B
chardonnay
87

Rochapel 2021 RD
garnacha
85

Rochapel Chardonnay 2021 B
chardonnay
87

Rochapel Garnacha 2021 T
garnacha
86

BODEGAS CAUDALIA
Cl. San Francisco Javier 14
31494 Lerga (Navarra)
☎: +34 670 833 340
info@bodegascaudalia.com
www.bodegascaudalia.com

Paal 01 100% Syrah 2020 T
syrah

90

Colour: deep cherry. Nose: ripe fruit, dried herbs, creamy oak. Palate: powerful, ripe fruit, spicy, round tannins.

Umea Garnacha 2021 T
garnacha

89

Fruity, herbal, jammy, tasty, kind finish.

Umea Garnacha Blanca 2021 B
garnacha blanca

89

Defined aromas, balanced, fruity, floral, wild, smooth.

Xi'Ipal Rosé El Olivo 2021 RD
garnacha

90

Pleasant, aromatic, floral, tasty.

BODEGAS CORELLANAS
Santa Bárbara, 29
31591 Corella (Navarra)
☎: +34 948 780 029
info@bodegascorellanas.com
www.bodegascorellanas.com

Blue Moscatel 2021 B SD
moscatel grano menudo

85

Sara Garnacha 2017 T
100% garnacha

90

Colour: deep cherry. Nose: creamy oak, fruit preserve, spicy. Palate: powerful, ripe fruit, spicy, round tannins.

Sarasate 2021 RD
garnacha

86

Sarasate Expresión Dulce Natural 2021 B MO D
moscatel grano menudo

84

Viña Rubicán Tempranillo 2017 T C
tempranillo

86

BODEGAS FERNÁNDEZ DE ARCAYA
La Serna, 31
31210 Los Arcos (Navarra)
☎: +34 948 640 811
info@fernandezdearcaya.com
www.fernandezdearcaya.com

Fernández de Arcaya Selección Privada 2017 T R
100% cabernet sauvignon

90

Colour: deep cherry. Nose: ripe fruit, dried herbs, creamy oak. Palate: powerful, ripe fruit, spicy, round tannins.

Viña Perguita 2018 T C
80% tempranillo, 15% cabernet sauvignon, 5% merlot

88

Age nuances, spicy, herbal, jammy, tasty, classic.

Viña Perguita 2019 T RB
85% tempranillo, 10% cabernet sauvignon, 5% merlot

86

BODEGAS GRAN FEUDO
Ribera, 34
31592 Cintruénigo (Navarra)
☎: +34 948 811 000
Fax: +34 948 811 407
info@granfeudo.com
www.granfeudo.com

Baluarte 2021 T RB
garnacha

87

Gran Feudo 2018 T C
41% tempranillo, 23% cabernet sauvignon, 19% merlot, 15% garnacha, 2% syrah

87

Gran Feudo 2021 RD
85% garnacha, 15% merlot

88

Pleasant, crisp, fruity, tasty.

Gran Feudo Chardonnay 2021 B
chardonnay

88

Standard, balanced, fruity, jammy, tasty.

Gran Feudo Dulce de Moscatel 2021 B MO D
moscatel grano menudo

91

Colour: bright straw. Nose: wild herbs, hint of anise, stone fruit, tropical fruit. Palate: flavourful, rich, balanced, ripe fruit.

DO NAVARRA / D.O.P.

DO NAVARRA / D.O.P.

Gran Feudo Edición Limitada Viñas Viejas 2017 T R
60% tempranillo, 40% garnacha

91

Colour: dark-red cherry. Nose: toasty, spicy, cocoa bean, ripe fruit, black fruit. Palate: flavourful, toasty, fine bitter notes.

Gran Feudo Garnacha Rosé 2021 RD
100% garnacha

89

Pleasant, aromatic, floral, fruity.

BODEGAS IÑAKI NÚÑEZ
San Prudencio, 13 4º Izda.
01005 Vitoria-Gasteiz (Araba/Álava)
☎: +34 945 140 126
ngarcia@arabafilms.com
www.bodegasiñakinuñez.es

Iñaki Núñez Selección Privada 2017 T
90% graciano, 10% garnacha

91

Colour: cherry, garnet rim. Nose: fruit preserve, powerful, dried herbs, toasty, tobacco. Palate: round tannins, balsamic, flavourful.

Iñaki Núñez Vendimia Seleccionada 2018 T
100% garnacha

88

Jammy, hot, spicy, tasty.

BODEGAS IRACHE
Avda. de Monasterio, 1
31240 Ayegui (Navarra)
☎: +34 948 551 608
irache@irache.com
www.irache.com

1891 2017 T C
88

Spicy, fruity, herbal, tasty.

1891 2021 RD
88

Pleasant, aromatic, fruity, crisp.

Irache 2016 T R
tempranillo, cabernet sauvignon, merlot

88

Defined aromas, balsamic herbs, standard, jammy, fruity, herbal, hot.

Irache 2017 T C
tempranillo, garnacha

88

Full-bodied, balanced, spicy, jammy, tasty, toasty.

Irache 2021 RD
88

Pleasant, aromatic, fruity, smooth, sweeties.

Irache Roble 2021 T RB
88

Fruity, exuberant, crisp, herbal, spicy.

BODEGAS LA CASA DE LÚCULO
Ctra. Larraga, s/n
31150 Mendigorria (Navarra)
☎: +34 948 343 148
bodega@luculo.es
www.luculo.es

Cátulo Ecológico 2021 RD
garnacha, tempranillo

89 🌿

Kind finish, smooth, tasty, fruity.

Cátulo Garnacha 2021 T
garnacha

87 🌿

Lúculo 2021 B FB
garnacha blanca

89 🌿

Aromatic, spicy, floral, fruity, jammy, smooth.

Luculo Origen 2019 T C
garnacha
88
Pruney, dried herbs, spicy, tasty.

BODEGAS LEZAUN
Egiarte, 1
31292 Lácar (Navarra)
☎: +34 948 541 339
info@lezaun.com
www.lezaun.com

Egiarte Rosado 2021 RD
100% garnacha
87

Lezaun 0,0 Sulfitos 2021 T
100% tempranillo
86

Lezaun 2014 T R
graciano, garnacha, cabernet sauvignon
89
Age nuances, spicy, fruity, pruney, tasty, coarse tannins.

Lezaun 2019 T C
tempranillo, graciano
88
Hot, spicy, fruity, herbal, tasty, coarse tannins.

Lezaun Gazaga 2020 T RB
tempranillo, graciano, cabernet sauvignon
87

Lezaun Tempranillo 2021 T MC
100% tempranillo
88
Fruity, aromatic, jammy, tasty, coarse tannins.

BODEGAS MALON DE ECHAIDE
Ctra. de Tarazona, 33
31520 Cascante (Navarra)
☎: +34 948 851 411
Fax: +34 948 844 504
info@malondeechaide.com
www.malondeechaide.com

Corazón de Malon 2021 RD
100% garnacha
86

Malón de Echaide 2018 T C
100% tempranillo
85

Malón de Echaide 2021 B
100% chardonnay
85

Malón de Echaide 2021 RD
100% garnacha
85

Malón de Echaide Chardonnay 2020 B FB
100% chardonnay
87

Malón de Echaide Garnacha 2020 T RB
100% garnacha
87

BODEGAS MANZANOS CAMPANAS
Avda. Zaragoza, 1
31398 Campanas (Navarra)
☎: +34 948 692 500
info@manzanos.com
www.bodegasmanzanos.com

Las Campanas 2009 T GR
91
Colour: dark-red cherry, garnet rim. Nose: ripe fruit, fruit preserve, aged wood nuances, tobacco, sweet spices, fine reductive notes. Palate: spicy, round tannins, long.

Las Campanas 2016 T R
cabernet sauvignon, merlot, tempranillo
90
Colour: deep cherry. Nose: dried herbs, creamy oak, red berry notes, black fruit, fine reductive notes. Palate: ripe fruit, spicy, round tannins.

Las Campanas 2018 T C
87

Las Campanas 2021 B
86

Las Campanas 2021 RD
100% garnacha
88
Kind finish, fruity, jammy, tasty.

Las Campanas Rosé 2021 RD PL
100% garnacha
89
Pleasant, aromatic, elegant, floral.

Primi Luis Gurpegui Chardonnay 2021 B
100% chardonnay
88
Defined aromas, fruity, floral, exuberant, jammy, simple.

DO NAVARRA / D.O.P.

BODEGAS MARCO REAL

Rua Romana 81
31390 Olite (Navarra)
☎: +34 948 712 193
Fax: +34 948 712 343
marcoreal@grupolanavarra.com
www.bodegasmarcoreal.com

Homenaje 2021 B
chardonnay
87

Marco Real Colección Privada 2018 T C
90
Colour: deep cherry. Nose: ripe fruit, dried herbs, creamy oak, dark chocolate. Palate: powerful, ripe fruit, spicy, round tannins.

Marco Real Finca La Pared Cuvée Especial 2019 T
91
Colour: deep cherry. Nose: ripe fruit, dried herbs, creamy oak. Palate: powerful, ripe fruit, spicy, round tannins.

Marco Real Finca la Pared Graciano 2019 T
graciano
92
Colour: very deep cherry. Nose: dried herbs, scrubland, black fruit, expressive, creamy oak, spicy, woody. Palate: flavourful, long, balanced, great length, round tannins.

Marco Real Flor de Chardonnay 2020 B
chardonnay
89
Kind finish, spicy, fruity, jammy, yeasty notes, tasty, toasty.

Marco Real Pequeñas Producciones Garnacha RD
garnacha
90
Colour: brilliant rose. Nose: balsamic herbs, red berry notes, wild herbs. Palate: fresh, fruity, good acidity.

BODEGAS MÁXIMO ABETE

Ctra. Tafalla s/n (NA-132)
31495 San Martín de Unx (Navarra)
☎: +34 948 386 525
info@bodegasmaximoabete.com
www.bodegasmaximoabete.com

Guerinda El Máximo 2020 T BA
90% garnacha, 10% cabernet sauvignon
89
Pruney, fruity, spicy, tasty.

Guerinda Navasentero 2020 T
graciano
89
Fruity, pruney, spicy, tasty, rustic.

Guerinda Parcelas de Garnacha "La Abejera" 2020 T
garnacha
92
Colour: Cherry. Nose: wild herbs, red berry notes, ripe fruit, scrubland, balanced, expressive. Palate: long, great length, varietal.

Guerinda Parcelas de Garnacha "Txirolas, Quitana y Vilarraga"" 2020 T BA
garnacha
91
Colour: cherry, garnet rim. Nose: ripe fruit, fragrant herbs, scrubland, spicy, varietal. Palate: fruity, flavourful, balanced.

Guerinda Tres Partes 2021 T
garnacha
89
Pleasant, wild, standard, fruity, varietally correct.

Guerinda+ La Viura 2020 B FB
viura
91
Colour: bright straw. Nose: expressive, ripe fruit, floral, fine lees. Palate: full-bodied, spicy, long, elegant.

BODEGAS OCHOA

Miranda de Arga, 35
31390 Olite (Navarra)
☎: +34 948 740 006
info@bodegasochoa.com
www.bodegasochoa.com

Ochoa 2014 T GR
60% tempranillo, 25% merlot, 15% cabernet sauvignon
91
Colour: dark-red cherry, garnet rim. Nose: fruit preserve, aged wood nuances, tobacco, sweet spices, dark chocolate. Palate: spicy, round tannins, long.

Ochoa 2014 T R
60% tempranillo, 25% merlot, 15% cabernet sauvignon
88
Pruney, balanced, spicy, age nuances, toasty.

Ochoa Rosado de Lágrima 2021 RD
40% garnacha, 40% merlot, 20% cabernet sauvignon
88
Pleasant, fruity, tasty.

Ochoa Moscatel Vendimia Tardía Dulce 2021 B MO D
moscatel grano menudo
90
Colour: bright yellow. Nose: candied fruit, honeyed notes, stone fruit, dried herbs. Palate: flavourful, unctuous, fruity, sweet, fruity aftertaste.

Ochoa 8A La Foto de 1938 2018 T C
87

BODEGAS PIEDEMONTE
Rúa Romana s/n
31390 Olite (Navarra)
☎: +34 948 712 406
bodega@piedemonte.com
www.piedemonte.com

Piedemonte 2016 T R
merlot, cabernet sauvignon, tempranillo
90
Colour: dark-red cherry, bright ochre rim. Nose: ripe fruit, aged wood nuances, tobacco, scrubland. Palate: spicy, round tannins, long.

Piedemonte Chardonnay 2021 B
chardonnay
88
Pleasant, fruity, standard, smooth, pronounced acidity.

Piedemonte Cuatro Tierras 2018 T C
merlot, cabernet sauvignon, tempranillo, garnacha
88
Classic, spicy, slight reduction, dried herbs, jammy, great length.

Piedemonte Gamma 2021 T
cabernet sauvignon, merlot, tempranillo
86

Piedemonte Moscatel 2021 B MO D
moscatel grano menudo
87

Piedemonte Old Vines Garnacha 2018 T C
garnacha
88
Balanced, pruney, tasty, spicy, toasty, herbaceous.

BODEGAS PRÍNCIPE DE VIANA
Mayor 191
31521 Murchante (Navarra)
☎: +34 948 838 640
Fax: +34 948 818 574
info@principedeviana.com
www.principedeviana.com

Príncipe de Viana 1423 2017 T R
78% tempranillo, 22% garnacha
91
Colour: dark-red cherry, garnet rim. Nose: fruit preserve, aged wood nuances, tobacco, sweet spices, dark chocolate. Palate: spicy, round tannins, long.

Príncipe de Viana Chardonnay 2021 B FB
100% chardonnay
89
Citrus fruit, fruity, dried flowers, dried herbs, tasty, crisp.

Príncipe de Viana Edición Blanca 2021 B
50% chardonnay, 50% sauvignon blanc
90
Colour: straw. Nose: expressive, white flowers, dried herbs, white fruit. Palate: flavourful, fruity, balanced, fresh.

Príncipe de Viana Edición Limitada 2018 T C
65% tempranillo, 29% syrah, 6% garnacha
89
Fruity, spicy, dried herbs, toasty, tasty, coarse tannins.

Príncipe de Viana Edición Rosa 2021 RD
100% garnacha
91
Colour: raspberry rose. Nose: elegant, red berry notes, floral, fragrant herbs. Palate: light-bodied, spicy, good acidity, fine bitter notes.

Príncipe de Viana Vendimia Seleccionada 2018 T C
54% tempranillo, 25% cabernet sauvignon, 19% merlot, 2% syrah
88
Spicy, smoky, fruity, dried herbs, mineral.

DO NAVARRA / D.O.P.

DO NAVARRA / D.O.P.

BODEGAS VALCARLOS
Ctra. Circunvalación, s/n
31210 Los Arcos (Navarra)
☎: +34 948 640 806
Fax: +34 948 640 866
info@bodegasvalcarlos.com
www.bodegasvalcarlos.com

Fortius 2015 T GR
tempranillo
89
Aromatic, spicy, dried herbs, jammy, tasty, slight reduction.

Fortius 2021 RD
tempranillo, garnacha, cabernet sauvignon, merlot
87

Fortius Chardonnay 2021 B
chardonnay
87

Fortius Chardonnay 2021 B FB
chardonnay
87

Fortius Crianza 2019 T C
tempranillo
87

Fortius Reserva 2018 T R
tempranillo
88
Standard, jammy, fruity, spicy, balanced.

Fortius Tempranillo 2020 T RB
tempranillo
88
Standard, smoky, fruity, spicy, jammy.

BODEGAS Y VIÑEDOS ARTAZU
Mayor, 3
31109 Artazu (Navarra)
☎: +34 945 600 119
comunicacion@artadi.com
www.artadi.com

Artazu Pasos de San Martín 2018 T
100% garnacha
92
Colour: bright cherry. Nose: balsamic herbs, scrubland, ripe fruit, red berry notes. Palate: spicy, balsamic, good acidity.

Artazu Santa Cruz de Artazu 2018 B
100% garnacha blanca
93
Colour: bright yellow. Nose: creamy oak, ripe fruit, spicy. Palate: good structure, toasty, fine bitter notes.

Artazu Santa Cruz de Artazu 2019 T
100% garnacha
94
Colour: bright cherry. Nose: complex, expressive, spicy, mineral, red berry notes. Palate: full-bodied, long, great length.

Artazuri 2021 RD
100% garnacha
90 🍷
Colour: deep cherry. Nose: ripe fruit, hot, faded flowers. Palate: fleshy, flavourful, powerful.

EMILIO VALERIO
Paraje de Argonga
31263 Dicastillo (Navarra)
☎: +34 667 753 497
bodega@bodegasemiliovalerio.es
www.bodegasemiliovalerio.com

Amburza 2017 T BA
cabernet sauvignon
89 🍷
Hot, classic, toasty, rustic, powerful, pruney.

Emilio Valerio 2019 T
garnacha, tempranillo, cabernet sauvignon, merlot, graciano
91 🍷
Colour: Cherry. Nose: balsamic herbs, sweet spices, scrubland, fruit liqueur notes, hot. Palate: spicy, balsamic, good acidity.

La Merced 2018 B FB
malvasía, garnacha blanca, viura
91 🍷
Colour: bright yellow. Nose: creamy oak, ripe fruit, sweet spices, candied fruit. Palate: rich, good structure, fruity, flavourful, balanced.

Usuaran 2016 T
graciano
92 🍷
Colour: cherry, garnet rim. Nose: creamy oak, hot, ripe fruit, black fruit, grassy, balsamic herbs. Palate: powerful, sweet tannins.

Viña de Aranbelza 2017 T
garnacha
91 🍷
Colour: cherry, garnet rim. Nose: overripe fruit, creamy oak, hot, scrubland, balsamic herbs. Palate: pruney, powerful, sweet tannins.

Viña de Leorin 2017 T
garnacha
89 🍷
Animal funk, dried herbs, jammy, tasty, great length.

FINCA ALBRET

Ctra. Cadreita-Villafranca, s/n
31515 Cadreita (Navarra)
☎: +34 948 406 806
Fax: +34 948 406 699
info@fincaalbret.com
www.fincaalbret.com

Albret El Alba Chardonnay 2021 B FB
100% chardonnay

90

Colour: bright straw. Nose: powerful, ripe fruit, stone fruit, aged wood nuances. Palate: rich, good structure, long.

Albret El Balcón 2018 T C
50% tempranillo, 40% cabernet sauvignon, 10% syrah

90

Colour: dark-red cherry. Nose: toasty, spicy, cocoa bean, grassy. Palate: flavourful, toasty, fine bitter notes.

Albret El Rocío 2021 RD
100% garnacha

90

Colour: rose, purple rim. Nose: red berry notes, floral, elegant. Palate: fruity, good acidity, easy to drink.

Albret La Loma Garnacha 2020 T RB
100% garnacha

92

Colour: bright cherry. Nose: balsamic herbs, sweet spices, scrubland, red berry notes, ripe fruit. Palate: spicy, balsamic, good acidity.

Albret La Viña de mi Madre 2017 T R
100% cabernet sauvignon

93

Colour: dark-red cherry, garnet rim. Nose: ripe fruit, aged wood nuances, tobacco, sweet spices, grassy. Palate: spicy, round tannins, long.

Albret Lastra 2017 T R
65% tempranillo, 25% cabernet sauvignon, 10% syrah

92

Colour: deep cherry. Nose: dried herbs, creamy oak, black fruit, ripe fruit, sweet spices. Palate: powerful, ripe fruit, spicy, round tannins.

FINCA LA CANTERA DE SANTA ANA

Cantera de Santa Ana
31521 Murchante (Navarra)
☎: +34 656 658 007
vinos@fincalacantera.com
www.fincalacantera.com

Finca la Cantera Garnacha 2020 T
garnacha

90

Colour: cherry, purple rim. Nose: powerful, ripe fruit, spicy. Palate: ripe fruit, flavourful, good structure.

Finca la Cantera Tempranillo 2021 T
tempranillo

89

Pleasant, toasty, jammy, fruity, spicy.

Nomeolvides Chardonnay 2021 B
chardonnay

89

Citrus fruit, balanced, spicy, tasty, toasty.

Nomeolvides Garnacha 2021 RD
garnacha

87

Nomeolvides Viura 2021 B
viura

90

Colour: bright straw. Nose: ripe fruit, fragrant herbs, fine lees, wild herbs. Palate: full-bodied, rich, good acidity.

SH 2019 T
syrah

88

Hot, overripe, toasty, powerful.

GALLINA DE PIEL WINES

Escoles, 43
17181 Aiguaviva (Girona/Gerona)
☎: +34 663 594 668
info@gallinadepielwines.com
www.gallinadepielwines.com

Pinkgall 2021 RD
95% garnacha, 3% garnacha roja, 2% garnacha blanca

86

DO NAVARRA / D.O.P.

GONZALO CELAYETA WINES

Barrandón, 6
31390 Olite (Navarra)
☎: +34 639 010 119
info@gonzalocelayetawines.com
www.gonzalocelayetawines.com

El Duende 2018 T
garnacha

91
Colour: bright cherry. Nose: red berry notes, fruit liqueur notes, faded flowers, dried herbs, sweet spices. Palate: flavourful, fruity, balanced, fruity aftertaste, slightly dry, soft tannins.

El Piano 2019 T
garnacha

90
Colour: bright cherry. Nose: ripe fruit, black fruit, red berry notes, spicy, black pepper. Palate: flavourful, fruity, balanced, slightly dry, soft tannins.

Huracán Daniela 2021 B FB
70% garnacha blanca, 30% viura, chardonnay, sauvignon blanc

90
Colour: bright straw. Nose: ripe fruit, spicy, dried herbs, white flowers, lactic notes. Palate: rich, fresh, flavourful, balanced.

Kimera 2019 T
garnacha

92
Colour: cherry, purple rim. Nose: floral, violets, red berry notes, ripe fruit, wild herbs. Palate: flavourful, fruity, good acidity, long, easy to drink.

La Huella de Aitana 2021 RD
garnacha

91
Colour: brilliant rose, purple rim. Nose: red berry notes, lactic notes, floral, sweet spices. Palate: balanced, fruity aftertaste.

La Huella de Aitana Cuvée Zen 2019 RD C
garnacha

92
Colour: salmon. Nose: powerful, ripe fruit, creamy oak, sweet spices. Palate: fleshy, flavourful, spicy.

J. CHIVITE FAMILY ESTATES

Ctra. NA-132, Km. 3,1
31132 Villatuerta (Navarra)
☎: +34 948 811 000
Fax: +34 948 811 407
info@bodegaschivite.com
www.chivite.com

🏆 **PODIUM**

Chivite Colección 125 1996 B FB
100% chardonnay

98
With personality, elegant, spicy. Colour: bright golden. Nose: spicy, toasted bread, saffron, chamomile, earthy notes, candied fruit, fine lees, petrol notes. Palate: rich, fruity, fleshy, long, toasty, good acidity, full of life.

Chivite Legardeta Garnacha 2019 T
garnacha

92
Taut. Colour: dark-red cherry. Nose: toasty, spicy, cocoa bean, red berry notes, ripe fruit, balsamic herbs. Palate: flavourful, toasty, fine bitter notes.

Chivite Legardeta Syrah 2019 T
syrah

91
Defined aromas. Colour: bright cherry. Nose: powerful, characterful, ripe fruit, floral, fresh fruit. Palate: good structure, flavourful, long, ripe fruit.

LA CALANDRIA. PURA GARNACHA
Calle Mayor, 81
31500 Murchante (Navarra)
☎: +34 609 476 387
javier@lacalandria.org
www.lacalandria.com

Cientruenos 2019 T BA
90
Colour: cherry, purple rim. Nose: spicy, black fruit, ripe fruit, scrubland. Palate: flavourful, fruity, good acidity, long.

Sonrojo 2021 RD
90
Colour: rose. Nose: ripe fruit, fruit preserve, hot, faded flowers. Palate: fleshy, flavourful, powerful, ripe fruit.

LMT WINES
Cerro Amurdi
31190 Cizur Menor (Navarra)
☎: +34 645 841 928
hola@lmtwines.com
www.lmtwines.com

Masusta 2020 T
garnacha
88
Balanced, floral, fruity, spicy.

NEKEAS
Las Huertas, s/n
31154 Añorbe (Navarra)
☎: +34 948 350 296
Fax: +34 948 350 300
nekeas@nekeas.com
www.nekeas.com

25 Vendimias 2021 B
garnacha blanca
91
Colour: bright straw. Nose: fruit expression, ripe fruit, floral, scrubland. Palate: flavourful, fresh, good acidity, fruity aftertaste, mineral.

El Chaparral de Vega Sindoa 2020 T
garnacha
89
Kind finish, dried flowers, fruity, jammy, hot, spicy, sweety finish.

El Rincón de Nekeas 2021 B
chardonnay
89
Fruity, dried flowers, dried herbs, tasty, jammy.

La Fuente de Nekeas 2019 T C
cabernet sauvignon, tempranillo
87

Los Olivos de Nekeas 2016 T R
merlot, cabernet sauvignon
90
Colour: cherry, garnet rim. Nose: fruit preserve, powerful, black fruit, smoky. Palate: flavourful, long, sweet tannins.

Nekeas Cepa x Cepa 2021 T
garnacha
89
Pleasant, floral, fruity, smooth.

NUEVOS VINOS CB
San Juan Bosco, 32 Bajo
03804 Alcoy (Alacant/Alicante)
☎: +34 965 549 172
Fax: +34 965 549 173
admon@nuevosvinos.es
www.nuevosvinos.es

Terraplén Garnacha 2021 T
100% garnacha
87

Terraplén Rosado Garnacha 2021 RD
100% garnacha
88
Kind finish, fruity, jammy, tasty.

Terraplén Viura 2021 B
viura
85

PAGO DE LARRÁINZAR
Camino de la Corona, s/n
31240 Ayegui (Navarra)
☎: +34 948 550 421
Fax: +34 948 556 120
info@pagodelarrainzar.com
www.pagodelarrainzar.com

Angel de Larrainzar 2019 T
45% tempranillo, 29% merlot, 16% garnacha, 10% cabernet sauvignon
88
Balanced, spicy, full-bodied, jammy, herbaceous, tasty.

Pago de Larrainzar Cabernet Sauvignon 2018 T
100% cabernet sauvignon
92
Colour: deep cherry. Nose: dried herbs, creamy oak, black fruit, ripe fruit, grassy. Palate: powerful, ripe fruit, spicy, round tannins.

DO NAVARRA / D.O.P.

DO NAVARRA / D.O.P.

Pago de Larrainzar Reserva Especial 2015 T R
38% merlot, 33% cabernet sauvignon, 17% garnacha, 12% tempranillo

93
Colour: deep cherry, garnet rim. Nose: aged wood nuances, ripe fruit, cocoa bean, cigar, toasty. Palate: flavourful, spicy, toasty, powerful tannins.

Raso de Larrainzar 2015 T R
47% tempranillo, 20% merlot, 21% cabernet sauvignon, 2% garnacha

91
Colour: dark-red cherry. Nose: toasty, spicy, cocoa bean, ripe fruit, black fruit. Palate: flavourful, toasty, fine bitter notes.

Rosado de Larrainzar 2021 RD
80% merlot, 20% tempranillo

89
Kind finish, dried flowers, fruity, jammy.

SEÑORÍO DE SARRÍA
Finca Señorío de Sarría, s/n
31100 Puente la Reina (Navarra)
☎: +34 948 202 202
rrpp@bornosbodegas.com
www.bodegadesarria.com

Señorío de Sarría 2010 T GR
merlot, cabernet sauvignon

91
Classic, full-bodied. Colour: deep cherry. Nose: ripe fruit, dried herbs, cocoa bean, spicy. Palate: ripe fruit, spicy, round tannins.

Señorío de Sarría 2016 T R
cabernet sauvignon, graciano

89
Pruney, balanced, spicy, dried herbs, toasty.

Señorío de Sarría 2018 T C
cabernet sauvignon, garnacha, graciano

89
Balanced, spicy, dried herbs, toasty, age nuances.

Señorío de Sarría 2021 RD
100% garnacha

89
Nose: red berry notes, floral. Palate: fruity, good acidity, easy to drink.

Señorío de Sarría Chardonnay 2021 B
100% chardonnay

86

Señorío de Sarría Rosé 2021 RD
garnacha, graciano

89
Kind finish, fruity, jammy, tasty.

Señorío de Sarría Viñedo Nº 1 2019 T C
100% garnacha

87

Señorío de Sarría Viñedo Nº 5 2021 RD
100% garnacha

89
Fruity, dried herbs, tasty, kind finish.

UBETA WINES
31523 Barillas (Navarra)
☎: +34 678 421 303
ubeta@ubetawines.com
www.ubetawines.com

Antón Aguirre 2021 T
100% garnacha

92 🍷
Colour: bright cherry. Nose: fresh fruit, red berry notes, wild herbs, floral. Palate: good acidity, spicy, fine tannins, easy to drink.

Berabal de Ubeta 2020 T

91
Colour: cherry, garnet rim. Nose: balsamic herbs, sweet spices, scrubland, red berry notes, ripe fruit. Palate: spicy, balsamic, good acidity.

Ubeta Rose 2021 RD
100% garnacha

89 🍀

Jammy, fruity, tasty, powerful.

Ubeta Garnacha 2019 T FB
100% garnacha

91 🍀

Colour: bright cherry. Nose: ripe fruit, wild herbs, dried flowers, sweet spices. Palate: fruity, spicy, round tannins, easy to drink.

Ubeta Garnacha Blanca 2021 B FB
100% garnacha blanca

90 🍀

Colour: bright yellow. Nose: ripe fruit, spicy, floral. Palate: rich, good structure, long, fine bitter notes.

UNSI
Rúa Alcalde Maillata, 3C
31390 Olite (Navarra)
☎: +34 689 482 741
unsi@unsiwines.com
www.unsiwines.com

Unsi "Finca El Boyeral" 2017 T BA
100% garnacha

93

Colour: deep cherry. Nose: ripe fruit, black fruit, wild herbs, spicy. Palate: fruity, flavourful, good structure, balanced, slightly dry, soft tannins.

Unsi "Finca Lasierra" 2016 T
100% garnacha

92

Colour: bright cherry. Nose: ripe fruit, black fruit, spicy, toasty, dried herbs. Palate: flavourful, fruity, balanced, slightly dry, soft tannins.

Unsi "Terrazas Blanco" 2021 B BA
garnacha blanca

91

Colour: bright straw. Nose: ripe fruit, white fruit, neat. Palate: flavourful, fresh, good acidity, fruity aftertaste, spicy.

Unsi "Terrazas" 2019 T
garnacha

90

Colour: cherry, purple rim. Nose: fruit expression, red berry notes, spicy, fragrant herbs. Palate: flavourful, fruity, good acidity, balanced, slightly dry, soft tannins.

Unsi Dulce Garnacha RF RC D
garnacha

91

Oxidativ. Colour: coppery red, hazy. Nose: varnish, candied fruit, faded flowers, honeyed notes. Palate: flavourful, sweet.

VALDELARES
Ctra. Eje del Ebro, km. 58
31579 Carcar (Navarra)
☎: +34 616 116 703
valdelares@valdelares.com
www.valdelares.com

Valdelares 2018 T C
87

Valdelares 2019 T C
87

Valdelares 2021 RD
merlot

89

Pleasant, tasty, jammy, fruity.

Valdelares 2021 T
tempranillo, merlot

87

Valdelares Chardonnay 2021 B
chardonnay

86

Valdelares Sauvignon Blanc 2021 B
sauvignon blanc

86

DO NAVARRA / D.O.P.

DO NAVARRA / D.O.P.

VINOS Y VIÑEDOS DOMINIO LASIERPE
Ribera, s/n
31592 Cintruénigo (Navarra)
☎: +34 948 811 033
comercial@dominiolasierpe.com
www.dominiolasierpe.com

Dominio Lasierpe 1920 Centenario 2020 T
100% garnacha
89
Jammy, wild, pleasant, standard, toasty.

Dominio Lasierpe 2019 T C
garnacha, tempranillo
88
Herbal, fruity, jammy, spicy, standard.

Finca Lasierpe Chardonnay 2021 B
100% chardonnay
86

Flor de Lasierpe Garnacha 2021 RD PL
100% garnacha
87

Flor de Lasierpe Tinto Selección Garnacha 2021 T
100% garnacha
88
Balanced, spicy, crisp, fruity.

VIÑA PALACIOS
Cerco de Fuera, 10
31390 Olite (Navarra)
☎: +34 616 055 414
bodega@vinapalacios.es
www.vinapalacios.es

El Arrebol de la Carra Cabra 2020 RD
garnacha
88
Pruney, fruity, sweety finish, balanced, tasty.

La Carra Cabra 2020 T
garnacha
90
Colour: bright cherry. Nose: black fruit, smoky, spicy, fruit preserve, wild herbs. Palate: fruity, flavourful, balanced, slightly dry, soft tannins.

VIÑA ZORZAL WINES
Ctra. del Villar, s/n
31591 Corella (Navarra)
☎: +34 948 780 617
xabi@vinazorzalwines.com
www.vinazorzalwines.com

Cuatro del Cuatro 2020 T
graciano
92
Defined aromas, herbal. Colour: bright cherry. Nose: wild herbs, dried herbs, expressive. Palate: varietal, balsamic.

Jirafas 2019 B
92
Colour: yellow. Nose: dried flowers, white flowers, dried herbs, fine lees. Palate: balanced, flavourful, fruity, good acidity, long.

Matias Michelini Garnache 2019 T
garnacha
93
Defined aromas. Colour: bright cherry. Nose: floral, red berry notes, fresh fruit, varietal, expressive. Palate: fruity, taut, balanced.

Punto de Fuga 2019 T
94
Nominado Vino Revelación
Colour: Cherry. Nose: complex, expressive, spicy, red berry notes. Palate: elegant, long, great length, easy to drink.

Viña Zorzal Corral de los Altos 2020 T
92
Floral, wild. Colour: light cherry. Nose: medium intensity, fresh, wild herbs, scrubland. Palate: fruity, easy to drink.

Viña Zorzal Garnacha 2020 T C
91
Colour: bright cherry. Nose: red berry notes, ripe fruit, floral, spicy. Palate: fruity, good acidity, long.

Viña Zorzal Garnacha 2021 T
garnacha
92
Colour: bright cherry, purple rim. Nose: fresh fruit, wild herbs, dried flowers. Palate: fresh, fruity, easy to drink.

Viña Zorzal Garnacha Blanca 2020 B
garnacha blanca
90
Colour: bright straw. Nose: fresh fruit, floral, medium intensity. Palate: fruity, balanced, easy to drink, fine bitter notes.

Viña Zorzal Graciano 2020 T
graciano

92

Defined aromas, herbal. Colour: bright cherry. Nose: balanced, fresh, fresh fruit. Palate: varietal, easy to drink, good acidity, fresh.

Viña Zorzal Lecciones de Vuelo 2020 T
garnacha

91

Colour: light cherry. Nose: ripe fruit, dried herbs, creamy oak, caramel, sweet spices. Palate: ripe fruit, spicy.

Viña Zorzal Señora de las Alturas 2020 T
94

Defined aromas, fruity. Colour: cherry, garnet rim. Nose: red berry notes, ripe fruit, wild herbs. Palate: balanced, flavourful, long.

Viñas Zorzal Malayeto 2020 T
93

Colour: bright cherry. Nose: red berry notes, ripe fruit, wild herbs, floral. Palate: fruity, long, full of life, flavourful, varietal.

DO. PENEDÈS

CONSEJO REGULADOR

Plaça Àgora. s/n. Pol. Ind. Domenys, II
08720 Vilafranca del Penedès (Barcelona)
☎: +34 938 904 811 - Fax: +34 938 904 754
@: dopenedes@dopenedes.cat
www.dopenedes.cat

LOCATION:

In the province of Barcelona, between the pre-coastal Catalonian mountain range and the plains that lead to the Mediterranean coast. There are three different areas: Penedès Superior, Penedès Central or Medio and Bajo Penedès.

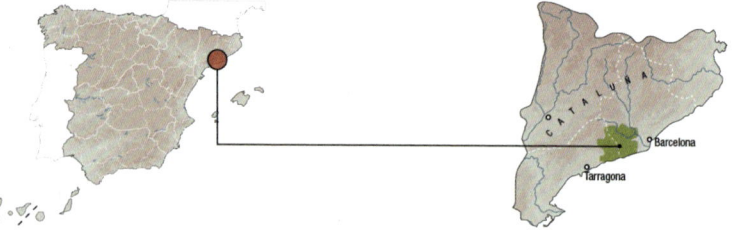

SUB-REGIONS:

Penedès Superior: The vineyards reach an altitude of 800 m; the traditional, characteristic variety is the Parellada, which is better suited to the cooler regions. **Penedès Central or Medio:** Cava makes up a large part of the production in this region; the most abundant traditional varieties are Macabeo and Xarel·lo. **Bajo Penedès:** This is the closest region to the sea, with a lower altitude and wines with a markedly Mediterranean character.

GRAPE VARIETIES:

WHITE: Macabeo, Xarel·lo, Parellada, Chardonnay, Riesling, Gewürztraminer, Chenin Blanc, Forcada, Moscatel de Alejandría, Garnacha Blanca, Viognier, Sumoi Blanc and Malvasía.

RED: Garnacha, Merlot, Samsó, Ull de Llebre (Tempranillo), Pinot Noir, Monastrell, Cabernet Sauvignon, Petit Verdot, Moneu, Syrah, Sumoll and Picapoll Negra.

FIGURES:

Vineyard surface: 2,500 – **Wine-Growers:** 2,100 – **Wineries:** 136 – **2021 Harvest rating:** N/A – **Production 21:** 13,425,000 litres – **Market percentages:** 75% National - 25% International.

SOIL:

There is deep soil, not too sandy or too clayey, permeable, which retains the rainwater well. The soil is poor in organic matter and not very fertile.

CLIMATE:

Mediterranean, in general warm and mild; warmer in the Bajo Penedès region due to the influence of the Mediterranean Sea, with slightly lower temperatures in Medio Penedès and Penedès Superior, where the climate is typically pre-coastal (greater contrasts between maximum and minimum temperatures, more frequent frosts and annual rainfall which at some places can reach 990 litres per square metre).

CLASSIFICATION OF YOUNG WINE HARVESTS PEÑÍNGUIDE

2017	2018	2019	2020	2021
GOOD	VERY GOOD	VERY GOOD	VERY GOOD	VERY GOOD

AGUSTI TORELLÓ MATA

Partida La Serra s/n
08770 Sant Sadurní d'Anoia (Barcelona)
☎: +34 938 911 173
info@agustitorellomata.com
www.agustitorellomata.com

Agustí Torelló Mata Espantallops 2019 B C
100% macabeo

91

Colour: bright yellow. Nose: dried flowers, candied fruit, fine lees, pattiserie, toasty. Palate: round, spicy, long, great length.

Agustí Torelló Mata XIC 2020 B
100% xarel.lo

90

Colour: straw. Nose: ripe fruit, dried herbs, faded flowers, dry nuts. Palate: powerful, ripe fruit, balanced.

Agustí Torelló Mata XV Xarel·lo Vermell 2021 RD

90

Colour: onion pink. Nose: red berry notes, wild herbs, dried flowers, fine lees. Palate: fresh, fruity, flavourful.

ALBET I NOYA

Can Vendrell de la Codina, s/n
08739 Sant Pau D'Ordal (Barcelona)
☎: +34 938 994 812
info@albetinoya.cat
www.albetinoya.cat

Albet i Noya Brut 21 2019 BE BR
chardonnay, parellada

91

Colour: bright straw. Nose: fine lees, floral, fragrant herbs, expressive. Palate: powerful, flavourful, good acidity, fine bead, balanced.

Albet i Noya El Fanio 2020 B
100% xarel.lo

92

Colour: bright straw. Nose: expressive, fine lees, dried flowers, balanced. Palate: full-bodied, spicy, long, elegant.

Albet i Noya La Milana 2018 T R
merlot, tempranillo, caladoc

93

Colour: dark-red cherry. Nose: toasty, spicy, cocoa bean, black fruit, ripe fruit, earthy notes. Palate: flavourful, toasty, fine bitter notes.

Albet i Noya Reserva Martí 2016 T GR
cabernet sauvignon, merlot, tempranillo, syrah

92

Colour: deep cherry. Nose: dried herbs, creamy oak, black fruit, spicy. Palate: powerful, ripe fruit, spicy, round tannins.

El Bosc Negre 2019 B
xarel.lo

92

Colour: bright yellow. Nose: powerful, creamy oak, ripe fruit, spicy. Palate: rich, good structure, long, toasty, fine bitter notes.

El Corral Cremat 2011 BE GR BR
100% xarel.lo

94

Colour: bright golden. Nose: fine lees, dry nuts, fragrant herbs, complex, toasty. Palate: powerful, flavourful, good acidity, fine bead, fine bitter notes.

ALEMANY I CORRIO

Melió, 78
08720 Vilafranca del Penedès (Barcelona)
☎: +34 938 180 949
sotlefriec@sotlefriec.com
www.alemany-corrio.com

Pas Curtei 2019 T

91

Colour: cherry, garnet rim. Nose: fruit preserve, powerful, dried herbs, hot, black fruit. Palate: flavourful, long.

Principia Mathematica 2021 B

91

Little interventionist, slight reduction. Colour: bright straw, greenish rim. Nose: fresh fruit, citrus fruit, wild herbs, fine lees, cereal notes. Palate: fresh, fruity, good acidity, fine bitter notes.

Sot Lefriec 2017 T

94

Colour: Cherry. Nose: balsamic herbs, ripe fruit, characterful, black fruit, dried herbs. Palate: spicy, balsamic, toasty, round tannins.

ALSINA & SARDÁ

Barrio Les Tarumbas, s/n
08733 El Pla del Penedès (Barcelona)
☎: +34 938 988 132
Fax: +34 938 988 132
alsina@alsinasarda.com
www.alsinasarda.com

Alsina & Sardá Boires 2018 T C
garnacha, cariñena

90

Colour: deep cherry. Nose: ripe fruit, dried herbs, spicy, red berry notes. Palate: ripe fruit, spicy, round tannins.

Alsina & Sardá Finca Cal Janes 2020 T C
merlot

87

DO PENEDÈS / D.O.P.

Alsina & Sardá Finca La Boltana 2021 B
xarel.lo
91
Colour: bright straw. Nose: fragrant herbs, fine lees, dried flowers, tropical fruit. Palate: full-bodied, rich, long, good acidity.

Cromàtic Chardonnay + Xarel.lo 2021 B
chardonnay, xarel.lo
87

Cromàtic Merlot 2021 RD
merlot
87

Cromàtic Muscat 2021 B
moscatel
88
Defined aromas, floral, varietally correct, smooth, simple.

AT ROCA
La Vinya, 15-17
08770 Sant Sadurní d'Anoia (Barcelona)
☎: +34 935 910 043
info@atroca.eu
www.atroca.eu

At Roca 2019 BE R BN
60% macabeo, 35% xarel.lo, 5% parellada
90
Colour: straw. Nose: expressive, fine lees, dried flowers, faded flowers. Palate: flavourful, fruity, fresh.

At Roca Rosat 2019 RE R BN
50% macabeo, 50% garnacha
90
Pleasant, aromatic. Nose: fresh fruit, floral, balanced. Palate: fresh, fruity.

Cantallops 2019 B
100% xarel.lo
92
Colour: bright yellow. Nose: dried herbs, faded flowers, candied fruit, stone fruit. Palate: powerful, ripe fruit, balanced.

Finca Els Gorgs 2013 BE GR
macabeo, xarel.lo
94
Colour: bright yellow. Nose: fine lees, fragrant herbs, characterful, ripe fruit, dry nuts. Palate: powerful, flavourful, good acidity, fine bead, fine bitter notes.

L'Esparter 2015 BE GR BN
100% macabeo
90
Colour: bright straw. Nose: fine lees, fragrant herbs, fresh fruit. Palate: good acidity, balanced, easy to drink, fresh.

Pedregar 2016 BE GR BN
85% garnacha, 15% macabeo
92

Pedregar 2016 RE R
garnacha
92
Colour: coppery red, bright. Nose: ripe fruit, bakery, brioche, dry nuts, dried flowers. Palate: flavourful, balanced.

AVGVSTVS FORVM
Ctra. Sant Vicenç, s/n
43700 El Vendrell (Tarragona)
☎: +34 977 666 910
avgvstvs@avgvstvsforvm.com
www.avgvstvsforvm.com

Avgvstvs Chardonnay 2021 B FB
100% chardonnay
91
Colour: bright yellow. Nose: powerful, creamy oak, spicy, white fruit, ripe fruit. Palate: rich, good structure, long, toasty, fine bitter notes.

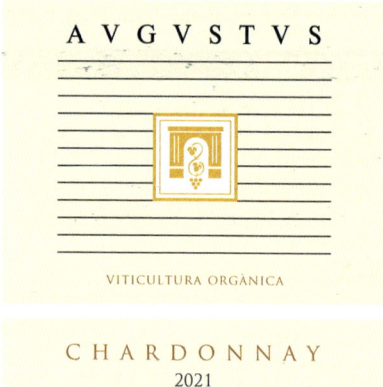

Avgvstvs Microvinificació Macabeo 2017 B FB
100% macabeo
91
Little interventionist. Colour: bright yellow. Nose: spicy, burnt matches, expressive, dried herbs. Palate: flavourful, fine bitter notes, balanced.

Avgvstvs Microvinificació Malvasia de Sitges 2021 B
100% malvasía de Sitges
91
Colour: straw. Nose: expressive, white flowers, jasmine, fresh, citrus fruit. Palate: flavourful, fruity, balanced.

DO PENEDÈS / D.O.P.

Avgvstvs Microvinificació Xarel.lo +100 2019 B FB
100% xarel.lo

91

Colour: bright yellow. Nose: ripe fruit, dried herbs, faded flowers, toasty. Palate: powerful, ripe fruit, balanced.

Avgvstvs Microvinificació Xarel.lo de Mar 2017 B FB
100% xarel.lo

91

Colour: bright yellow. Nose: dried flowers, candied fruit, fine lees, pattiserie, toasty. Palate: round, spicy, long, great length.

Avgvstvs Trajanvs 2017 T R
46% cabernet franc, 42% cabernet sauvignon, 12% merlot

90

Colour: cherry, garnet rim. Nose: creamy oak, hot, fruit preserve, dried herbs, sweet spices. Palate: sweet tannins, fruity, balanced, slightly dry, soft tannins.

AYMAR
Font de les Graus, 12-14
08720 Castellví de la Marca (Barcelona)
☎: +34 938 905 151
info@aymar.cat
www.aymar.cat

Aymar 2016 BE R BN
50% xarel.lo, 25% parellada, 25% macabeo

87

Aymar Rosé 2019 RE R EBR
100% garnacha

88

Pleasant, jammy, tasty.

Aymar Tranquil Blanc 2021 B
macabeo

87

Aymar Tranquil Negre 2021 T
cabernet sauvignon, tempranillo, garnacha

86

Castell de Pujades Xarel.lo 2021 B
100% xarel.lo

90

Colour: bright straw. Nose: fruit expression, ripe fruit, grassy, dried herbs. Palate: flavourful, fresh, good acidity, fruity aftertaste.

Castell de Pujades Xarel.lo Vermell 2021 RD
100% xarel.lo vermell

89

Smooth, wild, floral, balanced, dried flowers, fruity.

BLANCHER-CAPDEVILA PUJOL
Plaça Pont Romà, Edifici Blancher
08770 Sant Sadurní d'Anoia (Barcelona)
☎: +34 938 183 286
blancher@blancher.es
www.blancher.es

Capdevila Pujol Vica 2021 B
macabeo, moscatel de alejandría

89

Balanced, floral, fruity, herbal, full-bodied, jammy.

Mercè Jove 2021 T
merlot, tempranillo

87

BODEGA MIQUEL JANÉ
Masia Cal Costas, s/n
08736 Font-Rubí (Barcelona)
☎: +34 654 127 597
info@jmiqueljane.com
www.miqueljane.com

Clàssic Penedès Parellada 2018 BE R BN
100% parellada

92

Colour: bright straw. Nose: fresh fruit, citrus fruit, fine lees, fragrant herbs. Palate: fresh, fruity, good acidity.

Clàssic Sauvignon Blanc Miquel Jané 2017 BE BN
100% sauvignon blanc

90

Colour: bright straw. Nose: fine lees, floral, fragrant herbs, expressive. Palate: flavourful, good acidity, fine bead, balanced.

Masia Cal Costas Cabernet Sauvignon Syrah 2017 T
50% cabernet sauvignon, 50% syrah

88

Jammy, tasty, toasty, full-bodied, spicy.

Miquel Jané Baltana Garnacha 2021 RD
100% garnacha

89

Fruity, dried flowers, pleasant, jammy, tasty.

Miquel Jané Sauvignon Blanc 2021 B D
100% sauvignon blanc

89

Varietally correct, herbal, fruity, crisp.

Miquel Jané Xarel.lo 2021 B
100% xarel.lo

88

Citrus fruit, crisp, fruity, herbal.

BODEGA ST. JOANNES
Casa Gran S.Joan Samora s/n
08791 Sant Llorenç d'Hortons (Barcelona)
☎: +34 609 022 753
rosa@stjoannes.com
www.stjoannes.com

Matha by St Joannes 2019 T
90% syrah, 10% cabernet sauvignon
89
Spicy, dried flowers, balanced, jammy.

St Joannes 2019 T
85% cabernet sauvignon, 15% syrah
89
Wild, balsamic herbs, hot, jammy, tasty, herbal.

St Joannes 2021 B
xarel.lo
90
Oxidativ, tasty, jammy, herbal, yeasty notes.

St Joannes Rosé 2021 RD
syrah
87

BODEGAS CA N'ESTELLA
Masia Ca n'Estella, s/n
08635 Sant Esteve Sesrovires (Barcelona)
☎: +34 934 161 387
Fax: +34 934 161 620
a.vidal@fincacanestella.com
www.fincacanestella.com

Clot dels Oms Blanc 2021 B
89
Aromatic, floral, crisp, simple, wild, smooth.

Clot dels Oms Negre 2018 T R
87

Clot dels Oms Rosat 2021 RD
87

Gran Clot dels Oms Xarel.lo 2019 B BA
xarel.lo
91
Colour: bright yellow. Nose: ripe fruit, spicy, toasted bread. Palate: rich, long, fine bitter notes, good acidity.

BODEGAS CAPITÀ VIDAL
Ctra. Vilafranca- Igualada, C-15, km 21
08733 El Pla del Penedès (Barcelona)
☎: +34 938 988 630
Fax: +34 938 988 625
administracion@capitavidal.com
www.capitavidal.com

Clos Vidal Blanc de Blancs 2021 B
75% xarel.lo, 10% parellada, 10% moscatel de frontignan, 5% macabeo
88
Pleasant, standard, floral, crisp, fruity, smooth.

Clos Vidal Cabernet Sauvignon 2019 T RB
85% cabernet sauvignon, 15% syrah
90
Colour: deep cherry. Nose: ripe fruit, dried herbs, creamy oak. Palate: powerful, ripe fruit, spicy, round tannins.

Clos Vidal Merlot 2018 T C
85% merlot, 15% tempranillo
87

Clos Vidal Rosé Cuvée 2021 RD
40% syrah, 35% merlot, 25% garnacha
87

BODEGAS MASET
Ctra. Vilafranca-Igualada C-15 Km.19
08792 La Granada (Barcelona)
☎: +34 900 200 250
info@maset.com
www.maset.com

Maset Cabernet Sauvignon 2019 T R
cabernet sauvignon
88
Pleasant, toasty, jammy, spicy.

Maset Eufòria 2021 RD
merlot, garnacha
89
Balanced, floral, herbal, crisp.

Maset Foc Merlot 2018 T R
merlot
91
Colour: cherry, garnet rim. Nose: fruit preserve, fruit liqueur notes, powerful, dark chocolate, creamy oak. Palate: flavourful, long, slightly dry, soft tannins.

Maset La Sínia 2020 B FB
xarel.lo
89
Pleasant, standard, spicy, dried flowers, fruity, dried herbs, smooth.

Maset La Soledad 2020 B FB
chardonnay

91

Colour: bright yellow. Nose: dried flowers, candied fruit, fine lees, pattiserie. Palate: round, spicy, great length, flavourful, fresh.

Maset Singular Xarel·lo Biodinàmic 2021 B
xarel.lo

89

Fruity, dried herbs, tasty, jammy, pleasant.

BOLET AGRICULTURA ECOLÓGICA

Finca Mas Lluet s/n
08732 Castellví de la Marca (Barcelona)
☎: +34 938 918 153
cavasbolet@cavasbolet.com
www.cavasbolet.com

Bolet Apagallums Pinot Noir 2021 RD
pinot noir

86

Bolet Camagroc Xarel.lo 2021 B
xarel.lo

88

Balanced, spicy, herbal, jammy.

Bolet Cantarelus Ull de Llebre 2019 T
tempranillo

87

Bolet Fredolic (Sin Sulfitos) T

88

Rustic, herbaceous, jammy. Palate: slight fizz.

Bolet Sàpiens Merlot 2016 T C
merlot

88

Pleasant, herbal, jammy, standard, hot.

Bolet Vinya Sota Bosc 2021 B
moscatel, gewürztraminer

88

Kind finish, sweety finish, fruity, jammy, tasty.

CAN BAS DOMINI VINICOLA

Crtra. De Vilafranca km. 4
08739 Subirats (Barcelona)
☎: +34 938 994 173
info@can-bas.com
www.can-bas.com

Can Bas D'Origen P3 2021 B
100% xarel.lo

90

Colour: bright yellow. Nose: spicy, stone fruit, ripe fruit. Palate: rich, good structure, long, toasty, fine bitter notes.

Can Bas D'Origen P5 2021 B
moscatel grano menudo

91

Colour: straw. Nose: expressive, white flowers, jasmine, white fruit. Palate: fruity, balanced, good acidity.

Can Bas La Creu 2021 B
sauvignon blanc

91

Herbal. Colour: bright straw, greenish rim. Nose: fresh fruit, wild herbs. Palate: fresh, fruity, good acidity.

Can Bas La Romana 2018 B FB
55% chardonnay, 45% xarel.lo

92

Colour: yellow. Nose: toasted bread, spicy, smoky, ripe fruit, bakery. Palate: flavourful, long, toasty, great length, spicy.

Monreal 2015 T
cabernet sauvignon

90

Colour: bright cherry. Nose: fruit preserve, black fruit, dried herbs, dark chocolate, toasty, spicy. Palate: fruity, flavourful, balanced, roasted-coffee aftertaste, round tannins.

CAN RÀFOLS DELS CAUS

Can Rafols del Caus s/n
08793 Avinyonet del Penedès (Barcelona)
☎: +34 938 970 013
info@causgrup.com
www.canrafolsdelscaus.com

El Rocallís 2018 B FB
100% incrocio manzoni

92

Colour: bright yellow. Nose: powerful, creamy oak, ripe fruit, spicy. Palate: rich, good structure, toasty, fine bitter notes.

Gran Caus 2016 T R
cabernet franc, merlot, cabernet sauvignon

93

Colour: cherry, garnet rim. Nose: ripe fruit, scrubland, old leather, black liquorice. Palate: flavourful, spicy, soft tannins, balanced.

DO PENEDÈS / D.O.P.

DO PENEDÈS / D.O.P.

Gran Caus 2021 B
xarel.lo, chenin blanc, chardonnay

92

Colour: straw. Nose: expressive, white flowers, dried herbs, wild herbs, dry nuts. Palate: fruity, balanced, good acidity.

Gran Caus 2021 RD
merlot

91

Colour: brilliant rose. Nose: spicy, red berry notes, floral, ripe fruit, wild herbs. Palate: flavourful, good acidity, balanced.

La Calma 2019 B FB
100% chenin blanc

92

Colour: bright yellow. Nose: dried flowers, candied fruit, fine lees, pattiserie. Palate: round, spicy, long, great length.

Xarel.lo Pairal 2019 B FB
100% xarel.lo

93

Colour: straw. Nose: ripe fruit, dried herbs, faded flowers, fine lees. Palate: powerful, ripe fruit, balanced, fruity.

CAN SUMOI
Plaça del Roure s/n
08240 Sant Sadurní d'Anoia (Barcelona)
☎: +34 938 183 262
info@cansumoi.cat

Can Sumoi La Rosa 2021 RD
sumoll, montonega

90

Colour: rose. Nose: ripe fruit, fruit preserve, hot, faded flowers. Palate: fleshy, flavourful, ripe fruit.

Can Sumoi Perfum 2021 B
garnacha blanca, macabeo

88

Citrus fruit, herbal, floral, tasty, crisp.

Can Sumoi Xarel.lo 2021 B
xarel.lo

91

Colour: yellow. Nose: hint of anise, wild herbs, fine lees, lees reduction notes, characterful. Palate: flavourful, spicy, great length.

CANALS I MUNNÉ
Plaza Pau Casals, 6
08770 Sant Sadurní d'Anoia (Barcelona)
☎: +34 938 910 318
Fax: +34 938 911 945
marketing@canalsimunne.com
www.canalsimunne.com

Blanc Prínceps Muscat 2021 B
moscatel grano menudo

88

Standard, varietally correct, smooth, floral, balanced.

Gran Blanc Prínceps 2021 B

89

Aromatic, spicy, jammy, tasty, toasty, dried flowers.

Gran Prínceps 2016 T R
garnacha, samsó

90

Colour: Cherry. Nose: balsamic herbs, sweet spices, scrubland. Palate: spicy, good acidity, round tannins.

Noir Prínceps 2018 T C
50% cabernet sauvignon, 35% tempranillo, 15% syrah

88

Pleasant, tasty, jammy, spicy.

Rosé Prínceps 2021 RD
50% merlot, 50% syrah

88

Citrus fruit, standard, dried herbs, crisp.

VXVX Xarello Vermell 2021 RD

88

Citrus fruit, fruity, herbal, tasty.

CAVA & HOTEL MASTINELL
Ctra. Vilafranca a Sant Martí Sarroca, km. 0,5
08720 Vilafranca del Penedès (Barcelona)
☎: +34 938 170 586
info@mastinell.com
www.mastinell.com

Arte 2020 T
garnacha, cariñena, cabernet sauvignon

87

Eliane Chardonnay 2021 B
100% chardonnay

88

Hot, citrus fruit, standard, full-bodied, jammy, herbal.

Gisele 2021 B
100% xarel.lo

92

Colour: bright straw. Nose: ripe fruit, fragrant herbs, fine lees, bakery, spicy. Palate: full-bodied, rich, long, good acidity.

CAVA ORIOL ROSSELL

Propietat Cal Cassanyes
08732 Sant Marçal (Barcelona)
☎: +34 977 671 061
oriolrossell@oriolrossell.com
www.oriolrossell.com

Beach Rosé by Or 2021 RD
garnacha, syrah

88

Crisp, fruity, simple, pleasant, standard, floral, thin.

El Carro Gros 2016 T
40% syrah, 40% cabernet sauvignon, 20% merlot

91

Colour: cherry, garnet rim. Nose: fruit preserve, powerful, cereal notes, dried herbs. Palate: flavourful, coarse tannins, fruity, balanced.

Les Cerveres Xarel.lo 2019 B
100% xarel.lo

90

Colour: bright yellow. Nose: powerful, ripe fruit, spicy, toasty. Palate: rich, long, bitter, roasted-coffee aftertaste.

Ou de Paó 2020 B
xarel.lo

90

Colour: bright yellow. Nose: creamy oak, ripe fruit, spicy, dry nuts. Palate: rich, good structure, toasty, flavourful.

Rocaplana 2020 T
100% syrah

87

Virolet Xarel.lo 2021 B
8,5% xarel.lo

87

CAVAS HILL

Bonavista, 2
08734 Moja-Olérdola (Barcelona)
☎: +34 938 900 588
Fax: +34 938 170 246
cavashill@cavashill.com
www.cavashill.es

Blanc Bruc 2020 B

90

Colour: bright straw. Nose: fruit expression, ripe fruit, wild herbs, fine lees. Palate: flavourful, fresh, good acidity.

El Rei de Mambo 2020 T
cabernet sauvignon, syrah

87

Gran Toc 2016 T
syrah, merlot, cabernet sauvignon

88

Standard, herbal, jammy, tasty, dried herbs.

Kýma 2018 T R
cabernet sauvignon

89

Spicy, fruity, smoky, tasty.

Kýma Macabeo 2019 B
macabeo

90

Colour: straw. Nose: ripe fruit, dried herbs, faded flowers, white flowers. Palate: ripe fruit, balanced, flavourful, fruity.

La Clau de Volta 2020 T
merlot

88

Balanced, spicy, herbal, jammy.

CAVAS MIQUEL PONS

08792 La Granada (Barcelona)
☎: +34 938 974 541
Fax: +34 938 974 710
miquelpons@cavamiquelpons.com
www.cavamiquelpons.com

Miquel Pons 77 Veremas Garnacha 2020 T
garnacha, syrah, cabernet sauvignon

88

Spicy, fruity, herbal, jammy, tasty.

Miquel Pons 77 Veremas Xarel·lo 2021 B FB
xarel.lo

90

Colour: bright straw. Nose: ripe fruit, spicy, dried flowers. Palate: rich, good structure, fruity, fresh.

Miquel Pons Núria de Montargull 2021 B
moscatel de alejandría, chardonnay

88

Aromatic, fruity, floral, jammy.

Núria de Montargull 2021 RD
ull de llebre

89

Citrus fruit, dried flowers, crisp, dried herbs.

CAVES NAVERAN

Can Parellada
08735 Torrelavit (Barcelona)
☎: +34 938 988 274
naveran@naveran.com

Don Pablo 2018 T
cabernet sauvignon, merlot

90

Colour: deep cherry. Nose: dried herbs, creamy oak, black fruit. Palate: ripe fruit, spicy, round tannins.

Manuela de Naverán 2021 B
chardonnay

91

Colour: bright yellow. Nose: creamy oak, ripe fruit, spicy. Palate: rich, good structure, toasty, fine bitter notes.

Naveran Clos Antonia 2020 B FB
viognier

90

Colour: bright yellow. Nose: creamy oak, ripe fruit, spicy, dried flowers, waxy notes, fine lees. Palate: rich, good structure, fine bitter notes.

Naveran Clos dels Àngels 2020 T MC
syrah

90

Colour: deep cherry. Nose: dried herbs, creamy oak, black fruit. Palate: ripe fruit, spicy, round tannins.

Naveran Flor de Pinot 2021 RD
pinot noir

89

Defined aromas, crisp, fruity, balanced, floral, smooth.

CELLER CREDO

Tamarit, 10
08770 Sant Sadurní d'Anoia (Barcelona)
☎: +34 938 910 214
vins@cellercredo.cat
www.cellercredo.cat

Aloers 2019 B
100% xarel.lo

92

Little interventionist. Colour: straw. Nose: ripe fruit, dried herbs, faded flowers, dry nuts. Palate: powerful, ripe fruit, balanced.

Mirabelles 2018 B
100% malvasía de Sitges

91

Oxidativ. Colour: straw. Nose: ripe fruit, dried herbs, faded flowers. Palate: powerful, ripe fruit, balanced.

Miranius 2021 B
100% xarel.lo

90

Colour: bright straw. Nose: fruit expression, ripe fruit, fine lees, grassy. Palate: flavourful, fresh, good acidity.

Miranius Magnum 2019 B
100% xarel.lo

90

Oxidativ. Colour: bright straw. Nose: ripe fruit, fragrant herbs, fine lees. Palate: full-bodied, rich, good acidity.

Ratpenat 2019 B
100% macabeo

91

Colour: bright straw. Nose: fruit expression, ripe fruit, floral, fragrant herbs. Palate: flavourful, fresh, good acidity, fruity aftertaste.

Volaina 2019 B
100% parellada

91

Colour: bright straw, greenish rim. Nose: fresh fruit, citrus fruit, wild herbs, grassy. Palate: fresh, fruity, good acidity, fine bitter notes.

CELLER HOSPITAL DE SITGES

Plaza Joan Duran i Ferret s/n
08870 Sitges (Barcelona)
☎: +34 672 682 481
celler@hospitaldesitges.cat
www.hospitaldesitges.cat

Blanc Subur 2020 B
100% malvasía de Sitges

88

Little interventionist, yeasty notes, herbal, dried flowers, tasty.

Llegat Llopis 2017 B
100% malvasía de Sitges

91

Colour: bright yellow. Nose: citrus fruit, white fruit, faded flowers, honeyed notes. Palate: rich, good structure, long, toasty, fine bitter notes.

Monembasia 2017 BE BN
malvasía de Sitges

90

Colour: bright yellow. Nose: dried flowers, burnt matches, characterful. Palate: fruity, flavourful, balanced, sweetness.

COLET

Camino del Salinar, s/n
08796 Pacs del Penedès (Barcelona)
☎ +34 938 170 809
info@colet.cat
www.colet.cat

Colet A Priori 2019 BE R BR
macabeo, chardonnay, riesling, gewürztraminer, moscatel

91 🌱
Colour: bright straw. Nose: fresh fruit, citrus fruit, fine lees, fragrant herbs. Palate: fresh, fruity, good acidity.

Colet Assemblage 2016 RE EBR
pinot noir, chardonnay

91 🌱
Age nuances, weary. Colour: coppery red, bright. Nose: faded flowers, ripe fruit, spicy. Palate: flavourful, long, fine bead.

Colet Gran Cuveé 2018 BE R EBR
chardonnay, macabeo, xarel.lo

91 🌱
Colour: bright yellow. Nose: ripe fruit, fine lees, balanced, dried herbs, balsamic herbs. Palate: good acidity, fresh, easy to drink.

Colet Navazos (etiq.naranja) 2018 BE R EBR
xarel.lo

93 🌱
Colour: golden. Nose: ripe fruit, balanced, dried herbs, brioche, dry nuts. Palate: good acidity, flavourful, ripe fruit, fine solera notes.

Colet Navazos (etiq.verde) 2017 BE R EBR
chardonnay

93
Sweety finish. Colour: bright golden. Nose: ripe fruit, dry nuts, brioche, citrus fruit, caramel, faded flowers. Palate: ripe fruit, full-bodied, balanced.

Colet Tradicional 2018 BE R EBR
xarel.lo, macabeo

92 🌱
Colour: bright yellow. Nose: ripe fruit, fine lees, balanced, dried herbs. Palate: good acidity, flavourful, ripe fruit.

Colet Vatua! 2019 BE EBR
moscatel, gewürztraminer

91 🌱
Colour: bright straw. Nose: fine lees, floral, fragrant herbs, expressive. Palate: powerful, flavourful, good acidity, fine bead, balanced.

Colet Vatua! Rosé 2018 RE EBR
moscatel, gewürztraminer, garnacha

92 🌱
Colour: coppery red. Nose: fragrant herbs, wild herbs, medium intensity, balanced, expressive. Palate: fruity, flavourful, good acidity.

COVIDES VINYES - CELLERS

Finca Prunamala, Ctra.
St. Sadurní a Vilafranca, Km. 1
08770 Sant Sadurní D'Anoia (Barcelona)
☎ : +34 938 172 552
marketing@covides.com
www.covides.com

Duc de Foix Cabernet Sauvignon 2021 RD
cabernet sauvignon

87

Duc de Foix Cabernet Sauvignon Negre 2020 T C
cabernet sauvignon

88
Balanced, herbaceous, spicy, toasty, jammy.

Duc de Foix Merlot 2019 T
merlot

87

Duc de Foix Xarel.lo 2021 B
xarel.lo

87

Terra Terrae 2020 T
cabernet sauvignon, merlot, syrah

88 🌱
Dried herbs, jammy, standard, lactic, smooth, wild.

Terra Terrae Blanc 2020 B
chardonnay, xarel.lo, moscatel de alejandría

87 🌱

ESTEL D'ARGENT

Font-Rubí, 2 4º 1ª
08720 Vilafranca del Penedès (Barcelona)
☎ : +34 677 182 347
cava@esteldargent.com
www.esteldargent.com

Estel D'Argent 2019 T
cabernet sauvignon

89
Fruity, dried herbs, spicy, tasty, jammy.

Estel D'Argent 2021 RD
87

DO PENEDÈS / D.O.P.

DO PENEDÈS / D.O.P.

SusQuvat 2021 B BA
100% xarel.lo

90

Colour: bright straw. Nose: dry nuts, dried flowers, balanced, ripe fruit, spicy. Palate: flavourful, fruity.

ESTEVE I GIBERT VITICULTORS
Els Casots
08739 Subirats (Barcelona)
☎: +34 600 343 125
albert@esteveigibert.com
www.esteveigibert.com

Clot dels Eixams 2019 B FB
100% malvasía de Sitges

93 ♣

Colour: bright golden. Nose: wild herbs, candied fruit, ripe fruit, spicy. Palate: fresh, fruity, good acidity, balanced, flavourful, good finish.

L'Antana 2013 T R
100% merlot

90

Colour: cherry, garnet rim. Nose: fruit preserve, powerful, black fruit, old leather, sweet spices. Palate: flavourful, sweetness, full-bodied.

L'Oblit 2017 T FB
100% sumoll

92 ♣

Colour: cherry, purple rim. Nose: fruit expression, red berry notes, floral, spicy. Palate: flavourful, fruity, good acidity, long.

Les Vistes 2020 B
100% xarel.lo

92 ♣

Oxidativ. Colour: straw. Nose: ripe fruit, dried herbs, faded flowers, fine reductive notes, fine lees. Palate: powerful, ripe fruit, balanced.

Origen 2021 B
100% xarel.lo

88 ♣

Dried flowers, fruity, simple, wild, smooth.

FAMILIA AMETLLER
Barri La Rovira Roja, 10
08731 Sant Marti Sarroca (Barcelona)
☎: +34 938 904 735
info@familiaametller.com
www.familiaametller.com

Ametller Blanc Floral 2021 B
moscatel de frontignan, sauvignon blanc

89

Floral, crisp, fruity, herbal, tasty.

Ametller Garnatxa 2019 T
garnacha

89

Kind finish, toasty, tasty, smooth.

Ametller Garnatxa Negra i Carinyena 2019 T
garnacha, cariñena

89

Balanced, spicy, herbal, jammy.

Ametller Xarel.lo 2021 B
xarel.lo

87

FAMILIA TORRES
Miguel Torres i Carbó, 6
08720 Vilafranca del Penedès (Barcelona)
☎: +34 938 177 400
info@torres.es
www.torres.es

Clos Ancestral 2020 T
tempranillo, moneu, garnacha

91

Colour: bright cherry, garnet rim. Nose: balanced, red berry notes, ripe fruit, lactic notes, wild herbs. Palate: flavourful, long, easy to drink.

Forcada 2019 B
forcada

92

Colour: bright straw. Nose: ripe fruit, fine lees, wild herbs, waxy notes. Palate: full-bodied, rich, long, good acidity.

Fransola 2021 B
sauvignon blanc

91 ♣

Colour: bright straw, greenish rim. Nose: fresh fruit, citrus fruit, wild herbs, varietal. Palate: fresh, fruity, good acidity, flavourful.

Mas La Plana 2018 T R
cabernet sauvignon

93

Colour: deep cherry. Nose: ripe fruit, dried herbs, creamy oak, expressive, complex. Palate: powerful, ripe fruit, spicy, round tannins.

🏆 **PODIUM**

Reserva Real 2017 T
cabernet sauvignon, merlot, cabernet franc

96

Colour: Cherry. Nose: complex, expressive, spicy, cocoa bean, dried herbs, toasty, creamy oak. Palate: full-bodied, long, great length, good structure, flavourful.

Waltraud 2021 B
riesling

90

Colour: bright straw. Nose: floral, citrus fruit, fresh, medium intensity. Palate: fresh, good acidity, easy to drink, good finish.

FERRÉ I CATASÚS

Masía Gustems,
Crta. Sant Sadurní a Vilafranca, Km.8
08792 La Granada (Barcelona)
☎: +34 938 974 558
administracio@ferreicatasus.com
www.ferreicatasus.com

10 Sauvignon Blanc 2021 B
sauvignon blanc

88

Crisp, fruity, herbal, tropical.

Cap de Trons 2021 T
merlot, syrah, garnacha

87

Gall Negre 2017 T
merlot

90

Colour: bright cherry. Nose: fruit expression, red berry notes, spicy, ripe fruit, meaty notes. Palate: flavourful, fruity, balanced, round tannins.

Love is Vermell 2021 B
xarel.lo vermell

91

Colour: straw. Nose: expressive, white flowers, dried herbs, raspberry. Palate: flavourful, fruity, balanced, sweetness.

Somiatruites 2021 B
moscatel, garnacha blanca, chardonnay

88

Citrus fruit, fruity, herbal, crisp.

Sonat de l'Ala 2017 T
merlot

89

Fruity, herbal, jammy, tasty.

FINCA VILADELLOPS

Finca Viladellops
08734 Olérdola (Barcelona)
☎: +34 938 188 371
info@viladellops.com
www.viladellops.com

Finca Viladellops XXX Xarel.lo 2021 B FB
xarel.lo

90

Needs time. Colour: bright straw. Nose: fruit expression, ripe fruit, wild herbs. Palate: flavourful, fresh, good acidity.

Parany 2016 T C
cariñena

93

Colour: deep cherry. Nose: creamy oak, black fruit, scrubland, black pepper. Palate: ripe fruit, spicy, round tannins, elegant.

Turó de Les Abelles 2019 T
garnacha

91

Colour: cherry, garnet rim. Nose: ripe fruit, dried herbs, wild herbs, varietal, toasty. Palate: powerful, ripe fruit, spicy, round tannins.

Viladellops Selección Garnatxa 2019 T C
garnacha

91

Colour: dark-red cherry. Nose: toasty, spicy, cocoa bean, ripe fruit. Palate: flavourful, toasty, fine bitter notes.

GALLINA DE PIEL WINES

Escoles, 43
17181 Aiguaviva (Girona/Gerona)
☎: +34 663 594 668
info@gallinadepielwines.com
www.gallinadepielwines.com

Ikigall 2021 B
85% xarel.lo, 10% malvasía de Sitges, moscatel de alejandría

89

Jammy, fruity, dried flowers, aromatic, tasty.

DO PENEDÈS / D.O.P.

DO PENEDÈS / D.O.P.

GIRÓ DEL GORNER
Finca Giró del Gorner, s/n
08797 Puigdalber (Barcelona)
☎: +34 938 988 032
gorner@girodelgorner.com
www.girodelgorner.com

Giró del Gorner Blanc 2021 B
macabeo, xarel.lo, parellada, chardonnay

88

Pleasant, aromatic, standard, floral, fruity, simple.

Giró del Gorner Bru T RB
cabernet sauvignon, merlot

86

Giró del Gorner Les Saleres 2020 B FB
100% chardonnay

90 ♣

Colour: bright yellow. Nose: powerful, ripe fruit, roasted coffee, citrus fruit. Palate: rich, good structure, long, toasty, fine bitter notes.

Giró del Gorner Rosat 2021 RD
100% merlot

87

Giró del Gorner Vinya El Mocador 2020 B
100% moscatel de frontignan

89 ♣

Varietally correct, tasty, fruity, tropical.

Giró del Gorner Vinya La Serdalla 2020 B
xarel.lo

90 ♣

Oxidativ. Colour: straw. Nose: ripe fruit, dried herbs, faded flowers, dry nuts. Palate: ripe fruit, balanced, flavourful.

GIRÓ RIBOT
Finca El Pont, s/n
08792 Santa Fe del Penedès (Barcelona)
☎: +34 938 974 050
giroribot@giroribot.es
www.giroribot.com

Giró Ribot Mimat Blanc 2021 B
moscatel de frontignan, moscatel de alejandría, malvasía de Sitges

89 ♣

Citrus fruit, crisp, fruity, herbal, jammy.

Giró Ribot Solstici 2021 B
xarel.lo, chardonnay

89 ♣

Pleasant, defined aromas, floral, dried flowers, crisp, fruity, wild, smooth.

HUGUET DE CAN FEIXES
Finca Can Feixes, 1
08718 Cabrera D'Anoia (Barcelona)
☎: +34 937 718 227
canfeixes@canfeixes.com
www.canfeixes.com

Can Feixes Blanc Selecció 2021 B
parellada, macabeo, chardonnay, malvasía de Sitges

89 ♣

Dried flowers, crisp, dried herbs, standard, simple, smooth.

Can Feixes Blanc Tradició 2020 B
xarel.lo, malvasía de Sitges

90 ♣

Colour: yellow. Nose: fragrant herbs, fine lees, floral. Palate: full-bodied, rich, long, good acidity.

Can Feixes Chardonnay 2019 B FB
chardonnay

92 ♣

Colour: bright yellow. Nose: powerful, creamy oak, ripe fruit, spicy. Palate: rich, good structure, long, toasty, fine bitter notes.

Can Feixes Negre Selecció 2019 T
tempranillo, merlot

89 ♣

Pleasant, spicy, jammy, tasty.

Can Feixes Negre Tradició 2013 T C
merlot, petit verdot, cabernet sauvignon

89 ♣

Pruney, balanced, spicy, dried herbs.

Can Feixes Reserva Especial 2009 T R
cabernet sauvignon, merlot

91

Colour: dark-red cherry, garnet rim. Nose: ripe fruit, fruit preserve, tobacco, sweet spices, old leather. Palate: spicy, flavourful, balanced, slightly dry, soft tannins.

JANÉ VENTURA
Ctra. de Calafell, 2
43700 El Vendrell (Tarragona)
☎: +34 977 660 118
janeventura@janeventura.com
www.janeventura.com

Jané Ventura Finca Els Camps Macabeu 2019 B

92 ♣

Colour: bright yellow. Nose: powerful, ripe fruit, spicy, fine lees, sweet spices. Palate: rich, good structure, long, toasty, fine bitter notes, flavourful.

Jané Ventura
Finca Els Camps Negre 2018 T
90
Colour: cherry, garnet rim. Nose: fruit preserve, powerful, spicy, balanced. Palate: flavourful, long, round tannins.

Jané Ventura
Malvasía de Sitges 2020 B BA
91
Colour: straw. Nose: ripe fruit, dried herbs, faded flowers, brioche. Palate: powerful, ripe fruit, balanced.

Jané Ventura Sumoll 2018 T
92
Colour: deep cherry. Nose: ripe fruit, dried herbs, creamy oak, black fruit. Palate: powerful, ripe fruit, spicy, round tannins.

Jané Ventura Vinyes Blanques 2021 B
89
Aromatic, dried flowers, fruity, dried herbs, wild, balanced.

Jané Ventura Xarel.lo 2020 B
93
Colour: bright yellow. Nose: creamy oak, ripe fruit, spicy, toasted bread, brioche. Palate: rich, good structure, toasty, fine bitter notes.

JAUME LLOPART ALEMANY
Cl. Font Rubí, 9
08736 Font-Rubí (Barcelona)
☎: +34 938 979 133
info@jaumellopartalemany.com
www.jaumellopartalemany.com

Ani D'Anna 2021 B
moscatel de frontignan
87

Jaume Llopart Alemany 2021 B
87

Jaume Llopart Alemany 2021 RD
87

Jaume Llopart Alemany Merlot 2021 T
87

Vinya d'en Lluc 2021 B
88
Citrus fruit, dried herbs, floral, tasty.

JEAN LEON
Château Leon s/n
08775 Torrelavit (Barcelona)
☎: +34 938 995 512
jeanleon@jeanleon.com
www.jeanleon.com

Jean Leon 3055 Chardonnay 2021 B
chardonnay
90
Colour: bright straw. Nose: white fruit, citrus fruit, dried herbs. Palate: flavourful, fruity, balanced.

Jean Leon 3055 Merlot Petit Verdot 2020 T
petit verdot, merlot
90
Colour: deep cherry. Nose: ripe fruit, dried herbs, creamy oak. Palate: powerful, ripe fruit, round tannins.

Jean Leon 3055 Rosé 2021 RD
pinot noir, garnacha
90
Colour: brilliant rose. Nose: fruit expression, red berry notes, floral, fragrant herbs. Palate: fruity, easy to drink, flavourful.

Jean Leon Vinya Gigi Chardonnay 2019 B C
chardonnay
92
Colour: bright straw. Nose: fruit expression, ripe fruit, floral, dried herbs, balanced. Palate: flavourful, fresh, good acidity, fruity aftertaste.

Jean Leon Vinya Le Havre Cabernet Sauvignon Reserva 2017 T R
cabernet sauvignon, cabernet franc
90
Colour: deep cherry. Nose: ripe fruit, dried herbs, creamy oak, dark chocolate. Palate: powerful, spicy, round tannins.

Jean Leon Vinya Palau Merlot 2018 T C
merlot
91
Colour: Cherry. Nose: sweet spices, scrubland, dried herbs, fruit preserve, earthy notes. Palate: spicy, good structure, round tannins.

Jean Leon Xarel.lo 2021 B
100% xarel.lo
90
Colour: straw. Nose: dried herbs, faded flowers, white fruit, citrus fruit. Palate: ripe fruit, balanced, flavourful.

DO PENEDÈS / D.O.P.

LACRIMA BACCUS
Finca La Porxada s/n
08732 Castellet i la Gornal (Barcelona)
☎: +34 938 918 281
lacrimabaccus@bardinet.es
www.lacrimabaccus.com

Mar i Cel 2021 B
xarel.lo, chardonnay, sauvignon blanc
86

Més Feliç 2020 T
syrah
88
Pleasant, jammy, powerful, tasty.

Rosé & Clear 2021 RD
pinot noir, garnacha
88
Fruity, crisp, herbal, balanced.

Valent 2021 B
xarel.lo, chardonnay, sauvignon blanc
86

LLOPART – LA VIDA SECRETA DE LAS PLANTAS
Ctra. de Sant Sadurni - Ordal, Km. 4
08739 Subirats (Barcelona)
☎: +34 938 993 125
info@llopart.com
www.llopart.com

Llopart Clos dels Fòssils 2021 B
100% xarel.lo
90
Colour: bright straw. Nose: ripe fruit, fragrant herbs, fine lees. Palate: full-bodied, rich, good acidity.

MAS BERTRAN
Ctra. BP-2121 Km.7,7
08731 Sant Marti Sarroca (Barcelona)
☎: +34 938 990 859
info@masbertran.com
www.masbertran.com

Argila 2013 BE GR BN
92
Colour: bright golden. Nose: fine lees, characterful, ripe fruit, dry nuts, spicy. Palate: powerful, flavourful, good acidity, fine bead, fine bitter notes.

Balma 2017 BE R BR
50% xarel.lo, 40% macabeo, 10% parellada
90
Colour: bright yellow. Nose: ripe fruit, fine lees, dried herbs. Palate: good acidity, balanced, fine bitter notes.

Mas Bertran La Graua 2013 BE GR BN
50% xarel.lo, 50% macabeo
91
Colour: bright straw. Nose: fine lees, dry nuts, balanced, dried flowers, faded flowers. Palate: powerful, flavourful, good acidity, fine bead, balanced.

Mas Bertran X80 2012 BE GR BR
80% xarel.lo, 20% sumoll
90
Colour: bright straw. Nose: fine lees, fragrant herbs, expressive, white fruit. Palate: flavourful, good acidity, fine bead, balanced, fresh.

Nutt 2021 B
100% xarel.lo
89
Oxidativ, herbal, crisp, balanced.

Nutt Rosé 2021 RD
100% sumoll
88
Citrus fruit, fruity, herbal, balanced.

MAS ESCORPÍ
Industria, 36
08770 Sant Sadurní d'Anoia (Barcelona)
☎: +34 938 910 113
info@fincamasescorpi.com
www.fincamasescorpi.com/es/

Bru de Gramona 2018 T
88
Spicy, fruity, herbal, jammy, tasty.

Savinat Sauvignon Blanc 2019 B FB
100% sauvignon blanc
91
Colour: bright straw. Nose: ripe fruit, dried herbs, faded flowers, toasty. Palate: powerful, ripe fruit, balanced.

MAS RODÓ
Km. 2 Ctra. Sant Pere
Sacarrera a Sant Joan de Mediona
08773 Mediona (Barcelona)
☎: +34 932 385 780
Fax: +34 932 174 356
info@masrodo.com
www.masrodo.com

Mas Rodó Cabernet Sauvignon 2015 T
100% cabernet sauvignon
88
Spicy, fruity, dried herbs, varietally correct, jammy.

Mas Rodó Incògnit 2020 T
tempranillo, merlot, cabernet sauvignon
87

Mas Rodó Macabeo 2017 B
100% macabeo

90
Age nuances. Colour: bright golden. Nose: caramel, faded flowers, honeyed notes, dry nuts. Palate: rich, flavourful, bitter, toasty.

Mas Rodó Merlot 2015 T R
100% merlot

88
Fruity, spicy, jammy, tasty, wild.

Mas Rodó Montonega 2018 B
100% montonega

90
Oxidativ. Colour: straw. Nose: ripe fruit, dried herbs, faded flowers. Palate: ripe fruit, balanced, fresh.

Mas Rodó Riesling 2019 B FB
100% riesling

88
Pleasant, kind finish, jammy, tasty.

MONTRUBÍ
L'Avellà
08736 Font-Rubí (Barcelona)
☎: +34 938 979 066
xramos@montrubi.com
www.montrubi.com

Advent Samsó Dulce Natural 2019 T D
samsó

93
With personality, needs time. Colour: mahogany. Nose: characterful, balanced, varnish, candied fruit. Palate: flavourful, sweet, good acidity, unctuous.

Black 2021 T
garnacha

91
Colour: cherry, purple rim. Nose: fruit expression, red berry notes, floral. Palate: fruity, flavourful, balanced.

Gaintus Vertical 2017 T C
sumoll

92
Kind finish, smooth. Colour: bright cherry. Nose: burnt matches, expressive, balanced, wild herbs, red berry notes, ripe fruit. Palate: flavourful, varietal, easy to drink.

Serres Velles Macabeu 2021 B C
macabeo

90
Colour: bright yellow. Nose: creamy oak, ripe fruit, spicy, toasted bread. Palate: good structure, toasty, fine bitter notes.

White 2021 B
xarel.lo

90
Colour: bright straw. Nose: ripe fruit, floral, tropical fruit, fine lees. Palate: flavourful, fresh, good acidity, fruity aftertaste.

MUSCÀNDIA
Can Rosell de la Llena
08790 Gelida (Barcelona)
☎: +34 625 632 620
info@cavamuscandia.com
www.muscandia.com

Muscàndia Deliri Floral 2021 B
moscatel de frontignan, sauvignon blanc

90
Colour: bright straw. Nose: fruit expression, ripe fruit, floral, tropical fruit. Palate: flavourful, fresh, fruity aftertaste, fruity.

OLIVER VITICULTORS
Cal Xic de L'Agustí - Can Batista
08770 Subirats (Barcelona)
☎: +34 609 375 242
sadurni@oliverviticultors.com
www.oliverviticultors.com

10000 Hores Floral 2021 B
moscatel de frontignan, sauvignon blanc

88
Citrus fruit, floral, herbal, aromatic.

10000 Hores Negre Selecció 2020 T
tempranillo, syrah

88
Pleasant, defined aromas, standard, fruity, lactic, smooth.

10000 Hores Rosé 2021 RD
merlot, tempranillo

88
Kind finish, fruity, herbal, tasty.

10000 Hores Xarel.lo 2021 B
xarel.lo

88
Balanced, crisp, herbal.

Brots de Xarel.lo 2020 B
xarel.lo

89
Classic, spicy, crisp, full-bodied, tasty, herbal, toasty.

DO PENEDÈS / D.O.P.

DO PENEDÈS / D.O.P.

La Temptació 2019 B
malvasía de Sitges

91

Colour: bright yellow. Nose: creamy oak, ripe fruit, spicy, floral. Palate: rich, good structure, toasty, fine bitter notes.

PARATÓ VINÍCOLA
Can Respall de Renardes s/n
08733 El Pla del Penedès (Barcelona)
☎: +34 938 988 182
info@parato.es
www.parato.es

Finca Renardes 2019 T RB
65% tempranillo, 27% cabernet sauvignon, 8% cariñena

87

Parató Ática Tres x Tres 2020 B
65% xarel.lo, 7% macabeo, 28% chardonnay

89

Spicy, fruity, pruney, tasty.

Parató Xarel.lo 2021 B
100% xarel.lo

88

Crisp, fruity, herbal, oxidativ.

Santa Clara 2021 T
100% cabernet sauvignon

87

PARÉS BALTÀ
Masía Can Baltá, s/n
08796 Pacs del Penedès (Barcelona)
☎: +34 938 901 399
comunicacio@paresbalta.com
www.paresbalta.com

Parés Baltà Electio Xarel.lo 2020 B
100% xarel.lo

92

Colour: bright yellow. Nose: powerful, creamy oak, ripe fruit, spicy. Palate: rich, good structure, toasty, fine bitter notes.

Parés Baltà Hisenda Miret Garnatxa 2019 T R
100% garnacha

91

Colour: deep cherry. Nose: dried herbs, creamy oak, black fruit. Palate: ripe fruit, spicy, round tannins, sweetness.

Parés Baltà Indígena 2021 B
100% garnacha blanca

91

Colour: straw. Nose: ripe fruit, dried herbs, faded flowers. Palate: powerful, balanced, fruity, fresh.

Parés Baltà Indígena Negre 2020 T
100% garnacha

91

Colour: bright cherry. Nose: fruit expression, red berry notes, floral, spicy, wild herbs. Palate: flavourful, fruity, balanced.

Parés Baltà Satèl.lit 2019 B
cariñena blanca

92

Colour: bright yellow. Nose: powerful, spicy, toasted bread, creamy oak, ripe fruit, candied fruit, faded flowers. Palate: rich, good structure, long, toasty, fine bitter notes.

Parés Baltà Xarel.lo 2021 B
xarel.lo

90

Colour: straw. Nose: dried herbs, dry nuts, fresh, varietal. Palate: fruity, balanced, easy to drink.

PLANA D'EN JAN MICROVINIFICACIONES
Masia Benet, s/n
08796 Pacs del Penedès (Barcelona)
☎: +34 620 239 135
janmarrugat@planadenjan.com
www.planadenjan.com

Plana D'En Jan Bon Jan Negre 2018 T BA
90% ull de llebre, 5% merlot, 5% cabernet sauvignon

87

Plana D'En Jan Camp del Cuc 2019 B BA S
100% macabeo

93

With personality, little interventionist. Colour: bright yellow. Nose: balanced, expressive, fine reductive notes, burnt matches, dried flowers, spicy. Palate: long, balanced, taut.

Plana D'En Jan Clos de les Oliveres 2020 B C
100% chardonnay

88

Tropical, tasty, jammy, full-bodied.

Plana D'En Jan Els Ametllers 2017 B FB
100% xarel.lo

91

Colour: bright golden. Nose: dry nuts, candied fruit, honeyed notes, faded flowers, pattiserie. Palate: fine bitter notes, correct, light-bodied.

Plana D'En Jan Insólit
Blanc de Noir 2019 B S
garnacha

93
Colour: bright straw. Nose: fruit expression, ripe fruit, floral, dried herbs. Palate: flavourful, fresh, good acidity, fruity aftertaste.

Plana D'En
Jan La Sorrera 2018 T
100% merlot

92
Colour: deep cherry. Nose: dried herbs, creamy oak, black fruit, tobacco, tea leave. Palate: powerful, ripe fruit, spicy, round tannins.

Plana D'En Jan
Plana in Albis 2018 T R
100% cabernet sauvignon

93
Colour: Cherry. Nose: complex, expressive, spicy, mineral, ripe fruit. Palate: elegant, full-bodied, long, great length.

PLANAS ALBAREDA

Ctra. a Guardiola, km.3, s/n
08735 Vilobí del Penedès (Barcelona)
☎: +34 607 340 098
info@planasalbareda.com
www.planasalbareda.com

Planas Albareda Desclòs 2020 T
merlot

87

Planas Albareda L'Avenc 2021 B
xarel.lo

86

Rosat de Planas Albareda 2021 RD
merlot

87

PROPIETAT D'ESPIELLS

08770 Sant Sadurní d'Anoia (Barcelona)
☎: +34 938 911 000
Fax: +34 938 912 100
info@propietatdespiells.com
www.propietatdespiells.com

Casa Vella D'Espiells 2017 T R
cabernet sauvignon

92
Colour: bright cherry, garnet rim. Nose: fruit expression, red berry notes, spicy, ripe fruit. Palate: flavourful, fruity, balanced, slightly dry, soft tannins.

Flor de Xarel.lo D'Espiells 2021 B
100% xarel.lo

90
Colour: straw. Nose: dried herbs, faded flowers, fine lees. Palate: ripe fruit, balanced.

Iohannes 2015 T R
cabernet sauvignon, merlot

91
Colour: dark-red cherry, garnet rim. Nose: ripe fruit, fruit preserve, tobacco, sweet spices, black fruit. Palate: spicy, round tannins, flavourful, slightly dry, soft tannins.

Miranda D'Espiells 2021 B
100% chardonnay

90
Colour: bright straw. Nose: ripe fruit, fragrant herbs, fine lees, dried flowers. Palate: fresh, balanced, fine bitter notes.

Therasia 2020 B FB
xarel.lo, chardonnay, viognier

91
Colour: bright yellow. Nose: creamy oak, spicy, white fruit, ripe fruit. Palate: toasty, fine bitter notes, easy to drink.

ROS MARINA VITICULTORS

Camino Puigdàlber a Les Cases Noves, km. 1
08736 Font-Rubí (Barcelona)
☎: +34 938 988 185
rosmarina@rosmarina.es
www.rosmarina.es

Mas Uberni Blanc de Blanc 2021 B
xarel.lo, chardonnay

88
Citrus fruit, fruity, crisp, herbal.

Mas Uberni Chardonnay 2021 B
chardonnay

88
Citrus fruit, standard, creamy, herbal, crisp.

Ros Marina 2x 2021 B
xarel.lo

89
Oxidativ, crisp, fruity, herbal.

Ros Marina dCoster 2016 T C
cabernet sauvignon, merlot

88
Fruity, herbal, tasty, spicy, balanced.

DO PENEDÈS / D.O.P.

DO PENEDÈS / D.O.P.

Ros Marina Quatre Vinyes 2017 T
ull de llebre, cabernet sauvignon, merlot, samsó

88
Balanced, spicy, herbaceous, jammy, tasty.

Ros Marina Rosé 2021 RD
garnacha, cariñena, merlot

87

ROUND TABLE
Ramón Freixas, 40 Bx 2
08720 Vilafranca del Penedès (Barcelona)
☎: +34 637 716 723
hola@theroundtabletopics.com
www.roundtable.wine

Round Table 2021 B
gewürztraminer, sauvignon blanc, moscatel

88
Aromatic, kind finish, floral, smooth, simple.

ROVELLATS
Bº La Bleda
08731 Sant Marti Sarroca (Barcelona)
☎: +34 934 880 575
Fax: +34 934 880 819
rovellats@cavasrovellats.com
www.rovellats.com

Rovellats Blanc Chardonnay 2021 B
chardonnay

87

Rovellats Blanc Primavera 2021 B
chardonnay, macabeo, xarel.lo

87

Rovellats Merlot 2021 RD
merlot

87

SABATÉ I COCA - CASTELLROIG
Ctra. Sant Sadurní a Vilafranca, km. 1
08739 Subirats (Barcelona)
☎: +34 938 911 927
info@sabateicoca.com
www.SABATEICOCA.com

Castellroig So Blanc 2021 B
xarel.lo

91
Colour: bright straw, greenish rim. Nose: fresh fruit, citrus fruit, wild herbs, dried flowers. Palate: fresh, fruity, good acidity, fine bitter notes.

Castellroig So Blanc Magnum 2012 B
xarel.lo

93
Colour: bright yellow. Nose: dried flowers, candied fruit, fine lees, pattiserie, toasty. Palate: round, spicy, long, great length.

Castellroig So Negre 2019 T
tempranillo, cabernet sauvignon

90
Colour: cherry, purple rim. Nose: fruit expression, red berry notes, spicy, balsamic herbs. Palate: flavoured, fruity, balanced.

Castellroig So Serè 2018 B FB
xarel.lo

90
Colour: bright yellow. Nose: creamy oak, ripe fruit, spicy. Palate: rich, good structure, toasty, fine bitter notes.

Terroja de Sabaté i Coca 2016 B
xarel.lo

92
Colour: bright yellow. Nose: dried flowers, candied fruit, fine lees, pattiserie, toasty. Palate: round, spicy, long, great length.

SEGURA VIUDAS
Ctra. Sant Sadurní a St. Pere de Riudebitlles, Km. 5
08775 Torrelavit (Barcelona)
☎: +34 938 917 070
comunicacion@freixenet.es
www.seguraviudas.com

Creu de Lavit 2019 B
100% xarel.lo

89
Tasty, jammy, dried herbs, spicy.

SUMARROCA
Masia Molí Coloma. Barri el Rebato s/n
08739 Subirats (Barcelona)
☎: +34 938 911 092
sumarroca@sumarroca.com
www.sumarroca.com

Bòria Sumarroca 2017 T
syrah

92
Colour: deep cherry. Nose: ripe fruit, dried herbs, creamy oak. Palate: ripe fruit, spicy, grainy tannins.

Il·lògic Xarel·lo Orgànic Sumarroca 2021 B
xarel.lo

88
Fruity, herbal, tasty, kind finish.

Marger Sumarroca 2020 B FB
xarel.lo, macabeo

90

Colour: bright straw. Nose: ripe fruit, fragrant herbs, fine lees, toasted bread. Palate: full-bodied, rich, long, good acidity.

Sumarroca Chardonnay 2021 B
chardonnay

90

Colour: bright straw. Nose: fruit expression, ripe fruit, grassy. Palate: flavourful, fresh, good acidity, good finish.

Terral Sumarroca 2017 T
cabernet franc, syrah, merlot

91

Colour: deep cherry. Nose: creamy oak, black fruit, wild herbs, meaty notes. Palate: ripe fruit, spicy, round tannins.

Tuví or not to be Orgànic Sumarroca 2021 B
gewürztraminer, xarel.lo, riesling

87

TORELLÓ VITICULTORS
Finca Can Martí, Ctra. C-243b, km. 13.4
08790 Gelida (Barcelona)
☎: +34 938 910 793
Fax: +34 938 910 877
torello@torello.es
www.torello.com

Torelló 50 lliures 2021 B FB
100% xarel.lo

88

Balanced, dried flowers, fruity, jammy, standard, tasty.

Torelló Gran Crisalys 2019 B
xarel.lo, chardonnay

91

Colour: bright straw. Nose: fruit expression, ripe fruit, floral, dried herbs, fine lees. Palate: flavourful, fresh, good acidity, balanced.

Torelló Malvarel.lo 2021 B
60% xarel.lo, 40% malvasía

88

Oxidativ, herbal, tasty, citrus fruit.

Torelló Mas de la Torrevella 2021 B
100% chardonnay

89

Citrus fruit, sweety finish, full-bodied, jammy.

Torelló Raimonda 2017 T BA
cabernet sauvignon, merlot

90

Colour: bright cherry. Nose: sweet spices, black fruit, ripe fruit. Palate: fruity, spicy, round tannins.

TORRE DEL VEGUER
Urb. Torre de Veguer, s/n
08810 Sant Pere de Ribes (Barcelona)
☎: +34 938 963 190
torredelveguer@torredelveguer.com
www.torredelveguer.com

Raïms de la Inmortalitat Selecció 2016 T R
cabernet sauvignon, syrah

87

Raïms de la Inmortalitat Xarel·lo 2019 B FB
xarel.lo, xarel.lo vermell

91

Colour: bright straw. Nose: ripe fruit, fragrant herbs, fine lees, woody. Palate: full-bodied, rich, long, good acidity.

Torre del Veguer Abellerol 2021 B
moscatel grano menudo

87

Torre del Veguer Fonoll 2021 B
xarel.lo, xarel.lo vermell

90

Colour: straw. Nose: dried herbs, faded flowers, meaty notes, white fruit. Palate: ripe fruit, balanced, flavourful.

Torre del Veguer Maricel 2021 B
malvasía de Sitges

88

Citrus fruit, dried flowers, crisp, jammy.

Torre del Veguer Marta 2020 BE BN
moscatel grano menudo

90

Colour: bright straw. Nose: fine lees, floral, fragrant herbs, expressive. Palate: powerful, flavourful, good acidity, fine bead.

DO PENEDÈS / D.O.P.

TRIAS BATLLE
Pere El Gran, 24
08720 Vilafranca del Penedès (Barcelona)
☎: +34 677 497 892
peptrias@jtrias.com
www.jtrias.com

Trias Batlle 2020 T
tempranillo, merlot
88
Full-bodied, jammy, herbaceous, balanced.

Trias Batlle 2021 B
macabeo, xarel.lo, parellada, moscatel
88
Fruity, herbal, jammy, tasty.

Trias Batlle 2021 RD
merlot, syrah
87

Trias Batlle Cabernet Sauvignon 2017 T C
cabernet sauvignon
88
Defined aromas, pruney, spicy, dried herbs, jammy, tasty, standard.

Trias Batlle Xarel.lo 2019 B BA
xarel.lo
90
Smoky, spicy, jammy, toasty, floral, balanced. Nose: cocoa bean. Palate: flavourful, rich, spicy, easy to drink.

VALLDOLINA VITICULTORS I ELABORADORS
Pl. de la Creu 1 Masia Can Tutusaus
08795 Olesa de Bonesvalls (Barcelona)
☎: +34 938 984 181
info@valldolina.com
www.valldolina.com

Bivac 2021 B
xarel.lo, viognier
87

Bonesvalls Cabernet Sauvignon 2018 T BA
cabernet sauvignon
89
Herbaceous, jammy, spicy, great length, tasty.

ValLDolina Xarel.lo 2021 B
xarel.lo
90
Colour: bright straw. Nose: fruit expression, floral, white fruit, citrus fruit. Palate: flavourful, fresh, fruity.

ValLDolina Rosat 2021 RD
merlot
87

VARIAS
Plaça Manuel Raventós, 6
08770 Sant Sadurní d'Anoia (Barcelona)
☎: +34 938 912 763
info@cavavarias.es
www.cavavarias.com

Eureka Blanc 2020 B
xarel.lo
88
Roasted coffee, woody, citrus fruit, creamy, full-bodied, jammy.

Pere Punyetes Blanc 2021 B
xarel.lo, moscatel, chardonnay
87

Pere Punyetes Negre 2021 T
merlot, cabernet sauvignon, tempranillo
87

Pere Punyetes Rosat 2021 RD
merlot, tempranillo, cabernet sauvignon
88
Kind finish, fruity, balanced, sweeties, jammy.

Varias Selecció 2018 T C
cabernet sauvignon, tempranillo
90
Colour: deep cherry. Nose: ripe fruit, dried herbs, creamy oak, spicy. Palate: ripe fruit, spicy, round tannins, balanced.

VINS EL CEP
Can Llopart de Les Alzines, Ctra. Espiells
08770 Sant Sadurní d'Anoia (Barcelona)
☎: +34 938 912 353
info@vinselcep.com
www.vinselcep.com

Clot del Roure 2021 B
xarel.lo
92
Defined aromas, spicy. Colour: bright straw. Nose: white fruit, ripe fruit, dried herbs, scrubland, toasty. Palate: flavourful, full of life, full-bodied.

Clot del Roure Xarel.lo Brisat 2021 B
90
Colour: yellow, coppery red. Nose: characterful, dried herbs, hint of anise, spicy. Palate: flavourful, powerful, fine bitter notes.

GR 5 Senders 2020 B
xarel.lo

90

Colour: bright straw. Nose: ripe fruit, floral, neat, medium intensity. Palate: fruity aftertaste, easy to drink, long.

GR 5 Senders 2020 T

90

Colour: Cherry. Nose: balsamic herbs, scrubland, ripe fruit. Palate: good acidity, flavourful, fruity, bitter.

Pla del Bosc Xarel.lo Vermell 2021 B
xarel.lo vermell

91

Aromatic, wild. Colour: yellow, pale. Nose: ripe fruit, faded flowers, dried flowers. Palate: balanced, flavourful, dry, varietal.

VINS PER ESTIMAR EL VI
Calle Indústria, 36
08770 Sant Sadurní d'Anoia (Barcelona)
☎: +34 938 910 113
info@vinsperestimarelvi.com
www.vinsperestimarelvi.com/es/

Gessamí 2021 B
moscatel de frontignan, moscatel de alejandría, sauvignon blanc, gewürztraminer

90

Colour: straw. Nose: expressive, white flowers, jasmine, dried herbs. Palate: flavourful, fruity, balanced, sweetness.

Mart 2021 RD
xarel.lo vermell

91

Defined aromas, crisp, smooth. Colour: raspberry rose. Nose: expressive, dried flowers, rose petals, wild herbs. Palate: fresh, easy to drink.

Vi de Glass Gewürztraminer 0,375 2020 B D
100% gewürztraminer

93

Colour: bright yellow. Nose: spicy, dried flowers, tropical fruit, candied fruit, honeyed notes. Palate: flavourful, unctuous, fruity, sweet.

WINES RAMSURT BY MAGI RAVENTOS
Can Majó 18
08173 Sant Cugat (Barcelona)
☎: +34 652 439 922
comercialramsurt@gmail.com
www.ramsurt.com

Món Eixerit 2019 T
cabernet sauvignon, ull de llebre, merlot

85

DO PENEDÈS / D.O.P.

DO. PLA DE BAGES

CONSEJO REGULADOR

Casa de La Culla - La Culla, s/n
08240 Manresa (Barcelona)
☎: +34 938 748 236 - Fax: +34 938 748 094
@: info@dopladebages.com
www.dopladebages.com

LOCATION:

Covering one of the eastern extremes of the Central Catalonian Depression; it covers the natural region of Bages, of which the city of Manresa is the urban centre. To the south the region is bordered by the Montserrat mountain range, the dividing line which separates it from Penedés. It comprises the municipal areas of Fonollosa, Monistrol de Caldres, Sant Joan de Vilatorrada, Artés, Avinyó, Balsareny, Calders, Callús, Cardona, Castellgalí, Castellfollit del Boix, Castellnou de Bages, Manresa, Mura, Navarcles, Navàs, El Pont de Vilomara, Rajadell, Sallent, Sant Fruitós de Bages, Sant Mateu de Bages, Sant Salvador de Guardiola, Santpedor, Santa María d'Oló, Súria, Talamanca, Gaià, Sant Feliu Sasserra, Sant Vicenç de Castellet, Castellbell i el Vilar, Marganell, L'Estany and Moià i Monistrol de Montserrat.

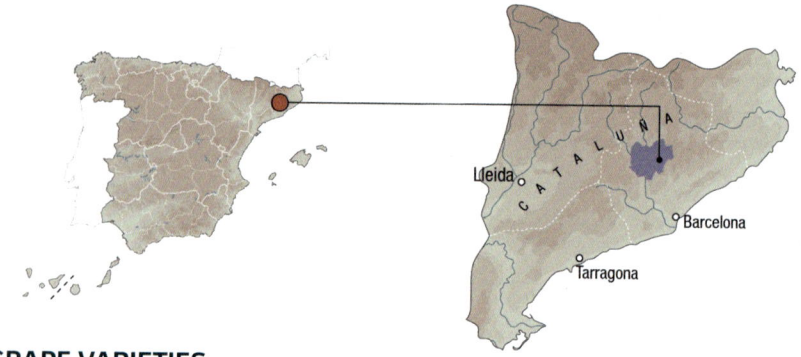

GRAPE VARIETIES:

WHITE: Chardonnay, Gewürztraminer, Macabeo, Picapoll, Parellada, Malvasía and Sauvignon Blanc.

RED: Sumoll, Ull de Llebre (Tempranillo), Merlot, Cabernet Franc, Cabernet Sauvignon, Syrah, Garnacha, Picapoll Negra, Cariñena and Mandó.

FIGURES:

Vineyard surface: 500 – **Wine-Growers:** 79 – **Wineries:** 15 – 20**20 Harvest rating:** N/A – **Production 20:** 605,000 litres – **Market percentages:** 80% National - 20% International.

SOIL:

The vineyards are situated at an altitude of about 400 m. The soil is franc-clayey, franc-sandy and franc-clayey-sandy.

CLIMATE:

Mid-mountain Mediterranean, with little rainfall (500 mm to 600 mm average annual rainfall) and greater temperature contrasts than in the Penedès.

CLASSIFICATION OF YOUNG WINE HARVESTS PEÑÍNGUIDE

2017	2018	2019	2020	2021
N/A	N/A	N/A	N/A	N/A

ABADAL

Masia Oliveras, s/n
08279 Santa María D'Horta D'Avinyó (Barcelona)
☎: +34 938 743 511
Fax: +34 938 737 204
info@abadal.net
www.abadal.net

Abadal 3.9 (Vi de Finca) 2018 T R
100% cabernet sauvignon

91

Colour: deep cherry. Nose: ripe fruit, dried herbs, creamy oak, aromatic coffee. Palate: powerful, ripe fruit, spicy, round tannins.

Abadal 5 Merlot 2018 T R
100% merlot

91

Colour: deep cherry. Nose: ripe fruit, dried herbs, fine reductive notes, spicy. Palate: ripe fruit, spicy, round tannins.

Abadal Mandó 2019 T
100% mandó

91

Colour: bright cherry. Nose: fresh fruit, raspberry, scrubland, fine reductive notes. Palate: good acidity, spicy, fine tannins.

Abadal Matis 2018 T C
45% cabernet sauvignon, 45% merlot, 10% morenillo

89

Balanced, spicy, herbal, age nuances, jammy, tasty.

Abadal Nuat 2019 B C
80% picapoll blanc, 20% macabeo

91

Colour: straw. Nose: dried herbs, faded flowers, white fruit, fine lees, mineral. Palate: powerful, ripe fruit, balanced.

Abadal Picapoll 2021 B
100% picapoll blanc

89

Citrus fruit, balanced, herbal, tasty.

CELLER GRAU I GRAU

Ctra. C-37, Km. 75,5
08255 Castellfollit del Boix (Barcelona)
☎: +34 938 356 002
info@cellergrauigrau.com
www.cellergrauigrau.com

Jaume Grau i Grau "Gratvs" 2014 T C
merlot, ull de llebre

89

Full-bodied, balsamic herbs, jammy, great length, balanced, tasty.

Jaume Grau i Grau Col.lecció Ull de Llebre 2014 T
ull de llebre

91

Colour: dark-red cherry, garnet rim. Nose: ripe fruit, fruit preserve, tobacco, sweet spices. Palate: spicy, round tannins, long.

Jaume Grau i Grau Selecció Especial 2017 T
ull de llebre, merlot, cabernet franc, syrah

89 ✿

Pruney, dried herbs, tasty, wild, jammy, balsamic herbs.

CELLER LES ACÀCIES

Ctra. B-431, km 56
08279 Avinyo (Barcelona)
☎: +34 933 620 010
info@lesacacies.com
www.lesacacies.com

Avinius Merlot i Syrah 2020 T
50% merlot, 50% syrah

89

Overripe, spicy, herbal, tasty.

Desbordant 2019 T
60% garnacha, 40% sumoll

89 ✿

Balanced, spicy, herbal, wild, tasty.

Instant de Flor 2021 B
100% chardonnay

88

Citrus fruit, fruity, herbal, smooth.

Opositor Noir 2019 T BA
picapoll negre, merlot, ull de llebre

89

Balanced, spicy, tasty, toasty, jammy.

CELLER SOLERGIBERT

Barquera, 40
08271 Artés (Barcelona)
☎: +34 938 305 084
josep@cellersolergibert.com
www.cellersolergibert.com

Pic de Solergibert 2021 B
picapoll blanc

89 ✿

Kind finish, aromatic, floral, fruity, tasty, balanced.

Sdm. Solergibert de Matacans 2019 T
cabernet sauvignon

91

Colour: Cherry. Nose: sweet spices, scrubland, earthy notes. Palate: balsamic, balanced, long, spicy.

DO PLA DE BAGES / D.O.P.

DO PLA DE BAGES / D.O.P.

Solergibert Cabernet 2011 T R
cabernet sauvignon, cabernet franc
89
Herbaceous, jammy, wild, tasty, mineral, spicy, slight reduction.

Solergibert Cabernet 2017 T R
cabernet sauvignon, cabernet franc
90
Colour: cherry, garnet rim. Nose: balsamic herbs, ripe fruit, scrubland. Palate: flavourful, spicy, long, good acidity.

Solergibert Merlot 2017 T R
merlot
89
Pruney, wild, herbaceous, slight reduction, full-bodied, tasty.

Sumoll de Solergibert 2019 T
sumoll
91
Dried herbs, wild, smooth. Nose: red berry notes, ripe fruit, wild herbs, earthy notes. Palate: good acidity, balanced, easy to drink, fruity.

COLLBAIX - CELLER EL MOLI
Camí de Rajadell, Km. 3
08242 Manresa (Barcelona)
☎: +34 931 021 965
collbaix@cellerelmoli.com
www.cellerelmoli.com

Collbaix 3 Nits 2019 T BA
merlot
90 ♣
Colour: Cherry. Nose: balsamic herbs, sweet spices, scrubland. Palate: spicy, good acidity, ripe fruit, easy to drink.

Collbaix Singular de Ánfora 2017 B BA
macabeo
92 ♣
Colour: bright yellow. Nose: creamy oak, ripe fruit, spicy, smoky. Palate: rich, good structure, toasty, fine bitter notes.

Collbaix Singular Negre 2016 T RB
cabernet sauvignon
90 ♣
Colour: deep cherry. Nose: dried herbs, creamy oak, black fruit, spicy. Palate: ripe fruit, spicy, round tannins.

Cor Valent 2016 T BA
sumoll, mandó, cabernet sauvignon, merlot
91 ♣
Classic, spicy, dried herbs, wild, tasty. Nose: red berry notes, ripe fruit, balanced. Palate: easy to drink, balanced, good acidity.

Gebrada 2021 B
picapoll blanc, macabeo
89
Dried flowers, pleasant, dried herbs, jammy, wild, simple.

Raw Mandó 2020 T FB
100% mandó
88 ♣
Spicy, balanced, herbal, toasty, jammy.

Raw Picapoll 2021 B
100% picapoll blanc
88 ♣
Citrus fruit, standard, crisp, herbal, yeasty notes.

U D'Urpina 2018 T R
cabernet sauvignon
91 ♣
Colour: Cherry. Nose: balsamic herbs, sweet spices, scrubland, ripe fruit, black liquorice. Palate: flavourful, good structure, spicy.

ENTREBOSC
Mas Pubill s/n
08255 Castellfollit del Boix (Barcelona)
☎: +34 640 379 227
ferranfc5@gmail.com
www.entrebosc.com

Entrebosc Negre 2019 T
cabernet franc
90 ♣
Colour: Cherry. Nose: scrubland, powerful, spicy, black pepper. Palate: spicy, balsamic, good acidity.

MÉS QUE PARAULES
Finca Jaumandreu, s/n
08259 Fonollosa (Barcelona)
☎: +34 936 556 057
hola@vinossinetiquetas.com
www.mesqueparaules.com

Més Que Paraules Blanc 2021 B
60% picapoll blanc, 30% sauvignon blanc, 10% mandó
88
Citrus fruit, floral, fruity, dried herbs.

Més Que Paraules Rosat 2021 RD
70% merlot, 30% sumoll
87

OLLER DEL MAS

Ctra. de Igualada a Manresa, km. 91 (C-37),
08241 Manresa (Barcelona)
☎: +34 938 768 315
admin@ollerdelmas.com
www.ollerdelmas.com

DO PLA DE BAGES / D.O.P.

Arnau Oller 2018 T R
94% merlot, 6% picapoll negre

91

Colour: Cherry. Nose: scrubland, tobacco, black fruit. Palate: spicy, round tannins, flavourful, varietal.

Bernat Oller 2018 T
70% merlot, 20% syrah, 5% picapoll negre, 5% picapoll blanc

88

Weary, age nuances, spicy, herbal, dried herbs, jammy.

Oller del Mas Especial Picapoll Negre 2019 T BA
100% picapoll negre

91

Colour: light cherry, garnet rim. Nose: wild herbs, dried herbs, ripe fruit, spicy. Palate: balanced, fine bitter notes, easy to drink, spicy.

DO. PLA I LLEVANT
CONSEJO REGULADOR

Canonge Barceló, 2
07200 Felanitx (Illes Balears)
☎: +34 971 168 569 - Fax: +34 971 184 49 34
@: correo@doplaillevant.com
www.doplaillevant.com

LOCATION:

The production region covers the eastern part of Majorca and consists of 18 municipal districts: Algaida, Ariany, Artá, Campos, Capdepera, Felanitx, Lluchamajor, Manacor, Mª de la Salud, Montuiri, Muro, Petra, Porreres, Sant Joan, Sant Llorens des Cardasar, Santa Margarita, Sineu and Vilafranca de Bonany.

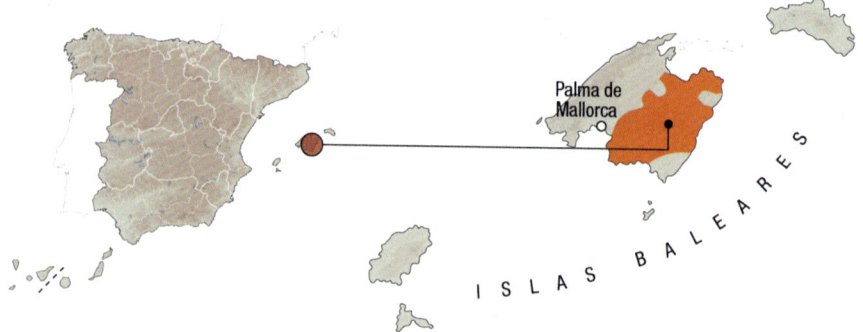

GRAPE VARIETIES:

WHITE: Prensal Blanc, Macabeo, Parellada, Moscatel, Chardonnay and Giró Blanc.

RED: Callet (majority), Manto Negro, Fogoneu, Tempranillo, Monastrell, Cabernet Sauvignon, Merlot, Syrah and Gorgollassa.

FIGURES:

Vineyard surface: 499 – **Wine-Growers:** 78 – **Wineries:** 14 – **2021 Harvest rating:** N/A – **Production 21:** 1,263,854 litres – **Market percentages:** 90% National - 10% International.

SOIL:

The soil is made up of limestone rocks, which give limy-clayey soils. The reddish Colour. of the terrain is due to the presence of iron oxide. The clays and calcium and magnesium carbonates, in turn, provide the whitish Colour. which can also be seen in the vineyards.

CLIMATE:

Mediterranean, with an average temperature of 16°C and with slightly cool winters and dry, hot summers. The constant sea breeze during the summer has a notable effect on these terrains close to the coast. The wet season is in autumn and the average annual rainfall is between 450 mm and 500 mm.

CLASSIFICATION OF YOUNG WINE HARVESTS PEÑÍNGUIDE

2017	2018	2019	2020	2021
GOOD	N/A	VERY GOOD	N/A	N/A

BODEGAS PERE SEDA

Cid Campeador, 22
07500 Manacor (Illes Balears/Islas Baleares)
☎: +34 971 605 087
Fax: +34 971 844 934
lucasreus@telefonica.net
www.pereseda.com

Gvivm Blanc de Blancs 2021 B
40% chardonnay, 35% moscatel, 21% prensal, 4% giró ros

88

Citrus fruit, crisp, fruity, herbal.

L'Arxiduc Rosat 2021 RD
60% merlot, 40% tempranillo

88

Citrus fruit, balanced, dried flowers, fruity.

Mossèn Alcover 2018 T
73% cabernet sauvignon, 27% callet

89

Balanced, dried flowers, toasty, jammy, slight reduction.

Pere Seda 2018 T C
39% merlot, 33% cabernet sauvignon, 21% syrah, 7% callet

89

Colour: deep cherry. Nose: dried herbs, creamy oak, black fruit, scrubland. Palate: powerful, ripe fruit, spicy, round tannins.

Pere Seda Chardonnay/Giro Ros 2021 B
80% chardonnay, 20% giró ros

89

Citrus fruit, spicy, floral, jammy, tasty, full-bodied.

Secret d´en Perico 2021 B FB
80% parellada, 20% giró ros

91

Colour: bright yellow. Nose: creamy oak, ripe fruit, spicy, roasted coffee, white flowers. Palate: rich, good structure, fine bitter notes.

MIQUEL OLIVER VINYES I BODEGUES

Ctra. Petra-Sta. Margarita km. 1,8
07520 Petra-Mallorca (Illes Balears/Islas Baleares)
☎: +34 971 561 117
Fax: +34 971 561 117
bodega@miqueloliver.com
www.miqueloliver.com

Aia 2016 T
100% merlot

90

Colour: dark-red cherry, garnet rim. Nose: fruit preserve, aged wood nuances, tobacco, sweet spices. Palate: spicy, round tannins, powerful, flavourful.

Muscat Miquel Oliver 2021 B S
moscatel

88

Citrus fruit, floral, varietally correct, sweety finish.

Orig 2021 B
100% giró ros

86

QBQ 2020 B FB
100% giró ros

91

Colour: straw. Nose: ripe fruit, dried herbs, white flowers. Palate: powerful, ripe fruit, balanced, rich, warm.

Ses Ferritges 2016 T C S
25% callet, 25% merlot, 25% syrah, cabernet sauvignon

91

Colour: deep cherry. Nose: dried herbs, creamy oak, black fruit, fruit preserve, tobacco. Palate: powerful, ripe fruit, spicy, round tannins.

Xperiment 2018 T FB S
callet

92

Colour: deep cherry. Nose: dried herbs, creamy oak, dark chocolate, black fruit. Palate: ripe fruit, spicy, round tannins, flavourful, elegant.

VINS MIQUEL GELABERT

Salas, 50
07500 Manacor (Illes Balears/Islas Baleares)
☎: +34 971 821 444
info@vinsmiquelgelabert.com
www.vinsmiquelgelabert.com

Chardonnay Roure 2020 B FB
chardonnay

90

Colour: bright yellow. Nose: powerful, creamy oak, ripe fruit, spicy. Palate: rich, good structure, long, toasty, fine bitter notes.

Gran Vinya Son Caules 2014 T
callet

91

Colour: pale ruby, brick rim edge. Nose: fruit preserve, aged wood nuances, tobacco, spicy. Palate: spicy, round tannins, balanced.

Sa Vall Selecció Privada 2016 B FB
giró ros, viognier, pinot noir

90

Colour: bright yellow. Nose: powerful, creamy oak, ripe fruit, roasted coffee. Palate: rich, good structure, fine bitter notes.

DO PLA I LEVANT / D.O.P.

DO PLA I LEVANT / D.O.P.

Son Moix Blanc 2017 B
giró ros, chardonnay

91

Colour: bright yellow. Nose: powerful, creamy oak, ripe fruit, sweet spices, woody. Palate: rich, good structure, long, toasty, fine bitter notes.

Torrent Negre 2015 T C
cabernet sauvignon, merlot, syrah

90

Classic. Colour: deep cherry. Nose: dried herbs, black fruit, spicy, tobacco. Palate: ripe fruit, spicy, round tannins.

Torrent Negre Selecció Privada Cabernet 2014 T C
cabernet sauvignon

92

Classic, balanced. Colour: deep cherry. Nose: dried herbs, creamy oak, black fruit, ripe fruit, cigar. Palate: powerful, ripe fruit, spicy, round tannins.

DO. Ca. PRIORAT
CONSEJO REGULADOR

Major, 2
43737 Torroja del Priorat (Tarragona)
☎: +34 977 839 495 - Fax. +34 977 839 472
@: info@doqpriorat.org
www.doqpriorat.org

LOCATION:

In the province of Tarragona. It is made up of the municipal districts of La Morera de Montsant, Scala Dei, La Vilella, Gratallops, Bellmunt, Porrera, Poboleda, Torroja, Lloá, Falset and Mola.

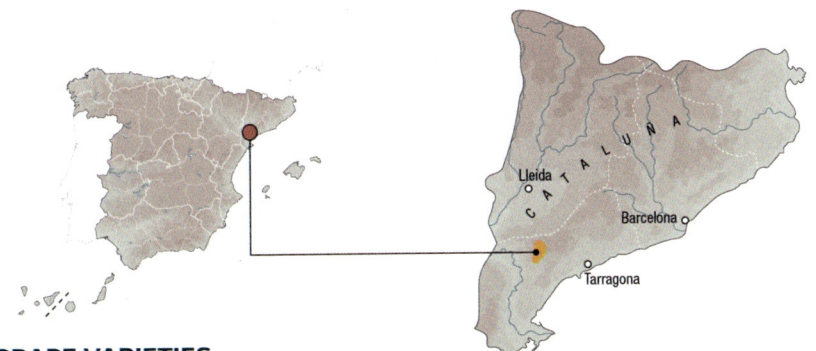

GRAPE VARIETIES:

WHITE: Chenin Blanc, Macabeo, Garnacha Blanca, Pedro Ximénez, Moscatel de Alejandría, Moscatel de grano menudo, Pansal, Picapoll Blanco.

RED: Cariñena, Garnacha, Garnacha Peluda, Cabernet Sauvignon, Merlot, Syrah, Cabernet Franc, Picapoll Negro, Pinot Noir, Tempranillo.

FIGURES:

Vineyard surface: 2,085 – **Wine-Growers:** 465 – **Wineries.** 116   **2021 Harvest rating:** Excellent – **Production 21:** 4,754,498 litres – **Market percentages:** 40% National - 60% International.

SOIL:

This is probably the most distinctive characteristic of the region and precisely what has catapulted it to the top positions in terms of quality, not only in Spain, but around the world. The soil, thin and volcanic, is composed of small pieces of slate (llicorella), which give the wines a markedly mineral character. The vineyards are located on terraces and very steep slopes.

CLIMATE:

Although with Mediterranean influences, it is temperate and dry. One of the most important characteristics is the practical absence of rain during the summer, which ensures very healthy grapes. The average rainfall is between 500 and 600 mm per year.

CLASSIFICATION OF YOUNG WINE HARVESTS PEÑÍNGUIDE

2017	2018	2019	2020	2021
VERY GOOD	VERY GOOD	EXCELLENT	EXCELLENT	EXCELLENT

DO Ca. PRIORAT / D.O.P.

ÁLVAREZ DURÁN PRIORAT
Avda. Cardenal Vidal i Barraquer, 2
43739 Porrera (Tarragona)
☎: +34 977 828 102
info@alvarezduranpriorat.com
www.alvarezduranpriorat.com

Dual 2019 T
40% garnacha, 30% cariñena, 15% syrah, 10% cabernet sauvignon, 5% merlot
91
Colour: deep cherry. Nose: dried herbs, creamy oak, black fruit, spicy. Palate: powerful, ripe fruit, spicy, round tannins.

Terroir X Blanco 2021 B
50% garnacha blanca, 30% viognier, 15% macabeo, 5% moscatel
90
Defined aromas. Colour: straw. Nose: white flowers, fragrant herbs, white fruit, ripe fruit. Palate: flavourful, fruity, fine bitter notes, long, easy to drink.

Terroir X La Tercera 2019 T
40% cariñena, 30% garnacha, 15% syrah, 10% cabernet sauvignon, 5% merlot
92
Colour: cherry, purple rim. Nose: red berry notes, spicy, ripe fruit, dried herbs, mineral. Palate: flavourful, fruity, good structure, good finish.

Terroir X El Segundo 2019 T
60% cariñena, 30% garnacha, 10% syrah
93
Colour: cherry, purple rim. Nose: fruit expression, spicy, smoky, balsamic herbs, ripe fruit. Palate: flavourful, fruity, long, fresh, good structure, slightly dry, soft tannins.

Terroir X La Viña Vieja 2019 T
95% cariñena, 5% garnacha
93
Colour: cherry, purple rim. Nose: fruit expression, spicy, ripe fruit, dried herbs, smoky. Palate: flavourful, fruity, good acidity, long, slightly dry, soft tannins.

ALVARO PALACIOS
Pol. 6, Parcela 26
43737 Gratallops (Tarragona)
☎: +34 977 839 195
info@alvaropalacios.com
www.alvaropalacios.com

Camins del Priorat 2021 T
40% garnacha, 15% cariñena, 21% cabernet sauvignon, 18% merlot, 6% syrah
92
Colour: deep cherry. Nose: dried herbs, red berry notes, ripe fruit, toasty, fresh, neat. Palate: powerful, ripe fruit, spicy, round tannins.

🏆 PODIUM

Finca Dofí 2020 T C
83% garnacha, 12% cariñena, 4% picapoll negre, 1% otras
96 🌿
Colour: bright cherry, light cherry. Nose: red berry notes, neat, expressive, elegant. Palate: fresh, fruity, flavourful, taut, easy to drink, long.

Gratallops Vi de la Vila 2020 T C
87% garnacha, 12% cariñena, 1% otras
94 🌿
Colour: bright cherry. Nose: balsamic herbs, sweet spices, scrubland, ripe fruit, black fruit. Palate: spicy, balsamic, good acidity, flavourful, complex.

🏆 PODIUM

L'Ermita 2020 T C
92% garnacha, 7% cariñena, 1% otras
98
Pleasant, complex. Colour: bright cherry. Nose: complex, expressive, spicy, mineral, earthy notes, red berry notes, balsamic herbs. Palate: elegant, full-bodied, long, great length.

🏆 PODIUM

La Baixada 2020 T
100% garnacha
96 🌿
Defined aromas, elegant. Colour: bright cherry. Nose: medium intensity, expressive, elegant, fresh fruit, wild herbs, citrus fruit. Palate: flavourful, round tannins, spicy, long.

🏆 PODIUM

Les Aubaguetes 2020 T C
70% garnacha, 27% cariñena, 3% otras
97 🌿
Colour: deep cherry. Nose: dried herbs, ripe fruit, black fruit, earthy notes, expressive, characterful. Palate: ripe fruit, spicy, round tannins, good structure, flavourful.

Les Terrasses 2020 T
72% garnacha, 28% cariñena
93
Colour: cherry, garnet rim. Nose: fruit preserve, characterful, powerful, sweet spices. Palate: flavourful, long, fruity, spicy, round tannins.

ATAVUS PRIORAT
Ctra. 710, Km. 8,3
43737 Gratallops (Tarragona)
☎: +34 977 020 350
info@atavusvines.com
www.atavusvines.com

Iura 2018 B
garnacha blanca, macabeo, pedro ximénez

91

Little interventionist, with personality. Colour: bright yellow. Nose: faded flowers, dried herbs, neat. Palate: flavourful, balanced, fine bitter notes, full of life.

Syrah D'Anfora 2019 T S
syrah

92 ♣

Colour: cherry, purple rim. Nose: fruit expression, red berry notes, spicy, mineral, dried herbs, black fruit. Palate: flavourful, fruity, long, fresh, round tannins.

BODEGA BRAVO ESCÓS
43737 Torroja del Priorat (Tarragona)
☎: +34 675 017 698
info@bodegabravoescospriorat.com
www.bodegabravoescospriorat.com

Bravo! 2021 RD
68% syrah, 32% cariñena

88

Pleasant, aromatic, fruity, tasty, wild, simple.

L'Escaleta 2020 T

92

Colour: cherry, garnet rim. Nose: balsamic herbs, scrubland, wild herbs. Palate: balsamic, spicy, soft tannins, fresh, easy to drink.

La Font del Mosquit 2021 B
72% macabeo, 28% garnacha blanca

91

Colour: bright straw. Nose: white fruit, wild herbs, dried herbs, waxy notes, balanced. Palate: flavourful, fruity, fine bitter notes.

La Roca de L'Abellar 2020 T
100% garnacha

93

Colour: bright cherry. Nose: ripe fruit, red berry notes, wild herbs, dried herbs, floral, sweet spices. Palate: fruity, flavourful, powerful, good finish, round tannins.

Les Camades 2020 T
50% cariñena, 35% garnacha, 10% syrah, 3% merlot, 2% cabernet sauvignon

91

Colour: deep cherry. Nose: ripe fruit, dried herbs. Palate: powerful, ripe fruit, spicy, round tannins.

Pas dels Caus 2021 T

91 ♣

Colour: bright cherry, purple rim. Nose: red berry notes, ripe fruit, wild herbs, mineral, spicy. Palate: flavourful, fruity, fresh, smoky aftertaste.

BODEGAS MASET
Ctra. Vilafranca-Igualada C-15 Km.19
08792 La Granada (Barcelona)
☎: +34 900 200 250
info@maset.com
www.maset.com

Maset Clos Viló 2019 T
60% cariñena, 40% garnacha

92

Taut, balanced. Colour: cherry, purple rim. Nose: fruit expression, floral, spicy, black fruit, red berry notes. Palate: flavourful, fruity, good acidity.

Maset Mas Viló 2019 T
60% garnacha, 40% cariñena

91

Colour: cherry, purple rim. Nose: spicy, ripe fruit, fruit preserve, dried herbs. Palate: flavourful, fruity, slightly dry, soft tannins, balanced.

Maset Mas Viló 2020 T
garnacha, cariñena

91

Colour: cherry, garnet rim. Nose: balsamic herbs, scrubland, ripe fruit, fruit preserve. Palate: flavourful, spicy.

BODEGAS PUIGGRÒS
Ctra. de Manresa, Km. 13
08711 Òdena (Barcelona)
☎: +34 629 853 587
info@bodegaspuiggros.com
www.bodegaspuiggros.com

Signes Del Priorat 2019 T

94

Colour: very deep cherry. Nose: complex, expressive, spicy, mineral, ripe fruit, black fruit. Palate: elegant, full-bodied, long, great length.

DO Ca. PRIORAT / D.O.P.

DO Ca. PRIORAT / D.O.P.

BODEGAS R & G
Ramón y Cajal 7, 1ºA
01007 Vitoria-Gasteiz (Araba/Álava)
☎: +34 945 150 589
araex@araex.com
www.araex.com

R&G Clos D'en Ferran 2019 T
62% cariñena, 38% garnacha
94
Colour: very deep cherry. Nose: complex, expressive, spicy, mineral, red berry notes, black fruit. Palate: full-bodied, long, great length.

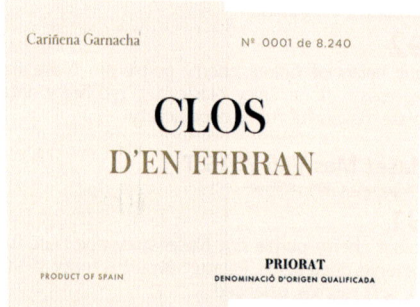

BODEGAS Y VIÑEDOS DE CAL GRAU
Ctra. del Molar a El Lloar, Finca "La Solana"
43736 El Molar (Tarragona)
☎: +34 977 054 851
info@vinosiberian.com
www.vinosiberian.com

Clos Badaceli de la Solana 2017 T C
91
Colour: deep cherry. Nose: dried herbs, creamy oak, black fruit, roasted coffee. Palate: powerful, ripe fruit, spicy, round tannins.

BUIL & GINÉ
Crta. Gratallops a la Vilella Baixa km 11.5
43737 Gratallops (Tarragona)
☎: +34 977 839 210
info@builgine.com
www.builgine.com

Angelia 2017 T
65% cabernet sauvignon, 35% garnacha, cariñena
92 🌿
Colour: deep cherry. Nose: ripe fruit, creamy oak, black fruit, scrubland, cocoa bean. Palate: powerful, ripe fruit, spicy, round tannins.

Giné Giné 2019 T
garnacha, cariñena
91
Jammy. Colour: Cherry. Nose: scrubland, fragrant herbs, ripe fruit. Palate: spicy, flavourful, balanced.

Giné Rose 2021 RD
90% garnacha, 10% merlot
89
Tasty, smooth, jammy, fruity, floral.

Joan Giné 2015 T
garnacha, cariñena, cabernet sauvignon
91
Colour: bright cherry. Nose: sweet spices, ripe fruit, dark chocolate, waxy notes, fine reductive notes, balsamic herbs. Palate: spicy, round tannins.

Joan Giné Blanc 2019 B FB
70% garnacha blanca, 20% macabeo, 10% pedro ximénez, viognier
90
Colour: bright yellow. Nose: spicy, white fruit, ripe fruit, balanced, dried flowers. Palate: flavourful, fruity, good structure.

Pleret 2013 T
garnacha, cariñena
92
Colour: dark-red cherry. Nose: ripe fruit, fruit preserve, aged wood nuances, tobacco, sweet spices. Palate: spicy, round tannins, long, good structure.

BURGOS PORTA - MAS SINÉS
Mas Sinén, s/n
43376 Poboleda (Tarragona)
☎: +34 696 094 509
burgosporta@massinen.com
www.massinen.com

Mas Sinén Clos 2017 T
garnacha, cariñena, cabernet sauvignon, syrah
93 🌿
Colour: deep cherry. Nose: dried herbs, creamy oak, black fruit, cocoa bean, fine reductive notes, mineral. Palate: powerful, ripe fruit, spicy, round tannins.

Mas Sinén Coster 2015 T C
garnacha, cariñena
92 🌿
Colour: cherry, garnet rim. Nose: powerful, black fruit, dried herbs, tobacco. Palate: flavourful, long, good structure.

Mas Sinén La Vall 2017 T BA
garnacha, cariñena, cabernet sauvignon, syrah
89 🌿
Full-bodied, balanced, spicy, dried herbs, jammy.

CASA GRAN DEL SIURANA

Ctra. de la Mina s/n
43738 Bellmunt del Priorat (Tarragona)
☎: +34 932 233 022
Fax: +34 932 231 370
info@casagrandelsiurana.com
www.casagrandelsiurana.com

Cruor 2018 T
48% cariñena, 42% garnacha, 10% syrah

93

Colour: deep cherry. Nose: dried herbs, creamy oak, black fruit, fine reductive notes. Palate: powerful, ripe fruit, spicy, round tannins.

La Fredat 2019 T
100% garnacha

90

Colour: cherry, purple rim. Nose: red berry notes, floral, spicy. Palate: flavourful, fruity, good acidity.

GR-174 2021 T
33% garnacha, 21% cariñena, 18% cabernet sauvignon, 15% syrah, 8% merlot, 3% cabernet franc

91

Colour: cherry, purple rim. Nose: fruit expression, red berry notes, floral. Palate: fruity, flavourful, balanced.

Gran Cruor 2014 T
100% syrah

92

Colour: black cherry. Nose: black fruit, fruit preserve, cigar, powerful. Palate: flavourful, long, reductive nuances, correct.

Gran Cruor Selecció Caranyena 2014 T
100% cariñena

92

Colour: dark-red cherry, garnet rim. Nose: ripe fruit, fruit preserve, aged wood nuances, tobacco, sweet spices, black fruit. Palate: spicy, round tannins, long.

CASTELL D'OR

Ctra. de Santes Creus, s/n
43814 Vila-Rodona (Tarragona)
☎: +34 977 459 860
castelldor@castelldor.com
www.castelldor.com

Abadía Mediterrània 2019 T C
garnacha, cariñena

89

Spicy, dried herbs, jammy, hot, great length, tasty, wild.

Esplugen 2019 T
garnacha, cariñena

90

Colour: bright cherry. Nose: fresh fruit, creamy oak, raspberry, scrubland. Palate: good acidity, spicy, fine tannins.

CELLER AIXALÀ I ALCAIT

Carrer Balandra, 8
43737 Torroja del Priorat (Tarragona)
☎: +34 629 507 807
pardelasses@gmail.com
www.pardelasses.com

Destrankis 2020 T BA
89 ♣

Spicy, smoky, fruity, jammy, tasty.

El Coster de L'Alzina 2017 T C
90 ♣

Colour: deep cherry. Nose: ripe fruit, dried herbs, cedar wood, spicy. Palate: powerful, ripe fruit, spicy, slightly dry, soft tannins.

DO Ca. PRIORAT / D.O.P.

DO Ca. PRIORAT / D.O.P.

Les Clivelles de l'Alzina 2019 T
91 🌱
Colour: cherry, purple rim. Nose: fruit expression, spicy, ripe fruit, black fruit, toasty, caramel. Palate: flavourful, fruity, good acidity, slightly dry, soft tannins.

Les Clivelles de Torroja 2020 T
90 🌱
Colour: cherry, purple rim. Nose: fruit preserve, black fruit, smoky, spicy, mineral. Palate: flavourful, fruity, balanced, round tannins.

Les Clivelles de Torroja Rosat 2021 RD
90
Colour: rose, purple rim. Nose: fruit expression, red berry notes, dried herbs, ripe fruit, mineral. Palate: fruity, good acidity, easy to drink, flavourful.

Pardelasses 2017 T
50% garnacha, 50% cariñena
89 🌱
Pruney, fruity, spicy, tasty.

CELLER BARTOLOMÉ
Major, 23
43738 Bellmunt del Priorat (Tarragona)
☎: +34 977 830 632
comercial@cellerbartolome.com
www.cellerbartolome.com

Clos Bartolomé Blanc 2021 B
macabeo, garnacha blanca
89
Jammy, needs time, tasty, yeasty notes, great length, spicy.

Primitiu de Bellmunt 2019 T
91
Colour: cherry, garnet rim. Nose: fruit preserve, fruit liqueur notes, powerful, dark chocolate. Palate: flavourful, sweetness, long.

CELLER CAL PLA
43739 Porrera (Tarragona)
☎: +34 977 828 125
Fax: +34 977 828 125
info@cellercalpla.com
www.cellercalpla.com

Celler Cal Pla 2019 T
garnacha, cariñena, cabernet sauvignon
88
Full-bodied, spicy, jammy, woody, roasted coffee.

La Carenyeta de Cal Pla 2016 T
100% cariñena
89
Overripe, fruity, dried herbs, tasty.

Mas D'en Compte 2017 B FB
garnacha blanca, xarel.lo, picapoll blanc
91
Colour: bright yellow. Nose: powerful, ripe fruit, spicy, toasted bread, smoky. Palate: rich, good structure, long, toasty.

Mas D'en Compte 2016 T R
garnacha, samsó, cabernet sauvignon
91
Colour: Cherry. Nose: sweet spices, scrubland, black fruit, ripe fruit, fine reductive notes. Palate: spicy, balsamic, good acidity.

Obac de L'Andreva 2017 T R
garnacha peluda, cariñena
91
Colour: deep cherry. Nose: ripe fruit, dried herbs, creamy oak, black fruit. Palate: powerful, ripe fruit, spicy, round tannins.

CELLER CASTELLET
Font de Dalt, 11
43739 Porrera (Tarragona)
☎: +34 630 849 874
info@cellercastellet.cat
www.cellercastellet.cat

Empit 2019 T C
garnacha peluda, cariñena, syrah
91
Colour: deep cherry. Nose: dried herbs, black fruit, dark chocolate. Palate: ripe fruit, spicy, round tannins.

Empit Selecció 2018 T R
100% cariñena
90
Colour: deep cherry. Nose: black fruit, cocoa bean, floral, scrubland. Palate: powerful, ripe fruit, spicy, round tannins.

Reblum 2019 T
garnacha

92

Dried flowers, wild. Colour: deep cherry, garnet rim. Nose: fruit preserve, black fruit, scrubland, dried flowers. Palate: fruity, easy to drink, balanced.

Terrotxa 2018 T
garnacha, cabernet sauvignon

91

Colour: deep cherry. Nose: creamy oak, black fruit, scrubland. Palate: powerful, ripe fruit, spicy, round tannins.

CELLER CLOS93 PRIORAT

Nou, 26
43737 El Lloar (Tarragona)
☎: +34 665 287 433
clos93@clos93.com
www.clos93.com

L'Exclamació 2019 T C
100% syrah

91 ♣

Colour: deep cherry. Nose: dried herbs, creamy oak, black fruit, meaty notes. Palate: powerful, ripe fruit, spicy, round tannins.

L'Interrogant 2020 T
garnacha, cariñena, cabernet sauvignon

91 ♣

Colour: cherry, purple rim. Nose: fruit expression, red berry notes, floral, spicy, fruit preserve. Palate: flavourful, fruity, good acidity.

Parentesis 2021 B
garnacha blanca, pedro ximénez

89 ♣

Dried flowers, dried herbs, aromatic, fruity, jammy, standard, balanced, spicy.

Vi de Vila del Lloar
"A la Memòria del Pé" 2017 T C
100% cariñena

91

Colour: very deep cherry. Nose: fruit preserve, dried fruit, dried herbs, fine reductive notes. Palate: flavourful, powerful, sweet tannins.

CELLER DE L'ABADÍA

Font, 38
43737 Gratallops (Tarragona)
☎: +34 627 032 134
jeroni@cellerabadia.com
www.cellerabadia.eu

Alice 2018 T R
30% cariñena, 30% garnacha, 20% monastrell, 10% syrah, 10% cabernet sauvignon

88

Spicy, herbaceous, jammy, powerful, tasty.

Clos Clara 2017 T GR
40% cariñena, 40% garnacha, 10% syrah, 10% cabernet sauvignon

89

Pruney, full-bodied, spicy, floral, dried herbs, full-bodied, tasty.

Sant Jeroni Hort 2019 T
70% garnacha, 30% syrah

90

Colour: deep cherry. Nose: ripe fruit, black fruit, scrubland, cocoa bean. Palate: ripe fruit, spicy, round tannins.

Mi Amor 2020 T
100% syrah

91

Colour: bright cherry. Nose: sweet spices, dark chocolate, scrubland, dried herbs, fruit preserve. Palate: fruity, spicy, round tannins, flavourful.

Sant Jeroni Aubada 2021 B
70% pedro ximénez, 30% garnacha blanca

90

Pleasant, aromatic. Colour: bright straw. Nose: spicy, white fruit, dried flowers, balanced, expressive. Palate: flavourful, spicy.

Sant Jeroni
Flor de Samsó 2021 RD
100% cariñena

86

Sant Jeroni Forn 2019 T
80% cariñena, 20% cabernet sauvignon

88

Pruney, spicy, toasty, hot.

CELLER DE L'ENCASTELL

Castell, 7
43739 Porrera (Tarragona)
☎: +34 630 941 959
roquers@roquers.com
www.roquers.com

Marge 2019 T
garnacha, merlot, cabernet sauvignon, cariñena, syrah

90

Colour: cherry, purple rim. Nose: floral, spicy, red berry notes, black fruit. Palate: flavourful, fruity, good acidity, long.

Roquers de Garnatxa 2019 T
garnacha

91

Colour: bright cherry. Nose: fresh fruit, creamy oak, black fruit, red berry notes, spicy. Palate: good acidity, spicy, fine tannins.

DO Ca. PRIORAT / D.O.P.

DO Ca. PRIORAT / D.O.P.

Roquers de Porrera 2018 T R
garnacha, cariñena, merlot

93

Colour: cherry, purple rim. Nose: fruit expression, floral, spicy, red berry notes, black fruit, fine reductive notes. Palate: flavourful, fruity, good acidity, long, balanced.

Roquers de Samsó 2019 T
cariñena

92

Representative, varietally correct. Colour: very deep cherry. Nose: fruit preserve, wild herbs, dried herbs, spicy. Palate: flavourful, round tannins, spicy, long.

CELLER ESCODA PALLEJÀ
La Font, 16
43737 Torroja del Priorat (Tarragona)
☎: +34 653 854 134
rescoda@hotmail.com
www.celleresccodapalleja.blogspot.com

Palet Most de Flor 2020 T
garnacha, cariñena, cabernet sauvignon, syrah

93

Colour: cherry, garnet rim. Nose: fruit preserve, fruit liqueur notes, powerful, black fruit, cocoa bean. Palate: flavourful, long, balanced.

Palet Vinya Tricolavet 2019 T C
garnacha, cariñena, cabernet sauvignon, syrah

90

Colour: cherry, garnet rim. Nose: fruit preserve, powerful, black fruit, scrubland. Palate: flavourful, sweetness.

Palet Vinyes Velles 2018 T C
cariñena

92

Colour: dark-red cherry. Nose: toasty, spicy, cocoa bean, scrubland, black fruit. Palate: flavourful, toasty, fine bitter notes, powerful, good structure, sweetness.

CELLER GRITELLES
Carrer Sant Andreu, 5
43360 Cornudella de Montsant (Tarragona)
☎: +34 637 407 184
celler@gritelles.com
www.gritelles.com

Gritelles Tros de la Serra 2019 B
macabeo

93

Oxidativ, with personality. Colour: golden, coppery red. Nose: white fruit, dry nuts, scrubland, hint of anise, balsamic herbs. Palate: fleshy, powerful, flavourful, fresh.

CELLER HIDALGO ALBERT
Pol. 14 Parcela 102
43376 Poboleda (Tarragona)
☎: +34 977 842 064
info@hidalgoalbert.com
www.hidalgoalbert.com

1270 a Vuit 2014 T R
garnacha, cariñena

92 🍷

Age nuances, pruney. Colour: cherry, garnet rim. Nose: old leather, black fruit, black pepper, dried herbs. Palate: flavourful, balanced.

Fina 1270 a Vuit 2019 T BA
garnacha, syrah, merlot, cabernet sauvignon, cabernet franc, cariñena

90 🍷

Colour: cherry, purple rim. Nose: red berry notes, floral, spicy, toasty. Palate: flavourful, fruity, good acidity.

Lo Petit Pau 2021 T
garnacha, syrah, merlot, cabernet sauvignon, cabernet franc, cariñena

88 🍷

Balanced, spicy, jammy, tasty, toasty.

CELLER JOAN AMETLLER
Ctra. La Morera de Monsant - Cornudella, km. 3,2
43361 La Morera de Monsant (Tarragona)
☎: +34 634 532 370
info@ametller.com
www.familiaametller.com

Clos Corriol 2021 B
100% garnacha blanca

88

Citrus fruit, herbal, crisp, balanced, not representative.

Clos Corriol Negre 2019 T
garnacha, merlot, syrah, cabernet sauvignon

88

Spicy, jammy, toasty, standard.

DO Ca. PRIORAT / D.O.P.

Clos Corriol Rosat 2021 RD
100% garnacha

88

Pleasant, floral, fruity, sweeties, dried herbs, tasty, simple.

Clos Mustardó 2016 T R
garnacha, merlot, cabernet sauvignon

89

Fruity, dried herbs, spicy, tasty, balanced.

Clos Mustardó 2018 B FB
100% garnacha blanca

89

Full-bodied, spicy, full-bodied, jammy, woody.

Els Igols "Vinyes de Coster" 2012 T GR
garnacha, cabernet sauvignon, merlot

91

Colour: dark-red cherry, garnet rim. Nose: ripe fruit, fruit preserve, aged wood nuances, tobacco. Palate: spicy, round tannins, long.

CELLER MAS BASTE
Font, 38
43737 Gratallops (Tarragona)
☎: +34 627 032 134
info@cellermasbaste.com
www.cellermasbaste.com

Peites 2009 T C
80% cariñena, 10% syrah, 10% cabernet sauvignon

91

Colour: cherry, garnet rim. Nose: red berry notes, spicy, meaty notes, dried flowers, dried herbs. Palate: flavourful, fruity, good acidity, long.

CELLER PASANAU
43361 La Morera de Montsant (Tarragona)
☎: +34 977 827 202
info@cellerpasanau.com
www.cellerpasanau.com

Pasanau Ceps Nous 2019 T
54% garnacha, 41% merlot, 4% cariñena, syrah

90 ♣

Pruney, full-bodied. Colour: deep cherry. Nose: black fruit, fruit preserve, dried herbs, black liquorice. Palate: flavourful, good structure.

Pasanau Finca La Planeta 2016 T
87% cabernet sauvignon, 13% garnacha

91

Colour: deep cherry. Nose: dried herbs, creamy oak, black fruit, fine reductive notes. Palate: powerful, ripe fruit, spicy, round tannins.

Pasanau Vi de Vila de La Morera de Montsant 2019 T
90% garnacha, 10% merlot

91 ♣

Colour: Cherry. Nose: balsamic herbs, scrubland, black fruit, ripe fruit. Palate: spicy, flavourful, round tannins, good structure.

CELLER VALL-LLACH
Del Pont, 9
43739 Porrera (Tarragona)
☎: +34 977 828 244
info@vallllach.com
www.vallllach.com

Aigua de Llum de Vall Llach 2021 B
100% viognier

91

Kind finish, flat. Colour: straw. Nose: white flowers, jasmine, dried herbs, balsamic herbs. Palate: flavourful, fruity, balanced, rich.

Embruix de Vall-Llach 2020 T
37% garnacha, 32% cariñena, 23% syrah, 8% cabernet sauvignon

92

Colour: deep cherry, purple rim. Nose: dried herbs, creamy oak, ripe fruit, black fruit. Palate: powerful, ripe fruit, spicy, round tannins.

🏆 PODIUM

Porrera Vi de Vila de Vall Llach 2020 T C
70% cariñena, 30% garnacha

95

Colour: deep cherry. Nose: ripe fruit, dried herbs, creamy oak, earthy notes, black fruit. Palate: powerful, ripe fruit, spicy, round tannins.

Porrera Vi de Vila de Vall Llach 2021 B
100% garnacha blanca

91

Oxidativ. Colour: straw. Nose: ripe fruit, dried herbs, faded flowers, cereal notes. Palate: ripe fruit, balanced, full-bodied, fine bitter notes.

Priorat Idus de Vall-Llach 2020 T
59% cariñena, 41% garnacha

92

Needs time. Colour: deep cherry. Nose: dried herbs, red berry notes, black fruit, fruit preserve, spicy. Palate: powerful, ripe fruit, spicy, slightly green tannins.

DO Ca. PRIORAT / D.O.P.

CELLERS DE SCALA DEI
Rambla de la Cartoixa, s/n
43379 Scala Dei (Tarragona)
☎: +34 977 827 027
info@cellersdescaladei.com
www.cellersdescaladei.com

Scala Dei Cartoixa 2018 T R
80% garnacha, 20% cariñena

93
Colour: bright cherry. Nose: balsamic herbs, sweet spices, scrubland, red berry notes. Palate: spicy, balsamic, good acidity.

Scala Dei L'Heretge 2018 T

94
Colour: bright cherry. Nose: complex, expressive, spicy, mineral, grassy. Palate: full-bodied, long, great length, good acidity.

Scala Dei Masdeu 2017 T
100% garnacha

94
Colour: cherry, garnet rim. Nose: overripe fruit, creamy oak, hot, fruit liqueur notes, scrubland. Palate: pruney, powerful, sweet tannins.

Scala Dei Massipa 2020 B
80% garnacha blanca, 20% chenin blanc

92
Colour: straw. Nose: creamy oak, dried flowers, white fruit, ripe fruit, waxy notes. Palate: flavourful, good structure.

Scala Dei Pla dels Àngels 2021 RD
100% garnacha

90
Colour: raspberry rose. Nose: red berry notes, floral, fragrant herbs. Palate: light-bodied, good acidity, fine bitter notes, easy to drink, fresh.

Scala Dei Prior 2018 T C
55% garnacha, 15% cariñena, 15% cabernet sauvignon, 15% syrah

93
Colour: Cherry. Nose: complex, expressive, spicy, mineral, red berry notes. Palate: elegant, full-bodied, long, great length.

🏆 PODIUM

Scala Dei Sant Antoni 2018 T
100% garnacha

95
Colour: light cherry. Nose: wild herbs, chamomile, fruit preserve, expressive. Palate: flavourful, fine bitter notes, easy to drink, long.

CELLERS TERRA I VINS
Av. Falset, 17 Baixos
43206 Reus (Tarragona)
☎: +34 658 567 409
celler@cellersterraivins.com
www.cellersterraivins.com

Brúixola 2016 B
garnacha blanca, macabeo, pedro ximénez

91
Colour: straw. Nose: spicy, toasty, burnt matches, ripe fruit, dried herbs. Palate: rich, flavourful, dry, long.

Brúixola 2017 T C
garnacha, samsó, syrah

91
Colour: cherry, garnet rim. Nose: fruit preserve, fruit liqueur notes, powerful, spicy, black pepper. Palate: flavourful, fruity, balanced, slightly dry, soft tannins.

Brúixola Carinyena Garnatxa Negra 2018 T
cariñena, garnacha

91
Colour: cherry, garnet rim. Nose: ripe fruit, dried herbs, mineral, spicy. Palate: fruity, flavourful, powerful, balanced, slightly dry, soft tannins.

CELLERS UNIÓ
Joan Oliver 16-24
43206 Reus (Tarragona)
☎: +34 977 330 055
Fax: +34 977 330 070
info@cellersunio.com
www.cellersunio.com

Roureda Llicorella Poboleda 2021 B
pedro ximénez

91
Colour: straw. Nose: expressive, white flowers, dried herbs. Palate: flavourful, fruity, balanced, long.

CELLERS UNIÓ - POBOLEDA
Plaça Portal, s/n
43376 Poboleda (Tarragona)
☎: +34 977 330 055
info@cellersunio.com
www.cellersunio.com

Vinyes de Poboleda 2021 T
garnacha, mazuelo

88
Pruney, spicy, powerful, dried herbs.

CLOS DEL PORTAL
43736 El Molar (Tarragona)
☎: +34 932 531 760
info@portaldelpriorat.com
www.portaldelpriorat.com

Gotes del Priorat 2020 T
garnacha, cariñena, syrah
91
Colour: cherry, purple rim. Nose: red berry notes, floral, spicy, black fruit. Palate: flavourful, fruity, good acidity.

La Solana dels Marges 2019 T R
cariñena
92
Colour: deep cherry. Nose: dried herbs, scrubland, fine reductive notes, spicy. Palate: spicy, long, ripe fruit.

Negre de Negres 2020 T
garnacha, cariñena, syrah, cabernet franc
93
Colour: deep cherry. Nose: dried herbs, dried flowers, cocoa bean, black fruit. Palate: ripe fruit, spicy, round tannins.

Negre de Negres Magnum 2019 T
92
Colour: dark-red cherry. Nose: ripe fruit, aged wood nuances, sweet spices, animal reductive notes, characterful. Palate: spicy, round tannins, long, balanced, fine bitter notes, elegant, fruity.

Somni 2018 T
syrah, cariñena
91
Colour: very deep cherry. Nose: roasted coffee, aromatic coffee, powerful, black fruit, woody. Palate: smoky aftertaste, great length, round tannins.

Tros de Clos Mas del Metge 2019 T
cariñena
91
Colour: bright cherry. Nose: red berry notes, black fruit, scrubland, toasty. Palate: good acidity, spicy, fine tannins.

CLOS FIGUERAS
Carrer La Font, 38
43737 Gratallops (Tarragona)
☎: +34 977 830 217
info@closfigueras.com
www.closfigueras.info

Clos Figueres 2020 T
93
Colour: cherry, purple rim. Nose: black fruit, ripe fruit, dried herbs, smoky, mineral. Palate: fruity, flavourful, balanced, smoky aftertaste, toasty, good structure.

Font de la Figuera 2020 T FB
92
Colour: cherry, purple rim. Nose: fruit expression, spicy, black fruit, smoky, dry stone. Palate: flavourful, fruity, balanced, slightly dry, soft tannins.

Font de la Figuera 2021 B
90
Kind finish, sweety finish. Colour: bright straw. Nose: floral, white fruit, ripe fruit. Palate: flavourful, rich.

Serras del Priorat 2021 B
garnacha blanca
91
Colour: bright straw. Nose: ripe fruit, floral, neat, dried herbs, wild herbs. Palate: flavourful, fruity aftertaste, varietal.

Serras del Priorat 2021 T
92
Colour: cherry, purple rim. Nose: fruit expression, spicy, grassy, black fruit. Palate: flavourful, fruity, balanced, good finish, smoky aftertaste, round tannins.

Sweet Clos Figueras 2021 TF D
garnacha
90
Colour: cherry, garnet rim. Nose: fruit preserve, ripe fruit. Palate: powerful, flavourful, fruity.

CLOS GALENA
Camino de la Solana, s/n
43736 El Molar (Tarragona)
☎: +34 607 421 822
laura@closgalena.com
www.closgalena.com

Clos Galena 2018 T R
93
Colour: bright cherry. Nose: creamy oak, raspberry, ripe fruit, petrol notes. Palate: good acidity, spicy, fine tannins, taut.

Clos Galena 2019 T R
93
Colour: bright cherry. Nose: ripe fruit, black fruit, spicy, cocoa bean, dried herbs, dry stone. Palate: good structure, fruity, flavourful, balanced, slightly dry, soft tannins.

Formiga de Seda 2021 B
89
Citrus fruit, spicy, balanced, floral, kind finish, jammy.

Formiga de Vellut 2019 T
90
Colour: deep cherry. Nose: dried herbs, scrubland, black fruit, dark chocolate. Palate: ripe fruit, spicy, round tannins.

DO Ca. PRIORAT / D.O.P.

DO Ca. PRIORAT / D.O.P.

CLOS MOGADOR
Camí Manyetes, s/n
43737 Gratallops (Tarragona)
☎: +34 977 839 171
Fax: +34 977 839 426
closmogador@closmogador.com
www.closmogador.com

Clos Mogador 2019 T C
94
Colour: deep cherry, bright cherry. Nose: complex, balanced, dried herbs, spicy. Palate: balanced, flavourful, spicy, ripe fruit.

🏆 **PODIUM**

Nelin 2019 B
96
Colour: straw. Nose: ripe fruit, dried herbs, faded flowers, cereal notes, spicy, dry stone, waxy notes. Palate: powerful, ripe fruit, balanced, full-bodied, flavourful.

CLOS PACHEM
Carrer de la Font 1D
43737 Gratallops (Tarragona)
☎: +34 621 229 185
admin@clospachem.com
www.clospachem.cat

Camí de la Mina 2020 T
garnacha, cariñena
92
Colour: Cherry. Nose: complex, expressive, spicy, mineral, wild herbs, floral. Palate: full-bodied, long, great length.

Pachem 2020 T S
garnacha
91
Colour: cherry, purple rim. Nose: red berry notes, floral, spicy, wild herbs, varietal. Palate: flavourful, fruity, good acidity.

Planassos 2019 T
100% cariñena
93
Colour: cherry, purple rim. Nose: fruit expression, red berry notes, spicy, black fruit, ripe fruit, dried flowers. Palate: flavourful, fruity, good acidity, long, full of life, balanced, slightly dry, soft tannins.

DE MULLER
Camí Pedra Estela, 34
43205 Reus (Tarragona)
☎: +34 977 757 473
promo@demuller.es
www.demuller.es

De Muller Carinyena 2018 T C
100% cariñena
90
Colour: bright cherry. Nose: aromatic coffee, dark chocolate, spicy, roasted coffee, ripe fruit. Palate: fruity, flavourful, balanced, slightly dry, soft tannins.

Legitim 2019 T C
35% garnacha, 35% cariñena, 15% merlot, 15% syrah
89
Full-bodied, balanced, spicy, dried herbs, toasty.

Les Pusses 2017 T C
50% merlot, 50% syrah
88
Balanced, spicy, jammy, full-bodied.

Lo Cabaló 2016 T R
75% garnacha, 15% cariñena, 10% syrah
89
Balanced, spicy, age nuances, dried herbs, jammy.

EDICIONES ILIMITADAS
Carrer Modolell, 56 Local A
08021 Barcelona (Barcelona)
☎: +34 932 531 760
info@edicionesi-limitadas.com

Fils de Vi 2020 T BA
cariñena
92
Colour: bright cherry. Nose: red berry notes, spicy, mineral, dried herbs, floral. Palate: fruity, flavourful, balanced, round tannins.

Lo MAS 2021 T
90
Colour: deep cherry. Nose: ripe fruit, dried herbs, creamy oak, wet leather. Palate: ripe fruit, spicy, round tannins.

ESPAIVI

Santa Ana, 13 Plaza Forum
43003 Tarragona (Tarragona)
☎: +34 672 266 894
espai.tgn@gmail.com

Trekan 2018 T BA
50% cariñena, 50% garnacha

91

Colour: bright cherry. Nose: fruit expression, red berry notes, floral, spicy, wild herbs. Palate: flavourful, fruity, good acidity, balanced, slightly dry, soft tannins.

FAMILIA NIN ORTIZ

Finca Planetes, Pol Partida Masis Parcela 288
Falset (Tarragona)
☎: +34 686 467 579
carlesov@gmail.com
http://fnovins.blogspot.com.es

Nit de Nin Coma d'en Romeu 2019 T
garnacha

94

Colour: cherry, garnet rim. Nose: black fruit, balsamic herbs, mineral, complex, smoky. Palate: concentrated, flavourful, powerful.

🏆 PODIUM

Nit de Nin La Rodeda 2019 T
garnacha peluda

96

Colour: deep cherry. Nose: black fruit, balsamic herbs, wild herbs, spicy, mineral, smoky. Palate: balanced, complex, flavourful.

Nit de Nin Mas d'en Caçador 2019 T
cariñena, garnacha, garnacha peluda, otras

94

Colour: deep cherry. Nose: red berry notes, black fruit, balsamic herbs, mineral, earthy notes, smoky. Palate: good structure, flavourful, full-bodied, warm.

Planetes Classic 2019 T
cariñena, garnacha, garnacha peluda

93

Colour: cherry, garnet rim. Nose: red berry notes, fresh fruit, scrubland, black liquorice, balsamic herbs, mineral. Palate: balanced, flavourful, full of life.

🏆 PODIUM

Selma de Nin 2018 B
parellada, roussanne, marsanne, chenin blanc

95

Taut. Colour: straw. Nose: white fruit, mineral, dry stone, smoky. Palate: good structure, great length, good acidity.

Terra Vermella de Nin 2018 B
parellada

94

Austere. Colour: straw. Nose: white fruit, citrus fruit, dry stone, petrol notes. Palate: flavourful, saline, good acidity.

FAMILIA TORRES

Finca La Solteta s/n
47311 El Lloar (Tarragona)
☎: +34 938 177 400
info@torres.es
www.torres.es

Mas de la Rosa 2019 T C
cariñena, garnacha

94

Taut, needs time. Colour: Cherry. Nose: complex, expressive, spicy, mineral, red berry notes, scrubland. Palate: elegant, full-bodied, long, great length.

Perpetual 2019 T C
cariñena, garnacha

93

Colour: bright cherry. Nose: ripe fruit, balanced, expressive, mineral, wild herbs. Palate: elegant, balanced, easy to drink.

Salmos 2019 T C
cariñena, garnacha

92

Colour: cherry, garnet rim. Nose: creamy oak, sweet spices, ripe fruit, dried herbs. Palate: good structure, flavourful.

FINCA TOBELLA

Les Aubagues
43737 Gratallops (Tarragona)
☎: +34 652 857 059
info@fincatobella.com
www.fincatobella.com

Bocins Blanc 2021 B
garnacha blanca, viognier

90

Colour: straw. Nose: ripe fruit, dried herbs, faded flowers, cereal notes. Palate: powerful, ripe fruit, balanced, flavourful, saline.

Tobella Negre 2020 T
garnacha, syrah, cabernet sauvignon, cariñena

90

Defined aromas, standard, jammy, tasty, fruity, spicy, dried herbs.

DO Ca. PRIORAT / D.O.P.

DO Ca. PRIORAT / D.O.P.

Tobella Negre 2021 T
garnacha, syrah, cabernet sauvignon, cariñena

90
Colour: cherry, purple rim. Nose: fruit expression, red berry notes, floral. Palate: fruity, flavourful, balanced.

GRAN CLOS DEL PRIORAT
Montsant, 2
43738 Bellmunt del Priorat (Tarragona)
☎: +34 977 830 675
office@granclos.com
www.granclos.com

Cartus 2007 T R
garnacha, cariñena

93
Colour: dark-red cherry, garnet rim. Nose: ripe fruit, fruit preserve, aged wood nuances, tobacco, sweet spices, black fruit. Palate: spicy, round tannins, long.

Finca El Puig 2019 T
garnacha, cariñena, syrah

90
Colour: cherry, purple rim. Nose: fruit expression, spicy, fruit preserve, fruit liqueur notes. Palate: flavourful, fruity, good finish, slightly dry, soft tannins.

Gran Clos 2018 T C
garnacha, cariñena

91
Colour: cherry, garnet rim. Nose: fruit preserve, powerful, creamy oak, dark chocolate. Palate: flavourful, long.

Gran Clos 2020 B FB
garnacha blanca, macabeo

91
Colour: bright straw. Nose: neat, white fruit, citrus fruit, balanced, floral. Palate: flavourful, easy to drink.

Les Mines 2019 T C
garnacha, merlot, cabernet sauvignon, cariñena

88
Overripe, spicy, fruity, hot, tasty.

Solluna 2019 T C
garnacha, merlot, cabernet sauvignon, cariñena

87

GRATAVINUM
Mas d'en Serres s/n
43737 Gratallops (Tarragona)
☎: +34 938 901 399
gratavinum@gratavinum.com
www.gratavinum.com

Gratavinum 2πr 2020 T
garnacha, cariñena, syrah

90 ✿
Colour: deep cherry. Nose: dried herbs, black fruit, spicy. Palate: ripe fruit, round tannins, balanced.

Gratavinum GV5 2016 T
cariñena, garnacha

93 ✿
Colour: cherry, garnet rim. Nose: balsamic herbs, ripe fruit, scrubland, old leather. Palate: flavourful, balsamic, spicy, balanced, round.

Gratavinum Silvestris 2020 T
cariñena, syrah

91 ✿
Colour: cherry, garnet rim. Nose: fruit preserve, powerful, red berry notes, toasty. Palate: flavourful, sweetness, long.

HAMMEKEN CELLARS

Llavador, 20
03700 Denia (Alacant/Alicante)
☎: +34 965 791 967
cellars@hammekencellars.com
www.hammekencellars.com

Tosalet Carignan Vinyes Velles 2017 T
100% carignan

92

Colour: Cherry. Nose: scrubland, burnt matches, wild herbs, spicy. Palate: balsamic, good acidity, round tannins, round.

Tosalet Vinyes Fins a 50 anys 2021 T
60% garnacha, 40% carignan

90

Colour: cherry, garnet rim. Nose: fruit preserve, powerful, scrubland, wild herbs. Palate: flavourful, long.

HODGKINSON PRIORAT

Mas del Habanero, Los Masos, km 2,7 T-710
43730 Falset (Tarragona)
☎: +34 687 565 731
caspar@hodgkinson-priorat.com

Hodgkinson Garnacha Peluda 2018 T R

91

Colour: cherry, purple rim. Nose: cereal notes, dried herbs, dry nuts, red berry notes. Palate: fruity, powerful, flavourful, powerful tannins.

Las Panzudas 2018 T

91

Colour: deep cherry. Nose: creamy oak, black fruit, cocoa bean, spicy, scrubland. Palate: powerful, ripe fruit, spicy, round tannins.

Mas del Habanero 2017 T
90% garnacha, 10% cariñena

92

Colour: bright cherry. Nose: scrubland, petrol notes, red berry notes, black fruit, cocoa bean, expressive. Palate: good acidity, spicy, fine tannins.

JOAN SIMÓ

11 de Setembre, 7
43739 Porrera (Tarragona)
☎: +34 977 830 993
info@cellerjoansimo.eu
www.cellerjoansimo.com

Les Eres 2016 T C
75% cariñena, 25% garnacha blanca

93

Colour: deep cherry. Nose: ripe fruit, dried herbs, creamy oak, spicy, fruit preserve. Palate: powerful, ripe fruit, spicy, round tannins, slightly dry, soft tannins.

Les Eres Especial dels Carners 2015 T C
75% garnacha, 25% cariñena

92

Nose: black liquorice, black fruit, ripe fruit, spicy. Palate: fruity, flavourful, good structure, balanced, slightly dry, soft tannins.

Tros Nou vi de vila Bellmunt 2018 T C
75% garnacha peluda, 25% cariñena

91

Colour: deep cherry. Nose: dried herbs, creamy oak, black fruit, floral. Palate: powerful, ripe fruit, spicy, round tannins.

JOSEP GRAU VITICULTOR

Paratge Les Taules
43776 Capçanes (Tarragona)
☎: +34 657 322 291
celler@josepgrauviticultor.com
www.josepgrauviticultor.com

Pedrabona 2020 T
garnacha, cariñena

92

Colour: Cherry. Nose: spicy, undergrowth, floral, red berry notes, ripe fruit. Palate: balanced, flavourful, fleshy.

DO Ca. PRIORAT / D.O.P.

L'INFERNAL

Polígon 8, Parcel·la 148
43737 Torroja del Priorat (Tarragona)
☎: +33 673 448 336
contact.linfernal@gmail.com
www.linfernalwine.com

Aguilera 2011 T GR
cariñena

92

Colour: cherry, garnet rim. Nose: sweet spices, tobacco, old leather, fruit preserve, wild herbs. Palate: fruity, balanced, classic aged character, soft tannins, smoky aftertaste.

Riu by Trío Infernal 2018 T R

92

Fruity, exuberant. Colour: bright cherry. Nose: red berry notes, floral, scrubland, black liquorice. Palate: good acidity, spicy, fine tannins, elegant.

Riu by Trío Infernal 2021 B
garnacha, macabeo

91

Oxidativ. Colour: straw. Nose: dried flowers, dried herbs, ripe fruit, white fruit. Palate: flavourful, powerful, long.

LA CONRERIA D'SCALA DEI

Carrer Mitja Galta, s/n (Finca Les Brugueres)
43379 Scala Dei (Tarragona)
☎: +34 977 827 055
laconreria@vinslaconreria.com
www.vinslaconreria.com

La Conreria Escaladei Vi de Vila 2017 T C
garnacha, cariñena

92

Colour: bright cherry. Nose: ripe fruit, dried herbs, creamy oak, spicy, wild herbs. Palate: powerful, spicy, round tannins, good finish, good structure.

Les Brugueres 2020 T
garnacha, syrah

90

Colour: bright cherry. Nose: fruit preserve, spicy, smoky, dried herbs. Palate: flavourful, fruity, good structure, slightly dry, soft tannins.

Les Brugueres 2021 B
garnacha blanca

89

Pleasant, fruity, jammy, balanced, standard, defined aromas, floral, tasty.

Nona 2020 T
garnacha, cariñena, syrah, merlot

90

Colour: deep cherry. Nose: fine reductive notes, ripe fruit, fruit preserve, dried herbs, wild herbs. Palate: powerful, ripe fruit, spicy, round tannins.

Primera Vinya Les Brugueres 2020 B
garnacha blanca

92

Colour: bright yellow. Nose: powerful, creamy oak, ripe fruit, spicy, pattiserie. Palate: rich, good structure, long, fine bitter notes.

Voltons 2018 T
cariñena, garnacha

90

Colour: cherry, purple rim. Nose: spicy, black fruit, red berry notes, scrubland. Palate: ripe fruit, flavourful, good structure.

LLICORELLA VINS

Carrer Vals, 6
43737 Gratallops (Tarragona)
☎: +34 968 435 022
Fax: +34 968 716 051
info@gilfamili.es
www.gilfaimly.es

Bluegray 2020 T
40% garnacha, 30% cariñena, 20% cabernet sauvignon, 10% merlot

90

Colour: deep cherry. Nose: ripe fruit, dried herbs, creamy oak. Palate: powerful, ripe fruit, spicy, round tannins.

Bluegray 2021 T
40% garnacha, 30% cariñena, 20% cabernet sauvignon, 10% merlot

92

Colour: cherry, purple rim. Nose: fruit expression, red berry notes, spicy, toasty, scrubland. Palate: flavourful, fruity, good acidity, long.

Clar del Bosc 2020 T
49% garnacha, 36% cariñena, 13% syrah, 2% cabernet sauvignon

93

Colour: deep cherry. Nose: dried herbs, creamy oak, black fruit, ripe fruit, toasty. Palate: powerful, ripe fruit, spicy, round tannins.

Minairo 2021 T
40% garnacha, 30% cariñena, 20% cabernet sauvignon, 10% merlot

92

Colour: cherry, purple rim. Nose: red berry notes, floral, sweet spices, earthy notes. Palate: flavourful, fruity, good acidity, long.

DO Ca. PRIORAT / D.O.P.

MAIUS VITICULTORS SCP
08172 St. Cugat del Valles (Barcelona)
☎: +34 696 998 575
jgomez@maiusviticultors.com
www.maiusviticultors.com

Maius Assemblage 2020 T
garnacha, cariñena

90

Colour: deep cherry. Nose: ripe fruit, dried herbs, creamy oak. Palate: ripe fruit, spicy, round tannins.

Maius Barranc de la Bruxa 2019 T C

90

Colour: deep cherry. Nose: ripe fruit, dried herbs, creamy oak. Palate: ripe fruit, spicy, round tannins.

Maius Garnatxa Blanca 2021 B
100% garnacha blanca

89

Standard, dried flowers, fruity, smooth, dried herbs, balanced.

MARCO ABELLA
Ctra. de Porrera a Cornudella de Montsant, Km. 1,2
43739 Porrera (Tarragona)
☎: +34 933 712 407
info@marcoabella.com
www.marcoabella.com

Altabella 2018 T

91

Colour: cherry, purple rim. Nose: fruit expression, floral, spicy, black fruit, scrubland. Palate: flavourful, fruity, fleshy, warm.

Clos Abella 2016 T
70% cariñena, 30% garnacha

90

Colour: bright cherry. Nose: creamy oak, red berry notes, scrubland, black fruit. Palate: good acidity, spicy, balanced.

Loidana 2020 T C
54% cariñena, 35% garnacha, 11% cabernet sauvignon

90

Colour: cherry, purple rim. Nose: fruit expression, red berry notes, floral, spicy, dried herbs. Palate: flavourful, fruity, balanced, smoky aftertaste.

Loidana Blanc 2021 B
25% garnacha blanca, 25% macabeo, 25% viognier, 25% picapoll blanc

90

Colour: straw. Nose: dried herbs, faded flowers, waxy notes, white fruit. Palate: ripe fruit, balanced, mineral.

Mas Mallola 2019 T R
70% garnacha, 30% cariñena

92

Colour: deep cherry. Nose: dried herbs, black fruit, violets, dark chocolate. Palate: powerful, ripe fruit, spicy, round tannins.

Òlbia 2018 B
48% garnacha blanca, 49% viognier, 2% pedro ximénez

92

Colour: straw. Nose: dried herbs, faded flowers, white fruit, waxy notes, balsamic herbs, dry stone. Palate: powerful, ripe fruit, balanced, rich.

MAS ALTA
Ctra. T-702, Km. 16,8
43375 La Vilella Alta (Tarragona)
☎: +34 977 054 151
info@bodegasmasalta.com
www.bodegasmasalta.com

Artigas 2019 B FB
70% garnacha blanca, 20% pedro ximénez, 10% macabeo

93

Colour: bright yellow. Nose: powerful, creamy oak, sweet spices, white fruit. Palate: rich, good structure, toasty, fine bitter notes.

Artigas 2019 T C
garnacha, cariñena, cabernet sauvignon

93

Colour: Cherry. Nose: expressive, spicy, mineral, scrubland. Palate: elegant, full-bodied, long, great length.

Cirerets 2019 T
60% cariñena, 40% garnacha

93

Colour: deep cherry. Nose: ripe fruit, dried herbs, spicy, complex, meaty notes. Palate: powerful, ripe fruit, spicy, round tannins.

Els Pics 2020 B
garnacha blanca, macabeo

91

Defined aromas. Colour: yellow, pale. Nose: white fruit, ripe fruit, wild herbs, citrus fruit, spicy. Palate: balanced, long, ripe fruit, easy to drink.

Els Pics 2020 T
50% garnacha, 40% cariñena, 5% merlot, 5% syrah

91

Colour: cherry, purple rim. Nose: fruit expression, spicy, ripe fruit, wild herbs, grassy. Palate: flavourful, fruity, fresh, balanced, good finish.

La Basseta 2018 T C
garnacha

94

Colour: bright cherry. Nose: sweet spices, dark chocolate, fruit preserve, dried herbs, wild herbs. Palate: fruity, spicy, round tannins.

DO Ca. PRIORAT / D.O.P.

🏆 PODIUM

La Creu Alta 2018 T BA
cariñena

96

Colour: Cherry. Nose: complex, expressive, spicy, tobacco, dried herbs, scrubland. Palate: full-bodied, long, great length, round tannins.

La Solana Alta 2019 B
garnacha blanca

94 🌿

Colour: bright yellow. Nose: powerful, spicy, dry stone, white fruit, waxy notes, dry nuts. Palate: rich, good structure, toasty, fine bitter notes, saline, long.

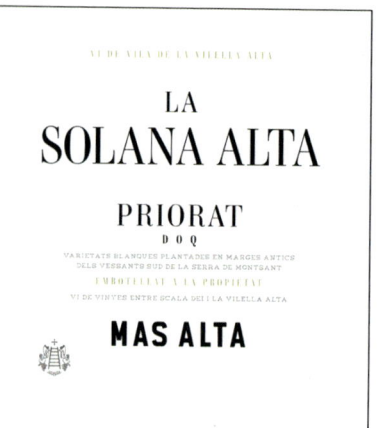

MAS DE L'A

Carrer Sort dels Capellans, 25
43730 Falset (Tarragona)
☎: +34 932 531 760
info@alfredoarribas.com

Les Margues 2019 B BA
garnacha blanca

94

Aromatic, reductive. Colour: bright straw. Nose: expressive, ripe fruit, floral, fine lees, mineral. Palate: full-bodied, complex, spicy, long, elegant.

Lo Noir 2019 T
garnacha, pinot noir

92

Colour: cherry, purple rim. Nose: black fruit, ripe fruit, dried herbs, spicy, smoky, mineral. Palate: fruity, flavourful, balanced, good finish, slightly dry, soft tannins.

Quars 2019 B
garnacha blanca

92

Reductive. Colour: yellow, pale. Nose: burnt matches, spicy, mineral. Palate: balanced, fine bitter notes, flavourful, complex.

Racorell 2019 T
garnacha

92

Colour: cherry, purple rim. Nose: tobacco, black fruit, ripe fruit, dried herbs, spicy. Palate: fruity, flavourful, good structure, balanced, fresh, slightly dry, soft tannins.

MAS DOIX

Ctra. T-702, Km. 4 - Partida Les Foreses
43376 Poboleda (Tarragona)
☎: +34 977 266 618
info@masdoix.com
www.masdoix.com

🏆 PODIUM

1902 Tossal d'en Bou
Gran Vinya Classificada 2019 T C
100% cariñena

97

Colour: Cherry. Nose: complex, expressive, spicy, mineral. Palate: elegant, full-bodied, long, great length.

🏆 PODIUM

1903 Coma de Cases
Garnatxa Velles Vinyes 2019 T C
100% garnacha

99

Colour: Cherry. Nose: complex, expressive, spicy, mineral. Palate: elegant, full-bodied, long, great length.

🏆 PODIUM

Doix Costers de Vinyes Velles 2019 T C
55% cariñena, 45% garnacha

95 🌱

Powerful, exuberant. Colour: cherry, garnet rim. Nose: complex, expressive, spicy, mineral, black fruit, ripe fruit. Palate: full-bodied, long, great length.

Les Crestes 2021 T
80% garnacha, 10% cariñena, 10% syrah

91

Colour: cherry, purple rim. Nose: fruit expression, red berry notes, spicy. Palate: flavourful, fruity, good acidity, long.

Murmuri 2021 B
95% garnacha blanca, 3% macabeo, 2% otras

90

Needs time. Colour: straw. Nose: ripe fruit, dried herbs, faded flowers, citrus fruit. Palate: ripe fruit, balanced, fruity.

Salanques 2020 T C
65% garnacha, 25% cariñena, 10% syrah

94

Colour: very deep cherry. Nose: complex, expressive, spicy, mineral, red berry notes, ripe fruit. Palate: elegant, full-bodied, long, great length, good acidity.

Salix 2021 B FB
65% garnacha blanca, 20% macabeo, 15% pedro ximénez

93

Colour: bright straw. Nose: expressive, fine lees, mineral, white fruit. Palate: spicy, long, elegant, good acidity.

MAS IGNEUS

Ctra.Falset- Vilella Baixa T-710 Km 11.1
43737 Gratallops (Tarragona)
☎: +34 676 293 435
daniel@masigneus.com
www.masigneus.com

Costers de L'Ermita 2018 T
100% cariñena

93 🌱

Colour: deep cherry. Nose: creamy oak, black fruit, ripe fruit, scrubland. Palate: powerful, ripe fruit, spicy, round tannins.

Costers de Pobleda 2013 T
95% cariñena, 5% syrah

94 🌱

Colour: deep cherry. Nose: ripe fruit, black fruit, balsamic herbs, dried herbs, spicy, black pepper, mineral. Palate: fruity, flavourful, powerful, balanced, long, slightly dry, soft tannins.

Fusió 2018 T
37% garnacha, 27% cariñena, 19% cabernet sauvignon, 10% merlot, 7% syrah

93 🌱

Colour: deep cherry. Nose: creamy oak, red berry notes, black fruit, scrubland. Palate: ripe fruit, spicy, round tannins.

La Capelleta 2019 B
100% garnacha blanca

92 🌱

Colour: yellow. Nose: ripe fruit, dried herbs, faded flowers, fine lees. Palate: ripe fruit, balanced, fine bitter notes, spicy.

M de Mas Igneus 2020 T
47% garnacha, 20% merlot, 14% syrah, 13% cabernet sauvignon, 6% cariñena

92 🌱

Colour: Cherry. Nose: woody, new oak, dark chocolate, spicy. Palate: powerful, ripe fruit, flavourful, round tannins, spicy.

V de Mas Igneus 2020 B
38% garnacha, 25% pedro ximénez, 23% merlot, 10% viognier, 4% garnacha blanca

90 🌱

Colour: straw. Nose: ripe fruit, dried herbs, faded flowers, spicy. Palate: ripe fruit, balanced, long.

MERITXELL PALLEJÀ

Carrer Piró, 10
43737 Gratallops (Tarragona)
☎: +34 670 960 735
info@nita.cat
www.nita.cat

Magran Partida Les Manyetes 2017 T C
garnacha

91

Colour: deep cherry. Nose: ripe fruit, dried herbs, fruit expression, black fruit, spicy. Palate: powerful, ripe fruit, good structure, slightly dry, soft tannins, good finish.

Nita 2020 T
garnacha, cariñena, syrah

90

Colour: cherry, purple rim. Nose: fruit expression, spicy, red berry notes, black fruit, wild herbs. Palate: flavourful, fruity, balanced, slightly dry, soft tannins.

Nita Blanc 2021 B
garnacha, chenin blanc, otras

90

Crisp, aromatic. Colour: straw. Nose: white flowers, dried herbs, white fruit, neat. Palate: flavourful, fruity, balanced, good acidity.

DO Ca. PRIORAT / D.O.P.

Ranci 2013-2018 T RC S
garnacha

91

Colour: old gold. Nose: varnish, candied fruit, honeyed notes. Palate: flavourful, balanced, classic aged character, old wood.

Torrent - Vi de Paratge 2018 T C
cariñena

92

Colour: deep cherry, purple rim. Nose: ripe fruit, dried herbs, creamy oak, black fruit, wild herbs, smoky. Palate: powerful, ripe fruit, spicy, round tannins, fresh.

MERUM PRIORATI
Ctra. de Falset, km. 9,3
43739 Porrera (Tarragona)
☎: +34 977 828 307
Fax: +34 977 828 324
comunicacio@pereventura.com
www.merumpriorati.com

Merum Priorati Desti 2019 T
60% garnacha, 30% cariñena, 10% syrah

92 🏆

Colour: Cherry. Nose: scrubland, dried herbs, characterful, black fruit, ripe fruit, waxy notes. Palate: spicy, balsamic, ripe fruit, easy to drink.

Merum Priorati El Cel 2019 T
35% garnacha, 25% cabernet sauvignon, 20% cariñena, syrah

93 🏆

Colour: deep cherry, bright cherry. Nose: ripe fruit, characterful, powerful, dried herbs, cocoa bean. Palate: good structure, flavourful, long, spicy.

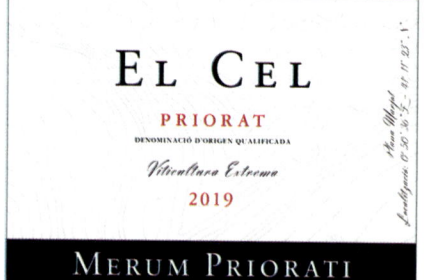

Merum Priorati Inici 2019 T
55% garnacha, 15% cariñena, 15% syrah, 15% cabernet sauvignon

92 🏆

Colour: bright cherry. Nose: red berry notes, floral, spicy, wild herbs, balanced. Palate: flavourful, good acidity, long, ripe fruit.

PERINET
Ctra de Poboleda T-702 PK | km 1,6
43376 Poboleda (Tarragona)
☎: +34 977 827 113
perinet@perinetwinery.com
www.perinetwinery.com

Merit 2018 T C
merlot, syrah, garnacha, cariñena

92

Colour: deep cherry. Nose: fruit expression, red berry notes, spicy, black fruit, grassy, dried herbs. Palate: flavourful, fruity, good acidity, good structure, balanced, slightly dry, soft tannins.

Perinet 1194 2017 T C
garnacha, cariñena, syrah

93

Colour: deep cherry. Nose: ripe fruit, black fruit, spicy, creamy oak, wild herbs, dried herbs. Palate: fruity, flavourful, powerful, good structure, round tannins.

Perinet 2017 T C
garnacha, cariñena, merlot, syrah

93

Colour: deep cherry. Nose: dried herbs, fruit preserve, violets, sweet spices, truffle notes. Palate: powerful, spicy, round tannins, good structure.

Vinya Mas Vell Cabernet Sauvignon 2017 T GR
cabernet sauvignon

91

Colour: bright cherry. Nose: fruit preserve, black fruit, wild herbs, spicy, black pepper. Palate: flavourful, fruity, balanced, slightly dry, soft tannins.

Vinya Mas Vell Garnatxa 2017 T GR
garnacha, cariñena, syrah

92

Colour: bright cherry. Nose: ripe fruit, red berry notes, black fruit, scrubland, wild herbs, spicy, fine reductive notes. Palate: good structure, flavourful, fruity, balanced.

PROYECTO GARNACHAS/VINTAE
Camí de la Pedra Estela, 34
43205 Reus (Tarragona)
☎: +34 941 483 557
marketing@vintae.com
www.vintae.com

La Garnatxa Fosca del Priorat 2020 T
garnacha

89

Pruney, spicy, powerful, tasty, kind finish.

RITME CELLER
Camí del Sindicat, s/n
43375 La Vilella Alta (Tarragona)
☎: +34 672 432 691
ritme@ritmeceller.com
www.ritmeceller.com

+ Ritme Blanc 2019 B
100% garnacha blanca

92
Colour: bright yellow. Nose: candied fruit, faded flowers, smoky, wild herbs. Palate: flavourful, fine bitter notes, spicy, long.

+ Ritme Blanc 2020 B
100% garnacha blanca

90
Colour: bright yellow. Nose: creamy oak, ripe fruit, spicy, woody. Palate: good structure, toasty, fine bitter notes.

+ Ritme Blanc 2021 B
100% garnacha blanca

92
Colour: bright straw. Nose: ripe fruit, fragrant herbs, fine lees, spicy. Palate: full-bodied, rich, flavourful, sweetness.

Etern 2018 T BA
75% cariñena, 25% garnacha

93
Colour: deep cherry. Nose: ripe fruit, spicy, toasty, dried flowers. Palate: fruity, flavourful, balanced, round tannins, good finish.

Etern 2020 T BA
75% cariñena, 25% garnacha

92
Colour: cherry, purple rim. Nose: ripe fruit, violets, sweet spices, dried herbs, expressive. Palate: fruity, flavourful, balanced, good finish, slightly dry, soft tannins.

Plaer 2018 T C
70% cariñena, 30% garnacha

92
Colour: deep cherry. Nose: ripe fruit, dried flowers, spicy, dried herbs. Palate: fruity, flavourful, balanced, round tannins.

Plaer 2020 T C
70% cariñena, 30% garnacha

92
Colour: cherry, purple rim. Nose: ripe fruit, dried flowers, faded flowers, spicy, expressive. Palate: fruity, flavourful, balanced, slightly dry, soft tannins.

Ritme Negre 2019 T
45% cariñena, 55% garnacha

93
Colour: deep cherry. Nose: red berry notes, ripe fruit, wild herbs, spicy. Palate: flavourful, powerful, good finish, fruity aftertaste.

Ritme Negre 2020 T
cariñena, garnacha

92
Colour: bright cherry. Nose: fresh fruit, creamy oak, red berry notes, spicy, toasty, scrubland. Palate: good acidity, spicy, round tannins.

SAMSARA PRIORAT
Ctra. T-702 Km 13
43379 Escaladei (Tarragona)
☎: +34 619 776 948
info@samsarapriorat.cat
www.samsarapriorat.cat

Samsara Priorat 2018 T
55% garnacha, 30% cariñena, 10% cabernet sauvignon, 5% syrah

91
Colour: bright cherry. Nose: fruit expression, red berry notes, spicy, grassy, ripe fruit. Palate: flavourful, fruity, long, fresh, balanced, good finish, slightly dry, soft tannins.

SANDRA DOIX CELLER
Carme, 115
43376 Poboleda (Tarragona)
☎: +34 605 241 851
sandra@sandradoix.com
www.sandradoix.com

MarLa vino de Paratge Les Salanques (Etiqueta Negra) 2019 T C
100% cariñena

93
Colour: cherry, purple rim. Nose: black fruit, ripe fruit, balsamic herbs, wild herbs, spicy, dry stone. Palate: fruity, good structure, flavourful, powerful, balanced, fruity aftertaste, long, round tannins.

MarLA vino de Vila de Poboleda (Etiqueta Roja) 2019 T C
70% cariñena, 30% garnacha

92
Colour: cherry, garnet rim. Nose: black fruit, balsamic herbs, ripe fruit, spicy, smoky, dried herbs. Palate: fruity, fresh, flavourful, powerful, balanced, good finish, slightly dry, soft tannins.

DO Ca. PRIORAT / D.O.P.

SANGENÍS I VAQUÉ

Pl. Catalunya, 3
43739 Porrera (Tarragona)
☎: +34 977 828 252
celler@sangenisivaque.com
www.sangenisivaque.com

Clos Monlleó 2011 T R
50% garnacha, 50% cariñena

92

Colour: cherry, garnet rim. Nose: balsamic herbs, scrubland, old leather, black fruit, creamy oak, dried herbs. Palate: flavourful, balsamic, spicy, round tannins.

Coranya 2015 T
50% garnacha, 50% cariñena

92

Colour: deep cherry. Nose: dried herbs, creamy oak, black fruit, fine reductive notes. Palate: powerful, ripe fruit, spicy, round tannins.

Dara 2019 T C
45% garnacha, 45% cariñena, 10% merlot

90

Colour: deep cherry. Nose: ripe fruit, dried herbs, toasty. Palate: powerful, spicy, round tannins.

Garbinada 2021 T
60% garnacha, 40% cariñena

91

Colour: cherry, purple rim. Nose: red berry notes, floral, spicy. Palate: flavourful, fruity, good acidity.

Lo Coster Blanc 2021 B
85% garnacha blanca, 15% macabeo

90

Colour: bright straw. Nose: ripe fruit, stone fruit, dried flowers. Palate: flavourful, fresh, good acidity, fruity aftertaste.

Vall Por 2017 T R
40% garnacha, 60% cariñena

89

Pruney, full-bodied, powerful, tasty, great length.

TERRES DE VIDALBA

43376 Poboleda (Tarragona)
☎: +34 616 413 722
info@terresdevidalba.com
www.terresdevidalba.com

No T'ho Diré 2019 B
garnacha blanca

88

Austere, dried herbs, full-bodied, jammy, smooth, flat.

No T'ho Diré Rosat 2020 RD
garnacha

90

Dried flowers, dried herbs, wild, spicy, tasty, jammy, fruity. Nose: scrubland, wild herbs. Palate: flavourful, fine bitter notes.

Tocs 2016 T R
garnacha, cariñena, syrah

92

Colour: Cherry. Nose: complex, spicy, mineral, wild herbs, scrubland, fruit preserve. Palate: full-bodied, long, great length, spicy, toasty.

Vidalba 2017 T
garnacha, syrah, cariñena

90

Colour: bright cherry. Nose: fruit expression, spicy, black fruit, toasted bread. Palate: flavourful, fruity, balanced, slightly dry, soft tannins.

TERROIR AL LIMIT

Carrer de Baixa Font, 10
43737 Torroja del Priorat (Tarragona)
☎: +34 977 839 391
vi@terroir-al-limit.com
www.terroir-al-limit.com

Arbossar 2020 T C
100% cariñena

94

Colour: Cherry. Nose: balsamic herbs, scrubland, wild herbs, ripe fruit, neat. Palate: balsamic, good acidity, taut, balanced, fruity.

Coreografía Priorat 2021 RD
85% garnacha, 15% garnacha blanca

92

Colour: raspberry rose. Nose: balsamic herbs, red berry notes, wild herbs. Palate: fresh, fruity, flavourful, ripe fruit.

🏆 PODIUM

Dits del Terra 2019 T C
100% cariñena

95

Colour: cherry, purple rim. Nose: fruit expression, floral, spicy, black fruit. Palate: flavourful, fruity, good acidity, long.

🏆 PODIUM

Les Manyes 2019 T
100% garnacha

97

Colour: light cherry. Nose: red berry notes, neat, floral, chamomile, expressive, elegant. Palate: fruity, full of life, long, easy to drink, good acidity, fine tannins.

🏆 PODIUM

Les Tosses 2019 T C
100% cariñena

96

Defined aromas. Colour: Cherry. Nose: balsamic herbs, scrubland, ripe fruit, spicy, mineral, expressive. Palate: balsamic, good acidity, balanced, fine bitter notes.

Pedra de Guix 2019 B C
33% garnacha blanca, 33% macabeo, 33% pedro ximénez

94

Wild, austere. Colour: yellow. Nose: dried flowers, faded flowers, chamomile, ripe fruit. Palate: taut, elegant, fine bitter notes, round.

Terra de Cuques Blanc 2021 B
50% pedro ximénez, 50% garnacha blanca

92

Colour: bright yellow. Nose: ripe fruit, dried herbs, faded flowers, white fruit. Palate: powerful, ripe fruit, balanced.

Terra de Cuques Negre 2019 T
50% garnacha, 50% cariñena

94

Wild. Colour: Cherry. Nose: complex, expressive, mineral, ripe fruit, dried herbs. Palate: elegant, full-bodied, long, great length.

Terroir Historic Blanc 2021 B
75% garnacha, 25% macabeo

90

Austere. Colour: straw. Nose: dried herbs, faded flowers, cereal notes. Palate: ripe fruit, balanced, flavourful.

Terroir Historic Negre 2019 T
75% garnacha, 25% cariñena

91

Colour: cherry, garnet rim. Nose: wild herbs, ripe fruit, neat. Palate: long, easy to drink, fruity.

Terroir Historic Negre 2021 T

92

Colour: Cherry. Nose: balsamic herbs, sweet spices, scrubland, fruit expression, black fruit, ripe fruit. Palate: spicy, balsamic, good acidity.

TROSSOS DEL PRIORAT

Ctra. Gratallops a La Vilella Baixa, Km. 10,65
43737 Gratallops (Tarragona)
☎: +34 670 590 788
celler@trossosdelpriorat.com
www.trossosdelpriorat.com

90 Minuts 2020 T
garnacha, cariñena, cabernet sauvignon

91

Nose: fruit expression, ripe fruit, black fruit, dried herbs, spicy. Palate: fruity, flavourful, balanced, good finish, round tannins.

Abracadabra 2019 B
garnacha blanca, macabeo

92

Colour: bright yellow. Nose: creamy oak, spicy, sweet spices, stone fruit, ripe fruit. Palate: rich, good structure, long, toasty, fine bitter notes.

L'Estaca 2017 T
100% garnacha

91 🌿

Colour: deep cherry. Nose: black fruit, scrubland, spicy, fine reductive notes. Palate: ripe fruit, spicy, round tannins.

Llum D'Alba 2021 B
garnacha blanca, macabeo, viognier

91

Colour: straw. Nose: dried herbs, faded flowers, white fruit, fine lees, dry stone. Palate: powerful, balanced, flavourful, saline.

Lo Món 2017 T
garnacha, cariñena, syrah, cabernet sauvignon

91

Colour: deep cherry. Nose: dried herbs, creamy oak, black fruit, fine reductive notes, meaty notes. Palate: powerful, ripe fruit, spicy, round tannins.

Pam de Nas 2017 T
garnacha, cariñena

91

Colour: cherry, purple rim. Nose: fruit expression, spicy, ripe fruit, black fruit, dried herbs. Palate: flavourful, fruity, powerful, good finish, slightly dry, soft tannins.

DO Ca. PRIORAT / D.O.P.

DO Ca. PRIORAT / D.O.P.

VINS DE LA MEMÒRIA

Montaner 320
08021 Barcelona (Barcelona)
vinsdelamemoria@gmail.com
www.vinsdelamemoria.com

Plom 2020 T
garnacha

93

Pronounced acidity, sweety finish. Colour: cherry, garnet rim. Nose: fruit preserve, powerful, dried herbs, black liquorice. Palate: flavourful, long, correct.

VITICULTORS COSTERS DEL PRIORAT

Mas dels Frares
43736 El Molar (Tarragona)
☎: +34 618 203 473
info@costersdelpriorat.com
www.costersdelpriorat.com

Blanc de Closos 2020 B

90

Little interventionist, oxidativ. Colour: straw. Nose: ripe fruit, dried herbs, faded flowers, cereal notes. Palate: powerful, ripe fruit, balanced.

Clos Alzina 2018 T
cariñena

91

Colour: cherry, garnet rim. Nose: fruit preserve, powerful, wild herbs. Palate: flavourful, long, balanced, fine bitter notes, round tannins.

Clos Cypres 2018 T
cariñena

92

Colour: cherry, purple rim. Nose: fruit expression, red berry notes, floral, spicy, toasty, earthy notes. Palate: flavourful, fruity, good acidity, long.

🏆 PODIUM

Memòries del Priorat Ranci Cal Marcelino TF RC
garnacha

95

Colour: dark mahogany. Nose: spicy, varnish, acetaldehyde, dry nuts. Palate: fine solera notes, bitter, flavourful, long, sweet.

Petit Pissarres 2020 T
garnacha, cariñena

90

Colour: cherry, purple rim. Nose: fruit expression, spicy, black fruit, dry stone. Palate: flavourful, fruity, good finish, smoky aftertaste, balanced.

VITICULTORS DEL PRIORAT

43738 Bellmunt del Priorat (Tarragona)
☎: +34 650 704 285
www.priorterrae.es

El Vol de L'Àliga 2020 B
garnacha, cariñena

90

With personality. Colour: bright straw. Nose: wild herbs, dried flowers, faded flowers. Palate: good acidity, fine bitter notes, long, balanced, flavourful.

Prior Terrae 2020 B

91

Reductive, jammy. Colour: yellow, pale. Nose: dried herbs, wild herbs. Palate: flavourful, balanced, fine bitter notes, spicy, long.

Prior Terrae Negre 2019 T

91

Colour: deep cherry. Nose: dried herbs, spicy, cocoa bean, black fruit, balanced. Palate: powerful, ripe fruit, spicy, round tannins.

Vol de L'Auga 2019 T

91

Colour: cherry, purple rim. Nose: spicy, wild herbs, black fruit, dry stone. Palate: flavourful, fruity, good structure, slightly dry, soft tannins, smoky aftertaste.

XAVI PALLEJÀ VITICULTOR

Les Aubagues
43730 Bellmunt del Priorat (Tarragona)
☎: +34 637 417 049
xavi.palleja@gmail.com
www.xavipalleja.cat

Buxus de les Aubagues 2020 T
cariñena, garnacha

91

Colour: deep cherry. Nose: ripe fruit, dried herbs, sweet spices. Palate: powerful, ripe fruit, spicy, round tannins.

DO. RÍAS BAIXAS
CONSEJO REGULADOR

Edif. Pazo de Mugartegui
36002 Pontevedra
☎: +34 986 854 850 - Fax: +34 986 864 546
@: consejo@doriasbaixas.eu
www.doriasbaixas.com

LOCATION:

In the southwest of the province of Pontevedra, covering five distinct sub-regions: Val do Salnés, O Rosal, Condado do Tea, Soutomaior and Ribeira do Ulla.

 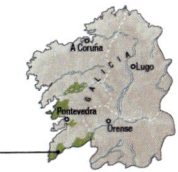

SUB-REGIONS:

Val do Salnés: This is the historic sub-region of the Albariño (in fact, here, almost all the white wines are produced as single-variety wines from this variety) and is centred around the municipal district of Cambados. It has the flattest relief of the four sub-regions. **Condado do Tea:** The furthest inland, it is situated in the south of the province on the northern bank of the Miño. It is characterized by its mountainous terrain. The wines must contain a minimum of 70% of Albariño and Treixadura. **O Rosal:** In the extreme southwest of the province, on the right bank of the Miño river mouth. The warmest sub-region, where river terraces abound. The wines must contain a minimum of 70% of Albariño and Loureira. **Soutomaior:** Situated on the banks of the Verdugo River, about 10 km from Pontevedra, it consists only of the municipal district of Soutomaior. It produces only single-varietals of Albariño. **Ribeira do Ulla:** A new sub-region along the Ulla River, which forms the landscape of elevated valleys further inland. It comprises the municipal districts of Vedra and part of Padrón, Deo, Boquixon, Touro, Estrada, Silleda and Vila de Cruce. Red wines predominate.

GRAPE VARIETIES:

WHITE: Albariño (majority), Loureira Blanca or Marqués, Treixadura and Caíño Blanco (preferred); Torrontés and Godello (authorized).

RED: Caíño Tinto, Espadeiro, Loureira Tinta and Sousón (preferred); Tempranillo, Mouratón, Garnacha Tintorera, Mencía, Brancellao (authorized) and Castañal.

FIGURES:

Vineyard surface: 4,1345 – **Wine-Growers:** 5,046 – **Wineries:** 178 – **2021 Harvest rating:** Good – **Production 21:** 30,171,614 litres – **Market percentages:** 67% National - 33% International.

SOIL:

Sandy, shallow and slightly acidic, which makes fine soil for producing quality wines. The predominant type of rock is granite, and only in the Concellos of Sanxenxo, Rosal and Tomillo is it possible to find a narrow band of metamorphous rock. Quaternary deposits are very common in all the sub-regions.

CLIMATE:

Atlantic, with moderate, mild temperatures due to the influence of the sea, high relative humidity and abundant rainfall (the annual average is around 1600 mm). There is less rainfall further downstream of the Miño (Condado de Tea), and as a consequence the grapes ripen earlier.

CLASSIFICATION OF YOUNG WINE HARVESTS PEÑÍNGUIDE

2017	2018	2019	2020	2021
VERY GOOD	VERY GOOD	VERY GOOD	VERY GOOD	VERY GOOD

DO RÍAS BAIXAS / D.O.P.

ADEGA CONDES DE ALBAREI
Lg. A Bouza, 1 Castrelo
36639 Cambados (Pontevedra)
☎: +34 986 543 535
inf@condesdealbarei.com
www.condesdealbarei.com

Albarei Áine 2016 B
100% albariño
92
Colour: bright yellow. Nose: powerful, creamy oak, ripe fruit, spicy. Palate: rich, good structure, long, toasty, fine bitter notes.

Albarei Orixe 2016 B
100% albariño
92
Colour: bright straw. Nose: ripe fruit, fragrant herbs, fine lees, toasted bread. Palate: full-bodied, rich, long, good acidity.

Condes de Albarei Albariño 2021 B
albariño
88
Citrus fruit, crisp, herbal, fruity.

Condes de Albarei Carballo Galego 2017 B FB
100% albariño
93
Colour: bright straw. Nose: ripe fruit, fragrant herbs, fine lees, ebb tide, spicy. Palate: full-bodied, rich, long, good acidity.

Condes de Albarei En Rama 2015 B C
100% albariño
92
Racy. Colour: bright straw. Nose: fine lees, dry stone, citrus fruit, wild herbs. Palate: fresh, ripe fruit, spicy.

Condes de Albarei Enxebre 2021 B MC
100% albariño
88
Citrus fruit, herbal, balanced, crisp.

ADEGA DOS EIDOS
Padriñán, 65
36960 Sanxenxo (Pontevedra)
☎: +34 986 690 009
info@adegaeidos.com
www.adegaeidos.com

Contraaparede 2017 B
albariño
93
Complex, with personality. Colour: bright yellow. Nose: fine lees, scrubland, wild herbs, bakery, floral. Palate: flavourful, long, fine bitter notes, good acidity.

Eidos de Padriñán 2019 B BA
100% albariño
91
Colour: bright yellow. Nose: creamy oak, ripe fruit, spicy, fine lees. Palate: rich, good structure, long, fine bitter notes.

Eidos de Padriñán 2021 B
100% albariño
90
Racy. Colour: bright straw, greenish rim. Nose: fresh fruit, citrus fruit, wild herbs. Palate: fresh, fruity, good acidity, fine bitter notes.

Veigas de Padriñán Selección de Añada 2021 B
100% albariño
91
Racy. Colour: bright straw. Nose: ripe fruit, fragrant herbs, fine lees, ebb tide. Palate: full-bodied, good acidity, flavourful.

ADEGAS CASTROBREY
Camanzo, s/n
36587 Vila de Cruces (Pontevedra)
☎: +34 986 583 643
bodegas@castrobrey.com
www.castrobrey.com

Nice to Meet You Madrid 2018 B
albariño, treixadura, godello
89
Fruity, dried flowers, jammy, tasty.

Sin Palabras 2018 B BA
100% albariño
91
Colour: bright yellow. Nose: ripe fruit, spicy, dried herbs. Palate: fruity, rich, flavourful, toasty.

Sin Palabras 2021 B MC
100% albariño
90
Colour: bright straw. Nose: fruit expression, ripe fruit, floral, dried herbs. Palate: flavourful, fresh, good acidity, fruity aftertaste.

Sin Palabras Edición Especial 2013 B
100% albariño
93
Colour: bright straw. Nose: ripe fruit, fragrant herbs, fine lees, spicy, mineral. Palate: full-bodied, rich, long, good acidity.

DO RÍAS BAIXAS / D.O.P.

Sin Palabras Mosto Flor 2018 B MC
100% albariño

91

Colour: bright yellow. Nose: ripe fruit, dried herbs, faded flowers, spicy. Palate: powerful, ripe fruit, balanced, rich.

ADEGAS GALEGAS
Lugar Meder, s/n
36457 Salvaterra do Miño (Pontevedra)
☎: +34 986 665 012
comunicacion@adegasgalegas.es
www.adegasgalegas.es

D. Pedro Soutomaior 2021 B
100% albariño

87

Danza Escumoso de Albariño BE BR
albariño

88

Citrus fruit, crisp, herbal, simple.

Veigadares 2018 B FB
albariño, treixadura, caíño blanco, loureiro

92

Colour: bright straw. Nose: ripe fruit, fragrant herbs, fine lees, cereal notes, ebb tide. Palate: full-bodied, rich, long, good acidity.

ADEGAS GRAN VINUM
Viñagrande 84B – San Miguel de Deiro
36627 Vilanova de Arousa (Pontevedra)
☎: +34 986 555 742
info@adegasgranvinum.com
www.granvinum.com

Esencia Diviña 2021 B
100% albariño

89

Pleasant, crisp, fruity, citrus fruit, simple, smooth, varietally correct.

Gran Vinum 2021 B
albariño

90

Colour: straw. Nose: white flowers, citrus fruit, fresh. Palate: flavourful, fruity, balanced, easy to drink.

Mar de Viñas 2021 B
100% albariño

88

Standard, crisp, fruity, floral, simple.

ADEGAS MORGADÍO
Albeos, s/n
36429 Albeos (Pontevedra)
☎: +34 988 261 212
Fax: +34 988 261 213
info@bodegasgrm.com
www.bodegasgrm.com

Morgadío 2021 B
100% albariño

91

Colour: bright straw, greenish rim. Nose: fresh fruit, citrus fruit, wild herbs, fine lees. Palate: fresh, fruity, good acidity, fine bitter notes.

Puerta Santa 2021 B
100% albariño

90

Colour: bright straw, greenish rim. Nose: fresh fruit, citrus fruit, wild herbs, fine lees. Palate: fresh, fruity, good acidity, fine bitter notes.

ADEGAS TERRA DE ASOREI
Autovía do Salnés (Salida 7) – San Martiño de Meis
36637 Meis (Pontevedra)
☎: +34 986 680 868
info@terradeasorei.com
www.terradeasorei.com

1953 Pazo Torrado 2018 BE BR
albariño

91

Colour: bright straw. Nose: fine lees, floral, expressive. Palate: flavourful, good acidity, balanced, fresh.

1953 Pazo Torrado 2021 B
100% albariño

88

Pleasant, defined aromas, fruity, very fruit-driven, standard.

Colección Nai E Señora 2020 B
100% albariño

92

Colour: straw. Nose: white flowers, wild herbs, expressive, fine lees. Palate: flavourful, fruity, easy to drink.

Nai E Señora 2021 B
100% albariño

89

Fruity, dried flowers, herbal, crisp.

Terra de Asorei 2020 T
espadeiro

91

Pleasant, herbal, representative. Colour: cherry, garnet rim. Nose: wild herbs, red berry notes. Palate: fruity, fresh, balanced.

DO RÍAS BAIXAS / D.O.P.

Terra de Asorei 2021 B
100% albariño
91
Pleasant, citrus fruit. Nose: fine lees, dried flowers, white fruit, neat, balanced. Palate: varietal, balanced, fine bitter notes, good finish.

Terra de Asorei Selección Privada 2020 B
100% albariño
92
Colour: straw. Nose: dried herbs, faded flowers, dry stone, fine lees, wild herbs, citrus fruit. Palate: powerful, ripe fruit, balanced.

ADEGAS TOLLODOURO
Ctra. de Tui a la Guardia, KM 55.5
36760 As Eiras (Pontevedra)
☎: +34 986 442 686
marketing@hgabodegas.com
www.hgabodegas.com

Pontellón Albariño 2021 B
100% albariño
88
Balanced, floral, fruity, jammy, standard.

Tollodouro 2021 B
85% albariño, 10% treixadura, 5% loureiro
90
Colour: straw. Nose: ripe fruit, floral, balanced, neat. Palate: powerful, ripe fruit, fine bitter notes.

ADEGAS VALMIÑOR
Estrada A Guarda – Tui nº245. Barrio Portela
36760 O'Rosal (Pontevedra)
☎: +34 986 609 060
valminor@valminorebano.com
www.valminorebano.com

Abade de Couto 2020 T
brancellao, sousón, caiño
89
Balsamic herbs, standard, herbal, jammy, dried flowers.

Davila 2021 B
loureiro, albariño, treixadura
91
Colour: bright straw. Nose: fruit expression, ripe fruit, dried herbs. Palate: flavourful, fresh, fruity aftertaste.

Serra da Estrela 2021 B
100% albariño
89
Citrus fruit, crisp, herbal, floral, yeasty notes.

Torroxal Albariño 2021 B
100% albariño
88
Citrus fruit, crisp, herbal, tasty.

Valmiñor 2021 B
100% albariño
88
Needs time, racy, herbal, tasty.

ADEGAS VALTEA
Lg. Portela, 14
36429 Crecente (Pontevedra)
☎: +34 986 666 344
Fax: +34 986 644 914
valtea@valtea.es
www.valtea.es

Finca Garabato Cepas Vellas 2020 B
100% albariño
91
Colour: bright straw. Nose: ripe fruit, dried herbs, faded flowers. Palate: powerful, ripe fruit, balanced.

Valtea 2021 B
100% albariño
90
Colour: bright straw. Nose: fruit expression, floral, citrus fruit. Palate: flavourful, fresh, fruity aftertaste, fruity.

Valtea BE BN
albariño
87

Valtea Cuvée Especial BE BN
100% albariño
88
Pleasant, standard, crisp, simple, smooth.

DO RÍAS BAIXAS / D.O.P.

ALDEA DE ABAIXO
Estrada de Novas, 62
36778 O'Rosal (Pontevedra)
☎: +34 986 626 121
Fax: +34 986 626 121
senoriodatorre@grannovas.com
www.bodegasorosal.com

Gran Novas Albariño 2021 B
100% albariño

88

Crisp, herbal, simple, standard.

Señorío da Torre 2021 B
70% albariño, 25% loureiro, 5% caiño

90

Colour: bright straw, greenish rim. Nose: fresh fruit, citrus fruit, wild herbs, fine lees. Palate: fresh, fruity, good acidity, fine bitter notes.

Señorío da Torre sobre Lías 2018 B
85% albariño, 10% loureiro, 5% caiño

92

Colour: bright straw. Nose: ripe fruit, fragrant herbs, fine lees, lactic notes, brioche. Palate: full-bodied, rich, good acidity.

ANÓNIMAS VITICULTORAS
Rua da Cruz, 25
36619 Vilagarcía de Arousa (Pontevedra)
☎: +34 678 561 175
info@anonimaswines.com
www.anonimaswines.com

Os Dunares Albariño 2021 B
albariño

90

Colour: straw. Nose: white flowers, dried herbs, citrus fruit, white fruit. Palate: flavourful, fruity, balanced.

Os Dunares Caiño 2020 T
caiño

87

ATLANTIC GALICIAN WINERIES
27334 A Pobra (Lugo)
info@atlanticgalicianwineries.com
www.altanticgalicianwineries.com

Marola & Mass 2021 B
albariño

89

Citrus fruit, balanced, herbal, yeasty notes, smoky, tasty.

ATTIS BODEGAS Y VIÑEDOS
Lg. Morouzos, 16D - Dena
36967 Meaño (Pontevedra)
☎: +34 986 744 790
administracion@attisbyv.com
www.attisbyv.com

Attis Atalante 2019 B
100% caiño blanco

91

Colour: bright straw. Nose: ripe fruit, fine lees, sweet spices, wild herbs. Palate: full-bodied, rich, good acidity.

Attis Brancellao 2020 T
100% brancellao

93

Colour: bright cherry. Nose: balanced, faded flowers, earthy notes, undergrowth, red berry notes, black fruit. Palate: flavourful, fruity, good acidity, balanced.

Attis Embaixador 2019 B
100% albariño

93

Colour: bright straw. Nose: fine lees, stone fruit, wild herbs, dry stone. Palate: full-bodied, good acidity, balanced.

Attis Lías Finas 2021 B
albariño

92

Colour: bright yellow. Nose: white fruit, wild herbs, white flowers, expressive. Palate: fruity, flavourful, balanced.

Genio y Figura 2021 B
100% albariño

90

Colour: bright straw. Nose: fruit expression, ripe fruit, floral, dried herbs. Palate: flavourful, fresh, good acidity, fruity aftertaste.

Nana 2019 B FB
100% albariño

92

Slight oxidation. Colour: bright yellow. Nose: powerful, creamy oak, ripe fruit, spicy, mineral. Palate: good structure, toasty, fine bitter notes.

Xión 2021 B
100% albariño

89

Tasty, tropical, fruity, herbal, crisp.

DO RÍAS BAIXAS / D.O.P.

AUTÉNTICOS VIÑADORES, VINOS DE TERROIR

24540 Cacabelos (León)
☎: +34 658 617 390
info@autenticosvinadores.com
www.autenticosvinadores.com

La Estela del Ciclohome 2018 B
100% albariño

91

Colour: bright straw. Nose: fruit expression, stone fruit, white flowers, dried herbs, mineral. Palate: fruity, flavourful, full-bodied, balanced.

BENJAMÍN MIGUEZ NOVAL

Porto de Abaixo, 10 - Porto
36458 Salvaterra de Miño (Pontevedra)
☎: +34 636 014 506
enoturismo@mariabargiela.com
www.mariabargiela.com

María Bargiela 2020 B
90% albariño, 8% treixadura, 2% loureiro

89

Aromatic, floral, fruity, tasty, smooth, standard.

BODEGA ELADIO PIÑEIRO

Sobran, 38
36611 Vilagarcía de Arousa (Pontevedra)
☎: +34 986 501 218
info@eladiopineiro.com
www.eladiopineiro.com

Amodiño 2018 B

92

Yeasty notes, age nuances. Colour: bright golden. Nose: bakery, fine lees, dried flowers, neat. Palate: flavourful, long, rich.

Envidiacochina 2020 B
albariño

92

Colour: bright straw. Nose: ripe fruit, fragrant herbs, fine lees, ebb tide, iodine notes, white flowers. Palate: full-bodied, rich, good acidity.

Frore de Carme 2017 B
albariño

92

Colour: bright yellow. Nose: ripe fruit, fine lees, bakery, floral, dried flowers. Palate: flavourful, rich, long.

Frore de Carme Millésime 2018 BE BN
albariño

91

Colour: bright yellow. Nose: dried flowers, faded flowers, citrus fruit, candied fruit. Palate: flavourful, fine bitter notes, fruity, good finish.

La Ola 2019 B
albariño

90

Colour: bright straw. Nose: ripe fruit, fragrant herbs, fine lees, white flowers, iodine notes. Palate: full-bodied, rich, long, good acidity, flavourful, saline.

Novoa 2017 T

87 🍷

BODEGA FILLABOA

Lugar de Fillaboa, s/n
36458 Salvaterra do Miño (Pontevedra)
☎: +34 986 658 132
info@bodegasfillaboa.masaveu.com
www.bodegasfillaboa.com

Fillaboa 2021 B
albariño

90

Colour: bright straw. Nose: fruit expression, ripe fruit, floral. Palate: flavourful, fresh, good acidity.

Fillaboa Selección Finca Monte Alto 2019 B
100% albariño

93

Colour: bright yellow. Nose: dried flowers, fine lees, stone fruit, complex, ebb tide. Palate: round, spicy, long, great length.

BODEGA GIL ARMADA

Pazo de Fefiñans, s/n
36630 Cambados (Pontevedra)
☎: +34 986 524 877
gilarmada@pazodefefinans.com
www.bodegagilarmada.com

Gil Armada (Viñedos propios da Torre de San Fardán) 2020 B
albariño

89

Citrus fruit, balanced, fruity, floral, tasty.

Gil Armada (Viñedos propios no Pazo de Fefiñáns) 2020 B
albariño

90

Colour: bright straw. Nose: white fruit, white flowers, wild herbs, ebb tide. Palate: fruity, flavourful, fresh, balanced.

BODEGA GRANBAZÁN
Lg. Tremoedo, 46
36628 Vilanova de Arousa (Pontevedra)
☎: +34 986 555 562
elena.sanchez@agrodebazan.com
www.bodegasgranbazan.com

🏆 PODIUM

Granbazán Don Álvaro de Bazán 2018 B
100% albariño

95

Colour: bright yellow. Nose: ripe fruit, fine lees, faded flowers, balanced. Palate: rich, good acidity, easy to drink, long.

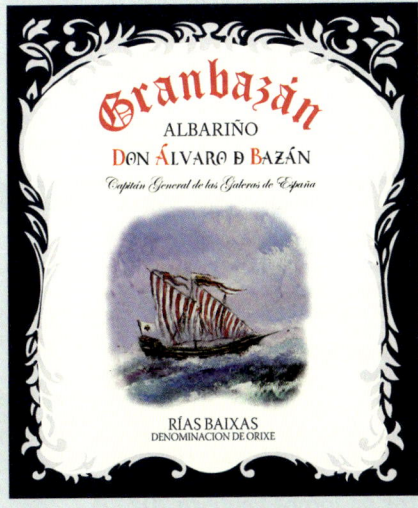

Granbazán Don Álvaro de Bazán 2019 B
albariño

93

Colour: bright straw. Nose: expressive, ripe fruit, floral, fine lees, mineral. Palate: full-bodied, spicy, long.

Granbazán Etiqueta Ámbar 2021 B
100% albariño

91

Colour: straw. Nose: expressive, white flowers, citrus fruit. Palate: fruity, balanced, easy to drink, varietal.

Granbazán Etiqueta Verde 2021 B
100% albariño

91

Colour: bright straw. Nose: floral, white fruit, citrus fruit. Palate: flavourful, fresh, good acidity, fruity aftertaste.

Granbazán Limousin 2019 B
100% albariño

93

Colour: bright yellow. Nose: spicy, white fruit, stone fruit, ripe fruit, lactic notes. Palate: rich, good structure, fine bitter notes.

BODEGA PAZO DE SAN MAURO
Pombal, 3 - Lugar de Porto
36458 Salvaterra de Miño (Pontevedra)
☎: +34 986 658 285
info@marquesdevargas.com
www.marquesdevargas.com

Pazo San Mauro Albariño 2021 B
albariño

88

Citrus fruit, herbal, crisp, yeasty notes.

BODEGA PAZO DE VILLAREI
Avda. de Peinador, 51
36416 Tameiga - Mos (Pontevedra)
☎: +34 986 441 732
marketing@hgabodegas.com
www.hgabodegas.com

Pazo Villarei 2021 B
100% albariño

90

Standard, kind finish. Colour: bright straw. Nose: tropical fruit, floral. Palate: fresh, easy to drink.

Villarei 2020 B
100% albariño

91

Colour: bright straw. Nose: citrus fruit, fine lees, wild herbs. Palate: fresh, flavourful, rich, balanced.

BODEGA Y VIÑEDOS VEIGA DA PRINCESA
Pol. Ind. Arbo, Parc. 2
32940 Arbo (Pontevedra)
☎: +34 988 261 256
info@pazodomar.com
www.pazodomar.com

Veiga da Princesa 2021 B

89

Standard, fruity, jammy, tropical, simple, pleasant, tasty.

DO RÍAS BAIXAS / D.O.P.

DO RÍAS BAIXAS / D.O.P.

BODEGAS AGUIUNCHO
Pedreiras, 1A Villalonga
36990 Sanxenxo (Pontevedra)
☎: +34 986 720 980
Fax: +34 986 727 063
info@aguiuncho.com
www.aguiuncho.com

Aguiuncho 2018 B BA
100% albariño
92
Colour: bright yellow. Nose: powerful, creamy oak, ripe fruit, spicy, woody, mineral, fine lees. Palate: rich, good structure, long, toasty, fine bitter notes.

Aguiuncho Selección 2018 B
100% albariño
91
Colour: bright straw. Nose: ripe fruit, fragrant herbs, fine lees, smoky. Palate: full-bodied, rich, long, good acidity.

Mar de Ons 2021 B
albariño
89
Citrus fruit, full-bodied, floral, fruity, full-bodied, jammy.

BODEGAS ALBAMAR
O Adro, 11 - Castrelo
36639 Cambados (Pontevedra)
☎: +34 660 292 750
xurxoalbamar@gmail.com

Albamar 2021 B
91
Colour: bright straw. Nose: ripe fruit, fragrant herbs, fine lees, wild herbs. Palate: full-bodied, good acidity, fresh.

Albamar 2021 RD
90
Colour: rose. Nose: faded flowers, raspberry, earthy notes, fine lees. Palate: fleshy, flavourful, ripe fruit.

Albamar Ancestral BE
91
Slight reduction, little interventionist. Colour: bright yellow. Nose: ripe fruit, fine lees, balanced, dried herbs, citrus fruit. Palate: good acidity, flavourful, ripe fruit.

Albamar O Esteiro 2020 T
91
Racy. Colour: cherry, purple rim. Nose: red berry notes, earthy notes, faded flowers. Palate: flavourful, fruity, good acidity.

Albamar Sesenta e Nove Arrobas 2020 B C
94
Colour: bright straw. Nose: ripe fruit, fragrant herbs, fine lees. Palate: full-bodied, rich, long, good acidity.

Alma de Mar 2020 B
92
Colour: bright straw. Nose: ripe fruit, fragrant herbs, fine lees, mineral, ebb tide. Palate: full-bodied, long, good acidity, flavourful.

PAI Edición Especial Albamar 2021 B
91
Colour: bright straw. Nose: ripe fruit, fine lees, citrus fruit. Palate: good acidity, flavourful, balanced.

Pepeluis 2020 B C
93
Colour: bright straw. Nose: ripe fruit, fragrant herbs, fine lees. Palate: full-bodied, rich, long, good acidity.

BODEGAS ALTOS DE TORONA
Vilachán, s/n
36740 As Eiras (Pontevedra)
☎: +34 986 442 073
marketing@hgabodegas.com
www.altosdetorona.com

Altos de Torona 2019 B BA
100% albariño
90
Colour: bright straw. Nose: ripe fruit, fragrant herbs, fine lees. Palate: full-bodied, rich, good acidity.

Altos de Torona Loureiro 2021 B
100% loureiro
89
Dried flowers, crisp, herbal, tasty.

Altos de Torona Rosal 2021 B
85% albariño, 10% caiño blanco, 5% loureiro
88
Citrus fruit, crisp, herbal, tasty.

Altos de Torona Caiño 2021 B
100% caiño blanco
88
Crisp, fruity, herbal, racy.

Altos de Torona Godello 2021 B
100% godello
89
Citrus fruit, crisp, herbal, standard.

Altos de Torona Albariño 2021 B
100% albariño

90

Colour: bright straw, greenish rim. Nose: fresh fruit, citrus fruit, wild herbs, fine lees. Palate: fresh, fruity, good acidity, fine bitter notes.

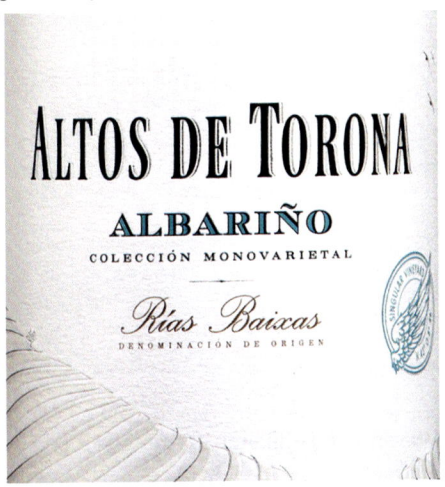

BODEGAS AQUITANIA
Bouza, 17 Castrelo
36639 Cambados (Pontevedra)
☎: +34 986 520 895
info@bodegasaquitania.com
www.bodegasaquitania.es

Aqvitania 2021 B
100% albariño

88

Fruity, crisp, herbal, smooth.

Bágoas Ledas 2021 B
100% albariño

86

Bernón 2021 B
100% albariño

87

BODEGAS AS LAXAS
As Laxas, 16
36430 Arbo (Pontevedra)
☎: +34 986 665 444
info@aslaxas.com
www.aslaxas.com

Bágoa do Miño 2020 B
100% albariño

91

Colour: bright straw. Nose: ripe fruit, fragrant herbs, fine lees, mineral. Palate: full-bodied, rich, good acidity.

Bágoa do Miño 2021 B
100% albariño

91

Colour: bright straw. Nose: ripe fruit, fragrant herbs, fine lees, mineral. Palate: full-bodied, rich, good acidity.

Laxas 2021 B
100% albariño

89

Citrus fruit, herbal, racy, tasty.

Val Do Sosego 2021 B
100% albariño

90

Colour: bright straw, greenish rim. Nose: citrus fruit, wild herbs, white fruit, stone fruit, fine lees. Palate: fresh, fruity, good acidity, fine bitter notes.

DO RÍAS BAIXAS / D.O.P.

DO RÍAS BAIXAS / D.O.P.

BODEGAS CHAVES
Condesa, 3, Barrantes
36636 Ribadumia (Pontevedra)
☎: +34 986 710 015
bodegaschaves@bodegaschaves.com
www.bodegaschaves.com

Castel de Fornos 2021 B
100% albariño

91

Colour: bright straw. Nose: white fruit, dried herbs, white flowers, citrus fruit. Palate: fresh, fruity, flavourful, balanced.

Cinco Islas Albariño Selección 2021 B
100% albariño

90

Colour: bright straw. Nose: white fruit, fruit expression, white flowers, ripe fruit. Palate: fruity, flavourful, balanced.

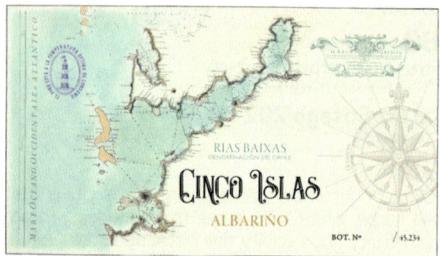

BODEGAS D. MATEOS
Camino de los Agudos, s/n
26559 Aldeanueva de Ebro (La Rioja)
☎: +34 941 261 897
Fax: +34 941 105 007
info@bodegasmateos.com
www.bodegasmateos.com

Lagar dos Mateos 2021 B
100% albariño

89

Citrus fruit, herbal, jammy, balanced.

BODEGAS DEL PALACIO DE
Pza. de Fefiñanes, s/n
36630 Cambados (Pontevedra)
☎: +34 986 542 204
fefinanes@fefinanes.com
www.fefinanes.com

1583 Albariño de Fefiñanes 2021 B
albariño

91

Needs time. Colour: bright straw. Nose: medium intensity, floral, citrus fruit, white fruit. Palate: fresh, good acidity.

Albariño de Fefiñanes 2021 B
100% albariño

91

Taut, very fruit-driven, needs time. Colour: bright straw. Nose: floral, fresh. Palate: easy to drink, good acidity.

Albariño de Fefiñanes III Año 2019 B
albariño

94

Elegant, dried flowers, crisp. Colour: yellow. Nose: expressive, fine lees, ebb tide, white fruit, elegant. Palate: rich, flavourful, long, saline.

Armas de Lanzós 2016 B

93

Dried flowers, crisp, citrus fruit. Colour: bright yellow. Nose: expressive, fine lees, characterful, balanced. Palate: flavourful, fresh.

BODEGAS EIDOSELA
Eidos de Abaixo, s/n - Sela
36494 Arbo (Pontevedra)
☎: +34 986 665 550
Fax: +34 986 665 299
info@bodegaseidosela.com
www.bodegaseidosela.com

Arbastrum 2021 B
albariño, treixadura, otras

89

Crisp, floral, fruity, wild, simple, smooth, balanced.

Eidosela 2018 T

88

Rustic, wild, fruity, spicy, balsamic herbs, smooth.

Eidosela 2021 B
100% albariño

89

Crisp, floral, fruity, smooth, standard.

DO RÍAS BAIXAS / D.O.P.

Eidosela
Burbujas del Atlántico BE BN
100% albariño
90
Colour: bright straw. Nose: fine lees, floral, fragrant herbs, spicy. Palate: flavourful, good acidity, balanced.

Eidosela
Burbujas del Atlántico BE EBR
100% albariño
90
Floral, varietally correct. Colour: bright straw. Nose: fresh, neat, fine lees. Palate: fresh, fruity, light-bodied.

BODEGAS FORJAS DEL SALNÉS
Pol. Ind. Salnes - Siete Pías Parc. 36
36630 Cambados (Pontevedra)
☎: +34 986 747 827
blancosdemar@gmail.com

Cies 2021 B
albariño
93
Floral, needs time. Colour: bright straw. Nose: expressive, ripe fruit, floral, fine lees, mineral. Palate: full-bodied, complex, spicy, long, elegant.

Finca Genoveva 2019 T
caiño
93
Colour: Cherry. Nose: expressive, spicy, mineral, red berry notes, wild herbs, dried herbs. Palate: elegant, full-bodied, long, great length, fresh.

Goliardo 2020 T
caiño, espadeiro, loureiro, sousón
93
With personality. Colour: bright cherry. Nose: ebb tide, scrubland, balsamic herbs, wild herbs, fresh fruit, red berry notes, black fruit. Palate: fresh, easy to drink, good acidity, good finish.

Goliardo Caiño 2019 T
caiño
93
Racy, herbal. Colour: light cherry. Nose: damp undergrowth, wild herbs, red berry notes, fresh fruit. Palate: racy, fresh, light-bodied.

Leirana 2021 B
albariño
93
Needs time. Colour: bright straw. Nose: ripe fruit, dried herbs, faded flowers, ebb tide, jasmine. Palate: powerful, ripe fruit, balanced.

Sálvora 2020 B FB
albariño
94
Taut, crisp. Colour: bright straw. Nose: ripe fruit, fragrant herbs, fine lees, fresh fruit. Palate: full-bodied, rich, long, good acidity.

BODEGAS GERARDO MÉNDEZ
Galiñans, 10 - Lores
36968 Meaño (Pontevedra)
☎: +34 986 747 046
info@bodegasgerardomendez.com
www.bodegasgerardomendez.com

Albariño Do Ferreiro 2021 B
92
Colour: bright straw, greenish rim. Nose: fresh fruit, citrus fruit, wild herbs, fine lees. Palate: fresh, fruity, good acidity, fine bitter notes.

BODEGAS LA CAÑA
Camiño Novo, 36
36600 Vilagarcía de Arousa (Pontevedra)
☎: +34 952 504 706
pedidos@jorgeordonez.es
www.jorgeordonez.es

La Caña 2021 B
albariño
91
Colour: bright straw. Nose: fruit expression, ripe fruit, floral, balsamic herbs. Palate: flavourful, fresh, good acidity, fruity aftertaste.

La Caña Navia 2019 B
albariño
93
Colour: bright yellow. Nose: dried flowers, candied fruit, fine lees, pattiserie, characterful, complex. Palate: round, spicy, long, great length.

DO RÍAS BAIXAS / D.O.P.

BODEGAS LA VAL
Lugar Muguiña, s/n - Arantei
36458 Salvaterra de Miño (Pontevedra)
☎: +34 986 610 728
ana@bodegaslaval.com
www.bodegaslaval.com

Finca Arantei 2020 B
100% albariño
91
Colour: bright straw. Nose: fragrant herbs, fine lees, stone fruit, spicy. Palate: full-bodied, rich, long, good acidity.

La Val Albariño 2021 B
100% albariño
90
Colour: bright straw, greenish rim. Nose: fresh fruit, citrus fruit, wild herbs, fine lees. Palate: fresh, fruity, good acidity, fine bitter notes.

🏆 PODIUM

La Val Crianza sobre Lías 2016 B C
100% albariño
95
Colour: bright straw. Nose: ripe fruit, fragrant herbs, fine lees, dry stone, mineral. Palate: full-bodied, rich, long, good acidity, fine bitter notes.

La Val Fermentado en Barrica 2016 B FB
100% albariño
94
Colour: bright straw. Nose: ripe fruit, fragrant herbs, fine lees, lees reduction notes, pattiserie. Palate: full-bodied, rich, long, good acidity.

La Val Sousón 2019 T RB
100% sousón
93
Defined aromas. Colour: cherry, purple rim. Nose: fruit expression, red berry notes, floral, spicy. Palate: flavourful, fruity, good acidity, elegant.

La Val Treixadura 2021 B
100% treixadura
91
Racy, austere. Colour: bright straw. Nose: fine lees, dry stone, grassy, white fruit, fresh fruit, citrus fruit. Palate: full-bodied, long, good acidity.

La Val Vendimia 2017 B
100% albariño
94
Colour: bright straw. Nose: ripe fruit, fragrant herbs, fine lees, smoky, mineral, ebb tide. Palate: full-bodied, rich, long, good acidity.

Mas que 2 2021 B
40% albariño, 40% treixadura, 20% loureiro
89
Citrus fruit, tasty, herbal, spicy, crisp.

Viña Ludy 2021 B
100% albariño
90
Colour: bright straw, greenish rim. Nose: fresh fruit, citrus fruit, wild herbs, fine lees. Palate: fresh, fruity, good acidity, fine bitter notes.

BODEGAS LAUREATUS
Lg. de Saramagoso, 13, San Martiño de Meis
36637 Meis (Pontevedra)
☎: +34 986 099 002
direccion@laureatus.es
www.laureatus.es

Laureatus 2021 B
100% albariño
90
Colour: bright straw. Nose: fruit expression, floral, wild herbs, citrus fruit. Palate: flavourful, fresh, good acidity, fruity.

Laureatus Dolium 2013 B C
100% albariño
91
With personality, age nuances. Colour: bright yellow. Nose: fine lees, dried flowers, white fruit, ripe fruit. Palate: spicy, bitter.

Laureatus Lías 2013 B C
100% albariño
92
Colour: bright yellow. Nose: ripe fruit, spicy, saffron, fine lees, faded flowers. Palate: rich, fine bitter notes, easy to drink.

BODEGAS MAR DE FRADES
Lg. Arosa, 16 - Finca Valiñas
36637 Meis (Pontevedra)
☎: +34 986 680 911
Fax: +34 986 680 926
info@mardefrades.es
www.mardefrades.es

**Finca Valiñas
"crianza sobre lías" 2018 B**
94
Colour: bright straw. Nose: ripe fruit, fragrant herbs, fine lees, stone fruit, ebb tide. Palate: full-bodied, rich, long, good acidity.

Mar de Frades Albariño 2021 B
albariño
90
Pleasant, crisp, fruity. Colour: bright yellow. Nose: medium intensity. Palate: easy to drink, light-bodied, fine bitter notes.

Mar de Frades BE BN
88
Pleasant, aromatic, standard, crisp, thin.

Mar de Frades Godello Atlántico 2020 B
91
Colour: bright straw. Nose: floral, neat. Palate: fresh, good acidity, fruity aftertaste, easy to drink.

Monteveiga 2018 B
94
Colour: bright straw. Nose: ripe fruit, fragrant herbs, fine lees, dry stone. Palate: full-bodied, long, good acidity.

BODEGAS MARQUÉS DE VIZHOJA
Finca La Moreira s/n
36438 Arbo (Pontevedra)
☎: +34 986 665 825
marquesdevizhoja@marquesdevizhoja.com
www.bodegasmarquesdevizhoja.com

Señor da Folla Verde 2020 B
70% albariño, 15% treixadura, 15% loureiro
91
Colour: straw. Nose: expressive, white flowers, jasmine, dried herbs. Palate: flavourful, fruity, balanced.

Torre La Moreira 2021 B
albariño
90
Colour: bright straw. Nose: expressive, white flowers, jasmine, dried herbs. Palate: flavourful, fruity, balanced.

BODEGAS MARTÍN CÓDAX
Lg. Burgáns 91, Vilariño
36633 Cambados (Pontevedra)
☎: +34 986 526 040
comunicacion@martincodax.com
www.martincodax.com

Martín Códax 2021 B
100% albariño
88
Citrus fruit, crisp, herbal, balanced.

Martín Códax Arousa 2019 B C
100% albariño
90
Colour: bright straw, greenish rim. Nose: fresh fruit, citrus fruit, wild herbs, fine lees. Palate: fresh, fruity, good acidity, fine bitter notes.

Martín Códax Finca Xieles 2020 B
100% albariño
91
Colour: bright straw, greenish rim. Nose: fresh fruit, citrus fruit, wild herbs, fine lees. Palate: fresh, fruity, good acidity, fine bitter notes.

Martín Códax Gallaecia 2017 B C
100% albariño
92
Colour: bright yellow. Nose: dried flowers, candied fruit, fine lees, pattiserie, stone fruit, dried fruit, spicy, saffron. Palate: spicy, fresh, racy.

Martin Códax Lías 2019 B C
100% albariño
91
Colour: bright straw. Nose: ripe fruit, fragrant herbs, fine lees. Palate: full-bodied, rich, good acidity.

DO RÍAS BAIXAS / D.O.P.

DO RÍAS BAIXAS / D.O.P.

Martín Códax Vindel 2018 B
100% albariño

91

Colour: bright yellow. Nose: creamy oak, ripe fruit, spicy, fine lees, wild herbs. Palate: rich, good structure, toasty, fine bitter notes.

BODEGAS PABLO PADÍN

Ameiro, 17 - Dena
36967 Meaño (Pontevedra)
☎: +34 986 743 231
Fax: +34 986 745 791
info@pablopadin.com
www.pablopadin.com

Eiral Albariño 2021 B
100% albariño

88

Pleasant, fruity, tasty, jammy.

Feitizo da Noite 2018 BE EBR
100% albariño

90

Colour: yellow. Nose: white fruit, white flowers, faded flowers, fine lees. Palate: fresh, easy to drink, good finish.

Feitizo da Noite BE BR
albariño

89

Dried flowers, pleasant. Nose: fine lees, ebb tide, fresh, medium intensity.

Segrel Albariño 2021 B
100% albariño

91

Colour: bright straw. Nose: white flowers, jasmine, dried herbs. Palate: flavourful, fruity, balanced.

Segrel Ámbar 2021 B
100% albariño

91

Colour: straw. Nose: ripe fruit, dried herbs, faded flowers. Palate: powerful, ripe fruit, balanced.

BODEGAS RODRÍGUEZ Y SANZO

Manuel Azaña, 9
47014 Valladolid (Valladolid)
☎: +34 983 150 150
comunicacion@valsanzo.com
www.rodriguezsanzo.com

María Sanzo 2021 B
100% albariño

89

Citrus fruit, herbal, crisp, tasty.

BODEGAS SANTIAGO ROMA

Catariño, 6 - Besomaño
36636 Ribadumia (Pontevedra)
☎: +34 679 469 218
bodega@santiagoroma.com
www.santiagoroma.com

Colleita de Martis Albariño 2021 B
albariño

90

Colour: bright straw, greenish rim. Nose: ripe fruit, fine lees, floral, white fruit. Palate: full-bodied, long, good acidity.

Metamorfose de Santiago Roma 2020 T
mencía

91

Colour: cherry, purple rim. Nose: fruit expression, red berry notes, spicy, dried flowers. Palate: flavourful, fruity, good acidity, ripe fruit.

Pedranai de Santiago Roma Albariño 2020 B
albariño

92

Colour: bright straw. Nose: expressive, ripe fruit, floral, fine lees, mineral, citrus fruit. Palate: easy to drink, fresh, balanced.

Santiago Roma Albariño 2021 B
albariño

90

Colour: bright straw. Nose: ripe fruit, fine lees, floral. Palate: good acidity, fine bitter notes, easy to drink.

Santiago Roma Albariño Selección 2020 B
albariño

92

Colour: bright straw. Nose: ripe fruit, fragrant herbs, fine lees, stone fruit. Palate: full-bodied, rich, long, good acidity.

Santiago Roma Burbulla 2020 BE BN
albariño

91

Colour: bright yellow. Nose: ripe fruit, fine lees, balanced, white flowers. Palate: good acidity, flavourful, ripe fruit.

BODEGAS TERRAS GAUDA

36760 O'Rosal (Pontevedra)
☎: +34 986 621 001
Fax: +34 986 621 084
terrasgauda@terrasgauda.com
www.terrasgauda.com

Abadía de San Campio 2021 B
albariño
89
Floral, crisp, fruity, dried herbs, tasty.

La Mar 2020 B
caiño
92
Colour: straw. Nose: ripe fruit, dried herbs, faded flowers. Palate: powerful, ripe fruit, balanced.

Terras Gauda 2021 B
albariño, loureiro, caiño
91
Colour: bright straw. Nose: white fruit, fruit expression, floral, expressive. Palate: fruity, fresh, flavourful, balanced.

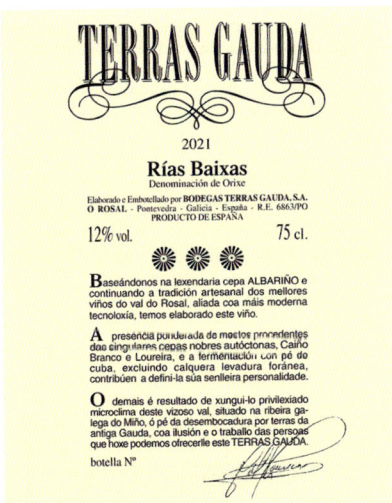

Terras Gauda Etiqueta Negra 2020 B FB
albariño, loureiro, caiño
91
Colour: bright yellow. Nose: powerful, ripe fruit, spicy, dry nuts. Palate: rich, long, fine bitter notes, balanced.

BODEGAS VIONTA

Lugar de Axis s/n - Simes
36636 Meaño (Pontevedra)
☎: +34 986 747 566
vionta@vionta.com
www.vionta.com

Agnusdei Albariño 2021 B
100% albariño
88
Racy, herbal, citrus fruit, balanced.

Depende 2021 B
100% albariño
89
Citrus fruit, crisp, herbal, racy, tasty.

Vionta 2021 B
100% albariño
90
Colour: bright straw, greenish rim. Nose: fresh fruit, citrus fruit, wild herbs. Palate: fresh, fruity, good acidity.

You & Me 2021 B
100% albariño
87

BOUZA DO REI

Rua Puxafeita, 21
36636 Ribadumia (Pontevedra)
☎: +34 986 710 257
Fax: +34 986 718 393
bouzadorei@bouzadorei.com
www.bouzadorei.com

Bouza Do Rei 2021 B
100% albariño
89
Racy, floral, herbal, tasty, yeasty notes.

Bouza do Rei Gran Selección 2020 B
100% albariño
91
Colour: bright straw. Nose: white fruit, dried flowers, white flowers, wild herbs, fine lees. Palate: fruity, fresh, flavourful, balanced.

Castel de Bouza 2021 B
100% albariño
89
Citrus fruit, fruity, crisp, herbal.

Gran Lagar de Bouza 2021 B
90
Colour: straw. Nose: expressive, white flowers, dried herbs. Palate: flavourful, fruity, balanced, mineral.

DO RÍAS BAIXAS / D.O.P.

DO RÍAS BAIXAS / D.O.P.

Pazo da Torre Albariño 2021 B
100% albariño
89
Kind finish, citrus fruit, crisp, fruity.

Xelmirez 2021 B
100% albariño
89
Floral, crisp, taut, citrus fruit.

CAMBADOS URBAN WINERY
36630 Cambados (Pontevedra)
☎: +34 652 885 545
comercial@cambadosurbanwinery.com
www.cambadosurbanwinery.com

Desconcierto 2021 B
albariño
91
Colour: bright straw. Nose: fruit expression, ripe fruit, floral, dried herbs. Palate: flavourful, fresh, good acidity, fruity aftertaste, good finish.

CAMINO DE CABRAS
Hermanos Maristas, 27
36700 Tui (Pontevedra)
☎: +34 698 145 790
info@caminodecabras.com
www.caminodecabras.com

Camino de Cabras Albariño 2021 B
albariño
89
Crisp, tasty, herbal, full-bodied.

Oro Valei 2021 B
albariño
89
Citrus fruit, herbal, jammy, full-bodied.

CARLOS REY LUSTRES
Axis, 11 - Simes
36969 Meaño (Pontevedra)
☎: +34 886 600 007
info@anadigna.com
www.anadigna.com

Anadigna 1932 2021 B
100% albariño
88
Crisp, floral, crisp, citrus fruit, smooth.

Anadigna Caiño 2020 T RB
100% caiño
90
Colour: bright cherry. Nose: fresh fruit, red berry notes, wild herbs, wild herbs. Palate: spicy, fruity, flavourful.

Anadigna Fudre 2019 B RB
100% albariño
91
Colour: bright yellow. Nose: spicy, expressive, dried flowers, white fruit, ripe fruit. Palate: rich, good structure, long, fine bitter notes, balanced.

Anadigna Fudre 2020 B RB
100% albariño
90
Pleasant, crisp. Colour: bright straw. Nose: fresh fruit, citrus fruit, floral, balanced, spicy. Palate: varietal, racy.

Anadigna sobre Lías 2020 B
100% albariño
91
Colour: bright straw. Nose: ripe fruit, fine lees, floral. Palate: full-bodied, rich, long, good acidity.

Anadigna Tradicional 2021 B
100% albariño
89
Pleasant, aromatic, floral, crisp, citrus fruit, smooth, wild.

CASA DO SOL
Laraño, 8 San Julián de Sales
15885 Vedra (A Coruña/La Coruña)
comercial@casadosol.es
www.casadosol.es

Casa Do Sol 2019 B
albariño
92
Colour: yellow. Nose: expressive, dried flowers, wild herbs, bakery, balanced, white fruit, ripe fruit. Palate: rich, full-bodied, spicy.

Froiña de Casa Do Sol 2021 B
albariño
89
Citrus fruit, crisp, herbal, balanced, yeasty notes.

CAZAPITAS
Salgosa, 25
36740 Tomiño (Pontevedra)
☎: +34 605 625 782
cazapitassl@gmail.com
www.cazapitas.com

Eido da Salgosa 2020 B
albariño
91
Colour: bright straw. Nose: white flowers, fresh fruit, dried herbs. Palate: fresh, flavourful, varietal.

Eido da Salgosa Rosal 2020 B
albariño, loureiro, caíño

90
Colour: bright straw. Nose: citrus fruit, white flowers, dried herbs. Palate: fresh, fruity.

La Sobrada 2019 B

91
Colour: bright straw. Nose: white flowers, jasmine, dried herbs, citrus fruit. Palate: flavourful, fruity, balanced.

CHAN DE ROSAS
Rua Escusa, 10
36636 Ribadumia (Pontevedra)
☎: +34 941 451 129
ervigio.adan@premiumfincas.com
www.premiumfincas.com

Chan de Rosas Clásico 2021 B
100% albariño

88
Pleasant, aromatic, fruity, tasty, simple.

Chan de Rosas Cuvée Especial 2021 B
100% albariño

90
Fruity, floral, balanced, standard, jammy, tasty, varietally correct. Palate: balanced, good acidity, fine bitter notes.

CORISCA
Bº Corisca, 7
36471 Entienza (Pontevedra)
☎: +34 615 430 296
info@bodegascorisca.com
www.bodegascorisca.com

Corisca 2021 B
albariño

91
Colour: bright straw. Nose: fragrant herbs, fine lees, white fruit, floral. Palate: full-bodied, long, good acidity, balanced.

DAVID MARTÍNEZ SOBRAL
Figueiro Lago, 14
36792 Tomiño (Pontevedra)
☎: +34 603 800 239
info@pedregales.es
www.pedregales.es

Lagar Pedregales Floración 2021 B
albariño

90
Colour: straw. Nose: expressive, white flowers, dried herbs, white fruit. Palate: fruity, balanced, good acidity, fine bitter notes.

DESTINOS CRUZADOS
Lg. Muguiña, s/n
36458 Salvaterra de Miño (Pontevedra)
☎: +34 600 536 154
infodestinoscruzados@gmail.com
www.destinoscruzados.es

Destinos Cruzados As Regadas 2018 T C
50% mencía, 35% brancellao, 15% espadeiro

91
Slight oxidation. Colour: cherry, purple rim. Nose: fruit expression, red berry notes, floral, spicy, faded flowers, earthy notes. Palate: flavourful, fruity, good acidity, long.

Destinos Cruzados Pousada 2019 B
75% albariño, 25% treixadura

92
Colour: bright straw. Nose: fragrant herbs, fine lees, white fruit, ripe fruit, white flowers. Palate: full-bodied, rich, balanced.

Destinos Cruzados Pousada 2019 B C
75% albariño, 25% treixadura

93
Colour: bright yellow. Nose: creamy oak, ripe fruit, spicy, toasted bread, fine lees. Palate: rich, good structure, long, toasty, fine bitter notes.

EIDO DA FONTE
Lugar A Fonte, 2 - Valeixe
36883 A Cañiza (Pontevedra)
☎: +34 986 654 242
info@eidodafonte.com
www.eidodafonte.com

Eido da Fonte Albariño 2021 B
100% albariño

87

Eido da Fonte Sousón 2017 T
sousón

88
Balanced, spicy, dried flowers, crisp, mineral.

ENEO
Paseo Virgen de la Vega 4 1ºI
26200 Haro (La Rioja)
☎: +34 941 310 494
soto@comercialeneo.com
www.comercialeneo.com

Rey Eneo 2020 B
albariño

89
Aromatic, floral, fruity, simple, pleasant, smooth.

DO RÍAS BAIXAS / D.O.P.

DO RÍAS BAIXAS / D.O.P.

Rey Eneo 2021 B
albariño

88

Citrus fruit, floral, jammy, thin, standard.

FAMILIA TORRES
Miguel Torres i Carbó, 6
08720 Vilafranca del Penedès (Barcelona)
☎: +34 938 177 400
info@torres.es
www.torres.es

Pazo das Bruxas 2021 B
albariño

88

Racy, citrus fruit, herbal, simple.

FAUSTINO RIVERO ULECIA
PO-400
36430 Arbo (Pontevedra)
☎: +34 941 380 057
www.faustinorivero.com

Faustino Rivero Ulecia Albariño 2021 B
albariño

89

Pleasant, aromatic, tasty, herbal, fruity.

FÉLIX SOLIS AVANTIS
Autovía del Sur, Km. 199
13300 Valdepeñas (Ciudad Real)
☎: +34 926 322 400
Fax: +34 926 322 417
fsa@felixsolisavantis.com
www.felixsolisavantis.com

Medusa Albariño 2021 B
100% albariño

88

Nose: white flowers, jasmine, dried herbs. Palate: flavourful, fruity.

FENTO WINES
Sisangándara, 22
36636 Ribadumia (Pontevedra)
☎: +34 986 099 486
Fax: +34 986 718 549
info@eulogiopomares.com
www.eulogiopomares.com

Castiñeiro Albariño 2019 B
albariño

93

Little interventionist. Colour: bright yellow. Nose: ripe fruit, dried herbs, faded flowers, fine lees. Palate: powerful, ripe fruit, balanced.

Eulogio Pomares Maceración con Pieles 2020 B

94

Colour: bright yellow. Nose: wild herbs, grassy, stone fruit, floral, petrol notes. Palate: fresh, fine bitter notes, flavourful.

FORTUNA WINES
Sanjurjo Badia 22, 3B
36207 Vigo (Pontevedra)
☎: +34 691 561 471
info@fortunawines.es
www.fortunawines.es

Catavento 2021 B
albariño

88

Pleasant, aromatic, floral, simple.

HAMMEKEN CELLARS
Llavador, 20
03700 Denia (Alacant/Alicante)
☎: +34 965 791 967
cellars@hammekencellars.com
www.hammekencellars.com

Gotas de Mar Albariño 2021 B
100% albariño

89

Aromatic, fruity, floral, varietally correct, tasty.

LAGAR DA CONDESA
Lugar de Maran s/n Arcos da Condesa
36650 Caldas de Reis (Pontevedra)
☎: +34 968 435 022
Fax: +34 968 716 051
info@gilfamily.es
www.gilfamily.es

Kentia 2021 B
albariño

92

Colour: bright yellow. Nose: ripe fruit, floral, fine lees, saline, ebb tide, white fruit. Palate: flavourful, fresh, good acidity, fruity aftertaste.

Lagar da Condesa 2021 B
100% albariño

93

Needs time. Colour: bright yellow. Nose: expressive, ripe fruit, floral, fine lees, mineral, stone fruit. Palate: full-bodied, complex, spicy, long, elegant.

O Fillo Da Condesa 2021 B
100% albariño

91

Colour: bright straw. Nose: ripe fruit, floral, saline, grassy. Palate: flavourful, fresh, good acidity, fruity aftertaste.

LAGAR DE BESADA

Pazo, 11 Xil
36968 Meaño (Pontevedra)
☎: +34 986 747 473
info@lagardebesada.com
www.lagardebesada.com

Añada de Baladiña 2011 B
100% albariño

92
Racy, full of life. Colour: bright straw. Nose: ripe fruit, fine lees, dry stone, ebb tide. Palate: full-bodied, good acidity, fine bitter notes.

Baladiña 2021 B
albariño

90
Colour: bright straw. Nose: fresh fruit, citrus fruit, wild herbs. Palate: fresh, fruity, good acidity, fine bitter notes, easy to drink.

Baladiña Barro 2014 B
100% albariño

93
Colour: bright yellow. Nose: elegant, fine lees, petrol notes, faded flowers, dried flowers, wild herbs. Palate: flavourful, good acidity, balanced, fine bitter notes.

Burbujas de Baladiña 2014 BE BN
100% albariño

90
Needs time. Colour: bright yellow. Nose: ripe fruit, fine lees, tropical fruit, wild herbs. Palate: good acidity, flavourful, ripe fruit.

Lagar de Besada 2021 B
albariño

89
Pleasant, aromatic, citrus fruit, crisp, fruity, simple.

LAGAR DE CERVERA

Estrada de Loureza, 86
36770 O'Rosal (Pontevedra)
☎: +34 986 625 875
Fax: +34 986 625 011
lagar@riojalta.com
www.riojalta.com

Lagar de Cervera 2021 B
100% albariño

91
Colour: bright straw, greenish rim. Nose: fresh fruit, citrus fruit, wild herbs. Palate: fresh, fruity, good acidity, fine bitter notes.

Pazo de Seoane O Rosal 2021 B
60% albariño, 40% loureiro, caiño blanco, treixadura

91
Colour: bright straw. Nose: ripe fruit, fragrant herbs, fine lees, white flowers, dried flowers, balsamic herbs. Palate: full-bodied, rich, good acidity.

LAGAR DE COSTA

Sartaxes, 8 - Castrelo
36639 Cambados (Pontevedra)
☎: +34 669 086 569
contacto@lagardecosta.com
www.lagardecosta.com

Calabobos 2018 B
100% albariño

92
Colour: bright straw. Nose: expressive, floral, fine lees. Palate: full-bodied, complex, spicy, long, balanced, fine bitter notes.

Lagar de Costa 2021 B
100% albariño

89
Pleasant, aromatic, standard, floral, fruity. Palate: easy to drink, good finish.

DO RÍAS BAIXAS / D.O.P.

DO RÍAS BAIXAS / D.O.P.

Lagar de Costa Tradición 2018 B BA
100% albariño

92

Colour: bright straw. Nose: fragrant herbs, fine lees, white fruit, ebb tide. Palate: full-bodied, rich, long, good acidity.

Maio 2018 B
100% albariño

93

Colour: bright straw. Nose: expressive, floral, fine lees, spicy. Palate: full-bodied, complex, long, rich.

MAIOR DE MENDOZA
Rúa de Xiabre, 58
36600 Vilagarcía de Arousa (Pontevedra)
☎: +34 986 508 896
maiordemendoza@hotmail.com
www.maiordemendoza.com

Maior de Mendoza 3 Crianzas 2018 B
100% albariño

93

Colour: bright yellow. Nose: dried flowers, candied fruit, fine lees, pattiserie. Palate: round, spicy, long.

Maior de Mendoza 3 Crianzas 2019 B
albariño

93

Colour: bright straw. Nose: white fruit, ripe fruit, faded flowers, dried herbs, fine lees. Palate: fruity, fresh, flavourful, good acidity, good finish.

🏆 PODIUM

Maior de Mendoza Finca Las Tablas 2017 B BA
albariño

98

Colour: bright yellow. Nose: ripe fruit, dried herbs, faded flowers, lactic notes, petrol notes. Palate: powerful, ripe fruit, balanced, good acidity, good finish.

Maior de Mendoza sobre Lías 2020 B
100% albariño

90

Colour: straw. Nose: expressive, jasmine, dried herbs, ripe fruit. Palate: flavourful, fruity, balanced, round.

Maior de Mendoza sobre Lías 2021 B
albariño

90

Colour: straw. Nose: expressive, white flowers, dried herbs, ripe fruit. Palate: flavourful, fruity, balanced.

MAR DE ENVERO
Lugar de Rarís, 30
15883 Teo (A Coruña/La Coruña)
☎: +34 981 195 202
bodega@mardeenvero.es
www.mardeenvero.es

Mar de Envero 2017 T BA
sousón, mencía, pedral

89

Herbal, fruity, slight reduction, animal funk, pronounced acidity, spicy, wild.

Troupe 2021 B
100% albariño

89

Pleasant, floral, crisp, fruity, saline, simple, smooth.

NOTAS FRUTALES DE ALBARIÑO
Ctra. Villar – Garabelos s/n
36429 Crecente (Pontevedra)
☎: +34 609 065 858
notasfrutales@gmail.com
www.notasfrutales.es

Finca Garabelos 2020 B
100% albariño

91

Colour: bright straw. Nose: ripe fruit, fragrant herbs, fine lees. Palate: full-bodied, rich, long, good acidity.

La Trucha 2021 B
100% albariño

88

Citrus fruit, crisp, herbal, balanced.

La Trucha Acero 2017 B
100% albariño

90

Colour: bright straw. Nose: fragrant herbs, fine lees, white fruit, lactic notes. Palate: full-bodied, rich, long, good acidity.

La Trucha Barrica 2020 B
100% albariño

92

Colour: yellow. Nose: ripe fruit, dried herbs, faded flowers, fine lees. Palate: fruity, flavourful, fresh.

La Trucha de Otoño 2017 B
100% albariño

93

Colour: bright straw. Nose: fragrant herbs, fine lees, stone fruit, dry nuts, brioche, caramel. Palate: full-bodied, rich, long, good acidity.

PACO & LOLA

Valdamor, 18 - Xil
36968 Meaño (Pontevedra)
☎: +34 986 747 779
comercial@pacolola.com
www.pacolola.com

Paco & Lola 2021 B
albariño

90

Colour: straw. Nose: white flowers, varietal, fresh, neat. Palate: fruity, balanced.

Paco & Lola BE
albariño

91

Colour: bright yellow. Nose: balanced, brioche, stone fruit. Palate: good acidity, flavourful, ripe fruit, balanced.

Paco & Lola Prime 2019 B

92

Colour: bright yellow. Nose: ripe fruit, fine lees, citrus fruit, floral, varietal. Palate: full-bodied, rich, long, good acidity.

Paco & Lola Vintage 2016 B

94

Age nuances. Colour: yellow. Nose: white fruit, ripe fruit, expressive, balanced, dried flowers, faded flowers. Palate: rich, fresh, fine bitter notes.

Nº12 by Paco & Lola 2021 B

88

Kind finish, floral, smooth.

PACO MULERO

Partida de la Hoya Torres s/n
30520 Jumilla (Murcia)
☎: +34 676 433 541
info@pacomulero.com
www.pacomulero.com

Paco Mulero Albariño 2021 B
100% albariño

89

Pleasant, balanced, fruity, jammy, smooth.

PAGOS DEL REY

Autovía del Sur, km.199
13300 Valdepeñas (Ciudad Real)
☎: +34 926 322 400
fsa@felixsolisavantis.com
www.pagosdelrey.com

Pulpo Albariño 2021 B

90

Colour: straw. Nose: expressive, white flowers, jasmine, dried herbs. Palate: flavourful, fruity, balanced.

PAZO BAIÓN

Lg. Abelleira 4, 5, 6 - Baión
36614 Vilanova de Arousa (Pontevedra)
☎: +34 636 800 234
inf@pazobaion.com
www.pazobaion.com

Pazo Baión Albariño 2021 B
100% albariño

92

Colour: bright straw. Nose: fruit expression, ripe fruit, floral, mineral. Palate: flavourful, fresh, good acidity, fruity aftertaste.

Pazo Baión Gran a Gran 2017 B
100% albariño
92
Colour: bright yellow. Nose: candied fruit, honeyed notes, dry nuts, petrol notes. Palate: flavourful, unctuous, fruity, sweet.

Pazo Baión Vides de Fontán 2018 B
100% albariño
93
Colour: bright straw. Nose: fragrant herbs, fine lees, white fruit, stone fruit, ebb tide. Palate: full-bodied, rich, long, good acidity, elegant.

PAZO DE BARRANTES
Finca Pazo de Barrantes
36636 Ribadumia (Pontevedra)
☎: +34 986 718 211
bodega@pazodebarrantes.com
www.nmarquesdemurrieta.com

🏆 PODIUM

La Comtesse 2018 B FB
100% albariño
96
Colour: bright yellow. Nose: dried flowers, candied fruit, fine lees, pattiserie, dry nuts. Palate: round, spicy, long, great length.

Pazo de Barrantes Albariño 2019 B
100% albariño
94
Briny. Colour: bright yellow. Nose: citrus fruit, wild herbs, mineral. Palate: fresh, fruity, good acidity, fine bitter notes.

Pazo de Barrantes Albariño 2020 B
100% albariño
93
Colour: bright yellow. Nose: ripe fruit, dried herbs, faded flowers, dried flowers. Palate: powerful, ripe fruit, balanced, spicy.

PAZO DE RUBIANES
Rúa do Pazo, 7 Rubianes
36619 Vilagarcía de Arousa (Pontevedra)
☎: +34 986 510 534
info@pazoderubianes.com
www.pazoderubianes.com

Pazo de Rubianes 1411 2017 B
albariño
93
Colour: bright yellow. Nose: powerful, creamy oak, ripe fruit, spicy, smoky, toasted bread. Palate: rich, good structure, toasty, fine bitter notes.

Pazo de Rubianes 1411 2019 B
albariño
92
Defined aromas, smooth. Colour: bright straw. Nose: ripe fruit, fine lees, spicy. Palate: rich, good acidity, good finish.

Pazo de Rubianes Albariño 2015 B
albariño
93
Colour: bright straw. Nose: ripe fruit, fragrant herbs, fine lees, pattiserie, caramel, wild herbs. Palate: full-bodied, rich, good acidity, balanced.

Pazo de Rubianes Albariño 2020 B
albariño
92
Colour: bright straw. Nose: ripe fruit, fragrant herbs, fine lees, white flowers. Palate: full-bodied, rich, balanced.

Pazo de Rubianes
García Caamaño 2015 B
100% albariño

92

Colour: bright straw. Nose: ripe fruit, fragrant herbs, fine lees, smoky. Palate: full-bodied, rich, good acidity.

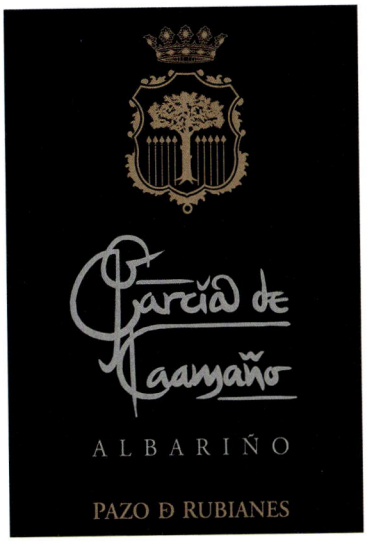

Pazo de Rubianes
García Caamaño 2018 B
albariño

92

Defined aromas. Colour: bright straw. Nose: fine lees, white fruit, ripe fruit, white flowers. Palate: easy to drink, good finish.

PAZO DE SEÑORANS

Vilanoviña, s/n
36616 Meis (Pontevedra)
☎: +34 986 715 373
info@pazodesenorans.com
www.pazodesenorans.com

Pazo Señorans 2021 B
100% albariño

92

Colour: bright straw, greenish rim. Nose: fresh fruit, wild herbs, fine lees. Palate: fresh, fruity, good acidity, fine bitter notes.

Pazo Señorans Colección 2018 B
100% albariño

94

Colour: bright straw. Nose: ripe fruit, fragrant herbs, fine lees, brioche, smoky, ebb tide. Palate: full-bodied, rich, long, good acidity.

🏆 PODIUM

Pazo Señorans
Selección de Añada 2013 B

98

Defined aromas, taut, crisp, original. Colour: bright straw. Nose: scrubland, fine lees, complex, varietal. Palate: full of life, flavourful, complex.

Tras los Muros 2018 B BA

94

Colour: bright straw. Nose: ripe fruit, fine lees, scrubland, creamy oak, toasty. Palate: full-bodied, rich, long, good acidity.

PAZO PONDAL

Lg. Coto, s/n - Cabeiras
36436 Arbo (Pontevedra)
☎: +34 986 665 551
Fax: +34 986 665 949
admin@pazopondal.com
www.pazopondal.com

Leira Pondal 2021 B
100% albariño

89

Balanced, crisp, herbal, tasty, yeasty notes.

Miña Vida 2021 B
70% albariño, 30% treixadura

89

Balanced, herbal, simple, citrus fruit.

Pazo Pondal Albariño 2020 B
albariño

90

Colour: bright straw. Nose: ripe fruit, fragrant herbs, fine lees, smoky. Palate: full-bodied, rich, long, good acidity.

DO RÍAS BAIXAS / D.O.P.

PAZO TORRE PENELAS
36657 Torre (Pontevedra)
☎: +34 938 177 400
info@torres.es
www.torres.es

Pazo Torre Penelas Blanco Granito 2019 B
albariño
90
Colour: bright straw. Nose: fragrant herbs, fine lees, white flowers. Palate: full-bodied, rich, good acidity.

PAZOS DE LUSCO
Grixó-Alxén s/n
36458 Salvaterra do Miño (Pontevedra)
☎: +34 986 659 102
prensa@gonzalezbyass.es
www.lusco.es

Lusco Albariño 2021 B
albariño
91
Colour: straw. Nose: expressive, white flowers, jasmine, dried herbs. Palate: flavourful, fruity, balanced, mineral.

Pazo de Piñeiro 2020 B
albariño
93
Colour: bright straw. Nose: ripe fruit, fragrant herbs, fine lees, ebb tide, white flowers. Palate: full-bodied, rich, long, good acidity.

PEKADO MORTAL
Rua Valdomiño, 35 Finca Fierro
32002 Barbadas (Ourense/Orense)
☎: +34 640 209 584
martinpadin9@hotmail.com

Pekado Mortal Albariño 2020 B
90
Colour: bright straw. Nose: ripe fruit, dried herbs, faded flowers, fine lees. Palate: fruity, flavourful, balanced, fresh.

PONTECABALEIROS
Chan da Ponte, 4
36430 Arbo (Pontevedra)
☎: +34 986 665 444
aslaxas@aslaxas.com
www.pontecabaleiros.com

Alvinte 2021 B
100% albariño
87

Ferrum 2020 B
100% albariño
88
Citrus fruit, herbal, simple.

Outón 2020 B
100% albariño
88
Citrus fruit, crisp, tasty.

Valdocea 2021 B
100% albariño
90
Colour: bright straw, greenish rim. Nose: fresh fruit, citrus fruit, wild herbs, fine lees. Palate: fresh, fruity, good acidity, fine bitter notes.

PRIVIOS
Soutelo, 3 Goián
36750 Tomiño (Pontevedra)
☎: +34 986 620 137
info@privios.com
www.privios.com

Goda 2021 B
albariño
88
Citrus fruit, balanced, herbal, thin.

QUINTA COUSELO
Barrio Couselo, 13
36770 O'Rosal (Pontevedra)
☎: +34 986 625 051
comercial@grandespagosgallegos.com
www.quintacouselo.com

Quinta de Couselo 2021 B
91
Colour: bright straw, greenish rim. Nose: citrus fruit, wild herbs, stone fruit. Palate: fresh, fruity, good acidity, fine bitter notes.

Quinta de Couselo Tradición 2015 B
90% albariño, 5% caíño, 5% loureiro
91
Age nuances. Colour: bright yellow. Nose: ripe fruit, spicy, dry nuts, toasted bread. Palate: rich, good structure, toasty, fine bitter notes.

Turonia 2021 B
91
Colour: bright straw, greenish rim. Nose: fresh fruit, citrus fruit, wild herbs, white fruit. Palate: fresh, fruity, good acidity, fine bitter notes.

RECTORAL DO UMIA

Rúa do Pan, Polígono Industrial do Salnés
36636 Ribadumia (Pontevedra)
☎: +34 986 716 360
vinos@bodegasgallegas.com
www.bodegasgallegas.com

Abellio 2021 B
100% albariño

87

Alectum 2021 B
100% albariño

89

Citrus fruit, herbal, racy, tasty.

Miudiño 2021 B
100% albariño

87

Rectoral do Umia 2021 B
100% albariño

88

Racy, citrus fruit, herbal, needs time.

Sentidiño 2021 B
100% albariño

87

Viñabade 2021 B
100% albariño

90

Colour: bright straw. Nose: ripe fruit, fragrant herbs, fine lees, white flowers. Palate: full-bodied, rich, good acidity.

SANTIAGO RUIZ

Rua do Viticultor Santiago Ruiz
36760 San Miguel de Tabagón (Pontevedra)
☎: +34 986 614 083
info@bodegasantiagoruiz.com
www.bodegasantiagoruiz.com

Rosa Ruiz 2021 B
albariño

90

Colour: bright straw. Nose: fruit expression, ripe fruit, dried herbs, white flowers. Palate: flavourful, fresh, good acidity, balanced.

Santiago Ruiz 2021 B
84% albariño, 6% godello, 4% loureiro, 3% caiño blanco, 3% treixadura

90

Colour: straw. Nose: expressive, white flowers, dried herbs, grassy, fresh fruit. Palate: flavourful, fruity, balanced, fresh.

SEÑORÍO DE RUBIÓS

Bouza do Rato, s/n - Rubiós
36449 As Neves (Pontevedra)
☎: +34 986 667 212
info@srubios.com
www.srubios.com

Manuel D'Amaro
Albariño Lías 2016 B
100% albariño

93

Colour: bright yellow. Nose: ripe fruit, fragrant herbs, fine lees. Palate: full-bodied, rich, long, good acidity.

DO RÍAS BAIXAS / D.O.P.

DO RÍAS BAIXAS / D.O.P.

Manuel d´Amaro Loureira Blanca 2018 B
loureiro
91
Colour: bright yellow. Nose: ripe fruit, white flowers, dried herbs. Palate: fruity, flavourful, balanced.

Señorío de Rubiós Albariño 2021 B
albariño
90
Colour: straw. Nose: ripe fruit, dried herbs. Palate: ripe fruit, balanced, fresh.

Señorío de Rubiós Condado Blanco BE BN
treixadura, albariño, loureiro, godello, torrontés
90
Colour: bright golden. Nose: fine lees, fragrant herbs, characterful, ripe fruit, dry nuts. Palate: powerful, flavourful, good acidity, fine bitter notes.

Señorío de Rubiós Condado do Tea Blanco 2021 B
treixadura, albariño, loureiro, godello, torrontés
90
Colour: bright straw. Nose: ripe fruit, tropical fruit, dried herbs. Palate: fruity, flavourful.

Señorío de Rubiós Condado Tinto 2018 T
brancellao, sousón, espadeiro, pedral, otras
92
Colour: bright cherry. Nose: fruit expression, floral, spicy, black fruit, dried herbs. Palate: flavourful, fruity, fresh.

TERRAS DE LANTAÑO
Baceiro, 1 - Lantaño
36658 Portas (Pontevedra)
☎: +34 615 646 442
bodega@terrasdelantano.es
www.terrasdelantano.com

Ruta 49 2021 B
albariño
89
Pleasant, aromatic, citrus fruit, floral, representative, crisp, smooth.

Terras de Lantaño 2021 B
albariño
89
Citrus fruit, balanced, fruity, crisp, floral, varietally correct.

Viña Cartin 2021 B
albariño
89
Aromatic, crisp, floral, varietally correct, smooth. Palate: good finish.

UVAS FELICES
Agullers, 7
08003 Barcelona (Barcelona)
☎: +34 902 327 777
www.vilaviniteca.es

El Jardín de Lucia 2021 B
90
Colour: bright straw, greenish rim. Nose: fresh fruit, citrus fruit, wild herbs. Palate: fruity, fine bitter notes, slightly acidic.

VAL DE MEIGAS
Travesía do Freixo, 3
36636 Ribadumia (Pontevedra)
☎: +34 675 600 102
info@valdemeigas.es
www.valdemeigas.es

Val de Meigas 2021 B
100% albariño
91
Colour: straw. Nose: ripe fruit, dried herbs, faded flowers. Palate: ripe fruit, balanced, flavourful.

VEIGA NAUM
Vilareis, 21
36967 Meaño (Pontevedra)
☎: +34 941 454 050
rrpp@bodegasriojanas.com
www.bodegasriojanas.com

Veiga Naúm 2021 B
albariño
87

VINTAE / ATLANTIS
As Laxas, 16
36430 Arbo (Pontevedra)
☎: +34 608 302 372
marketing@vintae.com
www.vintae.com

Atlantis Albariño 2021 B
albariño
90
Colour: bright straw. Nose: fruit expression, ripe fruit, floral. Palate: flavourful, fresh, good acidity, fruity aftertaste.

VIÑA ALMIRANTE

Peroxa, 5
36658 Portas (Pontevedra)
☎: +34 620 294 293
info@vinaalmirante.com
www.vinaalmirante.com

Adega Viña Almirante 2021 B
albariño

89

Citrus fruit, balanced, fruity, herbal, tasty.

Maccerato 2021 B
100% albariño

90

Colour: bright straw. Nose: floral, white fruit, fine lees, fresh. Palate: flavourful, good acidity.

Pionero 2021 B
100% albariño

89

Pleasant, crisp, smooth, simple, balanced, fruity.

Vanidade 2021 B
100% albariño

89

Pleasant, aromatic, floral, crisp, balanced, smooth.

VIÑA MORAIMA

Porráns 1, Baixo
36191 Barro (Pontevedra)
☎: +34 986 711 206
Fax: +34 986 711 206
contacto@adegamoraima.com
www.adegamoraima.com

Aba de Trasumia 2021 B
100% albariño

88

Racy, citrus fruit, herbal, tasty.

Moraima Albariño 2021 B
albariño

89

Racy, citrus fruit, herbal, tasty.

Moraima Caiño 2019 T
caiño

88

Herbal, fruity, dried flowers, spicy, pronounced acidity, balsamic herbs.

VIÑA NORA

Bruñeiras, 7
36440 As Neves (Pontevedra)
☎: +34 986 667 210
info@vinanora.com
www.vinanora.com

Nora 2021 B
100% albariño

91

Colour: bright straw, greenish rim. Nose: fresh fruit, citrus fruit, wild herbs. Palate: fresh, fruity, good acidity, fine bitter notes.

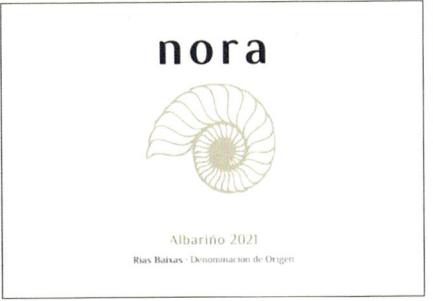

Nora da Neve 2019 B FB
100% albariño

93

Colour: straw. Nose: ripe fruit, dried herbs, faded flowers, pattiserie. Palate: powerful, ripe fruit, balanced.

Nora da Neve Encarnación Rodríguez 2019 B FB
100% albariño

94

Colour: bright yellow. Nose: dried flowers, candied fruit, fine lees, pattiserie. Palate: round, spicy, long, great length.

DO RÍAS BAIXAS / D.O.P.

DO RÍAS BAIXAS / D.O.P.

VIÑEDOS SINGULARES
Avda. de La Riera, 11 Nave 1
08960 Sant Just Desvern (Barcelona)
☎: +34 934 807 041
Fax: +34 934 807 076
info@vinedossingulares.com
www.vinedossingulares.com

Luna Creciente 2021 B
albariño
89
Standard, floral, fruity, smooth, thin.

ZÁRATE
Bouza, 23
36638 Padrenda (Pontevedra)
☎: +34 986 718 503
Fax: +34 986 718 549
info@zarate.es
www.albarino-zarate.com

Zárate Albariño 2021 B
albariño
92
Colour: bright straw. Nose: fine lees, white fruit, floral, varietal. Palate: flavourful, fruity, fresh, easy to drink.

Zárate Caiño Tinto 2020 T
caiño
93
Taut, exuberant. Colour: Cherry. Nose: balsamic herbs, sweet spices, scrubland, red berry notes. Palate: spicy, balsamic, good acidity, full-bodied.

Zárate El Balado 2021 B
93
Taut, needs time. Colour: bright straw. Nose: fruit expression, ripe fruit, floral. Palate: flavourful, fresh, good acidity, fruity aftertaste.

🏆 PODIUM

Zárate Tras da Viña 2020 B
95
With personality. Colour: bright yellow. Nose: spicy, yeasty notes, expressive, neat, smoky. Palate: full-bodied, spicy, fresh, fruity, good acidity.

DO. RIBEIRA SACRA
CONSEJO REGULADOR

Rúa do Comercio, 6-8
27400 Monforte de Lemos (Lugo)
☎: +34 982 410 968 - Fax: +34 982 411 265
@: info@ribeirasacra.org
www.ribeirasacra.org

LOCATION:

The region extends along the banks of the rivers Miño and Sil in the south of the province of Lugo and the northern region of the province of Orense; it is made up of 17 municipal districts in this region.

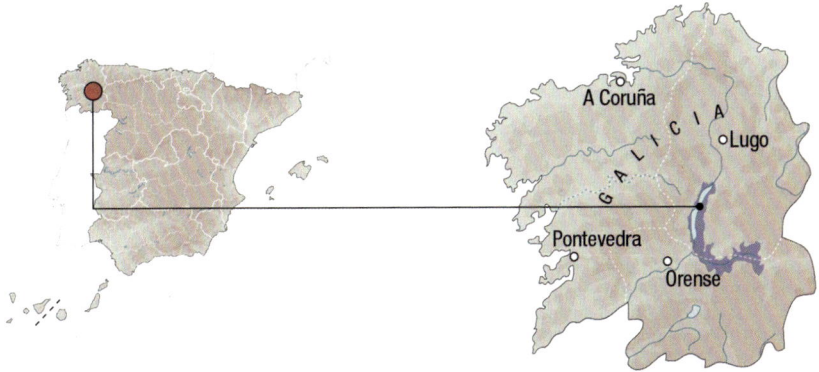

SUB-REGIONS:

Amandi, Chantada, Quiroga-Bibei, Ribeiras do Miño (in the province of Lugo) and Ribeiras do Sil.

GRAPE VARIETIES:

WHITE: Albariño, Loureira, Treixadura, Godello, Dona Blanca and Torrontés.

RED: Mencía, Brancellao Merenzao, Garnacha Tintorera, Tempranillo, Sousón, Caíño Tinto and Mouratón.

FIGURES:

Vineyard surface: 1,276 – **Wine-Growers:** 2,291 – **Wineries:** 98 – **2021 Harvest rating:** Very Good – **Production 21:** 4,329,390 litres – **Market percentages:** 94% National - 6% International.

SOIL:

In general, the soil is highly acidic, although the composition varies greatly from one area to another. The vineyards are located on steep terraces and are no higher than 400 m to 500 m above sea level.

CLIMATE:

Quite variable depending on the specific area. Less rain and slightly cooler climate and greater Continental influence in the Sil valley, and greater Atlantic character in the Miño valley. Altitude, on the other hand, also has an effect, with the vineyards closer to the rivers and with a more favourable orientation (south-southeast) being slightly warmer.

CLASSIFICATION OF YOUNG WINE HARVESTS PEÑÍNGUIDE

2017	2018	2019	2020	2021
VERY GOOD	VERY GOOD	VERY GOOD	EXCELLENT	VERY GOOD

DO RIBEIRA SACRA / D.O.P.

ABADÍA DA COVA
Avda. Buenos Aires, 12
27540 Escairón (Lugo)
☎: +34 982 452 031
abadiadacova@adegasmoure.com
www.adegasmoure.com

Abadía Da Cova 1124 Brancellao 2019 T BA
brancellao

93

Colour: bright cherry. Nose: complex, expressive, spicy, mineral, scrubland. Palate: elegant, full-bodied, long, great length.

Abadía Da Cova 1124 Caiño 2019 T
100% caiño

92

Colour: dark-red cherry. Nose: toasty, spicy, cocoa bean, dark chocolate. Palate: flavourful, toasty, fine bitter notes.

Abadía Da Cova 1124 Merenzao 2019 T BA
100% merenzao

93

Colour: cherry, purple rim. Nose: red berry notes, spicy, dried flowers, fine reductive notes, earthy notes. Palate: flavourful, fruity, good acidity, elegant.

Abadía Da Cova 2020 T
90% mencía, 10% otras

91

Colour: cherry, purple rim. Nose: red berry notes, floral, spicy, faded flowers. Palate: flavourful, fruity, good acidity.

Abadía Da Cova 2021 B
60% albariño, 30% godello, blanco lexitimo

90

Colour: bright straw. Nose: fresh fruit, floral, wild herbs, citrus fruit. Palate: fruity, correct, easy to drink.

Abadía Da Cova Mencía 2021 RD
100% mencía

90

Colour: salmon, bright. Nose: ripe fruit, red berry notes, rose petals, fragrant herbs. Palate: fresh, fruity, flavourful, balanced.

🏆 **PODIUM**

Abadía da Cova Penafión 2018 T BA
mencía, garnacha tintorera, otras

95

Colour: cherry, garnet rim. Nose: expressive, balanced, sweet spices, black pepper, red berry notes, black fruit, wild herbs, grassy. Palate: full-bodied, fruity, flavourful, round.

Abadía Da Cova Xuno 2019 T
85% mencía, 15% otras

92

Colour: Cherry. Nose: balsamic herbs, sweet spices, scrubland, ripe fruit. Palate: spicy, balsamic, good acidity.

ADEGA BARBADELO
Lugar de Doade, 55
27454 Sober (Lugo)
☎: +34 986 499 750
info@adegasterrae.com
www.adegasterrae.com

Barbadelo 2021 T
100% mencía

90

Colour: Cherry. Nose: balsamic herbs, sweet spices, scrubland, raspberry. Palate: spicy, balsamic, good acidity, fresh.

ADEGA MALCAVADA
Rosende, 67
27466 Sober (Lugo)
☎: +34 699 073 420
info@malcavada.com
www.malcavada.com

Malcavada 2021 RD

85

Malcavada Selección 2021 T
mencía

89

Floral, fruity, mineral, tasty, crisp, coarse tannins.

ADEGA PONTE DA BOGA
O Couto-Sampaio s/n
32764 Castro Caldelas (Ourense/Orense)
☎: +34 988 203 306
info@pontedaboga.es
www.pontedaboga.es

Bancales Olvidados 2019 T C
mencía

92

Colour: cherry, garnet rim. Nose: spicy, black fruit, neat, dried herbs. Palate: flavourful, long, good acidity, great length.

Porto de Lobos 2017 T
brancellao

92

Colour: light cherry, dark-red cherry. Nose: scrubland, wild herbs, ripe fruit. Palate: flavourful, fine bitter notes, balanced, reductive nuances.

Ponte da Boga Albariño 'A' 2021 B
albariño

91

Colour: straw. Nose: fresh, citrus fruit, floral, medium intensity. Palate: good acidity, easy to drink, fine bitter notes.

Capricho de Merenzao 2017 T
merenzao
92
Colour: cherry, garnet rim. Nose: ebb tide, red berry notes, ripe fruit, scrubland. Palate: flavourful, good acidity, taut.

Ponte da Boga Godello ´G´ 2021 B
godello
92
Colour: bright straw, greenish rim. Nose: fresh fruit, citrus fruit, wild herbs. Palate: fresh, fruity, good acidity, fine bitter notes.

Ponte Da Boga Pizarras y Esquistos 2020 T
mencía
90
Colour: bright cherry. Nose: ripe fruit, woody, spicy, dried herbs. Palate: fruity, spicy, flavourful.

ADEGA SAIÑAS
Espasantes
27450 Pantón (Lugo)
☎: +34 670 243 735
adega.sainas@gmail.com
www.sainas.com

Castro das Saíñas 2020 T
100% mencía
91
Colour: bright cherry. Nose: balsamic herbs, sweet spices, scrubland. Palate: spicy, balsamic, good acidity.

Saíñas 2020 T
95% mencía, 5% garnacha
89
Pleasant, fruity, tasty, simple, standard, dried flowers.

Saíñas - Areas 2020 T
100% mencía
90
Colour: Cherry. Nose: scrubland, red berry notes, floral. Palate: balsamic, good acidity.

Javier Fernández 2020 T
100% mencía
92
Colour: Cherry. Nose: balsamic herbs, sweet spices, scrubland, ripe fruit. Palate: spicy, balsamic, good acidity, flavourful.

Saíñas – O Boliño 2020 T BA
mencía, garnacha, mouratón, sousón, brancellao, merenzao
91
Colour: Cherry. Nose: balsamic herbs, sweet spices, scrubland. Palate: spicy, balsamic, good acidity.

ADEGA TOLO DO XISTO
Lugar Rubín, Rozavales, 1
27413 Monforte de Lemos (Lugo)
☎: +34 619 776 948
info@tolodoxisto.com
www.tolodoxisto.com

Tolo do Xisto 2018 T
100% mencía
89
Balanced, spicy, complex, jammy, tasty.

TX Os Conventos 2019 T R
100% mencía
93
Elegant. Colour: bright cherry. Nose: creamy oak, varietal, fresh, red berry notes, ripe fruit. Palate: good acidity, spicy, fine tannins, flavourful, long, great length.

ADEGAS GUIMARO
Sanmil, 43 Brosmos
27425 Sober (Lugo)
☎: +34 610 524 484
adegasguimaro@gmail.com
www.guimaro.es

🏆 PODIUM

A Ponte 2020 T
96
Colour: deep cherry. Nose: dried herbs, creamy oak, red berry notes, black fruit, earthy notes. Palate: powerful, ripe fruit, spicy, round tannins.

🏆 PODIUM

Finca Capeliños 2020 T
95
Colour: bright cherry. Nose: raspberry, ripe fruit, rose petals, earthy notes. Palate: good acidity, spicy, fine tannins, elegant, long.

DO RIBEIRA SACRA / D.O.P.

DO RIBEIRA SACRA / D.O.P.

🏆 PODIUM

Finca Meixeman 2020 T
96
Aromatic, defined aromas. Colour: cherry, purple rim. Nose: fruit expression, red berry notes, floral, spicy, balsamic herbs. Palate: flavourful, fruity, good acidity, long, elegant.

Finca Pombeiras 2020 T
94
Colour: deep cherry. Nose: ripe fruit, black fruit, dried flowers, spicy, scrubland, mineral. Palate: ripe fruit, spicy, round tannins.

Guimaro 2021 B
90
Needs time. Colour: bright straw, greenish rim. Nose: fresh fruit, citrus fruit, wild herbs, fine lees. Palate: fresh, fruity, good acidity, fine bitter notes.

Guimaro Mencía 91 T
91
Colour: cherry, purple rim. Nose: fruit expression, red berry notes, floral, dried flowers, earthy notes. Palate: fruity, flavourful, balanced.

ALGUEIRA
Doade s/n
27424 Sober (Lugo)
☎: +34 982 410 299
Fax: +34 982 410 299
info@algueira.com
www.algueira.com

Algueira Carravel 2017 T C
mencía
92
Colour: cherry, garnet rim. Nose: ripe fruit, dried herbs, wild herbs. Palate: flavourful, balsamic, spicy, great length.

Algueira Fincas 2017 T C
caiño, sousón
92
Racy, herbal. Colour: cherry, garnet rim. Nose: red berry notes, wild herbs, expressive, balanced, fresh. Palate: easy to drink, long, taut.

Algueira Serradelo 2018 T C
brancellao
93
Racy, defined aromas. Colour: light cherry. Nose: medium intensity, neat, elegant, complex. Palate: taut, fine bitter notes, elegant, racy.

Algueira Finca Cortezada 2021 B
godello, albarello, treixadura
92
Defined aromas, crisp. Colour: bright straw. Nose: fruit expression, floral, neat, balanced. Palate: fresh, fruity, easy to drink.

Algueira Risco 2018 T RB
94
Defined aromas. Colour: light cherry. Nose: wild herbs, dried flowers, wild herbs, chamomile. Palate: complex, elegant, easy to drink, long.

Algueira Escalada 2019 B FB
godello
93
Colour: bright straw. Nose: expressive, ripe fruit, floral, fine lees, spicy. Palate: full-bodied, spicy, long, rich.

ALMA DAS DONAS
Ribas de Sil, 1
27470 Pantón (Lugo)
☎: +34 988 200 045
info@almadasdonas.com
www.almadasdonas.com

Alma3once 2019 T
mencía
90
Pruney, spicy, jammy, tasty. Nose: balanced, spicy, fine reductive notes.

AlmaLarga 2019 B BA
godello
92
Colour: bright yellow. Nose: powerful, creamy oak, ripe fruit, spicy, pattiserie. Palate: rich, good structure, long, fine bitter notes, spicy.

AlmaLarga 2020 B
godello
91
Representative. Colour: bright straw. Nose: dried flowers, faded flowers, white fruit, ripe fruit, fine lees. Palate: fresh, good acidity, rich.

AlmaLola 2021 RD
mencía
85

DO RIBEIRA SACRA / D.O.P.

AlmaMadre 2017 T
mencía
90
Colour: deep cherry. Nose: ripe fruit, dried herbs, creamy oak. Palate: powerful, ripe fruit, spicy, round tannins.

AlmaNova 2020 T
mencía
91
Representative. Colour: cherry, garnet rim. Nose: balsamic herbs, ripe fruit, scrubland, neat. Palate: soft tannins, great length, easy to drink.

ATLANTIC GALICIAN WINERIES
A Pobra do Brollon
27334 Vilacha (Lugo)
info@atlanticgalicianwineries.com
www.atlanticgalicianwineries.com

Massimo Mencía & Sousón 2021 T
mencía, sousón
90
Representative, wild. Colour: cherry, garnet rim. Nose: scrubland, wild herbs. Palate: fruity, fine bitter notes, correct, easy to drink.

Massimo Selección Barrica Francesa 2019 T RB
mencía
91
Representative, balsamic herbs. Colour: cherry, garnet rim. Nose: fresh fruit, wild herbs, balsamic herbs. Palate: spicy, fine tannins, good acidity.

Massimo Selección Godello 2020 B
godello
90
Colour: bright yellow. Nose: powerful, creamy oak, ripe fruit, spicy. Palate: rich, good structure, toasty, fine bitter notes.

AUTÉNTICOS VIÑADORES, VINOS DE TERROIR
24540 Cacabelos (León)
☎: +34 658 617 390
info@autenticosvinadores.com
www.autenticosvinadores.com

Chuzos de Punta 2020 T
95% mencía, 5% merenzao
90
Defined aromas, representative. Colour: cherry, garnet rim. Nose: scrubland, red berry notes, wild herbs. Palate: fruity, easy to drink, good acidity.

BELESAR
Rua Monforte 9 local 210 Bajo
27003 Lugo (Lugo)
cecilia.flores@grupofosterlorca.com

Camino Empedrado 2020 T
mencía, garnacha
91
Colour: bright cherry. Nose: balsamic herbs, sweet spices, scrubland. Palate: spicy, balsamic, good acidity.

Camino Empedrado Blend de Fincas 2020 T RB
mencía, garnacha, tempranillo
92
Colour: Cherry. Nose: scrubland, spicy, red berry notes, black fruit, waxy notes. Palate: spicy, good acidity, fine bitter notes.

BODEGA A CARQUEIXA
Lugar A Carqueixa, 16
27460 Sober (Lugo)
☎: +34 610 765 472
info@bodegaacarqueixa.com
www.casasgallego.com

Coronín 2020 T
100% mencía
89
Herbal, jammy, wild, tasty, great length, fruity, bitter, reduced.

Garoubas 2020 T RB
mencía
89
Jammy, toasty, spicy, smoky, tasty.

BODEGA CRUCEIRO
Vilachá de Doade, 14
27424 Sober (Lugo)
☎: +34 609 183 352
info@adegacruceiro.es
www.adegacruceiro.es

Cruceiro 2021 T
mencía, merenzao, caiño, garnacha
89
Balanced, spicy, dried flowers, wild, tasty.

Cruceiro Rexio 2018 T
mencía, brancellao
91
Representative. Colour: Cherry. Nose: scrubland, wild herbs, red berry notes, ripe fruit. Palate: spicy, balsamic, good acidity.

Cruceiro Rexio 2021 B
godello
85

BODEGA GULLUFRE
Lobios
27423 Sober (Lugo)
☎: +34 639 843 457
Fax: +34 982 152 575
info@bodegagullufre.com
www.bodegagullufre.com

Don Xoán 2020 T
mencía
91
Colour: deep cherry. Nose: ripe fruit, wild herbs, spicy. Palate: ripe fruit, spicy, round tannins.

Romeo Domain 2020 T
mencía
88
Fruity, crisp, dried flowers, spicy.

BODEGA LAR DE RICOBAO
Parque Empresarial de Quiroga Parcela A-13
27320 Quiroga (Lugo)
☎: +34 662 677 215
info@lardericobao.com
www.lardericobao.com

Lar de Ricobao Godello - Branco Lexítimo - Treixadura 2020 B
godello, branco lexitimo, treixadura
89
Dried flowers, wild, smooth, yeasty notes, austere. Palate: easy to drink, good finish.

Lar de Ricobao Ouro do Val 2019 B BA
89
Smoky, spicy, fruity, jammy. Nose: dry nuts, toasted bread.

Lar de Ricobao Selección do Val 10 Lunas 2014 T
mencía
89
Age nuances, toasty, slight reduction, spicy, dried herbs, jammy.

Lar de Ricobao Selección do Val 2021 T
mencía
90
Pleasant, simple. Colour: cherry, purple rim. Nose: fruit expression, wild herbs, neat. Palate: fruity, easy to drink.

BODEGA MARCELINO I AMANDI
A Carqueixa, s/n
27460 Sober (Lugo)
☎: +34 647 164 040
contacto@dominiomarcelino.com
www.dominiomarcelino.com

Marcelino I 2020 T
86

Marcelino I 2021 T
87

BODEGA REGINA VIARUM
Doade, s/n
27424 Sober (Lugo)
☎: +34 986 442 686
tanialage@hgabodegas.com
www.reginaviarum.es

Origen 2019 T C
sousón, merenzao
90
Slight oxidation. Colour: cherry, purple rim. Nose: spicy, red berry notes, black fruit, faded flowers. Palate: flavourful, good acidity, spicy.

Regina Viarum Godello 2021 B
91
Colour: bright straw. Nose: neat, wild herbs, hint of anise, dried flowers. Palate: easy to drink, balanced, fine bitter notes.

Regina Viarum Mencía 2021 T
100% mencía
91
Wild. Colour: bright cherry, purple rim. Nose: fine reductive notes, red berry notes, ripe fruit, wild herbs, dried herbs, spicy. Palate: flavourful, varietal.

Regina Viarum Expresión 2018 T BA
mencía
90
Colour: deep cherry. Nose: dried herbs, creamy oak, spicy, black fruit. Palate: ripe fruit, spicy, round tannins.

Regina Viarum Rosae 2021 RD
mencía
88
Fruity, jammy, herbal, tasty, crisp.

BODEGA SOLLÍO
Lugar Os Vazquez, s/n
32765 A Teixeira (Ourense/Orense)
☎: +34 988 207 418
adegasollio@yahoo.es

Sollío 2020 T BA
mencía, brancellao, sousón
88
Full-bodied, spicy, jammy, roasted coffee, powerful.

Sollio Godello 2021 B
godello
87

Sollío Mencía 2020 T
mencía, brancellao, sousón
87

BODEGAS ALBAMAR
O Adro, 11 - Castrelo
36639 Cambados (Pontevedra)
☎: +34 660 292 750
xurxoalbamar@gmail.com

Albamar Nai 2019 T
93
Colour: cherry, purple rim. Nose: red berry notes, floral, spicy, earthy notes. Palate: flavourful, fruity, good acidity, mineral.

BODEGAS CASTRO CANDAZ
Bendollo s/n
27329 Quiroga (Lugo)
☎: +34 696 621 531
lidia@raulperez.com

Castro Candaz A Boca do Demo 2019 T
92
Colour: Cherry. Nose: balsamic herbs, sweet spices, scrubland, black fruit, ripe fruit, meaty notes. Palate: spicy, balsamic, good acidity, flavourful.

Castro Candaz La Vertical 2019 B
93
Colour: straw. Nose: ripe fruit, dried herbs, faded flowers, spicy. Palate: powerful, ripe fruit, balanced.

BODEGAS RECTORAL DE AMANDI
27423 Sober (Lugo)
☎: +34 988 384 200
vinos@bodegasgallegas.com
www.bodegasgallegas.com

Matilda Nieves Mencía 2021 T
85% mencía, 15% garnacha, sousón
89
Fruity, jammy, dried herbs, spicy, tasty.

Pasal de Esile Godello 2020 B
100% godello
90
Colour: bright straw. Nose: ripe fruit, fragrant herbs, fine lees, waxy notes, dried flowers. Palate: full-bodied, rich, good acidity.

Rectoral de Amandi 2021 T
mencía
90
Colour: cherry, purple rim. Nose: black fruit, wild herbs, spicy. Palate: balanced, fresh, smoky aftertaste.

Rectoral de Amandi Barrica Manolo Arnoya 2017 T
mencía
90
Reduced. Colour: Cherry. Nose: balsamic herbs, scrubland, ripe fruit. Palate: spicy, good acidity.

CARLOS CANEIRO NÚÑEZ
Dariz, s/n
32747 Parada de Sil (Ourense/Orense)
☎: +34 610 789 232
bodegacaneiro@gmail.com
bodega-carlos-caneiro.negocio.site

Viña Dariz 2019 T
mencía
91
Colour: cherry, purple rim. Nose: floral, spicy, damp earth, faded flowers, black fruit, fruit preserve. Palate: flavourful, fruity, good acidity.

Viña Dariz 2020 T
mencía
91
Defined aromas, fruity, crisp. Nose: red berry notes, wild herbs, balsamic herbs. Palate: flavourful, full of life.

DO RIBEIRA SACRA / D.O.P.

DO RIBEIRA SACRA / D.O.P.

CASA DE OUTEIRO
Parque Empresarial, Parcela B2
27320 Quiroga (Lugo)
☎: +34 637 895 831
hola@casadeouteiro.com
www.casadeouteiro.com

Casa de Outeiro Mencía Gama Escudo 2020 T
mencía
91
Rustic, tasty. Colour: dark-red cherry. Nose: black fruit, ripe fruit, spicy, characterful, waxy notes. Palate: flavourful, good structure.

CASA MOREIRAS
San Martín de Siós, s/n
27430 Pantón (Lugo)
☎: +34 680 545 830
Fax: +34 982 456 174
bodega@casamoreiras.com
www.casamoreiras.com

Campaza 2021 T
mencía
88
Pleasant, fruity, crisp, tasty.

Casa Moreiras 2021 B
88
Pleasant, dried herbs, fruity, aromatic, kind finish, balanced.

Casa Moreiras Mencía 2021 T
89
Nose: fruit expression, red berry notes, floral. Palate: fruity, flavourful.

Casa Moreiras Selección 2021 T
90
Colour: deep cherry. Nose: ripe fruit, dried herbs, black fruit. Palate: powerful, ripe fruit, spicy.

CASTROFIZ
Estrada LU 617 Km. 25,3 San Fiz de Asma
27516 Chantada (Lugo)
☎: +34 664 790 668
adegacastrofiz@gmail.com
www.castrofiz.es

Castrofiz 2019 T C
92
Colour: Cherry. Nose: balsamic herbs, sweet spices, scrubland, black fruit, ripe fruit, creamy oak. Palate: spicy, good acidity, concentrated.

Castrofiz 2020 T
mencía, 8% garnacha, 2% tempranillo
90
Colour: Cherry. Nose: balsamic herbs, scrubland, wild herbs. Palate: spicy, balsamic, good acidity.

Castrofiz Amenza 2021 B
55% godello, 30% treixadura, 15% branco lexitimo
87

Castrofiz Tradición 2020 T RB
82% mencía, 15% garnacha, 3% tempranillo
91
Defined aromas, wild. Colour: Cherry. Nose: scrubland, wild herbs, balanced, expressive. Palate: spicy, good acidity, fruity, taut.

CONDADO DE SEQUEIRAS
27500 Chantada (Lugo)
☎: +34 630 910 553
Fax: +34 944 128 502
condadodesequeiras@grupopeago.com
www.condadodesequeiras.com

Condado de Sequeiras 2017 T C
mencía
89
Toasty, jammy, classic, spicy.

Condado de Sequeiras Godello 2021 B
godello
87

Condado de Sequeiras Mencía 2020 T
mencía
89
Slight reduction, herbal, jammy, spicy, tasty.

CORZOÁS
Ribeiras de Miño, 51
27439 Pantón (Lugo)
☎: +34 628 228 276
info@adegacorzoas.com

Corzoás Godello 2021 B
godello
90
Colour: bright straw. Nose: ripe fruit, floral, spicy. Palate: flavourful, fresh, good acidity, fruity aftertaste, easy to drink.

Corzoás Mencía 2019 T BA
mencía
90
Colour: Cherry. Nose: balsamic herbs, scrubland, dried herbs. Palate: spicy, good acidity, ripe fruit, balanced.

DO RIBEIRA SACRA / D.O.P.

Corzoás Mencía 2020 T FB
mencía
90
Colour: cherry, garnet rim. Nose: black fruit, ripe fruit, scrubland, spicy, waxy notes. Palate: flavourful, dry, varietal.

Corzoás Mencía 2021 T
mencía
89
Pleasant, varietally correct, fruity, herbal, jammy, smooth.

DOMINIO DO BIBEI
Langullo, s/n
32781 Manzaneda (Ourense/Orense)
☎: +34 670 704 028
info@dominiodobibei.com
www.dominiodobibei.com

 PODIUM

Dominio do Bibei 2019 T
98
Defined aromas, wild, elegant. Colour: light cherry. Nose: wild herbs, scrubland, floral, fresh fruit, red berry notes. Palate: taut, balanced, full of life.

 PODIUM

Lacima 2019 T
95
Colour: Cherry. Nose: expressive, spicy, mineral. Palate: full-bodied, long, great length, ripe fruit, round, complex, good acidity.

Lalama 2019 T
94
Colour: Cherry. Nose: balsamic herbs, sweet spices, scrubland, red berry notes, wild herbs. Palate: spicy, balsamic, good acidity.

 PODIUM

Lapena 2019 B
96
Colour: yellow. Nose: white fruit, expressive, balanced, fine lees, mineral, spicy. Palate: elegant, balanced, good acidity, fine bitter notes, complex.

Lapola 2020 B
93
Colour: bright straw. Nose: fruit expression, ripe fruit, floral, stone fruit. Palate: flavourful, fresh, good acidity, fruity aftertaste.

DON BERNARDINO
Stª Cruz de Brosmos, s/n
27425 Sober (Lugo)
☎: +34 687 825 126
info@donbernardino.com
www.donbernardino.com

Don Bernardino 4ªGeneración 2018 T BA
90
Colour: dark-red cherry. Nose: toasty, spicy, cocoa bean. Palate: flavourful, toasty, fine bitter notes.

Don Bernardino Amandi 2020 T
mencía
90
Pleasant, fruity, jammy, spicy.

Don Bernardino Amandi 2021 T
mencía
89
Standard, jammy, balanced, wild, tasty, fruity.

Don Bernardino Finca Mezquita 2016 T C
mencía
88
Full-bodied, herbal, jammy, powerful, roasted coffee.

Don Bernardino Ibio 2017 T FB
mencía
90
Colour: dark-red cherry. Nose: toasty, spicy, cocoa bean, dark chocolate. Palate: flavourful, toasty, fine bitter notes.

Don Bernardino Ibio 2019 T FB
mencía
91
Colour: deep cherry. Nose: ripe fruit, dried herbs, balsamic herbs. Palate: powerful, ripe fruit, spicy, round tannins.

Don Bernardino Melanio 2017 T C
mencía
90
Colour: deep cherry. Nose: dried herbs, creamy oak, black fruit, roasted coffee, woody. Palate: powerful, ripe fruit, spicy, harsh oak tannins.

Don Bernardino Vacamulo 2017 T C
mencía
90
Colour: very deep cherry. Nose: roasted coffee, aromatic coffee, powerful, black fruit. Palate: smoky aftertaste, great length, round tannins.

DO RIBEIRA SACRA / D.O.P.

E.D.V. BODEGA
27320 Quiroga (Lugo)
☎: +34 650 983 654
info@bodegaedv.com

Don Cosme Godello 2020 B
godello
88
Pleasant, standard, fruity, jammy, tasty.

Don Cosme Mencía 2021 T
mencía
87

ENVINATE
Terrero 72
02630 La Roda (Albacete)
☎: +34 682 207 160
asesoria@envinate.es

🏆 PODIUM

Lousas Camiño Novo 2020 T
99
Elegant. Colour: cherry, purple rim. Nose: dried herbs, faded flowers, mineral, characterful, wild herbs, red berry notes, black fruit. Palate: full of life, flavourful, fruity, fresh, good structure, silky tannins, elegant.

FENTO WINES
Sisangándara, 22
36636 Ribadumia (Pontevedra)
☎: +34 986 099 486
Fax: +34 986 718 549
info@eulogiopomares.com
www.eulogiopomares.com

O Estranxeiro 2020 T
mencía
92
Rustic, representative. Colour: dark-red cherry. Nose: ripe fruit, expressive, wild herbs. Palate: fresh, fruity, easy to drink, balanced.

O Estranxeiro 2021 B
godello
91
Colour: bright straw. Nose: floral, white fruit, ripe fruit. Palate: flavourful, good acidity, fruity aftertaste.

FINCA CUARTA
Outeiro, Centeás, s/n
27460 Sober (Lugo)
☎: +34 982 178 852
info@priordepanton.com
www.fincacuarta.es

Finca Cuarta A Costa
por Rubén Moure 2019 T
100% mencía
92
Colour: cherry, garnet rim. Nose: scrubland, wild herbs, ripe fruit, expressive, fresh. Palate: fruity, easy to drink, good acidity.

Finca Cuarta Consentida
por Rubén Moure 2019 T
100% mencía
90
Colour: cherry, garnet rim. Nose: black fruit, fruit preserve, dried herbs, tobacco. Palate: long, correct.

Finca Cuarta Malcriado
por Rubén Moure 2018 T C
100% mencía
90
Colour: cherry, garnet rim. Nose: dried herbs, creamy oak, black fruit. Palate: ripe fruit, spicy.

Finca Cuarta Mencía
por Rubén Moure 2019 T BA
mencía
90
Colour: dark-red cherry. Nose: dried herbs, black fruit, neat, tobacco. Palate: spicy, correct, flavourful.

Finca Cuarta Mencía
por Rubén Moure 2021 T
mencía
91
Colour: deep cherry. Nose: ripe fruit, dried herbs, creamy oak. Palate: powerful, ripe fruit, spicy, round tannins.

FINCA MILLARA
Lugar de A Míllara, s/n
27430 Pantón (Lugo)
☎: +34 673 552 489
bodega@fincamillara.com
www.fincamillara.com

Finca Millara 2018 T C
mencía
92
With personality, wild. Nose: ripe fruit, dried flowers, spicy. Palate: long, easy to drink, balanced, taut.

DO RIBEIRA SACRA / D.O.P

Finca Millara Lagariza 2021 T
mencía
89
Pleasant, smooth, representative, fruity, herbal.

Finca Millara Selección Especial 2018 T
mencía
92
Colour: cherry, garnet rim. Nose: ripe fruit, balsamic herbs, spicy. Palate: flavourful, long, easy to drink, taut.

Ribera de los Naranjos 2019 T
mencía, sousón, garnacha tintorera, tempranillo
92
Balanced, herbal. Colour: deep cherry. Nose: wild herbs, citrus fruit, characterful, expressive. Palate: long, easy to drink.

JOSÉ MANUEL RODRÍGUEZ GONZÁLEZ
Vilachá - Doade
27424 Sober (Lugo)
☎: +34 982 460 613

Décima 2021 T
87

MIGUEL PAVÓN REINOSO
Lg. Valdomonde Parcela 53 - Pol. 54
28596 Taboada (Lugo)
☎: +34 638 800 747

Instinto Romano 2021 T
mencía, garnacha, merenzao, brancellao
87

OS CIPRESES
San Fiz - Tarrio
27500 Chantada (Lugo)
☎: +34 982 440 809
info@oscipreses.com
www.oscipreses.com

Cypressus 2019 T C
mencía, arauxa
91
Colour: cherry, garnet rim. Nose: balsamic herbs, ripe fruit, scrubland, characterful. Palate: flavourful, balsamic, spicy, soft tannins.

Os Cipreses Mencía 2020 T
mencía
90
Colour: deep cherry. Nose: ripe fruit, dried herbs, spicy. Palate: powerful, ripe fruit, spicy.

OSCAR PÉREZ RODRÍGUEZ
Naz de Abaixo, 55
27466 Sober (Lugo)
☎: +34 982 460 110
formarigo@yahoo.es

Naz 2019 T
mencía
89
Fruity, herbal, jammy, spicy, standard, tasty, balsamic herbs.

OTERO & PÉREZ BODEGUEROS
Mosqueiro, 22 Sisán
36636 Ribadumia (Pontevedra)
☎: +34 986 561 529
dismavi@dismavi.com
www.dismavi.com

Úber 760 2020 T
mencía
89
Fruity, wild, herbal, crisp, smooth.

RONSEL DO SIL
Sacardebois
32747 Parada do Sil (Ourense/Orense)
☎: +34 988 984 923
info@ronseldosil.com
www.ronseldosil.com

Al Pie del Cañón 2020 T
50% caiño, 25% caiño longo, 25% caiño bravo
92
Colour: bright cherry. Nose: balanced, faded flowers, red berry notes, black fruit, balsamic herbs. Palate: flavourful, fruity, good acidity.

Alpendre Merenzao 2020 T
merenzao
94
Colour: bright cherry. Nose: fresh fruit, expressive, faded flowers, earthy notes. Palate: balanced, good acidity, flavourful, balsamic, elegant.

Ourive Dona Branca 2020 B FB
dona blanca
92
Colour: straw. Nose: white fruit, ripe fruit, faded flowers, characterful. Palate: long, ripe fruit, flavourful.

Ourive Godello 2020 B
100% godello
90
Colour: bright straw. Nose: fine lees, floral, white fruit, balanced. Palate: rich, long, good acidity, easy to drink.

DO RIBEIRA SACRA / D.O.P.

Portico da Gloria Brancellao 2020 T
brancellao

93

Colour: cherry, purple rim. Nose: spicy, red berry notes, black fruit, balsamic herbs. Palate: flavourful, fruity, good acidity.

Vel'Uveyra Godello 2020 B
godello

89

Pleasant, spicy, tasty, simple. Nose: dried flowers, faded flowers.

Vel'Uveyra Mencía 2019 T
mencía

92

Colour: bright cherry. Nose: dried flowers, red berry notes, spicy, balsamic herbs. Palate: good acidity, spicy, fine tannins.

VÍA ROMANA ADEGAS E VIÑEDOS

A Ermida Belesar, s/n
27500 Chantada (Lugo)
☎: +34 982 469 069
marketing@viaromana.es
www.viaromana.es

Vía Romana do Camiño Brancellao 2019 T FB
100% brancellao

84

Vía Romana do Camiño Godello 2020 B
100% godello

90

Aromatic, balanced. Colour: bright straw. Nose: fine lees, faded flowers, dried flowers, white fruit. Palate: fine bitter notes, easy to drink, good finish.

VR 2018 T RB

90

Colour: dark-red cherry, light cherry. Nose: ripe fruit, scrubland, fine reductive notes. Palate: balsamic, spicy, soft tannins, easy to drink.

DO. RIBEIRO
CONSEJO REGULADOR

Rúa Redondela, 3 - 2º Andar
32400 Ribadavia (Ourense)
☎: +34 988 477 200 - Fax: +34 988 477 201
@: info@ribeiro.wine
www.ribeiro.wine

LOCATION:

In the west of the province of Ourense. The region comprises 13 municipal districts marked by the Miño and its tributaries.

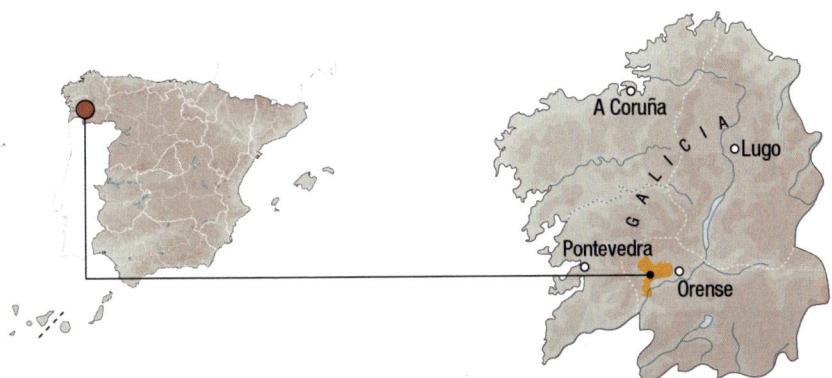

GRAPE VARIETIES:

WHITE: Treixadura, Torrontés, Palomino, Godello, Loureira, Albariño, Lado, Caíño Blanco, Albilla, Macabeo and Jerez.

RED: Caíño, Sousón, Ferrón, Mencía, Brancellao, Tempranillo and Garnacha Tintorera.

FIGURES:

Vineyard surface: 1,281 – **Wine-Growers:** 1,667 – **Wineries:** 98 – **2021 Harvest rating:** N/A – **Production 21:** 6,909,614 litres – **Market percentages:** 90% National - 10% International.

SOIL:

Predominantly granite, deep and rich in organic matter, although in some areas clayey soils predominate. The vineyards are on the slopes of the mountains (where higher quality wines are produced) and on the plains.

CLIMATE:

Atlantic, with low temperatures in winter, a certain risk of spring frosts, and high temperatures in the summer months. The average annual rainfall varies between 800 mm and 1,000 mm.

CLASSIFICATION OF YOUNG WINE HARVESTS PEÑÍNGUIDE

2017	2018	2019	2020	2021
VERY GOOD	VERY GOOD	VERY GOOD	VERY GOOD	VERY GOOD

DO RIBEIRO / D.O.P.

ADEGA DESEU
Ctra. Ribadavia – Carballiño, Km 6
32001 Leiro (Ourense/Orense)
☎: +34 986 499 750
info@adegasterrae.com
www.adegasterrae.com

Deseu 2021 B
100% treixadura
87

ADEGA DO VELEIRO
Laias, 1
32459 Cenlle (Ourense/Orense)
☎: +34 686 386 046
adegadoveleiro@gmail.com

Ana Veleiro 2020 B
treixadura, loureiro
88
Standard, crisp, floral, citrus fruit, smooth.

Cobiza 2020 B
treixadura, caiño blanco
91
Colour: bright straw. Nose: fresh fruit, citrus fruit, floral, fine lees. Palate: fresh, fruity, good acidity, fine bitter notes, flavourful.

ADEGA FRANCISCO FERNÁNDEZ SOUSA
Prado, 14
32430 Castrelo de Miño (Ourense/Orense)
☎: +34 678 530 898
info@terraminei.com
www.terraminei.com

Terra Minei 2021 B
100% treixadura
89
Dried flowers, fruity, jammy, tasty.

ADEGA JAVIER ESTÉVEZ ABELEDO
Porriñeira, 14
32431 Beade (Ourense/Orense)
☎: +34 616 121 375
estevezabeledo@gmail.com

Mal Raio te Parta 2021 B
albariño, godello, lado
91
Colour: bright straw. Nose: ripe fruit, fine lees, smoky, wild herbs. Palate: full-bodied, rich, long, good acidity.

Tinalla 2021 T
sousón, caiño longo, ferrón, brancellao
90
Colour: deep cherry. Nose: dried herbs, creamy oak, black fruit, mineral, spicy. Palate: powerful, ripe fruit, spicy, round tannins.

ADEGA MANUEL FORMIGO DE LA FUENTE
Ctra. Ribadavia Carballiño, km. 4,27
32431 Beade (Ourense/Orense)
☎: +34 627 569 885
info@fincateira.com
www.fincateira.com

Cholo 2020 B
loureiro
91
Colour: straw. Nose: expressive, citrus fruit, floral, fresh, white fruit. Palate: flavourful, balanced, spicy, easy to drink.

Cholo Agarda 2015 B
85% loureiro, treixadura, albariño
94

Oxidativ, with personality. Nose: fine lees, faded flowers, petrol notes, complex. Palate: full-bodied, rich, long, good acidity.

Finca Teira 2020 B
treixadura, godello, torrontés, caiño blanco
91
With personality, representative. Nose: dried flowers, burnt matches, medium intensity. Palate: flavourful, balanced, full-bodied.

Finca Teira 2020 T
caiño longo, sousón, brancellao
90
Representative, fruity, simple. Nose: wild herbs, scrubland, red berry notes, fresh fruit. Palate: balanced, easy to drink.

Finca Teira 2021 T
90
Dried flowers, fruity, simple, wild, standard, smooth. Colour: bright cherry.

Formigo 2020 B
treixadura, palomino, godello, torrontés, albariño, loureiro
88
Aromatic, floral, fruity, simple, tasty, standard.

Formigo 2021 B
treixadura, palomino, godello, torrontés, caiño blanco, loureiro
89
Dried flowers, fruity, dried herbs, tasty.

Teira X 2020 B
treixadura, alvilla, albariño, loureiro
93
Dried flowers, yeasty notes. Colour: bright straw. Nose: expressive, ripe fruit, fine lees, mineral. Palate: full-bodied, complex, long, elegant.

Tino Alvilla do Avia 2019 B
albillo

93

Colour: bright straw. Nose: floral, fine lees, mineral, fresh, medium intensity. Palate: complex, spicy, long, flavourful.

Tino Alvilla do Avia 2021 B

90

Defined aromas, balanced, simple. Colour: bright straw. Nose: white fruit, tropical fruit. Palate: correct, easy to drink.

ADEGA O COTARELO
O Barral, 22 Razamonde
32459 Cenlle (Ourense/Orense)
☎: +34 699 198 987
benirios22@gmail.com

O Cotarelo 2020 T BA
50% sousón, 50% brancellao

90

Colour: cherry, purple rim. Nose: fruit expression, red berry notes, floral, spicy, scrubland. Palate: flavourful, fruity, smoky aftertaste.

O Cotarelo 2021 B
60% treixadura, 20% godello, 10% loureiro, 10% albariño

90

Colour: bright yellow. Nose: ripe fruit, floral, dried herbs. Palate: flavourful, fresh, saline.

ADEGAS DO REXURDIR & RIBEIRO
Sobreira, 9 Rioboó - Osmo
32454 Cenlle (Ourense)
☎: +34 626 767 969
info@adegasdorexurdir.com
www.adegasdorexurdir.com

Fala de Min Brancellao 2019 T
brancellao

88

Fruity, jammy, tasty.

Fala de Min Godello 2021 B

88

Citrus fruit, crisp, fruity, herbal.

Louvre Lías 2021 B

89

Fruity, dried herbs, herbaceous, tasty, jammy.

Verbas 1917 2019 T

91

Wild, jammy. Colour: cherry, garnet rim. Nose: ripe fruit, wild herbs, scrubland. Palate: fresh, fruity, good acidity.

ADEGAS PARENTE GARCÍA
As Chavolas, 22
32454 Cenlle (Ourense/Orense)
☎: +34 639 648 173
adegasparente@gmail.com

Quinta do Avelino 1923 2020 B
treixadura, albariño, loureiro

90

Colour: yellow. Nose: dried herbs, dried flowers, faded flowers, expressive. Palate: balanced, long.

Quinta do Avelino 1923 2021 B
treixadura, albariño, loureiro

89

Pleasant, defined aromas, standard, wild, balanced, smooth.

Quinta do Avelino 2021 B
treixadura

88

Jammy, fruity, herbal, tasty.

Quinta do Avelino Rock na Viña 2021 B

88

Hot, sweety finish, jammy, fruity.

ADEGAS PAZO DO MAR
Ctra. Ourense-Castrelo, Km. 12,5
32940 Toén (Ourense/Orense)
☎: +34 988 261 256
Fax: +34 988 261 264
info@pazodomar.com
www.pazodomar.com

Pazo do Mar 2021 B
treixadura, torrontés, albariño, loureiro, godello

87

Torre do Olivar Expresión 2021 B
treixadura

90

Fruity, jammy, tasty, simple. Nose: neat, balanced. Palate: fruity, flavourful.

DO RIBEIRO / D.O.P.

DO RIBEIRO / D.O.P.

ADEGAS VALDAVIA
Lugar de Cuñas, 24
32454 Cenlle (Ourense/Orense)
☎: +34 669 892 681
info@adegasvaldavia.com
www.adegasvaldavia.com

Cuñas Davia 2020 B FB
treixadura, albariño, caiño
90
Colour: bright straw. Nose: fruit expression, ripe fruit, floral, dried herbs. Palate: flavourful, fresh, fruity.

Cuñas Davia 2020 T
sousón, brancellao, caiño longo
89
Balsamic herbs, fruity, jammy, tasty.

Cuñas Davia 2021 B
treixadura, albariño, godello, caiño
91
Colour: bright straw, greenish rim. Nose: fresh fruit, citrus fruit, wild herbs, burnt matches. Palate: fresh, fruity, good acidity, fine bitter notes.

Cuñas Davia A Xiada 2021 B
treixadura, albariño
92
Colour: bright straw, greenish rim. Nose: fresh fruit, citrus fruit, wild herbs, burnt matches. Palate: fresh, fruity, good acidity, fine bitter notes.

La Flor de Margot Mencía 2020 T
mencía
88
Spicy, dried flowers, jammy, rustic.

La Flor de Margot Treixadura 2021 B
treixadura
90
Colour: bright straw. Nose: fresh fruit, citrus fruit, wild herbs. Palate: fresh, fruity, good acidity, fine bitter notes.

AILALA-AILALELO
Pazo Lodeiro s/n - San Fiz Do Barón
32500 O Carballiño (Ourense/Orense)
☎: +34 610 602 672
export@ailalawine.com
www.ailalawine.com

Ailalá 2018 T
sousón
90
Colour: Cherry. Nose: balsamic herbs, scrubland, ripe fruit. Palate: spicy, balsamic, good acidity, easy to drink.

Ailalá 2021 B
treixadura
90
Defined aromas. Colour: yellow. Nose: white fruit, white flowers, balanced, neat. Palate: fruity, easy to drink, fine bitter notes.

ARCO DA VELLA A ADEGA DE ELADIO
Pza. de España, 1
32431 Beade (Ourense/Orense)
☎: +34 607 487 060
arcodavellaadegadeeladio@gmail.com
www.adegaarcodavella.com

Tarabelo 2019 T C
60% sousón, 15% caiño longo, 15% brancellao, 10% ferrón
88
Pleasant, tasty, jammy, slight reduction.

Tarabelo 2020 T C
sousón, caiño longo, brancellao, ferrón
88
Tasty, jammy, pruney, spicy.

Torques do Castro 2020 B
50% treixadura, 20% torrontés, 15% godello, 10% albariño, 5% loureiro
89
Fruity, jammy, tasty, dried herbs.

ATLANTIC GALICIAN WINERIES
Vilacha
27334 A Pobra do Brollón (Galicia)
info@atlanticgalicianwineries.com
www.altanticgalicianwineries.com

Mateo Colección Treixadura 2021 B
treixadura

91

Colour: bright straw, greenish rim. Nose: fresh fruit, citrus fruit, wild herbs, fine lees. Palate: fresh, fruity, good acidity, fine bitter notes.

Mateo Quintas 2021 B
treixadura, godello, torrontés

90

Colour: bright straw, greenish rim. Nose: fresh fruit, citrus fruit, wild herbs, fine lees. Palate: fresh, fruity, good acidity.

AUTÉNTICOS VIÑADORES, VINOS DE TERROIR
24540 Cacabelos (León)
☎: +34 658 617 390
info@autenticosvinadores.com
www.autenticosvinadores.com

Ciclohome Godello 2020 B
100% godello

89

Citrus fruit, dried herbs, tasty, aromatic, balanced, standard.

Ciclohome Treixadura 2017 B
95% treixadura, 5% godello

90

Colour: pale. Nose: white flowers, dried herbs, neat, expressive, balanced. Palate: fruity, easy to drink, good finish.

BODEGA ALANÍS
Santa Cruz de Arrabaldo
32990 Santa Cruz de Arrabaldo (Ourense/Orense)
☎: +34 988 384 200
vinos@bodegasgallegas.com
www.bodegasgallegas.com

Amavida 2021 B
100% treixadura

89

Dried flowers, fruity, herbal, tasty, briny.

Gran Alanís 2021 B
85% treixadura, 15% godello

89

Fruity, herbal, dried flowers, tasty.

Gran Campiño B
treixadura, torrontés

87

San Trocado 2021 B
100% treixadura

88

Dried herbs, fruity, jammy, tasty.

BODEGA CARLOS MORO
Camino Garugele, s/n
26338 San Vicente de la Sonsierra (La Rioja)
☎: +34 941 334 093
enoturismo@bodegacarlosmoro.com
www.bodegacarlosmoro.com

CM Viña Tenencia 2021 B
treixadura, godello, albariño

90

Aromatic, floral. Colour: bright straw. Nose: white fruit, white flowers. Palate: flavourful, correct, fine bitter notes.

BODEGA CASAR DE VIDE
Vide, 2
32430 Castrelo de Miño (Ourense/Orense)
www.casardevide.es

Casar de Vide 2021 B
treixadura

86

BODEGA COOP. SAN ROQUE DE BEADE S.C.G. TERRA DO CASTELO
Ctra. Ribadavia - Carballiño, Km. 4
32431 Beade (Ourense/Orense)
☎: +34 649 848 853
presidencia@terradocastelo.com
www.terradocastelo.com

Terra do Castelo "Sensación" 2021 B
treixadura, godello

88

Herbaceous, fruity, citrus fruit, simple.

Terra do Castelo Godello 2021 B
godello

89

Citrus fruit, fruity, herbal, tasty, crisp.

Terra do Castelo Tostado Dulce B D
100% treixadura

93

Colour: dark mahogany. Nose: candied fruit, honeyed notes, dried herbs, sweet spices, caramel, aromatic coffee. Palate: flavourful, unctuous, fruity, sweet.

DO RIBEIRO / D.O.P.

DO RIBEIRO / D.O.P.

Terra do Castelo Treixadura 2021 B
treixadura

90

Colour: bright yellow. Nose: expressive, varietal, neat, floral. Palate: flavourful, fruity, balanced.

Terra do Castelo Treixadura Selección 2013 B
100% treixadura

92

Colour: bright yellow. Nose: powerful, creamy oak, ripe fruit, spicy, fine lees, waxy notes. Palate: rich, good structure, long, toasty, fine bitter notes.

BODEGA SANCLODIO

Cubilledo-Gomariz
32429 Leiro (Ourense/Orense)
☎: +34 686 961 681
sanclodiovino@gmail.com
www.vinosanclodio.com

Sanclodio 2021 B
treixadura, godello, loureiro, albariño

89

Defined aromas, balanced, fruity, jammy, simple, standard.

BODEGA Y VIÑEDOS PAZO CASANOVA

Camiño Souto do Río, 1 Santa Cruz de Arrabaldo
32990 Ourense/Orense (Ourense/Orense)
☎: +34 988 384 186
Fax: +34 606 581 720
comercial@grandespagosgallegos.com
www.grandespagosgallegos.com

Casanova 2021 B
treixadura

92

Colour: bright straw. Nose: ripe fruit, fragrant herbs, fine lees, smoky. Palate: full-bodied, rich, long, good acidity.

Pazo Casanova 2021 B

91

Colour: bright straw, greenish rim. Nose: fresh fruit, citrus fruit, wild herbs, smoky, fine lees. Palate: fresh, fruity, good acidity, fine bitter notes.

BODEGAS CAMPANTE

Finca Reboreda (Puga)
32941 Toén (Ourense/Orense)
☎: +34 988 261 212
info@campante.com
www.bodegasgrm.com

A Telleira Godello 2021 B
godello

90

Colour: bright straw. Nose: floral, wild herbs, varietal. Palate: flavourful, fresh, good acidity, fruity aftertaste, easy to drink.

A Telleira Mencía 2020 T
mencía

88

Jammy, tasty, toasty, spicy.

A Telleira Parcelas 2021 B
treixadura, godello

90

Kind finish, wild. Nose: white fruit, wild herbs, stone fruit, neat, balanced. Palate: balanced, easy to drink, good finish.

Adeus 2021 B
treixadura

89

Floral, herbal, jammy, crisp, fruity, smooth.

BODEGAS EL PARAGUAS

Lugar de A Aldea de Cobas, 135
15594 Ferrol (A Coruña/La Coruña)
☎: +34 636 161 479
info@bodegaselparaguas.com
www.bodegaselparaguas.com

🏆 PODIUM

Fai un Sol de Carallo 2018 B
86% treixadura, 9% godello, 5% albariño

96

Colour: bright yellow. Nose: powerful, creamy oak, ripe fruit, spicy, cereal notes, waxy notes, white flowers. Palate: rich, good structure, long, toasty, fine bitter notes.

🏆 PODIUM

La Sombrilla 2019 B
92% treixadura, albariño

95

Full of life, austere. Colour: bright straw. Nose: fragrant herbs, fine lees, mineral, dry stone. Palate: full-bodied, rich, long, good acidity, great length.

DO RIBEIRO / D.O.P.

El Paraguas Atlántico 2020 B
88% treixadura, 9% godello, 3% albariño

93

Colour: bright yellow. Nose: creamy oak, ripe fruit, spicy, fine lees. Palate: rich, good structure, long, toasty, fine bitter notes.

BODEGAS MAURO ESTÉVEZ
A Ponte, 21
32417 Arnoia (Ourense/Orense)
☎: +34 617 090 616
joseestevezarnoia@gmail.com
www.mauroestevez.com

Mauro Estévez 2021 B
treixadura, albariño, loureiro, godello, lado

90

Colour: bright straw, greenish rim. Nose: fresh fruit, citrus fruit, wild herbs. Palate: fresh, fruity, fine bitter notes.

Uxía da Ponte 2020 B
100% lado

93

Colour: bright straw. Nose: fruit expression, citrus fruit, dried herbs, wild herbs, mineral, white flowers. Palate: fruity, fresh, flavourful, balanced.

BODEGAS NAIROA
A Ponte, 2
32417 Arnoia (Ourense/Orense)
☎: +34 988 492 867
info@bodegasnairoa.com
www.bodegasnairoa.com

Alberte Doble Lias 2020 B
treixadura, albariño, lado, loureiro

93

Colour: bright straw. Nose: ripe fruit, fragrant herbs, fine lees, stone fruit, sweet spices. Palate: full bodied, rich, long, good acidity.

Alberte Treixadura 2021 B
treixadura

90

Colour: bright straw. Nose: expressive, white flowers, jasmine, dried herbs. Palate: flavourful, fruity, balanced.

Val de Nairoa 2020 B

91

Colour: bright straw. Nose: ripe fruit, fragrant herbs, fine lees. Palate: good acidity, balanced, flavourful.

BODEGAS O VENTOSELA
Ctra. OU 504 Ribadavia-Carballiño, Km. 8.8
32420 Leiro (Ourense/Orense)
☎: +34 981 635 829
bodegasydestilerias@oventosela.com
www.oventosela.com

El Godello de Juan Miguez 2021 B
godello

89

Citrus fruit, vegetal, crisp, yeasty notes, tasty.

El Torrontés de Juan Miguez 2020 B

91

Colour: yellow. Nose: dried flowers, fresh fruit, white fruit, fine lees. Palate: flavourful, fine bitter notes, fresh.

O Ventosela 2021 B

88

Standard, dried herbs, wild, smooth, dried flowers, simple.

Viña Leiriña 2021 B

90

Colour: bright straw. Nose: ripe fruit, fragrant herbs, fine lees, citrus fruit. Palate: full-bodied, rich, long, good acidity.

BODEGAS SIAH
Ctra. Carballino, Km. 6
32419 Leiro (Ourense/Orense)
☎: +34 607 352 815
isabel.salgado@bodegas-siah.com
www.bodegas-siah.com

Siah Isabel Salgado 2020 B
treixadura, godello, albariño

91

Colour: bright straw. Nose: ripe fruit, fragrant herbs, fine lees, chamomile. Palate: full-bodied, rich, long, good acidity.

BODEGAS VAL DE SOUTO
Lugar Souto, 34
32430 Castrelo de Miño (Ourense/Orense)
☎: +34 636 024 205
valdesouto@gmail.com
www.valdesouto.com

Benedictus Fructus 2020 B BA
godello

90

Colour: straw. Nose: dried herbs, faded flowers, white fruit, fine lees. Palate: powerful, ripe fruit, balanced, flavourful.

DO RIBEIRO / D.O.P.

Benedictus Fructus 2021 B BA
godello

89

Varietally correct, fruity, jammy, tasty, balanced, standard.

Val de Souto Treixadura 2021 B
treixadura

89

Standard, floral, very fruit-driven, tasty, great length.

CASAL DE ARMÁN

San Andrés
32415 Ribadavia (Ourense/Orense)
☎: +34 680 979 763
info@casaldearman.net
www.casaldearman.net

Armán Finca Misenhora 2019 B
90% treixadura, 10% godello, albariño

94

Balanced, dried flowers. Colour: yellow. Nose: faded flowers, fine lees, white fruit, ebb tide. Palate: rich, flavourful, fine bitter notes, long, ripe fruit.

Arman Finca Os Loureiros 2019 B
100% treixadura

94

Colour: bright straw. Nose: ripe fruit, floral, spicy, fine lees. Palate: flavourful, fresh, good acidity, fruity aftertaste.

Casal de Armán 2020 T
caiño, sousón, brancellao

91

Colour: bright cherry. Nose: balsamic herbs, scrubland, wild herbs, ripe fruit. Palate: balsamic, good acidity, easy to drink.

Casal de Armán 2021 B
90% treixadura, 10% albariño, godello

92

Colour: bright straw. Nose: ripe fruit, fine lees, wild herbs, citrus fruit. Palate: full-bodied, rich, good acidity.

Pepe Carrasca 2020 B
treixadura

93

Colour: bright straw. Nose: fragrant herbs, fine lees, dry stone, white fruit. Palate: full-bodied, rich, long, good acidity, mineral.

COTO DE GOMARIZ

Barro de Gomariz s/n
32429 Leiro (Ourense/Orense)
☎: +34 610 602 672
info@losvinosdemiguel.com
www.cotodegomariz.com

Abadía de Gomariz 2017 T
sousón, brancellao, ferrol, mencía

92

Colour: cherry, garnet rim. Nose: balsamic herbs, ripe fruit, scrubland, old leather. Palate: flavourful, balsamic, spicy, soft tannins.

Coto de Gomariz 2020 B
treixadura, godello, albariño, loureiro

92

Colour: bright straw. Nose: fresh, floral, white fruit, expressive, neat. Palate: flavourful, fine bitter notes, easy to drink.

Coto de Gomariz
Finca O Figueiral 2019 B
treixadura, godello, albariño, loureiro

93

Colour: bright straw. Nose: expressive, ripe fruit, floral, fine lees, mineral, spicy. Palate: full-bodied, complex, spicy, long, elegant.

Gomariz X 2021 B

91

Colour: bright straw. Nose: fruit expression, ripe fruit, floral. Palate: flavourful, fresh, good acidity.

The Flower
and The Bee Treixadura 2021 B
treixadura

90

Aromatic, balanced. Colour: straw. Nose: expressive, white flowers, white fruit. Palate: flavourful, fruity, varietal.

The Flower and The Bee Sousón 2020 T
sousón
91
Colour: cherry, garnet rim. Nose: fine reductive notes, wild herbs, balanced, ripe fruit. Palate: flavourful, easy to drink.

DIEGO DONIZ DIÉGUEZ
Lg. O Pazo, s/n
32417 Arnoia (Ourense/Orense)
☎: +34 606 135 159
teardosdodi@gmail.com
www.diegodonizdieguez.com

Tear dos Dodi 2020 B
treixadura, torrontés
90
Colour: straw. Nose: expressive, white flowers, jasmine, dried herbs. Palate: flavourful, fruity, balanced, round.

DOMINIO DO BIBEI
Langullo, s/n
32781 Manzaneda (Ourense/Orense)
☎: +34 670 704 028
info@dominiodobibei.com
www.dominiodobibei.com

Lalume 2020 B
treixadura
93
Colour: bright straw, greenish rim. Nose: fresh fruit, citrus fruit, wild herbs, grassy, hint of anise. Palate: fresh, fruity, good acidity, fine bitter notes.

EDUARDO PEÑA
Ctra. de Cartelle, s/n
32430 Castrelo de Miño (Ourense/Orense)
☎: +34 629 872 130
bodega@bodegaeduardopenha.es
www.bodegaeduardopenha.es

Eduardo Peña 2021 B
60% treixadura, 20% albariño, 10% godello, 5% loureiro, 5% lado
90
Colour: straw. Nose: white flowers, wild herbs, fresh. Palate: fruity, balanced, easy to drink.

Sara Peña 2020 T RB
33% brancellao, 33% caíño, 33% sousón
90
Colour: dark-red cherry. Nose: toasty, spicy, cocoa bean, ripe fruit. Palate: flavourful, toasty, fine bitter notes.

EMILIO DOCAMPO DIÉGUEZ
San Andrés, 57
32415 Ribadavia (Ourense/Orense)
☎: +34 639 332 790
edocampodieguez@hotmail.com

Casal de Paula 2021 B
70% treixadura, 15% torrontés, 5% godello, 5% albariño, 5% loureiro
89
Fruity, herbal, tasty.

Casal de Paula 2021 T
50% mencía, 35% brancellao, 8% ferrón, 7% sousón
88
Fruity, herbal, simple, crisp.

Casal de Paula D.H. 2020 B
100% treixadura
90
Colour: bright yellow. Nose: ripe fruit, spicy, dried flowers, balanced. Palate: rich, good structure, long, fine bitter notes.

FINCA VIÑOA
Banga, s/n
32821 Carballino (Ourense/Orense)
☎: +34 606 581 720
comercial@grandespagosgallegos.com
www.fincavinoa.com

Finca Viñoa Embotellado Tardío 2019 B
treixadura, godello, albariño, loureiro
92
Kind finish. Colour: bright straw. Nose: ripe fruit, floral, fine lees, balanced. Palate: flavourful, fresh, good acidity, fruity aftertaste, rich.

DO RIBEIRO / D.O.P.

DO RIBEIRO / D.O.P.

Finca Viñoa Paraje Penaboa 2017 B
92
Colour: bright yellow. Nose: powerful, ripe fruit, spicy, toasty. Palate: rich, good structure, long, toasty, fine bitter notes.

Finca Viñoa Treixadura Sobre Lías 2021 B
treixadura
91
Colour: bright straw, greenish rim. Nose: fresh fruit, citrus fruit, wild herbs. Palate: fresh, fruity, good acidity, fine bitter notes.

GALLINA DE PIEL WINES
Escoles, 43
17181 Aiguaviva (Girona/Gerona)
☎: +34 663 594 668
info@gallinadepielwines.com
www.gallinadepielwines.com

Manar dos Seixas 2020 B
88% treixadura, 6% albariño, 4% godello, 2% loureiro
92
Colour: bright straw. Nose: ripe fruit, fragrant herbs, fine lees, smoky. Palate: full-bodied, rich, good acidity.

IVÁN VÁZQUEZ PATEIRO (PATEIRO VINOS DE GUARDA)
Ctra. de Francelos nº 34 2ºC
32400 Ribadavia (Ourense/Orense)
☎: +34 696 147 706
info@pateirovinosdeguarda.com
www.pateirovinosdeguarda.com

El Patito Feo Godello 2021 B BA
95% godello, 5% treixadura
89
Herbaceous, fruity, tasty, spicy.

El Patito Feo Treixadura Sobre Lías 2021 B
100% treixadura
89
Dried herbs, fruity, tasty, jammy.

Pateiro Anfora 2020 B
85% loureiro, 15% treixadura
91
With personality, oxidativ. Colour: yellow. Nose: dried flowers, faded flowers, expressive, neat. Palate: flavourful, bitter, long.

Pateiro Treixadura 2020 B BA
100% treixadura
91
Colour: bright straw. Nose: ripe fruit, floral, neat, spicy. Palate: flavourful, fresh, good acidity.

Saramusa Treixadura 2021 B
97% treixadura, 3% albariño
88
Fruity, tropical, dried herbs, overripe.

LEIVE ECOADEGA
Lg. a Torre s/n . Ctra. San Andres a Leiro
32415 Esposende - Cenlle (Ourense/Orense)
☎: +34 698 166 665
administracion@aurealux.com
www.aurealux.com

Leive Paradigma 2021 B
91
Colour: bright straw. Nose: fruit expression, ripe fruit, floral, citrus fruit. Palate: flavourful, fresh, good acidity, fruity aftertaste.

Leive Reliquia 2020 B FB
89
Citrus fruit, full-bodied, tasty, herbal, woody, roasted coffee.

Leive Treixadura 2020 B
87

Preto de Leive 2020 T
89 🌱
Pleasant, standard, fruity, herbal, jammy, great length, smooth, balanced.

LUIS A. RODRÍGUEZ VÁZQUEZ
Laxa, 7
32417 Arnoia (Ourense/Orense)
☎: +34 988 492 977
Fax: +34 988 492 977

Viña de Martín "Os Pasás" 2019 B
92
Colour: bright yellow. Nose: dried flowers, candied fruit, fine lees, pattiserie, waxy notes. Palate: round, spicy, long, great length.

MÓNICA ALBOR
Lg. Coedo, 4
32454 Cenlle (Ourense/Orense)
☎: +34 669 181 668
info@bodegasalbor.com
www.bodegasalbor.com

Mónica Albor Mencía Sousón Caiño 2021 T
85% mencía, 10% sousón, 5% caiño
88
Spicy, fruity, herbaceous, tasty.

Mónica Albor Treixadura 2020 B
100% treixadura

89

Kind finish, aromatic, dried flowers, fruity, jammy, tasty.

Mónica Albor Treixadura Godello Loureira 2020 B
85% treixadura, 10% godello, 5% loureiro

89

Kind finish, aromatic, floral, fruity, jammy, smooth, tropical.

O ALBOREXAR NO RIBEIRO

Cortiñas, 24
32430 Castrelo de Miño (Ourense/Orense)
☎: +34 617 400 595
oalborexarnoribeiro@gmail.com
www.oalborexarnoribeiro.com

O Alborexar 2019 T
25% brancellao, 25% caiño, 25% sousón, 25% ferrón

88

Jammy, tasty, spicy.

O Alborexar 4.0 2021 T
mencía

88

Balsamic herbs, herbaceous, jammy, tasty.

O Alborexar Frontón 2018 T
brancellao, caiño, ferrón

88

Roasted coffee, pruney, jammy, powerful.

O Alborexar Garnacha 2020 T
garnacha

88

Rustic, jammy, mineral, dried flowers, slight oxidation.

PABLO VIDAL - VINOS CON PERSONALIDAD

Rúa do Miradoiro 8
32004 Ourense/Orense (Ourense/Orense)
☎: +34 609 152 251
pablovidal@vinosconpersonalidad.com
www.vinosconpersonalidad.com

Big Bang 2019 T BA S
35% caiño longo, 30% sousón, 25% brachetto, 10% ferrón

92

Colour: cherry, garnet rim. Nose: overripe fruit, hot, black fruit, balsamic herbs, scrubland. Palate: pruney, powerful, sweet tannins.

Renacido 2019 B
80% treixadura, 10% godello, 7% albariño, 3% lado

92

Colour: straw. Nose: white flowers, ripe fruit, floral. Palate: fruity, balanced, fine bitter notes, good acidity.

PAZO DE VIEITE

Ctra. Ribadavia - Carballiño, Km. 6 Vieite
32419 Leiro (Ourense/Orense)
☎: +34 988 488 229
javier@javiervasallovinos.com
www.javiervasallovinos.com

1932 de Pazo Vieite 2021 B
100% treixadura

90

Colour: bright straw. Nose: ripe fruit, fragrant herbs, fine lees. Palate: full-bodied, rich, long, good acidity.

Farnadas 2021 B
94% treixadura, 3% albariño, 3% godello, loureiro

89

Pleasant, aromatic, lactic, dried herbs, floral, jammy.

JV Lías 2020 B
78% treixadura, 10% godello, 7% albariño, 5% loureiro

89

Fruity, sweety finish, jammy, tasty.

PRIORATO DE RAZAMONDE

Lugar de Razamonde s/n
32459 Cenlle (Ourense/Orense)
☎: +34 610 142 893
pedidos@prioratoderazamonde.com
www.prioratoderazamonde.com

Alter 2019 T
60% brancellao, 40% sousón

88

Tasty, jammy, spicy, powerful.

Alter 2021 B
85% treixadura, 10% godello, 5% loureiro

90

Colour: bright straw. Nose: fruit expression, ripe fruit, dried herbs, white flowers. Palate: flavourful, fresh, fruity aftertaste, fruity.

Priorato de Razamonde 2019 T
sousón, brancellao

88

Classic, toasty, jammy.

Priorato de Razamonde Treixadura 2021 B
100% treixadura

89

Pleasant, aromatic, simple, standard, floral, crisp, fruity.

DO RIBEIRO / D.O.P.

DO RIBEIRO / D.O.P.

RAMÓN DO CASAR
Prado de Miño, s/n
32430 Castrelo de Miño (Ourense/Orense)
☎: +34 988 036 097
adega@ramondocasar.es
www.ramondocasar.es

Ramón Do Casar Godello 2020 B FB
91
Colour: bright straw, greenish rim. Nose: fresh fruit, citrus fruit, wild herbs, white flowers. Palate: fresh, fruity, good acidity, fine bitter notes.

Ramón Do Casar Nobre 2020 B FB
92
Colour: bright yellow. Nose: wild herbs, fruit expression, stone fruit, ripe fruit. Palate: fresh, fruity, fine bitter notes, flavourful.

Ramón Do Casar Treixadura 2021 B
treixadura
90
Colour: bright yellow. Nose: fruit expression, ripe fruit, floral, dried herbs. Palate: flavourful, fresh, fruity.

Ramón Do Casar Varietal 2021 B
85% treixadura, 10% albariño, 5% godello
91
Colour: straw. Nose: expressive, white flowers, dried herbs, white fruit. Palate: flavourful, fruity, balanced.

SEÑORÍO DE BEADE
Piñeiros, s/n
32431 Beade (Ourense/Orense)
☎: +34 988 480 050
beade@beadeprimacia.com
www.senoriodebeade.com

Beade 25 Autor 2020 B
loureiro
90
Colour: bright straw. Nose: stone fruit, floral, neat. Palate: flavourful, good acidity, fruity aftertaste, long, easy to drink, fine bitter notes.

Beade Orixe 2016 B
treixadura, albariño, godello
91
Colour: bright yellow. Nose: powerful, creamy oak, ripe fruit, spicy. Palate: good structure, long, fine bitter notes.

Beade Orixe 2021 T C
mencía, caiño, sousón
90
Colour: deep cherry. Nose: dried herbs, creamy oak, black fruit, stone fruit, balsamic herbs. Palate: ripe fruit, spicy, round tannins.

Beade Primacía 2021 B
treixadura
89
Pleasant, simple, floral, crisp, aromatic, standard.

Señorío de Beade 2021 B
treixadura, godello, albariño, otras
88
Dried flowers, fruity, herbal, tasty.

Señorío de Beade 2021 T
mencía, caiño
88
Balsamic herbs, rustic, jammy.

VINOS ANTONIO MONTERO
Santa María, 7
32430 Castrelo do Miño (Ourense/Orense)
☎: +34 617 400 595
antoniomontero@antoniomontero.com
www.antoniomontero.com

Alejandrvs 2019 B
100% treixadura
91
Colour: bright straw. Nose: fragrant herbs, fine lees, spicy, stone fruit. Palate: full-bodied, rich, long, good acidity.

Antonio Montero "Autor" 2021 B
80% treixadura, 10% torrontés, 5% loureiro, 5% albariño
89
Aromatic, floral, balanced, crisp, great length, smooth.

Antonio Montero 50 Finca 2019 B BA
100% treixadura
91
Colour: bright yellow. Nose: creamy oak, ripe fruit, spicy, dried herbs. Palate: rich, long, fine bitter notes, good acidity.

Antonio Montero Colleita 2021 B
89
Citrus fruit, balanced, fruity, full-bodied, herbal.

VIÑA COSTEIRA
Valdepereira, 1
32400 Ribadavia (Ourense/Orense)
☎: +34 988 477 210
marketing@costeira.es
www.costeira.es

Colección 68 Costeira Treixadura 2021 B
91
Colour: straw. Nose: expressive, white flowers, dried herbs, white fruit, wild herbs. Palate: flavourful, fruity, balanced, good finish.

Modus Vivendi Ribeiro 2021 B S
treixadura, albariño, loureiro

88
Citrus fruit, fruity, herbal, jammy.

Tamborá 2021 B S
godello

88
Pleasant, standard, fruity, smooth, dried herbs.

Viña Costeira 2021 B
treixadura, torrontés, godello, albariño

88
Citrus fruit, dried herbs, floral, tasty.

VIÑA MEIN - EMILIO ROJO
Lugar de Mein, s/n
32420 San Clodio-Leiro (Ourense/Orense)
☎: +34 983 878 020
info@almacarraovejas.com
www.vinamein-emiliorojo.com

🏆 PODIUM

Emilio Rojo 2019 B
treixadura, lado, godello, caíño blanco, albariño, torrontés

96
Colour: bright straw. Nose: expressive, ripe fruit, floral, fine lees. Palate: full-bodied, complex, long, elegant, good acidity.

O Gran Mein 2020 B
treixadura, godello, albariño, torrontés, loureiro, lado

94
Balsamic herbs, complex. Colour: bright yellow. Nose: ripe fruit, dried herbs, faded flowers, citrus fruit. Palate: powerful, ripe fruit, balanced.

O Pequeno Mein 2020 T
caíño longo, brancellao, sousón, caíño redondo

91
Colour: cherry, purple rim. Nose: fruit expression, red berry notes, floral, spicy, violet drops. Palate: flavourful, fruity, good acidity, long.

O Pequeno Mein 2021 B
treixadura, albariño, torrontés, loureiro

92
Balsamic herbs, taut. Colour: bright straw. Nose: fruit expression, ripe fruit, floral. Palate: flavourful, fresh, good acidity, fruity aftertaste.

VIÑEDOS COTODEGAIO
Eirado, 1
32420 Beran (Leiro) (Ourense/Orense)
☎: +34 670 271 027
info@cotodegaio.com
www.cotodegaio.com

Terranegra 2021 B
treixadura, godello

89
Citrus fruit, balanced, floral, herbal, yeasty notes.

DO RIBEIRO / D.O.P.

DO. RIBERA DEL DUERO

CONSEJO REGULADOR

Hospital, 6
09300 Roa (Burgos)
☎: +34 947 541 221 - Fax: +34 947 541 116
@: info@riberadelduero.es
www.riberadelduero.es

LOCATION:

Between the provinces of Burgos, Valladolid, Segovia and Soria. This region comprises 19 municipal districts in the east of Valladolid, 5 in the north west of Segovia, 59 in the south of Burgos (most of the vineyards are concentrated in this province with 10,000 Ha) and 6 in the west of Soria.en la parte occidental de Soria.

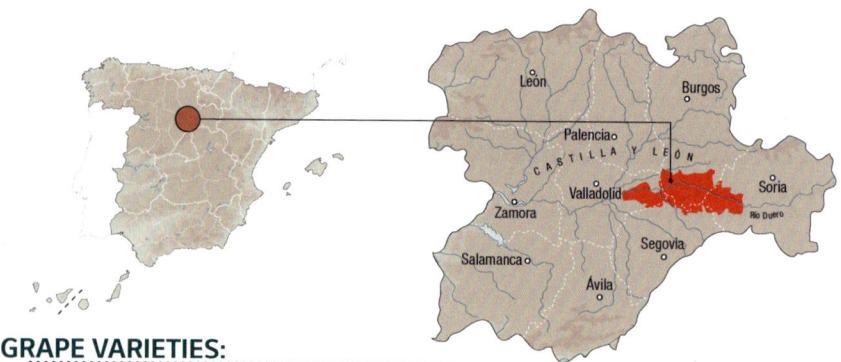

GRAPE VARIETIES:

WHITE: Albillo.

RED: Tinta del País (Tempranillo), Garnacha Tinta, Cabernet Sauvignon, Malbec and Merlot.

FIGURES:

Vineyard surface: 25,035– **Wine-Growers:** 7,877– **Wineries:** 307 – **2021 Harvest rating:** Excellent – **Production 21:** 76,900,000 litres. **Market percentages:** 82% National - 18% International.

SOIL:

In general, the soils are loose, not very fertile and with a rather high limestone content. Most of the sediment is composed of layers of sandy limestone or clay. The vineyards are located on the interfluvial hills and in the valleys at an altitude of between 700 and 850 m.

CLIMATE:

Continental in nature, with slight Atlantic influences. The winters are rather cold and the summers hot, although mention must be made of the significant difference in day-night temperatures contributing to the slow ripening of the grapes, enabling excellent acidity indexes to be achieved. The greatest risk factor in the region is the spring frosts, which are on many occasions responsible for sharp drops in production. The average annual rainfall is between 450 mm and 500 mm.

CLASSIFICATION OF YOUNG WINE HARVESTS PEÑÍNGUIDE

2017	2018	2019	2020	2021
VERY GOOD	GOOD	VERY GOOD	VERY GOOD	VERY GOOD

DO RIBERA DEL DUERO / D.O.P.

3 ASES BODEGAS Y VIÑEDOS
Camino de Pesquera, s/n
47360 Quintanilla de Arriba (Valladolid)
☎: +34 983 036 214
info@3asesvino.com
www.3asesvino.com

3 Ases 2018 T C
100% tempranillo

89
Full-bodied, creamy, balanced, spicy, tasty, roasted coffee.

3 Ases 2021 T
100% tempranillo

88
Aromatic, fruity, floral, tasty, sweety finish.

3 Ases 6 meses 2020 T RB
100% tempranillo

89
Balanced, spicy, dried herbs, tasty.

Hocicón 2021 RD
100% tempranillo

89
Balanced, racy, subtle, dried flowers, oxidativ.

AALTO BODEGAS Y VIÑEDOS
Paraje Vallejo de Carril, s/n
47360 Quintanilla de Arriba (Valladolid)
☎: +34 983 036 949
aalto@aalto.es
www.aalto.es

Aalto 2020 T
100% tempranillo

92
Colour: cherry, garnet rim. Nose: fruit preserve, powerful, toasty. Palate: flavourful, long, oaky, concentrated.

Aalto PS (Pagos Seleccionados) 2020 T R
100% tempranillo

94
Colour: deep cherry. Nose: black fruit, ripe fruit, creamy oak, spicy, roasted coffee. Palate: flavourful, balanced, round tannins, spicy.

ABADÍA SAN QUIRCE
Ctra. Madrid - Irun, Km. 171
09370 Gumiel De Izán (Burgos)
☎: +34 947 544 070
Fax: +34 947 525 759
adminis@abadiasanquirce.com
www.bodegasabadiasanquirce.com

Abadía de San Quirce 2016 T R
100% tempranillo

92
Colour: Cherry. Nose: scrubland, fruit preserve, old leather. Palate: spicy, balsamic, flavourful.

Abadía de San Quirce 2019 T C
92
Colour: dark-red cherry. Nose: toasty, spicy, cocoa bean, black fruit, ripe fruit. Palate: flavourful, toasty, fine bitter notes.

Abadía de San Quirce 6 meses 2021 T RB
100% tempranillo

89
Balanced, dried herbs, jammy, spicy.

Abadía de San Quirce Finca Helena 2019 T
100% tempranillo

92
Colour: dark-red cherry, garnet rim. Nose: ripe fruit, aged wood nuances, tobacco, sweet spices, wet leather. Palate: spicy, round tannins, long.

Abadía de San Quirce M9 2020 T
93
Colour: Cherry. Nose: spicy, black fruit, ripe fruit, earthy notes. Palate: elegant, full-bodied, long, great length.

ALEJANDRO FERNÁNDEZ TINTO PESQUERA
Real, 2
47315 Pesquera de Duero (Valladolid)
☎: +34 983 870 037
info@bodegapesquera.com
www.familiafernandezrivera.com

Tinto Pesquera 2019 T R
tempranillo

92
Colour: cherry, garnet rim. Nose: fruit preserve, powerful, black fruit, toasty. Palate: flavourful, sweetness, long.

Tinto Pesquera 2020 T C
tempranillo

91
Colour: cherry, purple rim. Nose: spicy, ripe fruit, creamy oak. Palate: flavourful, fruity, long.

DO RIBERA DEL DUERO / D.O.P.

ALTOS DEL ENEBRO
Ávila, 14
09460 Milagros (Burgos)
☎: +34 619 409 097
comercial@altosdelenebro.es
www.altosdelenebro.es

Altos del Enebro 2019 T C
92
Colour: dark-red cherry. Nose: toasty, spicy, cocoa bean, ripe fruit, truffle notes. Palate: flavourful, toasty, fine bitter notes.

Altos del Enebro Finca la Herradura 2019 T R
94
Colour: cherry, garnet rim. Nose: complex, expressive, spicy, mineral, black fruit, ripe fruit. Palate: full-bodied, long, great length, round tannins.

Tomás González 2020 T RB
89
Roasted coffee, jammy, spicy, toasty.

ALTOS DEL TERRAL
Barrio, 11 1ºB
09400 Aranda de Duero (Burgos)
☎: +34 670 706 947
bodega@altosdelterral.com
www.altosdelterral.com

Altos del Terral 2019 T C
tempranillo
91
Colour: deep cherry. Nose: dried herbs, creamy oak, black fruit, balsamic herbs. Palate: ripe fruit, spicy, round tannins.

Altos del Terral T1 2019 T
tempranillo
90
Colour: cherry, garnet rim. Nose: fruit preserve, fruit liqueur notes, powerful, dark chocolate, toasty. Palate: flavourful, sweetness, long.

Cuvée Julia Altos del Terral 2018 T R
tempranillo
91
Colour: cherry, garnet rim. Nose: fruit preserve, powerful, roasted coffee, dark chocolate. Palate: flavourful, sweetness, long.

ASTRALES
Ctra. Olmedillo, Km. 7
09313 Anguix (Burgos)
☎: +34 947 554 222
administracion@astrales.es
www.astrales.es

Astrales 2019 T
tempranillo
91
Colour: deep cherry. Nose: ripe fruit, dried herbs, creamy oak. Palate: powerful, ripe fruit, spicy, slightly dry, soft tannins.

Astrales Christina 2019 T C
tempranillo
93
Colour: cherry, garnet rim. Nose: ripe fruit, fruit preserve, aged wood nuances, creamy oak, new oak. Palate: spicy, round tannins, long, flavourful, ripe fruit.

AUSÀS BODEGAS Y VIÑEDOS
Síndico, 5
47008 Quintanilla de Onésimo (Valladolid)
☎: +34 669 653 217
info@ausasbodegas.com

🏆 **PODIUM**

Ausàs Interpretación 2020 T
100% tempranillo
96
Colour: cherry, garnet rim. Nose: complex, expressive, mineral, black fruit, ripe fruit, sweet spices. Palate: elegant, full-bodied, long, great length.

AVELINO VEGAS
Ctra. VA-101, km 3.7
47300 Peñafiel (Valladolid)
☎: +34 607 834 938
sevegarciavegas@avelinovegas.com
www.avelinovegas.com

Áureo 2018 T R
tempranillo
91
Colour: cherry, garnet rim. Nose: ripe fruit, aged wood nuances, sweet spices, wild herbs, dried herbs. Palate: spicy, round tannins, long.

Avelino Vegas 100 Aniversario 2019 T R
tempranillo
92
Colour: Cherry. Nose: sweet spices, dried herbs, neat. Palate: spicy, balsamic, round tannins.

BADEN NUMEN

Carreterilla, s/n
47359 San Bernardo (Valladolid)
☎: +34 615 995 552
bodega@badennumen.es
www.badennumen.es

Baden Numen "AU" 2019 T
tinto fino, cabernet sauvignon

88

Spicy, creamy, jammy, powerful, toasty.

Baden Numen "B" 2020 T
100% tinto fino

90

Colour: cherry, garnet rim. Nose: fruit preserve, powerful, toasty, aromatic coffee. Palate: flavourful, sweetness, long.

Baden Numen "N" 2019 T C
tinto fino, cabernet sauvignon

89

Jammy, tasty, toasty, full-bodied, roasted coffee.

BELA

Ctra. de Palencia-Aranda de Duero, km. 68
09443 Aranda de Duero (Burgos)
☎: +34 947 112 783
marketing@cvne.com
www.bodegabela.com

Finca Vallejo 2020 T RB
100% tinta del país

89

Spicy, fruity, jammy, tasty.

Bela – Ribera del Duero 2021 T
100% tinta del país

88

Spicy, jammy, toasty, balanced.

Finca Vallejo 2019 T C
100% tinta del país

90

Colour: cherry, garnet rim. Nose: ripe fruit, dried herbs, balanced, dried flowers. Palate: powerful, ripe fruit, spicy, round tannins.

Arano 2019 T C
100% tinta del país

92

Colour: dark-red cherry, garnet rim. Nose: ripe fruit, aged wood nuances, tobacco, sweet spices. Palate: spicy, round tannins, long.

BODEGA ASCENSIÓN REPISO BOCOS

Ctra. de Valbuena, 34
47315 Pesquera de Duero (Valladolid)
☎: +34 620 280 781
info@veronicasalgado.es
www.veronicasalgado.es

Verónica Salgado Capricho 2018 T C
100% tempranillo

92 🍃

Colour: deep cherry. Nose: ripe fruit, dried herbs, creamy oak, black fruit, violets. Palate: powerful, ripe fruit, spicy, round tannins.

Verónica Salgado Capricho Vino de Autor 2019 T

93 🍃

Colour: dark-red cherry. Nose: toasty, spicy, cocoa bean, ripe fruit. Palate: flavourful, toasty, fine bitter notes.

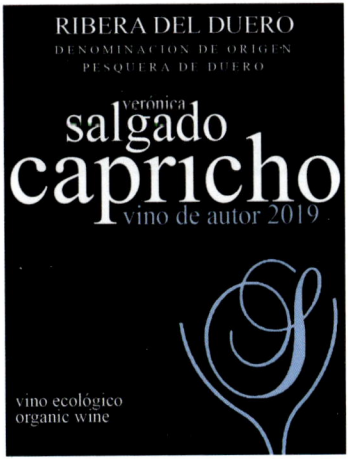

DO RIBERA DEL DUERO / D.O.P.

DO RIBERA DEL DUERO / D.O.P.

BODEGA BARDOS
Ctra. de Adrada, Km. 20
09315 Fuentemolinos (Burgos)
☎: +34 941 483 557
marketing@vintae.com
www.bardos.wine

Bardos 2018 T R
tinta del país, cabernet sauvignon

92

Colour: cherry, garnet rim. Nose: ripe fruit, creamy oak, dried herbs. Palate: ripe fruit, spicy, round tannins, flavourful, balsamic.

Bardos 2019 T RB
tinta del país

89

Pleasant, spicy, jammy, fruity, standard, tasty.

Bardos Romántica 2019 T C
tinta del país

91

Colour: cherry, purple rim. Nose: red berry notes, spicy, neat, balanced. Palate: flavourful, fruity, long, easy to drink.

Bardos Suprema 2018 T R
tinta del país

93

Colour: dark-red cherry. Nose: black fruit, complex, spicy, cocoa bean, characterful. Palate: flavourful, powerful, round tannins.

Clea 2019 T C
tinta del país

90

Colour: cherry, garnet rim. Nose: ripe fruit, red berry notes, wild herbs, spicy, balanced. Palate: ripe fruit, spicy, easy to drink.

BODEGA CONVENTO SAN FRANCISCO
Calvario, 22
47320 Peñafiel (Valladolid)
☎: +34 983 878 052
bodega@bodegaconvento.com
www.bodegaconvento.com

Convento San Francisco 2018 T C
tempranillo

92

Colour: deep cherry. Nose: ripe fruit, dried herbs, creamy oak, cocoa bean, black fruit. Palate: powerful, ripe fruit, spicy, round tannins.

Convento San Francisco La Zapatera 2018 T R
tempranillo

90

Colour: deep cherry, purple rim. Nose: ripe fruit, cocoa bean, roasted coffee. Palate: flavourful, spicy, toasty.

Convento San Francisco Selección Especial 2018 T BA
tempranillo

91

Colour: cherry, garnet rim. Nose: ripe fruit, toasty, cocoa bean. Palate: powerful, flavourful, balanced, toasty.

BODEGA COOP. SAN PEDRO REGALADO
Ctra. de Aranda, s/n
09370 La Aguilera (Burgos)
☎: +34 947 545 017
administracion@embocadero.com
www.embocadero.com

Embocadero 2018 T C
100% tempranillo

88

Spicy, smoky, dried herbs, toasty.

Embocadero 2020 T RB
100% tempranillo

90

Colour: Cherry. Nose: balsamic herbs, sweet spices, scrubland. Palate: spicy, good acidity, balanced, flavourful.

BODEGA COOP. SANTA ANA
Ctra. Aranda - Salas, km. 18,5
09410 Peñaranda de Duero (Burgos)
☎: +34 947 552 011
bodega@bodegasantaana.es
www.bodegasantaana.es

Castillo de Peñaranda 2017 T C
tempranillo

88

Fruity, herbal, spicy, jammy.

Cruz Sagra 2017 T
tempranillo

90

Colour: dark-red cherry, garnet rim. Nose: ripe fruit, fruit preserve, aged wood nuances, tobacco, sweet spices. Palate: spicy, round tannins, long.

Valdepisón 2021 RD
tempranillo

87

BODEGA COOP. VIRGEN DE LA ASUNCIÓN

Las Afueras, s/n
09311 La Horra (Burgos)
☎: +34 947 542 057
info@virgendelaasuncion.com
www.virgendelaasuncion.com

El Canto del Angel 2016 T RB
100% tinto fino

91

Colour: dark-red cherry, garnet rim. Nose: ripe fruit, fruit preserve, sweet spices, waxy notes, wild herbs. Palate: spicy, round tannins, long.

El Corazón de la Tierra 2016 T RB
100% tinto fino

91

Colour: Cherry. Nose: sweet spices, scrubland. Palate: spicy, good acidity, good structure, ripe fruit.

El Secreto de María 2018 T C
100% tinto fino

91

Colour: deep cherry. Nose: ripe fruit, dried herbs, creamy oak, spicy. Palate: powerful, ripe fruit, round tannins.

Viña Valera Gran Reserva 2015 T GR
100% tinto fino

92

Colour: deep cherry. Nose: ripe fruit, dried herbs, creamy oak. Palate: powerful, ripe fruit, spicy, harsh oak tannins.

BODEGA CUATRO RAYAS

Camino de la Fuentecilla, s/n
47491 La Seca (Valladolid)
☎: +34 983 816 320
info@cuatrorayas.es
www.cuatrorayas.es

Cuatro Rayas Cuarenta Vendimias Ribera del Duero 2020 T
100% tempranillo

89

Toasty, jammy, spicy, kind finish, standard. Nose: aromatic coffee, woody.

Cuatro Rayas Tempranillo 2020 T RB
tempranillo

88

Standard, simple, spicy, jammy, herbal.

BODEGA DOBLE R

Ctra. Valladolid, s/n
09315 Fuentecén (Burgos)
☎: +34 947 532 693
info@bodegasanmames.com
www.bodegasanmames.com

Doble R 2019 T C
tempranillo

89

Toasty, spicy, fruity, tasty.

Doble R 2021 T
tempranillo

87

Doble R 5 Meses 2020 T RB
tempranillo

87

Doble R 5 Meses 2021 T RB
89

Balanced, spicy, sweety finish, thin, full-bodied.

Doble R Vendimia Seleccionada 2017 T
tempranillo

88

Fruity, herbal, spicy, astringent.

BODEGA EMINA

Ctra. San Bernardo, s/n
47359 Valbuena de Duero (Valladolid)
☎: +34 983 683 315
www.emina.es

Emina 2019 T C
100% tempranillo

88

Pruney, dried herbs, lactic, standard.

Emina Emoción 2015 T R
100% tempranillo

91

Colour: deep cherry. Nose: dried herbs, creamy oak, black fruit, fine reductive notes. Palate: powerful, ripe fruit, spicy, round tannins.

Emina Pasión 2021 T
100% tempranillo

88

Jammy, very fruit-driven, fruity, tasty, balanced.

DO RIBERA DEL DUERO / D.O.P.

DO RIBERA DEL DUERO / D.O.P.

BODEGA HERMANOS DEL VILLAR
Cordel de las Merinas, s/n
47490 Rueda (Valladolid)
☎: +34 983 868 904
Fax: +34 983 868 905
pablo@orodecastilla.com
www.orodecastilla.com

Gaudeamus 2020 T RB
100% tinta del país
90
Colour: dark-red cherry. Nose: toasty, spicy, cocoa bean, ripe fruit. Palate: flavourful, toasty, fine bitter notes.

Oro de Castilla 2019 T C
100% tinta del país
91
Colour: very deep cherry. Nose: roasted coffee, aromatic coffee, powerful, black fruit. Palate: smoky aftertaste, great length, round tannins.

BODEGA MATARROMERA
Ctra. Renedo- Pesquera Km. 30
47359 Valbuena de Duero (Valladolid)
www.matarromera.es

Matarromera 2019 T C
100% tempranillo
90
Colour: cherry, garnet rim. Nose: fruit preserve, powerful, dark chocolate. Palate: flavourful, sweetness, long.

Matarromera Prestigio 2016 T
100% tempranillo
93
Colour: cherry, garnet rim. Nose: balsamic herbs, ripe fruit, scrubland, old leather. Palate: flavourful, spicy, round tannins.

Melior 9 Meses 2020 T
100% tempranillo
88
Fruity, spicy, pruney, tasty.

BODEGA S. ARROYO
Avda. del Cid, 99
09441 Sotillo de La Ribera (Burgos)
☎: +34 947 532 444
Fax: +34 947 532 499
info@tintoarroyo.com
www.tintoarroyo.com

Tinto Arroyo 2015 T GR
100% tempranillo
90
Colour: bright cherry. Nose: sweet spices, ripe fruit, dark chocolate. Palate: fruity, spicy, round tannins.

Tinto Arroyo 2017 T R
100% tempranillo
90
Colour: cherry, garnet rim. Nose: fruit preserve, fruit liqueur notes, powerful. Palate: flavourful, grainy tannins, fruity.

Tinto Arroyo 2020 T RB
100% tempranillo
88
Spicy, fruity, jammy, tasty.

Tinto Arroyo Vendimia Seleccionada 2015 T FB
100% tempranillo
91
Colour: deep cherry. Nose: ripe fruit, dried herbs, creamy oak. Palate: powerful, ripe fruit, spicy, round tannins.

Tinto Arroyo 2018 T C
100% tempranillo
91
Colour: deep cherry. Nose: ripe fruit, dried herbs, creamy oak. Palate: powerful, ripe fruit, spicy, round tannins.

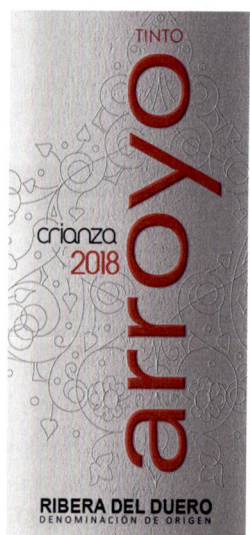

Viñarroyo 2021 RD
100% tempranillo
86

BODEGA SAN ROQUE DE LA ENCINA

San Roque, 73
09391 Castrillo de La Vega (Burgos)
☎: +34 671 486 273
exportacion@bodegasanroquedelaencina.com
www.pinadillo.com

299 de Monte Pinadillo 2018
89
Pruney, fruity, tasty.

El Notera by Monte Pinadillo 2019 T
100% tempranillo
88
Pleasant, aromatic, standard, jammy, wild, tasty.

Monte del Conde 2019 T RB
88
Pruney, fruity, herbal, tasty, toasty.

Monte Pinadillo 2018 T R
100% tempranillo
91
Colour: cherry, garnet rim. Nose: fruit preserve, powerful, sweet spices, toasty. Palate: flavourful, sweetness, long.

Monte Pinadillo 2020 T C
100% tempranillo
89
Balanced, spicy, herbal, jammy, tasty.

Monte Pinadillo 2021 T RB
100% tempranillo
88
Spicy, fruity, herbal, jammy.

Monte Pinadillo Rosado de Lágrima 2021 RD
100% tempranillo
88
Citrus fruit, balanced, crisp, sweeties.

BODEGA TIERRA ARANDA S. COOP.

San Francisco, 74
09400 Aranda de Duero (Burgos)
☎: +34 947 501 311
pedidos@vinotierraranda.es
www.vinotierraranda.es

Tierra Aranda 2016 T R
100% tempranillo
91
Colour: deep cherry. Nose: dried herbs, creamy oak, black fruit. Palate: powerful, ripe fruit, spicy, round tannins, warm.

Tierra Aranda 2019 T C
100% tempranillo
88
Fruity, herbal, spicy, tasty.

Tierra Aranda Edición Especial 2020 RD
100% tempranillo
88
Fruity, herbal, simple.

Tierra Aranda Edición Especial Viñedos Singulares 2020 T RB
100% tempranillo
88
Pleasant, jammy, tasty.

Tierra Aranda Vendimia Seleccionada 2020 T C
100% tempranillo
88
Nose: ripe fruit, creamy oak. Palate: ripe fruit, spicy.

Vega Valerio 2018 T C
100% tempranillo
88
Fruity, herbal, spicy.

BODEGA VALDRINAL

Camino de la Vega s/n, Polígono 4, Parcelas 130-133
40533 Aldehorno (Segovia)
☎: +34 634 578 839
general@valdrinal.com
www.valdrinal.com

Quinq 2016 T
92
Colour: cherry, garnet rim. Nose: spicy, ripe fruit, powerful. Palate: flavourful, fruity, good acidity, long.

DO RIBERA DEL DUERO / D.O.P.

DO RIBERA DEL DUERO / D.O.P.

Valdrinal 24 2018 T R
100% tempranillo

91

Colour: cherry, garnet rim. Nose: fruit preserve, powerful, creamy oak. Palate: flavourful, sweetness, long.

Valdrinal Entrega 2020 T RB
100% tempranillo

89

Dried flowers, jammy, dried herbs, fruity, great length, tasty.

Valdrinal SQR 2016 T

92

Colour: cherry, purple rim. Nose: fruit preserve, fruit liqueur notes, powerful, spicy. Palate: flavourful, slightly dry, soft tannins, toasty.

Valdrinal Tradición 2019 T C
100% tempranillo

91

Colour: deep cherry. Nose: ripe fruit, dried herbs, creamy oak. Palate: powerful, ripe fruit, spicy, round tannins.

BODEGA Y VIÑEDOS MARTÍN BERDUGO

Camino de la Colonia, s/n
09400 Aranda de Duero (Burgos)
☎: +34 947 506 331
bodega@martinberdugo.com
www.martinberdugo.com

Martín Berdugo 2015 T R
100% tempranillo

91

Colour: deep cherry. Nose: creamy oak, red berry notes, black fruit, wild herbs. Palate: ripe fruit, spicy, round tannins.

Martín Berdugo 2019 T C
100% tempranillo

89

Standard, spicy, dried herbs, jammy, kind finish, tasty.

Martín Berdugo 2020 T
tempranillo

88

Pleasant, tasty, spicy.

Martín Berdugo 2020 T BA
100% tempranillo

89

Kind finish, standard, jammy, toasty, tasty, balsamic herbs.

Martín Berdugo 2021 RD
tempranillo

87

Martín Berdugo 2021 T
tempranillo

89

Balanced, spicy, herbal, jammy.

MB Martín Berdugo T
tinto fino

91

Colour: deep cherry. Nose: ripe fruit, spicy, black pepper, tobacco. Palate: good structure, powerful, flavourful, fruity.

BODEGA Y VIÑEDOS NUNTIUM

Amsterdam, 61
47008 Valladolid (Valladolid)
☎: +34 983 870 037
rodrigo@nuntium.es
www.nuntium.es

De Rodrigo 2019 T
tempranillo

91

Colour: deep cherry. Nose: ripe fruit, dried herbs, fruit preserve. Palate: ripe fruit, spicy, round tannins.

Martin & Pons 2019 T
tempranillo

91

Colour: deep cherry. Nose: dried herbs, creamy oak, black fruit. Palate: ripe fruit, spicy, round tannins, balanced.

BODEGAS ABADÍA LA ARROYADA

La Tejera, s/n
09442 Terradillos de Esgueva (Burgos)
☎: +34 947 545 309
bodegas@abadialaarroyada.es
www.abadialaarroyada.es

Abadía la Arroyada 2017 T C
100% tempranillo

90

Colour: cherry, purple rim. Nose: fruit expression, red berry notes, floral, spicy, fruit preserve. Palate: flavourful, fruity, powerful, slightly dry, soft tannins.

Abadía la Arroyada 2019 T RB
100% tempranillo

88

Pleasant, standard, spicy, fruity, simple.

DO RIBERA DEL DUERO / D.O.P.

Abadía la Arroyada 2020 T
100% tempranillo
88
Balanced, spicy, fruity, herbal.

Abbatia Vendimia Seleccionada 2015 T
100% tempranillo
90
Colour: Cherry. Nose: sweet spices, scrubland, wild herbs. Palate: spicy, balsamic, flavourful, easy to drink.

Bocalaslobas 2016 T C
100% tempranillo
91
Colour: deep cherry. Nose: ripe fruit, dried herbs, creamy oak. Palate: powerful, ripe fruit, spicy, round tannins.

BODEGAS ANTÍDOTO
Ctra. a Atauta, 63B
42330 San Esteban de Gormaz (Soria)
☎: +34 975 350 493
bodegas@bodegasantidoto.com
www.bertrandsourdais.com/bodegas-antidoto

Antídoto 2020 T
tinto fino
89
Kind finish, fruity, jammy, tasty.

La Hormiga de Antídoto 2020 T
tinto fino
92
Colour: dark-red cherry. Nose: toasty, spicy, cocoa bean, black fruit, ripe fruit. Palate: flavourful, toasty, fine bitter notes.

Le Rosé de Antídoto 2020 RD
50% tinto fino, 50% albillo mayor
93
Colour: coppery red, bright. Nose: spicy, expressive, balanced, wild herbs. Palate: spicy, smoky aftertaste, great length, long.

Roselito 2021 RD
91
Colour: raspberry rose. Nose: fragrant herbs, dried flowers, citrus fruit. Palate: light-bodied, good acidity, fine bitter notes.

BODEGAS ARCANO
09317 San Martin de Rubiales (Burgos)
info@arcanowines.com
www.arcanowines.com

Arcano 2018 T
tempranillo
91 🌿
Colour: deep cherry. Nose: dried herbs, creamy oak, black fruit. Palate: powerful, ripe fruit, spicy, round tannins.

Arcano 2019 T
tempranillo
89 🌿
Standard, pruney, jammy, toasty, simple.

Carraroa 2019 T C
tempranillo
90 🌿
Colour: deep cherry, purple rim. Nose: ripe fruit, dried herbs, spicy. Palate: ripe fruit, round tannins, flavourful.

Eternal 2017 T
tempranillo
91 🌿
Colour: cherry, garnet rim. Nose: fruit preserve, fruit liqueur notes, powerful. Palate: flavourful, sweetness, long.

BODEGAS ARCO DE CURIEL
Calvario, s/n
47316 Curiel de Duero (Valladolid)
☎: +34 983 880 481
Fax: +34 983 881 766
info@arcocuriel.com
www.arcocuriel.com

Arcum 2018 T R
tempranillo
92
Colour: dark-red cherry, garnet rim. Nose: ripe fruit, aged wood nuances, tobacco, sweet spices. Palate: spicy, round tannins, long.

Arcum 2019 T C
tempranillo
89
Smoky, jammy, powerful, tasty, toasty.

Arcum 2020 T RB
87

Neptis Expresion 2017 T R
92
Colour: dark-red cherry, garnet rim. Nose: ripe fruit, fruit preserve, aged wood nuances, tobacco, sweet spices. Palate: spicy, round tannins, long.

DO RIBERA DEL DUERO / D.O.P.

Previus de Neptis 2020 T RB
91
Colour: deep cherry. Nose: ripe fruit, black fruit, dried herbs, spicy. Palate: fruity, flavourful, full-bodied.

Timonier 2019 T C
tempranillo
88
Balanced, jammy, tasty, toasty, spicy.

BODEGAS ARROCAL
Eras de Santa María, s/n
09443 Gumiel de Mercado (Burgos)
☎: +34 947 561 290
rodrigo@arrocal.com
www.arrocal.com

Arrocal 2019 T C
tinto fino
92
Colour: deep cherry. Nose: ripe fruit, creamy oak, spicy, expressive, balanced. Palate: ripe fruit, spicy, round tannins, varietal.

Arrocal Reserva de Familia 2018 T R
tinto fino
92
Colour: cherry, garnet rim. Nose: black fruit, ripe fruit, spicy, toasted bread, creamy oak. Palate: flavourful, long, round tannins, great length.

Arrocal Selección Especial 2020 T
tinto fino
91
Colour: deep cherry. Nose: ripe fruit, creamy oak, powerful. Palate: ripe fruit, spicy, round tannins, flavourful.

BODEGAS ARZUAGA NAVARRO
Ctra. Nac. 122, Km. 325
47350 Quintanilla de Onésimo (Valladolid)
☎: +34 983 681 146
bodeg@arzuaganavarro.com
www.arzuaganavarro.com

Arzuaga 2015 T GR
95% tempranillo, 5% cabernet sauvignon, merlot
92
Colour: dark-red cherry, garnet rim. Nose: fruit preserve, aged wood nuances, tobacco, sweet spices, roasted coffee. Palate: spicy, round tannins, long.

Arzuaga 2018 T R
93
Colour: dark-red cherry, garnet rim. Nose: ripe fruit, fruit preserve, aged wood nuances, tobacco, sweet spices. Palate: spicy, round tannins, long.

Arzuaga 2020 T C
94% tempranillo, 6% cabernet sauvignon
91
Colour: cherry, purple rim. Nose: ripe fruit, dried herbs, spicy, smoky. Palate: fruity, fresh, flavourful, slightly dry, soft tannins.

Arzuaga Ecológico 2018 T C
91 🌱
Colour: cherry, purple rim. Nose: floral, spicy, ripe fruit, dark chocolate. Palate: flavourful, fruity, good acidity, long.

Arzuaga Reserva Especial 2017 T R
97% tempranillo, 3% albillo
93
Colour: deep cherry. Nose: ripe fruit, dried herbs, creamy oak. Palate: powerful, ripe fruit, spicy, round tannins.

Gran Arzuaga 2015 T R
95% tempranillo, 5% albillo
92
Colour: deep cherry. Nose: dried herbs, creamy oak, black fruit. Palate: powerful, ripe fruit, spicy, round tannins.

La Planta 2021 T RB
tempranillo
90
Creamy. Colour: cherry, purple rim. Nose: fruit expression, floral, spicy, scrubland. Palate: flavourful, fruity, good acidity.

Laderas del Norte 2020 T
92 🌱
Colour: deep cherry. Nose: dried herbs, creamy oak, black fruit, lactic notes. Palate: ripe fruit, spicy, fine tannins, full-bodied.

Rosae Arzuaga 2021 RD
tempranillo
90
Crisp, floral. Colour: raspberry rose. Nose: fragrant herbs, red berry notes, ripe fruit, medium intensity. Palate: fine bitter notes, fresh, good finish.

BODEGAS ASENJO & MANSO

Ctra. Palencia, Km. 58,200
09311 La Horra (Burgos)
☎: +34 636 972 524
info@asenjo-manso.com

A&M 3 2020 T RB
tempranillo

87

A&M Autor 2015 T R
tempranillo

91

Colour: deep cherry. Nose: dried herbs, creamy oak, black fruit, woody. Palate: powerful, ripe fruit, spicy, round tannins.

Ceres 2019 T C
tempranillo

88

Fruity, jammy, spicy, toasty.

Silvanus 2019 T C
tempranillo

90

Colour: deep cherry. Nose: dried herbs, creamy oak, black fruit. Palate: ripe fruit, spicy, round tannins.

Silvanus Edición Limitada 2018 T
tempranillo

92

Colour: deep cherry. Nose: ripe fruit, dried herbs, creamy oak. Palate: powerful, ripe fruit, spicy, round tannins.

BODEGAS BALBÁS

La Majada, s/n
09311 La Horra (Burgos)
☎: +34 947 542 111
bodegas@balbas.es
www.balbas.es

Alitus 2015 T GR
tempranillo, cabernet sauvignon, merlot

93

Colour: dark-red cherry. Nose: toasty, spicy, cocoa bean, black fruit. Palate: flavourful, toasty, fine bitter notes, powerful, harsh oak tannins.

Balbás 2017 T R
tempranillo

90

Colour: dark-red cherry, cherry, purple rim. Nose: toasty, spicy, cocoa bean, ripe fruit. Palate: flavourful, toasty, powerful tannins.

Balbás 2019 T C
90% tempranillo, 10% cabernet sauvignon

90

Colour: deep cherry. Nose: ripe fruit, dried herbs, creamy oak. Palate: ripe fruit, spicy, harsh oak tannins.

Ritus 2018 T BA
75% tempranillo, 25% merlot

92

Colour: bright cherry. Nose: red berry notes, spicy, wild herbs, faded flowers. Palate: balanced, fine tannins, flavourful, fresh.

BODEGAS BRIEGO

Ctra. Cuellar, Km. 8,8
47311 Fompedraza (Valladolid)
☎: +34 983 892 156
Fax: +34 983 892 156
bodega@bodegasbriego.com
www.bodegasbriego.com

Ankal 2019 T C
100% tempranillo

90

Colour: cherry, purple rim. Nose: fruit expression, red berry notes, spicy, dried herbs. Palate: flavourful, fruity, slightly dry, soft tannins.

Ankal 2021 T RB
100% tempranillo

88

Jammy, pruney, spicy, powerful, tasty. Nose: dark chocolate.

Garabato 2021 T RB
100% tempranillo

89

Fruity, herbal, jammy, balanced, tasty.

Supernova 2019 T C
100% tempranillo

92

Colour: deep cherry. Nose: ripe fruit, dried herbs, spicy, toasty. Palate: powerful, slightly dry, soft tannins, flavourful.

Supernova Roble 2021 T RB
100% tempranillo

90

Colour: cherry, purple rim. Nose: black fruit, dried herbs, smoky, spicy. Palate: flavourful, fruity, balanced.

Tintito del Pueblo 2021 T
100% tempranillo

88

Balanced, spicy, creamy, dried herbs, jammy, woody.

DO RIBERA DEL DUERO / D.O.P.

BODEGAS BRIONES ABAD

Ctra. Fuentecén, Km. 1.2
09300 Roa (Burgos)
☎: +34 947 540 613
brionesabad@cantamuda.com
www.cantamuda.com

Cantamuda 2021 T RB
tempranillo

89

Pleasant, jammy, toasty, tasty, fruity.

Cantamuda Finca la Cebolla 2019 T
tempranillo

90

Colour: Cherry. Nose: woody, black fruit, ripe fruit. Palate: powerful, flavourful, balanced, long.

Cantamuda Parcela 64 2019 T
100% tempranillo

89

Full-bodied, spicy, jammy, creamy, coarse tannins.

BODEGAS CARRAMIMBRE

Ctra. N-122, Km. 311
47300 Peñafiel (Valladolid)
☎: +34 983 880 623
carramimbre@carramimbre.com
www.carramimbre.com

Altamimbre 2017 T
tinto fino

92

Colour: dark-red cherry. Nose: toasty, spicy, cocoa bean. Palate: flavourful, toasty, fine bitter notes.

Carramimbre 2019 T C
tinto fino

89

Pruney, fruity, spicy, tasty.

Carramimbre 2021 T RB
tinto fino

90

Colour: bright cherry. Nose: fresh fruit, creamy oak, red berry notes, black fruit. Palate: good acidity, spicy, fine tannins.

Torrepingón 2019 T C

87

Overripe, spicy, fruity, tasty.

Torrepingón 2021 T RB
88
Roasted coffee, fruity, smoky, tasty.

BODEGAS CASTILLEJO DE ROBLEDO
Ctra. Castillejo a Langa s/n
42328 Castillejo de Robledo (Soria)
☎: +34 975 355 062
Fax: +34 975 355 060
info@bodegascastillejo.com
www.bodegascastillejo.com

Silentium 2015 T R
tempranillo
91
Colour: cherry, garnet rim. Nose: balsamic herbs, ripe fruit, scrubland. Palate: flavourful, spicy, round tannins, ripe fruit, long.

Silentium 2020 T RB
100% tempranillo
89
Kind finish, defined aromas, spicy, fruity, jammy, tasty, balanced.

Silentium 2021 T RB
88
Balanced, spicy, crisp, fruity, dried herbs.

Silentium Expresión 2016 T C
100% tempranillo
92
Colour: deep cherry. Nose: creamy oak, black fruit, scrubland, toasty. Palate: powerful, ripe fruit, spicy, round tannins.

Silentium 2017 T C
tempranillo
90
Colour: Cherry. Nose: ripe fruit, sweet spices, characterful, varietal. Palate: powerful, ripe fruit, spicy, round tannins.

BODEGAS COMENGE
Camino del Castillo, s/n
47316 Curiel de Duero (Valladolid)
☎: +34 983 880 363
admin@comenge.com
www.comenge.com

Carmen by Comenge 2021 RD
albillo, tempranillo, garnacha, valenciana
91
Colour: rose. Nose: hot, faded flowers, red berry notes, ripe fruit. Palate: fleshy, flavourful, powerful, ripe fruit.

Comenge Biberius 2021 T RB
tempranillo
91
Colour: cherry, purple rim. Nose: fruit expression, red berry notes, floral, spicy. Palate: flavourful, fruity, good acidity.

Comenge El Origen 2019 T
tempranillo
93 ☘
Colour: very deep cherry. Nose: complex, expressive, spicy, mineral, ripe fruit, black fruit. Palate: elegant, full-bodied, long, great length.

Don Miguel Comenge 2017 T R
90% tempranillo, 10% cabernet sauvignon
93 ☘
Colour: dark-red cherry. Nose: toasty, spicy, cocoa bean, ripe fruit, black fruit. Palate: flavourful, toasty, fine bitter notes.

Familia Comenge Reserva 2018 T
tempranillo

93 🌱

Colour: deep cherry. Nose: dried herbs, creamy oak, black fruit, dark chocolate. Palate: powerful, ripe fruit, spicy, round tannins.

BODEGAS CONVENTO DE LAS CLARAS
Ctra. Valladolid, 0
47231 Serrada (Valladolid)
☎: +34 983 880 150
bodega@bodegasconventodelasclaras.com
www.bodegasconventodelasclaras.com

Convento Las Claras 2017 T C
100% tempranillo

92

Colour: deep cherry. Nose: ripe fruit, dried herbs, creamy oak, toasty. Palate: powerful, ripe fruit, spicy, round tannins.

BODEGAS COPABOCA
Autovía A-62, Salida 148
47100 Tordesillas (Valladolid)
☎: +34 983 395 655
comunicacion@copaboca.es
www.copaboca.com

Juan Galindo 2019 T C
tempranillo

89

Balanced, spicy, jammy, tasty, roasted coffee.

BODEGAS CRUZ DE ALBA
Camino de las Pozas, s/n
47350 Quintanilla de Onésimo (Valladolid)
☎: +34 941 310 295
info@cruzdealba.es
www.cruzdealba.es

Cruz de Alba 2019 T C

91

Colour: deep cherry. Nose: ripe fruit, dried herbs, creamy oak, cocoa bean. Palate: powerful, ripe fruit, spicy, balanced.

Finca Los Hoyales 2016 T
tempranillo

93

Colour: deep cherry. Nose: dried herbs, creamy oak, black fruit, roasted coffee. Palate: powerful, ripe fruit, spicy, round tannins.

Cruz de Alba Fuentelun 2017 T R

90

Colour: bright cherry. Nose: creamy oak, dried herbs, black fruit. Palate: spicy, correct.

DO RIBERA DEL DUERO / D.O.P.

BODEGAS CUEVAS JIMÉNEZ - FERRATUS
Ctra. Madrid-Irún, A-I km. 165
09370 Gumiel De Izán (Burgos)
☎: +34 638 007 140
bodega@ferratus.es
www.ferratus.es

Ferratus 2021 B
100% albillo mayor
90
Colour: bright straw. Nose: floral, white fruit, ripe fruit. Palate: flavourful, fresh, good acidity, fruity aftertaste.

Ferratus 2021 RD
90
Very fruit-driven. Colour: light cherry. Nose: red berry notes, floral, fine lees, rose petals. Palate: flavourful, good acidity, fruity aftertaste.

Ferratus AØ 2020 T RB
100% tempranillo
91
Colour: cherry, purple rim. Nose: spicy, ripe fruit, black fruit. Palate: flavourful, fruity, good acidity, long.

Ferratus Sensaciones 2017 T
100% tempranillo
93
Colour: dark-red cherry, garnet rim. Nose: ripe fruit, aged wood nuances, tobacco, sweet spices. Palate: spicy, round tannins, long.

BODEGAS DE LOS RÍOS PRIETO
Ctra. Pesquera-Renedo, km. 1
47315 Pesquera de Duero (Valladolid)
☎: +34 983 083 178
info@bodegasdelosriosprieto.com
www.bodegasdelosriosprieto.com

Prios Maximus 2016 T R
100% tempranillo
91
Colour: dark-red cherry, garnet rim. Nose: ripe fruit, aged wood nuances, tobacco, sweet spices. Palate: spicy, round tannins, long.

Prios Maximus 2019 T C
100% tempranillo
89
Fruity, spicy, herbal, tasty.

Prios Maximus 2021 T RB
100% tempranillo
88
Aromatic, spicy, fruity, dried herbs, tasty.

BODEGAS DOMINIO DE ATAUTA
Ctra. a Morcuera, s/n
42345 Atauta (Soria)
☎: +34 975 351 349
info@dominiodeatauta.com
www.dominiodeatauta.com

Dominio de Atauta 2019 T C
100% tinto fino
94
Colour: deep cherry. Nose: dried herbs, creamy oak, black fruit, spicy. Palate: powerful, ripe fruit, spicy, grainy tannins.

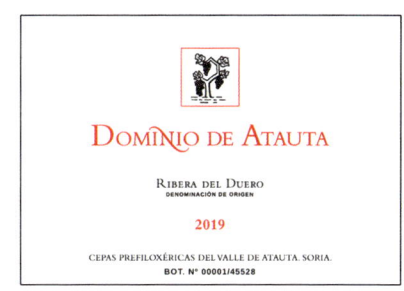

Dominio de Atauta Albillo Mayor Cepas Prefiloxérica 2020 B
albillo mayor
93
Colour: bright yellow. Nose: ripe fruit, fragrant herbs, fine lees, citrus fruit. Palate: rich, long, good acidity.

🏆 PODIUM

Dominio de Atauta La Mala 2017 T C
100% tinto fino
95
Colour: Cherry. Nose: expressive, spicy, mineral, scrubland. Palate: elegant, full-bodied, long, great length, good structure, taut.

🏆 PODIUM

Dominio de Atauta La Roza 2017 T
100% tinto fino
96
Colour: bright cherry. Nose: complex, expressive, spicy, mineral, ripe fruit, red berry notes, chalk. Palate: elegant, full-bodied, long, great length.

Dominio de Atauta Valdegatiles 2017 T
100% tinto fino
94
Colour: deep cherry. Nose: dried herbs, creamy oak, black fruit, woody. Palate: powerful, spicy, concentrated, pruney, warm, harsh oak tannins.

DO RIBERA DEL DUERO / D.O.P.

🏆 PODIUM

Dominio de Atauta
Llanos del Almendro 2017 T
100% tinto fino

97

Colour: Cherry. Nose: expressive, spicy, mineral, dark chocolate, meaty notes, red berry notes, black fruit. Palate: elegant, full-bodied, long, great length.

🏆 PODIUM

Dominio de Atauta San Juan 2017 T
100% tinto fino

95

Colour: bright cherry. Nose: complex, expressive, spicy, mineral, earthy notes, ripe fruit, black fruit. Palate: elegant, full-bodied, long, great length.

Parada de Atauta 2019 T
100% tinto fino

93

Colour: deep cherry. Nose: ripe fruit, dried herbs, creamy oak, black fruit. Palate: powerful, ripe fruit, spicy, round tannins.

BODEGAS DOMINIO DE CAIR

Ctra. Aranda - La Aguilera. km. 9
09370 La Aguilera (Burgos)
☎: +34 947 545 276
enoturismo@dominiodecair.com
www.dominiodecair.com

Cair Selección La Aguilera 2019 T
95% tempranillo, 5% merlot

91

Colour: deep cherry. Nose: ripe fruit, dried herbs, creamy oak, black fruit, dark chocolate. Palate: powerful, ripe fruit, spicy, round tannins.

Cruz del Pendón 2016 T
100% tempranillo

93

Colour: very deep cherry. Nose: complex, expressive, spicy, mineral. Palate: full-bodied, long, great length.

Las Matillas 2019 T
95% tempranillo, 5% albillo

93

Colour: deep cherry. Nose: complex, expressive, spicy, mineral, black fruit. Palate: elegant, full-bodied, long, great length.

Tierras de Cair 2018 T R
100% tempranillo

92

Colour: dark-red cherry. Nose: toasty, spicy, cocoa bean, ripe fruit, black fruit. Palate: flavourful, toasty, fine bitter notes.

BODEGAS EL LAGAR DE ISILLA

Camino Real, 1
09471 La Vid (Burgos)
☎: +34 947 530 434
administracion@lagarisilla.es
www.lagarisilla.es

El Lagar de Isilla Albillo Mayor 2020 B
100% albillo mayor

88

Citrus fruit, floral, fruity, tasty, tropical.

El Lagar de Isilla
Colección de La Familia 2016 T R
85% tinta del país, 13% cabernet sauvignon, 2% albillo mayor

93

Colour: Cherry. Nose: complex, expressive, spicy, mineral. Palate: full-bodied, long, great length, good structure, fruity, toasty.

El Lagar de Isilla Langa
de Duero 2018 T RB
100% tempranillo

92

Colour: bright cherry. Nose: expressive, balanced, dried herbs, ripe fruit, neat. Palate: flavourful, long, chalky, round tannins.

El Lagar de Isilla Matanza
de Soria 2019 T RB
100% tempranillo

91

Colour: deep cherry. Nose: ripe fruit, dried herbs, creamy oak, reduction notes. Palate: powerful, ripe fruit, spicy, round tannins.

El Lagar de Isilla
Paraje Peñalobos 2018 T RB
100% tempranillo

93

Colour: deep cherry. Nose: ripe fruit, dried herbs, creamy oak. Palate: powerful, ripe fruit, spicy, round tannins.

El Lagar de Isilla
San Juan del Monte 2019 T
100% tempranillo

93

Colour: cherry, purple rim. Nose: fruit expression, spicy, black fruit, red berry notes. Palate: flavourful, fruity, good acidity, balanced.

BODEGAS EMILIO MORO

Ctra. Peñafiel - Valoria, s/n
47315 Pesquera de Duero (Valladolid)
☎: +34 983 878 400
bodega@emiliomoro.com
www.emiliomoro.com

Clon de Familia 2016 T
tempranillo

93

Colour: deep cherry. Nose: ripe fruit, dried herbs, creamy oak. Palate: powerful, ripe fruit, spicy, round tannins.

Emilio Moro 2019 T
tempranillo

92

Colour: bright cherry. Nose: red berry notes, ripe fruit, wild herbs, spicy. Palate: flavourful, fruity, balanced, slightly dry, soft tannins.

Emilio Moro Vendimia Seleccionada 2019 T
tempranillo

92

Colour: dark-red cherry. Nose: black fruit, ripe fruit, spicy, creamy oak. Palate: flavourful, toasty, long.

Finca Resalso 2020 T
tempranillo

90

Colour: deep cherry. Nose: dried herbs, creamy oak, black fruit. Palate: ripe fruit, spicy, round tannins.

Finca Resalso 2021 T
tempranillo

90

Colour: cherry, purple rim. Nose: red berry notes, floral, spicy. Palate: flavourful, fruity, good acidity, long.

La Felisa 2020 T
tempranillo

92 ♣

Colour: deep cherry. Nose: ripe fruit, dried herbs, dried flowers. Palate: ripe fruit, spicy, slightly dry, soft tannins.

Malleolus 2019 T
100% tempranillo

92

Colour: bright cherry. Nose: ripe fruit, black fruit, dried flowers, sweet spices, creamy oak. Palate: flavourful, full-bodied, fruity, balanced, smoky aftertaste, fine tannins.

Malleolus de SanchoMartín 2018 T

93

Colour: dark-red cherry, garnet rim. Nose: ripe fruit, aged wood nuances, tobacco, sweet spices, black fruit. Palate: spicy, round tannins, long.

Malleolus de Valderramiro 2018 T
tempranillo

94

Powerful, tasty. Colour: cherry, garnet rim. Nose: characterful, powerful, creamy oak. Palate: round, flavourful, long.

BODEGAS EPIFANIO RIVERA/ERIAL

Onésimo Redondo, 45
47315 Pesquera de Duero (Valladolid)
☎: +34 983 870 109
info@epifaniorivera.com
www.epifaniorivera.com

Erial 2020 T
tinto fino

91

Colour: bright cherry. Nose: creamy oak, red berry notes, black fruit, scrubland. Palate: good acidity, spicy, fine tannins, balanced.

Erial TF (Tradición Familiar) 2019 T
tinto fino

91

Colour: deep cherry. Nose: ripe fruit, dried herbs, creamy oak, dark chocolate. Palate: powerful, ripe fruit, spicy, round tannins.

BODEGAS FÉLIX CALLEJO

Avda. del Cid, km. 16
09441 Sotillo de La Ribera (Burgos)
☎: +34 947 532 312
callejo@bodegasfelixcallejo.com
www.bodegasfelixcallejo.com

El Lebrero 2020 B
100% albillo mayor

92

Colour: bright straw. Nose: expressive, wild herbs, elegant, medium intensity. Palate: full-bodied, complex, spicy, long, elegant.

Félix Callejo Autor 2019 T
100% tinto fino

93

Colour: Cherry. Nose: complex, expressive, spicy, meaty notes. Palate: full-bodied, long, great length, taut.

Flores de Callejo 2021 T

90

Colour: cherry, purple rim. Nose: red berry notes, ripe fruit. Palate: flavourful, fruity, good acidity, long.

Majuelos de Callejo 2019 T
100% tinto fino

92

Colour: deep cherry. Nose: ripe fruit, dried herbs, creamy oak, fine reductive notes. Palate: ripe fruit, spicy, round tannins.

DO RIBERA DEL DUERO / D.O.P.

DO RIBERA DEL DUERO / D.O.P.

Parajes de Callejo 2019 T
95% tinto fino, 5% albillo mayor

93

Colour: dark-red cherry, garnet rim. Nose: ripe fruit, aged wood nuances, tobacco, sweet spices. Palate: spicy, round tannins, long.

Viña Pilar 2020 RD
50% tinto fino, 50% albillo mayor

90

Colour: onion pink. Nose: red berry notes, dried flowers, dried herbs. Palate: flavourful, fruity, fresh.

BODEGAS FUENTENARRO

Constitucion, 32
09311 La Horra (Burgos)
☎: +34 947 542 092
bodegas@fuentenarro.com
www.fuentenarro.com

Esenzias by Fuentenarro 2019 T RB
tinto fino

89

Reduced, jammy, tasty, wild. Nose: premature reduction notes.

Pérez Esteban by Fuentenarro 2018 T R
tinto fino

93

Colour: dark-red cherry, garnet rim. Nose: ripe fruit, aged wood nuances, tobacco, sweet spices. Palate: spicy, round tannins, long.

Viña Fuentenarro 2019 T C
tinto fino

88

Standard, dried herbs, jammy, roasted coffee.

Viña Fuentenarro 2020 T RB
tinto fino

88

Jammy, powerful, toasty.

BODEGAS FUENTESPINA

Camino Cascajo s/n
09470 Fuentespina (Burgos)
☎: +34 921 596 002
ana@avelinovegas.com
www.fuentespina.com

F de Fuentespina 2019 T R
tempranillo

92

Colour: dark-red cherry, garnet rim. Nose: fruit preserve, aged wood nuances, tobacco, sweet spices. Palate: spicy, round tannins, long.

Finca La Luna 2019 T R
tempranillo

88

Balanced, jammy, tasty, roasted coffee, woody.

Fuentespina 3 2021 T RB
tempranillo

86

Fuentespina C 2020 T C
tempranillo

88

Smoky, toasty, powerful, jammy.

Fuentespina R 2019 T R
tempranillo

90

Colour: dark-red cherry, garnet rim. Nose: ripe fruit, tobacco, sweet spices. Palate: spicy, round tannins.

Nicte 2020 T C
tempranillo

88

Smoky, powerful, toasty, jammy, fruity.

BODEGAS HACIENDA MONASTERIO

Ctra. Pesquera - Valbuena, s/n
47315 Pesquera de Duero (Valladolid)
☎: +34 983 484 002
bmonasterio@haciendamonasterio.com
www.haciendamonasterio.com

Hacienda Monasterio 2018 T R
80% tinto fino, 20% cabernet sauvignon

94

Colour: deep cherry. Nose: dried herbs, creamy oak, black fruit, spicy. Palate: powerful, ripe fruit, spicy, round tannins.

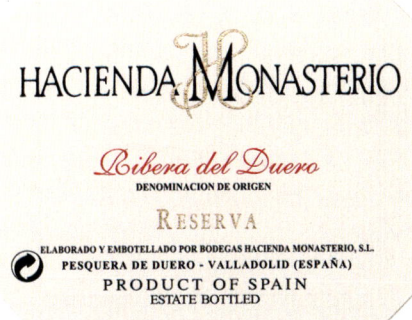

Hacienda Monasterio 2019 T

92

Colour: cherry, purple rim. Nose: black fruit, fruit preserve, spicy, dried herbs. Palate: flavourful, fruity, fresh, balanced.

Hacienda Monasterio Reserva Especial 2016 T R
75% tinto fino, 25% cabernet sauvignon

93

Colour: bright cherry. Nose: fruit preserve, fruit liqueur notes, powerful. Palate: flavourful, long, ripe fruit.

BODEGAS HEMAR

La Iglesia, 48
09315 Fuentecén (Burgos)
☎: +34 947 532 718
info@bodegashemar.com
www.bodegahemar.com

Hemar 2018 T C
100% tempranillo

90

Colour: deep cherry. Nose: dried herbs, creamy oak, black fruit. Palate: ripe fruit, spicy, round tannins.

Hemar 2020 T RB
100% tempranillo

89

Kind finish, fruity, jammy, spicy, smoky, smooth.

Llanum 2016 T R
100% tempranillo

91

Colour: dark-red cherry, garnet rim. Nose: ripe fruit, aged wood nuances, tobacco, sweet spices, black fruit. Palate: spicy, round tannins, long.

Los Jalones 2021 RD
100% tempranillo

87

Los Jalones 2021 T RB
100% tempranillo

86

BODEGAS HERCAL

Santo Domingo, 2
09300 Roa (Burgos)
☎: +34 947 541 281
ventas@somanilla.es
www.bodegashercal.com

Bocca 9 meses 2020 T RB

89

Balanced, spicy, jammy, tasty.

Corvul 20 meses Barrica 2018 T

91

Colour: dark-red cherry. Nose: toasty, spicy, cocoa bean, black fruit. Palate: flavourful, toasty, fine bitter notes.

Manvier 2021 T

88

Balanced, spicy, dried herbs, jammy.

Somanilla 16 meses Barrica 2019 T C

89

Spicy, smoky, fruity, tasty.

DO RIBERA DEL DUERO / D.O.P.

BODEGAS HERMANOS SASTRE

San Pedro, s/n
09311 La Horra (Burgos)
☎: +34 947 542 108
sastre@vinasastre.com
www.vinasastre.com

Regina Vides 2019 T
100% tempranillo

94

Colour: cherry, garnet rim. Nose: new oak, aromatic coffee, ripe fruit, powerful, roasted coffee. Palate: round tannins, balanced, fine bitter notes, good acidity, round.

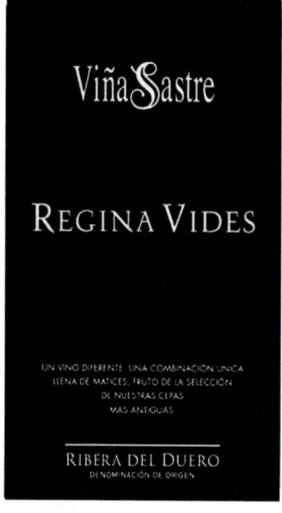

Viña Sastre 2019 T C
100% tinta del país

92

Colour: cherry, purple rim. Nose: toasty, creamy oak, ripe fruit. Palate: flavourful, balanced, toasty, smoky aftertaste, round tannins.

🏆 PODIUM

Viña Sastre Pago de Santa Cruz 2016 T GR
100% tinta del país

96

Colour: dark-red cherry, garnet rim. Nose: ripe fruit, aged wood nuances, tobacco, sweet spices. Palate: spicy, round tannins, long.

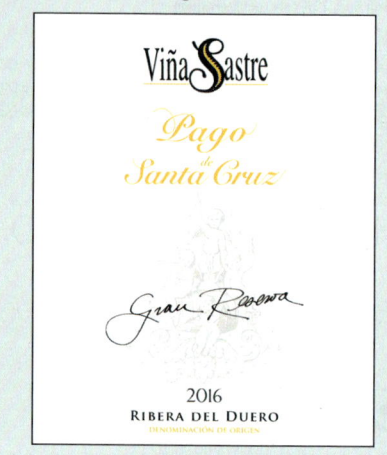

🏆 PODIUM

Viña Sastre Pesus 2016 T
80% tinta del país, 10% merlot, 10% cabernet sauvignon

95

Colour: very deep cherry. Nose: complex, expressive, spicy, mineral, black fruit. Palate: elegant, full-bodied, long, great length.

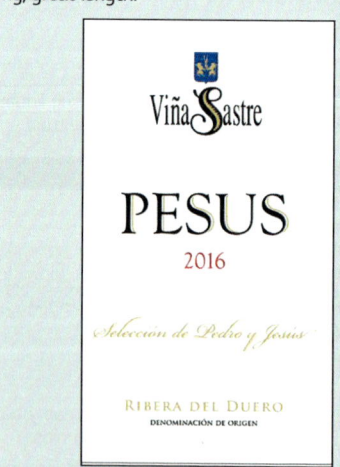

Viña Sastre 2020 T RB
100% tinta del país

91

Colour: dark-red cherry. Nose: sweet spices, black fruit, scrubland. Palate: spicy, round tannins, balanced.

Viña Sastre Marcelina Gómez 2021 RD
100% tinta del país

88

Colour: brilliant rose, purple rim. Nose: red berry notes, lactic notes, floral. Palate: balanced, pruney.

BODEGAS HESVERA
47315 Pesquera de Duero (Valladolid)
☎: +34 626 060 516
bodegashesvera@gmail.com
www.bodegashesvera.com

Hesvera 2019 T C
tinto fino

89

Pruney, balanced, spicy, jammy, toasty.

Hesvera 6 meses Barrica 2020 T RB
100% tinto fino

87

BODEGAS HNOS. PÉREZ PASCUAS – VIÑA PEDROSA
Avda. Ribera del Duero, 30
09314 Pedrosa de Duero (Burgos)
☎: +34 947 530 100
Fax: +34 947 530 002
vinapedrosa@perezpascuas.com
www.perezpascuas.com

Cepa Gavilán 2020 T C
100% tinto fino

90

Colour: very deep cherry. Nose: roasted coffee, aromatic coffee, powerful, fruit preserve. Palate: smoky aftertaste, great length, round tannins.

🏆 **PODIUM**

Pérez Pascuas Gran Selección 2015 T GR
tinto fino

95

Colour: dark-red cherry, garnet rim. Nose: ripe fruit, aged wood nuances, tobacco, sweet spices, black fruit, toasty. Palate: spicy, round tannins, long, rich, flavourful.

Viña Pedrosa 2016 T GR
100% tinto fino

94

Colour: deep cherry, garnet rim. Nose: ripe fruit, cocoa bean, cigar, toasty, fine reductive notes. Palate: flavourful, spicy, toasty, powerful tannins, sweetness.

Viña Pedrosa 2019 T R
100% tinto fino

93

Colour: deep cherry, garnet rim. Nose: aged wood nuances, ripe fruit, cocoa bean, cigar, toasty. Palate: flavourful, spicy, toasty, powerful tannins.

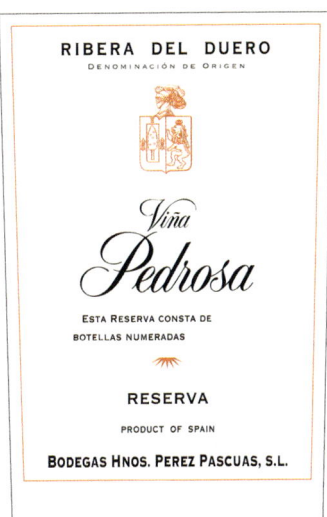

Viña Pedrosa 2020 T C
100% tinto fino

93

Colour: deep cherry. Nose: dried herbs, creamy oak, ripe fruit, red berry notes. Palate: powerful, ripe fruit, spicy, round tannins.

Viña Pedrosa Finca La Navilla 2019 T R
100% tinto fino

94

Colour: dark-red cherry. Nose: toasty, spicy, cocoa bean, chalk, ripe fruit, black fruit. Palate: flavourful, toasty, fine bitter notes.

BODEGAS ISMAEL ARROYO - VALSOTILLO
Los Lagares, 71
09441 Sotillo de La Ribera (Burgos)
☎: +34 947 532 309
bodega@valsotillo.com
www.valsotillo.com

ValSotillo 2017 T R
100% tinta del país

91

Colour: deep cherry. Nose: ripe fruit, dried herbs, creamy oak. Palate: powerful, ripe fruit, spicy, round tannins.

DO RIBERA DEL DUERO / D.O.P.

ValSotillo 2018 T C
100% tinta del país
91
Colour: deep cherry. Nose: dried herbs, creamy oak, black fruit. Palate: ripe fruit, spicy, round tannins.

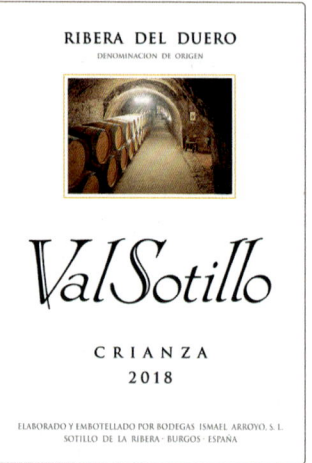

ValSotillo 2019 T
100% tinta del país
90
Colour: deep cherry. Nose: ripe fruit, dried herbs, creamy oak. Palate: ripe fruit, spicy, good structure.

ValSotillo Finca Buenavista 2018 T
100% tinta del país
92
Colour: deep cherry. Nose: ripe fruit, dried herbs, creamy oak, dark chocolate. Palate: powerful, ripe fruit, spicy, round tannins.

ValSotillo VS "40 Aniversario" 2016 T BA
100% tinta del país
93
Colour: dark-red cherry, garnet rim. Nose: ripe fruit, fruit preserve, aged wood nuances, tobacco, sweet spices. Palate: spicy, round tannins, long.

ValSotillo VS 2016 T R
100% tinta del país
93
Colour: deep cherry. Nose: ripe fruit, dried herbs, creamy oak. Palate: powerful, ripe fruit, spicy, powerful tannins.

BODEGAS IZQUIERDO
Paraje Pico El Otero, s/n Pol. 119 La Aguilera
09370 Aranda de Duero (Burgos)
☎: +34 620 941 470
bodega@bodegasizquierdo.com
www.bodegasizquierdo.com

Antonino Izquierdo Vendimia Seleccionada 2019 T
tempranillo
91
Colour: deep cherry. Nose: dried herbs, creamy oak, black fruit, roasted coffee. Palate: powerful, ripe fruit, spicy, round tannins.

Ninín 14 meses 2019 T RB
tempranillo
89
Spicy, dried herbs, jammy, pleasant, standard, fruity.

BODEGAS LA CELESTINA
Ctra. Morcuera s/n
42335 Atauta (Soria)

La Celestina 2019 T
100% tempranillo
91
Colour: deep cherry. Nose: ripe fruit, dried herbs, creamy oak. Palate: powerful, ripe fruit, spicy, round tannins.

La Celestina Vendimia Seleccionada 2018 T
100% tempranillo
90
Colour: cherry, purple rim. Nose: red berry notes, spicy, black fruit, dried herbs. Palate: fruity, good structure, toasty.

Viridiana 2021 T
100% tempranillo
88
Jammy, tasty, toasty.

BODEGAS LA HORRA
Camino de Anguix, s/n
09311 La Horra (Burgos)
☎: +34 947 613 963
rodarioja@roda.es
www.bodegaslahorra.es

Corimbo 2018 T
tinta del país
92
Colour: deep cherry. Nose: ripe fruit, dried herbs, creamy oak. Palate: powerful, ripe fruit, spicy, slightly dry, soft tannins.

Corimbo I 2016 T R
tinta del país

94
Colour: cherry, purple rim. Nose: fruit preserve, sweet spices, red berry notes. Palate: flavourful, slightly dry, soft tannins, good finish.

BODEGAS LLEIROSO
Ctra. Monasterio s/n
47359 Valbuena de Duero (Valladolid)
☎: +34 983 868 116
i.orduna@bornosbodegas.com
www.bodegaslleiroso.com

Lleiroso 2016 T R
100% tempranillo

91
Colour: cherry, garnet rim. Nose: meaty notes, fine reductive notes, balanced, characterful. Palate: flavourful, round tannins, spicy.

Lleiroso 2018 T C
100% tempranillo

88
Jammy, tasty, toasty, powerful.

Lvzmillar 2020 T RB
100% tempranillo

88
Fruity, jammy, powerful, smoky, tasty.

BODEGAS LÓPEZ CRISTÓBAL
Barrio Estación, s/n
09300 Roa (Burgos)
☎: +34 947 561 139
bodega@lopezcristobal.com
www.lopezcristobal.com

Bagús 2019 T
100% tempranillo

91
Colour: deep cherry. Nose: dried herbs, creamy oak, black fruit. Palate: ripe fruit, spicy, round tannins, balanced.

López Cristobal 2021 RD
100% tempranillo

90
Colour: rose. Nose: ripe fruit, hot, faded flowers. Palate: fleshy, flavourful, powerful, ripe fruit.

López Cristobal Albillo Mayor 2021 B S
100% albillo mayor

90
Colour: straw. Nose: ripe fruit, dried herbs, faded flowers, tropical fruit. Palate: powerful, ripe fruit, balanced.

López Cristobal La Colorada 2019 T C
100% tempranillo

91
Colour: deep cherry. Nose: dried herbs, creamy oak, black fruit. Palate: ripe fruit, spicy, round tannins, balanced.

López Cristobal Parcela 1 2018 T R
100% tinta del país

92
Colour: bright cherry. Nose: ripe fruit, balanced, varietal, spicy, smoky. Palate: fruity, spicy, round tannins.

Viracocha 2018 T BA
100% tinta del país

92
Colour: cherry, garnet rim. Nose: smoky, creamy oak, black fruit, ripe fruit, characterful. Palate: good structure, flavourful, long, great length.

BODEGAS MAESTE
09400 Aranda de Duero (Burgos)
☎: +34 603 834 379
info@bodegasmaeste.com
www.bodegasmaeste.com

Alma del Moral 2020 T
95% tempranillo, merlot, cabernet sauvignon, garnacha

92
Colour: very deep cherry, garnet rim. Nose: black fruit, ripe fruit, cocoa bean, dry stone, creamy oak, powerful. Palate: flavourful, round, round tannins, spicy, long

DO RIBERA DEL DUERO / D.O.P.

DO RIBERA DEL DUERO / D.O.P.

BODEGAS MARTA MATÉ
Camino de Caleruega s/n
09453 Tubilla del Lago (Burgos)
☎: +34 629 636 506
bodega@martamate.com
www.martamate.com

Marta Maté 2019 T C
tempranillo
92
Colour: bright cherry. Nose: complex, expressive, spicy, mineral. Palate: elegant, full-bodied, long, great length.

Viñas del Lago 2019 T
tempranillo, garnacha, albillo
89
Creamy, balanced, spicy, jammy, powerful.

BODEGAS MILVUS
Ctra. Salas de los Infantes, km 11
09490 Zazuar (Burgos)
☎: +34 662 477 974
info@bodegasmilvus.es
www.bodegasmilvus.es

Fuenconcejo 2019 T C
tempranillo
89
Aromatic, tasty, toasty, woody, full-bodied, spicy.

Fuenconcejo 2020 T RB
tempranillo
88
Jammy, spicy, toasty, smooth.

Fuenconcejo 2021 T
tempranillo
88
Pleasant, fruity, jammy, tasty.

Milvus 2021 RD
tempranillo, albillo mayor
88
Fruity, sweety finish, herbal.

Milvus 2021 T
tempranillo
90
Woody, smoky. Colour: deep cherry. Nose: powerful, toasty, black fruit. Palate: flavourful, toasty.

Milvus Edición Especial 2020 T
tempranillo
89
Fruity, roasted coffee, dried herbs, tasty.

BODEGAS NABAL
Ctra. Madrid-Irún, A-1, Salida 168
09370 Gumiel De Izán (Burgos)
☎: +34 947 544 218
info@bodegasnabal.com
www.bodegasnabal.com

Nabal 2018 T C
tempranillo
90
Colour: deep cherry. Nose: ripe fruit, dried herbs, creamy oak. Palate: powerful, ripe fruit, spicy, round tannins.

Nabal Albillo Mayor 2020 B
albillo mayor
91
Colour: bright yellow. Nose: dried flowers, candied fruit, fine lees, pattiserie. Palate: round, spicy, long, great length.

**Nabal Rosé
Rosado de Lágrima 2021 RD**
tempranillo, garnacha, albillo mayor
91
Colour: raspberry rose. Nose: red berry notes, floral, fragrant herbs, citrus fruit. Palate: light-bodied, good acidity, fine bitter notes.

Valle de Nabal 2020 T
tempranillo
89
Balanced, spicy, dried herbs, toasty.

Nabal Selección de la Familia 2016 T R
tempranillo

91

Classic. Colour: deep cherry. Nose: dried herbs, creamy oak, black fruit. Palate: powerful, ripe fruit, spicy, round tannins.

BODEGAS PAGOS DE MOGAR
Ctra. Pesquera, km. 0,2
47359 Valbuena de Duero (Valladolid)
☎: +34 983 683 011
comercial@bodegaspagosdemogar.com
www.bodegaspagosdemogar.com

Mogar 2020 T RB
100% tinta del país

87

Mogar Colección Privada 2016 T R
100% tinta del país

90

Colour: cherry, garnet rim. Nose: fruit preserve, fruit liqueur notes, powerful, sweet spices. Palate: flavourful, fruity, slightly dry, soft tannins.

Mogar Vendimia Seleccionada 2019 T C
100% tinta del país

88

Toasty, powerful, jammy.

BODEGAS PASCUAL
Ctra. de Aranda, Km. 5
09471 Fuentelcésped (Burgos)
☎: +34 947 557 351
export@bodegaspascual.com
www.bodegaspascual.com

Buró Vendimia Seleccionada 2018 T
100% tempranillo

88

Kind finish, jammy, toasty.

Diodoro Autor 2010 T
100% tempranillo

92

Colour: deep cherry, garnet rim. Nose: aged wood nuances, ripe fruit, cocoa bean, cigar, toasty. Palate: flavourful, spicy, toasty, powerful tannins.

Heredad de Peñalosa 2016 T R
100% tempranillo

89

Pleasant, tasty, wild, varietally correct, jammy, balsamic herbs, standard.

Heredad de Peñalosa 2019 T C
100% tempranillo

89

Standard, jammy, dried herbs, tasty, toasty.

Heredad de Peñalosa 2020 T RB
100% tempranillo

87

BODEGAS PEÑAFIEL
Ctra. N-122, Km. 311
47300 Peñafiel (Valladolid)
☎: +34 983 881 622
pedidos@bodegaspenafiel.com
www.bodegaspenafiel.com

Alma Serena 2020 T
tempranillo

87

Miros de Ribera 2018 T C
100% tempranillo

91

Colour: deep cherry. Nose: ripe fruit, balanced, spicy, smoky. Palate: powerful, spicy, round tannins.

Miros de Ribera 2020 T RB
tempranillo

89

Tasty, full-bodied, great length, fruity, smoky, toasty, coarse tannins.

DO RIBERA DEL DUERO / D.O.P.

Silencio de Miros 2018 T
100% tempranillo

92
Colour: bright cherry. Nose: fruit expression, red berry notes, floral, spicy. Palate: flavourful, fruity, long, roasted-coffee aftertaste.

BODEGAS PEÑALBA HERRÁIZ
Sol de las Moreras,3
09400 Aranda de Duero (Burgos)
☎: +34 617 331 609
Fax: +34 947 508 249
oficina@carravid.com
www.carravid.com

Aptus 2019 T RB
95% tempranillo, 5% otras

88
Pruney, toasty, powerful.

Aptus 2020 T RB
95% tempranillo, 5% otras

88
Fruity, spicy, jammy, tasty.

Carravid 2016 T C
100% tempranillo

88
Dried herbs, jammy, tasty, full-bodied, standard.

BODEGAS PINEA DEL DUERO
Camino de Valdeguzmán s/n
09314 Quintanamanvirgo (Burgos)
☎: +34 670 893 946
carmen@pinea.wine
www.pinea.wine

17 by Pinea 2019 T C
tempranillo

93
Varietally correct. Colour: very deep cherry. Nose: complex, expressive, spicy, ripe fruit, fruit preserve. Palate: elegant, full-bodied, long, great length.

Pinea 2017 T FB
100% tempranillo

94
Colour: bright cherry. Nose: sweet spices, ripe fruit, dark chocolate, creamy oak. Palate: spicy, round tannins, flavourful.

Pinea 2018 T FB
tempranillo

92
Colour: dark-red cherry. Nose: toasty, spicy, cocoa bean, ripe fruit, dark chocolate, aromatic coffee. Palate: flavourful, toasty, fine bitter notes.

BODEGAS PINNA FIDELIS
47300 Peñafiel (Valladolid)
☎: +34 983 878 034
info@pinnafidelis.com
www.pinnafidelis.com/es

Pinna Fidelis 2018 T R
tinta del país

90
Colour: dark-red cherry. Nose: black fruit, ripe fruit, spicy, creamy oak, tobacco, black pepper. Palate: flavourful, long.

Pinna Fidelis 2019 T C
tinta del país

89
Pruney, spicy, fruity, herbal, tasty.

Pinna Fidelis 2020 T RB
tinta del país

88
Toasty, jammy, kind finish.

Pinna Fidelis Roble Español 2019 T BA
tinta del país

90
Colour: cherry, purple rim. Nose: spicy, toasty, fruit preserve, smoky, dried herbs. Palate: fruity, flavourful, roasted-coffee aftertaste.

Pinna Fidelis Vendimia Seleccionada 2018 T
tinta del país

92
Colour: deep cherry. Nose: ripe fruit, dried herbs, creamy oak, expressive. Palate: powerful, ripe fruit, spicy, round tannins.

BODEGAS PORTIA
Antigua Ctra. N-I, km. 170
09370 Gumiel De Izán (Burgos)
☎: +34 947 102 700
Fax: +34 947 512 140
info@bodegasportia.com
www.bodegasportia.com

Portia 2019 T C
tempranillo

89
Spicy, toasty, fruity, jammy.

Portia 2020 T RB
tempranillo

88
Pleasant, jammy, fruity, toasty.

Portia Prima "La Encina" 2020 T
tempranillo
91
Nose: ripe fruit, black fruit, creamy oak, spicy, dried herbs. Palate: flavourful, fruity, powerful tannins, good finish.

Portia Summa 2019 T R
tempranillo
90
Needs time. Colour: Cherry. Nose: new oak, aromatic coffee, ripe fruit. Palate: powerful, harsh oak tannins, fruity.

Triennia de Bodegas Portia 2019 T
tempranillo
90
Colour: cherry, purple rim. Nose: black fruit, fruit preserve, black pepper, toasty. Palate: flavourful, fresh, powerful, slightly dry, soft tannins, good finish, roasted-coffee aftertaste.

Valpreciado 2019 T C
100% tempranillo
89
Spicy, toasty, fruity, jammy, tasty, balanced.

Valpreciado Prima 2020 T
91
Colour: deep cherry. Nose: dried herbs, creamy oak, black fruit, ripe fruit. Palate: powerful, ripe fruit, spicy, round tannins.

Valpreciado Roble 2020 T RB
88
Pleasant, fruity, jammy, toasty, simple, smooth.

BODEGAS PRADO DE OLMEDO
Paraje El Salegar, s/n
09370 Quintana del Pidío (Burgos)
☎: +34 947 546 960
Fax: +34 947 546 960
pradodeolmedo@pradodeolmedo.com
www.pradodeolmedo.com

Monasterio de San Miguel 2019 T C
tinta del país
90
Colour: dark-red cherry. Nose: toasty, spicy, cocoa bean. Palate: flavourful, toasty, fine bitter notes.

Monasterio de San Miguel 2020 T RB
tinta del país
88
Woody, tasty, great length, jammy, full-bodied.

Monasterio de San Miguel 2021 B
albillo mayor
90
Colour: straw. Nose: ripe fruit, dried herbs, faded flowers. Palate: powerful, ripe fruit, balanced.

BODEGAS R & G
Ramón y Cajal 7, 1ºA
01007 Vitoria-Gasteiz (Araba/Álava)
☎: +34 945 150 589
araex@araex.com
www.araex.com

R&G Ribera 2019 T
85% tempranillo, 15% merlot
93
Colour: dark-red cherry, garnet rim. Nose: ripe fruit, aged wood nuances, tobacco, sweet spices, dark chocolate. Palate: spicy, round tannins, long.

BODEGAS RAIZ DE GUZMÁN
Ctra. Circunvalación R-30 s/n
09300 Roa (Burgos)
☎: +34 947 541 191
info@raizyparamodeguzman.es
www.raizdeguzman.es

Raíz de Guzmán 2018 T R
91
Colour: deep cherry. Nose: ripe fruit, dried herbs, creamy oak, woody. Palate: powerful, ripe fruit, spicy, harsh oak tannins.

Raíz de Guzmán 2019 T C
100% tempranillo
92
Colour: deep cherry. Nose: ripe fruit, dried herbs, creamy oak, sweet spices, dark chocolate. Palate: powerful, ripe fruit, spicy, round tannins.

DO RIBERA DEL DUERO / D.O.P.

Raíz de Guzmán 2021 RD
100% tempranillo

90

Colour: brilliant rose, purple rim. Nose: fruit expression, red berry notes, lactic notes, floral. Palate: balanced, fruity aftertaste, easy to drink.

Raíz de Guzmán 9 meses 2020 T RB
100% tempranillo

90

Needs time. Colour: deep cherry. Nose: dried herbs, creamy oak, black fruit. Palate: ripe fruit, spicy, round tannins.

Raiz Profunda 2018 T
100% tempranillo

93

Colour: deep cherry, garnet rim. Nose: varietal, balanced, characterful, spicy, chalk. Palate: flavourful, long, good structure.

Raíz Voy Olé 2021 T
100% tempranillo

90

Colour: cherry, purple rim. Nose: powerful, ripe fruit, balanced. Palate: ripe fruit, flavourful.

BODEGAS RESALTE DE PEÑAFIEL
Ctra. N-122, Km. 312
47300 Peñafiel (Valladolid)
☎: +34 983 878 160
info@resalte.com
www.resalte.com

Albor de Resalte 2020 B
100% albillo

89

Fruity, herbal, tasty, citrus fruit.

Gran Resalte 2015 T GR
100% tempranillo

92

Colour: deep cherry. Nose: dried herbs, creamy oak, black fruit. Palate: powerful, ripe fruit, spicy, harsh oak tannins.

Origen de Resalte 2019 T
100% tempranillo

92

Colour: bright cherry. Nose: sweet spices, ripe fruit, dark chocolate, characterful, powerful. Palate: fruity, spicy, round tannins.

Resalte Expresión 2018 T R
100% tempranillo

92

Colour: cherry, garnet rim. Nose: ripe fruit, dried herbs, creamy oak, sweet spices. Palate: ripe fruit, round tannins, flavourful, varietal.

Resalte Vendimia Seleccionada 2020 T
100% tempranillo

91

Colour: bright cherry. Nose: creamy oak, red berry notes, black fruit, wild herbs. Palate: good acidity, spicy, fine tannins.

BODEGAS RODERO
Ctra. Boada, s/n
09314 Pedrosa de Duero (Burgos)
☎: +34 947 530 046
rodero@bodegasrodero.com
www.bodegasrodero.com

Carmelo Rodero 2019 T R
90% tempranillo, 10% cabernet sauvignon

92

Colour: dark-red cherry, garnet rim. Nose: fruit preserve, overripe fruit, waxy notes, toasty. Palate: warm, pruney, sweet tannins.

Carmelo Rodero TSM 2019 T
75% tempranillo, 10% cabernet sauvignon, 15% merlot

92

Colour: deep cherry. Nose: ripe fruit, dried herbs, creamy oak, dried flowers. Palate: powerful, ripe fruit, spicy, slightly dry, soft tannins.

Pago de Valtarreña 2019 T
100% tempranillo

92

Colour: very deep cherry. Nose: roasted coffee, aromatic coffee, powerful, black fruit, fruit preserve. Palate: smoky aftertaste, great length, round tannins.

BODEGAS RODRÍGUEZ Y SANZO
Manuel Azaña, 9
47014 Valladolid (Valladolid)
☎: +34 983 150 150
comunicacion@valsanzo.com
www.rodriguezsanzo.com

Valsanzo 2019 T C
100% tinto fino

89

Nose: fruit preserve, powerful, dark chocolate. Palate: flavourful, sweetness.

DO RIBERA DEL DUERO / D.O.P.

BODEGAS SANTA EULALIA
Malpica, s/n
09311 La Horra (Burgos)
☎: +34 983 586 868
Fax: +34 947 580 180
bodegasfrutosvillar@bodegasfrutosvillar.com
www.bodegasfrutosvillar.com

Conde de Siruela 2018 T C
100% tinta del país

90

Colour: cherry, purple rim. Nose: fruit expression, red berry notes, spicy. Palate: flavourful, fruity, taut.

Conde de Siruela 2020 T RB
100% tinta del país

87

Conde de Siruela 2021 T
100% tinta del país

86

La Horra 2019 T C
tinta del país

89

Spicy, herbal, fruity, tasty.

Riberal 2016 T R
tinta del país

90

Colour: dark-red cherry, garnet rim. Nose: fruit preserve, aged wood nuances, tobacco. Palate: spicy, round tannins.

BODEGAS SEÑORÍO DE ALDEA
Real, 68
42345 Aldea de San Esteban (Soria)
☎: +34 637 062 809
info@bodegasenoriodealdea.com
www.bodegasenoriodealdea.com

Agoris 2019 T RB
tempranillo, garnacha

88

Pleasant, jammy, tasty.

Agoris Viñedos Centenarios 2014 T R
tempranillo, garnacha, albillo mayor

91

Colour: dark-red cherry, garnet rim. Nose: fruit preserve, aged wood nuances, tobacco, sweet spices. Palate: spicy, round tannins, long.

Albus 2019 B RB
albillo mayor

91

Colour: bright straw. Nose: ripe fruit, fragrant herbs, fine lees. Palate: full-bodied, rich, long, good acidity.

Kamikaze de Aldea 2020 RD
tempranillo, albillo mayor

88

Citrus fruit, dried flowers, subtle, crisp.

Señor de Aldea 2015 T C
tempranillo

88

Rustic, coarse tannins, jammy, tasty.

Viejos Capos 2015 T R
tempranillo, garnacha, albillo mayor

92

Colour: cherry, garnet rim. Nose: fruit preserve, fruit liqueur notes, powerful, roasted coffee. Palate: flavourful, sweetness, long.

BODEGAS SEÑORÍO DE NAVA
Crta. Valladolid a Soria, km 63
09318 Nava de Roa (Burgos)
☎: +34 987 209 712
lafinca@lafinca.es
www.senoriodenava.es

Fuentenebro Tinto Fino 2020 T
100% tempranillo

91

Colour: bright cherry. Nose: red berry notes, black fruit, wild herbs, spicy. Palate: good acidity, fine tannins, flavourful, elegant.

Señorío de Nava 2016 T R
100% tempranillo

89

Pleasant, balsamic herbs, spicy, jammy, tasty.

Señorío de Nava 2017 T C
100% tempranillo

90

Colour: dark-red cherry, garnet rim. Nose: ripe fruit, aged wood nuances, tobacco, sweet spices. Palate: spicy, round tannins, long.

Señorío de Nava 2020 T RB
100% tempranillo

88

Toasty, jammy, tasty.

Señorío de Nava 2021 T
100% tempranillo

88

Balanced, spicy, herbal, jammy.

DO RIBERA DEL DUERO / D.O.P.

BODEGAS SERVILIO - ARRANZ
Onésimo Redondo, 39
47315 Pesquera de Duero (Valladolid)
☎: +34 983 870 062
bodega@bodegaservilio.com
www.bodegaservilio.com

Diego Rivera 2019 T C
100% tempranillo
90
Colour: deep cherry. Nose: ripe fruit, dried herbs, toasty, waxy notes. Palate: ripe fruit, spicy, round tannins, flavourful.

Diego Rivera 2020 T RB
100% tempranillo
88
Spicy, fruity, jammy, tasty, crisp.

Servilio 2019 T C
100% tempranillo
87

Servilio 2020 T RB
100% tempranillo
87

Servilio Vendimia Seleccionada 2016 T
100% tempranillo
89
Pruney, fruity, jammy.

BODEGAS TARSUS
Ctra. de Roa - Anguix, Km. 3
09313 Anguix (Burgos)
☎: +34 947 554 218
Teresa.Rodriguez@pernod-ricard.com
www.tarsusvino.com

Tarsus 2015 T R
98% tinta del país, 2% cabernet sauvignon
91
Colour: dark-red cherry. Nose: toasty, spicy, cocoa bean. Palate: flavourful, toasty, fine bitter notes.

Tarsus 2019 T C
100% tinta del país
91
Colour: deep cherry. Nose: ripe fruit, dried herbs, creamy oak, black fruit. Palate: powerful, ripe fruit, spicy, round tannins.

Tarsus Finca El Canto Vino de Autor 2018 T
100% tinta del país
93
Colour: deep cherry. Nose: dried herbs, creamy oak, black fruit, cocoa bean. Palate: powerful, ripe fruit, spicy, round tannins.

Tarsus La Despistada 2020 B
91
Colour: bright straw. Nose: ripe fruit, dried herbs, faded flowers. Palate: powerful, ripe fruit, balanced.

Tarsus Selección de Viñedos 2018 T
100% tinta del país
91
Colour: cherry, garnet rim. Nose: ripe fruit, spicy, dark chocolate. Palate: flavourful, good structure, round tannins, spicy, long.

Tarsus Terno Vino de Autor 2018 T
tinta del país
93
Colour: cherry, garnet rim. Nose: balanced, characterful, ripe fruit, sweet spices. Palate: flavourful, long, ripe fruit, spicy, great length, good structure, fruity.

BODEGAS TEÓFILO REYES
Ctra. Valladolid, s/n
47300 Peñafiel (Valladolid)
☎: +34 983 873 015
info@teofiloreyes.com

Tamiz de Teófilo Reyes 2021 T RB
100% tempranillo
89
Balanced, spicy, jammy, full-bodied, toasty.

Teófilo Reyes 15 meses 2020 T C
90
Colour: deep cherry. Nose: dried herbs, creamy oak, roasted coffee, cocoa bean, black fruit. Palate: powerful, ripe fruit, spicy, round tannins.

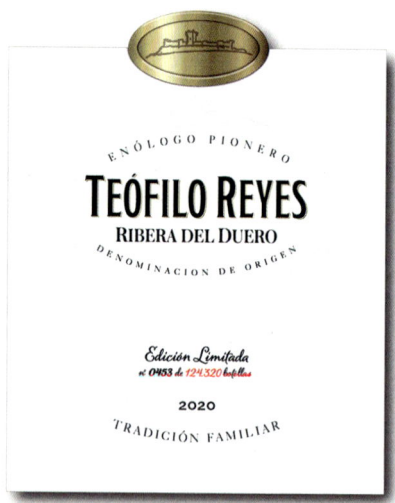

Teófilo Reyes 2017 T R
100% tempranillo
91
Colour: deep cherry. Nose: ripe fruit, dried herbs, smoky, spicy. Palate: ripe fruit, spicy, round tannins, balanced.

BODEGAS THESAURUS RIBERA DEL DUERO
Ctra. Cuellar, 17
47359 Olivares de Duero (Valladolid)
☎: +34 983 250 319
comercial@ciadevinos.com

Castillo de Peñafiel 2019 T C
100% tempranillo
88
Balanced, spicy, dried herbs, jammy.

Castillo de Peñafiel 2020 T RB
100% tempranillo
88
Balanced, spicy, dried herbs, tasty, toasty.

Castillo de Peñafiel Edición Limitada 2016 T R
100% tempranillo
90
Age nuances, pruney, balanced, dried herbs, tasty.

Dorivm 2019 T C
100% tempranillo
88
Full-bodied, balanced, spicy, dried herbs, jammy.

Dorivm 2020 T RB
100% tempranillo
88
Standard, spicy, dried herbs, toasty.

Dorivm Selección de la Familia 2016 T R
100% tempranillo
91
Colour: dark-red cherry, garnet rim. Nose: fruit preserve, aged wood nuances, tobacco, sweet spices. Palate: spicy, round tannins, balanced.

Thesaurus X 2019 T
100% tempranillo
89 ♣
Balanced, spicy, dried herbs, toasty.

BODEGAS TORREDEROS
Ctra. Valladolid, Km. 289,300
09318 Fuentelisendo (Burgos)
☎: +34 947 532 627
Fax: +34 947 532 731
administracion@torrederos.com
www.torrederos.com

Torrederos 2015 T R
100% tempranillo
91
Colour: deep cherry. Nose: ripe fruit, dried herbs, creamy oak, toasty. Palate: powerful, ripe fruit, spicy, harsh oak tannins.

Torrederos 2018 T C
tempranillo
91
Colour: deep cherry. Nose: ripe fruit, dried herbs, creamy oak. Palate: powerful, ripe fruit, spicy, round tannins.

Torrederos 2020 T RB
tempranillo
87

Torrederos 2021 RD
tempranillo
88
Pleasant, tasty, smooth, fruity.

Torrederos Selección 2016 T
91
Colour: cherry, garnet rim. Nose: fruit preserve, fruit liqueur notes, powerful. Palate: flavourful, sweetness, long.

BODEGAS TORREMORÓN
Ctra. Boada, s/n
09314 Quintanamanvirgo (Burgos)
☎: +34 947 554 075
administracion@bodegastorremoron.com
www.bodegastorremoron.com

Senderillo 2021 T
86

Tiberio 2021 T
90
Colour: cherry, purple rim. Nose: fruit expression, red berry notes, sweet spices. Palate: fruity, flavourful, balanced.

Torremorón 2019 T C
100% tempranillo
89
Spicy, fruity, tasty.

DO RIBERA DEL DUERO / D.O.P.

DO RIBERA DEL DUERO / D.O.P.

Torremorón 2020 T RB
100% tempranillo
87

Torremorón 2021 T
100% tempranillo
87

BODEGAS TRASLASCUESTAS
Ctra. Pedrosa a Mambrilla, s/n
09314 Valcavado de Roa (Burgos)
☎: +34 947 542 851
Fax: +34 947 542 861
administracion@bodegastraslascuestas.com

Traslascuestas 2019 T C
tempranillo
90
Reduced, great length. Colour: cherry, garnet rim. Nose: old leather, waxy notes, dried herbs. Palate: fruity, balanced, fine bitter notes, long, ripe fruit, spicy.

Traslascuestas 2020 RD FB
tempranillo
89
Fruity, dried flowers, tasty, balanced.

Traslascuestas 2020 T RB
tempranillo
88
Balanced, spicy, fruity, dried herbs, jammy.

BODEGAS TRENZA
Felix Mendelsohn, 8
03730 Jávea (Alacant/Alicante)
☎: +34 965 790 012
bodegas@bodegastrenza.com
www.bodegatrenza.com

Tofterup Brothers Tempranillo 2019 T
100% tempranillo
91
Colour: cherry, garnet rim. Nose: red berry notes, ripe fruit, dried flowers. Palate: powerful, ripe fruit, good acidity, fine bitter notes.

Viña Curvada Albillo Mayor 2019 B BA
100% albillo mayor
88
Balanced, herbal, toasty, spicy.

Viña Curvada Tempranillo 2019 T
100% tempranillo
88
Jammy, fruity, toasty. Palate: flavourful.

Viña Curvada Tempranillo 2017 T C
100% tempranillo
90
Colour: dark-red cherry. Nose: toasty, spicy, cocoa bean, ripe fruit, black fruit. Palate: flavourful, toasty, fine bitter notes.

BODEGAS TRUS
Ctra. Pesquera de Duero-Encinas km. 3
47316 Piñel de Abajo (Valladolid)
☎: +34 941 447 207
info@palaciosvinosdefinca.com
www.palaciosvinosdefinca.com/trus/

🏆 PODIUM

Pico de Luyas 2019 T
tinto fino
95
Colour: Cherry. Nose: complex, spicy, elegant, expressive, ripe fruit, neat. Palate: full-bodied, long, great length, flavourful, round tannins, fruity.

Punto Geodésico 2019 T
tinto fino
94
Colour: deep cherry. Nose: ripe fruit, dried herbs, balanced. Palate: powerful, ripe fruit, spicy, round tannins.

San Acislo 2019 T C
91
Colour: deep cherry. Nose: ripe fruit, dried herbs, creamy oak, fruit expression. Palate: powerful, ripe fruit, spicy, round tannins.

Trus 2017 T R
tinto fino
94
Colour: cherry, purple rim. Nose: fruit expression, red berry notes, floral, spicy. Palate: flavourful, fruity, long, fresh, powerful.

Trus 2019 T C
tinto fino
93
Colour: deep cherry. Nose: ripe fruit, dried herbs, creamy oak, sweet spices. Palate: powerful, ripe fruit, spicy, round tannins.

Trus 2021 T RB
tinto fino
91
Colour: cherry, purple rim. Nose: red berry notes, spicy, balanced. Palate: flavourful, fruity, good acidity, long.

BODEGAS VALDEMAR
Camino Viejo de Logroño, 24
01320 Oyón (Araba/Álava)
☎: +34 945 622 188
Fax: +34 945 622 111
adedios@valdemar.es
www.momentosvaldemar.com

Fincas de Valdemacuco 2019 T C
100% tempranillo
91
Colour: deep cherry. Nose: dried herbs, creamy oak, black fruit. Palate: powerful, ripe fruit, spicy, round tannins.

Fincas de Valdemacuco 2021 T RB
100% tempranillo
90
Colour: bright cherry. Nose: fresh fruit, creamy oak, black fruit, spicy. Palate: good acidity, spicy, fine tannins.

BODEGAS VALDUBÓN
Antigua Ctra. N-I, Km. 151
09400 Milagros (Burgos)
☎: +34 947 546 251
enoturismo@valdubon.com
www.valdubon.com

Honoris de Valdubón 2018 T
tempranillo, merlot
89
Spicy, jammy, slight reduction, standard, tasty, balanced, dried herbs.

Valdubón 2020 T RB
tempranillo
87

Valdubón 2021 RD
tempranillo
88
Kind finish, fruity, jammy, herbal.

BODEGAS VALLE DE MONZÓN
Paraje El Salegar, s/n
09370 Quintana del Pidío (Burgos)
☎: +34 947 545 694
Fax: +34 947 545 694
bodega@vallemonzon.com
www.vallemonzon.com

El Salegar 2018 T C
tinta del país
89
Full-bodied, balanced, spicy, jammy.

El Salegar 2020 T RB
tinta del país
85

Hoyo de la Vega 2015 T R
tinta del país
90
Colour: deep cherry. Nose: dried herbs, creamy oak, black fruit. Palate: powerful, ripe fruit, spicy, round tannins.

Hoyo de la Vega 2018 T C
tinta del país
88
Jammy, toasty, spicy.

Hoyo de la Vega 2019 RD
tinta del país, albillo
88
Aromatic, pruney, jammy, yeasty notes, tasty, bitter.

DO RIBERA DEL DUERO / D.O.P.

DO RIBERA DEL DUERO / D.O.P.

BODEGAS VALPARAISO
Paraje los Llanillos, s/n
09370 Quintana del Pidío (Burgos)
☎: +34 947 545 286
marketing@grupoeguizabal.com
www.bodegasvalparaiso.com

Raíces de Valparaiso 2019 T
100% tempranillo
92
Colour: deep cherry. Nose: ripe fruit, dried herbs, creamy oak. Palate: powerful, ripe fruit, spicy.

Valparaíso 2019 T C
100% tempranillo
88
Balanced, spicy, dried herbs, jammy.

Valparaíso 2020 T RB
100% tempranillo
88
Spicy, jammy, great length, toasty.

BODEGAS VEGA DE YUSO
Basilón 9 – Cañada Real s/n
47350 Quintanilla de Onésimo (Valladolid)
☎: +34 983 680 054
exportwine@vegadeyuso.com
www.vegadeyuso.com

Pozo de Nieve 2021 T
100% tempranillo
88
Kind finish, jammy, fruity, tasty, standard.

Tres Matas 2017 T R
100% tempranillo
90
Colour: very deep cherry. Nose: roasted coffee, aromatic coffee, powerful. Palate: smoky aftertaste, great length, round tannins.

Tres Matas 2019 T C
100% tempranillo
90
Colour: deep cherry. Nose: dried herbs, creamy oak, black fruit. Palate: ripe fruit, spicy, round tannins.

Tres Matas Vendimia Seleccionada 2018 T
100% tempranillo
90
Colour: cherry, purple rim. Nose: fruit expression, spicy, black fruit. Palate: flavourful, fruity, slightly dry, soft tannins.

Vegantigua 2020 T RB
100% tempranillo
88
Spicy, jammy, toasty, tasty.

BODEGAS VEGA SICILIA
Ctra. N-122, km 323
47359 Valbuena de Duero (Valladolid)
☎: +34 983 680 147
Fax: +34 983 680 263
vegasicilia@vega-sicilia.com
www.temposvegasicilia.com

🏆 **PODIUM**

Valbuena 5º 2018 T
96
Colour: cherry, garnet rim. Nose: fruit expression, black fruit, lactic notes, expressive, spicy, aromatic coffee. Palate: good structure, fruity, powerful, round tannins, balanced, long.

🏆 **PODIUM**

Vega Sicilia Único 2013 T
97% tempranillo, 3% cabernet sauvignon
96
Crisp, elegant. Colour: cherry, garnet rim. Nose: red berry notes, scrubland, balsamic herbs, chalk. Palate: fruity, easy to drink, good acidity, taut, chalky.

🏆 **PODIUM**

Vega Sicilia Único Reserva Especial T GR
95% tempranillo, 5% cabernet sauvignon
99
Classic, with personality, complex, defined aromas. Colour: dark-red cherry. Nose: spicy, black fruit, elegant, toasty, dark chocolate, fine reductive notes, waxy notes. Palate: spicy, ripe fruit, long, great length, round tannins, elegant, round.

BODEGAS VETUSTA
Avda. Portugal, 54
09400 Aranda de Duero (Burgos)
☎: +34 947 556 992
info@bodegasvetusta.com
www.bodegasvetusta.com

Vetusta Viñedo Especial Carrascalon Alto 2018 T
tempranillo
92
Colour: bright cherry. Nose: ripe fruit, black fruit, creamy oak, dried herbs. Palate: fruity, fresh, flavourful, balanced, silky tannins.

Vetusta 2018 T C
tempranillo
88
Spicy, jammy, hot, toasty.

Vetusta Viñas de Fuentenebro 2021 T
tempranillo
91 🌿
Fruity, taut. Colour: cherry, purple rim. Nose: red berry notes, floral, spicy, dried herbs. Palate: flavourful, fruity, good acidity, long.

BODEGAS VILANO
Avda. Tinta del País 2
09314 Pedrosa de Duero (Burgos)
☎: +34 947 530 029
Fax: +34 947 530 037
comunicacion@vilano.com
www.vilano.com

La Baraja 2019 T
tempranillo, cabernet sauvignon, merlot
90
Colour: deep cherry. Nose: ripe fruit, dried herbs, creamy oak. Palate: powerful, ripe fruit, spicy, round tannins.

Terra Incógnita 2019 T
tempranillo
92
Colour: deep cherry, garnet rim. Nose: aged wood nuances, ripe fruit, cocoa bean, cigar, toasty. Palate: flavourful, spicy, toasty, powerful tannins.

Vilano 2018 T R
90
Colour: deep cherry. Nose: ripe fruit, dried herbs, creamy oak, dark chocolate, woody. Palate: powerful, ripe fruit, spicy, harsh oak tannins.

Vilano 2019 T
tempranillo
92
Colour: cherry, garnet rim. Nose: fruit preserve, powerful, dark chocolate, toasty, new oak. Palate: flavourful, sweetness, long.

BODEGAS VIRTUS
Pago de Fuentecilla s/n
47313 Aldeayuso (Valladolid)
☎: +34 983 878 080
virtus@virtuswine.com
www.virtuswine.com

El Sueco 2018 T C
tempranillo
91
Colour: cherry, garnet rim. Nose: ripe fruit, scrubland, old leather. Palate: flavourful, balsamic, spicy, soft tannins.

El Sueco Albillo Mayor 2020 B
albillo mayor
88
Pleasant, floral, very fruit-driven, jammy, standard, crisp. Nose: tropical fruit.

Virtus 2015 T GR
tempranillo
92
Colour: dark-red cherry, garnet rim. Nose: ripe fruit, aged wood nuances, tobacco, sweet spices. Palate: spicy, round tannins, long.

BODEGAS VIYUELA
Ctra. de Quintanamanvirgo, s/n
09314 Boada de Roa (Burgos)
☎: +34 947 530 072
viyuela@bodegasviyuela.com
www.bodegasviyuela.com

Viyuela 2018 T R
100% tempranillo
88
Creamy, full-bodied, jammy, roasted coffee.

Viyuela 2019 T C
100% tempranillo
88
Pleasant, toasty, tasty, jammy.

Viyuela 2020 T RB
100% tempranillo
87

Viyuela Selección 2019 T
100% tempranillo
88
Powerful, hot, dried herbs, jammy, tasty.

Viyuela X Aniversario 2016 T
100% tempranillo
89
Balanced, spicy, creamy, roasted coffee, woody.

DO RIBERA DEL DUERO / D.O.P.

DO RIBERA DEL DUERO / D.O.P.

BODEGAS VIZCARRA
Finca Chirri, s/n
09317 Mambrilla de Castrejón (Burgos)
☎: +34 947 540 340
Fax: +34 947 540 340
bodegas@vizcarra.es
www.vizcarra.es

Alejandra Vizcarra 2020 B
albillo
91
Colour: straw. Nose: toasted bread, creamy oak, ripe fruit, woody. Palate: flavourful, rich, toasty, good acidity.

🏆 **PODIUM**

Celia Vizcarra 2018 T
93% tinto fino, 7% garnacha
95
Colour: deep cherry. Nose: complex, expressive, spicy, mineral, new oak, creamy oak. Palate: elegant, full-bodied, long, great length.

🏆 **PODIUM**

Inés Vizcarra 2018 T R
90% tinto fino, 10% merlot
95
Colour: cherry, garnet rim. Nose: balanced, powerful, ripe fruit, spicy. Palate: flavourful, round tannins, balanced, long.

Rosado Vizcarra 2021 RD
100% tinto fino
92
Colour: light cherry. Nose: red berry notes, ripe fruit, balanced. Palate: flavourful, fruity, full of life, good acidity.

Vizcarra 2019 T
100% tinto fino
93
Colour: deep cherry. Nose: dried herbs, creamy oak, ripe fruit, black fruit, sweet spices, dark chocolate. Palate: powerful, ripe fruit, spicy, round tannins.

Vizcarra Senda del Oro 2021 T
100% tinto fino
92
Colour: cherry, purple rim. Nose: fruit expression, red berry notes, floral, spicy, toasty, creamy oak. Palate: flavourful, fruity, good acidity, long.

Vizcarra Torralvo 2015 T GR
100% tinto fino
94
Colour: deep cherry. Nose: ripe fruit, creamy oak, spicy, balanced, varietal. Palate: ripe fruit, spicy, round tannins, round.

🏆 **PODIUM**

Vizcarra Torralvo 2019 T
100% tinto fino
95
Colour: dark-red cherry, garnet rim. Nose: ripe fruit, fruit preserve, aged wood nuances, tobacco, sweet spices. Palate: spicy, round tannins, long.

BODEGAS Y VIÑEDOS ACEÑA
Avda. Valladolid, 4
42330 San Esteban de Gormaz (Soria)
☎: +34 667 784 221
juanjo@terraesteban.com
www.terraesteban.com

Terraesteban 2017 T
tempranillo
91
Colour: dark-red cherry. Nose: toasty, spicy, cocoa bean, ripe fruit, black fruit. Palate: flavourful, toasty, fine bitter notes.

Terraesteban 2017 T C
tempranillo
90
Colour: dark-red cherry. Nose: toasty, spicy, cocoa bean. Palate: flavourful, toasty, fine bitter notes.

Terraesteban 2019 T RB
tempranillo
88
Fruity, pruney, herbal, spicy.

Terraesteban 2020 T
tempranillo
87

Terraesteban 2021 B
albillo mayor
88
Kind finish, floral, tasty.

Terraesteban 2021 RD
tempranillo
86

BODEGAS Y VIÑEDOS ALILIAN

Ctra. de la Aguilera km 3,5
09400 Aranda de Duero (Burgos)
☎: +34 947 506 659
info@bodegasalilian.es
www.bodegasalilian.es

Alilian Camino del Abuelo 2015 T
94
Colour: deep cherry. Nose: black fruit, ripe fruit, toasty, dark chocolate, spicy, dried herbs. Palate: fruity, flavourful, powerful, balanced, good finish, slightly dry, soft tannins.

BODEGAS Y VIÑEDOS ALIÓN

Ctra. N-122, Km. 312,4 Padilla de Duero
47300 Peñafiel (Valladolid)
☎: +34 983 881 236
Fax: +34 983 881 246
vegasicilia@vega-sicilia.com
www.tempovegasicilia.com

🏆 **PODIUM**

Alión 2019 T
tinto fino
95
Colour: bright cherry. Nose: complex, expressive, spicy, mineral, ripe fruit, black fruit. Palate: full-bodied, long, powerful, toasty, powerful tannins.

BODEGAS Y VIÑEDOS DEL CONDE DE SAN CRISTÓBAL

Ctra. Valladolid a Soria, Km. 303
47300 Peñafiel (Valladolid)
☎: +34 983 878 055
bodega@condesancristobal.com
www.marquesdevargas.com/es/bodegas/bodega-conde-san-cristobal

Conde de San Cristóbal 2019 T C
100% tinto fino
91
Colour: deep cherry. Nose: fruit preserve, dried herbs, spicy, creamy oak. Palate: flavourful, fruity, balanced, full-bodied.

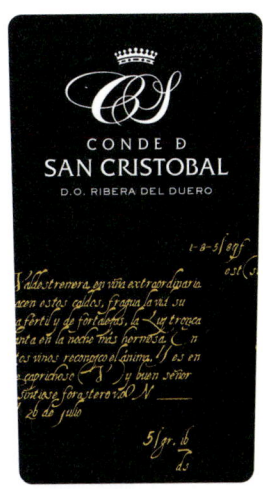

Conde de San Cristóbal Reserva Especial 2018 T R
100% tinto fino
92
Colour: deep cherry. Nose: dried herbs, creamy oak, black fruit. Palate: ripe fruit, spicy, round tannins.

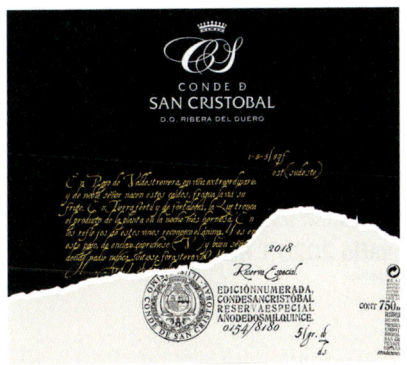

DO RIBERA DEL DUERO / D.O.P.

DO RIBERA DEL DUERO / D.O.P.

Conde de San Cristóbal Flamingo Rosé 2021 RD
100% tinta del país
91
Colour: salmon. Nose: floral, fragrant herbs, candied fruit. Palate: spicy, good acidity, fine bitter notes.

BODEGAS Y VIÑEDOS FRUTOS ARAGÓN
Paraje El Portillo, s/n
09300 Roa (Burgos)
☎: +34 600 735 508
info@frutosaragon.com
www.frutosaragon.com

Bravo 2018 T FB
tempranillo
91
Colour: deep cherry. Nose: dried herbs, creamy oak, black fruit. Palate: powerful, ripe fruit, spicy, round tannins, elegant.

Desafío 2017 T C
tempranillo
88
Toasty, reduced, spicy, jammy.

Desafío 2020 T RB
tempranillo
87

BODEGAS Y VIÑEDOS GALLEGO ZAPATERO
Segunda Travesía de la Olma, 4
09313 Anguix (Burgos)
☎: +34 648 180 777
bodega@bodegasgallegozapatero.com
www.bodegasgallegozapatero.com

Yotuel 2019 T BA
100% tinta del país
88
Pleasant, powerful, spicy, jammy, dried herbs.

Yotuel Finca La Nava 2016 T
100% tinta del país
91
Colour: deep cherry. Nose: ripe fruit, dried herbs, creamy oak. Palate: powerful, ripe fruit, spicy, round tannins.

Yotuel Finca Valdepalacios 2018 T
100% tinta del país
93
Colour: cherry, purple rim. Nose: fruit expression, red berry notes, floral, spicy. Palate: flavourful, fruity, long, round tannins.

Yotuel Selección 2017 T
100% tinta del país
91
Colour: deep cherry. Nose: ripe fruit, dried herbs, creamy oak. Palate: powerful, ripe fruit, spicy, round tannins.

BODEGAS Y VIÑEDOS JUAN MANUEL BURGOS
Aranda, 39
09471 Fuentelcésped (Burgos)
☎: +34 947 557 443
avan@avanvinos.com
www.avanvinos.com

Avan 2020 T C
tempranillo
90
Colour: deep cherry. Nose: ripe fruit, dried herbs, creamy oak, black fruit. Palate: powerful, ripe fruit, spicy, round tannins.

Avan Cepas Centenarias 2020 T
tempranillo
91
Colour: deep cherry. Nose: dried herbs, creamy oak, fruit preserve. Palate: powerful, ripe fruit, spicy, round tannins.

Avan Oak 2021 T BA
tempranillo
87

Avan Vinos de Viñedo de Los Cantillos 2018 T
100% tempranillo
92
Colour: cherry, purple rim. Nose: fruit expression, red berry notes, floral, spicy. Palate: flavourful, fruity, slightly dry, soft tannins.

Avan Vinos de Viñedo: Torrubio 2018 T C
100% tempranillo
91
Colour: cherry, garnet rim. Nose: sweet spices, dark chocolate, ripe fruit. Palate: flavourful, balanced, long, spicy.

Avan Viñedos Viejos 2020 T C
tempranillo
91
Colour: cherry, purple rim. Nose: ripe fruit, red berry notes, dried herbs, spicy. Palate: flavourful, fruity, balanced, slightly dry, soft tannins.

BODEGAS Y VIÑEDOS MONTEABELLÓN

Calvario, s/n
09318 Nava de Roa (Burgos)
☎: +34 947 550 000
Fax: +34 947 550 219
comunicacion@monteabellon.com
www.monteabellon.com

Monteabellón 5 meses 2021 T RB
100% tempranillo
88
Herbal, fruity, toasty, tasty.

Monteabellón Finca La Blanquera 2017 T GR
94
Colour: dark-red cherry, garnet rim. Nose: ripe fruit, fruit preserve, aged wood nuances, tobacco, sweet spices. Palate: spicy, round tannins, long.

Monteabellón Finca Matambres 2018 T
100% tempranillo
91
Colour: deep cherry. Nose: dried herbs, creamy oak, black fruit, toasty. Palate: ripe fruit, spicy, round tannins, balanced.

Monteabellón 14 meses 2019 T C
100% tempranillo
91
Colour: bright cherry. Nose: black fruit, ripe fruit, smoky, spicy, dried herbs. Palate: fruity, flavourful, balanced, fresh, round tannins.

DO RIBERA DEL DUERO / D.O.P.

DO RIBERA DEL DUERO / D.O.P.

BODEGAS Y VIÑEDOS MONTECASTRO

Ctra. VA-130, km. 12
47318 Castrillo de Duero (Valladolid)
☎: +34 983 484 013
contact@bodegasmontecastro.es
www.bodegasmontecastro.es

Montecastro 2018 T R
90% tinto fino, 5% cabernet sauvignon, 5% merlot

91

Colour: deep cherry. Nose: ripe fruit, dried herbs, creamy oak. Palate: ripe fruit, spicy, round tannins, toasty.

Montecastro 2019 T BA
95% tinto fino, 5% merlot

92

Complex, full-bodied. Colour: cherry, garnet rim. Nose: spicy, black fruit, dried herbs. Palate: flavourful, round tannins, toasty.

BODEGAS Y VIÑEDOS NEO

Ctra. N-122, Km. 274,5
09391 Castrillo de La Vega (Burgos)
☎: +34 947 514 393
ivan@bodegasneo.com
www.bodegasneo.com

El Arte de Vivir 2021 T
100% tempranillo

89

Jammy, tasty. Nose: black fruit, expressive.

Neo 2019 T BA
100% tempranillo

91

Colour: Cherry. Nose: woody, new oak, dark chocolate. Palate: powerful, good structure, flavourful, smoky aftertaste.

Neo Albillo Mayor 2020 B
100% albillo mayor

91

Colour: straw. Nose: ripe fruit, dried herbs, faded flowers. Palate: powerful, ripe fruit, balanced.

Sentido 2020 T
100% tempranillo

89

Full-bodied, pruney, powerful, tasty, spicy.

Vivir de Neo 2021 T
100% tempranillo

89

Aromatic, spicy, fruity, herbal, tasty.

Neo Punta Esencia 2019 T
100% tempranillo

92

Colour: cherry, garnet rim. Nose: fruit preserve, powerful, black fruit, new oak, toasty, dark chocolate. Palate: flavourful, sweetness, long.

El Buen Alfarero 2017 T
95% tempranillo, 5% albillo

94

Colour: very deep cherry. Nose: complex, expressive, spicy, mineral, ripe fruit, earthy notes. Palate: elegant, full-bodied, long, great length.

El Retablo III T
100% tempranillo

94

Colour: dark-red cherry, garnet rim. Nose: ripe fruit, fruit preserve, aged wood nuances, tobacco, sweet spices. Palate: spicy, round tannins, long, fruity.

DO RIBERA DEL DUERO / D.O.P.

🏆 PODIUM

Pradorey Élite 2019 T
100% tempranillo

96

Jammy, needs time. Colour: cherry, garnet rim. Nose: fruit preserve, powerful, toasty, black fruit. Palate: flavourful, long, round tannins.

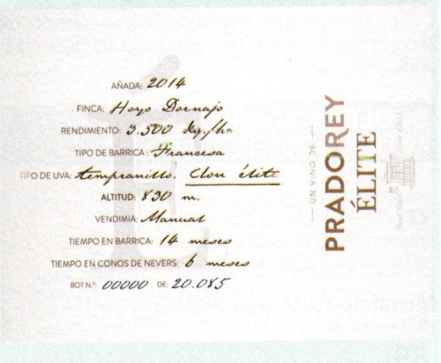

BODEGAS Y VIÑEDOS PRADOREY

Ctra. CL-619 (Magaz – Aranda) km.66, 1
09443 Gumiel de Mercado (Burgos)
☎: +34 947 546 900
info@pradorey.com
www.pradorey.com

Adaro 2019 T
100% tempranillo

93 🌿

Colour: cherry, garnet rim. Nose: ripe fruit, complex, creamy oak, black fruit. Palate: flavourful, balsamic, spicy.

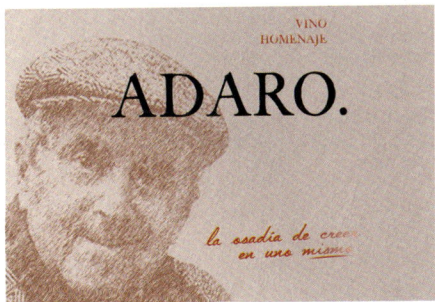

Pradorey Finca La Mina 2017 T R

93

Colour: deep cherry. Nose: dried herbs, creamy oak, black fruit, meaty notes, cocoa bean. Palate: powerful, ripe fruit, spicy, round tannins.

DO RIBERA DEL DUERO / D.O.P.

Lía de Pradorey 2021 RD
100% tempranillo
88
Kind finish, fruity, herbal, simple.

Pradorey Finca La Mina 2018 T R
tempranillo
93
Colour: deep cherry. Nose: dried herbs, creamy oak, spicy, black fruit, ripe fruit, smoky. Palate: ripe fruit, spicy, round tannins, flavourful.

Pradorey Finca Valdelayegua 2019 T
100% tempranillo
93
Colour: dark-red cherry. Nose: toasty, spicy, ripe fruit. Palate: flavourful, toasty, slightly dry, soft tannins.

Pradorey Origen 2021 T
tempranillo
90
Colour: bright cherry. Nose: floral, balanced, red berry notes. Palate: flavourful, fruity, good acidity.

Pradorey Rosado 2021 RD
50% tempranillo, 50% merlot
89
Pruney, full-bodied, sweety finish, jammy, great length, tasty.

Sr. Niño 2021 T
100% tempranillo
91
Colour: cherry, purple rim. Nose: fruit expression, red berry notes, floral. Palate: fruity, good acidity, easy to drink.

BODEGAS Y VIÑEDOS RAUDA S. COOP.

Ctra. de Pedrosa, s/n
09300 Roa (Burgos)
☎: +34 619 934 827
informacion@vinosderauda.com
www.vinosderauda.com

Musai de Tinto Roa 2015 T
100% tempranillo
89
Age nuances, spicy, balanced, dried herbs, jammy.

Tinto Roa 2014 T R
100% tempranillo
91
Colour: dark-red cherry, garnet rim. Nose: ripe fruit, aged wood nuances, tobacco, sweet spices. Palate: spicy, round tannins, long.

Tinto Roa 2019 T C
100% tempranillo
88
Spicy, jammy, herbaceous, toasty.

Tinto Roa 2020 T RB
tempranillo
87

BODEGAS Y VIÑEDOS TÁBULA

Ctra. de Valbuena, km. 2
47359 Olivares de Duero (Valladolid)
☎: +34 676 967 948
jlm@bodegastabula.es
www.bodegastabula.es

Clave de Tábula 2018 T
100% tempranillo
93
Colour: dark-red cherry, garnet rim. Nose: ripe fruit, aged wood nuances, tobacco, sweet spices, earthy notes. Palate: spicy, round tannins, long.

Damana 2019 T C
100% tempranillo
90
Colour: dark-red cherry. Nose: spicy, ripe fruit, smoky. Palate: flavourful, toasty, fine bitter notes, slightly dry, soft tannins.

Damana 5 2020 T
100% tempranillo
90
Colour: cherry, purple rim. Nose: red berry notes, ripe fruit, spicy, dried herbs. Palate: fruity, flavourful, balanced, slightly dry, soft tannins.

Gran Tábula 2017 T
100% tempranillo
94
Colour: dark-red cherry. Nose: spicy, cocoa bean, ripe fruit, dried herbs. Palate: flavourful, toasty, round tannins.

Tábula 2018 T
100% tempranillo
92
Colour: dark-red cherry, garnet rim. Nose: black fruit, ripe fruit, spicy, characterful. Palate: flavourful, good structure.

BODEGAS Y VIÑEDOS TAMARAL

Ctra. Nacional 122 Km 310,6
47314 Peñafiel (Valladolid)
☎: +34 983 878 017
club@tamaral.com
www.tamaral.com

Tamaral 2018 T C
100% tempranillo

90

Colour: cherry, garnet rim. Nose: fruit preserve, fruit liqueur notes, powerful, dark chocolate. Palate: flavourful, sweetness, long.

Tamaral Agricultura Ecológica 2019 T C
100% tempranillo

92 🌿

Colour: deep cherry. Nose: ripe fruit, dried herbs, creamy oak, dark chocolate. Palate: powerful, ripe fruit, spicy, round tannins.

Tamaral Finca la Mira 2018 T
100% tempranillo

92

Colour: cherry, garnet rim. Nose: expressive, spicy, wild herbs, creamy oak. Palate: flavourful, long, fleshy, balsamic.

BODEGAS Y VIÑEDOS VALDEBODEGA

Ctra. San Bernardo, s/n
47359 San Bernardo (Valladolid)
☎: +34 619 829 459
info@valdebodega.com
www.valdebodega.com

Montse Sola 2019 T
100% tempranillo

91

Colour: bright cherry. Nose: ripe fruit, black fruit, spicy, smoky, dried herbs. Palate: fruity, flavourful, balanced, fresh, slightly dry, soft tannins.

Valdebodega 2019 T C
100% tempranillo

90

Colour: bright cherry. Nose: ripe fruit, black fruit, spicy, toasty. Palate: flavourful, fruity, powerful, slightly dry, soft tannins.

Valdebodega 2020 T
100% tempranillo

89

Fruity, smoky, jammy, spicy.

BODEGAS Y VIÑEDOS VALTRAVIESO

Finca La Revilla, s/n
47316 Piñel de Arriba (Valladolid)
☎: +34 983 484 030
marketing@valtravieso.com
www.valtravieso.com

El Manifiesto 2018 B R
100% albillo mayor

91

Age nuances, oxidativ. Colour: bright golden. Nose: ripe fruit, dried herbs, faded flowers. Palate: ripe fruit, balanced, fine bitter notes, long, easy to drink.

Finca La Atalaya Valtravieso 2018 T R
75% tinto fino, 12,5% cabernet sauvignon, 12,5% merlot

92

Colour: dark-red cherry, garnet rim. Nose: ripe fruit, aged wood nuances, tobacco, sweet spices. Palate: spicy, round tannins, long.

Gran Valtravieso 2018 T R
100% tinto fino

93

Colour: dark-red cherry, garnet rim. Nose: ripe fruit, aged wood nuances, tobacco, sweet spices, dark chocolate. Palate: spicy, round tannins, long.

Valtravieso 2019 T C
93% tinto fino, 4% cabernet sauvignon, 3% merlot

91

Colour: deep cherry. Nose: ripe fruit, dried herbs, creamy oak, black fruit. Palate: powerful, ripe fruit, spicy, round tannins.

VT Tinto Fino Valtravieso 2018 T BA
tinto fino

92

Colour: cherry, garnet rim. Nose: powerful, black fruit, smoky, spicy, toasty. Palate: flavourful, long, ripe fruit, great length, round tannins.

VT Tinto Fino Valtravieso 2019 T BA
100% tinto fino

92

Colour: deep cherry. Nose: dried herbs, red berry notes, black fruit, dark chocolate. Palate: powerful, ripe fruit, spicy, round tannins.

VT Vendimia Seleccionada Valtravieso 2019 T
75% tinto fino, 17% merlot, 8% cabernet sauvignon

92

Colour: deep cherry. Nose: red berry notes, faded flowers, spicy, dried herbs. Palate: fruity, flavourful, balanced, slightly dry, soft tannins, smoky aftertaste.

DO RIBERA DEL DUERO / D.O.P.

BODEGAS Y VIÑEDOS VEGA REAL

Ctra. N-122, Km 298,6
47318 Castrillo de Duero (Valladolid)
☎: +34 679 180 532
visitas@vegareal.net
www.barbadillo.com/bodegas-vega-real

El Empecinado 2017 T C
tempranillo

89

Pruney, fruity, jammy.

Vaccayos 2017 T R
100% tempranillo

91

Colour: dark-red cherry. Nose: toasty, spicy, cocoa bean, toasted bread, black fruit. Palate: flavourful, toasty, fine bitter notes.

Vega Real 2019 T RB
100% tempranillo

87

BODEGAS ZIFAR

Afueras de D. Juan Manuel, 9-11
47300 Peñafiel (Valladolid)
☎: +34 983 873 147
bodegaszifar@zifar.com
www.zifar.com

Senda de los Olivos 2019 T C
100% tempranillo

92

Colour: dark-red cherry. Nose: toasty, spicy, cocoa bean, ripe fruit. Palate: flavourful, toasty, fine bitter notes.

Senda de los Olivos 2021 T RB
100% tempranillo

90

Taut. Colour: bright cherry. Nose: fresh fruit, creamy oak, black fruit, scrubland. Palate: good acidity, spicy, fine tannins.

BOSQUE DE MATASNOS

Avda. de Portugal, 18
09400 Aranda de Duero (Burgos)
☎: +34 947 458 375
administracion@bosquedematasnos.es
www.bosquedematasnos.es

Bosque de Matasnos 2019 T
tempranillo

92

Colour: dark-red cherry. Nose: toasty, spicy, cocoa bean, black fruit, ripe fruit. Palate: flavourful, toasty, fine bitter notes.

CASADETILIO

47013 Valladolid (Valladolid)
☎: +34 648 180 794
franmsj@hotmail.com
www.casadetilio.es

Detilio 2020 T
100% tempranillo

94

Colour: bright cherry. Nose: complex, expressive, spicy, mineral. Palate: full-bodied, long, great length.

CEPA 21

Ctra. N-122, Km. 297
47318 Castrillo de Duero (Valladolid)
☎: +34 983 484 083
bodega@cepa21.com
www.cepa21.com

Cepa 21 2018 T
tempranillo

91

Colour: deep cherry. Nose: ripe fruit, cocoa bean, toasty, spicy. Palate: flavourful, toasty, powerful tannins, fruity.

Cepa 21 2019 T
tempranillo

91

Colour: dark-red cherry. Nose: toasty, spicy, cocoa bean, cigar, waxy notes. Palate: flavourful, toasty, fine bitter notes.

Hito 2020 T
tempranillo

89

Balanced, spicy, dried herbs, toasty.

Hito 2021 RD
tempranillo

88

Fruity, jammy, herbal, tasty.

Malabrigo 2018 T
tempranillo

92

Colour: cherry, garnet rim. Nose: smoky, spicy, scrubland, wild herbs, ripe fruit. Palate: flavourful, powerful.

CILLAR DE SILOS

Paraje El Soto, s/n
09370 Quintana del Pidío (Burgos)
☎: +34 947 545 126
Fax: +34 947 545 605
bodega@cillardesilos.es
www.cillardesilos.es

Cillar 2021 T
tempranillo

90

Colour: cherry, purple rim. Nose: fruit expression, red berry notes, floral, spicy. Palate: flavourful, fruity, good acidity, long.

Cillar de Silos 2019 T C
tempranillo

91

Colour: dark-red cherry. Nose: toasty, spicy, cocoa bean, black fruit. Palate: flavourful, toasty, fine bitter notes.

Cillar Rosado de Silos 2021 RD
tempranillo, albillo mayor

90

Colour: raspberry rose. Nose: elegant, red berry notes, floral, fragrant herbs. Palate: light-bodied, spicy, good acidity, fine bitter notes.

Flor de Silos 2019 T
tempranillo

91

Colour: deep cherry. Nose: dried herbs, creamy oak, black fruit. Palate: ripe fruit, spicy, round tannins.

La Viña de Amalio 2019 T
tempranillo

91

Colour: cherry, garnet rim. Nose: fruit preserve, powerful, dark chocolate, new oak, toasty. Palate: flavourful, sweetness, long.

Torresilo 2019 T R
tempranillo

92

Colour: cherry, purple rim. Nose: fruit preserve, spicy, smoky, dried herbs. Palate: fruity, flavourful, balanced.

COMPAÑÍA DE VINOS TELMO RODRÍGUEZ

El Monte
01308 Lanciego (Araba/Álava)
☎: +34 945 628 315
contact@telmorodriguez.com
www.telmorodriguez.com

Matallana 2018 T

93

Colour: dark-red cherry. Nose: toasty, spicy, cocoa bean, ripe fruit, black fruit, earthy notes. Palate: flavourful, toasty, fine bitter notes.

CONDADO DE HAZA

Ctra. La Horra, s/n
09300 Roa (Burgos)
☎: +34 947 525 254
info@condadodehaza.com
www.familiafernandezrivera.com

Condado de Haza 2019 T R
100% tempranillo

92

Colour: cherry, purple rim. Nose: fruit expression, red berry notes, spicy, faded flowers. Palate: flavourful, fruity, balanced, slightly dry, soft tannins.

DO RIBERA DEL DUERO / D.O.P.

Condado de Haza 2020 T C
100% tempranillo

90

Colour: deep cherry. Nose: ripe fruit, dried herbs, creamy oak, sweet spices. Palate: ripe fruit, spicy, round tannins.

CONVENTO OREJA
Cl. de la Fuente s/n
47318 Mélida - Peñafiel (Valladolid)
☎: +34 601 363 197
jmvaquero@conventooreja.net
www.conventooreja.es

Convento Oreja 2019 T C
100% tempranillo

91

Colour: deep cherry. Nose: dried herbs, creamy oak, ripe fruit, black fruit. Palate: powerful, ripe fruit, spicy, round tannins.

Convento Oreja 2021 T RB
100% tempranillo

90

Colour: cherry, purple rim. Nose: powerful, ripe fruit, spicy. Palate: ripe fruit, flavourful, good structure.

Convento Oreja Memoria 2017 T R
100% tempranillo

92

Colour: very deep cherry. Nose: roasted coffee, aromatic coffee, powerful. Palate: smoky aftertaste, great length, round tannins.

Convento Oreja Selección de Familia 2018 T
100% tempranillo

93

Colour: cherry, purple rim. Nose: fruit expression, red berry notes, floral, spicy, wild herbs. Palate: flavourful, fruity, good acidity, long.

CUENTAVIÑAS
Vial B6 Peciña
26339 San Vicente de la Sonsierra (La Rioja)
☎: +34 686 498 183
info@cuentavinas.com

🏆 **PODIUM**

Cuentaviñas 2020 T
tinto fino

95

Aromatic, mineral. Colour: bright cherry. Nose: complex, expressive, spicy, mineral, fruit expression, red berry notes, chalk. Palate: elegant, full-bodied, long, great length.

CV SOLTERRA
Ctra. Roa a Pedrosa, km. 1,5
09300 Roa (Burgos)
☎: +34 915 196 651
info@cvsolterra.com
www.cvsolterra.com

AZ Alto de los Zorros 10 Meses 2019 T RB
tempranillo

90

Colour: cherry, garnet rim. Nose: faded flowers, dried flowers, ripe fruit, fruit preserve, black liquorice. Palate: balanced, toasty, correct.

AZ Alto de los Zorros 2017 T C

91

Hot, pruney, toasty, overripe.

AZ Alto de los Zorros Autor 2016 T R
tempranillo

92

Colour: dark-red cherry, garnet rim. Nose: ripe fruit, aged wood nuances, tobacco, sweet spices. Palate: spicy, round tannins, long.

CVNE
Barrio de la Estación, s/n
26200 Haro (La Rioja)
☎: +34 941 304 800
marketing@cvne.com
www.cvne.com

Cune Ribera del Duero 2020 T RB

87

Cune Ribera del Duero 2021 T RB
100% tinta del país

88

Balanced, spicy, dried herbs, toasty.

DEHESA DE LOS CANÓNIGOS
Ctra. Renedo - Pesquera, Km. 37
47315 Pesquera de Duero (Valladolid)
☎: +34 983 484 001
bodega@dehesadeloscanonigos.com
www.dehesadeloscanonigos.com

Dehesa de los Canónigos 2019 T BA
tempranillo, merlot, cabernet sauvignon
88
Standard, spicy, jammy, simple, herbal, tasty.

DEHESA VALDELAGUNA
47315 Pesquera de Duero (Valladolid)
☎: +34 619 460 308
montelaguna@montelaguna.es
www.montelaguna.es

Montelaguna 2019 T C
tempranillo
88
Jammy, toasty, spicy, standard, fruity, smoky.

Montelaguna 2020 T RB
tempranillo
88
Smoky, spicy, jammy, tasty, fruity, standard.

Montelaguna Selección 2018 T
90
Colour: very deep cherry. Nose: roasted coffee, aromatic coffee, powerful. Palate: smoky aftertaste, great length, round tannins.

DOMINIO BASCONCILLOS
Paraje Alto del Cura, Polígono 19 – Parcela 245
09370 Gumiel De Izán (Burgos)
☎: +34 947 473 300
comercial@dominiobasconcillos.com
www.dominiobasconcillos.com

Dominio Basconcillos Finca de Altura 2019 T RB
100% tempranillo
92
Colour: deep cherry. Nose: ripe fruit, dried herbs, creamy oak, red berry notes. Palate: powerful, ripe fruit, spicy, round tannins.

Dominio Basconcillos Viña Magna 2018 T C
90% tempranillo, 10% cabernet sauvignon
92
Colour: deep cherry. Nose: ripe fruit, dried herbs, creamy oak, black fruit. Palate: powerful, ripe fruit, spicy, round tannins.

Dominio Basconcillos Viña Magna 2018 T R
85% tempranillo, 10% cabernet sauvignon, 5% merlot
93
Colour: deep cherry. Nose: ripe fruit, dried herbs, creamy oak, dark chocolate. Palate: powerful, ripe fruit, spicy, round tannins.

Dominio Basconcillos Viña Magna 2019 T C
90% tempranillo, 10% cabernet sauvignon
93
Colour: Cherry. Nose: complex, expressive, spicy, mineral, red berry notes. Palate: elegant, full-bodied, long, great length.

DOMINIO DE BORNOS
Ctra. Monasterio, s/n
47359 Valbuena de Duero (Valladolid)
☎: +34 983 868 116
i.orduna@bornosbodegas.com
www.bodegaslleiroso.com

Dominio de Bornos 2017 T C
100% tempranillo
91
Colour: deep cherry. Nose: ripe fruit, dried herbs, creamy oak. Palate: powerful, ripe fruit, spicy, round tannins.

Dominio de Bornos 2019 T RB
100% tempranillo
87

DO RIBERA DEL DUERO / D.O.P.

DOMINIO DE CALOGÍA
Ctra. Pedrosa km 0,8
09300 Roa (Burgos)
☎: +34 947 124 360
calogia@calogia.com
www.calogia.com

**Dominio de Calogía
by José Manuel Pérez Ovejas 2020 T**
tinto fino
94
Colour: dark-red cherry. Nose: toasty, spicy, cocoa bean, ripe fruit, black fruit, dark chocolate. Palate: flavourful, toasty, fine bitter notes, spicy.

🏆 PODIUM

Dominio de Calogía by José Manuel Pérez Ovejas Cuveé S 2019 T
100% tinto fino
95
Colour: very deep cherry. Nose: complex, expressive, spicy, mineral, ripe fruit, black fruit. Palate: full-bodied, long, great length, flavourful, round tannins.

DOMINIO DE ES
Ctra. Atauta, 63B
42330 San Esteban de Gormaz (Soria)
☎: +34 676 536 390
bebervino@hotmail.com
www.bertrandsourdais.com

🏆 PODIUM

Dominio de Es Carravilla 2019 T
95
Colour: Cherry. Nose: complex, expressive, spicy, mineral. Palate: elegant, full-bodied, long, great length.

Dominio de Es Carravilla 2020 T
94
Colour: Cherry. Nose: complex, expressive, spicy, mineral, chalk, ripe fruit, red berry notes. Palate: elegant, full-bodied, long, great length.

Dominio de Es La Mata 2020 T
93 🌱
Colour: very deep cherry. Nose: roasted coffee, aromatic coffee, powerful, red berry notes, ripe fruit. Palate: smoky aftertaste, great length, round tannins.

**Dominio de Es
Viñas Viejas de Soria 2020 T**
93 🌱
Colour: dark-red cherry. Nose: toasty, spicy, cocoa bean, ripe fruit, black fruit. Palate: flavourful, toasty, fine bitter notes.

DOMINIO DE PINGUS
Millán Alonso, 49
47350 Quintanilla de Onésimo (Valladolid)
info@pingus.es
www.pingus.es

🏆 PODIUM

Flor de Pingus 2020 T
tempranillo
96
Creamy. Colour: cherry, garnet rim. Nose: red berry notes, ripe fruit, neat, expressive, dark chocolate. Palate: elegant, fine bitter notes, full of life, flavourful, varietal, full-bodied.

🏆 PODIUM

Pingus 2020 T
98
Elegant, complex. Colour: bright cherry. Nose: expressive, fresh, red berry notes, ripe fruit, wild herbs. Palate: fruity, full-bodied, full of life, fine tannins, spicy, long, fruity aftertaste.

PSI 2020 T
94
Colour: bright cherry. Nose: wild herbs, floral, expressive, spicy, red berry notes. Palate: flavourful, fruity, grainy tannins.

DOMINIO DEL ÁGUILA
Los Lagares, 42
09370 La Aguilera (Burgos)
☎: +34 638 899 236
administracion@gmail.com
www.dominiodelaguila.com

🏆 PODIUM

Canta la Perdiz 2017 T R
97 🌱
Defined aromas, complex, taut, great length. Colour: bright cherry. Nose: chalk, red berry notes, ripe fruit, wild herbs, mineral, complex, expressive. Palate: fruity, full of life, flavourful, round tannins.

🏆 PODIUM

Dominio del Aguila 2018 T R
tempranillo, otras
96 🌱
Colour: cherry, purple rim. Nose: fruit expression, red berry notes, floral, spicy, wild herbs. Palate: flavourful, fruity, good acidity, long.

🏆 PODIUM

Peñas Aladas 2013 T GR
98 🌿
With personality, fruity, original. Colour: bright cherry. Nose: fresh fruit, wild herbs, mineral, complex. Palate: fresh, fruity, balanced, taut, spicy, easy to drink, long.

Pícaro del Aguila 2020 T BA
tempranillo, otras
92 🌿
Colour: cherry, purple rim. Nose: fruit expression, red berry notes, sweet spices, toasty. Palate: flavourful, fruity, good acidity, long.

Pícaro del Aguila Clarete 2020 RD
94 🌿
Colour: rose. Nose: red berry notes, fragrant herbs, dried flowers, saffron, characterful. Palate: full-bodied, flavourful, spicy, sweetness, long.

DOMINIO DEL PIDIO
Lagares, 55-56
09370 Quintana del Pidío (Burgos)
☎: +34 947 545 126
Fax: +34 947 545 605
contacto@dominiodelpidio.com
www.dominiodelpidio.com

🏆 PODIUM

Dominio del Pidio 2019 T
95% tempranillo, 5% albillo mayor
95
Colour: Cherry. Nose: complex, expressive, spicy, mineral, fruit expression. Palate: elegant, full-bodied, long, great length.

Dominio del Pidio 2020 RD
50% tempranillo, 50% albillo mayor
93
Colour: raspberry rose. Nose: elegant, red berry notes, floral, fragrant herbs, fine lees, brioche. Palate: light-bodied, spicy, good acidity, fine bitter notes.

Dominio del Pidio Albillo 2020 B
100% albillo mayor
91
Very fruit-driven, needs time. Colour: straw. Nose: ripe fruit, dried herbs, faded flowers. Palate: powerful, ripe fruit, balanced.

DOMINIO DEL SOTO
Camino Real 28
09441 Sotillo de La Ribera (Burgos)
☎: +34 722 468 435
contact@dominiodelsoto.fr

Dominio Del Soto 2018 T C
92 🌿
Colour: bright cherry. Nose: creamy oak, expressive, fruit expression. Palate: ripe fruit, spicy, round tannins, balanced.

Dominio Del Soto 2018 T R
93
Colour: Cherry. Nose: complex, expressive, spicy, wild herbs. Palate: full-bodied, long, great length, round tannins.

Dominio Del Soto 2020 B
albillo mayor
90
Colour: bright straw. Nose: white fruit, ripe fruit, dried herbs, sweet spices. Palate: fruity, fresh, flavourful, balanced.

DOMINIO FOURNIER
Finca El Pinar, s/n
09316 Berlangas de Roa (Burgos)
☎: +34 947 533 006
prensa@gonzalezbyass.es
www.dominiofournier.com

Dominio Fournier 2019 T R
100% tinta del país
93
Colour: Cherry. Nose: complex, expressive, spicy, balanced, wild herbs. Palate: full-bodied, long, great length, round tannins.

Dominio Fournier 2020 T C
100% tinta del país
93
Colour: Cherry. Nose: complex, expressive, spicy, mineral, ripe fruit. Palate: elegant, full-bodied, long, great length.

ÉBANO VIÑEDOS Y BODEGAS
Ctra. N-122 Km., 299,6
47318 Castrillo de Duero (Valladolid)
☎: +34 986 609 060
ebano@valminorebano.com
www.valminorebano.com

Ébano 6 2020 T RB
100% tempranillo
90
Colour: deep cherry. Nose: ripe fruit, dried herbs, creamy oak. Palate: powerful, ripe fruit, spicy, round tannins.

Peñín Guide | SPANISH WINE

DO RIBERA DEL DUERO / D.O.P.

Ébano Salvaje 2018 T C
100% tempranillo
91
Colour: deep cherry. Nose: ripe fruit, dried herbs, creamy oak, black fruit. Palate: powerful, ripe fruit, spicy, round tannins.

FAMILIA BASTIDA
Canónigo Lozano, 11
30520 Jumilla (Murcia)
☎: +34 968 780 142
info@familiabastida.com
www.familiabastida.com

Vuela 2020 T
tempranillo
89
Fruity, spicy, dried herbs, tasty.

FELIZ COMPAÑÍA VINÍCOLA
Plaza Santa María, 2
09400 Aranda de Duero (Burgos)
☎: +34 685 822 275
conchi@lapicaragastroteca.com
www.vinofeliz.com

Más Feliz, Vides Olvidadas 2019 B
albillo mayor
89
Citrus fruit, dried herbs, balanced, needs time.

RQT Feliz, Cepas entre Viñas 2019 B C
albillo mayor
90
Colour: bright straw. Nose: ripe fruit, fragrant herbs, fine lees. Palate: full-bodied, rich, long, good acidity.

FINCA TORREMILANOS
Finca Torremilanos, s/n
09400 Aranda de Duero (Burgos)
☎: +34 947 510 377
Fax: +34 947 512 856
contacto@torremilanos.com
www.torremilanos.com

Los Cantos de Torremilanos 2019 T
89
Spicy, toasty, tasty, fruity, jammy.

Los Cantos de Torremilanos 2020 T
88
Fruity, pruney, toasty, spicy, slight oxidation.

Montecastrillo 2020 T RB
90
Colour: bright cherry. Nose: ripe fruit, red berry notes, black fruit, dried herbs, smoky, spicy. Palate: fruity, flavourful, balanced, slightly dry, soft tannins.

Montecastrillo 2021 T RB
89
Fruity, crisp, tasty, astringent, floral.

Torremilanos 2018 T C
91
Colour: deep cherry. Nose: dried herbs, creamy oak, black fruit, sweet spices. Palate: powerful, ripe fruit, spicy, round tannins.

FINCA VILLACRECES
Ctra. Soria N-122 Km 322
47350 Quintanilla de Onésimo (Valladolid)
☎: +34 983 680 437
Fax: +34 983 683 314
villacreces@villacreces.com
www.villacreces.com

Finca Villacreces 2019 T
tempranillo, cabernet sauvignon, merlot
94
Colour: deep cherry. Nose: dried herbs, creamy oak, black fruit, ripe fruit, sweet spices. Palate: powerful, ripe fruit, spicy.

🏆 PODIUM
Finca Villacreces Specimen Nº2 T
tinto fino, cabernet sauvignon
95
Colour: dark-red cherry, garnet rim. Nose: ripe fruit, aged wood nuances, tobacco, sweet spices, wet leather, cigar. Palate: spicy, round tannins, long.

Pruno 2020 T
90
Colour: cherry, purple rim. Nose: ripe fruit, dried herbs, herbaceous, sweet spices. Palate: powerful, ripe fruit, round tannins.

FINCA Y VIÑEDOS SAN COBATE

Ctra. Gumiel de Mercado – Oquillas Km 6,4
09443 Gumiel de Mercado (Burgos)
info@sancobate.com
www.sancobate.com

San Cobate La Finca 2018 T C
100% tinto fino

91

Colour: deep cherry. Nose: black liquorice, cocoa bean, red berry notes, black fruit, fine reductive notes. Palate: ripe fruit, spicy, round tannins, flavourful.

San Cucufate ¨Monasterio¨ 2018 T
100% tinto fino

92

Colour: deep cherry. Nose: ripe fruit, dried herbs, creamy oak, dark chocolate. Palate: powerful, ripe fruit, spicy, round tannins.

FRANCISCO BARONA

Ctra. Circunvalación s/n
09300 Roa (Burgos)
☎: +34 637 182 951
francisco@franciscobarona.com
www.franciscobarona.com

Francisco Barona 2020 T C
tinto fino, garnacha, bobal, albillo, jaen negro, otras

92

Colour: cherry, purple rim. Nose: ripe fruit, sweet spices, toasty, expressive. Palate: fruity, flavourful, balanced, slightly dry, soft tannins.

Francisco Barona
Finca Las Dueñas 2018 T R
tinto fino, bobal, jaen negro, otras

92

Colour: deep cherry. Nose: dried herbs, creamy oak, dark chocolate, black fruit. Palate: powerful, ripe fruit, spicy, round tannins.

GARMÓN CONTINENTAL

Camino de la Ribera s/n
47359 Olivares de Duero (Valladolid)
☎: +34 983 488 708
info@garmoncontinental.com
www.garmoncontinental.com

Garmón 2019 T
100% tempranillo

92

Colour: deep cherry. Nose: ripe fruit, dried herbs, creamy oak, aromatic coffee, woody. Palate: ripe fruit, spicy, good structure, harsh oak tannins.

HACIENDA MIGUEL SANZ

Ctra. BU-930, Km. 11
09491 Vadocondes (Burgos)
☎: +34 941 454 050
rrpp@bodegasriojanas.com
www.bodegasriojanas.com

Alacer 2019 T C
100% tempranillo

88

Kind finish, spicy, jammy, simple.

Alacer 2020 T RB
100% tempranillo

88

Fruity, spicy, herbal, smoky, coarse tannins, hot.

Alacer 2021 T
100% tempranillo

87

DO RIBERA DEL DUERO / D.O.P.

DO RIBERA DEL DUERO / D.O.P.

HACIENDA SOLANO
La Solana, 6
09370 La Aguilera (Burgos)
☎: +34 947 545 582
info@haciendasolano.com
www.haciendasolano.com

🏆 PODIUM

**Hacienda Solano
Finca Cascorrales 2018 T**
98% tempranillo, 2% albillo

95

Taut. Colour: deep cherry. Nose: complex, expressive, spicy, mineral, ripe fruit. Palate: full-bodied, long, great length, flavourful, mineral.

🏆 PODIUM

**Hacienda Solano
Finca Peñalobera 2019 T**
98% tempranillo, 2% albillo

95

Colour: cherry, garnet rim. Nose: expressive, spicy, wild herbs, elegant. Palate: full-bodied, long, great length, round tannins, complex.

Hacienda Solano Selección 2020 T BA
tempranillo

92

Colour: deep cherry. Nose: ripe fruit, creamy oak, fruit expression. Palate: ripe fruit, spicy, round tannins.

Hacienda Solano Viñas Viejas 2019 T BA
98% tempranillo, 2% albillo

93

Varietally correct. Colour: bright cherry. Nose: ripe fruit, black fruit, earthy notes, wild herbs. Palate: ripe fruit, spicy, fresh, easy to drink.

HAMMEKEN CELLARS
Llavador, 20
03700 Denia (Alacant/Alicante)
☎: +34 965 791 967
cellars@hammekencellars.com
www.hammekencellars.com

Oráculo 2019 T
100% tempranillo

89

Spicy, herbaceous, toasty, jammy.

HIJOS DE ANTONIO POLO
La Olma, 5
47300 Peñafiel (Valladolid)
☎: +34 983 873 183
info@pagopenafiel.com
www.pagopenafiel.com

**Pagos
de Peñafiel 2017 T R**
100% tempranillo

90

Colour: bright cherry. Nose: toasty, spicy, cocoa bean, fruit preserve. Palate: flavourful, toasty, fruity, slightly dry, soft tannins.

Pagos de Peñafiel 2018 T C
100% tempranillo

88

Pruney, fruity, spicy, toasty.

**Pagos
de Peñafiel 2020 T RB**
100% tempranillo

87

**Pagos de Peñafiel
Vendimia Selección 2019 T**
100% tempranillo

89

Pruney, fruity, spicy, tasty.

HORNILLOS BALLESTEROS
El Molino, 34
09300 Roa (Burgos)
☎: +34 636 282 170
Fax: +34 947 541 071
hornillosballesteros@telefonica.net
www.hornillosballesteros.info

MiBal 2019 T RB
100% tempranillo

88

Fruity, spicy, herbaceous, tasty.

MiBal 2021 T
100% tempranillo

88

Spicy, fruity, floral, herbal, tasty.

MiBal Selección 2014 T
100% tinta del país

92

Colour: deep cherry, garnet rim. Nose: aged wood nuances, cocoa bean, cigar, toasty. Palate: flavourful, spicy, toasty, powerful tannins.

Perfil 2014 T R
100% tinta del país

93

Representative. Colour: Cherry. Nose: complex, expressive, spicy, dried herbs. Palate: full-bodied, long, great length, round tannins.

MiBal 2018 T C
100% tempranillo

91

Colour: deep cherry. Nose: ripe fruit, dried herbs, creamy oak. Palate: powerful, ripe fruit, spicy, round tannins.

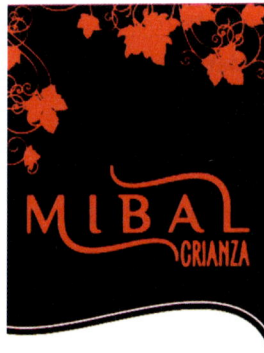

JESÚS DE MADRAZO WINES
San Ignacio de Loyola, 12 5ºG
26009 Logroño (La Rioja)
☎: +34 639 780 524
chus56madrazo@gmail.com

Selección Jesús Madrazo Selección Ribera del Duero 2019 T
88% tempranillo, 10% garnacha, 2% albillo

93

Colour: bright cherry, very deep cherry. Nose: ripe fruit, creamy oak, aromatic coffee, dried herbs, characterful. Palate: good structure, flavourful, long.

LA LOBA
42351 Matanza de Soria (Soria)
☎: +34 975 102 037
info@laloba.es

La Loba 2019 T

93

Colour: deep cherry. Nose: complex, expressive, spicy, mineral, ripe fruit. Palate: elegant, full-bodied, long, great length.

La Lobita 2020 T

92

Colour: cherry, purple rim. Nose: fruit expression, red berry notes, floral, spicy, balsamic herbs, scrubland. Palate: flavourful, fruity, good acidity, long.

LA QUINTA VENDIMIA
P.I. La Tapiada, Nave 5
42330 San Esteban de Gormaz (Soria)
☎: +34 627 744 125
n.ossa.alba@hotmail.es
www.laquintavendimia.com

La Quinta Vendimia 2018 T
tinta del país

90

Colour: cherry, purple rim. Nose: ripe fruit, dried herbs, creamy oak, smoky, black fruit. Palate: powerful, ripe fruit, spicy, slightly dry, soft tannins.

LEGARIS
Ctra. de Peñafiel - Encinas de Esgueva, km. 2,5
47316 Curiel de Duero (Valladolid)
☎: +34 983 878 088
info@legaris.com
www.legaris.com

Legaris 2019 T C
100% tinto fino

89

Pleasant, smoky, jammy, tasty, herbal, spicy.

Legaris 2020 T RB
100% tinto fino

89

Balanced, standard, fruity, jammy, wild, smooth.

Legaris Alcubilla de Avellaneda 2018 T BA
100% tinto fino

93

Colour: deep cherry. Nose: dried herbs, creamy oak, dark chocolate. Palate: powerful, ripe fruit, spicy, round tannins.

DO RIBERA DEL DUERO / D.O.P.

DO RIBERA DEL DUERO / D.O.P.

Legaris Moradillo de Roa 2018 T BA
100% tinto fino
92
Colour: cherry, garnet rim. Nose: overripe fruit, creamy oak, hot, black fruit, fruit preserve, dark chocolate. Palate: pruney, powerful, sweet tannins.

Legaris Olmedillo de Roa 2018 T
100% tinto fino
93
Colour: very deep cherry. Nose: roasted coffee, aromatic coffee, powerful, black fruit, ripe fruit. Palate: smoky aftertaste, great length, round tannins.

Páramos de Legaris 2018 T BA
100% tinto fino
91
Colour: dark-red cherry. Nose: toasty, spicy, cocoa bean, black fruit. Palate: flavourful, toasty, fine bitter notes.

LOESS VINOS
Plaza Cuartel Viejo, 7
49028 Zamora (Zamora)
☎: +34 983 664 898
loess@loess.es
www.loesscollection.com

Loess Blue Cap 2019 T C
tinta del país
91
Smoky, toasty. Colour: deep cherry. Nose: black fruit, dark chocolate, characterful. Palate: good structure, flavourful, round tannins.

Loess Collection 2018 T C
tinta del país
91
Colour: deep cherry. Nose: dried herbs, creamy oak, black fruit. Palate: powerful, ripe fruit, spicy, round tannins.

Loess Inspiration 2020 T
tinta del país
91
Colour: deep cherry. Nose: dried herbs, creamy oak, toasty, black fruit. Palate: powerful, ripe fruit, spicy, round tannins.

MARQUÉS DE BURGOS
Paraje de Buicio s/n
26360 Fuenmayor (La Rioja)
☎: +34 941 450 950
info@bodegaslan.com
www.lanencasa.com/ribera

8000 Marqués de Burgos 2018 T
89
Hot, pruney, overripe, toasty.

Marqués de Burgos 2018 T C
tempranillo
88
Toasty, jammy, spicy.

Marqués de Burgos 2020 T RB
tempranillo
87

MARQUÉS DE VELILLA
Ctra. de Sotillo de la Riibera , s/n
09311 La Horra (Burgos)
☎: +34 947 542 166
bodega@marquesdevelilla.com
www.marquesdevelilla.com

Doncel de Mataperras 2015 T
tinta del país
92
Colour: cherry, garnet rim. Nose: fruit preserve, powerful, complex, characterful. Palate: flavourful, long.

Finca La María 2019 T RB
90
Colour: deep cherry. Nose: ripe fruit, dried herbs, creamy oak, sweet spices. Palate: powerful, ripe fruit, spicy, round tannins.

Marqués de Velilla 2019 T C
tinta del país
89
Pleasant, spicy, jammy, toasty.

Marqués de Velilla 2020 T RB
tinta del país
88
Pleasant, smoky, toasty, tasty.

Marqués de Velilla 2021 T
tinta del país
86

MELIDA WINES
Pasadizo, 1
47318 Melida (Peñafiel) (Valladolid)
☎: +34 686 580 012
info@melidawines.com
www.melidawines.com

Dos Alas Rojas 2019 T C
tempranillo
92
Colour: Cherry. Nose: black fruit, ripe fruit, dried herbs, black pepper, neat, expressive, characterful. Palate: spicy, round tannins, long.

Eternauta 2020 T C
tempranillo
90
Colour: deep cherry. Nose: ripe fruit, dried herbs, creamy oak, woody. Palate: powerful, ripe fruit, spicy, harsh oak tannins.

Parpados 2020 T
100% tempranillo
92
Colour: deep cherry. Nose: dried herbs, creamy oak, ripe fruit, black fruit. Palate: powerful, ripe fruit, spicy, round tannins.

MILÉNICO
09317 San Martin de Rubiales (Burgos)
☎: +34 695 382 848
milenico@milenico.com
www.milenico.com

Valdepila 2018 T R
tempranillo
91 ☘
Colour: cherry, garnet rim. Nose: fruit preserve, fruit liqueur notes, powerful. Palate: flavourful, sweetness, long.

Valdepila 2019 T R
tempranillo
90 ☘
Colour: cherry, purple rim. Nose: fruit preserve, spicy, toasty, dried herbs. Palate: fruity, flavourful, balanced, slightly dry, soft tannins.

MILL WINES
Gremio del Cuero, 14
40195 Segovia (Segovia)
☎: +34 686 999 212
info@millwines.com
www.millwines.com

Mill 2019 T C
100% tempranillo
87

MILSETENTAYSEIS
Asturias 16 - Nave 18
09400 Aranda de Duero (Burgos)
☎: +34 983 878 020
info@almacarraovejas.com
www.milsetentayseis.com

Milsetentayseis 2019 T
tempranillo, 5% albillo
94 ☘
Colour: bright cherry. Nose: complex, expressive, spicy, mineral. Palate: elegant, full-bodied, long, great length.

Milsetentayseis La Peña 2020 RD
tinto fino, albillo
93 ☘
Colour: salmon. Nose: sweet spices, red berry notes, fragrant herbs, dried flowers. Palate: full-bodied, flavourful, spicy, sweetness, long.

MONTEBACO
Finca Montealto s/n
47300 Valbuena de Duero (Valladolid)
☎: +34 983 485 128
Fax: +34 983 485 033
montebaco@bodegasmontebaco.com
www.bodegasmontebaco.com

Montebaco Cara Norte 2020 T C
92 ☘
Colour: bright cherry. Nose: creamy oak, raspberry, wild herbs, black fruit, red berry notes. Palate: good acidity, spicy, fine tannins.

Montebaco de Finca 2020 T C
91
Colour: deep cherry. Nose: creamy oak, black fruit, scrubland. Palate: ripe fruit, spicy, round tannins.

Montebaco Selección Especial 2019 T
90
Colour: cherry, purple rim. Nose: black fruit, fruit preserve, black liquorice, spicy, smoky. Palate: fruity, flavourful, roasted-coffee aftertaste, slightly dry, soft tannins.

DO RIBERA DEL DUERO / D.O.P.

DO RIBERA DEL DUERO / D.O.P.

Semele 2020 T C
92
Colour: deep cherry. Nose: dried herbs, creamy oak, black fruit. Palate: powerful, ripe fruit, spicy, round tannins.

MONTEGAREDO
Ctra. Boada a Pedrosa, km. 1 Boada de Roa
09314 Pedrosa de Duero (Burgos)
☎: +34 600 300 636
Fax: +34 947 530 140
jlgaredo@gmail.com
www.montegaredo.com

Montegaredo 2020 T C
100% tinto fino
88
Jammy, toasty, spicy, fruity, balanced.

Montegaredo Gran Selección 2020 T C
100% tinto fino
91
Colour: cherry, purple rim. Nose: fruit expression, red berry notes, floral, spicy, scrubland. Palate: flavourful, fruity, good acidity.

Pirámide 2020 T C
100% tinto fino
89
Full-bodied, spicy, full-bodied, jammy, tasty, toasty.

NALUAR & ACEDIANO
Ctra. Pesquera-Encinas, Km. 3
47359 Piñel de Abajo (Valladolid)
34646215911

Acediano 2019 T C
tinto fino
93
Colour: deep cherry. Nose: dried herbs, creamy oak, black fruit, toasty. Palate: powerful, ripe fruit, spicy, grainy tannins.

Naluar 2020 T
tinto fino
93
Colour: Cherry. Nose: complex, expressive, spicy, mineral, ripe fruit. Palate: elegant, full-bodied, long, great length.

NEXUS BODEGAS
Ctra. Pesquera de Duero a Renedo, s/n
47315 Pesquera de Duero (Valladolid)
☎: +34 983 880 488
info@nexusfrontaura.com
www.nexusbodegas..com

Nexus + 2014 T
100% tempranillo
90
Weary. Colour: cherry, garnet rim. Nose: black fruit, spicy, dried herbs. Palate: good structure, grainy tannins.

Nexus 2017 T C
100% tempranillo
92
Colour: deep cherry. Nose: ripe fruit, dried herbs, creamy oak, toasty. Palate: powerful, ripe fruit, spicy, round tannins.

Nexus One 2011 T GR
93
Colour: deep cherry, garnet rim. Nose: aged wood nuances, ripe fruit, cocoa bean, cigar, toasty. Palate: flavourful, spicy, toasty, powerful tannins.

Nexus One 2019 T
91
Colour: deep cherry. Nose: ripe fruit, dried herbs, creamy oak, characterful. Palate: powerful, ripe fruit, spicy, round tannins.

Pisarrosas 2018 T C
90
Colour: deep cherry, purple rim. Nose: ripe fruit, dried herbs, creamy oak, spicy. Palate: powerful, ripe fruit, spicy, slightly dry, soft tannins.

NUESTRO DE DÍAZ BAYO
Camino de los Anarinos, s/n
09471 Fuentelcésped (Burgos)
☎: +34 941 451 129
info@premiumfincas.com
www.premiumfincas.com

Diaz Bayo 15 Meses 2020 T C
100% tinto fino
91
Colour: dark-red cherry. Nose: toasty, spicy, cocoa bean, black fruit. Palate: flavourful, toasty, fine bitter notes.

Diaz Bayo 4U 2021 T
100% tinto fino
90
Colour: deep cherry. Nose: ripe fruit, dried herbs, creamy oak. Palate: powerful, ripe fruit, spicy, round tannins.

Diaz Bayo 8 Meses 2021 T BA
100% tinto fino

89

Pleasant, balanced, spicy, jammy, tasty, fruity, great length.

Nuestro 15 meses 2020 T C
100% tinto fino

91

Colour: deep cherry. Nose: ripe fruit, dried herbs, waxy notes. Palate: powerful, ripe fruit, spicy, round tannins.

Nuestro 20 meses 2016 T BA
100% tinto fino

91

Colour: deep cherry. Nose: ripe fruit, dried herbs, creamy oak. Palate: powerful, ripe fruit, spicy, round tannins.

Nuestro 8 Meses 2021 T RB
100% tinto fino

89

Pleasant, spicy, jammy, great length, tasty, fruity, varietally correct.

PAGO DE CARRAOVEJAS

Camino de Carraovejas, s/n
47300 Peñafiel (Valladolid)
☎: +34 983 878 020
info@pagodecarraovejas.com
www.pagodecarraovejas.com

🏆 **PODIUM**

Pago de Carraovejas "Cuesta de las Liebres" 2018 T R
tinto fino

97

Colour: bright cherry. Nose: creamy oak, lactic notes, expressive, ripe fruit. Palate: ripe fruit, spicy, balanced, long, soft tannins.

Pago de Carraovejas 2020 T
tinto fino, merlot, cabernet sauvignon

93

Colour: deep cherry. Nose: dried herbs, creamy oak, ripe fruit, black fruit, toasty. Palate: powerful, ripe fruit, spicy, round tannins.

Pago de Carraovejas El Anejón 2018 T
tinto fino, merlot, cabernet sauvignon

94

Colour: very deep cherry. Nose: complex, expressive, spicy, mineral, powerful, roasted coffee. Palate: elegant, full-bodied, long, great length.

PAGO DE LOS CAPELLANES

Camino de la Ampudia, s/n
09314 Pedrosa de Duero (Burgos)
☎: +34 947 530 068
bodega@pagodeloscapellanes.com
www.pagodeloscapellanes.com

Pago de los Capellanes Crianza 2020 T C
100% tempranillo

92

Colour: very deep cherry. Nose: roasted coffee, aromatic coffee, powerful, ripe fruit, black fruit. Palate: smoky aftertaste, great length, round tannins.

Pago de los Capellanes Parcela El Nogal 2018 T FB
100% tempranillo

92

Pruney, overripe. Colour: dark-red cherry. Nose: characterful, ripe fruit, dark chocolate. Palate: good structure, flavourful, long.

🏆 **PODIUM**

Pago de los Capellanes Parcela El Picón 2016 T
100% tempranillo

95

Colour: cherry, purple rim. Nose: fruit expression, red berry notes, floral, spicy, ripe fruit. Palate: flavourful, fruity, long, slightly dry, soft tannins.

Pago de los Capellanes Reserva 2019 T R
100% tempranillo

94

Colour: dark-red cherry, garnet rim. Nose: ripe fruit, aged wood nuances, tobacco, sweet spices. Palate: spicy, round tannins, long.

PAGO DEL CIELO

Del Rosario, 56
47311 Fompedraza (Valladolid)
☎: +34 938 177 400
info@torres.es
www.torres.es

Celeste Crianza 2019 T C
100% tinto fino

90

Colour: cherry, purple rim. Nose: ripe fruit, dried herbs, spicy. Palate: fruity, flavourful, toasty.

Celeste Reserva 2018 T R
100% tinto fino

90

Colour: deep cherry. Nose: ripe fruit, dried herbs, creamy oak. Palate: powerful, ripe fruit, spicy, round tannins.

DO RIBERA DEL DUERO / D.O.P.

PAGOS DE ANGUIX
Camino de la Tejera, s/n
09313 Anguix (Burgos)
☎: +34 938 911 000
pagosdeanguix@pagosdeanguix.com
www.pagosdeanquix.com

Pagos de Anguix Barruecos 2019 T
100% tinto fino
93
Colour: bright cherry. Nose: ripe fruit, dark chocolate, waxy notes, tobacco. Palate: fruity, spicy, round tannins, flavourful.

Pagos de Anguix Costalara 2020 T
100% tinto fino
93
Colour: Cherry. Nose: expressive, spicy, mineral, black fruit, scrubland. Palate: elegant, full-bodied, long, great length.

Pagos de Anguix El Rosado 2021 RD
tinto fino, albillo mayor
90
Colour: rose. Nose: hot, faded flowers, ripe fruit, red berry notes. Palate: flavourful, powerful, ripe fruit.

Pagos de Anguix Prado Lobo 2018 T R
100% tinto fino
94
Colour: cherry, garnet rim. Nose: ripe fruit, chalk, expressive, elegant, neat. Palate: balanced, fruity, long, round, taut.

PAGOS DEL REY RIBERA DEL DUERO
Ctra. Palencia-Aranda, Km. 53
09311 Olmedillo de Roa (Burgos)
☎: +34 947 551 111
riberadelduero@pagosdelrey.com
www.pagosdelrey.com

409 2020 T C
tempranillo
89 ♣
Jammy, tasty, great length, dried herbs, full-bodied, toasty.

Altos de Tamarón 2017 T R
tempranillo
90
Colour: dark-red cherry, garnet rim. Nose: ripe fruit, fruit preserve, aged wood nuances, tobacco, sweet spices. Palate: spicy, round tannins, long.

Altos de Tamaron 2019 T C
tempranillo
88
Herbal, standard, spicy, fruity.

Altos de Tamarón 2021 T RB
tempranillo
87

Condado de Oriza 2021 T RB
tempranillo
88
Spicy, jammy, creamy, tasty.

Pago de Fuentecojo 2018 T
tempranillo
91
Colour: cherry, garnet rim. Nose: fruit preserve, powerful, dark chocolate. Palate: flavourful, sweetness, long.

PALACIO DE VILLACHICA
Ctra. N-122, Km. 433
49800 Toro (Zamora)
☎: +34 663 797 346
info@grupopalaciodevillachica.com
www.grupopalaciodevillachica.com

Palacio de Villachica VS Dehesa de Don Diego 2019 T C
0% tinto fino
92
Colour: deep cherry. Nose: ripe fruit, dried herbs, creamy oak, sweet spices, toasty. Palate: powerful, ripe fruit, spicy, round tannins.

PÁRAMO ARROYO
Ctra. de Roa - Pedrosa, km. 4
09314 Pedrosa de Duero (Burgos)
☎: +34 947 530 041
bodega@paramoarroyo.com
www.paramoarroyo.com

Eremus 2017 T C
tempranillo
89 ♣

Eremus 2020 T RB
tempranillo
88 ♣
Kind finish, spicy, tasty, jammy.

Páramo Arroyo 2014 T GR
tempranillo
90 ♣
Colour: deep cherry. Nose: ripe fruit, dried herbs, toasty. Palate: ripe fruit, round tannins.

Ser Vivo y Natural 2021 T
90 ♣
Colour: Cherry. Nose: scrubland, balanced, ripe fruit, meaty notes. Palate: spicy, good acidity, round tannins, flavourful.

PEKADO MORTAL
Rua Valdomiño, 35 Finca Fierro
32002 Barbadas (Ourense/Orense)
☎: +34 640 209 584
martinpadin9@hotmail.com

Pekado Mortal Tempranillo 2018 T
89
Spicy, full-bodied, woody, jammy, dried herbs.

Pekado Mortal Tempranillo 2019 T C
tempranillo
90
Colour: cherry, purple rim. Nose: ripe fruit, balsamic herbs, dried herbs. Palate: fruity, powerful, flavourful, balanced, slightly dry, soft tannins.

PICO CUADRO
Del Río, 22
47350 Quintanilla de Onésimo (Valladolid)
☎: +34 983 855 107
info@picocuadro.com
www.picocuadro.com

Pico Cuadro Original 2020 T
100% tempranillo
91
Colour: cherry, garnet rim. Nose: spicy, ripe fruit, balanced. Palate: flavourful, fruity, long, spicy.

Pico Cuadro Vendimia Seleccionada 2020 T
100% tempranillo
92
Colour: deep cherry. Nose: ripe fruit, dried herbs, creamy oak. Palate: powerful, ripe fruit, spicy, round tannins.

Pico Cuadro Wild 2020 T
100% tempranillo
91
Jammy, tasty, smooth, toasty.

PIEDRAS DE SAN PEDRO
Calle Arrabal (eras 40)
47315 Pesquera de Duero (Valladolid)
☎: +34 647 616 753
info@piedrasdesanpedro.com
www.piedrasdesanpedro.com

Loculto 2019 T C
100% tempranillo
88
Fruity, jammy, spicy, toasty, balanced.

Loculto Selección 2020 T RB
100% tempranillo
89
Jammy, spicy, dried herbs, bitter.

Piedras de San Pedro 2018 T C
100% tempranillo
91
Colour: deep cherry. Nose: dried herbs, creamy oak, black fruit, ripe fruit, sweet spices. Palate: powerful, ripe fruit, spicy, round tannins.

PROMETO SER INFIEL
Rododendro, 15
47008 Valladolid (Valladolid)
☎: +34 607 400 033
contacto@prometoserinfiel.com
www.prometoserinfiel.com

Arena y Caliza 2019 T C
tempranillo
91
Colour: cherry, purple rim. Nose: ripe fruit, red berry notes, dried flowers, spicy. Palate: fruity, flavourful, balanced, slightly dry, soft tannins.

Prometo ser Infiel 2019 T
85% tempranillo, 15% garnacha
93
Colour: deep cherry. Nose: ripe fruit, dried herbs, creamy oak, dark chocolate. Palate: powerful, ripe fruit, spicy, round tannins.

PROTOS BODEGAS RIBERA DUERO DE PEÑAFIEL
Bodegas Protos, 24-28
47300 Peñafiel (Valladolid)
☎: +34 983 878 011
bodega@bodegasprotos.com
www.bodegasprotos.com

Protos '27 2019 T
100% tempranillo
93
Colour: very deep cherry. Nose: aromatic coffee, powerful, ripe fruit, black fruit, new oak. Palate: smoky aftertaste, great length, round tannins.

Protos 2015 T GR
100% tempranillo
94
Classic, complex, representative. Nose: ripe fruit, cocoa bean, neat, balanced. Palate: good structure, slightly dry, soft tannins, complex, balanced.

Protos 2018 T C
100% tempranillo
92
Colour: deep cherry. Nose: ripe fruit, dried herbs, creamy oak, toasty. Palate: powerful, ripe fruit, spicy, round tannins.

DO RIBERA DEL DUERO / D.O.P.

DO RIBERA DEL DUERO / D.O.P.

Protos 5º Año 2016 T R
100% tempranillo

93

Colour: dark-red cherry, purple rim. Nose: ripe fruit, aged wood nuances, tobacco, sweet spices, meaty notes. Palate: spicy, round tannins, long.

Protos 9 meses 2020 T RB
100% tempranillo

90

Fruity, spicy, jammy. Colour: cherry, purple rim. Nose: ripe fruit, red berry notes, black fruit, spicy, smoky. Palate: fruity, flavourful, good structure, slightly dry, soft tannins.

🏆 **PODIUM**

Protos Selección Finca el Grajo Viejo 2018 T
100% tempranillo

95

Colour: cherry, purple rim. Nose: ripe fruit, sweet spices, dried herbs, dried flowers. Palate: spicy, round tannins, long, flavourful.

RUDELES
Trasterrera, 10
42345 Peñalba de San Esteban (Soria)
info@rudeles.com
www.rudeles.com

Rudeles "23" 2020 T
95% tempranillo, 5% garnacha

90

Colour: Cherry. Nose: scrubland, wild herbs. Palate: spicy, good acidity, ripe fruit, easy to drink.

Rudeles "23" 2021 B
100% albillo mayor

90

Colour: bright straw. Nose: fruit expression, ripe fruit, floral. Palate: flavourful, fresh, fruity aftertaste.

Rudeles 2021 RD
100% tempranillo

90

Colour: salmon. Nose: wild herbs, red berry notes, fresh, characterful. Palate: good acidity, balanced, fine bitter notes.

Rudeles Aire 2020 B FB
100% albillo mayor

90

Pleasant, defined aromas. Nose: wild herbs, white fruit, balanced, neat, fresh. Palate: spicy, easy to drink, fine bitter notes.

Rudeles Cerro El Cuberillo 2019 T
100% tempranillo

91

Colour: deep cherry. Nose: dried herbs, creamy oak, black fruit, woody. Palate: powerful, ripe fruit, spicy, harsh oak tannins.

Rudeles Finca La Nación 2018 T
100% tempranillo

92

Colour: deep cherry. Nose: dried herbs, creamy oak, black fruit. Palate: ripe fruit, spicy, round tannins.

SEÑORÍO DE BOCOS
Camino La Canaleja, s/n
47317 Bocos de Duero (Valladolid)
☎: +34 983 880 988
bodegas@senoriodebocos.com
www.bocoswine.com

Autor de Bocos 2019 T
tempranillo

89

Balanced, spicy, herbaceous, tasty, toasty.

Señorío de Bocos 2015 T R
tempranillo

87

Señorío de Bocos 2018 T C
tempranillo

88

Jammy, creamy, tasty, roasted coffee, woody.

Señorío de Bocos 2020 T RB
tempranillo

87

SOLAR DE SAMANIEGO
28036 Madrid (Madrid)
☎: +34 913 433 320
cofradia@solardesamaniego.es
www.solardesamaniego.com

Durón Cueva del Raposo 2018 T
80% tinta del país, 20% merlot

90

Colour: dark-red cherry, garnet rim. Nose: ripe fruit, aged wood nuances, tobacco, sweet spices. Palate: spicy, round tannins, long.

Durón Óptimo 2018 T R
75% tinta del país, 25% merlot

91

Pruney, jammy, tasty, rustic. Nose: grassy, dried herbs, black fruit. Palate: long, spicy, round tannins.

DO RIBERA DEL DUERO / D.O.P.

Maestro de Durón 2018 T C
95% tinta del país, 5% merlot

89

Jammy, creamy, pruney, full-bodied, smooth.

SPANISH PALATE
Avda. Carlos Latorre, 30
49800 Toro (Zamora)
☎: +34 980 690 643
export@spanishpalate.es
www.spanishpalate.es

Botas de Barro
Ribera del Duero 2019 T RB
100% tempranillo

88

Pleasant, fruity, jammy, tasty.

TR3SMANO
Pago de las Bodegas, s/n
47314 Padilla de Duero (Valladolid)
☎: +34 606 954 839
reservas@tresmano.com
www.tresmano.com

Tr3smano Albillo Mayor 2020 B
100% albillo

92

Colour: bright straw. Nose: expressive, ripe fruit, floral, fine lees, mineral. Palate: full-bodied, complex, spicy, long, elegant.

Tr3smano Tm 2018 T BA
100% tinta del país

94

Colour: deep cherry. Nose: ripe fruit, dried herbs, creamy oak, new oak, dark chocolate. Palate: powerful, ripe fruit, spicy, round tannins.

Tr3smano Vendimia 2019 T
100% tinta del país

92

Colour: very deep cherry. Nose: roasted coffee, aromatic coffee, powerful, black fruit, ripe fruit. Palate: smoky aftertaste, great length, round tannins.

TRESPIEDRAS
09315 Fuentecén (Burgos)
☎: +34 696 746 553
jorge@bodegastrespiedras.com
www.bodegastrespiedras.com

Nobbis 2020 T

90

Colour: dark-red cherry. Nose: toasty, spicy, cocoa bean, black fruit. Palate: flavourful, toasty, fine bitter notes.

Unanimous Finca La Tejera 2019 T C

94

Colour: very deep cherry. Nose: complex, expressive, spicy, mineral, black fruit. Palate: elegant, full-bodied, long, great length.

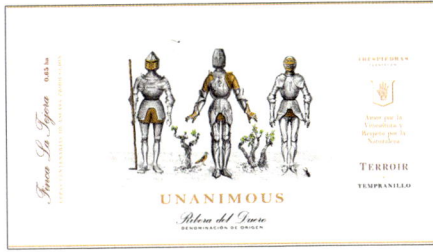

Unanimous Santa Cruz Albillo Mayor 2018 B
100% albillo mayor

91

Colour: bright straw. Nose: fruit expression, ripe fruit, floral, spicy. Palate: flavourful, fresh, good finish.

UVAS FELICES
Agullers, 7
08003 Barcelona (Barcelona)
☎: +34 902 327 777
www.vilaviniteca.es

La Cuartilleja 2018 T R

93

Colour: deep cherry. Nose: creamy oak, black fruit, scrubland, roasted coffee. Palate: powerful, ripe fruit, spicy, round tannins.

Venta Las Vacas 2020 T

93

Colour: bright cherry. Nose: fresh fruit, creamy oak, red berry notes, scrubland. Palate: good acidity, spicy, fine tannins, taut.

VALDEMONJAS
Ctra. N-122 Km. 322
47360 Quintanilla de Arriba (Valladolid)
☎: +34 983 248 294
info@valdemonjas.es
www.valdemonjas.es

Entre Palabras 2019 T
100% tempranillo

92 ♣

Colour: cherry, purple rim. Nose: fruit expression, red berry notes, floral, spicy. Palate: flavourful, fruity, good acidity, long.

DO RIBERA DEL DUERO / D.O.P.

VALDERIVERO
09318 Nava de Roa (Burgos)
☎: +34 948 379 994
info@marquesdelatrio.com
www.valderivero.es

Valderivero 2019 T C
tempranillo
90
Colour: dark-red cherry. Nose: toasty, spicy, cocoa bean, ripe fruit, varietal. Palate: flavourful, toasty, fine bitter notes.

Valderivero 2021 T RB
tempranillo
88
Toasty, powerful, jammy, spicy.

VEGA CLARA
Ctra. N-122, Km 328
47350 Quintanilla de Onésimo (Valladolid)
☎: +34 677 570 779
vegaclara@vegaclara.com
www.vegaclara.com

Diez Almendros 2020 T BA
91
Colour: deep cherry. Nose: ripe fruit, dried herbs, creamy oak. Palate: powerful, ripe fruit, spicy, round tannins.

Jilguerín 2021 T
tempranillo, cabernet sauvignon
87

Mario VC 2019 T
tempranillo, cabernet sauvignon
90
Colour: dark-red cherry. Nose: toasty, spicy, black fruit, ripe fruit, waxy notes, dried herbs. Palate: flavourful, toasty, fine bitter notes.

VELVETY WINES - DOMINIO LUBIANO
Ctra. Pesquera - Valbuena de Duero (Polig. 9 Parc. 66)
47315 Pesquera de Duero (Valladolid)
☎: +34 652 905 042
info@velvetywines.com
www.velvetywines.com

Dominio Lubiano 2019 T RB
100% tempranillo
92
Colour: deep cherry. Nose: ripe fruit, dried herbs, creamy oak, dark chocolate. Palate: powerful, ripe fruit, spicy, round tannins.

VICENTE GANDÍA
Ctra. Cheste a Godelleta, s/n
46370 Chiva (València/Valencia)
☎: +34 962 524 242
Fax: +34 962 524 243
info@vicentegandia.com
www.vicentegandia.es

Dolmo Tempranillo 2020 T RB
tempranillo
88
Balanced, spicy, toasty, tasty.

Nebla Tempranillo 2020 T RB
100% tempranillo
89
Full-bodied, balanced, spicy, dried herbs, toasty.

VIEJAS DE IZAN
Real, 14
09370 Gumiel De Izán (Burgos)
☎: +34 609 088 480
infovdi@viejasdeizan.com
www.viejasdeizan.com

🏆 PODIUM

Viejas de Izan 2019 T C
tempranillo
95
Colour: deep cherry. Nose: ripe fruit, black fruit, hint of anise, dried herbs, balsamic herbs. Palate: ripe fruit, spicy, round tannins, balanced.

Viejas de Izan 2020 T C
tempranillo
93
Colour: deep cherry. Nose: dried herbs, creamy oak, black fruit, red berry notes. Palate: powerful, ripe fruit, spicy, round tannins.

VINOS DE CULTO BANISIO
Paseo de Gracia, 8-10, 3ª planta
08007 Barcelona (Barcelona)
☎: +34 933 631 838
info@banisio.com
www.banisio.com

Banisio 2020 T RB
tempranillo
89
Full-bodied, creamy, balanced, spicy, jammy, tasty.

VINOS DE LA LUZ
Ctra. de Mélida, km. 3,5
47300 Peñafiel (Valladolid)
☎: +34 983 878 007
info@vinosdelaluz.com
www.vinosdelaluz.com

Cinema 2018 T C
100% tempranillo
90
Colour: deep cherry. Nose: dried herbs, creamy oak, black fruit, spicy. Palate: powerful, ripe fruit, spicy, round tannins.

Cinema 2020 T RB
100% tempranillo
90
Colour: deep cherry. Nose: ripe fruit, dried herbs, creamy oak. Palate: powerful, ripe fruit, spicy, round tannins.

Pagos de Valcerracin Crianza Vendimia Seleccionada 2018 T C
100% tempranillo
90
Colour: deep cherry. Nose: ripe fruit, dried herbs, toasty. Palate: ripe fruit, spicy, harsh oak tannins.

VINOS SANTOS ARRANZ (LÁGRIMA NEGRA)
47315 Pesquera de Duero (Valladolid)
Fax: +34 983 870 008
lagrimanegra82@hotmail.com
www.lagrima-negra.com

Lágrima Negra 2019 T C
88
Full-bodied, spicy, jammy, tasty, dried herbs.

Lágrima Negra 2020 T RB
87

Lágrima Negra Edición Limitada 2019 T
89
Creamy, spicy, balanced, jammy, toasty, great length.

VINOS Y VIÑEDOS FAMILIA FIEL
Camino de Valdeoliva, 9
28750 San Agustín de Guadalix (Madrid)
☎: +34 918 258 100
info@vinosfiel.es
www.vinosfiel.es

Colección 880 Vendimia Seleccionada 2020 T C
tempranillo
88
Jammy, pruney, toasty.

Cuesta Roa 940 2017 T C
tempranillo
91
Colour: dark-red cherry. Nose: toasty, spicy, cocoa bean. Palate: flavourful, toasty, fine bitter notes.

Cuesta Roa 940 Etiqueta Negra 2015 T C
tempranillo
90
Colour: deep cherry. Nose: dried herbs, creamy oak, fruit preserve, toasty. Palate: ripe fruit, spicy, round tannins.

VIÑA AGUILERA
Camino del Val s/n
09370 La Aguilera (Burgos)
☎: +34 947 504 192
vinaaguilerasl@hotmail.com

Torreval 6 meses Barricas 2021 T
90% tempranillo, 10% cabernet sauvignon
89
Jammy, spicy, tasty, toasty, smoky, fruity.

VIÑA ARNAIZ
Ctra. N-122, Km. 281
09463 Haza (Burgos)
☎: +34 947 536 227
Fax: +34 947 536 216
atcliente@jgc.es
www.familiagarciacarrion.com

Mayor de Castilla 2020 T RB
100% tempranillo
85

VIÑA BUENA
Avda. Portugal, 15
09400 Aranda de Duero (Burgos)
☎: +34 947 546 414
bodega@vinabuena.com
www.vinabuena.com

Barón de Santuy 2019 T C
87

Viña Buena 2020 T BA
88
Balanced, spicy, jammy, dried herbs.

Viña Buena 2021 T
tempranillo
88
Balanced, fruity, herbal, spicy.

DO RIBERA DEL DUERO / D.O.P.

Viña Buena 2019 T C
tempranillo
87
Jammy, toasty, powerful, pruney.

VIÑA VALDEMAZÓN
47359 Olivares de Duero (Valladolid)
☎: +34 983 680 220
info@valdemazon.com
www.valdemazon.com

Viña Valdemazón
Vendimia Seleccionada 2018 T
tempranillo
87

VIÑAS DEL JARO
Finca el Quiñón, Ctra. de Renedo a Pesquera de Duero, Km. 39
47359 Pesquera de Duero (Valladolid)
☎: +34 956 854 204
info@vinosiberian.com
www.vinosiberian.com

Chafandín 2018 T
tinto fino
91
Colour: deep cherry. Nose: dried herbs, creamy oak, black fruit, woody. Palate: powerful, ripe fruit, spicy, harsh oak tannins.

Jaros 2018 T
tinto fino
91
Colour: bright cherry. Nose: fresh fruit, creamy oak, black fruit. Palate: good acidity, spicy, fine tannins.

Sed de Caná 2018 T
tinto fino
94
Colour: deep cherry. Nose: ripe fruit, creamy oak, red berry notes, black fruit, dark chocolate. Palate: ripe fruit, spicy, round tannins, elegant.

Sembro 2021 T
tinto fino
89
Full-bodied, creamy, spicy, fruity.

Sembro
Edición Limitada 2020 T
tinto fino
89
Full-bodied, creamy, spicy, tasty, toasty.

VIÑEDOS ALONSO DEL YERRO
09300 Roa (Burgos)
☎: +34 610 789 997
Fax: +34 913 160 121
mariadelyerro@vay.es
www.alonsodelyerro.es

"María"
Alonso del Yerro 2018 T
tempranillo
91
Colour: deep cherry. Nose: ripe fruit, dried herbs, creamy oak. Palate: powerful, ripe fruit, spicy, round tannins.

Alonso del Yerro 2018 T
tempranillo
92
Colour: deep cherry. Nose: ripe fruit, dried herbs, creamy oak, black fruit. Palate: ripe fruit, spicy, round tannins.

VIÑEDOS LA NAVA
Isilla, 13
09400 Aranda de Duero (Burgos)
☎: +34 947 506 011
administracionvinos@grupotudanca.com
www.grupotudanca.com

Tudanca 2018 T C
tempranillo
90
Colour: deep cherry. Nose: ripe fruit, dried herbs, creamy oak. Palate: ripe fruit, spicy, round tannins.

Tudanca 2020 T RB
tempranillo
87

Tudanca Raytu 2015 T R
tempranillo
91
Colour: deep cherry. Nose: dried herbs, creamy oak, black fruit, woody. Palate: powerful, ripe fruit, spicy, round tannins.

Tudanca Vendimia Selección 2014 T
tempranillo

91

Colour: cherry, garnet rim. Nose: fruit preserve, fruit liqueur notes, powerful, toasty. Palate: flavourful, sweetness, long.

Tudanca Vicenta Mater 2016 T C
tempranillo

91

Colour: dark-red cherry, garnet rim. Nose: ripe fruit, aged wood nuances, tobacco, sweet spices. Palate: spicy, round tannins, long.

VIÑEDOS SINGULARES

Avda. de La Riera, 11 Nave 1
08960 Sant Just Desvern (Barcelona)
☎: +34 934 807 041
Fax: +34 934 807 076
info@vinedossingulares.com
www.vinedossingulares.com

Entrelobos 2020 T
tinto fino

89

Jammy, tasty, standard, dried herbs.

VIÑEDOS Y BODEGAS ÁSTER

Finca El Caño. Ctra. Palencia-Aranda, km. 54,9
09313 Anguix (Burgos)
☎: +34 947 522 700
aster@riojalta.com
www.riojalta.com

Áster Finca el Otero 2018 T
tinta del país

93

Colour: deep cherry. Nose: ripe fruit, dried herbs, creamy oak. Palate: powerful, ripe fruit, spicy, round tannins.

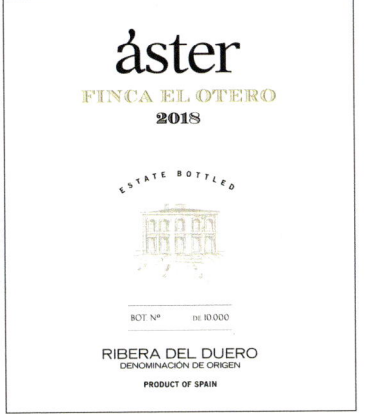

Áster 2019 T C
tinta del país

91

Colour: deep cherry. Nose: dried herbs, creamy oak, black fruit, stone fruit. Palate: powerful, ripe fruit, spicy, round tannins.

VIÑEDOS Y BODEGAS GARCÍA FIGUERO

Ctra. La Horra - Roa, Km. 2,2
09311 La Horra (Burgos)
☎: +34 686 008 952
cristina@tintofiguero.com
www.tintofiguero.com

🏆 **PODIUM**

Figuero Tinus 2018 T
100% tempranillo

95

Colour: cherry, purple rim. Nose: spicy, dried herbs, ripe fruit, red berry notes, black fruit, toasty. Palate: flavourful, fruity, long, slightly dry, soft tannins.

Milagros de Figuero 2018 T
100% tempranillo

94

Colour: deep cherry. Nose: ripe fruit, toasty, spicy, black fruit, dark chocolate. Palate: flavourful, spicy, slightly dry, soft tannins.

Tinto Figuero 12 2019 T C
100% tempranillo

92

Full-bodied, powerful. Nose: characterful, ripe fruit, powerful. Palate: flavourful, long, slightly dry, soft tannins.

Tinto Figuero 15 2018 T R
100% tempranillo

93

Colour: deep cherry. Nose: dried herbs, creamy oak, black fruit, cocoa bean. Palate: powerful, ripe fruit, spicy, round tannins.

Tinto Figuero Viñas Viejas 2019 T
100% tempranillo

93

Colour: deep cherry. Nose: creamy oak, black fruit, scrubland. Palate: powerful, ripe fruit, spicy, round tannins.

DO RIBERA DEL DUERO / D.O.P.

DO RIBERA DEL DUERO / D.O.P.

VIÑEDOS Y BODEGAS GORMAZ

Ctra. de Soria, s/n
42330 San Esteban de Gormaz (Soria)
☎: +34 913 982 303
Fax: +34 975 351 313
info@hispanobodegas.com
www.hispanobodegas.com

12 Linajes 2017 T R
tempranillo
91
Colour: Cherry. Nose: balsamic herbs, sweet spices, scrubland. Palate: spicy, flavourful, round tannins.

12 Linajes 2019 T C
tempranillo
91
Colour: deep cherry. Nose: ripe fruit, dried herbs, creamy oak. Palate: powerful, ripe fruit, spicy, round tannins.

12 Linajes 2021 T RB
tempranillo
90
Colour: deep cherry. Nose: dried herbs, creamy oak, ripe fruit, black fruit. Palate: powerful, ripe fruit, spicy, round tannins.

12 Linajes Grano a Grano 2018 T C
tempranillo
94
Colour: bright cherry, purple rim. Nose: red berry notes, spicy, wild herbs, dried flowers. Palate: balanced, fine tannins, flavourful.

12 Linajes Senda de la Estación 2019 T
tempranillo
92
Colour: Cherry. Nose: scrubland, fruit preserve, dried herbs. Palate: spicy, round tannins, long, flavourful.

YLLERA BODEGAS & VIÑEDOS

47490 Rueda (Valladolid)
☎: +34 983 868 097
grupoyllera@grupoyllera.com
www.grupoyllera.com

La Fleur Vivaltus 2017 T
94
Colour: dark-red cherry, garnet rim. Nose: ripe fruit, aged wood nuances, tobacco, sweet spices. Palate: spicy, round tannins, long.

Boada 2019 T RB
tempranillo
88
Spicy, jammy, tasty, toasty, smoky, pleasant, fruity.

Pepe Yllera 2019 T RB
91
Colour: cherry, garnet rim. Nose: ripe fruit, sweet spices, spicy, smoky. Palate: powerful, ripe fruit, spicy, round tannins.

🏆 PODIUM

Vivaltus 2017 T
95
Defined aromas, complex. Colour: dark-red cherry. Nose: chalk, ripe fruit, varietal, complex, meaty notes. Palate: flavourful, good structure, complex.

Boada 2015 T C
tempranillo
91
Colour: deep cherry. Nose: ripe fruit, dried herbs, creamy oak, meaty notes, fine reductive notes. Palate: powerful, ripe fruit, spicy, round tannins.

DO. RIBERA DEL GUADIANA
CONSEJO REGULADOR

Avda. Pte, Juan Carlos Rodríguez Ibarra, s/n. Apdo. 299
06200 Almendralejo (Badajoz)
☎: +34 924 671 302
Fax: +34 924 664 703
@: info@riberadelguadiana.eu
www.riberadelguadiana.eu

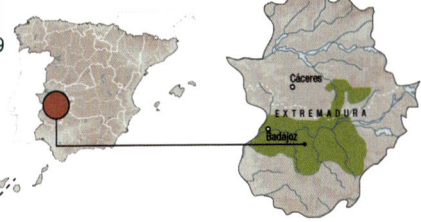

LOCATION:

Covering the 6 wine-growing regions of Extremadura, with a total surface of more than 87,000 Ha as described below.

SUB-REGIONS AND CLIMATE:

Cañamero: To the south east of the province of Cáceres, in the heart of the Sierra de Guadalupe. It comprises the municipal districts of Alia, Berzocana, Cañamero, Guadalupe and Valdecaballeros. The vineyards are located on the mountainside, at altitudes of between 600 m to 800 m. The terrain is rugged and the soil is slaty and loose. The climate is mild without great temperature contrasts, and the average annual rainfall is 750 mm to 800 mm. The main grape variety is the white Alarije.

Montánchez: Comprising 27 municipal districts. It is characterised by its complex terrain, with numerous hills and small valleys. The vineyards are located on brown acidic soil. The climate is Continental in nature and the average annual rainfall is between 500 mm and 600 mm. The white grape variety Borba occupies two thirds of the vineyards in the region.

Ribera Alta: This covers the Vegas del Guadiana and the plains of La Serena and Campo de Castuera and comprises 38 municipal districts. The soil is very sandy. The most common varieties are Alarije, Borba (white), Tempranillo and Garnacha (red).

Ribera Baja: Comprising 11 municipal districts. The vineyards are located on clayey-limy soil. The climate is Continental, with a moderate Atlantic influence and slight contrasts in temperature. The most common varieties are: Cayetana Blanca and Pardina among the whites, and Tempranillo among the reds.

Matanegra: Rather similar to Tierra de Barros, but with a milder climate. It comprises 8 municipal districts, and the most common grape varieties are Beba, Montua (whites), Tempranillo, Garnacha and Cabernet Sauvignon (reds).

Tierra de Barros: Situated in the centre of the province of Badajoz and the largest (4475 Ha and 37 municipal districts). It has flat plains with fertile soils which are rich in nutrients and have great water retention capacity (Rainfall is low: 350 mm to 450 mm per year). The most common varieties are the white Cayetana Blanca and Pardina, and the red Tempranillo, Garnacha and Cabernet Sauvignon.

GRAPE VARIETIES:

WHITE: alarije, borba, cayetana blanca, pardina, macabeo, chardonnay, chelva o montua, malvar, parellada, pedro ximénez, verdejo, eva, cigüente, perruno, moscatel de Alejandría, moscatel de grano menudo, sauvignon blanc, moscatel de Málaga, bobal blanco, sauvignon blanca, antão vaz, arinto, fernão pires, colombard and xarel.lo.

RED: garnacha tinta, tempranillo, bobal, cabernet sauvignon, garnacha tintorera, graciano, mazuela, merlot, monastrell, syrah, pinot noir, jaén tinto, touriga nacional, castelão, trincadeira and malbec.

FIGURES:

Vineyard surface: 42,672 - **Wine-Growers:** 3,546 - **Wineries:** 25 – **2021 Harvest rating:** Very Good – **Production 21:** 8,760,400 litres – **Market percentages:** 57% National - 43% International.

CLASSIFICATION OF YOUNG WINE HARVESTS PEÑÍNGUIDE

2017	2018	2019	2020	2021
AVERAGE	GOOD	AVERAGE	GOOD	N/A

DO RIBERA DEL GUADIANA / D.O.P.

BODEGAS CAÑALVA
Coto, 54
10136 Cañamero (Cáceres)
☎: +34 927 369 405
info@bodegascanalva.com
www.bodegascanalva.com

Cañalva Élégance 2019 T C
100% cabernet sauvignon
88
Spicy, fruity, jammy, smoky.

Cañalva Élégance 2020 T RB
100% garnacha
86

Cañalva Élégance 2021 B S
100% sauvignon blanc
88
Pleasant, varietally correct, jammy, fruity.

Cañalva Tempranillo 2019 T RB
100% tempranillo
87

Eburus 2018 T C
50% cabernet sauvignon, 50% tempranillo
90
Colour: bright cherry. Nose: ripe fruit, smoky, spicy, balsamic herbs. Palate: fruity, flavourful, slightly dry, soft tannins.

Fuente Cortijo 2019 T C
70% tempranillo, 30% cabernet sauvignon
89
Balsamic herbs, spicy, jammy, tasty, fruity.

BODEGAS LA CORTE
Ctra. Entrín Bajo, s/n
06196 Corte de Peleas (Badajoz)
☎: +34 924 693 014
Fax: +34 924 693 270
administracion@bodegaslacorte.com
www.bodegaslacorte.com

Conde de la Corte 2019 T C
100% tempranillo
89
Balsamic herbs, jammy, pruney, dried herbs, tasty, wild, smooth.

BODEGAS MARCELINO DÍAZ
Mecánica, s/n
06200 Almendralejo (Badajoz)
☎: +34 924 677 548
bodega@madiaz.com
www.madiaz.com

Puerta Palma 2020 T
85

Puerta Palma 2021 B
pardina
84

BODEGAS MARTÍNEZ PAIVA
Ctra. Gijón - Sevilla N-630, Km. 646
Apdo. Correos 87
06200 Almendralejo (Badajoz)
☎: +34 924 671 130
info@bodegasmartinezpaiva.com
www.bodegasmartinezpaiva.com

Paiva 56 Barricas 2018 T FB
tempranillo
88
Full-bodied, pruney, spicy, tasty.

Paiva Crianza 2019 T C
90% tempranillo, 10% cabernet sauvignon
88
Hot, jammy, tasty, toasty.

Paiva Solo Cayetana 2021 B
cayetana blanca
88
Citrus fruit, mineral, dried herbs, full-bodied, jammy.

Paiva Solo Tempranillo 2020 T S
tempranillo
87

BODEGAS ROMALE
Pol. Ind. Parc. 6, Manzana D - Mecánica s/n
06200 Almendralejo (Badajoz)
☎: +34 924 667 255
romale@romale.com
www.romale.com

Privilegio de Romale 2018 T C
tempranillo
87

BODEGAS RUIZ TORRES

Ctra. EX 116, km.33,8
10136 Cañamero (Cáceres)
☎: +34 927 369 027
info@ruiztorres.com
www.ruiztorres.com

Attelea 2018 T C
tempranillo, cabernet sauvignon

87

PAGO LOS BALANCINES

Paraje la Agraria, s/n
06475 Oliva de Mérida (Badajoz)
☎: +34 924 367 399
info@pagolosbalancines.com
www.pagolosbalancines.com

Balancines Blanco Sobre Lías B RB
chardonnay, viura

90

Colour: bright yellow. Nose: powerful, creamy oak, ripe fruit, spicy. Palate: long, toasty, fine bitter notes.

Balancines Garnacha & Garnacha 2017 T C
50% garnacha tintorera, 50% garnacha

91

Colour: Cherry. Nose: balsamic herbs, sweet spices, scrubland, earthy notes. Palate: spicy, balsamic, good acidity.

Balancines Gold 2018 T C
garnacha, tempranillo, garnacha tintorera

90

Colour: deep cherry. Nose: ripe fruit, dried herbs, creamy oak. Palate: powerful, ripe fruit, spicy, round tannins.

Haragán Reserva Especial 2018 T R
garnacha tintorera, tinta roriz

93

Colour: dark-red cherry, garnet rim. Nose: ripe fruit, fruit preserve, aged wood nuances, tobacco. Palate: spicy, round tannins, long.

Huno 2018 T R
garnacha tintorera, tempranillo, graciano

91

Colour: dark-red cherry. Nose: toasty, spicy, cocoa bean, fine reductive notes. Palate: flavourful, toasty, fine bitter notes.

PALACIO QUEMADO

Ctra. Almendralejo - Palomas, km 6,9
06840 Alange (Badajoz)
☎: +34 924 120 296
palacioquemado@alvear.es
www.palacioquemado.com

Palacio Quemado 2019 T C S

90

Colour: dark-red cherry. Nose: toasty, spicy, cocoa bean, powerful, ripe fruit. Palate: flavourful, toasty, fine bitter notes.

Palacio Quemado Viñedos Propios 2016 T R

91

Colour: deep cherry. Nose: ripe fruit, dried herbs, creamy oak, fine reductive notes. Palate: powerful, ripe fruit, spicy, round tannins.

S.C.A. SANTA MARTA VIRGEN

Cooperativa s/n
06150 Santa Marta de Los Barros (Badajoz)
☎: +34 924 690 218
Fax: +34 924 690 083
info@bodegasantamarta.com
www.cooperativasantamarta.com

Blasón del Turra Macabeo 2021 B
macabeo

86

Blasón del Turra Pardina 2021 B
pardina

85

Blasón del Turra Tempranillo 2020 T
tempranillo

84

Compass 2019 T RB
tempranillo

85

Valdeaurum 2018 T C
tempranillo

85

VIÑAOLIVA SOC. COOP.

Automoción, 1
06200 Almendralejo (Badajoz)
☎: +34 924 677 321
info@vinaoliva.com
www.zaleo.es

Grácil de Zaleo 2018 T C

87

DO RIBERA DEL GUADIANA / D.O.P.

DO RIBERA DEL GUADIANA / D.O.P.

Zaleo Pardina 2021 B
100% pardina
84

Zaleo Tempranillo 2021 T
100% tempranillo
88
Tasty, jammy, fruity.

VIÑEDOS POZANCO
Ctra. de BA-001 Km. 15,700
06800 Mérida (Badajoz)
☎: +34 924 143 249
info@bodegaspozanco.com
www.bodegaspozanco.com

10·12 (Diez Punto Doce) 2016 T R
60% tempranillo, 20% merlot, 20% graciano
88
Toasty, jammy, spicy, powerful, tasty.

Sinoble Macabeo 2020 B
100% macabeo
88
Citrus fruit, mineral, herbal, tasty.

Viñedos Pozanco 2017 T C
60% tempranillo, 20% graciano, 20% cabernet sauvignon
87

Viñedos Pozanco 2021 T
60% tempranillo, 40% merlot
83

DO. RIBERA DEL JÚCAR
CONSEJO REGULADOR

Deportes, 4.
16700 Sisante (Cuenca)
☎: +34 969 387 182 - Fax: +34 969 387 208
@: do@vinosriberadeljucar.com
www.vinosriberadeljucar.com

LOCATION:

The 7 wine producing municipal districts that make up the DO are located on the banks of the Júcar, in the south of the province of Cuenca. They are: Casas de Benítez, Casas de Guijarro, Casas de Haro, Casas de Fernando Alonso, Pozoamargo, Sisante and El Picazo. The region is at an altitude of between 650 and 750 m above sea level.

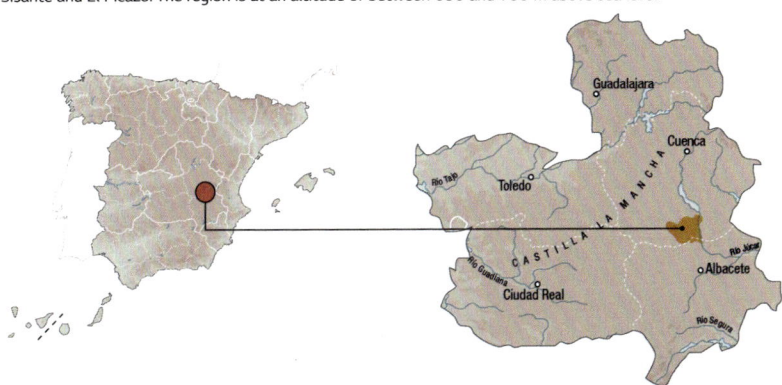

GRAPE VARIETIES:

RED: Bobal, Cencibel o Tempranillo, Cabernet Sauvignon, Merlot, Syrah, Petit Verdot, Cabernet Franc, Garnacha tinta, Garnacha Tintorera and Monastrell.

WHITE: Moscatel de grano menudo, Sauvignon Blanc, Airen, Macabeo, Verdejo, Pardillo (Marisancho) and Chardonnay.

FIGURES:

Vineyard surface: 6,700 – **Wine-Growers:** 850 – **Wineries:** 8 – **2021 Harvest rating:** Very Good – **Production 21:** 700,000 litres – **Market percentages:** 30% National - 70% International.

SOIL:

The most common type of soil consists of pebbles on the surface and a clayey subsoil, which provides good water retention capacity in the deeper levels.

CLIMATE:

Continental in nature, dry, and with very cold winters and very hot summers. The main factor contributing to the quality of the wine is the day-night temperature contrasts during the ripening season of the grapes, which causes the process to be carried out slowly.

CLASSIFICATION OF YOUNG WINE HARVESTS PEÑÍNGUIDE

2017	2018	2019	2020	2021
GOOD	GOOD	VERY GOOD	N/A	N/A

BODEGA LAS CALZADAS
Calle de la Virgen, 13
16708 Pozoamargo (Cuenca)
☎: +34 969 337 354
info@bodegalascalzadas.com
www.bodegalascalzadas.com

Tinácula El Santillo 2020 T
bobal, cencibel, pardilla

90

Colour: cherry, purple rim. Nose: ripe fruit, spicy, black pepper. Palate: fruity, flavourful, balanced, powerful tannins.

Tinácula Red 2020 T
bobal

90

Colour: deep cherry. Nose: ripe fruit, dried herbs, creamy oak, fruit expression. Palate: powerful, ripe fruit, spicy, round tannins.

Tinácula White 2021 B
pardilla

89

Floral, fruity, tasty.

Tinácula X 2019 T
bobal, cencibel

88

Fruity, spicy, rustic, simple.

BODEGAS TRENZA
Felix Mendelsohn, 8
03730 Jávea (Alacant/Alicante)
☎: +34 965 790 012
bodegas@bodegastrenza.com
www.bodegatrenza.com

Tofterup Brothers Tempranillo 2020 T

90

Colour: deep cherry. Nose: creamy oak, red berry notes, black fruit. Palate: ripe fruit, spicy, round tannins.

BODEGAS Y VIÑEDOS EL SOLEADO
Pozo Alfaro, 50
16237 Ledaña (Cuenca)
☎: +34 911 994 108
info@elsoleado.com
www.elsoleado.com

Billon Esperanza E316 2020 T
cabernet sauvignon, tempranillo

87

Billon Victoria V419 2020 T
syrah, merlot

87

BODEGAS Y VIÑEDOS ILLANA
Finca Buenavista, s/n
16708 Pozoamargo (Cuenca)
☎: +34 969 147 039
Fax: +34 969 147 057
info@bodegasillana.com
www.bodegasillana.com

Casa de Illana Alma 2021 B
100% sauvignon blanc

88

Balanced, herbal, spicy, crisp.

Casa de Illana Bobal Vino de Parcela 2019 T
100% bobal

92

Colour: Cherry. Nose: expressive, spicy, scrubland, dried herbs, waxy notes. Palate: full-bodied, long, great length, round tannins.

Casa de Illana Carmen 2019 B FB
100% sauvignon blanc

90

Colour: bright yellow. Nose: creamy oak, ripe fruit, spicy, dried herbs. Palate: good structure, toasty, fine bitter notes.

Casa de Illana Tresdecinco 2018 T C
31% cabernet sauvignon, 30% tempranillo, 24% merlot, 15% bobal

89

Defined aromas, herbal, jammy, tasty, balanced, spicy.

BODEGAS Y VIÑEDOS SUCRO
C. Tapias, 8
16708 Pozoamargo (Cuenca)
☎: +34 969 387 173
info@garciaperezgroup.com
www.parajesdelvalle.es

Sucro 2020 T
tempranillo

90

Colour: cherry, garnet rim. Nose: overripe fruit, creamy oak, hot. Palate: pruney, powerful, sweet tannins.

BODEGAS Y VIÑEDOS VALTRAVIESO

Finca La Revilla, s/n
47316 Piñel de Arriba (Valladolid)
☎: +34 983 484 030
marketing@valtravieso.com
www.valtravieso.com

Mil Cantos 2019 T
90% bobal, 10% airén

91

Colour: cherry, purple rim. Nose: floral, spicy, red berry notes, black fruit, ripe fruit, wild herbs. Palate: flavourful, fruity, good acidity, long, good structure, concentrated.

PURÍSIMA CONCEPCIÓN, S.C. DE CLM

Ctra. San Clemente, Km. 10
16610 Casas de Fernando Alonso (Cuenca)
☎: +34 969 383 043
info@vinoteatinos.com
www.vinoteatinos.com

Claros de Cuba Origen 2017 T
tempranillo

91

Colour: Cherry. Nose: woody, caramel, dark chocolate, black fruit, fruit preserve. Palate: powerful, balanced, flavourful, fruity, round tannins.

Teatinos 40 Barricas Tempranillo 2017 T R
tempranillo

89

Spicy, fruity, herbal, jammy, tasty.

Teatinos Claros de Cuba 2016 T R
tempranillo

89

Smoky, toasty, jammy, standard, tasty.

Teatinos Signvm 2016 T C
100% tempranillo

89

Spicy, fruity, smoky, tasty.

Teatinos Syrah 2021 T
100% syrah

89

Fruity, dried flowers, herbal, spicy, tasty.

Teatinos Tempranillo 2021 T
100% tempranillo

88

Spicy, fruity, jammy, toasty, tasty.

VIÑEDOS Y BODEGA LA MAGDALENA

Ctra. La Roda, s/n
16611 Casas de Haro (Cuenca)
☎: +34 969 380 722
vinos@vegamoragona.com
www.vegamoragona.com

Vega Moragona Cabernet Sauvignon 2020 T C
100% cabernet sauvignon

86

Vega Moragona La Duna 2019 T
100% tempranillo

89

Nose: overripe fruit, creamy oak, hot, aromatic coffee. Palate: pruney, powerful.

Vega Moragona Moscatel de Grano Menudo 2021 B D
100% moscatel grano menudo

88

Citrus fruit, balanced, floral, exuberant.

Vega Moragona Syrah 2020 T RB
100% syrah

88

Pleasant, kind finish, tasty, jammy.

DO RIBERA DEL JÚCAR / D.O.P.

DO. Ca. RIOJA
CONSEJO REGULADOR

Estambrera, 52
26006 Logroño (La Rioja)
☎: +34 941 500 400 - Fax: +34 941 500 672
@: info@riojawine.com
www.riojawine.com

LOCATION:

Occupying the Ebro valley. To the north it borders with the Sierra de Cantabria and to the south with the Sierra de la Demanda, and is made up of different municipal districts of La Rioja, the Basque Country and Navarra. The most western region is Haro and the easternmost, Alfaro, with a distance of 100 km between the two. The region is 40 km wide.es de 40 kilómetros.

SUB-REGIONS:

Rioja Alta: This has Atlantic influences; it is the most extensive with some 20,500 Ha and produces wines well suited for ageing. **Rioja Alavesa:** A mixture of Atlantic and Mediterranean influences, with an area under cultivation of some 11,500 Ha; both young wines and wines suited for ageing are produced. **Rioja Oriental:** With approximately 18,000 Ha, the climate is purely Mediterranean; white wines and rosés with a higher alcohol content and extract are produced.

GRAPE VARIETIES:

WHITE: Viura, Malvasía, Garnacha Blanca, Chardonnay, Sauvignon Blanc, Verdejo, Maturana Blanca, Tempranillo Blanco and Torrontés.

RED: Tempranillo, Garnacha, Graciano, Mazuelo and Maturana Tinta.

FIGURES:

Vineyard surface: 66,217– **Wine-Growers:** 14,300 – **Wineries:** 571– **2021 Harvest rating:** Very Good– **Production 21:** 281,640,000 litres – **Market percentages:** 56% National -44% International.

SOIL:

Various types: the clayey calcareous soil arranged in terraces and small plots which are located especially in Rioja Alavesa, la Sonsierra and some regions of Rioja Alta; the clayey ferrous soil, scattered throughout the region, with vineyards located on reddish, strong soil with hard, deep rock; and the alluvial soil in the area close to the rivers; these are the most level vineyards with larger plots; here the soil is deeper and has pebbles.

CLIMATE:

Quite variable depending on the different sub-regions. In general, there is a combination of Atlantic and Mediterranean influences, the latter becoming more dominant as the terrain descends from west to east, becoming drier and hotter. The average annual rainfall is slightly over 400 mm.

CLASSIFICATION OF YOUNG WINE HARVESTS PEÑÍNGUIDE

2017	2018	2019	2020	2021
GOOD	GOOD	VERY GOOD	VERY GOOD	VERY GOOD

ADRIÁN MORENO LLORENTE
Real, 57
26310 Badarán (La Rioja)
☎: +34 679 266 837
bodega@rulei.es
www.rulei.es

Rulei Viña Barracallo Renques de Chenin 2019 B FB
chenin blanc
91
Colour: bright yellow. Nose: creamy oak, ripe fruit, spicy. Palate: rich, good structure, long, toasty.

Rulei Viña Barracallo Tempranillo-Garnacha 2015 T
60% tempranillo, 40% garnacha
84

Rulei Viña El Moral Viñedo Singular 2019 T
garnacha
90
Colour: deep cherry. Nose: ripe fruit, fruit preserve, creamy oak, dried herbs. Palate: powerful, spicy, round tannins.

ALIAGA
Avda. de Navarra, 17
31591 Corella (Navarra)
☎: 948 401 321
sales@vinaaliaga.com
www.vinaaliaga.com

Gureaga 2015 T
70% tempranillo, 30% garnacha
86

ALTANZA
Ctra. Nacional 232, Km. 419,5
26360 Fuenmayor (La Rioja)
☎: +34 618 629 086
marketing@altanza.com
www.altanza.com

Alma Bohemia 2021 RD
89
Citrus fruit, dried flowers, herbal, balanced.

Altanza 2015 T GR
100% tempranillo
94
Colour: dark-red cherry, garnet rim. Nose: ripe fruit, fruit preserve, aged wood nuances, tobacco, sweet spices, scrubland. Palate: spicy, round tannins, long.

Altanza 2017 T R
100% tempranillo
92
Colour: bright cherry. Nose: ripe fruit, dried herbs, spicy, black fruit. Palate: fruity, flavourful, balanced, slightly dry, soft tannins.

Altanza Club 2015 T R
100% tempranillo
93
Colour: bright cherry. Nose: red berry notes, dried herbs, spicy, smoky. Palate: fruity, flavourful, balanced, smoky aftertaste, slightly dry, soft tannins.

Altanza Familia 2015 T R
100% tempranillo
92
Colour: deep cherry. Nose: dried herbs, creamy oak, black fruit, spicy. Palate: powerful, ripe fruit, spicy, round tannins.

Altanza Rosado 2021 RD
88
Overripe, fruity, dried herbs, simple.

Altanza Sauvignon Blanc 2021 B
sauvignon blanc
88
Citrus fruit, standard, herbal, simple.

Edulis 2019 T C
100% tempranillo
91
Colour: cherry, purple rim. Nose: fruit expression, red berry notes, spicy. Palate: flavourful, fruity, good acidity.

Uva por Uva 2019 T
tempranillo
94
Colour: dark-red cherry. Nose: toasty, spicy, cocoa bean, black fruit. Palate: flavourful, toasty, fine bitter notes.

Valvarés de Altanza 2018 T C
100% tempranillo
92
Colour: deep cherry. Nose: ripe fruit, dried herbs, creamy oak. Palate: powerful, ripe fruit, spicy, round tannins.

🏆 PODIUM

Velázquez Colección Artistas Españoles 2011 T R
tempranillo
95
Colour: deep cherry. Nose: dried herbs, creamy oak, black fruit, spicy. Palate: ripe fruit, spicy, round tannins.

DO Ca. RIOJA / D.O.P.

ALTOS DE RIOJA VITICULTORES Y BODEGUEROS
Solomillo, s/n
01309 Elvillar (Araba/Álava)
☎: +34 945 600 693
Fax: +34 945 600 692
altosderioja@altosderioja.com
www.altosderioja.com

Altos R 2019 T R
100% tempranillo
90
Colour: cherry, purple rim. Nose: red berry notes, grassy, spicy. Palate: fruity, flavourful, balanced, slightly dry, soft tannins.

Altos R 2020 T C
100% tempranillo
90
Colour: cherry, purple rim. Nose: red berry notes, ripe fruit, grassy, spicy. Palate: fruity, flavourful, balanced, slightly dry, soft tannins.

Altos R Pigeage 2019 T
80% tempranillo, 20% graciano
92
Colour: cherry, purple rim. Nose: red berry notes, ripe fruit, creamy oak, wild herbs, expressive. Palate: fruity, flavourful, balanced, powerful tannins.

Altos R Pigeage 2020 B FB
75% viura, 25% chardonnay
90
Colour: bright straw. Nose: white fruit, dried herbs, white flowers. Palate: fruity, flavourful, balanced.

Altos R Pigeage Graciano 2020 T
graciano
92
Colour: very deep cherry. Nose: fresh fruit, red berry notes, grassy, scrubland, spicy. Palate: good structure, fruity, flavourful, fresh, balanced, slightly dry, soft tannins.

Xai Alt 690M 2020 T
100% tempranillo
89
Fruity, overripe, dried flowers, tasty.

ARINAS GIL
Mentoste, 4
01330 Labastida (Araba/Álava)
☎: +34 678 939 064
arinasgilsc@gmail.com
www.zuzaran.es

Zuzarán 2019 T C
tempranillo
88
Toasty, classic, spicy, tasty.

Zuzarán 2020 B FB
viura
86 🌿

Zuzarán Fajero 2019 T C
50% tempranillo, 50% maturana
90
Colour: deep cherry. Nose: ripe fruit, dried herbs, creamy oak, black fruit. Palate: powerful, ripe fruit, spicy, round tannins.

Zuzarán Maturana 2019 T
maturana
88
Powerful, tasty, spicy, pruney.

Zuzarán Maturana Edición Limitada 2020 T
maturana
91 🌿
Colour: deep cherry. Nose: dried herbs, creamy oak, fruit preserve. Palate: powerful, ripe fruit, spicy, round tannins.

Zuzarán Viura 2020 B
viura
86 🌿

ARRIEZU VINEYARDS
Santa Gema, 28
31570 San Adrián (Navarra)
☎: +34 948 672 071
info@arriezuvineyards.com
www.arriezuvineyards.com

J.F. Arriezu 2016 T C
80% tempranillo, 10% graciano, 10% mazuelo
87

J.F. Arriezu 2020 T
50% tempranillo, 50% garnacha
90
Smooth, simple, tasty, jammy. Colour: bright cherry.

J.F. Arriezu Roble 2020 T
50% tempranillo, 50% garnacha
88
Pleasant, spicy, jammy, smooth.

J.F. Arriezu Rosado 2021 RD
45% tempranillo, 35% garnacha, 10% garnacha blanca, 10% viura
86

ARTUKE BODEGAS Y VIÑEDOS
La Serna, 24
01307 Baños de Ebro (Araba/Álava)
☎: +34 945 623 323
artuke@artuke.com
www.artuke.com

Artuke 2021 T MC
95% tempranillo, 5% viura
91
Colour: cherry, purple rim. Nose: fruit expression, red berry notes, floral, wild herbs. Palate: fruity, flavourful, balanced.

🏆 PODIUM
Artuke El Escolladero 2020 T
85% tempranillo, 15% graciano
95
Racy, fruity. Colour: Cherry. Nose: expressive, spicy, mineral, ripe fruit. Palate: full-bodied, long, great length, taut.

Artuke Finca de los Locos 2020 T
77% tempranillo, 20% graciano, 3% viura
93
Colour: cherry, purple rim. Nose: spicy, red berry notes, overripe fruit. Palate: flavourful, fruity, good acidity, long, correct, balanced.

🏆 PODIUM
Artuke La Condenada 2020 T
80% tempranillo, 20% graciano, garnacha, palomino
97
Colour: very deep cherry, purple rim. Nose: red berry notes, ripe fruit, spicy, balsamic herbs, grassy, faded flowers. Palate: fruity, flavourful, balanced, elegant, fine tannins.

Artuke Paso Las Mañas 2020 T
tempranillo
94
Colour: cherry, purple rim. Nose: red berry notes, dried herbs, spicy, balsamic herbs. Palate: fruity, flavourful, balanced, slightly dry, soft tannins, good acidity.

Artuke Pies Negros 2020 T C
90% tempranillo, 10% graciano
92
Colour: deep cherry. Nose: ripe fruit, black fruit, dried herbs, lactic notes. Palate: powerful, ripe fruit, spicy, round tannins.

AZPILICUETA
Avda. de la Estación, 30
26360 Fuenmayor (La Rioja)
☎: +34 941 279 900
elena.adell@pernod-ricard.com
www.azpilicueta.com

Azpilicueta 2018 T C
tempranillo, graciano, mazuelo
90
Colour: cherry, purple rim. Nose: fruit expression, red berry notes, floral, spicy. Palate: flavourful, fruity, good acidity.

Azpilicueta 2019 T C
91
Colour: cherry, garnet rim. Nose: ripe fruit, creamy oak, wild herbs, scrubland. Palate: ripe fruit, spicy, easy to drink.

Azpilicueta 2020 RD
viura, tempranillo
89
Balanced, spicy, dried flowers, fruity, full-bodied.

Azpilicueta 2021 B
viura
90
Colour: bright straw. Nose: stone fruit, jasmine, faded flowers, spicy. Palate: flavourful, fruity, balanced, toasty.

DO Ca. RIOJA / D.O.P.

Azpilicueta Selección de Barricas 2019 T C
tempranillo

90
Colour: deep cherry. Nose: dried herbs, creamy oak, black fruit, spicy, woody. Palate: powerful, ripe fruit, spicy, round tannins.

BARÓN DE LEY
Ctra. Mendavia - Lodosa, Km. 5,5
31587 Mendavia (Navarra)
☎: +34 948 694 303
Fax: +34 948 694 304
info@barondeley.com
www.barondeley.com

Barón de Ley 2017 T R
tempranillo, graciano, maturana

91
Colour: deep cherry. Nose: ripe fruit, dried herbs, creamy oak, roasted coffee, woody. Palate: powerful, ripe fruit, spicy, round tannins.

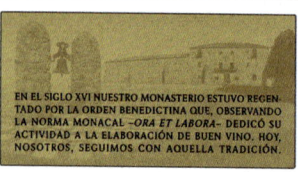

Baron de Ley 3 Viñas 2019 B R
viura, malvasía, garnacha blanca

90
Colour: bright yellow. Nose: creamy oak, ripe fruit, spicy, lactic notes, woody. Palate: rich, good structure, toasty.

Barón de Ley Finca Monasterio 2019 T BA
tempranillo, otras

91
Colour: deep cherry. Nose: ripe fruit, dried herbs, creamy oak, woody. Palate: powerful, ripe fruit, spicy, round tannins.

Barón de Ley Blanco sobre Lías 2020 B
garnacha blanca, tempranillo blanco

88
Tasty, jammy, aromatic.

Barón de Ley Rosado de Lágrima 2021 RD
garnacha

89
Pleasant, defined aromas, standard, crisp, floral, fruity, tasty.

Barón de Ley Varietal Graciano 2019 T
graciano

91
Colour: deep cherry. Nose: dried herbs, creamy oak, black fruit, ripe fruit, roasted coffee, woody. Palate: powerful, ripe fruit, spicy, round tannins.

Barón de Ley Varietal Maturana 2019 T BA
maturana

91
Colour: deep cherry. Nose: ripe fruit, dried herbs, creamy oak, tomato, roasted coffee. Palate: powerful, ripe fruit, spicy, round tannins.

BODEGA 202
01300 Laguardia (Araba/Álava)
☎: +34 667 852 704
luis.guemes@bodega202.com
www.bodega202.com

Aistear Vino de Autor 2017 T
100% tempranillo

92
Colour: dark-red cherry, garnet rim. Nose: ripe fruit, fruit preserve, aged wood nuances, sweet spices. Palate: spicy, round tannins, long.

Ansa 2017 T
100% tempranillo

92
Colour: bright cherry. Nose: sweet spices, ripe fruit, dark chocolate, creamy oak. Palate: fruity, spicy, round tannins.

Bodega 202 2019 T C
tempranillo

89
Smoky, spicy, jammy, tasty, coarse tannins, standard, balanced.

BODEGA 220 CÁNTARAS
Ctra. de Badarán, 36
26311 Cordovín (La Rioja)
☎: +34 681 132 258
comunicacion@bodega220cantaras.com
www.bodega220cantaras.com

Añadas por 220 Cántaras B
100% viura
91
Colour: bright yellow. Nose: ripe fruit, spicy, honeyed notes, faded flowers. Palate: rich, good structure, long, fine bitter notes.

Lías Finas por 220 Cántaras 2018 B C
viura
91
Colour: bright yellow. Nose: candied fruit, fine lees, lees reduction notes, faded flowers. Palate: flavourful, long, balanced, fine bitter notes.

Macerado por 220 Cántaras 2019 B
viura
89
With personality, dried flowers, dried herbs, wild, tasty, pleasant, bitter.

Senda de Haro 2019 T C
tempranillo, garnacha, maturana
91
Colour: cherry, garnet rim. Nose: ripe fruit, dried herbs, dried flowers. Palate: powerful, ripe fruit, spicy, round tannins.

Senda de Haro Garnacha 2020 T
garnacha
90
Colour: Cherry. Nose: scrubland, wild herbs, dried flowers, varietal. Palate: spicy, balsamic, good acidity.

Senda de Haro Vendimia Seleccionada 2019 T
tempranillo, garnacha, maturana
92
Colour: cherry, garnet rim. Nose: fruit preserve, powerful, dried herbs. Palate: flavourful, long, round tannins, spicy.

BODEGA ABEL MENDOZA MONGE
Paseo de Logroño, 7
26338 San Vicente de la Sonsierra (La Rioja)
☎: +34 941 308 010
jarrarte.abelmendoza@gmail.com

Abel Mendoza 5V 2021 B
malvasía, viura, garnacha blanca, tempranillo blanco, torrontés
92
Colour: bright straw, greenish rim. Nose: fresh fruit, citrus fruit, wild herbs, fine lees. Palate: fresh, fruity, good acidity, fine bitter notes, flavourful.

Abel Mendoza Garnacha Blanca 2021 B
garnacha blanca
92
Colour: bright yellow. Nose: ripe fruit, spicy, fine lees. Palate: rich, good structure, long, fine bitter notes, good acidity.

Abel Mendoza Guardaviñas 2020 T
tempranillo
92
Taut. Colour: bright cherry. Nose: fresh fruit, creamy oak, cocoa bean, black fruit. Palate: good acidity, spicy, fine tannins, flavourful.

Abel Mendoza Jarrarte 2018 T
tempranillo
91
Colour: cherry, purple rim. Nose: fruit expression, red berry notes, floral, spicy. Palate: flavourful, fruity, good acidity, elegant.

Abel Mendoza Risueño 2021 RD
graciano, garnacha, garnacha blanca
92
Colour: rose. Nose: ripe fruit, faded flowers, fine lees, sweet spices. Palate: fleshy, flavourful, ripe fruit.

Abel Mendoza Selección Personal 2019 T
tempranillo
92
Colour: cherry, purple rim. Nose: fruit expression, red berry notes, floral, spicy, dark chocolate. Palate: flavourful, fruity, good acidity, long.

DO Ca. RIOJA / D.O.P.

DO Ca. RIOJA / D.O.P.

BODEGA BIDEONA
Ctra. Samaniego, s/n.
01307 Villabuena de Álava (Araba/Álava)
☎: +34 945 609 408
hola@bideona.wine
www.bideona.wine

Bideona L3ZA 2019 T
tempranillo
93
Colour: deep cherry. Nose: ripe fruit, dried herbs, creamy oak, red berry notes. Palate: powerful, ripe fruit, spicy, round tannins.

Bideona L4GD4 (Laguardia) 2019 T
tempranillo
93
Colour: deep cherry. Nose: ripe fruit, dried herbs, creamy oak, sweet spices. Palate: powerful, ripe fruit, spicy, round tannins.

Bideona Las Parcelas 2019 B
viura
91
Colour: bright straw. Nose: ripe fruit, white fruit, wild herbs. Palate: flavourful, fresh, balanced.

Bideona Las Parcelas 2019 T
tempranillo
92
Colour: deep cherry. Nose: dried herbs, creamy oak, ripe fruit, red berry notes. Palate: powerful, ripe fruit, spicy, round tannins.

Bideona S4MG0 (Samaniego) 2019 T
tempranillo
94
Colour: dark-red cherry. Nose: toasty, cocoa bean, sweet spices, black fruit, ripe fruit. Palate: flavourful, toasty, fine bitter notes.

Bideona V1BN4 (Villabuena) 2019 T
tempranillo
94
Colour: cherry, purple rim. Nose: red berry notes, sweet spices, toasty. Palate: flavourful, fruity, good acidity, long.

BODEGA CARLOS MORO
Camino Garugele, s/n
26338 San Vicente de la Sonsierra (La Rioja)
☎: +34 941 334 093
enoturismo@bodegacarlosmoro.com
www.bodegacarlosmoro.com

CM 2018 T C
100% tempranillo
90
Colour: dark-red cherry. Nose: toasty, spicy, cocoa bean, ripe fruit. Palate: flavourful, toasty, fine bitter notes.

CM Prestigio 2017 T
100% tempranillo
91
Colour: very deep cherry. Nose: roasted coffee, aromatic coffee, powerful, ripe fruit. Palate: smoky aftertaste, great length, round tannins.

Oinoz 2017 T C
100% tempranillo
90
Colour: deep cherry. Nose: ripe fruit, dried herbs, creamy oak. Palate: powerful, ripe fruit, spicy, round tannins.

BODEGA CASA LA RAD
Ctra. Nacional 232 Km. 376-377
26513 Ausejo (La Rioja)
☎: +34 941 430 010
info@casalarad.com
www.casalarad.com

Casa La Rad 2018 B
chardonnay, malvasía, viura
90
Colour: bright straw. Nose: stone fruit, white fruit, dried herbs, sweet spices. Palate: fruity, flavourful, rich, balanced.

Casa La Rad 2018 T RB
cabernet sauvignon, tempranillo, garnacha
88
Balanced, spicy, dried herbs, jammy, toasty.

Solarce 2020 B
viura, tempranillo blanco, malvasía, chardonnay
88
Citrus fruit, herbal, crisp, floral, tasty.

Solarce 2020 T
tempranillo, graciano, maturana, mazuelo, garnacha
88
Standard, herbal, jammy, spicy.

Solarce 2021 RD
tempranillo, garnacha, graciano, mazuelo
88
Floral, fruity, herbal, tasty.

BODEGA CONTADOR
Ctra. Baños de Ebro, Km. 1
26338 San Vicente de la Sonsierra (La Rioja)
☎: +34 941 334 228
Fax: +34 941 334 537
delatierracontador@bodegacontador.com
www.bodegacontador.com

Carmen 2014 T GR
tempranillo
94
Colour: dark-red cherry, garnet rim. Nose: ripe fruit, aged wood nuances, tobacco, sweet spices, fine reductive notes. Palate: spicy, round tannins, long.

🏆 PODIUM

Contador 2020 T
tempranillo, garnacha
97
Colour: deep cherry, garnet rim. Nose: expressive, wild herbs, ripe fruit, black fruit, red berry notes, spicy, mineral. Palate: flavourful, fruity aftertaste, round tannins, long.

🏆 PODIUM

La Cueva del Contador 2020 T
tempranillo, garnacha
95
Colour: cherry, garnet rim. Nose: fruit expression, ripe fruit, spicy, neat. Palate: fruity, flavourful, good structure, round tannins, spicy, ripe fruit, long.

Predicador 2020 B
38% viura, 38% malvasía, 24% garnacha blanca
94
Colour: bright straw. Nose: ripe fruit, dried herbs, faded flowers, white flowers, mineral. Palate: powerful, ripe fruit, balanced.

🏆 PODIUM

Qué Bonito Cacareaba 2020 B
43% viura, 33% malvasía, 24% garnacha blanca
95
Colour: bright straw. Nose: expressive, ripe fruit, floral, fine lees, mineral, toasty. Palate: complex, spicy, long, elegant.

Predicador 2020 T
92% tempranillo, 3% garnacha, 3% mazuelo, 2% graciano
94
Colour: very deep cherry. Nose: complex, spicy, mineral, ripe fruit, fruit expression. Palate: elegant, full-bodied, long, soft tannins.

BODEGA CUATRO RAYAS
Camino de la Fuentecilla, S/N
47491 La Seca (Valladolid)
☎: +34 983 816 320
info@cuatrorayas.es
www.cuatrorayas.es

Cuatro Rayas Cuarenta Vendimias Rioja 2019 T C
tempranillo, garnacha, graciano
88
Pleasant, standard, fruity, herbal, wild, simple.

BODEGA DEL MONGE-GARBATI
Ctra. Rivas de Tereso, s/n
26338 San Vicente de la Sonsierra (La Rioja)
☎: +34 659 167 653
Fax: +34 941 311 870
bodegamg@yahoo.es
www.vinaane.com

El Laberinto de Viña Ane 2019 T C
100% tempranillo
91
Colour: very deep cherry. Nose: roasted coffee, aromatic coffee, powerful, ripe fruit, black fruit. Palate: smoky aftertaste, great length, round tannins.

Viña Ane Autor 2018 T C
100% tempranillo
91
Colour: deep cherry. Nose: creamy oak, black fruit, toasty. Palate: ripe fruit, spicy, round tannins.

Viña Ane Centenaria 2021 B FB
43% tempranillo blanco, 25% viura, 8% malvasía, 5% chardonnay, 4% sauvignon blanc, 4% torrontés
91
Colour: bright straw. Nose: jasmine, faded flowers, white fruit, dried herbs. Palate: fruity, fresh, flavourful, balanced, good finish.

Viña Ane Selección 2018 T
100% tempranillo
90
Toasty. Colour: deep cherry. Nose: ripe fruit, dried herbs, creamy oak. Palate: powerful, ripe fruit, spicy, round tannins.

DO Ca. RIOJA / D.O.P.

DO Ca. RIOJA / D.O.P.

BODEGA EL HOMBRE ORQUESTA
01330 Labastida (Araba/Álava)
☎: +34 606 526 649
elhombreorquestasl@gmail.com

El Hombre Orquesta 666 G 2019 T
100% garnacha
90
Colour: cherry, purple rim. Nose: red berry notes, spicy, violets. Palate: flavourful, fruity, good acidity.

El Hombre Orquesta 666 M 2019 T
mazuelo
92
Taut. Colour: bright cherry. Nose: fresh fruit, creamy oak, red berry notes, violets. Palate: good acidity, spicy, fine tannins.

El Hombre Orquesta 666 T 2019 T
tempranillo
89
Creamy, balanced, spicy, jammy, tasty.

El Hombre Orquesta 666 V 2020 B
100% viura
90
Colour: straw. Nose: ripe fruit, dried herbs, faded flowers, wild herbs, spicy. Palate: powerful, ripe fruit, balanced.

La Fuente de Mosito 2020 T
tempranillo, garnacha
86

Pindio 2019 T FB
tempranillo, garnacha, mazuelo
91
Colour: deep cherry. Nose: dried herbs, creamy oak, black fruit, dark chocolate. Palate: ripe fruit, spicy, round tannins.

BODEGA FINCA DE LOS ARANDINOS
Ctra. LR 137, km. 4,6
26375 Entrena (La Rioja)
☎: +34 941 446 065
comercial@fincadelosarandinos.com
www.fincadelosarandinos.com

Catay 2018 T C
93% tempranillo, 6% mazuelo, 1% garnacha
88
Pleasant, kind finish, tasty.

Catay 2021 RD
74% viura, 22% tempranillo, 4% garnacha
88
Fruity, herbal, tasty.

Catay Tempranillo Blanco 2021 B
100% tempranillo blanco
87

Catay Viñas Viejas 2019 B
100% viura
88
Citrus fruit, crisp, herbal, full-bodied.

El Conjuro 2017 T
80% tempranillo, 20% garnacha
88 🌱
Spicy, jammy, tasty, toasty.

BODEGA HACIENDA LÓPEZ DE HARO
Camino del Cementerio, s/n
23338 San Vicente de la Sonsierra (La Rioja)
☎: +34 941 483 557
marketing@vintae.com
www.haciendalopezdeharo.com

Classica Hacienda López de Haro 2004 T GR
tempranillo, garnacha
92
Classic. Colour: dark-red cherry, garnet rim. Nose: ripe fruit, aged wood nuances, cigar, spicy. Palate: long, flavourful, classic aged character.

Classica Hacienda López de Haro 2009 RD GR
viura, garnacha
90
Colour: onion pink. Nose: faded flowers, dried herbs, cigar. Palate: balanced, flavourful.

Classica Hacienda López de Haro 2013 B GR
viura, malvasía, garnacha blanca
92
Colour: bright yellow. Nose: powerful, creamy oak, ripe fruit, spicy, woody, roasted coffee. Palate: rich, good structure, long, toasty, fine bitter notes.

El Pacto de la Sonsierra 2019 T
tempranillo
91 🌱
Colour: cherry, garnet rim. Nose: ripe fruit, spicy, scrubland. Palate: ripe fruit, varietal, good acidity.

Hacienda López de Haro 2017 T R
tempranillo, graciano
89
Defined aromas, jammy, tasty, toasty, smooth, smoky.

Hacienda López de Haro 2021 B
viura, otras
88
Toasty, tasty, jammy.

BODEGA MANUEL QUINTANO
Avda. de Diputación, 22
01330 Labastida (Araba/Álava)
☎: +34 945 331 161
info@manuelquintano.com
www.manuelquintano.com

Manuel Quintano Cepas Viejas 2020 T
98% garnacha, 2% otras
92
Elegant, fruity. Colour: cherry, garnet rim. Nose: dried flowers, varietal, expressive, balanced, fruit expression. Palate: flavourful, fruity, good acidity, taut.

Manuel Quintano El Pionero 2020 T
95% tempranillo, 5% garnacha
90
Colour: cherry, garnet rim. Nose: red berry notes, floral, spicy. Palate: flavourful, fruity, good acidity, easy to drink.

Manuel Quintano Selección Particular 2019 T
91
Colour: bright cherry. Nose: sweet spices, ripe fruit, black fruit. Palate: fruity, spicy, round tannins.

BODEGA VICO
Polígono El Raposal, 80
26580 Arnedo (La Rioja)
☎: +34 941 380 257
comercial@bodegavico.com
www.bodegavico.com

Ciencuevas 2018 T C
100% tempranillo
87

Mocete 2021 T MC
100% tempranillo
89
Balanced, fruity, herbal, exuberant.

Ormus 2018 T C
100% tempranillo
88 🌱
Hot, pruney, fruity, spicy.

Ormus Edición Limitada 2021 T
89 🌱
Spicy, herbaceous, jammy, tasty, woody, toasty.

Ormus Viura 2020 B
100% viura
88 🌱
Fruity, thin, floral, jammy.

Pilares de Ciencuevas 2018 T RB
100% garnacha
91
Colour: bright cherry. Nose: red berry notes, faded flowers, fragrant herbs, sweet spices. Palate: fruity, fresh, balanced, good acidity, good finish, slightly dry, soft tannins.

BODEGA Y VIÑEDOS SOLABAL
Camino San Bartolomé, 6
26339 Ábalos (La Rioja)
☎: +34 941 334 492
solabal@solabal.es
www.solabal.es

Esculle de Solabal 2019 T C
tempranillo
89
Creamy, balanced, spicy, full-bodied, tasty.

Muñarrate de Solabal 2021 B
viura, sauvignon blanc
86

Muñarrate de Solabal 2021 T MC
tempranillo
88
Fruity, jammy, floral.

Solabal 2017 T R
tempranillo
89
Classic, pruney, fruity, spicy, tasty.

Solabal 2018 T C
tempranillo
90
Colour: deep cherry. Nose: dried herbs, creamy oak, black fruit, dark chocolate. Palate: ripe fruit, spicy, round tannins.

Vala de Solabal 2016 T
100% tempranillo
91
Colour: deep cherry. Nose: dried herbs, creamy oak, black fruit, woody. Palate: powerful, ripe fruit, spicy.

DO Ca. RIOJA / D.O.P.

DO Ca. RIOJA / D.O.P.

BODEGAS ALCONDE
Ctra. de Calahorra, s/n
31260 Lerin (Navarra)
☎: +34 948 530 058
Fax: +34 948 530 589
manuel@bodegasalconde.com
www.bodegasalconde.com

Marqués de Jubera 2019 T C
tempranillo, garnacha
87

Pórtico Mayor 2016 T R
tempranillo, mazuelo, graciano
89
Balanced, balsamic herbs, standard, jammy, tasty, spicy.

Pórtico Mayor 2017 T C
tempranillo, garnacha, graciano
88
Balanced, spicy, jammy, toasty.

Pórtico Mayor 2020 T
tempranillo, garnacha
87

BODEGAS ALCORTA
Camino de la Puebla, 50
26006 Logroño (La Rioja)
☎: +34 941 279 900
vinos@pernod-ricard.com
www.alcortavino.com

Audaz de Juan Alcorta 2017 T C
87

BODEGAS ALDONIA
Ctra. Fuenmayor, LR-137 Km 14
26370 Navarrete (La Rioja)
☎: +34 652 556 074
aldonia@aldonia.es
www.aldonia.es

Aldonia 2019 T
garnacha
88
Floral, crisp, fruity, spicy, toasty.

BODEGAS ALVIA
26371 Ventosa (La Rioja)
☎: +34 941 441 905
info@bodegasalvia.es
www.bodegasalvia.es

Livius Blanco 2018 B FB
70% viura, 30% malvasía
91
Aromatic, spicy, age nuances. Colour: bright yellow. Nose: ripe fruit, spicy, floral. Palate: rich, good structure, long, fine bitter notes, good acidity.

Livius Rosado 2018 RD FB
garnacha
90
Colour: salmon. Nose: sweet spices, red berry notes, fragrant herbs, dried flowers, brioche. Palate: full-bodied, flavourful, spicy, sweetness.

Marqués de Alvia Garnacha 2013 T GR
100% garnacha
91
Colour: bright cherry. Nose: ripe fruit, red berry notes, spicy, dried herbs. Palate: flavourful, fruity, balanced, slightly dry, soft tannins.

Marqués de Alvia Graciano 2014 T
100% graciano
91
Colour: bright cherry. Nose: red berry notes, ripe fruit, wild herbs, varietal. Palate: fresh, fruity, flavourful, good structure.

Marqués de Alvia Tempranillo 2014
100% tempranillo
91
Colour: bright cherry. Nose: ripe fruit, red berry notes, spicy, smoky, wet leather. Palate: fruity, flavourful, good structure, balanced.

Mileto Crianza 2018 T C
85% tempranillo, 15% garnacha, mazuelo
87

BODEGAS AMAREN
Ctra. Villabuena de Alava, 3
01307 Samaniego (Araba/Álava)
☎: +34 945 175 240
Fax: +34 945 174 566
info@bodegasamaren.com
www.bodegasamaren.com

Amaren 2019 B FB
viura, malvasía, tempranillo blanco
91
Colour: bright straw. Nose: stone fruit, ripe fruit, balanced, spicy. Palate: flavourful, fresh, good acidity, fruity aftertaste.

Amaren 60 Garnacha 2019 T
garnacha

92

Taut. Colour: cherry, purple rim. Nose: fruit expression, red berry notes, floral, spicy. Palate: flavourful, fruity, good acidity, long.

Amaren Selección de Viñedos 2019 T BA
90% tempranillo, 10% garnacha

89

Balanced, spicy, herbal, powerful, coarse tannins.

Carraquintana de Amaren 2018 T BA
tempranillo, garnacha, graciano, malvasía

93

Colour: cherry, purple rim. Nose: ripe fruit, creamy oak, spicy, dried herbs. Palate: flavourful, fruity, balanced.

El Cristo de Samaniego 2018 T
tempranillo, garnacha, viura, malvasía

93

Colour: deep cherry. Nose: fruit expression, fresh fruit, black fruit, ripe fruit, dried herbs, spicy, smoky. Palate: fruity, flavourful, balanced.

BODEGAS AMÉZOLA DE LA MORA
Paraje Viña Vieja, s/n
26359 Torremontalbo (La Rioja)
☎: +34 941 454 532
Fax: +34 941 454 537
info@bodegasamezola.es
www.bodegasamezola.es

Iñigo Amézola 2019 B FB
100% viura

90

With personality. Colour: yellow. Nose: candied fruit, ripe fruit, faded flowers, honeyed notes, sweet spices. Palate: flavourful, fruity, balanced.

Iñigo Amézola 2020 T
100% tempranillo

89

Pruney, fruity, balanced, tasty.

Señorío Amézola 2015 T R
85% tempranillo, 10% mazuelo, 5% graciano

85

Solar Amézola 2015 T GR
85% tempranillo, 10% mazuelo, 5% graciano

87

Viña Amézola 2018 T C
85% tempranillo, 10% mazuelo, 5% graciano

84

BODEGAS ANTONIO ALCARAZ
Ctra. Vitoria-Logroño, Km. 56
01300 Laguardia (Araba/Álava)
☎: +34 658 959 745
rioja@bodegasantonioalcaraz.com
www.bodegasantonioalcaraz.com

Antonio Alcaraz 2017 T C
100% tempranillo

89

Aromatic, smoky, spicy, tasty. Nose: black fruit, ripe fruit.

El Abuelo Pedro 2019 T
88% tempranillo, 7% viura, 3% garnacha, 2% mazuelo

91

Colour: Cherry. Nose: scrubland, smoky, sweet spices, dried herbs, dried flowers. Palate: spicy, balsamic, good acidity.

Gloria Antonio Alcaraz 2018 T
100% tempranillo

91

Colour: bright cherry. Nose: sweet spices, ripe fruit, smoky, spicy. Palate: fruity, round tannins, good acidity.

Madre Única Autor 2019 T
tempranillo

91

Powerful. Colour: cherry, garnet rim. Nose: fruit preserve, powerful, incense, sweet spices. Palate: flavourful, long.

BODEGAS ARAICO
La Hoya, 5
01307 Villabuena de Álava (Araba/Álava)
☎: +34 945 623 366
info@bodegasaraico.com
www.bodegasaraico.com

Araico 2018 T C
tempranillo, garnacha

88

Fruity, herbal, spicy, tasty.

Araico 2020 B FB
100% viura

86

Araico Autor 2020 T
100% tempranillo

87

El Orgullo de Julian 2019 T
100% tempranillo

90

Colour: cherry, purple rim. Nose: ripe fruit, red berry notes, dried herbs, expressive, spicy. Palate: fruity, flavourful, balanced, slightly dry, soft tannins.

DO Ca. RIOJA / D.O.P.

BODEGAS BAGORDI

Ctra. Estella s/n
31261 Andosilla (Navarra)
☎: +34 948 674 260
info@bagordi.com
www.bagordi.com

Bagordi 2013 T R
80% tempranillo, 20% graciano

90

Colour: dark-red cherry, garnet rim. Nose: fruit preserve, aged wood nuances, tobacco, sweet spices. Palate: spicy, round tannins, flavourful.

Bagordi 2018 T C
80% tempranillo, 10% garnacha, 10% graciano

90

Colour: cherry, purple rim. Nose: red berry notes, floral, spicy. Palate: flavourful, fruity, good acidity, long.

Bagordi 2021 B FB
50% sauvignon blanc, 50% garnacha blanca

88

Kind finish, fruity, floral, spicy, tropical.

Bagordi 2021 RD
100% garnacha

89

Floral, fruity, herbal, tasty, pleasant.

Bagordi Maturana 2020
100% maturana

88

Fruity, pruney, dried herbs, tasty.

Bagordi Vendimia Seleccionada 2017 T

87

BODEGAS BAIGORRI

Ctra. Vitoria-Logroño, Km. 53
01307 Samaniego (Araba/Álava)
☎: +34 945 609 420
elena.sanchez@bodegasbaigorri.com
www.bodegasbaigorri.com

Baigorri 2017 B FB
90% viura, 10% malvasía

92

Colour: straw. Nose: ripe fruit, dried herbs, faded flowers, toasted bread, sweet spices. Palate: ripe fruit, balanced, flavourful.

Baigorri Belus 2018 T
80% mazuelo, 20% otras

93

Colour: cherry, purple rim. Nose: red berry notes, floral, spicy, black fruit. Palate: flavourful, fruity, good acidity.

Baigorri de Garage 2017 T BA
100% tempranillo

93

Colour: deep cherry. Nose: ripe fruit, creamy oak, fine reductive notes, scrubland. Palate: ripe fruit, spicy, round tannins.

Baigorri Finca La Canoca 2020 T RB
100% tempranillo

94

Colour: Cherry. Nose: complex, expressive, spicy, mineral, black fruit, meaty notes. Palate: full-bodied, long, great length, slightly dry, soft tannins, elegant.

Baigorri Finca La Quintanilla 2019 T
100% tempranillo

91

Colour: deep cherry. Nose: dried herbs, creamy oak, black fruit, red berry notes, sweet spices. Palate: spicy, round tannins, balanced.

Baigorri Finca Las Navas 2021 T

93

Colour: deep cherry, bright cherry. Nose: chalk, ripe fruit, bakery, spicy. Palate: taut, flavourful, balanced, round tannins.

Baigorri Garnacha 2017 T
100% garnacha

92

Colour: deep cherry. Nose: ripe fruit, dried herbs, creamy oak. Palate: spicy, round tannins, fleshy, balanced.

Baigorri Maturana 2019 T
100% maturana

93

Colour: dark-red cherry. Nose: toasty, spicy, cocoa bean, fine reductive notes, black fruit. Palate: flavourful, toasty, fine bitter notes, fleshy.

BODEGAS BENJAMÍN DE ROTHSCHILD & VEGA SICILIA

Paraje de San Millán. Camino de las Cañas, s/n
01307 Samaniego (Araba/Álava)
☎: +34 945 567 508
macan@macan-brvs.com
www.temposvegasicilia.com

🏆 PODIUM

Macán 2018 T
100% tempranillo

96

Taut, balanced. Colour: deep cherry. Nose: ripe fruit, dried herbs, creamy oak, red berry notes, chalk, mineral. Palate: powerful, ripe fruit, spicy, round tannins.

Macán Clásico 2019 T
100% tempranillo

94

Balanced, defined aromas. Colour: bright cherry. Nose: balanced, characterful, powerful, red berry notes, ripe fruit. Palate: good structure, fruity, flavourful.

BODEGAS BERONIA

Ctra. Ollauri - Nájera, km. 1,8
26220 Ollauri (La Rioja)
☎: +34 941 338 000
prensa@gonzalezbyass.es
www.beronia.es

Alegra de Beronia 2021 RD
100% tempranillo

90

Colour: onion pink. Nose: red berry notes, dried herbs, floral. Palate: fruity, flavourful, balanced.

Beronia 198 Barricas 2013 T R
95% tempranillo, 5% graciano

91

Colour: dark-red cherry, garnet rim. Nose: ripe fruit, aged wood nuances, tobacco. Palate: spicy, round tannins.

Beronia 2013 T GR
95% tempranillo, 4% graciano, 1% mazuelo

93

Colour: dark-red cherry, garnet rim. Nose: ripe fruit, aged wood nuances, tobacco, sweet spices, elegant. Palate: spicy, round tannins, long.

Beronia 2017 T R
93% tempranillo, 6% graciano, 1% mazuelo

91

Colour: dark-red cherry, garnet rim. Nose: ripe fruit, aged wood nuances, sweet spices. Palate: spicy, round tannins.

Beronia III a.C. 2016 T
100% tempranillo

92

Colour: deep cherry. Nose: dried herbs, creamy oak, ripe fruit, black fruit, toasty. Palate: powerful, ripe fruit, spicy, round tannins.

Beronia Mazuelo 2017 T R
100% mazuelo

91

Colour: cherry, garnet rim. Nose: creamy oak, hot, ripe fruit, black fruit. Palate: pruney, powerful, sweet tannins.

Beronia Tempranillo Elaboración Especial 2020 T FB
100% tempranillo

88

Roasted coffee, jammy, spicy, smooth.

Beronia Viñas Viejas 2019 T
100% tempranillo

90

Colour: deep cherry. Nose: ripe fruit, dried herbs, creamy oak. Palate: powerful, ripe fruit, spicy, round tannins.

BODEGAS BETOLAZA

Cuesta Dulce, 12
26330 Briones (La Rioja)
☎: +34 650 862 104
betolaza@betolaza.es
www.betolaza.es

Betolaza 2016 T R
90% tempranillo, 10% garnacha

93

Colour: dark-red cherry, garnet rim. Nose: ripe fruit, fruit preserve, aged wood nuances, tobacco, sweet spices. Palate: spicy, round tannins, long.

Betolaza 2021 B
81% viura, 19% otras

88

Pruney, jammy, slight oxidation.

Betolaza Tempranillo 2021 T
100% tempranillo

86

Magadi 2018 B
100% viura

91

Colour: straw. Nose: ripe fruit, dried herbs, faded flowers. Palate: powerful, ripe fruit, balanced.

DO Ca. RIOJA / D.O.P.

Peñín Guide **SPANISH WINE** 517

DO Ca. RIOJA / D.O.P.

Resaco 2019 T
100% garnacha
92
Colour: cherry, purple rim. Nose: fruit expression, red berry notes, floral, sweet spices, scrubland. Palate: flavourful, fruity, good acidity, long.

BODEGAS BILBAÍNAS
Estación, 3
26200 Haro (La Rioja)
☎: +34 941 310 147
reservas@bodegasbilbainas.com
www.15bodegas.com/bodegas-bilbainas

Ederra 2016 T R
100% tempranillo
88
Age nuances, jammy, slight reduction, herbal, dried herbs.

Ederra 2018 T R
88
Pleasant, slight reduction, standard, dried herbs, jammy, simple.

Ederra Selección Especial 2020 B
viura
88
Pleasant, tasty, jammy, crisp.

Ederra Selección Tempranillo 2018 T C
100% tempranillo
89
Kind finish, jammy, tasty, wild, balanced.

La Vicalanda 2016 T R
100% tempranillo
93
Colour: dark-red cherry. Nose: ripe fruit, fruit preserve, tobacco, sweet spices, fine reductive notes. Palate: spicy, long, easy to drink.

Viña Pomal 2016 T R
100% tempranillo
91
Colour: cherry, garnet rim. Nose: sweet spices, dark chocolate, ripe fruit, dried herbs. Palate: balanced, ripe fruit, easy to drink.

Viña Pomal 2019 T C
100% tempranillo
89
Defined aromas, standard, spicy, jammy, tasty, smooth, classic.

Viña Pomal 2021 RD
70% garnacha, 30% viura
88
Floral, fruity, herbal, tasty.

Viña Pomal Ecológico 2019 T C
100% tempranillo
89 ♥
Defined aromas, balsamic herbs, dried flowers, spicy, jammy, tasty, pleasant.

Viña Pomal Selección 500 2018 T C
85% tempranillo, 15% garnacha
90
Colour: bright cherry, garnet rim. Nose: red berry notes, spicy, wild herbs. Palate: fresh, balanced, good acidity, easy to drink.

Viña Pomal Viura Malvasía 2021 B
70% viura, 30% malvasía
90
Colour: bright straw, greenish rim. Nose: fresh fruit, citrus fruit, wild herbs. Palate: fresh, fruity, good acidity, fine bitter notes, flavourful.

BODEGAS CAMPILLO
Ctra. Logroño, s/n
01300 Laguardia (Araba/Álava)
☎: +34 945 600 826
Fax: +34 945 600 837
info@bodegascampillo.es
www.bodegascampillo.es

Campillo 2019 T C
tempranillo
90
Colour: deep cherry. Nose: dried herbs, creamy oak, black fruit, sweet spices. Palate: ripe fruit, spicy, round tannins.

Campillo 57 Selección Especial 2015 T
tempranillo, graciano
90
Colour: deep cherry. Nose: dried herbs, creamy oak, black fruit, woody. Palate: powerful, ripe fruit, spicy, round tannins.

Campillo Rosé 2021 RD
garnacha
88
Pleasant, aromatic, standard, floral, simple.

BODEGAS CARLOS SAN PEDRO PEREZ DE VIÑASPRE

Páganos, 44
01300 Laguardia (Araba/Álava)
☎: +34 609 321 649
Fax: +34 945 621 111
info@bodegascarlossampedro.com
www.bodegascarlossampedro.com

Carlos San Pedro Perez de Viñaspre 2016 T
tempranillo

91

Colour: bright cherry. Nose: red berry notes, wild herbs, spicy. Palate: fruity, flavourful, balanced, slightly dry, soft tannins.

Viñasperi 2014 T GR
tempranillo

90

Colour: cherry, purple rim. Nose: fruit preserve, dried herbs, dried flowers, spicy. Palate: fruity, flavourful, balanced, slightly dry, soft tannins.

Viñasperi Blue Ocean 2019 T BA
tempranillo

89

Colour: cherry, purple rim. Nose: red berry notes, spicy, dried herbs, old leather. Palate: fruity, fresh, flavourful, slightly dry, soft tannins.

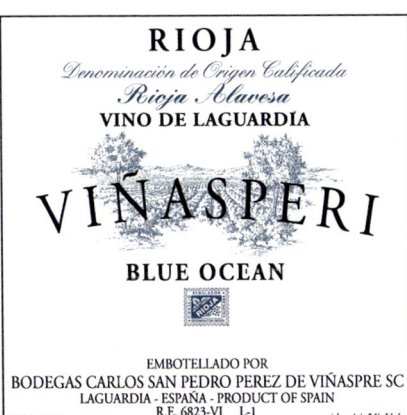

Viñasperi Selección 2016 T
tempranillo

90

Colour: bright cherry. Nose: ripe fruit, red berry notes, dried herbs, spicy. Palate: fruity, flavourful, balanced, toasty, slightly dry, soft tannins.

Viñasperi Selección Los Ojales 2019 T
tempranillo

90

Colour: cherry, purple rim. Nose: red berry notes, ripe fruit, spicy, dried herbs. Palate: fruity, flavourful, powerful tannins.

BODEGAS CERROLAZA

Ctra. Navarrete, 38
26372 Hornos de Moncalvillo (La Rioja)
☎: +34 941 286 728
Fax: +34 941 286 729
comercial@bodegascerrolaza.com
www.bodegascerrolaza.com

Altos del Marqués 2014 T R
100% tempranillo

88

Spicy, jammy, tasty, toasty, simple.

Altos del Marqués 2019 T BA
tempranillo, graciano, garnacha

87

Altos del Marqués 2019 T C
tempranillo

87

Altos del Marqués 2021 B
tempranillo blanco, viura

88

Fruity, dried flowers, herbal, jammy.

Aticus 2018 T C
tempranillo

88

Pruney, fruity, herbal, floral, tasty.

Aticus Vendimia Seleccionada 2019 T
tempranillo

89

Jammy, tasty, dried herbs, fruity, spicy, balanced, standard.

BODEGAS CORNELIO DINASTÍA

Carralaverde, 3 Nave 1
26370 Navarrete (La Rioja)
☎: +34 941 742 249
info@bodegascornelio.es
www.bodegascornelio.com

Cornelio Dinastía 2017 T C
tempranillo

88

Balanced, spicy, herbaceous, jammy.

DO Ca. RIOJA / D.O.P.

Cornelio Imperial Autor JM 2018 T
tempranillo
90
Jammy. Colour: deep cherry. Nose: dried herbs, creamy oak, black fruit. Palate: ripe fruit, spicy, round tannins.

Vega Vella 2017 T C
tempranillo, graciano, garnacha
87

Vega Vella 2018 B FB
garnacha blanca, sauvignon blanc
90
Colour: bright yellow. Nose: ripe fruit, spicy, white fruit, wild herbs, toasted bread. Palate: rich, good structure, long.

Vega Vella 2020 B
garnacha blanca, sauvignon blanc
90
Colour: bright straw, greenish rim. Nose: citrus fruit, wild herbs, stone fruit. Palate: fruity, good acidity, fine bitter notes, fleshy.

Vega Vella Graciano 2019 T
graciano
88
Spicy, herbal, full-bodied.

BODEGAS CORRAL
Ctra. de Logroño, Km. 10
26370 Navarrete (La Rioja)
☎: +34 941 440 193
Fax: +34 941 440 195
info@bodegascorral.com
www.bodegascorral.com

Altos de Corral Single Estate 2017 T C
100% tempranillo
91
Colour: cherry, garnet rim. Nose: fruit preserve, fruit liqueur notes, powerful. Palate: flavourful, sweetness, long.

Don Jacobo 2010 T GR
85% tempranillo, 10% graciano, 5% mazuelo
91
Colour: dark-red cherry, garnet rim. Nose: ripe fruit, aged wood nuances, tobacco, sweet spices. Palate: spicy, round tannins, long.

Don Jacobo Tempranillo Blanco 2021 B
100% tempranillo blanco
87

Los Corrales de Moncalvillo Maturana Blanca 2020 B
100% maturana blanca
93
Colour: bright yellow. Nose: dried flowers, candied fruit, fine lees, pattiserie. Palate: round, spicy, long, great length.

Los Corrales de Moncalvillo Maturana en Amphora 2020 T
100% maturana
92
Colour: Cherry, garnet rim. Nose: balsamic herbs, sweet spices, scrubland, ripe fruit. Palate: spicy, balsamic, good acidity.

Los Corrales de Moncalvillo Maturana Tinta 2019 T BA
100% maturana

92 ❦
Colour: cherry, garnet rim. Nose: fruit preserve, powerful, scrubland, toasty. Palate: flavourful, sweetness, long.

BODEGAS COSME PALACIO
San Lázaro, 1
01300 Laguardia (Araba/Álava)
☎: +34 945 600 151
rrpp@grupobodegaspalacio.es
www.entrecanalesdomecq.com

Cosme Palacio 1894 2018 B
93% viura, 7% malvasía

93
Colour: bright yellow. Nose: powerful, creamy oak, ripe fruit, spicy. Palate: rich, good structure, long, toasty, fine bitter notes.

Cosme Palacio 1894 2018 T
100% tempranillo

93
Colour: Cherry. Nose: complex, expressive, spicy, mineral, wild herbs. Palate: elegant, full-bodied, long, great length.

Cosme Palacio 2016 T R
100% tempranillo

93
Colour: cherry, purple rim. Nose: red berry notes, ripe fruit, balsamic herbs, spicy, creamy oak. Palate: flavourful, elegant, balanced, fine tannins.

Cosme Palacio 2018 B R
viura

93
Colour: yellow. Nose: spicy, smoky, dried flowers, faded flowers, characterful, expressive. Palate: spicy, long, rich, flavourful.

Cosme Palacio Vino de Laguardia 2019 T C
100% tempranillo

92
Colour: cherry, purple rim. Nose: red berry notes, floral, spicy, dark chocolate. Palate: flavourful, fruity, good acidity, long, full-bodied.

Glorioso 2018 T R
tempranillo

90
Colour: deep cherry. Nose: ripe fruit, dried herbs, creamy oak, cocoa bean. Palate: powerful, ripe fruit, spicy, round tannins.

BODEGAS COVILA
Avda. Soto, 26
01306 Lapuebla de Labarca (Araba/Álava)
☎: +34 945 627 232
exportacion@covila.es
www.covila.es

Covila 2016 T GR
100% tempranillo

89
Spicy, pruney, herbal, tasty.

Covila 2017 T R
tempranillo

88
Pruney, fruity, sweety finish, tasty.

Covila Aex 2019 T
100% tempranillo

90
Colour: cherry, purple rim. Nose: red berry notes, dried herbs, spicy, smoky. Palate: fresh, fruity, flavourful, balanced.

Covila 2019 T C
100% tempranillo

88
Fruity, floral, jammy, tasty.

Covila 2021 B
viura, sauvignon blanc, malvasía

85

Covila 2021 T
tempranillo

87

DO Ca. RIOJA / D.O.P.

DO Ca. RIOJA / D.O.P.

BODEGAS D. MATEOS
Camino de los Agudos, s/n
26559 Aldeanueva de Ebro (La Rioja)
☎: +34 941 261 897
Fax: +34 941 105 007
info@bodegasmateos.com
www.bodegasmateos.com

Colección de Familia La Mateo Tempranillo Blanco 2019 B
100% tempranillo blanco
89
Tropical, fruity, floral, spicy, jammy, tasty.

Colección de Familia La Mateo Garnacha Cepas Viejas 2018 T
100% garnacha
90
Age nuances, full-bodied, herbal, jammy. Nose: cigar, black fruit.

Colección de Familia La Mateo Parcelas Singulares 2017 T BA
50% tempranillo, 40% garnacha, 10% mazuelo
93
Colour: deep cherry. Nose: ripe fruit, dried herbs, creamy oak, earthy notes. Palate: powerful, ripe fruit, spicy, round tannins.

Colección de Familia La Mateo Vendimia 2018 T BA
70% tempranillo, 27% garnacha, 3% graciano
92
Colour: dark-red cherry. Nose: toasty, spicy, cocoa bean, earthy notes. Palate: flavourful, toasty, fine bitter notes.

La Vanidosa Nº1 2019 T
100% garnacha
91
Colour: deep cherry. Nose: ripe fruit, dried herbs, creamy oak, sweet spices. Palate: powerful, ripe fruit, spicy, round tannins.

Tío Martín 2019 T C
70% tempranillo, 22% garnacha, 5% mazuelo, 3% graciano
91
Colour: dark-red cherry. Nose: toasty, spicy, cocoa bean, black fruit, ripe fruit. Palate: flavourful, toasty, fine bitter notes.

BODEGAS DE FAMILIA BURGO VIEJO
Concordia, 8
26540 Alfaro (La Rioja)
☎: +34 941 183 405
Fax: +34 941 181 603
bodegas@burgoviejo.com
www.burgoviejo.com

Finca Vidales 2019 T C
tempranillo
88
Fruity, tasty, spicy, crisp, floral.

Finca Vidales 2021 B
100% viura
86

Licenciado 2017 T R
tempranillo
89
Balanced, spicy, fruity, dried herbs.

BODEGAS DE LOS HEREDEROS DEL MARQUÉS DE RISCAL
Torrea, 1
01340 Elciego (Araba/Álava)
☎: +34 945 606 000
Fax: +34 945 606 023
marquesderiscal@marquesderiscal.com
www.marquesderiscal.com

Arienzo de Marqués de Riscal 2018 T C
tempranillo, graciano
89
Pleasant, spicy, jammy, tasty, dried herbs, toasty, standard.

🏆 **PODIUM**

Barón de Chirel 2017 T
tempranillo
96
Colour: very deep cherry. Nose: complex, expressive, spicy, mineral, powerful, balsamic herbs. Palate: elegant, full-bodied, long, great length.

🏆 **PODIUM**

Finca Torrea 2018 T R
tempranillo, graciano
95
Colour: dark-red cherry, garnet rim. Nose: ripe fruit, aged wood nuances, tobacco, sweet spices, characterful, complex. Palate: spicy, round tannins, long.

DO Ca. RIOJA / D.O.P.

🏆 PODIUM

Marqués de Riscal 2016 T GR
tempranillo, otras

97

Colour: dark-red cherry, garnet rim. Nose: aged wood nuances, tobacco, sweet spices, ripe fruit, black fruit. Palate: spicy, round tannins, long.

Marqués de Riscal 2018 T R
95% tempranillo, 5% graciano

93

Colour: dark-red cherry, garnet rim. Nose: ripe fruit, aged wood nuances, tobacco, sweet spices. Palate: spicy, round tannins, long.

XR de Marqués de Riscal 2017 T R
tempranillo, graciano

94

Colour: deep cherry, garnet rim. Nose: ripe fruit, cocoa bean, cigar, toasty, tobacco. Palate: flavourful, spicy, toasty, soft tannins.

BODEGAS DE SANTIAGO

Avda. del Ebro, 50
01307 Baños de Ebro (Araba/Álava)
☎: +34 651 707 879
Fax: +34 945 609 201
info@bodegasdesantiago.es
www.bodegasdesantiago.es

Lagar de Santiago 2019 T C
87

Lagar de Santiago 2021 T MC
85

Lagar de Santiago Viura-Verdejo 2021 B
87

BODEGAS DEL MEDIEVO

Circunvalación San Roque, s/n
26559 Aldeanueva de Ebro (La Rioja)
☎: +34 941 163 141
info@bodegasdelmedievo.com
www.bodegasdelmedievo.com

Medievo 2017 T R
100% tempranillo

89

Classic, toasty, jammy, spicy.

Medievo 2019 T C
80% tempranillo, 10% garnacha, 5% graciano, 5% mazuelo

89

Spicy, toasty, tasty, powerful.

Monte Araya 2021 RD
garnacha

87

Tuercebotas Garnacha 2019 T C
100% garnacha

89

Pleasant, kind finish, toasty, tasty.

Tuercebotas Graciano 2019 T C
100% graciano

92

Powerful, rustic. Colour: cherry, garnet rim. Nose: fruit preserve, powerful, black fruit, toasty. Palate: flavourful, sweetness, long.

Tuercebotas Tempranillo Blanco 2021 B
100% tempranillo blanco

89

Pleasant, simple, fruity, standard, balanced, floral.

BODEGAS DIEZ DEL CORRAL

Avda. Príncipe de Asturias, 42
26210 Anguciana (La Rioja)
☎: +34 686 794 417
manuel@diezdelcorral.com
www.bodegasdiezdelcorral.com

La Piconada 2019 T
87

BODEGAS DOMECO DE JARAUTA

Camino Sendero Royal, 5
26559 Aldeanueva de Ebro (La Rioja)
☎: +34 941 163 078
info@bodegasdomecodejarauta.com
www.bodegasdomecodejarauta.com

Domeco de Jarauta 2015 T
85% tempranillo, 15% graciano

91

Colour: deep cherry. Nose: ripe fruit, dried herbs, creamy oak, scrubland. Palate: powerful, ripe fruit, spicy, round tannins.

Domeco de Jarauta 2020 B
100% garnacha blanca

89

Spicy, jammy, tasty, powerful, roasted coffee, woody.

Domeco de Jarauta 2020 T
100% garnacha

88

Balanced, fruity, herbaceous, jammy.

Sancho Barón 2020 T
100% tempranillo

87

Viña Marro 2019 T C
88
Fruity, pruney, spicy, herbal.

BODEGAS ESTRAUNZA
Avda. La Póbeda, 25
01306 Lapuebla de Labarca (Araba/Álava)
☎: +34 944 215 936
contacto@bodegasestraunza.com
www.bodegasestraunza.com

Blas de Lezo 2018 T C
tempranillo
86

Solar de Estraunza 2015 T R
tempranillo
87

Solar de Estraunza 2018 T C
tempranillo
87
Balanced, spicy, dried herbs, toasty.

Solar de Estraunza 2021 B
viura
85

Solar de Estraunza 2021 T
tempranillo
87

Solar de Estraunza Selección 2016 T
tempranillo
89
Kind finish, spicy, jammy, tasty. Palate: dry wood.

BODEGAS FAUSTINO
Ctra. de Logroño, s/n
01320 Oyón (Araba/Álava)
☎: +34 945 622 500
info@bodegasfaustino.es
www.bodegasfaustino.com

Faustino 2016 T R
tempranillo, mazuelo
90
Colour: cherry, garnet rim. Nose: ripe fruit, spicy. Palate: powerful, ripe fruit, spicy, round tannins.

Faustino 2019 T C
tempranillo
89
Aromatic, spicy, fruity, jammy, tasty.

Faustino Art Collection 2019 T C
tempranillo
89
Balanced, spicy, jammy, dried herbs.

Faustino Art Collection 2020 T
tempranillo
90
Colour: deep cherry. Nose: ripe fruit, dried herbs, sweet spices. Palate: powerful, ripe fruit, spicy, round tannins.

Faustino I 2011 T GR
92
Representative. Colour: dark-red cherry, garnet rim. Nose: ripe fruit, waxy notes, spicy, neat. Palate: spicy, long, good acidity.

Faustino I Magnum 2011 T GR
tempranillo, graciano, mazuelo
93
Colour: bright cherry, garnet rim. Nose: red berry notes, spicy, fine reductive notes, wild herbs. Palate: fresh, balanced, good acidity, fine tannins.

**Faustino Icon
Especial Selección 2017 T R**
tempranillo, graciano
89
Spicy, jammy, powerful, roasted coffee, woody.

Faustino V 2017 T R
tempranillo, mazuelo
88
Balanced, spicy, herbaceous, tasty, crisp.

Faustino VII 2021 T
tempranillo, mazuelo

87

Gran Faustino I 2004 T GR
tempranillo, graciano, mazuelo

94

Colour: dark-red cherry, garnet rim. Nose: ripe fruit, tobacco, sweet spices, dried herbs, meaty notes. Palate: spicy, round tannins, long.

BODEGAS FECO
El Prado 2A
26338 San Vicente de la Sonsierra (La Rioja)
☎: +54 651 103 444
info@bodegasfeco.com
www.bodegasfeco.com

Feco 2020 B MC
tempranillo blanco

91

Colour: bright straw. Nose: white fruit, ripe fruit, spicy. Palate: flavourful, fresh, good acidity, fruity aftertaste.

Feco 2020 T MC
tempranillo, viura

90

Colour: bright cherry. Nose: floral, ripe fruit, balanced. Palate: flavourful, fruity, good acidity.

Feco Piedra Miguel 2021 B
80% viura, 20% tempranillo blanco

89

Fruity, dried flowers, herbal, jammy.

Prettini 2021 B
80% viura, 20% tempranillo blanco

88

Dried flowers, fruity, sweety finish, tasty.

Prettini 2021 T
tempranillo, viura

88

Balsamic herbs, jammy, fruity, tasty, dried herbs, balanced.

BODEGAS FERNÁNDEZ EGUILUZ
Los Morales, 7 bajo
26339 Ábalos (La Rioja)
Fax: +34 659 146 888
p.larosa@hotmail.es

Cantarada de los Mozos San Prudencio 2018 T
70% tempranillo, 30% viura

92

Colour: dark-red cherry, garnet rim. Nose: ripe fruit, aged wood nuances, tobacco, sweet spices. Palate: spicy, round tannins, long.

Peña La Rosa 2021 B

86

Peña la Rosa 2021 T MC
tempranillo

89

Standard, floral, fruity, tasty.

Peña La Rosa El Secreto del Abuelo 2017 T
tempranillo

91

Colour: Cherry. Nose: balsamic herbs, sweet spices, scrubland, red berry notes. Palate: spicy, balsamic, good acidity.

Peña La Rosa Grano a Grano 2016 T
100% tempranillo

92

Colour: dark-red cherry, garnet rim. Nose: ripe fruit, aged wood nuances, tobacco, sweet spices. Palate: spicy, round tannins, long.

Peña la Rosa Vendimia Seleccionada 2016 T

88

Hot, spicy, balsamic herbs, smooth.

BODEGAS FOS
Término de Vialba, s/n
01340 Elciego (Araba/Álava)
☎: +34 945 606 681
fos@bodegasfos.com
www.bodegasfos.com

Finca Zuriena 2018 B
viura

94

Colour: bright yellow. Nose: powerful, creamy oak, ripe fruit, spicy, fruit preserve. Palate: good structure, long, toasty, fine bitter notes.

DO Ca. RIOJA / D.O.P.

DO Ca. RIOJA / D.O.P.

Fos Baranda 2018 T
90% tempranillo, 10% graciano

92

Colour: deep cherry. Nose: ripe fruit, dried herbs, creamy oak, black fruit, toasty. Palate: powerful, ripe fruit, spicy, round tannins.

🏆 **PODIUM**

Por los Cien 2018 T
100% tempranillo

95

Colour: deep cherry. Nose: ripe fruit, dried herbs, creamy oak, smoky, expressive. Palate: powerful, ripe fruit, spicy, round tannins, flavourful, soft tannins.

BODEGAS FRANCO ESPAÑOLAS

Cabo Noval, 2
26009 Logroño (La Rioja)
☎: +34 941 251 300
marketing@grupoeguizabal.com
www.bodegasfrancoespañolas.com

Bordón 2012 T GR
70% tempranillo, 20% graciano, 10% mazuelo

87

Bordón 2016 T R
80% tempranillo, 15% garnacha, 5% mazuelo

88

Fruity, spicy, herbal, tasty.

Bordón 2019 T C
80% tempranillo, 20% garnacha

88

Fruity, spicy, herbal, tasty.

Bordón 2021 B
100% viura

85

Bordón D'Anglade 2019 T C
tempranillo

90

Colour: dark-red cherry, garnet rim. Nose: ripe fruit, fruit preserve, aged wood nuances, tobacco. Palate: spicy, round tannins, long.

Talla de Diamante Semidulce 2021 B SD
75% viura, 20% tempranillo blanco, 5% chardonnay

87

BODEGAS FUIDIO

San Bartolome, 32
01322 Yecora (Araba/Álava)
☎: +34 679 255 045
bodegas@fuidio.com
www.fuidio.com

Fuidio 2021 T
tempranillo

86

Fuidio Iraley 2018 T C
tempranillo

88

Balanced, spicy, crisp, fruity, toasty.

BODEGAS GARCÍA DE OLANO

Ctra. Vitoria, s/n
01309 Páganos (Araba/Álava)
☎: +34 945 621 146
info@garciadeolano.com
www.garciadeolano.com

3 de Olano 2016 T C
100% tempranillo

88

Spicy, jammy, tasty, toasty.

3 de Olano Selección 2017 T
tempranillo

90

Colour: deep cherry. Nose: dried herbs, creamy oak, ripe fruit, black fruit. Palate: powerful, ripe fruit, spicy, round tannins.

Heredad García de Olano 2018 T C
100% tempranillo

88

Balanced, spicy, jammy, toasty, dried herbs.

Heredad García de Olano Tempranillo 2021 T MC
tempranillo

87

Mauleón 2014 T R
100% tempranillo

90

Colour: deep cherry. Nose: ripe fruit, dried herbs, creamy oak, fine reductive notes, cocoa bean. Palate: ripe fruit, spicy, round tannins.

Olanum Vendimia Seleccionada 2017 T BA
tempranillo

87

BODEGAS HEREDAD DE ADUNA

Matarredo, 39
01307 Samaniego (Araba/Álava)
☎: +34 945 623 343
bodegas@heredadaduna.com
www.heredadaduna.com

Aduna 2021 B
88
Aromatic, smooth, jammy, fruity.

Aduna T C
88
Toasty, tasty, jammy, spicy.

Aduna Vendimia Seleccionada 2016 T
100% tempranillo
91
Colour: very deep cherry. Nose: aromatic coffee, powerful, sweet spices. Palate: smoky aftertaste, great length, round tannins.

BODEGAS HERMANOS PECIÑA

Ctra. Vitoria Km. 47
26338 San Vicente de la Sonsierra (La Rioja)
☎: +34 941 334 366
info@bodegashermanospecina.com
www.bodegashermanospecina.com

Chobeo de Peciña 2012 T
100% tempranillo
91
Colour: dark-red cherry, garnet rim. Nose: fruit preserve, aged wood nuances, tobacco, sweet spices, wet leather. Palate: spicy, round tannins, long.

Chobeo de Peciña 2020 B FB
100% viura
88
Kind finish, dried flowers, fruity, herbal.

Señorío de P. Peciña 2011 T GR
95% tempranillo, 3% graciano, 2% garnacha
87

Señorío de P. Peciña 2016 T C
95% tempranillo, 3% graciano, 2% garnacha
88
Spicy, jammy, tasty, toasty.

Señorío de P. Peciña 2021 T
95% tempranillo, 3% graciano, 2% garnacha
87

BODEGAS HERMOSILLA

Avda. del Río Ebro, 36
01307 Baños de Ebro (Araba/Álava)
☎: +34 657 791 677
ventas@bodegashermosilla.com
www.bodegashermosilla.com

Hermosilla. Bodegas J. I. Hermosilla 2018 T C
100% tempranillo
87

Hermosilla. Bodegas J. I. Hermosilla 2021 B
viura
84

Hermosilla. Bodegas J. I. Hermosilla 2021 T MC
95% tempranillo, 5% viura
84

BODEGAS ISIDRO MILAGRO

Avda. del Ebro s/n
26540 Alfaro (La Rioja)
gerardo.export@bodegasisidromilagro.com
www.bodegasisidromilagro.com

Campo Burgo 2016 T R
tempranillo
88
Overripe, fruity, herbal, tasty.

Campo Burgo 2018 T C
tempranillo
88
Fruity, floral, herbal, spicy, balanced.

Campo Burgo 2021 T
tempranillo
87

Hacienda Susar 2018 T
tempranillo
90
Colour: cherry, purple rim. Nose: red berry notes, ripe fruit, spicy, dried herbs, toasty. Palate: flavourful, fruity, fresh, good structure.

DO Ca. RIOJA / D.O.P.

DO Ca. RIOJA / D.O.P.

BODEGAS IZADI
Herrería Travesía II, 5
01307 Villabuena de Álava (Araba/Álava)
☎: +34 945 609 086
Fax: +34 945 609 261
izadi@izadi.com
www.izadi.com

Izadi 2019 T C
tempranillo
92
Colour: deep cherry. Nose: ripe fruit, dried herbs, creamy oak, sweet spices, toasty. Palate: powerful, ripe fruit, spicy, round tannins.

🏆 **PODIUM**

Izadi El Regalo 2020 T
tempranillo
95
Colour: very deep cherry. Nose: complex, expressive, spicy, mineral, ripe fruit, red berry notes. Palate: full-bodied, long, great length.

Izadi Larrosa Blanca 2021 B
garnacha blanca
91
Colour: bright straw. Nose: fragrant herbs, fine lees, ripe fruit, stone fruit. Palate: full-bodied, rich, long, good acidity.

Izadi Larrosa Negra 2021 T
garnacha
92
Colour: bright cherry. Nose: balsamic herbs, sweet spices, scrubland, fresh fruit, red berry notes. Palate: spicy, balsamic, good acidity.

Izadi Larrosa Rosé 2021 RD
garnacha
90
Colour: raspberry rose. Nose: red berry notes, floral, fragrant herbs. Palate: light-bodied, spicy, good acidity.

Izadi Selección 2018 T
tempranillo, graciano, maturana, garnacha
93
Colour: bright cherry. Nose: balsamic herbs, sweet spices, scrubland, red berry notes, black fruit, complex. Palate: spicy, balsamic, good acidity.

Izadi Selección 2021 B
92
Needs time. Colour: bright straw. Nose: ripe fruit, dried herbs, faded flowers, sweet spices. Palate: powerful, ripe fruit, balanced.

BODEGAS JAVIER SAN PEDRO ORTEGA
Camino de La Hoya s/n
01300 Laguardia (Araba/Álava)
☎: +34 636 082 927
info@bodegasjaviersanpedro.com
www.bodegasjaviersanpedro.com

Cueva de Lobos Alpha 2020 T
garnacha
93
Varietally correct, herbal, wild. Colour: cherry, garnet rim. Nose: wild herbs, faded flowers, ripe fruit. Palate: flavourful, fruity, taut.

Viuda Negra Arca de Assa 2020 T BA
tempranillo
92
Colour: cherry, garnet rim. Nose: fruit preserve, powerful, dried herbs. Palate: flavourful, long, round tannins, good structure.

Viuda Negra La Taconera 2019 T
tempranillo
94
Colour: deep cherry. Nose: ripe fruit, dried herbs, creamy oak, fruit preserve. Palate: powerful, ripe fruit, spicy, round tannins.

Viuda Negra Nunca Jamás 2021 T
tempranillo
91
Colour: deep cherry. Nose: ripe fruit, dried herbs, creamy oak, toasty. Palate: powerful, ripe fruit, spicy, round tannins.

Viuda Negra Prado de las Almas 2021 RD FB
90% tempranillo, 10% garnacha
91
Colour: salmon. Nose: ripe fruit, red berry notes, fragrant herbs, faded flowers. Palate: fresh, fruity, flavourful.

Viuda Negra Villahuercos 2020 B FB
tempranillo blanco
93
Colour: bright straw. Nose: fruit expression, stone fruit, tropical fruit, dried herbs, spicy. Palate: fruity, flavourful, powerful, good finish.

BODEGAS JER

Ctra. Huércanos - Nájera
26004 Huércanos (La Rioja)
☎: +34 635 955 448
info@bodegasjer.es
www.bodegasjer.es

426 2019 T C
tempranillo
87

J. Cantera 2021 B
viura, malvasía
83

J. Cantera 2021 RD
garnacha, viura
85

J. Cantera 2021 T
tempranillo
86

Thaler de Plata 2020 T
88
Smoky, toasty, simple, jammy, standard.

BODEGAS LA CATEDRAL

Avda. de Mendavia, 30
26009 Logroño (La Rioja)
☎: +34 941 235 299
bodegasolarra@bodegasolarra.es

Epulum 2018 T R
90% tempranillo, 10% mazuelo, viura, garnacha
86

Epulum 2019 T C
90% tempranillo, 10% mazuelo, garnacha
86

Marqués de Rivallana 2015 T GR
90% tempranillo, 10% graciano, mazuelo
88
Fruity, classic, spicy, tasty.

Marqués de Rivallana 2018 T R
90% tempranillo, 10% garnacha, graciano, mazuelo
87

Marqués de Rivallana 2019 T C
90% tempranillo, 10% garnacha, graciano, mazuelo
87

BODEGAS LA ERALTA

Pol. Ind. El Sequero, Avda. de Cameros, 27
26150 Agoncillo (La Rioja)
☎: +34 941 395 092
sales@bodegaslaeralta.com
www.grupolaeralta.com

Altos del Bergasa 2016 T GR
tempranillo
90
Colour: deep cherry. Nose: dried herbs, creamy oak, black fruit, fine reductive notes. Palate: powerful, ripe fruit, spicy, round tannins.

Altos del Bergasa 2019 T C
88
Fruity, spicy, dried herbs, tasty.

Señorío de La Eralta 2015 T GR
tempranillo
89
Balanced, spicy, dried herbs, jammy, tasty.

Señorío de La Eralta 2016 T GR
tempranillo
88
Classic, spicy, fruity, jammy, tasty.

Señorío de La Eralta 2017 T GR
tempranillo
88
Hot, fruity, herbal, pruney, tasty.

Señorío de La Eralta 2019 T C
tempranillo
87

BODEGAS LAN

Paraje del Buicio, s/n
26360 Fuenmayor (La Rioja)
☎: +34 941 450 950
Fax: +34 941 450 567
info@bodegaslan.com
www.bodegaslan.com

LAN 7 metros 2019 T C
89
Balanced, spicy, jammy, tasty, toasty.

LAN a Mano 2019 T
tempranillo, graciano, mazuelo
92
Colour: deep cherry. Nose: dried herbs, creamy oak, black fruit, woody. Palate: powerful, ripe fruit, spicy, harsh oak tannins.

DO Ca. RIOJA / D.O.P.

DO Ca. RIOJA / D.O.P.

LAN D-12 2019 T C
tempranillo

91

Colour: deep cherry. Nose: dried herbs, spicy, black fruit, dark chocolate. Palate: ripe fruit, spicy, round tannins.

Culmen 2017 T R
tempranillo, graciano

94

Colour: deep cherry. Nose: dried herbs, creamy oak, black fruit, roasted coffee, meaty notes. Palate: powerful, ripe fruit, spicy, round tannins.

LAN 2017 T R
90

Colour: deep cherry. Nose: dried herbs, creamy oak, black fruit, woody. Palate: ripe fruit, spicy, round tannins.

LAN 2015 T GR
tempranillo, graciano, mazuelo, garnacha

92

Colour: dark-red cherry, garnet rim. Nose: ripe fruit, aged wood nuances, tobacco, sweet spices, cigar, old leather. Palate: spicy, round tannins, long.

BODEGAS LANDALUCE
Ctra. de los Molinos s/n
01300 Laguardia (Araba/Álava)
☎: +34 620 824 314
asier@bodegaslandaluce.es
www.bodegaslandaluce.es

Capricho de Landaluce 2019 T
100% tempranillo

91

Colour: cherry, purple rim. Nose: fruit expression, red berry notes, spicy. Palate: flavourful, fruity, good acidity, easy to drink.

Elle de Landaluce 2019 T
90% tempranillo, 10% graciano

90

Colour: bright cherry. Nose: wild herbs, dried flowers, ripe fruit, neat. Palate: flavourful, fruity, balanced.

Fincas de Landaluce 2018 T R
100% tempranillo

89

Pleasant, spicy, fruity, jammy, tasty, balanced, standard.

Fincas de Landaluce 2019 T C
100% tempranillo

88

Pleasant, fruity, jammy, tasty, spicy.

Fincas de Landaluce Graciano 2019 T
100% graciano

91

Representative. Colour: Cherry. Nose: scrubland, balanced, black liquorice, dried herbs. Palate: spicy, balsamic, good acidity.

BODEGAS LAR DE PAULA
Cercas Altas, 6
01309 Elvillar (Araba/Álava)
☎: +34 945 604 068
info@lardepaula.com
www.lardepaula.com

Gran Baroja 2011 T GR
tempranillo

92

Colour: deep cherry, garnet rim. Nose: aged wood nuances, cocoa bean, cigar, toasty. Palate: flavourful, spicy, toasty, powerful tannins.

Lar de Paula 2017 T C
100% tempranillo

88

Pleasant, tasty, jammy, spicy.

Lar de Paula 2018 B FB
viura, malvasía

89

Fruity, jammy, tasty.

Lar de Paula Edición Limitada 2015 T R
tempranillo

90

Colour: dark-red cherry, garnet rim. Nose: ripe fruit, aged wood nuances, tobacco, sweet spices. Palate: spicy, round tannins, long.

Lar de Paula Edición Limitada 2016 T C
tempranillo

91

Colour: dark-red cherry. Nose: toasty, spicy, cocoa bean, ripe fruit, black fruit. Palate: flavourful, toasty, fine bitter notes.

Lar de Paula Madurado en Bodega 2019 T
tempranillo

89

Spicy, jammy, tasty, toasty.

BODEGAS LAS CEPAS
Ctra Najera-Cenicero s/n
26313 Uruñuela (La Rioja)
☎: +34 625 259 216
ngarcia@lascepasriojawine.com
www.lascepasriojawine.com

Costalarbol 2020 T
60% graciano, 20% tempranillo, 20% garnacha

88

Standard, simple, wild, tasty, balanced, jammy, spicy.

Costalarbol 2021 B
100% maturana blanca

87

Costalarbol Graciano 2019 T
100% graciano

91

Colour: cherry, garnet rim. Nose: black fruit, fine reductive notes, balanced, varietal, wild herbs. Palate: good structure, flavourful, great length.

Garnacha 1921 2019 T
100% garnacha

88

Standard, dried herbs, jammy, wild, tasty, toasty.

Rebuzno 2020 T
100% maturana

89

Wild, smoky, toasty, jammy, balanced.

Serezhade 2020 B
80% maturana blanca, 10% sauvignon blanc, 10% verdejo

89

Fruity, dried flowers, tasty, jammy.

BODEGAS LAUNA
Ctra. Vitoria-Logroño, Km. 57
01300 Laguardia (Araba/Álava)
☎: +34 946 824 108
info@bodegaslauna.com
www.bodegaslauna.com

Amaita 2018 T
100% tempranillo

93

Colour: cherry, purple rim. Nose: tobacco, spicy, cocoa bean, smoky, black fruit, fruit preserve. Palate: full-bodied, fruity, flavourful, balanced, slightly dry, soft tannins.

Ikunus 2018 T
100% tempranillo

92

Colour: deep cherry. Nose: ripe fruit, black fruit, spicy, waxy notes. Palate: powerful, ripe fruit, spicy, round tannins, flavourful.

Launa 2019 T C
87% tempranillo, 13% otras

90

Colour: deep cherry. Nose: dried herbs, creamy oak, black fruit, dry nuts. Palate: ripe fruit, spicy, round tannins.

Launa 2021 B
40% chardonnay, 60% viura

89

Citrus fruit, balanced, spicy, crisp, dried herbs, jammy, woody.

Launa Selección Familiar 2018 T C
100% tempranillo

91

Smoky, standard. Nose: lactic notes, sweet spices, toasty. Palate: flavourful, spicy, ripe fruit.

Launa Selección Familiar 2021 B FB
100% viura

91

Colour: bright yellow. Nose: powerful, creamy oak, ripe fruit, spicy, sweet spices. Palate: rich, long, toasty, fine bitter notes.

DO Ca. RIOJA / D.O.P.

DO Ca. RIOJA / D.O.P.

Launa Selección Familiar 2017 T R
90% tempranillo, 10% graciano

91

Colour: bright cherry. Nose: sweet spices, ripe fruit, smoky. Palate: fruity, spicy, round tannins, powerful.

BODEGAS LECEA
Las Cuevas, 246
26340 San Asensio (La Rioja)
☎: +34 685 010 400
info@bodegaslecea.com
www.bodegaslecea.com

Corazón de Lago 2020 T MC
100% tempranillo

89

Pleasant, kind finish, fruity, floral, tasty, aromatic.

Cuevas de Lecea 2017 T BA
95% tempranillo, 5% mazuelo

91

Colour: cherry, purple rim. Nose: fruit expression, red berry notes, floral, spicy, dark chocolate. Palate: flavourful, fruity, good acidity, long.

Lecea 2010 T GR
tempranillo

90

Colour: cherry, garnet rim. Nose: wet leather, fine reductive notes, meaty notes, ripe fruit, sweet spices. Palate: flavourful, fruity, classic aged character, slightly dry, soft tannins.

Lecea 2014 T R
100% tempranillo

90

Colour: bright cherry. Nose: fruit preserve, red berry notes, spicy, smoky, dried herbs. Palate: fruity, flavourful, balanced, slightly dry, soft tannins.

Lecea 2016 T C
100% tempranillo

88

Fruity, herbal, spicy, balanced.

Lecea 2017 B C
chardonnay

87

BODEGAS LOLI CASADO
Avda. La Poveda, 46
01306 Lapuebla de Labarca (Araba/Álava)
☎: +34 678 041 484
loli@bodegaslolicasado.com
www.bodegaslolicasado.com

Jaun de Alzate 2015 T R
90% tempranillo, 5% graciano, 5% mazuelo

88

Toasty, smooth, jammy.

Polus 2015 T R
100% tempranillo

90

Colour: dark-red cherry, garnet rim. Nose: ripe fruit, aged wood nuances, tobacco, sweet spices. Palate: spicy, round tannins.

Polus 2019 T C
100% tempranillo

90

Colour: deep cherry. Nose: ripe fruit, dried herbs, creamy oak. Palate: powerful, ripe fruit, spicy, round tannins.

Polus 2021 T MC
89% tempranillo, 11% viura

87

Polus Tempranillo 2019 T
100% tempranillo

88

Toasty, powerful, jammy, spicy.

Polus Viura 2021 B
100% viura

88

Citrus fruit, balanced, herbal, yeasty notes.

BODEGAS LÓPEZ ORIA

Ctra. Elvillar, s/n
01300 Laguardia (Araba/Álava)
☎: +34 649 628 420
info@bodegaslopezoria.com
www.bodegaslopezoria.com

Pola 2019 T C
tempranillo

90

Colour: deep cherry. Nose: ripe fruit, dried herbs, creamy oak. Palate: powerful, ripe fruit, spicy, round tannins.

Pola 2021 T MC
tempranillo, viura

87

Pola Antonio López 2020 T
tempranillo

88

Toasty, tasty, spicy, reduced.

Pola Valecilla 2018 T
tempranillo

90

Colour: deep cherry. Nose: ripe fruit, dried herbs, creamy oak. Palate: powerful, ripe fruit, spicy, round tannins.

BODEGAS LUIS CAÑAS

Ctra. Samaniego, 10
01307 Villabuena de Álava (Araba/Álava)
☎: +34 945 623 373
Fax: +34 945 609 289
bodegas@luiscanas.com
www.luiscanas.com

Camino Leza 2019 T
tempranillo, viura, rojal

91

Colour: cherry, purple rim. Nose: fruit expression, red berry notes, floral, spicy. Palate: flavourful, fruity, good acidity, long, coarse tannins.

Luis Cañas 2015 T GR
95% tempranillo, 5% graciano

92

Colour: deep cherry. Nose: ripe fruit, dried herbs, creamy oak, meaty notes, sweet spices, scrubland. Palate: powerful, ripe fruit, spicy, round tannins.

Luis Cañas 2019 T C
95% tempranillo, 5% garnacha

90

Colour: dark-red cherry. Nose: toasty, spicy, cocoa bean, ripe fruit. Palate: flavourful, toasty, fine bitter notes.

Luis Cañas Selección de Familia 2017 T R
85% tempranillo, 15% otras

93

Colour: deep cherry. Nose: dried herbs, creamy oak, black fruit, red berry notes, roasted coffee. Palate: powerful, ripe fruit, spicy, round tannins.

Luis Cañas Viñas Viejas 2020 B
90% viura, 10% rojal

91

Colour: bright yellow. Nose: powerful, creamy oak, ripe fruit, spicy. Palate: rich, good structure, toasty, fine bitter notes.

Ribagaitas 2019 T
tempranillo, viura, rojal

92

Colour: cherry, purple rim. Nose: fruit expression, red berry notes, floral, spicy. Palate: flavourful, fruity, good acidity, long.

BODEGAS LUIS GURPEGUI MUGA

Luis Gurpegui, 3
31570 San Adrián (Navarra)
☎: +34 948 692 500
Fax: +34 948 692 700
info@manzanos.com
www.luisgurpeguimuga.com

Berceo 2018 T C
tempranillo, garnacha

89

Nose: ripe fruit, dried herbs, creamy oak. Palate: ripe fruit, round tannins.

Berceo Nueva Generación 2016 T R

87

Berceo Nueva Generación 2018 T C

90

Colour: dark-red cherry. Nose: toasty, spicy, cocoa bean, ripe fruit. Palate: flavourful, toasty, fine bitter notes.

Gonzalo de Berceo 2013 T GR

91

Colour: deep cherry, garnet rim. Nose: aged wood nuances, cocoa bean, cigar, toasty. Palate: flavourful, spicy, toasty, powerful tannins.

Gonzalo de Berceo 2016 T R

90

Colour: dark-red cherry, garnet rim. Nose: ripe fruit, fruit preserve, aged wood nuances, tobacco, sweet spices. Palate: spicy, round tannins, long.

Gonzalo de Berceo 2018 T C

89

Fruity, spicy, dried herbs, tasty.

DO Ca. RIOJA / D.O.P.

DO Ca. RIOJA / D.O.P.

Gonzalo de Berceo
Tempranillo Blanco C.V.C B FB
100% tempranillo blanco
89
Pleasant, toasty, tasty, jammy.

Los Dominios
de Berceo Prefiloxérico 2017 T
92
Colour: deep cherry. Nose: complex, expressive, spicy. Palate: elegant, full-bodied, long, great length.

Los Dominios
de Berceo Reserva 36 2006 T R
tempranillo
92
Colour: dark-red cherry, garnet rim. Nose: ripe fruit, aged wood nuances, tobacco, sweet spices. Palate: spicy, round tannins, long.

Luis Gurpegui 147 Aniversario 2018 T C
tempranillo, garnacha, graciano
88
Fruity, herbal, spicy, wild.

Primi Luis Gurpegui 2019 T C
tempranillo, garnacha
88
Classic, tasty, simple, spicy.

Siglo 2016 T R
tempranillo, garnacha
88
Balanced, spicy, fruity, herbal, jammy.

Siglo Familia
Fernandez de Manzanos 2015 T GR
90
Fruity, spicy, herbal, tasty. Colour: cherry, garnet rim. Nose: sweet spices, meaty notes, old leather, fruit preserve, red berry notes. Palate: flavourful, fruity, good structure, balanced.

Siglo Saco Tempranillo C.V.C T
tempranillo
90
Colour: dark-red cherry. Nose: toasty, spicy, cocoa bean, ripe fruit, black fruit. Palate: flavourful, toasty, fine bitter notes.

BODEGAS MANZANOS
Ctra. NA-134, km. 49
31560 Azagra (Navarra)
☎: +34 948 692 500
info@manzanos.com
www.bodegasmanzanos.com

Finca Manzanos 2015 T R
tempranillo, garnacha, graciano
87

Finca Manzanos Garnacha 2021 T
100% garnacha
88
Pleasant, tasty, jammy, fruity.

Manzanos 2014 T GR
tempranillo, garnacha, graciano
90
Colour: dark-red cherry, garnet rim. Nose: ripe fruit, tobacco, sweet spices, toasted bread. Palate: spicy, round tannins, long.

Manzanos 2016 T R
tempranillo, garnacha, graciano
88
Balanced, spicy, fruity, full-bodied, tasty.

Palacio de Manzanos Viñedo Singular
1890 Manzanos 2017 T
100% garnacha
90
Colour: dark-red cherry, garnet rim. Nose: aged wood nuances, tobacco, sweet spices, fruit preserve. Palate: spicy, round tannins, long.

Voché 5º Generación C.V.C. T
92
Colour: Cherry. Nose: balsamic herbs, sweet spices, scrubland, ripe fruit, black fruit. Palate: spicy, balsamic, good acidity.

Voché Selección Graciano 2017 T
graciano
90
Colour: cherry, purple rim. Nose: ripe fruit, red berry notes, spicy, smoky, dried herbs. Palate: fruity, flavourful, balanced, slightly dry, soft tannins.

BODEGAS MARQUÉS DE CÁCERES

Ctra. Logroño, s/n
26350 Cenicero (La Rioja)
☎: +34 941 454 000
comunicacion@marquesdecaceres.com
www.marquesdecaceres.com

Marqués de Cáceres 2016 T R
92
Colour: dark-red cherry, garnet rim. Nose: ripe fruit, aged wood nuances, tobacco, sweet spices, cigar. Palate: spicy, round tannins, long.

Marqués de Cáceres 2018 T C
89
Toasty, tasty, jammy, spicy.

BODEGAS MARQUÉS DE REINOSA

Ctra. Rincón de Soto, s/n
26560 Autol (La Rioja)
☎: +34 941 401 327
bodegas@marquesdereinosa.com
www.marquesdereinosa.com

Marqués de Reinosa 2018 T R
89
Toasty, jammy, spicy, classic.

Marqués de Reinosa 2021 B SD
tempranillo blanco
86

Marqués de Reinosa Maturana 2020 T
maturana
90
Colour: very deep cherry. Nose: roasted coffee, aromatic coffee, powerful, black fruit. Palate: smoky aftertaste, great length, round tannins.

Marqués de Reinosa Tempranillo 2021 T
tempranillo
88
Standard, balanced, fruity, jammy, simple.

Marqués de Reinosa Tempranillo Blanco 2021 B
tempranillo blanco
87

BODEGAS MARQUÉS DE TERÁN

Ctra. de Nájera, Km. 1
26220 Ollauri (La Rioja)
☎: +34 941 338 373
info@marquesdeteran.com
www.marquesdeteran.com

Marqués de Terán 2016 T R
tempranillo
90
Colour: deep cherry, garnet rim. Nose: aged wood nuances, cocoa bean, cigar, toasty. Palate: flavourful, spicy, toasty, powerful tannins.

Marqués de Terán 2018 T C
tempranillo
89
Toasty, jammy, spicy, balanced.

Marqués de Terán Edición Limitada 2015 T R
tempranillo
92
Colour: dark-red cherry, garnet rim. Nose: ripe fruit, fruit preserve, aged wood nuances, tobacco, sweet spices. Palate: spicy, round tannins, long.

Marqués de Terán Selección Especial 2018 T
tempranillo
91
Colour: deep cherry. Nose: ripe fruit, dried herbs, creamy oak. Palate: powerful, ripe fruit, spicy, round tannins.

Terán Versum 2018 T
92
Colour: dark-red cherry, garnet rim. Nose: ripe fruit, aged wood nuances, tobacco, sweet spices. Palate: spicy, round tannins, long.

BODEGAS MARQUÉS DEL ATRIO

Ctra. de Logroño NA-134, km. 86,200.
31587 Mendavia (Navarra)
☎: +34 948 379 994
visitas@marquesdelatrio.com
www.marquesdelatrio.com

2 Cepas Marqués del Atrio 2019 B BA
50% viura, 50% tempranillo blanco
91
Colour: bright yellow. Nose: ripe fruit, toasty, sweet spices, floral. Palate: rich, good structure, long, toasty, fine bitter notes.

Aullido 2021 T
tempranillo
88
Pleasant, crisp, fruity, tasty, simple.

DO Ca. RIOJA / D.O.P.

DO Ca. RIOJA / D.O.P.

Marqués del Atrio 2017 T R
tempranillo, graciano
90
Colour: dark-red cherry, garnet rim. Nose: ripe fruit, aged wood nuances, tobacco. Palate: spicy, round tannins, long.

Marqués del Atrio 2019 T C
tempranillo, graciano
90
Colour: deep cherry. Nose: ripe fruit, dried herbs, creamy oak, neat. Palate: powerful, ripe fruit, spicy, round tannins.

Marqués del Atrio Edición Limitada 2017 T
80% tempranillo, 20% graciano
91
Colour: dark-red cherry, garnet rim. Nose: ripe fruit, fruit preserve, aged wood nuances, cigar, dark chocolate. Palate: spicy, round tannins, long.

BODEGAS MARTÍNEZ ALESANCO
José García, 20
26310 Badarán (La Rioja)
☎: +34 941 367 075
info@bodegasmartinezalesanco.com
www.bodegasmartinezalesanco.com

Martínez Alesanco 2016 T R
90% tempranillo, 10% garnacha
90
Colour: cherry, garnet rim. Nose: waxy notes, fine reductive notes, ripe fruit, balanced. Palate: flavourful, long, good acidity.

Martínez Alesanco 2018 T C
80% tempranillo, 20% garnacha
88
Simple, jammy, standard, spicy, dried flowers.

Martínez Alesanco 2021 B FB
50% tempranillo blanco, 50% viura
88
Fruity, jammy, dried flowers, tasty.

Martínez Alesanco Rosado 2021 RD FB
100% garnacha
87

Nada que Ver 2017 T C
100% maturana
89
Pruney, full-bodied, herbal, jammy, hot, tasty, slight reduction.

Pedro Martínez Alesanco Selección 2015 T R
50% maturana, 25% tempranillo, 25% garnacha
91
Colour: deep cherry. Nose: creamy oak, black fruit, scrubland, floral. Palate: powerful, ripe fruit, spicy, round tannins.

BODEGAS MARTÍNEZ LACUESTA
Paraje de Ubieta, s/n
26200 Haro (La Rioja)
☎: +34 653 814 125
Fax: +34 941 303 748
luis@martinezlacuesta.com
www.martinezlacuesta.com

Campeador 2005 T GR
93
Colour: dark-red cherry. Nose: ripe fruit, fruit preserve, tobacco, incense. Palate: spicy, round tannins, long, classic aged character.

Hinia 2012 T
92
Colour: dark-red cherry, garnet rim. Nose: ripe fruit, fruit preserve, tobacco, animal reductive notes. Palate: spicy, round tannins, long.

Martínez Lacuesta 2011 T GR
91
Colour: dark-red cherry. Nose: ripe fruit, dried herbs, creamy oak, spicy. Palate: ripe fruit, spicy, round tannins, long.

Martínez Lacuesta 2014 B GR
92
Colour: yellow. Nose: toasted bread, smoky, bakery, white fruit, ripe fruit. Palate: full-bodied, flavourful, spicy, rich, smoky aftertaste.

Martínez Lacuesta 2018 B R
91
Colour: straw. Nose: ripe fruit, faded flowers, stone fruit, slightly evolved. Palate: powerful, ripe fruit, balanced, spicy, easy to drink.

Martínez Lacuesta 2019 T C
91
Colour: dark-red cherry, garnet rim. Nose: ripe fruit, aged wood nuances, tobacco, sweet spices. Palate: spicy, round tannins, long.

Martínez Lacuesta La Sucursal 2020 T BA
91
Colour: cherry, purple rim. Nose: fruit expression, floral, sweet spices. Palate: flavourful, fruity, good acidity, long, toasty.

BODEGAS MARTÍNEZ PALACIOS

Real, 48
26220 Ollauri (La Rioja)
☎: +34 941 338 023
bodega@bodegasmartinezpalacios.com
www.bodegasmartinezpalacios.com

Itran Tempranillo 2020 T
100% tempranillo
87

Martínez Palacios 2014 T R
90% tempranillo, 10% graciano
88
Overripe, toasty, creamy, tasty.

Martínez Palacios 2018 B C
90
Colour: bright yellow. Nose: powerful, creamy oak, ripe fruit, spicy. Palate: good structure, long, toasty, fine bitter notes.

Martínez Palacios 2018 T C
100% tempranillo
88
Woody, spicy, herbal, jammy, powerful.

Martínez Palacios Graciano 2019 T C
100% graciano
88
Balanced, wild, tasty, toasty.

Martínez Palacios Selección 30 Barricas 2018 T C
90
Colour: cherry, purple rim. Nose: red berry notes, floral, spicy, dark chocolate. Palate: flavourful, fruity, good acidity, long.

BODEGAS MASET RIOJA

Avda. Costa del Vino, 1
26200 Haro (La Rioja)
☎: +34 900 200 250
info@maset.com
www.maset.com/rioja

Maset Senderos del Molinero 2020 B
viura
88
Aromatic, floral, fruity, simple, jammy.

Maset Tempranillo 2018 T R
tempranillo
88
Pruney, spicy, herbaceous, jammy.

Maset Tempranillo 2019 T C
tempranillo
89
Pruney, fruity, full-bodied, jammy, tasty, spicy.

BODEGAS MAZUELA

Ctra. Elciego km 0,1
26350 Cenicero (La Rioja)
☎: +34 607 548 054
manuel@bodegasmazuela.com
www.bodegasmazuela.com

Corazón Indomable 2021 T MC
100% tempranillo
88
Jammy, herbal, tasty.

La Hoya 2016 T
100% tempranillo
91
Colour: bright cherry. Nose: ripe fruit, creamy oak, spicy, dried herbs. Palate: fruity, balanced, flavourful, good finish.

Liante 2020 T RB
100% tempranillo
88
Balanced, pruney, herbal, spicy.

Stelvio 2018 T C
100% tempranillo
88
Fruity, pruney, spicy, dried herbs.

Stelvio Blanco 2021 B
70% malvasía, 30% sauvignon blanc
88
Pruney, fruity, dried flowers, tasty.

Todo va a salir Bien 2021 RD FB
100% tempranillo
89
Creamy, spicy, sweety finish, jammy, toasty, woody.

BODEGAS MEDRANO IRAZU

San Pedro, 14
01309 Elvillar (Araba/Álava)
info@medranoirazu.com
www.medranoirazu.com

Amador Medrano Colecciòn Privada "Finca las Aguzaderas" 2018 T
93
Colour: cherry, garnet rim. Nose: expressive, spicy, mineral, ripe fruit, black fruit. Palate: elegant, full-bodied, long, great length.

DO Ca. RIOJA / D.O.P.

DO Ca. RIOJA / D.O.P.

Amador Medrano Graciano "Finca Valdegamarra" 2019 T
graciano

92

Colour: Cherry. Nose: balsamic herbs, sweet spices, scrubland, grassy. Palate: spicy, balsamic, good acidity, varietal.

Amador Medrano Parcela 14.8 2017 T FB
100% tempranillo

93

Slight reduction. Colour: dark-red cherry, garnet rim. Nose: ripe fruit, aged wood nuances, tobacco, sweet spices, black fruit. Palate: spicy, round tannins, long.

Amador Medrano Rosé "Finca El Encinal" 2021 RD PL S
50% garnacha, 30% tempranillo, 20% viura

89

Pleasant, aromatic, standard, floral, fruity, sweety finish.

Amador Medrano Tempranillo Blanco "Fincas Valdegamarra" 2020 B FB
100% tempranillo blanco

92

Colour: bright straw. Nose: ripe fruit, fragrant herbs, fine lees, expressive, complex. Palate: full-bodied, rich, long, good acidity.

Amador Medrano Terra Finca El Encinal 2019 T FB

93

Colour: deep cherry. Nose: complex, expressive, spicy, mineral, ripe fruit. Palate: elegant, full-bodied, long, great length.

Medrano Irazu 2017 T R

89

Defined aromas, spicy, balanced, fruity, crisp, wild, smooth.

Medrano Irazu 2019 T C
100% tempranillo

89

Pleasant, fruity, simple, smooth, balanced, standard.

Medrano Irazu 2021 T
100% tempranillo

87

BODEGAS MONTECILLO

Ctra. Fuenmayor, Km. 3
26370 Navarrete (La Rioja)
☎: +34 941 440 125
info@bodegasmontecillo.com
www.bodegasmontecillo.com

Montecillo 2015 T R
90% tempranillo, 10% mazuelo

90

Colour: dark-red cherry. Nose: dried herbs, wild herbs, ripe fruit, tobacco. Palate: flavourful, spicy, easy to drink.

Montecillo 2018 T C
90% tempranillo, 10% garnacha

89

Spicy, jammy, fruity, standard, smooth, balanced.

Montecillo 22 Barricas 2015 T GR
50% tempranillo, 25% graciano, 15% garnacha, 10% mazuelo

91

Colour: cherry, garnet rim. Nose: fruit preserve, fruit liqueur notes, powerful, dried herbs. Palate: flavourful, long, round tannins.

Montecillo Edición Limitada 2016 T
75% tempranillo, 25% graciano

90

Colour: dark-red cherry, garnet rim. Nose: ripe fruit, aged wood nuances, tobacco, sweet spices, cigar. Palate: spicy, round tannins, long.

Viña Cumbrero 2015 T R
100% tempranillo

87

Viña Cumbrero 2018 T C
95% tempranillo, 5% garnacha

86

BODEGAS MUGA

Avda. Vizcaya, s/n
26200 Haro (La Rioja)
☎: +34 941 311 825
info@bodegasmuga.com
www.bodegasmuga.com

🏆 PODIUM

Aro 2019 T
tempranillo, graciano

97

Colour: very deep cherry. Nose: complex, expressive, spicy, mineral, ripe fruit, fruit expression. Palate: full-bodied, long, great length, soft tannins.

El Andén 2019 T
70% tempranillo, 30% garnacha

91

Colour: deep cherry. Nose: ripe fruit, dried herbs, creamy oak, toasty. Palate: powerful, ripe fruit, spicy, round tannins.

Flor de Muga 2019 B R
viura, garnacha blanca, maturana blanca

93

Colour: bright yellow. Nose: powerful, creamy oak, spicy, ripe fruit. Palate: good structure, long, toasty, fine bitter notes.

Flor de Muga Rosé 2021 RD

91

Colour: raspberry rose. Nose: red berry notes, floral, fragrant herbs. Palate: light-bodied, spicy, good acidity, fine bitter notes.

Muga 2018 T C
tempranillo, garnacha, mazuelo, graciano

91

Colour: deep cherry. Nose: ripe fruit, dried herbs, wild herbs, meaty notes. Palate: powerful, ripe fruit, spicy, round tannins.

Muga 2019 T C
tempranillo, garnacha, graciano, mazuelo

92

Colour: dark-red cherry, garnet rim. Nose: ripe fruit, fruit preserve, aged wood nuances, tobacco, sweet spices. Palate: spicy, round tannins, long, toasty.

Muga 2021 B

90

Colour: bright straw. Nose: ripe fruit, stone fruit, dried herbs, white flowers. Palate: flavourful, fruity, balanced.

Muga 2021 RD

91

Colour: raspberry rose. Nose: ripe fruit, red berry notes, floral. Palate: fruity, fresh, flavourful.

Muga Selección Especial 2018 T R
tempranillo, garnacha, graciano, mazuelo

94

Colour: Cherry. Nose: complex, expressive, spicy, mineral, ripe fruit. Palate: elegant, full-bodied, long, great length.

🏆 PODIUM

Prado Enea 2015 T GR
tempranillo, garnacha, mazuelo, graciano

95

Colour: dark-red cherry, garnet rim. Nose: ripe fruit, fruit preserve, aged wood nuances, tobacco, sweet spices. Palate: spicy, round tannins, long.

🏆 PODIUM

Torre Muga 2019 T
tempranillo, mazuelo, graciano

95

Colour: very deep cherry. Nose: complex, expressive, spicy, mineral, black fruit, ripe fruit. Palate: full-bodied, long, great length, grainy tannins.

BODEGAS MURO

Avda. Gasteiz, 29
01306 Lapuebla de Labarca (Araba/Álava)
☎: +34 627 434 726
info@bodegasmuro.es
www.bodegasmuro.es

Amenital 2018 T BA
80% tempranillo, 20% graciano

91

Colour: Cherry, garnet rim. Nose: ripe fruit, red berry notes, balanced, expressive, wild herbs. Palate: balanced, fresh.

Apolinar´s Dream 2017 T BA
90% tempranillo, 5% maturana, 4% graciano, 1% viura

90

Colour: Cherry. Nose: sweet spices, scrubland, dried herbs. Palate: spicy, balsamic, good acidity.

Muro 2016 T R
90% tempranillo, 10% graciano

90

Colour: cherry, garnet rim. Nose: ripe fruit, scrubland, dried herbs, creamy oak, toasty. Palate: flavourful, spicy.

DO Ca. RIOJA / D.O.P.

DO Ca. RIOJA / D.O.P.

Muro 2018 T C
100% tempranillo

89

Pleasant, fruity, jammy, balsamic herbs, herbal, standard.

Retorno a los Palomares 2017 T BA
70% tempranillo, 30% graciano

91

Colour: dark-red cherry. Nose: balsamic herbs, sweet spices, scrubland, spicy. Palate: flavourful, ripe fruit, long.

BODEGAS MURUA

Ctra. Laguardia s/n
01340 Elciego (Araba/Álava)
☎: +34 945 606 260
info@bodegasmurua.masaveu.com
www.bodegasmurua.com

M de Murua 2019 T
100% tempranillo

91

Colour: deep cherry. Nose: dried herbs, creamy oak, black fruit. Palate: ripe fruit, spicy, round tannins.

Murua Blanco Fermentado en Barrica 2020 B FB
65% viura, 17% malvasía, 18% garnacha blanca

91

Colour: straw. Nose: ripe fruit, dried herbs, faded flowers, toasty. Palate: powerful, ripe fruit, balanced.

Murua Reserva 2015 T R
92% tempranillo, 8% graciano, mazuelo

92

Colour: deep cherry. Nose: ripe fruit, creamy oak, scrubland, aromatic coffee, sweet spices, fine reductive notes. Palate: powerful, ripe fruit, spicy, round tannins.

VS Murua 2019 T
90% tempranillo, 10% mazuelo

92

Colour: cherry, purple rim. Nose: red berry notes, floral, spicy, aromatic coffee. Palate: flavourful, good acidity, fleshy.

BODEGAS NAJERILLA, S.COOP.

Ctra. De Anguiano, s/n
26311 Arenzana de Abajo (La Rioja)
☎: +34 941 362 783
Fax: +34 941 410 369
najerilla@bodegasnajerilla.es
www.bodeganajerilla.es

Castezo Garnacha 2018 T

88

Balanced, spicy, herbal, jammy.

Castezo Tempranillo Blanco 2021 B

87

Cepa Negra 2015 T

89

Age nuances, balanced, spicy, jammy, tasty, toasty.

Cepa Negra Centenarium 2006 T R
95% tempranillo, 5% garnacha

88

Classic, weary. Colour: pale ruby, brick rim edge. Nose: old leather, fruit liqueur notes, cigar, spicy. Palate: light-bodied, classic aged character, bitter, reductive nuances.

Pura Cepa de Sarmiento 2021 T

89

Balanced, spicy, herbaceous, fruity, tasty.

Pura Cepa de Sarmiento Garnacha 2018 T C

89

Creamy, spicy, dried herbs, lactic, full-bodied.

BODEGAS NAVAJAS

Camino Balgarauz, 2
26370 Navarrete (La Rioja)
☎: +34 941 440 140
info@bodegasnavajas.com
www.bodegasnavajas.com

Cerro Los Curas Viñedo Singular 2019 B
100% viura

93

Colour: bright straw. Nose: ripe fruit, stone fruit, wild herbs, cocoa bean, sweet spices. Palate: fruity, flavourful, powerful, balanced, roasted-coffee aftertaste.

Navajas 2017 T C
100% tempranillo

87

Navajas 2018 B C
60% viura, 20% sauvignon blanc, 20% tempranillo blanco

88

Smoky, jammy, tasty, toasty.

Navajas Garnacha 2020 T
100% garnacha

86

BODEGAS NESTARES EGUIZÁBAL

26144 Galilea (La Rioja)
☎: +34 699 253 771
Fax: +34 941 480 351
bodegas@nestareseguizabal.com
www.nestareseguizabal.com

A Veredas 2018 B
sauvignon blanc, chardonnay

89

Spicy, tasty, herbal, roasted coffee, woody.

A Veredas 2019 RD
tempranillo, chardonnay

92

Colour: onion pink. Nose: fruit expression, wild herbs, red berry notes, wild herbs, dried flowers. Palate: fruity, flavourful, fresh, balanced.

Segares 2008 T GR
tempranillo

88

Smoky, toasty, jammy, spicy.

Segares 2010 T R
tempranillo

87

Segares 2015 T C
tempranillo

87

BODEGAS NIVARIUS

Ctra. de Nalda a Viguera, 46
26190 Nalda (La Rioja)
☎: +34 941 447 207
info@palaciosvinosdefinca.com
www.palaciosvinosdefinca.com

Nivarius 2021 B
tempranillo blanco

92

Colour: bright straw, greenish rim. Nose: fresh fruit, citrus fruit, wild herbs. Palate: fresh, fruity, good acidity, fine bitter notes.

Nivarius Edición Limitada 2018 B
viura, maturana blanca

93

Colour: bright straw. Nose: ripe fruit, fragrant herbs, fine lees, scrubland. Palate: full-bodied, rich, good acidity, powerful, flavourful.

🏆 **PODIUM**

Nivarius Finca La Nevera 2017 B
100% maturana blanca

95

Colour: bright yellow. Nose: powerful, creamy oak, ripe fruit, spicy. Palate: rich, good structure, long, toasty, fine bitter notes.

Valdesabril 2019 B
viura

94

Colour: bright straw. Nose: expressive, characterful, complex, fine lees, bakery. Palate: flavourful, fruity, rich, long, spicy.

BODEGAS OBALO

Ctra. N-232 A, Km. 26
26339 Ábalos (La Rioja)
☎: +34 941 744 056
esanz@terraselecta.com
www.bodegaobalo.com

Obalo 2021 RD
100% tempranillo

90

Austere. Colour: raspberry rose. Nose: red berry notes, fragrant herbs, citrus fruit. Palate: light-bodied, spicy, good acidity, fine bitter notes, elegant.

Obalo Las Arenas 2017 T R
100% tempranillo

92

Full-bodied, exuberant. Colour: cherry, garnet rim. Nose: ripe fruit, creamy oak, sweet spices. Palate: long, round tannins, ripe fruit, spicy.

DO Ca. RIOJA / D.O.P.

DO Ca. RIOJA / D.O.P.

Obalo San Roque 2020 T
100% tempranillo
89
Aromatic, fruity, spicy, standard, varietally correct, simple, balanced.

BODEGAS OLARRA
Avda. de Mendavia, 30
26009 Logroño (La Rioja)
☎: +34 941 235 299
Fax: +34 941 253 703
bodegasolarra@bodegasolarra.es
www.bodegasolarra.es

Cerro Añón 2015 T GR
85% tempranillo, 15% mazuelo, graciano
90
Colour: bright cherry. Nose: fruit preserve, old leather, meaty notes, dried herbs. Palate: balanced, flavourful, fruity, slightly dry, soft tannins.

Cerro Añón 2019 T C
90% tempranillo, 10% mazuelo, graciano
88
Full-bodied, jammy, toasty. Nose: woody.

El Rayo 2019 T C
85% tempranillo, 15% mazuelo, graciano
89
Balanced, spicy, fruity, jammy, tasty.

Olarra 2011 T GR
tempranillo, mazuelo, garnacha
91
Colour: cherry, garnet rim. Nose: fruit liqueur notes, red berry notes, meaty notes, wet leather, spicy, dried herbs. Palate: fruity, flavourful, balanced, slightly dry, soft tannins.

Otoñal T C
85

BODEGAS OLMAZA
Camino San Martín, s/n
26338 San Vicente de la Sonsierra (La Rioja)
☎: +34 647 682 032
info@bodegasolmaza.com
www.bodegasolmaza.com

Olmaza Autor 2019 T C
100% tempranillo
87

Osluna 2019 T C
100% tempranillo
88
Balanced, spicy, creamy, full-bodied, tasty.

Viña Olmaza 2021 B
65% viura, 20% tempranillo blanco, 10% garnacha blanca, 5% malvasía
87

Viña Olmaza 2021 T
40% tempranillo, 50% garnacha, 10% viura
88
Fruity, herbal, simple.

BODEGAS ONDALÁN
Ctra. de Logroño, 22
01230 Oyón (Araba/Álava)
☎: +34 945 622 537
Fax: +34 945 622 538
ondalan@ondalan.es
www.ondalan.es

Ondalán 100 Abades Graciano 2016 T
100% graciano
90
Colour: deep cherry. Nose: fruit preserve, fine reductive notes, wild herbs. Palate: ripe fruit, spicy, round tannins.

Ondalán 2016 T R
50% tempranillo, 50% graciano
90
Colour: bright cherry. Nose: red berry notes, fruit preserve, wild herbs, spicy, faded flowers. Palate: flavourful, powerful, balanced, slightly dry, soft tannins.

Ondalán 2019 T C
70% tempranillo, 30% graciano
86

Ondalán 2021 B
35% malvasía, 30% viura, 15% sauvignon blanc, 10% garnacha blanca, 10% tempranillo blanco
87

Ondalán Tempranillo Selección 2018 T
100% tempranillo
88
Balanced, spicy, jammy, toasty.

BODEGAS ONDARRE
Ctra. de Aras, s/n
31230 Viana (Navarra)
☎: +34 948 645 300
bodegasondarre@bodegasondarre.es
www.bodegasondarre.es

Mayor de Ondarre 2016 T R
90% tempranillo, 5% mazuelo, 5% otras
91
Colour: deep cherry. Nose: creamy oak, spicy, ripe fruit, dried herbs, dry nuts. Palate: flavourful, powerful, fresh, fruity, balanced, slightly dry, soft tannins.

Valdebarón 2019 T
tempranillo, mazuelo

90

Colour: bright cherry. Nose: red berry notes, ripe fruit, dried herbs. Palate: fruity, fresh, flavourful, balanced, slightly dry, soft tannins.

Valdebarón 2020 B
tempranillo blanco

91

Colour: bright yellow. Nose: creamy oak, ripe fruit, spicy. Palate: toasty, fine bitter notes, ripe fruit, long.

BODEGAS ORBEN

Ctra. Laguardia, Km. 60
01300 Laguardia (Araba/Álava)
☎: +34 945 609 086
Fax: +34 945 609 261
izadi@izadi.com
www.artevino.es/bodegas/orben

🏆 PODIUM

Malpuesto 2020 T
tempranillo

95

Colour: very deep cherry. Nose: complex, expressive, spicy, mineral, ripe fruit, black fruit. Palate: full-bodied, long, great length.

Orben 2020 T
tempranillo

94

Colour: deep cherry. Nose: dried herbs, creamy oak, ripe fruit, black fruit, toasty. Palate: powerful, ripe fruit, spicy, round tannins.

BODEGAS PALACIOS REMONDO

Avda. Zaragoza, 8
26540 Alfaro (La Rioja)
☎: +34 941 180 207
info@palaciosremondo.com
www.alvaropalacios.com

La Montesa 2020 T C
91% garnacha, 9% otras

93 🌿

Colour: bright cherry, garnet rim. Nose: wild herbs, grassy, fruit expression, red berry notes, floral, violets. Palate: fruity, easy to drink, taut, grainy tannins.

Plácet de Valtomelloso 2021 B
100% viura

94 🌿

Colour: bright straw. Nose: white fruit, balanced, expressive, fresh, spicy, floral. Palate: good structure, rich, flavourful, long.

🏆 PODIUM

Propiedad 2020 T
92% garnacha, 8% otras

96 🌿

Colour: bright cherry, cherry, garnet rim. Nose: red berry notes, ripe fruit, expressive, wild herbs, waxy notes. Palate: complex, good structure, flavourful, elegant.

🏆 PODIUM

Quiñón de Valmira 2020 T FB
90% garnacha, 10% otras

98 🌿

Aromatic, balsamic herbs. Colour: bright cherry. Nose: fresh fruit, red berry notes, earthy notes, floral, spicy, chalk. Palate: flavourful, fine tannins, fresh, taut, good acidity.

BODEGAS PROELIO

Camino Nalda a Viguera, 46
26190 Nalda (La Rioja)
☎: +34 941 447 207
info@palaciosvinosdefinca.com
www.palaciosvinosdefinca.com/proelio/

Proelio 2019 T C
tempranillo, garnacha, maturana

92

Colour: cherry, purple rim. Nose: spicy, black fruit, red berry notes, toasted bread. Palate: flavourful, fruity, good acidity.

🏆 PODIUM

Proelio La Canal del Rojo 2018 T
garnacha

95

Colour: Cherry. Nose: complex, expressive, spicy, mineral, wild herbs, floral, black pepper. Palate: elegant, full-bodied, long, good acidity.

Proelio Vendimia Seleccionada 2018 T R
tempranillo, garnacha, graciano

92

Colour: deep cherry. Nose: ripe fruit, dried herbs, creamy oak, toasty. Palate: powerful, ripe fruit, spicy, round tannins.

Proelio Viñedos Viejos 2019 T
garnacha

93

Colour: cherry, purple rim. Nose: fruit expression, red berry notes, floral, spicy. Palate: flavourful, fruity, good acidity.

DO Ca. RIOJA / D.O.P.

🏆 PODIUM

Puerto Rubio 2018 T
tempranillo

95

Colour: cherry, garnet rim. Nose: balsamic herbs, ripe fruit, scrubland, wild herbs, expressive. Palate: balsamic, spicy, taut, fruity, balanced.

BODEGAS R & G
Ramón y Cajal 7, 1ºA
01007 Vitoria-Gasteiz (Araba/Álava)
☎: +34 945 150 589
araex@araex.com
www.araex.com

R&G 2016 T
100% tempranillo

93

Colour: dark-red cherry, garnet rim. Nose: ripe fruit, aged wood nuances, tobacco, sweet spices, toasty, creamy oak. Palate: spicy, round tannins, long.

BODEGAS RAMÍREZ DE LA PISCINA
Ctra. Vitoria Laguardia s/n
26338 San Vicente de la Sonsierra (La Rioja)
☎: +34 665 746 500
Fax: +34 941 334 506
sonia@ramirezdelapiscina.com
www.ramirezdelapiscina.com

Selección Ramírez de la Piscina 2019 T C
100% tempranillo

92

Colour: very deep cherry. Nose: roasted coffee, aromatic coffee, powerful, black fruit. Palate: smoky aftertaste, great length, round tannins.

Ramírez de la Piscina 2020 B FB
70% viura, 30% chardonnay

89

Dried flowers, fruity, herbal, jammy.

BODEGAS RAMÓN BILBAO
Avda. Santo Domingo, 34
26200 Haro (La Rioja)
☎: +34 941 310 295
info@bodegasramonbilbao.es
www.bodegasramonbilbao.es

Mirto de Ramón Bilbao 2016 T

93

Colour: very deep cherry. Nose: roasted coffee, aromatic coffee, powerful, black fruit. Palate: smoky aftertaste, great length, round tannins.

Ramón Bilbao 2014 T GR

94

Kind finish, classic. Colour: deep cherry, garnet rim. Nose: aged wood nuances, ripe fruit, cocoa bean, cigar, toasty. Palate: flavourful, spicy, toasty, powerful tannins.

Ramón Bilbao 2016 T R

92

Colour: cherry, purple rim. Nose: red berry notes, floral, sweet spices, dark chocolate. Palate: flavourful, fruity, good acidity.

Ramón Bilbao 2021 RD

89

Defined aromas, kind finish, crisp, fruity, simple, balanced.

Ramón Bilbao Edición Limitada 2020 T

91

Colour: deep cherry. Nose: dried herbs, creamy oak, black fruit. Palate: ripe fruit, spicy, round tannins.

Ramón Bilbao Límite Norte 2019 B
91
Colour: bright straw, greenish rim. Nose: fresh fruit, citrus fruit, wild herbs. Palate: fresh, fruity, good acidity, fine bitter notes.

Ramón Bilbao 2020 T C
91
Colour: deep cherry. Nose: dried herbs, creamy oak, black fruit, cocoa bean. Palate: ripe fruit, spicy, round tannins.

Ramón Bilbao Viñedos de Altura 2020 T C
91
Colour: deep cherry. Nose: dried herbs, creamy oak, black fruit, roasted coffee. Palate: powerful, ripe fruit, spicy.

Ramón Bilbao Límite Sur 2019 T
91
Colour: cherry, purple rim. Nose: fruit expression, red berry notes, sweet spices. Palate: flavourful, fruity, good acidity.

Ramón Bilbao Organic 2020 T
87

Ramón Bilbao Reserva Original 2016 T
92
Colour: deep cherry. Nose: dried herbs, creamy oak, black fruit. Palate: ripe fruit, spicy, round tannins.

BODEGAS REMÍREZ DE GANUZA
Constitución, 1
01307 Samaniego (Araba/Álava)
☎: +34 945 609 022
bodegas@remirezdeganuza.com
www.remirezdeganuza.com

🏆 **PODIUM**

Remírez de Ganuza 2015 T R
88% tempranillo, 10% graciano, 2% otras
95
Colour: deep cherry. Nose: dried herbs, creamy oak, black fruit, ripe fruit, roasted coffee, woody. Palate: powerful, ripe fruit, spicy, round tannins.

BODEGAS RIOJANAS
Avda. Ricardo Ruiz Azcarraga, 1
26350 Cenicero (La Rioja)
☎: +34 941 454 050
Fax: +34 941 454 529
bodega@bodegasriojanas.com
www.bodegasriojanas.com

Monte Real 2015 T GR
100% tempranillo
92
Colour: dark-red cherry, garnet rim. Nose: ripe fruit, aged wood nuances, tobacco, sweet spices. Palate: spicy, round tannins, long.

DO Ca. RIOJA / D.O.P.

DO Ca. RIOJA / D.O.P.

Monte Real Crianza Familia 2020 T C
100% tempranillo

90

Colour: cherry, garnet rim. Nose: fruit preserve, powerful, characterful, sweet spices. Palate: flavourful, sweetness, long.

Monte Real Cuvée 2020 T C
100% garnacha

92

Colour: Cherry. Nose: expressive, spicy, mineral, ripe fruit, black fruit. Palate: elegant, full-bodied, long, great length.

Monte Real Garnacha 2021 T
100% garnacha

91

Colour: cherry, purple rim. Nose: red berry notes, fruit expression, floral. Palate: fruity, flavourful, powerful, fresh.

Monte Real Reserva de Familia 2019 T R
100% tempranillo

92

Colour: cherry, garnet rim. Nose: fruit preserve, fruit liqueur notes, powerful, sweet spices, black fruit. Palate: flavourful, sweetness, long.

Monte Real Tempranillo Blanco 2021 B
tempranillo blanco

91

Colour: bright straw. Nose: stone fruit, spicy, expressive. Palate: correct, ripe fruit, easy to drink, spicy.

Viña Albina Vendimia Seleccionada 2019 T R
80% tempranillo, 15% mazuelo, 5% graciano

89

Not representative. Colour: cherry, garnet rim. Nose: powerful, fruit expression, black fruit. Palate: flavourful, sweetness.

BODEGAS RODA
Avda. de Vizcaya, 5
26200 Haro (La Rioja)
☎: +34 941 303 001
rodarioja@roda.es
www.roda.es

🏆 PODIUM

Cirsion 2019 T R
88% tempranillo, 12% graciano

96

Colour: very deep cherry. Nose: roasted coffee, aromatic coffee, powerful, red berry notes, ripe fruit. Palate: smoky aftertaste, great length, round tannins.

Roda 2019 T R
91% tempranillo, 5% graciano, 4% garnacha

93

Colour: bright cherry. Nose: fruit expression, red berry notes, floral, spicy. Palate: flavourful, fruity, good acidity, long.

Roda I 2018 T R
92% tempranillo, 6% graciano, 2% garnacha

94

Colour: dark-red cherry, garnet rim. Nose: black fruit, dried herbs, spicy, characterful. Palate: good structure, flavourful, spicy, long.

🏆 PODIUM

Roda I 2019 B
viura, malvasía, garnacha blanca

95

Colour: bright straw. Nose: expressive, ripe fruit, floral, fine lees, mineral, dried flowers, spicy. Palate: full-bodied, long, complex, rich.

BODEGAS RODRÍGUEZ Y SANZO
Manuel Azaña, 9
47014 Valladolid (Valladolid)
☎: +34 983 150 150
comunicacion@valsanzo.com
www.rodriguezsanzo.com

La Senoba 2018 T C
70% tempranillo, 30% graciano

91

Colour: bright cherry. Nose: ripe fruit, red berry notes, black fruit, creamy oak, spicy, dried herbs. Palate: fruity, fresh, flavourful, good structure, slightly dry, soft tannins.

Lacrimus Crianza 2019 T C
tempranillo
89
Fruity, spicy, dried herbs, tasty.

Lacrimus Reserva 2018 T R
85% tempranillo, 10% graciano, 3% garnacha, 2% maturana
89
Fruity, spicy, herbal, tasty.

Lacrimus Rex 2020 T
75% garnacha, 25% graciano
90
Colour: cherry, purple rim. Nose: red berry notes, ripe fruit, wild herbs, spicy. Palate: fruity, flavourful, fresh, balanced, slightly dry, soft tannins.

BODEGAS SAN ESTEBAN
Ctra. Agoncillo s/n
26143 Murillo de Río Leza (La Rioja)
☎: +34 941 432 031
administracion@bodegassanesteban.com
www.bodegassanesteban.com

Kairos de San Esteban 2019 T
100% tempranillo
92
Colour: deep cherry. Nose: dried herbs, creamy oak, red berry notes, black fruit, ripe fruit. Palate: ripe fruit, spicy, round tannins.

Kairos de San Esteban 2020 RD
100% tempranillo
90
Colour: salmon. Nose: sweet spices, red berry notes, fragrant herbs, woody. Palate: full-bodied, flavourful, spicy.

Kairos de San Esteban Fermentado con sus Pieles 2020 B
100% sauvignon blanc
90
Colour: bright yellow. Nose: candied fruit, fine lees, pattiserie. Palate: spicy, great length.

Tierras de Murillo 2018 T C
100% tempranillo
88
Dried herbs, jammy, tasty, spicy.

Tierras de Murillo 2019 B FB
100% viura
91
Colour: bright straw. Nose: stone fruit, creamy oak, sweet spices. Palate: flavourful, fruity, balanced.

Tierras de Murillo Colección Privada 2019 T
100% tempranillo
91
Colour: cherry, purple rim. Nose: red berry notes, floral, spicy. Palate: flavourful, fruity, good acidity, long.

BODEGAS SANTALBA
Avda. de la Rioja, s/n
26221 Gimileo (La Rioja)
☎: +34 941 304 231
santalba@santalba.com
www.santalba.com

Ogga 2016 T R
90% tempranillo, 10% otras
91
Colour: dark-red cherry, garnet rim. Nose: ripe fruit, aged wood nuances, tobacco, sweet spices. Palate: spicy, round tannins, long.

Santalba Amaro 2017 T
100% tempranillo
92
Colour: cherry, garnet rim. Nose: fruit preserve, fruit liqueur notes, powerful, toasty, spicy. Palate: flavourful, sweetness, long.

Santalba Viña Hermosa 2015 T GR
100% tempranillo
92
Colour: dark-red cherry, garnet rim. Nose: ripe fruit, aged wood nuances, tobacco, sweet spices. Palate: spicy, round tannins, long.

BODEGAS SOLANA DE RAMIREZ RUIZ
Arana, 24
26339 Ábalos (La Rioja)
☎: +34 941 308 049
consultas@solanaderamirez.com
www.valsarte.com

Capricho de Valvanera 2021 B
viura, malvasía, sauvignon blanc, verdejo
86

Solana de Ramírez 2021 B
viura, malvasía, sauvignon blanc, verdejo
86

Solana de Ramírez 2021 RD
80% tempranillo, 20% garnacha
87

DO Ca. RIOJA / D.O.P.

DO Ca. RIOJA / D.O.P.

Pezón Negro 2019 T
100% garnacha

92

Colour: bright cherry. Nose: black fruit, fruit expression, red berry notes, dried herbs, spicy, woody. Palate: fruity, powerful, good structure, flavourful, balanced, long, round tannins.

Valsarte 2016 T R
91

Colour: bright cherry. Nose: ripe fruit, creamy oak, spicy, dried herbs. Palate: fruity, flavourful, good structure, balanced, slightly dry, soft tannins.

Valsarte 2019 T C
90% tempranillo, 10% graciano

90

Colour: deep cherry. Nose: red berry notes, ripe fruit, dried herbs, spicy, expressive. Palate: fruity, flavourful, good structure, balanced, slightly dry, soft tannins.

BODEGAS SOLAR VIEJO
Camino de la Hoya, s/n
01300 Laguardia (Araba/Álava)
☎: +34 945 600 113
solarviejo@solarviejo.com
www.solarviejo.com

Orube 2018 T R
50% tempranillo, 50% graciano

90

Creamy. Colour: deep cherry. Nose: dried herbs, creamy oak, black fruit. Palate: powerful, ripe fruit, spicy, round tannins.

Orube 2020 B FB
50% chardonnay, 50% tempranillo blanco

88

Jammy, toasty, tasty, powerful.

Orube 2021 RD
70% garnacha, 30% tempranillo

89

Dried flowers, dried herbs, fruity, tasty.

Orube Selección de Familia 2019 T C
100% tempranillo

91

Colour: deep cherry. Nose: dried herbs, creamy oak, black fruit, red berry notes. Palate: powerful, ripe fruit, spicy, round tannins.

Solar Viejo 2018 T C
100% tempranillo

87

Solar Viejo 2018 T R
70% tempranillo, 30% graciano

88

Balanced, herbaceous, fruity, jammy, toasty.

BODEGAS SONSIERRA
Paseo de Logroño, 3
26338 San Vicente de la Sonsierra (La Rioja)
☎: +34 941 334 031
sonsierra@sonsierra.com
www.sonsierra.com

Avior 2017 T R
89

Balanced, spicy, jammy, dried herbs, tasty.

Avior 2019 T C
100% tempranillo

88

Fruity, herbal, spicy, tasty.

Pagos de la Sonsierra 2015 T R
100% tempranillo

91

Colour: cherry, purple rim. Nose: black fruit, spicy, smoky, dried herbs. Palate: fruity, flavourful, good structure, slightly dry, soft tannins, balanced.

Perfume de Sonsierra 2015 T
100% tempranillo

90

Colour: cherry, purple rim. Nose: red berry notes, ripe fruit, meaty notes, old leather, sweet spices, dried herbs. Palate: fruity, flavourful, balanced, powerful tannins.

Sonsierra 2015 T GR
100% tempranillo

88

Classic, spicy, fruity, herbal, tasty.

Sonsierra 2017 T R
100% tempranillo

90

Colour: cherry, purple rim. Nose: black fruit, spicy, creamy oak, dried herbs. Palate: fruity, fresh, flavourful, balanced.

Sonsierra 2019 T C
100% tempranillo

88

Fruity, spicy, herbal, wild.

BODEGAS TARÓN

Ctra. de Miranda, s/n
26211 Tirgo (La Rioja)
☎: +34 941 301 650
Fax: +34 941 301 817
info@bodegastaron.com
www.bodegastaron.com

Tarón 2018 T C
95% tempranillo, 5% mazuelo

88

Full-bodied, creamy, spicy, full-bodied, jammy. Nose: woody.

Tarón 4M 2019 T
100% tempranillo

87

Pantocrator 2011 T R
100% tempranillo

91

Colour: deep cherry. Nose: dried herbs, creamy oak, woody, stone fruit, black fruit. Palate: ripe fruit, spicy, round tannins.

Patiens 2016 B R
100% viura

92

Colour: bright yellow. Nose: ripe fruit, dried herbs, faded flowers. Palate: powerful, ripe fruit, balanced.

Tarón 2015 T R
90% tempranillo, 10% mazuelo

90

Colour: deep cherry. Nose: ripe fruit, dried herbs, creamy oak, woody. Palate: powerful, ripe fruit, spicy, round tannins.

Tarón Tempranillo Blanco 2020 B C
100% tempranillo blanco

88

Fruity, herbal, full-bodied, tasty.

BODEGAS TIERRA

El Olmo, 16
01330 Labastida (Araba/Álava)
☎: +34 945 331 257
enoturismo@tierrayvino.com
www.tierrayvino.com

El Belisario 2019 T
92

Colour: cherry, garnet rim. Nose: smoky, spicy, toasty, ripe fruit. Palate: flavourful, fruity, powerful, round tannins.

Fernández Gómez 2021 T
88

Pleasant, fruity, jammy, dried herbs, simple.

DO Ca. RIOJA / D.O.P.

DO Ca. RIOJA / D.O.P.

La Abuela Visi 2020 B BA
91
Colour: bright straw. Nose: ripe fruit, fragrant herbs, fine lees, sweet spices. Palate: full-bodied, rich, long, good acidity.

Tierra 2019 T C
88
Pleasant, toasty, spicy, jammy.

Tierra 2021 B FB
87

Tierra de Fidel 2017 B FB
92
Colour: coppery red. Nose: scrubland, dried herbs, rose petals, red berry notes. Palate: flavourful, powerful, round.

BODEGAS TOBÍA – ÓSCAR TOBÍA
Paraje Senda Rutia, s/n
26214 Cuzcurrita de Rio Tirón (La Rioja)
☎: +34 941 301 789
tobia@bodegastobia.com
www.bodegastobia.com

Tobía Luz de Luna 2017 T R
100% tempranillo
91
Colour: deep cherry. Nose: ripe fruit, dried herbs, creamy oak, scrubland. Palate: powerful, ripe fruit, spicy, round tannins.

Tobía Luz de Luna 2019 T C
100% tempranillo
89
Spicy, jammy, tasty.

Tobía Selección de Autor 2019 T BA
77% tempranillo, 7% graciano, 6% garnacha, 10% otras
91
Colour: dark-red cherry, garnet rim. Nose: ripe fruit, aged wood nuances, tobacco, sweet spices, red berry notes. Palate: spicy, round tannins, long.

Tobía Selección de Autor 2020 B
50% chardonnay, 50% tempranillo blanco
92
Colour: bright yellow. Nose: powerful, creamy oak, ripe fruit, spicy. Palate: rich, good structure, long, toasty, fine bitter notes.

BODEGAS UBIDE
Ctra. Rio Hondillo, s/n
01300 Laguardia (Araba/Álava)
☎: +34 945 621 220
bodegasubide@bodegasubide.com
www.bodegasubide.com

Ubide Mar 2021 B
viura, malvasía
85

Ubide Reserva de la Familia 2012 T R
tempranillo
90
Colour: very deep cherry. Nose: roasted coffee, aromatic coffee, powerful, ripe fruit. Palate: smoky aftertaste, great length, round tannins.

Ubide Tierra 2017 T C
tempranillo
88
Toasty, jammy, powerful.

BODEGAS VALDELACIERVA
Ctra. Burgos, Km. 13
26370 Navarrete (La Rioja)
☎: +34 941 440 620
info@hispanobodegas.com
www.hispanobodegas.com

Valdelacierva 2017 T R
tempranillo
92
Colour: dark-red cherry, garnet rim. Nose: ripe fruit, aged wood nuances, sweet spices. Palate: spicy, round tannins.

Valdelacierva 2019 T C
tempranillo
90
Colour: cherry, purple rim. Nose: red berry notes, floral, sweet spices. Palate: flavourful, fruity, good acidity, long.

Valdelacierva Cantogordo 2018 T
tempranillo
93
Colour: very deep cherry. Nose: complex, expressive, spicy, mineral, black fruit. Palate: elegant, full-bodied, long, great length.

Valdelacierva Garnacha 2019 T
garnacha
92
Colour: deep cherry. Nose: ripe fruit, dried herbs, creamy oak, spicy. Palate: powerful, ripe fruit, spicy, round tannins.

Valdelacierva Grano a Grano 2018 T
tempranillo

94

Colour: dark-red cherry, garnet rim. Nose: ripe fruit, fruit preserve, aged wood nuances, tobacco, sweet spices. Palate: spicy, round tannins, long.

BODEGAS VALDELANA
Puente Barricuelo, 67-69
01340 Elciego (Araba/Álava)
☎: +34 653 298 659
export@bodegasvaldelana.com
www.bodegasvaldelana.com

Agnus de Valdelana Crianza 2019 T C
95% tempranillo, 5% graciano

90

Colour: deep cherry. Nose: fruit liqueur notes, red berry notes, fragrant herbs, spicy, smoky. Palate: flavourful, fruity, balanced, slightly dry, soft tannins.

Ladrón de Guevara de Autor 2019 T C
95% tempranillo, 5% graciano

91

Colour: deep cherry. Nose: ripe fruit, dried herbs, creamy oak, sweet spices, fine reductive notes. Palate: powerful, ripe fruit, spicy, round tannins.

Palador 2017 T R
95% tempranillo, 5% graciano

91

Colour: dark-red cherry, garnet rim. Nose: ripe fruit, aged wood nuances, tobacco, sweet spices. Palate: spicy, round tannins, long.

Palador 2019 T C
95% tempranillo, 5% mazuelo

90

Colour: deep cherry. Nose: ripe fruit, dried herbs, creamy oak. Palate: powerful, ripe fruit, spicy, round tannins.

BODEGAS VALDEMAR
Camino Viejo de Logroño, 24
01320 Oyón (Araba/Álava)
☎: +34 945 622 188
Fax: +34 945 622 111
adedios@valdemar.es
www.momentosvaldemar.com

Balcón de Pilatos by Valdemar 2019 T
100% maturana

93

Colour: cherry, purple rim. Nose: ripe fruit, wild herbs, dried herbs, sweet spices, roasted coffee, woody, black fruit. Palate: flavourful, fruity, balanced, slightly dry, soft tannins.

Conde de Valdemar 2018 T C
90% tempranillo, 5% garnacha, 5% mazuelo

90

Colour: deep cherry. Nose: dried herbs, creamy oak, black fruit, sweet spices. Palate: ripe fruit, spicy, round tannins.

Conde Valdemar Edición Limitada 2018 T
80% tempranillo, 10% maturana, 10% graciano

92

Colour: deep cherry. Nose: ripe fruit, dried herbs, creamy oak, scrubland. Palate: powerful, ripe fruit, spicy, round tannins.

Conde Valdemar Finca Alto Cantabria Blanco Viñedo Singular 2020 B FB
100% viura

92

Colour: bright yellow. Nose: candied fruit, fine lees, pattiserie, new oak. Palate: round, spicy, long, great length.

Conde Valdemar Finca Alto Cantabria Gran Añada Viñedo Singular Espumoso 2017 B

92

Colour: yellow. Nose: fine lees, dry nuts, toasted bread, spicy. Palate: flavourful, good acidity, fine bitter notes.

La Gargantilla Garnacha 2019 T
100% garnacha

93

Colour: cherry, purple rim. Nose: fruit expression, red berry notes, floral, spicy. Palate: flavourful, fruity, good acidity, long, fleshy.

BODEGAS VALLOBERA
Camino de la Hoya, 5
01300 Laguardia (Araba/Álava)
☎: +34 699 357 207
enologia@vallobera.com
www.vallobera.com

Caudalia de Vallobera 2006 B FB
100% viura

92

Colour: bright yellow. Nose: powerful, ripe fruit, spicy, toasted bread, smoky. Palate: rich, good structure, long, toasty, fine bitter notes.

Finca Vallobera 2019 T
100% tempranillo

91

Colour: cherry, garnet rim. Nose: ripe fruit, scrubland, earthy notes, sweet spices. Palate: flavourful, spicy, soft tannins.

DO Ca. RIOJA / D.O.P.

DO Ca. RIOJA / D.O.P.

Pensando en Tí 2020 B
100% garnacha blanca

91

Colour: bright yellow. Nose: creamy oak, ripe fruit, spicy, wild herbs, chamomile. Palate: rich, good structure, toasty, fine bitter notes.

Vallobera 2019 T C
100% tempranillo

88

Smoky, spicy, jammy, tasty, standard.

Vallobera Tempranillo 2021 T MC
100% tempranillo

88

Balanced, spicy, fruity, herbal.

BODEGAS VALORIA
Ctra. de Burgos, Km. 5
26006 Logroño (La Rioja)
☎: +34 941 204 059
jose.iraizoz@bvaloria.com
www.bvaloria.com

Viña Valoria 2011 T GR
tempranillo

90

Colour: bright cherry. Nose: ripe fruit, red berry notes, grassy, spicy, smoky, old leather. Palate: fruity, flavourful, powerful, fresh, good finish.

Viña Valoria 2014 T R
tempranillo

90

Colour: bright cherry. Nose: ripe fruit, red berry notes, sweet spices, grassy. Palate: fruity, flavourful, slightly dry, soft tannins, balanced.

Viña Valoria 2019 T C
tempranillo

89

Creamy, spicy, fruity, jammy, tasty.

Viña Valoria 2021 T
tempranillo

87

Viña Valoria Garnacha 2021 RD
garnacha

85

Viña Valoria Tempranillo Blanco 2021 B
tempranillo blanco

87

BODEGAS VINÍCOLA REAL
Ctra. Nalda, km. 9
26120 Albelda de Iregua (La Rioja)
☎: +34 941 444 233
info@vinicolareal.com
www.vinicolareal.com

200 Monges 2007 T GR
85% tempranillo, 10% graciano, 5% garnacha

92

Colour: dark-red cherry, garnet rim. Nose: aged wood nuances, tobacco, sweet spices, ripe fruit, black fruit. Palate: spicy, round tannins, long.

200 Monges 2009 B GR
viura

94

Colour: bright straw. Nose: expressive, ripe fruit, floral, fine lees, mineral, toasted bread. Palate: full-bodied, spicy, long, elegant.

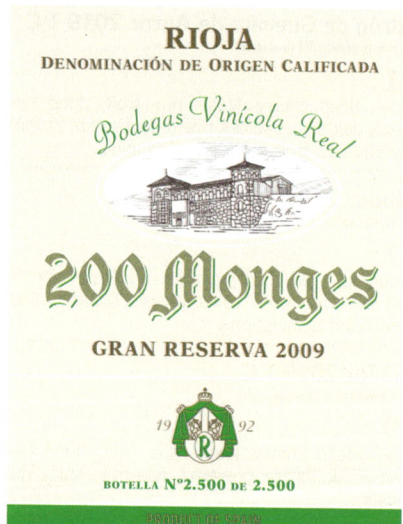

200 Monges 2011 B R
viura, malvasía

92

Colour: bright straw. Nose: stone fruit, sweet spices, creamy oak. Palate: flavourful, fruity, fresh, balanced.

200 Monges 2017 RD R
viura, garnacha

92

Colour: onion pink. Nose: red berry notes, dried herbs, dried flowers, sweet spices. Palate: fruity, flavourful, fresh, balanced.

DO Ca. RIOJA / D.O.P.

200 Monges Esencia 2012 B BA D
viura, garnacha, malvasía

94

Colour: golden. Nose: powerful, honeyed notes, candied fruit, fragrant herbs, acetaldehyde. Palate: flavourful, sweet, fresh, fruity, good acidity, long.

200 Monges Selección Especial 2006 T R
tempranillo

92

Colour: bright cherry. Nose: red berry notes, grassy, spicy. Palate: fruity, flavourful, good structure, fresh.

200 Monges Selección Especial 2011 B R S
malvasía, viura, garnacha blanca, moscatel

92

Colour: bright yellow. Nose: powerful, creamy oak, ripe fruit, spicy, woody. Palate: rich, good structure, toasty, fine bitter notes.

Cueva del Monge 2018 B FB
viura, malvasía, garnacha blanca

92

Colour: bright straw. Nose: stone fruit, wild herbs, dried flowers, white flowers. Palate: fruity, fresh, flavourful, balanced, spicy.

Cueva del Monge 2018 T
tempranillo

91

Colour: deep cherry. Nose: ripe fruit, dried herbs, creamy oak, black fruit. Palate: powerful, ripe fruit, spicy, round tannins.

Ondipuerko 2020 B
tempranillo blanco, maturana blanca, viura, chardonnay

89

Little interventionist, balsamic herbs, kind finish, pruney.

Ondipuerko 2020 RD
tempranillo, viura, garnacha

89

Pleasant, aromatic, little interventionist, tasty.

Ondipuerko 2020 T RB
tempranillo, graciano, garnacha, viura

90

Colour: cherry, purple rim. Nose: fruit expression, fruit preserve, red berry notes. Palate: flavourful, fruity, fresh, good structure, slightly dry, soft tannins.

Viña Los Valles 50 & 50 2018 T C
graciano, garnacha

89

Spicy, fruity, herbal, tasty.

BODEGAS VIÑA BERNEDA

Ctra. Somalo, 59
26313 Uruñuela (La Rioja)
☎: +34 941 371 304
Fax: +34 941 371 304
berneda@vinaberneda.com
www.vinaberneda.com

Berneda 2014 T R
100% tempranillo

88

Classic, fruity, dried herbs, jammy, tasty.

Berneda Vendimia Seleccionada 2018 T C
100% tempranillo

89

Fruity, spicy, tasty, toasty.

Viña Berneda 2018 T C
100% tempranillo

88

Fruity, herbaceous, spicy, tasty.

Viña Berneda 2021 B
100% viura

88

Citrus fruit, fruity, tasty, simple.

Viña Berneda 2021 T MC
100% tempranillo

86

BODEGAS VIÑA HERMINIA

Camino de los Agudos, 1
26559 Aldeanueva de Ebro (La Rioja)
☎: +34 941 142 305
vherminia@vherminia.es
www.viñaherminia.es

Herminia Vendimia Seleccionada 2019 T
100% tempranillo

89

Pleasant, fruity, toasty.

Viña Herminia 2018 T C
85% tempranillo, 15% garnacha

88

Jammy, toasty, spicy.

Viña Herminia Excelsus 2019 T
50% tempranillo, 50% garnacha

90

Colour: deep cherry. Nose: ripe fruit, dried herbs, creamy oak. Palate: powerful, ripe fruit, spicy, round tannins.

DO Ca. RIOJA / D.O.P.

BODEGAS VIVANCO
Ctra. N-232 s/n
26330 Briones (La Rioja)
☎: +34 941 322 360
Fax: +34 941 322 380
bodega@vivancoculturadevino.es
www.vivanculturadevino.es

Colección Vivanco 4 Varietales 2019 T D
70% tempranillo, 15% graciano, 10% garnacha, 5% mazuelo

91

Colour: cherry, purple rim. Nose: red berry notes, ripe fruit, dried herbs, sweet spices. Palate: fruity, flavourful, fresh, slightly dry, soft tannins.

Colección Vivanco Parcelas de Garnacha 2019 T
garnacha

93

Colour: cherry, purple rim. Nose: ripe fruit, red berry notes, dried herbs, dried flowers, spicy. Palate: flavourful, powerful, good structure, balanced, powerful tannins.

Colección Vivanco Parcelas de Maturana 2019 T C
maturana

93

Colour: deep cherry, purple rim. Nose: ripe fruit, black fruit, wild herbs, grassy, spicy. Palate: fruity, flavourful, good structure, balanced, slightly dry, soft tannins.

Vivanco 2019 T C
95% tempranillo, 3% graciano, 2% maturana

91

Colour: cherry, purple rim. Nose: red berry notes, ripe fruit, spicy, smoky, dried herbs, expressive. Palate: fruity, flavourful, balanced, slightly dry, soft tannins.

Vivanco Brunes 2019 T
90% tempranillo, 10% maturana

92 🌿

Colour: cherry, garnet rim. Nose: fruit preserve, fruit liqueur notes, powerful, toasty. Palate: flavourful, sweetness, long.

Vivanco Cuvée Inédita 2018 BE EBR
45% maturana blanca, 30% tempranillo blanco, 15% viura, 10% chardonnay

87

BODEGAS Y VIÑAS DEL CONDE
Avda. de la Póveda, 12
01306 Lapuebla de Labarca (Araba/Álava)
☎: +34 945 607 017
Fax: +34 945 063 173
info@casadomorales.es
www.condedealtava.com

Conde de Altava 2016 T R
tempranillo

87

Conde de Altava 2019 T C
tempranillo

87

Conde de Altava Tempranillo 2021 T
tempranillo

89

Colour: cherry, purple rim. Nose: red berry notes, floral, ripe fruit. Palate: fruity, flavourful, balanced.

BODEGAS Y VIÑEDOS ALVAREZ ALFARO
Ctra. Comarcal 384, Km. 0,8
26559 Aldeanueva de Ebro (La Rioja)
☎: +34 941 144 210
info@bodegasavarezalfaro.com
www.bodegasavarezalfaro.com

Alvarez Alfaro 2020 T C
100% tempranillo

87

Alvarez Alfaro 2021 B
100% garnacha blanca

88

Citrus fruit, crisp, fruity, jammy, tasty.

Alvarez Alfaro Selección Familiar 2017 T
100% tempranillo

90

Colour: deep cherry. Nose: black fruit, spicy, scrubland, toasty. Palate: ripe fruit, spicy, round tannins.

BODEGAS Y VIÑEDOS CASADO MORALES
Avda. La Póveda 12
01306 Lapuebla de Labarca (Araba/Álava)
☎: +34 945 607 017
Fax: +34 945 063 173
info@casadomorales.es
www.casadomorales.es

Casado Morales 2014 T GR
tempranillo, graciano

89

Age nuances, pruney, balanced, spicy, wild.

Casado Morales 2016 T R
tempranillo

91
Colour: bright cherry. Nose: fresh fruit, creamy oak, red berry notes, sweet spices. Palate: good acidity, spicy, balanced.

Casado Morales 2018 T C
tempranillo, graciano

91
Colour: deep cherry. Nose: black fruit, scrubland, cocoa bean. Palate: powerful, ripe fruit, spicy, round tannins.

Protesta 2021 T
tempranillo

87

BODEGAS Y VIÑEDOS GÓMEZ CRUZADO

Avda. Vizcaya, 6 Barrio de la Estación
26200 Haro (La Rioja)
☎: +34 941 312 502
bodega@gomezcruzado.com

Cerro Las Cuevas 2019 T
tempranillo, graciano, garnacha, mazuelo

94
Colour: cherry, purple rim. Nose: red berry notes, creamy oak, sweet spices, dried flowers, elegant. Palate: good structure, fruity, flavourful, full-bodied, good finish, fine tannins.

El Predilecto 2021 T
65% garnacha, 35% tempranillo

93
Colour: cherry, purple rim. Nose: fruit expression, fresh fruit, red berry notes, rose petals, violets, spicy. Palate: fruity, fresh, flavourful, balanced, powerful.

Gómez Cruzado 2º Año 2021 B
viura, tempranillo blanco, garnacha blanca, malvasía

91
Colour: straw. Nose: expressive, white flowers, jasmine, dried herbs. Palate: flavourful, fruity, balanced.

Montes Obarenes 2018 B
75% viura, 20% tempranillo blanco, 5% malvasía

93
Colour: bright yellow. Nose: dried flowers, candied fruit, fine lees, pattiserie, toasty. Palate: round, spicy, long, great length.

Pancrudo de Gómez Cruzado 2020 T
garnacha

94
Colour: Cherry. Nose: complex, expressive, wild herbs, dried flowers. Palate: elegant, full-bodied, long, great length, good acidity, balanced.

Viña Dorana 2017 T R
60% tempranillo, 40% garnacha

91
Colour: cherry, purple rim. Nose: ripe fruit, smoky, spicy, dried herbs. Palate: fruity, flavourful, powerful, slightly dry, soft tannins.

BODEGAS Y VIÑEDOS HERAS CORDÓN

Ctra. Lapuebla, Km. 2
26360 Fuenmayor (La Rioja)
☎: +34 941 451 413
Fax: +34 941 450 265
exportacion@herascordon.com
www.herascordon.com

Heras Cordón Vendimia Seleccionada 2019 T C

89
Pleasant, standard, aromatic, dried herbs, jammy, tasty, simple.

BODEGAS Y VIÑEDOS ILURCE

Ctra. Alfaro - Grávalos (LR-289), km. 23
26540 Alfaro (La Rioja)
☎: +34 941 180 829
info@ilurce.com
www.ilurce.com

Angel 2019 T
garnacha

89
Smoky, standard, spicy, jammy, tasty.

El Sueño de Amado 2017 T C
100% garnacha

90
Colour: cherry, garnet rim. Nose: fruit preserve, powerful, dried herbs. Palate: flavourful, long, round tannins.

Ilurce 2018 T C
38% tempranillo, 38% garnacha, 24% graciano

88
Spicy, balanced, fruity, jammy, tasty, dried herbs.

Ilurce 2020 T
100% tempranillo

87

Ilurce 2021 RD
100% garnacha

88
Fruity, floral, herbal, tasty, jammy.

DO Ca. RIOJA / D.O.P.

Sintauto 2017 T
graciano

91

Colour: Cherry. Nose: balsamic herbs, sweet spices, scrubland, black fruit. Palate: spicy, good acidity, long.

BODEGAS Y VIÑEDOS LA MALDITA

26330 Briones (La Rioja)
☎: +34 941 322 360
info@lamalditawines.com
www.lamalditawines.com

La Maldita Garnacha 2021 RD
garnacha

89

Fruity, herbal, tasty, crisp.

La Maldita Garnacha 2021 T
garnacha

87

La Maldita Garnacha Blanca 2021 B
garnacha blanca

87

La Maldita Revolution 2020 T
garnacha

90

Colour: deep cherry, bright cherry. Nose: red berry notes, ripe fruit, dried flowers, wild herbs. Palate: fruity, balanced.

BODEGAS Y VIÑEDOS LABASTIDA

Avda. Diputación, 53
01330 Labastida (Araba/Álava)
☎: +34 945 331 161
info@bodegaslabastida.com
www.bodegaslabastida.com

Solagüen 2016 T R
97% tempranillo, 3% otras

90

Colour: cherry, garnet rim. Nose: ripe fruit, spicy, black fruit. Palate: spicy, round tannins, easy to drink.

Solagüen 2019 T C
95% tempranillo, 5% garnacha

90

Colour: cherry, garnet rim. Nose: red berry notes, spicy, balanced. Palate: flavourful, fruity, good acidity, long.

BODEGAS Y VIÑEDOS LARRAZ

Paraje Ribarrey. Pol. 12- Parcela 50
26350 Cenicero (La Rioja)
☎: +34 639 728 581
info@bodegaslarraz.com
www.bodegaslarraz.com

Caudum Bodegas Larraz 2016 T
100% tempranillo

91

Colour: cherry, garnet rim. Nose: balsamic herbs, ripe fruit, scrubland, tobacco. Palate: flavourful, balsamic, spicy.

Caudum Bodegas Larraz 2018 T
100% tempranillo

90

Colour: cherry, garnet rim. Nose: ripe fruit, scrubland, fine reductive notes, tobacco. Palate: flavourful, balsamic, spicy.

Caudum Bodegas Larraz Selección Especial 2018 T BA
100% tempranillo

90

Colour: very deep cherry. Nose: dried herbs, scrubland, spicy, characterful, ripe fruit, fruit preserve. Palate: dry, flavourful.

BODEGAS Y VIÑEDOS LEZA GARCÍA

San Ignacio, 26
26313 Uruñuela (La Rioja)
☎: +34 941 371 142
bodegasleza@bodegasleza.com
www.bodegasleza.com

La Artesilla de Leza García 2019 T
tempranillo

93

Colour: deep cherry. Nose: dried herbs, creamy oak, ripe fruit, dried fruit. Palate: powerful, ripe fruit, spicy, round tannins.

La Artesilla de Leza García Viñedo Singular 2019 B
100% viura
90
Oxidativ. Colour: bright yellow. Nose: powerful, creamy oak, spicy, dry nuts, woody. Palate: rich, good structure, toasty, fine bitter notes.

Leza García Edición Garnacha 2019 T C
100% garnacha
89
Pleasant, toasty, jammy, spicy.

Leza García Tinto Familia 2018 T C
100% tempranillo
89
Pleasant, toasty, jammy, spicy.

LG de Leza García 2018 T
tempranillo
92
Colour: deep cherry. Nose: ripe fruit, dried herbs, creamy oak, aromatic coffee, dark chocolate. Palate: powerful, ripe fruit, spicy, round tannins.

Nube de Leza García Maturana 2021 T
100% maturana
89
Balsamic herbs, balanced, spicy, floral, fruity, pleasant.

BODEGAS Y VIÑEDOS MARQUÉS DE VARGAS
Ctra. Zaragoza, Km. 6
26006 Logroño (La Rioja)
☎: +34 941 261 401
info@marquesdevargas.com
www.marquesdevargas.com/es

Marqués de Vargas Hacienda Pradolagar 2017 T R
80% tempranillo, 20% mazuelo
94
Colour: Cherry. Nose: complex, spicy, creamy oak, dark chocolate. Palate: full-bodied, long, great length, spicy, ripe fruit, balanced.

Marqués de Vargas Selección Privada 2017 T R
72% tempranillo, 18% garnacha, 10% mazuelo
93
Jammy. Colour: Cherry. Nose: scrubland, dried flowers, faded flowers, sweet spices. Palate: spicy, flavourful, good structure.

Marqués de Vargas 2015 T GR
tempranillo, mazuelo, garnacha
92
Classic. Colour: deep cherry, garnet rim. Nose: black fruit, ripe fruit, old leather, spicy. Palate: fruity, flavourful, balanced, fresh, round tannins.

Marqués de Vargas 2017 T R
70% tempranillo, 17% garnacha, 13% mazuelo
92
Colour: deep cherry. Nose: ripe fruit, dried herbs, sweet spices, wild herbs. Palate: powerful, ripe fruit, spicy, round tannins.

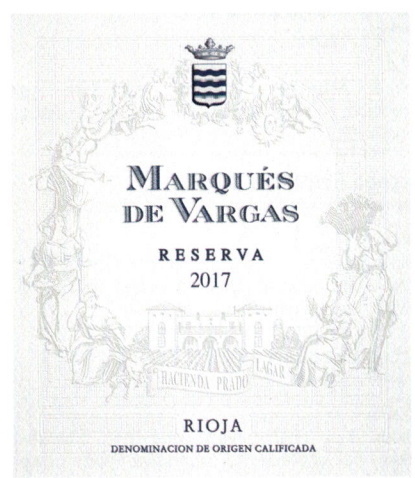

DO Ca. RIOJA / D.O.P.

BODEGAS Y VIÑEDOS MONTEABELLÓN

Calvario, s/n
09318 Nava de Roa (Burgos)
☎: +34 947 550 000
Fax: +34 947 550 219
comunicacion@monteabellon.com
www.monteabellon.com

Athus 2019 T C
90% tempranillo, 10% mazuelo

90

Colour: bright cherry. Nose: fresh fruit, wild herbs, balanced. Palate: good acidity, fine tannins, easy to drink.

Athus Vendimia 2021 T RB
100% tempranillo

88

Jammy, tasty, spicy.

BODEGAS Y VIÑEDOS ORTEGA EZQUERRO

Plaza del Frontón
26512 Tudelilla (La Rioja)
☎: +34 941 152 046
info@ortegaezquerro.com
www.ortegaezquerro.com

Don Quintin Ortega 2020 B FB
viura, malvasía, garnacha blanca

88

Citrus fruit, age nuances, spicy, crisp, full-bodied, woody.

OE Garnacha 2020 T
garnacha

92

Colour: Cherry. Nose: balsamic herbs, sweet spices, scrubland, red berry notes. Palate: spicy, balsamic, good acidity.

Ortega Ezquerro 2016 T R
tempranillo, garnacha, mazuelo

93

Colour: dark-red cherry, garnet rim. Nose: ripe fruit, aged wood nuances, tobacco, sweet spices. Palate: spicy, round tannins, long.

Ortega Ezquerro 2019 T C
tempranillo, garnacha

92

Colour: deep cherry. Nose: ripe fruit, dried herbs, creamy oak, characterful, expressive. Palate: powerful, ripe fruit, spicy, round tannins.

Ortega Ezquerro 2021 T MC
tempranillo, garnacha, viura

90

Colour: cherry, purple rim. Nose: powerful, ripe fruit, spicy. Palate: ripe fruit, flavourful, good structure.

BODEGAS Y VIÑEDOS PUENTE DEL EA

Camino Aguachal, s/n
26212 Sajazarra (La Rioja)
☎: +34 941 320 405
info@puentedelea.com
www.puentedelea.com

Coraz de Puente del Ea 2019 T
tempranillo

88

Spicy, coarse tannins, dried herbs, toasty.

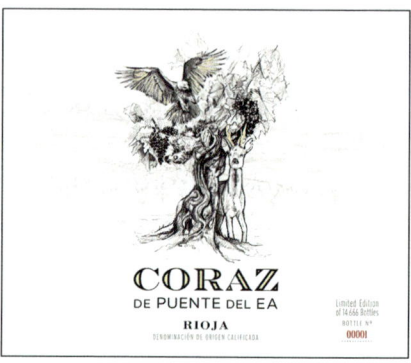

Obar de Puente del Ea 2020 B
viura

90

Colour: bright straw. Nose: ripe fruit, floral. Palate: flavourful, fresh, good acidity, fruity aftertaste.

Saiaz Puente del Ea 2017 T
tempranillo

89

Spicy, dried herbs, full-bodied, balanced.

BODEGAS Y VIÑEDOS QUIROGA DE PABLO

Antonio Pérez, 24
26323 Azofra (La Rioja)
☎: +34 941 379 334
Fax: +34 941 379 334
info@bodegasquiroga.com
www.bodegasquiroga.com

Lagar de Cayo 2021 B
86

Lagar de Cayo Clarete 2021 RD
tempranillo, viura, garnacha
88
Pleasant, defined aromas, floral, fruity, smooth, balanced, crisp.

Lagar de Cayo Tempranillo Blanco 2021 B
tempranillo blanco
88
Fruity, jammy, tasty, floral, standard.

Quirus 2021 B
viura, sauvignon blanc, verdejo, chardonnay
88
Aromatic, floral, fruity, jammy, very fruit-driven.

Quirus 2021 RD
tempranillo, viura, garnacha
88
Aromatic, jammy, fruity, balanced.

Quirus Tempranillo Blanco 2021 B
tempranillo blanco
89
Citrus fruit, fruity, jammy, tasty.

BODEGAS YSIOS

Camino de la Hoya, s/n
01300 Laguardia (Araba/Álava)
☎: +34 945 600 640
clara.canals@pernod-ricard.com
www.bodegasysios.com

Ysios 2019 B FB
viura
92
Colour: bright straw. Nose: white fruit, stone fruit, white flowers, dried herbs, expressive. Palate: fruity, flavourful, fresh, balanced.

Ysios Finca El Nogal 2017 T
tempranillo
94
Colour: cherry, garnet rim. Nose: complex, expressive, spicy, mineral, black fruit, ripe fruit. Palate: full-bodied, great length, ripe fruit.

🏆 PODIUM

Ysios Grano a Grano 2017 T
tempranillo
95
Colour: cherry, garnet rim. Nose: complex, expressive, spicy, mineral, ripe fruit, red berry notes. Palate: full-bodied, long, great length, balanced.

Ysios Las Naves 2017 T
tempranillo
94
Colour: dark-red cherry, garnet rim. Nose: ripe fruit, aged wood nuances, tobacco, sweet spices, chalk, truffle notes. Palate: spicy, round tannins, long.

Ysios Rosé 2021 RD
91
Colour: raspberry rose. Nose: elegant, red berry notes, floral, fragrant herbs. Palate: light-bodied, spicy, good acidity, fine bitter notes.

Ysios Selección 2016 T
tempranillo
91
Colour: deep cherry. Nose: dried herbs, creamy oak, red berry notes, ripe fruit. Palate: powerful, ripe fruit, spicy, round tannins.

BODEGAS ZINTZO

Buenavista, 2
01307 Villabuena de Álava (Araba/Álava)
☎: +34 945 609 186
info@bodegaszintzo.com
www.bodegaszintzo.com

5. Zintzo 2018 T C
100% tempranillo
87

Izena 2020 T
100% tempranillo
89
Hot, jammy, toasty.

Zintzo 2020 T
85% tempranillo, 15% viura
86

Zintzo 2021 B
100% viura
85

DO Ca. RIOJA / D.O.P.

SPANISH WINE

DO Ca. RIOJA / D.O.P.

BODEGAS ZURBAL
Camino de la Estación, 15
26330 Briones (La Rioja)
☎: +34 667 730 651
info@bodegaszurbal.com
www.bodegaszurbal.es

Zurbal 2017 T R
tempranillo
89
Pleasant, herbal, balsamic herbs, spicy, jammy, tasty.

Zurbal 2021 B
100% viura
85

Zurbal 2021 RD
tempranillo, garnacha
87

Zurbal 2021 T
tempranillo
87 🌿

BRUJO WINES
26008 Logroño (La Rioja)
☎: +34 941 046 155
comercial@brujowines.com
www.brujowines.com

Titania Edición Limitada 2017 T
100% tempranillo
91
Colour: bright cherry. Nose: sweet spices, ripe fruit, dark chocolate, creamy oak. Palate: fruity, spicy, round tannins.

Titania Edición Limitada 2018 B
100% tempranillo blanco
90
Colour: bright yellow. Nose: powerful, creamy oak, ripe fruit, spicy, caramel, woody. Palate: rich, good structure, toasty, fine bitter notes.

Titania Edición Limitada 2018 T
100% tempranillo
87

CAMPO VIEJO
Camino de la Puebla, 50
26007 Logroño (La Rioja)
☎: +34 941 279 900
elena.adell@pernod-ricard.com

Campo Viejo 2013 T GR
tempranillo
91
Colour: cherry, garnet rim. Nose: red berry notes, dried herbs, spicy, smoky. Palate: fruity, flavourful, fresh.

Campo Viejo 2017 T R
tempranillo
90
Colour: bright cherry. Nose: ripe fruit, spicy, smoky, dried herbs, meaty notes. Palate: fruity, flavourful, balanced.

Campo Viejo 2018 T C
tempranillo
88
Spicy, fruity, herbal, tasty.

Campo Viejo Tempranillo C.V.C. T
86

Campo Viejo Vendimia Seleccionada 2017 T C
tempranillo
91
Colour: dark-red cherry, garnet rim. Nose: ripe fruit, aged wood nuances, tobacco, sweet spices. Palate: spicy, round tannins, long.

Dominio Campo Viejo 2016 T
tempranillo
92
Colour: cherry, purple rim. Nose: red berry notes, dried herbs, spicy, ripe fruit. Palate: flavourful, balanced, fresh, slightly dry, soft tannins.

CARLOS SERRES
Avda. Santo Domingo, 40
26200 Haro (La Rioja)
☎: +34 941 310 279
Fax: +34 941 310 418
info@carlosserres.com
www.carlosserres.com

Carlos Serres 1896 Finca El Estanque Segundo Año 2018 RD R
60% mazuelo, 40% tempranillo
92
Defined aromas, spicy. Colour: coppery red. Nose: dried flowers, faded flowers, neat, expressive, spicy. Palate: rich, flavourful, balanced.

Carlos Serres 1896 Finca El Estanque Segundo Año 2019 T
90% tempranillo, 10% graciano
89
Full-bodied, pruney, wild, tasty, great length, spicy, standard.

Carlos Serres 2015 T GR
80% tempranillo, 10% graciano, 10% mazuelo
88
Kind finish, age nuances, animal funk, spicy, jammy.

Carlos Serres 2016 T R
90% tempranillo, 10% graciano
87

Carlos Serres 2019 T C
tempranillo
88
Pleasant, spicy, fruity, jammy, simple.

Carlos Serres Brut Rioja 2019 BE
47% viura, 35% chardonnay, tempranillo blanco
89
Balanced, taut, yeasty notes, tasty, herbal.

COMPAÑÍA DE VINOS HERACLIO
Crta. Corella s/n
26540 Alfaro (La Rioja)
☎: +34 941 181 570
heraclioalfaro@heraclioalfaro.com
www.terrasgauda.com

Heraclio Alfaro 2018 T C
garnacha, tempranillo, graciano, mazuelo
89
Spicy, dried herbs, jammy, tasty, wild, standard.

Heraclio Alfaro Finca Estarijo 2016 T
graciano, garnacha, tempranillo, mazuelo
90
Colour: dark-red cherry. Nose: toasty, spicy, cocoa bean, ripe fruit. Palate: flavourful, toasty, fine bitter notes.

COMPAÑÍA DE VINOS TELMO RODRÍGUEZ
El Monte
01308 Lanciego (Araba/Álava)
☎: +34 945 628 315
contact@telmorodriguez.com
www.telmorodriguez.com

🏆 PODIUM

El Velado 2019 T
95
Aromatic, complex, taut. Colour: Cherry. Nose: balsamic herbs, sweet spices, scrubland. Palate: spicy, balsamic, good acidity.

La Estrada 2019 T
93
Colour: bright cherry. Nose: balsamic herbs, sweet spices, scrubland, ripe fruit, red berry notes. Palate: spicy, balsamic, good acidity.

Lanzaga 2018 T
92
Colour: deep cherry. Nose: ripe fruit, creamy oak, scrubland, balanced. Palate: powerful, ripe fruit, spicy, round tannins.

🏆 PODIUM

Las Beatas 2019 T
97
Elegant, subtle. Colour: Cherry. Nose: complex, expressive, spicy, dried flowers, raspberry, black fruit, scrubland, tea leave. Palate: elegant, full-bodied, long, great length.

LZ 2021 T
92
Pleasant, defined aromas. Colour: cherry, purple rim. Nose: fruit expression, red berry notes, floral, powerful. Palate: fruity, flavourful, balanced.

Tabuerniga 2019 T
94
Colour: deep cherry. Nose: ripe fruit, dried herbs, sweet spices, mineral, expressive. Palate: powerful, ripe fruit, spicy, round tannins.

DO Ca. RIOJA / D.O.P.

CREACIONES EXEO
Costanilla del Hospital s/n
01330 Labastida (Araba/Álava)
☎: +34 945 331 257
Fax: +34 945 331 257
info@tierrayvino.com
www.bodegasexeo.com

Cifras 2020 B
91
Colour: straw. Nose: ripe fruit, dried herbs, faded flowers. Palate: powerful, ripe fruit, balanced.

Cifras 2020 T
91
Pleasant, smooth, tasty, jammy.

Enfudrecido 2019 T
91
Colour: deep cherry. Nose: ripe fruit, dried herbs, creamy oak. Palate: powerful, ripe fruit, spicy, round tannins.

Letras Minúsculas 2020 T
88
Kind finish, fruity, jammy, smooth.

CRIADORES DE RIOJA
26141 Alberite (La Rioja)
☎: +34 941 436 702
info@castilloclavijo.com
www.criadoresderioja.com

Alegro 2019 T C
tempranillo
86

Castillo Clavijo 2019 T C
tempranillo
87 🌱

Castillo San Lorenzo 2016 T R
tempranillo, graciano
86

El Guardián 2016 T R
tempranillo
88
Kind finish, toasty, jammy, spicy.

El Guardián 2019 T C
tempranillo
87

El Guardián sin Sulfitos 2021 T
tempranillo
87

CUENTAVIÑAS
Vial B6 Peciña
26339 San Vicente de la Sonsierra (La Rioja)
☎: +34 686 498 183
info@cuentavinas.com

Cuentaviñas Alomado 2020 T
tempranillo, viura, malvasía, calagraño
94
Pleasant, aromatic, complex. Colour: deep cherry. Nose: balsamic herbs, sweet spices, scrubland, red berry notes, floral. Palate: spicy, balsamic, good acidity.

🏆 **PODIUM**

Cuentaviñas El Tiznado 2020 T
tempranillo
96
Colour: deep cherry. Nose: dried herbs, creamy oak, sweet spices, ripe fruit, black fruit, earthy notes. Palate: powerful, ripe fruit, spicy, round tannins.

Cuentaviñas Garnacha CDVIN 2020 T
garnacha
94
Taut. Colour: bright cherry. Nose: complex, expressive, spicy, mineral, red berry notes, balsamic herbs. Palate: elegant, full-bodied, long, great length.

🏆 **PODIUM**

Cuentaviñas Los Yelsones 2020 T
tempranillo
95
Colour: deep cherry. Nose: complex, expressive, spicy, mineral, ripe fruit, red berry notes, black fruit. Palate: full-bodied, long, great length, flavourful, fresh, fruity.

CVNE

Barrio de la Estación, s/n
26200 Haro (La Rioja)
☎: +34 941 304 800
marketing@cvne.com
www.cvne.com

DO Ca. RIOJA / D.O.P.

Asúa 2019 T C
100% tempranillo
88
Tasty, spicy, jammy.

Corona 2015 B SD
100% viura
90
Colour: bright yellow. Nose: ripe fruit, candied fruit, honeyed notes. Palate: flavourful, unctuous, fruity, sweet.

Cune 2016 T GR
85% tempranillo, 10% graciano, 5% mazuelo
91
Colour: dark-red cherry, garnet rim. Nose: ripe fruit, aged wood nuances, tobacco, sweet spices. Palate: spicy, round tannins, long.

Cune 2018 T R
91
Colour: cherry, garnet rim. Nose: ripe fruit, sweet spices, balanced. Palate: powerful, ripe fruit, spicy, round tannins, easy to drink.

Cune 2019 B R
100% viura
88
Aromatic, jammy, tasty. Nose: toasty.

Cune 2019 T C
85% tempranillo, 15% garnacha, mazuelo
89
Toasty, tasty, jammy, spicy.

Cune Orgánico 2021 T
60% garnacha, 30% tempranillo, 10% graciano
89
Pleasant, tasty, jammy, herbal.

Cune Semidulce B SD
100% viura
88
Pleasant, smooth, tasty.

Imperial 2016 T GR
85% tempranillo, 10% garnacha, 5% mazuelo
94
Colour: dark-red cherry, garnet rim. Nose: ripe fruit, aged wood nuances, tobacco, sweet spices, toasty. Palate: spicy, round tannins, long.

Cune 2021 B
100% viura
87

Cune 2021 RD
100% tempranillo
87

Imperial 2018 T R
85% tempranillo, 15% garnacha, graciano, mazuelo
93
Colour: dark-red cherry, garnet rim. Nose: ripe fruit, aged wood nuances, tobacco, sweet spices. Palate: spicy, round tannins, long.

Monopole 2021 B
100% viura
89
Citrus fruit, standard, crisp, herbal.

Monopole Clásico 2019 B
viura
92
Colour: bright straw. Nose: ripe fruit, fragrant herbs, fine lees, pungent. Palate: full-bodied, rich, long, good acidity.

🏆 **PODIUM**

Real de Asúa 2019 T
95
Colour: very deep cherry. Nose: complex, expressive, spicy, mineral, red berry notes. Palate: elegant, full-bodied, long, great length.

DO Ca. RIOJA / D.O.P.

DIEZ-CABALLERO
Barrihuelo, 73
01340 Elciego (Araba/Álava)
☎: +34 944 807 295
diez-caballero@diez-caballero.es
www.diez-caballero.es

Díez-Caballero 2017 T R
tempranillo
90
Colour: cherry, purple rim. Nose: fruit expression, red berry notes, floral, spicy, burnt matches. Palate: flavourful, fruity, good acidity.

Díez-Caballero 2019 T C
tempranillo
91
Colour: deep cherry. Nose: dried herbs, creamy oak, red berry notes, black fruit. Palate: ripe fruit, spicy, round tannins.

Díez-Caballero 2021 B
viura
88
Standard, dried herbs, citrus fruit, balanced. Palate: good acidity, fine bitter notes.

Díez-Caballero Vendimia Seleccionada 2018 T R
tempranillo
91
Colour: deep cherry. Nose: dried herbs, creamy oak, black fruit, woody. Palate: powerful, ripe fruit, spicy, round tannins.

Victoria Díez-Caballero 2018 T R
tempranillo
92
Colour: deep cherry. Nose: dried herbs, creamy oak, black fruit, roasted coffee. Palate: ripe fruit, spicy, round tannins.

DOMINIO DE BERZAL
Término de Río Salado s/n
01307 Baños de Ebro (Araba/Álava)
☎: +34 945 623 368
Fax: +34 945 609 090
info@dominioberzal.com
www.dominioberzal.com

Dominio de Berzal 2019 T C
95% tempranillo, 5% graciano
91
Colour: deep cherry. Nose: dried herbs, creamy oak, black fruit. Palate: powerful, ripe fruit, spicy, round tannins.

Dominio de Berzal 2021 B
90% viura, 10% malvasía
89
Austere, citrus fruit, crisp, herbal, tasty.

Dominio de Berzal 2021 T MC
90% tempranillo, 10% viura
86

Dominio de Berzal 7 Varietales 2019 T
40% maturana, 10% graciano, garnacha, syrah, cabernet sauvignon, merlot
92
Colour: deep cherry. Nose: dried herbs, creamy oak, black fruit, spicy. Palate: powerful, ripe fruit, spicy, round tannins.

Dominio de Berzal Selección Privada 2019 T
tempranillo
91
Colour: cherry, purple rim. Nose: fruit expression, red berry notes, floral, spicy, violets, woody. Palate: flavourful, fruity, good acidity, long.

DOMINIO DE NOBLEZA
Bº Bodegas San Cristóbal, 79
26360 Fuenmayor (La Rioja)
☎: +34 941 450 507
bodegas@dominiodenobleza.com
www.dominiodenobleza.com

Dominio de Nobleza 2018 T C
tempranillo
86

Dominio de Nobleza Edición Limitada 2015 T R
tempranillo
90
Colour: cherry, garnet rim. Nose: ripe fruit, fruit preserve, smoky, spicy. Palate: flavourful, easy to drink.

Dominio de Nobleza Maturana 2020 T
maturana
89
Defined aromas, smoky, jammy, tasty, dried herbs. Nose: aromatic coffee.

Dominio de Nobleza Vendimia Seleccionada 2015 T R
tempranillo
89
Woody, full-bodied, spicy, jammy, tasty.

DON BALBINO

Avda. de la Póveda, 34
01306 Lapuebla de Labarca (Araba/Álava)
☎: +34 945 607 018
Fax: +34 945 607 018
administracion@donbalbino.com
www.donbalbino.com

Alto del Rincón, Viñedos de Altura 2019 T BA
85% tempranillo

91

Colour: deep cherry. Nose: dried herbs, creamy oak, black fruit. Palate: ripe fruit, spicy, round tannins.

Don Balbino 2020 B FB
100% viura

90

Dried flowers, spicy. Colour: bright straw. Nose: dried herbs, wild herbs, spicy, neat. Palate: flavourful, fruity, spicy, ripe fruit.

Don Balbino Vendimia Seleccionada 2019 T
100% tempranillo

90

Colour: very deep cherry. Nose: roasted coffee, aromatic coffee, powerful. Palate: smoky aftertaste, great length, round tannins.

El Capricho de Sofía 2021 B
viura

88

Tasty, toasty, jammy.

El Capricho de Sofía 2021 RD
tempranillo

88

Citrus fruit, herbal, tasty, simple.

Puentegrijos 2021 T MC
tempranillo, viura

88

Kind finish, fruity, herbal, tasty.

EGUREN UGARTE

Ctra. A-124, Km. 61
01309 Laguardia (Araba/Álava)
☎: +34 945 282 844
Fax: +34 945 271 319
info@egurenugarte.com
www.egurenugarte.com

Cedula Real 2015 T GR
90% tempranillo, 10% graciano

90

Colour: cherry, garnet rim. Nose: wet leather, fine reductive notes, creamy oak, red berry notes, ripe fruit. Palate: fruity, flavourful, balanced, slightly dry, soft tannins.

Cincuenta Eguren Ugarte 2017 T
tempranillo

89

Spicy, fruity, herbal, tasty.

Eguren Ugarte 2015 T R
95% tempranillo, 5% graciano

89

Aromatic, fruity, dried herbs, spicy, tasty.

Eguren Ugarte 2019 T C
tempranillo

88

Fruity, spicy, herbal, tasty.

Martín Cendoya 2016 T R
80% tempranillo, 20% graciano

90

Colour: deep cherry. Nose: ripe fruit, red berry notes, wild herbs, spicy. Palate: fruity, flavourful, fresh, balanced, slightly dry, soft tannins.

DO Ca. RIOJA / D.O.P.

EL COTO DE RIOJA

Camino Viejo de Logroño, 26
01320 Oyón (Araba/Álava)
☎: +34 945 622 216
info@elcoto.com
www.elcoto.com

875 m Chardonnay 2021 B FB
chardonnay

90

Colour: bright straw. Nose: stone fruit, dried herbs, sweet spices, faded flowers. Palate: fruity, flavourful, balanced.

Coto de Imaz 2017 T R
tempranillo

91

Colour: deep cherry. Nose: ripe fruit, dried herbs, creamy oak. Palate: powerful, ripe fruit, spicy, round tannins.

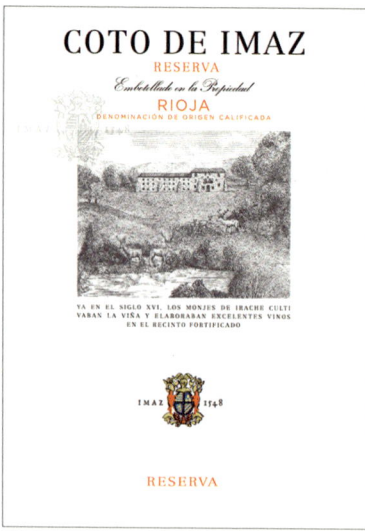

El Coto 2018 T C
tempranillo

88

Toasty, smooth, spicy, balanced.

El Coto Ecológico 2019 T C
tempranillo

89 🌱

Pleasant, toasty, tasty, fruity.

El Coto Selección Viñedos 2018 T C
tempranillo

89

Toasty, tasty, jammy, spicy.

EL INQUILINO WINES

Paseo del Prior, 3 – 5ºB
26004 Logroño (La Rioja)
☎: +34 948 780 617
xabi@vinazorzalwines.com

El Inquilino 2019 T C

90

Pleasant, crisp, balanced, fruity, smooth. Nose: red berry notes. Palate: easy to drink, good finish.

EL OTERO

Pza. Manuel Quintano, 1
01330 Labastida (Araba/Álava)
☎: +34 696 620 505
el.otero.sc@gmail.com
www.aimarez.com

Aimarez 2021 T MC
tempranillo, viura, garnacha

86

ELVIWINES

Ctra T-300 Falset-Marça, km 0.97
43775 Marça (Tarragona)
☎: +34 606 186 565
info@elviwines.com
www.elviwines.com

Herenza 2017 T R
97% tempranillo, 3% graciano

89

Balanced, spicy, dried herbs, lactic, jammy, tasty.

Herenza 2018 T C
100% tempranillo

88

Standard, spicy, balanced, tasty.

Herenza 2020 T
100% tempranillo

86

Herenza Rosé 2021 RD
73% tempranillo, 27% garnacha

87

ENEO
Paseo Virgen de la Vega 4 1ºI
26200 Haro (La Rioja)
☎: +34 941 310 494
soto@comercialeneo.com
www.comercialeneo.com

Rey Eneo 2015 T GR
tempranillo

87

Rey Eneo Edición Limitada 2010 T R
tempranillo

89
Creamy, spicy, jammy, tasty, smooth, toasty.

Rey Eneo Vendimia Seleccionada 2015 T R
tempranillo

88
Pleasant, standard, spicy, dried herbs, jammy, tasty, simple.

Rey Eneo Vendimia Seleccionada 2019 T C
tempranillo

87

EULOGIO & JAVIER WINES
Juan de Herrera, 23
47130 Simancas (Valladolid)
☎: +34 983 150 150
comunicacion@rodriguezysanzo.com
www.culogioyjavierwines.com

Capitán Gaona 2019 T
90% tempranillo, 10% garnacha

89
Floral, crisp, fruity, herbal, tasty.

FAMILIA BASTIDA
Canónigo Lozano, 11
30520 Jumilla (Murcia)
☎: +34 968 780 142
info@familiabastida.com
www.familiabastida.com

Bbastida 2019 T BA
100% tempranillo

88
Hot, pruney, toasty, powerful.

FAMILIA MONTAÑA
Pol. Ind. Lentiscares
26370 Navarrete (La Rioja)
info@premiumfincas.com
www.premiumfincas.com

Montaña Finca El Faraón 2019 T R
100% maturana

91
Reduced, pruney. Colour: deep cherry. Nose: black fruit, fruit preserve, wild herbs, scrubland, cigar. Palate: long, balanced, flavourful.

Montaña Finca La Valentina 2019 T C
90% tempranillo, 10% graciano

90
Colour: cherry, purple rim. Nose: red berry notes, floral, spicy. Palate: flavourful, fruity, good acidity.

FAUSTINO RIVERO ULECIA
Ctra. de Soria LR-115, km. 23
26580 Arnedo (La Rioja)
☎: +34 941 380 057
visitas@faustinorivero.com
www.faustinorivero.com

Faustino Rivero Ulecia 2017 T R
tempranillo, garnacha

90
Colour: deep cherry. Nose: aged wood nuances, ripe fruit, cigar, toasty. Palate: spicy, toasty.

Faustino Rivero Ulecia 2019 T C
tempranillo, garnacha

88
Pleasant, pruney, jammy, toasty, tasty.

Faustino Rivero Ulecia Viura 2021 B
viura

88
Aromatic, floral, fruity, tasty.

FINCA ALLENDE
Pza. Ibarra, 1
26330 Briones (La Rioja)
☎: +34 941 322 301
Fax: +34 941 322 302
sales@finca-allende.com
www.finca-allende.com

Allende 2016 T
100% tempranillo

93
Colour: dark-red cherry, garnet rim. Nose: ripe fruit, fruit preserve, aged wood nuances, tobacco, sweet spices. Palate: spicy, round tannins, long.

DO Ca. RIOJA / D.O.P.

DO Ca. RIOJA / D.O.P.

Allende 2018 B
95% viura, 5% malvasía
92
Colour: bright golden. Nose: ripe fruit, dried herbs, creamy oak, caramel. Palate: fruity, flavourful, powerful, fresh, balanced, good finish.

Allende 2018 RD
60% tempranillo, 40% garnacha
93
Colour: brilliant rose. Nose: ripe fruit, red berry notes, wild herbs, wild herbs, spicy. Palate: fruity, flavourful, fresh, complex, balanced, great length, soft tannins.

🏆 **PODIUM**

Avrvs 2016 T
85% tempranillo, 15% graciano
95
Colour: dark-red cherry. Nose: black fruit, dried herbs, wild herbs, creamy oak, characterful, smoky, aromatic coffee. Palate: ripe fruit, balsamic, round tannins, long, complex, silky tannins.

Calvario 2016 T
90% tempranillo, 8% garnacha, 2% graciano
94
Colour: cherry, garnet rim. Nose: ripe fruit, fruit preserve, varietal, dried herbs, spicy, black pepper. Palate: flavourful, round tannins, great length.

Gaminde 2016 T
100% tempranillo
94
Colour: deep cherry, garnet rim. Nose: black fruit, ripe fruit, dried herbs, spicy, sweet spices. Palate: flavourful, ripe fruit, round tannins.

Mártires 2020 B
100% viura
94
Colour: bright yellow. Nose: powerful, creamy oak, ripe fruit, spicy. Palate: rich, good structure, long, toasty, fine bitter notes.

Marure 2018 B
100% garnacha blanca
94
Colour: bright yellow. Nose: creamy oak, ripe fruit, spicy, toasty, smoky, cocoa bean. Palate: rich, good structure, long, fine bitter notes.

Mingortiz 2016 T
100% tempranillo
93
Colour: bright cherry, deep cherry. Nose: dried herbs, neat, spicy, expressive, characterful. Palate: varietal, flavourful, round tannins.

FINCA DE LA RICA
Las Cocinillas, s/n
01330 Labastida (La Rioja)
☎: +34 628 833 065
ignacio@fincadelarica.com
www.fincadelarica.com

El Buscador 2018 T C
90% tempranillo, 10% garnacha
90
Colour: cherry, purple rim. Nose: red berry notes, spicy, dried herbs. Palate: flavourful, fruity, slightly dry, soft tannins.

El Guía de Finca de la Rica 2021 T
88
Balanced, herbal, tasty, fruity.

El Nómada Selección de Parcelas 2019 T
90% tempranillo, 10% graciano
92
Colour: deep cherry. Nose: ripe fruit, dried herbs, creamy oak, mineral. Palate: powerful, ripe fruit, spicy, round tannins.

El Rincón de los Enebros 2020 T BA
70% tempranillo, 30% garnacha
92
Colour: bright cherry. Nose: spicy, mineral, ripe fruit, black fruit. Palate: full-bodied, long, great length.

Las Cabezadas de Matadula 2019 T
100% tempranillo
92
Colour: cherry, purple rim. Nose: black fruit, toasty, spicy, dried herbs. Palate: fruity, flavourful, balanced.

Las Candelera 2021 B FB
35% viura, 35% garnacha blanca, 30% malvasía
90
Colour: bright straw. Nose: fresh fruit, wild herbs, white fruit, wild herbs, floral. Palate: fruity, flavourful, balanced.

FINCA NUEVA
Ctra. de Fuenmayor, km. 1,5
26370 Navarrete (La Rioja)
☎: +34 941 322 301
Fax: +34 941 322 302
sales@fincanueva.com
www.fincanueva.com

Finca Nueva 2010 T GR
100% tempranillo
91
Colour: pale ruby, brick rim edge. Nose: old leather, creamy oak, spicy, fruit preserve, black fruit, dried herbs. Palate: fruity, flavourful, balanced, smoky aftertaste, round tannins.

Finca Nueva 2016 T R
100% tempranillo

91 Colour: deep cherry. Nose: sweet spices, incense, ripe fruit, red berry notes. Palate: flavourful, fruity, fresh, balanced, smoky aftertaste.

Finca Nueva 2018 T C
100% tempranillo

90 Colour: bright cherry. Nose: red berry notes, ripe fruit, spicy, dried herbs, dried flowers. Palate: fruity, flavourful, balanced, slightly dry, soft tannins.

Finca Nueva 2020 B FB
100% viura

89 Roasted coffee, pruney, spicy, tasty.

Finca Nueva 2021 RD
60% tempranillo, 40% garnacha

88 Fruity, sweety finish, jammy, tasty, herbal.

FINCA VALPIEDRA
El Montecillo s/n
26360 Fuenmayor (La Rioja)
☎: +34 941 450 876
Fax: +34 941 450 875
info@bujanda.com
www.fincavalpiedra.com

Cantos de Valpiedra 2018 T
100% tempranillo

90 Colour: deep cherry. Nose: ripe fruit, dried herbs, creamy oak. Palate: powerful, ripe fruit, spicy, round tannins.

Finca Valpiedra 2015 T R
92% tempranillo, 4% graciano, 4% maturana

94 Colour: deep cherry, garnet rim. Nose: aged wood nuances, ripe fruit, cocoa bean, cigar, toasty. Palate: flavourful, toasty, powerful tannins.

Finca Valpiedra 2017 B R
viura, malvasía, maturana

91 Colour: bright yellow. Nose: powerful, creamy oak, spicy, stone fruit. Palate: rich, good structure, toasty, fine bitter notes.

Petra de Valpiedra 2018 T
100% garnacha

92 Colour: Cherry, garnet rim. Nose: complex, spicy, mineral, scrubland. Palate: elegant, full-bodied, long, great length.

FINCAS DE AZABACHE
Avda. Juan Carlos I, 100
26559 Aldeanueva de Ebro (La Rioja)
☎: +34 941 163 039
Fax: +34 941 163 585
info@fincasdeazabache.com
www.fincasdeazabache.com

Azabache Vendimia Seleccionada 2019 T C
tempranillo, garnacha, graciano

89 Kind finish, spicy, jammy, herbal, tasty, wild, balanced.

Barón de Ebro 2019 T C
tempranillo, garnacha, graciano

87

Coscojares 2018 T
garnacha

91 Colour: dark-red cherry, garnet rim. Nose: fruit preserve, aged wood nuances, tobacco, sweet spices. Palate: spicy, round tannins, long.

DO Ca. RIOJA / D.O.P.

DO Ca. RIOJA / D.O.P.

Fincas de Azabache Tempranillo Blanco 2021 B
tempranillo blanco
88
Fruity, herbal, thin, tasty.

Tunante Tempranillo 2021 T
88
Tasty, fruity, smooth.

Fincas de Azabache Garnacha 2019 T C
garnacha
89
Varietally correct, wild, jammy, dried herbs, fruity, balanced.

FROM GALICIA GROUP
Cordelería, 39 Bajo
15003 A Coruña/La Coruña (A Coruña/La Coruña)
☎: +34 881 994 069
info@fromgaliciagroup.com
www.fromgaliciagroup.com

2 Kisses 2018 T C
tempranillo, graciano
88
Balsamic herbs, balanced, fruity, jammy, dried herbs, spicy.

2 Kisses 2021 T
tempranillo, garnacha
87

GÓMEZ DE SEGURA
El Campillar - Laguardia
01300 Laguardia (Araba/Álava)
☎: +34 615 929 828
info@gomezdesegura.com
www.gomezdesegura.com

Finca Ratón 2017 T R
100% tempranillo
90
Colour: dark-red cherry, garnet rim. Nose: aged wood nuances, tobacco, sweet spices. Palate: spicy, round tannins, long.

Gómez de Segura 2021 B
viura, malvasía
86

Gómez de Segura 2021 T MC
100% tempranillo
87

Gómez de Segura Reserva de la Familia 2018 T R
tempranillo
86

Gómez de Segura Vendimia Seleccionada 2020 T BA
100% tempranillo
87

HACIENDA EL TERNERO
Finca El Ternero, s/n
09200 Miranda de Ebro (Burgos)
☎: +34 941 320 021
info@elternero.com
www.elternero.com

Hacienda el Ternero 2014 T R
95% tempranillo, 5% mazuelo
90
Classic, toasty, spicy, jammy.

Hacienda el Ternero 2018 B FB
100% viura
91
Colour: bright straw. Nose: white fruit, dried herbs, floral, spicy. Palate: flavourful, fruity.

Hacienda el Ternero 2021 RD
tempranillo
87

Hacienda el Ternero Selección 2016 T C
100% tempranillo
91
Colour: dark-red cherry. Nose: toasty, spicy, cocoa bean, ripe fruit, tobacco. Palate: flavourful, toasty, fine bitter notes.

La Pera 2021 B
25% viura, 25% sauvignon blanc, 25% tempranillo blanco, 25% garnacha blanca

90

Colour: straw. Nose: expressive, white flowers, jasmine, dried herbs. Palate: flavourful, fruity, balanced.

Torno Hacienda el Ternero 2017 T C
100% tempranillo

90

Colour: deep cherry. Nose: ripe fruit, dried herbs, creamy oak. Palate: powerful, ripe fruit, spicy, round tannins.

HACIENDA GRIMÓN
Gallera, 6
26131 Ventas Blancas (La Rioja)
☎: +34 618 230 161
info@haciendagrimon.com
www.haciendagrimon.es

Desvelo Garnacha 2019 T
garnacha

91

Colour: cherry, purple rim. Nose: fruit expression, red berry notes, floral, spicy. Palate: flavourful, fruity, good acidity, elegant.

Finca La Oración 2020 T
tempranillo

92

Colour: cherry, purple rim. Nose: fruit expression, red berry notes, floral, spicy, scrubland. Palate: flavourful, fruity, elegant.

Hacienda Grimón 2019 T C
85% tempranillo, 10% garnacha, 5% graciano

90

Colour: cherry, purple rim. Nose: fruit expression, red berry notes, floral, spicy. Palate: flavourful, fruity, good acidity, fleshy.

Hacienda Grimón Chardonnay 2020 B

90

Colour: bright yellow. Nose: powerful, creamy oak, ripe fruit, spicy. Palate: rich, good structure, long, toasty, fine bitter notes.

Hacienda Grimón Viura 2019 B C
viura

91

Colour: bright straw. Nose: ripe fruit, fragrant herbs, fine lees, burnt matches. Palate: full-bodied, rich, long, good acidity.

Labarona 2018 T R
95% tempranillo, 5% graciano

91

Colour: cherry, purple rim. Nose: red berry notes, wild herbs, wild herbs, smoky, spicy. Palate: fruity, flavourful, good structure, balanced, slightly dry, soft tannins.

HACIENDA URBIÓN
Ctra. Nalda, km. 9
26120 Albelda de Iregua (La Rioja)
☎: +34 941 444 233
info@vinicolareal.com
www.vinicolareal.com

Urbión 2019 T C
tempranillo, garnacha

88

Fruity, crisp, herbal, tasty.

Urbión Cuvée 2019 B BA
viura, malvasía, garnacha blanca

87

Urbión Cuvée 2020 T RB
tempranillo, garnacha, viura

87

HERMANOS FRÍAS DEL VAL
Herrerías, 79
01307 Villabuena de Álava (Araba/Álava)
☎: +34 945 386 379
info@friasdelval.com
www.friasdelval.com

Don Peduz 2021 T
tempranillo

88

Kind finish, fruity, lactic, jammy, great length, simple, varietally correct.

Hermanos Frías del Val 2016 T R
tempranillo

88

Balanced, spicy, fruity, jammy, tasty.

Hermanos Frías del Val 2018 T C
tempranillo

89

Fruity, floral, dried herbs, spicy.

Hermanos Frías del Val Selección Personal 2015 T
tempranillo

89

Fruity, spicy, jammy, herbal.

DO Ca. RIOJA / D.O.P.

Viña El Flako 2020 B FB
60% malvasía, 40% viura

90

Colour: bright straw. Nose: powerful, creamy oak, ripe fruit, spicy. Palate: rich, good structure, long, toasty, fine bitter notes.

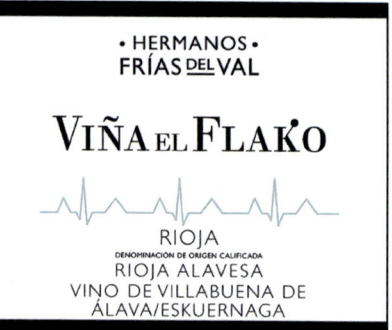

Viña El Flako 2019 B FB
60% malvasía, 40% viura

91

Defined aromas, floral. Nose: white flowers, white fruit, spicy. Palate: flavourful, rich, good structure, long.

IBAI VITICULTORES
Ctra. Navaridas s/n
01340 Elciego (Araba/Álava)
☎: +34 691 652 341
info@closibai.com
www.closibai.com

Clos Ibai 2018 B
90% viura, 10% garnacha blanca, malvasía, calagraño

89 ⚇

Citrus fruit, crisp, herbal, sweety finish.

Clos Ibai 2018 T
68% tempranillo, 17% graciano, 15% otras

91 ⚇

Exuberant, wild. Colour: cherry, purple rim. Nose: fruit expression, red berry notes, floral, spicy. Palate: flavourful, fruity, good acidity.

Clos Ibai Garnacha Blanca 2019 B
garnacha blanca

90 ⚇

Colour: straw. Nose: ripe fruit, dried herbs, faded flowers, stone fruit, lactic notes. Palate: powerful, ripe fruit, balanced, fruity, flavourful.

Clos Ibai Garnacha Tinta 2020 T
garnacha

90

Colour: bright cherry. Nose: fresh fruit, creamy oak, red berry notes. Palate: good acidity, spicy, fine tannins.

Clos Ibai Graciano 2020 T

87 ⚇

JAIME RUIZ DIAZ
Príncipes de Asturia, 4
26330 Briones (La Rioja)
☎: +34 670 306 209
jaimeru84@gmail.com

Troqueao 2019 T
50% garnacha, 50% tempranillo

92

Colour: cherry, purple rim. Nose: red berry notes, floral, creamy oak, sweet spices. Palate: flavourful, fruity, good acidity, long.

JESÚS DE MADRAZO WINES
San Ignacio de Loyola, 12 5ºG
26009 Logroño (La Rioja)
☎: +34 639 780 524
chus56madrazo@gmail.com

Jesús de Madrazo 2020 B R
95% viura, 5% rojal

90

Colour: bright yellow. Nose: creamy oak, ripe fruit, spicy, scrubland. Palate: good structure, toasty, fine bitter notes.

Jesús Madrazo 2019 T C
80% tempranillo, 8% garnacha, 4% graciano, 3% mazuelo

92

Colour: deep cherry. Nose: ripe fruit, dried herbs, creamy oak, wild herbs. Palate: powerful, ripe fruit, spicy, round tannins.

Selección Jesús Madrazo Rioja 2019 T
60% tempranillo, 28% graciano, 4% maturana

92

Colour: bright cherry. Nose: expressive, balanced, wild herbs, neat, lactic notes. Palate: fruity, easy to drink, spicy.

JOSÉ BASOCO BASOCO
Ctra. de Samaniego, s/n
01307 Villabuena de Álava (Araba/Álava)
☎: +34 657 794 964
info@fincabarronte.com
www.fincabarronte.com

Finca Barronte 2018 T C
tempranillo

90

Colour: cherry, purple rim. Nose: red berry notes, spicy, smoky, dried herbs. Palate: fruity, flavourful, fresh.

Finca Barronte Graciano 2019 T
100% graciano

90

Taut. Colour: bright cherry. Nose: fresh fruit, creamy oak, red berry notes, black fruit, spicy. Palate: good acidity, spicy, fine tannins.

Finca Matapaja 2019 T
tempranillo

90
Taut. Colour: bright cherry. Nose: fresh fruit, creamy oak, red berry notes, spicy. Palate: good acidity, spicy, fine tannins.

LA CARBONERA
Camino Los Arenales s/n
01330 Labastida (Araba/Álava)
☎: +34 938 177 400
info@torres.es
www.torres.es

Las Pisadas 2018 T
tempranillo

88
Pleasant, fruity, simple, smooth, balanced.

LA QUINTA
Castillo de Aledo, 47, ESC 5 Bajo A
30506 Molina de Segura (Murcia)
☎: +34 661 662 880
info@bodegalaquinta.com
www.bodegalaquinta.com

Ruido by La Quinta 2018 T
70% tempranillo, 22% garnacha, 5% mazuelo, 3% graciano

90
Colour: bright cherry. Nose: ripe fruit, dried herbs, spicy. Palate: ripe fruit, spicy, round tannins, easy to drink.

LA RIOJA ALTA
Avda. de Vizcaya, 8
26200 Haro (La Rioja)
☎: +34 941 310 346
Fax: +34 941 312 854
info@riojalta.com
www.riojalta.com

🏆 PODIUM

Gran Reserva 904 Selección Especial 2015 T GR
90% tempranillo, 10% graciano

97
Colour: cherry, garnet rim. Nose: balanced, complex, ripe fruit, spicy, fine reductive notes, meaty notes, earthy notes. Palate: good structure, flavourful, round tannins, balanced, great length.

Viña Alberdi 2018 T C
100% tempranillo

92
Colour: dark-red cherry. Nose: toasty, spicy, cocoa bean, black fruit, ripe fruit. Palate: flavourful, toasty, fine bitter notes.

🏆 PODIUM

La Rioja Alta Gran Reserva 890 2010 T GR
95% tempranillo, 5% graciano, mazuelo

99
Colour: cherry, garnet rim. Nose: balanced, complex, ripe fruit, spicy, fine reductive notes, cocoa bean, meaty notes. Palate: good structure, flavourful, round tannins, balanced, great length, elegant.

🏆 PODIUM

Viña Arana 2015 T GR
95% tempranillo, 5% graciano

95
Spicy, toasty, classic. Colour: dark-red cherry, garnet rim. Nose: ripe fruit, fruit preserve, aged wood nuances, tobacco, sweet spices. Palate: spicy, round tannins, long.

Viña Ardanza 2015 T R
80% tempranillo, 20% garnacha

94
Colour: dark-red cherry, garnet rim. Nose: ripe fruit, fruit preserve, tobacco, sweet spices, scrubland. Palate: spicy, round tannins, long, elegant.

🏆 PODIUM

Viña Ardanza 2016 T R
tempranillo, garnacha

95
Colour: dark-red cherry, garnet rim. Nose: ripe fruit, fruit preserve, aged wood nuances, tobacco, sweet spices, scrubland. Palate: spicy, round tannins, long, flavourful, elegant.

DO Ca. RIOJA / D.O.P.

LALOMBA
Avda. Santo Domingo, 34
26200 Haro (La Rioja)
☎: +34 941 310 295
alberto.saldon@zamoracompany.com
www.lalomba.es

Lalomba 2021 RD
92
Pleasant, floral. Colour: raspberry rose. Nose: floral, dried herbs, ripe fruit, fine lees. Palate: fruity, flavourful.

🏆 PODIUM

Lalomba Finca Iles 2019 T
95
Colour: bright cherry. Nose: complex, expressive, spicy, mineral, ripe fruit, red berry notes, toasty. Palate: full-bodied, long, great length.

Lalomba Finca Ladero 2017 T
94
Colour: dark-red cherry, garnet rim. Nose: tobacco, sweet spices, red berry notes, ripe fruit. Palate: spicy, round tannins, balanced, flavourful.

🏆 PODIUM

Lalomba Finca Valhonta 2018 T
95
Complex. Colour: deep cherry. Nose: complex, expressive, spicy, mineral, ripe fruit, black fruit. Palate: elegant, full-bodied, long, great length.

LECO PUNK WINES & VINEYARDS
26330 Briones (La Rioja)
☎: +34 941 322 360
Fax: +34 941 322 380
info@lecopunk.com
www.lecopunk.com

Leco Punk 2021 T
tempranillo
87 🌱

LMT WINES
Cerro Amurdi
31190 Cizur Menor (Navarra)
☎: +34 645 841 928
hola@lmtwines.com
www.lmtwines.com

La Tapada 2019 T
garnacha
91 🌱
Colour: cherry, purple rim. Nose: red berry notes, floral, spicy. Palate: flavourful, fruity, good acidity, long.

La Tapada 2020 T
garnacha
90 🌱
Colour: Cherry. Nose: balsamic herbs, sweet spices, scrubland. Palate: spicy, balsamic, good acidity.

LUBERRI MONJE AMESTOY
Camino de Rehoyos, s/n
01340 Elciego (Araba/Álava)
☎: +34 945 606 010
luberri@luberri.com
www.luberri.com

Biga de Luberri 2019 T C
100% tempranillo
90
Colour: bright cherry. Nose: red berry notes, spicy, balanced, neat. Palate: flavourful, fruity, good acidity, long.

Cepas Viejas de Luberri 2018 T
100% tempranillo
91
Colour: cherry, garnet rim. Nose: red berry notes, floral, spicy. Palate: fruity, good acidity, long, easy to drink.

Las Salinas Beltza 2019 T C
100% tempranillo
91
Colour: deep cherry. Nose: dried herbs, creamy oak, black fruit. Palate: powerful, ripe fruit, spicy, round tannins.

Las Salinas Zuri de Luberri 2020 B FB
80% viura, 20% malvasía
90
Colour: bright straw. Nose: dried flowers, medium intensity. Palate: flavourful, fruity, balanced, long, great length.

Luberri 2021 T MC
95% tempranillo, 5% viura
90
Colour: cherry, purple rim. Nose: fruit expression, ripe fruit, dried flowers. Palate: fruity, flavourful, balanced, easy to drink.

Monje Amestoy de Luberri 2016 T R
90% tempranillo, 10% cabernet sauvignon

91

Colour: Cherry. Nose: balsamic herbs, sweet spices, scrubland, toasty. Palate: spicy, good acidity, ripe fruit.

MACROBERT & CANALS
Soto Galo 12, Nave 1
26009 Logroño (La Rioja)
☎: +34 639 214 250
bryan@macrobertandcanals.com
www.macrobertandcanals.com

Barranco Del San Ginés 2017 T
80% tempranillo, 5% graciano, 5% garnacha, 5% mazuelo, 5% viura

92

Colour: deep cherry. Nose: ripe fruit, dried herbs, creamy oak, dark chocolate. Palate: powerful, ripe fruit, spicy, round tannins.

La Nave 2020 B
65% viura, 35% garnacha blanca

88

Pleasant, smooth, tasty.

La Nave 2020 T
35% garnacha, 35% tempranillo, 30% mazuelo

88

Slight reduction, fruity, herbal.

Laventura Tempranillo 2018 T
tempranillo

91

Colour: bright cherry. Nose: balsamic herbs, sweet spices, scrubland, ripe fruit. Palate: spicy, balsamic, good acidity.

Laventura Viura 2020 B
100% viura

92

Colour: bright straw. Nose: ripe fruit, fragrant herbs, fine lees, lactic notes. Palate: full-bodied, rich, good acidity.

Paraje de La Virgen 2020 T
80% tempranillo, 5% graciano, 5% garnacha, 5% mazuelo, 5% viura

92

Colour: dark-red cherry, garnet rim. Nose: ripe fruit, aged wood nuances, tobacco, sweet spices. Palate: spicy, round tannins, long.

MAISULAN
Camino Del Soto s/n
01309 Elvillar (Araba/Álava)
☎: +34 628 132 151
info@maisulan.com
www.maisulan.com

Maisulan 12 2018 T C
tempranillo, graciano

89 🌱

Balanced, fruity, dried herbs, jammy, tasty.

Maisulan El Hondón 2020 T
tempranillo, viura

90 🌱

Colour: deep cherry. Nose: dried herbs, creamy oak, black fruit. Palate: ripe fruit, spicy, round tannins.

Maisulan Los Magines 2020 T
tempranillo, garnacha, graciano, mazuelo, viura, malvasía

91 🌱

Colour: deep cherry. Nose: closed, black fruit, scrubland. Palate: powerful, ripe fruit, spicy, round tannins, saline.

Maisulan Sobremoro 2020 T BA
tempranillo, garnacha, viura, malvasía, calagraño

91 🌱

Colour: bright cherry. Nose: fresh fruit, creamy oak, red berry notes, floral. Palate: good acidity, spicy, flavourful.

Maisulan Txabola Graciano 2020 T
graciano

92 🌱

Colour: cherry, purple rim. Nose: fruit expression, red berry notes, scrubland, dried herbs, varietal. Palate: flavourful, fruity, fresh, balanced, fruity aftertaste, slightly dry, soft tannins.

Maisulan Txabola Tempranillo 2020 T
tempranillo

90 🌱

Colour: deep cherry. Nose: dried herbs, creamy oak, red berry notes. Palate: ripe fruit, spicy, round tannins.

MARQUÉS DE CARRIÓN
La Cadena Kalea, 14
01330 Labastida (Araba/Álava)
☎: +34 945 331 164
atcliente@jgc.es
www.garciacarrion.com

Marqués de Carrión 2018 T C
tempranillo, mazuelo, graciano

87

DO Ca. RIOJA / D.O.P.

DO Ca. RIOJA / D.O.P.

Marqués de Carrión T R
tempranillo, mazuelo, graciano
87

Pata Negra 2016 T R
tempranillo
87

Pata Negra 2017 T C
tempranillo
87

Pata Negra Selección 2019 T
86

MARQUÉS DE GRIÑÓN FAMILY ESTATES
Finca Casa de Vacas, CM-4015, Km. 3
45692 Malpica del Tajo (Toledo)
☎: +34 925 597 222
adiaz@marquesdegrinon.com

Marqués de Griñón Clásico 2019 T
100% tempranillo
88
Fruity, herbaceous, jammy.

Marqués de Griñón Selección Especial 2018 T C
100% tempranillo
88
Fruity, spicy, herbaceous, tasty.

MARQUÉS DE LA CONCORDIA FAMILY OF WINES
Ctra. El Ciego, s/n
26350 Cenicero (La Rioja)
www.marquesdelaconcordia.com

Marqués de la Concordia Rioja Santiago 2016 T R
100% tempranillo
88
Pruney, spicy, fruity, tasty.

Marqués de la Concordia Rioja Santiago 2018 T C
100% tempranillo
86

Marqués de la Concordia Rioja Santiago Segundo Año 2020 T
100% tempranillo
88
Fruity, herbal, spicy, tasty.

Paternina Banda Azul 2020 T C
80% tempranillo, 20% garnacha
86

MARQUÉS DE MURRIETA
Ctra. N-232-A, km. 402,365
26006 Logroño (La Rioja)
☎: +34 941 271 374
visitas@marquesdemurrieta.com
www.marquesdemurrieta.com

🏆 PODIUM

Capellania 2017 B R
100% viura
95
Colour: bright yellow. Nose: powerful, creamy oak, ripe fruit, spicy. Palate: rich, good structure, long, toasty, fine bitter notes.

🏆 PODIUM

Castillo Ygay 2011 T GR
84% tempranillo, 16% mazuelo
96
Colour: deep cherry, garnet rim. Nose: aged wood nuances, ripe fruit, cocoa bean, cigar, toasty. Palate: flavourful, spicy, toasty, powerful tannins.

🏆 PODIUM

Dalmau 2019 T R
86% tempranillo, 10% cabernet sauvignon, 4% graciano
96
Colour: deep cherry. Nose: ripe fruit, dried herbs, creamy oak, toasty, sweet spices. Palate: powerful, ripe fruit, spicy, round tannins.

Marqués de Murrieta 2018 T R
92
Colour: dark-red cherry, garnet rim. Nose: ripe fruit, aged wood nuances, tobacco, sweet spices. Palate: spicy, round tannins, long.

Marqués de Murrieta Primer Rosé 2021 RD
100% mazuelo
91
Colour: raspberry rose. Nose: red berry notes, floral, fragrant herbs. Palate: light-bodied, spicy, good acidity, fine bitter notes.

MARQUÉS DE TOMARES

Ctra. de Cenicero, s/n
26360 Fuenmayor (La Rioja)
☎: +34 941 451 129
info@premiumfincas.com
www.premiumfincas.com

Marqués de Tomares 2015 B GR
95% viura, 5% garnacha blanca

91

Colour: bright yellow. Nose: spicy, woody, white fruit, ripe fruit. Palate: rich, good structure, long, toasty, fine bitter notes.

Marqués de Tomares 2015 T GR
90% tempranillo, 7% graciano, 3% viura

91

Colour: bright cherry. Nose: sweet spices, ripe fruit, dark chocolate. Palate: fruity, spicy, round tannins, flavourful.

Marqués de Tomares 2016 T R
80% tempranillo, 10% graciano, 7% mazuelo, 3% viura

91

Colour: deep cherry. Nose: ripe fruit, dried herbs, creamy oak. Palate: powerful, ripe fruit, spicy, round tannins.

Marqués de Tomares 2018 B C
70% viura, 30% garnacha blanca

90

Colour: bright straw. Nose: white fruit, stone fruit, faded flowers, dried herbs. Palate: fruity, fresh, flavourful.

Marqués de Tomares 2019 T C
90% tempranillo, 10% graciano

90

Colour: cherry, garnet rim. Nose: balsamic herbs, ripe fruit, scrubland. Palate: flavourful, balsamic, spicy, soft tannins.

MARQUÉS DE VITORIA

Camino de Santa Lucía, s/n
01320 Oyón (Araba/Álava)
☎: +34 945 622 134
Fax: +34 945 601 496
info@marquesdevitoria.com
www.marquesdevitoria.com

Marqués de Vitoria 2015 T GR
tempranillo

91

Colour: cherry, purple rim. Nose: red berry notes, floral, spicy. Palate: flavourful, fruity, good acidity.

Marqués de Vitoria 2017 T R
tempranillo

90

Colour: deep cherry. Nose: dried herbs, creamy oak, red berry notes, black fruit. Palate: powerful, ripe fruit, spicy, round tannins.

Marqués de Vitoria 2019 T C
tempranillo

88

Balanced, spicy, fruity, herbaceous.

Marqués de Vitoria 2021 B
viura

87

Marqués de Vitoria Tempranillo 2021 T
tempranillo

88

Tasty, fruity, crisp.

MARQUÉS DEL PUERTO

Ctra. Logroño s/n
26360 Fuenmayor (La Rioja)
☎: +34 941 450 001
bodega@mbws.com
www.marquesdelpuerto.com

Marqués del Puerto 2016 T R
100% tempranillo

88

Balanced, spicy, creamy, jammy, roasted coffee.

Marqués del Puerto 2018 T C
88

Spicy, fruity, jammy, tasty.

MARTÍNEZ CORTA

Ctra. Cenicero, s/n
26313 Uruñuela (La Rioja)
☎: +34 948 202 200
info@bornosbodegas.com
www.bodegasmartinezcorta.com

Finca Iriarte 2019 T RB
100% tempranillo

87

Martínez Corta 2018 T C
100% tempranillo

89

Balanced, herbal, jammy, full-bodied.

DO Ca. RIOJA / D.O.P.

DO Ca. RIOJA / D.O.P.

MENDIETA OSABA WINES
Curillos, 36
01308 Lanciego (Araba/Álava)
☎: +34 945 608 140
mendi@mendietaosabawines.com
www.mendietaosabawines.com

El Camino Mendi 2020 T
mazuelo
91
Colour: cherry, purple rim. Nose: fresh fruit, red berry notes, dried flowers, spicy. Palate: flavourful, fruity, good structure, balanced, fresh, slightly dry, soft tannins.

Los Altos Mendi 2021 B
viura
90
Austere. Colour: bright straw. Nose: fragrant herbs, fine lees, dried herbs, white fruit. Palate: full-bodied, good acidity, mineral, saline.

Mendi by Mendieta Osaba 2021 T
tempranillo
89
Nose: red berry notes, floral. Palate: fruity, flavourful.

Osaba Mendi 2020 T
tempranillo
89
Fruity, dried herbs, floral, tasty.

Vascomendi V.S. 2019 T
tempranillo
89
Pruney, fruity, spicy, dried herbs, tasty.

Vascomendi V.S. 2021 B
viura, malvasía
90
Colour: bright yellow. Nose: powerful, creamy oak, ripe fruit, spicy. Palate: rich, good structure, long, toasty, fine bitter notes.

MONTES DE LEZA - LOZANO
Ctra A-124 Vitoria – Logroño, Km 60
01309 Leza (Araba/Álava)
☎: +34 945 605 197
admon.leza@bodegas-lozano.com
www.bodegas-lozano.com

Lozano Selección de Orígenes Viñas Viejas 2020 B
100% viura
90
Colour: bright straw. Nose: sweet spices, toasted bread, ripe fruit, stone fruit, dried herbs. Palate: fruity, fresh, flavourful, balanced.

Montes de Leza Edición Limitada 2019 T C
100% tempranillo
89
Balanced, spicy, dried herbs, tasty.

Montes de Leza Edición Limitada Lías 2019 B
100% viura
90
Colour: bright straw. Nose: ripe fruit, fragrant herbs, fine lees. Palate: full-bodied, long, good acidity.

Montes de Leza Tempranillo 2021 T
100% tempranillo
87

NUBORI
Camino Rio Hondillo, 1
01300 Laguardia (Araba/Álava)
☎: +34 941 183 502
administracion@nubori.es
www.bodegasnubori.com

Nubori 2015 T R
88
Spicy, dried flowers, fruity, jammy.

Nubori 2018 T C
tempranillo
85

Nubori Vendimia Seleccionada 2015 T R
tempranillo
89
Pruney, fruity, spicy, tasty.

OSTATU
Ctra. Vitoria, 1
01307 Samaniego (Araba/Álava)
☎: +34 945 609 133
ostatu@ostatu.com
www.ostatu.com

Gloria de Ostatu 2016 T BA
tempranillo
91
Colour: deep cherry. Nose: dried herbs, creamy oak, black fruit, roasted coffee, woody. Palate: ripe fruit, spicy, elegant.

Laderas Ostatu 2016 T
92% tempranillo, 8% viura
91
Colour: very deep cherry. Nose: roasted coffee, aromatic coffee, powerful, ripe fruit. Palate: smoky aftertaste, great length, round tannins.

Lore de Ostatu 2019 B FB
50% viura, 50% malvasía

91
Colour: straw. Nose: ripe fruit, dried herbs, faded flowers. Palate: powerful, ripe fruit, balanced.

Ostatu 2015 T GR
85% tempranillo, 15% graciano

93
Colour: bright cherry. Nose: red berry notes, ripe fruit, dried herbs, spicy. Palate: good structure, flavourful, balanced, slightly dry, soft tannins.

Ostatu 2016 T R
tempranillo

92
Colour: dark-red cherry, garnet rim. Nose: ripe fruit, aged wood nuances, tobacco, sweet spices. Palate: spicy, round tannins, long.

Ostatu 2019 T C
90% tempranillo, 10% graciano, mazuelo, garnacha

90
Colour: cherry, purple rim. Nose: meaty notes, ripe fruit, dried herbs, spicy. Palate: fruity, flavourful, balanced.

Selección de Ostatu 2017 T
90% tempranillo, 10% graciano

91
Colour: cherry, purple rim. Nose: red berry notes, floral, spicy, fine reductive notes. Palate: flavourful, good acidity, elegant.

Valdepedro de Ostatu 2020 T
tempranillo

89 ♣
Pruney, herbaceous, tasty, coarse tannins.

Ostatu 2021 B
85% viura, 15% malvasía

88
Fruity, dried flowers, herbal, tasty.

Ostatu 2021 RD
60% tempranillo, 30% garnacha, 10% viura

88 ♣
Racy, citrus fruit, herbal.

Ostatu 2021 T MC
90% tempranillo, 3% graciano, 3% mazuelo, 4% viura

89
Aromatic, fruity, jammy, tasty.

OXER WINES
Ctra. Navaridas, 15
01300 Laguardia (Araba/Álava)
☎: +34 616 984 118
oxerarnuak@gmail.com
www.oxerwines.com

Ahari 2020 T
60% tempranillo, 30% graciano, 10% viura

94
Colour: cherry, purple rim. Nose: red berry notes, floral, spicy, cocoa bean, earthy notes. Palate: flavourful, good acidity, long, round tannins.

🏆 PODIUM

Kalamity 2020 T
50% tempranillo, 47% garnacha, 3% viura

95
Colour: Cherry. Nose: expressive, spicy, mineral, red berry notes, hint of anise, cocoa bean, dried flowers. Palate: elegant, full-bodied, long, great length.

Suzzane 2020 T
97% garnacha, 3% viura

93
Colour: cherry, purple rim. Nose: spicy, ripe fruit, fruit preserve, scrubland. Palate: flavourful, fruity, good acidity, long.

Tartalo 2020 T
87% tempranillo, 10% viura, 3% graciano

94
Defined aromas. Colour: cherry, purple rim. Nose: red berry notes, floral, spicy, cocoa bean. Palate: flavourful, fruity, good acidity, long, grainy tannins.

PACO MULERO
Partida de la Hoya Torres s/n
30520 Jumilla (Murcia)
☎: +34 676 433 541
info@pacomulero.com
www.pacomulero.com

Prisma Tempranillo 2020 T
100% tempranillo

90
Simple, fruity, jammy, standard, varietally correct, spicy, dried flowers. Palate: ripe fruit, easy to drink.

DO Ca. RIOJA / D.O.P.

DO Ca. RIOJA / D.O.P.

PAGO DEL CAMINO
Ctra. de Nalda, Km. 9
26120 Albelda de Iregua (La Rioja)
☎: +34 941 444 233
info@vinicolareal.com
www.pagosdelcamino.com

Loriñón 2019 B
garnacha blanca, viura
88
Defined aromas, dried flowers, fruity, pleasant, standard.

Loriñón 2019 T C
87

Loriñón Cuvee 2020 T
88
Standard, simple, dried flowers, wild, smooth, kind finish.

Pagos del Camino Garnacha 2019 T
garnacha
90
Colour: Cherry. Nose: sweet spices, scrubland, dried flowers, balanced. Palate: spicy, balsamic, easy to drink.

PAGOS DEL REY
Ctra. N-232, PK 422,7
26360 Fuenmayor (La Rioja)
☎: +34 941 450 818
Fax: +34 941 450 818
rioja@pagosdelrey.com
www.pagosdelrey.com

Arnegui 2017 T R
tempranillo
88
Toasty, tasty, fruity, jammy.

Arnegui 2018 T C
tempranillo
87

Castillo de Albai 2017 T R
tempranillo
87

Castillo de Albai 2018 T C
tempranillo
87

Castillo de Albai 2021 B
viura
85

Castillo de Albai Tempranillo 2021 T
tempranillo
87

PAISAJES Y VIÑEDOS
Pza. Ibarra, 1
26330 Briones (La Rioja)
☎: +34 941 322 301
Fax: +34 941 322 302
comunicacio@vilaviniteca.es

Paisajes Cecias 2019 T
90
Colour: deep cherry. Nose: dried herbs, creamy oak, roasted coffee, black fruit. Palate: powerful, ripe fruit, spicy, round tannins.

Paisajes Valsalado 2019 T
90
Colour: cherry, purple rim. Nose: red berry notes, spicy, faded flowers. Palate: flavourful, fruity, good acidity.

QUEIRÓN
Nº 9 Barrio de Bodegas de Quel
26570 Quel (La Rioja)
☎: +34 941 234 200
info@queiron.es
www.queiron.es

Queirón de Gabriel 2011 T R
tempranillo, graciano
93
Representative. Colour: dark-red cherry, garnet rim. Nose: ripe fruit, aged wood nuances, tobacco, sweet spices, expressive, balanced. Palate: spicy, round tannins.

Queirón el Arca. Viñedo Singular 2018 T BA
garnacha
94
Colour: cherry, garnet rim. Nose: wild herbs, dried herbs, ripe fruit, fruit preserve, balanced, earthy notes. Palate: balanced, powerful, flavourful, varietal.

Queirón Ensayos Capitales Nº 3 Asoleao 2020 T
tempranillo
93
Colour: cherry, garnet rim. Nose: fruit preserve, powerful, dark chocolate, toasty, sweet spices. Palate: flavourful, sweetness, long.

Queirón mi Lugar Viñedo de Municipio 2018 T BA
tempranillo, garnacha
92
Colour: deep cherry. Nose: ripe fruit, dried herbs, creamy oak, sweet spices. Palate: powerful, ripe fruit, spicy, round tannins.

R. LÓPEZ DE HEREDIA VIÑA TONDONIA

Avda. Vizcaya, 3
26200 Haro (La Rioja)
☎: +34 941 310 244
Fax: +34 941 310 788
bodega@lopezdeheredia.com
www.tondonia.com

Viña Tondonia 2010 T R
94
Colour: deep cherry, garnet rim. Nose: aged wood nuances, ripe fruit, cocoa bean, cigar, toasty. Palate: flavourful, spicy, toasty, powerful tannins.

RAMÓN DE AYALA LETE E HIJOS

26290 Briñas (La Rioja)
☎: +34 941 310 575
bodegas@rayalaehijos.com
www.rayalaehijos.com

Deóbriga 2019 BE R BN
70% viura, 25% garnacha blanca, 5% malvasía
87

Deóbriga 2021 B FB
70% viura, 25% garnacha blanca, 5% malvasía
88
Balanced, spicy, fruity, standard, floral, great length.

Deóbriga 2021 T MC
100% tempranillo
86

Deóbriga Colección Privada 2017 T
90% tempranillo, 10% graciano
88
Pruney, spicy, balanced, toasty.

Deóbriga Selección Familiar 2018 T
90% tempranillo, 10% graciano
87

RAMÓN SAENZ ORGANIC WINES & VINEYARDS

Mayor, 12
01307 Baños de Ebro (Araba/Álava)
☎: +34 945 609 212
bodegasrs@hotmail.com
www.bodegasramonsaenz.com

Cimadago 2019 T C
tempranillo, garnacha, graciano
91
Colour: Cherry, garnet rim. Nose: balsamic herbs, sweet spices, scrubland. Palate: spicy, balsamic, good acidity.

Ramón Sáenz, Pasión de Vida 2021 T
tempranillo, viura
90
Colour: cherry, purple rim. Nose: ripe fruit, fresh fruit, violets. Palate: fruity, flavourful, balanced.

Ramón Sáenz, Pequeño Bastión 2020 T RB
tempranillo, garnacha
92
Colour: cherry, garnet rim. Nose: fruit preserve, powerful, toasty, red berry notes. Palate: flavourful, sweetness, long.

Ramón Sáenz, Piedras Rodantes 2020 T RB
tempranillo, graciano
92
Colour: deep cherry. Nose: ripe fruit, dried herbs, creamy oak. Palate: powerful, ripe fruit, spicy, round tannins.

REAL AGRADO

Camino de los Agudos s/n
26559 Aldeanueva de Ebro (La Rioja)
☎: +34 941 142 389
info@realagrado.com
www.realagrado.com

Las Planas 2017 B R
viura
91
Colour: bright golden. Nose: creamy oak, ripe fruit, stone fruit. Palate: fruity, rich, flavourful, fresh, old wood.

Real Agrado 2017 T R
tempranillo, graciano, mazuelo, garnacha
90
Colour: bright cherry. Nose: sweet spices, ripe fruit, dark chocolate. Palate: fruity, spicy, round tannins.

Real Agrado 2018 T C
garnacha, tempranillo
88
Jammy, toasty, tasty, spicy.

Real Agrado 2021 B
100% viura
85

Real Agrado 2021 RD
garnacha, viura
86

DO Ca. RIOJA / D.O.P.

DO Ca. RIOJA / D.O.P.

VA! Tempranillo 2021 T
tempranillo
89
Fruity, herbal, crisp, tasty.

RIOJA VEGA
Ctra. Logroño-Mendavia, Km. 92
31230 Viana (Navarra)
☎: +34 948 646 263
info@riojavega.com
www.riojavega.com

Rioja Vega 2018 T R
85% tempranillo, 10% graciano, 5% mazuelo
90
Colour: cherry, purple rim. Nose: ripe fruit, spicy, smoky, dried herbs. Palate: flavourful, fruity, balanced, slightly dry, soft tannins.

Rioja Vega 2019 T C
78% tempranillo, 20% graciano, 2% mazuelo
88
Spicy, jammy, fruity, tasty.

Rioja Vega Colección Tempranillo Blanco 2021 B
100% tempranillo blanco
90
Colour: bright straw. Nose: ripe fruit, stone fruit, floral, creamy oak, smoky. Palate: fruity, flavourful, toasty.

Rioja Vega Edición Limitada 2019 T C
70% tempranillo, 30% graciano
88
Fruity, spicy, herbal, tasty, crisp.

Rioja Vega Tempranillo Blanco 2019 B R
100% tempranillo blanco
91
Colour: bright straw. Nose: ripe fruit, white fruit, faded flowers, dried herbs. Palate: fruity, flavourful, fresh, good finish.

Rioja Vega Venta Jalón 2016 T R
75% graciano, 25% tempranillo
91
Colour: cherry, purple rim. Nose: ripe fruit, black fruit, creamy oak, spicy, dried herbs. Palate: fruity, flavourful, balanced, slightly dry, soft tannins.

ROSARIO VERA
Camino de la Hoya, 1
01300 Laguardia (Araba/Álava)
☎: +34 968 435 022
Fax: +34 968 453 051
info@gilfamily.es
www.gilfamily.es

Honoro Vera Rioja 2020 T
100% tempranillo
90
Colour: cherry, purple rim. Nose: red berry notes, floral, expressive. Palate: fruity, flavourful, balanced.

Honoro Vera Rioja 2021 T
100% tempranillo
91
Colour: cherry, purple rim. Nose: fruit expression, red berry notes, floral, balsamic herbs. Palate: fruity, flavourful, balanced.

Rosario Vera 2019 T
100% tempranillo
93
Colour: bright cherry. Nose: ripe fruit, dark chocolate, creamy oak, spicy. Palate: fruity, spicy, round tannins, fleshy.

Rosario Vera Amona 2021 T
100% tempranillo
92
Pleasant, fruity. Colour: cherry, purple rim. Nose: red berry notes, floral, spicy, toasty. Palate: flavourful, fruity, good acidity.

SEÑORÍO DE ARANA
La Cadena, 20
01330 Labastida (Araba/Álava)
☎: +34 945 331 150
Fax: +34 945 331 150
info@senoriodearana.com
www.senoriodearana.com

Sommelier 2015 T R
tempranillo
90
Colour: cherry, garnet rim. Nose: creamy oak, spicy, ripe fruit. Palate: flavourful, good structure, spicy, round tannins.

Sommelier 2017 T C
tempranillo
89
Pleasant, spicy, jammy, tasty, toasty, simple.

Viña del Oja 2014 T GR
tempranillo
90
Colour: cherry, garnet rim. Nose: floral, spicy, ripe fruit, smoky. Palate: flavourful, fruity, good acidity, long.

Viña del Oja 2016 T R
tempranillo
90
Balanced. Nose: red berry notes, ripe fruit, spicy, creamy oak. Palate: easy to drink, ripe fruit, spicy.

Viña del Oja 2018 T C
tempranillo
87

SEÑORÍO DE LÍBANO
Del Río, s/n
26212 Sajazarra (La Rioja)
☎: +34 941 320 066
bodega@castillodesajazarra.com
www.castillodesajazarra.com

Castillo de Sajazarra 2016 T R
97% tempranillo, 3% graciano
91
Colour: deep cherry. Nose: dried herbs, creamy oak, black fruit, woody. Palate: powerful, ripe fruit, spicy, round tannins.

Digma Graciano 2017 T R
100% graciano
91
Colour: cherry, purple rim. Nose: fruit preserve, creamy oak, smoky, sweet spices, dried herbs. Palate: fresh, flavourful, balanced, round tannins.

Digma Tempranillo 2016 T R
100% tempranillo
91
Colour: cherry, purple rim. Nose: fruit expression, red berry notes, floral, spicy. Palate: flavourful, fruity, good acidity, long.

Líbano 3 Generaciones 2018 T C
93% tempranillo, 7% graciano
87

SEÑORÍO DE SAN VICENTE
Los Remedios, 27
26338 San Vicente de la Sonsierra (La Rioja)
☎: +34 945 600 590
info@sierracantabria.com
www.sierracantabria.com

🏆 PODIUM

San Vicente 2018 T BA
tempranillo
95
Colour: cherry, purple rim. Nose: red berry notes, floral, spicy, toasty, expressive. Palate: flavourful, fruity, good acidity, grainy tannins.

SIERRA CANTABRIA
Amorebieta, 3
26338 San Vicente de la Sonsierra (La Rioja)
☎: +34 941 334 080
info@sierracantabria.com
www.sierracantabria.com

Murmurón 2021 T
90
Pleasant, fruity, jammy, great length, balanced. Nose: red berry notes, ripe fruit. Palate: easy to drink.

Sierra Cantabria 2012 T GR
tempranillo
93
Colour: dark-red cherry, garnet rim. Nose: ripe fruit, fruit preserve, spicy, fine reductive notes. Palate: round tannins, long, flavourful.

Sierra Cantabria 2016 T R
92
Colour: deep cherry. Nose: dried herbs, creamy oak, black fruit, dark chocolate. Palate: ripe fruit, spicy, round tannins.

Sierra Cantabria 2019 T C
92
Colour: deep cherry. Nose: ripe fruit, dried herbs, creamy oak, black fruit. Palate: powerful, ripe fruit, spicy, round tannins.

Sierra Cantabria 2021 B
89
Pleasant, smooth, tasty, jammy.

Sierra Cantabria 2021 RD
90
Colour: rose. Nose: fruit preserve, hot, faded flowers. Palate: fleshy, flavourful, powerful, ripe fruit.

DO Ca. RIOJA / D.O.P.

DO Ca. RIOJA / D.O.P.

Sierra Cantabria Garnacha 2018 T
92
Colour: cherry, purple rim. Nose: spicy, ripe fruit, dried flowers, faded flowers. Palate: flavourful, fruity, long, spicy, easy to drink.

Sierra Cantabria Selección 2020 T
89
Balanced, spicy, jammy, tasty, toasty.

XF Sierra Cantabria 2021 RD
garnacha, sauvignon blanc, viura, maturana blanca
91
Colour: raspberry rose. Nose: red berry notes, floral, fragrant herbs, dried herbs. Palate: spicy, good acidity, fine bitter notes.

SIERRA DE TOLOÑO
La Lleca s/n
01307 Villabuena de Álava (Araba/Álava)
info@sierradetolono.com
www.sierradetolono.com

La Dula 2020 T
garnacha
93
Colour: bright cherry. Nose: balsamic herbs, sweet spices, scrubland, stone fruit. Palate: spicy, balsamic, good acidity.

Nahikun 2021 B
viura, malvasía, otras
92
Colour: bright yellow. Nose: ripe fruit, dried herbs, faded flowers. Palate: powerful, ripe fruit, balanced.

Sierra de Toloño 2020 T
tempranillo
89
Tasty, jammy, fruity, spicy.

Sierra de Toloño 2021 B
viura
89
Pleasant, aromatic, balanced, dried flowers, dried herbs. Palate: easy to drink.

Sierra de Toloño Camino de Santa Cruz 2019 T
91 ♣
Colour: cherry, garnet rim. Nose: fruit preserve, powerful, sweet spices. Palate: flavourful, sweetness, long.

SIETE PASOS WINES
Ctra. Nac. 122, km. 311
47300 Peñafiel (Valladolid)
☎: +34 983 881 622
administracion@bodegaspenafiel.com
www.sietepasos.com

El Importante 2017 T
89
Colour: bright cherry. Nose: ripe fruit, sweet spices. Palate: fruity, spicy, round tannins, balanced.

SINODO VITIVINÍCOLA
Portillo, 8 Puerta B
26142 Villamediana de Iregua (La Rioja)
☎: +34 627 837 611
info@sinodovitivinicola.com
www.sinodovitivinicola.com

Sínodo 2020 B
85% viura, 15% sauvignon blanc
90
Colour: bright straw. Nose: stone fruit, wild herbs. Palate: fruity, flavourful, balanced.

Sínodo Centales Viñedo Singular 2020 T
85% tempranillo, 15% mazuelo, garnacha, viura, maturana
92
Colour: very deep cherry. Nose: roasted coffee, aromatic coffee, powerful. Palate: smoky aftertaste, great length, round tannins.

Sínodo Garnacha Graciano 2019 T
70% garnacha, 30% graciano
88
Toasty, spicy, pruney, jammy.

SOLAR DE SAMANIEGO
28036 Madrid (Madrid)
☎: +34 913 433 320
cofradia@solardesamaniego.es
www.solardesamaniego.com

Cabeza de Cuba 2018 T C
100% tempranillo
88
Fruity, balanced, herbal, tasty.

Carranavaridas 2019 T C
100% tempranillo
90
Colour: bright cherry. Nose: toasty, spicy, ripe fruit, dried herbs. Palate: fruity, flavourful, balanced, slightly dry, soft tannins.

Musco 2019 T C
95% tempranillo, 5% graciano
89
Fruity, spicy, dried herbs, tasty, toasty.

Solar de Samaniego 2019 T C
100% tempranillo

87

Solar de Samaniego
7 Cepas 2017 T R
90% tempranillo, 10% graciano

90

Colour: bright cherry. Nose: ripe fruit, sweet spices, creamy oak. Palate: flavourful, fruity, balanced, slightly dry, soft tannins.

Solar de Samaniego
Valcavada 2017 T R
80% tempranillo, 20% graciano

91

Colour: bright cherry. Nose: ripe fruit, sweet spices, spicy, toasty. Palate: fruity, flavourful, balanced, fresh, slightly dry, soft tannins.

TENTENUBLO WINES
La Fuente, 52-54 Sót. 2 Lis. 11-12
01308 Lanciego (Araba/Álava)
☎: +34 699 236 468
info@tentenublo.com
www.tentenublo.com

Custero 2020 T
60% garnacha, 20% tempranillo, viura, calagraño, garnacha gris, graciano

90

Colour: Cherry. Nose: ripe fruit, dried herbs, earthy notes, characterful. Palate: powerful, ripe fruit, spicy, round tannins.

Escondite del Ardacho:
Veriquete 2019 T
80% tempranillo, 10% garnacha, 5% viura, 5% malvasía

93

Crisp. Colour: deep cherry. Nose: creamy oak, black fruit, meaty notes, fine reductive notes, scrubland. Palate: powerful, ripe fruit, spicy, round tannins.

Tentenublo 2019 T
80% tempranillo, 5% garnacha, 5% viura

92

Colour: deep cherry. Nose: dried herbs, creamy oak, smoky, roasted coffee, meaty notes, black fruit. Palate: powerful, ripe fruit, spicy, round tannins.

Tentenublo 2020 B
20% malvasía, 30% Jaen blanca, 10% garnacha blanca, 40% viura

93

Oxidativ. Colour: straw. Nose: ripe fruit, dried herbs, dried flowers, lactic notes, cereal notes. Palate: powerful, ripe fruit, balanced, full-bodied.

Xérico 2018 T
85% tempranillo, 15% viura

91

Colour: cherry, purple rim. Nose: fruit expression, red berry notes, floral, spicy, toasty. Palate: flavourful, fruity, good acidity, long.

TERRAMAGNA
Ctra. Rincón del Moro, Km 10
02410 Lietor (Albacete)
☎: +34 965 211 955
earellano@grupoterramagna.com
www.grupoterramagna.com

Coso 2014 T R

90

Colour: dark-red cherry, garnet rim. Nose: overripe fruit, waxy notes, toasty, tobacco. Palate: warm, pruney, sweet tannins, good finish.

CT en Clave de DO 2018 T

89

Kind finish, toasty, tasty, spicy.

TOBELOS BODEGAS Y VIÑEDOS
Ctra. N 124, Km. 45
26290 Briñas (La Rioja)
☎: +34 941 305 630
tobelos@tobelos.com
www.tobelos.com

Quiñones de Tobelos
Viñedo Singular 2019 B BA
100% viura

92

Colour: bright straw. Nose: toasty, spicy, characterful, faded flowers. Palate: flavourful, good acidity, fine bitter notes.

Salinillas de Tobelos 2017 T
100% tempranillo

92

Colour: bright cherry. Nose: red berry notes, ripe fruit, spicy, balanced, elegant. Palate: fruity, good acidity, round tannins.

DO Ca. RIOJA / D.O.P.

Tahón de Tobelos 2015 T R
100% tempranillo

92

Colour: deep cherry. Nose: dried herbs, creamy oak, black fruit, scrubland, dark chocolate. Palate: powerful, ripe fruit, spicy, round tannins.

Tobelos 506 M. 2020 B
90% viura, 10% garnacha blanca

91

Colour: bright straw. Nose: neat, spicy, toasty, white fruit, dried flowers. Palate: rich, flavourful, spicy, easy to drink.

Tobelos Garnacha 2020 T BA
100% garnacha

90

Colour: deep cherry. Nose: dried herbs, creamy oak, black fruit, fruit preserve. Palate: powerful, ripe fruit, spicy, round tannins.

Tobelos Tempranillo 2018 T C
100% tempranillo

91

Colour: deep cherry. Nose: dried herbs, creamy oak, black fruit. Palate: ripe fruit, spicy, round tannins.

TORRE DE OÑA
Finca San Martín s/n
01309 Páganos (Araba/Álava)
☎: +34 945 621 154
info@torredeona.com
www.fincamartelo.com

Finca Martelo 2016 T R
95% tempranillo, 5% otras

94

Colour: dark-red cherry. Nose: toasty, spicy, fine reductive notes, woody. Palate: flavourful, toasty, fine bitter notes, elegant.

Finca San Martín 2019 T C
100% tempranillo

90

Colour: deep cherry. Nose: ripe fruit, dried herbs, creamy oak, aromatic coffee. Palate: ripe fruit, spicy, round tannins.

TRONADO WINES
El Olmo, 5
01330 Labastida (Araba/Álava)
☎: +34 672 255 142
info@tronadowines.com
www.tronadowines.com

Capitán Trueno 2020 T
95% tempranillo, 5% viura

91

Colour: deep cherry. Nose: ripe fruit, dried herbs, creamy oak. Palate: powerful, ripe fruit, spicy, round tannins.

UKAN WINERY
La Paz, 15
01300 Laguardia (Araba/Álava)
☎: +34 945 625 371
info@ukanwinery.com
www.ukanwinery.com

Senderos de Ukan 2019 T
100% tempranillo

90

Pleasant, crisp. Colour: cherry, garnet rim. Nose: red berry notes, spicy. Palate: easy to drink, fresh.

Ukan 2019 T
100% tempranillo

93

Colour: cherry, garnet rim. Nose: red berry notes, ripe fruit, dried flowers. Palate: fruity, fresh, balanced, good acidity, long, ripe fruit.

UVAS FELICES
Agullers, 7
08003 Barcelona (Barcelona)
☎: +34 902 327 777
www.vilaviniteca.es

La Locomotora Tempranillo 2020 T
88

Classic, dried herbs, jammy, spicy, toasty.

VICENTE GANDÍA

Ctra. Cheste a Godelleta, s/n
46370 Chiva (València/Valencia)
☎: +34 962 524 242
Fax: +34 962 524 243
info@vicentegandia.com
www.vicentegandia.es

Altos de Raiza 2020 T
100% tempranillo

87

Raiza Tempranillo 2013 T GR
100% tempranillo

89

Classic, spicy, fruity, jammy, tasty.

Raiza Tempranillo 2017 T R
tempranillo

88

Classic, fruity, smoky, spicy, tasty.

Raiza Tempranillo 2018 T C
tempranillo

87

VINOS DE CULTO BANISIO

Paseo de Gracia, 8-10, 3ª planta
08007 Barcelona (Barcelona)
☎: +34 933 631 838
info@banisio.com
www.banisio.com

Banisio 2016 T R
tempranillo

90

Colour: cherry, garnet rim. Nose: fruit preserve, powerful, toasty. Palate: flavourful, sweetness, long.

VIÑA BUJANDA

Ctra. Logroño, s/n
01320 Oyón (Araba/Álava)
☎: +34 941 450 876
Fax: +34 941 450 875
info@bujanda.com
www.vinabujanda.com

Viña Bujanda 2014 T GR
100% tempranillo

91

Colour: pale ruby, brick rim edge. Nose: old leather, fruit liqueur notes, cigar, spicy. Palate: classic aged character, bitter, reductive nuances.

Viña Bujanda 2016 T R
100% tempranillo

91

Colour: dark-red cherry, garnet rim. Nose: aged wood nuances, tobacco, sweet spices. Palate: spicy, round tannins, long.

Viña Bujanda 2019 T C
100% tempranillo

87

VIÑA DEL LENTISCO

Avda. de la Poveda, 16
01306 Lapuebla de Labarca (Araba/Álava)
☎: +34 648 117 198
info@vinovillota.com
www.vinovillota.com

Selvanevada Villota 2019 T RB
tempranillo, graciano, garnacha, mazuelo

90

Colour: Cherry. Nose: sweet spices, scrubland, ripe fruit. Palate: spicy, balsamic, good acidity, coarse tannins.

Selvanevada Villota 2020 B
viura, garnacha blanca

90

Colour: bright straw, greenish rim. Nose: fresh fruit, citrus fruit, wild herbs, fine lees. Palate: fresh, fruity, good acidity, fine bitter notes.

Villota 2018 B FB
100% viura, garnacha blanca, malvasía

91

Colour: bright yellow. Nose: dried flowers, candied fruit, fine lees, pattiserie. Palate: round, spicy, great length.

Villota 2019 T
tempranillo, garnacha, mazuelo, graciano

93

Colour: cherry, garnet rim. Nose: ripe fruit, dried herbs, expressive. Palate: good structure, flavourful, round tannins, spicy, long.

Villota Graciano 2019 T
graciano

91

Varietally correct, wild. Colour: cherry, garnet rim. Nose: meaty notes, spicy, wild herbs. Palate: flavourful, balanced, fine bitter notes.

DO Ca. RIOJA / D.O.P.

DO Ca. RIOJA / D.O.P.

VIÑA IJALBA
Ctra. Pamplona, Km. 1
26006 Logroño (La Rioja)
☎: +34 941 261 100
Fax: +34 941 261 128
vinaijalba@ijalba.com
www.ijalba.com

Ijalba 2018 B R
60% maturana blanca, 40% viura

92

Colour: bright yellow. Nose: fruit expression, white fruit, tropical fruit, dried herbs, creamy oak. Palate: flavourful, fruity, balanced.

Ijalba 2018 T R
80% tempranillo, 20% graciano

92 ♣

Colour: cherry, purple rim. Nose: fruit preserve, rose petals, faded flowers, dried herbs. Palate: fruity, flavourful, balanced, slightly dry, soft tannins.

Ijalba 2019 T C
90% tempranillo, 10% graciano

89 ♣

Animal funk, roasted coffee, jammy, dried herbs, tasty.

Ijalba 2020 B C
50% viura, 30% maturana blanca, 20% tempranillo blanco

91 ♣

Colour: bright straw. Nose: white fruit, white flowers, wild herbs, dried herbs, spicy. Palate: flavourful, balanced, round.

Ijalba Cuvèe 2020 T
70% tempranillo, 20% graciano, 10% maturana

91 ♣

Colour: deep cherry. Nose: dried herbs, creamy oak, fruit preserve. Palate: powerful, ripe fruit, spicy, round tannins.

Ijalba Maturana Blanca 2021 B
100% maturana blanca

92 ♣

Colour: bright straw. Nose: stone fruit, white fruit, wild herbs, wild herbs. Palate: fruity, fresh, flavourful.

VIÑA OLABARRI
Ctra. Haro - Anguciana, s/n
26200 Haro (La Rioja)
☎: +34 941 310 937
Fax: +34 941 311 602
info@bodegasolabarri.com
www.bodegasolabarri.com

Bikandi 2009 T R
100% tempranillo

90

Colour: dark-red cherry, garnet rim. Nose: ripe fruit, tobacco, cigar, old leather. Palate: spicy, round tannins, long.

Bikandi 2018 T C
100% tempranillo

87 ♣

Viña Olabarri 2015 T GR
tempranillo

90

Colour: dark-red cherry, garnet rim. Nose: ripe fruit, aged wood nuances, tobacco, sweet spices. Palate: spicy, round tannins.

Viña Olabarri 2016 T R
90% tempranillo, 10% graciano

89

Kind finish, toasty, tasty, jammy, classic.

Viña Olabarri 2019 T C
tempranillo

87

VIÑA REAL
Ctra. Logroño - Laguardia, Km. 4,8
01300 Laguardia (Araba/Álava)
☎: +34 945 625 255
marketing@cvne.com
www.cvne.com

Bakeder 2020 T

91

Colour: deep cherry. Nose: ripe fruit, dried herbs, creamy oak. Palate: powerful, ripe fruit, spicy, round tannins.

Pagos de Viña Real 2018 T
100% tempranillo

93

Colour: bright cherry. Nose: expressive, spicy, mineral, red berry notes. Palate: elegant, full-bodied, long, great length.

Viña Real 2016 T GR
95% tempranillo, 5% graciano
93
Colour: dark-red cherry, garnet rim. Nose: ripe fruit, aged wood nuances, tobacco, sweet spices. Palate: spicy, round tannins, long.

Viña Real 2021 RD
70% viura, 30% tempranillo, garnacha
88
Pleasant, floral, crisp, simple, wild, smooth.

VIÑA SALCEDA
Ctra. Cenicero, Km. 3
01340 Elciego (Araba/Álava)
☎: +34 945 606 125
Fax: +34 945 606 069
info@vinasalceda.com
www.vinasalceda.com

Conde de la Salceda 2017 T R
100% tempranillo
93
Colour: Cherry. Nose: complex, spicy, mineral, cocoa bean, hint of anise. Palate: elegant, full-bodied, good structure, flavourful.

El Mirador de la Salceda 2021 B
viura
88
Austere, herbal, jammy, tasty.

El Mirador de la Salceda 2021 T MC
100% tempranillo
91
Colour: cherry, purple rim. Nose: fruit expression, red berry notes, floral. Palate: fruity, flavourful, balanced.

La Salceda 2019 T C
100% tempranillo
90
Taut. Colour: ruby red. Nose: red berry notes, fresh fruit, wild herbs. Palate: fleshy, fresh, fruity.

Viña Salceda 2018 T R
95% tempranillo, 5% graciano
91
Colour: deep cherry. Nose: ripe fruit, dried herbs, creamy oak. Palate: powerful, ripe fruit, spicy, round tannins.

VIÑAS SILENCIOSAS
Crta Navaridas, s/n
01340 Elciego (Araba/Álava)
☎: +34 691 652 341
laulla@yahoo.com

Laulla 2021 T
tempranillo, viura
91
Colour: cherry, purple rim. Nose: fruit expression, floral, spicy, wild herbs, red berry notes, black fruit. Palate: flavourful, fruity, good acidity.

Viñas Silenciosas La Cuadrada 2019 T
tempranillo, graciano, garnacha, viura
91
Colour: cherry, garnet rim. Nose: fruit preserve, fruit liqueur notes, powerful, bakery, dried herbs, wild herbs. Palate: flavourful, long, round tannins.

Viñas Silenciosas Posadero 2019 B
viura, calagraño
93
With personality. Colour: bright golden. Nose: acetaldehyde, honeyed notes, wild herbs, dried herbs, sweet spices. Palate: fresh, fruity, flavourful, balanced, old wood, good finish.

DO Ca. RIOJA / D.O.P.

Viñas Silenciosas Posadero 2019 T
tempranillo, graciano

91

Colour: cherry, garnet rim. Nose: powerful, fruit preserve, black fruit, wild herbs, dried herbs. Palate: flavourful, toasty, good structure.

Viñas Silenciosas Regoyos 2019 T
tempranillo, viura

89

Pruney, full-bodied, rustic, tasty, wild.

VIÑEDOS DE PÁGANOS
Ctra. Navaridas, s/n
01309 Páganos (Araba/Álava)
☎: +34 945 600 590
info@sierracantabria.com
www.sierracantabria.com

Calados del Puntido 2018 T BA
tempranillo

93

Colour: dark-red cherry. Nose: toasty, spicy, cocoa bean, ripe fruit, black fruit. Palate: flavourful, toasty, fine bitter notes.

🏆 PODIUM

El Puntido 2019 T
tempranillo

96

Colour: cherry, garnet rim. Nose: complex, expressive, spicy, mineral, toasty, creamy oak. Palate: elegant, full-bodied, long, great length.

🏆 PODIUM

La Nieta 2019 T
tempranillo

98

Exuberant, fruity, mineral. Colour: bright cherry. Nose: complex, expressive, spicy, mineral, fruit expression, fresh fruit, red berry notes. Palate: elegant, full-bodied, long, great length.

VIÑEDOS DEL CONTINO
Finca San Rafael
01321 Laserna - Laguardia (Araba/Álava)
☎: +34 945 600 201
marketing@cvne.com
www.cvne.com

Contino 2017 T GR
82% tempranillo, 5% mazuelo, 10% graciano, 3% garnacha

94

Colour: dark-red cherry, garnet rim. Nose: ripe fruit, aged wood nuances, tobacco, sweet spices. Palate: spicy, round tannins, long.

Contino 2018 T R
85% tempranillo, 6% mazuelo, 6% graciano, 2% garnacha

92

Colour: light cherry, garnet rim. Nose: elegant, fine reductive notes, aged wood nuances, spicy. Palate: fine tannins, fine bitter notes.

Contino 2020 B
80% viura, 20% garnacha blanca

92

Colour: bright straw. Nose: ripe fruit, floral, fine lees, mineral, spicy. Palate: full-bodied, complex, spicy, long, elegant.

Contino Garnacha 2020 T
100% garnacha

92

Colour: bright cherry. Nose: balsamic herbs, sweet spices, scrubland, red berry notes. Palate: spicy, balsamic, good acidity.

Contino Mazuelo 2019 T
100% mazuelo

93

Colour: deep cherry. Nose: ripe fruit, dried herbs, creamy oak, red berry notes, sweet spices. Palate: powerful, ripe fruit, spicy, round tannins.

Contino Viña del Olivo 2019 T

94

Colour: dark-red cherry, garnet rim. Nose: ripe fruit, aged wood nuances, tobacco, sweet spices, red berry notes. Palate: spicy, round tannins, long.

VIÑEDOS HERMANOS HERNÁIZ
Ctra. Sto. Domingo – Haro Km 31,5 s/n
26241 Baños de Rioja (La Rioja)
☎: +34 941 300 105
correo@bodegaslaemperatriz.com
www.hermanoshernaiz.com

El Jardín de la Emperatriz 2019 T
91% tempranillo, 7% garnacha, 2% graciano

91

Colour: bright cherry. Nose: creamy oak, black fruit, spicy. Palate: good acidity, spicy, slightly green tannins.

El Jardín de la Emperatriz 2021 B
100% viura

89

Pleasant, kind finish, citrus fruit, floral, fruity.

El Pedal de Hermanos Hernáiz Tempranillo 2021 T
tempranillo

90

Colour: cherry, purple rim. Nose: fruit expression, red berry notes, floral. Palate: fruity, flavourful, balanced.

Finca La Emperatriz 2018 B
100% viura

91

Colour: bright straw. Nose: ripe fruit, fragrant herbs, fine lees, creamy oak. Palate: full-bodied, rich, long, good acidity.

Finca La Emperatriz 2018 T R
76% tempranillo, 20% garnacha, 4% viura

90

Colour: deep cherry. Nose: ripe fruit, dried herbs, creamy oak, dark chocolate. Palate: powerful, ripe fruit, spicy, round tannins.

Las Cenizas 2020 T
tempranillo

92

Colour: deep cherry. Nose: dried herbs, creamy oak, dark chocolate, black fruit. Palate: powerful, ripe fruit, spicy, round tannins.

VIÑEDOS REAL RUBIO

Avda. La Rioja s/n
26559 Aldeanueva de Ebro (La Rioja)
☎: +34 941 163 672
Fax: +34 941 163 672
export@realrubio.es
www.realrubio.es

Real Rubio 2010 T GR
tempranillo, graciano

91

Colour: dark-red cherry. Nose: ripe fruit, dried herbs, wild herbs. Palate: ripe fruit, spicy, round tannins, easy to drink.

Real Rubio 2013 T R
tempranillo, graciano

88

Crisp, herbaceous, fruity, spicy, balanced, great length.

Real Rubio 2018 T C
tempranillo, graciano, graciano

90

Colour: bright cherry. Nose: sweet spices, ripe fruit, fragrant herbs. Palate: fruity, spicy, round tannins.

Real Rubio 2021 B
sauvignon blanc, garnacha blanca

88

Balanced, herbal, dried flowers, standard, simple, smooth.

Real Rubio 2021 RD
garnacha

88

Fruity, herbal, floral, tasty.

Real Rubio Finca El Tordillo 2019 T
garnacha

92

Colour: Cherry. Nose: complex, expressive, spicy, floral, wild herbs. Palate: elegant, full-bodied, long, great length.

VIÑEDOS RUIZ JIMÉNEZ

Ctra. Comarcal LR-115, Km. 43
26559 Aldeanueva de Ebro (La Rioja)
☎: +34 941 163 577
Fax: +34 941 163 577
info@vinedosruizjimenez.es
www.vinedosruizjimenez.es

Osoti 2018 B
tempranillo blanco

88 🌱

Woody, spicy, tasty, powerful, toasty, age nuances.

Osoti 2021 T
tempranillo

88 🌱

Fruity, tasty, jammy.

Osoti La Era 2017 T RB
100% tempranillo

89 🌱

Pruney, classic, fruity.

Pago de Valcaliente 2018 B R
maturana blanca

90 🌱

Colour: bright yellow. Nose: powerful, creamy oak, ripe fruit, spicy, woody. Palate: rich, good structure, long, toasty, fine bitter notes.

Pago de Valcaliente Garnacha 2021 T C
garnacha

86 🌱

Paisajes Embotellados Viñedos Ruiz Jiménez 2021 T

87 🌱

DO Ca. RIOJA / D.O.P.

VIÑEDOS SIERRA CANTABRIA

Calle Fuente de la Salud s/n
26338 San Vicente de la Sonsierra (La Rioja)
☎: +34 941 334 080
info@sierracantabria.com
www.sierracantabria.com

DO Ca. RIOJA / D.O.P.

🏆 **PODIUM**

Amancio 2018 T
tempranillo

96

Colour: dark-red cherry, garnet rim. Nose: aged wood nuances, tobacco, sweet spices, toasty, new oak, dark chocolate, ripe fruit, black fruit. Palate: spicy, round tannins, long.

🏆 **PODIUM**

Finca El Bosque 2019 T
tempranillo

96

Powerful, jammy. Colour: deep cherry. Nose: dried herbs, creamy oak, ripe fruit, black fruit, toasty, earthy notes. Palate: powerful, ripe fruit, spicy, round tannins.

Sierra Cantabria Colección Privada 2020 T

94

Colour: dark-red cherry, garnet rim. Nose: ripe fruit, aged wood nuances, sweet spices, red berry notes, toasty. Palate: spicy, round tannins, long.

Sierra Cantabria Cuvèe 2018 T

93

Colour: deep cherry. Nose: ripe fruit, dried herbs, creamy oak, red berry notes, black fruit. Palate: powerful, ripe fruit, spicy, round tannins.

🏆 **PODIUM**

Sierra Cantabria Mágico 2018 T
tempranillo, graciano, mazuelo, garnacha

98

Kind finish, complex, floral, exuberant. Colour: deep cherry. Nose: complex, expressive, spicy, mineral, red berry notes, ripe fruit, black fruit. Palate: full-bodied, long, great length, flavourful.

Sierra Cantabria Organza 2020 B
viura, malvasía, garnacha blanca

93

Colour: bright yellow. Nose: ripe fruit, dried herbs, faded flowers, fine lees, brioche, stone fruit. Palate: powerful, ripe fruit, balanced.

VIÑEDOS Y BODEGAS LA MARQUESA

Herrería, 76
01307 Villabuena de Álava (Araba/Álava)
☎: +34 945 609 085
info@valserrano.com
www.valserrano.com

El Ribazo 2017 T

90

Colour: Cherry. Nose: smoky, spicy. Palate: fruity, good structure, flavourful, easy to drink, good acidity.

Finca Monteviejo 2017 T
95% tempranillo, 5% garnacha, graciano

90

Colour: Cherry, purple rim. Nose: fruit expression, red berry notes, floral, spicy. Palate: flavourful, fruity, good acidity, long.

Valserrano 2015 T GR
90% tempranillo, 10% graciano

91

Colour: dark-red cherry, garnet rim. Nose: ripe fruit, fruit preserve, aged wood nuances, tobacco, sweet spices. Palate: spicy, round tannins, long.

Valserrano 2017 B GR
viura, malvasía

92

Colour: bright yellow. Nose: dried flowers, candied fruit, fine lees, pattiserie. Palate: round, spicy, long, great length.

Valserrano 2018 T C
95% tempranillo, 5% mazuelo

89

Smoky, toasty, tasty, jammy, spicy, standard, kind finish.

Valserrano 2021 B FB
viura, malvasía

89
Fruity, herbal, tasty.

ZINIO BODEGAS
Ctra. Cenicero, s/n
26313 Uruñuela (La Rioja)
☎: +34 941 371 319
info@ziniobodegas.com
www.ziniobodegas.com

Lágrimas de María 2017 T R
100% tempranillo

88
Full-bodied, balanced, spicy, jammy.

Lágrimas de María 2018 T C
100% tempranillo

87

Sancho Garcés 2017 T R
100% tempranillo

89
Balanced, spicy, dried herbs, jammy, tasty, lactic.

Sancho Garcés 2018 T C
100% tempranillo

88 ♣
Balanced, spicy, dried herbs, toasty.

Zinio Street Art Collection 2020 T
90% tempranillo, 10% graciano

90
Colour: cherry, purple rim. Nose: spicy, red berry notes, ripe fruit, dried herbs. Palate: flavourful, fruity, good acidity, balanced.

Zinio Tempranillo 2017 T R
100% tempranillo

88
Full-bodied, balanced, jammy, roasted coffee.

ZUAZO GASTÓN BODEGAS Y VIÑEDOS
Las Norias, 2
01320 Oyón (Araba/Álava)
☎: +34 945 601 526
zuazogaston@zuazogaston.com
www.zuazogaston.com

ZG 2021 B
75% tempranillo, 25% viura

89
Citrus fruit, crisp, fruity, herbal, tasty.

ZG Edición Limitada 2020 T C
90
Colour: Cherry. Nose: ripe fruit, dried herbs, spicy. Palate: ripe fruit, spicy, round tannins, balanced.

Zuazo Gastón 2017 T R
90% tempranillo, 10% graciano

90
Colour: Cherry. Nose: balsamic herbs, sweet spices, scrubland, wild herbs. Palate: spicy, good acidity, round tannins, easy to drink.

Zuazo Gastón 2020 T C
90
Colour: cherry, purple rim. Nose: red berry notes, ripe fruit, dried flowers, faded flowers, dried herbs. Palate: flavourful, fruity, good acidity, long.

Zuazo Gastón Vendimia Seleccionada 2021 T
91
Colour: deep cherry. Nose: dried herbs, creamy oak, black fruit. Palate: powerful, ripe fruit, spicy, round tannins.

DO Ca. RIOJA / D.O.P.

DO. RUEDA
CONSEJO REGULADOR

Real, 8
47490 Rueda (Valladolid)
☎: +34 983 868 248 - Fax: +34 983 868 135
@: crdo.rueda@dorueda.com
www.dorueda.com

LOCATION:

In the provinces of Valladolid (53 municipal districts), Segovia (17 municipal districts) and Ávila (2 municipal districts). The vineyards are situated on the undulating terrain of a plateau and are conditioned by the influence of the river Duero that runs through the northern part of the region.

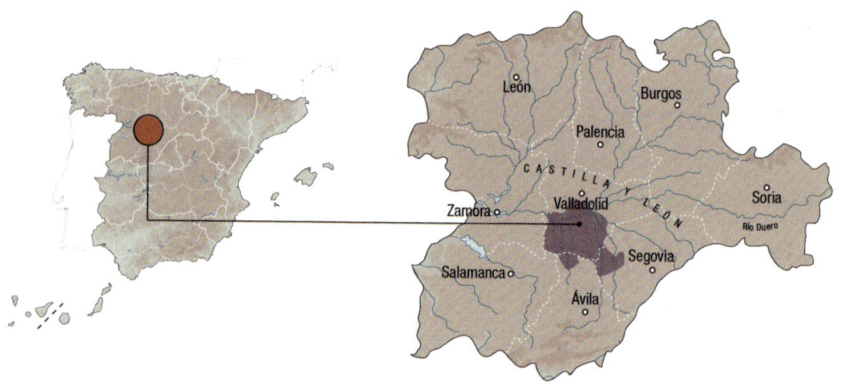

GRAPE VARIETIES:

WHITE: Verdejo, Viura, Sauvignon Blanc, Viognier, Chardonnay and Palomino Fino.

RED: Tempranillo, Cabernet Sauvignon, Syrah, Merlot and Garnacha.

FIGURES:

Vineyard surface: 20,727 – **Wine-Growers:** 1,648 – **Wineries:** 76 – **2021 Harvest rating:** Very Good – **Production 21:** 87,494,558 litres – **Market percentages:** 86% National - 14% International.

SOIL:

Many pebbles on the surface. The terrain is stony, poor in organic matter, with good aeration and drainage. The texture of the soil is variable although, in general, sandy limestone and limestone predominate.

CLIMATE:

Continental in nature, with cold winters and short hot summers. Rainfall is concentrated in spring and autumn. The average altitude of the region is between 600 m and 700 m, and only in the province of Segovia does it exceed 800 m.

CLASSIFICATION OF YOUNG WINE HARVESTS PEÑÍNGUIDE

2017	2018	2019	2020	2021
VERY GOOD	VERY GOOD	VERY GOOD	VERY GOOD	VERY GOOD

ALEGRE WINES & SPIRITS

Balmes, 345
08006 Barcelona (Barcelona)
☎: +34 935 641 262
info@alegrews.com
www.alegrews.com

Mirlo Blanco 2021 B
100% verdejo

91

Colour: yellow. Nose: expressive, ripe fruit, fine lees, spicy. Palate: full-bodied, long, balanced.

ALREGI

Pol. Ind. Empordà Internacional – C/Garrotxa, 8
17469 Vilamalla (Girona/Gerona)
☎: +34 972 526 061
alregi@alregi.es
www.alregi.es

Obsceno 2021 B
100% verdejo

88

Fruity, herbaceous, jammy, dried herbs, tropical.

ÁLVAREZ Y DÍEZ

Juan Antonio Carmona, 12
47500 Nava del Rey (Valladolid)
☎: +34 983 850 136
bodegas@alvarezydiez.com
www.alvarezydiez.com

Bento 2021 B
100% verdejo

88 ♣

Smooth, tasty, pleasant, kind finish, tropical.

Hacienda Alcaraz 2021 B
100% verdejo

88 ♣

Pleasant, kind finish, citrus fruit, floral.

Mantel Blanco Sauvignon Blanc 2021 B
100% sauvignon blanc

89

Aromatic, herbal, varietally correct, citrus fruit, great length, smooth.

Mantel Blanco Verdejo 2020 B FB
verdejo

92

Colour: bright yellow. Nose: powerful, creamy oak, ripe fruit, spicy. Palate: good structure, long, toasty, fine bitter notes.

Mantel Blanco Verdejo 2021 B
100% verdejo

91

Colour: bright straw, greenish rim. Nose: fresh fruit, citrus fruit, wild herbs, fine lees, hint of anise. Palate: fresh, fruity, good acidity, fine bitter notes.

Silga 2021 B
100% verdejo

90

Colour: bright straw, greenish rim. Nose: fresh fruit, citrus fruit, wild herbs. Palate: fresh, fruity, good acidity, fine bitter notes.

AVELINO VEGAS

Grupo Calvo Sotelo, 8
40460 Santiuste de San Juan Bautista (Segovia)
☎: +34 921 596 002
ana@avelinovegas.com
www.avelinovegas.com

Casa de la Vega Verdejo 2021 B
verdejo

88

Aromatic, kind finish, standard, balanced.

Montespina Verdejo 2021 B
verdejo

89

Pleasant, standard, herbal, simple, smooth.

Nicte Verdejo Eco B

89 ♣

Simple, tasty, herbal, crisp.

DO RUEDA / D.O.P.

Circe Verdejo 2021 B
verdejo
88
Pleasant, kind finish, crisp, fruity.

Montespina Sauvignon 2021 B
sauvignon blanc
88
Balanced, citrus fruit, herbal, crisp.

BARDOS
Zarzillo s/n
47491 Rueda (Valladolid)
☎: +34 941 843 557
marketing@vintae.com
www.bardos.wine/en

Bardos Verdejo 2021 B
verdejo
88
Fruity, dried herbs, jammy, tasty.

BELA
Ctra. de Palencia-Aranda de Duero, km. 68
09443 Aranda de Duero (Burgos)
☎: +34 947 112 783
marketing@cvne.com
www.bodegabela.com

Bela Gran Vino de Rueda 2020 B
100% verdejo
91
Colour: bright straw. Nose: ripe fruit, dried herbs, faded flowers. Palate: powerful, ripe fruit, balanced.

Bela Gran Vino de Rueda 2021 B
100% verdejo
91
Colour: bright straw. Nose: ripe fruit, dried herbs, faded flowers. Palate: powerful, ripe fruit, balanced.

Finca Vallejo 2020 B
100% verdejo
88
Citrus fruit, spicy, crisp, fruity, herbal.

Finca Vallejo 2021 B
100% verdejo
90
Colour: bright straw. Nose: fruit expression, ripe fruit, floral, hint of anise. Palate: flavourful, fresh, good acidity, fruity aftertaste.

BELONDRADE
Paraje de los Levantes. Quinta San Diego
47491 La Seca (Valladolid)
☎: +34 983 481 001
info@belondrade.com
www.belondrade.com

🏆 PODIUM

Belondrade y Lurton 2020 B FB
verdejo
95 🌱
Colour: bright golden. Nose: expressive, ripe fruit, fine lees, mineral, creamy oak. Palate: full-bodied, complex, spicy, long, elegant.

BERONIA RUEDA
Camino de la Peña, s/n. Finca La Perdiz
47490 Rueda (Valladolid)
☎: +34 983 664 460
prensa@gonzalezbyass.es
www.beronia.com

Beronia Rueda 2021 B
verdejo
89
Balanced, citrus fruit, kind finish, thin.

Las lias de Beronia 2020 B
92
Colour: straw, greenish rim. Nose: wild herbs, grassy, fresh fruit, hint of anise, spicy. Palate: flavourful, full-bodied, rich.

BODEAGAS PINDAL
Los Carriles, s/n
47220 Pozaldez (Valladolid)
☎: +34 617 194 404
pindal@pindalverdejo.com
www.pindalverdejo.com

Pindal Verdejo 2021 B
100% verdejo
86

Vardalón 2021 B
100% sauvignon blanc
87

BODEGA CAMPO ELISEO
Calle Nueva, 12
47391 La Seca (Valladolid)
☎: +34 983 034 030
bodega@francoislurton.es

Campo Eliseo Cuvée Alegre 2021 RD
90
Colour: light cherry. Nose: ripe fruit, creamy oak, sweet spices. Palate: flavourful, spicy.

Campo Eliseo Cuvée Alegre Verdejo 2021 B
91
Colour: yellow, greenish rim. Nose: wild herbs, white fruit, ripe fruit, balanced, neat. Palate: flavourful, fruity, rich.

Campo Eliseo Verdejo 2020 B FB
92
Spicy, jammy. Colour: bright yellow. Nose: powerful, creamy oak, ripe fruit, spicy. Palate: rich, good structure, long.

Hermanos Lurton Sauvignon Blanc 2021 B
sauvignon blanc
89
Herbaceous, defined aromas, fruity, wild, smooth, balanced.

Hermanos Lurton Valentín Rosé 2021 RD
tempranillo
90 🌱
Colour: raspberry rose. Nose: elegant, red berry notes, floral, fragrant herbs. Palate: light-bodied, good acidity, fine bitter notes, balanced, fresh.

Hermanos Lurton Verdejo 2021 B
verdejo
91
Colour: bright straw. Nose: ripe fruit, dried herbs, faded flowers. Palate: powerful, ripe fruit, balanced.

BODEGA CARLOS MORO
47400 Medina del Campo (Valladolid)
☎: +34 983 803 346
bodega@bodegacarlosmoro.com
www.bodegacarlosmoro.com

CM Verdejo 2018 B FB
100% verdejo
90
Colour: bright yellow. Nose: creamy oak, ripe fruit, spicy, toasty. Palate: rich, good structure, long, toasty, fine bitter notes.

BODEGA CUATRO RAYAS
Camino de la Fuentecilla, s/n
47491 La Seca (Valladolid)
☎: +34 983 816 320
info@cuatrorayas.es
www.cuatrorayas.es

🏆 **PODIUM**

61 Dorado en Rama 2021 BF Solera S
50% verdejo, 50% palomino
95
Colour: iodine, amber rim. Nose: complex, dry nuts, toasty, sweet spices. Palate: bitter, fine solera notes, long, spicy, saline.

Amador Diez Verdejo Cuvée 2017 B FB
100% verdejo
92
Colour: bright yellow. Nose: creamy oak, ripe fruit, spicy, candied fruit, grassy. Palate: good structure, long, toasty, fine bitter notes.

Amador Diez Verdejo Cuvée 2018 B FB
100% verdejo
93
Colour: bright yellow. Nose: toasty, citrus fruit, neat. Palate: rich, flavourful, fine bitter notes, good acidity, spicy, toasty.

Cuatro Rayas # Co. Organic Verdejo 2021 B
100% verdejo
90 🌱
Colour: bright straw. Nose: fresh fruit, citrus fruit, fine lees, hint of anise. Palate: fresh, fruity, good acidity, fine bitter notes.

DO RUEDA / D.O.P.

DO RUEDA / D.O.P.

Amador Diez Verdejo Cuvée 2015 B FB
100% verdejo

94

Colour: straw. Nose: dried herbs, sweet spices, candied fruit, dry nuts, wild herbs, dry stone. Palate: powerful, ripe fruit, balanced, flavourful.

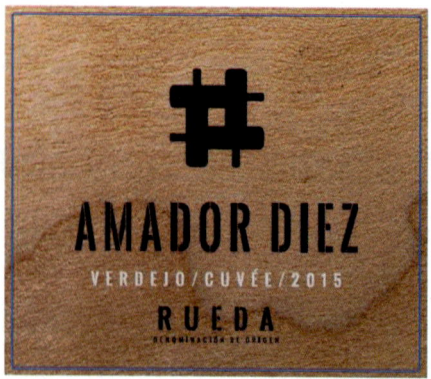

Cuatro Rayas 1935 Verdejo 2021 B
100% verdejo

90

Colour: bright straw. Nose: hint of anise, fine lees, fresh fruit. Palate: easy to drink, varietal, fresh.

Cuatro Rayas Cuarenta Vendimias Cuvée 2020 B
verdejo

93

Colour: bright yellow. Nose: fine lees, bakery, ripe fruit, stone fruit. Palate: flavourful, fruity, easy to drink, spicy.

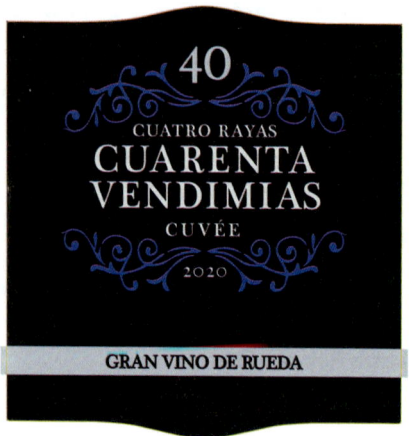

Cuatro Rayas Cuarenta Vendimias Verdejo 2021 B
verdejo

91

Colour: bright straw. Nose: fragrant herbs, fine lees, white fruit, ripe fruit. Palate: rich, good acidity, balanced, fine bitter notes.

Cuatro Rayas Longverdejo Viñedos Centenarios 2020 B
100% verdejo

93

Colour: yellow. Nose: expressive, ripe fruit, fine lees, dry stone. Palate: full-bodied, complex, spicy, long, elegant.

Cuatro Rayas Organic Verdejo 2021 B
100% verdejo

89 ❧

Balanced, herbal, aromatic, crisp, fruity, great length.

Cuatro Rayas Vendimia Nocturna Verdejo 2021 B
100% verdejo

89

Pleasant, smooth, tasty, varietally correct.

Cuatro Rayas Viñedos Centenarios 2021 B
100% verdejo

92

Colour: bright straw, greenish rim. Nose: wild herbs, white fruit, balanced. Palate: fruity, good acidity, fine bitter notes.

Green & Social Verdejo 2021 B
100% verdejo

88 ❧

Pleasant, dried herbs, crisp, wild, balanced.

BODEGA EMINA RUEDA

Ctra. Medina del Campo - Olmedo, Km. 1,4
47400 Medina del Campo (Valladolid)
☎: +34 983 803 346
eminarueda@emina.es
www.emina.es/bodega-emina-rueda

Emina Sauvignon 2021 B
100% sauvignon blanc

88

Aromatic, dried flowers, fruity, herbal, jammy.

Emina Verdejo 2018 B FB
100% verdejo

92

Colour: bright yellow. Nose: creamy oak, ripe fruit, spicy, wild herbs. Palate: rich, good structure, long, toasty, fine bitter notes.

Emina Verdejo 2021 B
100% verdejo

88

Citrus fruit, herbal, crisp, balanced.

BODEGA HERMANOS DEL VILLAR

Cordel de las Merinas, s/n
47490 Rueda (Valladolid)
☎: +34 983 868 904
Fax: +34 983 868 905
pablo@orodecastilla.com
www.orodecastilla.com

Crisol BE
100% verdejo

84

Oro de Castilla Finca Los Hornos 2019 B
100% verdejo

92

Colour: bright straw. Nose: ripe fruit, dried herbs, faded flowers. Palate: powerful, ripe fruit, balanced.

Oro de Castilla Sauvignon Blanc 2021 B
100% sauvignon blanc

89

Pleasant, aromatic, standard, balanced, herbal, wild.

Oro de Castilla Verdejo 2021 B
100% verdejo

90

Colour: bright straw. Nose: wild herbs, varietal, fresh, balanced. Palate: fine bitter notes, easy to drink, good finish.

Quivira Verdejo 2021 B
100% verdejo

88

Pleasant, kind finish, aromatic, dried flowers.

BODEGA HERMANOS LURTON

Calle Nueva 12
47491 La Seca (Valladolid)
☎: +34 983 034 030
Fax: +34 980 032 680
bodega@francoislurton.es
www.francoislurton.es

Hermanos Lurton Cuesta de Oro 2021 B FB

91 ❧

Colour: straw. Nose: stone fruit, ripe fruit, tropical fruit, spicy. Palate: ripe fruit, balanced, flavourful.

DO RUEDA / D.O.P.

DO RUEDA / D.O.P.

BODEGA LA GRANADILLA
Ctra. Tordesillas, km. 17
47500 Nava del Rey (Valladolid)
☎: +34 685 926 698
info@bodegaslagranadilla.es
www.bodegaslagranadilla.es

Dominio de La Granadilla Verdejo 2021 B
verdejo
88
Colour: straw. Nose: ripe fruit, dried herbs. Palate: powerful, ripe fruit, balanced.

Dominio La Granadilla Sauvignon 2021 B
sauvignon blanc
89
Aromatic, tropical, fruity, great length.

BODEGA MATARROMERA
Ctra. Renedo- Pesquera Km. 30
47359 Valbuena de Duero (Valladolid)
www.matarromera.es

Melior Sauvignon 2021 B
sauvignon blanc
89
Pleasant, kind finish, aromatic, tasty, jammy.

Melior Verdejo 2021 B
100% verdejo
90
Colour: bright straw. Nose: fruit expression, ripe fruit, floral. Palate: flavourful, fresh, good acidity, fruity aftertaste.

BODEGA REINA DE CASTILLA, S. COOP.
Cº de la Moya, s/n
47491 La Seca (Valladolid)
☎: +34 983 816 667
bodega@reinadecastilla.es
www.reinadecastilla.es

Divino Minino 2021 B
85% verdejo, 15% viura
86

El Bufón Verdejo 2021 B
100% verdejo
90
Colour: bright straw. Nose: fruit expression, ripe fruit, floral. Palate: flavourful, fresh, good acidity, fruity aftertaste.

Isabelino Verdejo 2021 B
100% verdejo
89
Pleasant, kind finish, aromatic, smooth.

La Loca Reina 2021 B
100% verdejo
91
Colour: bright straw. Nose: ripe fruit, fragrant herbs, fine lees, sweet spices. Palate: full-bodied, rich, long, good acidity.

Reina de Castilla Verdejo 2021 B
100% verdejo
91
Colour: bright straw, greenish rim. Nose: fresh fruit, citrus fruit, wild herbs, hint of anise, cedar wood. Palate: fresh, fruity, good acidity, fine bitter notes.

Reina de Castilla Verdejo Selección Cuvée 2020 B FB
100% verdejo
91
Colour: bright straw, greenish rim. Nose: fresh fruit, citrus fruit, wild herbs. Palate: fresh, fruity, good acidity, fine bitter notes.

BODEGA REJADORADA
Crta. De San Román a Morales Km. 0,9
47530 San Román de Hornija (Valladolid)
☎: +34 980 693 089
rejadorada@rejadorada.com
www.rejadorada.com

Rejadorada Verdejo 2021 B
100% verdejo
87

BODEGA TRES PILARES
El Rancho, 3
47491 La Seca (Valladolid)
☎: +34 983 816 682
bodega3pilares@gmail.com
www.bodega3pilares.com

Hermanos Fernández 2020 B
verdejo
91
Colour: bright straw, greenish rim. Nose: fresh fruit, citrus fruit, wild herbs. Palate: fresh, fruity, good acidity, fine bitter notes, flavourful.

BODEGA VALDEHERMOSO
Ctra. Nava del Rey - Rueda, km. 12,6
47500 Nava del Rey (Valladolid)
☎: +34 983 090 936
valdehermoso@valdehermoso.com
www.bodegavaldehermoso.com

Lagar del Rey 2019 B FB
verdejo
89
Toasty, citrus fruit, fruity, tasty.

Lagar del Rey
Sauvignon Blanc sobre Lías 2021 B
sauvignon blanc

88
Kind finish, dried flowers, tasty, tropical.

Lagar del Rey Verdejo sobre Lías 2021 B
verdejo

90
Colour: bright straw. Nose: fine lees, hint of anise, wild herbs. Palate: rich, long, good acidity.

Sotavento 100%
Sauvignon Blanc Lías 2021 B

90
Colour: straw. Nose: expressive, white flowers, jasmine, dried herbs. Palate: flavourful, fruity, balanced.

Sotavento 100% Verdejo Lías 2021 B
100% verdejo

89
Balsamic herbs, aromatic, smooth, tasty.

Viña Perez 100% Verdejo 2021 B
100% verdejo

87

BODEGA VALDRINAL
Camino de la Vega s/n, Polígono 4, Parcelas 130-133
40533 Aldehorno (Segovia)
☎: +34 634 578 839
general@valdrinal.com
www.valdrinal.com

Valdrinal de Santamaría 2021 B
100% verdejo

88
Citrus fruit, standard, crisp, slight reduction.

BODEGA Y VIÑEDOS MARTÍN BERDUGO
Camino de la Colonia, s/n
09400 Aranda de Duero (Burgos)
☎: +34 947 506 331
bodega@martinberdugo.com
www.martinberdugo.com

Martín Berdugo Verdejo 2021 B
verdejo

89
Pleasant, floral, herbal, jammy.

BODEGAS ALCONDE
Ctra. de Calahorra, s/n
31260 Lerin (Navarra)
☎: +34 948 530 058
Fax: +34 948 530 589
manuel@bodegasalconde.com
www.bodegasalconde.com

Tres Villas B
verdejo

89
Aromatic, varietally correct, smooth, tasty.

BODEGAS AURA
Ctra. Autovía del Noroeste, Km. 175
47490 Rueda (Valladolid)
☎: +34 683 588 222
info@bodegaslan.com
www.lanencasa.com/rueda

Aura Sauvignon Blanc B
sauvignon blanc

88
Tropical, powerful, simple, floral.

Aura Selección
Parcela Avutarda 2019 B FB
verdejo

91
Colour: bright yellow. Nose: powerful, creamy oak, ripe fruit, spicy. Palate: long, toasty, fine bitter notes, dry wood.

Aura Verdejo Vendimia Nocturna 2021 B
verdejo

91
Colour: bright straw, greenish rim. Nose: fresh fruit, citrus fruit, wild herbs. Palate: fresh, fruity, good acidity, fine bitter notes.

DO RUEDA / D.O.P.

BODEGAS CASTELO DE MEDINA

Ctra. CL-602, Km. 48
47465 Villaverde de Medina (Valladolid)
☎: +34 983 831 932
Fax: +34 983 831 857
info@castelodemedina.com
www.castelodemedina.com

Castelo de Medina 2020 B FB
100% verdejo

90

Colour: bright straw, greenish rim. Nose: fresh fruit, citrus fruit, wild herbs, spicy. Palate: fresh, fruity, good acidity, fine bitter notes.

Castelo de Medina Sauvignon Blanc 2020 B
100% sauvignon blanc

90

Colour: bright yellow. Nose: fragrant herbs, fine lees, citrus fruit. Palate: good acidity, easy to drink, good finish.

Castelo de Medina Verdejo 2021 B
100% verdejo

90

Colour: bright straw, greenish rim. Nose: fresh fruit, citrus fruit, wild herbs. Palate: fresh, fruity, good acidity, fine bitter notes.

Castelo de Medina Verdejo Vendimia Seleccionada 2020 B
100% verdejo

92

Colour: bright yellow. Nose: powerful, spicy, toasted bread, stone fruit, wild herbs. Palate: rich, good structure, long, toasty, fine bitter notes.

El Fisgón 2021 B
85% verdejo, 15% sauvignon blanc

88

Kind finish, crisp, citrus fruit, smooth.

BODEGAS CERROSOL

Camino Villagonzalo, s/n
40460 Santiuste de San Juan Bautista (Segovia)
☎: +34 921 596 326
Fax: +34 921 596 351
administracion@bodegascerrosol.com
www.bodegascerrosol.com

Doña Beatriz Verdejo 2021 B

87

Kind finish, fruity, tasty, simple.

Doña Beatriz Verdejo Cepas Viejas 2021 B
100% verdejo

89

With personality, jammy, dried herbs, wild, varietally correct, balanced.

BODEGAS COMENGE

Camino del Castillo, s/n
47316 Curiel de Duero (Valladolid)
☎: +34 983 880 363
admin@comenge.com
www.comenge.com

Colección Comenge Verdejo 2021 B
100% verdejo

90

Colour: bright straw. Nose: ripe fruit, dried herbs, faded flowers, grassy. Palate: powerful, ripe fruit, balanced.

Comenge Verdejo Gran Vino de Rueda 2020 B
verdejo

92

Colour: bright yellow. Nose: powerful, creamy oak, ripe fruit, spicy. Palate: rich, good structure, long, toasty, fine bitter notes.

BODEGAS CONVENTO DE LAS CLARAS

Ctra. Valladolid, 0
47231 Serrada (Valladolid)
☎: +34 983 880 150
bodega@bodegasconventodelasclaras.com
www.bodegasconventodelasclaras.com

Convento Las Claras Verdejo sobre Lías 2020 B FB
100% verdejo

92

Colour: straw. Nose: dried herbs, faded flowers, stone fruit, pattiserie. Palate: powerful, ripe fruit, balanced.

BODEGAS COPABOCA

Autovía A-62, Salida 148
47100 Tordesillas (Valladolid)
☎: +34 983 395 655
comunicacion@copaboca.es
www.copaboca.com

Juan Galindo 100% Verdejo 2020 B
100% verdejo

91

Colour: bright yellow. Nose: powerful, creamy oak, ripe fruit, sweet spices, wild herbs. Palate: good structure, toasty, fine bitter notes.

BODEGAS DE ALBERTO

Ctra. de Valdestillas, 2
47231 Serrada (Valladolid)
☎: +34 983 559 107
Fax: +34 983 559 084
info@dealberto.com
www.dealberto.com

🏆 PODIUM

De Alberto Dorado Verdejo 100% B Solera
verdejo

95

Colour: dark mahogany. Nose: candied fruit, fruit liqueur notes, spicy, varnish, dry nuts. Palate: fine solera notes, bitter, spicy, long.

De Alberto Edición Limitada 2020 B
verdejo

90

Colour: bright straw. Nose: white fruit, wild herbs, white flowers. Palate: fruity, flavourful, balanced.

De Alberto Pálido B PL

92

Colour: yellow. Nose: complex, dry nuts, macerated fruit, faded flowers, flor yeasts. Palate: rich, long, dry, great length, complex.

De Alberto Dorado Verdejo Dulce BF D
verdejo

93

Colour: bright golden. Nose: overripe fruit, cocoa bean, candied fruit, honeyed notes. Palate: complex, flavourful, fruity.

DO RUEDA / D.O.P.

DO RUEDA / D.O.P.

De Alberto Ecológico 100% Verdejo 2021 B
verdejo

89 🌱

Crisp, fruity, herbal, smooth, subtle.

De Alberto sobre Lías Verdejo 100% 2020 B
100% verdejo

91

Colour: bright straw. Nose: ripe fruit, fragrant herbs, fine lees. Palate: full-bodied, good acidity.

BODEGAS ERESMA - LA SOTERRAÑA

Ctra. N-601, Km. 151
47410 Olmedo (Valladolid)
☎: +34 983 601 026
info@bodegaslasoterrana.com
www.bodegaslasoterrana.com

Eresma + Cuvée Especial 2018 B
100% verdejo

92

Colour: bright yellow. Nose: powerful, creamy oak, ripe fruit, spicy, wild herbs. Palate: rich, good structure, long, toasty, fine bitter notes.

Eresma + Cuvée Especial 2019 B
100% verdejo

91

Colour: yellow. Nose: macerated fruit, candied fruit, spicy. Palate: flavourful, dry, balanced.

Eresma Fermentado Barrica 2020 B FB
100% verdejo

90

Colour: bright yellow. Nose: creamy oak, ripe fruit, spicy, wild herbs. Palate: good structure, toasty, fine bitter notes.

Eresma Sauvignon Blanc Vendimia Seleccionada 2021 B
100% sauvignon blanc

86

Eresma Verdejo sobre Lías 2021 B
100% verdejo

90

Colour: straw. Nose: ripe fruit, dried herbs, faded flowers. Palate: powerful, ripe fruit, balanced.

Eresma Verdejo Vendimia Seleccionada 2021 B
100% verdejo

90

Colour: bright straw, greenish rim. Nose: fresh fruit, citrus fruit, wild herbs. Palate: fresh, fruity, good acidity, fine bitter notes.

BODEGAS FÉLIX LORENZO CACHAZO

Ctra. Medina del Campo, Km. 9
47220 Pozaldez (Valladolid)
☎: +34 983 822 008
administracion@cachazo.com
www.cachazo.com

Carrasviñas 100% Verdejo 2021 B
verdejo

90

Colour: bright straw, greenish rim. Nose: fresh fruit, citrus fruit, wild herbs. Palate: fresh, fruity, good acidity, fine bitter notes.

Carrasviñas 2020 B FB
verdejo

90

Colour: bright yellow. Nose: powerful, ripe fruit, toasty, smoky. Palate: long, toasty, fine bitter notes.

Carrasviñas Dorado B
verdejo, palomino

92

Colour: old gold. Nose: spicy, flor yeasts, dry nuts. Palate: powerful, flavourful, rich.

BODEGAS FÉLIX SANZ

Santísimo Cristo, 28
47490 Rueda (Valladolid)
☎: +34 983 868 044
export@bodegasfelixsanz.es
www.bodegasfelixsanz.es

Viña Cimbrón Sauvignon 2021 B
100% sauvignon blanc

88

Kind finish, crisp, smooth, varietally correct. Nose: hint of anise.

Viña Cimbrón Verdejo 2021 B
100% verdejo

89

Citrus fruit, crisp, herbal, tasty.

Viña Cimbrón Verdejo Selección 2020 B
100% verdejo

90

Colour: bright straw. Nose: ripe fruit, fine lees, spicy. Palate: rich, long, good acidity, easy to drink.

BODEGAS FRONTAURA

Santiago, 17 5º
47001 Valladolid (Valladolid)
☎: +34 983 880 488
crobles@nexusfrontaura.com
www.bodegasfrontaura.com/es

Frontaura Verdejo 2019 B

92

Colour: bright yellow. Nose: dried flowers, candied fruit, fine lees, pattiserie. Palate: round, spicy, long, great length.

BODEGAS FRUTOS VILLAR

Ctra. Burgos - Portugal km. 113,7
47270 Cigales (Valladolid)
☎: +34 983 586 868
Fax: +34 983 580 180
bodegasfrutosvillar@bodegasfrutosvillar.com
www.bodegasfrutosvillar.com

María de Molina Verdejo 2021 B
100% verdejo

87

Muruve Verdejo 2021 B
100% verdejo

88

Citrus fruit, balanced, crisp, herbal.

Viña Cansina Verdejo 2021 B
100% verdejo

86

BODEGAS GARCÍAREVALO

Pza. San Juan, 4
47230 Matapozuelos (Valladolid)
☎: +34 983 832 914
enologo@garciarevalo.com
www.garciarevalo.com

Finca Tres Olmos Classic 2021 B
100% verdejo

90 🌱

Colour: bright straw, greenish rim. Nose: fresh fruit, citrus fruit, wild herbs, fine lees. Palate: fresh, fruity, good acidity, fine bitter notes.

Finca Tres Olmos Lías 2021 B
100% verdejo

90 🌱

Colour: bright straw, greenish rim. Nose: fresh fruit, citrus fruit, wild herbs, fine lees. Palate: fresh, fruity, good acidity, fine bitter notes.

Harenna - Tinaja 2020 B
100% verdejo

91

Colour: bright straw. Nose: fragrant herbs, fine lees, citrus fruit, bakery. Palate: long, good acidity, balanced.

Harenna 2020 B
100% verdejo

91

Colour: bright straw. Nose: ripe fruit, fine lees, dried herbs. Palate: full-bodied, rich, long, good acidity.

BODEGAS GARCIGRANDE

Aradillas, s/n
47490 Rueda (Valladolid)
☎: +34 913 982 303
info@hispanobodegas.com
www.hispanobodegas.com

12 Linajes Verdejo 2020 B
100% verdejo

91 🌱

Colour: bright straw. Nose: ripe fruit, fine lees, dried flowers. Palate: rich, long, good acidity.

12 Linajes Verdejo 2021 B
verdejo

90 🌱

Colour: bright straw, greenish rim. Nose: fresh fruit, citrus fruit, wild herbs. Palate: fresh, fruity, good acidity, fine bitter notes.

Señorío de Garcigrande Verdejo 2021 B
verdejo

90 🌱

Colour: yellow. Nose: fresh fruit, wild herbs. Palate: fresh, fruity, good acidity, fine bitter notes.

DO RUEDA / D.O.P.

DO RUEDA / D.O.P.

BODEGAS JOSÉ PARIENTE

Ctra. de Rueda, km. 2.5
47491 La Seca (Valladolid)
☎: +34 983 816 600
info@josepariente.com
www.josepariente.com

José Pariente Cuvée Especial 2019 B
100% verdejo

94 🏆

Colour: bright yellow. Nose: dried flowers, candied fruit, fine lees, pattiserie. Palate: round, spicy, long, great length.

José Pariente Cuvée Especial 2020 B
verdejo

93 🏆

Colour: bright straw. Nose: ripe fruit, fragrant herbs, fine lees, characterful. Palate: full-bodied, rich, long, good acidity.

🏆 PODIUM

José Pariente Finca Las Comas 2018 B
100% verdejo

95

Colour: yellow, greenish rim. Nose: dry stone, ripe fruit, fine lees, expressive, spicy. Palate: flavourful, elegant, fine bitter notes, long.

José Pariente Finca Las Comas 2019 B
verdejo

94

Colour: yellow. Nose: ripe fruit, creamy oak, bakery. Palate: flavourful, complex, fine bitter notes, spicy.

José Pariente Verdejo 2021 B
100% verdejo

91

Colour: bright straw. Nose: expressive, white flowers, jasmine, dried herbs. Palate: flavourful, fruity, balanced.

BODEGAS MARQUÉS DE RISCAL

Ctra. N-VI, km. 172,600
47490 Rueda (Valladolid)
☎: +34 983 868 029
Fax: +34 983 868 563
rherrero@marquesderiscal.com
www.marquesderiscal.com

Finca Montico 2021 B
verdejo

93 🏆

Colour: bright straw, greenish rim. Nose: fresh fruit, citrus fruit, wild herbs, fine lees. Palate: fresh, fruity, good acidity, fine bitter notes, balanced.

Marqués de Riscal Limousin 2021 B FB
verdejo

92

Colour: bright yellow. Nose: ripe fruit, spicy, creamy oak, wild herbs. Palate: rich, good structure, long, fine bitter notes.

Marqués de Riscal Sauvignon Blanc 2021 B
sauvignon blanc

91 🏆

Colour: bright straw, greenish rim. Nose: fresh fruit, citrus fruit, wild herbs, spicy. Palate: fresh, fruity, good acidity, fine bitter notes, flavourful.

Marqués de Riscal Verdejo Organic 2021 B
verdejo

90 🏆

Colour: bright straw, greenish rim. Nose: fresh fruit, citrus fruit, wild herbs, hint of anise. Palate: fresh, fruity, good acidity, easy to drink.

BODEGAS MOCÉN

Arribas, 7-9
47490 Rueda (Valladolid)
☎: +34 983 868 533
info@bodegasmocen.com
www.bodegasmocen.es

Alta Plata Verdejo 2021 B
100% verdejo

89

Pleasant, kind finish, tropical, smooth, tasty.

Hachón Verdejo Viura 2021 B
verdejo, viura

86

La Bien Pintá 2021 B

87

Mocén Sauvignon Blanc 2021 B
sauvignon blanc

87

Mocén Verdejo 2020 B
100% verdejo

89

Spicy, yeasty notes, tasty, jammy, great length.

Renacce 2019 BE BN
100% verdejo

87

Mocén Verdejo Selección Especial 2021 B
100% verdejo

89
Citrus fruit, standard, crisp, herbal.

Renacce 2020 B
70% verdejo, 30% chardonnay

91
Colour: straw. Nose: ripe fruit, dried herbs, faded flowers. Palate: powerful, ripe fruit, balanced.

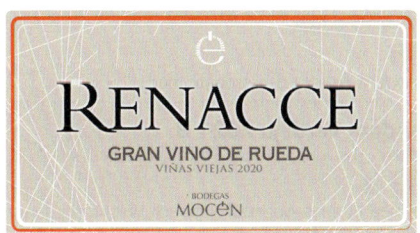

BODEGAS NAIA
Camino San Martín
47491 La Seca (Valladolid)
☎: +34 648 084 128
esanz@terraselecta.com
www.bodegasnaia.com

K-Naia 2021 B
85% verdejo, 15% sauvignon blanc

89
Varietally correct, crisp, dried herbs, bitter, balsamic herbs.

Naia 2021 B
100% verdejo

90
Colour: bright straw, greenish rim. Nose: fresh fruit, citrus fruit, wild herbs, fine lees. Palate: fresh, fruity, good acidity, fine bitter notes.

Naiades 2019 B FB
100% verdejo

94
Colour: bright yellow. Nose: expressive, fine lees, dry nuts, smoky, sweet spices, citrus fruit. Palate: full-bodied, complex, spicy, long.

S-Naia 2021 B
100% sauvignon blanc

89
Pleasant, aromatic. Nose: grassy, hint of anise.

DO RUEDA / D.O.P.

BODEGAS NIEVA

Camino Real, s/n
40447 Nieva (Segovia)
☎: +34 921 504 628
info@vinedosdenieva.com
www.martue.com

Blanco Nieva Pie Franco 2021 B
100% verdejo

92

Colour: bright straw. Nose: ripe fruit, dried herbs, faded flowers, fruit expression. Palate: powerful, ripe fruit, balanced.

Blanco Nieva Pie Franco Gran Vino de Rueda sobre Lías 2020 B
verdejo

93

Colour: bright yellow. Nose: powerful, ripe fruit, spicy, fine lees, lees reduction notes, earthy notes. Palate: rich, good structure, long, fine bitter notes.

Blanco Nieva Sauvignon Blanc 2021 B
sauvignon blanc

90

Slight reduction. Colour: bright straw, greenish rim. Nose: fresh fruit, citrus fruit, wild herbs. Palate: fresh, fruity, good acidity, fine bitter notes.

Blanco Nieva Verdejo 2021 B
verdejo

90

Colour: straw. Nose: expressive, white flowers, dried herbs, ripe fruit. Palate: flavourful, fruity, balanced.

BODEGAS PANDORA

Ctra. Nava del Rey, Km 1
47490 Rueda (Valladolid)
☎: +34 983 487 087
hablamos@bodegaspandora.com
www.bodegaspandora.com

Pandra Sauvignon Blanc 2019 B BA
100% sauvignon blanc

90 🌱

Colour: bright straw. Nose: expressive, white flowers, jasmine, dried herbs. Palate: flavourful, fruity, balanced.

Pandora Sauvignon Blanc 2021 B
100% sauvignon blanc

88

Standard, fruity, crisp, smooth, herbal.

Pandora Verdejo 2021 B
verdejo

89 🌱

Defined aromas, varietally correct, wild, tasty, jammy, balanced, citrus fruit.

Pandora Verdejo sobre Lías 2021 B
verdejo

89

Pleasant, jammy, wild, tasty, standard, citrus fruit.

Pandra Verdejo 2020 B FB
100% verdejo

91
Colour: bright yellow. Nose: powerful, creamy oak, ripe fruit, spicy. Palate: rich, good structure, long, toasty, fine bitter notes.

BODEGAS PITA
Camino Sendero del Monte s/n
47494 Rubi de Bracamonte (Valladolid)
☎: +34 983 824 240
info@bodegaspita.com
www.bodegaspita.com

Pita 2021 RD
100% garnacha

88 ♣
Kind finish, fruity, sweeties, smooth.

Pita Finca La Bonera Tempranillo 2018 T C
100% tempranillo

89 ♣
Pleasant, classic, toasty, jammy.

Pita Finca La Cantera 2018 B FB
100% verdejo

92 ♣
Colour: bright straw. Nose: expressive, ripe fruit, fine lees, brioche, bakery. Palate: full-bodied, complex, spicy, long, elegant.

Pita Sauvignon Blanc 2020 B
100% sauvignon blanc

88 ♣
Kind finish, yeasty notes, standard, smooth, citrus fruit.

Pita Verdejo (Dominio de Verderrubi) 2021 B
100% verdejo

90 ♣
Colour: bright straw. Nose: fragrant herbs, hint of anise, varietal. Palate: fresh, balanced.

BODEGAS PORTIA
Antigua Ctra. N-I, km. 170
09370 Gumiel De Izán (Burgos)
☎: +34 947 102 700
Fax: +34 947 512 140
info@bodegasportia.com
www.bodegasportia.com

Portia Verdejo 2021 B
verdejo

88
Kind finish, aromatic, floral, dried flowers.

BODEGAS PRIUS
47491 La Seca (Valladolid)
☎: +34 609 816 057
solar@solarmsancho.com
www.solarmsancho.com

Prius de Moraña Verdejo 2021 B
86

BODEGAS PROTOS
CL-610, Medina-La Seca, km 32,5
47491 La Seca (Valladolid)
☎: +34 983 816 608
fvillalba@bodegasprotos.com
www.bodegasprotos.com

Protos Verdejo 2019 B R
100% verdejo

92
Colour: bright yellow. Nose: creamy oak, ripe fruit, spicy, toasted bread, pattiserie. Palate: good structure, long, toasty, fine bitter notes.

Protos Verdejo 2021 B
100% verdejo

91
Colour: bright straw. Nose: fragrant herbs, wild herbs, hint of anise, balanced, varietal. Palate: flavourful, fine bitter notes, varietal.

Protos Verdejo Cuvée 2021 B
100% verdejo

91 ♣
Colour: bright straw, greenish rim. Nose: fresh fruit, citrus fruit, wild herbs, hint of anise. Palate: fresh, fruity, good acidity, fine bitter notes.

DO RUEDA / D.O.P.

DO RUEDA / D.O.P.

BODEGAS R & G
Ramón y Cajal 7, 1ºA
01007 Vitoria-Gasteiz (Araba/Álava)
☎: +34 945 150 589
araex@araex.com
www.araex.com

Rolland Galarreta Parcela 123 2020 B
100% verdejo
92 🌿
Colour: bright yellow. Nose: ripe fruit, dried herbs, faded flowers, spicy. Palate: powerful, ripe fruit, balanced.

BODEGAS RAMÓN BILBAO RUEDA
Autovía A6 Km. 168,
Paraje Las Amedias, Pol. 7, Parc. 14
47490 Rueda (Valladolid)
☎: +34 941 310 295
info@bodegasramonbilbao.es
www.ramonbilbao.es

Ramón Bilbao Verdejo 2021 B
89
Aromatic, standard, fruity, tasty, balanced.

Ramón Bilbao Organic Verdejo 2021 B
90 🌿
Kind finish, defined aromas. Nose: wild herbs, fresh fruit, varietal. Palate: balanced, balsamic, ripe fruit, flavourful.

Ramón Bilbao
Edición Limitada Lías 2019 B
verdejo
92
Colour: bright straw. Nose: fine lees, wild herbs, balanced, neat. Palate: rich, long, good acidity.

Ramón Bilbao Sauvignon Blanc 2021 B
88
Pleasant, crisp, herbal, varietally correct, defined aromas, simple.

BODEGAS RODRÍGUEZ Y SANZO
Manuel Azaña, 9
47014 Valladolid (Valladolid)
☎: +34 983 150 150
comunicacion@valsanzo.com
www.rodriguezsanzo.com

Rodríguez Sanzo Bajo Velo 2020 B
93
Creamy, kind finish. Colour: bright yellow. Nose: creamy oak, ripe fruit, spicy, pattiserie. Palate: rich, good structure, long, toasty, fine bitter notes.

Sanzo Viñas Viejas 2021 B FB
verdejo
90 🌿
Colour: bright straw. Nose: ripe fruit, floral, stone fruit. Palate: flavourful, fresh, good acidity, fruity aftertaste.

BODEGAS RUEDA PÉREZ

Boyón, 17
47220 Pozaldez (Valladolid)
☎: +34 650 454 657
info@bodegasruedaperez.es
www.bodegasruedaperez.es

José Galo Vendimia Seleccionada 2021 B
100% verdejo

88

Pleasant, varietally correct, tasty, jammy.

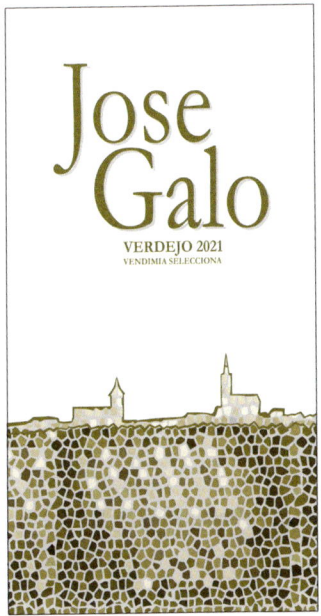

Viña Burón Verdejo 2021 B
verdejo

89

Citrus fruit, kind finish, balanced, herbal, tasty.

Zapadorado Verdejo 2021 B
100% verdejo

86

BODEGAS SEÑORÍO DE NAVA

Crta. Gumiel de Mercado – Orquillas Km 6.4
09443 Gumiel de Mercado (Burgos)
☎: +34 987 209 712
lafinca@lafinca.es
www.senoriodenava.es

Señorío de Nava Verdejo 2021 B
100% verdejo

86

BODEGAS TORREDEROS

Ctra. Valladolid, Km. 289,300
09318 Fuentelisendo (Burgos)
☎: +34 947 532 627
Fax: +34 947 532 731
administracion@torrederos.com
www.torrederos.com

Torrederos Verdejo 2021 B
100% verdejo

88

Balanced, herbal, tasty, wild, standard.

BODEGAS VAL DE VID

Ctra. de Valladolid - Medina del Campo, Km. 26.300
47231 Serrada (Valladolid)
☎: +34 983 559 914
administracion@valdevid.es
www.valdevid.es

Condesa Eylo Verdejo 2021 B
100% verdejo

91

Slight reduction. Colour: bright straw, greenish rim. Nose: fresh fruit, citrus fruit, wild herbs. Palate: fresh, fruity, good acidity, fine bitter notes.

Musgo Verdejo 2021 B

87

Val de Vid Verdejo 2021 B
100% verdejo

90

Colour: bright straw, greenish rim. Nose: fresh fruit, citrus fruit, wild herbs. Palate: fresh, fruity, good acidity, fine bitter notes.

Val de Vid Verdejo sobre Lías 2020 B BA
100% verdejo

92

Colour: bright yellow. Nose: powerful, creamy oak, ripe fruit, spicy, wild herbs, stone fruit. Palate: rich, good structure, toasty, fine bitter notes.

BODEGAS VATAN

Julio Romera de Torres, 12
29700 Vélez-Málaga (Málaga)
☎: +34 952 504 706
pedidos@jorgeordonez.es
www.jorgeordonez.es

Nisia 2021 B
100% verdejo

92

Colour: bright golden. Nose: ripe fruit, dried herbs, faded flowers, toasty, creamy oak. Palate: powerful, ripe fruit, balanced.

DO RUEDA / D.O.P.

DO RUEDA / D.O.P.

Nisia Las Suertes 2020 B
verdejo

94

Colour: bright yellow. Nose: fragrant herbs, fine lees, sweet spices, creamy oak, ripe fruit, stone fruit. Palate: full-bodied, rich, long, good acidity.

Nisia Las Suertes 2021 B
100% verdejo

93

Colour: bright yellow. Nose: powerful, creamy oak, spicy, ripe fruit, stone fruit. Palate: rich, good structure, long, toasty, fine bitter notes.

BODEGAS VERACRUZ
Juan Antonio Carmona, 1
47500 Nava del Rey (Valladolid)
☎: +34 670 581 157
bodegas@bodegasveracruz.com
www.bodegasveracruz.com

Ermita Veracruz Verdejo 2021 B
100% verdejo

92

Colour: bright straw, greenish rim. Nose: fresh fruit, citrus fruit, wild herbs, grassy. Palate: fresh, fruity, good acidity, fine bitter notes.

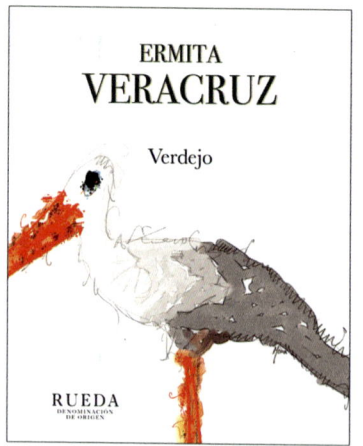

Veracruz Edición sobre Lías Verdejo 2021 B
100% verdejo

90

Colour: bright straw. Nose: fruit expression, ripe fruit, floral. Palate: flavourful, fresh, good acidity, fruity aftertaste.

BODEGAS VETUS
Ctra. Toro a Salamanca, Km. 9,5
49800 Toro (Zamora)
☎: +34 980 056 012
Fax: +34 980 056 012
vetus@bodegasvetus.com
www.bodegasvetus.com

Flor de Vetus Verdejo 2021 B
verdejo

89

Pleasant, aromatic, simple, tasty.

BODEGAS VIORE
Miguel Hernández, 31
47490 Rueda (Valladolid)
☎: +34 941 454 050
Fax: +34 941 454 529
bodega@bodegasriojanas.com
www.bodegasriojanas.com

Pregón 2021 B
100% verdejo

87

Viña Albina Verdejo 2021 B
100% verdejo

87

Viore Verdejo 2021 B
100% verdejo

88

Pleasant, varietally correct, smooth, wild.

Viore Verdejo sobre Lías B
verdejo

90

Pronounced acidity. Colour: bright straw. Nose: ripe fruit, white fruit, fine lees, neat. Palate: flavourful, rich.

BODEGAS Y VIÑEDOS ÁNGEL LORENZO CACHAZO

Estación, 53
47220 Pozaldez (Valladolid)
☎: +34 983 822 481
comercial@martivilli.com
www.martivilli.com

Martivillí Sauvignon Blanc 2021 B
100% sauvignon blanc

88
Pleasant, standard, herbal, tropical.

Martivillí Verdejo 2021 B
100% verdejo

90
Colour: straw. Nose: ripe fruit, dried herbs, faded flowers. Palate: powerful, ripe fruit, balanced.

BODEGAS Y VIÑEDOS MAYOR DE CASTILLA

Ctra. CL-610, km. 26,7
47491 La Seca (Valladolid)
☎: +34 968 758 190
www.garciacarrion.es

Mayor de Castilla Verdejo 2021 B
verdejo

87

Pata Negra Sauvignon Blanc 2021 B
sauvignon blanc

86

Pata Negra Verdejo 2021 B
verdejo

86

Solar de la Vega Verdejo 2021 B
verdejo

87

Viña Arnaiz Verdejo 2021 B
verdejo

88
Citrus fruit, balanced, spicy, herbal, tasty.

BODEGAS Y VIÑEDOS MONTEABELLÓN

Calvario, s/n
09318 Nava de Roa (Burgos)
☎: +34 947 550 000
Fax: +34 947 550 219
comunicacion@monteabellon.com
www.monteabellon.com

Monteabellón Verdejo 2021 B
100% verdejo

89
Citrus fruit, herbal, crisp, tasty.

BODEGAS Y VIÑEDOS NEO

Ctra. N-122, Km. 274,5
09391 Castrillo de La Vega (Burgos)
☎: +34 947 514 393
ivan@bodegasneo.com
www.bodegasneo.com

Neo Viñas Viejas 2021 B
100% verdejo

89
Spicy, full-bodied, roasted coffee, herbal.

BODEGAS Y VIÑEDOS SHAYA

Ctra. Aldeanueva del Codonal s/n
40642 Aldeanueva del Codonal (Segovia)
☎: +34 968 435 022
info@orowines.com

Arindo 2021 B
100% verdejo

90
Colour: bright straw, greenish rim. Nose: fresh fruit, citrus fruit, wild herbs. Palate: fresh, fruity, good acidity, fine bitter notes.

Honoro Vera Verdejo 2021 B
100% verdejo

89
Pleasant, smooth, tasty, herbal.

DO RUEDA / D.O.P.

Shaya 2021 B
100% verdejo

93 🌱

Colour: bright straw. Nose: ripe fruit, dried herbs, faded flowers, hint of anise, balsamic herbs. Palate: powerful, ripe fruit, balanced.

Shaya Habis 2020 B FB
100% verdejo

94

Colour: bright straw. Nose: expressive, ripe fruit, floral, fine lees, mineral. Palate: full-bodied, complex, spicy, long, elegant.

Vidilla 2021 B
100% verdejo

90

Colour: bright straw. Nose: fruit expression, ripe fruit, floral. Palate: flavourful, fresh, good acidity, fruity aftertaste.

BODEGAS Y VIÑEDOS VEGA REAL
Ctra. N-122, Km 298,6
47318 Castrillo de Duero (Valladolid)
☎: +34 679 180 532
visitas@vegareal.net
www.barbadillo.com/bodegas-vega-real

Vega Real Verdejo 2021 B
100% verdejo

87

BODEGUEROS QUINTA ESENCIA
Eras, 37
47520 Castronuño (Valladolid)
☎: +34 605 887 100
info@bodegueroquintaesencia.com
www.bodegueroquintaesencia.com

Silbon 2021 B
verdejo

88

Kind finish, herbal, varietally correct, wild, balanced.

CAMPOS DE SUEÑOS
Avda. Diagonal, 590, 5º 1ª
08021 Barcelona (Barcelona)
☎: +34 935 385 677
vinergia@vinergia.com
www.vinergia.com

Campos de Sueños 2021 B
100% verdejo

88

Jammy, tasty, tropical, thin.

CARRAMATA
Plaza San Agustín, 2 y 3
47400 Medina del Campo (Valladolid)
☎: +34 618 893 737
hola@carramata.es
www.carramata.es

Carramata 2021 B
100% verdejo

89

Aromatic, tasty, varietally correct, smooth.

CASERÍO DE DUEÑAS
Medina, 2
47465 Villaverde de Medina (Valladolid)
☎: +34 983 020 060
bibanez@entrecanalesdomecq.com
www.entrecanalesdomecq.com

La Poda Sauvignon Blanc 2021 B
100% sauvignon blanc

90

Colour: bright straw, greenish rim. Nose: fresh fruit, citrus fruit, wild herbs. Palate: fresh, fruity, good acidity, easy to drink.

Viña Mayor Verdejo 2021 B
verdejo

89

Pleasant, kind finish, floral, fruity.

COMPAÑÍA DE VINOS MIGUEL MARTÍN
Ctra. Valladolid, 14
47270 Cigales (Valladolid)
☎: +34 983 250 319
comercial@ciadevinos.com
www.bodegasthesaurus.com

Dòmine 2021 B
95% verdejo, 5% sauvignon blanc

87

Viña Goy Rueda 2021 B
100% verdejo

87

COMPAÑÍA DE VINOS TELMO RODRÍGUEZ
El Monte
01308 Lanciego (Araba/Álava)
☎: +34 945 628 315
contact@telmorodriguez.com
www.telmorodriguez.com

Basa 2021 B
verdejo
91
Colour: bright straw, greenish rim. Nose: fresh fruit, citrus fruit, wild herbs, hint of anise. Palate: fresh, fruity, good acidity, fine bitter notes.

El Transistor 2021 B
verdejo
93
Colour: bright straw. Nose: ripe fruit, dried herbs, faded flowers, hint of anise. Palate: powerful, ripe fruit, balanced.

CVNE
Barrio de la Estación, s/n
26200 Haro (La Rioja)
☎: +34 941 304 800
marketing@cvne.com
www.cvne.com

Cune Rueda 2021 B
100% verdejo
88
Pleasant, tropical, simple, tasty.

Monopole S. XXI 2021 B
100% verdejo
89
Pleasant, smooth, tasty, fruity.

DIEZ SIGLOS DE VERDEJO
Ctra. Valladolid Km. 24,5
47231 Serrada (Valladolid)
☎: +34 983 559 910
Fax: +34 983 559 470
info@diezsiglos.es
www.diezsiglos.es

Castillo de la Mota 2021 B
100% verdejo
87

Diez Siglos 2020 B FB
100% verdejo
90
Colour: bright yellow. Nose: creamy oak, spicy, aromatic coffee, grassy, stone fruit. Palate: good structure, toasty, fine bitter notes.

Diez Siglos Sauvignon Blanc 2021 B
100% sauvignon blanc
88
Citrus fruit, crisp, herbal, tasty.

Diez Siglos Verdejo 2021 B
100% verdejo
88 ♣
Citrus fruit, herbal, balanced, crisp.

Momento Diez 2020 B
100% verdejo
89
Citrus fruit, dried herbs, herbal, crisp, balanced.

Nekora Verdejo sobre Lías 2021 B
verdejo
88
Herbal, citrus fruit, crisp, balanced.

DOMINIO DEL BLANCO
Santísimo Cristo, 128
47490 Rueda (Valladolid)
☎: +34 699 726 469
botondegallo@botondegallo.com
www.botondegallo.com

Botón de Gallo 2020 B
90
Colour: straw. Nose: ripe fruit, dried herbs, faded flowers. Palate: powerful, ripe fruit, balanced.

ELÍAS MORA
Juan Mora s/n
47530 San Román de Hornija (Valladolid)
☎: +34 983 784 029
info@bodegaseliasmora.com
www.bodegaseliasmora.com

Elias Mora Contracorriente 2021 B
verdejo
88
Citrus fruit, crisp, herbal, standard.

DO RUEDA / D.O.P.

DO RUEDA / D.O.P.

EXPLOTACIONES VITIVINÍCOLAS LASECANAS C.B.
Cl. Saleta, 10
47491 La Seca (Valladolid)
☎: +34 605 876 524
info@lasecanasverdejo.es
www.lasecanasverdejo.es

Malvid Selección 2021 B
verdejo

90

Colour: bright straw. Nose: fruit expression, ripe fruit, floral. Palate: flavourful, fresh, good acidity, fruity aftertaste.

Malvid Verdejo 2021 B
verdejo

87

FINCA MONTEPEDROSO
Término La Morejona, s/n
47490 Rueda (Valladolid)
☎: +34 983 868 977
montepedroso@bujanda.com
www.fincamontepedroso.com

Finca Montepedroso Verdejo 2021 B
100% verdejo

90

Colour: bright straw. Nose: white flowers, jasmine, dried herbs. Palate: flavourful, fruity, balanced.

HERRERO BODEGA
Camino Real, 55
40447 Nieva (Segovia)
☎: +34 921 124 440
info@herrerobodega.com
www.herrerobodega.com

Erre de Herrero 2021 B
verdejo

87

Robert Vedel Cepas Viejas 2020 B

89

Balanced, spicy, yeasty notes, herbal.

J. FERNANDO VIÑEDOS Y BODEGAS
Camino Valverde s/n
47490 Rueda (Valladolid)
☎: +34 925 100 004
info@jfernandovb.com
www.jfernandofamilywines.com

J. Fernando Verdejo 2021 B
verdejo

88

Citrus fruit, floral, thin, tasty.

J. Fernando Verdejo Vendimia Seleccionada 2020 B
verdejo

92

Colour: bright straw. Nose: ripe fruit, fragrant herbs, fine lees, complex, spicy. Palate: full-bodied, rich, long, good acidity.

JAVIER SANZ VITICULTOR
San Judas, 2
47491 La Seca (Valladolid)
☎: +34 983 816 669
comunicaciones@bodegajaviersanz.com
www.bodegajaviersanz.com

Finca Saltamontes 2017 B
100% verdejo

94

Colour: bright yellow. Nose: dried flowers, candied fruit, fine lees, pattiserie. Palate: round, spicy, long, great length.

Javier Sanz 2019 B FB
100% verdejo

92

Colour: bright yellow. Nose: powerful, creamy oak, ripe fruit, spicy. Palate: good structure, long, toasty, fine bitter notes.

Javier Sanz Sauvignon Blanc 2021 B
100% sauvignon blanc

88

Citrus fruit, floral, tasty, slight reduction.

Javier Sanz Verdejo 2021 B
100% verdejo

91

Colour: bright straw. Nose: fruit expression, ripe fruit, floral, hint of anise. Palate: flavourful, fresh, good acidity, fruity aftertaste.

V Malcorta Verdejo 2020 B
100% verdejo

92

Colour: bright straw, greenish rim. Nose: citrus fruit, wild herbs, ripe fruit. Palate: fresh, fruity, good acidity, fine bitter notes.

LEGARIS

Ctra. de Peñafiel - Encinas de Esgueva, km. 2,5
47316 Curiel de Duero (Valladolid)
☎: +34 983 878 088
info@legaris.com
www.legaris.com

Legaris Sauvignon Blanc 2021 B
100% sauvignon blanc

88

Herbaceous, wild, varietally correct, aromatic, balsamic herbs, balanced, smooth.

Legaris Verdejo 2021 B
100% verdejo

89

Balanced, tasty, wild, standard, crisp, varietally correct.

LIBERSO CURIOSO VERDEJO

47331 Santibáñez de Valcorba (Valladolid)
☎: +34 983 507 439
hola@liberso.es
www.liberso.es

Liberso Curioso Verdejo 2016 B FB
100% verdejo

91

Colour: bright yellow. Nose: powerful, ripe fruit, spicy, fine lees. Palate: rich, good structure, long, bitter.

Liberso Curioso Verdejo 2020 B FB
100% verdejo

89

Pleasant, spicy, fruity, jammy, great length, tasty.

LLANO Y MONTE

Paraje de la Alquibla
30178 Mula (Murcia)
☎: +34 868 087 355
Fax: +34 868 087 388
info@bodegasllanoymonte.com
www.bodegasllanoymonte.com

Mar de Fondo 2020 B
100% verdejo

89

Citrus fruit, balanced, crisp, wild.

LOESS VINOS

Plaza Cuartel Viejo, 7
49028 Zamora (Zamora)
☎: +34 983 664 898
loess@loess.es
www.loesscollection.com

Loess Collection 2021 B FB
verdejo

91

Colour: bright straw. Nose: ripe fruit, floral, spicy, dried herbs. Palate: flavourful, good acidity, fruity aftertaste, rich, smoky aftertaste.

Loess Rueda 2021 B
verdejo

89

Crisp, fruity, very fruit-driven, great length, herbal, balanced, wild, simple.

MARQUÉS DE GRIÑÓN FAMILY ESTATES

Finca Casa de Vacas, CM-4015, Km. 23
45692 Malpica del Tajo (Toledo)
☎: +34 925 597 222
adiaz@marquesdegrinon.com

Marqués de Griñón Verdejo 2021 B
100% verdejo

87

MARQUÉS DE IRÚN

Albasanz, 12 Tercera Planta
28037 Madrid (Madrid)
☎: +34 667 495 225
carlos.ruiz@caballero.es
www.caballero.es/marcas/marques-de-irun

Marqués de Irún Verdejo 2021 B
100% verdejo

86

MARQUÉS DE LA CONCORDIA FAMILY OF WINES

Los Moros, 10
47490 Rueda (Valladolid)
☎: +34 915 767 327
www.marquesdelaconcordia.com

Vega de la Reina Verdejo 2021 B
100% verdejo

89

Pleasant, aromatic, dried flowers, fruity.

DO RUEDA / D.O.P.

DO RUEDA / D.O.P.

MARTINSANCHO BODEGA Y VIÑEDOS
Torcido, 1
47491 La Seca (Valladolid)
☎: +34 657 543 702
martinsancho@martinsancho.com
www.martinsancho.com

Martínsancho 2021 B
100% verdejo

89
Kind finish, smooth, herbal, aromatic, standard, crisp, simple.

MONTEBACO
Finca Montealto s/n
47300 Valbuena de Duero (Valladolid)
☎: +34 983 485 128
Fax: +34 983 485 033
montebaco@bodegasmontebaco.com
www.bodegasmontebaco.com

Montebaco Verdejo + Sauvignon 2021 B
verdejo, sauvignon blanc

90
Colour: bright straw. Nose: fruit expression, ripe fruit, floral. Palate: flavourful, fresh, good acidity, fruity aftertaste.

NUBORI
Camino Rio Hondillo, 1
01300 Laguardia (Araba/Álava)
☎: +34 941 183 502
administracion@nubori.es
www.bodegasnubori.com

Nubori Selección Familia 2018 T C
tempranillo

86

NUEVOS VINOS CB
San Juan Bosco, 32 Bajo
03804 Alcoy (Alacant/Alicante)
☎: +34 965 549 172
Fax: +34 965 549 173
admon@nuevosvinos.es
www.nuevosvinos.es

Perla Maris Verdejo 2021 B
100% verdejo

87

PACO MULERO
Partida de la Hoya Torres s/n
30520 Jumilla (Murcia)
☎: +34 676 433 541
info@pacomulero.com
www.pacomulero.com

Prisma Verdejo 2021 B
verdejo

89
Standard, balanced, tropical, smooth.

PAGO DEL CIELO
Del Rosario, 56
47311 Fompedraza (Valladolid)
☎: +34 938 177 400
info@torres.es
www.torres.es

Celeste Verdejo 2021 B
verdejo

88
Aromatic, herbaceous, citrus fruit, simple.

PAGO TRASLAGARES
Autovía Noroeste km 166,400
47490 Rueda (Valladolid)
☎: +34 671 006 565
export@traslagares.com
www.traslagares.com

Traslagares Sauvignon Blanc 2021 B
100% sauvignon blanc

86

Traslagares Verdejo 2021 B
100% verdejo

89
Kind finish, standard, balanced, wild, smooth.

Viña El Torreón Verdejo 2021 B
100% verdejo

87

PAGOS DEL REY
Avda. Morejona, 6
47490 Rueda (Valladolid)
☎: +34 926 322 400
rueda@pagosdelrey.com
www.pagosdelrey.com

Analivia Verdejo Selección 2021 B
verdejo

88
Pleasant, citrus fruit, tasty, smooth.

Blume Sauvignon Blanc 2021 B
100% sauvignon blanc

87

Blume Verdejo Selección 2021 B
verdejo
87

Blume Verdejo Viura 2021 B
50% verdejo, 50% viura
88
Citrus fruit, herbal, crisp, tasty.

PALACIO DE BORNOS

Ctra. Madrid - Coruña, km. 170,6
47490 Rueda (Valladolid)
☎: +34 983 868 116
info@bornosbodegas.com
www.palaciodebornos.com

Palacio de Bornos La Caprichosa 2021 B
100% verdejo
92
Colour: bright straw. Nose: fruit expression, ripe fruit, floral. Palate: flavourful, fresh, good acidity, fruity aftertaste.

Palacio de Bornos Sauvignon Blanc 2021 B
sauvignon blanc
88
Citrus fruit, simple, standard, floral.

Palacio de Bornos Sauvignon Blanc Semi-dulce 2021 B SD
100% sauvignon blanc
85

Palacio de Bornos Verdejo 2021 B
89
Pleasant, wild, smooth, herbal, crisp.

Palacio de Bornos Verdejo 2021 B FB
92
Colour: bright yellow. Nose: powerful, creamy oak, ripe fruit. Palate: good structure, long, toasty, fine bitter notes.

PERSEO 7 BODEGAS

Montero Calvo, 7
47001 Valladolid (Valladolid)
☎: +34 983 297 830
info@perseo7.com

Perseo 7 Verdejo sobre Lías 2021 B
100% verdejo
89
Tropical, tasty, jammy, crisp, floral.

SPANISH PALATE

Avda. Carlos Latorre, 30
49800 Toro (Zamora)
☎: +34 980 690 643
export@spanishpalate.es
www.spanishpalate.es

Botas de Barro Rueda 2021 B
88
Aromatic, tropical, citrus fruit, crisp.

UVAS FELICES

Agullers, 7
08003 Barcelona (Barcelona)
☎: +34 902 327 777
www.vilaviniteca.es

El Perro Verde 2021 B
90
Colour: bright straw, greenish rim. Nose: fresh fruit, citrus fruit, wild herbs, tropical fruit. Palate: fresh, fruity, good acidity, fine bitter notes.

Fenomenal 2021 B
88
Crisp, racy, citrus fruit, herbal.

VEGA DEL PAS

Ctra. CL-602, Km. 48
47465 Villaverde de Medina (Valladolid)
☎: +34 620 901 415
vinosvegadelpas@gmail.com

Vega del Pas Verdejo 2021 B
verdejo
89
Pleasant, aromatic, smooth, tasty.

VINITOR WINE GROUP

Santa Eugènia, 64
17005 Girona/Gerona (Girona/Gerona)
hola@vinitor.es
www.vinitorwinegroup.com

MOFO – The Free Soul 2021 B
100% verdejo
88
Herbal, fruity, jammy, tasty.

VINOS DE CULTO BANISIO

Paseo de Gracia, 8-10, 3ª planta
08007 Barcelona (Barcelona)
☎: +34 933 631 838
info@banisio.com
www.banisio.com

Banisio Sauvignon Blanc 2020 B
sauvignon blanc
88
Tropical, powerful, herbal, citrus fruit.

DO RUEDA / D.O.P.

DO RUEDA / D.O.P.

Banisio Sauvignon Blanc 2021 B
sauvignon blanc
88
Tropical, powerful, herbal, citrus fruit.

Banisio Verdejo 2020 B
verdejo
90
Colour: bright straw. Nose: fruit expression, ripe fruit, floral. Palate: flavourful, fresh, good acidity, fruity aftertaste.

Banisio Verdejo 2021 B
verdejo
90
Colour: bright straw. Nose: fruit expression, ripe fruit, floral. Palate: flavourful, fresh, good acidity, fruity aftertaste.

VINOS DE LA LUZ
Ctra. de Mélida, km. 3,5
47300 Peñafiel (Valladolid)
☎: +34 983 878 007
info@vinosdelaluz.com
www.vinosdelaluz.com

Cinema Verdejo 2021 B
100% verdejo
90
Colour: bright straw, greenish rim. Nose: fresh fruit, citrus fruit, wild herbs. Palate: fresh, fruity, good acidity, fine bitter notes.

Valcerracín Selección Limitada Verdejo 2021 B
100% verdejo
88
Pleasant, kind finish, fruity.

Valpincia Verdejo 2021 B
100% verdejo
89
Varietally correct, smooth, tasty, herbal.

VINOS DEL PASEANTE
Avda. Jaume Codorniu, s/n
08770 Sant Sadurní D'Anoia (Barcelona)
☎: +34 610 486 211
mj.lapuente@raventoscodorniu.com
www.vinosdelpaseante.com

La Charla 2021 B
85% verdejo, 15% sauvignon blanc
89
Citrus fruit, balanced, crisp, fruity, herbal, tasty.

VINOS SANZ
Ctra. Madrid - La Coruña, Km. 170,5
47490 Rueda (Valladolid)
☎: +34 916 408 730
vinossanz@vinossanz.com
www.vinossanz.com

El Loco de Finca La Colina 2021 B
97% verdejo, 3% sauvignon blanc
91
Colour: bright straw, greenish rim. Nose: fresh fruit, citrus fruit, wild herbs, hint of anise. Palate: fresh, fruity, good acidity, fine bitter notes.

Finca La Colina Sauvignon Blanc 2021 B
100% sauvignon blanc
91
Colour: bright straw, greenish rim. Nose: fresh fruit, citrus fruit, wild herbs, wild herbs. Palate: fresh, fruity, good acidity, fine bitter notes.

Finca La Colina Verdejo Cien x Cien 2021 B
100% verdejo
92
Colour: bright straw, greenish rim. Nose: fresh fruit, citrus fruit, wild herbs. Palate: fresh, fruity, fine bitter notes.

Sanz Clásico 2021 B
70% verdejo, 30% viura
89
Herbal, tasty, crisp, yeasty notes.

Sanz Sauvignon Blanc 2021 B
100% sauvignon blanc
88
Citrus fruit, dried herbs, crisp, balanced.

Sanz Verdejo 2021 B
100% verdejo
90
Colour: bright straw, greenish rim. Nose: fresh fruit, citrus fruit, wild herbs. Palate: fresh, fruity, good acidity, fine bitter notes.

VINS I LICORS GRAU
Torroella, 163
17200 Palafrugell (Girona/Gerona)
☎: +34 972 301 835
info@vinsilicorsgrau.com
www.grauonline.com

Qué no falte! 2021 B
100% verdejo
87

VIÑAS MURILLO
Polígono 1, Parcela 23
47238 Alcazarén (Valladolid)
☎: +34 652 054 177
administracion@vinasmurillo.es
www.vinasmurillo.es

Animoso 2021 B
100% verdejo
88
Citrus fruit, crisp, herbal, tasty.

Chapirete 2019 B FB
100% verdejo
93
Colour: bright yellow. Nose: dried flowers, candied fruit, fine lees, pattiserie. Palate: round, spicy, long, great length.

Chapirete Prefiloxérico 2020 B
100% verdejo
92
Colour: bright yellow. Nose: dried flowers, candied fruit, fine lees, pattiserie. Palate: round, spicy, long, great length.

Chapirete Selección 2021 B
100% verdejo
90
Colour: bright straw, greenish rim. Nose: fresh fruit, citrus fruit, wild herbs. Palate: fresh, fruity, good acidity, fine bitter notes.

Valdihuete 2021 B
valdihuete
87

VIÑEDOS ASTIGI
Calle Madrid, 1, 1º
28231 Las Rozas (Madrid)
☎: +34 661 528 004
info@econaturagourmet.com
www.econaturagourmet.com

Isabel I 2021 B
100% verdejo
87 🌱

VIÑEDOS SINGULARES
Avda. de La Riera, 11 Nave 1
08960 Sant Just Desvern (Barcelona)
☎: +34 934 807 041
Fax: +34 934 807 076
info@vinedossingulares.com
www.vinedossingulares.com

Afortunado 2021 B
verdejo
88
Aromatic, standard, fruity, jammy, simple, smooth.

YLLERA BODEGAS & VIÑEDOS
47490 Rueda (Valladolid)
☎: +34 983 868 097
grupoyllera@grupoyllera.com
www.grupoyllera.com

Yllera Verdejo Vendimia Nocturna Ecológico 2021 B
verdejo
90 🌱
Colour: bright straw, greenish rim. Nose: fresh fruit, citrus fruit, wild herbs, wild herbs. Palate: fresh, fruity, good acidity, fine bitter notes.

Boada Verdejo, Sauvignon Blanc, Chardonnay 2021 B
86

DO RUEDA / D.O.P.

DO RUEDA / D.O.P.

Cantosán Verdejo Viñas Viejas 2021 B
verdejo
90
Colour: bright straw. Nose: fruit expression, ripe fruit, floral, wild herbs. Palate: flavourful, fresh, fine bitter notes, balanced.

Yllera Sauvignon Blanc Vendimia Nocturna 2021 B
89
Varietally correct, tasty, herbaceous, fruity, crisp.

Yllera Verdejo Vendimia Nocturna 2021 B
verdejo
89
Dried flowers, fruity, dried herbs, tasty.

DO. SOMONTANO
CONSEJO REGULADOR

Avda. de la Merced, 64
22300 Barbastro (Huesca)
☎: +34 974 313 031 - Fax: +34 974 315 132
@: somontano@dosomontano.com
www.dosomontano.com

LOCATION:

In the province of Huesca, around the town of Barbastro. The region comprises 43 municipal districts, mainly centred round the region of Somontano and the rest of the neighbouring regions of Ribagorza and Monegros.

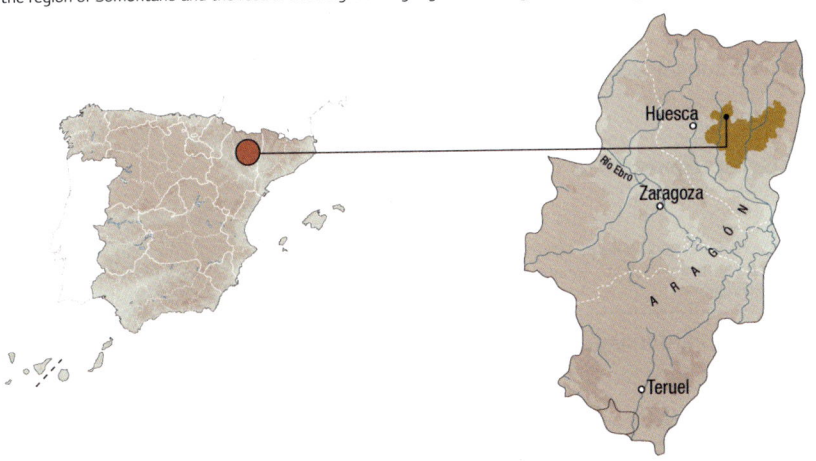

GRAPE VARIETIES:

WHITE: Macabeo, Garnacha Blanca, Alcañón, Chardonnay, Riesling, Sauvignon Blanc and Gewürztraminer.

RED: Tempranillo, Garnacha Tinta, Cabernet Sauvignon, Merlot, Moristel, Parraleta, Pinot Noir and Syrah.

FIGURES:

Vineyard surface: 3,832 – **Wine-Growers:** 324 – **Wineries:** 30 – 20**20 Harvest rating:** Very Good – **Production 21:** 14,422,600 litres – **Market percentages:** 75% domestic - 25% export.

SOIL:

The soil is mainly brownish limestone, not very fertile, with a good level of limestone and good permeability.

CLIMATE:

Characterised by cold winters and hot summers, with sharp contrasts in temperature at the end of spring and autumn. The average annual rainfall is 500 mm, although the rains are scarcer in the south and east

CLASSIFICATION OF YOUNG WINE HARVESTS PEÑÍNGUIDE

2017	2018	2019	2020	2021
GOOD	GOOD	VERY GOOD	VERY GOOD	VERY GOOD

DO SOMOTANO / D.O.P.

ALDAHARA
Carretera, 10
22423 Estadilla (Huesca)
☎: +34 620 309 217
bodega@aldahara.es
www.valdalferche.com

Aldahara 2018 T C
cabernet sauvignon, merlot, syrah
88
Spicy, herbal, jammy, tasty, great length, pleasant.

Aldahara 2021 RD
merlot
88
Pleasant, fruity, jammy, tasty, balanced, defined aromas.

Aldahara Chardonnay 2021 B
chardonnay
86

Aldahara Generaciones 2016 T
cabernet sauvignon, merlot
92
Colour: Cherry. Nose: sweet spices, scrubland, black fruit, fruit preserve, dark chocolate. Palate: spicy, good acidity, round tannins.

Aldahara Rasé Chardonnay 2019 B RB
chardonnay
88
Pleasant, smoky, spicy, dried flowers, jammy, bitter.

Aldahara Rasé Merlot 2019 T
merlot
86

BAL D'ISÁBENA BODEGAS
Ctra. A-1605, Km. 11,200
22587 Laguarres (Huesca)
☎: +34 605 785 178
info@baldisabena.com
www.baldisabena.com

Bal d'Isabena Garnacha 2021 T
garnacha
90
Colour: cherry, purple rim. Nose: fruit expression, red berry notes, floral. Palate: fruity, flavourful, balanced.

Bal d'Isabena Garnacha Blanca 2021 B
garnacha blanca
88
Kind finish, citrus fruit, floral, tasty.

Bal d'Isabena Gewürztraminer 2021 B
gewürztraminer
90
Colour: straw. Nose: expressive, white flowers, jasmine, dried herbs. Palate: flavourful, fruity, balanced.

Ixeia 2021 RD
88
Fruity, jammy, tasty.

Perrochico 2020 T
garnacha, tempranillo, syrah
87

Perrochico 2021 B
chardonnay, gewürztraminer
87

BLECUA
Ctra. de Naval, Km. 3,7
22300 Barbastro (Huesca)
☎: +34 974 302 216
Fax: +34 974 302 098
prensa@gonzalezbyass.es
www.bodegablecua.com

Blecua 2016 T R
cabernet sauvignon, syrah, merlot, garnacha
93
Colour: Cherry. Nose: sweet spices, scrubland, balanced, complex, characterful, ripe fruit. Palate: spicy, good acidity, round tannins.

BODEGA ENATE
Avda. de las Artes, 1
22314 Salas Bajas (Huesca)
☎: +34 974 302 580
Fax: +34 974 300 046
bodega.enate@grupoenate.es
www.enate.es

Enate Cabernet Sauvignon 2014 T R
cabernet sauvignon
91
Age nuances. Colour: Cherry. Nose: sweet spices, scrubland, smoky. Palate: spicy, balsamic, good acidity.

Enate Chardonnay 2020 B FB
chardonnay
92
Colour: bright yellow. Nose: creamy oak, ripe fruit, spicy, caramel, dry stone. Palate: rich, good structure, long, toasty, fine bitter notes.

Enate Merlot-Merlot 2017 T
merlot
91
Classic. Colour: deep cherry. Nose: dried herbs, creamy oak, black fruit, fruit liqueur notes, woody, earthy notes. Palate: powerful, ripe fruit, spicy, round tannins.

Enate Uno 2010 T
85% cabernet sauvignon, 15% merlot
94
Colour: dark-red cherry, garnet rim. Nose: fruit preserve, overripe fruit, waxy notes, toasty. Palate: warm, pruney, sweet tannins, good finish.

Enate Uno Chardonnay 2013 B GR
chardonnay
94
Colour: yellow, golden. Nose: toasted bread, smoky, faded flowers, ripe fruit, complex. Palate: rich, flavourful, smoky aftertaste, good structure, concentrated.

Enate Varietales 2019 T R
25% cabernet sauvignon, 25% syrah, 25% tempranillo, 25% merlot
92
Colour: bright cherry. Nose: sweet spices, ripe fruit, dark chocolate. Palate: spicy, round tannins, long, ripe fruit, round.

BODEGA LAUS
Ctra. N-240, km 154,8
22300 Barbastro (Huesca)
☎: +34 974 269 708
Fax: +34 974 269 715
bodega.laus@grupoenate.es
www.bodegaslaus.es

Laus 2017 T R
cabernet sauvignon
89
Age nuances, dried herbs, weary, classic, varietally correct.

Laus 2018 T C
50% merlot, 50% cabernet sauvignon
88
Jammy, tasty, full-bodied, balsamic herbs, spicy.

Laus 2021 RD
75% syrah, 25% garnacha
87

Laus 2021 T
50% merlot, 50% syrah
88
Balanced, fruity, jammy, tasty.

Laus Chardonnay 2021 B
chardonnay
86

BODEGA OTTO BESTUÉ
Ctra. A-138, Km. 0,5
22312 Enate (Huesca)
☎: +34 681 638 954
info@bodega-ottobestue.com
www.bodega-ottobestue.com

Bestué Chardonnay 2020 B FB
100% chardonnay
90
Woody. Colour: bright yellow. Nose: powerful, creamy oak, ripe fruit, spicy, aromatic coffee. Palate: rich, good structure, toasty, fine bitter notes.

Bestué Finca Santa Sabina 2019 T C
100% cabernet sauvignon
89
Powerful, spicy, tasty, jammy, great length.

DO SOMONTANO / D.O.P.

Bestué Marina 2021 B
100% gewürztraminer

88

Pleasant, aromatic, standard, floral, fruity, varietally correct, smooth, simple.

Bestué Monte Alicia 2021 B
100% chardonnay

89

Kind finish, floral, fruity, jammy, yeasty notes, tasty, balanced.

Bestué Viñadores 2018 T
80% garnacha, 20% cabernet sauvignon

90

Colour: Cherry. Nose: new oak, black fruit, ripe fruit, toasty, scrubland. Palate: powerful, harsh oak tannins.

BODEGA PIRINEOS

Ctra. Barbastro - Naval, Km. 3.5
22300 Barbastro (Huesca)
☎: +34 974 312 273
info@bodegapirineos.com
www.bodegapirineos.com

3404 Cabernet Garnacha Moristel 2021 T
cabernet sauvignon, garnacha, moristel

88

Kind finish, fruity, exuberant, herbal, tasty.

3404 Tuca D'Aneto 2019 T C
cabernet sauvignon, merlot, moristel

89

Full-bodied, balanced, spicy, dried herbs, full-bodied, jammy, tasty, toasty.

Marboré Cuvée 2019 T

90

Colour: deep cherry. Nose: dried herbs, creamy oak, black fruit. Palate: ripe fruit, spicy, round tannins.

Pirineos Chardonnay Viñedo Seleccionado 2021 B
chardonnay

89

Citrus fruit, full-bodied, herbal, tasty, smooth.

Pirineos Gewürztraminer 2021 B
gewürztraminer

89

Citrus fruit, balanced, floral, fruity, herbal.

Señorío de Lazán T C

86

BODEGA SOMMOS

Ctra. N-240, Km. 155
22300 Barbastro (Huesca)
☎: +34 974 269 900
diego.mur@bodegasommos.com
www.bodegasommos.com

Sommos Coleccción Sauvignon Blanc 2021 B
100% sauvignon blanc

89

Citrus fruit, crisp, herbal, balanced.

Sommos Colección Cabernet Sauvignon 2019 T R
100% cabernet sauvignon

91

Colour: dark-red cherry. Nose: characterful, hot, black fruit, dried herbs, scrubland. Palate: flavourful, round tannins.

Sommos Colección Chardonnay 2020 B
100% chardonnay

90

Colour: bright yellow. Nose: powerful, creamy oak, ripe fruit, spicy. Palate: rich, good structure, toasty, fine bitter notes.

Sommos Colección Merlot 2019 T BA
100% merlot

91

Colour: Cherry. Nose: sweet spices, scrubland, wild herbs, ripe fruit. Palate: flavourful, fruity, balsamic, good acidity.

Sommos Colección Syrah 2019 T R
syrah

90

Colour: cherry, garnet rim. Nose: fine reductive notes, wild herbs, dried flowers, ripe fruit. Palate: flavourful, toasty.

Sommos Colección Tempranillo 2019 T R
100% tempranillo

91

Colour: bright cherry. Nose: sweet spices, ripe fruit, dark chocolate. Palate: fruity, spicy, round tannins, flavourful.

BODEGAS ABINASA

Ctra. N-240, Km. 180
22124 Lascellas (Huesca)
☎: +34 685 806 425
info@bodegasabinasa.com
www.bodegasabinasa.com

Ana 2017 T C
85

Ana 2020 T RB
70% cabernet sauvignon, 20% merlot, 10% tempranillo
84

Ana 2021 RD
100% cabernet sauvignon
83

Ana Gewürztraminer 2020 B
gewürztraminer
86

Ana Selección Familia 2015 T
merlot
89
Age nuances, jammy, tasty, spicy, herbal, varietally correct.

BODEGAS CARLOS VALERO

Castillo de Capúa, 10 Nave 1
50197 Pol. Pla-Za (Zaragoza)
☎: +34 976 180 634
Fax: +34 976 186 326
info@bodegasvalero.com
www.bodegasvalero.com

Matarile 2018 T C
40% cabernet sauvignon, 30% merlot, 30% syrah
88
Age nuances, balanced, spicy, jammy.

Matarile 2020 B
70% chardonnay, 30% gewürztraminer
89
Citrus fruit, balanced, full-bodied, jammy, spicy.

BODEGAS FÁBREGAS

Cerler, 3
22300 Barbastro (Huesca)
☎: +34 974 310 498
info@bodegasfabregas.com
www.bodegasfabregas.com

Fábregas Garnacha Blanca 2020 B FB
garnacha blanca
90
Colour: bright straw. Nose: ripe fruit, fine lees, sweet spices, white flowers. Palate: full-bodied, rich, good acidity, good finish.

Fábregas Monteta 2019 T
garnacha
90
Jammy, rustic. Colour: bright cherry. Nose: balanced, neat, wild herbs, ripe fruit. Palate: fruity, easy to drink.

Fábregas Moristel 2020 T
moristel
90
Crisp. Colour: very deep cherry. Nose: wild herbs, animal reductive notes, characterful, spicy, toasty. Palate: fruity.

Fábregas Puro Syrah 2019 T C
syrah
91
Colour: deep cherry, purple rim. Nose: powerful, ripe fruit, balanced, dried flowers, dried herbs. Palate: fruity, balanced, easy to drink.

Mingua 2020 T
garnacha, moristel, syrah
89
Aromatic, wild, floral, fruity, simple, pleasant, kind finish.

Mingua 2021 B
garnacha blanca, chardonnay
87

BODEGAS MELER

Ctra. N-240, km. 154,2
22300 Barbastro (Huesca)
☎: +34 679 954 988
info@bodegasmeler.com
www.bodegasmeler.com

Andres Meler 2014 T R
cabernet sauvignon
90
Colour: deep cherry. Nose: dried herbs, creamy oak, black fruit. Palate: powerful, ripe fruit, spicy, round tannins.

Meler 10 Ballos de Garnacha 2019 T
garnacha
88
Balanced, spicy, fruity, jammy, creamy.

Meler 15 2015 T C
cabernet sauvignon, merlot
88
Balanced, spicy, jammy, tasty, age nuances.

Meler 6 2018 T C
cabernet sauvignon, merlot, syrah
88
Age nuances, creamy, spicy, jammy, tasty.

DO SOMONTANO / D.O.P.

DO SOMONTANO / D.O.P.

Meler 9 2016 T
cabernet sauvignon, garnacha
85

Meler Syrah 2020 T
syrah
85

BODEGAS OSCA
La Iglesia, 1
22124 Ponzano (Huesca)
☎: +34 974 319 017
bodega@bodegasosca.com
www.bodegasosca.com

Gran Eroles 2015 T R
cabernet sauvignon
91
Colour: dark-red cherry, garnet rim. Nose: ripe fruit, fruit preserve, aged wood nuances, tobacco, sweet spices. Palate: spicy, round tannins, flavourful.

Mascún Garnacha 2019 T C
87

Mascún Garnacha 2021 RD
garnacha
86

Mascún Garnacha Blanca 2021 B
86

Mascun Gewurztraminer 2021 B
86

Osca Merlot Reserva 2016 T R
88
Smoky, full-bodied, jammy, tasty, toasty.

BODEGAS VALDOVINOS
Camino de la Almunia, s/n
22133 Antillón (Huesca)
☎: +34 974 260 437
info@bodegasvaldovinos.com
www.bodegasvaldovinos.com

Berdá 2020 B
chardonnay
91
Colour: bright yellow. Nose: ripe fruit, spicy, waxy notes, dried flowers, dried herbs. Palate: good structure, flavourful, full-bodied.

Bresque Sauvignon Blanc 2021 B
sauvignon blanc
88
Citrus fruit, herbal, crisp, standard.

Bresque Syrah 2019 T RB
syrah
88
Balanced, spicy, herbaceous, jammy.

Valdovinos 2018 T C
cabernet sauvignon, tempranillo, garnacha
87

Valdovinos Selección Garnacha 2019 T
garnacha
88
Balanced, spicy, herbaceous, toasty.

Valdovinos Selección Syrah 2018 T
syrah
89
Balanced, spicy, full-bodied, age nuances, dried herbs, tasty.

DALCAMP
Pedania Monte Odina s/n
22415 Monesma de San Juan (Huesca)
☎: +34 973 760 018
Fax: +34 973 760 523
info@ramondalfo.com
www.castillodemonesma.com

Castillo de Monesma 2017 T C
cabernet sauvignon
86

Castillo de Monesma 2018 T R
tempranillo, syrah
88
Jammy, dried herbs, fruity, pruney, smoky.

Castillo de Monesma 2018 T RB
tempranillo, syrah
85

Castillo de Monesma 2019 T
syrah, merlot
86

IDRIAS
Ctra. Abiego 1229, Km 0,2
22124 Lascellas (Huesca)
☎: +34 974 340 671
info@bsdg.es
www.idrias.es

Idrias Chardonnay 2021 B
chardonnay
87 ♣

Idrias T RB
merlot, cabernet sauvignon, tempranillo
86

Idrias Tempranillo 2021 T
tempranillo
86

VINOS DIVERTIDOS
Avda. Jaime I El Conquistador, 90
03560 El Campello (Alicante)
☎: +34 966 105 325
Fax: +34 965 160 955
info@vinosdivertidos.es
www.vinosdivertidos.es

Cojón de Gato 2021 B
chardonnay, gewürztraminer
88
Aromatic, powerful, varietally correct, floral, crisp.

VIÑAS DEL VERO
Ctra. de Naval, Km. 3,7
22300 Barbastro (Huesca)
☎: +34 974 302 216
prensa@gonzalezbyass.es
www.vinasdelvero.es

Clarión de Viñas del Vero 2018 B
92
Colour: bright yellow. Nose: powerful, creamy oak, ripe fruit, spicy. Palate: rich, good structure, long, toasty, fine bitter notes.

Gran Vos de Viñas del Vero 2016 T R
92
Colour: Cherry. Nose: balsamic herbs, scrubland, spicy, black fruit, fruit preserve. Palate: flavourful, ripe fruit, balanced.

La Miranda de Secastilla 2019 T
85% garnacha, 12% syrah, 3% parraleta
89
Pruney, balanced, dried flowers, jammy, tasty, wild, pleasant.

La Miranda de Secastilla Garnacha 2020 RD
garnacha
88
Dried flowers, dried herbs, standard, jammy, great length.

La Miranda de Secastilla Garnacha Blanca 2020 B
garnacha blanca
90
Colour: bright straw. Nose: ripe fruit, dried herbs, faded flowers. Palate: powerful, ripe fruit, balanced, flavourful.

Secastilla 2018 T C
garnacha
91
Colour: cherry, garnet rim. Nose: powerful, dried herbs, characterful, fruit preserve, overripe fruit. Palate: flavourful, long, bitter.

Viñas del Vero Chardonnay 2021 B
chardonnay
90
Colour: bright straw. Nose: fruit expression, ripe fruit, floral. Palate: flavourful, fresh, good acidity, fruity aftertaste.

Viñas del Vero Gewürztraminer 2021 B
gewürztraminer
91
Colour: straw. Nose: ripe fruit, citrus fruit, jasmine. Palate: fruity, sweetness, easy to drink.

Viñas del Vero Pinot Noir 2021 RD
pinot noir
88
Fruity, dried flowers, tasty.

Viñas del Vero Sauvignon Blanc 2021 B
sauvignon blanc
87

DO SOMONTANO / D.O.P.

DO. TACORONTE - ACENTEJO

CONSEJO REGULADOR

Ctra. General del Norte, 97
38350 Tacoronte (Santa Cruz de Tenerife)
☎: +34 922 560 107 - Fax: +34 922 561 155
@: consejo@tacovin.com
www.tacovin.com

LOCATION:

Situated in the north of Tenerife, stretching for 23 km and is composed of 9 municipal districts: Tegueste, Tacoronte, El Sauzal, La Matanza de Acentejo, La Victoria de Acentejo, Santa Úrsula, La Laguna, Santa Cruz de Tenerife and El Rosario.

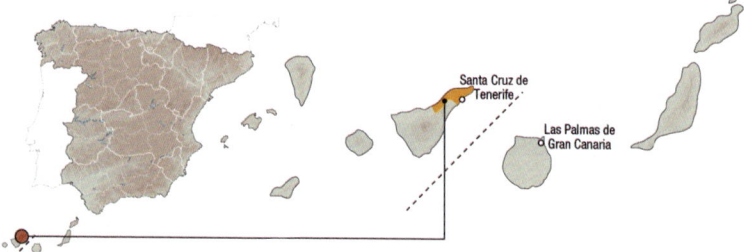

SUB-REGIONS:

Anaga (covering the municipal areas of La Laguna, Santa Cruz de Tenerife and Tegueste) which falls within the limits of the Anaga Rural Park.

GRAPE VARIETIES:

WHITE: PREFERRED: Güal, Malvasía, Listán Blanco and Marmajuelo.

AUTHORIZED: Pedro Ximénez, Moscatel, Verdello, Vijariego, Forastera Blanca, Albillo, Sabro, Bastardo Blanco, Breval, Burra Blanca and Torrontés.

RED: PREFERRED: Listán Negra and Negramoll.

AUTHORIZED: Tintilla, Moscatel Negro, Castellana Negra, Cabernet Sauvignon, Merlot, Pinot Noir, Ruby Cabernet, Syrah, Tempranillo, Bastardo Negro, Listán Prieto, Vijariego Negro and Malvasía Rosada.

FIGURES:

Vineyard surface: 740 – **Wine-Growers:** 1,123 – **Wineries:** 30 – **2021 Harvest rating:** Very Good – **Production 21:** 900,000 litres – **Market percentages:** 99% National -1% International.

SOIL:

The soil is volcanic, reddish, and is made up of organic matter and trace elements. The vines are cultivated both in the valleys next to the sea and higher up at altitudes of up to 1,000 m.

CLIMATE:

Typically Atlantic, affected by the orientation of the island and the relief which give rise to a great variety of microclimates. The temperatures are in general mild, thanks to the influence of the trade winds, which provide high levels of humidity, around 60%, although the rains are scarce.

CLASSIFICATION OF YOUNG WINE HARVESTS PEÑÍNGUIDE

2017	2018	2019	2020	2021
GOOD	GOOD	VERY GOOD	GOOD	VERY GOOD

AMBORA
Ctra. El Portezuelo Las Toscas, 197 La Padilla
38280 Tegueste (Santa Cruz de Tenerife)
☎: +34 629 955 639
amborabodegas@gmail.com

Ambora El Roquillo 2020 B
89
Balanced, spicy, dried herbs, standard.

Ambora Paraje San Ignacio 2020 T
91
Little interventionist. Colour: light cherry. Nose: red berry notes, balsamic herbs, floral. Palate: fruity, fresh, balanced.

Ambora Viña de Tegueste 2020 T
listán negro, negramoll, listán blanco
89
Balanced, spicy, fruity, smoky, tasty.

Negramoll Rosada 2020 T
90
Colour: bright cherry. Nose: fresh fruit, spicy, balsamic herbs, raspberry. Palate: good acidity, spicy, fine tannins.

BODEGA LA HIJUELA
Cº El Agua s/n Machado
38190 El Rosario (Santa Cruz de Tenerife)
☎: +34 696 050 759
bodegalahijuela@hotmail.com
www.bodegalahijuela.com

Híboro 2021 T
100% syrah
86

BODEGAS CRÁTER
San Nicolás, 122
38360 El Sauzal (Santa Cruz de Tenerife)
☎: +34 922 573 272
crater@craterbodegas.com
www.craterbodegas.com

Blanco de Cráter 2021 B
90% listán blanco, 10% albillo criollo
90
Colour: bright straw, greenish rim. Nose: fresh fruit, citrus fruit, dried herbs, mineral. Palate: fresh, fruity, good acidity.

Cráter 2017 T C
50% listán negro, 50% negramoll
91
Colour: Cherry, garnet rim. Nose: ripe fruit, fruit preserve, tobacco, fine reductive notes, tar. Palate: spicy, round tannins, long.

Cráter El Joven 2021 T
listán negro
89
Balanced, spicy, crisp, fruity, dried herbs.

Magma de Cráter 2018 T C
90% negramoll, 10% syrah
92
Colour: cherry, garnet rim. Nose: dried herbs, creamy oak, black fruit, fine reductive notes, tar. Palate: ripe fruit, spicy, round tannins.

BODEGAS INSULARES TENERIFE
Vereda del Medio, 48
38350 Tacoronte (Santa Cruz de Tenerife)
☎: +34 922 570 617
Fax: +34 922 570 043
contacto@bodegasinsularestenerife.es
www.bodegasinsularestenerife.es

🏆 PODIUM

Humboldt 1997 B D
listán blanco
95
Colour: bright cherry, garnet rim. Nose: candied fruit, sweet spices, dry nuts, toasty, stone fruit. Palate: fruity, flavourful, sweet, powerful.

Humboldt 2001 T D
100% listán negro
93
Colour: Cherry, brick rim edge. Nose: cigar, smoky, spicy, black fruit, dry nuts. Palate: full-bodied, powerful, flavourful, rich.

Viña Norte 2021 T
88
Balanced, spicy, crisp, mineral.

Viña Norte 2021 T MC
listán negro, negramoll
89
Crisp, fruity, floral, herbal, balanced.

Viña Norte Selección 2021 T
89
Spicy, crisp, fruity, mineral, balanced.

DO TACORONTE-ACENTEJO / D.O.P.

BODEGAS LOHER
Horno de La Teja, 26
38380 La Victoria de Acentejo
(Santa Cruz de Tenerife)
☎: +34 660 658 309
vinosloher@gmail.com

100% LN by LoHer 2020 T
listán negro
90
Colour: cherry, garnet rim. Nose: scrubland, black fruit, smoky, floral, ripe fruit. Palate: spicy, ripe fruit, round tannins, fleshy.

Cosechera Ensamblaje I 2020 T
91
Colour: dark-red cherry. Nose: toasty, spicy, cocoa bean, smoky. Palate: flavourful, toasty, fine bitter notes, fleshy.

Cosechera Ensamblaje II 2020 B
malvasía, gual, vijariego blanco, listán blanco
91
Colour: straw. Nose: expressive, white flowers, jasmine, dried herbs, waxy notes. Palate: flavourful, fruity, balanced, fresh, mineral.

Delirio 2019 T BA
listán negro
92
Colour: cherry, purple rim. Nose: fruit expression, floral, spicy, black fruit, tar. Palate: flavourful, fruity, good acidity, long.

Laboreo 2021 B BA
malvasía, listán blanco
91
Colour: straw. Nose: expressive, white flowers, dried herbs, waxy notes, sweet spices. Palate: flavourful, fruity, balanced.

Loher Tradicional 2020 T
listán negro, negramoll
90
Colour: deep cherry. Nose: ripe fruit, dried herbs, black fruit, smoky. Palate: powerful, ripe fruit, spicy, round tannins.

San Clemente 2020 B
baboso blanco, vijariego blanco, malvasía, marmajuelo
91
Colour: bright straw, greenish rim. Nose: fresh fruit, citrus fruit, wild herbs, sweet spices, creamy oak, honeyed notes. Palate: fresh, fruity, good acidity, fine bitter notes, flavourful.

San Clemente 2020 T BA
listán negro, tintilla
91
Colour: cherry, purple rim. Nose: fruit expression, red berry notes, spicy, smoky, rose petals. Palate: flavourful, fruity, good acidity, long.

BODEGAS MARBA
Ctra. Portezuelo - Las Toscas, 253
38280 Tegueste (Santa Cruz de Tenerife)
☎: +34 922 638 400
marba@bodegasmarba.es
www.bodegasmarba.com

Hugo Afrutado 2021 RD
listán negro, otras
87

Marba 2021 B BA
listán blanco, gual, albillo, forastera gomera
87

Marba 2021 RD
listán negro, otras
84

Marba 2021 T BA
listán negro, negramoll, tempranillo, ruby, syrah
86

Marba 2021 T MC
listán negro, otras
87

Marba Capricho 2021 T FB
syrah
87

DOMÍNGUEZ CUARTA GENERACIÓN
El Calvario, 79
38350 Tacoronte (Santa Cruz de Tenerife)
☎: +34 659 974 375
administracion@bodegasdominguez.es
www.bodegasdominguez.es

Domínguez 2020 T
listán negro, negramoll, castellana, listán blanco, albillo criollo, baboso
88
Standard, crisp, herbal, tasty, mineral.

Domínguez Antología 2017 T C
negramoll, castellana, baboso
89
Balanced, spicy, tasty, slight reduction.

**Domínguez Colección
El Marañón 2018 T C**
listán negro, negramoll, listán blanco, castellana

88
Balanced, spicy, jammy, tasty.

Domínguez con Firma 2019 T
55% listán negro, 35% negramoll, 10% castellana

87

Domínguez en Blanco 2020 B
albillo, negramoll, listán blanco, verdello

85

**Domínguez
Malvasía Clásico 2012 B D**
100% malvasía

91
Colour: bright yellow. Nose: ripe fruit, candied fruit, honeyed notes, stone fruit. Palate: flavourful, unctuous, fruity, sweet.

HACIENDA ACENTEJO

Pérez Díaz, 44 A
38380 La Victoria de Acentejo
(Santa Cruz de Tenerife)
☎: +34 650 974 598
jjgutierrez@haciendadeacentejo.com
www.haciendadeacentejo.com

Hacienda Acentejo 2021 B S
listán blanco

85

Hacienda Acentejo 2021 T S
listán negro, negramoll

87

Hacienda de Acentejo 2021 T BA
listán negro, negramoll, tintilla

85

PRESAS OCAMPO

Los Alamos de San Juan, 5
38350 Tacoronte (Santa Cruz de Tenerife)
☎: +34 922 571 689
Fax: +34 922 561 700
administracion@presasocampo.com
www.presasocampo.com

**Presas Ocampo
Gran Alysius 2020 T**
listán negro, syrah

90
Colour: deep cherry. Nose: dried herbs, creamy oak, black fruit, earthy notes. Palate: powerful, ripe fruit, spicy, round tannins.

**Presas Ocampo
Maceración Especial 2021 T MC**
listán negro

89
Balanced, spicy, floral, crisp, fruity, herbal.

Presas Ocampo Origen 2020 T
listán negro

91
Colour: cherry, purple rim. Nose: fruit expression, red berry notes, floral, spicy, rose petals. Palate: flavourful, fruity, good acidity, mineral.

**Presas Ocampo
Selección Especial 2021 T MC**
listán negro

88
Balanced, spicy, floral, herbal.

Presas Ocampo Vendimia Seleccionada 2020 T
listán negro, merlot, syrah

89
Balanced, spicy, crisp, fruity, herbal, tasty.

Presas Ocampo Viñedos Propios 2021 T
listán negro, negramoll, syrah

88
Balanced, spicy, floral, mineral, tasty.

TIERRA FUNDIDA

Camino Las Medianías 13. Finca Morales.
Los Baldios.
38291 San Cristóbal de la Laguna
(Santa Cruz de Tenerife)
☎: +34 647 989 081
vinosentandem@gmail.com
www.tierrafundida.es

Tierra Fundida 2020 T
negramoll, listán negro, castellana, baboso, verdello

91
Colour: deep cherry. Nose: dried herbs, black fruit, earthy notes, cocoa bean. Palate: ripe fruit, spicy, round tannins.

VIÑA ESTEVEZ

38380 La Victoria de Acentejo (Santa Cruz de Tenerife)
☎: +34 608 724 671
elena.vinaestevez@gmail.com

Viña Estévez 2020 T BA
50% baboso, 40% vijariego negro, 10% listán negro

88
Balanced, spicy, dried flowers, tasty, toasty.

DO TACORONTE-ACENTEJO / D.O.P.

DO TACORONTE-ACENTEJO / D.O.P.

WINERY BURGMANN TENERIFE

Ctra. Tacoronte Tejina, 78A
38350 Tacoronte (Santa Cruz de Tenerife)
☎: +34 610 750 437
propiedadesbcb@gmail.com
www.burgmannwinery.com

Burgmann Red Selection 2020 T
listán negro

88

Balsamic herbs, balanced, spicy, fruity, herbal, jammy.

Burgmann Rosé Selection 2020 RD
listán negro

86

Olivia by Burgmann 2020 B
listán blanco

92

Exuberant, floral, jammy, with personality. Colour: straw. Nose: ripe fruit, dried herbs, faded flowers, honeyed notes, balsamic herbs. Palate: powerful, ripe fruit, balanced, full-bodied.

DO. TARRAGONA
CONSEJO REGULADOR

Calle La Cort, 41
43800 Valls (Tarragona)
☎: +34 977 217 931 - Fax: +34 977 229 102
@: info@dotarragona.cat
www.dotarragona.cat

LOCATION:

The region is situated in the province of Tarragona. It comprises two different wine-growing regions: El Camp and Ribera d'Ebre, with a total of 72 municipal areas.

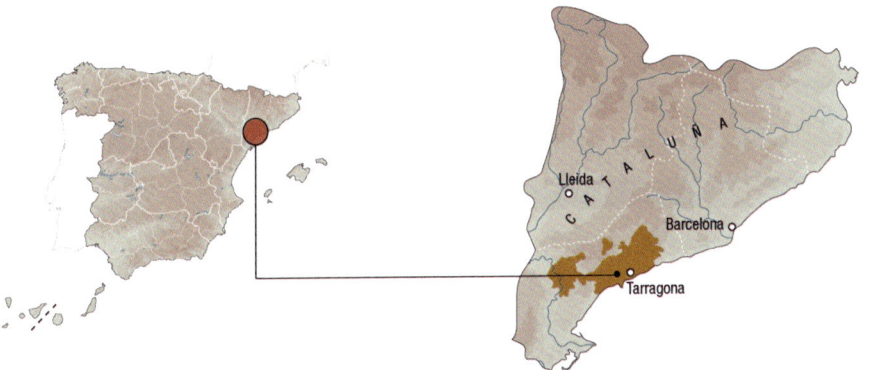

SUB-REGIONS:

El Camp and Ribera d'Ebre.

GRAPE VARIETIES:

WHITE: Chardonnay, Macabeo, Xarel·lo, Garnacha Blanca, Parellada, Moscatel de Alejandría, Moscatel de Frontignan, Sauvignon Blanc, Malvasía, Vinyater, Xarel.lo Vermell, Sumoi Blanc and Malvasia de Sitges.

RED: Samsó (Cariñena), Garnacha, Ull de Llebre (Tempranillo), Cabernet Sauvignon, Merlot, Monastrell, Pinot Noir, Syrah and Sumoll.

FIGURES:

Vineyard surface: 3,900 – **Wine-Growers:** 1,325 – **Wineries:** 37 – **2021 Harvest rating:** N/A – **Production 21:** 1,600,000 litres – **Market percentages:** 80% National - 20% International.

SOIL:

El Camp is characterized by its calcareous, light terrain, and the Ribera has calcareous terrain and also some alluvial terrain.

CLIMATE:

Mediterranean in the region of El Camp, with an average annual rainfall of 500 mm. The region of the Ribera has a rather harsh climate with cold winters and hot summers; it also has the lowest rainfall in the region (385 mm per year).

CLASSIFICATION OF YOUNG WINE HARVESTS PEÑÍNGUIDE

2017	2018	2019	2020	2021
AVERAGE	GOOD	GOOD	VERY GOOD	N/A

DO TARRAGONA / D.O.P.

ALREGI
Pol. Ind. Empordà Internacional – C/Garrotxa, 8
17469 Vilamalla (Girona/Gerona)
☎: +34 972 526 061
alregi@alregi.es
www.alregi.es

Lo Vy 2020 B
100% cartoixà
88
Little interventionist, slight oxidation, dried flowers, dried herbs, wild, smooth.

Lo Vy 2020 T
65% garnacha, 35% cabernet sauvignon
89
Balanced, spicy, herbaceous, rustic, tasty, crisp.

Lo Vy Ancestral 2020 B
100% cartoixà
88
Dried flowers, fruity, lactic, jammy, yeasty notes.

BIOPAUMERÀ
Plaza San Juan, 3
43513 Rasquera (Tarragona)
☎: +34 618 154 254
biopaumera@biopaumera.com
www.biopaumera.com

Adrià de Biopaumerà 2017 T RB
87

Erika de Paumera 2021 RD
86

Iuvenis de Biopaumerà 2020 T
88
Spicy, fruity, jammy, tasty.

CELLER MAS BELLA
La Font, 8 - Masmolets
43813 Valls (Tarragona)
☎: +34 600 269 786
info@cellermasbella.com
www.cellermasbella.com

Bella 2021 B
macabeo
86

Bella Blanc Cartoixa 2020 B
88
Fruity, jammy, dried herbs, spicy, crisp.

CELLER MAS D'EN BAIGET
Mas d'en Baiget s/n
43479 L'Albiol (Tarragona)
☎: +34 670 207 279
cellermasdenbaiget@gmail.com
www.cellermasdenbaiget.cat

Coster dels Rosers 2020 RD
ull de llebre, garnacha
86

Nuri 2020 B
moscatel grano menudo, macabeo
87

Pilanot Blanc 2020 B
garnacha blanca, parellada
89
Slight oxidation, little interventionist, with personality, jammy, tasty, bitter.

CELLER MAS DEL BOTÓ
Camí de Porrera a Alforja, s/n
43365 Alforja (Tarragona)
☎: +34 630 982 747
pep@masdelboto.cat
www.masdelboto.cat

Dolça de Ganagot 2017 T RC D
garnacha
90
Slight fizz. Colour. Cherry, bright ochre rim. Nose: spicy, toasty, aged wood nuances, dry nuts, dried fruit. Palate: powerful, flavourful, sweetness.

Ganagot 2007 T GR
65% garnacha, 25% cabernet sauvignon, 10% samsó
89
Pruney, full-bodied, spicy, jammy, tasty, herbal.

Ganagot 2009 T GR
garnacha, cariñena, cabernet sauvignon
90
Colour. light cherry. Nose: old leather, meaty notes, tobacco, ripe fruit, dried herbs. Palate: fruity, flavourful, powerful, slightly dry, soft tannins.

CELLER MAS VICENÇ
Mas Vicenç, s/n
43811 Cabra de Camp (Tarragona)
☎: +34 627 570 075
masvicens@masvicens.com
www.masvicens.com

Dent de Lleó 2019 B
60% chardonnay, 40% garnacha blanca
89
Toasty, jammy, tasty, spicy, full-bodied.

La Fonollosa 2019 T C
garnacha

90

Colour: bright cherry. Nose: sweet spices, ripe fruit, dark chocolate, roasted coffee. Palate: spicy, concentrated, flavourful.

CELLERS BLANCH

Avda. Catalunya 8
43812 Puigpelat (Tarragona)
☎: +34 649 993 294
info@cellersblanch.com
www.cellersblanch.com

Coordenades 1'19 2021 T
ull de llebre

87 🍷

Coordenades 41'17 2019 B
chardonnay, macabeo

84 🍷

Identitas 2019 B
subirat parent

89 🍷

Aromatic, varietally correct, dried flowers, wild, tasty, bitter.

Pont Fosc 2019 B
macabeo

91

Colour: bright straw. Nose: white fruit, balanced, wild herbs, dried herbs. Palate: easy to drink, balanced.

Promesa 2021 RD
ull de llebre

86 🍷

Subirat 2018 BE BN
macabeo, xarel.lo, subirat parent, parellada

88

Jammy, toasty, simple, standard.

DE MULLER

Camí Pedra Estela, 34
43205 Reus (Tarragona)
☎: +34 977 757 473
promo@demuller.es
www.demuller.es

De Muller Cabernet Sauvignon 2020 T C
100% cabernet sauvignon

85

De Muller Chardonnay 2020 B FB
chardonnay

90

Colour: bright yellow. Nose: ripe fruit, honeyed notes, sweet spices. Palate: fruity, full-bodied, flavourful, balanced, toasty.

De Muller Chardonnay 2021 B FB
100% chardonnay

89

Toasty, jammy, full-bodied, spicy, fruity.

De Muller Muscat 2021 B
100% moscatel

88

Aromatic, varietally correct, floral, standard.

De Muller Syrah 2020 T
100% syrah

87

Reina Violant BE R BN
50% chardonnay, 50% pinot noir

89

Aromatic, yeasty notes, dried flowers, smooth, balanced.

Solimar 2020 T C
60% cabernet sauvignon, 40% merlot

86

Solimar 2021 B
70% macabeo, 20% moscatel, 10% xarel.lo

86

Trilogía Pinot Noir Blanc de Noir BE R BN
100% pinot noir

90

Slight oxidation. Colour: yellow, pale. Nose: ripe fruit, dried flowers, faded flowers. Palate: flavourful, ripe fruit.

DO TARRAGONA / D.O.P.

DO TARRAGONA / D.O.P.

UNIVERSITAT ROVIRA I VIRGILI
Ctra TV 7211 km 7
43120 Constanti (Tarragona)
☎: +34 977 520 197
pedro.cabanillas@urv.cat
https://www.fe.urv.cat/es/facultad/bodega-mas-dels-frares

Reserva Lluis Arola i Ferrer 2017 T R
100% cabernet sauvignon

89
Colour: deep cherry. Nose: black fruit, wild herbs, black liquorice, spicy, ripe fruit. Palate: flavourful, full-bodied, slightly dry, soft tannins, balanced.

Universitat Rovira i Virgili Blanc de Blancs BE BR
60% macabeo, 8% chardonnay, 8% garnacha blanca, 12% parellada, 7% xarel.lo, 5% sauvignon blanc

88
Pleasant, fruity, jammy, dried flowers, tasty.

VINÍCOLA DE NULLES -ADERNATS
Raval de Sant Joan, 7
43887 Nulles (Tarragona)
☎: +34 977 602 622
botiga@vinicoladenulles.com
www.adernats.cat

100 Veremes 2019 B FB
100% macabeo

89 🌱
Colour: bright yellow. Nose: toasty, toasted bread, dry nuts, ripe fruit. Palate: fresh, full-bodied, flavourful, good finish.

Adernats Xarel.lo Vermell 2020 B
100% xarel.lo vermell

90 🌱
Colour: yellow, coppery red. Nose: ripe fruit, honeyed notes, dried flowers. Palate: dry, bitter.

VINYA JANINE (JOSEP M. SAUMELL)
Anselm Clavé, 1
43812 Rodonya (Tarragona)
☎: +34 629 014 231
vjanine@tinet.org
www.vinyajanine.com

Vinya Janine Syrah 2019 T C
syrah

87 🌱

VINYES DEL TIET PERE
Raval del Roser, 3
43886 Vilabella (Tarragona)
☎: +34 625 408 974
vinyesdeltietpere@gmail.com

Cami de la Font 2018 B
macabeo

91
Colour: old gold. Nose: ripe fruit, faded flowers, chamomile, characterful. Palate: rich, flavourful, long.

Cami de la Font 2019 B
macabeo

91
Colour: bright yellow. Nose: ripe fruit, spicy, wild herbs, dried flowers. Palate: rich, good structure, long, fine bitter notes.

Cami de la Font Brisat 2019 B
macabeo

92
With personality, exuberant. Colour: bright yellow. Nose: ripe fruit, dried flowers, characterful. Palate: bitter, flavourful.

Escabeces 2020 RD BA
xarel.lo vermell

90
With personality, little interventionist. Colour: coppery red. Nose: fresh, wild herbs, hint of anise. Palate: flavourful, dry, fine bitter notes.

Escabeces Blanques Primes 2020 B C
parellada

89
Little interventionist, balanced, dried flowers, wild, thin.

Escabeces Cartoixà Blanc 2020 B C
xarel.lo

89
Little interventionist, rustic, dried flowers, balanced, with personality.

VIVES AMBRÒS

Mayor, 39
43812 Montferri (Tarragona)
☎: +34 639 521 652
mail@vivesambros.com
www.vivesambros.com

Aïda de Vives Ambròs 2020 T
66% tempranillo, 33% garnacha

86

Aïda de Vives Ambròs 2021 B
80% macabeo, 20% xarel.lo

87

Ishii de Vives Ambròs 2021 B
50% moscatel de alejandría, 25% chardonnay, 25% xarel.lo

89

Fruity, herbal, dried flowers, tasty.

Jujol de Vives Ambròs 2018 B FB
100% xarel.lo vermell

87

DO TARRAGONA / D.O.P.

DO. TERRA ALTA
CONSEJO REGULADOR

Ctra. Vilalba, 31
43780 Gandesa (Tarragona)
☎: +34 977 421 278
@: info@doterraalta.cat
www.doterraalta.com

LOCATION:

In the southeast of Catalonia, in the province of Tarragona. It covers the municipal districts of Arnes, Batea, Bot, Caseres, Corbera d´Ebre, La Fatarella, Gandesa, Horta de Sant Joan, Pinell de Brai, La Pobla de Massaluca, Prat de Comte and Vilalba dels Arcs.

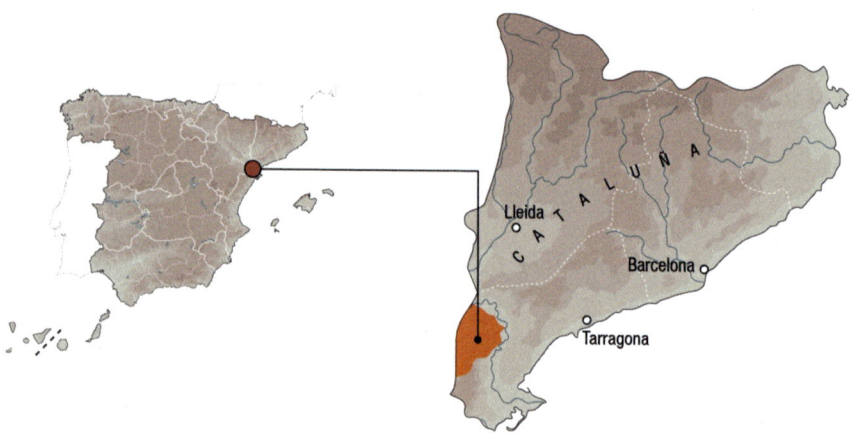

GRAPE VARIETIES:

WHITE: Chardonnay, Garnacha Blanca, Parellada, Macabeo, Moscatel, Sauvignon Blanc, Chenin, Pedro Ximénez and Viognier.

RED: Garnacha Tinta, Garnacha Peluda, Garnacha Tintorera, Cariñena, Syrah, Tempranillo (Ull De Llebre), Merlot, Cabernet Sauvignon, Cabernet Franc and Morenillo.

FIGURES:

Vineyard surface: 5,813 – **Wine-Growers:** 1,093 – **Wineries:** 62 – **2021 Harvest rating:** N/A – **Production 21:** 15,575,800 litres – **Market percentages:** 65% National - 35% International.

SOIL:

The vineyards are located on an extensive plateau at an altitude of slightly over 400 m. The soil is calcareous and the texture mainly clayey, poor in organic matter and with many pebbles.

CLIMATE:

Mediterranean, with continental influences. It is characterized by its hot, dry summers and very cold winters, especially in the higher regions in the east. The average annual rainfall is 400 mm. Another vital aspect is the wind: the 'Cierzo' and the 'Garbi' (Ábrego) winds.

CLASSIFICATION OF YOUNG WINE HARVESTS PEÑÍNGUIDE

2017	2018	2019	2020	2021
GOOD	GOOD	VERY GOOD	VERY GOOD	VERY GOOD

7 MAGNÍFICS
Miquel Torres i Carbó, 6
08720 Vilafranca del Penedès (Barcelona)
☎: +34 938 177 400
7magnifics@7magnifics.com
www.7magnifics.com

Rebels de Batea 2020 B
garnacha blanca
89
Citrus fruit, balanced, crisp, tasty, subtle, mineral.

Rebels de Batea 2020 T
garnacha
89
Jammy, spicy, tasty, powerful, standard, dried flowers, simple.

AGRÍCOLA CORBERA D'EBRE
Ponent, 21
43784 Corbera de Ebro (Tarragona)
☎: +34 977 420 432
bodega@agricolacorbera.com

L'Home Peix Negre 2021 T
100% garnacha
86

L'Home Peix Parellada 2021 B
100% parellada
84

L'Home Peix Rosat 2021 RD
100% garnacha
87

La Cisqueta de Corbera Blanc 2021 B
100% garnacha blanca
87

La Cisqueta de Corbera Negre 2021 T
65% garnacha, 35% cariñena
87

La Muntera 2019 T
100% cariñena
89
Fruity, crisp, spicy, herbal, tasty.

AGRÍCOLA I S. C. TERRA ALTA - CATERRA
Glorieta, s/n
43783 La Pobla de Massaluca (Tarragona)
☎: +34 608 590 780
info@caterra.es
www.caterra.es

Font Calenta 2021 B
macabeo, garnacha blanca
87

Font Calenta Negre 2021 T
84

Hereus Blanc 2021 B
100% garnacha blanca
88
Pleasant, floral, fruity, smooth, balanced.

Hereus Negre 2020 T
garnacha, cariñena
86

ALEGRE WINES & SPIRITS
Balmes, 345
08006 Barcelona (Barcelona)
☎: +34 935 641 262
info@alegrews.com
www.alegrews.com

La Dansada 2021 B
garnacha blanca
89
Citrus fruit, austere, crisp, herbal, mineral, tasty.

La Dansada 2021 T
garnacha
87

ALTAVINS VITICULTORS
Ctra. Vilalba dels Arcs s/n
43786 Batea (Tarragona)
☎: +34 977 430 596
altavins@altavins.com
www.altavins.com

Ilercavònia 2021 B
100% garnacha blanca
90
Colour: bright straw. Nose: dried flowers, white fruit, ripe fruit, balanced. Palate: fruity, easy to drink, good finish.

Selecció Carinyena 2018 T R
100% cariñena
92
Colour: cherry, garnet rim. Nose: fruit preserve, powerful, cocoa bean, dried herbs, varietal. Palate: flavourful, long, balanced, ripe fruit.

DO TERRA ALTA / D.O.P.

DO TERRA ALTA / D.O.P.

Selecció Garnatxa Blanca 2017 B
100% garnacha blanca

91

Age nuances, weary. Colour: bright yellow. Nose: characterful, roasted almonds, spicy, smoky. Palate: rich, flavourful.

Selecció Garnatxa Peluda 2017 T R
100% garnacha peluda

91

Colour: deep cherry. Nose: ripe fruit, dried herbs, creamy oak, wild herbs, earthy notes. Palate: powerful, ripe fruit, spicy, round tannins.

BERNAVÍ VINYATERS
Finca Mas Vernet
43782 Vilalba dels Arcs (Tarragona)
☎: +34 651 031 835
info@bernavi.com
www.bernavi.com

Bernaví Ca'Vernet 2018 T C
cabernet franc, cabernet sauvignon

89

Representative, balsamic herbs, dried herbs, jammy, powerful, tasty. Nose: earthy notes.

Bernaví Garnatxa Blanca 2021 B
garnacha blanca

88 🌿

Citrus fruit, balanced, simple, herbal.

Bernaví Morenillo 2021 RD
morenillo

87 🌿

Bernavi Tres de Tres 2020 T
garnacha, garnacha blanca, garnacha peluda

88

Kind finish, herbal, tasty, jammy, fruity, wild.

BIELSA RUANO VINS
Sant Isidre, 24
43782 Vilalba dels Arcs (Tarragona)
☎: +34 665 220 796
info@bielsaruano.com
www.bielsaruano.com

Lo Noi del Saxo 2021 B
garnacha blanca, macabeo

88

Citrus fruit, sweety finish, full-bodied, jammy, tropical.

Lo Noi del Saxo 2021 T
garnacha, syrah

89

Fruity, dried herbs, spicy, tasty.

Music de Carrer 2020 T C
garnacha

90

Colour: Cherry. Nose: ripe fruit, dried herbs, spicy. Palate: flavourful, spicy, ripe fruit.

Music de Carrer 2021 B
garnacha blanca

89

Pleasant, aromatic, dried flowers, fruity, jammy, standard.

CELLER ARRUFÍ
Avda. Terra Alta, 12
43786 Batea (Tarragona)
☎: +34 722 224 772
hola@cellerarrufi.com
www.cellerarrufi.com

Celler Arrufí Llicsó 2018 B BA
garnacha blanca

89 🌿

Citrus fruit, balanced, crisp, tasty, creamy.

Celler Arrufí Llicsó 2019 B BA
garnacha blanca

90 🌿

Pleasant. Colour: yellow. Nose: dry nuts, toasty, spicy, characterful, floral. Palate: long, flavourful, spicy.

Celler Arrufí Panical 2021 B
garnacha blanca

90 🌿

Colour: straw. Nose: dried herbs, faded flowers, white fruit, fine lees. Palate: ripe fruit, balanced, flavourful.

Celler Arrufí Panicort 2018 T
20% garnacha, 80% cariñena

89 🌿

Spicy, fruity, jammy, tasty.

Celler Arrufí Trepadella 2021 T

88 🌿

Crisp, fruity, floral, balanced.

CELLER BALART
Pol. La Plana, Parc. 20
43780 Gandesa (Tarragona)
☎: +34 600 484 900
info@germansbalart.com

Celler Balart 2021 B
100% garnacha blanca

89

Kind finish, aromatic, fruity, tropical, tasty, balanced.

Celler Balart Negre 2018 T
garnacha, cariñena
86

Llepolia Blanc 2021 B
garnacha, sauvignon blanc
87

Llepolia Negre 2019 T C
garnacha
86

Samerola Blanc 2021 B
garnacha, macabeo, moscatel
88
Citrus fruit, floral, crisp, tasty.

Samerola Negre 2021 T
garnacha, cariñena
88
Balanced, floral, crisp, fruity, herbal.

CELLER BÀRBARA FORÉS
Santa Anna, 28
43780 Gandesa (Tarragona)
☎: +34 977 420 160
info@cellerbarbarafores.com
www.cellerbarbarafores.com

Abrisa't Bàrbara Forés 2020 B C
garnacha blanca
93 ♣
With personality, defined aromas. Colour: old gold. Nose: faded flowers, dried flowers, chamomile. Palate: fruity, flavourful, balanced, fine bitter notes.

Bàrbara Forés Blanc 2021 B
garnacha blanca
91 ♣
Colour: bright straw. Nose: medium intensity, dried herbs, ripe fruit. Palate: flavourful, varietal, balanced.

Bàrbara Forés Rosat 2021 RD
75% garnacha, 20% syrah, 5% merlot
87 ♣

Coma d'En Pou Bàrbara Forés 2019 T C
garnacha, garnacha peluda
93 ♣
Colour: Cherry. Nose: balsamic herbs, sweet spices, scrubland. Palate: spicy, good acidity, varietal, easy to drink, taut.

El Quintà Bàrbara Forés 2020 B FB
92 ♣
Colour: yellow. Nose: dried flowers, faded flowers, spicy, expressive. Palate: ripe fruit, long, spicy, balanced.

El Templari Bàrbara Forés 2020 T C
morenillo
90 ♣
Colour: cherry, garnet rim. Nose: animal reductive notes, premature reduction notes, red berry notes, ripe fruit. Palate: fruity, easy to drink.

CELLER BATEA
Moli, 30
43786 Batea (Tarragona)
☎: +34 977 430 056
cellerbatea@cellerbatea.com
www.cellerbatea.com

Naturalis Mer Blanc 2021 B
100% garnacha blanca
88 ♣
Pleasant, aromatic, jammy, tropical, smooth, simple.

Naturalis Mer Roure 2020 T RB
garnacha
89 ♣
Varietally correct, wild, jammy, spicy, dried herbs, fruity, standard.

Primicia Blanc Bota 2021 B FB
100% garnacha blanca
87 ♣

Primicia Garnacha Syrah 2019 T C
garnacha, syrah
88
Balanced, spicy, fruity, dried herbs, jammy, toasty.

Vallmajor Blanc 2021 B
100% garnacha blanca
87

Vallmajor Negre 2021 T
100% garnacha
87

CELLER COMA D'EN BONET
Camí de Les Comes d'En Bonet s/n
43780 Gandesa (Tarragona)
☎: +34 678 036 487
pepefuster@comadenbonet.com
www.comadenbonet.com

Dardell Garnacha Blanca Viognier 2021 B
90% garnacha blanca, 10% viognier
87 ♣

ProHom Conceptia 2020 T
60% garnacha, 20% syrah, 20% cabernet sauvignon
88 ♣
Balanced, spicy, fruity, jammy, tasty.

DO TERRA ALTA / D.O.P.

DO TERRA ALTA / D.O.P.

ProHom Conceptia 2021 B
70% garnacha blanca, 30% viognier
88
Aromatic, fruity, simple, floral, jammy, standard.

ProHom Experientia 2018 T
garnacha, cariñena, syrah, cabernet sauvignon
91
Colour: bright cherry. Nose: fruit liqueur notes, black fruit, dried herbs, aromatic coffee, spicy. Palate: good structure, fruity, flavourful, balanced.

ProHom Experientia 2021 B FB
70% garnacha blanca, 30% viognier
89
Aromatic, floral, wild, smooth, standard, defined aromas.

ProHom Viognier 2021 B
100% viognier
90
Aromatic, varietally correct, kind finish, floral, fruity, jammy, tasty. Palate: flavourful, ripe fruit, easy to drink, varietal.

CELLER COOPERATIU GANDESA
Avda. Catalunya, 28
43780 Gandesa (Tarragona)
☎: +34 977 420 017
Fax: +34 977 420 403
cellercooperatiugandesa@gmail.com
www.coopgandesa.com

Puresa Garnatxa Blanca 2019 B FB
100% garnacha blanca
91
Colour: bright yellow. Nose: ripe fruit, dried herbs, faded flowers, spicy, honeyed notes. Palate: powerful, ripe fruit, balanced, flavourful, good finish.

Puresa Macabeu 2019 B FB
100% macabeo
92
Colour: bright yellow. Nose: powerful, ripe fruit, spicy, expressive. Palate: rich, good structure, long, fine bitter notes, toasty, flavourful.

Puresa Morenillo 2016 T C
100% morenillo
92
Defined aromas, jammy, wild. Colour: cherry, garnet rim. Nose: black fruit, ripe fruit, creamy oak, dried herbs, scrubland. Palate: flavourful, fruity, spicy.

Puresa Ranci BF RC
100% garnacha blanca
89
Toasty, tasty, spicy, oxidativ.

Puresa Samsó 2015 T BA
100% samsó
93
Colour: dark-red cherry, garnet rim. Nose: ripe fruit, sweet spices, waxy notes, fine reductive notes, dried herbs. Palate: spicy, round tannins, long, powerful.

CELLER LA BOTERA
Ctra. Maella, s/n
43786 Batea (Tarragona)
☎: +34 977 430 009
agrobotiga@labotera.com
www.labotera.com

Mudèfer Blanc 2019 B C
100% garnacha blanca
88
Jammy, full-bodied, exuberant, spicy, roasted coffee. Nose: woody.

Mudèfer Negre 2017 T C
75% cariñena, 25% garnacha
89
Pruney, balsamic herbs, dried herbs, jammy, great length, wild, tasty, spicy.

Vila Closa Blanc 2021 B
100% garnacha blanca
87

Vila Closa Chardonnay 2019 B FB
100% chardonnay
88
Dried flowers, standard, balanced, dried herbs, spicy, jammy.

Vila Closa Peluda 2020 T RB
100% garnacha peluda
91
Colour: bright cherry, light cherry. Nose: red berry notes, ripe fruit, dried flowers, wild herbs, expressive, neat. Palate: fruity, easy to drink, good acidity.

Vila Closa Rubor 2021 RD
100% garnacha
87

CELLER MARIOL
Les Forques, 2
43786 Batea (Tarragona)
☎: +34 977 430 303
oficina@cellermariol.com
www.casamariol.com

Casa Mariol Garnatxa Blanca 2021 B
garnacha blanca
87

Casa Mariol Garnatxa Negra 2020 T
garnacha
89
Spicy, floral, fruity, jammy, tasty.

Casa Mariol Garnatxa Negra 2021 T
garnacha
87 🌱

Casa Mariol Samsó 2018 T C
cariñena
85

Casa Mariol Syrah 2016 T R
syrah
87

CELLER PIÑOL
43786 Batea (Tarragona)
☎: +34 977 430 505
info@cellerpinol.com
www.casapinol.com

Anima L'Avi Arrufí 2020 B
100% garnacha blanca
92
Colour: bright straw. Nose: fine lees, fresh, white fruit, ripe fruit. Palate: full-bodied, rich, long, good acidity.

L'Avi Arrufí 2019 T
75% cariñena, 15% garnacha, 10% syrah
92
Colour: Cherry. Nose: ripe fruit, black fruit, wild herbs, spicy. Palate: flavourful, fruity, good finish, powerful.

L'Avi Arrufí 2020 B FB
100% garnacha blanca
93
Colour: yellow. Nose: spicy, dried flowers, faded flowers, balanced, expressive, white fruit, ripe fruit. Palate: rich, round, elegant.

Mather Teresina 2019 T
70% garnacha, 15% cariñena, 15% morenillo
93
Colour: bright cherry. Nose: black fruit, spicy, creamy oak, balsamic herbs. Palate: fruity, balanced, slightly dry, soft tannins, flavourful.

N. Sra. del Portal 2019 T C
70% garnacha, 15% cariñena, 15% syrah
90
Colour: bright cherry. Nose: ripe fruit, spicy, dried herbs, smoky. Palate: fruity, flavourful, slightly dry, soft tannins.

N. Sra. del Portal 2021 B
90% garnacha, 5% samsó blanc, 5% viognier
89
Aromatic, floral, fruity, jammy, balanced, tasty.

Sa Natura 2019 T
40% garnacha, 40% cariñena, 20% syrah
90 🌱
Colour: bright cherry. Nose: ripe fruit, black fruit, dried herbs, spicy. Palate: fruity, flavourful, spicy, slightly dry, soft tannins.

CELLER RIALLA
Saragossa, 10
43786 Batea (Tarragona)
☎: +34 679 200 774
rialla@riallavi.com
www.cellerrialla.com

Rialla Garnatxa Blanca 2021 B
100% garnacha blanca
90
Jammy, defined aromas. Colour: straw. Nose: dried flowers, dry nuts, white fruit, ripe fruit. Palate: flavourful, fine bitter notes, rich.

Rialla Garnatxa Peluda 2019 T
100% garnacha peluda
87

Rialla Garnatxa Peluda 2021 RD
100% garnacha peluda
85

Rialla Garnatxa Tintorera 2020 T
100% garnacha tintorera
87

DO TERRA ALTA / D.O.P.

DO TERRA ALTA / D.O.P.

CELLER XAVIER CLUA
Sant Isidre, 41
43782 Vilalba dels Arcs (Tarragona)
☎: +34 977 263 069
rosa@cellerclua.com
www.cellerclua.com

Clua Mil.lennium 2019 T
55% garnacha, 30% cabernet sauvignon, 15% syrah

89

Slight reduction, hot, balsamic herbs, jammy, tasty, powerful, spicy.

Il.lusió de Clua 2020 T
50% garnacha, 50% garnacha peluda

91

Colour: deep cherry. Nose: dried herbs, creamy oak, black fruit, undergrowth. Palate: powerful, ripe fruit, spicy, round tannins.

Il.lusió de Clua 2021 B
100% garnacha blanca

89

Pleasant, aromatic, floral, fruity, jammy, tasty.

Mas d'en Pol 2018 T C
60% garnacha, 20% syrah, 20% cabernet sauvignon

88

Slight reduction, classic, herbal, jammy, powerful, tasty.

Mas d'en Pol 2020 T
80% garnacha, 20% syrah

87

Mas d'en Pol 2021 B
65% garnacha blanca, 20% chardonnay, 15% sauvignon blanc

87

CELLERS TARRONÉ
Calvari, 22
43786 Batea (Tarragona)
☎: +34 977 430 109
info@cellerstarrone.com
www.cellerstarrone.com

A Part 2019 B
garnacha blanca

90

Colour: straw. Nose: ripe fruit, dried herbs, dried flowers, dry nuts. Palate: powerful, ripe fruit, balanced, spicy.

Merian Blanc 2021 B
100% garnacha blanca

88

Pleasant, dried flowers, fruity, jammy, tasty, standard.

Merian Negre 2021 T
100% garnacha

88

Dried herbs, jammy, rustic, tasty, standard.

Merian Rosat 2021 RD
100% garnacha

87

Punt i... 2020 T C
70% garnacha, 30% syrah

89

Jammy, wild, varietally correct, fruity, dried flowers.

Seguit 2019 T
100% garnacha

89

Jammy, toasty, wild, tasty, herbal, spicy.

CELLERS TERRA I VINS
Av. Falset, 17 Baixos
43206 Reus (Tarragona)
☎: +34 658 567 409
celler@cellersterraivins.com
www.cellersterraivins.com

Flor de Nit 2021 B
87

Flor de Nit 2021 RD
86

La Negra Flor 2020 T
garnacha, syrah

87

CELLERS UNIÓ
Joan Oliver 16-24
43206 Reus (Tarragona)
☎: +34 977 330 055
Fax: +34 977 330 070
info@cellersunio.com
www.cellersunio.com

Clos Dalian Garnacha 2021 T
garnacha

87

Clos Dalian Garnacha Blanca 2021 B
garnacha blanca

88

Citrus fruit, dried flowers, fruity, jammy, balanced.

Clos del Pinell Garnatxa 2021 T
garnacha

88

Balanced, spicy, jammy, dried herbs.

Clos del Pinell Garnatxa Blanca 2021 B
garnacha blanca
86

CLOS GALENA
Camino de la Solana, s/n
43736 El Molar (Tarragona)
☎: +34 607 421 822
laura@closgalena.com
www.closgalena.com

Secrets de Mar 2020 T RB
60% cariñena, 40% garnacha
89
Spicy, fruity, jammy, tasty.

Secrets de Mar 2021 B
70% garnacha blanca, 30% macabeo
88
Aromatic, pleasant, standard, fruity, floral, simple.

CLOS PACHEM
Carrer de la Font 1D
43737 Gratallops (Tarragona)
☎: +34 621 229 185
admin@clospachem.com
www.clospachem.cat

Licos 2020 B
100% garnacha blanca
91 ♣
Colour: bright yellow. Nose: powerful, creamy oak, ripe fruit, spicy. Palate: rich, good structure, long, toasty, fine bitter notes.

COCA I FITÓ
Avda. Onze de Setembre s/n
43736 El Masroig (Tarragona)
☎: +34 619 776 948
info@cocaifito.cat
www.cocaifito.cat

Coca i Fitó D'Or 2018 B
80% garnacha blanca, 20% macabeo
90
Colour: bright yellow. Nose: powerful, creamy oak, ripe fruit, spicy, toasty. Palate: rich, good structure, long, toasty, fine bitter notes.

Jaspi Blanc 2021 B
70% garnacha blanca, 30% macabeo
89
Citrus fruit, balanced, austere, dried flowers.

EDETÀRIA
Finca El Mas Ctra. Gandesa a Vilalba, Km. 2
43780 Gandesa (Tarragona)
☎: +34 977 421 534
export@edetaria.com
www.edetaria.com

Edetària Selecció 2019 B C
garnacha blanca
93
Colour: bright yellow. Nose: creamy oak, ripe fruit, spicy, white fruit, sweet spices, pattiserie, white flowers. Palate: good structure, long, toasty, fine bitter notes.

Edetària Selecció Vi de Finca El Mas 2018 T C
garnacha peluda, garnacha, cariñena
93
Colour: deep cherry. Nose: ripe fruit, dried herbs, creamy oak, dark chocolate. Palate: ripe fruit, flavourful, fruity.

Finca La Pedrissa 2018 T
94
Colour: cherry, purple rim. Nose: fruit expression, red berry notes, floral, spicy. Palate: flavourful, fruity, good acidity, long, spicy, easy to drink.

🏆 PODIUM

Finca La Terrenal 2018 B
garnacha blanca
95
Colour: bright yellow. Nose: powerful, creamy oak, ripe fruit, spicy, white fruit, white flowers, lactic notes. Palate: rich, good structure, long, toasty, fine bitter notes.

La Genuïna de Edetària 2018 T
garnacha
94
Colour: Cherry. Nose: dried flowers, wild herbs, scrubland, ripe fruit, creamy oak. Palate: good acidity, taut, easy to drink, long.

La Personal de Edetària 2018 T
93
Colour: dark-red cherry. Nose: spicy, cocoa bean, earthy notes, ripe fruit, toasty. Palate: flavourful, toasty, fine bitter notes.

🏆 PODIUM

Lo Mas D'Edetària 2019 T
garnacha peluda, cariñena
95
Colour: Cherry. Nose: expressive, mineral, red berry notes, ripe fruit. Palate: full-bodied, long, great length, flavourful, good structure.

DO TERRA ALTA / D.O.P.

DO TERRA ALTA / D.O.P.

Vía Edetana Blanc 2020 B
70% garnacha blanca, 30% viognier

91 🌿

Colour: straw. Nose: dried herbs, faded flowers, white fruit. Palate: powerful, ripe fruit, balanced, flavourful, fleshy.

Vía Edetana Blanc 2021 B
garnacha blanca

91 🌿

Colour: straw. Nose: dried herbs, white fruit, fine lees, white flowers. Palate: ripe fruit, balanced, fruity.

Vía Edetana Negre 2019 T BA
garnacha, garnacha peluda, syrah

91 🌿

Colour: cherry, garnet rim. Nose: ripe fruit, toasty, smoky, spicy. Palate: fruity, easy to drink, spicy.

FRANCK MASSARD
Calle de la Mota de Sant Pere, 26
08880 Cubelles (Barcelona)
☎: +34 938 956 541
info@franckmassard.com
www.franckmassard.com

El Mago 2020 B FB
garnacha blanca

91 🌿

Colour: straw. Nose: dried flowers, faded flowers, spicy, dry nuts. Palate: flavourful, ripe fruit, fine bitter notes.

El Mago 2020 T
100% garnacha

87 🌿

El Mago 2021 RD
garnacha

86 🌿

LAFOU CELLER
Plaça Catalunya, 34
43786 Batea (Tarragona)
☎: +34 938 743 511
Fax: +34 938 737 204
info@lafou.net
www.lafou.net

Lafou de Batea 2016 T R
85% garnacha, garnacha peluda, 15% samsó

92

Colour: deep cherry. Nose: creamy oak, black fruit, scrubland, dried fruit. Palate: powerful, ripe fruit, spicy, round tannins.

Lafou El Sender 2019 T C
70% garnacha, 20% syrah, 10% morenillo

91

Colour: deep cherry. Nose: dried herbs, creamy oak, black liquorice, red berry notes, black fruit. Palate: ripe fruit, spicy, round tannins.

Lafou Els Amelers 2021 B
100% garnacha blanca

91

Colour: straw. Nose: dried herbs, faded flowers, white fruit, ripe fruit, fine lees. Palate: ripe fruit, balanced.

MARCO ABELLA
Ctra. de Porrera
a Cornudella de Montsant, Km. 1,2
43739 Porrera (Tarragona)
☎: +34 933 712 407
info@marcoabella.com
www.marcoabella.com

Olbieta 2020 T
80% garnacha, 15% cariñena, 5% syrah

91

Colour: deep cherry. Nose: dried herbs, creamy oak, red berry notes, black fruit, ripe fruit, scrubland. Palate: ripe fruit, spicy, round tannins.

Olbieta Blanc 2020 B
70% garnacha blanca, 30% macabeo

89

Dried flowers, dried herbs, full-bodied, jammy, tasty, spicy.

MODERNISTA CELLER
Pilonet, 8
43594 El Pinell de Brai (Tarragona)
☎: +34 977 426 234
bodega@catedraldelvi.com
www.catedraldelvi.com

Gamberrillo Mistela Blanca 2017 BF Mistela
garnacha blanca

91

Colour: bright yellow. Nose: balsamic herbs, honeyed notes, sweet spices, expressive. Palate: rich, fruity, powerful, flavourful, elegant.

Gamberrillo Mistela Negra 2017 TF Mistela D
samsó

90

Sweety finish, overripe, full-bodied, tasty. Nose: dried fruit, characterful. Palate: sweet, long.

DO TERRA ALTA / D.O.P.

Gamberro Blanc 2017 B C
garnacha blanca

91
Colour: bright yellow. Nose: candied fruit, fine lees, pattiserie. Palate: round, spicy, long, great length.

Gamberro Negre de Guarda 2016 T C
garnacha, cariñena, syrah

92
Colour: dark-red cherry, garnet rim. Nose: ripe fruit, fruit preserve, aged wood nuances, tobacco, sweet spices. Palate: spicy, round tannins, long.

L'Indià Blanc 2019 B
garnacha blanca

90
Colour: bright straw. Nose: ripe fruit, fragrant herbs, fine lees. Palate: full-bodied, rich, long, good acidity.

SANT JOSEP VINS
Estació, 2
43785 Bot (Tarragona)
☎: +34 977 428 352
info@santjosepwines.com
www.santjosepvins.com

Clot D'Encís Mistela Blanca 2021 B Mistela D
100% garnacha blanca

86

Laquarta Grans Anyades Negre Vinyes Velles 2018 T
100% mazuelo

88
Balanced, spicy, jammy, dried herbs.

Laqvarta Blanc 2020 B
100% garnacha blanca

90
Colour: bright yellow. Nose: creamy oak, ripe fruit, spicy. Palate: rich, good structure, toasty, fine bitter notes.

Laqvarta Negre 2019 T
100% garnacha peluda

87

Llàgrimes de Tardor Garnatxa Blanca 2020 B FB
100% garnacha blanca

91
Colour: bright yellow. Nose: powerful, ripe fruit, spicy, toasted bread, dried flowers. Palate: rich, long, toasty, fine bitter notes, flavourful.

Llàgrimes de Tardor Rosat 2021 RD
45% garnacha, 33% garnacha blanca, 22% garnacha peluda

89
Wild, dried flowers, balanced, fruity, smooth, representative, great length.

SERRA DE CAVALLS
43594 El Pinell de Brai (Tarragona)
☎: +34 977 426 049
sat@serradecavalls.com
www.serradecavalls.com

Serra de Cavalls 2020 B FB
garnacha blanca

89
Jammy, pruney, spicy, toasty, tasty, great length.

Serra de Cavalls 2021 T
garnacha, merlot, syrah

86

Serra de Cavalls Garnacha Blanca 2021 B
garnacha blanca

88
Kind finish, wild, dried herbs, jammy, rustic, with personality.

Serra de Cavalls Garnatxa de Amfora 2019 T
garnacha

89
Pruney, dried herbs, hot. Nose: earthy notes, characterful.

VINS ALGARS
Algars, 68
43786 Batea (Tarragona)
☎: +34 635 189 982
info@vinsalgars.com
www.vinsalgars.com

Flor Trufes Blanc 2019 B FB
100% garnacha blanca

91 🍷
Colour: straw. Nose: ripe fruit, dried herbs, faded flowers, spicy. Palate: powerful, ripe fruit, balanced.

Flor Trufes Negre 2019 T
100% garnacha

90 🍷
Colour: deep cherry. Nose: dried herbs, creamy oak, black fruit, dark chocolate. Palate: ripe fruit, spicy, round tannins.

DO TERRA ALTA / D.O.P.

Trufes Blanc 2021 B
100% garnacha blanca

88
Standard, dried flowers, dried herbs, jammy, tasty. Nose: earthy notes.

Trufes Negre 2019 T
90% garnacha, 10% cabernet sauvignon

89
Pruney, full-bodied, hot, spicy, tasty, kind finish.

VINS DE LA MEMÒRIA
Montaner 320
08021 Barcelona (Barcelona)
vinsdelamemoria@gmail.com
www.vinsdelamemoria.com

La Bruixa 2020 B
garnacha blanca, macabeo

92
Dried flowers, wild. Colour: bright yellow. Nose: expressive, chamomile, fine lees. Palate: flavourful, full-bodied, fine bitter notes, long.

La Memòria 2019 B
100% garnacha blanca

93
Colour: bright yellow. Nose: white fruit, ripe fruit, dried flowers, faded flowers, spicy. Palate: rich, good structure, long, fine bitter notes, full-bodied.

Loebre "Blanc de Negres" 2020 B
cariñena, garnacha peluda

91
Slight reduction, with personality. Colour: yellow. Nose: characterful, powerful, dried herbs. Palate: flavourful, long, rich.

VINS DE MESIES
La Verge, 6
43782 Vilalba dels Arcs (Tarragona)
☎: +34 977 438 196
info@ecovitres.com
www.ecovitres.com

Mesies 2016 T C
60% garnacha, 30% samsó, 5% syrah, 5% merlot

87

Mesies Garnatxa Blanca 2019 B FB
garnacha blanca

89
With personality, spicy, balanced, tasty, defined aromas, pleasant, dried flowers.

Mesies Selecció 2015 T R
50% garnacha, 35% cariñena, 15% syrah

87

VINS DE RELAT
Gaudí, 4
43786 Batea (Tarragona)
☎: +34 938 743 511
Fax: +34 938 737 204
info@lapicossa.com
www.lapicossa.com

La Picossa Garnacha Blanca 2021 B
garnacha blanca

88
Citrus fruit, crisp, herbal, tasty.

La Picossa Garnacha Tinta 2021 T
100% garnacha

88
Balanced, spicy, jammy, smooth, roasted coffee.

Massaluca Garnatxa Blanca & Macabeo 2021 B
90% garnacha blanca, 10% macabeo

89
Kind finish, aromatic, dried flowers, fruity, tasty.

VINS DEL TROS
Major, 12
43782 Vilalba dels Arcs (Tarragona)
☎: +34 628 408 813
info@vinsdeltros.com
www.vinsdeltros.com

Cent x Cent Garnacha Blanca con Crianza en Anfora 2021 B C
garnacha blanca

89
Citrus fruit, crisp, fruity, saline, tasty.

VINS ESSÈNCIA DE LLUNA
Carrer Carnisseries, 5
43780 Gandesa (Tarragona)
☎: +34 680 594 381
hola@essenciadelluna.com
www.essenciadelluna.com

Essència de lluna 1925 2020 T C
samsó

91
Colour: cherry, purple rim. Nose: fruit liqueur notes, black fruit, spicy, smoky, dried herbs. Palate: fruity, good structure, powerful tannins.

Essència de lluna Blanc Cupatge 2021 B
70% garnacha, 30% macabeo

88
Citrus fruit, crisp, herbal, balanced.

Essència de Lluna Garnacha Blanca 2021 B
90
Colour: straw. Nose: dried herbs, white fruit, citrus fruit. Palate: balanced, saline, fresh.

Essència de lluna Rosat 2021 RD
garnacha
87

VITICULTORS BATEANS
Pol. 37 Par. 4
43786 Batea (Tarragona)
☎: +34 660 975 517
ashleyrosa10@hotmail.com
www.viticultorsbateans.com

Manyol 2020 T
garnacha
86

Manyol 2021 B
100% garnacha blanca
89
Citrus fruit, crisp, herbal, tasty, yeasty notes.

Manyol Eco 2021 T
60% garnacha, 20% cabernet sauvignon, 20% merlot
88 ❦
Fruity, jammy, spicy, smoky.

Manyol Selección 2019 T
60% garnacha, 40% syrah
89
Spicy, fruity, jammy, tasty.

DO TERRA ALTA / D.O.P.

DO. TIERRA DEL VINO DE ZAMORA

CONSEJO REGULADOR

Plaza Mayor, 1
49708 Villanueva de Campeán (Zamora)
☎: +34 980 560 055 - Fax: +34 980 560 055
@: info@tierradelvino.net
www.tierradelvino.net

LOCATION:

In the southeast part of Zamora, on the Duero river banks. This region comprises 46 municipal districts in the province Zamora and 10 in neighbouring Salamanca. Average altitude is 750 meters.

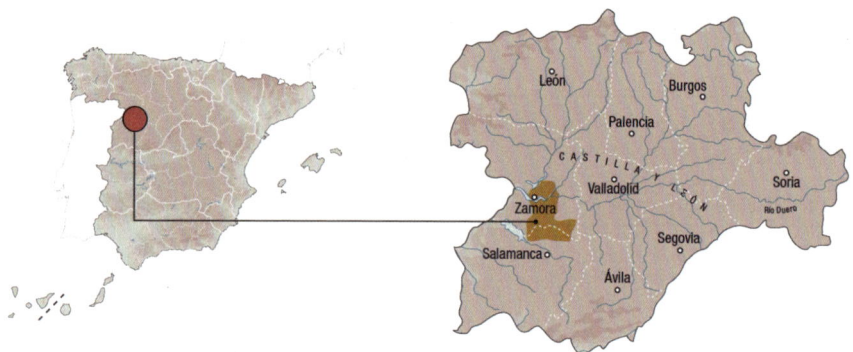

GRAPE VARIETIES:

WHITE: Malvasía, Moscatel de grano menudo and Verdejo (preferential); Albillo, Palomino and Godello (authorized).

RED: Tempranillo (main), Cabernet Sauvignon and Garnacha.

FIGURES:

Vineyard surface: 604 – **Wine-Growers:** 161 – **Wineries:** 11 – **2021 Harvest rating:** Very Good – **Production 21:** 436,959 litres – **Market percentages:** 88% National - 12% International.

SOIL:

The character of the territory derives from the river Duero tributaries, so it is predominantly alluvial and clay in the lower strata that might not allow great drainage, though they vary a lot depending on the altitude. There are also some sandy patches on the plain land and stony ones on the hill side.

CLIMATE:

Extreme temperatures as correspond to a dry continental pattern, with very hot summers and cold winters. It does not rain much and average annual rainfall hardly reaches 400 mm.

CLASSIFICATION OF YOUNG WINE HARVESTS PEÑÍNGUIDE

2017	2018	2019	2020	2021
GOOD	N/A	N/A	N/A	N/A

BODEGAS VALCABADINO

Ctra. N-122, Km. 463 Paraje Valcabadino
49026 Zamora (Zamora)
☎: +34 622 003 299
bodegasvalcabadino@bodegasvalcabadino.es
www.bodegasvalcabadino.es

Valcabadino 10 meses 2018 T RB
100% tempranillo

87

Valcabadino 2018 T C
100% tempranillo

88

Fruity, jammy, pruney, tasty.

BODEGAS VIÑAS DEL CÉNIT

Ctra. De Circunvalación, s/n
49708 Villanueva de Campeán (Zamora)
☎: +34 980 569 346

Cenit 2019 T C S
tempranillo

93

Colour: dark-red cherry. Nose: toasty, spicy, cocoa bean, roasted coffee, black fruit. Palate: flavourful, toasty, fine bitter notes.

Cenit 2020 B
dona blanca, albillo, godello, verdejo, palomino

92

Colour: bright yellow. Nose: creamy oak, ripe fruit, sweet spices, white flowers. Palate: rich, good structure, toasty, fine bitter notes.

Cenit Bonales 2019 T C
tempranillo

94

Colour: Cherry. Nose: complex, expressive, spicy, mineral. Palate: elegant, full-bodied, long, great length.

Via Cenit Colección 2019 B FB

91

Colour: bright straw. Nose: ripe fruit, fragrant herbs, fine lees, dry stone. Palate: full-bodied, rich, long, good acidity.

Via Cenit Colección 2019 T C
tempranillo

92

Colour: deep cherry. Nose: ripe fruit, dried herbs, creamy oak. Palate: powerful, ripe fruit, spicy, round tannins.

BODEGAS Y VIÑEDOS SEÑORÍO DE BOCOS

Ctra. Cubo del Vino, s/n
49719 Villamor de los Escuderos (Zamora)
☎: +34 980 690 240
bodegas@senoriodebocos.com
www.bocos.eu

Escudero de Bocos 2016 T C
tempranillo

89

Pleasant, jammy, fruity, tasty, spicy, standard, simple.

Escudero de Bocos 2017 T R
tempranillo

86

Escudero de Bocos 2020 T RB
tempranillo

87

Escudero de Bocos Verdejo 2019 B FB
verdejo

90

Colour: bright yellow. Nose: ripe fruit, spicy, toasted bread. Palate: rich, long, toasty, fine bitter notes, easy to drink.

Señorío de Bocos 2021 RD
tempranillo

87

Señorío de Bocos Verdejo 2021 B
verdejo

86

DO TIERRA DEL VINO DE ZAMORA / D.O.P.

DO. TORO

CONSEJO REGULADOR

Isaías Carrasco, s/n.
49800 Toro (Zamora)
☎: +34 980 690 335 - Fax: +34 980 693 201
@: consejo@dotoro.es
www.dotoro.es

LOCATION:

Comprising 12 municipal districts of the province of Zamora (Argujillo, Boveda de Toro, Morales de Toro, El Pego, Peleagonzalo, El Piñero, San Miguel de la Ribera, Sanzoles, Toro, Valdefinjas, Venialbo and Villanueva del Puente) and three in the province of Valladolid (San Román de la Hornija, Villafranca de Duero and the vineyards of Villaester de Arriba and Villaester de Abajo in the municipal district of Pedrosa del Rey), which practically corresponds to the agricultural region of Bajo Duero. The production area is to the south of the course of the Duero, which crosses the region from east to west.

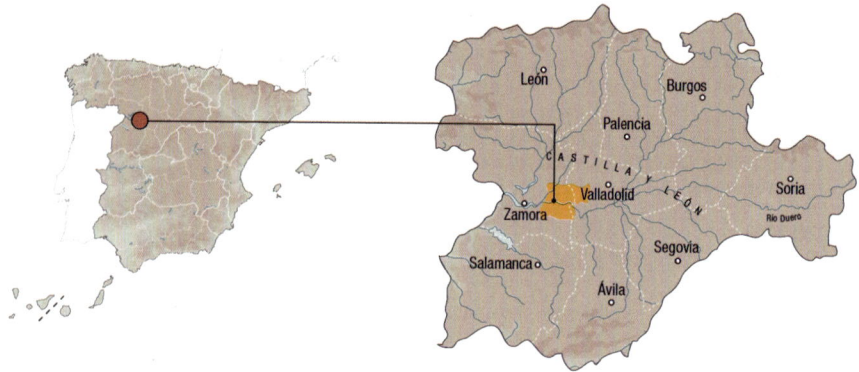

GRAPE VARIETIES:

WHITE: Malvasía, Verdejo, Moscatel de grano menudo and Albillo Real
RED: Tinta de Toro (majority) and Garnacha.

FIGURES:

Vineyard surface: 5,350 – **Wine-Growers:** 983 – **Wineries:** 62 – **2021 Harvest rating:** Excellent – **Production 21:** 15,821,000 litres – **Market percentages:** 65% National - 35% International.

SOIL:

The geography of the DO is characterised by a gently-undulating terrain. The vineyards are situated at an altitude of 620 m to 750 m and the soil is mainly brownish-grey limestone. However, the stony alluvial soil is better.

CLIMATE:

Extreme continental, with Atlantic influences and quite arid, with an average annual rainfall of between 350 mm and 400 mm. The winters are harsh (which means extremely low temperatures and long periods of frosts) and the summers short, although not excessively hot, with significant contrasts in day-night temperatures.

CLASSIFICATION OF YOUNG WINE HARVESTS PEÑÍNGUIDE

2017	2018	2019	2020	2021
GOOD	GOOD	VERY GOOD	VERY GOOD	VERY GOOD

ALGIL BODEGAS Y VIÑEDOS
Ctra. De la Estación, 1
47530 San Román de Hornija (Valladolid)
☎: +34 625 188 152
algil@algilbodegas.com
www.algilbodegas.com

Algil Crianza 2019 T C
100% tinta de Toro (Tinto)

90

Colour: dark-red cherry, purple rim. Nose: toasty, spicy, fruit preserve, faded flowers, dried herbs. Palate: flavourful, toasty, slightly dry, soft tannins, fruity.

Algil Expresión 2020 T C
89

Fruity, spicy, herbal, jammy.

Algil Garnacha 2020 T BA
100% garnacha

88

Kind finish, fruity, floral, sweety finish, jammy, tasty.

BODEGA CAMPO ELÍSEO TORO
Calle Nueva, 12
47491 La Seca (Valladolid)
☎: +34 983 034 030
Fax: +34 980 032 680
bodega@burdigala.es
www.campoeliseo.es

Campesino 2020 T
100% tinta de Toro (Tinto)

91

Colour: dark-red cherry. Nose: toasty, spicy, cocoa bean, ripe fruit, black fruit. Palate: flavourful, toasty, fine bitter notes.

Campo Eliseo 2018 T R
100% tinta de Toro (Tinto)

93

Colour: dark-red cherry, garnet rim. Nose: ripe fruit, fruit preserve, aged wood nuances, tobacco, sweet spices. Palate: spicy, round tannins, long.

Campo Eliseo Cuvée Alegre 2018 T
100% tinta de Toro

91 🌱

Colour: dark-red cherry. Nose: toasty, spicy, cocoa bean. Palate: flavourful, toasty, fine bitter notes.

Hermanos Lurton Cuesta Grande 2018 T
tinta de Toro (Tinto)

92 🌱

Colour: cherry, purple rim. Nose: fruit expression, red berry notes, floral, spicy, toasty, creamy oak. Palate: flavourful, fruity, good acidity, long.

Hermanos Lurton Orgánico sin sulfitos 2021 T
100% tinta de Toro (Tinto)

89 🌱

Pleasant, jammy, tasty.

Hermanos Lurton Tempranillo 2020 T
100% tinta de Toro (Tinto)

90

Colour: deep cherry. Nose: ripe fruit, creamy oak. Palate: powerful, ripe fruit, spicy, round tannins.

BODEGA CYAN
Ctra. Valdefinjas - Venialbo, Km. 9,2
49800 Toro (Zamora)
☎: +34 983 683 315
www.bodegacyan.es

Cyan 2018 T C
100% tinta de Toro (Tinto)

87 🌱

Cyan Pago de la Calera 2007 T
100% tinta de Toro (Tinto)

92

Defined aromas, age nuances. Colour: Rubí. Nose: black fruit, fruit liqueur notes, fine reductive notes, spicy, balanced, expressive. Palate: round tannins, spicy, long, round.

Cyan Prestigio 2016 T R
100% tinta de Toro (Tinto)

90 🌱

Colour: cherry, garnet rim. Nose: fruit preserve, powerful, spicy, incense, toasty. Palate: flavourful, long, slightly dry, soft tannins.

Cyan Tinta de Toro 2020 T RB
100% tinta de Toro (Tinto)

87 🌱

BODEGA LA PRESA DE SANZOLES
La Presa, 8
49152 Sanzoles (Zamora)
☎: +34 645 008 400
angel@bodegalapresa.com
www.bodegalapresa.com

El Cuco de Valdelaluna 2018 T RB
100% tinta de Toro (Tinto)

88

Pruney, spicy, dried herbs.

DO TORO / D.O.P.

BODEGA NUMANTHIA

Real s/n
49882 Valdefinjas (Zamora)
☎: +34 980 699 147
contact@numanthia.com
www.numanthia.com

Numanthia 2017 T
100% tinta de Toro (Tinto)

94

Colour: dark-red cherry, garnet rim. Nose: fruit preserve, overripe fruit, waxy notes, toasty, earthy notes, truffle notes. Palate: warm, pruney, sweet tannins, good finish.

 PODIUM

Termanthia 2015 T
100% tinta de Toro (Tinto)

95

Colour: cherry, garnet rim. Nose: balanced, complex, ripe fruit, spicy, fine reductive notes. Palate: good structure, flavourful, round tannins, balanced, great length.

Termes 2019 T
tinta de Toro (Tinto)

93

Colour: deep cherry. Nose: complex, expressive, spicy, mineral, ripe fruit, red berry notes. Palate: elegant, full-bodied, long, great length.

Termes 2020 B
malvasía castellana

91

Colour: bright straw. Nose: ripe fruit, floral, citrus fruit. Palate: flavourful, fresh, good acidity, fruity aftertaste.

BODEGA PAGO DE CUBAS

Ctra. Valdefinjas, km. 6,5
49882 Valdefinjas (Zamora)
☎: +34 626 410 524
direccion@bodegapagodecubas.com
www.bodegapagodecubas.com

Asterisco 2020 T
100% tinta de Toro (Tinto)

86

Incrédulo Blend 2019 T
100% tinta de Toro (Tinto)

91

Colour: dark-red cherry. Nose: toasty, spicy, fruit preserve. Palate: flavourful, toasty, fine bitter notes, balanced, slightly dry, soft tannins.

BODEGA REJADORADA

Crta. De San Román a Morales Km. 0,9
47530 San Román de Hornija (Valladolid)
☎: +34 980 693 089
rejadorada@rejadorada.com
www.rejadorada.com

Aier - Vino Cerámico 2020 T S
tinta de Toro (Tinto)

91

Colour: deep cherry. Nose: ripe fruit, black fruit, wild herbs, spicy, toasted bread. Palate: flavourful, fruity, spicy, slightly dry, soft tannins.

Antona García 2018 T C
tinta de Toro (Tinto)

91

Colour: deep cherry. Nose: ripe fruit, dried herbs, creamy oak, smoky, grassy. Palate: powerful, spicy, round tannins, fruity, good finish.

Bravo de Rejadorada 2016 T R
tinta de Toro (Tinto)

92

Colour: deep cherry. Nose: fruit preserve, fruit liqueur notes, black fruit, toasted bread, spicy, toasty, dried herbs. Palate: powerful, flavourful, good finish, slightly dry, soft tannins.

Novellum Temple 2018 T C
tinta de Toro (Tinto)

88

Fruity, smoky, jammy, toasty, tasty.

Rejadorada Roble 2020 T RB
tinta de Toro (Tinto)

89

Fruity, herbal, jammy, spicy, balanced, tasty.

Sango de Rejadorada 2015 T R
tinta de Toro (Tinto)

92

Colour: deep cherry. Nose: ripe fruit, dried herbs, creamy oak, smoky, spicy. Palate: powerful, ripe fruit, spicy, round tannins.

BODEGA VIÑAGUAREÑA

Ctra. Toro a Salamanca, Km. 12,5
49800 Toro (Zamora)
☎: +34 630 889 646
Fax: +34 980 568 013
jgarcia@vinotoro.com
www.vinotoro.com

Munia (6 meses en barrica) 2020 T RB
87

Munia Carácter 2020 T RB
100% tinta de Toro (Tinto)
89
Hot, pruney, fruity, spicy, tasty.

Munia Especial 2019 T R
100% tinta de Toro (Tinto)
92
Colour: deep cherry. Nose: ripe fruit, dried herbs, creamy oak, sweet spices. Palate: ripe fruit, spicy, round tannins, flavourful.

Tálamo 2018 T C
85% tinta de Toro (Tinto), 15% garnacha
90
Colour: deep cherry. Nose: ripe fruit, dried herbs, creamy oak, spicy. Palate: ripe fruit, spicy, round tannins.

Tres Julias 2019 T C
100% tinta de Toro (Tinto)
91 ♣
Colour: cherry, purple rim. Nose: spicy, ripe fruit, black fruit, smoky. Palate: flavourful, powerful, toasty, slightly dry, soft tannins.

BODEGA VOCARRAJE

49810 Morales de Toro (Zamora)
☎: +34 630 049 312
Fax: +34 980 698 172
info@vocarraje.es
www.vocarraje.es

Abdón Segovia 2018 T C
tinta de Toro (Tinto)
88
Balanced, spicy, jammy, powerful.

Abdón Segovia 2018 T RB
100% tinta de Toro (Tinto)
90
Colour: cherry, garnet rim. Nose: fruit preserve, powerful, dark chocolate. Palate: sweetness, flavourful, fleshy.

Abdón Segovia 2020 T
tinta de Toro (Tinto)
87

La Pasión Abdón Segovia 2012 T R
100% tinta de Toro (Tinto)
88
Slight oxidation. Colour: cherry, garnet rim. Nose: fruit preserve, fruit liqueur notes, powerful, dark chocolate. Palate: flavourful, sweetness, long.

BODEGA Y VIÑEDOS MAIRES

Camino los Llanos Km 0,4
49800 Toro (Zamora)
☎: +34 669 363 761
ademan@bodegamaires.com
www.bodegamaires.com

Ademán Adalia 2021 B
100% verdejo
89
Kind finish, fruity, jammy, tasty, toasty, herbal.

Ademán Carabizal 2021 T
100% tinta de Toro (Tinto)
90
Colour: deep cherry. Nose: ripe fruit, dried herbs, creamy oak. Palate: powerful, ripe fruit, spicy, round tannins.

Ademán Valdearanda 2019 T C
100% tinta de Toro (Tinto)
91
Colour: dark-red cherry. Nose: toasty, spicy, cocoa bean, ripe fruit, black fruit. Palate: flavourful, toasty, fine bitter notes.

Ademán Valdecarretas 2017 T FB
100% tinta de Toro (Tinto)
91
Colour: cherry, garnet rim. Nose: fruit preserve, fruit liqueur notes, powerful. Palate: flavourful, sweetness, long.

Cuzo 2021 T
100% tinta de Toro (Tinto)
87

BODEGAS A. VELASCO E HIJOS

Corredera, 23
49800 Toro (Zamora)
☎: +34 980 692 455
admon@bodegasvelascoehijos.com
www.bodegasvelascoehijos.com

Garabitas Viñas Viejas 2019 T
100% tinta de Toro (Tinto)
89 ♣
Colour: deep cherry. Nose: fruit preserve, fruit liqueur notes, dried herbs, spicy. Palate: flavourful, balanced, roasted-coffee aftertaste, slightly dry, soft tannins.

DO TORO / D.O.P.

DO TORO / D.O.P.

Peña Rejas 2021 T
tinta de Toro (Tinto)

87 🌿

BODEGAS CARODORUM
Ctra. Salamanca, ZA 605, Km. 1,5
49800 Toro (Zamora)
☎: +34 980 568 005
info@carodorum.com
www.carodorum.com

Carodorum Expresión 2021 T
91
Colour: dark-red cherry. Nose: toasty, spicy, cocoa bean, ripe fruit. Palate: flavourful, toasty, fine bitter notes.

Carodorum Issos 2020 T C
88
Spicy, toasty, jammy.

Carodorum Mantón de Manila 2021 T RB
91
Colour: deep cherry. Nose: ripe fruit, dried herbs, creamy oak. Palate: powerful, ripe fruit, spicy, round tannins.

BODEGAS COVITORO
Ctra. de Tordesillas, 13
49800 Toro (Zamora)
☎: +34 980 690 347
Fax: +34 980 690 143
comercial@covitoro.com
www.covitoro.com

Barbián 2020 T RB
88
Spicy, jammy, roasted coffee, powerful, tasty.

Cañus Verus 2019 T RB
100% tinta de Toro (Tinto)

88
Pruney, spicy, smoky, jammy, powerful.

Cañus Verus Malvasia 2020 B FB
100% malvasía

88
Toasty, jammy, dried flowers, standard. Nose: ripe fruit.

Cañus Verus Viñas Viejas 2018 T
100% tinta de Toro (Tinto)

90
Colour: bright cherry. Nose: sweet spices, dark chocolate, creamy oak, ripe fruit, characterful, smoky. Palate: fruity, spicy, round tannins.

Cermeño Vendimia Seleccionada 2021 T
100% tinta de Toro (Tinto)

87

Gran Cermeño 2018 T C
100% tinta de Toro (Tinto)

89
Jammy, tasty, spicy, great length, varietally correct. Colour: deep cherry. Nose: ripe fruit, creamy oak, wild herbs.

BODEGAS DIEZ GÓMEZ
Plaza Mayor, 3
49800 Toro (Zamora)
☎: +34 651 693 849
info@bodegasdiezgomez.com
www.bodegasdiezgomez.com

Americo 2019 T C
100% tinta de Toro (Tinto)

90
Jammy. Nose: sweet spices, ripe fruit, creamy oak. Palate: fruity, spicy, round tannins, toasty.

Americo Roble Español 2020 T RB
100% tinta de Toro (Tinto)

89
Smoky, toasty, spicy, tasty, powerful, jammy.

la JOTA de To V.R. (Viuda Rica) 2017 T RB
100% tinta de Toro (Tinto)

92
Full-bodied, woody. Colour: dark-red cherry. Nose: toasty, spicy, dark chocolate. Palate: flavourful, toasty, fine bitter notes.

BODEGAS ESTANCIA PIEDRA
Ctra. Toro a Salamanca (ZA-650), km. 5
49800 Toro (Zamora)
☎: +34 980 693 900
info@bodegaspiedra.com
www.bodegaspiedra.com

Cantadal 2021 T
100% tinta de Toro (Tinto)1

87 🌿

Piedra 2016 T R
91 🌿
Colour: dark-red cherry, garnet rim. Nose: ripe fruit, aged wood nuances, tobacco, sweet spices. Palate: spicy, round tannins, long.

Piedra 2019 T C
80% tinta de Toro (Tinto), 20% garnacha

89 🌿
Pruney, herbal, spicy, tasty, kind finish.

Piedra 2020 T RB
85% tinta de Toro (Tinto), 15% garnacha
88 🌿
Pruney, dried flowers, great length, tasty.

Piedra 2021 RD
100% tinta de Toro (Tinto)1
88 🌿
Pleasant, jammy, powerful, tasty.

Piedra Natural 2021 T
100% tinta de Toro (Tinto)1
88 🌿
Pruney, full-bodied, jammy, great length, tasty, fruity.

BODEGAS FARIÑA
Camino del Palo, s/n
49800 Toro (Zamora)
☎: +34 980 577 673
comercial@bodegasfarina.com
www.bodegasfarina.com

Dama de Toro 2020 T RB
88
Pruney, spicy, tasty, standard.

Fariña 2018 T C
tinta de Toro (Tinto)
90
Colour: bright cherry. Nose: sweet spices, ripe fruit, dark chocolate. Palate: fruity, spicy, round tannins.

Fariña Campus Gothorum 2019 T
tinta de Toro (Tinto)
94
Colour: cherry, garnet rim. Nose: fruit preserve, powerful, black fruit, creamy oak, cocoa bean. Palate: flavourful, round tannins, great length.

Fariña Lágrima 2020 T RB
tinta de Toro (Tinto)
89
Pruney, balanced, dried herbs, jammy, great length, spicy.

Gran Colegiata "Original" 2016 T R
tinta de Toro (Tinto)
92
Colour: Cherry. Nose: sweet spices, scrubland, black fruit, ripe fruit. Palate: spicy, good acidity, round tannins.

Gran Colegiata 2020 T RB
tinta de Toro (Tinto)
88
Pruney, spicy, tasty, standard.

Primero 2021 T
tinta de Toro (Tinto)
91
Colour: cherry, purple rim. Nose: fruit expression, red berry notes, floral. Palate: flavourful, fruity, good acidity, easy to drink.

BODEGAS FRANCISCO CASAS
Avda. de Los Comuneros, 67
49810 Morales de Toro (Zamora)
☎: +34 980 698 032
info@bodegascasas.com
www.bodegascasas.com

Camparrón 2017 T R
100% tinta de Toro (Tinto)
89
Toasty, spicy, fruity, tasty.

Camparrón 2018 T C
100% tinta de Toro (Tinto)
88
Fruity, smoky, toasty, jammy, spicy.

Camparrón Novum 2021 T
100% tinta de Toro (Tinto)
88
Aromatic, fruity, herbal, tasty, jammy.

Camparrón Seleccion 2021 T
100% tinta de Toro (Tinto)
87

Gamazo 2018 T C
100% tinta de Toro (Tinto)
88
Fruity, sweety finish, jammy, tasty.

Viña Abba 2017 T
100% tinta de Toro (Tinto)
89
Fruity, jammy, tasty, coarse tannins, spicy.

BODEGAS FRONTAURA
Santiago, 17 5º
47001 Valladolid (Valladolid)
☎: +34 983 880 488
crobles@nexusfrontaura.com
www.bodegasfrontaura.com/es

Aponte 2017 T R
93
Colour: dark-red cherry, garnet rim. Nose: aged wood nuances, tobacco, sweet spices, black fruit, ripe fruit, toasty. Palate: spicy, round tannins, long.

DO TORO / D.O.P.

Frontaura & Victoria Rosé Limited Edition 2021 RD

90

Colour: salmon. Nose: ripe fruit, fruit preserve, hot, dried herbs. Palate: flavourful, balanced, round, good finish.

Frontaura 2017 T R
tinta de Toro (Tinto)

92

Colour: dark-red cherry, garnet rim. Nose: fruit preserve, aged wood nuances, tobacco, sweet spices. Palate: spicy, round tannins, powerful, warm.

Frontaura 2018 T C

91

Colour: dark-red cherry. Nose: toasty, spicy, cocoa bean, black fruit, overripe fruit. Palate: flavourful, toasty, fine bitter notes.

BODEGAS ITURRIA

Camino de la Estación, s/n
47530 San Román de Hornija (Valladolid)
☎: +34 600 523 070
contact@bodegas-iturria.com
www.bodegas-iturria.com

La Viña de Segundo 2019 T RB
tinta de Toro (Tinto)

89 🌿

Fruity, jammy, spicy, tasty.

Tinto Iturria 2018 T
90% tinta de Toro (Tinto), 10% garnacha

90

Colour: deep cherry. Nose: dried herbs, creamy oak, black fruit, burnt matches. Palate: powerful, ripe fruit, spicy, round tannins.

Valdosan 2018 T
tinta de Toro (Tinto)

90 🌿

Colour: deep cherry. Nose: dried herbs, creamy oak, fruit preserve, toasty. Palate: powerful, ripe fruit, spicy, slightly dry, soft tannins.

BODEGAS MONTE LA REINA

Ctra. N-122, Km 436,7
49881 Toro (Zamora)
☎: +34 980 082 011
montelareina@montelareina.es
www.montelareina.es

Castillo de Monte la Reina 2018 T C
tinta de Toro (Tinto)

88

Fruity, spicy, tasty, pruney.

Castillo de Monte la Reina Cuvée Privée 2019 T
tinta de Toro (Tinto)

89

Fruity, floral, kind finish, jammy, tasty.

Castillo de Monte la Reina Vendimia Seleccionada 2016 T
tinta de Toro (Tinto)

90

Colour: dark-red cherry, garnet rim. Nose: fruit preserve, tobacco, sweet spices, dried herbs. Palate: spicy, flavourful, slightly dry, soft tannins.

BODEGAS RODRÍGUEZ Y SANZO

Manuel Azaña, 9
47014 Valladolid (Valladolid)
☎: +34 983 150 150
comunicacion@valsanzo.com
www.rodriguezsanzo.com

La Viña de Amaya 2020 T C
90% tinta de Toro (Tinto), 10% garnacha

90

Colour: bright cherry. Nose: fresh fruit, creamy oak, scrubland. Palate: good acidity, spicy, round tannins.

Las Tierras de Javier Rodríguez El Pego 2019 T C
100% tinta de Toro (Tinto)

92

Colour: cherry, purple rim. Nose: fruit expression, red berry notes, floral, spicy, black fruit. Palate: flavourful, fruity, good acidity, balanced, long.

🏆 **PODIUM**

Las Tierras de Javier Rodríguez El Teso Alto 2016 T
100% tinta de Toro (Tinto)

95

Colour: Cherry. Nose: spicy, mineral, cocoa bean, black fruit. Palate: full-bodied, long, great length, powerful tannins.

Las Tierras Garnacha 2021 T
garnacha

89

Pruney, balanced, fruity, tasty, great length, varietally correct.

Tras las Yesca 2021 T C

90

Kind finish, fruity, wild. Colour: bright cherry. Nose: neat, red berry notes, ripe fruit. Palate: correct, easy to drink.

Rodriguez & Sanzo
Gotas de Noche 2021 RD
85% tempranillo, 10% garnacha, 5% verdejo

92
Colour: brilliant rose, purple rim. Nose: red berry notes, lactic notes, floral, sweet spices. Palate: balanced, fruity aftertaste, easy to drink.

Finca Sobreño
Selección Especial 2019 T R
100% tinta de Toro (Tinto)

91
Colour: cherry, garnet rim. Nose: fruit preserve, powerful, sweet spices, toasty. Palate: flavourful, sweetness, long.

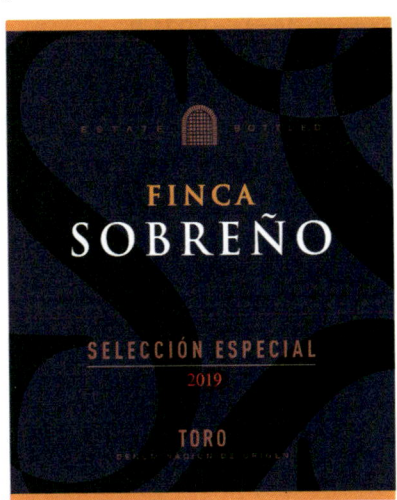

DO TORO / D.O.P.

BODEGAS SOBREÑO
Ctra. N-122, Km. 423
49800 Toro (Zamora)
☎: +34 980 693 417
Fax: +34 980 693 416
sobreno@sobreno.com
www.sobreno.com

Finca Sobreño 2020 T C
100% tinta de Toro (Tinto)

88
Pruney, toasty, powerful.

Finca Sobreño 2021 RD
100% tinta de Toro (Tinto)

88
Pleasant, aromatic, fruity, lactic, smooth, balanced.

Finca Sobreño Ecológico 2020 T
0% tinta de Toro (Tinto)

88 🌱
Tasty, jammy, fruity.

Finca Sobreño Ildefonso 2018 T R
100% tinta de Toro (Tinto)

92
Colour: cherry, garnet rim. Nose: fruit preserve, fruit liqueur notes, powerful, sweet spices, toasty. Palate: flavourful, sweetness, long.

La Cuadrilla 2021 T
100% tinta de Toro (Tinto)

87

BODEGAS TORREDUERO
Pol. Ind. Toro Norte, Parcela 5
49800 Toro (Zamora)
☎: +34 941 454 050
bodega@bodegasriojanas.com
www.bodegasriojanas.com

Marqués de Peñamonte
Colección Privada 2020 T
100% tinta de Toro (Tinto)

89
Spicy, fruity, pruney, dried herbs, tasty.

Peñamonte 2019 T C
100% tinta de Toro (Tinto)

87

Peñamonte 2021 RD
85% garnacha, 15% tinta de Toro (Tinto)

87

Peñamonte Cosecha 2021 T
100% tinta de Toro (Tinto)

87

DO TORO / D.O.P.

Peñamonte Garnacha 2021 T RB
garnacha
87

Peñamonte Roble 2021 T RB
100% tinta de Toro (Tinto)
85

Peñamonte Verdejo 2021 B
verdejo
85

BODEGAS VALBUSENDA
Ctra. Toro - Peleagonzalo s/n
49800 Toro (Zamora)
☎: +34 980 699 560
Fax: +34 980 699 566
bodega@valbusenda.com
www.valbusenda.com

Valbusenda Abios 2018 T C
100% tinta de Toro (Tinto)
88
Smoky, spicy, jammy, tasty, great length.

Valbusenda Abios Nude 2021 RD
100% tinta de Toro (Tinto)
85

Valbusenda Abios Tinta de Toro 2020 T RB
100% tinta de Toro (Tinto)
88
Creamy, fruity, jammy, full-bodied, toasty.

Valbusenda Verdejo 2019 B FB
100% verdejo
89
Colour: bright yellow. Nose: powerful, creamy oak, ripe fruit, spicy. Palate: rich, good structure, fine bitter notes, easy to drink.

BODEGAS VATAN
Pol. Norte, 1 Parcela 29B
49800 Toro (Zamora)
☎: +34 952 504 706
pedidos@jorgeordonez.es
www.jorgeordonez.es

Tritón Tinta Toro 2019 T
100% tinta de Toro (Tinto)
94
Colour: very deep cherry. Nose: complex, expressive, spicy, mineral, ripe fruit, black fruit. Palate: full-bodied, long, great length, flavourful.

🏆 **PODIUM**

Vatan 2019 T
100% tinta de Toro (Tinto)
95
Colour: cherry, garnet rim. Nose: complex, expressive, spicy, mineral, black fruit, ripe fruit, characterful. Palate: full-bodied, long, great length, toasty.

BODEGAS VEGA SAUCO
Avda. Comuneros, 108
49810 Morales de Toro (Zamora)
☎: +34 980 698 294
Fax: +34 980 698 294
comercial@vegasauco.es
www.vegasauco.es

Adoremus 2011 T R
100% tinta de Toro (Tinto)
88
Age nuances, spicy, jammy, tasty, creamy.

La Sonrisa del Nómada 2020 T RB
100% tinta de Toro (Tinto)
89
Full-bodied, spicy, fruity, powerful.

Piedras y Princesas 2018 T C
100% tinta de Toro (Tinto)
87

Vega Sauco El Beybi 2020 T RB
100% tinta de Toro (Tinto)
85

Vega Saúco Selección 2019 T
100% tinta de Toro (Tinto)

87

BODEGAS VETUS
Ctra. Toro a Salamanca, Km. 9,5
49800 Toro (Zamora)
☎: +34 980 056 012
Fax: +34 980 056 012
vetus@bodegasvetus.com
www.bodegasvetus.com

Celsus 2020 T
tinta de Toro (Tinto), garnacha

94

Colour: cherry, garnet rim. Nose: powerful, ripe fruit, black fruit, red berry notes, toasty. Palate: flavourful, long, full of life, ripe fruit.

Flor de Vetus 2019 T
tinta de Toro (Tinto), garnacha

92

Colour: cherry, purple rim. Nose: red berry notes, floral, toasty, sweet spices. Palate: flavourful, fruity, good acidity, long.

Vetus 2019 T
tinta de Toro (Tinto)

94

Colour: deep cherry. Nose: dried herbs, creamy oak, ripe fruit, black fruit, red berry notes, sweet spices. Palate: powerful, ripe fruit, spicy, round tannins.

BODEGAS VIRIATUS
Cl. Camino las Viñas, s/n
49622 Brime de Urz (Zamora)
☎: +34 649 876 187
vino@grupobarrero.com
www.viriatus.es

Alboca 2018 T C
100% tinta de Toro (Tinto)

89

Spicy, fruity, jammy, tasty.

Alboca 2020 T
100% tinta de Toro (Tinto)

88

Fruity, floral, dried herbs, tasty.

BODEGAS Y VIÑEDOS EL SOLEADO
Pozo Alfaro, 50
16237 Ledaña (Cuenca)
☎: +34 911 994 108
info@elsoleado.com
www.elsoleado.com

Biógrafo del Vino 2019 T
tinta de Toro (Tinto)

90

Colour: cherry, garnet rim. Nose: fruit preserve, fruit liqueur notes, powerful. Palate: flavourful, long, sweet tannins.

BODEGAS Y VIÑEDOS PINTIA
Ctra. San Román a Morales s/n
47530 San Román de Hornija (Valladolid)
☎: +34 983 784 178
Fax: +34 983 784 206
vegasicilia@vega-sicilia.com
www.temposvegasicilia.com

🏆 **PODIUM**

Pintia 2018 T

95

Aromatic, jammy, elegant. Colour: deep cherry. Nose: ripe fruit, sweet spices. Palate: powerful, ripe fruit, spicy, round tannins.

COMPAÑÍA DE VINOS TELMO RODRÍGUEZ
El Monte
01308 Lanciego (Araba/Álava)
☎: +34 945 628 315
contact@telmorodriguez.com
www.telmorodriguez.com

Gago 2019 T

92

Colour: dark-red cherry. Nose: toasty, spicy, cocoa bean, black fruit, ripe fruit, earthy notes. Palate: flavourful, toasty, fine bitter notes.

🏆 **PODIUM**

Pago La Jara 2018 T

95

Colour: very deep cherry. Nose: complex, expressive, spicy, mineral, black fruit, ripe fruit, toasty, creamy oak. Palate: elegant, full-bodied, long, great length.

DO TORO / D.O.P.

DO TORO / D.O.P.

CORAL DUERO
Calle Ascensión s/n
49154 El Pego (Zamora)
☎: +34 980 606 333
info@coralduero.com
www.coralduero.com

Las Parvas 2017 T
100% tinta de Toro (Tinto)

93

Colour: dark-red cherry, garnet rim. Nose: ripe fruit, fruit preserve, aged wood nuances, tobacco, sweet spices. Palate: spicy, round tannins, long.

Los Lastros 2019 T
100% tinta de Toro (Tinto)

91

Colour: deep cherry. Nose: ripe fruit, dried herbs, creamy oak, toasty. Palate: powerful, ripe fruit, spicy, round tannins.

R'sedas 2020 T
100% tinta de Toro (Tinto)

88

Nose: roasted coffee, aromatic coffee, powerful. Palate: smoky aftertaste.

Salgadero 2018 T BA
100% tinta de Toro (Tinto)

90

Colour: dark-red cherry. Nose: toasty, spicy, cocoa bean, ripe fruit, black fruit. Palate: flavourful, toasty, fine bitter notes.

DÍSCOLO
Ctra. El Pego – Guarrate, s/n
49154 El Pego (Zamora)
☎: +34 670 095 149
bodega@vinodiscolo.com
www.vinodiscolo.com

Cinco de Copas 2020 T BA S
100% tinta de Toro (Tinto)

89 🌿

Spicy, fruity, jammy, toasty, tasty.

Díscolo El Magnífico 2017 T BA S
100% tinta de Toro (Tinto)

92

Colour: deep cherry. Nose: aged wood nuances, ripe fruit, cocoa bean, toasty, dried herbs. Palate: flavourful, spicy, toasty, round tannins, good finish.

Díscolo Malvasía y Verdejo 2020 B FB S
40% malvasía, 60% verdejo

91 🌿

Colour: bright straw. Nose: ripe fruit, faded flowers, dried flowers, balanced, lees reduction notes. Palate: flavourful, rich.

Díscolo 2018 T BA S
100% tinta de Toro (Tinto)

90 🌿

Colour: dark-red cherry, cherry, purple rim. Nose: toasty, spicy, ripe fruit. Palate: flavourful, toasty, fruity, slightly dry, soft tannins.

DOMINIO DEL BENDITO
Llano La Silla s/n Pol.1 Parcela 4524
49800 Toro (Zamora)
☎: +34 980 667 010
info@bodegadominiodelbendito.es
www.bodegadominiodelbendito.com

Dominio del Bendito El Primer Paso 2020 T RB
100% tinta de Toro (Tinto)

91 🌿

Colour: deep cherry. Nose: ripe fruit, creamy oak, balanced, expressive. Palate: powerful, ripe fruit, spicy, round tannins.

Dominio del Bendito Las Sabias 2019 T
100% tinta de Toro (Tinto)

92 🌿

Colour: deep cherry, purple rim. Nose: ripe fruit, dried herbs, spicy, faded flowers, toasted bread. Palate: powerful, ripe fruit, spicy, slightly dry, soft tannins.

🏆 **PODIUM**

El Titán del Bendito 2019 T
100% tinta de Toro (Tinto)

95 🌿

Complex, full-bodied. Colour: Cherry. Nose: complex, expressive, spicy, creamy oak, cocoa bean. Palate: full-bodied, long, great length, round tannins.

Mi Verdadejo 2019 B FB
100% verdejo

92

Colour: pale. Nose: burnt matches, fine reductive notes, creamy oak, spicy, ripe fruit, woody. Palate: long, rich, good structure, flavourful.

DO TORO / D.O.P.

DUEBA VITIVINÍCOLA
Ctra. de Moraleja, s/n
49151 Casaseca de las Chanas (Zamora)
☎: +34 980 577 673
comercial@bodegasfarina.com
www.bodegasfarina.com

Astado 2017 T R
tinta de Toro (Tinto)

88

Toasty, pruney, spicy, jammy, tasty.

Astado 2018 T C
tinta de Toro (Tinto)

89

Standard, fruity, jammy, tasty, spicy. Palate: easy to drink.

Astado 2020 T RB
tinta de Toro (Tinto)

88

Standard, jammy, tasty, pruney.

Astado 2021 T
tinta de Toro (Tinto)

89

Fruity, jammy, full-bodied, great length. Nose: maceration notes.

Astado Viñas Viejas 2016 T RB
tinta de Toro

89

Pruney, exuberant, full-bodied, kind finish. Nose: dark chocolate.

ELÍAS MORA
Juan Mora s/n
47530 San Román de Hornija (Valladolid)
☎: +34 983 784 029
info@bodegaseliasmora.com
www.bodegaseliasmora.com

Descarte 2017 T
100% tinta de Toro (Tinto)

92

Colour: deep cherry. Nose: fruit liqueur notes, fruit preserve, black fruit, creamy oak, sweet spices, dried herbs. Palate: fruity, flavourful, balanced, good finish, round tannins.

Elías Mora 2014 T R
100% tinta de Toro (Tinto)

93

Colour: deep cherry. Nose: black fruit, fruit preserve, fruit liqueur notes, dried herbs, toasty. Palate: flavourful, powerful, spicy, good finish, round tannins.

Elías Mora 2019 T C
100% tinta de Toro (Tinto)

91

Colour: deep cherry. Nose: dried herbs, creamy oak, black fruit. Palate: powerful, ripe fruit, spicy, round tannins.

Gran Elías Mora 2015 T
100% tinta de Toro (Tinto)

92

Colour: bright cherry. Nose: ripe fruit, fruit preserve, tobacco, sweet spices. Palate: spicy, round tannins, long, flavourful.

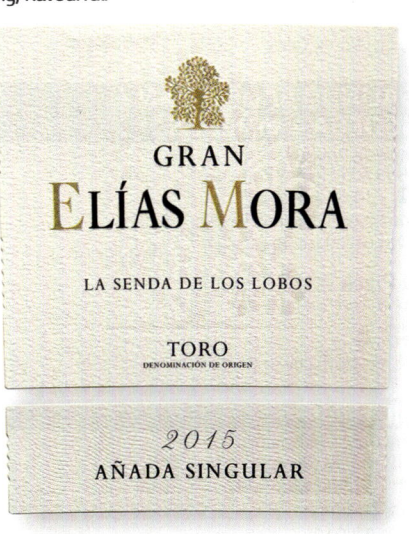

Viñas Elías Mora 2020 T RB
100% tinta de Toro (Tinto)

90

Colour: deep cherry, purple rim. Nose: ripe fruit, dried herbs, spicy, toasty. Palate: powerful, spicy, round tannins, flavourful.

FRUTOS VILLAR
Eras de Santa Catalina, s/n
49800 Toro (Zamora)
☎: +34 983 586 868
bodegasfrutosvillar@bodegasfrutosvillar.com
www.bodegasfrutosvillar.com

Muruve 2017 T R
100% tinta de Toro (Tinto)

91

Colour: dark-red cherry, garnet rim. Nose: ripe fruit, fruit preserve, aged wood nuances, tobacco, sweet spices. Palate: spicy, round tannins, long.

DO TORO / D.O.P.

Muruve 2018 T C
100% tinta de Toro (Tinto)

91

Colour: deep cherry. Nose: ripe fruit, dried herbs, creamy oak, mineral, earthy notes. Palate: powerful, ripe fruit, spicy, round tannins.

Muruve 2019 T RB
100% tinta de Toro (Tinto)

87

Muruve 2021 T
100% tinta de Toro (Tinto)

89

Fruity, jammy, tasty, smooth.

GALINDO SAN MILLÁN – BODEGA + VIÑEDOS
Travesía Prado, 1
49715 El Piñero (Zamora)
☎: +34 647 669 323
bodegagalindosanmillan@gmail.com
www.galindosanmillan.com

La Pizca 2017 T
tinta de Toro (Tinto)

88

Hot, powerful, overripe, tasty.

GIL LUNA
Ctra. Toro - Salamanca, Km. 2
49800 Toro (Zamora)
☎: +34 980 698 509
info@giluna.es
www.giluna.es

Lunas Nuevas 2018 T R
tinta de Toro (Tinto)

88

Fruity, pruney, spicy, smoky, tasty.

Sin Complejos 2020 T
100% tinta de Toro (Tinto)

87

Tres Lunas 2019 T
100% tinta de Toro (Tinto)

89

Full-bodied, pruney, toasty, tasty, great length.

Tres Lunas Verdejo 2021 B
100% verdejo

87

HACIENDA TERRA D'URO
Gamazo, 31
47004 Valladolid (Valladolid)
☎: +34 670 609 440
administracion@terraduro.com
www.terraduro.com

Finca La Rana 2020 T
100% tinta de Toro (Tinto)

91

Colour: cherry, purple rim. Nose: red berry notes, floral, spicy, black fruit, dried herbs. Palate: flavourful, fruity, good acidity, long.

Picio 2021 T
garnacha, tinta de Toro (Tinto)

90

Colour: cherry, purple rim. Nose: ripe fruit, spicy, red berry notes, woody. Palate: ripe fruit, flavourful, good structure.

Terra Duro La Enfermera 2021 T
100% tinta de Toro (Tinto)

88

Pruney, creamy, spicy, kind finish.

Terraduro Selección 2019 T
100% tinta de Toro (Tinto)

90

Colour: deep cherry. Nose: dried herbs, creamy oak, black fruit. Palate: ripe fruit, spicy, round tannins.

Uro 2019 T
100% tinta de Toro (Tinto)

90

Colour: deep cherry. Nose: dried herbs, creamy oak, black fruit. Palate: ripe fruit, spicy, round tannins.

LA VIÑA DEL ABUELO
Merced, 1
49800 Toro (Zamora)
☎: +34 617 329 426
bodega@abuelovino.com
www.abuelovino.com

La Viña del Abuelo Premium 2017 T
100% tinta de Toro (Tinto)

90

Colour: cherry, purple rim. Nose: ripe fruit, fruit preserve, sweet spices, dried herbs. Palate: spicy, round tannins, fruity, flavourful.

La Viña del Abuelo Selección Especial 2016 T C
100% tinta de Toro (Tinto)

92

Colour: cherry, purple rim. Nose: ripe fruit, sweet spices, faded flowers, dried herbs, wet leather, cigar. Palate: spicy, round tannins, fruity, flavourful.

La Viña del Abuelo Vendimia Seleccionada 2021 T
tinta de Toro (Tinto)
87

LEGADO DE ORNIZ
Ctra. San Román de Hornija-Toro, km 1
47530 San Román de Hornija (Valladolid)
☎: +34 669 545 976
info@legadodeorniz.com
www.legadodeorniz.com

Epitafio 2019 T RB
tinta de Toro (Tinto)
91
Colour: dark-red cherry. Nose: toasty, spicy, cocoa bean, fine reductive notes. Palate: flavourful, toasty, fine bitter notes.

Triens 2019 T RB
tinta de Toro (Tinto)
90
Colour: deep cherry. Nose: dried herbs, creamy oak, black fruit. Palate: ripe fruit, spicy, round tannins.

LUI & WILLIAM WINES
Puerta del Mercado, 9
49800 Toro (Zamora)
☎: +34 647 492 994
lui.william.wines@gmail.com
www.luiwilliamwines.com

Abelis Lui 2020 T C
90
Colour: dark-red cherry, garnet rim. Nose: fruit preserve, aged wood nuances, tobacco, sweet spices. Palate: spicy, round tannins, long.

Dominio de Taurum 2021 T RB
90
Colour: deep cherry. Nose: ripe fruit, dried herbs, creamy oak. Palate: powerful, ripe fruit, spicy, round tannins.

MATSU
Ctra. Tordesillas, 13
49800 Toro (Zamora)
☎: +34 941 483 557
marketing@vintae.com
www.vintae.com

Matsu El Pícaro 2021 T
tinta del país
90
Colour: bright cherry. Nose: sweet spices, ripe fruit. Palate: fruity, spicy, round tannins.

Matsu El Recio 2020 T
tinta de Toro (Tinto)
91
Colour: deep cherry. Nose: ripe fruit, dried herbs, creamy oak. Palate: powerful, ripe fruit, spicy, round tannins.

Matsu El Viejo 2020 T
tinta de Toro (Tinto)
91
Colour: cherry, garnet rim. Nose: fruit preserve, powerful, aromatic coffee, dark chocolate. Palate: flavourful, long, round tannins.

Matsu La Jefa 2019 B
malvasía, otras
91
Colour: bright straw. Nose: dried flowers, fine lees, spicy, white fruit, ripe fruit. Palate: rich, flavourful.

MOISÉS GRAN VINO
Avda. Comuneros, 102
49810 Morales de Toro (Zamora)
☎: +34 915 610 894
bodega@herediduruena.com

Moises Gran Vino 2014 T BA
100% tinta de Toro (Tinto)
92
Colour: deep cherry. Nose: ripe fruit, dried herbs, spicy, smoky. Palate: ripe fruit, spicy, round tannins, flavourful.

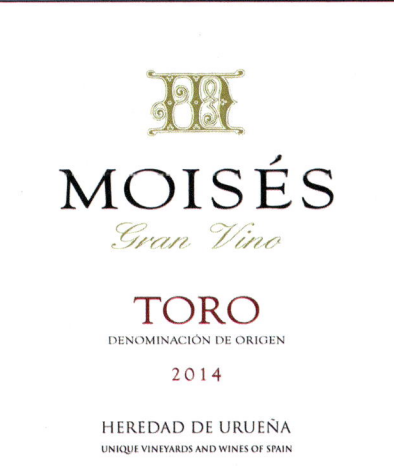

Toralto 2016 T BA
100% tinta de Toro (Tinto)
89

Toraltus 2016 T BA
100% tinta de Toro (Tinto)

92

Colour: deep cherry. Nose: dried herbs, wild herbs, black fruit, fruit liqueur notes, spicy. Palate: powerful, ripe fruit, round tannins, flavourful.

OROT
Ctra. Tordesillas, s/n
49800 Toro (Zamora)
☎: +34 983 868 116
info@bornosbodegas.com
www.orotbodega.com

Orot 2019 T C
100% tinta de Toro (Tinto)

88

Woody, full-bodied, jammy, powerful.

Orot 2019 T RB
100% tinta de Toro (Tinto)

87

Orot 2021 T
100% tinta de Toro (Tinto)

86

PAGOS DEL REY TORO
Avda. de los Comuneros, 90
49810 Morales de Toro (Zamora)
☎: +34 980 698 023
Fax: +34 980 698 020
toro@pagosdelrey.com
www.pagosdelrey.com

Bajoz 2018 T C
100% tinta de Toro (Tinto)

87

Bajoz 2021 T RB
100% tinta de Toro (Tinto)

85

Bajoz Tempranillo 2021 T
100% tinta de Toro (Tinto)

86

Gran Bajoz 2018 T
100% tinta de Toro (Tinto)

90

Colour: deep cherry. Nose: dried herbs, creamy oak, red berry notes, ripe fruit. Palate: ripe fruit, spicy, grainy tannins.

Sentero 2020 T RB
tinta de Toro (Tinto)

87

Sentero Expresión 2018 T
tinta de Toro (Tinto)

90

Colour: cherry, purple rim. Nose: red berry notes, floral, spicy, toasty, black liquorice. Palate: flavourful, fruity, good acidity, long.

PALACIO DE VILLACHICA
Ctra. N-122, Km. 433
49800 Toro (Zamora)
☎: +34 663 797 346
info@grupopalaciodevillachica.com
www.grupopalaciodevillachica.com

Palacio de Villachica Dehesa San Andrés Vendimia Seleccionada 2019 T

90

Colour: cherry, purple rim. Nose: fruit expression, spicy, black fruit, ripe fruit. Palate: flavourful, fruity, fresh, good finish.

SAN ROMÁN BODEGAS Y VIÑEDOS
Ctra. N-122, Km. 411
47112 Pedrosa del Rey (Valladolid)
☎: +34 983 784 118
info@bodegasanroman.com
www.bodegasanroman.com

Cartago 2017 T
90% tinta de Toro (Tinto), 10% otras

94 🌿

Colour: Cherry, orangey edge. Nose: wet leather, waxy notes, fruit liqueur notes, ripe fruit, black fruit. Palate: light-bodied, bitter, soft tannins, reductive nuances.

Prima 2020 T
85% tinta de Toro (Tinto), 15% garnacha, otras

91

Colour: cherry, purple rim. Nose: red berry notes, floral, spicy. Palate: flavourful, fruity, good acidity, long.

🏆 PODIUM

San Román 2019 T
100% tinta de Toro (Tinto)

95 🌿

Colour: Cherry. Nose: expressive, spicy, ripe fruit, characterful, complex. Palate: full-bodied, great length, round tannins, good structure, flavourful.

San Román Garnacha 2020 T
100% garnacha

93 🌿

Colour: bright cherry. Nose: balsamic herbs, sweet spices, scrubland, red berry notes, toasty. Palate: spicy, balsamic, good acidity.

DO TORO / D.O.P.

SIETE PASOS WINES

Ctra. Nac. 122, km. 311
47300 Peñafiel (Valladolid)
☎: +34 983 881 622
administracion@bodegaspenafiel.com
www.sietepasos.com

Cuentaovejas ZZZ 2018 T
100% tinta de Toro (Tinto)

89

Creamy, spicy, powerful, toasty, balanced.

SPANISH PALATE

Avda. Carlos Latorre, 30
49800 Toro (Zamora)
☎: +34 980 690 643
export@spanishpalate.es
www.spanishpalate.es

Botas de Barro Toro 2020 T
100% tempranillo

88

Toasty, powerful, jammy, tasty.

El Alma de Gildo 2020 T

90

Colour: cherry, purple rim. Nose: red berry notes, spicy, scrubland, black fruit. Palate: flavourful, fruity, good structure, balanced, powerful tannins.

Mi Tractor Azul 2021 T

88

Fruity, jammy, herbal, tasty.

TESO LA MONJA

Paraje Valdebuey s/n
49882 Valdefinjas (Zamora)
☎: +34 980 568 143
info@sierracantabria.com
www.sierracantabria.com

🏆 PODIUM

Alabaster 2019 T
tinta de Toro (Tinto)

97

Jammy, powerful. Colour: cherry, garnet rim. Nose: creamy oak, hot, ripe fruit, black fruit, overripe fruit, toasty, new oak. Palate: pruney, powerful, sweet tannins.

Almirez 2020 T
tinta de Toro (Tinto)

94

Colour: deep cherry. Nose: ripe fruit, dried herbs, creamy oak, red berry notes, black fruit. Palate: powerful, ripe fruit, spicy, round tannins.

Románico 2020 T
tinta de Toro (Tinto)

90

Colour: cherry, purple rim. Nose: red berry notes, spicy. Palate: flavourful, fruity, good acidity.

🏆 PODIUM

Victorino 2019 T
tinta de Toro (Tinto)

96

Colour: cherry, garnet rim. Nose: powerful, stone fruit, ripe fruit, black fruit, toasty, creamy oak, dark chocolate. Palate: flavourful, long, spicy.

TINTA ROSA

Alonso Cano, 41
47130 Simancas (Valladolid)
☎: +34 639 250 225
info@tintarosa.wine
www.tintarosa.wine

Tinta Rosa 2020 T
100% tinta de Toro (Tinto)

88

Fruity, smoky, jammy, toasty.

Tinta Rosa Selección 2019 T C
9,95% tinta de Toro (Tinto)

90

Colour: deep cherry. Nose: ripe fruit, dried herbs, creamy oak, sweet spices. Palate: powerful, ripe fruit, round tannins.

VINITOR WINE GROUP

Santa Eugènia, 64
17005 Girona/Gerona (Girona/Gerona)
hola@vinitor.es
www.vinitorwinegroup.com

MOFO - The Wild Child 2019 T
100% tinta de Toro (Tinto)

89

Fruity, jammy, spicy, tasty, coarse tannins.

MOFO – The Damn Master 2018 T
100% tinta de Toro (Tinto)

89

Fruity, pruney, tasty, toasty, coarse tannins.

DO TORO / D.O.P.

VINOS Y VIÑEDOS DE LA CASA MAGUILA

Ctra. El Piñero s/n Pol. 1 P. 715
49153 Venialbo (Zamora)
☎: +34 616 262 549
info@casamaguila.com
www.casamaguila.com

Angelitos Negros 2021 T
88
Pruney, jammy, tasty, fruity, sweety finish.

Cachito Mío Selección 14 de 24 2016 T
100% tinta de Toro (Tinto)
90
Colour: cherry, garnet rim. Nose: fruit preserve, powerful, dried herbs, black liquorice. Palate: flavourful, long.

Cachito Mío Selección 14 de 24 2017 T
100% tinta de Toro (Tinto)
91
Colour: deep cherry. Nose: ripe fruit, dried herbs, wild herbs. Palate: powerful, ripe fruit, spicy, round tannins.

Quizás Quizás Quizás 2017 T
92
Colour: cherry, garnet rim. Nose: fruit preserve, powerful, smoky, cocoa bean, creamy oak. Palate: flavourful, long, powerful tannins, spicy.

VIÑA ARNAIZ

Finca Villaester de Arriba, s/n
47112 Pedrosa del Rey (Valladolid)
☎: +34 983 784 421
orestes.garcia@jgc.es
www.familiagarciacarrion.com

Pata Negra 2021 T RB
tempranillo
86

VIÑEDOS ALONSO DEL YERRO

49810 Morales de Toro (Zamora)
☎: +34 610 789 997
mariadelyerro@vay.es
www.alonsodelyerro.es

Paydos 2017 T
tinta de Toro (Tinto)
93
Colour: Cherry. Nose: complex, spicy, toasty, black pepper, black fruit, ripe fruit. Palate: great length, round tannins, good structure, powerful.

Paydos 2018 T
tinta de Toro (Tinto)
92
Colour: Cherry. Nose: balsamic herbs, sweet spices, scrubland, black fruit. Palate: spicy, balsamic, good acidity.

DO. UCLÉS

CONSEJO REGULADOR

Avda. Miguel Cervantes, 93
16400 Tarancón (Cuenca)
☎: +34 969 135 056 - Fax: +34 969 135 421
@: info@vinosdeucles.com
www.vinosdeucles.com

LOCATION:

Midway between Cuenca (to the west) and Toledo (to the northwest), this DO is made up of 25 towns from the first province and three from the second. However, the majority of vineyards are situated in Tarancón and the neighbouring towns of Cuenca, as far as Huete - where La Alcarria starts - the largest stretch of border in the DO.

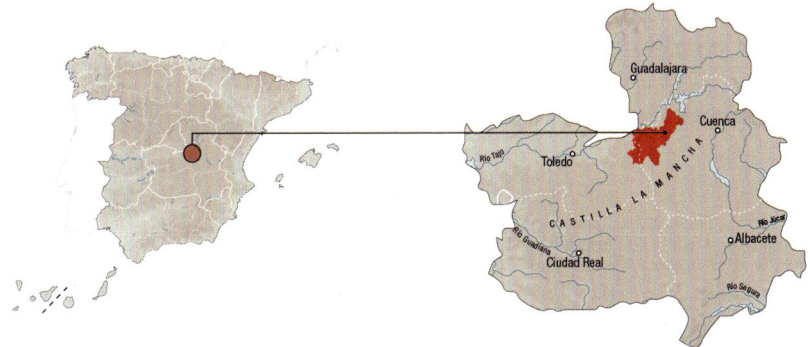

GRAPE VARIETIES:

RED: Tempranillo, Merlot, Cabernet Sauvignon, Garnacha and Syrah.

WHITE: Verdejo, Moscatel de Grano Menudo, Chardonnay, Airén, Sauvignon Blanc and Viura (macabeo).

FIGURES:

Vineyard surface: 1,700 – **Wine-Growers:** 532 – **Wineries:** 5 – **2021 Harvest rating:** N/A – **Production 21:** 3,800,300 litres – **Market percentages:** 79% National 21% International.

SOIL:

Despite spreading over two provinces with different soil components, the communal soils are deep and not very productive, of a sandy and consistent texture, becoming more clayey as you move towards the banks of the rivers Riansares and Bendija.

CLIMATE:

The Altamira sierra forms gentle undulations that rise from an average of 600 metres in La Mancha, reaching 1,200 metres. These ups and downs produce variations in the continental climate, which is less extreme, milder and has a Mediterranean touch. As such, rain is scarce, more akin to a semi-dry climate.

CLASSIFICATION OF YOUNG WINE HARVESTS PEÑÍNGUIDE

2017	2018	2019	2020	2021
N/A	N/A	N/A	N/A	N/A

DO UCLÉS / D.O.P.

BODEGAS & VIÑEDOS FONTANA
Extramuros s/n
16411 Fuente de Pedro Naharro (Cuenca)
☎: +34 969 125 433
o.moreno@peninsula.wine
www.peninsula.wine

Dominio de Fontana Sauvignon Blanc & Verdejo 2021 B
sauvignon blanc, verdejo

89

Herbaceous, fruity, wild, tasty, balanced, representative.

Dominio de Fontana Tempranillo & Cabernet Sauvignon 2019 T C
tempranillo, cabernet sauvignon

90

Colour: deep cherry. Nose: dried herbs, creamy oak, black fruit. Palate: powerful, ripe fruit, spicy, round tannins.

Dominio de Fontana Tempranillo & Syrah 2020 T RB
tempranillo, syrah

90

Colour: deep cherry. Nose: dried herbs, creamy oak, black fruit. Palate: ripe fruit, spicy, round tannins.

Mesta Tempranillo 2021 RD
tempranillo

87

Mesta Verdejo 2021 B
verdejo

87

Quinta de Quercus Single Vineyard 2018 T
tempranillo

90

Colour: deep cherry. Nose: dried herbs, creamy oak, black fruit. Palate: powerful, ripe fruit, spicy, round tannins.

BODEGAS FINCA LA ESTACADA
Ctra. N-400, Km. 103
16400 Tarancón (Cuenca)
☎: +34 969 327 099
Fax: +34 969 137 406
comunicacion@fincalaestacada.com
www.fincalaestacada.com

Finca la Estacada 12 meses 2018 T C
100% tempranillo

88

Standard, spicy, jammy, varietally correct, dried herbs, fruity.

Finca la Estacada 2021 B SD

86

Finca la Estacada 6 meses 2018 T RB
tempranillo

87

Finca La Estacada Chardonnay Sauvignon Blanc 2021 B

86

Finca la Estacada Varietales 2017 T R
40% tempranillo, 20% cabernet sauvignon, 20% syrah, 20% merlot

89

Spicy, jammy, tasty, toasty, herbal, balanced.

La Estacada Syrah Merlot 2018 T C
syrah, merlot

89

Jammy, fruity, herbal, spicy, balanced.

BODEGAS SOLEDAD
Ctra. Tarancón, s/n
16411 Fuente de Pedro Naharro (Cuenca)
☎: +34 969 125 039
Fax: +34 969 125 907
calidad@bodegasoledad.com
www.bodegasoledad.com

Bisiesto Cabernet Sauvignon 2017 T
cabernet sauvignon

88

Balsamic herbs, jammy, powerful, tasty.

Bisiesto Tempranillo 2016 T C
tempranillo

89

Kind finish, balsamic herbs, standard, balanced, jammy, tasty, toasty.

Bisiesto Verdejo 2020 B FB
verdejo

90

Colour: bright yellow. Nose: powerful, creamy oak, ripe fruit, spicy, wild herbs. Palate: rich, good structure, long, toasty, fine bitter notes.

Solmayor Airén 2021 B
airén

84

Solmayor Tempranillo 2021 T
tempranillo

87

Solmayor Verdejo-Sauvignon Blanc 2021 B
verdejo, sauvignon blanc

86

DO. UTIEL - REQUENA
CONSEJO REGULADOR

Sevilla, 12. Apdo. 61
46300 Utiel (Valencia)
☎: +34 962 171 062 - Fax: +34 962 172 185
@: info@utielrequena.org
www.utielrequena.org

LOCATION:

In thewest of the province of Valencia. It comprises the municipal districts of Camporrobles, Caudete de las Fuentes, Fuenterrobles, Requena, Siete Aguas, Sinarcas, Utiel, Venta del Moro and Villagordo de Cabriel.

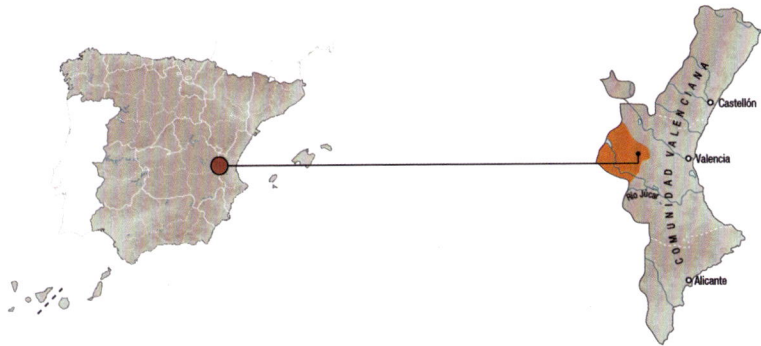

GRAPE VARIETIES:

RED: Bobal, Tempranillo, Garnacha, Cabernet Sauvignon, Merlot, Syrah, Pinot Noir, Garnacha Tintorera, Petit Verdot and Cabernet Franc.

WHITE: Tardana, Macabeo, Merseguera, Chardonnay, Sauvignon Blanc, Parellada, Verdejo and Moscatel de Grano Menudo.

FIGURES:

Vineyard surface: 32,013 – **Wine-Growers:** 4,579 – **Wineries:** 109 – **2021 Harvest rating:** Very Good – **Production 21:** 42,726,899 litres – **Market percentages:** 34% National - 66% International.

SOIL:

Mainly brownish-grey, almost red limestone, poor in organic matter andwith good permeability. The horizon of the vineyards are broken by the silhouette of the odd tree planted in the middle of the vineyards,which, bordered bywoods, offer a very attractive landscape.

CLIMATE:

Continental,with Mediterranean influences, coldwinters and slightly milder summers than in other regions of the province. Rainfall is quite scarcewith an annual average of 400 mm.

CLASSIFICATION OF YOUNG WINE HARVESTS PEÑÍNGUIDE

2017	2018	2019	2020	2021
GOOD	GOOD	VERY GOOD	VERY GOOD	VERY GOOD

DO UTIEL-REQUENA / D.O.P.

BODEGA & VIÑEDOS CARRES

Francho, 1
46352 Casas de Eufema (València/Valencia)
☎: +34 675 515 729
torrescarpiojl@gmail.com
www.bodegacarres.com

El Olivastro 2020 T
bobal

88

Spicy, fruity, herbal, smoky.

Malarado 2020 RD FB
bobal, garnacha

89

Kind finish, balanced, fruity, dried flowers, tasty.

Membrillera 2020 T C
bobal

90

Colour: light cherry. Nose: red berry notes, fruit expression, faded flowers, dried herbs, spicy. Palate: fruity, fresh, flavourful, balanced.

Pico d´Aliga 2020 T C
garnacha

88

Floral, fruity, jammy, tasty, crisp.

Sabinilla 2020 B FB
tardana

87

BODEGA ARANLEÓN

Ctra. Caudete, 3
46310 Los Marcos (València/Valencia)
☎: +34 963 631 640
Fax: +34 962 185 150
vinos@aranleon.com
www.aranleon.com

Aranleón Sólo 2020 T C
80% bobal, 20% cabernet sauvignon

88

Spicy, fruity, herbaceous, tasty.

BODEGA DE MOYA

Avda. de Las Bodegas, 14
46300 Utiel (València/Valencia)
☎: +34 665 330 991
yves@demoya.es
www.demoya.es

María 2019 T C
bobal

88

Overripe, fruity, spicy, tasty.

Sofía 2019 T
bobal

90

Colour: cherry, purple rim. Nose: black fruit, ripe fruit, spicy, smoky. Palate: fruity, flavourful, good structure, powerful tannins.

BODEGA DUSSART PEDRÓN

Saliente, 12
46355 Los Pedrones (València/Valencia)
☎: +34 722 270 944
bodegadussartpedron@gmail.com
www.bodegadussartpedron.com

Le Bobal 2020 T BA
bobal

90

Colour: cherry, purple rim. Nose: ripe fruit, black fruit, spicy, smoky, dried herbs. Palate: fruity, good structure, balanced, smoky aftertaste, slightly dry, soft tannins.

Le Cencibel 2020 T BA
tempranillo

89

Spicy, fruity, jammy, tasty.

Le Grenache 2019 T BA
garnacha

92

Colour: bright cherry. Nose: red berry notes, ripe fruit, floral, dried herbs, spicy. Palate: full-bodied, flavourful, powerful, balanced, good finish, slightly dry, soft tannins.

Le Rosé 2020 RD
45% bobal, 40% garnacha, 15% tempranillo

91

Slight oxidation. Colour: rose, pale. Nose: red berry notes, fruit preserve, fine lees, dried herbs. Palate: fleshy, flavourful, powerful, ripe fruit, great length.

BODEGA VERA DE ESTENAS

Ctra. N-III, km. 266 - Paraje La Cabezuela
46300 Utiel (València/Valencia)
☎: +34 962 171 141
estenas@veradeestenas.es
www.veradeestenas.es

Casa Don Ángel Bobal 2019 T
bobal

92

Colour: bright cherry. Nose: red berry notes, spicy, toasty, dried herbs. Palate: good acidity, grainy tannins, balanced.

El Bobal Estenas 2021 T
bobal
91
Colour: cherry, purple rim. Nose: red berry notes, ripe fruit, violets, wild herbs. Palate: flavourful, fruity, fresh, balanced.

Estenas 2018 T C
bobal, cabernet sauvignon
90
Colour: cherry, purple rim. Nose: black fruit, ripe fruit, smoky, spicy, dried herbs. Palate: flavourful, powerful, slightly dry, soft tannins.

Estenas 2021 B
macabeo, chardonnay
86

Estenas Bobal 2021 RD
bobal
87

Estenas Madurado en Barrica 2020 T
bobal, tempranillo, cabernet sauvignon, merlot
88
Spicy, smoky, fruity, pruney.

La Tardana Estenas 2021 B
tardana
90
Colour: bright straw. Nose: stone fruit, ripe fruit, dried herbs, dried flowers. Palate: fruity, flavourful, fresh, good finish.

BODEGAS ARRÁEZ
Pol. 6 Parcela 386 Paraje Ciscar
46630 La Font de la Figuera (València/Valencia)
☎: +34 962 290 031
info@bodegasarraez.com
www.bodegasarraez.com

Vividor 2020 T
bobal
88
Spicy, simple, standard.

BODEGAS COVIÑAS
Avda. Rafael Duyos, s/n
46340 Requena (València/Valencia)
☎: +34 962 300 680
covinas@covinas.com
www.covinas.com

Adnos Bobal 2014 T
bobal
92
Spicy, slight reduction. Colour: cherry, garnet rim. Nose: waxy notes, black fruit, ripe fruit. Palate: balanced, fine bitter notes, good acidity, round tannins.

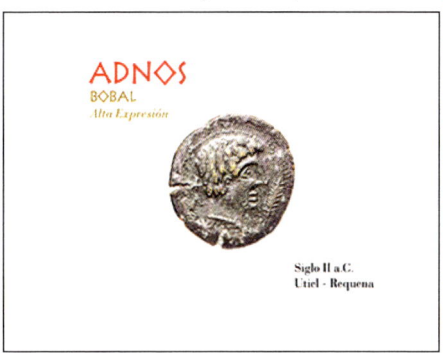

Aula Bobal Merlot 2019 T
bobal, merlot
88
Spicy, pruney, tasty, herbal.

Aula Bobal Tempranillo 2018 T C
bobal, tempranillo
86

Enterizo 2017 T R
bobal, tempranillo, cabernet sauvignon, garnacha
86

Enterizo Bobal 2021 RD
bobal
87

Veterum Vitium 2020 T
bobal
88
Aromatic, fruity, jammy, tasty, wild, smooth.

DO UTIEL-REQUENA / D.O.P.

BODEGAS DEL VALLE
Calle Molineta, 4
46354 Los Cojos (València/Valencia)
☎: +34 697 649 111
mireia@bodegasdelvalle.es

Avispero 2021 RD
bobal
84

Forcate 2019 T
bobal, tempranillo
89
Spicy, fruity, jammy, tasty.

Vera del Valle 2020 T
tempranillo, garnacha
88
Fruity, jammy, tasty, spicy.

BODEGAS EMILIO CLEMENTE
Camino de San Blas, s/n
46340 Requena (València/Valencia)
☎: +34 601 410 728
administracion@eclemente.es
www.eclemente.es

Chardo Day 2021 B
chardonnay
85

Emilio Clemente 2020 T
bobal
88
Spicy, fruity, jammy, simple.

Peñas Negras 2020 T
bobal
86

BODEGAS HISPANO SUIZAS
Ctra. N-322, Km. 451,7
46357 El Pontón (València/Valencia)
☎: +34 962 349 370
maria.salinas@bodegashispanosuizas.com
www.bodegashispanosuizas.com

Bassus Pinot Noir 2020 T
pinot noir
91
Colour: bright cherry, purple rim. Nose: red berry notes, fruit expression, spicy. Palate: fresh, fruity, flavourful, good acidity.

Bassus Pinot Noir Dulce 2021 RD D
pinot noir
89
Sweetish, fruity, tasty, citrus fruit.

Bobos Finca Casa la Borracha 2020 T
bobal
93
Colour: deep cherry. Nose: ripe fruit, black fruit, spicy, creamy oak. Palate: flavourful, good structure, balanced.

Impromptu 2021 B
sauvignon blanc
92
Colour: bright straw, greenish rim. Nose: citrus fruit, wild herbs, stone fruit. Palate: fresh, fine bitter notes, saline.

Quod Superius 2018 T
bobal, syrah, merlot, cabernet franc
93
Colour: deep cherry, garnet rim. Nose: fruit expression, ripe fruit, spicy, tobacco. Palate: flavourful, fruity, good structure, good finish, round tannins.

BODEGAS IRANZO
Ctra. de Madrid, 60
46315 Caudete de las Fuentes (València/Valencia)
☎: +34 962 319 282
comercial@bodegasiranzo.com
www.bodegasiranzo.com

0 por Ciento Tempranillo 2021 T
tempranillo
87 🌱

Finca Cañada Honda 2019 T BA
bobal, garnacha
87 🌱

Finca Cañada Honda 87 T C
tempranillo, cabernet sauvignon
87 🌱

BODEGAS JIMÉNEZ-VILA HNOS.
Construccion, 19 Pl. El Romeral
46190 Requena (València/Valencia)
☎: +34 662 118 746
export@jimenezvila.es
www.jimenezvila.es

La Novicia 2020 T RB
100% bobal
87

Nexo 2019 T
80% bobal, 20% syrah
89
Spicy, fruity, jammy, tasty.

Núcleo 2019 T
100% bobal

88

Fruity, overripe, spicy, astringent.

Terra de Tardor 2021 B FB
60% tardana, 40% sauvignon blanc

89

Spicy, toasty, jammy, citrus fruit, floral, tasty, great length.

BODEGAS LUIS CORBÍ - FINCA COR VI

Ctra. Nacional 322, Km 431 Los Isidros
46354 Requena (València/Valencia)
☎: +34 601 413 967
enologo@closcorvi.com
www.fincacorvi.com

Maloco 2021 T

90 🌱

Colour: bright cherry. Nose: fresh fruit, red berry notes, wild herbs. Palate: good acidity, spicy, fine tannins, fruity aftertaste.

BODEGAS MITOS

CV-450, Km.3 El Azagador
46357 Requena (València/Valencia)
☎: +34 962 300 703
admin@bodegasmitos.com
www.bodegasmitos.com

Mitos Macabeo 2020 B
macabeo

87

Mitos Tempranillo 2020 T
tempranillo

86

BODEGAS MURVIEDRO

Ampliación Pol. El Romeral, s/n
46340 Requena (València/Valencia)
☎: +34 962 329 003
murviedro@murviedro.es
www.murviedro.es

Expresion Reserva Bobal 2018 T R
100% bobal

90

Colour: bright cherry. Nose: black fruit, ripe fruit, spicy, dried herbs. Palate: fruity, flavourful, slightly dry, soft tannins.

Murviedro Colección Bobal 2020 T RB
100% bobal

88

Spicy, fruity, herbal.

La Casa de la Seda 2020 T BA
100% bobal

91

Colour: bright cherry. Nose: red berry notes, ripe fruit, dried herbs, spicy. Palate: fruity, powerful, balanced, slightly dry, soft tannins.

Murviedro Cepas Viejas Bobal 2019 T C
100% bobal

90

Colour: bright cherry, purple rim. Nose: black fruit, red berry notes, dried herbs, spicy. Palate: fruity, flavourful, slightly dry, soft tannins.

Sericis Cepas Viejas Bobal 2019 T
100% bobal

87

Vallejo Avenas 2020 B
100% chardonnay

90

Colour: bright yellow. Nose: ripe fruit, spicy, wild herbs, toasted bread. Palate: rich, good structure, fine bitter notes.

BODEGAS NODUS

Finca El Renegado, s/n
46315 Caudete de las Fuentes (València/Valencia)
☎: +34 962 174 029
andrea@bodegasnodus.com
www.bodegasnodus.com

Nodus Bobal 2020 T
bobal

87 🌱

Nodus Chardonnay 2021 B
100% chardonnay

89 🌱

Citrus fruit, spicy, herbal, lactic, tasty.

Nodus DP 2020 T
bobal

85 🌱

DO UTIEL-REQUENA / D.O.P.

DO UTIEL-REQUENA / D.O.P.

Nodus Merlot Delirium 2019 T
merlot

89

Balanced, spicy, dried herbs, toasty.

Nodus Sauvignon Blanc 2021 B
sauvignon blanc

88

Citrus fruit, crisp, herbal, balanced.

Nodus Tinto de Autor 2019 T C
45% merlot, 25% cabernet sauvignon, 15% syrah, 15% bobal

89

Balanced, spicy, herbal, full-bodied, tasty.

BODEGAS PASIEGO

Avda. Virgen de Tejeda, 28
46320 Sinarcas (València/Valencia)
☎: +34 609 076 575
bodega@bodegaspasiego.com
www.bodegaspasiego.com

Pasiego "CÆSAR" 2017 T C
cabernet sauvignon, merlot, bobal, syrah

89

Fruity, sweety finish, herbaceous, tasty, toasty.

Pasiego Bobal 2018 T C
100% bobal

89

Classic, spicy, smoky, fruity, jammy, toasty.

Pasiego de Autor 2017 T C
50% cabernet sauvignon, 25% merlot, 25% syrah

90

Colour: deep cherry. Nose: ripe fruit, black fruit, spicy, dried herbs. Palate: fruity, powerful, flavourful, smoky aftertaste, slightly dry, soft tannins.

Pasiego Julieta Dulce Natural 2017 B D
chardonnay, sauvignon blanc

94

Colour: bright yellow. Nose: honeyed notes, caramel, stone fruit, dried fruit, dried flowers. Palate: flavourful, unctuous, fruity, sweet.

Pasiego La Suertes 2021 B
50% sauvignon blanc, 50% macabeo

86

BODEGAS REBOLLAR ERNESTO CÁRCEL

Pago de Santana, s/n
46391 El Rebollar - Requena (València/Valencia)
☎: +34 607 436 362
bodegasrebollar@carceldecorpa.es
www.carceldecorpa.es

Carcel de Corpa 2013 T GR
tempranillo, garnacha, bobal

87

Carcel de Corpa 2018 T C
tempranillo, garnacha, bobal

86

Dame un Beso, Negro 2019 T BA
bobal

88

Spicy, fruity, jammy, tasty.

Seducción 2021 B
macabeo

85

Seducción 2021 RD
bobal

86

BODEGAS RODENO

Pol. 84 Parcela 207 Paraje Los Rodenos
46357 Hortunas (València/Valencia)
☎: +34 607 350 277
bodegasrodeno@bodegasrodeno.com

Rodeno 2020 T C
85% tempranillo, 15% bobal

86

Rodeno Superior 2020 B AROM S
80% macabeo, 20% chardonnay

85

BODEGAS UTIELANAS

Actor Rambal, 31
46300 Utiel (València/Valencia)
☎: +34 962 171 157
administracion@bodegasutielanas.com
www.bodegasutielanas.com

Castillo de Utiel 2018 T C
100% bobal

85

Castillo de Utiel 2021 T
100% bobal

85

Vega Infante 2021 T
90% bobal, 5% tempranillo, 5% garnacha

86

BODEGAS VEGALFARO
Ctra. Pontón - Utiel, Km. 3
46340 Requena (València/Valencia)
☎: +34 962 320 680
info@vegalfaro.com
www.vegalfaro.com

Caprasia 2021 B
macabeo, chardonnay

89 ♣

Citrus fruit, herbal, balanced, tasty.

Caprasia Bobal Ánfora 2020 T
bobal

91 ♣

Colour: cherry, purple rim. Nose: ripe fruit, spicy, dried herbs. Palate: fruity, balanced, flavourful, slightly dry, soft tannins.

Caprasia Rosé 2021 RD
bobal, merlot

87 ♣

Rebel.lia 2021 B
chardonnay, sauvignon blanc

87 ♣

Rebel.lia 2021 T
garnacha tintorera, tempranillo

88 ♣

Kind finish, standard, fruity, jammy.

Rebel.lia Selección Especial 2020 T RB

89 ♣

Spicy, fruity, jammy, tasty.

BODEGAS VIBE
Los Olmos, 1
46357 El Azagador - Requena (València/Valencia)
☎: +34 653 964 158
bodega@bodegasvibe.com
www.bodegasvibe.com

Parsimonia 2018 T C
merlot, cabernet sauvignon

88

Fruity, jammy, floral, tasty.

Parsimonia 2019 T FB
bobal

90

Colour: bright cherry. Nose: ripe fruit, black fruit, aromatic coffee. Palate: fruity, flavourful, good structure, slightly dry, soft tannins.

Parsimonia 2021 B
tardana

87

BRUNO MURCIANO
46315 Caudete de las Fuentes (València/Valencia)
☎: +34 663 924 485
bruno@brunomurciano.com
www.brunomurciano.com

Cambio de Tercio 2021 T

92 ♣

Colour: light cherry. Nose: fruit expression, red berry notes, ripe fruit, wild herbs, fragrant herbs. Palate: fruity, easy to drink.

El Sueño 2020 T
bobal

93 ♣

Rustic, herbal. Colour: cherry, garnet rim. Nose: ripe fruit, wild herbs. Palate: spicy, balanced, taut, fruity.

L'Alegria 2021 T

92 ♣

Colour: deep cherry. Nose: macerated fruit, ripe fruit, powerful. Palate: flavourful, ripe fruit, spicy, fruity aftertaste.

La Bruna 2020 T
bobal

91 ♣

Colour: dark-red cherry. Nose: fruit preserve, powerful, black pepper, creamy oak. Palate: flavourful, long.

Parajes del Cabriel 2021 T
bobal

90 ♣

Colour: cherry, garnet rim. Nose: fruit preserve, candied fruit, dried herbs. Palate: flavourful, fruity, correct, balanced.

CHOZAS CARRASCAL
Vereda Real
46390 San Antonio de Requena (València/Valencia)
☎: +34 963 410 395
chozas@chozascarrascal.es
www.chozascarrascal.com

Anma 2018 T BA
syrah, garnacha

92 ♣

Colour: cherry, purple rim. Nose: red berry notes, floral, spicy, scrubland. Palate: flavourful, fruity, good acidity, long.

Las Dos Ces 2020 T RB
bobal

91 ♣

Colour: cherry, purple rim. Nose: fruit expression, red berry notes, floral. Palate: fruity, flavourful, balanced.

DO UTIEL-REQUENA / D.O.P.

DO UTIEL-REQUENA / D.O.P.

Anma 2020 B MC
macabeo

91

Oxidativ. Colour: straw. Nose: dried herbs, dried flowers, white fruit. Palate: powerful, ripe fruit, balanced, flavourful.

Las Dos Ces 2021 B
macabeo

90

Colour: straw. Nose: white flowers, jasmine, dried herbs. Palate: flavourful, fruity, balanced.

Mudare 2019 B
100% bobal

93

Oxidativ, original. Colour: straw. Nose: ripe fruit, dried herbs, white flowers, fine lees. Palate: powerful, ripe fruit, balanced, flavourful, saline, great length.

Rose Marine 2021 RD
100% garnacha

91

With personality. Colour: raspberry rose. Nose: red berry notes, floral, fragrant herbs, dry nuts. Palate: spicy, good acidity, fine bitter notes, flavourful.

DOMINIO DE LA VEGA

Ctra. Madrid - Valencia, Km. 270,6
46390 San Antonio de Requena (València/Valencia)
☎: +34 962 320 570
dv@dominiodelavega.com
www.dominiodelavega.com

Blanco de Maria 2021 B
macabeo, sauvignon blanc

87

Blanco de Roberto 2021 B
sauvignon blanc

88

Standard, fruity, jammy, wild, kind finish.

Finca La Beata 2016 T GR
bobal

93

Colour: very deep cherry. Nose: fruit preserve, black fruit, creamy oak, spicy, black liquorice. Palate: good structure, powerful, flavourful, balanced, round tannins.

Paraje Tornel 2018 T R
bobal

90

Colour: bright cherry. Nose: black fruit, ripe fruit, spicy, dried herbs. Palate: fruity, flavourful, good structure, balanced, slightly dry, soft tannins.

Rosado de Fermin 2021 RD
bobal

87

Tinto de Abel 2021 T
bobal

88

Pleasant, fruity, jammy, tasty, simple.

ENVERO WINE COMPANY

Colón, 28
13700 Tomelloso (Ciudad Real)
☎: +34 630 565 000
info@allblackwines.com
www.allblackwines.com

Allblack 2018 T BA
100% bobal

91

Colour: cherry, purple rim. Nose: fruit preserve, black fruit, dried herbs, smoky, spicy. Palate: flavourful, good structure, powerful, balanced, slightly dry, soft tannins.

FAMILIA BASTIDA

Canónigo Lozano, 11
30520 Jumilla (Murcia)
☎: +34 968 780 142
info@familiabastida.com
www.familiabastida.com

Biftu 2018 T RB
bobal

89

Spicy, smoky, fruity, jammy, tasty.

FAUSTINO RIVERO ULECIA

Avda. Rafael Duyos, 6-8
46340 Requena (València/Valencia)
☎: +34 941 380 057
www.faustinorivero.com

Faustino Rivero Ulecia Bobal Tempranillo 2017 T R
bobal, tempranillo

88

Hot, toasty, smooth, jammy, spicy.

Faustino Rivero Ulecia Bobal Tempranillo 2019 T C
bobal, tempranillo

87

FINCA LA PICARAZA

Camino de La Picaraza s/n
46313 Las Casas, Utiel (València/Valencia)
☎: +34 675 043 456
info@bodegafincalapicaraza.com
www.bodegafincalapicaraza.com

Viña La Picaraza 2017 T C
bobal

87

LADRÓN DE LUNAS

Colón, 12
46357 La Portera (València/Valencia)
☎: +34 601 288 998
jorge.ferrer@ladrondelunas.es
www.ladrondelunas.es

La Ovejita Tinta 2021 T RB
100% bobal

86

Ladrón de Lunas Garnacha RE BR
garnacha

88

Fruity, floral, crisp, herbal, spicy.

Ladrón de Lunas Bobal 2021 T C
100% bobal

87

Ladron de Lunas Exclusive Vino de Autor 2015 T GR
100% bobal

88

Spicy, fruity, jammy, tasty, toasty.

Ladrón de Lunas Roble Bobal 2021 T RB
100% bobal

87

Ladrón de Lunas Sauvignon Macabeo 2021 B
80% sauvignon blanc, 20% macabeo

88

Aromatic, dried flowers, dried herbs, fruity, smooth.

LAS MERCEDES DEL CABRIEL

Ctra. Villargordo a Camporrobles, Km. 2
46317 Villargordo del Cabriel (València/Valencia)
☎: +34 659 954 310
jose.leon@bodegalasmercedes.com
www.bodegalasmercedes.com

Esencia de Las Mercedes 2018 T
100% bobal

91

Colour: deep cherry. Nose: ripe fruit, dried herbs, creamy oak, wet leather. Palate: powerful, ripe fruit, spicy, round tannins.

Las Mercedes Bobal al Límite 2019 T
100% bobal

91

Colour: cherry, purple rim. Nose: black fruit, ripe fruit, spicy, black pepper, dried herbs. Palate: fruity, powerful, fresh, slightly dry, soft tannins.

LATORRE AGROVINÍCOLA

Ctra. Requena, 2
46310 Venta del Moro (València/Valencia)
☎: +34 962 185 028
Fax: +34 962 185 422
bodega@latorreagrovinicola.com
www.latorreagrovinicola.com

Catamarán 2021 B
macabeo, verdejo

86

Duque de Arcas 2018 T C
bobal, tempranillo, cabernet sauvignon

87

Duque de Arcas 2018 T RB
tempranillo, cabernet sauvignon

85

Duque de Arcas Solo Bobal 2017 T R
bobal

89

Spicy, fruity, jammy, tasty.

Parreño 2021 B
viura, verdejo

84

Parreño 2021 RD
bobal

85

DO UTIEL-REQUENA / D.O.P.

DO UTIEL-REQUENA / D.O.P.

MONTESANCO
Casa de la Viña, Ctra. Utiel-Los Isidros km. 7
46340 Requena (València/Valencia)
☎: +34 962 121 626
vinos@montesanco.com
www.montesanco.com

Món Bobal 2018 T
100% bobal

92
Colour: deep cherry. Nose: dried herbs, black fruit, fruit preserve. Palate: powerful, ripe fruit, spicy, round tannins.

Món Macabeo 2021 B
100% macabeo

90
Colour: straw. Nose: dried herbs, faded flowers, spicy, white fruit. Palate: ripe fruit, balanced.

PAGO DE THARSYS
Ctra. Nacional III, km. 274
46340 Requena (València/Valencia)
☎: +34 962 303 354
bodega@pagodetharsys.com
www.pagodetharsys.com

Carlota Suria Organic Bobal 2019 T C
bobal

89
Spicy, fruity, herbal, jammy, tasty.

Carlota Suria Organic Chardonnay 2020 B FB
chardonnay

89
Creamy, spicy, herbal, roasted coffee, tasty.

Pago de Tharsys Cabernet Franc Sin Sulfitos 2021 T
cabernet franc

90
Colour: cherry, purple rim. Nose: red berry notes, floral, wild herbs. Palate: fruity, flavourful, balanced.

Pago de Tharsys Merseguera Sin Sulfitos 2021 B
merseguera

90
Colour: straw. Nose: ripe fruit, dried herbs, faded flowers, cereal notes. Palate: ripe fruit, balanced, full-bodied.

Tharsys Único 2018 BE R BR
bobal

90
Colour: bright yellow. Nose: ripe fruit, fine lees, balanced, dried herbs. Palate: good acidity, flavourful, ripe fruit.

RAÍCES IBÉRICAS
Avda. Mudejar, 61
50340 Maluenda (Zaragoza)
☎: +34 976 893 017
s.richard@raices.wine
www.raicesibericas.com

Raíces Bobal 2021 T
100% bobal

88
Fruity, herbaceous, floral, smoky, crisp.

TERRACOTA WINES CHI TAO JIU
Camí Tremolar
46012 Pinedo (València/Valencia)
☎: +34 629 013 515
info@terracottawines.es
www.terracottawines.es

Bobale Manneken Pis 2018 T C
100% bobal

89
Balanced, herbaceous, full-bodied, toasty, tasty.

Daniela 2017 T C
100% bobal

90
Colour: dark-red cherry. Nose: black fruit, ripe fruit, earthy notes, dried herbs. Palate: powerful, ripe fruit, concentrated, flavourful.

Dolia Amphorae Chardonnay 2021 B
chardonnay

88
Smoky, tasty, with personality, toasty, spicy, standard.

Dolia Bobal Amphorae 2021 T
bobal

89
Fruity, floral, sweety finish, jammy.

VICENTE GANDÍA
Ctra. Cheste a Godelleta, s/n
46370 Chiva (València/Valencia)
☎: +34 962 524 242
Fax: +34 962 524 243
info@vicentegandia.com
www.vicentegandia.es

BO - Bobal Único 2020 T
bobal

87

Bobal Blanco by Pepe Hidalgo 2021 B
bobal

89
Floral, fruity, smooth, jammy.

DO UTIEL-REQUENA / D.O.P.

Bobal Negro by Pepe Hidalgo 2020 T
bobal

90

Colour: cherry, garnet rim. Nose: creamy oak, hot, black fruit, overripe fruit, toasty. Palate: pruney, powerful, sweet tannins.

Bobal Rosa by Pepe Hidalgo 2021 RD
bobal

90

Colour: raspberry rose. Nose: elegant, floral, fragrant herbs, stone fruit. Palate: light-bodied, spicy, good acidity.

Con un Par Sauvignon Blanc 2021 B
sauvignon blanc

88

Citrus fruit, crisp, herbal, tasty.

Eora Chardonnay 2021 B
chardonnay

88

Citrus fruit, herbal, crisp, tasty.

Hoya de Cadenas Tempranillo 2018 T R
tempranillo

85

Marqués de Tena 2018 T R
tempranillo

87

VINOS SIERRA NORTE
Pol. Ind. El Romeral, Calle Transporte C2
46340 Requena (València/Valencia)
☎: +34 962 323 099
info@bodegasierranorte.com
www.bodegasierranorte.com

Bercial Ladera los Cantos 2018 T BA
bobal, cabernet sauvignon

91 🌿

Colour: Cherry. Nose: balsamic herbs, scrubland, balanced, spicy, black pepper. Palate: balsamic, good acidity, round tannins.

Bercial Selección 2020 B BA
chardonnay, sauvignon blanc, macabeo

90 🌿

Colour: bright yellow. Nose: spicy, stone fruit, ripe fruit. Palate: rich, long, toasty, fine bitter notes.

Fuenteseca 2021 B
macabeo, sauvignon blanc

88 🌿

Defined aromas, floral, fruity, smooth, balanced.

Fuenteseca 2021 RD
bobal, cabernet sauvignon

86 🌿

Pasion de Bobal 2020 T RB
bobal

89 🌿

Pleasant, wild, jammy, fruity, spicy. Palate: smoky aftertaste.

Pasion de Bobal 2021 RD
bobal

88 🌿

Balanced, fruity, herbal, crisp.

VIÑA MEMORIAS
Ctra. Madrid-Valencia, Km 270
46390 San Antonio de Requena (València/Valencia)
☎: +34 669 043 007
contact@vinamemorias.com
www.vinamemorias.com

Alenar Blanc Tardana 2021 B
100% tardana

88

Austere, citrus fruit, racy, herbal.

Yunikko 2020 T
100% bobal

91

Colour: deep cherry. Nose: black fruit, ripe fruit, wild herbs, spicy, toasty. Palate: flavourful, fruity, balanced, ripe fruit, toasty, round tannins.

VITICULTORES SAN JUAN BAUTISTA
Calle Vereda, 4
46340 San Juan (València/Valencia)
☎: +34 649 810 834
info@bobaldesanjuan.es
www.bobaldesanjuan.es

Bobal de Sanjuan 2020 T S
100% bobal

88

Fruity, spicy, herbaceous, crisp.

Bobal de Sanjuan 2021 RD
100% bobal

85

Camo 2020 T S
100% bobal

88

Fruity, crisp, herbaceous, spicy.

El Perdío 2020 T RB
100% bobal

87 🌿

DO. VALDEORRAS
CONSEJO REGULADOR

Ctra. Nacional 120, km. 463
32340 Vilamartín de Valdeorras (Ourense)
☎: +34 988 300 295 - Fax: +34 988 300 455
@: consello@dovaldeorras.com
www.dovaldeorras.tv

LOCATION:

The DO Valdeorras is situated in the northeast of the province of Orense. It comprises the municipal areas of Larouco, Petín, O Bolo, A Rua, Vilamartín, O Barco, Rubiá and Carballeda de Valdeorras.

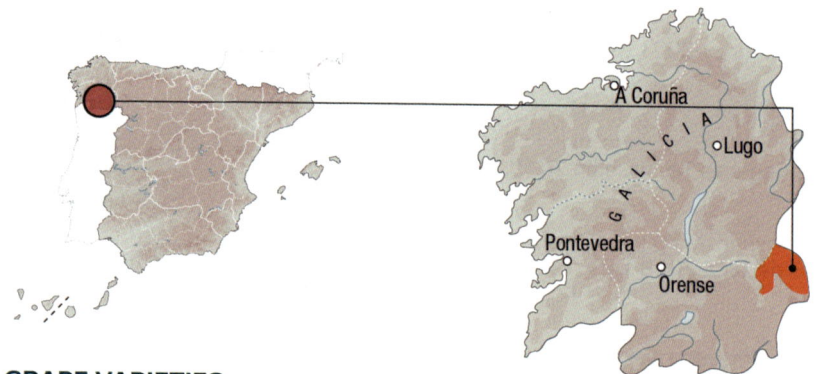

GRAPE VARIETIES:

WHITE: Godello, Dona Blanca, Palomino, Loureira, Treixadura, Albariño, Torrontes, Lado and Palomino.

RED: Mencía, Merenzao, Grao Negro, Garnacha, Tempranillo (Araúxa), Brancellao, Sousón, Caíño Tinto, Espadeiro, Ferrón, Garnacha Tintureira and Mouratón.

FIGURES:

Vineyard surface: 1,146 – **Wine-Growers:** 1,046 – **Wineries:** 42 – **2021 Harvest rating:** N/A – **Production 21:** 4,930,672 litres – **Market percentages:** 87% National - 13% International.

SOIL:

Quite varied. There are three types: the first type which is settled on shallow slate with many stones and a medium texture; the second type on deeper granite with a lot of sand and finally the type that lies on sediments and terraces, where there are usually a lot of pebbles.

CLIMATE:

Continental, with Atlantic influences. The average annual temperature is 11°C and the average annual rainfall ranges between 850 mm and 1,000 mm.

CLASSIFICATION OF YOUNG WINE HARVESTS PEÑÍNGUIDE

2017	2018	2019	2020	2021
VERY GOOD	VERY GOOD	EXCELLENT	VERY GOOD	VERY GOOD

ADEGA A COROA

A Coroa, s/n
32350 A Rúa (Ourense/Orense)
☎: +34 988 310 648
acoroa@acoroa.com
www.acoroa.com

A Coroa "Lías" 2020 B
godello

91

Colour: bright straw. Nose: balanced, medium intensity, floral, fine lees. Palate: rich, flavourful, fine bitter notes.

A Coroa 200 Cestos 2020 B FB
godello

92

Colour: bright straw. Nose: expressive, ripe fruit, floral, fine lees. Palate: full-bodied, long, rich, flavourful.

A Coroa Godello 2021 B
godello

90

Colour: bright straw. Nose: floral, white fruit, ripe fruit. Palate: flavourful, fresh, good acidity, fruity aftertaste, easy to drink.

Ladeira Vella 2018 T C
garnacha tintorera, mencía, otras

88

Toasty, jammy, powerful, hot.

ADEGA AGOREIRA

Valencia do Sil, s/n
32340 Vilamartín de Valdeorras (Ourense/Orense)
☎: +34 986 499 750
info@adegasterrae.com
www.adegasterrae.com

Agoreira 2021 B
100% godello

89

Pleasant, fruity, tropical, tasty, balanced, standard.

ADEGA ALAN DE VAL

Lugar de Pedrazais
32350 A Rúa (Ourense/Orense)
☎: +34 636 727 848
joaquin@alandeval.com

Alan de Val Caíño As Queimadas 2019 T
caíño longo

89

Smoky, spicy, fruity, jammy, tasty.

Alan de Val Castes Nobres 2019 T
brancellao, caíño, sousón

90

Colour: Cherry. Nose: ripe fruit, red berry notes, wild herbs, dried herbs, dried flowers. Palate: fruity, flavourful, fresh, slightly dry, soft tannins.

Alan de Val Godello 2021 B
100% godello

90

Dried flowers, jammy, wild, balanced, citrus fruit, aromatic. Nose: characterful, neat, varietal.

Alan de Val Mencía 2020 T
mencía

90

Colour: cherry, purple rim. Nose: red berry notes, wild herbs, grassy, spicy. Palate: flavourful, fruity, fresh, good structure, slightly dry, soft tannins.

Escada Garnacha Tintureira 2019 T
garnacha tintorera

90

Colour: cherry, garnet rim. Nose: ripe fruit, spicy, creamy oak, hot. Palate: flavourful, fruity, round tannins.

DO VALDEORRAS / D.O.P.

DO VALDEORRAS / D.O.P.

Pedrazais Godello 2020 B
100% godello

89

Pleasant, wild, dried flowers, fruity, balanced, kind finish, smooth.

ADEGA CEPADO
Patal, 11
32310 Rubiá (Ourense/Orense)
☎: +34 686 611 589
info@cepado.com
www.cepado.com

Cepado Black Edition 2020 B C
100% godello

91

Colour: bright straw. Nose: citrus fruit, white fruit, wild herbs, floral. Palate: fruity, fresh, flavourful, powerful.

Cepado Godello 2021 B
100% godello

89

Citrus fruit, herbal, tasty, crisp.

Cepado Mencía 2021 T
100% mencía

88

Jammy, fruity, tasty.

Finca a Coronela by Cepado 2019 T
garnacha tintorera

91

Colour: deep cherry. Nose: ripe fruit, dried herbs, creamy oak. Palate: powerful, ripe fruit, spicy, round tannins.

ADEGA O CASAL
Malladin, s/n
32310 Rubiá (Ourense/Orense)
☎: +34 663 563 079
casalnovo@casalnovo.es
www.casalnovo.es

Casal Novo Godello 2021 B
100% godello

89

Dried herbs, fruity, jammy, tasty.

Casal Novo Mencía 2020 T
95% mencía, 5% merenzao

88

Fruity, herbal, tasty, spicy.

BODEGA COOP. JESÚS NAZARENO S.C.G.
Avda. Florencio Delgado Gurriarán, 62
32300 O Barco de Valdeorras (Ourense/Orense)
☎: +34 988 320 262
Fax: +34 988 320 242
vinobarco@vinosbarco.com
www.vinosbarco.com

Menciño 2021 T
mencía

87

Menciño Summum 2020 T
mencía

87

Valdouro 2019 T BA
mencía, garnacha

88

Fruity, herbal, spicy, tasty.

Viña Abad Godello 2021 B
godello

89

Fruity, herbal, wild, tasty.

Viña Abad Sumum Godello 2020 B
godello

90

Colour: bright straw. Nose: fruit expression, floral, white fruit. Palate: flavourful, fresh, good acidity.

BODEGA GRANBAZÁN
Lg. Tremoedo, 46
36628 Vilanova de Arousa (Pontevedra)
☎: +34 986 555 562
elena.sanchez@agrodebazan.com
www.bodegasgranbazan.com

Quinta do Sil 2021 B
100% godello

88

Aromatic, floral, balanced, thin, standard.

BODEGA LA TAPADA
Finca La Tapada
32310 Rubiá (Ourense/Orense)
☎: +34 988 324 197
bodega.atapada@gmail.com
www.guitianvinos.com

Guitián Fermentado en Barrica de Acacia 2019 B
100% godello

92

Colour: bright yellow. Nose: ripe fruit, dried herbs, spicy, faded flowers. Palate: fruity, flavourful, balanced, smoky aftertaste.

Guitián Godello 2020 B FB
92
Colour: bright yellow. Nose: ripe fruit, dried herbs, spicy. Palate: fruity, flavourful, fresh, balanced, smoky aftertaste.

Guitián Godello 2021 B
100% godello
90
Colour: bright straw. Nose: white fruit, ripe fruit, dried flowers, jasmine, dried herbs. Palate: fruity, flavourful, balanced.

Guitián Godello sobre Lías 2020 B
100% godello
92
Colour: bright yellow. Nose: white fruit, fruit expression, scrubland, floral. Palate: fruity, fresh, good acidity, balanced.

Guitián Godello sobre Lías 2021 B
91
Colour: bright straw. Nose: fine lees, citrus fruit, white fruit, smoky, wild herbs. Palate: full-bodied, rich, long, good acidity.

BODEGA ROANDI
O Lagar, 1
32336 Entoma - O Barco (Ourense/Orense)
☎: +34 988 335 198
info@bodegaroandi.com
www.bodegaroandi.com

Alento 2019 B BA
godello
87

Bancales Moral 2020 T BA
87

Brinde de Godello 2014 BE R
100% godello
87

Domus de Roandi 2016 T C
90
Colour: bright cherry. Nose: ripe fruit, red berry notes, spicy, toasty, old leather. Palate: fruity, fresh, flavourful, powerful.

Roandi 2019 T
87

Roandi 2020 B
90
Colour: bright straw. Nose: ripe fruit, fine lees, balanced, floral. Palate: full-bodied, rich, long, great length.

BODEGAS ALBAMAR
O Adro, 11 - Castrelo
36639 Cambados (Pontevedra)
☎: +34 660 292 750
xurxoalbamar@gmail.com

Ceibo 2021 B
91
Colour: bright straw. Nose: ripe fruit, fragrant herbs, fine lees. Palate: full-bodied, rich, balanced.

BODEGAS AVANCIA
Parque Empresarial a Raña, 7 Parcela 135-136
32300 O Barco de Valdeorras (Ourense/Orense)
☎: +34 952 504 706
pedidos@jorgeordonez.es
www.jorgeordonez.es

Avancia Cuvee de O Godello 2021 B
100% godello
92
Colour: bright straw. Nose: fruit expression, ripe fruit, floral, stone fruit. Palate: flavourful, fresh, good acidity, fruity aftertaste.

Avancia Cuvée de O Mencía 2021 T
100% mencía
91
Colour: bright cherry. Nose: balsamic herbs, sweet spices, scrubland, ripe fruit, red berry notes. Palate: spicy, balsamic, good acidity.

Avancia Godello 2021 B
100% godello
93
Colour: bright yellow. Nose: dried herbs, faded flowers, ripe fruit, stone fruit, characterful. Palate: powerful, ripe fruit, balanced.

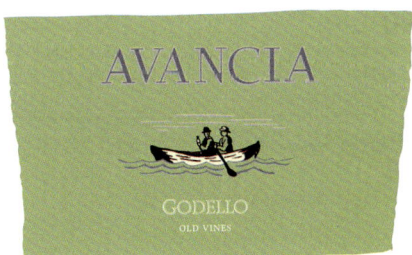

Avancia Nobleza Godello 2020 B
100% godello
94
Colour: bright yellow. Nose: dried flowers, candied fruit, fine lees, pattiserie. Palate: round, spicy, long, great length.

DO VALDEORRAS / D.O.P.

DO VALDEORRAS / D.O.P.

Avancia Nobleza Mencía 2020 T
90% mencía, 10% garnacha tintorera, mouratón, gran negro, sousón, cariñena

92

Colour: deep cherry. Nose: dried herbs, creamy oak, sweet spices, overripe fruit. Palate: powerful, ripe fruit, spicy, round tannins.

Avancia Nobleza Mencía 2021 T
90% mencía, 10% garnacha tintorera, mouratón, gran negro, sousón, cariñena

93

Colour: deep cherry. Nose: dried herbs, creamy oak, scrubland, balsamic herbs, ripe fruit, red berry notes. Palate: powerful, ripe fruit, spicy, round tannins.

BODEGAS D'BERNA
Corgomo, s/n
32340 Villamartín de Valdeorras (Ourense/Orense)
☎: +34 667 435 778
Fax: +34 988 324 557
info@bodegasdberna.com
www.bodegasdberna.com

D'Berna 2021 RD
87

D'Berna Garnacha Tintorera "ele" 2017 T
garnacha tintorera

89

Fruity, jammy, sweety finish, toasty.

D'Berna Godello 2021 B
godello

90

Colour: bright straw. Nose: stone fruit, dried herbs, dried flowers. Palate: fruity, flavourful, balanced, fresh.

D'Berna Mencía 2020 T
90

Colour: cherry, purple rim. Nose: ripe fruit, black fruit, scrubland, spicy. Palate: fruity, flavourful, fresh, balanced, round tannins.

D'Berna Souson Barrica "Juan" 2017 T
89

Spicy, crisp, fruity, toasty.

BODEGAS GODEVAL
32317 O Barco de Valdeorras (Ourense/Orense)
☎: +34 988 108 282
Fax: +34 988 325 309
godeval@godeval.com
www.godeval.com

Godeval 1986 2018 B
100% godello

93

Colour: bright yellow. Nose: ripe fruit, fragrant herbs, fine lees, faded flowers, dried flowers, dry nuts. Palate: full-bodied, rich, long, good acidity.

Godeval 2021 B
100% godello

91

Colour: bright straw. Nose: ripe fruit, wild herbs, faded flowers, mineral. Palate: fruity, flavourful, balanced, fresh.

Godeval Cepas Vellas 2020 B
100% godello

92

Colour: yellow. Nose: fine lees, brioche, dried flowers, balanced, expressive. Palate: fruity, good acidity.

Godeval Cepas Vellas 2021 B
100% godello

92

Colour: bright straw. Nose: ripe fruit, fine lees, wild herbs, floral. Palate: fruity, flavourful, full-bodied, great length.

Godeval Revival 2020 B
100% godello

92

Colour: bright straw. Nose: fresh fruit, citrus fruit, wild herbs, fresh, expressive. Palate: fruity, good acidity, fine bitter notes.

Godeval Revival 2021 B
100% godello

93

Defined aromas, citrus fruit, racy. Colour: straw. Nose: neat, fresh fruit. Palate: fruity, fresh, varietal, long.

BODEGAS SAMPAYOLO

Ctra. de Barxela, s/n
32356 Petín de Valdeorras (Ourense/Orense)
☎: +34 679 157 977
info@sampayolo.com
www.sampayolo.com

Garnacha Vella da Chaira do Ramiriño 2019 T
90
Colour: cherry, garnet rim. Nose: black fruit, fruit preserve, dried herbs, sweet spices, wild herbs. Palate: flavourful, round tannins.

Sampayolo Garnacha Tintorera 2020 T
89
Spicy, jammy, wild, toasty, tasty, standard.

Sampayolo Godello en Lágrimas de los Bancales de Olivedo 2021 B
godello
92
Colour: bright straw, greenish rim. Nose: fresh fruit, citrus fruit, wild herbs, dry stone. Palate: fresh, good acidity, fine bitter notes, flavourful, saline, great length.

Sampayolo Godello sobre Lías 2021 B
godello
90
Colour: straw. Nose: ripe fruit, dried herbs, faded flowers. Palate: powerful, ripe fruit, balanced.

Sampayolo Mencía 2021 T
mencía
88
Jammy, fruity, tasty, balanced, herbal.

CAMINO DE CABRAS

Hermanos Maristas, 27
36700 Tui (Pontevedra)
☎: +34 698 145 790
info@caminodecabras.com
www.caminodecabras.com

Camino de Cabras Godello 2021 B
100% godello
90
Colour: bright straw. Nose: white fruit, stone fruit, white flowers, dried herbs. Palate: fresh, fruity, flavourful, powerful.

Camino de Cabras Mencía 2020 T
100% mencía
88
Jammy, tasty, wild.

COMPAÑÍA DE VINOS TELMO RODRÍGUEZ

El Monte
01308 Lanciego (Araba/Álava)
☎: +34 945 628 315
contact@telmorodriguez.com
www.telmorodriguez.com

🏆 **PODIUM**

As Caborcas 2019 T
95
Colour: bright cherry. Nose: balsamic herbs, sweet spices, scrubland, fruit liqueur notes, red berry notes. Palate: spicy, balsamic, good acidity.

Branco de Santa Cruz 2019 B
godello
93
Colour: bright straw. Nose: ripe fruit, dried herbs, faded flowers, stone fruit, burnt matches. Palate: powerful, ripe fruit, balanced.

🏆 **PODIUM**

Falcoeira 2019 T
97
Complex, aromatic, with personality. Colour: bright cherry. Nose: complex, expressive, spicy, mineral, red berry notes, scrubland. Palate: elegant, full-bodied, long, great length.

Falcoeira Branco 2019 B
92
Colour: straw. Nose: ripe fruit, dried herbs, faded flowers, toasted bread, hint of anise, dry stone. Palate: powerful, ripe fruit, balanced, full-bodied.

Gaba do Xil Godello 2021 B
90
Balanced. Colour: bright straw. Nose: fruit expression, ripe fruit, floral. Palate: flavourful, fresh, good acidity, fruity aftertaste.

O Diviso 2019 T
94
Colour: bright cherry. Nose: balsamic herbs, sweet spices, scrubland, fruit liqueur notes, red berry notes, wild herbs, faded flowers. Palate: spicy, balsamic, good acidity.

Valbuxan 2019 T
93
Rustic. Colour: Cherry. Nose: balsamic herbs, sweet spices, scrubland, wild herbs, tomato. Palate: spicy, balsamic, good acidity.

DO VALDEORRAS / D.O.P.

DO VALDEORRAS / D.O.P.

ENTRECANALES DOMECQ E HIJOS
Anabel Segura, Ed D4
28108 Madrid (Madrid)
☎: +34 646 698 682
rrpp@entrecanalesdomecq.com
www.bodegadirecta.es

El Aeronauta 2020 B
100% godello
91
Colour: bright straw. Nose: fine lees, wild herbs, balanced. Palate: full-bodied, rich, long, good acidity.

La Poda Godello 2021 B
100% godello
89
Defined aromas, fruity, floral, varietally correct, tasty.

La Poda Mencía 2020 T
100% mencía
90
Colour: bright cherry. Nose: fresh fruit, wild herbs, spicy, wild herbs. Palate: fruity, flavourful, powerful, balanced, slightly dry, soft tannins.

HACIENDA UCEDIÑOS
Ucediños, s/n
32300 O Barco de Valdeorras (Ourense/Orense)
☎: +34 686 240 374
info@haciendaucediños.es
www.haciendaucediños.es

Hacienda Ucediños Don Eladio Expresión Mencía 2020 T RB
89
Fruity, dried herbs, tasty, wild.

Hacienda Ucediños Godello 2021 B
100% godello
89
Fruity, herbal, tasty, crisp.

Hacienda Ucediños Mencía 2020 T
mencía
90
Colour: cherry, purple rim. Nose: black fruit, ripe fruit, balsamic herbs, spicy. Palate: fruity, flavourful, slightly dry, soft tannins.

JOAQUÍN REBOLLEDO
San Roque, 20
32350 A Rúa (Ourense/Orense)
☎: +34 686 258 036
info@joaquinrebolledo.com
www.joaquinrebolledo.com

Finca Trasdairelas 2020 B
godello
89
Jammy, fruity, floral, balanced. Palate: fruity, flavourful, balanced, correct.

Joaquín Rebolledo 2020 T BA
90
Colour: deep cherry. Nose: dried herbs, creamy oak, black fruit. Palate: ripe fruit, spicy, round tannins.

Joaquín Rebolledo Godello 2021 B
godello
89
Pleasant, aromatic, fruity, balanced, varietally correct, tasty, standard.

Joaquín Rebolledo Mencía 2021 T
mencía
90
Colour: cherry, purple rim. Nose: red berry notes, floral, spicy. Palate: flavourful, fruity, good acidity, long.

Mª TERESA LÓPEZ FIDALGO (ADEGA O CABALIN)
Rúa Da Fonte, 15
32340 Vilamartín de Valdeorras (Ourense/Orense)
☎: +34 691 782 528
adegaocabalin@gmail.com
www.adegaocabalin.com

A Espedrada 2019 B FB
godello
91
Colour: bright yellow. Nose: fruit expression, stone fruit, wild herbs, spicy, characterful. Palate: fruity, flavourful, fresh, balanced, toasty.

O Cabalin 2019 T C
mencía, garnacha tintorera, merenzao, otras
91
Colour: deep cherry. Nose: fruit expression, ripe fruit, dried herbs, wild herbs, spicy, creamy oak. Palate: flavourful, fruity, good structure, slightly dry, soft tannins.

Viladequinta 2019 T C
mencía, merenzao, otras
91
Colour: deep cherry. Nose: ripe fruit, dried herbs, spicy, faded flowers. Palate: fruity, flavourful, balanced, good structure.

MANUEL CORZO RODRÍGUEZ
Chandoiro s/n
32372 O Bolo (Ourense/Orense)
☎: +34 689 978 094
info@manuelcorzo.es
www.manuelcorzo.es

Viña Corzo Godello 2021 B
godello
88
Sweety finish, fruity, jammy, tasty.

Viña Corzo Mencía 2021 T
mencía
89
Pleasant, fruity, jammy, tasty.

O LUAR DO SIL
Ctra. Petin - Seadur, Pol. 7, Parc. 569
32358 Seadur (Ourense/Orense)
☎: +34 988 343 661
bodega@pagodeloscapellanes.com
www.pagodeloscapellanes.com

O Luar do Sil Godello 2021 B
100% godello
90
Colour: bright straw. Nose: white fruit, ripe fruit, neat, balanced, tropical fruit. Palate: flavourful, fresh, good acidity.

O Luar do Sil Godello sobre Lías 2020 B
100% godello
92
Colour: yellow. Nose: ripe fruit, fragrant herbs, fine lees, balanced, expressive. Palate: full-bodied, rich, long, good acidity.

O Luar do Sil Vides de Córgomo 2019 B
100% godello
94
Colour: bright yellow. Nose: expressive, balanced, complex, neat, dried flowers, fine lees. Palate: rich, full-bodied, long, complex.

PABLO VIDAL - VINOS CON PERSONALIDAD
Rúa do Miradoiro 8
32004 Ourense/Orense (Ourense/Orense)
☎: +34 609 152 251
pablovidal@vinosconpersonalidad.com
www.vinosconpersonalidad.com

Maldito 2019 T
20% garnacha tintorera, 60% mencía, 20% brancellao, caíño longo, sousón
91
Colour: deep cherry. Nose: ripe fruit, dried herbs, creamy oak, balsamic herbs, scrubland. Palate: powerful, ripe fruit, spicy, round tannins.

PACO & LOLA
Valdamor, 18 - Xil
36968 Meaño (Pontevedra)
☎: +34 986 747 779
comercial@pacolola.com
www.pacolola.com

Paco & Lola Godello 2021 B
100% godello
89
Fruity, dried herbs, simple, crisp.

PEKADO MORTAL
Rua Valdomiño, 35 Finca Fierro
32002 Barbadas (Ourense/Orense)
☎: +34 640 209 584
martinpadin9@hotmail.com

Pekado Mortal Godello 2020 B
godello
90
Colour: bright golden. Nose: ripe fruit, fruit expression, wild herbs, dried herbs, spicy. Palate: fruity, flavourful, balanced, round, fine bitter notes.

RAFAEL PALACIOS
Avda. de Somoza, 22
32350 A Rúa (Ourense/Orense)
☎: +34 988 357 122
Fax: +34 988 310 643
bodega@rafaelpalacios.com
www.rafaelpalacios.com

🏆 PODIUM

As Sortes Val do Bibei 2020 B
100% godello
95
Colour: straw. Nose: ripe fruit, dried herbs, fine lees, dry stone, white flowers. Palate: powerful, ripe fruit, balanced, full-bodied, mineral.

DO VALDEORRAS / D.O.P.

DO VALDEORRAS / D.O.P.

Louro do Bolo Godello 2021 B
godello

93

Colour: straw. Nose: dried herbs, faded flowers, stone fruit, fine lees. Palate: powerful, ripe fruit, balanced, great length.

🏆 **PODIUM**

Sorte Antiga 2020 B
100% godello

95

Taut. Colour: straw. Nose: dried herbs, faded flowers, citrus fruit, white fruit. Palate: powerful, ripe fruit, balanced, elegant.

🏆 **PODIUM**

Sorte O Soro 2020 B
godello

99

Taut, mineral. Colour: bright straw. Nose: expressive, ripe fruit, floral, fine lees, mineral, wild herbs, sweet spices. Palate: full-bodied, complex, spicy, long, elegant, saline.

SANTA MARTA

Ctra. San Vicente, Km. 1.2
32348 Vilamartín de Valdeorras (Ourense/Orense)
☎: +34 988 324 559
Fax: +34 988 324 559
enologo@vinaredo.com
www.vinaredo.com

Fardelas de Viñaredo 2020 B
100% godello

91

Colour: bright yellow. Nose: white fruit, stone fruit, white flowers, mineral. Palate: fresh, fruity, good acidity, powerful.

Viñaredo Garnacha Tintorera 2018 T BA
100% garnacha tintorera

90

Colour: deep cherry. Nose: ripe fruit, black fruit, wild herbs, spicy. Palate: fresh, fruity, slightly dry, soft tannins.

Viñaredo Godello 2021 B
100% godello

90

Colour: bright straw. Nose: white fruit, ripe fruit, dried herbs, white flowers. Palate: fruity, flavourful, balanced, mineral, fine bitter notes.

Viñaredo Godello Barrica 2021 B
100% godello

89

Spicy, fruity, dried flowers, dried herbs, tasty.

Viñaredo Mencía 2021 T
85% mencía, 15% sousón

89

Fruity, jammy, sweety finish, spicy, crisp.

Viñaredo Sousón 2019 T BA
100% sousón

89

Fruity, jammy, tasty, balanced, crisp.

TERRIÑA

Estrada de Carballal s/n
32356 Petín de Valdeorras (Ourense/Orense)
☎: +34 982 454 005
marketing@mendezrojo.com
www.mendezrojo.com

Mil Ríos Garnacha 2019 T C
garnacha tintorera

91

Colour: cherry, purple rim. Nose: ripe fruit, red berry notes, black fruit, balsamic herbs, spicy. Palate: flavourful, fresh, fruity, balanced, round tannins.

Mil Ríos Godello 2019 B BA
100% godello

89

Slight oxidation, crisp, tasty, spicy, mineral.

Mil Ríos Godello 2020 B
100% godello

90

Colour: bright straw. Nose: ripe fruit, fragrant herbs, fine lees. Palate: rich, long, good acidity, balanced.

Mil Ríos Mencía 2020 T
100% mencía

90

Colour: deep cherry. Nose: ripe fruit, dried herbs, creamy oak. Palate: powerful, ripe fruit, spicy, round tannins.

Verdes Castros Godello 2021 B
godello

87

Verdes Castros Mencía 2020 T
100% mencía

88

Fruity, herbal, simple.

VALDESIL
Ctra. a San Vicente OU 807, km. 3
32348 Vilamartín de Valdeorras (Ourense/Orense)
☎: +34 988 337 900
raul.prada@valdesil.com
www.valdesil.com

Asadoira 2019 B
100% godello

93

Colour: bright straw. Nose: fragrant herbs, fine lees, white fruit, mineral. Palate: full-bodied, rich, long, good acidity, powerful.

🏆 PODIUM

Pedrouzos Magnum 2012 B FB
100% godello

96

Colour: bright yellow. Nose: candied fruit, fine lees, pattiserie, chamomile, caramel, dry nuts. Palate: round, spicy, long, great length, elegant.

Pezas da Portela 2019 B FB
100% godello

94

Taut. Colour: bright straw. Nose: expressive, ripe fruit, fine lees, mineral, toasted bread. Palate: full-bodied, complex, spicy, long.

Valdesil Godello sobre Lías 2020 B
100% godello

92

Colour: bright straw. Nose: ripe fruit, fragrant herbs, fine lees, dry stone. Palate: full-bodied, rich, long, good acidity.

Valdesil Parcela O Chao 2019 B FB
100% godello

94

Colour: bright straw. Nose: expressive, ripe fruit, floral, fine lees. Palate: full-bodied, spicy, long, elegant.

VINTAE / ATLANTIS
Av Florencio Delgado Gurriaran, 62
32300 O Barco (Ourense/Orense)
☎: +34 608 302 372
marketing@vintae.com
www.vintae.com

Atlantis Godello 2019 B
godello

91

Colour: bright straw. Nose: ripe fruit, fragrant herbs, fine lees, floral, pattiserie. Palate: full-bodied, rich, long, good acidity.

VIÑA COSTEIRA
Valdepereira, 1
32400 Ribadavia (Ourense/Orense)
☎: +34 988 477 210
marketing@costeira.es
www.costeira.es

Codos de Larouco Godello 2020 B
100% godello

91

Colour: bright yellow. Nose: powerful, ripe fruit, spicy, toasty. Palate: rich, good structure, long, fine bitter notes.

Vía Barrosa Godello 2021 B
godello

88

Aromatic, fruity, smooth, tropical, balanced.

Viña Costeira Mencía 2021 T
mencía, garnacha, sousón

88

Fruity, herbal, simple.

VIÑA SOMOZA
Rua do Pombar s/n
32350 A Rúa (Ourense/Orense)
☎: +34 915 130 180
bodega@vinosomoza.com
www.vinosomoza.com

Alma do Vello Tesouro 2020 T C
100% albarello

93

Colour: Cherry. Nose: balsamic herbs, sweet spices, scrubland, fruit expression, wild herbs. Palate: spicy, balsamic, flavourful, fresh, fruity.

As 2 Ladeiras 2020 B
100% godello

92

Colour: bright straw. Nose: ripe fruit, fine lees, white flowers, dried herbs. Palate: full-bodied, rich, long, good acidity, elegant.

Ededia 2020 B BA
100% godello

92

Elegant. Colour: bright straw. Nose: medium intensity, neat, balanced, dried flowers, wild herbs. Palate: balanced, fine bitter notes, flavourful.

Neno Viña Somoza Godello Sobre Lias 2020 B
100% godello

92

With personality. Colour: bright straw. Nose: fragrant herbs, fine lees, cereal notes, animal reductive notes. Palate: full-bodied, rich, good acidity.

DO VALDEORRAS / D.O.P.

DO VALDEORRAS / D.O.P.

Via XVIII 2020 T
mencía, garnacha tintorera, albarello

91

Colour: Cherry. Nose: balsamic herbs, ripe fruit, scrubland, spicy, wild herbs. Palate: spicy, good acidity, balanced.

Viña Somoza Taté 2020 T C
merenzao, garnacha tintorera, mencía, brancellao

92

Colour: Cherry. Nose: balsamic herbs, sweet spices, scrubland, ripe fruit. Palate: spicy, balsamic, good acidity, flavourful, good finish.

VIRGEN DEL GALIR
Las Escuelas, s/n Entoma
32336 O Barco de Valdeorras (Ourense/Orense)
☎: +34 988 335 600
www.cvne.com/bodegas/virgen-del-galir

A Villeira 2020 T
mencía, garnacha tintorera, brancellao, sousón, merenzao

92

Colour: cherry, purple rim. Nose: ripe fruit, red berry notes, wild herbs, faded flowers. Palate: fruity, flavourful, fresh, balanced.

Los Carismáticos 2019 T
100% merenzao

92

Colour: Cherry. Nose: balsamic herbs, sweet spices, scrubland, ripe fruit. Palate: spicy, balsamic, good acidity.

Maruxa Godello 2021 B
100% godello

88

Standard, crisp, fruity, simple, smooth.

Maruxa Mencía 2021 T
100% mencía

89

Pleasant, fruity, tasty.

Regueirón 2020 B
100% godello

92

Colour: bright straw. Nose: ripe fruit, spicy, toasted bread. Palate: rich, good structure, long, toasty, fine bitter notes.

Sede e Fame As Ermitas 2019 B
palomino

92

With personality. Colour: bright straw. Nose: pungent, fine lees, spicy, white fruit, ripe fruit. Palate: dry, balanced.

Val do Galir Godello 2021 B
100% godello

89

Fruity, balanced, standard, jammy, tasty, simple.

Val do Galir Mencía 2020 T RB
100% mencía

90

Colour: cherry, purple rim. Nose: ripe fruit, wild herbs, scrubland, creamy oak, spicy. Palate: flavourful, fruity, powerful, slightly dry, soft tannins.

DO. VALDEPEÑAS
CONSEJO REGULADOR

Constitución, 23
13300 Valdepeñas (Ciudad Real)
☎: +34 926 322 788 - Fax: +34 926 321 054
@: dovaldepenas@dovaldepenas.es
www.dovaldepenas.es

LOCATION:

On the southern border of the southern plateau, in the province of Ciudad Real. It comprises the municipal districts of Alcubillas, Moral de Calatrava, San Carlos del Valle, Santa Cruz de Mudela, Torrenueva and Valdepeñas and part of Alhambra, Granátula de Calatrava, Montiel and Torre de Juan Abad.

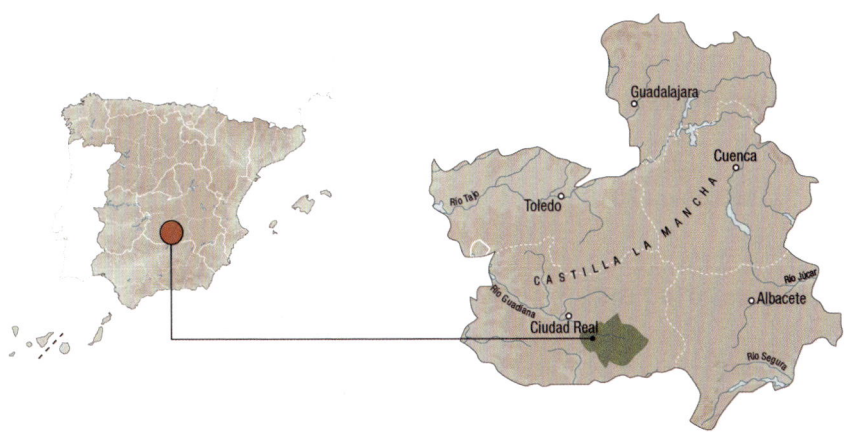

GRAPE VARIETIES:

WHITE: Airén, Macabeo, Chardonnay, Sauvignon Blanc, Moscatel de Grano Menudo and Verdejo.

RED: Cencibel (Tempranillo), Garnacha, Cabernet Sauvignon, Merlot, Syrah and Petit Verdot.

FIGURES:

Vineyard surface: 21,957 – **Wine-Growers:** 2,419 – **Wineries:** 21 – **20 Harvest rating:** N/A – **Production 20:** 82,080,000 litres – **Market percentages:** 70% National - 30% International.

SOIL:

Mainly brownish-red and brownish-grey limestone soil with a high lime content and quite poor in organic matter.

CLIMATE:

Continental in nature, with cold winters, very hot summers and little rainfall, which is usually around 250 and 400 mm per year.

CLASSIFICATION OF YOUNG WINE HARVESTS PEÑÍNGUIDE

2017	2018	2019	2020	2021
GOOD	GOOD	GOOD	GOOD	GOOD

DO VALDEPEÑAS / D.O.P.

BODEGA Y VIÑEDOS CASA DE LA NAVA

Avda. de España s/n
13300 Valdepeñas (Ciudad Real)
☎: +34 630 184 876
bodegacasadelanava@gmail.com
www.casadelanava.es

Media Legua 2021 B
airén
88
Pleasant, fruity, jammy, tropical, balanced, standard, thin.

BODEGAS FERNANDO CASTRO

Paseo Castelar, 70
13730 Santa Cruz de Mudela (Ciudad Real)
☎: +34 926 342 168
info@bodegasfernandocastro.com
www.bodegasfernandocastro.com

Caballo de Oro 2018 T R
tempranillo
84

Castillo Santa Bárbara 2018 T R
tempranillo
85

Finca Los Altos 2016 T GR
tempranillo
84

Finca Los Altos 2018 T R
tempranillo
85

Finca Los Altos Gran Selección 2021 T
tempranillo
86

Raíces 2016 T GR
tempranillo
84

Raíces Syrah 2018 T
syrah
83

BODEGAS MEGÍA E HIJOS -CORCOVO

Magdalena, 33
13300 Valdepeñas (Ciudad Real)
☎: +34 926 347 828
Fax: +34 926 347 829
comercial@corcovo.com
www.corcovo.com

Corcovo Airén 2021 B
airén
87

Corcovo Airen 24 Barricas 2020 B FB
airén
89
Balanced, dried herbs, jammy, yeasty notes, spicy, tasty.

Corcovo Syrah 24 Barricas 2020 T RB
syrah
89
Toasty, smoky, jammy, tasty, great length, spicy.

Corcovo Tempranillo 2017 T R
tempranillo
88
Fruity, sweeties, jammy, tasty, aromatic, balanced.

Corcovo Tempranillo 2019 T C
tempranillo
87

Corcovo Tempranillo 2020 T RB
tempranillo
87

BODEGAS NAVARRO LÓPEZ

Autovía Madrid - Cádiz, Km. 193
13300 Valdepeñas (Ciudad Real)
☎: +34 962 321 888
Fax: +34 926 320 977
info@navarrolopez.com
www.navarrolopez.com

Don Aurelio 2014 T GR
tempranillo
86

Don Aurelio 2015 T R
tempranillo
88
Balanced, spicy, jammy, smooth, fruity.

Don Aurelio 2016 T C
tempranillo
87

DO VALDEPEÑAS / D.O.P.

Don Aurelio 2021 RD
87

Don Aurelio Garnacha 2021 T
garnacha
88
Wild, herbal, fruity, smooth, standard, pleasant.

Don Aurelio Reserva de Familia 2017 T
88
Balanced, smooth, toasty, fruity.

Don Aurelio Syrah 2021 T
syrah
88
Aromatic, fruity, lactic, jammy, tasty, varietally correct.

Don Aurelio Tempranillo 2021 T
tempranillo
87

Don Aurelio Verdejo 2021 B
86

CORRALES ESPINOSA FAMILY WINES
Paraguay, 2
13300 Valdepeñas (Ciudad Real)
☎: 647 442 686
josemcorrales@hotmail.com
www.corralesesinosa.com

José Manuel Corrales 2020 T
tempranillo
91
Colour: dark-red cherry, purple rim. Nose: toasty, spicy, cocoa bean, fruit preserve. Palate: flavourful, toasty, fine bitter notes, fruity.

FÉLIX SOLIS AVANTIS
Autovía del Sur, Km. 199
13300 Valdepeñas (Ciudad Real)
☎: +34 926 322 400
Fax: +34 926 322 417
fsa@felixsolisavantis.com
www.felixsolisavantis.com

Los Molinos Airén Verdejo B
85

Los Molinos Tempranillo T
87

Viña Albali 2015 T GR
tempranillo
88
Classic, toasty, smooth, spicy, thin.

Viña Albali 2017 T R
tempranillo
87

Viña Albali 2018 T C
86

Viña Albali Gran Selección 2021 T
tempranillo
87

Viña Albali Tempranillo 2021 T
tempranillo
86

Viña Albali Verdejo Sauvignon Blanc 2021 B
50% verdejo, 50% sauvignon blanc
85

GRUPO DE BODEGAS VINARTIS
Autovía Madrid-Cádiz, Km. 200
13300 Valdepeñas (Ciudad Real)
☎: +34 926 320 300
atcliente@jgc.es
www.garciacarrion.com

Pata Negra 2016 T R
tempranillo
85

Pata Negra 2019 T C
tempranillo
85

Pata Negra 2020 T RB
tempranillo
85

Pata Negra Cepas Viejas 2016 T R
tempranillo
85

Pata Negra Tempranillo Cabernet Sauvignon 2020 T
tempranillo
84

Señorío de los Llanos 2016 T R
tempranillo
84

DO VALDEPEÑAS / D.O.P.

MUREDA ALIMENTACIÓN
Ctra. N-IV, Km. 184,1
13300 Valdepeñas (Ciudad Real)
☎: +34 926 318 058
ehorcajada@mureda.es
www.mureda.es

Mureda Cuvée BE R BN
airén
86 🌱

VINÍCOLA DE VALDEPEÑAS
Autovía Madrid - Andalucía, km. 198,3
13300 Valdepeñas (Ciudad Real)
☎: +34 926 347 074
coovival@gmail.com
www.coovival.com

Concejal Airén 2021 B
airén
85

Concejal Tempranillo 2018 T C
tempranillo
88
Pruney, powerful, tasty, spicy.

Concejal Tempranillo 2020 T
tempranillo
85

Concejal Verdejo 2021 B
verdejo
85

DO. VALENCIA
CONSEJO REGULADOR

Quart, 22
46001 Valencia
☎: +34 963 910 096 - Fax: +34 963 910 029
@: info@vinovalencia.org
www.vinovalencia.org

LOCATION:

In the province of Valencia. It comprises 66 municipal districts in 4 different sub-regions: Alto Turia, Moscatel de Valencia, Valentino and Clariano.

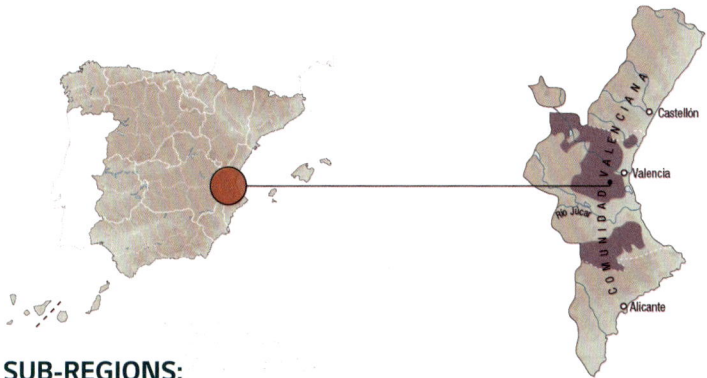

SUB-REGIONS:

There are four in total: Alto Turia, the highest sub-region (700 to 800 m above sea level) comprising 6 municipal districts; **Valentino** (23 municipal districts), in the centre of the province; the altitude varies between 250 m and 650 m; **Moscatel de Valencia** (9 municipal districts), also in the central region where the historical wine from the region is produced; and **Clariano** (33 municipal districts), to the south, at an altitude of between 400 m and 650 m.

GRAPE VARIETIES:

WHITE: Macabeo, Malvasía, Merseguera, Moscatel de Alejandría, Moscatel de Grano Menudo, Pedro Ximénez, Plantafina, Plantanova, Tortosí, Verdil, Chardonnay, Semillon Blanc, Sauvignon Blanc, Verdejo, Riesling, Viognier, Gewüztraminer, Albariño and Garnacha Blanca.

RED: Garnacha, Monastrell, Tempranillo, Tintorera, Forcallat Tinta, Bobal, Cabernet Sauvignon, Merlot, Pinot Noir, Syrah, Graciano, Malbec, Mandó, Marselan, Mencía, Merlot, Mazuelo, Petit Verdot, Miguel Arco, Mandó an Marselán.

FIGURES:

Vineyard surface: 13,069 – **Wine-Growers:** 6,150 – **Wineries:** 85 – **2021 Harvest rating:** Very Good – **Production 21:** 41,992,800 litres – **Market percentages:** 36% National - 64% International.

SOIL:

Mostly brownish-grey with limestone content; there are no drainage problems.

CLIMATE:

Mediterranean, marked by strong storms and downpours in summer and autumn. The average annual temperature is 15°C and the average annual rainfall is 500 mm.

CLASSIFICATION OF YOUNG WINE HARVESTS PEÑÍNGUIDE

2017	2018	2019	2020	2021
GOOD	GOOD	VERY GOOD	VERY GOOD	VERY GOOD

DO VALENCIA / D.O.P.

ANECOOP BODEGAS
Monforte, 1 – entlo.
46010 València/Valencia (València/Valencia)
☎: +34 963 938 500
Fax: +34 963 938 543
anecoopbodegas@anecoop.com
www.anecoopbodegas.com

Amatista 11.5 B MO D
verdosilla, moscatel de alejandría
87

El Enhebro 2020 T
garnacha tintorera, monastrell
87

El Enhebro 2021 B
50% verdil, 50% merseguera
87

Juan de Juanes Vendimia Bronce 2021 T
garnacha, tempranillo, syrah
87

Los Escribanos 2019 T
monastrell, garnacha tintorera
91
Colour: bright cherry. Nose: red berry notes, black fruit, spicy, dried flowers. Palate: good acidity, spicy, fine tannins.

Reymos Selección BE MO D
moscatel de alejandría
86

Reymos Versión Libre BE BN
moscatel de alejandría
86

Sol de Reymos B Mistela D
moscatel de alejandría
90
Colour: bright yellow. Nose: powerful, creamy oak, ripe fruit, spicy, dry nuts. Palate: rich, good structure, toasty, fine bitter notes.

Venta del Puerto Nº 12 2019 T BA
cabernet sauvignon, tempranillo, merlot, syrah
90
Colour: deep cherry. Nose: dried herbs, creamy oak, black fruit, spicy. Palate: ripe fruit, spicy, round tannins.

Venta del Puerto Nº 18 2017 T BA
cabernet sauvignon, tempranillo, merlot, syrah
90
Colour: cherry, garnet rim. Nose: fruit preserve, powerful, scrubland, dried herbs, dark chocolate. Palate: flavourful, long, spicy.

Vida Viña Tendida Moscato Bianco B MO D
moscatel de alejandría
85

BALDOVAR 923
Baldovar 17
46178 Alpuente (València/Valencia)
☎: +34 962 121 929
bodega@baldovar923.es
www.baldovar923.es

Arquela 2020 B FB
merseguera
91
Little interventionist, slight fizz. Colour: bright yellow. Nose: ripe fruit, spicy, cereal notes, toasted bread, dried herbs. Palate: good structure, toasty, fine bitter notes.

Rascaña 2020 B
merseguera, macabeo
92
Little interventionist. Colour: straw. Nose: ripe fruit, dried herbs, faded flowers. Palate: powerful, ripe fruit, balanced, full-bodied.

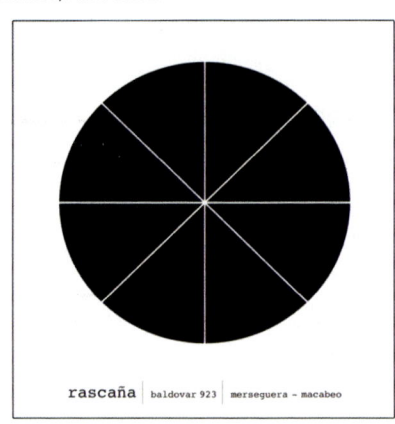

rascaña baldovar 923 merseguera – macabeo

Berandía 2019 T
bobal
90
Colour: cherry, purple rim. Nose: black fruit, red berry notes, spicy, smoky. Palate: fruity, spicy, smoky aftertaste.

Cañada París 2020 B C
merseguera
90
Colour: straw. Nose: dried herbs, faded flowers, white fruit, ripe fruit, dry stone. Palate: ripe fruit, balanced, rich, flavourful.

Cerro Negro 2020 T
mencía
90
Colour: cherry, purple rim. Nose: fine reductive notes, red berry notes, wild herbs, spicy, floral. Palate: fruity, flavourful, balanced.

Pieza la Moza 2020 RD
tempranillo, bobal
89
Dried flowers, weary, fruity, jammy, tasty, wild, with personality.

BODEGA AMANOVO
Finca Santa Rosa, Ctra. Fontanars, CV-660, K. 24.5
46870 Ontinyent (València/Valencia)
info@besvinos.com

Amanovo Verdejo Edición Especial 2020 B RB
verdejo
88
Spicy, toasty, jammy, smoky, great length.

BODEGA ARANLEÓN
Ctra. Caudete, 3
46310 Los Marcos (València/Valencia)
☎: +34 963 631 640
Fax: +34 962 185 150
vinos@aranleon.com
www.aranleon.com

Aranleón Sólo 2021 B
80% macabeo, 20% sauvignon blanc
89 🌱
Balanced, aromatic, yeasty notes, dried herbs, wild, jammy.

Blés 2020 T RB
bobal, tempranillo
86 🌱

Blés 2021 B
macabeo, sauvignon blanc
88 🌱
Standard, yeasty notes, pleasant, jammy, tasty, citrus fruit.

Blés Crianza de Aranleón 2019 T C
monastrell, tempranillo, cabernet sauvignon
88 🌱
Toasty, jammy, spicy, kind finish.

El Árbol de Aranleón 2019 T C
monastrell, bobal, tempranillo, cabernet sauvignon
88 🌱
Pruney, spicy, fruity, toasty.

BODEGA DE MOYA
Avda. de Las Bodegas, 14
46300 Utiel (València/Valencia)
☎: +34 665 330 991
yves@demoya.es
www.demoya.es

Gloria 2020 T
monastrell
89
Kind finish, jammy, pruney, dried herbs, spicy, tasty.

Julia 2019 T
100% monastrell
90
Colour: cherry, garnet rim. Nose: fruit preserve, fruit liqueur notes, powerful. Palate: flavourful, sweetness, long.

Tibó 2021 B
70% merseguera, 20% gewürztraminer, 10% moscatel
87

BODEGA EL ANGOSTO
Finca Sta. Rosa, Ctra. Fontanars CV-660, Km. 24.5
46870 Ontinyent (València/Valencia)
☎: +34 962 380 638
info@bodegaelangosto.com
www.bodegaelangosto.com

Almendros 2021 B
verdejo, sauvignon blanc, riesling, chardonnay
91
Colour: bright yellow. Nose: creamy oak, ripe fruit, spicy, white fruit, honeyed notes. Palate: good structure, toasty, fine bitter notes, full-bodied.

El Jefe de la Tribu 2019 T BA
bobal, malbec, syrah
92
Colour: cherry, garnet rim. Nose: overripe fruit, creamy oak, hot. Palate: pruney, powerful, sweet tannins.

Teuladí 2021 T RB
cabernet franc, garnacha, syrah
86

Vereda 2019 T BA
monastrell
89
Balanced, spicy, herbal, jammy, tasty.

DO VALENCIA / D.O.P.

DO VALENCIA / D.O.P.

BODEGA REYNA-NEPEFE
Finca El Churro
46357 La Portera (València/Valencia)
☎: +34 617 927 713
nepefesl@gmail.com

Calcetas 2021 B
viognier, sauvignon blanc

89 🌱
Jammy, dried herbs, dried flowers, full-bodied.

BODEGA SANT PERE DE MOIXENT
Plaza de la Hispanidad 4
46640 Moixent (València/Valencia)
☎: +34 962 260 020
info@coopmoixent.com
www.coopmoixent.com

Sant Pere 2020 T
monastrell, tempranillo, merlot

88
Spicy, fruity, dried herbs, tasty.

Sant Pere Blanc 2020 B
macabeo, malvasía, merseguera

89
Pleasant, dried flowers, balanced, fruity, jammy, age nuances.

Sant Pere Vinyes Velles Blanc 2019 B
pedro ximénez, macabeo, malvasía

90
Colour: bright straw. Nose: ripe fruit, fine lees, sweet spices, faded flowers. Palate: rich, good acidity, fine bitter notes, spicy.

Sant Pere Vinyes Velles Negre 2017 T
70% monastrell, 30% cariñena

90
Colour: light cherry. Nose: red berry notes, ripe fruit, wild herbs, spicy. Palate: fruity, fresh, balanced, slightly dry, soft tannins.

BODEGAS ARRÁEZ
Pol. 6 Parcela 386 Paraje Ciscar
46630 La Font de la Figuera (València/Valencia)
☎: +34 962 290 031
info@bodegasarraez.com
www.bodegasarraez.com

Los Arráez Lagares 2019 T RB
60% monastrell, 40% cabernet sauvignon

90
Colour: cherry, purple rim. Nose: ripe fruit, black fruit, balsamic herbs, spicy. Palate: fruity, flavourful, balanced, slightly dry, soft tannins.

Los Arráez Malvasía 2021 B RB
100% malvasía

90
Colour: bright straw. Nose: white fruit, floral, citrus fruit, expressive. Palate: long, good acidity, easy to drink, fine bitter notes.

Los Arráez Parcela 0 2018 T RB
40% garnacha tintorera, 30% monastrell, 15% cabernet sauvignon, 10% arco, 5% forcallat

90
Colour: cherry, purple rim. Nose: black fruit, ripe fruit, spicy, dried herbs. Palate: fruity, balanced, powerful tannins, good finish.

Los Arráez Verdil 2021 B
100% verdil

89
Balsamic herbs, fruity, citrus fruit, pleasant, thin. Palate: easy to drink, good acidity.

Mala Vida 2020 T RB
30% monastrell, 30% tempranillo, 20% syrah, 20% cabernet sauvignon

88
Spicy, jammy, tasty, powerful, toasty.

Mala Vida 2021 B

87

Mala Vida Edición Limitada 2020 T RB
100% monastrell

90
Colour: cherry, purple rim. Nose: black fruit, fruit preserve, spicy, toasty. Palate: fruity, balanced, slightly dry, soft tannins.

BODEGAS EL VILLAR

Avda. Agricultor, 1
46170 Villar del Arzobispo (València/Valencia)
☎: +34 962 720 050
Fax: +34 961 646 060
exportacion@elvillar.com
www.elvillar.com

Cantalares Garnacha 2021 RD
garnacha
88
Balanced, dried flowers, fruity, simple, wild, pleasant.

Cantalares Merlot 2020 T
merlot
86

Cantalares Merseguera 2021 B
merseguera
86

Viña Villar 2018 T C
tempranillo, merlot
83

BODEGAS ENGUERA

Ctra. CV - 590, Km. 51,5
46810 Enguera (València/Valencia)
☎: +34 962 224 318
oficina@bodegasenguera.com
www.bodegasenguera.com

Aliats 2021 B
100% verdil
88 🌿
Smoky, jammy, toasty, tasty, fruity. Nose: stone fruit.

Aliats 2021 T
50% monastrell, 50% marselan
89 🌿
Toasty, spicy, pruney, herbal, jammy.

Blanc d'Enguera Original 2021 B
70% verdil, 10% viognier, 10% sauvignon blanc, 10% chardonnay
89 🌿
Toasty, smoky, jammy, fruity, floral, aromatic, simple.

Megala 2019 T
60% monastrell, 20% tempranillo, 20% marselan
92 🌿
Colour: cherry, purple rim. Nose: black fruit, ripe fruit, fruit expression, balsamic herbs, smoky, spicy. Palate: fruity, balanced, spicy, round tannins, fine bitter notes.

Paradigma 2018 T BA
100% monastrell
90 🌿
Age nuances. Colour: Cherry, orangey edge. Nose: slightly evolved, fruit preserve, wet leather, spicy. Palate: fruity, flavourful, balanced, smoky aftertaste.

Sueño de Megala 2018 T BA
marselan
92 🌿
Nose: ripe fruit, dried herbs, spicy, smoky. Palate: fruity, flavourful, balanced, smoky aftertaste, round tannins.

BODEGAS GODELLETA

Mayor, 79
46388 Godelleta (València/Valencia)
☎: +34 961 800 017
bodega@godelleta.com
www.bodegagodelleta.com

Escala I Corda 2020 BF Mistela D
moscatel de alejandría
87

Silencio, es Música 2020 B
moscatel de alejandría
88
Aromatic, floral, dried flowers, fruity, smooth, standard.

BODEGAS HISPANO SUIZAS

Ctra. N-322, Km. 451,7
46357 El Pontón (València/Valencia)
☎: +34 962 349 370
maria.salinas@bodegashispanosuizas.com
www.bodegashispanosuizas.com

Bassus Finca Casilla Herrera 2018 T
90
Colour: deep cherry. Nose: dried herbs, creamy oak, black fruit, woody. Palate: powerful, ripe fruit, spicy, round tannins.

Finca Casa Julia 2021 B
90
Colour: bright straw. Nose: ripe fruit, floral, citrus fruit. Palate: flavourful, fruity aftertaste, fine bitter notes, easy to drink.

Impromptu Rosé 2021 RD
90
Colour: raspberry rose. Nose: dried herbs, floral, citrus fruit. Palate: light-bodied, good acidity, flavourful.

DO VALENCIA / D.O.P.

DO VALENCIA / D.O.P.

BODEGAS IRANZO
Ctra. de Madrid, 60
46315 Caudete de las Fuentes (València/Valencia)
☎: +34 962 319 282
comercial@bodegasiranzo.com
www.bodegasiranzo.com

Living Semillon 2020 B
86

Living Tempranillo 2019 T
tempranillo
85

BODEGAS LOS PINOS
Casa Los Pinos, s/n
46635 Fontanars dels Alforins (València/Valencia)
☎: +34 600 584 397
bodegaslospinos@bodegaslospinos.com
www.bodegaslospinos.com

Brote 2021 B
63,86% verdil, 36,14% viognier
89
Aromatic, wild, fruity, dried flowers, dried herbs, tasty.

Brote 2021 RD
garnacha
87

Dx de Dominio Los Pinos 2020 T RB
75% monastrell, 25% cabernet sauvignon
88
Toasty, powerful, pruney, hot.

Evento 2018 T C
50% monastrell, 50% garnacha
87

La Sort 2020 T
garnacha
90
Colour: cherry, purple rim. Nose: red berry notes, floral, spicy, scrubland. Palate: flavourful, fruity, good acidity, long.

Los Pinos 0 % Sulfito 2021 T
50% monastrell, 25% garnacha, 25% syrah
87

BODEGAS LUIS CORBÍ - CLOS COR VÍ
Camí de la Bodega km 1,5
46640 Moixent (València/Valencia)
☎: +34 601 413 967
enologo@closcorvi.com
www.closcorvi.com

Cimera Clos Cor Ví 2019 B C
50% riesling, 50% viognier
92
Colour: bright yellow. Nose: ripe fruit, dried herbs, faded flowers, dried flowers. Palate: powerful, ripe fruit, balanced, fresh, flavourful.

Clos Cor Ví Riesling 2020 B BA
riesling
91
Dried flowers, yeasty notes. Colour: yellow. Nose: ripe fruit, faded flowers, lees reduction notes. Palate: flavourful, long, ripe fruit.

Clos Cor Ví Viognier 2020 B S
viognier
91
Colour: yellow. Nose: dried flowers, faded flowers, dry nuts, fruit preserve. Palate: flavourful, ripe fruit, spicy.

Versat Clos Cor Ví 2021 B
verdil
90
Colour: yellow. Nose: wild herbs, balanced, dried flowers, fine lees. Palate: fruity, balanced, easy to drink.

BODEGAS LUIS CORBÍ - FINCA COR VI
Ctra. Nacional 322, Km 431 Los Isidros
46354 Requena (València/Valencia)
☎: +34 601 413 967
enologo@closcorvi.com
www.fincacorvi.com

Maloco 2020 T
bobal
90
Colour: cherry, purple rim. Nose: fruit preserve, spicy, dried herbs. Palate: fruity, powerful, slightly dry, soft tannins.

BODEGAS MURVIEDRO

Ampliación Pol. El Romeral, s/n
46340 Requena (València/Valencia)
☎: +34 962 329 003
murviedro@murviedro.es
www.murviedro.es

Audentia 2018 T R
40% tempranillo, 40% monastrell, 20% cabernet sauvignon

90

Colour: deep cherry. Nose: ripe fruit, black fruit, spicy, black pepper. Palate: powerful, fruity, toasty, smoky aftertaste, slightly dry, soft tannins.

Cañada La Torre 2021 B RB
100% malvasía

91 ♣

Aromatic, varietally correct. Colour: bright straw. Nose: neat, expressive, floral, fine lees. Palate: flavourful, fruity, balanced, fine bitter notes.

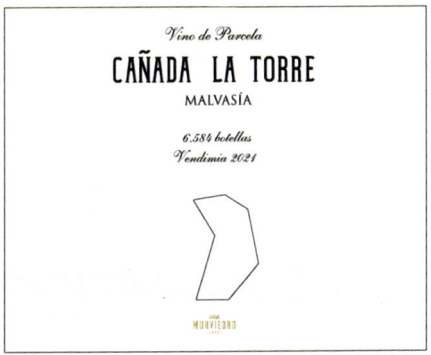

CV05 2020 T
100% cabernet sauvignon

91

Colour: cherry, purple rim. Nose: ripe fruit, black fruit, black pepper, smoky, dried herbs. Palate: flavourful, powerful, balanced, toasty, slightly dry, soft tannins.

Murviedro Colección 2018 T R
40% tempranillo, 40% monastrell, 20% cabernet sauvignon

90

Colour: bright cherry. Nose: ripe fruit, dried herbs, spicy. Palate: flavourful, fruity, balanced, slightly dry, soft tannins.

Murviedro Colección 2019 T C
50% tempranillo, 30% monastrell, 20% syrah

88

Herbal, fruity, spicy, wild.

BODEGAS NODUS

Finca El Renegado, s/n
46315 Caudete de las Fuentes (València/Valencia)
☎: +34 962 174 029
andrea@bodegasnodus.com
www.bodegasnodus.com

El Chaval 2021 RD
bobal

86 ♣

El Chaval 2021 T
bobal

88 ♣

Tasty, powerful, jammy, spicy.

El Chaval Macabeo 2021 B

87 ♣

El Renegado 2021 B
macabeo, moscatel

88 ♣

Aromatic, fruity, jammy, tropical, simple, smooth.

En la Parra 2021 B
chardonnay, moscatel

84 ♣

BODEGAS ONTINIUM

Avda. Almansa, 17
46870 Ontinyent (València/Valencia)
☎: +34 962 380 849
administracion@bodegasontinium.com
www.bodegasontinium.com

Capitan Julián 2020 T RB
100% tempranillo

88

Hot, jammy, toasty, powerful.

Capitan Julián 2021 B
75% chardonnay, 25% moscatel grano menudo

89

Floral, fruity, defined aromas, kind finish, balanced, tasty.

El Tesoro del Capitán 2020 T RB
100% monastrell

87

DO VALENCIA / D.O.P.

DO VALENCIA / D.O.P.

BODEGAS POLO MONLEÓN
Ctra. Valencia - Ademuz, Km. 86
46178 Titaguas (València/Valencia)
☎: +34 659 000 636
info@hoyadelcastillo.com
www.hoyadelcastillo.com

Hoya del Castillo 2021 B
merseguera
87

BODEGAS VEGAMAR
Garcesa, s/n
46175 Calles (València/Valencia)
☎: +34 962 781 443
info@vegamar.es
www.vegamar.es

Altos de la Muela 2019 T C
90
Colour: deep cherry. Nose: dried herbs, creamy oak, black fruit, cocoa bean. Palate: powerful, ripe fruit, spicy, round tannins.

Huella de Merseguera 2021 B
merseguera
89
Pleasant, aromatic, fruity, floral, jammy, balanced.

Huella de Syrah 2021 T
90
Colour: cherry, purple rim. Nose: powerful, ripe fruit, spicy. Palate: ripe fruit, flavourful, good structure.

Vegamar 2019 T C
90
Colour: very deep cherry. Nose: roasted coffee, aromatic coffee, powerful, ripe fruit. Palate: smoky aftertaste, great length, round tannins.

Vegamar 2021 B
89
Pleasant, aromatic, floral, fruity, tasty, simple.

BODEGAS VINIVAL
Ctra. Chiva a Montserrat, Km. 1
30520 Chiva (Valencia)
☎: +34 963 568 750
info@vinival.es
www.vinival.es

Dolçaina BF Mistela D
moscatel
90
Representative. Colour: old gold. Nose: honeyed notes, floral, sweet spices. Palate: powerful, flavourful.

BODEGAS VOLVER
Ctra de Pinoso a Fortuna, s/n
03658 Rodriguillo (Alacant/Alicante)
☎: +34 966 185 624
Fax: +34 965 075 376
export@bodegasvolver.com
www.bodegasvolver.com

Espeto Bobal 2020 T
bobal
88
Pleasant, herbal, fruity, jammy, standard.

BODEGAS Y DESTILERÍAS VIDAL
Pol. Ind. El Mijares. Valencia, 16
12550 Almazora (Castelló/Castellón)
☎: +34 964 503 300
Fax: +34 964 560 604
info@bodegasvidal.com
www.bodegasvidal.com

Moscatel Orgullo Vino de Licor B
moscatel
86

Uva D'Or Moscatel de Licor B D
moscatel
90
Colour: bright yellow. Nose: honeyed notes, floral, sweet spices, expressive. Palate: rich, fruity, powerful, flavourful.

BRUNO MURCIANO
46315 Caudete de las Fuentes (València/Valencia)
☎: +34 663 924 485
bruno@brunomurciano.com
www.brunomurciano.com

Las Blancas 2021 B
91 ❦
Colour: bright straw. Nose: ripe fruit, dried flowers, dried herbs, wild herbs. Palate: flavourful, fresh, good acidity.

Las Tintas 2021 T
89
Lactic, pruney, fruity, tasty, citrus fruit.

Pieles Doradas 2021 B
91 ❦
Colour: straw. Nose: expressive, white flowers, dried herbs. Palate: flavourful, fruity, balanced, good acidity, fine bitter notes.

CARMELITANO BODEGAS Y DESTILERÍAS
Bodolz, 12
12560 Benicassim (Castelló/Castellón)
☎: +34 964 300 849
carmelitano@carmelitano.com
www.carmelitano.com

Carmelitano Moscatel 2019 B MO D
moscatel de alejandría

88

Citrus fruit, creamy, floral, full-bodied, tasty.

CASA LO ALTO BODEGA Y VIÑEDOS
Crta. Caudete-Los Isidros, CV452
46310 Venta del Moro (València/Valencia)
☎: +34 962 329 003
info@casaloalto.com
www.casaloalto.com

Manzán 2020 T C
100% bobal

92

Colour: deep cherry. Nose: ripe fruit, dried herbs, creamy oak. Palate: powerful, ripe fruit, spicy, round tannins.

Rocha 2020 T
100% garnacha

90

Pleasant, fruity, jammy.

CASA LOS FRAILES
Casa Los Frailes, s/n
46635 Fontanars dels Alforins (València/Valencia)
☎: +34 962 222 220
info@bodegaslosfrailes.com
www.casalosfrailes.es

1771 Casa Los Frailes 2018 T C

93

Colour: Cherry. Nose: complex, expressive, spicy, mineral, earthy notes. Palate: elegant, full-bodied, long, great length.

Blanc de Trilogía 2021 B BA
60% sauvignon blanc, 20% moscatel, 20% verdil

89

Citrus fruit, herbal, balanced, crisp.

Los Frailes Caliza 2020 T
100% monastrell

92

Colour: deep cherry. Nose: ripe fruit, dried herbs, creamy oak, earthy notes, mineral. Palate: powerful, ripe fruit, spicy, round tannins.

Los Frailes Dolomitas 2020 T
100% monastrell

90

Colour: cherry, garnet rim. Nose: fruit preserve, fruit liqueur notes, powerful. Palate: flavourful, sweetness, long.

Los Frailes Rubificado 2020 T
100% garnacha

92

Colour: cherry, purple rim. Nose: red berry notes, floral, ripe fruit, sweet spices. Palate: flavourful, fruity, good acidity, long.

Trilogía 2018 T C
50% monastrell, 40% cabernet sauvignon, 10% tempranillo

91

Colour: dark-red cherry. Nose: toasty, spicy, cocoa bean. Palate: flavourful, toasty, fine bitter notes.

CELLER CATARUZ
Albacete, 48-11º
46007 València/Valencia (València/Valencia)
☎: +34 678 513 800
coque@cellercataruz.com
www.cellercataruz.com

Il.Lusiona't 2021 B
80% sauvignon blanc, 20% viognier

88

Defined aromas, floral, balanced, tasty, jammy.

Il.Lusiona't Rosé 2021 RD
90% marselan, 10% tempranillo

85

Malcriat 2020 T
80% garnacha tintorera, 20% merlot

89

Pleasant, kind finish, fruity, tasty.

Maneras de Vivir 2017 T
50% bobal, 50% syrah

86

Melic 2018 T
50% bobal, 30% cabernet sauvignon, 20% merlot

88

Smoky, toasty, jammy, tasty.

Xtrmo (Extremo) 2019 B FB
50% verdejo, 50% viognier

90

Colour: bright yellow. Nose: ripe fruit, spicy, stone fruit. Palate: rich, long, toasty, fine bitter notes.

DO VALENCIA / D.O.P.

CELLER DEL ROURE

Ctra. de Les Alcusses, Km. 11,1
46640 Moixent (València/Valencia)
☎: +34 962 295 020
info@cellerdelroure.es
www.alcusses.es

Cullerot 2021 B
30% pedro ximénez, 30% macabeo, 10% verdil, 10% merseguera, 10% malvasía, 10% chardonnay

90

Colour: bright straw. Nose: citrus fruit, white fruit, dry stone, dried flowers. Palate: fruity, flavourful, fresh, balanced.

Les Alcusses 2018 T
40% monastrell, 10% garnacha tintorera, 15% petit verdot, 15% syrah, 10% cabernet sauvignon, 10% merlot

90

Colour: light cherry. Nose: fruit preserve, red berry notes, spicy, meaty notes. Palate: fruity, balanced, slightly dry, soft tannins.

Maduresa 2020 T
80% monastrell, 20% cariñena

92

Colour: cherry, purple rim. Nose: black fruit, red berry notes, wild herbs, spicy, scrubland. Palate: fruity, flavourful, balanced, slightly dry, soft tannins.

Parotet 2020 T
70% arco, 30% mandó

91

Colour: cherry, purple rim. Nose: red berry notes, ripe fruit, wild herbs, spicy, dry stone. Palate: fruity, fresh, flavourful, balanced, good finish, slightly dry, soft tannins.

Safrà 2020 T
70% mandó, 30% arco

90

Colour: cherry, purple rim. Nose: red berry notes, wild herbs, grassy. Palate: fruity, light-bodied, flavourful, slightly dry, soft tannins.

Vermell 2020 T
70% garnacha tintorera, 30% mandó

90

Colour: bright cherry. Nose: red berry notes, ripe fruit, grassy, spicy. Palate: fruity, fresh, flavourful.

CLOS DE LÔM

Ctra. CV 655, km 6,8
46870 Ontinyent (València/Valencia)
☎: +34 963 349 777
info@closdelom.wine
www.closdelom.wine

Clos de Lôm Garnacha 2021 T
garnacha

91

Colour: cherry, garnet rim. Nose: fruit preserve, fruit liqueur notes, powerful, expressive. Palate: flavourful, sweetness, long.

Clos de Lôm Malvasía 2021 B
malvasía

89

Citrus fruit, floral, herbal, tasty, crisp.

Clos de Lôm Monastrell 2021 RD
monastrell

91

Colour: raspberry rose. Nose: red berry notes, wild herbs, floral. Palate: elegant, long, good acidity, fruity aftertaste.

Clos de Lôm Tempranillo 2021 T
tempranillo

92

Colour: Cherry. Nose: complex, expressive, spicy, mineral. Palate: elegant, full-bodied, long, great length.

FIL·LOXERA & CÍA.

Josep Renau, 53
46635 Fontanars dels Alforins (València/Valencia)
☎: +34 606 099 599
pilar@filoxeraycia.es

Beberás de la Copa de tu Hermana 2020 B
48% macabeo, 24% verdil, 16% monastrell, 12% subirat parent

90

Colour: bright golden. Nose: expressive, faded flowers, characterful, wild herbs. Palate: flavourful, dry, fine bitter notes.

Bienvenidos al Extraordinario Mundo de la Mujer Caballo mitad Mujer, mitad Caballo (Taronja-Orange) 2020 B C
64% valdihuete, 7% moscatel romano, 17% airén, 12% otras

92

With personality, little interventionist. Colour: old gold. Nose: wild herbs, grassy, characterful, animal reductive notes. Palate: flavourful, long.

Bienvenidos al Extraordinario Mundo de la Mujer Caballo mitad Mujer, mitad Caballo (Verd-Verde) 2020 T C
100% ullet de perdiu

93 🌱

Little interventionist. Colour: deep cherry. Nose: ripe fruit, dried herbs, creamy oak, toasty. Palate: powerful, ripe fruit, spicy, round tannins.

Bienvenidos al Extraordinario Mundo de la Mujer Caballo mitad Mujer, mitad Caballo 2020 T C
100% arco

92 🌱

Colour: very deep cherry. Nose: balsamic herbs, sweet spices, scrubland, ripe fruit. Palate: spicy, balsamic, good acidity.

El Cordero y las Vírgenes 2018 T R
36% garnacha tintorera, 8% graciano, 8% tempranillo, 22% monastrell, 24% garnacha, 2% malvasía

93 🌱

With personality, little interventionist. Colour: dark-red cherry. Nose: fine reductive notes, black fruit, ripe fruit, dried herbs, wild herbs. Palate: good acidity, balanced, easy to drink.

Sentada sobre La Bestia 2019 T BA
33% monastrell, 11% garnacha tintorera, 19% graciano, 27% garnacha, 8% tempranillo, 2% malvasía

89 🌱

Dried herbs, jammy, wild, smooth, great length. Palate: easy to drink.

LA BARONÍA DE TURIS
Avda. D. Bautista Soler Crespo, 22
46389 Turis (València/Valencia)
☎: +34 962 526 011
comercial@baroniadeturis.es
www.baroniadeturis.es

1000 Besos Merlot Selección 2020 T
100% merlot

88

Pruney, sweet, full-bodied, tasty, toasty, herbaceous, hot.

Henri Marc 01 Syrah 2021 T
100% syrah

88

Balanced, spicy, fruity, herbal.

Luna de Mar 2018 T C
merlot, tempranillo, syrah

87

Mistela Moscatel Turís Selección 2021 B D
100% moscatel de alejandría

89

Sweet, floral, fruity, herbal, full-bodied, hot.

Son 2 Días 2021 B

85

LA CASA DE LAS VIDES BODEGUES I VINYES
Corral el Galtero s/n
46890 Agullent (València/Valencia)
☎: +34 606 396 903
bodega@lacasadelasvides.com
www.lacasadelasvides.com

Acvlivs 2019 T C
syrah, tempranillo, merlot

85

CVP 2019 T RB
tempranillo, garnacha, merlot

86

Vallblanca 2021 B SS
gewürztraminer

86

RAFAEL CAMBRA
Casa Colaus, 1 Ctra. Fontanars a Moixent, Km. 1.8
46635 Fontanars dels Alforins (València/Valencia)
☎: +34 636 309 327
rafael@rafaelcambra.es
www.rafaelcambra.es

Casa Labor 2019 T

93

Colour: bright cherry. Nose: balsamic herbs, sweet spices, scrubland, chamomile. Palate: spicy, balsamic, good acidity.

Casa Sosegada 2020 T

91

Colour: cherry, purple rim. Nose: balsamic herbs, red berry notes, scrubland. Palate: balsamic, fruity, balanced.

Forcalla de Antonia 2020 T
forcallat

92

With personality, wild. Colour: light cherry. Nose: ripe fruit, fragrant herbs, characterful, expressive. Palate: balanced, fine bitter notes, easy to drink.

DO VALENCIA / D.O.P.

DO VALENCIA / D.O.P.

Rafael Cambra Dos 2020 T
cabernet sauvignon, cabernet franc, monastrell

92

Colour: cherry, purple rim. Nose: fruit expression, red berry notes, floral, spicy. Palate: flavourful, fruity, good acidity, long.

Rafael Cambra Uno 2020 T
monastrell

91

Colour: cherry, purple rim. Nose: fruit expression, red berry notes, floral, spicy. Palate: flavourful, fruity, good acidity, long.

Soplo 2020 T
90% garnacha, 10% forcallat

91

Colour: Cherry. Nose: balsamic herbs, sweet spices, scrubland. Palate: spicy, balsamic, good acidity.

RISKY GRAPES

El Terrerazo Ctra. N-330 Km 195
46300 Utiel (València/Valencia)
☎: +34 962 168 260
Fax: +34 962 168 259
info@riskygrapes.com
www.riskygrapes.com

Atance Bobal 2020 T
bobal

91 🌱

Colour: bright cherry. Nose: fresh fruit, wild herbs, floral, red berry notes, ripe fruit. Palate: good acidity, spicy, fine tannins.

Atance Cuvee 2021 B
75% merseguera, 25% malvasía

89 🌱

Standard, fruity, jammy, tasty.

SHUKHRAT KHAKIMOV & VITICULTORES

Plz Constitución, 7
03550 San Juan de Alicante (Alacant/Alicante)
☎: +34 965 943 090
Fax: +34 965 943 090
a.nogay@winexfood.com
wwww.winexfood.com

Gaspi Monastrell Gravas 2021 RD
100% monastrell

90

Dried flowers, with personality. Colour: coppery red. Nose: balanced, white fruit, ripe fruit, faded flowers. Palate: fruity, easy to drink.

Gaspi Monastrell Tierra Arenosa 2020 T C
100% monastrell

91

Little interventionist. Nose: animal reductive notes, wild herbs, black fruit, ripe fruit, characterful. Palate: flavourful, varietal, balsamic.

GGaspi Macabeo Tierra Blanca Orange 2021 B
macabeo

89

Herbal, with personality, bitter, balanced, dried herbs, original, simple.

GGaspi Macabeo Tierra Blanca/Roja 2021 B
100% macabeo

89

Dried flowers, yeasty notes, jammy. Palate: flavourful, fruity, balanced.

GGaspi Monastrell Tierra Blanca 2020 T C
100% monastrell

90

Slight oxidation, little interventionist. Colour: light cherry, hazy. Nose: cider notes, red berry notes, wild herbs, hint of anise. Palate: fleshy, fresh, fruity, flavourful.

VALSANGIACOMO

Ctra. Cheste - Godelleta, Km. 1
46370 Chiva (València/Valencia)
☎: +34 649 810 834
cherubino@cherubino.es
www.valsangiacomo.es

Malvasía de Sant Jaume 2021 B
85% malvasía, 15% merseguera

88

Fruity, jammy, dried flowers, simple, bitter, standard.

VICENTE GANDÍA

Ctra. Cheste a Godelleta, s/n
46370 Chiva (València/Valencia)
☎: +34 962 524 242
Fax: +34 962 524 243
info@vicentegandia.com
www.vicentegandia.es

Bahía Alta Afrutado 2021 B SS
moscatel

86

Ceramic Monastrell 2019 T
100% monastrell

92
Colour: dark-red cherry, garnet rim. Nose: ripe fruit, fruit preserve, aged wood nuances, tobacco, sweet spices, earthy notes. Palate: spicy, round tannins, long.

Clos de Gallur 2019 T RB
syrah, tempranillo, cabernet sauvignon

92
Colour: dark-red cherry. Nose: toasty, spicy, cocoa bean, ripe fruit, black fruit. Palate: flavourful, toasty, fine bitter notes.

Con un Par Monastrell Petit Verdot 2020 T BA
monastrell, petit verdot

87

El Miracle Nº1 2020 T
bobal, cabernet sauvignon

88
Spicy, herbaceous, fruity, tasty.

El Miracle Nº3 2021 B
macabeo, merseguera, malvasía, moscatel

87

El Miracle Nº5 2021 RD
bobal

88
Citrus fruit, crisp, fruity, herbal.

Fusta Nova Blanc 2021 B
moscatel, macabeo, chardonnay

85

Nebla Garnacha 2020 T S
garnacha

88
Spicy, fruity, jammy, tasty.

VINOS SIERRA NORTE
Pol. Ind. El Romeral, Calle Transporte C2
46340 Requena (València/Valencia)
☎: +34 962 323 099
info@bodegasierranorte.com
www.bodegasierranorte.com

Mariluna 2020 T BA
tempranillo, bobal

88 🌱
Kind finish, fruity, spicy, smoky, tasty.

Mariluna 2021 B
verdejo, macabeo

88 🌱
Pleasant, aromatic, dried herbs, fruity, jammy, balanced.

Pasion de Moscatel 2021 B
moscatel

88 🌱
Kind finish, aromatic, dried flowers, balanced, tropical.

VIÑAS DEL PORTILLO
F2 P4; Pol. Ind. El Llano
46360 Buñol (València/Valencia)
☎: +34 696 451 326
vinasdelportillo@vinasdelportillo.es
www.vinasdelportillo.com

Albufera 2020 T BA
86

Cañas y Barro 2021 B
malvasía, merseguera, moscatel

86

Cañas y Barro 2021 RD
85

VITICULTORES NELEMAN
San Vicente, 23
46310 Casas del Rey (València/Valencia)
☎: +34 634 794 119
hola@neleman.es
www.neleman.es

Neleman Bobal Hramony 2017 T
bobal

91 🌱
Colour: deep cherry. Nose: ripe fruit, dried herbs, creamy oak, cocoa bean, wild herbs. Palate: powerful, ripe fruit, spicy, round tannins.

Neleman Just Fucking Good Wine 2019 T
marselan

91 🌱
Colour: deep cherry. Nose: dried herbs, creamy oak, black fruit, scrubland. Palate: powerful, ripe fruit, spicy, round tannins, good finish.

Neleman Just Fucking Good Wine 2020 B
verdil, chardonnay, sauvignon blanc, viognier

88 🌱
Balanced, spicy, dried flowers, tasty, toasty.

Neleman Macabeo 2020 B
macabeo

87 🌱

DO VALENCIA / D.O.P.

DO VALENCIA / D.O.P.

Neleman Tempranillo Monastrell 2021 T
tempranillo, monastrell

90

Colour: deep cherry. Nose: dried herbs, red berry notes, black fruit, earthy notes. Palate: ripe fruit, spicy, round tannins.

Neleman Viognier Verdil 2021 B

88

Citrus fruit, balanced, floral, herbal, smooth.

The Swimmers 2019 B
riesling, viognier

90

Colour: straw. Nose: expressive, white flowers, jasmine, dried herbs, dry stone. Palate: flavourful, fruity, balanced, fresh.

WINES N' ROSES VITICULTORES

Arcediano Ros, 35
46630 La Font de la Foguera (Valencia)
☎: +34 677 388 186
hello@wnr.es
www.wnr.es

Born to Be Wild 2020 T RB
100% bobal

90

Colour: cherry, purple rim. Nose: red berry notes, ripe fruit, wild herbs, spicy. Palate: fruity, balanced, slightly dry, soft tannins.

Highway to Hell 2021 T RB
100% monastrell

90

Colour: deep cherry. Nose: ripe fruit, dried herbs, creamy oak. Palate: powerful, ripe fruit, spicy, round tannins.

Light My Fire 2020 T
100% garnacha tintorera

88

Spicy, fruity, jammy, tasty.

The Final Countdown 2020 T RB
100% monastrell

89

Fruity, spicy, tasty, crisp.

DO. VALLE DE GÜÍMAR
CONSEJO REGULADOR

Tafetana, 14
38500 Güímar (Santa Cruz de Tenerife)
☎: +34 922 514 709 - Fax: +34 922 514 485
@: consejo@vinosvalleguimar.com
www.vinosvalleguimar.com

LOCATION:

On the island of Tenerife. It practically constitutes a prolongation of the Valle de la Orotava region to the southeast, forming a valley open to the sea, with the Las Dehesas region situated in the mountains and surrounded by pine forests where the vines grow in an almost Alpine environment. It covers the municipal districts of Arafo, Candelaria and Güímar.

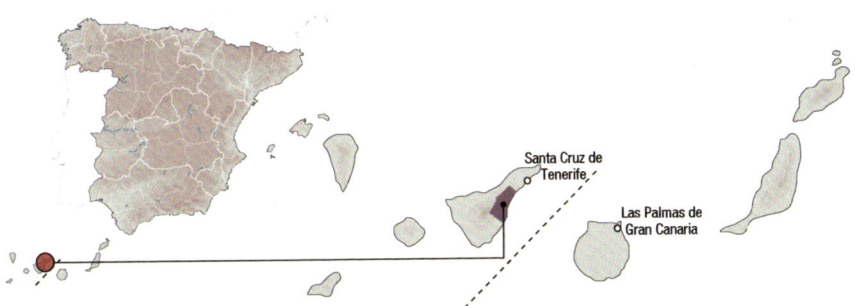

GRAPE VARIETIES:

WHITE: Gual, Listán Blanco, Malvasía, Moscatel, Verdello and Vijariego.

RED: Bastardo Negro, Listán Negro (15% of total), Malvasía Tinta, Moscatel Negro, Negramoll, Vijariego Negro, Cabernet Sauvignon, Merlot, Pinot Noir, Ruby Cabernet, Syrah and Tempranillo.

FIGURES:

Vineyard surface: 170 – **Wine-Growers:** 387 – **Wineries:** 12 **2021 Harvest rating:** N/A – **Production 21:** 269,095 litres – **Market percentages:** 99% National- 1% International.

SOIL:

Volcanic at high altitudes, there is a black tongue of lava crossing the area where the vines are cultivated on a hostile terrain with wooden frames to raise the long vine shoots.

CLIMATE:

Although the influence of the trade winds is more marked than in Abona, the significant difference in altitude in a much reduced space must be pointed out, which gives rise to different microclimates, and pronounced contrasts in day-night temperatures, which delays the harvest until 1st November.

CLASSIFICATION OF YOUNG WINE HARVESTS PEÑÍNGUIDE

2017	2018	2019	2020	2021
GOOD	GOOD	VERY GOOD	VERY GOOD	VERY GOOD

DO VALLE DE GÜIMAR / D.O.P.

BODEGA COMARCAL VALLE DE GÜIMAR
Ctra. a La Cumbre, Km. 4
38550 Arafo (Santa Cruz de Tenerife)
☎: +34 922 510 237
info@bodegacomarcalguimar.com
www.bodegavalledeguimar.com

Brumas de Ayosa 2019 BE R BN
listán negro
87

Brumas de Ayosa 2021 T
listán negro
87

Brumas de Ayosa Afrutado B SD
listán blanco, moscatel de alejandría
85

Brumas de Ayosa Malvasía Aromática 2021 B
malvasía
89
Crisp, balanced, herbal, tasty.

Brumas de Ayosa Seco 2021 B S
86

EL REBUSCO BODEGAS
La Punta, 75
38530 Candelaria (Santa Cruz de Tenerife)
☎: +34 670 540 401
info@elrebusco.es
www.elrebusco.es

Dis-Tinto 2020 T
merlot
85

El Rebusco Selección Syrah 2020 T
syrah
88
Pruney, sweet, full-bodied, dried herbs, spicy.

La Tentación 2019 B FB
listán blanco, albillo criollo, vijariego blanco
89
Citrus fruit, creamy, oxidativ, crisp, floral.

La Tentación Afrutado 2021 B SD
listán blanco, moscatel
87

La Tentación Seco 2021 B
listán blanco, moscatel
85

JUAN FRANCISCO FARIÑA PÉREZ
Subida Los Loros, km. 4,2
38550 Arafo (Santa Cruz de Tenerife)
☎: +34 636 824 919
jfcofarina@movistar.es

Los Loros "Tintilla Castellana" 2020 T RB
100% tintilla
91
With personality. Colour: cherry, purple rim. Nose: fruit expression, red berry notes, floral, spicy, mineral. Palate: flavourful, fruity, good acidity, balanced.

Los Loros "La Bota de Mateo" 2019 B
100% listán blanco
93
Colour: bright yellow. Nose: dried flowers, candied fruit, fine lees, pattiserie, caramel. Palate: round, spicy, long, great length, flavourful.

Los Loros "Listán Blanco de Canarias" 2021 B S
listán blanco
91
Austere. Colour: straw. Nose: dried herbs, dried flowers, dry stone, citrus fruit. Palate: flavourful, fruity, balanced, saline.

Los Loros "Siete Lomas" 2020 B FB
80% marmajuelo, 20% gual
90
Colour: bright yellow. Nose: creamy oak, ripe fruit, spicy, wild herbs. Palate: good structure, long, toasty.

Los Loros Albillo Criollo 2021 B
100% albillo criollo
91
Colour: straw. Nose: expressive, dried herbs, dried flowers, white fruit. Palate: flavourful, fruity, balanced, fresh.

Los Loros Listán Negro 2021 T BA
100% listán negro
92
Colour: light cherry. Nose: red berry notes, floral, spicy, wild herbs. Palate: flavourful, fruity, good acidity, balanced.

DO VALLE DE GÜÍMAR / D.O.P.

VIÑA GÓMEZ
Brisas de Chimisay, 1
38508 Güimar (Santa Cruz de Tenerife)
☎: +34 636 955 759
javiervinagomez@gmail.com
www.bodegavinagomez.com

1400M 2019 B C
90
Colour: bright yellow. Nose: powerful, creamy oak, ripe fruit, spicy. Palate: toasty, fine bitter notes, slightly evolved.

1400M 2020 B FB
92
Colour: bright yellow. Nose: creamy oak, ripe fruit, spicy, white flowers, honeyed notes. Palate: rich, good structure, long, toasty, fine bitter notes.

1400M 2021
87

1400M T
88
Crisp, fruity, herbal, tasty.

Viña Gómez Listán 2018 B D
listán blanco
89
Citrus fruit, spicy, full-bodied, jammy, tasty, herbal.

Viña Gómez Malvasia 2019 RD D
malvasía rosada
92
Colour: ruby red, bright ochre rim. Nose: citrus fruit, stone fruit, candied fruit, spicy, caramel. Palate: full-bodied, flavourful, balanced, round.

VIÑAS HERZAS
Morra del Estanque s/n
38550 Arafo (Santa Cruz de Tenerife)
☎: +34 639 157 290
morraherzas@yahoo.es
www.herzas.es

Viñas Herzas 2021 B
55% listán blanco, 25% marmajuelo, 15% albillo criollo, 5% moscatel de alejandría
88
Citrus fruit, floral, crisp, herbal.

Viñas Herzas 2021 T
50% tempranillo, 35% listán negro, 10% cabernet sauvignon, 5% merlot
88
Complex, with personality, floral, fruity, dried herbs, spicy.

DO. VALLE DE LA OROTAVA
CONSEJO REGULADOR

Parque Recreativo El Bosquito, nº1. Urb. La Marzagana II - La Perdona
38315 La Orotava (Santa Cruz de Tenerife)
☎: +34 922 309 922
@: tecnico@dovalleorotava.com
www.dovalleorotava.com

LOCATION:

In the north of the island of Tenerife. It borders to the west with the DO Ycoden-Daute-Isora and to the east with the DO Tacoronte-Acentejo. It extends from the sea to the foot of the Teide, and comprises the municipal districts of La Orotava, Los Realejos and El Puerto de la Cruz.

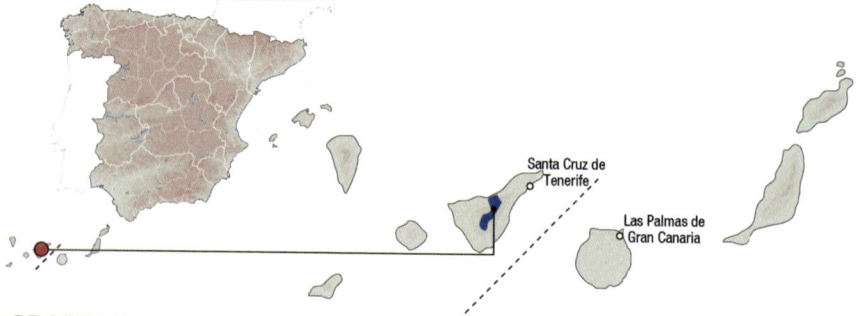

GRAPE VARIETIES:

WHITE: Main: Güal, Malvasía, Verdello, Vijariego, Albillo, Forastera Blanca o Doradilla, Sabro, Breval and Burrablanca.

Authorized: Bastardo Blanco, Forastera Blanca (Gomera), Listán Blanco, Marmajuelo, Moscatel, Pedro Ximénez and Torrontés.

RED: Main: Listán Negro, Malvasía Rosada, Negramoll, Castellana Negra, Mulata, Tintilla, Cabernet Sauvignon, Listán Prieto, Merlot, Pinot Noir, Ruby Cabernet, Syrah and Tempranillo.

Authorized: Bastardo Negro, Moscatel Negra, Tintilla and Vijariego Negra.

FIGURES:

Vineyard surface: 206 – **Wine-Growers:** 429 – **Wineries:** 18 – **2021 Harvest rating:** Excellent – **Production 21:** 560,000 litres – **Market percentages:** 70% National - 30% International.

SOIL:

Light, permeable, rich in mineral nutrients and with a slightly acidic pH due to the volcanic nature of the island. The vineyards are at an altitude of between 250 mm and 700 m.

CLIMATE:

As with the other regions on the islands, the weather is conditioned by the trade winds, which in this region result in wines with a moderate alcohol content and a truly Atlantic character. The influence of the Atlantic is also very important, in that it moderates the temperature of the costal areas and provides a lot of humidity. Lastly, the rainfall is rather low, but is generally more abundant on the north face and at higher altitudes.

CLASSIFICATION OF YOUNG WINE HARVESTS PEÑÍNGUIDE

2017	2018	2019	2020	2021
GOOD	GOOD	N/A	N/A	N/A

BODEGA ILLADA

Calle Nueva, 31, La Cruz Santa
38410 Los Realejos (Santa Cruz de Tenerife)
☎: +34 627 229 735
bodegaillada@gmail.com

El Reboso 2021 B
listán blanco
87

El Reboso 2021 T
listán negro
88
Balanced, smoky, tasty, fruity, crisp.

El Reboso Afrutado 2021 B
listán blanco
87

BODEGA TAFURIASTE

Las Candias Altas, 11
38312 La Orotava (Santa Cruz de Tenerife)
☎: +34 647 421 256
Fax: +34 922 336 027
vinos@bodegatafuriaste.com
www.bodegatafuriaste.com

Engazo Familia Tafuriaste 2019 T C
listán negro, tintilla, negramoll
89
Crisp, fruity, spicy, mineral, roasted coffee.

Engazo Familia Tafuriaste 2020 T RB
listán negro, tintilla, castellana
89
Balanced, spicy, dried flowers, smoky, crisp, tasty.

Prunet Esencia del Territorio 2020 T
listán negro, tintilla
89
Balanced, spicy, floral, crisp, fruity, toasty.

Tafuriaste 2021 T
listán negro
87

Tafuriaste Afrutado Semidulce 2021 RD SD
listán negro
86

BODEGAS EL PENITENTE

Camino La Habanera, 286
38300 La Orotava (Santa Cruz de Tenerife)
☎: +34 922 309 024
bodegas@elpenitentesl.es
www.elpenitentesl.es

Arautava 2020 T FB
87

Arautava Blanco Seco 2021 B S
87

Arautava Dulce 2002 B GR D
listán blanco
94
Colour: light mahogany. Nose: complex, dry nuts, toasty, sweet spices. Palate: long, fine solera notes, spicy, round.

Arautava Finca la Habanera 2020 T
listán negro
91
Colour: cherry, purple rim. Nose: red berry notes, floral, spicy, earthy notes, mineral. Palate: flavourful, fruity, good acidity.

Arautava Finca la Habanera Albillo Criollo 89 B
89
Balanced, spicy, crisp, tasty.

Arautava Finca la Habanera Listán Blanco 2020 B
listán blanco
91
Colour: bright straw, greenish rim. Nose: fresh fruit, citrus fruit, wild herbs, dry stone, mineral. Palate: fresh, fruity, good acidity, good structure, flavourful.

BODEGAS TAJINASTE

Ratiño 5
38315 La Orotava (Santa Cruz de Tenerife)
☎: +34 616 441 519
Fax: +34 922 105 080
bodega@tajinaste.net
www.bodegatajinaste.com

Can 2020 T
listán negro, vijariego negro
91
Classic. Colour: deep cherry. Nose: dried herbs, creamy oak, black fruit, earthy notes. Palate: powerful, ripe fruit, spicy, round tannins.

DO VALLE DE LA OROTAVA / D.O.P.

DO VALLE DE LA OROTAVA / D.O.P.

Tajinaste Naturalmente Dulce 2018 T D
90
Colour: cherry, garnet rim. Nose: fruit preserve, spicy, toasty, wild herbs, smoky. Palate: flavourful, fleshy, rich, balanced.

Tajinaste Vendimia Seleccionada 2020 T
listán negro, vijariego negro
90
Colour: deep cherry. Nose: dried herbs, creamy oak, black fruit, earthy notes. Palate: ripe fruit, spicy, round tannins.

LA HAYA
Calzadillas, 88
38413 Los Realejos (Santa Cruz de Tenerife)
☎: +34 629 051 413
lahayabodegas@gmail.com

La Haya 2019 B BA
90% listán blanco, 10% otras
90
Colour: bright yellow. Nose: creamy oak, ripe fruit, spicy, woody. Palate: rich, good structure, toasty, fine bitter notes.

La Haya Afrutado 2020 B SS
90% listán blanco, 10% otras
88
Citrus fruit, balanced, mineral, tasty, dried herbs, crisp.

La Haya Seco 2020 B
90% listán blanco, 10% otras
88
Balanced, spicy, dried flowers, full-bodied, mineral.

LA SUERTITA
Real de la Cruz Santa, 35-A
38413 Los Realejos (Santa Cruz de Tenerife)
☎: +34 669 408 761
bodegalasuertita@yahoo.es

Informal 2020 B
90
Standard. Colour: bright yellow. Nose: powerful, creamy oak, spicy, toasty. Palate: toasty, fine bitter notes, correct.

LA VIÑITA
Camino La Higuera, 9
38300 La Orotava (Santa Cruz de Tenerife)
☎: +34 639 369 330
daniari222@hotmail.com

Chivita 2021 B
86

Chivita Tinto Tradicional 2021 T
87

SUERTES DEL MARQUÉS
Cº Las Suertes Tercera, 10
38300 La Orotava (Santa Cruz de Tenerife)
☎: +34 922 501 300
Fax: +34 922 503 462
ventas@suertesdelmarques.com
www.suertesdelmarques.com

🏆 **PODIUM**

Las Suertes 2020 T
listán negro
95
Wild, elegant, representative. Colour: light cherry. Nose: complex, expressive, spicy, mineral, chamomile. Palate: elegant, full-bodied, long, great length, round.

Suertes del Marqués Cruz Santa 2020 T
vijariego negro
92
Colour: bright cherry. Nose: fresh fruit, red berry notes, wild herbs, wild herbs. Palate: good acidity, spicy, fine tannins, elegant.

Suertes del Marqués El Chibirique 2019 T
listán negro
94
Colour: bright cherry. Nose: fresh fruit, red berry notes, black pepper, rose petals, cocoa bean, wild herbs. Palate: good acidity, spicy, fine tannins, flavourful, fresh.

Suertes del Marqués El Esquilón 2019 T
listán negro
93
Herbal, crisp, with personality. Colour: cherry, garnet rim. Nose: damp undergrowth, fresh fruit, wild herbs. Palate: correct, balsamic.

Suertes del Marqués La Solana 2020 T
listán negro
94
Defined aromas, with personality. Colour: light cherry. Nose: ebb tide, wild herbs, expressive, red berry notes, ripe fruit. Palate: round, fruity, balanced, fine bitter notes.

Suertes del Marqués Los Pasitos 2020 T
baboso, vijariego blanco
93
Colour: bright cherry. Nose: fresh fruit, floral, wild herbs, dried flowers, defined aromas. Palate: balanced, good acidity, fresh, light-bodied.

Suertes del Marqués Trenzado 2020 B
100% listán blanco
93
Colour: bright straw. Nose: ripe fruit, fragrant herbs, fine lees, burnt matches. Palate: full-bodied, long, good acidity.

Suertes del Marqués Vidonia 2020 B
listán blanco

93

Dried flowers, wild. Colour: straw. Nose: burnt matches, neat, expressive. Palate: fine bitter notes, flavourful, long, balanced.

🏆 **PODIUM**

Suertes del Marqués Vidonia V.P. 2020 B C
listán blanco

97

Colour: golden, pale. Nose: white fruit, pattiserie, sweet spices, hint of anise, dry stone. Palate: elegant, good acidity, fresh.

Suertes del Marqués Vidueño 2020 T
listán negro, malvasía rosada, negramoll, baboso, castellana, otras

93

Colour: light cherry. Nose: wild herbs, fragrant herbs, balanced, medium intensity. Palate: flavourful, fruity, ripe fruit, long.

DO. VINOS DE MADRID

CONSEJO REGULADOR

Ronda de Atocha, 7
28012 Madrid
☎: +34 915 348 511 / Fax: +34 915 538 574
@: prensa@vinosdemadrid.es
www.vinosdemadrid.es

LOCATION:

In the south of the province of Madrid, it covers four distinct sub-regions: Arganda, Navalcarnero, San Martín de Valdeiglesias and El Molar.

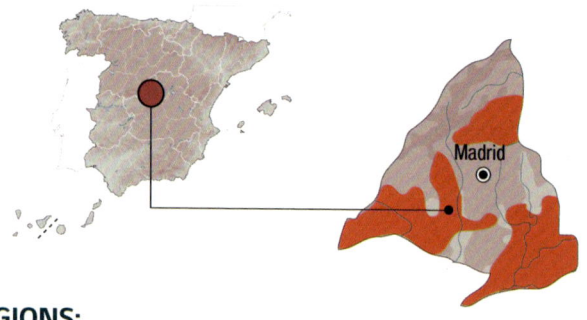

SUB-REGIONS:

San Martí: It comprises 9 municipal districts and has more than 3,821 Ha of vineyards, with mainly the Garnacha (red) and Albillo (white) varieties. **Navalcarnero:** It comprises 19 municipal districts with a total of about 2,107 Ha. The most typical wines are reds and rosés based on the Garnacha variety. **Arganda:** With 5,830 Ha and 26 municipal districts, it is the largest sub-region of the DO. The main varieties are the white Malvar and the red Tempranillo or Tinto Fino. **El Molar:** The new subzone in the DO. The main varieties are Malvar and Granacha Tinta.

GRAPE VARIETIES:

WHITE: Malvar, Airén, Albillo, Parellada, Macabeo, Torrontés, Moscatel de Grano Menudo and Sauvignon Blanc.

RED: Tinto Fino (Tempranillo), Garnacha, Garnacha Tintorera, Merlot, Cabernet Sauvignon, Syrah and Petit Verdot.

FIGURES:

Vineyard surface: 8,391 – **Wine-Growers:** 2,890 – **Wineries:** 50 – **2021 Harvest rating:** N/A – **Production 21:** 7,027,203 litres – **Market percentages:** 85% National - 15% International.

SOIL:

Rather unfertile soil and granite subsoil in the sub-region of San Martín de Valdeiglesias; in Navalcarnero the soil is brownish-grey, poor, with a subsoil of coarse sand and clay; in the sub-region of Arganda the soil is brownish-grey, with an acidic pH and granite subsoil.

CLIMATE:

Extreme continental, with cold winters and hot summers. The average annual rainfall ranges from 461 mm in Arganda to 658 mm in San Martín.

CLASSIFICATION OF YOUNG WINE HARVESTS PEÑÍNGUIDE

2017	2018	2019	2020	2021
GOOD	N/A	VERY GOOD	VERY GOOD	GOOD

ARGANDA BODEGAS

Camino de San Martín de la Vega, 16
28500 Arganda del Rey (Madrid)
☎: +34 918 710 201
vinicola@cvarganda.e.telefonica.net
www.vinicoladearganda.com

De Palabra 2019 T C
87

En tus Manos 2021 RD
tempranillo
84

Entre Silencios 2020 B FB
malvar
86

Siempre Contigo 2020 T RB
tempranillo
86

Sin Razón 2018 T R
tempranillo
89

Aromatic, spicy, jammy, tasty. Nose: black fruit, wild herbs.

Y Volarás 2018 T RB
syrah
89

Defined aromas, balanced, spicy, fruity, jammy, lactic, smoky.

BERNABELEVA

Ctra. Ávila-Toledo N-403, km 81,600
28680 San Martín de Valdeiglesias (Madrid)
☎: +34 915 091 909
bodega@bernabeleva.com
www.bernabeleva.com

Bernabeleva Arroyo de Tórtolas 2020 T
garnacha
94 🌱

Aromatic, balsamic herbs. Colour: bright cherry. Nose: balsamic herbs, sweet spices, scrubland, red berry notes, fruit expression. Palate: spicy, balsamic, good acidity.

🏆 **PODIUM**

Bernabeleva Carril del Rey 2020 T
garnacha
95 🌱

Colour: bright cherry. Nose: red berry notes, ripe fruit, floral, wild herbs, balanced, medium intensity. Palate: fruity, fresh, easy to drink, fine tannins.

Bernabeleva Viña Bonita 2020 T
garnacha
94 🌱

Colour: bright cherry. Nose: complex, expressive, spicy, mineral, red berry notes, fresh fruit. Palate: full-bodied, long, great length, flavourful.

Camino de Navaherreros 2020 B FB
albillo real, macabeo, moscatel grano menudo, malvar
90 🌱

Colour: yellow. Nose: dried herbs, dried flowers, white fruit, ripe fruit, dry nuts. Palate: fruity, balanced.

Camino de Navaherreros 2020 T
90% garnacha, 10% morenillo, tempranillo
92 🌱

Colour: light cherry. Nose: fruit expression, red berry notes, floral, spicy, wild herbs. Palate: flavourful, fruity, fresh, slightly dry, soft tannins.

Cantocuerdas Albillo 2020 B FB
albillo real
91 🌱

Defined aromas. Nose: dried flowers, faded flowers, ripe fruit, dried herbs. Palate: ripe fruit, easy to drink.

Manchomuelas
Blanco de Bernabeleva 2020 B FB
albillo real, macabeo, malvar
93 🌱

Colour: bright straw. Nose: citrus fruit, wild herbs, spicy, pattisserie, fine lees. Palate: fruity, good acidity, fine bitter notes, rich.

Navaherreros
Blanco de Bernabeleva 2020 B
albillo real, macabeo
92 🌱

Colour: bright straw. Nose: ripe fruit, fragrant herbs, fine lees, spicy. Palate: full-bodied, rich, long.

Navaherreros
Garnacha de Bernabeleva 2020 T
garnacha
93 🌱

Colour: light cherry. Nose: red berry notes, fruit expression, wild herbs, dry stone, dried flowers. Palate: fruity, flavourful, fresh, balanced, good finish, slightly dry, soft tannins.

DO VINOS DE MADRID / D.O.P.

DO VINOS DE MADRID / D.O.P.

BODEGA CUARTO LOTE
Travesía José de Churriguera, 1
28514 Nuevo Baztan (Madrid)
☎: +34 916 492 088
info@bodegacuartolote.com
www.bodegacuartolote.com

Arrabal del Conjuro 2019 T
60% cabernet sauvignon, 20% tempranillo, 20% merlot

90

Colour: bright cherry. Nose: sweet spices, black fruit, ripe fruit, scrubland, dried herbs. Palate: fruity, spicy, round tannins.

Bálsamo de Fierabrás 2018 T
70% tempranillo, 30% merlot

90

Nose: black fruit, fruit preserve, dried herbs, spicy. Palate: fruity, flavourful, powerful tannins, toasty.

Cuarto Lote 2019 T RB
70% tempranillo, 30% merlot

88

Colour: bright cherry. Nose: sweet spices, ripe fruit, wild herbs, dried herbs. Palate: spicy, slightly dry, soft tannins, toasty.

Cuarto Lote 2020 B
malvar

88

Aromatic, dried flowers, slight oxidation, balanced, fruity, smooth.

Cuarto Lote 2021 RD
33% tempranillo, 33% merlot, 33% syrah

87

BODEGA DEL NERO
Don Ramiro Ortíz de Zárate, 6
28370 Chinchón (Madrid)
☎: +34 679 499 695
bodegadelnero@gmail.com
www.bodegadelnero.com

Neri 2021 B
100% airén

86

Neri 2021 T
100% tempranillo

85

Neri Cepas Viejas 2020 T RB
100% tempranillo

88

Toasty, spicy, tasty, fruity.

Trapisondero 2021 T RB
100% tempranillo

86

Valdeliceda 2019 T BA
100% tempranillo

88

Toasty, spicy, fruity, tasty, jammy.

BODEGA ECOLÓGICA LUIS SAAVEDRA
Ctra. de Escalona, 5
28650 Cenicientos (Madrid)
☎: +34 629 124 729
info@bodegasaavedra.com
www.bodegasaavedra.com

100 y Cientos 2017 T RB
100% garnacha

90 🌱

Colour: deep cherry. Nose: ripe fruit, dried herbs, dried flowers, sweet spices, toasty. Palate: ripe fruit, spicy, round tannins.

100 y Cientos 2018 T RB
garnacha

90 🌱

Colour: bright cherry. Nose: dried herbs, creamy oak, fruit preserve, spicy. Palate: ripe fruit, spicy, round tannins, good finish.

Chulo 2020 T RB
garnacha

87 🌱

Corucho 2020 T RB
garnacha

89 🌱

Spicy, fruity, jammy, smoky, tasty.

Corucho 2021 RD
garnacha

86 🌱

Corucho Finca Peazo de la Encina 2018 T RB
garnacha, syrah, merlot, tempranillo

90 🌱

Colour: cherry, garnet rim. Nose: fruit preserve, wild herbs, dried flowers. Palate: flavourful, long.

Corucho Garnacha 2021 T
garnacha

87 🌱

Flor del Amanecer 2020 T RB
garnacha, tinto fino

87 🌱

Luis Saavedra 2019 T C
garnacha, syrah

87

Luis Saavedra Vendimia Nocturna 2017 T RB
garnacha, tempranillo, merlot, syrah, cabernet sauvignon

88

Defined aromas, balsamic herbs, dried herbs, jammy, spicy.

BODEGA EL REGAJAL

Antigua Ctra. de Andalucía, Km. 50.5
28300 Aranjuez (Madrid)
☎: +34 913 078 903
Fax: +34 913 576 312
reservas@elregajal.es
www.elregajal.es

El Regajal Selección Especial 2020 T
tempranillo, syrah, merlot, cabernet sauvignon, petit verdot

90

Colour: Cherry. Nose: balsamic herbs, sweet spices, scrubland, black fruit. Palate: spicy, round tannins, correct, easy to drink.

BODEGAS ANDRÉS MORATE

Camino del Horcajuelo, s/n
28390 Belmonte de Tajo (Madrid)
☎: +34 918 747 165
bodegas@andresmorate.com
www.andresmorate.com

Esther 2018 T C
tempranillo, syrah, cabernet sauvignon

86

Viña Bosquera 2021 B
airén, moscatel grano menudo

85

BODEGAS FIGUEROA

Convento, 19
28380 Colmenar de Oreja (Madrid)
☎: +34 608 046 227
info@bodegasfigueroa.com
www.bodegasfigueroa.com

Figueroa Blanco sobre Lías Finas 2021 B
moscatel, malvar

86

Figueroa Originem 2018 T C
100% tempranillo

88

Fruity, jammy, spicy, tasty, coarse tannins.

Señorío de Zafra Cabernet Sauvignon 2020 T RB

87

Señorío de Zafra Syrah 2020 T
syrah

89

Fruity, jammy, herbal, spicy, tasty.

Señorío de Zafra Tempranillo Merlot 2020 T RB
tempranillo, merlot

89

Fruity, tasty, balanced, spicy, coarse tannins.

VII Generación de Bodegas Figueroa 2019 T RB
tempranillo, cabernet sauvignon, merlot

88

Standard, spicy, herbal, jammy, tasty, balanced.

BODEGAS LA CASA DE MONROY

José Moya, 12
45940 Valmojado (Toledo)
☎: +34 699 124 752
info@bodegasmonroy.es
www.bodegasmonroy.es

El Gran Monroy Antheos 2019 T
60% merlot, 20% garnacha, 15% cabernet sauvignon, 5% graciano

89

Hot, pruney, toasty, powerful.

El Gran Monroy Theos 2018 T
85% merlot, 15% cabernet sauvignon

90

Colour: dark-red cherry. Nose: toasty, spicy, cocoa bean, black fruit, ripe fruit, dark chocolate. Palate: flavourful, toasty, fine bitter notes.

La Casa de Monroy "El Repiso" 2017 T C
tempranillo, syrah, merlot

88

Toasty, hot, spicy, jammy.

M de Monroy Garnacha Syrah 2020 T
95% garnacha, 5% syrah

88

Standard, herbal, jammy, wild, smooth, simple.

Monroy Malvar 2020 B
100% malvar

91

Colour: straw. Nose: expressive, white flowers, jasmine, dried herbs, sweet spices. Palate: flavourful, fruity, balanced.

DO VINOS DE MADRID / D.O.P.

BODEGAS PABLO MORATE - MUSEO DEL VINO
Avda. España, 34
28391 Valdelaguna (Madrid)
☎: +34 639 673 170
bodegasmorate@bodegasmorate.com
www.bodegasmorate.com

Arate Premium 2021 B
viura, malvar
83

Señorío de Morate 2020 T RB
tempranillo, syrah
85

Señorío de Morate Syrah 2015 T C
syrah
87

Señorío de Morate Syrah 2016 T
syrah
86

Viña Chozo 2020 T
tempranillo
86

BODEGAS TEJONERAS
Santo Domingo de los Silos, 6 1º
28660 Villa del Prado (Madrid)
☎: +34 915 649 495
info@tejoneras.com
www.tejoneras.com

750 2010 T GR
60% cabernet sauvignon, 40% merlot
89
Reduced, pruney, herbal, jammy, age nuances, spicy, tasty.

Lagraz 2019 T RB
garnacha, merlot, cabernet sauvignon
89
Defined aromas, balsamic herbs, dried herbs, jammy, spicy, toasty, tasty.

Tejoneras Alta Selección 2014 T R
merlot, cabernet sauvignon, syrah, garnacha
87

BODEGAS VIÑA BARDELA
De las Eras, 1
28729 Venturada (Madrid)
☎: +34 619 023 451
vinosbardela@yahoo.es
www.vinabardela.com

Sacasueños 2021 B
viura, malvar, airén
86

BODEGAS Y VIÑEDOS PEDRO GARCÍA
Soledad, 10
28380 Colmenar de Oreja (Madrid)
☎: +34 918 943 278
byvpedrogarcia@gmail.com
www.byvpedrogarcia.com

ISP (Isla de San Pedro) 2018 T C
89
Spicy, fruity, tasty, jammy, crisp.

La Romera T
merlot, tempranillo, syrah
86

Pedro García BE BR
85

Pedro García Malvar B
87

Pedro García Sauvignon y Malvar 2021 B
sauvignon blanc, malvar
89
Austere, citrus fruit, dried herbs, balanced.

SIP Barrica Selección 2019 T
merlot, tempranillo
89
Pruney, fruity, spicy, smoky, jammy, tasty.

BODEGAS Y VIÑEDOS VALLEYGLESIAS
Camino Fuente de los Huertos s/n
28680 San Martín de Valdeiglesias (Madrid)
☎: +34 606 842 636
bodega@valleyglesias.com
www.valleyglesias.com

A2 Albillo ValleYglesias 2021 B
albillo real
89
Fruity, herbal, jammy, tasty.

G2 Garnacha Centenaria 2020 T BA
100% garnacha

90

Colour: Cherry. Nose: scrubland, dried flowers, ripe fruit. Palate: spicy, flavourful, warm, balsamic.

Garnacha Rock 2019 T BA
100% garnacha

91

Colour: deep cherry. Nose: ripe fruit, dried herbs, creamy oak, scrubland. Palate: powerful, ripe fruit, spicy, round tannins.

La Pájara 2021 B FB
100% albillo real

90

Colour: bright straw. Nose: fruit expression, citrus fruit, dried herbs, dried flowers. Palate: fruity, flavourful, balanced.

Senderos de Valleyglesias 2020 T
garnacha, syrah, tempranillo

90

Colour: Cherry. Nose: woody, dried herbs, fruit preserve. Palate: spicy, balsamic, round tannins, concentrated.

Tete de la Course 2021 B
albillo real

89

Citrus fruit, fruity, herbal, tasty, wild.

CA' DI MAT
C. del Hilero, 7, nave 9
28696 Pelayos de la Presa (Madrid)
☎: +34 918 644 115
info@cadimat.wine
www.cadimat.wine

Fuente de Los Huertos 2019 T
garnacha

94

Colour: bright cherry. Nose: fruit expression, red berry notes, floral, spicy, fragrant herbs, scrubland. Palate: flavourful, fruity, good acidity, long.

Los Peros 2018 T

94

Colour: bright cherry. Nose: balsamic herbs, sweet spices, scrubland, red berry notes, mineral, chamomile. Palate: spicy, balsamic, good acidity.

Los Peros 2019 B
albillo real

93

With personality, little interventionist. Colour: bright straw, greenish rim. Nose: fresh fruit, citrus fruit, wild herbs, fine lees. Palate: fresh, fruity, good acidity, fine bitter notes.

Valautín 2019 B
albillo real

92

Colour: bright straw. Nose: white flowers, jasmine, dried herbs. Palate: flavourful, fruity, balanced, balsamic, ripe fruit.

Valautín 2019 T
garnacha

93

Colour: bright cherry. Nose: fruit preserve, fruit liqueur notes, powerful, earthy notes, mineral. Palate: flavourful, long, spicy, ripe fruit.

COMANDO G VITICULTORES
Avda. Constitución, 23
28640 Cadalso de Los Vidrios (Madrid)
☎: +34 918 640 602
info@comandog.es
www.comandog.es

La Bruja de Rozas 2020 T
garnacha

92

Colour: light cherry. Nose: floral, red berry notes, black fruit. Palate: fleshy, balanced, warm.

🏆 **PODIUM**

Las Umbrías 2019 T
garnacha

95

Colour: light cherry. Nose: red berry notes, scrubland, raspberry, balsamic herbs, expressive. Palate: fleshy, elegant, round tannins.

Rozas 1er Cru 2019 T
garnacha

93

Slight oxidation. Colour: light cherry. Nose: red berry notes, earthy notes, spicy, scrubland, floral. Palate: fleshy, flavourful, warm.

DO VINOS DE MADRID / D.O.P.

LAS MORADAS DE SAN MARTÍN

Pago de Los Castillejos Ctra. M-541, Km. 4,7
28680 San Martín de Valdeiglesias (Madrid)
☎: +34 687 457 235
bodega.lasmoradas@grupoenate.es
www.lasmoradasdesanmartin.es

Las Moradas de San Martín Albillo Real 2020 B
albillo real

91

Oxidativ, with personality. Colour: yellow. Nose: faded flowers, dried flowers, fine lees, lees reduction notes. Palate: flavourful, spicy, long.

Las Moradas de San Martín Initio 2017 T
garnacha

90

Colour: deep cherry. Nose: dried herbs, creamy oak, black fruit. Palate: ripe fruit, spicy, round tannins.

Las Moradas de San Martín La Sabina 2015 T R
garnacha

89

Weary, age nuances, dried herbs, vegetal, spicy.

Las Moradas de San Martín Las Luces 2011 T GR
garnacha

91

Colour: cherry, garnet rim. Nose: balanced, spicy, fine reductive notes, burnt matches, black fruit, tobacco. Palate: good structure, flavourful, round tannins, balanced, great length.

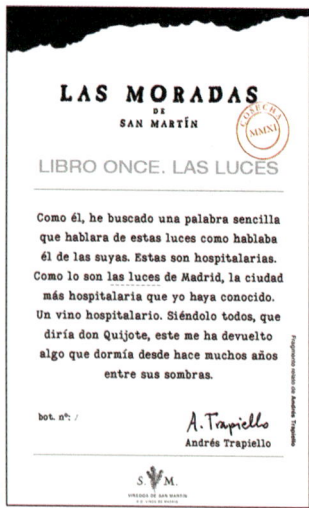

Las Moradas de San Martín Senda 2019 T
garnacha

91

Colour: cherry, purple rim. Nose: fruit expression, red berry notes, floral, spicy, black fruit, scrubland. Palate: flavourful, fruity, good acidity, fleshy.

LICINIA WINES

Soledad, 11
28530 Morata de Tajuña (Madrid)
☎: +34 696 448 050
valgoraj@liciniawines.com
www.liciniawines.com

Licinia 2016 T R
50% tempranillo, 24% syrah, 13% cabernet sauvignon, 13% merlot

92

Colour: deep cherry. Nose: fruit expression, red berry notes, sweet spices, dried herbs. Palate: flavourful, fruity, balanced, slightly dry, soft tannins.

Licinia 2020 B C
86% malvar, 14% torrontés

91

Colour: bright yellow. Nose: creamy oak, ripe fruit, spicy, stone fruit. Palate: rich, long, fine bitter notes.

Licinia MIRVS 2017 T R
100% tempranillo

92

Colour: cherry, garnet rim. Nose: fruit preserve, powerful, creamy oak, black fruit, dark chocolate. Palate: flavourful, long, sweet tannins, spicy, round.

Licinia Plus 2016 T R
65% cabernet sauvignon, 20% tempranillo, 8% syrah, 7% merlot

93

Colour: cherry, garnet rim. Nose: ripe fruit, dried herbs, scrubland, spicy. Palate: ripe fruit, spicy, round tannins, balanced, good acidity.

Licinia Plus 2019 B C
70% torrontés, 30% moscatel grano menudo

91

Yeasty notes, defined aromas. Colour: bright yellow. Nose: powerful, spicy, white fruit, ripe fruit, floral. Palate: rich, long, fine bitter notes.

MARAÑONES
28680 San Martín de Valdeiglesias (Madrid)
☎: +34 983 878 020
info@almacarraovejas.com
www.bodegamaranones.com

30.000 Maravedíes 2019 T
90% garnacha, 10% otras

93

Colour: deep cherry. Nose: dried herbs, red berry notes, fruit liqueur notes, spicy, balsamic herbs. Palate: powerful, ripe fruit, spicy, round tannins.

Marañones 2019 T
garnacha

93

Balsamic herbs, spicy, fruity. Colour: bright cherry. Nose: balsamic herbs, sweet spices, scrubland, red berry notes. Palate: spicy, balsamic, good acidity.

🏆 PODIUM

Peña Caballera 2019 T
garnacha

95

Colour: light cherry. Nose: red berry notes, ripe fruit, wild herbs, scrubland, hint of anise. Palate: good structure, fruity, flavourful, elegant.

Picarana 2020 B
albillo real

92

Colour: bright yellow. Nose: candied fruit, fine lees, pattiserie. Palate: round, spicy, great length.

MUJICA & DIAZ
Cº de Fuentecabrón, 10
40500 Miño de San Esteban (Soria)
☎: +34 620 214 755
mujicaydiazbodega@gmail.com

Matalba Malvar 2020 B
malvar

92

With personality, little interventionist. Colour: yellow, pale. Nose: expressive, balanced, fine lees, spicy. Palate: flavourful, long, rich.

PARAJES DE LOS VIDRIOS
Ronda de Madrid, 17
28640 Cadalso de Los Vidrios (Madrid)
☎: +34 913 020 247
comercial@parajesdelosvidrios.com
www.parajesdelosvidrios.com

Parajes de los Vidrios Garnacha 2018 T
100% garnacha

92 🌱

Colour: Cherry. Nose: scrubland, dried flowers, balanced, expressive, wild herbs. Palate: spicy, balsamic, good acidity.

Parajes de los Vidrios Pico del Mirlo 2018 T
100% garnacha

91 🌱

Colour: Cherry. Nose: balsamic herbs, dried flowers, balanced, red berry notes, ripe fruit. Palate: spicy, ripe fruit, easy to drink.

TIERRA CALMA
C. Carpinteros, 2
28680 San Martín de Valdeiglesias (Madrid)
☎: +34 609 099 024
comunicacion@tierracalma.com
www.tierracalma.com

Tierra Calma Albillo Real 2021 B
100% albillo real

91

Colour: straw. Nose: ripe fruit, dried herbs, faded flowers, toasted bread. Palate: powerful, ripe fruit, balanced, full-bodied, fleshy, warm.

Tierra Calma Cyster 2020 T
100% garnacha

88

Overripe, balanced, spicy, tasty.

Tierra Calma La Nava 2020 T
100% garnacha

85

Tierra Calma Las Cabreras 2020 T
100% garnacha

87

DO VINOS DE MADRID / D.O.P.

UVAS FELICES
Agullers, 7
08003 Barcelona (Barcelona)
☎: +34 902 327 777
www.vilaviniteca.es

Hombre Bala 2019 T
93
Colour: deep cherry. Nose: dried herbs, creamy oak, red berry notes, mineral. Palate: powerful, ripe fruit, spicy, round tannins.

Hombre Bala 2020 B
albillo real
92
Colour: bright yellow. Nose: creamy oak, ripe fruit, spicy, dried herbs. Palate: rich, good structure, toasty, fine bitter notes.

La Mujer Cañón 2019 T
94
Colour: bright cherry. Nose: complex, expressive, spicy, mineral, red berry notes, scrubland. Palate: elegant, long, great length.

VINOS JEROMÍN
Avda. Juan Carlos I Rey de España, 4
28590 Villarejo de Salvanés (Madrid)
☎: +34 918 742 030
Fax: +34 918 744 139
comercial@vinosjeromin.com
www.vinosjeromin.com

Dos de Mayo 2018 T C
tempranillo
89
Spicy, jammy, tasty, balanced, standard.

Félix Martínez Cepas Viejas 2018 T R
55% syrah, 45% tempranillo
91
Colour: deep cherry. Nose: ripe fruit, dried herbs, creamy oak, sweet spices, smoky. Palate: ripe fruit, spicy, round tannins, good finish.

Puerta de Alcalá 2018 T C
tempranillo, syrah, garnacha
88
Fruity, herbal, tasty, spicy, balanced.

Puerta del Sol Malvar 2021 B
malvar
87

Purificación Garnacha 2018 T C
garnacha
87

Purita Dynamita Garnacha 2018 T
garnacha
88
Wild, pruney, dried flowers, dried herbs, hot.

VINOS Y ACEITES LAGUNA
Illescas, 5
28360 Villaconejos (Madrid)
☎: +34 918 938 196
vyalaguna@gmail.com
www.lagunamadrid.com

Alma de Valdeguerra 2018 T C
80% tempranillo, 20% merlot
88
Defined aromas, balsamic herbs, fruity, jammy, dried herbs, herbal.

Alma de Valdeguerra 2020 RD SD
80% tempranillo, 20% garnacha
87

Alma de Valdeguerra 2020 T
tempranillo
88
Fruity, jammy, herbal, spicy, coarse tannins.

Alma de Valdeguerra 2020 T BA
tempranillo
85

Alma de Valdeguerra 2021 RD SD
80% tempranillo, 20% garnacha
85

Exun 2019 T
70% tempranillo, 20% merlot, 10% cabernet sauvignon
87

La Intrusa de Malasaña 2020 T BA
50% graciano, 50% tempranillo
90
Colour: cherry, purple rim. Nose: fruit expression, red berry notes, spicy, ripe fruit, black fruit. Palate: flavourful, fruity, good acidity, slightly dry, soft tannins.

DO. YCODEN-DAUTE-ISORA
CONSEJO REGULADOR

La Palmita, 10
38440 La Guancha (Sta. Cruz de Tenerife)
☎: +34 922 130 246 - Fax: +34 922 828 159
@: viticultura@ycoden.com
www.ycoden.com

LOCATION:

Occupying the northeast of the island of Tenerife and comprising the municipal districts of San Juan de La Rambla, La Guancha, Icod de los Vinos, Los Silos, El Tanque, Garachico, Buenavista del Norte, Santiago del Teide and Guía de Isora.

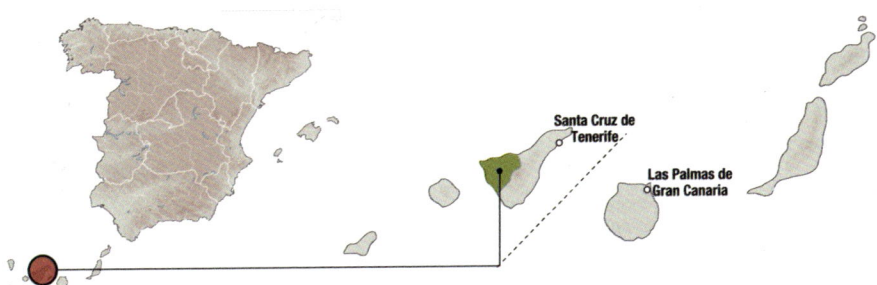

GRAPE VARIETIES:

WHITE: Bermejuela (or Marmajuelo), Güal, Malvasía, Moscatel, Pedro Ximénez, Verdello, Vijariego, Albillo, Bastardo Blanco, Forastera Blanca, Listán Blanco (majority), Sabro and Torrontés.

RED: Tintilla, Listán Negro (majority), Malvasía Rosada, Negramoll Castellana, Bastardo Negra, Moscatel Negra and Vijariego Negra.

FIGURES:

Vineyard surface: 115 – Wine-Growers: 350 – Wineries: 9 – 2021 Harvest rating: N/A – Production 21: 270,000 litres – Market percentages: 85% National - 15% International.

SOIL:

Volcanic ash and rock on the higher grounds, and clayey lower down. The vines are cultivated at very different heights, ranging from 50 to 1,400 m.

CLIMATE:

Mediterranean, characterised by the multitude of microclimates depending on the altitude and other geographical conditions. The trade winds provide the humidity necessary for the development of the vines. The average annual temperature is 19°C and the average annual rainfall is around 540 mm.

CLASSIFICATION OF YOUNG WINE HARVESTS PEÑÍNGUIDE

2017	2018	2019	2020	2021
N/A	N/A	N/A	N/A	N/A

DO YCODEN-DAUTE-ISORA / D.O.P.

BODEGAS INSULARES TENERIFE
Camino Cueva del Rey, 1
38430 Icod de los Vinos (Santa Cruz de Tenerife)
☎: +34 922 122 395
Fax: +34 922 814 688
contacto@bodegasinsularestenerife.es
www.bodegasinsularestenerife.es

Tágara 2021 B S
listán blanco, marmajuelo
88
Citrus fruit, standard, crisp, herbal.

Tágara 2021 B SD
listán blanco
87

ENVINATE
Terrero 72
02630 La Roda (Albacete)
☎: +34 682 207 160
asesoria@envinate.es

Benje 2020 B
92
Colour: yellow. Nose: meaty notes, burnt matches, wild herbs, dried herbs, hint of anise. Palate: flavourful, full of life.

Benje 2020 T
92
Pleasant. Colour: cherry, purple rim. Nose: red berry notes, floral, spicy, balsamic herbs, scrubland. Palate: flavourful, fruity, good acidity, long.

DO YECLA
CONSEJO REGULADOR

Poeta Francisco A. Jiménez, s/n - P.I. Urbayecla II
30510 Yecla (Murcia)
☎: +34 968 792 352 - Fax: +34 968 792 352
@: consejo@yeclavino.com
www.yeclavino.com

LOCATION:
In the northeast of the province of Murcia, within the plateau region, and comprising a single municipal district, Yecla.

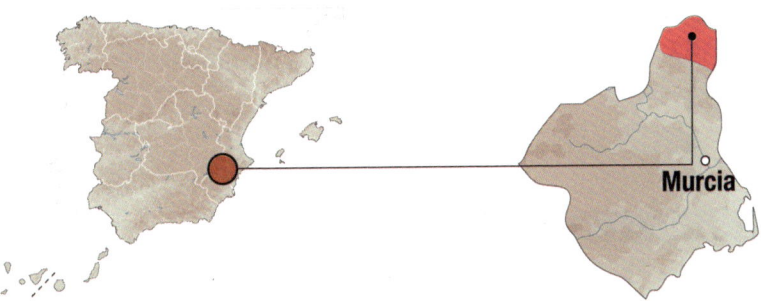

SUB-REGIONS:
Yecla Campo Arriba, with Monastrell as the most common variety and alcohol contents of up to 14°, and Yecla Campo Abajo, whose grapes produce a lower alcohol content (around 12° for reds and 11.5° for whites).

GRAPE VARIETIES:
WHITE: Merseguera, Airén, Macabeo, Malvasía, Chardonnay and Verdejo.

RED: Monastrell (majority 85% of total), Garnacha Tinta, Cabernet Sauvignon, Cencibel (Tempranillo), Merlot, Tintorera and Syrah.

FIGURES:
Vineyard surface: 4,349– Wine-Growers: 321 – Wineries: 9 – 2021 Harvest rating: Excellent– Production 21: 6,775,318 litres – Market percentages: 13% National - 87% International.

SOIL:
Fundamentally deep limestone, with good permeability. The vineyards are on undulating terrain at a height of between 400 m and 800 m above sea level.

CLIMATE:
Continental, with a slight Mediterranean influence, with hot summers and cold winters, and little rainfall, which is usually around 300 mm per annum.

CLASSIFICATION OF YOUNG WINE HARVESTS PEÑÍNGUIDE

2017	2018	2019	2020	2021
VERY GOOD	GOOD	VERY GOOD	VERY GOOD	N/A

DO YECLA / D.O.P.

ARTIGA FUSTEL
Progres, 19 Bajo
08720 Vilafranca del Penedès (Barcelona)
☎: +34 938 182 317
info@artiga-fustel.com
www.artiga-fustel.com

Gusto 2019 T BA
85% monastrell, 10% cabernet sauvignon, 5% garnacha tintorera

88

Spicy, jammy, full-bodied, roasted coffee.

ATLAN & ARTISAN
07010 Yecla (Murcia)
☎: +34 968 718 696
info@atlanandartisan.com
www.atlanandartisan.com

Epistem Nº 2 2018 T R
syrah

90 🌿

Colour: bright cherry, garnet rim. Nose: red berry notes, spicy, fine reductive notes, wild herbs. Palate: balanced, good acidity, fine tannins.

BODEGAS CASTAÑO
Ctra. Fuenteálamo, 3
30510 Yecla (Murcia)
☎: +34 968 791 115
Fax: +34 968 791 900
info@bodegascastano.com
www.bodegascastano.com

Detrás de la casa Monastrell 2016

93

Colour: deep cherry. Nose: dried herbs, creamy oak, tobacco, black fruit. Palate: powerful, ripe fruit, spicy, round tannins.

Altos de las Gateras Syrah 2017 T

94

Colour: deep cherry. Nose: dried herbs, creamy oak, black fruit, petrol notes. Palate: powerful, ripe fruit, spicy, round tannins, great length.

Casa Cisca 2017 T BA
100% monastrell

94

Colour: dark-red cherry, garnet rim. Nose: ripe fruit, fruit preserve, aged wood nuances, sweet spices, smoky, scrubland, tea leave. Palate: spicy, round tannins, long, powerful, flavourful.

Casa de la Cera 2017 T BA
50% monastrell, 50% garnacha tintorera, cabernet sauvignon, syrah, merlot

93

Colour: deep cherry. Nose: dried herbs, black fruit, meaty notes, tobacco. Palate: ripe fruit, spicy, round tannins, good structure, flavourful.

Castaño Colección 2018 T BA S
70% monastrell, 30% cabernet sauvignon

91

Colour: deep cherry. Nose: dried herbs, creamy oak, black fruit. Palate: powerful, ripe fruit, spicy, round tannins.

Castaño Monastrell Dulce 2018 T D
100% monastrell

89

Weary, overripe, spicy, tasty, wild.

Castaño Santa 2018 T BA
90% monastrell, 10% garnacha tintorera

92

Colour: cherry, garnet rim. Nose: dried herbs, creamy oak, black fruit, earthy notes, toasty. Palate: powerful, ripe fruit, spicy, round tannins.

Hécula Monastrell Organic 2020 T BA S
100% monastrell

90 🌿

Colour: deep cherry. Nose: ripe fruit, dried herbs, creamy oak. Palate: ripe fruit, spicy, round tannins.

Viña al lado de la Casa 2020 T

90

Smooth. Colour: deep cherry. Nose: dried herbs, black fruit, red berry notes. Palate: ripe fruit, spicy, round tannins, warm.

Viña Detrás de la Casa Garnacha Tintorera 2017 T C
garnacha tintorera
91
Colour: deep cherry. Nose: dried herbs, creamy oak, black fruit, earthy notes, dried flowers. Palate: powerful, ripe fruit, spicy, round tannins.

Viña Detrás de la Casa Syrah 2019 T
91
Colour: deep cherry. Nose: dried herbs, creamy oak, black fruit, violets. Palate: powerful, ripe fruit, spicy, round tannins.

BODEGAS LA PURÍSIMA
Ctra. de Pinoso, 3
30510 Yecla (Murcia)
☎: +34 968 751 257
Fax: +34 968 795 116
info@bodegaslapurisima.com
www.bodegaslapurisima.com

Consentido Monastrell Barrica 2019 T RB
100% monastrell
88
Colour: bright cherry. Nose: fruit preserve, black fruit, sweet spices. Palate: fruity, flavourful, fresh, slightly dry, soft tannins.

La Purísima Monastrell 2020 T
100% monastrell
88
Spicy, jammy, tasty, toasty.

La Purísima Old Vines Expression 2019 T RB
85% monastrell, 10% syrah, 5% garnacha
90
Colour: Cherry. Nose: new oak, black fruit, tea leave, scrubland. Palate: good structure, flavourful.

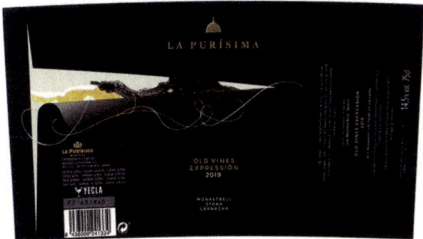

Old Hands 2019 T RB
100% monastrell
89
Spicy, fruity, jammy, tasty, toasty.

La Purísima Premium 2019 T BA
95% monastrell, 5% garnacha
90
Colour: cherry, purple rim. Nose: fruit preserve, smoky, caramel, expressive. Palate: flavourful, fresh, powerful tannins.

Trapío 2019 T RB
100% monastrell
90
Colour: very deep cherry. Nose: toasty, spicy, incense, dry nuts, herbaceous. Palate: flavourful, fresh, slightly green tannins.

BODEGAS TRENZA
Felix Mendelsohn, 8
03730 Jávea (Alacant/Alicante)
☎: +34 965 790 012
bodegas@bodegastrenza.com
www.bodegatrenza.com

La Nymphina Monastrell 2020 T
100% monastrell
90
Colour: bright cherry. Nose: ripe fruit, wild herbs, dried herbs, spicy. Palate: fruity, flavourful, balanced, good finish.

Tofterup Brothers Black Label Monastrell 2017 T
77% monastrell, 15% cabernet sauvignon, 8% garnacha tintorera
92
Colour: deep cherry. Nose: ripe fruit, black fruit, wild herbs, toasty, black liquorice. Palate: flavourful, fruity, good acidity, balanced, roasted-coffee aftertaste, slightly dry, soft tannins.

Tofterup Brothers Monastrell 2019 T
100% monastrell
90
Colour: bright cherry. Nose: ripe fruit, toasty, tomato, wild herbs. Palate: fruity, flavourful, slightly dry, soft tannins, balanced.

Tofterup Brothers Monastrell Barrel Select 2019 T
100% monastrell
91
Colour: bright cherry. Nose: ripe fruit, spicy, black fruit, wild herbs. Palate: flavourful, balanced, good acidity, good finish, harsh oak tannins.

Trenza Z-Strand 2020 T
monastrell, cabernet sauvignon, garnacha tintorera
88
Balanced, jammy, tasty, toasty, spicy.

DO YECLA / D.O.P.

DO YECLA / D.O.P.

Trenza Family Collection 2019 T
73% monastrell, 14% cabernet sauvignon, 13% garnacha tintorera

92

Colour: bright cherry. Nose: black fruit, ripe fruit, creamy oak, spicy, dried herbs, green pepper. Palate: powerful, flavourful, balanced, good acidity, smoky aftertaste, slightly dry, soft tannins.

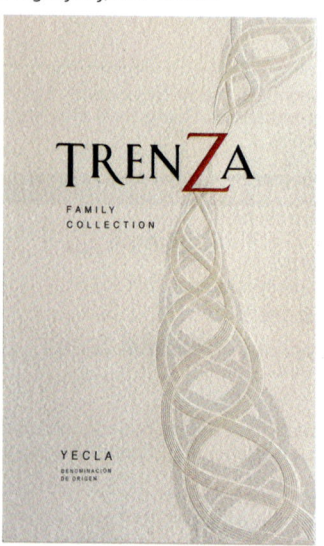

BOQUERA DEL CARCHE
Paraje Boquera del Carche, s/n
30510 Yecla (Murcia)
☎: +34 968 011 511
info@casaboquera.com
www.casaboquera.es

Casa Boquera 2017 T C
monastrell, petit verdot

90

Colour: cherry, purple rim. Nose: ripe fruit, black fruit, wild herbs, spicy, smoky. Palate: fruity, flavourful, fresh, powerful tannins.

Casa Boquera 2018 T RB
70% monastrell, 30% syrah

90

Colour: cherry, purple rim. Nose: ripe fruit, smoky, spicy, floral. Palate: flavourful, fruity, fresh, slightly dry, soft tannins.

Casa Boquera Selección 2017 T C
petit verdot

92

Colour: deep cherry. Nose: dried herbs, creamy oak, black fruit, old leather. Palate: powerful, ripe fruit, spicy, round tannins.

EVINE
Francisco Azorín, 18
30510 Yecla (Murcia)
☎: +34 653 997 673
bodegasevine@gmail.com
www.bodegasevine.es

Kyathos 2018 T C
monastrell

90

Colour: bright cherry. Nose: ripe fruit, black fruit, smoky, spicy. Palate: fruity, flavourful, slightly dry, soft tannins.

Llano Quintanilla 2018 T C
monastrell

86

María Sarmiento 2021 T
70% monastrell, 15% syrah, 15% garnacha tintorera

86

HAMMEKEN CELLARS
Llavador, 20
03700 Denia (Alacant/Alicante)
☎: +34 965 791 967
cellars@hammekencellars.com
www.hammekencellars.com

El Gringo Dark & Rich Monastrell 2020 T
monastrell

87

Gran Pasas Monastrell 2020 T
monastrell

90

Colour: bright cherry. Nose: fruit expression, fruit preserve, spicy, toasty. Palate: flavourful, balanced, fresh, good finish.

Pasas Sonrosado 2021 RD
monastrell

87

Pasas Uva Blanca 2021 B

87

Pasas Uva Tinta 2020 T
monastrell

88

Spicy, fruity, jammy, tasty.

Riurau Monastrell Apasionado 2020 T
monastrell

88

Balanced, jammy, floral, fruity.

VINOS DE PAGO

The "Vinos de Pago" are linked to a single winery, and it is a status given to that winery on the grounds of unique micro-climatic features and proven evidence of consistent high quality over the years, with the goal to produce wines of sheer singularity. So far, only 25 "Vinos de Pago" labels have been granted for different autonomous regions (Aragón, La Mancha, Comunidad Valenciana and Navarra). The "Vinos de Pago" category has the same status as a DO. This "pago" should not be confused with the other "pago" term used in the wine realm, which refers to a plot, a smaller vineyard within a bigger property. The "Pagos de España" association was formed in 2000 when a group of small producers of single estate wines got together to defend the singularity of their wines. In 2003, the association became Grandes Pagos de España, responding to the request of many colleagues in other parts of the country who wished to make the single-growth concept better known, and to seek excellence through the direct relationship between wines and their places of origin.

PAGO ABADÍA RETUERTA

560 hectares of vineyards in the municipality of Sardón de Duero, Valladolid, with the Duero River to the north and the „El Carrascal" mountain to the south as natural boundaries. It is located between two high moors, very close to each other, which provide cold air currents, while the river, at an altitude of 725 m, provides humidity. The area is characterised by a continental climate, with moderate rainfall (annual average of 450 mm) and dry summers. The soils have a sandy to clay loam texture, with sand predominating over almost the entire surface, although with high levels of clay as we approach the surface of the hillside. The estate has up to 10 red and 5 white grape varieties, the former being the most important to date due to their greater adaptation to the environment.

PAGO AYLES

Situated in the municipality of Mezalocha (Zaragoza), within the limits of the Cariñena appellation. The production area is located within the Ebro basin, principally around the depression produced by the River Huerva. The soils consist of limestone, marl and composites. The climate is temperate continental with low average annual rainfall figures of 350 to 550mm. The varieties authorized for the production of red and rosé wines are: garnacha, merlot, tempranillo and cabernet sauvignon.

PAGO CALZADILLA

Located in the Mayor river valley, in the part of the Alcarria region that belongs to the province of Cuenca, it enjoys altitude levels ranging between 845 and 1005 meters. The vines are mostly planted on limestone soils with pronounced slopes (with up to a 40% incline), so terraces and slant plots have become the most common feature, following the altitude gradients. The grape varieties planted are tempranillo, cabernet-sauvignon, garnacha and syrah.

PAGO CAMPO DE LA GUARDIA

The vineyards are in the town of La Guardia, to the northeast of the province of Toledo, on a high plateau known as Mesa de Ocaña. Soils are deep and with varying degrees of loam, clay and sand. The climate follows a continental pattern, with hot and dry summers and particularly dry and cold winters. The presence of the Tajo River to the north and the Montes de Toledo to the south promote lower rainfall levels than in neighbouring areas, and thus more concentration of aromas and phenolic compounds.

PAGO CASA DEL BLANCO

Its vineyards are located at an altitude of 617 metres in Campo de Calatrava, in the town of Manzanares, right in the centre of the province of Ciudad Real, and therefore with a mediterranean/continental climate. Soils have varying degrees of loam and sand, and are abundant in lithium, surely due to the ancient volcanic character of the region.

PAGO CHOZAS CARRASCAL

In San Antonio de Requena. This is the third Estate of the Community of Valencia, with just 31 hectares. Located at 720 metres above sea level. It has a continental climatology with Mediterranean influence. Low rainfall (and average of 350-400 litres annually), its soils are loam texture tending to clay and sandy. The varieties uses are: bobal, tempranillo, garnacha, cabernet sauvignon, merlot, syrah, cabernet franc and monastrell for red wines and chardonnay, sauvignon blanc and macabeo for white wines.

PAGO DE OTAZU

Its vineyards are located in Navarra, between two mountain ranges (Sierra del Perdón and Sierra de Echauri), and is probably the most northerly of all Spanish wine regions. It is a cool area with Atlantic climate and a high day-night temperature contrast. Soils in that part of the country, near the city of Pamplona, are limestone-based with abundant clay and stones, therefore with good drainage that allows vines to sink their roots deeper into the soil.

PAGO DE THARSYS

This Pago is located between the Castilian plateau and the Mediterranean, in the westernmost area of the province of Valencia. In the west and south it is bordered by the Cabriel river, and it's skirted on the east and north by the Pico Tejo and Juan Navarro mountain range, separating them from the Turia mountains. Its vines grow on clay loam soil and thrive in a Mediterranean climate with Atlantic influence. The varieties used are merlot, tempranillo, bobal, garnacha and cabernet franc for reds, and chardonnay and albariño for whites.

PAGO DEHESA DEL CARRIZAL

Property of Marcial Gómez Sequeira, Dehesa del Carrizal is located in the town of Retuerta de Bullaque, to the north of Ciudad Real. It enjoys a continental climate and high altitude (900 metres). The winemaker uses primarily foreign (French) varieties such as cabernet sauvignon.

PAGO DEHESA PEÑALBA

It is located in the municipality of Villabáñez, in the province of Valladolid, where it has an extension of 91 hectares. Its vineyard sits on a hot and poor soil with a high level of sand, pebbles and gravel. Only red wines are produced and the varieties allowed for their production are, in order of importance: tempranillo, syrah, cabernet sauvignon and merlot, the last three of which have a longer cycle and are able to ripen in an environment that is considerably more protected from frost and cold. The prevailing climate in the area is continental.

PAGO DOMINIO DE VALPEDUSA

Located in the town of Malpica de Tajo (Toledo), its owner, Carlos Falcó (Marqués de Griñón) pioneered the introduction in Spain of foreign grape varieties such as cabernet sauvignon.

PAGO EL TERRERAZO

El Terrerazo, property of Bodegas Mustiguillo, is the first "Vinos de Pago" label granted within the autonomous region of Valencia. It comprises 62 hectares at an altitude of 800 meters between Utiel and Sinarcas where an excellent clone of bobal –that yields small and loose berries– is grown. It enjoys a mediterranean-continental climate and the vineyard gets the influence of humid winds blowing from the sea, which is just 80 kilometres away from the property. Soils are characterized limestone and clay in nature, with abundant sand and stones.

PAGO EL VICARIO

Located in the municipality of Ciudad Real, among the first slopes of the Montes de Toledo, beside the Guadiana river at 638 meters above sea level. It has calcareous, light and shallow soils. The area covered by this appellation is 130.98 hectares where white varieties (chardonnay and sauvignon blanc) and red varieties (tempranillo, garnacha tinta, graciano, syrah, cabernet sauvignon, merlot and petit verdot) are grown.

PAGO FINCA BOLANDIN

140 hectares situated in the municipal area of Ablitas, in the southern limit of the province of Navarra, in the central area of the River Ebro Valley. This district is exposed to the influence of the Mediterranean which moves up the river valley. The vineyard is oriented to the south and has three types of soil; loam with abundant pebbles in the highest area, loam with silty clays with scarce stoniness in the upper half of the hillside and clay loam in the lower area. The authorised varieties are cabernet sauvignon, merlot, tempranillo and syrah for red wines and chardonnay, sauvignon blanc and small grain muscatel for white.

PAGO FINCA ÉLEZ

It became the first of all Vino de Pago designations of origin. Its owner is Manuel Manzaneque, and it is located at an altitude of 1000 metres in El Bonillo, in the province of Albacete. The winery became renown by its splendid chardonnay, but today also make a single-varietal syrah and some other red renderings.

PAGO FLORENTINO

Located in the municipality of Malagón (Ciudad Real), between natural lagoons to the south and the Sierra de Malagón to the north, at an altitude of some 630-670 metres. Soils are mainly siliceous with limestone and stones on the surface and a subsoil of slate and limestone. The climate is milder and dryer than that of neighbouring towns.

PAGO GUIJOSO

Finca El Guijoso is property of Bodegas Sánchez Muliterno, located in El Bonillo, between the provinces of Albacete and Ciudad Real. Surrounded by bitch and juniper woods, the vines are planted on stone (guijo in Spanish, from which it takes its name) soils at an altitude of 1000 metres. Wines are all made from French varieties, and have a clear French lean also in terms of style.

PAGO LA JARABA

Located in the municipalities of Villarrobledo (Albacete) and El Provencio (Cuenca), at 700 metres above sea level and with an extension of 75.18 hectares. Only red varieties are grown: Tempranillo, Cabernet Sauvignon, Merlot and Graciano. It offers sandy-loam and clay soils, with abundant alluvial elements.

PAGO LOS BALAGUESES

The "Pago de los Balagueses" is located to the south west of the Utiel-Requena wine region, just 20 kilometres away from Requena. At approximately 700 metres over the sea level, it enjoys a continental type of climate with mediterranean influence and an average annual rainfall of around 450 mm. The vines are planted on low hills –a feature that favours water drainage– surrounded by pines, almond and olive trees, thus giving shape to a unique landscape.

PAGO LOS CERRILLOS:

Located in the municipality of Argamasilla de Alba, Ciudad Real, where they grow tempranillo, cabernet sauvignon and syrah. It is an area that covers all of the high plain of the Guadiana River, in an area larger than 60 km2. It is located next to the Peñarroya reservoir, at an altitude of 695 metres, surrounded by hills and having primarily limestone soil. It has a continentalised Mediterranean climate.

PAGO PRADO DE IRACHE

Its vineyard is located in the municipality of Ayegui (Navarra) at an altitude of 450 metres. Climate is continental with strong Atlantic influence and soils are mainly of a loamy nature.

PAGO SEÑORIO DE ARINZANO

Sie befindet sich im Nordwesten Spaniens, genauer in Estella, Navarra. Ihr Weinstock wächst in einem Tal, das von den letzten Gebirgsausläufern der Pyrenäen gebildet wird, und das vom Fluss Ega, der die Rolle des Moderators der Temperaturen übernimmt, geteilt wird. Ihr Klima besitzt einen atlantischen Einfluss mit einem hohen thermischen Unterschied. Die Weinstöcke dieser Weinbergslagen befinden sich in einer komplexen geologischen Gegend mit unterschiedlichen Anteilen von Schlamm, Mergel, Ton und Degradierung von kalkigem Gestein.

PAGO URUEÑA

The first appellation of origin approved in Castilla León. Located in the municipality of Urueña, in the province of Valladolid, it has an area of 78 hectares where Tempranillo, Cabernet Sauvignon, Merlot and Syrah are grown. The vineyard is located on a wide plain at an altitude of 710 m.a.s.l, on sandy-clay-loam soils with good drainage capacity. The average annual rainfall here is around 410 mm.

PAGO VALLEGARCIA

It is located in the municipality of Retuerta del Bullaque, in the heart of the Montes de Toledo, in the province of Ciudad Real. The winery is integrated in one of the best examples of Mediterranean forest in the world, which preserves a flora and fauna of great ecological value and landscapes of untouched nature. It has an area of more than 1,500 hectares where both white varieties such as viognier, and reds (syrah, merlot, cabernet sauvignon, cabernet franc, petit verdot) are grown, as well as the recently incorporated red garnacha, cariñena and monastrell.

PAGO VERA DE ESTENAS

This is located in the area of Utiel-Requena, in the province of Valencia. It has a Mediterranean climate with a continental influence. Its soils are a dark chalky with a sandy clay loam texture. The average rainfall is 420 millimetres and the varieties planted are bobal, tempranillo, cabernet sauvignon and merlot for red wines and chardonnay for white wines.

PAGO ABADÍA RETUERTA

ABADÍA RETUERTA
Ctra. N-122, km. 332,5
47340 Sardón de Duero (Valladolid)
☎: +34 983 680 314
info@abadia-retuerta.es
www.abadia-retuerta.com

Abadía Retuerta Cuvée Palomar 2018 T R
40% garnacha, 40% tempranillo, 20% malbec

93

Colour: deep cherry. Nose: dried herbs, creamy oak, woody, black fruit, dark chocolate. Palate: powerful, ripe fruit, spicy, round tannins.

Abadía Retuerta Le Domaine 2021 B
80% sauvignon blanc, 20% verdejo

92 🌱

Colour: bright straw. Nose: ripe fruit, fine lees, wild herbs, hint of anise. Palate: full-bodied, good acidity, balanced, fine bitter notes.

Abadía Retuerta Pago Garduña 2018 T
100% syrah

93

Colour: deep cherry. Nose: dried herbs, creamy oak, black fruit, spicy, roasted coffee. Palate: ripe fruit, spicy, round tannins.

🏆 PODIUM

Abadía Retuerta Pago Negralada 2017 T BA
100% tempranillo

95

Colour: very deep cherry. Nose: complex, expressive, spicy, mineral, ripe fruit, black fruit. Palate: elegant, full-bodied, long, great length.

Abadía Retuerta Pago Valdebellón 2017 T R
100% cabernet sauvignon

93

Colour: cherry, garnet rim. Nose: powerful, ripe fruit, black fruit, balsamic herbs, grassy. Palate: flavourful, sweetness, long.

Abadía Retuerta Pago Valdebellón 2018 T R
93

Colour: deep cherry. Nose: dried herbs, creamy oak, black fruit, cocoa bean. Palate: powerful, ripe fruit, spicy, round tannins.

Abadía Retuerta Petit Verdot PV 2017 T
100% petit verdot

93

Rustic. Colour: very deep cherry. Nose: roasted coffee, aromatic coffee, powerful, grassy. Palate: smoky aftertaste, great length, round tannins, powerful.

Abadía Retuerta Selección Especial 2019 T
64% tempranillo, 15% cabernet sauvignon, 14% syrah, 6% merlot, 1% petit verdot

94

Colour: bright cherry. Nose: complex, expressive, spicy, mineral, red berry notes, ripe fruit. Palate: elegant, full-bodied, long, great length.

PAGO AYLÉS

BODEGA PAGO AYLÉS
Finca Aylés. Ctra. A-1101, Km. 24
50152 Mezalocha (Zaragoza)
☎: +34 976 140 473
pagoayles@pagoayles.com
www.pagoayles.com

Aylés "Tres de 3000" 2018 T
garnacha, merlot, cabernet sauvignon

93

Colour: dark-red cherry. Nose: toasty, spicy, cocoa bean, ripe fruit, black fruit. Palate: flavourful, toasty, fine bitter notes.

Cuesta del Herrero 2021 T BA
garnacha, tempranillo

92 🌱

Colour: cherry, purple rim. Nose: red berry notes, floral, spicy, sweet spices. Palate: flavourful, fruity, good acidity, long.

Patria Chica 2021 T
60% tempranillo, 26% garnacha, 12% merlot, 2% cabernet sauvignon

91 🌱

Colour: cherry, purple rim. Nose: red berry notes, floral, spicy. Palate: flavourful, fruity, good acidity.

Senda de los Leñadores 2020 T
garnacha, tempranillo

93

Colour: Cherry. Nose: complex, expressive, spicy, mineral, red berry notes, balsamic herbs. Palate: elegant, full-bodied, long, great length.

PAGO BOLANDIN

BODEGA PAGO DE CIRSUS
Ctra. de Ablitas a Ribafora, Km. 5
31523 Ablitas (Navarra)
☎: +34 948 386 427
info@pagodecirsus.com
www.pagodecirsus.com

Pago de Cirsus Chardonnay 2020 B FB
chardonnay

92

Colour: bright straw. Nose: ripe fruit, sweet spices, toasted bread. Palate: full-bodied, balanced, toasty, smoky aftertaste.

Pago de Cirsus Moscatel Vendimia Tardía 2017 B C D
moscatel

92

Colour: bright yellow. Nose: ripe fruit, honeyed notes, dried flowers, stone fruit. Palate: fruity, flavourful, spicy, sweet.

Pago de Cirsus Selección de Familia 2017 T C

93

Colour: deep cherry, garnet rim. Nose: aged wood nuances, ripe fruit, cocoa bean, cigar, toasty. Palate: flavourful, spicy, toasty, powerful tannins.

Pago de Cirsus La A T

92

Colour: Cherry. Nose: balsamic herbs, sweet spices, wild herbs, black fruit, ripe fruit. Palate: spicy, good acidity, round tannins.

Pago de Cirsus Cuvée Especial 2018 T

93

Colour: dark-red cherry, garnet rim. Nose: ripe fruit, aged wood nuances, tobacco, sweet spices. Palate: spicy, round tannins, long.

Pago de Cirsus Vendimia Seleccionada 2020 T C

91

Colour: deep cherry. Nose: ripe fruit, dried herbs, creamy oak, spicy. Palate: powerful, ripe fruit, spicy, round tannins.

PAGO CALZADILLA

BODEGA CALZADILLA
Ctra. Huete a Cuenca, Km. 3,3
16500 Huete (Cuenca)
☎: +34 969 143 020
info@pagocalzadilla.com
www.pagocalzadilla.com

Calzadilla Classic 2015 T
tempranillo, cabernet sauvignon, garnacha, syrah

93

Colour: dark-red cherry, garnet rim. Nose: ripe fruit, fruit preserve, aged wood nuances, tobacco, sweet spices. Palate: spicy, round tannins, long.

Gran Calzadilla 2012 T
cabernet sauvignon, tempranillo

93

Colour: deep cherry, garnet rim. Nose: aged wood nuances, ripe fruit, cocoa bean, cigar, toasty. Palate: flavourful, spicy, toasty, powerful tannins.

VINOS DE PAGO / D.O.P.

Opta Calzadilla 2017 T
tempranillo, syrah, garnacha
91
Colour: deep cherry. Nose: ripe fruit, dried herbs, creamy oak, toasty. Palate: powerful, ripe fruit, spicy, round tannins.

PAGO CAMPO DE LA GUARDIA

BODEGAS MARTUE
Campo de la Guardia, s/n
45760 La Guardia (Toledo)
☎: +34 658 915 812
admin@martue.com
www.martue.com

Martúe 20 Aniversario 2018 T
syrah, cabernet sauvignon, merlot, petit verdot, malbec
89
Jammy, herbal, balanced, tasty, smoky, spicy.

Martúe Chardonnay 2020 B BA
100% chardonnay
88
Pleasant, jammy, spicy, fruity, flat.

Martúe Especial 2017 T R
petit verdot, syrah, cabernet sauvignon, malbec
90
Colour: deep cherry. Nose: dried herbs, creamy oak, black fruit, fruit preserve, roasted coffee. Palate: powerful, ripe fruit, spicy, round tannins, flavourful.

Martúe Syrah 2018 T
syrah
90
Colour: deep cherry. Nose: ripe fruit, creamy oak, black fruit, meaty notes. Palate: powerful, ripe fruit, spicy, round tannins.

PAGO CASA DEL BLANCO

PAGO CASA DEL BLANCO
Ctra. Manzanares -Moral de Calatrava Km.23,200
13200 Manzanares (Ciudad Real)
☎: +34 619 306 251
comercial@pagocasadelblanco.com
www.pagocasadelblanco.com

Lithium 2019 T C
petit verdot
88
Roasted coffee, overripe, powerful.

Pilas Bonas 2021 B
chardonnay, sauvignon blanc
87

Quixote Cabernet Sauvignon Syrah 2018 T C
cabernet sauvignon, syrah
88
Roasted coffee, jammy, powerful, tasty.

Quixote Malbec Cabernet Franc 2018 T R
malbec, cabernet franc
90
Colour: dark-red cherry. Nose: toasty, spicy, cocoa bean, fruit preserve. Palate: flavourful, toasty, fine bitter notes.

Quixote Merlot Tempranillo Petit Verdot 2018 T
merlot, tempranillo, petit verdot
89
Overripe, powerful, spicy, kind finish.

Quixote Petit Verdot 2018 T C
petit verdot
89
Pruney, toasty, smoky, powerful.

PAGO CHOZAS CARRASCAL

CHOZAS CARRASCAL
Vereda Real
46390 San Antonio de Requena (València/Valencia)
☎: +34 963 410 395
chozas@chozascarrascal.es
www.chozascarrascal.com

🏆 **PODIUM**

El Cf de Chozas Carrascal 2017 T
cabernet franc
95 🌿
Colour: cherry, purple rim. Nose: black fruit, varietal, spicy, smoky, ripe fruit. Palate: flavourful, fruity, balanced.

Las Ocho 2018 T
bobal, monastrell, garnacha, tempranillo, syrah, cabernet sauvignon
94 🌿
Colour: dark-red cherry. Nose: black fruit, ripe fruit, black liquorice, wild herbs, spicy. Palate: flavourful, full-bodied, fresh, balanced, good finish, slightly dry, soft tannins.

Las Tres 2019 B
macabeo, chardonnay, sauvignon blanc

92 🌿

Colour: yellow. Nose: ripe fruit, stone fruit, scrubland, faded flowers, bakery. Palate: flavourful, full-bodied, balanced.

PAGO DE OTAZU

BODEGA OTAZU
Señorío de Otazu, s/n
31174 Otazu (Navarra)
☎: +34 948 329 200
Fax: +34 948 329 353
lboursier@bodegaotazu.es
www.otazu.com

Otazu 1 Ha
Una Historia Chardonnay 2018 B
100% chardonnay

92

Colour: bright yellow. Nose: ripe fruit, lactic notes, faded flowers, sweet spices, creamy oak. Palate: fruity, flavourful, fresh, good finish, smoky aftertaste.

Pago de Otazu 2019 T R
62% merlot, 38% cabernet sauvignon

92

Colour: dark-red cherry. Nose: toasty, spicy, cocoa bean, black fruit, ripe fruit. Palate: flavourful, toasty, fine bitter notes.

Pago de Otazu
Chardonnay con Crianza 2019 B
100% chardonnay

92

Colour: bright yellow. Nose: powerful, creamy oak, ripe fruit, spicy. Palate: rich, good structure, long, toasty, fine bitter notes.

PAGO DE THARSYS

PAGO DE THARSYS
Ctra. Nacional III, km. 274
46340 Requena (València/Valencia)
☎: +34 962 303 354
bodega@pagodetharsys.com
www.pagodetharsys.com

Pago de Tharsys Argila 2018 T
merlot

90 🌿

Colour: Cherry. Nose: balsamic herbs, sweet spices, scrubland, wild herbs, earthy notes. Palate: spicy, balsamic, good acidity.

Pago de Tharsys
Bobal Diana García 2019 T
bobal

89 🌿

Balsamic herbs, jammy, fruity, wild, balanced.

Pago de Tharsys
Vendimia Nocturna 2021 B
albariño

88 🌿

Balanced, citrus fruit, herbal, crisp, tasty.

Pago de Tharsys Vendimia
Nocturna Garnacha 2021 RD FB
garnacha

89 🌿

Dried flowers, full-bodied, jammy, smooth, lactic.

PAGO DEHESA DEL CARRIZAL

DEHESA DEL CARRIZAL
Ctra. Retuerta a Navas de Estena, Km 5
13194 Retuerta del Bullaque (Ciudad Real)
☎: +34 925 421 773
bodega@dehesadelcarrizal.com
www.dehesadelcarrizal.com

Dehesa del Carrizal
Cabernet Sauvignon 2019 T
cabernet sauvignon

93

Colour: cherry, purple rim. Nose: black fruit, black liquorice, green pepper, varietal, spicy. Palate: fruity, flavourful, full-bodied, balanced, powerful tannins.

Dehesa del Carrizal
Chardonnay 2020 B FB
chardonnay

91

Colour: bright yellow. Nose: powerful, ripe fruit, spicy, dried flowers. Palate: good structure, toasty, fine bitter notes, flavourful.

Dehesa del Carrizal
Colección Privada 2019 T
cabernet sauvignon, petit verdot, syrah

93

Colour: cherry, purple rim. Nose: black fruit, wild herbs, spicy, scrubland. Palate: fruity, flavourful, powerful, balanced.

Dehesa del Carrizal Syrah 2019 T
syrah

92

Colour: deep cherry. Nose: dried herbs, creamy oak, black fruit. Palate: ripe fruit, spicy, round tannins.

VINOS DE PAGO / D.O.P.

Dehesa del Carrizal MV 2019 T
cabernet sauvignon, syrah, merlot, tempranillo, petit verdot

91

Colour: cherry, purple rim. Nose: black fruit, ripe fruit, dried herbs, smoky, spicy. Palate: fruity, flavourful, fresh, slightly dry, soft tannins.

Dehesa del Carrizal Petit Verdot 2019 T
petit verdot

93

Colour: cherry, purple rim. Nose: black fruit, red berry notes, lactic notes, smoky, spicy. Palate: flavourful, fruity, balanced, slightly dry, soft tannins.

PAGO DEHESA PEÑALBA

BODEGAS VIZAR
Ctra. N-122. Km. 341
47329 Villabáñez (Valladolid)
☎: +34 983 682 690
vadalia@bodegasvizar.es
www.bodegasvizar.es

Vizar Selección Especial 2018 T R
tempranillo, syrah

91 🌱

Roasted coffee. Colour: cherry, purple rim. Nose: toasty, sweet spices, black fruit, ripe fruit, dried herbs. Palate: fruity, flavourful, slightly dry, soft tannins, powerful.

PAGO EL TERRERAZO

BODEGA MUSTIGUILLO
El Terrerazo Ctra. N-330 km. 195
46300 Utiel (València/Valencia)
☎: +34 962 168 260
info@bodegamustiguillo.com
www.bodegamustiguillo.com

Finca Terrerazo 2019 T
bobal

93 🌱

Colour: bright cherry. Nose: fruit expression, red berry notes, floral, spicy, scrubland, balsamic herbs. Palate: flavourful, fruity, good acidity, spicy, grainy tannins.

Mestizaje 2020 T
bobal, garnacha, syrah

91 🌱

Colour: bright cherry. Nose: ripe fruit, dried herbs. Palate: spicy, round tannins, fruity, flavourful, long.

🏆 **PODIUM**

Quincha Corral 2019 T
bobal

96 🌱

Colour: cherry, garnet rim. Nose: complex, expressive, spicy, mineral, red berry notes, elegant. Palate: full-bodied, long, great length, flavourful, fruity, complex, taut, easy to drink.

PAGO EL VICARIO

PAGO DEL VICARIO
Ctra. Ciudad Real - Porzuna, Km. 16
13196 Las Casas (Ciudad Real)
☎: +34 926 666 027
pedidos@pagodelvicario.com
www.pagodelvicario.com

Pago del Vicario 50-50 2018 T C
tempranillo, cabernet sauvignon

89

Fruity, herbal, jammy, tasty.

Pago del Vicario 6 meses 2020 T
tempranillo, garnacha, merlot, petit verdot

89

Pleasant, spicy, jammy, tasty.

Pago del Vicario Bancal del Río 2017 T
petit verdot

89

Smoky, tasty, powerful, jammy, hot.

Pago del Vicario Blanco de Tempranillo 2021 B
100% tempranillo

89
Kind finish, jammy, tasty, dried flowers.

Pago del Vicario Petit Verdot 2021 RD
100% petit verdot

91
Colour: brilliant rose, purple rim. Nose: red berry notes, lactic notes, floral. Palate: balanced, fruity aftertaste, easy to drink.

Pago del Vicario Talva 2019 B FB
chardonnay, sauvignon blanc, tempranillo, garnacha

91
Colour: bright yellow. Nose: ripe fruit, dried herbs, faded flowers. Palate: powerful, ripe fruit, balanced.

PAGO FINCA ÉLEZ

PAGO FINCA ÉLEZ
Ctra. Ossa de Montiel a El Bonillo, Km 11,5
02610 El Bonillo (Albacete)
☎: +34 626 882 250
administracion@fincaelez.com
www.pagofincaelez.com

Finca Élez Chardonnay Lías 2020 B
chardonnay

91 ♣
Colour: bright straw. Nose: fine lees, white fruit, ripe fruit, neat. Palate: full-bodied, rich, long, good acidity.

Finca Élez 2020 T
merlot, tempranillo, syrah

91 ♣
Colour: cherry, purple rim. Nose: fruit expression, red berry notes, floral, spicy. Palate: flavourful, fruity, good acidity, long.

PAGO FLORENTINO

BODEGAS LA SOLANA - PAGO FLORENTINO
Ctra. Porzuna -
Camino Cristo del Humilladero, km. 3
13420 Malagón (Ciudad Real)
☎: +34 983 681 146
Fax: +34 983 681 147
bodeg@arzuaganavarro.com
www.pagoflorentino.com

Pago Florentino 2019 T C
100% cencibel

89
Herbal, toasty, jammy, smoky, standard.

PAGO GUIJOSO

BODEGAS FAMILIA CONESA - PAGO GUIJOSO
Crta Ossa de Montiel - El Bonillo km 11
02610 El Bonillo (Albacete)
☎: +34 608 612 254
mvruiz@familiaconesa.com
www.familiaconesa.com

Finca La Sabina Cabernet 2016 T GR
100% cabernet sauvignon

89 ♣
Varietally correct, wild, jammy, fruity, spicy.

Finca La Sabina Merlot 2016 T
100% merlot

89 ♣
Fruity, jammy, spicy, balanced.

Finca La Sabina Syrah 2017 T C
100% syrah

89 ♣
Fruity, herbal, wild, pronounced acidity.

VINOS DE PAGO / D.O.P.

PAGO HEREDAD DE URUEÑA

PAGO HEREDAD DE URUEÑA
Ctra. Toro a Medina de Rioseco, km 21,400
47862 Urueña (Valladolid)
☎: +34 915 610 894
bodega@heredaduruena.com
www.heredaduruena.com

Santo Syrah 2016 T
100% syrah
92
Colour: dark-red cherry. Nose: toasty, spicy, cocoa bean, aromatic coffee, fruit liqueur notes. Palate: flavourful, toasty, fine bitter notes, fruity.

Santo Tempranillo 2015 T
100% tempranillo
92
Colour: dark-red cherry. Nose: toasty, spicy, fruit preserve, dried herbs. Palate: flavourful, toasty, concentrated, slightly dry, soft tannins.

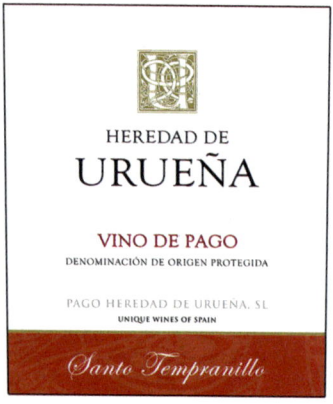

Selección Excelencia 2015 T C
tempranillo, syrah, merlot, cabernet sauvignon
91
Colour: deep cherry. Nose: ripe fruit, dried herbs, smoky, spicy, black fruit. Palate: ripe fruit, spicy, round tannins, flavourful.

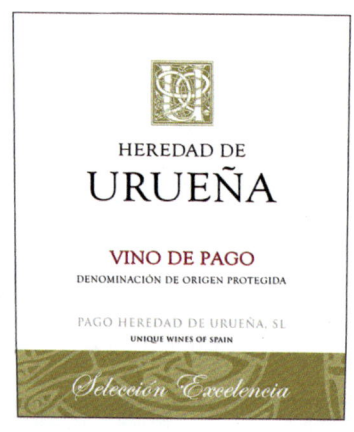

Santo Merlot 2016 T
90% merlot, 10% cabernet sauvignon
91
Colour: bright cherry. Nose: fruit expression, red berry notes, spicy, dried herbs. Palate: flavourful, fruity, good acidity, good finish.

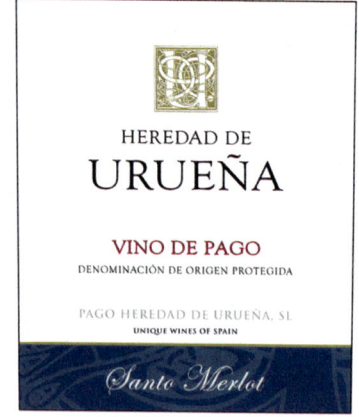

PAGO LA JARABA

PAGO DE LA JARABA

Ctra. Nacional 310, Km. 142,7
02600 Villarrobledo (Albacete)
☎: +34 967 138 250
Fax: +34 967 138 252
info@lajaraba.com
www.lajaraba.com

Azagador de la Jaraba Cosecha 2020 T BA
tempranillo, cabernet sauvignon, merlot

87

Azagador de la Jaraba Selección Especial 2019 T BA
70% tempranillo, 20% cabernet sauvignon, 10% malbec

89

Balanced, dried herbs, jammy, fruity, smoky.

Viña Jaraba 2019 T C
tempranillo, cabernet sauvignon, merlot

89

Dried herbs, jammy, kind finish, spicy, tasty.

Pago de la Jaraba 2020 T
tempranillo, cabernet sauvignon, merlot

91

Colour: bright cherry. Nose: sweet spices, ripe fruit, dark chocolate, dried herbs. Palate: spicy, round tannins.

Pago de la Jaraba Sauvignon Blanc 2021 B
sauvignon blanc

87

PAGO LOS BALAGUESES

BODEGAS VEGALFARO

Ctra. Pontón - Utiel, Km. 3
46340 Requena (València/Valencia)
☎: +34 962 320 680
info@vegalfaro.com
www.vegalfaro.com

Pago de los Balagueses Chardonnay 2021 B FB
chardonnay

91 ⚘

Colour: bright yellow. Nose: ripe fruit, spicy, powerful, toasted bread, stone fruit. Palate: rich, long, toasty, fine bitter notes.

Pago de los Balagueses Garnacha Tintorera 2020 T C
garnacha tintorera

93 ⚘

Rustic, tasty. Colour: Cherry. Nose: scrubland, wild herbs, spicy. Palate: flavourful, balanced, fine bitter notes.

Pago de los Balagueses Syrah 2020 T C
syrah

93 ⚘

Colour: deep cherry. Nose: ripe fruit, dried herbs, creamy oak, earthy notes. Palate: powerful, ripe fruit, spicy, round tannins.

Pasamonte Tintorera 2020 T
garnacha tintorera

92 ⚘

Colour: deep cherry. Nose: ripe fruit, spicy, dried flowers. Palate: ripe fruit, spicy, round tannins, balanced.

PAGO LOS CERRILLOS

BODEGAS MONTALVO WILMOT

Ctra. Ruidera, Km. 10,200 Finca los Cerrillos
13710 Argamasilla de Alba (Ciudad Real)
☎: +34 926 699 069
info@montalvowilmot.com
www.montalvowilmot.com

Montalvo Wilmot Petit Verdot 2019 T C
100% petit verdot

90

Colour: Cherry. Nose: balsamic herbs, sweet spices, scrubland, dried herbs. Palate: spicy, ripe fruit, round tannins.

Montalvo Wilmot Syrah 2020 T RB
100% syrah

89

Jammy, toasty, tasty, great length, spicy, hot.

Montalvo Wilmot Tempranillo Cabernet 2019 T RB
tempranillo, cabernet sauvignon

89

Exuberant, sweeties, powerful, coarse tannins. Nose: black fruit.

VINOS DE PAGO / D.O.P.

VINOS DE PAGO / D.O.P.

PAGO PRADO DE IRACHE

BODEGAS IRACHE
Avda. de Monasterio, 1
31240 Ayegui (Navarra)
☎: +34 948 551 608
irache@irache.com
www.irache.com

Prado Irache 2019 T BA
93
Colour: bright cherry. Nose: tobacco, toasty, black pepper, spicy, black fruit, ripe fruit, dried herbs. Palate: fruity, flavourful, full-bodied, balanced, roasted-coffee aftertaste, good finish, slightly dry, soft tannins.

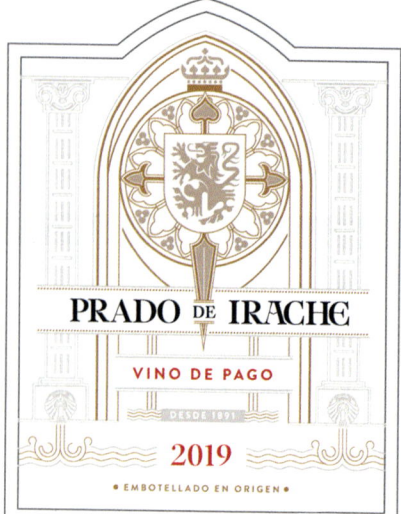

PAGO SEÑORIO DE ARÍNZANO

PROPIEDAD DE ARÍNZANO
Ctra. NA-132 Km. 3,1
31264 Aberin (Navarra)
☎: +34 689 934 018
cgarro@arinzano.com
www.arinzano.com

A de Arínzano 2021 RD
100% tempranillo
91
Colour: raspberry rose. Nose: elegant, red berry notes, floral, fragrant herbs. Palate: light-bodied, spicy, good acidity, fine bitter notes.

Arínzano Gran Vino 2017 B
100% chardonnay
93
Colour: bright yellow. Nose: dried flowers, candied fruit, fine lees, pattiserie. Palate: round, spicy, long, great length.

Arínzano Merlot Biológico 2018 T
100% merlot
92
Colour: dark-red cherry. Nose: toasty, spicy, cocoa bean, ripe fruit, black fruit. Palate: flavourful, toasty, fine bitter notes.

Hacienda de Arínzano 2018 T
85% tempranillo, 10% merlot, 5% cabernet sauvignon
92
Colour: deep cherry. Nose: ripe fruit, dried herbs, creamy oak, toasty, characterful, expressive. Palate: powerful, ripe fruit, spicy, round tannins.

Hacienda de Arínzano Chardonnay 2019 B
100% chardonnay
93
Colour: bright yellow. Nose: powerful, creamy oak, ripe fruit, spicy. Palate: good structure, toasty, fine bitter notes.

La Casona 2015 T
75% tempranillo, 25% merlot
93
Colour: cherry, garnet rim. Nose: overripe fruit, creamy oak, hot, toasty, dark chocolate. Palate: pruney, powerful, sweet tannins.

PAGO VALLEGARCÍA

PAGO DE VALLEGARCÍA
Finca Vallegarcía, s/n
13194 Retuerta del Bullaque (Ciudad Real)
☎: +34 925 421 407
comercial@vallegarcia.com
www.vallegarcia.com

Hipperia 2019 T
66% cabernet sauvignon, 21% cabernet franc, 10% merlot, 3% petit verdot

92

Colour: deep cherry. Nose: ripe fruit, dried herbs, creamy oak, dark chocolate. Palate: ripe fruit, spicy, round tannins, elegant.

Petit Hipperia 2020 T
35% cabernet franc, 34% syrah, 24% petit verdot, 6% merlot, 1% cabernet sauvignon

92

Colour: deep cherry. Nose: ripe fruit, dried herbs, creamy oak. Palate: powerful, ripe fruit, spicy, round tannins.

Vallegarcía Garnacha Cariñena 2020 T
garnacha, cariñena, syrah, monastrell

92

Colour: Cherry, garnet rim. Nose: complex, expressive, spicy, ripe fruit, black fruit, creamy oak. Palate: full-bodied, long, great length.

Vallegarcía Syrah 2019 T
100% syrah

91

Colour: very deep cherry. Nose: roasted coffee, aromatic coffee, powerful. Palate: smoky aftertaste, great length, round tannins

Vallegarcía Syrah 2020 T
syrah

93

Colour: deep cherry. Nose: ripe fruit, dried herbs, creamy oak, black fruit. Palate: powerful, ripe fruit, spicy, round tannins.

Vallegarcía Viognier 2020 B
100% viognier

92

Colour: bright straw. Nose: ripe fruit, dried herbs, faded flowers. Palate: powerful, ripe fruit, balanced.

PAGO VERA ESTENAS

BODEGA VERA DE ESTENAS
Ctra. N-III, km. 266 - Paraje La Cabezuela
46300 Utiel (València/Valencia)
☎: +34 962 171 141
estenas@veradeestenas.es
www.veradeestenas.es

Martínez Bermell Merlot 2019 T C
merlot

90

Colour: Cherry. Nose: sweet spices, scrubland, tobacco. Palate: spicy, balsamic, slightly dry, soft tannins.

Vera de Estenas 2018 T R

92

Colour: dark-red cherry, garnet rim. Nose: ripe fruit, fruit preserve, aged wood nuances, sweet spices, dried herbs. Palate: spicy, long, slightly dry, soft tannins.

Viña Lidón 2021 B FB
chardonnay

89

Jammy, spicy, floral, tasty, smoky.

VINOS DE PAGO / D.O.P.

VINOS DE CALIDAD

So far, there are only seven wine regions that have achieved the status "Vino de Calidad" ("Quality Wine Produced in Specified Regions"): Cangas, Cebreros, Lebrija, Valtiendas, Sierra de Salamanca, Valles de Benavente and Islas Canarias, regions that are allowed to label their wines with the VCPRD seal. This quality seal works as a sort of "training" session for the DO category, although it is still quite unknown for the average consumer.

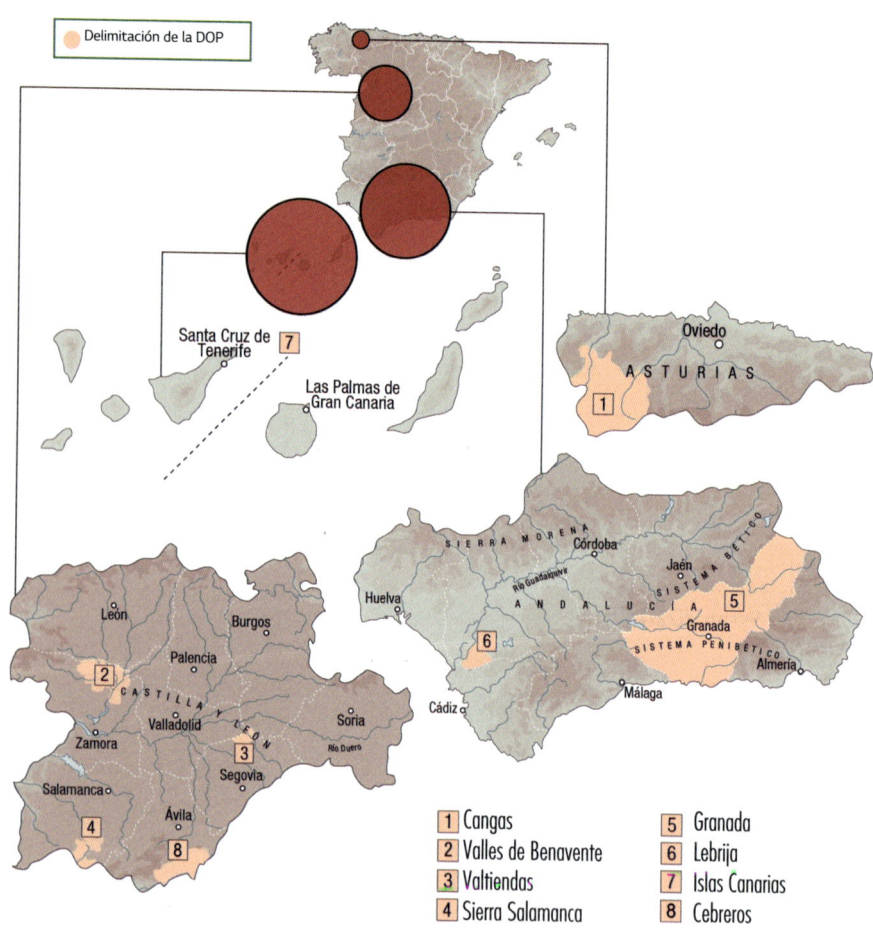

1. Cangas
2. Valles de Benavente
3. Valtiendas
4. Sierra Salamanca
5. Granada
6. Lebrija
7. Islas Canarias
8. Cebreros

CANGAS

Located to the south-eastern part of the province of Asturias, bordering with León, Cangas del Narcea has unique climatic conditions, completely different to the rest of the municipalities of Asturias; therefore, its wines have sheer singularity. With lower rainfall levels and more sunshine hours than the rest the province, vines are planted on slate, siliceous and sandy soils. The main varieties are albarín blanco and albillo (white), along with garnacha tintorera, mencía and verdejo negro (red).

CEBREROS

It's located to the south of the region of Castilla y León, in the province of Ávila, and it includes 35 municipalities in the province. The vineyard grows mainly on granite soil, with slate in certain areas. It is subject to a Mediterranean climate with continental influence, though in some parts it can have a more mountain climate in vines over 900m above sea level. The area is defined by the Sistema Central Ibérico (Central-Iberian mountain ranges), particularly by the Gredos mountain range and the rivers Alberche and Tiétar, both tributaries of the Tajo river. The main varieties are albillo real for whites and garnacha for reds, though tempranillo and garnacha tintorera are also permitted as secondary reds.

ISLAS CANARIAS

Approved in May 2011, the date of publication in the Boletín Oficial de Canarias (BOC), the constitution of its management board took place on 27 December, 2012. The production area covers the entire territory of the Canary Islands, allowing free movement of grapes in the Canary Islands. Its regulations cover broad grape varieties from the Canary Islands, as well as international ones.

LEBRIJA

Recognized by the Junta de Andalucía on March the 11th 2009. The production area includes the towns of Lebrija and El Cuervo, in the province of Sevilla.

The wines ascribed to the "Vino de Calidad de Lebrija" designation of quality will be made solely from the following grape varieties:

– **White varieties:** moscatel de Alejandría, palomino, palomino fino, sauvignon blanc and that traditionally known as vidueño (montúo de pilas, mollar cano, moscatel morisco, perruno).

– **Red varieties:** cabernet sauvignon, syrah, tempranillo, merlot and tintilla de Rota.

– **Types of wines:** white, red, generosos (fortified) and generosos de licor, naturally sweet and mistelas.

SIERRA DE SALAMANCA

The "Vino de Calidad" status was ratified to Sierra de Salamanca by the Junta de Castilla y León (Castilla y León autonomous government) in June 2010, becoming the third one to be granted within the region. Sierra de Salamanca lies in the south of the province of Salamanca, and includes 26 towns, all within the same province. Vines are planted mostly on terraces at the top of the hills and on clay soils based on limestone. Authorized varieties are viura, moscatel de grano menudo and palomino (white), as well as rufete, garnacha and tempranillo (red).

VALTIENDAS

An area to the north of the province of Segovia relatively known thanks to the brand name Duratón, also the name of the river that crosses a region that has mainly tempranillo planted, a grape variety known there also as tinta del país. The wines are fruitier and more acidic than those from Ribera del Duero, thanks to an altitude of some 900 metres and clay soils with plenty of stones.

VALLES DE BENAVENTE

Recognized by the Junta de Castilla y León in September 2000, the VCPRD comprises nowadays more than 50 municipalities and three wineries in Benavente, Santibáñez de Vidriales and San Pedro de Ceque. The production areas within the region are five (Valle Vidriales, Valle del Tera, Valle Valverde, La Vega and Tierra de Campos) around the city of Benavente, the core of the region. Four rivers (Tera, Esla, Órbigo and Valderadey, all of them tributary to the Duero river) give the region its natural borders.

VINO DE CALIDAD CANGAS

BODEGA MONASTERIO DE CORIAS
Monasterio de Corias, s/n
33800 Cangas del Narcea (Asturias)
☎: +34 985 810 493
bodega@monasteriodecorias.es
www.monasteriodecorias.es

Corias Guilfa 2020 B FB
91
Colour: bright yellow. Nose: creamy oak, ripe fruit, spicy, ebb tide, mineral. Palate: good structure, toasty, fine bitter notes.

Monasterio de Corias Finca los Frailes 2020 T
91
Colour: cherry, purple rim. Nose: fruit expression, red berry notes, floral, spicy. Palate: flavourful, fruity, good acidity, long.

Monasterio de Corias Pago del Narcea 2018 T
carrasquín, albarín negro, verdejo negro
91
Colour: deep cherry. Nose: dried herbs, black fruit, earthy notes, undergrowth. Palate: ripe fruit, spicy, round tannins.

Monasterio de Corias Viña Grandiella 2020 B
88
Citrus fruit, herbal, crisp, tasty, floral.

Valdemonje 2018 T
albarín negro
91
Herbal, representative, wild. Colour: cherry, purple rim. Nose: red berry notes, fresh fruit, spicy. Palate: fruity, fresh, easy to drink.

Valdemonje verdejo negro 2018 T
verdejo negro
89
Pronounced acidity, fruity, crisp, floral, herbal.

VINO DE CALIDAD DE CEBREROS

ARNACH
Dominio de Arroyo Quemao
05111 San Juan de La Nava (Ávila)
☎: +34 627 936 076
trescinco@live.com

Arnach 2019 T
garnacha
91
Colour: Cherry. Nose: scrubland, dried flowers, ripe fruit. Palate: spicy, good acidity, long.

BODEGA VETTONES S.C.
Plaza del Pilar, 4
05270 El Tiemblo (Ávila)
☎: +34 649 450 699
vetton16@gmail.com

Manliana 2020 T C
garnacha
90
Colour: Cherry. Nose: balsamic herbs, sweet spices, scrubland. Palate: spicy, balsamic, good acidity.

BODEGAS ARRAYÁN
Finca La Verdosa, s/n
28002 Santa Cruz del Retamar (Toledo)
☎: +34 916 633 131
comercial@arrayan.es
www.arrayan.es

Arrayán Albillo Real Granito 2020 B
albillo real
92
Colour: straw. Nose: ripe fruit, dried herbs, fine lees, smoky, red clay notes. Palate: ripe fruit, balanced, rich, warm.

BODEGAS AUSÍN
Cuesta de Santa María s/n
05460 Gavilanes (Ávila)
☎: +34 677 464 501
info@bodegasausin.com
www.bodegasausin.com

Fácil 2020 B
100% albillo real
91
Colour: bright golden. Nose: ripe fruit, fragrant herbs, fine lees. Palate: full-bodied, rich, long, good acidity.

Julia Esencia de Mmadre 2021 RD
100% garnacha
88
Hot, tasty, jammy, fruity.

Mmadre 2018 T
100% garnacha

90

Colour: bright cherry. Nose: toasty, spicy, cocoa bean, ripe fruit. Palate: flavourful, toasty, fine bitter notes.

COMPAÑIA DE VINOS TELMO RODRÍGUEZ

El Monte
01308 Lanciego (Araba/Álava)
☎: +34 945 628 315
contact@telmorodriguez.com
www.telmorodriguez.com

🏆 **PODIUM**

Arrebatacapas 2019 T
garnacha

95

Taut. Colour: Cherry. Nose: balsamic herbs, sweet spices, scrubland, red berry notes. Palate: spicy, balsamic, good acidity.

Pegaso "Zeta" 2020 T

92

Colour: bright cherry. Nose: balsamic herbs, sweet spices, scrubland, red berry notes, ripe fruit. Palate: spicy, balsamic, good acidity.

Pegaso "Barrancos de Pizarra" 2019 T
garnacha

93

Colour: cherry, garnet rim. Nose: fruit preserve, fruit liqueur notes, powerful, red berry notes, metalic. Palate: flavourful, sweetness, long.

Pegaso "Granito" 2019 T
garnacha

94

Colour: bright cherry. Nose: balsamic herbs, sweet spices, scrubland, red berry notes, fruit liqueur notes. Palate: spicy, balsamic, good acidity.

DANIEL RAMOS

San Pedro de Alcántara, 1
05270 El Tiemblo (Ávila)
☎: +34 687 410 952
dvrcru@gmail.com
www.danielramos.wine

Zerberos Clos Pepi 2020 T BA
garnacha, listán prieto, chelva

93 🍃

Personality. Colour: Cherry, bright ochre rim. Nose: fruit liqueur notes, dried herbs, black liquorice, faded flowers. Palate: powerful, flavourful, fruity, balanced, round tannins.

Zerberos El Altar 2020 T
garnacha, chelva

90 🍃

Wild. Colour: light cherry, orangey edge. Nose: fruit liqueur notes, wild herbs, black liquorice. Palate: flavourful, warm, long, sweetness.

Zerberos Llano Toledo 2020 T
garnacha, cariñena, jaen negro

92 🍃

Age nuances, oxidativ. Colour: light cherry, orangey edge. Nose: fruit preserve, fruit liqueur notes, dried flowers, faded flowers, caramel, characterful. Palate: sweetness, flavourful.

EME ERRE VIÑADORES

32950 Ferreiros (Ourense/Orense)
maite.maestre@videssingulares.com
www.videssingulares.com

Com_Pasión 2019 T
garnacha

91

Colour: bright cherry. Nose: balsamic herbs, scrubland, earthy notes. Palate: spicy, balsamic, good acidity.

Con_Sentido 2019 B
albillo real

92

Little interventionist. Colour: bright straw. Nose: ripe fruit, fragrant herbs, fine lees. Palate: full-bodied, rich, long, good acidity.

LAS PEDRERAS VIÑEDOS Y VINOS

Herguijuelas, 5
05114 Villanueva de Ávila (Ávila)
☎: +34 616 520 572
info.laspedreras@gmail.com

Arquitón 2021 RD
garnacha

90

Colour: raspberry rose. Nose: fruit expression, red berry notes, floral. Palate: fruity, good acidity, easy to drink.

Las Ánimas 2020 T
garnacha

94

Nominado Vino Revelación

Pleasant, balsamic herbs. Colour: bright cherry. Nose: red berry notes, fruit expression, wild herbs, spicy. Palate: flavourful, fruity, mineral.

Los Arroyuelos 2020 T
garnacha

93

Colour: bright cherry. Nose: sweet spices, raspberry, red berry notes, balsamic herbs, grassy. Palate: spicy, balsamic, good acidity.

PENÍNSULA VINICULTORES
Hermosilla 29, 1°A
28008 Madrid (Madrid)
☎: +34 916 039 363
o.moreno@peninsula.wine
www.peninsula.wine

Península Cebreros 2020 T
garnacha
91
Colour: bright cherry. Nose: balsamic herbs, sweet spices, scrubland, red berry notes. Palate: spicy, balsamic, good acidity.

RICO NUEVO VITICULTORES
Las Razuelas, 4
05113 Burgohondo (Ávila)
☎: +34 657 459 360
bodega@riconuevovinos.es
www.riconuevovinos.es

Al Raso 2021 B
garnacha
92 🌿
Colour: raspberry rose. Nose: red berry notes, floral, fragrant herbs. Palate: spicy, good acidity, fine bitter notes.

Barrera de Sol 2020 T BA
garnacha
93 🌿
Colour: deep cherry. Nose: dried herbs, spicy, ripe fruit, black fruit, violet drops. Palate: ripe fruit, spicy, flavourful. taut.

Flor de Albihar 2021 B
albillo real
93 🌿
Colour: bright straw. Nose: expressive, white flowers, jasmine, dried herbs, pattiserie. Palate: flavourful, balanced, ripe fruit.

Jirón de Niebla 2020 T C
garnacha
94 🌿
Colour: bright cherry. Nose: sweet spices, scrubland, red berry notes, fruit liqueur notes. Palate: spicy, balsamic, good acidity, rich, fruity.

La Quebrá 2019 T BA
92 🌿
Colour: deep cherry. Nose: dried herbs, stone fruit, scrubland, black fruit, fruit liqueur notes. Palate: powerful, ripe fruit, spicy.

Rico Nuevo Garnacha 2021 T
92 🌿
Colour: cherry, purple rim. Nose: balsamic herbs, red berry notes, scrubland, raspberry. Palate: balsamic, fruity, balanced.

Vereda de las Tórdigas 2020 T BA
garnacha
93 🌿
Colour: bright cherry. Nose: spicy, black fruit, wild herbs, floral, violet drops. ,Palate: flavourful, fresh, fruity, full of life. taut.

SOTO MANRIQUE BODEGA
Ctra. Avila Toledo s/n
05260 Cebreros (Ávila)
☎: +34 626 290 408
info@sotomanrique.com
www.sotomanrique.com

🏆 PODIUM

Alto de la Estrella 2019 T
garnacha
97
Colour: bright cherry. Nose: complex, expressive, spicy, floral, fruit expression, red berry notes, mineral. Palate: elegant, full-bodied, long, great length.

La Viña de Ayer Albillo Real 2019 B
93
Colour: bright yellow. Nose: ripe fruit, dried herbs, faded flowers, waxy notes. Palate: powerful, ripe fruit, balanced, bitter.

La Viña de Ayer Garnacha 2019 T
92
Taut. Colour: cherry, purple rim. Nose: floral, spicy, ripe fruit, red berry notes. Palate: flavourful, fruity, good acidity, long.

Naranjas Azules Garnacha 2021 RD
91
Colour: raspberry rose. Nose: red berry notes, floral, fragrant herbs. Palate: light-bodied, good acidity, fine bitter notes.

TIERRA DE CEBREROS
Ctra. de Cebreros a El Barraco, km. 3
05260 Cebreros (Ávila)
☎: +34 644 446 234
enoturismo@tierrasdecebreros.com
www.tierrasdecebreros.com

Panis Angelorum 2020 B
100% albillo real
91
Colour: bright yellow. Nose: dried flowers, candied fruit, fine lees, pattiserie. Palate: round, spicy, great length.

Peñín Guide **SPANISH WINE** 753

VINOS DE CALIDAD / D.O.P.

Sursum Corda 2021 T
100% garnacha

91

Colour: cherry, purple rim. Nose: red berry notes, floral, expressive. Palate: fruity, flavourful, balanced.

VIÑEDOS Y BODEGAS ALTO BUEN GRADO

Ctra. del Moral, Finca Alto Buen Grado
13200 Manzanares (Ciudad Real)
☎: +34 926 647 800
admon@bodegaaltobuengrado.es

La Cendra 2021 T
garnacha

91

Colour: Cherry. Nose: balsamic herbs, sweet spices, scrubland, red berry notes, ripe fruit. Palate: spicy, balsamic, good acidity.

La Cendra Selección de Familia 2020 T
garnacha

92

Colour: deep cherry. Nose: ripe fruit, dried herbs, creamy oak, earthy notes, red berry notes. Palate: powerful, ripe fruit, spicy, round tannins.

VINO DE CALIDAD DE LAS ISLAS CANARIAS

BODEGA EL LOMO

Ctra. El Lomo, 18
38280 Tegueste (Santa Cruz de Tenerife)
☎: +34 922 545 254
administracion@bodegaellomo.com
www.bodegaellomo.com

Barco de Tierra Adentro 2020 T FB
listán negro, negramoll, tintilla, listán blanco

88

Balanced, spicy, fruity, toasty.

Corazón Negro 2021 B SD
listán negro

87

El Lomo Listán Blanco 2021 B
listán blanco

87

El Lomo Listán Negro 2021 T
listán negro

88

Balanced, spicy, smoky, floral, thin.

Origen 1989 2020 T RB
listán negro, vijariego negro, castellana

88

Balanced, spicy, herbaceous, tasty.

BODEGA PIEDRA FLUIDA

Bencomo, 58
38390 Santa Úrsula (Santa Cruz de Tenerife)
☎: +34 608 385 544
gerencia@grupofeb.com
www.piedrafluida.com

Piedra Fluida 2021 B
80% listán blanco, 20% listán negro

91

Colour: bright straw. Nose: fragrant herbs, fine lees, fruit liqueur notes, white fruit, dry stone. Palate: full-bodied, long, good acidity, flavourful.

Piedra Fluida 2021 B SD

88

Sweet, citrus fruit, dried flowers, herbal.

Piedra Fluida Listán Negro 2020 T BA
100% listán negro

91

Colour: cherry, purple rim. Nose: fruit expression, floral, spicy, black fruit. Palate: flavourful, fruity, good acidity, long, fleshy.

Piedra Fluida Vidal 2020 T
100% listán negro

89

Floral, fruity, crisp, herbal, mineral.

BODEGA TAFURIASTE

Las Candias Altas, 11
38312 La Orotava (Santa Cruz de Tenerife)
☎: +34 647 421 256
Fax: +34 922 336 027
vinos@bodegatafuriaste.com
www.bodegatafuriaste.com

Ocho Islas 2020 T
listán negro, castellana, tintilla

90

Colour: cherry, purple rim. Nose: fruit expression, red berry notes, floral, spicy, fine reductive notes, mineral. Palate: flavourful, fruity, good acidity.

BODEGAS EL PENITENTE

Camino La Habanera, 286
38300 La Orotava (Santa Cruz de Tenerife)
☎: +34 922 309 024
bodegas@elpenitentesl.es
www.elpenitentesl.es

Cruz del Teide Afrutado Semidulce 2021 B SD
85

Cruz del Teide Seco 2020 T
listán negro
87

Tanganillo 2020 T
listán negro
86

Tanganillo Afrutado Semidulce 2021 B SD
85

BODEGAS MONJE

Camino Cruz de Leandro, 36
38359 El Sauzal (Santa Cruz de Tenerife)
☎: +34 922 585 027
Fax: +34 922 585 027
monje@bodegasmonje.com
www.bodegasmonje.com

Hollera Monje 2021 T MC
listán negro
87

BODEGAS TAJINASTE

Ratiño 5
38315 La Orotava (Santa Cruz de Tenerife)
☎: +34 616 441 519
Fax: +34 922 105 080
bodega@tajinaste.net
www.bodegatajinaste.com

Paisaje de las Islas Forastera 2021 B
forastera
91
Austere. Colour: bright straw. Nose: fragrant herbs, fine lees, white fruit, dried flowers, cereal notes. Palate: full-bodied, rich, long, good acidity.

Paisaje de Las Islas Malvasía Aromática Naturalmente Dulce 2019 B D
malvasía
90
Colour: bright yellow. Nose: honeyed notes, floral, sweet spices, stone fruit. Palate: fruity, flavourful, elegant.

Paisaje de las Islas Malvasia Marmajuelo 2021 B
marmajuelo, malvasía
90
Colour: straw. Nose: citrus fruit, white fruit, dried flowers, wild herbs. Palate: flavourful, fruity, balanced, mineral, saline.

Tajinaste 2020 T RB
listán negro, vijariego negro
88
Smoky, animal funk, powerful, tasty, toasty.

Tajinaste 2021 B S
listán blanco, albillo criollo
86

Tajinaste Tradicional 2020 T
listán negro
88
Balanced, spicy, crisp, mineral, dried herbs.

EL SITIO

Barranco San Juan, 47
38400 Tacoronte (Santa Cruz de Tenerife)
☎: +34 922 373 491
administracion@bodegaselsitio.com
www.bodegaselsitio.com

De Yanes Baboso Negro 2020 T
100% baboso
90
Colour: cherry, purple rim. Nose: fruit expression, floral, spicy, mineral, black fruit. Palate: flavourful, fruity, good acidity.

El Sitio Forastera 2021 B
100% forastera gomera
90
Colour: bright straw, greenish rim. Nose: fresh fruit, citrus fruit, wild herbs, mineral. Palate: fresh, fruity, good acidity, rich.

El Sitio Malvasía 2020 B
88
Balanced, floral, herbal, tasty, crisp.

El Sitio Orchilla 2021 B
100% vijariego blanco
90
Colour: bright straw, greenish rim. Nose: fresh fruit, citrus fruit, wild herbs. Palate: fresh, fruity, good acidity.

El Sitio Vijariego Negro 2020 T
100% vijariego negro
88
Balanced, fruity, herbal, thin.

VINOS DE CALIDAD / D.O.P.

La Finca 2020 T FB
85% vijariego negro, 15% syrah
89
Balanced, spicy, mineral, thin, floral, crisp.

LLANO DE EL PINO
Callejón la Hoyilla, 1
38280 Tegueste (Santa Cruz de Tenerife)
☎: +34 615 330 051
pjsantanah@hotmail.com

El Jardín de Abril Malvasía Aromática 2020 B
87

Jardín de Abril 2020 T BA
listán negro, negramoll, syrah, tempranillo
89
Crisp, fruity, floral, herbal, smoky.

Jardín de Abril Orange Wine 2020 B
albillo criollo
89
Citrus fruit, balanced, dried flowers, crisp, oxidativ, tasty, subtle.

Tasat 2021 T
listán negro, tempranillo, negramoll
87

Tasat Listán Blanco 2021 B
100% listán blanco
85

Tasat Listán Blanco Afrutado 2021 B SD
100% listán blanco
87

VIÑA ZANATA
El Sol, 3
38440 La Guancha (Santa Cruz de Tenerife)
☎: +34 922 828 166
zanata@zanata.net
www.zanata.net

Viña Zanata Afrutado 2021 B
listán blanco, moscatel, vijariego blanco, malvasía, marmajuelo
86

Viña Zanata Malvasía Seco 2021 B S
malvasía
87

Viña Zanata Marmajuelo 2021 B
marmajuelo
88
Citrus fruit, herbal, tasty, crisp.

Viña Zanata Tintilla 2020 T
tintilla
87

Viña Zanata Tradicional 2021 B S
listán blanco
86

Viña Zanata Vendimia Seleccionada 2021 B S
listán blanco, albillo, marmajuelo, vijariego blanco
87

VINO DE CALIDAD DE LOS VALLES DE BENAVENTE

BODEGAS OTERO
Avda. El Ferial, 22
49600 Benavente (Zamora)
☎: +34 980 631 600
info@bodegasotero.es
www.bodegasotero.es

Finca Valleoscuro 2021 B
verdejo
88
Dried flowers, jammy, kind finish.

Finca Valleoscuro Prieto Picudo 2021 RD
100% prieto picudo
90
Colour: brilliant rose, purple rim. Nose: red berry notes, lactic notes, floral. Palate: balanced, fruity aftertaste, easy to drink.

Finca Valleoscuro Prieto Picudo Tempranillo 2020 T
prieto picudo, tempranillo
89
Balanced, fruity, jammy, spicy, wild, smooth.

Finca Valleoscuro Prieto Picudo Tempranillo 2021 RD
prieto picudo, tempranillo
89
Pleasant, aromatic, elegant, floral.

Otero 2017 T C
100% prieto picudo
90
Colour: bright cherry. Nose: black fruit, ripe fruit, spicy, wild herbs. Palate: fruity, flavourful, balanced, slightly dry, soft tannins.

BODEGAS VERDES
Camino Benavente, 21
49610 Santibáñez de Vidriales (Zamora)
☎: +34 980 648 405
comercial@bodegasverdes.com
www.bodegasverdes.com

Carpurias Prieto Picudo 2021 T C
100% prieto picudo

89

Full-bodied, elegant, spicy, herbal, tasty.

Señorío de Vidriales 2021 RD
100% prieto picudo

87

Señorío de Vidriales 2021 T
50% prieto picudo, 50% tempranillo

87

Señorío de Vidriales Verdejo 2021 B
100% verdejo

86

HAMMEKEN CELLARS
Llavador, 20
03700 Denia (Alacant/Alicante)
☎: +34 965 791 967
cellars@hammekencellars.com
www.hammekencellars.com

Gotas de Mar Rosé 2021 RD
70% tempranillo, 30% prieto picudo

88

Kind finish, fruity, dried flowers, jammy, tasty.

VINO DE CALIDAD DE SIERRA DE SALAMANCA

AUTÉNTICOS VIÑADORES, VINOS DE TERROIR
24540 Cacabelos (León)
☎: +34 658 617 390
info@autenticosvinadores.com
www.autenticosvinadores.com

El Amante 2019 T C
100% rufete

91

Colour: cherry, garnet rim. Nose: ripe fruit, dried herbs, neat, tobacco, black liquorice. Palate: ripe fruit, spicy, round tannins.

Los Vientos 2019 T
100% rufete

90

Colour: Cherry. Nose: sweet spices, scrubland, ripe fruit. Palate: spicy, balsamic, good acidity, ripe fruit.

BODEGA DON CELESTINO
37671 San Esteban de la Sierra (Salamanca)
☎: +34 625 751 201
bodegadoncelestino@gmail.com

Don Celestino 2021 T
rufete

90

Colour: cherry, purple rim. Nose: red berry notes, floral, fresh, wild herbs. Palate: fruity, good acidity, fresh, easy to drink.

Don Celestino Rufete envejecido 2020 T
rufete

87

BODEGA EL ABUELO FLORES
La Mata, 5
37671 San Esteban de la Sierra (Salamanca)
☎: +34 653 151 694
bodegaelabueloflores@gmail.com
www.bodegaelabueloflores.es

El Notas Premium 2020 T
100% rufete

89

Spicy, dried herbs, jammy, tasty, standard, balanced.

Ilusión 2021 RD
tempranillo, palomino

87

Ilusión Rufete Serrano Blanco 2021 B
rufete blanco

89

Wild, dried flowers, crisp, fruity, herbal, smooth.

Kasiná 2020 T
tempranillo, rufete, garnacha

88

Kind finish, jammy, dried herbs, dried flowers, standard.

Notas 2020 T
100% rufete

89

Pleasant, dried herbs, wild, fruity, smooth.

Renegón 2020 T
tempranillo

87

VINOS DE CALIDAD / D.O.P.

VINOS DE CALIDAD / D.O.P.

BODEGAS Y VIÑEDOS EL ROBLEDO
Calle La Iglesia, 22
37650 Sequeros (Salamanca)
☎: +34 660 048 001
info@bodegaselrobledo.com
www.bodegaselrobledo.com

El Robledo Divino 2019 T
tempranillo
89
Pleasant, kind finish, fruity, jammy, tasty, spicy, defined aromas.

El Robledo Rufete 2019 T C
rufete
89
Jammy, toasty, powerful, spicy, great length, coarse tannins.

El Robledo Tempranillo Rufete 2019 T
tempranillo, rufete
89
Pleasant, standard, jammy, tasty, dried herbs, balanced.

CÁMBRICO
37658 Villanueva del Conde (Salamanca)
☎: +34 923 217 473
info@cambrico.com
www.cambrico.com

575 Uvas de Cámbrico 2019 T R
42% tempranillo, 37% garnacha, 21% rufete
91
Colour: cherry, garnet rim. Nose: ripe fruit, musky notes, fine reductive notes, wild herbs, dried herbs. Palate: spicy, correct, fruity, flavourful.

Cámbrico Rufete El Pocito 2018 T
100% rufete
94
Defined aromas, complex. Colour: cherry, garnet rim. Nose: spicy, waxy notes, ripe fruit, dried flowers, wild herbs. Palate: flavourful, long, balanced.

Cámbrico Rufete Valleoscuro 2018 T
100% rufete
92
Colour: Cherry. Nose: balsamic herbs, scrubland, spicy, ripe fruit. Palate: spicy, balsamic, good acidity, round tannins.

Viñas del Cámbrico Villanueva 2020 T
100% rufete
93
Pleasant, defined aromas, herbal. Colour: cherry, garnet rim. Nose: ripe fruit, wild herbs, dried herbs, complex, fine reductive notes. Palate: balanced, long, fine bitter notes, flavourful.

CUARTA GENERACIÓN BODEGAS Y VIÑEDOS
Castillo, 7
37658 Sotoserrano (Salamanca)
☎: +34 618 741 461
info@bodegasantonioaparicio.com
www.bodegasantonioaparicio.com

Cuarta Generación Sierrahonda 2017 T
88
Dried flowers, jammy, fruity, standard, spicy, tasty, toasty.

Cuarta Generación Valdeherreros 2020 T
100% rufete
89
Wild, spicy, jammy, fruity, herbal, coarse tannins, balanced.

IV Generación Rufete 2021 T
rufete
89
Pleasant, balanced, floral, fruity, jammy, tasty, great length.

Seis Décadas 2017 T
rufete, tempranillo
89
Classic, standard, jammy, fruity, spicy, smoky, dried herbs, tasty. Nose: waxy notes.

DOMINIO DE LA SIERRA
Chorrito s/n
37671 San Esteban de la Sierra (Salamanca)
☎: +34 630 030 348
info@dominiodelasierra.com
www.dominiodelasierra.com

Dominio de la Sierra 2021 B
rufete blanco, palomino, moscatel
89
Jammy, toasty, smoky, tasty, woody. Nose: stone fruit.

Dominio de la Sierra Dominivm 2019 T
rufete
91
Colour: deep cherry. Nose: ripe fruit, dried herbs, creamy oak. Palate: powerful, ripe fruit, spicy, round tannins.

Dominio de la Sierra Momentvm 2020 T
rufete, tempranillo
90
Colour: cherry, garnet rim. Nose: ripe fruit, black fruit, spicy, toasty. Palate: spicy, round tannins, long.

ROCHAL

Salas Pombo, 17
37670 Santibáñez de La Sierra (Salamanca)
☎: +34 923 435 260
Fax: +34 923 435 260
info@bodegasrochal.com
www.bodegasrochal.com

Calixto 2021 T
rufete
90
Colour: bright cherry, purple rim. Nose: red berry notes, ripe fruit, neat. Palate: flavourful, fruity.

Calixto Bolosea 2021 T
90
Colour: cherry, purple rim. Nose: fruit expression, red berry notes, floral, lactic notes. Palate: fruity, flavourful, balanced.

Calixto Nieto 2021 T
91
Colour: cherry, purple rim. Nose: fruit expression, red berry notes, floral, spicy, lactic notes. Palate: flavourful, fruity, good acidity, good structure.

Calixto Osiris 2019 T
tempranillo, rufete
89
Standard, jammy, tasty, powerful, toasty.

VINOS LA ZORRA

San Pedro, s/n
37610 Mogarraz (Salamanca)
☎: +34 609 392 591
estanverdes@vinoslazorra.es
www.vinoslazorra.es

La Zorra 8 Virgenes 2021 B
55% palomino, 30% rufete blanco, moscatel, 10% viura
91
Colour: bright straw, greenish rim. Nose: fresh fruit, wild herbs, white flowers. Palate: fresh, fruity, good acidity, fine bitter notes.

La Zorra Aragonés 2019 T
100% aragonés
91
Colour: bright cherry. Nose: ripe fruit, neat, balanced. Palate: good acidity, spicy, fine tannins, fruity, flavourful.

La Zorra La Novena Rufete Blanco 2020 B
100% rufete blanco
93
Colour: bright yellow. Nose: white flowers, fresh fruit, citrus fruit, balanced, expressive. Palate: fruity, flavourful, good acidity.

La Zorra Original 2019 T
60% rufete, aragonés
90
Colour: cherry, garnet rim. Nose: ripe fruit, sweet spices. Palate: spicy, long, ripe fruit, balanced.

La Zorra Rarísimo 2018 T
92
Racy, dried flowers, herbal. Nose: fresh fruit, ripe fruit, spicy, characterful, expressive. Palate: fresh, fruity, easy to drink.

La Zorra Rosada 2021 RD
100% rufete
90
Defined aromas, crisp, floral, wild, simple, tasty, fruity. Palate: fine bitter notes, balanced, good acidity, good finish.

VINOS RABILARGO

37660 Miranda del Castañar (Salamanca)
☎: +34 684 373 141
stmailan@yahoo.co.uk

Rabilargo 2020 T
70% rufete, 30% tempranillo
90
Colour: cherry, purple rim. Nose: floral, spicy, ripe fruit, red berry notes. Palate: flavourful, fruity, good acidity, long.

VINO DE CALIDAD VALTIENDAS

BODEGAS NAVALTALLAR

Calvario, s/n
40331 Navalilla (Segovia)
☎: +34 638 050 061
alejandro_costa@navaltallar.com
www.navaltallar.com

Navaltallar 2019 T C
tempranillo
86

Navaltallar Crianza Especial 2016 T C
88
Slight oxidation, pruney, creamy, spicy, fruity, jammy.

VINOS DE CALIDAD / D.O.P.

VINOS DE LA TIERRA

The number of "Vino de la Tierra" categories granted so far, 43, means the status is growing in importance, given that growers are only required to specify geographical origin, grape variety and alcohol content. For some, it means an easy way forward for their more experimental projects, difficult to be contemplated by the stern regulations of the designations of origin, as it is the case of vast autonomous regions such as La Mancha, Castilla y León or Extremadura. For the great majority, it is a category able to fostering vineyards with high quality potential, a broader varietal catalogue and therefore the opportunity to come up with truly singular wines, a sort of sideway entrance to the DO status.

The different "Vino de la Tierra" designations have been listed in alphabetical order.

In theory, the "Vino de la Tierra" status is one step below that of the DO, and it is the Spanish equivalent to the French "Vins de Pays", which pioneered worldwide this sort of category. In Spain, however, it has some unique characteristics. For example, the fact that the designation "Vino de la Tierra" is not always the ultimate goal, but it is rather used as a springboard to achieve the highly desired DO category. In addition, as it has happened in other countries, many producers have opted for this type of association with less stringent regulations that allow them greater freedom to produce wine. Therefore, in this section there is a bit of everything: from great wines to more simple and ordinary examples, a broad catalogue that works as a sort of testing (and tasting!) field for singularity as well as for new flavours and styles derived from the use of local, autochthonous varieties.

The new Spanish Ley del Vino (Wine Law) maintains the former status of "Vino de la Tierra", but establishes an intermediate step between this and the DO one. They are the so-called 'Vinos de Calidad con Indicación Geográfica' (Quality Wines with Geographical Indication), under which the region in question must remain for a minimum of five years.

In the light of the tasting carried out for this section, there is a steady improvement in the quality of these wines, as well as fewer misgivings on the part of the wineries about the idea of joining these associations.

3 RIBERAS

Granted by the administration at the end of 2008 for the wines produced and within the "3 Riberas" geographical indication. The different typologies are: rosé, white, red and noble wines.

ALTIPLANO DE SIERRA NEVADA

With the goal to free Granada's geographical indication exclusively for the "Vino de Calidad" category, in 2009 the VT I.G.P.Norte de Granada changed its name to VT I.G.P.Altiplano de Sierra Nevada. The new geographical indication comprises 43 municipalities in the north of the province of Granada. The authorized grape varieties for white wine production in the region are chardonnay, baladí verdejo, airen, torrontés, palomino, pedro ximénez, macabeo and sauvignon blanc; also tempranillo, monastrell, garnacha tinta, cabernet franc, cabernet sauvignon, pinot noir, merlot, and syrah for red wines.

BAILÉN

Bailén wine region comprises 350 hectares in some municipal districts within the province of Jaén but fairly close to La Mancha. Wines are made mainly from the grape variety known as "molinera de Bailén", that cannot be found anywhere else in the world, but also from other red grape varieties such as garnacha tinta, tempranillo and cabernet sauvignon, as well as the white pedro ximénez.

BAJO ARAGÓN

The most "mediterranean" region within Aragón autonomous community, it borders three different provinces (Tarragona, Castellón and Teruel) and is divided in four areas: Campo de Belchite, Bajo Martín, Bajo Aragón and Matarraña. Soils are mainly clay and limestone in nature, very rich in minerals with high potash content. The climate is suitable for the right maturation of the grapes, with the added cooling effect of the 'Cierzo' (northerly wind), together with the day-night temperature contrast, just the perfect combination for the vines. The main varieties are garnacha (both red and white), although foreign grapes like syrah, cabernet sauvignon, merlot and chardonnay are also present, as well as tempranillo and cariñena.

BARBANZA E IRIA

The last geographical indication to be granted to the autonomous region of Galicia back in 2007, Barbanza e Iria is located within the Ribera de la Ría de Arosa wine region, in the north of the province of Pontevedra. They make both red an white wines from with varieties such as albariño, caíño blanco, godello, loureiro blanco (also known as marqués), treixadura and torrontés (white); and brancellao, caíño tinto, espadeiro, loureiro tinto, mencía and susón (red).

VINOS DE LA TIERRA / I.G.P.

BETANZOS

Betanzos, in the province of La Coruña, became the second VT I.G.P.designation to be granted in Galicia. The vineyards is planted with local white varieties like blanco legítimo, Agudelo (godello) and jerez, as well as red grapes like garnacha, mencía and tempranillo.

CÁDIZ

Located in the south of the province of Cádiz, a vast region with a long history of wine production, the "Vinos de la Tierra de Cádiz" comprises 15 municipalities still under the regulations of the DO regarding grape production, but not winemaking. The authorised white varieties are: garrido, palomino, chardonnay, moscatel, mantúa, perruno, macabeo, sauvignon blanc y pedro ximénez; as well as the red tempranillo, syrah, cabernet sauvignon, garnacha tinta, monastrell, merlot, tintilla de rota, petit verdot and cabernet franc.

CAMPO DE CARTAGENA

Campo de Cartagena is a flatland region close to the Mediterranean Sea and surrounded by mountains of a moderate height. The vineyard surface ascribed to the VT I.G.P.is just 8 hectares. The climate is mediterranean bordering on an arid, desert type, with very hot summers, mild temperatures for the rest of the year, and low and occasional rainfall. The main varieties in the region are bonicaire, forcallat tinta, petit verdot, tempranillo, garnacha tintorera, crujidera, merlot, syrah and cabernet sauvignon (red); and chardonnay, malvasía, moravia dulce, moscatel de grano menudo and sauvignon blanc (white).

CASTELLÓ

Located in the eastern part of Spain, on the Mediterranean coast, the geographical indication Vinos de la Tierra de Castelló is divided in two different areas: Alto Palancia –Alto Mijares, Sant Mateu and Les Useres–, and Vilafamés. The climatic conditions in this wine region are good to grow varieties such as tempranillo, monastrell, garnacha, garnacha tintorera, cabernet sauvignon, merlot and syrah (red), along with macabeo and merseguera (white).

CASTILLA

Castilla-La Mancha, a region that has the largest vineyard surface in the planet (600.000 hectares, equivalent to 6% of the world's total vineyard surface, and to half of Spain's) has been using this Vino de la Tierra label since 1999 (the year the status was granted) for wines produced outside its designations of origin. The main grape varieties are airén, albillo, chardonnay, macabeo (viura), malvar, sauvignon blanc, merseguera, moscatel de grano menudo, pardillo (marisancho), Pedro Ximénez and torrontés (white);and bobal, cabernet sauvignon, garnacha tinta, merlot, monastrell, petit verdot, syrah, tempranillo, cencibel (jacivera), coloraíllo, frasco, garnacha tintorera, moravia agria, moravia dulce (crujidera), negral (tinto basto) and tinto velasco (red).

CASTILLA Y LEÓN

Another one of the regional 'macro-designations' for the wines produced in up to 317 municipalities within the autonomous region of Castilla y León. A continental climate with little rainfall, together with diverse soil patterns, are the most distinctive features of a region that can be divided into the Duero basin (part of the Spanish central high plateau) and the mountainous perimeter that surrounds it.

CÓRDOBA

It includes the wines produced in the province of Córdoba, with the exception of those bottled within the DO Montilla-Moriles label. All in all, we are talking of some 300 hectares and red and rosé wines made from cabernet sauvignon, merlot, syrah, tempranillo, pinot noir and tintilla de Rota grape varieties.

COSTA DE CANTABRIA

Wines produced in the Costa de Cantabria wine region as well as some inland valleys up to an altitude of 600 meters. The grape varieties used for white winemaking are godello, albillo, chardonnay, malvasía, ondarribi zuri, picapoll blanco and verdejo blanco; and just two for red wines: ondarribi beltza and verdejo negro. The region comprises some 8 hectares of vineyards.

CUMBRES DE GUADALFEO

Formerly known as "Vino de la Tierra de Contraviesa-Alpujarra", this geographical indication is used for wines made in the wine region located in the western part of the Alpujarras, in a border territory between two provinces (Granada and Almería), two rivers (Guadalfeo and Andarax), and very close to the Mediterranean Sea. The grape varieties used for white wine production are montúa, chardonnay, sauvignon blanc, moscatel, jaén blanca, Pedro Ximénez, vijirego y perruno; for red wines, they have garnacha tinta, tempranillo, cabernet sauvignon, cabernet franc, merlot, pinot noir and syrah.

DESIERTO DE ALMERÍA

Granted in the summer of 2003, the wine region comprises a diverse territory in the north of the province of Almería that includes the Tabernas Dessert as well as parts of the Sierra de Alhamilla, Sierra de Cabrera and the Cabo de Gata Natural Park. Harsh, climatic desert conditions follow a regular pattern of hot days and cooler nights that influence heavily the character of the resulting wines. The vineyard's average altitude is 525 meters. The varieties planted are chardonnay, moscatel, macabeo and sauvignon blanc (white); as well as tempranillo, cabernet sauvignon, monastrell, merlot, syrah and garnacha tinta (red).www.vinosdealmeria.es/zonas-viticolas/desierto-de-almeria

VINOS DE LA TIERRA / I.G.P.

EIVISSA

The production area includes the entire island of Ibiza (Eivissa), with the vineyards located in small valleys amongst the mountains –which are never higher than 500 meters– on clay-reddish soil covered by a thin limestone crust. Low rainfall levels and hot, humid summers are the most interesting climatic features. The authorized red varieties are monastrell, tempranillo, cabernet sauvignon, merlot and syrah; macabeo, parellada, malvasía, chardonnay and moscatel make up the white-grape catalogue.

EXTREMADURA

It comprises all the municipalities within the provinces of Cáceres and Badajoz, made up of six different wine regions. In December 1990, the regional government approved the regulations submitted by the Comisión Interprofesional de Vinos de la Tierra de Extremadura, and approved its creation. The varieties used for the production of white wines are alarije, borba, cayetana blanca, chardonnay, chelva, malvar, viura, parellada, Pedro Ximénez and verdejo; for red wines, they have bobal, mazuela, monastrell, tempranillo, garnacha, graciano, merlot, syrah and cabernet sauvignon.

FORMENTERA

This geographical indication comprises the wines produced in the island of Formentera. The dry, subtropical mediterranean climate, characterised by abundant sunshine hours and summers with high temperatures and humidity levels but little rainfall, evidently requires grape varieties well adapted to this type of weather. Red varieties are monastrell, fogoneu, tempranillo, cabernet sauvignon and merlot; malvasía, premsal blanco, chardonnay and viognier make up its white-grape catalogue.

ILLA DE MENORCA

The island of Menorca, a Biosphere Reserve, has a singular topography of gentle slopes; marl soils with a complex substratum of limestone, sandstone and slate, a mediterranean climate and northerly winter winds are the most significant features from a viticultural point of view. The wines produces in the island should be made exclusively from white grape varieties like chardonnay, macabeo, malvasía, moscatel, parellada or moll; as for the red renderings, cabernet sauvignon, merlot, monastrell, tempranillo and syrah get clearly the upper hand.

ILLES BALEARS

Includes all the municipalities of the Balearic Islands. It covers an area of 4,992 km2 and encompasses the islands of Mallorca, Menorca, Eivissa, Formentera and Cabrera. Accepted white grapes are: Moll, Chardonnay, Macabeo, Malvasia, Muscat of Alexandria, Muscat of Small Grain, Parellada, Riesling and Sauvignon Blanc. Accepted reds being: Callet, Manto Negro, Fogoneu, Monastrell, Cabernet Sauvignon, Merlot, Syrah, Tempranillo and Pinot Noir.

LADERAS DE GENIL

Formerly known (up to 2009) as VT I.G.P.Granada Suroeste, the label includes some 53 municipalities in the province of Granada. The region enjoys a unique microclimate very suitable for grape growing, given its low rainfall and the softening effect of the Mediterranean Sea. The white grape varieties used for wine production are vijiriego, macabeo, Pedro Ximénez, palomino, moscatel de Alejandría, chardonnay and sauvignon blanc; as well as the red garnacha tinta, perruna, tempranillo, cabernet sauvignon, merlot, syrah and pinot noir, predominantly.

LAUJAR-ALPUJARRA

This wine region is located at an altitude of 800 to 1500 meters between the Sierra de Gádor and the Sierra Nevada Natural Park. It has some 800 hectares of vines grown on terraces. Soils are chalk soils poor in organic matter, rocky and with little depth. The climate is moderately continental, given the sea influence and its high night-day temperature differential. The predominant grape varieties are jaén blanco, macabeo, vijiriego, Pedro Ximénez, chardonay and moscatel de grano menudo (white); and cabernet sauvignon, merlot, monastrell, tempranillo, garnachas tinta and syrah (red). www.vinosdealmeria.es/bodegas/vino-de-la-tierra-laujar-alpujarra

LIÉBANA

VT I.G.P.Liébana includes the municipalities of Potes, Pesagüero, Cabezón de Liébana, Camaleño, Castro Cillorigo y Vega de Liébana, all of them within the area of Liébana, located in the southwest of the Cantabria bordering with Asturias, León and Palencia. The authorized varieties are mencía, tempranillo, garnacha, garciano, merlot, syrah, pinot noir, albarín negro and cabernet sauvignon (red); and palomino, godello, verdejo, albillo, chardonnay and albarín blanco (white).

LOS PALACIOS

Los Palacios is located in the south-western part of the province of Sevilla, by the lower area of the Guadalquivir river valley. The wines included in this VT I.G.P.are white wines made from airén, chardonnay, colombard and sauvignon blanc.

MALLORCA

The production area of VT I.G.P.Mallorca includes all the municipalities within the island, which has predominantly limestone soils with abundant clay and sandstone, and a mediterranean climate with mild temperatures all-year-round. Red varieties present in the island are callet, manto negro, cabernet sauvignon, fogoneu, merlot, monastrell, syrah, tempranillo and pinot noir, along with the white prensal (moll), chardonnay, macabeo, malvasía, moscatel de Alejandría, moscatel de grano menudo, parellada, riesling and sauvignon blanc.

NORTE DE ALMERÍA

The Vinos de la Tierra Norte de Almería label comprises four municipalities in the Norte de Almería area, right in the north of the province. They produce white, red and rosé wines from grape varieties such as airén, chardonnay, macabeo and sauvignon blanc (white); as well as cabernet sauvignon, merlot, monastrell, tempranillo and syrah for red winemaking and tempranillo and monastrell for rosé.

POZOHONDO

Almost the entirety of surface dedicated to growing the vineyard is located in the coast (Morrazo peninsula and more inland areas of the estuaries of Pontevedra and Vigo). The wines under this new IGP must proceed from the municipalities of Bueu, Cangas, Marín, Moaña, Poio, Pontevedra, Redondela or Vilaboa. The area is exposed to a humid-oceanic climate, though protected by mountain ranges to the east and south that lessen rains during the summer. They posses a wide range of Galician varieties in white as well as red.

RIBEIRAS DO MORRAZO

The regulations for VT I.G.P.Pozoblanco were approved by the autonomous government of Castilla-La Mancha in the year 2000. It comprises the municipalities of Alcadozo, Peñas de San Pedro and Pozohondo, all of them in the province of Albacete.

RIBERA DEL ANDARAX

The Ribera del Andarax wine region is located in the middle area of the Andarax river valley at an altitude of 700 to 900 meters. Soils are varied in structure, with abundant slate, clay and sand. It enjoys an extreme mediterranean climate, with low occasional rainfall and high average temperatures. The grape varieties present in the region are predominantly macabeo, chardonnay and sauvignon blanc (white); and cabernet sauvignon, merlot, syrah, garnacha, tempranillo, monastrell and pinot noir (red). www.vinosdealmeria.es/zonas-viticolas/ribera-de-andarax

RIBERA DEL GÁLLEGO-CINCO VILLAS

Ribera del Gállego-Cinco Villas wine region is located in the territory along the Gállego river valley until it almost reaches the city of Zaragoza. Although small, its vineyards are shared between the provinces of Huesca and Zaragoza. Soils are mostly gravel in structure, which affords good drainage. The grape varieties used for wine production are garnacha, tempranillo, carbernet sauvignon and merlot (red), and mostly macabeo for white wines. www.vinosdelatierradearagon.es

RIBERA DEL JILOCA

Ribera del Jiloca, located in the south-eastern part of Aragón along the Jiloca river valley, is a wine region with a great winemaking potential, given its geo-climatic conditions. Vines are mostly planted on slate terraces perched on the slopes of the Sistema Ibérico mountain range, at high altitude, something that affords wines of great quality and singularity. Vines are planted mostly on alluvial limestone terraces of ancient river beds. Garnacha is the predominant grape, followed by macabeo. A dry climate, abundant sunlight hours and cold winters are the features that account for the excellent quality of the local grapes.www.vinosdelatierradearagon.es/empresas/ribera_del_jiloca.php

RIBERA DEL QUEILES

Up to sixteen municipalities from two different provinces (seven from Navarra and nine from Zaragoza) are part of the VT I.G.P.Ribera del Queiles. Wines are exclusively red, made from cabernet sauvignon, graciano, garnacha tinta, merlot, tempranillo and syrah. It has a regulating and controlling body (Comité Regulador de Control y Certificación) and so far just one winery. www.vinosdelatierradearagon.es

SERRA DE TRAMUNTANA-COSTA NORD

Currently, this VT I.G.P.comprises 41,14 hectares an up to eighteen municipal districts in the island of Mallorca, between the cape of Formentor and the southwest coast of Andratx, with mainly brownish-grey and limestone soils. Single-variety wines from malvasía, moscatel, moll, parellada, macabeo, chardonnay and sauvignon blanc (white), as well as cabernet sauvignon, merlot, syrah, monastrell, tempranillo, callet and manto negro (red) stand out.

SIERRA DE ALCARAZ

The Sierra del Alcaraz wine region comprises the municipal districts of Alcaraz, El Ballesteros, El Bonillo, Povedilla, Robledo, and Viveros, located in the western part of the province of Albacete, bordering with Ciudad Real. The VT I.G.P.status was granted by the autonomous government of Castilla-La Mancha in the year 2000. The red varieties planted in the region are cabernet sauvignon, merlot, bobal, monastrell, garnacha tinta and garnacha tintorera; along with white moravia dulce, chardonnay, chelva, eva, alarije, malvar, borba, parellada, cayetana blanca and Pedro Ximénez

SIERRA DE LAS ESTANCIAS Y LOS FILABRES

Located in the namesake mountain region in the province of Almería, this VT I.G.P. was approved along with its regulations in 2008. The grape varieties planted in the region are airén, chardonnay, macabeo, sauvignon blanc and moscatel de grano menudo –also known as morisco–, all of them white; and red cabernet sauvignon, merlot, monastrell, tempranillo, syrah, garnacha tinta, pinot noir and petit verdot.

SIERRA DEL NORTE DE SEVILLA

IThis region, located in the north of the province of Sevilla at the foothills of Sierra Morena, has a landscape of gentle hills and altitudes that range from 250 to almost 1000 metres. The climate in the region is mediterranean, with hot, dry summers, mild winters and a fairly high average rainfall. Since 1998, grape varieties such as tempranillo, garnacha tinta, cabernet sauvignon, cabernet franc, merlot, pinot noir, petit verdot and syrah (red); and chardonnay, Pedro Ximénez, colombard, sauvignon blanc, palomino and moscatel de Alejandría (white) have been planted in the region.

SIERRA SUR DE JAÉN

In this VT I.G.P. there are some 400 hectares planted with vines, although a minor percentage are table grapes. The label includes wines made in the Sierra Sur de Jaén wine region. White wines are made from jaén blanca and chardonnay, and red from garnacha tinta, tempranillo, cabernet sauvignon, merlot, syrah and pinot noir.

TORREPEROGIL

This geographical indication in the province of Jaén, whose regulations were approved in 2006, comprises 300 hectares in the area of La Loma, right in the centre of the province. The climate is mediterranean with continental influence, with cold winters and dry and hot summers. The wines are made mainly from garnacha tinta, syrah, cabernet sauvignon and tempranillo (red); and jaén blanco and Pedro Ximénez (white).

VALDEJALÓN

Established in 1998, it comprises 36 municipal districts in the mid- and lower-Jalón river valley. The vines are planted on alluvial, brownish-grey limestone soils, with low annual average rainfall of some 350 mm. They grape varieties planted are white (macabeo, garnacha blanca, moscatel and airén) and red (garnacha, tempranillo, cabernet sauvignon, syrah, monastrell and merlot). www.vinodelatierravaldejalon.com

VALLE DEL CINCA

Located in the southeast of the province of Huesca, almost bordering with Catalunya, Valle del Cinca is a traditional wine region that enjoys favourable climatic and soil conditions for vine growing: soils are mainly limestone and clay, and the average annual rainfall barely reaches 300 mm (irrigation is usually required). Grape varieties predominantly planted in the region are macabeo and chardonnay (white), along with garnacha tinta, tempranillo, cabernet sauvignon and merlot (red). www.vinosdelatierradearagon.es

VALLE DEL MIÑO-OURENSE

This wine region is located in the north of the province of Ourense, along the Miño river valley. The authorized grape varieties are treixadura, torrontés, godello, albariño, loureira and palomino –also known as xerez– for white wines, and mencía, brancellao, mouratón, sousón, caíno and garnacha for reds.

VALLES DE SADACIA

A designation created to include the wines made from the grape variety known as moscatel riojana, which was practically lost with the phylloxera bug and has been recuperated to produce both "vino de licor" and normal white moscatel. Depending on winemaking, the latter may either be dry, semi-dry or sweet. The vineyards that belong to this VT I.G.P. are mainly located in the south-western part of the region, in the Sadacia and Cidacos river valleys, overall a very suitable territory for vine growing purposes.

VILLAVICIOSA DE CÓRDOBA

One of the most recent geographical indications granted by the autonomous government of Andalucía back in 2008, it includes white and sweet wines made in the Villaviciosa wine region. The authorized varieties are baladí verdejo, moscatel de Alejandría, palomino fino, palomino, Pedro Ximénez, airén, calagraño Jaén, torrontés and verdejo.

VT 3 RIBERAS

J. CHIVITE FAMILY ESTATES

Ctra. NA-132, Km. 3,1
31132 Villatuerta (Navarra)
☎: +34 948 811 000
Fax: +34 948 811 407
info@bodegaschivite.com
www.chivite.com

Chivite Colección 125 2020 B FB
100% chardonnay

94

Aromatic, taut. Colour: bright straw. Nose: ripe fruit, dried herbs, faded flowers, stone fruit, white flowers. Palate: powerful, ripe fruit, balanced, toasty.

Chivite Colección 125 2021 RD FB
65% garnacha, 35% tempranillo

93

Colour: salmon. Nose: sweet spices, red berry notes, fragrant herbs, dried flowers. Palate: full-bodied, flavourful, spicy, great length, saline.

🏆 PODIUM

Chivite Colección 125 Vendimia Tardía 2020 B FB D
moscatel grano menudo

95

Colour: bright yellow. Nose: dried flowers, candied fruit, fine lees, pattiserie, brioche. Palate: round, spicy, long, great length.

Chivite Colección 125 Vino de Guarda 2017 T
100% tempranillo

94

Classic. Colour: cherry, garnet rim. Nose: elegant, varietal, ripe fruit, fine reductive notes. Palate: fruity, full of life, flavourful, round tannins.

Chivite La Zorrera Garnacha 2019 T
garnacha

94

Aromatic, fruity. Colour: very deep cherry. Nose: balsamic herbs, sweet spices, scrubland, red berry notes, floral. Palate: spicy, balsamic, good acidity.

Chivite Las Fincas 2 Garnachas 2020 B
51% garnacha, 49% garnacha blanca

90

Colour: bright straw. Nose: ripe fruit, floral, spicy. Palate: fresh, good acidity, easy to drink, good finish.

Chivite Las Fincas 2020 RD FB
100% garnacha

92

Colour: raspberry rose, bright. Nose: red berry notes, wild herbs, floral, mineral, sweet spices. Palate: elegant, flavourful, long, good acidity.

Chivite Las Fincas 2021 RD PL
70% garnacha, 30% tempranillo

91

Smooth, wild. Colour: raspberry rose. Nose: red berry notes, wild herbs, rose petals, medium intensity, neat. Palate: fruity, fine bitter notes, easy to drink.

VT ALTIPLANO DE SIERRA NEVADA

BODEGAS MUÑANA

Ctra. Graena a La Peza, s/n Finca Peñas Prietas
18517 Cortes Y Graena (Granada)
☎: +34 958 670 715
administracion@bodegasmunana.com
www.bodegasmunana.com

Muñana 3 Cepas 2018 T
tempranillo, cabernet sauvignon, petit verdot

90

Classic. Colour: Cherry. Nose: scrubland, spicy, waxy notes, black fruit, old leather. Palate: flavourful, round tannins.

VINO DE LA TIERRA - BAJO ARAGÓN / I.G.P.

Muñana Rojo 2020 T
tempranillo, cabernet sauvignon, petit verdot
89
Fruity, jammy, tasty, smooth.

Delirio Roble de Muñana 2018 T
88
Fruity, smoky, tasty.

VT BAJO ARAGÓN

AMPRIUS LAGAR
Los Enebros, 74 2ª Planta Edificio Galileo
44002 Teruel (Teruel)
☎: +34 978 623 077
info@ampriuslagar.es
www.ampriuslagar.es

Lagar d'Amprius 92/300 Syrah Noble 2016 T
100% syrah
92
Colour: cherry, garnet rim. Nose: powerful, black fruit, fruit preserve, meaty notes, characterful. Palate: flavourful, long, round tannins, ripe fruit, balanced.

Lagar d'Amprius Chardonnay 2019 B
90 🌱
Dried flowers, aromatic. Colour: yellow. Nose: dried flowers, wild herbs, hint of anise, white fruit, ripe fruit. Palate: balanced, ripe fruit, long.

Lagar d'Amprius Garnacha 2017 T
garnacha
91
Colour: dark-red cherry. Nose: fruit preserve, balanced, neat, dried flowers, dried herbs. Palate: flavourful, good structure, spicy, long.

Lagar d'Amprius Garnacha 2019 T
91
Colour: dark-red cherry. Nose: toasty, spicy, cocoa bean, ripe fruit, hot. Palate: flavourful, toasty, fine bitter notes.

Lagar d'Amprius Gewürztraminer 2018 B
90
Kind finish, aromatic, floral. Colour: bright yellow. Nose: varietal. Palate: flavourful, dry, balanced.

Lagar d'Amprius Kolenda 2018 B
gewürztraminer
88
Aromatic, slight oxidation, standard, dried flowers, jammy.

Lagar d'Amprius Nazarín 2019 T
89
Toasty, tasty, jammy, pruney.

BIOENOS
Mayor, 88 Bajo
50400 Cariñena (Zaragoza)
☎: +34 639 359 618
bioenos@bioenos.com
www.varietalesantiguos.es

Gorys Garnacha 2018 T
100% garnacha
88
Pleasant, hot, toasty, overripe.

BODEGA COOPERATIVA SAN PEDRO
Avda. Reino de Aragón, 10
44623 Cretas (Teruel)
☎: +34 978 850 309
info@cooperativasanpedro.es
www.cooperativasanpedro.es

Belví 2019 T C
garnacha, tempranillo
86

Belví 2021 RD
garnacha
86

Belví Ancestral 2021 BE BR
garnacha blanca
88
Pleasant, aromatic, floral, crisp, smooth, simple.

Belví Garnacha Blanca 2019 B
garnacha blanca
87

Belví Macabeo 2021 B
macabeo
83

Viñas Viejas 2016 T C
garnacha peluda
87

BODEGAS IGNACIO GUALLART
N-211, Km. 246,2
44600 Alcañiz (Teruel)
☎: +34 978 089 556
info@bodegasignacioguallart.com
www.bodegasignacioguallart.com

Ignacio Guallart Merlot 2019 T RB
100% merlot
89
Balanced, spicy, herbal, tasty, wild.

Pinzon 21 2021 RD
100% merlot
85 ♣

Trufado Garnacha 21 2021 T
100% garnacha
87 ♣

DOMINIO MAESTRAZGO
Royal III, B12
44550 Alcorisa (Teruel)
☎: +34 978 840 642
bodega@dominiomaestrazgo.com
www.dominiomaestrazgo.com

Dominio Maestrazgo 2018 T C
70% garnacha, 30% syrah
87

Dominio Maestrazgo Garnacha Blanca 2020 B RB
garnacha blanca
89
Tasty, toasty, dried flowers, jammy, fruity. Nose: caramel.

Dominio Maestrazgo Garnacha Blanca 2021 B
garnacha blanca
86

Dominio Maestrazgo Santolea 2020 T
60% garnacha, 40% syrah
88
Fruity, herbal, spicy, jammy.

Dominio Maestrazgo Syrah 2018 T BA
100% syrah
87

Rex Deus 2016 T R
80% garnacha, 20% syrah
91
Colour: cherry, garnet rim. Nose: black fruit, wet leather, sweet spices, dried herbs. Palate: fruity, flavourful, balanced, slightly dry, soft tannins.

MAS DE TORUBIO VITICULTORES
San Roque, 3
44623 Cretas (Teruel)
☎: +34 618 263 546
viticultores@masdetorubio.com
www.masdetorubio.com

Cloteta 2021 T
garnacha peluda
90
Colour: cherry, purple rim. Nose: fruit expression, red berry notes, floral, spicy. Palate: flavourful, fruity, good acidity, long.

La Clota 2019 T
garnacha peluda, merlot, cabernet sauvignon
92
Colour: Cherry. Nose: complex, expressive, spicy, mineral. Palate: elegant, full-bodied, long, great length.

Lo Pou 2020 B
garnacha blanca
92
Colour: bright yellow. Nose: powerful, creamy oak, ripe fruit, spicy. Palate: rich, good structure, long, toasty, fine bitter notes.

Lo Pou 2020 T
garnacha peluda
91
Colour: cherry, garnet rim. Nose: fruit preserve, fruit liqueur notes, powerful, earthy notes. Palate: flavourful, sweetness, long.

Nueve Rosas 2021 RD
garnacha peluda, merlot
89
Pleasant, kind finish, aromatic, floral.

Xado 2020 T
garnacha peluda, merlot, cabernet sauvignon
90
Colour: deep cherry. Nose: ripe fruit, dried herbs, creamy oak. Palate: powerful, ripe fruit, spicy, round tannins.

VINO DE LA TIERRA - BAJO ARAGÓN / I.G.P.

VINO DE LA TIERRA - BARBANZA E IRIA / I.G.P.

Xado 2021 B
garnacha blanca
89
Pleasant, aromatic, floral, jammy.

VENTA D'AUBERT
Ctra. Valderrobres a Arnes, Km. 28
44623 Cretas (Teruel)
☎: +34 978 769 021
info@ventadaubert.com
www.ventadaubert.com

Ventepico 2019 B SS
garnacha blanca, chardonnay
91 🌱
Colour: bright yellow. Nose: powerful, creamy oak, ripe fruit, spicy. Palate: rich, good structure, toasty, fine bitter notes.

VINOS TABERNER AMADO
Calle Extensión Agraria, 2
50794 Nonaspe (Zaragoza)
☎: +34 617 104 799
vinostaberneramado2020@gmail.com
www.vinostaberneramado.es

Entre dos Aguas 2019 T BA
garnacha, cabernet sauvignon, syrah, merlot
88
Age nuances, spicy, balanced, kind finish, jammy.

Entre dos Aguas 2020 T
garnacha, cabernet sauvignon, syrah, merlot
86

Entre dos Aguas Garnacha Blanca 2020 B
garnacha blanca
90
Colour: bright straw, greenish rim. Nose: citrus fruit, wild herbs, dried flowers. Palate: fresh, fruity, good acidity.

Entre dos Aguas Garnacha Blanca 2020 B BA
garnacha blanca
88
Citrus fruit, standard, dried herbs, spicy.

VITÍCOLA LA COSTALENA
Las Eras, 23
50710 Maella (Zaragoza)
☎: +34 626 072 761
joaquinemilio@hotmail.com

Arikai 2021 B
garnacha blanca
87

Arikai Tempranillo 2021 RD
tempranillo
87

VT BARBANZA E IRIA

ADEGA ENTREOSRIOS
Lugar de Entreosrios, 2
15948 Pobra do Caramiñal (A Coruña/La Coruña)
☎: +34 670 712 700
adega@entreosrios.com
www.adega.entreosrios.com

Komokabras Amarillo 2020 B BA
albariño
93
Colour: bright yellow. Nose: powerful, spicy, stone fruit, wild herbs, brioche. Palate: rich, good structure, long, toasty, fine bitter notes.

Komokabras Naranja Tinalla 2020 B
92
Oxidativ. Colour: bright golden. Nose: stone fruit, citrus fruit, wild herbs, fine lees, balsamic herbs. Palate: good structure, fruity, fleshy, fresh.

Komokabras Verde Lías 2020 B
albariño
93
Colour: bright straw. Nose: fragrant herbs, fine lees, stone fruit, citrus fruit, dry stone. Palate: full-bodied, rich, long, good acidity, balanced, mineral.

SaraS... 2020 B
albariño
90
Colour: straw. Nose: dried herbs, faded flowers, mineral, citrus fruit, white fruit. Palate: ripe fruit, balanced, fresh.

Vulpes Vulpes 2020 B
albarín
90
Slight reduction. Colour: bright straw. Nose: fine lees, white fruit, wild herbs, hint of anise. Palate: good acidity, flavourful, fleshy.

BODEGAS TORRES AUGUSTI
Forno 136-B Cordeiro
36647 Valga (Pontevedra)
☎: +34 603 425 302
torresaugusti@gmail.com
www.bodegastorresaugusti.com

Castellum Augusti 2020 B
albariño
90
Colour: bright straw, greenish rim. Nose: fresh fruit, citrus fruit, wild herbs, fine lees, spicy. Palate: fresh, fruity, good acidity, fine bitter notes.

Pepe Cabanas 2019 B
albariño

92
Colour: straw. Nose: expressive, white flowers, jasmine, dried herbs, stone fruit. Palate: flavourful, fruity, balanced, rich.

VT CÁDIZ

BODEGA TESALIA

Ctra. La Perdiz - Las Abiertas, CA 6106 km. 3,3
11630 Arcos de la Frontera (Cádiz)
☎: +34 608 883 292
comercial@bodegatesalia.com
www.bodegatesalia.com

ARX 2018 T C
60% syrah, 30% tintilla de rota, 8% petit verdot, 2% cabernet sauvignon

91
Colour: Cherry. Nose: sweet spices, scrubland, black fruit, ripe fruit. Palate: spicy, balsamic, good acidity.

Iceni 2020 T RB
50% tintilla de rota, 50% syrah

89
Kind finish, fruity, jammy, wild, smooth, dried herbs, smoky.

Tesalia 2017 T C
65% petit verdot, 30% syrah, 2,5% tintilla de rota, 2,5% cabernet sauvignon

92
Colour: Cherry. Nose: sweet spices, scrubland, black fruit, ripe fruit, meaty notes. Palate: spicy, balsamic, good acidity, toasty.

BODEGAS BARBADILLO

Luis de Eguilaz, 11
11540 Sanlúcar de Barrameda (Cádiz)
☎: +34 956 385 500
tamara@barbadillo.com
www.barbadillo.com

Ás Mirabrás 2020 B S
100% palomino

89
Aromatic, balsamic herbs, thin, dried flowers. Palate: fine bitter notes.

Barbadillo Blanco de Blancos 2021 B
sauvignon blanc

86

Castillo de San Diego 2021 B
100% palomino

86

Mirabrás 2019 B FB
100% palomino

91
Colour: straw. Nose: ripe fruit, dried herbs, faded flowers, flor yeasts, pungent. Palate: powerful, ripe fruit.

Quadis Envejecido 2019 T C
cabernet sauvignon, petit verdot, merlot, tintilla de rota, syrah

90
Kind finish, full-bodied. Colour: deep cherry. Nose: dried herbs, creamy oak, black fruit, scrubland, fruit preserve. Palate: powerful, ripe fruit, spicy, round tannins.

Sábalo 2021 B
100% palomino

89 🌱
Fruity, jammy, balanced, smooth, standard.

BODEGAS PÁEZ MORILLA

Avda. Medina Sidonia, 20
11406 Jerez de la Frontera (Cádiz)
☎: +34 956 181 717
Fax: +34 956 181 534
bodegas@paezmorilla.com
www.paezmorilla.com

Viñadero Roble 2018 T
87

BODEGAS PRIMITIVO COLLANTES

Calle Ancha, 51
11130 Chiclana de la Frontera (Cádiz)
☎: +34 956 400 150
administracion@bodegasprimitivocollantes.com
www.bodegaprimitivocollantes.es

Matalian 2021 B
palomino

90 🌱
Citrus fruit, floral, fruity, full-bodied, jammy, tasty. Colour: bright yellow.

Socaire 2019 B FB
palomino

92 🌱
Wild. Colour: straw, pale. Nose: citrus fruit, white fruit, spicy, burnt matches. Palate: full-bodied, flavourful, mineral, balanced.

Socaire Oxidativo 2016 B FB
palomino

93 🌱
With personality. Colour: straw, pale. Nose: white fruit, citrus fruit, spicy, dried flowers, burnt matches. Palate: fresh, full-bodied, flavourful, mineral.

VINO DE LA TIERRA - CÁDIZ / I.G.P.

VINO DE LA TIERRA - CÁDIZ / I.G.P.

Tivo 2018 B FB
uva rey

93 🌿

Flat. Colour: straw. Nose: white fruit, dry nuts, ripe fruit, dry stone. Palate: fleshy, full-bodied.

BODEGAS WILLIAMS & HUMBERT
Ctra. Madrid-Cádiz, PK. 641,75
11408 Jerez de la Frontera (Cádiz)
☎: +34 956 353 400
Fax: +34 956 353 408
williams@williams-humbert.com
www.williams-humbert.com

Estero Blanco 2020 B SD
84

CORTIJO DE JARA
11592 Jerez de la Frontera (Cádiz)
☎: +34 956 338 163
puertanueva.sl@cortijodejara.es
www.cortijodejara.es

Cortijo de Jara 6 meses 2020 T RB
syrah, merlot, tempranillo

87

COTA 45
Pórtico de Bajo de Guía 68
14540 Sanlúcar de Barrameda (Cádiz)
☎: +34 956 129 232
info@cota45.com

UBE Carrascal 2018 B S
88

Spicy, mineral, tasty, saline, complex. Nose: acetone.

FINCA MONCLOA
Manuel María González, 12
11403 Jerez de la Frontera (Cádiz)
☎: +34 956 357 000
prensa@gonzalezbyass.com
www.fincamoncloa.com

Finca Moncloa Tintilla de Rota 2019 TF D
tintilla de rota

93

Colour: cherry, garnet rim. Nose: overripe fruit, dried fruit, sweet spices, caramel. Palate: ripe fruit, powerful, long.

Finca Moncloa Tintilla de Rota Edición Limitada 2018 T BA S
tintilla de rota

90

Colour: deep cherry. Nose: dried herbs, creamy oak, black fruit, cedar wood, roasted coffee. Palate: powerful, ripe fruit, spicy, round tannins.

Finca Moncloa Tradicional 2018 T BA
syrah, tintilla de rota, cabernet sauvignon, petit verdot, cabernet franc, tempranillo

93

Colour: deep cherry. Nose: dried herbs, creamy oak, black fruit, meaty notes, fine reductive notes. Palate: ripe fruit, spicy, round tannins.

HUERTA DE ALBALÁ
Ctra. CA - 6105, Km. 4
11630 Arcos de la Frontera (Cádiz)
☎: +34 651 322 440
oscar.sobrino@huertadealbala.com
www.huertadealbala.com

Barbazul 2021 B
chardonnay

90

Colour: bright straw. Nose: ripe fruit, floral, dried herbs. Palate: flavourful, fresh, good acidity, fruity aftertaste.

Barbazul 2021 RD
syrah

88

Pleasant, aromatic, dried flowers, fruity.

Barbazul Selección Especial 2018 T
syrah, merlot, cabernet sauvignon

87

Taberner 2016 T
syrah

90

Colour: dark-red cherry, garnet rim. Nose: fruit preserve, aged wood nuances, tobacco, sweet spices, fine reductive notes. Palate: spicy, round tannins, long.

Taberner nº 1 2014 T
syrah, merlot

91

Colour: dark-red cherry, garnet rim. Nose: fruit preserve, aged wood nuances, tobacco, sweet spices. Palate: spicy, round tannins, long.

JOSÉ ESTÉVEZ

Ctra. N-IV Km. 640
11408 Jerez de la Frontera (Cádiz)
info@grupoestevez.com
www.grupoestevez.com

Albariza 2020 B S
85

MANUEL ARAGÓN

Olivo, 1
11130 Chiclana de la Frontera (Cádiz)
☎: +34 956 400 756
administracion@bodegamanuelaragon.com
www.bodegamanuelaragon.com

Campano 2019 T
tempranillo, syrah

84

La Batalla de la Barrosa 2021 B
sauvignon blanc

88
Aromatic, jammy, tasty.

MUCHADA-LÉCLAPART

Dorantes, 1
11540 Sanlúcar de Barrameda (Cádiz)
☎: +34 660 360 902
info@muchada-leclapart.com
www.muchada-leclapart.com

Elixir 2019 B
moscatel

93
With personality. Colour: yellow, hazy. Nose: stone fruit, wild herbs, white flowers, dry stone. Palate: saline, balanced, mineral.

🏆 PODIUM

Lumière 2019 B
palomino

96
Colour: straw. Nose: complex, cereal notes, dry stone, toasted bread. Palate: fleshy, flavourful, elegant.

REGANTÍO VIEJO

Ctra. de Alberite, Km. 9
11630 Arcos de la Frontera (Cádiz)
☎: +34 638 762 950
info@regantioviejo.com
www.regantioviejo.com

Duo Vites 2019 T
syrah, merlot

88
Hot, fruity, toasty, tasty.

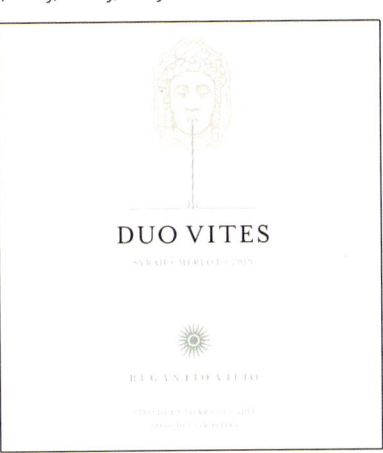

Relicta 2018 T C
syrah

89
Jammy, powerful, spicy, dried herbs, tasty, toasty.

VINO DE LA TIERRA - CÁDIZ / I.G.P.

VINO DE LA TIERRA - CASTELLÓN / I.G.P.

VALDESPINO
Ctra. Nacional IV, Km. 640
11404 Jerez de la Frontera (Cádiz)
☎: +34 956 321 004
Fax: +34 956 340 216
info@grupoestevez.com
www.grupoestevez.com

Ojo de Gallo 2019 B
palomino
91
Colour: straw. Nose: white flowers, jasmine, dried herbs, pungent. Palate: flavourful, fruity, balanced.

Viña Macharnudo alto Palomino 2020 B FB
92
Colour: bright yellow. Nose: powerful, ripe fruit, spicy, waxy notes, chalk, dry nuts. Palate: good structure, long, fine bitter notes, good acidity, balanced.

VIÑA CALLEJUELA
Camino del Reventón Chico, 27
11540 Sanlúcar de Barrameda (Cádiz)
☎: +34 617 492 483
info@callejuela.es
www.callejuela.es

Hacienda de Dª Francisca - Pago Callejuela B
palomino
90
Representative. Nose: dry nuts, roasted almonds, saline, dried flowers. Palate: flavourful, balanced.

La Choza Pago Macharnudo 2021 B
palomino
92
Yeasty notes. Nose: balanced, expressive, faded flowers, white fruit, ripe fruit. Palate: flavourful, long, fine bitter notes.

Las Mercedes Pago Añina 2021 B
palomino
89
Slight reduction, spicy, some off-flavours, tasty, great length.

Sobajanera Añina 2019 B
palomino
93
Complex. Nose: neat, expressive, flor yeasts, pungent, faded flowers. Palate: flavourful, long, full-bodied, saline.

Sobajanera Macharnudo 2019 B
palomino
91
Balanced, yeasty notes, representative, age nuances. Nose: lees reduction notes, faded flowers. Palate: flavourful, fine bitter notes.

VT CASTELLÓN

CLOS D' ESGARRACORDES
Partida Vilar La Call, 10
12118 Les Useres (Castelló/Castellón)
☎: +34 964 767 306
info@barondalba.com
www.barondalba.com

Clos D'Esgarracordes 2018 T C
cabernet sauvignon, tempranillo, monastrell, syrah
85

COOP. DE VIVER
Abadía, 4
12460 Viver (Castelló/Castellón)
☎: +34 964 141 050
info@cooperativaviver.es
www.cooperativaviver.es

La Perdición 2019 T C
70% tempranillo, 10% cabernet sauvignon, 10% merlot, 10% garnacha
88
Kind finish, fruity, sweety finish, full-bodied, jammy, toasty.

Nube 2021 B
chardonnay
88
Citrus fruit, herbal, crisp, balanced.

Viento sobre la Piel 2020 T BA
syrah
88
Balanced, spicy, herbal, thin.

VT CASTILLA

#GARAGEWINE
Camelia, 4
45800 Quintanar de la Orden (Toledo)
☎: +34 625 226 946
info@garagewine.es
www.garagewine.es

6 de 7 #garagewine 2020 T
100% cencibel
91
Colour: bright cherry, garnet rim. Nose: stone fruit, ripe fruit, spicy. Palate: flavourful, fruity, easy to drink.

Airén #garagewine 2021 B
airén
89
With personality, dried herbs, little interventionist, spicy, tasty. Nose: dry nuts. Palate: good finish.

Brujidera #garagewine 2020 T
100% brujidera
90
Colour: cherry, garnet rim. Nose: ripe fruit, fruit preserve, spicy. Palate: flavourful, fruity, easy to drink.

Cencibel #garagewine 2021 T
100% cencibel
91
Colour: cherry, purple rim. Nose: ripe fruit, stone fruit, neat, wild herbs. Palate: fresh, balanced, easy to drink, racy.

Garnacha Tintorera #garagewine 2020 T BA
100% garnacha tintorera
91
Colour: Cherry. Nose: black fruit, ripe fruit, wild herbs, faded flowers, varietal. Palate: powerful, ripe fruit, fine bitter notes.

Verdoncho #garagewine Orangewine 2021 B
100% verdoncho
90
With personality. Colour: coppery red. Nose: dried herbs, scrubland, dried flowers. Palate: correct, fine bitter notes.

ALDONZA
Ctra. N-430 km 462,3
02612 Munera (Albacete)
☎: +34 967 217 711
info@aldonzagourmet.com
www.aldonzagourmet.com

Aldonza Clásico 2015 T
88% tempranillo, 6% cabernet sauvignon, 4% merlot, 2% syrah
90
Colour: dark-red cherry, garnet rim. Nose: aged wood nuances, tobacco, sweet spices. Palate: spicy, round tannins, long.

Aldonza Navamarín 2013 T R
cabernet sauvignon, merlot, syrah, tempranillo
91
Colour: dark-red cherry, garnet rim. Nose: ripe fruit, fruit preserve, tobacco, sweet spices. Palate: spicy, round tannins, long.

Aldonza Selección 2015 T C S
cabernet sauvignon, merlot, tempranillo, syrah
89
Classic, toasty, tasty, jammy.

AÑADAS DEL QUIJOTE
CM 3102, Km. 14.6
13630 Socuéllamos (Ciudad Real)
☎: +34 692 110 292
atorres@quijotealimentacion.com

Torres Romero Ed. Limitada Syrah 2015 T
syrah
90
Colour: cherry, purple rim. Nose: black fruit, ripe fruit, wild herbs, toasty. Palate: flavourful, fruity, slightly dry, soft tannins.

Torres Romero Ed. Limitada Tempranillo 2014 T
tempranillo
89
Herbal, jammy, tasty, spicy.

Torres Romero Ed.Limitada Cabernet Sauvignon y Merlot 2012 T
60% cabernet sauvignon, 40% merlot
89
Overripe, spicy, fruity, tasty.

Torres Romero Petit Verdot Coleccion Privada 2013 T
petit verdot
90
Colour: deep cherry. Nose: dried herbs, creamy oak, black fruit, woody. Palate: powerful, ripe fruit, spicy, round tannins.

VINO DE LA TIERRA - CASTILLA / I.G.P.

VINO DE LA TIERRA - CASTILLA / I.G.P.

BODEGA CAMPOS DE DULCINEA
Garay, 1
45820 El Toboso (Toledo)
☎: +34 695 976 874
camposdedulcinea@camposdedulcinea.es
www.camposdedulcinea.es

Campos de Dulcinea Sauvignon Blanc 2021 B
100% sauvignon blanc
87 ♣

Campos de Dulcinea Selección de la Familia 2019 T
100% tempranillo
85

Campos de Dulcinea Tempranillo 2021 T
100% tempranillo
86

Unico 1926 2017 B FB
100% sauvignon blanc
91
Oxidativ. Colour: old gold, amber rim. Nose: candied fruit, dry nuts, sweet spices, varnish. Palate: flavourful, long.

Vale 2016 T R
100% tempranillo
89
Age nuances, pruney, kind finish, balanced, spicy, herbal.

Vale Serie Oro 2016 T GR
100% tempranillo
89
Slight oxidation, fruity, toasty, pruney, age nuances.

BODEGA CARRASCAS
Ctra. El Bonillo - Ossa de Montiel P.K. 11,4
02612 El Bonillo (Albacete)
☎: +34 967 965 880
info@carrascas.com
www.carrascas.com

Al Cobijo de una Gran Sabina 2018 T C
63% merlot, 37% cabernet sauvignon
91
Colour: deep cherry. Nose: dried herbs, black fruit, wild herbs, smoky. Palate: ripe fruit, spicy, round tannins, balanced.

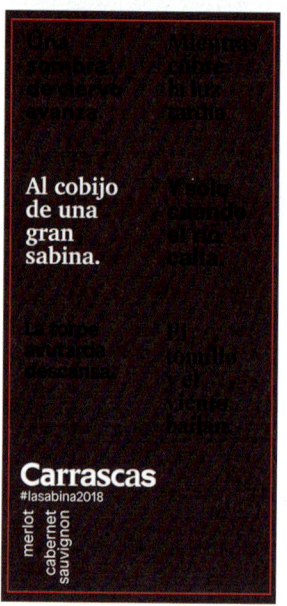

La Torpe Avutarda Descansa 2018 T C S
70% tempranillo, 30% syrah
90
Colour: deep cherry. Nose: dried herbs, creamy oak, black fruit, spicy. Palate: ripe fruit, round tannins, balanced.

Y solo cuando El Río Calla 2019 B
100% chardonnay
90
Colour: bright yellow. Nose: ripe fruit, spicy, wild herbs, citrus fruit. Palate: good structure, toasty, balanced.

El Tomillo y El Viento Bailan Viognier 2020 B
100% viognier

91

Colour: bright yellow. Nose: powerful, creamy oak, ripe fruit, spicy, dry nuts. Palate: rich, good structure, long, toasty, fine bitter notes.

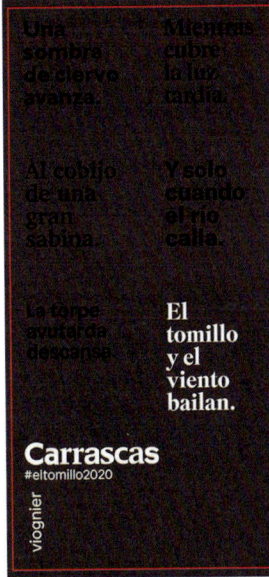

BODEGA DEHESA EL CARRASCAL
Ctra. de Alpera, km. 94.8 - CM-3201
02691 Bonete (Albacete)
☎: +34 967 240 458
Fax: +34 967 210 989
info@dehesaelcarrascal.com
www.vinoselcarrascal.com

Tudon´s Monastrell 2019 T
monastrell

89

Full-bodied, jammy, wild, herbal. Nose: caramel, sweet spices.

Tudon´s Petit Verdot 2019 T
petit verdot

89

Toasty, jammy, spicy, herbal, tasty, powerful.

BODEGA LA ERA
Estación, 6
19110 Mondéjar (Guadalajara)
☎: +34 609 684 689
laera@bodegalaera.com
www.bodegalaera.com

El Jardín de La Era 2020 T
tempranillo

86

La Era 2020 T
tempranillo

90

Colour: deep cherry. Nose: dried herbs, creamy oak, black fruit. Palate: ripe fruit, spicy, round tannins.

BODEGA LOS ALJIBES
Finca Los Aljibes
02520 Chinchilla de Montearagón (Albacete)
☎: +34 967 260 015
info@fincalosaljibes.com
www.fincalosaljibes.com

Aljibes 2018 T
cabernet sauvignon, merlot, merlot

91

Colour: dark-red cherry. Nose: toasty, spicy, cocoa bean, black fruit, ripe fruit. Palate: flavourful, toasty, fine bitter notes.

Aljibes Cabernet Franc 2018 T
cabernet franc

90

Colour: Cherry. Nose: scrubland, dried herbs, black fruit. Palate: spicy, flavourful, balsamic, round tannins.

Aljibes Garnacha Tintorera 2019 T C
garnacha tintorera

89

Pruney, dried herbs, tasty, great length, spicy, wild.

Aljibes Petit Verdot 2017 T
petit verdot

88

Smoky, toasty, tasty, powerful.

Enclave 2017 T
monastrell

91

Colour: dark-red cherry. Nose: toasty, spicy, cocoa bean, ripe fruit, black fruit. Palate: flavourful, toasty, fine bitter notes.

Viña Aljibes 2020 T RB
merlot, garnacha tintorera, cabernet sauvignon, petit verdot, syrah

87

VINO DE LA TIERRA - CASTILLA / I.G.P.

VINO DE LA TIERRA - CASTILLA / I.G.P.

BODEGA ROMAILA
45420 Almoncid de Toledo (Toledo)
☎: +34 915 416 561
administracion@romaila.es
www.romaila.es

100% Romaila 2018 T BA
100% garnacha
88
Kind finish, aromatic, floral, fruity.

Finca Romaila 2018 T C
cabernet sauvignon, syrah, tempranillo, cabernet franc
87

Oh! de Romaila T
89
Powerful, hot, toasty, jammy.

BODEGA Y VIÑEDOS TINEDO
Ctra. CM 3102, Km. 30
13630 Socuéllamos (Ciudad Real)
☎: +34 696 981 965
admin@tinedo.es
www.tinedo.es

Cala N 1 2019 T
77% tempranillo, 15% syrah, 6% cabernet sauvignon, 1% sauvignon blanc, 1% roussanne
89
Aromatic, toasty, tasty, jammy.

Cala N 2 2018 T
77% tempranillo, 13% graciano, 9% cabernet sauvignon, 1% sauvignon blanc
88
Spicy, overripe, toasty, kind finish.

JA! 2021 T
100% tempranillo
87

RunRun 2020 RD
33% moscatel grano menudo, 33% roussanne, 33% sauvignon blanc
88
Standard, fruity, simple, citrus fruit, floral.

Selección Parcela Moscatel 2021 B S
100% moscatel grano menudo
90
Colour: straw. Nose: expressive, white flowers, jasmine, dried herbs. Palate: flavourful, fruity, balanced.

Selección Parcela Syrah 2018 T
100% syrah
89
Toasty, tasty, jammy, spicy, complex.

BODEGAS ALCARDET / NTRA. SRA. DEL PILAR S.C. CLM
Mayor, 130
45810 Villanueva de Alcardete (Toledo)
☎: +34 925 166 375
Fax: +34 925 166 611
administracion@alcardet.com
www.alcardet.com

Alcardet 12 meses 2018 T S
tempranillo
88
Aromatic, toasty, spicy, jammy, tasty, standard.

Alcardet Cepas Viejas 2015 T BA
moravia
87

Alcardet Sauvignon Blanc 2021 B
sauvignon blanc
87

GEA Viña Galapagos 2021 B
verdejo
87

Pasaporte 2019 T
moravia
89
Jammy, spicy, dried herbs, tasty, wild. Nose: black fruit.

Pasaporte 2021 B
viognier
88
Aromatic, floral, balanced, fruity, jammy, smooth.

BODEGAS ALTOVELA
Ctra. Madrid - Alicante, Km. 100
45880 Corral de Almaguer (Toledo)
☎: +34 925 190 269
export@altovela.com
www.altovela.com

Cumplido 2021 B
sauvignon blanc, macabeo, airén
89
Balanced, spicy, fruity, jammy.

Cumplido 2021 T
88
Pleasant, tasty, smooth.

BODEGAS ANTONIO SERRANO

Galileo Galilei, 17
02600 Villarrobledo (Albacete)
☎: +34 627 029 943
administracion@bodegasantonioserrano.com
www.bodegasantonioserrano.com

Antonio Serrano Cencibel 2020 T
100% cencibel

87

Antonio Serrano Cencibel 2020 T RB
100% cencibel

88

Standard, pruney, jammy, spicy, fruity, varietally correct.

Antonio Serrano Coupage Premium 2017 T
50% cencibel, 25% monastrell, 25% garnacha

90

Colour: cherry, garnet rim. Nose: creamy oak, fruit liqueur notes, black fruit, waxy notes. Palate: pruney, powerful, sweet tannins.

BODEGAS ARÚSPIDE

Ciriaco Cruz, 2
13300 Valdepeñas (Ciudad Real)
☎: +34 926 347 075
export2@aruspide.com
www.aruspide.com

Ágora Ciento 69 Malbec 2020 RD
malbec

86

Ágora de Arúspide 2018 T RB
tempranillo

86

Ágora de Arúspide Lágrima 2021 B
airén, verdejo

84

Ágora Tempranillo 2021 T MC
tempranillo

86

Ágora Viognier 2021 B MC
viognier

85

Pura Savia 2019 T
tempranillo

88

Spicy, pruney, slight reduction, fruity.

BODEGAS BARREDA

Ramalazo, 2
45880 Corral de Almaguer (Toledo)
☎: +34 915 435 387
nacional@bodegas-barreda.com
www.bodegas-barreda.com

Torre de Barreda Amigos Multivarietal 2020 T
tempranillo, syrah, garnacha, graciano, cabernet sauvignon

89

Aromatic, fruity, jammy, lactic, dried herbs.

Torre de Barreda Amigos Multivarietal 2021 B
airén, viognier, sauvignon blanc

86

Torre de Barreda Cabernet Sauvignon 2020 T
100% cabernet sauvignon

88

Herbal, jammy, fruity, tasty, wild.

Torre de Barreda Garnacha 2019 T
100% garnacha

89

Pleasant, floral, aromatic, kind finish, fruity, smooth.

Torre de Barreda Graciano 2019 T
100% graciano

88

Balsamic herbs, spicy, herbal, jammy, wild, varietally correct.

Torre de Barreda PañoFino Viña Singular 2019 T S
tempranillo

90

Colour: deep cherry. Nose: ripe fruit, dried herbs, creamy oak. Palate: powerful, ripe fruit, spicy, round tannins.

BODEGAS BERBERANA

Ctra. Logroño Vitoria, Km. 423.700
26350 Fuenmayor (La Rioja)
☎: +34 607 604 413
jasanz@marquesdelaconcordia.com
www.berberana.com

Berberana Carta de Plata T BA
100% tempranillo

86

VINO DE LA TIERRA - CASTILLA / I.G.P.

Peñín Guide | SPANISH WINE | 777

VINO DE LA TIERRA - CASTILLA / I.G.P.

BODEGAS CAMPOS REALES
Castilla La Mancha, 4
16670 El Provencio (Cuenca)
☎: +34 967 166 066
tienda@bodegascamposreales.com
www.bodegascamposreales.com

Atril Bio Tempranillo Syrah 2021 T
tempranillo, syrah

89

Balanced, spicy, jammy, herbaceous, sweetish.

Canforrales Nature Tempranillo Syrah 2021 T
tempranillo, syrah

87

Canforrales Nature Viognier 2021 B
viognier

85

BODEGAS CORONADO
Ctra. San Isidro, s/n
16620 La Alberca de Záncara (Cuenca)
☎: +34 620 287 825
informacion@bodegascoronado.com
www.bodegascoronado.com

Charcón Alegría 2021 B
100% sauvignon blanc

87

Charcón Aromas 2020 B
100% gewürztraminer

88

Aromatic, varietally correct, fruity, floral, balanced, smooth.

Charcón Auténtico 2020 T
100% syrah

88

Pleasant, fruity, jammy, tasty, balanced, standard.

Charcón Diverso 2020 B FB
100% sauvignon blanc

86

Charcón Elegancia 2019 T
100% petit verdot

89

Spicy, fruity, jammy, tasty, toasty, dried herbs. Nose: dark chocolate.

Charcón Fortaleza 2019 T
100% petit verdot

88

Jammy, pruney, wild, dried herbs, standard, spicy.

BODEGAS CRIN ROJA
Paraje Mainetes
08651 Fuentealamo (Albacete)
☎: +34 938 743 511
Fax: +34 938 737 204
info@crinroja.com
www.crinroja.com

The Freaky Wines Tempranillo 2021 T
100% tempranillo

88

Fruity, tasty, aromatic.

The Freaky Wines Verdejo 2021 B
100% verdejo

85

BODEGAS DEL MUNI
Ctra. de Lillo, 48
45310 Villatobas (Toledo)
☎: +34 925 152 511
info@bodegasdelmuni.com
www.bodegasdelmuni.com

Corpus del Muni 2019 T RB
tempranillo, syrah, garnacha, petit verdot, merlot

87

Corpus del Muni 2021 RD
tempranillo, syrah

86

Corpus del Muni 60 Barricas 2020 T BA
tempranillo, syrah, garnacha

87

Corpus del Muni Blanca Selección 2021 B
chardonnay, sauvignon blanc, verdejo, riesling

87

Corpus del Muni Lucía Selección 2016 T C
tempranillo

88

Dried herbs, powerful, jammy, tasty, hot.

Corpus del Muni Sara Selección 2021 B SS
verdejo, riesling

84

BODEGAS EGUREN

Avda. del Cantábrico, s/n
01013 Vitoria-Gasteiz (Araba/Álava)
☎: +34 945 282 844
Fax: +34 945 271 319
info@egurenugarte.com
www.egurenugarte.com

Condado de Eguren Tempranillo 2020 T
tempranillo
84

Eguren Cabernet 2019 T
cabernet sauvignon
86

Eguren Sauvignon Blanc 2021 B
sauvignon blanc
84

Eguren Shiraz 2020 T
syrah
85

Eguren Verdejo 2021 B
verdejo
85

Reinares Tempranillo 2020 T BA
tempranillo
84

BODEGAS EL TANINO

Sol, s/n
02696 Hoya Gonzalo (Albacete)
☎: +34 696 946 071
dptotecnico@bodegaseltanino.com
www.bodegaseltanino.com

Altos de Santiago Chardonnay 2020 B
100% chardonnay
88
Kind finish, fruity, jammy, great length.

Mompichel 2020 B FB
chardonnay
89
Toasty, jammy, spicy, smoky, fruity, tasty.

BODEGAS ENTREMONTES

45800 Quintanar de la Orden (Toledo)
☎: +34 925 180 237
www.bodegasentremontes.com

Alto Blanco Merlot 2020 T AROM
merlot
87

Alto Blanco Verdejo 2020 B AROM
verdejo
89
Colour: bright straw. Nose: ripe fruit, fragrant herbs, fine lees. Palate: full-bodied, rich, long, good acidity.

BODEGAS FAMILIA CONESA - PAGO GUIJOSO

Crta Ossa de Montiel - El Bonillo km 11
02610 El Bonillo (Albacete)
☎: +34 608 612 254
mvruiz@familiaconesa.com
www.familiaconesa.com

El Beso Sauvignon Blanc 2019 B FB
100% sauvignon blanc
89 🌱
Balanced, toasty, jammy, herbal.

Finca La Sabina Tempranillo 2018 T
100% tempranillo
88
Fruity, jammy, balanced.

La Doncella de las Viñas 2021 RD
100% tempranillo
87 🌱

La Doncella de las Viñas Chardonnay 2021 B AROM
100% chardonnay
87 🌱

La Doncella de las Viñas Tempranillo 2020 T
100% tempranillo
87 🌱

La Doncella de las Viñas Tempranillo 2021 T RB
100% tempranillo
87 🌱

BODEGAS FAUSTINO RIVERO ULECIA (VINOS MONOVARIETALES)

Avda. de los Vinos, s/n
13600 Alcázar de San Juan (Ciudad Real)

Faustino Rivero Ulecia Afrutado 2021 B
moscatel
87

Faustino Rivero Ulecia Bobal 2021 RD
bobal
85

VINO DE LA TIERRA - CASTILLA / I.G.P.

VINO DE LA TIERRA - CASTILLA / I.G.P.

Faustino Rivero Ulecia Chardonnay 2021 B
chardonnay
86

Faustino Rivero Ulecia Macabeo 2021 B SD
macabeo
84

Faustino Rivero Ulecia Tempranillo 2021 T
tempranillo
86

Faustino Rivero Ulecia Verdejo 2021 B
verdejo
86

BODEGAS FERNÁNDEZ DE LA OSSA

San Antón, 42
02600 Villarrobledo (Albacete)
☎: +34 689 339 769
rubenossa99@hotmail.com
www.bodegasfernandezdelaossa.com

Casas de Peña 2021 RD
tempranillo
87

Casas de Peña Airén 2021 B
airén
85

Casas de Peña Chardonnay 2021 B
chardonnay
88
Fruity, jammy, lactic, tasty, herbal.

Casas de Peña Garnacha Tintorera 2021 T
garnacha tintorera
86

Casas de Peña Tempranillo 2021 T
tempranillo
87

BODEGAS FERNANDO CASTRO

Paseo Castelar, 70
13730 Santa Cruz de Mudela (Ciudad Real)
☎: +34 926 342 168
info@bodegasfernandocastro.com
www.bodegasfernandocastro.com

Mar de Flores 2021 T
pinot noir
84

Valdemonte Airén 2021 B
airén
84

Valdemonte Dark 2021 T
tempranillo
85

Valdemonte Pinot Noir 2021 T
pinot noir
85

Valdemonte Tempranillo 2021 T
tempranillo
86

Valdemonte Verdejo 2021 B
verdejo
87

BODEGAS FINCA LA ESTACADA

Ctra. N-400, Km. 103
16400 Tarancón (Cuenca)
☎: +34 969 327 099
Fax: +34 969 137 406
comunicacion@fincalaestacada.com
www.fincalaestacada.com

Hello World Petit Verdot 2021 T
petit verdot
87

Hello World Viognier 2021 B
viognier
87

Ochoymedio Tempranillo 2021 T
tempranillo
87

Ochoymedio Tinto Velasco 2021 T
tinto velasco
88
Balanced, spicy, herbaceous, jammy.

Secua Cabernet-Syrah 2017 T R
80% cabernet sauvignon, 20% syrah

91

Colour: deep cherry. Nose: dried herbs, creamy oak, black fruit, fine reductive notes. Palate: ripe fruit, spicy, round tannins.

Secua Crianza en lías 2021 B FB
viognier, sauvignon blanc

88

Balanced, spicy, herbal, dried flowers.

BODEGAS GARCÍA DE LARA
Del Cristo, 42
45360 Villarrubia de Santiago (Toledo)
☎: +34 648 733 298
info@bodegasgarciadelara.com

Finca Villalobillos 2020 B FB
100% airén

91

Colour: bright straw. Nose: fine lees, fresh fruit, balanced, neat. Palate: easy to drink, flavourful, balanced, round.

Finca Villalobillos 2021 B
100% airén

90

Colour: bright straw. Nose: ripe fruit, balanced, floral. Palate: flavourful, fresh, good acidity, balanced.

La Viña de La Cueva Colorá 2019 T C S
100% cencibel

90

Colour: deep cherry. Nose: ripe fruit, dried herbs, creamy oak. Palate: ripe fruit, spicy, round tannins.

BODEGAS GARDEL
Toledo, 2
16650 Las Mesas (Cuenca)
☎: +34 627 730 902
omar.gardel@hotmail.com
www.bodegasgardel.com

Poco a Poco 2021 B
87 🌱

Poco a Poco Envejecido en Barrica 2021 T C
87 🌱

Rosa de Alejandría 2020 B AG SD
moscatel

86 🌱

Tesón 2020 B FB
sauvignon blanc, verdejo

88 🌱

Citrus fruit, spicy, crisp, herbal, toasty.

BODEGAS LA SOLANA - PAGO FLORENTINO
Ctra. Porzuna -
Camino Cristo del Humilladero, km. 3
13420 Malagón (Ciudad Real)
☎: +34 983 681 146
Fax: +34 983 681 147
bodeg@arzuaganavarro.com
www.pagoflorentino.com

Pago Mota 2020 B
100% chardonnay

89

Kind finish, tropical, smooth, tasty.

Pago Mota 2021 B
100% chardonnay

87

BODEGAS LATÚE
Camino de la Esperilla, s/n
45810 Villanueva de Alcardete (Toledo)
☎: +34 925 166 350
web@latue.com
www.latue.com

Bohem 2021 B
87 🌱

Bohem 2021 T
tempranillo, garnacha

85 🌱

Clearly Organic 2021 B
airén, sauvignon blanc

85 🌱

Clearly Organic 2021 T
tempranillo

86 🌱

Clearly Organic sin sulfitos 2021 T
tempranillo

85 🌱

Latierra 2017 T RB
cabernet sauvignon, syrah

89 🌱

Pruney, dried herbs, reduced, spicy, great length. Nose: dark chocolate, balsamic herbs.

VINO DE LA TIERRA - CASTILLA / I.G.P.

BODEGAS LAZO

Finca La Zorrera
02436 Férez (Albacete)
☎: +34 622 766 900
info@lazotur.com
www.lazotur.com

Cabeza del Hierro 2019 T
tempranillo, cabernet sauvignon

88

Rustic, dried herbs, jammy, with personality, tasty.

Oriolus 2020 B
viognier

91

With personality, little interventionist, kind finish. Colour: yellow, pale. Nose: dried flowers, white fruit, wild herbs, stone fruit. Palate: flavourful, varietal, balanced.

Tarentola 2019 T
monastrell, otras

86

BODEGAS LEGANZA

Avda. IV Centenario, s/n
45800 Quintanar de la Orden (Toledo)
☎: +34 925 564 452
Fax: +34 925 564 021
info@bodegasleganza.es
www.bodegasleganza.com

Condesa de Leganza Rosado Selección de Familia 2021 RD
tempranillo

85

Condesa de Leganza Tempranillo 2019 T
tempranillo

86

Condesa de Leganza Verdejo 2021 B
verdejo

86

BODEGAS MANO A MANO

Ctra. CM-412, Km. 100
13248 Alhambra (Ciudad Real)
☎: +34 926 694 317
info@bodegamanoamano.com
www.manoamano.com

Venta la Ossa Cabernet Sauvignon 2019 T C
cabernet sauvignon

91

Colour: deep cherry. Nose: fruit preserve, powerful, dried herbs, scrubland. Palate: flavourful, long, round tannins.

Mano a Mano 2019 T C
100% tempranillo

89

Herbal, fruity, spicy, tasty.

Venta la Ossa 2021 T
tempranillo, cabernet sauvignon

89

Fruity, jammy, powerful, tasty, lactic.

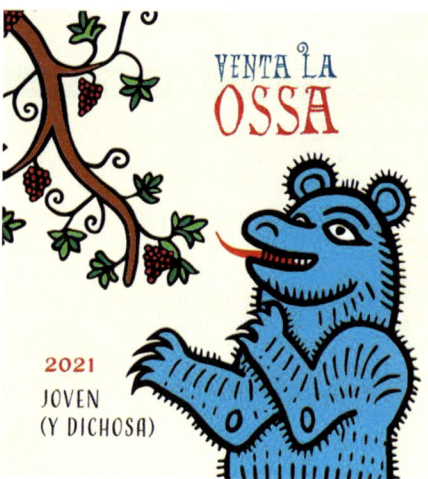

Venta la Ossa Syrah 2019 T
100% syrah

92

Colour: bright cherry, garnet rim. Nose: balanced, powerful, black fruit, ripe fruit. Palate: powerful, round tannins, long.

Venta la Ossa Tempranillo 2018 T C
100% tempranillo

92

Colour: Cherry. Nose: expressive, spicy, cocoa bean, sweet spices. Palate: full-bodied, long, great length, balanced.

Venta la Ossa TNT 2017 T
100% tempranillo

93

Colour: cherry, garnet rim. Nose: balsamic herbs, ripe fruit, scrubland, fine reductive notes, waxy notes, spicy. Palate: flavourful, spicy, round tannins.

BODEGAS MARISOL RUBIO
Goya, 54
45810 Villanueva de Alcardete (Toledo)
☎: +34 628 424 910
jorge@bodegasmarisolrubio.com
www.bodegasmarisolrubio.com

Marisol Rubio CIPMA I 2019 B
100% pedro ximénez

89

Citrus fruit, balanced, spicy, full-bodied, jammy, tasty, toasty.

Marisol Rubio CIPMA I 2020 B
100% pedro ximénez

89

Some off-flavours, balanced, spicy, tasty. Palate: sweetness.

Marisol Rubio CIPMA II 2020 B FB

91

Colour: bright golden. Nose: caramel, faded flowers, honeyed notes, toasted bread. Palate: flavourful, rich, long, spicy.

Marisol Rubio SON D SOL 2020 B
100% pedro ximénez

88

Pleasant, spicy, jammy, tasty.

BODEGAS MARTÍNEZ SÁEZ
Finca San José, Ctra. Barrax Km. 14,800
02600 Villarrobledo (Albacete)
☎: +34 967 443 088
Fax: +34 967 440 204
almacen@bodegasmartinezsaez.es
www.bodegasmartinezsaez.es

Martínez Saez Selección 2017 T C
88

Toasty, powerful, spicy, pruney.

Viña Orce 2019 T C
tempranillo, cabernet sauvignon

88

Kind finish, toasty, tasty, powerful.

Viña Orce 2021 RD
85

Viña Orce Chardonnay 2021 B
chardonnay

87

Viña Orce Macabeo Verdejo 2021 B
macabeo, verdejo

87

Viña Orce Tempranillo 2021 T RB
tempranillo

88

Smoky, fruity, jammy, tasty.

BODEGAS MARTUE
Campo de la Guardia, s/n
45760 La Guardia (Toledo)
☎: +34 658 915 812
admin@martue.com
www.martue.com

Dauco 2018 T
tempranillo

87

BODEGAS MEGÍA E HIJOS -CORCOVO
Magdalena, 33
13300 Valdepeñas (Ciudad Real)
☎: +34 926 347 828
Fax: +34 926 347 829
comercial@corcovo.com
www.corcovo.com

Corcovo 2021 RD
tempranillo

87

Corcovo Muscat 2021 B S
moscatel grano menudo

88

Aromatic, varietally correct, herbal, crisp, tasty.

Corcovo Verdejo 2021 B
verdejo

87

BODEGAS MIGUEL A. AGUADO
Cantalejos, 2
45165 San Martín de Montalbán (Toledo)
☎: +34 653 821 659
info@bodegasmiguelaguado.com
www.bodegasmiguelaguado.com

San Martineño 2014 T R
60% cabernet sauvignon, 40% garnacha

86

San Martineño Garnacha 2021 T BA
100% garnacha

87

San Martineño Tempranillo 2018 T
100% tempranillo

85

VINO DE LA TIERRA - CASTILLA / I.G.P.

VINO DE LA TIERRA - CASTILLA / I.G.P.

San Martineño Tempranillo 2021 T
tempranillo
85

BODEGAS MOISÉS CASAS (ECCE VINUM)
Victoria, 4
45830 Miguel Esteban (Toledo)
☎: +34 653 280 797
info@eccevinum.com
www.eccevinum.com

Ecce Vinum Tempranillo Organic (8 Months in oak barrel) 2015 T C
100% tempranillo
86 🌱

Ecce Vinum Tempranillo Organic 2021 T
100% tempranillo
86 🌱

Ecce Vinum Verdejo Organic 2021 B
100% verdejo
86 🌱

BODEGAS MONTALVO WILMOT
Ctra. Ruidera. Km. 10,200 Finca los Cerrillos
13710 Argamasilla de Alba (Ciudad Real)
☎: +34 926 699 069
info@montalvowilmot.com
www.montalvowilmot.com

Montalvo Wilmot Cabernet de Familia 2017 T
100% cabernet sauvignon
90
Colour: Cherry. Nose: scrubland, sweet spices, dark chocolate, ripe fruit. Palate: spicy, balsamic, round tannins, varietal.

Montalvo Wilmot Colección Privada 2017 T R
75% tempranillo, 25% cabernet sauvignon
91
Colour: light cherry, garnet rim. Nose: fine reductive notes, ripe fruit, aged wood nuances, spicy. Palate: balanced, complex, long, fine tannins, fine bitter notes.

Montalvo Wilmot Petit Verdot Selección 2017 T R
petit verdot
90
Colour: cherry, garnet rim. Nose: varietal, wild herbs, neat, dried herbs. Palate: flavourful, slightly dry, soft tannins.

Montalvo Wilmot Petit Verdot Tempranillo Syrah 2020 T BA
40% petit verdot, 30% tempranillo, 30% syrah
88
Wild, dried herbs, jammy, tasty. Palate: slightly dry, soft tannins.

BODEGAS NAVARRO LÓPEZ
Autovía Madrid - Cádiz, Km. 193
13300 Valdepeñas (Ciudad Real)
☎: +34 962 321 888
Fax: +34 926 320 977
info@navarrolopez.com
www.navarrolopez.com

Para Celsus 2021 T
88 🌱
Fruity, jammy, tasty, wild, balanced, great length.

Para Celsus Verdejo 2021 B
verdejo
86 🌱

Premium 1904 – Tempranillo Syrah 2020 T
tempranillo, syrah
89
Balanced, jammy, tasty, spicy, powerful, fruity.

Premium 1904 Sauvignon Blanc 2021 B
sauvignon blanc
88
Pleasant, herbal, fruity, smooth.

Rojo Tempranillo 2021 T
tempranillo
87

BODEGAS NOC
Orgaz, 12
45460 Manzaneque (Toledo)
☎: +34 925 344 727
info@bodegasnoc.com
www.bodegasnoc.es

Mernat de Noc 2017 T
41% cabernet sauvignon, 23% petit verdot, 20% syrah, 16% tempranillo
91
Colour: deep cherry. Nose: dried herbs, creamy oak, raspberry, floral, black fruit. Palate: ripe fruit, spicy, round tannins.

Mernat de Noc Viognier 2020 B
viognier
90
Colour: bright straw. Nose: ripe fruit, fragrant herbs, fine lees, tropical fruit. Palate: full-bodied, good acidity, fruity, saline.

Noc Chardonnay 2018 B
100% chardonnay
88
Citrus fruit, balanced, tasty, tropical, roasted coffee.

Noc Prestige BE BR
51% viognier, 49% chardonnay
87

Noc Rosé RE BR
100% tempranillo
88
Balanced, dried flowers, crisp, herbal.

Noc Tempranillo 2018 T
100% tempranillo
91
Colour: deep cherry. Nose: dried herbs, creamy oak, black fruit. Palate: powerful, ripe fruit, spicy, round tannins.

BODEGAS PUENTE DE RUS
Ctra. Almarcha, 50
16600 San Clemente (Cuenca)
☎: +34 969 300 155
exportassistant@puentederus.com
www.puentederus.com

Camino de Rus Selección 2015 T C
100% tempranillo
88
Spicy, jammy, smooth, toasty.

Camino de Rus Syrah 2020 T
100% syrah
87

Colmillo de Lobo T BA
40% cabernet sauvignon, 40% tempranillo, 10% merlot, 5% syrah, 5% petit verdot
87

Paso de Buey T
40% cabernet sauvignon, 40% tempranillo, 10% merlot, 5% syrah, 5% petit verdot
89
Aromatic, balsamic herbs, floral, jammy, sweety finish, spicy.

Pontevs 2017 T
60% tempranillo, 20% syrah, 10% cabernet sauvignon, 10% petit verdot
90
Colour: bright cherry. Nose: sweet spices, ripe fruit, dried herbs, characterful. Palate: fruity, spicy, round tannins.

Vista de Halcón 2020 B
80% sauvignon blanc, 20% verdejo
87

BODEGAS R & G
Ramón y Cajal 7, 1ºA
01007 Vitoria-Gasteiz (Araba/Álava)
☎: +34 945 150 589
araex@araex.com
www.araex.com

Eldoze 2015 T
100% syrah
90
Colour: cherry, garnet rim. Nose: fruit preserve, fruit liqueur notes, powerful, toasty, dark chocolate. Palate: flavourful, sweetness, long.

Eldoze 2017 T
91
Colour: very deep cherry. Nose: aromatic coffee, powerful, black fruit, ripe fruit. Palate: great length, round tannins, flavourful, spicy.

BODEGAS ROMERO DE ÁVILA SALCEDO
Avda. Constitución, 4
13240 La Solana (Ciudad Real)
☎: +34 926 631 426
media@bodegasromerodeavila.com
www.bodegasromerodeavila.com

Romero de Avila Salcedo 2011 T RB
50% tempranillo, 10% petit verdot, 10% merlot, 10% cabernet sauvignon, 20% syrah
89
Pruney, fruity, tasty, toasty, balanced.

Testigo 2016 T RB
60% tempranillo, 30% syrah, 10% cabernet sauvignon
89
Balsamic herbs, spicy, dried herbs, jammy, pruney, tasty. Palate: grainy tannins.

VINO DE LA TIERRA - CASTILLA / I.G.P.

VINO DE LA TIERRA - CASTILLA / I.G.P.

BODEGAS SOLEDAD

Ctra. Tarancón, s/n
16411 Fuente de Pedro Naharro (Cuenca)
☎: +34 969 125 039
Fax: +34 969 125 907
calidad@bodegasoledad.com
www.bodegasoledad.com

Honest GSM 2019 T
garnacha, syrah, monastrell

89

Balsamic herbs, standard, spicy, jammy, tasty, balanced.

Honest VVS 2020 B
verdejo, viura, sauvignon blanc

89

Citrus fruit, balanced, herbal, crisp, tasty.

BODEGAS TRENZA

Felix Mendelsohn, 8
03730 Jávea (Alacant/Alicante)
☎: +34 965 790 012
bodegas@bodegatrenza.com
www.bodegatrenza.com

Acentuado 2019 T
tempranillo

88

Pleasant, toasty, jammy, fruity.

Acentuado Rose Organic 2021 RD
garnacha

88 ♣

Pleasant, aromatic, fruity, sweeties, lactic, smooth.

Tofterup Brothers Organic Red T
tempranillo, syrah

88 ♣

Standard, fruity, spicy, jammy, tasty.

Tofterup Brothers Organic Rose 2021 RD
garnacha, bobal

88 ♣

Kind finish, aromatic, fruity, floral, lactic.

BODEGAS VILLAVID

Niño Jesús, 25
16280 Villarta (Cuenca)
☎: +34 649 361 942
export@villavid.com
www.villavid.com

Villavid Syrah 1952 2018 T RB
syrah

88

Spicy, jammy, varietally correct, full-bodied, smoky. Nose: meaty notes.

BODEGAS VIRGEN DE BELÉN

Paraje de Belén s/n Apdo. 540
02640 Almansa (Albacete)
☎: +34 655 624 850
bodega@bodegastiopio.com
www.bodegastiopio.com

Bastón 2017 T
100% petit verdot

86

BODEGAS VOLVER

Ctra de Pinoso a Fortuna, s/n
03658 Rodriguillo (Alacant/Alicante)
☎: +34 966 185 624
Fax: +34 965 075 376
export@bodegasvolver.com
www.bodegasvolver.com

Paso a Paso Tempranillo 2021 T
tempranillo

91

Colour: deep cherry. Nose: ripe fruit, dried herbs, creamy oak, red berry notes. Palate: powerful, ripe fruit, spicy, round tannins.

Volver 2020 T
100% tempranillo

90

Colour: very deep cherry. Nose: roasted coffee, aromatic coffee, powerful, ripe fruit, black fruit. Palate: smoky aftertaste, great length, round tannins.

Volver Cuvée 2018 T
80% tempranillo, 20% cabernet sauvignon

92

Colour: very deep cherry. Nose: roasted coffee, aromatic coffee, powerful, ripe fruit, black fruit. Palate: smoky aftertaste, great length, round tannins.

BODEGAS WINOL

Poeta García Carbonell, 8 - Bajo
02005 Albacete (Albacete)
☎: +34 686 136 446
aochoa@bodegaswinol.com
www.bodegaswinol.com

8Lgends Eterna Leyenda 2018 T
37% cabernet sauvignon, 25% syrah, 21% merlot, 17% tempranillo

86 ♣

8Lgends Leyenda del Caballero Cabernet Sauvignon 2020 T
100% cabernet sauvignon

87

8Lgends
Leyenda del Caballero Merlot 2020 T
100% merlot
86

8Lgends
Leyenda del Caballero Syrah 2020 T
100% syrah
84

BODEGAS Y VIÑEDOS CASA DEL VALLE

Ctra. de Yepes - Añover de Tajo, Km. 47,700
45313 Yepes (Toledo)
☎: +34 925 155 533
Fax: +34 925 147 019
casadelvalle@bodegaolarra.es
www.bodegascasadelvalle.es

Finca Valdelagua 2018 T
syrah, merlot, cabernet sauvignon
90
Colour: deep cherry. Nose: ripe fruit, dried herbs, wild herbs, sweet spices. Palate: ripe fruit, spicy, round tannins, flavourful.

Hacienda Casa del Valle 2020 T
tempranillo, cabernet sauvignon
87

Hacienda Casa del Valle Chardonnay 2020 B
100% chardonnay
86

Hacienda Casa del Valle Sauvignon Blanc 2021 B
sauvignon blanc
87

Hacienda Casa del Valle Selección Especial 2019 T
tempranillo, cabernet sauvignon
89
Balsamic herbs, jammy, wild, tasty, spicy, pleasant.

Hacienda Casa del Valle Syrah 2019 T
100% syrah
86

BODEGAS Y VIÑEDOS CASTIBLANQUE

Isaac Peral, 19
13610 Campo de Criptana (Ciudad Real)
☎: +34 926 589 147
administracion@bodegascastiblanque.com
www.bodegascastiblanque.com

Baldor Old Vines 2015 T
100% cabernet sauvignon
88
Pruney, spicy, balsamic herbs, great length. Palate: slightly dry, soft tannins.

Baldor Tradición Syrah 2017 T
100% syrah
86

Ilex 2020 T
syrah, tempranillo
86

Ilex Coupage 2019 T
syrah, tempranillo, garnacha
85

Ilex Verdejo 2021 B
100% verdejo
84

Manuel El Santero 2017 T RB
syrah, tempranillo
85

BODEGAS Y VIÑEDOS PINUAGA

Ctra. N-301 Km. 95,5
45880 Corral de Almaguer (Toledo)
☎: +34 629 058 900
bodega@bodegaspinuaga.com
www.bodegaspinuaga.com

La Senda de Pinuaga Premium Cuvée 2020 T RB
cencibel, garnacha, merlot
88 🌱
Spicy, pronounced acidity, toasty, jammy, dried herbs.

Pinuaga 200 Cepas 2019 T
cencibel
90 🌱
Colour: deep cherry. Nose: ripe fruit, dried herbs, creamy oak, lactic notes, woody. Palate: ripe fruit, spicy, round tannins.

Pinuaga Bianco 2021 B
sauvignon blanc
88 🌱
Citrus fruit, herbal, tasty, wild, crisp.

VINO DE LA TIERRA - CASTILLA / I.G.P.

Pinuaga Colección 2019 T C
cencibel

89

Spicy, lactic, jammy, roasted coffee, smooth.

Pinuaga Nature 2020 T RB
cencibel

88

Balanced, spicy, jammy, dried herbs.

Pinuaga Rosé 2021 RD
tempranillo, garnacha

85

BODEGAS Y VIÑEDOS VERUM

Ctra. N-310 Km. 228
13700 Tomelloso (Ciudad Real)
☎: +34 926 511 404
info@bodegasverum.com
www.bodegasverum.com

Ulterior Garnacha 2018 T
garnacha

91

Colour: light cherry. Nose: fruit preserve, fruit liqueur notes, faded flowers, spicy, wild herbs. Palate: correct, fine bitter notes, easy to drink.

Ulterior Mazuelo 2018 T
mazuelo

91

Wild, smooth. Colour: cherry, garnet rim. Nose: red berry notes, ripe fruit, characterful. Palate: flavourful, balanced, correct.

Verum Cabernet Franc Tempranillo 2021 RD
95% cabernet franc, 5% tempranillo

87

Verum Cencibel 2019 T RB
cencibel

90

Colour: deep cherry. Nose: dried herbs, creamy oak, ripe fruit, fruit preserve. Palate: powerful, ripe fruit, spicy, round tannins.

Verum Las Tinadas Airén de Pie Franco 2020 B
airén

89

Austere, smooth, dried flowers, yeasty notes, jammy, balanced.

Verum Terra Airén 2021 B
airén

88

Pleasant, crisp, fruity, smooth, simple.

CAPILLA DEL FRAILE

45170 San Martin de Pusa (Toledo)
☎: +34 675 217 334
info@gprocdf.com
www.fincacapilladelfraile.com

Capilla del Fraile Parcela Syrah 2018 T
syrah

89

Fruity, spicy, jammy, dried herbs, tasty, toasty.

Capilla del Fraile Parcela Syrah 2019 T
syrah

89

Fruity, sweety finish, jammy, toasty, tasty.

Capilla del Fraile Parcela Syrah 2020 T

88

Fruity, jammy, tasty, toasty.

Capilla del Fraile Parcela Syrah 2021 T

87

Capilla del Fraile Petit Verdot 2017 T
petit verdot

88

Spicy, fruity, dried herbs, smoky, astringent.

Capilla del Fraile Petit Verdot 2018 T

90

Fruity, sweety finish, herbal, spicy, tasty.

Capilla del Fraile Petit Verdot 2021 T

89

Fruity, spicy, herbal, toasty. Nose: fruit preserve.

Capilla del Fraile Selección Especial 2018 T
50% syrah, 50% petit verdot

88

Hot, spicy, fruity, smooth, tasty.

CARRIL CRUZADO

Ctra. Iniesta-Villagarcía del Llano km. 13
16236 Villagarcía del Llano (Cuenca)
☎: +34 616 960 992
bodega@carrilcruzado.com
www.carrilcruzado.es

Carril Cruzado Cabernet Sauvignon 2021 T
100% cabernet sauvignon

83

Carril Cruzado Multivarietal 2018 T
48,5% tempranillo, 27% syrah, 24,5% cabernet sauvignon

82

Carril Cruzado Selección Petit Verdot 9 Meses Barrica 2020 T
100% petit verdot

83

Carril Cruzado Selección Syrah 9 Meses Barrica 2020 T C
100% syrah

84

CASTILLO DE ARESAN
Ctra. N-310, Km.120
02600 Villarrobledo (Albacete)
☎: +34 679 981 697
irene.martinez@castillodearesan.es

Castillo de Aresan Cabernet Franc 2020 T
cabernet franc

87 🌱

Castillo de Aresan Sauvignon Blanc 2021 B
sauvignon blanc

88 🌱

Balanced, wild, varietally correct, herbal, smooth, pleasant.

Castillo de Aresan Verdejo 2021 B
86 🌱

El Rosé de Aresan 2021 RD
87 🌱

COOP. JESÚS DEL PERDÓN - BODEGAS YUNTERO
P.I. de Manzanares- Ctra. Alcazar s/n
13200 Manzanares (Ciudad Real)
☎: +34 926 610 309
Fax: +34 926 610 516
yuntero@yuntero.com
www.yuntero.com

Lazarillo 2021 T
100% tempranillo

85

Lazarillo Airén 2021 B
100% airén

84

COSECHEROS Y CRIADORES
Diputación, s/n
01320 Oyón (Araba/Álava)
☎: +34 941 450 876
info@bujanda.com
www.cosecherosycriadores.com

Copa Real 2021 T
garnacha, tempranillo

86

Mas Momentos Cabernet Sauvignon 2021 T
100% cabernet sauvignon

85 🌱

Mas Momentos Syrah 2021 T
100% syrah

86 🌱

Mas Momentos Tempranillo 2021 T
100% tempranillo

84 🌱

Sierra Almirón 2021 T
tempranillo, garnacha

86

Venta Vieja Malbec 2021 T
100% malbec

88

Balsamic herbs, spicy, jammy, tasty, wild.

DEHESA DE LOS LLANOS
Ctra. De Las Peñas de San Pedro, km. 5,5
02006 Albacete (Albacete)
☎: +34 967 243 100
a.granell@dehesadelosllanos.es
www.dehesadelosllanos.es

Cima Mazacruz Selección 2019 T
93% cabernet sauvignon, 7% merlot

88

Spicy, herbaceous, jammy, crisp, rustic.

Cima Selección 2021 B
verdejo

87

VINO DE LA TIERRA - CASTILLA / I.G.P.

VINO DE LA TIERRA - CASTILLA / I.G.P.

DOMINIO DE EGUREN
Camino de San Pedro, s/n
01309 Páganos (Araba/Álava)
☎: +34 945 600 590
info@sierracantabria.com
www.sierracamtabria.com

Códice 2020 T BA
100% tempranillo
88
Pleasant, fruity, smooth, tasty.

Protocolo 2020 T
100% tempranillo
86

Protocolo 2021 B
50% airén, 50% macabeo
86

Protocolo 2021 RD
50% bobal, 50% tempranillo
86

Protocolo Eco 2020 T
100% tempranillo
87

Protocolo Eco 2021 B
50% airén, 50% macabeo
84

Protocolo Eco 2021 RD
50% bobal, 50% tempranillo
85

DOMINIO DEL LINZE
Avda. Gregorio Prieto, 5
13300 Valdepeñas (Ciudad Real)
☎: +34 926 035 811
info@seleccionlucendo.com
www.seleccionlucendo.com

César Lucendo A. 2018 T R
tempranillo, cabernet sauvignon, merlot
89
Full-bodied, spicy, great length, tasty, dried herbs.

El Linze 2020 T
tinto velasco, syrah
89
Toasty, smoky, spicy, jammy, standard.

El Último Lobo 2020 T RB
tempranillo
87

Marta Cibelina 2021 B
chardonnay, viognier
87

EL SAUCERAL
Finca El Sauceral, Ctra. CM 403, km 19
45127 Las Ventas con Peña Aguilera (Toledo)
☎: +34 684 041 747
administracion@parajes-invest.com
www.elsauceral.com

La Peralosa 2020 T
cabernet sauvignon, syrah
89
Toasty, jammy, spicy, herbal, full-bodied.

Puerto del Milagro 2020 T
syrah
90
Kind finish, full-bodied. Colour. deep cherry. Nose: ripe fruit, dried herbs, creamy oak. Palate: powerful, ripe fruit, spicy, round tannins.

ENCOMIENDA DE CERVERA
13270 Almagro (Ciudad Real)
☎: +34 926 102 099
Fax: +34 902 222 297
info@ecervera.com
www.encomiendadecervera.com

1758 Reserva Familiar Petit Verdot 10 años 2011 T
petit verdot
90
Colour. deep cherry. Nose: dried herbs, creamy oak, black fruit, scrubland. Palate: powerful, ripe fruit, spicy, round tannins.

1758 Reserva Familiar Petit Verdot 5 años 2016 T
petit verdot
88
Standard, spicy, dried herbs, jammy.

1758 Selección Cencibel 2011 T C
cencibel
86

1758 Selección Reserva Familiar Cencibel 5 años 2016 T
tempranillo
90
Colour. cherry, purple rim. Nose: spicy, toasty, red berry notes, black fruit. Palate: flavourful, fruity, good acidity, long.

FÉLIX SOLIS AVANTIS

Autovía del Sur, Km. 199
13300 Valdepeñas (Ciudad Real)
☎: +34 926 322 400
Fax: +34 926 322 417
fsa@felixsolisavantis.com
www.felixsolisavantis.com

Viña Albali Cabernet Sauvignon* 2021 T
cabernet sauvignon

86

Viña Albali Chardonnay 2021 B
100% chardonnay

85

Viña Albali Garnacha Rosé 2021 RD
garnacha

84

Viña Albali Merlot 2021 T
100% merlot

86

Viña Albali Tempranillo Shiraz 2021 T
50% tempranillo, 50% syrah

86

FINCA CONSTANCIA

Camino del Bravo, s/n
45543 Otero (Toledo)
☎: +34 925 861 535
prensa@gonzalezbyass.com
www.fincaconstancia.es

Altos de la Finca 2017 T
60% petit verdot, 40% syrah

91

Colour: cherry, garnet rim. Nose: overripe fruit, creamy oak, hot, toasty, dark chocolate. Palate: pruney, powerful, sweet tannins.

Cosmológico Cabernet Sauvignon Biodinámico 2021 T
cabernet sauvignon

88 ♣

Varietally correct, herbal, fruity.

Finca Constancia Entre Lunas 2020 T BA
100% tempranillo

88 ♣

Fruity, herbaceous, simple.

Finca Constancia Graciano Parcela 12 2018 T
graciano

91

Colour: dark-red cherry, garnet rim. Nose: ripe fruit, tobacco, sweet spices. Palate: spicy, round tannins, long.

Finca Constancia Selección 2019 T BA
syrah, cabernet sauvignon, petit verdot, tempranillo, graciano, cabernet franc

89

Toasty, tasty, smooth, jammy.

Finca Constancia Tempranillo Parcela 23 2020 T
tempranillo

88

Kind finish, toasty, jammy, spicy.

Finca Constancia Verdejo Parcela 52 2020 B FB
verdejo

90

Colour: bright yellow. Nose: spicy, white fruit, ripe fruit. Palate: rich, good structure, long, fine bitter notes.

Universal Cabernet Sauvignon Biodinámico 2021 T
100% cabernet sauvignon

88 ♣

Aromatic, fruity, herbal, tasty.

Universal Sauvignon Blanc Biodinámico 2021 B
sauvignon blanc

88

Citrus fruit, varietally correct, tasty, fruity.

FINCA EL REFUGIO

Ctra. CM-3102, km. 14,6
13630 Socuéllamos (Ciudad Real)
☎: +34 629 512 478
info@fincaelrefugio.es
www.fincaelrefugio.es

Dominio del Prior Petit Verdot 2016 T BA
petit verdot

88 ♣

Fruity, jammy, dried herbs, powerful, spicy.

Dominio del Prior Syrah 2016 T BA
syrah

90 ♣

Colour: bright cherry. Nose: meaty notes, black fruit, ripe fruit, smoky. Palate: fruity, flavourful, slightly dry, soft tannins.

Legado Finca El Refugio Petit Verdot 2016 T
petit verdot

87 ♣

VINO DE LA TIERRA - CASTILLA / I.G.P.

VINO DE LA TIERRA - CASTILLA / I.G.P.

Legado Finca El Refugio Syrah 2016 T
syrah

88

Spicy, fruity, smoky.

Quorum – Le Blanc 2017 B FB
100% verdejo

90

Colour: yellow. Nose: toasted bread, yeasty notes, smoky, characterful. Palate: easy to drink, spicy.

Quorum de Finca El Refugio Private Collection 2012 T BA
75% tempranillo, 25% petit verdot

90

Colour: deep cherry. Nose: old leather, meaty notes, ripe fruit, spicy. Palate: flavourful, slightly dry, soft tannins, good structure.

FINCA LOS ALIJARES
Avda. de la Paz, 5
45180 Camarena (Toledo)
☎: +34 696 964 737
Fax: +34 918 174 364
gerencia@fincalosalijares.com
www.fincalosalijares.com

Finca Los Alijares Garnacha Petit Verdot 2016 T
garnacha, petit verdot

87

Finca Los Alijares Graciano 2015 T R
100% graciano

89

Pruney, spicy, varietally correct, tasty, powerful.

Finca Los Alijares Graciano Autor 2009 T GR
graciano

88

Colour: cherry, garnet rim. Nose: balsamic herbs, scrubland, old leather, earthy notes, fruit preserve. Palate: flavourful, balsamic, spicy, soft tannins.

Finca Los Alijares Moscatel 2019 B SD

87

Finca Los Alijares Syrah Graciano 2018 T C
50% syrah, 50% graciano

87

FINCA RÍO NEGRO
Ctra. CM 1001, Km. 37,400
19230 Cogolludo (Guadalajara)
☎: +34 913 022 648
Fax: +34 917 660 019
info@fincarionegro.com
www.fincarionegro.es

992 Finca Río Negro 2020 T
65% tempranillo, 20% syrah, 15% merlot

90

Colour: deep cherry, purple rim. Nose: ripe fruit, creamy oak, balanced, characterful. Palate: powerful, ripe fruit, spicy, round tannins.

Finca Río Negro 2018 T C
68% tempranillo, 12% cabernet sauvignon, 8% merlot, 12% syrah

91

Colour: bright cherry, garnet rim. Nose: scrubland, spicy, black fruit, ripe fruit. Palate: spicy, balsamic, round tannins.

Finca Río Negro 5º Año 2017 T GR
70% tempranillo, 30% cabernet sauvignon

93

Classic, jammy. Colour: cherry, garnet rim. Nose: ripe fruit, scrubland, fine reductive notes. Palate: flavourful, spicy.

Finca Río Negro Cerro del Lobo 2019 T
100% syrah

92

Wild. Colour: deep cherry. Nose: ripe fruit, dried herbs, dry nuts. Palate: powerful, ripe fruit, spicy, round tannins.

Finca Río Negro Gewürztraminer 2021 B
100% gewürztraminer

90

Aromatic, pleasant. Nose: white fruit, tropical fruit, floral, balanced. Palate: flavourful, varietal, rich.

FINCA VENTA DON QUIJOTE
Ctra. N-301, KM 131.4
45820 El Toboso (Toledo)
☎: +34 925 100 004
info@fincalaventa.com

J. Fernando Tempranillo 6 meses 2020 T
100% tempranillo

88

Balanced, spicy, dried herbs, toasty.

VQ Sauvignon Blanc 2021 B
100% sauvignon blanc

85

HACIENDA ALBAE

Ctra. de Argamasilla de Alba
a Cinco Casas, km. 25.500 CM
13710 Argamasilla de Alba (Ciudad Real)
☎: +34 667 109 422
administracion@haciendaalbae.com
www.haciendaalbae.com

Hacienda Albae Cabernet Sauvignon 2020 T
cabernet sauvignon

87

Hacienda Albae Chardonnay 2021 B
100% chardonnay

88

Aromatic, fruity, jammy, tropical, balanced.

Hacienda Albae Grand Viognier 2020 B
100% viognier

90

Colour: yellow. Nose: ripe fruit, sweet spices, fine lees, floral. Palate: rich, good structure, long, toasty, fine bitter notes.

Hacienda Albae Merlot 2020 T
100% merlot

85

Hacienda Albae Syrah 2020 T S
syrah

86

Hacienda Albae Top 888 2016 T R
90% cabernet sauvignon, 5% merlot, 5% syrah

87

HACIENDA REAL

Polígono Industrial Alces. Calle Mencía s/n
13600 Alcázar de San Juan (Ciudad Real)
☎: +34 926 547 404
sara.rodriguez@dcoop.es
www.dcoop.es

Hacienda Real 2021 B
airén

86

Hacienda Real 2021 T
100% cencibel

87

HACIENDA VILLARTA

Ctra. Nacional 403, km. 48
45910 Escalona (Toledo)
☎: +34 925 740 027
agrovillarta@gmail.com
www.haciendavillarta.com

Besanas 2020 T C
100% tempranillo

86

Tozara 2020 T RB
100% tempranillo

84

HAMMEKEN CELLARS

Llavador, 20
03700 Denia (Alacant/Alicante)
☎: +34 965 791 967
cellars@hammekencellars.com
www.hammekencellars.com

Allegranza Chardonnay 2021 B
100% chardonnay

86

Allegranza Tempranillo Vendimia Seleccionada 2021 T
100% tempranillo

86

Capa Vendimia Seleccionada Tempranillo 2021 T
100% tempranillo

90

Standard, pleasant. Nose: ripe fruit, dried herbs, creamy oak. Palate: powerful, ripe fruit, spicy.

El Gringo Dark Red Tempranillo 2021 T
100% tempranillo

88

Toasty, tasty, jammy, fruity.

El Paso del Lazo Tempranillo Shiraz 2021 T
85% tempranillo, 15% syrah

87

Sedosa Organic Rosé 2021 RD

88 ❀

Citrus fruit, standard, crisp, fruity, herbal.

VINO DE LA TIERRA - CASTILLA / I.G.P.

VINO DE LA TIERRA - CASTILLA / I.G.P.

HEREDAD DE ATENCIA
Acacio Moreno, 4
02600 Villarrobledo (Albacete)
☎: +34 967 138 457
bodegas@heredad-atencia.com
www.heredad-atencia.com

Atencia 2015 T BA
30% cabernet sauvignon, 20% syrah, 16% touriga nacional, 12% tempranillo, 12% tinta de Toro (Tinto), 5% petit verdot
87

N-A De Atencia 2014 T BA
65% cabernet franc, 17,5% petit verdot, 17,5% carmenère
92
Colour: cherry, garnet rim. Nose: overripe fruit, creamy oak, hot, expressive, powerful. Palate: pruney, powerful, sweet tannins.

LA BALLESTERA
Ctra. Torre de Juan Abad s/n, Km 26
13750 Castellar de Santiago (Ciudad Real)
☎: +34 916 585 065
imp@movistar.es
www.laballestera.com

La Ballestera Club Barrica 2019 T C
petit verdot, syrah, cabernet sauvignon
93
Classic. Colour: Cherry. Nose: spicy, black fruit, scrubland, cocoa bean. Palate: elegant, full-bodied, long.

La Ballestera Tinto Guarda Magnum 2019 T C
70% petit verdot, 15% cabernet sauvignon, 15% syrah
91
Colour: bright cherry. Nose: sweet spices, dark chocolate, ripe fruit, fruit preserve. Palate: spicy, round tannins.

MANUEL MANZANEQUE
Labradores, 2
02110 La Gineta (Albacete)
☎: +34 967 278 578
info@eavinos.com
www.eavinos.com

Manuel Manzaneque ¡Ea! 2019 T RB
cencibel
91
Colour: cherry, purple rim. Nose: fruit expression, floral, spicy, ripe fruit. Palate: flavourful, fruity, good acidity, long.

Mil Cepas Cencibel 2018 T BA
cencibel
88
Toasty, tasty, jammy, slight reduction.

MUREDA ALIMENTACIÓN
Ctra. N-IV, Km. 184,1
13300 Valdepeñas (Ciudad Real)
☎: +34 926 318 058
ehorcajada@mureda.es
www.mureda.es

Mureda 2021 RD
garnacha
85

Mureda Chardonnay 2021 B
chardonnay
86

Mureda Sauvignon Blanc Verdejo 2021 B S
50% sauvignon blanc, 50% verdejo
86

Mureda Syrah 2021 T
syrah
87

Mureda Tempranillo Syrah 2021 T
50% tempranillo, 50% syrah
87

PUNCTUM BIODYNAMIC FAMILY VINEYARDS
Finca El Fabián, s/n Apdo. Correos 71
16660 Las Pedroñeras (Cuenca)
☎: +34 912 918 326
exportsales@dominiodepunctum.com
www.dominiodepunctum.com

Lanzado Pet Nat 2021 BE
sauvignon blanc, viognier
88
Citrus fruit, herbal, jammy, tasty, tropical, floral.

Lanzado Pet Nat Rosé 2021 RE
garnacha
87

Pablo Claro Special Selection Chardonnay 2021 B
chardonnay
88
Citrus fruit, standard, herbal, simple, balanced.

Pablo Claro Special Selection Graciano Cabernet Sauvignon 2020 T
graciano, cabernet sauvignon
89
Jammy, herbaceous, tasty, wild, spicy.

Pomelado Vino Naranja 2021 B
chardonnay, viognier, viura

88 🌱

Citrus fruit, balanced, spicy, dried flowers, dried herbs, tasty.

Vaivén Blanc de Noir Tempranillo 2021 B
tempranillo

88 🌱

Citrus fruit, crisp, dried herbs, fruity, tasty.

TERRAMAGNA
Ctra. Rincón del Moro, Km 10
02410 Lietor (Albacete)
☎: +34 965 211 955
earellano@grupoterramagna.com
www.grupoterramagna.com

CT Cortijo de Trifillas Malbec 2019 T RB
87

Netón Garnacha Tintorera 2019 T C
garnacha tintorera

91

Colour: deep cherry. Nose: dried herbs, creamy oak, black fruit, dried flowers. Palate: powerful, ripe fruit, spicy, round tannins.

Netón Monastrell 2018 T C S
monastrell

90

Kind finish. Colour: deep cherry. Nose: scrubland, spicy, black fruit. Palate: ripe fruit, spicy, round tannins.

IP Isabel Peralta Garnacha Tintorera 2018 T
garnacha tintorera

90

Colour: deep cherry. Nose: dried herbs, dried flowers, black fruit, spicy. Palate: ripe fruit, spicy, round tannins.

IP Isabel Peralta Viogner 2020 B RB
viognier

90

Colour: straw. Nose: expressive, white flowers, dried herbs, sweet spices. Palate: flavourful, balanced, fleshy.

The Woods Of Tilo Cabernet Sauvignon 2017 T
cabernet sauvignon

88

Kind finish, balanced, spicy, fruity, creamy.

UNION CAMPESINA INIESTENSE
San Idefonso, 1
16235 Iniesta (Cuenca)
☎: +34 722 722 787
hans@senoriodeiniesta.com
www.señoriodeiniesta.com

Señorío de Iniesta Colección 2020 T RB
100% bobal

87 🌱

Señorío de Iniesta Rosé 2021 RD
100% bobal

85

Señorío de Iniesta Velvet 15 2020 T
100% tempranillo

87

UVAS FELICES
Agullers, 7
08003 Barcelona (Barcelona)
☎: +34 902 327 777
www.vilaviniteca.es

Sospechoso 2020 T
88

Balanced, spicy, herbal, toasty, jammy.

Sospechoso 2021 B
86

Sospechoso 2021 RD
88

Citrus fruit, fruity, herbal, balanced.

VINO DE LA TIERRA - CASTILLA / I.G.P.

VINÍCOLA DE CASTILLA

VINO DE LA TIERRA - CASTILLA / I.G.P.

Pol. Ind. Calle I, s/n
13200 Manzanares (Ciudad Real)
☎: +34 926 647 800
nacional@vinicoladecastilla.com
www.vinicoladecastilla.com

Olimpo 2020 T RB
50% tempranillo, 50% syrah

89

Spicy, jammy, wild, tasty, aromatic, balanced.

Olimpo Chardonnay 2021 B
100% chardonnay

88

Balanced, spicy, dried herbs, toasty.

Pago Peñuelas Garnacha 2021 T
100% garnacha

87

Pago Peñuelas Verdejo 2021 B
100% verdejo

84

VINÍCOLA DE TOMELLOSO

Ctra. Toledo - Albacete, Km. 130,8
13700 Tomelloso (Ciudad Real)
☎: +34 926 513 004
Fax: +34 926 538 201
vinicola@vinicoladetomelloso.com
www.vinicolatomelloso.com

El Viñedo de la Vida Tempranillo-Cabernet Sauvignon 2021 T

86 🌿

El Viñedo de la Vida Verdejo-Sauvignon Blanc 2021 B
verdejo, sauvignon blanc

85 🌿

VINÍCOLA DE VALDEPEÑAS

Autovía Madrid - Andalucía, km. 198,3
13300 Valdepeñas (Ciudad Real)
☎: +34 926 347 074
coovival@gmail.com
www.coovival.com

V5 by Consejal 2020 B BA
verdejo

88

Pleasant, smoky, jammy, toasty, powerful, lacks balance.

VINOS COLOMAN

Goya, 17
13620 Pedro Muñoz (Ciudad Real)
☎: +34 926 586 410
coloman@satcoloman.com
www.satcoloman.com

Pedroteño Airén 2021 B
airén

85

Pedroteño Tempranillo 2021 T
tempranillo

86

VINOS DE BERNARDO ORTEGA

Avda. de España, 17
02008 Albacete (Albacete)
☎: +34 627 465 228
bernardoenologo@gmail.com

Flor de Airén B AM

92

Colour: bright golden. Nose: acetaldehyde, cereal notes, dry nuts, dried fruit, dried herbs. Palate: fruity, flavourful, fine bitter notes, great length, smoky aftertaste.

Simbiosis Airén de Tinaja 2021 B

89

Fruity, dried flowers, dried herbs, wild, tasty.

Simbiosis Bobal Sincero 2019 T
100% bobal

91

Colour: cherry, purple rim. Nose: ripe fruit, black fruit, spicy, smoky, dried herbs. Palate: fruity, flavourful, balanced, slightly dry, soft tannins, smoky aftertaste.

VINOS LLÁMALO X

Madres, 35
02600 Villarrobledo (Albacete)
☎: +34 625 163 506
contacto@vinosllamalox.es
www.vinosllamalox.es

**Vinos Llámalo X Airén
sobre lías en Tinaja 2021 B**
100% airén

89

Dried herbs, citrus fruit, dried flowers, with personality. Nose: medium intensity, fresh fruit, fine lees. Palate: easy to drink.

**Vinos Llámalo X Crujidera
Madurado en Tinaja 2021 T**
100% moravia

91

Colour: cherry, purple rim. Nose: lactic notes, ripe fruit, red berry notes. Palate: flavourful, fruity, fruity aftertaste.

**Vinos Llámalo X Monastrell
madurado en Tinaja 2021 T**
100% monastrell

92

Colour: bright cherry. Nose: fruit expression, red berry notes, floral. Palate: flavourful, fruity, good acidity, long.

VINOS SIERRA NORTE

Ctra. de Munera La Roda, Km. 31,8
02630 La Roda (Albacete)
☎: +34 967 570 081
laroda@bodegasierranorte.com
www.bodegasierranorte.com

1564 Petit Verdot 2018 T BA
100% petit verdot

88

Overripe, roasted coffee, powerful. Nose: dried herbs.

1564 Syrah 2020 T BA
100% syrah

88

Spicy, fruity, tasty, wild.

1564 Viognier 2019 B BA
100% viognier

87

Olcaviana Chardonnay 2021 B
100% chardonnay

86

Olcaviana Sauvignon Blanc 2021 B
100% sauvignon blanc

86

Olcaviana Verdejo 2021 B
100% verdejo

87

VIÑEDOS Y BODEGA LA MAGDALENA

Ctra. La Roda, s/n
16611 Casas de Haro (Cuenca)
☎: +34 969 380 722
vinos@vegamoragona.com
www.vegamoragona.com

Vega Moragona Organic 2019 T
80% tempranillo, 20% syrah

88

Balanced, spicy, dried herbs, jammy, toasty.

VIÑEDOS Y BODEGAS MUÑOZ

Ctra. Villarrubia, 11
45350 Noblejas (Toledo)
☎: +34 925 140 070
info@bodegasmunoz.com
www.bodegasmunoz.com

**Blas Muñoz
Cepas Viejas 2016 T**
tempranillo

90

Colour: deep cherry. Nose: dried herbs, creamy oak, black fruit, toasty. Palate: ripe fruit, spicy, round tannins.

Blas Muñoz Chardonnay 2020 B
chardonnay

88

Citrus fruit, spicy, full-bodied, jammy, roasted coffee, woody.

**Finca Muñoz
Colección de la Familia 2019 T**
tempranillo, syrah, merlot

90

Colour: deep cherry. Nose: ripe fruit, dried herbs, creamy oak. Palate: powerful, ripe fruit, spicy, round tannins.

**Finca Muñoz
Colección de la Familia 2021 B BA**
chardonnay, sauvignon blanc

89

Pleasant, fruity, jammy, tasty, spicy, standard.

VINO DE LA TIERRA - CASTILLA / I.G.P.

VINO DE LA TIERRA - CASTILLA / I.G.P.

Legado Muñoz Chardonnay 2021 B
100% chardonnay
88
Citrus fruit, crisp, herbal, standard.

Legado Muñoz Garnacha 2020 T
garnacha
86

Legado Muñoz Macabeo Verdejo 2021 B
macabeo, verdejo
87

Legado Muñoz Merlot 2019 T
merlot
87

Legado Muñoz Tempranillo 2021 T
tempranillo
87

VIRGEN DE LAS VIÑAS BODEGA Y ALMAZARA

Ctra. Argamasilla de Alba, 1
13700 Tomelloso (Ciudad Real)
☎: +34 926 510 865
comercio.exterior@vinostomillar.com
www.vinostomillar.com

Lienzo Airén BE BN
airén
87
Pleasant, dried flowers, crisp, smooth, simple.

Sentir B
airén, sauvignon blanc
84 🌱

Sentir T
tempranillo, cabernet sauvignon
86

Lienzo Airén Pie Franco 2020 B
airén
89
Citrus fruit, balanced, tasty, herbal.

Lienzo Verdejo Sauvignon Blanc 2020 B
airén
87
Citrus fruit, herbal, slight fizz, jammy.

VT CASTILLA / CAMPO DE CALATRAVA

BODEGA AMANCIO MENCHERO
Legión, 27
13260 Bolaños de Calatrava (Ciudad Real)
☎: +34 680 402 931
amanciomenchero@hotmail.com
www.bodegamenchero.com

Cuba 38 2021 T
cencibel
86

Quarta Cabal 2021 B
82

BODEGAS NARANJO
Felipe II, 5
13150 Carrión de Calatrava (Ciudad Real)
☎: +34 687 045 574
comercial@bodegasnaranjo.com
www.bodegasnaranjo.com

Lahar de Calatrava 2021 B
83

Lahar de Calatrava 2021 B FB
85

Lahar de Calatrava 2021 T
86

BODEGAS QUINTA DE AVES
Ctra. CR-5222, Km. 11,200
13350 Moral de Calatrava (Ciudad Real)
☎: +34 915 716 514
hello@quintadeaves.es
www.quintadeaves.es

Quinta de Aves Cabernet Franc & Graciano Rosé 2021 RD
80% cabernet franc, 20% graciano
89
Balanced, fruity, citrus fruit, crisp, herbal, wild.

Quinta de Aves Coupage 2020 T C
50% tempranillo, 30% merlot, 18% graciano, 2% cabernet franc
91
Colour: bright cherry. Nose: sweet spices, ripe fruit, neat, dried herbs. Palate: fruity, spicy, round tannins.

Quinta de Aves Phoenix 2019 T C
100% tempranillo
92
Colour: dark-red cherry. Nose: ripe fruit, dried herbs, creamy oak, incense. Palate: powerful, ripe fruit, spicy, round tannins.

Quinta de Aves Sauvignon Blanc & Moscatel 2021 B
50% sauvignon blanc, 50% moscatel
89
Aromatic, floral, dried herbs, balanced, tasty. Nose: white fruit, stone fruit.

Quinta de Aves Syrah 2021 T
100% syrah
90
Colour: bright cherry. Nose: floral, ripe fruit, balanced. Palate: flavourful, fruity, good acidity.

Quinta de Aves Tempranillo 2021 T
100% tempranillo
90
Colour: cherry, purple rim. Nose: fruit expression, floral, lactic notes. Palate: fruity, flavourful, balanced.

BODEGAS WINOL
Poeta García Carbonell, 8 - Bajo
02005 Albacete (Albacete)
☎: +34 686 136 446
aochoa@bodegaswinol.com
www.bodegaswinol.com

8Lgends Leyenda del Volcán Moscatel de Alejandría Macabeo 2020 B
60% moscatel de alejandría, 40% macabeo
86

8Lgends Leyenda del Volcán Tempranillo 2019 T
100% tempranillo
80

SOTO DE ZEMTINAR
Ctra. CM-4111 Km. 17,500
13001 Aldea del Rey (Ciudad Real)
☎: +34 926 261 099
www.sotodezemtinar.com

Fábulo Sauvignon Blanc Airén Verdejo Volcánico 2021 B
sauvignon blanc, airén, verdejo
87

Fábulo Tempranillo Cabernet Sauvignon - Syrah Volcánico 2020 T
tempranillo, cabernet sauvignon, syrah
88
Colour: bright cherry. Nose: red berry notes, black fruit, balsamic herbs. Palate: good acidity, spicy, soft tannins, easy to drink.

Fábulo Tempranillo Graciano Volcánico 2020 T
tempranillo, graciano

89

Colour: bright cherry. Nose: ripe fruit, red berry notes, wild herbs, scrubland, spicy. Palate: fruity, flavourful, balanced, slightly dry, soft tannins.

VT CASTYLE

ABADÍA RETUERTA
Ctra. N-122, km. 332,5
47340 Sardón de Duero (Valladolid)
☎: +34 983 680 314
info@abadia-retuerta.es
www.abadia-retuerta.com

Abadía Retuerta Le Domaine 2020 B
80% sauvignon blanc, 20% verdejo

92

Colour: bright yellow. Nose: powerful, creamy oak, ripe fruit, spicy, waxy notes. Palate: rich, good structure, toasty, fine bitter notes.

ALTA PAVINA
Camino de Santibáñez, s/n
47328 La Parrilla (Valladolid)
☎: +34 983 681 521
bodegas@altapavina.com
www.altapavina.com

Alta Pavina Pinot Noir 2021 RD
pinot noir

89

Aromatic, balanced, dried flowers, wild, smooth, fruity, standard.

Alta Pavina Pinot Noir 2021 T RB
pinot noir

90

Defined aromas, dried flowers, fruity, jammy, smooth, wild, standard. Nose: varietal, balanced, fruit expression, floral.

Alta Pavina Verdejo 2021 B

89

Herbal, crisp, fruity, standard, balanced, simple.

Citius Pinot Noir 2018 T
pinot noir

90

Smooth. Colour: light cherry. Nose: red berry notes, ripe fruit, balanced, dried flowers, fine reductive notes. Palate: fruity, bitter.

Pavina Red 2019 T
pinot noir, tempranillo

89

Pleasant, standard, fruity, jammy, very fruit-driven, tasty, simple.

Pago La Pavina 2019 T
tempranillo, cabernet sauvignon

90

Colour: Cherry. Nose: balsamic herbs, scrubland, wild herbs, black fruit, ripe fruit. Palate: spicy, balsamic, good acidity.

ALVAR DE DIOS
Higinio Vázquez, 29
49154 El Pego (Zamora)
hola@alvardedios.com
www.alvardedios.com

Camino de los Arrieros 2019 T

92

Wild, defined aromas. Colour: light cherry. Nose: red berry notes, ripe fruit, spicy, dried herbs, wild herbs. Palate: balanced, fine bitter notes, easy to drink.

Tío Uco 2021 T
90% tinta de Toro (Tinto), 10% garnacha

91

Colour: cherry, purple rim. Nose: fruit expression, red berry notes, floral, spicy. Palate: flavourful, fruity, long, easy to drink.

ÁLVAREZ DE TOLEDO
Río Selmo, 8
24560 Toral de los Vados (León)
☎: +34 987 544 831
admon@bodegasalvarezdetoledo.com
www.bodegasalvarezdetoledo.com

Álvarez de Toledo Godello 2021 B

88

Pleasant, tropical, tasty, thin.

Álvarez de Toledo Mencía 2020 T RB
mencía

90

Colour: deep cherry. Nose: ripe fruit, dried herbs, creamy oak. Palate: powerful, ripe fruit, spicy, round tannins.

Álvarez de Toledo Mencía 2021 T RB
mencía

90

Colour: deep cherry. Nose: ripe fruit, dried herbs, creamy oak. Palate: powerful, ripe fruit, spicy, round tannins.

Álvarez de Toledo Colección Familia 2020 T
mencía

93

Colour: dark-red cherry. Nose: toasty, spicy, cocoa bean, fruit preserve, dark chocolate. Palate: flavourful, toasty, fine bitter notes.

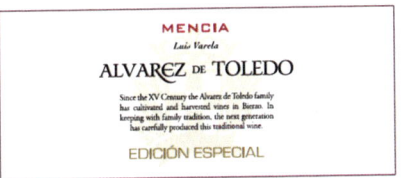

Marqués de Toro 2021 T
mencía

90

Colour: deep cherry. Nose: ripe fruit, dried herbs, creamy oak. Palate: powerful, ripe fruit, spicy, round tannins.

Señorío de la Antigua Mencía 2021 T
mencía

90

Colour: deep cherry. Nose: ripe fruit, dried herbs, creamy oak. Palate: powerful, ripe fruit, spicy, round tannins.

AVELINO VEGAS
Grupo Calvo Sotelo, 8
40460 Santiuste de San Juan Bautista (Segovia)
☎: +34 921 596 002
ana@avelinovegas.com
www.avelinovegas.com

Nicte Rosa Pálido 2021 RD
prieto picudo

89

Floral, tasty, fruity, smooth.

Tamina 2021 B
viognier

89

Citrus fruit, fruity, dried herbs, tasty.

BARCO DEL CORNETA
Carreventosa, 7
47491 La Seca (Valladolid)
☎: +34 648 454 958
info@barcodelcorneta.com
www.barcodelcorneta.com

Barco del Corneta 2020 B FB
100% verdejo

93 🌱

Needs time. Colour: bright straw. Nose: ripe fruit, dried herbs, faded flowers, sweet spices. Palate: powerful, ripe fruit, balanced.

Cucú (Cantaba la Rana) 2021 B
100% verdejo

90 🌱

Colour: bright straw. Nose: white flowers, jasmine, dried herbs, grassy. Palate: flavourful, fruity, balanced.

BARCOLOBO
Finca La Rinconada, Ctra. San Román de Hornija s/n
47520 Castronuño (Valladolid)
☎: +34 914 901 871
comercial@barcolobo.com
www.barcolobo.com

Barcolobo La Rinconada 2021 T
tempranillo

91

Colour: cherry, purple rim. Nose: red berry notes, floral, spicy, fruit expression. Palate: flavourful, fruity, good acidity, long.

Barcolobo Lacrimae Rerum 2021 RD
tempranillo

88

Dried flowers, fruity, jammy, tropical.

VINO DE LA TIERRA - CASTILLA Y LEÓN / I.G.P.

VINO DE LA TIERRA - CASTILLA Y LEÓN / I.G.P.

Barcolobo Verdejo 2021 B FB
verdejo
90
Colour: bright straw, greenish rim. Nose: fresh fruit, citrus fruit, grassy. Palate: fresh, fruity, good acidity.

BELONDRADE
Paraje de los Levantes. Quinta San Diego
47491 La Seca (Valladolid)
☎: +34 983 481 001
info@belondrade.com
www.belondrade.com

Belondrade Quinta Apolonia 2021 B
92
Colour: bright straw. Nose: expressive, ripe fruit, hint of anise, wild herbs, spicy. Palate: spicy, long, rich, balanced.

Belondrade Quinta Clarisa 2021 RD
95% tempranillo, 5% otras
89
Fruity, jammy, tasty, standard, kind finish, floral.

BODEGA CUATRO RAYAS
Camino de la Fuentecilla, s/n
47491 La Seca (Valladolid)
☎: +34 983 816 320
info@cuatrorayas.es
www.cuatrorayas.es

Cuatro Rayas Blush Rosé 2021 RD
50% tempranillo, 50% verdejo
87

Cuatro Rayas Organic Rosé Tempranillo-Verdejo 2021 RD
50% tempranillo, 50% verdejo
88
Pleasant, tasty, thin, fruity.

Cuatro Rayas Organic Tempranillo 2021 T
100% tempranillo
87

Dolce Bianco Verdejo Frizzante 2021 BE AG SD
100% verdejo
87

Green & Social Tempranillo 2021 T
100% tempranillo
88
Pleasant, fruity, tasty.

Vacceos Tempranillo 2021 T RB
100% tempranillo
87

BODEGA EL REGAJAL
Antigua Ctra. de Andalucía, Km. 50.5
28300 Aranjuez (Madrid)
☎: +34 913 078 903
Fax: +34 913 576 312
reservas@elregajal.es
www.elregajal.es

🏆 **PODIUM**

Galia Le Dean 2019 T
tempranillo
96
Colour: very deep cherry. Nose: complex, expressive, spicy, mineral, ripe fruit. Palate: full-bodied, long, great length, soft tannins.

Galia Villages 2019 T
94
Colour: cherry, purple rim. Nose: floral, spicy, fresh fruit, balanced, expressive. Palate: flavourful, fruity, good acidity, long.

Le Petit Galia 2018 T
tempranillo, garnacha
92
Colour: cherry, garnet rim. Nose: balsamic herbs, wild herbs, ripe fruit. Palate: flavourful, spicy, soft tannins, fresh, easy to drink.

BODEGA VOCARRAJE
49810 Morales de Toro (Zamora)
☎: +34 630 049 312
Fax: +34 980 698 172
info@vocarraje.es
www.vocarraje.es

Por Fin 2016 T R
100% garnacha
89
Pruney, lactic, full-bodied, jammy, tasty, toasty.

Por Fin 2020 B
malvasía
86

BODEGAS CANOPY
Herreros, 5
45180 Camarena (Toledo)
☎: +34 619 244 878
achacon@bodegascanopy.com
www.bodegascanopy.com

Kaos 2015 T R
100% garnacha
91
Colour: cherry, garnet rim. Nose: balsamic herbs, scrubland, meaty notes, fine reductive notes, fruit liqueur notes. Palate: flavourful, balsamic, spicy, soft tannins.

BODEGAS CASTELO DE MEDINA

Ctra. CL-602, Km. 48
47465 Villaverde de Medina (Valladolid)
☎: +34 983 831 932
Fax: +34 983 831 857
info@castelodemedina.com
www.castelodemedina.com

Castelo Nouveau 2021 T MC
100% tempranillo

89
Fruity, crisp, herbaceous, balanced.

Castelo Roble 2019 T RB S
60% syrah, 40% tempranillo

90
Colour: cherry, purple rim. Nose: spicy, scrubland, dry nuts, dark chocolate, red berry notes, black fruit. Palate: flavourful, fruity, good acidity, creamy.

Castelo Rosé 2021 RD
100% garnacha

88
Crisp, citrus fruit, fruity, herbal.

Darou Rosé 2021 RD
100% garnacha

87

Fisgón Rosé 2021 RD
100% garnacha

86

Sendero del Lobo Rosé 2021 RD
100% garnacha

87

BODEGAS CLUNIA

Ctra. BU-925, Km 27.25
09410 Coruña del Conde (Burgos)
☎: +34 948 838 640
info@bodegasclunia.com
www.bodegasclunia.com

Clunia Albillo 2021 B
100% albillo

91
Aromatic, wild. Colour: bright straw. Nose: dried flowers, dried herbs, spicy, stone fruit. Palate: rich, flavourful, good structure.

Clunia Malbec 2020 T
100% malbec

92
Colour: very deep cherry. Nose: black fruit, ripe fruit, wild herbs, dried herbs, waxy notes. Palate: flavourful, round tannins, spicy.

Clunia Syrah 2019 T
100% syrah

91
Colour: deep cherry. Nose: dried herbs, creamy oak, black fruit. Palate: ripe fruit, spicy, round tannins.

Clunia Tempranillo 2019 T
100% tempranillo

92
Colour: very deep cherry, bright cherry. Nose: varietal, sweet spices, creamy oak, ripe fruit. Palate: flavourful, good structure, round tannins, balanced.

Finca El Rincón de Clunia 2019 T
100% tempranillo

92
Woody. Colour: bright cherry. Nose: sweet spices, dark chocolate, black fruit, fruit preserve, creamy oak. Palate: fruity, spicy, round tannins.

Tesela by Clunia 2019 T
70% syrah, 30% tempranillo

91
Colour: deep cherry. Nose: dried herbs, creamy oak, black fruit, dark chocolate. Palate: powerful, ripe fruit, spicy, round tannins.

BODEGAS DE ALBERTO

Ctra. de Valdestillas, 2
47231 Serrada (Valladolid)
☎: +34 983 559 107
Fax: +34 983 559 084
info@dealberto.com
www.dealberto.com

De Alberto Selección 2018 T
80% tempranillo, 20% cabernet sauvignon

84

Finca Valdemoya 12 meses 2019 T C

83

Finca Valdemoya 2021 RD
100% tempranillo

88
Kind finish, aromatic, sweeties, lactic, simple, smooth.

BODEGAS EL LAGAR DE ISILLA

Camino Real, 1
09471 La Vid (Burgos)
☎: +34 947 530 434
administracion@lagarisilla.es
www.lagarisilla.es

La Casona de la Vid 4V 2018 T
25% syrah, 25% garnacha, 25% tempranillo, 25% cabernet sauvignon

89
Pleasant, spicy, dried herbs, jammy, tasty, toasty, standard.

VINO DE LA TIERRA - CASTILLA Y LEÓN / I.G.P.

VINO DE LA TIERRA - CASTILLA Y LEÓN / I.G.P.

La Casona de la Vid Cabernet 2017 T BA
cabernet sauvignon

93

Colour: deep cherry. Nose: ripe fruit, dried herbs, creamy oak, black fruit, spicy. Palate: powerful, ripe fruit, round tannins.

La Casona de la Vid Garnacha 2019 T
100% garnacha

90

Colour: cherry, purple rim. Nose: red berry notes, floral, spicy. Palate: fruity, easy to drink, balanced.

La Casona de la Vid Syrah 2019 T
100% syrah

89

Spicy, fruity, jammy, toasty.

BODEGAS ERESMA - LA SOTERRAÑA
Ctra. N-601, Km. 151
47410 Olmedo (Valladolid)
☎: +34 983 601 026
info@bodegaslasoterrana.com
www.bodegaslasoterrana.com

Eresma Tentazion BE AG
sauvignon blanc, verdejo

84

BODEGAS FARIÑA
Camino del Palo, s/n
49800 Toro (Zamora)
☎: +34 980 577 673
comercial@bodegasfarina.com
www.bodegasfarina.com

Águedas 2020 B BA
malvasía

91

Colour: bright straw. Nose: ripe fruit, floral, fine lees, spicy, dried flowers, faded flowers. Palate: full-bodied, complex, spicy, long.

Mascaradas 2019 T RB
tempranillo

87

Val de Reyes T D
tinta del país

91

Colour: cherry, garnet rim. Nose: fruit preserve, ripe fruit, spicy, aged wood nuances, black pepper, old leather. Palate: powerful, flavourful, dry wood.

BODEGAS GORDONZELLO
Alto de Santa Marina, s/n
24294 Gordoncillo (León)
☎: +34 987 758 030
info@gordonzello.com
www.gordonzello.com

Antojo Dulce 2020 B D
100% verdejo

85

BODEGAS LEDA
Mayor, 48
47320 Tudela de Duero (Valladolid)
☎: +34 983 520 682
info@bodegasleda.masaveu.com
www.bodegasleda.com

Leda Viñas Viejas 2018 T R
100% tempranillo

91

Colour: deep cherry. Nose: dried herbs, creamy oak, black fruit, woody. Palate: powerful, ripe fruit, spicy, round tannins.

Más de Leda 2018 T C
tempranillo

91

Colour: deep cherry. Nose: dried herbs, black fruit, burnt matches, fine reductive notes, cocoa bean. Palate: ripe fruit, spicy, round tannins, balanced.

BODEGAS MARQUÉS DE RISCAL
Ctra. N-VI, km. 172,600
47490 Rueda (Valladolid)
☎: +34 983 868 029
Fax: +34 983 868 563
rherrero@marquesderiscal.com
www.marquesderiscal.com

Barón de Chirel Viñas Centenarias Verdejo 2021 B
verdejo

93

Colour: bright yellow. Nose: powerful, creamy oak, ripe fruit, spicy. Palate: rich, good structure, long, toasty.

Marqués de Riscal 1860 Tempranillo 2021 T RB
tempranillo, syrah

90

Colour: cherry, purple rim. Nose: fruit expression, red berry notes, floral, raspberry. Palate: flavourful, fruity, balanced, slightly dry, soft tannins.

Marqués de Riscal Viñas Viejas 2021 RD
garnacha, tinta de Toro (Tinto)

90

Crisp, simple. Colour: raspberry rose. Nose: red berry notes, grassy. Palate: good acidity, fine bitter notes, fruity.

BODEGAS MAURO
Ctra. Villabañez, km. 1
47320 Tudela de Duero (Valladolid)
☎: +34 983 521 972
info@bodegasmauro.com
www.bodegasmauro.com

Mauro 2020 T
85% tempranillo, 15% syrah, cabernet sauvignon

93

Colour: deep cherry. Nose: ripe fruit, dried herbs, creamy oak, red berry notes, toasty. Palate: powerful, ripe fruit, spicy, grainy tannins.

Mauro Vendimia Seleccionada 2019 T
100% tempranillo

94 🌱

Colour: very deep cherry. Nose: roasted coffee, aromatic coffee, powerful, ripe fruit, black fruit. Palate: smoky aftertaste, great length, round tannins, flavourful.

BODEGAS MENADE
Ctra. Rueda - Nava del Rey, km. 1
47490 Rueda (Valladolid)
☎: +34 983 103 223
info@menade.es
www.menade.es

La Misión by Menade 2019 B
100% verdejo

93 🌱

Colour: bright yellow. Nose: ripe fruit, dried herbs, faded flowers, hint of anise. Palate: powerful, ripe fruit, balanced, flavourful, good acidity, smoky aftertaste, spicy.

Menade Sauvignon Blanc Dulce 2021 B D
sauvignon blanc

88 🌱

Citrus fruit, floral, fruity, full-bodied, sweetish, dried herbs.

Menade Verdejo 2021 B
100% verdejo

90 🌱

Colour: bright straw, greenish rim. Nose: fresh fruit, citrus fruit, wild herbs. Palate: fresh, fruity, good acidity, fine bitter notes.

Nossa de Menade 2021 T
100% tempranillo

90 🌱

Tasty, rustic. Colour: bright cherry. Nose: fresh fruit, creamy oak, red berry notes. Palate: good acidity, spicy.

Nosso by Menade 2021 B
100% verdejo

90 🌱

Oxidativ. Colour: bright straw, greenish rim. Nose: citrus fruit, wild herbs, white fruit, waxy notes. Palate: fresh, fruity, good acidity, fine bitter notes.

Sobrenatural by Menade 2016 B C
100% verdejo

93 🌱

Age nuances, oxidativ. Colour: yellow. Nose: ripe fruit, stone fruit, white fruit, honeyed notes, dried flowers. Palate: fruity, fresh, balanced, good finish.

BODEGAS MOCÉN
Arribas, 7-9
47490 Rueda (Valladolid)
☎: +34 983 868 533
info@bodegasmocen.com
www.bodegasmocen.es

Arlequín 2021 B
verdejo

88

Citrus fruit, herbal, balanced, crisp.

Arlequín 2021 RD
100% tempranillo

87

Arlequín 2021 T
100% tempranillo

89

Kind finish, fruity, lactic, wild, tasty, simple.

Cobranza Tempranillo 2018 T BA
100% tempranillo

88

Kind finish, jammy, toasty, spicy, dried herbs, tasty.

Cobranza Vendimia Seleccionada 2015 T
100% tempranillo

92

Colour: Cherry. Nose: complex, expressive, spicy, tobacco. Palate: full-bodied, long, great length, round tannins.

Renacce Blush Rosé 2021 RD
100% tempranillo

86

VINO DE LA TIERRA - CASTILLA Y LEÓN / I.G.P.

BODEGAS NIDIA

Camino de la Casa de La Cabaña
47410 Olmedo (Valladolid)
☎: +34 616 155 501
hola@bodegasnidia.com
www.bodegasnidia.com

Nidia 2020 RD
merlot, verdejo

88

Citrus fruit, dried flowers, herbaceous, weary.

Nidia Verdejo Lías 2020 B
verdejo

90

Colour: bright straw, greenish rim. Nose: citrus fruit, wild herbs, ripe fruit, fine lees. Palate: fresh, fruity, good acidity, fine bitter notes.

BODEGAS PRIETO PARIENTE

Ctra. de Rueda, km. 2,5
47491 La Seca (Valladolid)
☎: +34 983 816 600
info@prietopariente.com
www.prietopariente.com

El Origen de Prieto Pariente 2018 T C
95% tempranillo, 5% garnacha

93 🌱

Colour: deep cherry. Nose: ripe fruit, creamy oak, wild herbs, characterful, expressive. Palate: powerful, ripe fruit, spicy, round tannins.

La Provincia de Prieto Pariente 2019 T C
55% tempranillo, 45% garnacha

92 🌱

Colour: deep cherry. Nose: dried herbs, creamy oak, red berry notes, black fruit. Palate: ripe fruit, spicy, round tannins, balanced.

Los Confines 2019 T C
100% garnacha

93 🌱

Colour: cherry, purple rim. Nose: fruit expression, red berry notes, floral, spicy, earthy notes. Palate: flavourful, fruity, good acidity, long.

Viognier de Prieto Pariente 2020 B
100% viognier

90 🌱

Colour: bright straw, greenish rim. Nose: fresh fruit, citrus fruit, wild herbs, floral. Palate: fresh, fruity, good acidity, fine bitter notes.

BODEGAS RAMIRO'S

Camino Viejo de Simancas, km. 3,5
47008 Valladolid (Valladolid)
☎: +34 639 306 279
bodegasramiros@hotmail.com
www.bodegasramiros.com

Ramiro's 2018 T
100% tempranillo

92

Colour: very deep cherry. Nose: aromatic coffee, powerful, ripe fruit, black fruit, dark chocolate, new oak. Palate: smoky aftertaste, great length, round tannins.

BODEGAS REBROTAR

Real, 7
47238 Hornillos de Eresma (Valladolid)
estherslv@gmail.com

Rebrotar 2020 B FB
100% verdejo

92

Colour: bright yellow. Nose: powerful, creamy oak, ripe fruit, spicy. Palate: rich, good structure, toasty, fine bitter notes.

BODEGAS RODRÍGUEZ Y SANZO

Manuel Azaña, 9
47014 Valladolid (Valladolid)
☎: +34 983 150 150
comunicacion@valsanzo.com
www.rodriguezsanzo.com

Rodriguez & Sanzo WhisBa 2019 T
tempranillo

93

Colour: dark-red cherry, purple rim. Nose: toasty, spicy, cocoa bean, ripe fruit, black fruit. Palate: flavourful, toasty, slightly dry, soft tannins, roasted-coffee aftertaste.

VINO DE LA TIERRA - CASTILLA Y LEÓN / I.G.P.

Rodríguez Sanzo
Orange Wine 2020 B BA
91
With personality. Colour: yellow. Nose: faded flowers, dried flowers, ripe fruit, expressive. Palate: long, flavourful.

BODEGAS SINFORIANO
47194 Mucientes (Valladolid)
☎: +34 983 663 008
sinfo@sinforianobodegas.com
www.sinforianobodegas.com

Raimun Garnacha 2020 T
95% garnacha, 5% tempranillo
87

Raimun Verdejo Natural 2020 B FB
100% verdejo
89
Colour: bright straw. Nose: ripe fruit, floral, tropical fruit, stone fruit. Palate: fruity, balanced, easy to drink, flavourful.

BODEGAS TRIDENTE
Pol.1 Parc. 146/148 Paraje Cantagrillos
49708 Villanueva de Campeán (Zamora)
☎: +34 968 435 022
info@gilfamily.es
www.gilfamily.es

Entresuelos 2019 T
100% tempranillo
91
Colour: deep cherry. Nose: dried herbs, creamy oak, ripe fruit, black fruit, toasty. Palate: powerful, ripe fruit, spicy, round tannins.

Entresuelos 2020 T
100% tempranillo
91
Colour: deep cherry. Nose: ripe fruit, dried herbs, creamy oak. Palate: powerful, ripe fruit, spicy, round tannins.

Rejón 2019 T
100% tempranillo
94
Colour: dark-red cherry, garnet rim. Nose: aged wood nuances, tobacco, sweet spices, black fruit, ripe fruit, new oak, black pepper. Palate: spicy, round tannins, long.

Tridente Malvasía 2021 B
malvasía castellana
91
Little interventionist. Colour: bright yellow. Nose: dried flowers, candied fruit, fine lees, pattiserie. Palate: round, spicy, long, great length.

Tridente Prieto Picudo 2019 T
100% prieto picudo
92
Colour: dark-red cherry. Nose: toasty, spicy, cocoa bean, cigar, black fruit. Palate: flavourful, toasty, fine bitter notes.

Tridente Prieto Picudo 2020 T
100% prieto picudo
91
Colour: deep cherry. Nose: ripe fruit, dried herbs, creamy oak. Palate: powerful, ripe fruit, spicy, round tannins.

Tridente Tempranillo 2019 T
100% tempranillo
93
Colour: dark-red cherry. Nose: toasty, spicy, cocoa bean, ripe fruit, black fruit. Palate: flavourful, toasty, fine bitter notes, ripe fruit, good structure.

Tridente Tempranillo 2020 T
100% tempranillo
92
Colour: deep cherry. Nose: ripe fruit, dried herbs, creamy oak, toasty. Palate: powerful, ripe fruit, spicy, round tannins.

BODEGAS VERDES
Camino Benavente, 21
49610 Santibáñez de Vidriales (Zamora)
☎: +34 980 648 405
comercial@bodegasverdes.com
www.bodegasverdes.com

Lyrius One from Prieto Picudo 2021 RD
100% prieto picudo
86

VINO DE LA TIERRA - CASTILLA Y LEÓN / I.G.P.

Lyrius One from Verdejo 2021 B S
100% verdejo
86

Lyrius One from Verdejo 2021 B SD
100% verdejo
84

BODEGAS VIÑAS DEL CÉNIT
Ctra. De Circunvalación, s/n
49708 Villanueva de Campeán (Zamora)
☎: +34 980 569 346

Venta Mazarrón 2020 T
100% tempranillo
89
Tasty, powerful, jammy.

BODEGAS VIZAR
Ctra. N-122. Km. 341
47329 Villabáñez (Valladolid)
☎: +34 983 682 690
vadalia@bodegasvizar.es
www.bodegasvizar.es

Vizar 2018 T BA
tempranillo, syrah, merlot
89
Fruity, spicy, herbal, jammy.

Vizar Prestigio 2018 T
tempranillo, syrah, cabernet sauvignon, merlot
87

Vizar Verdejo 2018 B FB
verdejo
90
Colour: bright straw, greenish rim. Nose: fresh fruit, citrus fruit, wild herbs, sweet spices, woody. Palate: fresh, fruity, good acidity, fine bitter notes.

BODEGAS Y VIÑEDOS LA MEJORADA
Monasterio de La Mejorada, s/n
47410 Olmedo (Valladolid)
☎: +34 983 483 057
info@lamejorada.es
www.lamejorada.es

La Mejorada Las Cercas 2018 T RB
60% tempranillo, 40% syrah
92
Colour: deep cherry. Nose: ripe fruit, dried herbs, creamy oak, meaty notes. Palate: powerful, ripe fruit, spicy, round tannins.

La Mejorada Las Norias 2018 T RB
tempranillo
93
Colour: Cherry. Nose: expressive, spicy, smoky, cocoa bean, ripe fruit. Palate: full-bodied, long, great length, round tannins, spicy.

Palomar de la Reina 2019 T
syrah
93
Colour: deep cherry. Nose: dried herbs, creamy oak, black fruit. Palate: spicy, round tannins, pruney.

Tiento La Mejorada 2016 T
tempranillo, syrah, merlot, malbec
93
Colour: deep cherry. Nose: ripe fruit, dried herbs, creamy oak, meaty notes. Palate: powerful, ripe fruit, spicy, round tannins, sweetness.

Villalar 2019 T RB
tempranillo, cabernet sauvignon
90
Colour: deep cherry. Nose: dried herbs, creamy oak, black fruit. Palate: ripe fruit, spicy, grainy tannins.

BODEGAS Y VIÑEDOS PRADOREY
Ctra. CL-619 (Magaz – Aranda) km.66, 1
09443 Gumiel de Mercado (Burgos)
☎: +34 947 546 900
Fax: +34 947 546 999
info@pradorey.com
www.pradorey.com

Pradorey Blanco 2020 B
90
Colour: bright yellow. Nose: powerful, creamy oak, ripe fruit, spicy. Palate: good structure, long, toasty, fine bitter notes.

BODEGAS Y VIÑEDOS VALTRAVIESO

Finca La Revilla, s/n
47316 Piñel de Arriba (Valladolid)
☎: +34 983 484 030
marketing@valtravieso.com
www.valtravieso.com

Valtravieso Rupture 2018 T
70% merlot, 30% cabernet sauvignon

91

Colour: Cherry. Nose: balsamic herbs, sweet spices, scrubland, black fruit. Palate: spicy, powerful, round tannins, balsamic.

CANTALAPIEDRA VITICULTORES

Ctra. VP-9901 km. 3.7
47491 La Seca (Valladolid)
☎: +34 665 235 709
info@isaaccantalapiedra.com
www.cantalapiedraviticultores.com

Majuelo El Espejo 2020 B
verdejo

93

Little interventionist. Colour: bright yellow. Nose: expressive, ripe fruit, floral, fine lees, mineral, burnt matches. Palate: full-bodied, complex, spicy, long, elegant.

COMPAÑÍA DE VINOS MIGUEL MARTÍN

Ctra. Valladolid, 14
47270 Cigales (Valladolid)
☎: +34 983 250 319
comercial@ciadevinos.com
www.bodegasthesaurus.com

Martín Verástegui 2019 T BA
100% tempranillo

90 ✿

Colour: deep cherry. Nose: ripe fruit, dried herbs, creamy oak, spicy. Palate: ripe fruit, spicy, round tannins.

CONDADO DE HAZA

Ctra. La Horra, s/n
09300 Roa (Burgos)
☎: +34 947 525 254
info@condadodehaza.com
www.familiafernandezrivera.com

20 Aldeas 2019 T
tempranillo

89 ✿

Creamy, full-bodied, spicy, coarse tannins, roasted coffee.

DANIEL LANDI VITICULTOR

Avda de la Constitución, 23
28640 Cadalso de Los Vidrios (Madrid)
☎: +34 918 640 602
info@danilandi.com
www.danilandi.com

🏆 PODIUM

El Reventón 2019 T
garnacha

95

Colour: light cherry. Nose: red berry notes, scrubland, balsamic herbs, floral, spicy. Palate: elegant, fleshy, fine tannins.

Las Iruelas 2019 T
garnacha

94

Colour: light cherry. Nose: red berry notes, balsamic herbs, spicy. Palate: fleshy, elegant, warm.

DEHESA LA GRANJA

Finca La Granja
49420 Vadillo de la Guareña (Zamora)
☎: +34 980 566 009
www.familiafernandezrivera.com

Dehesa La Granja 2019 T
tempranillo

87 ✿

DOMINIO DOS TARES

Los Barredos, 4
24318 San Román de Bembibre (León)
☎: +34 987 514 550
Fax: +34 987 514 570
info@dominiodetares.com
www.dominiodostares.com

Cumal 2018 T
prieto picudo

92

Colour: deep cherry. Nose: dried herbs, creamy oak, black fruit, faded flowers. Palate: powerful, ripe fruit, spicy, round tannins.

Estay 2018 T RB
prieto picudo

90

Colour: Cherry. Nose: balsamic herbs, scrubland, red berry notes, ripe fruit. Palate: spicy, good acidity.

Tombú 2021 RD
prieto picudo

89

Aromatic, jammy, tasty, smooth.

VINO DE LA TIERRA - CASTILLA Y LEÓN / I.G.P.

VINO DE LA TIERRA - CASTILLA Y LEÓN / I.G.P.

EULOGIO & JAVIER WINES
Juan de Herrera, 23
47130 Simancas (Valladolid)
☎: +34 983 150 150
comunicacion@rodriguezysanzo.com
www.eulogioyjavierwines.com

Clavius Verdejo 2018 B
100% verdejo
90
Colour: yellow. Nose: faded flowers, white fruit, ripe fruit, spicy, dry nuts. Palate: flavourful, toasty, rich.

El Quinto Paraje Verdejo 2021 B
verdejo
90
Colour: bright straw, greenish rim. Nose: citrus fruit, wild herbs, white fruit. Palate: fresh, fruity, fine bitter notes.

FINCA TORREMILANOS
Finca Torremilanos, s/n
09400 Aranda de Duero (Burgos)
☎: +34 947 510 377
Fax: +34 947 512 856
contacto@torremilanos.com
www.torremilanos.com

Peñalba-López 2020 B
92
Colour: bright straw. Nose: ripe fruit, fragrant herbs, fine lees, cereal notes. Palate: full-bodied, rich, long, good acidity, flavourful.

FRUTOS VILLAR
Ctra. Burgos-Portugal, Km. 113,7
47270 Cigales (Valladolid)
☎: +34 983 586 868
Fax: +34 983 580 180
bodegasfrutosvillar@bodegasfrutosvillar.com
www.bodegasfrutosvillar.com

Atalaya de Barahona 2021 T
100% tempranillo
86

Blanco Polar Verdejo 2021 B
100% verdejo
86

Don Frutos Tempranillo 2021 T
100% tempranillo
87

Don Frutos Verdejo 2021 B
100% verdejo
85

FUENTES DEL SILENCIO
Plaza Mayor, 2
24767 Herrreros de Jamuz (León)
☎: +34 987 688 861
bcabero@fuentesdelsilencio.com
www.fuentesdelsilencio.com

Fuentes del Silencio Las Jaras 2019 T
mencía, prieto picudo, alicante bouchet (Tinto)
91
Colour: cherry, garnet rim. Nose: ripe fruit, scrubland, grassy. Palate: flavourful, balsamic, spicy, round tannins.

Fuentes del Silencio Mataperezosa 2019 B FB
palomino, dona blanca
93
With personality. Colour: yellow. Nose: burnt matches, candied fruit, expressive, neat, spicy, faded flowers. Palate: flavourful, long, spicy, complex.

Fuentes del Silencio Prieto Picudo Viejo 2019 T
prieto picudo
92
Colour: bright cherry. Nose: fresh fruit, fruit expression, wild herbs, grassy, spicy. Palate: spicy, fruity, fresh, smoky aftertaste.

JAVIER SANZ VITICULTOR
San Judas, 2
47491 La Seca (Valladolid)
☎: +34 983 816 669
comunicaciones@bodegajaviersanz.com
www.bodegajaviersanz.com

Javier Sanz Paraje la Encina 2020 T RB
100% bruñal
92
Colour: bright cherry. Nose: wild herbs, expressive, neat, dry stone. Palate: flavourful, fine bitter notes, fruity, easy to drink, long, balsamic.

LA FURGONETA VINOS
Arriba, 9
24393 Santa Marina del Rey (León)
☎: +34 609 793 812
leon@sarahselections.com

La Furgoneta que miraba al Órbigo 2020 T
90
Rustic. Colour: deep cherry. Nose: dried herbs, creamy oak, black fruit, faded flowers. Palate: ripe fruit, spicy, round tannins.

LAR DE MAÍA

Pº de Castilviejo, 2
47290 Cubillas de Santa Marta (Valladolid)
☎: +34 650 986 098
export@lardemaia.com
www.lardemaia.com

Lar de Maía 5º 2020 T BA
100% tempranillo

87

Lar de Maía 7º Autor 2020 T BA
100% tempranillo

90

Colour: deep cherry. Nose: ripe fruit, dried herbs, spicy, smoky. Palate: powerful, ripe fruit, round tannins.

Lar de Maía 8º 2021 RD
100% tempranillo

89

Aromatic, dried flowers, jammy, fruity.

LEYENDA DEL PÁRAMO

Ctra. de León s/n, Paraje El Cueto
24230 Valdevimbre (León)
☎: +34 987 050 039
info@leyendadelparamo.com
www.leyendadelparamo.com

El Aprendiz 2019 T RB
100% prieto picudo

90

Colour: cherry, purple rim. Nose: fruit expression, red berry notes, spicy, dried herbs, black liquorice. Palate: flavourful, fruity, slightly dry, soft tannins, toasty, balanced.

El Aprendiz 2021 B
100% albarín

90

Aromatic, jammy, dried herbs. Nose: dry nuts, balanced, dried flowers. Palate: ripe fruit, long.

El Aprendiz 2021 RD
100% prieto picudo

88

Kind finish, pruney, sweeties, fruity, sweety finish.

El Médico 2015 T RB
100% prieto picudo

90

Colour: deep cherry. Nose: ripe fruit, dried herbs, creamy oak, meaty notes. Palate: ripe fruit, spicy, round tannins, fruity.

El Músico 2014 T
100% prieto picudo

88

Fruity, dried herbs, spicy, smoky.

El Rescatado 2021 RD
100% prieto picudo

84 🌿

MEDINA AGRICULTURA ECOLÓGICA (FINCA LAS CARABALLAS)

47400 Medina del Campo (Valladolid)
☎: +34 678 552 943
info@lascaraballas.com
www.lascaraballas.es

Finca Las Caraballas Sector 2.8 2020 B
viognier

90 🌿

Dried flowers. Colour: yellow. Nose: dried flowers, faded flowers, white fruit, ripe fruit, fine lees. Palate: rich, fine bitter notes, easy to drink.

Finca Las Caraballas Verdejo 2019 B
100% verdejo

91 🌿

Colour: yellow. Nose: white fruit, ripe fruit, fine lees, faded flowers. Palate: flavourful, ripe fruit, long.

Finca Las Caraballas Verdejo 2021 B
100% verdejo

90 🌿

Colour: bright straw, greenish rim. Nose: fresh fruit, citrus fruit, wild herbs. Palate: fruity, fine bitter notes, easy to drink.

OSSIAN VIDES Y VINOS

Cordel de las Merinas s/n
40447 Nieva (Segovia)
☎: +34 983 878 020
info@almacarraovejas.com
www.ossianvinos.com

🏆 PODIUM

Ossian Capitel 2019 B FB
verdejo

95 🌿

Colour: bright yellow. Nose: ripe fruit, spicy, burnt matches, balanced, sweet spices. Palate: rich, good structure, long, fine bitter notes, good acidity, toasty.

Ossian Quintaluna 2020 B
verdejo

92

Colour: bright straw, greenish rim. Nose: fresh fruit, citrus fruit, wild herbs, balsamic herbs, grassy. Palate: fresh, fruity, good acidity, fine bitter notes

VINO DE LA TIERRA - CASTILLA Y LEÓN / I.G.P.

VINO DE LA TIERRA - CASTILLA Y LEÓN / I.G.P.

🏆 PODIUM

Ossian 2019 B
verdejo

95 🌱

Colour: bright yellow. Nose: dried flowers, candied fruit, fine lees, pattiserie, characterful, complex. Palate: round, spicy, long, great length.

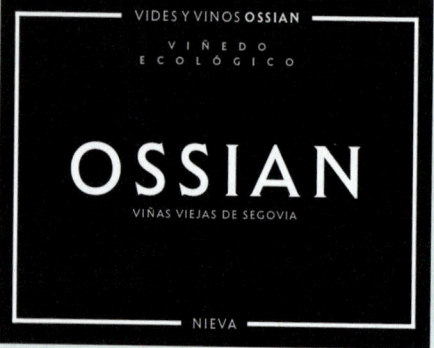

PACO MULERO
Partida de la Hoya Torres s/n
30520 Jumilla (Murcia)
☎: +34 676 433 541
info@pacomulero.com
www.pacomulero.com

Paco Mulero Tempranillo 2019 T
100% tempranillo

90

Colour: cherry, garnet rim. Nose: fruit preserve, powerful, creamy oak, dried herbs. Palate: flavourful, long, round tannins.

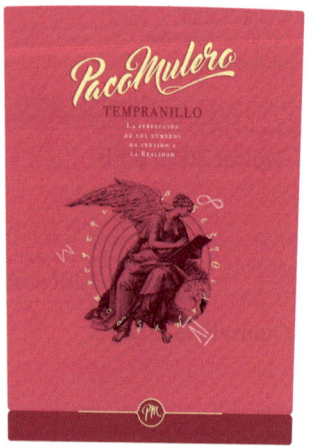

QUINTA SARDONIA
Casas Blancas s/n, Granja Sardón
47340 Sardón de Duero (Valladolid)
☎: +34 983 032 883
info@quintasardonia.com
www.quintasardonia.com

🏆 PODIUM

Quinta Sardonia QS 2019 T
tempranillo, cabernet sauvignon, merlot, malbec, syrah

95 🌱

Colour: bright cherry. Nose: expressive, spicy, mineral, ripe fruit, black fruit, complex. Palate: elegant, full-bodied, long, great length.

Quinta Sardonia QS2 2019 T
tempranillo, petit verdot, cabernet sauvignon, syrah, malbec

93 🌱

Colour: cherry, garnet rim. Nose: fruit preserve, powerful, sweet spices, creamy oak. Palate: flavourful, long.

Sardón 2019 T
95% tempranillo, 3% garnacha, 2% otras

92

Colour: deep cherry. Nose: ripe fruit, dried herbs, creamy oak. Palate: powerful, ripe fruit, spicy, round tannins.

SPANISH PALATE

Avda. Carlos Latorre, 30
49800 Toro (Zamora)
☎: +34 980 690 643
export@spanishpalate.es
www.spanishpalate.es

Botas de Barro León 2019 T RB
100% prieto picudo

86

VDB BODEGAS VALLE DEL BOTIJAS

Ctra. Valoria, 45
47315 Pesquera de Duero (Valladolid)
☎: +34 616 998 323
info@bodegasvdb.com
www.bodegasvdb.com

Valle del Botijas Sauvignon Blanc 2021 B
sabro

87

Vdb Valle del Botijas 14 meses 2016 T C
tempranillo, merlot, cabernet sauvignon, syrah

91

Colour. Cherry. Nose: balsamic herbs, sweet spices, scrubland. Palate: spicy, balsamic, good acidity, balanced, flavourful.

Vdb Valle del Botijas Angela 2021 B
verdejo

84

VICENTE GANDÍA

Ctra. Cheste a Godelleta, s/n
46370 Chiva (València/Valencia)
☎: +34 962 524 242
Fax: +34 962 524 243
info@vicentegandia.com
www.vicentegandia.es

Nebla Verdejo 2021 B
verdejo

88

Aromatic, fruity, herbal, crisp, standard.

Vicente Gandía Godello Edición Limitada 2021 B
godello

88

Fruity, jammy, tasty, simple, bitter, herbal.

VILE LA FINCA, BODEGAS Y VIÑEDOS

Crta. Cembranos-Valdevimbre km 6,2
24230 León (León)
☎: +34 987 209 712
lafinca@lafinca.es
www.vilelafinca.es

Real Arbás 2021 B D
100% albarín

86

VINOS DE ARGANZA

Río Ancares, 2
24560 Toral de los Vados (León)
☎: +34 987 544 831
admon@vinosdearganza.com
www.vinosdearganza.com

Castillo Colina 2021 T
89

Overripe, toasty, powerful, hot.

Encanto Godello 2021 B
88

Pleasant, smooth, tasty, tropical.

Encanto Mencía 2021 T RB
89

Hot, tasty, overripe.

Encanto Selección 2021 T
mencía

91

Colour. deep cherry. Nose: ripe fruit, dried herbs, creamy oak, sweet spices. Palate: powerful, ripe fruit, spicy, round tannins.

Flavium Premium 2021 T
91

Colour. deep cherry. Nose: ripe fruit, dried herbs, creamy oak, characterful. Palate: powerful, ripe fruit, spicy, round tannins.

VINO DE LA TIERRA - CASTILLA Y LEÓN / I.G.P.

Flavium Selección 2021 T
89
Toasty, overripe, powerful.

Lagar de Robla Colección Cuatro Hermanos Finca Valdelebre 2020 T
mencía
93
Colour: deep cherry. Nose: spicy, mineral, ripe fruit, expressive, characterful. Palate: elegant, full-bodied, long, great length.

Lagar de Robla Premium 2021 T
89
Toasty, overripe, hot, powerful.

Lagar de Robla Selección 2021 T
89
Overripe, pruney, tasty, spicy.

Legado de Farro Godello 2021 B
88
Tropical, smooth, tasty, thin.

Legado de Farro Mencía 2021 T RB
91
Colour: deep cherry. Nose: ripe fruit, dried herbs, creamy oak, characterful. Palate: powerful, ripe fruit, spicy, round tannins.

Século Cepas Viejas 2021 T RB
91
Colour: deep cherry. Nose: ripe fruit, dried herbs, creamy oak, characterful, complex. Palate: powerful, ripe fruit, spicy, round tannins.

VIÑADESGRACIA
Camino La Cueva, s/n
47320 Valladolid (Valladolid)
☎: +34 615 218 638
vinadesgracia.00@gmail.com
www.vinadesgracia.com

Viña Desgracia 2019 T
100% tempranillo
88
Fruity, jammy, toasty, dried herbs.

VIÑAS SERRANAS
Ctra. Coria s/n
37656 Cepeda (Salamanca)
☎: +34 634 555 355
info@vserranas.com

El Canchorral 2020 T
92
Wild, herbal. Colour: cherry, garnet rim. Nose: wild herbs, dried herbs, red berry notes, ripe fruit, balanced, fresh. Palate: fruity, fresh.

El Ciclón Serrano 2019 T
rufete
91
Colour: light cherry. Nose: red berry notes, ripe fruit, fine reductive notes, spicy. Palate: flavourful, fruity, correct.

El Helechal Rufete Blanca 2020 B
rufete blanco
92
Colour: yellow. Nose: dried flowers, wild herbs, characterful, hint of anise. Palate: flavourful, long, complex.

Fuente Grulla 2020 T
93
Kind finish, defined aromas. Colour: light cherry. Nose: red berry notes, fresh, neat, expressive. Palate: fruity, taut, easy to drink, fresh.

🏆 PODIUM

Renvivas 2019 T
95
Colour: Cherry. Nose: complex, expressive, spicy, mineral, raspberry, fresh fruit, wild herbs. Palate: elegant, full-bodied, long, great length, taut.

VIÑEDOS DE LAS ACACIAS
Rio Selmo, 10
24560 Toral de los Vados (León)
☎: +34 987 544 831
bodegaslasacacias@gmail.com
www.palaciodearganza.com

Marqués de Montejos Selección 2021 T
mencía

90

Colour: dark-red cherry. Nose: toasty, spicy, cocoa bean, ripe fruit. Palate: flavourful, toasty, fine bitter notes.

Palacio de Arganza 2021 T
mencía

90

Colour: dark-red cherry. Nose: toasty, spicy, cocoa bean, ripe fruit. Palate: flavourful, toasty, fine bitter notes.

Palacio de Arganza Cabernet Sauvignon Mencía 2020 T
mencía, cabernet sauvignon

88

Overripe, toasty, powerful, pruney.

Señorío de Peñalba Selección 2021 T
mencía

90

Colour: dark-red cherry. Nose: toasty, spicy, cocoa bean, ripe fruit. Palate: flavourful, toasty, fine bitter notes.

WEIN & VINOS GMBH
Hardenbergstr. 9A
10715 Berlin (Berlin)
☎: +49 303 150 6080
info@vinos.de
www.vinos.de

Intuición Verdejo 2021 B
100% verdejo

90 ♣

Colour: bright straw. Nose: fruit expression, ripe fruit, floral. Palate: flavourful, fresh, good acidity, fruity aftertaste.

YLLERA BODEGAS & VIÑEDOS
47490 Rueda (Valladolid)
☎: +34 983 868 097
grupoyllera@grupoyllera.com
www.grupoyllera.com

Yllera 12 meses 2019 T RB
tempranillo, syrah, cabernet sauvignon, merlot

89

Jammy, tasty, fruity, spicy, simple.

Yllera 9 Meses 2019 T RB
tempranillo, cabernet sauvignon, syrah, merlot

88

Pleasant, spicy, jammy, tasty.

Yllera Chardonnay Vendimia Nocturna 2021 B
chardonnay

88

Citrus fruit, herbal, jammy, balanced.

VT COSTA DE CANTABRIA

CASONA MICAELA
Barrio Henales, s/n
39880 Valle de Villa Verde (Cantabria)
☎: +34 635 503 059
casonamicaela@hotmail.com
www.casonamicaela.com

Micaela Selección de Añada 2020 B
albariño, riesling

90

Colour: straw. Nose: dried herbs, fresh fruit, citrus fruit. Palate: fruity, good acidity, easy to drink, good finish.

VINO DE LA TIERRA - EIVISSA / I.G.P.

PAGO CASA DEL BLANCO - CANTABRIA

Barrio de la Llamosa, s/n
39761 Nates (Cantabria)
☎: +34 619 306 251
comercial@pagocasadelblanco.com
www.pagocasadelblanco.com

Palacio de Treto 2020 B C
albariño

90

Colour: yellow. Nose: citrus fruit, white fruit, dried flowers, mineral, sweet spices. Palate: fresh, fruity, flavourful, good finish.

ViñaMar 2021 B
albariño

90

Colour: bright straw. Nose: white fruit, dried herbs, mineral. Palate: fresh, fruity, flavourful, balanced.

VT EIVISSA

CAN RICH

Camí de Sa Vorera, s/n
07820 San Antonio (Illes Balears/Islas Baleares)
☎: +34 971 803 377
info@bodegascanrich.com
www.bodegascanrich.com

BES Can Rich 2021 RD OL
monastrell

86 🌱

Can Rich Blanc D'Amfora 2021 B
moscatel

90 🌱

With personality. Colour: straw. Nose: expressive, white flowers, jasmine, yeasty notes, wild herbs. Palate: flavourful, fruity, balanced.

Can Rich Negre D'Amfora 2020 T
monastrell

87 🌱

Lausos 2018 T
cabernet sauvignon, monastrell

89 🌱

Balanced, spicy, jammy, herbal.

IBIZKUS WINES

Ctra. Eivissa – Santa Eulària, Km. 9,5
07849 Santa Eulària des Riu
(Illes Balears/Islas Baleares)
☎: +34 971 807 330
david@ibizkus.com
www.ibizkus.com

Ibizkus Syrah 2019 T
syrah

88

Hot, pruney, spicy, jammy, powerful.

Totem Red 2019 T
monastrell

92

Colour: dark-red cherry, garnet rim. Nose: ripe fruit, fruit preserve, aged wood nuances, tobacco, sweet spices. Palate: spicy, round tannins, long.

Totem Rosé Sant Josep 2019 RD FB
monastrell

88

Hot, jammy, tasty, powerful.

Totem White 2020 B FB
malvasía

91

Colour: bright yellow. Nose: ripe fruit, dried herbs, faded flowers, sweet spices. Palate: powerful, ripe fruit, balanced.

VT EXTREMADURA

BODEGAS CARLOS PLAZA

Sol s/n
06196 Cortegana (Badajoz)
☎: +34 924 687 932
info@bodegascarlosplaza.com
www.bodegascarlosplaza.com

Carlos Plaza 2019 T RB
merlot

86

Carlos Plaza 2020 T
tempranillo, syrah

86

Carlos Plaza 2021 B
macabeo, sauvignon blanc

85

Carlos Plaza Selección 2018 T
tempranillo, merlot, syrah

88

Balanced, spicy, jammy, classic, toasty.

La Llave Roja 2020 T
tempranillo, merlot, syrah
87

La Llave Roja 8 meses 2018 T C
tempranillo, merlot, syrah
87

BODEGAS DE OCCIDENTE
Mecánica, 40
06200 Almendralejo (Badajoz)
☎: +34 662 952 801
info@bodegasoran.com
www.bodegasoran.com

Buche Valle El Raposo 2019 T C
tempranillo
87

Buche Garnacha Tempranillo 2016 T RB
garnacha tintorera, tempranillo
86

Gran Buche 2015 T R
89
Jammy, full-bodied, fruity, lactic, spicy, roasted coffee.

BODEGAS LENEUS
Ctra. Alange, km. 2
06200 Almendralejo (Badajoz)
☎: +34 691 340 843
contacto@vinosleneus.com
www.vinosleneus.com

Leneus 2017 T BA
100% tempranillo
85

Leneus Cayetana 2021 B
100% cayetana blanca
87

Leneus Reishi 2019 T
tempranillo, reishi
86

Leneus Tempranillo 2021 T MC
100% tempranillo
87

BODEGAS MARCELINO DÍAZ
Mecánica, s/n
06200 Almendralejo (Badajoz)
☎: +34 924 677 548
bodega@madiaz.com
www.madiaz.com

Theodosivs 2018 T C
87

VINO DE LA TIERRA - EXTREMADURA / I.G.P.

VINO DE LA TIERRA - EXTREMADURA / I.G.P.

BODEGAS RUIZ TORRES
Ctra. EX 116, km.33,8
10136 Cañamero (Cáceres)
☎: +34 927 369 027
info@ruiztorres.com
www.ruiztorres.com

Ruiz Torres Syrah 2018 T
100% syrah
86

Trampal 2018 T C
100% tempranillo
85

Verdejo de Bodegas Ruiz Torres 2021 B
100% verdejo
85

BODEGAS SANI PRIMAVERA
La Zarza, s/n
06200 Almendralejo (Badajoz)
☎: +34 924 677 917
tienda@bodegassani.com
www.bodegasani.com

Corte Real 2018 B FB
moscatel, sauvignon blanc
88
Pleasant, aromatic, tropical, tasty.

Magia de Adventus 2015 T FB
cabernet sauvignon, tempranillo, syrah
91
Colour: dark-red cherry, garnet rim. Nose: ripe fruit, aged wood nuances, tobacco, sweet spices. Palate: spicy, round tannins, long.

Sani Viña Extremeña 2018 T FB
cabernet sauvignon, tempranillo
86

ENCINA BLANCA DE ALBURQUERQUE
Ctra. Ex 302, Km. 85,3
06510 Alburquerque (Badajoz)
☎: +34 679 807 326
bodega@encinablanca.com
www.encinablancadealburquerque.es

Blanco 12 Cepas 2021 B
cayetana blanca, pardina, cigüente, otras
89 🌱
Fruity, tasty, jammy, sweety finish, full-bodied.

Espumoso Edición Especial 2017 BE R BN
cayetana blanca, pardina, cigüente, otras
92
Oxidativ. Colour: bright straw. Nose: brioche, dry nuts, ripe fruit, chalk. Palate: flavourful, balanced, fine bead, full-bodied.

Tinto 9 Cepas 2018 T R
tempranillo, cabernet sauvignon, petit verdot, merlot, syrah, garnacha
87

MARQUÉS DE VALDUEZA
Autovía A-5, km. 360
06800 Mérida (Badajoz)
☎: +34 913 191 508
contact@marquesdevaldueza.com
www.marquesdevaldueza.com

Marqués de Valdueza Etiqueta Roja 2018 T C
70% cabernet sauvignon, 30% syrah
87

Marqués de Valdueza Gran Vino de Guarda 2017 T GR
94% syrah, 6% cabernet sauvignon
89
Balanced, spicy, dried flowers, age nuances, toasty.

Valdueza 2019 T BA
75% syrah, 20% petit verdot, 5% cabernet sauvignon
87

PALACIO QUEMADO
Ctra. Almendralejo - Palomas, km 6,9
06840 Alange (Badajoz)
☎: +34 924 120 296
palacioquemado@alvear.es
www.palacioquemado.com

Palacio Quemado La Raya 2018 T
91
Colour: dark-red cherry, garnet rim. Nose: ripe fruit, aged wood nuances, tobacco, sweet spices. Palate: spicy, round tannins, long.

Palacio Quemado La Zarcita 2020 T
touriga nacional, garnacha tintorera, syrah
91
Colour: Cherry, garnet rim. Nose: balsamic herbs, sweet spices, scrubland, ripe fruit. Palate: spicy, good acidity, easy to drink.

Palacio Quemado Los Acilates 2019 T C S
90
Colour: deep cherry. Nose: dried herbs, creamy oak, overripe fruit. Palate: powerful, ripe fruit, round tannins.

Palacio Quemado Vendimia Seleccionada 2020 T RB
tempranillo, garnacha tintorera
89
Jammy, toasty, hot, herbal.

S.C.A. SANTA MARTA VIRGEN
Cooperativa s/n
06150 Santa Marta de Los Barros (Badajoz)
☎: +34 924 690 218
Fax: +34 924 690 083
info@bodegasantamarta.com
www.cooperativasantamarta.com

Calamón RD AG SD
tempranillo
84

VIÑA SANTA MARINA
Ctra. N-630, Km. 634
06800 Mérida (Badajoz)
☎: +34 824 607 001
administracion@vsantamarina.com
www.vsantamarina.com

Torremayor 12 meses 2019 T
tempranillo
86

Viña Santa Marina Cabernet Sauvignon-Syrah 2019 T C
cabernet sauvignon, syrah
87

VIÑAOLIVA SOC. COOP.
Automoción, 1
06200 Almendralejo (Badajoz)
☎: +34 924 677 321
info@vinaoliva.com
www.zaleo.es

Tinaja de Zaleo 2021 T S
100% syrah
88
Vegetal, full-bodied, fruity.

Z de Zaleo de Aguja Natural 2021 BE AG SD
85

Z de Zaleo de Aguja Natural 2021 RE AG SD
85

VIÑEDOS POZANCO
Ctra. de BA-001 Km. 15,700
06800 Mérida (Badajoz)
☎: +34 924 143 249
info@bodegaspozanco.com
www.bodegaspozanco.com

10·12 (Diez Punto Doce) 2021 T
50% tempranillo, 50% merlot, graciano
84

10·12 (Diez Punto Doce) BE AG SD
80% moscatel, 20% cayetana blanca
84

10·12 Selección (Diez Punto Doce) 2018 T C
60% tempranillo, 20% merlot, 20% cabernet sauvignon
88
Herbal, jammy, toasty, spicy, wild, balanced.

Viñedos Pozanco Verdejo 2021 B
100% verdejo
86

VINO DE LA TIERRA - EXTREMADURA / I.G.P.

VT FORMENTERA

CAP DE BARBARIA
Ctra. de Cap de Barbaria, km. 5,8
07860 Formentera (Illes Balears/Islas Baleares)
☎: +34 647 707 572
info@capdebarbaria.com
www.capdebarbaria.com

Cap de Barbaria 2017 T C
cabernet sauvignon, merlot, monastrell, fogoneu

93

Colour: garnet rim, light cherry. Nose: fruit preserve, tobacco, sweet spices, dried herbs. Palate: spicy, round tannins, balanced.

🏆 PODIUM

Cap de Barbaria Pansit de Formentera B Solera D
prensal

96

Colour: light mahogany. Nose: powerful, complex, dry nuts, toasty, acetaldehyde. Palate: rich, long, fine solera notes, spicy, round.

Ophiusa 2019 T GR
cabernet sauvignon, merlot, monastrell, fogoneu

90

Colour: deep cherry. Nose: dried herbs, creamy oak, fruit preserve. Palate: spicy, good structure, grainy tannins.

TERRAMOLL
Ctra. de La Mola Km. 15.2 El Pilar de La Mola
07872 Formentera (Illes Balears/Islas Baleares)
☎: +34 971 327 293
info@terramoll.es
www.terramoll.es

Es Monestir 2018 T R
monastrell

92 🍃

Colour: cherry, purple rim. Nose: spicy, red berry notes, black fruit, undergrowth, scrubland. Palate: flavourful, good acidity, long, spicy, mineral.

Rosa de Mar 2021 RD
merlot, cabernet sauvignon, monastrell

89 🍃

Kind finish, balanced, exuberant, fruity, yeasty notes, tasty.

Savina 2021 B
viognier, malvasía, garnacha blanca, moscatel grano menudo

89 🍃

Citrus fruit, balanced, dried herbs, yeasty notes.

Terramoll Natural 2021 T BA

86 🍃

VT ILLA DE MENORCA

BINITORD
Camí Lloc de Monges s/n
07760 Ciutadella de Menorca (Illes Balears/Islas Baleares)
☎: +34 665 560 224
clarasalord@binitord.com
www.binitord.com

Binitord Blanc 2021 B
chardonnay, malvasía

89

Citrus fruit, floral, crisp, herbal.

Binitord Ciutat de Parella 2017 T R
cabernet sauvignon, merlot, syrah

87

Binitord Negre 2020 T
cabernet sauvignon, merlot, syrah

89

Balanced, spicy, dried flowers, herbaceous.

Binitord Rosat 2021 RD
tempranillo, syrah

87

BODEGAS TORRALBENC VELL
Ctra de Mahon a Cala en Porter, km 10
07730 Alaior (Illes Balears/Islas Baleares)
☎: +34 660 806 145
info@bodegastorralbenc.com
www.bodegastorralbenc.com

Torralbenc 2021 RD
monastrell, syrah

88

Citrus fruit, balanced, dried flowers, tasty.

Torralbenc Chardonnay 2019 B
chardonnay

90

Colour: bright yellow. Nose: powerful, creamy oak, ripe fruit, spicy, woody. Palate: rich, good structure, toasty, fine bitter notes.

Torralbenc Coupage Tinto 2019 T
merlot, syrah

92

Colour: cherry, purple rim. Nose: red berry notes, spicy, animal reductive notes, scrubland, dried flowers. Palate: flavourful, fruity, good acidity.

VINO DE LA TIERRA - FORMENTERA / I.G.P.

Torralbenc Merlot 2019 T
merlot

89

Overripe, spicy, dried herbs, full-bodied, tasty.

Torralbenc Syrah 2019 T RB
syrah

92

Colour: deep cherry. Nose: dried herbs, creamy oak, black fruit. Palate: ripe fruit, spicy, round tannins.

SA FORANA
Cugullonet Nou
07712 Sant Climent - Mahón (Illes Balears/Islas Baleares)
☎: +34 607 242 510
saforana@saforana.com
www.saforana.com

600 Metros Sa Forana 2021 T
40% cabernet sauvignon, 30% merlot, 15% ull de llebre, 15% syrah

88

Balanced, fruity, herbal, spicy, crisp.

600 Metros Sa Forana Blanc 2021 B
60% prensal, 40% chardonnay

87

Sa Forana 2019 T
42% cabernet sauvignon, 34% ull de llebre, 24% syrah

88

Spicy, balanced, jammy, dried herbs.

Sa Forana Blanc 2021 B FB
75% chardonnay, 25% prensal

89

Citrus fruit, hot, spicy, jammy, full-bodied.

VT ILLES BALEARS

ÀNIMA NEGRA VITICULTORS
3ª Volta, 18
07200 Faianitx (Illes Balears/Islas Baleares)
☎: +34 971 584 481
admin@annegra.com
www.animanegra.com

Àn 2019 T
callet

94

Colour: cherry, garnet rim. Nose: toasty, spicy, cocoa bean, ripe fruit. Palate: flavourful, toasty, fine bitter notes.

Àn/2 2020 T
callet, mantonegro, fogoneu, syrah

93

Colour: bright cherry. Nose: balsamic herbs, sweet spices, scrubland, wild herbs, ripe fruit. Palate: spicy, balsamic, good acidity.

ATLAN & ARTISAN
07579 Colonia de San Pere (Illes Balears/Islas Baleares)
info@atlanandartisan.com
www.atlanandartisan.com

8 Vents 2019 T C
callet, cabernet sauvignon, merlot

93

Colour: deep cherry. Nose: dried herbs, creamy oak, black fruit, fine reductive notes. Palate: powerful, ripe fruit, spicy, round tannins.

8 Vents Gran 2019 T
cabernet sauvignon, merlot

92

Colour: deep cherry. Nose: dried herbs, creamy oak, black fruit, meaty notes. Palate: powerful, ripe fruit, spicy, round tannins.

BODEGAS TORRALBENC VELL
Ctra de Mahon a Cala en Porter, km 10
07730 Alaior (Illes Balears/Islas Baleares)
☎: +34 660 806 145
info@bodegastorralbenc.com
www.bodegastorralbenc.com

Torralbenc Coupage Blanc 2020 B
sauvignon blanc, parellada, viognier

91

Colour: straw. Nose: ripe fruit, dried herbs, faded flowers. Palate: ripe fruit, balanced, full-bodied, rich.

TERRA DE FALANIS
Ctra. Campos Felanitx 10.3 Km.
07200 Felanitx (Illes Balears/Islas Baleares)
☎: +34 679 314 406
contactoterradefalanis@gmail.com
www.terradefalanis.com

Castell de Santueri Rouge 2020 T RB
65% callet, 20% mantonegro, 15% cabernet sauvignon

88

Balanced, spicy, herbal, crisp.

VINO DE LA TIERRA - ILLES BALEARS / I.G.P.

VINO DE LA TIERRA - LADERAS DEL GENIL / I.G.P.

VT LADERAS DEL GENIL

BODEGAS SEÑORÍO DE NEVADA
Ctra. de Cónchar, s/n
18659 Villamena (Granada)
☎: +34 958 777 092
Fax: +34 958 107 367
info@senoriodenevada.es
www.senoriodenevada.es

Señorío de Nevada 2021 B
vijariego blanco, viognier
86

Señorío de Nevada 2021 RD
garnacha
86

VT LIÉBANA

BODEGA PICOS DE CABARIEZO
39571 Cabezón de Liébana (Cantabria)
☎: +34 942 735 177
info@vinosylicorespicos.es
www.vinosylicorespicos.com

Finca Morillas 2018 T RB
mencía, syrah
90
Colour: bright cherry. Nose: fruit expression, wild herbs, wild herbs, spicy. Palate: fruity, fresh, balanced.

Picos de Cabariezo Selección 2020 T
garnacha, tempranillo, mencía
88
Spicy, smoky, fruity, jammy.

DESTILERÍA Y BODEGA CAYO
Mesasinpan s/n
39584 Cabezón de Liébana (Cantabria)
☎: +34 942 730 689
info@bodegacayo.com
www.bodegacayo.com

Enza 2019 B RB
palomino
89
Kind finish, elegant, dried flowers, fruity, thin.

Lusia 2019 T RB SD
75% mencía, 25% tempranillo
92
Colour: deep cherry. Nose: ripe fruit, dried herbs, creamy oak, scrubland. Palate: powerful, ripe fruit, spicy, round tannins.

VT MALLORCA

4 KILOS VINÍCOLA
1ª Volta, 168 Puigverd
07200 Felanitx (Illes Balears/Islas Baleares)
☎: +34 971 580 523
Fax: +34 971 580 523
fgrimalt@4kilos.com
www.4kilos.com

12 volts 2020 T
cabernet sauvignon, callet, fogoneu, syrah, merlot
94
Crisp, pleasant. Colour: light cherry. Nose: red berry notes, fresh fruit, floral, wild herbs. Palate: flavourful, fruity, good acidity.

🏆 **PODIUM**

4 Kilos 2020 T
callet
95
Colour: Rubí, purple rim. Nose: defined aromas, raspberry, floral, toasted bread, wild herbs, earthy notes. Palate: flavourful, fresh, round tannins, great length.

🏆 **PODIUM**

Gallinas & Focas 2018 T
mantonegro
96
Crisp. Colour: bright cherry. Nose: defined aromas, red berry notes, raspberry, floral, violets, earthy notes. Palate: fresh, flavourful, elegant, great length.

7103 PETIT CELLER
Pol. 6 Parc. 151
07320 Santa María del Camí
(Illes Balears/Islas Baleares)
☎: +34 690 101 647
pladebuc@gmail.com
www.7103-petitceller.com

40 Braces Negre 2019 T
100% mantonegro
91
Colour: cherry, purple rim. Nose: red berry notes, floral, spicy, scrubland. Palate: flavourful, fruity, good acidity, elegant, fleshy.

7103 Blanc de Mantonegro 2021 B
100% mantonegro
89
Balanced, spicy, dried herbs, jammy, smooth.

7103 Blanc Premsal Blanc 2021 B
100% prensal

87

7103 Giró 2021 B
100% giró ros

90

Colour: straw. Nose: ripe fruit, dried herbs, faded flowers, citrus fruit. Palate: powerful, ripe fruit, full-bodied, warm, bitter.

7103 Negre Bóta 2019 T BA
70% mantonegro, 30% merlot

90

Colour: cherry, purple rim. Nose: red berry notes, floral, spicy, balsamic herbs. Palate: flavourful, fruity, good acidity.

7103 Rosat 2021 RD
100% mantonegro

88

Citrus fruit, dried flowers, fruity, smooth.

BINIGRAU
Fiol, 33
07143 Biniali (Illes Balears/Islas Baleares)
☎: +34 971 512 023
info@binigrau.es
www.binigrau.es

Binigrau Bi-Blanc 2020 B FB
chardonnay

92

Colour: bright straw. Nose: ripe fruit, fragrant herbs, fine lees, sweet spices. Palate: full-bodied, rich, long, good acidity.

Binigrau Bi-Negre 2017 T BA
mantonegro, callet, merlot

93

Colour: dark-red cherry, garnet rim. Nose: ripe fruit, fruit preserve, aged wood nuances, tobacco, sweet spices. Palate: spicy, round tannins, long.

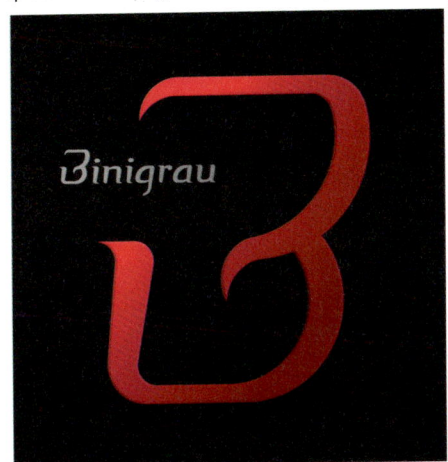

Nou Nat 2021 B
60% prensal, 40% chardonnay

92

Colour: bright straw. Nose: ripe fruit, dried herbs, faded flowers. Palate: powerful, ripe fruit, balanced.

Binigrau E-Negre 2020 T
mantonegro, merlot

92 🌿

Colour: Cherry. Nose: balsamic herbs, sweet spices, scrubland. Palate: spicy, balsamic, good acidity.

VINO DE LA TIERRA - MALLORCA / I.G.P.

VINO DE LA TIERRA - MALLORCA / I.G.P.

Binigrau E-Rosat 2021 RD
mantonegro, merlot

91 🌿

Colour: raspberry rose. Nose: red berry notes, floral, fragrant herbs. Palate: light-bodied, spicy, good acidity, fine bitter notes.

Obac de Binigrau 2018 T BA
50% mantonegro, callet, 30% merlot, 10% syrah, 10% cabernet sauvignon

93

Colour: Cherry. Nose: balsamic herbs, sweet spices, scrubland, ripe fruit, characterful. Palate: spicy, balsamic, good acidity.

BODEGA AVA VI MALLORCA
Camino de Muro, Polig. 9 Parcela 40
07141 Sencelles (Illes Balears/Islas Baleares)
☎: +34 687 789 932
juliosumiller@gmail.com
www.ava-vi.com

ANAVA…a cercar un somni Rosat 2020 RD BA
mantonegro

90

Colour: coppery red, bright. Nose: wild herbs, dried herbs, spicy, white fruit, ripe fruit. Palate: flavourful, rich, fruity.

Ava Vi Blanc 2021 B
70% prensal, 30% chardonnay

87

AVA Vi Negre 2019 T BA
mantonegro, callet, cabernet sauvignon, syrah, merlot

88

Pruney, full-bodied, spicy, dried herbs, powerful.

AVA Selecció 2019 T C
70% mantonegro, 30% otras

91

Colour: deep cherry. Nose: red berry notes, black fruit, scrubland, spicy. Palate: ripe fruit, spicy, round tannins, balanced.

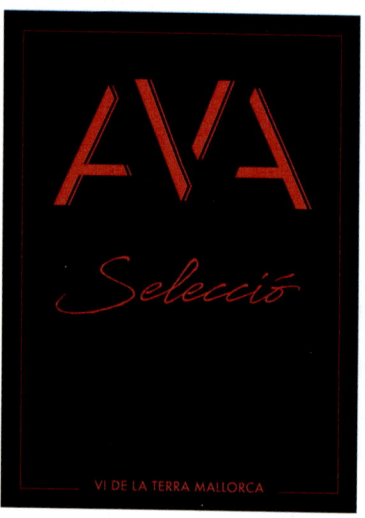

TRIAVA Blanc "Vino de Guarda" 2020 B FB

91

Colour: straw. Nose: powerful, creamy oak, ripe fruit, spicy. Palate: good structure, long, toasty, fine bitter notes.

TRIAVA Heritage "Vino de Guarda" 2019 T R
50% mantonegro, 25% merlot, 25% cabernet sauvignon

91

Colour: deep cherry. Nose: dried herbs, creamy oak, black fruit, woody. Palate: powerful, ripe fruit, spicy, round tannins.

Avanero Giro Ros "Vino de Parcela" 2020 B FB
100% giró ros

92

Colour: straw. Nose: ripe fruit, dried herbs, faded flowers, waxy notes. Palate: powerful, ripe fruit, balanced.

BODEGA BLANCA TERRA
Ctra. Palma Manacor, Km. 30
07260 Montuïri (Illes Balears/Islas Baleares)
☎: +34 971 267 163
info@blancaterra.com
www.blancaterra.com

Arrelat 2020 B
60% malvasía, 40% giró ros

87

Febrer 2018 T C
60% merlot, 30% monastrell, 10% cabernet sauvignon

89

Balanced, spicy, age nuances, tasty.

Foravila 2020 B
80% chardonnay, 20% riesling

88

Citrus fruit, herbal, tasty, wild.

Passió 2020 RD
50% merlot, 50% malvasía

88

Balanced, dried flowers, fruity, crisp.

Son Roca 2020 T
100% syrah

88

Balanced, spicy, fruity, full-bodied.

BODEGA CAP ANDRITXOL
07140 Sencelles (Illes Balears/Islas Baleares)
☎: +34 628 883 171
info@capandritxol.com
www.capandritxol.com

1580 2017 T
50% callet, 50% cabernet sauvignon

91

Colour: deep cherry. Nose: dried herbs, black fruit, burnt matches, balsamic herbs, spicy. Palate: powerful, ripe fruit, spicy, round tannins.

Corsari 2021 B
prensal, giró ros, chardonnay

89

Dried flowers, dried herbs, standard, jammy, wild, smooth. Nose: earthy notes.

Defensor 2021 B
prensal, giró ros

89

Citrus fruit, herbal, crisp, tasty.

Matacá 2017 T
merlot, syrah, callet, mantonegro, tempranillo

91

Colour: deep cherry. Nose: dried herbs, black fruit, balsamic herbs, fine reductive notes. Palate: ripe fruit, spicy, round tannins, elegant.

Sa Talaia 2017 T BA
merlot, cabernet sauvignon, syrah, gorgollassa, mantonegro, callet

90

Colour: deep cherry. Nose: dried herbs, creamy oak, fine reductive notes, black fruit, meaty notes. Palate: ripe fruit, spicy, round tannins, balanced.

Vigilia 2021 RD
mantonegro

87

BODEGA CASTELL MIQUEL
Ctra. Alaró-Lloseta, Km. 8,7
07340 Alaró (Illes Balears/Islas Baleares)
☎: +34 971 510 698
info@castellmiquel.com
www.castellmiquel.com

Bodega Castell Miquel Cabernet Sauvignon 2019 T R
cabernet sauvignon

89

Full-bodied, balanced, spicy, jammy, powerful, tasty.

Bodega Castell Miquel Syrah 2019 T R

88

Full-bodied, balanced, spicy, dried herbs, jammy.

Rosado Stairway To Heaven 2021 RD
syrah, tempranillo

87

Sauvignon Blanc Owner's Edition 2021 B
sauvignon blanc

88

Citrus fruit, balanced, herbal, yeasty notes, smoky, sweety finish.

VINO DE LA TIERRA - MALLORCA / I.G.P.

VINO DE LA TIERRA - MALLORCA / I.G.P.

Sauvignon Blanc Stairway To Heaven 2021 B
sauvignon blanc
87

Stairway to Heaven Cuvée 2019 T R
cabernet sauvignon, syrah, merlot
90
Colour: cherry, purple rim. Nose: fruit expression, spicy, red berry notes, black fruit, scrubland. Palate: flavourful, fruity, good acidity.

BODEGA SON MAYOL
Camí Can Mallol, 9
07010 Establiments (Illes Balears/Islas Baleares)
☎: +34 871 600 026
info@bodegasonmayol.es
www.bodegasonmayol.es

Grand Vin Son Mayol 2017 T GR
cabernet sauvignon, merlot, cabernet franc, petit verdot
93
Classic. Colour: deep cherry. Nose: dried herbs, creamy oak, black fruit, ripe fruit. Palate: powerful, ripe fruit, spicy, round tannins, good structure.

Grand Vin Son Mayol 2018 T GR
cabernet sauvignon, cabernet franc, petit verdot, merlot
92
Classic. Colour: deep cherry. Nose: dried herbs, creamy oak, black fruit. Palate: powerful, ripe fruit, spicy, good structure.

Premier Vin Son Mayol 2017 T R
merlot, cabernet sauvignon, malbec
90
With personality. Colour: deep cherry. Nose: dried herbs, creamy oak, black fruit, scrubland, lactic notes. Palate: ripe fruit, spicy, flavourful.

Premier Vin Son Mayol 2018 T R
merlot, cabernet sauvignon, malbec
91
Colour: bright cherry. Nose: creamy oak, ripe fruit, scrubland. Palate: good acidity, spicy, fine tannins.

Vin Blanc Son Mayol 2020 B
roussanne, marsanne, viognier, sauvignon blanc
91
Colour: straw. Nose: expressive, white flowers, dried herbs, waxy notes, sweet spices. Palate: flavourful, balanced, full-bodied.

Vin Rose Son Mayol 2021 RD AG
merlot, cabernet franc, malbec, touriga
89
Balanced, dried flowers, fruity, herbal, tasty.

BODEGAS ÁNGEL
Ctra. Sta María - Sencelles, km. 4,8
07320 Santa María del Camí
(Illes Balears/Islas Baleares)
☎: +34 971 180 118
info@bodegasangel.com
www.bodegasangel.com

Ángel Blanc de Blanca 2021 B
33% prensal, 30% malvasía, 17% chardonnay, 20% viognier
88
Citrus fruit, dried flowers, herbal, jammy, balanced.

Ángel Cabernet Sauvignon 2016 T
cabernet sauvignon
89
Balanced, spicy, fruity, jammy, tasty.

Ángel Manto Negro 500 2019 T C
95% mantonegro, 5% syrah
91
Colour: cherry, purple rim. Nose: fruit expression, red berry notes, floral, spicy, wild herbs. Palate: flavourful, fruity, good acidity.

Ángel Rosat 2021 RD
37% merlot, 35% prensal, 28% mantonegro
86

Moni Dolç Orange Wine 2021 B D
88% prensal, 4% viognier, 4% chardonnay, 2% moscatel
85

BODEGAS CAN VIDALET
Ctra. Alcudia - Pollença Ma 2201, Km. 4,85
07460 Pollença (Illes Balears/Islas Baleares)
☎: +34 971 531 719
Fax: +34 971 535 395
info@canvidalet.com
www.canvidalet.com

Barros de Cecili 2019 B
100% prensal
91
With personality. Colour: bright straw. Nose: ripe fruit, fragrant herbs, fine lees, mineral, spicy. Palate: full-bodied, rich, long, good acidity.

Can Vidalet Blanc de Blancs 2021 B
prensal, chardonnay, moscatel, sauvignon blanc
87 🌿

Can Vidalet Blanc de Negres 2021 B
merlot, callet, syrah
89 🌿
Citrus fruit, austere, herbal, saline, crisp.

Can Vidalet Ses Pedres 2018 B
chardonnay
91
Colour: bright yellow. Nose: powerful, creamy oak, spicy, candied fruit, caramel. Palate: good structure, long, toasty, fine bitter notes.

Can Vidalet So del Xiprer 2017 T
cabernet sauvignon, syrah, merlot
88
Full-bodied, spicy, tasty, toasty, pruney.

BODEGUES MACIÀ BATLE
Camí Coanegra
07320 Santa María del Camí (Illes Balears/Islas Baleares)
☎: +34 971 140 014
info@maciabatle.com
www.maciabatle.com

Macià Batle 1856 2018 T BA
89
Balanced, spicy, dried herbs, jammy, toasty.

Macià Batle 2020 T
87

Macià Batle 2021 B MC
89
Citrus fruit, herbal, fruity, crisp, balsamic herbs.

Macià Batle 2021 T MC
manto negro (Tinto)
90
Colour: cherry, purple rim. Nose: fruit expression, red berry notes, floral, black liquorice. Palate: fruity, flavourful, balanced, varietal.

Macià Batle Blanc de Blancs 2021 B
89
Floral, jammy, kind finish, great length, sweeties, tasty.

Macià Batle Blanc de Blancs Dolç 2019 B D
90
Colour: bright yellow. Nose: ripe fruit, candied fruit, honeyed notes. Palate: flavourful, unctuous, fruity, sweet.

Macià Batle Blanc de Blancs Orange 2021 B
89
Defined aromas, standard, floral, jammy, kind finish, great length.

Macià Batle Col.lecció Privada 2018 T
90
Colour: deep cherry. Nose: dried herbs, creamy oak, black fruit, tobacco, fruit preserve, roasted coffee. Palate: powerful, spicy, toasty.

Macià Batle Negre Dolç 2018 T D
90
Colour: cherry, garnet rim. Nose: fruit preserve, spicy, toasty, aged wood nuances, dried herbs. Palate: flavourful, balanced.

Macià Batle Rosat 2021 RD
90
Colour: rose. Nose: faded flowers, scrubland, earthy notes, raspberry. Palate: fleshy, flavourful, ripe fruit.

Macià Batle Sauvignon Blanc 2020 B FB
sauvignon blanc
89
Citrus fruit, creamy, herbal, jammy, full-bodied, toasty.

Macià Batle Sauvignon Blanc 2021 B
sauvignon blanc
89
Citrus fruit, crisp, herbal, yeasty notes.

Macià Batle Xeremia Blanc 2020 B
prensal
87

Macià Batle Xeremia Blanc 2021 B
89
Austere, citrus fruit, dried herbs, tasty, fruity.

Margalida Llompart Blanc 2019 B
prensal, chardonnay
89
Woody, standard, spicy, jammy, tasty. Palate: rich.

Margalida Llompart Rosat 2021 RD
mantonegro, cabernet sauvignon
89
Kind finish, dried flowers, dried herbs, jammy, tasty, fruity.

Margarida Llompart Negre 2017 T
89
Pruney, creamy, woody, tasty, spicy.

Santa Clara 2021 T
mantonegro, merlot, cabernet sauvignon, syrah
88
Balanced, spicy, herbal, roasted coffee.

VINO DE LA TIERRA - MALLORCA / I.G.P.

VINO DE LA TIERRA - MALLORCA / I.G.P.

....
Santa Clara Blanc de Blancs 2021 B
88
Aromatic, floral, jammy, tasty. Palate: rich.

BODEGUES VIDAL SERRA
Ca ses Monges, 15
07142 Santa Eugènia (Illes Balears/Islas Baleares)
☎: +34 673 466 992
bodeguesvidalserra@gmail.com
....
Dièresi 2020 T
50% merlot, 50% mantonegro
89
Pruney, full-bodied, balanced, spicy, powerful.
....
María Serra 2021 B
prensal, moscatel, chardonnay, sauvignon blanc
89
Citrus fruit, balanced, herbal, crisp, tasty.
....
Xafarder 2021 RD
syrah, merlot
87

CAN GELAT
07316 Moscari (Illes Balears/Islas Baleares)
☎: +31 633 321 964
info@cangelat.com
www.cangelat.com
....
Can Gelat Callet 2021 RD
callet
90 🍇
Standard. Colour: coppery red. Nose: lees reduction notes, wild herbs, red berry notes, ripe fruit. Palate: fruity, balanced, flavourful.
....
Can Gelat Giró 2021 B
giró ros
91 🍇
Colour: yellow, pale. Nose: wild herbs, dried herbs, faded flowers, ripe fruit. Palate: balanced, fine bitter notes, long, flavourful.

CELLER TIANNA NEGRE
Cami des Mitjans s/n, Parc. 67 – Pol. 7
07350 Binissalem (Illes Balears/Islas Baleares)
☎: +34 971 886 826
Fax: +34 971 226 201
info@tiannanegre.com
www.tiannanegre.com
....
Tianna 1 Negre 2020 T
100% callet
93 🍇
Subtle, slight reduction. Colour: light cherry. Nose: scrubland, red berry notes, ripe fruit, petrol notes. Palate: flavourful, balsamic, spicy, soft tannins.

....
Véloblanc 2021 B
100% mantonegro
87 🍇
....
Tianna
Bocchoris Negre 2020 T
mantonegro, syrah, merlot, callet, cabernet sauvignon
90 🍇
Colour: light cherry. Nose: ripe fruit, dried herbs, creamy oak, sweet spices, woody. Palate: ripe fruit, spicy, round tannins.
....
Tianna Negre 2020 T
syrah, merlot, mantonegro, callet
91 🍇
Colour: light cherry. Nose: ripe fruit, dried herbs, creamy oak, sweet spices. Palate: ripe fruit, spicy, round tannins.
....
Vélonegre 2020 T
100% mantonegro
91 🍇
Colour: cherry, purple rim. Nose: fruit expression, red berry notes, floral, spicy. Palate: flavourful, fruity, good acidity, balanced.

DALT TURÓ
Camí es Figueral, Pol. 14 par 65
07630 Campos (Illes Balears/Islas Baleares)
☎: +34 971 583 657
daltturo@daltturo.com
www.daltturo.com
....
Dalt Turó Acopinyat 2021 B
40% callet, 36% malvasía, 24% mantonegro
90
Colour: bright straw, greenish rim. Nose: citrus fruit, wild herbs, ripe fruit, white flowers, mineral. Palate: fresh, fruity, good acidity.
....
Dalt Turó Brescat 2019 T
35% callet, 25% cabernet sauvignon, 15% syrah, 20% merlot, 5% mantonegro
91
Colour: cherry, purple rim. Nose: spicy, red berry notes, black fruit, faded flowers, undergrowth. Palate: flavourful, good acidity, balanced.
....
Dalt Turó Sauló 2019 T
60% callet, 5% mantonegro, 30% cabernet sauvignon, 3% syrah
92
With personality, smooth. Colour: cherry, purple rim. Nose: fruit expression, red berry notes, floral, spicy, chamomile, honeyed notes. Palate: flavourful, fruity, good acidity, long, elegant.
....
Dalt Turó Granat 2019 T C
61% cabernet sauvignon, 11% callet, 8% syrah, 20% merlot
89
Balanced, spicy, dried herbs, jammy, tasty, smoky.

Dalt Turó Roget 2021 RD
52% mantonegro, 48% callet

89
Balanced, dried flowers, smooth, fruity.

DUNORD VITÍCOLA
El Cano, 28
07470 Port de Pollença (Illes Balears/Islas Baleares)
☎: +34 670 645 978
dunordviticola@gmail.com

Curolla 2020 T
100% gargallosa

92
Colour: cherry, purple rim. Nose: fruit expression, red berry notes, floral, sweet spices, scrubland. Palate: flavourful, fruity, good acidity.

Insomni 2021 B
50% giró ros, 50% malvasía

90
Colour: straw. Nose: white flowers, jasmine, dried herbs, white fruit. Palate: flavourful, fruity, balanced.

FINCA CAN AXARTELL
Ctra. Pollença – Campanet km 1,5
07460 Pollença (Illes Balears/Islas Baleares)
☎: +34 871 870 353
info@canaxartell.es
www.canaxartell.com

Aurorum 2021 RD C
92 ♣
Colour: raspberry rose. Nose: red berry notes, citrus fruit, fine lees, spicy. Palate: good acidity, fine bitter notes, great length.

The Artist 2019 T C
54% callet, 19% syrah, 27% merlot

91 ♣
Colour: deep cherry. Nose: dried herbs, creamy oak, black fruit, sweet spices. Palate: powerful, ripe fruit, round tannins.

Ventum 2017 T C
merlot, syrah, callet

92 ♣
Colour: deep cherry. Nose: dried herbs, creamy oak, cocoa bean, black fruit. Palate: powerful, ripe fruit, spicy, round tannins.

Terrum 2019 T
callet

92 ♣
Colour: cherry, purple rim. Nose: fruit expression, red berry notes, floral, rose petals, balsamic herbs. Palate: fruity, flavourful, balanced.

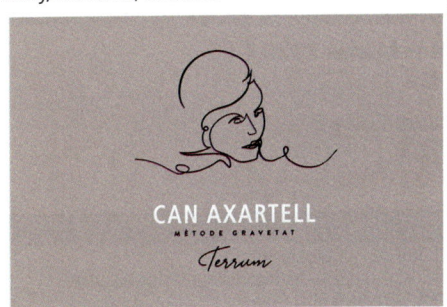

Ventum 2018 T C
merlot, syrah, callet

93 ♣
Colour: deep cherry. Nose: creamy oak, black fruit, scrubland, spicy. Palate: powerful, ripe fruit, round tannins, great length.

VINO DE LA TIERRA - MALLORCA / I.G.P.

JAUME DE PUNTIRÓ

Pza. Nova, 23
07320 Santa María del Camí
(Illes Balears/Islas Baleares)
☎: +34 606 429 023
pere@vinsjaumedepuntiro.com
www.vinsjaumedepuntiro.com

Brisat de Puntiró 2021 B
giró ros

88 🌱

Standard, spicy, dried flowers, herbaceous, rustic.

MURGUIALDI 3 DE BALEARS

Dos Marias 2020 T RB
87

Llum 2021 B
88

Crisp, citrus fruit, herbal, balanced.

SA CABANA

Camí de Pou des Torrent
07350 Binissalem (Illes Balears/Islas Baleares)
☎: +34 650 800 080
info@bodegasacabana.com
www.bodegasacabana.com

Sa Cabana Chardonnay 2021 B
100% chardonnay

89

Crisp, fruity, herbal, slight reduction, tasty.

Sa Cabana Girò Ros 2021 B
100% giró ros

90

Colour: bright straw, greenish rim. Nose: fresh fruit, citrus fruit, wild herbs, hint of anise. Palate: fresh, fruity, good acidity.

Sa Cabana Merlot 2020 T BA
100% merlot

89

Herbal, tasty, crisp, fruity.

SANTA CATARINA

Ctra. Inca – Sencelles, Km. 3
07140 Sencelles (Illes Balears/Islas Baleares)
☎: +34 971 137 115
info@bodegasantacatarina.com
www.bodegasantacatarina.com

'Nguany Blanc 2021 B
55% giró ros, 45% viognier

89

Balanced, dried flowers, crisp, jammy, herbal.

'Nguany Negre 2020 T
60% syrah, 20% callet, 20% mantonegro

91

Colour: cherry, purple rim. Nose: fruit expression, red berry notes, floral, spicy. Palate: flavourful, fruity, good acidity.

Sta Callet 2021 T
callet

90

Colour: cherry, purple rim. Nose: red berry notes, floral, spicy, scrubland. Palate: flavourful, fruity, good acidity.

Sta Giró Ros 2021 B
100% giró ros

87

SON GRAU GRAN

Ctra. Alaró – Lloseta km 9,1
07340 Alaró (Illes Balears/Islas Baleares)
☎: +34 649 038 389
adm@songrau.com
www.songrau.com

Son Grau
Gran Escursac 2021 T
90% escursac, 10% syrah

92 🌱

Colour: cherry, purple rim. Nose: fruit expression, red berry notes, floral, spicy, wild herbs. Palate: flavourful, fruity, good acidity, long.

Son Grau
Gran Gargollassa 2021 T BA
100% gargallosa

89 🌱

Floral, fruity, spicy, tasty, pronounced acidity.

Son Grau
Gran Gargollassa 2021 RD
100% gargallosa

90
Colour: salmon. Nose: fruit expression, red berry notes, floral, dried herbs, dry stone. Palate: fruity, good acidity, balanced, flavourful.

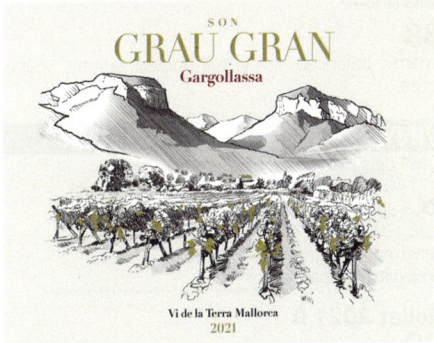

SON JULIANA
Ctra. Santa Maria - Sencelles KM 7.2
07142 Santa Eugènia (Illes Balears/Islas Baleares)
☎: +34 615 930 852
ventas.sonjuliana@gmail.com
www.sonjuliana.es

Cuvée #1 Son Juliana 2019 B
prensal, giró ros, chardonnay

89
Citrus fruit, herbal, tasty, wild.

Cuvée #2 Son Juliana 2017 T
55% cabernet sauvignon, 45% merlot

90
Colour: deep cherry. Nose: dried herbs, creamy oak, red berry notes, black fruit. Palate: ripe fruit, spicy, round tannins.

Cuvée #3
Son Juliana 2019 RD
mantonegro, callet, cabernet sauvignon

90
Colour: raspberry rose. Nose: citrus fruit, white fruit, fine lees, dried herbs. Palate: spicy, good acidity, fine bitter notes.

Mantonegro
Blanco Son Juliana 2019 B
100% mantonegro

89
Citrus fruit, crisp, fruity, herbal, balanced.

Mantonegro Tinto Son Juliana 2019 T
mantonegro

90
Colour: cherry, purple rim. Nose: red berry notes, floral, scrubland. Palate: flavourful, fruity, ripe fruit, balanced.

SON PRIM PETIT
Ctra. Inca - Sencelles, Km. 4,9
07014 Sencelles (Illes Balears/Islas Baleares)
☎: +34 971 872 758
jaime@sonprim.com
www.sonprim.com

El Peligro 2018 T

92
Colour: cherry, purple rim. Nose: fruit expression, red berry notes, black fruit, fruit preserve, scrubland. Palate: flavourful, fruity, good acidity, long.

SON RAMON VINS I VINYES
Ctra Muro-Inca, Km 4,5
07430 Llubí (Illes Balears/Islas Baleares)
☎: +34 651 870 085
bodega@sonramon.com
www.sonramon.com

Cabernet de Son Ramon 2017 T
cabernet sauvignon

89
Full-bodied, balanced, spicy, dried herbs, roasted coffee.

Sirà de Son Ramon 2017 T
syrah

90
Colour: deep cherry. Nose: dried herbs, creamy oak, black fruit, meaty notes. Palate: ripe fruit, spicy, round tannins.

Son Ramon Blanc de Blancs 2021 B
prensal, chardonnay

87

Son Ramon Chardonnay 2021 B
chardonnay

86

Son Ramon Negre 2019 T
syrah, cabernet sauvignon, merlot

87

Son Ramon Selecció Especial 2017 T
syrah, cabernet sauvignon

91
Colour: deep cherry. Nose: creamy oak, wild herbs, black fruit, meaty notes. Palate: powerful, ripe fruit, spicy, round tannins.

VINO DE LA TIERRA - MALLORCA / I.G.P.

VINOS Y VIÑEDOS TRAMUNTANA
07003 Palma de Mallorca
(Illes Balears/Islas Baleares)
☎: +34 619 302 880
info@canxanet.com
www.canxanet.com

Cadmo 2018 T
50% syrah, 25,8% mantonegro, 24,2% gorgollassa

93

Colour: Cherry. Nose: balsamic herbs, scrubland, red berry notes, fine reductive notes. Palate: spicy, balsamic, good acidity, flavourful, great length.

Cumas 2018 T
100% mantonegro

93

Colour: light cherry. Nose: balsamic herbs, scrubland, undergrowth, fine reductive notes, red berry notes. Palate: spicy, balsamic, good acidity, flavourful.

🏆 PODIUM

Sibila 2018 T
100% gorgollassa

95

Colour: light cherry. Nose: balsamic herbs, sweet spices, scrubland, red berry notes, earthy notes. Palate: spicy, balsamic, good acidity.

Siurell 2021 B
76% giró ros, 24% malvasía

90

Colour: bright straw, greenish rim. Nose: fresh fruit, citrus fruit, wild herbs, balsamic herbs. Palate: fresh, fruity, good acidity, fine bitter notes.

Xanet 2018 T
43,2% merlot, 35,2% syrah, 11,9% gorgollassa, 9,3% mantonegro

92

Colour: cherry, purple rim. Nose: floral, red berry notes, black fruit, scrubland. Palate: fruity, flavourful, balanced, spicy.

Xanet Rosé 2021 RD
40% mantonegro, 19% callet, 18% syrah, 14% gorgollassa, 9% merlot

88

Slight reduction, citrus fruit, dried herbs, crisp, tasty.

VINS NADAL
Ramón Llull, 2
07350 Binissalem (Illes Balears/Islas Baleares)
☎: +34 971 511 058
vinsnadal@vinsnadal.es
www.vinsnadal.es

Blanc 110 Vins Nadal 2021 B
sauvignon blanc

88

Creamy, balanced, spicy, full-bodied, jammy, yeasty notes.

VINYES I VINS CA SA PADRINA
Camí dels Horts, s/n
07140 Sencelles (Illes Balears/Islas Baleares)
☎: +34 646 318 600
Fax: +34 971 874 370
cellermantonegro@gmail.com
www.vinscasapadrina.com

Mollet 2021 B

90

Colour: straw. Nose: dried herbs, faded flowers, citrus fruit, white fruit. Palate: ripe fruit, bitter, full-bodied.

Rosat 2021 RD

88

Hot, fruity, dried herbs, full-bodied, jammy.

VINYES MORTITX
Ctra. Pollença Lluc, Km. 10,9
07315 Escorca (Illes Balears/Islas Baleares)
☎: +34 971 533 889
info@vinyesmortitx.com
www.vinyesmortitx.com

L'U de Mortitx 2019 T R
syrah

90

Classic. Colour: deep cherry. Nose: dried herbs, creamy oak, black fruit. Palate: ripe fruit, spicy, round tannins, balanced.

Mortitx Callet - Gorgollassa 2021 T
gorgollassa, callet

89

Crisp, fruity, balsamic herbs, balanced.

Racó Fred
de Mortitx 2018 B FB
malvasía, giró ros, chardonnay

91

Colour: bright yellow. Nose: creamy oak, ripe fruit, spicy, waxy notes. Palate: rich, good structure, long, toasty, fine bitter notes.

Rodal Pla de Mortitx 2019 T
merlot, syrah, cabernet sauvignon
90
Colour: deep cherry. Nose: ripe fruit, creamy oak, scrubland. Palate: ripe fruit, spicy, round tannins, balanced.

VT RIBEIRAS DO MORRAZO

BODEGA OS AREEIROS
Lg. Areeiros, 30 Santa Cristina de Cobres
36142 Vilaboa (Pontevedra)
☎: +34 986 672 349
contacto@osareeiros.com
www.osareeiros.com

Os Areeiros 2021 B
albariño
87

Os Areeiros 2021 T
mencía, caiño, sousón
88
Aromatic, tasty, simple, herbal.

VT RIBERA DEL GÁLLEGO-CINCO VILLAS

BODEGAS EJEANAS
Avda. Cosculluela, 23
50600 Ejea de Los Caballeros (Zaragoza)
☎: +34 976 663 770
info@bodegasejeanas.com
www.bodegasejeanas.com

Un Garnacha Blanc de Noir 2018 B FB
100% garnacha
89
Balanced, crisp, herbal, tasty, mineral, spicy.

Un Uva Nocturna Garnacha + 2018 T C
garnacha
88
Balanced, spicy, fruity, herbal, jammy.

Un Uva Nocturna Garnacha Syrah 2019 T
60% garnacha, 40% syrah
87

Un Uva Nocturna Merlot 2019 T C
merlot
85

Un Uva Nocturna Tempranillo Garnacha 2019 T
60% tempranillo, 40% garnacha
87

Vega de Luchán 2021 RD
garnacha
84

VT RIBERA DEL QUEILES

GUELBENZU
Finca La Lombana
50513 Vierlas (Zaragoza)
☎: +34 948 202 200
info@bornosbodegas.com
www.guelbenzu.com

Guelbenzu Azul 2018 T
tempranillo, merlot
89
Spicy, jammy, tasty, toasty.

Guelbenzu Evo 2015 T
cabernet sauvignon, graciano, syrah
89
Wild, jammy, dried herbs, fruity, balanced, spicy.

Guelbenzu Lautus 2010 T
tempranillo, cabernet sauvignon, graciano, merlot
91
Colour: deep cherry. Nose: black fruit, black liquorice, dried herbs, spicy. Palate: fruity, powerful, flavourful, slightly dry, soft tannins.

Guelbenzu Vierlas 2017 T RB
merlot, syrah
87

VT SIERRA NORTE DE SEVILLA

BODEGAS DE FUENTE REINA
Yedra, 2
41450 Constantina (Sevilla)
☎: +34 640 573 891
Fax: +34 955 880 017
oz@bodegasfuentereina.com

Fuente Reina 2018 T
70% merlot, 30% tempranillo
88
Spicy, fruity, herbal, wild.

Fundus Roble 2020 T
tempranillo, merlot, cabernet sauvignon
86

VINO DE LA TIERRA - SIERRA NORTE DE SEVILLA / I.G.P.

Peñín Guide **SPANISH WINE 833**

VINO DE LA TIERRA - VAL DO MIÑO-OURENSE / I.G.P.

Fundus Tinaja 2021 T
merlot, garnacha, mollar
87

Torre Beraun 2015 T
100% merlot
90
Colour: cherry, purple rim. Nose: ripe fruit, spicy, toasty, wild herbs. Palate: fruity, flavourful, good finish.

COLONIAS DE GALEÓN
Polígono industrial Los Manantiales Fase II Nave 16
41370 Cazalla de La Sierra (Sevilla)
☎: +34 638 438 396
m.angeles@coloniasdegaleon.com
www.coloniasdegaleon.com

Cantueso 2020 T
syrah, garnacha, viognier, chardonnay
90 🌿
Colour: cherry, purple rim. Nose: ripe fruit, red berry notes, wild herbs, sweet spices, dried herbs. Palate: fruity, balanced, slightly dry, soft tannins.

Ermita del Monte 2020 T C
tempranillo, syrah, merlot
88 🌿
Spicy, floral, fruity, jammy.

Marrurro 2020 T C
100% cabernet franc
88 🌿
Colour: cherry, purple rim. Nose: toasted bread, black fruit, ripe fruit, wild herbs. Palate: fruity, flavourful, balanced.

Pinchaperas 2020 T C
garnacha, tempranillo, cabernet sauvignon
87 🌿

Silente 2020 B BA
100% viognier
89 🌿
Oxidativ, tasty, herbal, spicy, floral, hot.

Soplagaitas 2020 B
chardonnay, viognier
90 🌿
Oxidativ. Colour: straw, pale. Nose: citrus fruit, white fruit, wild herbs, white flowers. Palate: fleshy, fresh, flavourful, warm.

VT VAL DO MIÑO-OURENSE

REMIGIO ENRIQUEZ
Rua Dos Chaos, 20
32001 Valdo Regueiro (Ourense/Orense)
☎: +34 659 327 453
cotodesantamarta@gmail.com
www.vinosremigioenriquez.com

Remigio Enriquez Godello 2018 B
godello
87

Remigio Enriquez Mencía 2019 T
mencía
86

VT VALDEJALÓN

BODEGAS FRONTONIO
Cº de las Bodegas, 22
50109 Alpartir (Zaragoza)
☎: +34 638 961 395
sales@bodegasfrontonio.com
www.bodegasfrontonio.com

Botijo Blanco 2021 B
100% garnacha blanca
91 🌿
Colour: bright straw. Nose: expressive, white flowers, jasmine, dried herbs. Palate: flavourful, fruity, balanced.

Botijo Rojo 2021 T
100% garnacha
92 🌿
Colour: bright cherry. Nose: balsamic herbs, sweet spices, scrubland, raspberry. Palate: spicy, balsamic, good acidity.

🏆 **PODIUM**

El Jardín de las Iguales 2020 T
garnacha
96
Aromatic. Colour: bright cherry. Nose: complex, expressive, spicy, mineral, red berry notes, faded flowers. Palate: elegant, full-bodied, long, great length.

VT VALLES DE SADACIA

LIBALIS
Ctra. de Murillo, s/n
26500 Calahorra (La Rioja)
☎: +34 941 483 557
marketing@vintae.com
www.vintae.com

Libalis Rosé 2020 RD
garnacha, moscatel grano menudo

87

Libalis White 2020 B
moscatel grano menudo, viura, malvasía

87

TABLE WINES / WINE

Just outside the "Vino de Calidad" status, we find the "Vino de Mesa" ("Table Wine") category, which are those not included in any of the other categories (not even in the "Vino de la Tierra" one, regarded as "Vino de Mesa" by the Ley del Vino ("Wine Law"). The present editions of our Guide has up to 287 table wines rated as excellent, something which is quite telling, and force us to a change of mind in regard to the popular prejudice against this category, traditionally related –almost exclusively– to bulk, cheap wines.

In this section we include wines made in geographical areas that do not belong to any designation of origin (DO as such) or association of Vino de la Tierra, although most of them come indeed from wines regions with some vine growing and winemaking tradition.

We do not pretend to come up with a comprehensive account of the usually overlooked vinos de mesa (table wines), but to enumerate here some Spanish wines that were bottled with no geographic label whatsoever.

The wineries are listed alphabetically within their Autonomous regions. The reader will discover some singular wines of –in some cases– excellent quality that could be of interest to those on the look out for novelties or alternative products to bring onto their tables.

ADEGA DO RICÓN
Vieiro, 12 Mourentan
36437 Arbo (Pontevedra)
☎: +34 670 495 163
adrian@adegadoricon.com
www.adegadoricon.com

Adega do Ricón 2019 T
sousón, mencía, caíño
88
Spicy, jammy, smooth, fruity, wild, simple.

Adega do Ricón 2021 B
albariño
90
Colour: straw. Nose: white flowers, fresh fruit, fragrant herbs, citrus fruit. Palate: fruity, balanced, good acidity.

Adega do Ricón Ancestral 2021 BE
88
Kind finish, fruity, smooth, dried flowers, simple.

Anne do Ricón 2020 B BA
93
Colour: bright yellow. Nose: spicy, faded flowers, white fruit, ripe fruit. Palate: rich, good structure, long, fine bitter notes, balanced, good acidity.

Anne do Ricón Laranxa 2020 B
90
With personality, little interventionist. Colour: golden. Nose: faded flowers, chamomile, white fruit. Palate: correct, bitter.

Anne do Ricón Velo de Flor 2018 B
91
Colour: yellow, pale. Nose: flor yeasts, faded flowers, white fruit, roasted almonds. Palate: balanced, dry, old wood.

ADEGA ENTREOSRIOS
Lugar de Entreosrios, 2
15948 Pobra do Caramiñal (A Coruña/La Coruña)
☎: +34 670 712 700
adega@entreosrios.com
www.adega.entreosrios.com

Komokabras Rose D-mencial 2020 RD S
89
Little interventionist, fruity, tasty, wild, dried flowers.

ADEGA JAVIER ESTÉVEZ ABELEDO
Porriñeira, 14
32431 Beade (Ourense/Orense)
☎: +34 616 121 375
estevezabeledo@gmail.com

Duastempas 2019 T
brancellao, torrontés
88
Balanced, spicy, tasty, toasty, crisp.

ALEMANY I CORRIO
Melió, 78
08720 Vilafranca del Penedès (Barcelona)
☎: +34 938 180 949
sotlefriec@sotlefriec.com
www.alemany-corrio.com

El Microscopi 2019 T
89
Spicy, dried flowers, fruity, sweety finish, tasty.

ALKIMIA WINES
Rambla de Prim, 248 Local 2
08020 Barcelona (Barcelona)
☎: +34 669 486 715
info@alkimiapriorat.com
www.alkimiawines.com

Alkimia 2019 B
garnacha blanca
91 🍷
Colour: bright golden. Nose: stone fruit, wild herbs, faded flowers, expressive. Palate: fruity, flavourful, fresh, balanced, spicy.

Mistik Moscatell 2019 B D
moscatel
90
Colour: straw. Nose: expressive, white flowers, jasmine, dried herbs, dry nuts. Palate: flavourful, fruity, balanced, rich.

ALTA ALELLA
Camí Baix de Tiana s/n
08328 Alella (Barcelona)
☎: +34 934 693 720
info@altaalella.wine
www.altaalella.wine

AA Dolç de Neu 2021 B BA D
pansa blanca
94 🍷
Colour: bright yellow. Nose: ripe fruit, chamomile, floral, honeyed notes. Palate: fruity, fresh, flavourful, powerful, long.

TABLE WINES / WINES

AA Dolç Mataró 2020 T D
mataró

91 🌿

Colour: dark-red cherry. Nose: dried fruit, honeyed notes, faded flowers. Palate: flavourful, balanced.

ALTA ALELLA - CELLER DE LES AUS
Cami Baix de Tiana, s/n
08328 Alella (Barcelona)
☎: +34 934 693 720
info@altaalella.wine
www.altaalella.wine

AUS Merla 2021 T C
100% mataró

87 🌿

ALTANZA
Ctra. Nacional 232, Km. 419,5
26360 Fuenmayor (La Rioja)
☎: +34 618 629 086
marketing@altanza.com
www.altanza.com

La Niña de mis Ojos 2021 B SD
60% verdejo, 40% sauvignon blanc

87

ALTO DE INAZARES
Camino de Majarazan, s/n Finca El Altico
30413 Moratalla (Murcia)
☎: +34 654 782 541
apina@altodeinazares.com
www.altodeinazares.com

Alto de Inazares Chardonnay 2020 B
100% chardonnay

92 🌿

Colour: bright straw. Nose: white fruit, ripe fruit, dried flowers, faded flowers, fine lees. Palate: flavourful, full-bodied, fine bitter notes.

Alto de Inazares Monastrell 2020 T
monastrell

93 🌿

Colour: Cherry. Nose: scrubland, red berry notes, ripe fruit, wild herbs. Palate: spicy, balsamic, good acidity, flavourful, easy to drink.

Alto de Inazares Pinot Noir 2019 T C
pinot noir

91 🌿

Colour: cherry, garnet rim. Nose: characterful, fine reductive notes, spicy, black liquorice. Palate: flavourful, long, good acidity, balanced.

Alto de Inazares Syrah 2020 T
syrah

92 🌿

Colour: deep cherry, bright cherry. Nose: meaty notes, ripe fruit, varietal, powerful. Palate: flavourful, balanced, long.

Alto de Inazares Viognier 2020 B
viognier

90 🌿

Colour: straw. Nose: ripe fruit, dried herbs, dried flowers. Palate: powerful, ripe fruit, balanced.

AMADIS DE YÉBENES
Jardín, 58
45830 Miguel Esteban (Toledo)
☎: +34 925 172 730
info@amadisdeyebenes.com
www.amadisdeyebenes.com

Amadis de Yébenes 2020 T
tempranillo, cabernet sauvignon

86

ÀNIMA NEGRA VITICULTORS
3ª Volta, 18
07200 Faianitx (Illes Balears/Islas Baleares)
☎: +34 971 584 481
admin@annegra.com
www.animanegra.com

Quíbia 2021 B
callet, prensal, giró ros

91

Colour: straw. Nose: expressive, white flowers, jasmine, dried herbs. Palate: flavourful, fruity, balanced.

AT ROCA
La Vinya, 15-17
08770 Sant Sadurní d'Anoia (Barcelona)
☎: +34 935 165 043
info@atroca.eu
www.atroca.eu

Anima Mundi Gres 2021 B
xarel.lo

91 🌿

With personality, little interventionist. Nose: dried herbs, wild herbs, ripe fruit, neat. Palate: fresh, long, good acidity.

Anima Mundi Pells 2021 B
macabeo

91 🌿

Colour: pale. Nose: faded flowers, ripe fruit, powerful, characterful. Palate: flavourful, good acidity, balanced.

Anima Mundi Xarel.lo 2020 B
xarel.lo

92

Taut, wild, full of life. Nose: expressive, white fruit, citrus fruit, floral. Palate: varietal, balanced, good acidity, easy to drink.

ATLAN & ARTISAN
07010 Yecla (Murcia)
☎: +34 968 718 696
info@atlanandartisan.com
www.atlanandartisan.com

Epistem 2018 T
monastrell, syrah, garnacha tintorera

91

Colour: deep cherry. Nose: dried herbs, creamy oak, black fruit, mineral. Palate: powerful, ripe fruit, spicy, round tannins, concentrated.

Epistem Nº 5 2018 T
monastrell

91

Colour: deep cherry. Nose: dried herbs, creamy oak, black fruit. Palate: powerful, ripe fruit, spicy, round tannins.

Five Miles 2017 T
monastrell

90

Colour: deep cherry. Nose: black fruit, scrubland, spicy, earthy notes. Palate: powerful, ripe fruit, round tannins.

ATTIS BODEGAS Y VIÑEDOS
Lg. Morouzos, 16D - Dena
36967 Meaño (Pontevedra)
☎: +34 986 744 790
administracion@attisbyv.com
www.attisbyv.com

Sitta 2021 RD
caiño, pedral, espadeiro

90

Colour: rose. Nose: fresh fruit, red berry notes, faded flowers, mineral. Palate: fresh, fruity, good finish.

Sitta Doliola 2018 B
100% albariño

91

With personality. Nose: ebb tide, faded flowers, complex, characterful, red clay notes. Palate: concentrated, bitter, ripe fruit.

Sitta Finca Molinero 2018 T

90

Rustic, slight oxidation, tasty. Colour: light cherry. Nose: red berry notes, fine reductive notes, scrubland, balsamic herbs. Palate: fine bitter notes, easy to drink.

Sitta Maceración 2021 B
100% albariño

90

With personality, crisp. Colour: straw, amber. Nose: faded flowers, citrus fruit. Palate: dry, fresh, easy to drink, bitter.

Sitta Pereiras 2021 B D
albariño

90

Colour: straw. Nose: floral, citrus fruit, fresh. Palate: sweet, good finish, easy to drink.

AUTÓCTON CELLER
Camí del Cementiri, s/n
43717 La Bisbal del Penedès (Tarragona)
☎: +34 672 432 691
autocton@autoctonceller.com
www.autoctonceller.com

Autócton Blanc 2020 B FB
xarel.lo

90

Colour: bright straw. Nose: fragrant herbs, white fruit, fine lees. Palate: flavourful, easy to drink.

Autócton Blanc 2021 B FB
100% xarel.lo

93

Colour: straw. Nose: white flowers, dried herbs, dry stone, expressive. Palate: flavourful, fruity, balanced, good acidity.

Autócton Negre 2016 T
100% sumoll

91

Colour: light cherry, garnet rim. Nose: red berry notes, ripe fruit, fragrant herbs, dried flowers. Palate: fruity, easy to drink, long, good finish.

Gran Autócton Blanc 2020 B
90% xarel.lo, 10% malvasía de Sitges

92

Slight reduction. Colour: yellow. Nose: faded flowers, burnt matches, dried flowers. Palate: flavourful, spicy, long.

Gran Autócton Blanc 2021 B
90% xarel.lo, 10% malvasía de Sitges

92

Colour: straw. Nose: balanced, medium intensity, dried herbs, hint of anise, floral. Palate: easy to drink, fine bitter notes.

TABLE WINES / WINES

Gran Autócton Negre 2017 T
100% sumoll
93
Colour: Cherry. Nose: complex, expressive, spicy, wild herbs. Palate: elegant, long, great length, easy to drink.

AVELINO VEGAS
Grupo Calvo Sotelo, 8
40460 Santiuste de San Juan Bautista (Segovia)
☎: +34 921 596 002
ana@avelinovegas.com
www.avelinovegas.com

Vegas B
verdejo, sauvignon blanc, viognier
87

Vegas Colección Privada 2020 T
cabernet sauvignon, petit verdot
88
Balanced, spicy, tasty, full-bodied, jammy.

Vegas RD
88
Aromatic, sweeties, lactic, jammy, simple.

AVINYÓ
Masia Can Fontanals
08793 Avinyonet del Penedès (Barcelona)
☎: +34 938 970 055
avinyo@avinyo.com
www.avinyo.com

Avinyó Dolç
de Cabernet 2018 B Mistela D
100% cabernet sauvignon
91
Colour: golden. Nose: honeyed notes, candied fruit, fragrant herbs, dry nuts. Palate: flavourful, sweet, fruity, full-bodied.

La Mirona Blanc 2020 B
malvasía de Sitges, moscatel de frontignan, xarel.lo
88
Balanced, dried flowers, smooth, standard.

La Mirona Rosat 2021 RD
100% merlot
89
Fruity, wild, floral, balanced, dried herbs, tasty.

Petillant Blanc 2021 BE AG
60% moscatel de frontignan, 15% xarel.lo, 25% macabeo
86

BARCO DEL CORNETA
Carreventosa, 7
47491 La Seca (Valladolid)
☎: +34 648 454 958
info@barcodelcorneta.com
www.barcodelcorneta.com

🏆 **PODIUM**

Parajes del Infierno "El Judas" 2019 B FB
100% viura
95
Colour: straw. Nose: dried herbs, stone fruit, cereal notes, fine lees. Palate: powerful, ripe fruit, balanced.

Parajes del Infierno
"La Sillería" 2019 B FB
100% verdejo
94
Colour: bright yellow. Nose: ripe fruit, floral, fine lees, mineral, sweet spices, toasty. Palate: full-bodied, complex, spicy, long, elegant.

Parajes del Infierno
"Las Envidias" 2019 B
100% palomino
94
Taut, little interventionist. Colour: bright straw. Nose: ripe fruit, floral, fine lees, burnt matches, cereal notes. Palate: complex, spicy, long, elegant.

Prapetisco 2019 T
juan garcía
91
Colour: deep cherry. Nose: dried herbs, black fruit, wild herbs, earthy notes. Palate: powerful, ripe fruit, spicy, round tannins.

BARONÍA DEL MONTSANT
Comte de Rius, 1
43360 Cornudella de Montsant (Tarragona)
☎: +34 977 821 483
englora@baronia-m.com
www.baronia-m.com

Clos D'Englora AV 14 2017 T
garnacha, cariñena, syrah
90
Colour: cherry, garnet rim. Nose: overripe fruit, creamy oak, hot. Palate: pruney, powerful, sweet tannins.

Com Gat i Gos Seleccio 2015 T C
garnacha
88
Overripe, jammy, toasty.

Englora 603.8 2018 T C
garnacha, cariñena

90

Colour: cherry, garnet rim. Nose: fruit preserve, fruit liqueur notes, powerful, dark chocolate. Palate: flavourful, sweetness, long.

BERNAVÍ VINYATERS
Finca Mas Vernet
43782 Vilalba dels Arcs (Tarragona)
☎: +34 651 031 835
info@bernavi.com
www.bernavi.com

Bernaví Akrònim 2019 T C
morenillo, montepulciano

89

With personality, dried flowers, dried herbs, jammy, wild, yeasty notes, slight oxidation. Nose: characterful, overripe fruit.

Coster de la Devea 2018 B
garnacha blanca, macabeo

92

Colour: bright yellow. Nose: ripe fruit, spicy, toasted bread, dry stone. Palate: rich, good structure, toasty, fine bitter notes.

BICHO PALO WINE
Barón del Solar, 6
30520 Jumilla (Murcia)
☎: +34 616 983 509
pedromartinez.vsb@gmail.com
www.bichopalowine.com

Bicho Palo 2020 T C
petit verdot

88

Hot, jammy, powerful, pruney, full-bodied.

BODEGA ALISTE
Plaza de España, 4
49520 Figueruela de Abajo (Zamora)
☎: +34 676 986 570
javier@hacedordevino.com
www.vinosdealiste.com

Geijo 2020 B FB
91 🌱

Colour: straw. Nose: ripe fruit, dried herbs, faded flowers, toasty, sweet spices. Palate: powerful, ripe fruit, balanced.

Mar de Aliste 2020 T
89 🌱

Toasty, tasty, powerful, jammy.

Marina de Aliste 2020 T
90% tempranillo, 10% syrah

90

Colour: cherry, garnet rim. Nose: ripe fruit, dried herbs, hot, sweet spices. Palate: powerful, ripe fruit, spicy, round tannins.

BODEGA COOPERATIVA CIGALES
Las Bodegas, s/n
47270 Cigales (Valladolid)
☎: +34 983 580 135
administracion@bodegacigales.com
www.bodegacigales.com

Castillo de Robles T
tempranillo

87

Cosechero B
verdejo

87

Cosechero RD
87

Cosechero T
tempranillo

84

BODEGA DUSSART PEDRÓN
Saliente, 12
46355 Los Pedrones (València/Valencia)
☎: +34 722 270 944
bodegadussartpedron@gmail.com
www.bodegadussartpedron.com

Le Vermentino 2020 B
vermentino

91

Oxidativ. Colour: bright straw. Nose: citrus fruit, wild herbs, ripe fruit, dried flowers. Palate: fresh, fruity, good acidity, fine bitter notes, flavourful.

BODEGA KIENINGER
Paraje Rural los Frontones, 67
29400 Ronda (Málaga)
☎: +34 630 161 156
martin@bodegakieninger.com
www.bodegakieninger.com

Rosara 2021 RD
100% blaufraenkisch

88 🌱

Fruity, herbal, simple, wild.

TABLE WINES / WINES

BODEGA LA DOLORES
Juan José Lorente, 19 bajo
50005 Zaragoza (Zaragoza)
☎: +34 609 251 412
pregunta@bodegaladolores.com
www.bodegaladolores.com

La Dolores Afrutado 2021 B
macabeo, sauvignon blanc
87

BODEGA LA ENCINA
Pedro Más, 23
03408 La Encina (Alacant/Alicante)
☎: +34 610 410 945
Fax: +34 962 387 808
vinosnaturaleslaencina@gmail.com
www.bodegalaencina.com

Forcallat by Bodega La Encina 2021 T
forcallat
87 🌿

Garnacha Rosé by Bodega La Encina 2021 RD
1% garnacha
84 🌿

Monastrell by Bodega La Encina 2021 T
monastrell
82 🌿

Tierra de Forcallat 2021 B
forcallat
84 🌿

Vino Viejo del Abuelo Monastrell 2012 T RC
monastrell
92 🌿
Colour: light cherry, bright ochre rim. Nose: fruit preserve, dry nuts, sweet spices, aged wood nuances. Palate: good structure, powerful, flavourful, fleshy.

Zero Garnacha 2021 T
garnacha
84 🌿

BODEGA MADRID ROMERO
Ctra. de El Carche, Km 8,3.
30520 Jumilla (Murcia)
☎: +34 611 043 352
info@bodegamadridromero.com
www.bodegamadridromero.com

Madrid Romero Blanco BF Mistela D
moscatel
90
Colour: bright yellow. Nose: white fruit, fruit liqueur notes, balanced, neat, varietal, floral. Palate: sweet, correct.

Madrid Romero Tinto TF D
monastrell
89
Age nuances, weary, pruney, tasty.

BODEGA MALVAJIO
Camino de Campanales, s/n Nave 9
29649 Mijas (Málaga)
☎: +34 622 017 228
bodegamalvajio@outlook.com
www.bodegamalvajio.com

Flor de Malvajio 2020 T
syrah, merlot
90
Colour: bright cherry. Nose: red berry notes, black fruit, scrubland, mineral. Palate: good acidity, spicy, fine tannins, ripe fruit.

Malaje 2021 RD
90
Colour: coppery red, bright. Nose: faded flowers, dried flowers, ripe fruit. Palate: rich, flavourful, long.

Malapipa 2021 B
moscatel de alejandría
88
Little interventionist, slight oxidation, balanced. Nose: white fruit, ripe fruit, cereal notes.

Malavia 2021 B
89
Fruity, jammy, floral, tasty, dried flowers.

Malvajio Vino de Garaje 2018 T
syrah
90
Pruney. Colour: deep cherry. Nose: ripe fruit, dried herbs, creamy oak, black fruit, meaty notes. Palate: ripe fruit, spicy, round tannins, concentrated.

BODEGA MUSTIGUILLO

El Terrazo Ctra. N-330 km. 195
46300 Utiel (València/Valencia)
☎: +34 962 168 260
info@bodegamustiguillo.com
www.bodegamustiguillo.com

Finca Calvestra Margas 2016 B
100% merseguera

93 🍷

Colour: yellow, pale. Nose: expressive, floral, fine lees, mineral, white fruit, ripe fruit. Palate: full-bodied, complex, spicy, long, elegant, flavourful.

Finca Calvestra Merseguera 2020 B
merseguera

93 🍷

Colour: bright straw. Nose: medium intensity, white fruit, dried flowers, balanced. Palate: easy to drink, good acidity, balanced, fine bitter notes, spicy, elegant.

La Garnacha de Mustiguillo 2020 T
garnacha

93 🍷

Colour: bright cherry. Nose: scrubland, fresh fruit, red berry notes, wild herbs. Palate: spicy, balsamic, good acidity, fine tannins, easy to drink.

Mestizaje 2020 B
merseguera, malvasía, xarel.lo, viognier

91 🍷

Standard, very fruit-driven, defined aromas. Colour: straw. Nose: white flowers, white fruit, balanced. Palate: fruity, easy to drink.

Pela Roques de Mustiguillo 2020 T
syrah

91 🍷

Rustic, with personality. Colour: bright cherry. Nose: ripe fruit, red berry notes, scrubland, balsamic herbs, burnt matches, fine reductive notes. Palate: easy to drink, flavourful, grainy tannins.

BODEGA NAVE ROVER

Pol. 24, Par 416. Camino de Castellitx
07210 Algaida (Illes Balears/Islas Baleares)
☎: 610 468 132
admin@naverover.com
www.naverover.com

Rover Nº 1 2019 T
75% syrah, 25% mantonegro

92 🍷

Full-bodied. Colour: black cherry. Nose: black fruit, fruit preserve, toasty, creamy oak, tobacco, dried herbs, wet leather. Palate: good structure, flavourful.

BODEGA PARDO TOLOSA

Villatoya, 26
02215 Alborea (Albacete)
☎: +34 963 517 067
Fax: +34 961 668 213
ventas@bodegapardotolosa.com
www.bodegapardotolosa.com

Señorío de Pardo 2020 B
macabeo, moscatel

85

Señorío de Pardo 2020 T
tempranillo

85

BODEGA PRIVILEGIO DEL CONDADO

San José, 2
21710 Bollullos Par del Condado (Huelva)
☎: +34 959 410 261
comercial@vinicoladelcondado.com
www.vinicoladelcondado.com

Carámbano Ice Wine B D

92

Colour: bright yellow. Nose: ripe fruit, candied fruit, honeyed notes, citrus fruit, dried flowers. Palate: flavourful, unctuous, fruity, sweet.

BODEGA ROMAILA

45420 Almoncid de Toledo (Toledo)
☎: +34 915 416 561
administracion@romaila.es
www.romaila.es

Romaila 2021 RD

88 🍷

Fruity, jammy, sweeties, balanced, pleasant, smooth.

BODEGA SIESTO

Calle La Presa, 40
49152 Sanzoles (Zamora)
☎: +34 657 689 542
vsiesto@hotmail.com
www.bodegasiesto.com

Siesto 2018 T C
50% bruñal, 50% tinta de Toro (Tinto)

90

Colour: cherry, garnet rim. Nose: fruit preserve, fruit liqueur notes, powerful, dark chocolate, aromatic coffee. Palate: flavourful, sweetness, long.

TABLE WINES / WINES

BODEGA SIGUÍN

Hilero 7 Bj 2
28696 Pelayos de la Presa (Madrid)
☎: +34 629 816 225
info@bodegasiguin.com
www.bodegasiguin.es

Flytrap 2018 T D
100% garnacha

91

Colour: dark-red cherry. Nose: overripe fruit, pattiserie, fruit liqueur notes. Palate: flavourful, sweet, balanced.

BODEGA SOMMOS GARNACHA

Ctra. Murero – Atea s/n
50366 Murero (Zaragoza)
☎: +34 976 174 740
info@bodegasommosgarnacha.com
www.bodegasommosgarnacha.com

Alquéz de Sommos 2019 T
garnacha

90

Colour: bright cherry. Nose: sweet spices, ripe fruit, balanced, wild herbs, toasty. Palate: spicy, round tannins.

Lamin de Sommos 2019 T
garnacha

92

Colour: Cherry. Nose: expressive, spicy, cocoa bean. Palate: full-bodied, long, great length, round tannins, balanced.

BODEGA TINAHA

DS El Reguero, 17-18
30550 Abarán (Murcia)
☎: +34 639 414 174
contacto@bodegatinaha.com
www.bodegatinaha.com

Lorenzo Boris Tinaha 2020 T
95% monastrell, 5% forcallat, tardana, airén, rojal

90

Colour: deep cherry. Nose: ripe fruit, dried herbs, creamy oak. Palate: powerful, ripe fruit, spicy, round tannins.

BODEGA Y VIÑEDOS 13 VIÑAS

Campo del Obispo, 13
24492 Cubillos del Sil (León)
Fax: +34 649 312 360
trecevinas@gmail.com
www.13viñas.com

A Ponte Vella 2019 T RB
80% mencía, 20% jerez, godello

89

Spicy, dried herbs, pruney, tasty, rustic, coarse tannins. Nose: waxy notes.

Alto de San Esteban 2019 T C

88

Jammy, pruney, dried herbs, coarse tannins, tasty.

BODEGAS 7LINDES

Calle Los Bonillas, 4
16270 Villalpardo (Cuenca)
☎: +34 606 514 936
bodegas7lindes@gmail.com

Clemencia 2019 T
bobal, rojal

91

Colour: cherry, garnet rim. Nose: fruit preserve, powerful, lactic notes, dried herbs, wild herbs. Palate: flavourful, long, easy to drink.

BODEGAS 9C+

46167 Chulilla (València/Valencia)
☎: +34 628 883 973
l.entiscar@hotmail.com

Purita 2020 T
100% merlot

90

Colour: deep cherry. Nose: ripe fruit, dried herbs, creamy oak. Palate: ripe fruit, spicy, round tannins.

Purita 2020 T BA
merlot

91

Colour: deep cherry. Nose: creamy oak, black fruit, red berry notes, scrubland. Palate: powerful, ripe fruit, spicy, round tannins.

Purita Blanc de Noirs 2021 B
16% mersegera, 8% garnacha, 76% merlot

86

BODEGAS ALCONDE

Ctra. de Calahorra, s/n
31260 Lerin (Navarra)
☎: +34 948 530 058
Fax: +34 948 530 589
manuel@bodegasalconde.com
www.bodegasalconde.com

Alconde Óptimo 2018 T C S
cabernet sauvignon, merlot

89

Smoky, jammy, herbal, powerful, tasty.

Alconde Óptimo 2021 RD SS
merlot

85

Pink Pearl 2021 RD PL S
garnacha

86

BODEGAS ARRÁEZ

Pol. 6 Parcela 386 Paraje Ciscar
46630 La Font de la Figuera (València/Valencia)
☎: +34 962 290 031
info@bodegasarraez.com
www.bodegasarraez.com

Los Arráez Arcos 2020 T
arco

90

Smoky, toasty. Colour: cherry, garnet rim. Nose: characterful, smoky, incense. Palate: flavourful, ripe fruit.

Los Arráez Arcos 2021 T
100% arco

91

Colour: cherry, purple rim. Nose: red berry notes, floral, spicy, toasty. Palate: flavourful, fruity, good acidity.

Los Arráez Syrah 2020 T
100% syrah

90

Colour: cherry, garnet rim. Nose: ripe fruit, dried herbs, faded flowers, varietal. Palate: powerful, round tannins, flavourful.

BODEGAS ARZUAGA NAVARRO

Ctra. Nac. 122, Km. 325
47350 Quintanilla de Onésimo (Valladolid)
☎: +34 983 681 146
bodeg@arzuaganavarro.com
www.arzuaganavarro.com

Fan D.Oro 2020 B FB S
100% chardonnay

90

Colour: bright yellow. Nose: powerful, creamy oak, ripe fruit, spicy. Palate: rich, long, toasty.

BODEGAS BIGARDO

Ctra, Toro Alaejos, km. 3.3 N-602
49800 Toro (Zamora)
☎: +34 651 999 917
vinobigardo@gmail.com
www.bigardo.es

Bigarda 2021 RD

89

Dried flowers, fruity, sweety finish, jammy, tasty.

Bigardo 2020 T
100% tinta de Toro (Tinto)

87

Maldito Parné 2020 T C
100% tinta de Toro (Tinto)

91

Colour: Cherry. Nose: balsamic herbs, sweet spices, scrubland, ripe fruit. Palate: spicy, round tannins, flavourful.

Pellejo 2020 T
100% tinta de Toro (Tinto)

90

Powerful, dried herbs, full-bodied. Nose: fruit liqueur notes, overripe fruit. Palate: good structure, flavourful, balanced, round tannins, great length.

Satélite Boarding Wine 2020 T
100% tinta de Toro (Tinto)

90

Colour: cherry, garnet rim. Nose: powerful, sweet spices, black fruit, fruit preserve. Palate: flavourful, long, powerful, slightly dry, soft tannins.

BODEGAS BORSAO

50540 Borja (Zaragoza)
☎: +34 976 867 116
contacto@bodegasborsao.com
www.bodegasborsao.com

Viña Borgia by Borsao 2020 T
garnacha

87 🌱

BODEGAS BRECA

Ctra. Monasterio de Piedra, s/n
50219 Munébrega (Zaragoza)
☎: +34 952 504 706
pedidos@jorgeordonez.es
www.jorgeordonez.es

Breca El Nacido 2021 T
garnacha

92

Colour: deep cherry. Nose: ripe fruit, dried herbs, creamy oak, sweet spices. Palate: powerful, ripe fruit, spicy, round tannins.

Breca Rosé 2021 RD
100% garnacha

90

Colour: raspberry rose. Nose: floral, fragrant herbs, ripe fruit, red berry notes. Palate: spicy, good acidity, fine bitter notes.

Garnacha de Fuego 2021 T
100% garnacha

90

Colour: cherry, purple rim. Nose: balsamic herbs, red berry notes, scrubland. Palate: balsamic, fruity, grainy tannins.

TABLE WINES / WINES

Breca 2019 T FB
100% garnacha

92

Colour: bright cherry. Nose: balsamic herbs, sweet spices, scrubland, black fruit. Palate: spicy, balsamic, good acidity, fruity.

BODEGAS CABALLERO
San Francisco, 32
11500 El Puerto de Santa María (Cádiz)
☎: +34 956 751 851
marketing1@caballero.es
www.caballero.es

Abulaga B SS
100% moscatel de alejandría

86

BODEGAS CAMPESTRAL
Ctra. Arcos-Algar, km. 7
11630 Arcos de la Frontera (Cádiz)
☎: +34 670 586 035
info@campestral.es
www.campestral.es

Campestral Clarete 2020 RD
palomino, otras

91

Colour: salmon. Nose: ripe fruit, hot, faded flowers, spicy. Palate: fleshy, flavourful, ripe fruit.

Campestral Red 2020 T
syrah, merlot, cabernet sauvignon, tintilla de rota, petit verdot

85

Campestral Red 2020 T RB
syrah, merlot, cabernet sauvignon, tintilla de rota, petit verdot

84

Campestral White envejecido bajo Velo 2020 B
palomino

91

Little interventionist, original. Colour: amber. Nose: ripe fruit, dried herbs, faded flowers, flor yeasts. Palate: ripe fruit, balanced, fleshy, bitter.

BODEGAS CAN VIDALET
Ctra. Alcudia - Pollença Ma 2201, Km. 4,85
07460 Pollença (Illes Balears/Islas Baleares)
☎: +34 971 531 719
Fax: +34 971 535 395
info@canvidalet.com
www.canvidalet.com

Barros de Cecili Submarino 2019 B
100% prensal

93

Colour: straw. Nose: ripe fruit, dried herbs, faded flowers, ebb tide, dry stone. Palate: ripe fruit, balanced, flavourful, great length.

Port de Cecili Blanc BF D
moscatel, prensal

93

Colour: amber. Nose: acetaldehyde, candied fruit, dry nuts, sweet spices, caramel. Palate: flavourful, sweet, fleshy, full-bodied.

Port de Cecili Blanc BF S
moscatel, prensal

93

Colour: amber. Nose: macerated fruit, dry nuts, sweet spices, scrubland, faded flowers. Palate: fleshy, powerful, flavourful, warm.

Port de Cecili T D
cabernet sauvignon, merlot

93

Colour: light cherry, brick rim edge. Nose: dried fruit, dry nuts, complex, sweet spices, creamy oak, varnish. Palate: rich, flavourful, old wood, warm, great length.

BODEGAS CERRO DEL ÁGUILA
Avda. de Ciudad Real, 18
45127 Las Ventas con Peña Aguilera (Toledo)
bodegascerrodelaguila@gmail.com

Puerto Carbonero Vino de Parcela 2019 T S
100% garnacha

87

BODEGAS COVIÑAS

Avda. Rafael Duyos, s/n
46340 Requena (València/Valencia)
☎: +34 962 300 680
covinas@covinas.com
www.covinas.com

Aula Verdejo 2021 B
verdejo

87

Voramar 2021 B S
chardonnay

86

Voramar 2021 RD
garnacha

86

Voramar 2021 T
tempranillo

88

Kind finish, sweety finish, fruity, jammy.

BODEGAS FRONTONIO

Cº de las Bodegas, 22
50109 Alpartir (Zaragoza)
☎: +34 638 961 395
sales@bodegasfrontonio.com
www.bodegasfrontonio.com

El Jardín de las Iguales 2019 B
macabeo

94

Colour: bright straw. Nose: ripe fruit, fragrant herbs, fine lees, black pepper, toasty, bakery. Palate: full-bodied, rich, long, good acidity.

Frontonio La Cerqueta 2020 T
100% garnacha

93

Colour: bright cherry. Nose: balsamic herbs, sweet spices, scrubland, fruit expression, citrus fruit. Palate: spicy, balsamic, good acidity, fruity, grainy tannins.

Frontonio La Loma y Los Santos 2020 B
garnacha blanca, macabeo

94

Colour: bright yellow. Nose: ripe fruit, fragrant herbs, fine lees, burnt matches, chalk. Palate: full-bodied, rich, long, good acidity.

Frontonio La Tejera 2020 T FB
garnacha, macabeo

94

Colour: bright cherry. Nose: complex, expressive, spicy, mineral, red berry notes. Palate: elegant, full-bodied, long, great length.

Frontonio Telescópico 2020 T
garnacha, garnacha peluda, mazuelo

93

Colour: deep cherry. Nose: ripe fruit, dried herbs, creamy oak, red berry notes, floral. Palate: powerful, ripe fruit, spicy, round tannins.

BODEGAS GRATIAS. FAMILIA Y VIÑEDOS

Calle A, 12
02200 Casas Ibáñez (Albacete)
☎: +34 646 700 966
silvia@bodegasgratias.com
www.bodegasgratias.com

¿Y tú de quién eres? 2020 T

90 🌱

Colour: Cherry. Nose: balsamic herbs, sweet spices, scrubland. Palate: spicy, balsamic, good acidity.

¿Y tú de quién eres? 2021 B

86 🌱

Gratias Got 2020 T
bobal

88 🌱

Jammy, tasty, balsamic herbs, aromatic.

Gratias Máximas 2018 T
bobal

88 🌱

Toasty, tasty, jammy.

Gratias Rosé 2021 RD
bobal

87 🌱

Gratias Sol 2020 B S
tardana

87 🌱

BODEGAS GUTIÉRREZ DE LA VEGA

03792 Parcent (Alacant/Alicante)
☎: +34 966 403 871
info@bodegasgutierrezdelavega.es
www.bodegasgutierrezdelavega.es

Casta Diva Cosecha Dorada 2021 B
moscatel

91

Aromatic. Colour: yellow. Nose: white flowers, white fruit, ripe fruit, varietal. Palate: flavourful, fruity, balanced, rich.

Casta Diva Cosecha Miel Dulce 2019 B D

94

Colour: bright yellow. Nose: balsamic herbs, honeyed notes, floral, sweet spices, expressive. Palate: rich, fruity, powerful, flavourful, elegant.

🏆 PODIUM

Casta Diva Esencia Dulce 2018 B D
96
Oxidativ. Colour: golden. Nose: powerful, honeyed notes, candied fruit, fragrant herbs, acetaldehyde, complex, sweet spices. Palate: flavourful, sweet, fruity, good acidity, long.

Casta Diva Monte Diva 2021 B
moscatel
94
With personality, floral, balsamic herbs. Colour: bright golden. Nose: honeyed notes, scrubland, fine lees, fruit expression. Palate: balanced, easy to drink, fruity.

🏆 PODIUM

Casta Diva Reserva Real Dulce 2002 B R D
moscatel de alejandría
96
Colour: mahogany. Nose: caramel, overripe fruit, cocoa bean, aromatic coffee, dry nuts. Palate: sweetness, complex, flavourful, powerful.

Casta Diva Selección Especial 2021 B
merseguera, moscatel
91
Colour: bright straw. Nose: fruit expression, ripe fruit, dried flowers. Palate: flavourful, good acidity, fruity aftertaste, balanced.

Mr. Tambourine Wine 2018 T
giró, merlot
90
Defined aromas. Colour: light cherry. Nose: wild herbs, red berry notes, neat, expressive. Palate: easy to drink, slightly dry, soft tannins.

🏆 PODIUM

Recóndita Armonía 2010 2010 T Solera D
98
Colour: cherry, garnet rim. Nose: overripe fruit, dried fruit, sweet spices, toasty, tar, dark chocolate. Palate: ripe fruit, warm, powerful, good acidity, round.

Recóndita Armonía Dulce 2019 T D
92
Slight oxidation. Colour: bright cherry. Nose: candied fruit, balanced, dried flowers. Palate: flavourful, balanced, correct.

Tío Raimundo 2017 B
moscatel
93
Defined aromas, original, full of life, smooth. Colour: bright yellow. Nose: floral, flor yeasts, dry nuts, balsamic herbs. Palate: varietal, balanced, easy to drink.

V. Alejandría 2021 RD
moscatel de alejandría
90
Standard, slight oxidation, with personality. Colour: coppery red. Nose: expressive, wild herbs, floral, neat. Palate: flavourful, varietal, easy to drink, fine bitter notes.

BODEGAS HIDALGO-LA GITANA
Banda de la Playa, 42
11540 Sanlúcar de Barrameda (Cádiz)
☎: +34 956 385 304
bodegashidalgo@lagitana.es
www.lagitana.es

Las 30 del Cuadrado 2020 B FB
100% palomino
91
Colour: straw. Nose: ripe fruit, dry nuts, dried flowers, faded flowers, yeasty notes. Palate: flavourful, fine bitter notes.

BODEGAS LA CASA DE LÚCULO
Ctra. Larraga, s/n
31150 Mendigorria (Navarra)
☎: +34 948 343 148
bodega@luculo.es
www.luculo.es

Chloss 2021 T
88
Defined aromas, yeasty notes, fruity, bitter, with personality, little interventionist.

BODEGAS LATÚE
Camino de la Esperilla, s/n
45810 Villanueva de Alcardete (Toledo)
☎: +34 925 166 350
web@latue.com
www.latue.com

By Latúe Tempranillo 2021 T
tempranillo
86 🌱

La Casa de Bio 2021 B
sauvignon blanc
87 🌱

Latúe Método tradicional BE BR
airén
87 🌱

Oye 2021 RD
tempranillo
87

Really Awesome Wine 2021 T
cabernet sauvignon, tempranillo, merlot
86

Wine by Nature 2021 B
airén, sauvignon blanc
85

BODEGAS LOCUAZ
Avda. Reyes Católicos, 88
02600 Villarrobledo (Albacete)
☎: +34 670 304 827
bodegaslocuaz@gmail.com
www.bodegaslocuaz.com

Locuaz Viura Tinaja 2021 B
viura
88
Floral, oxidativ, little interventionist, jammy.

BODEGAS LOZANO
Avda. Reyes Católicos, 156
02600 Villarrobledo (Albacete)
☎: +34 967 141 907
info@bodegas-lozano.com
www.bodegas-lozano.com

Marqués de Toledo Verdejo 2021 B
verdejo
87

BODEGAS LUZÓN
Ctra. Jumilla-Calasparra, Km. 3,1
30520 Jumilla (Murcia)
☎: +34 968 784 135
info@bodegasluzon.com
www.bodegasluzon.com

Mindoro 2020 B FB
viognier
88
Toasty, powerful, jammy, spicy.

BODEGAS MÁLAGA VIRGEN
Autovía A-92, Km. 132
29520 Fuente de Piedra (Málaga)
☎: +34 952 319 454
info@bodegasmalagavirgen.com
www.bodegasmalagavirgen.com

Barón de Rivero Chardonnay 2021 B
chardonnay
85

Barón de Rivero Semidulce 2021 B SD
pedro ximénez
86

Barón de Rivero Syrah 2021 T
syrah
85

Barón de Rivero Verdejo 2021 B
verdejo
86

BODEGAS MARQUÉS DE VIZHOJA
Finca La Moreira s/n
36438 Arbo (Pontevedra)
☎: +34 986 665 825
marquesdevizhoja@marquesdevizhoja.com
www.bodegasmarquesdevizhoja.com

Marqués de Vizhoja 2021 B
88
Kind finish, aromatic, floral, fruity, smooth, simple.

BODEGAS MENADE
Ctra. Rueda - Nava del Rey, km. 1
47490 Rueda (Valladolid)
☎: +34 983 103 223
info@menade.es
www.menade.es

Adorado by Menade Crianza de 1967 B Solera
verdejo, palomino
94
Colour: old gold. Nose: sweet spices, acetaldyhyde, dry nuts, flor yeasts. Palate: full-bodied, dry, spicy, long, fine bitter notes, complex.

TABLE WINES / WINES

BODEGAS MUÑANA
Ctra. Graena a La Peza, s/n Finca Peñas Prietas
18517 Cortes Y Graena (Granada)
☎: +34 958 670 715
administracion@bodegasmunana.com
www.bodegasmunana.com

Muñana 2019 B
chardonnay, sauvignon blanc, moscatel

86

BODEGAS MURVIEDRO
Ampliación Pol. El Romeral, s/n
46340 Requena (València/Valencia)
☎: +34 962 329 003
murviedro@murviedro.es
www.murviedro.es

Audentia Petit Verdot 2021 T
100% petit verdot

87

Audentia Sauvignon Blanc 2021 B
100% sauvignon blanc

87

Sericis Cepas Viejas Viognier 2021 B FB
viognier

89 🌿

Aromatic, floral, jammy, balanced, standard, pleasant.

BODEGAS ONEROM
Colombia s/n, P.I La Serreta
30500 Molina de Segura (Murcia)
☎: +34 968 237 124
info@bodegasonerom.com
www.bodegasonerom.com

Onerom Monastrell 2021 T
monastrell

90

Colour: cherry, purple rim. Nose: black fruit, spicy, smoky, dried herbs. Palate: full-bodied, flavourful, fruity, good structure, slightly dry, soft tannins.

Onerom Verdejo 2021 B
verdejo

90

Colour: bright straw, greenish rim. Nose: fresh fruit, citrus fruit, wild herbs, spicy. Palate: fresh, fruity, good acidity, fine bitter notes.

Pinadel T
monastrell, syrah, cabernet sauvignon

89

Fruity, jammy, spicy, smoky, tasty.

BODEGAS OTERO
Avda. El Ferial, 22
49600 Benavente (Zamora)
☎: +34 980 631 600
info@bodegasotero.es
www.bodegasotero.es

El Cubeto Reserva Familia Otero B
verdejo, otras

93

Colour: iodine. Nose: flor yeasts, cereal notes, ripe fruit, smoky, dry nuts. Palate: flavourful, powerful, balanced, saline, fresh.

BODEGAS PÁEZ MORILLA
Avda. Medina Sidonia, 20
11406 Jerez de la Frontera (Cádiz)
☎: +34 956 181 717
Fax: +34 956 181 534
bodegas@paezmorilla.com
www.paezmorilla.com

Tierra Blanca 2021 B
85

Tierra Blanca Semidulce 2021 B SD
84

BODEGAS PERE SEDA
Cid Campeador, 22
07500 Manacor (Illes Balears/Islas Baleares)
☎: +34 971 605 087
Fax: +34 971 844 934
lucasreus@telefonica.net
www.pereseda.com

Pere Seda 2018 BE BN
95% parellada, 5% chardonnay

85

BODEGAS PUIGGRÒS
Ctra. de Manresa, Km. 13
08711 Òdena (Barcelona)
☎: +34 629 853 587
info@bodegaspuiggros.com
www.bodegaspuiggros.com

Impressionant 2021 B
garnacha blanca

92

With personality, little interventionist. Colour: bright golden, hazy. Nose: dried flowers, candied fruit, fine lees, pattiserie. Palate: spicy, long, great length.

BODEGAS ROSALÍO ALONSO & CO
Bosque, 21
45350 Noblejas (Toledo)
☎: +34 677 460 225
ralopec@ciccp.es

Casa de Isaac 2019 T
100% merlot

91

Full-bodied. Colour: cherry, garnet rim. Nose: fruit preserve, powerful, dark chocolate, black fruit. Palate: flavourful, long, sweet tannins.

Va por Ustedes 2019 T
tempranillo, syrah, merlot

90

Colour: deep cherry. Nose: ripe fruit, dried herbs. Palate: ripe fruit, spicy, round tannins, long.

BODEGAS SANI PRIMAVERA
La Zarza, s/n
06200 Almendralejo (Badajoz)
☎: +34 924 677 917
tienda@bodegassani.com
www.bodegasani.com

Flamenco de Sani Árabe 2021 B AG S
sauvignon blanc

85

BODEGAS SILVANO GARCÍA
Avda. de Murcia, 29
30520 Jumilla (Murcia)
☎: +34 968 780 767
bodegas@silvanogarcia.es
www.silvanogarcia.es

Silvano García Blue 2021 B
moscatel

87

BODEGAS TAGONIUS
Ctra. de Tielmes a Carabaña Km 4,4
28550 Tielmes (Madrid)
☎: +34 918 737 505
Fax: +34 918 746 161
info@tagonius.com
www.bodegastagonius.com

Tagonius T
cabernet sauvignon, syrah, merlot, tempranillo

86

BODEGAS TIERRA SAVIA
41370 Cazalla de La Sierra (Sevilla)
☎: +34 623 068 615
bodegastierrasavia@gmail.com

Piu Ánfora 2019 B
parrona

89 ♣

Citrus fruit, dried flowers, jammy, little interventionist, wild, tasty.

Piu Ánfora 2019 T FB
garnacha

90 ♣

Colour: light cherry. Nose: fruit preserve, fruit liqueur notes, powerful. Palate: light-bodied, easy to drink.

Tierra Savia Mirlo 2019 B
viognier

89 ♣

Citrus fruit, floral, jammy, smooth, dried herbs.

Tierra Savia Zaranda El Viejo Francés T
100% tempranillo

91 ♣

Colour: deep cherry. Nose: creamy oak, black fruit, ripe fruit, wild herbs. Palate: powerful, ripe fruit, spicy, round tannins.

BODEGAS TOTÓ MARQUÉS
Avda. Verge de l'Abellera, 18
43364 Prades (Tarragona)
☎: +34 678 184 754
guillermo@totomarques.com
www.totomarques.com

Prades 950 2016 T
garnacha, syrah, merlot, pinot noir

91

Colour: cherry, garnet rim. Nose: balsamic herbs, ripe fruit, scrubland, tobacco, fine reductive notes, earthy notes. Palate: flavourful, balsamic, spicy.

BODEGAS TRENZA

Felix Mendelsohn, 8
03730 Jávea (Alacant/Alicante)
☎: +34 965 790 012
bodegas@bodegastrenza.com
www.bodegatrenza.com

La Orphica Monastrell Iluminada 2020 T BA SS
91
Colour: dark-red cherry. Nose: toasty, spicy, cocoa bean, black fruit, fruit preserve. Palate: flavourful, toasty, fine bitter notes.

La Orphica Monastrell Selección Tardia 2021 T SS
monastrell
90
Colour: cherry, garnet rim. Nose: overripe fruit, creamy oak, hot. Palate: pruney, powerful, sweet tannins.

La Orphica Selección Aurora 2021 B SS
87

La Orphica Sintonía 2021 RD
86

BODEGAS Y VIÑEDOS ARTADI

Ctra. de Logroño, s/n
01300 Laguardia (Araba/Álava)
☎: +34 945 600 119
comunicacion@artadi.com
www.artadi.com

Artadi Canales 2020 T
100% tempranillo
94
Colour: dark-red cherry. Nose: toasty, spicy, cocoa bean, ripe fruit, black fruit. Palate: flavourful, toasty, fine bitter notes.

🏆 **PODIUM**

Artadi El Carretil 2020 T
100% tempranillo
98
Taut, complex. Colour: bright cherry. Nose: complex, expressive, spicy, mineral, fruit expression, chalk. Palate: elegant, full-bodied, long, great length.

Artadi La Hoya 2020 T
100% tempranillo
94
Colour: very deep cherry. Nose: balsamic herbs, sweet spices, scrubland, fruit expression, red berry notes. Palate: spicy, balsamic, good acidity, full of life, mineral.

🏆 **PODIUM**

Artadi La Poza de Ballesteros 2020 T
100% tempranillo
96
Colour: very deep cherry. Nose: complex, expressive, spicy, mineral, ripe fruit, black fruit. Palate: elegant, full-bodied, long, great length.

🏆 **PODIUM**

Artadi Majadales 2020 T
100% tempranillo
95
Colour: bright cherry. Nose: complex, expressive, spicy, mineral, red berry notes, ripe fruit. Palate: full-bodied, long, great length, good acidity.

🏆 **PODIUM**

Artadi Quintanilla 2020 T
100% tempranillo
96
Colour: cherry, purple rim. Nose: fruit expression, red berry notes, floral, spicy, mineral, creamy oak. Palate: flavourful, fruity, good acidity, long.

🏆 PODIUM

Artadi San Lázaro 2020 T
100% tempranillo

96

Tasty, powerful. Colour: deep cherry. Nose: ripe fruit, dried herbs, creamy oak, black fruit, earthy notes, truffle notes. Palate: powerful, ripe fruit, spicy, round tannins.

Artadi Terreras 2020 T
100% tempranillo

94

Colour: dark-red cherry. Nose: toasty, spicy, cocoa bean, creamy oak, ripe fruit, black fruit. Palate: flavourful, toasty, grainy tannins.

🏆 PODIUM

Artadi Valdeginés 2020 T
100% tempranillo

95

Taut, fruity. Colour: deep cherry. Nose: dried herbs, creamy oak, powerful, ripe fruit, red berry notes. Palate: powerful, ripe fruit, spicy, round tannins.

Artadi Viñas de Gain 2018 B
100% viura

93

Colour: bright yellow. Nose: ripe fruit, dried herbs, faded flowers, spicy. Palate: powerful, ripe fruit, balanced.

Artadi Viñas de Gain 2019 T
100% tempranillo

93

Colour: deep cherry. Nose: dried herbs, creamy oak, ripe fruit, black fruit. Palate: powerful, ripe fruit, spicy, round tannins.

Artadi Viñas de Gain 2020 T
tempranillo

93

Taut. Colour: deep cherry. Nose: complex, expressive, spicy, mineral. Palate: long, great length, grainy tannins.

BODEGAS Y VIÑEDOS CASTIBLANQUE

Isaac Peral, 19
13610 Campo de Criptana (Ciudad Real)
☎: +34 926 589 147
administracion@bodegascastiblanque.com
www.bodegascastiblanque.com

Baldor Chardonnay 2020 B FB
chardonnay

87

Señorío de Mareste T
tempranillo

82

Silicon Red T
tempranillo

84

Solamente B
100% airén

83

Solamente T
tempranillo, syrah, cabernet sauvignon

84

BODEGAS Y VIÑEDOS PRADOREY

Ctra. CL-619 (Magaz – Aranda) km.66, 1
09443 Gumiel de Mercado (Burgos)
☎: +34 947 546 900
info@pradorey.com
www.pradorey.com

El Cuentista 2019 B
100% tempranillo

91

Colour: straw. Nose: dried herbs, faded flowers, lactic notes, bakery, white fruit. Palate: powerful, ripe fruit, balanced, flavourful.

BODEGAS ZIRÍES

Menasalbas, 18
45120 San Pablo de los Montes (Toledo)
☎: +34 639 502 147
javier@ziries.es
www.ziries.es

Con Viento Fresco 2018 T
garnacha

88

Slight oxidation, balsamic herbs, spicy, dried flowers, wild, tasty.

Melé 2014 T
garnacha

91

Colour: dark-red cherry, garnet rim. Nose: tobacco, sweet spices, red berry notes, black fruit, ripe fruit. Palate: spicy, round tannins, elegant.

Ziries 2014 T
garnacha

88

Weary, age nuances, jammy, slight reduction, dried herbs.

TABLE WINES / WINES

BRUGAROL
Camino de Bell-Lloc, 63
17230 Palamòs (Girona/Gerona)
☎: +34 666 763 540
rrius@brugarol.com
www.brugarol.com

Ancestral 2019 RE
88
Citrus fruit, dried flowers, fruity, herbal, tasty.

BSI - BODEGAS SAN ISIDRO
Ctra. Murcia, s/n
30520 Jumilla (Murcia)
☎: +34 968 780 700
info@bsi.es
www.bsi.es

Celia Rosé 2021 RD
100% monastrell
88
Kind finish, floral, fruity, jammy.

Celia Verdejo 2021 B
100% verdejo
88
Aromatic, fruity, jammy.

CÁMBRICO
37658 Villanueva del Conde (Salamanca)
☎: +34 923 217 473
info@cambrico.com
www.cambrico.com

Cámbrico Rufete Blanca Pizarra 2019 B
rufete blanco
93 ♣
Colour: bright straw. Nose: expressive, characterful, dried flowers, meaty notes, lees reduction notes. Palate: sweetness, fruity, flavourful.

CAMINO DE CABRAS
Hermanos Maristas, 27
36700 Tui (Pontevedra)
☎: +34 698 145 790
info@caminodecabras.com
www.caminodecabras.com

Almaviño 2021 B
89
Pleasant, dried herbs, tasty, wild, simple, balanced.

CAMINO DEL NORTE, VINOS BY RAÚL PÉREZ
24540 Cacabelos (León)
☎: +34 658 617 390
info@caminodelnortevinos.com
www.caminodelnortevinos.com

El Soradal 2016 T BA
95% mencía, 5% petit verdot
91
Colour: dark-red cherry. Nose: meaty notes, waxy notes, ripe fruit, characterful, dried herbs. Palate: ripe fruit, long, spicy.

El Tesón 2015 T
90% mencía, 10% pinot noir
90
Colour: cherry, garnet rim. Nose: black fruit, wild herbs, black liquorice, tobacco. Palate: flavourful, bitter.

El Tesón 2016 T
93% mencía, 7% pinot noir
91
Slight reduction, little interventionist, rustic. Colour: cherry, garnet rim. Nose: black fruit, dried herbs, characterful. Palate: flavourful, fine bitter notes.

CAN SUMOI
Plaça del Roure s/n
08240 Sant Sadurní d'Anoia (Barcelona)
☎: +34 938 183 262
info@cansumoi.cat

Can Sumoi Sumoll – Garnatxa 2020 T
garnacha, sumoll
91
Colour: deep cherry. Nose: dried herbs, raspberry, ripe fruit. Palate: ripe fruit, spicy, round tannins.

CANTALAPIEDRA VITICULTORES
Ctra. VP-9901 km. 3.7
47491 La Seca (Valladolid)
☎: +34 665 235 709
info@isaaccantalapiedra.com
www.cantalpiedraviticultores.com

El Parvon 2020 T
94
Colour: Cherry. Nose: balsamic herbs, sweet spices, scrubland, mineral, dry stone, fruit expression, red berry notes. Palate: spicy, balsamic, good acidity.

CARLOS CANEIRO NÚÑEZ
Dariz, s/n
32747 Parada de Sil (Ourense/Orense)
☎: +34 610 789 232
bodegacaneiro@gmail.com
bodega-carlos-caneiro.negocio.site

Cocoloco 2021 B
godello
86

CASA CORREDOR
Autovía Alicante/Albacete, Salida 168 La Encina
02660 Caudete (Albacete)
☎: +34 966 842 064
info@mgwinesgroup.com
www.mgwinesgroup.com

Alagu Alicante Bouschet 2018 T
91
Colour: cherry, garnet rim. Nose: overripe fruit, creamy oak, hot, grassy. Palate: pruney, powerful, sweet tannins.

Alagu Forcallat 2018 T
forcallat
91
Colour: deep cherry. Nose: balsamic herbs, sweet spices, scrubland. Palate: spicy, balsamic, good acidity.

Alagu Rosé 2020 RD
89
Pleasant, aromatic, kind finish, tasty.

CASA SICILIA 1707
Paraje Alcaydias, 4
03660 Novelda (Alacant/Alicante)
☎: +34 965 605 385
administracion@casasicilia1707.es
www.casasicilia1707.es

Ad Gaude 2021 B
100% albariño
90 🌿
Colour: bright yellow. Nose: floral, white fruit, ripe fruit, characterful. Palate: flavourful, rich.

CASAL DE ARMÁN
San Andrés
32415 Ribadavia (Ourense/Orense)
☎: +34 680 979 763
info@casaldearman.net
www.casaldearman.net

Arman Doce Dulce B D
92
Colour: old gold. Nose: expressive, floral, candied fruit, honeyed notes. Palate: fresh, good acidity, long, sweet.

CAZAPITAS
Salgosa, 25
36740 Tomiño (Pontevedra)
☎: +34 605 625 782
cazapitassl@gmail.com
www.cazapitas.com

Cazapitas 2020 B
87

Falares 2020 T
sousón, caíño redondo, castañal, brancellao
90
With personality, wild. Colour: cherry, garnet rim. Nose: balsamic herbs, red berry notes. Palate: fresh, fruity, easy to drink.

CELLER ARRUFÍ
Avda. Terra Alta, 12
43786 Batea (Tarragona)
☎: +34 722 224 772
hola@cellerarrufi.com
www.cellerarrufi.com

77 dies Celler Arrufi 2021 B
garnacha blanca
89 🌿
Pleasant, floral, fruity, simple, jammy, crisp.

77 Nits Celler Arrufi 2021 T
garnacha
88 🌿
Little interventionist, pruney, fruity, yeasty notes, rustic.

CELLER HOSPITAL DE SITGES
Plaza Joan Duran i Ferret s/n
08870 Sitges (Barcelona)
☎: +34 672 682 481
celler@hospitaldesitges.cat
www.hospitaldesitges.cat

Malvasia de Sitges 2011 B Mistela D
100% malvasía de Sitges
94
Colour: old gold. Nose: candied fruit, honeyed notes, aged wood nuances, faded flowers. Palate: complex, varietal, flavourful, balanced.

Malvasia de Sitges Rancio BF RB S
100% malvasía de Sitges
93
Colour: iodine, amber rim. Nose: dry nuts, citrus fruit, sweet spices. Palate: spicy, old wood, good acidity.

TABLE WINES / WINES

Peñín Guide | SPANISH WINE

TABLE WINES / WINES

CELLER JOC
Ctra. Vilajuiga, s/n
17750 Capmany (Girona/Gerona)
☎: +34 607 222 002
info@vinojoc.com
www.vinojoc.com

Amunt Negre 2020 T
80% syrah, 10% garnacha, 10% monastrell
89
Hot, powerful, jammy, toasty.

Peligru 2021 T
garnacha, merlot
87

CELLER LES SOQUES
Partida de Asprillas, pol.1, 136
03292 Elche (Alacant/Alicante)
☎: +34 646 364 848
info@cellerlessoques.es
www.cellerlessoques.es

Atabalat 2018 T
75% monastrell, 25% cabernet franc
91
Colour: Cherry. Nose: scrubland, tobacco, waxy notes, black fruit, ripe fruit. Palate: spicy, balsamic, coarse tannins.

Atabalat 2019 RD
100% monastrell
90
Colour: onion pink. Nose: red berry notes, characterful, spicy, dried herbs, faded flowers. Palate: fruity, fresh, balanced.

Atabalat 2019 T
86% monastrell, 13% syrah, 1% moscatel de alejandría
92
Colour: cherry, garnet rim. Nose: ripe fruit, red berry notes, expressive, neat. Palate: long, ripe fruit, round tannins, good acidity.

Atabalat 2020 T
100% monastrell
93
Colour: deep cherry. Nose: ripe fruit, dried herbs, smoky, waxy notes. Palate: powerful, ripe fruit, spicy, round tannins, flavourful, balanced, long.

Rebombori 2020 B
50% macabeo, 0% moscatel
92
Little interventionist. Colour: bright straw, greenish rim. Nose: fresh fruit, citrus fruit, wild herbs, spicy. Palate: fresh, fruity, good acidity, fine bitter notes, good structure.

Rebombori Moscatel 2019 B
100% moscatel de alejandría
91
Colour: bright yellow. Nose: creamy oak, spicy, white fruit, candied fruit, dried flowers. Palate: balanced, fresh, flavourful.

CELLER MAR DE VINS
Avda. de Benidorm, 48
03530 La Nucía (Alicante)
☎: +34 686 829 739
info@cellermardevins.com
www.cellermardevins.com

Mar de Vins Alguer Vinyes Velles 2020 B
100% malvasía
91
Colour: bright straw. Nose: expressive, white flowers, jasmine, dried herbs. Palate: flavourful, fruity, balanced.

Mar de Vins Els Fustals Vinyes Velles 2020 B
100% malvasía
90
Colour: yellow. Nose: faded flowers, fine lees, dry nuts. Palate: flavourful, toasty, rich.

Mar de Vins Ermità Vinyes Velles 2020 B
100% malvasía
91
Colour: bright yellow. Nose: ripe fruit, dried herbs, faded flowers. Palate: powerful, ripe fruit, balanced.

CELLER MARIOL
Les Forques, 2
43786 Batea (Tarragona)
☎: +34 977 430 303
oficina@cellermariol.com
www.casamariol.com

Casa Mariol Verdejo 2021 B
verdejo
85

CELLER MAS DEL BOTÓ
Camí de Porrera a Alforja, s/n
43365 Alforja (Tarragona)
☎: +34 630 982 747
pep@masdelboto.cat
www.masdelboto.cat

Mas del Botó 2016 T
garnacha, cabernet sauvignon
87

Mas del Botó 2019 T
garnacha, cabernet sauvignon, cariñena
85

CELLER RUBIÓ DE SÒLS
25600 Foradada (Lleida/Lérida)
☎: +34 690 872 356
juditsogas@gmail.com
www.rubiodesols.cat

Rubiòls 2019 B
100% xarel.lo
91
Colour: bright yellow. Nose: spicy, toasty, stone fruit, ripe fruit. Palate: rich, good structure, fine bitter notes.

Sòls Xarel 2019 B BA
100% xarel.lo
89
Aromatic, spicy, jammy, tasty, fruity, balanced.

CELLER TIANNA NEGRE
Cami des Mitjans s/n, Parc. 67 – Pol. 7
07350 Binissalem (Illes Balears/Islas Baleares)
☎: +34 971 886 826
Fax: +34 971 226 201
info@tiannanegre.com
www.tiannanegre.com

Tianna 3 Negre 2020 T
100% escursac
93
Colour: cherry, purple rim. Nose: balsamic herbs, red berry notes, scrubland, spicy. Palate: balsamic, fruity, flavourful, elegant, great length.

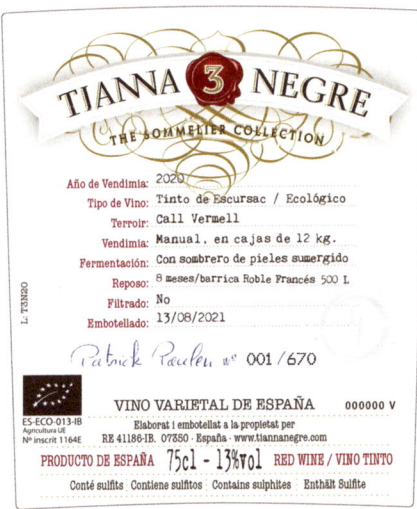

CÍA. DE VINOS ENTRE DOS AGUAS
Finca la Venta-Mesas de Asta
11590 Jerez de la Frontera (Cádiz)
☎: +34 651 923 577
cveda.vinos@gmail.com
www.e2a.es

La Viña de Robin 2018 T R
malbec, syrah, petit verdot
90
Colour: cherry, garnet rim. Nose: balsamic herbs, ripe fruit, scrubland, fine reductive notes. Palate: flavourful, balsamic, spicy.

CLOS DELS GUARANS
08730 Santa Margarida i Els Monjos (Barcelona)
☎: +34 675 040 848
info@closdelsguarans.com
www.closdelsguarans.cat

El Guarà 2019 B
xarel.lo vermell
92
Colour: bright straw. Nose: ripe fruit, dried herbs, faded flowers. Palate: powerful, ripe fruit, balanced.

L'Abellar 2020 B
subirat parent
89
Aromatic, dried herbs, little interventionist, tasty.

Les Someres 2021 B
malvasía, alarije, vinyater, xarel.lo vermell, llimonera
88
Pronounced acidity, herbal, oxidativ.

Les Someres 2021 T
monastrell, garnacha, malvasía rosada
91
Colour: bright cherry. Nose: balsamic herbs, sweet spices, scrubland. Palate: spicy, balsamic, good acidity.

Tardatio Blanc de Noir 2018 B
malvasía rosada
93
Colour: bright yellow. Nose: dried flowers, candied fruit, fine lees, pattiserie. Palate: round, spicy, long, great length.

Tardatio Subtil Rosé 2020 RD
malvasía rosada
92
Colour: raspberry rose. Nose: red berry notes, floral, fragrant herbs. Palate: light-bodied, spicy, good acidity, fine bitter notes.

TABLE WINES / WINES

CLOS PISSARRA
Pol. 14, Parcelas 106-108
43730 Falset (Tarragona)
☎: +34 626 656 730
maruxa.roel@gmail.com
www.clospissarra.com

Monje Terruño 2016 T R
67% merlot, 33% syrah

88
Age nuances, balanced, spicy, jammy, toasty.

COOP. SANTÍSIMO CRISTO DE LAS INJURIAS
Ctra. Villarrubia, 27
45350 Noblejas (Toledo)
☎: +34 925 140 291
gestion@coopereativanoblejas.es
www.cooperativanoblejas.es

Hijón del Santo 2020 T BA
cencibel, syrah, otras

89
Balanced, spicy, dried herbs, jammy, full-bodied.

La Zaranda Chardonnay Viognier 2021 B
chardonnay, viognier

88 🌱
Toasty, tropical, jammy, citrus fruit.

La Zaranda Syrah 2021 T
syrah

84 🌱

CORTIJO DE JARA
11592 Jerez de la Frontera (Cádiz)
☎: +34 956 338 163
puertanueva.sl@cortijodejara.es
www.cortijodejara.es

Cortijo de Jara Gewürztraminer 2021 B
gewürztraminer, sauvignon blanc

85

COTO DE GOMARIZ
Barro de Gomariz s/n
32429 Leiro (Ourense/Orense)
☎: +34 610 602 672
info@losvinosdemiguel.com
www.cotodegomariz.com

Gomariz 12 (Doce) Dulce 2020 B D
treixadura

91
Colour: straw. Nose: ripe fruit, citrus fruit, jasmine, stone fruit, candied fruit. Palate: fruity, sweetness, easy to drink, flavourful.

VX Cuvée Caco 2018 T
sousón, caiño longo, caiño da Terra, carabuñeira, mencía

92
Colour: black cherry. Nose: balsamic herbs, sweet spices, scrubland, black fruit. Palate: spicy, balsamic, good acidity.

CUME DO AVIA
Santo André, s/n
32415 Ribadavia (Ourense/Orense)
☎: +34 630 727 037
diego@cumedoavia.com

Arraiano 2020 T BA
caiño longo, brancellao, sousón, mencía, garnacha, mouratón

92
Colour: bright cherry. Nose: balsamic herbs, scrubland, red berry notes, floral. Palate: spicy, balsamic, good acidity.

Brancellao Dos Canotos 2020 T
brancellao

93 🌱
Colour: bright cherry. Nose: fruit expression, red berry notes, balsamic herbs, scrubland, rose petals, chamomile. Palate: fruity, fresh, flavourful.

Caiño dos Canotos 2019 T
caiño

91 🌱
Pronounced acidity. Colour: Cherry. Nose: balsamic herbs, sweet spices, scrubland, fruit expression. Palate: spicy, balsamic.

Nº8 Colletia 2020 T
caiño, brancellao, sousón, carabuñeira, ferrón, merenzao

92 🌱
Colour: Cherry. Nose: balsamic herbs, sweet spices, scrubland, floral, red berry notes. Palate: spicy, balsamic, good acidity.

Nº8 Colletia Branco 2020 B
treixadura, albariño, lado, loureiro, caiño blanco

91 🌱
Little interventionist. Colour: bright straw, greenish rim. Nose: fresh fruit, citrus fruit, wild herbs. Palate: fresh, fruity, good acidity, fine bitter notes.

Treixadura Dos Canotos 2019 B BA
treixadura

89 🌱
Aromatic, kind finish, dried flowers, simple.

DANIEL RAMOS

San Pedro de Alcántara, 1
05270 El Tiemblo (Ávila)
☎: +34 687 410 952
dvrcru@gmail.com
www.danielramos.wine

El Berrakin 2021 T
garnacha, jaen negro, cariñena

89

Fruity, sweety finish, dried herbs, floral, tasty.

Kπ Rosé 2019 RD
garnacha

93

Colour: rose. Nose: fruit preserve, hot, faded flowers, dry nuts, lees reduction notes, caramel. Palate: fleshy, flavourful, powerful, ripe fruit.

Kπ Iluminati 2020 T
garnacha

93

Defined aromas. Colour: light cherry. Nose: red berry notes, fruit liqueur notes, earthy notes, characterful, expressive, wild herbs. Palate: long, balanced, ripe fruit.

DE MULLER

Camí Pedra Estela, 34
43205 Reus (Tarragona)
☎: +34 977 757 473
promo@demuller.es
www.demuller.es

De Muller Avreo Dulce Solera 1954 BF RC D
70% garnacha, 30% garnacha blanca

94

Colour: bright cherry, garnet rim. Nose: acetaldehyde, varnish, candied fruit. Palate: fruity, flavourful, sweet.

De Muller Avreo Seco Solera 1954 BF Añejo S
70% garnacha, 30% garnacha blanca

93

Colour: iodine, amber rim. Nose: dry nuts, creamy oak, caramel, sweet spices. Palate: rich, long, spicy.

🏆 PODIUM

De Muller Garnacha Solera 1926 BF Solera D
garnacha

95

Colour: light mahogany. Nose: powerful, complex, dry nuts, toasty, acetaldehyde. Palate: rich, long, fine solera notes, spicy, round.

De Muller Misa Dulce Superior B D
50% macabeo, 50% garnacha blanca

92

Colour: amber. Nose: stone fruit, dried flowers, sweet spices, fine lees. Palate: fleshy, balanced.

De Muller Moscatel Añejo Vino de Licor BF
moscatel de alejandría

92

Colour: golden. Nose: powerful, honeyed notes, candied fruit, acetaldehyde. Palate: flavourful, sweet, fresh, fruity, good acidity, long.

De Muller Rancio Seco BF Añejo S
40% garnacha blanca, 40% garnacha blanca, 20% cariñena

91

Colour: iodine, amber rim. Nose: dry nuts, creamy oak, woody. Palate: spicy, good structure, dry.

Dom Berenguer Vino de Licor 1918 BF Solera

94

Colour: light mahogany. Nose: powerful, complex, dry nuts, toasty, caramel. Palate: rich, long, fine solera notes, spicy.

Dom Juncosa Solera 1939 BF Solera
garnacha

93

With personality, age nuances. Colour: iodine, amber rim. Nose: candied fruit, pattiserie, varnish. Palate: flavourful, long, old wood.

DOBLEDEPEREZ MICROBODEGA

Plaza del Ayuntamiento, 7 Bajo A
02653 Albatana (Albacete)
☎: +34 627 588 119
info@dobledeperez.es
www.dobledeperez.es

La Jarana 2021 B
100% moscatel grano menudo

90

Subtle, with personality. Colour: straw. Nose: ripe fruit, dried herbs, ebb tide, dried flowers. Palate: ripe fruit, balanced, flavourful.

TABLE WINES / WINES

DUNORD VITÍCOLA
El Cano, 28
07470 Port de Pollença
(Illes Balears/Islas Baleares)
☎: +34 670 645 978
dunordviticola@gmail.com

Trabucat 2019 T
65% syrah, 35% escursac
94
Colour: deep cherry. Nose: creamy oak, black fruit, scrubland. Palate: powerful, ripe fruit, spicy, round tannins, balanced, good acidity.

ECCOCIVI CELLER
Paratge Montrodó, 3
17462 San Martí Vell (Girona/Gerona)
☎: +34 872 000 015
info@eccociwine.com
www.eccocivi.com

Ca L'Elsa 2016 T R
cabernet sauvignon
92
Colour: Cherry. Nose: spicy, dark chocolate, creamy oak, balanced. Palate: full-bodied, long, great length, round tannins.

Can Noves Blanc 2020 B
garnacha blanca, xarel.lo
90
Colour: bright straw. Nose: spicy, toasty, white fruit, ripe fruit. Palate: rich, flavourful.

Montrodó Blanc 2020 B
chardonnay, garnacha blanca, viognier, xarel.lo
89
Aromatic, floral, dried flowers, jammy, fruity, tasty.

Montrodó Blanc 2021 B
chardonnay, garnacha blanca, viognier, xarel.lo
87

Montrodó Negre 2020 T S
merlot, cabernet sauvignon, garnacha
90
Colour: deep cherry. Nose: ripe fruit, dried herbs, creamy oak. Palate: ripe fruit, spicy, round tannins.

Montrodó Rosat 2021 RD
merlot, garnacha
89
Aromatic, with personality, floral, fruity, kind finish, jammy.

EDRA
Ctra A - 132, km 26
22800 Ayerbe (Huesca)
☎: +34 679 420 455
edra@bodega-edra.com
www.bodega-edra.com

Edra Blancoluz 2020 B
100% viognier
90 ♣
Full-bodied, oxidativ. Colour: straw. Nose: dried herbs, faded flowers, stone fruit. Palate: powerful, ripe fruit, balanced.

EGO BODEGAS
Paraje Hoya de Torres s/n
30520 Jumilla (Murcia)
☎: +34 868 680 939
info@egobodegas.com
www.egobodegas.com

Ego Bodegas Sauvignon Blanc 2021 B
sauvignon blanc
86

El Goru 2021 RD
garnacha
86

Goru El Blanco 2021 B
moscatel de alejandría, chardonnay
89
Citrus fruit, dried flowers, dried herbs, tasty.

EL CELLER DE LA IBOLA
Ctra. CV-223 km. 15
12222 Ain (Castelló/Castellón)
☎: +34 674 503 087
rrivasnavarro@gmail.com
www.elcellerdelaibola.com

El Celler de la Ibola 2020 B
xarel.lo
89
Smoky, dried flowers, jammy, toasty, great length.

GR 36 2018 T
garnacha
82

ELÍAS MORA
Juan Mora s/n
47530 San Román de Hornija (Valladolid)
☎: +34 983 784 029
info@bodegaseliasmora.com
www.bodegaseliasmora.com

Dulce Benavides T D
100% tinta de Toro (Tinto)
91
Colour: cherry, garnet rim. Nose: fruit preserve, ripe fruit, spicy. Palate: flavourful, balanced, long.

VINATE
Terrero 72
02630 La Roda (Albacete)
☎: +34 682 207 160
asesoria@envinate.es

Albahra 2020 T
91
Rustic. Colour: deep cherry. Nose: dried herbs, creamy oak, overripe fruit, black fruit. Palate: powerful, ripe fruit, spicy.

La Santa de Úrsula 2020 T
94
Little interventionist, needs time. Colour: cherry, purple rim. Nose: fruit expression, red berry notes, floral, spicy, burnt matches, wild herbs. Palate: flavourful, fruity, good acidity, long.

🏆 PODIUM

Lousas Seoane 2020 T
95
Colour: bright cherry. Nose: red berry notes, wild herbs, fruit expression, wild herbs, floral. Palate: fresh, fruity, good acidity, taut, elegant.

Lousas Viñas de Aldea 2020 T
94
Colour: cherry, purple rim. Nose: fruit expression, red berry notes, floral, violets, fragrant herbs, earthy notes. Palate: fruity, fresh, full of life, flavourful, balanced, spirituous, spicy.

Migan 2020 T
93
Colour: bright cherry. Nose: balsamic herbs, sweet spices, scrubland, burnt matches, red berry notes. Palate: spicy, balsamic, good acidity.

🏆 PODIUM

Palo Blanco 2020 B
95
Colour: bright straw. Nose: cereal notes, burnt matches, citrus fruit, fresh fruit, dry stone, grassy. Palate: powerful, fresh, flavourful, mineral, saline.

Táganan 2020 B
93
Little interventionist, rustic. Colour: bright straw, golden. Nose: fragrant herbs, fine lees, cereal notes, dry stone, stone fruit. Palate: full-bodied, balanced, saline, flavourful.

Táganan 2020 T
93
Colour: cherry, purple rim. Nose: burnt matches, toasty, spicy, ripe fruit, dried herbs. Palate: fruity, flavourful, fresh, good finish.

Táganan Margalagua 2020 T
94
Colour: cherry, purple rim. Nose: red berry notes, wild herbs, dry stone, dried herbs, spicy, dried flowers. Palate: fresh, fruity, full of life, balanced.

EQUIPO NAVAZOS
11403 Jerez de la Frontera (Cádiz)
equipo@navazos.com
www.equiponavazos.com

La Bota de Manzanilla 107 Florpower MMXX B
93
Colour: yellow, golden. Nose: faded flowers, ripe fruit, candied fruit, dry nuts, flor yeasts. Palate: flavourful, dry.

Navazos Niepoort 2020 B
94
Colour: bright yellow. Nose: pungent, flor yeasts, neat, expressive. Palate: complex, saline, long, balanced, fine bitter notes.

FERRÉ I CATASÚS
Masía Gustems, Crta. Sant Sadurní a Vilafranca, Km.8
08792 La Granada (Barcelona)
☎: +34 938 974 558
administracio@ferreicatasus.com
www.ferreicatasus.com

Ferré i Catasús Dos 2019 T C
cabernet sauvignon, cabernet franc, garnacha
89 🌱
Colour: bright cherry. Nose: fruit expression, ripe fruit, black fruit, black liquorice, spicy, creamy oak. Palate: flavourful, powerful, balanced, slightly dry, soft tannins.

TABLE WINES / WINES

FINCA BACARA
Calle Rio Taibilla, 13 - 7ºA
30110 Churra (Murcia)
☎: +34 868 680 939
Fax: +34 868 580 285
info@fincabacara.com
www.fincabacara.com

Finca Bacara Sauvignon Blanc 2021 B
sauvignon blanc

89
Herbal, dried herbs, fruity, tasty.

Yeya 2021 B
moscatel de alejandría, chardonnay

90
Colour: bright straw. Nose: stone fruit, tropical fruit, faded flowers. Palate: fresh, flavourful, balanced.

FINCA CAN AXARTELL
Ctra. Pollença – Campanet km 1,5
07460 Pollença (Illes Balears/Islas Baleares)
☎: +34 871 870 353
info@canaxartell.es
www.canaxartell.com

Selecció Familiar sin sulfitos 2019 T
petit verdot

93 ☘
Colour: deep cherry. Nose: dried herbs, black fruit, spicy, scrubland. Palate: powerful, ripe fruit, spicy, round tannins.

FRANCISCO JAVIER SANZ SOGUERO
Barranco, 60
50108 Almonacid de La Sierra (Zaragoza)
☎: +34 696 453 234
franciscojavier.sanzsoguero@gmail.com
www.vignius.com

IGnius (Etiqueta Morada) 2015 T
garnacha

88 ☘
Age nuances, reduced, full-bodied, pruney.

GERMÁN R BLANCO
El Olivo s/n
24310 Albares de la Ribera (León)
☎: +34 635 432 351
ad@lively-wines.com
www.germanrblanco.com

Clos Pepin Casa Aurora 2020 T
garnacha tintorera, palomino, portuguesa, garnacha, mencía

93
Colour: cherry, purple rim. Nose: red berry notes, floral, spicy, balsamic herbs, scrubland. Palate: flavourful, fruity, good acidity, long.

La Nave Casa Aurora 2020 T
90% mencía, 10% garnacha tintorera, palomino

91
Colour: deep cherry. Nose: ripe fruit, dried herbs, creamy oak. Palate: powerful, ripe fruit, spicy, round tannins.

Poula Casa Aurora 2020 T
mencía, garnacha tintorera, portuguesa, palomino, godello, dona blanca

94
Colour: Cherry. Nose: balsamic herbs, sweet spices, scrubland, ripe fruit, earthy notes. Palate: spicy, balsamic, good acidity.

Valle del Río Casa Aurora 2020 T
60% garnacha tintorera, 10% garnacha, 10% mencía, 17% godello

93
Colour: very deep cherry. Nose: complex, expressive, spicy, mineral. Palate: elegant, full-bodied, long, great length.

GIRÓ RIBOT
Finca El Pont, s/n
08792 Santa Fe del Penedès (Barcelona)
☎: +34 938 974 050
giroribot@giroribot.es
www.giroribot.es

Giró Ribot Giró2 2017 B FB
100% giró blanc

91
Colour: yellow. Nose: spicy, toasted bread, smoky, yeasty notes. Palate: rich, good structure, toasty, bitter.

Giró Ribot Mimat 2018 T BA
50% marselan, 30% petit verdot, 20% cabernet franc

92
Classic. Colour: deep cherry. Nose: dried herbs, creamy oak, black fruit, old leather. Palate: ripe fruit, spicy, round tannins.

GRIFOLL DECLARA GRUP
La Font, 7
43736 El Molar (Tarragona)
☎: +34 977 825 149
info@cgrifolldeclara.com
www.cgrifolldeclara.com

Declara Selección Especial 2018 T
70% cariñena, 30% garnacha

91
Colour: deep cherry. Nose: red berry notes, black fruit, scrubland, balsamic herbs, spicy. Palate: powerful, ripe fruit, spicy, round tannins.

Tossals Selecció 2019 T
50% garnacha, 50% cariñena

91
Colour: cherry, purple rim. Nose: red berry notes, floral, spicy. Palate: flavourful, fruity, good acidity.

HATORI
Las Damas, 30
40500 Riaza (Segovia)
☎: +34 620 214 755
hatorihanzosl@yahoo.es

Artico B

91
Colour: bright yellow. Nose: faded flowers, fine reductive notes, burnt matches, wild herbs, characterful. Palate: flavourful, dry, fine bitter notes.

IDRIAS
Ctra. Abiego 1229, Km 0,2
22124 Lascellas (Huesca)
☎: +34 974 340 671
info@bsdg.es
www.idrias.es

Idrias Merlot 2021 RD
merlot

88
Pleasant, dried herbs, fruity, jammy, tasty, simple.

JAVIER SANZ VITICULTOR
San Judas, 2
47491 La Seca (Valladolid)
☎: +34 983 816 669
comunicaciones@bodegajaviersanz.com
www.bodegajaviersanz.com

V Colorado 2016 T FB
100% tardillo

93
Wild. Colour: Cherry. Nose: balsamic herbs, scrubland, ripe fruit, dried herbs, fine reductive notes. Palate: spicy, balsamic, good acidity.

🏆 PODIUM

V Dulce de Invierno 2019 B D
100% verdejo

95
Colour: bright yellow. Nose: dried fruit, honeyed notes, fragrant herbs, faded flowers. Palate: fruity, rich, flavourful, balanced, unctuous.

JORGE ORDÓÑEZ MÁLAGA
Bartolome Esteban Murillo, 11
29700 Vélez (Málaga)
☎: +34 952 504 706
pedidos@jorgeordonez.es
www.jorgeordonez.es

Botani Sparkling Muscat 2021 BE
100% moscatel de alejandría

91
Colour: bright yellow. Nose: dried herbs, fruit expression, white fruit, floral. Palate: correct, flavourful.

🏆 PODIUM

Ordóñez & Co. Nº4 Esencia (sin fortificar) 2016 B D
100% moscatel de alejandría

99
Colour: mahogany. Nose: caramel, cocoa bean, aromatic coffee, toasty, overripe fruit, dried fruit. Palate: sweetness, complex, flavourful.

LA ALQUERÍA DE PRUNA
Camino del Rancho s/n
41670 Pruna (Sevilla)
☎: +34 617 551 575
info@alqueriadepruna.com
www.alqueriadepruna.com

Alqueria de Pruna (etiqueta blanca) 2019 T BA
petit verdot, syrah

89
With personality, smoky, tasty, dried herbs, spicy. Nose: meaty notes, spicy.

Alqueria de Pruna (etiqueta roja) 2020 T S
petit verdot, syrah, tintilla de rota

88
Kind finish, jammy, rustic, herbal, smooth, fruity.

Alqueria de Pruna (etiqueta roja) 2021 T S
petit verdot, syrah

89
Jammy, dried flowers, dried herbs, wild, fruity.

LA CALANDRIA. PURA GARNACHA
Calle Mayor, 81
31500 Murchante (Navarra)
☎: +34 609 476 387
javier@lacalandria.org
www.lacalandria.com

Bardalera 2019 B
100% garnacha
88
Standard, jammy, tasty, balanced, kind finish.

Tierga T
garnacha
88
Pruney, overripe, full-bodied, dried herbs, tasty.

LADRÓN DE LUNAS
Colón, 12
46357 La Portera (València/Valencia)
☎: +34 601 288 998
jorge.ferrer@ladrondelunas.es
www.ladrondelunas.es

Bum Bum Ciao Verdejo 2021 B SD
85

Bum Bum Rosé 2021 RD SD
86

Obejita Pink 2021 RD
86

OBejita Verde 2020 B
86

LAGAR DE COSTA
Sartaxes, 8 - Castrelo
36639 Cambados (Pontevedra)
☎: +34 669 086 569
contacto@lagardecosta.com
www.lagardecosta.com

Viva la Vida 2018 T
100% espadeiro
89
Balanced, spicy, dried flowers, crisp, age nuances, pruney.

LAGAR DE LA SALUD
Ctra. Córdoba_Málaga, km. 41
14550 Montilla (Córdoba)
☎: +34 659 467 525
info@lagardelasalud.com
www.lagardelasalud.com

Dulas del Lagar de la Salud "un americano en lagar de la salud" 2020 T C
cabernet sauvignon
88
Full-bodied, jammy, tasty, spicy, slight oxidation, overripe.

Dulas del Lagar de la Salud "un francés en lagar de la salud" 2020 T
cabernet sauvignon
88
With personality, herbaceous, rustic, tasty, spicy, elegant, overripe, slight oxidation.

LAOSA
Las Cuevas, 8
24232 Ardón (León)
☎: +34 666 217 032
noelia@laosavinos.com
www.laosavinos.com

La Voz del Viñador 2019 T
88
Balsamic herbs, astringent, vegetal, jammy.

La Voz del Viñador 2020 T
prieto picudo
91
Colour: bright cherry. Nose: balsamic herbs, sweet spices, scrubland, red berry notes. Palate: spicy, balsamic, good acidity.

LAR DE MAÍA
Pº de Castilviejo, 2
47290 Cubillas de Santa Marta (Valladolid)
☎: +34 650 986 098
export@lardemaia.com
www.lardemaia.com

Gntx Lar de Maía 2020 T C
100% garnacha
89
Pruney, dried herbs, jammy, great length, tasty, coarse tannins.

LAS PEDRERAS VIÑEDOS Y VINOS
Herguijuelas, 5
05114 Villanueva de Ávila (Ávila)
☎: +34 616 520 572
info.laspedreras@gmail.com

La Coronela 2020 T
tempranillo
93
Colour: deep cherry. Nose: dried herbs, creamy oak, dark chocolate, ripe fruit, black fruit. Palate: powerful, ripe fruit, spicy, round tannins.

LMT WINES
Cerro Amurdi
31190 Cizur Menor (Navarra)
☎: +34 645 841 928
hola@lmtwines.com
www.lmtwines.com

Artaxo 2019 T
100% garnacha
93
Colour: Cherry. Nose: expressive, spicy, wild herbs, varietal, neat. Palate: elegant, full-bodied, long, great length.

Artaxo 2020 T
garnacha
92
Balanced, jammy. Colour: cherry, garnet rim. Nose: red berry notes, ripe fruit, dried flowers. Palate: flavourful, varietal, ripe fruit.

Bikote 2020 T
91
Rustic, jammy. Colour: cherry, garnet rim. Nose: ripe fruit, scrubland, dried herbs. Palate: flavourful, fruity.

Cerro Amurdi 2020 B
garnacha blanca
92
Colour: golden. Nose: ripe fruit, wild herbs, macerated fruit, mineral. Palate: fruity, balanced.

El Yesal 2019 T
garnacha
93
With personality, wild. Colour: cherry, garnet rim. Nose: red berry notes, chalk, expressive, neat, fragrant herbs. Palate: flavourful, varietal, spicy, long.

Korteta 2020 T S
garnacha, graciano
92
Colour: bright cherry, purple rim. Nose: ripe fruit, faded flowers, dried flowers. Palate: ripe fruit, spicy, flavourful, round.

MANUEL ARAGÓN
Olivo, 1
11130 Chiclana de la Frontera (Cádiz)
☎: +34 956 400 756
administracion@bodegamanuelaragon.com
www.bodegamanuelaragon.com

Retallo Blanco 2020 B
80% pedro ximénez, 20% moscatel
90
Jammy, tasty, kind finish, aromatic. Colour: bright yellow.

Señorío de Galdón 2021 B
moscatel
88
Aromatic, tropical, varietally correct.

MAS COMTAL
Mas Comtal, 1
08793 Avinyonet del Penedès (Barcelona)
☎: +34 938 970 052
mascomtal@mascomtal.com
www.mascomtal.com

Mas Comtal Incrocio Manzoni 2018 B
incrocio manzoni
91 🌿
Colour: bright yellow. Nose: ripe fruit, dry nuts, dried herbs, spicy. Palate: fruity, flavourful, fresh.

Mas Comtal Lyric Solera 1993 TF Solera D
merlot
93
With personality, age nuances. Nose: smoky, roasted almonds, dark chocolate. Palate: flavourful, long, old wood, great length.

Mas Comtal Pétrea Chardonnay 2017 B FB
chardonnay
91
Colour: bright golden. Nose: ripe fruit, toasty, spicy, dried herbs. Palate: full-bodied, fruity, fresh, spicy, smoky aftertaste.

Mas Comtal Pétrea Merlot 2017 T
merlot
91
Colour: deep cherry. Nose: dried herbs, creamy oak, fruit liqueur notes, spicy. Palate: powerful, ripe fruit, spicy, round tannins.

Mas Comtal Pizzicato 2021 RD
moscatel de hamburgo
86

TABLE WINES / WINES

SPANISH WINE

MAS VILELLA

Masia Mas Vilella – Camí del Cementiri, s/n
43717 La Bisbal del Penedès (Tarragona)
☎: +34 672 432 691
masvilella@masvilella.net

Mas Vilella Blanc 2020 B FB
malvasía de Sitges

92

Colour: yellow. Nose: expressive, ripe fruit, floral. Palate: full-bodied, complex, long, balanced, good acidity.

Mas Vilella Blanc 2021 B FB
100% malvasía de Sitges

92

Colour: bright straw. Nose: ripe fruit, jasmine, white flowers, dried herbs. Palate: fruity, flavourful, rich, fresh, balanced.

Mas Vilella Negre 2020 T
95% cabernet sauvignon, 5% sumoll

93

Colour: deep cherry. Nose: ripe fruit, dried herbs, creamy oak, scrubland. Palate: powerful, ripe fruit, spicy, round tannins.

Tros de Mas Vilella 2019 T
50% sumoll, 50% cabernet sauvignon

91

Colour: Cherry. Nose: balsamic herbs, sweet spices, scrubland, earthy notes, ripe fruit. Palate: spicy, flavourful.

MIXTURA

Lugar de Esposende, 40
32415 Esposende (Ourense)
www.mixturaiw.com

Mix 2020 B C
treixadura, albariño

93

Colour: bright yellow. Nose: powerful, creamy oak, ripe fruit, spicy. Palate: rich, good structure, long, toasty, fine bitter notes.

Mix 2020 T C
mencía, caíño

91

Colour: cherry, garnet rim. Nose: overripe fruit, creamy oak, hot, red berry notes. Palate: pruney, powerful, sweet tannins.

Mixtura 2020 B C
treixadura, albariño

94

Colour: bright straw. Nose: ripe fruit, fragrant herbs, fine lees, sweet spices, citrus fruit. Palate: full-bodied, rich, long, good acidity.

🏆 PODIUM

Mixtura Etiqueta Dorada 2018 B
treixadura, albariño

98

Colour: bright straw. Nose: expressive, ripe fruit, floral, fine lees, mineral. Palate: full-bodied, complex, spicy, long, elegant.

Mixtura Etiqueta Roja 2018 T
caíño longo

94

Rustic. Colour: deep cherry. Nose: ripe fruit, creamy oak, scrubland, balsamic herbs, grassy, meaty notes. Palate: powerful, ripe fruit, spicy, round tannins.

Mixtura Etiqueta Verde 2020 B
albariño

94

Colour: bright yellow. Nose: dried flowers, candied fruit, fine lees, pattiserie. Palate: round, spicy, long, great length.

MUJICA & DIAZ

Cº de Fuentecabrón, 10
40500 Miño de San Esteban (Soria)
☎: +34 620 214 755
mujicaydiazbodega@gmail.com

Matalba Colección Privada 2020 B
albillo mayor

89

Aromatic, spicy, jammy, smooth, balanced, pleasant.

Matalba Colección Privada 2020 T
tempranillo

88

Balanced, spicy, herbaceous, rustic.

Matalba Garnacha 2020 T

88

Spicy, herbaceous, dried herbs, astringent, rustic.

Matalba Velo 2020 T
tempranillo

87

Matalba Velo de Flor Edición Limitada 2020 T
tempranillo

91

Colour: cherry, purple rim. Nose: dried herbs, creamy oak, black fruit, flor yeasts. Palate: powerful, ripe fruit, spicy, round tannins.

Matalba Vino Blanco 2020 B
albillo mayor

88

Kind finish, spicy, toasty, jammy, standard.

Matalba Viñas Viejas 2020 T
tempranillo
89
Powerful, herbaceous, spicy, complex.

OLLER DEL MAS
Ctra. de Igualada a Manresa, km. 91 (C-37),
08241 Manresa (Barcelona)
☎: +34 938 768 315
admin@ollerdelmas.com
www.ollerdelmas.com

Càndia 2018 T BA
garnacha, sumoll, samsó, syrah
90
Colour: deep cherry. Nose: ripe fruit, dried herbs, cigar, meaty notes. Palate: ripe fruit, spicy, round tannins.

Oller del Mas Especial Carinyena 2018 T BA
cariñena
93
Colour: Cherry. Nose: complex, expressive, spicy, varietal, neat. Palate: elegant, full-bodied, long, great length.

Ròmia 2018 T
cariñena, garnacha, mandó, sumoll, picapoll negre
91
Colour: Cherry. Nose: balsamic herbs, scrubland, black fruit, toasted bread, burnt matches. Palate: flavourful, long.

OXER WINES
Ctra. Navaridas, 15
01300 Laguardia (Araba/Álava)
☎: +34 616 984 118
oxerarnuak@gmail.com
www.oxerwines.com

🏆 PODIUM
Premio al Vino Revelación

Kuusu 2020 T
92% tinta de Toro (Tinto), 8% albillo mayor
95
Defined aromas. Colour: cherry, purple rim. Nose: fruit expression, red berry notes, floral, spicy, wild herbs. Palate: flavourful, fruity, good acidity, long, elegant, round tannins.

PABLO VIDAL - VINOS CON PERSONALIDAD
Rúa do Miradoiro 8
32004 Ourense/Orense (Ourense/Orense)
☎: +34 609 152 251
pablovidal@vinosconpersonalidad.com
www.vinosconpersonalidad.com

Rock & Roll 2020 T RB
90% mencía, 5% caíño, 5% garnacha tintorera
91
Colour: Cherry. Nose: balsamic herbs, sweet spices, scrubland. Palate: spicy, balsamic, good acidity.

PAGO DE ALMARAES
Ctra. de Fonelas, Km. 1,5
18510 Benalua de Guadix (Granada)
☎: +34 958 348 752
info@bodegaspagodealmaraes.es
www.bodegaspagodealmaraes.es

Elvira B
moscatel grano menudo
88
Aromatic, varietally correct, jammy, floral, tasty.

PAGO DE LA BOTICARIA
Diseminados, 31
50360 Daroca (Zaragoza)
☎: +34 976 800 494
info@pagodelaboticaria.com
www.pagodelaboticaria.com

Trilo-Vites 2019 T
garnacha
92
Colour: dark-red cherry. Nose: toasty, spicy, cocoa bean, dark chocolate, ripe fruit. Palate: flavourful, toasty, fine bitter notes.

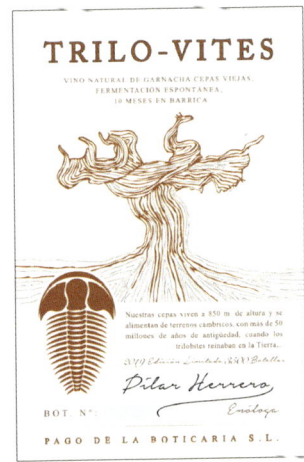

TABLE WINES / WINES

Viña Satoshi 2019 T BA
85% garnacha, 15% tempranillo

89

Colour: deep cherry. Nose: black fruit, ripe fruit, dark chocolate, dried herbs, toasty. Palate: powerful, ripe fruit, spicy, round tannins.

PAGO DEL MARE NOSTRUM
Ctra. A-348
04460 Fondón (Almería)
☎: +34 926 666 027
Fax: +34 926 666 029
pedidos@pagodelvicario.com
www.pagodelvicario.com

1500 H Coupage 2010 T C S
pinot noir, tempranillo, merlot, cabernet sauvignon

89

Age nuances, dried herbs, spicy, slight reduction, jammy, tasty.

PAGO FINCA ÉLEZ
Ctra. Ossa de Montiel a El Bonillo, Km 11,5
02610 El Bonillo (Albacete)
☎: +34 626 882 250
administracion@fincaelez.com
www.pagofincaelez.com

El Secreto de LZ 2020 B
viognier

91 🌿

Colour: bright yellow. Nose: powerful, creamy oak, stone fruit, toasty. Palate: good structure, long, toasty, fine bitter notes.

PAZO PONDAL
Lg. Coto, s/n - Cabeiras
36436 Arbo (Pontevedra)
☎: +34 986 665 551
Fax: +34 986 665 949
admin@pazopondal.com
www.pazopondal.com

Olivia Rosé 2021 RD
40% albariño, 30% treixadura, 20% sousón, 10% pedral

91

Colour: coppery red, bright. Nose: spicy, red berry notes, wild herbs, floral, mineral. Palate: elegant, flavourful, long, good acidity.

PENÍNSULA VINICULTORES
Hermosilla 29, 1ºA
28008 Madrid (Madrid)
☎: +34 916 039 363
o.moreno@peninsula.wine
www.peninsula.wine

Península Skin Contact 2019 B

92 🌿

With personality, little interventionist. Colour: bright straw, greenish rim. Nose: citrus fruit, wild herbs, macerated fruit, faded flowers, chamomile. Palate: good acidity, fine bitter notes.

Península Vino de Montaña Sierras Gata & Gredos 2020 T
garnacha, rufete, piñuela

90

Colour: Cherry. Nose: balsamic herbs, sweet spices, scrubland, ripe fruit. Palate: spicy, balsamic, good acidity.

PLA DE MOREI
Cami de la Garça s/n
08789 La Torre de Claramunt (Barcelona)
☎: +34 931 313 454
enoturisme@plademorei.com
www.plademorei.com

Riu de Gost Garnacha Blanca 2020 B RB
100% garnacha blanca

92 🌿

Colour: bright straw. Nose: dry nuts, faded flowers, spicy. Palate: flavourful, rich, balanced.

PRIMITIVO QUILES
Mayor, 4
03640 Monovar (Alacant/Alicante)
☎: +34 965 470 099
Fax: +34 966 960 235
info@primitivoquiles.com
www.primitivoquiles.com

Primitivo Quiles Moscatel Extra BF Mistela D
moscatel

91

Colour: mahogany. Nose: stone fruit, dried fruit, faded flowers. Palate: fruity, sweet, flavourful, balanced, spicy.

Primitivo Quiles Moscatel Laurel BF Mistela D
moscatel

88

Kind finish, varietally correct, balanced, floral, sweet, representative.

PROYECTO GARNACHAS/VINTAE

Ctra. de Murillo
26500 Calahorra (La Rioja)
☎: +34 941 483 557
marketing@vintae.com
www.vintae.com

La Garnacha Perdida del Pirineo 2019 T C
garnacha

90

Colour: cherry, purple rim. Nose: black fruit, dried herbs, toasty, spicy. Palate: fruity, flavourful, balanced.

La Garnacha Salvaje del Moncayo 2019 T

90

Pleasant, balsamic herbs. Colour: light cherry. Nose: ripe fruit, wild herbs. Palate: flavourful, balanced, spicy.

RAÍCES IBÉRICAS

Avda. Mudejar, 61
50340 Maluenda (Zaragoza)
☎: +34 976 893 017
s.richard@raices.wine
www.raicesibericas.com

Antonio & Antonia 2021 T
100% garnacha

90

Colour: cherry, garnet rim. Nose: fruit preserve, powerful, black liquorice, aromatic coffee. Palate: flavourful, long.

Doble Cuerpo 2019 T
100% garnacha

88

Jammy, dried herbs, tasty, toasty.

La Cuna 2021 T RB
100% garnacha

89

Pruney, full-bodied, tasty, great length, sweety finish, dried herbs, varietally correct.

Montuno 2019 T RB
50% garnacha tintorera, 20% monastrell, 20% bobal, 10% graciano

87

Raíces Monastrell 2021 T
100% monastrell

89

Jammy, wild, varietally correct, balanced, great length, tasty.

Ramito Tempranillo Syrah 2021 T
70% tempranillo, 30% syrah

88 ❦

Jammy, fruity, tasty, wild, balanced.

RAÚL PÉREZ BODEGAS Y VIÑEDOS

Bulevar Rey Juan Carlos 1º Rey de España, 11 B
24400 Ponferrada (León)
contacto@raulperez.com
www.raulperez.com

Raúl Pérez El Rapolao 2020 T

93

Colour: bright cherry. Nose: balsamic herbs, sweet spices, scrubland, red berry notes, fruit expression. Palate: spicy, balsamic, good acidity.

Raúl Pérez Pan y Carne 2020 T

94

Colour: Cherry. Nose: balsamic herbs, sweet spices, scrubland, red berry notes, chamomile. Palate: spicy, balsamic, good acidity.

REMIGIO ENRIQUEZ

Rua Dos Chaos, 20
32001 Valdo Regueiro (Ourense/Orense)
☎: +34 659 327 453
cotodesantamarta@gmail.com
www.vinosremigioenriquez.com

Remigio Enriquez (Yayos) 2016 T BA
80% tempranillo, 20% garnacha

84

Remigio Enríquez Restlessness 2016 T BA
90% tempranillo, 10% cabernet sauvignon, merlot

88

Fruity, spicy, toasty, jammy.

Remigio Enríquez Restlessness 2017 T BA
tempranillo, cabernet sauvignon, merlot

89

Age nuances, spicy, jammy, balanced, dried herbs. Nose: fine reductive notes, tobacco.

Remigio Enríquez Restlessness 2018 T BA
tempranillo, cabernet sauvignon, merlot

90

Colour: deep cherry. Nose: ripe fruit, spicy, dried herbs, smoky. Palate: powerful, ripe fruit, round tannins, good acidity.

SEÑORÍO DE LOS ARCOS
Ctra.. Caboalles, 332
24191 Villlabalter (León)
☎: +34 987 226 594
info@senoriodelosarcos.es
www.senoriodelosarcos.es

Seda B SD
85

SEÑORÍO DEL JÚCAR
Pol. Ind. Parc 64-70
02200 Casas Ibáñez (Albacete)
☎: +34 967 460 564
administracion@bsjucar.es
www.senoriodeljucar.com

Parajes del Valle Maceración Macabeo 2021 B
macabeo
90 🌿
Little interventionist. Colour: coppery red, ochre. Nose: citrus fruit, white fruit, medium intensity, fragrant herbs. Palate: easy to drink, good finish, fresh.

SHUKHRAT KHAKIMOV & VITICULTORES
Plz Constitución, 7
03550 San Juan de Alicante (Alacant/Alicante)
☎: +34 965 943 090
Fax: +34 965 943 090
a.nogay@winexfood.com
wwww.winexfood.com

Orange Sk & V 2020 B
100% semillón
87

Teo Viognier 2020 B S
100% viognier
89
Little interventionist, slight oxidation, citrus fruit, fruity, dried flowers, herbal, spicy.

Teo Viognier 2021 B RB
viognier
91
Little interventionist. Colour: straw. Nose: dried herbs, faded flowers, white fruit, stone fruit, toasted bread. Palate: ripe fruit, balanced, rich.

Teo Viognier Maceración 2020 B S
100% viognier
90
Little interventionist, slight oxidation. Colour: straw. Nose: dried herbs, faded flowers, stone fruit, macerated fruit. Palate: ripe fruit, balanced, rich, fruity, flavourful.

Villana Llévame al Huerto 2020 B
100% tardana
91 🌿
Little interventionist. Colour: straw. Nose: dried herbs, faded flowers, white fruit. Palate: powerful, ripe fruit, balanced.

SILICE VITICULTORES
Travesía de Pardavila, 17 - 2º A
36205 Vigo (Pontevedra)
☎: +34 986 312 137
info@sciliceviticultores.com

Sílice 2019 T
mencía, albarello, merenzao, garnacha tintorera
93
Colour: Cherry. Nose: red berry notes, raspberry, floral, wild herbs, balsamic herbs, musky notes. Palate: fleshy, fresh, round tannins.

Sílice Klarete 2019 RD
mencía, palomino
90
Colour: rose. Nose: red berry notes, ripe fruit, dried herbs, smoky. Palate: fleshy, ripe fruit, fresh.

Sílice 2020 B
treixadura, godello, palomino
92
Colour: bright straw. Nose: fragrant herbs, fine lees, citrus fruit, burnt matches. Palate: good acidity, balanced, flavourful.

Sílice Rosado 2019 RD
mencía, merenzao, palomino
90
Colour: onion pink. Nose: lactic notes, dried herbs, red berry notes, fine lees. Palate: fresh, balanced, fine bitter notes.

Sílice Xabrega 2019 T
mencía, garnacha tintorera
93
Colour: light cherry. Nose: red berry notes, sweet spices, wild herbs, smoky. Palate: fresh, round tannins, balanced, flavourful.

SOLAR DE URBEZO
San Valero, 14
50600 Cariñena (Zaragoza)
☎: +34 976 621 968
urbezoecologico@solardeurbezo.es
www.solardeurbezo.es

Nostre 2021 T
88
Pleasant, fruity, sweety finish, tasty.

SOPLA LEVANTE
Pérez Galdós, 3.
02003 Albacete (Albacete)
☎: +34 696 168 873
jose@rodriguezdevera.com
www.rodriguezdevera.com

Sopla Levante La Horca 2020 B
85% merseguera, 15% otras
86

Sopla Levante Lomas del Polo 2020 T
85% parrell, 15% otras
85

Sopla Levante Pinomar Orange Wine 2020 B
85% merseguera, 15% otras
83

SOTO DE OÑATIL
Pol. Ind. Lentiscares
26370 Navarrete (La Rioja)
☎: +34 941 451 129
info@premiumfincas.com

C0001 Numbered Edition B
87

Celler i Vinyes Nostre Blanc 2021 B
86

Celler i Vinyes Nostre Rosat 2021 RD
86

SOTO MANRIQUE BODEGA
Ctra. Avila Toledo s/n
05260 Cebreros (Ávila)
☎: +34 626 290 408
info@sotomanrique.com
www.sotomanrique.com

La Orquesta 2019 B
93
Colour: bright yellow. Nose: expressive, white flowers, jasmine, dried herbs, mineral. Palate: flavourful, fruity, balanced.

SUERTES DEL MARQUÉS
Cº Las Suertes Tercera, 10
38300 La Orotava (Santa Cruz de Tenerife)
☎: +34 922 501 300
Fax: +34 922 503 462
ventas@suertesdelmarques.com
www.suertesdelmarques.com

Sortevera 2020 B
listán blanco
94
Colour: bright yellow. Nose: citrus fruit, white flowers, dried flowers, dry stone, chalk, sweet spices. Palate: good structure, long, toasty, fine bitter notes, saline.

Sortevera 2020 T
listán negro, negramoll, listán rosado, vijariego negro, tintilla, moscatel negro
94
Crisp, herbal, elegant, representative. Colour: cherry, garnet rim. Nose: red berry notes, neat, expressive, wild herbs. Palate: fruity, easy to drink, long.

TEO LEGIDO
05229 Castellanos de Zapardiel (Ávila)
☎: +34 605 619 723
teolegido@icloud.com
www.teolegido.com

El Rosal 2019 T
garnacha
93
Colour: light cherry. Nose: balsamic herbs, ripe fruit, scrubland, dried flowers, red berry notes. Palate: flavourful, balsamic, spicy, soft tannins.

🏆 **PODIUM**

La Bovila 2019 T
tempranillo, syrah, verdejo
95 🍷
Colour: deep cherry. Nose: balsamic herbs, sweet spices, ripe fruit, red berry notes, scrubland. Palate: spicy, balsamic, good acidity.

Las Galgas 2020 B
verdejo
93 🍷
Briny, little interventionist. Colour: bright yellow. Nose: powerful, creamy oak, ripe fruit, spicy. Palate: rich, good structure, long, fine bitter notes.

TERRA DE FALANIS
Ctra. Campos Felanitx 10.3 Km.
07200 Felanitx (Illes Balears/Islas Baleares)
☎: +34 679 314 406
contactoterradefalanis@gmail.com
www.terradefalanis.com

Castell de Santueri Blanc 2021 B
50% callet, 40% prensal, 10% giró ros

88
Fruity, tropical, dried herbs, pleasant.

GTX Pirate Wine 2016 T
100% garnacha

91
Colour: light cherry. Nose: ripe fruit, dried flowers, wild herbs. Palate: balanced, easy to drink, spicy, fine bitter notes.

THUNDERWINEMAKERS & AVGVSTVS FORVM
Sicilia, 376.
08025 Barcelona (Barcelona)
☎: +34 625 652 276
info@winethunder.com
www.winethunder.com

Nvde MCB (Macabeu Ánfora) 2020 B
macabeo

89 🌿
Little interventionist, dried flowers, dried herbs, smooth, wild.

Nvde MCB (Macabeu Ciment) 2020 B

88 🌿
Standard, dried flowers, great length, wild, balanced, little interventionist.

Nvde MCB (Macabeu Ciment) 2021 B
macabeo

88 🌿
Standard, little interventionist, wild, smooth.

Nvde Sauvignon Blanc 2020 B
sauvignon blanc

91 🌿
Colour: yellow. Nose: characterful, expressive, dried herbs, wild herbs. Palate: flavourful, full-bodied.

Nvde Sauvignon Blanc 2021 B
sauvignon blanc

90 🌿
Colour: bright straw. Nose: white fruit, ripe fruit, hint of anise. Palate: flavourful, easy to drink.

TIERRA FUNDIDA
Camino Las Medianías 13.
Finca Morales. Los Baldíos.
38291 San Cristóbal de la Laguna
(Santa Cruz de Tenerife)
☎: +34 647 989 081
vinosentandem@gmail.com
www.tierrafundida.es

Tierra Fundida Blanco Cercado del Pino 2020 B S
verdello

90
Oxidativ. Colour: bright straw. Nose: fresh fruit, citrus fruit, wild herbs. Palate: fresh, fruity, good acidity.

Tierra Fundida Clarete Los Topes 2021 T
vijariego negro, gual

89
Citrus fruit, dried herbs, mineral, crisp.

Tierra Fundida Los Topes 2020 B
albillo criollo

90
Colour: straw. Nose: white flowers, dried herbs, white fruit. Palate: flavourful, fruity, balanced, mineral.

TOMÁS TORRESANO
Calle Colón, 67
45840 La Puebla de Almoradiel (Toledo)
☎: +34 655 646 353
info@tomastorresano.com
www.tomastorresano.com

Vivo del Aire Airén 2021 B
airén

89
Jammy, simple, rustic, tasty.

TORRE DEL VEGUER
Urb. Torre de Veguer, s/n
08810 Sant Pere de Ribes (Barcelona)
☎: +34 938 963 190
torredelveguer@torredelveguer.com
www.torredelveguer.com

Torre del Veguer Llum del Cadí 2019 T
pinot noir

92
Colour: light cherry, garnet rim. Nose: balsamic herbs, sweet spices, scrubland, toasty, earthy notes. Palate: spicy, balsamic, good acidity.

Torre del Veguer Pur 2021 B
malvasía de Sitges

90 🌿
With personality, little interventionist. Nose: characterful, chamomile, dried flowers. Palate: flavourful, bitter.

UNGLES NEGRES

Farigola, 24
08023 Barcelona (Barcelona)
☎: +34 606 804 025
ferran@ferrannoquera.com

Ungles Negres 2021 T
100% syrah

92

Colour: deep cherry, bright cherry. Nose: black fruit, dried flowers, wild herbs. Palate: ripe fruit, spicy, round tannins, flavourful, easy to drink.

VALSANGIACOMO

Ctra. Cheste - Godelleta, Km. 1
46370 Chiva (València/Valencia)
☎: +34 649 810 834
cherubino@cherubino.es
www.valsangiacomo.es

Inverso Godello 2021 B S
100% godello

90

Colour: straw. Nose: ripe fruit, dried herbs, faded flowers, fine lees. Palate: ripe fruit, balanced, rich, flavourful.

Ulises Valsangiacomo Chardonnay 2021 B FB
100% chardonnay

91

Colour: bright straw. Nose: ripe fruit, fragrant herbs, fine lees. Palate: full-bodied, rich, long, good acidity.

VENTA D'AUBERT

Ctra. Valderrobres a Arnes, Km. 28
44623 Cretas (Teruel)
☎: +34 978 769 021
info@ventadaubert.com
www.ventadaubert.com

Dionus 2016 T R
cabernet sauvignon, merlot

91 ♣

Colour: bright cherry. Nose: ripe fruit, black fruit, spicy, smoky, wild herbs. Palate: fruity, flavourful, round tannins, balanced.

Divertus 2018 T
garnacha, syrah, monastrell

90 ♣

Colour: cherry, garnet rim. Nose: toasty, spicy, black fruit, scrubland. Palate: flavourful, toasty, fine bitter notes.

Venta D'Aubert Viognier 2021 B
viognier

90 ♣

Colour: bright straw. Nose: fragrant herbs, fine lees, stone fruit, floral. Palate: full-bodied, rich, good acidity, balanced.

Ventus 2016 T R
cabernet sauvignon, garnacha, merlot, syrah, monastrell, cabernet franc

89 ♣

Spicy, dried herbs, jammy, age nuances, toasty.

VICENTE GANDÍA

Ctra. Cheste a Godelleta, s/n
46370 Chiva (València/Valencia)
☎: +34 962 524 242
Fax: +34 962 524 243
info@vicentegandia.com
www.vicentegandia.es

Avento Malvasía Moscatel 2021 B SS
malvasía, moscatel

86

Finca del Mar Verdejo 2021 B
verdejo

86

Nebla Verdejo Rosé 2021 RD
verdejo, garnacha

88

Creamy, balanced, herbaceous, fruity.

Uva Pirata 2021 B
garnacha blanca

89

Pleasant, dried flowers, balanced, wild, tasty, kind finish.

Uva Pirata Garnacha 2021 T
garnacha tintorera

87

Vicente Gandía Chardonnay sobre Lías 2021 B
chardonnay

88

Standard, floral, fruity, jammy, tasty, balanced.

TABLE WINES / WINES

VINS PEPE RAVENTOS

Plaça del Roure s/n
08240 Sant Sadurní d'Anoia (Barcelona)
☎: +34 938 183 262
info@raventos.com
www.vinspeperaventos.com

Bastard Negre de les Terrasses del Serral 2015 T
bastardo

93

Colour: cherry, garnet rim. Nose: balsamic herbs, ripe fruit, scrubland, old leather. Palate: flavourful, balsamic, spicy.

Bastard Negre de les Terrasses del Serral 2017 T
bastardo

93

Colour: deep cherry. Nose: ripe fruit, creamy oak, balsamic herbs, grassy, complex, characterful. Palate: powerful, ripe fruit, spicy, round tannins.

Xarel.lo Vinya del Noguer 2018 B
xarel.lo

93

Balsamic herbs, little interventionist. Colour: bright straw, greenish rim. Nose: citrus fruit, wild herbs, candied fruit, ripe fruit. Palate: fresh, fruity, good acidity, fine bitter notes.

Xarel.lo Vinya del Noguer 2019 B
xarel.lo

91

Little interventionist, needs time. Colour: bright straw, greenish rim. Nose: fresh fruit, citrus fruit, wild herbs. Palate: fresh, fruity, good acidity, fine bitter notes.

VINYES MORTITX

Ctra. Pollença Lluc, Km. 10,9
07315 Escorca (Illes Balears/Islas Baleares)
☎: +34 971 533 889
info@vinyesmortitx.com
www.vinyesmortitx.com

Giró Ros de Mortitx 2021 B
giró ros

87

VIÑA BALBAINA

Viña La Torre, Ctra. Jerez-Rota, Km. 8
11408 Jerez de la Frontera (Cádiz)
☎: +34 617 071 349
mgalan@vbalbaina.es
www.vbalbaina.es

Esencia de la Torre 2017 T BA
60% petit verdot, 40% tempranillo

90

Herbal, spicy, wild. Colour: cherry, garnet rim. Nose: black fruit, dried herbs, characterful. Palate: flavourful, round tannins.

Esencia de la Torre Chardonnay 2021 B
chardonnay

85

Esencia de la Torre Petit Verdot 2018 T BA
100% petit verdot

88

Fruity, jammy, spicy, standard, tasty.

VIÑA CALLEJUELA

Camino del Reventón Chico, 27
11540 Sanlúcar de Barrameda (Cádiz)
☎: +34 617 492 483
info@callejuela.es
www.callejuela.es

Blanco de Hornillos 2021 B
palomino

89

Oxidativ, herbal, yeasty notes, tasty.

VIÑA EL PISÓN

Santa Engracia, 11
01300 Laguardia (Araba/Álava)
☎: +34 945 600 119

🏆 **PODIUM**

Viña El Pisón 2020 T
100% tempranillo

97 ♣

Complex, exuberant. Colour: bright cherry. Nose: complex, expressive, spicy, mineral, ripe fruit, black fruit, creamy oak. Palate: elegant, full-bodied, long, great length.

VIÑOS DE ENCOSTAS

Lugar de O Pazo, 4 As Viñas Gomariz
32429 Leiro (Ourense/Orense)
☎: +34 630 862 953
sebio@xlsebio.es
www.xlsebio.es

As Viñas 2019 B
albariño, treixadura, otras

93

Colour: bright yellow. Nose: dried flowers, candied fruit, fine lees, pattiserie. Palate: round, spicy, long, great length.

Heaven & Hell 2020 B
treixadura, godello, albariño

93

Colour: straw. Nose: ripe fruit, dried herbs, faded flowers, sweet spices, mineral. Palate: powerful, ripe fruit, balanced.

Máis Alá 2020 B
godello

94

Colour: bright straw. Nose: expressive, ripe fruit, floral, fine lees, mineral. Palate: full-bodied, complex, spicy, long, elegant.

O Con 2020 B
albariño

94

Colour: bright straw. Nose: fruit expression, ripe fruit, floral, spicy, dried herbs. Palate: flavourful, fresh, good acidity, fruity aftertaste.

Salvaxe 2019 B
lado, treixadura, albariño, godello, silveiriña

93

Colour: bright yellow. Nose: dried flowers, candied fruit, fine lees, toasted bread, roasted almonds. Palate: round, spicy, long, great length.

Village 2020 B
albariño, treixadura, otras

93

Colour: bright straw. Nose: dried herbs, faded flowers, stone fruit. Palate: powerful, ripe fruit, balanced.

VIRGEN DE LAS VIÑAS BODEGA Y ALMAZARA

Ctra. Argamasilla de Alba, 1
13700 Tomelloso (Ciudad Real)
☎: +34 926 510 865
comercio.exterior@vinostomillar.com
www.vinostomillar.com

Octavo Arte Cabernet Sauvignon T
cabernet sauvignon

85

Octavo Arte Merlot T
merlot

88

Balanced, crisp, fruity, herbal, toasty.

Octavo Arte Sauvignon Blanc B
sauvignon blanc

87

VITICULTOR Y ELABORADOR JUAN BERNAL

Avda. América 8. 1ºA
11520 Rota (Cádiz)
☎: +34 607 813 154
enología@bodegajf.es

Alumbra 2019 T C
malbec

90

Slight reduction. Colour: deep cherry. Nose: characterful, waxy notes, wild herbs, violets. Palate: bitter, flavourful, fleshy.

Oceánidas 2019 T C
syrah, pinot noir, malbec

91

Colour: Cherry. Nose: scrubland, red berry notes, black fruit, dried flowers, waxy notes. Palate: spicy, balsamic, good acidity.

Oceánidas Tintilla 2019 T C
tintilla de rota

90

Colour: cherry, garnet rim. Nose: ripe fruit, aged wood nuances, sweet spices. Palate: spicy, long, good acidity, grainy tannins.

XOLAYR

Avda. Andalucía, 1
18659 Cozvijar (Granada)
☎: +34 620 126 514
lopezdelacasa@gmail.com

Elvira Moscatel Grano Menudo 2020 B
moscatel grano menudo

86

Elvira Vigiriega 2020 B
vijariego blanco

88

Citrus fruit, herbal, toasty, tasty.

SPARKLING WINES

Under this heading you will find all the table wines that have some relation with fizz and that do not have any specific mention such as Denominación de Origen, Vinos de calidad, Vinos de pago, etc.

The reader will find here some of the best second fermentation sparkling wines in Spain today, which coexist in these pages together with other wines whose relationship with carbon dioxide comes from other production methods such as the pet nat, semi-sparkling, Charmat, etc.

SPARKLING WINES

3VVV SINGULAR & WINES
Avda. Tarragona, s/n
25300 Tárrega (Lleida/Lérida)
☎: +34 630 179 962
isidre.ribalta@carviresa.com

Trans Trepat Ancestral 2020 RE
88
Kind finish, jammy, fruity, spicy, toasty.

ALTA ALELLA - CELLER DE LES AUS
Cami Baix de Tiana, s/n
08328 Alella (Barcelona)
☎: +34 934 693 720
info@altaalella.wine
www.altaalella.wine

AUS Pét-Nat 2021 BE AG
pansa blanca
90
Colour: bright yellow. Nose: ripe fruit, fine lees, balanced, dried herbs. Palate: flavourful, ripe fruit, full-bodied.

AUS Rosé Pét-Nat 2021 RE AG
89
Fruity, jammy, kind finish, floral, smooth, pleasant.

AT ROCA
La Vinya, 15-17
08770 Sant Sadurní d'Anoia (Barcelona)
☎: +34 935 165 043
info@atroca.eu
www.atroca.eu

Anima Mundi Ancestral Cami dels Xops 2021 BE
xarel.lo, macabeo
90
Aromatic, kind finish. Nose: floral, white fruit. Palate: fruity, easy to drink, good finish.

Anima Mundi Ancestral Noguer Baix 2016 BE
macabeo
92
Colour: yellow. Nose: spicy, dry nuts, dried flowers, fine lees, bakery, wild herbs. Palate: good acidity, taut, fine bitter notes.

BODEGA CASTILLO DE MONJARDÍN
Viña Rellanada, s/n
31242 Villamayor de Monjardin (Navarra)
☎: +34 948 537 412
contacto@monjardin.es
www.monjardin.es

Monjardín Chardonnay Selección Familiar 2016 BE GR BN
chardonnay
89
Aromatic, spicy, smooth. Nose: caramel. Palate: easy to drink, good finish.

BODEGA ROMAILA
45420 Almoncid de Toledo (Toledo)
☎: +34 915 416 561
administracion@romaila.es
www.romaila.es

Romaila Pet Nat 2021 RE
100% garnacha
86

Romaila Pet Nat BE
88
Pleasant, aromatic, smooth, wild, dried flowers, sweety finish.

BODEGAS ALCEÑO
Barrio Iglesias, 34
30520 Jumilla (Murcia)
☎: +34 968 780 142
info@alceno.com
www.alceno.com

Alceño Rosé RE BN
monastrell
88
Kind finish, fruity, jammy, tasty, balanced.

BODEGAS CAMPESTRAL
Ctra. Arcos-Algar, km. 7
11630 Arcos de la Frontera (Cádiz)
☎: +34 670 586 035
info@campestral.es
www.campestral.es

Campestral Red Ancestral 2020 TE
syrah, merlot, cabernet sauvignon, tintilla de rota, petit verdot
82

Campestral White Ancestral 2020 BE
palomino
89
Citrus fruit, balanced, spicy, dried flowers, yeasty notes, tasty, original.

BODEGAS DELGADO
Cosano, 2
14500 Puente Genil (Córdoba)
☎: +34 957 600 085
fino@bodegasdelgado.com
www.bodegasdelgado.com

Lemonier 2021 BE BN
pedro ximénez
91
Colour: bright yellow. Nose: ripe fruit, fine lees, balanced, dried herbs, flor yeasts. Palate: good acidity, flavourful, ripe fruit, long.

BODEGAS MIGUEL A. AGUADO
Cantalejos, 2
45165 San Martín de Montalbán (Toledo)
☎: +34 653 821 659
info@bodegasmiguelaguado.com
www.bodegasmiguelaguado.com

Pasión de Castillo de Montalban 2021 BE BN
100% macabeo
79

Pasión de Castillo de Montalban BE R
81

BODEGAS MUÑANA
Ctra. Graena a La Peza, s/n Finca Peñas Prietas
18517 Cortes Y Graena (Granada)
☎: +34 958 670 715
administracion@bodegasmunana.com
www.bodegasmunana.com

Muñana 1188 BE BN
chardonnay
85

BODEGAS TIERRA SAVIA
41370 Cazalla de La Sierra (Sevilla)
☎: +34 623 068 615
bodegastierrasavia@gmail.com

Piu Ancestral 2020 BE
parrona
88
Citrus fruit, yeasty notes, dried herbs, spicy, kind finish.

CAN DESCREGUT
Masia Can Descregut s/n
08735 Vilobí del Penedès (Barcelona)
☎: +34 938 978 273
info@descregut.com
www.descregut.com

Descregut 2019 BE BN
60% xarel.lo, 30% macabeo, 10% parellada
90 🌱
Colour: bright straw. Nose: fresh fruit, dried herbs, fine lees, floral. Palate: fresh, fruity, flavourful, good acidity.

Fumissola 2018 BE BN
50% sumoll, 50% xarel.lo vermell
93 🌱
With personality. Colour: bright straw. Nose: burnt matches, expressive, neat, fragrant herbs. Palate: flavourful, long, easy to drink.

Memòria 2016 BE BN
60% xarel.lo, 40% macabeo
92 🌱
Colour: bright straw. Nose: fine lees, floral, fragrant herbs, expressive, burnt matches. Palate: flavourful, good acidity, fine bead, balanced.

CAN GRAU VELL
Can Grau Vell, s/n
08781 Hostalets de Pierola (Barcelona)
☎: +34 676 586 933
Fax: +34 932 684 965
info@grauvell.cat
www.grauvell.cat

Y la Nave Va RE
88
Little interventionist, wild, dried herbs, standard. Nose: cereal notes.

CAN RICH
Camí de Sa Vorera, s/n
07820 San Antonio (Illes Balears/Islas Baleares)
☎: +34 971 803 377
info@bodegascanrich.com
www.bodegascanrich.com

Can Rich Blanc de Blancs 2020 BE
malvasía
84 🌱

SPARKLING WINES

SPARKLING WINES

CAN SUMOI
Plaça del Roure s/n
08240 Sant Sadurní d'Anoia (Barcelona)
☎: +34 938 183 262
info@cansumoi.cat

Ancestral Montonega 2021 BE
montonega
90
Creamy, crisp. Colour: bright yellow. Nose: fine lees, balanced, dried herbs, citrus fruit. Palate: good acidity, flavourful.

CODORNÍU
Avda. Jaume Codorníu, s/n
08770 Sant Sadurní d'Anoia (Barcelona)
☎: +34 938 183 232
info@codorniu.com
www.codorniu.es

Codorníu 150 Aniversario Edición Limitada BE BR
91
Colour: bright yellow. Nose: ripe fruit, fine lees, balanced, dried herbs. Palate: good acidity, flavourful, ripe fruit.

Codorníu Zero BE
86

GRAMONA
Industria, 36
08770 Sant Sadurní d'Anoia (Barcelona)
☎: +34 938 910 113
info@gramona.com
www.gramona.com

🏆 PODIUM
Enoteca Gramona 2005 BE BN
98
Colour: bright golden. Nose: fine lees, dry nuts, fragrant herbs, complex, toasty, fine reductive notes. Palate: powerful, flavourful, good acidity, fine bead, fine bitter notes.

🏆 PODIUM
Enoteca Gramona 2006 BE BN
60% xarel.lo, 40% macabeo
99
Colour: bright golden. Nose: dry nuts, fragrant herbs, complex, fine lees, sweet spices, expressive. Palate: powerful, flavourful, good acidity, fine bead, fine bitter notes, elegant.

Gramona Argent 2017 BE BR
100% chardonnay
92 🌱
Colour: bright golden. Nose: fine lees, fragrant herbs, ripe fruit, dry nuts, wild herbs. Palate: powerful, flavourful, good acidity, fine bead, fine bitter notes.

Gramona Argent Rosé 2018 RE BN
90% pinot noir, 10% chardonnay
91 🌱
Colour: raspberry rose. Nose: elegant, red berry notes, floral, fragrant herbs. Palate: light-bodied, spicy, good acidity, fine bitter notes.

🏆 PODIUM
Gramona Celler Batlle 2012 BE BR
60% xarel.lo, 40% macabeo
98
Complex, spicy. Colour: bright yellow. Nose: elegant, expressive, spicy, balanced. Palate: rich, full-bodied, flavourful, full of life, long, spicy, fine bitter notes.

🏆 PODIUM
Gramona III Lustros 2013 BE BN
70% xarel.lo, 30% macabeo
95
Colour: bright golden. Nose: fragrant herbs, characterful, ripe fruit, dry nuts. Palate: powerful, flavourful, good acidity, fine bead.

Gramona III Lustros 2014 BE BN
94
Colour: bright yellow. Nose: fine lees, bakery, complex, balanced, neat. Palate: flavourful, long, fine bitter notes, fine bead.

Gramona Imperial 2015 BE BR
xarel.lo, 30% macabeo, 15% chardonnay, 5% parellada
92 🌱
Colour: bright straw. Nose: fine lees, floral, fragrant herbs, expressive, spicy. Palate: good acidity, fine bead, balanced.

Gramona Innoble BE BN
100% xarel.lo
93
Colour: bright straw. Nose: fine lees, floral, fragrant herbs, expressive, sweet spices, toasted bread. Palate: powerful, flavourful, good acidity, fine bead, balanced.

Gramona La Cuvée 2017 BE
60% xarel.lo, 30% macabeo, 10% parellada
91 🌱
Colour: bright yellow. Nose: ripe fruit, fine lees, balanced, stone fruit. Palate: good acidity, flavourful, ripe fruit, fleshy, sweetness.

🏆 PODIUM

Gramona TLN 2009 BE BN
60% xarel.lo, 40% macabeo

96
Colour: bright golden. Nose: fine lees, dry nuts, fragrant herbs, toasty, bakery. Palate: flavourful, good acidity, fine bead, fine bitter notes.

HATORI
Las Damas, 30
40500 Riaza (Segovia)
☎: +34 620 214 755
hatorihanzosl@yahoo.es

Malevaje 2019 BE
albillo mayor

91
Colour: straw, pale. Nose: cereal notes, burnt matches, dry nuts, citrus fruit. Palate: flavourful, fresh.

HUGUET DE CAN FEIXES
Finca Can Feixes, 1
08718 Cabrera D'Anoia (Barcelona)
☎: +34 937 718 227
canfeixes@canfeixes.com
www.canfeixes.com

Huguet de Can Feixes 2014 BE GR BN
parellada, macabeo, pinot noir

92 🌱
Colour: straw. Nose: expressive, dry nuts, fragrant herbs, spicy, white fruit. Palate: flavourful, fine bead, long, great length.

Huguet de Can Feixes Classic 2014 BE GR BR
parellada, macabeo, pinot noir

92 🌱
Colour: bright straw. Nose: fine lees, floral, fragrant herbs, expressive. Palate: powerful, flavourful, good acidity, fine bead, balanced.

Huguet Magnum 2010 BE R BN

92
Colour: bright golden. Nose: fine lees, fragrant herbs, ripe fruit, dry nuts, faded flowers. Palate: powerful, flavourful, good acidity, fine bead, fruity.

JÚLIA BERNET
Avda.de Barcelona, 24
08739 El Pago (Barcelona)
☎: +34 639 273 965
info@juliabernet.com
www.juliabernet.com

Júlia Bernet 130 2017 BE BN
xarel.lo

93 🌱
Colour: bright yellow. Nose: ripe fruit, fine lees, balanced, dried herbs, brioche. Palate: good acidity, flavourful, ripe fruit, long, elegant.

Júlia Bernet Exsum 2018 BE BN
xarel.lo

94 🌱
Defined aromas, balanced. Colour: straw. Nose: neat, expressive, elegant, spicy, pattiserie. Palate: rich, flavourful, bitter, good acidity.

🏆 PODIUM

María Bernet 2013 BE BN
xarel.lo

95 🌱
Colour: bright golden. Nose: fine lees, dry nuts, fragrant herbs, complex, toasty, brioche. Palate: powerful, flavourful, good acidity, fine bead, fine bitter notes, full of life, good finish.

María Bernet 2015 BE BN
xarel.lo

92 🌱
Colour: bright golden. Nose: fine lees, fragrant herbs, ripe fruit, dry nuts, lactic notes. Palate: flavourful, good acidity, fine bead, fine bitter notes.

LLOPART
Ctra. de Sant Sadurni - Ordal, Km. 4
08739 Subirats (Barcelona)
☎: +34 938 993 125
info@llopart.com
www.llopart.com

Llopart 2018 BE R BN
45% xarel.lo, 35% macabeo, 20% parellada

91 🌱
Austere, representative. Colour: bright straw. Nose: fine lees, fragrant herbs, fresh. Palate: powerful, flavourful, good acidity, fine bead, balanced.

Llopart Ex·Vite Viñas Singulares Les Flandes 2012 BE BR
60% xarel.lo, 40% macabeo

94 🌱
Colour: bright golden. Nose: fine lees, dry nuts, fragrant herbs, complex, wild herbs. Palate: powerful, flavourful, good acidity, fine bead, fine bitter notes.

SPARKLING WINES

SPARKLING WINES

Llopart Leopardi 2015 BE BN
45% xarel.lo, 40% macabeo, 15% parellada
92
Colour: bright straw. Nose: fine lees, fragrant herbs, expressive, candied fruit. Palate: powerful, flavourful, good acidity, fine bead.

Llopart Microcosmos Rosé 2019 RE BN
85% pinot noir, 15% monastrell
91
Floral, crisp, fruity, kind finish. Colour: salmon. Nose: fresh, balanced, rose petals. Palate: fruity, fine bitter notes.

🏆 PODIUM

Llopart Original 1887 Viñas Singulares Les Flandes 2011 BE BN
50% montonega, 25% xarel.lo, 25% macabeo
96
Colour: bright golden. Nose: dry nuts, fragrant herbs, complex, fine lees, sweet spices. Palate: powerful, flavourful, good acidity, fine bead, fine bitter notes.

Llopart Panoramic 2016 BE BR
50% xarel.lo, 40% macabeo, 10% parellada
93
Colour: bright straw. Nose: fine lees, fragrant herbs, expressive, stone fruit, toasted bread. Palate: powerful, flavourful, good acidity, fine bead.

MAS CANDÍ
Mas Candí, s/n
08793 Les Gunyoles d'Avinyonet (Barcelona)
☎: +34 636 621 510
info@mascandi.com
www.mascandi.com

Indomable 2016 BE BN
xarel.lo, sumoll
91
Colour: bright straw. Nose: fine lees, floral, fragrant herbs, expressive. Palate: flavourful, good acidity, fine bead, balanced, fresh.

Mas Candí 2019 BE BN
xarel.lo, macabeo, parellada, sumoll
91
Colour: bright straw. Nose: ripe fruit, fine lees, dried herbs, faded flowers. Palate: flavourful, good acidity, fine bead, fruity.

Prohibit 2018 RE BN
91
Colour: raspberry rose. Nose: elegant, red berry notes, floral, fragrant herbs. Palate: spicy, good acidity, fine bitter notes, powerful.

Segunyola 2017 BE BN
100% xarel.lo
93
Dried flowers, rustic, with personality. Colour: bright straw. Nose: wild herbs, fragrant herbs, dried herbs. Palate: flavourful, fine bitter notes, balanced.

NADAL
Finca Nadal de la Boadella, s/n
08775 Torrelavit (Barcelona)
☎: +34 938 988 011
comunicacio@nadal.com
www.nadal.com

Ramón Nadal Giró 2004 BE GR BR
62% xarel.lo, 32% parellada
92
Colour: bright golden. Nose: fine lees, fragrant herbs, ripe fruit, dry nuts, dried flowers. Palate: powerful, flavourful, good acidity, fine bead, fine bitter notes.

🏆 PODIUM

RNG 20 Magnum 1997 BE BN
97
Colour: bright golden. Nose: fragrant herbs, complex, fine lees, sweet spices, toasted bread. Palate: powerful, flavourful, good acidity, fine bead, fine bitter notes.

RNG 2014 BE BR
91
Colour: bright straw. Nose: fine lees, floral, fragrant herbs, expressive. Palate: powerful, flavourful, good acidity, fine bead.

Salvatge Edició Limitada 2013 BE BN
macabeo, xarel.lo, parellada
93
Colour: bright straw. Nose: fine lees, fragrant herbs, expressive, sweet spices, dry nuts. Palate: powerful, flavourful, good acidity, fine bead.

Salvatge Edició Limitada 2015 BE BR
91
Colour: bright straw. Nose: medium intensity, dried herbs, fine lees, candied fruit. Palate: fruity, flavourful, good acidity.

Salvatge Rosé 2018 RE BR
54% pinot noir, 46% macabeo
89
Pleasant, standard, balanced, floral, fruity, smooth, kind finish.

RAVENTÓS I BLANC
Plaça del Roure, s/n
08770 Sant Sadurní d'Anoia (Barcelona)
☎: +34 938 183 262
info@raventos.com
www.raventos.com

🏆 PODIUM

Enoteca Personal Manuel Raventós Negra Magnum 2013 BE BN
sumoll, xarel.lo

95

Nose: fine lees, dry nuts, fragrant herbs, toasty. Palate: flavoured, good acidity, fine bead, fine bitter notes.

🏆 PODIUM

Manuel Raventós 2011 BE BN
100% xarel.lo

97

Colour: bright golden. Nose: dry nuts, fragrant herbs, complex, fine lees, sweet spices, expressive. Palate: powerful, flavourful, good acidity, fine bead, fine bitter notes, elegant.

Manuel Raventós 2014 BE BN
sumoll, xarel.lo

94

Colour: bright golden. Nose: fine lees, characterful, ripe fruit, dry nuts, pattiserie, toasted bread. Palate: flavourful, fine bead, fine bitter notes, balanced.

🏆 PODIUM

Manuel Raventós 2015 BE BN
xarel.lo, macabeo

95

Colour: bright golden. Nose: fine lees, dry nuts, fragrant herbs, complex, toasty. Palate: powerful, flavourful, good acidity, fine bead, fine bitter notes.

Raventós i Blanc Blanc de Blancs 2020 BE

90

Colour: bright yellow. Nose: dried herbs, fresh fruit, white flowers. Palate: good acidity, flavourful, ripe fruit.

Raventós i Blanc De La Finca 2019 BE BN
xarel.lo, macabeo, parellada

92 🏆

Colour: bright straw. Nose: fine lees, fragrant herbs, expressive, toasted bread. Palate: powerful, flavourful, good acidity, fine bead, balanced.

Raventós i Blanc De Nit 2020 RE BR
xarel.lo, macabeo, parellada, monastrell

91

Colour: raspberry rose. Nose: red berry notes, floral, fragrant herbs. Palate: light-bodied, spicy, good acidity.

Raventos i Blanc Textures de Pedra 2017 BE BN
30% xarel.lo vermell, xarel.lo, sumoll, bastardo, parellada

93 🏆

Colour: bright golden. Nose: fine lees, dry nuts, fragrant herbs, complex. Palate: powerful, flavourful, good acidity, fine bead.

RECAREDO
Tamarit, 10
08770 Sant Sadurní d'Anoia (Barcelona)
☎: +34 938 910 214
info@recaredo.com
www.recaredo.com

🏆 PODIUM

Recaredo Serral del Vell 2011 BE BN
50% xarel.lo, 50% macabeo

95 🏆

Colour: bright golden. Nose: fine lees, dry nuts, fragrant herbs, complex, toasty. Palate: powerful, flavourful, good acidity, fine bead, fine bitter notes.

Recaredo Serral del Vell 2016 BE BN
75% xarel.lo, 25% macabeo

94 🏆

Oxidativ. Colour: bright yellow. Nose: ripe fruit, fine lees, balanced, brioche. Palate: good acidity, flavourful, ripe fruit, long.

Recaredo Subtil 2017 BE BN
100% xarel.lo

92 🏆

Colour: bright golden. Nose: fine lees, dry nuts, fragrant herbs, complex, dried herbs. Palate: flavourful, good acidity, fine bead, fine bitter notes, fruity.

Recaredo Terrers 2018 BE BN
65% xarel.lo, 18% macabeo, 17% parellada, 65% xarel.lo, 18% macabeo, 17% parellada

92 🏆

Little interventionist, wild. Colour: bright straw. Nose: dried flowers, expressive. Palate: flavourful, fresh, balanced.

SPARKLING WINES

SPARKLING WINES

🏆 PODIUM

Reserva Particular de Recaredo 2012 BE BN
59% xarel.lo, 41% macabeo

97

Colour: yellow. Nose: candied fruit, dry nuts, dried herbs, wild herbs, fragrant herbs, spicy, fine lees. Palate: rich, flavourful, long.

🏆 PODIUM

Turo d'en Mota 2008 BE BN
96

Colour: bright golden. Nose: dry nuts, fragrant herbs, complex, toasty, caramel, faded flowers. Palate: powerful, flavourful, good acidity, fine bead, fine bitter notes, complex, elegant.

SABATÉ I COCA - CASTELLROIG
Ctra. Sant Sadurní a Vilafranca, km. 1
08739 Subirats (Barcelona)
☎: +34 938 911 927
info@sabateicoca.com
www.sabateicoca.com

Castellroig 2019 BE R BR
xarel.lo, macabeo, parellada, chardonnay

89

Crisp, smooth, very fruit-driven, balanced, tasty. Palate: fine bitter notes.

Castellroig Xarel.lo Vermell 2019 BE R BR
xarel.lo vermell, garnacha

91

Colour: bright straw. Nose: fine lees, floral, expressive, stone fruit, grassy. Palate: powerful, flavourful, good acidity, fine bead.

Sabaté i Coca Josep Coca 2015 BE GR BN
xarel.lo, macabeo

92

Colour: bright straw. Nose: fine lees, floral, fragrant herbs, expressive, roasted almonds. Palate: powerful, flavourful, good acidity, fine bead.

Sabaté i Coca Josep Coca Magnum 2013 BE GR BN
xarel.lo, macabeo

93

Colour: bright golden. Nose: fine lees, fragrant herbs, characterful, ripe fruit, dry nuts. Palate: powerful, flavourful, good acidity, fine bead, fine bitter notes.

Sabaté i Coca Mosset 2017 BE GR BN
xarel.lo, parellada, macabeo

91

Colour: bright straw. Nose: fresh fruit, citrus fruit, fine lees. Palate: fresh, fruity, good acidity.

TORELLÓ VITICULTORS
Finca Can Martí, Ctra. C-243b, km. 13.4
08790 Gelida (Barcelona)
☎: +34 938 910 793
Fax: +34 938 910 877
torello@torello.es
www.torello.com

Gran Torelló 2015 BE BN
xarel.lo, macabeo, parellada

93

Colour: bright golden. Nose: fine lees, dry nuts, fragrant herbs, complex, toasty. Palate: powerful, flavourful, good acidity, fine bead, fine bitter notes.

Torelló 225 2017 BE BN
macabeo, xarel.lo, parellada

92

Colour: bright golden. Nose: fine lees, fragrant herbs, characterful, ripe fruit, dry nuts, lactic notes. Palate: powerful, flavourful, good acidity, fine bead, fine bitter notes, fruity.

🏆 PODIUM

Torelló Collection 2011 BE BN
macabeo, xarel.lo, parellada

95

Colour: bright straw. Nose: fresh fruit, citrus fruit, fine lees, fragrant herbs, brioche. Palate: fresh, good acidity, flavourful, fleshy.

Torelló Finca Can Martí 2011 BE BR
32% chardonnay, 31% xarel.lo, 21% macabeo, 16% parellada

93

Colour: bright yellow. Nose: ripe fruit, fine lees, balanced, dried herbs, bakery, dry nuts. Palate: good acidity, flavourful, ripe fruit, long.

Torelló Finca Can Martí 2017 BE BR
xarel.lo, chardonnay, macabeo, parellada

93

Colour: bright straw. Nose: wild herbs, dried herbs, expressive, balanced, fine lees. Palate: fresh, good acidity, fine bitter notes.

Torelló Special Edition 2017 BE BR
90 🏆
Colour: bright straw. Nose: medium intensity, fresh fruit, dried herbs, fine lees. Palate: fresh, fruity, good acidity, easy to drink.

Torelló Tradicional 2016 BE BN
xarel.lo, macabeo, parellada

91 🏆
Colour: bright straw. Nose: fine lees, floral, fragrant herbs, expressive. Palate: powerful, flavourful, good acidity, fine bead.

VICENTE GANDÍA
Ctra. Cheste a Godelleta, s/n
46370 Chiva (València/Valencia)
☎: +34 962 524 242
Fax: +34 962 524 243
info@vicentegandia.com
www.vicentegandia.es

Sandara Blanco BE
viura, verdejo, sauvignon blanc

84

Sandara Premium Moscato BE D
moscatel

84

Sandara Rosado RE D
bobal

84

VINÍCOLA DE TOMELLOSO
Ctra. Toledo - Albacete, Km. 130,8
13700 Tomelloso (Ciudad Real)
☎: +34 926 513 004
Fax: +34 926 538 001
vinicola@vinicoladetomelloso.com
www.vinicolatomelloso.com

Xtales! Moscatel Frizzante BE SD
100% moscatel

87

Xtales! Tempranillo Rosé Frizzante RE SD
100% tempranillo

85

VINOS COLOMAN
Goya, 17
13620 Pedro Muñoz (Ciudad Real)
☎: +34 926 586 410
coloman@satcoloman.com
www.satcoloman.com

Manchegal 2021 BE SD
airén

83

Manchegal Rosado 2021 RE SD
tempranillo

84

VINS PEPE RAVENTOS
Plaça del Roure s/n
08240 Sant Sadurní d'Anoia (Barcelona)
☎: +34 938 183 262
info@raventos.com
www.vinspeperaventos.com

🏆 PODIUM

Mas del Serral 2010 BE BN
xarel.lo, bastardo

97
Colour: bright yellow. Nose: fine lees, fragrant herbs, expressive, stone fruit, ripe fruit, toasted bread. Palate: powerful, flavourful, good acidity, fine bead, balanced.

🏆 PODIUM

Mas del Serral 2011 BE BN
xarel.lo, bastardo

97
Colour: bright golden. Nose: fragrant herbs, characterful, ripe fruit, dry nuts, complex, expressive, fine lees. Palate: powerful, flavourful, good acidity, fine bead, fine bitter notes.

VIÑA BALBAINA
Viña La Torre, Ctra. Jerez-Rota, Km. 8
11408 Jerez de la Frontera (Cádiz)
☎: +34 617 071 349
mgalan@vbalbaina.es
www.vbalbaina.es

Galantería 2019 BE C BR
palomino, moscatel de alejandria

90
Colour: bright straw. Nose: fresh fruit, citrus fruit, fine lees, fragrant herbs, floral. Palate: fresh, fruity, good acidity.

BODEGAS NARANJO
Felipe II, 5
13150 Carrión de Calatrava (Ciudad Real)
☎: +34 687 045 574
comercial@bodegasnaranjo.com
www.bodegasnaranjo.com

Lahar BE BN
60% moscatel de alejandría, 40% macabeo

87

Lahar Tempranillo RE BN
100% tempranillo

83

INDEX

 ORGANIC WINES

 WINERIES

 WINE

WINE	PAGE
¿Y TÚ DE QUIÉN ERES? 2020 T	847
¿Y TÚ DE QUIÉN ERES? 2021 B	847
■ DISSIDENT 2019 T	42
+ RITME BLANC 2019 B	373
+ RITME BLANC 2020 B	373
+ RITME BLANC 2021 B	373
0 POR CIENTO TEMPRANILLO 2021 T	676
10 SAUVIGNON BLANC 2021 B	335
100 VEREMES 2019 B FB	638
100 X 100 MONASTRELL 2020 T	233
100 X 100 SYRAH 2020 T	233
100 Y CIENTOS 2017 T RB	722
100 Y CIENTOS 2018 T RB	722
100% ROMAILA 2018 T BA	776
10000 HORES FLORAL 2021 B	339
10000 HORES NEGRE SELECCIÓ 2020 T	339
10000 HORES ROSÉ 2021 RD	339
10000 HORES XARELLO 2021 B	339
11 PINOS BOBAL 2020 T RB	271
12 LINAJES VERDEJO 2020 B	605
12 LINAJES VERDEJO 2021 B	605
12 VOLTS 2020 T	822
1270 A VUIT 2014 T R	360
1564 PETIT VERDOT 2018 T BA	797
1564 SYRAH 2020 T BA	797
1564 VIOGNIER 2019 B BA	797
1771 CASA LOS FRAILES 2018 T C	707
20 ALDEAS 2019 T	809
37 BARRICAS 2018 T C	90
4 KILOS 2020 T	822
409 2020 T C	488
575 UVAS DE CÁMBRICO 2019 T R	758

WINE	PAGE
7 VIN 2016 T BA	260
7 VIN 2017 T BA	260
77 DIES CELLER ARRUFI 2021 B	855
77 NITS CELLER ARRUFI 2021 T	855
8LGENDS ETERNA LEYENDA 2018 T	786

A

WINE	PAGE
A PART 2019 B	646
AA CAU D'EN GENIS 2020 B	32
AA CAU D'EN GENIS 2021 B	32
AA DOLÇ DE NEU 2021 B BA D	837
AA DOLÇ MATARÓ 2020 T D	838
AA LANIUS 2020 B S	32
AA ORBUS 2019 T	32
AA PARVUS CHARDONNAY 2021 B	32
AA PARVUS SYRAH 2019 T	32
ABADÍA RETUERTA LE DOMAINE 2021 B	738
ABRISA'T BÀRBARA FORÉS 2020 B C	643
ACENTUADO ROSE ORGANIC 2021 RD	786
ACINIPO 2017 T C	259
ACORDE 2020 T	233
ACÚSTIC BLANC 2019 B	291
ACÚSTIC BLANC 2021 B	291
ACÚSTIC ROSAT 2021 RD	291
AD GAUDE 2021 B	855
ADARAS CALIZO GARNACHA TINTORERA 2021 T BA	49
ADARAS HUELLA GARNACHA TINTORERA - MONASTRELL 2020 T SS	49
ADARAS KALIZO SIN SULFITOS 2021 T S	49
ADARAS LLUVIA 2021 B SS	49
ADARO 2019 T	471
ADERNATS XARELLO VERMELL 2020 B	638

WINE	PAGE
ADRIÀ DE BIOPAUMERÀ 2017 T RB	636
ADVENT SAMSÓ DULCE NATURAL 2019 T D	339
AGALIU 2020 B FB	175
ÁGORA CIENTO 69 MALBEC 2020 RD	777
ÁGORA TEMPRANILLO 2021 T MC	777
ÁGORA VIOGNIER 2021 B MC	777
AGUILERA 2011 T GR	368
AGUSTÍ TORELLÓ MATA 2016 BE GR BN	121
AGUSTÍ TORELLÓ MATA 2019 BE R BR	121
AGUSTÍ TORELLÓ MATA BARRICA GRAN RESERVA 2017 BE GR BN	121
AGUSTÍ TORELLÓ MATA ESPANTALLOPS 2019 B C	325
AGUSTÍ TORELLO MATA KRIPTA 2014 BE GR BN	121
AGUSTÍ TORELLO MATA KRIPTA GRAN ANYADA 2008 BE GR BN	121
AGUSTÍ TORELLÓ MATA ROSAT TREPAT 2020 RE R BR	121
AGUSTÍ TORELLÓ MATA XIC 2020 B	325
AGUSTÍ TORELLÓ MATA XV XAREL·LO VERMELL 2021 RD	325
AIXARTA BLANC 2020 B	117
AL RASO 2021 B	753
ALBET I NOYA BRUT 21 2019 BE BR	325
ALBET I NOYA EL FANIO 2020 B	325
ALBET I NOYA LA MILANA 2018 T R	325
ALBET I NOYA RESERVA MARTÍ 2016 T GR	325
ALDEA DE ADARAS 2021 T BA	49
ALIATS 2021 B	703
ALIATS 2021 T	703
ALKIMIA 2019 B	837
ALMAVID 2018 T R	273
ALMAVID GARNACHA 2020 T RB	273

ORGANIC WINES

WINE	PAGE
ALMAVID ROSÉ 2021 RD	273
ALMUDÍ 2018 T	90
ALOERS 2019 B	332
ALSINA & SARDÁ 2020 BE R BR	121
ALSINA & SARDÁ BOIRES 2018 T C	325
ALSINA & SARDÁ FINCA LA BOLTANA 2021 B	326
ALSINA & SARDÁ SELLO 2018 BE GR BN	121
ALTA ALELLA 10 2010 BE BN	122
ALTA ALELLA 10 2012 BE BN	122
ALTA ALELLA LAIETÀ 2017 BE GR BN	122
ALTA ALELLA LAIETÀ ROSÉ 2018 RE GR BN	122
ALTA ALELLA MIRGIN 2018 BE GR BN	122
ALTA ALELLA MIRGIN EXEO EVOLUCIÓ + 2005 BE GR BN	122
ALTA ALELLA MIRGIN EXEO PARAJE CALIFICADO VALL-CIRERA 2016 BE GR BN	122
ALTA ALELLA MIRGIN OPUS PARAJE CALIFICADO VALL-CIRERA 2017 BE BN	122
ALTA ALELLA MIRGIN ROSÉ 2018 RE GR BN	122
ALTAMENTE MONASTRELL 2020 T	222
ALTO DE INAZARES CHARDONNAY 2020 B	838
ALTO DE INAZARES MONASTRELL 2020 T	838
ALTO DE INAZARES PINOT NOIR 2019 T C	838
ALTO DE INAZARES SYRAH 2020 T	838
ALTO DE INAZARES VIOGNIER 2020 B	838
ALTOLANDON BY ROSALÍA 2019 T C	266
ALTOLANDÓN CUENCAME 2020 T C	266
ALTOLANDÓN RED 2017 T R	266
ALTOS DE LUZÓN 2020 T	226
ALTOS DEL CABRIEL 2020 T	267
ALTOS DEL CABRIEL 2021 B	267
ALTOS DEL CABRIEL 2021 RD	267
AMBURZA 2017 T BA	316
AMETLLER 2020 BE BR	138
AMETLLER 2020 BE R BN	138
AMETLLER BLANC DE NOIRS 2020 BE BN	138
ANGELIA 2017 T	356
ANIMA MUNDI ANCESTRAL CAMI DELS XOPS 2021 BE	878
ANIMA MUNDI ANCESTRAL NOGUER BAIX 2016 BE	878
ANIMA MUNDI GRES 2021 B	838
ANIMA MUNDI PELLS 2021 B	838
ANIMA MUNDI XARELLO 2020 B	839
ANIMA NUA COR VIU 2020 B	164
ANIMA NUA COR VIU 2020 T C	164
ANMA 2018 T BA	679
ANMA 2020 B MC	680
ANNA DE CODORNÍU ECOLÓGICO RESERVA BE R BR	135
ANTÓN AGUIRRE 2021 T	320
ANTONIO SERRANO CENCIBEL 2020 T RB	777
ARANLEÓN SÓLO 2020 T C	674
ARANLEÓN SÓLO 2021 D	701
ARCANO 2018 T	439
ARCANO 2019 T	439
ARESTEL BE BR	153
ARÍNZANO MERLOT BIOLÓGICO 2018 T	746
ARNAU DE RENDÉ MASDEU 2020 T C S	164
ARNAU OLLER 2018 T R	349
AROA 2018 T	189
AROA JAUNA 2018 T C	305
AROA LAIA 2021 B	305
AROA LARROSA 2021 RD	305
AROA MUTIKO 2020 T	305
ARRAYÁN SELECCIÓN 2019 T C	273
ARROYUELO EN RAMA BF FI	211
ARTADI CANALES 2020 T	852
ARTADI EL CARRETIL 2020 T	852
ARTADI LA HOYA 2020 T	852
ARTADI LA POZA DE BALLESTEROS 2020 T	852
ARTADI MAJADALES 2020 T	852
ARTADI QUINTANILLA 2020 T	852
ARTADI SAN LÁZARO 2020 T	853
ARTADI TERRERAS 2020 T	853
ARTADI VALDEGINÉS 2020 T	853
ARTADI VIÑAS DE GAIN 2018 B	853
ARTADI VIÑAS DE GAIN 2019 T	853
ARTADI VIÑAS DE GAIN 2020 T	853
ARTAZURI 2021 RD	316
ARTUKE EL ESCOLLADERO 2020 T	507
ARTUKE FINCA DE LOS LOCOS 2020 T	507
ARTUKE LA CONDENADA 2020 T	507
ARTUKE PASO LAS MAÑAS 2020 T	507
ARTZAI 2019 B	84
ARTZAI 2020 B	84
ARZUAGA ECOLÓGICO 2018 T C	440
AT ROCA 2019 BE R BN	326
AT ROCA ROSAT 2019 RE R BN	326
ATANCE BOBAL 2020 T	710
ATANCE CUVEE 2021 B	710
ÁTICA 2017 BE GR EBR	146
ÁTICA ROSÉ 2018 RE R EBR	146
ATRIL BIO TEMPRANILLO SYRAH 2021 T	778
AUDITORI 2019 T C	291
AUDITORI BLANC 2021 B	291

Peñín Guide | SPANISH WINE

ORGANIC WINES

WINE	PAGE
AURORUM 2021 RD C	829
AUS BRUANT 2020 BE R BN	122
AUS CAPSIGRANY SIN SULFITOS 2020 BE R BN	122
AUS MERLA 2021 T C	838
AUS PÉT-NAT 2021 BE AG	878
AUS ROSÉ PÉT-NAT 2021 RE AG	878
AUZELLS 2021 B	176
AVENT BE BR	140
AVENT ROSÉ RE BR	140
AVGVSTVS CHARDONNAY 2021 B FB	326
AVGVSTVS MICROVINIFICACIÓ MACABEO 2017 B FB	326
AVGVSTVS MICROVINIFICACIÓ MALVASIA DE SITGES 2021 B	326
AVGVSTVS MICROVINIFICACIÓ XARELLO +100 2019 B FB	327
AVGVSTVS MICROVINIFICACIÓ XARELLO DE MAR 2017 B FB	327
AVGVSTVS TRAJANVS 2017 T R	327
AVINYÓ 2017 BE R BN	123
AVINYÓ 2019 BE R BR	123
AVINYÓ BLANC DE NOIRS BE R BN	123
AVINYÓ DOLÇ DE CABERNET 2018 B MISTELA D	840
AVINYÓ ROSÉ SUBLIM 2019 RE R BR	123
AYMAR 2016 BE R BN	327
AYMAR ROSÉ 2019 RE R EBR	327
AYMAR TRANQUIL BLANC 2021 B	327

B

WINE	PAGE
BAGORDI 2013 T R	516
BAGORDI 2018 T C	516
BAGORDI 2021 B FB	516
BAGORDI 2021 RD	516
BAGORDI MATURANA 2020	516
BAGORDI VENDIMIA SELECCIONADA 2017 T	516
BALMA 2017 BE R BR	338
BÀLSAM BE BR	145
BÀRBARA FORÉS BLANC 2021 B	643
BÀRBARA FORÉS ROSAT 2021 RD	643
BARCO DEL CORNETA 2020 B FB	801
BARRERA DE SOL 2020 T BA	753
BEACH ROSÉ BY OR 2021 RD	331
BEBERÁS DE LA COPA DE TU HERMANA 2020 B	708
BELONDRADE Y LURTON 2020 B FB	596
BENTO 2021 B	595
BERCIAL LADERA LOS CANTOS 2018 T BA	683
BERCIAL SELECCIÓN 2020 B BA	683
BERNABELEVA ARROYO DE TÓRTOLAS 2020 T	721
BERNABELEVA CARRIL DEL REY 2020 T	721
BERNABELEVA VIÑA BONITA 2020 T	721
BERNAT OLLER 2018 T	349
BERNAVÍ GARNATXA BLANCA 2021 B	642
BERNAVÍ MORENILLO 2021 RD	642
BERTHA 2019 BE R BN	131
BERTHA CARDÚS 2018 BE GR BN	131
BERTHA LOUNGE 2019 BE BR	131
BES CAN RICH 2021 RD OL	816
BIENVENIDOS AL EXTRAORDINARIO MUNDO DE LA MUJER CABALLO MITAD MUJER, MITAD CABALLO (TARONJA-ORANGE) 2020 B C	708
BIENVENIDOS AL EXTRAORDINARIO MUNDO DE LA MUJER CABALLO MITAD MUJER, MITAD CABALLO (VERD-VERDE) 2020 T C	709
BIENVENIDOS AL EXTRAORDINARIO MUNDO DE LA MUJER CABALLO MITAD MUJER, MITAD CABALLO 2020 T C	709

WINE	PAGE
BIKANDI 2018 T C	588
BINIGRAU E-NEGRE 2020 T	823
BINIGRAU E-ROSAT 2021 RD	824
BIODIVERSO 2020 B	266
BIVAC 2021 B	344
BLACK 2021 T	339
BLANC D'ENGUERA ORIGINAL 2021 B	703
BLANC DE CLOSOS 2020 B	376
BLANC DE GRESA 2020 B FB	192
BLANC DE TRILOGÍA 2021 B BA	707
BLANC SUBUR 2020 B	332
BLANCO 12 CEPAS 2021 B	818
BLANCO DE MARIA 2021 B	680
BLANCO DE ROBERTO 2021 B	680
BLAUVERD 2021 T	295
BLÉS 2020 T RB	701
BLÉS 2021 B	701
BLÉS CRIANZA DE ARANLEÓN 2019 T C	701
BOBAL ICON 2020 T RB	271
BOHEM 2021 B	781
BOHEM 2021 T	781
BOLET APAGALLUMS PINOT NOIR 2021 RD	329
BOLET CAMAGROC XARELLO 2021 B	329
BOLET CANTARELUS ULL DE LLEBRE 2019 T	329
BOLET CARTOIXÀ 2014 BE GR BN	128
BOLET CLASSIC ECO 2019 BE BR	128
BOLET ECO 2015 BE GR BN	128
BOLET ECO 2019 BE R BN	128
BOLET ECO 2019 BE R BR	128
BOLET FREDOLIC (SIN SULFITOS) T	329
BOLET PINOT NOIR ROSAT 2017 RE R BR	128
BOLET SÀPIENS MERLOT 2016 T C	329

ORGANIC WINES

WINE	PAGE
BOLET VINYA SOTA BOSC 2021 B	329
BONESVALLS CABERNET SAUVIGNON 2018 T BA	344
BOTIJO BLANCO 2021 B	834
BOTIJO ROJO 2021 T	834
BRANCELLAO DOS CANOTOS 2020 T	858
BRAÓ 2018 T C	291
BRAÓ 2019 T C	291
BRI NEGRE 2015 T R	294
BRISAT DE PUNTIRÓ 2021 B	830
BRISAT DEL COSTER 2021 B	165
BROTE 2021 B	704
BROTE 2021 RD	704
BROTS DE XAREL.LO 2020 B	339
BRU DE GRAMONA 2018 T	338
BUC 2017 T C	81
BY LATÚE TEMPRANILLO 2021 T	848

C

WINE	PAGE
CAIÑO DOS CANOTOS 2019 T	858
CALA N 1 2019 T	776
CALA N 2 2018 T	776
CALCETAS 2021 B	702
CALZADILLA CLASSIC 2015 T	739
CAMBIO DE TERCIO 2021 T	679
CÁMBRICO RUFETE BLANCA PIZARRA 2019 B	854
CÁMBRICO RUFETE EL POCITO 2018 T	758
CAMÍ DE CORMES 2019 T C	191
CAMÍ DE CORMES 2021 B	191
CAMINO DE NAVAHERREROS 2020 B FB	721
CAMINO DE NAVAHERREROS 2020 T	721
CAMPO ELISEO CUVÉE ALEGRE 2018 T	655
CAMPOS DE DULCINEA SAUVIGNON BLANC 2021 B	774

WINE	PAGE
CAMPOS DE RISCA 2020 T	230
CAN BAS D'ORIGEN P3 2021 B	329
CAN BAS D'ORIGEN P5 2021 B	329
CAN BAS LA CREU 2021 B	329
CAN BAS LA ROMANA 2018 B FB	329
CAN FEIXES BLANC SELECCIÓ 2021 B	336
CAN FEIXES BLANC TRADICIÓ 2020 B	336
CAN FEIXES CHARDONNAY 2019 B FB	336
CAN FEIXES NEGRE SELECCIÓ 2019 T	336
CAN FEIXES NEGRE TRADICIÓ 2013 T C	336
CAN GELAT CALLET 2021 RD	828
CAN GELAT GIRÓ 2021 B	828
CAN RICH BLANC D'AMFORA 2021 B	816
CAN RICH BLANC DE BLANCS 2020 BE	879
CAN RICH NEGRE D'AMFORA 2020 T	816
CAN SUMOI PERFUM 2021 B	330
CAN SUMOI XARELLO 2021 B	330
CAN VIDALET BLANC DE BLANCS 2021 B	826
CAN VIDALET BLANC DE NEGRES 2021 B	826
CANALS NADAL CN 1986 BLANC DE NOIRS 2017 BE R BR	129
CANALS NADAL ROSÉ 2019 RE BR	129
CÀNDIA 2018 T BA	867
CANFORRALES NATURE TEMPRANILLO SYRAH 2021 T	778
CANFORRALES NATURE VIOGNIER 2021 B	778
CANTA LA PERDIZ 2017 T R	478
CANTADAL 2021 T	658
CANTALLOPS 2019 B	326
CANTOCUERDAS ALBILLO 2020 B FB	721
CANTUESO 2020 T	834
CAÑADA LA TORRE 2021 B RB	705

WINE	PAGE
CAP DE TRONS 2021 T	335
CAPRASIA 2018 BE R BN	127
CAPRASIA 2021 B	679
CAPRASIA BOBAL ÁNFORA 2020 T	679
CAPRASIA ROSÉ 2021 RD	679
CAPRICHO DIVINO 2019 BE BN	271
CAPRICHO DIVINO SAUVIGNON BLANC 2021 B	271
CARA NORD TREPAT ROSAT 2021 RD	164
CARCHELO ROSÉ 2021 RD	224
CARLES ANDREU 12@ 2021 T	164
CARLOTA SURIA ORGANIC BE R BN	146
CARLOTA SURIA ORGANIC BE R BR	146
CARLOTA SURIA ORGANIC BOBAL 2019 T C	682
CARLOTA SURIA ORGANIC CHARDONNAY 2020 B FB	682
CARRAROA 2019 T C	439
CARREDUEÑAS 2019 T C	160
CARTAGO 2017 T	668
CASA BOQUERA 2017 T C	734
CASA BOQUERA 2018 T RB	734
CASA BOQUERA SELECCIÓN 2017 T C	734
CASA CASTILLO EL MOLAR 2021 T	224
CASA CASTILLO LA TENDIDA 2020 T	224
CASA CASTILLO LA TENDIDA 2021 T	224
CASA CASTILLO LAS GRAVAS 2020 T	224
CASA CASTILLO MONASTRELL 2021 T	224
CASA CASTILLO PIE FRANCO 2020 T C	224
CASA MARIOL GARNATXA BLANCA 2021 B	644
CASA MARIOL GARNATXA NEGRA 2021 T	645
CASA VELLA D'ESPIELLS 2017 T R	341
CASAR DE BURBIA 2020 T RB	71
CASAR DE BURBIA GODELLO 2021 B	71

ORGANIC WINES

WINE	PAGE
CASAR GODELLO 2020 B FB	71
CASTELL D'OR ORGÀNIC BE BR	129
CASTELL DE PUJADES XARELLO 2021 B	327
CASTELL DE PUJADES XARELLO VERMELL 2021 RD	327
CASTELLROIG 2019 BE R BR	884
CASTELLROIG SO BLANC 2021 B	342
CASTELLROIG SO BLANC MAGNUM 2012 B	342
CASTELLROIG SO NEGRE 2019 T	342
CASTELLROIG SO SERÈ 2018 B FB	342
CASTELLROIG XARELLO VERMELL 2019 BE R BR	884
CASTILLO CLAVIJO 2019 T C	562
CASTILLO DE ARESAN CABERNET FRANC 2020 T	789
CASTILLO DE ARESAN SAUVIGNON BLANC 2021 B	789
CASTILLO DE ARESAN TERRUÑO 2019 T	243
CASTILLO DE ARESAN VERDEJO 2021 B	789
CASTILLO DE LIRIA ORGANIC 2021 T	44
CÁTULO ECOLÓGICO 2021 RD	312
CÁTULO GARNACHA 2021 T	312
CAVA ROXANNE 2020 BE BR	135
CELLER ARRUFÍ LLICSÓ 2018 B BA	642
CELLER ARRUFÍ LLICSÓ 2019 B BA	642
CELLER ARRUFÍ PANICAL 2021 B	642
CELLER ARRUFÍ PANICORT 2018 T	642
CELLER ARRUFÍ TREPADELLA 2021 T	642
CENT X CENT GARNACHA BLANCA CON CRIANZA EN ANFORA 2021 B C	650
CERRO DE LA RETAMA 2016 T C	203
CÉRVOLES BLANC VINYES ALTES DE LES GARRIGUES 2020 B FB	174
CÉRVOLES COLORS BLANC 2021 B	174
CÉRVOLES COLORS NEGRE 2020 T C	174
CÉRVOLES NEGRE 2019 T	174
CÉSAR LUCENDO A. 2018 T R	790
CESILIA 2020 T RB	41
CESILIA BLANC 2021 B	41
CESILIA LA GARNACHA 2020 T	41
CESILIA ROSÉ 2021 RD	41
CESILIA ROSÉ LA RÉSERVE 2021 RD	41
CHAVEO 2019 T C	90
CHULO 2020 T RB	722
CIMADAGO 2019 T C	581
CIMERA CLOS COR VÍ 2019 B C	704
CINCO DE COPAS 2020 T BA S	664
CIRERETS 2019 T	369
CLAVIUS VERDEJO 2018 B	810
CLEARLY ORGANIC 2021 B	781
CLEARLY ORGANIC 2021 T	781
CLEARLY ORGANIC SIN SULFITOS 2021 T	781
CLOS ALZINA 2018 T	376
CLOS COR VÍ RIESLING 2020 B BA	704
CLOS COR VÍ VIOGNIER 2020 B S	704
CLOS CYPRES 2018 T	376
CLOS FIGUERES 2020 T	363
CLOS GALENA 2018 T R	363
CLOS GALENA 2019 T R	363
CLOS IBAI 2018 B	572
CLOS IBAI 2018 T	572
CLOS IBAI GARNACHA BLANCA 2019 B	572
CLOS IBAI GRACIANO 2020 T	572
CLOS LOJEN 2021 T	267
CLOS PONS ALGES 2016 T C	174
CLOS PONS PLA DEL TET 2018 T C	174
CLOS PONS ROC DE FOC 2014 B	174
CLOS PONS ROC NU 2012 T R	174
CLOS PONS SISQUELLA 2018 B C	174
CLOS VIDAL BLANC DE BLANCS 2021 B	328
CLOS VIDAL CABERNET SAUVIGNON 2019 T RB	328
CLOS VIDAL ROSÉ CUVÉE 2021 RD	328
CLOT DEL ROURE 2021 B	344
CLOT DEL ROURE XARELLO BRISAT 2021 B	344
CLOT DELS EIXAMS 2019 B FB	334
CLOT DELS OMS BLANC 2021 B	328
COCA I FITÓ NATURA 2021 T	297
CODORNIU NON PLUS ULTRA 2018 BE R BR	135
COLECCIÓN COMENGE VERDEJO 2021 B	603
COLET A PRIORI 2019 BE R BR	333
COLET ASSEMBLAGE 2016 RE EBR	333
COLET GRAN CUVÉE 2018 BE R EBR	333
COLET NAVAZOS (ETIQ.NARANJA) 2018 BE R EBR	333
COLET TRADICIONAL 2018 BE R EBR	333
COLET VATUA! 2019 BE EBR	333
COLET VATUA! ROSÉ 2018 RE EBR	333
COLLBAIX 3 NITS 2019 T BA	348
COLLBAIX SINGULAR DE ÁNFORA 2017 B BA	348
COLLBAIX SINGULAR NEGRE 2016 T RB	348
COMA D'EN POU BÀRBARA FORÉS 2019 T C	643
COMENGE EL ORIGEN 2019 T	443
CON VIENTO FRESCO 2018 T	853
CONCEJO 2017 T R	160
COORDENADES 1'19 2021 T	637
COORDENADES 41'17 2019 B	637
COR VALENT 2016 T BA	348
CORET 2019 B	293
CORISCA 2021 B	393
CORRAL DE CASTRO 2014 T R	203
CORRAL DE CASTRO 2017 T	203

WINE	PAGE
CORUCHO 2020 T RB	722
CORUCHO 2021 RD	722
CORUCHO FINCA PEAZO DE LA ENCINA 2018 T RB	722
CORUCHO GARNACHA 2021 T	722
COSMOLÓGICO CABERNET SAUVIGNON BIODINÁMICO 2021 T	791
COSTALARBOL 2020 T	531
COSTALARBOL 2021 B	531
COSTALARBOL GRACIANO 2019 T	531
COSTER DELS ROSERS 2020 RD	636
COSTERS DE CORNUDELLA 2019 T R	296
COSTERS DE L'ERMITA 2018 T	371
COSTERS DE POBLEDA 2013 T	371
CRAZY GRAPES 2021 T	231
CROMÀTIC CHARDONNAY + XARELLO 2021 B	326
CROMÀTIC MERLOT 2021 RD	326
CROMÀTIC MUSCAT 2021 B	326
CRUZ CONDE 1902 BF PX D	286
CTÒNIA 2019 B FB	189
CUATRO RAYAS = CO. ORGANIC VERDEJO 2021 B	597
CUATRO RAYAS ORGANIC ROSÉ TEMPRANILLO-VERDEJO 2021 RD	802
CUATRO RAYAS ORGANIC TEMPRANILLO 2021 T	802
CUATRO RAYAS ORGANIC VERDEJO 2021 B	599
CUCÚ (CANTABA LA RANA) 2021 B	801
CUESTA DEL HERRERO 2021 T BA	738
CUEVA DE CHAMÁN ROBLE MONASTRELL 2020 T RB	50
CUEVA DEL CHAMÁN GARNACHA TINTORERA 2021 T	50
CUEVA DEL CHAMÁN VERDEJO 2021 B	50
CUEVA LLANA BOBAL 2021 T	270
CUEVA LLANA MACABEO 2021 B	270
CUEVA LLANA ROSÉ 2021 RD	270
CUEVA LLANA SYRAH 2021 T	270
CUMPLIDO 2021 B	776
CUMPLIDO 2021 T	776
CUNE ORGÁNICO 2021 T	563
CUSCÓ BERGA 2013 BE GR BN	136
CUSCÓ BERGA BE GR BR	136
CUSCÓ BERGA BE R BN	136
CUSTERO 2020 T	585
CUVÉE #1 SON JULIANA 2019 B	831
CUVÉE #3 SON JULIANA 2019 RD	831
CUVILA 2021 T	301
CYAN 2018 T C	655
CYAN PRESTIGIO 2016 T R	655
CYAN TINTA DE TORO 2020 T RB	655

D

WINE	PAGE
DARDELL GARNACHA BLANCA VIOGNIER 2021 B	643
DE ALBERTO ECOLÓGICO 100% VERDEJO 2021 B	604
DE BUENA JERA 2018 T C	59
DECLARA SELECCIÓN ESPECIAL 2018 T	863
DECORUS 2020 T RA S	55
DEHESA LA GRANJA 2019 T	809
DESBORDANT 2019 T	347
DESCREGUT 2019 BE BN	879
DESTRANKIS 2020 T BA	357
DIDO 2020 T	301
DIDO BLANC 2020 B	301
DIEZ SIGLOS VERDEJO 2021 B	615
DIONUS 2016 T R	873
DÍSCOLO 2018 T BA S	664
DÍSCOLO MALVASÍA Y VERDEJO 2020 B FB S	664
DISFRUTAR 2019 T R	228

WINE	PAGE
DIVERTUS 2018 T	873
DOIX COSTERS DE VINYES VELLES 2019 T C	371
DOMINIO DE ES LA MATA 2020 T	478
DOMINIO DE ES VIÑAS VIEJAS DE SORIA 2020 T	478
DOMINIO DE FONTANA SAUVIGNON BLANC & VERDEJO 2021 B	672
DOMINIO DE FONTANA TEMPRANILLO & CABERNET SAUVIGNON 2019 T C	672
DOMINIO DE FONTANA TEMPRANILLO & SYRAH 2020 T RB	672
DOMINIO DE LA SIERRA DOMINIVM 2019 T	758
DOMINIO DE LA SIERRA MOMENTVM 2020 T	758
DOMINIO DE LA VEGA NO.1 2020 BE BR	137
DOMINIO DEL AGUILA 2018 T R	478
DOMINIO DEL BENDITO EL PRIMER PASO 2020 T RB	664
DOMINIO DEL BENDITO LAS SABIAS 2019 T	664
DOMINIO DEL PRIOR PETIT VERDOT 2016 T BA	791
DOMINIO DEL PRIOR SYRAH 2016 T BA	791
DOMINIO DEL SOTO 2018 T C	479
DON CARLOS 2018 T RB	233
DON JACOBO TEMPRANILLO BLANCO 2021 B	520
DON MIGUEL COMENGE 2017 T R	443
DOÑA LEO ALTOLANDÓN 2021 B	266
DX DE DOMINIO LOS PINOS 2020 T RB	704

E

WINE	PAGE
ECCE VINUM TEMPRANILLO ORGANIC (8 MONTHS IN OAK BARREL) 2015 T C	784
ECCE VINUM TEMPRANILLO ORGANIC 2021 T	784
ECCE VINUM VERDEJO ORGANIC 2021 B	784
ECOS 2018 T C	259
EDRA BLANCOLUZ 2020 B	860
EGIARTE ROSADO 2021 RD	313

ORGANIC WINES

ORGANIC WINES

WINE	PAGE
EL ÁRBOL DE ARANLEÓN 2019 T C	701
EL BADIU 2020 B	33
EL BERRAKIN 2021 T	859
EL BESO SAUVIGNON BLANC 2019 B FB	779
EL BOSC NEGRE 2019 B	325
EL CARRO GROS 2016 T	331
EL CAVA DE CHOZAS CARRASCAL 2018 BE R BN	135
EL CF DE CHOZAS CARRASCAL 2017 T	740
EL CHAVAL 2021 RD	705
EL CHAVAL 2021 T	705
EL CHAVAL MACABEO 2021 B	705
EL CONJURO 2017 T	512
EL CONVENTO DE LA VEGA 2018 T	199
EL CORDERO Y LAS VÍRGENES 2018 T R	709
EL CORRAL CREMAT 2011 BE GR BR	325
EL COSTER DE L'ALZINA 2017 T C	357
EL COTO ECOLÓGICO 2019 T C	566
EL FOLLET ROSAT 2021 RD	164
EL FUNDAMENTALISTA 2020 T	268
EL FUNDAMENTALISTA 2021 T	269
EL LINZE 2020 T	790
EL MAGO 2020 B FB	648
EL MAGO 2020 T	648
EL MAGO 2021 RD	648
EL MIRACLE ORGANIC BE BR	152
EL OLIVASTRO 2020 T	674
EL ORIGEN DE PRIETO PARIENTE 2018 T C	806
EL PACTO DE LA SONSIERRA 2019 T	512
EL PERDÍO 2020 T RB	683
EL PLANTONAL 2020 T	233
EL PLANTONAL 2021 T	233
EL QUINTÀ BÀRBARA FORÉS 2020 B FB	643

WINE	PAGE
EL QUINTO PARAJE VERDEJO 2021 B	810
EL RENEGADO 2021 B	705
EL RESCATADO 2021 RD	811
EL ROCALLÍS 2018 B FB	329
EL ROSÉ DE ARESAN 2021 RD	789
EL SECRETO DE LZ 2020 B	868
EL SEQUÉ 2020 T	40
EL SEQUÉ DULCE 2017 TF C D	40
EL SUEÑO 2020 B	679
EL TEMPLARI BÀRBARA FORÉS 2020 T C	643
EL TITÁN DEL BENDITO 2019 T	664
EL VELADO 2019 T	561
EL VIÑEDO DE LA VIDA TEMPRANILLO-CABERNET SAUVIGNON 2021 T	796
EL VIÑEDO DE LA VIDA VERDEJO-SAUVIGNON BLANC 2021 B	796
EL VOL DE L'ÀLIGA 2020 B	376
ELIANE CHARDONNAY 2021 B	330
ELITIA CARIÑENAS VIEJAS 2018 T	181
ELVIRA 2019 B	293
EMILIO VALERIO 2019 T	316
EN LA PARRA 2021 B	705
ENTRE PALABRAS 2019 T	491
ENTREBOSC NEGRE 2019 T	348
ENTRELIMITES LIMITE NATURAL 2016 T	60
ENTREMONTES AIRÉN 2021 B	239
ENTREMONTES TEMPRANILLO 2021 T	239
EPISTEM 2018 T	839
EPISTEM Nº 2 2018 T R	732
EPISTEM Nº 5 2018 T	839
EQUILIBRIO 2021 T	233
EQUILIBRIO SAUVIGNON BLANC 2021 B	233

WINE	PAGE
EQUILIBRIO-4 2020 T	233
EQUILIBRIO-9 2020 T BA	233
ERA 2019 T C	294
EREMUS 2017 T C	488
EREMUS 2020 T RB	488
ERIKA DE PAUMERA 2021 RD	636
ERMITA DEL MONTE 2020 T C	834
ES MONESTIR 2018 T R	820
ESCONDITE DEL ARDACHO: VERIQUETE 2019 T	545
ESENCIA DE LA TORRE 2017 T BA	874
ESENCIA DE LA TORRE CHARDONNAY 2021 B	874
ESENCIA DE LA TORRE PETIT VERDOT 2018 T BA	874
ESPECTACLE 2019 T C	300
ESPELT QUINZE ROURES 2021 B FB	188
ESPELT SAULÓ 2021 T	188
ESPELT TERRES NEGRES 2019 T	188
ESPORRERES 2019 T	296
ESTELA DE ARRAYÁN 2018 T R	273
ESTHER 2018 T C	723
ETERN 2018 T BA	373
ETERN 2020 T BA	373
ETERNAL 2017 T	439
EVENTO 2018 T C	704
EXEDRA 2021 B	115
EXEDRA 2021 T	115
EXTREMARIUM DE MONT MARÇAL 2019 BE R BN	144
EXTREMARIUM DE MONT MARÇAL 2019 BE R BR	144
EYA 2021 B	224
EYA 2021 RD	224

WINE	PAGE	WINE	PAGE	WINE	PAGE
EYA 2021 T	224	FINCA LOS ALIJARES MOSCATEL 2019 B SD	792	FORMIGA DE VELLUT 2019 T	363
F		FINCA LOS ALIJARES SYRAH GRACIANO 2018 T C	792	FOSSI BF AM	211
		FINCA LOS FRUTALES 2019 RD RB	260	FRANSOLA 2021 B	334
FAMILIA COMENGE RESERVA 2018 T	444	FINCA LOS FRUTALES CABERNET SAUVIGNON 2019 T C	260	FRUTO NOBLE JOVEN 2021 T	36
FAMILIA PACHECO MONASTRELL ORGÁNICO 2020 T	229			FRUTO NOBLE ROBLE 2021 T RB	36
FERRÉ I CATASÚS DOS 2019 T C	861	FINCA LOS FRUTALES GARNACHA 2017 T C	260	FRUTO NOBLE ROSADO 2021 RD	36
FILIGRANA 2020 T	118	FINCA LOS FRUTALES IGUALADO 2017 T	260	FRUTO NOBLE SAUVIGNON BLANC 2021 B	36
FILIGRANA 2021 B	118	FINCA LOS FRUTALES MERLOT SYRAH 2019 T C	260	FUCHS DE VIDAL 2017 BE GR BN	124
FINA 1270 A VUIT 2019 T BA	360	FINCA LOS HALCONES BOBAL 2019 T	271	FUCHS DE VIDAL ROSÉ PINOT NOIR RE R EBR	124
FINCA CALVESTRA MARGAS 2016 B	843	FINCA LOS HALCONES VIOGNIER 2021 B FB	271	FUCHS DE VIDAL UNIC BE R BN	124
FINCA CALVESTRA MERSEGUERA 2020 B	843	FINCA MONTICO 2021 B	606	FUENTES DEL SILENCIO PRIETO PICUDO VIEJO 2019 T	810
FINCA CAÑADA HONDA 2019 T BA	676	FINCA OLIVARDOTS GROC D'ANFORA 2020 B	192		
FINCA CAÑADA HONDA 87 T C	676	FINCA RENARDES 2019 BE R BN	146	FUENTESECA 2021 B	683
FINCA COMABARRA 2019 T C	176	FINCA RENARDES 2019 T RB	340	FUENTESECA 2021 RD	683
FINCA CONSTANCIA ENTRE LUNAS 2020 T BA	791	FINCA ROMAILA 2018 T C	776	FUMISSOLA 2018 BE BN	879
FINCA DOFÍ 2020 T C	354	FINCA SANDOVAL 2020 T	269	FURVUS 2020 T BA	302
FINCA EL MOLAR 2020 T RB	268	FINCA SANGUIJUELA 2014 T C	259	FUSIÓ 2018 T	371
FINCA EL MOLAR BOBAL 2020 T	268	FINCA SOBREÑO ECOLÓGICO 2020 T	661	**G**	
FINCA EL MOLAR MORAVIA AGRIA 2020 RD	268	FINCA TERRERAZO 2019 T	742	GALEAM MONASTRELL 2021 T	37
FINCA ÉLEZ 2020 T	743	FINCA TRES OLMOS CLASSIC 2021 B	605	GALLINAS & FOCAS 2018 T	822
FINCA ÉLEZ CHARDONNAY LÍAS 2020 B	743	FINCA TRES OLMOS LÍAS 2021 B	605	GARABITAS VIÑAS VIEJAS 2019 T	657
FINCA ELS GORGS 2013 BE GR	326	FINCA VILADELLOPS XXX XARELLO 2021 B FB	335	GARCIANO DE AZUL Y GARANZA 2020 T BA	305
FINCA LA SABINA CABERNET 2016 T GR	743	FINCA VILLACRECES 2019 T	480	GARNACHA 1921 2019 T	531
FINCA LA SABINA MERLOT 2016 T	743	FIVE MILES 2017 T	839	GARNACHA ROSÉ BY BODEGA LA ENCINA 2021 RD	842
FINCA LA SABINA SYRAH 2017 T C	743	FLOR DE ALBIHAR 2021 B	753		
FINCA LAS CARABALLAS SECTOR 2.8 2020 B	811	FLOR DE ENYA 2020 T RB	44	GEA VIÑA GALAPAGOS 2021 B	776
FINCA LAS CARABALLAS VERDEJO 2019 B	811	FLOR DE XARELLO D'ESPIELLS 2021 B	341	GEIJO 2020 B FB	841
FINCA LAS CARABALLAS VERDEJO 2021 B	811	FLOR DEL AMANECER 2020 T RB	722	GEMMA 2017 BE GR BN	145
FINCA LOS ALIJARES GARNACHA PETIT VERDOT 2016 T	792	FLOR TRUFES BLANC 2019 B FB	649	GESSAMÍ 2021 B	345
		FLOR TRUFES NEGRE 2019 T	649	GG 2020 T	42
FINCA LOS ALIJARES GRACIANO 2015 T R	792	FORCALLAT BY BODEGA LA ENCINA 2021 T	842	GIRÓ DEL GORNER LES SALERES 2020 B FB	336
FINCA LOS ALIJARES GRACIANO AUTOR 2009 T GR	792	FORMIGA DE SEDA 2021 B	363	GIRÓ DEL GORNER ROSAT 2020 RE BR	139

ORGANIC WINES

ORGANIC WINES

WINE	PAGE
GIRÓ DEL GORNER VINYA EL MOCADOR 2020 B	336
GIRÓ DEL GORNER VINYA LA SERDALLA 2020 B	336
GIRÓ RIBOT MIMAT BLANC 2021 B	336
GIRÓ RIBOT SOLSTICI 2021 B	336
GISELE 2021 B	330
GOTES DEL PRIORAT 2020 T	363
GR 5 SENDERS 2020 B	345
GRAMONA ARGENT 2017 BE BR	880
GRAMONA ARGENT ROSÉ 2018 RE BN	880
GRAMONA IMPERIAL 2015 BE BR	880
GRAMONA LA CUVÉE 2017 BE	880
GRAN AUTÓCTON BLANC 2020 B	839
GRAN AUTÓCTON BLANC 2021 B	839
GRAN AUTÓCTON NEGRE 2017 T	840
GRAN CALZADILLA 2012 T	739
GRAN CAUS 2016 T R	329
GRAN CAUS 2021 B	330
GRAN CAUS 2021 RD	330
GRAN FUCHS DE VIDAL BE R BN	124
GRAN JUVÉ CAMPS 2016 BE GR BN	141
GRANIT 2020 B	299
GRANIT 2021 B	299
GRATALLOPS VI DE LA VILA 2020 T C	354
GRATAVINUM 2πR 2020 T	366
GRATAVINUM GV5 2016 T	366
GRATAVINUM SILVESTRIS 2020 T	366
GRATIAS GOT 2020 T	847
GRATIAS MÁXIMAS 2018 T	847
GRATIAS ROSÉ 2021 RD	847
GRATIAS SOL 2020 B S	847
GREEN & SOCIAL VERDEJO 2021 B	599
GRESA 2015 T R	192
GROSS 2021 RD	259
GROSS CABERNET SAUVIGNON 2018 T R	260
GROSS PETIT VERDOT 2018 T C	260
GUERINDA PARCELAS DE GARNACHA "LA ABEJERA" 2020 T	314
GUILLEM CAROL 2017 BE GR BN	134

H

WINE	PAGE
HACIENDA ALCARAZ 2021 B	595
HACIENDA DE LA VIZCONDESA 2020 T RB	260
HACIENDA MONASTERIO 2018 T R	449
HACIENDA MONASTERIO 2019 T	449
HACIENDA MONASTERIO RESERVA ESPECIAL 2016 T R	449
HARVEYS PEDRO XIMÉNEZ VORS BF PX D	217
HÉCULA MONASTRELL ORGANIC 2020 T BA S	732
HERETAT D'LÁCRIMA BACCUS 2019 BE R BN	142
HERETAT D'LÁCRIMA BACCUS 2019 BE R BR	142
HERMANOS LURTON CUESTA DE ORO 2021 B FB	599
HERMANOS LURTON CUESTA GRANDE 2018 T	655
HERMANOS LURTON ORGÁNICO SIN SULFITOS 2021 T	655
HERMANOS LURTON VALENTÍN ROSÉ 2021 RD	597
HI 2019 T C	231
HIBEU 2019 T	274
HIBEU 2020 T	274
HIBEU FINCA LA MINERAL 2019 T	274
HIGHWAY TO HELL 2021 T RB	712
HOMBROS 2020 T BA	71
HONORO VERA ORGÁNICO 2021 T	226
HUGUET DE CAN FEIXES 2014 BE GR BN	881
HUGUET DE CAN FEIXES CLASSIC 2014 BE GR BR	881

I

WINE	PAGE
IDENTITAS 2019 B	637
IDRIAS CHARDONNAY 2021 B	628
IDRIAS MERLOT 2021 RD	863
IDRIAS T RB	629
IDRIAS TEMPRANILLO 2021 T	629
IGNIUS (ETIQUETA MORADA) 2015 T	862
IJALBA 2018 T R	588
IJALBA 2019 T C	588
IJALBA 2020 B C	588
IJALBA CUVÈE 2020 T	588
IJALBA MATURANA BLANCA 2021 B	588
IKIGALL 2021 B	335
IL·LUSIONAT 2021 B	707
IL·LÒGIC XAREL·LO ORGÀNIC SUMARROCA 2021 B	342
IMAGINAR 2020 T C	228
INANNAT 2021 RD	118
INDOMABLE 2016 BE BN	882
INQUIET DE RENDÉ MASDEU 2021 T	164
INTUICIÓN VERDEJO 2021 B	815
IO MASIA SERRA 2018 T R	189
IOHANNES 2015 T R	341
IRVING SYRAH 2020 T	202
ISABEL I 2021 B	621
IUVENIS DE BIOPAUMERÀ 2020 T	636
IVETTE 2019 BE R BR	131
IXENT 2021 RD	182

J

WINE	PAGE
J.F. ARRIEZU 2016 T C	507
J.F. ARRIEZU 2020 T	507

WINE	PAGE
J.F. ARRIEZU ROBLE 2020 T	507
J.F. ARRIEZU ROSADO 2021 RD	507
JA! 2021 T	776
JANÉ VENTURA FINCA ELS CAMPS MACABEU 2019 B	336
JANÉ VENTURA MALVASÍA DE SITGES 2020 B BA	337
JAUME DE PUNTIRÓ BLANC 2021 B	81
JAUME DE PUNTIRÓ CARMESÍ 2018 T	81
JAUME DE PUNTIRÓ ROSAT 2021 RD	81
JAUME GRAU I GRAU SELECCIÓ ESPECIAL 2017 T	347
JAUME SERRA ORGÁNICO BE BR	123
JEAN LEON 3055 CHARDONNAY 2021 B	337
JEAN LEON 3055 MERLOT PETIT VERDOT 2020 T	337
JEAN LEON 3055 ROSÉ 2021 RD	337
JEAN LEON VINYA GIGI CHARDONNAY 2019 B C	337
JEAN LEON VINYA LE HAVRE CABERNET SAUVIGNON RESERVA 2017 T R	337
JEAN LEON VINYA PALAU MERLOT 2018 T C	337
JEAN LEON XARELLO 2021 B	337
JIMÉNEZ-LANDI EL CORRALÓN 2021 T	275
JIMÉNEZ-LANDI PIÉLAGO 2019 T	275
JIMÉNEZ-LANDI SOTORRONDERO 2020 T	275
JIMÉNEZ-LANDI VALDINIEBLA 2021 RD	275
JIRÓN DE NIEBLA 2020 T C	753
JOSÉ PARIENTE CUVÉE ESPECIAL 2019 B	606
JOSÉ PARIENTE CUVÉE ESPECIAL 2020 B	606
JOSEP FORASTER BLANC SELECCIÓ 2021 B	165
JOSEP FORASTER TREPAT 2021 T	165
JUAN GIL ETIQUETA PLATA/SILVER LABEL 2020 T	226
JÚLIA BERNET 130 2017 BE BN	881
JÚLIA BERNET EXSUM 2018 BE BN	881
JULIETA 2020 T	165

WINE	PAGE
JUMENTA MERLOT SYRAH GARNACHA TINTORERA 2019 T	47
JUVÉ & CAMPS LA SIBERIA 2013 RE GR BN	141
JUVÉ & CAMPS MILESIMÉ 2016 BE R BR	141
JUVÉ & CAMPS RESERVA DE LA FAMILIA 2018 BE GR BN	141

K

KYATHOS 2018 T C	734
Kπ ILUMINATI 2020 T	859
Kπ ROSÉ 2019 RD	859

L

L´HERAVI SELECCIÓ 2021 T	302
L'ALEGRIA 2021 T	679
L'ENTRADA P.7 2020 B	49
L'ESPARTER 2015 BE GR BN	326
L'ESTACA 2017 T	375
L'EXCLAMACIÓ 2019 T C	359
L'HERAVI 2021 T	301
L'HERAVI BLANC DE NOIR 2021 B	302
L'INTERROGANT 2020 T	359
L'INTRÍS 2019 T	191
L'OBLIT 2017 T FB	334
LA BAIXADA 2020 T	354
LA BASSETA 2018 T C	369
LA BELLA ANCESTRAL PET NAT 2020 RE C BN	270
LA BELLA ANCESTRAL PET NAT 2021 RE BN	270
LA BOTERA 2018 T C	191
LA BOVILA 2019 T	871
LA BRUIXA 2020 B	650
LA BRUNA 2020 T	679
LA CALMA 2019 B FB	330
LA CAPELLETA 2019 B	371

WINE	PAGE
LA CASA DE BIO 2021 B	848
LA CASILLA 2020 T RB	268
LA CEPA DE PELAYO 2020 B	269
LA CEPA DE PELAYO BOBAL 2019 T	269
LA DONCELLA DE LAS VIÑAS 2021 RD	779
LA DONCELLA DE LAS VIÑAS CHARDONNAY 2021 B AROM	779
LA DONCELLA DE LAS VIÑAS TEMPRANILLO 2020 T	779
LA DONCELLA DE LAS VIÑAS TEMPRANILLO 2021 T RB	779
LA ESTRADA 2019 T	561
LA ESTRECHA 2020 T	268
LA FARAONA 2020 T BA	73
LA FELISA 2020 T	447
LA FLORENS 2019 T	299
LA FLORENS 2020 T	299
LA GARNACHA DE MUSTIGUILLO 2020 T	843
LA MEMÒRIA 2019 B	650
LA MERCED 2018 B FB	316
LA MIRONA BLANC 2020 B	840
LA MIRONA ROSAT 2021 RD	840
LA MISIÓN BY MENADE 2019 B	805
LA MONTESA 2020 T C	543
LA NIMFA BLANC 2021 B	164
LA PIZCA 2017 T	666
LA PRINCESA 2020 T BA	232
LA PROVINCIA DE PRIETO PARIENTE 2019 T C	806
LA PRUNERA 2016 T	293
LA QUEBRÁ 2019 T BA	753
LA SENDA DE PINUAGA PREMIUM CUVÉE 2020 T RB	787
LA SIMA 2020 T	266

ORGANIC WINES

Peñín Guide SPANISH WINE

ORGANIC WINES

WINE	PAGE
LA SIRVIENTA 2020 B	232
LA SOLANA ALTA 2019 B	370
LA SOLANA DELS MARGES 2019 T R	363
LA SORT 2020 T	704
LA TAPADA 2019 T	574
LA TAPADA 2020 T	574
LA TEMPTACIÓ 2019 B	340
LA VIÑA DE SEGUNDO 2019 T RB	660
LA ZARANDA CHARDONNAY VIOGNIER 2021 B	858
LA ZARANDA SYRAH 2021 T	858
LÁCRIMA BACCUS ROSÉ 2019 RE R BR	142
LADERAS DE SEDELLA 2018 T	262
LADERAS DEL NORTE 2020 T	440
LAGAR D'AMPRIUS CHARDONNAY 2019 B	766
LANZADO PET NAT 2021 BE	794
LANZADO PET NAT ROSÉ 2021 RE	794
LANZAGA 2018 T	561
LAS BEATAS 2019 T	561
LAS BLANCAS 2021 B	706
LAS DOS CES 2020 T RB	679
LAS DOS CES 2021 B	680
LAS GALGAS 2020 B	871
LAS LAMAS 2020 T BA	73
LAS MORADAS DE SAN MARTÍN ALBILLO REAL 2020 B	726
LAS MORADAS DE SAN MARTÍN INITIO 2017 T	726
LAS OCHO 2018 T	740
LAS TRES 2019 B	741
LATIERRA 2017 T RB	781
LATÚE 2021 RD	240
LATÚE AIRÉN 2021 B	240
LATÚE CABERNET SAUVIGNON & SYRAH 2019 T	240
LATÚE MÉTODO TRADICIONAL BE BR	848
LAUDUM 2020 T RB	35
LAUSOS 2018 T	816
LAVIA FINCA PASO MALO 2019 T C	92
LAVIA VALLE DEL ACENICHE 2019 T C	92
LAVIA VALLE VENTA DEL PINO 2019 T C	92
LE NATUREL 2021 B	305
LE NATUREL 2021 T	305
LECO PUNK 2021 T	574
LEGADO FINCA EL REFUGIO PETIT VERDOT 2016 T	791
LEGADO FINCA EL REFUGIO SYRAH 2016 T	792
LENEUS 2017 T BA	817
LENEUS CAYETANA 2021 B	817
LENEUS REISHI 2019 T	817
LENEUS TEMPRANILLO 2021 T MC	817
LES ALCUSSES 2018 T	708
LES ARGILES D'ORTO VINS BLANC 2021 B	300
LES ARGILES D'ORTO VINS NEGRE 2021 T	300
LES ARGILES D'ORTO VINS ROSAT 2021 RD	300
LES AUBAGUETES 2020 T C	354
LES BRUGUERES 2020 T	368
LES CASETES 2020 T	299
LES CERVERES XARELLO 2019 B	331
LES CLIVELLES DE L'ALZINA 2019 T	358
LES CLIVELLES DE TORROJA 2020 T	358
LES ELIES 2019 T BA	188
LES GALLINETES 2021 T	165
LES VISTES 2020 B	334
LEZAUN 0,0 SULFITOS 2021 T	313
LEZAUN 2014 T R	313
LEZAUN 2019 T C	313
LEZAUN GAZAGA 2020 T RB	313
LEZAUN TEMPRANILLO 2021 T MC	313
LIBER 2021 B	183
LICOS 2020 B	647
LIVING SEMILLON 2020 B	704
LIVING TEMPRANILLO 2019 T	704
LLANO QUINTANILLA 2018 T C	734
LLOPART 2018 BE R BN	881
LLOPART CLOS DELS FÒSSILS 2021 B	338
LLOPART EX-VITE VIÑAS SINGULARES LES FLANDES 2012 BE BR	881
LLOPART LEOPARDI 2015 BE BN	882
LLOPART MICROCOSMOS ROSÉ 2019 RE BN	882
LLOPART ORIGINAL 1887 VIÑAS SINGULARES LES FLANDES 2011 BE BN	882
LLOPART PANORAMIC 2016 BE BR	882
LO PETIT PAU 2021 T	360
LO VY 2020 B	636
LO VY 2020 T	636
LO VY ANCESTRAL 2020 B	636
LOEBRE "BLANC DE NEGRES" 2020 B	650
LOMA DE LOS FELIPES 2018 T C	203
LOS CABOS DE URBEZO 2021 T	112
LOS CANTOS DE TORREMILANOS 2019 T	480
LOS CANTOS DE TORREMILANOS 2020 T	480
LOS CONFINES 2019 T C	806
LOS CORRALES DE MONCALVILLO MATURANA BLANCA 2020 B	520
LOS CORRALES DE MONCALVILLO MATURANA TINTA 2019 T BA	521
LOS FRAILES CALIZA 2020 T	707
LOS FRAILES DOLOMITAS 2020 T	707

WINE	PAGE
LOS FRAILES RUBIFICADO 2020 T	707
LOS PINOS 0 % SULFITO 2021 T	704
LOVE IS VERMELL 2021 B	335
LÚCULO 2021 B FB	312
LUIS SAAVEDRA 2019 T C	723
LUIS SAAVEDRA VENDIMIA NOCTURNA 2017 T RB	723
LUNAS NUEVAS 2018 T R	666
LUZÓN COLECCIÓN MONASTRELL 2021 T	227
LZ 2021 T	561

M

WINE	PAGE
M DE MAS IGNEUS 2020 T	371
MABAL 2020 T	90
MABAL MACABEO DE BALCONA 2019 B	90
MADURESA 2020 T	708
MAINETES VERDEJO 2021 B FB	227
MAISULAN 12 2018 T C	575
MAISULAN EL HONDÓN 2020 T	575
MAISULAN LOS MAGINES 2020 T	575
MAISULAN SODREMORO 2020 T BA	575
MAISULAN TXABOLA GRACIANO 2020 T	575
MAISULAN TXABOLA TEMPRANILLO 2020 T	575
MAIUS GARNATXA BLANCA 2021 B	369
MALARADO 2020 RD FB	674
MALCRIAT 2020 T	707
MALLERENGA 2017 BE GR BN	135
MALLOLET 2021 B	191
MALOCO 2020 T	704
MALOCO 2021 T	677
MANCHOMUELAS BLANCO DE BERNABELEVA 2020 B FB	721
MANERAS DE VIVIR 2017 T	707

WINE	PAGE
MANTONEGRO BLANCO SON JULIANA 2019 B	831
MANTONEGRO TINTO SON JULIANA 2019 T	831
MANYOL ECO 2021 T	651
MAR DE ALISTE 2020 T	841
MAR GARCÍA 2019 T	47
MAR I CEL 2021 B	338
MARCO VALERO MARCIAL 2019 T	98
MAREVIA 2019 BE BR	151
MARGER SUMARROCA 2020 B FB	343
MARÍA BERNET 2013 BE BN	881
MARÍA BERNET 2015 BE BN	881
MARÍA CASANOVAS 2019 BE GR BN	142
MARÍA CASANOVAS XP 2019 BE GR BN	142
MARIA RIGOL ORDI 2015 BE GR BN	142
MARÍA RIGOL ORDI 2017 BE R BN	142
MARIA RIGOL ORDI MÀGNUM CUPATGE DOS MIL DISSET 2017 BE R BN	143
MARÍA RIGOL ORDI MICROTIRATGE #6: PARELLADA 2016 BE GR BN	143
MARÍA RIGOL ORDI MICROTIRATGE #7: TREPAT 2020 BE GR BN	143
MARIA RIGOL ORDI MIL·LENNI 2017 BE R BN	143
MARÍA SARMIENTO 2021 T	734
MARILUNA 2020 T BA	711
MARILUNA 2021 B	711
MARIOLA T	112
MARQUÉS DE RISCAL SAUVIGNON BLANC 2021 B	606
MARQUÉS DE RISCAL VERDEJO ORGANIC 2021 B	606
MARRURRO 2020 T C	834
MART 2021 RD	345
MARTA CIBELINA 2021 B	790
MARTÍN BERDUGO 2020 T	438

WINE	PAGE
MARTÍN VERÁSTEGUI 2019 T BA	809
MARTÍNEZ ROSÉ RIMARTS 2020 RE BN	148
MAS CANDÍ 2019 BE BN	882
MAS COMTAL INCROCIO MANZONI 2018 B	865
MAS D'ALSINA & SARDÁ 2018 BE R BN	122
MAS DE LA PANSA PARELLADA 2018 B	118
MAS MOMENTOS CABERNET SAUVIGNON 2021 T	789
MAS MOMENTOS SYRAH 2021 T	789
MAS MOMENTOS TEMPRANILLO 2021 T	789
MAS OLLER BLAUNIT 2021 T	189
MAS OLLER MALVASÍA 2019 B MISTELA D	189
MAS OLLER MAR 2021 B	189
MAS OLLER PLUS 2018 T	189
MAS OLLER PUR 2020 T C	189
MAS OLLER SYRAH DE LA MUNTANYA 2019 T C	189
MAS PICOSA BLANC 2021 B	117
MAS PICOSA NEGRE 2021 T	117
MAS SINÉN CLOS 2017 T	356
MAS SINÉN COSTER 2015 T C	356
MAS SINÉN LA VALL 2017 T BA	356
MAS TORTÓ 2020 T C	293
MAS TORTÓ BLANC 2021 B	294
MAS VILELLA BLANC 2020 B FB	866
MAS VILELLA BLANC 2021 B FB	866
MAS VILELLA NEGRE 2020 T	866
MASET L'AVI PAU 2018 BE GR BN	126
MASET NU BRUT ROSÉ 2020 RE BR	126
MATALIAN 2021 B	769
MATHA BY ST JOANNES 2019 T	328
MAURO VENDIMIA SELECCIONADA 2019 T	805
MAXX 2016 T C	260
MAXX 2017 T C	260

ORGANIC WINES

ORGANIC WINES

WINE	PAGE
MEGALA 2019 T	703
MELÉ 2014 T	853
MELIC 2018 T	707
MEMBRILLERA 2020 T C	674
MEMÒRIA 2016 BE BN	879
MEMÒRIES DEL PRIORAT RANCI CAL MARCELINO TF RC	376
MENADE SAUVIGNON BLANC DULCE 2021 B D	805
MENADE VERDEJO 2021 B	805
MERIAN BLANC 2021 B	646
MERIAN NEGRE 2021 T	646
MERIAN ROSAT 2021 RD	646
MERUM PRIORATI DESTI 2019 T	372
MERUM PRIORATI EL CEL 2019 T	372
MERUM PRIORATI INICI 2019 T	372
MÉS FELIÇ 2020 T	338
MESIES 2016 T C	650
MESIES GARNATXA BLANCA 2019 B FB	650
MESIES SELECCIÓ 2015 T R	650
MESTA TEMPRANILLO 2021 RD	672
MESTA VERDEJO 2021 B	672
MESTIZAJE 2020 B	843
MESTIZAJE 2020 T	742
MESTRE VILA VELL 2019 T	115
MIL HISTORIAS BOBAL 2021 T RB	266
MIL HISTORIAS MALBEC 2021 T	266
MILIARIO 2019 T RB	69
MILSETENTAYSEIS 2019 T	485
MILSETENTAYSEIS LA PEÑA 2020 RD	485
MIM 2013 T R	294
MIM NATURA PINOT NOIR ROSADO 2019 RE R BR	154
MIQUEL JANÉ XARELLO 2021 B	327

WINE	PAGE
MIQUEL PONS 2019 BE R BN	132
MIQUEL PONS 77 VEREMAS GARNACHA 2020 T	331
MIQUEL PONS 77 VEREMAS XAREL·LO 2021 B FB	331
MIQUEL PONS NÚRIA DE MONTARGULL 2021 B	331
MIRABELLES 2018 B	332
MIRANDA D'ESPIELLS 2021 B	341
MIRANIUS 2021 B	332
MIRANIUS MAGNUM 2019 B	332
MIZARAN 2020 B	266
MIZARAN BOBAL 2021 RD	267
MIZARAN TEMPRANILLO 2020 T RB	267
MOCANAL 2020 T	199
MÓN BOBAL 2018 T	682
MÓN MACABEO 2019 BE R BN	144
MÓN MACABEO 2021 B	682
MONASTRELL BY BODEGA LA ENCINA 2021 T	842
MONCERBAL 2020 T	73
MONREAL 2015 T	329
MONT MARÇAL 2019 BE R BN	144
MONT MARÇAL 2019 BE R BR	144
MONT MARÇAL ROSÉ 2020 RE R BR	144
MONTEBACO CARA NORTE 2020 T C	485
MONTECASTRILLO 2020 T RB	480
MONTECASTRILLO 2021 T RB	480
MONTESQUIUS NATURELOVERS 2018 BE R BN	145
MOSST 2021 B	189
MUDARE 2019 B	680
MUNDO DE YUNTERO VERDEJO-SAUVIGNON BLANC 2021 B	243
MUREDA 2021 RD	794
MUREDA CHARDONNAY 2021 B	794
MUREDA CUVÉE BE R BN	698

WINE	PAGE
MUREDA SAUVIGNON BLANC VERDEJO 2021 B S	794
MUREDA SYRAH 2021 T	794
MUREDA TEMPRANILLO SYRAH 2021 T	794
MURVIEDRO ARTS DE LUNA ORGANIC BE BN	126
MURVIEDRO ARTS DE LUNA ORGANIC BE BR	126
MURVIEDRO ARTS DE LUNA ORGANIC ROSÉ RE BR	126
MURVIEDRO EKO 2021 T	37
MUSCÀNDIA 2018 BE R EBR	145
MUSCÀNDIA ANHEL BLANC DE NOIRS 2017 BE GR BN	145
MUSCÀNDIA DELIRI FLORAL 2021 B	339
MUSCÀNDIA ROSÉ PINOT NOIR 2020 RE R EBR	145

N

WINE	PAGE
NATURALEZA SALVAJE GARNACHA 2020 T	305
NATURALIS MER BLANC 2021 B	643
NATURALIS MER ROURE 2020 T RB	643
NAVAHERREROS BLANCO DE BERNABELEVA 2020 B	721
NAVAHERREROS GARNACHA DE BERNABELEVA 2020 T	721
NAVERAN CLOS ANTONIA 2020 B FB	332
NAVERAN CLOS DELS ÀNGELS 2020 T MC	332
NAVERAN FLOR DE PINOT 2021 RD	332
NAVERÁN ODISEA 2019 BE BR	133
NAVERAN PERLES BLANQUES 2017 BE GR BR	133
NAVERÁN PERLES D'OR 2017 BE GR BR	133
NAVERÁN PERLES ROSES PINOT NOIR 2019 RE BR	133
NEGRE DE NEGRES 2020 T	363
NEGRE DE NEGRES MAGNUM 2019 T	363
NELEMAN BOBAL HRAMONY 2017 T	711
NELEMAN JUST FUCKING GOOD WINE 2019 T	711
NELEMAN JUST FUCKING GOOD WINE 2020 B	711

WINE	PAGE
NELEMAN MACABEO 2020 B	711
NELEMAN TEMPRANILLO MONASTRELL 2021 T	712
NEREUS GARNACHA NEGRA 2018 T	181
NEREUS SELECCIÓ 2019 T C	181
NICTE VERDEJO ECO B	595
NIMI GERRA 2019 B	42
NIMI TOSSAL 2017 B R	42
N°8 COLLETIA 2020 T	858
N°8 COLLETIA BRANCO 2020 B	858
NODUS BOBAL 2020 T	677
NODUS CHARDONNAY 2021 B	677
NODUS DP 2020 T	677
NODUS MERLOT DELIRIUM 2019 T	678
NODUS SAUVIGNON BLANC 2021 B	678
NODUS TINTO DE AUTOR 2019 T C	678
NONA 2020 T	368
NOSSA DE MENADE 2021 T	805
NOSSO BY MENADE 2021 B	805
NOVOA 2017 T	382
NURI 2020 B	636
NUTT 2021 B	338
NUTT ROSÉ 2021 RD	338
NVDE MCB (MACABEU ÁNFORA) 2020 B	872
NVDE MCB (MACABEU CIMENT) 2020 B	872
NVDE MCB (MACABEU CIMENT) 2021 B	872
NVDE SAUVIGNON BLANC 2020 B	872
NVDE SAUVIGNON BLANC 2021 B	872

O

WINE	PAGE
OCHOA ROSADO DE LÁGRIMA 2021 RD	314
OH! DE ROMAILA T	776
OLCAVIANA CHARDONNAY 2021 B	797
OLCAVIANA SAUVIGNON BLANC 2021 B	797
OLCAVIANA VERDEJO 2021 B	797
OLD HANDS 2019 T RB	733
OLE DE AROMAS 2019 T	269
OLE DE PASSION 2020 B	269
OLIVER VITICULTORS 2020 BE BN	145
OLIVER VITICULTORS ROSÉ 2019 RE BN	145
OLLER DEL MAS ESPECIAL CARINYENA 2018 T BA	867
OLLER DEL MAS ESPECIAL PICAPOLL NEGRE 2019 T BA	349
ONTINAR MERLOT 2021 RD	304
ONTINAR MERLOT 2021 T	304
ONTINAR TEMPRANILLO 2021 T	304
ORIGEN 2021 B	334
ORIOL ROSSELL 2019 BE R BN	131
ORIOL ROSSELL ARIADNA 2016 BE GR BN	131
ORIOL ROSSELL MITIC 2017 BE GR BN	131
ORIOL ROSSELL RESERVA DE LA PROPIETAT 2015 BE GR BN	131
ORIOL ROSSELL RESERVA DE LA PROPIETAT ROSÉ 2016 RE GR BR	131
ORMUS 2018 T C	513
ORMUS EDICIÓN LIMITADA 2021 T	513
ORMUS VIURA 2020 B	513
OROVELO 2019 B	269
OSOTI 2018 B	591
OSOTI 2021 T	591
OSOTI LA ERA 2017 T RB	591
OSSIAN 2019 B	812
OSSIAN CAPITEL 2019 B FB	811
OSTATU 2021 RD	579
OTRO CUENTO 2019 B	59
OU DE PAÓ 2020 B	331
OYE 2021 RD	849

P

WINE	PAGE
P.F. 2020 T	268
PABLO CLARO SPECIAL SELECTION CHARDONNAY 2021 B	794
PABLO CLARO SPECIAL SELECTION GRACIANO CABERNET SAUVIGNON 2020 T	794
PAGO DE LOS BALAGUESES CHARDONNAY 2021 B FB	745
PAGO DE LOS BALAGUESES GARNACHA TINTORERA 2020 T C	745
PAGO DE LOS BALAGUESES SYRAH 2020 T C	745
PAGO DE THARSYS 2017 BE GR BN	146
PAGO DE THARSYS ARGILA 2018 T	741
PAGO DE THARSYS BOBAL DIANA GARCÍA 2019 T	741
PAGO DE THARSYS CABERNET FRANC SIN SULFITOS 2021 T	682
PAGO DE THARSYS MILLESIME 2018 BE R BR	146
PAGO DE THARSYS MILLESIME BARRICA 2019 BE R BN	146
PAGO DE THARSYS MILLÉSIME ROSÉ RESERVA 2018 RE R BR	146
PAGO DE THARSYS VENDIMIA NOCTURNA 2021 B	741
PAGO DE THARSYS VENDIMIA NOCTURNA GARNACHA 2021 RD FB	741
PAGO DE VALCALIENTE 2018 B R	591
PAGO DE VALCALIENTE GARNACHA 2021 T C	591
PAGOS DE REVERÓN 2020 T BA	30
PAGOS DE REVERÓN 2021 B S	30
PAGOS DE REVERÓN 2021 T S	30
PAISAJES EMBOTELLADOS VIÑEDOS RUIZ JIMÉNEZ 2021 T	591

ORGANIC WINES

ORGANIC WINES

WINE	PAGE
PANDORA VERDEJO 2021 B	608
PANDRA SAUVIGNON BLANC 2019 B BA	608
PARA CELSUS 2021 T	784
PARA CELSUS VERDEJO 2021 B	784
PARADIGMA 2018 T BA	703
PARAJES DE LOS VIDRIOS GARNACHA 2018 T	727
PARAJES DE LOS VIDRIOS PICO DEL MIRLO 2018 T	727
PARAJES DEL CABRIEL 2021 T	679
PARAJES DEL INFIERNO "EL JUDAS" 2019 B FB	840
PARAJES DEL INFIERNO "LA SILLERÍA" 2019 B FB	840
PARAJES DEL INFIERNO "LAS ENVIDIAS" 2019 B	840
PARAJES DEL VALLE 2021 RD	270
PARAJES DEL VALLE MACABEO 2021 B	270
PARAJES DEL VALLE MACERACIÓN MACABEO 2021 B	870
PARAJES DEL VALLE MONASTRELL 2020 T	232
PÁRAMO ARROYO 2014 T GR	488
PARAÑY 2016 T C	335
PARATÓ 2018 BE R BN	146
PARATÓ 2018 BE R BR	146
PARATÓ ÁTICA TRES X TRES 2020 B	340
PARATÓ XAREL.LO 2021 B	340
PARDELASSES 2017 T	358
PARDELLS 2019 B	188
PARENTESIS 2021 B	359
PARÉS BALTÀ CUVÉE DE CAROL 2013 BE GR BN	146
PARÉS BALTÀ ELECTIO XAREL.LO 2020 B	340
PARÉS BALTÀ HISENDA MIRET GARNATXA 2019 T R	340
PARÉS BALTÀ HISTORIC 2018 BE GR BN	147
PARÉS BALTÀ INDÍGENA 2021 B	340
PARÉS BALTÀ INDÍGENA NEGRE 2020 T	340
PARÉS BALTÀ ROSA CUSINÉ 2017 RE GR BN	147

WINE	PAGE
PARÉS BALTÀ SATÈL.LIT 2019 B	340
PARÉS BALTÀ XAREL.LO 2021 B	340
PAROTET 2020 T	708
PARTAL CEPAS VIEJAS 2017 T	90
PAS CURTEI 2019 T	325
PAS DELS CAUS 2021 T	355
PASAMONTE TINTORERA 2020 T	745
PASANAU CEPS NOUS 2019 T	361
PASANAU VI DE VILA DE LA MORERA DE MONTSANT 2019 T	361
PASION DE BOBAL 2020 T RB	683
PASION DE BOBAL 2021 RD	683
PASIÓN DE MONASTRELL 2020 T	44
PASION DE MOSCATEL 2021 B	711
PATA NEGRA APASIONADO ORGANIC T	222
PATA NEGRA ORGANIC BE BR	123
PATOJO 2020 T	35
PATRIA CHICA 2021 T	738
PEDREGAR 2016 BE GR BN	326
PEDREGAR 2016 RE R	326
PELA ROQUES DE MUSTIGUILLO 2020 T	843
PENÍNSULA SKIN CONTACT 2019 B	868
PEÑA HIBEU 2018 T	274
PEÑA REJAS 2021 T	658
PEÑALBA-LÓPEZ 2020 B	810
PEÑAS ALADAS 2013 T GR	479
PERAFITA GARNATXA NEGRA 2021 T	186
PERAFITA PICAPOLL 2021 B	186
PERAFITA ROSAT 2021 RD	186
PERE VENTURA GRAN VINTAGE PARAJE CALIFICADO CAN BAS 2015 BE GR BR	147
PERELADA CUVÉE ESPECIAL 2020 BE BN	147

WINE	PAGE
PERELADA STARS TOUCH OF ROSÉ 2020 RE BR	148
PERNALES MERLOT SYRAH 2018 T C	262
PETILLANT BLANC 2021 BE AG	840
PETIT BLANC SAÓ 2021 B	175
PETIT CUPADA DE CANALS NADAL 2019 BE R BN	129
PETIT PISSARRES 2020 T	376
PETIT SAÓ 2019 T	175
PETIT SIÓS 2021 T RB	174
PIC DE SOLERGIBERT 2021 B	347
PÍCARO DEL AGUILA 2020 T BA	479
PÍCARO DEL AGUILA CLARETE 2020 RD	479
PICO D´ALIGA 2020 T C	674
PIEDRA 2016 T R	658
PIEDRA 2019 T C	658
PIEDRA 2020 T RB	659
PIEDRA 2021 RD	659
PIEDRA NATURAL 2021 T	659
PIELES DORADAS 2021 B	706
PINCHAPERAS 2020 T C	834
PINO 2020 T RB	268
PINOSO CLÁSICO 2019 T C	38
PINUAGA 200 CEPAS 2019 T	787
PINUAGA BIANCO 2021 B	787
PINUAGA COLECCIÓN 2019 T C	788
PINUAGA NATURE 2020 T RB	788
PINUAGA ROSÉ 2021 RD	788
PINZON 21 2021 RD	767
PIQUERAS VERDEJO WILD FERMENTED 2021 B FB	48
PITA 2021 RD	609
PITA FINCA LA BONERA TEMPRANILLO 2018 T C	609
PITA FINCA LA CANTERA 2018 B FB	609
PITA SAUVIGNON BLANC 2020 B	609

WINE	PAGE
PITA VERDEJO (DOMINIO DE VERDERRUBI) 2021 B	609
PIU ÁNFORA 2019 B	851
PIU ÁNFORA 2019 T FB	851
PLA DE MOREI 2019 T	118
PLÁCET DE VALTOMELLOSO 2021 B	543
PLAER 2018 T C	373
PLAER 2020 T C	373
PLANAS ALBAREDA 2019 BE R BN	148
PLANAS ALBAREDA 2020 BE BN	148
PLANAS ALBAREDA 2020 BE BR	148
PLANAS ALBAREDA DESCLÒS 2020 T	341
PLANAS ALBAREDA GRAN RESERVA DE L'AVI 2018 BE GR BN	148
PLANAS ALBAREDA L'AVENC 2021 B	341
PLANAS ALBAREDA ROSAT 2020 RE BR	148
PLOM 2020 T	376
POCO A POCO 2021 B	781
POCO A POCO ENVEJECIDO EN BARRICA 2021 T C	781
PÓLVORA 2020 T	166
POMELADO VINO NARANJA 2021 B	795
PORPRAT 2016 T	82
PRAPETISCO 2019 T	840
PRETO DE LEIVE 2020 T	426
PRIMERA VINYA LES BRUGUERES 2020 B	368
PRIMICIA BLANC BOTA 2021 B FB	643
PRINCIPIA MATHEMATICA 2021 B	325
PRIOR TERRAE NEGRE 2019 T	376
PRISMA ORGÁNICO 2021 T	232
PRIVAT 2020 BE R BN	134
PRIVAT 2020 BE R RR	134
PRIVAT ROSAT 2020 RE BN	134
PROHIBIT 2018 RE BN	882

WINE	PAGE
PROHOM CONCEPTIA 2020 T	643
PROHOM CONCEPTIA 2021 B	644
PROHOM EXPERIENTIA 2018 T	644
PROHOM EXPERIENTIA 2021 B FB	644
PROHOM VIOGNIER 2021 B	644
PROMESA 2021 RD	637
PROPIEDAD 2020 T	543
PROTOCOLO ECO 2020 T	790
PROTOCOLO ECO 2021 B	790
PROTOCOLO ECO 2021 RD	790
PROTOS ROSÉ 2021 RD	158
PROTOS VERDEJO CUVÉE 2021 B	609
PUEBLO DE LAVIA 2019 T	92
PUNT I... 2020 T C	646
PURA SANGRE 2014 T R	233
PURA SAVIA 2019 T	777

Q

WINE	PAGE
QUINCHA CORRAL 2019 T	742
QUINTA DE QUERCUS SINGLE VINEYARD 2018 T	672
QUINTA SARDONIA QS 2019 T	812
QUINTA SARDONIA QS2 2019 T	812
QUIÑÓN DE VALMIRA 2020 T FB	543
QUORUM – LE BLANC 2017 B FB	792
QUORUM DE FINCA EL REFUGIO PRIVATE COLLECTION 2012 T BA	792

R

WINE	PAGE
RAIMAT EL MOLÍ 2018 T C	176
RAÏMS DE LA INMORTALITAT SELECCIÓ 2016 T R	343
RAÏMS DE LA INMORTALITAT XAREL·LO 2019 B FB	343
RAMITO TEMPRANILLO SYRAH 2021 T	869
RAMÓN BILBAO ORGANIC 2020 T	545

WINE	PAGE
RAMÓN BILBAO ORGANIC VERDEJO 2021 B	610
RAMÓN SÁENZ, PASIÓN DE VIDA 2021 T	581
RAMÓN SÁENZ, PEQUEÑO BASTIÓN 2020 T RB	581
RAMÓN SÁENZ, PIEDRAS RODANTES 2020 T RB	581
RATPENAT 2019 B	332
RAVENTÓS I BLANC DE LA FINCA 2019 BE BN	883
RAVENTOS I BLANC TEXTURES DE PEDRA 2017 BE BN	883
RAW MANDÓ 2020 T FB	348
RAW PICAPOLL 2021 B	348
RAYUELO 2018 T	266
REALLY AWESOME WINE 2021 T	849
REBEL·LIA 2021 B	679
REBEL·LIA 2021 T	679
REBEL·LIA SELECCIÓN ESPECIAL 2020 T RB	679
REBUZNO 2020 T	531
RECAREDO SERRAL DEL VELL 2011 BE BN	883
RECAREDO SERRAL DEL VELL 2016 BE BN	883
RECAREDO SUBTIL 2017 BE BN	883
RECAREDO TERRERS 2018 BE BN	883
RESERVA PARTICULAR DE RECAREDO 2012 BE BN	884
RICO NUEVO GARNACHA 2021 T	753
RIMARTS 2017 BE GR BN	148
RIMARTS 2018 BE R BN	148
RIMARTS 2019 BE R BR	148
RIMARTS GRAN RESERVA ESPECIAL CHARDONNAY 2017 BE GR BN	148
RITA 2021 B	302
RIU BY TRÍO INFERNAL 2018 T R	368
RIU BY TRÍO INFERNAL 2021 B	368
RIU DE GOST GARNACHA BLANCA 2020 B RB	868
ROCAPLANA 2020 T	331

ORGANIC WINES

WINE	PAGE
RODRÍGUEZ DE VERA 2020 RD	47
RODRÍGUEZ DE VERA CHARDONNAY PEQUEÑAS PARCELAS 2017 B FB	47
RODRÍGUEZ DE VERA SERIE LIMITADA 2017 T C	47
RODRÍGUEZ DE VERA SERIE LIMITADA CHARDONNAY 2019 B	47
ROLLAND GALARRETA PARCELA 123 2020 B	610
ROMAILA 2021 RD	843
ROMAILA PET NAT 2021 RE	878
ROMAILA PET NAT BE	878
RÒMIA 2018 T	867
ROS MARINA 2X 2021 B	341
ROS MARINA DCOSTER 2016 T C	341
ROS MARINA QUATRE VINYES 2017 T	342
ROS MARINA ROSÉ 2021 RD	342
ROSA DE ALEJANDRÍA 2020 B AG SD	781
ROSA DE MAR 2021 RD	820
ROSADO DE FERMIN 2021 RD	680
ROSARA 2021 RD	841
ROSAT DE PLANAS ALBAREDA 2021 RD	341
ROSÉ & CLEAR 2021 RD	338
ROSE MARINE 2021 RD	680
ROSMARINUS 2018 T RB	92
ROSMARINUS 2020 T	92
ROSSINYOL 2017 BE GR BN	135
ROVER Nº 1 2019 T	843
RUBIÒLS 2019 B	857
RUNRUN 2020 RD	776

S

WINE	PAGE
SA NATURA 2019 T	645
SÁBALO 2021 B	769
SABATÉ I COCA JOSEP COCA 2015 BE GR BN	884
SABATÉ I COCA JOSEP COCA MAGNUM 2013 BE GR BN	884
SABATÉ I COCA MOSSET 2017 BE GR BN	884
SABINILLA 2020 B FB	674
SADURNÍ OLIVER 2018 BE R BN	145
SADURNÍ OLIVER BARRICA 2017 BE R BN	145
SADURNÍ OLIVER ROSAT PINOT NOIR 2020 RE BN	146
SAFRÀ 2020 T	708
SAIAL 2019 B BA	118
SALIA 2020 T R	269
SALTO DEL USERO 2021 T	90
SALVATGE ROSÉ 2018 RE BR	882
SAMAS 2019 T	183
SAN ROMÁN 2019 T	668
SAN ROMÁN GARNACHA 2020 T	668
SANCHO GARCÉS 2018 T C	593
SANTA CLARA 2021 T	340
SANZO VIÑAS VIEJAS 2021 B FB	610
SAÓ ABRIVAT 2019 T C	175
SAÓ BLANC 2020 B FB	175
SAÓ EXPRESSIU 2019 T R S	175
SAVINA 2021 B	820
SAVINAT SAUVIGNON BLANC 2019 B FB	338
SCHATZ CHARDONNAY 2021 B	259
SCHATZ PETIT VERDOT 2016 T C	259
SCHATZ PINOT NOIR 2016 T C	259
SCHATZ ROSADO 2021 RD	259
SEDELLA 2018 T	262
SEDOSA ORGANIC ROSÉ 2021 RD	793
SEGUIT 2019 T	646
SEGUNYOLA 2017 BE BN	882
SEIS DE AZUL Y GARANZA 2019 T	305
SELECCIÓ FAMILIAR SIN SULFITOS 2019 T	862
SELECCIÓN PARCELA MOSCATEL 2021 B S	776
SELECCIÓN PARCELA SYRAH 2018 T	776
SENDA DE LAS ROCHAS BOBAL 2014 T RB	267
SENTADA SOBRE LA BESTIA 2019 T BA	709
SENTIR B	798
SENTITS BLANCS 2020 B FB	115
SENTITS NEGRES GARNATXA NEGRA 2017 T	115
SEÑORÍO DE FUENTEÁLAMO MONASTRELL 2021 RD	228
SEÑORÍO DE FUENTEÁLAMO MONASTRELL 2021 T	228
SEÑORÍO DE FUENTEÁLAMO VERDEJO 2021 B	228
SEÑORÍO DE GARCIGRANDE VERDEJO 2021 B	605
SEÑORÍO DE INIESTA COLECCIÓN 2020 T RB	795
SER VIVO Y NATURAL 2021 T	488
SEREZHADE 2020 B	531
SERICIS CEPAS VIEJAS VIOGNIER 2021 B FB	850
SERRES VELLES MACABEU 2021 B C	339
SHAYA 2021 B	614
SIERRA DE TOLOÑO CAMINO DE SANTA CRUZ 2019 T	584
SIGNES 2016 T	116
SIGNO BOBAL 2020 T BA	269
SILENTE 2020 B BA	834
SILEO 2020 T	296
SILVANO GARCÍA VERDEJO 2021 B	228
SIN BLANCA 2018 T C	60
SIN COMPLEJOS 2020 T	666
SIÓS CAU DEL GAT 2020 T C	174
SIÓS PLA DEL LLADONER 2021 B	175
SO DE MASIA CAN RODA PANSA BLANCA 2021 B	32

ORGANIC WINES

WINE	PAGE
SOBRENATURAL BY MENADE 2016 B C	805
SOCAIRE 2019 B FB	769
SOCAIRE OXIDATIVO 2016 B FB	769
SOGAS MASCARÓ BE R BN	150
SÒLS XAREL 2019 B BA	857
SOMIATRUITES 2021 B	335
SOMNI 2018 T	363
SON GRAU GRAN ESCURSAC 2021 T	830
SON GRAU GRAN GARGOLLASSA 2021 RD	831
SON GRAU GRAN GARGOLLASSA 2021 T BA	830
SOPLAGAITAS 2020 B	834
ST JOANNES 2019 T	328
ST JOANNES 2021 B	328
ST JOANNES ROSÉ 2021 RD	328
SUEÑO DE MEGALA 2018 T BA	703
SUMARROCA 2018 BE GR BN	151
SUMARROCA CHARDONNAY 2021 B	343
SUMARROCA ECOLÒGIC 2019 BE R BR	151
SUMMUM LÁCRIMA BACCUS 2018 BE R BN	142
SUMMUM LÁCRIMA BACCUS 2018 BE R BR	142
SUMMUM LÁCRIMA BACCUS 2019 BE R BN	142
SUMOLL AMFORA 2017 T R	118
SUNEUS BLANC 2021 B	181
SYRAH D'ANFORA 2019 T S	355

T

WINE	PAGE
TABÁ 2020 T C	231
TABUERNIGA 2019 T	561
TAMARAL AGRICULTURA ECOLÓGICA 2019 T C	473
TEBAIDA 2020 T RB	71
TEBAIDA NEMESIO 2020 T RB	72
TEBAIDA Nº5 2020 T RB	72
TEIXAR 2017 T	302
TENTENUBLO 2019 T	585
TENTENUBLO 2020 B	585
TERRA FITER 2015 T GR	42
TERRA TERRAE 2020 T	333
TERRA TERRAE BE BR	136
TERRA TERRAE BE R BR	136
TERRA TERRAE BLANC 2020 B	333
TERRAJE 2018 T BA	232
TERRAL SUMARROCA 2017 T	343
TERRAMOLL NATURAL 2021 T BA	820
TERRITORI 2019 T R	299
TERROJA DE SABATÉ I COCA 2016 B	342
TERRUM 2019 T	829
TESÓN 2020 B FB	781
TEULERA 2020 B	118
THARSYS ÚNICO 2018 BE R BR	682
THE ARTIST 2019 T C	829
THE FINAL COUNTDOWN 2020 T RB	712
THE SWIMMERS 2019 B	712
THERASIA 2020 B FB	341
THESAURUS X 2019 T	461
TIANNA 1 NEGRE 2020 T	828
TIANNA 3 NEGRE 2020 T	857
TIANNA BOCCHORIS NEGRE 2020 T	828
TIANNA NEGRE 2020 T	828
TIERRA DE FORCALLAT 2021 B	842
TIERRA SAVIA MIRLO 2019 B	851
TIERRA SAVIA ZARANDA EL VIEJO FRANCÉS T	851
TIME WAITS FOR NO ONE 6M 2021 T RB	231
TIME WAITS FOR NO ONE BLACK 2020 T	231
TIME WAITS FOR NO ONE DOUBLE 2021 T	231
TINÁCULA EL SANTILLO 2020 T	502
TINÁCULA RED 2020 T	502
TINÁCULA WHITE 2021 B	502
TINÁCULA X 2019 T	502
TINTO DE ABEL 2021 T	680
TIVO 2018 B FB	770
TOFTERUP BROTHERS ORGANIC RED T	786
TOFTERUP BROTHERS ORGANIC ROSE 2021 RD	786
TORELLÓ 225 2017 BE BN	884
TORELLÓ 50 LLIURES 2021 B FB	343
TORELLÓ GRAN CRISALYS 2019 B	343
TORELLÓ MALVARELLO 2021 B	343
TORELLÓ SPECIAL EDITION 2017 BE BR	885
TORELLÓ TRADICIONAL 2016 BE BN	885
TORRE DEL VEGUER ABELLEROL 2021 B	343
TORRE DEL VEGUER FONOLL 2021 B	343
TORRE DEL VEGUER MARICEL 2021 B	343
TORRE DEL VEGUER MARTA 2020 BE BN	343
TORRE DEL VEGUER PUR 2021 B	872
TORREMILANOS 2018 T C	480
TOSSALS SELECCIÓ 2019 T	863
TRADICIÓN DE ARESAN TEMPRANILLO - CABERNET SAUVIGNON - MERLOT 2020 T	243
TRANCO DEL LOBO 2019 T C	270
TREIXADURA DOS CANOTOS 2019 B BA	858
TREPAT DEL JORDIET DE RENDÉ MASDEU 2020 T	164
TRES JULIAS 2019 T	657
TRES LUNAS 2019 T	666
TRES LUNAS VERDEJO 2021 B	666
TREVEJOS ORGANIC WINES LISTÁN BLANCO 2020 B FB	29
TRIAS BATLLE 2020 T	344
TRIAS BATLLE 2021 B	344

» ORGANIC WINES

WINE	PAGE
TRIAS BATLLE 2021 RD	344
TRILOGÍA 2018 T C	707
TROBALLA 2019 T	175
TROS BLANC NOTARIA 2016 B FB	292
TROS BLANC SALERES 2016 B	292
TROS DE CLOS MAS DEL METGE 2019 T	363
TROS DE MAS VILELLA 2019 T	866
TROS NEGRE NOTARIA 2017 T	292
TRUFADO GARNACHA 21 2021 T	767
TRUFES BLANC 2021 B	650
TRUFES NEGRE 2019 T	650
TUDANCA 2020 T RB	494
TURÓ DE LES ABELLES 2019 T	335
TURONS VALLCORBA 2019 T C	176
TUTUSAUS 2017 BE GR BN	151
TUVÍ OR NOT TO BE ORGÀNIC SUMARROCA 2021 B	343

U

WINE	PAGE
U D'URPINA 2018 T R	348
UBETA GARNACHA 2019 T FB	321
UBETA GARNACHA BLANCA 2021 B FB	321
UBETA ROSE 2021 RD	321
ULIBARRI 2019 B	84
ULIBARRI 2020 B	84
ULTERIOR GARNACHA 2018 T	788
ULTERIOR MAZUELO 2018 T	788
UNIVERSAL CABERNET SAUVIGNON BIODINÁMICO 2021 T	791
URBEZO 2019 T C	112
URBEZO CHARDONNAY 2021 B	112
USUARAN 2016 T	316

V

WINE	PAGE
V DE MAS IGNEUS 2020 B	371
V89 2018 B FB	175
VAIVÉN BLANC DE NOIR TEMPRANILLO 2021 B	795
VALCHÉ 2019 T C	90
VALDEPEDRO DE OSTATU 2020 T	579
VALDEPILA 2018 T R	485
VALDEPILA 2019 T R	485
VALDOSAN 2018 T	660
VALENT 2021 B	338
VALLDOLINA 2018 BE GR BR	151
VALLDOLINA 2019 BE R BN	152
VALLDOLINA ROSAT 2021 RD	344
VALLDOLINA XARELLO 2021 B	344
VALLE DEL BOTIJAS SAUVIGNON BLANC 2021 B	813
VALTOSCA 2021 T	224
VD'O 1.15 2015 T	192
VD'O 2.15 2015 T R	192
VD'O 6.19 2019 B	192
VEGA MEDIEN ECOLÓGICO BE BR	151
VEGA MEDIEN ROSÉ RE BR	151
VEGA MORAGONA ORGANIC 2019 T	797
VEGA TOLOSA CHARDONNAY 2018 BE BN	271
VEGA VELLA 2017 T C	520
VEGA VELLA 2018 B FB	520
VEGA VELLA 2020 B	520
VEGA VELLA GRACIANO 2019 T	520
VEGALFARO 2017 BE GR BN	127
VÉLOBLANC 2021 B	828
VÉLONEGRE 2020 T	828
VENTA D'AUBERT VIOGNIER 2021 B	873
VENTEPICO 2019 B SS	768

WINE	PAGE
VENTUM 2017 T C	829
VENTUM 2018 T C	829
VENTUS 2016 T R	873
VENUS 2018 T C	301
VENUS DE CARTOIXÀ 2018 B	301
VENUS DE LA FIGUERA 2018 T	301
VEREDA DE LAS TÓRDIGAS 2020 T BA	753
VERMELL 2020 T	708
VERÓNICA SALGADO CAPRICHO 2018 T C	433
VERÓNICA SALGADO CAPRICHO VINO DE AUTOR 2019 T	433
VERSAT CLOS COR VÍ 2021 B	704
VERUM CABERNET FRANC TEMPRANILLO 2021 RD	788
VERUM CENCIBEL 2019 T RB	788
VERUM LAS TINADAS AIRÉN DE PIE FRANCO 2020 B	788
VESPRES 2020 T	299
VESPRES BLANC 2021 B	299
VETUSTA VIÑAS DE FUENTENEBRO 2021 T	465
VI D'AMFORA 2021 T	302
VI DE FANG 2020 T C	164
VÍA EDETANA BLANC 2020 B	648
VÍA EDETANA BLANC 2021 B	648
VÍA EDETANA NEGRE 2019 T BA	648
VIDAL I FERRÉ 2018 BE R BN	133
VIDAL I FERRÉ 2020 BE BR	133
VIDAL I FERRÉ BLANC DE NOIRS 2020 BE R BN	133
VIDAL I FERRÉ ROSAT 2020 RE BR	133
VIDUEÑOS DE SEDELLA 2020 B	262
VIENTO ALISEO 2021 RD	245
VIENTO ALISEO GRACIANO CABERNET SAUVIGNON 2020 T BA S	245

WINE	PAGE
VIENTO ALISEO TEMPRANILLO PETIT VERDOT 2021 T S	245
VIENTO ALISEO VIOGNIER 2021 B	246
VILADELLOPS SELECCIÓN GARNATXA 2019 T C	335
VILARNAU 2019 BE R BN	153
VILARNAU BRUT ROSÉ DELICAT 2020 RE R BR	153
VILLA DE CORULLÓN 2020 T	73
VILLANA LLÉVAME AL HUERTO 2020 B	870
VILOSELL 2019 T	177
VINANA 2017 T	260
VINO VIEJO DEL ABUELO MONASTRELL 2012 T RC	842
VINS DE TALLER BASEIA 2019 B	119
VINS DE TALLER GEUM 2020 T C	119
VINS DE TALLER GRIS 2021 RD	119
VINS DE TALLER LARIX 2021 B	119
VINS DE TALLER PHLOX 2021 B	119
VINYA JANINE SYRAH 2019 T C	638
VIÑA BORGIA BY BORSAO 2020 T	845
VIÑA BOSQUERA 2021 B	723
VIÑA CORRALES PAGO BALBAINA BF FI	206
VIÑA DE ARANBELZA 2017 T	316
VIÑA DE LEORIN 2017 T	316
VIÑA EL PISÓN 2020 T	875
VIÑA LOS VALLES 50 & 50 2018 T C	553
VIÑA POMAL ECOLÓGICO 2019 T C	518
VIÑA URBEZO 2021 T	112
VIÑAHONDA ORGANIC + 2019 T BA	228
VIÑAS DEL CÁMBRICO VILLANUEVA 2020 T	758
VIOGNIER DE PRIETO PARIENTE 2020 B	806
VIROLET XARELLO 2021 B	331
VIVANCO BRUNES 2019 T	554
VIVIR SIN DORMIR 2020 T RB	223

WINE	PAGE
VIZAR 2018 T BA	808
VIZAR PRESTIGIO 2018 T	808
VIZAR SELECCIÓN ESPECIAL 2018 T R	742
VIZAR VERDEJO 2018 B FB	808
VOL D'ANIMA DE RAIMAT BLANC 2021 B	176
VOL D'ANIMA DE RAIMAT ROSAT 2021 RD	176
VOL DE L'AUGA 2019 T	376
VOLAINA 2019 B	332
VOLALTO 2020 T BA	222

W

WINE	PAGE
WALTRAUD 2021 B	335
WHITE 2021 B	339
WINE BY NATURE 2021 B	849

X

WINE	PAGE
XARELLO PAIRAL 2019 B FB	330
XARELLO VINYA DEL NOGUER 2018 B	874
XARELLO VINYA DEL NOGUER 2019 B	874
XÉRICO 2018 T	585
XTRMO (EXTREMO) 2019 B FB	707

Y

WINE	PAGE
YEMANUEVA AIRÉN PIE FRANCO 2021 B	237
YEMANUEVA TEMPRANILLO 2019 T	237
YEMASERENA 2014 T C	237
YEMASERENA 2016 T RB	237
YEMASERENA 2018 B FB	237
YLLERA VERDEJO VENDIMIA NOCTURNA ECOLÓGICO 2021 B	621
YO SOLO 2020 T FB	261

Z

WINE	PAGE
ZERBEROS CLOS PEPI 2020 T BA	752
ZERBEROS EL ALTAR 2020 T	752

WINE	PAGE
ZERBEROS LLANO TOLEDO 2020 T	752
ZERO GARNACHA 2021 T	842
ZIRIES 2014 T	853
ZURBAL 2021 T	560
ZUZARÁN 2020 B FB	506
ZUZARÁN MATURANA EDICIÓN LIMITADA 2020 T	506
ZUZARÁN VIURA 2020 B	506

ORGANIC WINES

WINERIES

WINERIES	PAGE
#GARAGEWINE	773
3 ASES BODEGAS Y VIÑEDOS	431
3VVV SINGULAR & WINES	171, 878
4 KILOS VINÍCOLA	822
7 MAGNÍFICS	181, 291, 641
7103 PETIT CELLER	822

A

WINERIES	PAGE
AALTO BODEGAS Y VIÑEDOS	431
ABADAL	347
ABADÍA DA COVA	406
ABADÍA DE POBLET	163
ABADÍA RETUERTA	738, 800
ABADÍA SAN QUIRCE	431
ACÚSTIC CELLER	291
ADEGA A COROA	685
ADEGA ABELEDOS	279
ADEGA AGOREIRA	685
ADEGA ALAN DE VAL	685
ADEGA BARBADELO	406
ADEGA CEPADO	686
ADEGA CONDES DE ALBAREI	378
ADEGA DESEU	418
ADEGA DO RICÓN	837
ADEGA DO VELEIRO	418
ADEGA DOS EIDOS	378
ADEGA ENTREOSRIOS	768, 837
ADEGA FRANCISCO FERNÁNDEZ SOUSA	418
ADEGA JAVIER ESTÉVEZ ABELEDO	418, 837
ADEGA MALCAVADA	406
ADEGA MANUEL FORMIGO DE LA FUENTE	418
ADEGA O CASAL	686
ADEGA O COTARELO	419

WINERIES	PAGE
ADEGA PONTE DA BOGA	406
ADEGA SAIÑAS	407
ADEGA TOLO DO XISTO	407
ADEGAS CASTROBREY	378
ADEGAS DO REXURDIR & RIBEIRO	419
ADEGAS GALEGAS	379
ADEGAS GRAN VINUM	379
ADEGAS GUIMARO	407
ADEGAS MORGADÍO	379
ADEGAS PARENTE GARCÍA	419
ADEGAS PAZO DAS TAPIAS	279
ADEGAS PAZO DO MAR	419
ADEGAS TERRA DE ASOREI	379
ADEGAS TOLLODOURO	380
ADEGAS VALDAVIA	420
ADEGAS VALMIÑOR	279, 380
ADEGAS VALTEA	380
ADRIÁN MORENO LLORENTE	505
ADUR TXAKOLINA	195
AGRÍCOLA CORBERA D'EBRE	641
AGRÍCOLA DE ULLDEMOLINS	291
AGRÍCOLA I S. C. TERRA ALTA - CATERRA	641
AGUSTI TORELLÓ MATA	121, 325
AILALA-AILALELO	420
AITAREN	195
AIZPURUA	195
ALBANTO WAIÑES	255
ALBET I NOYA	325
ALDAHARA	624
ALDEA DE ABAIXO	381
ALDONZA	121, 773
ALEGRE WINES & SPIRITS	181, 595, 641

WINERIES	PAGE
ALEJANDRO FERNÁNDEZ TINTO PESQUERA	431
ALEMANY I CORRIO	325, 837
ALEX	304
ALFREDO ARRIBAS	292
ALGIL BODEGAS Y VIÑEDOS	655
ALGUEIRA	408
ALIAGA	304, 505
ALKIMIA WINES	837
ALMA DAS DONAS	408
ALMÁZCARA MAJARA	63
ALREGI	181, 595, 636
ALSINA & SARDÁ	121, 325
ALTA ALELLA	32, 122, 837
ALTA ALELLA - CELLER DE LES AUS	122, 838, 878
ALTA PAVINA	800
ALTAMENTE VINOS	222
ALTANZA	505, 838
ALTANZA - COLECCIÓN R. AMILLO	205
ALTAVINS VITICULTORS	641
ALTO DE INAZARES	838
ALTOLANDÓN	266
ALTOS DE RIOJA VITICULTORES Y BODEGUEROS	506
ALTOS DE TREVEJOS	29
ALTOS DEL ENEBRO	432
ALTOS DEL TERRAL	432
ALVAR DE DIOS	800
ÁLVAREZ DE TOLEDO	800
ÁLVAREZ DURÁN PRIORAT	354
ÁLVAREZ Y DÍEZ	595
ÁLVARO DOMECQ	205
ÁLVARO PALACIOS	354
ALVEAR	285

WINERIES	PAGE
AMADIS DE YÉBENES	838
AMBORA	631
AMPRIUS LAGAR	766
ANECOOP BODEGAS	304, 700
ANGUERA DOMENECH	292
ÀNIMA NEGRA VITICULTORS	821, 838
ANÓNIMAS VITICULTORAS	381
AÑADAS DEL QUIJOTE	773
ARCO DA VELLA A ADEGA DE ELADIO	420
ARGANDA BODEGAS	721
ARINAS GIL	506
ARLESE NEGOCIOS	55
ARNACH	751
AROA BODEGAS	305
AROMAS EN MI COPA	222
ARRIEZU VINEYARDS	507
ARTIGA FUSTEL	101, 222, 732
ARTOMAÑA TXAKOLINA	52
ARTUKE BODEGAS Y VIÑEDOS	507
ARTURO GARCÍA VIÑEDOS Y BODEGAS	63
ASTRALES	432
AT ROCA	326, 838, 878
ATAVUS PRIORAT	292, 355
ATLAN & ARTISAN	732, 821, 839
ATLANTIC GALICIAN WINERIES	381, 409, 421
ATTIS BODEGAS Y VIÑEDOS	63, 381, 839
AURELIO FEO VITICULTOR	64
AUSÀS BODEGAS Y VIÑEDOS	432
AUTÉNTICOS VIÑADORES, VINOS DE TERROIR	382, 409, 421, 757
AUTÓCTON CELLER	839
AV BODEGUERS	181

WINERIES	PAGE
AVELINO VEGAS	156, 432, 595, 801, 840
AVGVSTVS FORVM	326
AVINYÓ	123, 840
AXPE SAGARDOTEGIA TXAKOLINDEGIA	84
AYMAR	327
AYSV. MENASA	279
AZPILICUETA	507
AZUL PERDIDO	250
AZUL Y GARANZA	305

B

BADEN NUMEN	433
BAL D'ISÁBENA BODEGAS	624
BALDOVAR 923	700
BARCO DEL CORNETA	801, 840
BARCOLOBO	801
BARDOS	596
BARÓN DE LEY	508
BARONÍA DEL MONTSANT	840
BELA	433, 596
BELESAR	409
BFI ONDRADE	596, 802
BENJAMÍN MIGUEZ NOVAL	382
BERMEJO	252
BERNABELEVA	721
BERNAVÍ VINYATERS	642, 841
BERONIA RUEDA	596
BICHO PALO WINE	841
BIELSA RUANO VINS	642
BIEN DE ALTURA	199
BIMBACHE VINÍCOLA	179
BINIGHAU	823
BINITORD	820

WINERIES	PAGE
BIOENOS	108, 766
BIOPAUMERÀ	636
BLANCHER-CAPDEVILA PUJOL	123, 327
BLECUA	624
BODEGAS PINDAL	597
BODEGA & VIÑEDOS CARRES	674
BODEGA 202	508
BODEGA 220 CÁNTARAS	509
BODEGA A CARQUEIXA	409
BODEGA ABEL MENDOZA MONGE	509
BODEGA ALANÍS	421
BODEGA ALISTE	841
BODEGA AMANCIO MENCHERO	237, 799
BODEGA AMANOVO	701
BODEGA AMEZTOI	195
BODEGA ANDRÉS INIESTA	266
BODEGA ARANLEÓN	674, 701
BODEGA ARRIBES DEL DUERO	58
BODEGA ASCENSIÓN REPISO BOCOS	433
BODEGA ASTOBIZA	52
BODEGA AVA VI MALLORCA	824
BODEGA BALCONA	90
BODEGA BARDOS	434
BODEGA BATGARA	52
BODEGA BERROJA TXAKOLI	84
BODEGA BIDEONA	510
BODEGA BLANCA TERRA	825
BODEGA BRAVO ESCÓS	355
BODEGA CALZADILLA	739
BODEGA CAMPO ELISEO	597
BODEGA CAMPO ELÍSEO TORO	655
BODEGA CAMPOS DE DULCINEA	774

WINERIES	PAGE	WINERIES	PAGE	WINERIES	PAGE
BODEGA CAP ANDRITXOL	825	BODEGA DON CELESTINO	757	BODEGA LA ENCINA	842
BODEGA CARLOS MORO	421, 510, 597	BODEGA DUSSART PEDRÓN	674, 841	BODEGA LA ERA	775
BODEGA CARRASCAS	774	BODEGA ECOLÓGICA LUIS SAAVEDRA	722	BODEGA LA FLORIDA	252
BODEGA CASA LA RAD	510	BODEGA EL ABUELO FLORES	757	BODEGA LA GRANADILLA	600
BODEGA CASAR DE VIDE	421	BODEGA EL ANGOSTO	701	BODEGA LA HIJUELA	631
BODEGA CASTELL MIQUEL	825	BODEGA EL HOMBRE ORQUESTA	512	BODEGA LA MELONERA	260
BODEGA CASTILLO DE ENÉRIZ	305	BODEGA EL LOMO	754	BODEGA LA PRESA DE SANZOLES	655
BODEGA CASTILLO DE MONJARDÍN	306, 878	BODEGA EL REGAJAL	723, 802	BODEGA LA TAPADA	686
BODEGA CÉSAR PRÍNCIPE	156	BODEGA ELADIO PIÑEIRO	382	BODEGA LA TERCIA - ORGANIC WINES	237
BODEGA COMARCAL VALLE DE GÜIMAR	714	BODEGA EMINA	435	BODEGA LACASMI	29
BODEGA CONTADOR	511	BODEGA EMINA RUEDA	599	BODEGA LAR DE RICOBAO	410
BODEGA CONVENTO SAN FRANCISCO	434	BODEGA ENATE	624	BODEGA LAS CALZADAS	502
BODEGA COOP. JESÚS NAZARENO S.C.G.	686	BODEGA F. SCHATZ	259	BODEGA LAS VIRTUDES	35
BODEGA COOP. SAN PEDRO REGALADO	434	BODEGA FABIO COULLET	259	BODEGA LAUS	625
BODEGA COOP. SAN ROQUE DE BEADE S.C.G. TERRA DO CASTELO	421	BODEGA FILLABOA	382	BODEGA LÓPEZ CAMPOS	273
		BODEGA FINCA DE LOS ARANDINOS	512	BODEGA LOS ALJIBES	775
BODEGA COOP. SANTA ANA	434	BODEGA FRONTIO	58	BODEGA MADRID ROMERO	222, 842
BODEGA COOP. VIRGEN DE LA ASUNCIÓN	435	BODEGA GIL ARMADA	382	BODEGA MAGALARTE ZAMUDIO	84
BODEGA COOPERATIVA CIGALES	156, 841	BODEGA GRANBAZÁN	383, 686	BODEGA MALVAJIO	842
BODEGA COOPERATIVA SAN PEDRO	766	BODEGA GROSS HERMANOS	259	BODEGA MANUEL QUINTANO	513
BODEGA COSECHEROS REUNIDOS S.COOP	306	BODEGA GULLUFRE	410	BODEGA MARCELINO I AMANDI	410
BODEGA CRUCEIRO	409	BODEGA HACIENDA LÓPEZ DE HARO	512	BODEGA MARQUÉS MONTECIERZO	308
BODEGA CUARTO LOTE	722	BODEGA HERMANOS DEL VILLAR	436, 599	BODEGA MATARROMERA	436, 600
BODEGA CUATRO RAYAS	435, 511, 597, 802	BODEGA HERMANOS LURTON	599	BODEGA MENCEY CHASNA	29
BODEGA CYAN	655	BODEGA HIRIART	156	BODEGA MIQUEL JANÉ	327
BODEGA DE LIÉDENA	306	BODEGA ILLADA	717	BODEGA MONASTERIO DE CORIAS	751
BODEGA DE MOYA	674, 701	BODEGA INURRIETA	307	BODEGA MONASTRELL	90
BODEGA DEHESA EL CARRASCAL	47, 775	BODEGA JAUME SERRA	123	BODEGA MONTORO, MARIO RODRÍGUEZ MENDOZA	235
BODEGA DEL ABAD	64	BODEGA JOAQUÍN FERNÁNDEZ	260		
BODEGA DEL MONGE-GARBATI	511	BODEGA K5	195	BODEGA MUNICIPAL DE AGÜIMES	199
BODEGA DEL NERO	722	BODEGA KIENINGER	260, 841	BODEGA MUSEUM	156
BODEGA DOBLE R	435	BODEGA LA DOLORES	94, 842	BODEGA MUSTIGUILLO	742, 843

WINERIES	PAGE
BODEGA NAVE ROVER	843
BODEGA NUMANTHIA	656
BODEGA OS AREEIROS	833
BODEGA OTAZU	308, 741
BODEGA OTTO BESTUÉ	625
BODEGA PAGO AYLÉS	738
BODEGA PAGO DE CIRSUS	308, 739
BODEGA PAGO DE CUBAS	656
BODEGA PAGOS DE ARAIZ	309
BODEGA PARDO TOLOSA	266, 843
BODEGA PAZO DE SAN MAURO	383
BODEGA PAZO DE VILLAREI	383
BODEGA PÉREZ CARAMÉS	64
BODEGA PICOS DE CABARIEZO	822
BODEGA PIEDRA FLUIDA	754
BODEGA PIRINEOS	626
BODEGA PRIVILEGIO DEL CONDADO	168, 843
BODEGA QUINTA LAS VELAS	58
BODEGA REGINA VIARUM	410
BODEGA REINA DE CASTILLA, S. COOP.	600
BODEGA REJADORADA	600, 656
BODEGA REYNA-NEPEFE	702
BODEGA ROANDI	687
BODEGA RODRÍGUEZ DE VERA	47
BODEGA ROMAILA	776, 843, 878
BODEGA S. ARROYO	436
BODEGA SAN ANTONIO ABAD COOPERATIVA DE CLM	267
BODEGA SAN FRANCISCO JAVIER	206
BODEGA SAN GREGORIO	94
BODEGA SAN ISIDRO. BULLAS	90
BODEGA SAN JUAN	199

WINERIES	PAGE
BODEGA SAN MARTÍN	309
BODEGA SAN ROQUE DE LA ENCINA	437
BODEGA SANCLODIO	422
BODEGA SANSTRAVÉ	163
BODEGA SANT PERE DE MOIXENT	702
BODEGA SANTO DOMINGO DE GUZMÁN	273
BODEGA SIESTO	843
BODEGA SIGUÍN	844
BODEGA SOLLÍO	411
BODEGA SOMMOS	626
BODEGA SOMMOS GARNACHA	844
BODEGA SON MAYOL	826
BODEGA SOPLA PONIENTE (RODRÍGUEZ DE VERA)	285
BODEGA ST. JOANNES	328
BODEGA TAFURIASTE	717, 754
BODEGA TERCIA DE ULEA	90
BODEGA TESALIA	769
BODEGA TIERRA ARANDA S. COOP.	437
BODEGA TINAHA	844
BODEGA TRES PILARES	600
BODEGA ULIBARRI	84
BODEGA VALDEHERMOSO	600
BODEGA VALDELOSFRAILES	157
BODEGA VALDRINAL	437, 601
BODEGA VANDAMA	199
BODEGA VEGA AIXALÁ	163
BODEGA VEGA DE GÁLDAR	199
BODEGA VENTURA	200
BODEGA VERA DE ESTENAS	124, 674, 747
BODEGA VERÓNICA ORTEGA	65
BODEGA VETTONES S.C.	751
BODEGA VICO	513

WINERIES	PAGE
BODEGA VIÑAGUAREÑA	657
BODEGA VIRGEN DE LA SIERRA S. COOP.	94
BODEGA VOCARRAJE	657, 802
BODEGA Y VIÑAS ALDOBA	237
BODEGA Y VIÑEDOS 13 VIÑAS	65, 844
BODEGA Y VIÑEDOS CASA DE LA NAVA	696
BODEGA Y VIÑEDOS HEREDAD MORÁN & LÓPEZ	65
BODEGA Y VIÑEDOS HIJA DE ANÍBAL	66
BODEGA Y VIÑEDOS MAIRES	657
BODEGA Y VIÑEDOS MARTÍN BERDUGO	438, 601
BODEGA Y VIÑEDOS NUNTIUM	438
BODEGA Y VIÑEDOS PAZO CASANOVA	422
BODEGA Y VIÑEDOS SOLABAL	513
BODEGA Y VIÑEDOS TINEDO	776
BODEGA Y VIÑEDOS VEIGA DA PRINCESA	383
BODEGAS	#N/D
BODEGAS & VIÑEDOS FONTANA	672
BODEGAS 1890	222
BODEGAS 7LINDES	844
BODEGAS 9C+	844
BODEGAS A. VELASCO E HIJOS	657
BODEGAS ABADÍA I A ARROYADA	438
BODEGAS ABINASA	627
BODEGAS ADRIÁ	66
BODEGAS AESSIR	124
BODEGAS AGUIUNCHO	384
BODEGAS AINZÓN	101
BODEGAS AL ZAGAL	202
BODEGAS ALBAMAR	384, 411, 687
BODEGAS ALCARDET / NTRA. SRA. DEL PILAR S.C. CLM	776
BODEGAS ALCEÑO	223, 878

WINERIES	PAGE
BODEGAS ALCONDE	309, 514, 601, 844
BODEGAS ALCORTA	514
BODEGAS ALDONIA	514
BODEGAS ALONSO CUESTA	273
BODEGAS ALTO MONCAYO	101
BODEGAS ALTOS DE TORONA	384
BODEGAS ALTOVELA	776
BODEGAS ALVIA	514
BODEGAS AMAREN	514
BODEGAS AMÉZOLA DE LA MORA	515
BODEGAS ANDRÉS MORATE	723
BODEGAS ÁNGEL	826
BODEGAS ANTÍDOTO	439
BODEGAS ANTONIO ALCARAZ	515
BODEGAS ANTONIO SERRANO	777
BODEGAS AQUITANIA	385
BODEGAS ARAGONESAS	102
BODEGAS ARAICO	515
BODEGAS ARCANO	439
BODEGAS ARCO DE CURIEL	439
BODEGAS ARFE	206
BODEGAS ARGÜESO	206
BODEGAS ARMENDÁRIZ	310
BODEGAS ARRÁEZ	35, 124, 223, 675, 702, 845
BODEGAS ARRAYÁN	273, 751
BODEGAS ARROCAL	440
BODEGAS ARÚSPIDE	777
BODEGAS ARZUAGA NAVARRO	440, 845
BODEGAS AS LAXAS	385
BODEGAS ASENJO & MANSO	441
BODEGAS ATALAQUE (RODRÍGUEZ DE VERA)	274
BODEGAS ATALAYA	47

WINERIES	PAGE
BODEGAS ATECA	95
BODEGAS AUGUSTA BILBILIS	95
BODEGAS AURA	601
BODEGAS AUSÍN	751
BODEGAS AVANCIA	687
BODEGAS AYUSO	237
BODEGAS BAGORDI	516
BODEGAS BAIGORRI	516
BODEGAS BALBÁS	441
BODEGAS BARBADILLO	206, 769
BODEGAS BARREDA	777
BODEGAS BENJAMÍN DE ROTHSCHILD & VEGA SICILIA	517
BODEGAS BERAMENDI	310
BODEGAS BERBERANA	777
BODEGAS BERNARDO ÁLVAREZ	66
BODEGAS BERONIA	517
BODEGAS BETOLAZA	517
BODEGAS BIGARDO	845
BODEGAS BILBAÍNAS	518
BODEGAS BLEDA	223
BODEGAS BOCOPA	35
BODEGAS BORSAO	102, 845
BODEGAS BRECA	96, 845
BODEGAS BRIEGO	441
BODEGAS BRIONES ABAD	442
BODEGAS CA N'ESTELLA	124, 328
BODEGAS CABALLERO	846
BODEGAS CAMPANTE	422
BODEGAS CAMPESTRAL	846, 878
BODEGAS CAMPILLO	518
BODEGAS CAMPOS DE ENANZO S. COOP.	310

WINERIES	PAGE
BODEGAS CAMPOS REALES	238, 778
BODEGAS CAN VIDALET	826, 846
BODEGAS CANO	47
BODEGAS CANOPY	274, 802
BODEGAS CANTALOBOS	66
BODEGAS CAÑALVA	498
BODEGAS CAPITÀ VIDAL	124, 328
BODEGAS CARCHELO	224
BODEGAS CARE	108
BODEGAS CARLOS PLAZA	816
BODEGAS CARLOS SAN PEDRO PEREZ DE VIÑASPRE	519
BODEGAS CARLOS VALERO	104, 108, 627
BODEGAS CARODORUM	658
BODEGAS CARRAMIMBRE	442
BODEGAS CARREÑO	91
BODEGAS CASA ANTONETE	238
BODEGAS CASA CASTILLO	224
BODEGAS CASTAÑO	732
BODEGAS CASTELO DE MEDINA	602, 803
BODEGAS CASTILLEJO DE ROBLEDO	443
BODEGAS CASTRO CANDAZ	411
BODEGAS CAUDALIA	311
BODEGAS CENTRO ESPAÑOLAS	238
BODEGAS CERRO DEL ÁGUILA	846
BODEGAS CERROLAZA	519
BODEGAS CERROSOL	602
BODEGAS CÉSAR FLORIDO	207
BODEGAS CHAVES	279, 386
BODEGAS CLOS D'AGON	115, 181
BODEGAS CLUNIA	803
BODEGAS COMENGE	443, 603

WINERIES	PAGE	WINERIES	PAGE	WINERIES	PAGE
BODEGAS CONTRERAS	91	BODEGAS DEL PALACIO DE FEFIÑANES	386	BODEGAS ESTEFANIA	67
BODEGAS CONVENTO DE LAS CLARAS	444, 603	BODEGAS DEL ROSARIO	91	BODEGAS ESTRAUNZA	524
BODEGAS COPABOCA	444, 603	BODEGAS DEL SEÑORÍO	108	BODEGAS FÁBREGAS	627
BODEGAS CORELLANAS	311	BODEGAS DEL VALLE	676	BODEGAS FAELO	36
BODEGAS CORNELIO DINASTÍA	519	BODEGAS DELGADO	286, 879	BODEGAS FAMILIA CONESA - PAGO GUIJOSO	743, 779
BODEGAS CORONADO	778	BODEGAS DIEZ DEL CORRAL	523	BODEGAS FARIÑA	659, 804
BODEGAS CORRAL	520	BODEGAS DIEZ GÓMEZ	658	BODEGAS FAUSTINO	125, 524
BODEGAS COSME PALACIO	521	BODEGAS DIMOBE	261	BODEGAS FAUSTINO RIVERO ULECIA (VINOS MONOVARIETALES)	779
BODEGAS COVILA	521	BODEGAS DIOS BACO	208		
BODEGAS COVIÑAS	124, 675, 847	BODEGAS DOMECQ DE JARAUTA	523	BODEGAS FECO	525
BODEGAS COVITORO	658	BODEGAS DOMINIO DE ATAUTA	445	BODEGAS FÉLIX CALLEJO	447
BODEGAS CRAPULA & LANENA	225	BODEGAS DOMINIO DE CAIR	446	BODEGAS FÉLIX LORENZO CACHAZO	604
BODEGAS CRÁTER	631	BODEGAS E. MENDOZA	35	BODEGAS FÉLIX SANZ	605
BODEGAS CRIN ROJA	778	BODEGAS EGUREN	779	BODEGAS FERNÁNDEZ DE ARCAYA	311
BODEGAS CRUZ CONDE	286	BODEGAS EIDOSELA	386	BODEGAS FERNÁNDEZ DE LA OSSA	780
BODEGAS CRUZ DE ALBA	444	BODEGAS EJEANAS	833	BODEGAS FERNÁNDEZ EGUILUZ	525
BODEGAS CUATRO PASOS	67	BODEGAS EL LAGAR DE ISILLA	446, 803	BODEGAS FERNÁNDEZ HERRERO	202
BODEGAS CUEVAS JIMÉNEZ - FERRATUS	445	BODEGAS EL NIDO	225	BODEGAS FERNANDO CASTRO	109, 239, 696, 780
BODEGAS D. MATEOS	386, 522	BODEGAS EL PARAGUAS	422	BODEGAS FIGUEROA	723
BODEGAS D'BERNA	688	BODEGAS EL PENITENTE	717, 755	BODEGAS FINCA LA ESTACADA	672, 780
BODEGAS DE ALBERTO	603, 803	BODEGAS EL TANINO	47, 779	BODEGAS FONTEDEI	202
BODEGAS DE FAMILIA BURGO VIEJO	522	BODEGAS EL VILLAR	703	BODEGAS FORJAS DEL SALNÉS	387
BODEGAS DE FUENTE REINA	833	BODEGAS EL VINCULO	239	BODEGAS FOS	525
BODEGAS DE GALDAMES	85	BODEGAS EMILIO CLEMENTE	125, 676	BODEGAS FRANCISCO CASAS	659
BODEGAS DE LOS HEREDEROS DEL MARQUÉS DE RISCAL	195, 522	BODEGAS EMILIO MORO	67, 447	BODEGAS FRANCISCO GÓMEZ	36
		BODEGAS EMINA	157	BODEGAS FRANCISCO RODRÍGUEZ GARROTE	58
BODEGAS DE LOS RÍOS PRIETO	445	BODEGAS ENGUERA	703	BODEGAS FRANCO ESPAÑOLAS	526
BODEGAS DE OCCIDENTE	817	BODEGAS ENTREMONTES	239, 779	BODEGAS FRONTAURA	605, 659
BODEGAS DE SANTIAGO	523	BODEGAS EPIFANIO RIVERA/ERIAL	447	BODEGAS FRONTONIO	834, 847
BODEGAS DECORUS	55	BODEGAS ERESMA - LA SOTERRAÑA	604, 804	BODEGAS FRUTOS VILLAR	605
BODEGAS DEL MEDIEVO	523	BODEGAS ESTANCIA PIEDRA	658	BODEGAS FUENTENARRO	448
BODEGAS DEL MUNI	778	BODEGAS ESTEBAN MARTÍN	109	BODEGAS FUENTESPINA	448

WINERIES	PAGE
BODEGAS FUIDIO	526
BODEGAS GARCÍA DE LARA	781
BODEGAS GARCÍA DE OLANO	526
BODEGAS GARCÍAREVALO	605
BODEGAS GARCIGRANDE	605
BODEGAS GARDEL	239, 781
BODEGAS GERARDO MÉNDEZ	387
BODEGAS GODELIA	68
BODEGAS GODELLETA	703
BODEGAS GODEVAL	688
BODEGAS GORDONZELLO	255, 804
BODEGAS GRAN FEUDO	311
BODEGAS GRATIAS. FAMILIA Y VIÑEDOS	847
BODEGAS GUTIÉRREZ COLOSÍA	208
BODEGAS GUTIÉRREZ DE LA VEGA	847
BODEGAS HABLA	817
BODEGAS HACIENDA MOLLEDA	109
BODEGAS HACIENDA MONASTERIO	449
BODEGAS HEMAR	449
BODEGAS HERCAL	449
BODEGAS HEREDAD ANSÓN	109
BODEGAS HEREDAD DE ADUNA	527
BODEGAS HERMANOS PECIÑA	527
BODEGAS HERMANOS SASTRE	450
BODEGAS HERMOSILLA	527
BODEGAS HESVERA	451
BODEGAS HIBÉU	274
BODEGAS HIDALGO-LA GITANA	209, 848
BODEGAS HIJOS DE FÉLIX SALAS	157
BODEGAS HISPANO SUIZAS	125, 676, 703
BODEGAS HNOS. PÉREZ PASCUAS – VIÑA PEDROSA	451

WINERIES	PAGE
BODEGAS IGLESIAS	168
BODEGAS IGNACIO GUALLART	767
BODEGAS INFANTE	168
BODEGAS INSULARES TENERIFE	631, 730
BODEGAS IÑAKI NÚÑEZ	312
BODEGAS IRACHE	312, 746
BODEGAS IRANZO	676, 704
BODEGAS ISIDRO MILAGRO	240, 527
BODEGAS ISLA	240
BODEGAS ISMAEL ARROYO - VALSOTILLO	451
BODEGAS ITSASMENDI	85
BODEGAS ITURRIA	660
BODEGAS IZADI	528
BODEGAS IZQUIERDO	452
BODEGAS JAVIER SAN PEDRO ORTEGA	528
BODEGAS JER	529
BODEGAS JIMÉNEZ LANDI	275
BODEGAS JIMÉNEZ-VILA HNOS.	676
BODEGAS JOSÉ L. FERRER	81
BODEGAS JOSÉ PARIENTE	606
BODEGAS JUAN GIL	225
BODEGAS LA CAÑA	387
BODEGAS LA CASA DE LÚCULO	312, 848
BODEGAS LA CASA DE MONROY	723
BODEGAS LA CATEDRAL	529
BODEGAS LA CELESTINA	452
BODEGAS LA CIGARRERA	210
BODEGAS LA CORTE	498
BODEGAS LA ERALTA	529
BODEGAS LA HORRA	452
BODEGAS LA PURÍSIMA	733
BODEGAS LA REMEDIADORA	240

WINERIES	PAGE
BODEGAS LA SOLANA - PAGO FLORENTINO	743, 781
BODEGAS LA VAL	388
BODEGAS LADAIRO	279
BODEGAS LAN	529
BODEGAS LANDALUCE	530
BODEGAS LAR DE PAULA	530
BODEGAS LAS CEPAS	531
BODEGAS LAS TIRAJANAS	200
BODEGAS LATÚE	240, 781, 848
BODEGAS LAUNA	531
BODEGAS LAUREATUS	388
BODEGAS LAVIA	92
BODEGAS LAZO	782
BODEGAS LECEA	532
BODEGAS LEDA	804
BODEGAS LEGANZA	782
BODEGAS LENEUS	125, 817
BODEGAS LERMA	55
BODEGAS LEZAUN	313
BODEGAS LEZCANO-LACALLE	158
BODEGAS LLEIROSO	453
BODEGAS LOCUAZ	849
BODEGAS LOHER	632
BODEGAS LOLI CASADO	532
BODEGAS LÓPEZ CRISTÓBAL	453
BODEGAS LÓPEZ ORIA	533
BODEGAS LOS PINOS	704
BODEGAS LOZANO	241, 849
BODEGAS LUIS CAÑAS	533
BODEGAS LUIS CORBÍ - CLOS COR VÍ	704
BODEGAS LUIS CORBÍ - FINCA COR VI	677, 704
BODEGAS LUIS GURPEGUI MUGA	533

WINERIES	PAGE
BODEGAS LUNARES DE RONDA	261
BODEGAS LUZÓN	226, 849
BODEGAS MAESTE	453
BODEGAS MÁLAGA VIRGEN	261, 286, 849
BODEGAS MALON DE ECHAIDE	313
BODEGAS MANO A MANO	782
BODEGAS MANZANOS	534
BODEGAS MANZANOS CAMPANAS	313
BODEGAS MAR DE FRADES	388
BODEGAS MARBA	632
BODEGAS MARCELINO DÍAZ	125, 498, 817
BODEGAS MARCO REAL	314
BODEGAS MARISOL RUBIO	783
BODEGAS MARQUÉS DE CÁCERES	535
BODEGAS MARQUÉS DE REINOSA	535
BODEGAS MARQUÉS DE RISCAL	606, 804
BODEGAS MARQUÉS DE TERÁN	535
BODEGAS MARQUÉS DE VIZHOJA	389, 849
BODEGAS MARQUÉS DEL ATRIO	535
BODEGAS MARTA MATÉ	454
BODEGAS MARTÍN CÓDAX	380
BODEGAS MARTÍNEZ ALESANCO	536
BODEGAS MARTÍNEZ LACUESTA	536
BODEGAS MARTÍNEZ PAIVA	126, 498
BODEGAS MARTÍNEZ PALACIOS	537
BODEGAS MARTÍNEZ SÁEZ	783
BODEGAS MARTUE	740, 783
BODEGAS MASET	115, 126, 328, 355
BODEGAS MASET RIOJA	537
BODEGAS MAURO	805
BODEGAS MAURO ESTÉVEZ	423
BODEGAS MÁXIMO ABETE	314
BODEGAS MAZUELA	537
BODEGAS MEDRANO IRAZU	537
BODEGAS MEGÍA E HIJOS -CORCOVO	696, 783
BODEGAS MELER	627
BODEGAS MENADE	805, 849
BODEGAS MIGUEL A. AGUADO	783, 879
BODEGAS MILVUS	454
BODEGAS MITOS	677
BODEGAS MOCÉN	606, 805
BODEGAS MOISÉS CASAS (ECCE VINUM)	784
BODEGAS MONJE	755
BODEGAS MONÓVAR	36
BODEGAS MONTALVO WILMOT	745, 784
BODEGAS MONTE AMÁN	55
BODEGAS MONTE LA REINA	660
BODEGAS MONTECILLO	538
BODEGAS MORCA	104
BODEGAS MUCY	158
BODEGAS MUGA	126, 539
BODEGAS MUÑANA	202, 765, 850, 879
BODEGAS MURO	539
BODEGAS MURUA	540
BODEGAS MURVIEDRO	37, 126, 677, 705, 850
BODEGAS NABAL	454
BODEGAS NAIA	607
BODEGAS NAIROA	423
BODEGAS NAJERILLA, S.COOP.	540
BODEGAS NARANJO	241, 799, 885
BODEGAS NAVAJAS	540
BODEGAS NAVALTALLAR	759
BODEGAS NAVARRO LÓPEZ	696, 784
BODEGAS NESTARES EGUIZÁBAL	541
BODEGAS NIDIA	806
BODEGAS NIEVA	608
BODEGAS NIVARIUS	541
BODEGAS NOC	784
BODEGAS NODUS	677, 705
BODEGAS O VENTOSELA	423
BODEGAS OBALO	541
BODEGAS OCHOA	314
BODEGAS OLARRA	542
BODEGAS OLIVARES	227
BODEGAS OLIVEROS	169
BODEGAS OLMAZA	542
BODEGAS ONDALÁN	542
BODEGAS ONDARRE	542
BODEGAS ONEROM	850
BODEGAS ONTINIUM	705
BODEGAS ORBEN	543
BODEGAS ORTIGOSA	37
BODEGAS OSBORNE	210
BODEGAS OSCA	628
BODEGAS OTERO	756, 850
BODEGAS PABLO MORATE - MUSEO DEL VINO	724
BODEGAS PABLO PADÍN	390
BODEGAS PÁEZ MORILLA	769, 850
BODEGAS PAGOS DE MOGAR	455
BODEGAS PALACIOS REMONDO	543
BODEGAS PANDORA	608
BODEGAS PANIZA	110
BODEGAS PARCENT C.B.	37
BODEGAS PASCUAL	455
BODEGAS PASCUAL FERNÁNDEZ - FRONTERA NATURAL	59

WINERIES	PAGE	WINERIES	PAGE	WINERIES	PAGE
BODEGAS PASIEGO	678	BODEGAS RAMÓN BILBAO RUEDA	610	BODEGAS SANTALBA	547
BODEGAS PASTRANA	59	BODEGAS REBOLLAR ERNESTO CÁRCEL	678	BODEGAS SANTIAGO APÓSTOL	228
BODEGAS PEDROHERAS	241	BODEGAS REBROTAR	806	BODEGAS SANTIAGO ROMA	390
BODEGAS PEIQUE	68	BODEGAS RECTORAL DE AMANDI	411	BODEGAS SAUCI	169
BODEGAS PEÑAFIEL	455	BODEGAS REMÍREZ DE GANUZA	545	BODEGAS SEDELLA	262
BODEGAS PEÑALBA HERRÁIZ	456	BODEGAS RESALTE DE PEÑAFIEL	458	BODEGAS SEÑORÍO DE ALDEA	459
BODEGAS PERE SEDA	351, 850	BODEGAS REVERÓN	30	BODEGAS SEÑORÍO DE NAVA	459, 611
BODEGAS PIEDEMONTE	315	BODEGAS RIKO	38	BODEGAS SEÑORÍO DE NEVADA	202, 822
BODEGAS PINCERNA	255	BODEGAS RIOJANAS	545	BODEGAS SERVILIO - ARRANZ	460
BODEGAS PINEA DEL DUERO	456	BODEGAS RODA	546	BODEGAS SIAH	423
BODEGAS PINNA FIDELIS	456	BODEGAS RODENO	678	BODEGAS SIERRA DE CABRERAS	38
BODEGAS PINOSO	38	BODEGAS RODERO	458	BODEGAS SIERRA SALINAS	39
BODEGAS PIQUERAS	48	BODEGAS RODRÍGUEZ Y SANZO	69, 390, 458, 546, 610, 660, 806	BODEGAS SILVANO GARCÍA	228, 851
BODEGAS PITA	609			BODEGAS SIMBOLO	242
BODEGAS POLO MONLEÓN	706	BODEGAS ROMALE	126, 498	BODEGAS SINFORIANO	159, 807
BODEGAS PORTIA	456, 609	BODEGAS ROMÁN	104	BODEGAS SOBREÑO	661
BODEGAS PRADO DE OLMEDO	457	BODEGAS ROMERO DE ÁVILA SALCEDO	242, 785	BODEGAS SOLANA DE RAMIREZ RUIZ	547
BODEGAS PRIETO PARIENTE	806	BODEGAS ROSALÍO ALONSO & CO	851	BODEGAS SOLAR VIEJO	548
BODEGAS PRIMITIVO COLLANTES	211, 769	BODEGAS ROURA, J.A. PEREZ ROURA	32, 127	BODEGAS SOLEDAD	672, 786
BODEGAS PRÍNCIPE DE VIANA	315	BODEGAS RUBERTE	104	BODEGAS SONSIERRA	548
BODEGAS PRIUS	609	BODEGAS RUBICÓN	252	BODEGAS TAGONIUS	851
BODEGAS PROELIO	543	BODEGAS RUEDA PÉREZ	611	BODEGAS TAJINASTE	717, 755
BODEGAS PROTOS	158, 609	BODEGAS RUIZ TORRES	499, 818	BODEGAS TAMPESTA	255
BODEGAS PUENTE DE RUS	242, 785	BODEGAS SALVUEROS	158	BODEGAS TARÓN	549
BODEGAS PUIGGRÒS	115, 355, 850	BODEGAS SALZILLO	227	BODEGAS TARSUS	460
BODEGAS QUINTA DE AVES	799	BODEGAS SAMPAYOLO	689	BODEGAS TEJONERAS	724
BODEGAS QUITAPENAS	262	BODEGAS SAN ALEJANDRO	96	BODEGAS TENEGUÍA	250
BODEGAS R & G	211, 356, 457, 544, 610, 785	BODEGAS SAN DIONISIO, S. COOP.	227	BODEGAS TEÓFILO REYES	460
BODEGAS RAIZ DE GUZMÁN	457	BODEGAS SAN ESTEBAN	547	BODEGAS TERRAS GAUDA	391
BODEGAS RAMÍREZ DE LA PISCINA	544	BODEGAS SAN VALERO	110	BODEGAS THESAURUS CIGALES	159
BODEGAS RAMIRO'S	806	BODEGAS SANI PRIMAVERA	127, 818, 851	BODEGAS THESAURUS RIBERA DEL DUERO	461
BODEGAS RAMÓN BILBAO	544	BODEGAS SANTA EULALIA	459	BODEGAS TIERRA	549

WINERIES	PAGE
BODEGAS TIERRA SAVIA	851, 879
BODEGAS TINTO CORAZÓN	228
BODEGAS TOBÍA – ÓSCAR TOBÍA	550
BODEGAS TORRALBENC VELL	820, 821
BODEGAS TORREDEROS	461, 611
BODEGAS TORREDUERO	661
BODEGAS TORREMORÓN	461
BODEGAS TORRES AUGUSTI	768
BODEGAS TOTÓ MARQUÉS	851
BODEGAS TRADICIÓN	211
BODEGAS TRASLASCUESTAS	462
BODEGAS TRENZA	462, 502, 733, 786, 852
BODEGAS TRIAY	280
BODEGAS TRIDENTE	807
BODEGAS TROBAT	127
BODEGAS TRUS	462
BODEGAS UBIDE	550
BODEGAS UTIELANAS	678
BODEGAS VAL DE SOUTO	423
BODEGAS VAL DE VID	611
BODEGAS VALBUSENDA	662
BODEGAS VALCABADINO	653
BODEGAS VALCARLOS	316
BODEGAS VALDELACIERVA	550
BODEGAS VALDELANA	551
BODEGAS VALDEMAR	463, 551
BODEGAS VALDEORITE	127
BODEGAS VALDESNEROS	55
BODEGAS VALDOVINOS	628
BODEGAS VALDUBÓN	463
BODEGAS VALLE DE MONZÓN	463
BODEGAS VALLOBERA	551

WINERIES	PAGE
BODEGAS VALORIA	552
BODEGAS VALPARAISO	464
BODEGAS VATAN	611, 662
BODEGAS VEGA DE YUSO	464
BODEGAS VEGA SAUCO	662
BODEGAS VEGA SICILIA	464
BODEGAS VEGALFARO	127, 679, 745
BODEGAS VEGAMAR	127, 706
BODEGAS VERACRUZ	612
BODEGAS VERDES	757, 807
BODEGAS VERDÚGUEZ	242
BODEGAS VETUS	612, 663
BODEGAS VETUSTA	464
BODEGAS VIBE	679
BODEGAS VILANO	465
BODEGAS VILLA CONCHI	128
BODEGAS VILLAVID	267, 786
BODEGAS VINÍCOLA REAL	552
BODEGAS VINIVAL	706
BODEGAS VIÑA BARDELA	724
BODEGAS VIÑA BERNEDA	553
BODEGAS VIÑA ELENA	228
BODEGAS VIÑA HERMINIA	553
BODEGAS VIÑA ROMANA	59
BODEGAS VIÑAS DE VIÑALES	69
BODEGAS VIÑAS DEL CÉNIT	653, 808
BODEGAS VIONTA	391
BODEGAS VIORE	69, 612
BODEGAS VIRGEN DE BELÉN	48, 786
BODEGAS VIRGEN DE LOREA	85
BODEGAS VIHIATUS	663
BODEGAS VIRTUS	465

WINERIES	PAGE
BODEGAS VITALIS	256
BODEGAS VITIVINOS	267
BODEGAS VIVANCO	554
BODEGAS VIVANZA	39
BODEGAS VIYUELA	465
BODEGAS VIZAR	742, 808
BODEGAS VIZCARRA	466
BODEGAS VOLVER	39, 48, 706, 786
BODEGAS WILLIAMS & HUMBERT	212, 770
BODEGAS WINOL	49, 786, 799
BODEGAS XALO	40
BODEGAS Y DESTILERÍAS VIDAL	706
BODEGAS Y VIÑAS DEL CONDE	554
BODEGAS Y VIÑEDOS A.A.	69
BODEGAS Y VIÑEDOS ACEÑA	466
BODEGAS Y VIÑEDOS ALFREDO SANTAMARÍA	159
BODEGAS Y VIÑEDOS ALILIAN	467
BODEGAS Y VIÑEDOS ALIÓN	467
BODEGAS Y VIÑEDOS ALVAREZ ALFARO	554
BODEGAS Y VIÑEDOS ÁNGEL LORENZO CACHAZO	613
BODEGAS Y VIÑEDOS ARTADI	852
BODEGAS Y VIÑEDOS ARTAZU	316
BODEGAS Y VIÑEDOS CASA DEL VALLE	787
BODEGAS Y VIÑEDOS CASADO MORALES	554
BODEGAS Y VIÑEDOS CASTIBLANQUE	787, 853
BODEGAS Y VIÑEDOS DE CAL GRAU	356
BODEGAS Y VIÑEDOS DEL CONDE DE SAN CRISTÓBAL	467
BODEGAS Y VIÑEDOS DIEZ MERITO	213
BODEGAS Y VIÑEDOS EL ROBLEDO	758
BODEGAS Y VIÑEDOS EL SEQUÉ	40
BODEGAS Y VIÑEDOS EL SOLEADO	267, 502, 663

WINERIES	PAGE
BODEGAS Y VIÑEDOS FRUTOS ARAGÓN	468
BODEGAS Y VIÑEDOS GALLEGO ZAPATERO	468
BODEGAS Y VIÑEDOS GANCEDO	69
BODEGAS Y VIÑEDOS GÓMEZ CRUZADO	555
BODEGAS Y VIÑEDOS HERAS CORDÓN	555
BODEGAS Y VIÑEDOS ILLANA	502
BODEGAS Y VIÑEDOS ILURCE	555
BODEGAS Y VIÑEDOS JOSÉ ANTONIO GARCÍA	70
BODEGAS Y VIÑEDOS JUAN MANUEL BURGOS	468
BODEGAS Y VIÑEDOS LA MALDITA	556
BODEGAS Y VIÑEDOS LA MEJORADA	808
BODEGAS Y VIÑEDOS LABASTIDA	556
BODEGAS Y VIÑEDOS LADERO	243
BODEGAS Y VIÑEDOS LARRAZ	556
BODEGAS Y VIÑEDOS LEZA GARCÍA	556
BODEGAS Y VIÑEDOS LUNA BEBERIDE	70
BODEGAS Y VIÑEDOS MARQUÉS DE VARGAS	557
BODEGAS Y VIÑEDOS MAYOR DE CASTILLA	613
BODEGAS Y VIÑEDOS MERAYO	70
BODEGAS Y VIÑEDOS MONTEABELLÓN	469, 558, 613
BODEGAS Y VIÑEDOS MONTECASTRO	470
BODEGAS Y VIÑEDOS NEO	470, 613
BODEGAS Y VIÑEDOS ORTEGA EZQUERRO	558
BODEGAS Y VIÑEDOS PEDRO GARCÍA	724
BODEGAS Y VIÑEDOS PINTIA	663
BODEGAS Y VIÑEDOS PINUAGA	787
BODEGAS Y VIÑEDOS PONCE	267
BODEGAS Y VIÑEDOS PRADOREY	471, 808, 853
BODEGAS Y VIÑEDOS PUENTE DEL EA	558
BODEGAS Y VIÑEDOS QUINTA DA MURADELLA	280
BODEGAS Y VIÑEDOS QUIROGA DE PABLO	559
BODEGAS Y VIÑEDOS RAUDA S. COOP.	472

WINERIES	PAGE
BODEGAS Y VIÑEDOS ROSAN	160
BODEGAS Y VIÑEDOS SEÑORIO DE BOCOS	653
BODEGAS Y VIÑEDOS SHAYA	613
BODEGAS Y VIÑEDOS SUCRO	502
BODEGAS Y VIÑEDOS TÁBULA	472
BODEGAS Y VIÑEDOS TAMARAL	473
BODEGAS Y VIÑEDOS TAVERA	275
BODEGAS Y VIÑEDOS VALDEBODEGA	473
BODEGAS Y VIÑEDOS VALLEYGLESIAS	724
BODEGAS Y VIÑEDOS VALTRAVIESO	56, 229, 473, 503, 809
BODEGAS Y VIÑEDOS VEGA REAL	474, 614
BODEGAS Y VIÑEDOS VENTA LA VEGA	49
BODEGAS Y VIÑEDOS VENTUA	71
BODEGAS Y VIÑEDOS VERUM	788
BODEGAS YSIOS	559
BODEGAS YUSTE	213
BODEGAS ZIFAR	474
BODEGAS ZINTZO	559
BODEGAS ZIRÍES	853
BODEGAS ZURBAL	560
BODEGUEROS QUINTA ESENCIA	614
BODEGUES MACIÀ BATLE	827
BODEGUES VIDAL SERRA	828
BODEGUES VISENDRA	116
BODEM BODEGAS	110
BOLET AGRICULTURA ECOLÓGICA	128, 329
BOQUERA DEL CARCHE	734
BOSQUE DE MATASNOS	474
BOUZA DO REI	391
BROTONS V & A.	40
BRUGAROL	182, 854

WINERIES	PAGE
BRUJO WINES	560
BRUNO MURCIANO	679, 706
BRUTO	229
BSI - BODEGAS SAN ISIDRO	229, 854
BUIL & GINÉ	356
BURGOS PORTA - MAS SINÉS	356

C

WINERIES	PAGE
CA N'ESTRUC	116
CA' DI MAT	725
CAL SERRADOR	128
CAMBADOS URBAN WINERY	392
CÁMBRICO	758, 854
CAMINO DE CABRAS	392, 689, 854
CAMINO DEL NORTE, VINOS BY RAÚL PÉREZ	854
CAMPO VIEJO	560
CAMPOS DE LUZ	111
CAMPOS DE RISCA	230
CAMPOS DE SUEÑOS	614
CAN BAS DOMINI VINICOLA	329
CAN DESCREGUT	879
CAN GELAT	828
CAN GRAU VELL	116, 879
CAN RÀFOLS DELS CAUS	329
CAN RICH	816, 879
CAN SUMOI	330, 854, 880
CANALS I MUNNÉ	128, 330
CANALS NADAL	129
CANTALAPIEDRA VITICULTORES	809, 854
CANTARIÑA	71
CAP DE BARBARIA	820
CAPILLA DEL FRAILE	788
CAR VINÍCOLAS REUNIDAS	171

WINERIES	PAGE	WINERIES	PAGE	WINERIES	PAGE
CARA NORD CELLER	163, 292	CAVA REVERTÉ	131	CELLER CREDO	332
CARLOS CANEIRO NÚÑEZ	411, 855	CAVAS BERTHA	131	CELLER DE CAPÇANES	117, 293
CARLOS REY LUSTRES	392	CAVAS HILL	132, 331	CELLER DE L'ABADÍA	359
CARLOS SERRES	560	CAVAS JANÉ SANTACANA	132	CELLER DE L'ENCASTELL	359
CARMELITANO BODEGAS Y DESTILERÍAS	707	CAVAS MIQUEL PONS	132, 331	CELLER DE L'ERA	294
CARRAMATA	614	CAVES BOHIGAS	116, 133	CELLER DE SANUI	172
CARRASCALEJO	92	CAVES NAVERAN	133, 332	CELLER DEL ROURE	708
CARRIL CRUZADO	268, 788	CAVES ROMAGOSA TORNÉ	117	CELLER ESCODA PALLEJÀ	360
CASA CORREDOR	41, 855	CAVES VIDAL I FERRÉ	133	CELLER GERISENA	184
CASA DE OUTEIRO	412	CAZAPITAS	392, 855	CELLER GRAU I GRAU	117, 347
CASA DO SOL	392	CELLER AIXALÀ I ALCAIT	357	CELLER GRITELLES	294, 360
CASA GRAN DEL SIURANA	357	CELLER ARCHÉ PAGÈS	182	CELLER HIDALGO ALBERT	360
CASA LO ALTO BODEGA Y VIÑEDOS	707	CELLER ARRUFÍ	642, 855	CELLER HOSPITAL DE SITGES	332, 855
CASA LOS FRAILES	707	CELLER BALART	642	CELLER HUGAS DE BATLLE	184
CASA MOREIRAS	412	CELLER BÀRBARA FORÉS	643	CELLER JOAN AMETLLER	360
CASA SICILIA 1707	41, 855	CELLER BARTOLOMÉ	358	CELLER JOC	184, 856
CASADETILIO	474	CELLER BATEA	643	CELLER JORDI LLUCH	134
CASAL DE ARMÁN	424, 855	CELLER BATLLIU DE SORT	171	CELLER LA BOTERA	644
CASAR DE BURBIA	71	CELLER CAL BESSÓ	293	CELLER LES ACÀCIES	347
CASERÍO DE DUEÑAS	614	CELLER CAL PLA	358	CELLER LES FRESES	41
CACONA MICAELA	015	CELLER CARLES ANDREU	133, 164	CELLER LES SOQUES	856
CASTELL D'OR	129, 293, 357	CELLER CASTELLET	358	CELLER MAR DE VINS	856
CASTELL DEL REMEI	171	CELLER CASTELLÓ MURPHY	182	CELLER MARIÀ PAGÈS	185
CASTELL SANT ANTONI	129	CELLER CATARUZ	707	CELLER MARIOL	644, 856
CASTELO DE PEDREGOSA	130	CELLER CEDÓ ANGUERA	293	CELLER MARTÍ FABRA	185
CASTILLO DE ARESAN	243, 789	CELLER CERCAVINS	172	CELLER MARTÍN FAIXÓ	186
CASTRO DE LOBARZÁN	280	CELLER CLOS93 PRIORAT	359	CELLER MAS BASTE	361
CASTRO VENTOSA	72	CELLER COMA D'EN BONET	643	CELLER MAS BELLA	636
CASTROFIZ	412	CELLER COOPERATIU D'ESPOLLA	183	CELLER MAS D'EN BAIGET	636
CAVA & HOTEL MASTINELL	130, 330	CELLER COOPERATIU D'ESPOLLA – VINS DE POSTAL	183	CELLER MAS DEL BOTÓ	636, 856
CAVA ANGEL	130			CELLER MAS VICENÇ	636
CAVA ORIOL ROSSELL	131, 331	CELLER COOPERATIU GANDESA	644	CELLER MASROIG	294

WINERIES	PAGE	WINERIES	PAGE	WINERIES	PAGE
CELLER MASSIS DE L'ALBERA	186	CHOZAS CARRASCAL	135, 679, 740	CONVENTO OREJA	476
CELLER MATALLONGA	172	CÍA. DE VINOS ENTRE DOS AGUAS	857	COOP. ALBARIZAS DE TREBUJENA	214
CELLER MONTSEC	173	CÍA. VINÍCOLA DEL SUR - TOMÁS GARCÍA	286	COOP. DE VIVER	772
CELLER PASANAU	361	CÍA. VITÍCOLA SILEO	296	COOP. JESÚS DEL PERDÓN - BODEGAS YUNTERO	243, 789
CELLER PASCONA	295	CIEN Y PICO WINE	268		
CELLER PIÑOL	645	CILLAR DE SILOS	475	COOP. SANTA QUITERIA - TINTORALBA	49
CELLER PRIVAT	134	CINGLES BLAUS	296	COOP. SANTÍSIMO CRISTO DE LAS INJURIAS	858
CELLER PURGATORI	173	CLOS D' ESGARRACORDES	772	COOPERATIVA CONDES DE FUENSALIDA	275
CELLER RENDÉ MASDÉU	164	CLOS DE LÔM	708	COOPERATIVA DE GARRIGUELLA	187
CELLER RIALLA	645	CLOS DEL PORTAL	363	COOPERATIVA FALSET MARÇA	297
CELLER RONADELLES	295	CLOS DELS GUARANS	857	CORAL DUERO	664
CELLER RUBIÓ DE SÒLS	857	CLOS FIGUERAS	297, 363	CORISCA	393
CELLER SOLERGIBERT	347	CLOS GALENA	363, 647	CORRALES ESPINOSA FAMILY WINES	697
CELLER TIANNA NEGRE	828, 857	CLOS MOGADOR	297, 364	CORTIJO DE JARA	770, 858
CELLER VALL-LLACH	361	CLOS PACHEM	364, 647	CORTIJO LOS AGUILARES	263
CELLER VILA CORONA	173	CLOS PISSARRA	858	CORZOÁS	412
CELLER VINÍRIC	186	CLOS PONS	174	COSECHEROS Y CRIADORES	789
CELLER XAVIER CLUA	646	COCA I FITÓ	297, 647	COSTERS DEL SIÓ	174
CELLERS BLANCH	637	CODORNÍU	135, 880	COTA 45	770
CELLERS CAN BLAU	295	COLECCIÓN DE TONELES CENTENARIOS	41	COTO DE GOMARIZ	424, 858
CELLERS CAROL VALLÈS	134	COLET	333	COVIDES VINYES - CELLERS	136, 333
CELLERS D'EN GUILLA	186	COLLBAIX - CELLER EL MOLI	348	CREACIONES EXEO	562
CELLERS DE SCALA DEI	362	COLONIAS DE GALEÓN	834	CREGO E MONAGUILLO	281
CELLERS SANT RAFEL	295	COMANDO G VITICULTORES	725	CRIADORES DE RIOJA	562
CELLERS TARRONÉ	646	COMPAÑÍA DE VINOS HERACLIO	561	CUARTA GENERACIÓN BODEGAS Y VIÑEDOS	758
CELLERS TERRA I VINS	296, 362, 646	COMPAÑÍA DE VINOS MIGUEL MARTÍN	614, 809	CUENTAVIÑAS	476, 562
CELLERS UNIÓ	296, 362, 646	COMPAÑÍA DE VINOS PEÑA SERRANO (PE ESE)	73	CUEVAS DE AROM	96
CELLERS UNIÓ - POBOLEDA	362	COMPAÑÍA DE VINOS TELMO RODRÍGUEZ	262, 475, 561, 615, 663, 689, 752	CUME DO AVIA	858
CEPA 21	475			CUSCÓ BERGA	136
CÉRVOLES CELLER	174	CONCEJO BODEGAS	160	CV SOLTERRA	476
CÉSAR MÁRQUEZ BODEGAS Y VIÑEDOS	72	CONDADO DE HAZA	475, 809	CVNE	476, 563, 615
CHAN DE ROSAS	393	CONDADO DE SEQUEIRAS	412		

WINERIES	PAGE	WINERIES	PAGE	WINERIES	PAGE
D		DOMINIO DE BORNOS	477	EDUARDO PEÑA	425
DALCAMP	628	DOMINIO DE CALOGÍA	478	EGO BODEGAS	230, 860
DALT TURÓ	828	DOMINIO DE EGUREN	790	EGUREN UGARTE	565
DANIEL FERNÁNDEZ (ALBA AL-BAR)	281	DOMINIO DE ES	478	EIDO DA FONTE	393
DANIEL LANDI VITICULTOR	276, 809	DOMINIO DE LA SIERRA	758	EL CELLER D'EN MARC	187
DANIEL RAMOS	752, 859	DOMINIO DE LA VEGA	137, 680	EL CELLER DE LA IBOLA	860
DAVID LANA GARCÍA VERDUGO	250	DOMINIO DE NOBLEZA	564	EL COTO DE RIOJA	566
DAVID MARTÍNEZ SOBRAL	393	DOMINIO DE PINGUS	478	EL GRIFO	252
DE BRINGAS	86	DOMINIO DE TARES	73	EL HATO Y EL GARABATO	59
DE LA RIVA	214	DOMINIO DEL ÁGUILA	478	EL INQUILINO WINES	566
DE MULLER	364, 637, 859	DOMINIO DEL BENDITO	664	EL OTERO	566
DEHESA DE LOS CANÓNIGOS	477	DOMINIO DEL BLANCO	615	EL PROGRESO SDAD. COOP. CLM	244
DEHESA DE LOS LLANOS	789	DOMINIO DEL LINZE	790	EL REBUSCO BODEGAS	714
DEHESA DEL CARRIZAL	741	DOMINIO DEL PIDIO	479	EL SAUCERAL	790
DEHESA LA GRANJA	809	DOMINIO DEL SOTO	479	EL SITIO	755
DEHESA VALDELAGUNA	477	DOMINIO DO BIBEI	413, 425	ELÍAS MORA	615, 665, 861
DELAMPA	230	DOMINIO DOS TARES	809	ELVIWINES	32, 298, 566
DELGADO ZULETA	214	DOMINIO FOURNIER	479	EME ERRE VIÑADORES	752
DESCENDIENTES DE J. PALACIOS	73	DOMINIO MAESTRAZGO	767	EMILIO DOCAMPO DIÉGUEZ	425
DESTILERÍA Y BODEGA CAYO	822	DON BALBINO	565	EMILIO VALERIO	316
DESTINOS CRUZADOS	393	DON BERNARDINO	413	EMPORDÀLIA	187
DIEGO DONIZ DIÉGUEZ	425	DON PEDRONES	74	ENCIMA WINES	74
DIEZ SIGLOS DE VERDEJO	615	DONIENE GORRONDONA TXAKOLINA	86	ENCINA BLANCA DE ALBURQUERQUE	818
DIEZ-CABALLERO	564	DUEBA VITIVINÍCOLA	665	ENCOMIENDA DE CERVERA	790
DÍSCOLO	664	DUNORD VITÍCOLA	829, 860	ENEO	393, 567
DOBLEDEPEREZ MICROBODEGA	230, 859	**E**		ENTREBOSC	348
DOMENIO WINES BY CELLERS DOMENYS	117, 136, 164	E.D.V. BODEGA	414	ENTRECANALES DOMECQ E HIJOS	690
		ÉBANO VIÑEDOS Y BODEGAS	479	ENVERO WINE COMPANY	680
DOMÍNGUEZ CUARTA GENERACIÓN	632	ECCOCIVI CELLER	860	ENVINATE	414, 730
DOMINIO BASCONCILLOS	477	EDETÀRIA	647	EQUIPO NAVAZOS	215, 287, 861
DOMINIO DE BACO	244	EDICIONES ILIMITADAS	298, 364	ESENCIA WINES CELLARS	231
DOMINIO DE BERZAL	564	EDRA	860	ESPAIVI	365

WINERIES	PAGE
ESPELT VITICULTORS	188
ESTEBAN CASTEJÓN	97
ESTEL D'ARGENT	137, 333
ESTEVE I GIBERT VITICULTORS	334
ESTÉVEZ BODEGAS Y VIÑEDOS	74
ESTONES VINS	298
EULOGIO & JAVIER WINES	567, 810
EVINE	734
EXPLOTACIONES VITIVINÍCOLAS LASECANAS C.B.	616

F

WINERIES	PAGE
FAMILIA AMETLLER	138, 334
FAMILIA BASTIDA	244, 480, 567, 680
FAMILIA MONTAÑA	567
FAMILIA NIN ORTIZ	365
FAMILIA TORRES	165, 334, 365, 394
FAUSTINO RIVERO ULECIA	394, 567, 680
FAUSTO RIVERO PARDO	281
FÉLIX SOLÍS	244
FÉLIX SOLIS AVANTIS	394, 697, 791
FELIZ COMPAÑIA VINÍCOLA	480
FENTO WINES	394, 414
FERNANDO DE CASTILLA	215
FERRÉ I CATASÚS	138, 335, 861
FIL·LOXERA & CÍA.	708
FILTRO	#N/D
FINCA ALBRET	317
FINCA ALLENDE	567
FINCA ANTIGUA	245
FINCA BACARA	231, 862
FINCA BINIAGUAL	81
FINCA CAN AXARTELL	829, 862
FINCA COLLADO	42

WINERIES	PAGE
FINCA CONSTANCIA	791
FINCA CUARTA	414
FINCA DE LA RICA	568
FINCA EL MOLAR	268
FINCA EL REFUGIO	791
FINCA LA CANTERA DE SANTA ANA	317
FINCA LA PICARAZA	681
FINCA LOS ALIJARES	792
FINCA MILLARA	414
FINCA MONCLOA	770
FINCA MONTEPEDROSO	616
FINCA NUEVA	568
FINCA RÍO NEGRO	792
FINCA SANDOVAL	268
FINCA TOBELLA	365
FINCA TORREMILANOS	480, 810
FINCA VALLDOSERA	138
FINCA VALPIEDRA	569
FINCA VENTA DON QUIJOTE	792
FINCA VILADELLOPS	335
FINCA VILLACRECES	480
FINCA VIÑOA	425
FINCA Y VIÑEDOS SAN COBATE	481
FINCAS DE AZABACHE	569
FORTUNA WINES	394
FRAGAS DO LECER	281
FRANCISCO BARONA	481
FRANCISCO JAVIER SANZ SOGUERO	862
FRANCK MASSARD	648
FREIXA RIGAU	138
FREIXENET	117, 138
FROM GALICIA GROUP	570

WINERIES	PAGE
FRUTOS VILLAR	160, 665, 810
FUENTES DEL SILENCIO	810

G

WINERIES	PAGE
GAINTZA	196
GALINDO SAN MILLÁN – BODEGA + VIÑEDOS	666
GALLINA DE PIEL WINES	97, 188, 317, 335, 426
GARAIKOETXEA TXAKOLINDEGIA	196
GARENA TXAKOLINA	86
GARGALO	281
GARKALDE TXAKOLINA	86
GARMÓN CONTINENTAL	481
GATELL	139
GERMÁN R BLANCO	862
GIL LUNA	666
GIL PEJENAUTE	105
GIRÓ DEL GORNER	139, 336
GIRÓ RIBOT	139, 336, 862
GÓMEZ DE SEGURA	570
GONZÁLEZ BYASS JEREZ	216
GONZALO CELAYETA WINES	318
GORKA IZAGIRRE	86
GRACIA	287
GRAMONA	880
GRAN CLOS DEL PRIORAT	366
GRANDES VINOS	111
GRATAVINUM	366
GRIFOLL DECLARA GRUP	863
GRUPO DE BODEGAS VINARTIS	697
GUELBENZU	833
GURE AHALEGINAK	87

WINERIES	PAGE
H	
HACIENDA ACENTEJO	633
HACIENDA ALBAE	793
HACIENDA EL TERNERO	570
HACIENDA GRIMÓN	571
HACIENDA MIGUEL SANZ	481
HACIENDA REAL	793
HACIENDA SOLANO	482
HACIENDA TERRA D'URO	666
HACIENDA UCEDIÑOS	690
HACIENDA URBIÓN	571
HACIENDA VILLARTA	793
HAMMEKEN CELLARS	42, 140, 231, 245, 367, 394, 482, 734, 757, 793
HARVEYS	217
HASIBERRIAK WINES	87
HATORI	863, 881
HEREDAD DE ATENCIA	794
HERETAT MASCORRUBÍ	117
HERMANOS FRÍAS DEL VAL	571
HERRERO BODEGA	616
HIJOS DE ANTONIO POLO	482
HIJOS DE CRESCENCIA MERINO	160
HIJOS DE RAINERA PÉREZ MARÍN. "LA GUITA"	217
HIJOS DE RUFINO IGLESIAS	160
HIRUZTA BODEGA	196
HODGKINSON PRIORAT	367
HORNILLOS BALLESTEROS	482
HUERTA DE ALBALÁ	770
HUERTO DE LA CONDESA	263
HUGUET DE CAN FEIXES	336, 881

WINERIES	PAGE
I	
IBAI VITICULTORES	572
IBIZKUS WINES	816
IDRIAS	628, 863
IVÁN VÁZQUEZ PATEIRO (PATEIRO VINOS DE GUARDA)	426
J	
J. CHIVITE FAMILY ESTATES	318, 765
J. FERNANDO VIÑEDOS Y BODEGAS	616
J. GARCIA CARRION LA MANCHA	245
JAIME RUIZ DIAZ	572
JANÉ VENTURA	140, 336
JAUME DE PUNTIRÓ	81, 830
JAUME LLOPART ALEMANY	140, 337
JAVIER SANZ VITICULTOR	616, 810, 863
JEAN LEON	337
JESÚS DE MADRAZO WINES	483, 572
JOAN DE LA CASA. VITICULTOR	42
JOAN SIMÓ	367
JOAQUÍN REBOLLEDO	690
JON ANDER REKALDE	87
JORGE ORDÓÑEZ MÁLAGA	264, 863
JOSÉ BASOCO BASOCO	572
JOSÉ ESTÉVEZ	771
JOSÉ MANUEL RODRÍGUEZ GONZÁLEZ	415
JOSEP FORASTER	165
JOSEP GRAU VITICULTOR	299, 367
JOTA JIMÉNEZ WINES	231
JUAN FRANCISCO FARIÑA PÉREZ	714
JÚLIA BERNET	881
JUVÉ & CAMPS	141

WINERIES	PAGE
L	
L'INFERNAL	368
L'OLIVERA	175
L'ORIGAN	141
LA ALQUERÍA DE PRUNA	863
LA BALLESTERA	794
LA BARONÍA DE TURIS	709
LA CALANDRIA. PURA GARNACHA	319, 864
LA CARBONERA	573
LA CASA DE LAS VIDES BODEGUES I VINYES	709
LA CEPA DE PELAYO	269
LA CONRERIA D'SCALA DEI	368
LA COVA DELS VINS	299
LA FURGONETA VINOS	810
LA GAVACHA	97
LA GENERAL DE VINOS	105
LA HAYA	718
LA LOBA	483
LA NIÑA DE CUENCA, BODEGA Y VIÑEDOS PROPIOS	269
LA QUINTA	573
LA QUINTA VENDIMIA	483
LA RIOJA ALTA	573
LA SETERA	60
LA SUERTITA	718
LA VIÑA DEL ABUELO	666
LA VIÑITA	718
LACRIMA BACCUS	142, 338
LADRÓN DE LUNAS	142, 681, 864
LAFOU CELLER	648
LAGAR DA CONDESA	394
LAGAR DE BESADA	395

WINERIES	PAGE
LAGAR DE CERVERA	395
LAGAR DE COSTA	395, 864
LAGAR DE DEUSES	281
LAGAR DE LA SALUD	287, 864
LAGAR DEL DUQUE	160
LALOMBA	574
LANGA	97
LAOSA	256, 864
LAR DE MAÍA	811, 864
LAS MERCEDES DEL CABRIEL	681
LAS MORADAS DE SAN MARTÍN	726
LAS PEDRERAS VIÑEDOS Y VINOS	752, 865
LATORRE AGROVINÍCOLA	681
LECO PUNK WINES & VINEYARDS	574
LEGADO DE ORNIZ	667
LEGARIS	483, 617
LEIVE ECOADEGA	426
LEYENDA DEL PÁRAMO	811
LIBALIS	835
LIBERSO CURIOSO VERDEJO	617
LIBRE Y SALVAJE	111
LICINIA WINES	726
LLANO DE EL PINO	756
LLANO Y MONTE	92, 617
LLANOS NEGROS	250
LLICORELLA VINS	368
LLOPART	881
LLOPART – LA VIDA SECRETA DE LAS PLANTAS	338
LMT WINES	319, 574, 865
LOCOS POR EL VINO	105
LOESS VINOS	484, 617
LOS BARRANCOS	203

WINERIES	PAGE
LOSADA VINOS DE FINCA	75
LUBERRI MONJE AMESTOY	574
LUI & WILLIAM WINES	667
LUIS A. RODRÍGUEZ VÁZQUEZ	426
LUSTAU	217

M

Mª TERESA LÓPEZ FIDALGO (ADEGA O CABALIN)	690
MACROBERT & CANALS	575
MAGALARTE LEZAMA	87
MAIOR DE MENDOZA	396
MAISULAN	575
MAIUS VITICULTORS SCP	369
MANUEL ARAGÓN	218, 771, 865
MANUEL CORZO RODRÍGUEZ	691
MANUEL MANZANEQUE	794
MAR DE ENVERO	396
MARAÑONES	727
MARCO ABELLA	369, 648
MARÍA CASANOVAS	142
MARIA RIGOL ORDI	142
MARIA ZAMARREÑO	75
MARQUÉS DE BURGOS	484
MARQUÉS DE CARRIÓN	575
MARQUÉS DE GRIÑÓN FAMILY ESTATES	143, 576, 617
MARQUÉS DE IRÚN	617
MARQUÉS DE LA CONCORDIA FAMILY OF WINES	576, 617
MARQUÉS DE LA CONCORDIA FAMILY OF WINES CAVA	143
MARQUÉS DE MURRIETA	576
MARQUÉS DE TOMARES	577
MARQUÉS DE VALDUEZA	818

WINERIES	PAGE
MARQUÉS DE VELILLA	484
MARQUÉS DE VITORIA	577
MARQUÉS DEL PUERTO	577
MARQUÉS DEL REAL TESORO	219
MARTÍNEZ CORTA	577
MARTINSANCHO BODEGA Y VIÑEDOS	618
MAS ALTA	369
MAS BERTRAN	338
MAS BLANCH I JOVÉ	175
MAS CANDÍ	882
MAS COMTAL	865
MAS DE L'A	370
MAS DE LA PANSA	118, 165
MAS DE LES VINYES	300
MAS DE TORUBIO VITICULTORES	767
MAS DOIX	370
MAS ESCORPÍ	338
MAS IGNEUS	371
MAS LLUNES	188
MAS OLLER	189
MAS RAMONEDA	175
MAS RODÓ	338
MAS VILELLA	866
MASCARÓ	143
MASIA CAN RODA	32
MASIA SERRA	189
MATSU	667
MEDINA AGRICULTURA ECOLÓGICA (FINCA LAS CARABALLAS)	811
MELIDA WINES	485
MENDIETA OSABA WINES	578
MERCÈ SANGUESA I MILLAN	118

WINERIES	PAGE
MERITXELL PALLEJÀ	371
MERRUTXU	88
MERUM PRIORATI	372
MÉS QUE PARAULES	348
MIGUEL PAVÓN REINOSO	415
MILÉNICO	485
MILL WINES	485
MILSETENTAYSEIS	485
MIQUEL OLIVER VINYES I BODEGUES	351
MIXTURA	866
MODERNISTA CELLER	648
MOISÉS GRAN VINO	667
MÓNICA ALBOR	426
MONT MARÇAL	144
MONTEBACO	485, 618
MONTEGAREDO	486
MONTES DE LEZA - LOZANO	578
MONTESANCO	144, 682
MONTESQUIUS	144
MONTRUBÍ	339
MUCHADA-LÉCLAPART	771
MUJICA & DIAZ	727, 866
MUREDA ALIMENTACIÓN	698, 794
MURGUIALDI 3 DE BALEARS	830
MUSCÀNDIA	145, 339

N

WINERIES	PAGE
NADAL	145, 882
NALUAR & ACEDIANO	486
NEKEAS	319
NEXUS BODEGAS	486
NOTAS FRUTALES DE ALBARIÑO	396

WINERIES	PAGE
NTRA. SRA. DE LA CABEZA DE CASAS IBÁÑEZ SOC. COOP. DE CLM	269
NUBORI	578, 618
NUESTRO DE DÍAZ BAYO	486
NUEVOS VINOS CB	319, 618

O

WINERIES	PAGE
O ALBOREXAR NO RIBEIRO	427
O LUAR DO SIL	691
OCELLUM DURII	60
OLIVEDA	190
OLIVER VITICULTORS	145, 339
OLLER DEL MAS	349, 867
OROT	668
ORTO VINS	300
OS CIPRESES	415
ÓSCAR OLMOS VINOS	232
OSCAR PÉREZ RODRÍGUEZ	415
OSSIAN VIDES Y VINOS	811
OSTATU	578
OTERO & PÉREZ BODEGUEROS	415
OVIDIO GARCÍA	161
OXER WINES	579, 867

P

WINERIES	PAGE
PABLO VIDAL - VINOS CON PERSONALIDAD	282, 427, 691, 867
PACO & LOLA	397, 691
PACO MULERO	49, 98, 232, 300, 397, 579, 618, 812
PAGO CASA DEL BLANCO	740
PAGO CASA DEL BLANCO - CANTABRIA	816
PAGO DE ALMARAES	867
PAGO DE CARRAOVEJAS	487

WINERIES	PAGE
PAGO DE LA BOTICARIA	867
PAGO DE LA JARABA	745
PAGO DE LARRÁINZAR	319
PAGO DE LOS ABUELOS	75
PAGO DE LOS CAPELLANES	487
PAGO DE THARSYS	146, 682, 741
PAGO DE VALLEGARCÍA	747
PAGO DEL CAMINO	580
PAGO DEL CIELO	487, 618
PAGO DEL MARE NOSTRUM	868
PAGO DEL VICARIO	742
PAGO FINCA ÉLEZ	743, 868
PAGO HEREDAD DE URUEÑA	744
PAGO LOS BALANCINES	499
PAGO TRASLAGARES	618
PAGOS ALTOS DE ACERED	98
PAGOS DE ANGUIX	488
PAGOS DEL MONCAYO	105
PAGOS DEL REY	397, 580, 618
PAGOS DEL REY RIBERA DEL DUERO	488
PAGOS DEL REY TORO	668
PAISAJES Y VIÑEDOS	580
PALACIO DE BORNOS	619
PALACIO DE VILLACHICA	488, 668
PALACIO QUEMADO	499, 819
PALMERI SICILIA	106
PARAJES DE LOS VIDRIOS	727
PARAJES DEL VALLE BODEGAS Y VIÑEDOS	232, 270
PÁRAMO ARROYO	488
PARATÓ VINÍCOLA	146, 340
PARDAI Y PUNTO	60
PARDEVALLES	256

WINERIES	PAGE
PARÉS BALTÀ	146, 340
PARXET	147
PAZO BAIÓN	397
PAZO DE BARRANTES	398
PAZO DE RUBIANES	398
PAZO DE SEÑORANS	399
PAZO DE VIEITE	427
PAZO PONDAL	399, 868
PAZO TORRE PENELAS	400
PAZOS DE LUSCO	400
PAZOS DEL REY	282
PEKADO MORTAL	400, 489, 691
PENÍNSULA VINICULTORES	88, 753, 868
PEÑOS MARTIN MARCOS	61
PEPE MENDOZA CASA AGRÍCOLA	43
PERE GUARDIOLA	190
PERE VENTURA	147
PERELADA	147, 190
PÉREZ BARQUERO	287
PERINET	372
PERSEO 7 BODEGAS	619
PICO CUADRO	489
PIEDRAS DE SAN PEDRO	489
PITUCO VITICULTOR	232
PLA DE MOREI	118, 868
PLANA D'EN JAN MICROVINIFICACIONES	340
PLANAS ALBAREDA	148, 341
PONTECABALEIROS	400
PONY FOODS	148
PRESAS OCAMPO	633
PRIMA VINIA	282
PRIMITIVO QUILES	43, 868

WINERIES	PAGE
PRIORATO DE RAZAMONDE	427
PRIVIOS	400
PROMETO SER INFIEL	489
PROPIEDAD DE ARÍNZANO	746
PROPIETAT D'ESPIELLS	341
PROTOS BODEGAS RIBERA DUERO DE PEÑAFIEL	489
PROYECTO GARNACHAS/VINTAE	98, 372, 869
PUNCTUM BIODYNAMIC FAMILY VINEYARDS	245, 794
PURÍSIMA CONCEPCIÓN, S.C. DE CLM	503
PURO ROFE	253
Q	
QUEIRÓN	580
QUINTA COUSELO	400
QUINTA SARDONIA	812
R	
R. LÓPEZ DE HEREDIA VIÑA TONDONIA	581
RAFAEL CAMBRA	709
RAFAEL PALACIOS	691
RAÍCES IBÉRICAS	76, 98, 682, 869
RAIMAT	176
RAMÓN DE AYALA LETE E HIJOS	581
RAMÓN DO CASAR	428
RAMÓN ROQUETA	118
RAMÓN SAENZ ORGANIC WINES & VINEYARDS	581
RAUL PÉREZ BODEGAS Y VIÑEDOS	76, 869
RAUL PÉREZ BODEGAS Y VIÑEDOS	257
RAVENTÓS I BLANC	883
REAL AGRADO	581
RECAREDO	883
RECTORAL DO UMIA	401
REGANTÍO VIEJO	771
REMIGIO ENRIQUEZ	834, 869

WINERIES	PAGE
REZABAL	196
RICO NUEVO VITICULTORES	753
RIMARTS	148
RINCÓN DEL GUINIGUADA	200
RIOJA VEGA	582
RISKY GRAPES	710
RITME CELLER	373
ROBERT J. MUR	149
ROCHAL	759
ROGER GOULART (CVNE)	149
ROIG PARALS	191
RONSEL DO SIL	415
ROS MARINA VITICULTORS	341
ROSARIO VERA	582
ROSELL GALLART	149
ROUND TABLE	300, 342
ROVELLATS	150, 342
RUDELES	490
S	
S. COOP. NIÑO JESÚS DE ANIÑÓN	99
S.C.A. SANTA MARTA VIRGEN	499, 819
SA CABANA	830
SA FORANA	821
SABATÉ I COCA - CASTELLROIG	342, 884
SAM'S WINE	232
SAMSARA PRIORAT	373
SAN ANTONIO DE PADUA	270
SAN ISIDRO DE ALBOREA	270
SAN ROMÁN BODEGAS Y VIÑEDOS	668
SÁNCHEZ ROMATE HNOS.	219
SANDRA DOIX CELLER	373
SANGENÍS I VAQUÉ	374

WINERIES	PAGE
SANT JOSEP VINS	118, 649
SANTA CATARINA	82, 830
SANTA CRUZ DE ALPERA SOC. COOP. DE C-L-M	50
SANTA MARTA	692
SANTIAGO RUIZ	401
SDAD. COOP. CRISTO DE LA VEGA	246
SEGURA VIUDAS	119, 150, 342
SEÑORÍO DE ARANA	582
SEÑORÍO DE BEADE	428
SEÑORÍO DE BOCOS	490
SEÑORÍO DE CABRERA	200
SEÑORÍO DE LÍBANO	33, 583
SEÑORÍO DE LOS ARCOS	76, 257, 870
SEÑORÍO DE RUBIÓS	401
SEÑORÍO DE SAN VICENTE	583
SEÑORÍO DE SARRÍA	320
SEÑORÍO DEL JÚCAR	270, 870
SERRA DE CAVALLS	649
SHUKHRAT KHAKIMOV & VITICULTORFS	43, 270, 710, 870
SIERRA CANTABRIA	583
SIERRA DE TOLOÑO	584
SIETE PASOS WINES	282, 584, 669
SILICE VITICULTORES	870
SILVIA MARRAO BARREIRO	77
SINODO VITIVINÍCOLA	584
SOC. COOP. AGRARIA DE CLM SAN ISIDRO	271
SOC. COOP. VITIVINÍCOLA DE LONGARES - COVINCA	112
SOGAS MASCARÓ	150
SOLAR DE SAMANIEGO	490, 584
SOLAR DE URBEZO	112, 870

WINERIES	PAGE
SON GRAU GRAN	830
SON JULIANA	831
SON PRIM PETIT	831
SON RAMON VINS I VINYES	831
SOPLA LEVANTE	871
SOTO DE OÑATIL	871
SOTO DE ZEMTINAR	799
SOTO DEL VICARIO	77
SOTO MANRIQUE BODEGA	753, 871
SPANISH PALATE	50, 232, 491, 619, 669, 813
SPECTACLE VINS	300
SUERTES DEL MARQUÉS	718, 871
SUMARROCA	150, 342

T

WINERIES	PAGE
TALAI BERRI TXAKOLINA	196
TALLERI BERRIA UPATEGIA ETA MAHASTIAK	88
TANTAKA WINES	53
TENTENUBLO WINES	585
TEO LEGIDO	871
TERRA DE FALANIS	151, 300, 821, 872
TERRACOTA WINES CHI TAO JIU	682
TERRAMAGNA	232, 585, 795
TERRAMOLL	820
TERRAS DE LANTAÑO	402
TERRAS DO CIGARRÓN	282
TERRER DE PALLARS	176
TERRES DE VIDALBA	374
TERRIÑA	692
TERROIR AL LIMIT	374
TERROIR SENSE FRONTERES	301
TESO LA MONJA	669
THUNDERWINEMAKERS & AVGVSTVS FORVM	872

WINERIES	PAGE
TIERRA CALMA	727
TIERRA DE CEBREROS	753
TIERRA FUNDIDA	633, 872
TINTA ROSA	669
TOBELOS BODEGAS Y VIÑEDOS	585
TOCAT DE L'ALA	191
TOMÁS CUSINÉ	176
TOMÁS TORRESANO	872
TORELLÓ VITICULTORS	343, 884
TORO ALBALÁ	289
TORRE DE OÑA	586
TORRE DEL VEGUER	343, 872
TOTAL RESULTADO	#N/D
TR3SMANO	491
TRASLANZAS BODEGAS Y VIÑEDOS	161
TRES GENERACIONES	264
TRESPIEDRAS	491
TRIAS BATLLE	151, 344
TRONADO WINES	586
TROSSOS DEL PRIORAT	375
TXAKOLI TXABARRI	88
TXOMIN ETXANIZ	196

U

WINERIES	PAGE
UBETA WINES	320
UKAN WINERY	586
UNGLES NEGRES	873
UNION CAMPESINA INIESTENSE	795
UNIÓN VINÍCOLA DEL ESTE	151
UNIVERSITAT ROVIRA I VIRGILI	119, 638
UNSI	321
UVAS FELICES	402, 491, 586, 619, 728, 795

WINERIES	PAGE
V	
VAL DE MEIGAS	402
VALDELARES	321
VALDEMONJAS	491
VALDERIVERO	492
VALDESIL	693
VALDESPINO	219, 772
VALL DE BALDOMAR	177
VALLDOLINA VITICULTORS I ELABORADORS	151, 344
VALLFORMOSA	119
VALSANGIACOMO	710, 873
VARIAS	152, 344
VDB BODEGAS VALLE DEL BOTIJAS	813
VEGA CLARA	492
VEGA DEL PAS	619
VEGA TOLOSA	271
VEIGA NAUM	402
VELVETY WINES - DOMINIO LUBIANO	492
VENTA D'AUBERT	768, 873
VENUS LA UNIVERSAL	301
VÍA ROMANA ADEGAS E VIÑEDOS	416
VICENTE GANDÍA	44, 152, 492, 587, 682, 710, 813, 873, 885
VID VICA	153
VIEJAS DE IZAN	492
VILARNAU	153
VILE LA FINCA, BODEGAS Y VIÑEDOS	257, 813
VINATE	861
VINÍCOLA DE CASTILLA	246, 796
VINÍCOLA DE NULLES -ADERNATS	153, 638
VINÍCOLA DE TOMELLOSO	247, 796, 885
VINÍCOLA DE VALDEPEÑAS	698, 796

WINERIES	PAGE
VINÍCOLA SARRAL I SECCIÓ DE CRÈDIT	165
VINITOR WINE GROUP	619, 669
VINOS ANTONIO MONTERO	428
VINOS COLOMAN	247, 796, 885
VINOS DE ALGUEÑA	44
VINOS DE ARGANZA	813
VINOS DE BERNARDO ORTEGA	796
VINOS DE CULTO BANISIO	492, 587, 619
VINOS DE LA LUZ	493, 620
VINOS DEL PASEANTE	620
VINOS DIVERTIDOS	629
VINOS GUERRA	77
VINOS JEROMÍN	728
VINOS LA ZORRA	759
VINOS LLÁMALO X	797
VINOS RABILARGO	759
VINOS SANTOS ARRANZ (LÁGRIMA NEGRA)	493
VINOS SANZ	620
VINOS SIERRA NORTE	44, 233, 683, 711, 797
VINOS TABERNER AMADO	768
VINOS VALTUILLE	78
VINOS Y ACEITES LAGUNA	728
VINOS Y VIÑEDOS DE LA CASA MAGUILA	670
VINOS Y VIÑEDOS DOMINIO LASIERPE	322
VINOS Y VIÑEDOS FAMILIA FIEL	493
VINOS Y VIÑEDOS TRAMUNTANA	832
VINS ALGARS	649
VINS DE LA MEMÒRIA	33, 166, 376, 650
VINS DE MESIES	650
VINS DE RELAT	192, 650
VINS DE TALLER	119
VINS DEL COMTAT	44

WINERIES	PAGE
VINS DEL TROS	650
VINS EL CEP	154, 344
VINS ESSÈNCIA DE LLUNA	650
VINS I LICORS GRAU	112, 192, 621
VINS MIQUEL GELABERT	351
VINS NADAL	82, 832
VINS PEPE RAVENTOS	874, 885
VINS PER ESTIMAR EL VI	345
VINTAE / ATLANTIS	53, 402, 693
VINYA ELS VILARS	177
VINYA JANINE (JOSEP M. SAUMELL)	638
VINYA TAUJANA	82
VINYES D'EN GABRIEL	301
VINYES D'OLIVARDOTS	192
VINYES DEL TIET PERE	638
VINYES DELS ASPRES	193
VINYES DOMÈNECH	302
VINYES I VINS CA SA PADRINA	82, 832
VINYES MORTITX	832, 874
VIÑA AGUILERA	493
VIÑA ALMIRANTE	283, 403
VIÑA ARNAIZ	493, 670
VIÑA BALBAINA	874, 885
VIÑA BUENA	493
VIÑA BUJANDA	587
VIÑA CALLEJUELA	772, 875
VIÑA COSTEIRA	428, 693
VIÑA DEL LENTISCO	587
VIÑA EL PISÓN	875
VIÑA ESTEVEZ	633
VIÑA GÓMEZ	715
VIÑA IJALBA	588

WINERIES	PAGE
VIÑA MEIN - EMILIO ROJO	429
VIÑA MEMORIAS	683
VIÑA MORAIMA	403
VIÑA NORA	403
VIÑA OLABARRI	588
VIÑA PALACIOS	322
VIÑA REAL	588
VIÑA SALCEDA	589
VIÑA SANTA MARINA	819
VIÑA SOMOZA	693
VIÑA VALDEMAZÓN	494
VIÑA ZANATA	756
VIÑA ZORZAL WINES	322
VIÑADESGRACIA	814
VIÑAOLIVA SOC. COOP.	499, 819
VIÑAS DEL BIERZO	78
VIÑAS DEL JARO	494
VIÑAS DEL PORTILLO	711
VIÑAS DEL RÓDAN	233
VIÑAS DEL VERO	629
VIÑAS HERZAS	715
VIÑAS MURILLO	621
VIÑAS SERRANAS	814
VIÑAS SILENCIOSAS	589
VIÑEDOS ALONSO DEL YERRO	494, 670
VIÑEDOS ASTIGI	621
VIÑEDOS COTODEGAIO	429
VIÑEDOS DE CAMARENA, SDAD. COOP. DE CLM	276
VIÑEDOS DE LAS ACACIAS	815
VIÑEDOS DE PÁGANOS	590
VIÑEDOS DEL CONTINO	590
VIÑEDOS HERMANOS HERNÁIZ	590

WINERIES	PAGE
VIÑEDOS LA NAVA	494
VIÑEDOS POZANCO	500, 819
VIÑEDOS REAL RUBIO	591
VIÑEDOS RUIZ JIMÉNEZ	591
VIÑEDOS SIERRA CANTABRIA	592
VIÑEDOS SINGULARES	79, 302, 404, 495, 621
VIÑEDOS Y BODEGA LA MAGDALENA	503, 797
VIÑEDOS Y BODEGAS ALTO BUEN GRADO	754
VIÑEDOS Y BODEGAS ASENSIO CARCELÉN	233
VIÑEDOS Y BODEGAS ÁSTER	495
VIÑEDOS Y BODEGAS GARCÍA FIGUERO	495
VIÑEDOS Y BODEGAS GONZÁLEZ	276
VIÑEDOS Y BODEGAS GORMAZ	496
VIÑEDOS Y BODEGAS LA MARQUESA	592
VIÑEDOS Y BODEGAS MUÑOZ	247, 797
VIÑEDOS Y BODEGAS PABLO	113
VIÑEDOS Y BODEGAS PITTACUM	79
VIÑOS DE ENCOSTAS	875
VIRGEN DE LAS VIÑAS BODEGA Y ALMAZARA	248, 798, 875
VIRGEN DEL GALIR	694
VITÍCOLA LA COSTALENA	768
VITICULTOR Y ELABORADOR JUAN BERNAL	876
VITICULTORES NELEMAN	711
VITICULTORES SAN JUAN BAUTISTA	683
VITICULTORS BATEANS	651
VITICULTORS COSTERS DEL PRIORAT	376
VITICULTORS DEL PRIORAT	376
VIVES AMBRÒS	154, 639

W

WINERIES	PAGE
WEIN & VINOS GMBH	99, 233, 815
WINERY BURGMANN TENERIFE	634

WINERIES	PAGE
WINES N' ROSES VITICULTORES	712
WINES RAMSURT BY MAGI RAVENTOS	177, 193, 345

X

XAVI PALLEJÀ VITICULTOR	376
XOLAYR	203, 876

Y

YLLERA BODEGAS & VIÑEDOS	496, 621, 815

Z

ZÁRATE	404
ZINIO BODEGAS	593
ZUAZO GASTÓN BODEGAS Y VIÑEDOS	593

WINES

WINE	PAGE
'NGUANY NEGRE 2020 T	830
¿Y TÚ DE QUIÉN ERES? 2020 T	847
¿Y TÚ DE QUIÉN ERES? 2021 B	847
'NGUANY BLANC 2021 B	830
"MARÍA" ALONSO DEL YERRO 2018 T	494
= DISSIDENT 2019 T	42
% UZ CIEN X CIEN UVA ZALEMA 2021 B	168
+ RITME BLANC 2019 B	373
+ RITME BLANC 2020 B	373
+ RITME BLANC 2021 B	373
0 POR CIENTO TEMPRANILLO 2021 T	676
031 BARRICA ALICANTE BOUSCHET 2018 T BA	44
031 BARRICA CABERNET SAUVIGNON 2018 T C	43
10 SAUVIGNON BLANC 2021 B	335
10-12 (DIEZ PUNTO DOCE) 2016 T R	500
10-12 (DIEZ PUNTO DOCE) 2021 T	819
10-12 (DIEZ PUNTO DOCE) BE AG SD	819
10-12 SELECCIÓN (DIEZ PUNTO DOCE) 2018 T C	819
100 VEREMES 2019 B FB	638
100 X 100 MONASTRELL 2020 T	233
100 X 100 SYRAH 2020 T	233
100 Y CIENTOS 2017 T RB	722
100 Y CIENTOS 2018 T RB	722
100% LN BY LOHER 2020 T	632
100% ROMAILA 2018 T BA	776
1000 BESOS MERLOT SELECCIÓN 2020 T	709
10000 HORES FLORAL 2021 B	339
10000 HORES NEGRE SELECCIÓ 2020 T	339
10000 HORES ROSÉ 2021 RD	339
10000 HORES XARELLO 2021 B	339
11 PINOS BOBAL 2020 T RB	271
12 LINAJES 2017 T R	496

WINE	PAGE
12 LINAJES 2019 T C	496
12 LINAJES 2021 T RB	496
12 LINAJES GRANO A GRANO 2018 T C	496
12 LINAJES SENDA DE LA ESTACIÓN 2019 T	496
12 LINAJES VERDEJO 2020 B	605
12 LINAJES VERDEJO 2021 B	605
12 VOLTS 2020 T	822
1270 A VUIT 2014 T R	360
1400M 2019 B C	715
1400M 2020 B FB	715
1400M 2021	715
1400M T	715
1428 GARNACHA SYRAH 2020 T	99
1500 H COUPAGE 2010 T C S	868
1564 PETIT VERDOT 2018 T BA	797
1564 SYRAH 2020 T BA	797
1564 VIOGNIER 2019 B BA	797
1580 2017 T	825
1583 ALBARIÑO DE FEFIÑANES 2021 B	386
17 BY PINEA 2019 T C	456
1730 BF PX D	205
1730 FINO EN RAMA BF FI S	205
1730 VORS BF AM S	205
1730 VORS BF OL S	205
1730 VORS BF PC S	205
1752 GARNACHA TINTORERA ALTA EXPRESIÓN 2017 T C	47
1758 RESERVA FAMILIAR PETIT VERDOT 10 AÑOS 2011 T	790
1758 RESERVA FAMILIAR PETIT VERDOT 5 AÑOS 2016 T	790
1758 SELECCIÓN CENCIBEL 2011 T C	790
1758 SELECCIÓN RESERVA FAMILIAR CENCIBEL 5 AÑOS 2016 T	790
1771 CASA LOS FRAILES 2018 T C	707
180 NOCHES 2021 B	97

WINE	PAGE
1822 BF PC S	206
1822 FINO EN RAMA BF FI S	206
1822 MAGNUM BF MZ	206
1822 SOLERA FUNDACIONAL BOTA NO BF AM S	206
1860 SELECCIÓN 2018 T R	47
1864 CASTILLO DE OLITE 2016 T R	305
1864 CASTILLO DE OLITE 2018 T C	305
1864 CASTILLO DE OLITE CHARDONNAY 2021 B FB	305
1891 2017 T C	312
1891 2021 RD	312
1902 TOSSAL D'EN BOU GRAN VINYA CLASSIFICADA 2019 T C	370
1903 COMA DE CASES GARNATXA VELLES VINYES 2019 T C	370
1913 2019 B	306
1913 2021 RD	306
1913 2021 T	306
1921 MONASTRELL 2019 T	44
1932 DE PAZO VIEITE 2021 B	427
1953 PAZO TORRADO 2018 BE BR	379
1953 PAZO TORRADO 2021 B	379
1955 AMONTILLADO SOLERA CINCUENTENARIO B AM S	287
1955 OLOROSO SOLERA CINCUENTENARIO B OL S	288
1955 PALO CORTADO CINCUENTENARIO B PC S	288
1955 PEDRO XIMENEZ SOLERA CINCUENTENARIO BF PX D	288
1962 ORIGEN 2018 T	40
2 CEPAS MARQUÉS DEL ATRIO 2019 B BA	535
2 KISSES 2018 T C	570
2 KISSES 2021 T	570
20 ALDEAS 2019 T	809
200 MONGES 2007 T GR	552
200 MONGES 2009 B GR	552
200 MONGES 2011 B R	552
200 MONGES 2017 RD R	552

WINE	PAGE
200 MONGES ESENCIA 2012 B BA D	553
200 MONGES SELECCIÓN ESPECIAL 2006 T R	553
200 MONGES SELECCIÓN ESPECIAL 2011 B R S	553
25 VENDIMIAS 2021 B	319
299 DE MONTE PINADILLO 2018	437
3 ASES 2018 T C	431
3 ASES 2021 T	431
3 ASES 6 MESES 2020 T RB	431
3 DE OLANO 2016 T C	526
3 DE OLANO SELECCIÓN 2017 T	526
3 MIRADAS "CERRO MACHO" 2019 B	285
3 MIRADAS PARAJE DE RÍO FRÍO ALTO 2018 B	285
3 MIRADAS VINO DE PUEBLO 2020 B	285
3 SETMANES 2021 B	171
3 SETMANES 2021 RD	171
30.000 MARAVEDÍES 2019 T	727
30.70 2021 B	184
3404 CABERNET GARNACHA MORISTEL 2021 T	626
3404 TUCA D'ANETO 2019 T C	626
37 BARRICAS 2018 T C	90
4 KILOS 2020 T	822
40 BRACES NEGRE 2019 T	822
409 2020 T C	488
426 2019 T C	529
5 PULGARES 2018 T C	66
5. ZINTZO 2018 T C	559
50 VENDIMIAS DE SINFORIANO 2015 T	159
575 UVAS DE CÁMBRICO 2019 T R	758
6 DE 7 #GARAGEWINE 2020 T	773
600 METROS SA FORANA 2021 T	821
600 METROS SA FORANA BLANC 2021 B	821
61 DORADO EN RAMA 2021 BF SOLERA S	597

WINE	PAGE
7 VIN 2016 T BA	260
7 VIN 2017 T BA	260
7103 BLANC DE MANTONEGRO 2021 B	822
7103 BLANC PREMSAL BLANC 2021 B	823
7103 GIRÓ 2021 B	823
7103 NEGRE BÓTA 2019 T BA	823
7103 ROSAT 2021 RD	823
750 2010 T GR	724
77 DIES CELLER ARRUFI 2021 B	855
77 NITS CELLER ARRUFI 2021 T	855
8 VENTS 2019 T C	821
8 VENTS GRAN 2019 T	821
8.0.1 T R	110
8000 MARQUÉS DE BURGOS 2018 T	484
875 M CHARDONNAY 2021 B FB	566
8LGENDS ETERNA LEYENDA 2018 T	786
8LGENDS LEYENDA DEL CABALLERO CABERNET SAUVIGNON 2020 T	786
8LGENDS LEYENDA DEL CABALLERO MERLOT 2020 T	787
8LGENDS LEYENDA DEL CABALLERO SYRAH 2020 T	787
8LGENDS I FYFNDA DEL JINETE 2017 T C	49
8LGENDS LEYENDA DEL VOLCÁN MOSCATEL DE ALEJANDRÍA MACABEO 2020 B	799
8LGENDS LEYENDA DEL VOLCÁN TEMPRANILLO 2019 T	799
9 SET 2 2020 T	184
90 MINUTS 2020 T	375
99 PUNTS 2019 T	192
992 FINCA RÍO NEGRO 2020 T	792

A

WINE	PAGE
A COROA "LÍAS" 2020 B	685
A COROA 200 CESTOS 2020 B FB	685
A COROA GODELLO 2021 B	685

WINE	PAGE
A DE ARÍNZANO 2021 RD	746
A ESPEDRADA 2019 B FB	690
A PART 2019 B	646
A PONTE 2020 T	407
A PONTE VELLA 2019 T RB	844
A TELLEIRA GODELLO 2021 B	422
A TELLEIRA MENCÍA 2020 T	422
A TELLEIRA PARCELAS 2021 B	422
A VEREDAS 2018 B	541
A VEREDAS 2019 RD	541
A VILLEIRA 2020 T	694
A&M 3 2020 T RB	441
A&M AUTOR 2015 T R	441
A2 ALBILLO VALLEYESIGNAS 2021 B	724
AA CAU D'EN GENIS 2020 B	32
AA CAU D'EN GENIS 2021 B	32
AA DOLÇ DE NEU 2021 B BA D	837
AA DOLÇ MATARÓ 2020 T D	838
AA LANIUS 2020 B S	32
AA ORBUS 2019 T	32
AA PARVUS CHARDONNAY 2021 B	32
AA PARVUS SYRAH 2019 T	32
AALTO 2020 T	431
AALTO PS (PAGOS SELECCIONADOS) 2020 T R	431
ABA DE TRASUMIA 2021 B	403
ABAD DOM BUENO 2021 RD	64
ABAD DOM BUENO ESENCIA 2021 B	64
ABAD DOM BUENO GODELLO 2021 B	64
ABAD DOM BUENO LADERAS DEL NORTE 2020 T RB	64
ABAD DOM BUENO MENCÍA 2021 T	64
ABADAL 3.9 (VI DE FINCA) 2018 T R	347
ABADAL 5 MERLOT 2018 T R	347

WINE	PAGE	WINE	PAGE	WINE	PAGE
ABADAL MANDÓ 2019 T	347	ABADÍA RETUERTA PAGO VALDEBELLÓN 2018 T R	738	ACÚSTIC NEGRE 2020 T RB	291
ABADAL MATIS 2018 T C	347	ABADÍA RETUERTA PETIT VERDOT PV 2017 T	738	ACÚSTIC ROSAT 2021 RD	291
ABADAL NUAT 2019 B C	347	ABADÍA RETUERTA SELECCIÓN ESPECIAL 2019 T	738	ACVLIVS 2019 T C	709
ABADAL PICAPOLL 2021 B	347	ABBATIA VENDIMIA SELECCIONADA 2015 T	439	AD GAUDE 2021 B	855
ABADE DE COUTO 2020 T	380	ABDÓN SEGOVIA 2018 T C	657	ADARAS CALIZO GARNACHA TINTORERA 2021 T BA	49
ABADÍA DA COVA 1124 BRANCELLAO 2019 T BA	406	ABDÓN SEGOVIA 2018 T RB	657	ADARAS HUELLA GARNACHA TINTORERA - MONASTRELL 2020 T SS	49
ABADÍA DA COVA 1124 CAIÑO 2019 T	406	ABDÓN SEGOVIA 2020 T	657		
ABADÍA DA COVA 1124 MERENZAO 2019 T BA	406	ABEL MENDOZA 5V 2021 B	509	ADARAS KALIZO SIN SULFITOS 2021 T S	49
ABADÍA DA COVA 2020 T	406	ABEL MENDOZA GARNACHA BLANCA 2021 B	509	ADARAS LLUVIA 2021 B SS	49
ABADÍA DA COVA 2021 B	406	ABEL MENDOZA GUARDAVIÑAS 2020 T	509	ADARO 2019 T	471
ABADÍA DA COVA MENCÍA 2021 RD	406	ABEL MENDOZA JARRARTE 2018 T	509	ADEGA DO RICÓN 2019 T	837
ABADÍA DA COVA PENAFIÓN 2018 T BA	406	ABEL MENDOZA RISUEÑO 2021 RD	509	ADEGA DO RICÓN 2021 B	837
ABADÍA DA COVA XUNO 2019 T	406	ABEL MENDOZA SELECCIÓN PERSONAL 2019 T	509	ADEGA DO RICÓN ANCESTRAL 2021 BE	837
ABADÍA DE GOMARIZ 2017 T	424	ABELEDOS 2021 B	279	ADEGA VIÑA ALMIRANTE 2021 B	403
ABADÍA DE POBLET BLANC 2019 B	163	ABELEDOS 2021 T	279	ADEMÁN ADALIA 2021 B	657
ABADÍA DE POBLET NEGRE 2018 T	163	ABELIS LUI 2020 T C	667	ADEMÁN CARABIZAL 2021 T	657
ABADÍA DE SAN CAMPIO 2021 B	391	ABELLIO 2021 B	401	ADEMÁN VALDEARANDA 2019 T C	657
ABADÍA DE SAN QUIRCE 2016 T R	431	ABRACADABRA 2019 B	375	ADEMÁN VALDECARRETAS 2017 T FB	657
ABADÍA DE SAN QUIRCE 2019 T C	431	ABRISAT BÀRBARA FORÉS 2020 B C	643	ADERNATS 2018 BE R BN	153
ABADÍA DE SAN QUIRCE 6 MESES 2021 T RB	431	ABUELO PACO 2012 T R	242	ADERNATS 2018 BE R BR	153
ABADÍA DE SAN QUIRCE FINCA HELENA 2019 T	431	ABULAGA B SS	846	ADERNATS PURN GRAN RESERVA 2012 BE GR BN	154
ABADÍA DE SAN QUIRCE M9 2020 T	431	ACACIA MADRID ROMERO 2020 B FB	222	ADERNATS XARELLO VERMELL 2020 B	638
ABADÍA LA ARROYADA 2017 T C	438	ACECHO, SELECCIÓN DEL COLLADO T	111	ADERNATS XC 2015 BE GR BN	154
ABADÍA LA ARROYADA 2019 T RB	438	ACEDIANO 2019 T C	486	ADEUS 2021 B	422
ABADÍA LA ARROYADA 2020 T	439	ACENTUADO 2019 T	786	ADN CANALS 2017 BE GR BN	128
ABADÍA MEDITERRÀNIA 2019 T C	357	ACENTUADO ROSE ORGANIC 2021 RD	786	ADNOS BOBAL 2014 T	675
ABADÍA RETUERTA CUVÉE PALOMAR 2018 T R	738	ACINIPO 2017 T C	259	ADORADO BY MENADE CRIANZA DE 1967 B SOLERA	849
ABADÍA RETUERTA LE DOMAINE 2020 B	800	ACORDE 2020 T	233	ADOREMUS 2011 T R	662
ABADÍA RETUERTA LE DOMAINE 2021 B	738	ACUMA 2021 T	230	ADRIÀ DE BIOPAUMERÀ 2017 T RB	636
ABADÍA RETUERTA PAGO GARDUÑA 2018 T	738	ACÚSTIC BLANC 2019 B	291	ADUNA 2021 B	527
ABADÍA RETUERTA PAGO NEGRALADA 2017 T BA	738	ACÚSTIC BLANC 2021 B	291	ADUNA T C	527
ABADÍA RETUERTA PAGO VALDEBELLÓN 2017 T R	738	ACÚSTIC NEGRE 2019 T RB	291	ADUNA VENDIMIA SELECCIONADA 2016 T	527

WINE	PAGE
ADUR SOBRE LÍAS 2019 B	195
ADUR SOBRE LÍAS 2020 B	195
ADUR XO 2017 B	195
ADVENT SAMSÓ DULCE NATURAL 2019 T D	339
AFORTUNADO 2021 B	621
AGALIU 2020 B FB	175
AGAN 2020 T	199
AGNUS DE VALDELANA CRIANZA 2019 T C	551
AGNUSDEI ALBARIÑO 2021 B	391
ÁGORA CIENTO 69 MALBEC 2020 RD	777
ÁGORA DE ARÚSPIDE 2018 T RB	777
ÁGORA DE ARÚSPIDE LÁGRIMA 2021 B	777
ÁGORA TEMPRANILLO 2021 T MC	777
ÁGORA VIOGNIER 2021 B MC	777
AGOREIRA 2021 B	685
AGORIS 2019 T RB	459
AGORIS VIÑEDOS CENTENARIOS 2014 T R	459
ÁGUEDAS 2020 B BA	804
AGUILERA 2011 T GR	368
AGUIRREBEKO LASAL 2019 B C	84
AGUIRREBEKO LASAL 2020 B C	84
AGUIUNCHO 2018 B BA	384
AGUIUNCHO SELECCIÓN 2018 B	384
AGUSTÍ TORELLÓ MATA 2016 BE GR BN	121
AGUSTÍ TORELLÓ MATA 2019 BE R BR	121
AGUSTÍ TORELLÓ MATA BARRICA GRAN RESERVA 2017 BE GR BN	121
AGUSTÍ TORELLÓ MATA ESPANTALLOPS 2019 B C	325
AGUSTÍ TORELLÓ MATA KRIPTA 2014 BE GR BN	121
AGUSTÍ TORELLÓ MATA KRIPTA GRAN ANYADA 2008 BE GR BN	121
AGUSTÍ TORELLÓ MATA ROSAT TREPAT 2020 RE R BR	121
AGUSTÍ TORELLÓ MATA XIC 2020 B	325
AGUSTÍ TORELLÓ MATA XV XAREL·LO VERMELL 2021 RD	325
AHARI 2020 T	579
AIA 2016 T	351
AÏDA DE VIVES AMBRÒS 2020 T	639
AÏDA DE VIVES AMBRÒS 2021 B	639
AIER - VINO CERÁMICO 2020 T S	656
AIGUA DE LLUM DE VALL LLACH 2021 B	361
AILALÁ 2018 T	420
AILALÁ 2021 B	420
AIMAREZ 2021 T MC	566
AINA JAUME LLOPART ALEMANY ROSADO RE R BR	140
AIRE DE L'O DE L'ORIGAN BE BN	141
AIRE DE L'ORIGAN BE BN	141
AIRE DE L'ORIGAN MAGNUM 2019 BE BN	141
AIRE DE L'ORIGAN ROSÉ MAGNUM 2019 RE BN	141
AIRE DE PROTOS 2021 RD	158
AIRÉN =GARAGEWINE 2021 B	773
AISTEAR VINO DE AUTOR 2017 T	508
AITAKO 2019 B AG	196
AITAREN 2019 B FB	195
AITU! 2020 T	87
AIXA 2020 RD	202
AIXARTA BLANC 2020 B	117
AL COBUJO DE UNA GRAN SABINA 2018 T C	774
AL PIE DEL CAÑÓN 2020 T	415
AL RASO 2021 B	753
ALABASTER 2019 T	669
ALACER 2019 T C	481
ALACER 2020 T RB	481
ALACER 2021 T	481
ALAGU ALICANTE BOUSCHET 2018 T	855
ALAGU FORCALLAT 2018 T	855
ALAGU ROSÉ 2020 RD	855
ALAMEDA BF CRM	209
ALAN DE VAL CAÍÑO AS QUEIMADAS 2019 T	685
ALAN DE VAL CASTES NOBRES 2019 T	685
ALAN DE VAL GODELLO 2021 B	685
ALAN DE VAL MENCÍA 2020 T	685
ALAYA TIERRA 2020 T	47
ALBACEA MERLOT 2021 T	225
ALBADA FINCA GEMELO 2020 T	94
ALBADA FINCA SANTOS 2020 T	94
ALBADA GARNACHA VIÑAS VIEJAS SOBRE LÍAS 2021 T	94
ALBADA MACABEO VIÑAS VIEJAS SOBRE LÍAS 2021 B	94
ALBADA PARAJE LA CAÑADILLA 2019 T	94
ALBADA PARAJE LA CAÑADILLA 2020 T	94
ALBADA PARAJE LLANO HERRERA 2020 T	94
ALBAFLOR BLANC 2021 B	82
ALBAHRA 2020 T	861
ALBALEIA COLOMBARD 2021 B	168
ALBAMAR 2021 B	384
ALBAMAR 2021 RD	384
ALBAMAR ANCESTRAL BE	384
ALBAMAR NAI 2019 I	411
ALBAMAR O ESTEIRO 2020 T	384
ALBAMAR SESENTA E NOVE ARROBAS 2020 B C	384
ALBANTO LA VIÑA DE LAS FLORES 2021 RD	255
ALBANTO LA VIÑA DE LAS FLORES LÍAS FINAS 2021 B	255
ALBAREI ÁINE 2016 B	378
ALBAREI ORIXE 2016 B	378
ALBARIÑO DE FEFIÑANES 2021 B	386
ALBARIÑO DE FEFIÑANES III AÑO 2019 B	386
ALBARIÑO DO FERREIRO 2021 B	387
ALBARIZA 2020 B S	771

WINE	PAGE
ALBAYDA 2020 B FB	202
ALBÉITAR 2021 T	160
ALBERT DE VILARNAU CHARDONNAY PINOT NOIR 2015 BE GR BN	153
ALBERT DE VILARNAU XARELLO CASTANYER 2016 BE GR BN	153
ALBERTE DOBLE LÍAS 2020 B	423
ALBERTE TREIXADURA 2021 B	423
ALBET I NOYA BRUT 21 2019 BE BR	325
ALBET I NOYA EL FANIO 2020 B	325
ALBET I NOYA LA MILANA 2018 T R	325
ALBET I NOYA RESERVA MARTÍ 2016 T GR	325
ALBOCA 2018 T C	663
ALBOCA 2020 T	663
ALBOR DE RESALTE 2020 B	458
ALBRET EL ALBA CHARDONNAY 2021 B FB	317
ALBRET EL BALCÓN 2018 T C	317
ALBRET EL ROCÍO 2021 RD	317
ALBRET LA LOMA GARNACHA 2020 T RB	317
ALBRET LA VIÑA DE MI MADRE 2017 T R	317
ALBRET LASTRA 2017 T R	317
ALBUFERA 2020 T BA	711
ALBUREJO BF OL S	205
ALBUS 2019 B RB	459
ALCARDET 12 MESES 2018 T S	776
ALCARDET CEPAS VIEJAS 2015 T BA	776
ALCARDET SAUVIGNON BLANC 2021 B	776
ALCEÑO 2021 T	223
ALCEÑO DULCE 2017 T D	223
ALCEÑO MONASTRELL 12 MESES 2019 T	223
ALCEÑO ROSÉ RE BN	878
ALCEÑO SAUVIGNON BLANC 2020 B FB	223
ALCEÑO SELECCIÓN 2017 T C	223

WINE	PAGE
ALCEÑO SYRAH 50 BARRICAS 2020 T	223
ALCEO 2020 T RB	244
ALCONDE 2016 T R	309
ALCONDE ÓPTIMO 2018 T C S	844
ALCONDE ÓPTIMO 2021 RD SS	844
ALCOR 2015 T	116
ALDAHARA 2018 T C	624
ALDAHARA 2021 RD	624
ALDAHARA CHARDONNAY 2021 B	624
ALDAHARA GENERACIONES 2016 T	624
ALDAHARA RASÉ CHARDONNAY 2019 B RB	624
ALDAHARA RASÉ MERLOT 2019 T	624
ALDEA DE ADARAS 2021 T BA	49
ALDOBA 2021 T MC	237
ALDOBA VERDEJO 2021 B	237
ALDONIA 2019 T	514
ALDONZA BE BN	121
ALDONZA BE BR	121
ALDONZA BE R BR	121
ALDONZA CLÁSICO 2015 T	773
ALDONZA NAVAMARÍN 2013 T R	773
ALDONZA ROSÉ RE BR	121
ALDONZA SELECCIÓN 2015 T C S	773
ALECTUM 2021 B	401
ALEGRA DE BERONIA 2021 RD	517
ALEGRO 2019 T C	562
ALEJANDRA VIZCARRA 2020 B	466
ALEJANDRVS 2019 B	428
ALENAR BLANC TARDANA 2021 B	683
ALENTO 2019 B BA	687
ALEX 2018 T C	304
ALEX GARNACHA 2021 RD	304

WINE	PAGE
ALEX VIURA 2021 B	304
ALGAIREN 2020 T	113
ALGIL CRIANZA 2019 T C	655
ALGIL EXPRESIÓN 2020 T C	655
ALGIL GARNACHA 2020 T BA	655
ALGUEIRA CARRAVEL 2017 T C	408
ALGUEIRA ESCALADA 2019 B FB	408
ALGUEIRA FINCA CORTEZADA 2021 B	408
ALGUEIRA FINCAS 2017 T C	408
ALGUEIRA RISCO 2018 T RB	408
ALGUEIRA SERRADELO 2018 T C	408
ALIAGA DOSCARLOS SAUVIGNON BLANC 2021 B	304
ALIAGA HELENA SYRAH SYRAH 2019 T	304
ALIAGA LÁGRIMA DE GARNACHA 2021 RD	304
ALIAGA LÁGRIMA DE LUNA 2021 RD	304
ALIAGA MOSCATEL VENDIMIA TARDÍA 2019 B D	304
ALIAGA RESERVA DE LA FAMILIA 2014 T R	304
ALIATS 2021 B	703
ALIATS 2021 T	703
ALICANTE BOUSCHET BY TARIMA 2019 T BA	39
ALICE 2018 T R	359
ALILIAN CAMINO DEL ABUELO 2015 T	467
ALIÓN 2019 T	467
ALITUS 2015 T GR	441
ALIBES 2018 T	775
ALIBES CABERNET FRANC 2018 T	775
ALIBES GARNACHA TINTORERA 2019 T C	775
ALIBES PETIT VERDOT 2017 T	775
ALKIMIA 2019 B	837
ALLBLACK 2018 T BA	680
ALLEGRANZA CHARDONNAY 2021 B	793
ALLEGRANZA TEMPRANILLO VENDIMIA SELECCIONADA 2021 T	793

WINE	PAGE
ALLEGRANZA TEMPRANILLO-SHIRAZ 2021 T	245
ALLENDE 2016 T	567
ALLENDE 2018 B	568
ALLENDE 2018 RD	568
ALLOZO 2014 T GR	238
ALLOZO 2017 T R	238
ALLOZO 2019 T C	239
ALLOZO TEMPRANILLO 2021 T	239
ALLOZO VERDEJO 2021 B	239
ALMA BOHEMIA 2021 RD	505
ALMA DE AUTOR GODELLO 2020 B	279
ALMA DE AUTOR TINTO 2020 T	279
ALMA DE BLANCO 2021 B	279
ALMA DE LUZON 2019 T	226
ALMA DE MAR 2020 B	384
ALMA DE TINTO 2020 T	279
ALMA DE UNX 2016 T	309
ALMA DE VALDEGUERRA 2018 T C	728
ALMA DE VALDEGUERRA 2020 RD SD	728
ALMA DE VALDEGUERRA 2020 T	728
ALMA DE VALDEGUERRA 2020 T BA	728
ALMA DE VALDEGUERRA 2021 RD SD	728
ALMA DEL MORAL 2020 T	453
ALMA DO VELLO TESOURO 2020 T C	693
ALMA SERENA 2020 T	455
ALMA3ONCE 2019 T	408
ALMALARGA 2019 B BA	408
ALMALARGA 2020 B	408
ALMALOLA 2021 RD	408
ALMAMADRE 2017 T	409
ALMANAQUE 2017 T C	55
ALMANAQUE 2017 T RB	55

WINE	PAGE
ALMANAQUE 2020 T	55
ALMANAQUE 2021 RD	55
ALMANOVA 2020 T	409
ALMAVID 2018 T R	273
ALMAVID GARNACHA 2020 T RB	273
ALMAVID ROSÉ 2021 RD	273
ALMAVIÑO 2021 B	854
ALMÁZCARA MAJARA 2018 T C	63
ALMENDROS 2021 B	701
ALMIRANTE BF OL S	219
ALMIREZ 2020 T	669
ALMUDÍ 2018 T	90
ALMUDÍ UNO 2017 T C	90
ALOERS 2019 B	332
ALONSO CUESTA CÁLLATE 2019 T RB	273
ALONSO CUESTA CUVÉE 2018 T BA	273
ALONSO CUESTA LA GARNACHA DE LOLA 2020 T BA S	273
ALONSO DEL YERRO 2018 T	494
ALPAIRO 2021 RD	158
ALPENDRE MERENZAO 2020 T	415
ALQUERIA DE PRUNA (ETIQUETA BLANCA) 2019 T BA	863
ALQUERIA DE PRUNA (ETIQUETA ROJA) 2020 T S	863
ALQUERIA DE PRUNA (ETIQUETA ROJA) 2021 T S	863
ALQUÉZ DE SOMMOS 2019 T	844
ALSINA & SARDÁ 2020 BE R BR	121
ALSINA & SARDÁ BOIRES 2018 T C	325
ALSINA & SARDÁ FINCA CAL JANES 2020 T C	325
ALSINA & SARDÁ FINCA LA BOLTANA 2021 B	326
ALSINA & SARDÁ GRAN RESERVA ESPECIAL 2016 BE GR BN	121
ALSINA & SARDA PINOIR 2020 RE R BN	121
ALSINA & SARDÁ SELLO 2018 BE GR BN	121
ALSINA & SARDÁ VESTIGIS GRAN CUVÉE 2016 BE GR BN	122

WINE	PAGE
ALTA ALELLA 10 2010 BE BN	122
ALTA ALELLA 10 2012 BE BN	122
ALTA ALELLA LAIETÀ 2017 BE GR BN	122
ALTA ALELLA LAIETÀ ROSÉ 2018 RE GR BN	122
ALTA ALELLA MIRGIN 2018 BE GR BN	122
ALTA ALELLA MIRGIN EXEO EVOLUCIÓ + 2005 BE GR BN	122
ALTA ALELLA MIRGIN EXEO PARAJE CALIFICADO VALLCIRERA 2016 BE GR BN	122
ALTA ALELLA MIRGIN OPUS PARAJE CALIFICADO VALLCIRERA 2017 BE BN	122
ALTA ALELLA MIRGIN ROSÉ 2018 RE GR BN	122
ALTA PAVINA PINOT NOIR 2021 RD	800
ALTA PAVINA PINOT NOIR 2021 T RB	800
ALTA PAVINA VERDEJO 2021 B	800
ALTA PLATA VERDEJO 2021 B	606
ALTABELLA 2018 T	369
ALTAMENTE MONASTRELL 2020 T	222
ALTAMIMBRE 2017 T	442
ALTANZA 2015 T GR	505
ALTANZA 2017 T R	505
ALTANZA CLUB 2015 T R	505
ALTANZA COLECCIÓN ROBERTO AMILLO AMONTILLADO BF AM S	205
ALTANZA COLECCIÓN ROBERTO AMILLO OLOROSO BF OL S	205
ALTANZA COLECCIÓN ROBERTO AMILLO PALO CORTADO BF PC S	205
ALTANZA COLECCIÓN ROBERTO AMILLO PEDRO XIMÉNEZ BF PX D	205
ALTANZA FAMILIA 2015 T R	505
ALTANZA ROSADO 2021 RD	505
ALTANZA SAUVIGNON BLANC 2021 B	505
ALTER 2019 T	427

WINE	PAGE
ALTER 2021 B	427
ALTERÓN 2018 T C	270
ALTERÓN 2021 RD	270
ALTERÓN MACABEO 2019 B	270
ALTO BLANCO MERLOT 2020 T AROM	779
ALTO BLANCO VERDEJO 2020 B AROM	779
ALTO CARMONA 2017 T C	55
ALTO CARMONA 2018 T R	55
ALTO DE INAZARES CHARDONNAY 2020 B	838
ALTO DE INAZARES MONASTRELL 2020 T	838
ALTO DE INAZARES PINOT NOIR 2019 T C	838
ALTO DE INAZARES SYRAH 2020 T	838
ALTO DE INAZARES VIOGNIER 2020 B	838
ALTO DE LA ESTRELLA 2019 T	753
ALTO DE SAN ESTEBAN 2019 T C	844
ALTO DEL RINCÓN, VIÑEDOS DE ALTURA 2019 T BA	565
ALTO LAS PIZARRAS 2021 T BA	98
ALTO MONCAYO 2018 T	101
ALTO MONCAYO VERATÓN 2019 T	102
ALTO SIÓS 2018 T R	174
ALTOCIELO 2016 T R	261
ALTOLANDON BY ROSALÍA 2019 T C	266
ALTOLANDÓN CUENCAME 2020 T C	266
ALTOLANDÓN RED 2017 T R	266
ALTOS DE CORRAL SINGLE ESTATE 2017 T C	520
ALTOS DE INURRIETA 2017 T R	307
ALTOS DE LA FINCA 2017 T	791
ALTOS DE LA HOYA 2020 T BA	227
ALTOS DE LA MUELA 2019 T C	706
ALTOS DE LAS GATERAS SYRAH 2017 T	732
ALTOS DE LOSADA EL CEPÓN 2020 T	75
ALTOS DE LOSADA LA BIENQUERIDA 2020 T	75

WINE	PAGE
ALTOS DE LOSADA LA CHANA 2020 T	75
ALTOS DE LOSADA MENCÍA 2020 T	75
ALTOS DE LUZÓN 2020 T	226
ALTOS DE RAIZA 2020 T	587
ALTOS DE SANTIAGO CHARDONNAY 2020 B	779
ALTOS DE TAMARÓN 2017 T R	488
ALTOS DE TAMARON 2019 T C	488
ALTOS DE TAMARÓN 2021 T RB	488
ALTOS DE TORONA 2019 B BA	384
ALTOS DE TORONA ALBARIÑO 2021 B	385
ALTOS DE TORONA CAIÑO 2021 B	384
ALTOS DE TORONA GODELLO 2021 B	384
ALTOS DE TORONA LOUREIRO 2021 B	384
ALTOS DE TORONA ROSAL 2021 B	384
ALTOS DEL BERGASA 2016 T GR	529
ALTOS DEL BERGASA 2019 T C	529
ALTOS DEL CABRIEL 2020 T	267
ALTOS DEL CABRIEL 2021 B	267
ALTOS DEL CABRIEL 2021 RD	267
ALTOS DEL ENEBRO 2019 T C	432
ALTOS DEL ENEBRO FINCA LA HERRADURA 2019 T R	432
ALTOS DEL MARQUÉS 2014 T R	519
ALTOS DEL MARQUÉS 2019 T BA	519
ALTOS DEL MARQUÉS 2019 T C	519
ALTOS DEL MARQUÉS 2021 B	519
ALTOS DEL TERRAL 2019 T C	432
ALTOS DEL TERRAL T1 2019 T	432
ALTOS R 2019 T R	506
ALTOS R 2020 T C	506
ALTOS R PIGEAGE 2019 T	506
ALTOS R PIGEAGE 2020 B FB	506
ALTOS R PIGEAGE GRACIANO 2020 T	506

WINE	PAGE
ALUMBRA 2019 T C	876
ALVAREZ ALFARO 2020 T C	554
ALVAREZ ALFARO 2021 B	554
ALVAREZ ALFARO SELECCIÓN FAMILIAR 2017 T	554
ÁLVAREZ DE TOLEDO COLECCIÓN FAMILIA 2020 T	801
ÁLVAREZ DE TOLEDO GODELLO 2021 B	800
ÁLVAREZ DE TOLEDO MENCÍA 2020 T RB	800
ÁLVAREZ DE TOLEDO MENCÍA 2021 T RB	800
ALVEAR PALO CORTADO Nº 7 BF PC	285
ALVEAR PEDRO XIMÉNEZ 1927 BF PX D	285
ALVEAR PEDRO XIMÉNEZ DE SACRISTÍA 1998 BF PX D	285
ALVEAR PEDRO XIMÉNEZ SOLERA 1830 BF PX D	285
ALVINTE 2021 B	400
AMA DE GORKA IZAGIRRE 2019 B	86
AMA DE GORKA IZAGIRRE MAGNUM 2015 B	86
AMADIS DE YÉBENES 2020 T	838
AMADOR DIEZ VERDEJO CUVÉE 2015 B FB	598
AMADOR DIEZ VERDEJO CUVÉE 2017 B FB	597
AMADOR DIEZ VERDEJO CUVÉE 2018 B FB	597
AMADOR MEDRANO COLECCIÓN PRIVADA "FINCA LAS AGUZADE-RAS" 2018 T	537
AMADOR MEDRANO GRACIANO "FINCA VALDEGAMARRA" 2019 T	538
AMADOR MEDRANO PARCELA 14.8 2017 T FB	538
AMADOR MEDRANO ROSÉ "FINCA EL ENCINAL" 2021 RD PL S	538
AMADOR MEDRANO TEMPRANILLO BLANCO "FINCAS VALDEGAMARRA" 2020 B FB	538
AMADOR MEDRANO TERRA FINCA EL ENCINAL 2019 T FB	538
AMAITA 2018 T	531
AMALIA MALVASÍA SECO 2021 B	252
AMANCIO 2018 T	592
AMANOVO VERDEJO EDICIÓN ESPECIAL 2020 B RB	701

WINE	PAGE	WINE	PAGE	WINE	PAGE
AMAREN 2019 B FB	514	ÀN/2 2020 T	821	ANGEL NOIR ROSAT RE BN	130
AMAREN 60 GARNACHA 2019 T	515	ANA 2017 T C	627	ÁNGEL ROSAT 2021 RD	826
AMAREN SELECCIÓN DE VIÑEDOS 2019 T BA	515	ANA 2020 T RB	627	ANGELIA 2017 T	356
AMATISTA 11.5 B MO D	700	ANA 2021 RD	627	ANGELITOS NEGROS 2021 T	670
AMAVIDA 2021 B	421	ANA GEWÜRZTRAMINER 2020 B	627	ANHEL D'EMPORDÀ 2017 T BA	190
AMBORA EL ROQUILLO 2020 B	631	ANA SELECCIÓN FAMILIA 2015 T	627	ANHEL D'EMPORDÀ 2021 B	190
AMBORA PARAJE SAN IGNACIO 2020 T	631	ANA VELEIRO 2020 B	418	ANHEL D'EMPORDÀ 2021 RD	190
AMBORA VIÑA DE TEGUESTE 2020 T	631	ANA VENDIMIA TARDÍA 2020 B	87	ANI D'ANNA 2021 B	337
AMBURZA 2017 T BA	316	ANADIGNA 1932 2021 B	392	ANÍBAL DE OTERO LOS FORNOS 2016 T	66
AMENITAL 2018 T BA	539	ANADIGNA CAIÑO 2020 T RB	392	ANÍBAL DE OTERO VILLA OTERO 2017 T	66
AMERICO 2019 T C	658	ANADIGNA FUDRE 2019 B RB	392	ANIMA L'AVI ARRUFÍ 2020 B	645
AMERICO ROBLE ESPAÑOL 2020 T RB	658	ANADIGNA FUDRE 2020 B RB	392	ANIMA MUNDI ANCESTRAL CAMI DELS XOPS 2021 BE	878
AMETLLER 2016 BE GR BN	138	ANADIGNA SOBRE LÍAS 2020 B	392	ANIMA MUNDI ANCESTRAL NOGUER BAIX 2016 BE	878
AMETLLER 2020 BE BR	138	ANADIGNA TRADICIONAL 2021 B	392	ANIMA MUNDI GRES 2021 B	838
AMETLLER 2020 BE R BN	138	ANALIVIA VERDEJO SELECCIÓN 2021 B	618	ANIMA MUNDI PELLS 2021 B	838
AMETLLER BLANC DE NOIRS 2020 BE BN	138	ANAVA..A CERCAR UN SOMNI ROSAT 2020 RD BA	824	ANIMA MUNDI XARELLO 2020 B	839
AMETLLER BLANC FLORAL 2021 B	334	ANAYÓN GARNACHA 2019 T	111	ANIMA NUA COR VIU 2020 B	164
AMETLLER GARNATXA 2019 T	334	ANAYÓN SELECCIÓN 2019 T BA	111	ANIMA NUA COR VIU 2020 T C	164
AMETLLER GARNATXA NEGRA I CARINYENA 2019 T	334	ANCESTRAL 2019 RE	854	ANIMOSO 2021 B	621
AMETLLER XARELLO 2021 B	334	ANCESTRAL MONTONEGA 2021 BE	880	ANINIUS 2020 T	99
AMIC DE CLOS D'AGON 2019 I	181	ANDRÉS ALONSO GARNACHA SYRAH 2021 T	98	ANKAL 2019 T C	441
AMIC DE CLOS D'AGON 2021 B	182	ANDRES MELER 2014 T R	627	ANKAL 2021 T RD	441
AMIC DE CLOS D'AGON 2021 RD	182	ANEXE 2021 T	293	ANMA 2018 T BA	679
AMODIÑO 2018 B	382	ANEXE SYRAH 2020 T	293	ANMA 2020 B MC	680
AMON BF AM S	286	ANEXE VINYES VELLES DE CARINYENA 2020 T	293	ANNA DE CODORNÍU BE BN	135
AMONTILLADO 51-1ª VORS BF AM S	210	ANGEL 2019 T	555	ANNA DE CODORNÍU BE BR	135
AMONTILLADO DEL DUQUE VORS BF AM S	216	ANGEL BE R BN	130	ANNA DE CODORNÍU BLANC DE NOIRS BE BR	135
AMONTILLADO FINO IMPERIAL BF AM	213	ÁNGEL BLANC DE BLANCA 2021 B	826	ANNA DE CODORNÍU ECOLÓGICO RESERVA BE R BR	135
AMONTILLADO TRADICIÓN VORS BF AM S	211	ÁNGEL CABERNET SAUVIGNON 2016 T	826	ANNA DE CODORNÍU ICE EDITION BE SS	135
AMORANY 2016 BE GR BN	153	ANGEL CUPATGE BE R BN	130	ANNA DE CODORNÍU ICE EDITION BE SS	135
AMUNT NEGRE 2020 T	856	ANGEL DE LARRAINZAR 2019 T	319	ANNA DE CODORNÍU ROSÉ RE BR	135
ÀN 2019 T	821	ÁNGEL MANTO NEGRO 500 2019 T C	826	ANNE DO RICÓN 2020 B BA	837

WINE	PAGE	WINE	PAGE	WINE	PAGE
ANNE DO RICÓN LARANXA 2020 B	837	APTUS 2020 T RB	456	ARETXABALETA 2019 B	84
ANNE DO RICÓN VELO DE FLOR 2018 B	837	AQUIANA 2019 T C	70	ARETXABALETA 2020 B	84
ANSA 2017 T	508	AQUILÓN 2016 T	101	ARGILA 2013 BE GR BN	338
ANTÍDOTO 2020 T	439	AQVITANIA 2021 B	385	ARIENZO DE MARQUÉS DE RISCAL 2018 T C	522
ANTIMA 2018 T C	187	ARAGONIA SELECCIÓN ESPECIAL 2019 T	102	ARIKAI 2021 B	768
ANTOJO DULCE 2020 B D	804	ARAICO 2018 T C	515	ARIKAI TEMPRANILLO 2021 RD	768
ANTÓN AGUIRRE 2021 T	320	ARAICO 2020 B FB	515	ARINDO 2021 B	613
ANTONA GARCÍA 2018 T C	656	ARAICO AUTOR 2020 T	515	ARÍNZANO GRAN VINO 2017 T	746
ANTONI CANALS NADAL CUPADA SELECCIÓ 2018 BE GR BR	129	ARANDA CREAM BF CRM	205	ARÍNZANO MERLOT BIOLÓGICO 2018 T	746
ANTONINO IZQUIERDO VENDIMIA SELECCIONADA 2019 T	452	ARANLEÓN SÓLO 2020 T C	674	ARLEQUÍN 2021 B	805
ANTONIO & ANTONIA 2021 T	869	ARANLEÓN SÓLO 2021 B	701	ARLEQUÍN 2021 RD	805
ANTONIO ALCARAZ 2017 T C	515	ARANO 2019 T C	433	ARLEQUÍN 2021 T	805
ANTONIO MONTERO "AUTOR" 2021 B	428	ARATE PREMIUM 2021 B	724	ARMAN DOCE DULCE B D	855
ANTONIO MONTERO 50 FINCA 2019 B BA	428	ARAUTAVA 2020 T FB	717	ARMÁN FINCA MISENHORA 2019 B	424
ANTONIO MONTERO COLLEITA 2021 B	428	ARAUTAVA BLANCO SECO 2021 B S	717	ARMAN FINCA OS LOUREIROS 2019 B	424
ANTONIO SERRANO CENCIBEL 2020 T	777	ARAUTAVA DULCE 2002 B GR D	717	ARMANTES VENDIMIA SELECCIONADA 2019 T	94
ANTONIO SERRANO CENCIBEL 2020 T RB	777	ARAUTAVA FINCA LA HABANERA 2020 T	717	ARMAS DE GUERRA 2021 B	77
ANTONIO SERRANO COUPAGE PREMIUM 2017 T	777	ARAUTAVA FINCA LA HABANERA ALBILLO CRIOLLO 89 B	717	ARMAS DE GUERRA GODELLO 2021 B	77
AÑADA 1975 BF AM	216	ARAUTAVA FINCA LA HABANERA LISTÁN BLANCO 2020 B	717	ARMAS DE GUERRA MENCÍA 2016 T C	77
AÑADA DE BALADIÑA 2011 B	395	ARBASTRUM 2021 B	386	ARMAS DE GUERRA MENCÍA 2021 T	78
AÑADAS POR 220 CÁNTAROS B	509	ARBOSSAR 2020 T C	374	ARMAS DE LANZÓS 2016 B	386
AÑIL FRESH 2021 B	247	ARCANO 2018 T	439	ARMENDÁRIZ 2018 T C	310
AÑORANZA 2021 RD	241	ARCANO 2019 T	439	ARMENDÁRIZ 2020 T	310
AÑORANZA CABERNET SHIRAZ 2021 T	241	ARCOS DE MOCLINEJO BF SOLERA D	261	ARMENDÁRIZ 2021 B	310
AÑORANZA SAUVIGNON BLANC 2021 B	241	ARCOS DE MOCLINEJO BF SOLERA S	261	ARMENDÁRIZ 2021 RD	310
APAGA Y VÁMONOS 2020 T	29	ARCUM 2018 T R	439	ARMENDÁRIZ DULCE NATURAL 2021 B D	310
APAGA Y VÁMONOS 2021 RD SD	29	ARCUM 2019 T C	439	ARNACH 2019 T	751
APAGA Y VÁMONOS AFRUTADO 2021 B SD	29	ARCUM 2020 T RB	439	ARNAU DE RENDÉ MASDEU 2020 T C S	164
APOLINAR´S DREAM 2017 T BA	539	ARENA Y CALIZA 2019 T C	489	ARNAU OLLER 2018 T R	349
APONTE 2017 T R	659	ARESTEL BE BR	153	ARNEGUI 2017 T R	580
APÓSTOLES VORS BF PC MED	216	ARESTEL BE SS	153	ARNEGUI 2018 T C	580
APTUS 2019 T RB	456	ARESTEL VINTAGE 2019 BE EBR	153	ARO 2019 T	539

WINE	PAGE
AROA 2018 T	189
AROA JAUNA 2018 T C	305
AROA LAIA 2021 B	305
AROA LARROSA 2021 RD	305
AROA MUTIKO 2020 T	305
AROTZ 2020 B BA	87
ARQUELA 2020 B FB	700
ARQUITÓN 2021 RD	752
ARRABAL DEL CONJURO 2019 T	722
ARRAIANO 2020 T BA	858
ARRAYÁN ALBILLO REAL 2020 B	273
ARRAYÁN ALBILLO REAL GRANITO 2020 B	751
ARRAYAN ROSADO DE GARNACHA 2021 RD	273
ARRAYÁN SELECCIÓN 2019 T C	273
ARREBATACAPAS 2019 T	752
ARRELAT 2020 B	825
ARRIBES DE VETTONIA 2018 T R	58
ARRIBES DE VETTONIA 2019 T C	58
ARRIBES DE VETTONIA 2021 B	58
ARRIBES DE VETTONIA 2021 RD	58
ARRIBES DE VETTONIA VENDIMIA SELECIONADA 2016 T R	58
ARROCAL 2019 T C	440
ARROCAL RESERVA DE FAMILIA 2018 T R	440
ARROCAL SELECCIÓN ESPECIAL 2020 T	440
ARROYO DE ARRAYÁN 2020 B	273
ARROYUELO EN RAMA BF FI	211
ARS COLLECTA 459 2010 BE GR BR	135
ARS COLLECTA BLANC DE BLANCS 2018 BE GR BR	135
ARS COLLECTA EL TROS NOU PINOT NOIR 2010 BE GR BR	135
ARS COLLECTA JAUME CODORNÍU 2018 BE GR BR	135
ARS COLLECTA LA FIDEUERA XARELLO 2011 BE GR BR	135
ARS COLLECTA LA PLETA CHARDONNAY 2011 BE GR BR	135

WINE	PAGE
ART LUNA BEBERIDE 2020 T C	70
ARTADI CANALES 2020 T	852
ARTADI EL CARRETIL 2020 T	852
ARTADI LA HOYA 2020 T	852
ARTADI LA POZA DE BALLESTEROS 2020 T	852
ARTADI MAJADALES 2020 T	852
ARTADI QUINTANILLA 2020 T	852
ARTADI SAN LÁZARO 2020 T	853
ARTADI TERRERAS 2020 T	853
ARTADI VALDEGINÉS 2020 T	853
ARTADI VIÑAS DE GAIN 2018 B	853
ARTADI VIÑAS DE GAIN 2019 T	853
ARTADI VIÑAS DE GAIN 2020 T	853
ARTAXO 2019 T	865
ARTAXO 2020 T	865
ARTAZU PASOS DE SAN MARTÍN 2018 T	316
ARTAZU SANTA CRUZ DE ARTAZU 2018 B	316
ARTAZU SANTA CRUZ DE ARTAZU 2019 T	316
ARTAZURI 2021 RD	316
ARTE 2020 T	330
ARTERO 2016 T R	247
ARTERO 2017 T C	247
ARTERO 2021 RD	248
ARTERO MACABEO VERDEJO 2021 B	248
ARTERO TEMPRANILLO 2021 T	248
ARTICO B	863
ARTIGA GARNACHA SELECCIÓN 2021 T	101
ARTIGAS 2019 B FB	369
ARTIGAS 2019 T C	369
ARTUKE 2021 T MC	507
ARTUKE EL ESCOLLADERO 2020 T	507
ARTUKE FINCA DE LOS LOCOS 2020 T	507

WINE	PAGE
ARTUKE LA CONDENADA 2020 T	507
ARTUKE PASO LAS MAÑAS 2020 T	507
ARTUKE PIES NEGROS 2020 T C	507
ARTXANDA 2020 B	87
ARTZAI 2019 B	84
ARTZAI 2020 B	84
ARX 2018 T C	769
ARZUAGA 2015 T GR	440
ARZUAGA 2018 T R	440
ARZUAGA 2020 T C	440
ARZUAGA ECOLÓGICO 2018 T C	440
ARZUAGA RESERVA ESPECIAL 2017 T R	440
AS 2 LADIEIRAS 2020 B	693
AS CABORCAS 2019 T	689
ÁS MIRABRÁS 2020 B S	769
AS SORTES VAL DO BIBEI 2020 B	691
AS VIÑAS 2019 B	875
ASADOIRA 2019 B	693
ASTADO 2017 T R	665
ASTADO 2018 T C	665
ASTADO 2020 T RB	665
ASTADO 2021 T	665
ASTADO VIÑAS VIEJAS 2016 T RB	665
ÁSTER 2019 T C	495
ÁSTER FINCA EL OTERO 2018 T	495
ASTERISCO 2020 T	656
ASTOBIZA ROSÉ 2021 RD	52
ASTRALES 2019 T	432
ASTRALES CHRISTINA 2019 T C	432
ASÚA 2019 T C	563
AT ROCA 2019 BE R BN	326
AT ROCA ROSAT 2019 RE R BN	326

WINE	PAGE	WINE	PAGE	WINE	PAGE
ATABALAT 2018 T	856	AUDENTIA 2018 T R	705	AUZELLS 2021 B	176
ATABALAT 2019 RD	856	AUDENTIA PETIT VERDOT 2021 T	850	AVA SELECCIÓ 2019 T C	824
ATABALAT 2019 T	856	AUDENTIA SAUVIGNON BLANC 2021 B	850	AVA VI BLANC 2021 B	824
ATABALAT 2020 T	856	AUDITORI 2019 T C	291	AVA VI NEGRE 2019 T BA	824
ATALAQUE GARNACHA DEL HORCAJO 2018 T C	274	AUDITORI BLANC 2021 B	291	AVAN 2020 T C	468
ATALAQUE GARNACHA LA PERALEDA 2019 T	274	AULA BE BN	124	AVAN CEPAS CENTENARIAS 2020 T	468
ATALAQUE MOSCATEL DEL HORCAJO 2018 B D	274	AULA BE BR	124	AVAN OAK 2021 T BA	469
ATALAYA DE BARAHONA 2021 T	810	AULA BE SS	124	AVAN VINOS DE VIÑEDO DE LOS CANTILLOS 2018 T	469
ATANCE BOBAL 2020 T	710	AULA BOBAL MERLOT 2019 T	675	AVAN VINOS DE VIÑEDO: TORRUBIO 2018 T C	469
ATANCE CUVEE 2021 B	710	AULA BOBAL TEMPRANILLO 2018 T C	675	AVAN VIÑEDOS VIEJOS 2020 T C	469
ATENCIA 2015 T BA	794	AULA CHARDONNAY BE R BN	125	AVANCIA CUVEE DE O GODELLO 2021 B	687
ATHUS 2019 T C	558	AULA MACABEO BE R BN	125	AVANCIA CUVÉE DE O MENCÍA 2021 T	687
ATHUS VENDIMIA 2021 T RB	558	AULA ROSÉ RE BR	125	AVANCIA GODELLO 2021 B	687
ÁTICA 2017 BE GR EBR	146	AULA VERDEJO 2021 B	847	AVANCIA NOBLEZA GODELLO 2020 B	687
ÁTICA ROSÉ 2018 RE R EBR	146	AULLIDO 2021 T	535	AVANCIA NOBLEZA MENCÍA 2020 T	688
ATICUS 2018 T C	519	AURA SAUVIGNON BLANC B	601	AVANCIA NOBLEZA MENCÍA 2021 T	688
ATICUS VENDIMIA SELECCIONADA 2019 T	519	AURA SELECCIÓN PARCELA AVUTARDA 2019 B FB	601	AVANERO GIRO ROS "VINO DE PARCELA" 2020 B FB	825
ATLANTIS ALBARIÑO 2021 B	402	AURA VERDEJO VENDIMIA NOCTURNA 2021 B	601	AVELINO VEGAS 100 ANIVERSARIO 2019 T R	432
ATLANTIS GODELLO 2019 B	693	ÁUREO 2018 T R	432	AVENT BE BR	140
ATLANTIS HONDARRABI ZURI 2020 B	53	AURO 2021 B	38	AVENT ROSÉ RE BR	140
ATRIL BIO TEMPRANILLO SYRAH 2021 T	778	AURORA EN RAMA MAGNUM BF MZ	213	AVENTO MALVASÍA MOSCATEL 2021 B SS	873
ATTECA 2020 T	95	AURORUM 2021 RD C	829	AVGVSTVS CHARDONNAY 2021 B FB	326
ATTECA ARMAS 2019 T	95	AUS BRUANT 2020 BE R BN	122	AVGVSTVS MICROVINIFICACIÓ MACABEO 2017 B FB	326
ATTECA ARMAS 2020 T	95	AUS CAPSIGRANY SIN SULFITOS 2020 BE R BN	122	AVGVSTVS MICROVINIFICACIÓ MALVASIA DE SITGES 2021 B	326
ATTECA SELECCIÓN DE LA FAMILIA 2019 T	95	AUS MERLA 2021 T C	838	AVGVSTVS MICROVINIFICACIÓ XARELLO +100 2019 B FB	327
ATTECA SELECCIÓN DE LA FAMILIA 2020 T	95	AUS PÉT-NAT 2021 BE AG	878	AVGVSTVS MICROVINIFICACIÓ XARELLO DE MAR 2017 B FB	327
ATTELEA 2018 T C	499	AUS ROSÉ PÉT-NAT 2021 RE AG	878	AVGVSTVS TRAJANVS 2017 T R	327
ATTIS ATALANTE 2019 B	381	AUSÁS INTERPRETACIÓN 2020 T	432	AVINIUS MERLOT I SYRAH 2020 T	347
ATTIS BRANCELLAO 2020 T	381	AUTÓCTON BLANC 2020 B FB	839	AVINYÓ 2017 BE R BN	123
ATTIS EMBAIXADOR 2019 B	381	AUTÓCTON BLANC 2021 B FB	839	AVINYÓ 2019 BE R BR	123
ATTIS LÍAS FINAS 2021 B	381	AUTÓCTON NEGRE 2016 T	839	AVINYÓ BLANC DE NOIRS BE R BN	123
AUDAZ DE JUAN ALCORTA 2017 T C	514	AUTOR DE BOCOS 2019 T	490	AVINYÓ DOLÇ DE CABERNET 2018 B MISTELA D	840

WINE	PAGE
AVINYÓ LA TICOTA 2015 BE GR BN	123
AVINYÓ ROSÉ SUBLIM 2019 RE R BR	123
AVIOR 2017 T R	548
AVIOR 2019 T C	548
AVISPERO 2021 RD	676
AVRVS 2016 T	568
AXPE TXAKOLI 2020 B	84
AYLÉS "TRES DE 3000" 2018 T	738
AYMAR 2016 BE R BN	327
AYMAR ROSÉ 2019 RE R EBR	327
AYMAR TRANQUIL BLANC 2021 B	327
AYMAR TRANQUIL NEGRE 2021 T	327
AZ ALTO DE LOS ZORROS 10 MESES 2019 T RB	476
AZ ALTO DE LOS ZORROS 2017 T C	476
AZ ALTO DE LOS ZORROS AUTOR 2016 T R	476
AZABACHE VENDIMIA SELECCIONADA 2019 T C	569
AZACÁN GARNACHA SHIRAZ 2018 T	105
AZAGADOR DE LA JARABA COSECHA 2020 T BA	745
AZAGADOR DE LA JARABA SELECCIÓN ESPECIAL 2019 T BA	745
AZPILICUETA 2018 T C	507
AZPILICUETA 2019 T C	507
AZPILICUETA 2020 RD	507
AZPILICUETA 2021 B	507
AZPILICUETA SELECCIÓN DE BARRICAS 2019 T C	508
AZUA BOBAL 2018 T C	267
AZUA BOBAL 2018 T RB	267
AZUA BOBAL 2020 T	267
AZUA VERDEJO 2021 B	267
AZUL PERDIDO MALVASÍA 2019 B	250

B

WINE	PAGE
BABALÀ - VI NEGRE EIXERIT 2021 T	183
BABALÀ VI BLANC SIMPÀTIC 2021 B	183
BABALÀ VI ROSAT ALEGRE 2021 RD	183
BABU 2021 B	65
BAC DE LES GINESTERES B RC D	193
BADEN NUMEN "AU" 2019 T	433
BADEN NUMEN "B" 2020 T	433
BADEN NUMEN "N" 2019 T C	433
BÁGOA DO MIÑO 2020 B	385
BÁGOA DO MIÑO 2021 B	385
BÁGOAS LEDAS 2021 B	385
BAGORDI 2013 T R	516
BAGORDI 2018 T C	516
BAGORDI 2021 B FB	516
BAGORDI 2021 RD	516
BAGORDI MATURANA 2020	516
BAGORDI VENDIMIA SELECCIONADA 2017 T	516
BAGÚS 2019 T	453
BAHÍA ALTA AFRUTADO 2021 B SS	710
BAHÍA DE DENIA 2021 B	40
BAIGORRI 2017 B FB	516
BAIGORRI BELUS 2018 T	516
BAIGORRI DE GARAGE 2017 T BA	516
BAIGORRI FINCA LA CANOCA 2020 T RB	516
BAIGORRI FINCA LA QUINTANILLA 2019 T	516
BAIGORRI FINCA LAS NAVAS 2021 T	516
BAIGORRI GARNACHA 2017 T	516
BAIGORRI MATURANA 2019 T	516
BAJOZ 2018 T C	668
BAJOZ 2021 T RB	668
BAJOZ TEMPRANILLO 2021 T	668
BAKEDER 2020 T	588
BAL D'ISABENA GARNACHA 2021 T	624
BAL D'ISABENA GARNACHA BLANCA 2021 B	624

WINE	PAGE
BAL D'ISABENA GEWÜRZTRAMINER 2021 B	624
BALA PERDIDA 2020 T	35
BALADIÑA 2021 B	395
BALADIÑA BARRO 2014 B	395
BALANCINES BLANCO SOBRE LÍAS B RB	499
BALANCINES GARNACHA & GARNACHA 2017 T C	499
BALANCINES GOLD 2018 T C	499
BALBÁS 2017 T R	441
BALBÁS 2019 T C	441
BALCÓN DE PILATOS BY VALDEMAR 2019 T	551
BALDOMÀ SELECCIÓ 2019 T	177
BALDOR CHARDONNAY 2020 B FB	853
BALDOR OLD VINES 2015 T	787
BALDOR TRADICIÓN SYRAH 2017 T	787
BALDÚS 2017 BE GR BN	132
BALDÚS 2018 BE R BN	132
BALMA 2017 BE R BR	338
BALMETA 2019 T	187
BÀLSAM BE BR	145
BÁLSAMO DE FIERABRÁS 2018 T	722
BALTASAR GRACIÁN ARTE DE INGENIO 2018 T R	96
BALTASAR GRACIAN ARTE DE PRUDENCIA 2019 T C	96
BALTASAR GRACIÁN EL CRITICÓN 2021 RD	96
BALTASAR GRACIÁN EL DISCRETO 2021 B	96
BALTASAR GRACIÁN EL POLÍTICO 2021 T	96
BALTASAR GRACIÁN VIÑAS VIEJAS EL HÉROE 2020 T	96
BALUARTE 2021 T RB	311
BANCAL DEL BOSC BLANC 2021 B	302
BANCALES MORAL 2020 T BA	687
BANCALES OLVIDADOS 2019 T C	406
BANISIO 2016 T R	587
BANISIO 2020 T RB	492

WINE	PAGE	WINE	PAGE	WINE	PAGE
BANISIO SAUVIGNON BLANC 2020 B	619	BARÓN DE LEY BLANCO SOBRE LÍAS 2020 B	508	BEADE 25 AUTOR 2020 B	428
BANISIO SAUVIGNON BLANC 2021 B	620	BARÓN DE LEY FINCA MONASTERIO 2019 T BA	508	BEADE ORIXE 2016 B	428
BANISIO VERDEJO 2020 B	620	BARÓN DE LEY ROSADO DE LÁGRIMA 2021 RD	508	BEADE ORIXE 2021 T C	428
BANISIO VERDEJO 2021 B	620	BARÓN DE LEY VARIETAL GRACIANO 2019 T	508	BEADE PRIMACÍA 2021 B	428
BANZAO 2018 T	77	BARÓN DE LEY VARIETAL MATURANA 2019 T BA	508	BÉBEME 2020 T RB	58
BARBADELO 2021 T	406	BARÓN DE RIVERO CHARDONNAY 2021 B	849	BEBERÁS DE LA COPA DE TU HERMANA 2020 B	708
BARBADILLO BLANCO DE BLANCOS 2021 B	769	BARÓN DE RIVERO SEMIDULCE 2021 B SD	849	BEGASTRI 2018 T C	91
BARBADILLO CUCO BF OL S	206	BARÓN DE RIVERO SYRAH 2021 T	849	BEGASTRI 2021 B	91
BARBADILLO LA CILLA BF PX D	206	BARÓN DE RIVERO VERDEJO 2021 B	849	BEGASTRI 2021 RD	91
BARBADILLO SAN RAFAEL BF MED	207	BARÓN DE SANTUY 2019 T C	493	BEGOA 2019 B	58
BÀRBARA FORÉS BLANC 2021 B	643	BARRANCO DEL SAN GINÉS 2017 T	575	BELA - RIBERA DEL DUERO 2021 T	433
BÀRBARA FORÉS ROSAT 2021 RD	643	BARRERA DE SOL 2020 T BA	753	BELA GRAN VINO DE RUEDA 2020 B	596
BARBAZUL 2021 B	770	BARROS DE CECILI 2019 B	826	BELA GRAN VINO DE RUEDA 2021 B	596
BARBAZUL 2021 RD	770	BARROS DE CECILI SUBMARINO 2019 B	846	BELLA 2021 B	636
BARBAZUL SELECCIÓN ESPECIAL 2018 T	770	BASA 2021 B	615	BELLA BLANC CARTOIXA 2020 B	636
BARBIÁN 2020 T RB	658	BASSUS FINCA CASILLA HERRERA 2018 T	703	BELONDRADE QUINTA APOLONIA 2021 B	802
BARCO DE TIERRA ADENTRO 2020 T FB	754	BASSUS PINOT NOIR 2020 T	676	BELONDRADE QUINTA CLARISA 2021 RD	802
BARCO DEL CORNETA 2020 B FB	801	BASSUS PINOT NOIR DULCE 2021 RD D	676	BELONDRADE Y LURTON 2020 B FB	596
BARCOLOBO LA RINCONADA 2021 T	801	BASTARD NEGRE DE LES TERRASSES DEL SERRAL 2015 T	874	BELVÍ 2019 T C	766
BARCOLOBO LACRIMAE RERUM 2021 RD	801	BASTARD NEGRE DE LES TERRASSES DEL SERRAL 2017 T	874	BELVÍ 2021 RD	766
BARCOLOBO VERDEJO 2021 B FB	802	BASTIÓN 2021 RD	276	BELVÍ ANCESTRAL 2021 BE BR	766
BARDALERA 2019 B	864	BASTIÓN DE CAMARENA 2021 B	276	BELVÍ GARNACHA BLANCA 2019 B	766
BARDOS 2018 T R	434	BASTIÓN GARNACHA + SYRAH 2021 T	276	BELVÍ MACABEO 2021 B	767
BARDOS 2019 T RB	434	BASTIÓN GARNACHA 2019 T	276	BEMBIBRE 2018 T R	73
BARDOS ROMÁNTICA 2019 T C	434	BASTIÓN SELECCIÓN 2020 T	276	BENEDICTUS FRUCTUS 2020 B BA	423
BARDOS SUPREMA 2018 T R	434	BASTÓN 2017 T	786	BENEDICTUS FRUCTUS 2021 B BA	424
BARDOS VERDEJO 2021 B	596	BASTÓN 2019 T	48	BENIZAR SEMIDULCE 2021 B SD	237
BARÓN DE CHIREL 2017 T	522	BATEC 2014 BE GR BR	138	BENJE 2020 B	730
BARÓN DE CHIREL VIÑAS CENTENARIAS VERDEJO 2021 B	804	BATGARA 18 MESES 2018 B BA	52	BENJE 2020 T	730
BARÓN DE EBRO 2019 T C	569	BATGARA AROMAS DEL SUR 2019 B BA	52	BENTO 2021 B	595
BARÓN DE LEY 2017 T R	508	BBASTIDA 2019 T BA	567	BERABAL DE UBETA 2020 T	320
BARON DE LEY 3 VIÑAS 2019 B R	508	BEACH ROSÉ BY OR 2021 RD	331	BERAMENDI 3 FLORES GARNACHA 2021 RD	310

WINE	PAGE
BERAMENDI CHARDONNAY 2021 B	310
BERAMENDI EDIC. ESPECIAL GRACIANO 2019 T	310
BERAMENDI EDIC. ESPECIAL TEMPRANILLO 2019 T RB	310
BERAMENDI GARNACHA 2021 RD	310
BERAMENDI MERLOT 2017 T C	310
BERANDÍA 2019 T	700
BERBERANA CARTA DE PLATA T BA	777
BERBERANA GRAN TRADICIÓN BE BR	143
BERBERANA GRAN TRADICIÓN BE SS	143
BERBERANA GRAN TRADICIÓN RE BR	143
BERCEO 2018 T C	533
BERCEO NUEVA GENERACIÓN 2016 T R	533
BERCEO NUEVA GENERACIÓN 2018 T C	533
BERCIAL LADERA LOS CANTOS 2018 T BA	683
BERCIAL SELECCIÓN 2020 B BA	683
BERDÁ 2020 B	628
BERMEJO DIEGO SECO 2021 B	252
BERMEJO LISTÁN NEGRO MACERACIÓN TRADICIONAL 2018 T BA	252
BERMEJO MALVASIA BE BN	252
BERMEJO MALVASIA NATURALMENTE DULCE B D	252
BERMEJO MALVASÍA SECO 2021 B FB	252
BERNABELEVA ARROYO DE TÓRTOLAS 2020 T	721
BERNABELEVA CARRIL DEL REY 2020 T	721
BERNABELEVA VIÑA BONITA 2020 T	721
BERNAT OLLER 2018 T	349
BERNAVÍ AKRÒNIM 2019 T C	841
BERNAVÍ CAVERNET 2018 T C	642
BERNAVÍ GARNATXA BLANCA 2021 B	642
BERNAVÍ MORENILLO 2021 RD	642
BERNAVI TRES DE TRES 2020 T	642
BERNEDA 2014 T R	553
BERNEDA VENDIMIA SELECCIONADA 2018 T C	553
BERNÓN 2021 B	385
BERONIA 198 BARRICAS 2013 T R	517
BERONIA 2013 T GR	517
BERONIA 2017 T R	517
BERONIA III A.C. 2016 T	517
BERONIA MAZUELO 2017 T R	517
BERONIA RUEDA 2021 B	596
BERONIA TEMPRANILLO ELABORACIÓN ESPECIAL 2020 T FB	517
BERONIA VIÑAS VIEJAS 2019 T	517
BERTHA 2019 BE R BN	131
BERTHA CARDÚS 2018 BE GR BN	131
BERTHA LOUNGE 2019 BE BR	131
BERTHA SEGLE XXI 2009 BE GR BN	131
BERTHA SEGLE XXI ROSÉ 2017 RE GR BR	131
BERTOLA 12 AÑOS BF AM S	213
BERTOLA 12 AÑOS BF OL S	213
BERTOLA 12 AÑOS BF PC S	213
BERTOLA BF CRM	213
BES CAN RICH 2021 RD OL	816
BESANA REAL MACABEO 2021 B	247
BESANA REAL MACABEO 2021 B FB	247
BESANA REAL TEMPRANILLO 2018 T C	247
BESANA REAL TEMPRANILLO 2021 T	247
BESANA REAL VERDEJO 2021 B	247
BESANAS 2020 T C	793
BESO DE VINO GARNACHA VIÑAS VIEJAS 2020 T	111
BESTUÉ CHARDONNAY 2020 B FB	625
BESTUÉ FINCA SANTA SABINA 2019 T C	625
BESTUÉ MARINA 2021 B	626
BESTUÉ MONTE ALICIA 2021 B	626
BESTUÉ VIÑADORES 2018 T	626
BETOLAZA 2016 T R	517
BETOLAZA 2021 B	517
BETOLAZA TEMPRANILLO 2021 T	517
BICHO PALO 2020 T C	841
BIDEONA L3ZA 2019 T	510
BIDEONA L4GD4 (LAGUARDIA) 2019 T	510
BIDEONA LAS PARCELAS 2019 B	510
BIDEONA LAS PARCELAS 2019 T	510
BIDEONA S4MG0 (SAMANIEGO) 2019 T	510
BIDEONA V1BN4 (VILLABUENA) 2019 T	510
BIENVENIDOS AL EXTRAORDINARIO MUNDO DE LA MUJER CABALLO MITAD MUJER, MITAD CABALLO (TARONJA-ORANGE) 2020 B C	708
BIENVENIDOS AL EXTRAORDINARIO MUNDO DE LA MUJER CABALLO MITAD MUJER, MITAD CABALLO (VERD-VERDE) 2020 T C	709
BIENVENIDOS AL EXTRAORDINARIO MUNDO DE LA MUJER CABALLO MITAD MUJER, MITAD CABALLO 2020 T C	709
BIFTU 2018 T RB	680
BIG BANG 2019 T BA S	427
DIGA DE LUDEIRI 2019 T C	574
BIGARDA 2021 RD	845
BIGARDO 2020 T	845
BIKANDI 2009 T R	588
BIKANDI 2018 T C	588
BIKOTE 2020 T	865
BILLON ESPERANZA E316 2020 T	502
BILLON VICTORIA V419 2020 T	502
BIMBACHE 2020 B	179
BINIGRAU BI-BLANC 2020 B FB	823
BINIGRAU BI-NEGRE 2017 T BA	823

WINE	PAGE	WINE	PAGE	WINE	PAGE
BINIGRAU E-NEGRE 2020 T	823	BLANC SUBUR 2020 B	332	BLUEGRAY 2020 T	368
BINIGRAU E-ROSAT 2021 RD	824	BLANCA 2021 B	295	BLUEGRAY 2021 T	368
BINITORD BLANC 2021 B	820	BLANCHER 2016 BE GR BN	123	BLUME SAUVIGNON BLANC 2021 B	618
BINITORD CIUTAT DE PARELLA 2017 T R	820	BLANCHER BE R BN	123	BLUME VERDEJO SELECCIÓN 2021 B	619
BINITORD NEGRE 2020 T	820	BLANCHER BE R BR	123	BLUME VERDEJO VIURA 2021 B	619
BINITORD ROSAT 2021 RD	820	BLANCHER ROSAT 2018 RE R BR	123	BO - BOBAL ÚNICO 2020 T	682
BIODIVERSO 2020 B	266	BLANCO 12 CEPAS 2021 B	818	BOADA 2015 T C	496
BIÓGRAFO DEL VINO 2019 T	663	BLANCO DE CRÁTER 2021 B	631	BOADA 2019 T RB	496
BISIESTO CABERNET SAUVIGNON 2017 T	672	BLANCO DE HORNILLOS 2021 B	875	BOADA VERDEJO, SAUVIGNON BLANC, CHARDONNAY 2021 B	621
BISIESTO TEMPRANILLO 2016 T C	672	BLANCO DE MARIA 2021 B	680	BOBAL BLANCO BY PEPE HIDALGO 2021 B	682
BISIESTO VERDEJO 2020 B FB	672	BLANCO DE ROBERTO 2021 B	680	BOBAL DE SANJUAN 2020 T S	683
BISILA ROSÉ BRUT NATURE PRESTIGE BE BN	142	BLANCO NIEVA PIE FRANCO 2021 B	608	BOBAL DE SANJUAN 2021 RD	683
BITXIA BERRIA 2021 B	88	BLANCO NIEVA PIE FRANCO GRAN VINO DE RUEDA SOBRE LÍAS 2020 B	608	BOBAL ICON 2020 T RB	271
BIU FINCA DE LA BORDA 2018 B	171			BOBAL NEGRO BY PEPE HIDALGO 2020 T	683
BIU PINOT NOIR 2020 T	171	BLANCO NIEVA SAUVIGNON BLANC 2021 B	608	BOBAL ROSA BY PEPE HIDALGO 2021 RD	683
BIVAC 2021 B	344	BLANCO NIEVA VERDEJO 2021 B	608	BOBALE MANNEKEN PIS 2018 T C	682
BK! 2020 B	292	BLANCO POLAR VERDEJO 2021 B	810	BOBOS FINCA CASA LA BORRACHA 2020 T	676
BK! 2020 T BA	292	BLANEO CHARDONNAY 2021 B FB	309	BOCA GROSSA 2019 T	181
BLACK 2021 T	339	BLANEO SYRAH 2019 T FB	309	BOCA GROSSA 2021 RD	181
BLANC 110 VINS NADAL 2021 B	832	BLAS DE LEZO 2018 T C	524	BOCA NEGRA DULCE 2016 T D	36
BLANC BRUC 2020 B	331	BLAS MUÑOZ CEPAS VIEJAS 2016 T	797	BOCA PETITA 2021 B	181
BLANC D'ENGUERA ORIGINAL 2021 B	703	BLAS MUÑOZ CHARDONNAY 2020 B	797	BOCA PETITA 2021 RD	181
BLANC DE CLOSOS 2020 B	376	BLASÓN DEL TURRA MACABEO 2021 B	499	BOCA PETITA 2021 T	181
BLANC DE GERISENA 2021 B RB	184	BLASÓN DEL TURRA PARDINA 2021 B	499	BOCALASLOBAS 2016 T C	439
BLANC DE GRESA 2020 B FB	192	BLASÓN DEL TURRA TEMPRANILLO 2020 T	499	BOCATINTA 2018 T	259
BLANC DE TRILOGÍA 2021 B BA	707	BLAU 2020 T	295	BOCATINTA 2018 T C	259
BLANC DELS ASPRES 2020 B FB	193	BLAUVERD 2021 T	295	BOCCA 9 MESES 2020 T RB	449
BLANC MOSCATEL SEC LES FRESES DE JESÚS POBRE 2021 B MO	41	BLECUA 2016 T R	624	BOCINS BLANC 2021 B	365
BLANC PRÍNCEPS MUSCAT 2021 B	330	BLÉS 2020 T RB	701	BODEGA 202 2019 T C	508
BLANC SEC ÁMFORA LES FRESES DE JESÚS POBRE 2021 B MO	41	BLÉS 2021 B	701	BODEGA CASTELL MIQUEL CABERNET SAUVIGNON 2019 T R	825
BLANC SEC L'HORABONA LES FRESES DE JESÚS POBRE 2021 B MO	41	BLÉS CRIANZA DE ARANLEÓN 2019 T C	701	BODEGA CASTELL MIQUEL SYRAH 2019 T R	825
		BLUE MOSCATEL 2021 B SD	311	BODEGAS ADRIA GODELLO 2021 B	66

WINE	PAGE
BODEGAS ADRIA MENCÍA 2021 T	66
BODEGAS ADRIA SILK 2020 T RB	66
BODEGAS ADRIA VELVET 2019 T	66
BODEGAS ADRIA VILLA EL TOLEIRO 2019 B C	66
BOHEM 2021 B	781
BOHEM 2021 T	781
BOHIGAS 2020 BE GR EBR	133
BOHIGAS BE R BN	133
BOHIGAS BE R BR	133
BOHIGAS GARNATXA NEGRA 2019 T BA	116
BOHIGAS ROSAT RE BR	133
BOHIGAS XARELLO 2021 B	116
BOLET APAGALLUMS PINOT NOIR 2021 RD	329
BOLET CAMAGROC XARELLO 2021 B	329
BOLET CANTARELUS ULL DE LLEBRE 2019 T	329
BOLET CARTOIXÀ 2014 BE GR BN	128
BOLET CLASSIC ECO 2019 BE BR	128
BOLET ECO 2015 BE GR BN	128
BOLET ECO 2019 BE R BN	128
BOLET ECO 2019 BE R BR	128
BOLET FREDOLIC (SIN SULFITOS) T	329
DOLET PINOT NOIR ROSAT 2017 RE R DR	128
BOLET SÀPIENS MERLOT 2016 T C	329
BOLET VINYA SOTA BOSC 2021 B	329
BONA NIT 2019 T	296
BONESVALLS CABERNET SAUVIGNON 2018 T BA	344
BONFILL 2019 T C	182
BORDÓN 2012 T GR	526
BORDÓN 2016 T R	526
BORDÓN 2019 T C	526
BORDÓN 2021 B	526
BORDÓN D'ANGLADE 2019 T C	526

WINE	PAGE
BÒRIA SUMARROCA 2017 T	342
BORN TO BE WILD 2020 T RB	712
BORRASCA 2018 T	36
BORSAO BEROLA 2018 T	103
BORSAO BOLÉ 2018 T RB	103
BORSAO CABRIOLA 2018 T	102
BORSAO CLÁSICO 2021 T	103
BORSAO SELECCIÓN 2021 B	103
BORSAO SELECCIÓN 2021 RD	103
BORSAO SELECCIÓN 2021 T	103
BORSAO TRES PICOS 2020 T	103
BORSAO ZARIHS 2018 T	103
BOSQUE DE MATASNOS 2019 T	474
BOTANI 2021 B	264
BOTANI GARNACHA 2021 T	264
BOTANI NOBLEZA 2021 B	264
BOTANI SPARKLING MUSCAT 2021 BE	863
BOTAS DE BARRO ALMANSA 2019 T RB	50
BOTAS DE BARRO JUMILLA 2020 T RB	232
BOTAS DE BARRO LEÓN 2019 T RB	813
BOTAS DE BARRO RIBERA DEL DUERO 2019 T RB	491
BOTAS DE BARRO RUEDA 2021 B	619
BOTAS DE BARRO TORO 2020 T	669
BOTIJO BLANCO 2021 B	834
BOTIJO ROJO 2021 T	834
BOTÓN DE GALLO 2020 B	615
BOUZA DO REI 2021 B	391
BOUZA DO REI GRAN SELECCIÓN 2020 B	391
BRANCELLAO DOS CANOTOS 2020 T	858
BRANCO DE SANTA CRUZ 2019 B	689
BRAÓ 2018 T C	291
BRAÓ 2019 T C	291

WINE	PAGE
BRAVO 2018 T FB	468
BRAVO DE REJADORADA 2016 T R	656
BRAVO! 2021 RD	355
BRECA 2019 T FB	846
BRECA EL NACIDO 2021 T	845
BRECA ROSÉ 2021 RD	845
BREGA 2018 T	96
BRESQUE SAUVIGNON BLANC 2021 B	628
BRESQUE SYRAH 2019 T RB	628
BRI NEGRE 2015 T R	294
BRINDE DE GODELLO 2014 BE R	687
BRISAT DE PUNTIRÓ 2021 B	830
BRISAT DEL COSTER 2021 B	165
BRISÉE 2021 B	182
BROTE 2021 B	704
BROTE 2021 RD	704
BROTONS GRAN FONDILLON RESERVA 1964 T FO	40
BROTONS GRAN FONDILLON RESERVA 1970 T FO	41
BROTONS GRAN FONDILLON RESERVA 1978 T FO	40
BROTS DE XARELLO 2020 B	339
BRU DE GRAMONA 2018 T	338
BRU DE VERDÚ 14 2017 T	172
BRU DE VERDÚ 2019 B	172
BRU DE VERDÚ 2019 T	172
BRÚIXOLA 2016 B	362
BRÚIXOLA 2017 T C	362
BRÚIXOLA CARINYENA GARNATXA NEGRA 2018 T	362
BRUJIDERA #GARAGEWINE 2020 T	773
BRUMA DEL ESTRECHO DE MARÍN "PARAJE MARÍN" 2020 T	228
BRUMA DEL ESTRECHO DE MARÍN "PARCELA MANDILES" 2020 T	229
BRUMA DEL ESTRECHO DE MARÍN FINCA CQ 2020 T	229

WINE	PAGE
BRUMAS DE AYOSA 2019 BE R BN	714
BRUMAS DE AYOSA 2021 T	714
BRUMAS DE AYOSA AFRUTADO B SD	714
BRUMAS DE AYOSA MALVASÍA AROMÁTICA 2021 B	714
BRUMAS DE AYOSA SECO 2021 B S	714
BRUÑAL QUINTA LAS VELAS 2019 T C	58
BRUTO 2019 T	229
BUC 2017 T C	81
BUCHE GARNACHA TEMPRANILLO 2016 T RB	817
BUCHE VALLE EL RAPOSO 2019 T C	817
BUENCOMIEZO MENCIA SELECCIÓN 2017 T	64
BUFARUT 2019 T C	181
BULEZA 2021 RD	65
BUM BUM CIAO VERDEJO 2021 B SD	864
BUM BUM ROSÉ 2021 RD SD	864
BURBUJAS DE BALADIÑA 2014 BE BN	395
BURGMANN RED SELECTION 2020 T	634
BURGMANN ROSÉ SELECTION 2020 RD	634
BURÓ VENDIMIA SELECCIONADA 2018 T	455
BUXUS DE LES AUBAGUES 2020 T	376
BY LATÚE TEMPRANILLO 2021 T	848

C

WINE	PAGE
C0001 NUMBERED EDITION B	871
CA L'ELSA 2016 T R	860
CA N'ESTRUC BLANC 2021 B	116
CA N'ESTRUC NEGRE 2021 T	116
CA N'ESTRUC ROSAT 2021 RD	116
CA N'ESTRUC XARELLO 2021 B	116
CABALLERO HIDALGO 2017 T R	248
CABALLERO HIDALGO 2018 T C	248
CABALLO DE ORO 2018 T R	696
CABANELAS 2018 T	78

WINE	PAGE
CABERNET DE SON RAMON 2017 T	831
CABEZA DE CUBA 2018 T C	584
CABEZA DEL HIERRO 2019 T	782
CABRIDA 2020 T C	293
CACHITO MÍO SELECCIÓN 14 DE 24 2016 T	670
CACHITO MÍO SELECCIÓN 14 DE 24 2017 T	670
CADMO 2018 T	832
CAÍÑO DOS CANOTOS 2019 T	858
CAIR SELECCIÓN LA AGUILERA 2019 T	446
CALA MARQUESA 2021 B	181
CALA N 1 2019 T	776
CALA N 2 2018 T	776
CALABOBOS 2018 B	395
CALADOS DEL PUNTIDO 2018 T BA	590
CALAMÓN RD AG SD	819
CALCETAS 2021 B	702
CALIBER 2019 T BA	99
CALIXTO 2021 T	759
CALIXTO BOLOSEA 2021 T	759
CALIXTO NIETO 2021 T	759
CALIXTO OSIRIS 2019 T	759
CALIZA MERLOT SYRAH TEMPRANILLO 2021 T	244
CALIZA SAUVIGNON BLANC AIRÉN VIURA 2021 B	244
CALIZA TEMPRANILLO 2021 RD	244
CALVARIO 2016 T	568
CALZADILLA CLASSIC 2015 T	739
CAMARILLAS 2019 T C	38
CAMBIO DE TERCIO 2021 T	679
CÁMBRICO RUFETE BLANCA PIZARRA 2019 B	854
CÁMBRICO RUFETE EL POCITO 2018 T	758
CÁMBRICO RUFETE VALLEOSCURO 2018 T	758
CAMELOT DULCE MONASTRELL 2019 T D	227

WINE	PAGE
CAMÍ D'EN POCA SANG 2021 RD	184
CAMÍ DE CORMES 2019 T C	191
CAMÍ DE CORMES 2021 B	191
CAMÍ DE LA FONT 2018 B	638
CAMÍ DE LA FONT 2019 B	638
CAMÍ DE LA FONT BRISAT 2019 B	638
CAMÍ DE LA MINA 2020 T	364
CAMÍ DE TARDOR 2021 RD	163
CAMINO DE CABRAS ALBARIÑO 2021 B	392
CAMINO DE CABRAS GODELLO 2021 B	689
CAMINO DE CABRAS MENCÍA 2020 T	689
CAMINO DE LOS ARRIEROS 2019 T	800
CAMINO DE NAVAHERREROS 2020 B FB	721
CAMINO DE NAVAHERREROS 2020 T	721
CAMINO DE RUS SELECCIÓN 2015 T C	785
CAMINO DE RUS SYRAH 2020 T	785
CAMINO DE SEDA 2021 T	222
CAMINO DEL BOSQUE T	111
CAMINO EMPEDRADO 2020 T	409
CAMINO EMPEDRADO BLEND DE FINCAS 2020 T RB	409
CAMINO LEZA 2019 T	533
CAMINS DEL PRIORAT 2021 T	354
CAMO 2020 T S	683
CAMPANO 2019 T	771
CAMPARRÓN 2017 T R	659
CAMPARRÓN 2018 T C	659
CAMPARRÓN NOVUM 2021 T	659
CAMPARRÓN SELECCION 2021 T	659
CAMPAZA 2021 T	412
CAMPEADOR 2005 T GR	536
CAMPESINO 2020 T	655
CAMPESTRAL CLARETE 2020 RD	846

WINE	PAGE
CAMPESTRAL RED 2020 T	846
CAMPESTRAL RED 2020 T RB	846
CAMPESTRAL RED ANCESTRAL 2020 TE	878
CAMPESTRAL WHITE ANCESTRAL 2020 BE	878
CAMPESTRAL WHITE ENVEJECIDO BAJO VELO 2020 B	846
CAMPILLO 2019 T C	518
CAMPILLO 57 SELECCIÓN ESPECIAL 2015 T	518
CAMPILLO ROSÉ 2021 RD	518
CAMPO BURGO 2016 T R	527
CAMPO BURGO 2018 T C	527
CAMPO BURGO 2021 T	527
CAMPO ELISEO 2018 T R	655
CAMPO ELISEO CUVÉE ALEGRE 2018 T	655
CAMPO ELISEO CUVÉE ALEGRE 2021 RD	597
CAMPO ELISEO CUVÉE ALEGRE VERDEJO 2021 B	597
CAMPO ELISEO VERDEJO 2020 B FB	597
CAMPO REDONDO 2018 T RB	66
CAMPO REDONDO GODELLO 2021 B	66
CAMPO VIEJO 2013 T GR	560
CAMPO VIEJO 2017 T R	560
CAMPO VIEJO 2018 T C	560
CAMPO VIEJO TEMPRANILLO C.V.C. T	560
CAMPO VIEJO VENDIMIA SELECCIONADA 2017 T C	560
CAMPOS DE DULCINEA SAUVIGNON BLANC 2021 B	774
CAMPOS DE DULCINEA SELECCIÓN DE LA FAMILIA 2019 T	774
CAMPOS DE DULCINEA TEMPRANILLO 2021 T	774
CAMPOS DE LUZ 2018 T R	111
CAMPOS DE LUZ 2021 B	111
CAMPOS DE LUZ 2021 RD	111
CAMPOS DE LUZ GARNACHA 2021 T	111
CAMPOS DE LUZ REVELACIÓN 2019 T C	111
CAMPOS DE RISCA 2020 T	230
CAMPOS DE SUEÑOS 2021 B	614
CAMPOS REALES CABERNET SAUVIGNON 2019 T RB	238
CAMPOS REALES SYRAH 2019 T	238
CAN 2020 T	717
CAN BAS D'ORIGEN P3 2021 B	329
CAN BAS D'ORIGEN P5 2021 B	329
CAN BAS LA CREU 2021 B	329
CAN BAS LA ROMANA 2018 B FB	329
CAN BLAU 2020 T	295
CAN FEIXES BLANC SELECCIÓ 2021 B	336
CAN FEIXES BLANC TRADICIÓ 2020 B	336
CAN FEIXES CHARDONNAY 2019 B FB	336
CAN FEIXES NEGRE SELECCIÓ 2019 T	336
CAN FEIXES NEGRE TRADICIÓ 2013 T C	336
CAN FEIXES RESERVA ESPECIAL 2009 T R	336
CAN GELAT CALLET 2021 RD	828
CAN GELAT GIRÓ 2021 B	828
CAN NOVES BLANC 2020 B	860
CAN RICH BLANC D'AMFORA 2021 B	816
CAN RICH BLANC DE BLANCS 2020 BE	879
CAN RICH NEGRE D'AMFORA 2020 T	816
CAN SUMOI LA ROSA 2021 RD	330
CAN SUMOI PERFUM 2021 B	330
CAN SUMOI SUMOLL - GARNATXA 2020 T	854
CAN SUMOI XARELLO 2021 B	330
CAN VIDALET BLANC DE BLANCS 2021 B	826
CAN VIDALET BLANC DE NEGRES 2021 B	826
CAN VIDALET SES PEDRES 2018 B	827
CAN VIDALET SO DEL XIPRER 2017 T	827
CANALS & MUNNÉ ICE SWEET WHITE 2020 BE R D	128
CANALS & MUNNÉ INSUPERABLE 2019 BE R BR	128
CANALS & MUNNÉ X10 2011 BE GR BN	128
CANALS NADAL 2017 BE GR BN	129
CANALS NADAL 2019 BE R BN	129
CANALS NADAL CN 1986 BLANC DE NOIRS 2017 BE R BR	129
CANALS NADAL ROSÉ 2019 RE BR	129
CANASTA CREAM 20 AÑOS BF CRM	212
CANASTA CREAM BF CRM	212
CÀNDIA 2018 T BA	867
CANDILES DE HIRIART 2020 T C	156
CÁNFORA PIE FRANCO 2016 T R	238
CANFORRALES CHARDONNAY 2021 B	238
CANFORRALES CLÁSICO TEMPRANILLO 2021 T	238
CANFORRALES LUCÍA AIRÉN 2021 B	238
CANFORRALES NATURE TEMPRANILLO SYRAH 2021 T	778
CANFORRALES NATURE VIOGNIER 2021 B	778
CANFORRALES PETIT VERDOT 2021 RD	238
CANFORRALES SELECCIÓN 2019 T RB	238
CANTA LA PERDIZ 2017 T R	478
CANTADAL 2021 T	658
CANTALARES GARNACHA 2021 RD	703
CANTALARES MERLOT 2020 T	703
CANTALARES MERSEGUERA 2021 B	703
CANTALLOPS 2019 B	326
CANTALOBOS 2019 T C	66
CANTALOBOS 2021 B	67
CANTALOBOS 2021 T	67
CANTAMUDA 2021 T RB	442
CANTAMUDA FINCA LA CEBOLLA 2019 T	442
CANTAMUDA PARCELA 64 2019 T	442
CANTARADA DE LOS MOZOS SAN PRUDENCIO 2018 T	525
CANTARIÑA 2 VIÑA DE LOS PINOS 2018 T	71
CANTARIÑA 3 EL TRIÁNGULO 2018 T	71
CANTARIÑA 5 VALDEOBISPO 2018 T	71

WINES

WINE	PAGE
CANTARIÑA 6 MERENZAO 2020 T	71
CANTOCUERDAS ALBILLO 2020 B FB	721
CANTOS DE VALPIEDRA 2018 T	569
CANTOS DEL DIABLO 2019 T	276
CANTOSÁN VERDEJO VIÑAS VIEJAS 2021 B	622
CANTUESO 2020 T	834
CAÑADA DEL SOTO 2018 T C	47
CAÑADA LA TORRE 2021 B RB	705
CAÑADA PARÍS 2020 B C	700
CAÑALVA ÉLÉGANCE 2019 T C	498
CAÑALVA ÉLÉGANCE 2020 T RB	498
CAÑALVA ÉLÉGANCE 2021 B S	498
CAÑALVA TEMPRANILLO 2019 T RB	498
CAÑAS Y BARRO 2021 B	711
CAÑAS Y BARRO 2021 RD	711
CAÑUS VERUS 2019 T RB	658
CAÑUS VERUS MALVASIA 2020 B FB	658
CAÑUS VERUS VIÑAS VIEJAS 2018 T	658
CAP DE BARBARIA 2017 T C	820
CAP DE BARBARIA PANSIT DE FORMENTERA B SOLERA D	820
CAP DE CREUS NACRE 2021 B	192
CAP DE TRONS 2021 T	335
CAP SENTIT ORANGE WINE 2021 B	117
CAPA VENDIMIA SELECCIONADA TEMPRANILLO 2021 T	793
CAPDEVILA PUJOL BE R BN	123
CAPDEVILA PUJOL VICA 2021 B	327
CAPELLANIA 2017 B R	576
CAPILLA DEL FRAILE PARCELA SYRAH 2018 T	788
CAPILLA DEL FRAILE PARCELA SYRAH 2019 T	788
CAPILLA DEL FRAILE PARCELA SYRAH 2020 T	788
CAPILLA DEL FRAILE PARCELA SYRAH 2021 T	788
CAPILLA DEL FRAILE PETIT VERDOT 2017 T	788

WINE	PAGE
CAPILLA DEL FRAILE PETIT VERDOT 2018 T	788
CAPILLA DEL FRAILE PETIT VERDOT 2021 T	788
CAPILLA DEL FRAILE SELECCIÓN ESPECIAL 2018 T	788
CAPITÁN GAONA 2019 T	567
CAPITAN JULIÁN 2020 T RB	705
CAPITAN JULIÁN 2021 B	705
CAPITÁN TRUENO 2020 T	586
CAPRASIA 2018 BE R BN	127
CAPRASIA 2021 B	679
CAPRASIA BOBAL ÁNFORA 2020 T	679
CAPRASIA ROSÉ 2021 RD	679
CAPRICHO DE LANDALUCE 2019 T	530
CAPRICHO DE MERENZAO 2017 T	407
CAPRICHO DE VALVANERA 2021 B	547
CAPRICHO DIVINO 2019 BE BN	271
CAPRICHO DIVINO SAUVIGNON BLANC 2021 B	271
CAPRICHO VAL DE PAXARIÑAS 2021 B	69
CAPRICHO VAL DE PAXARIÑAS 2021 RD	69
CAPTIROT 2018 T C	117
CAPVESPRE ORIGEN 2021 B	117
CAPVESPRE ORIGEN 2021 RD	117
CAPVESPRE ORIGEN 2021 T	117
CAPVESPRE SUNSET 2021 B	117
CAPVESPRE SUNSET 2021 RD	117
CAPVESPRE SUNSET 2021 T	117
CARA NORD BLANC 2021 B	163
CARA NORD NEGRE 2020 T C	164
CARA NORD TREPAT NEGRE 2021 T	164
CARA NORD TREPAT ROSAT 2021 RD	164
CARABIBAS 640 NOCHES 21 MESES 2015 T R S	38
CARABIBAS LA VIÑA DEL CARPINTERO 2019 T C	38
CARABIBAS VS VENDIMIA SELECCIONADA 2019 T C S	38

WINE	PAGE
CARÁMBANO ICE WINE B D	843
CARCEL DE CORPA 2013 T GR	678
CARCEL DE CORPA 2018 T C	678
CARCHELO 2020 T RB	224
CARCHELO ROSÉ 2021 RD	224
CARCHELO SELECTO 2015 T C	224
CARDENAL VORS BF PC S	219
CARE CARIÑENA NATIVA 2020 T C	108
CARE FINCA BANCALES 2018 T R	108
CARE FINCA MARIMÚ 2020 T BA	108
CARE GARNACHA BLANCA NATIVA 2021 B	108
CARE GARNACHA NATIVA 2020 T	108
CARE SOBRE LÍAS 2021 T	108
CARE SOLIDARITY ROSÉ 2021 RD	108
CARLES ANDREU 12@ 2021 T	164
CARLES ANDREU 2018 BE GR BN	133
CARLES ANDREU 2019 BE BN	134
CARLES ANDREU BARRICA 2017 BE R BN	134
CARLES ANDREU PARELLADA 2020 B	164
CARLES ANDREU ROSAT 2019 RE BR	134
CARLES ANDREU	
ROSAT BARRICA 2019 RE R BR	134
CARLES ANDREU TREPAT 2019 T BA	164
CARLOS PLAZA 2019 T RB	816
CARLOS PLAZA 2020 T	816
CARLOS PLAZA 2021 B	816
CARLOS PLAZA SELECCIÓN 2018 T	816
CARLOS RUBÉN PARCELA ÚNICA 2021 T	98
CARLOS RUBÉN SIN DUDA 2021 T	98
CARLOS SAN PEDRO PEREZ DE VIÑASPRE 2016 T	519
CARLOS SERRES 1896 FINCA EL ESTANQUE SEGUNDO AÑO 2018 RD R	560

WINE	PAGE
CARLOS SERRES 1896 FINCA EL ESTANQUE SEGUNDO AÑO 2019 T	561
CARLOS SERRES 2015 T GR	561
CARLOS SERRES 2016 T R	561
CARLOS SERRES 2019 T C	561
CARLOS SERRES BRUT RIOJA 2019 BE	561
CARLOS VII B AM S	285
CARLOTA SURIA ORGANIC BE R BN	146
CARLOTA SURIA ORGANIC BE R BR	146
CARLOTA SURIA ORGANIC BOBAL 2019 T C	682
CARLOTA SURIA ORGANIC CHARDONNAY 2020 B FB	682
CARMELITANO MOSCATEL 2019 B MO D	707
CARMELO RODERO 2019 T R	458
CARMELO RODERO TSM 2019 T	458
CARMEN 2014 T GR	511
CARMEN BY COMENGE 2021 RD	443
CÁRMINE 2019 T BA	225
CARODORUM EXPRESIÓN 2021 T	658
CARODORUM ISSOS 2020 T C	658
CARODORUM MANTÓN DE MANILA 2021 T RB	658
CARPURIAS PRIETO PICUDO 2021 T C	757
CARRAMATA 2021 B	614
CARRAMIMBRE 2019 T C	442
CARRAMIMBRE 2021 T RB	442
CARRANAVARIDAS 2019 T C	584
CARRAQUINTANA DE AMAREN 2018 T BA	515
CARRAROA 2019 T C	439
CARRASCALEJO 2017 T C	92
CARRASCALEJO 2020 T	92
CARRASCALEJO 2021 RD	92
CARRASVIÑAS 100% VERDEJO 2021 B	604
CARRASVIÑAS 2020 B FB	604

WINE	PAGE
CARRASVIÑAS DORADO B	604
CARRATRAVIESA 2021 RD	160
CARRAVID 2016 T C	456
CARREDUEÑAS 2019 T C	160
CARREDUEÑAS 2020 RD FB	160
CARREDUEÑAS 2021 RD	160
CARRIL CRUZADO CABERNET SAUVIGNON 2021 T	788
CARRIL CRUZADO MULTIVARIETAL 2018 T	788
CARRIL CRUZADO PETIT VEDOT EDICIÓN LIMITADA 2019 T C	268
CARRIL CRUZADO SAUVIGNON BLANC 2021 B	268
CARRIL CRUZADO SELECCIÓN PETIT VERDOT 9 MESES BARRICA 2020 T	789
CARRIL CRUZADO SELECCIÓN SYRAH 9 MESES BARRICA 2020 T C	789
CARTAGO 2017 T	668
CARTESIUS NEGRE 2019 T	182
CARTESIUS ROSAT 2021 RD	182
CARTUS 2007 T R	366
CASA ANTONETE MACABEO 2021 B	238
CASA ANTONETE TEMPRANILLO 2017 T C	238
CASA ANTONETE TEMPRANILLO 2021 T MC	238
CASA BOQUERA 2017 T C	734
CASA BOQUERA 2018 T RB	734
CASA BOQUERA SELECCIÓN 2017 T C	734
CASA CASTILLO EL MOLAR 2021 T	224
CASA CASTILLO LA TENDIDA 2020 T	224
CASA CASTILLO LA TENDIDA 2021 T	224
CASA CASTILLO LAS GRAVAS 2020 T	224
CASA CASTILLO MONASTRELL 2021 T	224
CASA CASTILLO PIE FRANCO 2020 T C	224
CASA CISCA 2017 T BA	732
CASA DE ILLANA ALMA 2021 B	502

WINE	PAGE
CASA DE ILLANA BOBAL VINO DE PARCELA 2019 T	502
CASA DE ILLANA CARMEN 2019 B FB	502
CASA DE ILLANA TRESDECINCO 2018 T C	502
CASA DE ISAAC 2019 T	851
CASA DE LA CERA 2017 T BA	732
CASA DE LA DEHESA 2014 T C	241
CASA DE LA ERMITA 2020 T C	231
CASA DE LA ERMITA PETIT VERDOT 2019 T C	231
CASA DE LA VEGA VERDEJO 2021 B	595
CASA DE OUTEIRO MENCÍA GAMA ESCUDO 2020 T	412
CASA DO SOL 2019 B	392
CASA DON ÁNGEL BOBAL 2019 T	674
CASA JIMÉNEZ 2019 T C	44
CASA LA RAD 2018 B	510
CASA LA RAD 2018 T RB	510
CASA LABOR 2019 T	709
CASA MARIOL GARNATXA BLANCA 2021 B	644
CASA MARIOL GARNATXA NEGRA 2020 T	645
CASA MARIOL GARNATXA NEGRA 2021 T	645
CASA MARIOL SAMSÓ 2018 T C	645
CASA MARIOL SYRAH 2016 T R	645
CASA MARIOL VERDEJO 2021 B	856
CASA MOREIRAS 2021 B	412
CASA MOREIRAS MENCÍA 2021 T	412
CASA MOREIRAS SELECCIÓN 2021 T	412
CASA RITAS 2021 T	35
CASA SOSEGADA 2020 T	709
CASA VELLA D'ESPIELLS 2017 T R	341
CASADO MORALES 2014 T GR	554
CASADO MORALES 2016 T R	555
CASADO MORALES 2018 T C	555
CASAL DE ARMÁN 2020 T	424

WINE	PAGE
CASAL DE ARMÁN 2021 B	424
CASAL DE PAULA 2021 B	425
CASAL DE PAULA 2021 T	425
CASAL DE PAULA D.H. 2020 B	425
CASAL NOVO GODELLO 2021 B	686
CASAL NOVO MENCÍA 2020 T	686
CASANOVA 2021 B	422
CASAR DE BURBIA 2020 T RB	71
CASAR DE BURBIA GODELLO 2021 B	71
CASAR DE VALDAIGA PARAJE EL TOLEIRO 2020 T	64
CASAR DE VIDE 2021 B	421
CASAR GODELLO 2020 B FB	71
CASAS DE PEÑA 2021 RD	780
CASAS DE PEÑA AIRÉN 2021 B	780
CASAS DE PEÑA CHARDONNAY 2021 B	780
CASAS DE PEÑA GARNACHA TINTORERA 2021 T	780
CASAS DE PEÑA TEMPRANILLO 2021 T	780
CASTA DIVA COSECHA DORADA 2021 B	847
CASTA DIVA COSECHA MIEL DULCE 2019 B D	847
CASTA DIVA ESENCIA DULCE 2018 B D	848
CASTA DIVA MONTE DIVA 2021 B	848
CASTA DIVA RESERVA REAL DULCE 2002 B R D	848
CASTA DIVA SELECCIÓN ESPECIAL 2021 B	848
CASTAÑO COLECCIÓN 2018 T BA S	732
CASTAÑO MONASTRELL DULCE 2018 T D	732
CASTAÑO SANTA 2018 T BA	732
CASTEL DE BOUZA 2021 B	391
CASTEL DE FORNOS 2021 B	386
CASTELL D'AIXA 2021 B	40
CASTELL D'OR BE BN	129
CASTELL D'OR BE C BR	129
CASTELL D'OR BE GR BN	129

WINE	PAGE
CASTELL D'OR BRUT ROSAT RE BR	129
CASTELL D'OR ORGÀNIC BE BR	129
CASTELL D'OR RESERVA IMPERIAL BE R BR	129
CASTELL DE FALSET 2014 T R	297
CASTELL DE PUJADES XARELLO 2021 B	327
CASTELL DE PUJADES XARELLO VERMELL 2021 RD	327
CASTELL DE SANTUERI BLANC 2021 B	872
CASTELL DE SANTUERI ROUGE 2020 T RB	821
CASTELL DEL REMEI COMTAT D'URGELL 2017 B	171
CASTELL DEL REMEI GOTIM BLANC 2021 B	171
CASTELL DEL REMEI GOTIM BRU 2019 T	171
CASTELL DEL REMEI ODA NEGRE 2020 T C	171
CASTELL DEL REMEI VI DOLÇ 2019 B D	171
CASTELL SANT ANTONI CAMÍ DEL SOT 2014 BE GR BR	129
CASTELL SANT ANTONI GRAN BARRICA 2015 BE GR BN	129
CASTELL SANT ANTONI GRAN RESERVA 2012 BE GR BN	130
CASTELL SANT ANTONI JAZZ NATURE BE R BN	130
CASTELL SANT ANTONI JAZZ NATURE ROSÉ RE R BR	130
CASTELL SANT ANTONI TORRE DE L'HOMENATGE 2005 BE GR BN	130
CASTELLROIG 2019 BE R BR	884
CASTELLROIG SO BLANC 2021 B	342
CASTELLROIG SO BLANC MAGNUM 2012 B	342
CASTELLROIG SO NEGRE 2019 T	342
CASTELLROIG SO SERÉ 2018 B FB	342
CASTELLROIG XARELLO VERMELL 2019 BE R BR	884
CASTELLUM AUGUSTI 2020 B	768
CASTELO DE MEDINA 2020 B FB	602
CASTELO DE MEDINA SAUVIGNON BLANC 2020 B	602
CASTELO DE MEDINA VERDEJO 2021 B	602
CASTELO DE MEDINA VERDEJO VENDIMIA SELECCIONADA 2020 B	602

WINE	PAGE
CASTELO NOUVEAU 2021 T MC	803
CASTELO ROBLE 2019 T RB S	803
CASTELO ROSÉ 2021 RD	803
CASTEZO GARNACHA 2018 T	540
CASTEZO TEMPRANILLO BLANCO 2021 B	540
CASTILLO CLAVIJO 2019 T C	562
CASTILLO COLINA 2021 T	813
CASTILLO DE ALBAI 2017 T R	580
CASTILLO DE ALBAI 2018 T C	580
CASTILLO DE ALBAI 2021 B	580
CASTILLO DE ALBAI TEMPRANILLO 2021 T	580
CASTILLO DE ARESAN CABERNET FRANC 2020 T	789
CASTILLO DE ARESAN SAUVIGNON BLANC 2021 B	789
CASTILLO DE ARESAN TERRUÑO 2019 T	243
CASTILLO DE ARESAN VERDEJO 2021 B	789
CASTILLO DE AYUD 2020 T R	97
CASTILLO DE BELARFONSO 2020 T RB	274
CASTILLO DE BENIZAR CABERNET SAUVIGNON 2021 RD	237
CASTILLO DE BENIZAR MACABEO 2021 B	237
CASTILLO DE BENIZAR TEMPRANILLO 2020 T	237
CASTILLO DE ENERIZ COLECCIÓN 2020 T C	305
CASTILLO DE GUZMÁN BF AM	214
CASTILLO DE GUZMÁN BF FI	214
CASTILLO DE GUZMÁN BF OL	214
CASTILLO DE GUZMÁN BF PC	214
CASTILLO DE JUMILLA 2018 T C	223
CASTILLO DE JUMILLA MONASTRELL 2020 T	223
CASTILLO DE LA MOTA 2021 B	615
CASTILLO DE LIRIA ORGANIC 2021 T	44
CASTILLO DE MONESMA 2017 T C	628
CASTILLO DE MONESMA 2018 T R	628
CASTILLO DE MONESMA 2018 T RB	628

WINE	PAGE
CASTILLO DE MONESMA 2019 T	628
CASTILLO DE MONJARDÍN CHARDONNAY 2020 B FB	306
CASTILLO DE MONJARDÍN DEYO MERLOT DE AUTOR 2018 T C	306
CASTILLO DE MONTE LA REINA 2018 T C	660
CASTILLO DE MONTE LA REINA CUVÉE PRIVÉE 2019 T	660
CASTILLO DE MONTE LA REINA VENDIMIA SELECCIONADA 2016 T	660
CASTILLO DE PEÑAFIEL 2019 T C	461
CASTILLO DE PEÑAFIEL 2020 T RB	461
CASTILLO DE PEÑAFIEL EDICIÓN LIMITADA 2016 T R	461
CASTILLO DE PEÑARANDA 2017 T C	434
CASTILLO DE ROBLES T	841
CASTILLO DE SAJAZARRA 2016 T R	583
CASTILLO DE SAN DIEGO 2021 B	769
CASTILLO DE UTIEL 2018 T C	678
CASTILLO DE UTIEL 2021 T	678
CASTILLO SAN LORENZO 2016 T R	562
CASTILLO SANTA BÁRBARA 2017 T C	239
CASTILLO SANTA BÁRBARA 2018 T C	239
CASTILLO SANTA BÁRBARA 2018 T R	696
CASTILLO YGAY 2011 T GR	576
CASTIÑEIRO ALBARIÑO 2019 B	394
CASTRO CANDAZ A BOCA DO DEMO 2019 T	411
CASTRO CANDAZ LA VERTICAL 2019 B	411
CASTRO DAS SAÍÑAS 2020 T	407
CASTRO DE LOBARZÁN 2021 B	280
CASTRO DE LOBARZÁN 2021 T	280
CASTROFIZ 2019 T C	412
CASTROFIZ 2020 T	412
CASTROFIZ AMENZA 2021 B	412
CASTROFIZ TRADICIÓN 2020 T RB	412
CATAMARÁN 2021 B	681
CATAVENTO 2021 B	394
CATAY 2018 T C	512
CATAY 2021 RD	512
CATAY TEMPRANILLO BLANCO 2021 B	512
CATAY VIÑAS VIEJAS 2019 B	512
CATÓN BF OL	285
CÁTULO ECOLÓGICO 2021 RD	312
CÁTULO GARNACHA 2021 T	312
CAUDALIA DE VALLOBERA 2006 B FB	551
CAUDUM BODEGAS LARRAZ 2016 T	556
CAUDUM BODEGAS LARRAZ 2018 T	556
CAUDUM BODEGAS LARRAZ SELECCIÓN ESPECIAL 2018 T BA	556
CAVA ÁRABE 2020 BE BN	127
CAVA BALLBÉ BE BN	138
CAVA ESENCIA VEGAMAR BE BN	127
CAVA ESTENAS 2020 BE BN	124
CAVA FAUSTINO ART COLLECTION BE R BR	125
CAVA FAUSTINO BE R BR	125
CAVA FAUSTINO ROSE RE R BR	125
CAVA LENEUS SELECCIÓN 2020 BE SS	125
CAVA REVERTÉ "ELECTE" 2017 BE R BN	131
CAVA REVERTÉ 2016 BE R BN	131
CAVA ROXANNE 2020 BE BR	135
CAVAS HILL COLE-LECCIÓ PRIVADA BE BN	132
CAVAS HILL CUVÉE 1887 BE BN	132
CAVAS HILL PANOT GAUDÍ BE BN	132
CAVAS HILL PANOT GAUDÍ BE BR	132
CAVAS HILL PANOT GAUDÍ CORAL RE BR	132
CAZAPITAS 2020 B	855
CEDULA REAL 2015 T GR	565
CEIBO 2021 B	687
CELESTE CRIANZA 2019 T C	487
CELESTE RESERVA 2018 T R	487
CELESTE VERDEJO 2021 B	618
CELIA ROSÉ 2021 RD	854
CELIA VERDEJO 2021 B	854
CELIA VIZCARRA 2018 T	466
CELLER ARRUFÍ LLICSÓ 2018 B BA	642
CELLER ARRUFÍ LLICSÓ 2019 B BA	642
CELLER ARRUFÍ PANICAL 2021 B	642
CELLER ARRUFÍ PANICORT 2018 T	642
CELLER ARRUFÍ TREPADELLA 2021 T	642
CELLER BALART 2021 B	642
CELLER BALART NEGRE 2018 T	643
CELLER CAL PLA 2019 T	358
CELLER I VINYES NOSTRE BLANC 2021 B	871
CELLER I VINYES NOSTRE ROSAT 2021 RD	871
CELLER MARIÀ PAGÈS GARNATXA D'EMPORDÀ DULCE 2020 B D	185
CELLER MARIÀ PAGÈS MOSCAT D'EMPORDÀ DULCE 2020 B D	185
CELLER MARIÀ PAGÈS ROSA-T 2021 RD	185
CELLER MARIÀ PAGÈS VINYA DE L'HORT 2021 B	185
CELLER MATALLONGA M 2021 RD	172
CELLER TROBAT 2018 BE R BN	127
CELSUS 2020 T	663
CENCIBEL =GARAGEWINE 2021 T	773
CENIT 2019 T C S	653
CENIT 2020 B	653
CENIT BONALES 2019 T C	653
CENT X CENT GARNACHA BLANCA CON CRIANZA EN ANFORA 2021 B C	650
CEPA 21 2018 T	475
CEPA 21 2019 T	475
CEPA GAVILÁN 2020 T C	451

WINE	PAGE
CEPA NEGRA 2015 T	540
CEPA NEGRA CENTENARIUM 2006 T R	540
CEPADO BLACK EDITION 2020 B C	686
CEPADO GODELLO 2021 B	686
CEPADO MENCÍA 2021 T	686
CEPAS DEL ZORRO 2019 B FB	90
CEPAS DEL ZORRO 2019 T C	90
CEPAS DEL ZORRO 2021 RD	90
CEPAS DEL ZORRO MACABEO 2021 B	90
CEPAS DEL ZORRO MONASTRELL 2020 T	90
CEPAS VIEJAS DE LUBERRI 2018 T	574
CERAMIC MONASTRELL 2019 T	711
CERCIUM 2019 T	188
CEREMONIA MONASTRELL BLANC DE NOIRS 2021 B	44
CERES 2019 T C	441
CERMEÑO VENDIMIA SELECCIONADA 2021 T	658
CERRO AMURDI 2020 B	865
CERRO AÑÓN 2015 T GR	542
CERRO AÑÓN 2019 T C	542
CERRO CEREZO 2019 T	56
CERRO DE LA RETAMA 2016 T C	203
CERRO LAS CUEVAS 2019 T	555
CERRO LOS CURAS VIÑEDO SINGULAR 2019 B	540
CERRO NEGRO 2020 T	701
CÉRVOLES BLANC VINYES ALTES DE LES GARRIGUES 2020 B FB	174
CÉRVOLES COLORS BLANC 2021 B	174
CÉRVOLES COLORS NEGRE 2020 T C	174
CÉRVOLES ESTRATS 2019 T	174
CÉRVOLES NEGRE 2019 T	174
CÉRVOLES TOSSALETS 2019 T	174
CÉSAR FLORIDO MOSCATEL DORADO BF MO D	207
CÉSAR FLORIDO MOSCATEL PASAS BF MO D	207
CÉSAR LUCENDO A. 2018 T R	790
CÉSAR PRÍNCIPE 2019 T C	156
CESILIA 2020 T RB	41
CESILIA BLANC 2021 B	41
CESILIA LA GARNACHA 2020 T	41
CESILIA ROSÉ 2021 RD	41
CESILIA ROSÉ LA RÉSERVE 2021 RD	41
CHAFANDÍN 2018 T	494
CHAN DE ROSAS CLÁSICO 2021 B	393
CHAN DE ROSAS CUVÉE ESPECIAL 2021 B	393
CHAPIRETE 2019 B FB	621
CHAPIRETE PREFILOXÉRICO 2020 B	621
CHAPIRETE SELECCIÓN 2021 B	621
CHARCÓN ALEGRÍA 2021 B	778
CHARCÓN AROMAS 2020 B	778
CHARCÓN AUTÉNTICO 2020 T	778
CHARCÓN DIVERSO 2020 B FB	778
CHARCÓN ELEGANCIA 2019 T	778
CHARCÓN FORTALEZA 2019 T	778
CHARDO DAY 2021 B	676
CHARDONNAY ROURE 2020 B FB	351
CHASNERO LISTÁN NEGRO 2021 T	29
CHAVEO 2019 T C	90
CHIBUSQUE 2020 B	253
CHIVITA 2021 B	718
CHIVITA TINTO TRADICIONAL 2021 T	718
CHIVITE COLECCIÓN 125 1996 B FB	318
CHIVITE COLECCIÓN 125 2020 B FB	765
CHIVITE COLECCIÓN 125 2021 RD FB	765
CHIVITE COLECCIÓN 125 VENDIMIA TARDÍA 2020 B FB D	765
CHIVITE COLECCIÓN 125 VINO DE GUARDA 2017 T	765
CHIVITE LA ZORRERA GARNACHA 2019 T	765
CHIVITE LAS FINCAS 2 GARNACHAS 2020 B	765
CHIVITE LAS FINCAS 2020 RD FB	765
CHIVITE LAS FINCAS 2021 RD PL	765
CHIVITE LEGARDETA GARNACHA 2019 T	318
CHIVITE LEGARDETA SYRAH 2019 T	318
CHIVO 2020 B	179
CHLOSS 2021 T	848
CHOBEO DE PECIÑA 2012 T	527
CHOBEO DE PECIÑA 2020 B FB	527
CHOLO 2020 B	418
CHOLO AGARDA 2015 B	418
CHOZAS CARRASCAL CH 2016 BE GR BN	135
CHULO 2020 T RB	722
CHUPADERO 2020 B	253
CHUZOS DE PUNTA 2020 T	409
CICLOHOME GODELLO 2020 B	421
CICLOHOME TREIXADURA 2017 B	421
CIEN Y PICO DOBLE PASTA 2019 T	268
CIEN Y PICO EN VASO 2019 T	268
CIENCUEVAS 2018 T C	513
CIENTRUENOS 2019 T BA	319
CIES 2021 B	387
CIFRAS 2020 B	562
CIFRAS 2020 T	562
CILLAR 2021 T	475
CILLAR DE SILOS 2019 T C	475
CILLAR ROSADO DE SILOS 2021 RD	475
CIMA MAZACRUZ SELECCIÓN 2019 T	789
CIMA SELECCIÓN 2021 B	789
CIMADAGO 2019 T C	581
CIMERA CLOS COR VÍ 2019 B C	704
CINCO DE COPAS 2020 T BA S	664

WINE	PAGE	WINE	PAGE	WINE	PAGE
CINCO ISLAS ALBARIÑO SELECCIÓN 2021 B	386	CLÍO 2019 T	225	CLOS DE LÔM MONASTRELL 2021 RD	708
CINCUENTA EGUREN UGARTE 2017 T	565	CLON DE FAMILIA 2016 T	447	CLOS DE LÔM TEMPRANILLO 2021 T	708
CINEMA 2018 T C	493	CLÒNIC 2019 T	293	CLOS DE SANUI 2019 T	172
CINEMA 2020 T RB	493	CLÒNIC CARINYENA VINYAS VIEJAS 2018 T C	293	CLOS DEL GOS 2020 T	296
CINEMA VERDEJO 2021 B	620	CLOS ABELLA 2016 T	369	CLOS DEL PINELL GARNATXA 2021 T	646
CINGLES BLAUS MAS DE LES MORERES 2019 T	296	CLOS ALZINA 2018 T	376	CLOS DEL PINELL GARNATXA BLANCA 2021 B	647
CINGLES BLAUS MAS DE LES MORERES 2020 B	297	CLOS ANCESTRAL 2020 T	334	CLOS FIGUERES 2020 T	363
CINGLES BLAUS OCTUBRE 2020 T	297	CLOS BADACELI DE LA SOLANA 2017 T C	356	CLOS GALENA 2018 T R	363
CINGLES BLAUS OCTUBRE 2021 B	297	CLOS BALTASAR 2020 T	96	CLOS GALENA 2019 T R	363
CINGLES BLAUS SELECCIÓ 2019 T C	297	CLOS BARTOLOMÉ BLANC 2021 B	358	CLOS GELIDA 4 HERETATS 2017 BE GR BN	154
CIRCE VERDEJO 2021 B	596	CLOS CLARA 2017 T GR	359	CLOS IBAI 2018 B	572
CIRERETS 2019 T	369	CLOS COR VÍ RIESLING 2020 B BA	704	CLOS IBAI 2018 T	572
CIRSION 2019 T R	546	CLOS COR VÍ VIOGNIER 2020 B S	704	CLOS IBAI GARNACHA BLANCA 2019 B	572
CITIUS PINOT NOIR 2018 T	800	CLOS CORRIOL 2021 B	360	CLOS IBAI GARNACHA TINTA 2020 T	572
CLAR DEL BOSC 2020 T	368	CLOS CORRIOL NEGRE 2019 T	360	CLOS IBAI GRACIANO 2020 T	572
CLARET DE TARDOR 2020 T	193	CLOS CORRIOL ROSAT 2021 RD	361	CLOS LOJEN 2021 T	267
CLARIÓN DE VIÑAS DEL VERO 2018 B	629	CLOS CYPRES 2018 T	376	CLOS MESORAH 2019 T R	298
CLAROR PARAJE CALIFICADO CAN PRATS 2014 BE GR BN	154	CLOS D'AGON 2019 T	115	CLOS MOGADOR 2019 T C	364
CLAROS DE CUBA ORIGEN 2017 T	503	CLOS D'AGON 2020 B	115	CLOS MONLLEÓ 2011 T R	374
CLÀSSIC PENEDÈS PARELLADA 2018 BE R BN	327	CLOS D'AGON SELECCIÓN ESPECIAL 2019 T	115	CLOS MUSTARDÓ 2016 T R	361
CLÀSSIC SAUVIGNON BLANC MIQUEL JANÉ 2017 BE BN	327	CLOS D'AGON SYRAH MAS PALET 2019 T	182	CLOS MUSTARDÓ 2018 B FB	361
CLASSICA HACIENDA LÓPEZ DE HARO 2004 T GR	512	CLOS D'AGON VALMANA 2019 T C	182	CLOS PEPIN CASA AURORA 2020 T	862
CLASSICA HACIENDA LÓPEZ DE HARO 2009 RD GR	512	CLOS D'AGON VIOGNIER 2020 B FB	115	CLOS PONS ALGES 2016 T C	174
CLASSICA HACIENDA LÓPEZ DE HARO 2013 B GR	512	CLOS D'ENGLORA AV 14 2017 T	840	CLOS PONS PLA DEL TET 2018 T C	174
CLAVE DE TÁBULA 2018 T	472	CLOS D'ESGARRACORDES 2018 T C	772	CLOS PONS ROC DE FOC 2014 B	174
CLAVIS 2014 T R	245	CLOS DALIAN GARNACHA 2021 T	646	CLOS PONS ROC NU 2012 T R	174
CLAVIUS VERDEJO 2018 B	810	CLOS DALIAN GARNACHA BLANCA 2021 B	646	CLOS PONS SISQUELLA 2018 B C	174
CLEA 2019 T C	434	CLOS DE GALLUR 2019 T RB	711	CLOS VIDAL BLANC DE BLANCS 2021 B	328
CLEARLY ORGANIC 2021 B	781	CLOS DE LES DÒMINES 2018 T R	183	CLOS VIDAL CABERNET SAUVIGNON 2019 T RB	328
CLEARLY ORGANIC 2021 T	781	CLOS DE LES DÒMINES BLANC 2021 B FB	183	CLOS VIDAL MERLOT 2018 T C	328
CLEARLY ORGANIC SIN SULFITOS 2021 T	781	CLOS DE LÔM GARNACHA 2021 T	708	CLOS VIDAL ROSÉ CUVÉE 2021 RD	328
CLEMENCIA 2019 T	844	CLOS DE LÔM MALVASÍA 2021 B	708	CLOT D'ENCÍS MISTELA BLANCA 2021 B MISTELA D	649

WINE	PAGE	WINE	PAGE	WINE	PAGE
CLOT DEL ROURE 2021 B	344	CODORNÍU ZERO BE	880	COM GAT I GOS SELECCIO 2015 T C	840
CLOT DEL ROURE XARELLO BRISAT 2021 B	344	CODOS DE LAROUCO GODELLO 2020 B	693	COM TU 2019 T C	297
CLOT DELS EIXAMS 2019 B FB	334	COJÓN DE GATO 2021 B	629	COM_PASIÓN 2019 T	752
CLOT DELS OMS BLANC 2021 B	328	COLECCIÓN 68 COSTEIRA TREIXADURA 2021 B	428	COMA D'EN POU BÀRBARA FORÉS 2019 T C	643
CLOT DELS OMS NEGRE 2018 T R	328	COLECCIÓN 880 VENDIMIA SELECCIONADA 2020 T C	493	COMA FREDOSA 2014 T BA	184
CLOT DELS OMS ROSAT 2021 RD	328	COLECCIÓN COMENGE VERDEJO 2021 B	603	COMENGE BIBERIUS 2021 T RB	443
CLOTETA 2021 T	767	COLECCIÓN DE FAMILIA LA MATEO GARNACHA CEPAS VIEJAS 2018 T	522	COMENGE EL ORIGEN 2019 T	443
CLUA MILLENNIUM 2019 T	646			COMENGE VERDEJO GRAN VINO DE RUEDA 2020 B	603
CLUNIA ALBILLO 2021 B	803	COLECCIÓN DE FAMILIA LA MATEO PARCELAS SINGULARES 2017 T BA	522	COMOLOCO 2021 T	225
CLUNIA MALBEC 2020 T	803			COMPASS 2019 T RB	499
CLUNIA SYRAH 2019 T	803	COLECCIÓN DE FAMILIA LA MATEO TEMPRANILLO BLANCO 2019 B	522	CON UN PAR MONASTRELL PETIT VERDOT 2020 T BA	711
CLUNIA TEMPRANILLO 2019 T	803			CON UN PAR SAUVIGNON BLANC 2021 B	683
CM 2018 T C	510	COLECCIÓN DE FAMILIA LA MATEO VENDIMIA 2018 T BA	522	CON VIENTO FRESCO 2018 T	853
CM PRESTIGIO 2017 T	510	COLECCIÓN NAI E SEÑORA 2020 B	379	CON_SENTIDO 2019 B	752
CM VERDEJO 2018 B FB	597	COLECCIÓN VIVANCO 4 VARIETALES 2019 T D	554	CONCA DE TREMP 2019 T C	176
CM VIÑA TENENCIA 2021 B	421	COLECCIÓN VIVANCO PARCELAS DE GARNACHA 2019 T	554	CONCA DE TREMP BLANC 2020 B	176
COBIJA DEL POBRE 2021 B	63	COLECCIÓN VIVANCO PARCELAS DE MATURANA 2019 T C	554	CONCEJAL AIRÉN 2021 B	698
COBIZA 2020 B	418	COLET A PRIORI 2019 BE R BR	333	CONCEJAL TEMPRANILLO 2018 T C	698
COBRANA 2020 T	65	COLET ASSEMBLAGE 2016 RE EBR	333	CONCEJAL TEMPRANILLO 2020 T	698
COBRANZA TEMPRANILLO 2018 T BA	805	COLET GRAN CUVEÉ 2018 BE R EBR	333	CONCEJAL VERDEJO 2021 B	698
COBRANZA VENDIMIA SELECCIONADA 2015 T	805	COLET NAVAZOS (ETIQ.NARANJA) 2018 BE R EBR	333	CONCEJO 2017 T R	160
COCA I FITÓ CARINYENA 2015 T	297	COLET NAVAZOS (ETIQ.VERDE) 2017 BE R EBR	333	CONDADO DE EGUREN TEMPRANILLO 2020 T	779
COCA I FITÓ D'OR 2018 B	647	COLET TRADICIONAL 2018 BE R EBR	333	CONDADO DE FERMOSEL 2015 T C	60
COCA I FITÓ MARAGDA 2016 T	297	COLET VATUA! 2019 BE EBR	333	CONDADO DE FERMOSEL 2016 T C	60
COCA I FITÓ NATURA 2021 T	297	COLET VATUA! ROSÉ 2018 RE EBR	333	CONDADO DE HAZA 2019 T R	475
COCA I FITÓ NU 2020 T	297	COLISEO VORS BF AM	219	CONDADO DE HAZA 2020 T C	476
COCA I FITÓ ROSA 2020 RD	297	COLLAGE 2021 B	64	CONDADO DE ORIZA 2021 T RB	488
COCOLOCO 2021 B	855	COLLBAIX 3 NITS 2019 T BA	348	CONDADO DE SEQUEIRAS 2017 T C	412
CODI/02 2021	37	COLLBAIX SINGULAR DE ÁNFORA 2017 B BA	348	CONDADO DE SEQUEIRAS GODELLO 2021 B	412
CÓDICE 2020 T BA	790	COLLBAIX SINGULAR NEGRE 2016 T RB	348	CONDADO DE SEQUEIRAS MENCÍA 2020 T	412
CODORNÍU 150 ANIVERSARIO EDICIÓN LIMITADA BE BR	880	COLLEITA DE MARTIS ALBARIÑO 2021 B	390	CONDE ANSÚREZ 2021 RD	160
CODORNIU NON PLUS ULTRA 2018 BE R BR	135	COLMILLO DE LOBO T BA	785	CONDE DE ALDAMA "BOTA NO" BF AM	213

WINE	PAGE
CONDE DE ALDAMA 1/3 BF OL	213
CONDE DE ALDAMA 1/8 BF PC S	214
CONDE DE ALDAMA MAGNUM MANZANILLA PASADA BF MZ	214
CONDE DE ALTAVA 2016 T R	554
CONDE DE ALTAVA 2019 T C	554
CONDE DE ALTAVA TEMPRANILLO 2021 T	554
CONDE DE CARALT BE BN	132
CONDE DE HARO 2019 BE R BR	126
CONDE DE HARO ROSÉ 2016 RE BR	126
CONDE DE LA CORTE 2019 T C	498
CONDE DE LA SALCEDA 2017 T R	589
CONDE DE MONTERROSO 2013 T R	240
CONDE DE MONTERROSO 2018 T C	240
CONDE DE MONTERROSO 2020 T	240
CONDE DE SAN CRISTÓBAL 2019 T C	467
CONDE DE SAN CRISTÓBAL FLAMINGO ROSÉ 2021 RD	468
CONDE DE SAN CRISTÓBAL RESERVA ESPECIAL 2018 T R	467
CONDE DE SIRUELA 2018 T C	459
CONDE DE SIRUELA 2020 T RB	459
CONDE DE SIRUELA 2021 T	459
CONDE DE VALDEMAR 2018 T C	551
CONDE VALDEMAR EDICIÓN LIMITADA 2010 T	551
CONDE VALDEMAR FINCA ALTO CANTABRIA BLANCO VIÑEDO SINGULAR 2020 B FB	551
CONDE VALDEMAR FINCA ALTO CANTABRIA GRAN AÑADA VIÑEDO SINGULAR ESPUMOSO 2017 B	551
CONDES DE ALBAREI ALBARIÑO 2021 B	378
CONDES DE ALBAREI CARBALLO GALEGO 2017 B FB	378
CONDES DE ALBAREI EN RAMA 2015 B C	378
CONDES DE ALBAREI ENXEBRE 2021 B MC	378
CONDES DE FUENSALIDA 100 AÑOS 2020 T	275
CONDES DE FUENSALIDA 2019 T C	275
CONDES DE FUENSALIDA 2020 T	275
CONDES DE FUENSALIDA 2021 RD	275
CONDES DE FUENSALIDA 2021 RE D	275
CONDESA DE LEGANZA ROSADO S ELECCIÓN DE FAMILIA 2021 RD	782
CONDESA DE LEGANZA TEMPRANILLO 2019 T	782
CONDESA DE LEGANZA VERDEJO 2021 B	782
CONDESA EYLO VERDEJO 2021 B	611
CONSENTIDO MONASTRELL BARRICA 2019 T RB	733
CONTADOR 2020 T	511
CONTINO 2017 T GR	590
CONTINO 2018 T R	590
CONTINO 2020 B	590
CONTINO GARNACHA 2020 T	590
CONTINO MAZUELO 2019 T	590
CONTINO VIÑA DEL OLIVO 2019 T	590
CONTRAAPAREDE 2017 B	378
CONVENTO LAS CLARAS 2017 T C	444
CONVENTO LAS CLARAS VERDEJO SOBRE LÍAS 2020 B FB	603
CONVENTO OREJA 2019 T C	476
CONVENTO OREJA 2021 T RB	476
CONVENTO OREJA MEMORIA 2017 T R	476
CONVENTO OREJA SELECCIÓN DE FAMILIA 2018 T	476
CONVENTO SAN FRANCISCO 2018 T C	434
CONVENTO SAN FRANCISCO LA ZAPATERA 2018 T R	434
CONVENTO SAN FRANCISCO SELECCIÓN ESPECIAL 2018 T BA	434
COOP 1958 2021 B	173
COOP 1958 T	173
COORDENADES 1'19 2021 T	637
COORDENADES 41'17 2019 B	637
COPA REAL 2021 T	789
COQUINERO EN RAMA BF FI S	210
COR VALENT 2016 T BA	348
CORANYA 2015 T	374
CORAZ DE PUENTE DEL EA 2019 T	558
CORAZÓN DE LAGO 2020 T MC	532
CORAZÓN DE MALON 2021 RD	313
CORAZÓN INDOMABLE 2021 T MC	537
CORAZÓN LOCO BOBAL 2018 T	266
CORAZÓN LOCO PREMIUM 2016 T R	266
CORAZÓN LOCO SELECCIÓN 2016 T C	266
CORAZÓN NEGRO 2021 B SD	754
CORCOVO 2021 RD	783
CORCOVO AIRÉN 2021 B	696
CORCOVO AIREN 24 BARRICAS 2020 B FB	696
CORCOVO MUSCAT 2021 B S	783
CORCOVO SYRAH 24 BARRICAS 2020 T RB	696
CORCOVO TEMPRANILLO 2017 T R	696
CORCOVO TEMPRANILLO 2019 T C	696
CORCOVO TEMPRANILLO 2020 T RB	696
CORCOVO VERDEJO 2021 B	783
COREOGRAFÍA MONTSANT 2021 T	301
COREOGRAFÍA PRIORAT 2021 RD	374
CORET 2019 B	293
CORIAS GUILFA 2020 B FB	751
CORIMBO 2018 T	452
CORIMBO I 2016 T R	453
CORISCA 2021 B	393
CORNELIO DINASTÍA 2017 T C	519
CORNELIO IMPERIAL AUTOR JM 2018 T	520
CORNEO 2020 RD RB	58
CORNITERO 2021 T MC	55
CORONA 2015 B SD	563
CORONA DE ARAGÓN 2019 T C	111

WINE	PAGE	WINE	PAGE	WINE	PAGE
CORONA DE ARAGÓN SAUVIGNON BLANC 2021 B	111	COSECHERO B	841	CRÁPULA GOLD 2019 T	225
CORONÍN 2020 T	409	COSECHERO RD	841	CRÁPULA GOLD 2020 T	225
CORPUS DEL MUNI 2019 T RB	778	COSECHERO T	841	CRÁPULA SOUL 2019 T C	225
CORPUS DEL MUNI 2021 RD	778	COSME PALACIO 1894 2018 B	521	CRÁPULA SOUL EDICIÓN LIMITADA 2019 T	225
CORPUS DEL MUNI 60 BARRICAS 2020 T BA	778	COSME PALACIO 1894 2018 T	521	CRÁTER 2017 T C	631
CORPUS DEL MUNI BLANCA SELECCIÓN 2021 B	778	COSME PALACIO 2016 T R	521	CRÁTER EL JOVEN 2021 T	631
CORPUS DEL MUNI LUCÍA SELECCIÓN 2016 T C	778	COSME PALACIO 2018 B R	521	CRAZY GRAPES 2021 T	231
CORPUS DEL MUNI SARA SELECCIÓN 2021 B SS	778	COSME PALACIO VINO DE LAGUARDIA 2019 T C	521	CREAM TRADICIÓN VOS BF CRM	211
CORRAL DE CASTRO 2014 T R	203	COSMOLÓGICO CABERNET SAUVIGNON BIODINÁMICO 2021 T	791	CREGO E MONAGUILLO GODELLO 2021 B	281
CORRAL DE CASTRO 2017 T	203	COSO 2014 T R	585	CREGO E MONAGUILLO MENCÍA 2021 T	281
CORRAL DEL OBISPO 2020 T RB	79	COSTALARBOL 2020 T	531	CREGO E MONAGUILLO TREIXADURA ALBARIÑO 2020 B	281
CORSARI 2021 B	825	COSTALARBOL 2021 B	531	CREU DE LAVIT 2019 B	342
CORTE REAL 2018 B FB	818	COSTALARBOL GRACIANO 2019 T	531	CRISOL BE	599
CORTEO 2019 T	225	COSTER D'EN FORNÓS 2019 T	298	CRISTALÍ DULCE NATURAL BF D	44
CORTIJO DE JARA 6 MESES 2020 T RB	770	COSTER DE LA DEVEA 2018 B	841	CRISTIARI 2021 B	177
CORTIJO DE JARA GEWÜRZTRAMINER 2021 B	858	COSTER DELS ROSERS 2020 RD	636	CRISTIARI 2021 RD	177
CORTIJO LOS AGUILARES 2021 RD	263	COSTERS DE CORNUDELLA 2019 T R	296	CRISTIARI D'ALÒS MERLOT 2019 T BA	177
CORTIJO LOS AGUILARES PAGO EL ESPINO 2019 T BA	263	COSTERS DE L'ERMITA 2018 T	371	CROMÀTIC CHARDONNAY + XARELLO 2021 B	326
CORTIJO LOS AGUILARES PINOT NOIR 2020 T C	263	COSTERS DE POBLEDA 2013 T	371	CROMÀTIC MERLOT 2021 RD	326
CORTIJO LOS AGUILARES PINOT NOIR 2021 T C	263	COSTERS DEL GRAVET 2020 T C	293	CROMÀTIC MUSCAT 2021 B	326
CORUCHO 2020 T RB	722	COTO DE GOMARIZ 2020 B	424	CRUCEIRO 2021 T	409
CORUCHO 2021 RD	722	COTO DE GOMARIZ FINCA O FIGUEIRAL 2019 B	424	CRUCEIRO REXIO 2018 T	409
CORUCHO FINCA PEAZO DE LA ENCINA 2018 T RB	722	COTO DE HAYAS 2019 T C	102	CRUCEIRO REXIO 2021 B	409
CORUCHO GARNACHA 2021 T	722	COTO DE IMAZ 2017 T R	566	CRUOR 2018 T	357
CORVUL 20 MESES BARRICA 2018 T	449	COVILA 2016 T GR	521	CRUZ 2020 T C	200
CORZOÁS GODELLO 2021 B	412	COVILA 2017 T R	521	CRUZ CONDE 1902 BF PX D	286
CORZOÁS MENCÍA 2019 T BA	412	COVILA 2019 T C	521	CRUZ CONDE B FI S	286
CORZOÁS MENCÍA 2020 T FB	413	COVILA 2021 B	521	CRUZ CONDE BF PX D	286
CORZOÁS MENCÍA 2021 T	413	COVILA 2021 T	521	CRUZ DE ALBA 2019 T C	444
COSCOJARES 2018 T	569	COVILA AEX 2019 T	521	CRUZ DE ALBA FUENTELUN 2017 T R	444
COSECHERA ENSAMBLAJE I 2020 T	632	CR MONTELIOS 2015 T	64	CRUZ DE PIEDRA SELECCIÓN ESPECIAL GARNACHA 2020 T	94
COSECHERA ENSAMBLAJE II 2020 T	632	CRÁPULA BASADO EN HECHOS REALES 2019 T C	225	CRUZ DE SAN ANDRÉS 2020 T RB	64

WINE	PAGE
CRUZ DEL MAR BF AM S.	207
CRUZ DEL PENDÓN 2016 T	446
CRUZ DEL TEIDE AFRUTADO SEMIDULCE 2021 B SD	755
CRUZ DEL TEIDE SECO 2020 T	755
CRUZ SAGRA 2017 T	434
CT CORTIJO DE TRIFILLAS MALBEC 2019 T RB	795
CT EN CLAVE DE DO 2018 T	585
CT EN CLAVE DE DO 2019 T RB	232
CTÒNIA 2019 B FB	189
CUARTA GENERACIÓN SIERRAHONDA 2017 T	758
CUARTA GENERACIÓN VALDEHERREROS 2020 T	758
CUARTO LOTE 2019 T RB	722
CUARTO LOTE 2020 B	722
CUARTO LOTE 2021 RD	722
CUATRO DEL CUATRO 2020 T	322
CUATRO PASOS 2020 T	67
CUATRO PASOS BLACK 2019 T	67
CUATRO PASOS ROSÉ 2021 RD	67
CUATRO RAYAS = CO. ORGANIC VERDEJO 2021 B	597
CUATRO RAYAS 1935 VERDEJO 2021 B	598
CUATRO RAYAS BLUSH ROSÉ 2021 RD	802
CUATRO RAYAS CUARENTA VENDIMIAS CUVÉE 2020 B	598
CUATRO RAYAS CUARENTA VENDIMIAS RIBERA DEL DUERO 2020 T	435
CUATRO RAYAS CUARENTA VENDIMIAS RIOJA 2019 T C	511
CUATRO RAYAS CUARENTA VENDIMIAS VERDEJO 2021 B	598
CUATRO RAYAS LONGVERDEJO VIÑEDOS CENTENARIOS 2020 B	598
CUATRO RAYAS ORGANIC ROSÉ TEMPRANILLO-VERDEJO 2021 RD	802
CUATRO RAYAS ORGANIC TEMPRANILLO 2021 T	802
CUATRO RAYAS ORGANIC VERDEJO 2021 B	599
CUATRO RAYAS TEMPRANILLO 2020 T RB	435
CUATRO RAYAS VENDIMIA NOCTURNA VERDEJO 2021 B	599
CUATRO RAYAS VIÑEDOS CENTENARIOS 2021 B	599
CUBA 38 2021 T	799
CUCOS DE LA ALBERQUILLA 2020 T	229
CUCÚ (CANTABA LA RANA) 2021 B	801
CUENTAOVEJAS ZZZ 2018 T	669
CUENTAVIÑAS 2020 T	476
CUENTAVIÑAS ALOMADO 2020 T	562
CUENTAVIÑAS EL TIZNADO 2020 T	562
CUENTAVIÑAS GARNACHA CDVIN 2020 T	562
CUENTAVIÑAS LOS YELSONES 2020 T	562
CUESTA DEL HERRERO 2021 T BA	738
CUESTA ROA 940 2017 T C	493
CUESTA ROA 940 ETIQUETA NEGRA 2015 T C	493
CUEVA DE CHAMÁN ROBLE MONASTRELL 2020 T RB	50
CUEVA DE LOBOS ALPHA 2020 T	528
CUEVA DEL CHAMÁN GARNACHA TINTORERA 2021 T	50
CUEVA DEL CHAMÁN VERDEJO 2021 B	50
CUEVA DEL MONGE 2018 B FB	553
CUEVA DEL MONGE 2018 T	553
CUEVA LLANA BOBAL 2021 T	270
CUEVA LLANA MACABEO 2021 B	270
CUEVA LLANA ROSÉ 2021 RD	270
CUEVA LLANA SYRAH 2021 T	270
CUEVAS DE AROM ALTAS PARCELAS 2020 T	96
CUEVAS DE AROM AS LADIERAS 2020 T C	97
CUEVAS DE AROM OS CANTALS 2020 T	97
CUEVAS DE AROM TUCA NEGRA 2020 T	97
CUEVAS DE LECEA 2017 T BA	532
CULLEROT 2021 B	708
CULMEN 2017 T R	530
CUM LAUDE BE R BN	130
CUMAL 2018 T	809
CUMAS 2018 T	832
CUMPLIDO 2021 B	776
CUMPLIDO 2021 T	776
CUNE 2016 T GR	563
CUNE 2018 T R	563
CUNE 2019 B R	563
CUNE 2019 T C	563
CUNE 2021 B	563
CUNE 2021 RD	563
CUNE ORGÁNICO 2021 T	563
CUNE RIBERA DEL DUERO 2020 T RB	476
CUNE RIBERA DEL DUERO 2021 T RB	476
CUNE RUEDA 2021 B	615
CUNE SEMIDULCE B SD	563
CUÑAS DAVIA 2020 B FB	420
CUÑAS DAVIA 2020 T	420
CUÑAS DAVIA 2021 B	420
CUÑAS DAVIA A XIADA 2021 B	420
CUPIDO BOBAL 2018 T	269
CUPIDO MACABEO 2018 B BA	269
CUROLLA 2020 T	829
CUSCÓ BERGA 2013 BE GR BN	136
CUSCÓ BERGA 2019 BE R BR	136
CUSCÓ BERGA BE GR BR	136
CUSCÓ BERGA BE R BN	136
CUSCÓ BERGA ROSÉ RE R BR	136
CUSTERO 2020 T	585
CUVÉE #1 SON JULIANA 2019 B	831
CUVÉE #2 SON JULIANA 2017 T	831
CUVÉE #3 SON JULIANA 2019 RD	831
CUVÉE ANTONIO MASCARÓ 2014 BE GR BN	143

WINE	PAGE
CUVÉE JULIA ALTOS DEL TERRAL 2018 T R	432
CUVILA 2021 T	301
CUZO 2021 T	657
CV05 2020 T	705
CVP 2019 T RB	709
CYAN 2018 T C	655
CYAN PAGO DE LA CALERA 2007 T	655
CYAN PRESTIGIO 2016 T R	655
CYAN TINTA DE TORO 2020 T RB	655
CYPRESSUS 2019 T C	415

D

WINE	PAGE
D. PEDRO SOUTOMAIOR 2021 B	379
D'BERNA 2021 RD	688
D'BERNA GARNACHA TINTORERA "ELE" 2017 T	688
D'BERNA GODELLO 2021 B	688
D'BERNA MENCÍA 2020 T	688
D'BERNA SOUSON BARRICA "JUAN" 2017 T	688
DAINA 2021 RD	188
DALMAU 2019 T R	576
DALT TURÓ ACOPINYAT 2021 B	828
DALT TURÓ BRESCAT 2019 T	828
DALT TURÓ GRANAT 2019 T C	828
DALT TURÓ ROGET 2021 RD	829
DALT TURÓ SAULÓ 2019 T	828
DAMA DE TORO 2020 T RB	659
DAMANA 2019 T C	472
DAMANA 5 2020 T	472
DAME UN BESO, NEGRO 2019 T BA	678
DANIELA 2017 T C	682
DANZA ESCUMOSO DE ALBARIÑO BE BR	379
DARA 2019 T C	374
DARDELL GARNACHA BLANCA VIOGNIER 2021 B	643

WINE	PAGE
DAROU ROSÉ 2021 RD	803
DAUCO 2018 T	783
DAVILA 2021 B	380
DE ALBERTO DORADO VERDEJO 100% B SOLERA	603
DE ALBERTO DORADO VERDEJO DULCE BF D	603
DE ALBERTO ECOLÓGICO 100% VERDEJO 2021 B	604
DE ALBERTO EDICIÓN LIMITADA 2020 B	603
DE ALBERTO PÁLIDO B PI	603
DE ALBERTO SELECCIÓN 2018 T	803
DE ALBERTO SOBRE LÍAS VERDEJO 100% 2020 B	604
DE BRINGAS 2020 B	86
DE BRINGAS 2021 B	86
DE BUENA JERA 2018 T C	59
DE CAP A PEUS 2021 B	185
DE LA CRUZ DE 1767 BF PC S	206
DE LOS ABUELOS TEIRÓ GODELLO 2020 B	75
DE LOS ABUELOS VIÑAS CENTENARIAS 2021 RD	75
DE LOS ABUELOS VIÑEDO BARREIROS GODELLO 2020 B FB	75
DE LOS ABUELOS VIÑEDO BARREIROS MENCÍA 2020 T	75
DE LOS ABUELOS VIÑEDO SATURNO 2021 T	75
DE MULLER AVREO DULCE SOLERA 1954 BF RC D	859
DE MULLER AVREO SECO SOLERA 1954 BF AÑEJO S	859
DE MULLER CABERNET SAUVIGNON 2020 T C	637
DE MULLER CARINYENA 2018 T C	364
DE MULLER CHARDONNAY 2020 B FB	637
DE MULLER CHARDONNAY 2021 B FB	637
DE MULLER GARNACHA SOLERA 1926 BF SOLERA D	859
DE MULLER MISA DULCE SUPERIOR B D	859
DE MULLER MOSCATEL AÑEJO VINO DE LICOR BF	859
DE MULLER MUSCAT 2021 B	637
DE MULLER RANCIO SECO BF AÑEJO S	859
DE MULLER SYRAH 2020 T	637

WINE	PAGE
DE PALABRA 2019 T C	721
DE RODRIGO 2019 T	438
DE YANES BABOSO NEGRO 2020 T	755
DÉCIMA 2021 T	415
DECLARA SELECCIÓN ESPECIAL 2018 T	863
DECORUS 2020 T BA S	55
DEFENSOR 2021 B	825
DEHESA DE LOS CANÓNIGOS 2019 T BA	477
DEHESA DEL CARRIZAL CABERNET SAUVIGNON 2019 T	741
DEHESA DEL CARRIZAL CHARDONNAY 2020 B FB	741
DEHESA DEL CARRIZAL COLECCIÓN PRIVADA 2019 T	741
DEHESA DEL CARRIZAL MV 2019 T	742
DEHESA DEL CARRIZAL PETIT VERDOT 2019 T	742
DEHESA DEL CARRIZAL SYRAH 2019 T	741
DEHESA LA GRANJA 2019 T	809
DEL PRÍNCIPE BF AM S	219
DELAMPA 50 AÑOS 2018 T C	230
DELAMPA MONASTRELL 2021 T	230
DELAMPA SELECCIÓN 2021 T	230
DELER 2021 B	299
DELER 2021 RD	299
DELER VINYES SELECCIONADES 2021 T	299
DELGADO 1874 B AM S	286
DELGADO 1874 B OL S	286
DELGADO 1874 BF PX D	286
DELICIOSA EN RAMA SACA PRIMAVERA (375CC) BF MZ S	219
DELIRIO 2019 T BA	632
DELIRIO ROBLE DE MUÑANA 2018 T	766
DELIT 2019 T	42
DEMASIADO CORAZÓN 2019 B FB	63
DENT DE LLEÓ 2019 B	636
DEÓBRIGA 2019 BE R BN	581

WINE	PAGE
DEÓBRIGA 2021 B FB	581
DEÓBRIGA 2021 T MC	581
DEÓBRIGA COLECCIÓN PRIVADA 2017 T	581
DEÓBRIGA SELECCIÓN FAMILIAR 2018 T	581
DEPAULA PONCE 2021 T RB	268
DEPENDE 2021 B	391
DESAFÍO 2017 T C	468
DESAFÍO 2020 T RB	468
DESBORDANT 2019 T	347
DESCARTE 2017 T	665
DESCONCIERTO 2021 B	392
DESCREGUT 2019 BE BN	879
DESEU 2021 B	418
DESTINOS CRUZADOS AS REGADAS 2018 T C	393
DESTINOS CRUZADOS POUSADA 2019 B	393
DESTINOS CRUZADOS POUSADA 2019 B C	393
DESTRANKIS 2020 T BA	357
DESVELO GARNACHA 2019 T	571
DETILIO 2020 T	474
DETRÁS DE LA CASA MONASTRELL 2016	732
DIAPIRO 2020 B BA	38
DIAPIRO 2020 T RB	38
DIAZ BAYO 15 MESES 2020 T C	486
DIAZ BAYO 4U 2021 T	486
DIAZ BAYO 8 MESES 2021 T BA	487
DIDO 2020 T	301
DIDO BLANC 2020 B	301
DIDO LA SOLUCIÓ ROSA 2020 RD	301
DIEGO RIVERA 2019 T C	460
DIEGO RIVERA 2020 T RB	460
DIERESI 2020 T	828
DIEZ ALMENDROS 2020 T BA	492

WINE	PAGE
DIEZ SIGLOS 2020 B FB	615
DIEZ SIGLOS SAUVIGNON BLANC 2021 B	615
DIEZ SIGLOS VERDEJO 2021 B	615
DÍEZ-CABALLERO 2017 T R	564
DÍEZ-CABALLERO 2019 T C	564
DÍEZ-CABALLERO 2021 B	564
DÍEZ-CABALLERO VENDIMIA SELECCIONADA 2018 T R	564
DIGMA GRACIANO 2017 T R	583
DIGMA TEMPRANILLO 2016 T R	583
DIMOBE PAJARETE BF TRASAÑEJO D	261
DIMOBE SECO BF TRASAÑEJO S	261
DIODORO AUTOR 2010 T	455
DIONUS 2016 T R	873
DIOS BACO BF AM S	208
DIOS BACO BF CRM	208
DIOS BACO BF FI	208
DIOS BACO BF PX	208
DIOS BACO MEDIUM BF OL	208
DIOSA DE LAS 3 CABEZAS 2021 B	168
DIS-TINTO 2020 T	714
DÍSCOLO 2018 T BA S	664
DÍSCOLO EL MAGNÍFICO 2017 T BA S	664
DÍSCOLO MALVASÍA Y VERDEJO 2020 B FB S	664
DISFRUTAR 2019 T R	228
DITS DEL TERRA 2019 T C	374
DIVERTUS 2018 T	873
DIVINO MINIÑO 2021 B	600
DIVUS 2018 T RB	223
DOBIÑON 2020 T	76
DOBLE CUERPO 2019 T	869
DOBLE R 2019 T C	435
DOBLE R 2021 T	435

WINE	PAGE
DOBLE R 5 MESES 2020 T RB	435
DOBLE R 5 MESES 2021 T RB	435
DOBLE R VENDIMIA SELECCIONADA 2017 T	435
DOIX COSTERS DE VINYES VELLES 2019 T C	371
DOLÇ D'ART SELECCIÓN DE LICOR 2017 BF D	37
DOLÇ LES FRESES DE JESÚS POBRE 2021 B MO D	41
DOLÇA DE GANAGOT 2017 T RC D	636
DOLÇAINA BF MISTELA D	706
DOLCE BIANCO VERDEJO FRIZZANTE 2021 BE AG SD	802
DOLIA AMPHORAE CHARDONNAY 2021 B	682
DOLIA BOBAL AMPHORAE 2021 T	682
DOLMO TEMPRANILLO 2020 T RB	492
DOM BERENGUER VINO DE LICOR 1918 BF SOLERA	859
DOM JUNCOSA SOLERA 1939 BF SOLERA	859
DOMECO DE JARAUTA 2015 T	523
DOMECO DE JARAUTA 2020 B	523
DOMECO DE JARAUTA 2020 T	523
DOMENIO SUMOI NEGRE 2017 T	117
DOMENIO TREPAT 2019 T	164
DOMENIO ULL DE LLEBRE 2019 T	165
DÒMINE 2021 B	614
DOMINE RD S	159
DOMÍNGUEZ 2020 T	632
DOMÍNGUEZ ANTOLOGÍA 2017 T C	632
DOMÍNGUEZ COLECCIÓN EL MARAÑÓN 2018 T C	633
DOMÍNGUEZ CON FIRMA 2019 T	633
DOMÍNGUEZ EN BLANCO 2020 B	633
DOMÍNGUEZ MALVASÍA CLÁSICO 2012 B D	633
DOMINIO BASCONCILLOS FINCA DE ALTURA 2019 T RB	477
DOMINIO BASCONCILLOS VIÑA MAGNA 2018 T C	477
DOMINIO BASCONCILLOS VIÑA MAGNA 2018 T R	477
DOMINIO BASCONCILLOS VIÑA MAGNA 2019 T C	477

WINE	PAGE
DOMINIO CAMPO VIEJO 2016 T	560
DOMINIO DE ATAUTA 2019 T C	445
DOMINIO DE ATAUTA ALBILLO MAYOR CEPAS PREFILOXÉRICA 2020 B	445
DOMINIO DE ATAUTA LA MALA 2017 T C	445
DOMINIO DE ATAUTA LA ROZA 2017 T	445
DOMINIO DE ATAUTA LLANOS DEL ALMENDRO 2017 T	446
DOMINIO DE ATAUTA SAN JUAN 2017 T	446
DOMINIO DE ATAUTA VALDEGATILES 2017 T	445
DOMINIO DE BACO 2020 T RB	244
DOMINIO DE BACO AIRÉN 2021 B	244
DOMINIO DE BACO CABERNET 2021 T	244
DOMINIO DE BACO SYRAH 2021 T	244
DOMINIO DE BACO TEMPRANILLO 2021 T	244
DOMINIO DE BACO VERDEJO 2021 B	244
DOMINIO DE BERZAL 2019 T C	564
DOMINIO DE BERZAL 2021 B	564
DOMINIO DE BERZAL 2021 T MC	564
DOMINIO DE BERZAL 7 VARIETALES 2019 T	564
DOMINIO DE BERZAL SELECCIÓN PRIVADA 2019 T	564
DOMINIO DE BORNOS 2017 T C	477
DOMINIO DE BORNOS 2019 T RB	477
DOMINIO DE CALOGÍA BY JOSÉ MANUEL PÉREZ OVEJAS 2020 T	478
DOMINIO DE CALOGÍA BY JOSÉ MANUEL PÉREZ OVEJAS CUVÉE S 2019 T	478
DOMINIO DE ES CARRAVILLA 2019 T	478
DOMINIO DE ES CARRAVILLA 2020 T	478
DOMINIO DE ES LA MATA 2020 T	478
DOMINIO DE ES VIÑAS VIEJAS DE SORIA 2020 T	478
DOMINIO DE FONTANA SAUVIGNON BLANC & VERDEJO 2021 B	672
DOMINIO DE FONTANA TEMPRANILLO & CABERNET SAUVIGNON 2019 T C	672
DOMINIO DE FONTANA TEMPRANILLO & SYRAH 2020 T RB	672
DOMINIO DE GARDEL 2017 T C	239
DOMINIO DE GARDEL 2020 T RB	240
DOMINIO DE LA GRANADILLA VERDEJO 2021 B	600
DOMINIO DE LA SIERRA 2021 B	758
DOMINIO DE LA SIERRA DOMINIVM 2019 T	758
DOMINIO DE LA SIERRA MOMENTVM 2020 T	758
DOMINIO DE LA VEGA AUTHENTIQUE 2019 BE R BN	137
DOMINIO DE LA VEGA CUVÉE PRESTIGE 2017 BE R BN	137
DOMINIO DE LA VEGA EXPRESSION 2019 BE R BR	137
DOMINIO DE LA VEGA NO.1 2020 BE BR	137
DOMINIO DE LA VEGA RESERVA ESPECIAL 2018 BE R BR	137
DOMINIO DE LA VEGA RESERVA ESPECIAL ROSÉ 2018 RE R BR	137
DOMINIO DE NOBLEZA 2018 T C	564
DOMINIO DE NOBLEZA EDICIÓN LIMITADA 2015 T R	564
DOMINIO DE NOBLEZA MATURANA 2020 T	564
DOMINIO DE NOBLEZA VENDIMIA SELECCIONADA 2015 T R	564
DOMINIO DE TARES CEPAS VIEJAS 2018 T C	73
DOMINIO DE TARES GODELLO 2021 B FB	73
DOMINIO DE TAURUM 2021 T RB	667
DOMINIO DE UNX CHARDONNAY 2021 B	304
DOMINIO DE UNX GARNACHA BLANCA SUR LIE 2021 B	304
DOMINIO DE UNX GARNACHA OLD VINES 2020 T	304
DOMINIO DE UNX GARNACHA ROSADO DE LÁGRIMA 2021 RD	304
DOMINIO DEL AGUILA 2018 T R	478
DOMINIO DEL BENDITO EL PRIMER PASO 2020 T RB	664
DOMINIO DEL BENDITO LAS SABIAS 2019 T	664
DOMINIO DEL PIDIO 2019 T	479
DOMINIO DEL PIDIO 2020 RD	479
DOMINIO DEL PIDIO ALBILLO 2020 B	479
DOMINIO DEL PRIOR PETIT VERDOT 2016 T BA	791
DOMINIO DEL PRIOR SYRAH 2016 T BA	791
DOMINIO DEL SOTO 2018 T C	479
DOMINIO DEL SOTO 2018 T R	479
DOMINIO DEL SOTO 2020 B	479
DOMINIO DO BIBEI 2019 T	413
DOMINIO FOURNIER 2019 T R	479
DOMINIO FOURNIER 2020 T C	479
DOMINIO LA GRANADILLA SAUVIGNON 2021 B	600
DOMINIO LASIERPE 1920 CENTENARIO 2020 T	322
DOMINIO LASIERPE 2019 T C	322
DOMINIO LUBIANO 2019 T RB	492
DOMINIO MAESTRAZGO 2018 T C	767
DOMINIO MAESTRAZGO GARNACHA BLANCA 2020 B RB	767
DOMINIO MAESTRAZGO GARNACHA BLANCA 2021 B	767
DOMINIO MAESTRAZGO SANTOLEA 2020 T	767
DOMINIO MAESTRAZGO SYRAH 2018 T BA	767
DOMUS DE ROANDI 2016 T C	687
DON AURELIO 2014 T GR	696
DON AURELIO 2015 T R	696
DON AURELIO 2016 T C	696
DON AURELIO 2021 RD	697
DON AURELIO GARNACHA 2021 T	697
DON AURELIO RESERVA DE FAMILIA 2017 T	697
DON AURELIO SYRAH 2021 T	697
DON AURELIO TEMPRANILLO 2021 T	697
DON AURELIO VERDEJO 2021 B	697
DON BAFFO 2021 T	230
DON BALBINO 2020 B FB	565
DON BALBINO VENDIMIA SELECCIONADA 2019 T	565
DON BERNARDINO 4ªGENERACIÓN 2018 T BA	413
DON BERNARDINO AMANDI 2020 T	413
DON BERNARDINO AMANDI 2021 T	413
DON BERNARDINO FINCA MEZQUITA 2016 T C	413

WINE	PAGE
DON BERNARDINO IBIO 2017 T FB	413
DON BERNARDINO IBIO 2019 T FB	413
DON BERNARDINO MELANIO 2017 T C	413
DON BERNARDINO VACAMULO 2017 T C	413
DON CARLOS 2018 T RB	233
DON CELESTINO 2021 T	757
DON CELESTINO RUFETE ENVEJECIDO 2020 T	757
DON COSME GODELLO 2020 B	414
DON COSME MENCÍA 2021 T	414
DON FRUTOS TEMPRANILLO 2021 T	810
DON FRUTOS VERDEJO 2021 B	810
DON GUIDO SOLERA ESPECIAL 20 AÑOS VOS BF PX D	212
DON JACOBO 2010 T GR	520
DON JACOBO TEMPRANILLO BLANCO 2021 B	520
DON JUAN PEDRO XIMÉNEZ TRASAÑEJO 30 AÑOS BF PX D	261
DON LUCIANO 2018 T C	245
DON LUCIANO 2021 RD	245
DON MIGUEL COMENGE 2017 T R	443
DON P.X. 1999 BF PX D	289
DON P.X. 2019 BF PX	289
DON PABLO 2018 T	332
DON PEDUZ 2021 T	571
DON QUIEN 2020 B FB	171
DON QUIEN 2020 T BA	171
DON QUINTIN ORTEGA 2020 B FB	558
DON SALVADOR MOSCATEL TRASAÑEJO 30 AÑOS BF MO D	261
DON SUERO 2018 T C	257
DON SUERO 2020 T RB	257
DON XOÁN 2020 T	410
DON ZOILO 12 AÑOS BF OL S	212
DON ZOILO BF AM S	212
DON ZOILO BF FI S	212
DON ZOILO BF PC	212
DON ZOILO PX BF PX D	212
DONCEL DE MATAPERRAS 2015 T	484
DONCELES CRUZ CONDE B FI	286
DONIENE 2021 B	86
DONIENE APARDUNE 2020 BE BN	86
DONIENE XX 2019 B BA	86
DOÑA BEATRIZ VERDEJO 2021 B	602
DOÑA BEATRIZ VERDEJO CEPAS VIEJAS 2021 B	602
DOÑA LEO ALTOLANDÓN 2021 B	266
DORIVM 2019 T C	461
DORIVM 2020 T RB	461
DORIVM SELECCIÓN DE LA FAMILIA 2016 T R	461
DOS ALAS ROJAS 2019 T C	485
DOS CORTADOS 20 AÑOS BF PC S	212
DOS DE MAYO 2018 T C	728
DOS MARIAS 2020 T RB	830
DRY SACK MEDIUM 15 AÑOS BF MED	212
DUAL 2019 T	354
DUASTEMPAS 2019 T	837
DUC DE FOIX BE BN	136
DUC DE FOIX BE R BN	136
DUC DE FOIX BE R BR	136
DUC DE FOIX CABERNET SAUVIGNON 2021 RD	333
DUC DE FOIX CABERNET SAUVIGNON NEGRE 2020 T C	333
DUC DE FOIX MERLOT 2019 T	333
DUC DE FOIX RESERVA ESPECIAL BE R BR	136
DUC DE FOIX XARELLO 2021 B	333
DULAS DEL LAGAR DE LA SALUD "UN AMERICANO EN LAGAR DE LA SALUD" 2020 T C	864
DULAS DEL LAGAR DE LA SALUD "UN FRANCÉS EN LAGAR DE LA SALUD" 2020 T	864
DULAS DEL LAGAR DE LA SALUD 2020 B FB	287
DULCE BENAVIDES T D	861
DUO VITES 2019 T	771
DUQUE DE ARCAS 2018 T C	681
DUQUE DE ARCAS 2018 T RB	681
DUQUE DE ARCAS SOLO BOBAL 2017 T R	681
DURÓN CUEVA DEL RAPOSO 2018 T	490
DURÓN ÓPTIMO 2018 T R	490
DX DE DOMINIO LOS PINOS 2020 T RB	704

E

WINE	PAGE
E-GALA 2021 B	86
ÉBANO 6 2020 T RB	479
ÉBANO SALVAJE 2018 T C	480
EBURUS 2018 T C	498
ECCE VINUM TEMPRANILLO ORGANIC (8 MONTHS IN OAK BARREL) 2015 T C	784
ECCE VINUM TEMPRANILLO ORGANIC 2021 T	784
ECCE VINUM VERDEJO ORGANIC 2021 B	784
ECLÉCTICO - PUESTA EN CRUZ LÍAS B	59
ECOS 2018 T C	259
EDEDIA 2020 B BA	693
EDERRA 2016 T R	518
EDERRA 2018 T R	518
EDERRA SELECCIÓN ESPECIAL 2020 B	518
EDERRA SELECCIÓN TEMPRANILLO 2018 T C	518
EDETÀRIA SELECCIÓ 2019 B C	647
EDETÀRIA SELECCIÓ VI DE FINCA EL MAS 2018 T C	647
EDRA BLANCOLUZ 2020 B	860
EDUARDO PEÑA 2021 B	425
EDULIS 2019 T C	505
EGIARTE ROSADO 2021 RD	313
EGO BODEGAS SAUVIGNON BLANC 2021 B	860

WINE	PAGE
EGUREN CABERNET 2019 T	779
EGUREN SAUVIGNON BLANC 2021 B	779
EGUREN SHIRAZ 2020 T	779
EGUREN UGARTE 2015 T R	565
EGUREN UGARTE 2019 T C	565
EGUREN VERDEJO 2021 B	779
EIDAN 2021 B SD	200
EIDAN 2021 T	200
EIDO DA FONTE ALBARIÑO 2021 B	393
EIDO DA FONTE SOUSÓN 2017 T	393
EIDO DA SALGOSA 2020 B	392
EIDO DA SALGOSA ROSAL 2020 B	393
EIDOS DE PADRIÑÁN 2019 B BA	378
EIDOS DE PADRIÑÁN 2021 B	378
EIDOSELA 2018 T	386
EIDOSELA 2021 B	386
EIDOSELA BURBUJAS DEL ATLÁNTICO BE BN	387
EIDOSELA BURBUJAS DEL ATLÁNTICO BE EBR	387
EIRAL ALBARIÑO 2021 B	390
EJE MONASTRELL 2020 T	39
EL ABUELO PEDRO 2019 T	515
EL ABUELO SELECCIÓN 2017 T	48
EL AERONAUTA 2020 B	690
EL ALMA DE GILDO 2020 T	669
EL AMANTE 2019 T C	757
EL ANDÉN 2019 T	539
EL APRENDIZ 2019 T RB	811
EL APRENDIZ 2021 B	811
EL APRENDIZ 2021 RD	811
EL ÁRBOL DE ARANLEÓN 2019 T C	701
EL ARQUITECTO 2018 T BA	264
EL ARREBOL DE LA CARRA CABRA 2020 RD	322

WINE	PAGE
EL ARTE DE VIVIR 2021 T	470
EL BADIU 2020 B	33
EL BELISARIO 2019 T	549
EL BERRAKIN 2021 T	859
EL BESO SAUVIGNON BLANC 2019 B FB	779
EL BOBAL ESTENAS 2021 T	675
EL BOSC NEGRE 2019 B	325
EL BUEN ALFARERO 2017 T	471
EL BUFÓN VERDEJO 2021 B	600
EL BUSCADOR 2018 T C	568
EL CAIRE MONASTRELL 2019 T	36
EL CAMINO MENDI 2020 T	578
EL CANCHORRAL 2020 T	814
EL CANDADO BF PX D	219
EL CANTO DEL ANGEL 2016 T RB	435
EL CAPRICHO DE SOFÍA 2021 B	565
EL CAPRICHO DE SOFÍA 2021 RD	565
EL CARRIEGO 2021 RD	257
EL CARRO GROS 2016 T	331
EL CASTRO DE VALTUILLE 2021 T	72
EL CASTRO DE VALTUILLE GODELLO 2020 B BA	72
EL CAVA DE CHOZAS CARRASCAL 2018 BE R BN	135
EL CELLER DE LA IBOLA 2020 B	860
EL CF DE CHOZAS CARRASCAL 2017 T	740
EL CHAPARRAL DE VEGA SINDOA 2020 T	319
EL CHAVAL 2021 RD	705
EL CHAVAL 2021 T	705
EL CHAVAL MACABEO 2021 B	705
EL CICLÓN SERRANO 2019 T	814
EL CONJURO 2017 T	512
EL CONVENTO DE LA VEGA 2018 T	199
EL CORAZÓN DE LA TIERRA 2016 T RB	435

WINE	PAGE
EL CORDERO Y LAS VÍRGENES 2018 T R	709
EL CORRAL CREMAT 2011 BE GR BR	325
EL COSTER DE L'ALZINA 2017 T C	357
EL COTO 2018 T C	566
EL COTO ECOLÓGICO 2019 T C	566
EL COTO SELECCIÓN VIÑEDOS 2018 T C	566
EL CRISTO DE SAMANIEGO 2018 T	515
EL CUBETO RESERVA FAMILIA OTERO B	850
EL CUCO DE VALDELALUNA 2018 T RB	655
EL CUENTISTA 2019 B	853
EL DUENDE 2018 T	318
EL EMPECINADO 2017 T C	474
EL ENHEBRO 2020 T	700
EL ENHEBRO 2021 B	700
EL ERIAL DE VALDECAÑADA 2018 B	76
EL FISGÓN 2021 B	602
EL FOLLET ROSAT 2021 RD	164
EL FUNDAMENTALISTA 2020 T	268
EL FUNDAMENTALISTA 2021 T	269
EL GODELLO DE JUAN MIGUEZ 2021 B	423
EL GORU 2021 RD	860
EL GRAN MONROY ANTHEOS 2019 T	723
EL GRAN MONROY THEOS 2018 T	723
EL GRIFO ARIANA 2019 T	252
EL GRIFO MALVASÍA LÍAS 2019 B	252
EL GRIFO MALVASÍA SECO COLECCIÓN 2021 B	252
EL GRIFO MOSCATEL DE ANA B D	252
EL GRINGO DARK & RICH MONASTRELL 2020 T	734
EL GRINGO DARK RED TEMPRANILLO 2021 T	793
EL GUARÁ 2019 B	857
EL GUARDIÁN 2016 T R	562
EL GUARDIÁN 2019 T C	562

WINE	PAGE
EL GUARDIÁN SIN SULFITOS 2021 T	562
EL GUÍA DE FINCA DE LA RICA 2021 T	568
EL HELECHAL RUFETE BLANCA 2020 B	814
EL HIJO DE LA DOLORES 2021 T	94
EL HOMBRE ORQUESTA 666 G 2019 T	512
EL HOMBRE ORQUESTA 666 M 2019 T	512
EL HOMBRE ORQUESTA 666 T 2019 T	512
EL HOMBRE ORQUESTA 666 V 2020 B	512
EL IMPORTANTE 2017 T	584
EL INQUILINO 2019 T C	566
EL JARDÍN DE ABRIL MALVASÍA AROMÁTICA 2020 B	756
EL JARDÍN DE LA EMPERATRIZ 2019 T	590
EL JARDÍN DE LA EMPERATRIZ 2021 B	591
EL JARDÍN DE LA ERA 2020 T	775
EL JARDÍN DE LAS IGUALES 2019 B	847
EL JARDÍN DE LAS IGUALES 2020 T	834
EL JARDÍN DE LUCÍA 2021 B	402
EL JEFE DE LA TRIBU 2019 T BA	701
EL LABERINTO DE VIÑA ANE 2019 T C	511
EL LAGAR DE ISILLA ALBILLO MAYOR 2020 B	446
EL LAGAR DE ISILLA COLECCIÓN DE LA FAMILIA 2016 T R	446
EL LAGAR DE ISILLA LANGA DE DUERO 2018 T RB	446
EL LAGAR DE ISILLA MATANZA DE SORIA 2019 T RB	446
EL LAGAR DE ISILLA PARAJE PEÑALOBOS 2018 T RB	446
EL LAGAR DE ISILLA SAN JUAN DEL MONTE 2019 T	446
EL LEBRERO 2020 B	447
EL LERO 2018 T BA	264
EL LINZE 2020 T	790
EL LOCO DE FINCA LA COLINA 2021 B	620
EL LOMO LISTÁN BLANCO 2021 B	754
EL LOMO LISTÁN NEGRO 2021 T	754
EL MAGO 2020 B FB	648
EL MAGO 2020 T	648
EL MAGO 2021 RD	648
EL MAGO CHALUPA 2021 B	74
EL MAGO CHALUPA 2021 T	74
EL MANIFIESTO 2018 B R	473
EL MÉDICO 2015 T RB	811
EL MICROSCOPI 2019 T	837
EL MIRACLE ART 2020 T	44
EL MIRACLE BE BR	152
EL MIRACLE Nº1 2020 T	711
EL MIRACLE Nº3 2021 B	711
EL MIRACLE Nº5 2021 RD	711
EL MIRACLE ORGANIC BE BR	152
EL MIRACLE ROSÉ RE BR	152
EL MIRADOR DE LA SALCEDA 2021 B	589
EL MIRADOR DE LA SALCEDA 2021 T MC	589
EL MÚSICO 2014 T	811
EL NIDO 2019 T	225
EL NÓMADA SELECCIÓN DE PARCELAS 2019 T	568
EL NOTAS PREMIUM 2020 T	757
EL NOTERA BY MONTE PINADILLO 2019 T	437
EL OLIVASTRO 2020 T	674
EL ORGULLO DE JULIAN 2019 T	515
EL ORIGEN DE PRIETO PARIENTE 2018 T C	806
EL PACTO DE LA SONSIERRA 2019 T	512
EL PARAGUAS ATLÁNTICO 2020 B	423
EL PARVON 2020 T	854
EL PASO DEL LAZO TEMPRANILLO SHIRAZ 2021 T	793
EL PASO OLD VINES 2020 T	231
EL PASO TEMPRANILLO SHIRAZ 2021 T	245
EL PATITO FEO GODELLO 2021 B BA	426
EL PATITO FEO TREIXADURA SOBRE LÍAS 2021 B	426
EL PEDAL DE HERMANOS HERNÁIZ TEMPRANILLO 2021 T	591
EL PELIGRO 2018 T	831
EL PERDÍO 2020 T RB	683
EL PERRO VERDE 2021 B	619
EL PETIT DE SANUI BLANC 2021 B	172
EL PETIT DE SANUI NEGRE 2019 T	172
EL PIANO 2019 T	318
EL PLANTONAL 2020 T	233
EL PLANTONAL 2021 T	233
EL POLVORETE 2021 B	67
EL PREDILECTO 2021 T	555
EL PRESUMIT DEL PALLARS 2020 T	176
EL PUNTIDO 2019 T	590
EL QUINTÀ BÀRBARA FORÉS 2020 B FB	643
EL QUINTO PARAJE VERDEJO 2021 T	810
EL RAPOLAO 2020 T	76
EL RAPOLAO VINO DE PARAJE 2020 T	72
EL RAYO 2019 T C	542
EL REBOSO 2021 B	717
EL REBOSO 2021 T	717
EL REBOSO AFRUTADO 2021 B	717
EL REBUSCO SELECCIÓN SYRAH 2020 T	714
EL REGAJAL SELECCIÓN ESPECIAL 2020 T	723
EL REI DE MAMBO 2020 T	331
EL RENEGADO 2021 B	705
EL RESCATADO 2021 RD	811
EL RETABLO III T	471
EL REVENTÓN 2019 T	809
EL RIBAZO 2017 T	592
EL RINCÓN DE LOS ENEBROS 2020 T BA	568
EL RINCÓN DE NEKEAS 2021 B	319
EL ROBLEDO DIVINO 2019 T	758

WINE	PAGE
EL ROBLEDO RUFETE 2019 T C	758
EL ROBLEDO TEMPRANILLO RUFETE 2019 T	758
EL ROCALLÍS 2018 B FB	329
EL ROSAL 2019 T	871
EL ROSÉ DE ARESAN 2021 RD	789
EL SALEGAR 2018 T C	463
EL SALEGAR 2020 T RB	463
EL SALZE 2019 T	44
EL SECRETO DE LZ 2020 B	868
EL SECRETO DE MARÍA 2018 T C	435
EL SECRETO DEL ABUELO 2018 T C	92
EL SECRETO DEL ABUELO 2020 T RB	92
EL SENAT DE MONTSANT 2020 T	291
EL SEQUÉ 2020 T	40
EL SEQUÉ DULCE 2017 TF C D	40
EL SITIO FORASTERA 2021 B	755
EL SITIO MALVASÍA 2020 B	755
EL SITIO ORCHILLA 2021 B	755
EL SITIO VIJARIEGO NEGRO 2020 T	755
EL SORADAL 2016 T BA	854
EL SUECO 2018 T C	465
EL SUECO ALBILLO MAYOR 2020 B	465
EL SUEÑO 2020 T	679
EL SUEÑO DE AMADO 2017 T C	555
EL TEMPLARI BÀRBARA FORÉS 2020 T C	643
EL TESÓN 2015 T	854
EL TESÓN 2016 T	854
EL TESORO DEL CAPITÁN 2020 T RB	705
EL TITÁN DEL BENDITO 2019 T	664
EL TOMILLO Y EL VIENTO BAILAN VIOGNIER 2020 B	775
EL TORRONTÉS DE JUAN MIGUEZ 2020 B	423
EL TOSSAL DE LA SALUT 2019 T	163

WINE	PAGE
EL TRANSISTOR 2021 B	615
EL ÚLTIMO LOBO 2020 T RB	790
EL VALÍN MENCÍA 2020 T	77
EL VEÏNAT 2021 T	302
EL VELADO 2019 T	561
EL VÍNCULO 2018 T C	239
EL VÍNCULO ALEJAIREN 2019 B C	239
EL VÍNCULO PARAJE LA GOLOSA 2016 T GR	239
EL VIÑEDO DE LA VIDA TEMPRANILLO-CABERNET SAUVIGNON 2021 T	796
EL VIÑEDO DE LA VIDA VERDEJO-SAUVIGNON BLANC 2021 B	796
EL VISIONARI 2021 RD	186
EL VOL DE L'ÀLIGA 2020 B	376
EL YESAL 2019 T	865
ELDOZE 2015 T	785
ELDOZE 2017 T	785
ELIANE CHARDONNAY 2021 B	330
ELIAS I TERNS 2009 BE R BN	146
ELÍAS MORA 2014 T R	665
ELÍAS MORA 2019 T C	665
ELIAS MORA CONTRACORRIENTE 2021 B	615
ELITIA CARIÑENAS VIEJAS 2018 T	181
ELITIA GARNATXA D'EMPORDÀ B SOLERA D	181
ELIXIR 2019 B	771
ELLE DE LANDALUCE 2019 T	530
ELS IGOLS "VINYES DE COSTER" 2012 T GR	361
ELS PÁJAROS 2020 T	293
ELS PICS 2020 B	369
ELS PICS 2020 T	369
ELVIRA 2019 B	293
ELVIRA B	867
ELVIRA MOSCATEL GRANO MENUDO 2020 B	876

WINE	PAGE
ELVIRA VIGIRIEGA 2020 B	876
ELYSSIA GRAN CUVÉE 2020 BE R BR	138
ELYSSIA PINOT NOIR ROSÉ 2020 RE BR	138
EMBOCADERO 2018 T C	434
EMBOCADERO 2020 T RB	434
EMBRUIX DE VALL-LLACH 2020 T	361
EMERGENTE 2019 T C	308
EMERGENTE CHARDONNAY 2021 B	308
EMERGENTE MOSCATEL DE GRANO MENUDO 2021 B SD	308
EMERGENTE ROSADO DE LÁGRIMA 2021 RD	308
EMERGENTE VINO DE AUTOR GARNACHA 2016 T R	308
EMERGENTE VINO DE AUTOR GARNACHA 2019 T R	308
EMILIO CLEMENTE 2020 T	676
EMILIO MORO 2019 T	447
EMILIO MORO VENDIMIA SELECCIONADA 2019 T	447
EMILIO ROJO 2019 B	429
EMILIO VALERIO 2019 T	316
EMINA 2019 T C	435
EMINA EMOCIÓN 2015 T R	435
EMINA PASIÓN 2021 T	435
EMINA ROSÉ 2021 RD	157
EMINA ROSÉ PRESTIGIO 2021 RD	157
EMINA SAUVIGNON 2021 B	599
EMINA VERDEJO 2018 B FB	599
EMINA VERDEJO 2021 B	599
EMPELTS 2019 T	302
EMPIT 2019 T C	358
EMPIT SELECCIÓ 2018 T R	358
EMPÓRION 2018 T	188
EN CONTACTO 2021 B	230
EN LA PARRA 2021 B	705
EN TUS MANOS 2021 RD	721

WINE	PAGE
ENATE CABERNET SAUVIGNON 2014 T R	624
ENATE CHARDONNAY 2020 B FB	624
ENATE MERLOT-MERLOT 2017 T	625
ENATE UNO 2010 T	625
ENATE UNO CHARDONNAY 2013 B GR	625
ENATE VARIETALES 2019 T R	625
ENCANTERI 2020 T C	186
ENCANTO GODELLO 2021 B	813
ENCANTO MENCÍA 2021 T RB	813
ENCANTO SELECCIÓN 2021 T	813
ENCINA DEL INGLÉS 2021 T	260
ENCLAVE 2017 T	775
ENCONTRADO BF OL S	219
ENFUDRECIDO 2019 T	562
ENGAZO FAMILIA TAFURIASTE 2019 T	717
ENGAZO FAMILIA TAFURIASTE 2020 T RB	717
ENGLORA 603.8 2018 T C	841
ENOTECA GRAMONA 2005 BE BN	880
ENOTECA GRAMONA 2006 BE BN	880
ENOTECA PERSONAL MANUEL RAVENTÓS NEGRA MAGNUM 2013 BE BN	883
ENRIQUE MENDOZA CHARDONNAY 2021 B	35
ENRIQUE MENDOZA CHARDONNAY 2021 B FB	35
ENRIQUE MENDOZA LAS QUEBRADAS 2020 T C	36
ENRIQUE MENDOZA MOSCATEL DE LA MARINA DULCE 2021 B D	36
ENRIQUE MENDOZA SANTA ROSA 2019 T FB	36
ENTERIZO 2017 T R	675
ENTERIZO BOBAL 2021 RD	675
ENTRE DOS AGUAS 2019 T BA	768
ENTRE DOS AGUAS 2020 T	768
ENTRE DOS AGUAS GARNACHA BLANCA 2020 B	768
ENTRE DOS AGUAS GARNACHA BLANCA 2020 B BA	768

WINE	PAGE
ENTRE PALABRAS 2019 T	491
ENTRE SILENCIOS 2020 B FB	721
ENTREBOSC NEGRE 2019 T	348
ENTRELIMITES LA BALANZA BRUÑAL 2015 T	60
ENTRELIMITES LIMITE NATURAL 2016 T	60
ENTRELIMITES LIMITE NATURAL 2019 T	60
ENTRELIMITES TRANSITIUM DURII 2009 T GR	60
ENTRELOBOS 2020 T	495
ENTREMONTES 2017 T C	239
ENTREMONTES AIRÉN 2021 B	239
ENTREMONTES SELECCIÓN 2021 B	239
ENTREMONTES TEMPRANILLO 2021 T	239
ENTRESUELOS 2019 T	807
ENTRESUELOS 2020 T	807
ENVIDIACOCHINA 2020 B	382
ENZA 2019 B RB	822
EORA CHARDONNAY 2021 B	683
EPÍLOGO CHARDONNAY 2019 B BA	243
EPÍLOGO SAUVIGNON BLANC 2021 B	243
EPISTEM 2018 T	839
EPISTEM Nº 2 2018 T R	732
EPISTEM Nº 5 2018 T	839
EPITAFIO 2019 T RB	667
EPULUM 2018 T R	529
EPULUM 2019 T C	529
EQUILIBRIO 2021 T	233
EQUILIBRIO SAUVIGNON BLANC 2021 B	233
EQUILIBRIO-4 2020 T	233
EQUILIBRIO-9 2020 T BA	233
ERA 2019 T C	294
EREMUS 2017 T C	488
EREMUS 2020 T RB	488

WINE	PAGE
ERESMA + CUVÉE ESPECIAL 2018 B	604
ERESMA + CUVÉE ESPECIAL 2019 B	604
ERESMA FERMENTADO BARRICA 2020 B FB	604
ERESMA SAUVIGNON BLANC VENDIMIA SELECCIONADA 2021 B	604
ERESMA TENTAZION BE AG	804
ERESMA VERDEJO SOBRE LÍAS 2021 B	604
ERESMA VERDEJO VENDIMIA SELECCIONADA 2021 B	604
ERIAL 2020 T	447
ERIAL TF (TRADICIÓN FAMILIAR) 2019 T	447
ERIKA DE PAUMERA 2021 RD	636
ERMITA DEL MONTE 2020 T C	834
ERMITA VERACRUZ VERDEJO 2021 B	612
ERNIO SOBRE LÍAS 2020 B	196
ERRE DE HERRERO 2021 B	616
ERUELO 2016 T C	55
ES MONESTIR 2018 T R	820
ESCABECES 2020 RD BA	638
ESCABECES BLANQUES PRIMES 2020 B C	638
ESCABECES CARTOIXÀ BLANC 2020 B C	638
ESCADA GARNACHA TINTUREIRA 2019 T	685
ESCALA I CORDA 2020 BF MISTELA D	703
ESCONDITE DEL ARDACHO: VERIQUETE 2019 T	585
ESCORÇA 2020 B	172
ESCUDERO DE BOCOS 2016 T C	653
ESCUDERO DE BOCOS 2017 T R	653
ESCUDERO DE BOCOS 2020 T RB	653
ESCUDERO DE BOCOS VERDEJO 2019 B FB	653
ESCULLE DE SOLABAL 2019 T C	513
ESCUMOLL 2017 BE GR BN	117
ESENCIA DE LA TORRE 2017 T BA	874

WINE	PAGE	WINE	PAGE	WINE	PAGE
ESENCIA DE LA TORRE CHARDONNAY 2021 B	874	ESTENAS BOBAL 2021 RD	675	EXTREM DE BONAVAL BE R BR	127
ESENCIA DE LA TORRE PETIT VERDOT 2018 T BA	874	ESTENAS MADURADO EN BARRICA 2020 T	675	EXTREM DE BONAVAL ROSADO RE R BR	127
ESENCIA DE LAS MERCEDES 2018 T	681	ESTERO BLANCO 2020 B SD	770	EXTREMARIUM DE MONT MARÇAL 2019 BE R BN	144
ESENCIA DIVIÑA 2021 B	379	ESTHER 2018 T C	723	EXTREMARIUM DE MONT MARÇAL 2019 BE R BR	144
ESENCIA RUPESTRE GARNACHA TINTORERA 2016 T C	50	ESTOLA 2014 T GR	237	EXUN 2019 T	728
ESENZIAS BY FUENTENARRO 2019 T RB	448	ESTOLA 2018 T C	237	EYA 2021 B	224
ESPECTACLE 2019 T C	300	ESTOLA 2020 RD	237	EYA 2021 RD	224
ESPELT COMABRUNA 2017 T	188	ESTOLA 2020 T RB	238	EYA 2021 T	224
ESPELT QUINZE ROURES 2021 B FB	188	ESTOLA SELECCIÓN 2018 T	238	F	
ESPELT SAULÓ 2021 T	188	ESTOLA VERDEJO 2021 B	238		
ESPELT TERRES NEGRES 2019 T	188	ESTONES GS 2017 T	298	F DE FUENTESPINA 2019 T R	448
ESPETO BOBAL 2020 T	706	ETAPA 24 2020 B SS	66	FÁBREGAS GARNACHA BLANCA 2020 B FB	627
ESPINAPURA BF FI ES	169	ETERN 2018 T BA	373	FÁBREGAS MONTETA 2019 T	627
ESPLUGEN 2019 T	357	ETERN 2020 T BA	373	FÁBREGAS MORISTEL 2020 T	627
ESPORRERES 2019 T	296	ETERNAL 2017 T	439	FÁBREGAS PURO SYRAH 2019 T C	627
ESPUMOSO EDICIÓN ESPECIAL 2017 BE R BN	818	ETERNAUTA 2020 T C	485	FABULA DE PANIZA SYRAH 2021 T	110
ESSÈNCIA DE LLUNA 1925 2020 T C	650	ÈTIM DOLÇA CARINYENA 2020 TF D	298	FÁBULO SAUVIGNON BLANC AIRÉN VERDEJO VOLCÁNICO 2021 B	799
ESSÈNCIA DE LLUNA BLANC CUPATGE 2021 B	650	ÈTIM EL VIATGE 2021 T	298		
ESSÈNCIA DE LLUNA GARNACHA BLANCA 2021 B	651	ÈTIM L'ANTULL 2021 B	298	FÁBULO TEMPRANILLO CABERNET SAUVIGNON - SYRAH VOLCÁNICO 2020 T	799
ESSÈNCIA DE LLUNA ROSAT 2021 RD	651	ÈTIM LA PAUSA 2021 RD	298		
ESTAY 2018 T RB	809	ÈTIM SINESTÈSICS TARDANA NEGRA 2018 TF D	298	FÁBULO TEMPRANILLO GRACIANO VOLCÁNICO 2020 T	800
ESTEBAN MARTÍN 2021 T	109	EUKENI TXAKOLI 2021 B	52	FÁCIL 2020 B	751
ESTECILLO MACABEO 2021 B	99	EULÀLIA DE PONS CUVÉE 2018 BE R BR	132	FAGUS DE COTO DE HAYAS 2019 T BA	102
ESTEL D'ARGENT 2018 BE R BN	137	EULOGIO POMARES MACERACIÓN CON PIELES 2020 B	394	FAGUS DE COTO DE HAYAS 2020 T BA	102
ESTEL D'ARGENT 2019 T	333	EUREKA BLANC 2020 B	344	FAI UN SOL DE CARALLO 2018 B	422
ESTEL D'ARGENT 2021 RD	333	EVENTO 2018 T C	704	FALA DE MIN BRANCELLAO 2019 T	419
ESTEL D'ARGENT ESPECIAL 2017 BE R EBR	137	EVOL 2021 T BA	222	FALA DE MIN GODELLO 2021 B	419
ESTEL D'ARGENT ESPECIAL 2018 BE GR EBR	137	EXEDRA 2021 B	115	FALARES 2020 T	855
ESTEL D'ARGENT ROSÉ 2019 RE R BN	137	EXEDRA 2021 T	115	FALCOEIRA 2019 T	689
ESTELA DE ARRAYÁN 2018 T R	273	ÉXODO 2019 T RB	230	FALCOEIRA BRANCO 2019 B	689
ESTENAS 2018 T C	675	ÉXODO AUTOR 2018 T C	230	FALGUERA 2012 T C	184
ESTENAS 2021 B	675	EXPRESION RESERVA BOBAL 2018 T R	677	FAMILIA COMENGE RESERVA 2018 T	444
				FAMILIA PACHECO MONASTRELL ORGÁNICO 2020 T	229

WINE	PAGE	WINE	PAGE	WINE	PAGE
FAMILIA PACHECO SELECCIÓN PARA PACO 2019 T C	229	FECO 2020 B MC	525	FILOXERA 2020 T BA	87
FAN D.ORO 2020 B FB S	845	FECO 2020 T MC	525	FILS DE VI 2020 T BA	364
FARAÓN 30 AÑOS VORS 50 CL. BF OL S	209	FECO PIEDRA MIGUEL 2021 B	525	FINA 1270 A VUIT 2019 T BA	360
FARAÓN BF OL S	209	FEITIZO DA NOITE 2018 BE EBR	390	FINCA A CORONELA BY CEPADO 2019 T	686
FARDELAS DE VIÑAREDO 2020 B	692	FEITIZO DA NOITE BE BR	390	FINCA ANTIGUA 2015 T R	245
FARIÑA 2018 T C	659	FÉLIX CALLEJO AUTOR 2019 T	447	FINCA ANTIGUA PETIT VERDOT 2019 T	245
FARIÑA CAMPUS GOTHORUM 2019 T	659	FÉLIX MARTÍNEZ CEPAS VIEJAS 2018 T R	728	FINCA ANTIGUA SYRAH 2019 T C	245
FARIÑA LÁGRIMA 2020 T RB	659	FÉLIX SALAS 2018 T C	157	FINCA ANTIGUA ÚNICO 2016 T C	245
FARNADAS 2021 B	427	FENOMENAL 2021 B	619	FINCA ANTIGUA VIURA 2021 B	245
FATHER 1943 2019 T BA	281	FERNÁNDEZ DE ARCAYA SELECCIÓN PRIVADA 2017 T R	311	FINCA ARANTEI 2020 B	388
FAUSTINO 2016 T R	524	FERNÁNDEZ GÓMEZ 2021 T	549	FINCA BACARA SAUVIGNON BLANC 2021 B	862
FAUSTINO 2019 T C	524	FERNANDO DE CASTILLA "AMONTILLADO ANTIQUE" BF AM S	215	FINCA BARRONTE 2018 T C	572
FAUSTINO ART COLLECTION 2019 T C	524	FERNANDO DE CASTILLA "FINO ANTIQUE" BF FI S	215	FINCA BARRONTE GRACIANO 2019 T	572
FAUSTINO ART COLLECTION 2020 T	524	FERNANDO DE CASTILLA "PALO CORTADO ANTIQUE" BF PC S	215	FINCA BINIAGUAL MANTONEGRO 2020 T FB	81
FAUSTINO I 2011 T GR	524	FERNANDO DE CASTILLA FINO CLASSIC BF FI S	215	FINCA BINIAGUAL VERÁN 2019 T BA	81
FAUSTINO I MAGNUM 2011 T GR	524	FERNANDO DE CASTILLA OLOROSO SINGULAR BF OL S	215	FINCA BUTARÓS 2016 T	188
FAUSTINO ICON ESPECIAL SELECCIÓN 2017 T R	524	FERNANDO DE CASTILLA PX CLASSIC BF PX D	216	FINCA CALVESTRA MARGAS 2016 B	843
FAUSTINO RIVERO ULECIA 2017 T R	567	FERRATUS 2021 B	445	FINCA CALVESTRA MERSEGUERA 2020 B	843
FAUSTINO RIVERO ULECIA 2019 T C	567	FERRATUS 2021 RD	445	FINCA CAÑADA HONDA 2019 T BA	676
FAUSTINO RIVERO ULECIA AFRUTADO 2021 B	779	FERRATUS AØ 2020 T RB	445	FINCA CAÑADA HONDA 87 T C	676
FAUSTINO RIVERO ULECIA ALDARIÑO 2021 B	394	FERRATUS SENSACIONES 2017 T	445	FINCA CAPELIÑOS 2020 T	407
FAUSTINO RIVERO ULECIA BOBAL 2021 RD	779	FERRÉ I CATASÚS DOS 2019 T C	861	FINCA CASA JULIA 2021 B	703
FAUSTINO RIVERO ULECIA BOBAL TEMPRANILLO 2017 T R	680	FERRERET MANTONEGRO CALLET 2021 RD	81	FINCA COLLADO MESSEGUERA 2019 B	42
FAUSTINO RIVERO ULECIA BOBAL TEMPRANILLO 2019 T C	680	FERRUM 2020 B	400	FINCA COMABARRA 2019 T C	176
FAUSTINO RIVERO ULECIA CHARDONNAY 2021 B	780	FET A MÀ 2019 T	42	FINCA CONSTANCIA ENTRE LUNAS 2020 T BA	791
FAUSTINO RIVERO ULECIA MACABEO 2021 B SD	780	FIGUERO TINUS 2018 T	495	FINCA CONSTANCIA GRACIANO PARCELA 12 2018 T	791
FAUSTINO RIVERO ULECIA TEMPRANILLO 2021 T	780	FIGUEROA BLANCO SOBRE LÍAS FINAS 2021 B	723	FINCA CONSTANCIA SELECCIÓN 2019 T BA	791
FAUSTINO RIVERO ULECIA VERDEJO 2021 B	780	FIGUEROA ORIGINEM 2018 T C	723	FINCA CONSTANCIA TEMPRANILLO PARCELA 23 2020 T	791
FAUSTINO RIVERO ULECIA VIURA 2021 B	567	FILIGRANA 2020 T	118	FINCA CONSTANCIA VERDEJO PARCELA 52 2020 B FB	791
FAUSTINO V 2017 T R	524	FILIGRANA 2021 B	118	FINCA CUARTA A COSTA POR RUBÉN MOURE 2019 T	414
FAUSTINO VII 2021 T	525	FILLABOA 2021 B	382	FINCA CUARTA CONSENTIDA POR RUBÉN MOURE 2019 T	414
FEBRER 2018 T C	825	FILLABOA SELECCIÓN FINCA MONTE ALTO 2019 B	382	FINCA CUARTA MALCRIADO POR RUBÉN MOURE 2018 T C	414

WINE	PAGE
FINCA CUARTA MENCÍA POR RUBÉN MOURE 2019 T BA	414
FINCA CUARTA MENCÍA POR RUBÉN MOURE 2021 T	414
FINCA DEL MAR VERDEJO 2021 B	873
FINCA DOFÍ 2020 T C	354
FINCA EL BOSQUE 2019 T	592
FINCA EL CARRIL HECHICERO 2016 T C	266
FINCA EL CARRIL PAOLO ANDREA 2016 T	266
FINCA EL CARRIL VALERIA 2019 B	266
FINCA EL MOLAR 2020 T RB	268
FINCA EL MOLAR BOBAL 2020 T	268
FINCA EL MOLAR MORAVIA AGRIA 2020 RD	268
FINCA EL PUIG 2019 T	366
FINCA EL RINCÓN DE CLUNIA 2019 T	803
FINCA ÉLEZ 2020 T	743
FINCA ÉLEZ CHARDONNAY LÍAS 2020 B	743
FINCA ELS GORGS 2013 BE GR	326
FINCA GARABATO CEPAS VELLAS 2020 B	380
FINCA GARABELOS 2020 B	396
FINCA GENOVEVA 2019 T	387
FINCA IRIARTE 2019 T RB	577
FINCA JAKUE TXAKOLINA 2021 B	196
FINCA LA ATALAYA VALTRAVIESO 2018 T R	473
FINCA LA BEATA 2016 T GR	680
FINCA LA CANTERA GARNACHA 2020 T	317
FINCA LA CANTERA TEMPRANILLO 2021 T	317
FINCA LA COLINA SAUVIGNON BLANC 2021 B	620
FINCA LA COLINA VERDEJO CIEN X CIEN 2021 B	620
FINCA LA EMPERATRIZ 2018 B	591
FINCA LA EMPERATRIZ 2018 T R	591
FINCA LA ESTACADA 12 MESES 2018 T C	672
FINCA LA ESTACADA 2021 B SD	672
FINCA LA ESTACADA 6 MESES 2018 T RB	672

WINE	PAGE
FINCA LA ESTACADA CHARDONNAY SAUVIGNON BLANC 2021 B	672
FINCA LA ESTACADA VARIETALES 2017 T R	672
FINCA LA LUNA 2019 T R	448
FINCA LA MARÍA 2019 T RB	484
FINCA LA MATEA GARNACHA 2017 T C	109
FINCA LA ORACIÓN 2020 T	571
FINCA LA PEDRISSA 2018 T	647
FINCA LA RANA 2020 T	666
FINCA LA SABINA CABERNET 2016 T GR	743
FINCA LA SABINA MERLOT 2016 T	743
FINCA LA SABINA SYRAH 2017 T C	743
FINCA LA SABINA TEMPRANILLO 2018 T	779
FINCA LA SERRA DEL VENT 2021 B	176
FINCA LA TERRENAL 2018 B	647
FINCA LAS CARABALLAS SECTOR 2.8 2020 B	811
FINCA LAS CARABALLAS VERDEJO 2019 B	811
FINCA LAS CARABALLAS VERDEJO 2021 B	811
FINCA LASIERPE CHARDONNAY 2021 B	322
FINCA LES LLERES 2016 B BA	172
FINCA LES ROQUES 2020 B RB	184
FINCA LOS ALIJARES GARNACHA PETIT VERDOT 2016 T	792
FINCA LOS ALIJARES GRACIANO 2015 T R	792
FINCA LOS ALIJARES GRACIANO AUTOR 2009 T GR	792
FINCA LOS ALIJARES MOSCATEL 2019 B SD	792
FINCA LOS ALIJARES SYRAH GRACIANO 2018 T C	792
FINCA LOS ALTOS 2016 T GR	696
FINCA LOS ALTOS 2018 T R	696
FINCA LOS ALTOS GRAN SELECCIÓN 2021 T	696
FINCA LOS AZARES TEMPRANILLO CABERNET MERLOT 2017 T	238
FINCA LOS FRUTALES 2019 RD RB	260
FINCA LOS FRUTALES CABERNET SAUVIGNON 2019 T C	260

WINE	PAGE
FINCA LOS FRUTALES GARNACHA 2017 T C	260
FINCA LOS FRUTALES IGUALADO 2017 T	260
FINCA LOS FRUTALES MERLOT SYRAH 2019 T C	260
FINCA LOS HALCONES BOBAL 2019 T	271
FINCA LOS HALCONES VIOGNIER 2021 B FB	271
FINCA LOS HOYALES 2016 T	444
FINCA LUNA BEBERIDE 2020 T RB	70
FINCA LUZÓN 2019 T C	226
FINCA LUZÓN 2021 T RB	227
FINCA LUZÓN MONASTRELL SYRAH 2021 T	227
FINCA MANZANOS 2015 T R	534
FINCA MANZANOS GARNACHA 2021 T	534
FINCA MARTELO 2016 T R	586
FINCA MATAPAJA 2019 T	573
FINCA MEIXEMAN 2020 T	408
FINCA MILLARA 2018 T C	414
FINCA MILLARA LAGARIZA 2021 T	415
FINCA MILLARA SELECCIÓN ESPECIAL 2018 T	415
FINCA MONCLOA TINTILLA DE ROTA 2019 TF D	770
FINCA MONCLOA TINTILLA DE ROTA EDICIÓN LIMITADA 2018 T BA S	770
FINCA MONCLOA TRADICIONAL 2018 T BA	770
FINCA MONTEPEDROSO VERDEJO 2021 B	616
FINCA MONTEVIEJO 2017 T	592
FINCA MONTICO 2021 B	606
FINCA MORIANA 2018 T C	237
FINCA MORILLAS 2018 T RB	822
FINCA MUÑOZ COLECCIÓN DE LA FAMILIA 2019 T	797
FINCA MUÑOZ COLECCIÓN DE LA FAMILIA 2021 B BA	797
FINCA NUEVA 2010 T GR	568
FINCA NUEVA 2016 T R	569
FINCA NUEVA 2018 T C	569

WINE	PAGE
FINCA NUEVA 2020 B FB	569
FINCA NUEVA 2021 RD	569
FINCA OLIVARDOTS GROC D'ANFORA 2020 B	192
FINCA POMBEIRAS 2020 T	408
FINCA RACONS 2017 B	177
FINCA RATÓN 2017 T R	570
FINCA RENARDES 2019 BE R BN	146
FINCA RENARDES 2019 T RB	340
FINCA RESALSO 2020 T	447
FINCA RESALSO 2021 T	447
FINCA RÍO NEGRO 2018 T C	792
FINCA RÍO NEGRO 5º AÑO 2017 T GR	792
FINCA RÍO NEGRO CERRO DEL LOBO 2019 T	792
FINCA RÍO NEGRO GEWÜRZTRAMINER 2021 B	792
FINCA ROMAILA 2018 T C	776
FINCA SALTAMONTES 2017 B	616
FINCA SAN MARTÍN 2019 T C	586
FINCA SANDOVAL 2020 T	269
FINCA SANGUIJUELA 2014 T C	259
FINCA SOBREÑO 2020 T C	661
FINCA SOBREÑO 2021 RD	661
FINCA SOBREÑO ECOLÓGICO 2020 T	661
FINCA SOBREÑO ILDEFONSO 2018 T R	661
FINCA SOBREÑO SELECCIÓN ESPECIAL 2019 T R	661
FINCA TEIRA 2020 B	418
FINCA TEIRA 2020 T	418
FINCA TEIRA 2021 T	418
FINCA TERRERAZO 2019 T	742
FINCA TORREA 2018 T R	522
FINCA TRASDAIRELAS 2020 B	690
FINCA TRES OLMOS CLASSIC 2021 B	605
FINCA TRES OLMOS LÍAS 2021 B	605
FINCA VALDELAGUA 2018 T	787
FINCA VALDEMOYA 12 MESES 2019 T C	803
FINCA VALDEMOYA 2021 RD	803
FINCA VALIÑAS "CRIANZA SOBRE LÍAS" 2018 B	388
FINCA VALLDOSERA 2013 BE GR BN	138
FINCA VALLDOSERA 2014 BE R BN	138
FINCA VALLDOSERA MONASTRELL 2018 RE R BN	138
FINCA VALLEJO 2019 T C	433
FINCA VALLEJO 2020 B	596
FINCA VALLEJO 2020 T RB	433
FINCA VALLEJO 2021 B	596
FINCA VALLEOSCURO 2021 B	756
FINCA VALLEOSCURO PRIETO PICUDO 2021 RD	756
FINCA VALLEOSCURO PRIETO PICUDO TEMPRANILLO 2020 T	756
FINCA VALLEOSCURO PRIETO PICUDO TEMPRANILLO 2021 RD	756
FINCA VALLOBERA 2019 T	551
FINCA VALPIEDRA 2015 T R	569
FINCA VALPIEDRA 2017 B R	569
FINCA VIDALES 2019 T C	522
FINCA VIDALES 2021 B	522
FINCA VILADELLOPS XXX XARELLO 2021 B FB	335
FINCA VILLACRECES 2019 T	400
FINCA VILLACRECES SPECIMEN Nº 2 T	480
FINCA VILLALOBILLOS 2020 B FB	781
FINCA VILLALOBILLOS 2021 B	781
FINCA VIÑOA EMBOTELLADO TARDÍO 2019 B	425
FINCA VIÑOA PARAJE PENABOA 2017 B	426
FINCA VIÑOA TREIXADURA SOBRE LÍAS 2021 B	426
FINCA ZURIENA 2018 B	525
FINCAS DE AZABACHE GARNACHA 2019 T C	570
FINCAS DE AZABACHE TEMPRANILLO BLANCO 2021 B	570
FINCAS DE LANDALUCE 2018 T R	530
FINCAS DE LANDALUCE 2019 T C	530
FINCAS DE LANDALUCE GRACIANO 2019 T	530
FINCAS DE VALDEMACUCO 2019 T C	463
FINCAS DE VALDEMACUCO 2021 T RB	463
FINO C.B. BF FI S	285
FINO CORREDERA B FI S	287
FINO GRANERO EN RAMA BF FI	218
FINO LOS AMIGOS B FI S	288
FINO TRADICIÓN BF FI S	212
FINQUES INCANSABLES GARNATXA NEGRE 2019 T BA	186
FINQUES INCANSABLES MONASTRELL 2020 T BA	186
FISGÓN ROSÉ 2021 RD	803
FIVE MILES 2017 T	839
FLAMENCO DE SANI ÁRABE 2021 B AG S	851
FLAVIUM PREMIUM 2021 T	813
FLAVIUM SELECCIÓN 2021 T	814
FLOR D'ALBERA 2019 B FB	185
FLOR D'EMPORDÀ 2018 T	192
FLOR DE AIRÉN B AM	796
FLOR DE ALBIHAR 2021 B	753
FLOR DE ALLOZO TEMPRANILLO GARNACHA 2019 T	239
FLOR DE CAYUS 2019 T BA	101
FLOR DE ENYA 2020 T RB	44
FLOR DE LASIERPE GARNACHA 2021 RD PL	322
FLOR DE LASIERPE TINTO SELECCIÓN GARNACHA 2021 T	322
FLOR DE MALVAJÍO 2020 T	842
FLOR DE MORCA 2021 T	104
FLOR DE MUGA 2019 B R	539
FLOR DE MUGA ROSÉ 2021 RD	539
FLOR DE NIT 2021 B	646
FLOR DE NIT 2021 RD	646
FLOR DE PINGUS 2020 T	478

Peñín Guide | SPANISH WINE

WINE	PAGE
FLOR DE SILOS 2019 T	475
FLOR DE VETUS 2019 T	663
FLOR DE VETUS VERDEJO 2021 B	612
FLOR DE XARELLO D'ESPIELLS 2021 B	341
FLOR DEL AMANECER 2020 T RB	722
FLOR MALVÉS 2020 B	42
FLOR MEDITERRÁNEA 2020 T	295
FLOR TRUFES BLANC 2019 B FB	649
FLOR TRUFES NEGRE 2019 T	649
FLORES DE CALLEJO 2021 T	447
FLYTRAP 2018 T D	844
FOLLACO 2020 T R	58
FONDILLÓN ED. LIMITADA 1959 T FO	37
FONDILLÓN LUIS XIV 25 AÑOS T FO	41
FONDILLÓN LUIS XIV 50 AÑOS (TONEL LUNA) T FO	42
FONDONET SELECCIÓN 5 AÑOS 2010 T BA D	44
FONT CALENTA 2021 B	641
FONT CALENTA NEGRE 2021 T	641
FONT DE LA FIGUERA 2020 T FB	363
FONT DE LA FIGUERA 2021 B	363
FORAVILA 2020 B	825
FORCADA 2019 B	334
FORCALLA DE ANTONIA 2020 T	709
FORCALLAT BY BODEGA LA ENCINA 2021 T	842
FORCATE 2019 T	676
FORMIGA DE SEDA 2021 B	363
FORMIGA DE VELLUT 2019 T	363
FORMIGO 2020 B	418
FORMIGO 2021 B	418
FORTIUS 2015 T GR	316
FORTIUS 2021 RD	316
FORTIUS CHARDONNAY 2021 B	316

WINE	PAGE
FORTIUS CHARDONNAY 2021 B FB	316
FORTIUS CRIANZA 2019 T C	316
FORTIUS RESERVA 2018 T R	316
FORTIUS TEMPRANILLO 2020 T RB	316
FOS BARANDA 2018 T	526
FOSSI BF AM	211
FRAGA DO CORVO GODELLO 2021 B	281
FRAGA DO CORVO MENCÍA 2021 T	281
FRAGAS DO LECER 2021 B	281
FRANCISCO BARONA 2020 T C	481
FRANCISCO BARONA FINCA LAS DUEÑAS 2018 T R	481
FRANSOLA 2021 B	334
FREIXENET CARTA ROSÉ 2020 RE ES	138
FREIXENET CORDÓN NEGRO 2020 BE BR	139
FREIXENET SELECCIÓN ESPECIAL 2021 B	117
FREIXENET SELECCIÓN ESPECIAL 2021 RD	117
FRESQUITO VINO DE TINAJA 2020 B	288
FROIÑA DE CASA DO SOL 2021 B	392
FRONTAURA & VICTORIA ROSÉ LIMITED EDITION 2021 RD	660
FRONTAURA 2017 T R	660
FRONTAURA 2018 T C	660
FRONTAURA VERDEJO 2019 B	605
FRONTONIO LA CERQUETA 2020 T	847
FRONTONIO LA LOMA Y LOS SANTOS 2020 B	847
FRONTONIO LA TEJERA 2020 T FB	847
FRONTONIO TELESCÓPICO 2020 T	847
FRORE DE CARME 2017 B	382
FRORE DE CARME MILLÉSIME 2018 BE BN	382
FRUIT D'AUTOR 2017 RF	38
FRUTO NOBLE JOVEN 2021 T	36
FRUTO NOBLE ROBLE 2021 T RB	36
FRUTO NOBLE ROSADO 2021 RD	36

WINE	PAGE
FRUTO NOBLE SAUVIGNON BLANC 2021 B	36
FUCHS DE VIDAL 2017 BE GR BN	124
FUCHS DE VIDAL BE R BN	124
FUCHS DE VIDAL ROSÉ PINOT NOIR RE R EBR	124
FUCHS DE VIDAL UNIC BE R BN	124
FUEGO LENTO 2017 T S	35
FUEGO LENTO MONASTRELL EXTREMO 2019 T BA	35
FUENCONCEJO 2019 T C	454
FUENCONCEJO 2020 T RB	454
FUENCONCEJO 2021 T	454
FUENTE CORTIJO 2019 T C	498
FUENTE DE LOS HUERTOS 2019 T	725
FUENTE GRULLA 2020 T	814
FUENTE REINA 2018 T	833
FUENTENEBRO TINTO FINO 2020 T	459
FUENTES DEL SILENCIO LAS JARAS 2019 T	810
FUENTES DEL SILENCIO MATAPEREZOSA 2019 B FB	810
FUENTES DEL SILENCIO PRIETO PICUDO VIEJO 2019 T	810
FUENTESECA 2021 B	683
FUENTESECA 2021 RD	683
FUENTESPINA 3 2021 T RB	448
FUENTESPINA C 2020 T C	448
FUENTESPINA R 2019 T R	448
FUERZA BY EGO 2020 T	230
FUIDIO 2021 T	526
FUIDIO IRALEY 2018 T C	526
FUMISSOLA 2018 BE BN	879
FUNDUS ROBLE 2020 T	833
FUNDUS TINAJA 2021 T	834
FURO GARNACHA 2019 T	105
FUROT DEL MONESTIR 2017 T C	190
FURVUS 2020 T BA	302

WINE	PAGE
FUSIÓ 2018 T	371
FUSTA NOVA BLANC 2021 B	711

G

WINE	PAGE
G2 GARNACHA CENTENARIA 2020 T BA	725
G22 DE GORKA IZAGIRRE 2020 B	87
G22 DE GORKA IZAGIRRE MAGNUM 2019 B	87
GABA DO XIL GODELLO 2021 B	689
GABRIEL MARTÍNEZ. PEQUEÑOS PASOS, GRANDES ILUSIONES T	225
GAGO 2019 T	663
GAINTUS VERTICAL 2017 T C	339
GAINTZA 2021 B	196
GALANTERÍA 2019 BE C BR	885
GALEAM 2018 T C	37
GALEAM DRY MUSCAT 2021 B	37
GALEAM MONASTRELL 2021 T	37
GALIA LE DEAN 2019 T	802
GALIA VILLAGES 2019 T	802
GALL NEGRE 2017 T	335
GALLINAS & FOCAS 2018 T	822
GALVÁN 2019 B	281
GALVÁN 2020 B	281
GAMAZO 2018 T C	659
GAMBERRILLO MISTELA BLANCA 2017 BF MISTELA	648
GAMBERRILLO MISTELA NEGRA 2017 TF MISTELA D	648
GAMBERRO BLANC 2017 B C	649
GAMBERRO NEGRE DE GUARDA 2016 T C	649
GAMELLÓN 2021 T	227
GAMELLÓN VIÑAS VIEJAS 2019 T BA	227
GAMINDE 2016 T	568
GANADERO 2021 B	274
GANADERO 2021 T	274
GANAGOT 2007 T GR	636

WINE	PAGE
GANAGOT 2009 T GR	636
GANCEDO 2020 T RB	69
GARABATO 2021 T RB	441
GARABITAS VIÑAS VIEJAS 2019 T	657
GARBINADA 2021 T	374
GARBUIX ORIGEN 2021 RD	173
GARBUIX VEREMA VERMELLA 2021 RD	173
GARCIANO DE AZUL Y GARANZA 2020 T BA	305
GARENA 2021 B	86
GARGALO GODELLO 2021 B	281
GARKALDE TXAKOLINA 2021 B	86
GARMÓN 2019 T	481
GARNACHA 1921 2019 T	531
GARNACHA BLANCA Y RADIANTE HEREDAD CARLOS VALERO 2020 B	104
GARNACHA CENTENARIA DE COTO DE HAYAS 2020 T	102
GARNACHA DE FUEGO 2021 T	845
GARNACHA ROCK 2019 T BA	725
GARNACHA ROSÉ BY BODEGA LA ENCINA 2021 RD	842
GARNACHA TINTORERA =GARAGEWINE 2020 T BA	773
GARNACHA VELLA DA CHAIRA DO RAMIRIÑO 2019 T	689
GARNATA 2016 T R	202
GARNATXA CASTELL DEL REMEI 2021 T	171
GARNATXA DE CELLERS D'EN GUILLA BF D	186
GAROUBAS 2020 T RB	409
GARRIGUELLA GARNATXA D'EMPORDÁ AMBRÉ DULCE RD BA D	187
GARRIGUELLA GARNATXA D'EMPORDÁ ROBÍ DULCE NATURAL T D	187
GARRIGUELLA MOSCATEL D'EMPORDÁ DULCE 2021 B D	187
GASPI MONASTRELL GRAVAS 2021 RD	710
GASPI MONASTRELL TIERRA ARI-NOSA 2020 T C	710
GATELL AMBROSÍA 2017 BE GR BN	139

WINE	PAGE
GATELL HERITAGE 2017 BE GR BN	139
GATELL INITIAL 2017 BE GR BN	139
GATELL ROSÉ 2017 RE GR BN	139
GAUDEAMUS 2020 T RB	436
GAUDIR 2021 BE BN	148
GEA VIÑA GALAPAGOS 2021 B	776
GEBRADA 2021 B	348
GEIJO 2020 B FB	841
GÉMINA CUVÉE SELECCIÓN 2019 T C	229
GÉMINA FINCA EL VOLCÁN 2018 T	229
GÉMINA FINCA LA CABRA 2018 T C	229
GÉMINA FINCA LOS TOMILLARES 2018 T	229
GÉMINA SELECCIÓN MONASTRELL 2020 T	230
GEMMA 2017 BE GR BN	145
GENIO ESPAÑOL 12 MESES MONASTRELL 2019 T RB	223
GENIO ESPAÑOL 2020 T RB	223
GENIO Y FIGURA 2021 B	381
GERARD 2018 T R	177
GEROA LÍAS 2020 B	86
GESSAMÍ 2021 B	345
GG 2020 T	42
GGASPI MACABEO TIERRA BLANCA ORANGE 2021 B	710
GGASPI MACABEO TIERRA BLANCA/ROJA 2021 B	710
GGASPI MONASTRELL TIERRA BLANCA 2020 T C	710
GHM GARNACHA + GARNACHA 2019 T C	109
GIL ARMADA (VIÑEDOS PROPIOS DA TORRE DE SAN FARDÁN) 2020 B	382
GIL ARMADA (VIÑEDOS PROPIOS NO PAZO DE FEFIÑÁNS) 2020 B	382
GINÉ GINÉ 2019 T	356
GINÉ ROSE 2021 RD	356
GIRÓ DEL GORNER 2013 BE GR BN	139

WINE	PAGE	WINE	PAGE	WINE	PAGE
GIRÓ DEL GORNER 2013 BE GR BR	139	GODELIA SELECCIÓN GODELLO 2018 B	68	GOTES DEL PRIORAT 2020 T	363
GIRÓ DEL GORNER 2017 BE R BN	139	GODELIA SELECCIÓN MENCÍA 2016 T	68	GOYA XL BF MZ S	214
GIRÓ DEL GORNER 2018 BE R BR	139	GODEVAL 1986 2018 B	688	GR 36 2018 T	860
GIRÓ DEL GORNER BLANC 2021 B	336	GODEVAL 2021 B	688	GR 5 SENDERS 2020 B	345
GIRÓ DEL GORNER BRU T RB	336	GODEVAL CEPAS VELLAS 2020 B	688	GR 5 SENDERS 2020 T	345
GIRÓ DEL GORNER LES SALERES 2020 B FB	336	GODEVAL CEPAS VELLAS 2021 B	688	GR-174 2021 T	357
GIRÓ DEL GORNER ROSAT 2020 RE BR	139	GODEVAL REVIVAL 2020 B	688	GRÀ D'OR BLANCO SECO 2021 B	38
GIRÓ DEL GORNER ROSAT 2021 RD	336	GODEVAL REVIVAL 2021 B	688	GRACIA PEDRO XIMÉNEZ DULCE VIEJO BF PX D	287
GIRÓ DEL GORNER VINYA EL MOCADOR 2020 B	336	GODINA 2020 T	104	GRÁCIL DE ZALEO 2018 T C	499
GIRÓ DEL GORNER VINYA LA SERDALLA 2020 B	336	GOLD RUPESTRE GARNACHA TINTORERA 2019 T C	50	GRAMONA ARGENT 2017 BE BR	880
GIRÓ RIBOT AVANT 2016 BE R BR	139	GOLIARDO 2020 T	387	GRAMONA ARGENT ROSÉ 2018 RE BN	880
GIRÓ RIBOT EXCELSUS 100 MAGNUM 2011 BE GR BR	139	GOLIARDO CAIÑO 2019 T	387	GRAMONA CELLER BATLLE 2012 BE BR	880
GIRÓ RIBOT GIRÓ2 2017 B FB	862	GOMARIZ 12 (DOCE) DULCE 2020 B D	858	GRAMONA III LUSTROS 2013 BE BN	880
GIRÓ RIBOT MARE 2017 BE GR BN	139	GOMARIZ X 2021 B	424	GRAMONA III LUSTROS 2014 BE BN	880
GIRÓ RIBOT MIMAT 2018 T BA	862	GÓMEZ CRUZADO 2º AÑO 2021 T	555	GRAMONA IMPERIAL 2015 BE BR	880
GIRÓ RIBOT MIMAT BLANC 2021 B	336	GÓMEZ DE SEGURA 2021 B	570	GRAMONA INNOBLE BE BN	880
GIRÓ RIBOT SOLSTICI 2021 B	336	GÓMEZ DE SEGURA 2021 T MC	570	GRAMONA LA CUVÉE 2017 BE	880
GIRÓ RIBOT SPUR 2017 BE GR BN	140	GÓMEZ DE SEGURA RESERVA DE LA FAMILIA 2018 T R	570	GRAMONA TLN 2009 BE BN	881
GIRÓ RIBOT UNPLUGGED ROSADO 2018 RE R BR	140	GÓMEZ DE SEGURA VENDIMIA SELECCIONADA 2020 T BA	570	GRAN ALANÍS 2021 B	421
GIRÓ ROS DE MORTITX 2021 B	874	GONZALO DE BERCEO 2013 T GR	533	GRAN ALLEGRANZA 2021 T	42
GISELE 2021 B	330	GONZALO DE BERCEO 2016 T R	533	GRAN AMAT 2019 BE BN	127
GLADIUM VIÑAS VIEJAS 2019 T C	238	GONZALO DE BERCEO 2018 T C	533	GRAN ARZUAGA 2015 T R	440
GLORIA 2020 T	701	GONZALO DE BERCEO TEMPRANILLO BLANCO C.V.C B FB	534	GRAN AUTÓCTON BLANC 2020 B	839
GLORIA ANTONIO ALCARAZ 2018 T	515	GORKA IZAGIRRE 2021 B	87	GRAN AUTÓCTON BLANC 2021 B	839
GLORIA DE OSTATU 2016 T BA	578	GORRONDONA 2021 B	86	GRAN AUTÓCTON NEGRE 2017 T	840
GLORIOSO 2018 T R	521	GORU 18 M 2018 T	230	GRAN BAJOZ 2018 T	668
GNEIS 2015 T	189	GORU 38 BARRELS 2020 T BA	230	GRAN BAROJA 2011 T GR	530
GNTX LAR DE MAÍA 2020 T C	864	GORU EL BLANCO 2021 B	860	GRAN BARQUERO B AM S	288
GODA 2021 B	400	GORYS GARNACHA 2018 T	766	GRAN BARQUERO B FI S	288
GODELIA GODELLO 2021 B	68	GOTAS DE MAR ALBARIÑO 2021 B	394	GRAN BARQUERO B OL S	288
GODELIA MENCÍA 2017 T RB	68	GOTAS DE MAR ROSÉ 2021 RD	757	GRAN BARQUERO B PC	288
GODELIA MENCÍA ROSÉ 2021 RD	68	GOTES DEL MONTSANT 2019 T	292	GRAN BARQUERO BF PX D	288

WINE	PAGE
GRAN BARQUERO EN RAMA B FI S	288
GRAN BIERZO EL CULEBRAL 2021 T	78
GRAN BIERZO MENCÍA 2020 T	78
GRAN BIERZO ORIGEN 2020 T	78
GRAN BLANC PRÍNCEPS 2021 B	330
GRAN BUCHE 2015 T R	817
GRAN CALZADILLA 2012 T	739
GRAN CAMPIÑO B	421
GRAN CAUS 2016 T R	329
GRAN CAUS 2021 B	330
GRAN CAUS 2021 RD	330
GRAN CERMEÑO 2018 T C	658
GRAN CLOS 2018 T C	366
GRAN CLOS 2020 B FB	366
GRAN CLOT DELS OMS XARELLO 2019 B BA	328
GRAN COLEGIATA "ORIGINAL" 2016 T R	659
GRAN COLEGIATA 2020 T RB	659
GRAN CRUOR 2014 T	357
GRAN CRUOR SELECCIÓ CARANYENA 2014 T	357
GRAN CRUZ DEL CALVARIO 2020 B	179
GRAN ELÍAS MORA 2015 T	665
GRAN EROLES 2015 T R	628
GRAN FAUSTINO I 2004 T GR	525
GRAN FEUDO 2018 T C	311
GRAN FEUDO 2021 RD	311
GRAN FEUDO CHARDONNAY 2021 B	311
GRAN FEUDO DULCE DE MOSCATEL 2021 B MO D	311
GRAN FEUDO EDICIÓN LIMITADA VIÑAS VIEJAS 2017 T	312
GRAN FEUDO GARNACHA ROSÉ 2021 RD	312
GRAN FUCI IS DE VIDAL BE R BN	124
GRAN HACIENDA MOLLEDA GHM CARIÑENA + CARIÑENA 2016 T C	109

WINE	PAGE
GRAN HACIENDA MOLLEDA GHM CARIÑENA + GARNACHA 2019 T C	109
GRAN JUVÉ CAMPS 2016 BE GR BN	141
GRAN LAGAR DE BOUZA 2021 B	391
GRAN LERMA VINO DE AUTOR 2016 T R	55
GRAN MINGUET 2017 BE R BN	32
GRAN NOVAS ALBARIÑO 2021 B	381
GRAN PADRINA 2020 T	82
GRAN PASAS MONASTRELL 2020 T	734
GRAN PRÍNCEPS 2016 T R	330
GRAN RESALTE 2015 T GR	458
GRAN RESERVA 904 SELECCIÓN ESPECIAL 2015 T GR	573
GRAN RESERVA FAMILIAR MILLENIUM 2013 BE GR BR	134
GRAN RIGAU CHARDONNAY BE R BN	138
GRAN TÁBULA 2017 T	472
GRAN TOC 2016 T	331
GRAN TORELLÓ 2015 BE BN	884
GRAN TORONDOS 2020 T	156
GRAN TORONDOS 2021 RD	156
GRAN VALTRAVIESO 2018 T R	473
GRAN VERÁN 2019 T BA	81
GRAN VINUM 2021 B	379
GRAN VINYA SON CAULES 2014 T	351
GRAN VOS DE VIÑAS DEL VERO 2016 T R	629
GRANBAZÁN DON ÁLVARO DE BAZÁN 2018 B	383
GRANBAZÁN DON ÁLVARO DE BAZÁN 2019 B	383
GRANBAZÁN ETIQUETA ÁMBAR 2021 B	383
GRANBAZÁN ETIQUETA VERDE 2021 B	383
GRANBAZÁN LIMOUSIN 2019 B	383
GRAND VIN SON MAYOL 2017 T GR	826
GRAND VIN SON MAYOL 2018 T GR	826
GRANIT 2020 B	299

WINE	PAGE
GRANIT 2021 B	299
GRANS MURALLES 2018 T R	165
GRATALLOPS VI DE LA VILA 2020 T C	354
GRATAVINUM 2πR 2020 T	366
GRATAVINUM GV5 2016 T	366
GRATAVINUM SILVESTRIS 2020 T	366
GRATIANVS 2018 T	306
GRATIAS GOT 2020 T	847
GRATIAS MÁXIMAS 2018 T	847
GRATIAS ROSÉ 2021 RD	847
GRATIAS SOL 2020 B S	847
GREDAS VIEJAS 2017 T RB	267
GREEN & SOCIAL TEMPRANILLO 2021 T	802
GREEN & SOCIAL VERDEJO 2021 B	599
GRESA 2015 T R	192
GRITELLES MANOU 2020 T	294
GRITELLES SIURANA BRISAT 2020 B BA	294
GRITELLES SIURANA NEGRE 2018 T BA	294
GRITELLES TROS DE LA SERRA 2019 B	360
GRITELLES VINYES VELLES CARINYENA 2019 T C	294
GRITELLES VINYES VELLES GARNATXA 2019 T C	294
GRIZZLY 2019 T	256
GROSS 2021 RD	259
GROSS CABERNET SAUVIGNON 2018 T R	260
GROSS PETIT VERDOT 2018 T C	260
GRUÑÓN 2017 T	105
GTX PIRATE WINE 2016 T	872
GUADIANEJA PARAJE ALTO HUNGRAO 2020 T RB	246
GUADIANEJA PARAJE ALTO HUNGRAO 2021 B	246
GUARDIANES DEL FONDILLÓN 1955 T FO	44
GUEI BENZU AZUL 2018 T	833
GUELBENZU EVO 2015 T	833

WINE	PAGE
GUELBENZU LAUTUS 2010 T	833
GUELBENZU VIERLAS 2017 T RB	833
GUERINDA EL MÁXIMO 2020 T BA	314
GUERINDA NAVASENTERO 2020 T	314
GUERINDA PARCELAS DE GARNACHA "LA ABEJERA" 2020 T	314
GUERINDA PARCELAS DE GARNACHA "TXIROLAS, QUITANA Y VILARRAGA"" 2020 T BA	314
GUERINDA TRES PARTES 2021 T	314
GUERINDA+ LA VIURA 2020 B FB	314
GÜERTANA SAUVIGNON BLANC 2021 B	227
GÜERTANO MONASTRELL 2019 T	227
GUILLAMINA 2020 B	172
GUILLEM CAROL 2017 BE GR BN	134
GUIMARO 2021 B	408
GUIMARO MENCÍA 91 T	408
GUITIÁN FERMENTADO EN BARRICA DE ACACIA 2019 B	686
GUITIÁN GODELLO 2020 B FB	687
GUITIÁN GODELLO 2021 B	687
GUITIÁN GODELLO SOBRE LÍAS 2020 B	687
GUITIÁN GODELLO SOBRE LÍAS 2021 B	687
GURDOS 2021 RD	255
GURE ABERRIA 2021 B	88
GURE AHALEGINAK 2021 B	87
GURE NATURA 2020 B	88
GUREAGA 2015 T	505
GUSTO 2019 T BA	732
GUTIÉRREZ COLOSÍA BF AM S	208
GUTIÉRREZ COLOSÍA BF FI S	208
GUTIÉRREZ COLOSÍA BF OL	208
GUTIÉRREZ COLOSÍA BF PX D	208
GUTIÉRREZ COLOSÍA FINO EN RAMA 15 GRADOS BF FI S	208
GUTIÉRREZ COLOSÍA FINO EN RAMA 16 GRADOS BF FI S	208
GVIVM BLANC DE BLANCS 2021 B	351

H

WINE	PAGE
HACHÓN VERDEJO VIURA 2021 B	606
HACIENDA ACENTEJO 2021 B S	633
HACIENDA ACENTEJO 2021 T S	633
HACIENDA ALBAE CABERNET SAUVIGNON 2020 T	793
HACIENDA ALBAE CHARDONNAY 2021 B	793
HACIENDA ALBAE GRAND VIOGNIER 2020 B	793
HACIENDA ALBAE MERLOT 2020 T	793
HACIENDA ALBAE SYRAH 2020 T S	793
HACIENDA ALBAE TOP 888 2016 T R	793
HACIENDA ALCARAZ 2021 B	595
HACIENDA CASA DEL VALLE 2020 T	787
HACIENDA CASA DEL VALLE CHARDONNAY 2020 B	787
HACIENDA CASA DEL VALLE SAUVIGNON BLANC 2021 B	787
HACIENDA CASA DEL VALLE SELECCIÓN ESPECIAL 2019 T	787
HACIENDA CASA DEL VALLE SYRAH 2019 T	787
HACIENDA DE ACENTEJO 2021 T BA	633
HACIENDA DE ARÍNZANO 2018 T	746
HACIENDA DE ARÍNZANO CHARDONNAY 2019 B	746
HACIENDA DE Dª FRANCISCA - PAGO CALLEJUELA B	772
HACIENDA DE LA VIZCONDESA 2020 T RB	260
HACIENDA DEL CARCHE CEPAS VIEJAS 2019 T C	231
HACIENDA EL TERNERO 2014 T R	570
HACIENDA EL TERNERO 2018 B FB	570
HACIENDA EL TERNERO 2021 RD	570
HACIENDA EL TERNERO SELECCIÓN 2016 T C	570
HACIENDA ELSA GODELLO 2021 B	63
HACIENDA ELSA MENCÍA 2020 T	63
HACIENDA GRIMÓN 2019 T C	571
HACIENDA GRIMÓN CHARDONNAY 2020 B	571
HACIENDA GRIMÓN VIURA 2019 B C	571

WINE	PAGE
HACIENDA LÓPEZ DE HARO 2017 T R	512
HACIENDA LÓPEZ DE HARO 2021 B	513
HACIENDA MOLLEDA 2021 RD	109
HACIENDA MOLLEDA CARIÑENA 2021 T	109
HACIENDA MOLLEDA GARNACHA 2018 T C	109
HACIENDA MOLLEDA GARNACHA 2020 T RB	109
HACIENDA MOLLEDA GARNACHA 2021 T	109
HACIENDA MOLLEDA GARNACHA BLANCA 2019 B	109
HACIENDA MONASTERIO 2018 T R	449
HACIENDA MONASTERIO 2019 T	449
HACIENDA MONASTERIO RESERVA ESPECIAL 2016 T R	449
HACIENDA REAL 2021 B	793
HACIENDA REAL 2021 T	793
HACIENDA SAEL GODELLO 2021 B	63
HACIENDA SAEL MENCÍA 2020 T	63
HACIENDA SOLANO FINCA CASCORRALES 2018 T	482
HACIENDA SOLANO FINCA PEÑALOBERA 2019 T	482
HACIENDA SOLANO SELECCIÓN 2020 T BA	482
HACIENDA SOLANO VIÑAS VIEJAS 2019 T BA	482
HACIENDA SUSAR 2018 T	527
HACIENDA UCEDIÑOS DON ELADIO EXPRESIÓN MENCÍA 2020 T RB	690
HACIENDA UCEDIÑOS GODELLO 2021 B	690
HACIENDA UCEDIÑOS MENCÍA 2020 T	690
HARAGÁN RESERVA ESPECIAL 2018 T R	499
HARENNA - TINAJA 2020 B	605
HARENNA 2020 B	605
HARVEYS AMONTILLADO PREMIUM BF AM S	217
HARVEYS AMONTILLADO VORS BF AM S	217
HARVEYS BRISTOL CREAM BF CRM	217
HARVEYS FINO PREMIUM BF FI S	217
HARVEYS PALO CORTADO VORS BF PC MED	217

WINE	PAGE	WINE	PAGE	WINE	PAGE
HARVEYS PEDRO XIMÉNEZ VORS BF PX D	217	HERENZA 2021 B	32	HINIA 2012 T	536
HEAVEN & HELL 2020 B	875	HERENZA ROSÉ 2021 RD	567	HIPPERIA 2019 T	747
HECHANZA REAL 2019 T C	58	HERETAT D'LÁCRIMA BACCUS 2019 BE R BN	142	HIRIART 2016 T C	156
HÉCULA MONASTRELL ORGANIC 2020 T BA S	732	HERETAT D'LÁCRIMA BACCUS 2019 BE R BR	142	HIRIART 2021 RD	156
HELLO WORLD PETIT VERDOT 2021 T	780	HEREUS BLANC 2021 B	641	HIRIART ÉLITE 2021 RD	156
HELLO WORLD VIOGNIER 2021 B	780	HEREUS NEGRE 2020 T	641	HIRUZTA BEREZIA 2020 B	196
HEMAR 2018 T C	449	HERMANOS FERNÁNDEZ 2020 B	600	HIRUZTA ROSÉ 2021 RD	196
HEMAR 2020 T RB	449	HERMANOS FRÍAS DEL VAL 2016 T R	571	HIRUZTA TXAKOLINA 2021 B	196
HENRI MARC 01 SYRAH 2021 T	709	HERMANOS FRÍAS DEL VAL 2018 T C	571	HITO 2020 T	475
HERACLIO ALFARO 2018 T C	561	HERMANOS FRÍAS DEL VAL SELECCIÓN PERSONAL 2015 T	571	HITO 2021 RD	475
HERACLIO ALFARO FINCA ESTARIJO 2016 T	561	HERMANOS LURTON CUESTA DE ORO 2021 B FB	599	HOCICÓN 2021 RD	431
HERAS CORDÓN VENDIMIA SELECCIONADA 2019 T C	555	HERMANOS LURTON CUESTA GRANDE 2018 T	655	HODGKINSON GARNACHA PELUDA 2018 T R	367
HEREDAD 26 2020 T RB	65	HERMANOS LURTON ORGÁNICO SIN SULFITOS 2021 T	655	HOLLERA MONJE 2021 T MC	755
HEREDAD 26 GODELLO 2020 B	65	HERMANOS LURTON SAUVIGNON BLANC 2021 B	597	HOMBRE BALA 2019 T	728
HEREDAD 26 MENCÍA 2020 T	65	HERMANOS LURTON TEMPRANILLO 2020 T	655	HOMBRE BALA 2020 B	728
HEREDAD 26 MENCÍA 2021 T	65	HERMANOS LURTON VALENTÍN ROSÉ 2021 RD	597	HOMBROS 2020 T BA	71
HEREDAD ALTOS DE TALAN 2020 B FB	65	HERMANOS LURTON VERDEJO 2021 B	597	HOMENAJE 2021 B	314
HEREDAD DE ANSÓN 2017 T C	109	HERMINIA VENDIMIA SELECCIONADA 2019 T	553	HONEST GSM 2019 T	786
HEREDAD DE ANSÓN 2021 B	110	HERMOSILLA. BODEGAS J. I. HERMOSILLA 2018 T C	527	HONEST VVS 2020 B	786
HEREDAD DE ANSÓN 2021 RD	110	HERMOSILLA. BODEGAS J. I. HERMOSILLA 2021 B	527	HONORIS DE VALDUBÓN 2018 T	463
HEREDAD DE ANSÓN MERLOT SYRAH 2021 T	110	HERMOSILLA. BODEGAS J. I. HERMOSILLA 2021 T MC	527	HONORO VERA GARNACHA 2021 T	95
HEREDAD DE PEÑALOSA 2016 T R	455	HESVERA 2019 T C	451	HONORO VERA MONAGTRIELL 2021 T	226
HEREDAD DE PEÑALOSA 2019 T C	455	HESVERA 6 MESES BARRICA 2020 T RB	451	HONORO VERA ORGÁNICO 2021 T	226
HEREDAD DE PEÑALOSA 2020 T RB	455	HI 2019 T C	231	HONORO VERA RIOJA 2020 T	582
HEREDAD DEL VIEJO IMPERIO HOMENAJE 2016 T	59	HIBEU 2019 T	274	HONORO VERA RIOJA 2021 T	582
HEREDAD GARCÍA DE OLANO 2018 T C	526	HIBEU 2020 T	274	HONORO VERA VERDEJO 2021 B	613
HEREDAD GARCÍA DE OLANO TEMPRANILLO 2021 T MC	526	HIBEU FINCA LA MINERAL 2019 T	274	HOYA DE CADENAS BE BN	152
HEREDAD X CARLOS VALERO 2020 T	108	HÍBORO 2021 T	631	HOYA DE CADENAS BE BR	152
HERENCIA DEL CAPRICHO 2020 B FB	69	HIDALGO CASTILLA 2016 T GR	242	HOYA DE CADENAS ROSADO RE BR	152
HERENZA 2017 T R	566	HIDALGO CASTILLA 2017 T R	242	HOYA DE CADENAS TEMPRANILLO 2018 T R	683
HERENZA 2018 T C	566	HIGHWAY TO HELL 2021 T RB	712	HOYA DEL CASTILLO 2021 B	706
HERENZA 2020 T	566	HIJÓN DEL SANTO 2020 T BA	858	HOYA MONTÉS 2021 RD	270

WINE	PAGE
HOYO DE LA VEGA 2015 T R	463
HOYO DE LA VEGA 2018 T C	463
HOYO DE LA VEGA 2019 RD	463
HOYO MEMBRILLO MEDIUM BF MED	218
HUELLA DE MERSEGUERA 2021 B	706
HUELLA DE SYRAH 2021 T	706
HUERTO DE LA CONDESA LA HIEDRA 2020 T RB	263
HUERTO DE LA CONDESA LA PALMERA 2019 T C	263
HUERTO DE LA CONDESA LOS CIPRESES 2019 T C	263
HUERTO DE LA CONDESA LOS CIPRESES 2021 RD	263
HUERTO DE LA CONDESA PAMPANEANDO 2021 T	263
HUGO AFRUTADO 2021 RD	632
HUGUET DE CAN FEIXES 2014 BE GR BN	881
HUGUET DE CAN FEIXES CLASSIC 2014 BE GR BR	881
HUGUET MAGNUM 2010 BE R BN	881
HUMBOLDT 1997 B D	631
HUMBOLDT 2001 T D	631
HUNO 2018 T R	499
HURACÁN DANIELA 2021 B FB	318
I	
IBERO II 2018 T	110
IBIZKUS SYRAH 2019 T	816
ICENI 2020 T RB	769
ICNOS 2018 T C	188
IDENTITAS 2019 B	637
IDOIA 2018 T	116
IDOIA BLANC 2020 B FB	116
IDRIAS CHARDONNAY 2021 B	628
IDRIAS MERLOT 2021 RD	863
IDRIAS T RB	629
IDRIAS TEMPRANILLO 2021 T	629
IEUP 2020 B	87

WINE	PAGE
IEUP BARRIKAN 2020 B FB	87
IEUP SOBRE LÍAS MAGNUM 2019 B	88
IGNACIO GUALLART MERLOT 2019 T RB	767
IGNIUS (ETIQUETA MORADA) 2015 T	862
IJALBA 2018 B R	588
IJALBA 2018 T R	588
IJALBA 2019 T C	588
IJALBA 2020 B C	588
IJALBA CUVÉE 2020 T	588
IJALBA MATURANA BLANCA 2021 B	588
IKEWEN 2 AÑOS VELO DE FLOR 2019 B	199
IKEWEN 2020 T	199
IKIGALL 2021 B	335
IKUNUS 2018 T	531
ILLUSIÓ DE CLUA 2020 T	646
ILLUSIÓ DE CLUA 2021 B	646
ILLUSIONAT 2021 B	707
ILLUSIONAT ROSÉ 2021 RD	707
ILAGARES 2021 B	309
ILAGARES 2021 RD	309
ILAGARES GARNACHA 2021 T	309
ILDANIA 2017 T	269
ILERCAVÒNIA 2021 B	641
ILEX 2020 T	787
ILEX COUPAGE 2019 T	787
ILEX VERDEJO 2021 B	787
IL·LÒGIC XAREL·LO ORGÀNIC SUMARROCA 2021 B	342
ILURCE 2018 T C	555
ILURCE 2020 T	555
ILURCE 2021 RD	555
ILUSIÓN 2021 RD	757
ILUSIÓN RUFETE SERRANO BLANCO 2021 B	757

WINE	PAGE
IMAGINAR 2020 T C	228
IMPERIAL 2016 T GR	563
IMPERIAL 2018 T R	563
IMPRESSIONANT 2021 B	850
IMPROMPTU 2021 B	676
IMPROMPTU ROSÉ 2021 RD	703
INANNAT 2021 RD	118
INAUDITA 2021 T	230
INCRÉDULO BLEND 2019 T	656
INDOMABLE 2016 BE BN	882
INÉS VIZCARRA 2018 T R	466
INFILTRADO 2021 T	231
INFINITO 2018 T C	230
INFORMAL 2020 B	718
INGÉNITO 2021 T RB	259
INMORTALIS 2019 T	91
INNOCENCE 2021 B FB	168
INO GARNATXA DE L'EMPORDÀ 2020 RF SOLERA D	189
INQUIET DE RENDÉ MASDEU 2021 T	164
INSOMNI 2021 B	829
INSTANT DE FLOR 2021 B	347
INSTINTO ROMANO 2021 T	415
INTUICIÓN VERDEJO 2021 B	815
INURRIETA CUATROCIENTOS 2019 T C	307
INURRIETA MIMAO 2020 T	307
INURRIETA MIMAO 2021 B BA	307
INURRIETA PURO VICIO 2019 T	307
INVERSO GODELLO 2021 B S	873
INVITA 2021 B	33
INVITA ROSÉ 2021 RD	33
IÑAKI NÚÑEZ SELECCIÓN PRIVADA 2017 T	312
IÑAKI NÚÑEZ VENDIMIA SELECCIONADA 2018 T	312

WINE	PAGE
IÑIGO AMÉZOLA 2019 B FB	515
IÑIGO AMÉZOLA 2020 T	515
IO MASIA SERRA 2018 T R	189
IOHANNES 2015 T R	341
IP ISABEL PERALTA GARNACHA TINTORERA 2018 T	795
IP ISABEL PERALTA VIOGNER 2020 B RB	795
IRACHE 2016 T R	312
IRACHE 2017 T C	312
IRACHE 2021 RD	312
IRACHE ROBLE 2021 T RB	312
IRVING SYRAH 2020 T	202
ISABEL I 2021 B	621
ISABELINO VERDEJO 2021 B	600
ISHII DE VIVES AMBRÒS 2021 B	639
ISLA ORO AIRÉN 2021 B	240
ISLA ORO CABERNET SAUVIGNON 2021 T	240
ISLA ORO GARNACHA 2021 RD	240
ISLA ORO MACABEO 2021 B	240
ISLA ORO TEMPRANILLO 2018 T C	240
ISLA ORO TEMPRANILLO 2021 T	240
ISP (ISLA DE SAN PEDRO) 2018 T C	724
ITRAN TEMPRANILLO 2020 T	537
ITSASMENDI ADOS 2020 B	85
ITSASMENDI ARTIZAR (BOTELLA BORGOÑA) 2017 B	85
ITSASMENDI Nº 7 2020 B	85
ITSASMENDI UREZTI 2018 B	85
IURA 2018 B	355
IUVENIS DE BIOPAUMERÀ 2020 T	636
IV GENERACIÓN RUFETE 2021 T	758
IVETTE 2019 BE R BR	131
IXEIA 2021 RD	624
IXENT 2021 RD	182

WINE	PAGE
IZADI 2019 T C	528
IZADI EL REGALO 2020 T	528
IZADI LARROSA BLANCA 2021 B	528
IZADI LARROSA NEGRA 2021 T	528
IZADI LARROSA ROSÉ 2021 RD	528
IZADI SELECCIÓN 2018 T	528
IZADI SELECCIÓN 2021 B	528
IZENA 2020 T	559

J

WINE	PAGE
J DE SAMITIER 2020 T	95
J. BELMONTE 2021 T	112
J. CANTERA 2021 B	529
J. CANTERA 2021 RD	529
J. CANTERA 2021 T	529
J. FERNANDO TEMPRANILLO 6 MESES 2020 T	792
J. FERNANDO VERDEJO 2021 B	616
J. FERNANDO VERDEJO VENDIMIA SELECCIONADA 2020 B	616
J.F. ARRIEZU 2016 T C	507
J.F. ARRIEZU 2020 T	507
J.F. ARRIEZU ROBLE 2020 T	507
J.F. ARRIEZU ROSADO 2021 RD	507
J.L. VILELA LADAIRO 2019 T C	279
JAI 2021 T	776
JABALÍ GARNACHA & SYRAH 2021 T	110
JABALÍ TEMPRANILLO & CABERNET 2021 T	110
JALIFA VORS "30 YEARS" BF AM S	213
JANE SANTACANA 2017 BE GR BN	132
JANE SANTACANA ETIQUETA BLANCA 2018 BE R BN	132
JANE SANTACANA ETIQUETA COBRE 2018 BE R BR	132
JANE SANTACANA ETIQUETA DORADA 2018 BE R BN	132
JANÉ VENTURA 1914 2012 BE GR RN	140
JANÉ VENTURA DO MAJOR VINYES VELLES 2016 BE GR BN	140

WINE	PAGE
JANÉ VENTURA DO MAJOR VINYES VELLES 2017 BE GR BN	140
JANÉ VENTURA FINCA ELS CAMPS MACABEU 2019 B	336
JANÉ VENTURA FINCA ELS CAMPS NEGRE 2018 T	337
JANÉ VENTURA MALVASÍA DE SITGES 2020 B BA	337
JANÉ VENTURA RESERVA DE LA MÚSICA 2019 BE R BN	140
JANÉ VENTURA RESERVA DE LA MÚSICA MAGNUM 2017 BE R BN	140
JANÉ VENTURA RESERVA DE LA MÚSICA ROSÉ 2019 RE R BR	140
JANÉ VENTURA SUMOLL 2018 T	337
JANÉ VENTURA VINYES BLANQUES 2021 B	337
JANÉ VENTURA XARELLO 2020 B	337
JARABE DE ALMÁZCARA MAJARA 2019 T	63
JARDÍN DE ABRIL 2020 T BA	756
JARDÍN DE ABRIL ORANGE WINE 2020 B	756
JAROS 2018 T	494
JASPI BLANC 2021 B	647
JASPI NEGRE 2018 T	297
JAUME DE PUNTIRÓ BLANC 2021 B	81
JAUME DE PUNTIRÓ CARMESÍ 2018 T	81
JAUME DE PUNTIRÓ ROSAT 2021 RD	81
JAUME GRAU I GRAU "GRATVS" 2014 T C	347
JAUME GRAU I GRAU COLLECCIÓ UII DE LLEBRE 2014 T	347
JAUME GRAU I GRAU SELECCIÓ ESPECIAL 2017 T	347
JAUME LLOPART ALEMANY 2015 BE GR BN	140
JAUME LLOPART ALEMANY 2021 B	337
JAUME LLOPART ALEMANY 2021 RD	337
JAUME LLOPART ALEMANY BE R BN	140
JAUME LLOPART ALEMANY BE R BN	140
JAUME LLOPART ALEMANY MERLOT 2021 T	337
JAUME SERRA BE R BN	123
JAUME SERRA CHARDONNAY BE GR BN	123
JAUME SERRA ORGÁNICO BE BR	123

WINE	PAGE
JAUME SERRA PINOT NOIR ROSÉ RE BR	123
JAUME SERRA VINTAGE 2019 BE BN	123
JAUN DE ALZATE 2015 T R	532
JAVIER FERNÁNDEZ 2020 T	407
JAVIER SANZ 2019 B FB	616
JAVIER SANZ PARAJE LA ENCINA 2020 T RB	810
JAVIER SANZ SAUVIGNON BLANC 2021 B	616
JAVIER SANZ VERDEJO 2021 B	616
JEAN LEON 3055 CHARDONNAY 2021 B	337
JEAN LEON 3055 MERLOT PETIT VERDOT 2020 T	337
JEAN LEON 3055 ROSÉ 2021 RD	337
JEAN LEON VINYA GIGI CHARDONNAY 2019 B C	337
JEAN LEON VINYA LE HAVRE CABERNET SAUVIGNON RESERVA 2017 T R	337
JEAN LEON VINYA PALAU MERLOT 2018 T C	337
JEAN LEON XARELLO 2021 B	337
JESÚS DE MADRAZO 2020 B R	572
JESÚS MADRAZO 2019 T C	572
JILGUERÍN 2021 T	492
JIMÉNEZ-LANDI BLANCO MALVAR 2020 B	275
JIMÉNEZ-LANDI EL CORRALÓN 2021 T	275
JIMÉNEZ-LANDI NATURAL 2021 T	275
JIMÉNEZ-LANDI PIÉLAGO 2019 T	275
JIMÉNEZ-LANDI SOTORRONDERO 2020 T	275
JIMÉNEZ-LANDI VALDINIEBLA 2021 RD	275
JIRAFAS 2019 B	322
JIRÓN DE NIEBLA 2020 T C	753
JOAN GINÉ 2015 T	356
JOAN GINÉ BLANC 2019 B FB	356
JOANA 2020 T	296
JOANA RIGAU ROS 2021 B	190
JOAQUÍN REBOLLEDO 2020 T BA	690

WINE	PAGE
JOAQUÍN REBOLLEDO GODELLO 2021 B	690
JOAQUÍN REBOLLEDO MENCÍA 2021 T	690
JONCÀRIA GARNACHA 2015 T	190
JORGE ORDÓÑEZ & CO Nº 1 SELECCIÓN ESPECIAL DULCE (SIN FORTIFICAR) 2020 B D	264
JORGE ORDÓÑEZ & CO Nº 2 VICTORIA DULCE (SIN FORTIFICAR) 2021 B D	264
JORGE ORDÓÑEZ & CO. Nº3 VIÑAS VIEJAS (SIN FORTIFICAR) 2018 B D	264
JOSÉ GALO VENDIMIA SELECCIONADA 2021 B	611
JOSÉ L FERRER 2016 T R	81
JOSÉ L FERRER 2018 T C	81
JOSÉ MANUEL CORRALES 2020 T	697
JOSÉ PARIENTE CUVÉE ESPECIAL 2019 B	606
JOSÉ PARIENTE CUVÉE ESPECIAL 2020 B	606
JOSÉ PARIENTE FINCA LAS COMAS 2018 B	606
JOSÉ PARIENTE FINCA LAS COMAS 2019 B	606
JOSÉ PARIENTE VERDEJO 2021 B	606
JOSEP FORASTER BLANC SELECCIÓ 2021 B	165
JOSEP FORASTER PEP 2020 T	165
JOSEP FORASTER TREPAT 2021 T	165
JUAN DE JUANES VENDIMIA BRONCE 2021 T	700
JUAN GALINDO 100% VERDEJO 2020 B	603
JUAN GALINDO 2019 T C	444
JUAN GIL 2021 RD	226
JUAN GIL ETIQUETA AMARILLA/YELLOW LABEL 2021 T	226
JUAN GIL ETIQUETA AZUL/BLUE LABEL 2020 T	226
JUAN GIL ETIQUETA PLATA/SILVER LABEL 2020 T	226
JUAN GIL MOSCATEL SECO 2021 B	226
JUAN GIL PETIT VERDOT 2021 T	226
JUAN GIL SELECCIÓN BARTOLOMÉ ABELLÁN 2021 T BA	226
JUJOL DE VIVES AMBRÒS 2018 B FB	639

WINE	PAGE
JULIA 2019 T	701
JÚLIA BERNET 130 2017 BE BN	881
JÚLIA BERNET EXSUM 2018 BE BN	881
JULIA ESENCIA DE MMADRE 2021 RD	751
JULIÁN SANTOS MARTÍNEZ EDICIÓN CENTENARIO 2019 T C	230
JULIETA 2020 T	165
JUMENTA MERLOT SYRAH GARNACHA TINTORERA 2019 T	47
JUVÉ & CAMPS BLANC DE NOIRS 2010 BE GR BR	141
JUVÉ & CAMPS LA CAPELLA 2009 BE GR BN	141
JUVÉ & CAMPS LA SIBERIA 2013 RE GR BN	141
JUVÉ & CAMPS MILESIMÉ 2008 BE GR BR	141
JUVÉ & CAMPS MILESIMÉ 2016 BE R BR	141
JUVÉ & CAMPS RESERVA DE LA FAMILIA 2018 BE GR BN	141
JV LÍAS 2020 B	427

K

WINE	PAGE
K PILOTA 2020 B	195
K-NAIA 2021 B	607
K-THARSIS 2018 T C	173
KÝMA 2018 T R	331
KÝMA MACABEO 2019 B	331
K5 2016 B	195
K5 2019 B	195
KAIROS DE SAN ESTEBAN 2019 T	547
KAIROS DE SAN ESTEBAN 2020 RD	547
KAIROS DE SAN ESTEBAN FERMENTADO CON SUS PIELES 2020 B	547
KALAMITY 2020 T	579
KAMIKAZE DE ALDEA 2020 RD	459
KAOS 2015 T R	802
KASINÁ 2020 T	757
KENTIA 2021 B	394
KIMERA 2019 T	318

WINE	PAGE
KINKI 2021 T	65
KIRKILLA AMEZTOI 2017 B	195
KOMOKABRAS AMARILLO 2020 B BA	768
KOMOKABRAS NARANJA TINALLA 2020 B	768
KOMOKABRAS ROSE D-MENCIAL 2020 RD S	837
KOMOKABRAS VERDE LÍAS 2020 B	768
KORTETA 2020 T S	865
KUUSU 2020 T	867
KYATHOS 2018 T C	734
KYRA 2020 B FB	255
Kπ ILUMINATI 2020 T	859
Kπ ROSÉ 2019 RD	859

L

WINE	PAGE
L´HERAVI SELECCIÓ 2021 T	302
L'ABELLAR 2020 B	857
L'ALBA 2019 T	36
L'ALBA DE FAELO 2021 RD	36
L'ALBA DEL MAR 2021 B	36
L'ALEGRIA 2021 T	679
L'ANICETO 2019 T C	291
L'ANTANA 2013 T R	334
L'APHRODISIAQUE GODELLO 2021 B	63
L'APHRODISIAQUE MENCÍA 2021 T	63
L'ARXIDUC ROSAT 2021 RD	351
L'AVI ARRUFÍ 2019 T	645
L'AVI ARRUFÍ 2020 B FB	645
L'ENTRADA P.7 2020 B	49
L'EQUILIBRISTA GARNATXA 2018 T	116
L'EQUILIBRISTA NEGRE 2018 T	116
I'FRA DEL CELDONI 2012 BE GR BN	134
L'ERMITA 2020 T C	354
L'ESCALETA 2020 T	355

WINE	PAGE
L'ESPARTER 2015 BE GR BN	326
L'ESTACA 2017 T	375
L'ESTACIÓ BLANC 2019 B	118
L'ESTACIÓ NEGRE 2016 T C	118
L'EXCLAMACIÓ 2019 T C	359
L'HERAVI 2021 T	301
L'HERAVI BLANC DE NOIR 2021 B	302
L'HOME PEIX NEGRE 2021 T	641
L'HOME PEIX PARELLADA 2021 B	641
L'HOME PEIX ROSAT 2021 RD	641
L'INDIÀ BLANC 2019 B	649
L'INTERROGANT 2020 T	359
L'INTRÚS 2019 T	191
L'O DE L'ORIGAN BE BN	141
L'O DE L'ORIGAN ROSAT RE BN	141
L'OBLIT 2017 T FB	334
L'OLIVERA RESERVA SUPERIOR 2018 BE BN	175
L'ORATORI 2020 T	185
L'U DE MORTITX 2019 T R	832
L'ÚNIC 2020 T	292
LA ABUELA VISI 2020 B BA	550
LA ARTESILLA DE LEZA GARCÍA 2019 T	556
LA ARTESILLA DE LEZA GARCÍA VIÑEDO SINGULAR 2019 B	557
LA ATALAYA DEL CAMINO 2020 T	47
LA BAIXADA 2020 T	354
LA BALLESTERA CLUB BARRICA 2019 T C	794
LA BALLESTERA TINTO GUARDA MAGNUM 2019 T C	794
LA BARAJA 2019 T	465
LA BASSETA 2018 T C	369
LA BATALLA DE LA BARROSA 2021 B	771
LA BELLA ANCESTRAI PET NAT 2020 RE C BN	270
LA BELLA ANCESTRAL PET NAT 2021 RE BN	270

WINE	PAGE
LA BELLA GARNACHA 2020 T	101
LA BESTIA BLANCA 2021 B	190
LA BESTIA GARNACHA 2019 T RB	101
LA BESTIA MONASTRELL 2019 T BA	222
LA BESTIA NEGRA 2020 T	190
LA BIEN PINTÁ 2021 B	606
LA BOTA DE AMONTILLADO (BOTA Nº 109) "BOTA PUNTA" BF AM	215
LA BOTA DE MANZANILLA 107 FLORPOWER MMXX B	861
LA BOTA DE MANZANILLA PASADA Nº103 MAGNUM BF MZ	215
LA BOTA DE OLOROSO (BOTA Nº 108) "BOTA NO" BF	215
LA BOTA DE PALO CORTADO (BOTA Nº 102) BF PC	215
LA BOTERA 2018 T C	191
LA BOVILA 2019 T	871
LA BRUIXA 2020 B	650
LA BRUJA DE ROZAS 2020 T	725
LA BRUNA 2020 T	679
LA CABEZA DE PERRO 2020 B	71
LA CABEZA DE PERRO 2021 RD	71
LA CALMA 2019 B FB	330
LA CALMA MÁGICA 2018 T	304
LA CALMA MÁGICA 2019 B	304
LA CAÑA 2021 B	387
LA CAÑA NAVIA 2019 B	387
LA CAÑADA BF PX D	288
LA CAPELLETA 2019 B	371
LA CARENYETA DE CAL PLA 2016 T	358
LA CARRA CABRA 2020 T	322
LA CASA DE BIO 2021 B	848
LA CASA DE LA SEDA 2020 T BA	677
LA CASA DE LAS LOCAS GODELLO 2021 B	282
LA CASA DE MONROY "EL REPISO" 2017 T C	723

WINE	PAGE
LA CASA DEL VOLCÁN 2019 B	250
LA CASILLA 2020 T RB	268
LA CASONA 2015 T	746
LA CASONA DE LA VID 4V 2018 T	803
LA CASONA DE LA VID CABERNET 2017 T BA	804
LA CASONA DE LA VID GARNACHA 2019 T	804
LA CASONA DE LA VID SYRAH 2019 T	804
LA CELESTINA 2019 T	452
LA CELESTINA VENDIMIA SELECCIONADA 2018 T	452
LA CENDRA 2021 T	754
LA CENDRA SELECCIÓN DE FAMILIA 2020 T	754
LA CEPA DE PELAYO 2020 B	269
LA CEPA DE PELAYO BOBAL 2019 T	269
LA CHARLA 2021 B	620
LA CHOZA PAGO MACHARNUDO 2021 B	772
LA CIGARRERA BF AM ES	210
LA CIGARRERA BF MO D	210
LA CIGARRERA BF MZ ES	210
LA CIGARRERA BF OL S	210
LA CIGARRERA BF PX D	210
LA CIGARRERA MANZANILLA PASADA BF MZ ES	210
LA CIGÜEÑA CLARETE 2021 RD	74
LA CIGÜEÑA DOÑA BLANCA 2021 B	74
LA CIGÜEÑA GODELLO 2021 B	74
LA CIGÜEÑA MALVASÍA 2021 B	74
LA CISQUETA DE CORBERA BLANC 2021 B	641
LA CISQUETA DE CORBERA NEGRE 2021 T	641
LA CLAU DE VOLTA 2020 T	331
LA CLOTA 2019 T	767
LA COMTESSE 2018 B FB	398
LA CONRERIA ESCALADEI VI DE VILA 2017 T C	368
LA CORONELA 2020 T	865

WINE	PAGE
LA COSTANA 2018 T C	255
LA CREU ALTA 2018 T BA	370
LA CRUSSET BE R BN	124
LA CUADRILLA 2021 T	661
LA CUARTILLEJA 2018 T R	491
LA CUEVA DEL CONTADOR 2020 T	511
LA CUNA 2021 T RB	869
LA DAMA 2017 T C	36
LA DANSADA 2021 B	641
LA DANSADA 2021 T	641
LA DEL VIVO 2020 B	76
LA DEPA 2020 T	264
LA DEPA ROSÉ 2021 RD	264
LA DOLORES AFRUTADO 2021 B	842
LA DOLORES TINAJA 2020 T	94
LA DOLORES VIÑAS VIEJAS 2021 T	94
LA DONCELLA DE LAS VIÑAS 2021 RD	779
LA DONCELLA DE LAS VIÑAS CHARDONNAY 2021 B AROM	779
LA DONCELLA DE LAS VIÑAS TEMPRANILLO 2020 T	779
LA DONCELLA DE LAS VIÑAS TEMPRANILLO 2021 T RB	779
LA DULA 2020 T	584
LA E ARGÜESO EN RAMA MAGNUM BF MZ	206
LA ERA 2020 T	775
LA ESTACADA SYRAH MERLOT 2018 T C	672
LA ESTELA DEL CICLOHOME 2018 B	382
LA ESTRADA 2019 T	561
LA ESTRECHA 2020 T	268
LA FARAONA 2020 T BA	73
LA FELISA 2020 T	447
LA FINCA 2020 T FB	756
LA FLEUR VIVALTUS 2017 T	496
LA FLOR DE MARGOT MENCÍA 2020 T	420

WINE	PAGE
LA FLOR DE MARGOT TREIXADURA 2021 B	420
LA FLORENS 2019 T	299
LA FLORENS 2020 T	299
LA FLORIDA 2021 RD	252
LA FLORIDA MALVASÍA VOLCÁNICA 2021 B	252
LA FONOLLOSA 2019 T C	637
LA FONT DEL MOSQUIT 2021 B	355
LA FONT VOLTADA 2019 T C	163
LA FREDAT 2019 T	357
LA FUENTE DE MOSITO 2020 T	512
LA FUENTE DE NEKEAS 2019 T C	319
LA FURGONETA QUE MIRABA AL ÓRBIGO 2020 T	810
LA GALBANA 2019 T R	70
LA GARGANTILLA GARNACHA 2019 T	551
LA GARNACHA DE MUSTIGUILLO 2020 T	843
LA GARNACHA OLVIDADA DE ARAGÓN 2020 T	98
LA GARNACHA PERDIDA DEL PIRINEO 2019 T C	869
LA GARNACHA SALVAJE DEL MONCAYO 2019 T	869
LA GARNATXA D'EN PITU RD GR D	186
LA GARNATXA FOSCA DEL PRIORAT 2020 T	372
LA GAVACHA 2020 T BA	97
LA GENUÏNA DE EDETÀRIA 2018 T	647
LA GERMANA DE PASCONA 2021 B	295
LA GITANA ANIVERSARIO BF MZ S	209
LA GITANA BF MZ S	209
LA GITANA EN RAMA BF MZ S	209
LA GUERRERA FINCA CENTENARIA 2017 T C	158
LA GUERRERA FINCA CENTENARIA 2019 T MC	158
LA GUITA BF MZ S	217
LA GUITA EN RAMA BF MZ S	217
LA HAYA 2019 B BA	718
LA HAYA AFRUTADO 2020 B SS	718

WINE	PAGE
LA HAYA SECO 2020 B	718
LA HONDA AMONTILLADO EN RAMA BF AM	210
LA HONDA BF FI	210
LA HORMIGA DE ANTÍDOTO 2020 T	439
LA HORRA 2019 T C	459
LA HOYA 2016 T	537
LA HUELLA DE AITANA 2021 RD	318
LA HUELLA DE AITANA CUVÉE ZEN 2019 RD C	318
LA INA BF FI S	217
LA INTRUSA DE MALASAÑA 2020 T BA	728
LA JACA BF MZ S	206
LA JANDA BF FI S	206
LA JARANA 2021 B	859
LA JOTA DE TO V.R. (VIUDA RICA) 2017 T RB	658
LA KIKA MAGNUM SACA PRIMAVERA BF MZ	214
LA LLAVE ROJA 2020 T	817
LA LLAVE ROJA 8 MESES 2018 T C	817
LA LLORONA 2021 B	65
LA LOBA 2019 T	483
LA LOBITA 2020 T	483
LA LOCA REINA 2021 B	600
LA LOCOMOTORA TEMPRANILLO 2020 T	586
LA MALDITA GARNACHA 2021 RD	556
LA MALDITA GARNACHA 2021 T	556
LA MALDITA GARNACHA BLANCA 2021 B	556
LA MALDITA REVOLUTION 2020 T	556
LA MALQUERIDA TEMPRANILLO- BOBAL- SYRAH 2016 T BA	267
LA MAR 2020 B	391
LA MARE DE PASCONA 2019 T	295
LA MATACALVA 2019 T	309
LA MEJORADA LAS CERCAS 2018 T RB	808
LA MEJORADA LAS NORIAS 2018 T RB	808
LA MEMÒRIA 2019 B	650
LA MERCED 2018 B FB	316
LA MIRANDA DE SECASTILLA 2019 T	629
LA MIRANDA DE SECASTILLA GARNACHA 2020 RD	629
LA MIRANDA DE SECASTILLA GARNACHA BLANCA 2020 B	629
LA MIRONA BLANC 2020 B	840
LA MIRONA ROSAT 2021 RD	840
LA MISIÓN BY MENADE 2019 B	805
LA MONTESA 2020 T C	543
LA MUJER CAÑON 2019 T	728
LA MUNTERA 2019 T	641
LA NAVE 2020 B	575
LA NAVE 2020 T	575
LA NAVE CASA AURORA 2020 T	862
LA NEGRA FLOR 2020 T	646
LA NIETA 2019 T	590
LA NIMFA BLANC 2021 B	164
LA NIÑA DE MIS OJOS 2021 B SD	838
LA NIT DE LES GARNATXES LIMESTONE 2021 T	293
LA NOVICIA 2020 T RB	676
LA NYMPHINA MONASTRELL 2020 T	733
LA OLA 2019 B	382
LA ORPHICA MONASTRELL ILUMINADA 2020 T BA SS	852
LA ORPHICA MONASTRELL SELECCIÓN TARDIA 2021 T SS	852
LA ORPHICA SELECCIÓN AURORA 2021 B SS	852
LA ORPHICA SINTONÍA 2021 RD	852
LA ORQUESTA 2019 B	871
LA OVEJITA TINTA 2021 T RB	681
LA PÁJARA 2021 B FB	725
LA PASIÓN ABDÓN SEGOVIA 2012 T R	657
LA PERA 2021 B	571
LA PERALOSA 2020 T	790
LA PERDICIÓN 2019 T C	772
LA PERSONAL DE EDETÀRIA 2018 T	647
LA PICONADA 2019 T	523
LA PICOSSA GARNACHA BLANCA 2021 B	650
LA PICOSSA GARNACHA TINTA 2021 T	650
LA PITXOTXA CABERNET SAUVIGNON 2017 T C	37
LA PITXOTXA MOSCATEL DE ALEJANDRÍA 2020 B SD	37
LA PITXOTXA ROSÉ 2020 RD	37
LA PIZCA 2017 T	666
LA PLANTA 2021 T RB	440
LA PODA GODELLO 2021 B	690
LA PODA MENCÍA 2020 T	690
LA PODA SAUVIGNON BLANC 2021 B	614
LA POULOSA LA VIZCAÍNA 2019 T	76
LA PRINCESA 2020 T BA	232
LA PROHIBICIÓN 2019 T RB	79
LA PROVINCIA DE PRIETO PARIENTE 2019 T C	806
LA PRUNERA 2016 T	293
LA PURÍSIMA MONASTRELL 2020 T	733
LA PURISIMA OLD VINES EXPRESSION 2019 T RB	733
LA PURÍSIMA PREMIUM 2019 T BA	733
LA QUEBRÁ 2019 T BA	753
LA QUINTA DE RAFA 2021 T	48
LA QUINTA VENDIMIA 2018 T	483
LA RENACIDA 2020 T	156
LA RIOJA ALTA GRAN RESERVA 890 2010 T GR	573
LA RIVA MANZANILLA FINA MIRAFLORES BAJA BF MZ	214
LA RIVA MANZANILLA PASADA BALBAÍNA ALTA BF MZ	214
LA ROCA DE L'ABELLAR 2020 T	355
LA ROMERA T	724
LA SACA DE ALTANZA BF PC S	205
LA SALCEDA 2019 T C	589

WINE	PAGE	WINE	PAGE	WINE	PAGE
LA SALVACIÓN 2020 B	72	LA VAL CRIANZA SOBRE LÍAS 2016 B C	388	LA ZORRA ORIGINAL 2019 T	759
LA SANTA DE ÚRSULA 2020 T	861	LA VAL FERMENTADO EN BARRICA 2016 B FB	388	LA ZORRA RARÍSIMO 2018 T	759
LA SENDA DE PINUAGA PREMIUM CUVÉE 2020 T RB	787	LA VAL SOUSÓN 2019 T RB	388	LA ZORRA ROSADA 2021 RD	759
LA SENOBA 2018 T C	546	LA VAL TREIXADURA 2021 B	388	LABARONA 2018 T R	571
LA SETERA 2019 T C	60	LA VAL VENDIMIA 2017 B	388	LABOREO 2021 B BA	632
LA SETERA 2021 B	60	LA VANIDOSA Nº1 2019 T	522	LACIMA 2019 T	413
LA SETERA 2021 T	60	LA VICALANDA 2016 T R	518	LÁCRIMA BACCUS ROSÉ 2019 RE R BR	142
LA SETERA SELECCIÓN ESPECIAL 2014 T C	60	LA VILLA REAL CABERNET SAUVIGNON 2018 T C S	240	LACRIMUS CRIANZA 2019 T C	547
LA SETERA TINAJA VARIETALES 2015 T RB	60	LA VILLA REAL MACABEO 2021 B	240	LACRIMUS RESERVA 2018 T R	547
LA SIMA 2020 T	266	LA VILLA REAL MOSCATEL 2021 B D	240	LACRIMUS REX 2020 T	547
LA SIRVIENTA 2020 B	232	LA VILLA REAL SAUVIGNON BLANC 2021 B	240	LADAIRO 2019 T BA	280
LA SOBRADA 2019 B	393	LA VILLA REAL TEMPRANILLO 2020 T BA S	240	LADAIRO 2020 B FB	280
LA SOLANA ALTA 2019 T	370	LA VILLA REAL TEMPRANILLO SYRAH 2021 T	240	LADAIRO COLECCIÓN FAMILIA GODELLO TREIXADURA 2021 B	280
LA SOLANA DELS MARGES 2019 T R	363	LA VIÑA DE AMALIO 2019 T	475	LADAIRO COLECCIÓN FAMILIA MENCÍA Y ARAÚXA 2021 T	280
LA SOMBRILLA 2019 B	422	LA VIÑA DE AMAYA 2020 T C	660	LADEIRA VELLA 2018 T C	685
LA SONRISA DEL NÓMADA 2020 T RB	662	LA VIÑA DE AYER ALBILLO REAL 2019 B	753	LADERAS DE INURRIETA 2019 T	307
LA SORT 2020 T	704	LA VIÑA DE AYER GARNACHA 2019 T	753	LADERAS DE MONTORO 2020 T	235
LA SUERTE DE ARRAYÁN 2019 T	274	LA VIÑA DE LA CUEVA COLORÁ 2019 T C S	781	LADERAS DE SEDELLA 2018 T	262
LA TAPADA 2019 T	574	LA VIÑA DE ROBIN 2018 T R	857	LADERAS DEL NORTE 2020 T	440
LA TAPADA 2020 T	574	LA VIÑA DE SEGUNDO 2019 T RB	660	LADERAS OSTATU 2016 T	578
LA TARDANA ESTENAS 2021 B	675	LA VIÑA DEL ABUELO PREMIUM 2017 T	666	LADERO 2016 T C	243
LA TEMPTACIÓ 2019 B	340	LA VIÑA DEL ABUELO SELECCIÓN ESPECIAL 2016 T C	666	LADERO 2021 RD	243
LA TENTACIÓN 2019 B FB	714	LA VIÑA DEL ABUELO VENDIMIA SELECCIONADA 2021 T	667	LADERO AIRÉN VERDEJO 2021 B	243
LA TENTACIÓN AFRUTADO 2021 B SD	714	LA VIÑA ESCONDIDA 2018 T	274	LADERO TEMPRANILLO 2021 T	243
LA TENTACIÓN SECO 2021 B	714	LA VITORIANA 2020 T	76	LADRÓN DE GUEVARA DE AUTOR 2019 T C	551
LA TERCIA B PX D	286	LA VOZ DEL VIÑADOR 2019 T	864	LADRÓN DE LUNAS BE BN	142
LA TORPE AVUTARDA DESCANSA 2018 T C S	774	LA VOZ DEL VIÑADOR 2020 T	864	LADRÓN DE LUNAS BOBAL 2021 T C	681
LA TRUCHA 2021 B	396	LA ZARANDA CHARDONNAY VIOGNIER 2021 B	858	LADRON DE LUNAS EXCLUSIVE VINO DE AUTOR 2015 T GR	681
LA TRUCHA ACERO 2017 B	396	LA ZARANDA SYRAH 2021 T	858	LADRÓN DE LUNAS GARNACHA RE BR	681
LA TRUCHA BARRICA 2020 B	396	LA ZORRA 8 VIRGENES 2021 B	759	LADRÓN DE LUNAS ROBLE BOBAL 2021 T RB	681
LA TRUCHA DE OTOÑO 2017 B	396	LA ZORRA ARAGONÉS 2019 T	759	LADRÓN DE LUNAS SAUVIGNON MACABEO 2021 B	681
LA VAL ALBARIÑO 2021 B	388	LA ZORRA LA NOVENA RUFETE BLANCO 2020 B	759	LAFOU DE BATEA 2016 T R	648

WINE	PAGE
LAFOU EL SENDER 2019 T C	648
LAFOU ELS AMELERS 2021 B	648
LAGAR D'AMPRIUS 92/300 SYRAH NOBLE 2016 T	766
LAGAR D'AMPRIUS CHARDONNAY 2019 B	766
LAGAR D'AMPRIUS GARNACHA 2017 T	766
LAGAR D'AMPRIUS GARNACHA 2019 T	766
LAGAR D'AMPRIUS GEWÜRZTRAMINER 2018 B	766
LAGAR D'AMPRIUS KOLENDA 2018 B	766
LAGAR D'AMPRIUS NAZARÍN 2019 T	766
LAGAR DA CONDESA 2021 B	394
LAGAR DE BENAVIDES B FI S	286
LAGAR DE BESADA 2021 B	395
LAGAR DE CAYO 2021 B	559
LAGAR DE CAYO CLARETE 2021 RD	559
LAGAR DE CAYO TEMPRANILLO BLANCO 2021 B	559
LAGAR DE CERVERA 2021 B	395
LAGAR DE COSTA 2021 B	395
LAGAR DE COSTA TRADICIÓN 2018 B BA	396
LAGAR DE DEUSES GODELLO 2021 B	281
LAGAR DE DEUSES MENCÍA 2020 T	282
LAGAR DE ROBLA COLECCIÓN CUATRO HERMANOS FINCA VALDELEBRE 2020 T	814
LAGAR DE ROBLA PREMIUM 2021 T	814
LAGAR DE ROBLA SELECCIÓN 2021 T	814
LAGAR DE SANTIAGO 2019 T C	523
LAGAR DE SANTIAGO 2021 T MC	523
LAGAR DE SANTIAGO VIURA-VERDEJO 2021 B	523
LAGAR DEL DUQUE 2021 RD	160
LAGAR DEL REY 2019 B FB	600
LAGAR DEL REY SAUVIGNON BLANC SOBRE LÍAS 2021 B	601
LAGAR DEL REY VERDEJO SOBRE LÍAS 2021 B	601
LAGAR DOS MATEOS 2021 B	386
LAGAR PEDREGALES FLORACIÓN 2021 B	393
LAGRAZ 2019 T RB	724
LÁGRIMA DE VITALIS 2021 RD	256
LÁGRIMA DE VITALIS ALBARÍN 2021 B	256
LÁGRIMA NEGRA 2019 T C	493
LÁGRIMA NEGRA 2020 T RB	493
LÁGRIMA NEGRA EDICIÓN LIMITADA 2019 T	493
LÁGRIMAS DE MARÍA 2017 T R	593
LÁGRIMAS DE MARÍA 2018 T C	593
LAHAR BE BN	885
LAHAR DE CALATRAVA 2021 B	799
LAHAR DE CALATRAVA 2021 B FB	799
LAHAR DE CALATRAVA 2021 T	799
LAHAR TEMPRANILLO RE BN	885
LAJAS "FINCA EL PEÑISCAL" 2016 T	98
LAJAS "FINCA EL PEÑISCAL" 2019 T	98
LALAMA 2019 T	413
LALOMBA 2021 RD	574
LALOMBA FINCA ILES 2019 T	574
LALOMBA FINCA LADERO 2017 T	574
LALOMBA FINCA VALJONTA 2018 T	574
LALUME 2020 B	425
LALUT BLANC DE NOIR 2021 B	187
LALUT CREACIÓN MARCUS 2017 T C	187
LALUT D'EN MARC MAGNIFICO 2017 T	187
LALUT SELECCIÓN ESPECIAL 2017 T	187
LALUT SUPERIOR 2018 T C	187
LAMIN DE SOMMOS 2019 T	844
LAN 2015 T GR	530
LAN 2017 T R	530
LAN 7 METROS 2019 T C	529
LAN A MANO 2019 T	529
LAN D-12 2019 T C	530
LANGA CLASSIC 2019 T	98
LANZADO PET NAT 2021 BE	794
LANZADO PET NAT ROSÉ 2021 RE	794
LANZAGA 2018 T	561
LAPENA 2019 B	413
LAPILLI 2019 B	250
LAPOLA 2020 B	413
LAQUARTA GRANS ANYADES NEGRE VINYES VELLES 2018 T	649
LAQVARTA BLANC 2020 B	649
LAQVARTA NEGRE 2019 T	649
LAR DE MAÍA 5º 2020 T BA	811
LAR DE MAÍA 7º AUTOR 2020 T BA	811
LAR DE MAÍA 8º 2021 RD	811
LAR DE PAULA 2017 T C	531
LAR DE PAULA 2018 B FB	531
LAR DE PAULA EDICIÓN LIMITADA 2015 T R	531
LAR DE PAULA EDICIÓN LIMITADA 2016 T C	531
LAR DE PAULA MADURADO EN BODEGA 2019 T	531
LAR DE RICOBAO GODELLO - BRANCO LEXÍTIMO - TREIXADURA 2020 B	410
LAR DE RICOBAO OURO DO VAL 2019 B RA	410
LAR DE RICOBAO SELECCIÓN DO VAL 10 LUNAS 2014 T	410
LAR DE RICOBAO SELECCIÓN DO VAL 2021 T	410
LAS 30 DEL CUADRADO 2020 B FB	848
LAS ÁNIMAS 2020 T	752
LAS BEATAS 2019 T	561
LAS BLANCAS 2021 B	706
LAS CABEZADAS DE MATADULA 2019 T	568
LAS CAMPANAS 2009 T GR	313
LAS CAMPANAS 2016 T R	313
LAS CAMPANAS 2018 T C	313

WINE	PAGE
LAS CAMPANAS 2021 B	313
LAS CAMPANAS 2021 RD	313
LAS CAMPANAS ROSÉ 2021 RD PL	313
LAS CANDELERA 2021 B FB	568
LAS CENIZAS 2020 T	591
LAS CHOVAS 2019 T	231
LAS DOS CES 2020 T RB	679
LAS DOS CES 2021 B	680
LAS GALGAS 2020 B	871
LAS GUNDIÑAS LA VIZCAÍNA 2019 T	76
LAS IRUELAS 2019 T	809
LAS LAMAS 2020 T BA	73
LAS MAMBLAS 2019 T	56
LAS MARGAS SIERRA DE ALGAIREN GARNACHA BLANCA 2021 B	110
LAS MARGAS SIERRA DE ALGAIREN GARNACHA TINTA 2020 T	110
LAS MATILLAS 2019 T	446
LAS MERCEDES BOBAL AL LÍMITE 2019 T	681
LAS MERCEDES PAGO AÑINA 2021 B	772
LAS MORADAS DE SAN MARTÍN ALBILLO REAL 2020 B	726
LAS MORADAS DE SAN MARTÍN INITIO 2017 T	726
LAS MORADAS DE SAN MARTÍN LA SABINA 2015 T R	726
LAS MORADAS DE SAN MARTÍN LAS LUCES 2011 T GR	726
LAS MORADAS DE SAN MARTÍN SENDA 2019 T	726
LAS MUSAS 2021 RD	157
LAS OCHO 2018 T	740
LAS PANZUDAS 2018 T	367
LAS PARADAS 2020 T	105
LAS PARVAS 2017 T	664
LAS PISADAS 2018 T	573
LAS PIZARRAS FABLA 2021 T BA	98
LAS PIZARRAS VIÑA ALARBA 2020 T BA	98

WINE	PAGE
LAS PLANAS 2017 B R	581
LAS REÑAS SELECCIÓN MONASTRELL SYRAH 2018 T C	91
LAS ROCAS GARNACHA 2020 T	96
LAS ROCAS GARNACHA VIÑAS VIEJAS 2020 T	96
LAS SALINAS BELTZA 2019 T C	574
LAS SALINAS ZURI DE LUBERRI 2020 B FB	574
LAS SEÑORAS MEDIUM BF OL MED	214
LAS SUERTES 2020 T	718
LAS TIERRAS DE JAVIER RODRÍGUEZ EL PEGO 2019 T C	660
LAS TIERRAS DE JAVIER RODRÍGUEZ EL TESO ALTO 2016 T	660
LAS TIERRAS GARNACHA 2021 T	660
LAS TINTAS 2021 T	706
LAS TIRAJANAS 2019 T RB	200
LAS TIRAJANAS MALVASÍA VOLCÁNICA 2019 B	200
LAS TIRAJANAS MALVASÍA VOLCÁNICA 2021 B FB	200
LAS TIRAJANAS PARAJE CALDERA DE TIRAJANA 2021 T RB	200
LAS TIRAJANAS TINAMAR 2021 T RB	200
LAS TIRAJANAS VERIJADIEGO 2019 B	200
LAS TRES 2019 B	741
LAS TRES FILAS 2020 T RB	70
LAS UMBRÍAS 2019 T	725
LAS UVAS DE LA IRA 2020 T	276
LAS VIRTUDES FORTALEZA 2021 B FB	35
LASCALA 2021 B	39
LASLÍAS DE BERONIA 2020 B	596
LATIERRA 2017 T RB	781
LATÚE 2021 RD	240
LATÚE AIRÉN 2021 B	240
LATÚE CABERNET SAUVIGNON & SYRAH 2019 T	240
LATÚE MÉTODO TRADICIONAL BE BR	848
LAUDUM 2020 T RB	35
LAUDUM MONASTRELL 2020 T RB	35

WINE	PAGE
LAUDUM XII PLUS 2017 T C	35
LAULLA 2021 T	589
LAUNA 2019 T C	531
LAUNA 2021 B	531
LAUNA SELECCIÓN FAMILIAR 2017 T R	532
LAUNA SELECCIÓN FAMILIAR 2018 T C	531
LAUNA SELECCIÓN FAMILIAR 2021 B FB	531
LAUREATUS 2021 B	388
LAUREATUS DOLIUM 2013 B C	388
LAUREATUS LÍAS 2013 B C	388
LAUS 2017 T R	625
LAUS 2018 T C	625
LAUS 2021 RD	625
LAUS 2021 T	625
LAUS CHARDONNAY 2021 B	625
LAUSOS 2018 T	816
LAVENTURA TEMPRANILLO 2018 T	575
LAVENTURA VIURA 2020 B	575
LAVIA FINCA PASO MALO 2019 T C	92
LAVIA VALLE DEL ACENICHE 2019 T C	92
LAVIA VALLE VENTA DEL PINO 2019 T C	92
LAVIÑA 2021 B	119
LAVIÑA 2021 RD	119
LAVIÑA TEMPRANILLO MERLOT 2019 T BA	119
LAXAS 2021 B	385
LAYA 2021 T	47
LAZARILLO 2021 T	789
LAZARILLO AIRÉN 2021 B	789
LE BOBAL 2020 T BA	674
LE CENCIBEL 2020 T BA	674
LE GRENACHE 2019 T BA	674
LE NATUREL 2021 B	305

WINE	PAGE
LE NATUREL 2021 T	305
LE PETIT GALIA 2018 T	802
LE ROSÉ 2020 RD	674
LE ROSÉ DE ANTÍDOTO 2020 RD	439
LE VERMENTINO 2020 B	841
LECEA 2010 T GR	532
LECEA 2014 T R	532
LECEA 2016 T C	532
LECEA 2017 B C	532
LECO PUNK 2021 T	574
LEDA VIÑAS VIEJAS 2018 T R	804
LEDEA 2019 T C	306
LEDEA GARNACHA 2021 RD	306
LEGADO DE FARRO GODELLO 2021 B	814
LEGADO DE FARRO MENCÍA 2021 T RB	814
LEGADO ESTECILLO GARNACHA & SYRAH 2020 T BA	99
LEGADO FINCA EL REFUGIO PETIT VERDOT 2016 T	791
LEGADO FINCA EL REFUGIO SYRAH 2016 T	792
LEGADO GARNACHA 2020 T BA	99
LEGADO GARNACHA BLANCA MACABEO 2021 B BA	99
LEGADO MACABEO 2020 B FB	99
LEGADO MUÑOZ CHARDONNAY 2021 B	798
LEGADO MUÑOZ GARNACHA 2020 T	798
LEGADO MUÑOZ MACABEO VERDEJO 2021 B	798
LEGADO MUÑOZ MERLOT 2019 T	798
LEGADO MUÑOZ TEMPRANILLO 2021 T	798
LEGARIS 2019 T C	483
LEGARIS 2020 T RB	483
LEGARIS ALCUBILLA DE AVELLANEDA 2018 T BA	483
LEGARIS MORADILLO DE ROA 2018 T BA	484
LEGARIS OLMEDILLO DE ROA 2018 T	484
LEGARIS SAUVIGNON BLANC 2021 B	617
LEGARIS VERDEJO 2021 B	617
LEGITIM 2019 T C	364
LEGUM 2018 T BA	110
LEIRA PONDAL 2021 B	399
LEIRANA 2021 B	387
LEIVE PARADIGMA 2021 B	426
LEIVE RELIQUIA 2020 B FB	426
LEIVE TREIXADURA 2020 B	426
LEIX 2019 T	177
LEMONIER 2021 BE BN	879
LENEUS 2017 T BA	817
LENEUS CAYETANA 2021 B	817
LENEUS REISHI 2019 T	817
LENEUS TEMPRANILLO 2021 T MC	817
LEONOR BF PC S	216
LERMA 2018 T C	55
LES ALCUSSES 2018 T	708
LES ARGILES D'ORTO VINS BLANC 2021 B	300
LES ARGILES D'ORTO VINS NEGRE 2021 T	300
LES ARGILES D'ORTO VINS ROSAT 2021 RD	300
LES AUBAGUETES 2020 T C	354
LES BRUGUERES 2020 T	368
LES BRUGUERES 2021 B	368
LES CAMADES 2020 T	355
LES CASETES 2020 T	299
LES CERVERES XARELLO 2019 B	331
LES CLIVELLES DE L'ALZINA 2019 T	358
LES CLIVELLES DE TORROJA 2020 T	358
LES CLIVELLES DE TORROJA ROSAT 2021 RD	358
LES COMES D'ORTO 2019 T	300
LES CRESTES 2021 T	371
LES ELIES 2019 T BA	188
LES ERES 2016 T C	367
LES ERES ESPECIAL DELS CARNERS 2015 T C	367
LES GALLINETES 2021 T	165
LES MANYES 2019 T	375
LES MARGUES 2019 B BA	370
LES MINES 2019 T C	366
LES PEDRENYERES 2020 B	291
LES PUSSES 2017 T C	364
LES SOMERES 2021 B	857
LES SOMERES 2021 T	857
LES SORTS BLANC 2021 B FB	294
LES SORTS JOVE 2021 T	294
LES SORTS SYCAR 2019 T	294
LES SORTS VINYES VELLES 2018 T C	294
LES TALLADES DE CAL NICOLAU 2017 T C	300
LES TERRASSES 2020 T	354
LES TOSSES 2019 T C	375
LES VISTES 2020 B	334
LETRADO SOLERA 1992 BF SOLERA	168
LETRAS MINÚSCULAS 2020 T	562
LEZA GARCÍA EDICIÓN GARNACHA 2019 T C	557
LEZA GARCÍA TINTO FAMILIA 2018 T C	557
LEZAUN 0,0 SULFITOS 2021 T	313
LEZAUN 2014 T R	313
LEZAUN 2019 T C	313
LEZAUN GAZAGA 2020 T RB	313
LEZAUN TEMPRANILLO 2021 T MC	313
LEZCANO-LACALLE 2017 T R	158
LEZCANO-LACALLE DÚ 2015 T	158
LG DE LEZA GARCÍA 2018 T	557
LÍA DE PRADOREY 2021 RD	472
LIALA 2020 RD FB	159

WINE	PAGE	WINE	PAGE	WINE	PAGE
LIANTE 2020 T RB	537	LIVIUS BLANCO 2018 B FB	514	LO CABALÓ 2016 T R	364
LÍAS FINAS POR 220 CÁNTARAS 2018 B C	509	LIVIUS ROSADO 2018 RD FB	514	LO CARNISSER 2020 B	291
LIASON GARNACHA 2021 T	110	LLABUSTES CHARDONNAY 2021 B	173	LO COSTER BLANC 2021 B	374
LIBALIS ROSÉ 2020 RD	835	LLABUSTES MERLOT 2018 T C	173	LO MAS 2021 T	364
LIBALIS WHITE 2020 B	835	LLABUSTES RIESLING 2020 B	173	LO MAS D'ETDETÀRIA 2019 T	647
LÍBANO 3 GENERACIONES 2018 T C	583	LLABUSTES ULL DE LLEBRE 2019 T C	173	LO MÓN 2017 T	375
LIBER 2021 B	183	LLÀGRIMES DE TARDOR GARNATXA BLANCA 2020 B FB	649	LO NOI DEL SAC DE PASCONA 2021 B	295
LIBERSO CURIOSO VERDEJO 2016 B FB	617	LLÀGRIMES DE TARDOR ROSAT 2021 RD	649	LO NOI DEL SAXO 2021 B	642
LIBERSO CURIOSO VERDEJO 2020 B FB	617	LLANO QUINTANILLA 2018 T C	734	LO NOI DEL SAXO 2021 T	642
LIBERTARIO 2021 T	240	LLANOS NEGROS LA BATISTA 2020 B FB	250	LO NOIR 2019 T	370
LIBRE Y SALVAJE CARIÑENA 2020 T	111	LLANOS NEGROS LA TIME 2000 B	250	LO PETIT PAU 2021 T	360
LIBRE Y SALVAJE CLARETE 2020 RD	111	LLANOS NEGROS LOS GRILLOS 2020 T	250	LO POU 2020 B	767
LIBRE Y SALVAJE GARNACHA 2020 T	111	LLANUM 2016 T R	449	LO POU 2020 T	767
LIBRE Y SALVAJE GARNACHA BLANCA 2020 B	111	LLEDA COUPAGE 2020 T	109	LO VY 2020 B	636
LIBRE Y SALVAJE NARANCHA 2019 B	112	LLEGAT LLOPIS 2017 B	332	LO VY 2020 T	636
LIBRE Y SALVAJE TINAJAS ANTIGUAS 2019	112	LLEIROSO 2016 T R	453	LO VY ANCESTRAL 2020 B	636
LICENCIADO 2017 T R	522	LLEIROSO 2018 T C	453	LOBARZÁN ISAURA 2020 B FB	280
LICINIA 2016 T R	726	LLENCA PLANA 2019 T RB	300	LOBARZÁN ISAURA 2020 T	280
LICINIA 2020 B C	726	LLEPOLIA BLANC 2021 B	643	LOCUAZ VIURA TINAJA 2021 B	849
LICINIA MIRVS 2017 T R	726	LLEPOLIA NEGRE 2019 T C	643	LOCULTO 2019 T C	489
LICINIA PLUS 2016 T R	726	LLOPART 2018 BE R BN	881	LOCULTO SELECCIÓN 2020 T RB	489
LICINIA PLUS 2019 B C	727	LLOPART CLOS DELS FÒSSILS 2021 B	338	LOEBRE "BLANC DE NEGRES" 2020 B	650
LICOS 2020 B	647	LLOPART EXVITE VIÑAS SINGULARES LES FLANDES 2012 BE BR	881	LOESS BLUE CAP 2019 T C	484
LIENZO AIRÉN BE BN	798	LLOPART LEOPARDI 2015 BE BN	882	LOESS COLLECTION 2018 T C	484
LIENZO AIRÉN PIE FRANCO 2020 B	798	LLOPART MICROCOSMOS ROSÉ 2019 RE BN	882	LOESS COLLECTION 2021 B FB	617
LIENZO BLEND 2017 T R	248	LLOPART ORIGINAL 1887 VIÑAS SINGULARES LES FLANDES 2011 BE BN	882	LOESS INSPIRATION 2020 T	484
LIENZO VERDEJO SAUVIGNON BLANC 2020 B	798			LOESS RUEDA 2021 B	617
LIGHT MY FIRE 2020 T	712	LLOPART PANORAMIC 2016 BE BR	882	LOHER TRADICIONAL 2020 T	632
LINDARAJA 2019 T RB	202	LLUM 2021 B	830	LOIDANA 2020 T C	369
LITHIUM 2019 T C	740	LLUM D'ALBA 2021 B	375	LOIDANA BLANC 2021 B	369
LIVING SEMILLON 2020 B	704	LLUM DE VI BLANC 2021 B	163	LOLA ROSÉ PINOT NOIR 2019 RE R BR	128
LIVING TEMPRANILLO 2019 T	704	LLUOR NEGRE 2019 T	82	LOMA DE LOS FELIPES 2018 T C	203

WINE	PAGE
LÓPEZ CRISTOBAL 2021 RD	453
LÓPEZ CRISTOBAL ALBILLO MAYOR 2021 B S	453
LÓPEZ CRISTOBAL LA COLORADA 2019 T C	453
LÓPEZ CRISTOBAL PARCELA 1 2018 T R	453
LORE DE OSTATU 2019 B FB	579
LORENZO BORIS TINAHA 2020 T	844
LORIÑÓN 2019 B	580
LORIÑÓN 2019 T C	580
LORIÑÓN CUVEE 2020 T	580
LOS ALTOS MENDI 2021 B	578
LOS ARRÁEZ ARCOS 2020 T	845
LOS ARRÁEZ ARCOS 2021 T	845
LOS ARRÁEZ LAGARES 2019 T RB	702
LOS ARRÁEZ MALVASÍA 2021 B RB	702
LOS ARRÁEZ PARCELA 0 2018 T RB	702
LOS ARRÁEZ SYRAH 2020 T	845
LOS ARRÁEZ VERDIL 2021 B	702
LOS ARROTOS DEL PENDÓN 2019 B	257
LOS ARROYUELOS 2020 T	752
LOS CABOS DE URBEZO 2021 T	112
LOS CANTOS DE TORREMILANOS 2019 T	400
LOS CANTOS DE TORREMILANOS 2020 T	480
LOS CARISMÁTICOS 2019 T	694
LOS CONFINES 2019 T C	806
LOS CORRALES DE MONCALVILLO MATURANA BLANCA 2020 B	520
LOS CORRALES DE MONCALVILLO MATURANA EN AMPHORA 2020 T	520
LOS CORRALES DE MONCALVILLO MATURANA TINTA 2019 T BA	521
LOS DOMINIOS DE BERCEO PREHILOXÉRICO 2017 T	534
LOS DOMINIOS DE BERCEO RESERVA 36 2006 T R	534
LOS ESCRIBANOS 2019 T	700

WINE	PAGE
LOS FRAILES CALIZA 2020 T	707
LOS FRAILES DOLOMITAS 2020 T	707
LOS FRAILES RUBIFICADO 2020 T	707
LOS JALONES 2021 RD	449
LOS JALONES 2021 T RB	449
LOS LASTROS 2019 T	664
LOS LOROS "LA BOTA DE MATEO" 2019 B	714
LOS LOROS "LISTÁN BLANCO DE CANARIAS" 2021 B S	714
LOS LOROS "SIETE LOMAS" 2020 B FB	714
LOS LOROS "TINTILLA CASTELLANA" 2020 T RB	714
LOS LOROS ALBILLO CRIOLLO 2021 B	714
LOS LOROS LISTÁN NEGRO 2021 T BA	714
LOS LOSARES GARNACHA TINTORERA 2019 T	48
LOS LOSARES MONASTRELL 2019 T	48
LOS MOLINOS AIRÉN VERDEJO B	697
LOS MOLINOS TEMPRANILLO T	697
LOS OLIVOS DE NEKEAS 2016 T R	319
LOS PEROS 2018 T	725
LOS PEROS 2019 B	725
LOS PINOS 0 % SULFITO 2021 T	704
LOS TABLEROS AFRUTADO 2021 B SS	29
LOS VIENTOS 2019 T	757
LOSADA 2020 T	75
LOSADA GODELLO 2021 B	75
LOURO DO BOLO GODELLO 2021 B	692
LOUSAS CAMIÑO NOVO 2020 T	414
LOUSAS SEOANE 2020 T	861
LOUSAS VIÑAS DE ALDEA 2020 T	861
LOUVRE LÍAS 2021 B	419
LOVE IS VERMELL 2021 B	335
LOZANO SELECCIÓN DE ORÍGENES VIÑAS VIEJAS 2020 B	578
LUBERRI 2021 T MC	574

WINE	PAGE
LÚCULO 2021 B FB	312
LUCULO ORIGEN 2019 T C	313
LUIS CAÑAS 2015 T GR	533
LUIS CAÑAS 2019 T C	533
LUIS CAÑAS SELECCIÓN DE FAMILIA 2017 T R	533
LUIS CAÑAS VIÑAS VIEJAS 2020 B	533
LUIS GURPEGUI 147 ANIVERSARIO 2018 T C	534
LUIS PEIQUE 2017 T RB	68
LUIS SAAVEDRA 2019 T C	723
LUIS SAAVEDRA VENDIMIA NOCTURNA 2017 T RB	723
LUIS XIV ÁNFORAS 2021 T	41
LUMIÈRE 2019 B	771
LUNA BEBERIDE GODELLO 2021 B	70
LUNA BEBERIDE MENCÍA 2021 T	70
LUNA CRECIENTE 2021 B	404
LUNA DE MAR 2018 T C	709
LUNARES 2018 T	261
LUNARES 2020 RD MC	261
LUNAS NUEVAS 2018 T R	666
LUNÁTICO MONASTRELL 2020 T	231
LUNO 2020 T	298
LUNO BLANC 2020 R	298
LURRETIK 2021 B	195
LUSCO ALBARIÑO 2021 B	400
LUSIA 2019 T RB SD	822
LUSTAU ALMACENISTA CAYETANO DEL PINO BF PC S	217
LUSTAU ALMACENISTA MANZANILLA PASADA MANUEL CUEVAS JURADO BF MZ	218
LUSTAU ALMACENISTAS PATA DE GALLINA GARCÍA JARANA BF OL S	218
LUSTAU AMONTILLADO DEL CASTILLO ANTONIO CABALLERO Y SOBRINOS BF AM S	217

WINE	PAGE	WINE	PAGE	WINE	PAGE
LUSTAU AMONTILLADO VORS BF AM S	218	MACIÀ BATLE NEGRE DOLÇ 2018 T D	827	MÁIS ALÁ 2020 B	875
LUSTAU AÑADA VINTAGE SHERRY 2003 BF OL D	218	MACIÀ BATLE ROSAT 2021 RD	827	MAISULAN 12 2018 T C	575
LUSTAU EMPERATRIZ EUGENIA BF OL S	218	MACIÀ BATLE SAUVIGNON BLANC 2020 B FB	827	MAISULAN EL HONDÓN 2020 T	575
LUSTAU JARANA BF FI	218	MACIÀ BATLE SAUVIGNON BLANC 2021 B	827	MAISULAN LOS MAGINES 2020 T	575
LUSTAU PEDRO XIMÉNEZ VORS BF PX D	218	MACIÀ BATLE XEREMIA BLANC 2020 B	827	MAISULAN SOBREMORO 2020 T BA	575
LUSTAU SAN EMILIO BF PX D	218	MACIÀ BATLE XEREMIA BLANC 2021 B	827	MAISULAN TXABOLA GRACIANO 2020 T	575
LUXURIA 2020 B	282	MADRE ÚNICA AUTOR 2019 T	515	MAISULAN TXABOLA TEMPRANILLO 2020 T	575
LUZÓN COLECCIÓN MONASTRELL 2021 T	227	MADRID ROMERO BLANCO BF MISTELA D	842	MAIUS ASSEMBLAGE 2020 T	369
LVZMILLAR 2020 T RB	453	MADRID ROMERO ROBLE 12 MESES 2017 T C	222	MAIUS BARRANC DE LA BRUXA 2019 T C	369
LYRIUS ONE FROM PRIETO PICUDO 2021 RD	807	MADRID ROMERO ROBLE 6 MESES 2019 T RB	222	MAIUS GARNATXA BLANCA 2021 B	369
LYRIUS ONE FROM VERDEJO 2021 B S	808	MADRID ROMERO TINTO TF D	842	MAJUELO EL ESPEJO 2020 B	809
LYRIUS ONE FROM VERDEJO 2021 B SD	808	MADURESA 2020 T	708	MAJUELOS DE CALLEJO 2019 T	447
LZ 2021 T	561	MAESTRO DE DURÓN 2018 T C	491	MAL RAIO TE PARTA 2021 B	418
M		MAGADI 2018 B	517	MALA VIDA 2020 T RB	702
M DE MAS IGNEUS 2020 T	371	MAGALARTE LEZAMA 2020 B	88	MALA VIDA 2021 B	702
M DE MONROY GARNACHA SYRAH 2020 T	723	MAGALARTE ZAMUDIO 2019 B FB	84	MALA VIDA EDICIÓN LIMITADA 2020 T RB	702
M DE MURUA 2019 T	540	MAGALARTE ZAMUDIO 2020 B FB	84	MALABRIGO 2018 T	475
MABAL 2020 T	90	MAGALARTE ZAMUDIO 2021 B	84	MÁLAGA ORO VIEJO BF TRASAÑEJO D	262
MABAL MACABEO DE BALCONA 2019 B	90	MAGENC 2021 B	186	MÁLAGA PX NOBLE QUITAPENAS BF PX D	262
MABRE 2021 B	188	MAGIA DE ADVENTUS 2015 T FB	818	MÁLAGA VIRGEN DUNKEL BF PX D	261
MACÁN 2018 T	517	MAGMA DE CRÁTER 2018 T C	631	MÁLAGA VIRGEN SWEET BF PX D	262
MACÁN CLÁSICO 2019 T	517	MAGNUM DE BARBIANA BF MZ S	214	MALAJE 2021 RD	842
MACCERATO 2021 B	403	MAGNUM DE LA GOYA BF MZ S	214	MALAPIPA 2021 B	842
MACERADO POR 220 CÁNTARAS 2019 B	509	MAGRAN PARTIDA LES MANYETES 2017 T C	371	MALARADO 2020 RD FB	674
MACIÀ BATLE 1856 2018 T BA	827	MAINETES 12 MESES 2018 T C	227	MALAVIA 2021 B	842
MACIÀ BATLE 2020 T	827	MAINETES VERDEJO 2021 B FB	227	MALCAVADA 2021 RD	406
MACIÀ BATLE 2021 B MC	827	MAIO 2018 B	396	MALCAVADA SELECCIÓN 2021 T	406
MACIÀ BATLE 2021 T MC	827	MAIOR DE MENDOZA 3 CRIANZAS 2018 B	396	MALCRIAT 2020 T	707
MACIÀ BATLE BLANC DE BLANCS 2021 B	827	MAIOR DE MENDOZA 3 CRIANZAS 2019 B	396	MALCUAT 2021 T	192
MACIÀ BATLE BLANC DE BLANCS DOLÇ 2019 B D	827	MAIOR DE MENDOZA FINCA LAS TABLAS 2017 B BA	396	MALDITO 2019 T	691
MACIÀ BATLE BLANC DE BLANCS ORANGE 2021 B	827	MAIOR DE MENDOZA SOBRE LÍAS 2020 B	396	MALDITO PARNÉ 2020 T C	845
MACIÀ BATLE COLLECCIÓ PRIVADA 2018 T	827	MAIOR DE MENDOZA SOBRE LÍAS 2021 B	396	MALEVAJE 2019 BE	881

WINE	PAGE	WINE	PAGE	WINE	PAGE
MALKOA PRIVATE COLLECTION 2017 B BA S	52	MANTEL BLANCO SAUVIGNON BLANC 2021 B	595	MAR DE FLORES 2021 T	780
MALKOA PRIVATE COLLECTION 2018 B BA S	52	MANTEL BLANCO VERDEJO 2020 B FB	595	MAR DE FONDO 2020 B	617
MALKOA TXAKOLI EDICIÓN LIMITADA 2018 B	52	MANTEL BLANCO VERDEJO 2021 B	595	MAR DE FRADES ALBARIÑO 2021 B	389
MALLEOLUS 2019 T	447	MANTOLÁN BE BN	247	MAR DE FRADES BE BN	389
MALLEOLUS DE SANCHOMARTÍN 2018 T	447	MANTONEGRO BLANCO SON JULIANA 2019 B	831	MAR DE FRADES GODELLO ATLÁNTICO 2020 B	389
MALLEOLUS DE VALDERRAMIRO 2018 T	447	MANTONEGRO TINTO SON JULIANA 2019 T	831	MAR DE LLUNA MOSCAT 2021 B	185
MALLERENGA 2017 BE GR BN	135	MANUEL ARAGÓN BF AM S	218	MAR DE ONS 2021 B	384
MALLOLET 2021 B	191	MANUEL ARAGÓN BF OL S	218	MAR DE VINS ALGUER VINYES VELLES 2020 B	856
MALOCO 2020 T	704	MANUEL ARAGÓN BF PC S	218	MAR DE VINS ELS FUSTALS VINYES VELLES 2020 B	856
MALOCO 2021 T	677	MANUEL ARAGÓN BF PX D	218	MAR DE VINS ERMITÀ VINYES VELLES 2020 B	856
MALÓN DE ECHAIDE 2018 T C	313	MANUEL D´AMARO LOUREIRA BLANCA 2018 B	402	MAR DE VIÑAS 2021 B	379
MALÓN DE ECHAIDE 2021 B	313	MANUEL D'AMARO ALBARIÑO LÍAS 2016 B	401	MAR GARCÍA 2019 T	47
MALÓN DE ECHAIDE 2021 RD	313	MANUEL EL SANTERO 2017 T RB	787	MAR I CEL 2021 B	338
MALÓN DE ECHAIDE CHARDONNAY 2020 B FB	313	MANUEL MANZANEQUE ¡EA! 2019 T RB	794	MARA MARTIN GODELLO 2021 B	282
MALÓN DE ECHAIDE GARNACHA 2020 T RB	313	MANUEL QUINTANO CEPAS VIEJAS 2020 T	513	MARA MOURA 2020 B	282
MALPASO 2020 T	274	MANUEL QUINTANO EL PIONERO 2020 T	513	MARAÑONES 2019 T	727
MALPUESTO 2020 T	543	MANUEL QUINTANO SELECCIÓN PARTICULAR 2019 T	513	MARBA 2021 B BA	632
MALVAJIO VINO DE GARAJE 2018 T	842	MANUEL RAVENTÓS 2011 BE BN	883	MARBA 2021 RD	632
MALVASÍA 2020 B	182	MANUEL RAVENTÓS 2014 BE BN	883	MARBA 2021 T BA	632
MALVASÍA DE SANT JAUME 2021 B	710	MANUEL RAVENTÓS 2015 BE BN	883	MARBA 2021 T MC	632
MALVASIA DE SITGES 2011 B MISTELA D	855	MANUELA DE NAVERÁN 2021 R	332	MARBA CAPRICHO 2021 T FB	632
MALVASIA DE SITGES RANCIO BF RB S	855	MANVIER 2021 T	449	MARBELLA BLUSH ROSÉ 2020 RD	262
MALVID SELECCIÓN 2021 B	616	MANYOL 2020 T	651	MARBORÉ CUVÉE 2019 T	626
MALVID VERDEJO 2021 B	616	MANYOL 2021 B	651	MARCELINO I 2020 T	410
MANAR DOS SEIXAS 2020 B	426	MANYOL ECO 2021 T	651	MARCELINO I 2021 T	410
MANCHEGAL 2021 BE SD	885	MANYOL SELECCIÓN 2019 T	651	MARCENCA 2021 T	301
MANCHEGAL ROSADO 2021 RE SD	885	MANZÁN 2020 T C	707	MARCO REAL COLECCIÓN PRIVADA 2018 T C	314
MANCHOMUELAS BLANCO DE BERNABELEVA 2020 B FB	721	MANZANOS 2014 T GR	534	MARCO REAL FINCA LA PARED CUVÉE ESPECIAL 2019 T	314
MANERAS DE VIVIR 2017 T	707	MANZANOS 2016 T R	534	MARCO REAL FINCA LA PARED GRACIANO 2019 T	314
MANLIANA 2020 T C	751	MAR ADENTRO 2021 B	283	MARCO REAL FLOR DE CHARDONNAY 2020 B	314
MANO A MANO 2019 T C	782	MAR DE ALISTE 2020 T	841	MARCO REAL PEQUEÑAS PRODUCCIONES GARNACHA RD	314
MANS DE SAMSÓ 2020 T	302	MAR DE ENVERO 2017 T BA	396	MARCO VALERO MARCIAL 2019 T	98

WINE	PAGE
MARESIA DEL ATLÁNTICO 2019 B	250
MARESIA DEL ATLÁNTICO 2020 B	250
MAREVIA 2019 BE BR	151
MARGALIDA LLOMPART BLANC 2019 B	827
MARGALIDA LLOMPART ROSAT 2021 RD	827
MARGARIDA LLOMPART NEGRE 2017 T	827
MARGE 2019 T	359
MARGER SUMARROCA 2020 B FB	343
MARÍA 2019 T C	674
MARÍA BARGIELA 2020 B	382
MARÍA BERNET 2013 BE BN	881
MARÍA BERNET 2015 BE BN	881
MARÍA CASANOVAS 2019 BE GR BN	142
MARÍA CASANOVAS XP 2019 BE GR BN	142
MARIA CATASÚS BE R BN	138
MARÍA DE MOLINA VERDEJO 2021 B	605
MARIA GANXA DE PASCONA 2021 T	295
MARIA RIGOL ORDI 2015 BE GR BN	142
MARÍA RIGOL ORDI 2017 BE R BN	142
MARIA RIGOL ORDI MÀGNUM CUPATGE DOS MIL DISSET 2017 BE R BN	143
MARÍA RIGOL ORDI MICROTIRATGE #6: PARELLADA 2016 BE GR BN	143
MARÍA RIGOL ORDI MICROTIRATGE #7: TREPAT 2020 BE GR BN	143
MARIA RIGOL ORDI MIL·LENNI 2017 BE R BN	143
MARÍA SANZO 2021 B	390
MARÍA SARMIENTO 2021 T	734
MARÍA SERRA 2021 B	828
MARILUNA 2020 T BA	711
MARILUNA 2021 B	711
MARINA ALTA 2021 B	35
MARINA DE ALISTE 2020 T	841

WINE	PAGE
MARIO VC 2019 T	492
MARIOLA T	112
MARISOL RUBIO CIPMA I 2019 B	783
MARISOL RUBIO CIPMA I 2020 B	783
MARISOL RUBIO CIPMA II 2020 B FB	783
MARISOL RUBIO SON D SOL 2020 B	783
MARLA VINO DE PARATGE LES SALANQUES (ETIQUETA NEGRA) 2019 T C	373
MARLA VINO DE VILA DE POBOLEDA (ETIQUETA ROJA) 2019 T C	373
MARMALLEJO 2018 T C	91
MAROLA & MASS 2021 B	381
MARQUÉS DE ALVIA GARNACHA 2013 T GR	514
MARQUÉS DE ALVIA GRACIANO 2014 T	514
MARQUÉS DE ALVIA TEMPRANILLO 2014	514
MARQUÉS DE BURGOS 2018 T C	484
MARQUÉS DE BURGOS 2020 T RB	484
MARQUÉS DE CÁCERES 2016 T R	535
MARQUÉS DE CÁCERES 2018 T C	535
MARQUÉS DE CARRIÓN 2018 T C	575
MARQUÉS DE CARRIÓN T R	576
MARQUÉS DE CASTILLA TEMPRANILLO SYRAH MERLOT 2021 T	246
MARQUÉS DE GRIÑÓN CLÁSICO 2019 T	576
MARQUÉS DE GRIÑÓN ROSÉ "LA VIE EN ROSE" RE BR	143
MARQUÉS DE GRIÑÓN SELECCIÓN ESPECIAL 2018 T C	576
MARQUÉS DE GRIÑÓN VERDEJO 2021 B	617
MARQUÉS DE IRÚN VERDEJO 2021 B	617
MARQUÉS DE JUBERA 2019 T C	514
MARQUÉS DE LA CONCORDIA MM SELECCIÓN ESPECIAL BE BN	143
MARQUÉS DE LA CONCORDIA RESERVA DE LA FAMILIA BLANC DE BLANCS 2018 BE BN	143
MARQUÉS DE LA CONCORDIA RESERVA DE LA FAMILIA ROSÉ RE BR	143

WINE	PAGE
MARQUÉS DE LA CONCORDIA RIOJA SANTIAGO 2016 T R	576
MARQUÉS DE LA CONCORDIA RIOJA SANTIAGO 2018 T C	576
MARQUÉS DE LA CONCORDIA RIOJA SANTIAGO SEGUNDO AÑO 2020 T	576
MARQUÉS DE LA CONCORDIA SELECCIÓN ESPECIAL ROSÉ 2019 RE BR	143
MARQUÉS DE MONTECIERZO SELECCIÓN MERLOT 2016 T RB	308
MARQUÉS DE MONTEJOS SELECCIÓN 2021 T	815
MARQUÉS DE MURRIETA 2018 T R	576
MARQUÉS DE MURRIETA PRIMER ROSÉ 2021 RD	576
MARQUÉS DE PEÑAMONTE COLECCIÓN PRIVADA 2020 T	661
MARQUÉS DE REINOSA 2018 T R	535
MARQUÉS DE REINOSA 2021 B SD	535
MARQUÉS DE REINOSA MATURANA 2020 T	535
MARQUÉS DE REINOSA TEMPRANILLO 2021 T	535
MARQUÉS DE REINOSA TEMPRANILLO BLANCO 2021 B	535
MARQUÉS DE RISCAL 1860 TEMPRANILLO 2021 T RB	804
MARQUÉS DE RISCAL 2016 T GR	523
MARQUÉS DE RISCAL 2018 T R	523
MARQUÉS DE RISCAL LIMOUSIN 2021 B FB	606
MARQUÉS DE RISCAL SAUVIGNON BLANC 2021 B	606
MARQUÉS DE RISCAL TXAKOLI 2020 B	195
MARQUÉS DE RISCAL VERDEJO ORGANIC 2021 B	606
MARQUÉS DE RISCAL VIÑAS VIEJAS 2021 RD	805
MARQUÉS DE RIVALLANA 2015 T GR	529
MARQUÉS DE RIVALLANA 2018 T R	529
MARQUÉS DE RIVALLANA 2019 T C	529
MARQUÉS DE TENA 2018 T R	683
MARQUÉS DE TERÁN 2016 T R	535
MARQUÉS DE TERÁN 2018 T C	535
MARQUÉS DE TERÁN EDICIÓN LIMITADA 2015 T R	535
MARQUÉS DE TERÁN SELECCIÓN ESPECIAL 2018 T	535

WINE	PAGE	WINE	PAGE	WINE	PAGE
MARQUÉS DE TOLEDO VERDEJO 2021 B	849	MARTIN & PONS 2019 T	438	MARTÍNEZ PALACIOS SELECCIÓN 30 BARRICAS 2018 T C	537
MARQUÉS DE TOMARES 2015 B GR	577	MARTÍN BERDUGO 2015 T R	438	MARTÍNEZ ROSÉ RIMARTS 2020 RE BN	148
MARQUÉS DE TOMARES 2015 T GR	577	MARTÍN BERDUGO 2019 T C	438	MARTÍNEZ SAEZ SELECCIÓN 2017 T C	783
MARQUÉS DE TOMARES 2016 T R	577	MARTÍN BERDUGO 2020 T	438	MARTÍNSANCHO 2021 B	618
MARQUÉS DE TOMARES 2018 B C	577	MARTÍN BERDUGO 2020 T BA	438	MÁRTIRES 2020 B	568
MARQUÉS DE TOMARES 2019 T C	577	MARTÍN BERDUGO 2021 RD	438	MARTIVILLÍ SAUVIGNON BLANC 2021 B	613
MARQUÉS DE TORO 2021 T	801	MARTÍN BERDUGO 2021 T	438	MARTIVILLÍ VERDEJO 2021 B	613
MARQUÉS DE VALDUEZA ETIQUETA ROJA 2018 T C	818	MARTÍN BERDUGO VERDEJO 2021 B	601	MARTÚE 20 ANIVERSARIO 2018 T	740
MARQUÉS DE VALDUEZA GRAN VINO DE GUARDA 2017 T GR	818	MARTÍN CENDOYA 2016 T R	565	MARTÚE CHARDONNAY 2020 B BA	740
MARQUÉS DE VARGAS 2015 T GR	557	MARTÍN CÓDAX 2021 B	389	MARTÚE ESPECIAL 2017 T R	740
MARQUÉS DE VARGAS 2017 T R	557	MARTÍN CÓDAX AROUSA 2019 B C	389	MARTÚE SYRAH 2018 T	740
MARQUÉS DE VARGAS HACIENDA PRADOLAGAR 2017 T R	557	MARTÍN CÓDAX FINCA XIELES 2020 B	389	MARURE 2018 B	568
MARQUÉS DE VARGAS SELECCIÓN PRIVADA 2017 T R	557	MARTÍN CÓDAX GALLAECIA 2017 B C	389	MARUXA GODELLO 2021 B	694
MARQUÉS DE VELILLA 2019 T C	484	MARTÍN CÓDAX LÍAS 2019 B C	389	MARUXA MENCÍA 2021 T	694
MARQUÉS DE VELILLA 2020 T RB	484	MARTÍN CÓDAX VINDEL 2018 B	390	MAS BERTRAN LA GRAUA 2013 BE GR BN	338
MARQUÉS DE VELILLA 2021 T	485	MARTÍN SARMIENTO 2017 T	67	MAS BERTRAN X80 2012 BE GR BR	338
MARQUÉS DE VITORIA 2015 T GR	577	MARTÍN VERÁSTEGUI 2019 T BA	809	MAS CANDÍ 2019 BE BN	882
MARQUÉS DE VITORIA 2017 T R	577	MARTÍNEZ ALESANCO 2016 T R	536	MAS COMTAL INCROCIO MANZONI 2018 B	865
MARQUÉS DE VITORIA 2019 T C	577	MARTÍNEZ ALESANCO 2018 T C	536	MAS COMTAL LYRIC SOLERA 1993 TF SOLERA D	865
MARQUÉS DE VITORIA 2021 B	577	MARTÍNEZ ALESANCO 2021 B FB	536	MAS COMTAL PÉTREA CHARDONNAY 2017 B FB	865
MARQUÉS DE VITORIA TEMPRANILLO 2021 T	577	MARTÍNEZ ALESANCO ROSADO 2021 RD FB	536	MAS COMTAL PÉTREA MERLOT 2017 T	865
MARQUÉS DE VIZHOJA 2021 B	849	MARTÍNEZ BERMELL MERLOT 2019 T C	747	MAS COMTAL PIZZICATO 2021 RD	865
MARQUÉS DEL ATRIO 2017 T R	536	MARTÍNEZ CORTA 2018 T C	577	MAS D'ALSINA & SARDÁ 2018 BE R BN	122
MARQUÉS DEL ATRIO 2019 T C	536	MARTÍNEZ LACUESTA 2011 T GR	536	MAS D'EN COMPTE 2016 T R	358
MARQUÉS DEL ATRIO EDICIÓN LIMITADA 2017 T	536	MARTÍNEZ LACUESTA 2014 B GR	536	MAS D'EN COMPTE 2017 B FB	358
MARQUÉS DEL PUERTO 2016 T R	577	MARTÍNEZ LACUESTA 2018 B R	536	MAS D'EN POL 2018 T C	646
MARQUÉS DEL PUERTO 2018 T C	577	MARTÍNEZ LACUESTA 2019 T C	536	MAS D'EN POL 2020 T	646
MARRURRO 2020 T C	834	MARTÍNEZ LACUESTA LA SUCURSAL 2020 T BA	536	MAS D'EN POL 2021 B	646
MART 2021 RD	345	MARTÍNEZ PALACIOS 2014 T R	537	MAS DE LA PANSA DOLÇ DE TREPAT 2019 T D	165
MARTA CIBELINA 2021 B	790	MARTÍNEZ PALACIOS 2018 B C	537	MAS DE LA PANSA ESCUMÓS ROSAT 2018 RE S	165
MARTA MATÉ 2019 T C	454	MARTÍNEZ PALACIOS 2018 T C	537	MAS DE LA PANSA PARELLADA 2018 B	118
MARTÍ FABRA SELECCIÓ VINYES VELLES 2019 T RB	185	MARTÍNEZ PALACIOS GRACIANO 2019 T C	537	MAS DE LA PANSA TREPAT 2018 T	165

WINE	PAGE
MAS DE LA ROSA 2019 T C	365
MÁS DE LEDA 2018 T C	804
MAS DEL BOTÓ 2016 T	856
MAS DEL BOTÓ 2019 T	856
MAS DEL HABANERO 2017 T	367
MAS DEL SERRAL 2010 BE BN	885
MAS DEL SERRAL 2011 BE BN	885
MAS DELS METS T	296
MÁS FELIZ, VIDES OLVIDADAS 2019 B	480
MAS LA PLANA 2018 T R	335
MAS MALLOLA 2019 T R	369
MAS MOMENTOS CABERNET SAUVIGNON 2021 T	789
MAS MOMENTOS SYRAH 2021 T	789
MAS MOMENTOS TEMPRANILLO 2021 T	789
MAS OLLER BLAUNIT 2021 T	189
MAS OLLER MALVASÍA 2019 B MISTELA D	189
MAS OLLER MAR 2021 B	189
MAS OLLER PLUS 2018 T	189
MAS OLLER PUR 2020 T C	189
MAS OLLER SYRAH DE LA MUNTANYA 2019 T C	189
MAS PICOSA BLANC 2021 B	117
MAS PICOSA NEGRE 2021 T	117
MAS QUE 2 2021 B	388
MAS RAMONEDA BOIRA BLANC DE NOIR 2021 B	175
MAS RAMONEDA ERIÇÓ 2018 T C	176
MAS RAMONEDA PERDIU 2018 T RB	176
MAS RAMONEDA VOLA VOLA PAPALLONA 2021 B	176
MAS RODÓ CABERNET SAUVIGNON 2015 T	338
MAS RODÓ INCÒGNIT 2020 T	338
MAS RODÓ MACABEO 2017 B	339
MAS RODÓ MERLOT 2015 T R	339
MAS RODÓ MONTONEGA 2018 B	339

WINE	PAGE
MAS RODÓ RIESLING 2019 B FB	339
MAS SINÉN CLOS 2017 T	356
MAS SINÉN COSTER 2015 T C	356
MAS SINÉN LA VALL 2017 T BA	356
MAS TORTÓ 2020 T C	293
MAS TORTÓ BLANC 2021 B	294
MAS UBERNI BLANC DE BLANC 2021 B	341
MAS UBERNI CHARDONNAY 2021 B	341
MAS VILELLA BLANC 2020 B FB	866
MAS VILELLA BLANC 2021 B FB	866
MAS VILELLA NEGRE 2020 T	866
MASCARADAS 2019 T RB	804
MASCARÓ "AMBROSIA" 2019 BE R SS	143
MASCARÓ MAGNUM 2019 BE BN	143
MASCARÓ NIGRUM 2019 BE R BR	144
MASCARÓ PURE 2019 BE R BN	144
MASCARÓ RUBOR AURORAE 2019 RE BR	144
MASCÚN GARNACHA 2019 T C	628
MASCÚN GARNACHA 2021 RD	628
MASCÚN GARNACHA BLANCA 2021 B	628
MASCUN GEWURZTRAMINER 2021 B	628
MASDACHE 2020 B	253
MASET 1917 2018 BE GR BN	126
MASET CABERNET SAUVIGNON 2019 T R	328
MASET CLOS VILÓ 2019 T	355
MASET DEL LLEÓ SYRAH 2018 T R	115
MASET EUFÒRIA 2021 RD	328
MASET FOC MERLOT 2018 T R	328
MASET L'AVI PAU 2018 BE GR BN	126
MASET LA SÍNIA 2020 B FB	328
MASET LA SOLEDAD 2020 B FB	329
MASET MAS VILÓ 2019 T	355

WINE	PAGE
MASET MAS VILÓ 2020 T	355
MASET NU BRUT ROSÉ 2020 RE BR	126
MASET SENDEROS DEL MOLINERO 2020 B	537
MASET SINGULAR XAREL·LO BIODINÀMIC 2021 B	329
MASET TEMPRANILLO 2018 T R	537
MASET TEMPRANILLO 2019 T C	537
MASET VINTAGE 2018 BE R BN	126
MASIA CAL COSTAS CABERNET SAUVIGNON SYRAH 2017 T	327
MASÍA CARRERAS BLANC 2019 B FB	185
MASÍA CARRERAS NEGRE 2019 T	185
MASÍA PAIRAL CAN CARRERAS GARNATXA DE L'EMPORDÀ B D	185
MASSALUCA GARNATXA BLANCA & MACABEO 2021 B	650
MASSIMO MENCÍA & SOUSÓN 2021 T	409
MASSIMO SELECCIÓN BARRICA FRANCESA 2019 T RB	409
MASSIMO SELECCIÓN GODELLO 2020 B	409
MASTINELL CARPE DIEM 2016 BE GR BN	130
MASTINELL CRISTINA 2016 BE GR EBR	130
MASTINELL NATURE 2014 BE GR BN	130
MASUSTA 2020 T	319
MATACÁ 2017 T	825
MATALBA COLECCIÓN PRIVADA 2020 B	866
MATALBA COLECCIÓN PRIVADA 2020 T	866
MATALBA GARNACHA 2020 T	866
MATALBA MALVAR 2020 B	727
MATALBA VELO 2020 T	866
MATALBA VELO DE FLOR EDICIÓN LIMITADA 2020 T	866
MATALBA VINO BLANCO 2020 B	866
MATALBA VIÑAS VIEJAS 2020 T	867
MATALIAN 2021 B	769
MATALLANA 2018 T	475
MATALLONGA SELECCIÓ 2017 T C	173
MATARILE 2018 T C	627

WINE	PAGE
MATARILE 2020 B	627
MATARROMERA 2019 T C	436
MATARROMERA PRESTIGIO 2016 T	436
MATEO COLECCIÓN TREIXADURA 2021 B	421
MATEO QUINTAS 2021 B	421
MATHA BY ST JOANNES 2019 T	328
MATHER TERESINA 2019 T	645
MATIAS MICHELINI GARNACHE 2019 T	322
MATILDA NIEVES MENCÍA 2021 T	411
MATIUS MONASTRELL 2018 T C	227
MATSU EL PÍCARO 2021 T	667
MATSU EL RECIO 2020 T	667
MATSU EL VIEJO 2020 T	667
MATSU LA JEFA 2019 B	667
MATUSALEM VORS BF OL D	216
MAUDES 2018 T C	158
MAULEÓN 2014 T R	526
MAURO 2020 T	805
MAURO ESTÉVEZ 2021 B	423
MAURO VENDIMIA SELECCIONADA 2019 T	805
MAXX 2016 T C	260
MAXX 2017 I C	260
MAYOR DE CASTILLA 2020 T RB	493
MAYOR DE CASTILLA VERDEJO 2021 B	613
MAYOR DE ONDARRE 2016 T R	542
MAYORAL 2016 T C	222
MAYORAL 2020 T	222
MAYORAL RESERVADO T	222
MB MARTÍN BERDUGO T	438
MEDIA LEGUA 2021 B	696
MEDIEVO 2017 T R	523
MEDIEVO 2019 T C	523

WINE	PAGE
MEDRANO IRAZU 2017 T R	538
MEDRANO IRAZU 2019 T C	538
MEDRANO IRAZU 2021 T	538
MEDUSA ALBARIÑO 2021 B	394
MEGALA 2019 T	703
MELÉ 2014 T	853
MELER 10 BALLOS DE GARNACHA 2019 T	627
MELER 15 2015 T C	627
MELER 6 2018 T C	627
MELER 9 2016 T	628
MELER SYRAH 2020 T	628
MELIC 2018 T	707
MELIOR 9 MESES 2020 T	436
MELIOR SAUVIGNON 2021 B	600
MELIOR VERDEJO 2021 B	600
MEMBRILLERA 2020 T C	674
MEMÒRIA 2016 BE BN	879
MEMÒRIES DE BINIAGUAL NEGRE 2019 T C	81
MEMÒRIES DEL PRIORAT RANCI CAL MARCELINO TF RC	376
MENADE SAUVIGNON BLANC DULCE 2021 B D	805
MENADE VERDEJO 2021 B	805
MENCEY CHASNA SECO 2021 B	29
MENCEY CHASNA SEMISECO 2021 B SS	29
MENCEY DE CHASNA AFRUTADO 2021 RD	29
MENCEY DE CHASNA AFRUTADO SEMIDULCE 2021 B SD	30
MENCEY DE CHASNA VIJARIEGO NEGRO 2020 T	30
MENCIÑO 2021 T	686
MENCIÑO SUMMUM 2020 T	686
MENDI BY MENDIETA OSABA 2021 T	578
MENGUANTE CARIÑENA SELECCIÓN 2019 T BA	113
MENGUANTE GARNACHA 2021 T	113
MENGUANTE GARNACHA BLANCA 2020 B	113

WINE	PAGE
MENGUANTE GARNACHA SELECCIÓN 2019 T	113
MENGUANTE VIDADILLO 2019 T	113
MENTIDERO 2020 B	253
MERAYO GARNACHA TINTORERA 2018 T	70
MERAYO GODELLO 2021 B	70
MERAYO MENCÍA 2021 T	71
MERCÈ JOVE 2021 T	327
MERIAN BLANC 2021 B	646
MERIAN NEGRE 2021 T	646
MERIAN ROSAT 2021 RD	646
MERIT 2018 T C	372
MERITUM 2016 BE BN	139
MERNAT DE NOC 2017 T	784
MERNAT DE NOC VIOGNIER 2020 B	784
MERRUTXU 2021 B	88
MERUM PRIORATI DESTI 2019 T	372
MERUM PRIORATI EL CEL 2019 T	372
MERUM PRIORATI INICI 2019 T	372
MÉS FELIÇ 2020 T	338
MÉS QUE PARAULES BLANC 2021 B	348
MÉS QUE PARAULES ROSAT 2021 RD	348
MESIES 2016 T C	650
MESIES GARNATXA BLANCA 2019 B FB	650
MESIES SELECCIÓ 2015 T R	650
MESTA TEMPRANILLO 2021 RD	672
MESTA VERDEJO 2021 B	672
MESTÍS 2021 B	182
MESTIZAJE 2020 B	843
MESTIZAJE 2020 T	742
MESTRE VILA VELL 2019 T	115
METÁFORA 2019 T	64
METAMORFOSE DE SANTIAGO ROMA 2020 T	390

Peñín Guide SPANISH WINE 993

WINE	PAGE	WINE	PAGE	WINE	PAGE
MF 2013 T	186	MIMETIC 2020 T	97	MISSENYORA 2020 B FB S	175
MI AMOR 2020 T	359	MINAIRO 2021 T	368	MISTELA MOSCATEL TURÍS SELECCIÓN 2021 B D	709
MI TRACTOR AZUL 2021 T	669	MINDORO 2020 B FB	849	MISTERIO CONDADO VIEJO BF S	168
MI VERDADEJO 2019 B FB	664	MINERAL 2020 T C	292	MISTERIO ORANGE NARANJA BF SOLERA D	168
MIBAL 2018 T C	483	MINGORTIZ 2016 T	568	MISTIK MOSCATELL 2019 B D	837
MIBAL 2019 T RB	482	MINGUA 2020 T	627	MITOS MACABEO 2020 B	677
MIBAL 2021 T	482	MINGUA 2021 B	627	MITOS TEMPRANILLO 2020 T	677
MIBAL SELECCIÓN 2014 T	482	MINGUS 2021 T	65	MIUDIÑO 2021 B	401
MICAELA SELECCIÓN DE AÑADA 2020 B	815	MINIUS GODELLO 2021 B	279	MIUT EL JABONERO 2020 B FB	289
MIGAN 2020 T	861	MIÑA VIDA 2021 B	399	MIX 2020 B C	866
MIL CANTOS 2019 T	503	MIQUEL JANÉ BALTANA GARNACHA 2021 RD	327	MIX 2020 T C	866
MIL CEPAS CENCIBEL 2018 T BA	794	MIQUEL JANÉ SAUVIGNON BLANC 2021 B D	327	MIXTURA 2020 B C	866
MIL HISTORIAS BOBAL 2021 T RB	266	MIQUEL JANÉ XARELLO 2021 B	327	MIXTURA ETIQUETA DORADA 2018 B	866
MIL HISTORIAS MALBEC 2021 T	266	MIQUEL PONS 2019 BE R BN	132	MIXTURA ETIQUETA ROJA 2018 T	866
MIL RÍOS GARNACHA 2019 T C	692	MIQUEL PONS 77 VEREMAS GARNACHA 2020 T	331	MIXTURA ETIQUETA VERDE 2020 B	866
MIL RÍOS GODELLO 2019 B BA	692	MIQUEL PONS 77 VEREMAS XAREL·LO 2021 B FB	331	MIZARAN 2020 B	266
MIL RÍOS GODELLO 2020 B	692	MIQUEL PONS GRAN RESERVA VINTAGE 2015 BE GR BN	132	MIZARAN BOBAL 2021 RD	267
MIL RÍOS MENCÍA 2020 T	692	MIQUEL PONS MONTARGULL 2015 BE GR BN	132	MIZARAN TEMPRANILLO 2020 T RB	267
MILAGROS DE FIGUERO 2018 T	495	MIQUEL PONS MONTARGULL XARELLO 2015 BE GR BR	133	MMADRE 2018 T	752
MILETO CRIANZA 2018 T C	514	MIQUEL PONS NÚRIA DE MONTARGULL 2020 B	331	MO SALINAS 2019 T FB	39
MILIARIO 2019 T RB	69	MIRA SALINAS 2013 T	39	MO SALINAS MOSCATEL MACABEO 2021 B	39
MILL 2019 T C	485	MIRABELLES 2018 B	332	MOCANAL 2020 T	199
MILLATOS 2020 T	160	MIRABRÁS 2019 B FB	769	MOCÉN SAUVIGNON BLANC 2021 B	606
MILMANDA 2019 B C	165	MIRANDA D'ESPIELLS 2021 B	341	MOCÉN VERDEJO 2020 B	606
MILSETENTAYSEIS 2019 T	485	MIRANIUS 2021 B	332	MOCÉN VERDEJO SELECCIÓN ESPECIAL 2021 B	607
MILSETENTAYSEIS LA PEÑA 2020 RD	485	MIRANIUS MAGNUM 2019 B	332	MOCETE 2021 T MC	513
MILVUS 2021 RD	454	MIRENE 2020 B	87	MODERNITXEN VINO NOBLE DE ALICANTE 2017 TF CRM	37
MILVUS 2021 T	454	MIRLO BLANCO 2021 B	595	MODUS VIVENDI RIBEIRO 2021 B S	429
MILVUS EDICIÓN ESPECIAL 2020 T	454	MIROS DE RIBERA 2018 T C	455	MOFO - THE WILD CHILD 2019 T	669
MIM 2013 T R	294	MIROS DE RIBERA 2020 T RB	455	MOFO – THE DAMN MASTER 2018 T	669
MIM NATURA BLANC DE NOIRS 2016 BE GR BN	154	MIRTHYA 2020 T	90	MOFO – THE FREE SOUL 2021 B	619
MIM NATURA PINOT NOIR ROSADO 2019 RE R BR	154	MIRTO DE RAMÓN BILBAO 2016 T	544	MOGAR 2020 T RB	455

WINE	PAGE
MOGAR COLECCIÓN PRIVADA 2016 T R	455
MOGAR VENDIMIA SELECCIONADA 2019 T C	455
MOISES GRAN VINO 2014 T BA	667
MOLINO REAL 2018 B	262
MOLLET 2021 B	832
MOMENTO DIEZ 2020 B	615
MOMPICHEL 2020 B FB	779
MÓN BOBAL 2018 T	682
MÓN EIXERIT 2019 T	345
MÓN MACABEO 2019 BE R BN	144
MÓN MACABEO 2021 B	682
MONASTERIO DE CORIAS FINCA LOS FRAILES 2020 T	751
MONASTERIO DE CORIAS PAGO DEL NARCEA 2018 T	751
MONASTERIO DE CORIAS VIÑA GRANDIELLA 2020 B	751
MONASTERIO DE LAS VIÑAS 2018 T R	111
MONASTERIO DE SAN MIGUEL 2019 T C	457
MONASTERIO DE SAN MIGUEL 2020 T RB	457
MONASTERIO DE SAN MIGUEL 2021 B	457
MONASTERIO REAL 2019 T BA	117
MONASTRELL BY BODEGA LA ENCINA 2021 T	842
MONCERBAL 2020 T	73
MONDEO SELECCIÓN ESPECIAL 2020 T BA	233
MONEMBASIA 2017 BE BN	332
MONI DOLÇ ORANGE WINE 2021 B D	826
MÓNICA ALBOR MENCÍA SOUSÓN CAIÑO 2021 T	426
MÓNICA ALBOR TREIXADURA 2020 B	427
MÓNICA ALBOR TREIXADURA GODELLO LOUREIRA 2020 B	427
MONJARDÍN CHARDONNAY ESPECIAL SUPERIOR 2009 B GR	306
MONJARDÍN CHARDONNAY SELECCIÓN FAMILIAR 2016 BE GR BN	878
MONJARDÍN TEMPRANILLO ESPECIAL SUPERIOR 2017 T R	306
MONJE AMESTOY DE LUBERRI 2016 T R	575

WINE	PAGE
MONJE TERRUÑO 2016 T R	858
MONÓCULO 2019 T RB	48
MONOPOLE 2021 B	563
MONOPOLE CLÁSICO 2019 B	563
MONOPOLE S. XXI 2021 B	615
MONREAL 2015 T	329
MONROY MALVAR 2020 B	723
MONT MARÇAL 2019 BE R BN	144
MONT MARÇAL 2019 BE R BR	144
MONT MARÇAL ROSÉ 2020 RE R BR	144
MONTALVO WILMOT CABERNET DE FAMILIA 2017 T	784
MONTALVO WILMOT COLECCIÓN PRIVADA 2017 T R	784
MONTALVO WILMOT PETIT VERDOT 2019 T C	745
MONTALVO WILMOT PETIT VERDOT SELECCIÓN 2017 T R	784
MONTALVO WILMOT PETIT VERDOT TEMPRANILLO SYRAH 2020 T BA	784
MONTALVO WILMOT SYRAH 2020 T RB	745
MONTALVO WILMOT TEMPRANILLO CABERNET 2019 T RB	745
MONTAÑA FINCA EL FARAÓN 2019 T R	567
MONTAÑA FINCA LA VALENTINA 2019 T C	567
MONTCABRER 2017 T C	44
MONTE AMÁN 2017 T C	55
MONTE AMÁN 2018 T RB	55
MONTE AMÁN 2020 T	55
MONTE AMÁN 2021 RD	55
MONTE ARAYA 2021 RD	523
MONTE DE LAS MOZAS BOBAL 2021 RD	271
MONTE DE LAS MOZAS MACABEO 2021 B	271
MONTE DEL CONDE 2019 T RB	437
MONTE OTON 2020 T	103
MONTE PINADILLO 2018 T R	437
MONTE PINADILLO 2020 T C	437

WINE	PAGE
MONTE PINADILLO 2021 T RB	437
MONTE PINADILLO ROSADO DE LÁGRIMA 2021 RD	437
MONTE REAL 2015 T GR	545
MONTE REAL CRIANZA FAMILIA 2020 T C	546
MONTE REAL CUVÉE 2020 T C	546
MONTE REAL GARNACHA 2021 T	546
MONTE REAL RESERVA DE FAMILIA 2019 T R	546
MONTE REAL TEMPRANILLO BLANCO 2021 B	546
MONTEABELLÓN 14 MESES 2019 T C	469
MONTEABELLÓN 5 MESES 2021 T RB	469
MONTEABELLÓN FINCA LA BLANQUERA 2017 T GR	469
MONTEABELLÓN FINCA MATAMBRES 2018 T	469
MONTEABELLÓN VERDEJO 2021 B	613
MONTEAGUDO BF OL S	215
MONTEARRUIT VIEJÍSIMO 75 AÑOS BOTA 1/1 B AM	287
MONTEBACO CARA NORTE 2020 T C	485
MONTEBACO DE FINCA 2020 T C	485
MONTEBACO SELECCIÓN ESPECIAL 2019 T	485
MONTEBACO VERDEJO + SAUVIGNON 2021 B	618
MONTECASTRILLO 2020 T RB	480
MONTECASTRILLO 2021 T RB	480
MONTECASTRO 2018 T R	470
MONTECASTRO 2019 T BA	470
MONTECIERZO ROSÉ 2021 RD	308
MONTECILLO 2015 T R	538
MONTECILLO 2018 T C	538
MONTECILLO 22 BARRICAS 2015 T GR	538
MONTECILLO EDICIÓN LIMITADA 2016 T	538
MONTECRUZ 2014 T GR	109
MONTECRUZ 2015 T GR	109
MONTEGAREDO 2020 T C	486
MONTEGAREDO GRAN SELECCIÓN 2020 T C	486

WINE	PAGE
MONTELAGUNA 2019 T C	477
MONTELAGUNA 2020 T RB	477
MONTELAGUNA SELECCIÓN 2018 T	477
MONTEPLOGAR 2018 T C	111
MONTES DE LEZA EDICIÓN LIMITADA 2019 T C	578
MONTES DE LEZA EDICIÓN LIMITADA LÍAS 2019 B	578
MONTES DE LEZA TEMPRANILLO 2021 T	578
MONTES OBARENES 2018 B	555
MONTESPINA SAUVIGNON 2021 B	596
MONTESPINA VERDEJO 2021 B	595
MONTESQUIUS 1918 MAGNUM 2004 BE GR BN	144
MONTESQUIUS 2010 BE GR BN	144
MONTESQUIUS COLECCIÓN PRIVADA 2013 BE GR BN	144
MONTESQUIUS EDICIÓN ESPECIAL BLANC DE BLANCS 2015 BE GR BN	144
MONTESQUIUS LA ESENCIA ROSÉ 2011 RE GR BN	145
MONTESQUIUS NATURELOVERS 2018 BE R BN	145
MONTESQUIUS VINTAGE 2018 BE R EBR	145
MONTEVEIGA 2018 B	389
MONTNEGRE 2019 T	82
MONTORO 2021 B	235
MONTORO DE FORASTERA 2021 B FB	235
MONTRODÓ BLANC 2020 B	860
MONTRODÓ BLANC 2021 B	860
MONTRODÓ NEGRE 2020 T S	860
MONTRODÓ ROSAT 2021 RD	860
MONTSE SOLA 2019 T	473
MONTSEC MIRAPALLARS 2018 T	173
MONTSEC SAUVIGNON BLANC 2019 B	173
MONTUNO 2019 T RB	869
MORAIMA ALBARIÑO 2021 B	403
MORAIMA CAIÑO 2019 T	403

WINE	PAGE
MORCA 2019 T	104
MORGADÍO 2021 B	379
MORTITX CALLET - GORGOLLASSA 2021 T	832
MOSCATEL ORGULLO VINO DE LICOR B	706
MOSSÈN ALCOVER 2018 T	351
MOSST 2021 B	189
MOUNTAIN 2019 B	263
MR DULCE 2020 B D	263
MR. TAMBOURINE WINE 2018 T	848
MUCHAS MANOS 2021 RD	244
MUCHAS MANOS AIRÉN 2021 B	244
MUCHAS MANOS TEMPRANILLO 2021 T	244
MUCY 2018 T C	158
MUCY 2021 RD	158
MUDARE 2019 B	680
MUDÈFER BLANC 2019 B C	644
MUDÈFER NEGRE 2017 T C	644
MUGA 2018 T C	539
MUGA 2019 T C	539
MUGA 2021 B	539
MUGA 2021 RD	539
MUGA SELECCIÓN ESPECIAL 2018 T R	539
MUNDO DE YUNTERO VERDEJO- SAUVIGNON BLANC 2021 B	243
MUNIA (6 MESES EN BARRICA) 2020 T RB	657
MUNIA CARÁCTER 2020 T RB	657
MUNIA ESPECIAL 2019 T R	657
MUNIADONA 2019 B FB	56
MUÑANA 1188 BE BN	879
MUÑANA 1188 CABERNET SAUVIGNON 2018 T	202
MUÑANA 2019 B	850
MUÑANA 3 CEPAS 2018 T	765
MUÑANA ROJO 2020 T	766

WINE	PAGE
MUÑARRATE DE SOLABAL 2021 B	513
MUÑARRATE DE SOLABAL 2021 T MC	513
MUREDA 2021 RD	794
MUREDA CHARDONNAY 2021 B	794
MUREDA CUVÉE BE R BN	698
MUREDA SAUVIGNON BLANC VERDEJO 2021 B S	794
MUREDA SYRAH 2021 T	794
MUREDA TEMPRANILLO SYRAH 2021 T	794
MURI VETERES 2017 T C	224
MURMURI 2021 B	371
MURMURÓN 2021 T	583
MURO 2016 T R	539
MURO 2018 T C	540
MURRI 2021 B	192
MURUA BLANCO FERMENTADO EN BARRICA 2020 B FB	540
MURUA RESERVA 2015 T R	540
MURUVE 2017 T R	665
MURUVE 2018 T C	666
MURUVE 2019 T RB	666
MURUVE 2021 T	666
MURUVE VERDEJO 2021 B	605
MURVIEDRO ARTS DE LUNA ORGANIC BE BN	126
MURVIEDRO ARTS DE LUNA ORGANIC BE BR	126
MURVIEDRO ARTS DE LUNA ORGANIC ROSÉ RE BR	126
MURVIEDRO CEPAS VIEJAS BOBAL 2019 T C	677
MURVIEDRO CEPAS VIEJAS MONASTRELL 2018 T R SS	37
MURVIEDRO COLECCIÓN 2018 T R	705
MURVIEDRO COLECCIÓN 2019 T C	705
MURVIEDRO COLECCIÓN BOBAL 2020 T RB	677
MURVIEDRO EKO 2021 T	37
MUSAI DE TINTO ROA 2015 T	472
MUSCÀNDIA 2016 BE GR BN	145

WINE	PAGE
MUSCÀNDIA 2018 BE R EBR	145
MUSCÀNDIA ANHEL BLANC DE NOIRS 2017 BE GR BN	145
MUSCÀNDIA DELIRI FLORAL 2021 B	339
MUSCÀNDIA MAGNUM 2014 BE GR BN	145
MUSCÀNDIA ROSÉ PINOT NOIR 2020 RE R EBR	145
MUSCAT MIQUEL OLIVER 2021 B S	351
MUSCO 2019 T C	584
MUSEUM 2018 T R	157
MUSGO VERDEJO 2021 B	611
MUSIC DE CARRER 2020 T C	642
MUSIC DE CARRER 2021 B	642

N

WINE	PAGE
N-A DE ATENCIA 2014 T BA	794
N. SRA. DEL PORTAL 2019 T C	645
N. SRA. DEL PORTAL 2021 B	645
NABAL 2018 T C	454
NABAL ALBILLO MAYOR 2020 B	454
NABAL ROSÉ ROSADO DE LÁGRIMA 2021 RD	454
NABAL SELECCIÓN DE LA FAMILIA 2016 T R	455
NABULÉ 2019 T	102
NADA QUE VER 2017 T C	536
NADAL ESPECIAL 2015 BE GR BN	145
NAHIKUN 2021 B	584
NAI E SEÑORA 2021 B	379
NAIA 2021 B	607
NAIADES 2019 B FB	607
NALUAR 2020 T	486
NANA 2019 B FB	381
NAPOLEÓN 30 AÑOS VORS 50 CL. BF AM S	209
NAPOLEÓN BF AM S	209
NARANJAS AZULES GARNACHA 2021 RD	753
NATURALEZA SALVAJE GARNACHA 2020 T	305

WINE	PAGE
NATURALIS MER BLANC 2021 B	643
NATURALIS MER ROURE 2020 T RB	643
NAVAHERREROS BLANCO DE BERNABELEVA 2020 B	721
NAVAHERREROS GARNACHA DE BERNABELEVA 2020 T	721
NAVAJAS 2017 T C	540
NAVAJAS 2018 B C	540
NAVAJAS GARNACHA 2020 T	541
NAVALTALLAR 2019 T C	759
NAVALTALLAR CRIANZA ESPECIAL 2016 T C	759
NAVAZOS NIEPOORT 2020 B	861
NAVERAN CLOS ANTONIA 2020 B FB	332
NAVERAN CLOS DELS ÀNGELS 2020 T MC	332
NAVERAN FLOR DE PINOT 2021 RD	332
NAVERÁN ODISEA 2019 BE BR	133
NAVERAN PERLES BLANQUES 2017 BE GR BR	133
NAVERÁN PERLES D'OR 2017 BE GR BR	133
NAVERÁN PERLES ROSES PINOT NOIR 2019 RE BR	133
NAZ 2019 T	415
NDQ SELECCIÓN 2019 T	225
NEBLA GARNACHA 2020 T S	711
NEBLA TEMPRANILLO 2020 T RB	492
NEBLA VERDEJO 2021 B	813
NEBLA VERDEJO ROSÉ 2021 RD	873
NÉGORA CHARDONNAY 2021 B	238
NÉGORA SAUVIGNON BLANC 2021 B	238
NÉGORA VERDEJO 2021 B	238
NEGRAMOLL ROSADA 2020 T	631
NEGRE DE GERISENA 2020 T	184
NEGRE DE NEGRES 2020 T	363
NEGRE DE NEGRES MAGNUM 2019 T	363
NEGRE DELS ASPRES 2018 T C	193
NEKAZARI 2020 B C	87

WINE	PAGE
NEKEAS CEPA X CEPA 2021 T	319
NEKORA VERDEJO SOBRE LÍAS 2021 B	615
NELEMAN BOBAL HRAMONY 2017 T	711
NELEMAN JUST FUCKING GOOD WINE 2019 T	711
NELEMAN JUST FUCKING GOOD WINE 2020 B	711
NELEMAN MACABEO 2020 B	711
NELEMAN TEMPRANILLO MONASTRELL 2021 T	712
NELEMAN VIOGNIER VERDIL 2021 B	712
NELIN 2019 B	364
NENA 2021 RD	177
NENO VIÑA SOMOZA GODELLO SOBRE LIAS 2020 B	693
NEO 2019 T BA	470
NEO ALBILLO MAYOR 2020 B	470
NEO PUNTA ESENCIA 2019 T	471
NEO VIÑAS VIEJAS 2021 B	613
NEPTIS EXPRESION 2017 T R	439
NEREUS GARNACHA NEGRA 2018 T	181
NEREUS SELECCIÓ 2019 T C	181
NERI 2021 B	722
NERI 2021 T	722
NERI CEPAS VIEJAS 2020 T RB	722
NERO DE SORT 2020 T	172
NEROS 2021 B	55
NEROS 2021 RD	56
NETÓN GARNACHA TINTORERA 2019 T C	795
NETÓN MONASTRELL 2018 T C S	795
NEXO 2019 T	676
NEXUS + 2014 T	486
NEXUS 2017 T C	486
NEXUS ONE 2011 T GR	486
NEXUS ONE 2019 T	486
NICE TO MEET YOU MADRID 2018 B	378

WINE	PAGE	WINE	PAGE	WINE	PAGE
NICTE 2020 T	448	NOC PRESTIGE BE BR	785	NUESTRO 15 MESES 2020 T C	487
NICTE ROSA PÁLIDO 2021 RD	801	NOC ROSÉ RE BR	785	NUESTRO 20 MESES 2016 T BA	487
NICTE VERDEJO ECO B	595	NOC TEMPRANILLO 2018 T	785	NUESTRO 8 MESES 2021 T RB	487
NIDIA 2020 RD	806	NODUS BOBAL 2020 T	677	NUEVE ROSAS 2021 RD	767
NIDIA VERDEJO LÍAS 2020 B	806	NODUS CHARDONNAY 2021 B	677	NUMANTHIA 2017 T	656
NIMI GERRA 2019 B	42	NODUS DP 2020 T	677	NUMERUS CLAUSUS 2019 T	157
NIMI NATURALMENT DOLÇ 2017 B FB D	42	NODUS MERLOT DELIRIUM 2019 T	678	NURI 2020 B	636
NIMI TOSSAL 2017 B R	42	NODUS SAUVIGNON BLANC 2021 B	678	NÚRIA CLAVEROL ALLIER 2015 BE GR BR	150
NINÍN 14 MESES 2019 T RB	452	NODUS TINTO DE AUTOR 2019 T C	678	NÚRIA CLAVEROL BLANC DE NOIRS 2015 BE GR BR	150
NIÑO DE LAS UVAS MONASTRELL 2020 T RB	91	NOÉ VORS BF PX D	216	NÚRIA CLAVEROL HOMENATGE 2015 BE GR BR	151
NIÑO DE LAS UVAS MONASTRELL 2021 RD	91	NOIR PRÍNCEPS 2018 T C	330	NÚRIA DE MONTARGULL 2021 RD	331
NIÑO MIMADO CHARDONNAY 2021 B FB	109	NOMEOLVIDES CHARDONNAY 2021 B	317	NURIA ROSÉ 2019 RE R BR	133
NIÑO MIMADO GARNACHA 2016 T C	109	NOMEOLVIDES GARNACHA 2021 RD	317	NUTT 2021 B	338
NISIA 2021 B	611	NOMEOLVIDES VIURA 2021 B	317	NUTT ROSÉ 2021 RD	338
NISIA LAS SUERTES 2020 B	612	NONA 2020 T	368	NVDE MCB (MACABEU ÁNFORA) 2020 B	872
NISIA LAS SUERTES 2021 B	612	NORA 2021 B	403	NVDE MCB (MACABEU CIMENT) 2020 B	872
NIT DE NIN COMA D'EN ROMEU 2019 T	365	NORA DA NEVE 2019 B FB	403	NVDE MCB (MACABEU CIMENT) 2021 B	872
NIT DE NIN LA RODEDA 2019 T	365	NORA DA NEVE ENCARNACIÓN RODRÍGUEZ 2019 B FB	403	NVDE SAUVIGNON BLANC 2020 B	872
NIT DE NIN MAS D'EN CAÇADOR 2019 T	365	NOSSA DE MENADE 2021 T	805	NVDE SAUVIGNON BLANC 2021 B	872
NITA 2020 T	371	NOSSO BY MENADE 2021 B	805		
NITA BLANC 2021 B	371	NOSTRE 2021 T	870	O	
NIVARIUS 2021 B	541	NOTAS 2020 T	757	O ALBOREXAR 2019 T	427
NIVARIUS EDICIÓN LIMITADA 2018 B	541	NOU NAT 2021 B	823	O ALBOREXAR 4.0 2021 T	427
NIVARIUS FINCA LA NEVERA 2017 B	541	NOVELLUM TEMPLE 2018 T C	656	O ALBOREXAR FRONTÓN 2018 T	427
NIVIA 2020 B FB	189	NOVOA 2017 T	382	O ALBOREXAR GARNACHA 2020 T	427
NO T'HO DIRÉ 2019 B	374	NUBE 2021 B	772	O CABALIN 2019 T C	690
NO T'HO DIRÉ ROSAT 2020 RD	374	NUBE DE LEZA GARCÍA MATURANA 2021 T	557	O CON 2020 B	875
Nº12 BY PACO & LOLA 2021 B	397	NUBORI 2015 T R	578	O CON DA MOURA 2019 B C	282
Nº8 COLLETIA 2020 T	858	NUBORI 2018 T C	578	O COTARELO 2020 T BA	419
Nº8 COLLETIA BRANCO 2020 B	858	NUBORI SELECCIÓN FAMILIA 2018 T C	618	O COTARELO 2021 B	419
NOBBIS 2020 T	491	NUBORI VENDIMIA SELECCIONADA 2015 T R	578	O DIVISO 2019 T	689
NOC CHARDONNAY 2018 B	785	NÚCLEO 2019 T	677	O ESTRANXEIRO 2020 T	414
				O ESTRANXEIRO 2021 B	414

WINE	PAGE
O FILLO DA CONDESA 2021 B	394
O GRAN MEIN 2020 B	429
O LUAR DO SIL GODELLO 2021 B	691
O LUAR DO SIL GODELLO SOBRE LÍAS 2020 B	691
O LUAR DO SIL VIDES DE CÓRGOMO 2019 B	691
O PEQUENO MEIN 2020 T	429
O PEQUENO MEIN 2021 B	429
O VENTOSELA 2021 B	423
OBAC DE BINIGRAU 2018 T BA	824
OBAC DE L'ANDREVA 2017 T R	358
OBALO 2021 RD	541
OBALO LAS ARENAS 2017 T R	541
OBALO SAN ROQUE 2020 T	542
OBAR DE PUENTE DEL EA 2020 B	558
OBEJITA PINK 2021 RD	864
OBEJITA VERDE 2020 B	864
OBSCENO 2021 B	595
OBSESIÓN 2020 T BA	233
OCEÁNIDAS 2019 T C	876
OCEÁNIDAS TINTILLA 2019 T C	876
OCHO ISLAS 2020 T	754
OCHOA 2014 T GR	314
OCHOA 2014 T R	314
OCHOA 8A LA FOTO DE 1938 2018 T C	315
OCHOA MOSCATEL VENDIMIA TARDÍA DULCE 2021 B MO D	315
OCHOA ROSADO DE LÁGRIMA 2021 RD	314
OCHOYMEDIO TEMPRANILLO 2021 T	780
OCHOYMEDIO TINTO VELASCO 2021 T	780
OCTAVO ARTE CABERNET SAUVIGNON T	875
OCTAVO ARTE MERLOT T	875
OCTAVO ARTE SAUVIGNON BLANC B	875
OE GARNACHA 2020 T	558

WINE	PAGE
OFERENTE 2018 T C	228
OFERENTE 2019 B BA	228
OFERENTE 2020 B	228
OFERENTE SELECCIÓN 2020 T	228
OGGA 2016 T R	547
OH! DE ROMAILA T	776
OINOZ 2017 T C	510
OJO DE GALLO 2019 B	772
OJOS DEL GUADIANA 2018 T R	244
OJOS DEL GUADIANA AIRÉN 2021 B	244
OJOS DEL GUADIANA CHARDONNAY 2021 B	244
OJOS DEL GUADIANA SELECCIÓN 2020 T BA	244
OJOS DEL GUADIANA TEMPRANILLO 2021 T	244
OJOS DEL GUADIANA VERDEJO 2021 B	244
OLANUM VENDIMIA SELECCIONADA 2017 T BA	526
OLARRA 2011 T GR	542
ÒLBIA 2018 B	369
OLBIETA 2020 T	648
OLBIETA BLANC 2020 B	648
OLCAVIANA CHARDONNAY 2021 B	797
OLCAVIANA SAUVIGNON BLANC 2021 B	797
OLCAVIANA VERDEJO 2021 B	797
OLD & PLUS AMONTILLADO VORS BF AM S	219
OLD & PLUS OLOROSO BF OL S	219
OLD & PLUS P.X. BF PX D	219
OLD HANDS 2019 T RB	733
OLE DE AROMAS 2019 T	269
OLE DE PASSION 2020 B	269
OLIMPO 2020 T RB	796
OLIMPO CHARDONNAY 2021 B	796
OLIVARES 2020 T	227
OLIVARES 2021 RD	227

WINE	PAGE
OLIVER VITICULTORS 2020 BE BN	145
OLIVER VITICULTORS ROSÉ 2019 RE BN	145
OLIVEROS PEDRO XIMÉNEZ BF PX D	169
OLIVEROS VINO NARANJA BF D	169
OLIVIA BY BURGMANN 2020 B	634
OLIVIA ROSÉ 2021 RD	868
OLLER DEL MAS ESPECIAL CARIÑENA 2018 T BA	867
OLLER DEL MAS ESPECIAL PICAPOLL NEGRE 2019 T BA	349
OLMAZA AUTOR 2019 T C	542
OLOROSO TRADICIÓN VORS BF OL S	212
OLVIDADO BF AM S	219
OLVIDAR 2021 B MC	228
OMBRA 2019 T BA	299
OMBRA 2021 B	299
ONDALÁN 100 ABADES GRACIANO 2016 T	542
ONDALÁN 2016 T R	542
ONDALÁN 2019 T C	542
ONDALÁN 2021 B	542
ONDALÁN TEMPRANILLO SELECCIÓN 2018 T	542
ONDIPUERKO 2020 B	553
ONDIPUERKO 2020 RD	553
ONDIPUERKO 2020 T RB	553
ONEROM MONASTRELL 2021 T	850
ONEROM VERDEJO 2021 B	850
ONTINAR MERLOT 2021 RD	304
ONTINAR MERLOT 2021 T	304
ONTINAR TEMPRANILLO 2021 T	304
OPERA PRIMA CABERNET SAUVIGNON 2021 T	245
OPERA PRIMA CHARDONNAY 2021 B	245
OPERA PRIMA SYRAH 2021 T	245
OPERA PRIMA VERDEJO 2021 B	245
OPHIUSA 2019 T GR	820

WINE	PAGE	WINE	PAGE	WINE	PAGE
OPOSITOR NOIR 2019 T BA	347	OROT 2019 T C	668	OSTATU 2021 RD	579
OPTA CALZADILLA 2017 T	740	OROT 2019 T RB	668	OSTATU 2021 T MC	579
ORÁCULO 2019 T	482	OROT 2021 T	668	OTAZU 1 HA UNA HISTORIA CHARDONNAY 2018 B	741
ORANGE 2021 B	259	OROVELO 2019 B	269	OTAZU CHARDONNAY 2021 B	308
ORANGE SK & V 2020 B	870	ORTEGA EZQUERRO 2016 T R	558	OTAZU MERLOT 2021 RD	308
ORBEN 2020 T	543	ORTEGA EZQUERRO 2019 T C	558	OTAZU PREMIUM CUVÉE 2019 T C	308
ORDÓÑEZ & CO. Nº4 ESENCIA (SIN FORTIFICAR) 2016 B D	863	ORTEGA EZQUERRO 2021 T MC	558	OTAZU RESERVA CLÁSICO 2016 T R	308
ORIG 2021 B	351	ORUBE 2018 T R	548	OTAZU TEMPRANILLO ROSÉ 2021 RD	308
ORIGEN 1989 2020 T RB	754	ORUBE 2020 B FB	548	OTERO 2017 T C	756
ORIGEN 2019 T C	410	ORUBE 2021 RD	548	OTOÑAL T C	542
ORIGEN 2021 B	334	ORUBE SELECCIÓN DE FAMILIA 2019 T C	548	OTRO CUENTO 2019 B	59
ORIGEN BRUÑAL QUINTA LAS VELAS 2017 T R	58	OS AREEIROS 2021 B	833	OU DE PAÓ 2020 B	331
ORIGEN DE RESALTE 2019 T	458	OS AREEIROS 2021 T	833	OURIVE DONA BRANCA 2020 B FB	415
ORIOL ROSSELL 2019 BE R BN	131	OS CIPRESES MENCÍA 2020 T	415	OURIVE GODELLO 2020 B	415
ORIOL ROSSELL ARIADNA 2016 BE GR BN	131	OS DUNARES ALBARIÑO 2021 B	381	OUTÓN 2020 B	400
ORIOL ROSSELL GRAN PROPIETAT 2009 BE GR BN	131	OS DUNARES CAIÑO 2020 T	381	OVIDIO GARCÍA DE AUTOR 2016 T R	161
ORIOL ROSSELL MITIC 2017 BE GR BN	131	OSABA MENDI 2020 T	578	OVIDIO GARCÍA ESENCIA 2018 T C	161
ORIOL ROSSELL RESERVA DE LA PROPIETAT 2015 BE GR BN	131	OSBORNE SOLERA INDIA BF OL S	210	OVIDIO GARCÍA SELECCIÓN 2020 T RB	161
ORIOL ROSSELL RESERVA DE LA PROPIETAT ROSÉ 2016 RE GR BR	131	OSCA MERLOT RESERVA 2016 T R	628	OVNI PEDRO XIMENEZ 2021 B	287
		OSCAR MESTRE GIRÓ 2019 T	38	OYE 2021 RD	849
ORIOLUS 2020 B	782	OSCAR MESTRE INSURRECTE 2020 B C	38	**P**	
ORISTÁN 2018 T R	241	OSLUNA 2019 T C	542	P.F. 2020 T	268
ORISTÁN 2019 T C	241	OSOTI 2018 B	591	PAAL 01 100% SYRAH 2020 T	311
ORISTÁN VERDEJO 2021 B	241	OSOTI 2021 T	591	PABLO CLARO SPECIAL SELECTION CHARDONNAY 2021 B	794
ORMUS 2018 T C	513	OSOTI LA ERA 2017 T RB	591	PABLO CLARO SPECIAL SELECTION GRACIANO CABERNET SAUVIGNON 2020 T	794
ORMUS EDICIÓN LIMITADA 2021 T	513	OSSIAN 2019 B	812		
ORMUS VIURA 2020 B	513	OSSIAN CAPITEL 2019 B FB	811	PACHEM 2020 T S	364
ORO DE CASTILLA 2019 T C	436	OSSIAN QUINTALUNA 2020 B	811	PACO & LOLA 2021 B	397
ORO DE CASTILLA FINCA LOS HORNOS 2019 B	599	OSTATU 2015 T GR	579	PACO & LOLA BE	397
ORO DE CASTILLA SAUVIGNON BLANC 2021 B	599	OSTATU 2016 T R	579	PACO & LOLA GODELLO 2021 B	691
ORO DE CASTILLA VERDEJO 2021 B	599	OSTATU 2019 T C	579	PACO & LOLA PRIME 2019 B	397
ORO VALEI 2021 B	392	OSTATU 2021 B	579	PACO & LOLA VINTAGE 2016 B	397

WINE	PAGE
PACO MULERO ALBARIÑO 2021 B	397
PACO MULERO QUINCE MESES 2020 T	50
PACO MULERO TEMPRANILLO 2019 T	812
PACO MULERO VEINTE MESES 2019 T	232
PAGO DE CARRAOVEJAS "CUESTA DE LAS LIEBRES" 2018 T R	487
PAGO DE CARRAOVEJAS 2020 T	487
PAGO DE CARRAOVEJAS EL AÑEJÓN 2018 T	487
PAGO DE CIRSUS CHARDONNAY 2020 B FB	739
PAGO DE CIRSUS CHARDONNAY 2021 B	308
PAGO DE CIRSUS CUVÉE ESPECIAL 2018 T	739
PAGO DE CIRSUS LA A T	739
PAGO DE CIRSUS MOSCATEL VENDIMIA TARDÍA 2017 B C D	739
PAGO DE CIRSUS ROSÉ GRAN CUVÉE ESPECIAL 2021 RD FB	309
PAGO DE CIRSUS SELECCIÓN DE FAMILIA 2017 T C	739
PAGO DE CIRSUS VENDIMIA SELECCIONADA 2020 T C	739
PAGO DE FUENTECOJO 2018 T	488
PAGO DE LA JARABA 2020 T	745
PAGO DE LA JARABA SAUVIGNON BLANC 2021 B	745
PAGO DE LARRAINZAR CABERNET SAUVIGNON 2018 T	319
PAGO DE LARRAINZAR RESERVA ESPECIAL 2015 T R	320
PAGO DE LOS BALAGUESES CHARDONNAY 2021 B FB	745
PAGO DE LOS BALAGUESES GARNACHA TINTORERA 2020 T C	745
PAGO DE LOS BALAGUESES SYRAH 2020 T C	745
PAGO DE LOS CAPELLANES CRIANZA 2020 T C	487
PAGO DE LOS CAPELLANES PARCELA EL NOGAL 2018 T FB	487
PAGO DE LOS CAPELLANES PARCELA EL PICÓN 2016 T	487
PAGO DE LOS CAPELLANES RESERVA 2019 T R	487
PAGO DE OTAZU 2019 T R	741
PAGO DE OTAZU CHARDONNAY CON CRIANZA 2019 B	741
PAGO DE THARSYS 2017 BE GR BN	146
PAGO DE THARSYS ARGILA 2018 T	741
PAGO DE THARSYS BOBAL DIANA GARCÍA 2019 T	741
PAGO DE THARSYS CABERNET FRANC SIN SULFITOS 2021 T	682
PAGO DE THARSYS MERSEGUERA SIN SULFITOS 2021 B	682
PAGO DE THARSYS MILLESIME 2018 BE R BR	146
PAGO DE THARSYS MILLESIME BARRICA 2019 BE R BN	146
PAGO DE THARSYS MILLÉSIME ROSÉ RESERVA 2018 RE R BR	146
PAGO DE THARSYS VENDIMIA NOCTURNA 2021 B	741
PAGO DE THARSYS VENDIMIA NOCTURNA GARNACHA 2021 RD FB	741
PAGO DE VALCALIENTE 2018 B R	591
PAGO DE VALCALIENTE GARNACHA 2021 T C	591
PAGO DE VALDONEJE 2021 T	78
PAGO DE VALDONEJE EL VALAO 2019 T BA	78
PAGO DE VALDONEJE GODELLO 2021 B	78
PAGO DE VALDONEJE LA TELLERÍA 2019 T	78
PAGO DE VALTARREÑA 2019 T	458
PAGO DEL VICARIO 50-50 2018 T C	742
PAGO DEL VICARIO 6 MESES 2020 T	742
PAGO DEL VICARIO BANCAL DEL RÍO 2017 T	742
PAGO DEL VICARIO BLANCO DE TEMPRANILLO 2021 B	743
PAGO DEL VICARIO PETIT VERDOT 2021 RD	743
PAGO DEL VICARIO TALVA 2019 B FB	743
PAGO EL CORDONERO GARNACHA ALBILLO 2020 RD FB	159
PAGO EL CORDONERO TEMPRANILLO 12 MESES 2018 T BA	159
PAGO EL CORDONERO TEMPRANILLO 9 MESES 2018 T	159
PAGO EL CORDONERO VERDEJO 2020 B FB	159
PAGO FLORENTINO 2019 T C	743
PAGO FRANCISCO GÓMEZ FONDILLÓN 1988 T FO	36
PAGO LA JARA 2018 T	663
PAGO LA PAVINA 2019 T	800
PAGO MOTA 2020 B	781
PAGO MOTA 2021 B	781
PAGO PEÑUELAS GARNACHA 2021 T	796
PAGO PEÑUELAS VERDEJO 2021 B	796
PAGOS DE ANGUIX BARRUECOS 2019 T	488
PAGOS DE ANGUIX COSTALARA 2020 T	488
PAGOS DE ANGUIX EL ROSADO 2021 RD	488
PAGOS DE ANGUIX PRADO LOBO 2018 T R	488
PAGOS DE ARAIZ CRIANZA 2019 T C	309
PAGOS DE ARAIZ ROBLE 2020 T RB	309
PAGOS DE ARAIZ ROSADO 2021 RD	309
PAGOS DE ARAIZ ROSÉ 2021 RD	309
PAGOS DE LA SONSIERRA 2015 T R	548
PAGOS DE PEÑAFIEL 2017 T R	482
PAGOS DE PEÑAFIEL 2018 T C	482
PAGOS DE PEÑAFIEL 2020 T RB	482
PAGOS DE PEÑAFIEL VENDIMIA SELECCIÓN 2019 T	482
PAGOS DE REVERÓN 2020 T BA	30
PAGOS DE REVERÓN 2021 B S	30
PAGOS DE REVERÓN 2021 T S	30
PAGOS DE REVERÓN AFRUTADO 2021 B SD	30
PAGOS DE VALCERRACIN CRIANZA VENDIMIA SELECCIONADA 2018 T C	493
PAGOS DE VIÑA REAL 2018 T	588
PAGOS DEL CAMINO GARNACHA 2019 T	500
PAGOS REVERÓN AFRUTADO 2021 RD SD	30
PAGOS REVERÓN MALVASIA 2019 B S	30
PAI EDICIÓN ESPECIAL ALBAMAR 2021 B	384
PAISAJE DE LAS ISLAS FORASTERA 2021 B	755
PAISAJE DE LAS ISLAS MALVASÍA AROMÁTICA NATURALMENTE DULCE 2019 B D	755
PAISAJE DE LAS ISLAS MALVASIA MARMAJUELO 2021 B	755
PAISAJES CECIAS 2019 T	580
PAISAJES EMBOTELLADOS VIÑEDOS RUIZ JIMÉNEZ 2021 T	591
PAISAJES VALSALADO 2019 T	580

WINE	PAGE	WINE	PAGE	WINE	PAGE
PAIVA 2019 BE R BN	126	PALADOR 2019 T C	551	PARAJE DE LOS BANCALES 2018 T R	59
PAIVA 2020 BE BN	126	PALAU SOLÁ BE BN	124	PARAJE DE TITOS 2020 T	244
PAIVA 56 BARRICAS 2018 T FB	498	PALET MOST DE FLOR 2020 T	360	PARAJE TORNEL 2018 T R	680
PAIVA CRIANZA 2019 T C	498	PALET VINYA TRICOLAVET 2019 T C	360	PARAJES DE CALLEJO 2019 T	448
PAIVA SOLO CAYETANA 2021 B	498	PALET VINYES VELLES 2018 T C	360	PARAJES DE LOS VIDRIOS GARNACHA 2018 T	727
PAIVA SOLO TEMPRANILLO 2020 T S	498	PALMA BLANCA DULCE 2019 B D	36	PARAJES DE LOS VIDRIOS PICO DEL MIRLO 2018 T	727
PAIXAR MENCÍA 2020 T	70	PALMERI ADÁN 2015 T GR	106	PARAJES DEL CABRIEL 2021 T	679
PAIXÓN 2021 T	283	PALMERI NAVALTA 2017 T C	106	PARAJES DEL INFIERNO "EL JUDAS" 2019 B FB	840
PAJARETE NOBLE QUITAPENAS 2018 BF SD	262	PALO BLANCO 2020 B	861	PARAJES DEL INFIERNO "LA SILLERÍA" 2019 B FB	840
PÁJARO LOCO GODELLO 2021 B	282	PALO CORTADO TRADICIÓN VORS BF PC S	212	PARAJES DEL INFIERNO "LAS ENVIDIAS" 2019 B	840
PÁJARO LOCO MENCÍA 2021 T	282	PALOMAR DE LA REINA 2019 T	808	PARAJES DEL VALLE 2021 RD	270
PALACIO DE ARGANZA 2021 T	815	PAM DE NAS 2017 T	375	PARAJES DEL VALLE MACABEO 2021 B	270
PALACIO DE ARGANZA CABERNET SAUVIGNON MENCÍA 2020 T	815	PANCRUDO DE GÓMEZ CRUZADO 2020 T	555	PARAJES DEL VALLE MACERACIÓN MACABEO 2021 B	870
PALACIO DE BORNOS LA CAPRICHOSA 2021 B	619	PANDORA SAUVIGNON BLANC 2021 B	608	PARAJES DEL VALLE MONASTRELL 2020 T	232
PALACIO DE BORNOS SAUVIGNON BLANC 2021 B	619	PANDORA VERDEJO 2021 B	608	PARAJES VINO DE REGIÓN 2020 T	72
PALACIO DE BORNOS SAUVIGNON BLANC SEMIDULCE 2021 B SD	619	PANDORA VERDEJO SOBRE LÍAS 2021 B	608	PÁRAMO ARROYO 2014 T GR	488
		PANDRA SAUVIGNON BLANC 2019 B BA	608	PÁRAMOS DE LEGARIS 2018 T BA	484
PALACIO DE BORNOS VERDEJO 2021 B	619	PANDRA VERDEJO 2020 B FB	609	PARANY 2016 T C	335
PALACIO DE BORNOS VERDEJO 2021 B FB	619	PANIS ANGELORUM 2020 B	753	PARATÓ 2018 BE R BN	146
PALACIO DE MANZANOS VIÑEDO SINGULAR 1890 MANZANOS 2017 T	534	PANTIGANA 2020 B	172	PARATÓ 2018 BE R BR	146
		PANTOCRATOR 2011 T R	549	PARATÓ ÁTICA TRES X TRES 2020 B	340
PALACIO DE TRETO 2020 B C	816	PAÑO DE LÁGRIMAS 2020 T RB	158	PARATÓ XARELLO 2021 B	340
PALACIO DE VILLACHICA DEHESA SAN ANDRÉS VENDIMIA SELECCIONADA 2019 T	668	PAR VINO NARANJA BF D	168	PARCENT GIRÓ 2021	38
		PARA CELSUS 2021 T	784	PARDELASSES 2017 T	358
PALACIO DE VILLACHICA VS DEHESA DE DON DIEGO 2019 T C	488	PARA CELSUS VERDEJO 2021 B	784	PARDELLS 2019 B	188
PALACIO QUEMADO 2019 T C S	499	PARADA DE ATAUTA 2019 T	446	PARDEVALLES 2021 RD	256
PALACIO QUEMADO LA RAYA 2018 T	819	PARADIGMA 2018 T BA	703	PARDEVALLES ALBARÍN 2021 B	256
PALACIO QUEMADO LA ZARCITA 2020 T	819	PARADISUAK JANEO 2020 RD	85	PARDEVALLES CARROLEÓN 2018 T	256
PALACIO QUEMADO LOS ACILATES 2019 T C S	819	PARADISUAK JAUREGI - ABADIÑO 2018 B	85	PARDEVALLES CARROLEÓN 2020 B FB	256
PALACIO QUEMADO VENDIMIA SELECCIONADA 2020 T RB	819	PARADISUAK LEIOA 2019 B	85	PARDEVALLES GAMONAL 2018 T C	256
PALACIO QUEMADO VIÑEDOS PROPIOS 2016 T R	499	PARADISUAK OTXOLARRE - MORGA 2019 B	85	PARDEVALLES PRIETO PICUDO 2021 T	257
PALADOR 2017 T R	551	PARAJE DE LA VIRGEN 2020 T	575	PARENTESIS 2021 B	359

WINE	PAGE
PARÉS BALTÀ CUVÉE DE CAROL 2013 BE GR BN	146
PARÉS BALTÀ ELECTIO XARELLO 2020 B	340
PARÉS BALTÀ HISENDA MIRET GARNATXA 2019 T R	340
PARÉS BALTÀ HISTORIC 2018 BE GR BN	147
PARÉS BALTÀ INDÍGENA 2021 B	340
PARÉS BALTÀ INDÍGENA NEGRE 2020 T	340
PARÉS BALTÀ ROSA CUSINÉ 2017 RE GR BN	147
PARÉS BALTÀ SATÈLLIT 2019 B	340
PARÉS BALTÀ XARELLO 2021 B	340
PAROTET 2020 T	708
PARPADOS 2020 T	485
PARRADEZ 2021 RD	156
PARREÑO 2021 B	681
PARREÑO 2021 RD	681
PARSIMONIA 2018 T C	679
PARSIMONIA 2019 T FB	679
PARSIMONIA 2021 B	679
PARTAL CEPAS VIEJAS 2017 T	90
PARTAL DE AUTOR 2006 T	90
PARTICULAR CHARDONNAY 2019 B FB	110
PARTICULAR GARNACHA BLANCA 2021 B	110
PARTICULAR GARNACHA OLD VINE 2018 T C	110
PARTICULAR GARNACHA VIÑAS CENTENARIAS 2015 T R	110
PARXET 2019 BE BN	147
PAS CURTEI 2019 T	325
PAS DELS CAUS 2021 T	355
PASAL DE ESILE GODELLO 2020 B	411
PASAMONTE TINTORERA 2020 T	745
PASANAU CEPS NOUS 2019 T	361
PASANAU FINCA LA PLANETA 2016 T	361
PASANAU VI DE VILA DE LA MORERA DE MONTSANT 2019 T	361
PASAPORTE 2019 T	776
PASAPORTE 2021 B	776
PASAS SONROSADO 2021 RD	734
PASAS UVA BLANCA 2021 B	734
PASAS UVA TINTA 2020 T	734
PASIEGO "CÆSAR" 2017 T C	678
PASIEGO BOBAL 2018 T C	678
PASIEGO DE AUTOR 2017 T C	678
PASIEGO JULIETA DULCE NATURAL 2017 B D	678
PASIEGO LA SUERTES 2021 B	678
PASION DE BOBAL 2020 T RB	683
PASION DE BOBAL 2021 RD	683
PASIÓN DE CASTILLO DE MONTALBAN 2021 BE BN	879
PASIÓN DE CASTILLO DE MONTALBAN BE R	879
PASIÓN DE MONASTRELL 2020 T	44
PASION DE MOSCATEL 2021 B	711
PASO A PASO TEMPRANILLO 2021 T	786
PASO DE BUEY T	785
PASSIÓ 2020 RD	825
PASTORA PASADA BF MZ S	207
PASTRANA 50 CL BF AM	209
PASTRANA MANZANILLA PASADA BF MZ S	209
PATA NEGRA 2016 T R	576
PATA NEGRA 2016 T R	697
PATA NEGRA 2017 T C	576
PATA NEGRA 2019 T C	697
PATA NEGRA 2020 T RB	697
PATA NEGRA 2021 T RB	670
PATA NEGRA APASIONADO ORGANIC T	222
PATA NEGRA APASIONADO SIN SULFITOS 2020 T	222
PATA NEGRA APASIONADO T	222
PATA NEGRA CEPAS VIEJAS 2016 T R	697
PATA NEGRA ORGANIC BE BR	123
PATA NEGRA SAUVIGNON BLANC 2021 B	613
PATA NEGRA SELECCIÓN 2019 T	576
PATA NEGRA TEMPRANILLO CABERNET SAUVIGNON 2020 T	697
PATA NEGRA VERDEJO 2021 B	613
PATEIRO ANFORA 2020 B	426
PATEIRO TREIXADURA 2020 B BA	426
PATERNINA BANDA AZUL 2020 T C	576
PATIENS 2016 B R	549
PATOJO 2020 T	35
PATRIA CHICA 2021 T	738
PAUL CHENEAU 2017 BE R BR	140
PAUL CHENEAU BLANC DE BLANCS BE R BR	140
PAVINA RED 2019 T	800
PAYDOS 2017 T	670
PAYDOS 2018 T	670
PAYOYA NEGRA 2019 T R	260
PAZO BAIÓN ALBARIÑO 2021 B	397
PAZO BAIÓN GRAN A GRAN 2017 B	398
PAZO BAIÓN VIDES DE FONTÁN 2018 B	398
PAZO CASANOVA 2021 B	422
PAZO DA TORRE ALBARIÑO 2021 B	392
PAZO DAS BRUXAS 2021 B	394
PAZO DE BARRANTES ALBARIÑO 2019 B	398
PAZO DE BARRANTES ALBARIÑO 2020 B	398
PAZO DE MONTERREY GODELLO 2021 B	282
PAZO DE MONTERREY MENCÍA 2020 T	282
PAZO DE MONTERREY RAÚL BOO MENCÍA 2020 T	282
PAZO DE PIÑEIRO 2020 B	400
PAZO DE RUBIANES 1411 2017 B	398
PAZO DE RUBIANES 1411 2019 B	398
PAZO DE RUBIANES ALBARIÑO 2015 B	398
PAZO DE RUBIANES ALBARIÑO 2020 B	398

WINE	PAGE	WINE	PAGE	WINE	PAGE
PAZO DE RUBIANES GARCÍA CAAMAÑO 2015 B	399	PEGASO "BARRANCOS DE PIZARRA" 2019 T	752	PEÑA LA ROSA GRANO A GRANO 2016 T	525
PAZO DE RUBIANES GARCÍA CAAMAÑO 2018 B	399	PEGASO "GRANITO" 2019 T	752	PEÑA LA ROSA VENDIMIA SELECCIONADA 2016 T	525
PAZO DE SEOANE O ROSAL 2021 B	395	PEGASO "ZETA" 2020 T	752	PEÑA REJAS 2021 T	658
PAZO DO MAR 2021 B	419	PEIQUE GODELLO 2021 B	68	PEÑALBA-LÓPEZ 2020 B	810
PAZO PONDAL ALBARIÑO 2020 B	399	PEIQUE MENCÍA 2021 T	69	PEÑAMONTE 2019 T C	661
PAZO SAN MAURO ALBARIÑO 2021 B	383	PEIQUE RAMÓN VALLE 2020 T	69	PEÑAMONTE 2021 RD	661
PAZO SEÑORANS 2021 B	399	PEIQUE SELECCIÓN FAMILIAR 2018 T	68	PEÑAMONTE COSECHA 2021 T	661
PAZO SEÑORANS COLECCIÓN 2018 B	399	PEIQUE VIÑEDOS VIEJOS 2019 T RB	69	PEÑAMONTE GARNACHA 2021 T RB	662
PAZO SEÑORANS SELECCIÓN DE AÑADA 2013 B	399	PEITES 2009 T C	361	PEÑAMONTE ROBLE 2021 T RB	662
PAZO TORRE PENELAS BLANCO GRANITO 2019 B	400	PEKADO MORTAL ALBARIÑO 2020 B	400	PEÑAMONTE VERDEJO 2021 B	662
PAZO VILLAREI 2021 B	383	PEKADO MORTAL GODELLO 2020 B	691	PEÑAS ALADAS 2013 T GR	479
PE ESE 2019 T	73	PEKADO MORTAL TEMPRANILLO 2018 T	489	PEÑAS NEGRAS 2020 T	676
PEDRA DE GUIX 2019 B C	375	PEKADO MORTAL TEMPRANILLO 2019 T C	489	PEÑAZUELA VENDIMIA SELECCIONADA GARNACHA 2020 T RB	101
PEDRABONA 2020 T	367	PELA ROQUES DE MUSTIGUILLO 2020 T	843		
PEDRANAI DE SANTIAGO ROMA ALBARIÑO 2020 B	390	PELIGRU 2021 T	856	PEÑAZUELA VENDIMIA SELECCIONADA GARNACHA BLANCA 2021 B	101
PEDRAZAIS GODELLO 2020 B	686	PELL DE GERRES 2020 B	298		
PEDREGAR 2016 BE GR BN	326	PELLEJO 2020 T	845	PEPE CABANAS 2019 B	769
PEDREGAR 2016 RE R	326	PEMARTIN BF FI S	213	PEPE CARRASCA 2020 B	424
PEDRO GARCÍA BE BR	724	PENÍNSULA CEBREROS 2020 T	753	PEPE MENDOZA CASA AGRÍCOLA 2020 T	43
PEDRO GARCÍA MALVAR B	724	PENÍNSULA SKIN CONTACT 2019 B	868	PEPE MENDOZA CASA AGRÍCOLA 2021 B	43
PEDRO GARCÍA SAUVIGNON Y MALVAR 2021 B	724	PENÍNSULA VINO ATLÁNTICO 2021 B	88	PEPE MENDOZA CASA AGRÍCOLA PUREZA 2021 B	43
PEDRO MARTÍNEZ ALESANCO SELECCIÓN 2015 T R	536	PENÍNSULA VINO DE MONTAÑA SIERRAS GATA & GREDOS 2020 T	868	PEPE MENDOZA EL VENENO 2020 T BA	43
PEDRO XIMÉNEZ RESERVA DE FAMILIA BF PX D	262			PEPE MENDOZA FIERROCA 2020 T	43
PEDRO XIMÉNEZ TRADICIÓN VOS BF PX D	212	PENSANDO EN TÍ 2020 B	552	PEPE MENDOZA GIRÓ DE ABARGUES 2020 T C	43
PEDROHERAS 2015 T R	241	PEÑA CABALLERA 2019 T	727	PEPE MENDOZA MOSCATEL PASA ORIGEN 2016 B D	43
PEDROHERAS AIRÉN 2021 B	241	PEÑA CADIELLA 2019 T RB	45	PEPE YLLERA 2019 T RB	496
PEDROHERAS SYRAH TEMPRANILLO 2021 T S	241	PEÑA DEL AGUILA BF PC S	207	PEPELUIS 2020 B C	384
PEDROHERAS TEMPRANILLO 2021 T	241	PEÑA DEL AGUILA FINO EN RAMA BF FI S	207	PERAFITA GARNATXA NEGRA 2021 T	186
PEDROHERAS VERDEJO 2021 B	241	PEÑA HIBEU 2018 T	274	PERAFITA PICAPOLL 2021 B	186
PEDROTEÑO AIRÉN 2021 B	796	PEÑA LA ROSA 2021 B	525	PERAFITA ROSAT 2021 RD	186
PEDROTEÑO TEMPRANILLO 2021 T	796	PEÑA LA ROSA 2021 T MC	525	PERE PUNYETES BLANC 2021 B	344
PEDROUZOS MAGNUM 2012 B FB	693	PEÑA LA ROSA EL SECRETO DEL ABUELO 2017 T	525	PERE PUNYETES NEGRE 2021 T	344

WINE	PAGE
PERE PUNYETES ROSAT 2021 RD	344
PERE SEDA 2018 BE BN	850
PERE SEDA 2018 T C	351
PERE SEDA CHARDONNAY/GIRO ROS 2021 B	351
PERE VENTURA GRAN VINTAGE PARAJE CALIFICADO CAN BAS 2015 BE GR BR	147
PERE VENTURA TRESOR 2018 BE GR BR	147
PERE VENTURA TRESOR ANNIVERSARY 2017 BE GR BR	147
PERE VENTURA TRESOR CUVÉE BARRIQUE 2018 BE GR BR	147
PERE VENTURA VINTAGE 2015 BE GR BR	147
PEREGRINO 14 2016 T C	255
PEREGRINO ALBARÍN 2021 B	255
PEREGRINO MIL 100 2011 T C	255
PEREGRINO ROSÉ 2021 RD	255
PERELADA 5 FINQUES 2018 T R	190
PERELADA AIRES DE GARBET 2019 T R	190
PERELADA CUVÉE ESPECIAL 2020 BE BN	147
PERELADA EX EX 14 2019 T C	190
PERELADA FINCA GARBET 2016 T R	191
PERELADA FINCA GARBET 2017 T R	191
PERELADA FINCA MALAVEÏNA 2019 T	191
PERELADA GARNATXA DE L'EMPORDÀ 12 ANYS DULCE NATURAL BF SOLERA D	191
PERELADA GRAN CLAUSTRO 2015 T C	190
PERELADA GRAN CLAUSTRO 2017 BE GR BN	148
PERELADA GRAN CLAUSTRO CUVÉE ESPECIAL 2017 BE GR BN	148
PERELADA STARS 2020 BE R BN	148
PERELADA STARS TOUCH OF ROSÉ 2020 RE BR	148
PÉREZ ESTEBAN BY FUENTENARRO 2018 T R	448
PÉREZ PASCUAS GRAN SELECCIÓN 2015 T GR	451
PERFIL 2014 T R	483
PERFUME DE SONSIERRA 2015 T	548

WINE	PAGE
PERINET 1194 2017 T C	372
PERINET 2017 T C	372
PERLA MARIS VERDEJO 2021 B	618
PERLAT 2020 T RB	296
PERLAT BLANC 2021 B	296
PERLAT GARNATXA 2020 T BA	296
PERLAT SYRAH 2020 T RB	296
PERNALES MERLOT SYRAH 2018 T C	262
PERPETUAL 2019 T C	365
PERROCHICO 2020 T	624
PERROCHICO 2021 B	624
PERSEO 7 VERDEJO SOBRE LÍAS 2021 B	619
PÉTALOS DEL BIERZO 2020 T	73
PÉTALOS DEL BIERZO 2021 T	73
PETARDO 2020 T	185
PETHIOX 2019 T C	74
PETHIOX 2021 T	74
PETHIOX CIELO 2021 B	74
PETHIOX CLARETE 2021 RD	74
PETHIOX DISTINCIÓN 2018 T	74
PETILLANT BLANC 2021 BE AG	840
PETIT BALDOMA 2020 T	177
PETIT BALDOMÀ 2021 B	177
PETIT BLANC SAÓ 2021 B	175
PETIT CUPADA DE CANALS NADAL 2019 BE R BN	129
PETIT HIPPERIA 2020 T	747
PETIT OBSESIÓN 2020 T BA	233
PETIT PISSARRES 2020 T	376
PETIT PITTACUM 2021 T	79
PETIT SAÓ 2019 T	175
PETIT SIÓS 2021 T RB	174
PETITES ESTONES NEGRE 2021 T	298

WINE	PAGE
PETRA DE VALPIEDRA 2018 T	569
PEZAS DA PORTELA 2019 B FB	693
PEZÓN NEGRO 2019 T	548
PIAMATER NATURALMENTE DULCE 2018 B MO D	261
PIC DE SOLERGIBERT 2021 B	347
PICARANA 2020 B	727
PÍCARO DEL AGUILA 2020 T BA	479
PÍCARO DEL AGUILA CLARETE 2020 RD	479
PICIO 2021 T	666
PICO CUADRO ORIGINAL 2020 T	489
PICO CUADRO VENDIMIA SELECCIONADA 2020 T	489
PICO CUADRO WILD 2020 T	489
PICO D´ALIGA 2020 T C	674
PICO DE LUYAS 2019 T	462
PICO FERREIRA 2020 T	72
PICOS DE CABARIEZO SELECCIÓN 2020 T	822
PIE FIRME 2019 T R	229
PIEDEMONTE 2016 T R	315
PIEDEMONTE CHARDONNAY 2021 B	315
PIEDEMONTE CUATRO TIERRAS 2018 T C	315
PIEDEMONTE GAMMA 2021 T	315
PIEDEMONTE MOSCATEL 2021 B MO D	315
PIEDEMONTE OLD VINES GARNACHA 2018 T C	315
PIEDRA 2016 T R	658
PIEDRA 2019 T C	658
PIEDRA 2020 T RB	659
PIEDRA 2021 RD	659
PIEDRA FLUIDA 2021 B	754
PIEDRA FLUIDA 2021 B SD	754
PIEDRA FLUIDA LISTÁN NEGRO 2020 T BA	754
PIEDRA FLUIDA VIDAL 2020 T	754
PIEDRA NATURAL 2021 T	659

WINE	PAGE	WINE	PAGE	WINE	PAGE
PIEDRAS DE SAN PEDRO 2018 T C	489	PINO DONCEL 12 MESES 2019 T C	223	PIU ÁNFORA 2019 T FB	851
PIEDRAS Y PRINCESAS 2018 T C	662	PINO DONCEL 24 MESES 2017 T C	223	PIZARRAS DE OTERO 2021 T	67
PIELES DORADAS 2021 B	706	PINO DONCEL BLACK 2020 T RB	224	PLA DE MOREI 2019 T	118
PIEZA LA MOZA 2020 RD	701	PINOSO CLÁSICO 2019 T C	38	PLA DEL BOSC XARELLO VERMELL 2021 B	345
PILANOT BLANC 2020 B	636	PINTIA 2018 T	663	PLÁCET DE VALTOMELLOSO 2021 B	543
PILARES DE CIENCUEVAS 2018 T RB	513	PINUAGA 200 CEPAS 2019 T	787	PLAER 2018 T C	373
PILAS BONAS 2021 B	740	PINUAGA BIANCO 2021 B	787	PLAER 2020 T C	373
PILLET BLANC 2021 B	112	PINUAGA COLECCIÓN 2019 T C	788	PLANA D'EN FONOLL CABERNET SAUVIGNON 2020 T	118
PILLET NEGRE 2020 T	112	PINUAGA NATURE 2020 T RB	788	PLANA D'EN FONOLL CHARDONNAY 2020 B BA	118
PILLET ROSAT 2021 RD	112	PINUAGA ROSÉ 2021 RD	788	PLANA D'EN FONOLL SAUVIGNON BLANC 2021 B	119
PINADEL T	850	PINYERES BLANC 2021 B	294	PLANA D'EN JAN BON JAN NEGRE 2018 T BA	340
PINCERNA ALBARÍN 2021 B	255	PINYERES NEGRE 2019 T	295	PLANA D'EN JAN CAMP DEL CUC 2019 B BA S	340
PINCERNA PRIETO PICUDO 2021 RD	255	PINZON 21 2021 RD	767	PLANA D'EN JAN CLOS DE LES OLIVERES 2020 B C	340
PINCERNA PRIETO PICUDO 2021 T	255	PIONERO 2021 B	403	PLANA D'EN JAN ELS AMETLLERS 2017 B FB	340
PINCERNA SUMILLER 2020 T	255	PIQUERAS VERDEJO WILD FERMENTED 2021 B FB	48	PLANA D'EN JAN INSÒLIT BLANC DE NOIR 2019 B S	341
PINCHAPERAS 2020 T C	834	PIQUERAS VS 2017 T	48	PLANA D'EN JAN LA SORRERA 2018 T	341
PINDAL VERDEJO 2021 B	597	PIRÁMIDE 2020 T C	486	PLANA D'EN JAN PLANA IN ALBIS 2018 T R	341
PINDIO 2019 T FB	512	PIRINEOS CHARDONNAY VIÑEDO SELECCIONADO 2021 B	626	PLANAS ALBAREDA 2019 BE R BN	148
PINEA 2017 T FB	456	PIRINEOS GEWÜRZTRAMINER 2021 B	626	PLANAS ALBAREDA 2020 BE BN	148
PINEA 2018 T FB	456	PISARROSAS 2018 T C	486	PLANAS ALBAREDA 2020 BE BR	148
PINGOROTE SAUVIGNON BLANC 2021 B	240	PITA 2021 RD	609	PLANAS ALBAREDA DESCLÒS 2020 T	341
PINGOROTE TEMPRANILLO 2016 T C	241	PITA FINCA LA BONERA TEMPRANILLO 2018 T C	609	PLANAS ALBAREDA GRAN RESERVA DE L'AVI 2018 BE GR BN	148
PINGOROTE TEMPRANILLO 2016 T R	241	PITA FINCA LA CANTERA 2018 B FB	609	PLANAS ALBAREDA L'AVENC 2021 B	341
PINGUS 2020 T	478	PITA SAUVIGNON BLANC 2020 B	609	PLANAS ALBAREDA ROSAT 2020 RE BR	148
PINK PEARL 2021 RD PL S	844	PITA VERDEJO (DOMINIO DE VERDERRUBI) 2021 B	609	PLANASSOS 2019 T	364
PINKGALL 2021 RD	317	PITTACUM 2019 T RB	79	PLANETES CLASSIC 2019 T	365
PINNA FIDELIS 2018 T R	456	PITTACUM AUREA 2018 T RB	79	PLEAMAR EN RAMA BF MZ S	208
PINNA FIDELIS 2019 T C	456	PITTACUM VAL DE LA OSA 2017 T RB	79	PLERET 2013 T	356
PINNA FIDELIS 2020 T RB	456	PITUCO MST 2020 T	232	PLOM 2020 T	376
PINNA FIDELIS ROBLE ESPAÑOL 2019 T BA	456	PITUCO PARAJE LAS ZORRERAS 2017 T R	232	POBLETS DEL MONTSANT 2020 T	297
PINNA FIDELIS VENDIMIA SELECCIONADA 2018 T	456	PIU ANCESTRAL 2020 BE	879	POBLETS DEL MONTSANT 2021 B	297
PINO 2020 T RB	268	PIU ÁNFORA 2019 B	851	POCO A POCO 2021 B	781

WINE	PAGE
POCO A POCO ENVEJECIDO EN BARRICA 2021 T C	781
POLA 2019 T C	533
POLA 2021 T MC	533
POLA ANTONIO LÓPEZ 2020 T	533
POLA VALECILLA 2018 T	533
POLEY AMONTILLADO EN RAMA BF AM	289
POLEY BF CRM	289
POLEY FINO DEL LAGAR EN RAMA B FI S	289
POLEY FINO PASADO EN RAMA B FI	289
POLEY OLOROSO EN RAMA BF OL	289
POLEY PALO CORTADO EN RAMA BF PC	289
POLUS 2015 T R	532
POLUS 2019 T C	532
POLUS 2021 T MC	532
POLUS TEMPRANILLO 2019 T	532
POLUS VIURA 2021 B	532
PÓLVORA 2020 T	166
POMELADO VINO NARANJA 2021 B	795
PONT FOSC 2019 B	637
PONTE DA BOGA ALBARIÑO 'A' 2021 B	406
PONTE DA BOGA GODELLO ´G´ 2021 B	407
PONTE DA BOGA PIZARRAS Y ESQUISTOS 2020 T	407
PONTELLÓN ALBARIÑO 2021 B	380
PONTEVS 2017 T	785
POR FIN 2016 T R	802
POR FIN 2020 B	802
POR LOS CIEN 2018 T	526
POR TÍ 2019 T	227
PORPRAT 2016 T	82
PORRERA VI DE VILA DE VALL LLACH 2020 T C	361
PORRERA VI DE VILA DE VALL LLACH 2021 B	361
PORT DE CECILI BLANC BF D	846

WINE	PAGE
PORT DE CECILI BLANC BF S	846
PORT DE CECILI T D	846
PORTAL DE MONCAYO ILUSIÓN GARNACHA 2020 T	104
PORTAL DE MONCAYO PASIÓN 2019 T	104
PORTAL DE MONCAYO ROSÉ 2021 RD	104
PORTELL 2016 T C	165
PORTELL 2016 T R	166
PORTELL BLANC DE BLANCS 2021 B S	166
PORTELL BLANC SEMI DOLÇ 2021 BE SD	166
PORTELL MERLOT 2019 T BA	166
PORTELL ROSAT TREPAT 2021 RD	166
PORTENTO 2012 T R	242
PORTENTO 2015 T C	242
PORTENTO SAUVIGNON BLANC 2021 B	242
PORTENTO TEMPRANILLO 2021 T	242
PORTIA 2019 T C	456
PORTIA 2020 T RB	456
PORTIA PRIMA "LA ENCINA" 2020 T	457
PORTIA SUMMA 2019 T R	457
PORTIA VERDEJO 2021 B	609
PORTICO DA GLORIA BRANCELLAO 2020 T	416
PÓRTICO MAYOR 2016 T R	514
PÓRTICO MAYOR 2017 T C	514
PÓRTICO MAYOR 2020 T	514
PORTO DE LOBOS 2017 T	406
POULA CASA AURORA 2020 T	862
POZO DE NIEVE 2021 T	464
PRADES 950 2016 T	851
PRADO ENEA 2015 T GR	539
PRADO IRACHE 2019 T BA	746
PRADO NEGRO 2016 T C	202
PRADOREY BLANCO 2020 B	808

WINE	PAGE
PRADOREY ÉLITE 2019 T	471
PRADOREY FINCA LA MINA 2017 T R	471
PRADOREY FINCA LA MINA 2018 T R	472
PRADOREY FINCA VALDELAYEGUA 2019 T	472
PRADOREY ORIGEN 2021 T	472
PRADOREY ROSADO 2021 RD	472
PRADOS COLECCIÓN GARNACHA 2020 T BA	105
PRADOS COLECCIÓN SYRAH 2020 T	105
PRADOS FUSION GARNACHA SYRAH 2021 T	105
PRADOS PRIVÉ 2020 T C	105
PRAPETISCO 2019 T	840
PRAT D'HORES BLANC DE SANUI 2019 B FB	172
PRAT D'HORES NEGRE DE SANUI 2019 T	172
PREDICADOR 2020 B	511
PREDICADOR 2020 T	511
PREGÓN 2021 B	612
PREMIER VIN SON MAYOL 2017 T R	826
PREMIER VIN SON MAYOL 2018 T R	826
PREMIUM 1904 – TEMPRANILLO SYRAH 2020 T	784
PREMIUM 1904 SAUVIGNON BLANC 2021 B	784
PRESAS OCAMPO GRAN ALYSIUS 2020 T	633
PRESAS OCAMPO MACERACIÓN ESPECIAL 2021 T MC	633
PRESAS OCAMPO ORIGEN 2020 T	633
PRESAS OCAMPO SELECCIÓN ESPECIAL 2021 T MC	633
PRESAS OCAMPO VENDIMIA SELECCIONADA 2020 T	633
PRESAS OCAMPO VIÑEDOS PROPIOS 2021 T	633
PRETO DE LEIVE 2020 T	426
PRETTINI 2021 B	525
PRETTINI 2021 T	525
PREVIUS DE NEPTIS 2020 T RB	440
PRIDE 2017 BE GR BR	129
PRIMA 2020 T	668

WINE	PAGE	WINE	PAGE	WINE	PAGE
PRIMERA VINYA LES BRUGUERES 2020 B	368	PRISMA TEMPRANILLO 2020 T	579	PROTOS '27 2019 T	489
PRIMERO 2021 T	659	PRISMA VERDEJO 2021 B	618	PROTOS 2015 T GR	489
PRIMI LUIS GURPEGUI 2019 T C	534	PRIUS DE MORAÑA VERDEJO 2021 B	609	PROTOS 2018 T C	489
PRIMI LUIS GURPEGUI CHARDONNAY 2021 B	313	PRIVAT 2020 BE R BN	134	PROTOS 5º AÑO 2016 T R	490
PRIMICIA BLANC BOTA 2021 B FB	643	PRIVAT 2020 BE R BR	134	PROTOS 9 MESES 2020 T RB	490
PRIMICIA GARNACHA SYRAH 2019 T C	643	PRIVAT ROSAT 2020 RE BN	134	PROTOS CLARETE 2021 RD	158
PRIMITIU DE BELLMUNT 2019 T	358	PRIVILEGIO DE ROMALE 2018 T C	498	PROTOS ROSÉ 2021 RD	158
PRIMITIVO QUILES FONDILLÓN 1948 T FO	43	PRIVILEGIO DE ROMALE BE R BN	126	PROTOS SELECCIÓN FINCA EL GRAJO VIEJO 2018 T	490
PRIMITIVO QUILES GRAN IMPERIAL 1892 BF SOLERA D	43	PRIVILEGIO DEL CONDADO 2021 B S	168	PROTOS VERDEJO 2019 B R	609
PRIMITIVO QUILES MONASTRELL 2017 T C	43	PROELIO 2019 T C	543	PROTOS VERDEJO 2021 B	609
PRIMITIVO QUILES MOSCATEL EXTRA BF MISTELA D	868	PROELIO LA CANAL DEL ROJO 2018 T	543	PROTOS VERDEJO CUVÉE 2021 B	609
PRIMITIVO QUILES MOSCATEL LAUREL BF MISTELA D	868	PROELIO VENDIMIA SELECCIONADA 2018 T R	543	PRUNET ESENCIA DEL TERRITORIO 2020 T	717
PRIMUS AMEZTOI 2020 B	195	PROELIO VIÑEDOS VIEJOS 2019 T	543	PRUNO 2020 T	480
PRÍNCIPE DE BARBADILLO BF AM S	207	PROHIBIT 2018 RE BN	882	PSI 2020 T	478
PRÍNCIPE DE VIANA 1423 2017 T R	315	PROHOM CONCEPTIA 2020 T	643	PUEBLO DE LAVIA 2019 T	92
PRÍNCIPE DE VIANA CHARDONNAY 2021 B FB	315	PROHOM CONCEPTIA 2021 B	644	PUENTE DE RUS 2020 T RB	242
PRÍNCIPE DE VIANA EDICIÓN BLANCA 2021 B	315	PROHOM EXPERIENTIA 2018 T	644	PUENTE DE RUS 3 VARIEDADES 2020 T	242
PRÍNCIPE DE VIANA EDICIÓN LIMITADA 2018 T C	315	PROHOM EXPERIENTIA 2021 B FB	644	PUENTE DE RUS LA GUACHA BLANCO 2021 B SD	242
PRÍNCIPE DE VIANA EDICIÓN ROSA 2021 RD	315	PROHOM VIOGNIER 2021 B	644	PUENTE DE RUS SAUVIGNON BLANC 2021 B	242
PRÍNCIPE DE VIANA VENDIMIA SELECCIONADA 2018 T C	315	PROMESA 2021 RD	637	PUENTE DE RUS TEMPRANILLO 2015 T C	242
PRINCIPIA MATHEMATICA 2021 B	325	PROMESA BF MO D	219	PUENTE DE RUS VERDEJO 2021 B	242
PRIOR TERRAE 2020 B	376	PROMETO SER INFIEL 2019 T	489	PUENTEGRIJOS 2021 T MC	565
PRIOR TERRAE NEGRE 2019 T	376	PROPIEDAD 2020 T	543	PUERTA DE ALCALÁ 2018 T C	728
PRIORAT IDUS DE VALL-LLACH 2020 T	361	PROPÒSIT 2019 T	186	PUERTA DEL SOL MALVAR 2021 B	728
PRIORATO DE RAZAMONDE 2019 T	427	PROPÒSIT BLANC 2021 B	186	PUERTA PALMA 2020 T	498
PRIORATO DE RAZAMONDE TREIXADURA 2021 B	427	PROTESTA 2021 T	555	PUERTA PALMA 2021 B	498
PRIOS MAXIMUS 2016 T R	445	PROTOCOLO 2020 T	790	PUERTA PALMA BE EBR	125
PRIOS MAXIMUS 2019 T C	445	PROTOCOLO 2021 B	790	PUERTA PALMA BE R BN	125
PRIOS MAXIMUS 2021 T RB	445	PROTOCOLO 2021 RD	790	PUERTA PALMA ROSÉ 2021 RE BR	125
PRISMA GARNACHA 2020 T	98	PROTOCOLO ECO 2020 T	790	PUERTA SANTA 2021 B	379
PRISMA GARNACHA TINTORERA MONASTRELL 2021 T	49	PROTOCOLO ECO 2021 B	790	PUERTO ALICANTE AROMÁTICO 2021 B S	44
PRISMA ORGÁNICO 2021 T	232	PROTOCOLO ECO 2021 RD	790	PUERTO ALICANTE SELECCIÓN 2020 T	44

WINE	PAGE
PUERTO CARBONERO VINO DE PARCELA 2019 T S	846
PUERTO DEL MILAGRO 2020 T	790
PUERTO RUBIO 2018 T	544
PUERTO SALINAS 2013 T R SS	39
PULCHRUM CRESPIELLO EDICIÓN LIMITADA 2012 T R	108
PULCHRUM CRESPIELLO EDICIÓN LIMITADA 2016 T R	108
PULPO ALBARIÑO 2021 B	397
PUNT I. 2020 T C	646
PUNTES DE CALNEGRE 2020 T	300
PUNTILS 2018 T C	187
PUNTILS BLANC 2021 B	187
PUNTILS NEGRE 2021 T	187
PUNTO DE FUGA 2019 T	322
PUNTO GEODÉSICO 2019 T	462
PURA CEPA DE SARMIENTO 2021 T	540
PURA CEPA DE SARMIENTO GARNACHA 2018 T C	540
PURA SANGRE 2014 T R	233
PURA SAVIA 2019 T	777
PURESA GARNATXA BLANCA 2019 B FB	644
PURESA MACABEU 2019 B FB	644
PURESA MORENILLO 2016 T C	644
PURESA RANCI BF RC	644
PURESA SAMSÓ 2015 T BA	644
PURGATORI 2019 T BA	173
PURIFICACIÓN GARNACHA 2018 T C	728
PURITA 2020 T	844
PURITA 2020 T BA	844
PURITA BLANC DE NOIRS 2021 B	844
PURITA DYNAMITA GARNACHA 2018 T	728

Q

WINE	PAGE
QBQ 2020 B FB	351
QUADIS ENVEJECIDO 2019 T C	769
QUARS 2019 B	370
QUARTA CABAL 2021 B	799
QUÉ BONITO CACAREABA 2020 B	511
QUÉ NO FALTE! 2021 B	621
QUEIRÓN DE GABRIEL 2011 T R	580
QUEIRÓN EL ARCA. VIÑEDO SINGULAR 2018 T BA	580
QUEIRÓN ENSAYOS CAPITALES Nº 3 ASOLEAO 2020 T	580
QUEIRÓN MI LUGAR VIÑEDO DE MUNICIPIO 2018 T BA	580
QUELÍAS ROSÉ 2021 RD	159
QUÍBIA 2021 B	838
QUIKE 2019 RD	116
QUIM 2021 B	177
QUINCHA CORRAL 2019 T	742
QUINQ 2016 T	437
QUINTA DE AVES CABERNET FRANC & GRACIANO ROSÉ 2021 RD	799
QUINTA DE AVES COUPAGE 2020 T C	799
QUINTA DE AVES PHOENIX 2019 T C	799
QUINTA DE AVES SAUVIGNON BLANC & MOSCATEL 2021 B	799
QUINTA DE AVES SYRAH 2021 T	799
QUINTA DE AVES TEMPRANILLO 2021 T	799
QUINTA DE COUSELO 2021 B	400
QUINTA DE COUSELO TRADICIÓN 2015 B	400
QUINTA DE QUERCUS SINGLE VINEYARD 2018 T	672
QUINTA DEL 67 2020 T	48
QUINTA DO AVELINO 1923 2020 B	419
QUINTA DO AVELINO 1923 2021 B	419
QUINTA DO AVELINO 2021 B	419
QUINTA DO AVELINO ROCK NA VIÑA 2021 B	419
QUINTA DO SIL 2021 B	686
QUINTA LAS VELAS TEMPRANILLO 2019 T C	58
QUINTA REGIA BOBAL 2020 T	271
QUINTA SARDONIA QS 2019 T	812
QUINTA SARDONIA QS2 2019 T	812
QUINTA SOUTULLO 2019 T	281
QUINTA SOUTULLO 2020 B	281
QUIÑÓN DE VALMIRA 2020 T FB	543
QUIÑONES DE TOBELOS VIÑEDO SINGULAR 2019 B BA	585
QUIRUS 2021 B	559
QUIRUS 2021 RD	559
QUIRUS TEMPRANILLO BLANCO 2021 B	559
QUITAPENAS MÁLAGA DULCE 2020 BF PX D	262
QUITAPENAS MOSCATEL DORADO 2021 BF D	262
QUITAPENAS MOSCATEL PLATA 2021 BF D	262
QUITE 2021 T	65
QUIVIRA VERDEJO 2021 B	599
QUIXOTE CABERNET SAUVIGNON SYRAH 2018 T C	740
QUIXOTE MALBEC CABERNET FRANC 2018 T R	740
QUIXOTE MERLOT TEMPRANILLO PETIT VERDOT 2018 T	740
QUIXOTE PETIT VERDOT 2018 T C	740
QUIZÁS QUIZÁS QUIZÁS 2017 T	670
QUO VADIS FONDILLÓN 1972 T FO	36
QUO VADIS VORS BF AM S	215
QUOD SUPERIUS 2018 T	676
QUORUM – LE BLANC 2017 B FB	792
QUORUM DE FINCA EL REFUGIO PRIVATE COLLECTION 2012 T BA	792

R

WINE	PAGE
R'SEDAS 2020 T	664
R&G 2016 T	544
R&G CLOS D'EN FERRAN 2019 T	356
R&G RIBERA 2019 T	457
R&G SOLERAS OLVIDADAS BF AM	211
R&G SOLERAS OLVIDADAS BF FI	211

WINE	PAGE
R&G SOLERAS OLVIDADAS BF MZ	211
RABETLLAT I VIDAL 2018 BE R BN	124
RABETLLAT I VIDAL ROSAT 2019 RE R BR	124
RABILARGO 2020 T	759
RACÓ FRED DE MORTITX 2018 B FB	832
RACORELL 2019 T	370
RAFAEL CAMBRA DOS 2020 T	710
RAFAEL CAMBRA UNO 2020 T	710
RAÍCES 2016 T GR	696
RAÍCES BOBAL 2021 T	682
RAÍCES DE VALPARAISO 2019 T	464
RAÍCES GODELLO 2021 B	76
RAÍCES MENCÍA 2021 T	76
RAÍCES MONASTRELL 2021 T	869
RAÍCES SYRAH 2018 T	696
RAIMAT EL MOLÍ 2018 T C	176
RAIMAT EL NIU DE LA CIGONYA 2020 B	176
RAÏMS DE LA INMORTALITAT SELECCIÓ 2016 T R	343
RAÏMS DE LA INMORTALITAT XAREL·LO 2019 B FB	343
RAIMUN GARNACHA 2020 T	807
RAIMUN VERDEJO NATURAL 2020 B FB	807
RAÍZ DE GUZMÁN 2018 T R	457
RAÍZ DE GUZMÁN 2019 T C	457
RAÍZ DE GUZMÁN 2021 RD	458
RAÍZ DE GUZMÁN 9 MESES 2020 T RB	458
RAIZ PROFUNDA 2018 T	458
RAÍZ VOY OLÉ 2021 T	458
RAIZA TEMPRANILLO 2013 T GR	587
RAIZA TEMPRANILLO 2017 T R	587
RAIZA TEMPRANILLO 2018 T C	587
RALTOLLO 24 M 2019 T C	233
RAMBLA DE ULEA 2021 T	91

WINE	PAGE
RAMÍREZ DE LA PISCINA 2020 B FB	544
RAMIRO'S 2018 T	806
RAMITO TEMPRANILLO SYRAH 2021 T	869
RAMÓN BILBAO 2014 T GR	544
RAMÓN BILBAO 2016 T R	544
RAMÓN BILBAO 2020 T C	545
RAMÓN BILBAO 2021 RD	544
RAMÓN BILBAO EDICIÓN LIMITADA 2020 T	544
RAMÓN BILBAO EDICIÓN LIMITADA LÍAS 2019 B	610
RAMÓN BILBAO LÍMITE NORTE 2019 B	545
RAMÓN BILBAO LÍMITE SUR 2019 T	545
RAMÓN BILBAO ORGANIC 2020 T	545
RAMÓN BILBAO ORGANIC VERDEJO 2021 B	610
RAMÓN BILBAO RESERVA ORIGINAL 2016 T	545
RAMÓN BILBAO SAUVIGNON BLANC 2021 B	610
RAMÓN BILBAO VERDEJO 2021 B	610
RAMÓN BILBAO VIÑEDOS DE ALTURA 2020 T C	545
RAMÓN DO CASAR GODELLO 2020 B FB	428
RAMÓN DO CASAR NOBRE 2020 B FB	428
RAMÓN DO CASAR TREIXADURA 2021 B	428
RAMÓN DO CASAR VARIETAL 2021 B	428
RAMÓN NADAL GIRÓ 2004 BE GR BR	882
RAMÓN ROQUETA GARNACHA 2021 T	118
RAMÓN ROQUETA GARNACHA BLANCA 2021 B	118
RAMÓN ROQUETA TEMPRANILLO 2021 T	118
RAMÓN SÁENZ, PASIÓN DE VIDA 2021 T	581
RAMÓN SÁENZ, PEQUEÑO BASTIÓN 2020 T RB	581
RAMÓN SÁENZ, PIEDRAS RODANTES 2020 T RB	581
RANCI 2013-2018 T RC S	372
RANCI ORTO VINS B RC	300
RANGUELOS 2021 T	239
RAPOLAO 2019 T C	78

WINE	PAGE
RASCAÑA 2020 B	700
RASIM VI PANSIT NATURALMENTE DULCE 2018 B D	175
RASIM VIMADUR 2018 T D	175
RASO DE LARRAINZAR 2015 T R	320
RATPENAT 2019 B	332
RAÚL PÉREZ EL RAPOLAO 2020 T	869
RAÚL PÉREZ PAN Y CARNE 2020 T	869
RAVENTÓS I BLANC BLANC DE BLANCS 2020 BE	883
RAVENTÓS I BLANC DE LA FINCA 2019 BE BN	883
RAVENTÓS I BLANC DE NIT 2020 RE BR	883
RAVENTOS I BLANC TEXTURES DE PEDRA 2017 BE BN	883
RAW MANDÓ 2020 T FB	348
RAW PICAPOLL 2021 B	348
RAYUELO 2018 T	266
RCR 2019 T C	190
REAL AGRADO 2017 T R	581
REAL AGRADO 2018 T C	581
REAL AGRADO 2021 B	581
REAL AGRADO 2021 RD	581
REAL ARBÁS 2021 B D	813
REAL DE ASÚA 2019 T	563
REAL RUBIO 2010 T GR	591
REAL RUBIO 2013 T R	591
REAL RUBIO 2018 T C	591
REAL RUBIO 2021 B	591
REAL RUBIO 2021 RD	591
REAL RUBIO FINCA EL TORDILLO 2019 T	591
REALLY AWESOME WINE 2021 T	849
REBELLIA 2021 B	679
REBELLIA 2021 T	679
REBELLIA SELECCIÓN ESPECIAL 2020 T RB	679
REBELS DE BATEA 2020 B	641

WINE	PAGE
REBELS DE BATEA 2020 T	641
REBLUM 2019 T	359
REBOMBORI 2020 B	856
REBOMBORI MOSCATEL 2019 B	856
REBROTAR 2020 B FB	806
REBUZNO 2020 T	531
RECAREDO SERRAL DEL VELL 2011 BE BN	883
RECAREDO SERRAL DEL VELL 2016 BE BN	883
RECAREDO SUBTIL 2017 BE BN	883
RECAREDO TERRERS 2018 BE BN	883
RECLOT 2021 T	292
RECLOT ROSAT 2021 RD	292
RECÓNDITA ARMONÍA 2010 2010 T SOLERA D	848
RECÓNDITA ARMONÍA DULCE 2019 T D	848
RECTORAL DE AMANDI 2021 T	411
RECTORAL DE AMANDI BARRICA MANOLO ARNOYA 2017 T	411
RECTORAL DO UMIA 2021 B	401
REGENTE BF PC S	219
REGINA VIARUM EXPRESIÓN 2018 T BA	411
REGINA VIARUM GODELLO 2021 B	410
REGINA VIARUM MENCÍA 2021 T	410
REGINA VIARUM ROSAE 2021 RD	411
REGINA VIDES 2019 T	450
REGUEIRÓN 2020 B	694
REGULUS BE BR	125
REINA DE CASTILLA VERDEJO 2021 B	600
REINA DE CASTILLA VERDEJO SELECCIÓN CUVÉE 2020 B FB	600
REINA VIOLANT BE R BN	637
REINARES TEMPRANILLO 2020 T BA	779
REINO DE ALTUZARRA 2016 T R	305
REINO DE ALTUZARRA 2018 T C	305
REJADORADA ROBLE 2020 T RB	656

WINE	PAGE
REJADORADA VERDEJO 2021 B	600
REJÓN 2019 T	807
RELICTA 2018 T C	771
REMIGIO ENRIQUEZ (YAYOS) 2016 T BA	869
REMIGIO ENRIQUEZ GODELLO 2018 B	834
REMIGIO ENRIQUEZ MENCÍA 2019 T	834
REMIGIO ENRÍQUEZ RESTLESSNESS 2016 T BA	869
REMIGIO ENRÍQUEZ RESTLESSNESS 2017 T BA	869
REMIGIO ENRÍQUEZ RESTLESSNESS 2018 T BA	869
REMÍREZ DE GANUZA 2015 T R	545
REMOLÓN ROSÉ 2021 RD PL S	161
REMONTE 2017 T C	310
REMONTE 2021 RD	310
REMONTE CHARDONNAY 2021 B	310
RENACCE 2019 BE BN	606
RENACCE 2020 B	607
RENACCE BLUSH ROSÉ 2021 RD	805
RENACIDO 2019 B	427
RENAIX DE GIRÓ 2020 T	38
RENAIX LA PASSIÓ 2021 B	38
RENEGÓN 2020 T	757
RENVIVAS 2019 T	814
RESACO 2019 T	518
RESALTE EXPRESIÓN 2018 T R	458
RESALTE VENDIMIA SELECCIONADA 2020 T	458
RESERVA LLUIS AROLA I FERRER 2017 T R	638
RESERVA PARTICULAR DE RECAREDO 2012 BE BN	884
RESERVA REAL 2017 T	335
RETALLO BLANCO 2020 B	865
RETORNO A LOS PALOMARES 2017 T BA	540
REX DEUS 2016 T R	767
REY ENEO 2015 T GR	567

WINE	PAGE
REY ENEO 2020 B	393
REY ENEO 2021 B	394
REY ENEO EDICIÓN LIMITADA 2010 T R	567
REY ENEO VENDIMIA SELECCIONADA 2015 T R	567
REY ENEO VENDIMIA SELECCIONADA 2019 T C	567
REY ZAGAL 2011 T C	202
REY ZAGAL 2013 T R	202
REY ZAGAL CASCAMORRAS 2016 T	202
REY ZAGAL SAUVIGNON BLANC 2020 B	202
REYES DE ARAGÓN PREMIUM 2021 T BA	98
REYMOS SELECCIÓN BE MO D	700
REYMOS VERSIÓN LIBRE BE BN	700
RHODES 2019 T C	189
RIALLA GARNATXA BLANCA 2021 B	645
RIALLA GARNATXA PELUDA 2019 T	645
RIALLA GARNATXA PELUDA 2021 RD	645
RIALLA GARNATXA TINTORERA 2020 T	645
RIBAGAITAS 2019 T	533
RIBERA DE LOS NARANJOS 2019 T	415
RIBERAL 2016 T R	459
RICAHEMBRA SOLERA 1980 BF SOLERA D	168
RICO NUEVO GARNACHA 2021 T	753
RIESLING 2021 B	37
RIGAU ROS 2015 T R	190
RIMARTS 2017 BE GR BN	148
RIMARTS 2018 BE R BN	148
RIMARTS 2019 BE R BR	148
RIMARTS GRAN RESERVA ESPECIAL CHARDONNAY 2017 BE GR BN	148
RIMARTS UVAE 2011 BE BN	149
RINCÓN DEL GUINIGUADA 2020 B	200
RINCÓN DEL GUINIGUADA 2020 T	200

WINE	PAGE
RIOJA VEGA 2018 T R	582
RIOJA VEGA 2019 T C	582
RIOJA VEGA COLECCIÓN TEMPRANILLO BLANCO 2021 B	582
RIOJA VEGA EDICIÓN LIMITADA 2019 T C	582
RIOJA VEGA TEMPRANILLO BLANCO 2019 B R	582
RIOJA VEGA VENTA JALÓN 2016 T R	582
RITA 2021 B	302
RITME NEGRE 2019 T	373
RITME NEGRE 2020 T	373
RITUS 2018 T BA	441
RIU BY TRÍO INFERNAL 2018 T R	368
RIU BY TRÍO INFERNAL 2021 B	368
RIU DE GOST GARNACHA BLANCA 2020 B RB	868
RIU RAU DULCE 2019 B MISTELA D	40
RIURAU MONASTRELL APASIONADO 2020 T	734
RNG 20 MAGNUM 1997 BE BN	882
RNG 2014 BE BR	882
ROANDI 2019 T	687
ROANDI 2020 B	687
ROBERT J. MUR ED. ESPECIAL 2012 BE GR BN	149
ROBERT J. MUR ESPECIAL TRADICIÓ 2018 BE R BN	149
ROBERT J. MUR ESPECIAL TRADICIÓ 2018 BE R BR	149
ROBERT J. MUR ESPECIAL TRADICIÓ ROSÉ 2018 RE R BN	149
ROBERT J. MUR ESPECIAL TRADICIÓ ROSÉ 2018 RE R BR	149
ROBERT J. MUR MILLÉSIMÉ 2016 BE R BN	149
ROBERT VEDEL CEPAS VIEJAS 2020 B	616
ROC 2020 T	65
ROCA DEL CRIT 2020 T	188
ROCAPLANA 2020 T	331
ROCHA 2020 T	707
ROCHAPEL 2021 RD	310
ROCHAPEL CHARDONNAY 2021 B	310
ROCHAPEL GARNACHA 2021 T	310
ROCK & ROLL 2020 T RB	867
ROCOSO 2021 T RB	58
RODA 2019 T R	546
RODA I 2018 T R	546
RODA I 2019 B	546
RODAL PLA DE MORTITX 2019 T	833
RODENO 2020 T C	678
RODENO SUPERIOR 2020 B AROM S	678
RODRIGUEZ & SANZO GOTAS DE NOCHE 2021 RD	661
RODRIGUEZ & SANZO WHISBA 2019 T	806
RODRÍGUEZ DE VERA 2020 RD	47
RODRÍGUEZ DE VERA CHARDONNAY PEQUEÑAS PARCELAS 2017 B FB	47
RODRÍGUEZ DE VERA SERIE LIMITADA 2017 T C	47
RODRÍGUEZ DE VERA SERIE LIMITADA CHARDONNAY 2019 B	47
RODRÍGUEZ SANZO BAJO VELO 2020 B	610
RODRÍGUEZ SANZO ORANGE WINE 2020 B BA	807
ROFE 2020 B	253
ROFE 2020 T	253
ROGER GOULART 2019 BE R BN	149
ROGER GOULART CORAL ROSÉ 2019 RE BR	149
ROGER GOULART JOSEP VALLS 2017 BE EBR	149
ROGER GOULART MILLESIMÉ 2020 BE BR	149
ROGER GOULART ROSÉ MILLÉSIME 2020 RE BR	149
ROGER GOULART SEMISECO RESERVA 2020 BE R SS	149
ROJO TEMPRANILLO 2021 T	784
ROLLAND GALARRETA PARCELA 123 2020 B	610
ROMAILA 2021 RD	843
ROMAILA PET NAT 2021 RE	878
ROMAILA PET NAT BE	878
ROMÁN CEPAS VIEJAS 2017 T	104
ROMÁNICO 2020 T	669
ROMANORUM 2016 T R	61
ROMANORUM 2018 T C	61
ROMÉ 2021 T RB	259
ROMEO DOMAIN 2020 T	410
ROMERO DE AVILA SALCEDO 2011 T RB	785
RÒMIA 2018 T	867
ROQUERS DE GARNATXA 2019 T	359
ROQUERS DE PORRERA 2018 T R	360
ROQUERS DE SAMSÓ 2019 T	360
ROS MARINA 2X 2021 B	341
ROS MARINA DCOSTER 2016 T C	341
ROS MARINA QUATRE VINYES 2017 T	342
ROS MARINA ROSÉ 2021 RD	342
ROSA DE ALEJANDRÍA 2020 B AG SD	781
ROSA DE MAR 2021 RD	820
ROSA RUIZ 2021 B	401
ROSADO DE FERMIN 2021 RD	680
ROSADO DE LARRAINZAR 2021 RD	320
ROSADO STAIRWAY TO HEAVEN 2021 RD	825
ROSADO VIZCARRA 2021 RD	466
ROSAE ARZUAGA 2021 RD	440
ROSAN 2021 RD	160
ROSARA 2021 RD	841
ROSARIO VERA 2019 T	582
ROSARIO VERA AMONA 2021 T	582
ROSAT 110 VINS NADAL 2021 RD	82
ROSAT 2021 RD	832
ROSAT DE GERISENA 2021 RD	184
ROSAT DE PLANAS ALBAREDA 2021 RD	341
ROSÉ & CLEAR 2021 RD	338
ROSE MARINE 2021 RD	680

WINE	PAGE
ROSÉ PRÍNCEPS 2021 RD	330
ROSELITO 2021 RD	439
ROSELL GALLART BE R BN	149
ROSMARINUS 2018 T RB	92
ROSMARINUS 2020 T	92
ROSSINYOL 2017 BE GR BN	135
ROUND TABLE 2021 B	342
ROUND TABLE 2021 T MC	300
ROURA 5* BE EBR	127
ROURA BE BN	127
ROURA BE BR	127
ROURA COUPAGE 2017 T BA	32
ROURA GARNATXA SYRAH 2017 T C	32
ROURA SAUVIGNON BLANC 2021 B	32
ROURA TRES CEPS 2017 T C	32
ROURA XARELLO 2021 B	32
ROUREDA LLICORELLA POBOLEDA 2021 B	362
ROVELLATS BLANC CHARDONNAY 2021 B	342
ROVELLATS BLANC PRIMAVERA 2021 B	342
ROVELLATS COL·LECCIÓ 2016 BE GR EBR	150
ROVELLATS CUVÉE ESPECIAL 2019 BE R BN	150
ROVELLATS GRAN RESERVA 2016 BE GR BN	150
ROVELLATS MASIA S. XV 2012 BE GR BN	150
ROVELLATS MERLOT 2021 RD	342
ROVELLATS RESERVA IMPERIAL 2019 BE BR	150
ROVELLATS RESERVA IMPERIAL ROSÉ 2019 RE R	150
ROVER Nº 1 2019 T	843
ROZAS 1ER CRU 2019 T	725
RQT FELIZ, CEPAS ENTRE VIÑAS 2019 B C	480
RUBERTE TR3SOR 2020 T	104
RUBICÓN 2018 I	252
RUBICÓN MALVASÍA SECO 2021 B	252

WINE	PAGE
RUBICÓN MOSCATEL DULCE 2020 B D	252
RUBICÓN ROSADO 2020 RD	252
RUBICÓN SEMIDULCE 2021 B SD	252
RUBIÒLS 2019 B	857
RUDELES "23" 2020 T	490
RUDELES "23" 2021 B	490
RUDELES 2021 RD	490
RUDELES AIRE 2020 B FB	490
RUDELES CERRO EL CUBERILLO 2019 T	490
RUDELES FINCA LA NACIÓN 2018 T	490
RUIDO BY LA QUINTA 2018 T	573
RUIZ TORRES SYRAH 2018 T	818
RUJAQ ANDALUSÍ BF TRASAÑEJO D	261
RULEI VIÑA BARRACALLO RENQUES DE CHENIN 2019 B FB	505
RULEI VIÑA BARRACALLO TEMPRANILLO-GARNACHA 2015 T	505
RULEI VIÑA EL MORAL VIÑEDO SINGULAR 2019 T	505
RUNRUN 2020 RD	776
RUPESTRE DE ALPERA GARNACHA TINTORERA 2018 T R	50
RUTA 49 2021 B	402

S

WINE	PAGE
S-NAIA 2021 B	607
S. ROBERTO BOTA ÚNICA 1/1 BF PC	207
S. ROBERTO BOTA ÚNICA 1/2 BF AM	207
S. ROBERTO BOTA ÚNICA 2/2 BF AM	207
S'ALOU 2018 T C	193
SA CABANA CHARDONNAY 2021 B	830
SA CABANA GIRÒ ROS 2021 B	830
SA CABANA MERLOT 2020 T BA	830
SA FORANA 2019 T	821
SA FORANA BLANC 2021 B FB	821
SA ILLA 2017 T R	187
SA NATURA 2019 T	645

WINE	PAGE
SA TALAIA 2017 T BA	825
SA VALL SELECCIÓ PRIVADA 2016 B FB	351
SÁBALO 2021 B	769
SABARIA 2019 T C	60
SABATACHA 2019 T C	230
SABATÉ I COCA JOSEP COCA 2015 BE GR BN	884
SABATÉ I COCA JOSEP COCA MAGNUM 2013 BE GR BN	884
SABATÉ I COCA MOSSET 2017 BE GR BN	884
SABINILLA 2020 B FB	674
SACASUEÑOS 2021 B	724
SADURNÍ OLIVER 2018 BE R BN	145
SADURNÍ OLIVER BARRICA 2017 BE R BN	145
SADURNÍ OLIVER ROSAT PINOT NOIR 2020 RE BN	146
SAFRÀ 2020 T	708
SAGASTIBELTZA 2020 B	88
SAIAL 2019 B BA	118
SAIAZ PUENTE DEL EA 2017 T	558
SAIÑAS - AREAS 2020 T	407
SAIÑAS – O BOLIÑO 2020 T BA	407
SAIÑAS 2020 T	407
SALANQUES 2020 T C	371
SALETA MONASTRELL 2020 T RB	222
SALGADERO 2018 T BA	664
SALIA 2020 T R	269
SALINILLAS DE TOBELOS 2017 T	585
SALIX 2021 B FB	371
SALMOS 2019 T C	365
SALSIPUEDES 2019 T C	60
SALTO DEL USERO 2021 B	90
SALTO DEL USERO 2021 T	90
SALUD DE SANI PRIMAVERA CAVA 2020 BE BR	127
SALVATGE EDICIÓ LIMITADA 2013 BE BN	882

WINE	PAGE	WINE	PAGE	WINE	PAGE
SALVATGE EDICIÓ LIMITADA 2015 BE BR	882	SAN ROMÁN 2019 T	668	SANTA CLARA BLANC DE BLANCS 2021 B	828
SALVATGE ROSÉ 2018 RE BR	882	SAN ROMÁN GARNACHA 2020 T	668	SANTALBA AMARO 2017 T	547
SALVAVIDES 2020 T	172	SAN SALVADOR GODELLO 2019 B FB	64	SANTALBA VIÑA HERMOSA 2015 T GR	547
SALVAXE 2019 B	875	SAN TROCADO 2021 B	421	SANTIAGO ROMA ALBARIÑO 2021 B	390
SÁLVORA 2020 B FB	387	SAN VICENTE 2018 T BA	583	SANTIAGO ROMA ALBARIÑO SELECCIÓN 2020 B	390
SALVUEROS 2020 T RB	158	SANCHO BARÓN 2020 T	523	SANTIAGO ROMA BURBULLA 2020 BE BN	390
SALVUEROS 2021 RD	158	SANCHO GARCÉS 2017 T R	593	SANTIAGO RUIZ 2021 B	401
SALVUEROS GARNACHA GRIS 2021 RD	159	SANCHO GARCÉS 2018 T C	593	SANTO MERLOT 2016 T	744
SAM EDICIÓN LIMITADA JUMILLA 2018 T	232	SANCLODIO 2021 B	422	SANTO SYRAH 2016 T	744
SAMAS 2019 T	183	SANDARA BLANCO BE	885	SANTO TEMPRANILLO 2015 T	744
SAMEROLA BLANC 2021 B	643	SANDARA PREMIUM MOSCATO BE D	885	SANZ CLÁSICO 2021 B	620
SAMEROLA NEGRE 2021 T	643	SANDARA ROSADO RE D	885	SANZ SAUVIGNON BLANC 2021 B	620
SAMITIER 2020 T RB	95	SANGARIDA GODELLO 2021 B	63	SANZ VERDEJO 2021 B	620
SAMITIER GARNACHA 2019 T	95	SANGARIDA LA GUIANA 2020 B	63	SANZO VIÑAS VIEJAS 2021 B FB	610
SAMITIER GARNACHA BLANCA 2021 B	95	SANGARIDA LA YEGUA 2020 B	63	SAÓ ABRIVAT 2019 T C	175
SAMITIER SYRAH 2020 T	95	SANGARIDA MENCÍA 2021 T	63	SAÓ BLANC 2020 B FB	175
SAMPAYOLO GARNACHA TINTORERA 2020 T	689	SANGARIDA PICO TUERTO 2020 T	64	SAÓ EXPRESSIU 2019 T R S	175
SAMPAYOLO GODELLO EN LÁGRIMAS DE LOS BANCALES DE OLIVEDO 2021 B	689	SANGO DE REJADORADA 2015 T R	656	SARA GARNACHA 2017 T	311
		SANGRE Y TRABAJADERO BF OL S	208	SARA PEÑA 2020 T RB	425
SAMPAYOLO GODELLO SOBRE LÍAS 2021 B	689	SANI VIÑA EXTREMEÑA 2018 T FB	818	SARAMUSA TREIXADURA 2021 B	426
SAMPAYOLO MENCÍA 2021 T	689	SANSOFÍ 2020 T	199	SARAS... 2020 B	768
SAMSARA PRIORAT 2018 T	373	SANT JERONI AUBADA 2021 B	359	SARASATE 2021 RD	311
SAN ACISLO 2019 T C	463	SANT JERONI FLOR DE SAMSÓ 2021 RD	359	SARASATE EXPRESIÓN DULCE NATURAL 2021 B MO D	311
SAN CLEMENTE 2020 B	632	SANT JERONI FORN 2019 T	359	SARDASOL 2021 RD	309
SAN CLEMENTE 2020 T BA	632	SANT JERONI HORT 2019 T	359	SARDASOL SAUVIGNON BLANC 2021 B	309
SAN COBATE LA FINCA 2018 T C	481	SANT PERE 2020 T	702	SARDASOL SELECCIÓN GARNACHA 2021 T	309
SAN CUCUFATE "MONASTERIO" 2018 T	481	SANT PERE BLANC 2020 B	702	SARDASOL SELECCIÓN GARNACHA BLANCA 2021 B	309
SAN LEÓN RESERVA DE FAMILIA BF MZ S	206	SANT PERE VINYES VELLES BLANC 2019 B	702	SARDASOL TEMPRANILLO 2019 T C	310
SAN MARTINEÑO 2014 T R	783	SANT PERE VINYES VELLES NEGRE 2017 T	702	SARDÓN 2019 T	812
SAN MARTINEÑO GARNACHA 2021 T BA	783	SANTA BÁRBARA DE ALICANTE 2019 T RB	45	SARGAS DE IDUES GARNACHA 2020 T	97
SAN MARTINEÑO TEMPRANILLO 2018 T	783	SANTA CLARA 2021 T	340	SARGAS DE IDUES GARNACHA BLANCA 2020 B	97
SAN MARTINEÑO TEMPRANILLO 2021 T	784	SANTA CLARA 2021 T	827	SATÉLITE BOARDING WINE 2020 T	845

WINE	PAGE
SÀTIRS NEGRE 2019 T C	182
SAUCI 2019 T C	169
SAUCI 2021 B	169
SAULÓ PANSA BLANCA 2019 B C	33
SAUVIGNON BLANC OWNER'S EDITION 2021 B	825
SAUVIGNON BLANC STAIRWAY TO HEAVEN 2021 B	826
SAVINA 2021 B	820
SAVINAT SAUVIGNON BLANC 2019 B FB	338
SCALA DEI CARTOIXA 2018 T R	362
SCALA DEI L'HERETGE 2018 T	362
SCALA DEI MASDEU 2017 T	362
SCALA DEI MASSIPA 2020 B	362
SCALA DEI PLA DELS ÀNGELS 2021 RD	362
SCALA DEI PRIOR 2018 T C	362
SCALA DEI SANT ANTONI 2018 T	362
SCHATZ CHARDONNAY 2021 B	259
SCHATZ PETIT VERDOT 2016 T C	259
SCHATZ PINOT NOIR 2016 T C	259
SCHATZ ROSADO 2021 RD	259
SDM. SOLERGIBERT DE MATACANS 2019 T	347
SECASTILLA 2018 T C	629
SECRET D'EN PERICO 2021 B FB	351
SECRETS DE MAR 2020 T RB	647
SECRETS DE MAR 2021 B	647
SECUA CABERNET-SYRAH 2017 T R	781
SECUA CRIANZA EN LÍAS 2021 B FB	781
SECUENCIAL 2021 B	259
SÉCULO CEPAS VIEJAS 2021 T RB	814
SED DE CANÁ 2018 T	494
SEDA B SD	870
SEDE E FAME AS ERMITAS 2019 B	694
SEDELLA "LAS JACINTAS" 2019 T	262

WINE	PAGE
SEDELLA 2018 T	262
SEDOSA ORGANIC ROSÉ 2021 RD	793
SEDUCCIÓN 2021 B	678
SEDUCCIÓN 2021 RD	678
SEGARES 2008 T GR	541
SEGARES 2010 T R	541
SEGARES 2015 T C	541
SEGEDA DE SAMITIER 2018 T BA	95
SEGREL ALBARIÑO 2021 B	390
SEGREL ÁMBAR 2021 B	390
SEGUIT 2019 T	646
SEGUNDA BOTA B FI	286
SEGUNYOLA 2017 BE BN	882
SEGURA VIUDAS BE R BR	150
SEGURA VIUDAS LAVIT BRUT NATURE 2020 BE R BN	150
SEGURA VIUDAS MAS D'ARANYÓ 2016 T C	119
SEGURA VIUDAS VINTAGE 2015 BE GR BR	150
SEGURAS VIUDAS 2021 RD	119
SEIS DE AZUL Y GARANZA 2019 T	305
SEIS DÉCADAS 2017 T	758
SELECCIÓ 259 2013 T C	119
SELECCIÓ CARINYENA 2018 I R	641
SELECCIÓ DE GERISENA 2018 T C	184
SELECCIÓ FAMILIAR SIN SULFITOS 2019 T	862
SELECCIÓ GARNATXA BLANCA 2017 B	642
SELECCIÓ GARNATXA PELUDA 2017 T R	642
SELECCIÓN DE OSTATU 2017 T	579
SELECCIÓN EXCELENCIA 2015 T C	744
SELECCIÓN JESÚS MADRAZO RIOJA 2019 T	572
SELECCIÓN JESÚS MADRAZO SELECCIÓN RIBERA DEL DUERO 2019 T	483
SELECCIÓN PARCELA MOSCATEL 2021 B S	776

WINE	PAGE
SELECCIÓN PARCELA SYRAH 2018 T	776
SELECCIÓN RAMÍREZ DE LA PISCINA 2019 T C	544
SELMA DE NIN 2018 B	365
SELVANEVADA VILLOTA 2019 T RB	587
SELVANEVADA VILLOTA 2020 B	587
SEMBRO 2021 T	494
SEMBRO EDICIÓN LIMITADA 2020 T	494
SEMELE 2020 T C	486
SEMSUM 2 2021 B SD	41
SENDA DE HARO 2019 T C	509
SENDA DE HARO GARNACHA 2020 T	509
SENDA DE HARO VENDIMIA SELECCIONADA 2019 T	509
SENDA DE HOYAS 2021 T	104
SENDA DE HOYAS ORÍGENES 2020 T RB	104
SENDA DE LAS ROCHAS BOBAL 2014 T RB	267
SENDA DE LAS ROCHAS TEMPRANILLO 2018 T C	267
SENDA DE LOS LEÑADORES 2020 T	738
SENDA DE LOS OLIVOS 2019 T C	474
SENDA DE LOS OLIVOS 2021 T RB	474
SENDA LASARDA 2019 T C	108
SENDERILLO 2021 T	461
SENDERO DEL LOBO ROSÉ 2021 RD	003
SENDEROS DE UKAN 2019 T	586
SENDEROS DE VALLEYGLESIAS 2020 T	725
SENSE SENTITS 2019 T	115
SENTADA SOBRE LA BESTIA 2019 T BA	709
SENTERO 2020 T RB	668
SENTERO EXPRESIÓN 2018 T	668
SENTIDIÑO 2021 B	401
SENTIDO 2020 T	470
SENTIR B	798
SENTIR T	798

WINE	PAGE
SENTITS BLANCS 2020 B FB	115
SENTITS NEGRES GARNATXA NEGRA 2017 T	115
SEÑOR DA FOLLA VERDE 2020 B	389
SEÑOR DE ALDEA 2015 T C	459
SEÑORÍO AMÉZOLA 2015 T R	515
SEÑORÍO DA TORRE 2021 B	381
SEÑORÍO DA TORRE SOBRE LÍAS 2018 B	381
SEÑORÍO DE AGÜIMES 2021 B D	199
SEÑORÍO DE AGÜIMES 2021 RD	199
SEÑORÍO DE BEADE 2021 B	428
SEÑORÍO DE BEADE 2021 T	428
SEÑORÍO DE BOCOS 2015 T R	490
SEÑORÍO DE BOCOS 2018 T C	490
SEÑORÍO DE BOCOS 2020 T RB	490
SEÑORÍO DE BOCOS 2021 RD	653
SEÑORÍO DE BOCOS VERDEJO 2021 B	653
SEÑORÍO DE CABRERA 2019 T	200
SEÑORÍO DE CABRERA 2020 T	200
SEÑORÍO DE CABRERA 2021 B	200
SEÑORÍO DE FUENTEÁLAMO MONASTRELL 2020 T RB	227
SEÑORÍO DE FUENTEÁLAMO MONASTRELL 2021 RD	228
SEÑORÍO DE FUENTEÁLAMO MONASTRELL 2021 T	228
SEÑORÍO DE FUENTEÁLAMO VERDEJO 2021 B	228
SEÑORÍO DE GALDÓN 2021 B	865
SEÑORÍO DE GARCIGRANDE VERDEJO 2021 B	605
SEÑORÍO DE GUADIANEJA MACABEO 2021 B	246
SEÑORÍO DE GUADIANEJA SYRAH 2021 T	246
SEÑORÍO DE GUADIANEJA TEMPRANILLO 2021 T S	246
SEÑORÍO DE INIESTA COLECCIÓN 2020 T RB	795
SEÑORÍO DE INIESTA ROSÉ 2021 RD	795
SEÑORÍO DE INIESTA VELVET 15 2020 T	795
SEÑORÍO DE LA ANTIGUA MENCÍA 2021 T	801

WINE	PAGE
SEÑORÍO DE LA ERALTA 2015 T GR	529
SEÑORÍO DE LA ERALTA 2016 T GR	529
SEÑORÍO DE LA ERALTA 2017 T GR	529
SEÑORÍO DE LA ERALTA 2019 T C	529
SEÑORÍO DE LAZÁN T C	626
SEÑORÍO DE LOS LLANOS 2016 T R	697
SEÑORÍO DE MARESTE T	853
SEÑORÍO DE MORATE 2020 T RB	724
SEÑORÍO DE MORATE SYRAH 2015 T C	724
SEÑORÍO DE MORATE SYRAH 2016 T	724
SEÑORÍO DE NAVA 2016 T R	459
SEÑORÍO DE NAVA 2017 T C	459
SEÑORÍO DE NAVA 2020 T RB	459
SEÑORÍO DE NAVA 2021 T	459
SEÑORÍO DE NAVA VERDEJO 2021 B	611
SEÑORÍO DE NEVADA 2021 B	822
SEÑORÍO DE NEVADA 2021 RD	822
SEÑORÍO DE NEVADA BRONCE 2019 T	202
SEÑORÍO DE NEVADA CLUB DE LA BARRICA 2018 T	202
SEÑORÍO DE NEVADA ORO 2019 T	202
SEÑORÍO DE NEVADA PLATA 2019 T	202
SEÑORÍO DE OTXARAN 2017 B	85
SEÑORÍO DE OTXARAN 2019 B	85
SEÑORÍO DE P. PECIÑA 2011 T GR	527
SEÑORÍO DE P. PECIÑA 2016 T C	527
SEÑORÍO DE P. PECIÑA 2021 T	527
SEÑORÍO DE PARDO 2020 B	843
SEÑORÍO DE PARDO 2020 T	843
SEÑORÍO DE PEÑALBA SELECCIÓN 2021 T	815
SEÑORÍO DE RUBIÓS ALBARIÑO 2021 B	402
SEÑORÍO DE RUBIÓS CONDADO BLANCO BE BN	402
SEÑORÍO DE RUBIÓS CONDADO DO TEA BLANCO 2021 B	402

WINE	PAGE
SEÑORÍO DE RUBIÓS CONDADO TINTO 2018 T	402
SEÑORÍO DE SARRÍA 2010 T GR	320
SEÑORÍO DE SARRÍA 2016 T R	320
SEÑORÍO DE SARRÍA 2018 T C	320
SEÑORÍO DE SARRÍA 2021 RD	320
SEÑORÍO DE SARRÍA CHARDONNAY 2021 B	320
SEÑORÍO DE SARRÍA ROSÉ 2021 RD	320
SEÑORÍO DE SARRÍA VIÑEDO Nº 1 2019 T C	320
SEÑORÍO DE SARRÍA VIÑEDO Nº 5 2021 RD	320
SEÑORÍO DE UNX 2018 T C	309
SEÑORÍO DE VIDRIALES 2021 RD	757
SEÑORÍO DE VIDRIALES 2021 T	757
SEÑORÍO DE VIDRIALES VERDEJO 2021 B	757
SEÑORÍO DE ZAFRA CABERNET SAUVIGNON 2020 T RB	723
SEÑORÍO DE ZAFRA SYRAH 2020 T	723
SEÑORÍO DE ZAFRA TEMPRANILLO MERLOT 2020 T RB	723
SEÑORÍO DEL BISPO 2019 T RB	276
SEÑORÍO DEL BISPO 2021 T RB	276
SER VIVO Y NATURAL 2021 T	488
SERÉ 2019 T C	117
SEREZHADE 2020 B	531
SERICIS CEPAS VIEJAS BOBAL 2019 T	677
SERICIS CEPAS VIEJAS MONASTRELL 2018 T R	37
SERICIS CEPAS VIEJAS VIOGNIER 2021 B FB	850
SERRA D'OR CHARDONNAY BE R BN	128
SERRA D'OR COUPAGE CLASSIC BE R BN	128
SERRA D'OR PINOT NOIR BE R BN	128
SERRA DA ESTRELA 2021 B	380
SERRA DE CAVALLS 2020 B FB	649
SERRA DE CAVALLS 2021 T	649
SERRA DE CAVALLS GARNACHA BLANCA 2021 B	649
SERRA DE CAVALLS GARNATXA DE AMFORA 2019 T	649

WINE	PAGE
SERRAS DEL PRIORAT 2021 B	363
SERRAS DEL PRIORAT 2021 T	363
SERRES VELLES MACABEU 2021 B C	339
SERVILIO 2019 T C	460
SERVILIO 2020 T RB	460
SERVILIO VENDIMIA SELECCIONADA 2016 T	460
SES FERRITGES 2016 T C S	351
SET TOTA LA VIDA T	298
SEULALIA GODELLO 2021 B	69
SEULALIA MENCÍA 2020 T	69
SEULALIA PALOMINO 2021 B	69
SH 2019 T	317
SHAYA 2021 B	614
SHAYA HABIS 2020 B FB	614
SIAH ISABEL SALGADO 2020 B	423
SIBILA 2018 T	832
SIEMPRE CONTIGO 2020 T RB	721
SIERRA ALMIRÓN 2021 T	789
SIERRA CANTABRIA 2012 T GR	583
SIERRA CANTABRIA 2016 T R	583
SIERRA CANTABRIA 2019 T C	583
SIERRA CANTABRIA 2021 B	583
SIERRA CANTABRIA 2021 RD	583
SIERRA CANTABRIA COLECCIÓN PRIVADA 2020 T	592
SIERRA CANTABRIA CUVÉE 2018 T	592
SIERRA CANTABRIA GARNACHA 2018 T	584
SIERRA CANTABRIA MÁGICO 2018 T	592
SIERRA CANTABRIA ORGANZA 2020 B	592
SIERRA CANTABRIA SELECCIÓN 2020 T	584
SIERRA DE TOLOÑO 2020 T	584
SIERRA DE TOLOÑO 2021 B	584
SIERRA DE TOLOÑO CAMINO DE SANTA CRUZ 2019 T	584
SIESTO 2018 T C	843
SIETE PELDAÑOS 2020 B BA	59
SIETE PELDAÑOS BRUÑAL 2020 T	59
SIETE PELDAÑOS DOÑA BLANCA 2020 B	59
SIETE PELDAÑOS GARNACHA 2020 T C	59
SIETE PELDAÑOS MALVASÍA 2020 B	59
SIETE PELDAÑOS TEMPRANILLO 2020 T	59
SIGLO 2016 T R	534
SIGLO FAMILIA FERNANDEZ DE MANZANOS 2015 T GR	534
SIGLO SACO TEMPRANILLO C.V.C T	534
SIGNES 2016 T	116
SIGNES DEL PRIORAT 2019 T	355
SIGNO BOBAL 2020 T BA	269
SILBON 2021 B	614
SILENCIO DE MIROS 2018 T	456
SILENCIO, ES MÚSICA 2020 B	703
SILENTE 2020 B BA	834
SILENTIUM 2015 T R	443
SILENTIUM 2017 T C	443
SILENTIUM 2020 T RB	443
SILENTIUM 2021 T RB	443
SILENTIUM EXPRESIÓN 2016 T C	443
SILEO 2020 T	296
SILGA 2021 B	595
SÍLICE 2019 T	870
SÍLICE 2020 B	870
SÍLICE KLARETE 2019 RD	870
SÍLICE ROSADO 2019 RD	870
SÍLICE XABREGA 2019 T	870
SILICON RED T	853
SILVA DAPONTE GODELLO 2021 B	279
SILVA DAPONTE MENCÍA 2021 T	279
SILVANO GARCÍA 4 MESES 2019 T BA	228
SILVANO GARCÍA BLUE 2021 B	851
SILVANO GARCÍA DULCE MONASTRELL 2020 T D	228
SILVANO GARCÍA ETIQUETA NEGRA 2019 T	228
SILVANO GARCÍA VERDEJO 2021 B	228
SILVANUS 2019 T C	441
SILVANUS EDICIÓN LIMITADA 2018 T	441
SIMBIOSIS AIRÉN DE TINAJA 2021 B	796
SIMBIOSIS BOBAL SINCERO 2019 T	796
SÍMBOLO AIRÉN 2021 B	242
SÍMBOLO CHARDONNAY SELECCIÓN 2021 B	242
SÍMBOLO PETIT VERDOT 2021 T	242
SÍMBOLO TEMPRANILLO 2017 T RB	242
SÍMBOLO TEMPRANILLO 2021 T	242
SÍMBOLO VERDEJO 2021 B	242
SIN BLANCA 2018 T C	60
SIN COMPLEJOS 2020 T	666
SIN PALABRAS 2018 B BA	378
SIN PALABRAS 2021 B MC	378
SIN PALABRAS EDICIÓN ESPECIAL 2013 B	378
SIN PALABRAS MOSTO FLOR 2018 B MC	379
SIN RAZÓN 2018 T R	721
SINFO 2020 RD FB	159
SINFORIANO 2016 T R	159
SINFORIANO 2018 T C	159
SINGULARS GARNATXA ROJA 2020 B FB	189
SINOBLE MACABEO 2020 B	500
SÍNODO 2020 B	584
SÍNODO CENTALES VIÑEDO SINGULAR 2020 T	584
SÍNODO GARNACHA GRACIANO 2019 T	584
SINOLS GARNATXA RD SOLERA	188
SINTAUTO 2017 T	556

WINE	PAGE	WINE	PAGE	WINE	PAGE
SIÓS CAU DEL GAT 2020 T C	174	SOLAR DE ESTRAUNZA 2021 B	524	SOLLÍO MENCÍA 2020 T	411
SIÓS PLA DEL LLADONER 2021 B	175	SOLAR DE ESTRAUNZA 2021 T	524	SOLLUNA 2019 T C	366
SIP BARRICA SELECCIÓN 2019 T	724	SOLAR DE ESTRAUNZA SELECCIÓN 2016 T	524	SOLMAYOR AIRÉN 2021 B	672
SIRÀ DE SON RAMON 2017 T	831	SOLAR DE LA VEGA VERDEJO 2021 B	613	SOLMAYOR TEMPRANILLO 2021 T	672
SITTA 2021 RD	839	SOLAR DE SAEL MENCÍA 2018 T C	63	SOLMAYOR VERDEJO-SAUVIGNON BLANC 2021 B	672
SITTA DOLIOLA 2018 B	839	SOLAR DE SAMANIEGO 2019 T C	585	SOLPOST CARINYENA 2017 T	296
SITTA FINCA MOLINERO 2018 T	839	SOLAR DE SAMANIEGO 7 CEPAS 2017 T R	585	SOLPOST GARNATXA 2016 T C	296
SITTA MACERACIÓN 2021 B	839	SOLAR DE SAMANIEGO VALCAVADA 2017 T R	585	SOLPOST ORIGEN 2018 T	296
SITTA PEREIRAS 2021 B D	839	SOLAR VIEJO 2018 T C	548	SÒLS XAREL 2019 B BA	857
SIURELL 2021 B	832	SOLAR VIEJO 2018 T R	548	SOMANILLA 16 MESES BARRICA 2019 T C	449
SO DE MASIA CAN RODA PANSA BLANCA 2021 B	32	SOLARCE 2020 B	510	SOMIADORS 2019 T	181
SOBAJANERA AÑINA 2019 B	772	SOLARCE 2020 T	510	SOMIATRUITES 2021 B	335
SOBAJANERA MACHARNUDO 2019 B	772	SOLARCE 2021 RD	511	SOMMELIER 2015 T R	582
SOBRENATURAL BY MENADE 2016 B C	805	SOLEAR EDICIÓN BICENTENARIO BF MZ	207	SOMMELIER 2017 T C	582
SOCAIRE 2019 B FB	769	SOLEAR EN RAMA SACA DE PRIMAVERA BF MZ S	207	SOMMOS COLECCCIÓN SAUVIGNON BLANC 2021 B	626
SOCAIRE OXIDATIVO 2016 B FB	769	SOLEAR EN RAMA SACA DE VERANO 2022 BF MZ	207	SOMMOS COLECCIÓN CABERNET SAUVIGNON 2019 T R	626
SOFÍA 2019 T	674	SOLERA 1842 MEDIUM SWEET VOS BF OL MED	220	SOMMOS COLECCIÓN CHARDONNAY 2020 B	626
SOGAS MASCARÓ BE BN	150	SOLERA 1931 ESPOLLA GARNATXA D'EMPORDÀ VINO DULCE NATURAL BF SOLERA D	183	SOMMOS COLECCIÓN MERLOT 2019 T BA	626
SOGAS MASCARÓ BE BR	150			SOMMOS COLECCIÓN SYRAH 2019 T R	626
SOGAS MASCARÓ BE R BN	150	SOLERA FAMILIAR GUTIÉRREZ COLOSÍA BF AM S	208	SOMMOS COLECCIÓN TEMPRANILLO 2019 T R	626
SOL DE REYMOS B MISTELA D	700	SOLERA FAMILIAR GUTIÉRREZ COLOSÍA BF OL S	209	SOMNI 2018 T	363
SOLABAL 2017 T R	513	SOLERA FAMILIAR GUTIÉRREZ COLOSÍA BF PC S	209	SOMNIS DE GERISENA - SOL I RESENA RD AÑEJO D	184
SOLABAL 2018 T C	513	SOLERA FINA MARÍA DEL VALLE B FI S	287	SON 2 DÍAS 2021 B	709
SOLAGÜEN 2016 T R	556	SOLERA FINA MARÍA DEL VALLE EN RAMA B FI S	287	SON GRAU GRAN ESCURSAC 2021 T	830
SOLAGÜEN 2019 T C	556	SOLERGIBERT CABERNET 2011 T R	348	SON GRAU GRAN GARGOLLASSA 2021 RD	831
SOLAMENTE B	853	SOLERGIBERT CABERNET 2017 T R	348	SON GRAU GRAN GARGOLLASSA 2021 T BA	830
SOLAMENTE T	853	SOLERGIBERT MERLOT 2017 T R	348	SON MOIX BLANC 2017 B	352
SOLANA DE RAMÍREZ 2021 B	547	SOLIMAR 2020 T C	637	SON RAMON BLANC DE BLANCS 2021 B	831
SOLANA DE RAMÍREZ 2021 RD	547	SOLIMAR 2021 B	637	SON RAMON CHARDONNAY 2021 B	831
SOLAR AMÉZOLA 2015 T GR	515	SOLISERENA ESPOLLA GARNACHA D'EMPORDÀ BF SOLERA D	183	SON RAMON NEGRE 2019 T	831
SOLAR DE ESTRAUNZA 2015 T R	524	SOLLÍO 2020 T BA	411	SON RAMON SELECCIÓ ESPECIAL 2017 T	831
SOLAR DE ESTRAUNZA 2018 T C	524	SOLLÍO GODELLO 2021 B	411	SON ROCA 2020 T	825

WINE	PAGE	WINE	PAGE	WINE	PAGE
SONAT DE L'ALA 2017 T	335	SOTO DEL VICARIO EL ORIGEN 2017 T	77	SUMARROCA CHARDONNAY 2021 B	343
SONROJO 2021 RD	319	SOTO DEL VICARIO GO DE GODELLO 2021 B FB	77	SUMARROCA ECOLÓGIC 2019 BE R BR	151
SONS DE PRADES 2019 B	165	SOTO DEL VICARIO GODELLO ESPECIAL CUVÉE 2020 B	77	SUMMUM LÁCRIMA BACCUS 2018 BE R BN	142
SONSIERRA 2015 T GR	549	SOTO DEL VICARIO MEN DE MENCÍA 2017 T C	77	SUMMUM LÁCRIMA BACCUS 2018 BE R BR	142
SONSIERRA 2017 T R	549	SOTO DEL VICARIO MEN DE MENCÍA SELECCIÓN DE VIÑEDOS 2014 T C	77	SUMMUM LÁCRIMA BACCUS 2019 BE R BN	142
SONSIERRA 2019 T C	549			SUMOLL AMFORA 2017 T R	118
SOÑADOR CRUZ CONDE B FI S	286	SR. NIÑO 2021 T	472	SUMOLL DE SOLERGIBERT 2019 T	348
SOPLA LEVANTE LA HORCA 2020 B	871	SSSSSHHHHHH ULLDEMOLINS 2021 T	291	SUNEUS BLANC 2021 B	181
SOPLA LEVANTE LOMAS DEL POLO 2020 T	871	ST JOANNES 2019 T	328	SUNEUS NEGRE 2020 T RB	181
SOPLA LEVANTE PINOMAR ORANGE WINE 2020 B	871	ST JOANNES 2021 B	328	SUPER TRAMP 2017 T	116
SOPLA PONIENTE AMONTILLADO VIEJÍSIMO SAN ROQUE 30 B AM	285	ST JOANNES ROSÉ 2021 RD	328	SUPERNOVA 2019 T C	441
		STA CALLET 2021 T	830	SUPERNOVA ROBLE 2021 T RB	441
SOPLA PONIENTE FINO EN RAMA CERRO DEL MAJUELO 20 B FI	285	STA GIRÓ ROS 2021 B	830	SURSUM CORDA 2021 T	754
SOPLA PONIENTE PX VIEJÍSIMO EL COSCOJAL 70 BF PX	285	STA MANTONEGRO 2020 T	82	SUSQUVAT 2021 B BA	334
SOPLAGAITAS 2020 B	834	STA PRENSAL 2021 B	82	SUTRA BY TONI ARRAEZ BE BR	124
SOPLO 2020 T	710	STAIRWAY TO HEAVEN CUVÉE 2019 T R	826	SUTRA BY TONI ARRAEZ BE R BR	124
SOQUES 2018 T R	193	STELVIO 2018 T C	537	SUZZANE 2020 T	579
SORTE ANTIGA 2020 B	692	STELVIO BLANCO 2021 B	537	SWEET CLOS FIGUERAS 2021 TF D	363
SORTE O SORO 2020 B	692	SUBIRAT 2018 BE BN	637	SYRAH D'ANFORA 2019 T S	355
SORTEVERA 2020 B	871	SUCRO 2020 T	502		
SORTEVERA 2020 T	871	SUEÑO DE MEGALA 2018 T BA	703	T	
SORTIUS MONASTRELL 2020 T	91	SUERTES DEL MARQUÉS CRUZ SANTA 2020 I	718	TABÁ 2020 T C	231
SORTIUS MONASTRELL 2021 T	91	SUERTES DEL MARQUÉS EL CHIBIRIQUE 2019 T	718	TABERNER 2016 T	770
SORTIUS SYRAH 2020 T RB	91	SUERTES DEL MARQUÉS EL ESQUILÓN 2019 T	718	TABERNER Nº 1 2014 T	770
SOSPECHOSO 2020 T	795	SUERTES DEL MARQUÉS LA SOLANA 2020 T	718	TABUCA 2020 T	105
SOSPECHOSO 2021 B	795	SUERTES DEL MARQUÉS LOS PASITOS 2020 T	718	TABUERNIGA 2019 T	561
SOSPECHOSO 2021 RD	795	SUERTES DEL MARQUÉS TRENZADO 2020 B	718	TÁBULA 2018 T	472
SOT LEFRIEC 2017 T	325	SUERTES DEL MARQUÉS VIDONIA 2020 B	719	TACA NEGRA 2017 T C	185
SOT NERAL MACABEU 2021 B	171	SUERTES DEL MARQUÉS VIDONIA V.P. 2020 B C	719	TADEO CORTIJO LOS AGUILARES 2019 T C	263
SOT NERAL SYRAH 2020 T	171	SUERTES DEL MARQUÉS VIDUEÑO 2020 T	719	TADEO TINAJA CORTIJO LOS AGUILARES 2019 T	263
SOTAVENTO 100% SAUVIGNON BLANC LÍAS 2021 B	601	SUFREIRAL 2020 T	72	TADEO TINAJA CORTIJO LOS AGUILARES 2020 T	263
SOTAVENTO 100% VERDEJO LÍAS 2021 B	601	SUMARROCA 2018 BE GR BN	151	TAFURIASTE 2021 T	717
				TAFURIASTE AFRUTADO SEMIDULCE 2021 RD SD	717

WINE	PAGE	WINE	PAGE	WINE	PAGE
TÁGANAN 2020 B	861	TANTAKA 2019 T	53	TAUROMAQUIA OLOROSO VIEJÍSIMO B OL S	287
TÁGANAN 2020 T	861	TANTAKA 2020 T	53	TAUROMAQUIA PEDRO XIMÉNEZ SUPERIOR BF PX D	287
TÁGANAN MARGALAGUA 2020 T	861	TANTAKA DIAPIRO (LACRE CALABAZA) 2019 B	53	TAUS FUSION 2020 T	231
TÁGARA 2021 B S	730	TANTAKA DIAPIRO (LACRE VERDE) 2019 B	53	TAVERA ANTIGUOS VIÑEDOS 2021 RD	275
TÁGARA 2021 B SD	730	TANTUM ERGO CHARDONNAY PINOT NOIR 2019 BE BN	125	TAVERA ANTIGUOS VIÑEDOS GARNACHA 2019 T RB	275
TAGONIUS T	851	TANTUM ERGO EXCLUSIVE MAGNUM 2011 BE GR BN	125	TAVERA EDICIÓN NUMERADA SYRAH 2019 T C	275
TAHÓN DE TOBELOS 2015 T R	586	TANTUM ERGO PINOT NOIR ROSÉ 2019 RE BN	125	TAVERA SYRAH TEMPRANILLO 2019 T RB	275
TAJINASTE 2020 T RB	755	TANTUM ERGO VINTAGE 2018 BE BN	125	TAVERA VENDIMIA SELECCIONADA 2017 T C	275
TAJINASTE 2021 B S	755	TARABELO 2019 T C	420	TEAR DOS DODI 2020 B	425
TAJINASTE NATURALMENTE DULCE 2018 T D	718	TARABELO 2020 T C	420	TEATINOS 40 BARRICAS TEMPRANILLO 2017 T R	503
TAJINASTE TRADICIONAL 2020 T	755	TARDATIO BLANC DE NOIR 2018 B	857	TEATINOS CLAROS DE CUBA 2016 T R	503
TAJINASTE VENDIMIA SELECCIONADA 2020 T	718	TARDATIO SUBTIL ROSÉ 2020 RD	857	TEATINOS SIGNVM 2016 T C	503
TALAI BERRI ROSÉ 2021 RD	196	TARENTOLA 2019 T	782	TEATINOS SYRAH 2021 T	503
TALAI BERRI TXAKOLINA 2021 B	196	TARES P. 3 2017 T R	73	TEATINOS TEMPRANILLO 2021 T	503
TÁLAMO 2018 T C	657	TARIMA HILL 2020 T	39	TEBAIDA 2020 T RB	71
TALLA DE DIAMANTE SEMIDULCE 2021 B SD	526	TARIMA HILL 2021 B FB	40	TEBAIDA NEMESIO 2020 T RB	72
TALLAT DE LLUNA 2019 T	177	TARIMA SELECCIÓN 2021 T	40	TEBAIDA Nº5 2020 T RB	72
TAMARAL 2018 T C	473	TARÓN 2015 T R	549	TEIRA X 2020 B	418
TAMARAL AGRICULTURA ECOLÓGICA 2019 T C	473	TARÓN 2018 T C	549	TEIXAR 2017 T	302
TAMARAL FINCA LA MIRA 2018 T	473	TARÓN 4M 2019 T	549	TEJONERAS ALTA SELECCIÓN 2014 T R	724
TAMBORÁ 2021 B S	429	TARÓN TEMPRANILLO BLANCO 2020 B C	549	TEMPLER 2019 T C	293
TAMINA 2021 B	801	TARSUS 2015 T R	460	TEMPLER 2021 T	293
TAMIZ DE TEÓFILO REYES 2021 T RB	460	TARSUS 2019 T C	460	TENEGUÍA 2021 T	250
TAMPESTA 2021 RD	255	TARSUS FINCA EL CANTO VINO DE AUTOR 2018 T	460	TENEGUÍA LA GOTA 2021 B	250
TAMPESTA ALBARÍN 2021 B	255	TARSUS LA DESPISTADA 2020 B	460	TENEGUÍA MALVASÍA AROMÁTICA NATURALMENTE DULCE 2020 B D	250
TAMPESTA GOLÁN 2020 RD	256	TARSUS SELECCIÓN DE VIÑEDOS 2018 T	460		
TAMPESTA GOLÁN EDICIÓN LIMITADA 2015 T GR	256	TARSUS TERNO VINO DE AUTOR 2018 T	460	TENEGUÍA MALVASÍA DULCE 2000 B R	250
TAMPESTA MANEKI ED. ESPECIAL 2019 B FB	255	TARTALO 2020 T	579	TENTENUBLO 2019 T	585
TAMPESTA NEKO GODELLO 2020 B	256	TASAT 2021 T	756	TENTENUBLO 2020 B	585
TANGANILLO 2020 T	755	TASAT LISTÁN BLANCO 2021 B	756	TEO VIOGNIER 2020 B S	870
TANGANILLO AFRUTADO SEMIDULCE 2021 B SD	755	TASAT LISTÁN BLANCO AFRUTADO 2021 B SD	756	TEO VIOGNIER 2021 B RB	870
TANTAKA 2019 B	53	TAUROMAQUIA AMONTILLADO VIEJÍSIMO B AM S	287		

WINE	PAGE
TEO VIOGNIER MACERACIÓN 2020 B S	870
TEÓFILO REYES 15 MESES 2020 T C	460
TEÓFILO REYES 2017 T R	461
TERÁN VERSUM 2018 T	535
TERCIA DE ULEA 2020 T C	91
TERESA BLANCHER DE LA TIETA 2012 BE GR BN	123
TERESA MATA GARRIGA BE R BN	128
TERMANTHIA 2015 T	656
TERMES 2019 T	656
TERMES 2020 B	656
TERRA DE ASOREI 2020 T	379
TERRA DE ASOREI 2021 B	380
TERRA DE ASOREI SELECCIÓN PRIVADA 2020 B	380
TERRA DE CUQUES BLANC 2021 B	375
TERRA DE CUQUES NEGRE 2019 T	375
TERRA DE TARDOR 2021 B FB	677
TERRA DO CASTELO "SENSACIÓN" 2021 B	421
TERRA DO CASTELO GODELLO 2021 B	421
TERRA DO CASTELO TOSTADO DULCE B D	421
TERRA DO CASTELO TREIXADURA 2021 B	422
TERRA DO CASTELO TREIXADURA SELECCIÓN 2013 B	422
TERRA DURO LA ENFERMERA 2021 T	666
TERRA FITER 2015 T GR	42
TERRA INCÓGNITA 2019 T	465
TERRA MINEI 2021 B	418
TERRA RUBIA 3 UVAS 2021 B	281
TERRA TERRAE 2020 T	333
TERRA TERRAE BE BR	136
TERRA TERRAE BE R BR	136
TERRA TERRAE BLANC 2020 B	333
TERRA VERMELLA DE NIN 2018 T	365
TERRADURO SELECCIÓN 2019 T	666

WINE	PAGE
TERRAESTEBAN 2017 T	466
TERRAESTEBAN 2017 T C	466
TERRAESTEBAN 2019 T RB	466
TERRAESTEBAN 2020 T	466
TERRAESTEBAN 2021 B	466
TERRAESTEBAN 2021 RD	466
TERRAI OVC OLD VINE CARIÑENA 2020 T RB	112
TERRAI OVG 2021 T BA	112
TERRAI OVG VENDIMIA SELECCIONADA 2020 T BA	112
TERRAJE 2018 T BA	232
TERRAL SUMARROCA 2017 T	343
TERRAMOLL NATURAL 2021 T BA	820
TERRANEGRA 2021 B	429
TERRAPLÉN GARNACHA 2021 T	319
TERRAPLÉN ROSADO GARNACHA 2021 RD	319
TERRAPLÉN VIURA 2021 B	319
TERRAS DE LANTAÑO 2021 B	402
TERRAS DO CIGARRÓN 2021 B	282
TERRAS GAUDA 2021 B	391
TERRAS GAUDA ETIQUETA NEGRA 2020 B FB	391
TERRAZAS DEL MONCAYO GARNACHA 2018 T BA	101
TERRÍCOLA 2019 T	298
TERRÍCOLA BLANC 2020 B	298
TERRITORI 2019 T R	299
TERROIR HISTORIC BLANC 2021 B	375
TERROIR HISTORIC NEGRE 2019 T	375
TERROIR HISTORIC NEGRE 2021 T	375
TERROIR SENSE FRONTERES BRISAT 2021 B	301
TERROIR SENSE FRONTERES NEGRE 2021 T	301
TERROIR X BLANCO 2021 B	354
TERROIR X EL SEGUNDO 2019 T	354
TERROIR X LA TERCERA 2019 T	354

WINE	PAGE
TERROIR X LA VIÑA VIEJA 2019 T	354
TERROJA DE SABATÉ I COCA 2016 B	342
TERRÒS 2019 T	299
TERROTXA 2018 T	359
TERRUM 2019 T	829
TESALIA 2017 T C	769
TESELA BY CLUNIA 2019 T	803
TESÓN 2020 B FB	781
TESORO DE VILLENA FONDILLÓN 1972 T FO	35
TESTIGO 2016 T RB	785
TETE DE LA COURSE 2021 B	725
TEULADÍ 2021 T RB	701
TEULERA 2020 B	118
THALER DE PLATA 2020 T	529
THARSYS ÚNICO 2018 BE R BR	682
THE ARTIST 2019 T C	829
THE FINAL COUNTDOWN 2020 T RB	712
THE FLOWER AND THE BEE SOUSÓN 2020 T	425
THE FLOWER AND THE BEE TREIXADURA 2021 B	424
THE FREAKY WINES TEMPRANILLO 2021 T	778
THE FREAKY WINES VERDEJO 2021 B	778
THE SWIMMERS 2019 B	712
THE WOODS OF TILO CABERNET SAUVIGNON 2017 T	795
THEODOSIVS 2018 T C	817
THERASIA 2020 B FB	341
THESAURUS X 2019 T	461
TIANNA 1 NEGRE 2020 T	828
TIANNA 3 NEGRE 2020 T	857
TIANNA BOCCHORIS NEGRE 2020 T	828
TIANNA NEGRE 2020 T	828
TIBERIO 2021 T	461
TIBÓ 2021 B	701

WINE	PAGE	WINE	PAGE	WINE	PAGE
TIENTO LA MEJORADA 2016 T	808	TILAMA 2020 B	253	TINTO FIGUERO VIÑAS VIEJAS 2019 T	495
TIERGA T	864	TILENUS ENTRECUESTAS GODELLO 2020 B FB	67	TINTO ITURRIA 2018 T	660
TIERRA 2019 T C	550	TILENUS ENVEJECIDO EN ROBLE 2019 T RB	67	TINTO PESQUERA 2019 T R	431
TIERRA 2021 B FB	550	TILENUS GODELLO MONTESEIROS 2021 B	67	TINTO PESQUERA 2020 T C	431
TIERRA ARANDA 2016 T R	437	TILENUS LA FLORIDA 2019 T C	67	TINTO ROA 2014 T R	472
TIERRA ARANDA 2019 T C	437	TILENUS LAS LADERAS MENCÍA 2019 T BA	67	TINTO ROA 2019 T C	472
TIERRA ARANDA EDICIÓN ESPECIAL 2020 RD	437	TILENUS PAGOS DE POSADA 2017 T BA	67	TINTO ROA 2020 T RB	472
TIERRA ARANDA EDICIÓN ESPECIAL VIÑEDOS SINGULARES 2020 T RB	437	TILENUS VENDIMIA 2021 T	68	TINTORALBA 2018 T C	49
		TIME WAITS FOR NO ONE 6M 2021 T RB	231	TINTORALBA 2021 T RB	49
TIERRA ARANDA VENDIMIA SELECCIONADA 2020 T C	437	TIME WAITS FOR NO ONE BLACK 2020 T	231	TINTORALBA GARNACHA TINTORERA 2021 T	49
TIERRA BLANCA 2021 B	850	TIME WAITS FOR NO ONE DOUBLE 2021 T	231	TINTORALBA SAUVIGNON BLANC - VERDEJO 2021 B	49
TIERRA BLANCA SEMIDULCE 2021 B SD	850	TIMONIER 2019 T C	440	TINTORALBA SELECCIÓN 2017 T C	49
TIERRA CALMA ALBILLO REAL 2021 B	727	TINÁCULA EL SANTILLO 2020 T	502	TINTORALBA SYRAH 2021 RD	49
TIERRA CALMA CYSTER 2020 T	727	TINÁCULA RED 2020 T	502	TÍO DIEGO BF AM S	220
TIERRA CALMA LA NAVA 2020 T	727	TINÁCULA WHITE 2021 B	502	TÍO MARTÍN 2019 T C	522
TIERRA CALMA LAS CABRERAS 2020 T	727	TINÁCULA X 2019 T	502	TÍO MATEO BF FI S	219
TIERRA DE FIDEL 2017 B FB	550	TINAJA DE ZALEO 2021 T S	819	TÍO PEPE BF FI S	216
TIERRA DE FORCALLAT 2021 B	842	TINALLA 2021 T	418	TÍO PEPE CUATRO PALMAS BF AM S	216
TIERRA FUNDIDA 2020 T	633	TINO ALVILLA DO AVIA 2019 B	419	TÍO PEPE DOS PALMAS BF FI S	216
TIERRA FUNDIDA BLANCO CERCADO DEL PINO 2020 B S	872	TINO ALVILLA DO AVIA 2021 B	419	TÍO PEPE EN RAMA BF FI S	216
TIERRA FUNDIDA CLARETE LOS TOPES 2021 T	872	TINTA ROSA 2020 T	669	TÍO PEPE TRES PALMAS BF FI S	216
TIERRA FUNDIDA LOS TOPES 2020 B	872	TINTA ROSA SELECCIÓN 2019 T C	669	TÍO PEPE UNA PALMA BF FI S	216
TIERRA IMPERIAL 2016 T GR	242	TINTITO DEL PUEBLO 2021 T	441	TÍO RAIMUNDO 2017 B	848
TIERRA IMPERIAL 2017 T R	243	TINTO 9 CEPAS 2018 T R	818	TÍO UCO 2021 T	800
TIERRA IMPERIAL 2019 T C	243	TINTO ARROYO 2015 T GR	436	TIRABEQUE 2020 T	279
TIERRA IMPERIAL OAKED SELECTION 2020 T	243	TINTO ARROYO 2017 T R	436	TIRABEQUE 2021 B	279
TIERRA SAVIA MIRLO 2019 B	851	TINTO ARROYO 2018 T C	436	TITANIA EDICIÓN LIMITADA 2017 T	560
TIERRA SAVIA ZARANDA EL VIEJO FRANCÉS T	851	TINTO ARROYO 2020 T RB	436	TITANIA EDICIÓN LIMITADA 2018 B	560
TIERRAS DE CAIR 2018 T R	446	TINTO ARROYO VENDIMIA SELECCIONADA 2015 T FB	436	TITANIA EDICIÓN LIMITADA 2018 T	560
TIERRAS DE MURILLO 2018 T C	547	TINTO DE ABEL 2021 T	680	TIVO 2018 B FB	770
TIERRAS DE MURILLO 2019 B FB	547	TINTO FIGUERO 12 2019 T C	495	TOBELLA NEGRE 2020 T	365
TIERRAS DE MURILLO COLECCIÓN PRIVADA 2019 T	547	TINTO FIGUERO 15 2018 T R	495	TOBELLA NEGRE 2021 T	366

WINE	PAGE
TOBELOS 506 M. 2020 B	586
TOBELOS GARNACHA 2020 T BA	586
TOBELOS TEMPRANILLO 2018 T C	586
TOBÍA LUZ DE LUNA 2017 T R	550
TOBÍA LUZ DE LUNA 2019 T C	550
TOBÍA SELECCIÓN DE AUTOR 2019 T BA	550
TOBÍA SELECCIÓN DE AUTOR 2020 B	550
TOCAT DE L'ALA 2015 T	191
TOCAT DE L'ALA BLANC 2021 B	192
TOCS 2016 T R	374
TODO VA A SALIR BIEN 2021 RD FB	537
TOFTERUP BROTHERS BLACK LABEL MONASTRELL 2017 T	733
TOFTERUP BROTHERS MONASTRELL 2019 T	733
TOFTERUP BROTHERS MONASTRELL BARREL SELECT 2019 T	733
TOFTERUP BROTHERS ORGANIC RED T	786
TOFTERUP BROTHERS ORGANIC ROSE 2021 RD	786
TOFTERUP BROTHERS TEMPRANILLO 2019 T	462
TOFTERUP BROTHERS TEMPRANILLO 2020 T	502
TOLLODOURO 2021 B	380
TOLO DO XISTO 2018 T	407
TOMÁS GONZÁLEZ 2020 T RB	432
TOMBÚ 2021 RD	809
TOMILLAR 2017 T R	248
TOMILLAR 2018 T C	248
TOMILLAR AIRÉN 2021 B	248
TOMILLAR CHARDONNAY 2021 B	248
TOMILLAR MERLOT 2021 RD	248
TOMILLAR TEMPRANILLO 2021 T	248
TORALTO 2016 T BA	667
TORALTUS 2016 T BA	668
TORELLÓ 225 2017 BE BN	884
TORELLÓ 50 LLIURES 2021 B FB	343
TORELLÓ COLLECTION 2011 BE BN	884
TORELLÓ FINCA CAN MARTÍ 2011 BE BR	884
TORELLÓ FINCA CAN MARTÍ 2017 BE BR	884
TORELLÓ GRAN CRISALYS 2019 B	343
TORELLÓ MALVARELLO 2021 B	343
TORELLÓ MAS DE LA TORREVELLA 2021 B	343
TORELLÓ RAIMONDA 2017 T BA	343
TORELLÓ SPECIAL EDITION 2017 BE BR	885
TORELLÓ TRADICIONAL 2016 BE BN	885
TORNO HACIENDA EL TERNERO 2017 T C	571
TORONDOS CLARETE 2021 RD	156
TORONDOS ROSÉ 2021 RD	156
TORQUES DO CASTRO 2020 B	420
TORRALBENC 2021 RD	820
TORRALBENC CHARDONNAY 2019 B	820
TORRALBENC COUPAGE BLANC 2020 B	821
TORRALBENC COUPAGE TINTO 2019 T	820
TORRALBENC MERLOT 2019 T	821
TORRALBENC SYRAH 2019 T RB	821
TORRE BERAUN 2015 T	834
TORRE DE BARREDA AMIGOS MULTIVARIETAL 2020 T	777
TORRE DE BARREDA AMIGOS MULTIVARIETAL 2021 B	777
TORRE DE BARREDA CABERNET SAUVIGNON 2020 T	777
TORRE DE BARREDA GARNACHA 2019 T	777
TORRE DE BARREDA GRACIANO 2019 T	777
TORRE DE BARREDA PAÑOFINO VIÑA SINGULAR 2019 T S	777
TORRE DE CAPMANY MOSCATEL 2021 B	190
TORRE DE CAPMANY VIEJAS SOLERAS B D	190
TORRE DE GAZATE 2017 T C	247
TORRE DE GAZATE AIRÉN 2021 B	247
TORRE DE GAZATE CABERNET SAUVIGNON 2021 RD	247
TORRE DE GAZATE VERDEJO 2021 B	247
TORRE DE LOIZAGA BIGARREN 2021 B	85
TORRE DE LOIZAGA SELECCIÓN 2020 B	85
TORRE DEL VEGUER ABELLEROL 2021 B	343
TORRE DEL VEGUER FONOLL 2021 B	343
TORRE DEL VEGUER LLUM DEL CADÍ 2019 T	872
TORRE DEL VEGUER MARICEL 2021 B	343
TORRE DEL VEGUER MARTA 2020 BE BN	343
TORRE DEL VEGUER PUR 2021 B	872
TORRE DO OLIVAR EXPRESIÓN 2021 B	419
TORRE LA MOREIRA 2021 B	389
TORRE MUGA 2019 T	539
TORREDEROS 2015 T R	461
TORREDEROS 2018 T C	461
TORREDEROS 2020 T RB	461
TORREDEROS 2021 RD	461
TORREDEROS SELECCIÓN 2016 T	461
TORREDEROS VERDEJO 2021 B	611
TORRELONGARES 2018 T C	112
TORRELONGARES 2021 RD	112
TORRELONGARES GARNACHA 2021 T	112
TORREMAYOR 12 MESES 2019 T	819
TORREMILANOS 2018 T C	480
TORREMORÓN 2019 T C	461
TORREMORÓN 2020 T RB	462
TORREMORÓN 2021 T	462
TORRENT - VI DE PARATGE 2018 T C	372
TORRENT FALS 2019 T C	82
TORRENT NEGRE 2015 T C	352
TORRENT NEGRE SELECCIÓ PRIVADA CABERNET 2014 T C	352
TORREPINGÓN 2019 T C	442
TORREPINGÓN 2021 T RB	443
TORRES ROMERO ED. LIMITADA SYRAH 2015 T	773

WINE	PAGE
TORRES ROMERO ED. LIMITADA TEMPRANILLO 2014 T	773
TORRES ROMERO ED.LIMITADA CABERNET SAUVIGNON Y MERLOT 2012 T	773
TORRES ROMERO PETIT VERDOT COLECCION PRIVADA 2013 T	773
TORRESILO 2019 T	475
TORREVAL 6 MESES BARRICAS 2021 T	493
TORROXAL ALBARIÑO 2021 B	380
TOSALET CARIGNAN VINYES VELLES 2017 T	367
TOSALET VINYES FINS A 50 ANYS 2021 T	367
TOSSALS SELECCIÓ 2019 T	863
TOTEM RED 2019 T	816
TOTEM ROSÉ SANT JOSEP 2019 RD FB	816
TOTEM WHITE 2020 B FB	816
TOURAN 2019 T	104
TOZARA 2020 T RB	793
TR3SMANO ALBILLO MAYOR 2020 B	491
TR3SMANO TM 2018 T BA	491
TR3SMANO VENDIMIA 2019 T	491
TRABUCAT 2019 T	860
TRACA I MOCADOR BLANC 2020 B	300
TRADICIÓN DE ARESAN TEMPRANILLO - CABERNET SAUVIGNON - MERLOT 2020 T	243
TRAGANTIA ZALEMA 2021 B	169
TRAMP 2017 T	116
TRAMPAL 2018 T C	818
TRANCO DEL LOBO 2019 T C	270
TRANS TREPAT ANCESTRAL 2020 RE	878
TRAPELLA 2021 RD	192
TRAPÍO 2019 T RB	733
TRAPISONDERO 2021 T RB	722
TRAS LAS YESCA 2021 T C	660
TRAS LOS MUROS 2018 B BA	399

WINE	PAGE
TRASLAGARES SAUVIGNON BLANC 2021 B	618
TRASLAGARES VERDEJO 2021 B	618
TRASLANZAS 2018 T	161
TRASLANZAS 2021 RD	161
TRASLANZAS V+A 2020 B	161
TRASLASCUESTAS 2019 T C	462
TRASLASCUESTAS 2020 RD FB	462
TRASLASCUESTAS 2020 T RB	462
TRASTO 2020 B BA	256
TRASTO 2020 T	256
TRASTO FINCA EL BARRANCO 2020 T	256
TRAVESURA CABERNET SAUVIGNON 2021 T	91
TRAVESURA SHIRAZ 2021 T	91
TREIXADURA DOS CANOTOS 2019 B BA	858
TREKAN 2018 T BA	365
TRENCACLOSQUES DE PASCONA 2021 RD	295
TRENZA FAMILY COLLECTION 2019 T	734
TRENZA Z-STRAND 2020 T	733
TREPAT DEL JORDIET DE RENDÉ MASDEU 2020 T	164
TRES AMIGOS 2018 T BA S	273
TRES CUESTAS 2018 T BA S	161
TRES JULIAS 2019 T C	657
TRES LEONES NATURALMENTE DULCE B D	262
TRES LUNAS 2019 T	666
TRES LUNAS VERDEJO 2021 B	666
TRES MATAS 2017 T R	464
TRES MATAS 2019 T C	464
TRES MATAS VENDIMIA SELECCIONADA 2018 T	464
TRES MULLERES GODELLO 2020 B	280
TRES MULLERES TREIXADURA 2020 B	280
TRES NAUS BE BR	136
TRES NAUS BE C BN	136

WINE	PAGE
TRES NAUS BE R BN	137
TRES NAUS BE R BR	137
TRES NAUS ROSÉ INTENSO RE C BR	136
TRES NAUS ROSÉ PÁLIDO RE BR	137
TRES OJOS GARNACHA 2020 T	94
TRES PATAS 2018 T	274
TRES VILLAS B	601
TREVEJOS BLANC DE NOIRS 2017 BE BN	29
TREVEJOS MOUNTAIN WINES BABOSO NEGRO 2020 T	29
TREVEJOS MOUNTAIN WINES LISTÁN BLANCO & MALVASÍA 2020 B	29
TREVEJOS MOUNTAIN WINES LISTÁN PRIETO 2018 T FB	29
TREVEJOS MOUNTAIN WINES VIJARIEGO NEGRO 2020 T	29
TREVEJOS ORGANIC WINES LISTÁN BLANCO 2020 B FB	29
TREVEJOS VOLCANIC WINES BABOSO NEGRO & SYRAH 2018 T	29
TREVEJOS VOLCANIC WINES BLANCO ALBILLO & VERDELLO 2020 B	29
TRIANA 30 AÑOS VORS 50 CL BF PX D	209
TRIANA HIDALGO LA GITANA BF PX D	209
TRIAS BATLLE 2020 BE R BN	151
TRIAS BATLLE 2020 BE R BR	151
TRIAS BATLLE 2020 T	344
TRIAS BATLLE 2021 B	344
TRIAS BATLLE 2021 RD	344
TRIAS BATLLE BLAUÉ 2018 BE GR BN	151
TRIAS BATLLE CABERNET SAUVIGNON 2017 T C	344
TRIAS BATLLE ROSÉ 2020 RE BR	151
TRIAS BATLLE XARELLO 2019 B BA	344
TRIAVA BLANC "VINO DE GUARDA" 2020 B FB	824
TRIAVA HERITAGE "VINO DE GUARDA" 2019 T R	825
TRIAY 2021 B	280
TRIAY 38 2019 T	280

WINE	PAGE	WINE	PAGE	WINE	PAGE
TRIAY MENCÍA 2021 T	280	TUDANCA 2018 T C	494	TXAKOLI LEYES DE ABELLANEDA BEREZIA MAGNUM 2021 B	86
TRIDENTE MALVASÍA 2021 B	807	TUDANCA 2020 T RB	494	TXAKOLI REZABAL 2021 B	196
TRIDENTE PRIETO PICUDO 2019 T	807	TUDANCA RAYTU 2015 T R	494	TXAKOLI REZABAL ROSÉ 2021 RD	196
TRIDENTE PRIETO PICUDO 2020 T	807	TUDANCA VENDIMIA SELECCIÓN 2014 T	495	TXAKOLI TXABARRI EXTRA 2021 B	88
TRIDENTE TEMPRANILLO 2019 T	807	TUDANCA VICENTA MATER 2016 T C	495	TXAKOLI UNO 2020 B	52
TRIDENTE TEMPRANILLO 2020 T	807	TUDON´S MONASTRELL 2019 T	775	TXOMIN ETXANIZ 2021 B	196
TRIENNIA DE BODEGAS PORTIA 2019 T	457	TUDON´S PETIT VERDOT 2019 T	775	TXOMIN ETXANIZ 2021 RD	197
TRIENS 2019 T RB	667	TUERCEBOTAS GARNACHA 2019 T C	523	TXOMIN ETXANIZ BEREZIA 2020 B	197
TRIGA MAGNUM 2019 T	39	TUERCEBOTAS GRACIANO 2019 T C	523		
TRILO-VITES 2019 T	867	TUERCEBOTAS TEMPRANILLO BLANCO 2021 B	523	U	
TRILOGÍA 2018 T C	707	TUNANTE TEMPRANILLO 2021 T	570	U D'URPINA 2018 T R	348
TRILOGÍA PINOT NOIR BLANC DE NOIR BE R BN	637	TURO D'EN MOTA 2008 BE BN	884	UBE CARRASCAL 2018 B S	770
TRITÓN TINTA TORO 2019 T	662	TURÓ DE LES ABELLES 2019 T	335	ÚBER 760 2020 T	415
TROBALLA 2019 T	175	TURONIA 2021 B	400	UBETA GARNACHA 2019 T FB	321
TROBAT ROSAT RE BR	127	TURONS DE LA PLETA 2020 B	176	UBETA GARNACHA BLANCA 2021 B FB	321
TROQUEAO 2019 T	572	TURONS VALLCORBA 2019 T C	176	UBETA ROSE 2021 RD	321
TROS BLANC NOTARIA 2016 B FB	292	TUTUM BA 2019 BE BN	151	UBIDE MAR 2021 B	550
TROS BLANC SALERES 2016 B	292	TUTUSAUS 2017 BE GR BN	151	UBIDE RESERVA DE LA FAMILIA 2012 T R	550
TROS DE CLOS MAS DEL METGE 2019 T	363	TUVÍ OR NOT TO BE ORGÀNIC SUMARROCA 2021 B	343	UBIDE TIERRA 2017 T C	550
TROS DE MAS VILELLA 2019 T	866	TX BERRIA 2020 B	88	UCEDO 2020 T RB	70
TROS NEGRE NOTARIA 2017 T	292	TX OS CONVENTOS 2019 T R	407	UKAN 2019 T	586
TROS NOU VI DE VILA BELLMUNT 2018 T C	367	TX TXOMIN ETXANIZ 2020 B BA	196	ULIBARRI 2019 B	84
TROSSOS SANTS 2019 B	292	TXABARRI 2021 RD	88	ULIBARRI 2020 R	84
TROSSOS VELLS 2019 T BA	292	TXABARRI 2021 T	88	ULISES VALSANGIACOMO CHARDONNAY 2021 B FB	873
TROUPE 2021 B	396	TXABARRI ABEITXA 2020 B	88	ULL DE SERP LA CLOSA CARINYENA 2019 T C	182
TRUFADO GARNACHA 21 2021 T	767	TXAKOLI AGUIRREBEKO 2021 B	84	ULL DE SERP LA CLOSA MACABEU 2019 B FB	182
TRUFES BLANC 2021 B	650	TXAKOLI AIALLE 2020 B	195	ULTERIOR GARNACHA 2018 T	788
TRUFES NEGRE 2019 T	650	TXAKOLI AIZPURUA 2021 B	195	ULTERIOR MAZUELO 2018 T	788
TRUS 2017 T R	463	TXAKOLI ARETXAGA 2021 B	85	ULTREIA "LA CLAUDINA" 2020 B	76
TRUS 2019 T C	463	TXAKOLI BERROJA 2019 B	84	ULTREIA "SAINT JACQUES" 2020 T	76
TRUS 2021 T RB	463	TXAKOLI LAÍNOA MAGNUM 2020 B	85	ULTREIA GODELLO 2020 B	76
TU RAI 2019 T BA	173	TXAKOLI LAÍNOA MAGNUM 2021 B	85	ULTREIA VALTUILLE 2020 T	76
				ULULA 2020 T	109

WINE	PAGE
UMEA GARNACHA 2021 T	311
UMEA GARNACHA BLANCA 2021 B	311
UN CULÍN MENCÍA DE VALTUILLE 2020 T	70
UN GARNACHA BLANC DE NOIR 2018 B FB	833
UN UVA NOCTURNA GARNACHA + 2018 T C	833
UN UVA NOCTURNA GARNACHA SYRAH 2019 T	833
UN UVA NOCTURNA MERLOT 2019 T C	833
UN UVA NOCTURNA TEMPRANILLO GARNACHA 2019 T	833
UNA NIT EN GLOBUS 2020 T	299
UNANIMOUS FINCA LA TEJERA 2019 T C	491
UNANIMOUS SANTA CRUZ ALBILLO MAYOR 2018 B	491
UNGLES NEGRES 2021 T	873
UNICO 1926 2017 B FB	774
UNIVERSAL CABERNET SAUVIGNON BIODINÁMICO 2021 T	791
UNIVERSAL SAUVIGNON BLANC BIODINÁMICO 2021 B	791
UNIVERSITAT ROVIRA I VIRGILI 2021 B	119
UNIVERSITAT ROVIRA I VIRGILI BLANC DE BLANCS BE BR	638
UNSI "FINCA EL BOYERAL" 2017 T BA	321
UNSI "FINCA LASIERRA" 2016 T	321
UNSI "TERRAZAS BLANCO" 2021 B BA	321
UNSI "TERRAZAS" 2019 T	321
UNSI DULCE GARNACHA RF RC D	321
URBEZO 2019 T C	112
URBEZO CHARDONNAY 2021 B	112
URBIÓN 2019 T C	571
URBIÓN CUVÉE 2019 B BA	571
URBIÓN CUVÉE 2020 T RB	571
URO 2019 T	666
URTARAN 2020 B FB	52
USUARAN 2016 T	316
UVA D'OR MOSCATEL DE LICOR B D	706
UVA PIRATA 2021 B	873
UVA PIRATA GARNACHA 2021 T	873
UVA POR UVA 2019 T	505
UVIO C 2019 T C	91
UXÍA DA PONTE 2020 B	423
UZT TARDÍA B	168

V

WINE	PAGE
V COLORADO 2016 T FB	863
V DE MAS IGNEUS 2020 B	371
V DULCE DE INVIERNO 2019 B D	863
V MALCORTA VERDEJO 2020 B	617
V. ALEJANDRÍA 2021 RD	848
V5 BY CONSEJAL 2020 B BA	796
V89 2018 B FB	175
VA DE BO 2019 T C	42
VA POR USTEDES 2019 T	851
VA! TEMPRANILLO 2021 T	582
VACCAYOS 2017 T R	474
VACCEOS TEMPRANILLO 2021 T RB	802
VAIVÉN BLANC DE NOIR TEMPRANILLO 2021 B	795
VAL DE MEIGAS 2021 B	402
VAL DE NAIROA 2020 B	423
VAL DE REYES T D	804
VAL DE SOUTO TREIXADURA 2021 B	424
VAL DE VID VERDEJO 2021 B	611
VAL DE VID VERDEJO SOBRE LÍAS 2020 B BA	611
VAL DO GALIR GODELLO 2021 B	694
VAL DO GALIR MENCÍA 2020 T RB	694
VAL DO SOSEGO 2021 B	385
VALA DE SOLABAL 2016 T	513
VALAUTÍN 2019 B	725
VALAUTÍN 2019 T	725
VALBUENA 5º 2018 T	464

WINE	PAGE
VALBUSENDA ABIOS 2018 T C	662
VALBUSENDA ABIOS NUDE 2021 RD	662
VALBUSENDA ABIOS TINTA DE TORO 2020 T RB	662
VALBUSENDA VERDEJO 2019 B FB	662
VALBUXAN 2019 T	689
VALCABADINO 10 MESES 2018 T RB	653
VALCABADINO 2018 T C	653
VALCERRACÍN SELECCIÓN LIMITADA VERDEJO 2021 B	620
VALCHÉ 2019 T C	90
VALDEAURUM 2018 T C	499
VALDEBARÓN 2019 T	543
VALDEBARÓN 2020 B	543
VALDEBODEGA 2019 T C	473
VALDEBODEGA 2020 T	473
VALDEJUANA SYRAH 2020 T	273
VALDELACIERVA 2017 T R	550
VALDELACIERVA 2019 T C	550
VALDELACIERVA CANTOGORDO 2018 T	550
VALDELACIERVA GARNACHA 2019 T	550
VALDELACIERVA GRANO A GRANO 2018 T	551
VALDELARES 2018 T C	321
VALDELARES 2019 T C	321
VALDELARES 2021 RD	321
VALDELARES 2021 T	321
VALDELARES CHARDONNAY 2021 B	321
VALDELARES SAUVIGNON BLANC 2021 B	321
VALDELICEDA 2019 T BA	722
VALDELOSFRAILES 2016 T R	157
VALDELOSFRAILES 2017 T C	157
VALDELOSFRAILES CLARETE 2021 RD	157
VALDEMILANOS 2020 T	102
VALDEMONJE 2018 T	751

WINE	PAGE
VALDEMONJE VERDEJO NEGRO 2018 T	751
VALDEMONTE AIRÉN 2021 B	780
VALDEMONTE DARK 2021 T	780
VALDEMONTE PINOT NOIR 2021 T	780
VALDEMONTE TEMPRANILLO 2021 T	780
VALDEMONTE VERDEJO 2021 B	780
VALDEPEDRO DE OSTATU 2020 T	579
VALDEPILA 2018 T R	485
VALDEPILA 2019 T R	485
VALDEPISÓN 2021 RD	434
VALDERICA MENCÍA 2021 T	63
VALDERIVERO 2019 T C	492
VALDERIVERO 2021 T RB	492
VALDESABRIL 2019 B	541
VALDESALAS 2021 B	65
VALDESIL GODELLO SOBRE LÍAS 2020 B	693
VALDESIL PARCELA O CHAO 2019 B FB	693
VALDESNEROS 2021 RD	56
VALDIHUETE 2021 B	621
VALDOCEA 2021 B	400
VALDOSAN 2018 T	660
VALDOURO 2019 T BA	686
VALDOVINOS 2018 T C	628
VALDOVINOS SELECCIÓN GARNACHA 2019 T	628
VALDOVINOS SELECCIÓN SYRAH 2018 T	628
VALDRINAL 24 2018 T R	438
VALDRINAL DE SANTAMARÍA 2021 B	601
VALDRINAL ENTREGA 2020 T RB	438
VALDRINAL SQR 2016 T	438
VALDRINAL TRADICIÓN 2019 T C	438
VALDUBÓN 2020 T RB	463
VALDUBÓN 2021 RD	463
VALDUEZA 2019 T BA	818
VALE 2016 T R	774
VALE SERIE ORO 2016 T GR	774
VALENT 2021 B	338
VALERNA 2020 B FB	171
VALERNA 2020 T BA	171
VALJUNCO 2021 RD	257
VALJUNCO ALBARÍN 2020 B	257
VALL DE MOLINÀS 2013 T RB	184
VALL DE MOLINÀS 2021 B	184
VALL DE XALÓ 2021 T MC	40
VALL DE XALÓ GIRÓ VINO DE LICOR TF D	40
VALL POR 2017 T R	374
VALLBLANCA 2021 B SS	709
VALLDOLINA 2018 BE GR BR	151
VALLDOLINA 2019 BE R BN	152
VALLDOLINA ROSAT 2021 RD	344
VALLDOLINA XARELLO 2021 B	344
VALLE DE NABAL 2020 T	454
VALLE DEL BOTIJAS SAUVIGNON BLANC 2021 B	813
VALLE DEL RÍO CASA AURORA 2020 T	862
VALLEGARCÍA GARNACHA CARIÑENA 2020 T	747
VALLEGARCÍA SYRAH 2019 T	747
VALLEGARCÍA SYRAH 2020 T	747
VALLEGARCÍA VIOGNIER 2020 B	747
VALLEJO AVENAS 2020 B	677
VALLMAJOR BLANC 2021 B	643
VALLMAJOR NEGRE 2021 T	643
VALLOBERA 2019 T C	552
VALLOBERA TEMPRANILLO 2021 T MC	552
VALMAGAZ MENCÍA 2021 T	78
VALMIÑOR 2021 B	380
VALPARAÍSO 2019 T C	464
VALPARAÍSO 2020 T RB	464
VALPINCIA VERDEJO 2021 B	620
VALPRECIADO 2019 T C	457
VALPRECIADO PRIMA 2020 T	457
VALPRECIADO ROBLE 2020 T RB	457
VALSANZO 2019 T C	458
VALSARTE 2016 T R	548
VALSARTE 2019 T C	548
VALSERRANO 2015 T GR	592
VALSERRANO 2017 B GR	592
VALSERRANO 2018 T C	592
VALSERRANO 2021 B FB	593
VALSOTILLO 2017 T R	451
VALSOTILLO 2018 T C	452
VALSOTILLO 2019 T	452
VALSOTILLO FINCA BUENAVISTA 2018 T	452
VALSOTILLO VS "40 ANIVERSARIO" 2016 T BA	452
VALSOTILLO VS 2016 T R	452
VALTEA 2021 B	380
VALTEA BE BN	380
VALTEA CUVÉE ESPECIAL BE BN	380
VALTOSCA 2021 T	224
VALTRAVIESO 2019 T C	473
VALTRAVIESO RUPTURE 2018 T	809
VALTUILLE CEPAS CENTENARIAS 2020 T BA	72
VALTUILLE RAPOLAO 2020 T C	72
VALTUILLE VINO DE VILLA 2020 T	72
VALTUILLE VINO DE VILLA 2020 T BA	72
VALVARÉS DE ALTANZA 2018 T C	505
VALVINOSO TEMPRANILLO 2021 RD	159
VAN GUS VANA 2014 T	75

WINE	PAGE	WINE	PAGE	WINE	PAGE
VANDAMA HOYA OSCURA 2021 T	199	VEGA DE LUCÍA GODELLO 2021 B	279	VEIGA NAÚM 2021 B	402
VANDAMA VINO DE FINCA 2021 T	199	VEGA DE LUCÍA MENCÍA 2020 T	279	VEIGADARES 2018 B FB	379
VANIDADE 2021 B	403	VEGA DEL PAS VERDEJO 2021 B	619	VEIGAS DE PADRIÑÁN SELECCIÓN DE AÑADA 2021 B	378
VARDALÓN 2021 B	597	VEGA INFANTE 2021 T	679	VEL'UVEYRA GODELLO 2020 B	416
VARIAS AL·LEGORIA 2015 BE R BN	152	VEGA MEDIEN ECOLÓGICO BE BR	151	VEL'UVEYRA MENCÍA 2019 T	416
VARIAS AL·LEGORIA 2015 BE R BR	152	VEGA MEDIEN ROSÉ RE BR	151	VELÁZQUEZ COLECCIÓN ARTISTAS ESPAÑOLES 2011 T R	505
VARIAS CUVÉE CLÀSSIC 2008 BE GR BN	152	VEGA MORAGONA CABERNET SAUVIGNON 2020 T C	503	VELLA LOLA 2021 B	186
VARIAS CUVÉE IMPERIAL 2009 BE GR BN	152	VEGA MORAGONA LA DUNA 2019 T	503	VELLA LOLA 2021 T	186
VARIAS EDICIÓ LIMITADA XARELLO 2008 BE GR BN	152	VEGA MORAGONA MOSCATEL DE GRANO MENUDO 2021 B D	503	VÉLOBLANC 2021 B	828
VARIAS GENUÍ 2019 BE BN	152	VEGA MORAGONA ORGANIC 2019 T	797	VÉLONEGRE 2020 T	828
VARIAS SELECCIÓ 2018 T C	344	VEGA MORAGONA SYRAH 2020 T RB	503	VENERABLE VORS BF PX D	210
VASCOMENDI V.S. 2019 T	578	VEGA REAL 2019 T RB	474	VENTA D'AUBERT VIOGNIER 2021 B	873
VASCOMENDI V.S. 2021 B	578	VEGA REAL VERDEJO 2021 B	614	VENTA DEL PUERTO Nº 12 2019 T BA	700
VATAN 2019 T	662	VEGA SAUCO EL BEYBI 2020 T RB	662	VENTA DEL PUERTO Nº 18 2017 T BA	700
VD'O 1.15 2015 T	192	VEGA SAÚCO SELECCIÓN 2019 T	663	VENTA LA OSSA 2021 T	782
VD'O 2.15 2015 T R	192	VEGA SICILIA ÚNICO 2013 T	464	VENTA LA OSSA CABERNET SAUVIGNON 2019 T C	782
VD'O 6.19 2019 B	192	VEGA SICILIA ÚNICO RESERVA ESPECIAL T GR	464	VENTA LA OSSA SYRAH 2019 T	782
VDB VALLE DEL BOTIJAS 14 MESES 2016 T C	813	VEGA TOLOSA CHARDONNAY 2018 BE BN	271	VENTA LA OSSA TEMPRANILLO 2018 T C	782
VDB VALLE DEL BOTIJAS ANGELA 2021 B	813	VEGA VALERIO 2018 T C	437	VENTA LA OSSA TNT 2017 T	782
VDM ORANGE DULCE BF AROM D	168	VEGA VELLA 2017 T C	520	VENTA LAS VACAS 2020 T	491
VEGA AIXALÀ CALIU 2015 T C S	163	VEGA VELLA 2018 B FB	520	VENTA MAZARRÓN 2020 T	808
VEGA AIXALÀ CARINYENA 2015 T C	163	VEGA VELLA 2020 B	520	VENTA VIEJA MALBEC 2021 T	789
VEGA AIXALÀ GARNATXA VILANOVA 2015 T C S	163	VEGA VELLA GRACIANO 2019 T	520	VENTEPICO 2019 B SS	768
VEGA AIXALÀ LA BAUMA 2019 B	163	VEGALFARO 2017 BE GR BN	127	VENTUA 2021 RD	71
VEGA AIXALÀ LA FONT DELS AUBACS 2018 T BA	163	VEGAMAR 2019 T C	706	VENTUA MOURELO 2018 T C	71
VEGA AIXALÀ VIERN 2012 T R S	163	VEGAMAR 2021 B	706	VENTUM 2017 T C	829
VEGA CARRIEGOS 2014 T C	257	VEGAMAR PRIVÉE 18 BE R BN	127	VENTUM 2018 T C	829
VEGA CARRIEGOS 2016 T MC	257	VEGANTIGUA 2020 T RB	464	VENTUS 2016 T R	873
VEGA CARRIEGOS 2021 RD	257	VEGAS B	840	VENUS 2018 T C	301
VEGA DE GÁLDAR 2020 T	199	VEGAS COLECCIÓN PRIVADA 2020 T	840	VENUS DE CARTOIXÀ 2018 B	301
VEGA DE LA REINA VERDEJO 2021 B	617	VEGAS RD	840	VENUS DE LA FIGUERA 2018 T	301
VEGA DE LUCHÁN 2021 RD	833	VEIGA DA PRINCESA 2021 B	383	VERA DE ESTENAS 2018 T R	747

WINE	PAGE	WINE	PAGE	WINE	PAGE
VERA DEL VALLE 2020 T	676	VETUSTA 2018 T C	465	VIDUEÑOS DE SEDELLA 2020 B	262
VERACRUZ EDICIÓN SOBRE LÍAS VERDEJO 2021 B	612	VETUSTA VIÑAS DE FUENTENEBRO 2021 T	465	VIEJAS DE IZAN 2019 T C	492
VERÁN BLANC 2021 B	81	VETUSTA VIÑEDO ESPECIAL CARRASCALON ALTO 2018 T	464	VIEJAS DE IZAN 2020 T C	492
VERBAS 1917 2019 T	419	VI D'AMFORA 2021 T	302	VIEJO CP BF PC S	220
VERBENERA B FI S	286	VI DE FANG 2020 T C	164	VIEJO ZULETA VOS BF AM S	215
VERDEJO DE BODEGAS RUIZ TORRES 2021 B	818	VI DE GLASS GEWÜRZTRAMINER 0,375 2020 B D	345	VIEJOS CAPOS 2015 T R	459
VERDELLO ANTIGUO 2019 B	280	VI DE VILA DEL LLOAR "A LA MEMÒRIA DEL PÉ" 2017 T C	359	VIENTO ALISEO 2021 RD	245
VERDES CASTROS GODELLO 2021 B	692	VI DEL BANYA 2020 T	173	VIENTO ALISEO GRACIANO CABERNET SAUVIGNON 2020 T BA S	245
VERDES CASTROS MENCÍA 2020 T	692	VÍA BARROSA GODELLO 2021 B	693	VIENTO ALISEO TEMPRANILLO PETIT VERDOT 2021 T S	245
VERDONCHO #GARAGEWINE ORANGEWINE 2021 B	773	VIA CENIT COLECCIÓN 2019 B FB	653	VIENTO ALISEO VIOGNIER 2021 B	246
VEREDA 2019 T BA	701	VIA CENIT COLECCIÓN 2019 T C	653	VIENTO SOBRE LA PIEL 2020 T BA	772
VEREDA DE LAS TÓRDIGAS 2020 T BA	753	VÍA EDETANA BLANC 2020 B	648	VIERNES 2021 T	68
VEREMA MONASTRELL CRIADO EN ÁNFORA 2020 T	44	VÍA EDETANA BLANC 2021 B	648	VIGILIA 2021 RD	825
VERITAS 2019 BE BN	81	VÍA EDETANA NEGRE 2019 T BA	648	VII GENERACIÓN DE BODEGAS FIGUEROA 2019 T RB	723
VERITAS BLANC 2020 B	81	VÍA ROMANA DO CAMIÑO BRANCELLAO 2019 T FB	416	VILA CLOSA BLANC 2021 B	644
VERITAS ROIG 2021 RD	81	VÍA ROMANA DO CAMIÑO GODELLO 2020 B	416	VILA CLOSA CHARDONNAY 2019 B FB	644
VERMELL 2020 T	708	VIA XVIII 2020 T	694	VILA CLOSA PELUDA 2020 T RB	644
VERÓNICA SALGADO CAPRICHO 2018 T C	433	VIARIL CABERNET SAUVIGNON 2020 T	269	VILA CLOSA RUBOR 2021 RD	644
VERÓNICA SALGADO CAPRICHO VINO DE AUTOR 2019 T	433	VIARIL MACABEO 2021 B	269	VILADELLOPS SELECCIÓN GARNATXA 2019 T C	335
VERSAT CLOS COR VÍ 2021 B	704	VIARIL VERDEJO SAUVIGNON BLANC 2021 B	269	VILADEQUINTA 2019 T C	690
VERSOS DE VALTUILLE F 15 MESES 2020 T	74	VICENTE GANDÍA CHARDONNAY SOBRE LÍAS 2021 B	873	VILANO 2018 T R	465
VERSOS DE VALTUILLE CEPAS CENTENARIAS 2020 T RB	74	VICENTE GANDÍA GODELLO EDICIÓN LIMITADA 2021 B	813	VILANO 2019 T	465
VERSOS DE VALTUILLE GODELLO 2020 B FB	74	VICTORIA DÍEZ-CABALLERO 2018 T R	564	VILARNAU 2019 BE R BN	153
VÉRTEBRA DE LA FIGUERA 2021 T	301	VICTORINO 2019 T	669	VILARNAU BRUT ROSÉ DELICAT 2020 RE R BR	153
VERUM CABERNET FRANC TEMPRANILLO 2021 RD	788	VIDA VIÑA TENDIDA MOSCATO BIANCO B MO D	700	VILARNAU VINTAGE 2014 BE GR BN	153
VERUM CENCIBEL 2019 T RB	788	VIDAL I FERRÉ 2017 BE GR BN	133	VILARS 2019 T RB	177
VERUM LAS TINADAS AIRÉN DE PIE FRANCO 2020 T	788	VIDAL I FERRÉ 2018 BE R BN	133	VILE LA FINCA 2017 T	257
VERUM TERRA AIRÉN 2021 B	788	VIDAL I FERRÉ 2020 BE BR	133	VILLA CONCHI IMPERIAL 2018 BE EBR	128
VESPRES 2020 T	299	VIDAL I FERRÉ BLANC DE NOIRS 2020 BE R BN	133	VILLA DE CORULLÓN 2020 T	73
VESPRES BLANC 2021 B	299	VIDAL I FERRÉ ROSAT 2020 RE BR	133	VILLA DE SAN LORENZO GODELLO 2019 B	75
VETERUM VITIUM 2020 T	675	VIDALBA 2017 T	374	VILLAGE 2020 B	875
VETUS 2019 T	663	VIDILLA 2021 B	614	VILLALAR 2019 T RB	808

WINE	PAGE	WINE	PAGE	WINE	PAGE
VILLANA LLÉVAME AL HUERTO 2020 B	870	VINS DE TALLER PHLOX 2021 B	119	VIÑA ALBALI TEMPRANILLO SHIRAZ 2021 T	791
VILLAREI 2020 B	383	VINYA D'EN FERRAN JAUME LLOPART ALEMANY 2012 BE GR BN	141	VIÑA ALBALI VERDEJO SAUVIGNON BLANC 2021 B	697
VILLAVID BOBAL 2021 RD	267	VINYA D'EN LLUC 2021 B	337	VIÑA ALBERDI 2018 T C	573
VILLAVID SYRAH 1952 2018 T RB	786	VINYA DEL METGE 2021 RD	187	VIÑA ALBINA VENDIMIA SELECCIONADA 2019 T R	546
VILLAVID TEMPRANILLO 2021 T	267	VINYA ESCUDÉ 523 2018 BE R EBR	134	VIÑA ALBINA VERDEJO 2021 B	612
VILLAVID VERDEJO 2021 B	267	VINYA ESCUDÉ DAURAT 2019 BE R BN	134	VIÑA ALIJIBES 2020 T RB	775
VILLAZO 2021 B	259	VINYA ESCUDÉ SINGLE VINEYARD 2019 BE R BN	134	VIÑA AMABLE 12 MESES 2017 T	199
VILLOTA 2018 B FB	587	VINYA GASÓ 2017 T C	292	VIÑA AMABLE 4 MESES 2020 T RB	199
VILLOTA 2019 T	587	VINYA JANINE SYRAH 2019 T C	638	VIÑA AMALIA 2021 B SS	288
VILLOTA GRACIANO 2019 T	587	VINYA MAG RAVEN 2018 T BA	177	VIÑA AMÉZOLA 2018 T C	515
VILOSELL 2019 T	177	VINYA MAS VELL CABERNET SAUVIGNON 2017 T GR	372	VIÑA ANE AUTOR 2018 T C	511
VIN BLANC SON MAYOL 2020 B	826	VINYA MAS VELL GARNATXA 2017 T GR	372	VIÑA ANE CENTENARIA 2021 B FB	511
VIN ROSE SON MAYOL 2021 RD AG	826	VINYA TAUJANA BLANC DE BLANC 2021 B	82	VIÑA ANE SELECCIÓN 2018 T	511
VINALOPÓ 2020 T RB	35	VINYA TAUJANA ROSAT 2021 RD	82	VIÑA ARANA 2015 T GR	573
VINALOPÓ PETIT VERDOT 2019 T RB	35	VINYES DE POBOLEDA 2021 T	362	VIÑA ARDANZA 2015 T R	573
VINANA 2017 T	260	VIÑA 98 BF PX D	206	VIÑA ARDANZA 2016 T R	573
VINEA 2020 T C	157	VIÑA AB BF AM S	217	VIÑA ARNAIZ VERDEJO 2021 B	613
VINEA 2021 RD	157	VIÑA ABAD GODELLO 2021 B	686	VIÑA AZENICHE PETIT VERDOT 2018 T	91
VINO VIEJO DEL ABUELO MONASTRELL 2012 T RC	842	VIÑA ABAD SUMUM GODELLO 2020 B	686	VIÑA AZENICHE SYRAH 2019 T C	91
VINOS LLÁMALO X AIRÉN SOBRE LÍAS EN TINAJA 2021 B	797	VIÑA ABBA 2017 T	659	VIÑA BERNEDA 2018 T C	553
VINOS LLÁMALO X CRUJIDERA MADURADO EN TINAJA 2021 T	797	VIÑA AINZÓN 2018 T R	101	VIÑA BERNEDA 2021 B	553
VINOS LLÁMALO X MONASTRELL MADURADO EN TINAJA 2021 T	797	VIÑA AINZÓN 2019 T C	101	VIÑA BERNEDA 2021 T MC	553
		VIÑA AL LADO DE LA CASA 2020 T	732	VIÑA BISPO 2021 B	276
VINS DE POSTAL - COLL DE RIBERA 2014 T BA	183	VIÑA ALBALI 2015 T GR	697	VIÑA BISPO 2021 RD	276
VINS DE POSTAL – LA CREU 2018 B	183	VIÑA ALBALI 2017 T R	697	VIÑA BISPO 2021 T	276
VINS DE POSTAL – LA SUREDA 2014 T	183	VIÑA ALBALI 2018 T C	697	VIÑA BORGIA BY BORSAO 2020 T	845
VINS DE POSTAL – LES AULEDES 2017 B	183	VIÑA ALBALI CABERNET SAUVIGNON* 2021 T	791	VIÑA BOSQUERA 2021 B	723
VINS DE POSTAL – LES MUNTADES 2013 T	183	VIÑA ALBALI CHARDONNAY 2021 B	791	VIÑA BOTIAL 2020 T RB	91
VINS DE TALLER BASEIA 2019 B	119	VIÑA ALBALI GARNACHA ROSÉ 2021 RD	791	VIÑA BUENA 2019 T C	494
VINS DE TALLER GEUM 2020 T C	119	VIÑA ALBALI GRAN SELECCIÓN 2021 T	697	VIÑA BUENA 2020 T BA	493
VINS DE TALLER GRIS 2021 RD	119	VIÑA ALBALI MERLOT 2021 T	791	VIÑA BUENA 2021 T	493
VINS DE TALLER LARIX 2021 B	119	VIÑA ALBALI TEMPRANILLO 2021 T	697	VIÑA BUJANDA 2014 T GR	587

WINE	PAGE
VIÑA BUJANDA 2016 T R	587
VIÑA BUJANDA 2019 T C	587
VIÑA BURÓN VERDEJO 2021 B	611
VIÑA CALDERONA 2021 RD	160
VIÑA CANSINA 2021 RD	160
VIÑA CANSINA VERDEJO 2021 B	605
VIÑA CARTIN 2021 B	402
VIÑA CATAJARROS 2021 RD	160
VIÑA CHOZO 2020 T	724
VIÑA CIMBRÓN SAUVIGNON 2021 B	605
VIÑA CIMBRÓN VERDEJO 2021 B	605
VIÑA CIMBRÓN VERDEJO SELECCIÓN 2020 B	605
VIÑA CORRALES PAGO BALBAINA BF FI	206
VIÑA CORZO GODELLO 2021 B	691
VIÑA CORZO MENCÍA 2021 T	691
VIÑA COSTEIRA 2021 B	429
VIÑA COSTEIRA MENCÍA 2021 T	693
VIÑA CUERVA 2016 T R	241
VIÑA CUERVA 2018 T C	241
VIÑA CUERVA 2021 RD	241
VIÑA CUERVA 2021 T	241
VIÑA CUERVA AIRÉN 2021 B	241
VIÑA CUMBRERO 2015 T R	538
VIÑA CUMBRERO 2018 T C	538
VIÑA CURVADA ALBILLO MAYOR 2019 B BA	462
VIÑA CURVADA TEMPRANILLO 2017 T C	462
VIÑA CURVADA TEMPRANILLO 2019 T	462
VIÑA DARIZ 2019 T	411
VIÑA DARIZ 2020 T	411
VIÑA DE ARANBELZA 2017 T	316
VIÑA DE LEORIN 2017 T	316
VIÑA DE MARTÍN "OS PASÁS" 2019 B	426

WINE	PAGE
VIÑA DEL OJA 2014 T GR	583
VIÑA DEL OJA 2016 T R	583
VIÑA DEL OJA 2018 T C	583
VIÑA DESGRACIA 2019 T	814
VIÑA DETRÁS DE LA CASA GARNACHA TINTORERA 2017 T C	733
VIÑA DETRÁS DE LA CASA SYRAH 2019 T	733
VIÑA DORANA 2017 T R	555
VIÑA EL FLAKO 2019 B FB	572
VIÑA EL FLAKO 2020 B FB	572
VIÑA EL PISÓN 2020 T	875
VIÑA EL TORREÓN VERDEJO 2021 B	618
VIÑA ESTÉVEZ 2020 T BA	633
VIÑA FUENTENARRO 2019 T C	448
VIÑA FUENTENARRO 2020 T RB	448
VIÑA GÓMEZ LISTÁN 2018 B D	715
VIÑA GÓMEZ MALVASIA 2019 RD D	715
VIÑA GOY 2021 RD	159
VIÑA GOY RUEDA 2021 B	614
VIÑA HERMINIA 2018 T C	553
VIÑA HERMINIA EXCELSUS 2019 T	553
VIÑA JARABA 2019 T C	745
VIÑA JUGUERA SELECCIÓN 2018 T C	306
VIÑA LA CEJA 2019 T	268
VIÑA LA PICARAZA 2017 T C	681
VIÑA LEIRIÑA 2021 B	423
VIÑA LIDÓN 2021 B FB	747
VIÑA LOS VALLES 50 & 50 2018 T C	553
VIÑA LUDY 2021 B	388
VIÑA MACHARNUDO ALTO PALOMINO 2020 B FB	772
VIÑA MARRO 2019 T C	524
VIÑA MAYOR VERDEJO 2021 B	614
VIÑA MIGARRÓN 2018 T C	66

WINE	PAGE
VIÑA MIGARRÓN 2020 T	66
VIÑA MIGARRÓN 2021 RD	66
VIÑA NORTE 2021 T	631
VIÑA NORTE 2021 T MC	631
VIÑA NORTE SELECCIÓN 2021 T	631
VIÑA OLABARRI 2015 T GR	588
VIÑA OLABARRI 2016 T R	588
VIÑA OLABARRI 2019 T C	588
VIÑA OLMAZA 2021 B	542
VIÑA OLMAZA 2021 T	542
VIÑA ORCE 2019 T C	783
VIÑA ORCE 2021 RD	783
VIÑA ORCE CHARDONNAY 2021 B	783
VIÑA ORCE MACABEO VERDEJO 2021 B	783
VIÑA ORCE TEMPRANILLO 2021 T RB	783
VIÑA PEDROSA 2016 T GR	451
VIÑA PEDROSA 2019 T R	451
VIÑA PEDROSA 2020 T C	451
VIÑA PEDROSA FINCA LA NAVILLA 2019 T R	451
VIÑA PEREZ 100% VERDEJO 2021 B	601
VIÑA PERGUITA 2018 T C	311
VIÑA PERGUITA 2019 T RB	311
VIÑA PICOTA 2021 RD	158
VIÑA PILAR 2020 RD	448
VIÑA POMAL 2016 T R	518
VIÑA POMAL 2019 T C	518
VIÑA POMAL 2021 RD	518
VIÑA POMAL ECOLÓGICO 2019 T C	518
VIÑA POMAL SELECCIÓN 500 2018 T C	518
VIÑA POMAL VIURA MALVASÍA 2021 B	518
VIÑA REAL 2016 T GR	589
VIÑA REAL 2021 RD	589

WINE	PAGE	WINE	PAGE	WINE	PAGE
VIÑA ROMALE XARELLO 2021 BE BN	126	VIÑA ZORZAL GRACIANO 2020 T	323	VIÑASPERI 2014 T GR	519
VIÑA RUBICÁN TEMPRANILLO 2017 T C	311	VIÑA ZORZAL LECCIONES DE VUELO 2020 T	323	VIÑASPERI BLUE OCEAN 2019 T BA	519
VIÑA SALCEDA 2018 T R	589	VIÑA ZORZAL SEÑORA DE LAS ALTURAS 2020 T	323	VIÑASPERI SELECCIÓN 2016 T	519
VIÑA SANTA MARINA CABERNET SAUVIGNON-SYRAH 2019 T C	819	VIÑABADE 2021 B	401	VIÑASPERI SELECCIÓN LOS OJALES 2019 T	519
VIÑA SASTRE 2019 T C	450	VIÑADERO ROBLE 2018 T	769	VIÑAVERDE 2021 B SS	287
VIÑA SASTRE 2020 T RB	451	VIÑAHONDA 2017 T C	228	VIÑEDOS POZANCO 2017 T C	500
VIÑA SASTRE MARCELINA GÓMEZ 2021 RD	451	VIÑAHONDA ORGANIC + 2019 T BA	228	VIÑEDOS POZANCO 2021 T	500
VIÑA SASTRE PAGO DE SANTA CRUZ 2016 T GR	450	VIÑAMALEA 2018 T C	267	VIÑEDOS POZANCO VERDEJO 2021 B	819
VIÑA SASTRE PESUS 2016 T	450	VIÑAMAR 2021 B	816	VIOGNIER DE PRIETO PARIENTE 2020 B	806
VIÑA SATOSHI 2019 T BA	868	VIÑAREDO GARNACHA TINTORERA 2018 T BA	692	VIOLETES DE FANG 2018 T	298
VIÑA SOMOZA TATÉ 2020 T C	694	VIÑAREDO GODELLO 2021 B	692	VIONTA 2021 B	391
VIÑA TONDONIA 2010 T R	581	VIÑAREDO GODELLO BARRICA 2021 B	692	VIORE MENCÍA 2021 T	69
VIÑA URBEZO 2021 T	112	VIÑAREDO MENCÍA 2021 T	692	VIORE VERDEJO 2021 B	612
VIÑA VALDEMAZÓN VENDIMIA SELECCIONADA 2018 T	494	VIÑAREDO SOUSÓN 2019 T BA	692	VIORE VERDEJO SOBRE LÍAS B	612
VIÑA VALERA GRAN RESERVA 2015 T GR	435	VIÑARROYO 2021 RD	436	VIRACOCHA 2018 T BA	453
VIÑA VALORIA 2011 T GR	552	VIÑAS DE PANIZA SYRAH 2020 T	110	VIRIDIANA 2021 T	452
VIÑA VALORIA 2014 T R	552	VIÑAS DEL CÁMBRICO VILLANUEVA 2020 T	758	VIROLET XARELLO 2021 B	331
VIÑA VALORIA 2019 T C	552	VIÑAS DEL LAGO 2019 T	454	VIRTUS 2015 T GR	465
VIÑA VALORIA 2021 T	552	VIÑAS DEL VERO CHARDONNAY 2021 B	629	VISENDRA 2020 T R	116
VIÑA VALORIA GARNACHA 2021 RD	552	VIÑAS DEL VERO GEWÜRZTRAMINER 2021 B	629	VISTA DE HALCÓN 2020 B	785
VIÑA VALORIA TEMPRANILLO BLANCO 2021 B	552	VIÑAS DEL VERO PINOT NOIR 2021 RD	629	VITALIS 2016 T C	256
VIÑA VILLAR 2018 T C	703	VIÑAS DEL VERO SAUVIGNON BLANC 2021 B	629	VITALIS SELECCIÓN 2014 T C	256
VIÑA ZANATA AFRUTADO 2021 B	756	VIÑAS ELÍAS MORA 2020 T RB	665	VITIS EXTREMA 2019 T	69
VIÑA ZANATA MALVASÍA SECO 2021 B S	756	VIÑAS HERZAS 2021 B	715	VIUDA NEGRA ARCA DE ASSA 2020 T BA	528
VIÑA ZANATA MARMAJUELO 2021 B	756	VIÑAS HERZAS 2021 T	715	VIUDA NEGRA LA TACONERA 2019 T	528
VIÑA ZANATA TINTILLA 2020 T	756	VIÑAS SILENCIOSAS LA CUADRADA 2019 T	589	VIUDA NEGRA NUNCA JAMÁS 2021 T	528
VIÑA ZANATA TRADICIONAL 2021 B S	756	VIÑAS SILENCIOSAS POSADERO 2019 B	589	VIUDA NEGRA PRADO DE LAS ALMAS 2021 RD FB	528
VIÑA ZANATA VENDIMIA SELECCIONADA 2021 B S	756	VIÑAS SILENCIOSAS POSADERO 2019 T	590	VIUDA NEGRA VILLAHUERCOS 2020 B FB	528
VIÑA ZORZAL CORRAL DE LOS ALTOS 2020 T	322	VIÑAS SILENCIOSAS REGOYOS 2019 T	590	VIVA LA VIDA 2018 T	864
VIÑA ZORZAL GARNACHA 2020 T C	322	VIÑAS VIEJAS 2016 T C	767	VIVALTUS 2017 T	496
VIÑA ZORZAL GARNACHA 2021 T	322	VIÑAS VIEJAS DE PANIZA 2020 T	110	VIVANCO 2019 T C	554
VIÑA ZORZAL GARNACHA BLANCA 2020 B	322	VIÑAS ZORZAL MALAYETO 2020 T	323	VIVANCO BRUNES 2019 T	554

WINE	PAGE
VIVANCO CUVÉE INÉDITA 2018 BE EBR	554
VIVANZA ELITE 2017 T C	39
VIVANZA GOLD 2016 T C	39
VIVES AMBRÒS 2018 BE GR BN	154
VIVES AMBRÒS 2018 BE R BR	154
VIVES AMBRÒS JUJOL 2017 BE GR BN	154
VIVES AMBRÒS ROSAT 2019 RE R BR	154
VIVIDOR 2020 T	675
VIVIR DE NEO 2021 T	470
VIVIR SIN DORMIR 2020 T RB	223
VIVO DEL AIRE AIRÉN 2021 B	872
VIYUELA 2018 T R	465
VIYUELA 2019 T C	465
VIYUELA 2020 T RB	465
VIYUELA SELECCIÓN 2019 T	465
VIYUELA X ANIVERSARIO 2016 T	465
VIZAR 2018 T BA	808
VIZAR PRESTIGIO 2018 T	808
VIZAR SELECCIÓN ESPECIAL 2018 T R	742
VIZAR VERDEJO 2018 B FB	808
VIZCARRA 2019 T	466
VIZCARRA SENDA DEL ORO 2021 T	466
VIZCARRA TORRALVO 2015 T GR	466
VIZCARRA TORRALVO 2019 T	466
VOCHÉ 5º GENERACIÓN C.V.C. T	534
VOCHÉ SELECCIÓN GRACIANO 2017 T	534
VOL D'ANIMA DE RAIMAT BLANC 2021 B	176
VOL D'ANIMA DE RAIMAT ROSAT 2021 RD	176
VOL DE L'AUGA 2019 T	376
VOLAINA 2019 B	332
VOLALTO 2020 T BA	222
VOLTONS 2018 T	368

WINE	PAGE
VOLVER 2020 T	786
VOLVER CUVÉE 2018 T	786
VORAMAR 2021 B S	847
VORAMAR 2021 RD	847
VORAMAR 2021 T	847
VQ SAUVIGNON BLANC 2021 B	792
VR 2018 T RB	416
VS MURUA 2019 T	540
VT TINTO FINO VALTRAVIESO 2018 T BA	473
VT TINTO FINO VALTRAVIESO 2019 T BA	473
VT VENDIMIA SELECCIONADA VALTRAVIESO 2019 T	473
VUELA 2020 T	480
VULPES VULPES 2020 B	768
VX CUVÉE CACO 2018 T	858
VXVX XARELLO VERMELL 2021 RD	330

W

WINE	PAGE
WALTRAUD 2021 B	335
WELLINGTON 20 AÑOS VOS 50 CL BF PC S	209
WELLINGTON 30 AÑOS VORS 50 CL BF PC S	210
WHITE 2021 B	339
WILLIAMS & HUMBERT 2001 BF AM S	213
WILLIAMS & HUMBERT 2001 BF OL S	213
WILLIAMS & HUMBERT 2009 BF OL S	213
WILLIAMS & HUMBERT AÑADA 2002 BF PC S	213
WINE BY NATURE 2021 B	849
WINNER PREMIUM 2016 T C	59

X

WINE	PAGE
XADO 2020 T	767
XADO 2021 B	768
XAFARDER 2021 RD	828
XAI AIT 690M 2020 T	506
XANET 2018 T	832

WINE	PAGE
XANET ROSÉ 2021 RD	832
XARELLO PAIRAL 2019 B FB	330
XARELLO VINYA DEL NOGUER 2018 B	874
XARELLO VINYA DEL NOGUER 2019 B	874
XARMANT TXAKOLI 2021 B	52
XAVI 2018 T	296
XELMIREZ 2021 B	392
XÉRICO 2018 T	585
XESTAL 2015 T RB	70
XF SIERRA CANTABRIA 2021 RD	584
XI'IPAL ROSÉ EL OLIVO 2021 RD	311
XIÓN 2021 B	381
XOLAYR 2019 T BA	203
XOT BLANC 2020 B	193
XPERIMENT 2018 T FB S	351
XR DE MARQUÉS DE RISCAL 2017 T R	523
XTALESI MOSCATEL FRIZZANTE BE SD	885
XTALESI TEMPRANILLO ROSÉ FRIZZANTE RE SD	885
XTRMO (EXTREMO) 2019 B FB	707

Y

WINE	PAGE
Y LA NAVE VA RE	879
Y SOLO CUANDO EL RÍO CALLA 2019 B	774
Y VOLARÁS 2018 T RB	721
YAÑORIA 2016 T R	108
YAÑORIA 2017 T C	108
YEMANUEVA AIRÉN PIE FRANCO 2021 B	237
YEMANUEVA TEMPRANILLO 2019 T	237
YEMASERENA 2014 T C	237
YEMASERENA 2016 T RB	237
YEMASERENA 2018 B FB	237
YEYA 2021 B	862
YLLERA 12 MESES 2019 T RB	815

WINE	PAGE
YLLERA 9 MESES 2019 T RB	815
YLLERA CHARDONNAY VENDIMIA NOCTURNA 2021 B	815
YLLERA SAUVIGNON BLANC VENDIMIA NOCTURNA 2021 B	622
YLLERA VERDEJO VENDIMIA NOCTURNA 2021 B	622
YLLERA VERDEJO VENDIMIA NOCTURNA ECOLÓGICO 2021 B	621
YNOCENTE BF FI S	220
YO SOLO 2020 T FB	261
YOTUEL 2019 T BA	468
YOTUEL FINCA LA NAVA 2016 T	468
YOTUEL FINCA VALDEPALACIOS 2018 T	468
YOTUEL SELECCIÓN 2017 T	468
YOU & ME 2021 B	391
YSIOS 2019 B FB	559
YSIOS FINCA EL NOGAL 2017 T	559
YSIOS GRANO A GRANO 2017 T	559
YSIOS LAS NAVES 2017 T	559
YSIOS ROSÉ 2021 RD	559
YSIOS SELECCIÓN 2016 T	559
YUGO 2018 T C	246
YUGO 2021 T	246
YUGO AIRÉN 2021 B	246
YUGO GARNACHA TEMPRANILLO 2021 RD	246
YUGO VERDEJO 2021 B	246
YUNIKKO 2020 T	683
YUNTERO 2015 T R	243
YUNTERO 68 VENDIMIA 2015 T R	243
YUNTERO MACABEO – SAUVIGNON BLANC 2021 B	244

Z

WINE	PAGE
Z DE ZALEO DE AGUJA NATURAL 2021 BE AG SD	819
Z DE ZALEO DE AGUJA NATURAL 2021 RE AG SD	819
ZABALONDO 2021 B	84
ZACATIN SOBRE LÍAS 2019 B	202
ZAÍNO SYRAH TEMPRANILLO 2020 T	271
ZAÍNO TEMPRANILLO 2020 T C	271
ZALEO PARDINA 2021 B	500
ZALEO TEMPRANILLO 2021 T	500
ZAPADORADO VERDEJO 2021 B	611
ZÁRATE ALBARIÑO 2021 B	404
ZÁRATE CAIÑO TINTO 2020 T	404
ZÁRATE EL BALADO 2021 B	404
ZÁRATE TRAS DA VIÑA 2020 B	404
ZARZALES 2021 RD	156
ZENIZATE CABERNET SAUVIGNON 2020 T	227
ZENIZATE MONASTRELL 2020 T	227
ZERBEROS CLOS PEPI 2020 T BA	752
ZERBEROS EL ALTAR 2020 T	752
ZERBEROS LLANO TOLEDO 2020 T	752
ZERO GARNACHA 2021 T	842
ZG 2021 B	593
ZG EDICIÓN LIMITADA 2020 T C	593
ZINIO STREET ART COLLECTION 2020 T	593
ZINIO TEMPRANILLO 2017 T R	593
ZINTZO 2020 T	559
ZINTZO 2021 B	559
ZIRIES 2014 T	853
ZISMERO 2021 T	105
ZUAZO GASTÓN 2017 T R	593
ZUAZO GASTÓN 2020 T C	593
ZUAZO GASTÓN VENDIMIA SELECCIONADA 2021 T	593
ZURA 2019 B	87
ZURBAL 2017 T R	560
ZURBAL 2021 B	560
ZURBAL 2021 RD	560
ZURBAL 2021 T	560
ZUZARÁN 2019 T C	506
ZUZARÁN 2020 B FB	506
ZUZARÁN FAJERO 2019 T C	506
ZUZARÁN MATURANA 2019 T	506
ZUZARÁN MATURANA EDICIÓN LIMITADA 2020 T	506
ZUZARÁN VIURA 2020 B	506

MAPS

MAP OF THE DO'S IN SPAIN AND VINOS DE PAGO

ANDALUCÍA
1 DO/DOP Málaga- Sierras de Málaga
2 DO/DOP Montilla-Moriles
3 DO/DOP Condado de Huelva y Vino Naranja del Condado de Huelva
4 DO/DOP Jerés-Xérèz-Sherry Manzanilla de SanLlucar
5 DO/DOP Granada

ARAGÓN
1 DO/DOP Somontano
2 DO/DOP Cariñena
 P1. Pago de Aylés /DOP
3 DO/DOP Campo de Borja
4 DO/DOP Calatayud

CASTILLA LA MANCHA
1 DO/DOP La Mancha
 P1.Pago Guijoso /DOP
 P2.Pago Finca Élez /DOP
 P3.Pago Calzadilla /DOP
 P4.Pago Campo de la Guardia /DOP
 P5.Pago de Valdepusa /DOP
 P6.Pago Dehesa del Carrizal /DOP
 P7.Pago Casa del Blanco /DOP
 P8.Pago Florentino /DOP
 P9.Pago El Vicario /DOP
 P10.Pago Los Cerillos /DOP
 P11.Pago Vallegarcía /DOP
 P12. Pago La Jaraba /DOP
3 DO/DOP Almansa
4 DO/DOP Ribera de Júcar
5 DO/DOP Manchuela
6 DO/DOP Uclés
7 DO/DOP Méntrida
8 DO/DOP Valdepeñas

CASTILLA Y LEÓN
1 DO/DOP Rueda
2 DO/DOP Ribera del Duero
3 DO/DOP Arlanza
4 DO/DOP Cigales
5 DO/DOP León
6 DO/DOP Bierzo
7 DO/DOP Toro
8 DO/DOP Tierra del Vino de Zamora
9 DO/DOP Arribes
 P1. Pago Urueña /DOP
 P2. Pago Abadía Retuerta /DOP
 P3. Pago Dehesa Peñalba /DOP

CATALUÑA
1 DO/DOP Terra Alta
2 DO/DOP Montsant
3 DO/DOP Ca. Piorat
4 DO/DOP Tarragona
5 DO/DOP Conca de Barberà
6 DO/DOP Penedès
7 DO/DOP Alella
8 DO/DOP Catalunya
9 DO/DOP Empordà
10 DO/DOP Pla de Bages
11 DO/DOP Costers del Segre

EXTREMADURA
1 DO/DOP Ribera del Guadiana

COMUNIDAD VALENCIANA
1 DO/DOP Alicante
2 DO/DOP Valencia
3 DO/DOP Utiel-Requena
 P1.Pago El Terrerazo /DOP
 P2.Pago Los Balagueses /DOP
 P3. Pago de Tharsys /DOP
 P4 Pago Chozas Carrascal /DOP

GALICIA
1 DO/DOP Monterrei
2 DO/DOP Valdeorras
3 DO/DOP Ribeira Sacra
4 DO/DOP Rías Baixas
5 DO/DOP Ribeiro

ILLES BALEARS
1 DO/DOP Pla i Llevant
2 DO/DOP Binissalem Mallorca

ISLAS CANARIAS
1 DO/DOP Lanzarote
2 DO/DOP Gran Canaria
3 DO/DOP Valle de Güimar
4 DO/DOP Tacoronte-Acentejo
5 DO/DOP Ycoden-Daute-Isora
6 DO/DOP Abona
7 DO/DOP Valle de la Orotava
8 DO/DOP La Gomera
9 DO/DOP El Hierro
10 DO/DOP La Palma

MADRID
1 DO/DOP Vinos de Madrid

MURCIA
1 DO/DOP Bullas
2 DO/DOP Yecla

NAVARRA
1 DO/DOP Navarra
 P1.Pago Otazu /DOP
 P2.Pago Señorío de Arínzano /DOP
 P3.Pago Prado de Irache /DOP

PAÍS VASCO
1 DO/DOP Getariako Txakolina
2 DO/DOP Bizkaiko Txakolina
3 DO/DOP Arabako Txakolina

CROSS-BORDER APPELLATIONS
1 DO Ca Rioja
2 DO Jumilla

MAP OF THE DO'S IN SPAIN AND VINOS DE PAGO

ILLES BALEARS

ISLAS CANARIAS

DO CAVA

① Indicator of the appellation of origin in each province.

P8 Indicator of "Vinos de Pago" in each province

Peñín Guide | SPANISH WINE 1037

MAP OF VINOS DE LA TIERRA AND VINOS DE CALIDAD

ANDALUCÍA
1 Norte de Almería
2 Sierra de las Estancias
3 Desierto de Almería
4 Ribera de Andarax
5 Laujar-Alpujarra
6 Contraciesa-Alpujar/Cumbres de Guadalfeo
7 Laderas de Genil
8 Altiplano de Sierra Nevada
9 Sierra Sur de Jaén
10 Bailén
11 Torreperogil
12 Córdoba
13 Villaviciosa de Córdoba
14 Sierra Norte de Sevilla
15 Los Palacios
16 Cádiz

ARAGÓN
17 Ribera del Gállego-Cinco Villas
18 Ribera del Jiloca
19 Valdejalón
20 Bajo Aragón
21 Valle del Cinca

CANTABRIA
22 Liébana
23 Costa de Cantabria

CASTILLA-LA MANCHA
24 Castilla
25 Pozohondo
26 Sierra de Alcaraz
27 Gálvez

CASTILLA Y LEÓN
28 Castilla y León

EXTREMADURA
29 Extremadura

GALICIA
30 Betanzos
31 Barbanza e Iria
32 Val DO Miño-Ourense
33 Riberias do Morrazo

ILLES BALEARS
34 Illa de Menorca
35 Mallorca
36 Serra de Tramuntana-Costa Nord
37 Eivissa
38 Formentera
39 Illes Balears

LA RIOJA
40 Valles de Sadacia

MURCIA
41 Murcia
42 Campo de Cartagena
43 Abanilla

NAVARRA-ARAGÓN
44 Ribera del Queiles
45 3 Riberas

COMUNIDAD VALENCIAN
46 Castelló

VINOS DE CALIDAD
47 V.C. Cangas / DOP
48 V.C. Lebrija / DOP
49 V.C. Sierra de Salamanca / DOP
50 V.C. Valles de Benavente / DOP
51 V.C. Valtiendas / DOP
52 V.C. Islas Canarias / DOP
53 V.C. de Cebreros / DOP

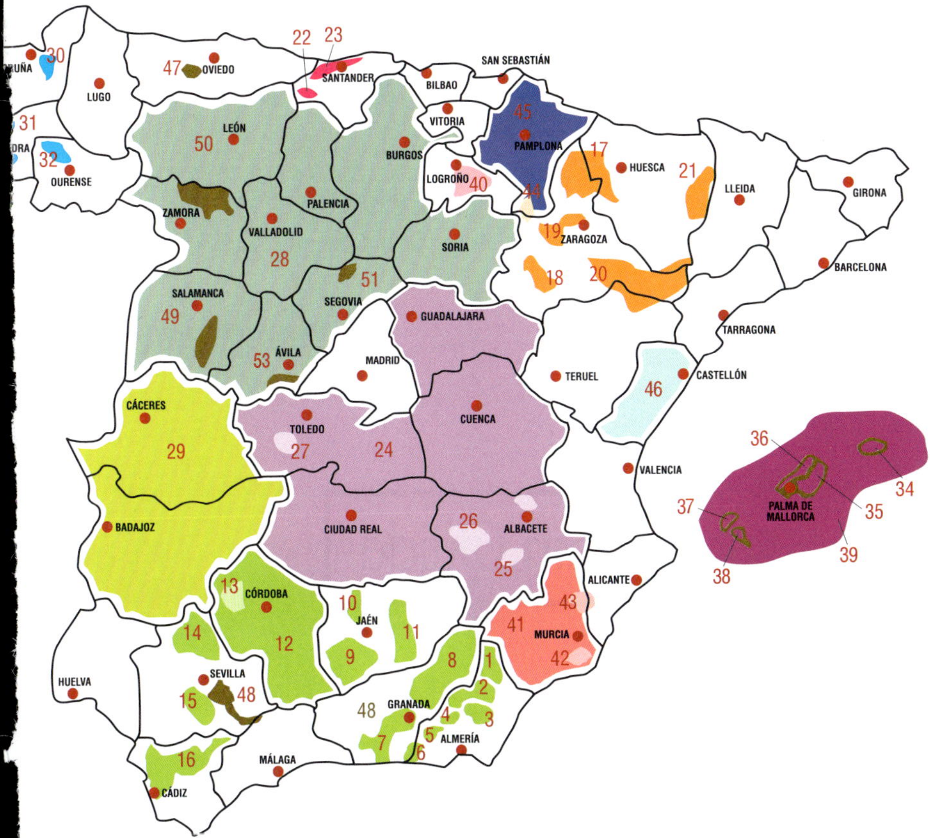

MAP OF VINOS DE LA TIERRA AND VINOS DE CALIDAD

ISLAS CANARIAS

Santa Cruz de Tenerife

Las Palmas de Gran Canaria

Peñín Guide | SPANISH WINE

Peñín Guide